Greenland
(Kalatdlit-Nunat)
Alaska
Yukon
N.W.T.
Nunavut
[Labrador]
B.C.
Alta.
Sask.
Man.
Ont.
Que.
Nfld.
St.Pierre
and Miquelon
P.E.I.
N.B.
N.S.
Vt.
Maine
N.H.
Mass.
R.I.
Conn.
N.J.
Del.
Md.
Wash.
Oreg.
Idaho
Mont.
Wyo.
N.Dak.
S.Dak.
Minn.
Wis.
Mich.
N.Y.
Pa.
Iowa
Nebr.
Nev.
Utah
Colo.
Calif.
Ill.
Ind.
Ohio
W.Va.
Va.
Kans.
Mo.
Ky.
N.C.
Tenn.
S.C.
Okla.
Ark.
Ariz.
N.Mex.
Ala.
Ga.
Miss.
Tex.
La.
Fla.
Mexico

Flora of North America

Contributors to Volume 9

Crinan Alexander
Lawrence A. Alice
Stéphane M. Bailleul
Luc Brouillet
Anne Bruneau
Michael B. Burgess
Christopher S. Campbell
Paul M. Catling
Kevin R. Cushman
Alison C. Dibble
Elizabeth E. Dickson
Eric T. Doucette
Martin Dubé
Reidar Elven
Barbara Ertter
Peter Frost-Olsen
Christopher T. Frye
Jeanette Fryer
Douglas H. Goldman
Arthur Haines
Alice L. Heikens
James Henrickson
William J. Hess
Bertil Hylmö†
William V. Jones
Genevieve J. Kline
Ronald W. Lance
Walter H. Lewis
Richard A. Lis
James A. Macklin
Bryony Macmillan
Charles T. Mason Jr.†
John McNeill
T. Lawrence Mellichamp
Gisèle Mitrow
Gerry Moore
David F. Murray
Guy L. Nesom
Richard J. Pankhurst†
Bruce D. Parfitt†
James B. Phipps
James L. Reveal
Joseph R. Rohrer
Ivan A. Schanzer
Paul D. Sørensen
Joshua C. Springer
Günter Staudt†
John L. Strother
W. Wayt Thomas
Brian D. Vanden Heuvel
Alan S. Weakley
Robert A. S. Wright
George Yatskievych
Peter F. Zika

Editors for Volume 9

Luc Brouillet
Lead Editor, Taxon Editor for Crossosomataceae, Picramniaceae, and Staphyleaceae, Co-Taxon Editor for Rosaceae

Kanchi Gandhi
Nomenclatural Editor

Cassandra L. Howard
Technical Editor (2011–2014)

Helen Jeude
Technical Editor (2006–2010)

Robert W. Kiger
Bibliographic Editor

James B. Phipps
Co-Taxon Editor for Rosaceae

Andrew C. Pryor
Assisting Technical Editor

Heidi H. Schmidt
Managing Editor

John L. Strother
Reviewing Editor

James L. Zarucchi
Editorial Director

Volume Composition

Cassandra L. Howard
Compositor and Technical Editor

Kristin Pierce
Compositor and Editorial Assistant

Heidi H. Schmidt
Production Coordinator and Managing Editor

Crataegus okanaganensis
var. okanaganensis

Flora of North America

North of Mexico

Edited by FLORA OF NORTH AMERICA EDITORIAL COMMITTEE

VOLUME 9

Magnoliophyta: Picramniaceae to Rosaceae

NEW YORK OXFORD • OXFORD UNIVERSITY PRESS • 2014

Oxford University Press is a department of the University of Oxford.
It furthers the University's objective of excellence in research,
scholarship, and education by publishing worldwide.

Oxford New York
Auckland Cape Town Dar es Salaam Hong Kong Karachi
Kuala Lumpur Madrid Melbourne Mexico City Nairobi
New Delhi Shanghai Taipei Toronto

With offices in
Argentina Austria Brazil Chile Czech Republic France Greece
Guatemala Hungary Italy Japan Poland Portugal Singapore
South Korea Switzerland Thailand Turkey Ukraine Vietnam

Published by Oxford University Press, Inc.
198 Madison Avenue, New York, New York 10016
www.oup.com

Library of Congress Cataloging-in-Publication Data
(Revised for Volume 9)
Flora of North America North of Mexico
edited by Flora of North America Editorial Committee.
Includes bibliographical references and indexes.
Contents: v. 1. Introduction—v. 2. Pteridophytes and gymnosperms—
v. 3. Magnoliophyta: Magnoliidae and Hamamelidae—
v. 22. Magnoliophyta: Alismatidae, Arecidae, Commelinidae (in part), and Zingiberidae—
v. 26. Magnoliophyta: Liliidae: Liliales and Orchidales—
v. 23. Magnoliophyta: Commelinidae (in part): Cyperaceae—
v. 25. Magnoliophyta: Commelinidae (in part): Poaceae, part 2—
v. 4. Magnoliophyta: Caryophyllidae (in part): part 1—
v. 5. Magnoliophyta: Caryophyllidae (in part): part 2—
v. 19, 20, 21. Magnoliophyta: Asteridae (in part): Asteraceae, parts 1–3—
v. 24. Magnoliophyta: Commelinidae (in part): Poaceae, part 1—
v. 27. Bryophyta, part 1—
v. 8. Magnoliophyta: Paeoniaceae to Ericaceae—
v. 7. Magnoliophyta: Salicaceae to Brassicaceae—
v. 28. Bryophyta, part 2—
v. 9. Magnoliophyta: Picramniaceae to Rosaceae

ISBN: 978-0-19-534029-7 (v. 9)
1. Botany—North America.
2. Botany—United States.
3. Botany—Canada.
I. Flora of North America Editorial Committee.
QK110.F55 2002 581.97 92-30459

1 2 3 4 5 6 7 8 9
Printed in the United States of America
on acid-free paper

Contents

Dedication vii
Founding Member Institutions viii
Donors ix
Art Sponsorship x
Editorial Committee xi
Project Staff xii
Contributors xiii
Taxonomic Reviewers xv
Regional Reviewers xv
Preface xviii

Introduction xix

Picramniaceae 3
Staphyleaceae 6
Crossosomataceae 9
Rosaceae 18

Literature Cited 663
Index 690

Courtesy of The University of British Columbia

This volume of the

Flora of North America North of Mexico

is dedicated to

Walter Hepworth Lewis

enthusiastic student of *Rosa*,

early supporter of and author for the *Flora* effort,

respected professor, and esteemed colleague

FOUNDING MEMBER INSTITUTIONS

Flora of North America Association

Agriculture and Agri-Food Canada
Ottawa, Ontario

Arnold Arboretum
Jamaica Plain, Massachusetts

Canadian Museum of Nature
Ottawa, Ontario

Carnegie Museum of Natural History
Pittsburgh, Pennsylvania

Field Museum of Natural History
Chicago, Illinois

Fish and Wildlife Service
United States Department of the Interior
Washington, D.C.

Harvard University Herbaria
Cambridge, Massachusetts

Hunt Institute for Botanical Documentation
Carnegie Mellon University
Pittsburgh, Pennsylvania

Jacksonville State University
Jacksonville, Alabama

Jardin Botanique de Montréal
Montréal, Québec

Kansas State University
Manhattan, Kansas

Missouri Botanical Garden
St. Louis, Missouri

New Mexico State University
Las Cruces, New Mexico

The New York Botanical Garden
Bronx, New York

New York State Museum
Albany, New York

Northern Kentucky University
Highland Heights, Kentucky

Université de Montréal
Montréal, Québec

University of Alaska
Fairbanks, Alaska

University of Alberta
Edmonton, Alberta

The University of British Columbia
Vancouver, British Columbia

University of California
Berkeley, California

University of California
Davis, California

University of Idaho
Moscow, Idaho

University of Illinois
Urbana-Champaign, Illinois

University of Iowa
Iowa City, Iowa

The University of Kansas
Lawrence, Kansas

University of Michigan
Ann Arbor, Michigan

University of Oklahoma
Norman, Oklahoma

University of Ottawa
Ottawa, Ontario

University of Southwestern Louisiana
Lafayette, Louisiana

The University of Texas
Austin, Texas

University of Western Ontario
London, Ontario

University of Wyoming
Laramie, Wyoming

Utah State University
Logan, Utah

For their support of the preparation of this volume, we gratefully acknowledge and thank:

The Philecology Foundation

Franklinia Foundation

The Andrew W. Mellon Foundation

The David and Lucile Packard Foundation

an anonymous foundation

Chanticleer Foundation

The Fairweather Foundation

ChevronTexaco

Agriculture and Agri-Food Canada

ESRI

Chris Davidson and Sharon Christoph

Botanical Research Foundation of Idaho

For sponsorship of illustrations included in this volume, we express sincere appreciation to:

Lawrence A. Alice, Bowling Green, Kentucky
Rubus repens, Rosaceae

American Rose Society
all *Rosa* species, Rosaceae

Anita Cholewa, Blaine, Minnesota
Kelseya uniflora, Rosaceae

Marshall R. Crosby & Shelley L. Pryor, St. Louis, Missouri
Malus ioensis, Rosaceae

Timothy A. Dickinson, Toronto, Ontario
Crataegus douglasii, Rosaceae – in honor of Rhoda M. Love

Timothy A. Dickinson & Deborah Metsger, Toronto, Ontario
Crataegus brachyacantha, Rosaceae – in memory of Jane M. Bowles

Field Botanists of Ontario, Ontario
Waldsteinia fragarioides, Rosaceae

Douglas Goldman, Greensboro, North Carolina
Rubus hispidus, Rosaceae – in memory of Hobbes Goldman

Haverford College Arboretum, Haverford, Pennsylvania
Exochorda racemosa, Rosaceae

Alice Heikens, Franklin, Indiana
Neviusia alabamensis, Rosaceae

William and Betsy Hess, Monterey, Tennessee
Oemleria cerasiformis, Rosaceae

James Macklin, Ottawa, Ontario
Rubus allegheniensis, Rosaceae

Maryland Native Plant Society, Silver Spring, Maryland
Amelanchier laevis, Rosaceae

Nancy R. Morin, Point Arena, California
Chamaebatiaria millefolium, Rosaceae – in appreciation of the Arizona Native Plant Society
Lyonothamnus floribundus, Rosaceae – in honor of Sherwin Carlquist

Western Kentucky University, Department of Biology, Bowling Green, Kentucky
Rubus pubescens, Rosaceae

Project Staff — past and present involved with the preparation of Volume 9

Barbara Alongi, *Illustrator*
Burgund Bassüner, *Information Coordinator*
Ariel S. Buback, *Editorial Assistant*
Trisha K. Distler, *GIS Analyst*
Pat Harris, *Editorial Assistant*
Cassandra L. Howard, *Technical Editor and Compositor*
Helen Jeude, *Senior Technical Editor*
Ruth T. King, *Editorial Assistant*
Marjorie C. Leggitt, *Illustrator*
John Myers, *Illustrator and Illustration Compositor*
Kristin Pierce, *Editorial Assistant and Compositor*
Andrew C. Pryor, *Editorial Assistant*
Hong Song, *Programmer*

Contributors to Volume 9

Crinan Alexander
Royal Botanic Garden
Edinburgh, Scotland

Lawrence A. Alice
Western Kentucky University
Bowling Green, Kentucky

Stéphane M. Bailleul
Jardin Botanique de Montréal
Montréal, Québec

Luc Brouillet
Université de Montréal
Montréal, Québec

Anne Bruneau
Université de Montréal
Montréal, Québec

Michael B. Burgess
Morris Arboretum
University of Pennsylvania
Philadelphia, Pennsylvania

Christopher S. Campbell
The University of Maine
Orono, Maine

Paul M. Catling
Agriculture and Agri-Food Canada
Ottawa, Ontario

Kevin R. Cushman
The University of Maine
Orono, Maine

Alison C. Dibble
The University of Maine
Orono, Maine

Elizabeth E. Dickson
University of Calgary
Calgary, Alberta

Eric T. Doucette
The University of Maine
Orono, Maine

Martin Dubé
Université de Moncton
Edmundston, New Brunswick

Reidar Elven
Natural History Museum
University of Oslo
Oslo, Norway

Barbara Ertter
University of California
Berkeley, California

Peter Frost-Olsen
Aarhus University
Aarhus, Denmark

Christopher T. Frye
Natural Heritage Program
Maryland Department of Natural Resources
Wye Mills, Maryland

Jeanette Fryer
Froxfield, England

Douglas H. Goldman
U.S.D.A. Natural Resources Conservation Service
Greensboro, North Carolina

Arthur Haines
Bowdoin, Maine

Alice L. Heikens
Franklin College
Franklin, Indiana

James Henrickson
The University of Texas at Austin
Austin, Texas

William J. Hess
Monterey, Tennessee

Bertil Hylmö†
Bjuv, Sweden

William V. Jones
Glendora, California

Genevieve J. Kline
Northern Illinois University
DeKalb, Illinois

Ronald W. Lance
Chimney Rock Park
Chimney Rock, North Carolina

Walter H. Lewis
Washington University
St. Louis, Missouri

Richard A. Lis
Department of Fish and Game
Redding, California

James A. Macklin
Agriculture and Agri-Food Canada
Ottawa, Ontario

Bryony Macmillan
Landcare Research
Lincoln, New Zealand

Charles T. Mason Jr.†
University of Arizona
Tucson, Arizona

John McNeill
Royal Botanic Garden
Edinburgh, Scotland

T. Lawrence Mellichamp
University of North Carolina
Charlotte, North Carolina

Gisèle Mitrow
Agriculture and Agri-Food Canada
Ottawa, Ontario

Gerry Moore
U.S.D.A. Natural Resources Conservation Service
Greensboro, North Carolina

David F. Murray
University of Alaska Museum of the North
University of Alaska Fairbanks
Fairbanks, Alaska

Guy L. Nesom
Fort Worth, Texas

Richard J. Pankhurst†
Royal Botanic Garden
Edinburgh, Scotland

Bruce D. Parfitt†
University of Michigan–Flint
Flint, Michigan

James B. Phipps
University of Western Ontario
London, Ontario

James L. Reveal
Cornell University
Ithaca, New York

Joseph R. Rohrer
University of Wisconsin
Eau Claire, Wisconsin

Ivan A. Schanzer
Main Botanical Garden
Russian Academy of Sciences
Moscow, Russia

Paul D. Sørensen
Northern Illinois University
DeKalb, Illinois

Joshua C. Springer
Michigan State University
East Lansing, Michigan

Günter Staudt†
Merzhausen, Germany

John L. Strother
University of California
Berkeley, California

W. Wayt Thomas
The New York Botanical Garden
Bronx, New York

Brian D. Vanden Heuvel
Colorado State University–Pueblo
Pueblo, Colorado

Alan S. Weakley
University of North Carolina
Chapel Hill, North Carolina

Robert A. S. Wright
Blacksburg, Virginia

George Yatskievych
Missouri Botanical Garden
St. Louis, Missouri

Peter F. Zika
University of Washington
Seattle, Washington

Taxonomic Reviewers

Fred R. Barrie
Missouri Botanical Garden
St. Louis, Missouri

Christopher S. Campbell
The University of Maine
Orono, Maine

Timothy A. Dickinson
Royal Ontario Museum
University of Toronto
Toronto, Ontario

Barbara Ertter
University of California
Berkeley, California

James Henrickson
The University of Texas at Austin
Austin, Texas

James A. Macklin
Agriculture and Agri-Food Canada
Ottawa, Ontario

David F. Murray
University of Alaska
Museum of the North
University of Alaska Fairbanks
Fairbanks, Alaska

James L. Reveal
Cornell University
Ithaca, New York

Regional Reviewers

ALASKA / YUKON

Bruce Bennett
Yukon Department of Environment
Whitehorse, Yukon

Robert Lipkin
Alaska Natural Heritage Program
University of Alaska
Anchorage, Alaska

David F. Murray
University of Alaska
Museum of the North
University of Alaska Fairbanks
Fairbanks, Alaska

Carolyn Parker
University of Alaska
Museum of the North
University of Alaska Fairbanks
Fairbanks, Alaska

Mary Stensvold
U.S.D.A. Forest Service
Sitka, Alaska

PACIFIC NORTHWEST

Edward R. Alverson
The Nature Conservancy
Eugene, Oregon

Adolf Ceska
British Columbia Conservation Data Centre
Victoria, British Columbia

Richard R. Halse
Oregon State University
Corvallis, Oregon

Eugene N. Kozloff
Shannon Point Marine Center
Anacortes, Washington

Jim Pojar
British Columbia Forest Service
Smithers, British Columbia

Cindy Roché
Talent, Oregon

Peter F. Zika
University of Washington
Seattle, Washington

SOUTHWESTERN UNITED STATES

H. David Hammond†
Northern Arizona University
Flagstaff, Arizona

G. Frederic Hrusa
Courtland, California

Charles T. Mason Jr.†
University of Arizona
Tucson, Arizona

James D. Morefield
Nevada Natural Heritage Program
Carson City, Nevada

Donald J. Pinkava
Arizona State University
Tempe, Arizona

Jon P. Rebman
San Diego Natural History Museum
San Diego, California

Margriet Wetherwax
University of California
Berkeley, California

WESTERN CANADA

William J. Cody†
Agriculture and Agri-Food Canada
Ottawa, Ontario

Lynn Gillespie
Canadian Museum of Nature
Ottawa, Ontario

A. Joyce Gould
Alberta Tourism, Parks and Recreation
Edmonton, Alberta

Vernon L. Harms
University of Saskatchewan
Saskatoon, Saskatchewan

Elizabeth Punter
University of Manitoba
Winnipeg, Manitoba

ROCKY MOUNTAINS

Curtis R. Björk
University of Idaho
Moscow, Idaho

Bonnie Heidel
University of Wyoming
Laramie, Wyoming

B. E. Nelson
University of Wyoming
Laramie, Wyoming

NORTH CENTRAL UNITED STATES

William T. Barker
North Dakota State University
Fargo, North Dakota

Anita F. Cholewa
University of Minnesota
St. Paul, Minnesota

Neil A. Harriman
University of Wisconsin
Oshkosh, Wisconsin

Bruce W. Hoagland
University of Oklahoma
Norman, Oklahoma

Robert B. Kaul
University of Nebraska
Lincoln, Nebraska

Deborah Q. Lewis
Iowa State University
Ames, Iowa

Ronald L. McGregor†
The University of Kansas
Lawrence, Kansas

Lawrence R. Stritch
U.S.D.A. Forest Service
Shepherdstown, West Virginia

George Yatskievych
Missouri Botanical Garden
St. Louis, Missouri

SOUTH CENTRAL UNITED STATES

David E. Lemke
Southwest Texas State University
San Marcos, Texas

EASTERN CANADA

Sean Blaney
Atlantic Canada Conservation Data Centre
Sackville, New Brunswick

Jacques Cayouette
Agriculture and Agri-Food Canada
Ottawa, Ontario

Frédéric Coursol
Mirabel, Québec

William J. Crins
Ontario Ministry of Natural Resources
Peterborough, Ontario

John K. Morton†
University of Waterloo
Waterloo, Ontario

Marian Munro
Nova Scotia Museum of Natural History
Halifax, Nova Scotia

Michael J. Oldham
Natural Heritage Information Centre
Peterborough, Ontario

GREENLAND

Geoffrey Halliday
University of Lancaster
Lancaster, England

NORTHEASTERN UNITED STATES

Ray Angelo
New England Botanical Club
Cambridge, Massachusetts

Tom S. Cooperrider
Kent State University
Kent, Ohio

Arthur Haines
Bowdoin, Maine

Edward G. Voss†
University of Michigan
Ann Arbor, Michigan

Kay Yatskievych
Missouri Botanical Garden
St. Louis, Missouri

SOUTHEASTERN UNITED STATES

Mac H. Alford
University of Southern Mississippi
Hattiesburg, Mississippi

J. Richard Carter Jr.
Valdosta State University
Valdosta, Georgia

L. Dwayne Estes
Austin Peay State University
Clarksville, Tennessee

Paul J. Harmon
West Virginia Natural Heritage Program
Charleston, West Virginia

W. John Hayden
University of Richmond
Richmond, Virginia

John B. Nelson
University of South Carolina
Columbia, South Carolina

Bruce A. Sorrie
University of North Carolina
Chapel Hill, North Carolina

R. Dale Thomas
Seymour, Tennessee

Lowell E. Urbatsch
Louisiana State University
Baton Rouge, Louisiana

Thomas F. Wieboldt
Virginia Polytechnic Institute and State University
Blacksburg, Virginia

B. Eugene Wofford
University of Tennessee
Knoxville, Tennessee

FLORIDA

Loran C. Anderson
Florida State University
Tallahassee, Florida

Bruce F. Hansen
University of South Florida
Tampa, Florida

Richard P. Wunderlin
University of South Florida
Tampa, Florida

Preface for Volume 9

Since the publication of *Flora of North America* volume 7 (the sixteenth volume and the last one covering vascular plants published in the *Flora* series) in late 2010, the membership of the Flora of North America Association [FNAA] Board of Directors has undergone changes. Guy Baillargeon and James S. Miller have retired from the board. New board members include: Kerry A. Barringer (Taxon Editor), Gerry Moore, and Elizabeth H. Zacharias. As a result of a reorganization finalized in 2003, the Board of Directors of FNAA succeeded the former Editorial Committee; for the sake of continuity of citation, authorship of *Flora* volumes is still to be cited as "Flora of North America Editorial Committee, eds."

Prior to the completion of volume 9, the tome containing the second half of mosses for the flora area – volume 28 and seventeenth in the series – was published earlier in 2014. In total, volumes 27 and 28 treat 1402 species in 333 genera classified in 81 families for the flora area.

The initial editorial process was done at the Canada center, Université de Montréal. The final editorial processing for this volume of the *Flora* was undertaken at the Missouri Botanical Garden in St. Louis. Pre-press production for the volume, including typesetting and layout, plus coordination for all aspects of planning, executing, and scanning the illustrations, was done at the St. Louis center. Other aspects of production, such as art panel composition plus labeling and occurrence map generation, were carried out in Gaston, Oregon, and Miami, Florida, respectively.

Line drawings published in this volume were executed by three very talented artists: Barbara Alongi prepared illustrations for Crossosomataceae and *Mespilus* (Rosaceae); Marjorie C. Leggett illustrated Picramniaceae, Rosaceae (except those for *Mespilus* and *Rosa*), and Staphyleaceae plus she prepared the color frontispiece depicting *Crataegus okanaganensis* var. *okanaganensis* (Rosaceae); and John Myers illustrated species of *Rosa* (Rosaceae), plus he composed and labeled all of the line drawings in this volume.

Starting with volume 8 published in 2009, the circumscription and ordering of some families within the *Flora* have been modified so they mostly reflect that of the Angiosperm Phylogeny Group [APG] rather than the previously followed Cronquist organizational structure. The groups of families found in this and future volumes in the series are mostly ordered following E. Haston et al. (2007); since APG views of relationships and circumscriptions have evolved, and will certainly change further through time, some discrepancies in organization will occur. Volume 30 of the *Flora of North America* will contain an overall index to the published volumes.

Support from many institutions and by individuals has enabled the *Flora* to be produced. The team that authored the account of *Amelanchier* (Rosaceae) extends special thanks to the Maine Agricultural and Forest Experiment Station and the National Science Foundation.

The Flora of North America Association remains deeply thankful to the many people who continue to help create, encourage, and sustain the *Flora*.

Introduction

Scope of the Work

Flora of North America North of Mexico is a synoptic account of the plants of North America north of Mexico: the continental United States of America (including the Florida Keys and Aleutian Islands), Canada, Greenland (Kalâtdlit-Nunât), and St. Pierre and Miquelon. The *Flora* is intended to serve both as a means of identifying plants within the region and as a systematic conspectus of the North American flora.

The *Flora* will be published in 30 volumes. Volume 1 contains background information that is useful for understanding patterns in the flora. Volume 2 contains treatments of ferns and gymnosperms. Families in volumes 3–26, the angiosperms, were first arranged according to the classification system of A. Cronquist (1981) with some modifications, and starting with Volume 8, the circumscriptions and ordering of families generally follow those of the Angiosperm Phylogeny Group [APG] (see E. Haston et al. 2007). Bryophytes are being covered in volumes 27–29. Volume 30 will contain the cumulative bibliography and index.

The first two volumes were published in 1993, Volume 3 in 1997, and Volumes 22, 23, and 26, the first three of five volumes covering the monocotyledons, appeared in 2000, 2002, and 2002, respectively. Volume 4, the first part of the Caryophyllales, was published in late 2003. Volume 25, the second part of the Poaceae, was published in mid 2003, and Volume 24, the first part, was published in January 2007. Volume 5, completing the Caryophyllales plus Polygonales and Plumbaginales, was published in early 2005. Volumes 19–21, treating Asteraceae, were published in early 2006. Volume 27, the first of two volumes treating mosses in North America, was published in late 2007. Volume 8, Paeoniaceae to Ericaceae, was published in September 2009, and Volume 7, Salicaceae to Brassicaceae appeared in 2010. Earlier this year, Volume 28 was published, completing the treatment of mosses for the flora area. The correct bibliographic citation for the *Flora* is: Flora of North America Editorial Committee, eds. 1993+. Flora of North America North of Mexico. 18+ vols. New York and Oxford.

Volume 9 treats 691 species in 74 genera contained in four families. For additional statistics please refer to Table 1 on p. xx.

Contents · General

The *Flora* includes accepted names, selected synonyms, literature citations, identification keys, descriptions, phenological information, summaries of habitats and geographic ranges, and other biological observations. Each volume contains a bibliography and an index to the taxa included in that volume. The treatments, written and reviewed by experts from throughout the systematic botanical community, are based on original observations of herbarium specimens and, whenever possible, on living plants. These observations are supplemented by critical reviews of the literature.

Table 1. *Statistics for Volume 9 of Flora of North America.*

Family	Total Genera	Endemic Genera	Introduced Genera	Total Species	Endemic Species	Introduced Species	Conservation Taxa
Picramniaceae	2	0	0	2	0	0	0
Staphyleaceae	1	0	0	2	2	0	0
Crossosomataceae	3	1	0	7	4	0	6
Rosaceae	68	12	15	680	416	149	156
Totals	**74**	**13**	**15**	**691**	**422**	**149**	**162**

Basic Concepts

Our goal is to make the *Flora* as clear, concise, and informative as practicable so that it can be an important resource for both botanists and nonbotanists. To this end, we are attempting to be consistent in style and content from the first volume to the last. Readers may assume that a term has the same meaning each time it appears and that, within groups, descriptions may be compared directly with one another. Any departures from consistent usage will be explicitly noted in the treatments (see References).

Treatments are intended to reflect current knowledge of taxa throughout their ranges worldwide, and classifications are therefore based on all available evidence. Where notable differences of opinion about the classification of a group occur, appropriate references are mentioned in the discussion of the group.

Documentation and arguments supporting significantly revised classifications are published separately in botanical journals before publication of the pertinent volume of the *Flora*. Similarly, all new names and new combinations are published elsewhere prior to their use in the *Flora*. No nomenclatural innovations will be published intentionally in the *Flora*.

Taxa treated in full include extant and recently extinct native species, hybrids that are well established (or frequent), and waifs or cultivated plants that are found frequently outside cultivation and give the appearance of being naturalized. Taxa mentioned only in discussions include waifs or naturalized plants now known only from isolated old records and some non-native, economically important or extensively cultivated plants, particularly when they are relatives of native species. Excluded names and taxa are listed at the ends of appropriate sections, for example, species at the end of genus, genera at the end of family.

Treatments are intended to be succinct and diagnostic but adequately descriptive. Characters and character states used in the keys are repeated in the descriptions. Descriptions of related taxa at the same rank are directly comparable.

With few exceptions, taxa are presented in taxonomic sequence. If an author is unable to produce a classification, the taxa are arranged alphabetically and the reasons are given in the discussion.

Treatments of hybrids follow that of one of the putative parents. Hybrid complexes are treated at the ends of their genera, after the descriptions of species.

We have attempted to keep terminology as simple as accuracy permits. Common English equivalents usually have been used in place of Latin or Latinized terms or other specialized terminology, whenever the correct meaning could be conveyed in approximately the same space, for example, "pitted" rather than "foveolate," but "striate" rather than "with fine longitudinal lines." See *Categorical Glossary for the Flora of North America Project* (R. W. Kiger and D. M. Porter 2001; also available online at http://huntbot.andrew.cmu.edu) for standard definitions of generally used terms. Very specialized terms are defined, and sometimes illustrated, in the relevant family or generic treatments.

References

Authoritative general reference works used for style are *The Chicago Manual of Style,* ed. 14 (University of Chicago Press 1993); *Webster's New Geographical Dictionary* (Merriam-Webster 1988); and *The Random House Dictionary of the English Language*, ed. 2, unabridged (S. B. Flexner and L. C. Hauck 1987). *B-P-H/S. Botanico-Periodicum-Huntianum/Supplementum* (G. D. R. Bridson and E. R. Smith 1991), *BPH-2: Periodicals with Botanical Content* (Bridson 2004), and *BPH Online* [http://fmhibd.library.cmu.edu/fmi/iwp/cgi?-db=BPH_Online&-loadframes] (Bridson and D. W. Brown) have been used for abbreviations of serial titles, and *Taxonomic Literature*, ed. 2 (F. A. Stafleu and R. S. Cowan 1976–1988) and its supplements by Stafleu et al. (1992–2009) have been used for abbreviations of book titles.

Graphic Elements

All genera and approximately 28 percent of the species in this volume are illustrated. The illustrations may show diagnostic traits or complex structures. Most illustrations have been drawn from herbarium specimens selected by the authors. Data on specimens that were used and parts that were illustrated have been recorded. This information, together with the archivally preserved original drawings, is deposited in the Missouri Botanical Garden Library and is available for scholarly study.

Specific Information in Treatments

Keys

Dichotomous keys are included for all ranks below family if two or more taxa are treated. More than one key may be given to facilitate identification of sterile material.

Nomenclatural Information

Basionyms of accepted names, with author and bibliographic citations, are listed first in synonymy, followed by any other synonyms in common recent use, listed in alphabetical order, without bibliographic citations.

The last names of authors of taxonomic names have been spelled out. The conventions of *Authors of Plant Names* (R. K. Brummitt and C. E. Powell 1992) have been used as a guide for including first initials to discriminate individuals who share surnames.

If only one infraspecific taxon within a species occurs in the flora area, nomenclatural information (literature citation, basionym with literature citation, relevant other synonyms) is given for the species, as is information on the number of infraspecific taxa in the species and their distribution worldwide, if known. A description and detailed distributional information are given only for the infraspecific taxon.

Descriptions

Character states common to all taxa are noted in the description of the taxon at the next higher rank. For example, if sexual condition is dioecious for all species treated within a genus, that character state is given in the generic description. Characters used in keys are repeated in the descriptions. Characteristics are given as they occur in plants from the flora area. Characteristics that occur only in plants from outside the flora area may be given within square brackets, or instead may be noted in the discussion following the description. In families with one genus and one or more species, the family description is given as usual, the genus description is condensed, and the species are described as usual. Any special terms that may be used when describing members of a genus are presented and explained in the genus description or discussion.

Starting with volumes of the *Flora* published in 2014, some phrases in descriptions may appear within chevrons (often called angled brackets). Chevrons are used when information is included for one or some taxa at the same rank within the next higher level taxon but parallel information is missing in one or more of the remaining ones. An example where chevrons are used in this volume is in the description provided of *Alvaradoa* [see p. 4], where additional information about the seeds — **Seeds** 1, <ellipsoid, ± ¼–⅓ length of samara> — is given; for *Picramnia*, the other genus of Picramniaceae treated in the *Flora*, the description has just "**Seeds** 1–3," lacking parallel information.

In reading descriptions, the reader may assume, unless otherwise noted, that: the plants are green, photosynthetic, and reproductively mature; woody plants are perennial; stems are erect; roots are fibrous; leaves are simple and petiolate; flowers are bisexual, radially symmetric, and pediceled; perianth parts are hypogynous, distinct, and free; and ovaries are superior. Because measurements and elevations are almost always approximate, modifiers such as "about," "circa," or "±" are usually omitted.

Unless otherwise noted, dimensions are length × width. If only one dimension is given, it is length or height. All measurements are given in metric units. Measurements usually are based on dried specimens but these should not differ significantly from the measurements actually found in fresh or living material.

Chromosome numbers generally are given only if published, documented counts are available from North American material or from an adjacent region. No new counts are published intentionally in the *Flora*. Chromosome counts from nonsporophyte tissue have been converted to the $2n$ form. The base number (x =) is given for each genus. This represents the lowest known haploid count for the genus unless evidence is available that the base number differs.

Flowering time and often fruiting time are given by season, sometimes qualified by early, mid, or late, or by months. Elevations over 200 m generally are rounded to the nearest 100 m; those 100 m and under are rounded to the nearest 10 m. Mean sea level is shown as 0 m, with the understanding that this is approximate. Elevation often is omitted from herbarium specimen labels, particularly for collections made where the topography is not remarkable, and therefore precise elevation is sometimes not known for a given taxon.

The term "introduced" is defined broadly to refer to plants that were released deliberately or accidentally into the flora and that now exist as wild plants in areas in which they were not recorded as native in the past. The distribution of non-native plants is often poorly documented and presence of the plants in the flora may be ephemeral.

If a taxon is globally rare or if its continued existence is threatened in some way, the words "of conservation concern" appear before the statements of elevation and geographic range.

Criteria for taxa of conservation concern are based on NatureServe's (formerly The Nature Conservancy)—see http://www.natureserve.org—designations of global rank (G-rank) G1 and G2:

G1 Critically imperiled globally because of extreme rarity (5 or fewer occurrences or fewer than 1000 individuals or acres) or because of some factor(s) making it especially vulnerable to extinction.

G2 Imperiled globally because of rarity (5–20 occurrences or fewer than 3000 individuals or acres) or because of some factor(s) making it very vulnerable to extinction throughout its range.

The occurrence of species and infraspecific taxa within political subunits of the *Flora* area is depicted by dots placed on the outline map to indicate occurrence in a state or province. The Nunavut boundary on the maps has been provided by the GeoAccess Division, Canada Centre for Remote Sensing, Earth Science. Authors are expected to have seen at least one specimen documenting each geographic unit record (except in rare cases when undoubted literature reports may be used) and have been urged to examine as many specimens as possible from throughout the range of each taxon. Additional information about taxon distribution may be presented in the discussion.

Distributions are stated in the following order: Greenland; St. Pierre and Miquelon; Canada (provinces and territories in alphabetic order); United States (states in alphabetic order); Mexico (11 northern states may be listed specifically, in alphabetic order); West Indies; Bermuda; Central America (Belize, Costa Rica, El Salvador, Guatemala, Honduras, Nicaragua, Panama); South America; Europe, or Eurasia; Asia (including Indonesia); Africa; Atlantic Islands; Indian Ocean Islands; Pacific Islands; Australia; Antarctica.

Discussion

The discussion section may include information on taxonomic problems, distributional and ecological details, interesting biological phenomena, and economic uses.

Selected References

Major references used in preparation of a treatment or containing critical information about a taxon are cited following the discussion. These, and other works that are referred to in discussion or elsewhere, are included in Literature Cited at the end of the volume.

CAUTION

The Flora of North America Editorial Committee **does not encourage, recommend, promote, or endorse** any of the folk remedies, culinary practices, or various utilizations of any plant described within this volume. Information about medicinal practices and/or ingestion of plants, or of any part or preparation thereof, has been included only for historical background and as a matter of interest. Under no circumstances should the information contained in these volumes be used in connection with medical treatment. Readers are strongly cautioned to remember that many plants in the flora are toxic or can cause unpleasant or adverse reactions if used or encountered carelessly.

Key to boxed codes following accepted names:

- [C] of conservation concern
- [E] endemic to the flora area
- [F] illustrated
- [I] introduced to the flora area
- [W] weedy, based mostly on R. H. Callihan et al. (1995) and/or D. T. Patterson et al. (1989)

Flora of North America

PICRAMNIACEAE Fernando & Quinn

• Bitterbush or Picramnia Family

W. Wayt Thomas

Shrubs or trees, rarely subshrubs, dioecious. **Stems** usually simple, wood dense. **Leaves** usually persistent, cauline, alternate, imparipinnately compound; stipules absent; petiole present; blade usually chartaceous, margins entire, <leaflets alternate or subopposite (often darkening upon drying), bases of lateral leaflets often oblique>. **Inflorescences** terminal or lateral, simple racemes (either of separate flowers or glomerules of flowers), pendent, or compound thyrses. **Flowers** unisexual, perianth and androecium hypogynous; epicalyx bractlet absent; hypanthium absent; sepals [3–]5, basally connate or distinct; petals [3–]5, sometimes 0, distinct, <stramineous to maroon or red, usually ligulate>; stamens or staminodia (if present) (3–)5, distinct, free, antipetalous, anthers longitudinally dehiscent; staminate flowers: (filaments unappendaged) intrastaminal discs or, rarely, column present, pistillodia usually conic and hairy or vestigial; pistillate flowers: staminodia present or absent; pistils 1, 2 or 3(or 4)-carpellate, carpels connate, free, ovary borne on disc or gynophore, placentation apical, styles 2 or 3(or 4), usually ± absent, stigmas sessile and divergent on apices of ovaries; ovules 2 or 3 per locule. **Fruits** berries or samaras; <sepals often persistent; stigmas persistent>. **Seeds** 1, not arillate.

Genera 3, species 52 (2 genera, 2 species in the flora): Florida, Mexico, West Indies, Central America, South America.

The length of the pedicel is taken from fruiting material for *Alvaradoa* and *Picramnia*. The third genus, *Nothotalisia* W. W. Thomas, is known from Panama and northwestern South America with three species.

SELECTED REFERENCE Thomas, W. W. 1990. The American genera of Simaroubaceae and their distribution. Acta Bot. Brasil. 4: 11–18.

1. Fruits samaras; leaflets elliptic, 8–33 × 3–9 mm, apices rounded . . . . 1. *Alvaradoa*, p. 4
1. Fruits berries; leaflets ovate to narrowly elliptic, 14–300 × 8–160 mm, apices usually acute to acuminate . . . . 2. *Picramnia*, p. 4

1. ALVARADOA Liebmann, Vidensk. Meddel. Dansk Naturhist. Foren. Kjøbenhavn 1853: 100. 1854 • [For Pedro de Alvarado y Contreras, ca. 1485–1541, member of Cortez's expedition to Mexico]

Shrubs or trees, 10–70[–150] dm. **Leaves** imparipinnate (mimosoid in appearance); leaflets [14–]17–35[–40], alternate, elliptic, 8–33 × 3–9 mm, base rounded to oblique. **Inflorescences** lateral to subterminal, racemes. **Flowers:** sepals 5; petals 5 or 0; stamens 5, anthers dorsifixed with enlarged connective, thecae parallel; pistils 3-carpellate, 2 lateral carpels sterile, central fertile, in staminate flowers pistils vestigial. **Fruits** samaras, flattened or recurved, usually hairy; sepals not persistent. **Seeds** 1, <ellipsoid, ± $\frac{1}{4}$–$\frac{1}{3}$ length of samara>. x = 8.

Species 6 (1 in the flora): Florida, Mexico, West Indies, Central America, South America (Argentina, Bolivia).

A chromosome count for *Alvaradoa subovata* Cronquist of Argentina is $2n$ = 16 (A. M. Sanso and M. N. Seo 2005). The samara is derived from a dry, indehiscent capsule; the wings are formed from the two lateral, sterile carpels.

1. Alvaradoa amorphoides Liebmann, Vidensk. Meddel. Dansk Naturhist. Foren. Kjøbenhavn 1853: 101. 1854 • Mexican alvaradoa [F]

Subspecies 2 (1 in the flora): Florida, Mexico, West Indies, Central America.

Subspecies *amorphoides* is found in Mexico, West Indies (Grand Cayman), and Central America.

1a. Alvaradoa amorphoides Liebmann subsp. psilophylla (Urban) Cronquist, Brittonia 5: 135. 1944 [F]

Alvaradoa psilophylla Urban, Repert. Spec. Nov. Regni Veg. 20: 304. 1924

Shrubs or trees, 10–70 dm. **Stems** of current year puberulent. **Leaves** glabrous except for puberulent rachis; rachis (including petiole) 9.5–21 cm; petiolules 1–1.5 mm; leaflets 17–35, abaxially pale green, adaxially olive, oblong to elliptic, 5–8(–10) × 14–19 mm, apex rounded. **Racemes** pendent; staminate-flowered racemes 7–15 cm, puberulent; pistillate-flowered racemes 7–10 cm, pubescent. **Pedicels:** staminate 1.5–2 mm; pistillate 1–2 mm. **Staminate flowers:** sepals 5, ovate, 1.5 mm, puberulent; petals 5, filiform, 3–3.5 mm, narrower than filaments; intrastaminal discs lobed; stamens 5, filaments 2–5 mm; pistils vestigial. **Pistillate flowers:** sepals 5, ovate, puberulent; petals 0; intrastaminal discs circular, thin, unlobed; stamens 0; pistils 3-carpellate, ovate, flattened, 2–2.2 × 1–1.4 mm, pubescent, stigmas unequal, 2 at apex of ovary, divergent, 3d slightly more proximal along medial line of pistil, sometimes only distal 2 stigmas present, one larger than other. **Samaras** stramineous or purplish tinged, flattened, (8–)10–12 × 2.7–3.2(–4) mm, dry, margins ciliate, puberulent; pedicels 6–10 mm.

Flowering and fruiting year-round. Hammocks, pine woodlands, on limestone; 0–20 m; Fla.; West Indies (Bahamas, Cuba).

In the flora area, subsp. *psilophylla* is found in Miami-Dade County. Specimens of subsp. *amorphoides* are, in general, more robust with longer, narrower fruits. Plants of subsp. *amorphoides* occur in a wider variety of habitats and are not obligate on limestone formations as are those of subsp. *psilophylla*.

2. PICRAMNIA Swartz, Prodr., 2, 27. 1788, name conserved • Bitterbush [Greek *picros*, bitter, and *thamnos*, shrub, alluding to bitterness of whole plant]

Shrubs or trees, rarely subshrubs, (2–)15–80(–220) dm. **Leaves** imparipinnate [paripinnate (with 1 of distal pair of leaflets oriented terminally)]; leaflets [1–]5–11[–33], alternate or subopposite, ovate to elliptic, 14–300 × 8–160 mm, base of terminal usually symmetric, of lateral ones usually oblique. **Inflorescences** terminal or axillary, rarely cauliflorous, [racemes or simple] compound

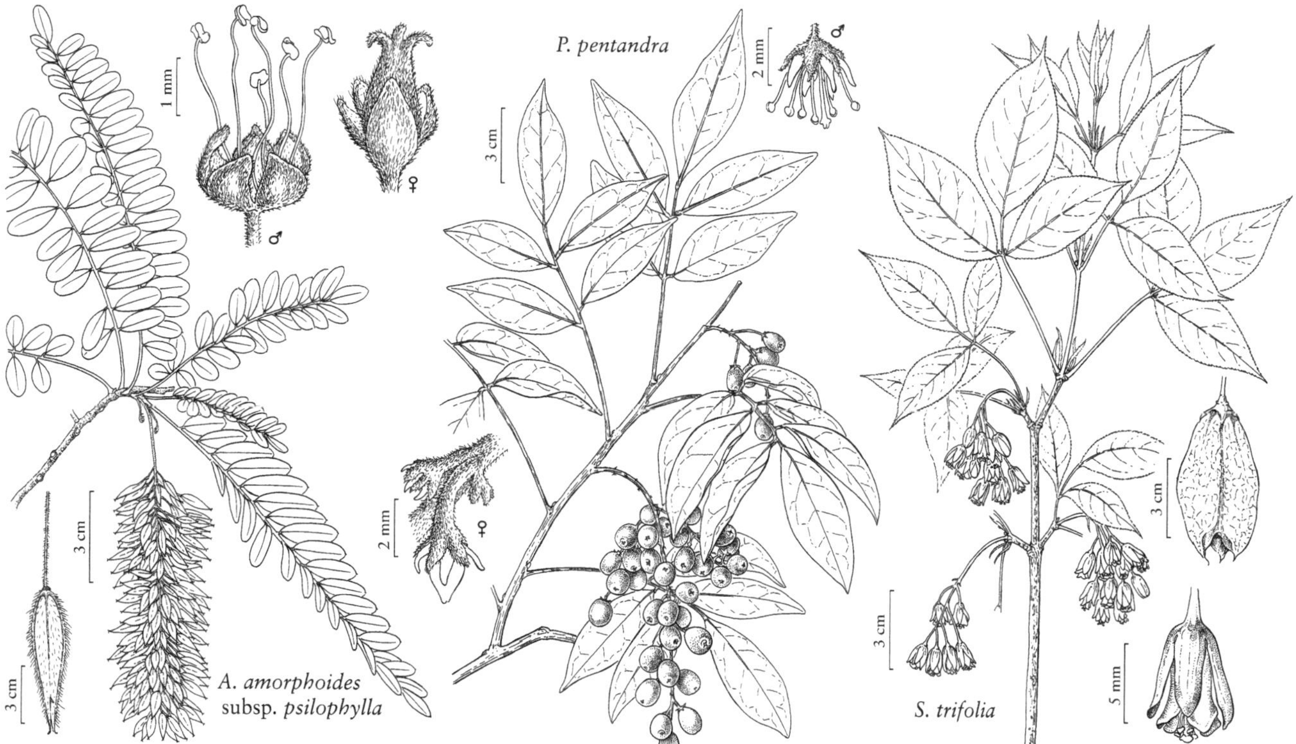

ALVARADOA ◦ PICRAMNIA ◦ STAPHYLEA

thyrses. **Flowers:** sepals 5; petals 5; staminate flowers: anthers usually with enlarged connective, thecae parallel; pistils reduced to cylindric or lobed pistillodia or tufts of hairs; pistillate flowers: stamens reduced to staminodia or 0; pistils 2- or 3-carpellate, all carpels potentially fertile. **Fruits** berries; sepals persistent. **Seeds** 1–3.

Species 43 (1 in the flora): Florida, Mexico, West Indies, Central America, South America.

Picramnia is a neotropical genus, extending from Florida (Miami-Dade County) and Mexico to subtropical South America (northern Argentina and southern Brazil); members of the genus are found primarily in moist and semideciduous forests.

SELECTED REFERENCE Thomas, W. W. 1988. A conspectus of Mexican and Central American *Picramnia* (Simaroubaceae). Brittonia 40: 89–105.

1. Picramnia pentandra Swartz, Fl. Ind. Occid. 1: 220. 1797 • Florida bitterbush [F]

Shrubs or trees, 15–40 dm. **Stems:** young puberulent. **Leaves** imparipinnate; petiole ± puberulent; larger leaves 8–14 cm, sparsely puberulent; petiolules 1.5–3 mm, wrinkled, ± puberulent; leaflets 5–11, ovate to narrowly elliptic, larger 6–12.5 × 1.8–5 cm, apex usually acute to acuminate, usually shiny adaxially, secondary and tertiary veins salient on both surfaces, glabrous except along midveins adaxially and at base near petiolule. **Thyrses** 6–20 cm. **Pedicels:** staminate 0.5–0.7 mm; pistillate 1–2 mm. **Staminate flowers:** sepals lanceolate, 1–1.2 mm, puberulent; petals lanceolate to narrowly oblong, 1.5–1.6 mm, glabrous; intrastaminal discs circular; stamens 5, filaments 2.2–2.8 mm, anthers dorsifixed; pistils vestigial. **Pistillate flowers:** sepals triangular, 1 mm, puberulent; petals narrowly oblong, 1.3–1.6 mm, glabrous; stamens 0; pistils ovoid, 1–1.5 mm, glabrous, stigmas 2, slightly divergent. **Berries** becoming black, spheric to ellipsoid, 6–9 mm diam., glabrous; pedicels 7–10 mm.

Flowering and fruiting year-round. Hammocks, pine woodlands, on limestone; 0–20 m; Fla.; West Indies (Bahamas, Cuba); Central America; n South America (Venezuela).

In the flora area, *Picramnia pentandra* is known from Miami-Dade County.

STAPHYLEACEAE Martinov • Bladdernut Family

Luc Brouillet

Shrubs, rarely trees. Stems branched [simple]. **Leaves** deciduous [persistent], cauline, opposite, usually ternate [odd-pinnate, 1-foliolate]; stipules caducous, extrapetiolar [intrapetiolar], distinct [connate]; petiole present; blade herbaceous, margins serrulate to serrate or crenulate-serrulate; <leaflets: articulations or petiolule bases with 2 glands or stipels>. **Inflorescences** terminal [axillary], thyrses, drooping. **Flowers** bisexual [unisexual, plants dioecious], perianth and androecium hypogynous; epicalyx bractlet absent; hypanthium short-campanulate; sepals 5, distinct or basally connate; petals 5, distinct; nectar disc lobed, <surrounding bases of ovaries>; stamens 5 in 1 series (alternating with nectary lobes), distinct, free, alternipetalous, anthers <X-shaped or sagittate>, longitudinally dehiscent; pistils 1[or 2], [2 or]3(or 4)-carpellate, carpels connate [distinct], free, ovary superior, placentation axile or axile-basal, styles (2 or)3(or 4), basally connate, distally postgenitally connate, stigmas expanded, flattened; ovules [1–]3–12 per locule. **Fruits** capsules [berries], <inflated, usually 3-lobed, carpels becoming distinct distally, tardily so adaxially, and apically dehiscent; styles persistent, splitting>. **Seeds** 1 or 2 per locule, not arillate [arillate].

Genera 2, species ca. 45 (1 genus, 2 species in the flora): North America, Mexico, South America, Europe, Asia (to New Guinea).

A. Cronquist (1981) classified Staphyleaceae within Sapindales. His circumscription of the family included Tapiscioideae (*Huertea* Ruiz & Pavón and *Tapiscia* Oliver), now placed in Tapisciaceae (Huerteales), an order unrelated to Crossosomatales but close to Brassicales, Malvales, and Sapindales (V. Savolainen et al. 2000b; D. E. Soltis et al. 2000). Staphyleaceae are thus restricted here to the genera *Dalrympelea* Roxburgh (*Turpinia* Ventenat) and *Staphylea* (including *Euscaphis* Siebold & Zuccarini) (Staphyleoideae Arnott in Cronquist; generic delimitation according to S. L. Simmons 2007). Based on molecular phylogenetic data (V. Sosa and M. W. Chase 2003), the Angiosperm Phylogeny Group (2003) placed Staphyleaceae in the strict sense within Crossosomatales, an order remote from Sapindales. Sosa and Chase, and Oh S. H. and D. Potter (2006), showed that Staphyleaceae are sister to Crossosomataceae, Guamatelaceae, and Stachyuraceae (Crossosomatales). Detailed morphologic information on the family was provided by W. C. Dickison (1986, 1987), M. L. Matthews and P. K. Endress

(2005), and Simmons (2007), among others. Matthews and Endress plus Simmons indicated the presence of a floral cup in the family; the cup is here called hypanthium. The inflorescence is usually described as a panicle or raceme; it appears to be a thyrse. The fruit is usually interpreted as capsular; Matthews and Endress described it as follicular.

SELECTED REFERENCES Dickison, W. C. 1986. Floral morphology and anatomy of Staphyleaceae. Bot. Gaz. 147: 312–326. Dickison, W. C. 1987. Leaf and nodal anatomy and systematics of Staphyleaceae. Bot. Gaz. 148: 475–489. Simmons, S. L. 2007. Staphyleaceae. In: K. Kubitzki et al., eds. 1990+. The Families and Genera of Vascular Plants. 10+ vols. Berlin etc. Vol. 9, pp. 440–445. Spongberg, S. 1971b. The Staphyleaceae in the southeastern United States. J. Arnold Arbor. 52: 196–203.

1. STAPHYLEA Linnaeus, Sp. Pl. 1: 270. 1753; Gen. Pl. ed. 5, 130. 1754 • Bladdernut, staphylier [Abridged from Greek *Staphylodendron*, ancient name for the genus; Greek *staphyle*, cluster of grapes, alluding to disposal of flowers and fruits]

Shrubs, rarely trees. Stems 1–10+. **Leaves** 3(–5)-foliolate [odd-pinnate]; stipules linear, papery; terminal leaflets long-petiolulate, laterals short-petiolulate or sessile; leaflet blades ovate to elliptic or obovate, base oblique, rounded, or cuneate, lateral ± asymmetric, apex ± acuminate to apiculate, glabrous or sparsely villous, densely when young; hairs unicellular. **Thyrses** 5–15[–25+]-flowered, on 2d-year wood, pedunculate, bracteate. **Pedicels** basally 2-bracteolate, with swollen point of disarticulation mid length. **Flowers** appearing before or with leaves, campanulate or cylindro-campanulate; sepals white or pale greenish pink to pale pink, petaloid, equal to or shorter than petals; petals white [pink]; stamens ± exserted, equal to or longer than petals; pistils short-stalked, stigmas (2 or)3(or 4)-lobed. **Capsules** brownish, ellipsoid or obovoid, trigonous, papery; carpels becoming distinct distally; sepals persistent; styles ± persistent. **Seeds** light brown, ± spheric. $x = 13$.

Species 10 (2 in the flora): North America, Mexico, Eurasia.

Staphylea species are cultivated as garden ornamentals.

The two North American species of *Staphylea* differ in their floral biology: the flowers of *S. trifolia* are pinkish green (sepals) and campanulate with barely exserted stamens and styles; those of *S. bolanderi* are white and more tubular (cylindro-campanulate) with well-exserted stamens and styles. These differences appear to correlate with pollinators available in the eastern deciduous forest and in California, respectively.

1. Leaves abaxially ± villous; flowers campanulate; sepals pale greenish, greenish pink to pale pink to whitish; stamen filaments equal to or slightly longer than petals; filaments (basally) and pistils villous; capsules obovoid; e North America .... 1. *Staphylea trifolia*
1. Leaves glabrous; flowers cylindro-campanulate; sepals white; stamen filaments longer than petals; filaments and pistils glabrous; capsules ellipsoid; California .... 2. *Staphylea bolanderi*

1. Staphylea trifolia Linnaeus, Sp. Pl. 1: 270. 1753 • American bladdernut, staphylier à trois folioles [E] [F]

Plants <suckering>, 2–4(–5) m. **Leaves:** leaflet blades elliptic to widely obovate, 4.5–13 cm, margins serrulate to serrate, apex acuminate, abaxial surfaces ± villous, densely when young, adaxial glabrous or sparsely hairy. **Flowers** campanulate; sepals pale greenish, greenish pink to pale pink to whitish, 4–7 mm; petals spatulate, 6–8 mm, apex recurved; filaments not or barely exserted, 6–8 mm, equal to or slightly longer than petals, basally villous; pistils villous, styles ± equal to stamens. **Capsules** obovoid, 3–5 cm. $2n = 72$.

Flowering Apr–Jun. Deciduous forests, often humid, often at bases of rocky slopes, floodplain woods, thickets, stream banks; 0–1000 m; Ont., Que.; Ala., Ark., Conn., Del., D.C., Fla., Ga., Ill., Ind., Iowa, Kans., Ky., La., Md., Mass., Mich., Minn., Miss., Mo., Nebr., N.H., N.J., N.Y., N.C., Ohio, Okla., Pa., S.C., Tenn., Vt., Va., W.Va., Wis.

Leaves of *Staphylea trifolia* are sometimes 5-foliolate.

The relationship of *Staphylea trifolia* with the northern Mexican *S. pringlei* S. Watson remains to be determined.

2. Staphylea bolanderi A. Gray, Proc. Amer. Acad. Arts 10: 69. 1874 • Sierra bladdernut [E]

Plants 2–6 m. **Leaves:** leaflet blades widely ovate to widely elliptic or ± round, (2–)2.5–8(–15) cm, margins crenulate-serrulate, apex short-acuminate to apiculate, glabrous. **Flowers** cylindro-campanulate; sepals white, 5–7(–9) mm; petals narrowly oblong or oblanceolate, 7–8(–12) mm, apex not recurved; filaments exserted, 9–10(–15) mm, longer than petals, glabrous; pistils glabrous, styles equal to or surpassing stamens. **Capsules** ellipsoid, (2.5–)3.5–4(–5.5) cm.

Flowering Mar–May. Chaparral, foothill woodlands, yellow pine forests, slopes, canyon sides; (200–)300–1400 m; Calif.

Staphylea bolanderi occurs in the foothills and at the lower elevations of the Cascades and Klamath ranges, the Sierra Nevada, and the Tehachapi Mountains.

CROSSOSOMATACEAE Engler

• Crossosoma Family

Charles T. Mason Jr.†
George Yatskievych

Shrubs, rarely trees. Stems intricately branched, often spinescent; <bark becoming finely and irregularly grooved, usually peeling in thin longitudinal plates>. **Leaves** deciduous or drought-deciduous, cauline, alternate or opposite, sometimes appearing fasciculate on short shoots, simple; stipules absent or relatively small; petiole absent or short, <base usually slightly expanded, thickened and persistent (in *Glossopetalon*)>; blade herbaceous to slightly coriaceous, margins entire or apically 2- or 3-lobed (in *Apacheria*). **Inflorescences** terminal on short shoots, sometimes axillary or appearing so, usually flowers solitary, occasionally 2- or 3-flowered clusters (in *Glossopetalon*), <short shoots or pedicels usually with minute, scarious or hardened bractlike structures at base>. **Flowers** bisexual, perianth and androecium perigynous; epicalyx bractlet absent; hypanthium saucer-shaped to shallowly cup-shaped [narrowly funnelform (in *Velascoa*)]; sepals (3 or)4 or 5(or 6), distinct; petals (3 or)4 or 5(or 6), distinct, <base often short-clawed>; nectar disc absent or crenately lobed, <fleshy or thin>; stamens 4–50, distinct, free, <sometimes unequal in length>, anthers attached at or near bases, yellow, dehiscing by longitudinal slits; pistils 1–9, distinct, free, short-stipitate or sessile, styles usually ± oblique, often indistinct, stigmas capitate or relatively short, sometimes linear, ventrally decurrent; ovules 1–22+. **Fruits** follicles, <angled or short-tapered at both ends, coriaceous>. **Seeds** 1–22 per follicle, arillate.

Genera 4, species ca. 8 (3 genera, 7 species in the flora): w United States, Mexico.

For years, Crossosomataceae were thought to comprise only the type genus. In 1975, the circumscription was expanded with the discovery of *Apacheria*, and three years later it was enlarged further with the transfer of *Glossopetalon* from Celastraceae (R. F. Thorne and R. Scoggin 1978). A fourth genus, the monospecific *Velascoa* Calderón & Rzedowski, endemic to northeastern Querétaro, Mexico, was described in 1997. *Velascoa recondita* Calderón & Rzedowski is unique in the family in its elongate, narrowly funnelform hypanthium and nearly sessile stamens.

The affinities of the family have been controversial. A. Cronquist (1981) outlined the problems, noting that the presence of an aril (and some embryologic features) suggested a placement in the Dilleniales, but he chose instead to classify the family tentatively in his concept of Rosales, near Rosaceae, based on superficial floral similarities such as perigyny. V. Sosa and M. W. Chase (2003), who studied DNA sequence variation, and M. L. Matthews and P. K. Endress (2005), who studied developmental floral morphology, independently concluded that the Crossosomatales should be segregated as a distinct order (including such morphologically diverse families as Stachyuraceae and Staphyleaceae). The circumscription and placement of this order in relation to other Rosids are still uncertain.

SELECTED REFERENCE Sosa, V. and M. W. Chase. 2003. Phylogenetics of Crossosomataceae based on *rbc*L sequence data. Syst. Bot. 28: 96–105.

1. Stamens 15–50; sepals and petals 5; petals 9–18 mm; follicles finely transversely verrucose or indistinctly reticulate; seeds 4–22 per follicle; arils deeply fimbriate . . . . 2. *Crossosoma*, p. 11
1. Stamens 8 or 4–10; sepals and petals 4 or 3–5(or 6); petals 4–5 mm or 2–9 mm; follicles longitudinally striate; seeds 1 or 2 per follicle; arils irregularly discoid or ± fimbrillate.
    2. Leaves opposite, sometimes appearing fasciculate on short shoots, blade margins entire or apically 2- or 3-lobed; stigmas linear; sepals and petals 4 . . . . 1. *Apacheria*, p. 10
    2. Leaves alternate, blade margins entire; stigmas capitate, oblique; sepals and petals 3–5(or 6) . . . . 3. *Glossopetalon*, p. 12

1. APACHERIA C. T. Mason, Madroño 23: 105, figs. 1, 2a. 1975 • [For Apache Indians] [C] [E]

Shrubs. Stems <main often ± spreading>, irregularly and usually densely branched; twigs tan or orangish tinged, often with pair of opposite grooves, sometimes beginning to exfoliate during 1st year, not punctate, glabrous; bark grayish tan to gray or black. **Leaves** opposite, sometimes appearing fasciculate on short shoots; stipules absent or minute, narrowly triangular; sessile, those on long shoots decurrent as lines or grooves; blade oblanceolate to cuneate or spatulate, base tapered, margins entire or apically 2- or 3-lobed, <occasionally minutely ciliate near base>, apex tip angled or tapered to sharply pointed, often mucronate (or lobe tips), venation obscure, not glaucous, glabrous. **Inflorescences** terminal on short shoots, solitary flowers. **Pedicels** absent or short. **Flowers:** hypanthium without nectar disc; sepals 4, lanceolate to ovate, ± concave, not cucullate, margins usually narrowly hyaline at least proximally, not glaucous, glabrous; petals 4, white, sometimes cream or ± reddish tinged with age, oblanceolate to oblanceolate-spatulate, 4–5 mm; stamens 8, usually in 2 series, filaments inserted on hypanthium outside narrow disc, longer than anthers; carpels 4, usually distinct, sessile; stigmas linear, ventrally decurrent. **Follicles** 1–4, asymmetrically turbinate to narrowly obovoid, turgid to slightly flattened, adaxial surface straight or nearly so prior to dehiscence, longitudinally striate with prominent venation, hispidulous, sometimes glabrescent; adaxial suture cartilaginous. **Seeds** 1 or 2 per follicle, brown to yellowish brown, reniform, not or only slightly flattened, shiny, finely papillose (appearing pebbled); arils white (drying tan), irregularly discoid or ± fimbrillate.

Species 1: sw United States.

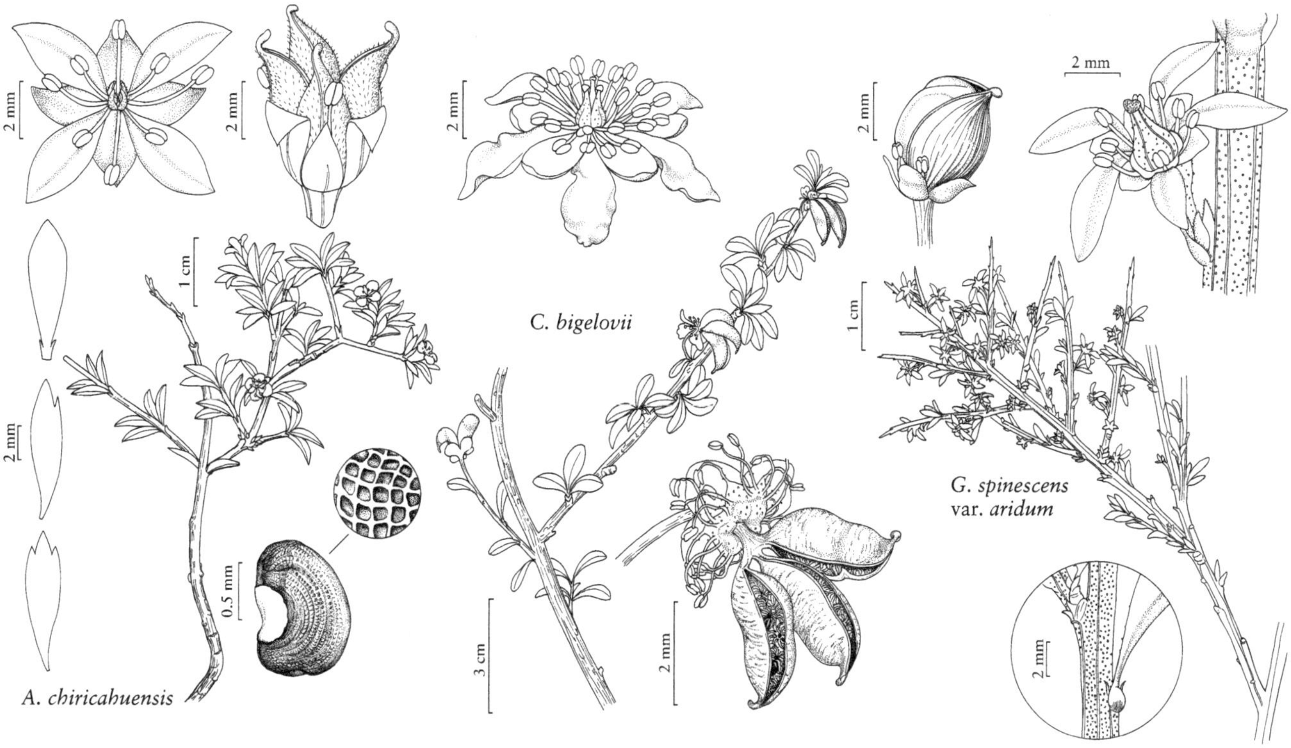

APACHERIA ◦ CROSSOSOMA ◦ GLOSSOPETALON

1. **Apacheria chiricahuensis** C. T. Mason, Madroño 23: 105, figs. 1, 2a. 1975 • Chiricahua rockflower [C] [E] [F]

Shrubs 7–50 cm (often wider than tall). **Leaves** 3.5–7.5 × 1–2.3 mm. **Sepals** 3–3.5 mm, usually 2 opposing sepals slightly narrower and longer than other pair. **Follicles** 4–5 mm. **Seeds** 1–1.5 × 0.8–1 mm.

Flowering Apr–Nov. Rock crevices, cliff faces, on igneous or volcanic substrates (mostly rhyolite); of conservation concern; 1700–2100 m; Ariz., N.Mex.

Apacheria chiricahuensis has not been documented to occur outside the flora area; its presence in poorly botanized portions of adjacent northwestern Mexico is likely.

2. CROSSOSOMA Nuttall, Proc. Acad. Nat. Sci. Philadelphia 4: 7. 1848 • [Greek *krossos*, fringe or tassel, and *soma*, body, alluding to aril]

Shrubs, occasionally trees. Stems densely branched; twigs tan or reddish tinged, <± angled>, not decurrently ridged, smooth, not punctate, glabrous; bark gray to grayish tan. **Leaves** alternate, sometimes appearing fasciculate on short shoots; stipules absent; petiole absent or indistinctly short, not decurrent; blade narrowly elliptic to elliptic or narrowly oblanceolate to obovate, occasionally spatulate, base attenuate, sometimes cuneate, margins entire, apex tip rounded to blunt, usually mucronate, <midvein usually apparent (at least abaxially)>, sometimes 3–5 pairs of arched-ascending secondary veins and fine network of tertiary veins also apparent (especially upon drying), glaucous, glabrous. **Inflorescences** terminal on short shoots or apparently axillary, flowers solitary. **Pedicels** short to long. **Flowers:** hypanthium with fleshy, crenately lobed nectar disc; sepals 5, ovate to nearly circular, ± cucullate, margins narrowly hyaline, glaucous, glabrous; petals 5, usually white, sometimes ± purplish tinged with age, narrowly lanceolate, elliptic, or oblanceolate, 9–18 mm; stamens 15–50 in 2–4 series, filaments inserted on disc, usually longer than anthers; carpels 2–9, distinct, short-stipitate; stigmas oblique, capitate. **Follicles** (1 or)2–9, ± cylindric, turgid, adaxial surface mostly curved prior to dehiscence, <apex sometimes ± beaked>, finely transversely verrucose or indistinctly reticulate with rounded ridges, glabrous; adaxial suture cartilaginous. **Seeds** 4–22 per follicle, brown to nearly black (finely mottled), usually reniform, sometimes nearly spheric, slightly to moderately flattened, shiny, smooth to faintly striate, rugulose, or verruculose; arils white or pale yellow, deeply fimbriate, <seeds and arils often ± coherent>. *x* = 12.

Species 2 (2 in the flora): sw United States, nw Mexico.

1. Sepals 4–5 mm; leaves 5–15(–25) mm, mostly fasciculate . . . . 1. *Crossosoma bigelovii*
1. Sepals 8–12 mm; leaves 20–90 mm, mostly well spaced. . . . . .2. *Crossosoma californicum*

1. Crossosoma bigelovii S. Watson, Proc. Amer. Acad. Arts 11: 122. 1876 • Rockflower [F]

Crossosoma bigelovii var. *glaucum* (Small) Kearney & Peebles; *C. bigelovii* var. *parviflorum* (B. L. Robinson & Fernald) K. C. Davis; *C. glaucum* Small; *C. parviflorum* B. L. Robinson & Fernald

Shrubs, 0.7–2 m. **Leaves** mostly fasciculate on short shoots, 5–15 mm (–25 mm on long shoots). **Pedicels** short to long. **Flowers:** sepals 4–5 mm; petals 9–12(–15) mm. **Follicles** (1 or)2 or 3(or 4), (5–)7–15 mm. **Seeds** 4–8, 2–2.5 mm diam. **2*n*** = 24.

Flowering Jan–May. Dry rocky slopes and cliff faces, often in canyons, most commonly on igneous or volcanic substrates, mostly in the Sonoran and Mojave deserts; 50–1700 m; Ariz., Calif., Nev.; Mexico (Baja California, Chihuahua, Sonora).

The name *Crossosoma parviflorum* was based on specimens from the north side of the Grand Canyon and Sonora, Mexico, and applied to plants with somewhat smaller flowers and fruits than those in most other populations. Further taxonomic research may determine that this constitutes a valid infrataxon; it has not been accepted for the present treatment. Although it is available through specialty nurseries, this species is seldom seen in southwestern gardens as a specimen plant and ornamental.

2. Crossosoma californicum Nuttall, Proc. Acad. Nat. Sci. Philadelphia 4: 8. 1848 (as californica) • Catalina rockflower [C]

Shrubs or trees, 1–5 m. **Leaves** mostly well spaced, 20–90 mm. **Pedicels** mostly long. **Flowers:** sepals 8–12 mm; petals 12–18 mm. **Follicles** 2–9, 15–20(–32) mm. **Seeds** 18–22, 2–3 mm diam. **2*n*** = 24.

Flowering Feb–May. Dry rocky slopes, often in canyons, coastal scrub and chaparral, on igneous or volcanic substrates; of conservation concern; 50–500 m; Calif.; Mexico (Baja California).

Crossosoma californicum is known only from the Palos Verdes Peninsula of Los Angeles County and the islands of San Clemente and Santa Catalina, as well as on Guadalupe Island (Baja California); it occasionally is cultivated as an ornamental.

3. GLOSSOPETALON A. Gray, Smithsonian Contr. Knowl. 5(6): 29, plate 12, fig. B. 1853 • Greasebush [Greek *glossa*, tongue, and *petalon*, spreading, alluding to outspreading tonguelike petals]

Shrubs. Stems densely branched, <branches ascending or divaricate>; twigs yellowish green to green or grayish green, eventually turning yellowish brown to orangish brown, with fine decurrent ridges, punctate (sometimes inconspicuously), glabrous or puberulent; bark becoming gray to dark gray, pale yellow, or dark brown, sometimes with black patches. **Leaves** alternate; stipules absent or usually relatively small (often adnate proximally to petiole and appearing as a pair of minute lobes), triangular to subulate or filiform; petiole usually short, persistent after leaves are shed, base sometimes glandular-thickened, becoming darkened, decurrent as fine ridges proximal to nodes; blade narrowly elliptic to oblanceolate, base angled or tapered, margins entire, apex tip blunt to sharply pointed, sometimes rounded, sometimes mucronate (spinescent in *G. pungens*), <midvein sometimes apparent as low keel proximally on abaxial surfaces>, venation obscure, sometimes inconspicuously punctate, usually glaucous, glabrous or puberulent. **Inflorescences** terminal or axillary (then terminating short shoots), flowers solitary, sometimes 2- or 3-flowered clusters. **Pedicels** mostly short. **Flowers:** hypanthium lined with fleshy, crenately lobed nectar disc; sepals 3–5(or 6), ovate to nearly circular, sometimes slightly concave at base, not cucullate, margins hyaline, glaucous, glabrous; petals <persistent at fruiting>, 3–5(or 6), white, <spirally twisted with age>, narrowly elliptic-oblanceolate to oblanceolate, 2–9 mm; stamens 4–10 in 1 or 2 series, filaments inserted along disc margin, equal to or longer than anthers; carpels 1(–3), distinct, sessile; stigmas oblique, capitate. **Follicles** 1(–3), asymmetrically ovoid to obovoid or ellipsoid, turgid or slightly angular-flattened, adaxial surface curved or ± contorted prior to dehiscence, <apex often appearing ± beaked>, longitudinally striate with prominent venation, glabrous, occasionally sparsely puberulent; adaxial suture <narrow>, ± cartilaginous. **Seeds** 1 or 2 per follicle, usually tan or yellowish, sometimes brownish white or cream, spheric or somewhat reniform, shiny, smooth or minutely reticulate/pitted; arils white (drying tan), irregularly discoid or ± fimbrillate.

Species 4 (4 in the flora): w, c United States, Mexico.

Glossopetalon is treated here as comprising one relatively widespread and morphologically variable species and three narrowly endemic, more morphologically homogeneous relatives. The plants occur mostly on calcareous substrates. Taxa within *Glossopetalon* are relatively character-poor, especially on herbarium specimens, and the characters used to distinguish taxa in past treatments are variable among populations. The genus is in need of more detailed study using biosystematic and molecular methods. In navigating the keys, characters of leaf shape, stipule development, and numbers and shapes of perianth parts should be studied closely by observing multiple examples of a given structure from a specimen in order to facilitate determinations. Rapidly expanding extension shoots sometimes produce atypical leaves that are larger and broader than the limits stated in the keys and descriptions.

Forsellesia Greene, often included as a synonym of *Glossopetalon*, is an illegitimate name.

SELECTED REFERENCES Ensign, M. R. 1942. A revision of the celastraceous genus *Forsellesia* (*Glossopetalon*). Amer. Midl. Naturalist 27: 501–511. Holmgren, N. H. 1988. *Glossopetalon* (Crossosomataceae) and a new variety of *G. spinescens* from the Great Basin, U.S.A. Brittonia 40: 269–274. Thorne, R. F. and R. Scoggin. 1978. *Forsellesia* Greene (*Glossopetalon* Gray), a third genus in the Crossosomataceae, Rosineae, Rosales. Aliso 9: 171–178.

1. Plants 5–25 cm, forming relatively low mounds or mats.
    2. Leaf blade apices short-acuminate to sharply acute, sometimes mucronate, mucro 0.1–0.4 mm; petals 3–5, 2–4 × 0.4–1 mm; stamens 4–6, in 1 equal series . . . . . . 1. *Glossopetalon clokeyi*
    2. Leaf blade apices acute or short-acuminate, mucronate, mucro 0.6–1.2 mm; petals 5, 6–8 × 1.5–2.2 mm; stamens 10, in 2 unequal series . . . . 2. *Glossopetalon pungens*
1. Plants 25–300 cm, forming relatively tall mounds or upright.
    3. Stem tips mostly strongly spinescent; leaf blade margins usually not, sometimes slightly and evenly, thickened; stipules absent (var. *spinescens*) or triangular to subulate . . . . 3. *Glossopetalon spinescens*
    3. Stem tips not or weakly spinescent; leaf blade margins (and midveins) ± thickened; stipules absent . . . . 4. *Glossopetalon texense*

1. **Glossopetalon clokeyi** (Ensign) H. St. John, Proc. Biol. Soc. Wash. 55: 112. 1942 • Clokey's greasebush C E

Forsellesia clokeyi Ensign, Amer. Midl. Naturalist 27: 504, fig. 2. 1942

Shrubs forming relatively low mounds or mats, (10–) 15–25 cm. **Stems:** branchlets ascending, green to yellowish green during 1st (and sometimes 2d) year, eventually turning yellowish brown, very slender, less than 0.5 mm diam.; bark of older branches pale yellow to gray; tips <sharply pointed, flexible>, not or only weakly spinescent; glabrous or sparsely puberulent. **Leaves:** stipules triangular to narrowly triangular or filiform, <0.2–0.5 mm>, fused basal portion not or only slightly thickened, not darkened or glandular, free tips minute; blade oblanceolate to narrowly elliptic-oblanceolate, 4–6 × 1–1.5 mm, margins not thickened, apex short-acuminate to sharply acute, usually non-mucronate, sometimes mucronate, mucro 0.1–0.4 mm, glabrous. **Inflorescences** appearing axillary, usually flowers solitary. **Flowers:** sepals 3–5, obovate, 1.2–1.5 mm, often slightly unequal, <blunt to sharply acute>, tips not spinescent; petals 3–5 (in different flowers on same plant), oblanceolate, 2–4 × 0.4–1 mm; stamens 4–6, in 1 equal series. **Follicles** 1, broadly ovoid, 3–4 mm, moderately striate, glabrous.

Flowering May–Jul. Crevices and ledges of limestone cliffs; of conservation concern; 2100–3000 m; Nev.

Glossopetalon clokeyi is known from the Spring (Charleston) Mountains in Clark County, where it occurs mostly at relatively inaccessible sites high on cliffs.

2. **Glossopetalon pungens** Brandegee, Bot. Gaz. 27: 445. 1899 • Dwarf greasebush C E

Forsellesia pungens (Brandegee) A. Heller; *F. pungens* var. *glabra* Ensign; *Glossopetalon pungens* var. *glabrum* (Ensign) H. St. John

Shrubs forming relatively low mounds or mats, 5–20 cm. **Stems:** branchlets usually ascending, sometimes divaricate, green to yellowish green during 1st year, soon turning yellowish brown, slender, 0.5–1 mm diam.; bark of older branches dark brown to black; tips not or weakly spinescent; glabrous or sparsely to ± densely puberulent. **Leaves:** stipules absent; <petiole base ± persistent, not or only slightly thickened, not darkened or glandular>; blade oblanceolate to narrowly elliptic, 6–10 × 3–5 mm, margins thickened, apex acute or short-acuminate, mucronate, mucro 0.6–1.2 mm, glabrous or scabridulous. **Inflorescences** terminal, usually flowers solitary. **Flowers:** sepals 5, ovate, 2–3 mm, slightly unequal, tips spinescent; petals 5, narrowly to broadly oblanceolate, 6–8 × 1.5–2.2 mm; stamens 10, in 2 unequal series, <longer opposite sepals>. **Follicles** 1(–3), ovoid to broadly ellipsoid, 2.5–4 mm, moderately striate, usually sparsely puberulent, sometimes glabrous.

Flowering Apr–Jul. Crevices and ledges of limestone cliffs and outcrops; of conservation concern; 1300–2400 m; Calif., Nev.

Populations of *Glossopetalon pungens* in the mountains of San Bernardino County, California, tend to occur at lower elevations than do those in Clark, Lincoln, and Nye counties, Nevada. The California plants also have glabrous leaves (the basis of var. *glabrum*); most of the Nevada plants have densely scabridulous leaves. Both glabrous and hairy plants sometimes have strongly veined leaves that appear somewhat corrugated, at least on herbarium specimens.

3. **Glossopetalon spinescens** A. Gray, Smithsonian Contr. Knowl. 5(6): 29, plate 12, fig. B. 1853 • Spiny greasebush F

Forsellesia spinescens (A. Gray) Greene

Shrubs forming relatively tall mounds or upright, 25–300 cm. **Stems:** branchlets ascending to divaricate, green to grayish green or yellowish green during at least 1st year, soon or tardily turning yellowish brown to orangish brown, slender to relatively stout, 0.5–2 mm diam.; bark of older branches dark gray; tips mostly strongly spinescent (except on new growth); usually glabrous, sometimes puberulent (especially near nodes). **Leaves:** stipules absent (var. *spinescens*) or triangular to subulate, fused basal portion usually thickened, darkened, glandular, free tips 0.2–1 mm; blade usually narrowly to broadly oblanceolate, sometimes narrowly elliptic, 3–17 × 1.2–6 mm, margins usually not, sometimes slightly and evenly, thickened, apex acute, sometimes minutely mucronate, mucro 0.1–0.4 mm, usually glabrous, sometimes sparsely to moderately puberulent. **Inflorescences** axillary, usually flowers solitary, rarely 2- or 3-flowered. **Flowers:** sepals 4 or 5(or 6), narrowly ovate to ovate, 1–3 mm, equal or slightly unequal, tips not spinescent; petals 4 or 5(or 6), narrowly elliptic-oblanceolate to oblanceolate, 4–9 × 0.4–1.5 mm; stamens 5–10, in 1 equal or 2 unequal series, <longer opposite sepals>. **Follicles** 1 or 2, ovoid to broadly ovoid or broadly ellipsoid, 3–5 mm, moderately to strongly striate, glabrous.

Varieties 6 (5 in the flora): w United States, Mexico.

Glossopetalon spinescens is here treated as a widespread complex of intergrading varieties showing some geographic localization. This follows N. H. Holmgren (1988, 1997) and differs from M. R. Ensign (1942) and H. St. John (1942), who had fewer specimens available for study. The sixth infraspecific taxon, var. *mexicanum* (Ensign) H. St. John, is known only from limestone and gypsum outcrops in the mountains of Coahuila and Nuevo León, Mexico. Variety *mexicanum* is a small-leaved variant similar to var. *microphyllum*; it differs from that taxon in lacking stipules. Isolated disjunct occurrences of the species in more southern portions of Mexico (Guanajuato, Oaxaca, Tlaxcala, Veracruz) have not been studied.

1. Stipules absent .... 3a. *Glossopetalon spinescens* var. *spinescens*
1. Stipules present (sometimes not observable when leaves are fasciculate on short shoots).
   2. Largest leaf blades 3–7(–8) mm; branchlets usually turning from grayish green to dull orangish brown toward end of 1st year, sometimes later .... 3d. *Glossopetalon spinescens* var. *microphyllum*
   2. Largest leaf blades 6–17 mm; branchlets usually turning from grayish green to yellowish brown toward end of 2d or 3d year (sometimes earlier in var. *meionandrum*).
      3. Stipule free portion relatively well developed, mostly narrowly triangular to subulate, 0.5–1.7 mm; leaves often still developing at flowering and early deciduous, branches often appearing nearly leafless during much of growing season .... 3b. *Glossopetalon spinescens* var. *aridum*
      3. Stipule free portion relatively poorly developed, usually triangular to narrowly triangular, sometimes subulate, 0.2–0.5 (–0.8) mm; leaves mostly well developed at flowering and tardily deciduous, branches appearing leafy during most of growing season.
         4. Stipule adnate portions yellowish or brownish tinged, often poorly developed, slightly thickened; petals mostly widest near apex, apices rounded or abruptly acute to short-acuminate; stamens 5–7, equal or subequal...... .... 3c. *Glossopetalon spinescens* var. *meionandrum*
         4. Stipule adnate portions dark reddish purple to nearly black, usually well developed, noticeably thickened; petals mostly widest proximal to apex (sometimes nearly oblong), apices rounded to gradually angled or acuminate; stamens usually 8, in 2 unequal series .... 3e. *Glossopetalon spinescens* var. *planitierum*

3a. Glossopetalon spinescens A. Gray var. **spinescens**

Shrubs (25–)50–300 cm. **Branchlets** ascending to divaricate, turning from grayish green to dull yellowish brown toward end of 2d year or later, usually glabrous, sometimes sparsely puberulent (around nodes). **Leaves** usually well developed at flowering and early deciduous, branches often appearing nearly leafless during much of growing season; stipules absent, <persistent petiole bases usually becoming whitened with brown basal band, sometimes turning dark reddish purple to nearly black, ± expanded and thickened, glandular>; largest blade 6–12 × 1.5–3 mm, margins glabrous or sparsely ciliolate proximally, glabrous. **Petals** 5–8 mm, mostly widest proximal to apex <sometimes nearly oblong>, apex acute or acuminate, tips usually sharply pointed. **Stamens** 6–10, in 1 equal or 2 unequal series.

Flowering Apr–Sep. Rocky margins of washes, steep rocky hillsides, crevices and ledges of cliffs and outcrops, mostly on limestone substrates; 1300–2300 m; Ariz., N.Mex., Okla., Tex.; Mexico (Chihuahua).

3b. Glossopetalon spinescens A. Gray var. **aridum** M. E. Jones, Contr. W. Bot. 8: 28. 1898 E F

Forsellesia arida (M. E. Jones) A. Heller; *F. nevadensis* (A. Gray) Greene; *F. stipulifera* (H. St. John) Ensign; *Glossopetalon nevadense* A. Gray; *G. nevadense* var. *stipuliferum* (H. St. John) C. L. Hitchcock; *G. stipuliferum* H. St. John

Shrubs (25–)50–200(–300) cm. **Branchlets** often ± ascending, turning from grayish green to dull yellowish brown toward end of 2d or 3d year, glabrous. **Leaves** often still developing at flowering and early deciduous, branches often appearing nearly leafless during much of growing season; stipules present, sometimes difficult to observe, adnate portion dark reddish purple to nearly black, usually well developed, noticeably thickened, free portion relatively well developed, mostly narrowly triangular to subulate, 0.5–1.7 mm; largest blade 7–12(–17) × (1.5–)2–6 mm, margins ciliolate, glabrous or sparsely to moderately puberulent. **Petals** 4–9 mm, mostly widest proximal to apex <sometimes nearly oblong>, apex rounded to acute or acuminate, tips blunt or sharp. **Stamens** 6–10, in 2 unequal series.

Flowering Mar–Jun. Rocky slopes, crevices and ledges of limestone, igneous cliffs and outcrops; 1100–2500 m; Ariz., Calif., Idaho, Mont., Nev., Oreg., Utah, Wash.

N. H. Holmgren (1988, 1997) noted apparent intergrades of var. *aridum* with var. *meionandrum* in southeastern Utah and with var. *microphyllum* in northern Arizona, southern Nevada, and southwestern Utah.

3c. Glossopetalon spinescens A. Gray var. **meionandrum** (Koehne) Trelease in A. Gray et al., Syn. Fl. N. Amer. 1: 401. 1897 [E]

Glossopetalon meionandrum Koehne, Gartenflora 43: 237, fig. 52. 1894; *Forsellesia meionandra* (Koehne) A. Heller

Shrubs 25–65 cm. **Branchlets** often ± ascending, turning from grayish green to yellowish brown toward end of 2d or 3d year (sometimes earlier), glabrous or sparsely to moderately puberulent (especially near nodes). **Leaves** mostly well developed at flowering and tardily deciduous, branches appearing leafy during most of growing season; stipules present, sometimes difficult to observe, adnate portion yellowish or brownish tinged, often poorly developed, slightly thickened, free portion relatively poorly developed, usually triangular to narrowly triangular, sometimes subulate, 0.2–0.5 (–0.8) mm; largest blade 7–15 × 2–4 mm, margins ciliolate or eciliolate, glabrous. **Petals** 4–6 mm, mostly widest near apex, apex rounded or abruptly acute to short-acuminate, tips blunt or sharp. **Stamens** 5–7, equal or subequal.

Flowering Mar–Jun. Rocky slopes, crevices and ledges of limestone, shale cliffs and outcrops; 1600–2300 m; Ariz., Colo., Utah, Wyo.

3d. Glossopetalon spinescens A. Gray var. **microphyllum** N. H. Holmgren, Brittonia 40: 272, fig. 1i–l. 1988 [E]

Shrubs 25–60 cm. **Branchlets** divaricate to ± ascending, usually turning from grayish green to dull orangish brown toward end of 1st year, (sometimes later), glabrous or ciliolate. **Leaves** usually well developed at flowering, sometimes early deciduous, branches sometimes appearing nearly leafless later in growing season; stipules present, sometimes difficult to observe, adnate portion dark reddish purple to nearly black, usually well developed, noticeably thickened, free portion usually relatively well developed, mostly narrowly triangular to subulate, 0.4–1.2 mm; largest blade 3–7(–8) × 1.2–2.7 mm, margins glabrous or ciliolate, glabrous. **Petals** 4–8 mm, mostly widest proximal to apex <sometimes nearly oblong>, apex rounded to gradually acute or acuminate, tips blunt or sharply pointed. **Stamens** 6–10 in 2 unequal series.

Flowering Apr–Jun. Rocky margins of washes, steep rocky hillsides, crevices and ledges of cliffs and outcrops, mostly on limestone substrates; 1300–2400 m; Ariz., Nev., Utah.

A single collection from San Bernardino County, California (*Davidson, Gustafson & Reynolds 7853*, MO, RSA), compares well with specimens from the main range of var. *microphyllum* in its small leaves, divaricate branching, and orangish brown older branchlets, but appears to differ in being a somewhat taller plant with the twigs changing color more tardily and the leaves moderately puberulent (versus glabrous) abaxially. Further studies may confirm a disjunct range extension for this variety in eastern California.

3e. Glossopetalon spinescens A. Gray var. **planitierum** (Ensign) Yatskievych, Novon 17: 530. 2007 • Plains greasebush [E]

Forsellesia planitierum Ensign, Amer. Midl. Naturalist 27: 509, fig. 8. 1942; *Glossopetalon planitierum* (Ensign) H. St. John

Shrubs 40–120 cm. **Branchlets** usually ascending, usually turning from grayish green to yellowish brown toward end of 2d or 3d year, glabrous or nearly so. **Leaves** mostly well developed at flowering and tardily deciduous, branches appearing leafy during most of growing season; stipules present, sometimes difficult to observe, adnate portion dark reddish purple to nearly black, usually well developed, noticeably thickened, free portion relatively poorly developed, triangular to narrowly triangular or subulate, 0.2–0.4 mm; largest blade 6–14 × 2–4 mm, margins ciliolate, moderately puberulent, sometimes glabrescent. **Petals** 4–6 mm, mostly widest proximal to apex <sometimes nearly oblong>, apex rounded to gradually angled or acuminate, tips blunt or sharply pointed. **Stamens** usually 8 in 2 unequal series.

Flowering Mar–May. Upland prairie slopes, crevices, ledges, cliff tops, drainages, outcrops, on limestone, gypsum, and sandstone substrates; 800–1600 m; Colo., N.Mex., Okla., Tex.

The range of var. *planitierum* is centered on the Texas Panhandle; it does not overlap with those of other members of the *Glossopetalon spinescens* complex. M. R. Ensign (1942) noted an unusual collection from near the Louisiana border in Marion County, Texas (*Harris 470*, US). This disjunct occurrence, upon which Ensign equivocated, would extend the distribution of the taxon significantly eastward and significantly lower its elevational limit (and that of the entire genus) to about 50 m. Plants have not been rediscovered in eastern Texas since 1930.

4. Glossopetalon texense (Ensign) H. St. John, Proc. Biol. Soc. Wash. 55: 112. 1942 (as texensis) • Texas greasebush [C] [E]

Forsellesia texensis Ensign, Amer. Midl. Naturalist 27: 510, fig. 6. 1942

Shrubs forming relatively tall mounds or upright, 25–200 cm. **Stems:** branchlets mostly loosely ascending, green to yellowish green or grayish green through 2d year, slender to relatively stout, 0.5–1.2 mm diam.; bark of older branches dark gray, sometimes with black patches; tips not or weakly spinescent; glabrous. **Leaves:** stipules absent; <petiole base persistent, usually becoming whitened or light brown after leaf has been shed, occasionally turning dark reddish purple to nearly black, ± expanded and thickened, sometimes glandular, sometimes ciliolate>; blade oblanceolate to broadly oblanceolate, 6–20 × 3–5 mm, margins (and midvein) ± thickened, apex rounded to acutely pointed, sometimes mucronulate, mucro 0.1–0.4 mm, glabrous. **Inflorescences** axillary, usually of solitary flowers, rarely 2- or 3-flowered. **Flowers:** sepals (4 or)5, ovate, 1.5–2.5 mm, equal, tips not spinescent; petals (4 or)5, narrowly elliptic-oblanceolate, sometimes oblanceolate, 5–7 × 0.4–0.9(–1.5) mm; stamens 7–9, in 1 equal series. **Follicles** 1, broadly ovoid to broadly obovoid, 4–5 mm, moderately to strongly striate, glabrous.

Flowering Apr–Jun(–Oct). Crevices and ledges of limestone and chalk cliffs, narrow ridge tops, and adjacent steep slopes; of conservation concern; 300–500 m; Tex.

Glossopetalon texense is known only from along stretches of the Devil's (Val Verde County) and Nueces (Uvalde County) rivers in Texas. Although M. R. Ensign (1942) described the petals as lanceolate and 1.5 mm wide, specimens examined during the present study had flowers with more slender petals; it is difficult to separate from *G. spinescens*.

ROSACEAE Jussieu

• Rose Family

James B. Phipps

Herbs (annual or perennial), shrubs, or trees. Stems simple or branched. **Leaves** persistent or deciduous, basal and/or cauline, usually alternate, rarely opposite, simple or compound (palmate or imparipinnate); stipules usually present, sometimes absent; petiole present or absent; blade thin to coriaceous, margins ± lobed or unlobed, usually toothed. **Inflorescences** terminal, sometimes axillary, panicles with terminal flower (that is, determinate) or reductions of this: 1-flowered, glomerules, fascicles, spikes, racemes, corymbs, umbels, or cymes. **Flowers** usually bisexual, rarely unisexual, perianth and androecium perigynous or epigynous; epicalyx bractlet sometimes present; hypanthium flat to hemispheric, or cylindric to funnelform or urceolate; sepals (0–)4 or 5(–10), distinct, free; petals (0–)4 or 5(–12, rarely more in double ornamentals), distinct, free; nectar disc sometimes absent; stamens 0–130(–220), distinct, free, anthers usually longitudinally dehiscent; torus well developed, inconspicuous, or absent; pistils 1–250 (–450), distinct or ± connate, free or ± adnate to hypanthium, ovary superior or inferior (then 2–5-carpellate and -locular and ± connate with axile placentation), styles terminal, subterminal, lateral, or ± basal, sometimes basally connate, stigmas usually capitate; ovules 1 or 2(–5+), <basal, marginal, or apical, collateral, superposed, biseriate, or clustered, integuments 2, crassinucellate, with or without obturator>. **Fruits** achenes aggregated or not, follicles aggregated or not, drupes aggregated or not, aggregated nutlets, pomes, aggregated drupelets, or capsules; <sometimes involving accessory organs, for example, hypanthium, torus>. **Seeds** 1 or 2(–12+), not arillate.

Genera 88, species ca. 3000 (68 genera, 680 species, including 22 hybrids, in the flora): North America, Mexico, West Indies, Bermuda, Central America, South America, Eurasia, Africa, Atlantic Islands, Indian Ocean Islands, Pacific Islands (Hawaii, New Zealand), Australia.

Three subfamilies and 16 tribes are recognized for the family with representatives of all tribes found in the flora area.

Rosaceae grow most commonly in north-temperate regions and are more or less absent from hot deserts and high-rainfall, low-altitude tropics. The family is large and diverse, characterized by radially symmetric flowers with a fundamentally saucer-shaped hypanthium and peripheral calyx, corolla, androecium, and, usually, superior gynoecium. Considerable variation occurs in important details of flower and fruit, this enhanced by a very adaptable hypanthium, which is discussed in more detail below.

Rosaceous inflorescences vary in the number of flowers from one to about 500. A great variation occurs also in inflorescence form, though a pattern is perceived when they are viewed as reduction series from a terminal, determinate panicle that is fundamentally bracteate, thus generating a range of panicles, racemes, corymbs, cymes, solitary flowers, or other forms. In this treatment, appendages on the inflorescence are distinguished by the terms bract and bracteole. Bract is used for the larger, laminate, usually chlorophyllous leaf homologues that subtend axes and may be indistinguishable from foliage leaves. Bracteoles are scalelike, often membranous, often caducous, and of uncertain homology, in part due to not being restricted to axis-subtending positions.

The floral architecture in Rosaceae is radially symmetric around a disc-shaped to urceolate hypanthium. Flowers normally have a four- or five-merous corolla and calyx; great variation is found in the numbers of stamens and carpels. Pollination is usually entomophilous, the flowers having normally green sepals and showy, often more or less clawed, usually white, yellow, or pink, less commonly red or green, petals. The flowers in some genera are relatively small and anemophilous and may lack one or two of the principal whorls. An unusual feature of some rosaceous flowers is the torus, a pad of receptacular, usually spongy, tissue, sometimes relatively large, in the center of the hypanthium that, when present, bears the gynoecium. Another unusual feature of some genera is an epicalyx that comprises a ring of sepaloid bractlets, usually of the same number as sepals, which is located on the hypanthium proximal to the calyx.

Fruit types are particularly significant both in rosaceous identification, taxonomy, and diversity, as well as for successful dispersal. Fruit in Rosaceae may be either dry, then dehiscent or not, or succulent and indehiscent; it sometimes involves accessory organs such as the torus or a persistent hypanthium.

Dry indehiscent fruits are achenaceous, in certain cases involving modifications to the style. In anemochorous (adapted for dispersal by wind) situations, this may involve a plumose style (for example, *Cercocarpus*); in epizoochorous (distributed on the outside of animals) cases, the styles may bear stiff hairs or barbs (for example, *Geum*). Alternatively, achenaceous fruits may lack styles (sometimes due to abscission) or, more commonly, have a relatively short one (for example, *Potentilla*). In situations where there appears to be no significant post-flowering function for the style, dispersal may be myrmecochorous (by ants), anemochorous, or in the rare torus-borne cases (for example, *Fragaria*), the fruit may be endozoochorous (eaten by animals and passed through the gut). Sometimes, fruits with such styles are aggregated in acheneta that rely on barbed, persistent hypanthia for epizoochorous dispersion (for example, *Acaena* and *Agrimonia*). Dry dehiscent fruits are follicular with seeds normally distributed by air after splitting of the ripe follicle (for example, *Gillenia*).

Succulent, endozoochorous fruits exhibit a similar range of variation. The most common types are: multiple drupelets on the surface of a more or less conic torus (for example, *Rubus*); individual and sometimes large drupes on a flat receptacular apex (*Prunus*); pomes, which are berrylike fruits in which a fleshy hypanthium more or less completely surrounds and is generally more or less fully adnate to the carpels (for example, *Amelanchier* and *Crataegus*). The result is that a terminal orifice may remain open, often bearing floral remnants around its rim. Pomes with hard pyrenes are sometimes distinguished as polypyrenous drupes, a term not used in this treatment due to antonymic confusion caused by combining the roots for pyrene (hardened mesocarp) and drupe (fleshy mesocarp), although the relevant distinction is well recognized (J. R. Rohrer et al. 1991).

Some important temperate fruits are members of Rosaceae: apples (*Malus*), pears (*Pyrus*), almonds, apricots, cherries, peaches, and plums (*Prunus*), blackberries and raspberries (*Rubus*), strawberries (*Fragaria*), loquat (*Eriobotrya*); minor fruits include those of *Amelanchier*, *Crataegus*,

Cydonia, *Mespilus*, and others. Some genera are popular in ornamental horticulture in North America, for example, most of the above as well as, especially, *Chaenomeles*, *Cotoneaster*, *Pyracantha*, *Rhaphiolepis* Lindley, *Sorbus*, *Photinia*, *Physocarpus*, *Rosa*, and *Spiraea* among trees and shrubs; and *Filipendula*, *Geum*, *Potentilla*, and *Spiraea* among plants suitable for flower borders.

Apomixis is a feature of some rosaceous genera and may make taxonomic decisions difficult or equivocal in genera such as *Alchemilla*, *Amelanchier*, *Crataegus*, *Rubus*, and others. Apomixis always seems to be associated with polyploidy and often with hybridization; in some apomictic genera, sexual species and apomicts are facultatively sexual.

Rosaceae lack alkaloids and blue anthocyanic pigments. Some have foliage or seed that becomes toxic due to hydrolysis of benzaldehyde cyanohydrins.

The great diversity of Rosaceae reflects the age of the family (since at least the mid Eocene) and its evolutionary success. The family has needed no fundamental change in circumscription since Jussieu first recognized it, the sometimes included outgroups (for example, Chrysobalanaceae and Neuradaceae), already being confidently excluded in more recent pre-molecular classifications, while segregate families (for example, Malaceae) have received little currency. This reflects a lack both of close neighbors and of large internal discontinuities. Species limits are often debatable, especially where apomixis is present; generic limits are much more fixed, apart from the familiar historic trend to recognizing smaller, more discrete units. What had principally been fluid in rosaceous taxonomy until modern molecular research were the phylogenetic relationships, both to neighboring families and within the family. Rosales once contained families such as Crassulaceae and Saxifragaceae, which have many morphological similarities to Rosaceae, but Angiosperm Phylogeny Group (2003) indicated instead that in Rosales, Rosaceae is sister to a group of families including Moraceae, Rhamnaceae, Ulmaceae, and Urticaceae.

The internal relationships of Rosaceae have received much study, including morphological, anatomical, cytological, phytochemical, breeding system, fungal pathogenicity (especially rusts), and molecular. Among these studies, that of D. Potter et al. (2007) resulted in the first comprehensive molecular phylogeny of Rosaceae, and the treatment in this volume reflects it. Potter et al. recognized three subfamilies with 15 tribes worldwide. Their classification used the ranks supertribe and subtribe, not used here. Subfamily Dryadoideae, with one tribe and four genera, diverged early and is unique in Rosaceae for its actinorhizal symbiosis; it has achenaceous fruits. Subfamily Rosoideae has six tribes and 29 genera and is only slightly altered from its traditional circumscription by removing Dryadoideae and three small genera. Most Rosoideae have achenaceous fruit; some have multiple drupelets or achenes borne on a torus. The remainders of Rosaceae are found in the large and heterogeneous subfam. Amygdaloideae of nine tribes and 54 genera. This subfamily has the largest diversity of fruit types. Amygdaloideae now contains the traditional spiraeoid genera with follicular fruit, Maleae with pomaceous fruit, and Amygdaleae with drupaceous fruit.

SELECTED REFERENCES Kalkman, C. 1988. The phylogeny of the Rosaceae. Bot. J. Linn. Soc. 98: 37–59. Kalkman, C. 2004. Rosaceae. In: K. Kubitzki et al., eds. 1990+. The Families and Genera of Vascular Plants. 10+ vols. Berlin etc. Vol. 6, pp. 343–386. Morgan, D. R., D. E. Soltis, and K. R. Robertson. 1994. Systematic and evolutionary implications of *rbc*L sequence variation in Rosaceae. Amer. J. Bot. 81: 890–903. Potter, D. et al. 2002. Phylogenetic relationships in Rosaceae inferred from chloroplast *mat*K and *trn*L-*trn*F nucleotide sequence data. Pl. Syst. Evol. 231: 77–89. Potter, D. et al. 2007. Phylogeny and classification of Rosaceae. Pl. Syst. Evol. 266: 5–43. Robertson, K. R. 1974. The genera of Rosaceae in the southeastern United States. J. Arnold Arbor. 55: 303–332, 344–401, 611–662. Rydberg, P. A. 1908–1918. Rosaceae. In: N. L. Britton et al., eds. 1905+. North American Flora.... 47+ vols. New York. Vol. 22, pp. 239–533.

Key to Subfamilies and Tribes of Rosaceae

Luc Brouillet

1. Leaves simple, sometimes pinnately, rarely palmately compound, stipules deciduous or absent, rarely persistent, if pinnately compound and stipules persistent, then fruits follicles (rarely achenes); flowers: perianth and androecium perigynous or epigynous, tori absent or minute; ovules 2+ (rarely 1, sometimes by abortion); fruits usually aggregated follicles or drupes or nutlets, pomes, ± woody capsules, follicles, or drupes, rarely achenes [c. Rosaceae subfam. Amygdaloideae, p. 345].
   2. Leaves opposite; carpels 2(or 3), distinct, free; fruits aggregated follicles .... c1. Lyonothamneae, p. 345
   2. Leaves alternate, rarely opposite; carpels 1–8, distinct or connate, free or ± adnate to hypanthium; fruits achenes, drupes, aggregated achenes, drupes, nutlets, or follicles, or capsules or pomes.
      3. Shrubs or herbs; stipules present; leaves pinnately compound, rarely simple.
         4. Shrubs; carpels 1–5, basally connate or distinct .... c6. Sorbarieae, p. 392
         4. Herbs; carpels 5, basally connate .... c8. Gillenieae, p. 424
      3. Trees or shrubs, rarely herbs (then stipules absent); stipules present or absent; leaves simple, rarely pinnately compound (then herbs or trees).
         5. Carpels 1; fruits drupes .... c3. Amygdaleae, p. 352
         5. Carpels 1–5; fruits achenes, aggregated drupes, aggregated nutlets, aggregated follicles, capsules, or pomes.
            6. Flowers: perianth and androecium epigynous (perigynous in *Vauquelinia*); carpels ± connate or distinct, adnate to hypanthium (free in *Vauquelinia*); fruits pomes or woody capsules (surrounded by persistent hypanthia and splitting into 5 follicles in *Vauquelinia*) .... c9. Maleae, p. 427
            6. Flowers: perianth and androecium perigynous; carpels distinct, sometimes ± connate, free, rarely adnate to hypanthium base; fruits achenes, aggregated follicles, aggregated nutlets, aggregated drupes, or capsules splitting into 5 follicles (not surrounded by hypanthia).
               7. Shrubs; leaf venation palmate (stipules present); carpels basally connate .... c2. Neillieae, p. 347
               7. Shrubs, sometimes subshrubs, trees, or herbs; leaf venation pinnate (sometimes palmate in Spiraeeae, then stipules absent); carpels distinct (sometimes connate in Exochordeae).
                  8. Fruits capsules or aggregated drupes.................. c4. Exochordeae, p. 383
                  8. Fruits aggregated follicles or nutlets, or achenes.
                     9. Shrubs; stipules persistent or deciduous; leaves alternate or opposite; carpels 2–8 or 1 (in *Coleogyne*); ovules 1 or 2, marginal; fruits achenes or aggregated nutlets .... c5. Kerrieae, p. 386
                     9. Shrubs, subshrubs, or herbs; stipules absent; leaves alternate; carpels 3–5(or 6); ovules 2–5, apical; fruits aggregated follicles or achenes (in *Holodiscus*) .... c7. Spiraeeae, p. 398

1. Leaves pinnately, sometimes palmately, compound, sometimes simple, stipules persistent and adnate to petioles, sometimes deciduous and free, rarely absent; flowers: perianth and androecium perigynous, tori enlarged, sometimes small or absent; ovules 1(or 2, then fruits achenes or drupes); fruits achenes, or aggregated drupelets or achenes (within fleshy, urn-shaped hypanthia [hips] in Roseae).
   10. Shrubs or trees, unarmed; leaves simple, sometimes pinnately compound; styles persistent, elongate in fruit [b. Rosaceae subfam. Dryadoideae] .... b1. Dryadeae, p. 326
   10. Herbs, shrubs, or subshrubs, these sometimes armed; leaves pinnately or palmately compound, rarely simple; styles deciduous, if persistent in fruit, then either not elongate, or rarely elongate (in Colurieae) [a. Rosaceae subfam. Rosoideae, p. 23].

[11. Shifted to left margin.—Ed.]

11. Shrubs or subshrubs.
 12. Hypanthia ovoid, oblong, globose to cupulate or urceolate, rarely obovoid or hemispheric, tori absent (conic if present); fruits aggregated achenes within hypanthium (hips) . . . . a4. Roseae, p. 74
 12. Hypanthia saucer-, cup-, or funnel-shaped, hemispheric to flat, or obconic to obcampanulate, tori enlarged; fruits achenes, or aggregated achenes or drupelets.
  13. Plants armed, sometimes unarmed; stipules free or adnate to petiole base; fruits aggregated drupelets . . . . a2. Rubeae (in part), p. 27
  13. Plants unarmed; stipules absent or ± free or ± adnate to petiole; fruits aggregated achenes, sometimes achenes.
   14. Epicalyx bractlets absent, sometimes present; styles terminal, sometimes subterminal, elongate in fruit . . . . a3. Colurieae (in part), p. 57
   14. Epicalyx bractlets present, sometimes absent; styles basal or lateral to subterminal, not elongate in fruit . . . . a5. Potentilleae (in part), p. 119

11. Herbs.
 15. Leaves pinnately compound, sometimes simple with palmate venation; stipules free; fruits aggregated drupelets; ovules 2, collateral . . . . a2. Rubeae (in part), p. 27
 15. Leaves pinnately or palmately compound, rarely simple with pinnate or palmate venation; stipules adnate to petiole, sometimes free; fruits achenes or aggregated achenes; ovules 1, if 2 then superposed.
  16. Tori inconspicuous or absent, sometimes enlarged in fruit; carpels 1–12.
   17. Ovules 2; fruits aggregated achenes, not enveloped by hypanthia; tori enlarged in fruit . . . . a1. Ulmarieae, p. 23
   17. Ovules 1; fruits achenes, enclosed within enlarged, often hardened, hypanthia; tori not enlarged in fruit . . . . a6. Agrimonieae, p. 312
  16. Tori usually enlarged; carpels 1–260(–450).
   18. Epicalyx bractlets absent, sometimes present; styles terminal, sometimes subterminal, elongate in fruit . . . . a3. Colurieae (in part), p. 57
   18. Epicalyx bractlets present, sometimes absent; styles basal or lateral to subterminal, not elongate in fruit . . . . a5. Potentilleae (in part), p. 119

a. Rosaceae Jussieu subfam. Rosoideae Arnott, Botany, 107. 1832 (as Roseae)

Luc Brouillet

Herbs, shrubs, or subshrubs. Leaves alternate, rarely opposite, pinnately compound, sometimes simple or palmately compound; stipules present, rarely absent. **Flowers:** torus usually enlarged, sometimes small or absent; carpels 1–260(–450), distinct, free, styles distinct, rarely connate (Roseae); ovules 1(or 2), collateral (Rubeae) or superposed (*Fallugia*, *Filipendula*). **Fruits** achenes or aggregated achenes <sometimes with fleshy, urn-shaped hypanthium or enlarged torus>, sometimes aggregated drupelets; styles persistent or deciduous, not elongate (elongate <but not plumose> in *Geum*). $x = 7(8)$.

Genera 28–35, species ca. 1600 (26 genera, 302 species, including 1 hybrid, in the flora): North America, Mexico, West Indies, Bermuda, Central America, South America, Eurasia, Africa, Atlantic Islands, Indian Ocean Islands, Pacific Islands, Australia.

Representative species in all six tribes recognized in subfam. Rosoideae occur in the flora area.

Variation in the number of genera in subfam. Rosoideae is due to differences in generic delimitation between D. Potter et al. (2007) and the authors of some Potentilleae genera. Cyanogenic glycosides and sorbitol are absent in the subfamily.

SELECTED REFERENCES Eriksson, T. et al. 1998. Phylogenetic analysis of *Potentilla* using DNA sequences of nuclear ribosomal internal transcribed spacers (ITS), and implications for the classification of Rosoideae (Rosaceae). Pl. Syst. Evol. 211: 155–179. Eriksson, T. et al. 2003. The phylogeny of Rosoideae (Rosaceae) based on sequences of the internal transcribed spacers (ITS) of nuclear ribosomal DNA and the *trn*L/F region of chloroplast DNA. Int. J. Pl. Sci. 164: 197–211.

a1. Rosaceae Jussieu (subfam. Rosoideae) tribe Ulmarieae Lamarck & de Candolle, Syn. Pl. Fl. Gall., 338. 1806 (as Ulmariae)

Luc Brouillet

Herbs, perennial; unarmed. **Leaves** alternate, pinnately compound; stipules persistent, free or adnate to petiole; venation pinnate to nearly palmate basally. **Flowers:** perianth and androecium perigynous; epicalyx bractlets absent; hypanthium flat to concave; torus inconspicuous, sometimes enlarged in fruit; carpels 5–12, distinct, free, styles terminal to subterminal, distinct; ovules 2, apical, superposed. **Fruits** aggregated achenes; styles persistent, not elongate.

Genus 1, species 15 (4 in the flora): North America, Eurasia, n Africa; introduced in Pacific Islands (New Zealand), Australia.

The base chromosome number for Ulmarieae is $x = 7$.

1. FILIPENDULA Miller, Gard. Dict. Abr. ed. 4, vol. 1. 1754 • Meadowsweet, dropwort, filipendule [Latin *filum*, thread, and *pendulus*, hanging, alluding to root tubers of *Filipendula vulgaris* hanging together with threads]

Ivan A. Schanzer

Thecanisia Rafinesque; *Ulmaria* Miller

Herbs, perennial, 5–25 dm; rhizomatous. **Stems** 1–10, erect, simple, leafy, glabrous or hairy. **Leaves** deciduous by or persistent to flowering, basal and cauline; stipules ovate to elliptic or round, <herbaceous, base auriculate or not>, margins dentate; petiole present; blade pinnate, 10–40 cm, herbaceous, lateral leaflets 1–17 pairs (with smaller, interspersed leaflets between them in *F. ulmaria*), sometimes absent (in *F. occidentalis*), lanceolate or ovate to elliptic, unlobed or palmately lobed, terminal leaflets palmately 3–9-lobed, margins flat, dentate, serrate, or doubly serrate, abaxial surface glabrous or sparsely short-hairy to tomentose, adaxial glabrous or sparsely short-hairy. **Inflorescences** terminal, usually 100+-flowered, dense to lax, corymbiform panicles <with nearly erect proximal branches overtopping central axis (anthela)>, glabrous or hairy; bracts absent; bracteoles absent. **Pedicels** present. **Flowers** 8–15 mm diam.; hypanthium 1 mm, glabrous; sepals (4 or)5 or 6(–9), reflexed, spatulate to triangular; petals (4 or)5 or 6(–9), white to cream or pink to purple, oblanceolate or obovate to orbiculate; stamens 20–30, shorter to longer than petals; carpels <stipitate or sessile>, glabrous or ciliate to hirsute, styles <persistent>. **Fruits** aggregated achenes, 3–18, ± flattened, lanceolate to oblanceolate, 3–14 mm, margins ciliate or glabrous, faces hairy or glabrous; hypanthium persistent, in some species becoming convex in fruit; sepals persistent, reflexed. $x = 7$.

Species 15 (4 in the flora): North America, Eurasia, n Africa; introduced in Pacific Islands (New Zealand), Australia.

According to the results of molecular phylogenetic researches (D. R. Morgan et al. 1994; T. Eriksson et al. 2003), *Filipendula* is phylogenetically the basal genus of subfam. Rosoideae. This readily explains its previously controversial and long-disputed position in Rosaceae; for a long time *Filipendula* was regarded as part of *Spiraea* (C. J. Maximowicz 1879; D. B. O. Savile 1968). No taxon closely related to *Filipendula* is known.

The inflorescence shape is one of the most characteristic features of the genus; it is called an anthela and represents a corymbiform panicle with the proximal branches nearly upright and long overriding the central axis of the inflorescence (A. A. Fedorov and Z. Artjuschenko 1979).

A number of species of *Filipendula* are used as ornamentals. They have attractive, dense inflorescences of pink to purplish or white flowers and rather decorative, deeply dissected foliage. Among ornamental species, the Japanese *F.* ×*purpurea* Maximowicz, the North American *F. rubra*, and the Eurasian *F. ulmaria* are noteworthy; these are plants of wet habitats and are mostly used in gardens to fringe ponds. In *F. ulmaria*, cultivars with double flowers are known. Both *F. ulmaria* and *F. vulgaris* were introduced to North America by early colonists, probably as ornamentals.

Filipendula ulmaria is of interest as a medicinal plant; its flowers contain tannins, salicylates, and heparin-like compounds, and were tested for anti-inflammatory and antiulcer activities (O. D. Barnaulov and P. P. Denisenko 1980; B. A. Kudriashov et al. 1990, 1991). *Filipendula rubra* has been reported to have similar properties (S. Foster and J. A. Duke 1990).

SELECTED REFERENCES Savile, D. B. O. 1968. Parasite relationships and disposition of *Filipendula*. Brittonia 20: 230–231. Schanzer, I. A. 1994. Taxonomic revision of the genus *Filipendula* Mill. (Rosaceae). J. Jap. Bot. 69: 290–319.

1. Root tubers present; basal leaves 3–6(or 7), persistent to flowering; lateral leaflets in 7–17 pairs; petals (5 or)6 or 7(–9); achenes densely appressed-hairy throughout . . . . 4. *Filipendula vulgaris*
1. Root tubers absent; basal leaves 1 or 2, deciduous by flowering; lateral leaflets in 1–5 pairs, sometimes absent in *F. occidentalis*; petals (4 or)5(or 6); achenes glabrous or sutures ciliate.
    2. Petals pink to purple, margins unevenly toothed; plants strongly rhizomatous, forming irregular patches; lateral leaflets 2- or 3-lobed; terminal leaflets 7–9-lobed . . . . 2. *Filipendula rubra*
    2. Petals white to cream, margins entire; plants cespitose or weakly rhizomatous; lateral leaflets not lobed; terminal leaflets 3–7-lobed.

[3. Shifted to left margin.—Ed.]

3. Petals not clawed; lateral leaflets many times smaller than terminal, in 1–3 pairs or absent; terminal leaflets 5–7-lobed; plants weakly rhizomatous; hypanthia nearly flat; achenes straight, stipitate, sutures densely ciliate . . . . 1. *Filipendula occidentalis*
3. Petals clawed; lateral leaflets comparable in size with terminal, in 2–4 pairs; terminal leaflets 3–5-lobed; plants cespitose; hypanthia concave, becoming slightly convex in fruit; achenes twisted, sessile, glabrous . . . . 3. *Filipendula ulmaria*

1. Filipendula occidentalis (S. Watson) Howell, Fl. N.W. Amer., 185. 1898 • Queen of the forest C E

Spiraea occidentalis S. Watson, Proc. Amer. Acad. Arts 18: 192. 1883; *S. camtschatica* Pallas var. *occidentalis* (S. Watson) Wenzig

Plants weakly rhizomatous, 7–15 dm. **Rhizomes** horizontal, stout, 10–15 mm wide, internodes 1–4 cm; root tubers absent. **Stems** finely puberulent distally. **Leaves:** basal 1 or 2, deciduous by flowering; stipules round, 1–1.5 cm diam., base not auriculate; lateral leaflets in 1–3 pairs or absent, remote, ovate to elliptic, to 1 cm, margins serrate; terminal leaflets round, 6–16 cm diam., palmately 5–7-lobed, lobes ovate to oblong-lanceolate, margins doubly serrate, apex acute to acuminate, surfaces short-appressed hairy at least on veins. **Inflorescences** 100+-flowered; branches and pedicels densely short-appressed hairy, hairs straight. **Flowers:** hypanthium nearly flat, saucer-shaped; sepals (4 or)5(or 6), green, narrowly triangular, 4–6 mm, margins <often serrate>, usually with midrib, abaxially puberulent, adaxially glabrous; petals (4 or)5(or 6), white, in buds sometimes marginally pink, oblanceolate, 10–15 mm, not clawed, base broad, margins entire; stamens white, about equal to petals. **Achenes** 7–12, flattened, lanceolate, straight, 6–7.5 mm, sutures densely ciliate, faces sparsely hairy; stipes 1–2 mm; styles 1.5–2 mm.

Flowering summer (Jun–Aug). Wet, mossy rock along forest streams, at or slightly above water level, wet rock on mountain slopes, riverbanks, rocky summits; of conservation concern; 30–1000 m; Oreg., Wash.

Filipendula occidentalis is known only from the valleys of several small rivers in the coast ranges of northwestern Oregon and southwestern Washington. The species is threatened by logging and associated forest-management activities. The species is most probably related to *F. camtschatica* (Pallas) Maximowicz from the Pacific coast of Asia.

2. Filipendula rubra (Hill) B. L. Robinson, Rhodora 8: 204. 1906 • Queen of the prairie, filipendule rouge E F

Ulmaria rubra Hill, Hort. Kew., 214, plate 7. 1768; *Filipendula lobata* (Gronovius ex Jacquin) Maximowicz; *Spiraea rubra* (Hill) Britton; *Thecanisia angustifolia* Rafinesque; *T. lobata* (Gronovius ex Jacquin) Rafinesque; *T. purpurea* Rafinesque

Plants strongly rhizomatous, forming irregular patches, 12–25 dm. **Rhizomes** horizontal, thin, 2.5–3 mm wide, internodes 3–7 cm; root tubers absent. **Stems** glabrous. **Leaves:** basal 1 or 2, deciduous by flowering; stipules ovate to elliptic, 1–1.5 cm, base auriculate; lateral leaflets in 1–5 pairs, remote, ovate to elliptic, to 9 cm, <palmately 2- or 3-lobed, lobes lanceolate>, margins serrate or doubly serrate; terminal leaflets round, 10–15 cm diam., palmately 7–9-lobed, lobes oblanceolate to lanceolate, margins doubly serrate, apex acute to acuminate, surfaces hairy at least on veins, <hairs appressed, straight, short, 0.5 mm>, or adaxial glabrous. **Inflorescences** 100+-flowered; branches and pedicels glabrous. **Flowers:** hypanthium slightly concave; sepals (4 or)5 (or 6), purplish, spatulate to triangular, 1–3 mm, margins without midrib, abaxially glabrous, adaxially finely appressed-hairy or glabrous; petals (4 or)5(or 6), pink to purple, orbiculate, 2.5–7 mm, claws distinct, short, base narrow, margins unevenly toothed; stamens pink to purplish, about equal to or longer than petals. **Achenes** 3–7, flattened, oblanceolate, straight, 8–14 mm, glabrous; stipes 0.5–1.5 mm; styles 1–2 mm.

Flowering summer (Jun–Jul). Moist meadows and bogs, roadsides, ditches, often persistent in abandoned gardens; 0–1000 m; N.B., N.S., Ont., Que.; Conn., Ill., Ind., Iowa, Maine, Md., Mass., Mich., Miss., Mo., N.H., N.J., N.Y., N.C., Ohio, Pa., Vt., Va., W.Va., Wis.

Filipendula rubra is widely planted as an ornamental and is known only in cultivation or as an escape in the northern part of the listed range (eastern Canada, Maine, New York) as well as in West Virginia. The species is related to the east Asian *F. angustiloba* (Turczaninow) Maximowicz and *F. palmata* (Pallas) Maximowicz; it was used by Native Americans for heart troubles and love potions (D. E. Moerman 1998). The plant probably

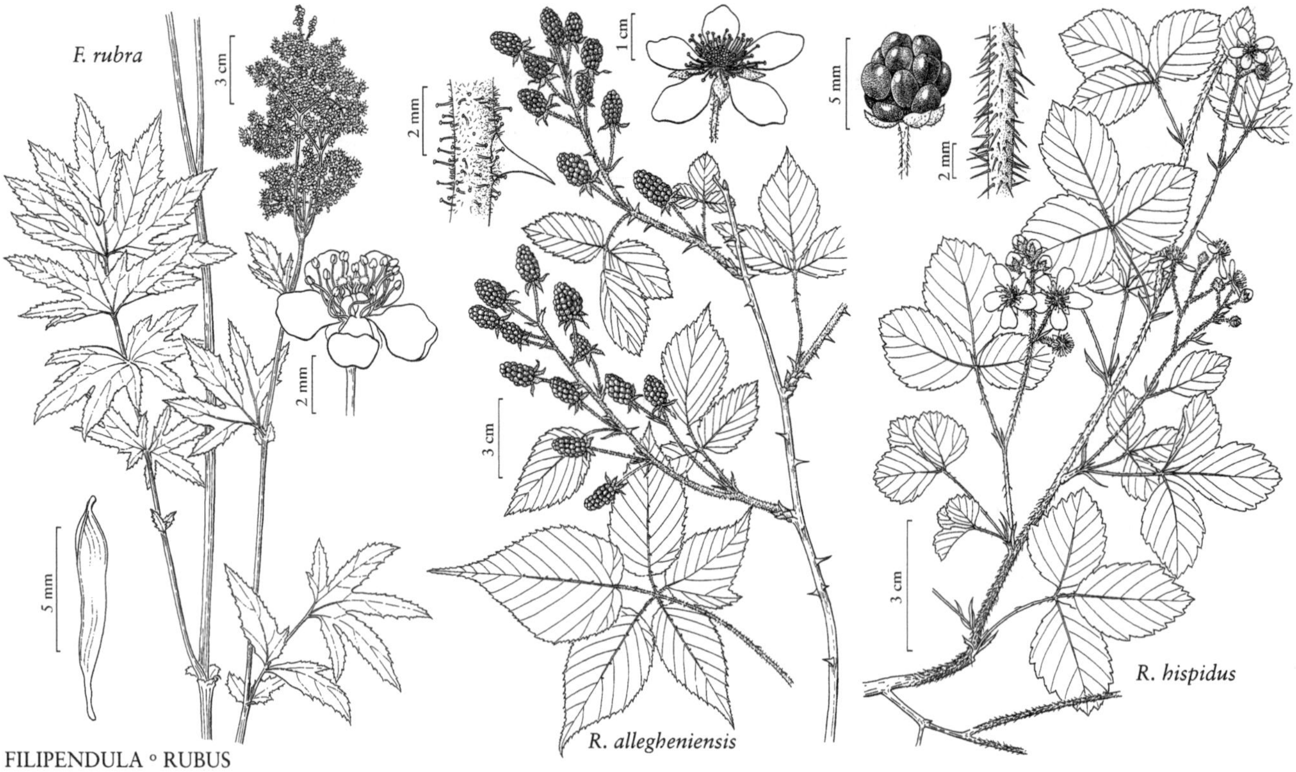

FILIPENDULA ◦ RUBUS

contains salicylic acid (natural precursor to aspirin), which has anti-inflammatory and analgesic effects. It has been used to treat gout, influenza, rheumatism, arthritis, fever, and kidney and bladder problems. The root is rich in tannins; it is used as an astringent in the treatment of, for example, diarrhea, dysentery, and bleeding (S. Foster and J. A. Duke 1990).

3. Filipendula ulmaria (Linnaeus) Maximowicz, Trudy Imp. S.-Peterburgsk. Bot. Sada 6: 251. 1879 • Reine-des-prés [I]

Spiraea ulmaria Linnaeus, Sp. Pl. 1: 490. 1753; *Filipendula denudata* (J. Presl & C. Presl) Fritsch; *F. ulmaria* var. *denudata* (J. Presl & C. Presl) Maximowicz; *S. denudata* J. Presl & C. Presl; *Thecanisia ulmaria* (Linnaeus) Rafinesque

Plants cespitose, 12–17 dm. **Rhizomes** horizontal, stout, 5–10 mm wide, internodes 1–2 cm; root tubers absent. **Stems** sparsely curly-puberulous distally or glabrous. **Leaves:** basal 1 or 2, deciduous by flowering; stipules ovate, 0.6–1 cm, base auriculate; lateral leaflets in 2–4 pairs, remote, with smaller intermediate leaflets between them, ovate to elliptic, 2–6 cm, margins doubly serrate; terminal leaflets round, 3–7 cm diam., palmately 3–5-lobed, lobes ovate to lanceolate, margins doubly serrate, apex acute to acuminate, surfaces hairy at least on nerves, <hairs appressed, straight to curly, short>, or abaxial grayish tomentose. **Inflorescences** 100+-flowered; branches and pedicels glabrous or tomentose to tomentulose. **Flowers:** hypanthium concave, becoming slightly convex in fruit; sepals (4 or)5(or 6), green, spatulate to triangular, 2–3 mm, margins without or with weak midrib, abaxially tomentose puberulent, adaxially glabrous; petals (4 or)5(or 6), white to cream, orbiculate to obovate, 2.5–5 mm, claws distinct, short, base narrow, margins entire; stamens white, longer than petals. **Achenes** 6–8, flattened, elongate, twisted, 3–6 mm, glabrous; sessile; styles 1–1.5 mm. $2n$ = 14, 14+2B (Europe).

Flowering summer (Jul). Moist meadows, roadside and railway ditches, near abandoned houses; 0–500 m; introduced; St. Pierre and Miquelon; N.B., Nfld. and Labr. (Nfld.), N.S., Ont., P.E.I., Que.; Colo., Conn., Ill., Ind., Maine, Mass., Mich., Minn., N.H., N.J., N.Y., Ohio, Pa., Vt., W.Va., Wis.; Eurasia.

Filipendula ulmaria is cultivated as an ornamental and sometimes escapes. The species is fairly common and relatively persistent or locally spreading from old gardens and thoroughly established well into the wild at some locations in the Maritimes. The native range stretches from Atlantic Europe to eastern Siberia (basin of Lena River), and from the Arctic Circle to the Altai Mountains of southern Siberia.

Two or three varieties or subspecies are often recognized within Europe and Russia in *Filipendula ulmaria* in the broad sense. Among them, subsp. *picbaueri* (Podpěra) Smejkal represents a separate species, *F. stepposa* Juzepczuk, and does not occur in North America. Two other frequently used infraspecific names are subsp. *denudata* and subsp. *ulmaria*, the former sometimes also regarded as a separate species. The latter has abaxially grayish tomentose leaves; the former has leaves only minutely hairy along the nerves. The full spectrum of intermediates usually occurs within a single population, and both forms clearly belong to the same species (I. A. Schanzer 1994). Both of them may occur in North America as escapes from cultivation.

The flowers contain tannins and salicylates and are thought to reduce pain and fever, mildly. They have also been used to treat stomach complaints, such as heartburn. Some research in laboratory animals has been conducted to investigate their effectiveness for ulcers, but results are not conclusive. Some laboratory studies appear to show that the flowers and seeds contain a chemical similar to heparin. That and the salicylate component may have some inhibiting effect on blood clotting (O. D. Barnaulov and P. P. Denisenko 1980; S. Foster and J. A. Duke 1990; B. A. Kudriashov et al. 1990, 1991). None of these properties have been documented by well-controlled clinical studies.

4. Filipendula vulgaris Moench, Methodus, 663. 1794 • Filipendule vulgaire [I]

Spiraea filipendula Linnaeus, Sp. Pl. 1: 490. 1753

Plants cespitose, 5–10 dm. **Rhizomes** horizontal to nearly vertical, relatively stout, 5–10 mm wide, internodes 0.5–1 cm; root tubers round to elongate, 1–3 cm. **Stems** glabrous. **Leaves:** basal 3–6(or 7), persistent to flowering; stipules ovate, 0.8–1.2 cm, <apiculate>, base auriculate; lateral leaflets in 7–17 pairs, close, lanceolate, to 3 cm, margins deeply dentate; terminal leaflets lanceolate, to 3 cm, 3-lobed, lobes lanceolate, margins deeply dentate, apex acute, abaxial surface glabrous or hairy on nerves, <hairs appressed, straight to curly>. **Inflorescences** usually less than 100-flowered; branches and pedicels glabrous. **Flowers:** hypanthium concave, becoming convex in fruit; sepals (5 or)6 or 7(–9), green, spatulate to triangular, 2–3 mm, margins without midrib, glabrous; petals (5 or)6 or 7(–9), white to cream, obovate, 5–8 mm, clawed, base narrow, margins entire; stamens white, about equal to petals. **Achenes** 10–18, slightly flattened, ovate, straight, 3–5 mm, densely appressed-hairy throughout, <hairs straight, short>; sessile (attached to enlarged globose torus); styles 0.5–1 mm. **2*n*** = 14, 14+2B (Europe, w Asia).

Flowering summer (Jun–Jul). Abandoned gardens and cemeteries, well-drained short grasslands; 0–1000 m; introduced; Nfld. and Labr. (Nfld.), N.S., Ont.; Calif., Conn., Maine, Mass., N.Y., Vt.; Eurasia; n Africa.

Filipendula vulgaris has been cultivated as an ornamental in North America and probably escapes. The native range stretches from Atlantic Europe to southwestern Siberia (Altai Mountains), and from southern Scandinavia to northern Africa. Despite strong morphologic differences from the other species of *Filipendula*, it is probably related to *F. ulmaria* (I. A. Schanzer 1994). Two additional names have been associated with this species: *Filipendula hexapetala* Gilibert is a rejected name, and *F. hexapetala* Gilibert ex Maximowicz is illegitimate.

a2. ROSACEAE Jussieu (subfam. ROSOIDEAE) tribe RUBEAE Dumortier, Anal. Fam. Pl., 39. 1829

Luc Brouillet

Shrubs, subshrubs, or herbs, perennial; armed or unarmed. **Leaves** alternate, imparipinnately or palmately compound or simple; stipules persistent [deciduous], free or adnate to petiole; venation pinnate or palmate. **Flowers:** perianth and androecium perigynous; epicalyx bractlets absent; hypanthium flat to hemispheric; torus flat or convex to conic; carpels 5–150, styles apical, distinct; ovules 2, apical, collateral, <only 1 maturing>. **Fruits** aggregated drupelets; styles persistent, not elongate.

Genus 1, species 250–700 (37 in the flora): North America, Mexico, West Indies, Central America, South America, Eurasia, Africa, Pacific Islands, Australia; introduced nearly worldwide.

The base chromosome number for Rubeae is $x = 7$. The tribe is host to *Phragmidium* rusts.

2. RUBUS Linnaeus, Sp. Pl. 1: 492. 1753; Gen. Pl. ed. 5, 218. 1754 • Bramble [Ancient Latin name for bramble, from *ruber*, red]

Lawrence A. Alice Douglas H. Goldman
James A. Macklin Gerry Moore

Dalibarda Linnaeus

Shrubs, subshrubs, or herbs, perennial, 0.5–30(–50) dm; fibrous, ± woody in species with larger plants. **Stems** 1–several, <biennial or perennial, rarely annual (*R. illecebrosus*)>, erect, arching, mounding, or creeping, rarely decumbent, ascending, or scrambling, <rooting or not at nodes or tips, terete or angled>; <prickles absent or sparse to dense, erect to retrorse, weak to stout, broad based or not; bristles absent or sparse to dense, erect to slightly retrorse, weak to stiff>; glabrous or hairy, eglandular or stipitate-glandular, sometimes sessile-glandular, <pruinose or not>. **Leaves** winter-persistent to deciduous, cauline; stipules filiform or elliptic to ovate, margins entire; petiole present; blade reniform to orbiculate, 2–30 cm, herbaceous to ± coriaceous, leaflets 0 or 3, 5, 7, or 9, terminal ovate to elliptic to obovate, <1.7–15 cm, base cuneate to rounded or cordate, sometimes truncate, rarely tapered or obtuse, unlobed or lobed>, margins flat or revolute, finely to coarsely crenate, dentate to doubly dentate, or serrate to doubly serrate, abaxial surface <unarmed or with prickles on midvein consistent with those on stems>, glabrous or ± densely hairy, eglandular or ± densely stipitate-glandular, sometimes sessile-glandular, along veins. **Inflorescences** axillary or terminal, 1–35(–100)-flowered, cymiform, racemiform, umbelliform, thyrsiform, or paniculiform, glabrous or sparsely to densely pubescent, <eglandular or sparsely to densely glandular, armed or unarmed>; bracts usually present; bracteoles absent. **Pedicels** present, <unarmed or sparsely armed with prickles similar to those of stems, glabrous or sparsely to densely hairy, eglandular or sparsely to densely stipitate-glandular, sometimes sessile-glandular>. **Flowers** bisexual (unisexual in *R. chamaemorus*, *R. ursinus*, and subg. *Micranthobatus* [in the sense of Kalkman]), 5–80 mm diam.; hypanthium 3–10 mm diam., glabrous or sparsely to densely pubescent, <eglandular or sparsely to densely glandular>; sepals 5, erect or spreading to reflexed, lanceolate to long-caudate, <unarmed or armed, glabrous or hairy, eglandular or sparsely to densely stipitate-glandular, sometimes sessile-glandular>; petals (0–)5(or 6), white to pink or magenta, suborbiculate to elliptic, obovate, or spatulate; stamens 20–100+, shorter to longer than petals, <filaments filiform or laminar>; carpels glabrous or hairy, <styles slender or clavate>. **Fruits** aggregated drupelets, (1–)5–100[–150], <not or weakly to strongly coherent>, separating with or without torus attached, golden yellow to red or black, globose to hemispheric or cylindric, 5–20 mm, <fleshy or dryish>, glabrous or finely hairy, <sometimes pruinose>; hypanthium usually persistent; sepals usually persistent, usually reflexed. **Seeds** 1 per drupelet. $x = 7$.

Species 250–700 (37 in the flora): North America, Mexico, West Indies, Central America, South America, Eurasia, Africa, Pacific Islands (Hawaii, New Zealand), Australia; introduced widely.

Molecular data (L. A. Alice and C. S. Campbell 1999) have shown *Rubus* to be monophyletic when including *Dalibarda* (*R. repens*). These data also show that long-recognized infrageneric taxa, such as the raspberries (subg. *Idaeobatus* Focke) and dwarf raspberries [subg. *Cylactis* (Rafinesque) Focke], may not be monophyletic. Although species are not organized here according to the subgeneric treatment by L. H. Bailey (1941–1945), they would belong to his subgenera as follows: flowering raspberries or thimbleberries [subg. *Anoplobatus* (Focke) Focke, species 3, 9, 18, 22, 23], snow raspberries [subg. *Chamaebatus* (Focke) Focke, species 19], cloudberries or baked-apple berries [subg. *Chamaemorus* (Hill) Focke, species 7], dwarf raspberries or plumboys (subg. *Cylactis*, species 2, 16, 26, 29, 31), raspberries (subg. *Idaeobatus*, species 11, 13, 14, 17, 20, 21, 24, 28, 33), and blackberries (subg. *Rubus*, species 1, 4–6, 8, 10, 12, 15, 25, 27, 32, 34, 35, 36, 37). Bailey did not include *D. repens* (30. *R. repens*) in *Rubus*, but W. O. Focke (1910) included it in subg. *Dalibarda* (under the illegitimate name *R. dalibarda* Linnaeus).

Rubus, especially the blackberries, presents some of the most difficult species-level problems, because of polyploidy, apomixis, and hybridization. As a result, differences of opinion on the number of species to be recognized from a given region can vary tremendously (for example, treatments in M. L. Fernald 1950 and H. A. Gleason and A. Cronquist 1991 for the northern half of the eastern United States plus parts of southeastern Canada, the former recognizing 205 species, the latter 25). A relatively conservative approach is taken here, one with much precedent (for example, A. E. Radford et al. 1968; R. G. Brown and M. L. Brown 1972; R. K. Godfrey and J. W. Wooten 1981; E. G. Voss 1972–1996, vol. 2; Great Plains Flora Association 1986; Godfrey 1988; Gleason and Cronquist; A. F. Rhoads and W. M. Klein 1993; E. B. Smith 1994; R. S. Mitchell and G. C. Tucker 1997; R. P. Wunderlin 1998; D. W. Magee and H. E. Ahles 1999; T. S. Cooperrider et al. 2001; R. L. Jones 2005; A. S. Weakley 2012; Weakley et al. 2012).

The decision to treat *Rubus* conservatively herein is due in part to the broad geographic distribution of many species in which quantitative characters are highly variable and the general absence of empirical data demonstrating that minor morphological variants are genetically based and not environmentally plastic. There is an overall lack of comprehensive herbarium specimens (for example, primocane, floricane with flowers and/or fruits, and habit data) making study difficult. Regarding *Rubus*, R. K. Godfrey (1988) wrote, "oversimplification appears to be the only way to achieve a practicable solution to the dilemma."

Other genera present a similar predicament by containing a large number of microspecies, such as *Alchemilla*, *Crepis*, *Hieracium*, *Oenothera*, and *Taraxacum*. W. Dietrich et al. (1997) stated for *Oenothera* taxonomy, and equally applicable to *Rubus*, that "immense splitting would result in a taxonomic system where even the specialist would lose the overview." See P. A. Rydberg (1913c), L. H. Bailey (1941–1945), P. D. Strausbaugh and E. L. Core (1978), H. A. Davis (1990), and M. P. Widrlechner (1998) for *Rubus* treatments that recognize many species.

Definitions for two morphologic terms primarily associated with *Rubus* are: primocane, a biennial or perennial stem prior to being developed enough for flowering, with flowering not occurring until at least its second season of growth, and floricane, a biennial or perennial stem after it has begun flowering, which does not occur until at least its second season of growth (R. W. Kiger and D. M. Porter 2001). Parcifrond is an elongate, sparsely leafy, sterile branch arising from a floricane proximal to its flowering branches. Novirame is a fertile (flowering, fruiting) branch on a primocane. Descriptions in this treatment are based on primocanes and floricanes, and exclude parcifronds and noviramEs.

In key leads and descriptions in this treatment, references to leaves are for those on primocanes, unless otherwise indicated. Also, for stems unsupported by other vegetation, measurements of vertical distances above the ground are given, not stem lengths. For example, stems of *Rubus flagellaris* may be as much as 3 dm above the ground even though the stems may be several meters long. *Rubus* inflorescences have been described as cymose or racemose. The inflorescences are determinate (thus cymose), with the terminal flower opening first. Flowering then proceeds acropetally (J. C. Guillard 1857; H. W. Rickett 1944), a pattern that is more typical of racemes. In the treatment here, inflorescences are described as cymiform, umbelliform, paniculiform, racemiform, or thyrsiform, these characterizations based on what the inflorescence most closely resembles. Also, inflorescence descriptions here refer to the flowering structures that generally terminate a branch distal to the leaves on that branch, including undivided, smaller, bractlike leaves. In plants with racemiform, paniculiform, or thyrsiform inflorescences, smaller axillary inflorescences containing one to few flowers are common and typically occur just proximal to the main inflorescences. Although most *Rubus* species in the flora area bear inflorescences that terminate a branch, such inflorescences often can appear to be axillary if terminating very short, leafy, axillary branches. Especially in subg. *Rubus*, terminal inflorescences almost always form on lateral branches, but occasionally form on a distal extension of the primocane of the previous year. In the flora area, native species in subg. *Rubus* have shorter floricanes, generally less than 30 cm; most non-native members in the flora area can have floricane branches exceeding 1 m.

Rubus shows diversity in armature and glands. Prickles, which protect the plant and aid some species in climbing, range from stout and broad-based to weak and narrow-based. Bristles may be stiff or weak. Prickles and bristles range from erect to retrorse. Generally, plants with sparse armature have spaces between the prickle bases or bristles that are more than two to three times the length of the prickles or bristles; plants with dense armature have spaces that are equal to or less than lengths. Glands vary significantly in size and shape and can be stalked or sessile. Glandularity is not always obvious; glands are sometimes hidden by other hairs or can be very small. Sessile glands can be less than 0.2 mm long or wide, requiring a good lens to be viewed; they are spheric to obovoid, whereas stipitate glands can be spheric to obovoid or discoid to cupulate.

Cultivars of blackberries and raspberries are grown for their edible fruit, and other *Rubus* taxa are grown for their ornamental value. Cultivars of blackberries have parentages or origins that include species native to North America and Europe, as well as red raspberry (for example, *R. idaeus* subsp. *idaeus* and subsp. *strigosus*) (J. R. Clark et al. 2007). Cultivars of red and black raspberries are primarily derived from *R. idaeus* and *R. occidentalis*, respectively. Purple raspberry cultivars are derived from the crossing of red and black raspberries. In both black and red raspberries, a lack of anthocyanin pigments can lead to the production of yellowish fruit. Cultivars of *R. idaeus* with yellowish fruit are sold as golden or amber raspberries. The Logan and Boysen cultivars are widely grown for their excellent fruits. Both cultivars are polyploid (6*x* and 8*x*, respectively) involving *R. ursinus* as the maternal parent. In Loganberry, the paternal parent is *R. idaeus* subsp. *idaeus*, while Boysenberry has a more complicated ancestry.

Due to horticultural or agricultural interest, it can be expected that more *Rubus* species will be introduced into the flora, with some naturalizing. *Rubus tomentosus* Borkhausen var. *canescens* (de Candolle) Wirtgen, a European species similar to *R. bifrons* and *R. vestitus* but much more pubescent than either, was reported growing wild in a small area in Randolph County, West Virginia (E. E. Hutton and R. B. Clarkson 1961), although more recently it has not been confirmed as naturalized there (P. J. Harmon et al. 2006). *Rubus hirsutus* Thunberg, native of China, Japan, and the Korean peninsula (L. Lingdi and D. E. Boufford 2003), was found naturalized in a woodland in Chapel Hill, North Carolina (D. H. Goldman, pers. obs.).

SELECTED REFERENCES Alice, L. A. and C. S. Campbell. 1999. Phylogeny of *Rubus* (Rosaceae) based on nuclear ribosomal DNA internal transcribed spacer region sequences. Amer. J. Bot. 86: 81–97. Bailey, L. H. 1941–1945. Species batorum. The genus *Rubus* in North America (north of Mexico) I–X. Gentes Herbarum 5: 1–918. Davis, H. A. 1990. Studies in *Rubus*. Castanea 55: 22–30. Davis, H. A., A. M. Fuller, and T. Davis. 1967–1970. Contributions towards the revision of eubati of eastern North America. Castanea 32: 20–37; 33: 50–76, 206–241; 34: 157–179, 235–266; 35: 176–194. Widrlechner, M. P. 1998. The genus *Rubus* L. in Iowa. Castanea 63: 415–465.

1. Leaves simple.
  2. Plants armed.
    3. Stipules broadly elliptic to ovate . . . . . 19. *Rubus nivalis* (in part)
    3. Stipules filiform to linear . . . . . 36. *Rubus ursinus* (in part)
  2. Plants unarmed.
    4. Herbs (stems sometimes woody basally); styles filiform.
      5. Flowers unisexual . . . . . 7. *Rubus chamaemorus*
      5. Flowers bisexual.
        6. Leaves unlobed; (sterile petaliferous flowers present) . . . . . 30. *Rubus repens*
        6. Leaves 3–5-lobed.
          7. Ovaries glabrous or sparsely hairy . . . . . 2. *Rubus arcticus* (in part)
          7. Ovaries densely hairy . . . . . 16. *Rubus lasiococcus* (in part)
    4. Shrubs; styles clavate.
      8. Leaf blades (5–)10–25(–30) cm wide; styles glabrous.
        9. Petals usually magenta, rarely white; stems, abaxial leaf surfaces, and pedicels: stipitate glands dark purple . . . . . 22. *Rubus odoratus*
        9. Petals white; stems, abaxial leaf surfaces, and pedicels: stipitate glands yellowish to reddish . . . . . 23. *Rubus parviflorus*
      8. Leaf blades (2.5–)3–5.5(–8) cm wide; styles villous.
        10. Leaf blades orbiculate to reniform; fruits dark purple . . . . . 9. *Rubus deliciosus*
        10. Leaf blades cordate to broadly ovate; fruits red or deep red.
          11. Abaxial leaf surfaces glabrous or sparsely hairy; Idaho, Oregon . . . 3. *Rubus bartonianus*
          11. Abaxial leaf surfaces moderately hairy; Arizona, Colorado, New Mexico, Utah . . . . . 18. *Rubus neomexicanus*
1. Leaves compound.
  12. Stems strongly pruinose.
    13. Stems creeping; plants to 3 dm.
      14. Leaflet abaxial surfaces densely white-tomentose; drupelets separating from tori . . . . . 11. *Rubus glaucifolius*
      14. Leaflet abaxial surfaces sparsely to densely hairy, but not white-tomentose; drupelets separating with tori attached.
        15. Flowers bisexual; fruits pruinose . . . . . 5. *Rubus caesius*
        15. Flowers usually functionally unisexual; fruits not pruinose . . . . . 36. *Rubus ursinus* (in part)
    13. Stems erect, ascending, or arching, sometimes climbing over other vegetation; plants 5+ dm.
      16. Leaves pinnately compound.
        17. Leaflet abaxial surfaces unarmed or with prickles; petals white to greenish white; fruits red to whitish, rarely amber; not Florida . . . . . 13. *Rubus idaeus*
        17. Leaflet abaxial surfaces with scattered, reflexed, broad-based prickles on midveins; petals pink to magenta; fruits purple-black; Florida . . . . . 20. *Rubus niveus*
      16. Leaves palmately compound or ternate.
        18. Stems usually moderately, rarely sparsely, hairy, hairs stellate; inflorescences 10–60-flowered, paniculiform or thyrsiform; drupelets separating with tori attached . . . . . 35. *Rubus ulmifolius*
        18. Stems glabrous or sparsely puberulent; inflorescences (1–)3–10(–20)-flowered, flat-topped cymiform, cymiform, or umbelliform; drupelets separating from tori.
          19. Lateral leaflets stalked (at least 2 larger ones); leaflet apices acuminate; w North America . . . . . 17. *Rubus leucodermis*
          19. Lateral leaflets sessile or subsessile; leaflet apices acute; e North America, Great Plains . . . . . 21. *Rubus occidentalis*

[12. Shifted to left margin.—Ed.]

12. Stems not pruinose or some stems weakly pruinose.
  20. Plants to 3 dm (rarely to 7 dm in *R. trivialis*, but then falling); stems usually creeping, sometimes erect but low growing, or higher only when using other vegetation for support.
    21. Stems unarmed, without prickles or bristles.
      22. Stems erect, not creeping . . . . 2. *Rubus arcticus* (in part)
      22. Stems creeping.
        23. Leaflets 5, rarely 3; (lateral leaflets deeply lobed, sinuses extending nearly to leaflet base) . . . . 26. *Rubus pedatus*
        23. Leaflets 3(–5); (lateral leaflets not or shallowly lobed).
          24. Ovaries densely hairy . . . . 16. *Rubus lasiococcus* (in part)
          24. Ovaries glabrous . . . . 29. *Rubus pubescens*
    21. Stems armed, with prickles and/or bristles.
      25. Stipules lanceolate or elliptic to ovate.
        26. Stipules adnate to petioles; leaflet abaxial surfaces with prickles along midveins . . . . 19. *Rubus nivalis* (in part)
        26. Stipules free from petioles; leaflet abaxial surfaces without prickles along midveins . . . . 31. *Rubus saxatilis*
      25. Stipules filiform to linear or lanceolate.
        27. Leaflet abaxial surfaces whitish-tomentose or short-velutinous.
          28. Leaves palmately compound; fruits black; drupelets separating with tori attached. . . . . 4. *Rubus bifrons* (in part)
          28. Leaves ternate or pinnately compound; fruits red; drupelets separating from tori . . . . 24. *Rubus parvifolius*
        27. Leaflet abaxial surfaces glabrous or sparsely to densely hairy, not strongly whitened.
          29. Leaflets (at least some) usually deeply or rarely shallowly lobed; inflorescences terminal, thyrsiform . . . . 15. *Rubus laciniatus* (in part)
          29. Leaflets usually unlobed, rarely shallowly lobed; inflorescences terminal on short shoots, (sometimes appearing axillary) usually racemiform or cymiform, sometimes solitary flowers or thyrsiform.
            30. Flowers usually functionally unisexual; stems usually strongly pruinose; Pacific Northwest, West Coast . . . . 36. *Rubus ursinus* (in part)
            30. Flowers bisexual; stems not pruinose; e, c, (sw) North America.
              31. Stems: prickles broad-based; bristles usually absent, sometimes sparse to dense.
                32. Stems: bristles absent; leaves deciduous, some occasionally semipersistent, not lustrous; inflorescences 1–3(–8) -flowered; petals white . . . . 10. *Rubus flagellaris*
                32. Stems: bristles absent or gland-tipped, red to purple, rarely green, slender; leaves persistent or semipersistent, lustrous; inflorescences 1(–3)-flowered; petals white to pink . . . . 34. *Rubus trivialis*
              31. Stems: prickles absent or narrow-based; bristles sparse to dense.
                33. Stems tip-rooting; leaves persistent; inflorescences 1–7 (–10)-flowered . . . . 12. *Rubus hispidus*
                33. Stems not node- or tip-rooting; leaves deciduous; inflorescences 5–15(–20)-flowered . . . . 32. *Rubus setosus* (in part)
  20. Plants 3+ dm; stems erect or arching.
    34. Leaflet abaxial surfaces usually closely, densely white-hairy or gray-hairy.
      35. Leaves palmately compound.

36. Terminal leaflets cuneate to obovate, margins serrate, rarely doubly serrate, apices broadly rounded to subtruncate, often cuspidate; petals white . . . . 8. *Rubus cuneifolius* (in part)
36. Terminal leaflets ovate to broadly elliptic or suborbiculate to orbiculate, margins moderately to coarsely, usually doubly, rarely singly, serrate, apices acute or acuminate to cuspidate or short-attenuate, short-acuminate, or cuspidate; petals white, pink, or magenta.
37. Terminal leaflets usually suborbiculate to orbiculate, sometimes broadly elliptic; pedicels moderately to densely stipitate-glandular; petals pink to magenta . . . . 37. *Rubus vestitus* (in part)
37. Terminal leaflets elliptic or ovate to suborbiculate; pedicels eglandular or sparsely to moderately sessile-glandular; petals white to pink.
38. Inflorescences thyrsiform, elongate, (projected well beyond subtending leaves), 10–60(–100)-flowered . . . . 4. *Rubus bifrons* (in part)
38. Inflorescences cymiform to thyrsiform, compact, (not projected well beyond subtending leaves), 3–15(–25)-flowered . . . 25. *Rubus pascuus* (in part)
35. Leaves (of primocanes) ternate.
39. Stems sparsely to moderately hairy, densely bristly . . . . 28. *Rubus phoenicolasius*
39. Stems glabrate or sparsely to densely hairy, bristles absent.
40. Bark usually papery in age, peeling (especially towards base); inflorescences 1- or 2-flowered; petals pink to magenta; filaments laminar; fruits yellow, orange, or red; drupelets separating from tori . . . . 33. *Rubus spectabilis* (in part)
40. Bark not papery, peeling; inflorescences (1–)3–15(–25)-flowered; petals white to pale pink; filaments filiform; fruits black; drupelets separating with tori attached.
41. Terminal leaflets cuneate to obovate, margins serrate, rarely doubly serrate, apices broadly rounded to subtruncate, often cuspidate . . . . 8. *Rubus cuneifolius* (in part)
41. Terminal leaflets broadly elliptic or ovate, margins moderately to coarsely, usually doubly, rarely singly, serrate, apices acute or short-attenuate. . . . . 25. *Rubus pascuus* (in part)

[34. Shifted to left margin.—Ed.]

34. Leaflet abaxial surfaces usually glabrous or sparsely to densely hairy or puberulent, (not white-hairy or gray-hairy).
42. Leaflets (at least some) usually deeply, rarely shallowly, lobed.
43. Bark not papery, peeling; inflorescences usually 5–25-flowered; petals white to pink; filaments filiform; fruits black; drupelets separating with tori attached . . . . 15. *Rubus laciniatus* (in part)
43. Bark usually papery in age, peeling (especially towards base); inflorescences usually 1–2-flowered; petals pink to magenta; filaments laminar; fruits yellow, orange, or red; drupelets separating from tori . . . . 33. *Rubus spectabilis* (in part)
42. Leaflets shallowly lobed or not lobed.
44. Leaves pinnately compound, (leaflets 3–7) . . . . 14. *Rubus illecebrosus*
44. Leaves ternate or palmately compound.
45. Bark usually papery in age, peeling (especially towards base); inflorescences 1–2-flowered; petals pink to magenta; filaments laminar; fruits yellow, orange, or red; drupelets separating from tori. . . . . 33. *Rubus spectabilis* (in part)
45. Bark not papery, peeling; inflorescences (1–)5–25(–35)-flowered; petals white, pink, or magenta; filaments filiform; fruits black; drupelets separating with tori attached.
46. Stems: prickles absent or widely scattered and narrow-based.
47. Stems: bristles present . . . . 32. *Rubus setosus* (in part)
47. Stems: bristles absent.

48. Pedicels densely stipitate-glandular; inflorescences often elongate, racemiform . . . . 1. *Rubus allegheniensis* (in part)
48. Pedicels eglandular or sparsely to moderately sessile- or stipitate-glandular; inflorescences short to elongate, racemiform, cymiform, or thyrsiform.
    49. Stems: prickles absent or widely scattered, narrow-based; leaves lustrous; inflorescences racemiform . . . . 6. *Rubus canadensis*
    49. Stems: prickles widely scattered, narrow-based; leaves not lustrous; inflorescences cymiform, racemiform, or thyrsiform . . . . 27. *Rubus pensilvanicus* (in part)

[46. Shifted to left margin.—Ed.]

46. Stems: prickles broad-based.
    50. Pedicels eglandular or sparsely to moderately sessile-glandular.
        51. Terminal leaflets cuneate to obovate, 2–6 × 3–4 cm . . . . 8. *Rubus cuneifolius* (in part)
        51. Terminal leaflets ovate to lanceolate, 5–15 × 3–13 cm . . . . 27. *Rubus pensilvanicus* (in part)
    50. Pedicels sparsely to densely stipitate-glandular.
        52. Terminal leaflets usually suborbiculate to orbiculate, sometimes broadly elliptic, apex acute to cuspidate; inflorescences cymiform or thyrsiform; petals pink to magenta . . . . 37. *Rubus vestitus* (in part)
        52. Terminal leaflets ovate to lanceolate, apex acuminate to long-attenuate; inflorescences racemiform, cymiform, or thyrsiform; petals white.
            53. Inflorescences (5–)15–25-flowered, racemiform, often elongate; pedicels densely short- to long-stipitate-glandular . . . . 1. *Rubus allegheniensis* (in part)
            53. Inflorescences (2–)5–12(–16)-flowered, cymiform, thyrsiform, or short-racemiform; pedicels sparsely to moderately sessile- to short-stipitate-glandular . . . . 27. *Rubus pensilvanicus* (in part)

1. Rubus allegheniensis Porter, Bull. Torrey Bot. Club 23: 153. 1896 • Allegheny or common blackberry, sow-teat berry, ronce des Alléghanys [E] [F]

Rubus villosus Aiton var. *montanus* Porter, Bull. Torrey Bot. Club 17: 15. 1890; *R. acadiensis* L. H. Bailey; *R. allegheniensis* var. *calycosus* (Fernald) Fernald; *R. allegheniensis* var. *gravesii* (Fernald) Fernald; *R. allegheniensis* var. *neoscoticus* (Fernald) L. H. Bailey; *R. allegheniensis* var. *plausus* L. H. Bailey; *R. allegheniensis* var. *populifolius* Fernald; *R. alumnus* L. H. Bailey; *R. apianus* L. H. Bailey; *R. attractus* L. H. Bailey; *R. auroralis* L. H. Bailey; *R. campester* L. H. Bailey; *R. concameratus* H. A. Davis & T. Davis; *R. fryei* H. A. Davis & T. Davis; *R. glandicaulis* Blanchard; *R. impos* L. H. Bailey; *R. inclinis* L. H. Bailey; *R. irregularis* L. H. Bailey; *R. licitus* L. H. Bailey; *R. montpelierensis* Blanchard ex L. H. Bailey; *R. nigrobaccus* L. H. Bailey; *R. nuperus* L. H. Bailey; *R. paulus* L. H. Bailey; *R. pugnax* L. H. Bailey; *R. rosa* L. H. Bailey; *R. saltuensis* L. H. Bailey; *R. sceleratus* Brainerd ex Fernald

Shrubs, 10–30 dm, armed. **Stems** <biennial>, erect-arching, sparsely to densely hairy, sparsely to abundantly stipitate-glandular, <glands usually flattened to cupulate>, not pruinose; prickles sparse to dense, erect or slightly retrorse, stout, 4–10 mm, broad-based. **Leaves** deciduous, palmately compound, <not lustrous>; stipules filiform to narrowly lanceolate, (2–)5–15(–20) mm; leaflets (3–)5(–7), terminal ovate to lanceolate, 7–16 × 4–9 cm, base rounded to shallowly cordate, unlobed, margins finely to coarsely serrate or doubly serrate, apex acuminate to long-attenuate, abaxial surfaces with hooked prickles on midveins, moderately hairy, moderately to densely stipitate-glandular on veins. **Inflorescences** terminal <on short shoots, sometimes appearing axillary>, (5–)15–25-flowered, racemiform, often elongate. **Pedicels** unarmed or prickles sparse, hooked, densely hairy, densely short- to long-stipitate-glandular. **Flowers** bisexual; petals white, obovate to elliptic, rarely suborbiculate, 8–20 mm; filaments filiform; ovaries glabrous or glabrate. **Fruits** black, globose to cylindric, 1–2 cm; drupelets 20–100, strongly coherent, separating with torus attached. $2n$ = 14, 21, 28.

Flowering May–Jul. Woodlands, savannas, prairies, meadows, rock outcrops, disturbed areas, dry to damp soil; 0–1600 m; B.C., N.B., N.S., Ont., P.E.I., Que.; Ala., Ark., Calif., Conn., Del., D.C., Ga., Ill., Ind., Iowa, Kans., Ky., Maine, Md., Mass., Mich., Minn., Mo., Nebr., N.H., N.J., N.Y., N.C., Ohio, Okla., Pa., R.I., S.C., Tenn., Vt., Va., W.Va., Wis.

Rubus allegheniensis consists of erect-arching, usually well-armed plants, usually densely stipitate-glandular with flattened to cupulate glands, particularly about the inflorescence.

The Menominee, Meskwaki, and Potawatomi used root extracts of *Rubus allegheniensis* to treat eye irritation (Hu. H. Smith 1923, 1928, 1933). The Ojibwa used boiled extracts of the canes as a diuretic and a root extract for treating diarrhea (Smith 1932). *Rubus allegheniensis* is introduced in British Columbia and California, where rare and in at least California, probably no more than a waif.

Rubus montanus (Porter) Porter, a name that has been used for this species, is a later homonym of *R. montanus* Libert ex Lejeune.

The following nothospecies names are based on putative hybrids involving *Rubus allegheniensis* and: *R. canadensis* (*R.* ×*forestalis* L. H. Bailey, *R.* ×*immanis* Ashe); *R. flagellaris* (*R.* ×*boyntonii* Ashe, *R.* ×*fraternalis* L. H. Bailey [based on *R.* ×*fraternus* Brainerd & Peitersen (not *R. fraternus* Gremli)], *R.* ×*licens* L. H. Bailey, *R.* ×*ostryifolius* Rydberg); *R. pensilvanicus* (*R.* ×*avipes* L. H. Bailey, *R.* ×*floricomus* Blanchard, *R.* ×*orarius* Blanchard); *R. setosus* (*R.* ×*abbrevians* Blanchard, *R.* ×*aculiferus* Fernald, *R.* ×*adenocaulis* Fernald, *R.* ×*angustifoliatus* L. H. Bailey, *R.* ×*ascendens* Blanchard, *R.* ×*atwoodii* L. H. Bailey, *R.* ×*clausenii* L. H. Bailey, *R.* ×*flavinanus* Blanchard, *R.* ×*frondisentis* Blanchard, *R.* ×*perspicuus* L. H. Bailey, *R.* ×*ravus* L. H. Bailey).

2. Rubus arcticus Linnaeus, Sp. Pl. 1: 494. 1753 • Arctic raspberry

Herbs, 0.5–3 dm, unarmed. **Stems** erect, <thin, weak, ± woody basally>, glabrous or sparsely hairy, eglandular, not pruinose. **Leaves** deciduous, simple and usually 3-lobed or ternate; stipules ovate to lanceolate, 4–8 mm; blade margins dentate, serrate, or doubly serrate, abaxial surfaces sparsely to moderately pubescent, eglandular. **Inflorescences** axillary, 1–3-flowered. **Pedicels** glabrous or sparsely to moderately pubescent, eglandular or stipitate-glandular. **Flowers** bisexual; petals pink to rose or magenta, obovate to oblanceolate, 8–25 mm; outer filaments dilated basally, inner filiform; ovaries glabrous or sparsely hairy, <styles filiform>. **Fruits** reddish to dark purple, globose, to 1 cm diam., rarely larger; drupelets (10–)15–30, weakly coherent, separating with torus attached.

The type specimen of *Rubus alaskensis* L. H. Bailey has long glandular hairs on the stems and has unusually large leaflets. It may represent a hybrid between *R. arcticus* and *R. idaeus* subsp. *strigosus* or some other species of *Rubus*.

Subspecies 3 (3 in the flora): North America, Eurasia.

1. Leaves simple, usually 3-lobed . . . . 2c. *Rubus arcticus* subsp. *stellatus*
1. Leaves usually ternate.
   2. Central leaflets ovate to rhombic, bases rounded to attenuate, not cuneate, apices usually acute; flowers 1–3; petals obovate, apex entire or slightly emarginate . . . . 2a. *Rubus arcticus* subsp. *arcticus*
   2. Central leaflets ovate to obovate, bases cuneate, apices usually rounded; flowers solitary; petals oblanceolate to obovate, apex slightly to strongly emarginate. . . . . 2b. *Rubus arcticus* subsp. *acaulis*

2a. Rubus arcticus Linnaeus subsp. **arcticus** • Arctic raspberry

Rubus arcticus var. *pentaphylloides* Hultén

Plants 0.5–3 dm. **Stems** glabrous or sparsely hairy. **Leaves** ternate; petiole hairy; central leaflets ovate to rhombic, to 6 cm (lateral leaflets shorter), base rounded to attenuate, not cuneate, margins dentate to serrate (sometimes doubly), apex usually acute. **Inflorescences:** flowers 1–3. **Pedicels** eglandular. **Flowers:** sepals deltate, hairy, eglandular or glandular; petals pink, obovate, 8–20 mm, apex entire or emarginate. **2*n*** = 21.

Flowering Jun–Aug. Subalpine meadows, bogs, thickets, tundra, peaty soil; 0–2000 m; Alta., B.C., N.W.T., Yukon; Alaska, Wash.; Eurasia.

2b. Rubus arcticus Linnaeus subsp. **acaulis** (Michaux) Focke, Biblioth. Bot. 72: 24. 1910 • Dwarf raspberry, ronce acaule [E]

Rubus acaulis Michaux, Fl. Bor.-Amer. 1: 298. 1803; *R. arcticus* var. *acaulis* (Michaux) B. Boivin; *R. arcticus* var. *grandiflorus* Ledebour

Plants 0.5–2 dm. **Stems** glabrous or hairy. **Leaves** usually ternate; petiole glabrous or hairy; central leaflets ovate to obovate, to 6 cm (lateral leaflets shorter), base cuneate, margins dentate to serrate (often doubly), apex usually rounded. **Inflorescences:** flowers solitary. **Pedicels** stipitate-glandular. **Flowers:** sepals caudate, glabrous or hairy, eglandular or glandular; petals pink to rose, oblanceolate to obovate, 15–25 mm, apex slightly to strongly emarginate. **2*n*** = 14, 21.

Flowering May–Aug. Muskegs, boggy woods, fens, bogs, swamps, thickets, moist tundra; 0–3000 m; St. Pierre and Miquelon; Alta., B.C., Man., Nfld. and Labr., N.W.T., Nunavut, Ont., Que., Sask., Yukon; Alaska, Colo., Maine, Mich., Minn., Mont., Oreg., Wash., Wyo.

Subspecies *acaulis* hybridizes with *Rubus pubescens* (*R.* ×*paracaulis* L. H. Bailey [= *R. pubescens* var. *paracaulis* (L. H. Bailey) B. Boivin]) in areas of sympatry.

2c. Rubus arcticus Linnaeus subsp. **stellatus** (Smith) B. Boivin, Bull. Soc. Bot. France 102: 235. 1955

Rubus stellatus Smith, Pl. Icon. Ined. 3: plate 64. 1791

Plants to 1.6 dm. **Stems** often with soft, spreading hairs. **Leaves** simple, usually 3-lobed, rarely unlobed; petiole hairy; blade ± rotund, 2–4.5 cm (usually broader than long), base cordate to subcordate, margins doubly serrate, lobe apices usually obtuse. **Inflorescences:** flowers solitary. **Pedicels** sparsely short-stipitate-glandular, sometimes eglandular. **Flowers:** sepals caudate, hairy, sparsely to moderately stipitate-glandular; petals pink to magenta, oblanceolate to obovate, 10–22 mm, apex emarginate. $2n = 14$.

Flowering Jun–Aug. Alpine stream banks and meadows, lake shores; 0–1500 m; B.C., N.W.T., Yukon; Alaska; e Asia (Russian Far East).

3. Rubus bartonianus M. Peck, Rhodora 36: 267. 1934 • Barton's raspberry [C] [E]

Shrubs, 8–25 dm, unarmed. **Stems** erect, sparsely short-hairy, glabrescent, eglandular, not pruinose. **Leaves** deciduous, simple; stipules lanceolate, 4–6 mm; blade cordate to broadly ovate, (2–)2.5–4(–5) × (2.5–)3.5–4.5(–5.5) cm, base deeply cordate, 3–5-lobed, lobe apices acute to obtuse, margins coarsely doubly dentate, abaxial surfaces glabrous or sparsely hairy, eglandular or sparsely stipitate-glandular. **Inflorescences** 1-flowered. **Pedicels** moderately hairy, eglandular or sparsely stipitate-glandular. **Flowers** bisexual; petals white, obovate, (15–)20–25 mm; filaments filiform; ovaries glabrous, <styles clavate, villous>. **Fruits** deep red, hemispheric, to 1 cm; drupelets 10–30, coherent, separating from torus.

Flowering Mar–May. Dry, rocky slopes; of conservation concern; 300–400 m; Idaho, Oreg.

Rubus bartonianus is distinguished from the other flowering raspberries within its geographic range by its erect, unarmed stems, relatively small, simple leaves with acute to obtuse lobes, deeply cordate bases, sparsely hairy or glabrous abaxial surfaces, relatively large flowers with white petals, and densely long-hairy, clavate styles. The leaves superficially resemble those of *Acer glabrum* or some species of *Ribes*.

Rubus bartonianus is most similar to *R. neomexicanus* but especially *R. deliciosus*. The species is known only from the Snake River Canyon of Idaho and Oregon.

4. Rubus bifrons Vest, Steyermärk. Z. 3: 163. 1821 • European or Himalayan blackberry [I] [W]

Rubus armeniacus Focke; *R. linkianus* Seringe

Shrubs, (3–)10–50 dm, armed. **Stems** <biennial>, arching, sometimes creeping early or with age, sparsely to densely hairy, eglandular or sparsely sessile-glandular, not or weakly pruinose; prickles moderately dense, hooked, retrorse, or erect, stout, 4–10 mm, broad- or narrow-based. **Leaves** deciduous to semievergreen, palmately compound; stipules filiform to linear, 7–15 mm; leaflets 3–5, terminal elliptic or ovate to suborbiculate, 6–15 × 4–9 cm, base rounded to shallowly cordate, unlobed, margins moderately to coarsely serrate, apex acute or acuminate to short-attenuate, abaxial surfaces white (gray-green in shade), with hooked prickles on largest veins, short-velutinous to tomentose, eglandular, rarely sparsely sessile-glandular along midveins. **Inflorescences** terminal, sometimes also axillary, 10–60(–100)-flowered, thyrsiform, <projected well beyond subtending leaves>. **Pedicels:** prickles moderate to dense, hooked to retrorse or erect, densely hairy, eglandular or sparsely to moderately sessile-glandular. **Flowers** bisexual; petals white to pink, obovate or elliptic to orbiculate, 10–15 mm; filaments filiform; ovaries apically hairy. **Fruits** black, globose to subcylindric, 1–2 cm; drupelets 15–40(–50), strongly coherent, separating with torus attached. $2n = 28, 48$.

Flowering (Apr–)May–Aug(–Sep). Woodland edges, dry or mixed woods, open areas, often disturbed, roadsides, moist soil; 0–1200(–1800) m; introduced; B.C., Ont.; Ala., Alaska, Ariz., Ark., Calif., Colo., Conn., D.C., Ga., Idaho, Ill., Kans., Ky., La., Md., Mass., Miss., Mo., Mont., Nev., N.J., N.Mex., N.Y., N.C., Ohio, Okla., Oreg., Pa., R.I., S.C., Tenn., Tex., Utah, Va., Wash.; Europe.

Rubus bifrons lacks the strongly pruinose stems of the related *R. ulmifolius* and has usually much larger leaves. The species also often has much broader inflorescences than *R. ulmifolius* or the related *R. vestitus*. It is less densely glandular and has longer and more open inflorescences than *R. pascuus*, and unlike the latter it has spreading, not erect, primocane leaves. *Rubus bifrons* also lacks the stipitate-glandular hairs and nearly round primocane terminal leaflets of *R. vestitus*; seedlings of *R. bifrons* can be densely stipitate-glandular.

L. H. Bailey (1941–1945), M. L. Fernald (1950), Y. Heslop-Harrison (1968), and H. A. Gleason and A. Cronquist (1991) distinguished between *Rubus bifrons* and another species (referred to as either

R. discolor or *R. procerus* P. J. Müller ex Boulay, both often considered synonyms of *R. armeniacus*), based on stem shape and indument, prickle shape and angle, leaf shape and margins, inflorescence shape, and petal color. Plants in North America identified as either *R. bifrons* or the other species can vary broadly in any of these characters, even within individual stems, making it extremely difficult or impossible to distinguish between these species in the flora area. Recent research addressing the genetics or species distinctiveness in this complex in Europe (J. Kollmann et al. 2000) and Australia (K. J. Evans et al. 2007) did not sample plants from North America, and recent work within the flora area (L. V. Clark et al. 2013) addressed this complex only in California, Oregon, and Washington. Until a much-needed genetic study is performed that includes a broad selection of plants of ser. *Discolores* from throughout North America, Europe, and elsewhere, the precise identities of all of our material in this complex will be elusive. With this degree of uncertainty, a conservative approach is taken here, which is consistent with the treatment herein of the native blackberries. The results of more comprehensive DNA research may show that a different name than *R. bifrons* is to be used in the flora area. The name *R. discolor* often has been misattributed as a synonym of *R. armeniacus*; it is a synonym of *R. ulmifolius* (H. E. Weber 1985).

In the flora area, *Rubus bifrons* is most abundant from northern California northward to British Columbia, where it can grow in great density over large areas, often to the exclusion of all other vegetation (for example, K. Williams et al. 2006). There are reports of *R. bifrons* (as *R. armeniacus* or *R. procerus*) hybridizing with *R. ursinus* (T. S. Mallah 1954; L. V. Clark and M. Jasieniuk 2012; B. Sutherland and L. Alice, unpubl.) and *R. laciniatus* (R. K. Bammi and H. P. Olmo 1966).

Rubus thyrsoideus Wimmer is an illegitimate name that applies here.

5. **Rubus caesius** Linnaeus, Sp. Pl. 1: 493. 1753
• European dewberry [I]

Shrubs, to 3[–10] dm, armed. **Stems** <biennial>, creeping or arching then creeping, glabrous or hairy, eglandular or sparsely sessile- to short-stipitate-glandular, strongly pruinose; prickles sparse to dense, erect to retrorse, slender, 1–3 mm, narrow- or broad-based. **Leaves** deciduous, ternate; stipules linear-lanceolate to ovate, 5–15 mm; terminal leaflets ovate, 4–14 × 3–10 cm, base rounded to shallowly cordate, unlobed or shallowly lobed, margins coarsely serrate to doubly serrate, apex acute to acuminate, abaxial surfaces with prickles on midveins or unarmed, sparsely to moderately hairy, eglandular or sparsely short-stipitate-glandular along larger veins. **Inflorescences** terminal <on short shoot, usually appearing axillary>, 1–6-flowered, solitary flowers, cymiform, or racemiform. **Pedicels** unarmed or prickles sparse to dense, erect to retrorse, densely hairy, sparsely to densely sessile- to short- or long-stipitate-glandular. **Flowers** bisexual; petals white, obovate to elliptic or suborbiculate, 7–13 mm; filaments filiform; ovaries glabrous. **Fruits** appearing bluish, black beneath bloom, <pruinose>, globose, 0.5–1.5 cm; drupelets 1–5(–20), weakly coherent, separating with torus attached. $2n = 28$.

Flowering May–Oct. Open woodlands, meadows, roadsides, disturbed areas; 0–400 m; introduced; Ont.; Iowa, Ky., Mich., N.Y., Ohio, Pa.; Europe; w, c Asia (to w China); introduced also in South America (Argentina).

Within the flora area, *Rubus caesius* usually produces fruit with few, relatively large drupelets; in Eurasia, fruit with more drupelets are typical, and weakly pruinose fruit have been observed.

Rubus caesius is vegetatively similar to *R. flagellaris* and is introduced into parts of the geographic range of the latter; see M. P. Widrlechner (1998) for a comparison of these two species. *Rubus caesius* shows striking vegetative similarity to pruinose, trifoliate-leaved plants of *R. ursinus*, with which it may have a close relationship (L. A. Alice, unpubl.). In the absence of flowers or fruit, these two species could be mistaken for one another. *Rubus caesius* generally has broader middle primocane leaflets than *R. ursinus*, but that is not a consistent difference. *Rubus caesius* may have been introduced into the geographic range of *R. ursinus*; most herbarium specimens surveyed have been inadequate to make a proper determination. On specimen labels, herbarium collectors should describe at least the color of the fruit. In the last decade, *R. caesius* apparently has become much more aggressive around Ithaca, New York, despite occurring there since the early twentieth century (F. R. Wesley, pers. comm.).

6. **Rubus canadensis** Linnaeus, Sp. Pl. 1: 494. 1753
• Canadian or smooth highbush blackberry, ronce du Canada [E]

Rubus argutus Link var. *randii* (L. H. Bailey) L. H. Bailey; *R. besseyi* L. H. Bailey; *R. canadensis* var. *imus* L. H. Bailey; *R. illustris* L. H. Bailey; *R. kennedyanus* Fernald; *R. laetabilis* L. H. Bailey; *R. montensis* L. H. Bailey

Shrubs, 5–30 dm, unarmed or armed. **Stems** <biennial>, arching, glabrous, eglandular or sparsely sessile- to short stipitate-glandular, not pruinose; prickles absent or sparse, erect or slightly retrorse, slender, 2–5 mm, narrow- to broad-based. **Leaves** deciduous, usually palmately compound, <lustrous>; stipules filiform to narrowly lanceolate, (5–)8–15(–22) mm; leaflets (3–)5, terminal ovate to elliptic, (3–)7–9(–11) × (3–)4–5(–6) cm, base rounded to shallowly cordate, unlobed, margins finely to coarsely serrate or doubly serrate, apex acuminate to attenuate, abaxial surfaces sometimes with hooked prickles on midvein, glabrous or puberulent, eglandular. **Inflorescences** terminal <on short shoots, sometimes appearing axillary>, (5–)15–25-flowered, racemiform. **Pedicels** usually unarmed, hairy, eglandular or sparsely sessile- to short-stipitate-glandular. **Flowers** bisexual; petals white, obovate to elliptic, 8–22 mm; filaments filiform; ovaries glabrous. **Fruits** black, globose to short-cylindric, 1–2 cm; drupelets 10–75, strongly coherent, separating with torus attached. $2n$ = 14, 21.

Flowering May–Jul. Mountainous highlands, cool, shaded intermediate regions within deciduous and boreal forests, margins of lakes and woods; 0–2000 m; St. Pierre and Miquelon; N.B., Nfld. and Labr. (Nfld.), N.S., Ont., P.E.I., Que.; Ala., Conn., D.C., Ga., Ky., Maine, Md., Mass., Mich., Minn., N.H., N.J., N.Y., N.C., Ohio, Pa., R.I., S.C., Tenn., Vt., Va., W.Va., Wis.; introduced in Europe (Poland).

Rubus canadensis has a wide range and is the most northerly of all the highbush blackberries. The species is morphologically similar to *R. allegheniensis* except that the plants have little armature, are not as hairy, and have no glands. It also has characteristically lustrous leaves. The open and edge habitats in which it grows can greatly influence the phenotype, as with most highbush blackberries. At higher elevations or in poorer soil conditions, the plants are smaller; when partially shaded, leaves of primocanes may be larger. Leaves of floricanes vary in size and shape and are unreliable for identification. Because of the variation, multiple names have been erected, especially locally; it is felt that a stricter approach to defining it is prudent. Some of the names traditionally associated with sect. *Canadenses* (L. H. Bailey) L. H. Bailey are here associated with other *Rubus* species or are considered to refer to putative hybrids.

The following nothospecies names are based on putative hybrids involving *Rubus canadensis* and: *R. allegheniensis* (*R.* ×*forestalis* L. H. Bailey, *R.* ×*immanis* Ashe); *R. flagellaris* (*R.* ×*lepagei* L. H. Bailey, *R.* ×*rixosus* L. H. Bailey); *R. hispidus* (*R.* ×*novanglicus* L. H. Bailey); *R. pensilvanicus* (*R.* ×*amabilis* Blanchard [not Focke], *R.* ×*amicalis* Blanchard, *R.* ×*crux* Ashe, *R.* ×*elegantulus* Blanchard [= *R. canadensis* var. *elegantulus* (Blanchard) Farwell], *R.* ×*multilicius* L. H. Bailey, *R.* ×*noveboracus* L. H. Bailey, *R.* ×*pergratus* Blanchard [= *R. canadensis* var. *pergratus* (Blanchard) L. H. Bailey]); *R. setosus* (*R.* ×*miscix* L. H. Bailey [based on *R.* ×*peculiaris* Blanchard (not *R. peculiaris* Sampaio)]).

7. **Rubus chamaemorus** Linnaeus, Sp. Pl. 1: 494. 1753
• Baked-apple berry, cloudberry, chicouté, plaquebière

Chamaemorus anglica Greene; *C. norwegica* Greene

Herbs, (0.5–)1–2.5(–3) dm, unarmed. **Stems** <from rhizomes>, erect, <rarely woody basally>, sparsely hairy, sparsely stipitate-glandular distally, not pruinose. **Leaves** deciduous, simple; stipules broadly ovate to oblong, 3–6 mm; blade reniform or orbiculate, 2.5–6 × (3–)5–8(–12) cm, base cordate, 3–7-lobed, lobe apices rounded, margins finely doubly serrate or dentate, abaxial surfaces sparsely hairy, sparsely to moderately stipitate-glandular along veins. **Inflorescences** 1-flowered. **Pedicels** pubescent, sparsely to densely short- to long-stipitate-glandular. **Flowers** unisexual; petals white, obovate, (8–)10–15(–18) mm; filaments filiform; ovaries glabrous, <styles filiform, glabrous>. **Fruits** golden yellow to orange or reddish, globose, 1.2–1.5 cm; drupelets (5–)10–20, moderately coherent, separating from torus. $2n$ = 56.

Flowering May–Aug. Bogs, muskeg, wet meadows, mossy woods, boreal forests, peaty tundra, lichen peat polygons, exposed sandy areas, lakesides, gravelly stream beds, rock outcrops; 0–1300 m; Greenland; St. Pierre and Miquelon; Alta., B.C., Man., N.B., Nfld. and Labr., N.W.T., N.S., Nunavut, Ont., P.E.I., Que., Sask., Yukon; Alaska, Maine, Minn., N.H., N.Y.; Eurasia.

Rubus chamaemorus is circumboreal; it is presumed to be extirpated from New York, where it was known only from a single population on Long Island, far removed from the next southernmost localities in northern New Hampshire.

Rubus chamaemorus is defined by its unarmed stems, simple, coriaceous, and plicate leaves, unisexual flowers, and golden yellow to reddish fruits. *Rubus chamaemorus* is likely an ancient allopolyploid (K. V. Ambrose 2006).

The edible fruits of *Rubus chamaemorus* are highly valued in northern regions for consumption fresh and in yogurt, jams, syrups, and juices. An alcoholic beverage is made from the fruits and sold commercially in Newfoundland, Quebec, and Finland.

8. Rubus cuneifolius Pursh, Fl. Amer. Sept. 1: 347. 1813 • Sand blackberry [E]

Rubus audax L. H. Bailey; *R. chapmanii* L. H. Bailey; *R. cuneifolius* var. *angustior* L. H. Bailey; *R. cuneifolius* var. *austrifer* L. H. Bailey; *R. cuneifolius* var. *spiniceps* L. H. Bailey; *R. cuneifolius* var. *subellipticus* Fernald; *R. georgiensis* L. H. Bailey; *R. probabilis* L. H. Bailey; *R. probativus* L. H. Bailey; *R. randolphiorum* L. H. Bailey; *R. sejunctus* L. H. Bailey

Shrubs, 5–10(–30) dm, armed. **Stems** <biennial>, erect, rarely arching, sparsely to densely hairy (especially young), eglandular, not pruinose; <bark not papery, peeling>; prickles usually dense, hooked to retrorse, usually stout, 3–6 mm, broad-based. **Leaves** deciduous to semievergreen, usually ternate, sometimes palmately compound; stipules filiform or linear to lanceolate, 3–15 mm; leaflets 3–5, terminal cuneate to obovate, 2–6 × 3–4 cm, base cuneate, unlobed, margins serrate, rarely doubly serrate, apex broadly rounded to subtruncate, often cuspidate, abaxial surfaces often with prickles on midveins, densely gray- to white-hairy, sparsely to densely sessile-glandular. **Inflorescences** terminal <on short shoots, sometimes appearing axillary>, (1–)3–5 (–12)-flowered, cymiform. **Pedicels** unarmed or prickles sparse to moderate, hooked to retrorse, moderately to densely hairy, eglandular or sparsely to moderately sessile-glandular. **Flowers** bisexual; petals white, elliptic to obovate, 5–15 mm; filaments filiform; ovaries glabrous or glabrate. **Fruits** black, globose to cylindric, 0.6–2 cm; drupelets 15–50, strongly coherent, separating with torus attached. $2n = 21, 28$.

Flowering May–Jul. Dry to damp open areas, sandy or rocky soil; 0–300 m; Ala., Conn., Del., D.C., Fla., Ga., La., Md., Mass., Miss., N.H., N.J., N.Y., N.C., Pa., S.C., Tenn., Va.

Rubus cuneifolius is characteristic of the Coastal Plain and Piedmont. The species is distinguished from most other blackberries in the flora area by its usually strongly erect and shrub-forming habit and abaxial leaflet surfaces often with dense white to gray indument; it may be confused with *R. pascuus*. *Rubus cuneifolius* has white petals and cuneate to obovate leaflets with usually revolute margins, the proximal third entire; *R. pascuus* has pale pink to white petals and ovate to broadly elliptic leaflets with the margins flat and entire only at base.

Rubus longii Fernald may represent hybridization between *R. cuneifolius* and another, yet-undetermined, blackberry (perhaps *R. pascuus*). Hybrids of *R. cuneifolius* and *R. hispidus* can be found around abandoned commercial cranberry bogs in New Jersey (G. Moore, pers. obs.). *Rubus cuneifolius* is introduced in South Africa and is officially recognized there as a noxious weed (L. Henderson 1995; T. Olkers and M. P. Hill 1999; H. Klein 2002). Preliminary examination of plants identified as *R. cuneifolius* from South Africa has shown that the plants are not *R. cuneifolius* but may be hybrids involving *R. cuneifolius* and another species.

The following nothospecies names are based on putative hybrids involving *Rubus cuneifolius* and: *R. pensilvanicus* (*R.* ×*acer* L. H. Bailey, *R.* ×*acer* var. *subacer* L. H. Bailey, *R.* ×*argutinus* L. H. Bailey, *R.* ×*floridensis* L. H. Bailey); *R. trivialis* (*R.* ×*inferior* L. H. Bailey).

9. Rubus deliciosus Torrey, Ann. Lyceum Nat. Hist. New York 2: 196. 1827 • Delicious raspberry [E]

Rubus medius Kuntze; *R. roezlii* Regel

Shrubs, 5–15(–20) dm, unarmed. **Stems** decumbent to erect, glabrous or sparsely short-hairy, eglandular or sparsely stipitate-glandular, not pruinose. **Leaves** deciduous, simple; stipules lanceolate to narrowly ovate, 5–8(–14) mm; blade orbiculate to reniform, (1.5–) 2–4(–5) × (2.5–)3–4.5(–7) cm, base shallowly cordate, 3–5(–7)-lobed, lobe apices broadly rounded to obtuse, margins doubly dentate, abaxial surfaces sparsely hairy, mostly along veins, sparsely stipitate-glandular. **Inflorescences** 1-flowered. **Pedicels** pubescent, sparsely to densely short- to long-stipitate-glandular. **Flowers** bisexual; petals white, obovate to ovate, 15–25(–30) mm; filaments filiform; ovaries glabrous, <styles clavate, villous>. **Fruits** dark purple, hemispheric, to 1 cm; drupelets 10–40, coherent, separating from torus. $2n = 14, 21$.

Flowering May–Aug. Rocky canyons, outcrops, stream banks; 1800–2900 m; Colo., N.Mex., Okla., Wyo.

Rubus deliciosus is recognized by its decumbent to erect, unarmed stems, relatively small simple leaves with obtuse to broadly rounded lobes, shallowly cordate bases, sparsely hairy abaxial surfaces, large flowers, white petals, glabrous or sparsely hairy sepals, and densely villous clavate styles. The distinction between *R. deliciosus* and *R. neomexicanus* is not always clear, especially when the former has more narrowly obtuse lobes and denser leaf abaxial hairs. *Rubus deliciosus* is also similar to *R. bartonianus* of Idaho and Oregon, as well as the Mexican *R. trilobus* Seringe. It might be reasonable to recognize a broader concept of *R. deliciosus*, one containing multiple infraspecific taxa. *Rubus deliciosus* and its close relatives require a revisionary study.

Rubus deliciosus has long been planted for its ornamental value because of its large, showy flowers.

10. Rubus flagellaris Willdenow, Enum. Pl., 549. 1809 • Northern dewberry, ronce à flagelles

Rubus aboriginum Rydberg; *R. alacer* L. H. Bailey; *R. alius* L. H. Bailey; *R. almus* (L. H. Bailey) L. H. Bailey; *R. apogaeus* L. H. Bailey; *R. aptatus* L. H. Bailey; *R. arenicola* Blanchard; *R. arizonensis* Focke; *R. armatus* (Fernald) L. H. Bailey; *R. arundelanus* Blanchard; *R. arundelanus* var. *jecklyanus* (Blanchard) L. H. Bailey; *R. ashei* L. H. Bailey; *R. austrinus* L. H. Bailey; *R. baileyanus* Britton; *R. bollianus* L. H. Bailey; *R. bonus* L. H. Bailey; *R. botruosus* L. H. Bailey; *R. bretonis* L. H. Bailey; *R. brevipedalis* L. H. Bailey; *R. cacaponensis* H. A. Davis & T. Davis; *R. camurus* L. H. Bailey; *R. cathartium* Fernald; *R. celer* L. H. Bailey; *R. census* L. H. Bailey; *R. centralis* L. H. Bailey; *R. clairbrownii* L. H. Bailey; *R. clandestinus* L. H. Bailey; *R. clarus* L. H. Bailey; *R. coloniatus* L. H. Bailey; *R. complex* L. H. Bailey; *R. conabilis* L. H. Bailey; *R. connixus* L. H. Bailey; *R. cordialis* L. H. Bailey; *R. cordifrons* (L. H. Bailey) L. H. Bailey; *R. currulis* L. H. Bailey; *R. curtipes* L. H. Bailey; *R. deamii* L. H. Bailey; *R. decor* L. H. Bailey; *R. depavitus* L. H. Bailey; *R. dives* L. H. Bailey; *R. eflagellaris* L. H. Bailey; *R. enslenii* Trattinnick; *R. exemptus* L. H. Bailey; *R. exsularis* L. H. Bailey; *R. fandus* L. H. Bailey; *R. fecundus* L. H. Bailey; *R. felix* L. H. Bailey; *R. ferrofluvius* H. A. Davis, A. M. Fuller & T. Davis; *R. flagellaris* var. *almus* L. H. Bailey; *R. flagellaris* var. *occidualis* L. H. Bailey; *R. foliaceus* L. H. Bailey; *R. folioflorus* L. H. Bailey; *R. grimesii* L. H. Bailey; *R. hancinianus* L. H. Bailey; *R. housei* L. H. Bailey; *R. ignarus* L. H. Bailey; *R. imperiorum* Fernald; *R. indianensis* L. H. Bailey; *R. iniens* L. H. Bailey; *R. injunctus* L. H. Bailey; *R. inobvius* L. H. Bailey; *R. invisus* (L. H. Bailey) Britton; *R. ithacanus* L. H. Bailey; *R. jactus* L. H. Bailey; *R. jaysmithii* L. H. Bailey; *R. jaysmithii* var. *angustior* L. H. Bailey; *R. kentuckiensis* L. H. Bailey; *R. lassus* L. H. Bailey; *R. leviculus* L. H. Bailey; *R. mainensis* L. H. Bailey; *R. maltei* L. H. Bailey; *R. maniseesensis* L. H. Bailey; *R. meracus* L. H. Bailey; *R. michiganensis* (F. W. Card ex L. H. Bailey) L. H. Bailey; *R. minnesotanus* L. H. Bailey; *R. multifer* L. H. Bailey; *R. mundus* L. H. Bailey; *R. nefrens* L. H. Bailey; *R. obsessus* L. H. Bailey; *R. obvius* L. H. Bailey; *R. occidualis* (L. H. Bailey) L. H. Bailey; *R. particeps* L. H. Bailey; *R. particularis* L. H. Bailey; *R. pernagaeus* Fernald; *R. pityophilus* S. J. Smith ex L. H. Bailey; *R. plexus* Fernald; *R. plicatifolius* Blanchard; *R. profusiflorus* L. H. Bailey; *R. pronus* L. H. Bailey; *R. redundans* L. H. Bailey; *R. rhodophyllus* Rydberg; *R. roribaccus* (L. H. Bailey) Rydberg; *R. rosagnetis* L. H. Bailey; *R. russeus* L. H. Bailey; *R. sailori* L. H. Bailey; *R. sanfordii* L. H. Bailey; *R. satis* L. H. Bailey; *R. scambens* L. H. Bailey; *R. schoolcraftianus* L. H. Bailey; *R. sewardianus* Fernald; *R. steelei* L. H. Bailey; *R. subinnoxius* Fernald; *R. temerarius* L. H. Bailey; *R. tetricus* L. H. Bailey; *R. uvidus* L. H. Bailey; *R. vixalacer* L. H. Bailey; *R. whartoniae* L. H. Bailey

Shrubs, to 3 dm, armed. **Stems** <biennial>, usually creeping, sometimes low-arching and then creeping <flowering branches usually erect>, glabrous or densely hairy, eglandular or sparsely sessile- to short-stipitate-glandular, not pruinose; prickles sparse to dense, hooked, sometimes distally slender, 1–4 mm, broad-based; <bristles absent>. **Leaves** deciduous, some sometimes semipersistent, ternate or palmately compound, <not lustrous>; stipules filiform or linear to lanceolate, 3–20 mm; leaflets 3–5, terminal ovate or elliptic to suborbiculate, 3–11 × 2–7.5 cm, base broadly cuneate or rounded to shallowly cordate, usually unlobed, rarely shallowly lobed, margins moderately to coarsely serrate to doubly serrate or serrate-dentate, apex acute or acuminate to short-attenuate, abaxial surfaces with prickles on midvein or unarmed, sparsely to moderately hairy, eglandular or sessile- or short-stipitate-glandular along largest veins. **Inflorescences** terminal <on short shoots, usually appearing axillary>, 1–3(–8)-flowered, racemiform. **Pedicels** unarmed or prickles sparse to moderate, retrorse to hooked, moderately to densely hairy, usually sparsely to densely sessile- or short-stipitate-glandular, rarely eglandular. **Flowers** bisexual; petals white, elliptic, obovate, or oblanceolate, 8–20 mm; filaments filiform; ovaries glabrous. **Fruits** black, sometimes dark red, globose to cylindric, 1–2 cm; drupelets 10–40, strongly coherent, separating with torus attached. $2n$ = 14, 21, 28, 35, 42, 49, 56, 63.

Flowering Mar–Jun. Woodlands, savannas, pine barrens, prairies, meadows, rock outcrops, disturbed areas, dry to seasonally wet soil; 0–1000 m; N.B., N.S., Ont., P.E.I., Que.; Ala., Ariz., Ark., Conn., Del., D.C., Fla., Ga., Ill., Ind., Iowa, Kans., Ky., La., Maine, Md., Mass., Mich., Minn., Miss., Mo., Nebr., N.H., N.J., N.Mex., N.Y., N.C., Ohio, Okla., Pa., R.I., S.C., Tenn., Tex., Vt., Va., W.Va., Wis.; Mexico (Coahuila, Hidalgo, Nuevo León, San Luis Potosí, Sonora).

Rubus flagellaris is extremely polymorphic, ranging from plants with low-arching (and later creeping) stems and relatively few prickles to low, creeping plants with abundant prickles. Individual plants in some years will produce abundant, arching, poorly armed stems, and in others creeping, well-armed stems. Prickle shape also varies in these plants both within a year and among different years. Local variants seem to readily intergrade with other variants; over the entire North American range of what would be known as sect. *Procumbentes* (Rydberg) L. H. Bailey [= sect. *Flagellares* (L. H. Bailey) L. H. Bailey], a continuum of variation seems to be common.

Apparent consistent features of *Rubus flagellaris* are terete primocanes to 7 mm diam. near the base and presence of rigid, hooked primocane prickles to 4 mm. Primocanes that tip-root and are low and long-running are nearly consistent features of *R. flagellaris*. Flower number per inflorescence throughout most of the geographic range of *R. flagellaris* is one to three or, rarely, five; in the far western part of the range (as *R. arizonensis* [= *R. oligospermus* Thornber ex Rydberg, name illegitimate]) up to eight flowers per inflorescence have been reported.

Rubus flagellaris lacks pruinose fruits and stems and is similar in general appearance to the introduced *R. caesius*.

Rubus vagus L. H. Bailey is an illegitimate later homonym, and *R. longipes* Fernald is an illegitimate name; they both apply here.

The following nothospecies names are based on putative hybrids involving *Rubus flagellaris* and: *R. allegheniensis* (*R.* ×*boyntonii* Ashe, *R.* ×*fraternalis* L. H. Bailey [based on *R.* ×*fraternus* Brainerd & Peitersen (not *R. fraternus* Gremli)], *R.* ×*licens* L. H. Bailey, *R.* ×*ostryifolius* Rydberg); *R. canadensis* (*R.* ×*lepagei* L. H. Bailey, *R.* ×*rixosus* L. H. Bailey); *R. hispidus* (*R.* ×*ambigens* Fernald, *R.* ×*distinctus* L. H. Bailey, *R.* ×*elongatus* Brainerd & Peitersen [not *R. elongatus* Smith], *R.* ×*emeritus* L. H. Bailey, *R.* ×*furtivus* L. H. Bailey, *R.* ×*kalamazoensis* L. H. Bailey, *R.* ×*permixtus* Blanchard, *R.* ×*rosendahlii* L. H. Bailey, *R.* ×*segnis* L. H. Bailey, *R.* ×*varus* L. H. Bailey, *R.* ×*vigoratus* L. H. Bailey); *R. pensilvanicus* (*R.* ×*akermanii* Fernald, *R.* ×*darlingtonii* L. H. Bailey, *R.* ×*dissitiflorus* Fernald, *R.* ×*hypolasius* Fernald, *R.* ×*janssonii* L. H. Bailey, *R.* ×*largus* L. H. Bailey, *R.* ×*recurvicaulis* Blanchard, *R.* ×*rossbergianus* Blanchard); *R. setosus* (*R.* ×*alter* L. H. Bailey, *R.* ×*arcuans* Fernald & H. St. John, *R.* ×*biformispinus* Blanchard, *R.* ×*bigelovianus* L. H. Bailey, *R.* ×*bracteolifer* Fernald, *R.* ×*jacens* Blanchard var. *specialis* L. H. Bailey, *R.* ×*laevior* (L. H. Bailey) Fernald [based on *R. permixtus* var. *laevior* L. H. Bailey], *R.* ×*multiformis* Blanchard, *R.* ×*setospinosus* L. H. Bailey, *R.* ×*severus* Brainerd ex Fernald).

11. Rubus glaucifolius Kellogg, Proc. Calif. Acad. Sci. 1(ed. 2): 70. 1873 • San Diego raspberry [E]

Rubus ganderi L. H. Bailey; *R. glaucifolius* subsp. *ganderi* (L. H. Bailey) R. M. Beauchamp; *R. glaucifolius* var. *ganderi* (L. H. Bailey) Munz; *R. leucodermis* Douglas ex Torrey & A. Gray var. *glaucifolius* (Kellogg) Jepson

Shrubs, to 3 dm, weakly armed. **Stems** <biennial>, creeping, glabrous, eglandular, strongly pruinose; prickles sparse, erect or slightly curved, weak, slender, 2–3 mm, narrow-based. **Leaves** deciduous, ternate; stipules filiform to linear, 3–10 mm; terminal leaflets ovate to elliptic, 4–8 × 3–7 cm, base tapered or subcordate, often 2-lobed, margins coarsely dentate, apex acute to rounded, abaxial surfaces unarmed, densely white-tomentose, eglandular. **Inflorescences** terminal and axillary, (1 or) 2–4(–10)-flowered, cymiform. **Pedicels** unarmed, finely hairy, stipitate-glandular. **Flowers** bisexual; petals white, oblong to oblanceolate, 4–8 mm; filaments laminar; ovaries white-tomentose. **Fruits** reddish purple, hemispheric to conic, 0.4–1 cm diam.; drupelets 10–40, coherent, separating from torus.

Flowering Apr–Jul. Semiopen montane forests; 800–2100 m; Calif., Oreg.

Rubus glaucifolius is found in the Klamath and Sierra Nevada mountains as well as the Peninsular and northern Coastal ranges in California, and in adjacent Oregon only in Jackson County.

12. Rubus hispidus Linnaeus, Sp. Pl. 1: 493. 1753 • Bristly or swamp dewberry, ronce hispide [E] [F]

Rubus blanchardianus (L. H. Bailey) L. H. Bailey; *R. compos* L. H. Bailey; *R. cubitans* Blanchard; *R. davisiorum* L. H. Bailey; *R. fassettii* L. H. Bailey; *R. hispidus* var. *blanchardianus* L. H. Bailey; *R. hispidus* var. *cupulifer* L. H. Bailey; *R. hispidus* var. *obovalis* (Michaux) Fernald; *R. huttonii* L. H. Bailey; *R. paganus* L. H. Bailey; *R. pervarius* (L. H. Bailey) L. H. Bailey; *R. plus* L. H. Bailey; *R. porteri* L. H. Bailey; *R. provincialis* L. H. Bailey; *R. pudens* L. H. Bailey; *R. rowleei* L. H. Bailey; *R. signatus* L. H. Bailey; *R. trifrons* Blanchard var. *pudens* (L. H. Bailey) Fernald; *R. vagulus* L. H. Bailey; *R. vigil* L. H. Bailey

Shrubs, sometimes herblike, to 2 dm, armed. **Stems** <biennial>, creeping, <tip-rooting, flowering branches often erect>, glabrous or sparsely hairy, sparsely to densely short- to long-stipitate-glandular, rarely eglandular, not pruinose; prickles sparse to dense, erect to retrorse, weak, slender, 1–3(–5) mm, narrow-based; <bristles sparse to dense, erect or retrorse, narrow, flexible and weak, not gland-tipped>. **Leaves** persistent, ternate or, rarely, palmately compound, <lustrous>; stipules linear to narrowly lanceolate, (2–)5–15(–20) mm; leaflets 3(–5), terminal obovate to suborbiculate, 1.7–6.5 × 1–5.2 cm, base cuneate to rounded, unlobed, margins moderately to coarsely doubly serrate, apex acute to rounded, abaxial surfaces with bristles on midvein or unarmed, glabrous or sparsely hairy, sparsely to densely short-stipitate-glandular along largest veins. **Inflorescences** terminal <on short shoots, usually appearing axillary>, 1–7(–10)-flowered, racemiform. **Pedicels** unarmed or bristles moderate to sometimes dense, erect to retrorse, moderately to densely hairy, sparsely to densely short- to long-stipitate-glandular. **Flowers** bisexual; petals white, obovate to oblanceolate, 6–10 mm; filaments filiform; ovaries glabrous. **Fruits** black, globose to cylindric, to 1 cm; drupelets (5–)10–15(–20), strongly coherent, separating with torus attached. $2n$ = 14, 21, 28, 35, 56.

Flowering May–Jul(–Aug). Moist to wet woodlands, meadows, openings, bogs, fens, disturbed areas, wet to dry soil; 0–1500 m; N.B., Nfld. and Labr. (Nfld.), N.S., Ont., P.E.I., Que.; Conn., Del., D.C., Ga., Ill., Ind., Ky., Maine, Md., Mass., Mich., N.H., N.J., N.Y., N.C., Ohio, Pa., R.I., S.C., Tenn., Vt., Va., Wash., W.Va., Wis.

Rubus hispidus is identified mostly by its creeping primocanes with somewhat lustrous, coriaceous, and relatively small leaves that persist for nearly a year, and an armature of stiff hairs or, rarely, weak, slender prickles. Plants with few persistent leaves may be evidence of introgression with *R. flagellaris* or *R. setosus*. Hybrids with *R. cuneifolius* have been seen around abandoned commercial cranberry bogs in New Jersey (G. Moore, pers. obs.). The superficially similar *R. caesius* differs from *R. hispidus* in its pruinose stems, deciduous and non-lustrous leaves, and pruinose fruit.

Rubus hispidus is introduced in southwestern Washington State.

The following nothospecies names are based on putative hybrids involving *Rubus hispidus* and: *R. canadensis* (*R.* ×*novanglicus* L. H. Bailey); *R. flagellaris* (*R.* ×*ambigens* Fernald, *R.* ×*distinctus* L. H. Bailey, *R.* ×*elongatus* Brainerd & Peitersen [not *R. elongatus* Smith], *R.* ×*emeritus* L. H. Bailey, *R.* ×*furtivus* L. H. Bailey, *R.* ×*kalamazoensis* L. H. Bailey, *R.* ×*permixtus* Blanchard, *R.* ×*rosendahlii* L. H. Bailey, *R.* ×*segnis* L. H. Bailey, *R.* ×*varus* L. H. Bailey, *R.* ×*vigoratus* L. H. Bailey); *R. setosus* (*R.* ×*adjacens* Fernald, *R.* ×*grandidens* L. H. Bailey, *R.* ×*harmonicus* L. H. Bailey, *R.* ×*jacens* Blanchard, *R.* ×*parlinii* L. H. Bailey, *R.* ×*spiculosus* Fernald, *R.* ×*tardatus* Blanchard, *R.* ×*tholiformis* Fernald, *R.* ×*trifrons* Blanchard, *R.* ×*zaplutus* L. H. Bailey).

13. Rubus idaeus Linnaeus, Sp. Pl. 1: 492. 1753
• Red raspberry, framboisier

Shrubs, 5–25 dm, armed or unarmed. **Stems** <biennial>, erect, glabrescent, eglandular or stipitate-glandular, strongly pruinose; prickles absent or sparse to dense, erect, weak, 1–4 mm, broad-based; <bristles present or absent>. **Leaves** deciduous, pinnately compound; stipules filiform, 5–10 mm; <petiole unarmed or with prickles, strigose or glabrous, glandular or eglandular; terminal leaflets petiolulate, lateral sessile or subsessile>; leaflets 3–5(–7), terminal ovate to lanceolate, 7–15 × 4–11 cm, base rounded to cordate, unlobed or 2-lobed, margins serrate to doubly serrate, apex acute to attenuate, abaxial surfaces unarmed or with prickles, strongly white-tomentose, eglandular or glandular. **Inflorescences** terminal and axillary, (1–)3–7(–20) -flowered, racemiform. **Pedicels** unarmed or prickles sparse, hooked, glabrous or sparsely pubescent, stipitate-glandular or eglandular. **Flowers** bisexual; petals white to greenish white, spatulate to obovate, 5–10 mm; filaments filiform or slightly dilated basally; ovaries pubescent. **Fruits** usually red to whitish, rarely amber, globose to conic, 0.5–2 cm; drupelets 10–60, coherent, separating from torus.

Subspecies 2 (2 in the flora): North America, Mexico, Eurasia, Africa.

Rubus idaeus is the source of most of the cultivated red and amber raspberries. Plants were used by the Cherokee, Cree, and others as a parturient; an infusion of dried leaves in water was used to prepare the uterus for childbirth and help women recover after childbirth (D. E. Moerman 1998; D. Hoffman 2003).

1. Stems eglandular . . . . . . 13a. *Rubus idaeus* subsp. *idaeus*
1. Stems stipitate-glandular . . . . 13b. *Rubus idaeus* subsp. *strigosus*

13a. Rubus idaeus Linnaeus subsp. **idaeus**
• European raspberry, framboisier d'Europe I

Stems to 1 cm diam., eglandular; bark usually not peeling with age. **Leaves:** petiole eglandular; leaflets 3–5(–7), terminal broadly ovate to oblong-ovate, base cordate, unlobed or lobed in larger leaflets, apex acute to acuminate, <laterals acute to attenuate>, abaxial surfaces eglandular. **Flowers:** sepals eglandular. **Fruits:** drupelets 20–60, coherent, separating without torus attached. $2n = 14$.

Flowering May–Jul. Fields, roadsides, disturbed areas, dry soil; 0–2000 m; introduced; N.B., N.S., Ont., P.E.I., Que.; Alaska, Ark., Conn., Idaho, Ill., Ind., Iowa, Maine, Mass., Mich., Minn., Mont., N.J., N.Y., N.C., N.Dak., Ohio, R.I., S.Dak., Tenn., Utah, Vt., Va., Wash., Wyo.; Eurasia; Africa.

In the flora area, subsp. *idaeus* primarily is an escape from cultivation. M. L. Fernald (1919b) reported a thornless form (forma *inermis* Kaufmann) as apparently native to the Magdalen Islands, Quebec, and Minnesota. L. H. Bailey (1941–1945) concluded that typical *Rubus idaeus* might be native to the northern regions of North America. Subspecies *idaeus* usually has narrow prickles but rarely can have some bristles.

13b. Rubus idaeus Linnaeus subsp. **strigosus** (Michaux) Focke, Abh. Naturwiss. Vereins Bremen 13: 473. 1896 • American red raspberry, framboisier sauvage

Rubus strigosus Michaux, Fl. Bor.-Amer. 1: 297. 1803; *R. aliceae* L. H. Bailey; *R. arizonicus* (Greene) Rydberg; *R. carolinianus* Rydberg; *R. greeneanus* L. H. Bailey; *R. idaeus* var. *aculeatissimus* Regel & Tiling; *R. idaeus* var. *canadensis* Richardson ex Fernald; *R. idaeus* var. *caudatus* (B. L. Robinson & H. Schrenk) Fernald & B. G. Schubert; *R. idaeus* var. *egglestonii* (Blanchard) Fernald; *R. idaeus* var. *eucyclus* Fernald & Weatherby; *R. idaeus* var. *gracilipes* M. E. Jones; *R. idaeus* var. *heterolasius* Fernald; *R. idaeus* subsp. *melanolasius* (Dieck) Focke; *R. idaeus* var. *melanolasius* (Dieck) R. J. Davis; *R. idaeus* var. *peramoenus* (Greene) Fernald; *R. idaeus* subsp. *sachalinensis* (H. Léveillé) Focke; *R. idaeus* var. *strigosus* (Michaux) Maximowicz; *R. melanolasius* Dieck; *R. strigosus* var. *acalyphaceus* (Greene) L. H. Bailey; *R. strigosus* var. *arizonicus* (Greene) Kearney & Peebles

Stems to 0.6 cm diam., stipitate-glandular; bark usually peeling with age. **Leaves:** petiole stipitate-glandular; leaflets 3–5, terminal ovate to lanceolate, base rounded to cordate, unlobed or lobed (less deeply so in ternate leaves), apex acuminate to attenuate, abaxial surfaces stipitate-glandular. **Flowers:** sepals stipitate-glandular. **Fruits:** drupelets 10–60, coherent, separating without torus attached or drupelets not coherent and falling separately. **2*n*** = 14, 21, 28.

Flowering May–Jul. Fields, woodlands, roadsides, disturbed areas, dry to moist soil; 0–3400 m; St. Pierre and Miquelon; Alta., B.C., Man., N.B., Nfld. and Labr., N.W.T., N.S., Nunavut, Ont., P.E.I., Que., Sask., Yukon; Alaska, Ariz., Calif., Colo., Conn., Del., D.C., Idaho, Ill., Ind., Iowa, Maine, Md., Mass., Mich., Minn., Mo., Mont., Nebr., Nev., N.H., N.J., N.Mex., N.Y., N.C., N.Dak., Ohio, Okla., Oreg., Pa., R.I., S.Dak., Tenn., Utah, Vt., Va., Wash., W.Va., Wis., Wyo.; Mexico (Chihuahua, Nuevo León, Sonora).

Hybrids between subsp. *strigosus* and *Rubus occidentalis*, including the cultivated purple raspberries, have usually gone under the name *R.* ×*neglectus* Peck. Peck did not treat *R. neglectus* as a hybrid, although his description, based on plants from northeastern New York, acknowledged the plants to be intermediate between the two taxa. L. H. Bailey (1941–1945) suspected *R. neglectus* to be a distinct species of local range. The fruit of these plants has a strikingly distinct taste; Peck noted that the locals referred to it as cream berries. M. L. Fernald (1900) misapplied the name var. *anomalus* Arrhenius to reduced, unarmed sterile plants of subsp. *strigosus* with simple leaves on the floricanes; such plants were later described by Blanchard as *R. egglestonii*. *Rubus viburnifolius* (Greene) Rydberg (not Franchet) is an illegitimate name that applies here.

14. Rubus illecebrosus Focke, Abh. Naturwiss. Vereine Bremen 16: 278. 1899 • Strawberry raspberry [I]

Shrubs, 5–10(–15) dm, armed. **Stems** <annual>, arching to mounding, glabrous, eglandular, not pruinose; prickles sparse, slightly hooked, stout, 3–4 mm, broad-based. **Leaves** deciduous, pinnately compound; stipules lanceolate to oblanceolate, 8–13 mm; leaflets (3–)5–7, terminal lanceolate to narrowly ovate, (5.5–)6.2–8.5(–9.7) × (1.8–)2.2–3(–4.2) cm, base usually rounded, sometimes truncate to subcordate, unlobed, margins finely to coarsely doubly serrate, apex acuminate, abaxial surfaces with prickles on midvein, glabrous, eglandular. **Inflorescences** solitary flowers or 2- or 3-flowered clusters. **Pedicels:** prickles sparse, slightly hooked, glabrous, eglandular. **Flowers** bisexual; petals white, broadly obovate to suborbiculate, 12–15(–18) mm; filaments laminar; ovaries glabrous, <styles glabrous>. **Fruits** red, oblong, 1.4–2 cm; drupelets 50–100, strongly coherent, separating from torus. **2*n*** = 14.

Flowering Jun–Sep. Disturbed sites; 0–1200 m; introduced; N.S., Ont.; Conn., Ky., Maine, Mass., Minn., N.Y., N.C., Oreg., Pa., Va., W.Va.; e Asia; introduced also in South America (Colombia, Paraguay, Venezuela).

Rubus illecebrosus is distinguished from other raspberries by its prickly, glabrous, eglandular stems, pinnately compound leaves with typically seven lanceolate to narrowly ovate leaflets, relatively large flowers, white petals, and 50–100 drupelets.

15. Rubus laciniatus Willdenow, Hort. Berol. 2: plate 82. 1806 • Cutleaf blackberry [I]

Shrubs, 1–20 dm, armed. **Stems** <biennial>, arching, sometimes creeping, glabrate to moderately hairy, sparsely to densely sessile- to short-stipitate-glandular, not or weakly pruinose; prickles moderately dense to dense, suberect to retrorse or hooked, stout, 2–8 mm, broad-based. **Leaves** evergreen to late-deciduous, ternate or palmately compound; stipules filiform, 5–15 mm; leaflets 3–5, terminal ovate to elliptic, 6–12 × 7–12 cm, base rounded to cordate, usually deeply, rarely shallowly, lobed, margins coarsely serrate to doubly serrate, apex acuminate to short-attenuate, abaxial surfaces with hooked prickles primarily on midvein, moderately to densely hairy, eglandular or moderately sessile- to short-stipitate-glandular primarily on midvein. **Inflorescences** terminal, 5–25-flowered, thyrsiform. **Pedicels:** prickles moderate to dense, erect to reflexed, densely hairy, moderately to densely sessile- to short-stipitate-glandular. **Flowers** bisexual; petals white to pink, elliptic to obovate, 8–15 mm; filaments filiform; ovaries usually apically hairy. **Fruits** black, globose to subcylindric, 1–1.5 cm; drupelets 25–60, strongly coherent, separating with torus attached. $2n$ = 28.

Flowering Jun–Sep(–Oct). Woodland edges, semi-open woods, rocky and/or moist soil, open areas, often disturbed, roadsides; 0–1800 m; introduced; B.C., Ont.; Calif., Colo., Conn., Del., D.C., Idaho, Ill., Ind., Ky., Md., Mass., Mich., Mo., Mont., N.J., N.Y., N.C., Ohio, Oreg., Pa., R.I., S.C., Tenn., Vt., Va., Wash., W.Va., Wyo.; Europe.

Rubus laciniatus is the only blackberry in the flora area that has leaflets so deeply lobed and dissected or has 3-lobed petals. Rare individuals bearing relatively unlobed leaflets and perianth parts may be part of the natural variation of this species or be hybrids with some other species. Putative natural hybrids have been reported with *R. bifrons* (as *R. procerus*; R. K. Bammi and H. P. Olmo 1966). Such unusual plants of *R. laciniatus* would be distinguished from the native blackberries by their thyrsiform and notably stipitate-glandular inflorescences and evergreen to late-deciduous leaves.

Although widely introduced in the flora area, *Rubus laciniatus* is a less aggressive weed than *R. bifrons*.

16. Rubus lasiococcus A. Gray, Proc. Amer. Acad. Arts 17: 201. 1882 • Roughfruit berry [E]

Herbs, 1–1.5 dm, unarmed. **Stems** creeping, <flowering branches erect>, sparsely hairy, eglandular or sparsely to moderately short-stipitate-glandular, not pruinose. **Leaves** deciduous, simple or ternate; stipules broadly elliptic to ovate, 5–10 mm; leaflets 3(–5), blade reniform or orbiculate, terminal lobe or leaflets broadly obovate to obovate-cuneate, 1.5–4(–6) × 1.5–4(–6) cm, base cordate, 3(–5)-lobed, margins coarsely doubly dentate, apex rounded to acute, abaxial surfaces sparsely hairy on veins, eglandular. **Inflorescences** 1- or 2-flowered. **Pedicels** sparsely to moderately hairy, short-stipitate-glandular. **Flowers** bisexual; petals white, broadly elliptic to obovate to orbiculate, (5–)8–10(–12) mm; filaments filiform; ovaries densely hairy, <styles filiform, glabrous>. **Fruits** red, hemispheric, 1 cm; drupelets 5–15, weakly coherent, falling separately or as loose units free from torus. $2n$ = 14.

Flowering May–Aug. Moist to dry, semiopen forests, wet meadows, roadsides, dry sand; 800–2000 m; B.C.; Calif., Oreg., Wash.

Rubus lasiococcus is recognized by its creeping, unarmed stems, simple 3-lobed to 3-foliate leaves, small flowers, white petals, and densely hairy ovaries.

17. Rubus leucodermis Douglas ex Torrey & A. Gray, Fl. N. Amer. 1: 454. 1840

Rubus occidentalis Linnaeus subsp. *leucodermis* (Douglas ex Torrey & A. Gray) Focke

Shrubs, 10–30 dm, armed. **Stems** <biennial>, ascending, often arching, glabrous or sparsely puberulent, eglandular or sparsely to densely short-stipitate-glandular, strongly pruinose; prickles moderate to dense, erect or curved, slender to stout, 4–8 mm, usually broad-based, sometimes narrow-based. **Leaves** deciduous, palmately compound or ternate; stipules filiform, 5–10 mm; <petiole with hooked or erect prickles, glabrous, eglandular or sparsely to densely sessile- to short-stipitate-glandular; lateral leaflets stalked, at least 2 larger ones, often with prickles, glabrous, eglandular or stipitate-glandular>; leaflets 3–5(–7), terminal ovate to lanceolate, 6–10 × 3–6 cm, base cordate to truncate, unlobed or shallowly 3-lobed, margins serrate to doubly serrate, apex acuminate, abaxial surface sometimes with prickles on midrib and some lateral veins, glabrous to strongly white-tomentose, usually eglandular. **Inflorescences** axillary and terminal, (1–)3–10(–12)-flowered, flat-topped cymiform or umbelliform. **Pedicels:** prickles sparse to dense, hooked or erect, puberulent, eglandular or sparsely to densely sessile- or short-stipitate-glandular. **Flowers** bisexual; petals <erect, then ascending>, white, oblong to oblanceolate, 3–6(–8) mm; filaments laminar; ovaries pubescent, rarely glabrous. **Fruits** reddish purple to nearly black, depressed globose to conic, 1–2 cm; drupelets 20–60, coherent, separating from torus. $2n$ = 14.

Varieties 2 (2 in the flora): w North America, n Mexico.

Rubus leucodermis is similar to *R. occidentalis*, differing primarily by its leaflets having acuminate apices and lateral leaflets that are distinctly stalked. Prickles on the pedicels of *R. occidentalis* are erect; those of *R. leucodermis* tend to be hooked.

1. Leaves abaxially white-tomentose; drupelets (20–)30–60 . . . . 17a. *Rubus leucodermis* var. *leucodermis*
1. Leaves abaxially glabrous, rarely sparsely hairy; drupelets 20–40 . . . . 17b. *Rubus leucodermis* var. *nigerrimus*

17a. Rubus leucodermis Douglas ex Torrey & A. Gray var. **leucodermis** • Whitebark raspberry

Rubus bernardinus (Greene) Fedde; *R. leucodermis* subsp. *bernardinus* (Greene) Thorne; *R. leucodermis* var. *bernardinus* (Greene) Jepson; *R. leucodermis* var. *trinitatis* A. Berger

Plants 10–30 dm, moderately to densely armed. **Leaves** abaxially white-tomentose. **Inflorescences** (2 or)3–10(–12)-flowered. **Flowers:** petals oblong to oblanceolate, 3–6(–8) mm. **Fruits:** drupelets (20–)30–60, pubescent. $2n$ = 14.

Flowering Apr–Aug. Moist to dry, open, often rocky areas, shaded roadsides, sheltered openings, thickets, open forests; 0–2700 m; B.C.; Alaska, Ariz., Calif., Idaho, Mont., Nev., N.Mex., Oreg., Utah, Wash.; Mexico (Chihuahua).

17b. Rubus leucodermis Douglas ex Torrey & A. Gray var. **nigerrimus** (Greene) H. St. John, Fl. S.E. Washington, 213. 1937 • Dark or northwest raspberry [C] [E]

Melanobatus nigerrimus Greene, Leafl. Bot. Observ. Crit. 1: 244. 1906, based on *Rubus hesperius* Piper, Erythea 5: 103. 1897, not Rogers 1896; *R. nigerrimus* (Greene) Rydberg

Plants 10–20 dm, densely armed. **Leaves** abaxially glabrous, rarely sparsely hairy. **Inflorescences** 1–5-flowered. **Flowers:** petals oblong, 4–7 mm. **Fruits:** drupelets 20–40, glabrous, rarely puberulent.

Flowering Apr–Aug. Moist to dry shrubby thickets and open forests in canyons; of conservation concern; 200–700 m; Wash.

Variety *nigerrimus* can be distinguished from the typical variety by its essentially glabrous foliage. This variety is endemic to southeastern Washington near the Snake River in Garfield and Whitman counties. M. E. Peck (1941) reported it (as *Rubus hesperius*) from Oregon; no collection from there has been seen as part of this review. The taxonomic status of this variant remains questionable as characters such as pubescence are not discrete and may reflect the effect of environment.

18. Rubus neomexicanus A. Gray, Smithsonian Contr. Knowl. 5(6): 55. 1853 (as neo-mexicanus) • New Mexico raspberry [E]

Rubus deliciosus Torrey var. *neomexicanus* (A. Gray) Kearney; *R. exrubicundus* L. H. Bailey

Shrubs, (10–)20–30(–35) dm, unarmed. **Stems** erect, sparsely to moderately short-hairy, sparsely stipitate-glandular, not pruinose. **Leaves** deciduous, simple; stipules lanceolate to narrowly ovate, (4–)6–8(–10) mm; blade cordate to broadly ovate, (2.5–)3.5–5.5(–6.5) × (2.5–)3.5–5.5(–8) cm, base shallowly to deeply cordate, palmately, shallowly to deeply 3–5(–7)-lobed, margins coarsely doubly dentate, apex acute to broadly obtuse, abaxial surfaces moderately hairy, sparsely stipitate-glandular. **Inflorescences** 1(or 2)-flowered. **Pedicels** sparsely to moderately hairy, sparsely to densely short- to long-stipitate-glandular. **Flowers** bisexual; petals white, broadly obovate to suborbiculate, (15–)20–30(–35) mm; filaments filiform; ovaries glabrous, <styles clavate, villous>. **Fruits** red, hemispheric, to 1 cm; drupelets 20, coherent, separating from torus. $2n$ = 14.

Flowering May–Sep. Mountain slopes, canyons, streams; 1400–2600 m; Ariz., Colo., N.Mex., Utah.

Rubus neomexicanus is recognized by its erect, unarmed stems, relatively small, simple leaves with acute to obtuse lobes, the terminal prominent, deeply cordate base, moderately hairy abaxially, large flowers, white petals, and densely long-hairy, clavate styles. The species is similar to *R. bartonianus* and *R. deliciosus*, and treating it as a variety of the latter species would be reasonable, as W. O. Focke (1910) considered the two conspecific.

19. Rubus nivalis Douglas in W. J. Hooker, Fl. Bor.-Amer. 1: 181. 1832 • Snow raspberry [E]

Shrubs, to 1.5 dm, armed. **Stems** <perennial>, creeping, sparsely hairy, glabrescent, eglandular, not pruinose; prickles sparse, strongly retrorse, stout, to 1 mm, broad-based. **Leaves** evergreen, simple or ternate; stipules <adnate to petioles>, broadly elliptic to ovate, (6–)8–10 mm; blade ovate to cordate, (2.5–)3–5(–8) cm, lobe or leaflet base cordate, shallowly 3-lobed, margins coarsely, singly or doubly dentate, apex acute to obtuse, abaxial surfaces with prickles along midveins, glabrous or sparsely hairy, eglandular, <both surfaces lustrous fresh>. **Inflorescences** 1- or 2-flowered. **Pedicels:** prickles scattered, retrorse, moderately to densely hairy, eglandular. **Flowers** bisexual; petals magenta to pink, elliptic to oblanceolate or spatulate, (5–)8–10 mm; filaments filiform; ovaries moderately hairy, <styles glabrous>. **Fruits** red, hemispheric, 0.4–1 cm; drupelets 3–10, not coherent, separating from torus. **2*n*** = 14.

Flowering Jun–Sep. Moist, semishaded forests, glades, moist soil, logged areas; 0–1700 m; B.C.; Calif., Idaho, Oreg., Wash.

Rubus nivalis is recognized by its creeping, prickly stems, simple to ternate, evergreen leaves, broadly elliptic to ovate stipules, two leaflets, relatively small flowers, and magenta to pink petals. Its closest relative is likely the Mexican *R. pumilus* Focke. Asian species previously classified in subg. *Chamaebatus* (Focke) Focke are hexaploid (M. M. Thompson 1997) and not phylogenetically close; *R. nivalis* appears to be sister to all blackberries of subg. *Rubus* (L. A. Alice and C. S. Campbell 1999; Alice et al. 2008).

The fruits of *Rubus nivalis* are eaten fresh, stewed, and canned by the Hoh and Quileute Indians (A. B. Reagan 1936).

The only known specimen of *Rubus nivalis* from California was collected in 1961 from Del Norte County at 1250 m near the Oregon border.

20. Rubus niveus Thunberg, Rubo, 9, fig. 3. 1813
• Snowpeaks, Mysore, or hill raspberry [I]

Shrubs, 10–30 dm, armed. **Stems** <biennial>, erect to arching, often scrambling, sparsely hairy, glabrescent, eglandular, strongly pruinose; prickles sparsely to moderately dense, hooked to reflexed, stout, 3–9 mm, broad-based. **Leaves** deciduous, pinnately compound; stipules linear-lanceolate, 4–8 mm; leaflets (3–)5–7(–9), terminal ovate to broadly ovate, 4–6.7 × (2.4–)3.4–5.4 cm, base shallowly cordate, sometimes shallowly 3-lobed, margins coarsely serrate to doubly serrate, apex acute to short-acuminate, abaxial surfaces with scattered, reflexed, broad-based prickles on midveins, densely white-tomentose, eglandular. **Inflorescences** terminal and axillary, 1–11-flowered, cymiform. **Pedicels:** prickles absent or sparse, erect to reflexed, pubescent, eglandular. **Flowers** bisexual; petals pink to magenta, broadly obovate to orbiculate, 4 mm; filaments laminar; ovaries densely hairy. **Fruits** purple-black, hemispheric to globose, 0.5–1.5 cm; drupelets 50–75, coherent, separating from torus. **2*n*** = 14.

Flowering year-round. Disturbed sites adjacent to tropical or subtropical woodlands; 0–30 m; introduced, Fla.; Africa; introduced also in Central America, South America, Pacific Islands (Hawaii).

Rubus niveus is distinguished from other raspberries by its pinnately compound leaves with 5–7 leaflets, ovate to broadly ovate terminal leaflet, leaflets with prickles on the midvein, and pink to magenta, broadly obovate to orbiculate petals. The species was introduced into southern Florida around 1949 from seeds obtained in Kenya and quickly became popular for its fruits, which are edible fresh or frozen and are often used in preserves, pies, and tarts (J. F. Morton 1987).

21. Rubus occidentalis Linnaeus, Sp. Pl. 1: 493. 1753
• Black raspberry, framboisier noir [E] [F]

Shrubs, 5–25 dm, armed. **Stems** <biennial>, erect, <primocanes and floricanes later over-arching>, glabrous or sparsely puberulent, eglandular, strongly pruinose; prickles sometimes sparse, erect or hooked, narrow to stout, 4–8 mm, narrow- to broad-based. **Leaves** deciduous, ternate or palmately compound; stipules filiform, 5–10 mm; <petiole usually armed with prickles; petiolules of terminal leaflets also sometimes armed; lateral leaflets sessile or subsessile>; leaflets 3(–5), terminal ovate to lanceolate, 5–14 × 3–11 cm, base rounded to cordate, unlobed or laterals sometimes lobed, margins finely to doubly serrate, apex acute, abaxial surfaces sometimes with hooked prickles on midveins, densely white-canescent to tomentose, eglandular. **Inflorescences** (2 or)3–7(–20)-flowered, cymiform or umbelliform. **Pedicels:** prickles erect, hooked, puberulent to pubescent, eglandular. **Flowers** bisexual; petals <initially erect, later ascending>, white, narrowly obovate to elliptic, 5–10 mm, <apex sometimes emarginate>; filaments laminar; ovaries tomentose. **Fruits** usually dark purplish to black, rarely amber, <pruinose, aromatic>, depressed-globose, 1–1.5 cm; drupelets 20–50, coherent, separating from torus. **2*n*** = 14.

Flowering late Apr–Jul. Woodlands, fields, prairies, meadows, savannas, in disturbed areas, dry to moist soil; 0–1000 m; N.B., Ont., Que.; Ala., Ark., Colo., Conn., Del., D.C., Ga., Ill., Ind., Iowa, Kans., Ky., Maine, Md., Mass., Mich., Minn., Miss., Mo., Nebr., N.H., N.J., N.Y., N.C., N.Dak., Ohio, Okla., Pa., R.I., S.C., S.Dak., Tenn., Vt., Va., W.Va., Wis.

Rubus occidentalis is the source of most of the black raspberries in cultivation. See 13b. *R. idaeus* subsp. *strigosus* for discussion of application of the name *R. neglectus*. Also, see 17. *R. leucodermis* for a discussion of its taxonomic closeness. The morphologically similar species *R. eriocarpus* Liebmann is known from southern Mexico and Central America, and *R. pringlei* Rydberg is known from Mexico and Guatemala.

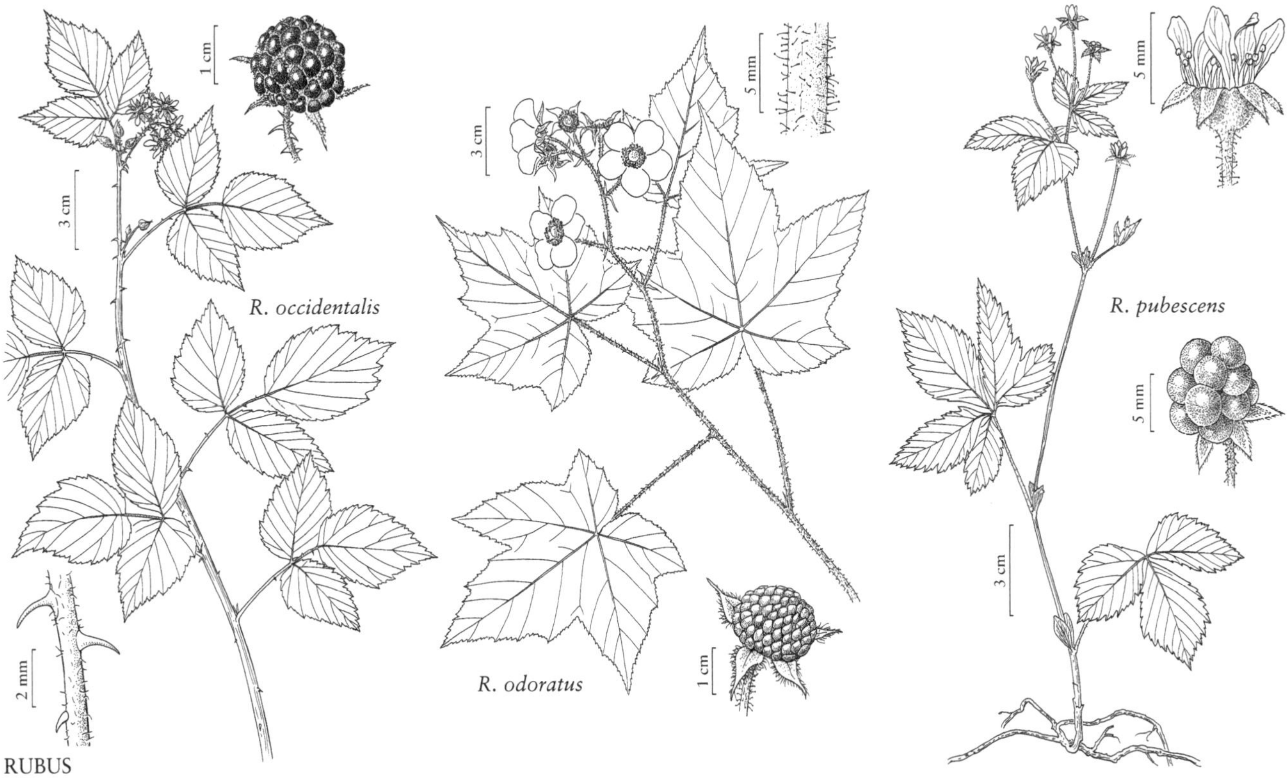

RUBUS

22. Rubus odoratus Linnaeus, Sp. Pl. 1: 494. 1753 • Purple flowering raspberry, ronce odorante [E] [F]

Rubacer odoratum (Linnaeus) Rydberg; *Rubus odoratus* var. *albidus* L. H. Bailey; *R. odoratus* var. *columbianus* Millspaugh

Shrubs, 10–20 dm, unarmed. **Stems** erect, sparsely to moderately hairy, moderately to densely stipitate-glandular, <glands dark purple>, not pruinose. **Leaves** deciduous, simple; stipules lanceolate to ovate, 5–15 mm; blade subrotund to reniform, 9–20(–30) × (10–)15–25(–30) cm, base cordate, palmately, ± deeply, (3–)5-lobed, margins finely, irregularly serrate to doubly serrate, apex acute to acuminate, abaxial surfaces sparsely to moderately hairy, sparsely to densely stipitate-glandular, <glands dark purple>. **Inflorescences** terminal and axillary, 4–7(–22)-flowered, cymiform to thyrsiform. **Pedicels** pubescent, moderately to densely stipitate-glandular, <glands dark purple>. **Flowers** bisexual; petals usually magenta, rarely white, broadly obovate to suborbiculate, (12–)17–25(–30) mm; filaments filiform; ovaries distally densely hairy, <styles clavate, glabrous>. **Fruits** pale to dark red, hemispheric, 0.7–1.5 cm; drupelets 30–60, coherent, separating from torus. $2n = 14$.

Flowering May–Jul. Moist shady sites in deciduous forests, margins of woods, rocky slopes, wooded talus, stream banks, roadsides; 10–1500 m; N.B., Ont., Que.; Ala., Conn., Del., D.C., Ga., Ill., Ind., Ky., Maine, Md., Mass., Mich., Mo., N.H., N.J., N.Y., N.C., Ohio, Pa., R.I., Tenn., Vt., Va., Wash., W.Va., Wis.; introduced in Europe.

Rubus odoratus is introduced in Washington State. The species is distinguished from other flowering raspberries by its erect, unarmed stems, simple leaves, large flowers, magenta petals, glabrous, clavate styles, and purple stipitate glands densely covering most plant parts. A hybrid with *R. parviflorus* (*R.* ×*fraseri* Rehder) is thought to occur in areas of overlap in northern Michigan (E. G. Voss 1972–1996, vol. 2).

The Cherokee used leaf infusions for labor pains and the Iroquois used plant infusions for miscarriage (P. Bergner 1997). *Rubus odoratus* is grown as an ornamental for its relatively large flowers and magenta petals and its edible fruits that are somewhat dry and insipid.

23. Rubus parviflorus Nuttall, Gen. N. Amer. Pl. 1: 308. 1818 • Thimbleberry

Rubus parviflorus subsp. *velutinus* (W. H. Brewer) Roy L. Taylor & MacBryde; *R. parviflorus* var. *velutinus* (W. H. Brewer) Greene

Shrubs, 5–20(–30) dm, unarmed. **Stems** <biennial>, erect, sparsely hairy, glabrescent, moderately stipitate-glandular, <glands yellowish to reddish>, not pruinose. **Leaves** deciduous, simple; stipules lanceolate to ovate, 5–15 mm; blade orbiculate to reniform, 5–20 × 5–25 cm, base cordate, palmately, shallowly to moderately deeply, (3–)5(–7) -lobed, margins coarsely, irregularly serrate to doubly serrate, apex shortly acuminate to obtuse, abaxial surfaces glabrate to densely hairy, sparsely to moderately stipitate-glandular, <glands yellowish to reddish>. **Inflorescences** terminal and axillary, (1–)3–7(–15), cymiform to thyrsiform. **Pedicels** sparsely to moderately hairy, moderately to densely stipitate-glandular, <glands yellowish to reddish>. **Flowers** bisexual; petals white, broadly obovate, (10–)14–22(–28) mm; filaments filiform; ovaries distally densely hairy, <styles clavate, glabrous>. **Fruits** red, hemispheric, 1–1.8 cm; drupelets 50–60, coherent, separating from torus. **2*n*** = 14.

Flowering May–Aug. Moist areas, open woods, thickets, clearings, stream banks, canyons, grassy meadows, rocky cliffs, sand dunes, upper beaches, dry sandy areas, roadsides; 0–3000 m; Alta., B.C., Ont.; Alaska, Ariz., Calif., Colo., Idaho, Ill., Iowa, Mich., Minn., Mont., Nev., N.Mex., Oreg., S.Dak., Utah, Wash., Wis., Wyo.; Mexico (Chihuahua).

Rubus parviflorus is distinguished from other flowering raspberries by its erect, unarmed stems, simple leaves, large flowers, white petals, glabrous, clavate styles, and yellowish to reddish stipitate glands covering most plant parts. A hybrid with *R. odoratus* (*R.* ×*fraseri* Rehder) is thought to occur in areas of overlap in northern Michigan (E. G. Voss 1972–1996, vol. 2). Varieties and forms have been described based on indument and glandularity of stems, leaves, petioles, pedicels, and sepals (for example, M. L. Fernald 1950). Different variants often occur mixed in the same population and generally do not correspond to definite geographic regions (N. C. Fassett 1941). Plants from coastal California (considered as var. *velutinus*) tend to have densely hairy leaf abaxial surfaces; this variation may be environmentally induced; other *Rubus* species show the same pattern (for example, *R. spectabilis* var. *franciscanus*) in that area. Wild thimbleberries are collected in Keweenaw Peninsula, Michigan, and are used in beverages and especially for jam.

Rubus parviflorus is typically associated with western North America but was originally described from plants at its eastern range limit in northern Michigan. The Great Lakes populations are disjunct from western mountain populations by around 600 miles.

The name *Rubus parviflorus* var. *grandiflorus* Farwell, which applies here, is illegitimate.

24. Rubus parvifolius Linnaeus, Sp. Pl. 2: 1197. 1753 • Japanese bramble, Australian raspberry or bramble [I]

Rubus triphyllus Thunberg

Shrubs, 1–3 dm, armed. **Stems** <biennial>, low-mounding to creeping, sparsely to moderately hairy, glabrescent, eglandular, not pruinose; prickles sparse to moderate, slightly to strongly hooked, weak to stout, 0.8–1.5(–2) mm, broad-based. **Leaves** deciduous, ternate or pinnately compound; stipules filiform to linear, (4–)6–11 mm; leaflets 3(–5), terminal broadly ovate-rhombic to obovate, (2–)3–5(–9) × 2.2–5(–7.5) cm, base cuneate to obtuse, slightly 3-lobed, margins coarsely serrate to doubly serrate, apex acute to obtuse, abaxial surfaces with slightly to strongly hooked prickles on veins, densely whitish-tomentose, eglandular or sparsely short-stipitate-glandular. **Inflorescences** terminal and axillary, 2–20-flowered, cymiform to thyrsiform. **Pedicels:** prickles moderate, slightly to strongly hooked, moderately hairy, eglandular or sparsely short-stipitate-glandular. **Flowers** bisexual; petals pinkish to magenta, oblanceolate to obovate, 4–7 mm; filaments filiform; ovaries moderately to densely hairy, <styles glabrous>. **Fruits** red, round, 0.6–0.9 cm; drupelets 10–50, strongly coherent, separating from torus. **2*n*** = 14.

Flowering May–Jun. Disturbed sites; 0–400 m; introduced; Del., Ill., Iowa, Mass., Mo., Nebr., N.J., Ohio, Va.; e Asia; Australia.

Rubus parvifolius is distinguished from other raspberries by its broadly ovate-rhombic to obovate leaflets and relatively small flowers with pinkish to magenta petals. This species has the potential to become a very significant weed (P. M. Drobney and M. P. Widrlechner 2012).

25. Rubus pascuus L. H. Bailey, Gentes Herbarum 5: 440, fig. 200. 1943 • Topsy, Nanticoke or Chesapeake blackberry [E]

Rubus serissimus L. H. Bailey; *R. uliginosus* Fernald

Shrubs, 7–30 dm, armed. **Stems** <biennial>, arching, sparsely to densely hairy, sparsely to densely sessile-glandular, rarely short-stipitate-glandular, not pruinose; <bark not papery, peeling>; prickles moderate to dense, hooked, retrorse, or erect, stout, 4–10 mm, broad- or narrow-based. **Leaves** deciduous to semievergreen, palmately compound, rarely ternate; stipules filiform to linear, rarely narrowly lanceolate, 5–16(–20) mm; leaflets 3–5, terminal broadly elliptic or ovate, (3.5–)4.5–8.5 × 2.5–6 cm, base rounded, unlobed, margins moderately to coarsely doubly, rarely singly, serrate, apex acute or short-acuminate, abaxial surfaces with hooked prickles on larger veins, short-velutinous to tomentose, sparsely to densely sessile-glandular, rarely short-stipitate-glandular. **Inflorescences** terminal and axillary, 3–15(–25)-flowered, cymiform to thyrsiform, <not projected well beyond subtending leaves>. **Pedicels:** prickles moderate to dense, hooked, retrorse, or erect, densely hairy, eglandular or sparsely to moderately sessile-glandular. **Flowers** bisexual; petals white to pale pink, obovate or elliptic to orbiculate, (8–)10–15 mm; filaments filiform; ovaries glabrous or apically hairy. **Fruits** black, globose to subcylindric, 1–2 cm; drupelets (15–)20–40(–50), strongly coherent, separating with torus attached. **2*n*** = 21, 28.

Flowering (Apr–)May–Jun. Woodland edges, swamp margins, dry thickets, open and often disturbed areas, roadsides; 0–300(–700) m; Ala., Ark., Del., Ga., Ill., Ind., Ky., Md., Miss., Mo., N.J., N.C., Okla., Pa., S.C., Tenn., Tex., Va.

Rubus pascuus is probably more widespread than indicated here. The species is most likely to be mistaken for *R. bifrons*, *R. cuneifolius*, or *R. pensilvanicus*. Unlike *R. bifrons*, *R. pascuus* is generally more densely glandular, has much shorter, compact inflorescences that usually barely surpass the leaves, and tends to hold its primocane leaves relatively erect (versus spreading in *R. bifrons*). Although both can be semievergreen, *R. pascuus* will lose its leaves more readily than *R. bifrons* in areas where they co-occur. When compared with *R. cuneifolius*, *R. pascuus* has pale pink to white petals and ovate to broadly elliptic leaflets with flat margins, the entire part restricted to the base; *R. cuneifolius* has white petals and cuneate to obovate leaflets with usually revolute margins, the proximal third entire. When compared with *R. pensilvanicus*, *R. pascuus* has pale pink to white petals. *Rubus pensilvanicus* only has white petals as well as abaxial leaf surfaces, which are white to gray-green in *R. pascuus* with dense, short-velutinous to tomentose indument, while those of *R. pensilvanicus* can be hairy but never whitish to gray-green. A notable aspect of *R. pascuus* is how vigorously armed it is, more so than most other *Rubus* species in the flora area.

Rubus pascuus may represent a stabilized hybrid between a European and a native blackberry species, suggested by whitish pubescence on the abaxial leaf surface and pale pink petals. L. H. Bailey placed *R. pascuus* in his sect. *Cuneifolii* (L. H. Bailey) L. H. Bailey, likely because of the whitish hairs on the abaxial leaf surface.

The use of the name *Rubus pascuus* here is tentative; it is the best fit to the material observed. This treatment follows H. A. Davis (1990) for the choice of the name to use and its associated synonymy.

26. Rubus pedatus Smith, Pl. Icon. Ined. 3: plate 63. 1791 • Strawberryleaf raspberry

Herbs, to 1 dm, unarmed. **Stems** creeping, <flowering branches erect>, sparsely hairy, eglandular or sparsely stipitate-glandular, not pruinose. **Leaves** deciduous, pedately compound; stipules broadly ovate to suborbiculate, 2–3 mm; leaflets 3 (often appearing as if 5), <lateral leaflets deeply lobed, sinuses nearly to leaflet base>, terminal obovate to obovate-rhombic, (1–)1.5–2.5(–3.4) × (0.8–)1.2–2(–2.5) cm, base cuneate, margins coarsely singly or doubly serrate, apex rounded to obtuse, abaxial surfaces glabrous or sparsely hairy on midvein, eglandular or sparsely stipitate-glandular. **Inflorescences** 1-flowered. **Pedicels** sparsely hairy, eglandular or sparsely stipitate-glandular. **Flowers** bisexual; petals white, oblong to narrowly obovate, 6–10 mm; filaments filiform; ovaries glabrous, <styles glabrous>. **Fruits** red, 0.6–1 cm; drupelets 3–6, weakly coherent, separating from torus. **2*n*** = 14.

Flowering Jun–Aug. Coniferous woods, forest edges, glades, meadows, bogs, stream banks, streambeds, roadsides; 20–2300 m; Alta., B.C., Yukon; Alaska, Idaho, Mont., Oreg., Wash.; ne Asia.

Rubus pedatus is recognized by its creeping, unarmed stems, pedately 3–5-foliate leaves, relatively small flowers, white petals, and glabrous ovaries.

27. Rubus pensilvanicus Poiret in J. Lamarck et al., Encycl. 6: 246. 1804 • Pennsylvania blackberry, ronce de Pennsylvanie [E]

Rubus abactus L. H. Bailey; *R. ablatus* L. H. Bailey; *R. abundiflorus* L. H. Bailey; *R. amnicola* Blanchard; *R. andrewsianus* Blanchard; *R. argutus* Link; *R. arvensis* L. H. Bailey; *R. associus* Hanes ex L. H. Bailey; *R. barbarus* L. H. Bailey; *R. bellobatus* L. H. Bailey; *R. betulifolius* Small; *R. blakei* L. H. Bailey; *R. bractealis* L. H. Bailey; *R. brainerdii* Rydberg; *R. burnhamii* L. H. Bailey; *R. bushii* L. H. Bailey; *R. cardianus* L. H. Bailey; *R. condensiflorus* L. H. Bailey; *R. congruus* L. H. Bailey; *R. cupressorum* Fernald; *R. defectionis* Fernald; *R. densipubens* L. H. Bailey; *R. densissimus* H. A. Davis & T. Davis; *R. facetus* L. H. Bailey; *R. fatuus* L. H. Bailey; *R. floridus* Trattinnick; *R. frondosus* Bigelow; *R. gnarus* L. H. Bailey; *R. heterophyllus* Willdenow; *R. humilior* L. H. Bailey; *R. impar* L. H. Bailey; *R. independens* L. H. Bailey; *R. insons* L. H. Bailey; *R. insulanus* L. H. Bailey; *R. interioris* L. H. Bailey; *R. jennisonii* L. H. Bailey; *R. jugosus* L. H. Bailey; *R. kelloggii* L. H. Bailey; *R. latifoliolus* L. H. Bailey; *R. laudatus* A. Berger; *R. leggii* H. A. Davis & T. Davis; *R. libratus* L. H. Bailey; *R. louisianus* A. Berger; *R. missouricus* L. H. Bailey; *R. mollior* L. H. Bailey; *R. multispinus* Blanchard; *R. oklahomus* L. H. Bailey; *R. originalis* L. H. Bailey; *R. paludivagus* Fernald; *R. pauper* L. H. Bailey; *R. pauxillus* L. H. Bailey; *R. penetrans* L. H. Bailey; *R. perfoliosus* L. H. Bailey; *R. perpauper* L. H. Bailey; *R. persistens* Rydberg; *R. philadelphicus* Blanchard; *R. praepes* L. H. Bailey; *R. pratensis* L. H. Bailey; *R. prestonensis* H. A. Davis & T. Davis; *R. pubifolius* L. H. Bailey; *R. recurvans* Blanchard; *R. saepescandens* L. H. Bailey; *R. sativus* (L. H. Bailey) Brainerd; *R. subsolanus* L. H. Bailey; *R. subtractus* L. H. Bailey; *R. suus* L. H. Bailey; *R. tygartensis* H. A. Davis & T. Davis; *R. ucetanus* L. H. Bailey; *R. virilis* L. H. Bailey; *R. vixargutus* L. H. Bailey; *R. wahlii* L. H. Bailey; *R. wiegandii* L. H. Bailey

Shrubs, 10–30 dm, armed. **Stems** <biennial>, erect to arching, glabrous or sparsely to densely hairy, eglandular or sparsely to moderately, rarely densely, sessile- to short-stipitate-glandular, not pruinose; prickles sparse to dense, erect or slightly retrorse, stout, 4–10 mm, broad-based. **Leaves** deciduous, palmately compound, <not lustrous>; stipules filiform to narrowly lanceolate, (3–)5–15(–20) mm; leaflets (3–)5(–7), terminal ovate to lanceolate, 5–15 × 3–13 cm, base rounded to shallowly cordate, unlobed, margins finely to coarsely singly or doubly serrate, apex acuminate to long-attenuate, abaxial surfaces <green>, usually with retrorse prickles on midveins, moderately hairy, eglandular or sparsely to moderately sessile- to short-stipitate-glandular along veins. **Inflorescences** terminal <on short shoots, usually appearing axillary>, (2–)5–12(–16)-flowered, cymiform, racemiform, or thyrsiform. **Pedicels** unarmed or prickles sparse, erect, glabrous or sparsely to densely hairy, eglandular or sparsely to moderately sessile- to short-stipitate-glandular. **Flowers** bisexual; petals white, usually obovate to elliptic, rarely suborbiculate, 8–40 mm; filaments filiform; ovaries glabrous. **Fruits** black, globose to cylindric, 1–2 cm; drupelets 10–100, strongly coherent, separating with torus attached. **2*n*** = 14, 21, 28, 35, 36, 42.

Flowering May–Jul. Woodlands, savannas, prairies, fields, meadows, swamps, rock outcrops, sand dunes, sandy soil, disturbed areas, dry to wet soil; 0–1400 m; N.B., Nfld. and Labr. (Nfld.), N.S., Ont., P.E.I., Que.; Ala., Ark., Calif., Conn., Del., D.C., Fla., Ga., Idaho, Ill., Ind., Iowa, Kans., Ky., La., Maine, Md., Mass., Mich., Minn., Miss., Mo., Nebr., N.H., N.J., N.Y., N.C., Ohio, Okla., Pa., R.I., S.C., Tenn., Tex., Vt., Va., Wash., W.Va., Wis.; introduced in Pacific Islands (Hawaii).

As circumscribed here, *Rubus pensilvanicus* represents sect. *Arguti* (L. H. Bailey) L. H. Bailey, in which Bailey (1941–1945) recognized 109 species.

More strongly glandular plants of *Rubus pensilvanicus* may represent introgression with *R. allegheniensis* or other species, especially when exhibiting long-stipitate glands. Thyrsiform inflorescences are present in plants that have been called *R. missouricus*. There are reports (T. S. Mallah 1954; L. V. Clark and M. Jasieniuk 2012; B. Sutherland and L. A. Alice, unpubl.) of *R. pensilvanicus* hybridizing with *R. ursinus*.

Rubus pensilvanicus is introduced in California, Idaho, and Washington.

The following nothospecies names are based on putative hybrids involving *Rubus pensilvanicus* and: *R. allegheniensis* (*R.* ×*avipes* L. H. Bailey, *R.* ×*floricomus* Blanchard, *R.* ×*orarius* Blanchard); *R. canadensis* (*R.* ×*amabilis* Blanchard [not Focke], *R.* ×*amicalis* Blanchard, *R.* ×*crux* Ashe, *R.* ×*elegantulus* Blanchard [= *R. canadensis* var. *elegantulus* (Blanchard) Farwell], *R.* ×*multilicius* L. H. Bailey, *R.* ×*noveboracus* L. H. Bailey, *R.* ×*pergratus* Blanchard [= *R. canadensis* var. *pergratus* (Blanchard) L. H. Bailey]); *R. cuneifolius* (*R.* ×*acer* L. H. Bailey, *R.* ×*acer* var. *subacer* L. H. Bailey, *R.* ×*argutinus* L. H. Bailey, *R.* ×*floridensis* L. H. Bailey); *R. flagellaris* (*R.* ×*akermanii* Fernald, *R.* ×*darlingtonii* L. H. Bailey, *R.* ×*dissitiflorus* Fernald, *R.* ×*hypolasius* Fernald, *R.* ×*janssonii* L. H. Bailey, *R.* ×*largus* L. H. Bailey, *R.* ×*recurvicaulis* Blanchard, *R. rossbergianus* Blanchard); *R. setosus* (*R.* ×*wisconsinensis* L. H. Bailey).

28. Rubus phoenicolasius Maximowicz, Bull. Acad. Imp. Sci. Saint-Pétersbourg 17: 160. 1872 • Wineberry I W

Shrubs, 10–20(–30) dm, armed. **Stems** <biennial>, arching, sparsely to moderately hairy, long-stipitate-glandular, not pruinose; prickles sparse to dense, erect to hooked, slender, 2–8(–10) mm, slender-based; <bristles dense, 2–6 mm, usually gland-tipped and reddish purple, glands ellipsoid to narrowly cup-shaped>. **Leaves** deciduous, ternate; stipules filiform, 5–12 mm; terminal leaflets broadly deltate or ovate to suborbiculate, 4–15 × 3.5–14 cm, base truncate, rounded, or shallowly cordate, unlobed or dentate to shallowly, sharply or bluntly 3-lobed, margins coarsely serrate or doubly serrate, apex acute to acuminate, abaxial surfaces with scattered, erect to curved prickles on moderate- to large-sized veins, densely white-hairy, moderately to densely long-stipitate-glandular along veins. **Inflorescences** terminal and axillary, 5–30-flowered, cymiform to racemiform. **Pedicels** usually unarmed or prickles sparse, erect, moderately to densely hairy, densely long-stipitate-glandular. **Flowers** bisexual; petals white to pink, ovate to suborbiculate, 4–6 mm; filaments laminar; ovaries glabrous. **Fruits** red to maroon, globose, 1–1.5 cm; drupelets 15–40, strongly coherent, separating from torus. **2*n*** = 14.

Flowering May–Jul. Woodlands, roadsides, disturbed open areas, moist soil; 0–1200 m; introduced; B.C.; Ala., Ark., Conn., Del., D.C., Ga., Ill., Ind., Ky., Md., Mass., Mo., N.J., N.Y., N.C., Ohio, Pa., R.I., S.C., Tenn., Vt., Va., W.Va.; e Asia (China, Japan, Korea).

Rubus phoenicolasius was introduced to North America for edible fruit, breeding stock, and for ornament. Attractive for its typically reddish purple glandular hairs, *R. phoenicolasius* nonetheless can be invasive.

29. Rubus pubescens Rafinesque, Med. Repos., hexade 3, 2: 333. 1811 • Dwarf red blackberry or raspberry, red raspberry, ronce pubescente E F

Rubus saxatilis Linnaeus var. *canadensis* Michaux, Fl. Bor.-Amer. 1: 298. 1803; *R. pubescens* var. *pilosifolius* A. F. Hill

Herbs, to 0.4 dm, unarmed. **Stems** creeping, <flowering branches erect>, glabrous or sparsely to moderately hairy, eglandular, not pruinose. **Leaves** deciduous, usually ternate, sometimes pedately 5-foliate; stipules lance-ovate to oblanceolate, (5–)8–12(–20) mm; leaflets 3(–5), <petiolule of terminal leaflet 2–7 mm>, terminal rhombic to obovate, (2–)4–8(–10) × 2–4(–6) cm, base cuneate, unlobed or 2-lobed, margins deeply serrate to doubly serrate, apex acute or acuminate, abaxial surfaces glabrous or sparsely hairy, eglandular. **Inflorescences** 1–3-flowered, sometimes umbelliform. **Pedicels** moderately to densely long-hairy, eglandular to densely stipitate-glandular. **Flowers** bisexual; petals white to pink, oblanceolate to obovate, (4–)6–8(–10) mm; filaments laminar; ovaries glabrous, <styles glabrous>. **Fruits** red, globose to conical, 0.5–1.4 cm diam., drupelets (5–)10–25, loosely coherent, falling separately or as a unit with torus attached. **2*n*** = 14.

Flowering May–Jul. Swamps, bogs, fens, stream banks, moist woods, bluffs, gravel sites, sandy soil; 0–2200 m; St. Pierre and Miquelon; Alta., B.C., Man., N.B., Nfld. and Labr., N.W.T., N.S., Ont., P.E.I., Que., Sask., Yukon; Alaska, Colo., Conn., Idaho, Ill., Ind., Iowa, Maine, Mass., Mich., Minn., Mont., Nebr., N.H., N.J., N.Y., N.Dak., Ohio, Pa., R.I., S.Dak., Vt., Wash., W.Va., Wis., Wyo.

Rubus pubescens is recognized by its creeping, unarmed stems, oblanceolate to lance-ovate stipules, relatively small flowers, and white to pink petals. It hybridizes with *R. arcticus* subsp. *acaulis* in areas of sympatry, and the hybrids are referable to *R.* ×*paracaulis* L. H. Bailey [= *R. pubescens* var. *paracaulis* (L. H. Bailey) B. Boivin]. The hybrids are similar to *R. arcticus* subsp. *acaulis* in having more obovate, rounded leaflets, and larger, pink to magenta petals; they are larger, creeping, and have hairy and stipitate-glandular pedicels and sepals like *R. pubescens*.

30. Rubus repens (Linnaeus) Kuntze, Revis. Gen. Pl. 1: 223. 1891 • Robin runaway, false violet, dalibarde rampante E F

Dalibarda repens Linnaeus, Sp. Pl. 1: 491. 1753

Herbs, 0.5–1 dm, unarmed. **Stems** creeping, moderately appressed-hairy, eglandular or sparsely to moderately stipitate-glandular, not pruinose. **Leaves** deciduous, simple; stipules lanceolate-laciniate, (2–)3–5(–8) mm; blade ovate-orbiculate, 1.5–2.7(–3) × 1.5–3.5(–4) cm, base deeply cordate, unlobed, margins crenate to broadly dentate, apex rounded, abaxial surfaces sparsely to moderately hairy, sparsely to moderately short-stipitate-glandular. **Inflorescences** 1-flowered. **Pedicels** moderately and retrorsely long-hairy, eglandular. **Flowers** bisexual; petals <absent in fertile flowers, present in sterile flowers>, white, narrowly elliptic to oblanceolate, 5–8 mm; filaments filiform; ovaries densely hairy, <styles glabrous>. **Fruits** whitish, 0.3–0.5 cm, <dry>; drupelets 5–10(–15), not coherent, separating from torus, <enclosed by converging sepals>. **2*n*** = 14.

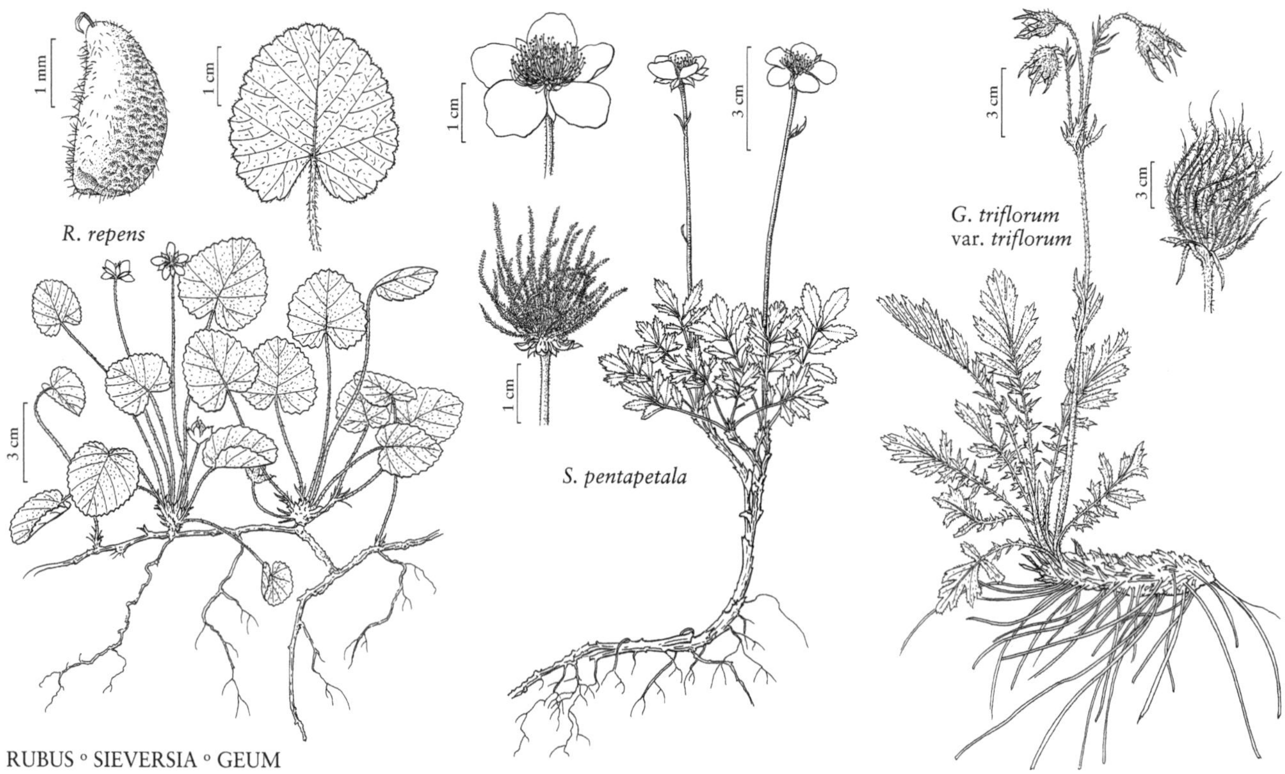

RUBUS ◦ SIEVERSIA ◦ GEUM

Flowering Jun–Sep. Moist woods, swamps; 0–1000 m; N.B., N.S., Ont., P.E.I., Que.; Conn., Maine, Md., Mass., Mich., Minn., N.H., N.J., N.Y., N.C., Ohio, Pa., R.I., Vt., Va., W.Va.

Rubus repens is recognized by its creeping, unarmed stems, simple, ovate-orbiculate leaves, long petioles with spreading hairs, sterile petaliferous flowers on long pedicels and fertile apetalous flowers on short pedicels, and essentially dry fruits. Phylogenetic analysis of DNA sequence data supports inclusion of *Dalibarda repens* in *Rubus* (L. A. Alice and C. S. Campbell 1999), among other basal species to *R. lasiococcus* (K. V. Ambrose 2006).

The Iroquois use a decoction of powdered plants of *Rubus repens* as a blood purifier and for venereal disease (J. W. Herrick 1977).

31. Rubus saxatilis Linnaeus, Sp. Pl. 1: 494. 1753
• Stone bramble

Herbs or subshrubs, to 4 dm, armed or unarmed. **Stems** usually creeping, <± woody basally>, sparsely to densely hairy, eglandular, not pruinose; prickles absent or sparse, erect, weak, 1–3 mm, narrow-based. **Leaves** deciduous, ternate; stipules <free from petioles>, lanceolate to ovate or elliptic, 6–11 mm; <petiole 2–11 cm; petiolule of terminal leaflet 8–20 mm; lateral leaflet sessile or subsessile>; terminal leaflets ovate to elliptic, 4–8 × 3–7 cm, base tapered, unlobed, <lateral leaflets sometimes shallowly lobed>, margins coarsely doubly serrate, apex acute, abaxial surfaces unarmed, sparsely hairy, eglandular. **Inflorescences** axillary, 1–6-flowered, cymiform or umbelliform. **Pedicels** unarmed or prickles weak, erect, hairy, eglandular or sparsely stipitate-glandular. **Flowers** bisexual; petals greenish white, narrowly obovate, 5.5–7 mm; filaments laminar; ovaries glabrous. **Fruits** red, globose, 0.5–1.5 cm; drupelets 1–10, not or weakly coherent, separating with torus attached. $2n = 28$.

Flowering May–Jul. Birch scrub, talus slopes; 1300–2100 m; Greenland; Eurasia.

Rubus saxatilis is defined by its creeping, unarmed or weakly armed stems, ternate leaves, and relatively small white petals. It has very weak prickles that detach easily, sometimes giving plants an unarmed appearance. The species is known in the flora area from only two sites in southern Greenland (T. W. Böcher 1938; D. B. Jensen and K. D. Christensen 2003).

32. Rubus setosus Bigelow, Fl. Boston. ed. 2, 198. 1824 • Bristly blackberry, ronce sétuleuse [E]

Rubus apparatus L. H. Bailey; *R. beatus* L. H. Bailey; *R. benneri* L. H. Bailey; *R. bicknellii* L. H. Bailey; *R. boottianus* L. H. Bailey; *R. condignus* L. H. Bailey; *R. deaneanus* L. H. Bailey; *R. discretus* L. H. Bailey; *R. dissensus* L. H. Bailey; *R. dissimilis* L. H. Bailey; *R. electus* L. H. Bailey; *R. exter* L. H. Bailey; *R. fulleri* L. H. Bailey; *R. groutianus* Blanchard; *R. gulosus* L. H. Bailey; *R. hispidoides* L. H. Bailey; *R. hispidus* Linnaeus var. *suberectus* Peck; *R. jejunus* L. H. Bailey; *R. junceus* Blanchard; *R. junior* L. H. Bailey; *R. lawrencei* L. H. Bailey; *R. mediocris* L. H. Bailey; *R. navus* L. H. Bailey; *R. nocivus* L. H. Bailey; *R. notatus* L. H. Bailey; *R. ortivus* (L. H. Bailey) L. H. Bailey; *R. perinvisus* L. H. Bailey; *R. racemiger* L. H. Bailey; *R. regionalis* (L. H. Bailey) L. H. Bailey; *R. reravus* L. H. Bailey; *R. rotundior* (L. H. Bailey) L. H. Bailey; *R. schneideri* L. H. Bailey; *R. semisetosus* Blanchard; *R. spectatus* L. H. Bailey; *R. stipulatus* L. H. Bailey; *R. uniformis* L. H. Bailey; *R. univocus* L. H. Bailey; *R. vermontanus* Blanchard; *R. viridifrons* L. H. Bailey; *R. wheeleri* (L. H. Bailey) L. H. Bailey

Shrubs, 2–10(–15) dm, armed. **Stems** <biennial>, erect to arching, rarely creeping, <not node- or tip-rooting>, sparsely to moderately hairy, sparsely to densely long-, rarely short-stipitate-glandular, <glands rounded to flattened>, not pruinose; prickles absent or sparsely to moderately dense, erect to retrorse, weak, slender, 1–3(–5) mm, narrow-based; <bristles sparse to dense, erect to retrorse, green to reddish, narrow, semirigid, sometimes flexible, not gland-tipped>. **Leaves** deciduous, ternate or palmately compound, <sometimes ± lustrous>; stipules filiform to lanceolate, 5–20(–38) mm; leaflets (3–)5, terminal obovate or elliptic to rhombic, 4.5–11.5 × 2–7 cm, base cuneate to narrowly rounded, unlobed, margins moderately, coarsely serrate to doubly serrate or serrate-dentate, apex acute or acuminate to short-attenuate, abaxial surfaces unarmed or with bristles on midvein, sparsely to moderately hairy, eglandular or sparsely to moderately sessile- to long-stipitate-glandular along larger veins. **Inflorescences** terminal <on short shoots, sometimes appearing axillary>, 5–15(–20)-flowered, racemiform, cymiform, or thyrsiform. **Pedicels:** prickles or bristles sparse to dense, erect to retrorse, moderately to densely hairy, moderately to densely stipitate-glandular. **Flowers** bisexual; petals white, obovate to oblanceolate, 6–13 mm; filaments filiform; ovaries glabrous. **Fruits** black, globose, 0.7–1.5 cm; drupelets 5–25, strongly coherent, separating with torus attached. $2n$ = 14, 21, 28, 35.

Flowering Jun–Jul(–Aug). Open woodlands, savannas, prairies, meadows, disturbed areas, dry to wet soil; 0–1000 m; N.B., Nfld. and Labr. (Nfld.), N.S., Ont., P.E.I., Que.; Conn., Del., D.C., Ill., Ind., Iowa, Maine, Md., Mass., Mich., Minn., N.H., N.J., N.Y., Ohio, Pa., R.I., Vt., Va., W.Va., Wis.

Rubus setosus is identified by erect to arching, rarely creeping, not tip-rooting, terete primocanes, always setose but variable in density, and deciduous primocane leaves that are often palmately compound with five leaflets. Cultivated plants sometimes alternate from erect to creeping (or vice-versa) in successive years, depending on changing soil conditions.

The following nothospecies names are based on putative hybrids involving *Rubus setosus* and: *R. allegheniensis* (*R.* ×*abbrevians* Blanchard, *R.* ×*aculiferus* Fernald, *R.* ×*adenocaulis* Fernald, *R.* ×*angustifoliatus* L. H. Bailey, *R.* ×*ascendens* Blanchard, *R.* ×*atwoodii* L. H. Bailey, *R.* ×*clausenii* L. H. Bailey, *R.* ×*flavinanus* Blanchard, *R.* ×*frondisentis* Blanchard, *R.* ×*perspicuus* L. H. Bailey, *R.* ×*ravus* L. H. Bailey); *R. canadensis* (*R.* ×*miscix* L. H. Bailey [based on *R.* ×*peculiaris* Blanchard (not *R. peculiaris* Sampaio)]); *R. flagellaris* (*R.* ×*alter* L. H. Bailey, *R.* ×*arcuans* Fernald & H. St. John, *R.* ×*biformispinus* Blanchard, *R.* ×*bigelovianus* L. H. Bailey, *R.* ×*bracteolifer* Fernald, *R.* ×*jacens* Blanchard var. *specialis* L. H. Bailey, *R.* ×*laevior* (L. H. Bailey) Fernald [based on *R. permixtus* var. *laevior* L. H. Bailey], *R.* ×*multiformis* Blanchard, *R.* ×*setospinosus* L. H. Bailey, *R.* ×*severus* Brainerd ex Fernald); *R. hispidus* (*R.* ×*adjacens* Fernald, *R.* ×*grandidens* L. H. Bailey, *R.* ×*harmonicus* L. H. Bailey, *R.* ×*jacens* Blanchard, *R.* ×*parlinii* L. H. Bailey, *R.* ×*spiculosus* Fernald, *R.* ×*tardatus* Blanchard, *R.* ×*tholiformis* Fernald, *R.* ×*trifrons* Blanchard, *R.* ×*zaplutus* L. H. Bailey); *R. pensilvanicus* (*R.* ×*wisconsinensis* L. H. Bailey).

33. Rubus spectabilis Pursh, Fl. Amer. Sept. 1: 348, plate 16. 1813 • Salmonberry

Rubus franciscanus Rydberg; *R. spectabilis* var. *franciscanus* (Rydberg) J. T. Howell

Shrubs, 10–40 dm, usually armed. **Stems** erect to arching, glabrate or sparsely to densely hairy, eglandular or sparsely short-stipitate-glandular, rarely densely long-stipitate-glandular, not pruinose; <bark usually papery with age, peeling (especially toward base)>; prickles absent or sparse to dense, erect, slender, 1–5 mm, broad- to narrow-based. **Leaves** deciduous, ternate; stipules filiform to linear, 3–10 mm; terminal leaflets ovate, 4–15 × 3.5–15 cm, base truncate, rounded to shallowly cordate, shallowly, sharply lobed, margins coarsely serrate to doubly serrate, apex acute to acuminate, abaxial surfaces unarmed or with erect prickles on midvein, moderately to densely hairy, eglandular, rarely stipitate-glandular along midvein. **Inflorescences** terminal and axillary, 1- or 2-flowered. **Pedicels** unarmed or prickles sparse, erect, moderately to densely hairy, eglandular, rarely short-stipitate-glandular. **Flowers** bisexual; petals pink to magenta, broadly to narrowly obovate, 10–30 mm; filaments laminar; ovaries glabrous. **Fruits** yellow, orange, or red, globose to ovoid, 1–2 cm; drupelets 20–80, strongly coherent, separating from torus. $2n = 14$.

Flowering (Feb–)Mar–Jul. Woodlands, woodland edges, bogs, shorelines, roadsides, disturbed areas, moist to wet soil; 0–2000 m; B.C.; Alaska, Calif., Idaho, Oreg., Wash.; e Asia (Japan); introduced in Europe.

Rubus spectabilis is a thicket-forming shrub that has relatively large and desirably edible fruit. The species is used as an ornamental primarily for its robust, showy flowers and is naturalized in parts of western Europe. It is sister to the Hawaiian endemic *R. hawaiiensis* A. Gray. See discussion under 36. *R. ursinus* for the uncertain application of the name *R. menziesii* Hooker.

34. Rubus trivialis Michaux, Fl. Bor.-Amer. 1: 296. 1803 • Southern dewberry

Rubus agilis L. H. Bailey; *R. continentalis* (Focke) L. H. Bailey; *R. duplaris* Shinners; *R. ictus* L. H. Bailey; *R. lucidus* Rydberg; *R. macvaughii* L. H. Bailey; *R. mirus* L. H. Bailey; *R. nessianus* L. H. Bailey; *R. riograndis* L. H. Bailey; *R. rubrisetus* Rydberg; *R. sons* L. H. Bailey; *R. tallahasseanus* L. H. Bailey

Shrubs, to 3(–7) dm, <sometimes climbing higher through other vegetation>, moderately to densely armed. **Stems** <biennial>, initially low-arching, then falling and creeping (or climbing through other vegetation), glabrous or moderately hairy, sparsely to densely short- to long-stipitate-glandular, not pruinose; prickles moderate to dense, recurved, sometimes distally slender, 1–4 mm, broad-based; <bristles absent or sparse to dense, erect to retrorse, red to purple, rarely green, slender, weak, gland-tipped>. **Leaves** persistent or semipersistent, ternate to palmately compound, <lustrous>; stipules filiform, linear, or lanceolate, 2–12(–15) mm; leaflets 3–5, terminal narrowly elliptic or ovate to obovate, 2–8.5 × 0.7–4.5 cm, base rounded to cuneate, unlobed, margins moderately to coarsely serrate to doubly serrate, apex acute to acuminate, abaxial surfaces with hooked prickles on midvein, glabrous or sparsely to moderately hairy, eglandular or sparsely short-stipitate-glandular along central vein. **Inflorescences** terminal <on short shoots, usually appearing axillary>, 1(–3)-flowered. **Pedicels:** prickles and, often, bristles moderate to dense, recurved, moderately to densely hairy, sparsely to moderately sessile- to short-stipitate-glandular. **Flowers** bisexual; petals white to pink, elliptic to obovate, 10–16(–25) mm; filaments filiform; ovaries glabrous. **Fruits** black, globose to ovoid, 1–1.5(–2) cm; drupelets 10–50, strongly coherent, separating with torus attached. $2n = 14$.

Flowering Jan–Jun. Open woodlands, savannas, prairies, meadows, sand dunes, disturbed areas, dry to seasonally wet soil; 0–200 m; Ala., Ark., Fla., Ga., Ill., Kans., Ky., La., Miss., Mo., N.C., Okla., S.C., Tenn., Tex., Va.; Mexico (Coahuila, Nuevo León, Tamaulipas).

Rubus trivialis is distinguished from other species of *Rubus* by its frequently glandular-bristly and generally creeping stems, abundant recurved prickles, and typically persistent or semipersistent, lustrous primocane leaves with relatively narrow leaflets. Although emerging primocanes typically reach to 30 cm above the ground, vigorous plants can have new primocanes standing erect to 70 cm that later fall to the ground or onto adjacent vegetation as they continue to enlarge. L. H. Bailey (1941–1945) believed that there was no clear separation between members of sect. *Persistentes* Fernald (referred to by Bailey as sect. *Verotriviales* [illegitimate sectional name for *R. trivialis* and associated species]) and sect. *Procumbentes* (by Bailey as sect. *Flagellares*; including *R. flagellaris* and related species). All characteristics that he recognized for distinguishing sect. *Persistentes* from sect. *Procumbentes* are often present in the latter, with the exception of hispid-setose stems, and glabrous plants of *R. trivialis* are difficult to distinguish from some plants of *R. flagellaris*; to include the latter species within the former would be reasonable.

Rubus ×*inferior* L. H. Bailey is a putative hybrid of *R. trivialis* and *R. cuneifolius*.

35. Rubus ulmifolius Schott, Isis (Oken) 1818: 821. 1818 • Elm-leaf blackberry I

Rubus discolor Weihe & Nees; *R. ulmifolius* var. *anoplothyrsus* Sudre

Shrubs, 30–50 dm, unarmed or armed. **Stems** <biennial>, erect, then arching, usually moderately, rarely sparsely, hairy <hairs stellate>, eglandular, strongly pruinose; prickles absent or sparse to dense, erect to hooked, stout, 3–10 mm, broad-based. **Leaves** persistent or nearly so, ternate to palmately compound; stipules filiform to linear, 4–12 mm; leaflets 3–5, terminal ovate, elliptic, or suborbiculate to obovate, 4–10 × 2–6(–9) cm, base rounded to cuneate, unlobed, margins finely to moderately serrate, apex acute or acuminate to attenuate, abaxial surfaces with hooked prickles on larger veins, densely white short-velutinous, hairs usually minute, eglandular. **Inflorescences** terminal, 10–60-flowered, paniculiform or thyrsiform. **Pedicels** unarmed or prickles moderate to dense, erect to hooked, densely short-hairy, eglandular. **Flowers** bisexual; petals usually pink, sometimes white, obovate or elliptic to suborbiculate, 5–12 mm; filaments filiform; ovaries apically hairy. **Fruits** black, globose to ovoid, 1–1.5 cm; drupelets 10–40, strongly coherent, separating with torus attached. **2*n*** = 14.

Flowering May–Jul. Woodland edges, open areas, disturbed areas, moist soil; 0–500 m; introduced; Calif., Nev., N.J., Oreg., Wash.; s, c Europe; n Africa; introduced also in s South America.

Rubus ulmifolius can be distinguished, especially from the closely related *R. bifrons* and *R. vestitus*, by its strongly pruinose stems, finely serrate leaflets, and lack of glands throughout. Unlike *R. ulmifolius*, strongly pruinose native *Rubus* species lack relatively large and showy pink petals. Some new stems developing from tip-rooting, and early leaves on such stems (especially in shade), are not whitened abaxially and are tomentose. Such unusual stems develop typical leaves and surfaces in parts formed later.

There has been nomenclatural confusion over unarmed plants of *Rubus ulmifolius* (see E. Monasterio-Huelin and H. E. Weber 1996). *Rubus inermis* Willdenow is an illegitimate later homonym of *R. inermis* Pourret, and the type specimens for both names are of poor quality. If an unarmed variety of *R. ulmifolius* is to be recognized, such as occurs in California, the name var. *anoplothyrsus* Sudre should be used instead of var. *inermis* (Willdenow) Focke.

Rubus ulmifolius is attractive and is potentially weedy but relatively uncommon, found primarily in coastal California from the San Francisco Bay region southward. It probably no longer persists in New Jersey, last collected there in 1897, and it is rare in Nevada and Oregon. *Rubus ulmifolius* likely hybridizes with *R. bifrons* where they co-occur; it is one of relatively few diploid and sexually reproducing species of European blackberries.

36. Rubus ursinus Chamisso & Schlechtendal, Linnaea 2: 11. 1827 • California or Pacific or creeping blackberry

Rubus eastwoodianus Rydberg; *R. macropetalus* Douglas ex Hooker; *R. sirbenus* L. H. Bailey; *R. ursinus* var. *eastwoodianus* (Rydberg) J. T. Howell; *R. ursinus* subsp. *macropetalus* (Douglas ex Hooker) Roy L. Taylor & MacBryde; *R. ursinus* var. *macropetalus* (Douglas ex Hooker) S. W. Brown; *R. ursinus* var. *sirbenus* (L. H. Bailey) J. T. Howell; *R. vitifolius* Chamisso & Schlechtendal; *R. vitifolius* var. *eastwoodianus* (Rydberg) Munz; *R. vitifolius* subsp. *ursinus* (Chamisso & Schlechtendal) Abrams

Shrubs, <usually dioecious>, to 2 dm, armed. **Stems** <biennial>, creeping (often growing over other vegetation), sparsely to densely hairy, eglandular or sparsely to densely sessile- to stipitate-glandular, usually strongly pruinose; prickles moderate to dense, erect to retrorse or hooked, weak to moderately stout, 4–10 mm, narrow- to broad-based. **Leaves** deciduous or persistent, usually ternate, sometimes simple or pinnately compound; stipules filiform to linear, 7–17 mm; leaflets 3(–5), terminal ovate to lanceolate, 3.5–12 × 3–10 cm, base cuneate or rounded to cordate, shallowly sharp-lobed or unlobed, margins coarsely serrate to doubly serrate, apex acute to acuminate, abaxial surfaces with slender, erect prickles on larger veins, moderately to densely hairy, eglandular, rarely sparsely sessile-glandular along midvein. **Inflorescences** terminal <on short shoots, usually appearing axillary>, 1–5-flowered, cymiform or racemiform. **Pedicels:** prickles moderate to dense, erect, densely hairy, eglandular or sparsely to densely stipitate-glandular. **Flowers** usually functionally unisexual; petals white, in staminate flowers ovate or obovate to narrowly elliptic, 7–18 mm, in pistillate ovate to elliptic, 6–11 mm; filaments filiform; ovaries glabrous or hairy. **Fruits** black, sometimes red or purple, rarely white, <not pruinose>, globose to cylindric, 1–2.5 cm; drupelets 20–50, strongly coherent, separating with torus attached. **2*n*** = 42, 56, 63, 70, 77, 84, 91.

Flowering (Jan–)Mar–Aug(–Sep). Woodlands, shrublands, open or disturbed areas, dry to damp soil; 0–1600 m; B.C.; Calif., Idaho, Mont., Oreg., Wash.; Mexico (Baja California).

In habit, *Rubus ursinus* is often reminiscent of the predominantly eastern North American, non-sympatric *R. flagellaris*; it is also polymorphic. *Rubus ursinus* comprises a polyploid spectrum dominated by octoploid and dodecaploid plants. It is an allopolyploid involving phylogenetically distant ancestors with its closest relative being the Hawaiian endemic *R. macraei* A. Gray. See S. W. Brown (1943) for a discussion of morphological and chromosome number variation in *R. ursinus*. The widespread, glaucous-stemmed, trifoliate-leaved forms of *R. ursinus* are vegetatively nearly identical to *R. caesius*. Of agricultural significance, *R. ursinus* is a parent of some important cultivars, including loganberry and boysenberry.

There has been uncertainty whether the name *Rubus menziesii* Hooker and subsequent combinations [*Parmena menziesii* (Hooker) Greene, *R. spectabilis* var. *menziesii* (Hooker) S. Watson, and *R. ursinus* var. *menziesii* (Hooker) Focke] pertain to *R. spectabilis* or *R. ursinus*. The protologue by Hooker describes a hairy, relatively small, procumbent plant, which is congruent with *R. ursinus*. The type specimen at Kew also looks like *R. ursinus*. In the protologue Hooker also claimed the species has red petals and suggested that it may have an affinity with *R. spectabilis*.

Specimens examined for this study of North American collections attributed to *Rubus macrophyllus* Weihe & Nees (synonym *R. amplificatus* Lees) are *R. ursinus*. If *R. macrophyllus* was once present in the flora area it is likely absent now; some reports of *R. macrophyllus* could represent hybrids. There are reports of *R. ursinus* hybridizing with *R. bifrons* and *R. pensilvanicus* in California (T. S. Mallah 1954; L. V. Clark and M. Jasieniuk 2012; L. A. Alice, unpubl.).

37. Rubus vestitus Weihe & Nees in M. J. Bluff et al., Comp. Fl. German. 1: 684. 1825 • European blackberry [I] [W]

Shrubs, 10–20 dm, armed. **Stems** <biennial>, arching, usually densely, sometimes moderately, hairy, usually eglandular, sometimes sessile- to short-stipitate-glandular, not pruinose; prickles moderate to dense, recurved to erect, slender, 3–10 mm, broad-based. **Leaves** deciduous or ± persistent, palmately compound; stipules filiform to linear, 5–15 mm; leaflets 3–5, terminal usually suborbiculate to orbiculate, sometimes broadly elliptic, 4–9 × 4–8 cm, base rounded to shallowly cordate, unlobed, margins moderately to coarsely serrate, apex acute to cuspidate, abaxial surfaces with hooked prickles on midvein, densely white-gray to gray-green tomentose, eglandular or obscurely short-stipitate- to sessile-glandular along midveins. **Inflorescences** terminal, 7–35-flowered, cymiform or thyrsiform. **Pedicels:** prickles moderate to dense, recurved to erect, densely hairy, moderately to densely stipitate-glandular. **Flowers** bisexual; petals pink to magenta, elliptic to suborbiculate, 6–12(–15) mm; filaments filiform; ovaries apically hairy. **Fruits** black, globose, 1–1.5 cm; drupelets 15–40, coherent, separating with torus attached. $2n = 28$.

Flowering Jul–Sep. Woodland edges, open areas, disturbed areas, in moist soil; 0–600 m; introduced; B.C.; Oreg.; Europe.

Rubus vestitus can be distinguished, especially from the closely related *R. bifrons* and *R. ulmifolius*, by its long-stipitate-glandular hairs, particularly in the inflorescence, and terminal primocane leaflets that are typically suborbiculate and abaxially densely tomentose. *Rubus allegheniensis* can possess similar glandular hairs; it does not have pink petals, thyrsiform inflorescences, nor rounded and abaxially densely tomentose, often whitened terminal leaflets.

Rubus vestitus has not yet been found in Washington, but it is likely there. Because of the overall similarity of *R. vestitus* to *R. bifrons*, it is possible that this potentially weedy species is widespread, but rare and possibly overlooked, occurring as far south as California. In Oregon, *R. vestitus* prefers shady habitats (B. Wilson, pers. comm.).

a3. ROSACEAE Jussieu (subfam. ROSOIDEAE) tribe COLURIEAE Rydberg in N. L. Britton et al., N. Amer. Fl. 22: 240. 1908

Luc Brouillet

Shrubs [trees], subshrubs, or herbs, perennial; unarmed. **Leaves** alternate, sometimes opposite (*Geum*), pinnately compound or simple; stipules persistent, ± free or ± adnate to petiole; venation pinnate or palmate. **Flowers:** perianth and androecium perigynous; epicalyx bractlets absent or present; hypanthium funnelform, saucer- to cup-shaped, or obconic to obcampanulate; torus usually enlarged; carpels (2 or)3–250(–450), styles terminal (subterminal in *Waldsteinia*), distinct; ovules 1 or 2, basal, superposed. **Fruits** aggregated achenes; styles persistent, sometimes deciduous (*Waldsteinia*), elongate, <sometimes hooked, glabrous or hairy, rarely plumose>.

Genera 4, species ca. 60 (4 genera, 22 species in the flora): North America, Mexico, South America, Eurasia, Africa, Pacific Islands (New Zealand), Australia.

Waldsteinia is recognized as distinct from *Geum* in this treatment, while D. Potter et al. (2007) included the former in the latter. Names are provided for users who would prefer to include *Waldsteinia* within *Geum*, as supported by molecular data (J. E. E. Smedmark and T. Eriksson 2002).

The base chromosome number for Colurieae is $x = 7$. Except for *Fallugia*, the genera of Colurieae are host to *Phragmidium* rusts. C. Kalkman (2004) stated that two ovules had been reported for *Fallugia* but that he observed only one, which would accord better with membership in Colurieae.

SELECTED REFERENCE Smedmark, J. E. E. and T. Eriksson. 2002. Phylogenetic relationships of *Geum* (Rosaceae) and relatives inferred from the nrITS and *trn*L-*trn*F regions. Syst. Bot. 27: 303–317.

1. Shrubs, 10–20(–35) dm; leaves simple; flowers all pistillate, all staminate, or staminate with terminal one bisexual. . . . . 6. *Fallugia*, p. 73
1. Herbs or subshrubs, 0.2–12 dm; leaves simple, sometimes lobed, or ternately or pinnately compound; flowers bisexual.
    2. Subshrubs; inflorescences: flowers solitary. . . . . 3. *Sieversia*, p. 57
    2. Herbs; inflorescences 1–18-flowered, open cymes or panicles.
        3. Plants subscapose or leafy-stemmed; tori hemispheric to cylindric; carpels (2–)20–250(–450); fruiting styles persistent, elongating, hooked or not. . . . . 4. *Geum*, p. 58
        3. Plants scapose; tori not evident; carpels 3–7; fruiting styles deciduous, not hooked . . . . 5. *Waldsteinia*, p. 71

3. SIEVERSIA Willdenow, Ges. Naturf. Freunde Berlin Mag. Neuesten Entdeck. Gesammten Naturk. 5: 397. 1811 • [For Johann August Carl Sievers, 1762–1795, German-born apothecary who explored eastern Russia in search of medicinal rhubarb]

Joseph R. Rohrer

Subshrubs (dwarf), <subscapose>, 0.2–0.8[–1.5] dm; rhizomatous, <rhizomes branching, slender, woody; fibrous-rooted>. **Stems** 1–5+, decumbent to ascending, simple, glabrous. **Leaves** deciduous, cauline (often crowded near stem apices), pinnately compound; stipules adnate to petiole base, linear, margins entire; petiole present; blade elliptic to obovate, 1–3.5 cm, herbaceous, leaflets 3–11, narrowly elliptic to obovate or rhombic, margins flat, entire or sharply lobed or toothed apically, venation pinnate, surfaces glabrous. **Inflorescences** terminal, flowers solitary; bracts 1 or 2. **Pedicels** present. **Flowers** 15–25 mm diam.; epicalyx bractlets 5; hypanthium saucer-shaped, 2 mm, puberulent; sepals 5, spreading, ovate; petals 5, white, some yellow at base or fading yellowish, obovate to elliptic; stamens 100–120, shorter than petals; torus hemispheric, <short-hairy>; carpels 30–70, hairy, <styles not geniculate-jointed>; ovule 1. **Fruits** aggregated achenes, 30–70, ovoid to fusiform, tapered apically, 2 mm, hairy; hypanthium persistent; sepals persistent, spreading to reflexed; styles persistent, elongating to 20[–27] mm, not hooked, <pilose except distal 1 mm>. $x = 7$.

Species 1 or 2 (1 in the flora): Alaska, e Asia.

Sieversia is circumscribed here in its original narrow sense and follows the treatment by J. E. E. Smedmark (2006). Late nineteenth- and early twentieth-century authors often included within *Sieversia* all members of Colurieae with fruits bearing relatively long, slender, plumose styles that lack a geniculate joint. Smedmark and her colleagues have shown that this fruit type is probably ancestral in Colurieae and have argued for placement of all North American species, except for *S. pentapetala*, in *Geum* (see discussion under 4. *Geum*). The Asian endemic *S. pusilla* Hultén differs from *S. pentapetala* largely in its greater degree of blade division (the leaflets are deeply divided resulting in nearly bipinnate blades) and may or may not be a distinct species.

1. Sieversia pentapetala (Linnaeus) Greene, Pittonia 4: 49. 1899 • Aleutian or low avens [F]

Dryas pentapetala Linnaeus, Sp. Pl. 1: 501. 1753; *Geum pentapetalum* (Linnaeus) Makino

Leaves: stipules 2–6 mm; petiole 4–15 mm, proximally expanded and sheathing stems, sheath margins ciliate; leaflets proximally narrowly elliptic, 2–6 mm, margins entire or remotely toothed, distally obovate to rhombic, 8–12 mm, margins sharply lobed or toothed apically. **Inflorescences:** bracts 3-fid or linear, 6–10 mm; stipule present; petiole reduced to sheathing bases. **Pedicels** 2.5–8 cm, glabrate to puberulent proximally, puberulent to downy distally. **Flowers** erect; epicalyx bractlets sometimes 2-fid distally or divided nearly in 2, 3–4 mm; sepals 4–7 mm, margins entire; petals spreading, 7–11 mm, distinctly longer than sepals. $2n = 14$.

Flowering summer. Moist tundra; 0–200[–2700] m; Alaska; e Asia (Japan, Russian Far East).

In North America, *Sieversia pentapetala* is known only from the central Aleutian Islands. Japanese plants are somewhat larger overall than those of the Aleutians.

4. GEUM Linnaeus, Sp. Pl. 1: 500. 1753; Gen. Pl. ed. 5, 220. 1754 • Avens, benoîte [Greek *geuo* or *geyo*, to give relish, alluding to quality of roots of St. Benedict's herb, *Geum urbanum*]

Joseph R. Rohrer

Acomastylis Greene; *Erythrocoma* Greene; *Novosieversia* F. Bolle; *Stylypus* Rafinesque

Herbs, perennial, <subscapose or leafy-stemmed>, 1–12 dm, glabrous or hairy; from stout caudices or rhizomatous, <fibrous-rooted>. **Stems** 1–5, erect, simple. **Leaves** winter-persistent (basal), basal and cauline, simple, 3-foliolate, lyrate-pinnate, or odd-pinnate (commonly with smaller leaflets intermixed with larger ones); stipules <often conspicuous>, ± free or adnate to petiole or blade, linear, lanceolate, or ovate, margins entire, toothed, or lobed; petiole present, sometimes absent on cauline leaves; blade oblanceolate, obovate, elliptic, ovate, or orbiculate, herbaceous, major leaflets 1–26, rhombic, elliptic, oblong, obovate, oblanceolate, orbiculate, reniform, or cordate, margins flat, 2–7-lobed and/or laciniate, serrate, dentate, or crenate, venation pinnate or palmate. **Inflorescences** terminal, 1–18-flowered, cymes, open; bracts present as reduced cauline leaves; bracteoles present or absent. **Pedicels** present. **Flowers** 4–46 mm diam.; epicalyx bractlets, if present, 5 (10 in *G. glaciale*); hypanthium saucer-shaped to cup-shaped, 2–6 mm; sepals 5(–10 in *G. glaciale*), erect to erect-spreading, reflexed or not, deltate, deltate-ovate, or deltate-lanceolate; petals 5(–9 in *G. glaciale*), white to yellow, sometimes suffused with pink or purple, or purple-veined, obovate, orbiculate, ovate, obcordate, obdeltate, spatulate, suborbiculate, elliptic, or oblong; stamens [10–]25–120, shorter than petals; torus hemispheric to cylindric; carpels (2–)20–250(–450), <styles entire or geniculate-jointed, distal portions then deciduous>; ovule 1. **Fruits** aggregated achenes, (2–)20–250(–450), ovoid to fusiform, tapered apically into style, 2–4.5 mm; hypanthium persistent; sepals persistent, erect, spreading, recurved, or reflexed; styles persistent, becoming accrescent, hooked and elongating to 10 mm, or not hooked and elongating to 70 mm. $x = 7$.

Species ca. 45 (16 in the flora): North America, Mexico, South America, Eurasia, Africa, Pacific Islands (New Zealand), Australia; temperate and arctic-alpine regions.

The most distinctive feature of many members of *Geum* is their geniculate-jointed style. The style has two parts, a persistent proximal segment and a deciduous distal segment. Where the two segments join, each bends to form a hook, with the hook of the proximal segment joined to the hook of the distal one. Seven of the 16 North American species of *Geum* lack this distinctive feature. Differences in style structure have led to differences in opinion as to generic delimitation and subgeneric classification of *Geum*.

During the early twentieth century, species traditionally included in *Geum* were divided among an increasing number of segregate genera. F. Bolle (1933), in his worldwide monograph, recognized five genera. After conducting extensive cytogenetic studies on *Geum* for two decades, W. Gajewski (1957) concluded that the most reasonable approach was to classify all the species in one genus. Most flora writers since Gajewski have maintained the species in a single heteromorphic genus. J. E. E. Smedmark and her collaborators have utilized DNA sequences from ITS, chloroplast *trn*L-*trn*F region, and nuclear low-copy gene GBSSI (Smedmark and T. Eriksson 2002; Smedmark et al. 2003), along with scanning electron microscopy of gynoecial development (Smedmark and Eriksson 2006), to test hypotheses by Gajewski of fruit evolution and allopolyploidy in *Geum* and to explore the phylogenetic relationships among the species. Their data support none of the previous circumscriptions of *Geum* as a monophyletic group, nor most of the suggested segregate genera favored by taxonomists such as Bolle, E. L. Greene (1906), and P. A. Rydberg (1913b). The taxonomic solution chosen here for North American species is traditional yet consistent with the molecular data and recognizes two genera: *Geum* for the polyploid ($2n$ = 28–116), herbaceous species and *Sieversia* for the diploid ($2n$ = 14), suffrutescent species. Smedmark (2006) found that *Waldsteinia* might best be treated within *Geum*. It is also possible that *Coluria* R. Brown, *Taihangia* T. T. Yu & C. L. Li, and *Waldsteinia* are sister to *Geum* (Smedmark and Eriksson 2002). *Waldsteinia* ($2n$ = 14) is retained here as a separate genus, with its traditional circumscription.

Although most species of *Geum* in the flora area are reasonably distinct morphologically, their distinctive characteristics are not well correlated so as to be combined in a key using several characters per couplet, as was pointed out by E. G. Voss (1972–1996, vol. 2), whose key to *Geum* in Michigan served as a model for the key below. In particular, the characters that are useful in identifying specimens in flower do not correlate well with characters helpful in the identification of fruiting specimens. Specimens with both flowers and fruits are the easiest to identify with certainty, and both leads at couplet 11 in the key should be followed if possible. Petal color in *Geum* fades upon drying. Both white and yellow petals ultimately turn cream to pale yellow, thus rendering an important key character for fresh plants unreliable with most herbarium specimens. Collectors should note petal color on their labels.

For taxonomists familiar with *Geum*, leaves provide valuable characters for distinguishing the species, especially when flower color is not known. The complex morphology of leaves and their variability do not suit them for wide use in keys. Variability occurs both among plants of one species and on one plant. In many of the species with geniculate-jointed styles, the basal leaf form changes over the growing season. Spring-developing leaves are often pinnate followed by lyrate-pinnate or 3-foliolate leaves, and in late summer by simple leaves that are winter-persistent. For instance, the basal leaves of *G. laciniatum* may appear like those of typical *G. aleppicum* in late spring, but look more like those of *G. macrophyllum* var. *macrophyllum* in late summer. Cauline leaves may vary from pinnate proximally to 3-foliolate at mid stem to simple just proximal to the inflorescence. Because variability of leaves within a single species is nearly as great as between some species, leaf descriptions are kept short. Measurements of leaf length include the petiole, if present.

SELECTED REFERENCES Bolle, F. 1933. Eine Übersicht über die Gattung *Geum* L. und die ihr nahestehenden Gattungen. Repert. Spec. Nov. Regni Veg. Beih. 72: 1–119. Gajewski, W. 1957. A cytogenetic study on the genus *Geum* L. Monogr. Bot. 4: 1–416. Smedmark, J. E. E. 2006. Recircumscription of *Geum* (Colurieae: Rosaceae). Bot. Jahrb. Syst. 126: 409–417.

1. Styles not geniculate-jointed, wholly persistent, 2–70 mm, apices not hooked (*G. triflorum* var. *ciliatum* often inconspicuously geniculate-jointed, distal segments tardily deciduous, remainder elongating at least 15 mm; *G. schofieldii* apices sometimes ± hooked); plants subscapose; stems 2–40(–60) cm; cauline leaves 0.7–5 cm, (not resembling basal).
    2. Basal leaves pinnate, terminal leaflets slightly larger than laterals.
        3. Flowers nodding (erect in fruit); hypanthia maroon, purple, or greenish mottled with purple (may turn pale brown in fruit); petals erect, cream to yellowish suffused with pink or purple, or purple-veined; cauline leaves opposite . . . . 2. *Geum triflorum*
        3. Flowers erect; hypanthia green, or green and purple-tinged to strongly purple; petals spreading, yellow; cauline leaves alternate.
            4. Stems densely pilose, hairs 2–5 mm; fruiting styles 18–30 mm, pilose except distal 1–3 mm . . . . 1. *Geum glaciale*
            4. Stems glabrous or downy, hairs to 1 mm; fruiting styles 2–5(–10) mm, glabrous throughout or pilose only at base . . . . 3. *Geum rossii*
    2. Basal leaves lyrate-pinnate, sometimes simple, terminal leaflets larger than laterals.
        5. Basal leaves interruptedly lyrate-pinnate, major leaflets 5–7, interspersed with 2–8 minor ones . . . . 4. *Geum schofieldii*
        5. Basal leaves strongly lyrate-pinnate or simple, major leaflet 1, proximally 0–10 minor ones.
            6. Stems densely hirsute proximally; inflorescences 3–10-flowered; pedicels glandular-hairy; fruiting styles glandular in basal 1/3 . . . . 6. *Geum radiatum*
            6. Stems glabrate, sparsely downy, sparsely hirsute, or pilose proximally; inflorescences 1 or 2(–4)-flowered; pedicels downy and/or hirsute, usually eglandular; fruiting styles eglandular.
                7. Cauline leaves 1.5–4.5 cm; fruiting styles 9–14 mm, pilose in basal 3/4; British Columbia, Alaska . . . . 5. *Geum calthifolium*
                7. Cauline leaves 0.8–1.7(–2.5) cm; fruiting styles 6–9 mm, pilose in basal 1/3; Nova Scotia, New Hampshire . . . . 7. *Geum peckii*
1. Styles geniculate-jointed, distal segments eventually deciduous, proximal segments persistent, 1.5–9 mm, apices hooked; plants leafy-stemmed; stems (20–)40–120 cm; cauline leaves (2–)5–23 cm, (proximal resembling basal, distal smaller).
    8. Flowers nodding; sepals erect to erect-spreading, not reflexed; petals spatulate-obovate, apices rounded, truncate, or shallowly emarginate.
        9. Hypanthia greenish maroon to maroon; styles with distal segments 3–4.5 mm, shorter than mature proximal segments; fruiting tori on 4–10 mm stipes. . . . . 8. *Geum rivale*
        9. Hypanthia green to greenish maroon; styles with distal segments 5–8 mm, longer than mature proximal segments; fruiting tori sessile . . . . 9. *Geum geniculatum*

[8. Shifted to left margin.—Ed.]

8. Flowers erect; sepals erect to spreading but soon reflexed; petals oblong, elliptic, suborbiculate, ovate, or obovate, apices obtuse to rounded.
   10. Fruiting tori on 3–7 mm stipes; epicalyx bractlets absent; sepals 1–3 mm; petals 1–2 mm . . . . . 16. *Geum vernum*
   10. Fruiting tori sessile or on less than 3 mm stipes; epicalyx bractlets usually present (often absent in *G. macrophyllum*); sepals (2–)3–10 mm; petals 2–9 mm (rarely 1.5 mm in *G. virginianum*).
      11. Plants with flowers.
         12. Petals white or cream.
            13. Petals (3–)4–8 mm, ± equal to or longer than sepals; stems glabrate to downy, hairs to 1.5 mm . . . . . 13. *Geum canadense* (in part)
            13. Petals (1.5–)2–5 mm, shorter than sepals; stems hirsute or puberulent and hirsute to densely hirsute, some hairs 2–2.5 mm.
               14. Pedicels puberulent, densely hirsute; cauline leaves with stipules 4–14 mm . . . . . 12. *Geum laciniatum* (in part)
               14. Pedicels puberulent (sometimes with scattered hairs); cauline leaves with stipules 11–48 mm . . . . . 14. *Geum virginianum* (in part)
         12. Petals yellow.
            15. Styles with distal segments glabrous or with short hairs, hairs shorter than diam. of style; cauline leaves with stipules 10–40 × 5–35 mm . . . . . 15. *Geum urbanum* (in part)
            15. Styles with distal segments pilose at base, hairs much longer than diam. of style; cauline leaves with stipules 7–28 × 3–22 mm.
               16. Epicalyx bractlets often absent; styles with proximal segments sparsely to densely stipitate-glandular; basal leaves interruptedly lyrate-pinnate, terminal leaflets usually much larger than laterals . . . . 10. *Geum macrophyllum* (in part)
               16. Epicalyx bractlets present; styles with proximal segments eglandular; basal leaves interruptedly pinnate, terminal leaflets usually only slightly larger than laterals. . . . . . 11. *Geum aleppicum* (in part)
      11. Plants with fruits.
         17. Fruiting tori glabrous, puberulent, or with ring of bristles at base and tuft at apex.
            18. Fruiting styles with proximal segments sparsely to densely stipitate-glandular; epicalyx bractlets often absent; pedicels densely puberulent (sometimes with scattered longer hairs), sometimes stipitate-glandular . . . . . 10. *Geum macrophyllum* (in part)
            18. Fruiting styles with proximal segments glabrous, sometimes 1 or 2 eglandular bristles at base; epicalyx bractlets present; pedicels densely puberulent, hirsute, eglandular . . . . . 12. *Geum laciniatum* (in part)
         17. Fruiting tori densely downy or bristly.
            19. Fruiting styles with proximal segments with bristles on basal ⅓; fruiting tori densely downy (hairs soft, 0.3–0.7 mm); proximal cauline leaves mostly pinnately compound. . . . . . 11. *Geum aleppicum* (in part)
            19. Fruiting styles with proximal segments glabrous or glabrate, sometimes sparsely hairy or stipitate-glandular in *G. canadense*; fruiting tori densely bristly (hairs 1–2.3 mm); proximal cauline leaves usually 3-foliolate or simple and 3-lobed to unlobed, sometimes pinnate in *G. urbanum*.
               20. Fruiting styles with distal segments glabrous or with short hairs, hairs shorter than diam. of style. . . . . . 15. *Geum urbanum* (in part)
               20. Fruiting styles with distal segments pilose in basal ½, hairs much longer than diam. of style.
                  21. Cauline leaves with stipules 4–13 × 1–7 mm; stems glabrate to downy . . . . . 13. *Geum canadense* (in part)
                  21. Cauline leaves with stipules 11–48 × 6–35 mm; stems puberulent and hirsute to densely hirsute . . . . . 14. *Geum virginianum* (in part)

1. Geum glaciale Adams ex Fischer, Mém. Soc. Imp. Naturalistes Moscou 2: 187, plate 11, fig. 20. 1809 • Glacier avens

Novosieversia glacialis (Adams ex Fischer) F. Bolle; *Sieversia glacialis* (Adams ex Fischer) Sprengel

Plants subscapose. **Stems** 2–30 cm, densely pilose, <hairs 2–5 mm>. **Leaves:** basal 3–10 cm, blade pinnate-pinnatifid, leaflets/lobes 11–23, terminal leaflet slightly larger; cauline 1.4–2.5 cm, stipules adnate to leaf, indistinguishable from lobes, blade <bractlike, not resembling basal, alternate>, simple, 3-fid. **Inflorescences** 1-flowered. **Pedicels** densely pilose, eglandular. **Flowers** erect; epicalyx bractlets 6–9 mm; hypanthium green; sepals spreading-erect, 7–14 mm; petals <5–9>, spreading, yellow, obovate, orbiculate, or ovate, 10–20 mm, longer than sepals, apex rounded to emarginate. **Fruiting tori** sessile, silky. **Fruiting styles** wholly persistent, not geniculate-jointed, 18–30 mm, apex not hooked, pilose except distal 1–3 mm. **2*n*** = 28.

Flowering early summer. Tundra, rocky slopes, heaths; 0–1400 m; N.W.T., Yukon; Alaska; Asia (n Russia).

Flowers of *Geum glaciale* have from five to nine petals with seven being the most common. Other species of *Geum* nearly always have five petals, which is standard in most rosaceous genera. The sepals of *G. glaciale* are usually in two whorls for a total of about ten, and the bractlets are in two to three whorls. The exact number of epicalyx bractlets and sepals is made difficult to determine by the thick covering of hairs and because individual sepals and bractlets are sometimes deeply divided into two equal lobes.

2. Geum triflorum Pursh, Fl. Amer. Sept. 2: 736. 1813 • Prairie smoke, old man's whiskers, torchflower, three-sisters F

Erythrocoma triflora (Pursh) Greene; *Sieversia triflora* (Pursh) R. Brown

Plants subscapose. **Stems** 10–45 cm, downy to pilose, <hairs 0.1–3 mm, sometimes septate-glandular>. **Leaves:** basal 4–30 cm, blade interruptedly pinnate, major leaflets 10–18, alternating with 6–16 minor ones gradually increasing in size distally, terminal leaflet slightly larger than major laterals; cauline 1–5 cm, stipules adnate to leaf, indistinguishable from leaflets/lobes, blade <bractlike, not resembling basal, opposite>, pinnate-pinnatifid. **Inflorescences** (1–)3–5(–7)-flowered. **Pedicels** densely woolly, sometimes glandular. **Flowers** nodding, <erect in fruit>; epicalyx bractlets 6–15 mm; hypanthium maroon, purple, or greenish mottled with purple, <may turn pale brown in fruit>; sepals erect, 7–14 mm; petals erect, cream to yellowish suffused with pink or purple, or purple-veined, elliptic, 7–13 mm, shorter to longer than sepals, apex rounded to obtuse. **Fruiting tori** sessile, densely puberulent. **Fruiting styles** wholly persistent or distal 3–7 mm tardily deciduous, not or inconspicuously geniculate-jointed, 15–70 mm, apex not or occasionally ± hooked, pilose to apex or nearly so.

Varieties 2 (2 in the flora): North America, Mexico.

At the beginning of the twentieth century, E. L. Greene described over a dozen species belonging to the *Geum triflorum* complex based on differences in leaf form and indument, the relative length and shape of the epicalyx bractlets and sepals, and petal length and shape. Most of these species were quickly reduced to synonymy by other botanists. When specimens are examined from across the continent, most of the characters used to separate species in the *G. triflorum* complex show nearly continuous variation. It seems best to treat these variants as belonging to one species. Whether and how to classify the variation within the species will remain controversial. Some character expressions correlate reasonably well with dividing the species into two varieties, as was first proposed by N. C. Fassett (1928). Variety *triflorum* occurs east of the Rocky Mountains and is typical of the grasslands covering the northern Great Plains; it also is the variety found in the mountains of Arizona and New Mexico. Variety *ciliatum* is common throughout the rest of the Rocky Mountains, the Sierra Nevada, and Cascade Range. Intermediate specimens occur here and there, particularly in Alberta and British Columbia.

1. Basal leaves with larger leaflets toothed or cleft less than ½ their lengths; fruiting styles persistent, 28–70 mm . . . . 2a. *Geum triflorum* var. *triflorum*
1. Basal leaves with larger leaflets cleft more than ½ their lengths; fruiting styles: proximal segments persistent, 15–40 mm, distal segments tardily deciduous, 3–7 mm. . . .2b. *Geum triflorum* var. *ciliatum*

2a. Geum triflorum Pursh var. **triflorum** F

Basal leaves: larger leaflets toothed or cleft less than ½ their lengths. **Inflorescences** (1–)3(–7)-flowered. **Fruiting styles** persistent, 28–70 mm, apex not hooked, pilose except for apical 3–5 mm. **2*n*** = 42.

Flowering spring–early summer. Prairies, grassy openings in dry forests, bluffs, open coniferous forests, thin soil over limestone; 200–1800(–2900) m; Alta., B.C., Man.,

N.W.T., Ont., Sask., Yukon; Ariz., Colo., Ill., Iowa, Mich., Minn., Mont., N.Mex., N.Y., N.Dak., S.Dak., Wis., Wyo.; Mexico.

2b. Geum triflorum Pursh var. **ciliatum** (Pursh) Fassett, Rhodora 30: 207. 1928 [E]

Geum ciliatum Pursh, Fl. Amer. Sept. 1: 352. 1813; *G. canescens* (Greene) Munz; *G. triflorum* var. *campanulatum* (Greene) C. L. Hitchcock; *G. triflorum* var. *canescens* (Greene) Kartesz & Gandhi; *Sieversia campanulata* (Greene) Rydberg; *S. canescens* (Greene) Rydberg; *S. ciliata* (Pursh) G. Don

Basal leaves: larger leaflets cleft more than ½ their lengths. **Inflorescences** (1–)3–5(–7)-flowered. **Fruiting styles:** proximal segment persistent, 15–40 mm, pilose, distal segment tardily deciduous, 3–7 mm, apex not or ± hooked, glabrous or pilose on base. **2*n*** = 42.

Flowering late spring–early summer. Mountain meadows, grassy and rocky slopes, open coniferous and aspen woodlands, sagebrush scrub; 60–3300 m; Alta., B.C., Sask.; Calif., Colo., Idaho, Mont., Nev., Oreg., Utah, Wash., Wyo.

3. Geum rossii (R. Brown) Seringe in A. P. de Candolle and A. L. P. P. de Candolle, Prodr. 2: 553. 1825 • Ross's avens

Sieversia rossii R. Brown, Chlor. Melvill., 18, plate C. 1823; *Acomastylis rossii* (R. Brown) Greene; *Geum rossii* var. *depressum* (Greene) C. L. Hitchcock; *G. rossii* var. *turbinatum* (Rydberg) C. L. Hitchcock; *G. turbinatum* Rydberg; *S. gracilipes* (Piper) Greene

Plants subscapose. **Stems** 4–28 cm, glabrous or downy, <hairs to 1 mm, sometimes septate-glandular>. **Leaves:** basal 3–13 cm, blade pinnate to interruptedly pinnate, major leaflets 13–26, alternating with 0–14 minor ones, terminal leaflet slightly larger than major laterals; cauline 0.7–2 cm, stipules adnate to leaf, indistinguishable from pair of lobes, blade <bractlike, not resembling basal, alternate>, simple, pinnatifid to 3-fid. **Inflorescences** 1–3(or 4)-flowered. **Pedicels** woolly, sometimes glandular. **Flowers** erect; epicalyx bractlets 1.5–7 mm; hypanthium green, slightly purple-tinged to strongly purple; sepals erect to erect-spreading, 3–10 mm; petals spreading, yellow, obovate to nearly orbiculate, 5–12 (–17) mm, longer than sepals, apex broadly rounded to irregularly emarginate. **Fruiting tori** sessile, glabrous. **Fruiting styles** wholly persistent, not geniculate-jointed, 2–5(–10) mm, apex not hooked, glabrous throughout or pilose only at base. **2*n*** = 56.

Flowering summer. Alpine and arctic tundra, rocky slopes, often in gravelly or peaty soil; 0–4000 m; Greenland; B.C., N.W.T., Nunavut, Yukon; Alaska, Ariz., Colo., Idaho, Mont., Nev., N.Mex., Oreg., Utah, Wash., Wyo.; e Asia (Russian Far East).

The variability accommodated here in *Geum rossii* was distributed by earlier monographers such as P. A. Rydberg (1913b) and F. Bolle (1933) among a half dozen species. W. Gajewski (1957) reduced them to two species, *G. rossii* and *G. turbinatum*; most recent taxonomists have recognized the two taxa as subspecies or varieties of a single species. The large geographic discontinuity between the Rocky Mountain and arctic ranges makes it easy for those wishing to follow this tradition. No one morphologic character or combination of characters neatly separates the arctic plants from those of the Rockies.

Where their ranges overlap in Alaska, *Geum rossii* hybridizes with *G. calthifolium* to form sterile plants known as *G.* ×*macranthum* (Kearney ex Rydberg) B. Boivin; see discussion under 4. *G. schofieldii*.

4. Geum schofieldii Calder & Roy L. Taylor, Canad. J. Bot. 43: 1394. 1965 • Queen Charlotte avens [C] [E]

Plants subscapose. **Stems** 13–30 cm, puberulent, <hairs scattered, 0.1–1 mm>. **Leaves:** basal 5–25 cm, blade interruptedly lyrate-pinnate, major leaflets 5–7, interspersed with 2–8 minor ones increasing in size distally, terminal leaflet larger than major laterals; cauline 1.5–4.5 cm, stipules adnate to leaf, indistinguishable from pair of lobes, blade <bractlike, not resembling basal>, simple, 3-fid. **Inflorescences** 1-flowered. **Pedicels** densely puberulent, sometimes glandular. **Flowers** erect; epicalyx bractlets 4–8 mm; hypanthium green; sepals erect to erect-spreading, 7–12 mm; petals spreading, yellow, broadly obovate to nearly orbiculate, 14–16 mm, distinctly longer than sepals, apex broadly rounded to irregularly emarginate. **Fruiting tori** sessile, glabrous. **Fruiting styles** wholly persistent, not geniculate-jointed, 7–10 mm, apex not or ± hooked, pilose only on bases. **2*n*** = 112.

Flowering summer. Rocky runnels, cliffs, rock outcrops; of conservation concern; 500–1000 m; B.C.

Geum schofieldii is one of the more attractive yet seldom seen species of avens in North America; it is known from the rocky wet mountains of the Queen Charlotte Islands and northern Vancouver Island.

Because *Geum schofieldii* appears to be intermediate morphologically between *G. calthifolium* and *G. rossii*, some botanists have considered it to be a hybrid and included it in synonymy with *G.* ×*macranthum* [*G. calthifolium* × *G. rossii*]. The 16-ploid chromosome number ($2n$ = 112) of *G. schofieldii* suggests an allopolyploid origin. *Geum rossii* is not known to occur in the Queen Charlotte Islands or on Vancouver Island; the plants cannot be of recent hybrid origin. *Geum* ×*macranthum* is found in the Aleutian Islands and the Alaska Peninsula where the two parental species overlap. Compared to *G. schofieldii*, its leaves are more pinnate than lyrate-pinnate, with nine to 13 major leaflets instead of five to seven. Also, *G.* ×*macranthum* appears to be sterile; *G. schofieldii* appears to be fertile, suggesting that the latter is a valid species. The distinctness of *G. schofieldii* pollen also supports its specific status (R. J. Hebda and C. C. Chinnappa 1990).

5. **Geum calthifolium** Smith in A. Rees, Cycl. 16: Geum no. 13. 1810 • Caltha-leaved avens

Plants subscapose. **Stems** 8–40 cm, sparsely hirsute proximally, puberulent throughout. **Leaves:** basal 5–25 cm, blade strongly lyrate-pinnate, sometimes simple, major leaflet 1, minor leaflets 1–7, terminal leaflet much larger than minor laterals; cauline 1.5–4.5 cm, stipules not evident, blade <bractlike, not resembling basal>, simple. **Inflorescences** 1 or 2(–4)-flowered. **Pedicels** downy, hirsute, eglandular. **Flowers** erect; epicalyx bractlets 2–6 mm; hypanthium green or green with slight maroon tinge; sepals erect-spreading, 6–11 mm; petals spreading, yellow, obcordate-obdeltate to nearly orbiculate, 9–13 mm, longer than sepals, apex broadly and shallowly emarginate. **Fruiting tori** sessile, glabrous. **Fruiting styles** wholly persistent, not geniculate-jointed, 9–14 mm, apex not hooked, pilose in basal 3/4, eglandular. $2n$ = 42.

Flowering summer. Muskegs, moist meadows, heathlands, moist sites on rocky slopes, ledges, and cliffs; 0–1500 m; B.C.; Alaska; e Asia (Russian Far East).

Geum calthifolium hybridizes with *G. rossii* [= *G.* ×*macranthum* (Kearney ex Rydberg) B. Boivin]; see discussion under 4. *G. schofieldii.*

6. **Geum radiatum** Michaux, Fl. Bor.-Amer. 1: 300. 1803 • Appalachian or spreading avens [C] [E]

Sieversia radiata (Michaux) G. Don

Plants subscapose. **Stems** 10–45 (–60) cm, densely hirsute proximally to puberulent distally. **Leaves:** basal 10–30 cm, blade strongly lyrate-pinnate, sometimes simple, major leaflet 1, minor leaflets 1–6, terminal leaflet much larger than minor laterals; cauline 1.2–4 cm, stipules not evident, blade <bractlike, not resembling basal>, simple. **Inflorescences** 3–10 -flowered. **Pedicels** glandular-hairy. **Flowers** erect; epicalyx bractlets 2–5 mm; hypanthium green; sepals spreading in flower, erect in fruit, 6–10 mm; petals spreading, yellow, orbiculate to obcordate, 9–16 mm, longer than sepals, apex emarginate. **Fruiting tori** sessile, glabrous. **Fruiting styles** wholly persistent, not geniculate-jointed, 8–12 mm, apex not hooked, pilose and stipitate-glandular in basal 1/3. $2n$ = 42.

Flowering summer. Rocky cliffs and ledges, montane balds; of conservation concern; 1500–1900 m; N.C., Tenn.

Geum radiatum is closely related to *G. calthifolium* and *G. peckii*. The morphologic differences among them are slight compared to the discontinuities separating other *Geum* species. On the basis of morphology it would be possible to combine all three in a single species. Yet they occupy distinct ranges separated by a minimum of 1200 km. Traditionally, they have been treated as separate species, and recently I. G. Paterson and M. Snyder (1999) reported molecular genetic evidence for continuing to recognize *G. peckii* and *G. radiatum* as separate species.

Geum radiatum is listed as an endangered species by the U.S. Fish and Wildlife Service.

7. **Geum peckii** Pursh, Fl. Amer. Sept. 1: 352. 1813 • Mountain-avens [C] [E]

Plants subscapose. **Stems** 7–40 cm, glabrate, sparsely downy, or pilose proximally, downy or pilose distally. **Leaves:** basal 5–15 cm, blade strongly lyrate-pinnate, major leaflet 1, minor leaflets 4–10, terminal leaflet much larger than minor laterals; cauline 0.8–1.7(–2.5) cm, stipules not evident, blade <bractlike, not resembling basal>, simple, 3-fid. **Inflorescences** 1–2(–4)-flowered. **Pedicels** densely downy, usually eglandular. **Flowers** erect; epicalyx bractlets 2–3 mm; hypanthium green; sepals spreading, 4–7 mm; petals spreading, yellow, obcordate, nearly orbiculate, or broadly ovate, 9–13 mm, nearly 2 times sepals, apex broadly rounded, emarginate, or irregularly erose. **Fruiting tori** sessile, glabrous. **Fruiting styles** wholly persistent, not geniculate-jointed, 6–9 mm, apex not hooked, pilose in basal 1/3, eglandular. $2n$ = 42.

Flowering summer. Alpine meadows, wet spots on rocky cliffs and slopes, montane streamsides, coastal bogs; of conservation concern; 0–1900 m; N.S.; N.H.

Geum peckii occurs at 0–10 m in Nova Scotia and at 1500–1900 m in New Hampshire.

8. Geum rivale Linnaeus, Sp. Pl. 1: 501. 1753 • Purple or water avens, chocolate-root, benoîte des ruisseaux

Plants leafy-stemmed. **Stems** 30–85 cm, sparsely downy to scattered-hirsute proximally, downy distally. **Leaves:** basal 6–40 cm, blade interruptedly pinnate, major leaflets 5–7, mixed with 7–14 minor ones, terminal leaflet slightly larger than major laterals; cauline 2–10 cm, stipules ± free, 7–18 × 5–9 mm, blade pinnate to 3-foliolate. **Inflorescences** 2–8-flowered. **Pedicels** densely downy, some hairs glandular. **Flowers** nodding; epicalyx bractlets 2–4 mm; hypanthium greenish maroon to maroon; sepals erect, 7–10 mm; petals erect, yellow, suffused with purple and purple-veined, spatulate-obovate, 8–10 mm, ± equal to sepals, apex rounded, truncate, or shallowly emarginate. **Fruiting tori** on 4–10 mm stipes, densely bristly. **Fruiting styles** geniculate-jointed, proximal segment persistent, 5–9 mm, apex hooked, stipitate-glandular, bristles on basal ½, distal segment deciduous, 3–4.5 mm, pilose except apical 1 mm. $2n = 42$.

Flowering mid spring–summer. Swamps, fens, bogs, wet meadows, along streams and lakes, moist rich woods, in circumneutral to alkaline soil; 0–3000 m; Greenland; St. Pierre and Miquelon; Alta., B.C., Man., N.B., Nfld. and Labr., N.S., Nunavut, Ont., P.E.I., Que., Sask.; Colo., Conn., Idaho, Ill., Ind., Maine, Md., Mass., Mich., Minn., Mont., N.H., N.J., N.Mex., N.Y., N.Dak., Ohio, Pa., R.I., S.Dak., Vt., Wash., W.Va., Wis., Wyo.; Eurasia.

Geum rivale forms hybrids with three other species of the genus in North America: *G.* ×*aurantiacum* Fries ex Scheutz [*G. aleppicum* × *G. rivale*] is reported from Alberta, Ontario, Montana, and New York; *G.* ×*pervale* B. Boivin [*G. macrophyllum* var. *perincisum* × *G. rivale*] is known from Alberta, Ontario, Quebec, and Saskatchewan; and *G.* ×*pulchrum* Fernald [*G. macrophyllum* var. *macrophyllum* × *G. rivale*] is reported from Ontario, Quebec, Idaho, New Hampshire, and Vermont. The hybrids generally have the habit and foliage of *G. rivale* but exhibit shallower hypanthia, spreading sepals, and pure yellow, clawed, obovate to suborbiculate petals. The glandular hairs on the pedicels usually lack the dark tips seen in *G. rivale*. Given that the morphologic differences among the three named hybrids are vague, suspected hybrid specimens are best determined while in the field where collectors can note the proximity of the hybrids to their supposed parental species.

9. Geum geniculatum Michaux, Fl. Bor.-Amer. 1: 300. 1803 • Bent avens [C] [E]

Plants leafy-stemmed. **Stems** 50–84 cm, sparsely hirsute proximally, hirsute distally. **Leaves:** basal 14–30 cm, blade interruptedly lyrate-pinnate, major leaflets 1–5, mixed with 2–5 minor ones, terminal leaflet slightly to much larger than major laterals; cauline 4–12 cm, stipules ± free, 7–20 × 4–8 mm, blade pinnate, 3-foliolate, or simple and 3-lobed. **Inflorescences** 5–15-flowered. **Pedicels** densely pilose, some hairs septate-glandular. **Flowers** nodding; epicalyx bractlets 2.5–3 mm; hypanthium green with purple tinge at base to greenish maroon; sepals erect to erect-spreading, 5–8 mm; petals erect, white, pink, or lavender, spatulate-obovate, 5–7 mm, ± equal to sepals, apex rounded, truncate, or shallowly emarginate. **Fruiting tori** sessile, densely bristly. **Fruiting styles** geniculate-jointed, proximal segment persistent, 2–4.5 mm, apex hooked, septate-glandular hairs on basal ½, distal segment very tardily deciduous, 5–8 mm, pilose except apical 1–2 mm, with some glandular hairs.

Flowering summer. Mountain balds, seeps, stream banks, cliffs, wooded boulder fields; of conservation concern; 1600–1900 m; N.C., Tenn.

Geum geniculatum is enigmatic and is currently known from 15 populations in western North Carolina and eastern Tennessee. As a rare plant, it has attracted far less attention than *G. radiatum*, which is also known from the same region. Although *G. geniculatum* is not listed federally as endangered or threatened, it is listed as threatened by North Carolina and as endangered by Tennessee.

W. Gajewski (1957) speculated that *Geum geniculatum* is a hybrid, with one of the parents being *G. rivale*. K. R. Robertson (1974) pointed out that the nearest locality for *G. rivale* is in West Virginia and that pollen stainability of *G. geniculatum* is very high. *Geum geniculatum* may be of hybrid origin; it cannot be a first-generation hybrid. North American taxonomists have treated it as a valid species.

10. Geum macrophyllum Willdenow, Enum. Pl., 557. 1809 • Large-leaved avens, benoîte à grandes feuilles

Plants leafy-stemmed. **Stems** 30–110 cm, puberulent and hirsute or sparsely hirsute. **Leaves:** basal 10–45 cm, blade interruptedly lyrate-pinnate, major leaflets 5–9, alternating with 4–15 minor ones, terminal leaflet usually much larger than major laterals; cauline 2–12 cm, stipules ± free, 7–23 × 3–12 mm, blade lyrate-pinnate, pinnate, 3-foliolate, or simple and 3-lobed. **Inflorescences** 3–16-flowered. **Pedicels** densely puberulent, <sometimes with scattered longer hairs>, sometimes stipitate-glandular. **Flowers** erect; epicalyx bractlets often absent, 0.5–2 mm; hypanthium green; sepals erect-spreading but soon reflexed, 2.5–5.5 mm; petals spreading, yellow, obovate, broadly elliptic, or suborbiculate, 3.5–7 mm, longer than sepals, apex rounded. **Fruiting tori** sessile or on less than 1 mm stipes, puberulent. **Fruiting styles** geniculate-jointed, proximal segment persistent, 2.5–6 mm, apex hooked, sparsely to densely stipitate-glandular, distal segment deciduous, 1–2 mm, pilose in basal 1/3, <hairs much longer than diam. of style>.

Varieties 2 (2 in the flora): North America, Eurasia.

Characters useful in recognizing specimens of *Geum macrophyllum* are yellow petals, epicalyx bractlets often absent, proximal style segment sparsely to densely stipitate-glandular, and fruiting receptacles puberulent. Across its broad North American range from Alaska to California and Nova Scotia, *G. macrophyllum* exhibits considerable variation. Based largely on the shape of the terminal leaflet of the basal leaves and the degree of dissection and shape of the divisions of the distal cauline leaves, P. A. Rydberg (1913b) distinguished three species within the range of variation treated here as one species. Basal leaves of *G. macrophyllum* in the strict sense (var. *macrophyllum* in this treatment) have relatively large reniform to rounded terminal leaflets and the distal cauline leaves are three-cleft into rhombic or cuneate lobes. The basal-leaf terminal leaflets of *G. perincisum* (= var. *perincisum*) are only slightly larger than the laterals and are deeply lobed into rhombic-obovate segments; the distal cauline leaves are dissected into oblanceolate divisions. Both basal and cauline leaves of *G. oregonense* (= var. *perincisum*) are intermediate between those of *G. macrophyllum* and *G. perincisum*. W. Gajewski (1955) crossed all three taxa under discussion and examined leaf morphology and cytology of the F_1 and F_2 hybrids. He concluded that they were distinct but not yet completely separated species. Fairly well correlated with the more dissected leaves of *G. oregonense* and *G. perincisum* is the presence of minute stalked glands on the pedicels. The treatment here follows H. M. Raup (1931) in recognizing two varieties of *G. macrophyllum* based more on pedicel glandularity than leaf morphology.

SELECTED REFERENCE Gajewski, W. 1955. Cytogenetic relations of *Geum macrophyllum* Willd. with *G. perincisum* Rydb. and *G. oregonense* Rydb. Acta Soc. Bot. Poloniae 24: 311–334.

1. Pedicels not glandular-puberulent; cauline leaves with distal simple and 3-lobed (divided less than 3/4 to base), lobes rhombic-oblong . . . . 10a. *Geum macrophyllum* var. *macrophyllum*
1. Pedicels glandular-puberulent; cauline leaves with distal 3-foliolate or simple and 3-lobed (divided almost to base), lobes oblanceolate to obovate . . . . 10b. *Geum macrophyllum* var. *perincisum*

10a. Geum macrophyllum Willdenow var. **macrophyllum**

Cauline leaves: proximal strongly lyrate-pinnate, distal simple and 3-lobed (divided less than 3/4 to base), lobes rhombic-oblong. **Pedicels** densely puberulent with scattered longer hairs, eglandular. **2*n*** = 42.

Flowering late spring–summer. Wet meadows, stream banks, thickets, moist woods, clearings, coastal terraces, forest edges, roadsides; 0–1500 m; St. Pierre and Miquelon; B.C., N.B., Nfld. and Labr., N.S., Ont., P.E.I., Que.; Alaska, Calif., Idaho, Maine, Mass., Mich., Mont., Nebr., N.H., N.Y., N.Dak., Oreg., Vt., Wash., Wis.; Eurasia.

Alaskan specimens of var. *macrophyllum* often have relatively few stalked glands on the styles and some appear almost glandless.

Variety *macrophyllum* hybridizes with *Geum rivale* (= *G.* ×*pulchrum* Fernald); see discussion under 8. *G. rivale*.

10b. Geum macrophyllum Willdenow var. **perincisum** (Rydberg) Raup, Rhodora 33: 176. 1931 [E]

Geum perincisum Rydberg in N. L. Britton et al., N. Amer. Fl. 22: 405. 1913; *G. macrophyllum* subsp. *perincisum* (Rydberg) Hultén; *G. macrophyllum* var. *rydbergii* Farwell; *G. oregonense* (Scheutz) Rydberg; *G. perincisum* var. *intermedium* B. Boivin

Cauline leaves: proximal pinnate or lyrate-pinnate, distal 3-foliolate or simple and 3-lobed (divided almost to base), lobes oblanceolate to obovate. **Pedicels** densely puberulent, sometimes also with scattered longer hairs, stipitate-glandular. **2*n*** = 42.

Flowering late spring–summer. Wet meadows, stream banks, thickets, muskegs, clearings, forest edges, roadsides; 100–3100 m; Alta., B.C., Man., N.W.T., Nunavut, Ont., Que., Sask., Yukon; Alaska, Ariz., Calif., Colo., Idaho, Mich., Minn., Mont., Nev., N.Mex., N.Dak., Oreg., S.Dak., Utah, Wash., Wis., Wyo.

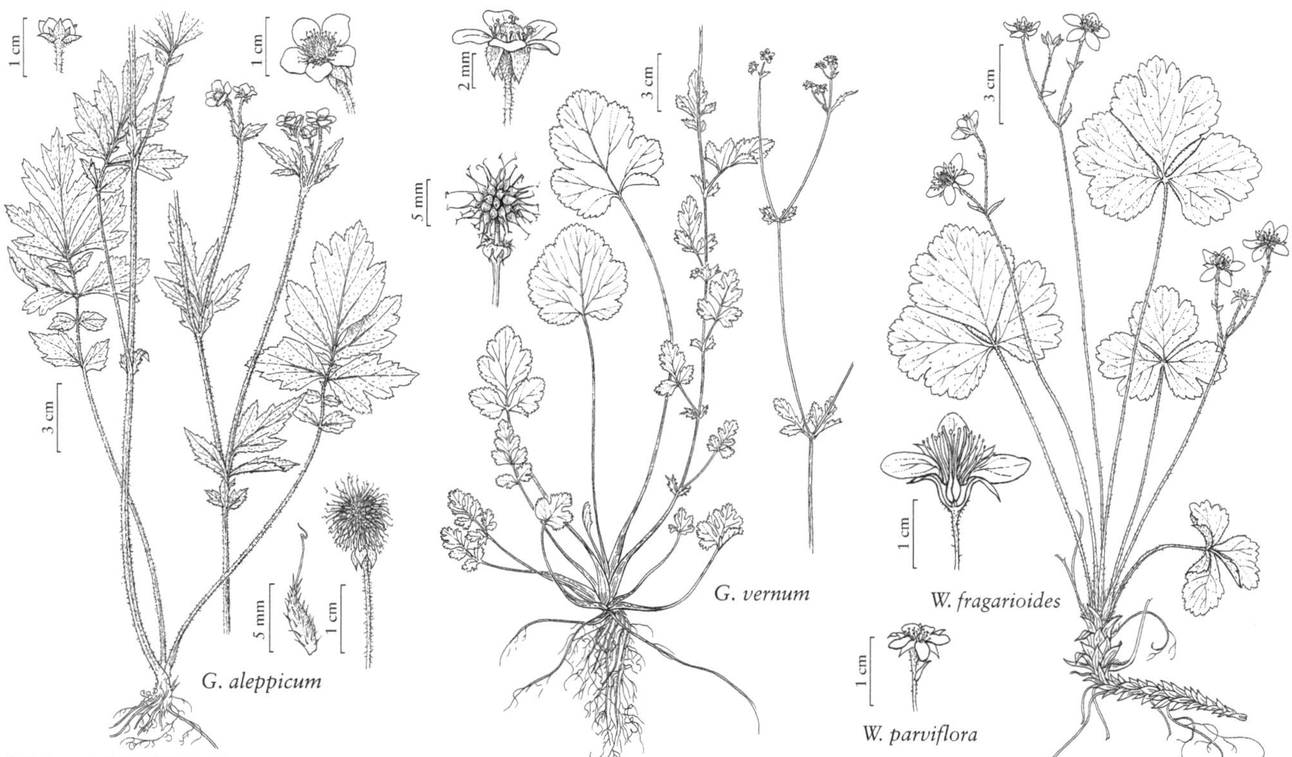

GEUM ◦ WALDSTEINIA

Most specimens of *Geum macrophyllum* with glands on the pedicels also have the more deeply dissected leaves typical of var. *perincisum*; exceptions occur. Plants collected in the San Bernardino Mountains and the southern Sierra Nevada of California are so densely glandular as to fit this variety, yet their leaves are like those of var. *macrophyllum*. More troublesome are specimens from the Upper Peninsula of Michigan with sparsely glandular pedicels and leaves of var. *macrophyllum*. These plants are likely hybrids between the two varieties as their ranges overlap at the western end of the Great Lakes region. Occasional intermediate specimens were also seen from northern California to British Columbia where the ranges of the two varieties meet. Specimens of var. *perincisum* from around James Bay have deeply dissected leaves, but some have glandless pedicels and occasionally the styles are completely glabrous.

Variety *perincisum* hybridizes with *Geum rivale* (= *G.* ×*pervale* B. Boivin); see discussion under 8. *G. rivale*.

11. Geum aleppicum Jacquin, Collectanea 1: 88, plate 127. 1787 • Yellow avens, benoîte d'Alep [F]

Geum aleppicum subsp. *strictum* (Fernald) R. T. Clausen; *G. aleppicum* var. *strictum* Fernald; *G. decurrens* Rydberg; *G. strictum* Aiton var. *decurrens* (Rydberg) Kearney & Peebles

Plants leafy-stemmed. **Stems** 30–120 cm, hirsute to sparsely hirsute. **Leaves:** basal 10–35 cm, blade interruptedly pinnate, major leaflets 5–9, alternating with 4–6 minor ones, terminal leaflet usually only slightly larger than major laterals; cauline 4–19 cm, stipules ± free, 8–28 × 5–22 mm, blade pinnate to 3-foliolate. **Inflorescences** 3–7-flowered. **Pedicels** puberulent, hirsute, eglandular. **Flowers** erect; epicalyx bractlets 2–3.5 mm; hypanthium green; sepals erect but soon reflexed, 4–8 mm; petals spreading, yellow, nearly orbiculate to broadly ovate or obovate, 5–9 mm, equal to or slightly longer than sepals, apex rounded, occasionally irregularly notched. **Fruiting tori** on 2–3 mm stipes, densely downy, <hairs soft, 0.3–0.7 mm>. **Fruiting styles** geniculate-jointed, proximal segment persistent, 3–6 mm, apex hooked, bristles on basal $\frac{1}{3}$, eglandular, distal segment deciduous, 1–2 mm, pilose in basal $\frac{1}{2}$, <hairs much longer than diam. of style>. **2*n*** = 42.

Flowering summer. Moist habitats, meadows, stream banks, alluvial thickets, swamps, forests, clearings, ditches, roadsides; 0–2600 m; Alta., B.C., Man., N.B., N.W.T., N.S., Ont., P.E.I., Que., Sask., Yukon; Alaska, Ariz., Calif., Colo., Conn., Idaho, Ill., Ind., Iowa, Maine, Md., Mass., Mich., Minn., Mont., Nebr., N.H., N.J., N.Mex., N.Y., N.C., N.Dak., Ohio, Oreg., Pa., R.I., S.Dak., Tenn., Utah, Vt., Va., Wash., W.Va., Wis., Wyo.; Mexico; Eurasia.

In the broad area where their ranges overlap, *Geum aleppicum* is often confused with *G. macrophyllum* var. *perincisum*. The two taxa are similar in habit and leaf form, and they occupy the same habitats. Mixed collections mounted on one herbarium sheet are sometimes seen. One macroscopic clue useful in quickly sorting herbarium specimens is the color of the dried plants. *Geum aleppicum* often, but not always, darkens considerably upon drying and appears blackish green; *G. macrophyllum* is lighter and greener. In *G. aleppicum*, the epicalyx bractlets are always present and conspicuous; the bractlets are relatively small or absent in *G. macrophyllum*. In fruiting specimens, the styles of *G. aleppicum* have bristles at the base and are without glands; the styles of *G. macrophyllum* lack bristles and are nearly always stipitate-glandular. The bristles at the base of the style are a consistent and useful characteristic in fruiting specimens of *G. aleppicum*. Other species of avens may have bristles on the achenes; no others (except for occasional specimens of *G. laciniatum*) have bristles on the styles.

Geum aleppicum hybridizes with *G. rivale* (= *G. ×aurantiacum* Fries ex Scheutz); see discussion under 8. *G. rivale*.

Geum strictum Aiton is a superfluous and illegitimate name for *G. aleppicum*.

12. Geum laciniatum Murray, Novi Comment. Soc. Regiae Sci. Gott. 5: 30, plate 2. 1775 • Rough avens, benoîte lacinicée [E]

Geum laciniatum var. *trichocarpum* Fernald

Plants leafy-stemmed. **Stems** 30–100 cm, hirsute, <some hairs 2–2.5 mm>. **Leaves:** basal 15–32 cm, blade simple, lyrate-pinnate, or pinnate, major leaflets 1–7, alternating with 0–10 minor leaflets, terminal leaflet slightly to much larger than major laterals; cauline 3.5–18 cm, stipules ± free, 4–14 × 7–17 mm, blade pinnate, 3-foliolate, or simple and unlobed. **Inflorescences** 2–9-flowered. **Pedicels** densely puberulent, hirsute, eglandular. **Flowers** erect; epicalyx bractlets 1–2 mm; hypanthium green; sepals spreading but soon reflexed, 3–10 mm; petals spreading, white, oblong to elliptic, 2–5 mm, shorter than sepals, apex obtuse to rounded. **Fruiting tori** sessile, glabrous except for ring of bristles at base and tuft at apex. **Fruiting styles** geniculate-jointed, proximal segment persistent, 2.5–5 mm, apex hooked, glabrous, sometimes 1 or 2 eglandular bristles at base, distal segment deciduous, 1–2 mm, short hairs on basal ½. $2n = 42$.

Flowering early summer. Wet woods and thickets, flood plains, wet woods around lakes, stream banks, boggy meadows; 0–1000 m; N.B., N.S., Ont., P.E.I., Que.; Conn., Del., D.C., Ill., Ind., Iowa, Kans., Ky., Maine, Md., Mass., Mich., Minn., Mo., Nebr., N.H., N.J., N.Y., N.C., Ohio, Pa., R.I., S.Dak., Tenn., Vt., Va., W.Va., Wis.

B. L. Robinson and M. L. Fernald (1908) and J. K. Small (1933) misapplied the name *Geum virginianum* to this species. As a consequence, older specimens of *G. laciniatum* are often labeled (and filed in herbaria) as *G. virginianum*.

Geum laciniatum is unique among members of the genus in having some of the heads of achenes pop off the stem and disperse as a unit. The heads disarticulate where the torus joins the hypanthium, leaving the hypanthium inverted at the tip of the stem.

Fernald described var. *trichocarpum* based on the presence of bristles on the summits of the achenes. Although this morphology occurs to the near exclusion of the glabrous one in the western part of the species range, it also occurs as far east as the Carolinas, Maine, New Jersey, and Virginia. The glabrous condition has a smaller range, and it is the prominent one in Nova Scotia, Kentucky, Maine, and Ohio. The two are well mixed in Quebec, New York, and Pennsylvania. Although they are easily distinguished (hairs absent versus present), they seem to have little phytogeographic significance, and presence of achene hairs is not correlated with any other characteristics. It appears to be a trivial variation.

Geum laciniatum hybridizes with *G. urbanum* (= *G. ×macneillii* J.-P. Bernard & R. Gauthier); see discussion under 15. *G. urbanum*.

13. Geum canadense Jacquin, Hort. Bot. Vindob. 2: 82, plate 175. 1772–1773 • White avens, benoîte du Canada

Geum camporum Rydberg; *G. canadense* var. *brevipes* Fernald; *G. canadense* var. *camporum* (Rydberg) Fernald & Weatherby; *G. canadense* var. *grimesii* Fernald & Weatherby; *G. canadense* var. *texanum* Fernald & Weatherby

Plants leafy-stemmed. **Stems** 30–100 cm, glabrate to downy, <hairs to 1.5 mm, sometimes glandular>. **Leaves:** basal 10–25 cm, blade simple or pinnate, major leaflets 3–5, plus 0–4 minor basal ones, terminal leaflet larger than major laterals; cauline 3–8 cm, stipules ± free, 4–13 × 1–7 mm, blade 3-foliolate or simple and 3-lobed to unlobed. **Inflorescences** 3–15-flowered. **Pedicels** densely hairy, <hairs of varying lengths, few long stiff ones>, sometimes glandular. **Flowers** erect; epicalyx bractlets 0.5–1.5 mm; hypanthium green; sepals spreading but soon reflexed, 3–6 mm; petals spreading, white, obovate to oblong, (3–)4–8 mm, ± equal to or slightly longer than sepals, apex rounded. **Fruiting tori** sessile, densely bristly, <hairs 1–2.3 mm>. **Fruiting styles** geniculate-jointed, proximal segment persistent, 2–8 mm, apex hooked, usually glabrous, sometimes sparsely hairy or stipitate-glandular, distal segment deciduous, 1–2 mm, pilose in basal ½, <hairs much longer than diam. of style>. **2*n*** = 42.

Flowering spring–summer. Lowlands and upland forests, meadows, along streams, thickets, bottomland hardwoods, swamps; 0–600 m; Man., N.B., N.S., Ont., P.E.I., Que.; Ala., Ark., Conn., Del., D.C., Ga., Ill., Ind., Iowa, Kans., Ky., La., Maine, Md., Mass., Mich., Minn., Miss., Mo., Mont., Nebr., N.H., N.J., N.Y., N.C., N.Dak., Ohio, Okla., Pa., R.I., S.C., S.Dak., Tenn., Tex., Vt., Va., W.Va., Wis., Wyo.; Mexico (Chiapas).

Varieties and forms have been described in an effort to classify the variation encompassed in *Geum canadense*. In the eastern half of the United States, it is by far the most common, widespread, and variable of the *Geum* species. Nearly all writers of recent floras have not found it worthwhile to apply names to the variants. Perhaps the most distinctive and worthy of further consideration are plants from the southwestern corner of the range in Arkansas, Oklahoma, and Texas. These plants, some of which fit Fernald and Weatherby's description of var. *texanum*, bloom from late March through May, significantly earlier than the rest of the species, which typically flowers after June first, even in the other southern states.

Geum canadense hybridizes with *G. urbanum* (= *G.* ×*catlingii* J.-P. Bernard & R. Gauthier); see discussion under 15. *G. urbanum*.

Geum album J. F. Gmelin is a superfluous name that pertains here.

SELECTED REFERENCE Fernald, M. L. and C. A. Weatherby. 1922. Varieties of *Geum canadense*. Rhodora 24: 47–50.

14. Geum virginianum Linnaeus, Sp. Pl. 1: 500. 1753
• Cream or Virginia or pale avens [E]

Geum flavum (Porter) E. P. Bicknell; *G. hirsutum* Muhlenberg ex Link

Plants leafy-stemmed. **Stems** 25–110 cm, puberulent and hirsute to densely hirsute, <some hairs 2–2.5 mm>. **Leaves:** basal 12–25 cm, blade simple or pinnate, major leaflets 3–5, plus 0–4 minor basal ones, terminal leaflet slightly larger than major laterals; cauline 4–15(–23) cm, stipules ± free, 11–48 × 6–35 mm, blade 3-foliolate or simple and 3-lobed to unlobed. **Inflorescences** 3–14-flowered. **Pedicels** puberulent, <sometimes with scattered hairs>, eglandular. **Flowers** erect; epicalyx bractlets 1–1.5 mm; hypanthium green; sepals spreading but soon reflexed, 3–6 mm; petals spreading, cream, oblong to elliptic, (1.5–)2–3.5 mm, shorter than sepals, apex rounded. **Fruiting tori** sessile, densely bristly, <hairs 1–2.3 mm>. **Fruiting styles** geniculate-jointed, proximal segment persistent, (3–)4.5–7 mm, apex hooked, glabrate, distal segment deciduous, 1–2 mm, pilose in basal ½, <hairs much longer than diam. of style>.

Flowering summer. Mostly forests, river bottoms to dry uplands, rocky slopes, oak-hickory woods; 0–700 m; Ont.; Ala., Ark., Conn., Del., D.C., Ga., Ill., Ind., Ky., Md., Mass., Mich., Mo., N.J., N.Y., N.C., Ohio, Okla., Pa., R.I., S.C., Tenn., Vt., Va., W.Va.

Geum virginianum has been confused nomenclaturally with *G. laciniatum*. B. L. Robinson and M. L. Fernald (1908) and J. K. Small (1933) misapplied the name *G. virginianum* to the species correctly named *G. laciniatum*, and they used the names *G. flavum* (Robinson and Fernald) and *G. hirsutum* (Small) for what is correctly named *G. virginianum*. Much of what is named *G. virginianum* in older herbarium collections is actually *G. laciniatum*.

In habit and leaf form, *Geum virginianum* is similar to *G. canadense*. *Geum virginianum* differs in having cream petals shorter than the sepals (versus white and usually equal to or longer), the largest stipules of the cauline leaves 20–48 × 10–35 mm (versus smaller), and the stems hirsute to densely hirsute (versus glabrate to downy with only scattered longer hairs). The variability within *G. canadense* is so great that for some specimens it is difficult to determine whether they belong to *G. canadense* or *G. virginianum*.

L. A. Raynor (1952) believed *Geum virginianum* to be an F_1 hybrid between *G. aleppicum* and *G. canadense*. K. R. Robertson (1974) acknowledged that some herbarium specimens assignable to *G. virginianum* from the area where *G. aleppicum* and *G. canadense* are sympatric have mostly aborted pollen and may represent natural hybrids. He pointed out that farther south, where *G. aleppicum* is absent, specimens of *G. virginianum* appear fully fertile and seem to represent a valid species.

15. Geum urbanum Linnaeus, Sp. Pl. 1: 501. 1753 • Wood avens, herb bennet, clover-root, benoîte commune [I]

Plants leafy-stemmed. **Stems** 25–85 cm, puberulent and sparsely hirsute. **Leaves:** basal 12–25 cm, blade strongly lyrate-pinnate to pinnate, major leaflets 1–9, sometimes alternating with minor ones, terminal leaflet slightly to much larger than major laterals; cauline 4–15 cm, stipules ± free, 10–40 × 5–35 mm, blade pinnate, 3-foliolate, or simple and 3-lobed. **Inflorescences** 3–18-flowered. **Pedicels** puberulent, less densely hirsute, eglandular. **Flowers** erect; epicalyx bractlets 0.5–2 mm; hypanthium green; sepals spreading but soon reflexed, 2–5 mm; petals spreading, yellow, suborbiculate, elliptic, or obovate, 2–6 mm, shorter to longer than sepals, apex rounded. **Fruiting tori** sessile, densely bristly, <hairs 1–2.3 mm>. **Fruiting styles** geniculate-jointed, proximal segment persistent, 3–7 mm, apex hooked, glabrous, distal segment deciduous, 1–1.5 mm, glabrous or with short hairs, <hairs shorter than diam. of style>. **2*n*** = 42.

Flowering summer. Waste places, hedgerows, gardens, edge of sidewalks, disturbed woods, along roads and trails in mesic woods; 0–1400 m; introduced; N.B., Ont., Que.; Idaho, Ill., Mass., Mich., N.Y., Ohio, Oreg., Pa., R.I., Utah, Vt., Wash., Wis.; Eurasia; n Africa.

Specimens from Chicago, Illinois, and Cambridge, Massachusetts, indicate that *Geum urbanum* has been present in North America for at least 100 years; herbarium specimens are scarce. The species seems to be spreading in ruderal habitats of the northeastern United States and adjacent Canada. The plants have the habit and foliage of *G. canadense* and *G. virginianum*, with stipules as large as or even larger than those of the latter; the petals are bright yellow versus white or cream, and the distal segments of the styles are glabrous or nearly so versus partially pilose.

Geum urbanum is often found growing near *G. canadense*, and the two occasionally form hybrids; *G.* ×*catlingii* J.-P. Bernard & R. Gauthier is known from New Brunswick, Ontario, Quebec, and Michigan. Less common is *G.* ×*macneillii* J.-P. Bernard & R. Gauthier [*G. laciniatum* × *G. urbanum*], known from Quebec. Both hybrids closely resemble *G. urbanum* in vegetative form, with little hint of their other parent; it is likely that hybrid plants often go undetected. Most herbarium specimens of these hybrids were collected along with their parental species at the same locality. The hybrids are often robust, apparently showing hybrid vigor. Unlike *G. urbanum*, they have cream petals and heads with mostly sterile ovaries.

SELECTED REFERENCE Bernard, J.-P. and R. Gauthier. 1986. Observations sur le *Geum urbanum* L. dans la région de Québec et description de deux hybrides. Naturaliste Canad. 113: 317–324.

16. Geum vernum (Rafinesque) Torrey & A. Gray, Fl. N. Amer. 1: 422. 1840 • Spring avens [E] [F]

Stylypus vernus Rafinesque, Neogenyton, 3. 1825

Plants leafy-stemmed. **Stems** 20–70 cm, pilose or sparsely pilose, <hairs septate>. **Leaves:** basal 4–27 cm, blade simple or pinnate, leaflets 3–11, terminal leaflet larger; cauline 2–7 cm, stipules ± free, 8–25 × 7–12 mm, blade pinnate to 3-foliolate. **Inflorescences** 3–13-flowered. **Pedicels** glandular-downy, becoming glabrate in fruit. **Flowers** erect; epicalyx bractlets absent; hypanthium green; sepals reflexed, 1–3 mm; petals spreading, yellow to cream, oblong to elliptic, sometimes obovate, 1–2 mm, equal to or shorter than sepals, apex rounded. **Fruiting tori** on 3–7 mm stipes, glabrous. **Fruiting styles** geniculate-jointed, proximal segment persistent, 1.5–3 mm, apex hooked, nearly glabrous except for few septate-glandular hairs, distal segment deciduous, 0.7–1 mm, nearly glabrous except for short hairs. **2*n*** = 42.

Flowering spring. Moist woods, disturbed moist areas, flood plains, openings; 0–600 m; Ont.; Ark., Del., D.C., Ga., Ill., Ind., Iowa, Kans., Ky., Md., Mass., Mich., Mo., Nebr., N.J., N.Y., N.C., Ohio, Okla., Pa., Tenn., Va., W.Va., Wis.

Geum vernum is distinctive and is recognized by its early flowering, minute petals, lack of epicalyx bractlets, and heads of achenes soon elevated well beyond the recurved hypanthia and sepals. Recently, it has been expanding its range northward and has been collected with increasing frequency in southern Ontario, Massachusetts, Michigan, New York, and Wisconsin. No specimens could be located to confirm reports of *G. vernum* from Mississippi or Texas.

5. WALDSTEINIA Willdenow, Ges. Naturf. Freunde Berlin Neue Schriften 2: 105, plate 4, fig. 1. 1799 • Barren-strawberry, waldsteinie [For Franz de Paula Adam von Waldstein, 1759–1823, Austrian soldier and botanist]

James B. Phipps

Herbs, perennial, <scapose>, often mat-forming, 1–4 dm, strigose; rhizomatous. **Stems** 1–20+, ± erect. **Leaves** deciduous, basal, <sometimes clustered around scape>, simple or compound; stipules adnate to petiole base, narrow, <scarious>, margins entire; petiole present; blade ± rounded, deeply cordate, <ternate or 3–5(–7)-lobed>, 3–5 cm, herbaceous, margins flat, when lobes usually shallow, dentate, <often with large teeth distally>, venation palmate, surfaces strigose or hirsute. **Inflorescences** terminal, 3–8-flowered, panicles, open, <± as long as petioles, elongating somewhat in fruit>; bracts present; bracteoles absent. **Pedicels** present. **Flowers** 8–20 mm diam.; epicalyx bractlets 0 [5, minute]; hypanthium broadly obconic to obcampanulate, slightly contracted at mouth, 2–4 mm, strigose or glabrous; sepals 5, spreading, triangular; petals 5, yellow, obovate to elliptic, elliptic-lanceolate, or suborbiculate; stamens 50–70, ± equal to petals, <filaments persistent>; torus not evident, <receptacle base villous>; carpels 3–7, short-villous; ovule 1. **Fruits** aggregated achenes, 3–7, broadly crescent-shaped, 1.5–2.5 mm, short-villous; hypanthium persistent; sepals persistent, ± reflexed; styles deciduous, not hooked. $x = 7$.

Species 6 or 7 (4 in the flora): North America, Eurasia; temperate regions.

Plants of *Waldsteinia* are distinctive, small, rhizomatous, and often mat-forming herbs of woodlands with ternate or three- to five-lobed leaves, differing from *Geum* by their long, straight, deciduous styles. J. E. E. Smedmark (2006) showed that the type species of *Waldsteinia* is nested in *Geum*.

SELECTED REFERENCE Teppner, H. 1968. Zur Kenntniss der Gattung *Waldsteinia*. Graz.

1. Leaves mostly ternately compound, rarely merely deeply lobed.
    2. Petals obovate to broadly elliptic, 3.5–10 mm, much longer than sepals . . . . . . 1. *Waldsteinia fragarioides*
    2. Petals elliptic to elliptic-lanceolate, 2.5–4.5 mm, shorter than to barely exceeding sepals . . . . . . 2. *Waldsteinia parviflora*
1. Leaves 3–5(–7)-lobed, shallowly so or to ± ⅔ cleft.
    3. Flowers 8–10 mm diam.; petals narrowly elliptic, barely exceeding sepals, bright yellow . . . . . . 3. *Waldsteinia lobata*
    3. Flowers 15–18 mm diam.; petals broadly elliptic to suborbiculate, much longer than sepals, creamy yellow . . . . . . 4. *Waldsteinia idahoensis*

1. Waldsteinia fragarioides (Michaux) Trattinnick, Rosac. Monogr. 3: 107. 1823 • Common barren-strawberry, faux-fraisier [E] [F]

Dalibarda fragarioides Michaux, Fl. Bor.-Amer. 1: 300, plate 28. 1803 (as fragaroides); *Geum donianum* (Trattinnick) Weakley & Gandhi; *G. fragarioides* (Michaux) Smedmark; *Waldsteinia doniana* Trattinnick; *W. fragarioides* subsp. *doniana* (Trattinnick) Teppner

Plants 10–20 cm. **Leaves** mostly ternately compound, rarely merely deeply lobed; leaflets broadly cuneate-obovate, 3–5 cm, lateral ones ± asymmetric, margins commonly shallowly and irregularly lobed, surfaces sparsely strigose, <principal veins often densely strigose>. **Inflorescences** 3–8 -flowered. **Flowers** 14–20 mm diam.; hypanthium obcampanulate, 2–4 mm, externally strigose or glabrous; petals obovate to broadly elliptic, 3.5–10 mm, much longer than sepals, apex obtuse or rounded; carpels 3–5.

Flowering Apr–May. Moist or dry hardwood forests or mixed conifer-hardwood forests; 0–600 m; N.B., Ont., Que.; Conn., Ill., Ind., Ky., Maine, Md., Mass., Mich., Minn., Mo., N.H., N.J., N.Y., N.C., Ohio, Pa., Tenn., Vt., Va., W.Va., Wis.

Waldsteinia doniana is usually synonymized with *W. parviflora*; according to M. L. Fernald (1935) it is based on an aberrant cultivated form of *W. fragarioides*.

Waldsteinia fragarioides and *W. parviflora* are often combined, then often distinguished as varieties. The flower size of *W. fragarioides* is very variable, and it seems that larger flowers are more usual in the north of the range. The best differentiator from *W. parviflora* is sepal length relative to petal length, and all specimens with adequate material seen by the author were distinguishable. *Waldsteinia fragarioides* is used as a mat-forming groundcover in ornamental horticulture.

2. Waldsteinia parviflora Small, Bull. Torrey Bot. Club 25: 137. 1898 • Small-petaled barren-strawberry [E] [F]

Waldsteinia fragarioides (Michaux) Trattinnick var. *parviflora* (Small) Fernald

Plants 10–20 cm. **Leaves** mostly ternately compound, rarely merely deeply lobed; leaflets broadly cuneate-obovate, 3–5 cm, lateral ones asymmetric, margins commonly shallowly and irregularly lobed, surfaces sparsely hirsute. **Inflorescences** 3–7-flowered. **Flowers** 8–10 mm diam.; hypanthium obcampanulate, 2–4 mm, strigose; petals elliptic to elliptic-lanceolate, 2.5–4.5 mm, shorter than to barely exceeding sepals, apex often acute; carpels 3 or 4.

Flowering Mar–Apr. Woods, mountains, and piedmont; 100–400 m; Ala., Ga., Ky., N.C., S.C., Tenn., Va.

Waldsteinia parviflora closely resembles *W. fragarioides* but has generally smaller flowers with petals shorter than to subequal to the sepals. The possibility that *W. parviflora* originated as a hybrid between *W. fragarioides* and *W. lobata* should be investigated.

3. Waldsteinia lobata (Baldwin) Torrey & A. Gray, Fl. N. Amer. 1: 426. 1840 • Piedmont barren-strawberry [E]

Dalibarda lobata Baldwin in S. Elliott, Sketch Bot. S. Carolina 1: 571. 1821; *Geum lobatum* (Baldwin) Smedmark

Plants 15–30 cm. **Leaves** 3-lobed, sometimes to ca. ⅔ cleft, 4–8 cm wide, surfaces sparsely strigose. **Inflorescences** 3–7 -flowered. **Flowers** 8–10 mm diam.; hypanthium broadly obconic, 2–3 mm, strigose; petals <bright yellow>, narrowly elliptic, 4 mm, barely exceeding sepals, apex subacute; carpels 3 or 4.

Flowering Mar–Apr. Moist woods near streams in piedmont; 100–500 m; Ga., N.C., S.C.

Waldsteinia lobata is very similar to *W. parviflora* except in leaf form.

4. Waldsteinia idahoensis Piper, Bull. Torrey Bot. Club 30: 180. 1903 • Idaho barren-strawberry [C] [E]

Geum idahoense (Piper) Smedmark

Plants 15–40 cm. **Leaves** shallowly 3–5(–7)-lobed, 3–7 cm wide, surfaces hirsute. **Inflorescences** 3–8-flowered. **Flowers** 15–18 mm diam.; hypanthium broadly obconic, 2–3 mm, glabrous; petals <creamy yellow>, broadly elliptic to suborbiculate, 5–8 mm, much longer than sepals, apex obtuse; carpels 4–7.

Flowering Jun. Mesic forest floors, under open canopy of grand fir or western red-cedar; of conservation concern; 500–1500 m; Idaho, Mont.

Waldsteinia idahoensis is easily distinguished and highly disjunct from other species of the genus. The species usually proliferates after clear-cuts. Most of the records are from the Clearwater Region of north-central Idaho.

6. FALLUGIA Endlicher, Gen. Pl. 16: 1246. 1840 • Apache plume, yerba del pasmo, barba de chivo [For Virgilio Fallugi, 1627–1707, Italian abbot]

James Henrickson

Bruce D. Parfitt†

Shrubs, 10–20(–35) dm, polygamodioecious; rhizomatous in sandy soil. **Stems** 1–20+, spreading-ascending; bark brown, <with age separating in papery sheets>; long and short shoots present; young stems whitish, hirtellous to villous, also stellate-lepidote with white to rust orange trichomes. **Leaves** deciduous, cauline, <abscising distal to persistent, clasping bases, crowded on short shoots>, simple; stipules adnate, acicular, reduced on short-shoot leaves, margins entire; petiole present; blade obovate to oblanceolate or linear, <1- or 2-pinnately lobed>, 0.4–3 cm, leathery, margins revolute, entire, venation pinnate, surfaces hirtellous, villous, orange-lepidote throughout, abaxially densely so, glabrescent. **Inflorescences** terminal <on long shoots of season>, 1–7-flowered, loose corymbose racemes; bracts present, reduced; bracteoles present <often linear-acicular or with reduced, paired, basal lobes borne at pedicel base>. **Pedicels** present. **Flowers** all pistillate, all staminate, or staminate with terminal one bisexual, 20–35 (–42) mm diam.; epicalyx bractlets 5, <entire or 3-toothed>; hypanthium broadly funnelform, 2.5–3.5 mm, exterior rusty lepidote-stellate and villose-pilose, interior densely hirsute; sepals 5, ascending, broadly ovate to suborbiculate; petals 5, white (to pinkish), oblong-obovate to suborbiculate; stamens (24–)50–95(–120), shorter than petals, <anthers fertile in staminate flowers, sterile in pistillate flowers>; torus ovoid-cylindric; carpels (24–)50–95(–120), reduced in staminate flowers, sericeous; ovules 2. **Fruits** aggregated achenes, (24–)50–95(–120), compressed, fusiform, 1–2.5 mm, sericeous; hypanthium persistent; sepals persistent, spreading; <torus elongating>; styles persistent, greatly elongating, <filiform, plumose>. $x = 14$.

Species 1: sw United States, n Mexico.

Fallugia has long been considered closely related to *Cercocarpus*, *Cowania*, and *Dryas*. Cytologic (E. D. McArthur et al. 1983), morphologic (J. Henrickson 2001), and molecular (D. R. Morgan et al. 1994; D. Potter et al. 2007) data substantiate its close relationship with *Geum* (tribe Colurieae).

SELECTED REFERENCE Henrickson, J. 2001. Systematics and relationships of *Fallugia* (Rosoideae–Rosaceae). Aliso 20: 1–15.

1. Fallugia paradoxa (D. Don) Endlicher ex Torrey in W. H. Emory, Not. Milit. Reconn., 139. 1848 [F]

Sieversia paradoxa D. Don, Trans. Linn. Soc. London 14: 576, plate 22, figs. 7–10. 1825; *Fallugia micrantha* Cockerell; *F. paradoxa* var. *acuminata* Wooton

Stems: long-shoot internodes 2.5–10(–15) mm, those proximal to flowers 15–30 (–50) mm; short-shoot spurs 1–3.5(–7) mm. **Leaves:** long-shoot leaf bases 1.5–4.5 mm, short-shoot leaf bases much reduced; stipules 1–2 mm; blade (4–)7–18(–30) × (0.8–)3–12(–23) mm, lobes 3–7, linear-lanceolate, 1–7(–13) × 0.7–1.4(–2.2) mm. **Flowers:** epicalyx bractlets leaflike, subulate to linear, 2–5(–8) mm, margins entire or apically 2- or 3-toothed; hypanthium 2.5–3.5 × (4–)5–6(–7) mm; sepals (3.2–)4–8 (–10) × 2.2–5(–6.5) mm, subterminal leafy apiculations 1(–3), terminal, linear-subulate, 1–3, 1–2.5(–3.5) mm; petals (6–)10–18(–21) × (6–)7–14(–17) mm, ± larger in staminate flowers; staminate flower filaments 2.2–4.3 mm, anthers 0.7–1.2 mm; pistillate flower filaments 1.5–2.5 mm, anthers 0.3–0.4 mm (sterile); stigmas terminal. **Achenes** 2-veined; styles ± curved, 2.5–5 cm, hairs straight, spreading-ascending, white to purplish. $2n = 28$.

Flowering Apr–Oct. Desert scrub in sandy to rocky (rarely clay) drainages, or sandy-loamy to rocky uplands in grasslands, chaparral, pinyon-oak, juniper, or ponderosa pine woodlands; (200–)700–2800 m; Ariz., Calif., Colo., Nev., N.Mex., Okla., Tex., Utah; Mexico (Baja California, Chihuahua, Coahuila, Durango, Zacatecas).

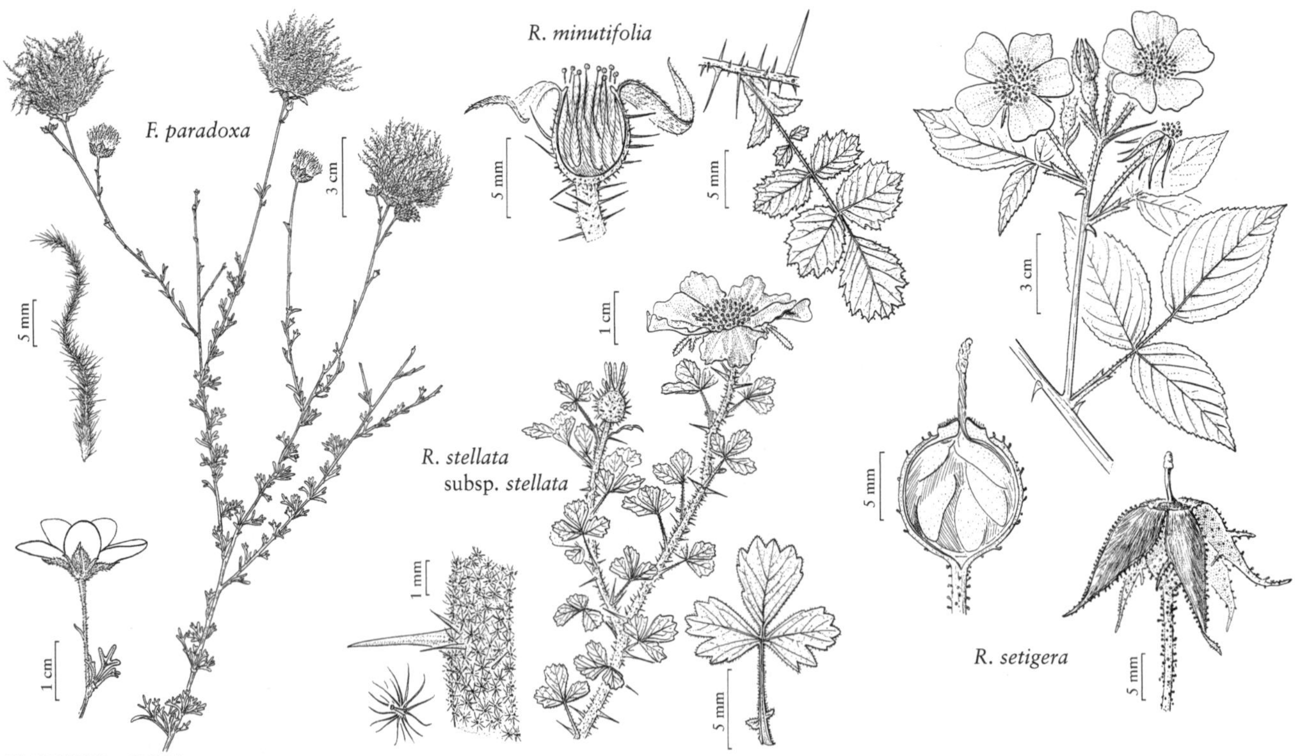

FALLUGIA ◦ ROSA

Fallugia paradoxa grows in diverse habitats, ranging from the Chihuahuan, Great Basin, and Mojave Desert scrub to woodlands. Throughout its range, plants exhibit considerable variation in underground structures, and lobing and indument of leaves, bracts, and epicalyx. In sandy drainages, it is commonly rhizomatous and colonial, especially in California. In upland sites, it is not rhizomatous. Lobing of sepal tips and epicalyx bracts shows considerable variation on individuals and to some extent regionally. Lobes are more consistently found in plants from Texas and northern Mexico.

a4. ROSACEAE Jussieu (subfam. ROSOIDEAE) tribe ROSEAE Lamarck & de Candolle, Syn. Pl. Fl. Gall., 331. 1806 (as Rosae)

Luc Brouillet

Shrubs or subshrubs; usually armed, rarely unarmed. **Leaves** alternate, imparipinnately compound or 3-foliate [simple]; stipules persistent, sometimes caducous [absent], adnate to petiole base or free, auricles free; venation pinnate. **Flowers:** perianth and androecium perigynous; epicalyx bractlets absent; hypanthium ovoid, oblong, globose to cupulate or urceolate, rarely obovoid or hemispheric, <with or without distinct necks>; torus usually absent, conic if present; carpels 3–110+, styles terminal, distinct, rarely connate; ovule 1, apical. **Fruits** accessory; aggregated achenes enclosed within hypanthium (hip); styles persistent, not elongate.

Genus 1, species ca. 140 (33 in the flora): North America, Mexico, Eurasia, Africa, Pacific Islands, Australia; introduced nearly worldwide.

The base chromosome number for Roseae is $x = 7$.

7. ROSA Linnaeus, Sp. Pl. 1: 491. 1753; Gen. Pl. ed. 5, 217. 1754 • Rose, brier, rosier [Latin name for rose]

Walter H. Lewis

Barbara Ertter

Anne Bruneau

Shrubs or subshrubs, 1–50(–100) dm; rhizomatous or stoloniferous. **Stems** 1–20, erect to spreading, arching, climbing, or procumbent, <simple or branched>; bark brown, red, and gray, <often exfoliating>; long and short shoots present; glabrous, rarely densely puberulent to tomentose. **Leaves** deciduous, rarely persistent or semipersistent, cauline; stipules present [absent], narrow, rarely broad, margins entire or serrate, sometimes pectinate, crenate, or lacinulose, rarely serrulate, undulate, or sinuate; petiole present; blade cordate, elliptic, lanceolate, ovate-lanceolate, oblong, obovate, oval, ovoid, orbiculate, or suborbiculate, 1–3 cm, membranous or leathery, <rugose or smooth, dull or lustrous>; leaflets (3–)5–11(–13), <lateral subsessile, terminal petiolulate>, elliptic to obovate, ± oblong, obovate-elliptic, lanceolate-elliptic, or obovate-oblong, margins flat, serrate, crenate, or incised, surfaces glabrous, puberulent, pubescent, or tomentose, <eglandular or glandular>. **Inflorescences** terminal <usually on lateral branches, sometimes on primary stems and shoots>, (1–)4–30(–50)-flowered, usually panicles, sometimes corymbs, glabrous; bracts absent or present, <(0 or)1–3(–8)>; bracteoles absent. **Pedicels** present. **Flowers** rarely unisexual (dioecious in *R. setigera* or monoecious), 10–90(–100) mm diam.; hypanthium 2–5(–10) mm, glabrous, puberulent, tomentose, or setose, <eglandular or glandular>; sepals 5, erect or spreading to reflexed, ovate-lanceolate or lanceolate, rarely ovate-acuminate or deltate, <margins entire or pinnatifid, apex acute, often prolonged tip>; petals [4]5 (sometimes numerous if "double"), pink to red, sometimes white, obovate, <apex usually emarginate>; stamens 35–220, shorter than petals; <carpels usually borne on inner hypanthial walls, rarely on basal tori, styles glabrous or pilose, sometimes lanate or villous, exsert>. **Fruits** (hips), 1–50, globose, ovoid-obovoid, ellipsoid, oblong, pyriform, or urceolate, 7–11(–20) diam., glabrous, <glandular or eglandular>; hypanthium persistent, <red or orange-red, purplish red, or purplish black, fleshy or leathery>; sepals persistent or deciduous, erect to spreading-erect. $x = 7$.

Species ca. 140 (33 in the flora): North America, Mexico, Europe, Asia, n Africa; introduced nearly worldwide.

Most species of *Rosa* occur in the cooler parts of the northern hemisphere. Only three or four species extend south of the Tropic of Cancer in the Old World, none do so in the New World.

In an infrafamilial classification of the Rosaceae by D. Potter et al. (2007), *Rosa* is included in subfam. Rosoideae, supertribe Rosodae (no tribe). Distinctive features of *Rosa* include shrubby habit, concave or urn-shaped hypanthia, relatively numerous pistils, and fruits aggregated achenes (achenetum) inside fleshy hypanthia (hips).

Character states often are part of a continuum instead of being discrete. When addressed by differing taxonomic philosophies, radically different treatments of *Rosa* in North America have resulted. P. A. Rydberg (1918) recognized 129 native species and dozens of hybrids among them. E. W. Erlanson (1934) accepted 15 species, plus seven ecotype species and four putative hybrids based on cytogenetics, experimental hybridizations, common garden experiments, and field observations. Although no consensus treatment has been adopted by subsequent regional (for example, B. Ertter 1993e; A. Cronquist and N. H. Holmgren 1997) and national floras (for example, H. J. Scoggan 1978–1979, vol. 1) or monographic treatments (W. H. Lewis 1958, 1959, 1959b, 1962, 1965), a unifying feature has been a conservative approach to

Rosa taxonomy, with relatively few species recognized (Ertter and Lewis 2008). Caution is advised in relying on names in synonymy for updating the identification of old collections; synonymy given here often differs from previous applications.

The approach of R. Elven et al. (1999) is followed here in using subspecies as the main infraspecific rank. The rank of variety also is used, either within subspecies or as subunits of species.

The following summary describes the methods of obtaining data. Information on introduced species was obtained chiefly from North American plants with only supplemental use of specimens from Eurasia.

Prickles, including aciculi, of all *Rosa* are epidermal outgrowths of stems and not thorns, which are modified stems. Acicular prickles are shaped like aciculi, needle-shaped and relatively short. Declining prickles are retrorse, being turned inward toward the axis. Infrastipular prickles are proximal to stipules, single or paired in contrast to internodal. Setaceous prickles are shaped like bristles or stiff hairs (setae). Pricklets are miniature prickles on petioles and rachises closely resembling infrastipular/internodal prickles found on stems except that they are smaller.

Herbarium specimens often do not have information on plant height, habit, or bark; field research and observations in gardens have helped provide information for these and other characters. First-year rhizomatous shoots (suckers, turions) are densely covered with mixed internodal prickles and aciculi; leaves of these often develop an extra pair of lateral leaflets when compared to older stems. Identifications using these shoots are difficult because specific differentiation at an early stage of development is limited; no attempt was made to incorporate such data in this treatment. Whenever possible, first-year rhizomatous shoots should be collected, particularly if they flower the first year, as *Rosa arkansana*, or when mowing or otherwise broken shoots stimulate early flowering. Data for prickles and aciculi were obtained using distal stems and distal, usually fertile branches at roughly the third or fourth internode from branch tips; measurements of infrastipular prickles (or, if none, internodal) consisted of prickle lengths to base × base lengths; relatively few aciculi or setae were measured. *Leaves:* stipules were measured for length of adnation to petioles, width across both sides of petioles near distal regions, and length of distally free auricles on one side. Petioles and rachises often have pricklets resembling main prickles of stems. Terminal leaflet measurements of serrations per side exclude terminal serration, and for double- or multi-serrations only each main serration is counted. *Inflorescences* include measurements of pedicel lengths; pedicels are slender to stout, usually varying between 0.5–1 mm diam. *Flowers:* sepal base width measurements at junctures with hypanthia were made of the three outer sepals; sepal margins are characteristically tomentose or pubescent and are not described (varying glandularity is), nor are induments of sepal adaxial surfaces (also tomentose or pubescent); petal numbers are either single and five or double and ten or more; available stamen number averages are from E. W. Erlanson (1934). *Achenes* are bony walled, round or angular (not in descriptions unless different); when unusually small, shriveled, or darkened, they are considered abortive and, if counted, are provided separately from fertile achenes. *Cytology:* ploidy levels obtained from flow cytometry calculations are based on nuclear DNA amounts using fresh leaflets; these are provided with the ploidy level followed by DNA only in species discussions (all research by Anne Bruneau).

When using the keys, be aware that most species known to possess prickles also have some stems and branches lacking armature.

Phenotypic variability in *Rosa* may be due to seasonal or annual environmental factors, varying gene expressions, hybridization, and polyploidy; in sect. *Caninae*, hemisexual reproduction and such features represent confounding factors in rose morphology and taxonomy.

All species of the gall-inducing cynipid wasps *Diplolepis* are restricted to native and introduced roses and induce structurally distinct galls. For example, in Ontario, *D. polita* induces a single-chambered gall on leaves of *Rosa acicularis* subsp. *sayi*; *D. spinosa* induces a multichambered gall on stems of *R. blanda* (J. D. Shorthouse 1991).

Most *Rosa* species have general usages, such as fleshy hips long and widely used for eating raw and as prepared foods, such as jams, jellies, and preserves, and for food supplements due to high vitamin C content. Ethnobotanical uses provided for species are based on D. E. Moerman (1998) and W. H. Lewis and M. P. F. Elvin-Lewis (2003), together with original articles and personal communications.

North American rose species have contributed modestly to the development of hybrids for the horticulture industry. Such contributions are reported in publications associated with the American Rose Society and Royal Horticultural Society (J. H. McFarland 1938; P. Harkness 2003; C. Quest-Ritson and B. Quest-Ritson 2003).

The infrageneric structure used here is relatively well established in floral and taxonomic works (with some ambivalent results from molecular analyses). The main difference is that sect. *Rosa* replaces the better-known sect. *Cinnamomeae*. This results from conservation of the type of *Rosa* on *R. cinnamomea* (J. McNeill et al. 2006, App. IIIA) over the hybrid *R. ×centifolia* Linnaeus in sect. *Gallicae*.

Two species of sect. *Chinenses* de Candolle ex Seringe, both cultivated in warmer areas, are reported as occasionally escaping in North America. Key distinguishing features of the section include shrubs erect or climbing, stems with curved prickles, leaflets usually three to five with relatively narrow, adnate stipules, sepals reflexed and deciduous after anthesis, and styles free, exsert, nearly equaling stamens. *Rosa chinensis* Jacquin, China rose, has been confirmed as occurring at some time in Arkansas, Florida, Georgia, Louisiana, Mississippi, South Carolina, and Virginia. The closely allied *R. ×odorata* (Andrews) Sweet, tea rose, was also found at some time in Florida, Georgia, Louisiana, and South Carolina, and unconfirmed in Prince Edward Island, Pennsylvania, and Utah. *Rosa chinensis* is distinguished from *R. ×odorata* by its shrubby habit (versus climbing or scrambling lianas to 20 m), leaves with three to five (to seven) leaflets (versus five to nine), stipules and auricles marginally stipitate-glandular (versus eglandular or glandular only on auricle margins), flowers usually four or five, rarely solitary (versus one to three), slightly or not fragrant (versus fragrant), 3–5 cm diam. (versus 5–10 cm diam.), with often pinnately lobed sepals (versus entire or slightly incised), and hips ovoid or pyriform (versus depressed-globose). Both species are of major importance in breeding new garden roses, especially the hybrid teas, by incorporating continuous flowering into roses.

SELECTED REFERENCES Bruneau, A., J. R. Starr, and S. Joly. 2007. Phylogenetic relationships in the genus *Rosa*: New evidence from chloroplast DNA sequences and an appraisal of current knowledge. Syst. Bot. 32: 366–378. Erlanson, E. W. 1934. Experimental data for a revision of the North American wild roses. Bot. Gaz. 96: 197–259. Joly, S. et al. 2006. Polyploid and hybrid evolution in roses east of the Rocky Mountains. Amer. J. Bot. 93: 412–425. Rydberg, P. A. 1918. *Rosa*. In: N. L. Britton et al., eds. 1905+. North American Flora.... 47+ vols. New York. Vol. 22, pp. 482–533. Wissemann, V. and C. M. Ritz. 2005. The genus *Rosa* (Rosoideae, Rosaceae) revisited: Molecular analysis of nrITS-1 and *atp*B-*rbc*L intergenic spacer (IGS) versus conventional taxonomy. Bot. J. Linn. Soc. 147: 275–290.

1. Styles: hypanthial discs absent; leaves 1.5–3 cm, leaflet margins deeply incised, broadly crenate, sometimes serrate, terminal petiolules 0.5–4 mm; s California to w Texas [7a. *Rosa* subg. *Hesperhodos*, p. 78] . . . . 7a.1. *Rosa* sect. *Minutifoliae*, p. 79
1. Styles: hypanthial discs present; leaves (2.5–)3.5–15(–18) cm, leaflet margins usually serrate, rarely deeply incised or broadly crenate, terminal petiolules (1–)5–20(–40) mm; throughout North America [7b. *Rosa* subg. *Rosa*, p. 82].
   2. Stipules ± free from petioles or adnate at bases; petals white, rarely rose; leaves persistent or semipersistent, leathery.

3. Bracts 6–8 (covering ½ or all of hypanthia), margins deeply laciniate; pedicels 3–6 mm, densely white sericeous-tomentose, setae dense; leaflets (5–)7–9(–11), terminal petiolules 3–5 mm, blades 14–30 × 5–15 mm . . . . 7b.2. *Rosa* sect. *Bracteatae*, p. 86
3. Bracts 1(or 2), margins serrate; pedicels 12–30 mm, glabrous, setae apical or absent; leaflets 3(–5), terminal petiolules 9–13 mm, blades 40–65 × 30–40 mm . . . . 7b.3. *Rosa* sect. *Laevigatae*, p. 87

[2. Shifted to left margin.—Ed.]

2. Stipules adnate to petioles; petals rose to pink or white, rarely crimson; leaves deciduous, rarely semipersistent or persistent, membranous, sometimes leathery.
   4. Styles connate (exsert 3–6 mm), hypanthial discs conic; shrubs stoloniferous; stipule margins pinnatifid or lacinulose, sometimes entire; hips usually globose, 4–10 × 5–9 mm . . . . .7b.1. *Rosa* sect. *Systylae*, p. 82
   4. Styles free (exsert 0.5–4 mm in sect. *Rosa*), hypanthial discs conic or flat; shrubs rhizomatous, rhizomes sometimes absent; stipule margins usually entire, slightly crenate, or serrate; hips varying in shape, 6–18(–24) × 5–22(–25) mm.
      5. Rhizomes absent or relatively short; stems erect, arching, or spreading; leaflets adaxially usually tomentulose or pubescent . . . . 7b.5. *Rosa* sect. *Caninae*, p. 89
      5. Rhizomes relatively long; stems erect, sometimes arching; leaflets adaxially usually glabrous.
         6. Distal branches: infrastipular prickles usually present, internodal prickles sometimes intermixed with aciculi, rarely absent; pedicel bracts persistent . . . . 7b.7. *Rosa* sect. *Rosa*, p. 94
         6. Distal branches: infrastipular prickles usually absent, rarely present (in sect. *Gallicae*), internodal prickles usually intermixed with aciculi; pedicel bracts caducous or absent.
            7. Leaflets 3–5(–7), terminal petiolules 6–8 mm; flowers 6–9 cm diam.; inflorescences 1–3(–8)-flowered; pedicel bracts caducous; hips bright to dark red; hypanthial discs conic, 4–5 mm diam.; sepals of hips deciduous, erect to reflexed . . . . 7b.4. *Rosa* sect. *Gallicae*, p. 88
            7. Leaflets (5–)9–11(–13), terminal petiolules 2–4 mm; flowers 2–4 cm diam.; inflorescences 1-flowered; pedicel bracts absent; hips blackish purple; hypanthial discs flat, 4–4.5 mm diam.; sepals of hips persistent, erect . . . . .7b.6. *Rosa* sect. *Pimpinellifoliae*, p. 93

7a. Rosa Linnaeus subg. Hesperhodos (Cockerell) Cockerell ex Rehder, Man. Cult. Trees ed. 2, 451. 1940

Hesperhodos Cockerell, Nature 90: 571. 1913

Leaves 1.5–3 cm; leaflets 3–7, margins deeply incised or broadly crenate, sometimes serrate, terminal: petiolule 0.5–4 mm, blade oval, suborbiculate, obovate, or deltate, base cuneate. **Inflorescences** 1(–3)-flowered. **Pedicels:** bracts present or absent. **Flowers:** hypanthium subglobose to globose or cupulate, densely pubescent or glabrous, setose or not, glandular or eglandular; carpels 10–110 (average 65), styles free, pilose, stylar orifice 2–5 mm diam., rims 0.5–1 mm wide, hypanthial disc absent. **Hips** leathery, setose.

Species 2 (2 in the flora): sw, sc United States, nw Mexico.

A phylogenetic analysis using plastid DNA sequences suggests that subg. *Hesperhodos* is sister to the rest of *Rosa*, but with weak support (A. Bruneau et al. 2007). This contrasts with analyses indicating that *R. persica* Michaux of subg. *Hulthemia* (Dumortier) Focke is sister to the remainder of the genus (S. Wu et al. 2001; V. Wissemann and C. M. Ritz 2005). A more compelling resolution might be possible once the related and more variable *R. stellata* is examined phylogenetically.

SELECTED REFERENCE Lewis, W. H. 1965. Monograph of *Rosa* in North America. V. Subgenus *Hesperhodos*. Ann. Missouri Bot. Gard. 52: 99–113.

7a.1. Rosa Linnaeus (subg. Hesperhodos) sect. Minutifoliae Crépin, J. Roy. Hort. Soc. 11: 226. 1889

Shrubs, forming dense thickets, <xerophytic>; rhizomatous. **Stems** erect, rarely arching, 2.5–12(–15) dm; distal branches densely pubescent or glabrous, with or without stellate hairs; prickles erect, rarely curved, infrastipular prickles present, mixed with internodal prickles and aciculi. **Leaves** deciduous, 1.5–3 × 1.5–2.5 cm; stipule margins entire or dentate, glandular or eglandular, auricles flared; leaflets 3–7, terminal: petiolule 0.5–4 mm, blade oval, suborbiculate, obovate, or deltate, 6–18 × 4–13 mm, abaxial surfaces usually pubescent or tomentulose, rarely glabrous. **Inflorescences** 1(–3)-flowered. **Pedicels** erect, stout, 2–16 mm, mostly setose, often pubescent, rarely glabrous; bracts 0 or 1 or 2, margins serrate to deeply incised. **Flowers** 2.5–5 cm; hypanthium thick-walled, setose or not; sepals ovate-lanceolate, 8–20 × (1–)2–5 mm, margins pinnatifid, abaxial surfaces glabrous or pubescent, sessile- or stipitate-glandular, sometimes eglandular; petals single, pink to rose pink, sometimes white; carpels 10–110, styles free, stylar orifice 2–5 mm diam., rims 0.5–1 mm wide, hypanthial disc absent. **Hips** dull red to dark reddish purple, subglobose or cupulate, 5–13(–18) × 5–16(–18) mm, hard, leathery, pubescent or glabrous, setose, stipitate-glandular or eglandular; sepals persistent, erect. **Achenes** basal.

Species 2 (2 in the flora): sw, sc United States, nw Mexico.

1. Leaflets 5–7, terminal blades 3–7 × 2–6 mm; bracts 1 or 2; flowers 2.5–3(–4) cm diam.; hypanthia subglobose to globose, densely pubescent and setose; distal branches pubescent or glabrous, without stellate hairs; sepals 8–12 × 2–4 mm, abaxial surfaces pubescent; hips eglandular, densely setose . . . . . 1. *Rosa minutifolia*
1. Leaflets 3–5, terminal blades 8–18 × 5–13 mm; bracts 0; flowers 4–5 cm diam.; hypanthia cupulate, puberulent or glabrous; distal branches densely pubescent, with or without stellate hairs; sepals 12–20 × 4–5 mm, abaxial surfaces glabrous, sometimes puberulent; hips stipitate-glandular, densely or sparsely setose . . . . . 2. *Rosa stellata*

1. Rosa minutifolia Engelmann, Bull. Torrey Bot. Club 9: 97. 1882 • Ensenada or small-leaved rose C F

Hesperhodos minutifolius (Engelmann) Hurst

Shrubs forming dense, low thickets. **Stems** usually erect, (3–)5–12(–15) dm; distal branches pubescent or glabrous, without stellate hairs; infrastipular prickles absent, internodal prickles sparse to common, erect, (2–)6–10(–12) × 1–3 mm, pubescent at least basally, mixed with dense aciculi, to 3 mm. **Leaves** 1.5–2.5 cm; stipules 3.5–4 × 1.5–2 mm, margins entire or dentate with glands, surfaces pubescent, eglandular, auricles acute, (0.5–)2–3 mm, surfaces pubescent, eglandular; petiole and rachis puberulent, stipitate glands and pricklets sparse; leaflets 5–7, terminal: petiolule 1–4 mm, blade oval, suborbiculate, or obovate, 3–7 × 2–6 mm, margins deeply 1- or multi-lobed, usually glandular, teeth <lobelike>, (3 or)4 or 5(or 6) per side on distal ½ of blades, some multi-serrate, apex acute to obtuse, abaxial surfaces pubescent (especially on veins), adaxial dull, sometimes pubescent. **Inflorescences** 1(–3)-flowered. **Pedicels** 2–9 mm, setae sparse, eglandular; bracts 1 or 2. **Flowers** 2.5–3 cm diam.; hypanthium subglobose to globose, 3–4 × 2.5–3.5 mm, densely pubescent and setose; sepals spreading, 8–12 × 2–4 mm, tip 4–6 × 1–1.5 mm, <lobes 3 or 4>, margins usually gland-tipped, abaxial surfaces pubescent, setae sparse, eglandular or sparsely glandular; petals usually pink to light rose pink, sometimes white, 10–15(–20) × 9–14 mm; stamens 45; carpels 10–25(–30), styles exsert 1–2 mm beyond stylar orifice (2–4 mm diam.), rims 0.5–0.8 mm wide. **Hips** dark reddish purple, subglobose, 5–7 × 5–7 mm, setae 1–4 mm, pubescent, eglandular. **Achenes** 6–16, dark, ± terete-elongate, 3.5–4 × 1.5–2 mm. $2n = 14$.

Flowering Mar–Jul. Dry washes, brush, grasslands, sagebrush, rocky hillsides; of conservation concern; 0–200 m; Calif.; Mexico (Baja California).

Rosa minutifolia was first collected in the United States in 1985 (San Diego County; G. A. Levin 1986). Before being extirpated by a development project, the single population was re-established in a nearby protected site, where the transplants are reportedly doing well (C. Burrascano, pers. comm.). Existence of *R. minutifolia* in the United States remains of conservation concern. Although common where found in coastal scrub of Baja California, Mexico, the habitat there is considered threatened by development.

2. Rosa stellata Wooton, Bull. Torrey Bot. Club 25: 152, plate 335. 1898 E F

Hesperhodos minutifolius (Engelmann) Hurst subsp. *stellatus* (Wooton) Hurst; *H. stellatus* (Wooton) Boulenger

Shrubs forming dense, low thickets. **Stems** erect, rarely arching, 2.5–8(–15) dm; distal branches densely pubescent, usually with, rarely without, stellate hairs, rarely glabrous, sometimes densely or sparsely sessile- or stipitate-glandular; infrastipular prickles paired or single, erect, rarely curved, 10–13 × 2.5–6 mm, sometimes with shorter internodal prickles and aciculi 0.5–5 mm. **Leaves** 2–3 cm; stipules 4–8 × 1–2 mm, margins ± entire, surfaces glabrous or tomentulose, sessile- or stipitate-glandular, sometimes eglandular; auricles obtuse, 2–3 × 1.5–2 mm, surfaces tomentulose, eglandular or glandular; petiole and rachis glabrous or pubescent, sparsely sessile- and stipitate-glandular, aciculi sparse, with or without glands; leaflets 3–5, terminal: petiolule 0.5–2 mm, blade obovate or deltate, 8–18 × 5–13 mm, margins broadly 1- or, rarely, multi-crenate, sometimes 1-serrate, glandular, teeth 3–8 per side on distal $\frac{1}{2}$–$\frac{2}{3}$ of blades, apex acute or rounded, abaxial surfaces pubescent or tomentulose, rarely glabrous, <eglandular>, adaxial dull to lustrous, glabrous or tomentulose. **Inflorescences** 1(or 2)-flowered. **Pedicels** 2–16 mm, setae absent or dense or few, <to 3 mm, pubescent, rarely glabrous>, eglandular or stipitate- or sessile-glandular; bracts 0. **Flowers** 4–5 cm diam.; hypanthium cupulate, 2–5 × 3.5–6(–8) mm, puberulent or glabrous, setae <subulate>, 1–6 mm, eglandular, rarely glandular; sepals erect, 12–20 × 4–5 mm, tip 3–5 × 1–2 mm, <lobes lanceolate>, margins glandular, abaxial surfaces glabrous, basally setose, glandular; petals pink or dark pink, <with strong almond fragrance>, 15–25 × 14–25 mm; stamens 175; carpels 60–110, styles exsert 1–2.5 mm beyond stylar orifice (4–5 mm diam.), rims 1 mm wide. **Hips** dull red (darkening with age), broadly subglobose or cupulate, 7–13(–18) × (7–)9–16(–18) mm, setae to 4 mm, puberulent or glabrous, stipitate-glandular. **Achenes** 10–15, dark, ± terete-elongate, 3.5–4.5 × 2 mm.

Subspecies 3 (3 in the flora): sw, sc United States.

1. Leaflets (3–)5; distal branches usually glabrous, rarely with stellate hairs (usually stipitate-glandular); petioles and rachises glabrous or puberulent; se New Mexico, w Texas .... 2c. *Rosa stellata* subsp. *mirifica*
1. Leaflets 3; distal branches tomentose-woolly, with stellate hairs; petioles and rachises pubescent or puberulent, sometimes glabrous; nw Arizona, sc New Mexico.
   2. Hypanthia: setae 1–3 mm; terminal leaflets abaxially tomentulose on midveins, otherwise glabrous; sc New Mexico.................. 2a. *Rosa stellata* subsp. *stellata*
   2. Hypanthia: setae 4–6 mm; terminal leaflets abaxially tomentulose or glabrous; nw Arizona .... 2b. *Rosa stellata* subsp. *abyssa*

2a. Rosa stellata Wooton subsp. **stellata** • Desert or gooseberry or star rose E F

Stems erect, 4–15 dm; distal branches tomentose-woolly, with stellate hairs, sessile- and stipitate-glandular; infrastipular prickles single or paired, <largest flattened>, 11–13 × 2.5–6 mm, internodal prickles and aciculi sparse, larger, eglandular. **Leaves:** stipules 4–8 × 1–2 mm, margins ± entire, eglandular or glands sparse, auricles foliaceous; petiole and rachis pubescent, sometimes glabrous; leaflets 3, terminal: petiolule 0.5–1.5 mm, blade obovate, margins broadly 1- or multi-crenate, teeth 3–6 per side, abaxial surfaces tomentulose on midveins, otherwise glabrous. **Flowers** 4.5 cm diam.; hypanthium densely setose, setae 1–3 mm; sepals 12–17 × 5 mm; petals pink. **Hips** subglobose or cupulate, 7–13 mm diam., sparsely or densely setose. $2n = 14$.

Flowering Apr–Sep. Streams, roadsides, openings in pine-juniper woods, dry rocky hillsides and slopes, overhangs in rocky canyons; 500–700 m; N.Mex.

Subspecies *stellata* is known from the Organ and San Andres mountains in Doña Ana County of south-central New Mexico (W. H. Lewis 1965).

2b. Rosa stellata Wooton subsp. **abyssa** A. M. Phillips, Madroño 39: 31, fig. 1. 1992 • Grand Canyon rose E

Rosa stellata var. *abyssa* (A. M. Phillips) N. H. Holmgren

Stems erect or arching, 2.5–10(–15) dm; distal branches tomentose-woolly, with stellate hairs, stipitate-glandular; infrastipular prickles paired, <stout>, 10–12 × 3.5–6 mm, internodal prickles and aciculi sparse, smaller, eglandular, rarely glandular. **Leaves:** stipules 4–8 × 1–2 mm, margins entire, often sessile- or stipitate-glandular, glands sparse to common, auricles foliaceous; petiole and rachis puberulent, sometimes glabrous; leaflets 3, terminal: petiolule 1 mm, blade obovate, margins broadly 1-crenate, teeth 4–8 per side, abaxial surfaces tomentulose or glabrous, <adaxial lustrous>. **Flowers** 5 cm diam.; hypanthium: densely setose, setae 4–6 mm; sepals 12–20 × 1–3(–4) mm; petals dark pink. **Hips** subglobose, 10–18 mm diam., densely setose.

Flowering May–Jun. Canyon rims and drainages, tops of cliffs, edges of mesas or plateaus, sandy to gravelly soil with limestone chips; 1400–2300 m; Ariz.

Subspecies *abyssa* is known only from Coconino and Mohave counties approximately 750 km from the nearest populations of subsp. *stellata* in New Mexico (A. M. Phillips 1992).

2c. Rosa stellata Wooton subsp. **mirifica** (Greene) W. H. Lewis, Ann. Missouri Bot. Gard. 52: 111. 1965 • Wonderful rose E

Rosa mirifica Greene, Leafl. Bot. Observ. Crit. 2: 62. 1910; *Hesperhodos mirificus* (Greene) Cockerell; *R. stellata* var. *mirifica* (Greene) Cockerell; *R. vernonii* Greene

Stems erect, 10–15 dm; distal branches usually glabrous, rarely with stellate hairs, usually stipitate-glandular; infrastipular prickles paired, 10–12 × 3–4 mm, internodal prickles and aciculi mixed with sessile and stipitate glands. **Leaves:** stipules 8 × 2 mm, margins entire, sessile-glandular, glands sparse, auricles not notably foliaceous; petiole and rachis glabrous or puberulent; leaflets (3–)5, terminal: petiolule 1.5–2 mm, blade deltate, margins 1-serrate, teeth 4–6 per side, abaxial surfaces glabrous or tomentulose on midveins and elsewhere, adaxial dull. **Flowers** 4 cm diam.; hypanthium setose, setae 1.5–3 mm; sepals 16–20 × 5 mm; petals pink. **Hips** broadly subglobose, 9–15 mm diam., densely setose.

Flowering May–Sep. Igneous or limestone cliffs, arroyos, edges of pinyon pine woods, along streams, roads, openings in pine-juniper woods, rocky north slopes; 1700–2300 m; N.Mex., Tex.

Subspecies *mirifica* occurs to the east of subsp. *stellata* in the Sacramento and White mountains of Otero County, New Mexico, to Culberson and Hudspeth counties in the Guadalupe and Eagle mountains of western Texas, where it is found on Permian limestone.

Within subsp. *mirifica* is var. *erlansoniae* W. H. Lewis (Erlanson rose) having floral branches flexuous at stem nodes, devoid of internodal prickles or nearly so, sepals entire rather than pinnatifid as in var. *mirifica*, leaflets and stems glabrous, and pedicels eglandular and glabrous; var. *erlansoniae* is known only from the Guadalupe Mountains of western Texas (W. H. Lewis 1965).

7b. ROSA Linnaeus subg. ROSA

Leaves (2–)3.5–15(–18) cm; leaflets 3–11(–13), margins usually serrate, rarely deeply incised or broadly crenate, terminal: petiolule (1–)5–20(–40) mm, blade elliptic, lanceolate, oblong, ovate, suborbiculate, obovate, and/or cordate, base not cuneate. **Inflorescences** corymbs or panicles, commonly solitary or few, 1–16(–50)-flowered. **Pedicels:** bracts present (absent in sect. *Pimpinellifoliae*). **Flowers:** hypanthium globose to depressed-globose, ovoid, oblong, or urceolate, rarely cupulate, obovoid, or hemispheric, glabrous, rarely setose, eglandular or stipitate-glandular; carpels 3–35(–68), styles free or connate, pubescent, sometimes glabrous, stylar orifice 0.5–3(–5) mm diam., rims 0.5 mm wide, hypanthial disc flat or conic. **Hips** fleshy, rarely leathery, ± setose or not.

Species ca. 135 (31 in the flora): North America, Mexico, Eurasia, n Africa; introduced nearly worldwide.

Subgenus *Rosa* includes seven sections, all represented in the flora area, five only by introduced species.

7b.1. ROSA Linnaeus (subg. ROSA) sect. SYSTYLAE de Candolle, Cat. Pl. Hort. Monsp., 137. 1813 (as Synstylae)

Shrubs, sometimes forming thickets; stoloniferous. **Stems** erect to procumbent, climbing, vinelike, (10–)15–30(–100) dm; distal branches glabrous, eglandular; prickles infrastipular and/or internodal, single or paired, usually curved or declined, sometimes erect, flattened, ± stout, sometimes mixed with aciculi, rarely absent. **Leaves** deciduous or semipersistent, 5–12 cm, membranous or leathery; stipules persistent, adnate to petiole, <auricles flared or erect>, margins lacinulose, sometimes entire, eglandular or stipitate-glandular; leaflets 3–9(–11), terminal: petiolule 5–16 mm, blade elliptic, ovate, elliptic-ovate, or obovate, 10–48(–70) × 8–27(–40) mm, abaxial surfaces glabrous or pubescent to tomentose, eglandular, sometimes glandular, adaxial dull or lustrous. **Inflorescences** panicles, 1–6(–30+)-flowered. **Pedicels** erect, slender, 5–25 mm, glabrous or pubescent, eglandular or stipitate-glandular; bracts persistent, 1–3, margins lacinulose, stipitate-glandular. **Flowers** 1.5–5 cm diam.; hypanthium ovoid, oblong, or urceolate, glabrous, eglandular or stipitate-glandular; sepals persistent, reflexed, ovate-acuminate or -lanceolate, 6–18 × 1–4 mm, margins entire or (outer) pinnatifid, abaxial surfaces glabrous or pubescent, eglandular or stipitate-glandular; petals single or double, pink, rose-purple, or white; carpels 6–25, styles connate <in columns, exsert 3–6 mm, sometimes ± free with age>, glabrous or pilose, stylar orifice 0.5–2 mm diam., hypanthial disc conic, 2–4 mm diam. **Hips** red or orange-red, usually globose, sometimes ovoid or subglobose, (4–)5–10 × 5–9 mm, glabrous, eglandular or sparsely stipitate-glandular; sepals late deciduous, reflexed. **Achenes** basiparietal.

Species ca. 35 (3 in the flora): North America, Europe, Asia (primarily China), n Africa; introduced in Mexico, West Indies, Central America, South America, Pacific Islands (New Zealand, Philippines).

Other species of sect. *Systylae*, hybrids, and cultivars that spread by stoloniferous shoots and are mostly sterile have escaped from cultivation and are occasionally found. One is *Rosa sempervirens* Linnaeus, evergreen rose, native to southern Europe, Asiatic Turkey, and northern Africa and found in California (B. Ertter and W. H. Lewis 2008); introduced in 1827 in France, it may also be the basis of a collection from Massachusetts (*B. N. Gates 15713*, 8 Nov. 1942, CONN), noted as doubtless an escape and spreading by rooting freely at tips of canes, and in Oklahoma, where a scrambling vine was collected (*P. Kirtley 68*, 6 Oct. 1935, MO). Such occasional findings of *R. sempervirens* can be expected in mostly warmer areas of the continent where it is readily recognized by its sprawling stems to 60 dm, and usually persistent, leathery, and lustrous leaflets with terminal leaflets distinctly longer than laterals. The species is introduced sporadically in the West Indies, Central America, and South America.

Rosa ×*moschata* Herrmann, musk rose, an escape from mostly southern latitude gardens, is found in Alabama, Illinois (not confirmed), and Louisiana. It is identified by long climbing or prostrate stoloniferous stems reaching 90 dm with erect or slightly curved prickles, leaflets ovate to lanceolate-elliptic, 5–50(–70) mm, mostly glabrous, finely serrate, and adaxially not lustrous, flowers to 5 cm diam. in clusters having a musky fragrance, petals white, or less commonly, pink, often double, sepals narrowly lanceolate and long-acuminate to 20 mm, strongly reflexed, and dark red or purple-brown hips.

1. Leaflets 3–5, terminal: petiolules 10–16 mm, blade teeth (18–)35(–42) per side; stipule margins entire, sometimes fimbriate; flowers 3–5 cm diam.; sepal margins entire . . . . 3. *Rosa setigera*
1. Leaflets (5–)7–9(–11), terminal: petiolules 5–13 mm, blade teeth 12–20 per side; stipule margins fimbriate; flowers 1.5–2.5 cm diam.; sepal margins pinnatifid.
   2. Pedicels 5–12 mm, stipitate-glandular or eglandular, usually pubescent; sepals 6–10 × 1.5–2 mm, abaxial surfaces usually stipitate-glandular; flowers 1.5–2.5 cm diam.; petals single or double; leaflet adaxial surfaces dull, terminal blade membranous . . . . . 4. *Rosa multiflora*
   2. Pedicels 18–25 mm, eglandular, glabrous; sepals 6–8 × 1–1.5 mm, abaxial surfaces eglandular; flowers 2–2.5 cm diam.; petals usually double; leaflet adaxial surfaces lustrous, terminal blade leathery . . . . 5. *Rosa lucieae*

3. Rosa setigera Michaux, Fl. Bor.-Amer. 1: 295. 1803 • Climbing prairie rose [E] [F]

Rosa rubifolia R. Brown; *R. setigera* var. *elatior* Persoon; *R. setigera* var. *glabra* Torrey & A. Gray; *R. setigera* var. *tomentosa* Torrey & A. Gray

Stems erect to procumbent and vinelike, 10–20(–60) dm; bark of canes green to light brown; prickles infrastipular and internodal, single or paired, declined, usually curved, sometimes erect, stout, 3–4 × 7–9 mm, <broad-based>, sometimes mixed with aciculi, rarely absent. **Leaves** deciduous, 8–12 cm; stipules narrowly lanceolate, 12–15 × 0.5–3 mm, auricles flared, 3–4 mm, margins entire, sometimes fimbriate, stipitate-glandular, surfaces glabrous, sparsely glandular; petiole and rachis with or without pricklets, usually pubescent, stipitate-glandular; leaflets 3 (mostly younger stems)–5 (older stems), terminal: petiolule 10–16 mm, blade ovate to elliptic-ovate, (30–)48(–70) × (20–)27(–40) mm, membranous or leathery, base rounded or obtuse, margins 1(or 2)-serrate, teeth (18–)35(–42) per side, coarse, gland-tipped, apex usually acuminate, abaxial surfaces <pale green>, glabrous or pubescent to tomentose, sometimes sessile- and/or stipitate-glandular, adaxial darker green, dull, glabrous. **Panicles** (1–)6(–15+)-flowered. **Pedicels** 15–25 mm, glabrous, stipitate-glandular; bracts 1 or 2, narrowly lanceolate, 10–30 × 1–2 mm, margins short stipitate-glandular, surfaces glabrous, eglandular. **Flowers** <functionally unisexual or monoecious, plants dioecious>, 3–5 cm diam.; hypanthium ovoid, 4–6 × 4–5 mm, stipitate-glandular; sepals narrowly to broadly ovate-lanceolate, 10–18 × 2–4 mm, margins entire, tip 3–4 × 0.5–1 mm, abaxial surfaces pubescent, stipitate-glandular; petals single, rose-purple to pink, fading to white, 18–25 × 16–25 mm; <stamens 212>; carpels 20–25, styles glabrous, exsert 5–6 mm beyond stylar orifice rims (0.5 mm diam.), hypanthial disc 2–3 mm diam. **Hips** bright red, subglobose to globose, 6–10 × 6–9 mm, <firm>, sparsely stipitate-glandular <often undeveloped because of dioecy, then early deciduous>. **Achenes** 17–22, fawn, 4–5 × 2.5–3 mm. $2n = 14$.

Flowering May–Aug. Prairies, savannas, woodland borders, clearings, open fields, abandoned pastures, waste areas, roadsides, fence rows; 100–500 m; Ont.; Ala., Ark., Conn., Del., D.C., Fla., Ga., Ill., Ind., Iowa, Kans., Ky., La., Maine, Md., Mass., Mich., Miss., Mo., Nebr., N.H., N.J., N.Y., N.C., Ohio, Okla., Pa., S.C., Tenn., Tex., Vt., Va., W.Va., Wis.; introduced in Europe (Channel Islands).

In the eastern United States *Rosa setigera* has been introduced from the Midwest or escaped from cultivation (W. H. Lewis 1959b). Based primarily on herbarium records, *R. setigera* is introduced in Connecticut, Delaware, District of Columbia, Florida, Louisiana, Maine, Maryland, Massachusetts, Mississippi, New Hampshire, New Jersey, North Carolina, South Carolina, Vermont, and Virginia; other states (Alabama, Georgia, New York, Pennsylvania, West Virginia) probably have both native and introduced populations.

Rosa setigera is the only native rose with procumbent or climbing stems to 60 dm with three leaflets on younger stems and five on older stems, and with unisexual flowers and caducous sepals.

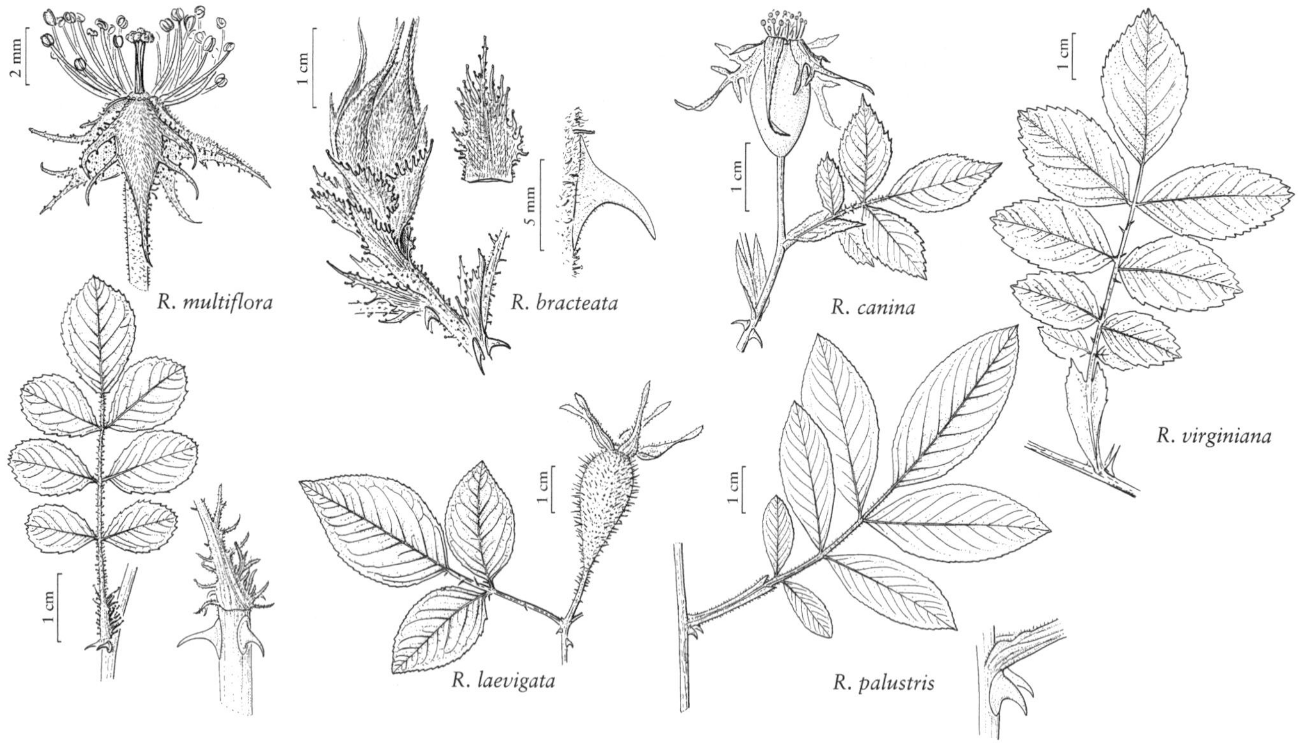

ROSA

Rosa setigera is the only species of sect. *Systylae* native to North America. The species is distinct from other members of the section in its flavonoid patterns, which show linkages to sect. *Cinnamomeae* [= sect. *Rosa*] (C. Grossi et al. 1998); it is also the only dioecious species of the genus. Microscopic floral characters are detailed elsewhere (W. H. Lewis 1959b; P. G. Kevan et al. 1990; J. R. Kemp et al. 1993, 1993b).

Since 1886, *Rosa setigera* has been used also as one parent in climbing hybrid cultivars produced in central Europe to increase hardiness and vigorous growth.

SELECTED REFERENCE Lewis, W. H. 1959b. A monograph of the genus *Rosa* in North America. III. *R. setigera*. SouthW. Naturalist 3: 154–174.

4. Rosa multiflora Thunberg in J. A. Murray, Syst. Veg. ed. 14, 474. 1784 • Japanese or multiflora rose, rosier multiflore [F] [I] [W]

Stems erect, arching, procumbent, or climbing, 15–30(–100) dm; bark of canes green to reddish brown; infrastipular prickles paired, curved, rarely erect, stout, 4–6 × 3 mm, internodal prickles fewer, similar, aciculi absent. **Leaves** deciduous, 5–12 cm; stipules narrowly lanceolate, 8–13 × 1.5–2 mm, auricles flared, 4–8 mm, margins fimbriate, <cleft to 3 mm>, usually stipitate-glandular or gland-tipped, surfaces glabrous, sometimes puberulent, eglandular or sessile- or stipitate-glandular; petiole and rachis with pricklets, puberulent, pubescent, or tomentose, usually stipitate-glandular, sometimes eglandular; leaflets (5–)7–9(–11), terminal: petiolule 7–13 mm, blade obovate to elliptic, 10–45 × 8–25 mm, membranous, base cuneate, margins 1(or 2)-serrate, teeth 12–20 per side, sharp, rarely gland-tipped, apex acute to long-acuminate, abaxial surfaces glabrous or pubescent (especially midveins), eglandular, adaxial green, dull, glabrous. **Panicles** 5–30+-flowered. **Pedicels** 5–12 mm, pubescent (especially proximally), sometimes glabrous, stipitate-glandular or eglandular; bracts 1 or 2, narrowly lanceolate, 5–10 × 1–2 mm, margins short stipitate-glandular, surfaces glabrous, eglandular. **Flowers** <sweet-scented>, 1.5–2.5 cm diam.; hypanthium oblong, 2 × 1–1.5 mm, eglandular or stipitate-glandular; sepals ovate-acuminate, 6–10 × 1.5–2 mm, <shorter than petals>, margins pinnatifid, tip 1–2 × 0.5 mm, abaxial surfaces glabrous, usually stipitate-glandular; petals single or double, white, rarely pale pink, 7–13 × 5–9 mm; carpels 6–11, styles glabrous, exsert 3–4 mm beyond stylar orifice rims (0.5–1 mm diam.), hypanthial disc 2–3 mm diam. **Hips** orange-red to red, ovoid to globose, (4–)5–7(–10) × 5–7 mm, eglandular or sparsely stipitate-glandular. **Achenes** 6–11, tan, 3.8–5 × 2–2.5 mm. **2*n*** = 14.

Flowering Apr–Jun. Disturbed areas, thickets, borders of woods, fences, streamsides, roadsides, old pastures, fields, acidic soil; 0–1700 m; introduced; B.C., N.B., Nfld. and Labr. (Nfld.), N.S., Ont., Que.; Ala., Ark., Calif., Colo., Conn., Del., D.C., Fla., Ga., Ill., Ind., Iowa, Kans., Ky., La., Maine, Md., Mass., Mich., Minn., Miss., Mo., Nebr., N.H., N.J., N.Mex., N.Y., N.C., Ohio, Okla., Oreg., Pa., R.I., S.C., Tenn., Tex., Utah, Vt., Va., Wash., W.Va., Wis.; Asia (China, Japan); introduced also in Mexico, Pacific Islands (Philippines).

Rosa multiflora was introduced in North America before 1811 as an ornamental. In the twentieth century, seedlings were widely provided by the U.S. Soil Conservation Service for growing hedgerows and similar plantings where it thrived in more acidic soil east of the Rocky Mountains. The species invaded pastures, degraded forage quality, reduced grazing area and agricultural productivity, and was widely reported as a noxious weed, causing, for instance, severe eye and skin irritations in cattle. Domestic sheep and goats feed on leaves, new buds, and shoots, and goats in pastures with severe *R. multiflora* infestations can virtually eliminate this rose within four seasons. The USDA Forest Service reported in 2006 that in 31 states the species had become a major conservation issue (S. E. Banasiak and S. J. Meiners 2009).

Vigorous climbers known as the "hard-luck" roses because of their great tolerance to frost, neglect, and varying habitats have been developed from *Rosa multiflora*, which has long served as a major root stock for budding roses in the rose industry worldwide.

Achenes of *Rosa multiflora* have long been used as a purgative in Japanese traditional medicine.

5. **Rosa lucieae** Franchet & Rochebrune ex Crépin, Bull. Soc. Roy. Bot. Belgique 10: 324. 1871 (as luciae) • Lucie or memorial rose I

Rosa wichuraiana Crépin

Stems procumbent or climbing, 10+ dm, <rooting at nodes>; bark of canes green or brown; prickles infrastipular and/or internodal, single or paired, curved or declined, ± stout, 4–5 × 2 mm, aciculi absent. **Leaves** deciduous (north) or semipersistent (south), 8–10 cm; stipules narrowly lanceolate, 10–12 × 2–3 mm, auricles erect, sometimes flared, 2–4 mm, margins fimbriate, stipitate-glandular, surfaces glabrous, eglandular; petiole and rachis with pricklets, glabrous, stipitate- or, sometimes, sessile-glandular; leaflets 7–9, terminal: petiolule 5–9 mm, blade broadly ovate to obovate, 15–30 × 12–20 mm, leathery, base cuneate, margins 1(or 2)-serrate, teeth 12–16 per side, acute, gland-tipped, apex acute to acuminate, abaxial surfaces glabrous, eglandular except for glands on midveins, adaxial green, lustrous, glabrous. **Panicles** (1–)5–20+-flowered. **Pedicels** 18–25 mm, glabrous, eglandular; bracts 1–3, lanceolate, 9–16 × 2.5–5 mm, margins sparsely stipitate-glandular, surfaces glabrous, eglandular. **Flowers** <scent of apple or clover>, 2–2.5 cm diam.; hypanthium urceolate, 4–6.5 × 2–3 mm, eglandular, <neck (0–)1 × 4–5 mm>; sepals ovate-acuminate, 6–8 × 1–1.5 mm, margins pinnatifid, tip 2 × 0.5 mm, abaxial surfaces glabrous, eglandular; petals double, sometimes single, white, pink to rose distally, 13–15 × 11–15 mm; carpels 12–21, styles pilose, exsert 3.5–5 mm beyond stylar orifice (1.5–2 mm diam.), hypanthial disc <appearing flattened with age>, 3–4 mm diam. **Hips** red, globose, 5–10 × 5–9 mm, eglandular. **Achenes** 1–11, dark tan, 4–4.5 × 2–2.5 mm. $2n$ = 14(28).

Flowering Jun–Aug. Disturbed areas, roadsides, flood plains, old homesteads, dry woods, highway verges; 0–500 m; introduced; Ala., Ark., Calif., Conn., Fla., Ga., Ill., Iowa, Ky., La., Md., Mass., Miss., Mo., N.C., Ohio, Okla., Pa., S.C., Tenn., Tex., Va., W.Va.; Asia (China, Japan, Korea); introduced also in Mexico, Pacific Islands (New Zealand, Philippines).

Introductions of *Rosa lucieae* in North America are largely from cultivars. The species is most common in southern and mid-latitudes of eastern and south-central United States. With spreading stems to 60 dm, *R. lucieae* is a typical stoloniferous plant forming carpets of colorful groundcover.

Rosa lucieae cultivars and hybrids have been widely used in breeding rambler and climbing roses. Its creeping stems extend to 60 dm and root readily, forming a white- or pink-petaled ground cover for gardens and cemeteries, and along highways where it also serves to stabilize verges. The most commonly found cultivar in North America is 'Dorothy Perkins'.

7b.2. Rosa Linnaeus (subg. Rosa) sect. Bracteatae Thory, Prodr. Monogr. Rosier, 125. 1820 [I]

Shrubs, forming dense clumps and thickets; stoloniferous. **Stems** ± climbing, spreading, or vinelike, sometimes decumbent, 30–50 dm; distal branches densely puberulent to tomentose, stipitate-glandular; infrastipular prickles usually paired, curved or hooked, flattened, stout, internodal prickles curved or hooked, sometimes erect and slender, usually mixed with aciculi and stipitate glands. **Leaves** persistent or semipersistent, 4.5–6.5 cm, leathery; stipules persistent, adnate to petiole at bases, margins pectinate <cleft into segments>, sparsely stipitate-glandular; leaflets (5–)7–9(–11), terminal: petiolule 3–5 mm, blade narrowly obovate, sometimes elliptic, 14–30 × 5–15 mm, abaxial surfaces glabrous, sometimes pubescent, stipitate-glandular, <adaxial lustrous>. **Inflorescences** panicles, 1 or 2(–8)-flowered. **Pedicels** erect, stout, 3–6 mm, densely white sericeous-tomentose, eglandular, <setae dense>; bracts caducous, 6–8, <closely subtending short pedicels and covering ½ or all of hypanthia>, margins deeply laciniate, eglandular. **Flowers** 5–7(–10) cm diam.; hypanthium urceolate, densely white sericeous-tomentose, eglandular; sepals persistent, spreading or reflexed, ovate, 12–17 × 3–6 mm, margins (outer) sometimes pinnatifid, abaxial surfaces densely white-sericeous, eglandular; petals single, white; carpels 60–68, styles free, pilose, stylar orifice 2.5–3 mm diam., hypanthial disc flat, 7–9 mm diam. **Hips** orange-red, subglobose or globose, rarely oblong, 7–30 × 8–30 mm, sparsely to densely white sericeous-tomentose, not setose, eglandular; sepals tardily deciduous, spreading or reflexed. **Achenes** basiparietal.

Species 2 (1 in the flora): introduced; Asia; introduced also in West Indies, Europe, Australia.

6. Rosa bracteata J. C. Wendland, Bot. Beob., 50. 1798 • Macartney rose [F] [I] [W]

Stems climbing or vinelike, sometimes decumbent, 30–50 dm; branches flexuous, brown; infrastipular prickles usually paired, broad-based, internodal prickles paired or single, reddish brown, 4.5–9 × 6–9 mm. **Leaves:** stipules 3–4 × 2.5–3.5 mm, auricles 3–6 × 1.5–2 mm, surfaces tomentose; petiole and rachis with pricklets, glabrous or pubescent, usually stipitate-glandular; leaflets: base cuneate, margins 1-serrulate-crenulate, teeth 12–16 per side, broad-based, apex obtuse to acute, often apiculate, gland-tipped, abaxial surfaces green with scattered pricklets and stipitate glands along prominent midveins, sometimes with hairs along midveins, eglandular, sometimes glabrous, adaxial glabrous, eglandular. **Panicles** lateral and terminal. **Pedicels:** bracts ovate-acuminate, (6–)8–10 × 5–7 mm, margins densely white sericeous-tomentose. **Flowers** 5–7(–10) cm diam.; hypanthium 5–7 × 8–9 mm; sepal margins sparsely glandular-ciliate, tip 2–5 × 1–2 mm; petals 20–35 × 20–30 mm; carpels apically pilose, exsert 2–2.5 mm beyond stylar orifice of hypanthial disc (1–2 mm diam.). **Hips** leathery. **Achenes** 30–40+, tan, 4–5 × 2–3 mm. **2*n*** = 14.

Flowering May–Oct. Disturbed areas, roadside hedges, open fields, pastures, bayou edges, ditches, along railways, woods, swampy thickets, pinelands; 0–500 m; introduced; Ala., Ark., Fla., Ga., La., Md., Miss., N.C., S.C., Tex., Va.; Asia (China, s Japan, Taiwan); introduced also in West Indies, Europe, Australia.

Rosa bracteata was introduced to Europe in 1795, and later to the United States, where it became widely naturalized and invasive in parts of the southeast, and in the West Indies. The species forms dense thickets of puberulent to tomentose stems; bracts, pedicels, and hips are densely sericeous-tomentose. The six to eight bracts closely subtend the short pedicels and tightly cover about half or all of the hypanthia.

7b.3. Rosa Linnaeus (subg. Rosa) sect. Laevigatae Thory, Prodr. Monogr. Rosier, 37. 1820 [I]

Rosa sect. *Sinicae* Crépin

Shrubs, <evergreen>, forming dense thickets; rhizomatous, <rhizomes relatively long>. **Stems** climbing, 20–75(–100) dm; distal branches glabrous, eglandular; infrastipular and internodal prickles paired or single, curved or hooked, ± flattened, stout. **Leaves** persistent, 4–8 cm, leathery; stipules caducous, ± free from petiole, margins fimbriate, some teeth gland-tipped; leaflets 3(–5), terminal: petiolule 9–13 mm, blade ovate-elliptic or -lanceolate, 40–65 × 30–40 mm, abaxial surfaces glabrous, eglandular, adaxial lustrous. **Inflorescences** 1(or 2)-flowered. **Pedicels** erect, stout, 12–30 mm, glabrous, eglandular, <setae apical or absent>; bracts caducous, 0 or 1(or 2), margins serrate, most teeth gland-tipped. **Flowers** 5.5–8(–10) cm diam.; hypanthium cupulate, densely setose, eglandular; sepals ± persistent, erect, lanceolate or ovate-lanceolate, 18–35 × 4 mm, margins entire, abaxial surfaces glabrous or sparsely pubescent, stipitate-glandular; petals single, usually white, rarely rose; carpels 45–50, styles free, glabrous or pilose, hypanthium orifice 1.5–2 mm diam., hypanthial disc conic, 6–8 mm diam. **Hips** dark red, ovoid to oblong, 20–35 × 9–20 mm, glabrous, densely setose (with age only setae remnants), eglandular; sepals tardily deciduous, spreading to reflexed. **Achenes** basiparietal.

Species 1: introduced; Asia; introduced also in West Indies, s Africa, Atlantic Islands, Pacific Islands.

7. Rosa laevigata Michaux, Fl. Bor.-Amer. 1: 295. 1803 • Cherokee rose [F] [I]

Rosa cherokeensis Donn ex Small

Stems: distal branches dark brown to purplish red; infrastipular prickles paired, internodal prickles paired or single, reddish brown, rarely deltate or erect, 6–9 × 6–11 mm. **Leaves:** stipules 4–6 × 1–2 mm, auricles 5–7(–10) mm, surfaces glabrous; petiole and rachis sometimes with pricklets, glabrous, eglandular; leaflets: terminal larger than laterals, base cuneate, margins 1-serrate, teeth 20–30 per side, tipped with dark glands, apex acute or acuminate, abaxial surfaces usually with pricklets on midveins. **Pedicels** densely setose; bracts attached near base, lanceolate, 5 × 2–3 mm, margins not ciliate, surfaces glabrous, eglandular. **Flowers** 1(or 2); hypanthium 5–9 × 4–6 mm, densely setose, setae 2–3 mm, some gland-tipped; sepal tip 5–10 × 2 mm; petals 25–35 × 25–30 mm; styles exsert 1–2 mm beyond hypanthium orifice. **Hips** leathery, base elongate, 7–12 mm, neck 1–2 × 6–9 mm. **Achenes** 20–30, light tan, 6–8 × 3–4 mm. $2n = 14$.

Flowering (Feb–)Apr–Jun. Swamp edges, pastures, fence and hedge rows, abandoned fields, roadsides, pine barrens, disturbed areas; 0–500 m; introduced; Ala., Fla., Ga., La., Miss., N.C., S.C., Tex.; Asia; introduced also in West Indies, s Africa, Atlantic Islands (Madeira), Pacific Islands (Hawaii).

By the nineteenth century, *Rosa laevigata* was widely found in the southern United States as a hedging shrub, being commonly distributed by stem cuttings and rhizomes, representing relatively few clones (C. A. Walker 1996; Walker and D. J. Werner 1997). The species was named and described from plants collected in Georgia in the 1790s by A. Michaux, who was unaware that his newly discovered species was native to Asia. This suggests a very old introduction into the United States, where it occurs only in the southeast to eastern Texas and the Carolinas. Shrubs form dense, rhizomatous thickets with stems capable of climbing to 100 dm. Leaves are persistent and leathery with mostly three leaflets and stipules free for most of their lengths, then caducous. Inflorescences consist of one (rarely two) white-petaled flowers and have densely setose hypanthia and hips.

In China, where the species is considered one of the most beautiful and cherished of roses, *Rosa laevigata* has a long history of use by herbal practitioners to treat bronchitis, dysentery, and urogenital disorders, and to restore hair color (J. A. Duke and E. S. Ayensu 1985, vol. 2).

7b.4. ROSA Linnaeus (subg. ROSA) sect. GALLICAE Crépin, J. Roy. Hort. Soc. 11: 221. 1889 [I]

Shrubs, single or forming patches; rhizomatous, <rhizomes relatively long>. **Stems** erect or arching, (5–)10–15(–20) dm; distal branches glabrous, sparsely setose- and stipitate-glandular; infrastipular prickles absent, rarely present, internodal prickles curved, sometimes erect, rarely hooked or declined, mixed with aciculi, stipitate glands, and smaller prickles <to 3 mm>. **Leaves** deciduous, rarely persistent, 10.5–12 cm, leathery; stipules caducous, adnate, <auricles flared or erect>, margins entire or serrate, gland-tipped; leaflets 3–5(–7), terminal: petiolule 6–8 mm, blade suborbiculate or ovate to narrowly elliptic, 20–26(–60) × 17–30 mm, abaxial surfaces pubescent, sometimes sessile-glandular, adaxial dull, <glabrous>. **Inflorescences** panicles, 1–3(–8)-flowered. **Pedicels** erect, stout, (10–)20–60 mm, pubescent, stipitate-glandular; bracts caducous, 1 or 2, margins entire, stipitate-glandular. **Flowers** 6–9 cm diam.; hypanthium cupulate, glabrous, stipitate-glandular; sepals ± caducous, reflexed, spreading, or erect, deltate, 16–23 × 2.5–3.5 mm, margins usually pinnatifid, abaxial surfaces glabrous or pubescent, glandular; petals usually double, deep pink to crimson; carpels 30–47, styles free, pilose, rarely glabrous, hypanthium orifice 2 mm diam., hypanthial disc conic, 4–5 mm diam. **Hips** bright to dark red, globose or subglobose, 8–9 × 10–12 mm, glabrous, setose, densely stipitate-glandular; sepals deciduous, erect to reflexed. **Achenes** basiparietal.

Species 1: introduced; c, s Europe, w Asia; introduced also in Mexico, Central America, South America, s Africa, Australia.

8. Rosa gallica Linnaeus, Sp. Pl. 1: 492. 1753 • French or Gallic rose, rose of Provence, rosier de France [I]

Stems: distal branches green to dull red; prickles internodal, curved, sometimes erect, rarely hooked, declined, 3–7 × 2–5 mm. **Leaves:** stipules subulate, 14–24 × 3–5 mm, auricles 4–10 mm, surfaces pubescent, eglandular; petiole and rachis with sparse pricklets, puberulent, densely stipitate-glandular; leaflets 5(–7), terminal blade slightly rugose, base obtuse to subcordate, margins shallowly 1(or 2)-dentate-crenate, teeth 14–23 per side, apex acute to acuminate, abaxial surfaces pale, gray-green, sessile-glandular particularly on midveins, adaxial bluish green or dark green. **Inflorescences** 1–3(–8) -flowered. **Pedicels:** bracts 1 or 2, caducous, lanceolate, 8–14 × 2–4 mm, margins ciliate, surfaces glabrous or pubescent, eglandular or stipitate-glandular. **Flowers:** hypanthium 5–7 × 3–5(–7) mm, neck (0–)1 × 3 mm; sepal tip erect or spreading, 7 × 2 mm; petals 27–35 × 20–30 mm [or larger]; styles exsert 2–4 mm beyond hypanthium orifice. **Hips** leathery. **Achenes** 3, 5 × 4–5 mm. $2n = 28$.

Flowering May–Jul. Waste areas, roadside thickets, railways, former house sites; 0–500 m; introduced; N.S., Ont., Que.; Conn., Ill., Ind., Ky., La., Maine, Mass., Mich., Miss., Mo., N.H., N.J., N.Y., N.C., Ohio, Pa., R.I., S.C., Tenn., Tex., Vt., Va., Wis.; c, s Europe; w Asia (Caucasus, Iraq); introduced also in Mexico (Oaxaca), Central America, South America, w Europe (Guernsey), s Africa, Australia.

Rosa ×*centifolia* Linnaeus (cabbage rose), long cultivated with about 500 ornamental cultivars known, is a probable hybrid of *R. gallica* × *R.* ×*moschata* Herrmann; it is not known to produce seed (P. V. Heath 1992). A. Bruneau et al. (2007) found that plastid DNA markers of *R. gallica* and *R.* ×*centifolia* are identical, indicating that *R. gallica* is the maternal parent. *Rosa damascena* Miller (Damask rose) also is close to *R.* ×*centifolia* and is thought to have the same parentage; if this proves true, then *R.* ×*centifolia* is the earlier name.

Rosa gallica is used as a tonic, mild astringent, and eye wash, and to treat bowel complaints and excessive mucous discharges. Petals of the closely related *R.* ×*centifolia* are collected for the distillation of rose water, a laxative used also to treat infantile diseases (J. Lindley 1838).

7b.5. Rosa Linnaeus (subg. Rosa) sect. Caninae de Candolle ex Seringe, Mus. Helv. Bot. 1: 3. 1818 [I]

Shrubs, <compact>, usually forming thickets; rhizomatous, <rhizomes absent or relatively short>. **Stems** erect, arching, or spreading, 7–30(–50) dm; distal branches glabrous, eglandular; infrastipular prickles paired or single, curved or appressed, rarely erect, flattened, stout, internodal prickles rare, smaller, or aciculi, sometimes absent. **Leaves** deciduous, rarely persistent, 4–11(–18) cm, membranous to leathery; stipules persistent, adnate to petiole, auricles flared, margins entire, rarely slightly crenate, <sometimes densely glandular-ciliate>, stipitate-glandular or eglandular; leaflets 5–7(–9), terminal: petiolule 5–17(–40) mm, blade ovate, obovate, orbiculate, or elliptic, sometimes oval or ovate-lanceolate, 10–45(–65) × 8–30(–50) mm, abaxial surfaces pubescent or tomentose (sometimes on midveins) or glabrous, eglandular or glandular, <with or without resinous glands>, adaxial <pale, light, or dark green, sometimes glaucous>, lustrous to dull, <tomentulose or pubescent, sometimes glabrous>. **Inflorescences** panicles, sometimes corymbs, 1–4(–7)-flowered. **Pedicels** erect or reflexed, stout, (5–)11–20(–35) mm, glabrous, eglandular or stipitate-glandular; bracts persistent, 1 or 2, margins glandular-serrate or stipitate- or ciliate-glandular. **Flowers** 2.5–5 cm diam.; hypanthium globose, ovoid, obovoid, oblong, or urceolate, glabrous, usually eglandular, <base sometimes stipitate- or setose-glandular>; sepals persistent or deciduous, appressed-reflexed, spreading, or erect, lanceolate or ovate-lanceolate, 10–25 × 2–5 mm, margins (outer) often deeply pinnatifid, abaxial surfaces glabrous, eglandular, densely glandular, or stipitate-glandular; petals single, pink or rose, rarely white; carpels 25–65, styles free, lanate or villous, rarely glabrous, stylar orifice 1–2.5(–3.5) mm diam., hypanthial disc flat or conic, 2–5 mm diam. **Hips** red to purplish or orange, globose, depressed-globose, obovoid, ovoid, oblong, urceolate, ellipsoid, or pyriform, 10–25 × 6–22 mm, glabrous, sometimes setose, eglandular or stipitate-glandular; sepals deciduous or persistent, spreading, reflexed, or erect. **Achenes** basiparietal.

Species ca. 20 (5 in the flora): introduced; Europe, Asia, n Africa; introduced also in Mexico, Central America, South America, Pacific Islands (New Zealand), Australia.

All species of sect. *Caninae* are polyploid, mostly pentaploid ($2n = 35$), but also triploid ($2n = 21$) and tetraploid ($2n = 28$) or all three polyploid levels for a single species, with a hemisexual, matroclinal (maternal) reproduction unlike that known for any other flowering plant (H. Nybom et al. 2000). In addition, unknown elsewhere in *Rosa*, apomixis appears to occur to a limited extent among species of *Caninae*, which complicates further their reproductive anomaly (G. Werlemark 2000). Even though over 350 taxa have been described in the section, the relatively few species now recognized are highly complex groups of microspecies, and the authors agree with the editors of W. J. Bean (1970–1988, vol. 4) that no attempt should be made in modern floristic works to take account of every character-state combination. Further, those editors believed that the species of *Caninae* are cryptohybrids, that is, they are all products of ancient hybridizations between species that no longer exist, at least not in their primitive states.

In addition to the species treated here, *Rosa obtusifolia* Desvaux is marginally naturalized in California, New Mexico, and Virginia; it differs from *R. canina* in having hairy leaflets that are usually glandular abaxially.

1. Sepals abaxially eglandular; prickle lengths ± uniform; leaflets abaxially glabrous, rarely pubescent or tomentose on midveins, eglandular; bracts 6–18 × 4–5 mm, surfaces glabrous . . . . 9. *Rosa canina*
1. Sepals abaxially densely glandular or stipitate-glandular; prickle lengths varying or ± uniform; leaflets abaxially usually tomentose, rarely glabrous or pubescent, usually glandular; bracts 13–21 × 4–7 mm, surfaces usually tomentose, rarely glabrous or pubescent.

[2. Shifted to left margin.—Ed.]

2. Leaflets abaxially usually densely viscid-glandular, glabrous or pubescent, glands with ripe apple scent; prickles curved or falcate . . . . 10. *Rosa rubiginosa*
2. Leaflets abaxially usually resinous-glandular, tomentose, glands resin-scented; prickles all erect or some curved.
   3. Hips: sepals spreading or reflexed, deciduous after anthesis; stylar orifices 1 mm diam. . . . . 11. *Rosa tomentosa*
   3. Hips: sepals usually erect, persistent after anthesis; stylar orifices 2.5–3.5 mm diam.
      4. Stipules 10–14 × 3–5 mm, auricles 3–4 mm; bracts 16–21 mm; sepals lanceolate, 15–20 × 3–3.5 mm; stylar orifices 2.5 mm diam. . . . . 12. *Rosa sherardii*
      4. Stipules 15–20 × 5–10 mm, auricles 5–8 mm; bracts 10–12 mm; sepals ovate-lanceolate, 20–25 × (4–)5 mm; stylar orifices 3.5 mm diam. . . . . 13. *Rosa mollis*

9. Rosa canina Linnaeus, Sp. Pl. 1: 491. 1753
• Dog rose, rosier des chiens F I W

Rosa corymbifera Borkhausen; *R. dumetorum* Thuillier; *R. montezumae* Humboldt & Bonpland ex Thory

Shrubs, arching, not rhizomatous. **Stems** usually erect to sprawling, 10–25(–50) dm; distal branches arching, bark green; infrastipular prickles paired, curved or appressed, 6–7 × 4–9 mm, lengths ± uniform, internodal prickles rare, single, rarely absent. **Leaves** deciduous, 6–11 cm; stipules 10–22 × 3–5 mm, auricles 3–5 mm, margins stipitate-glandular or eglandular, surfaces glabrous, eglandular or sparsely stipitate-glandular; petiole and rachis sometimes with pricklets, glabrous, eglandular; leaflets 5–7, terminal: petiolule 5–11 mm, blade ovate, obovate, or elliptic, 15–40 × 12–20 mm, base obtuse to slightly cuneate, margins 1- or multi-serrate, teeth 20–30 per side, apex acute, sometimes acuminate, abaxial surfaces glabrous, rarely pubescent or tomentose on midveins, eglandular, adaxial dark green to green, lustrous to dull, glabrous, rarely tomentose. **Inflorescences** panicles, sometimes corymbs, solitary, sometimes 2 or 3(–7)-flowered. **Pedicels** erect to reflexed as hips mature, 8–20 mm, eglandular or stipitate-glandular; bracts 2, ovate-lanceolate, 6–18 × 4–5 mm, margins glandular-serrate, abaxial surfaces puberulent, adaxial surfaces glabrous, eglandular. **Flowers** 3.5–5 cm diam.; hypanthium narrowly urceolate, 7–9 × 3–6 mm, eglandular, neck 2–3 × 1–2 mm; sepals appressed-reflexed, spreading, or erect, ovate-lanceolate, 10–17 × 3–5 mm, margins deeply pinnatifid, tip 4–6 × 0.5 mm, abaxially eglandular; petals rose, pink, or white, <sweet-scented>, 18–25 × 15–18 mm; carpels 26–36, styles villous, exsert 1–2 mm beyond stylar orifice (0.7–1.5 mm diam.) of hypanthial disc (4–5 mm diam.). **Hips** red, globose, ovoid, urceolate, or ellipsoid, 10–16(–24) × 6–16 mm, glabrous, eglandular; sepals deciduous as hips mature, reflexed. **Achenes** 14–23, tan, 5–6 × 3–3.5 mm. $2n = 35$.

Flowering May–Jul. Foothills, woodlands, along roads and railways, abandoned homesteads, riparian habitats; 0–700 m; introduced; B.C., N.B., N.S., Ont., Que.; Ala., Ariz., Ark., Calif., Conn., Idaho, Ind., Kans., Ky., Maine, Md., Mass., Mich., Mo., Nebr., N.Y., N.C., Ohio, Oreg., Pa., R.I., Tenn., Utah, Vt., Va., Wash., W.Va., Wis.; Europe; wc Asia (Turkmenistan); n Africa; introduced also in Mexico, Central America, South America, Pacific Islands (New Zealand), Australia.

Rosa canina occurs sporadically throughout Canada and the United States in disturbed areas. Shrubs lack rhizomes and have arching stems with paired or single curved infrastipular prickles all more or less uniform in length. Sepals (abaxially), stipules, petioles, rachises, pedicels, blades (abaxially, except some veins), hypanthia, and hips are all glabrous and most are also eglandular.

W. J. Bean (1970–1988) cited *Rosa canina* var. *corymbifera* (Borkhousen) Rouy, *R. canina* var. *dumetorum* (Thuillier) Baker, and *R. dumetorum* as synonyms of *R. corymbifera*, and that treatment is followed here.

Moderate evidence exists for the efficacy of powdered achenes and hip receptacles of *Rosa canina* to treat patients suffering from osteoarthritis, specifically of the knee and hip (K. Winther et al. 2005; C. Chrubasik et al. 2006).

10. Rosa rubiginosa Linnaeus, Mant. Pl. 2: 564. 1771
• Eglantine or sweet briar rose, rosier églantier I W

Shrubs, erect; not rhizomatous. **Stems** 10–30 dm; distal branches arching, bark dark brownish red; infrastipular prickles single or paired, curved, <falcate>, 6–12 × 3–7 mm, lengths varying or ± uniform, internodal prickles sometimes mixed with aciculi and glandular setae. **Leaves** persistent, 4–6.5 cm; stipules 6–10 × 2–4 mm, auricles 3–5 mm, margins stipitate-glandular, surfaces glabrous, eglandular; petiole and rachis with pricklets, puberulent, stipitate-glandular; leaflets 5–7(–9), <viscid glands with ripe apple scent>, terminal: petiolule 5–10 mm, blade mostly suborbiculate or broadly oval, 10–25 × 8–15 mm, base obtuse, margins 2- or multi-serrate, teeth 10–18 per side, apex acute to obtuse, abaxial surfaces glabrous or pubescent, usually densely viscid-glandular, adaxial green, lustrous to dull, puberulent or glabrous. **Inflorescences** panicles, 1–3(–7)-flowered. **Pedicels** erect, 6–9 mm, densely stipitate-glandular, sometimes mixed with aciculi [and setae]; bracts 2, ovate-lanceolate to lanceolate, 15 × 5 mm, margins stipitate-glandular, surfaces glabrous, eglandular. **Flowers** 2–4 cm diam.; hypanthium obovoid or broadly oblong, 5–6 × 3–4 mm, eglandular, neck (0–)1–1.5 × 3–4 mm; sepals erect or spreading, rarely reflexed, lanceolate or ovate-lanceolate, 14–18 × 2 mm, margins mostly pinnatifid, tip 3–5 × 0.5–1 mm, abaxially densely stipitate-glandular; petals bright or deep pink, 11–20 × 11–18 mm; carpels 25–45, styles villous or glabrous, exsert 1–2 mm beyond stylar orifice (1.2–2 mm diam.) of hypanthial disc (2.5–4 mm diam.). **Hips** dark red, subglobose to broadly ovoid, ellipsoid, or pyriform, 10–25 × 7–22 mm, glabrous, sometimes setose, eglandular; sepals tardily deciduous, mostly erect. **Achenes** 15–25, tan, 3.5–4(–5) × 2–2.5(–3) mm. **2*n*** = 35, 42.

Varieties 2 (2 in the flora): introduced; Europe, w Asia, n Africa; introduced widely worldwide.

Rosa rubiginosa has been introduced throughout Canada and the United States except the desert southwest. Plants are compact, upright shrubs without rhizomes. Stems have stout, falcate infrastipular prickles mixed with internodal prickles, aciculi, and glandular setae. Leaflet blades are densely viscid-glandular with ripe apple scent and margins two- or multi-serrate with stipitate glands.

Rosa eglanteria Linnaeus is a formally rejected name that pertains here.

1. Distal branches: prickle lengths varying, aciculi and setae sometimes present; hips 10–25 × 10–22 mm; flowers 2.5–4 cm diam., sepals deciduous as or after hips mature, styles usually villous, stylar orifices 1/3 diam. of rims 4 mm diam. . . . . 10a. *Rosa rubiginosa* var. *rubiginosa*
1. Distal branches: prickle lengths ± uniform, aciculi and setae absent; hips 10–12 × 7–9 mm; flowers 2–3.5 cm diam., sepals deciduous before or as hips mature, styles usually glabrous, stylar orifices 1/6–1/5 diam. of rims 2.5–4 mm diam. . . . . 10b. *Rosa rubiginosa* var. *nemoralis*

10a. Rosa rubiginosa Linnaeus var. **rubiginosa** [I]

Rosa suaveolens Pursh

Prickles infrastipular and internodal, lengths varying, aciculi and setae sometimes present on distal or flowering branches. **Leaves:** terminal blades mostly suborbiculate or broadly oval, 10–25 × 8–15 mm, apex usually acute or short-acuminate, abaxial surfaces densely glandular. **Flowers** 2.5–4 cm diam.; sepals deciduous as or after hips mature, styles usually villous, stylar orifice 1/3 diam. of rims 4 mm diam. **Hips** 10–25 × 10–22 mm. **2*n*** = 35.

Flowering Apr–Aug. Thickets, clearings, roadsides; 0–1100 m; introduced; B.C., N.B., Nfld. and Labr. (Nfld.), N.S., Ont., P.E.I., Que.; Ala., Ark., Calif., Colo., Conn., Del., D.C., Ga., Idaho, Ill., Ind., Iowa, Kans., Ky., La., Maine, Mass., Mich., Minn., Miss., Mo., Mont., Nebr., N.H., N.J., N.Y., N.C., Ohio, Okla., Oreg., Pa., R.I., S.C., Tenn., Tex., Utah, Vt., Va., Wash., W.Va., Wis., Wyo.; Europe; w Asia (Iran); n Africa; introduced also in Central America, South America, elsewhere in Africa, Pacific Islands (New Zealand), Australia.

10b. Rosa rubiginosa Linnaeus var. **nemoralis** (Léman) Thory in P. J. Redouté, Roses 2: 23. 1821 [I]

Rosa nemoralis Léman, Bull. Sci. Soc. Philom. Paris 1818: 94. 1818; *R. micrantha* Borrer

Prickles infrastipular and internodal, lengths ± uniform, aciculi and setae absent on flowering branches. **Leaves:** terminal blades mostly oval, 12–20 × 8–14 mm, apex usually obtuse, abaxial surfaces seldom densely glandular. **Flowers** 2–3.5 cm diam.; sepals deciduous before or as hips mature, styles usually glabrous, stylar orifice 1/6–1/5 diam. of rims 2.5–4 mm diam. **Hips** 10–12 × 7–9 mm. **2*n*** = 35, 42.

Flowering May–Aug. Thickets, clearings, roadsides; 0–700 m; introduced; N.S., Ont., P.E.I., Que.; Conn., D.C., Ill., Ind., Ky., Maine, Md., Mass., Mo., N.J., N.Y., N.C., Ohio, Oreg., Pa., R.I., S.C., Tenn., Tex., Va., Wash., W.Va., Wis.; Europe; introduced also in Pacific Islands (New Zealand), Australia.

Variety *nemoralis* is often found near var. *rubiginosa* with apparently little or no ecological distinctions between them. They differ in morphological characters by degrees of expression rather than complete distinctions. The rank of variety follows, for example, the recommendation of H. J. Scoggan (1978–1979).

11. Rosa tomentosa Smith, Fl. Brit. 2: 539. 1800 • Harsh downy-rose [I]

Shrubs, climbing; rhizomatous or not. **Stems** arching, 7–20(–30) dm, <internodes long>; distal branches spreading, bark pale green; infrastipular prickles single or paired, curved or ± erect, 3–6 × 2.5–3.5 mm, lengths ± uniform, internodal prickles sometimes mostly smaller, with aciculi, rarely absent. **Leaves** deciduous, 5–7.5(–11) cm; stipules 10–16(–20) × 3–4 mm, auricles 2–5 mm, margins densely glandular-ciliate, surfaces tomentose, often glandular abaxially; petiole and rachis with sparse pricklets, tomentulose, glandular; leaflets 5–7, terminal: petiolule 6–8 mm, blade orbiculate to ovate-lanceolate or elliptic, 15–30(–60) × 8–22(–50) mm, base obtuse, margins 1- or multi-serrate, teeth 13–20 per side, apex acute, abaxial surfaces tomentose, usually resinous-glandular, <glands resin-scented>, adaxial pale or grayish green, dull, glabrous or pubescent. **Inflorescences** panicles, 1–6-flowered. **Pedicels** erect, (11–)20–35 mm, densely stipitate-glandular; bracts 2, ovate-lanceolate, 13–14 × 5–7 mm, margins ciliate-glandular, surfaces tomentose, glandular or eglandular. **Flowers** 3.5–5 cm diam.; hypanthium ovoid, 6–7 × 4 mm, stipitate-glandular, setose, rarely eglandular, neck 1.5 × 2 mm; sepals spreading or erect, ovate-lanceolate, 15–25 × 2–3 mm, margins pinnatifid, tip 4–5 × 1–1.5 mm, abaxially densely stipitate-glandular; petals pink, rarely white, 8–15 × 10–15 mm; carpels 26–34, styles glabrous or villous, exsert 2–2.5 mm beyond stylar orifice (1 mm diam.) of hypanthial disc (2–4 mm diam.). **Hips** dark red, oblong, ovoid, or globose, 10–15 × 10–12 mm, glabrous, stipitate-glandular at least proximally, <neck 1 × 5–6 mm>; sepals deciduous after anthesis, spreading or reflexed. **Achenes** ca. 23, tan, 4–4.5 × 2 mm. **2*n*** = 35.

Flowering Jun–Jul. Roadside thickets, hedgerows, woodland edges; 0–300 m; introduced; P.E.I.; N.Y., N.C., Tenn., Tex., Vt.; Europe.

A rare introduction from Europe, *Rosa tomentosa* was reported by P. A. Rydberg (1918) from North Carolina and Texas and collected by Fernald (GH) in Queens County, Prince Edward Island. The species is also reported to occur in disturbed areas in New York, Tennessee, and Vermont (USDA PLANTS database). Whether the species has persisted at any of these occurrences is unknown; it is not currently included in the flora of Canada (H. G. Scoggan 1978–1979; VASCAN).

Rosa tomentosa is best recognized by its climbing habit, leaflets that are abaxially, and sometimes adaxially, tomentose or pubescent, and relatively long petioles. Sepals are spreading or reflexed after anthesis and deciduous when hips mature. Prickles are usually erect, and the stylar orifices are small (1 mm diam.).

12. Rosa sherardii Davies, Welsh Botanol., 49. 1813 (as sherardi) • Sherard's downy-rose [I]

Rosa omissa Déséglise; *R. subglobosa* Smith

Shrubs, erect, <forming small thickets>; rhizomatous or not. **Stems** erect, <glaucous, stout>, 10–20 dm; distal branches often flexuous, bark reddish, <± pruinose or glaucous>; infrastipular prickles single or paired, curved or erect, 6–9 × 3–4 mm, lengths ± uniform, internodal prickles sometimes present, aciculi absent. **Leaves** deciduous, 6–10(–18) cm; stipules 10–14 × 3–5 mm, auricles 3–4 mm, margins short stipitate-glandular, surfaces usually glabrous, sometimes pubescent, glandular; petiole and rachis with pricklets, pubescent, glandular; leaflets 5–7, terminal: petiolule 6–12(–40) mm, blade ovate to narrowly or broadly elliptic, 30–45(–65) × 15–30(–45) mm, base rounded, sometimes slightly cuneate, margins 2- or multi-serrate, teeth 17–26 per side, apex acute or rounded, abaxial surfaces <light green, ± rugose>, tomentose, densely resinous-glandular, <glands reddish brown, resin-scented>, adaxial green, dull, tomentulose or glabrous, eglandular. **Inflorescences** panicles, 1–3(or 4)-flowered. **Pedicels** erect to reflexed as hips mature, 12–17 mm, sparsely stipitate-glandular; bracts 2, ovate-lanceolate, 16–21 × 4–6 mm, margins glandular, surfaces tomentose to puberulent or glabrous, glandular or eglandular. **Flowers** 2.5–3.5 cm diam.; hypanthium globose, 5–6 × 5 mm, eglandular or sparsely stipitate-glandular, neck absent; sepals spreading to erect, lanceolate, 15–20 × 3–3.5 mm, margins sometimes pinnatifid, tip 4–8 × 1.5–2 mm, abaxially densely stipitate-glandular; petals deep rose pink, rarely white, 15–19 × 13–15 mm; carpels 52, styles lanate or villous, exsert 1–2 mm beyond stylar orifice (2.5 mm diam.) of hypanthial disc (4.3 mm diam.). **Hips** orange to bright red, globose or obovoid to ellipsoid, 13–18 × 11–18 mm, glabrous, sparsely stipitate-glandular; sepals persistent, usually spreading-erect. **Achenes** 40, tan, 5–5.5 × 2.5–3 mm. **2*n*** = 28, 35, 42.

Flowering Jun–Jul. Thickets of woodland margins, overgrown pastures, open scrub, wastelands, roadsides; 300–400 m; introduced; Vt.; Europe.

Known from northeastern Vermont, *Rosa sherardii*, a Scottish introduction in the nineteenth century, is an erect, compact shrub often forming rhizomatous thickets. Sepals are spreading to erect and persistent after anthesis, auricles are 3–4 mm, bracts 16–21 mm, and stylar orifices 2.5 mm diam.

Rosa sherardii occurs frequently in northern England and Scotland, as does the closely allied *R. mollis*. *Rosa sherardii* and, undoubtedly, *R. mollis*, both recently identified in the flora area by A. V. Gilman, were brought to Vermont by Scottish immigrants who settled and farmed there (Gilman 2012).

SELECTED REFERENCE Gilman, A. V. 2012. *Rosa sherardii* Davies, an overlooked species in Vermont, and new to North America. Phytoneuron 2012-31: 1–6.

13. Rosa mollis Smith in J. E. Smith et al., Engl. Bot. 35: plate 2459. 1812 • Soft downy-rose [I]

Shrubs, <forming large thickets>; rhizomatous. **Stems** usually spreading and arching, erect, 8–15(–20) dm; distal branches erect or spreading, bark reddish when exposed, <pruinose when young or shaded>; infrastipular prickles paired, erect, 5–6 × 3–4 mm, lengths ± uniform, internodal prickles similar or smaller, mixed with aciculi. **Leaves** deciduous, 8–11(–14) cm; stipules 15–20 × 5–10 mm, auricles 5–8 mm, margins <glabrous or ciliate>, surfaces usually puberulent to tomentulose, sometimes glabrous, eglandular; petiole and rachis without pricklets, pubescent, eglandular; leaflets (5–)7, <glands resin-scented when crushed>, terminal: petiolule 10–17 mm, blade elliptic to ovate, 12–35 × 10–18 mm, base cuneate, sometimes rounded, margins 1- or 2-serrate, teeth 14–18 per side, apex acute, sometimes rounded, abaxial surfaces tomentose, resinous-glandular, sometimes eglandular, adaxial light green, dull, glabrous or mostly tomentulose. **Inflorescences** panicles, 1 or 2(–5)-flowered. **Pedicels** erect or slightly reflexed, (5–)15–35 mm, sparingly stipitate-glandular or eglandular; bracts (1 or)2, ovate-lanceolate, 10–12 × 4–5 mm, margins irregularly glandular-serrate, surfaces glabrous or pubescent, eglandular. **Flowers** 3–4.5 cm diam.; hypanthium globose, 5–7 × 5–7 mm, stipitate-glandular or setose, neck absent; sepals spreading, ovate-lanceolate, 20–25 × (4–)5 mm, margins entire, tip 8–10 × 0.5–1 mm, abaxially densely glandular or stipitate-glandular; petals deep pink, rarely white, 9–16 × 10–16 mm; carpels 50–65, styles lanate, exsert 1.5 mm beyond stylar orifice (3.5 mm diam.) of hypanthial disc (5 mm diam.). **Hips** red to purplish, globose to depressed-globose, 11–16 × 12–15(–20) mm, glabrous, sometimes setose, stipitate-glandular; sepals persistent, erect. **Achenes** 60, dark tan to black, 4–5 × 1.5–2 mm. $2n$ = 28, 35, 42.

Flowering Jun–Jul. Thickets, stream banks, roadsides, overgrown pastures; 300–400 m; introduced, Vt.; n, c Europe.

Rosa mollis is introduced from Scotland and is closely allied to *R. villosa* Linnaeus; the two were treated as distinct by W. J. Bean (1970–1988) and A. V. Gilman (2012).

7b.6. Rosa Linnaeus (subg. Rosa) sect. Pimpinellifoliae de Candolle ex Seringe, Mus. Helv. Bot. 1: 3. 1818 [I]

Rosa sect. *Luteae* Crépin; *Rosa* sect. *Sericeae* Crépin

Subshrubs, forming dense patches; rhizomatous, <rhizomes long>. **Stems** erect, 2–10(–15) dm; distal branches glabrous, eglandular; infrastipular prickles absent, internodal prickles erect or ± curved, usually intermixed with aciculi. **Leaves** persistent, 2.5–8 cm, membranous; stipules persistent, adnate to petiole, <auricles flared>, margins entire or serrate, eglandular or glandular; leaflets (5–)9–11(–13), terminal: petiolule 2–4 mm, blade oblong-ovate, suborbiculate, or broadly elliptic, 5–22 × 5–12 mm, abaxial surfaces glabrous, eglandular, adaxial <green>, dull, <glabrous>. **Inflorescences:** flowers solitary. **Pedicels** erect, slender, 15–35 mm, glabrescent, eglandular, sometimes with aciculi; bracts absent. **Flowers** 2–4 cm diam.; hypanthium subglobose, glabrous, eglandular; sepals persistent, erect, ovate-lanceolate, 10–12 × 2.5–3 mm, margins entire, abaxial surfaces glabrous, eglandular; petals single, white [rarely pink]; carpels 18–23, styles free, lanate, stylar orifice 2.5–3 mm diam., hypanthial disc flat, 4–4.5 mm diam. **Hips** blackish purple, globose or subglobose, 8–13 × 8–16 mm, <lustrous>, glabrous, eglandular; sepals persistent, erect. **Achenes** basiparietal.

Species ca. 25 (1 in the flora): introduced; Eurasia; introduced also in Atlantic Islands (Iceland), Pacific Islands (New Zealand).

In addition to the species treated here, yellow-flowered *Rosa hugonis* Hemsley has been reported from Ontario.

14. Rosa spinosissima Linnaeus, Sp. Pl. 1: 491. 1753 • Burnet or Scots rose, rosier pimprenelle [I]

Rosa illinoiensis Baker; *R. lutescens* Pursh; *R. pimpinellifolia* Linnaeus

Prickles internodal, dense on main stems, sparser on some flowering branches, paired or single, erect or ± curved, terete, 5–6 × 1–3 mm, mixed with shorter prickles and usually aciculi. **Leaves:** stipules 9–14 × 2–4 mm, auricles 2.5–5 mm, margins entire or serrate-glandular, surfaces glabrous; petiole and rachis with pricklets, pubescent, glandular; leaflets: blade oblong-ovate or suborbiculate, (5–)7–9 (–11) × 5–8 mm, or broadly elliptic, 5–22 × 5–12 mm, base obtuse or rounded to broadly cuneate, margins 1-serrate, sometimes multi-serrate, teeth 8–12 per side, gland-tipped, apex obtuse to acute. **Pedicels:** bracts absent. **Flowers:** hypanthium 3–5 × 3–4 mm; sepal tip 2–3 mm, apex acuminate or caudate-acuminate; petals white, 15–25 × 14–23 mm; styles exsert 2 mm beyond stylar orifice of hypanthial disc. **Achenes** 8–12, dark tan, 4 × 2.5 mm. $2n = 28$.

Flowering May–Jun. Disturbed areas; 0–500 m; introduced; N.B., Ont., Que.; Conn., Ill., Ind., Kans., Maine, Mass., Mo., N.H., N.J., N.Y., Ohio, Tenn., Vt., Va., Wis.; Europe; Asia (China, Japan); introduced also in Atlantic Islands (Iceland), Pacific Islands (New Zealand).

Rosa spinosissima hybridizes with other species of the genus. Some garden cultivars are more robust and glandular than wild types.

Rosa spinosissima is an erect, relatively low subshrub characterized by erect, relatively short stems with relatively long rhizomes forming dense patches. Stems extend to 10 dm with dense, intermixed internodal prickles and aciculi and lacking infrastipular prickles. Leaflets are relatively small and mostly nine to eleven; flowers are solitary and petals are white; inflorescences lack bracts; and hips are lustrous, blackish purple with erect, persistent sepals.

Using *Rosa spinosissima* as one parent, George Harison crossed *R.* ×*foetida* (Austrian brier rose) in his New York City garden in about 1830 to produce Harison's yellow rose, *R.* ×*harisonii* Rivers. This unique yellow-flowering rose was planted widely as pioneers moved west across the plains where even today it has been found in Arizona, Colorado, Idaho, Iowa, Nebraska, Nevada, Texas, Utah, and other states.

Of a group of medicinal plants tested for antioxidant properties, *Rosa spinosissima*, with the largest amounts of phenolic compounds, proved to have the highest radical scavenging activity and provided the highest peroxidation inhibition (A. Mavi et al. 2004).

7b.7. Rosa Linnaeus (subg. Rosa) sect. Rosa

Rosa sect. *Cinnamomeae* de Candolle ex Seringe

Shrubs or subshrubs, forming dense, clumped thickets, colonies, or hedge clusters, miniatures forming small, low-clustered, or open habits; usually rhizomatous. **Stems** erect, flexuous, spreading, ascending, sprawling, or procumbent, 1–50 dm; distal <and fertile> branches glabrous, rarely puberulent to tomentose, eglandular; prickles infrastipular, single or paired (rarely to 3 in *R. californica*, to 4 in *R. gymnocarpa*), erect, curved, hooked, declined, introrse, appressed, thin, terete, deltate, subulate, stout, sparse, internodal similar, smaller, sometimes mixed with aciculi, rarely absent. **Leaves** deciduous, (2–)4–11(–17) cm, leathery to membranous; stipules persistent, adnate to petiole, <auricles usually flared or erect>, margins entire or serrate, sometimes serrulate, undulate, or sinuate, usually glandular; leaflets (3–)5–9(–11), terminal: petiolule 1–15(–20) mm, blade elliptic, lanceolate, obovate, or ovate, sometimes cordate, ovoid, oblong, or orbiculate, (4–)10–40(–63) × (3–)5–40 mm, <rugose in *R. rugosa*>, abaxial surfaces <sometimes glaucous>, glabrous or pubescent, eglandular, sessile-, or stipitate-glandular, adaxial dull or lustrous, <glabrous, rarely pubescent, eglandular>. **Inflorescences** corymbs, panicles, sometimes solitary, 1–30(–50)-flowered. **Pedicels** erect or recurved, sometimes curved or reflexed as hips mature, slender or stout, 2–20(–35) mm, glabrous or pubescent, eglandular or usually stipitate-, rarely sessile- or setose-glandular; bracts persistent, 1–3, margins entire or serrate, sometimes undulate. **Flowers** 1–9 cm diam.; hypanthium globose to depressed-globose, ovoid, or ovoid-urceolate, sometimes cupulate, oblong, or hemispheric, glabrous or sparsely hairy, eglandular, stipitate-, or setose-glandular; sepals persistent or deciduous, erect, spreading, or reflexed, rarely ascending, lanceolate or ovate-lanceolate, 5–30(–40) × 1.5–6 mm, margins entire or pinnatifid, abaxial surfaces glabrous, rarely pubescent, sessile-, stipitate-, or setose-glandular, sometimes eglandular; petals single or double, pink to rose or purplish pink, rarely white or crimson; carpels (3–)20–40(–65), styles free <exsert (0.5–)1–2.5(–4) mm>, pilose, stylar orifice 1–3 mm diam., hypanthial disc flat, 2–5(–10) mm diam. **Hips** scarlet or red to orange-red, rarely purplish, brownish, or crimson or blue-purple, globose to depressed-globose, ovoid, obovoid, or ellipsoid, rarely pyriform, urceolate, elongate, or oblong, 6–18(–24) × 5–15(–25) mm, glabrous, sometime setose, eglandular or stipitate-, rarely setose-, glandular; sepals persistent or deciduous, erect, reflexed, or spreading. **Achenes** basal or basiparietal.

Species ca. 50 (19 in the flora): North America, Mexico, Eurasia; introduced in Pacific Islands (New Zealand), Australia.

Molecular phylogenetic and multivariate analyses of *Rosa* (S. Joly et al. 2006; Joly and A. Bruneau 2006, 2007; Bruneau et al. 2007) determined that certain eastern diploid species form a clade distinct from more western diploids; that *R. arkansana* is derived from the *R. blanda–R. woodsii* group; that *R. carolina* is likely of hybrid origin from one of the latter groups and *R. palustris*; that *R. virginiana* is derived either from *R. nitida*, *R. palustris*, or a hybrid between the two; and that the three polyploid species, *R. arkansana*, *R. carolina*, and *R. virginiana*, evolved multiple times.

Rosa ×*dulcissima* Lunell (*R. blanda* × *R. woodsii*) (W. H. Lewis 1962) occurs in Manitoba, southeastern Saskatchewan, Minnesota, North Dakota, South Dakota, and Wisconsin. In its Manitoba epicenter, B. Ziola and J. R. Dugle (1970) recognized numbers of the hybrid in excess of either parent. However, the validity of the hybrid has been questioned, as has the distinctiveness of *R. woodsii* from *R. blanda* (S. Joly and A. Bruneau 2007). The two species and their hybrid are retained here based on morphological and ecogeographic distinctions.

Rosa collaris Rydberg ($2n$ = 14, diploid) (D. Zlesak and W. H. Lewis, unpubl.) probably represents the hybrid between *R. gymnocarpa* and *R. woodsii* subsp. *ultramontana*. Plants annotated with this name differ from the former in having more persistent sepals and/or hairy leaflets with eglandular teeth, and from the latter in having solitary flowers, small hips, or other features more characteristic of *R. gymnocarpa*.

SELECTED REFERENCE Joly, S. and A. Bruneau. 2007. Delimiting species boundaries in *Rosa* sect. *Cinnamomeae* (Rosaceae) in eastern North America. Syst. Bot. 32: 819–836.

1. Distal branches usually unarmed, rarely with prickles and aciculi, glabrous.
    2. Stems sparsely branched; stamens 115 (unknown in *R. glauca*); stylar orifices 1.5–2 mm diam.; e Rocky Mountains to e North America.
        3. Stipules 1.5–2 mm wide; pedicels stipitate-glandular; flowers 2–3 cm diam.; leaflets abaxially glaucous, pale green, sometimes dull red, adaxially green-red or purplish, glaucous . . . . 16. *Rosa glauca*
        3. Stipules 4–6 mm wide; pedicels eglandular, rarely sparsely stipitate-glandular; flowers 3–6(–7) cm diam.; leaflets abaxially not glaucous, abaxially pale green, adaxially green, dull to ± lustrous . . . . 24. *Rosa blanda*
    2. Stems densely or openly branched; stamens 65–100; stylar orifices 2–2.5 mm diam.; w North America from Cascades to w Rocky Mountains.

4. Inflorescences (1–)3–10(–25+)-flowered, usually panicles; sepals 8–15(–21) × 1.5–2.5 mm; hypanthia 3–5 mm diam., necks 1.5–3.5 mm diam.; usually in full sun near water, centered in Great Basin and Columbia Plateau . . . . 25. *Rosa woodsii* (in part)
4. Inflorescences 1–3(–9)-flowered, corymbs; sepals 14–25(–35) × (2–)3–4 mm; hypanthia 5–8.5 mm diam., necks 3–6 mm diam.; forested areas, from Cascades to Rocky Mountains, skirting Great Basin . . . . 28. *Rosa nutkana* (in part)

1. Distal branches armed, usually glabrous, rarely tomentose or puberulent (in *R. rugosa*).
   5. Distal branches without or with relatively few infrastipular prickles, usually with mixed internodal prickles and aciculi.
      6. Sepals deciduous; hips irregularly ellipsoid or ellipsoid to nearly globose; terminal leaflet margins 2- or multi-serrate, teeth 7–13 per side; flowers 1.5–3 cm diam., stylar orifices 1 mm diam.; carpels 3–12(–16). . . . . .30. *Rosa gymnocarpa*
      6. Sepals persistent; hips subglobose, globose, ellipsoid, oblong, or urceolate; terminal leaflet margins 1-serrate or 1-, 2-, or multi-dentate-serrate, teeth 8–25 per side; flowers 3–6 cm diam., stylar orifices 1.5–2 mm diam.; carpels 18–33 or 26–43.
         7. Inflorescences 1–6(–16)-flowered on current year stems; leaflets (5–)7–9(–11), contiguous, terminal blade margins serrate, teeth 8–16 per side; pedicels erect . . . . 23. *Rosa arkansana*
         7. Inflorescences 1 or 2(or 3)-flowered on current and older stems; leaflets 5–7, not contiguous, terminal blade margins dentate-serrate, teeth 11–25 per side; pedicels reflexed (as hips mature) . . . . 29. *Rosa acicularis*
   5. Distal branches with paired, sometimes single, infrastipular prickles (rare in *R. nitida*), with or without internodal prickles and aciculi.
      8. Prickle bases pubescent; flowers 6–9 cm diam.; leaflets abaxially deeply veined, adaxially rugose (rarely smooth in *R.* ×*hollandica* [see p. 99]); hypanthial discs 5–10 mm diam.. . . . . .17. *Rosa rugosa*
      8. Prickle bases glabrous; flowers (1–)2.5–5.5(–9) cm diam.; leaflets not deeply veined or rugose; hypanthial discs (2–)3–5 mm diam.
         9. Sepals deciduous as hips mature, usually reflexed or widely spreading after anthesis; pedicels and hypanthia stipitate-glandular, rarely eglandular; achenes mostly basal, rarely basiparietal (in *R. virginiana*).
            10. Distal branches: internodal prickles (red to reddish purple) mixed with aciculi, infrastipular prickles rare or absent; stems usually procumbent (sometimes relatively short) . . . . .21. *Rosa nitida*
            10. Distal branches: internodal prickles and aciculi rare or absent, infrastipular prickles not mixed with aciculi; stems erect, sometimes spreading.
                11. Terminal leaflet margins 1- or 2-serrulate, teeth 20–30 per side; auricles erect, rarely flared; swamps and wetlands . . . . .18. *Rosa palustris*
                11. Terminal leaflet margins serrate, teeth 8–18(–23) per side; auricles flared; rarely swamps or wetlands.
                    12. Pedicels 2–8 mm; petals white, rarely pink; terminal blades 3–7 mm wide; hips leathery; mostly Oklahoma, n Texas . . . . 22. *Rosa foliolosa*
                    12. Pedicels 5–19(–25) mm; petals pink to deep rose, rarely white; terminal blades 6–28 mm wide; hips fleshy; mostly e North America, rarely e Texas.
                        13. Leaflets adaxially lustrous; stipules 4–9 mm wide; sepals 20–40 mm; stems erect, stout; infrastipular prickles stout, erect, curved, or declined, rarely hooked, bases relatively broad . . . . 19. *Rosa virginiana*
                        13. Leaflets adaxially dull, rarely slightly lustrous; stipules 2–3 mm wide; sepals 10–22 mm; stems spreading, weak, sometimes erect; infrastipular prickles subulate, erect, sometimes declined, rarely curved, bases relatively narrow. . . . . .20. *Rosa carolina*

[9. Shifted to left margin.—Ed.]

9. Sepals persistent as hips mature, usually erect or ± spreading after anthesis; pedicels and hypanthia eglandular, sometimes stipitate-glandular (in *R. spithamea*, *R. nutkana*); achenes mostly basiparietal, rarely basal (in *R. cinnamomea*).
    14. Subshrubs, forming open colonies; stems 1–8(–15) dm; California, Oregon.
        15. Hypanthia stipitate- to setose-glandular, very rarely eglandular . . . . 32. *Rosa spithamea*
        15. Hypanthia eglandular.
            16. Hypanthia ovoid to depressed-globose; infrastipular prickles single or paired, internodal prickles and aciculi dense; inflorescences 1–5(+)-flowered; coastal . . . . 31. *Rosa pinetorum*
            16. Hypanthia ovoid-globose to widely urceolate; infrastipular prickles paired, internodal prickles and aciculi rarely present; inflorescences 1 or 2(–7)-flowered; Sierra Nevada and Cascade mountains. . . . . .33. *Rosa bridgesii*
    14. Shrubs, loosely clustered, or forming thickets; stems 2–25(–50) dm.
        17. Leaflets abaxially ± glaucous; petals double, rarely single; stipules often convolute; achenes basal; introduced . . . . 15. *Rosa cinnamomea*
        17. Leaflets abaxially usually not glaucous; petals single; stipules not convolute; achenes basiparietal; native.
            18. Inflorescences 1–3(–9)-flowered, corymbs; petals (15–)20–32 mm; sepals 14–25 (–35) mm, tips 5–10(–20) × 1–3 mm; hypanthia 5–8.5 mm diam., necks 3–6 mm diam.; hypanthial discs 5 mm diam.; hips 10–20(+) mm diam.. . . . . . . 28. *Rosa nutkana* (in part)
            18. Inflorescences 1–30(–50)-flowered, panicles or corymbs; petals (10–)15–20 (–25) mm; sepals 8–17(–21) mm, tips 1–7(–10) × 0.3–2.5 mm; hypanthia 2.5–5.5 mm diam., necks 1.5–4.5 mm diam.; hypanthial discs 3–4(–5) mm diam.; hips 5–15(–20) mm diam.
                19. Petioles and rachises usually puberulent-velutinous, hairs 0.1(–0.5) mm; terminal leaflet blades usually widest at or above middle, teeth (5–)7–14 per side on distal ½+ of margin; internodal prickles usually sparse to dense (except subsp. *ultramontana*); mostly e of Sierra-Cascade axis. . . . . 25. *Rosa woodsii* (in part)
                19. Petioles and rachises glabrous or hairy, hairs to 1 mm; terminal leaflet blades usually widest at or below middle, teeth 7–22 per side on distal ¾+ of margin; internodal prickles absent or rare; mostly Sierra-Cascade axis and west.
                    20. Infrastipular prickles erect, rarely curved, usually subulate, 2–10 × 2–4 mm; sepals abaxially stipitate-glandular (except subsp. *ahartii*); pedicels usually glabrous, sometimes finely pubescent; hypanthia 2.5–3.5 mm diam., necks 2 mm diam. . . . . .26. *Rosa pisocarpa*
                    20. Infrastipular prickles strongly curved to nearly erect, ± flattened, stout, subulate, 3–15(–20) × 2.5–8 mm (to 10–15 mm); sepals usually abaxially eglandular; pedicels hairy, rarely glabrous; hypanthia 3–5.5 mm diam., necks 2–4.5 mm diam.. . . . . 27. *Rosa californica*

15. **Rosa cinnamomea** Linnaeus, Sp. Pl. 1: 491. 1753, name conserved • Cinnamon rose, rosier canelle I

Shrubs, forming dense thickets. **Stems** ± flexuous, 10–20 dm, densely branched; bark dark red to brown, glabrous; infrastipular prickles paired, erect, sometimes curved, stout, 4–7 × 2.5–3 mm, <base glabrous>, internodal prickles absent, aciculi rare. **Leaves** 4–8 cm; stipules <often convolute>, 8–15 × 3–8 mm, auricles flared, (1.5–)3–5(–7) mm, margins entire or crenate, sparsely to densely stipitate-glandular, surfaces glabrous, sometimes pubescent, eglandular; petiole and rachis without pricklets, pubescent, eglandular, sometimes sparsely or densely sessile-glandular; leaflets 5(–7), terminal: petiolule 4–9 mm, blade elliptic to ovate, 20–40 × 12–18 mm, membranous, margins 1-serrate, teeth 9–18 per side, eglandular or gland-tipped, apex acute, abaxial surfaces pale green, ± glaucous, pubescent (especially on veins) or glabrous, glandular, adaxial green, dull, sometimes sparsely pubescent. **Inflorescences** corymbs, 1 or 2(or 3)-flowered. **Pedicels** erect, slender, 7–17 mm, glabrous, eglandular; bracts 2, ovate or ovate-lanceolate to obovate, 6–15 × 6–10 mm, margins ciliate, stipitate-glandular, surfaces mostly pubescent, sometimes glabrous. **Flowers** 3–5 cm diam.; hypanthium hemispheric, 4–5 × 3–4 mm, glabrous, eglandular, neck absent; sepals spreading, lanceolate or ovate-lanceolate, 18–25 × 2.5–4 mm, tip 5–8 × 1–2 mm, margins entire, rarely pinnatifid, <lobes 1 or 2>, abaxial surfaces pubescent, eglandular or sparsely sessile- or, rarely, stipitate-glandular; petals double, rarely single, pinkish purple, 9–15 × 9–11 mm; carpels 25–35, styles exsert 4 mm beyond stylar orifice (2.5–3 mm diam.) of hypanthial disc (4–4.5 mm diam.). **Hips** scarlet, globose, 9–11 × 8–10 mm, fleshy, glabrous, eglandular, neck absent; sepals persistent, erect or spreading, <hips rarely maturing>. **Achenes** basal, 17, dark tan, 3–3.5 × 1.5 mm. **2*n*** = 14.

Flowering Jun–Jul. Roadsides, old pastures, fencerows, fields, stream banks, near dwellings; 0–400 m; introduced; N.B., Nfld. and Labr. (Nfld.), N.S., Ont., P.E.I., Que.; Conn., Ill., Maine, Mass., Mich., N.H., N.Y., Ohio, Pa., R.I., Vt., Va., Wis.; Europe; Asia (Japan, Siberia).

The double-flowered form of *Rosa cinnamomea* is often found in eastern North America, where hip formation and achene development are infrequent. The species is distinguished by its mostly convoluted stipules, abaxially glaucous leaflets, and basal achenes.

16. Rosa glauca Pourret, Hist. & Mém. Acad. Roy. Sci. Toulouse 3: 326. 1788 • Redleaf rose, rosier glauque [I]

Rosa rubrifolia Villars

Shrubs, erect and tall. **Stems** ± flexuous, erect, to 20 dm, sparsely branched, <distal branches and leaves glaucous, with purplish bloom>; bark becoming red-brown, glabrous; infrastipular prickles absent, internodal prickles sparse or absent, aciculi absent or erect, curved, or declined, subulate, 2–5 × 1.5–4.5 mm, base mostly broad, eglandular. **Leaves** 6–10.5 cm; stipules 13–16 × 1.5–2 mm, auricles flared, 2.5–5 mm, margins entire, eglandular or sparsely gland-fringed, surfaces glabrous, eglandular; petiole and rachis usually with pricklets, glabrous, eglandular; leaflets 5–7(–9), terminal: petiolule 10–20 mm, blade narrowly elliptic to ovate, 20–45 × 15–25 mm, leathery, base cuneate, sometimes rounded, margins 1-serrate, <eglandular or few gland-tipped>, teeth 12–18 per side, eglandular, apex acute, abaxial surfaces pale green, sometimes dull red, glaucous, glabrous, eglandular, adaxial green-red or purplish, glaucous, dull, glabrous. **Inflorescences** corymbs, rarely panicles, 1–5-flowered. **Pedicels** erect, slender, (10–)15–25 mm, glabrous, stipitate-glandular; bracts 1 or 2, narrowly ovate-lanceolate, 8–23 × 3–6 mm, margins entire, eglandular, surfaces glabrous, eglandular. **Flowers** 2–3 cm diam.; hypanthium narrowly ovoid, 5.5–7 × 3.5–5.5 mm, glabrous, eglandular, neck <purplish>, absent or 1 × 2 mm; sepals spreading, lanceolate, 15–25 × 1.5–2 mm, tip 10–15 × 1 mm, margins entire, sometimes pinnatifid, abaxial surfaces glabrous, mostly stipitate-glandular; petals single, deep pink to crimson-red, sometimes white basally, 8–14 × 5–6 mm; carpels 30–32, styles exsert 1–2 mm beyond stylar orifice (1.5–2 mm diam.) of hypanthial disc (3 mm diam.). **Hips** dark brownish red to crimson red, globose, ovoid, or obovoid, 10–13 × 9–11 mm, fleshy, glabrous, eglandular, neck absent; sepals deciduous as hips mature, erect to spreading. **Achenes** basiparietal, 15–23, light tan to tan, 4–4.5 × 2.5–3 mm. **2*n*** = 28.

Flowering May–Jul. Roadsides, wastelands, neglected areas; 0–2000 m; introduced; N.B., Nfld. and Labr. (Nfld.), N.S., Que.; Colo., Ill., Maine, Mass., Minn., N.Y., S.C.; c, s Europe; introduced also in n Europe.

Rosa glauca, included by A. J. Rehder (1940) in sect. *Caninae*, belongs in sect. *Rosa* (I. Klášterský 1968).

Rosa glauca is an infrequent introduction to the flora area from the mountains of Europe. As its garden cultivation increases, the species will continue to spread and become naturalized. The following collection label is a perfect illustration: *W. A. Weber 15003* (COLO, NY), 19 June 1973, at Boulder, Colorado, found at the bottom of Gregory Gulch at West end of Baseline Rd., north slopes at base of talus slides, 2000 m, thoroughly naturalized cultivar, probably spread by jays from adjacent city. The shrubs usually produce abundant hips and achenes as well as rhizomatous shoots; once established the plants reproduce vegetatively and sexually.

Rosa glauca consists of erect shrubs reaching 20 dm; more or less flexuous stems and glaucous distal branches have a purplish bloom with sparse or no broad-based internodal prickles. Leaflet blades are glaucous abaxially and often crimson-red; adaxially, blades are purplish-glaucous.

The taxonomic affiliation of the name *Rosa ferruginea* Villars is uncertain.

17. Rosa rugosa Thunberg in J. A. Murray, Syst. Veg. ed. 14, 473. 1784 • Japanese, rugose, or Sitka rose, rosier rugueux [I] [W]

Shrubs, forming thickets. **Stems** erect, sometimes arching, 10–25 dm, densely branched; bark greenish white when young, purplish black with age, densely tomentose or puberulent; infrastipular prickles paired, erect, stout, subulate, 10 × 4 mm, <base pubescent>, internodal prickles similar, densely mixed with aciculi and stipitate glands. **Leaves** 7–11 cm; stipules 20–30 × 4–7 mm, auricles flared, 4–6 mm, margins entire or unevenly serrate, sessile-glandular, surfaces <rugose>, glabrous or pubescent, glandular; petiole and rachis with pricklets <usually curved, subequal>, pubescent, usually sparsely glandular; leaflets 5–9, terminal: petiolule 8–18 mm, blade widely elliptic to ovate, rarely obovate, 20–55 × 10–35 mm, leathery, <rugose>, base cuneate to obtuse, margins usually 1-crenate-serrate, <eglandular, sometimes gland-tipped>, teeth 11–17 per side, sometimes gland-tipped, apex obtuse to acute, abaxial surfaces gray-green, <deeply veined>, pubescent, eglandular, adaxial dark green, lustrous, <rugose>, glabrous. **Inflorescences** corymbs, 1(–4)-flowered. **Pedicels** erect, sometimes reflexed (as hip matures), stout, 10–15(–30) mm, pubescent, sometimes setose, sparsely sessile- or stipitate-glandular; bracts 2, ovate-lanceolate, 12(–15) × 2(–5) mm, margins entire with few hairs, eglandular or gland-tipped, surfaces pubescent, glandular. **Flowers** 6–9 cm diam.; hypanthium depressed-globose, 6–8 × 5–6 mm, glabrous, sometimes setose overall or only in neck area, eglandular, rarely glandular, neck 1–2 × 4–5 mm; sepals erect, ovate-lanceolate, 20–37 × 4–6 mm, tip 4–5 × 1–2 mm, margins entire, rarely pinnatifid, <tomentose, glandular>, abaxial surfaces pubescent, sessile- and stipitate-glandular, often setose; petals single, sometimes double (cultivars), purplish pink or white, 35–50 × 30–45 mm; carpels 48–60, styles exsert 1–2 mm beyond stylar orifice of hypanthial disc (5–10 mm diam.). **Hips** scarlet, depressed-globose, 18–20 × 20–25 mm, leathery, glabrous, sometimes setose, eglandular, rarely glandular, neck 1–2 × 4–5 mm; sepals persistent, erect. **Achenes** basiparietal, 40, tan, 4–6 × 2–4.5 mm. **2*n*** = 14.

Flowering Jun–Jul(–Oct). Disturbed areas, roadsides, fields, maritime dunes and bluffs, shores, riparian sites; 0–1000 m; introduced; St. Pierre and Miquelon; B.C., N.B., Nfld. & Labr. (Nfld.), N.S., Ont., P.E.I., Que.; Alaska, Conn., Del., Ill., Maine, Md., Mass., Mich., Minn., Mo., N.H., N.J., N.Y., Ohio, Pa., R.I., Utah, Vt., Va., Wash., W.Va.; Asia (n China, Japan, Siberia); introduced also in Europe.

Rosa rugosa is possibly indigenous to the western Aleutian Islands, Alaska; such nativity requires further study. Elsewhere in Alaska and other northern habitats, particularly near coastal estuaries and dunes, the species is widely naturalized in North America. It is readily identified by its rugose, leathery leaflets with deep veins, distal branches that are densely tomentose, and flowers 6–9 cm diam., consisting of mostly purplish pink double petals. It can be weedy and is invasive in eastern parts of the United States and Canada, especially along coastlines and waterways.

Some *Rosa rugosa* hybrids are recognized by their reticulated, deeply veined, rugose, and dark green leaves. The most widespread naturalized hybrid is *R.* ×*hollandica* Persoon ex Steudel (Dutch rose), found throughout central and northern Europe, where it has been widely used as rootstock for grafting and budding. In 2010, A. V. Gilman discovered and identified the hybrid in five Vermont localities and, together with D. Werier, also found robust shrubs in an open pasture in Schuyler County, New York.

Gallic acid derivatives and polysaccharides, found in aqueous extracts of *Rosa rugosa* flowers, demonstrated major antioxidant activity (T. B. Ng et al. 2004). In experiments with animals, extracts of *R. rugosa* alleviated oxidative stress associated with diabetes by inhibiting lipid peroxidation (E. J. Cho et al. 2004).

18. Rosa palustris Marshall, Arbust. Amer., 135. 1785 • Swamp rose, rosier palustre [E] [F]

Rosa floridana Rydberg; *R. gemella* Willdenow; *R. lancifolia* Small; *R. obtusiuscula* Rydberg; *R. palustris* var. *dasistema* (Rafinesque) E. J. Palmer & Steyermark

Shrubs, forming thickets. **Stems** erect, 10–25(–30) dm, sparsely branched; bark reddish brown, glabrous; infrastipular prickles paired, curved, rarely erect, stout, 3.5–8 × 2–5(–10) mm, <base glabrous>, internodal prickles and aciculi rare, sometimes absent. **Leaves** 8–11 cm; stipules <narrow>, 10–22(–35) × 2.5–4 mm, auricles erect, rarely flared, 2.5–4.5(–8) mm, margins serrulate, eglandular or stipitate-glandular, surfaces glabrous, eglandular; petiole and rachis usually with pricklets, puberulent to pubescent, sometimes glabrous, eglandular or sparsely stipitate-glandular; leaflets 5–7, terminal: petiolule 5–10 mm, blade ovate-lanceolate, rarely broadly lanceolate or narrowly elliptic, 23–45 × 10–18 mm, membranous, base cuneate, margins 1- or 2-serrulate, <eglandular>, teeth 20–30 per side, acute to ± obtuse, eglandular, apex acute to subacute, abaxial surfaces pale green, glabrous or pubescent, eglandular, adaxial green, dull, glabrous. **Inflorescences** corymbs, (1 or)2–10(–40)-flowered. **Pedicels** erect, slender, 6–15 mm, glabrous, densely stipitate glandular; bracts 2, lanceolate, 6–15 × 3–4 mm, <margins and central veins pubescent>, eglandular, surfaces pubescent, eglandular. **Flowers** 2.5–5 cm diam.; hypanthium cupulate, 2–4 × 2–4 mm, glabrous, sparsely to densely stipitate-glandular, neck absent or 1–3 mm; sepals spreading to reflexed, rarely erect, lanceolate to narrowly ovate-lanceolate, 15–30(–40) × 2–3.5 mm, tip 2.5–3.5 × 0.5–1 mm, margins entire, rarely pinnate, abaxial surfaces glabrous, densely, sometimes sparsely, stipitate-glandular; petals single, pink to deep pink, 14–28 × 13–28 mm; <stamens 200>; carpels 24–50, styles exsert 0.5–1 m beyond stylar orifice (1.5 mm diam.) of hypanthial disc (3.5–4.5 mm diam.). **Hips** deep red, usually globose to subglobose, rarely elongate, 7–11 × 7–11 mm, fleshy, glabrous, sparsely or densely stipitate-glandular, neck absent or 3 × 1 mm; sepals deciduous, spreading. **Achenes** basal, 26, tan, 3 × 1.5–2 mm. **2*n*** = 14.

Flowering (Jun–)Jul(–Aug). Swampy woods and pastures, marshes, edges of ponds, springs, lakes, backwaters, sloughs, streams, ditches; 0–700 m; N.B., N.S., Ont., Que.; Ala., Ark., Conn., Del., D.C., Fla., Ga., Ill., Ind., Iowa, Ky., Maine, Md., Mass., Mich., Miss., Mo., N.H., N.J., N.Y., N.C., Ohio, Pa., R.I., S.C., Tenn., Vt., Va., W.Va., Wis.; introduced in Europe.

The single 1952 collection of *Rosa palustris* from Lake of Three Fires State Park (ISC), Taylor County, southwestern Iowa, is about 430 km northwest of the nearest known collection of the species, in Missouri. The species was probably introduced. It has the most serrulate leaflet margins of all roses in North America; the stems usually have short, stout, curved infrastipular prickles, rarely without armature.

Rosa ×*palustriformis* (Rydberg) Voss (*R. carolina* var. *aculeata* Schuette, *R. michiganensis* Erlanson, *R. schuetteana* Erlanson) refers to putative hybrids between *R. blanda* × *R. palustris* from Maine, Michigan, Ohio, and Wisconsin. *Rosa schuetteana* is morphologically intermediate between the parental species; *R.* ×*palustriformis* is more similar to *R. palustris*. *Rosa blanda* and *R. palustris* differ by: branch armature (*R. blanda* unarmed or with sparse prickles or aciculi, *R. palustris* with stout, curved infrastipular prickles or unarmed); length of each auricle (*R. blanda* average 4.8 mm, *R. palustris* average 2.6 mm); petioles and rachises with pricklets (*R. blanda* rare, *R. palustris* common); leaflet serrations (*R. blanda* serrate, acute, teeth ten to 26 per blade side, *R. palustris* serrulate, slightly blunt, teeth 20–30 per blade side); pedicels stipitate-glandular (*R. blanda* eglandular, *R. palustris* almost always); hypanthia stipitate-glandular (*R. blanda* eglandular, *R. palustris* almost always); inflorescences corymbs (*R. blanda* rare, *R. palustris* common).

Root decoctions of *Rosa palustris* were drunk by Cherokee to treat diarrhea (W. H. Lewis and M. P. F. Elvin-Lewis 2003). In Maine, *R. palustris* hips, including their achenes, are gathered about February, flattened, dried, and ground into flour for use with ground wheat to make leavened bread. The bread has a red color and a fine taste reminiscent of tomatoes (A. Haines, pers. comm.).

19. Rosa virginiana Miller, Gard. Dict. ed. 8, Rosa no. 10. 1768, name conserved • Virginia rose, rosier de Virginie E F

Rosa lucida Ehrhart; *R. lucida* var. *lamprophylla* (Rehder) P. V. Heath; *R. nanella* Rydberg

Shrubs, forming dense thickets and hedge clusters. **Stems** erect to ascending, (2–)10–30 dm, densely branched; bark red to purplish red, glabrous; infrastipular prickles paired or single, usually curved, sometimes erect, or declined, appressed, stout, 6–10 × 4–10 mm, <base glabrous>, internodal prickles or aciculi rare, smaller, sometimes absent. **Leaves** 5–8(–11) cm; stipules 14–25 × 4–9 mm, auricles flared, 3–5 mm, margins undulate, irregularly glandular-serrate, surfaces glabrous, eglandular; petiole and rachis sometimes with pricklets and aciculi, glabrous, puberulent, or sparsely pubescent, stipitate-glandular; leaflets 5–7(–9), terminal: petiolule 6–14 mm, blade narrowly elliptic to ovate, 17–32 × 6–16 mm, membranous, base cuneate, margins 1- or 2-serrate, teeth 10–18(–23) per side, gland-tipped or eglandular, apex acute, sometimes obtuse, abaxial surfaces pale green, glabrous or pubescent, eglandular, adaxial deep green, turning purplish red in fall, lustrous, glabrous. **Inflorescences** corymbs, 1–6(–15)-flowered. **Pedicels** erect, slender to stout, 7–14(–25) mm, glabrous, sparsely to densely stipitate-glandular; bracts 2, broadly lanceolate, 16–25 × 4–6 mm, margins entire, sometimes serrate, gland-tipped, surfaces glabrous with few hairs, eglandular. **Flowers** 4.3–5.5 cm diam.; hypanthium subglobose or depressed-globose, sometimes globose, 3.5–5.5 × 5.5–6.5 mm, glabrous, stipitate-glandular, neck absent; sepals spreading or reflexed, lanceolate, 20–40 × 2.5–4 mm, tip 6–12 × 0.5–2 mm, margins usually pinnatifid, rarely entire, inner 2 usually entire, abaxial surfaces glabrous, densely stipitate- or setose-glandular; petals single, pink to deep rose, rarely white, 22–26 × 25–30 mm; <stamens 140>; carpels 26–40(–65), styles exsert 1–2.5 mm beyond stylar orifice (1.5–3 mm diam.) of hypanthial disc (3–5 mm diam.). **Hips** orange-red to red or red-black, globose to depressed-globose, 8–12 × 9–13 mm, fleshy, glabrous, stipitate-glandular, neck absent; sepals deciduous, erect. **Achenes** mostly basal, fewer basiparietal, 8–14, tan, 3–4 × 1.5–3.5 mm. **2*n*** = 28.

Flowering Jun–early Aug. Grasslands, woods, cliffs, maritime heathlands and grasslands, ditches, old fields, edges of wet spruce woods, rocky ledges, damp thickets, swamps, streams, shores; 0–200 m; St. Pierre and Miquelon; N.B., Nfld. and Labr. (Nfld.), N.S., Ont., P.E.I., Que.; Conn., Del., D.C., Maine, Md., Mass., Mich., N.H., N.J., N.Y., Pa., R.I., Va.; introduced in Europe (Austria, France, Great Britain).

Rosa virginiana is primarily coastal from Newfoundland to New Jersey. Distribution extends inland along estuaries and streams for relatively short distances or, rarely, outlying populations as in central New York (glacial lakes of Green Lakes State Park). Some collections from the District of Columbia and adjacent Virginia are *R. carolina* × *R. virginiana*, indicating that at one time, the hybrid and both parents existed in the region.

Disjunct introductions of *Rosa virginiana* are found along railroads, highways, and ports. The species is introduced in Ontario, eastern Michigan, and Virginia.

Rosa virginiana Miller was conserved over *R. virginiana* Herrmann in 2011 (W. H. Lewis 2008b).

Shrubs of *Rosa virginiana* form thickets and hedge clusters having erect, stout stems that densely branch, and in Newfoundland reach to six feet tall. These are armed with stout and relatively long infrastipular prickles, erect or curved and broad-based. Leaflets are lustrous adaxially, stipule widths are 4–9 mm, and sepal lengths are 20–40 mm; these traits differentiate *R. virginiana* from the closely allied *R. carolina*.

The majority of plants determined as *Rosa virginiana* from the Allegheny and Appalachian mountains (for example, in West Virginia) and from the Midwest (for example, Indiana) are *R. carolina* subsp. *subserrulata*. In the eastern United States, putative hybrids and their introgressants with *R. carolina* subsp. *carolina* occur from Massachusetts to New Jersey and, rarely, south or north of these states. These are the nothospecies *R.* ×*novae-angliae* W. H. Lewis.

20. Rosa carolina Linnaeus, Sp. Pl. 1: 492. 1753
• Carolina or pasture rose, rosier de Caroline [F]

Shrubs or subshrubs, forming scattered populations. **Stems** spreading and weak, sometimes erect, slender, 3–10(–13) dm, openly branched; bark dull reddish brown, glabrous; infrastipular prickles usually paired, erect, sometimes declined, rarely curved, flat, subulate, (2–)3–9 × 1.5–3 mm, <base glabrous>, internodal prickles sparsely or densely mixed with aciculi, stipitate glands and smaller internodal prickles. **Leaves** 5–10(–16) cm; stipules 10–18(–23) × 2–3 mm, auricles flared, 2–4 mm, margins entire or subserrate, eglandular or finely stipitate-glandular, surfaces glabrous, rarely puberulent, eglandular; petiole and rachis sometimes with pricklets <3 mm>, aciculi few, glabrous, rarely pubescent, rarely stipitate-glandular; leaflets (3–)5–7(–9), terminal: petiolule 4–11 mm, blade ovate, elliptic, or lanceolate, 18–50 × 9–28 mm, membranous, base cuneate, margins 1- or 2+-serrate, teeth 8–14(–18) per side, eglandular or gland-tipped, apex acute to acuminate, rarely obtuse, abaxial surfaces pale green, glabrous, rarely pubescent, eglandular or glandular, adaxial usually green, dull, rarely slightly lustrous, glabrous. **Inflorescences** corymbs, 1–3(–6)-flowered. **Pedicels** erect, slender, 5–19 mm, glabrous, sparsely to ± densely stipitate-glandular, sometimes eglandular; bracts 2, lanceolate, 10–17 × 2–4 mm, margins entire, usually eglandular, surfaces with sparse hairs, stipitate-glandular. **Flowers** 3–5.5 cm diam.; hypanthium globose or ovoid, 4–6(–8) × 3.5–5(–8) mm, glabrous, ± densely to sparsely stipitate-glandular, sometimes eglandular, neck (0–)0.5–1 × 2 mm; sepals reflexed, sometimes spreading, lanceolate, 10–22 × 2–3 mm, tip 2–10 × 0.5–1 mm, margins pinnatifid or entire, abaxial surfaces rarely puberulent, stipitate-glandular, rarely eglandular; petals single, pink, 15–24 × 13–19 mm; <stamens 105>; carpels 32–46, styles exsert 1 mm beyond stylar orifice (1.5–2 mm diam.) of hypanthial disc (4–5 mm diam.). **Hips** red or orange-red, globose or depressed-globose, rarely ellipsoid, 7–14 × 6–15 mm, fleshy, glabrous, densely to sparsely stipitate-glandular, sometimes eglandular, neck 0–0.5 × 5–6 mm; sepals early deciduous, spreading to reflexed. **Achenes** basal, 2–6(–10), tan, 4–5 × 2.5–3 mm.

Subspecies 3 (2 in the flora): North America, ne Mexico.

Rosa carolina is a polymorphic allotetraploid derived from diploids found in eastern North America (S. Joly et al. 2006). Two nothospecies representing presumptive secondary hybridization and introgression are *R.* ×*medioccidentis* W. H. Lewis (*R. arkansana* × *R. carolina*) in Iowa, eastern Kansas, and western Missouri, and *R.* ×*novae-angliae* W. H. Lewis (*R. carolina* × *R. virginiana*) of New England, infrequently south to the District of Columbia, nearby Virginia, and New Jersey (W. H. Lewis 2008).

The Menominee of Wisconsin once ate hips of *Rosa carolina* to treat gastrointestinal problems (Hu. H. Smith 1928).

Subspecies *mexicoensis* W. H. Lewis is found in the Sierra Madre Oriental of northeastern Mexico (Coahuila, Nuevo León, Tamaulipas).

SELECTED REFERENCE Lewis, W. H. 2008. *Rosa carolina* (Rosaceae) subspecies and hybrids in eastern and midwestern United States, Canada, and Mexico. Novon 18: 192–198.

1. Stems spreading, sometimes erect; prickles (rarely absent) infrastipular, infrequently mixed with sparse, shorter internodal prickles and aciculi not to apices, eglandular; most common in e United States . . . . 20a. *Rosa carolina* subsp. *carolina*
1. Stems erect; prickles infrastipular, mixed with dense, shorter internodal, aciculi, and sparse stipitate glands to apices; epicenter in the Ozark Plateau and surrounding states . . . . 20b. *Rosa carolina* subsp. *subserrulata*

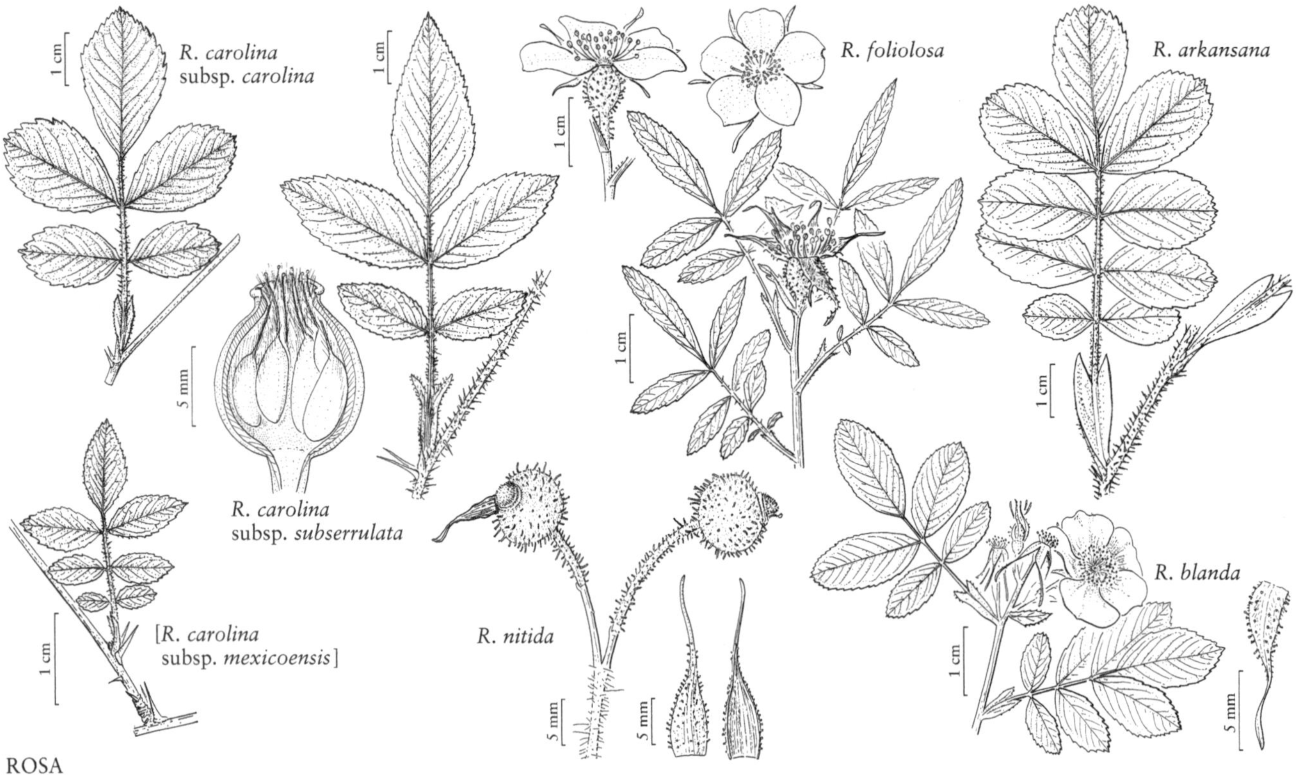

ROSA

20a. Rosa carolina Linnaeus subsp. **carolina** E F

Rosa aucuparia Rydberg; *R. carolina* var. *grandiflora* (Baker) Rehder; *R. carolina* var. *villosa* (Best) Rehder; *R. humilis* Marshall; *R. lyonii* Pursh; *R. parviflora* Ehrhart; *R. serrulata* Rafinesque

Stems spreading, sometimes erect, usually weak; infrastipular prickles (rarely absent), subulate, rarely stout, infrequently mixed with sparse, shorter internodal prickles and aciculi not to apices, eglandular. $2n = 28$.

Flowering late May–Aug [turions Sep]. Pastures, woodlands, glades, openings in woodlands, dry forests and woodlands, roadsides, stream banks, rocky ravines, ledges, bluffs, power line rights-of-way; 0–800 m; N.B., N.S., Ont., P.E.I., Que.; Ala., Ark., Conn., Del., D.C., Fla., Ga., Ill., Ind., Iowa, Kans., Ky., La., Maine, Md., Mass., Mich., Miss., Mo., Nebr., N.H., N.J., N.Y., N.C., Ohio, Okla., Pa., R.I., S.C., Tenn., Tex., Vt., Va., W.Va., Wis.

Subspecies *carolina* is characterized by short, weak, spreading or erect stems armed with erect, flat or, infrequently, subulate, paired infrastipular prickles, and is common in the eastern and southeastern United States, corresponding to the Austroriparian Subprovincial Element (S. P. McLaughlin 2007). To the west, it occurs less frequently in lowland areas and is replaced elsewhere in drier areas by subsp. *subserrulata*; intergrades are found in eastern Midwest regions. Subspecies *carolina* is relatively rare in Arkansas, Kansas, Louisiana, Oklahoma, and Texas.

20b. Rosa carolina Linnaeus subsp. **subserrulata** (Rydberg) W. H. Lewis, Novon 18: 195. 2008 E F

Rosa subserrulata Rydberg in N. L. Britton et al., N. Amer. Fl. 22: 500. 1918; *R. palmeri* Rydberg; *R. rudiuscula* Greene; *R. texarkana* Rydberg

Stems erect, robust; infrastipular prickles subulate or stout, mixed with dense, shorter internodal prickles, aciculi, and sparse stipitate glands to apices. $2n = 28$.

Flowering May–Jul. Dry wooded slopes, hillside thickets, rocky hills, upland woods, cherty upper slopes, sandstone glades, prairies; 100–1100 m; Ont.; Ala., Ark., Ga., Ill., Ind., Iowa, Kans., La., Md., Mich., Miss., Mo., N.C., Okla., Pa., S.C., Tenn., Tex., Va., W.Va., Wis.

Subspecies *subserrulata* is distinguished by relatively short, stout, erect stems armed with paired, subulate or stout, long, infrastipular prickles together with shorter prickles, aciculi, and stipitate glands to stem apices. It is most common in the Midwest with an epicenter in drier habitats of the Ozark Plateau, where it is particularly common in Arkansas, eastern Kansas, Missouri, eastern Oklahoma, and northeastern Texas. This range corresponds to the Illinoian Floristic Subprovince and, in part, the Appalachian Subprovince, where both subspecies occur.

21. Rosa nitida Willdenow, Enum. Pl., 544. 1809 • Red-spined or shining rose, rosier brillant E F

Shrubs or subshrubs, some forming clusters. **Stems** spreading, procumbent, or erect, 2–8(–10) dm, openly branched; bark dark brown, glabrous; infrastipular prickles sometimes present, paired, erect or declined, rarely curved, subulate, 6 × 6 mm, <base glabrous>, internodal prickles usually dense, mixed with aciculi, erect or declined, subulate, terete, 2–7 × 2 mm, eglandular. **Leaves** 3–10 cm; stipules 10–14 × 4–5.5 mm, auricles flared, 4–5 mm, margins entire or glandular-serrate, eglandular or gland-tipped, surfaces glabrous, eglandular; petiole and rachis with pricklets, glabrous, rarely puberulent, eglandular; leaflets (5–)7–9, terminal: petiolule 3–5(–9) mm, blade narrowly elliptic or lanceolate, rarely ovate, 13–27(–40) × 7–17 mm, membranous, margins 1-serrate, eglandular, rarely gland-tipped, teeth 12–17 per side, sometimes gland-tipped, apex acute to subacuminate, abaxial surfaces green, glabrous or pubescent, eglandular, adaxial deep green, purplish red in fall, lustrous, glabrous. **Inflorescences** corymbs, 1–3-flowered. **Pedicels** erect or recurved (as hips mature), slender, 13–25 mm, glabrous, densely long stipitate-glandular; bracts 1 or 2, lanceolate, 9–18 × 3.5–7 mm, margins entire, rarely serrate, few gland-tipped, <apically cleft 2–3.5 mm>, surfaces glabrous, eglandular. **Flowers** 4–5 cm diam.; hypanthium globose or subglobose, 3–4 × 3–5 mm, glabrous, densely stipitate-glandular, neck absent; sepals spreading to reflexed, lanceolate, 14–22 × 2–3 mm, tip 5–10 × 0.5–1 mm, margins entire, abaxial surfaces glabrous, densely stipitate-glandular; petals single, pink to rose, 19–23 × 20–23 mm; carpels 20–24, styles exsert 0.5–1 mm beyond stylar orifice (1.5–2 mm diam.) of hypanthial disc (4 mm diam.). **Hips** red to dark red, globose to depressed-globose, 8–10 × 7–10 mm, fleshy, glabrous, densely stipitate-glandular, neck absent; sepals deciduous, erect or spreading to reflexed. **Achenes** basal, 10–14, tan, darkening with age, 2.6–2.8 × 1.6–1.8 mm. **2*n*** = 14.

Flowering Jul–Aug. Wet edges of spruce woods, bogs, swamps, rocky ledges, wet thickets, margins of ponds and streams, shores, rocky and grassy hills and bluffs; 0–400 m; St. Pierre and Miquelon; N.B., Nfld. and Labr. (Nfld.), N.S., Ont., P.E.I., Que.; Conn., Maine, Mass., N.H., N.Y., R.I., Vt.

Rosa nitida is typical of the eastern Canadian Provincial Element (S. P. McLaughlin 2007) and is often found near or intermixed with *R. virginiana*. Putative hybrids between *R. nitida* (2*x*) and both *R. palustris* (2*x*) and *R. virginiana* (4*x*) are reported from Nova Scotia and New England. Because the armature of both *R. palustris* and *R. virginiana* is predominately or exclusively of infrastipular prickles, hybridity might explain the infrequent occurrences of *R. nitida* with infrastipular prickles. In a sample of 23 sheets of *R. nitida* having abundant prickles, two possess infrastipular prickles.

The most diagnostic feature of these shrubs with weak, sprawling stems is their distal branches, which are densely covered with mixed red to reddish purple internodal prickles and aciculi with infrastipular prickles relatively few.

22. Rosa foliolosa Nuttall ex Torrey & A. Gray, Fl. N. Amer. 1: 460. 1840 • White prairie rose E F

Rosa ignota Shinners

Shrubs, forming thickets or not. **Stems** erect to deflexed, 3–6 dm, densely branched; bark dull reddish brown, sometimes green, glabrous; infrastipular prickles paired, erect, subulate, 2–5 × 0.5–1 mm, <base glabrous>, internodal prickles or aciculi rare, sometimes absent. **Leaves** 3–7 cm; stipules 6–14 × 1.5–4 mm, auricles flared, 2–3 mm, margins entire or glandular-serrate, eglandular, surfaces glabrous, rarely puberulent, eglandular; petiole and rachis with pricklets, usually glabrous, stipitate-glandular; leaflets (5–)9(–11), terminal: petiolule 1–4(–6) mm, blade lanceolate to narrowly elliptic, 14–25 × 3–7 mm, membranous, base long-cuneate, margins 1-serrate, <gland-tipped or eglandular>, teeth 9–15 per side, gland-tipped, apex acute, abaxial surfaces light green, glabrous, eglandular, <exserted midveins sometimes with sparse pricklets, pubescent, stipitate-glandular>, adaxial deep green, lustrous, glabrous. **Inflorescences** corymbs, 1(–5)-flowered. **Pedicels** erect, slender, 2–8 mm, glabrous, stipitate-glandular; bracts 2, lanceolate, 8–12 × 3–4 mm, margins entire, few stipitate-glandular, surfaces glabrous, eglandular. **Flowers** 3.5–4.5 cm diam.; hypanthium subglobose, globose, or ovoid, 4–4.5 × 2.5–4 mm, glabrous, usually densely stipitate-glandular, neck absent; sepals reflexed or spreading, lanceolate, 13–20 × 2–3 mm, tip 3 × 0.5 mm, margins pinnatifid, abaxial surfaces glabrous or puberulent, stipitate-glandular; petals single, white, rarely pink, 17–28 × (6–)10–18 mm; carpels 20–32, styles exsert 1–1.5 mm beyond stylar orifice (1.5–2 mm diam.) of hypanthial disc (3–4 mm diam.). **Hips** dull red, globose to depressed-globose, rarely urceolate, 9–10 × 7–9 mm, leathery, glabrous, densely stipitate-glandular, neck absent or insignificant; sepals deciduous, spreading to erect. **Achenes** mostly basal, 8–12, tan, 4(–5) × 1.5–2 mm. **2*n*** = 14.

Flowering May–Jul. Blackland prairies, dry hillsides and woods, roadside and railroad verges, slopes and ravines, limestone and sandstone hills; 200–500 m; Ark., Kans., Okla., Tex.

The Kansas distribution of *Rosa foliolosa* is limited to Cherokee and Neosho counties, bordering Missouri and Oklahoma.

Rosa foliolosa is the only white rose native to North America; a pink form is known from Wise County, Texas (W. H. Lewis 1959). *Rosa foliolosa* is the most distinctive and geographically limited native *Rosa* in eastern North America, readily recognized by its white or, rarely, pink, narrow (6–18 mm) petals, lustrous, deep green and narrow (9 mm) leaflets, short pedicels (2–8 mm), leathery, dull red hips, and mostly basal achenes.

SELECTED REFERENCE Lewis, W. H. 1959. A monograph of the genus *Rosa* in North America. II. *R. foliolosa*. SouthW. Naturalist 3: 145–153.

23. Rosa arkansana Porter in T. C. Porter and J. M. Coulter, Syn. Fl. Colorado, 38. 1874 • Prairie rose [E] [F]

Rosa alcea Greene; *R. arkansana* var. *suffulta* (Greene) Cockerell; *R. conjuncta* Rydberg; *R. suffulta* Greene

Shrubs, forming hedge clusters. **Stems** erect, slender or stout, 6–15 dm, openly branched; bark dull red to purplish red, glabrous; infrastipular prickles rarely present, internodal prickles densely mixed with aciculi <to stem apices>, erect, terete, 1–3(–4) × 0.5–2 mm, base rarely extending to 3 mm, smallest often gland-tipped, aciculi rarely absent. **Leaves** 5–10(–16) cm; stipules 18–24 × 4–7 mm, auricles flared, 2.5–4(–7) mm, margins undulate, coarsely or shallowly glandular-serrate, surfaces glabrous or puberulent, eglandular; petiole and rachis sometimes with pricklets, sometimes with sparse aciculi, pubescent, sometimes glabrous, rarely stipitate-glandular; leaflets (5–)7–9(–11), terminal: petiolule 4–12 mm, blade obovate, sometimes elliptic, 15–40 × 8–20 mm, membranous, margins 1(or 2+)-serrate, teeth 8–16 per side, eglandular, rarely gland-tipped, apex acute, abaxial surfaces pale green, pubescent, sometimes glabrous, eglandular, adaxial green, ± glaucous, dull, sometimes pubescent (especially along midveins). **Inflorescences** corymbs, 1–6(–16)-flowered. **Pedicels** erect, slender, 10–20 mm, glabrous, eglandular; bracts 1 or 2(or 3), broadly lanceolate, 11–20 × 5–8 mm, margins entire, eglandular, surfaces glabrous, eglandular. **Flowers** 3.3–4 cm diam.; hypanthium globose, 5–6.5 × 4–5.5 mm, glabrous, eglandular, neck (0–)0.5–1.5 × 2 mm; sepals spreading to erect, lanceolate, 11–20 (–30) × (1.5–)3–4 mm, tip 3–7 × 0.5–1 mm, margins pinnatifid or entire, abaxial surfaces glabrous, stipitate-glandular or eglandular; petals single, rarely double, pink or rose, sometimes fading white, rarely white, 22–26 × 21–30 mm; <stamens 120>; carpels 26–43, styles exsert 1.5–2 mm beyond stylar orifice (1.5 mm diam.) of hypanthial disc (3 mm diam.). **Hips** dull orange-red, globose, subglobose, or oblong, 10–11 × 7.5–13 mm, fleshy, glabrous, eglandular, rarely stipitate-glandular, neck 0–2 mm; sepals persistent, erect at hip maturity. **Achenes** basiparietal, 12–15, dark buff, <ellipsoid>, 4.5–5 × 2.5 mm. **2*n*** = 28.

Flowering May–Jul. Rocky slopes, dry hillsides, prairies, bluffs, open woods, grassy roadsides; 200–2100 m; Alta., B.C., Man., Ont., Sask.; Ark., Colo., Ill., Ind., Iowa, Kans., Maine, Mich., Minn., Mo., Mont., Nebr., N.Mex., N.Y., N.Dak., Ohio, Okla., S.Dak., Tex., Wis., Wyo.

Rosa arkansana is one of the more invasive indigenous roses in North America. Collections from Ontario, Quebec, Maine, Massachusetts, New York, North Carolina, and Vermont, where nativity is unknown, are introductions that by and large are ephemeral. In other states where *R. arkansana* is native, disjunct populations may be either introduced or ephemeral, including those in Illinois, Indiana, Michigan, Ohio, and north-central Texas in disturbed areas. It is indigenous from northeastern British Columbia to Manitoba, east to Ohio, western Missouri, northeastern New Mexico, and Colorado in prairies and plains, and within the eastern Rocky Mountains of the North American Prairies Province (A. Cronquist 1982).

Rosa arkansana possibly arose from the diploid *R. blanda* and *R. woodsii* complex as an autopolyploid (S. Joly et al. 2006). It hybridizes with *R. carolina*; the hybrids are known as *R.* ×*medioccidentis* W. H. Lewis.

The Chippewa use root infusions or decoctions of *Rosa arkansana* as anticonvulsants, to treat bleeding wounds, and as stimulants and tonics; Omahas (Nebraska) use roots as an eye medicine and petals as a perfume for hair oil (D. E. Moerman 1998).

In a breeding program initiated by Agriculture-Canada to establish winter hardy roses for the Canadian prairies, the best results were obtained by crossing *Rosa arkansana* and, sometimes, *R. spinosissima* with floribundas and hybrid teas to produce new cultivars such as 'Prairie Joy' (L. M. Collicutt 1992) and 'Winnipeg Parks' (Collicutt 1992b).

24. Rosa blanda Aiton, Hort. Kew. 2: 202. 1789 • Meadow or smooth rose, rosier inerme [E] [F]

Rosa johannensis Fernald; *R. rousseauiorum* B. Boivin; *R. subblanda* Rydberg; *R. williamsii* Fernald

Shrubs, forming thickets. **Stems** erect or arching, (5–)10–20 dm, sparsely branched; bark green when young, red to orange-red with age, proximally glaucous, glabrous; infrastipular prickles absent, internodal prickles rare, sparse, distal aciculi rare. **Leaves** 8.5–11 cm; stipules (9–)15–25 × 4–6 mm, auricles usually flared, (2.5–)4–6 mm, margins entire or serrate, sparsely to densely stipitate-glandular, surfaces glabrous, sometimes pubescent, eglandular or glandular; petiole and rachis with pricklets rare or absent, glabrous or puberulent to pubescent, sessile- or stipitate-glandular, sometimes eglandular; leaflets 5–7(–9), terminal: petiolule 5–9(–13) mm, blade elliptic or ovate, sometimes obovate, (15–)25–40(–55) × (8–)12–20(–30) mm, membranous, margins 1-serrate, teeth 10–26 per side, acute, sometimes black gland-tipped, apex acute, sometimes obtuse, abaxial surfaces pale green, glabrous, sometimes pubescent or tomentulose, eglandular, adaxial green, dull to ± lustrous, glabrous, sometimes marginally hairy. **Inflorescences** corymbs, 1–5(–10)-flowered. **Pedicels** erect, slender, 14–25 mm, glabrous, eglandular, rarely sparsely stipitate-glandular; bracts 1 or 2, ovate-lanceolate, 14–20 × 2–8 mm, margins entire, sometimes undulate, irregularly serrate, short stipitate-glandular, surfaces glabrous, glandular. **Flowers** 3–6(–7) cm diam.; hypanthium ovoid or globose, 4–5 × 4.5–5 mm, glabrous, eglandular, rarely stipitate-glandular, neck (0–)0.5–0.8 × 3–3.5 mm; sepals usually erect <usually forming a beak-cap> or ± reflexed, ovate-lanceolate, (12–)20–30 × 2.5–3.5 mm, tip 2.5–4 × 0.5–1.5 mm, margins entire or pinnatifid, abaxial surfaces glabrous, stipitate-glandular, rarely eglandular; petals single, pink or rose, 13–26 × 12–26 mm; <stamens 115>; carpels 32–55, styles exsert 1–2 mm beyond stylar orifice (1.5 mm diam.) of hypanthial disc (4 mm diam.). **Hips** red, subglobose to globose, sometimes ellipsoid, urceolate, or pyriform, 8–11 × 8–11 mm, fleshy, glabrous, eglandular, rarely sparsely stipitate-glandular, neck (0–)0.5–1(–1.5) × 3.5–4.5 mm; sepals persistent, erect or reflexed. **Achenes** basiparietal, 26, tan, 3.5–4 × 1.5–3 mm. $2n = 14$.

Flowering Jun–Jul. Thickets, grassy verges, edges of woods, ditches, stream banks, gravelly and sandy flats; 0–700 m; Man., N.B., N.W.T., N.S., Ont., Que., Sask.; Ill., Ind., Iowa, Maine, Mass., Mich., Minn., Mo., Nebr., N.H., N.Y., N.Dak., Ohio, Pa., R.I., S.Dak., Vt., W.Va., Wis.; introduced in c Europe.

Rosa blanda is best recognized by its erect, slender distal stems and branches without armature or, rarely, with internodal prickles and/or aciculi. In addition, auricles are flared, 4–6 mm; abaxial surfaces of sepals are stipitate-glandular, usually forming an erect beak-cap, and the average stamen number is 115, nearly double that of *R. woodsii* with 65.

M. Mercure and A. Bruneau (2008) described and genetically confirmed spontaneous hybrids between *Rosa blanda* and *R. rugosa* forming colonies along the St. Lawrence Estuary, Quebec. Analyses indicated that most of the hybrids were first generation; second generation and backcrosses were also detected. This is a recently established hybrid zone with the potential for introgression of the aggressive *R. rugosa* into native *R. blanda* and thereby a serious threat to the integrity of the native species.

A. Bruneau et al. (2007) found that, within *Rosa blanda*, previously recognized segregate species (*R. johannensis*, *R. rousseauiorum*, *R. subblanda*, *R. williamsii*) should not be considered specifically distinct from *R. blanda*, although populations representing *R. williamsii* might be worthy of subspecific recognition.

25. Rosa woodsii Lindley, Ros. Monogr., 21. 1820 [F]

Shrubs, forming thickets or ± open stands. **Stems** erect, slender to stout, 2–20(–50) dm, densely or openly branched; bark dark red, glabrous; infrastipular prickles usually paired, sometimes absent, erect or curved to hooked, declined, or introrse, usually subulate, sometimes terete or flattened, stout, (0.5–)2–7(–13) × 1–5 mm, <base glabrous>, internodal prickles usually sparse to dense, often intermixed with aciculi. **Leaves** 2.5–8(–12) cm; stipules (6–)9–16(–25) × 2–5 mm, auricles usually flared, 2–6 mm, margins usually entire, sometimes undulate, sparsely serrate, eglandular or glandular, surfaces glabrous, eglandular, rarely glandular; petiole and rachis with pricklets sparse or absent, usually puberulent-velutinous <hairs 0.1(–0.5) mm>, rarely glabrate, eglandular, rarely stipitate-glandular; leaflets 5–7(–9), terminal: petiolule 3–12 mm, blade obovate, elliptic, or ovate, rarely cordate, (6–)12–35(–40) × 6–20(–26) mm, <usually widest at or above middle>, membranous, base cuneate, rarely obtuse, margins 1(or 2+)-serrate, teeth (5–)7–14 per side, <on distal ½–⅘ of margin>, acute, usually eglandular, apex acute or obtuse, abaxial surfaces pale green, glabrous, sometimes pubescent or puberulent, eglandular, sometimes glandular, adaxial green, rarely glaucous, dull, glabrous. **Inflorescences** usually panicles, sometimes corymbs or solitary flowers, 1–10(–25+)-flowered.

Pedicels erect, slender, 10–20(–33) mm, glabrous, eglandular, rarely stipitate-glandular; bracts 2, ovate or lanceolate, (6–)9–20 × 4–9 mm, margins entire or ciliate, eglandular, sometimes sparsely glandular, surfaces pubescent, eglandular. **Flowers** (2–)3–3.5(–5) cm diam.; hypanthium ovoid, rarely oblong or globose, 3–6 × 3–5 mm, glabrous, eglandular, rarely stipitate-glandular, neck (0–)0.5–1 × 1.5–3.5 mm; sepals spreading, ovate-lanceolate, 8–15(–21) × 1.5–2.5 mm, tip 4–6 × 0.3–1(–2) mm, margins usually entire, abaxial surfaces glabrous, eglandular, sometimes sessile- or stipitate-glandular; petals single, pink to deep rose, 15–20(–25) × 15–20(–25) mm; <stamens 65>; carpels (16–)20–40(–50), styles exsert 1–2 mm beyond stylar orifice (2–2.5 mm diam.) of hypanthial disc (3–5 mm diam.). **Hips** red, orange-red, or purplish red, globose, depressed-globose, ovoid, oblong, or urceolate, 6–13(–16) × 5–12(–15) mm, fleshy, usually glabrous, eglandular, neck (0–)1–2 × 3–4(–7) mm; sepals persistent, erect to spreading. **Achenes** basiparietal, 15–40, tan to dark tan, (3.5–)4–5(–6) × 2–4 mm.

Subspecies 6 (6 in the flora): c, w North America, n Mexico.

Rosa woodsii is the most common and most variable rose species in central and western North America (W. H. Lewis and B. Ertter 2007, 2010). Among its diagnostic features are relatively slender prickles, usually 1-serrate, eglandular leaflets, and a finely velutinous indument of 0.2–0.5 mm hairs on petioles and rachises. Its range extends from the prairies of Canada and the United States to inland Alaska and to north-central New Mexico, California east of the Cascade Mountains and Sierra Nevada, and northern Mexico.

At least 25 species names have been proposed to accommodate variation encompassed here within *Rosa woodsii*; some of these names have been used extensively (for example, *R. fendleri* Crépin, *R. macounii* Greene). The six subspecies recognized here represent significant morphological tendencies occurring in well-defined ecogeographic settings (W. H. Lewis and B. Ertter 2007, 2010). Localized varieties are recognized within two of the subspecies. Plants with intermediate characteristics often occur in transitional zones, and occasional anomalous collections display features not otherwise characteristic of a given ecogeographic region.

Leaves of *Rosa woodsii* inhibit HIV-1 replication in acutely infected cells. Both oleanolic and pomolic acids were identified as anti-HIV agents (Y. Kashiwada et al. 1998). Root decoctions of *R. woodsii* were drunk by the Shoshoni as a blood tonic for general debility and to treat diarrhea, and also by the Paiute for treating diarrhea (W. H. Lewis and M. P. F. Elvin-Lewis 2003).

1. Infrastipular prickles usually erect (erect or curved in subsp. *gratissima*, rarely curved in subsp. *woodsii*), sometimes absent; sepals abaxially eglandular, rarely stipitate-glandular (except subsp. *woodsii* n of 49° parallel); mostly w of Wasatch Range or e of Continental Divide.
    2. Shrubs 2–10(–20) dm; stems usually densely branched; terminal leaflets usually obovate; centered in prairies and plains e of Continental Divide, occasionally extending w through passes . . . . 25a. *Rosa woodsii* subsp. *woodsii*
    2. Shrubs usually 10–30(–50) dm; stems openly to densely branched; terminal leaflets ovate, ovate-elliptic, obovate, or elliptic; mostly within or w of Wasatch Range and n Rocky Mountains.
        3. Prickles sparse or absent on distal stems and branches; stems openly branched; inflorescences (1–)3–10(–25)-flowered; centered in n Great Basin and Columbia Plateau . . . . 25e. *Rosa woodsii* subsp. *ultramontana*
        3. Prickles usually abundant on distal stems and branches; stems densely branched; inflorescences 1–3(–5+)-flowered; centered in Mojave Desert . . . . 25f. *Rosa woodsii* subsp. *gratissima*
1. Infrastipular prickles usually curved or hooked, rarely absent; sepals often abaxially stipitate-glandular (except subsp. *puberulenta*); centered in Colorado Plateau Subprovince.
    4. Shrubs 2–10(+) dm; stems usually densely branched; terminal leaflets obovate, sometimes ovate or elliptic; high montane forests to scrub oak woodlands, 2100–3300 m. . . . . 25b. *Rosa woodsii* subsp. *manca*
    4. Shrubs 10–25+ dm; stems openly branched; terminal leaflets elliptic, rarely cordate or ovate; riparian sites in bottomlands, extending into mountains along waterways, 1300–2400 m.
        5. Sepals usually eglandular abaxially; prickles usually falcate, sometimes erect, declined, or introrse; terminal leaflets 20–40 mm; center of Colorado Plateau Subprovince . . . . 25c. *Rosa woodsii* subsp. *puberulenta*
        5. Sepals usually densely stipitate-glandular abaxially; prickles usually strongly curved or hooked, sometimes introrse; terminal leaflets 10–30 mm; Apachian and s edge of Colorado Plateau Subprovinces . . . . 25d. *Rosa woodsii* subsp. *arizonica*

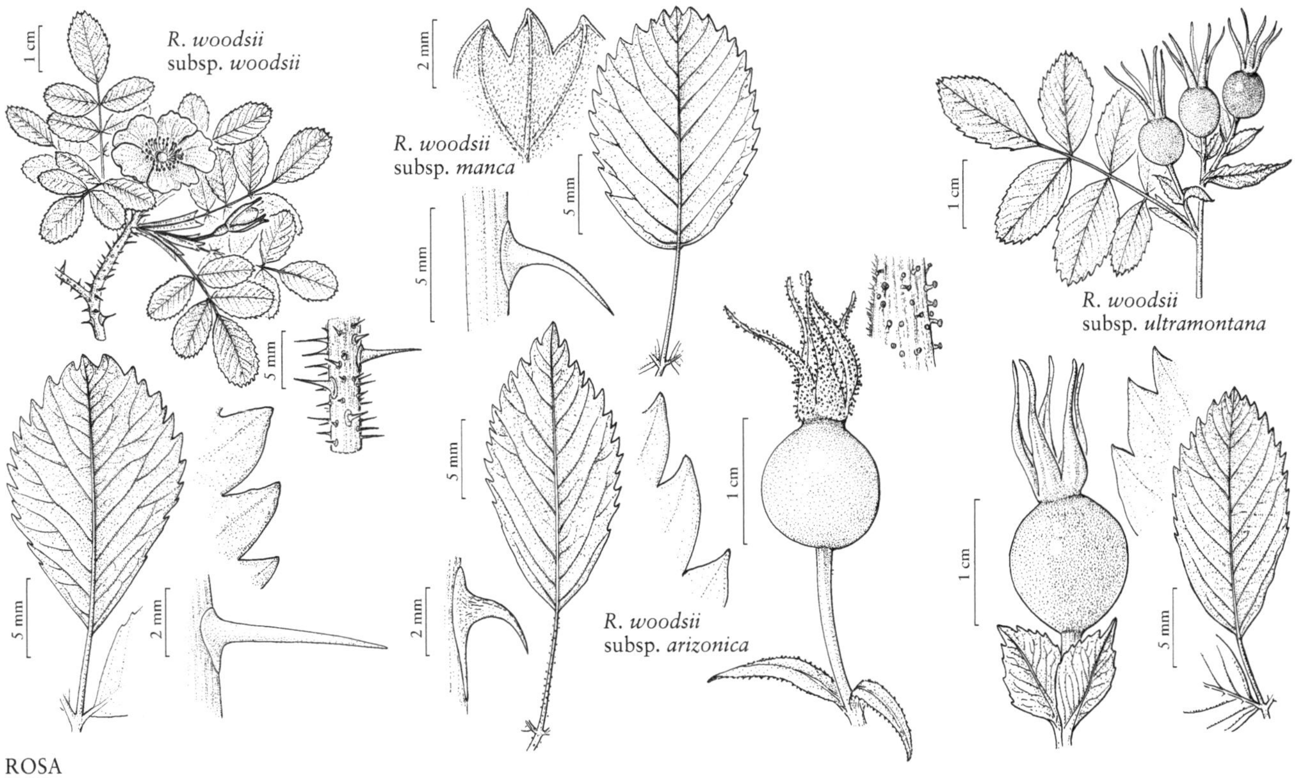

ROSA

25a. Rosa woodsii Lindley subsp. **woodsii** • Woods's rose, rosier de Woods [F]

Rosa adenosepala Wooton & Standley; *R. demareei* E. J. Palmer; *R. fendleri* Crépin; *R. macounii* Greene; *R. woodsii* var. *fendleri* (Crépin) Rydberg; *R. woodsii* var. *terrens* (Lunell) Breitung

Shrubs, 2–10(–20) dm. **Stems** usually densely branched; prickles usually present on distal stems and branches, infrastipular erect, rarely appressed or curved, terete or subulate, 2–8 mm, internodal dense, sometimes sparse or absent. **Terminal leaflets** obovate, sometimes ovate or elliptic, 13–30(–35) mm. **Inflorescences** 1–3(+)-flowered. **Sepals:** abaxial surfaces eglandular, rarely stipitate-glandular (north of 49° parallel). **2*n*** = 14.

Flowering Jun–Jul(–Aug). Prairies, plains, riparian and woodland areas; 100–2200(–3500) m; Alta., B.C., Man., N.W.T., Ont., Sask., Yukon; Alaska, Ariz., Colo., Idaho, Iowa, Kans., Minn., Mont., Nebr., N.Mex., N.Dak., Okla., S.Dak., Tex., Wyo.; Mexico (Coahuila).

Subspecies *woodsii* is the only subspecies of *Rosa woodsii* occurring east of the Rocky Mountains, where it is found from the prairies of northern Canada and central Alaska to the mountains of northern Mexico. The subspecies also extends west in valleys of the Rocky Mountains (for example, barely entering Idaho at Monida Pass), where it intergrades with other subspecies. The subspecies is sometimes introduced in Quebec, Illinois, Missouri, and Wisconsin; such introductions are mostly transient. Pending further study, var. *maderensis* Henrickson of northern Mexico is tentatively included in this subspecies. Within this large geographic range, significant variation occurs in glandularity and other features (V. L. Harms 1974; W. H. Lewis and B. Ertter 2010); usually plants are relatively compact, with abundant erect prickles, obovate leaflets, and relatively few flowers.

In central Canada, particularly in Alberta, western Manitoba, and Saskatchewan, populations of subsp. *woodsii* often express gigas (enlarged) features of leaves, stem heights, floral diameters, and petals at extreme morphologic ranges of the subspecies. A tetraploid (2*n* = 28) having these characteristics (W. H. Lewis, unpubl.) has been found in populations near Edmonton, Alberta; such diversity may be due, at least in part, to autopolyploidy or, alternately, hybridization with *Rosa acicularis* subsp. *sayi* (2*n* = 42), as B. Ziola and J. R. Dugle (1970) concluded in Manitoba and V. L. Harms (1974) in Saskatchewan. *Rosa acicularis* usually completes flowering about ten days before *R. woodsii*, although in some years overlaps occur.

25b. Rosa woodsii Lindley subsp. **manca** (Greene) W. H. Lewis & Ertter, Novon 17: 351. 2007 • Mancos rose [E] [F]

Rosa manca Greene, Pittonia 4: 11. 1899

Shrubs, 2–10(+) dm. **Stems** usually densely branched; prickles usually present on distal stems and branches, infrastipular curved or hooked, rarely erect, terete, 3–6 mm, internodal usually dense, rarely sparse or absent. **Terminal leaflets** obovate, sometimes ovate or elliptic, (6–)13–17(–35) mm. **Inflorescences** 1–3(+)-flowered. **Sepals:** abaxial surfaces usually stipitate-glandular.

Flowering Jun–Jul. Rocky meadows, high mountain forests to aspen and scrub oak woodlands, along streams, moist ditch banks; 2100–3300 m; Ariz., Colo., N.Mex., Utah, Wyo.

Subspecies *manca* [2*x* (DNA)] is the high-elevation representative of *Rosa woodsii* in the southern Rocky Mountains and outlying peaks and ridges of the Colorado Plateau, particularly characteristic of Gambel oak woodlands. Plants are relatively short and tend to have curved prickles, obovate leaflets, and stipitate-glandular sepals. As so circumscribed, this is not the *R. manca* of E. W. Erlanson (1934), which was regarded as a hexaploid (2*n* = 42) ecotype of *R. nutkana*. Neither is it the *R. manca* of A. Cronquist and N. H. Holmgren (1997), which represented a mixture of subsp. *manca* and *R. nutkana*, probably subsp. *melina*.

25c. Rosa woodsii Lindley subsp. **puberulenta** (Rydberg) W. H. Lewis & Ertter, Novon 20: 48. 2010 • Plateau rose [E]

Rosa puberulenta Rydberg, Fl. Rocky Mts., 443, 1062. 1917; *R. chrysocarpa* Rydberg; *R. neomexicana* Cockerell

Shrubs, 10–25+ dm. **Stems** openly branched; prickles usually present on distal stems and branches, infrastipular usually falcate, sometimes erect, declined, or introrse, usually subulate, 3–10 mm, internodal usually sparse or absent. **Terminal leaflets** usually broadly elliptic, 20–40 mm. **Inflorescences** 1 or 2(–5)-flowered. **Sepals:** abaxial surfaces eglandular, rarely sessile-glandular.

Flowering May–Jul. Edges of streams, flats, riparian woodlands dominated by cottonwood and pine; 1300–2400 m; Ariz., Colo., Nev., N.Mex., Utah, Wyo.

Subspecies *puberulenta* is the predominant *Rosa* occurring on bottomlands throughout the Colorado Plateau, extending at least as far as Lincoln and White counties, Nevada. Populations with comparable morphology that occur in northeastern California, Idaho, southeastern Oregon, and southeastern Washington might also belong here. The distribution of subsp. *puberulenta* overlaps those of all five other subspecies of *R. woodsii*. Diagnostic features include riparian habitat, relatively tall open habit, relatively prominent curved, declined, or introrse infrastipular prickles, and eglandular sepals.

25d. Rosa woodsii Lindley subsp. **arizonica** (Rydberg) W. H. Lewis & Ertter, Novon 17: 350. 2007 • Arizona rose [F]

Rosa arizonica Rydberg in N. L. Britton et al., N. Amer. Fl. 22: 516. 1918; *R. granulifera* Rydberg; *R. woodsii* var. *arizonica* (Rydberg) W. H. Lewis

Shrubs, 10–25+ dm, <usually forming thickets>. **Stems** openly branched; prickles usually present on distal stems and branches, infrastipular strongly curved or hooked, sometimes introrse, stout, sometimes flattened, 2–8 mm, internodal sparse or absent. **Terminal leaflets** usually elliptic, rarely cordate or narrowly ovate, 10–30 mm. **Inflorescences** 1 or 2(–4)-flowered. **Sepals:** abaxial surfaces usually stipitate-glandular, sometimes eglandular. 2***n*** = 14.

Flowering May–Jun. Sandy soil of oak, juniper, and pine forests, alluvial flats, brushy hillsides, sandstone cliff bases, sunny road banks; 1300–2400 m; Ariz., Nev., N.Mex., Utah; Mexico (Chihuahua).

Subspecies *arizonica* occurs primarily along streams in forested areas of Arizona, southern Nevada (Spring Mountains), northwestern New Mexico, and southern Utah with extensions into mountainous regions. Tall open habit, strongly curved prickles, and stipitate-glandular sepals are diagnostic, although not always present in combination. Plants with relatively few or no prickles, endemic to the West Fork of Oak Creek Canyon in Coconino County, Arizona, are var. *ertterae* W. H. Lewis, the Ertter rose; conservation attention to this variety is warranted. The remainder of the subspecies is var. *arizonica*.

25e. Rosa woodsii Lindley subsp. **ultramontana** (S. Watson) Roy L. Taylor & MacBryde, Canad. J. Bot. 56: 189. 1978 • Ultramontane or interior rose [E] [F]

Rosa californica Chamisso & Schlechtendal var. *ultramontana* S. Watson in W. H. Brewer et al., Fl. California 1: 187. 1876; *R. grosseserrata* E. E. Nelson; *R. lapwaiensis* H. St. John; *R. pisocarpa* A. Gray var. *ultramontana* (S. Watson) M. Peck; *R. pyrifera* Rydberg; *R. rotundata* Rydberg; *R. salictorum* Rydberg; *R. ultramontana* (S. Watson) A. Heller

Shrubs, usually 10–30(–50) dm. **Stems** openly branched; prickles absent or sparse on distal stems and branches, infrastipular prickles usually erect, usually subulate, 2–7 mm, internodal usually sparse or absent. **Terminal leaflets** ovate to elliptic, sometimes obovate, (10–)20–32(–40) mm. **Inflorescences** (1–)3–10(–25)-flowered. **Sepals:** abaxial surfaces eglandular, rarely stipitate-glandular. $2n = 14$.

Flowering May–Jul. Rocky banks and hillsides, steep hillsides, shady riparian banks, wooded stream bottoms, roadsides, fence rows, sagebrush hills; 150–2500 m; B.C.; Calif., Colo., Idaho, Mont., Nev., Oreg., Utah, Wash., Wyo.

Subspecies *ultramontana* occurs from the northern Rocky Mountains and Wasatch Front across the northern Great Basin and Columbia Plateau to the Sierra-Cascade axis, most often in full sun near water. Plants are relatively tall and open, generally with sparse, erect prickles.

Achene extracts of subsp. *ultramontana* exhibited strong antioxidant and antimicrobial activities, which were greater than such activities from *Rosa nutkana* hip extracts. Extracts from both were strongly active, the first roses from British Columbia known to express such activities (O. Yi et al. 2007).

25f. Rosa woodsii Lindley subsp. **gratissima** (Greene) W. H. Lewis & Ertter, Novon 17: 351. 2007 • Mojave or Tehachapi rose [E]

Rosa gratissima Greene, Fl. Francisc., 73. 1891; *R. pisocarpa* A. Gray var. *gratissima* (Greene) Jepson; *R. woodsii* var. *gratissima* (Greene) D. Cole

Shrubs, usually 10–30(+) dm. **Stems** densely branched; prickles present on distal stems and branches, infrastipular erect or curved, subulate or stout, 2–10(–13) mm, internodal usually abundant. **Terminal leaflets** ovate-elliptic, sometimes obovate, 9–28 mm. **Inflorescences** 1–3(–5+)-flowered, sometimes 10(–15)-flowered. **Sepals:** abaxial surfaces eglandular, rarely stipitate-glandular.

Flowering May–Aug. Mountain and desert springs, along streams, moist canyons, slopes, yellow pine forests; 800–3400 m; Calif., Nev.

Subspecies *gratissima* is diploid [2*x* (DNA)]. It occurs on mountains in and adjacent to the Mojave Desert of southeastern California and southern Nevada, US Highway 50 across central Nevada approximately coinciding with the boundary with subsp. *ultramontana*. Plants tend to be densely branched with abundant, relatively stout prickles extending into the inflorescence.

Within subsp. *gratissima*, localized clusters of glabrous populations on the northern foot of the San Bernardino Mountains represent var. *glabrata* (Parish) D. Cole (basionym *Rosa californica* Chamisso & Schlechtendal var. *glabrata* Parish); *R. mohavensis* Parish is a homotypic synonym and *R. woodsii* var. *mohavensis* (Parish) Jepson is superfluous and illegitimate. Conservative attention to var. *glabrata* is warranted. The remainder of subsp. *gratissima* is var. *gratissima*. A. Cronquist and N. H. Holmgren (1997) identified the NY syntype of *R. gratissima* as *R. californica*; the holotype (NDG 23620) clearly falls within the circumscription here.

26. Rosa pisocarpa A. Gray, Proc. Amer. Acad. Arts 8: 382. 1872 [E] [F]

Shrubs, loosely clustered or in dense thickets. **Stems** ascending to erect, (2–)4–20(–25) dm, openly branched; bark ± glaucous when young, dark reddish brown or dull red with age <outer layer may exfoliate as thin ash gray peel>, glabrous; infrastipular prickles usually paired, erect, rarely curved, usually subulate, 2–10 × 2–4 mm, <base glabrous>, internodal prickles rare or absent. **Leaves** 5–10(–13) cm; stipules 8–22 × 2–5 mm, auricles flared, 2–5 mm, margins usually entire, sometimes erose or lobed, <finely ciliolate to ciliate>, eglandular, surfaces glabrous or sparsely pubescent, sparsely stipitate-glandular or eglandular; petiole and rachis sometimes with pricklets, glabrous or hairy <hairs to 1 mm>, sometimes stipitate-glandular; leaflets 5–7(–9), terminal: petiolule 8–12 mm, blade elliptic-ovate, (15–)20–45(–60) × 9–16(–20) mm, <widest at or below middle>, membranous, base cuneate to obtuse, margins 1(or 2)-serrate, teeth 12–22 per side, <on distal $\frac{3}{4}$–$\frac{4}{5}$ of margin>, acute, eglandular, apex acute, sometimes obtuse, abaxial surfaces pale green, usually sparsely pubescent, eglandular, adaxial green, dull, glabrous, rarely puberulent. **Inflorescences** corymbs, sometimes panicles or solitary flowers, 1–12-flowered.

Pedicels erect, sometimes recurved, slender, 10–22 mm, usually glabrous, sometimes finely puberulent, eglandular, rarely stipulate-glandular; bracts 2 or 3, lanceolate, 8–14 × 3–6 mm, margins entire, sometimes serrate, irregularly stipitate-glandular and/or erose, surfaces glabrous or pubescent, eglandular. **Flowers** 2.4–3.8 cm diam.; hypanthium ovoid-urceolate, 3–5 × 2.5–3.5 mm, glabrous, rarely setose, eglandular, neck (0–)0.5–1 × 2 mm; sepals spreading, ovate-lanceolate, 10–17 × 1.5–3 mm, tip 3–7(–10) × 1.5–2.5 mm, margins entire, abaxial surfaces sometimes puberulent, densely or sparsely stipitate-glandular, rarely eglandular; petals single, pink to deep pink, 12–18 × 10–18 mm; <stamens 75>; carpels 22–35, styles exsert 1–2 mm beyond stylar orifice (1 mm diam.) of hypanthial disc (3.5 mm diam.). **Hips** scarlet, globose, sometimes subglobose or ovoid, 7–15 × 7–13 mm, fleshy, glabrous, eglandular, rarely setose- or stipitate-glandular, neck (0–)1–1.5 × 1.5–3.5 mm; sepals persistent, erect. **Achenes** basiparietal, 5–35, tan, 3–4 × 1.5–2.5 mm. **2*n*** = 14, 28.

Subspecies 2 (2 in the flora): w North America.

1. Inflorescences (1–)3–12-flowered; sepal abaxial surfaces usually stipitate-glandular, tips to 10 mm; infrastipular prickles (1 or)2; hips usually globose, rarely subglobose, abruptly narrowed to necks 1.5–3 mm diam.; leaflets most commonly 7, terminal blade 15–35 mm . . . . 26a. *Rosa pisocarpa* subsp. *pisocarpa*
1. Inflorescences 1–3(–10+)-flowered; sepal abaxial surfaces usually eglandular, rarely stipitate-glandular, tips to 7 mm; infrastipular prickles 0 or 1(or 2); hips subglobose to ovoid, gradually to abruptly narrowed to necks 2.5–3.5 mm diam.; leaflets most commonly 5, terminal blade 20–45(–60) mm . . . . 26b. *Rosa pisocarpa* subsp. *ahartii*

26a. Rosa pisocarpa A. Gray subsp. **pisocarpa** • Cluster or pea rose [E] [F]

Rosa rivalis Eastwood

Plants often forming thickets. **Stems** loosely to densely clustered, 10–25 dm; infrastipular prickles primarily (1 or)2, 2–10 mm. **Leaves** 5–10 cm; leaflets 5–7(–9), most commonly 7, terminal blade 15–35 mm. **Inflorescences** (1–)3–12-flowered. **Flowers:** sepal tip to 10 mm, abaxial surfaces usually stipitate-glandular; carpels 22–35. **Hips** usually globose, rarely subglobose, 7–10 mm diam., abruptly narrowed to neck 1.5–3 mm diam. **Achenes** 25–35. **2*n*** = 14.

Flowering Jun–Aug. Coasts, stream banks, riparian areas, open and low places, sedge meadows, swamps, roadside hedges, thickets, montane oak belts; 30–2100 m; B.C.; Calif., Oreg., Wash.

Subspecies *pisocarpa* encompasses traditional circumscriptions of *Rosa pisocarpa*. It occurs in coastal regions from southwestern British Columbia to northern California, west of and within the Cascade Mountains. Stems tend to have paired erect prickles, stipitate-glandular sepals with elongate tips, and clusters of pea-shaped hips. Limited populations intermediate with *R. woodsii* subsp. *ultramontana* occur in Cascade Mountain passes along the Fraser River (British Columbia), the Columbia River (Oregon and Washington), and the Klamath River (southern Oregon). Reports of *R. pisocarpa* from farther east (for example, R. J. Davis 1952) were based on *R. woodsii* or other species.

The Snohomish and Squaxin of Puget Sound, Washington, used subsp. *pisocarpa* root decoctions to treat sore throats, and bark infusions for after child-birth, respectively (D. E. Moerman 1998). Southeastern Vancouver Island Saanich have used young shoots for food in the spring; ripe hips were eaten raw in the fall. They also boiled branches to make a decoction, which was used to flush the eyes of those not able to see well; this preparation served to treat any eye problem, including cataracts (N. J. Turner, pers. comm.).

26b. Rosa pisocarpa A. Gray subsp. **ahartii** Ertter & W. H. Lewis, Madroño 55: 171, fig. 1. 2008 • Ahart's rose [E]

Plants rarely forming thickets. **Stems** usually solitary or loosely clustered, (2–)4–14(–18) dm; infrastipular prickles 0 or 1 (or 2), 2–5 mm. **Leaves** (5–)6–11(–13) cm; leaflets 5–7, most commonly 5, terminal blade 20–45(–60) mm. **Inflorescences** 1–3(–10+)-flowered. **Flowers:** sepal tip to 7 mm, abaxial surfaces usually eglandular, rarely stipitate-glandular; carpels 32–34. **Hips** subglobose to ovoid, 8–13 mm diam., gradually to abruptly narrowed to neck 2.5–3.5 mm diam. **Achenes** 5–20. **2*n*** = 28.

Flowering Jun–Aug. Streamsides, meadow margins, roadsides in woodlands, seasonally moist areas in openings of midmontane forests; 100–1700 m; Calif., Oreg.

Subspecies *ahartii* comprises populations from the northern Sierra Nevada and southern Cascade Range (B. Ertter and W. H. Lewis 2008). Stems tend to have relatively few (or no) prickles, relatively large ovate leaves, and relatively few flowers.

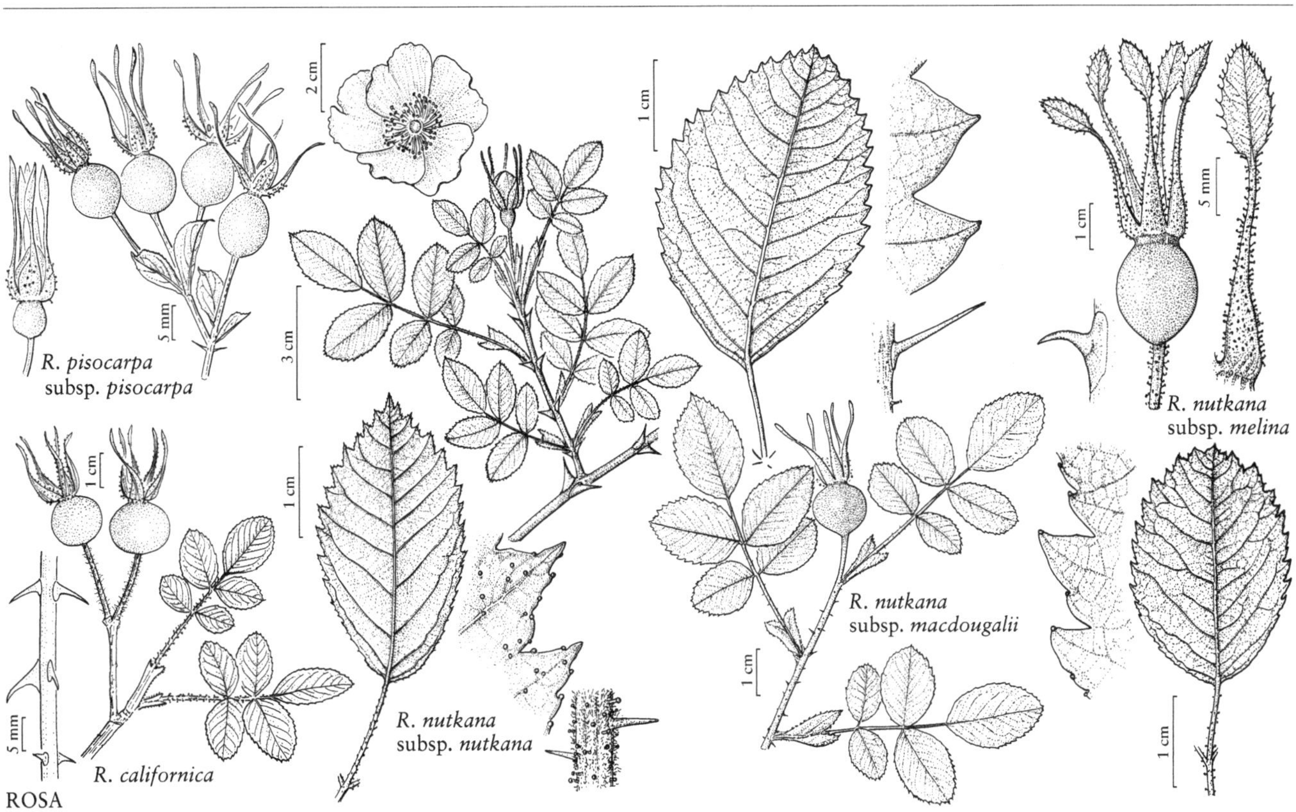

27. Rosa californica Chamisso & Schlechtendal, Linnaea 2: 35. 1827 • California rose [F]

Rosa aldersonii Greene

Shrubs, usually forming thickets. **Stems** ascending to erect, 8–25(–30) dm, densely to openly branched; bark often glaucous when young, dark reddish brown with age <sometimes exfoliating as thin ash-gray sheets>, glabrous; infrastipular prickles 1 or 2(or 3), strongly curved to nearly erect, ± flattened, stout, subulate, 3–15(–20) × 2.5–8 mm (to 10–15 mm), <base glabrous>, internodal prickles usually absent, sometimes sparse, smaller, aciculi usually absent, rarely stipitate-glandular. **Leaves** (2–)4–14 cm; stipules 5–20 × 2–5 mm, auricles flared to erect, 2–5 mm, margins entire or serrate, stipitate-glandular, surfaces glabrous, eglandular; petiole and rachis sometimes with pricklets, usually sparsely to densely hairy <hairs to 1 mm>, rarely glabrate, sometimes stipitate-glandular; leaflets 5–7(–9), terminal: petiolule 5–15(–20) mm, blade ovate to elliptic, sometimes obovate, (10–)15–40(–60) × 10–25(–30) mm, <usually widest at or below middle>, membranous to ± leathery, margins 1- or 2-serrate, teeth 7–20 per side <on distal ¾–⅘ of margin>, obtuse to acute, eglandular or gland-tipped, apex ± obtuse, sometimes acute, abaxial surfaces pale green, sparsely to abundantly shaggy-hairy, sometimes more finely hairy and/or sparsely glandular, adaxial green, dull, sparsely hairy to subglabrous. **Inflorescences** panicles, few corymbs, (1–)3–30(–50)-flowered. **Pedicels** erect, slender, 3–20 mm, hairy, rarely glabrous, eglandular, rarely glandular; bracts 1–3, broadly lanceolate, 8–20 × 2–10 mm, margins entire or serrate, eglandular or stipitate-glandular, surfaces ± hairy, eglandular. **Flowers** 2.5–5 cm diam.; hypanthium ± ovoid, 4–6 × 3–5.5 mm, glabrous or sparsely hairy, eglandular, neck 1 × 2–4.5 mm; sepals spreading, ovate-lanceolate, 10–15 × 2–4.5 mm, tip 1–5 × 0.3–1 mm, margins entire, <eglandular, sometimes glandular>, abaxial surfaces ± hairy, eglandular, sometimes stipitate-glandular; petals single, pink, (10–)15–25 × (10–)15–25 mm; <stamens 90>; carpels (20–)25–40, styles exsert 1–2.5 mm beyond stylar orifice (1.5–2 mm diam.) of hypanthial disc (3–4 mm diam.). **Hips** scarlet, ± ovoid, sometimes globose to obovoid, 10–18 × 7–20 mm, fleshy, glabrous or ± hairy, eglandular, neck 0.5–2 × (2.5–)3–5(–6) mm; sepals persistent, usually erect. **Achenes** basiparietal, (1–)5–20, cream to pale brown, (3–)3.5–4.5(–5) × 2–3 mm. $2n$ = 28.

Flowering Feb–Nov. Sunny streamsides, slough banks, mesic draws, moist areas in open woodlands, brushlands, and grasslands; 0–1800 m; Calif., Oreg.; Mexico (Baja California).

Rosa californica is the common thicket-forming rose in the California Floristic Province, barely entering the Mojave Desert along the Mojave River at Victorville;

Oregon occurrences are problematic. Characteristics that distinguish *R. californica* include large compressed, often strongly curved to nearly erect prickles, shaggy leaf vestiture, leaflets most commonly ovate with obtuse apices and cuneate to rounded to subcordate bases, numerous flowers with hairy pedicels, and large ovoid hips.

Some patterns of intraspecific variation in *Rosa californica* correlate with the ecogeographic complexity of the California Floristic Province, and they gain significance given the extent to which the plants are used in restoration efforts. Among those ecogeographic variants that may merit taxonomic recognition are: the *hartwegiana* phase, comprising exceptionally robust populations in the Central Valley with consistently hooked prickles, elliptic, short-hairy leaflets with acute apices, and relatively numerous flowers; the *nutkanoid* phase, consisting of scattered coastal populations that combine characteristics of *R. californica* (for example, hip size) and *R. nutkana* (for example, exceptionally heavy armature); and the *orthacantha* phase occurring in brushlands and open woodlands around Monterey and San Francisco bays and tending to have prominent pairs of erect infrastipular prickles, 1–3 flowers on elongate pedicels that exceed the subtending bracts, and ellipsoid hips containing relatively large achenes, possibly resulting from introgression with *R. gymnocarpa*. Variety *orthacantha* C. Presl probably represents this last phase.

The Costanoans of Californian Santa Clara Valley used hip decoctions of *Rosa californica* for treating indigestion, sore throat, fever, colds, rheumatism, and kidney ailments, and as a wash for scabs and sores (B. R. Bocek 1984).

28. Rosa nutkana C. Presl, Abh. Königl. Böhm. Ges. Wiss., ser. 5, 6: 563. 1851 [E] [F]

Shrubs, loosely clustered or forming dense thickets. **Stems** erect or spreading, stout or lax, 4–25(–40) dm, densely or openly branched; bark dull reddish brown, older wood with gray exfoliate, glabrous; infrastipular prickles 0–2, erect, curved, or hooked, often flattened, stout, subulate, sometimes deltoid, (2–)3–15(–20) × 2–10(–15) mm, <base glabrous>, internodal prickles usually sparse or absent, smaller, mixed with aciculi, sometimes absent. **Leaves** (3–)5–12(–17) cm; stipules 12–20(–28) × 3–5 (–10) mm, auricles flared, 3–6(–10) mm, margins sinuate or shallowly serrate, sparsely to ± densely stipitate-glandular, surfaces glabrous, rarely pubescent, eglandular; petiole and rachis with pricklets absent or sparse, glabrous or pubescent, sessile- or stipitate-glandular; leaflets 5–7(–9), terminal: petiolule (3–)5–17 mm, blade ovate, elliptic, or suborbiculate, rarely cordate or oblong, (14–)20–55(–63) × (6–)10–30(–40) mm, firm, base cuneate to obtuse, margins 1- or 2+-serrate, teeth 8–20 per side, <often broad-based 2–4 mm>, acute to obtuse, eglandular or gland-tipped, apex acute or obtuse, rarely acuminate, abaxial surfaces green, glabrous or sparsely pubescent, eglandular or sessile-glandular, adaxial green, dull, glabrous. **Inflorescences** corymbs, 1–3(–9)-flowered. **Pedicels** erect or recurved, slender to stout, 10–20(–33) mm, glabrous, eglandular or sparsely stipitate-glandular; bracts 1 or 2(or 3), ovate, 13–20 × 4–9 mm, margins entire, surfaces glabrous, glandular or eglandular. **Flowers** (3.5–)4.5–7(–8.5) cm diam.; hypanthium globose to subglobose or ovoid, 4–10 × 5–8.5 mm, glabrous, eglandular, sometimes setose-glandular, neck (0–)0.5–1.5 × 3–6 mm; sepals spreading, lanceolate, 14–25(–35) × (2–)3–4 mm, tip 5–10(–20) × 1–3 mm, margins entire, abaxial surfaces glabrous or slightly hairy, stipitate-glandular and sessile-glandular or eglandular; petals single, pink to deep rose, (15–) 20–32 × (15–)20–32 mm; <stamens 100>; carpels 28–58, styles exsert 1 mm beyond stylar orifice (2–2.5 mm diam.) of hypanthial disc (5 mm diam.). **Hips** red, orange-red, or purplish red, globose, depressed-globose, ovoid, oblong, or urceolate, (8–)10–20(–24) × 10–20(+) mm, fleshy, glabrous, sometimes setose, eglandular or glandular, neck (0–)1–2 × (4–)5–8 mm; sepals persistent, erect, sometimes spreading. **Achenes** basiparietal, (1–)16–40, tan to dark tan, 4–6 × 2–4 mm.

Subspecies 3 (3 in the flora): w North America.

Rosa nutkana is a widespread, polymorphic, hexaploid species that occurs from coastal Alaska to northern California, and eastward in the Rocky Mountains from Montana to New Mexico (W. H. Lewis and B. Ertter 2007). Flowers and hips are usually solitary and significantly larger than in most sympatric species, and sepal tips are usually conspicuously prolonged and expanded. Eglandular and stipitate-glandular or setose hips can occur in the same population.

A presumed hybrid between subsp. *nutkana* (6*x*) and *Rosa rugosa* (2*x*) has been reported from Washington state. In Colorado, at the southeasternmost range of *R. nutkana*, a putative hybrid between subsp. *melina* and *R. acicularis* subsp. *sayi* has been recognized as *R.* ×*engelmannii* S. Watson. It is based on such intermediate characteristics as few infrastipular prickles with interspersed aciculi on distal stems and fertile branches, sparsely stipitate-glandular abaxial sepal surfaces, petioles and rachises with few pricklets, and blade margins one- or multi-serrate.

Among the Lillooet and Thompson tribes of southwestern British Columbia, *Rosa nutkana* is considered of high value for use in traditional medicine (N. J. Turner 1988). They do not distinguish *R. acicularis* and *R. woodsii* from *R. nutkana* and use them similarly; *R. gymnocarpa* is distinct in name and use. Root extracts of *R. nutkana*, probably subsp. *nutkana*, were found to be active against an enteric coronavirus in antiviral screening (A. R. McCutcheon et al. 1995).

1. Stems 4–10(–15) dm, internodes 2–2.5 cm; prickles (fertile branches) usually curved or hooked, rarely erect, rarely absent; sepal abaxial surfaces sessile- or stipitate-glandular; 2500–3800 m, s Rocky Mountains to Wasatch Mountains .... 28c. *Rosa nutkana* subsp. *melina*
1. Stems (6–)10–30(–40) dm, internodes 2–6.5 cm; prickles (fertile branches) erect, rarely curved, sometimes absent; sepal abaxial surfaces stipitate-glandular or eglandular; 0–2300 m, coast to n Rocky Mountains.
   2. Shrubs densely branched; internodes 2–4 cm; infrastipular prickles rarely absent; terminal leaflets abaxially glandular or eglandular, margins usually 2+-serrate, teeth gland-tipped; Alaska to n California w of Cascade Mountains...... 28a. *Rosa nutkana* subsp. *nutkana*
   2. Shrubs openly branched; internodes 5–6.5 cm; infrastipular prickles usually absent; terminal leaflets abaxially usually eglandular, margins 1-serrate, teeth eglandular; e Cascade Mountains to n Rocky Mountains.................... 28b. *Rosa nutkana* subsp. *macdougalii*

28a. Rosa nutkana C. Presl subsp. **nutkana** • Nootka rose [E] [F]

Rosa aleutensis Crépin; *R. durandii* Crépin; *R. nutkana* var. *muriculata* (Greene) G. N. Jones; *R. nutkana* var. *setosa* G. N. Jones

Shrubs, densely branched. **Stems** 10–30(–40) dm, internodes 2–4 cm; prickles (distal stems and fertile branches) infrastipular, paired, erect, stout, rarely absent. **Terminal leaflet** margins usually 2+-serrate, teeth gland-tipped, abaxial surfaces glandular or eglandular. **Inflorescences** usually 1(or 2)-flowered. **Sepals:** abaxial surfaces eglandular or (outer) stipitate-glandular. **2*n*** = 42.

Flowering Apr–Jul. Rocky coasts and bluffs, stream banks, edges of woods, roadsides; 0–300 m; B.C., N.W.T.; Alaska, Calif., Oreg., Wash.

The predominant rose of coastal western North America, subsp. *nutkana* is found in the Vancouverian Floristic Subprovince west of the Coast-Cascade Mountains from Alaska and Northwest Territories to California. In stream passes through the coastal mountains, intergrades occur between the coastal subspecies and the interior subsp. *macdougalii*, particularly along the Fraser and Skeena rivers in British Columbia and the Columbia River between Oregon and Washington. Populations directly on the coast are often heavily armed, with exceptionally large, flattened prickles.

28b. Rosa nutkana C. Presl subsp. **macdougalii** (Holzinger) Piper, Contr. U.S. Natl. Herb. 11: 335. 1906 • Macdougal's or Spalding rose [E] [F]

Rosa macdougalii Holzinger, Bot. Gaz. 21: 36. 1896 (as macdougali); *R. nutkana* var. *hispida* Fernald; *R. rainierensis* G. N. Jones; *R. spaldingii* Crépin ex Rydberg

Shrubs, openly branched. **Stems** (6–)10–25(–30) dm, internodes 5–6.5 cm; prickles (distal stems and fertile branches) usually absent, infrastipular rarely present, usually erect or subulate. **Terminal leaflet** margins 1-serrate, teeth eglandular, abaxial surfaces eglandular. **Inflorescences** 1–3(–9)-flowered. **Sepals:** abaxial surfaces usually eglandular, rarely setose or stipitate-glandular at bases. **2*n*** = 42.

Flowering May–Aug. Rocky dry slopes, stream banks and bottoms, moist meadows, forest edges, sheltered slopes, roadside banks, fences; 200–2300 m; B.C.; Calif., Colo., Idaho, Mont., Oreg., Utah, Wash., Wyo.

The interior subsp. *macdougalii* occurs primarily in forested areas in central British Columbia, Oregon, and Washington east of the Cascade Mountains to the northern Rocky Mountains, barely entering northern Utah and northwestern Colorado, mostly west of the continental divide. Shrubs are more openly branched and sparsely armed than those of subsp. *nutkana*, with eglandular, usually one-serrate leaflets. Isolated populations from north-central California (including *Rosa brownii* Rydberg, an illegitimate name) have more uniformly ovate or obovate leaflets and more compact branches than in northern populations.

28c. Rosa nutkana C. Presl subsp. **melina** (Greene) W. H. Lewis & Ertter, Novon 17: 345. 2007 • Rocky Mountain rose [E] [F]

Rosa melina Greene, Pittonia 4: 10. 1899; *R. oreophila* Rydberg; *R. pandorana* Greene; *R. underwoodii* Rydberg

Shrubs, openly branched. **Stems** 4–10(–15) dm, internodes 2–2.5 cm; prickles (distal stems and fertile branches) infrastipular, curved or hooked, often stout, internodal sometimes present, smaller, rarely erect, sometimes absent. **Terminal leaflet** margins usually 2+-serrate, teeth gland-tipped, abaxial surfaces glandular or eglandular. **Inflorescences** usually 1(–5)-flowered. **Sepals:** abaxial surfaces sessile- or stipitate-glandular.

Flowering late Jun–Jul. Aspen, fir, spruce, and pine forest edges and clearings, stream edges and banks, brushy hillsides, peaks, ridges, grassy rocky slopes, meadows, canyon slopes; 2500–3800 m; Ariz., Colo., Idaho, N.Mex., Utah, Wyo.

Subspecies *melina* [6x (DNA)] is endemic to high elevations of the Rocky Mountains Biogeographic Province, where it is often associated with *Populus tremuloides*. Curved prickles, glandular sepals, and two- or three-serrate leaflets help distinguish subsp. *melina* where it overlaps with subsp. *macdougalii* in southeastern Idaho and northern Utah. Significant distributional overlap occurs also with *Rosa woodsii* subsp. *manca*, which can occur in mixed or adjacent populations. Subspecies *melina* is generally at higher elevations, flowers somewhat earlier than subsp. *manca*, and is hexaploid (the latter is diploid).

29. Rosa acicularis Lindley, Ros. Monogr., 44, plate 8. 1820 • Acicular rose F

Shrubs, forming dense thickets. **Stems** erect, stout, (3–)10–20(–25) dm, sparsely or densely branched distally; bark pale brown with tips dull red, glabrous; infrastipular prickles absent, internodal prickles dense, erect, subulate, terete, ± flattened, 9 × 4 mm, sparsely stipitate-glandular, mixed with dense aciculi. **Leaves** 5.5–15 cm; stipules 20–25 × 3–8 mm, auricles flared or erect, 5–7 mm, margins usually entire, stipitate-glandular, surfaces glabrous, sometimes puberulent, sparsely sessile-glandular or eglandular; petiole and rachis usually (rarely) with pricklets, glabrous or puberulent to pubescent, sessile-glandular; leaflets 5–7 (on annual shoots), terminal: petiolule 6–20 mm, blade elliptic, ovoid, or ovate-lanceolate, 20–60 × 13–32 mm, firm, margins 1(or 2+)-dentate-serrate, teeth 11–25 per side acute or obtuse, gland-tipped or eglandular, apex acute or obtuse, abaxial surfaces light green, glabrous or puberulent (on main veins), eglandular or sparsely to densely sessile-glandular, adaxial green, dull, glabrous or sparsely hairy. **Inflorescences** corymbs, 1 or 2(or 3)-flowered. **Pedicels** reflexed as hips mature, slender, (13–)20–28(–35) mm, glabrous, stipitate-glandular or eglandular; bracts 1–3, ovoid, 18–22 × 4–14 mm, margins entire, sessile- or short stipitate-glandular, surfaces glabrous, sometimes hairy, eglandular. **Flowers** 3–6 cm diam.; hypanthium ovoid to oblong, 5–8 × 4–6 mm, glabrous, eglandular, neck 1 × 2.5–3 mm; sepals spreading, often beak-capped, lanceolate, 20–33 × (2.5–)3–3.5 mm, tip 7–12 × 0.5–0.8 mm, margins entire, sometimes pinnatifid, abaxial surfaces glabrous, sometimes pubescent, eglandular or stipitate-glandular (north); petals single, rose pink to pale pink, (13–)22–25 × (11–)20–25 mm; <stamens 75–100>; carpels 18–33, styles exsert 1 mm beyond stylar orifice (2 mm diam.) of ± flat hypanthial disc (3.5–4.5 mm diam.). **Hips** orange-red to bright red or blue-purple, globose to ellipsoid or urceolate, 10–23 × 9–11 mm, fleshy, glabrous, eglandular, neck 1–2 × 2–2.5 mm; sepals persistent, erect, often beak-capped. **Achenes** basiparietal, 14–25, tan, 4 × 2–2.5 mm.

Subspecies 2 (2 in the flora): North America, Eurasia.

Rosa acicularis is circumpolar in forests of Eurasia and North America. Octoploid subsp. *acicularis* occurs from Sweden across northern Russia to Outer Mongolia, northern China, Japan, and Korea; hexaploid subsp. *sayi* is found throughout northern North America. The two subspecies meet in an intergradation zone from eastern Siberia to Alaska, and perhaps Yukon (W. H. Lewis 1958), where ploidy level is required for subspecies or hybrid confirmation.

SELECTED REFERENCE Lewis, W. H. 1958. A monograph of the genus *Rosa* in North America. I. *R. acicularis*. Brittonia 11: 1–24.

1. Pedicels densely stipitate-glandular; leaflets 5 (or 7), margins 1(or 2+)-dentate-serrate; sepal abaxial surfaces usually stipitate-glandular . . . . 29a. *Rosa acicularis* subsp. *acicularis*
1. Pedicels usually eglandular, if stipitate-glandular, not to apex or stipitate glands mostly sparse; leaflets 5–7, margins 1- or 2-dentate-serrate; sepal abaxial surfaces usually eglandular . . . . 29b. *Rosa acicularis* subsp. *sayi*

29a. Rosa acicularis Lindley subsp. **acicularis** • Eurasian acicular rose

Leaves: stipule auricles 4–8 mm wide; petiole and rachis rarely with pricklets; leaflets 5(or 7), margins 1(or 2+)-dentate-serrate, teeth 11–25 per ½ blade, eglandular, abaxial surfaces eglandular. **Pedicels** densely stipitate-glandular. **Sepals** usually to 3 mm wide at bases, abaxial surfaces usually stipitate-glandular, sometimes eglandular. $2n = 56$.

Flowering May–Jun. Timberlines, larch-birch and pine forests, stony glades, grassy meadows, watersheds; 1500–2800 m; Alaska; Eurasia.

At least one collection from central Alaska has been confirmed as subsp. *acicularis* on both morphological and cytological evidence (W. H. Lewis 1958). Other reports from Alaska, and Yukon, need confirmation as to whether they are octoploid ($2n = 56$) subsp. *acicularis* plants or hybrids with hexaploid subsp. *sayi* individuals, forming $2n = 49$ septaploids.

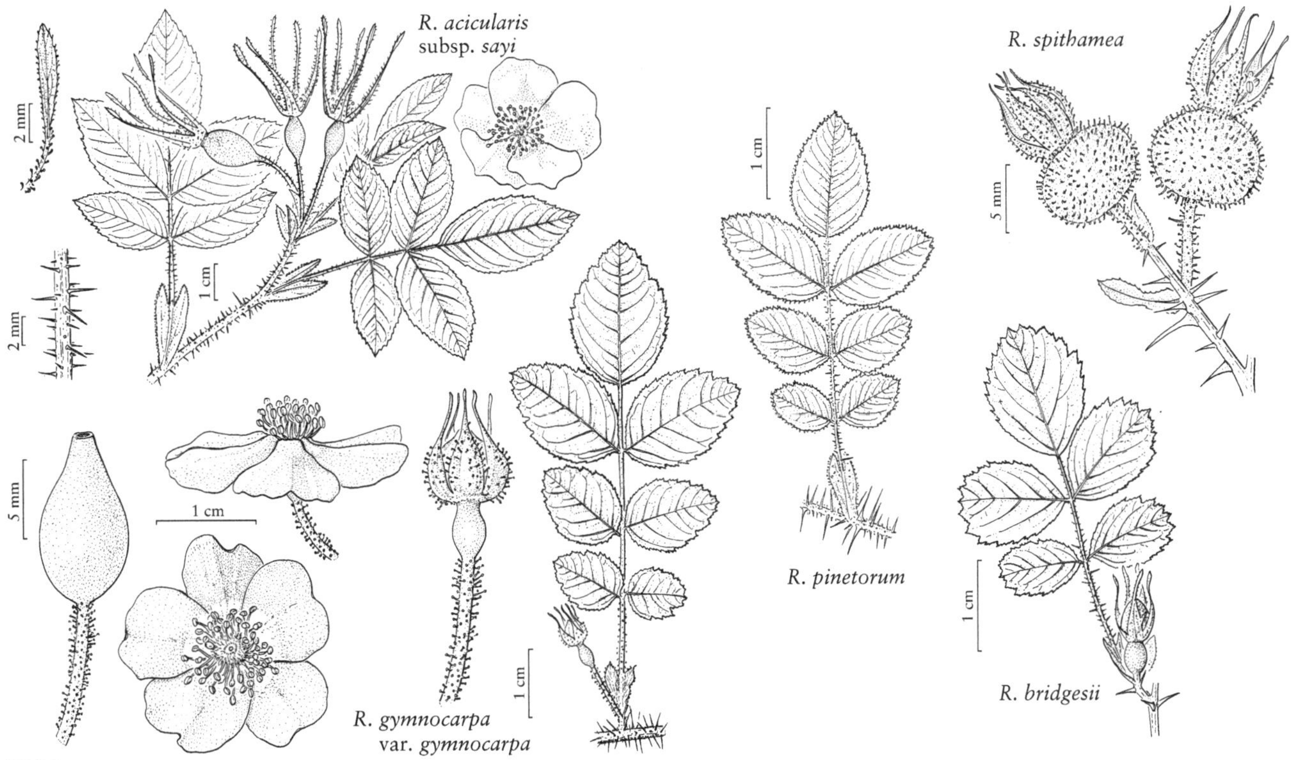

ROSA

29b. Rosa acicularis Lindley subsp. **sayi** (Schweinitz) W. H. Lewis, Brittonia 11: 19. 1959 • Say's acicular rose, rosier aciculaire [E] [F]

Rosa sayi Schweinitz in W. H. Keating, Narrat. Exp. St. Peter's River 2: 388. 1824; *R. acicularis* var. *bourgeauiana* Crépin; *R. butleri* Rydberg

Leaves: stipule auricles 4–7 mm wide; petiole and rachis usually with pricklets; leaflets 5–7, margins 1- or 2-dentate-serrate, teeth 10–25 per ½ blade, gland-tipped, abaxial surfaces usually eglandular, sometimes glandular. **Pedicels** usually eglandular, if stipitate-glandular, not to apex or mostly sparse. **Sepals** usually 3+ mm wide at bases, abaxial surfaces eglandular, sometimes sparsely stipitate-glandular. $2n = 42$.

Flowering Jun–Jul. Meadows, moist forest edges, flood plains, forested uplands, sandy aspen slopes, hillsides, calcareous ridges, banks of bogs and streams, dry sandy ridges, clearings in woods, roadsides, ravines; 100–2500 m; Alta., B.C., Man., N.B., N.W.T., N.S., Nunavut, Ont., Que., Sask., Yukon; Alaska, Colo., Iowa, Maine, Mass., Mich., Minn., Mont., N.H., N.Y., N.Dak., S.Dak., Vt., W.Va., Wis., Wyo.

Southern limits of subsp. *sayi* occur in the Rocky Mountains of Colorado, Black Hills of South Dakota, and algific slopes of talus or broken rock covering ice caves in northwestern Iowa, southwestern Wisconsin, Ice Mountain of West Virginia, and mountainous regions of New England where habitats similar to more northern ecosystems exist. Among these outlying populations, the density of stem prickles and aciculi often decreases and fertile branches rarely may be devoid of most armature. In central Canada and northward to Alaska, sepals have relatively small, short stipitate glands on abaxial surfaces more commonly than farther south, where they are more or less eglandular.

Plants with the morphological features of subsp. *sayi* also exist in eastern Siberia; confirmation of their identity depends on ploidy levels (W. H. Lewis 1958).

Branches of subsp. *sayi*, as well as of *Rosa nutkana* subsp. *macdougalii* and *R. woodsii* subsp. *ultramontana*, were used as cleansing agents and disinfectants, especially at times of illness and death, by the Shuswap people of interior British Columbia. The branches were placed around houses to prevent diseases in corpses from spreading to others; they were also used to disinfect rooms, clean air, and keep bad spirits away. People also used solutions made from rose bushes for steaming and cleansing in sweathouses. Root solutions were used to treat burns by soaking the burned area, which healed well without blistering or scarring. Hips and branches were boiled and decoctions drunk as medicine for treating diarrhea and upset stomach; teas made from branches were drunk to protect oneself from anything bad (N. J. Turner, pers. comm.).

30. Rosa gymnocarpa Nuttall in J. Torrey and A. Gray, Fl. N. Amer. 1: 461. 1840 [E] [F]

Shrubs or subshrubs, usually loosely clustered. **Stems** erect, sometimes spreading, slender, (1–)3–15(–25) dm, sparsely or densely branched; bark sometimes glaucous, reddish brown with age, glabrous; infrastipular prickles 0–2, erect, subulate, 2–8(–10) × 1.5 mm, terete, internodal prickles similar or smaller, sparse to dense, sometimes absent on distal stems, mixed with aciculi, base terete, eglandular. **Leaves** (2–)4–10(–17) cm; stipules 5–15 × 2–5 mm, auricles flared, 2–3 mm, margins entire, stipitate-glandular, surfaces glabrous, eglandular; petiole and rachis with pricklets, glabrous, rarely finely puberulent, sparsely stipitate-glandular; leaflets 5–9(–11), terminal: petiolule (2–)5–12(–20) mm, blade elliptic to obovate or ovate to nearly orbiculate, (4–)10–40(–60) × (4–)10–20(–40) mm, membranous to ± leathery, margins 2+-serrate, teeth 7–13 per side, obtuse to acute, gland-tipped, apex obtuse, sometimes nearly acute, rounded, or truncate, abaxial surfaces pale green, glabrous, eglandular, adaxial green, dull, glabrous. **Inflorescences** corymbs, usually 1–3-flowered, rarely in multi-flowered candelabras. **Pedicels** erect, ± curved as hips mature, slender, 10–25(–35) mm, glabrous, stipitate-glandular, rarely eglandular (except var. *serpentina*); bracts <some early caducous>, 1 or 2, attached near pedicel bases, ovate or lanceolate, 4–12 × 2–8 mm, margins entire, short stipitate-glandular, <apex acute or rounded>, surfaces glabrous, eglandular. **Flowers** 1.5–3 cm diam.; hypanthium narrowly ovoid-urceolate, 2–4 × 1.5–2 mm, glabrous, eglandular, neck 0–1 × 1.5 mm; sepals ascending to reflexed, lanceolate, 5–10 × 2–3 mm, tip 0.1–5 × 1 mm, margins entire, abaxial surfaces glabrous, eglandular, sometimes stipitate-glandular; petals single, deep pink, 8–15 × 6–13 mm; <stamens 57>; carpels 3–12(–16), styles exsert 1–1.5 mm beyond stylar orifice (1 mm diam.) of hypanthial disc (2–4 mm diam.). **Hips** scarlet, irregularly ellipsoid or ellipsoid to nearly globose, 7–15 × 5–13 mm, fleshy, glabrous, eglandular, neck 0–2 × 1.5–2.5 mm; sepals, styles, and distal receptacle collectively deciduous at fruit maturity along well-defined, circumscissile line, erect to reflexed. **Achenes** basiparietal, (1–)4–10(–12), cream to pale brown, (3–)4.5–7 × 2–4 mm.

Varieties 2 (2 in the flora): w North America.

Rosa gymnocarpa is most easily distinguished in mature fruit, in that the sepals collectively and cleanly separate from the hip together with styles and, sometimes, associated hypanthia. Other diagnostic features include solitary to few, small flowers, stipitate-glandular pedicels, hypanthia extremely small in bud, and glabrous leaflets doubly glandular-toothed. The species is also one of the relatively few roses that flourishes in partial shade. *Rosa gymnocarpa* occurs in forested areas from British Columbia and Montana to central California, with disjunct populations in southern California; it skirts the Great Basin, with the possible exception of the type of *R. leucopsis* Greene, a likely synonym, purportedly from central Oregon's sagebrush steppe, which is otherwise devoid of the species (A. Cronquist and N. H. Holmgren 1997).

The distinctive features of *Rosa gymnocarpa* have been used as the basis for sect. *Gymnocarpae* Crépin. Recognition of that section is not supported by the molecular analysis by A. Bruneau et al. (2007), in which *R. gymnocarpa* occurs in a clade with *R. pinetorum* and one sample of *R. californica*, separate from a distinct clade that groups diverse specimens of *R. bridgesii*, *R. spithamea*, and some Asian species of sect. *Rosa*.

1. Leaflets (5–)7–9(–11), terminal blades elliptic to narrowly obovate or ovate, (5–)10–30(–60) mm, apices usually obtuse, sometimes nearly acute or rounded; pedicels (10–)15–25(–35) mm; stems 3–15(–25) dm; light shade (rarely full sun) in open woodlands, forest edges, often near streams or rocky sites; s British Columbia to s California, e to Montana . . . . .30a. *Rosa gymnocarpa* var. *gymnocarpa*
1. Leaflets 5(–7), terminal blades broadly elliptic to obovate or ovate to nearly orbiculate, 4–20 mm, apices usually broadly obtuse to rounded, sometimes nearly truncate; pedicels 10–15 mm; stems (1–)3–6(–13) dm; full sun in roadsides, ridges, and other openings in chaparral and stunted forests on ultramafic substrates; Siskiyou Mountains, sw Oregon and nw California . . . . 30b. *Rosa gymnocarpa* var. *serpentina*

30a. Rosa gymnocarpa Nuttall var. **gymnocarpa** • Wood or baldhip rose [E] [F]

Rosa abietorum Greene; *R. helleri* Greene; *R. leucopsis* Greene; *R. piscatoria* Greene

Stems 3–15(–25) dm. **Leaves** (2–)4–10(–17) × (1.5–)3–6(–8) cm; stipules 6–15 mm; leaflets (5–)7–9(–11), usually bright green, thin, terminal: petiolule (3–)5–12(–20) mm, blade elliptic to narrowly obovate or ovate, (5–)10–40(–60) × (5–)10–20(–40) mm, apex usually obtuse, sometimes nearly acute or rounded. **Pedicels** (10–)15–25(–35) mm, stipitate-glandular, rarely eglandular. **Flowers** 1.5–3 cm diam.; petals 8–15 × 6–13 mm. **Hips** irregularly ellipsoid or ellipsoid to nearly globose. **Achenes** (1–)4–10(–13). $2n = 14$.

Flowering (Feb–)Apr–Jul. Light shade (rarely full sun) in open woodlands, forest edges, often near streams or rocky sites; 0–2000 m; B.C.; Calif., Idaho, Mont., Oreg., Wash.

The Saanich of southern Vancouver Island consider var. *gymnocarpa* a spiritual plant, used for bringing strength and ritual protection: by young people at puberty, by new dancers, and for the recently bereaved who have lost close relatives (N. J. Turner, pers. comm.).

30b. Rosa gymnocarpa Nuttall var. **serpentina** Ertter & W. H. Lewis, Madroño 55: 174, fig. 2. 2008 • Gasquet rose C E

Stems (1–)3–6(–13) dm. **Leaves** 2–6(–8) × 1.5–3(–4) cm; stipules 5–11 mm; leaflets 5(–7), often bluish green or red-tinged, ± leathery, terminal: petiolule 2–10 mm, blade broadly elliptic to obovate or ovate to nearly orbiculate, 4–20 × 4–20 mm, apex usually broadly obtuse to rounded, sometimes nearly truncate. **Pedicels** 10–15 mm, eglandular or stipitate-glandular (in same inflorescence). **Flowers** 2 cm diam.; petals 8–10 × 6–10 mm. **Hips** irregularly ovoid to elongate-ellipsoid. **Achenes** 1–4.

Flowering Apr–Jun. Full sun in roadsides, ridges, and other openings in chaparral and stunted forests on ultramafic substrates; of conservation concern; 400–1500(–2300) m; Calif., Oreg.

Variety *serpentina* is the form that usually occurs in full sun on ultramafic substrates in the Siskiyou Mountains of northwestern California and southwestern Oregon.

31. Rosa pinetorum A. Heller, Muhlenbergia 1: 53. 1904 • Pine rose C E F

Rosa gymnocarpa Nuttall var. *pinetorum* (A. Heller) Jepson

Subshrubs, forming open colonies. **Stems** erect, 1–8(–15) dm, openly branched; bark dark reddish brown, glabrous; infrastipular prickles single or paired, erect, subulate, 3–10 × 0.5–1.5 mm, <base glabrous>, internodal prickles dense, smaller, sometimes mixed with aciculi. **Leaves** (2–)3.5–7(–8.5) cm; stipules 5–10 (–30) × 2–5 mm, auricles slightly flared, 1–3 mm, margins entire, sometimes stipitate-glandular, surfaces glabrous, eglandular; petiole and rachis usually with pricklets, glabrous or finely hairy, stipitate-glandular; leaflets 5(–7), terminal: petiolule 5–12 mm, blade ± elliptic, 12–25(–32) × 6–15(–20) mm, membranous to ± leathery, base rounded, sometimes obtuse to cuneate, margins multi-serrate to nearly 1-serrate, teeth (5–)7–12 per side, acute to obtuse, usually gland-tipped, apex obtuse to rounded, abaxial surfaces ± pale green, glabrous or sparsely hairy, glandular, adaxial green, dull, glabrous. **Inflorescences** corymbs, 1–5(+)-flowered. **Pedicels** erect, mostly slender, 2–20(–35) mm, glabrous or sparsely hairy, sometimes stipitate-glandular; bracts 1–3, lanceolate or ovate-lanceolate, 6–25 × 2–7 mm, margins entire or serrate, short stipitate-glandular, surfaces glabrous or sparsely pubescent, glandular. **Flowers** 2–4 cm diam.; hypanthium ovoid to depressed-globose, 3.5–4 × 3.5–4 mm, glabrous, eglandular, neck 0.5 × 2.5–3.5 mm; sepals spreading to reflexed, ovate-lanceolate, 7–12 × 3 mm, tip 2–8 × 0.5–2 mm, margins entire or slightly toothed, abaxial surfaces finely hairy, stipitate-glandular, sometimes eglandular; petals single, deep pink, 10–20 × 10–20 mm; carpels 10–40, styles exsert 1.5 mm beyond stylar orifice (1.5–2 mm diam.) of hypanthial disc (3.5–5 mm diam.). **Hips** scarlet, subglobose to ovoid, 10 × 10–12 mm, fleshy, glabrous, eglandular, neck 1 × 4–5 mm; sepals persistent, erect. **Achenes** basiparietal, 1–10, pale tan, 3–4 × 2.5–3 mm. $2n = 14$.

Flowering May–Jun. Seasonally moist areas, openings in pine forests; of conservation concern; 0–200 m; Calif.

The present circumscription of *Rosa pinetorum*, restricted to the central California coast, is narrower than that of historical usage (for example, W. L. Jepson 1909–1943, vol. 2; L. Abrams and R. S. Ferris 1923–1960, vol. 2; P. A. Munz 1959), in which the name was applied to dwarf roses throughout the Coast Ranges and Sierra Nevada of California that lacked the conspicuous stipitate hypanthial glands that characterize *R. spithamea*. Here, the Sierra Nevada populations are treated as *R. bridgesii*.

Coast Range populations are more problematic; the eglandular characteristic occurs sporadically throughout the range of *Rosa spithamea*, possibly as a result of hybridization with other species; R. F. Hoover (1966) called such plants forma *pinetorum* (A. Heller) Hoover. Here, however, the epithet is restricted to relatively uniform populations occurring on coastal terraces in open forests of *Pinus muricata* and *P. radiata* on the Monterey Peninsula in central California. Given that these populations are well removed from the nearest known populations of *R. spithamea* and that Monterey Peninsula has other highly localized endemic species, *R. pinetorum* is tentatively retained as a distinct species. Plants assigned to this reduced circumscription clustered with *R. gymnocarpa* in a molecular phylogenetic analysis (A. Bruneau et al. 2007); whether this is indicative of close relationship or introgression is unresolved.

Based on pollen size and shape from the type specimen of *Rosa pinetorum*, E. W. Erlanson (1931) concluded that it is tetraploid ($2n = 28$). This is consistent with flow cytometry data, which indicated either triploid or tetraploid status (A. Bruneau et al., unpubl.), but not with chromosome counts using the same plant, which clearly showed a diploid $2n = 14$ (D. Zlesak and W. H. Lewis, unpubl.). Erlanson (1934) used the name *R. yainacensis* to encompass *R. pinetorum* in the strict sense and putatively tetraploid populations from Fort Ross, California, distinct from the diploid *R. calvaria*, a synonym of *R. bridgesii.*

32. Rosa spithamea S. Watson in W. H. Brewer et al., Bot. California 2: 444. 1880 • Coast ground rose [E] [F]

Rosa granulata Greene; *R. sonomensis* Greene; *R. spithamea* var. *sonomensis* (Greene) Jepson

Subshrubs, forming open colonies. **Stems** erect, 1–5(–10) dm, openly branched; bark dark reddish brown, glabrous; infrastipular prickles paired, erect to slightly curved, largest sometimes ± flattened, subulate, 2–8(–12) × 2–5 mm, <base glabrous>, internodal prickles absent or sparse to dense, mixed with aciculi, aciculi sometimes absent, stipitate-glandular or eglandular. **Leaves** 3–10 cm; stipules 6–15 × 1.5–2.5(–3.5) mm, auricles flared to ± erect, 1.5–6 mm, margins entire, stipitate-glandular, surfaces glabrous, eglandular; petiole and rachis with pricklets, usually glabrous, rarely finely hairy, stipitate-glandular; leaflets 5–7(–9), terminal: petiolule 3–13 mm, blade ovate-elliptic, sometimes obovate-elliptic, 10–30 × 8–20 mm, membranous to ± leathery, base obtuse, margins ± 2-serrate, teeth 8–20 per side, acute to obtuse, gland-tipped, apex obtuse to ± truncate, abaxial surfaces pale green to green, glabrous, sometimes sparsely hairy, sessile-glandular or eglandular, adaxial green, dull, rarely sparsely hairy. **Inflorescences** corymbs, 1–10-flowered. **Pedicels** erect, slender to stout, 2–15 mm, glabrous, rarely finely pubescent, stipitate-glandular, sometimes setose-glandular; bracts 1 or 2, lanceolate, 2–15 × 1.5–5 mm, margins entire, stipitate-glandular, surfaces glabrous, eglandular. **Flowers** 2.5–3.8 cm diam.; hypanthium ± ovoid, 3–6 × 2.5–5 mm, glabrous, sparsely to densely stipitate- to setose-glandular, rarely eglandular, neck 0.5–1 × 2.5–4 mm; sepals spreading, ovate-lanceolate, 8–15 × 2.5–3.5 mm, tip 0–4 × 1.3 mm, margins entire, abaxial surfaces glabrous, stipitate-glandular; petals single, deep pink, 12–23 × 15 mm; carpels 15–30, styles exsert 1 mm beyond stylar orifice (2 mm diam.) of hypanthial disc (3.5 mm diam.). **Hips** scarlet, ± depressed-subglobose, 7–15 × 7–15 mm, fleshy, glabrous, sparsely to densely stipitate- to setose-glandular, rarely eglandular, neck 0.5–2 × 2–5 mm; sepals persistent, erect. **Achenes** basiparietal, 4–17, cream to pale brown, 3.5–5 × 2.5–3.5 mm. $2n = 14$.

Flowering May–Aug. Edges and understory of chaparral and mixed forests, post-burn openings; 200–2000 m; Calif., Oreg.

The combination of miniature size and stipitate- to setose-glandular hips makes *Rosa spithamea* one of the more easily and reliably identified species in the genus, although the density of hypanthial glands can vary. The species occurs from Douglas County, Oregon, to San Luis Obispo County, California. *Rosa spithamea* is fire-adapted and blooms profusely only after fires or equivalent disturbances; at other times, it persists in the understory in a vegetative state.

E. W. Erlanson (1934) referred to *Rosa spithamea* as tetraploid ($2n = 28$); twice in her key to *R. spithamea* she questioned that count.

33. Rosa bridgesii Crépin ex Rydberg, Bull. Torrey Bot. Club 44: 83. 1917, name conserved • Sierran ground rose [E] [F]

Rosa gymnocarpa Nuttall var. *pubescens* S. Watson; *R. spithamea* S. Watson var. *solitaria* L. F. Henderson; *R. spithamea* var. *subinermis* Engelmann

Subshrubs, forming open colonies. **Stems** erect, (1–)2–4(–8) dm, openly branched; bark often glaucous when young, gray to dark brown with age, glabrous; infrastipular prickles paired, erect, largest ± flattened, subulate, 3–10 × 2–7 mm, <base glabrous>, internodal rare or absent, smaller, mixed with aciculi, eglandular. **Leaves** 2.5–11 cm; stipules 5–10 × 2–4 mm, auricles flared, 1–3 mm, margins entire, usually sessile-glandular, surfaces glabrous, stipitate-glandular; petiole and rachis with sparse pricklets, glabrous or densely fine hairy <hairs to 0.2 mm>, densely stipitate-glandular; leaflets (3–)5–7, terminal: petiolule 3–20 mm, blade obovate or widely elliptic-cuneate to ± orbiculate, 10–30(–50) × 8–25(–30) mm, membranous to ± leathery, base rounded-obtuse, margins 2+-crenate, teeth 8–12 per side, obtuse, gland-tipped, apex obtuse to truncate, abaxial surfaces gray-green, finely puberulent, sometimes sparsely stipitate-glandular, adaxial green, dull, finely puberulent to glabrate. **Inflorescences** corymbs, 1 or 2(–7)-flowered. **Pedicels** erect, relatively slender, 5–17 mm, glabrous, eglandular or stipitate-glandular; bracts 1–3, lanceolate, 3–10 × 1–6 mm, margins entire, stipitate-glandular, surfaces glabrous, eglandular. **Flowers** 2.5–3.5 cm diam.; hypanthium ovoid-globose to widely urceolate, 4–5 × 3–4 mm, glabrous, rarely sparsely setose, usually eglandular, neck 0.5–1(–2) × 1.5–3 mm; sepals spreading, lanceolate, 6–14 × 2–3 mm, tip 0–4 × 0.5 mm, margins entire, abaxial surfaces puberulent, usually stipitate-glandular;

petals single, deep pink, 10–20 × 10–20 mm; carpels 10–30, styles exsert 0.5–1 mm beyond stylar orifice (1–1.5 mm diam.) of hypanthial disc (2–4 mm diam.). **Hips** scarlet, ovoid to depressed-globose to pyriform, (8–)10–18 × 7–14 mm, fleshy, glabrous, rarely sparsely setose distally, eglandular, neck 1 × 2.5–4.5 mm; sepals persistent, erect. **Achenes** basiparietal, 1–11, cream to pale brown, 3.5–5(–6.5) × 2.5–4 mm. $2n$ = 14, 28.

Flowering May–Aug. Open forest floors, meadow edges, rocky outcrops, midmontane forests, shade to sun; 700–2500 m; Calif., Oreg.

As here circumscribed, *Rosa bridgesii* is the Sierran-Cascade counterpart of *R. spithamea*, encompassing previous references to *R. pinetorum* and *R. spithamea* from the Sierra Nevada and most applications of *R. yainacensis*. The species is the low-growing, openly rhizomatous rose of midmontane forests in the Sierra Nevada and southern Cascade Range of California and Oregon. Finely puberulent leaflets help distinguish non-fruiting *R. bridgesii* from sympatric *R. gymnocarpa*, which has glabrous leaflets. Plants from Oregon differ from Sierran plants in that prickles are less likely to be infrastipular only.

In the phylogenetic analysis by A. Bruneau et al. (2007), *Rosa bridgesii* and *R. spithamea* comprise a distinct clade more closely related to Asian species than to others from western North America.

Rosa bridgesii has been conserved against *R. calvaria* Greene, *R. covillei* Greene, *R. crenulata* Greene, *R. myriadenia* Greene, and *R. yainacensis* Greene (B. Ertter 2007b). Only *R. calvaria* and *R. crenulata* are unambiguous synonyms of *R. bridgesii*; the taxonomic identity of the other names remains unresolved and may involve hybridization with other species.

On the basis of pollen size, E. W. Erlanson (1931) believed that material falling within the current circumscription of *Rosa bridgesii* was tetraploid ($2n$ = 28). In a subsequent paper (1934), she reported a diploid somatic count ($2n$ = 14) for *R. calvaria*, a synonym of *R. bridgesii*. Recent flow cytometry indicates a tetraploid condition for the sampled specimen of *R. bridgesii* (A. Bruneau et al., unpubl. data). It is possible that both ploidy levels are present within the species.

a5. Rosaceae Jussieu (subfam. Rosoideae) tribe Potentilleae Sweet, Brit. Fl. Gard. 2: sub plate 124. 1825 (as Potentillae)

Luc Brouillet

Herbs, perennial, rarely annual or biennial, shrubs, or subshrubs; unarmed. **Leaves** alternate, rarely opposite, pinnately (palmately) compound (simple in *Alchemilla*, *Aphanes*, and *Chamaerhodos*); stipules persistent (absent in *Chamaerhodos*), adnate to petiole; venation pinnate or palmate. **Flowers:** perianth and androecium perigynous; epicalyx bractlets present, sometimes absent; hypanthium usually patelliform, cupulate, or campanulate, sometimes turbinate, saucer-shaped, flat-bottomed, or subglobose to ellipsoid or ovoid; torus flat to conic or turbinate, enlarged (absent or reduced in *Alchemilla*, *Aphanes*, and *Chamaerhodos*); carpels 1–260, styles basal or lateral to subterminal, distinct; ovules 1(or 2), basal. **Fruits** aggregated achenes (achenes in *Alchemilla* and *Aphanes*); <torus sometimes fleshy>; styles deciduous or persistent, not elongate.

Genera 14–22, species ca. 860 (14 genera, 189 species, including 1 hybrid, in the flora): North America, Mexico, West Indies, Bermuda, Central America, South America, Eurasia, Africa, Atlantic Islands, Pacific Islands, Australia.

The base chromosome number for Potentilleae is mostly x = 7 (8 in *Alchemilla* and *Aphanes*; 14 in *Comarum*).

Variation in the number of genera recognized in Potentilleae is due to differences in generic delimitation between D. Potter et al. (2007) and the authors of *Potentilla* and segregates here (see 9. *Ivesia* and 8. *Potentilla* for discussion). In the former, *Duchesnea*, *Horkelia*, *Horkeliella*, and *Ivesia* are included within *Potentilla*. Likewise, *Aphanes* is included within *Alchemilla* by Potter et al. while it is kept distinct here.

Potentilla and its segregates and *Fragaria* are host to *Phragmidium* rusts, but not the other genera of the tribe.

SELECTED REFERENCES Keck, D. D. 1938[1939]. Revision of *Horkelia* and *Ivesia*. Lloydia 1: 75–142. Rydberg, P. A. 1898. A Monograph of the North American Potentilleae. New York. [Mem. Dept. Bot. Columbia Coll. 2.] Soják, J. 2008. Notes on *Potentilla* XXI. A new division of the tribe Potentilleae (Rosaceae) and notes on generic delimitations. Bot. Jahrb. Syst. 127: 349–358.

1. Shrubs; leaf lobe margins entire; achenes hirsute . . . . 16. *Dasiphora*, p. 295
1. Herbs, perennial, sometimes annual or biennial, or subshrubs; leaf lobe margins or apices ± toothed, sometimes entire; achenes glabrous (sometimes ± hairy).
    2. Petals 0, sepals 4; achenes 1, enclosed in dry, urceolate or subglobose to ellipsoid or ovoid hypanthia.
        3. Herbs perennial; leaves basal, blades reniform to orbiculate, palmately lobed, sometimes palmately compound; stamens 4 . . . . 20. *Alchemilla*, p. 302
        3. Herbs annual; leaves cauline, blades cuneate, deeply divided into segments, each lobed; stamens 1(or 2) . . . . 21. *Aphanes*, p. 309
    2. Petals and sepals usually 5; achenes 1–260, usually aggregated (sometimes on elongating tori), usually in (± open) patelliform, cupulate, campanulate, or turbinate hypanthia (not enclosed in dry hypanthium).
        4. Leaves all or mostly basal or proximal (if cauline, deeply pinnatifid), ternate or 2–4-ternate (sometimes simple and coarsely toothed apically in *Sibbaldia*).
            5. Tori becoming red and fleshy in fruit; leaf margins serrate to crenate.
                6. Leaves ± doubly serrate or crenate; stolons leafy; inflorescences: flowers solitary, axillary at stolon nodes; petals yellow . . . . 12. *Duchesnea*, p. 272
                6. Leaves serrate to crenate; stolons not leafy; inflorescences 1–10-flowered cymes, axillary from leaf rosettes; petals usually white . . . . 13. *Fragaria*, p. 274
            5. Tori hemispheric (not enlarged or fleshy) in fruit or absent; leaf margins entire or (2 or)3(–5)-toothed apically.
                7. Leaves pinnately compound or simple and deeply pinnatifid, margins entire, stipules absent . . . . 14. *Chamaerhodos*, p. 279
                7. Leaves ternate, margins toothed apically, stipules persistent.
                    8. Petals ± yellow; stamens 5 . . . . 17. *Sibbaldia*, p. 297
                    8. Petals usually white; stamens 20(–30) . . . . 18. *Sibbaldiopsis*, p. 299
        4. Leaves basal or cauline, the latter usually reduced distally, odd-pinnate to palmate, rarely ternate or ± bipinnate.
            9. Petals deep red to purple, rarely pink, shorter than sepals; tori enlarged and spongy at maturity; horizontal stems sometimes floating, wetland habitats . . . . 19. *Comarum*, p. 300
            9. Petals yellow to white, rarely pink or red (then equal to or longer than sepals); tori not enlarged and spongy at maturity; stems erect to decumbent, not horizontal or floating even if in wetlands.
                10. Anthers dehiscing by continuous marginal slit (with a single theca); styles sub-basal . . . . 15. *Drymocallis*, p. 280
                10. Anthers dehiscing longitudinally; styles subterminal to lateral.
                    11. Hypanthium patelliform to campanulate or cupulate to turbinate (not flat-bottomed); filaments not forming tube; petals white to yellow, sometimes reddish or pink tinged.

12. Plants not aromatic; leaves ± cordate or reniform to narrowly elliptic in outline, leaflets 3–15(–41); petals oblanceolate or obovate to obcordate to nearly round, rarely elliptic; carpels 3–260 .... 8. *Potentilla*, p. 121
12. Plants often aromatic; leaves planar to cylindric, leaflets (3–) 7–161; petals linear or narrowly oblanceolate to obovate, sometimes obcordate; carpels 1–20(–40) .... 9. *Ivesia*, p. 219

[11. Shifted to left margin.—Ed.]

11. Hypanthium ± cupulate or bluntly campanulate and flat-bottomed; filaments forming tube; petals usually white, sometimes pink-tinged, rose-veined, or cream.
    13. Stamens 10; leaflets (3–)5–41 .... 10. *Horkelia*, p. 246
    13. Stamens 20; leaflets 30–70 .... 11. *Horkeliella*, p. 270

8. POTENTILLA Linnaeus, Sp. Pl. 1: 495. 1753; Gen. Pl. ed. 5, 219. 1754 • Cinquefoil, potentille [Latin *potens*, powerful, and *-illa*, diminutive, alluding to supposed medicinal qualities of some species]

Barbara Ertter Reidar Elven

James L. Reveal David F. Murray

Herbs, perennial, rarely annual or biennial, rosetted to densely matted, 0.1–7(–10) dm, sparsely to densely hairy <hairs straight, crisped, and/or cottony>, sometimes glabrate, sometimes glandular; usually short-rhizomatous, sometimes stoloniferous. **Stems** 1–many, prostrate to erect, sometimes rooting at nodes, <usually ± green, sometimes reddish>. **Leaves** winter marcescent or persistent, primarily basal or cauline, <cauline (0–)1–9, reduced or well developed>, alternate, rarely opposite, ternate, palmate, subpalmate, or odd-pinnate, sometimes ± bipinnate; stipules persistent, basally adnate to petiole, linear to ovate, margins entire or lobed, sometimes toothed; petiole present; blade ± cordate or reniform to narrowly elliptic in outline, 0.5–25(–30) cm, foliaceous, rarely ± coriaceous; leaflets 3–15(–41), <terminal sometimes confluent with distalmost lateral ones, separate to overlapping>, narrowly oblanceolate to obovate, elliptic, oblong, cuneate, or flabellate in outline, margins flat or revolute, toothed to divided nearly to base into linear to oblanceolate lobes, sometimes entire, venation pinnate or palmate. **Inflorescences** terminal, sometimes axillary <on stolons>, 1–100+-flowered, often cymose, sometimes solitary or racemiform, open to ± congested; bracts present, usually ± reduced; bracteoles absent. **Pedicels** present, <straight or becoming ± recurved in fruit>. **Flowers** (5–)8–20(–26) mm diam.; epicalyx bractlets (4 or)5(–10); hypanthium patelliform to cupulate, rarely turbinate, 0.5–2.5(–5) × (1.5–)2–7(–10) mm; sepals (4 or)5(–10), spreading at anthesis, lanceolate to broadly ovate or deltate; petals (4 or)5(–10), pale to bright yellow, less often dark reddish, reddish orange, or white, oblanceolate or obovate to most commonly obcordate to nearly round, rarely elliptic (sect. *Pentaphylloides*); stamens (5–)20(–30), shorter than petals; <filaments not forming tube>; torus ± conic; carpels 3–260, glabrous, rarely sparsely hairy, styles usually subapical, rarely lateral; ovule 1. **Fruits** aggregated achenes, individually deciduous, 1–260, obliquely ± ovoid, 0.5–2.6 mm, glabrous or rarely ± hairy apically or on style scar; hypanthium persistent; sepals persistent, erect; styles tardily deciduous, <jointed, filiform to conic, sometimes clavate, often rough-thickened basally to nearly whole length>. $x = 7$.

Species ca. 400 (98, including 1 hybrid, in the flora): North America, Mexico, West Indies, Bermuda, Central America, South America, Eurasia, Africa, Atlantic Islands (Azores, Macaronesia), Pacific Islands (New Guinea, New Zealand), Australia; arctic, temperate, and montane in Northern Hemisphere, mostly montane in Southern Hemisphere and equatorial regions.

Potentilla as presented here has a circumscription significantly reduced from treatments in most recent floras, since the convergence of morphologic (J. Soják 1985[1989], 2008) and molecular (T. Eriksson et al. 1998, 2003; C. Dobeš and J. Paule 2010; M. H. Töpel et al. 2011) interpretations has justified the segregation of *P. fruticosa*, *P. glandulosa* and relatives, *P. palustris*, and *P. tridentata* into *Dasiphora*, *Drymocallis*, *Comarum*, and *Sibbaldiopsis*, respectively (A. Kurtto and Eriksson 2003; Ertter 2007). The generic assignment of *P. anserina* is problematic (Soják 2010); the species is retained here within *Potentilla*. The senior author has chosen to maintain *Duchesnea*, *Horkelia*, *Horkeliella*, and *Ivesia* as distinct genera. As a result, the generic arrangement used here nearly recapitulates that used by P. A. Rydberg (1898, 1908d), representing the last comprehensive treatment of Potentilleae in North America.

Species circumscriptions and nomenclature are also significantly revised, reflecting not only the first continental assessment of the genus since Rydberg but also the need to synthesize American and Eurasian approaches to the genus. Whereas the revision by P. A. Rydberg (1908d) has served as the basis for subsequent floristic works in North America, subject to significant lumping of species and realignment of genera (for example, D. D. Keck 1938; J. Clausen et al. 1940), Eurasians generally have used as their starting point a monograph by T. Wolf (1908), a contemporary of Rydberg. Rydberg was handicapped by limited access to type material in Europe, resulting in the misapplication of several names.

Even with the reduced generic circumscription and improved global perspective, *Potentilla* remains challenging. Key contributing factors are facultative pseudogamous agamospermy and hybridization (summarized by J. Clausen et al. 1940; B. Eriksen 1996), which result in everything from ephemeral F_1 hybrids to fully stabilized species of hybrid origin, sometimes across sectional boundaries. In the face of such biosystematic complexity, this treatment uses the operational species concept for arctic species outlined by R. Elven et al. (1999), with the exception that variety has been used as the primary infraspecific rank for temperate species.

Several species have been recently described or resurrected from synonymy for use in the current treatment, and anomalous collections examined during this preparation probably represent additional novelties. The isolated mountain ranges and complex geology of the North American Cordillera in particular contains a rich trove of unresolved diversity. Collectors should be alert for sympatric species in these situations, and the possible hybrid origin of aberrant individuals or clones.

Sections that have been established for North American *Potentilla* [for example, by A. Nelson (in J. M. Coulter and Nelson 1909), O. A. Stevens 1959, B. C. Johnston 1985] are mostly based on a subset of validly published but unranked groups by P. A. Rydberg (1898, 1908). Species numbers for primarily Eurasian sections have been provided by J. Soják (pers. comm.). The sectional arrangement used here is both pragmatic and tentative; it also represents an implicit phylogenetic hypothesis. The arrangement reflects molecular analyses (C. Dobeš and J. Paule 2010; M. H. Töpel et al. 2011), in which the first four sections correspond to well-supported clades that diverge basal to the ivesioid clade and the remaining sections. The majority of species beyond the first six sections are mingled in a largely unresolved crown clade, informally termed the *Potentilla* core group by Dobeš and Paule, and the *Argentea* clade by Töpel et al. Their results also suggest an Asian origin for *Potentilla* in the Oligocene, multiple colonizations in North America, and rapid radiation of the core group beginning in the Pliocene.

Detailed morphology and anatomy in *Potentilla* are provided by P. A. Rydberg (1898) and T. Wolf (1908). With some exceptions, floral morphology is of less taxonomic value than are leaf division and leaflet incision, particularly of basal leaves, which renders specimens without them often impossible to identify correctly. Habit and inflorescence architecture are also important features, such that specimens with fully developed inflorescences are typically more confidently identifiable than those in early anthesis.

Vestiture, especially of petioles and abaxial leaflet surfaces, is of critical diagnostic importance. Four primary hair groups are recognized by B. Eriksen and B. A. Jurtzev (1999): straight (or curved), crispate (short, twisted hairs, called crisped in the current treatment), floccose (corresponding to cottony in FNA terminology), and glandular (mostly short-stipitate). We further divide straight hairs into short hairs 0.2 mm, which sometimes form an even underlayer to other hair types, and long hairs for all other straight to curved hairs. Because these five groups can occur in any combination, all are addressed separately in descriptions. Hair density is categorized as sparse, common (or moderately), abundant, or dense if obscuring the surface; rigidity of long hairs is categorized as soft, weak, or stiff.

In keys and descriptions, stem length includes inflorescence, hairs are eglandular, and leaves are basal unless otherwise indicated. Rosetted plants have a basal rosette of leaves from a largely unbranched caudex; tufted plants have multiple stems from a short-branched caudex; and matted plants have dense, spreading, multi-branched caudices. Stolons and stoloniferous are used for plants that sprawl on the ground with apparently solitary flowers arising from nodes of flagelliform stems, even though these may be more homologous to prostrate cymose inflorescences than true stolons. Flagelliform stems form short shoots at the nodes but usually are not otherwise branched. In plants with stems arising centrally from ephemeral basal rosettes, new basal rosettes for the following year often appear in fall and might be confused with persistent rosettes. The diagnostic character of marcescent whole leaves (in sects. *Niveae* and *Rubricaules*) is in addition to sheathing stipular leaf bases and petioles, which are also commonly present on plants lacking marcescent whole leaves. Transitions between the extremes of strictly palmate and strongly pinnate leaves are roughly quantified as a fraction of the leaf axis (petiole plus rachis) bearing leaflets. Cauline leaves consist of foliar and bractlike structures on the stem proximal to the proximalmost branch. Leaflet dimensions and teeth number are of the larger leaflets (central in palmate leaves, terminal or distal in pinnate leaves). Leaflet tooth length is measured on the distal edge of the largest and medium-sized teeth. The central portion of the leaflet between opposite tooth sinuses is called the undivided medial blade. Hairs projecting beyond the tips of the leaflet teeth are referred to as apical tufts. Pedicel length is measured from the flower to the subtending bract, even if this appears to be a mid-pedicel bract. Because pedicels at the proximal nodes are often significantly longer than elsewhere, pedicel length is primarily for the nonproximal nodes, with the proximalmost extremes noted separately. Curved pedicels are those that commonly become sigmoidally recurved in fruit; this is characteristic of sections with predominately prostrate inflorescences. Measurements of pedicels, epicalyx bractlets, hypanthia, and sepals are taken at flowering and fruiting but not in bud. Hypanthium diameter is an external measurement of pressed flowers. Although not otherwise indicated, petals of most species commonly have a basal patch of more intense color, which often coincides with ultraviolet nectar guides (N. Naruhashi and H. Ikeda 1999). Style features are based primarily on dried mature styles that are tardily deciduous. Rugose achenes are often smooth until mature.

SELECTED REFERENCES Dobeš, C. and J. Paule. 2010. A comprehensive chloroplast DNA-based phylogeny of the genus *Potentilla* (Rosaceae): Implications for its geographic origin, phylogeography and generic circumscription. Molec. Phylogen. Evol. 56: 156–175. Rydberg, P. A. 1908d. *Potentilla*. In: N. L. Britton et al., eds. 1905+. North American Flora.... 47+ vols. New York. Vol. 22, 293–388.

1. Petals white or dark reddish to reddish orange.
   2. Petals dark reddish to reddish orange (see also 19. *Comarum*, *Ivesia multifoliolata* [see p. 244]) . . . . 8j. *Potentilla* sect. *Rubrae*, p. 148
   2. Petals white (see also 18. *Sibbaldiopsis*).
      3. Leaves ternate; perennials . . . . 8c. *Potentilla* sect. *Lupinoides*, p. 131
      3. Leaves ± pinnate, sometimes ± bipinnate; biennials, sometimes annuals or short-lived perennials . . . . 8e. *Potentilla* sect. *Arenicolae*, p. 136
1. Petals pale to bright yellow, rarely cream.
   4. Plants stoloniferous; stems flagelliform, becoming prostrate, rooting at some nodes; inflorescences solitary flowers at stolon nodes (see also 12. *Duchesnea*).
      5. Leaves pinnate, usually white at least abaxially, sometimes green . . . . 8a. *Potentilla* sect. *Pentaphylloides*, p. 126
      5. Leaves ternate or palmate, usually ± green on both surfaces, sometimes silvery white abaxially . . . . 8d. *Potentilla* sect. *Potentilla* (in part), p. 132
   4. Plants not stoloniferous; stems not flagelliform, usually decumbent to erect, sometimes prostrate or pendent, not rooting at nodes; inflorescences usually cymose, sometimes racemiform or solitary flowers.
      6. Leaves ternate, appearing ± palmate, central leaflets incised nearly to base into 3 lobes, lateral ones incised nearly to base into 2 lobes, margins strongly revolute; Beringia . . . . 8b. *Potentilla* sect. *Biflorae*, p. 130
      6. Leaves ternate or palmate to pinnate, leaflets incised 1/10 (or less) to nearly to midrib into (0 or)1–15+ teeth, margins flat or revolute; North America including Beringia.
         7. Styles 0.5–1.3(–1.5) mm, conic-tapered to columnar-tapered or columnar-clavate, rarely ± filiform.
            8. Stems from centers of ephemeral basal rosettes, primary leaves cauline; primarily ± disturbed or seasonally wet habitats.
               9. Annuals, biennials, or short-lived perennials; leaves ternate or palmate to pinnate; petals (1–)1.5–5 mm; styles slightly to strongly papillate-swollen ± whole length, 0.5–0.8 mm; primarily seasonally wet habitats . . . . 8f. *Potentilla* sect. *Rivales*, p. 137
               9. Perennials; leaves usually palmate, rarely ternate; petals (2–)2.5–10 (–13) mm; styles not or ± papillate-swollen in proximal 1/5–1/2, 0.6–1.5 mm; primarily ± disturbed habitats.
                  10. Flowers 4(or 5)-merous; leaflets green abaxially . . . . 8d. *Potentilla* sect. *Potentilla* (in part), p. 132
                  10. Flowers 5-merous; leaflets green to white abaxially.
                      11. Leaflets green abaxially, cottony and/or crisped hairs absent; petioles: long hairs stiff, spreading; hypanthia 5–9 mm diam. . . . . 8g. *Potentilla* sect. *Rectae*, p. 141
                      11. Leaflets green to white abaxially, crisped and/or cottony hairs absent or sparse to dense; petioles: long hairs soft to weak, loosely appressed to spreading; hypanthia 2–5 mm diam. . . . . 8h. *Potentilla* sect. *Terminales*, p. 143
            8. Stems lateral to persistent basal rosettes, primary leaves basal; primarily arctic, alpine, and prairie habitats.
               12. Leaves pinnate to subpalmate.
                   13. Styles papillate-swollen only at very base if at all; inflorescences ± congested to very open, sometimes solitary flowers.
                       14. Stems (0.2–)0.4–1(–1.5) dm; leaflets incised 1/2–3/4 to midvein, teeth 2–4(or 5) per side; inflorescences (1 or)2–5(–10)-flowered; -flowered; Rocky Mountains . . . . 8m. *Potentilla* sect. *Subjugae* (in part), p. 165
                       14. Stems 1–5 dm; leaflets incised nearly to midvein, teeth (3 or)4–7 per side; inflorescences 2–40-flowered; Alaska to Quebec . . . . 8u. *Potentilla* sect. *Pensylvanicae* (in part), p. 211

13. Styles papillate-swollen in proximal ⅕–¾+; inflorescences congested to ± open, sometimes solitary flowers.
    15. Leaflets on up to distal ⅙(–¼) of leaf axis, usually grayish white to white abaxially, sometimes gray, cottony (or cottony-crisped) hairs abundant to dense ..... 8t. *Potentilla* sect. *Rubricaules* (in part), p. 205
    15. Leaflets on distal ⅙–½+, rarely less, of leaf axis, green to white abaxially, cottony hairs absent or sparse to dense..... 8u. *Potentilla* sect. *Pensylvanicae* (in part), p. 211

12. Leaves ternate or palmate.
    16. Leaflets gray to white abaxially, sometimes yellowish white or reddish, cottony(/crisped) hairs abundant to dense.
        17. Leaflets 3, rarely more, on all basal leaves ....8s. *Potentilla* sect. *Niveae*, p. 195
        17. Leaflets 3–5(–7), usually more than 3 on at least some leaves ..... 8t. *Potentilla* sect. *Rubricaules* (in part), p. 205
    16. Leaflets pale to dark green abaxially, sometimes reddish, grayish, or brownish, cottony hairs absent.
        18. Leaves ternate; leaflets 3 .... 8r. *Potentilla* sect. *Aureae* (in part), p. 188
        18. Leaves usually palmate; leaflets (3–)5–11.
            19. Stem lengths usually 2–5 times basal leaves, rarely shorter than leaves; leaflet teeth 1–10(–13) per side; glands sometimes red .... 8i. *Potentilla* sect. *Chrysanthae*, p. 146
            19. Stem lengths 1–3 times basal leaves; leaflet teeth 2–4 (or 5) per side; glands not red ... 8r. *Potentilla* sect. *Aureae* (in part), p. 188

[7. Shifted to left margin.—Ed.]

7. Styles (1–)1.5–3(–3.5) mm, filiform to tapered, rarely columnar.
    20. Stems usually prostrate to decumbent, sometimes ascending (especially in supporting vegetation) or pendent, lengths ½–3(–5) times basal leaves; pedicels often recurved in fruit, proximal usually not much longer than distal; plants usually rosetted, sometimes tufted or ± matted; taproots sometimes fleshy-thickened.
        21. Cottony hairs sparse to dense, at least on abaxial leaflet surfaces.
            22. Leaves pinnate; leaflets on distal ½–⅔(–¾) of leaf axis, surfaces ± similar, adaxial hairs soft .... 8n. *Potentilla* sect. *Multijugae* (in part), p. 167
            22. Leaves palmate to pinnate; leaflets usually on tip or to distal ½ of leaf axis, or, if more, then surfaces ± to strongly dissimilar, adaxial hairs stiff ..... 8o. *Potentilla* sect. *Concinnae* (in part), p. 176
        21. Cottony hairs absent.
            23. Leaves pinnate .... 8n. *Potentilla* sect. *Multijugae* (in part), p. 167
            23. Leaves ternate or palmate to subpalmate.
                24. Petioles: straight hairs ± appressed, glands usually absent or sparse, rarely abundant .... 8o. *Potentilla* sect. *Concinnae* (in part), p. 176
                24. Petioles: straight hairs usually spreading to ascending, sometimes loosely appressed, glands usually abundant, sometimes sparse ..... 8p. *Potentilla* sect. *Subviscosae*, p. 182
    20. Stems decumbent to erect, rarely prostrate, lengths (1–)2–4(–5) times basal leaves; pedicels straight in fruit, proximal often much longer than distal; plants usually tufted, sometimes cushion-forming or matted, rarely rosetted; taproots not fleshy-thickened.
        25. Leaves ternate; leaflets 3 .... 8r. *Potentilla* sect. *Aureae* (in part), p. 188
        25. Leaves usually palmate to pinnate; leaflets (3–)5–15(–41).
            26. Leaves usually palmate to subpalmate, leaflets on tip or less than distal ⅒ of leaf axis.
                27. Styles (1–)1.5–3 mm; basal leaves 2-ranked or not, surfaces green to white, sometimes blue-green, sometimes glaucous; North America and Greenland, especially mid-montane w North America ..... 8k. *Potentilla* sect. *Graciles* (in part), p. 150
                27. Styles 1–1.6 mm; basal leaves 2-ranked, surfaces green, not glaucous; e Canada and Greenland, mostly coastal .... 8r. *Potentilla* sect. *Aureae* (in part), p. 188

[26. Shifted to left margin.—Ed.]

26. Leaves pinnate to subpinnate, or palmate with additional lateral leaflets; leaflets on distal 1/10 to nearly whole leaf axis.
    28. Leaves usually palmate with additional lateral leaflets, sometimes pinnate; leaflets (3–)5 at tip of leaf axis plus 1(or 2) additional pairs, distal leaflets distinct, surfaces strongly dissimilar; petioles: long hairs spreading on first-formed leaves, tightly appressed to ascending on later-formed leaves; 3400–4000 m . . . . 8m. *Potentilla* sect. *Subjugae* (in part), p. 165
    28. Leaves pinnate to subpinnate with 3 apical leaflets and usually more than 2 additional pairs, distal leaflets often confluent, surfaces similar to strongly dissimilar; petioles: long hairs absent or weak to stiff, tightly appressed to spreading; 300–3800 m.
        29. Leaflets (3–)5–7(–9) on distal 1/10–1/2(–3/4) of leaf axis, surfaces ± similar, cottony hairs usually absent.
            30. Stems (1–)1.5–5(–6) dm; glands usually absent or sparse, rarely abundant; basal leaves (2–)5–25(–28) cm; leaflets ± obovate to cuneate; petals 5–12 mm . . . . 8k. *Potentilla* sect. *Graciles* (in part), p. 150
            30. Stems (0.2–)0.5–1.5(–2.5) dm; glands common to abundant; basal leaves 2–7 cm; leaflets ± flabellate; petals 3.5–6.5 mm . . . . 8q. *Potentilla* sect. *Brevifoliae*, p. 187
        29. Leaflets (5–)7–21(–41) on distal (1/6–)1/5–2/3 to whole rachis, surfaces similar to strongly dissimilar, cottony hairs often present.
            31. Leaflets usually ± evenly incised 1/4–1/2 or less to midvein, sometimes entire; stems ascending to erect, (0.3–)1.5–7(–8) dm; proximal pedicels usually much longer than distal; Canadian prairies to s Nevada, Arizona, and New Mexico . . . . 8l. *Potentilla* sect. *Leucophyllae*, p. 160
            31. Leaflets evenly or unevenly incised 1/2 to completely to midvein, sometimes entire; stems usually prostrate to decumbent, sometimes ascending, (0.3–)0.5–3(–5) dm; proximal pedicels often not much longer than distal; Rocky Mountains w to California, Oregon, and Washington. . . . . 8n. *Potentilla* sect. *Multijugae* (in part), p. 167

8a. POTENTILLA Linnaeus sect. PENTAPHYLLOIDES Tausch, Hort. Canal. 1: sub *P. ornithopoda*. 1823

Reidar Elven

David F. Murray

Argentina Hill; *Potentilla* sect. *Anserina* (Gaudin) Pfeiffer; *Potentilla* ser. *Anserina* (Gaudin) T. T. Yu & C. L. Li; *Potentilla* subg. *Argentina* (Hill) Jepson

Perennials, openly matted, stoloniferous [to ± tufted, not stoloniferous]; taproots not fleshy-thickened, <often absent and replaced by ± thickened lateral roots>; vestiture primarily of long and crisped-cottony hairs, glands absent or sparse, not red. **Stems** prostrate [to erect], flagelliform [or not], rooting at nodes [or not], lateral to ± persistent basal rosettes, 1–10+ dm, lengths (0.5–)1–5+ times basal leaves. **Leaves:** basal not or ± 2-ranked; cauline (between flowering and/or rosette-forming nodes) 0 or 1; primary leaves pinnate (distal leaflets sometimes confluent), (1–)3–50(–75) cm; petiole: long hairs absent or ascending to weakly appressed, sometimes spreading, soft to weak, glands absent or very sparse; primary leaflets (5–)7–21(–31[–51]) plus additional small leaflets interspersed, on distal (1/2–)2/3–4/5+ of leaf axis, overlapping or not, elliptic to obovate, margins revolute or nearly flat, distal (1/2–)2/3–3/4+ evenly incised 1/4–1/2 to midvein, teeth (2–)5–12(–16) per side, surfaces similar to strongly dissimilar, abaxial usually white, sometimes green, cottony(/crisped) hairs usually dense, sometimes sparse or absent, adaxial green to white, not glaucous, long hairs absent or weak to soft. **Inflorescences** solitary flowers <at stolon nodes>

[± cymose]. **Pedicels** straight in fruit, (1–)2–45(–55) cm. **Flowers** 5(–7)-merous; hypanthium 3–7 mm diam.; petals bright yellow, elliptic to broadly elliptic, (4–)5–15(–20) mm, longer than sepals, apex rounded or slightly retuse; stamens [5–]20(–30); styles ± lateral [subapical or subbasal], narrowly columnar-filiform, not papillate-swollen proximally, 1.5–2.5 mm. **Achenes** slightly rugose and papillate.

Species ca. 50 (1 in the flora): North America, Mexico, South America (Argentina, Chile), Eurasia, Pacific Islands (New Guinea, New Zealand), s Australia.

The use here of sect. *Pentaphylloides* is contrary to the argument by J. Soják (2007) that the use of *Pentaphylloides* Tournefort by G. R. Boehmer (in C. G. Ludwig 1760) qualified as an unranked infrageneric taxon, making sect. *Pentaphylloides* Tausch a later homonym. In that Boehmer was citing synonyms, not infrataxa, sect. *Pentaphylloides* Tausch is the correct name for this section. The names sect. *Chenopotentilla* Focke and sect. *Leptostylae* (Th. Wolf) Guşuleac are superfluous and illegitimate.

Potentilla anserina is the atypical and only representative in North America of a primarily southeastern Asian section. Although provisionally retained within *Potentilla*, the morphological distinctiveness of the section has led to the recognition of *Argentina* as a segregate genus (for example, P. A. Rydberg 1908d; Á Löve and D. Löve 1975b; B. C. Johnston 1985; J. Soják 2010). Molecular analyses (C. Dobeš and J. Paule 2010; M. H. Töpel et al. 2011) have confirmed the section is a strongly supported monophyletic clade, albeit including some other Asian species treated by J. Soják (1994) as the segregate genera *Piletophyllum* (Soják) Soják and *Tylosperma* Botschantzev. Chloroplast markers (Dobeš and Paule; Töpel et al.) place this clade sister to all other Potentilleae; nuclear markers (Töpel et al.) indicate instead a sister relation to Fragariinae.

SELECTED REFERENCES Rousi, A. 1965. Biosystematic studies on the species aggregate *Potentilla anserina* L. Ann. Bot. Fenn. 2: 47–112. Soják, J. 2010. *Argentina* Hill, a genus distinct from *Potentilla* (Rosaceae). Thaiszia 20: 91–97.

1. Potentilla anserina Linnaeus, Sp. Pl. 1: 495. 1753 • Silverweed [F]

Argentina anserina (Linnaeus) Rydberg

Basal leaves: petiole (0.5–)1–15(–25) cm, long hairs absent or sparse to dense, 1.5–3.5 mm; larger leaflets (0.4–)0.5–5(–7) × (0.2–)0.3–2(–3) cm, surfaces: abaxial with long hairs absent or sparse to abundant, not restricted to veins, 0.5–2(–2.5) mm, cottony-crisped hairs usually dense or usually absent (in subsp. *groenlandica*), adaxial with long hairs absent or sparse to abundant, cottony-crisped hairs usually absent, sometimes sparse to common (especially subsp. *yukonensis*). **Flowers:** epicalyx bractlets narrowly to broadly ovate-triangular or linear to elliptic, (2–)2.5–7(–8) × (0.3–)0.5–3(–3.5) mm, often 2-fid or dentate; sepals (3–)3.5–7(–9) mm, apex subacute to acuminate; petals (4–)5–15(–20) × (2.5–)3–10(–12) mm; filaments (1–)2.5–3.5(–4.5) mm, anthers 0.7–1.3 mm; carpels (10–)20–200(–250). **Achenes** 2 mm.

Subspecies 4 or 5 (4 in the flora): North America, Mexico, South America (Argentina, Chile), Eurasia, Pacific Islands (New Guinea, New Zealand), s Australia.

Potentilla anserina is polymorphic in most features, especially in hairiness, but also in size and in degree of dissection of leaflets, epicalyx bractlets, and sepals. While most of the variation described by A. G. Blytt (1906) is taxonomically insignificant, A. Rousi (1965) found support for three northern races: subspp. *anserina*, *egedei*, and *pacifica*. He also suggested that *P. yukonensis* Hultén might qualify as a separate subspecies. This treatment follows J. Soják (1994) in accepting four northern subspecies of *P. anserina*, all present in North America.

A. Rousi (1965) demonstrated partial interfertility between the races of *Potentilla anserina*, which form intermediates in all zones of contact. The distinctness of the four races, in spite of intermediates, is upheld by being partly allopatric and occupying different habitats where they are sympatric. As partly interfertile parapatric entities, they are best treated as subspecies.

The vast majority of chromosome counts are tetraploid. Tetraploid plants ($2n$ = 28) are fully fertile; hexaploids ($2n$ = 42) are largely pollen and seed sterile; pentaploids ($2n$ = 35) are probably occasional hybrids (S. Erlandsson 1942, 1942b; A. Rousi 1965). A. Kurtto et al. (in J. Jalas et al. 1972+, vol. 13) considered numbers above tetraploid level as cases of occasional autopolyploidy.

1. Epicalyx bractlets equal to or longer than sepals, often 2-fid or dentate; leaflet surfaces: abaxial with cottony-crisped hairs dense, long hairs common to abundant, on and between veins, adaxial glabrous or sparsely to densely hairy; achenes with dorsal groove; inland and seashore plants.
   2. Hypanthium patelliform (wider than deep) in fruit; epicalyx bractlets narrowly to broadly ovate-triangular, equal to sepals; most of North America. . . . . 1a. *Potentilla anserina* subsp. *anserina*
   2. Hypanthium turbinate (± as deep as wide) in fruit; epicalyx bractlets usually linear to elliptic, longer than sepals, rarely subequal; interior w Canada and Alaska . . . . 1b. *Potentilla anserina* subsp. *yukonensis*
1. Epicalyx bractlets shorter than sepals, usually entire, rarely 2-fid or dentate; leaflet surfaces: abaxial with cottony-crisped hairs absent or sparse to dense, long hairs absent or sparse, on veins, adaxial usually glabrous, rarely sparsely to densely hairy; achenes without dorsal groove; seashore or near-coastal plants.
   3. Carpels (20–)50–200(–250); leaves (3–)10–50 (–75) cm; leaflets (4 or)5–10(–15) per side, abaxial surfaces densely hairy, teeth (4–) 6–12(–16) per side, teeth apices acute to acuminate, rarely subacute; flowers 1–2.5 (–3.5) cm diam. . . . . 1c. *Potentilla anserina* subsp. *pacifica*
   3. Carpels 25–60; leaves (1–)2–10 cm, rarely longer; leaflets 2–4(or 5) per side, abaxial surfaces glabrous, sometimes sparsely to densely hairy, teeth 2–6(–10) per side, teeth apices rounded to subacute; flowers 0.8–1.5 cm diam. . . . . 1d. *Potentilla anserina* subsp. *groenlandica*

1a. Potentilla anserina Linnaeus subsp. **anserina** • Common silverweed, potentille ansérine

Argentina argentea Rydberg; *Potentilla anserina* var. *sericea* Hayne; *P. pratincola* B. Boivin

Stems densely hairy. **Leaves** horizontal to erect, 3–20 cm, rarely longer; leaflets (5 or) 6–11(–12) per side, overlapping, rarely separate, teeth (4 or)5–8 (–11+) per side, teeth apex acute to acuminate, surfaces: abaxial with long hairs common to abundant, on and between veins, cottony-crisped hairs dense, adaxial glabrous or sparsely to densely hairy. **Flowers** 1–1.5(–2) cm diam.; epicalyx bractlets narrowly to broadly ovate-triangular, equal to sepals, often 2-fid or dentate; hypanthium patelliform (much wider than deep) in fruit; petals overlapping or not, elliptic to broadly elliptic; carpels (10–)20–60. **Achenes** with dorsal groove. **$2n$** = 28, 35, 42.

Flowering spring–fall. Dry meadows, pastures, open dry pine and aspen forests, dry sandy and gravelly stream-, lake-, and seashores, sand dunes, inland alkaline habitats, dry ruderal habitats; 0–3000 m; Greenland; St. Pierre and Miquelon; Alta., B.C., Man., N.B., Nfld. and Labr., N.W.T., N.S., Ont., P.E.I., Que., Sask., Yukon; Alaska, Ariz., Calif., Colo., Conn., Idaho, Ill., Ind., Iowa, Maine, Mass., Mich., Minn., Mont., Nebr., Nev., N.H., N.Mex., N.Y., N.Dak., Ohio, Oreg., R.I., S.Dak., Utah, Vt., Wash., Wis., Wyo.; Mexico; South America (Chile); Eurasia; Australia.

Subspecies *anserina* is the most widespread subspecies, occurring both on seashores and well inland. It is presumed introduced in Greenland and Alaska.

A. Rousi (1965) compared American and European plants of subsp. *anserina* and found no consistent morphologic differences between the continents. An ecologic difference does exist: whereas the European plants are common both in coastal and inland habitats, the American plants are mainly inland. Both native and introduced plants are probably present in North America, as the European plant is a widely established, aggressive weed. Plants with a dense sericeous vestiture on both leaf surfaces (var. *sericea*) are especially characteristic of inland saline habitats.

POTENTILLA

1b. Potentilla anserina Linnaeus subsp. **yukonensis** (Hultén) Soják ex Elven & D. F. Murray, J. Bot. Res. Inst. Texas 2: 441. 2008 • Yukon silverweed [E] [F]

Potentilla yukonensis Hultén, Fl. Alaska Yukon 6: 1033. 1946, based on *Argentina subarctica* Rydberg in N. L. Britton et al., N. Amer. Fl. 22: 354. 1908, not *P. subarctica* Rydberg 1908; *P. anserina* var. *yukonensis* (Hultén) B. Boivin; *P. egedei* Wormskjöld ex Hornemann subsp. *yukonensis* (Hultén) Hultén

Stems sparsely to densely hairy, rarely glabrous. **Leaves** horizontal to erect, 4–20 cm, rarely longer; leaflets (4 or)5–8(or 9) per side, separate, rarely overlapping, teeth (3–)5–9 per side, teeth apex acute to acuminate (more sharply than in other subspp.), surfaces: abaxial with long hairs common to abundant, on and between veins, cottony-crisped hairs dense, adaxial sparsely to densely hairy, rarely glabrous. **Flowers** 1.2–2.2 cm diam.; epicalyx bractlets usually linear to elliptic, longer than sepals, rarely subequal, often 2-fid or dentate; hypanthium turbinate (as deep as wide) in fruit; petals not overlapping, elliptic; carpels 20–60. **Achenes** with dorsal groove.

Flowering summer. Dry sandy and gravelly stream- and lakeshores, dry pastures, open dry pine forests, inland alkaline habitats, road verges and gravelly sites; 0–800 m; Alta., Man., N.W.T., Sask., Yukon; Alaska.

Subspecies *yukonensis* seems to be the predominant subspecies in interior northwestern Canada and Alaska and is perhaps the only native race in interior Northwest Territories, Yukon, and Alaska. The range is probably larger than reported here. While the flowers of other subspecies generally remain erect, those of subsp. *yukonensis* tend to nod in fruit.

A. Rousi (1965) suggested that there might be an additional, densely hairy subspecies (undescribed) in western North America. The northern plants observed by the authors mostly share the distinctive hypanthium character with subsp. *yukonensis* and are assigned here.

1c. Potentilla anserina Linnaeus subsp. **pacifica** (Howell) Rousi, Ann. Bot. Fenn. 2: 104. 1965 • Pacific silverweed [E] [F]

Potentilla pacifica Howell, Fl. N.W. Amer., 179. 1898; *Argentina occidentalis* Rydberg; *A. pacifica* (Howell) Rydberg; *P. anserina* var. *grandis* Torrey & A. Gray; *P. egedei* Wormskjold ex Hornemann subsp. *grandis* (Torrey & A. Gray) Hultén; *P. egedei* subsp. *pacifica* (Howell) L. A. Sergienko; *P. rolandii* B. Boivin

Stems glabrous or sparsely hairy. **Leaves** erect, rarely horizontal or ascending, (3–)10–50(–75) cm; leaflets

(4 or)5–10(–15) per side, separate, rarely overlapping, teeth (4–)6–12(–16) per side, teeth apex acute to acuminate, rarely subacute, surfaces: abaxial with long hairs absent or sparse, on veins, cottony-crisped hairs dense, adaxial glabrous, rarely sparsely to densely hairy. **Flowers** 1–2.5(–3.5) cm diam.; epicalyx bractlets narrowly to broadly ovate-triangular, shorter than sepals, entire, rarely 2-fid or dentate; hypanthium patelliform (wider than deep) in fruit; petals not or rarely overlapping, elliptic; carpels (20–)50–200(–250). **Achenes** without dorsal groove. **2*n*** = 28.

Flowering summer. Salt and brackish marshes, clay and sandy seashores, maritime drift deposits; 0–150 m; B.C., N.B., Nfld. and Labr., N.S., P.E.I., Que.; Alaska, Calif., Conn., Maine, Mass., N.H., N.Y., Oreg., R.I., Wash.

Pacific and Atlantic populations of subsp. *pacifica* are widely separated geographically but are essentially the same morphologically, except for the high number of achenes in some west coast populations.

1d. Potentilla anserina Linnaeus subsp. **groenlandica** Trattinnick, Rosac. Monogr. 4: 13. 1824 [F]

Argentina anserina (Linnaeus) Rydberg subsp. *egedei* (Wormskjold ex Hornemann) Á. Löve & Ritchie; *A. anserina* subsp. *groenlandica* (Trattinnick) Á. Löve; *A. egedei* (Wormskjold ex Hornemann) Rydberg; *Potentilla anserina* subsp. *egedei* (Wormskjold ex Hornemann) Hiitonen; *P. egedei* Wormskjold ex Hornemann; *P. egedei* var. *groenlandica* (Trattinnick) Polunin

Stems glabrous. **Leaves** horizontal or ascending, rarely erect, (1–)2–10 cm, rarely longer; leaflets 2–4(or 5) per side, separate, teeth 2–6(–10) per side, teeth apex rounded to subacute, surfaces: abaxial with long hairs absent or sparse, on veins, cottony-crisped hairs absent, sometimes sparse to dense, adaxial glabrous. **Flowers** 0.8–1.5 cm diam.; epicalyx bractlets narrowly to broadly ovate-triangular, shorter than sepals, entire; hypanthium patelliform (wider than deep) in fruit; petals not overlapping, elliptic; carpels 25–60. **Achenes** without dorsal groove. **2*n*** = 28, 35, 42.

Flowering summer. Salt and brackish marshes, clayey and sandy seashores, driftwalls; 0 m; Greenland; Man., Nfld. and Labr. (Labr.), N.W.T., Nunavut, Ont., Que., Yukon; Alaska; Eurasia.

The northern arctic plants have glabrous leaves and are compatible with *Potentilla egedei*. Some of the southern arctic and northern boreal plants in the Atlantic regions have leaves with white-hairy abaxial surfaces, more leaflets, often more pointed teeth, and slightly larger flowers. These are compatible with subsp. *groenlandica* as originally described. The variation between these extremes is continuous. Both names are based on plants from Greenland, far north of the native range of subsp. *anserina*, and are probably the same entity. Following the argument by J. Soják (1969) that trinomials by Trattinnick are subspecies, the epithet *groenlandica* has priority over *egedei* at this rank.

8b. Potentilla Linnaeus sect. Biflorae (Rydberg) O. Stevens in N. L. Britton et al., N. Amer. Fl. 22(7): 11. 1959

Reidar Elven

David F. Murray

Barbara Ertter

Potentilla [unranked] *Biflorae* Rydberg, Bull. Torrey Bot. Club 24: 13. 1897; *Potentilla* ser. *Biflorae* (Rydberg) Soják; *Potentilla* subg. *Micropogon* Bunge

Perennials, tufted, not stoloniferous; taproots not fleshy-thickened; vestiture of long and crisped hairs, glands absent or sparse, not red. **Stems** ± erect, not flagelliform, not rooting at nodes, lateral to persistent basal rosettes, (0.1–)0.5–1(–1.5) dm, lengths 1.5–3 times basal leaves. **Leaves:** basal sometimes 2-ranked; cauline leaves 0(or 1); primary leaves ternate, appearing ± palmate, 2–7 cm; petiole: long hairs absent or spreading to ascending, weak, glands usually absent; leaflets 3, at tip of leaf axis, separate, cuneate-ovate in outline, margins strongly revolute, central leaflets divided nearly to base into 3 equal lobes, lateral divided nearly to base into 2 lobes, lobes entire, rarely again 2-fid, surfaces ± dissimilar, abaxial whitish with thick waxy bloom, cottony hairs absent, adaxial light green, often purple-tinged, not glaucous, long hairs weak or absent. **Inflorescences** solitary flowers or 2- or 3(or 4)-flowered, ± cymose, open. **Pedicels** straight in fruit, 0.5–1.5(–2) cm, proximal not much longer than distal. **Flowers** 5-merous; hypanthium 2–2.5 mm diam.; petals pale yellow to yellow, ± obcordate, 6–10 mm, longer than sepals, apex retuse; stamens ca. 20; styles subapical, filiform-tapered, not papillate-swollen proximally, 2.5–4 mm. **Achenes** smooth.

Species 1: nw North America, Asia.

Although sect. *Biflorae* deviates from the main body of *Potentilla* and superficially resembles the mainly Asian genus *Dasiphora* in shoot and leaf architecture and in some floral features, molecular analyses place the species unequivocally in the *Potentilla* clade, not in the *Fragaria* clade with *Dasiphora* (C. Dobeš and J. Paule 2010; M. H. Töpel et al. 2011).

2. Potentilla biflora D. F. K. Schlechtendal, Ges. Naturf. Freunde Berlin Mag. Neuesten Entdeck. Gesammten Naturk. 7: 297. 1816 • Two-flower cinquefoil F

Basal leaves: petiole 1–4 cm, long hairs absent or sparse, 1.5–2 mm, crisped hairs absent or sparse; leaflet lobes linear, 10–20 × 1–2 mm, surfaces with long hairs sparse or absent. **Flowers:** epicalyx bractlets oblong, ovate, or lanceolate, 2.5–4 × 0.4–1.3 mm; sepals 3.5–5 mm, apex acute to apiculate; petals 6–10 × 4–10 mm; filaments 2.8–4 mm, anthers 0.6 mm; carpels 15–25. **Achenes** 1.5–2 mm. $2n$ = 14.

Flowering summer. *Dryas* fellfields, gravel terraces, well-drained soil, often with seasonal seepage, on calcareous substrates; 0–2000 m; B.C., N.W.T., Nunavut, Yukon; Alaska; Asia.

In North America, *Potentilla biflora* is mostly Beringian, barely reaching northern British Columbia and western Northwest Territories, with one locality in Nunavut (southern Victoria Island). The amphi-Beringian range is strongly isolated from other Asian populations, the closest being in the Altai Mountains of south-central Siberia. Two varieties are accepted in China (Li C. L. et al. 2003c), with North American plants falling in var. *biflora*.

8c. Potentilla Linnaeus sect. Lupinoides Tausch, Hort. Canal. 1: sub *P. ornithopoda*. 1823 I

Barbara Ertter

James L. Reveal

Potentilla sect. *Fragariastrum* (Heister ex Fabricus) Seringe; *Potentilla* subg. *Fragariastrum* (Heister ex Fabricus) Reichenbach; *Potentilla* sect. *Fragarioides* (Th. Wolf) Juzepczuk; *Potentilla* sect. *Quinquefolium* Poeverlein

Perennials, rosetted or ± tufted to openly matted, sometimes ± stoloniferous; taproots not fleshy-thickened; vestiture of long hairs, glands absent or sparse, not red. **Stems:** mostly ± prostrate, not flagelliform, not rooting at nodes, intermixed in persistent basal rosettes, 0.4–1(–1.5) dm, lengths 0.5–1 times basal leaves, <longer non-flowering often developing, sometimes rooting at nodes and forming new basal rosettes>. **Leaves:** basal not in ranks; cauline (flowering stems) 0–2; primary leaves ternate [palmate], (2–)6–10(–20) cm; petiole: long hairs spreading, weak, glands absent or sparse; leaflets 3[–7], at tip of leaf axis, overlapping, broadly elliptic to obovate, margins flat, distal ½–¾ evenly incised ¼ or less to midvein [entire], teeth (3 or)4–7 per side, surfaces ± similar, green (abaxial often paler), not glaucous, long hairs weak, cottony hairs absent. **Inflorescences** solitary flowers or 2- or 3-flowered, ± cymose, open. **Pedicels** often recurved in fruit, 1–3 cm, proximal not much longer than distal. **Flowers** 5-merous; hypanthium 3–5 mm diam.; petals white, usually obcordate, 4–7 mm, equal to or slightly longer than sepals, apex usually ± retuse; stamens ca. 20; styles subapical, filiform, not papillate-swollen proximally, 1.5–2 mm. **Achenes** faintly rugose.

Species 6 (1 in the flora): introduced, Newfoundland; Eurasia.

Section *Lupinoides* belongs to the *Alba* clade of M. H. Töpel et al. (2011), which is an early radiation of *Potentilla* in southern Europe and southwestern Asia (C. Dobeš and J. Paule 2010). In addition to the species included here, *P. alba* Linnaeus has been reported in Massachusetts (B. A. Sorrie 2005) and Connecticut; it has apparently not become naturalized. Plants of *P. alba* are not stoloniferous and the five leaflets are nearly entire.

3. **Potentilla sterilis** (Linnaeus) Garcke, Fl. N. Mitt.-Deutschland ed. 4, 112. 1858 • Strawberry leaved cinquefoil, Barren-strawberry [I]

Fragaria sterilis Linnaeus, Sp. Pl. 1: 495. 1753

Basal leaves: petiole 2–7(–17) cm, long hairs common to abundant, 1–2 mm; central leaflets (0.8–)1.5–3(–5) × (0.6–)1–2(–3) cm, straight hairs common to abundant (sparser adaxially), glands absent or sparse. **Flowers:** epicalyx bractlets ± lanceolate, 2.5–4 × 0.8–1.3 mm; sepals 4–6.5 mm, apex ± acute; petals 4–7 × 3–5 mm; filaments 0.8–2 mm, anthers 0.6–0.9 mm; carpels numerous. **Achenes** 1.5 mm. $2n$ = 28 (Europe).

Flowering spring. Dry to moist, often rocky slopes; 0 m; introduced, Nfld. and Labr. (Nfld.); Europe.

Potentilla sterilis was collected from the coast of Newfoundland in 1928 (*A. M. Ayre s.n.*, GH); no recent documentation of occurrence is known (J. Maunder, pers. comm. to L. Brouillet). Although previously treated as native (M. L. Fernald 1950; E. Hultén and M. Fries 1986), this status has been challenged (for example, A. Kurtto et al. in J. Jalas et al. 1972+, vol. 13). According to H. A. Gleason and A. Cronquist (1991), the species is also rarely introduced in the eastern United States; no voucher has been seen.

Potentilla sterilis is superficially similar to strawberry (*Fragaria*) but lacks the fleshy fruit; other differences include stolon structure, anther morphology, and style length and attachment.

8d. Potentilla Linnaeus sect. Potentilla

Barbara Ertter

James L. Reveal

Potentilla sect. *Tormentilla* (Linnaeus) Tausch; *Tormentilla* Linnaeus

Perennials, openly matted or ± tufted, often stoloniferous; taproots usually replaced by thick rootstocks; vestiture mostly of long hairs, glands absent or sparse, rarely common, sometimes reddish. **Stems** usually becoming ± prostrate, sometimes ascending to erect, often flagelliform, often rooting at nodes, lateral or central to persistent or ephemeral basal rosettes, 0.3–12+ dm, lengths (1–)2–10+ times basal leaves. **Leaves:** basal and cauline not in ranks; cauline (proximal to flowering and/or branching nodes) 0–3(–7); primary leaves ternate or palmate, 2–20(–30) cm; petiole: long hairs appressed to spreading, weak to stiff, glands absent or sparse, sometimes common; leaflets 3–5(–7), at tip of leaf axis, separate to slightly overlapping, obovate to narrowly elliptic, cuneate, or oblanceolate, margins flat or slightly revolute, distal ½–¾ evenly incised

$\frac{1}{4}$–$\frac{1}{2}$ to midvein, teeth 2–13 per side, surfaces similar to ± dissimilar, abaxial usually green, sometimes silvery white, cottony hairs absent, adaxial green, not glaucous, long hairs usually ± stiff, sometimes weak or absent. **Inflorescences** solitary flowers <at stolon nodes> or 3–30 -flowered, cymose, open. **Pedicels** straight or slightly curved in fruit, (1–)2–12(–17) cm, proximal not longer than distal. **Flowers** 4 or 5(–10)-merous; hypanthium 1.5–5(–7) mm diam.; petals usually bright yellow, rarely cream, ± obcordate or obovate to round, (2–)4–9(–12) mm, usually longer than sepals, apex rounded to retuse; stamens 15–20; styles subapical, columnar-clavate to ± filiform, not papillate-swollen proximally, 0.6–1.5 mm. **Achenes** smooth or rugose.

Species 7 or 8 (5 in the flora): e North America, Eurasia, n Africa, Atlantic Islands; introduced in w North America, Mexico, West Indies, Bermuda, Central America, South America, c Africa (Ethiopia), Pacific Islands (New Zealand), Australia.

The species of sect. *Potentilla* comprise a monophyletic clade that includes *Duchesnea*, diverging basally to the species placed here in *Horkelia*, *Horkeliella*, and *Ivesia* (C. Dobeš and J. Paule 2010; M. H. Töpel et al. 2011). There are at least some morphological features by which this and other clades basal to the core *Potentilla* could be treated as separate genera. However, since the type of the genus (*P. reptans*) is in this section, doing so would require either a massive renaming of most *Potentilla*, or else conserving the type of the genus on a different species. If the latter course were taken, the species in this section would comprise *Tormentilla*.

Among the distinctive features of sect. *Potentilla* are the high percentage of stoloniferous species (shared with *Duchesnea*), a tuberous rootstock in many species, and the presence of tetramerous-flowered species (*P. anglica*, *P. erecta*). The section is also distinctive in having an amphi-Atlantic distribution, with native species in both eastern North America and Europe. The native material has sometimes been treated as a single species, or with confused nomenclature (M. L. Fernald 1931), leading to much unreliability in older herbarium annotations.

When leaves are palmate, the lateral leaflet pairs are usually more or less fused at the base, suggesting a ternate origin. Distal cauline leaves and inflorescence bracts are sometimes opposite. For comparison with other sections, counts of cauline leaves are restricted to nodes proximal to the first flowering stolon node, but descriptions of cauline leaves otherwise include all well-developed foliar structures at stolon nodes (until such time as these root and form new basal rosettes).

SELECTED REFERENCE Fernald, M. L. 1931. *Potentilla canadensis* and *P. simplex*. Rhodora 33: 180–191.

1. Flowers 4(or 5)-merous; stems usually openly branched; leaflets 3–5; hypanthia 1.5–4 mm diam.
    2. Stems soon becoming prostrate, ± flagelliform, eventually rooting at some nodes; inflorescences mostly solitary flowers at stolon nodes; cauline leaves: petioles 0.3–4(–8) cm; basal leaves ± persistent . . . . 7. *Potentilla anglica*
    2. Stems mostly ascending to erect, not flagelliform, not rooting at nodes; inflorescences cymose, 3–30-flowered; cauline leaves: petioles usually 0 cm; basal leaves ephemeral . . . . 8. *Potentilla erecta*
1. Flowers 5(–10)-merous; stems not branched; leaflets (3–)5(–7); hypanthia 2.5–7 mm diam.
    3. Petals 7–9(–12) × 6–9(–11) mm; anthers (1–)1.3–2 mm; epicalyx bractlets elliptic or oblong to ovate, 4–10 × 1.5–3.5 mm, often much larger than sepals (especially in fruit); petiole hairs tightly to loosely appressed, 0.5–1.5 mm, leaflet surfaces similar, abaxial green, sparsely to moderately hairy; cauline leaves 2 or 3(or 4) proximal to 1st flowering node; achenes ± rugose . . . . 6. *Potentilla reptans*
    3. Petals 4–7(–8) × 3–6.5(–8) mm; anthers 0.4–1 mm; epicalyx bractlets linear to narrowly lanceolate or lanceolate-elliptic, 2–5(–6) × 0.8–1.5(–2) mm, slightly smaller than to larger than sepals (especially in bud); petiole hairs appressed to spreading, (0.5–)1–3 mm, leaflet surfaces similar to ± dissimilar, abaxial green to silvery white, sparsely to densely hairy; cauline leaves 0 or 1 proximal to 1st flowering node; achenes smooth or faintly rugose.

[4. Shifted to left margin.—Ed.]

4. Cauline leaves not fully expanded at anthesis, 0 or 1 proximal to 1st flowering node, leaflets ± resembling those of basal leaves, apex rounded to obtuse, teeth 2–7 per side on distal ½ of leaflet; rootstocks erect, stout, 0.5–2 cm . . . . 4. *Potentilla canadensis*
4. Cauline leaves well expanded at anthesis, (0 or)1 proximal to 1st flowering node, leaflets often more elongate than those of basal leaves, apex acute to obtuse, teeth 4–8(–13) per side on distal ½–¾ of leaflet; rootstocks horizontal, irregularly thickened or moniliform, 1–8 cm . . . . .5. *Potentilla simplex*

4. Potentilla canadensis Linnaeus, Sp. Pl. 1: 498. 1753 • Dwarf cinquefoil [E]

Potentilla canadensis var. *pumila* (Poiret) Torrey & A. Gray; *P. canadensis* var. *villosissima* Fernald; *P. caroliniana* Poiret; *P. pumila* Poiret

Rootstocks erect, stout, 0.5–2 cm. **Stems** initially ascending to erect, soon becoming prostrate, flagelliform, not branched, eventually rooting at some nodes, (0.3–)0.5–12 dm. **Basal leaves** ± persistent, usually palmate, 2–9(–11) cm; petiole 1–7(–8) cm, long hairs abundant to dense, appressed to spreading, 1.5–3 mm, mostly ± weak, glands absent or sparse; leaflets (3–)5(–7), central mostly ± obovate to cuneate, sometimes narrowly so, (0.5–)1–4(–6) × (0.5–)0.8–2(–3) cm, distal ½ of margin incised ¼–⅓ to midvein, teeth 2–7 per side, surfaces similar to ± dissimilar, abaxial green to silvery white, sparsely to densely hairy, adaxial green, sparsely to moderately hairy. **Cauline leaves** 0 or 1 proximal to 1st flowering node, not fully expanded at anthesis, usually palmate, 2–7(–9) cm; petiole (0.5–)1–6(–7.5) cm; leaflets (3–)5, ± resembling those of basal leaves, apex rounded to obtuse. **Inflorescences** solitary flowers at stolon nodes. **Pedicels** (1–)2–5(–9) cm. **Flowers** 5-merous; epicalyx bractlets linear to lanceolate-elliptic, (2–)3–5(–6) × 0.8–1.5 mm, slightly smaller than to ± equal to sepals; hypanthium 2.5–5 mm diam.; sepals (2–)3–5(–6) mm, apex acute; petals 4–6(–8) × 3.5–6.5(–8) mm, apex rounded to slightly retuse; stamens ca. 20, filaments 0.5–2 mm, anthers 0.4–1 mm; carpels 20–40, styles 0.8–1.4 mm. **Achenes** 1.2–1.4 mm, smooth. **2*n*** = 28.

Flowering late Mar–early Jun. Dry flats and slopes in lawns, pastures, roadsides, cherty slopes, dry meadows, edges of oak and conifer woodlands, often on acidic soil; 0–1500 m; N.S., Ont.; Conn., Del., D.C., Ky., Maine, Md., Mass., Mo., N.H., N.J., N.Y., N.C., Ohio, Pa., R.I., S.C., Tenn., Vt., Va., W.Va.

A cream-colored petaled form (*ochroleuca*) was described from a now-obliterated site in Massachusetts (M. L. Fernald 1931).

5. Potentilla simplex Michaux, Fl. Bor.-Amer. 1: 303. 1803 • Simple or old-field cinquefoil, potentille simple [E]

Potentilla canadensis Linnaeus var. *simplex* (Michaux) Torrey & A. Gray; *P. simplex* var. *argyrisma* Fernald; *P. simplex* var. *calvescens* Fernald

Rootstocks horizontal, irregularly thickened or moniliform, 1–8 cm. **Stems** initially erect to ascending <to 5 dm>, then arching above ground-level, becoming prostrate, flagelliform, not branched, rooting at distal nodes, (0.5–)1.5–12 dm. **Basal leaves** persistent or ephemeral, usually palmate, 3.5–20 cm; petiole 1–15 cm, long hairs ± abundant, appressed to spreading, (0.5–)1–3 mm, weak to ± stiff, glands absent or sparse, sometimes common; leaflets 5(–7), central narrowly elliptic to obovate, 1.5–5 × 0.5–2.5 cm, distal ½–¾ of margin incised ¼–⅓ to midvein, teeth 4–8(–13) per side, surfaces similar to ± dissimilar, abaxial green to silvery white, sparsely to abundantly hairy, adaxial green, glabrate or sparsely to moderately hairy. **Cauline leaves** (0 or)1 proximal to 1st flowering node, well expanded at anthesis, usually palmate, 2.5–6(–9) cm; petiole 0–2(–4) cm; leaflets (3–)5, often more elongate than those of basal leaves, <± elliptic to oblanceolate, sometimes obovate, 1.5–6(–7.5) × 0.5–2(–3) cm>, apex acute to obtuse. **Inflorescences** solitary flowers at stolon nodes. **Pedicels** 1–5(–8.5) cm. **Flowers** 5(or 6)-merous; epicalyx bractlets linear to narrowly lanceolate, 2–5 × 1–1.5(–2) mm, often larger than sepals (especially in bud); hypanthium 3–5 mm diam.; sepals 4–6 mm, apex broadly acute; petals 4–7 × 3–6 mm, apex rounded to ± retuse; stamens ca. 20, filaments 0.5–2 mm, anthers 0.6–1 mm; carpels 20–50, styles 0.8–1.5 mm. **Achenes** 0.9–1.2 mm, faintly rugose.

Flowering (Mar–)Apr–Jul(–Aug). Dry or infrequently moist flats and slopes in pastures, dry meadows, roadsides, old fields, edges of mixed oak and conifer woodlands, often on acidic soil; 0–800 m; St. Pierre and Miquelon; N.B., Nfld. and Labr. (Nfld.), N.S., Ont., P.E.I., Que.; Ala., Ark., Conn., Del., D.C., Ga., Ill., Ind., Iowa, Kans., Ky., La., Maine, Md., Mass., Mich., Minn., Miss., Mo., Nebr., N.H., N.J., N.Y., N.C., Ohio, Okla., Pa., R.I., S.C., Tenn., Tex., Vt., Va., W.Va., Wis.

The distinction between *Potentilla simplex* and *P. canadensis* is subtle, and the former is often misidentified as the latter; additional features to distinguish the two species are provided by M. L. Fernald (1931).

6. Potentilla reptans Linnaeus, Sp. Pl. 1: 499. 1753 • Creeping cinquefoil, potentille rampante [I]

Rootstocks erect, slender to stout, 2–6+ cm. **Stems** soon becoming prostrate, flagelliform, not branched, rooting at some nodes, 1.5–10+ dm. **Basal leaves** persistent, usually palmate, 3–12(–30) cm; petiole 2–10(–25) cm, long hairs sparse to abundant, tightly to loosely appressed, 0.5–1.5 mm, usually stiff, sometimes weak, glands absent; leaflets (3–)5(–7), central oblanceolate to obovate, (0.5–)2–4(–7) × 0.3–1.5(–2.5) cm, distal ± ¾ of margin incised ¼–⅓ to midvein, teeth (3 or)4–12 per side, surfaces similar, green, sparsely to moderately hairy. **Cauline leaves** 2 or 3(or 4) proximal to 1st flowering node, usually well expanded at anthesis, usually palmate, 2–8(–20) cm; petiole 0.5–6(–15) cm; leaflets (3–)5, ± resembling those of basal leaves, apex rounded to obtuse. **Inflorescences** solitary flowers at stolon nodes. **Pedicels** (2.5–)4–12(–15) cm. **Flowers** 5(–10)-merous; epicalyx bractlets elliptic or oblong to ovate, 4–10 × 1.5–3.5 mm, often much larger than sepals (especially in fruit); hypanthium 4–7 mm diam.; sepals (3–)5–7 mm, apex broadly acute to obtuse; petals 7–9 (–12) × 6–9(–11) mm, apex usually ± retuse; stamens ca. 20, filaments (0.5–)1–2.5(–2.8) mm, anthers (1–)1.3–2 mm; carpels 60–120, styles 0.6–1.3 mm. **Achenes** 1.3–1.6 mm, ± rugose. **2*n*** = 28 (Eurasia).

Flowering late Apr–Aug. Dry to moist lawns, roadsides, waste places, on non-acidic soil; 0–500 m; introduced; N.S., Ont., Que.; Calif., Colo., D.C., Fla., Ga., Ill., Ky., La., Md., Mass., Mich., Minn., N.J., N.Y., Ohio, Oreg., Pa., Va., Wash., Wis.; Europe; w Asia; n Africa; Atlantic Islands (Azores, Macaronesia); introduced also in West Indies, Bermuda, South America, Africa (Ethiopia), Pacific Islands (New Zealand), Australia.

Potentilla reptans is becoming widely introduced in North America, though not fully naturalized in some of the states indicated here. It may be confused with *P. canadensis* and *P. simplex*, but, in addition to the characters highlighted in the key, *P. reptans* tends to have more crenately toothed leaves than the more sharply toothed leaves of the native species. The plant has a history of medicinal uses similar to that of *P. erecta*. The cultivated double-flowered form is sometimes found established in weedy places.

7. Potentilla anglica Laicharding, Veg. Europ. 1: 475. 1790 • English or wood cinquefoil, trailing tormentil, potentille d'Angleterre [I]

Rootstocks ± erect, stout to slender, 1–4 cm. **Stems** soon becoming prostrate, ± flagelliform, usually openly branched, eventually rooting at some nodes, (0.3–)1.5–7+ dm. **Basal leaves** ± persistent, ternate or palmate, 2–10(–12) cm; petiole 1–7(–8) cm, long hairs sparse to abundant, appressed, 0.5–1 mm, stiff, glands absent; leaflets 3–5, central ± obovate to cuneate, 1–3 (–3.5) × 0.8–2(–2.5) cm, distal ± ½ of margin incised ¼–⅓(–½) to midvein, teeth 2–4 per side, surfaces similar, green, sparsely to moderately hairy. **Cauline leaves** 2 or 3(or 4) proximal to 1st flowering or branching node, usually well expanded at anthesis, usually ternate, 2–6 (–10) cm; petiole 0.3–4(–8) cm; leaflets (3–)5, ± resembling or narrower than those of basal leaves, <narrowly cuneate, 1–2(–3.5) × 0.8–2(–2.5) cm>, apex rounded to obtuse. **Inflorescences** solitary flowers at stolon nodes. **Pedicels** (1–)3–10(–17) cm. **Flowers** 4(–5)-merous; epicalyx bractlets narrowly elliptic to oblong or ovate, 3–4(–7) × 1–1.5 mm, smaller than to slightly larger than sepals; hypanthium 2–4 mm diam.; sepals (3–)4–6 mm, apex broadly acute or acuminate; petals 6–9 × 5–9 mm, apex ± retuse; stamens 15–20, filaments 0.8–1.2 mm, anthers 0.8–1.2 mm; carpels 20–50, styles 0.9–1.5 mm. **Achenes** 1–1.5(–1.8) mm, ± smooth. **2*n*** = 28, 56 (Europe).

Flowering May–Aug(–Nov). Moist flats and slopes, mainly on acidic soil; 0–1200 m; introduced; St. Pierre and Miquelon; B.C., Nfld. and Labr., N.S., Que.; Calif., Colo., Maine, N.Y., Oreg., Pa., Utah, Wash.; Europe; introduced also in Atlantic Islands (Azores, Madeira), Pacific Islands (New Zealand), Australia.

Potentilla anglica apparently is a product of hybridization involving *P. erecta* and *P. reptans* (B. Matfield and J. R. Ellis 1972) that has become stabilized and distinct from both parents. In Europe, it forms back-cross hybrids with both *P. erecta* (*P.* ×*suberecta* Zimmeter) and *P. reptans* (*P.* ×*mixta* Nolte); these hybrids are not known from North America. A garden hybrid supposedly between *P. anglica* and *P. nepalensis* Hooker (known as *P.* ×*tonguei* Mallett) was found in Allegany State Park, New York, but this was likely cultivated rather than naturalized.

Potentilla anglica probably was introduced even in Newfoundland (A. Kurtto et al. in J. Jalas et al. 1972+, vol. 13), contrary to the view expressed by M. L. Fernald (1950).

The name *Potentilla procumbens* Sibthorp was previously used for this species; it is a superfluous and illegitimate name.

8. Potentilla erecta (Linnaeus) Raeuschel, Nomencl. Bot, 152. 1797 • Erect cinquefoil, tormentil [I]

Tormentilla erecta Linnaeus, Sp. Pl. 1: 500. 1753

Rootstocks erect to horizontal, irregularly thickened, 1–8 cm. **Stems** mostly ascending to erect, not flagelliform, openly branched, not rooting at nodes, (0.5–)1–5(–7) dm. **Basal leaves** ephemeral, usually ternate, 2–15 cm; petiole (1–)3–13 cm, long hairs sparse to common, appressed, 0.5–1 mm, stiff, glands absent; leaflets 3(–5), central ± obovate to cuneate, (0.5–)1–2 × (0.3–)1–1.5 cm, distal ½–⅔ of margin incised ± ⅓ to midvein, teeth 2–4 per side, surfaces similar, green (abaxial sometimes paler), glabrate or sparsely to moderately hairy. **Cauline leaves** (0–)1–3(–7) proximal to 1st flowering node, expanded at anthesis, usually ternate, rarely palmate (or appearing so because of leafletlike stipules), (0.5–) 1–3 cm; petiole usually 0 cm; leaflets 3(–5), usually much narrower than those of basal leaves, <narrowly cuneate to oblanceolate, (0.5–)1–3 × (0.2–)0.5–1 cm, distal ⅓–⅔ of margin incised ± ½ to midvein, teeth 2–4(–6) per side>, apex usually obtuse to acute. **Inflorescences** 3–30-flowered, cymose, open. **Pedicels** (1–)2–3(–5) cm. **Flowers** 4(or 5)-merous; epicalyx bractlets ovate to narrowly elliptic, 1.5–4 × 0.5–1 mm, much smaller than to sometimes equal to sepals; hypanthium 1.5–2.5 mm diam.; sepals (2–)3–5 mm, apex broadly acute; petals (2–)4–6 × (1.5–)3–6 mm, apex ± retuse; stamens 15–20, filaments 2–3.8 mm, anthers 0.4–0.6 mm; carpels 4–8 (–20), styles 0.9–1.4 mm. **Achenes** 1.2–2 mm, rugose. $2n = 28$ (Eurasia).

Flowering unknown. Moist, mossy flats and slopes, acidic soil; 0–100 m; introduced; N.S.; Mass.; Europe; w Asia; n Africa; Atlantic Islands (Azores).

Although included here, *Potentilla erecta* is questionably extant as a naturalized species in North America and is not currently known from Newfoundland where historically reported. It is a medicinal plant called tormentil, high in tannins and strongly astringent. A tea made from the tuberous rootstock is widely used in Europe and parts of Asia for diarrhea, dysentery, sore throats, and related problems. A decoction is a mild antibiotic used for cuts and minor infections, especially as an anti-inflammatory for gingivitis and other gum problems. The rootstock is used as a red dye.

Another synonym with a long history of use is *Potentilla tormentilla* Necker, illegitimate by virtue of being superfluous.

8e. Potentilla Linnaeus sect. Arenicolae (Rydberg) O. Stevens in N. L. Britton et al., N. Amer. Fl. 22(7): 11. 1959 [E]

Barbara Ertter

James L. Reveal

Potentilla [unranked] *Arenicolae* Rydberg, Monogr. N. Amer. Potentilleae, 31. 1898

Biennials, sometimes annuals or short-lived perennials, ± rosetted, not stoloniferous; taproots not fleshy-thickened; vestiture mostly of long hairs, glands ± abundant, not red. **Stems** prostrate to decumbent, not flagelliform, not rooting at nodes, lateral to weakly persistent or ephemeral basal rosettes, (0.2–)0.5–4(–6) dm, lengths (1–)1.5–5 times basal leaves. **Leaves** basal not in ranks; cauline leaves 1–4; primary leaves ± pinnate (with distal leaflets confluent), sometimes ± bipinnate, 2–9 cm; petiole: long hairs ± spreading, weak to ± stiff, glands ± abundant; leaflets (5–)7–15(–21) (or appearing more because of deep lobes), on distal ⅓–⅔ of leaf axis, overlapping, obovate or elliptic to flabellate, margins flat, ± whole length irregularly incised ½ to completely to midvein (proximal leaflets sometimes pinnately so), sometimes entire (especially distal leaflets), teeth (or lobes) 0–4 per side, surfaces similar, grayish green, not glaucous, long hairs ± stiff, cottony and/or crisped hairs absent. **Inflorescences** (2–)6–50 (–100+)-flowered, irregularly cymose, sometimes racemiform, ± open. **Pedicels** recurved in fruit, 0.7–2(–2.5) cm, proximal not much longer than distal. **Flowers** 5-merous; hypanthium 2.5–4.5 mm diam.; petals white, broadly obovate to obcordate or nearly round, (3–)4–6 mm, longer than sepals, apex usually ± retuse; stamens ca. 20; styles subapical, filiform-tapered, papillate-swollen in proximal ⅕–⅓(–½), 0.8–2 mm. **Achenes** rugose.

Species 1: w United States.

While E. Crum (1936) followed T. Wolf (1908) in associating *Potentilla newberryi* with sect. *Rivales*, recent phylogenetic analysis using chloroplast DNA places *P. newberryi* as basal or sister to all other species in the *Potentilla* core group (C. Dobeš and J. Paule 2010). The two sections share similar life histories and habitats, sometimes growing intermixed, and the abundant glands of *P. newberryi* are also characteristic of *P. biennis*. In other characteristics, *P. newberryi* differs from sect. *Rivales*, for example, in its finely pinnate (to bipinnate) leaves, prominently recurved pedicels, large white petals, and longer styles. Here P. A. Rydberg (1898, 1908d) and B. C. Johnston (1985) are followed in viewing the combination of distinctive features as sufficient to justify a monospecific section.

An intriguing comparison can be made to *Ivesia*, which shares many diagnostic morphological characters; *Potentilla newberryi* was described as *I. gracilis*, one of the two species of the genus *Ivesia* as originally circumscribed by A. Gray.

9. Potentilla newberryi A. Gray, Proc. Amer. Acad. Arts 6: 532. 1865 • Newberry's cinquefoil [E] [F]

Ivesia gracilis Torrey & A. Gray in War Department [U.S.], Pacif. Railr. Rep. 6(3): 72, plate 11. 1858, not *Potentilla gracilis* Douglas ex Hooker 1830

Basal leaves: petiole 1–3.5 cm, long hairs ± abundant, 0.5–1.5 mm, short hairs absent or sparse, rarely common; leaflets 0.2–1 cm, lobes oblanceolate to narrowly elliptic, (0.5–)1–2 mm wide, long hairs ± abundant, short hairs absent or sparse, glands sparse to abundant. **Flowers:** epicalyx bractlets broadly lanceolate to elliptic, 1.5–4(–5) × (0.4–)0.8–1.3 mm; sepals 2–4(–5) mm, apex ± acute; petals (3–)4–6 × (2–)3–5 mm; filaments 1–2(–2.5) mm, anthers 0.4–0.6 mm; carpels 20–50. **Achenes** 0.9–1.2 mm.

Flowering summer. Moist, sandy to clayey, more or less alkaline soil, especially where seasonally inundated near streams, ponds, and lakes; 1300–1800 m; Calif., Nev., Oreg., Wash.

Potentilla newberryi grows in valley bottoms in south-central Oregon, northeastern California, and northwestern Nevada. The only specimen supposedly collected in south-central Washington (*W. N. Suksdorf 2718*, WTU) was gathered in 1898.

8f. Potentilla Linnaeus sect. Rivales Poeverlein in P. F. A. Ascherson et al., Syn. Mitteleur. Fl. 6(1): 669. 1904

Barbara Ertter

James L. Reveal

Potentilla sect. *Supinae* (Lehmann) A. Nelson; *Potentilla* ser. *Supinae* Lehmann; *Tridophyllum* Necker ex Greene

Annuals, biennials, or short-lived perennials, 1-stemmed to ± tufted, not stoloniferous; taproots not fleshy-thickened; vestiture of long hairs, sometimes ± crisped hairs present, glands absent or sparse to abundant, not red (sometimes red-septate). **Stems** decumbent to erect, sometimes prostrate, not flagelliform, not rooting at nodes, from centers of ephemeral basal rosettes, (1–)2–6(–9) dm, lengths (1–)2–5(–12) times basal leaves. **Leaves:** basal not in ranks; cauline (0–)1–9(–14); basal and proximal cauline ternate or palmate to pinnate (with distal leaflets

± confluent), (2–)3–15(–25) cm; petiole: long hairs ascending to spreading, rarely appressed, weak to stiff, glands absent or sparse to abundant; leaflets 3–9, at tip to distal ⅔ of leaf axis, separate to overlapping, oblanceolate or elliptic to obovate, sometimes oval or nearly round, margins flat, distal ½ to ± whole length evenly to unevenly incised ¼–½ to midvein, teeth (2 or)3–8(–15) per side, surfaces ± similar, green (abaxial often paler), not glaucous, long hairs weak to stiff, cottony hairs absent. **Inflorescences** (5–)20–100+-flowered, ± cymose, sometimes racemiform, compact to open. **Pedicels** straight to ± recurved in fruit, 0.2–2(–3) cm, proximal often ± longer than distal. **Flowers** 5-merous; hypanthium (2–)3–6(–7) mm diam.; petals pale yellow to yellow, broadly oblanceolate to oblong or broadly obovate, (1–)1.5–5 mm, usually shorter than sepals, apex rounded to ± retuse; stamens (5–)10–20(–25); styles subapical, conic-tapered, slightly to strongly papillate-swollen ± whole length, 0.5–0.8 mm. **Achenes** smooth to strongly rugose.

Species 18–21 (4 in the flora): North America, Mexico, Central America, South America, Eurasia, Africa; introduced in Pacific Islands (New Zealand), Australia.

The North American representatives of sect. *Rivales* are morphologically distinctive in being annuals, biennials, or short-lived perennials (especially in cold regions) with very short, conic-tapered styles. Plants grow primarily in wet habitats, especially where periodically inundated. These distinctions led E. L. Greene (1906c) to adopt *Tridophyllum*. However, the gross morphological cohesiveness is not matched by molecular monophyly. Instead, chloroplast data scatter *Potentilla norvegica*, *P. rivalis*, and *P. supina* among the core *Potentilla*, while *P. biennis* is sister species to the monophyletic *Ivesia-Horkelia-Horkeliella* clade (C. Dobeš and J. Paule 2010; M. H. Töpel et al. 2011). Nuclear markers, on the other hand, place *P. norvegica* as sister to the *Ivesia-Horkelia-Horkeliella* clade (Töpel et al.), suggesting a possible hybrid origin and/or multiple origins of the biennial habit.

Existing herbarium annotations of *Potentilla biennis*, *P. norvegica*, and *P. rivalis* are not reliable, but the three species can be readily distinguished by vestiture of proximal petioles and stems. Only *P. biennis* has prominently septate gland-tipped trichomes; *P. norvegica* is characterized by relatively stiff, spreading, tubercle-based hairs to 3 mm, resembling those of *P. recta*. In addition, the tiny, smooth, pale achenes of *P. biennis* and *P. rivalis* contrast sharply with the larger, darker, strongly ridged ones of *P. norvegica*.

Basal leaves usually wither by anthesis, and cauline leaves (that is, those proximal to the first flowering and/or branching node) often wither by mid-anthesis as well. Unless otherwise indicated, leaves include both basal and cauline leaves. If leaves are ephemeral, and/or where stems branch at the proximal nodes, proximal inflorescence bracts can be used as leaf-equivalents in the keys and descriptions.

1. Leaves pinnate to subpalmate, leaflets 5–9.
   2. Leaflets 5–9, on distal (¼–)½–⅔ of leaf axis; achenes usually strongly rugose, developing a smooth, corky protuberance often as large as body of achene .... 10. *Potentilla supina*
   2. Leaflets 5(–7), on less than distal ⅕ (basal leaves) or ½ (cauline leaves) of leaf axis; achenes ± smooth, without a corky protuberance .... 11. *Potentilla rivalis* (in part)
1. Leaves ternate or palmate, leaflets 3–5.
   3. Petiole and stem base hairs usually ± stiff, 1–2.5(–3) mm, glands absent or sparse, inconspicuous; achenes usually strongly rugose, tan to brown, 0.8–1.3 mm; hypanthia 4–7 mm diam.; stamens 15 or 20, anthers 0.3–0.5 mm..................... 13. *Potentilla norvegica*
   3. Petiole and stem base hairs usually ± weak, 0.5–1.5 mm (very weak if longer), glands absent or sparse to abundant, sometimes conspicuous; achenes ± smooth, white to yellowish, 0.5–0.9 mm; hypanthia (2–)3–4(–5.5) mm diam.; stamens (5–)10 or 15, anthers 0.2–0.3 mm.

[4. Shifted to left margin.—Ed.]

4. Petiole and stem base glands absent or sparse, inconspicuous; leaflets oblanceolate-elliptic to obovate, separate to ± overlapping; stems decumbent to erect, sometimes prostrate . . . . 11. *Potentilla rivalis* (in part)
4. Petiole and stem base glands sparse to abundant, conspicuous (to 1 mm, septate); leaflets mostly obovate or oval to nearly round, usually overlapping; stems ascending to erect . . . . 12. *Potentilla biennis*

10. Potentilla supina Linnaeus, Sp. Pl. 1: 497. 1753 [F]

Subspecies 7 (1 in the flora): North America, Mexico, Eurasia, Africa; introduced in Australia.

The status of *Potentilla supina* in North America is open to question, given the otherwise Eurasian and northern African distribution of the species and ease of transport by migratory waterfowl. Although North American material has commonly been treated as a distinct species (*P. paradoxa*), E. Hultén and M. Fries (1986) went so far as to consider *P. supina* introduced in North America. The use of subspecies within a broadly defined *P. supina* follows J. Soják (1987, 1993) and is also adopted by A. Kurtto et al. (in J. Jalas et al. 1972+, vol. 13).

10a. Potentilla supina Linnaeus subsp. **paradoxa** (Nuttall) Soják, Folia Geobot. Phytotax. 4: 207. 1969 • Bushy cinquefoil [F]

Potentilla paradoxa Nuttall in J. Torrey and A. Gray, Fl. N. Amer. 1: 437. 1840; *P. nicolletii* (S. Watson) E. Sheldon; *P. supina* var. *nicolletii* S. Watson; *P. supina* var. *paradoxa* (Nuttall) Th. Wolf

Stems prostrate to ascending, sometimes erect, (1–)2–4(–6) dm, hairs at base not particularly stiff, not tubercle-based, glands absent or sparse, inconspicuous. **Leaves** pinnate (with distal leaflets ± confluent), 5–12(–15) cm; petiole 1–6(–8) cm, long hairs sparse to common, ± ascending, sometimes erect or appressed, 0.5–1(–1.5) mm, usually ± weak, crisped hairs absent or sparse to common, glands sparse to common, inconspicuous; leaflets 5–9, on distal (¼–)½–⅔ of leaf axis, ± separate, largest ones oblanceolate to elliptic or obovate, 1–2(–4) × 0.6–1.2 (–2.5) cm, distal ⅔ to ± whole margin ± evenly incised ¼–½ to midvein, teeth (2–)4–6 per side, surfaces glabrate or sparsely hairy, glands mostly absent. **Inflorescences** (10–)20–100+-flowered. **Pedicels** 0.5–2(–2.5) cm. **Flowers:** epicalyx bractlets ± elliptic to narrowly ovate, 3–5(–7) × 1–2(–3) mm; hypanthium 3–5 mm diam.; sepals 3–5(–6) mm, apex acute to apiculate; petals yellow, obovate, 2.5–5 × 2–4.5 mm; stamens 20(–25), filaments 0.7–1.8(–2) mm, anthers 0.3–0.5 mm; carpels ca. 100, styles 0.5–0.6 mm. **Achenes** brown, 0.7–1.3 mm, usually strongly rugose, developing a smooth, corky protuberance on ventral suture, often as large as body of achene. **2*n*** = 28.

Flowering late spring–summer. Sandy shorelines of lakes, reservoirs, and streams; 200–1600 m; Alta., B.C., Man., N.W.T., Ont., Sask.; Colo., Idaho, Ill., Iowa, Kans., La., Mich., Minn., Mo., Mont., Nebr., N.Mex., N.Y., N.Dak., Ohio, Okla., Oreg., Pa., S.Dak., Tex., Utah, Wash., Wyo.; n Mexico; Eurasia.

Although subsp. *paradoxa* is relatively common and widespread in central North America, occurrences are more sporadic and possibly ephemeral in the far western states. The diagnostic corky protuberance on the achene is well developed only on fully mature achenes, appearing merely as a flap of membranous tissue in earlier stages.

11. Potentilla rivalis Nuttall in J. Torrey and A. Gray, Fl. N. Amer. 1: 437. 1840 • Brook or river cinquefoil

Potentilla millegrana Engelmann ex Lehmann; *P. pentandra* Engelmann; *P. rivalis* var. *millegrana* (Engelmann ex Lehmann) S. Watson; *P. rivalis* var. *pentandra* (Engelmann) S. Watson

Stems decumbent to erect, sometimes prostrate, (0.5–)1–4(–7) dm, hairs at base not stiff, not tubercle-based, glands absent or sparse, inconspicuous. **Leaves** ternate, palmate, or subpalmate (with distal leaflets ± confluent), basal 3–15(–25) cm, cauline 2–7(–12) cm; petiole: basal 1–9(–16) cm, cauline 0.5–4(–8) cm, long hairs usually common to abundant, ascending to spreading, 0.5–1.5 mm, usually ± weak, ± crisped hairs common to abundant, glands absent or sparse, inconspicuous; leaflets 3–5(–7), at tip to distal ⅕ (basal) or ½ (cauline) of leaf axis, separate to ± overlapping, largest ones oblanceolate-elliptic to obovate, (0.5–)1–5(–6) × 0.5–2 (–2.5) cm, distal ½–¾ of margin evenly to unevenly incised ⅓–½ to midvein, <sometimes cleft nearly to base>, teeth 3–8 per side, surfaces moderately to abundantly hairy, glands absent or sparse. **Inflorescences** (5–)20–100+-flowered. **Pedicels** 0.2–1(–2) cm. **Flowers:** epicalyx bractlets narrowly elliptic to ovate, (1.5–)2–4(–6) × 0.8–1.5 mm; hypanthium (2–)3–5 mm diam.; sepals 3–5 mm, apex broadly acute to obtuse; petals pale yellow to yellow, broadly oblong-obovate,

1.5–2 × 1 mm; stamens (5–)10(–15), filaments 0.4–0.9 mm, anthers 0.2–0.3 mm; carpels 40–100, styles 0.5–0.6 mm. **Achenes** yellowish, 0.7–0.9 mm, ± smooth, without a corky protuberance. 2*n* = 14, 70.

Flowering spring–summer. Moist meadows, stream banks, lakeshores, gravel bars in flood plains, drying marshes, open areas in river-bottom forests; 200–2400 m; Alta., B.C., Man., Ont., Sask.; Ariz., Calif., Colo., Idaho, Ill., Iowa, Kans., Maine, Md., Mass., Minn., Mo., Mont., Nebr., Nev., N.Mex., N.Y., N.Dak., Okla., Oreg., S.Dak., Tex., Utah, Va., Wash., Wis., Wyo.; Mexico (Baja California).

Potentilla rivalis is most abundant in central and western North America; reports of occurrences in more eastern states (including those listed here) need confirming, as *P. rivalis* and *P. norvegica* often have been confused. *Potentilla leucocarpa* Rydberg was provided as a superfluous replacement name for *P. millegrana*; specimens annotated by Rydberg with this name include both *P. biennis* and *P. rivalis*.

Potentilla rivalis is sometimes divided into three species or varieties (for example, H. J. Scoggan 1978–1979). In a strict sense, var. *rivalis* has subpalmately compound (5-foliate) leaves. Variety *milligrana*, the most common phase, has 3-foliate leaves. Plants with both 3- and 5-foliolate leaves are var. *pentandra*, which also tends to have five stamens, though this latter character is not correlated with the leaf features. On the Great Plains, where var. *pentandra* is confined, all three expressions are often found in a single population (R. L. McGregor 1986b).

12. Potentilla biennis Greene, Fl. Francisc., 65. 1891 • Biennial or Greene's cinquefoil [E]

Stems ascending to erect, (1–)2–6(–8.5) dm, hairs at base not stiff, not tubercle-based, glands sparse to abundant, conspicuous (to 1 mm, septate). **Leaves** ternate, (2–)4–12(–25) cm; petiole (1–)2–8(–20) cm, long hairs sparse to common, spreading to ascending, 0.5–1.5(–3) mm, weak, crisped hairs usually absent, glands sparse to abundant, conspicuous (to 1 mm, septate); leaflets 3, at tip of leaf axis, usually overlapping, largest ones mostly obovate or oval to nearly round, (0.5–)1–3(–4) × 0.3–3(–3.5) cm, distal 3/4 to ± whole margin evenly to unevenly incised 1/4–1/3(–1/2) to midvein, teeth 3–8(–13) per side, surfaces sparsely to moderately hairy, rarely glabrate, glands sparse to abundant at least abaxially. **Inflorescences** (10–)20–70+-flowered. **Pedicels** 0.2–1(–2) cm. **Flowers:** epicalyx bractlets narrowly oblanceolate to narrowly obovate or elliptic, (1.5–)2–3.5(–4.5) × 0.5–1.5(–2) mm; hypanthium (2–)3–4(–5.5) mm diam.; sepals (2–)3–5 mm, apex broadly acute to obtuse; petals pale yellow to yellow, broadly oblanceolate to obovate, (1–)2–2.7 × 1.5 mm; stamens 10 or 15, filaments 0.5–1.2 mm, anthers 0.2–0.3 mm; carpels 40–100, styles 0.5–0.7 mm. **Achenes** whitish or pale yellowish, 0.5–0.9 mm, smooth, without a corky protuberance.

Flowering summer. Moist meadows, stream banks, ditches, seepages; 400–3100 m; Alta., B.C., Sask., Yukon; Alaska, Ariz., Calif., Colo., Idaho, Mont., Nev., N.Mex., N.Dak., Oreg., Utah, Wash., Wyo.

Potentilla biennis may be recognized by its septate hairs with a round, terminal gland, which dominate the vestiture especially at the base of the plant. Stems are more commonly erect and leaflets rounder than those of *P. rivalis*. The species is most common along the eastern side of the Sierra Nevada of California, across the Intermountain Region to the Rocky Mountains. All specimens seen by the authors from Baja California, Mexico, which were previously referred to *P. biennis* are now assigned to *P. rivalis*.

J. Soják (1996) lectotypified *Potentilla millegrana* Engelmann ex Lehmann upon a collection of *P. rivalis*, thus preventing *P. biennis* from being a later taxonomic synonym.

13. Potentilla norvegica Linnaeus, Sp. Pl. 1: 499. 1753 • Norwegian or rough cinquefoil, potentille de Norvège

Potentilla flexuosa Rafinesque; *P. labradorica* Lehmann; *P. monspeliensis* Linnaeus; *P. norvegica* subsp. *hirsuta* (Michaux) Hylander; *P. norvegica* var. *hirsuta* (Michaux) Lehmann; *P. norvegica* var. *labradorica* (Lehmann) Fernald; *P. norvegica* subsp. *monspeliensis* (Linnaeus) Ascherson & Graebner

Stems ascending to erect, (0.5–)2–5(–9) dm, hairs at base ± stiff, tubercle-based, glands absent or sparse, inconspicuous. **Leaves** ternate, rarely palmate, 3–15(–20) cm; petiole 1–6(–10) cm, long hairs sparse to abundant, spreading to ascending, 1–2.5(–3) mm, usually ± stiff, ± crisped hairs absent or sparse to common, glands absent or sparse, inconspicuous; leaflets 3(–5), at tip of leaf axis, separate to ± overlapping, largest ones broadly oblanceolate or elliptic to obovate, 1–6(–10) × 0.7–4(–5) cm, distal (1/2–)2/3–3/4+ of margin usually ± evenly incised 1/4–1/3 to midvein, teeth (3 or)4–8(–15) per side, surfaces sparsely to moderately hairy, sometimes glabrate or abundantly hairy, glands mostly absent. **Inflorescences** 5–40+-flowered. **Pedicels** (0.2–)0.5–2.5(–3) cm. **Flowers:** epicalyx bractlets ± elliptic to narrowly ovate, (3–)4–8(–13) × 1.5–3(–5) mm; hypanthium 4–7 mm diam.; sepals 5–8 mm, apex acute to obtuse; petals yellow, broadly obovate, (2–)3–5 × 2–4 mm; stamens 15 or 20, filaments 0.7–2 mm, anthers 0.3–0.5 mm; carpels 60–150, styles 0.7–0.8 mm. **Achenes** tan to brown, 0.8–1.3 mm, usually strongly rugose, without a corky protuberance. **2*n*** = 42, 56, 63, 70.

Flowering summer. Moist meadows, stream banks, lakeshores, roadsides, grasslands, hardwood and conifer woodlands, tundra; 300–2700 m; Greenland; St. Pierre and Miquelon; Alta., B.C., Man., N.B., Nfld. and Labr., N.W.T., N.S., Nunavut, Ont., P.E.I., Que., Sask., Yukon; Ala., Alaska, Ariz., Ark., Calif., Colo., Conn., Del., D.C., Ga., Idaho, Ill., Ind., Iowa, Kans., Ky., Maine, Md., Mass., Mich., Minn., Mo., Mont., Nebr., Nev., N.H., N.J., N.Mex., N.Y., N.C., N.Dak., Ohio, Okla., Oreg., Pa., R.I., S.C., S.Dak., Tenn., Tex., Utah, Vt., Va., Wash., W.Va., Wis., Wyo.; Mexico; Central America; Eurasia; introduced in South America, Pacific Islands (New Zealand).

Potentilla norvegica is considered native in both North America and Eurasia, with the American race occasionally recognized as subsp. *hirsuta* for the stiff hairs on the stems, petioles, and pedicels of most plants in the flora area. This feature and other purported differences between the two races, in addition to being relatively subtle and inconstant, can be found in both America and Eurasia, though undoubtedly at least in part as introductions. Some populations in eastern Canada with glabrous stems have been distinguished as var. *labradorica* (for example, M. L. Fernald 1950), but such plants typically are intermixed with hairy individuals. The Löves (Á Löve 1954; Á Löve and D. Löve 1966) have argued that all three variants should be treated as distinct species due to chromosomal differences (*hirsuta* 2*n* = 56; *labradorica* 2*n* = 42; *norvegica* in the narrow sense 2*n* = 70) and obligate apomixis. Further research is needed to determine the taxonomic validity and rank of these expressions. If treated as species, *P. flexuosa* antedates *P. labradorica* (as noted by J. Soják 1969), while *P. monspeliensis*, although commonly used for the American race, is based on a European type (as summarized by A. Kurtto et al. in J. Jalas et al. 1972+, vol. 13).

An even greater challenge is determining native versus introduced ranges, especially given the likelihood that both native and Eurasian populations are widespread in North America. Achenes are produced prolifically and easily dispersed, to the extent that *Potentilla norvegica* is a contaminant in clover and hay fields and considered a weed at least in Canada (P. A. Werner and J. D. Soule 1976).

8g. Potentilla Linnaeus sect. Rectae (Th. Wolf) Juzepczuk in V. L. Komarov et al., Fl. URSS 10: 160. 1941 [I]

Barbara Ertter

James L. Reveal

Potentilla [unranked] *Rectae* Th. Wolf in P. F. A. Ascherson et al., Syn. Mitteleur. Fl. 6(1): 671, 750. 1904

Perennials, 1-stemmed to ± tufted, not stoloniferous; taproots not fleshy-thickened, often giving rise to ± thickened lateral roots; vestiture of <well-differentiated> long and short hairs, glands common to abundant, not red. **Stems** erect, not flagelliform, not rooting at nodes, from centers of ephemeral basal rosettes, 1.5–7 dm, lengths (2–)3–5 times basal <or proximal cauline> leaves. **Leaves:** basal not in ranks; cauline (3–)5–9; primary leaves palmate, proximal ones 5–18 cm; petiole: long hairs spreading, stiff, glands common to abundant; leaflets 5–7, at tip of leaf axis, separate, oblanceolate, margins flat or ± revolute, ± whole length ± evenly incised ⅓–½ to midvein, teeth 7–17 per side, surfaces ± similar, green, not glaucous, long hairs stiff, cottony and/or crisped hairs absent. **Inflorescences** (3–)7–60+-flowered, cymose, ± open. **Pedicels** straight in fruit, 0.5–2(–3) cm, proximal often much longer than distal. **Flowers** 5-merous; hypanthium 5–9 mm diam.; petals pale [to bright] yellow, obcordate, (4–)7–10(–13) mm, equal to or longer than sepals, apex retuse, rarely rounded; stamens 25–30; styles subapical, columnar-tapered, not or slightly papillate-swollen in proximal ⅓, 0.8–1.3 mm. **Achenes** strongly rugose.

Species 8–14 (1 in the flora): introduced; Eurasia, n Africa; introduced also in South America, Pacific Islands (New Zealand), Australia.

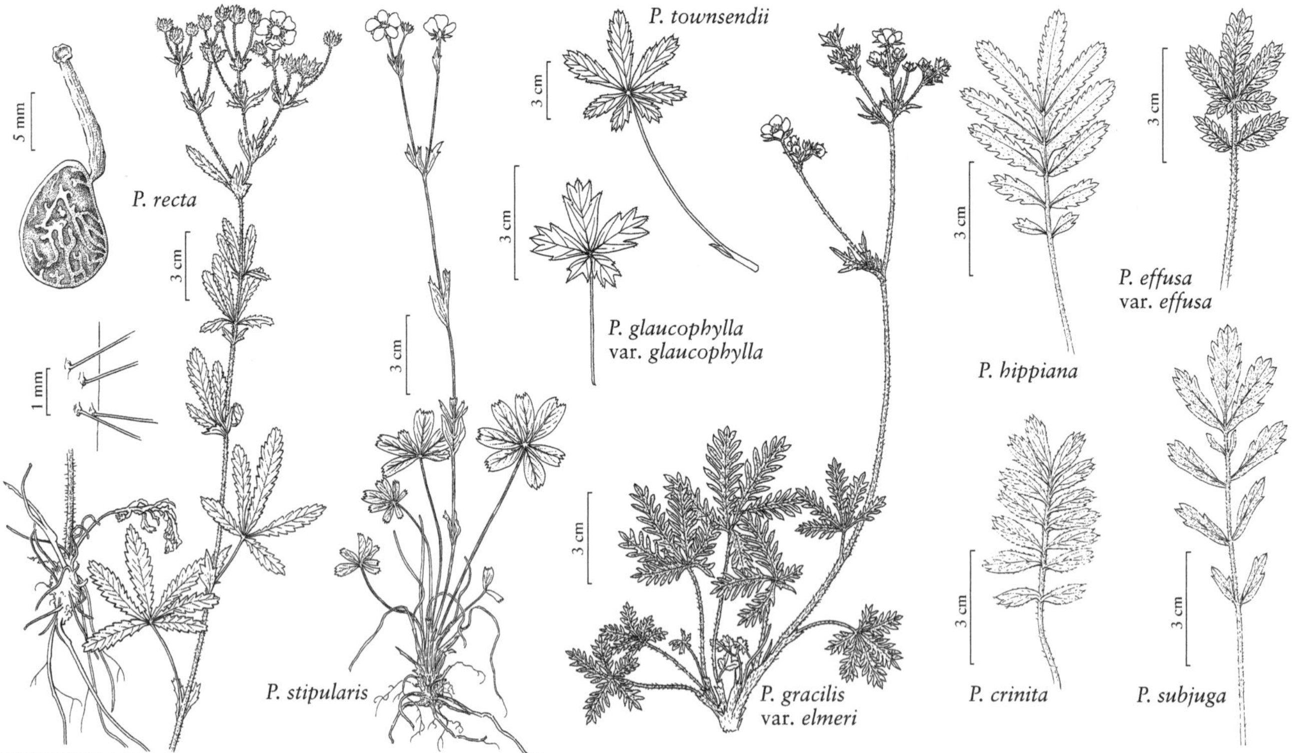

POTENTILLA

14. Potentilla recta Linnaeus, Sp. Pl. 1: 497. 1753 • Sulphur cinquefoil, potentille dressée F I W

Potentilla recta subsp. *obscura* (Willdenow) Arcangeli; *P. recta* var. *obscura* (Willdenow) Poiret; *P. recta* subsp. *pilosa* (Willdenow ex Poiret) Reichenbach f. ex Rothmaler; *P. recta* var. *pilosa* Willdenow ex Poiret; *P. recta* var. *sulphurea* (Lamarck) de Candolle; *P. sulphurea* Lamarck

Cauline leaves: petiole (3–)4–8 cm, long hairs common, 2–4 mm, short hairs common to abundant; leaflets 5–7, 1.5–10 × 0.5–3.5 cm, long hairs common, 0.5–2 mm abaxially, 1 mm and stiffly appressed adaxially, short hairs common on primary veins abaxially, glands sparse to common abaxially. **Flowers:** epicalyx bractlets narrowly lanceolate to lanceolate, 5–12 × 1–2 mm; sepals 4–10(–12) mm, apex acute to acuminate; petals (4–)7–10(–13) × 7–10 mm; filaments 0.5–2.8 mm, anthers 0.8–1.2 mm; carpels 80–140. **Achenes** 1.2–1.8 mm. **2*n*** = 28, 42 (Eurasia).

Flowering late spring–summer. Roadsides, waste places, fields, grasslands, shrublands, forests; 0–2100 m; introduced; Alta., B.C., Man., N.B., Nfld. and Labr. (Nfld.), N.S., Ont., P.E.I., Que., Sask.; Ala., Alaska, Ariz., Ark., Calif., Colo., Conn., Del., D.C., Fla., Ga., Idaho, Ill., Ind., Iowa, Kans., Ky., La., Maine, Md., Mass., Mich., Minn., Miss., Mo., Mont., Nebr., Nev., N.H., N.J., N.Mex., N.Y., N.C., N.Dak., Ohio, Okla., Oreg., Pa., R.I., S.C., S.Dak., Tenn., Tex., Utah, Vt., Va., Wash., W.Va., Wis., Wyo.; Eurasia; n Africa; introduced also in South America (Argentina), Pacific Islands (New Zealand), Australia.

The *Potentilla recta* species group is among the most complex in Europe and western Asia, with diverse taxonomic circumscriptions. The most conservative is adopted here (also used by P. W. Ball et al. 1968; J. Soják 2004), in which most variation is included in a broadly defined *P. recta.* An alternative is provided by A. Kurtto et al. (in J. Jalas et al. 1972+, vol. 13), who recognized six subspecies of *P. recta* and three additional species. The phase found in North America is rather uniform, corresponding to the pale-flowered *sulphurea* variant that Kurtto et al. included in subsp. *recta.* However, B. Boivin (1952, 1966b) accepted two additional varieties in Canada, locally naturalized in Ontario and Saskatchewan. Nomenclaturally these correspond to subsp. *obscura* and subsp. *pilosa* as recognized by Kurtto et al., but other than 5-foliate leaves in var. *obscura* (Boivin 1967–1979), the morphological basis for the taxonomic conclusions by Boivin is unclear.

The earliest North American collection of *Potentilla recta* seen by the authors was gathered by C. W. Short as a weed in a garden in Montreal in 1852 (MO). It was encountered initially as a naturalized plant in New York in 1879, and by 1900 it was fairly widespread in

New England before being found in Ontario, Canada. From there it spread rapidly throughout much of the northern United States and southern Canada (K. Zouhar, http://www.fs.fed.us/database/feis/). It was in British Columbia by 1914, and in a Forest Service pasture in eastern Utah by 1924. The species is listed as a noxious weed in much of North America, spreading aggressively in both natural and disturbed habitats.

The species is often confused with the native, widespread *Potentilla gracilis* (sect. *Graciles*), but differs from the latter in having ephemeral basal leaves (often replaced by a new flush in later summer), pale yellow petals, and shorter styles. The long, stiff, spreading, tubercle-based hairs on stems and petioles of *P. recta*, overtopping a layer of short hairs and glands, also serve to distinguish this aggressive species from native *P. gracilis* at any stage of the life cycle. The leaves are also sometimes mistaken for *Cannabis*. The name *P. recta* is sometimes confused with *P. erecta*, a much different plant in sect. *Potentilla*.

8h. Potentilla Linnaeus sect. Terminales (Döll) Grenier in J. C. M. Grenier and D. A. Godron, Fl. France 1: 532. 1848–1849 [I]

Barbara Ertter

James L. Reveal

Potentilla [unranked] *Terminales* Döll, Rhein. Fl., 772. 1843; *Potentilla* [unranked] *Argenteae* Rydberg; *Potentilla* sect. *Argenteae* (Rydberg) Juzepczuk

Perennials, rosetted or tufted, not stoloniferous; taproots not fleshy-thickened; vestiture of long, short-crisped, and cottony or crisped-cottony hairs, glands absent or sparse to common, not red. **Stems** decumbent to erect, not flagelliform, not rooting at nodes, from centers of ephemeral basal rosettes, 1–6 dm, lengths (2–)3–5(–10) times basal <or proximal cauline> leaves. **Leaves:** basal not in ranks; cauline 2–9; primary leaves usually palmate, sometimes ternate, proximal ones 2–14 cm; petiole: long hairs loosely appressed to spreading, soft to weak, glands absent or sparse to common; leaflets 5–7, at tip of leaf axis, ± overlapping or not, oblanceolate to obovate, margins flat or revolute, distal ½–¾+ evenly to unevenly incised ⅓–¾+ to midvein, teeth 2–10 per side, surfaces similar to strongly dissimilar, abaxial green to white, cottony and/or crisped hairs absent or sparse to dense, adaxial green, not glaucous, long hairs weak to stiff. **Inflorescences** 10–100+-flowered, cymose, ± open. **Pedicels** usually straight in fruit, 0.3–1.5(–3) cm, proximal ± longer than distal. **Flowers** 5-merous; hypanthium 2–5 mm diam.; petals yellow, obovate to cuneate-obcordate, (2–)2.5–7(–8) mm, slightly shorter to ± longer than sepals, apex rounded to truncate or retuse; stamens ca. 20; styles subapical, columnar-tapered, scarcely to strongly papillate-swollen in proximal ⅕–½, 0.6–1.2 mm. **Achenes** smooth to rugose.

Species 20–30 (3 in the flora): introduced; Eurasia; also introduced in Pacific Islands (New Zealand).

As summarized by A. Kurtto et al. (in J. Jalas et al. 1972+, vol. 13), the species comprising sect. *Terminales* (including sect. *Argenteae*) consist of both sexual and apomictic populations of various ploidy levels that can be subdivided into more or less consistent species. The three species adventive in North America are relatively distinct, representing only a subset of European variation. Collections of *Potentilla inclinata* and *P. intermedia* are sometimes confused; the former often has petals smaller than the European average, and anthers are often intermediate in size. The distribution of these species in North America may need adjusting from what is presented here.

Another species complex, the *Potentilla collina* Wibel group (as addressed by A. Kurtto et al. in J. Jalas et al. 1972+, vol. 13), may be present in North America, at least as an occasional waif. The specimens underlying the citation of this species by P. A. Rydberg (1898, 1908d) are here identified as *P. argentea* (New York) and *P. inclinata* (Minnesota); however, variation of traits distinguishing members of sect. *Terminales* can be subtle and difficult to interpret out of their European context.

Potentilla intermedia is considered to be of hybrid origin involving *P. argentea* and *P. norvegica*; it appears to reproduce by both sexual and apomictic means. Some authors consider *P. inclinata* to be the hybrid derivative of *P. argentea* and *P. recta* (A. Kurtto et al. in J. Jalas et al. 1972+, vol. 13). The *P. collina* group is likewise thought to have a hybrid origin, involving members of sections *Aureae* and *Terminales*. Placement of these species in sect. *Terminales* is made on the basis of key morphologic characters.

1. Leaflet surfaces strongly dissimilar, abaxial white, cottony hairs dense, distal ½–⅔ of margin incised ½–¾+ to midvein, teeth 2 or 3 per side (more if lobed or secondarily toothed) . . . . 15. *Potentilla argentea*
1. Leaflet surfaces similar or ± dissimilar, abaxial green, grayish, or gray-green, cottony hairs absent, distal (½–)¾+ of margin incised ⅓–⅔ to midvein, teeth 4–10 per side.
   2. Anthers (0.5–)0.8–1.2 mm; petals 4–7(–8) mm; leaflets grayish to gray-green abaxially, short or crisped hairs usually ± abundant, sometimes sparse, margins usually evenly incised; epicalyx bractlets: lengths ± 1 times sepals . . . . 16. *Potentilla inclinata*
   2. Anthers 0.3–0.5 mm; petals 3–5 mm; leaflets green to grayish green abaxially, short or crisped hairs ± sparse, margins usually unevenly, sometimes evenly, incised (often with 1 or 2 incisions nearly to midvein); epicalyx bractlets: lengths usually ⅔, sometimes 1, times sepals . . . . 17. *Potentilla intermedia*

15. Potentilla argentea Linnaeus, Sp. Pl. 1: 497. 1753
• Silver or hoary cinquefoil, potentille argentée [I]

Stems mostly decumbent to ascending, 1–6 dm. **Basal leaves** palmate. **Cauline leaves** 2–9, proximal ones 2–7(–10) cm; proximal petioles 1–4(–7) cm, long hairs absent or sparse, ± ascending, 1–2 mm, soft, short and crisped hairs absent or obscured, cottony hairs dense, glands absent or obscured; leaflets 5(–7), central one ± oblanceolate, (0.5–)1–3 × 0.3–1(–1.3) cm, margins revolute, distal ½–⅔ evenly or unevenly incised ½–¾+ to midvein, teeth 2 or 3 per side (more if lobed or secondarily toothed), surfaces strongly dissimilar, abaxial white, long hairs ± sparse (mostly on veins), 1–2 mm, weak, short and crisped hairs absent or obscured, cottony hairs dense, glands absent or obscured, adaxial long hairs absent or sparse to common, short or crisped hairs absent or sparse, cottony hairs absent, glands absent or sparse. **Inflorescences** 10–80-flowered. **Pedicels** 0.3–1.5(–2) cm. **Flowers:** epicalyx bractlets oblong to narrowly ovate, 1.5–3 × 0.7–1.2 mm, lengths ⅔–1 times sepals; sepals 2–4.5 mm, apex ± acute; petals (2–)2.5–4 × 1.5–3 mm; filaments 0.8–1.5 mm, anthers 0.3–0.6(–0.8) mm; carpels 30–60, styles 0.6–0.9 mm, often strongly papillate-swollen proximally. **Achenes** 0.8–1.1 mm, smooth to lightly rugose. **2*n*** = 14, 28, 35, 42, 56, 62 (Eurasia).

Flowering spring–summer. Dry flats and slopes, roadsides, dry ditches, other open sites, in grasslands, oak and conifer woodlands; 0–2000 m; introduced; St. Pierre and Miquelon; Alta., B.C., Man., N.B., Nfld. and Labr. (Nfld.), N.S., Ont., P.E.I., Que., Sask.; Colo., Conn., D.C., Idaho, Ill., Ind., Iowa, Ky., Maine, Md., Mass., Mich., Minn., Mo., Mont., N.H., N.J., N.Y., N.C., N.Dak., Ohio, Oreg., Pa., R.I., S.Dak., Tenn., Vt., Va., Wash., W.Va., Wis., Wyo.; Eurasia; introduced also in Pacific Islands (New Zealand).

The bright, silvery vestiture of the leaves quickly distinguishes *Potentilla argentea* from *P. inclinata* and *P. intermedia*. Although treated here as introduced, some (P. A. Rydberg 1898; P. A. Werner and J. D. Soule 1976) considered *P. argentea* to be probably native in eastern North America.

The *Potentilla argentea* species group is an amphi- or apomictic polyploid complex that has been the subject of numerous cytological, sexual, and molecular studies (for example, A. Müntzing and G. Müntzing 1945; A. Müntzing 1958; J. Paule et al. 2011). Taxonomic treatments have ranged from a highly polymorphic single species to an abundance of species, varieties, and forms (T. Wolf 1908). This treatment is at the conservative end of the spectrum. Some New England references (R. C. Bean et al. 1967; F. C. Seymour 1969) distinguish plants with densely white-hairy adaxial leaflet surfaces as var. *pseudocalabra* Th. Wolf, a name based on plants from southeastern Europe. Molecular analysis by Paule et al. supports the conclusion by Wolf that the variety consists of intermediates between *P. argentea* and *P. calabra* Tenore. The optimum taxonomic disposition of var. *pseudocalabra*, and whether any North American plants are properly included, remains to be determined.

Potentilla argentea has been reported as an active agent in treating cirrhosis of the liver (N. T. Starostenko and V. N. Starostenko 1971).

16. Potentilla inclinata Villars, Hist. Pl. Dauphiné 3: 567, plate 45 [bottom left]. 1788, name conserved • Ashy cinquefoil, potentille inclinée [I]

Potentilla canescens Besser; *P. intermedia* Linnaeus var. *canescens* (Besser) Ruprecht

Stems ascending to erect, (1–)1.5–5 dm. **Basal leaves** palmate. **Cauline leaves** 3–9, proximal ones 5–10(–14) cm; proximal petioles 2–6(–8) cm, long hairs sparse to common, spreading to loosely appressed, 1.5–3 mm, soft to weak, short or crisped hairs abundant, cottony hairs mostly absent, glands sparse to common, usually obscured; leaflets 5–7, central one usually ± oblanceolate, (1.5–)2–5 × 0.5–1.5 cm, margins flat or ± revolute, distal (½–)¾+ usually evenly incised ½ to midvein, teeth 4–6(–10) per side, surfaces ± dissimilar, abaxial grayish to gray-green, long hairs sparse to common (especially on veins), 1–2.5 mm, soft to weak, short or crisped hairs ± abundant, sometimes sparse, cottony hairs absent, glands ± sparse (usually obscured), adaxial long hairs usually sparse, sometimes absent, 1–2 mm, short hairs absent or sparse, crisped and cottony hairs absent, glands absent or sparse. **Inflorescences** 10–50+-flowered. **Pedicels** 0.3–1.5(–3) cm. **Flowers:** epicalyx bractlets lanceolate or elliptic to oblong, 3–6 × 1–1.5 mm, length ± 1 times sepals; sepals 3.5–6.5 mm, apex acute to acuminate; petals 4–7(–8) × 4–5(–6) mm; filaments 0.8–2.5 mm, anthers (0.5–)0.8–1.2 mm; carpels 80–100+, styles 0.8–1.1 mm, scarcely papillate-swollen proximally. **Achenes** 1 mm, lightly rugose. **2*n*** = 14, 28, 35, 42, 84 (Eurasia).

Flowering summer. Dry waste places, along roadsides, ditches, other open sites, in grasslands, oak and conifer woodlands; 0–2200 m; introduced; B.C., Ont., P.E.I., Que.; Ariz., Conn., Ind., Mass., Mich., Minn., N.H., N.J., N.Y., N.C., Ohio, Pa., R.I., Vt., Va., Wash., Wis., Wyo.; Eurasia.

In addition to the characters given in the key, *Potentilla inclinata* in the flora area tends to have narrower leaflets, more compact inflorescences with fewer flowers, and more coarsely hairy epicalyx bractlets and sepals than *P. intermedia*. The species is currently becoming more common in western North America.

The name *Potentilla inclinata* has been conserved over *P. assurgens* Villars (T. Gregor et al. 2009).

17. Potentilla intermedia Linnaeus, Syst. Nat. ed. 12, 2: 351. 1767; Mant. Pl. 1: 76. 1767 • Downy cinquefoil, potentille intermédiaire [I]

Potentilla heidenreichii Zimmeter

Stems ascending to erect, 2–5 dm. **Basal leaves** palmate or ternate. **Cauline leaves** 3–8, proximal ones (3–)5–14 cm; proximal petioles (1–)2–10 cm, long hairs sparse to common, spreading to loosely appressed, 1–2 mm, weak, short or crisped hairs sparse to common, cottony hairs absent, glands absent or sparse; leaflets usually 5, central one oblanceolate to obovate, 2–4.5 × 1–2.5 cm, margins ± flat, distal ¾+ usually unevenly, sometimes evenly, incised ⅓–⅔ to midvein (often with 1 or 2 incisions nearly to midvein), teeth 5–10 per side, surfaces similar or ± dissimilar, abaxial green to grayish green, long hairs sparse to common (especially on veins), 0.5–1.5 mm, soft to weak, short or crisped hairs ± sparse, cottony hairs absent, glands absent or sparse, adaxial long hairs absent or sparse, 1 mm, short hairs absent or sparse, crisped and cottony hairs absent, glands absent or sparse. **Inflorescences** 20–100+-flowered. **Pedicels** 0.5–1.5(–2.5) cm. **Flowers:** epicalyx bractlets lanceolate to ovate or elliptic, 2–5 × 0.8–1.5 mm, lengths usually ⅔, sometimes 1, times sepals; sepals 3.5–6.5 mm, apex ± acute; petals 3–5 × 2–3 mm; filaments 0.8–2 mm, anthers 0.3–0.5 mm; carpels 40–70, styles 1–1.2 mm, scarcely papillate-swollen proximally. **Achenes** 1 mm, rugose. **2*n*** = 28, 56 (Eurasia).

Flowering late spring–summer. Dry waste places along roadsides, ditches, other open sites, in grasslands, oak and conifer woodlands; 0–300 m; introduced; St. Pierre and Miquelon; N.B., Nfld. and Labr. (Nfld.), N.S., Ont., P.E.I., Que.; Conn., Ind., Iowa, Maine, Mass., Mich., N.H., N.J., N.Y., Ohio, Pa., R.I., Vt., Wis.; Eurasia.

Potentilla intermedia is more common than *P. inclinata* at least in eastern Canada and New England.

8i. Potentilla Linnaeus sect. Chrysanthae (Th. Wolf) Juzepczuk in V. L. Komarov et al., Fl. URSS 10: 180. 1941

Reidar Elven

Barbara Ertter

James L. Reveal

Potentilla [unranked] *Chrysanthae* Th. Wolf in P. F. A. Ascherson et al., Syn. Mitteleur. Fl. 6(1): 671, 775. 1904

Perennials, rosetted or tufted, not stoloniferous; taproots not fleshy-thickened; vestiture mostly of long hairs (and crisped hairs in inflorescence), glands absent or sparse to abundant, sometimes red (*P. rubella*). **Stems** ascending to erect, not flagelliform, not rooting at nodes, lateral to persistent basal rosettes, (0.2–)1–5(–7) dm, lengths usually 2–5 times basal leaves, rarely shorter than leaves. **Leaves:** basal not in ranks; cauline 1–3; primary leaves palmate, 3–25(–35) cm; petiole: long hairs spreading to appressed, weak to stiff, glands absent or sparse to common; leaflets (4–)5–11, at tip of leaf axis, ± overlapping or not, narrowly oblanceolate or oblanceolate-elliptic to obovate, margins flat, distal ½–¾+ evenly incised ¼–⅓(–½) to midvein, teeth 1–10(–13) per side, or distal ¼ or less incised less than ⅒(–¼) to midvein, teeth 1 or 2(–5) per side (*P. stipularis*), surfaces ± similar, green to pale green or reddish, sometimes glaucous, long hairs absent or weak to stiff, cottony hairs absent. **Inflorescences** (1 or)2–20-flowered, usually cymose, open. **Pedicels** usually ± straight in fruit, sometimes ± curved (*P. thuringiaca*), 1–3(–6) cm, proximal often much longer than distal. **Flowers:** 5-merous; hypanthium 2.5–5 mm diam.; petals yellow, ± obcordate, (4–)5–10 mm, longer than sepals, apex retuse; stamens ca. 20; styles subapical, ± columnar, rarely columnar-filiform, to tapered, not or ± papillate-swollen proximally, 0.7–1.3 mm. **Achenes** smooth to faintly rugose.

Species ca. 8 (3 in the flora): n North America, Eurasia.

1. Cauline leaves: stipules fused with less than ⅓ of petiole, free portion longer than fused portion; basal leaves: distal ⅔–¾+ of leaflet margins with (4–)6–10(–13) teeth per side; petiole hairs ± spreading, 1–2.5 mm; introduced; Quebec and Rhode Island . . . . 18. *Potentilla thuringiaca*
1. Cauline leaves: stipules fused with all or most of petiole, free portion shorter than fused portion; basal leaves: distal ⅒–½ or less of leaflet margins with 1 or 2(–5) teeth per side; petiole hairs absent or appressed to spreading, 0.5–1.5 mm; native; Alaska and Greenland.
   2. Leaflets 7–11, margins incised in distal ¼ or less; glands mostly absent . . . . 19. *Potentilla stipularis*
   2. Leaflets (4 or)5(or 6), margins incised in distal ± ½; glands sparse to abundant (especially on pedicels, epicalyx bractlets, and sepals) . . . . 20. *Potentilla rubella*

18. Potentilla thuringiaca Bernhardi ex Link, Enum. Hort. Berol. Alt. 2: 64. 1822 • German cinquefoil, potentille de Thuringe I

Potentilla chrysantha Treviranus subsp. *thuringiaca* (Bernhardi ex Link) Ascherson & Graebner; *P. goldbachii* Ruprecht; *P. nestleriana* Trattinnick

Glands sparse, uncolored, minute. **Stems** 1.5–5(–7) dm. **Basal leaves** 4–25(–35) cm; petiole 2–20(–25) cm, hairs sparse to common, ± spreading, 1–2.5 mm, weak, glands absent or sparse; leaflets 5–7(–9), central one narrowly to broadly oblanceolate, 1–6(–9) × 0.5–2 cm, distal ⅔–¾+ of margin incised ¼–⅓(–½) to midvein, teeth (4–)6–10(–13) per side, surfaces ± similar, abaxial green to pale green, hairs sparse to common on primary veins, ascending to spreading, 0.5–1.5 mm, weak, adaxial green, hairs more appressed and shorter. **Cauline leaves** 1–3; stipules fused with less than ⅓ of petiole, free portion longer than fused portion. **Inflorescences** 5–20-flowered. **Pedicels** 1–3 cm (proximalmost to 5 cm). **Flowers:** epicalyx bractlets linear to lanceolate, 3–8 × 1–1.5(–2) mm; hypanthium 3–5 mm diam.; sepals 3–5 mm, apex ± acute; petals (5–)8–10 × 4–10 mm; filaments 0.3–1.5 mm, anthers 0.7–0.9 mm; carpels 35–50, styles 1.1–1.3 mm. **Achenes** 1–1.2 mm, faintly rugose. $2n$ = 42, 56 (Eurasia).

Flowering late spring–summer. Edges of and openings in hardwood and conifer woodlands; 0–100 m; introduced; Que.; R.I.; Eurasia.

Potentilla thuringiaca might be mistaken for the more widely established *P. intermedia* of sect. *Terminales.* In addition to having flowering stems lateral to persistent basal leaves, *P. thuringiaca* has much larger petals (mostly 8–10 mm versus 4–5 mm), five to nince leaflets, and tubercles at the bases of the long, spreading hairs.

To the Quebec records (and subsequent collections) reported by R. Cayouette (1966) is added Block Island, Rhode Island (*Fernald et al. 9650*, GH).

19. Potentilla stipularis Linnaeus, Sp. Pl. 1: 498. 1753 • Stipulate cinquefoil [F]

Potentilla stipularis var. *groenlandica* T. J. Sørensen

Glands mostly absent. **Stems** (0.2–)1–2.5(–3.5) dm. **Basal leaves** 3–7(–10) cm; petiole 2–5(–7.5) cm, hairs absent or sparse, appressed to spreading, 0.5–0.8 mm, stiff, glands absent or nearly so; leaflets 7–11, central one oblanceolate-elliptic, 1–2(–3) × 0.3–0.8 cm, distal ¼ or less of margin incised less than 1/10(–¼) to midvein, teeth 1 or 2(–5) per side, surfaces similar, abaxial pale green to reddish, often glaucous, hairs absent or sparse to common on primary veins, appressed to ascending, 0.6–1 mm, stiff, adaxial green or reddish, glabrous or nearly so. **Cauline leaves** 1–3; stipules fused with all or most of petiole, free portion shorter than fused portion. **Inflorescences** 2–5-flowered. **Pedicels** 1–3 cm (proximalmost to 6 cm). **Flowers:** epicalyx bractlets lanceolate to narrowly lanceolate, 3–5 × 0.7–1(–1.5) mm; hypanthium 2.5–3 mm diam.; sepals 4–6 mm, apex acute; petals (4–)6–8 × 4–6 mm; filaments 1.6–2 mm, anthers 0.4–0.6 mm; carpels 40–50, styles 0.7–0.9 mm. **Achenes** 1.1–1.3 mm, smooth. $2n$ = 28 (Eurasia).

Flowering summer. Herb meadows, alluvial meadows and flats, open shrub thickets, *Salix-Dryas* tundra; 10–600 m; Greenland; Alaska; Eurasia.

The occurrences of *Potentilla stipularis* in Alaska are a continuation of the range from Asia. The six known occurrences in eastern and northeastern Greenland are disjunct (A. E. Porsild 1964) and were assigned to var. *groenlandica* by Sørensen, differing from Asian and Alaskan counterparts mainly in quantitative features: nine to eleven leaflets, seven to 11 teeth per leaflet, and not conspicuously glaucous (G. Halliday, pers. comm.). Here var. *groenlandica* is considered a high-arctic ecotype that provisionally is not accepted taxonomically.

20. Potentilla rubella T. J. Sørensen, Meddel. Grønland 101(2): 106, plates 1–3. 1934

Potentilla rubelloides V. V. Petrovsky

Glands sparse to abundant, red, stipitate (at least on calyx and hypanthium). **Stems** 0.5–2 dm. **Basal leaves** 3–6 cm; petiole 2–4 cm, hairs common to abundant, ± ascending to spreading, 1–1.5 mm, weak to stiff, glands sparse to common; leaflets (4 or)5(or 6), central one obovate, 0.5–1.5 × 0.3–0.9 cm, distal ca. ½ of margin incised ca. ¼ to midvein, teeth (2 or)3(or 4) per side, surfaces ± similar, abaxial pale reddish green, hairs sparse to common on primary veins, spreading, 0.6–1 mm, ± stiff, adaxial green or reddish, glabrous or sparsely hairy. **Cauline leaves** 1 or 2; stipules fused with all or most of petiole, free portion shorter than fused portion. **Inflorescences** (1 or)2–5-flowered. **Pedicels** 1–3 cm (proximalmost to 5 cm). **Flowers:** epicalyx bractlets linear to narrowly lanceolate, 4–6 × 0.9–1.2 mm; hypanthium 3–4 mm diam.; sepals 5–7 mm, apex subacute or acute; petals 7–9 × 4–6 mm; filaments 0.7–0.9 mm, anthers 0.4 mm; carpels 40–50, styles 0.9–1.1 mm. **Achenes** not known. $2n$ = 28, 42 (Siberia).

Flowering summer. Herb-*Salix* meadows, herb slopes, open tundra, among mosses; 0–500 m; Greenland; Asia.

Potentilla rubella combines characteristics of *P. hyparctica* (sect. *Aureae*) and *P. stipularis.* The species is strongly supported as an intersectional hybrid with polytopic origins. It reproduces by seed, shows no transitions to its presumed parental species, and has significant ranges in eastern and northeastern Greenland and northern Asia.

8j. Potentilla Linnaeus sect. Rubrae (Rydberg) O. Stevens in N. L. Britton et al., N. Amer. Fl. 22(7): 11. 1959

Barbara Ertter

James L. Reveal

Potentilla [unranked] *Rubrae* Rydberg in N. L. Britton et al., N. Amer. Fl. 22(4): 293, 323. 1906; *Potentilla* sect. *Haematochrus* (Rydberg) Schiman-Czeika; *Potentilla* [unranked] *Haematochrus* Rydberg

Perennials, ± tufted, not stoloniferous; taproots not fleshy-thickened; vestiture of long, short, crisped, and/or cottony hairs, glands sparse to abundant, rarely reddish. **Stems** ascending to nearly erect, not flagelliform, not rooting at nodes, lateral to persistent basal rosettes, (2–)3–7 (–10) dm, lengths (2–)3–5 times basal leaves. **Leaves:** basal not 2-ranked; cauline 1–3; primary leaves palmate to ± subpalmate, (2–)4–15(–30) cm; petiole: long hairs spreading to weakly appressed, weak to ± stiff, glands sparse to abundant, rarely absent; leaflets 5–7, on tip or less than distal 1/10 of leaf axis, ± overlapping or not, oblanceolate or oblong to elliptic or obovate, margins flat, distal 1/2 to whole length evenly incised 1/5–1/4 or less to midvein, teeth (0–)4–15(–19) per side, surfaces similar or slightly to strongly dissimilar, abaxial green to white, cottony hairs absent or sparse to dense, adaxial green, not glaucous, long hairs weak to stiff. **Inflorescences** (4–)10–40(–70)-flowered, cymose, open. **Pedicels** straight in fruit, 0.3–4 cm, proximal ± longer than distal. **Flowers** 5-merous; hypanthium 4–6 mm diam.; petals dark reddish to reddish orange, ± obcordate, (3.5–)5–10 mm, equal to or longer than sepals, apex retuse; stamens 20–30; styles subapical, tapered-filiform, papillate-swollen in proximal 1/5, 2–3.5 mm or less. **Achenes** ± rugose.

Species 9 (2 in the flora): sw United States, Mexico, Asia.

Red petals distinguish species of sect. *Rubrae* from the remaining species assigned here to *Potentilla.* Most species in the section occur in Mexico. As traditionally defined, sect. *Rubrae* also includes the southeastern Asian species *P. atrosanguinea* Loddiges, G. Loddiges & W. Loddiges (three leaflets), *P. argyrophylla* Wallich ex Lehmann (three leaflets), and *P. nepalensis* Hooker (five leaflets). The first is available in the horticultural trade and was reported in a strawberry field in New Brunswick, Canada (B. Boivin 1966b; H. J. Scoggan 1978–1979, part 3); it did not become naturalized. The name *P. argyrophylla* var. *atrosanguinea* (Loddiges, G. Loddiges & W. Loddiges) Hooker f., used in the Flora of China (Li C.-L. et al. 2003c) is incorrect; *P. argyrophylla* (1831) is a name later than *P. atrosanguinea* (1823).

1. Basal leaves palmate; leaflets 5(–7), distal 3/4 to whole margin incised, surfaces slightly to strongly dissimilar; petals ± dark reddish throughout; cauline leaves: stipules usually toothed, sometimes entire; widespread in Arizona, New Mexico . . . . 21. *Potentilla thurberi*
1. Basal leaves usually ± subpalmate, sometimes palmate; leaflets (5–)7, distal 1/2–3/4 (sometimes nearly whole) margin incised, surfaces ± similar; petals dark reddish proximally, reddish orange distally; cauline leaves: stipules usually entire; Coconino County, Arizona. . . . 22. *Potentilla sanguinea*

21. Potentilla thurberi A. Gray, Pl. Nov. Thurb., 318. 1854 • Scarlet or Thurber's cinquefoil

Stems (2–)3–7(–10) dm. **Basal leaves** palmate, (2–)4–15(–30) cm; petiole (2–)3–10(–25) cm, long hairs sparse to common, spreading, 1–3.5 mm, weak, glands sparse to abundant, sometimes absent; leaflets 5(–7), at tip of leaf axis, central one (1–)2–6(–8) × (0.5–)1–2.5(–4) cm, petiolules 0(–10) mm, distal 3/4 to whole margin incised 1/5–1/4 to midvein, teeth (4–)7–15(–19) per side (sometimes secondarily toothed), (1–)2–3(–5) mm, teeth apex acute to ± obtuse, surfaces slightly to strongly dissimilar, abaxial pale green to white, straight hairs sparse to abundant, often dense on veins, 0.5–3 mm, cottony hairs absent or sparse to dense, glands ± sparse or absent, sometimes obscured, adaxial green, straight hairs sparse to abundant, 0.2–1 mm, cottony hairs absent, glands ± sparse or absent. **Cauline leaves:** stipules usually toothed, sometimes entire. **Inflorescences** (4–)10–35 -flowered. **Pedicels** 0.5–4 cm. **Flowers:** epicalyx bractlets narrowly to broadly lanceolate-elliptic (rarely toothed or lobed), 4–9(–14) × (1–)1.5–2 mm; sepals 4–10(–15) mm, apex acute to acuminate; petals ± dark reddish throughout, (3.5–)6–10 × 6–10 mm; filaments 1.5–3 mm, anthers 0.8–1 mm; carpels 30–70, styles 2.5–3.5 mm. **Achenes** 1.5 mm, ± rugose.

Varieties 2 (2 in the flora): sw United States, nw Mexico.

The abaxially white-cottony leaflets of var. *atrorubens* usually make it easily distinguished from var. *thurberi*, which lacks cottony hairs and is usually less hairy in general. The two varieties sometimes occur together, and intermediates between the extremes are common.

The degree of glandularity and the lengths of the central leaflets are greater in the southern populations of both varieties. The glands are less widely distributed on the plants northward and are fewer in northern Arizona and New Mexico. Leaflets are shorter and broader north of Graham County, Arizona, and Otero County, New Mexico.

1. Leaflets pale green abaxially, cottony hairs absent . . . . 21a. *Potentilla thurberi* var. *thurberi*
1. Leaflets ± white to white abaxially, cottony hairs sparse to dense . . . . 21b. *Potentilla thurberi* var. *atrorubens*

21a. Potentilla thurberi A. Gray var. **thurberi**

Potentilla rubida L. O. Williams

Leaflets: teeth apex ± obtuse, abaxial surfaces pale green, straight hairs sparse to common on veins, cottony hairs absent.

Flowering summer. Moist meadows, dry grassy slopes, in oak and conifer woodlands; 1800–2900 m; Ariz., N.Mex.; Mexico (Chihuahua).

21b. Potentilla thurberi A. Gray var. **atrorubens** (Rydberg) Kearney & Peebles, J. Wash. Acad. Sci. 29: 481. 1939

Potentilla atrorubens Rydberg, Bull. Torrey Bot. Club 24: 11, plate 288. 1897

Leaflets: teeth apex acute, abaxial surfaces ± white to white, straight hairs usually dense on veins, cottony hairs sparse to dense.

Flowering summer. Dry grassy slopes, moist meadows and forests, oak and conifer woodlands; 1800–2900 m; Ariz., N.Mex.; Mexico (Chihuahua, Sonora).

22. Potentilla sanguinea Rydberg in N. L. Britton et al., N. Amer. Fl. 22: 324. 1908 • Flagstaff cinquefoil [C] [E]

Potentilla thurberi A. Gray var. *sanguinea* (Rydberg) Kearney & Peebles

Stems 3–7(–10) dm. **Basal leaves** ± subpalmate, sometimes palmate, (4–)6–15(–21) cm; petiole (2–)4–10(–15) cm, long hairs sparse to abundant, spreading to weakly appressed, 0.5–2(–3) mm, weak to ± stiff, glands sparse to abundant; leaflets (5–)7, on less than distal 1/10 (1–5[–10] mm) of leaf axis, central one 1–6(–7) × 1–2(–3) cm, petiolules 0(–6) mm, distal 1/2–3/4 (sometimes nearly whole), margin incised 1/5 or less to midvein, teeth 4–10(–15) per side, 1–2(–3) mm, teeth apex ± obtuse, surfaces ± similar, abaxial green, straight hairs sparse to abundant, especially on veins, 0.5–1.5 mm, cottony hairs absent, glands ± sparse or absent, adaxial slightly paler green, hairs shorter and sparser. **Cauline leaves:** stipules usually entire. **Inflorescences** (4–)8–40(–70)-flowered. **Pedicels** 0.3–2 cm. **Flowers:** epicalyx bractlets lanceolate-elliptic, 3.5–8 × 1–1.5(–2) mm; sepals 4–8(–10) mm, apex acuminate; petals dark reddish proximally, reddish orange distally, (3.5–)5–9 × 5–10 mm; filaments 1.5–2.5 mm, anthers 0.8 mm; carpels 30–40, styles 2–3 mm. **Achenes** 1.5–2 mm, lightly rugose.

Flowering summer. Moist forest settings in oak and conifer woodlands; of conservation concern; 2100–2200 m; Ariz.

Potentilla sanguinea is known only from south and east of Flagstaff, Coconino County. The species may be a hybrid between *P. thurberi* var. *thurberi* and *P. hippiana* or another member of sect. *Leucophyllae* that occurs sympatrically.

8k. POTENTILLA Linnaeus sect. GRACILES (Rydberg) A. Nelson, Bull. Torrey Bot. Club 26: 480. 1899

Barbara Ertter

James L. Reveal

Potentilla [unranked] *Graciles* Rydberg, Bull. Torrey Bot. Club 24: 4. 1897; *Potentilla* sect. *Candidae* (Rydberg) O. Stevens; *Potentilla* sect. *Heptaphyllae* (Rydberg) O. Stevens; *Potentilla* sect. *Longipedunculatae* (Rydberg) O. Stevens; *Potentilla* sect. *Nuttallianae* (Rydberg) O. Stevens; *Potentilla* sect. *Pectinisectae* (Rydberg) O. Stevens; *Potentilla* sect. *Permolles* (Rydberg) O. Stevens

Perennials, ± tufted, not stoloniferous; taproots not fleshy-thickened; vestiture of long, short-crisped, and sometimes cottony(/crisped) hairs, glands absent or sparse to abundant, sometimes red (*P. pulcherrima*). **Stems** decumbent to erect, not flagelliform, not rooting at nodes, lateral to persistent basal rosettes, (0.5–)2–7(–10) dm, lengths (1.5–)2–4 times basal leaves. **Leaves:** basal sometimes 2-ranked; cauline (0 or)1–4(or 5); primary leaves palmate to subpalmate, sometimes subpinnate (with distal leaflets ± confluent), not flagelliform, 2–40 cm; petiole: long hairs spreading to appressed, weak to stiff, glands absent or sparse to abundant; leaflets 5–9, at tip or on up to distal ¼(–¾) of leaf axis, overlapping or not, narrowly oblanceolate, elliptic, or cuneate to obovate, margins flat to ± revolute, distal ¼ to nearly whole length evenly to unevenly incised ¼ to nearly to midvein, teeth (1 or)2–10(–12) per side, surfaces similar to strongly dissimilar, abaxial green to white, cottony hairs absent or sparse to dense, adaxial green to grayish, sometimes blue-green and glaucous, long hairs mostly weak, sometimes stiff, soft, or absent. **Inflorescences** (2–)5–50(–70)-flowered, cymose, ± to very open. **Pedicels** straight in fruit, 0.2–3(–6.5) cm, proximal often much longer than distal. **Flowers** 5-merous; hypanthium (3–)3.5–6(–8) mm diam.; petals bright yellow, ± obcordate, (3–)4–10(–12) mm, usually longer than sepals, apex retuse; stamens ca. 20; styles subapical, filiform to tapered, papillate-swollen in proximal ⅒–⅓(–⅔), (1–)1.5–3 mm. **Achenes** smooth.

Species 7 or more (7 in the flora): North America, Mexico.

Section *Graciles* consists primarily of decumbent to erect, palmate-leaved, yellow-petaled perennials that are the most abundant *Potentilla* in mid-montane western North America. Most of the diversity treated here as five varieties of the most common species, *P. gracilis*, was divided by P. A. Rydberg (1908d) into 32 species arranged in six of his unranked groups: *Candidae*, *Graciles*, *Longipedunculatae*, *Nuttallianae*, *Pectinisectae*, and *Permolles*. The number of species in sect. *Graciles* as now defined is problematic. Although understanding of variation in the continental United States has benefited from the biosystematic analyses of J. Clausen et al. (1940), the potential Mexican representatives have not received equivalent attention. It is likely that some Mexican species placed by B. C. Johnston (1985) in sect. *Ranunculoides* (Th. Wolf) Juzepczuk are better accommodated in sect. *Graciles*, as already done here with an expanded circumscription of *P. townsendii*. *Potentilla leptophylla* Rydberg, the sole representative of group *Heptaphyllae* Rydberg, probably belongs in this section as well. The relationship of the North American sect. *Graciles* to the Eurasian sect. *Chrysanthae* remains to be determined.

The section contains some of the most taxonomically challenging groups among the North American cinquefoils, made all the more complex because its members often hybridize with species within and beyond sect. *Graciles*. Facultative apomixis is highly probable, resulting in populations of uncertain taxonomic status. The approach used here is to key and describe only the core representatives of the entities treated here, with some of the additional variation accounted for in the discussion.

The circumscriptions adopted here approximate those of J. Clausen et al. (1940), albeit with some species treated as varieties. Because the studies by Clausen et al. were centered in California, variation in the Rocky Mountains, Colorado Plateau, and northern prairies remains poorly analyzed.

1. Basal leaves subpalmate to pinnate; leaflets on distal 1/10–1/2(–3/4) of leaf axis.
    2. Leaflet surfaces usually strongly dissimilar, abaxial grayish to white, cottony (or crisped-cottony) hairs abundant to dense; mostly Great Basin and east . . . . . . 27. *Potentilla pulcherrima* (in part)
    2. Leaflet surfaces ± similar, abaxial green or ± blue-green to grayish, cottony hairs usually absent (short-crisped hairs sometimes sparse to common); mostly w, n of the Great Basin.
        3. Leaflets usually glaucous, ± blue-green, ± evenly incised on distal 1/3–1/2(–2/3), teeth 1–3(–5) per side; anthers 0.4–0.7(–0.9) mm . . . . 23. *Potentilla glaucophylla* (in part)
        3. Leaflets not glaucous, green to grayish, evenly to unevenly incised on distal 1/2–3/4, teeth (2 or)3–7(or 8) per side (sometimes secondarily toothed); anthers 0.7–1.2 mm.
            4. Leaflets usually gray-green to grayish, long hairs ± abundant, short-crisped hairs usually sparse to common; petioles: long hairs abundant, weak, appressed to ascending-spreading . . . . 28. *Potentilla bruceae*
            4. Leaflets green, long hairs nearly absent or sparse to common (sometimes restricted to abaxial veins), short and crisped hairs usually absent; petioles: long hairs absent or sparse to abundant, usually stiff, tightly appressed . . . . 29. *Potentilla drummondii*
1. Basal leaves palmate, rarely subpalmate; leaflets on tip or less than distal 1/10 of leaf axis.
    5. Leaflet surfaces usually strongly dissimilar, abaxial grayish to white, not glaucous, cottony (or crisped-cottony) hairs abundant to dense.
        6. Glands usually absent or inconspicuous, uncolored; leaflets evenly to unevenly incised 1/4–3/4+ to midvein; styles ± tapered, papillate-swollen proximally . . . . 25. *Potentilla gracilis* (in part)
        6. Glands usually conspicuous, red-tipped; leaflets evenly incised 1/4–1/2 to midvein; styles filiform to tapering above papillate-swollen base. . . . . 27. *Potentilla pulcherrima* (in part)
    5. Leaflet surfaces similar to ± dissimilar, abaxial green, ± blue-green, or pale green to grayish, rarely whitish, sometimes glaucous, crisped hairs sparse or abundant, sometimes absent, cottony hairs usually absent.
        7. Leaflets usually glaucous, ± blue-green, incised on distal 1/3–1/2(–2/3), teeth 1–3(–5) per side; anthers 0.4–0.7(–0.9) mm; stems 0.5–3(–4.5) dm; inflorescences 2–10 (–20)-flowered . . . . 23. *Potentilla glaucophylla* (in part)
        7. Leaflets not glaucous, dark green to grayish, rarely whitish, incised on distal (1/4–)1/2 to nearly whole length, teeth (2–)5–10(or 11) per side; anthers (0.5–)0.6–1.2(–1.6) mm; stems (0.5–)2–7(–10) dm; inflorescences (4–)10–50(–70)-flowered.

[8. Shifted to left margin.—Ed.]

8. Leaflets narrowly oblanceolate-elliptic, incised ¼–⅓ to midvein, teeth 1–2 mm, abaxially green, long hairs usually absent or limited to veins and margins, short and crisped hairs absent; Arizona and New Mexico . . . . 24. *Potentilla townsendii*
8. Leaflets oblanceolate to elliptic or obovate, incised ¼–¾+ to midvein, teeth 2–17(–20) mm, abaxially green or pale green to grayish, rarely whitish, long hairs sparse to abundant, sometimes mostly on veins, short-crisped hairs absent or sparse to abundant; mostly n, w of Arizona and New Mexico.
    9. Glands usually absent or inconspicuous, uncolored; leaflets dark green to grayish, rarely whitish; epicalyx bractlets: hairs sparse to abundant, rarely dense, ± appressed to ascending, rarely spreading . . . . 25. *Potentilla gracilis* (in part)
    9. Glands conspicuous, usually golden; leaflets green; epicalyx bractlets: hairs sparse to common, loosely appressed to spreading. . . . . 26. *Potentilla brunnescens*

23. Potentilla glaucophylla Lehmann, Index Seminum (Hamburg) 1836: 7. 1836 • Blueleaf or vari-leaved cinquefoil [E] [F]

Potentilla diversifolia Lehmann var. *glaucophylla* (Lehmann) Lehmann

Glands absent or inconspicuous, uncolored. **Stems** mostly ± ascending, 0.5–3(–4.5) dm. **Basal leaves** often 2-ranked, palmate, sometimes subpalmate, 2–20 cm; petiole (0.5–)1–8(–14) cm, long hairs often absent, sometimes sparse to abundant, usually appressed, 1–2 mm, weak to ± stiff, short, crisped, and cottony hairs absent, glands absent or sparse; leaflets 5 or 6(or 7), sometimes with 1 or 2(–4) additional, smaller, entire leaflets, on tip or to distal 1/10(–¼) of leaf axis, separate, largest ones oblanceolate or cuneate to obovate, 1–4(–6) × 0.5–1.5(–2) cm, margins flat, distal ⅓–½(–⅔) ± evenly incised ½ or nearly to midvein, undivided medial blade 1.5–7(–10) mm wide, teeth 1–3(–5) per side, broadly lanceolate or linear to narrowly oblong, 1–5(–8) mm, surfaces similar <adaxial ± less hairy>, ± blue-green, usually glaucous, long hairs sparse to abundant (often restricted to veins, distal teeth, and margins), short, crisped, and cottony hairs absent, glands absent or sparse. **Cauline leaves** (0 or)1 or 2(or 3). **Inflorescences** 2–10(–20)-flowered. **Pedicels** (0.5–)1–3(–6.5) cm. **Flowers:** epicalyx bractlets lanceolate to elliptic, 2–5 × 0.8–1.2(–1.6) mm, hairs sparse to common, rarely abundant, ± ascending, glands absent or inconspicuous; hypanthium 3.5–5 mm diam.; sepals (2–)2.5–4.5(–5) mm, apex acute to short acuminate; petals (4–)5–10(–12) × 4–9(–10) mm; filaments 1.5–2.5(–3) mm, anthers 0.4–0.7(–0.9) mm; carpels 25–40, styles filiform above papillate-swollen base, 1.5–2.5(–3) mm. **Achenes** 1.2–1.6 mm.

Varieties 2 (2 in the flora): North America.

Potentilla glaucophylla replaces *P. diversifolia*; an examination of the lectotype of the latter confirms J. Soják's (1996) conclusion that *P. diversifolia* applies to a hybrid involving *P. glaucophylla* and *P. hippiana*. Historically, this species was often termed *P. dissecta* Pursh, a misapplied name that has been rejected (B. Ertter et al. 2008).

In addition to the nomenclatural change, the circumscription of *Potentilla glaucophylla* is here restricted to plants with usually glaucous, blue-green, distally toothed leaflets that are usually glabrate (at least in var. *glaucophylla*). Plants from the Colorado Plateau and southern Rocky Mountains formerly included in this species and having dark green leaves, more leaflet teeth, and larger anthers are transferred either to *P. townsendii* or to a currently undescribed entity.

The two-ranked leaves of *Potentilla glaucophylla* suggest a possible connection to *P. crantzii* in sect. *Aureae*.

1. Leaflets incised ⅓–½ to midvein, teeth broadly lanceolate; range of species . . . . 23a. *Potentilla glaucophylla* var. *glaucophylla*
1. Leaflets incised ¾+ to midvein, teeth linear to narrowly oblong; northern Rocky Mountains . . . . 23b. *Potentilla glaucophylla* var. *perdissecta*

23a. Potentilla glaucophylla Lehmann var. **glaucophylla** [E] [F]

Potentilla diversifolia Lehmann subsp. *ranunculus* (Lange) A. E. Porsild; *P. diversifolia* var. *ranunculus* (Lange) B. Boivin; *P. ranunculus* Lange

Stems (0.5–)1–3(–4.5) dm. **Basal leaves** palmate, rarely subpalmate; distal ⅓–½(–⅔) of leaflet margins incised ⅓–½ to midvein, undivided medial blade 3–7(–10) mm wide, teeth broadly lanceolate. **2*n*** = 42, 90, 91, 101.

Flowering summer. Edges of meadows, dry gravelly flats and slopes in montane to subalpine conifer woodlands, mixed grassland communities, alpine tundra; (0–)100–3500 m; Greenland; Alta., B.C., Nfld. and Labr. (Labr.), N.W.T., Sask., Yukon; Alaska, Calif., Colo., Idaho, Mont., Nev., Oreg., Utah, Wash., Wyo.

Variety *glaucophylla* is widespread through much of western North America from southern Alaska, Yukon, and Northwest Territories south mainly in the high mountains to California, central Colorado, and Utah. The variety is also found on scattered mountain ranges in the Great Basin and is a provincial rarity in southwestern Saskatchewan. Greenland plants that have been called var. *ranunculus* are also included here, since the authors can find no morphologic differences in spite of the significant geographic disjunction. The most common phase is glabrate; hairy populations are primarily found in the Rocky Mountains and the Mount Angeles area of Clallam County, Washington.

The lower elevation is provided by the Greenland populations. The variety otherwise tends to occur at higher elevations, and to bloom somewhat earlier, than sympatric *Potentilla gracilis*, which is the species most often confused with var. *glaucophylla*.

23b. Potentilla glaucophylla Lehmann var. **perdissecta** (Rydberg) Soják, Thaiszia 16: 49. 2006 C E

Potentilla perdissecta Rydberg in N. L. Britton et al., N. Amer. Fl. 22: 327. 1908; *P. diversifolia* Lehmann var. *perdissecta* (Rydberg) C. L. Hitchcock

Stems 0.5–2 dm. **Basal leaves** palmate to subpalmate; distal 1/2–2/3 of leaflet margins incised 3/4+ to midvein, undivided medial blade 1.5–3 mm wide, teeth linear to narrowly oblong.

Flowering summer. Edges of meadows, dry gravelly flats and slopes, in montane to subalpine conifer woodlands, mixed grassland communities in alpine tundra; of conservation concern; 1500–3000 m; Alta.; Idaho, Mont., Wash., Wyo.

Variety *perdissecta* is confined to the northern Rocky Mountains of extreme southwestern Alberta, south into eastern Idaho, western Montana, and western Wyoming. The variety is also reported from Okanogan County, Washington. Apparent intergradation occurs with pinnate-leaved *Potentilla ovina* var. *decurrens* (sect. *Multijugae*).

24. Potentilla townsendii Rydberg, Bull. Torrey Bot. Club 28: 174. 1901 • Townsend's cinquefoil F

Glands absent or inconspicuous, uncolored. **Stems** ascending to nearly erect, (1–)2–5(–7.5) dm. **Basal leaves** usually not in ranks, palmate, rarely subpalmate, (2–)5–20(–30) cm; petiole (1–)2–15(–25) cm, long hairs usually absent or sparse and appressed, sometimes common and spreading, 1–1.5 mm, stiff, short, crisped, and cottony hairs absent, glands absent; leaflets 5–7(or 8), usually on tip, rarely on less than distal 1/10, of leaf axis, separate, largest ones narrowly oblanceolate-elliptic, (1.5–)2–5(–6) × 0.5–1.2(–1.5) cm, margins flat, distal (1/4–)1/2–3/4 evenly incised 1/4–1/3 to midvein, undivided medial blade 4–11 mm wide, teeth (2–)4–8(or 9) per side, mostly broadly lanceolate, 1–2 mm, surfaces similar, green, not glaucous, long hairs usually absent or limited to veins and margins, short, crisped, and cottony hairs absent, glands absent. **Cauline leaves** 1–3. **Inflorescences** (5–)10–20(–25)-flowered. **Pedicels** 0.7–3 (–3.5) cm. **Flowers:** epicalyx bractlets lanceolate to elliptic, (2–)2.5–5(–6) × 0.5–1(–1.4) mm, hairs sparse, ± appressed, glands absent; hypanthium 3.5–5 mm diam.; sepals 4–8(–9) mm, apex acute to long acuminate; petals 5–10 × 5–10(–12) mm; filaments 1–3 mm, anthers (0.5–) 0.6–1 mm; carpels 15–30, styles filiform above papillate-swollen base, (1–)1.5–2 mm. **Achenes** 1.1–1.6 mm.

Flowering summer. Edges of meadows, dry gravelly flats and slopes, in montane to subalpine conifer woodlands, mixed grassland communities; (2300–)2700–3700 m; Ariz., N.Mex.; Mexico (Chihuahua).

Although previously applied only to collections from Chihuahua, Mexico, *Potentilla townsendii* is here used to accommodate most plants from Arizona and New Mexico previously placed in *P. glaucophylla* (as *P. diversifolia* Lehmann). Plants have narrow, shallowly toothed leaflets; stems, petioles, and leaflet veins that are often reddish; and larger anthers than *P. glaucophylla*. As now circumscribed, *P. townsendii* is known in the United States primarily from the White Mountains of east-central Arizona and the high mountains of northern New Mexico. Whether these plants are truly conspecific with those from Chihuahua, Mexico, remains to be confirmed. The species might also occur in Colorado and Utah, but the best disposition of these problematic populations is uncertain. Somewhat hairy plants with subpalmate leaves are interpreted as introgression from *P. hippiana* or other species.

25. Potentilla gracilis Douglas ex Hooker, Bot. Mag. 57: plate 2984. 1830 • Slender cinquefoil, potentille grêle E F

Glands usually absent or inconspicuous, uncolored. **Stems** ascending to erect, (0.5–) 2–7(–12) dm. **Basal leaves** usually not in ranks, palmate, rarely subpalmate, 4–40(–55) cm; petiole (2–)3–25(–45) cm, long hairs sparse to dense, appressed to spreading, 0.5–2(–3) mm, weak to stiff, short hairs absent, crisped hairs usually absent, cottony hairs absent or sparse to abundant (var. *owyheensis*), glands absent or sparse; leaflets 5–9, on tip, rarely less than distal 1/10 of leaf axis, separate to overlapping, largest ones oblanceolate to elliptic or obovate, (1.5–)2–9(–11) × (0.7–)1.5–4 cm, margins flat to revolute, distal (2/3–)3/4 to nearly whole length evenly to unevenly incised 1/4–3/4+ to midvein, undivided medial blade 2–20 mm wide, teeth (4–) 6–10(–11) per side (sometimes secondarily toothed), linear to broadly lanceolate, 2–20(–35) mm, surfaces ± similar to strongly dissimilar, abaxial green or pale green to white, usually not glaucous, long hairs sparse to abundant (sometimes mostly on veins), short-crisped hairs absent or sparse to abundant, cottony hairs absent or sparse to dense, glands usually absent or obscured to sparse, sometimes common, adaxial dark green to grayish, long hairs sparse to ± abundant, rarely absent or dense, short-crisped hairs absent or sparse, cottony hairs absent or sparse to abundant (var. *owyheensis*), glands usually absent or sparse to sometimes common. **Cauline leaves** 1–4(or 5). **Inflorescences** (4–)10–50(–60) -flowered. **Pedicels** 0.3–3 cm. **Flowers:** epicalyx bractlets linear to narrowly lanceolate or elliptic, 2–6(–8) × 0.5–1.5(–2.5) mm, hairs sparse to abundant, rarely dense, ± appressed to ascending, rarely spreading, glands usually absent or inconspicuous; hypanthium 3.5–6 mm diam.; sepals 4–8(–10) mm, apex acute to long acuminate; petals (3–)4–10(–11) × (3–)4–10(–12) mm; filaments (1–)1.5–2.5(–3.5) mm, anthers 0.6–1.2(–1.6) mm; carpels (15–)20–50, styles ± tapered, papillate-swollen proximally, (1–)1.5–2.5(–3) mm. **Achenes** 1–1.8 mm.

Varieties 5 (5 in the flora): w North America.

Because variation within *Potentilla gracilis* and other members of sect. *Graciles* is still far from resolved, taxonomic recognition is given to only the most distinctive extremes and that mostly at the varietal level. The goal is to call attention to areas that need study rather than to provide a full resolution, and even this leaves out some large, widespread populations in both keys and descriptions. The taxonomic placement of some available names, such as *P. glomerata* A. Nelson, *P. longipedunculata* Rydberg, and *P. macropetala* Rydberg, likewise remains unresolved.

The vouchers for the report by M. L. Fernald (1950) of *Potentilla gracilis* and *P. pulcherrima* from New Hampshire are presumably specimens collected in Coos County by A. S. Pease (NEBC). The three collections differ from one another, corresponding most closely to var. *fastigiata*, var. *flabelliformis*, and *P. pulcherrima*. The presence of such diversity in the same meadow, all as disjunct occurrences, suggests an artificial, non-persisting introduction.

1. Petioles: cottony hairs usually abundant, rarely sparse . . . . 25d. *Potentilla gracilis* var. *owyheensis*
1. Petioles: cottony hairs usually absent.
   2. Leaflets incised 1/4–1/2 to midvein, undivided medial blade 5–20 mm wide, teeth narrowly lanceolate to ovate.
      3. Leaflet surfaces usually strongly dissimilar, abaxial white, sometimes grayish, crisped-cottony hairs abundant to dense, rarely sparse, adaxial green; near coastal n California to Alaska . . . . 25a. *Potentilla gracilis* var. *gracilis*
      3. Leaflet surfaces ± similar to ± dissimilar, abaxial pale green to greenish gray, crisped hairs absent or sparse to abundant, cottony hairs usually absent, adaxial green to grayish green; widespread. . . . . .25b. *Potentilla gracilis* var. *fastigiata*

   [2. Shifted to left margin.—Ed.]
2. Leaflets incised 3/4+ to midvein, undivided medial blade 2–6 mm wide, teeth linear.
   4. Leaflet surfaces usually ± dissimilar, abaxial grayish to whitish, long hairs abundant to dense, crisped hairs sparse to abundant, cottony hairs usually absent, adaxial green to grayish green, teeth 3–13(–20) mm, not secondarily toothed; petioles 2–15(–25) cm . . . . 25c. *Potentilla gracilis* var. *elmeri*
   4. Leaflet surfaces strongly dissimilar, abaxial white, long hairs sparse to common (mostly on veins), cottony or crisped-cottony hairs dense, straight hairs dense, mostly on veins, adaxial usually dark green, sometimes grayish green, teeth (8–)10–20(–35) mm, often secondarily toothed; petioles (7–)10–25(–30) cm . . . . .25e. *Potentilla gracilis* var. *flabelliformis*

25a. **Potentilla gracilis** Douglas ex Hooker var. **gracilis** E

Potentilla alaskana Rydberg; *P. blaschkeana* Turczaninow ex Lehmann

Stems (2–)4–10(–12) dm. **Basal leaves:** petiole (7–)10–25 cm, long hairs abundant, usually spreading, sometimes appressed, 0.5–2(–3) mm, stiff, crisped and cottony hairs usually absent, glands absent; leaflets (5–)7–9, ± separate, largest oblanceolate to narrowly obovate, 3–9 × (1.3–) 1.5–3(–3.5) cm, margins flat to ± revolute, nearly whole length ± evenly incised 1/3–1/2 to midvein, undivided medial blade 5–12 mm wide, teeth 7–10(or 11) per side, sometimes secondarily toothed, narrowly lanceolate to ovate, 3–8 mm, surfaces usually strongly dissimilar, abaxial white, sometimes grayish, long hairs abundant (mostly on veins), crisped-cottony hairs abundant to dense, rarely sparse, glands absent or sparse, adaxial green, straight hairs sparse, rarely absent, crisped and cottony hairs absent, glands absent. **Flowers:** epicalyx bractlets 3–5(–6) × 0.5–1.1 mm; hypanthium 4–6 mm diam.; petals (5–)7–10 × 5–12 mm; carpels 20–40, styles 2–3 mm. **Achenes** 1.2–1.8 mm.

Flowering summer. Moist meadows, usually in oak and conifer woodlands; 100–500(–1100) m; B.C.; Alaska, Calif., Oreg., Wash.

Variety *gracilis* is a relatively tall plant with serrate, narrow, strongly bicolor leaves, confined mainly to meadows west of the Cascade Mountains, from northwestern California northward through western Oregon to southern Washington. From there, an expression with increasingly broad leaflets transitional to var. *fastigiata*, represented by *Potentilla alaskana* and *P. blaschkeana*, extends in the coastal mountains to southern Alaska.

References to var. *gracilis* from the Canadian prairie provinces (B. Boivin 1967–1979) correspond to the *Potentilla filipes* Rydberg extreme of *P. pulcherrima*. The two are superficially similar; var. *gracilis* has more sharply and deeply serrate leaflets and lacks the conspicuous glands and contrasting leaflet veins of the latter. The inclusion of Saskatchewan in the range of var. *gracilis* by H. J. Scoggan (1978–1979, part 3) is most likely based on var. *flabelliformis* or its intergrades with other varieties.

25b. Potentilla gracilis Douglas ex Hooker var. **fastigiata** (Nuttall) S. Watson, Proc. Amer. Acad. Arts 8: 557. 1873 • Nuttall's or Hall's cinquefoil, potentille de Nuttall [E]

Potentilla fastigiata Nuttall in J. Torrey and A. Gray, Fl. N. Amer. 1: 440. 1840; *P. angustata* Rydberg; *P. blaschkeana* Turczaninow ex Lehmann var. *permollis* (Rydberg) Th. Wolf; *P. etomentosa* Rydberg; *P. etomentosa* subsp. *hallii* (Rydberg) Abrams; *P. fastigiata* var. *jucunda* (A. Nelson) Soják; *P. fastigiata* var. *permollis* (Rydberg) Soják; *P. gracilis* var. *glabrata* (Lehmann) C. L. Hitchcock; *P. gracilis* var. *hallii* (Rydberg) Th. Wolf; *P. gracilis* subsp. *nuttallii* (Lehmann) D. D. Keck; *P. gracilis* var. *nuttallii* (Lehmann) E. Sheldon; *P. gracilis* var. *permollis* (Rydberg) C. L. Hitchcock; *P. gracilis* var. *rigida* S. Watson; *P. indiges* M. Peck; *P. jucunda* A. Nelson; *P. nuttallii* Lehmann; *P. nuttallii* var. *glabrata* Lehmann

Stems (0.5–)1.5–7(–9) dm. **Basal leaves:** petiole (2–)3–20(–45) cm, long hairs common to dense, appressed to spreading, 0.5–2(–3) mm, stiff, crisped hairs usually absent, sometimes sparse to common, cottony hairs absent, glands absent or sparse; leaflets 5–7(–9), separate to ± overlapping, largest oblanceolate to elliptic or narrowly obovate, (1.5–)2–7(–10) × (0.7–)1.5–3.5 cm, margins flat to slightly revolute, distal (2/3–)3/4 to nearly whole length ± evenly incised 1/4–1/2 to midvein, undivided medial blade 6–20 mm wide, teeth 5–9(–11) per side, usually not secondarily toothed, narrowly lanceolate to ovate, 2–10 mm, surfaces ± similar to ± dissimilar, abaxial pale green to greenish gray, long hairs sparse to abundant (mostly on veins), crisped hairs absent or sparse to abundant, cottony hairs usually absent, glands absent or sparse, rarely common, adaxial green to grayish green, long hairs sparse to common, sometimes absent, crisped and cottony hairs absent, glands absent or sparse, rarely common. **Flowers:** epicalyx bractlets 2–6.5(–7) × 0.5–1.5(–2.5) mm; hypanthium 4–6 mm diam.; petals (3–)4–8(–10) × (3–)4–10(–12) mm; carpels (15–)30–50, styles 1.5–2.5(–3) mm. **Achenes** 1.4–1.8 mm.

Flowering summer. Vernally moist to dry meadows or gravelly flats and slopes, in grassland and sagebrush communities, oak, aspen, or conifer woodlands; (10–)100–3500 m; Alta., B.C., Man., N.W.T., Ont., P.E.I., Que., Sask., Yukon; Alaska, Ariz., Calif., Colo., Idaho, Minn., Mont., Nev., N.Dak., Oreg., S.Dak., Utah, Wash., Wyo.

Variety *fastigiata* was commonly known as var. *nuttallii* until N. H. Holmgren (1997b) noted that var. *fastigiata* had priority at varietal rank, due to the publication of *Potentilla holopetala* Turczaninow var. *fastigiata* (Nuttall) Lehmann. As used here, the variety encompasses most collections of *P. gracilis* with moderately and evenly toothed leaflets that have at most somewhat dissimilar leaflet surfaces that usually lack cottony hairs. As such, the variety is essentially the catch-all category for most variation within *P. gracilis* that is not assigned elsewhere.

Within this broad circumscription are at least three candidates for taxonomic recognition. First is var. *fastigiata* in the narrow sense, consisting of relatively large, hairy plants found throughout the range given above. The *hallii* phase, found mostly in the mountains of California, is shorter on average, and leaflets tend to be broader, less hairy, and less deeply incised. Glabrate populations that are common from eastern Oregon to the northern Rocky Mountains, some possibly indicating introgression from *Potentilla glaucophylla*, represent the *nuttallii* or *glabrata* phase.

Occurrences in Quebec and Prince Edward Island, and at least some other occurrences in eastern provinces and states, are most likely introduced. Eastern populations of var. *fastigiata* in general tend to be relatively hairy and to intergrade with *Potentilla pulcherrima*.

25c. Potentilla gracilis Douglas ex Hooker var. **elmeri** (Rydberg) Jepson, Fl. Calif. 2: 189. 1936 • Elmer's or combleaf cinquefoil [E] [F]

Potentilla elmeri Rydberg in N. L. Britton et al., N. Amer. Fl. 22: 315. 1908; *P. bakeri* Rydberg; *P. flabelliformis* Lehmann var. *inyoensis* Jepson; *P. pectinisecta* Rydberg; *P. pectinisecta* var. *comosa* (Rydberg) Soják

Stems (1.5–)2–6(–8) dm. **Basal leaves:** petiole 2–15 (–25) cm, long hairs common to dense, usually tightly appressed, sometimes ascending to spreading, 0.5–2.5 mm, stiff, crisped hairs absent or sparse, cottony hairs absent, glands absent or sparse; leaflets 5–7, usually ± overlapping, largest mostly oblanceolate to narrowly obovate, 2–6(–9) × 1.5–2.5 cm, margins flat to ± revolute, distal 3/4 to nearly whole length evenly incised 3/4+ to midvein, undivided medial blade 2–4 mm wide, teeth (5 or)6–10 per side, not secondarily toothed, linear, 3–13(–20) mm, surfaces usually ± dissimilar, abaxial grayish to whitish, long hairs abundant to dense, crisped hairs sparse to abundant, cottony hairs usually absent, glands absent or sparse, adaxial green to grayish, long hairs ± abundant, rarely dense, crisped hairs absent or sparse, cottony hairs absent, glands absent or sparse. **Flowers:** epicalyx bractlets 2.5–5 × 0.5–1.2(–1.5) mm; hypanthium 3.5–5 mm diam.; petals 4–7(–8) × (3–)5–10 mm; carpels (20–)30–50, styles 1.5–3 mm. **Achenes** 1–1.3 mm. **2*n*** = 42.

Flowering summer. Vernally moist meadows, often semialkaline, dry gravelly flats and slopes, in sagebrush communities, aspen and conifer woodlands; 1300–3300 m; Calif., Colo., Idaho, Mont., Nev., Oreg., Utah, Wyo.

Variety *elmeri* is the characteristic representative of *Potentilla gracilis* in the Great Basin, distinguished by deeply and evenly pectinate-toothed leaflets with a very narrow undivided medial blade. Leaflet surfaces are only somewhat dissimilar, most often grayish with abundant crisped (versus cottony) hairs abaxially. Basal leaves tend to be shorter than in other varieties, and petioles are most often tightly strigose. Intergradation with other varieties is common; populations beyond the Great Basin are problematic.

25d. Potentilla gracilis Douglas ex Hooker var. **owyheensis** Ertter & Mansfield, Madroño 54: 181, fig. 1. 2007 • Owyhee cinquefoil [E]

Stems (1.5–)3–6 dm. **Basal leaves:** petiole (3–)5–15(–19) cm, long hairs weak, sparse to abundant, loosely appressed to ascending, 1–2 mm, crisped and/or cottony hairs usually abundant, rarely sparse, glands absent or obscured; leaflets 5–7, ± overlapping, largest oblanceolate to obovate, (2–)3.5–7 × (1–)2–4 cm, margins flat, distal 3/4 to nearly whole length evenly to unevenly incised (1/2–)3/4+ to midvein, undivided medial blade 4–10 mm wide, teeth 4–8 per side, often secondarily toothed, linear to lanceolate, 7–17 mm, surfaces ± similar to ± dissimilar, abaxial grayish white, long hairs sparse to common, crisped and/or cottony hairs abundant to dense, rarely sparse, glands absent or obscured, adaxial greenish gray, long hairs sparse to common, crisped and/or cottony hairs sparse to abundant, glands absent or obscured to sparse. **Flowers:** epicalyx bractlets 2–5(–6) × 0.5–1.5 mm; hypanthium 3.5–5 mm diam.; petals (3–)4–6(–11) × 4–12 mm; carpels 25–40, styles 1.5–2.5 mm. **Achenes** 1.4–1.5 mm.

Flowering summer. Vernally moist meadows, in sagebrush communities, aspen and conifer woodlands; 1200–2700 m; Idaho, Oreg.

Variety *owyheensis* ranges from the Owyhee uplands of Owyhee County, Idaho, to the Ochoco Mountains of central and eastern Oregon. It is readily distinguished from other varieties in the abundance of loose cottony hairs (grading to long soft hairs) on petioles and both leaflet surfaces. The grayish white leaflets tend to be unevenly incised and secondarily toothed. Where sharing meadows with other varieties, var. *owyheensis* generally grows closest to the dry margins.

Comparable plants from the northern Sierra Nevada of California, represented by the type of *Potentilla subvillosa* Rydberg, might prove to belong here as well.

25e. Potentilla gracilis Douglas ex Hooker var. **flabelliformis** (Lehmann) Nuttall in J. Torrey and A. Gray, Fl. N. Amer. 1: 440. 1840 [E]

Potentilla flabelliformis Lehmann, Nov. Stirp. Pug. 2: 12. 1830

Stems (3–)5–8(–10) dm. **Basal leaves:** petiole (7–)10–25(–30) cm, long hairs abundant, usually ± appressed, 0.5–2(–3) mm, stiff, crisped and cottony hairs usually absent, glands usually absent; leaflets 5–7, separate to overlapping, largest oblanceolate to obovate, (3.5–)5–9(–11) × 2–4 cm, margins revolute, ± whole length evenly or unevenly incised ¾+ to midvein, undivided medial blade 3–6 mm wide, teeth 6–10 per side, often secondarily toothed, linear, (8–)10–20(–35) mm, surfaces strongly dissimilar, abaxial white, long hairs sparse to common (mostly on veins), cottony or crisped-cottony hairs dense, glands absent or obscured, adaxial usually dark green, sometimes grayish green, long hairs sparse to common (mostly on veins), crisped hairs absent or sparse, cottony hairs and glands usually absent. **Flowers:** epicalyx bractlets 4–8 × 0.8–2 mm; hypanthium 3.5–5.5 mm diam.; petals 5–10 × 5–12 mm; carpels 20–50, styles (1–)1.5–2 mm. **Achenes** 1.1–1.5 mm. **2*n*** = 56–65.

Flowering summer. Montane meadows in aspen and conifer woodlands, sagebrush communities, grasslands; 200–2000 m; Alta., B.C., Man., Ont., Sask.; Calif., Idaho, Mich., Minn., Mont., Oreg., Wash., Wis., Wyo.

Variety *flabelliformis* is relatively distinctive, perhaps on par with members of sect. *Graciles* treated here as species. Plants are relatively large, with deeply incised leaves that are notably bicolored with dense cottony hairs abaxially. There are two morphs with some geographic underpinning, possibly deserving taxonomic recognition. One morph, characteristic of the northern prairies and including the type, has simply pectinate leaves with a very narrow undivided medial blade and narrow teeth lacking secondary toothing. The second morph tends to have unevenly incised leaves with a wider undivided medial blade and teeth that are often tapering and secondarily toothed. This morph is centered in the montane valleys of the northern Rocky Mountains, extending as far south as Modoc County in northeastern California.

Collections of var. *flabelliformis* from eastern locations (Michigan, Minnesota, Wisconsin) are probably introductions, possibly of uncertain persistence. Historical reports from Alaska, Colorado, Nebraska, and Nevada are unconfirmed as this variety.

Although *Potentilla flabelliformis* var. *ctenophora* Rydberg [= *P. gracilis* var. *ctenophora* (Rydberg) B. Boivin] is commonly treated as a synonym of *P. flabelliformis*, P. A. Rydberg (1908d) had placed it in synonymy under *P. blaschkeana* (addressed here under var. *gracilis*). Rydberg (1897, 1898) did not designate a type or syntypes, so the proper disposition of var. *ctenophora* awaits lectotypification.

26. Potentilla brunnescens Rydberg, Bull. Torrey Bot. Club 28: 173. 1901 [E]

Potentilla gracilis Douglas ex Hooker var. *brunnescens* (Rydberg) C. L. Hitchcock

Glands conspicuous, usually golden. **Stems** ascending to nearly erect, (1.5–)3–7 dm. **Basal leaves** sometimes 2-ranked, palmate, (5–)10–30 cm; petiole (1.5–)5–25(–50) cm, long hairs sparse to abundant, appressed to spreading, 1–2 mm, ± stiff, short, crisped, and cottony hairs absent, glands ± abundant; leaflets (5–)7–9, at tip of leaf axis, separate to slightly overlapping, largest ones oblanceolate-elliptic to narrowly obovate, (2–)5–10 × 1–4 cm, margins flat to ± revolute, distal ¾ to nearly whole length ± evenly incised (¼–)½–¾+ to midvein, undivided medial blade 2–15(–20) mm wide, teeth (5–)7–10 per side, linear to broadly lanceolate, 3–15 mm, surfaces ± similar, <abaxial paler>, green, not glaucous, long hairs sparse to common, short-crisped hairs absent or sparse, cottony hairs absent, glands ± abundant. **Cauline leaves** 1–3(–4). **Inflorescences** 12–30(–70)-flowered. **Pedicels** 0.2–1(–4) cm. **Flowers:** epicalyx bractlets lanceolate to elliptic, 3–6 × 1–2 mm, hairs sparse to common, loosely appressed to spreading, glands common to abundant; hypanthium 4–6(–8) mm diam.; sepals 4–8(–9) mm, apex acute to acuminate; petals 6–9 × 6–10 mm; filaments 1–2 mm, anthers 0.6–1 mm; carpels 30–60, styles filiform-tapered, papillate-swollen proximally, 1.5–2 mm. **Achenes** 1.5 mm.

Flowering summer. Vernally moist rocky flats, sagebrush benchlands and slopes, usually basaltic; 400–3100 m; Alaska, Colo., Idaho, Mont., Oreg., Utah, Wash., Wyo.

The conspicuous golden glands, dark green color, and relatively broad leaflets make *Potentilla brunnescens* one of the better-defined species of sect. *Graciles*, though it has often been treated as a variety of *P. gracilis*. The glands, which are often on septate stalks, are particularly conspicuous on the epicalyx bractlets, hypanthium, and sepals, and at the junction of petiole and leaflets. The species occurs from the Blue Mountains of Oregon and Washington to western Montana, northern Utah, and north-central Colorado; collections from Alaska are probably not native.

As presented here, *Potentilla brunnescens* encompasses two morphological extremes that might merit varietal status. Leaflets of the more common and widespread extreme are incised about halfway to the midvein; in contrast, some populations in western Wyoming and adjacent Idaho have leaflets incised three-fourths or more to the midvein. The latter have often been assigned to *P. gracilis* var. *elmeri* while reserving *P. gracilis* var. *brunnescens* for the former; this is untenable because the type of *P. brunnescens* is the latter form.

27. Potentilla pulcherrima Lehmann, Nov. Stirp. Pug. 2: 10. 1830 • Beautiful cinquefoil [E]

Potentilla filipes Rydberg; *P. gracilis* Douglas ex Hooker var. *filipes* (Rydberg) B. Boivin; *P. gracilis* var. *pulcherrima* (Lehmann) Fernald; *P. pulcherrima* var. *condensata* Th. Wolf; *P. pulcherrima* var. *filipes* (Rydberg) Th. Wolf; *P. pulcherrima* var. *wardii* (Greene) Soják

Glands usually conspicuous, red-tipped (palmate-leaved plants), sometimes absent or inconspicuous, colorless (subpalmate-leaved plants). **Stems** ascending to erect, (1.5–)3–6(–8) dm. **Basal leaves** sometimes 2-ranked, palmate to subpalmate, 3–25(–35) cm; petiole 1–20 (–30) cm, long hairs sparse to abundant, appressed to spreading, 2–3 mm, weak to stiff, short hairs usually absent, crisped hairs absent or sparse, cottony hairs usually absent, glands sparse to abundant; leaflets 5–7, at tip or on distal 1/10 of leaf axis, separate to slightly overlapping, largest ones ± oblanceolate, 1–6(–9) × 1–4 cm, margins flat, distal 2/3 to nearly whole length evenly incised 1/4–1/2 to midvein, undivided medial blade 6–15 mm wide, teeth 6–12 per side, ± broadly lanceolate, 1–5 mm, surfaces usually strongly dissimilar, abaxial grayish to white, long hairs sparse to common (mostly on veins), short hairs absent, cottony or crisped-cottony hairs abundant to dense, glands usually ± abundant (at least on veins), adaxial green, rarely grayish green, not glaucous, long hairs sparse to common, short, crisped, and cottony hairs usually absent, glands usually sparse to abundant. **Cauline leaves** 1–3. **Inflorescences** (4–)10–50(–60)-flowered. **Pedicels** 0.5–3(–4.5) cm. **Flowers:** epicalyx bractlets narrowly lanceolate to narrowly elliptic, 3–5(–6) × 0.8–2 mm, hairs sparse to common, ± appressed to spreading, glands absent or sparse to abundant; hypanthium (3–)4–6 mm diam.; sepals 4–8(–9) mm, apex acuminate; petals 6–9(–11) × 6–13 mm; filaments 1–3.5 mm, anthers 0.8–1.1 mm; carpels 20–40, styles filiform to tapering above papillate-swollen base, (1–)1.5–2 mm. **Achenes** 1.1–1.5(–1.6) mm. **2*n*** = 70, 71, 108.

Flowering summer. Dry meadows in grasslands, sagebrush, scrub oak, aspen and conifer woodlands; (300–)800–3600 m; Alta., B.C., Man., Ont., Que., Sask.; Ariz., Calif., Colo., Conn., Idaho, Mont., Nev., N.Mex., N.Dak., S.Dak., Utah, Wyo.

Potentilla pulcherrima is here retained provisionally in the broad sense, encompassing intergrades between morphological extremes. The most distinctive extreme has strictly palmate leaves and conspicuous red-tipped glands. Eglandular hairs are relatively sparse and spreading on epicalyx bractlets and abaxial leaflet veins, such that the veins tend to contrast with the cottony white abaxial surface. At the other extreme are plants with subpalmate leaves, no or inconspicuous glands, and densely strigose abaxial leaflet veins. The former is understood to be the core species, towards which the description is accordingly weighted. The latter in turn represents introgression with *P. hippiana* (sect. *Leucophyllae*) or other species. The subpalmate extreme includes the lectotype of *P. pulcherrima* (designated by J. Soják 1996), so if a narrower circumscription of the species is adopted then either *P. filipes* would become the accepted name of the core species or predominant current usage of *P. pulcherrima* would have to be conserved by designating a different type.

By whatever circumscription, *Potentilla pulcherrima* is one of the more common members of sect. *Graciles* in the Rocky Mountains, extending east into the northern prairies. The species grows mainly in the mountains from southeastern British Columbia and southern Alberta to northern and eastern Arizona and southern New Mexico. The native range extends at least to Manitoba and North Dakota, with outlying populations in the Black Hills of South Dakota. Populations are also scattered in the mountains of eastern Nevada and disjunctly in the White Mountains of California. Adventive occurrences in fields and disturbed sites are reported from as far east as Connecticut; see discussion of 25. *P. gracilis* for reports from New Hampshire. All known specimens from Minnesota and Oregon have been reidentified as particularly hairy variants of *P. gracilis* in the broad sense; reports from Washington are suspect.

Potentilla pulcherrima has often been treated as a variety of *P. gracilis*, sometimes with the *filipes* extreme included in var. *gracilis* (B. Boivin 1967–1979). The current decision to treat *P. pulcherrima* as a species is in part because it has a monsoonal Rocky Mountain rather than a Pacific Northwest center of distribution.

Plants combining the vestiture of *Potentilla pulcherrima*, including abundant glandularity, with subpalmate leaves approaching *P. hippiana* (leaflets on about one-fifth of leaf axis) have been named *P. gracilis* var. *hippianoides* S. L. Welsh & N. D. Atwood.

28. Potentilla bruceae Rydberg in N. L. Britton et al., N. Amer. Fl. 22: 342. 1908 • Bruce's cinquefoil [E]

Potentilla drummondii Lehmann subsp. *bruceae* (Rydberg) D. D. Keck; *P. drummondii* var. *bruceae* (Rydberg) N. H. Holmgren

Glands absent or inconspicuous, uncolored. **Stems** ascending to nearly erect, (1–)2–5(–6) dm. **Basal leaves** not in ranks, ± subpalmate to subpinnate (proximal leaflets sometimes doubled, distal leaflets confluent and/or decurrent), (2–)5–20(–28) cm; petiole (1–)3–15(–20) cm, long hairs abundant, appressed to ascending-spreading, 1–2.5 mm, weak <often flattened and twisted>, sometimes grading to crisped or cottony hairs, short hairs absent, glands absent or sparse; leaflets 5–7, on distal 1/10–1/4 of leaf axis, ± overlapping, largest ones usually ± obovate, (1–)1.5–5(–6) × 1–3.5(–4) cm, margins flat, distal 2/3–3/4 evenly to unevenly incised ± 1/2 to midvein (often with additional incisions nearly to midvein), undivided medial blade 4–20 mm wide, teeth (2 or)3–6(–8) per side (sometimes secondarily toothed), ± lanceolate, 4–11 mm, surfaces ± similar, <abaxial ± lighter and hairier>, gray-green to grayish, rarely green, not glaucous, long hairs ± abundant, short-crisped hairs absent or sparse to common, cottony hairs usually absent, glands absent or sparse. **Cauline leaves** 1 or 2. **Inflorescences** 5–50-flowered. **Pedicels** 0.5–3(–4) cm. **Flowers:** epicalyx bractlets broadly lanceolate to elliptic, 3–6(–7.5) × 1.5–2.5(–3) mm, hairs ± common to abundant, loosely appressed to ascending, glands absent or inconspicuous; hypanthium 3.5–5.5 mm diam.; sepals 5–10 mm, apex acute to acuminate; petals 5–10 × 5–10 mm; filaments 1–2.5 mm, anthers 0.7–1.2 mm; carpels 15–30, styles filiform-tapered, papillate-swollen proximally, 2–3 mm. **Achenes** 1.4–1.6 mm. **2*n*** = 63–73, 79, 87, 98, 129.

Flowering summer. Dry montane meadows and adjacent slopes, in conifer woodlands; 1200–3700 m; Calif., Nev., Oreg.

As redefined by B. Ertter and D. Mansfield (2007), *Potentilla bruceae* is characterized by irregularly subpalmate, weakly villose leaves. Such plants form relatively consistent and extensive populations in dry meadows centered near Lake Tahoe in California and Nevada, extending to the central Sierra Nevada and to the Warner Mountains in Oregon. This interpretation differs from that previously suggested by Ertter (1992), in which *P. bruceae* was interpreted as the catchall category for hybrids between *P. breweri* (sect. *Multijugae*) and *P. drummondii*. All three taxa were accordingly treated as varieties of *P. drummondii*. D. D. Keck (in J. Clausen et al. 1940) treated *P. bruceae* as a subspecies of *P. drummondii*; B. C. Johnston (1980) concluded that its placement as a variety of *P. breweri* was more justified. The three entities unquestionably intergrade (with the type of *P. anomalofolia* M. Peck possibly one result), but not beyond what is the norm for facultatively apomictic species of *Potentilla*.

Chromosome numbers reported by J. Clausen et al. (1940) require verification as this species; the irregular behavior observed in meiotic material suggests that at least some of the individuals were hybrids.

See discussion of 29. *Potentilla drummondii* regarding leaflet dissection and the inclusion of *P. bruceae* in sect. *Graciles*.

29. Potentilla drummondii Lehmann, Nov. Stirp. Pug. 2: 9. 1830 • Drummond's cinquefoil [E]

Potentilla cascadensis Rydberg; *P. drummondii* var. *cascadensis* (Rydberg) Th. Wolf

Glands usually absent or inconspicuous, rarely conspicuous, uncolored. **Stems** decumbent to nearly erect, 1.5–4.5(–6) dm. **Basal leaves** sometimes 2-ranked, subpinnate to pinnate (proximal leaflets often doubled, distal leaflets confluent and/or decurrent), (4–)10–25 cm; petiole 1–10(–15) cm, long hairs absent or sparse to abundant, appressed, 1–1.5 mm, usually stiff, short, crisped, and cottony hairs usually absent, glands absent or (regionally) abundant; leaflets 5–9, on distal 1/10–1/2(–3/4) of leaf axis, separate to ± overlapping, largest ones obovate to cuneate, (1–)2–5 × (0.7–)1–3(–3.5) cm, margins flat, distal 1/2–3/4 unevenly, sometimes evenly, incised ± 1/2 to midvein (often with additional incisions nearly to midvein), undivided medial blade 3–15 mm wide, teeth 3–7 per side (sometimes secondarily toothed), linear to lanceolate, 3–10 mm, surfaces similar, green, not glaucous, long hairs nearly absent or sparse to common (sometimes restricted to abaxial veins), short, crisped, and cottony hairs usually absent, glands usually absent, sometimes regionally abundant. **Cauline leaves** 1–3. **Inflorescences** 3–15-flowered. **Pedicels** 1–3(–6.5) cm. **Flowers:** epicalyx bractlets lanceolate to narrowly elliptic, 3–5 × 1–1.6 mm, hairs usually sparse, appressed to ascending, glands usually absent, sometimes regionally sparse to common; hypanthium (3–)4–6(–7) mm diam.; sepals 5–10 mm, apex acute to acuminate; petals (6–)7–12 × 5–10 mm; filaments 2–3.5 mm, anthers 0.7–1 mm; carpels (20–)25–40, styles filiform above papillate-swollen base, 2–3 mm. **Achenes** 1.5–2 mm. **2*n*** = 64, 92, 96, 98–108.

Flowering summer. Moist to dry meadows and adjacent slopes, in conifer woodlands, alpine tundra communities; 300–3000 m; Alta., B.C.; Alaska, Calif., Mont., Oreg., Wash.

The present circumscription of *Potentilla drummondii* differs from that of B. Ertter (1992, 1993), which encompassed *P. breweri* and *P. bruceae* as varieties (see 28. *P. bruceae* discussion for details). The description above focuses on large, eglandular, subpinnate plants that occur in mountains from the central Sierra Nevada and northern Coast Ranges of California to the Olympic Peninsula of Washington, east to southern Alberta. At least some collections previously identified as *P. glaucophylla* (as *P. diversifolia* Lehmann) from the Kenai Peninsula in Alaska can be readily accommodated within this concept, as can collections from the Cabinet Mountains in Montana. The distinction between *P. drummondii* and subpalmate *P. glaucophylla* can be vague; in general, the proximalmost leaflets of *P. glaucophylla* are relatively small and often entire; in *P. drummondii* proximalmost leaflets are similar in size and dissection to other leaflets.

Pinnate-leaved regional variants merit further attention; in particular the smaller form described as *Potentilla cascadensis*, and glandular plants in the northern Coast Ranges of California. Relatively small plants forming uniform populations in the southern Sierra Nevada may represent stabilized hybrids with *P. breweri*.

The inclusion of *Potentilla bruceae* and *P. drummondii* in sect. *Graciles* differs from that of J. Clausen et al. (1940) and B. C. Johnston (1980, 1985), who allied both species with *P. breweri* in sect. *Multijugae*. Although subpalmate to subpinnate leaves are anomalous in primarily palmate sect. *Graciles*, doubled proximal leaflets and an indistinct transition between the rachis and axis of the pinnatisect terminal leaflet in these two species suggest that this state is derived from a palmate progenitor. Habit and petioles also are more compatible with sect. *Graciles* than sect. *Multijugae*.

81. POTENTILLA Linnaeus sect. LEUCOPHYLLAE (Rydberg) A. Nelson in J. M. Coulter and A. Nelson, New Man. Bot. Rocky Mt., 255. 1909 [E]

Barbara Ertter

James L. Reveal

Potentilla [unranked] *Leucophyllae* Rydberg, Monogr. N. Amer. Potentilleae, 31. 1898; *Potentilla* [unranked] *Hippianae* Rydberg; *Potentilla* sect. *Hippianae* (Rydberg) Ertter & Reveal

Perennials, ± tufted, not stoloniferous; taproots not fleshy-thickened; vestiture primarily of long and cottony or sometimes crisped-cottony hairs, glands usually absent or sparse, sometimes common, not red. **Stems** ascending to erect, not flagelliform, not rooting at nodes, lateral to persistent basal rosettes, (0.3–)1.5–7(–8) dm, lengths (1–)1.5–4(–5) times basal leaves. **Leaves:** basal not 2-ranked; cauline (0 or)1–6+; primary leaves pinnate to subpinnate (with distal leaflets often confluent), 3–45(–50) cm; petiole: long hairs usually ± to tightly appressed, sometimes ascending, usually stiff, sometimes weak, glands absent or sparse or obscured; leaflets (5–)7–15, on distal (1/6–)1/5–2/3 of leaf axis, separate to ± overlapping distally, oblanceolate to narrowly obovate, oblong, or cuneate, margins flat, distal 1/4 to whole length, rarely less, usually ± evenly incised 1/4–1/2 or less to midvein, sometimes entire, teeth (0–)2–18 per side, surfaces similar to strongly dissimilar, abaxial white to ± green, cottony hairs absent or sparse to dense, adaxial green to white, not glaucous, long hairs mostly weak, sometimes stiff. **Inflorescences** (5–)7–60 -flowered, cymose, open. **Pedicels** straight in fruit, 0.3–3(–6) cm, proximal usually much longer than distal. **Flowers** 5-merous; hypanthium 2–7 mm diam.; petals yellow, ± obcordate, (3–) 4–10 mm, usually longer than sepals, retuse; stamens 20–25; styles subapical, tapered-filiform, papillate-swollen in proximal 1/5 or less, (1–)1.5–3 mm. **Achenes** smooth to slightly rugose.

Species 4 (4 in the flora): w, c North America.

Section *Leucophyllae* is an apparent radiation of four species in monsoonal regions of the American Southwest, ranging from the Colorado Plateau, Rocky Mountains, and western Great Plains to the prairies of Canada. The plants share features with sect. *Graciles* but have consistently pinnate leaves, distal leaflets often decurrent, and petioles commonly densely strigose. Plants are often locally abundant in relatively dry forest openings and open grasslands, tending to be more xeric-tolerant than plants of sect. *Graciles*.

Descriptions and keys focus on the most distinctive expressions, which are blurred by a propensity for the species to hybridize wherever they grow sympatrically. This is counteracted by at least some level of habitat partitioning; for example, *Potentilla hippiana* tends to occur on deeper soils and *P. crinita* on rockier slopes than other species in this section.

As a further complication, key diagnostic characters sometimes sort independently in different parts of the relatively large ranges of the species. The prairie expression described as *Potentilla argyrea* is particularly problematic, undermining the distinction between *P. effusa* and *P. hippiana* in the northern part of their ranges. In the southern Rocky Mountains, the two species are relatively distinct on the basis of correlated vestiture and leaflet toothing. In contrast, the prairie expression combines vestiture closer to *P. hippiana* with leaflet features of *P. effusa*. Pending further analysis, B. Boivin (1952) and B. C. Johnston (1980) are followed in retaining this variant in *P. hippiana*, though without formal taxonomic recognition. Inclusion in *P. effusa* might prove more justified, as done by J. Soják (2006).

Since *Potentilla subjuga* (sect. *Subjugae*) and *P. ovina* (sect. *Multijugae*) are sometimes identified as members of sect. *Leucophyllae*, they are included herein and key out in the third and fifth couplets, respectively.

1. Leaflets: abaxial surfaces white to gray, crisped or cottony hairs abundant to dense.
   2. Epicalyx bractlets: abaxial vestiture often much sparser than that of sepals, often glabrate or glabrescent distally, straight hairs absent or sparse, cottony hairs usually abundant (at least proximally); lateral leaflets evenly to unevenly paired, teeth (1–)4–9 per side on distal (½–)⅔–¾ (rarely more) of margin, surfaces similar, cottony hairs abundant to dense abaxially, abundant adaxially; Alberta to Manitoba, to Colorado, possibly disjunct in nc New Mexico and ne Utah . . . . 31. *Potentilla effusa* (in part)
   2. Epicalyx bractlets: abaxial vestiture similar to or ± sparser than that of sepals, usually not glabrescent, straight hairs ± abundant, crisped or sometimes ± cottony hairs absent or sparse to abundant; lateral leaflets evenly (to unevenly in *argyrea* phase) paired, teeth (5–)7–12(–18) per side on distal (⅔–)¾ to whole margin, surfaces ± to strongly dissimilar, crisped-cottony hairs abundant to dense abaxially, absent or sparse to common adaxially; c Canada to Nevada, Arizona, and New Mexico, introduced farther east.
      3. Leaflets (2 or)3–6(or 7) per side, distal pairs ± decurrent, often confluent with terminal leaflet, surfaces ± to strongly dissimilar; 500–3400 m . . . . 30. *Potentilla hippiana*
      3. Leaflets 2(or 3) per side, distal pairs not decurrent or confluent with terminal leaflet, surfaces strongly dissimilar; 3400–4000 m . . . . 34. *Potentilla subjuga* (sect. *Subjugae*)
1. Leaflets: abaxial surfaces ± green to grayish or silvery, crisped or cottony hairs usually absent or sparse.
   4. Lateral leaflets often unevenly paired, distal pairs usually not decurrent or confluent with terminal leaflet, surfaces ± green; epicalyx bractlets with cottony hairs sometimes abundant (at least proximally).
      5. Stems ascending to erect; basal leaves: leaflets 2 or 3 per side, incised ± ½ to midvein, teeth 2–5 per side; cauline leaves 2–6+; cottony hairs often present (at least on epicalyx bractlets); nc Colorado . . . . 31. *Potentilla effusa* (in part)
      5. Stems prostrate to ascending; basal leaves: leaflets 3–5 per side, incised ½–⅔ (–¾) to midvein, teeth (0 or)1 or 2(or 3) per side; cauline leaves 1 or 2; cottony hairs absent; Uinta and middle Rocky mountains, to Nevada and Colorado . . . . 39. *Potentilla ovina* (sect. *Multijugae*)
   4. Lateral leaflets evenly paired, distal pairs ± decurrent, often confluent with terminal leaflet, surfaces greenish to silvery or grayish; epicalyx bractlets with cottony hairs usually absent.

[6. Shifted to left margin.—Ed.]

6. Basal leaves 15–45(–50) cm; larger leaflets (2–)3–7 cm, not conduplicate, teeth 6–18 per side on distal 3/4+ of margin; se Wyoming, e slope Rocky Mountains of Colorado to Sacramento Mountains, New Mexico . . . . 32. *Potentilla ambigens*
6. Basal leaves 3–15(–20) cm; larger leaflets 1–3(–4) cm, often conduplicate, teeth (0 or)1–5(–9) per side on distal 1/4–1/2(–2/3) or less of margin; s Nevada to sw Colorado, n Arizona, and nw New Mexico . . . . 33. *Potentilla crinita*

30. Potentilla hippiana Lehmann, Nov. Stirp. Pug. 2: 7. 1830 • Woolly or Hipp's cinquefoil, potentille de Hipp [E] [F]

Potentilla leucophylla Torrey, Ann. Lyceum Nat. Hist. New York 2: 197. 1827, not Pallas 1773; *P. argyrea* Rydberg; *P. effusa* Douglas ex Lehmann var. *argyrea* (Rydberg) Soják; *P. hippiana* var. *argyrea* (Rydberg) B. Boivin; *P. hippiana* var. *diffusa* Lehmann; *P. propinqua* Rydberg

Stems (0.3–)2–4(–5) dm, lengths (1–)2–4 times basal leaves. **Basal leaves** pinnate to subpinnate, (3–)5–15 (–25) cm; petiole 1–10(–15) cm, long hairs abundant to dense, tightly appressed, 1–2.5 mm, stiff, short hairs absent or obscured, crisped-cottony hairs absent or sparse to common, glands absent or obscured; leaflets not conduplicate, lateral ones evenly (to unevenly in *argyrea* phase) paired, (2 or)3–6(or 7) per side on distal (1/6–)1/5–1/2 of leaf axis, distal pairs ± decurrent, often confluent with terminal leaflet, larger leaflets oblanceolate or narrowly obovate to oblong, 1–5(–6) × 0.3–1.5 cm, distal (2/3–)3/4 to whole margin incised 1/2 or less (rarely more) to midvein, teeth (5–)7–12(–18) per side, 1–4 mm, surfaces ± to strongly dissimilar, abaxial white, long hairs abundant (mostly on, but not limited to, veins), 1–2 mm, stiff, short hairs absent or obscured, crisped-cottony hairs abundant to dense, glands absent or obscured, adaxial green to grayish, long hairs sparse to common, rarely absent, short or crisped to, sometimes, cottony hairs absent or sparse to common, glands sparse. **Cauline leaves** 1 or 2(or 3). **Inflorescences** 10–30-flowered. **Pedicels** 0.3–3(–5) cm. **Flowers:** epicalyx bractlets narrowly elliptic to lanceolate, rarely linear, 2–5 × 0.5–1.5 mm, (1/2–)2/3 as long as sepals, abaxial vestiture similar to or ± sparser than sepals, usually not glabrescent, straight hairs ± abundant, crisped to sometimes ± cottony hairs absent or sparse to abundant; hypanthium 3–7 mm diam.; sepals 4–5.5(–6.5) mm, apex acute to acuminate; petals 4–8 × 4–7 mm; filaments 0.5–2.5 mm, anthers 0.6–1.1 mm; carpels (5–)10–30, styles 1.7–2.5 mm. **Achenes** 1.4–1.8 mm, smooth to faintly rugose. **2*n*** = 42, 70, 77, 84, 98.

Flowering summer. Dry grasslands and meadows, in aspen and conifer woodlands or alpine tundra, disturbed sites; 500–3400 m; Alta., B.C., Man., N.W.T., N.S., Ont., Que., Sask.; Ariz., Colo., Idaho, Mich., Minn., Mont., Nebr., Nev., N.Mex., N.Dak., S.Dak., Utah, Wyo.

Potentilla hippiana is most common and distinctive in the Colorado Plateau and southern Rocky Mountains, with outliers in the mountains of Arizona, Nevada, and New Mexico. Occurrences are more scattered in the northern Rocky Mountains and Great Plains, where intergradation with *P. effusa* is common. Among these intermediates are the *argyrea* phase and the single report from Idaho (B. C. Johnston 1980). *Potentilla hippiana* is probably adventive in Nova Scotia, eastern Ontario, Quebec, and Michigan, and possibly the Northwest Territories. The reported occurrence in central Alaska (E. Hultén 1968) is based on a specimen of *P. pensylvanica* (CAN).

Significant variation exists throughout the range of *Potentilla hippiana*, including dwarfism, leaf division, leaflet dissection, adaxial leaflet vestiture, and pedicel length. Some of this variation might merit taxonomic recognition upon further analysis, especially if correlated with ploidy level (6x through 12x). At present, the differences used to distinguish var. *argyrea* are not found to be sufficiently correlated with distribution to merit formal taxonomic recognition.

Complicating the infraspecific variation is a tendency for *Potentilla hippiana* to intergrade with other species, in particular *P. effusa*. Beyond sect. *Leucophyllae*, the frequency of hybrids with *P. pulcherrima* blurs the distinction between the two species; *P. gracilis* var. *hippianoides* S. L. Welsh & N. D. Atwood is probably one of the resultant intermediates.

31. Potentilla effusa Douglas ex Lehmann, Nov. Stirp. Pug. 2: 8. 1830 [E] [F]

Potentilla hippiana Lehmann subsp. *effusa* (Douglas ex Lehmann) Dorn; *P. hippiana* var. *effusa* (Douglas ex Lehmann) Dorn

Stems (0.5–)1.5–4(–5.5) dm, lengths 2–3(–4) times basal leaves. **Basal leaves** pinnate, (3–)5–15(–25) cm; petiole 0.5–5(–8) cm, long hairs absent to abundant,

± appressed, 1–3 mm, stiff to weak, short and crisped hairs absent or obscured, cottony hairs abundant (at least on first-formed leaves), glands sparse or obscured; leaflets not conduplicate, lateral ones evenly to unevenly paired, 2–5(–7) per side (secondary leaflets sometimes interspersed) on distal $\frac{1}{3}$–$\frac{1}{2}$ of leaf axis, distal pairs usually not decurrent or confluent with terminal leaflet, larger leaflets oblanceolate to narrowly obovate, 1–2.5(–3) × 0.3–1.3(–1.8) cm, distal ($\frac{1}{2}$–)$\frac{2}{3}$–$\frac{3}{4}$ (rarely more) of margin incised ± $\frac{1}{2}$ to midvein, teeth (1 or)2–9 per side, 1–4 mm, surfaces similar, gray to white or ± green, abaxial: long hairs sparse to common (at least on veins), 0.5–1.5 mm, weak to stiff, short or crisped hairs absent or sparse, sometimes obscured, cottony hairs absent or sparse to dense, glands absent, sparse, or obscured, adaxial: long hairs absent or sparse to common, short or crisped hairs absent or sparse, cottony hairs absent or sparse to abundant, glands absent or sparse. **Cauline leaves** 2–6+. **Inflorescences** 7–30-flowered. **Pedicels** 0.3–3 cm. **Flowers:** epicalyx bractlets linear to narrowly lanceolate, 1–2.5(–3) × 0.3–1 mm, $\frac{1}{2}$–$\frac{2}{3}$ as long as sepals, abaxial vestiture often much sparser than sepals, often glabrate or glabrescent distally, straight hairs absent or sparse, cottony hairs usually abundant (at least proximally), sometimes absent or nearly so (var. *rupincola*); hypanthium 2–6 mm diam.; sepals 3–6(–7) mm, apex long acuminate; petals 4–6.5 × 4–6 mm; filaments 1–3.5 mm, anthers 0.5–1 mm; carpels 3–15, styles 1.5–2.1 mm. **Achenes** 1.8–2.1 mm, smooth.

Varieties 2 (2 in the flora): w, c North America.

In the southern Rocky Mountains, *Potentilla effusa* is relatively distinct from *P. hippiana* both morphologically and geographically, with *P. effusa* mostly east of the Continental Divide and *P. hippiana* mostly west. In general, *P. hippiana* differs from *P. effusa* in having leaflets more evenly paired, more tightly serrate with teeth often occurring on the whole margin, and more bicolored with abaxial vestiture of crisped rather than cottony hairs. Stems of *P. effusa* have more cauline leaves on average than *P. hippiana*, and vestiture of epicalyx bractlets of *P. effusa* (as here circumscribed) is never sericeous.

The distinction between the two species is marred by intermediate populations throughout their shared range in the northern Rocky Mountains and Great Plains. Most regional floras focused on these areas have accordingly included *Potentilla effusa* within *P. hippiana*, sometimes without infraspecific distinction. An alternate solution has been the recognition of *P. hippiana* subsp. *effusa*, with the subspecies further subdivided into the two varieties recognized here. The varieties themselves intergrade; *P. effusa* var. ×*coloradensis* (Rydberg) Th. Wolf is available for the hybrid.

1. Leaflets gray to white, surfaces moderately long-hairy and moderately to densely cottony. . . . . 31a. *Potentilla effusa* var. *effusa*
1. Leaflets ± green, surfaces glabrous or sparsely to moderately long-hairy (at least on veins) and not or sparsely cottony . . . . 31b. *Potentilla effusa* var. *rupincola*

31a. Potentilla effusa Douglas ex Lehmann var. **effusa** • Branched cinquefoil [E] [F]

Potentilla effusa var. *filicaulis* Nuttall; *P. hippiana* Lehmann var. *filicaulis* (Nuttall) B. Boivin

Leaflets gray to white, teeth (1–)4–9 per side, abaxial surfaces moderately long-hairy especially on veins, moderately to densely cottony, adaxial moderately long-hairy and abundantly cottony. **Flowers:** epicalyx bractlets and sepals usually abundantly cottony at least proximally; anthers (0.6–)0.7–1 mm. $2n = 56$.

Flowering summer. Dry rocky slopes, slightly moist meadows, grasslands, oak, aspen, and montane conifer woodlands (limestone pavement); 700–2600 m; Alta., Man., Sask.; Colo., Idaho, Minn., Mont., Nebr., N.Mex., N.Dak., S.Dak., Utah, Wyo.

Variety *effusa* occurs from southern Alberta to Manitoba southward on plains and in the mountains to central Colorado. Disjunct populations from northern foothills of the Uinta Mountains in Utah are included in var. *effusa* by B. C. Johnston (1980), but relegated to *Potentilla hippiana* by N. H. Holmgren (1997b). Verification is needed for single occurrences of the variety mapped by Johnston in eastern Idaho, northwestern Nebraska, and north-central New Mexico, without citation of specimens. A single collection from Valley County, Idaho, (*E. A. Christenson* 42, CONN) is almost certainly an introduction of unknown persistence.

The small, narrow, cottony-floccose epicalyx bractlets tend to be glabrescent at least distally, contrasting with the sharply acuminate sepals. Variety *effusa*, in its most distinctive form, has unevenly pinnate leaves with coarsely toothed gray leaflets that are equally cottony-floccose on both surfaces.

Collections of otherwise typical var. *effusa* from central Wyoming and adjacent Montana are uniformly smaller than average with only one to four teeth per side of the leaflets. These occur mainly on limestone outcrops (1300–2600 m elevation) and might qualify as a distinct variety.

31b. Potentilla effusa Douglas ex Lehmann var. **rupincola** (Osterhout) Th. Wolf, Biblioth. Bot. 71: 199. 1908 • Rockloving cinquefoil [C] [E]

Potentilla rupincola Osterhout, Bull. Torrey Bot. Club 26: 256. 1899

Leaflets ± green, teeth 2–5 per side, abaxial surfaces sparsely to moderately long-hairy at least on veins, not or sparsely cottony, adaxial glabrous or sparsely long-hairy and/or cottony. **Flowers:** epicalyx bractlets and sepals mostly glabrate or glabrous except proximally; anthers 0.5–0.6 (–0.7) mm.

Flowering summer. Granitic outcrops or thin, gravelly soil, montane conifer woodlands; of conservation concern; 2100–3300 m; Colo.

Besides being bright green (rather than grayish), var. *rupincola* plants are generally less robust and somewhat shorter than plants of var. *effusa*, which also tends to be more upright. The variety is known only from the Rocky Mountains of north-central Colorado.

32. Potentilla ambigens Greene, Erythea 1: 5. 1893 • Silkyleaf or giant or southern Rocky Mountain cinquefoil [C] [E]

Stems (2–)3–7(–8) dm, lengths (1–)1.5–2.5 times basal leaves. **Basal leaves** pinnate, 15–45 (–50) cm; petiole (5–)8–20 cm, long hairs abundant to dense, ascending to appressed, (1–)2–3 mm, ± stiff, rarely weak, short or crisped hairs absent or sparse, cottony hairs absent, glands sparse, often obscured; leaflets not conduplicate, lateral ones evenly paired, (3 or)4–7 per side on distal ± ½ of leaf axis, most pairs ± decurrent, <rachis winged>, distal sometimes confluent with terminal leaflet, larger leaflets narrowly obovate, (2–)3–7 × 1–2.5 cm, distal ¾+ of margin incised ⅓ to midvein, teeth 6–18 per side, 1–4 mm, surfaces ± similar, abaxial grayish, long hairs sparse to common (mostly on veins), 0.5–2 mm, ± stiff, rarely weak, short-crisped hairs absent or sparse to common, cottony hairs absent, glands sparse to common, adaxial green to grayish, long and short hairs not differentiated, sparse to common, crisped and cottony hairs absent, glands absent or sparse. **Cauline leaves** (0 or)1–3. **Inflorescences** (10–)20–60-flowered. **Pedicels** (0.5–) 1–3(–6) cm. **Flowers:** epicalyx bractlets lanceolate to narrowly ovate or elliptic, 3–6 × 1–1.5 mm, ⅔ as long as to nearly equal to sepals, abaxial vestiture ± similar to sepals, not glabrescent, straight hairs sparse to common, crisped or cottony hairs usually absent; hypanthium 3–6 mm diam.; sepals 4–7 mm, apex acute to acuminate; petals (6–)7–10 × 5–9 mm; filaments (1–)1.5–4 mm, anthers 0.7–0.9(–1) mm; carpels 20–30, styles (1–)1.8–3 mm. **Achenes** 1.4–1.6 mm, smooth. **2*n*** = 82.

Flowering summer. Dry meadows and adjacent slopes, in conifer woodlands; of conservation concern; 2000–2700 m; Colo., N.Mex., Wyo.

Potentilla ambigens is one of the more distinctive species in sect. *Leucophyllae*, comprising exceptionally large plants with large, coarsely hairy leaves. Although sometimes confused with *P. hippiana*, *P. ambigens* lacks the crisped-cottony vestiture of that species and is usually morphologically distinctive even where both species occur together (D. G. Anderson, www.fs.fed.us/r2/projects/scp/assessments/potentillaambigens.pdf).

Potentilla ambigens occurs in Wyoming (at least historically), along the eastern slope of the Rocky Mountains to northern Colorado, and in scattered locations in the high mountains of New Mexico as far south as the Sacramento Mountains, Otero County. The species is of conservation concern in Colorado and has not been documented in Wyoming since 1900 (D. G. Anderson, www.fs.fed.us/r2/projects/scp/assessments/potentillaambigens.pdf).

33. Potentilla crinita A. Gray, Mem. Amer. Acad. Arts, n. s. 4: 41. 1849 • Bearded or Lemmon's cinquefoil [E] [F]

Ivesia lemmonii S. Watson; *Potentilla crinita* var. *lemmonii* (S. Watson) Kearney & Peebles; *P. lemmonii* (S. Watson) Greene

Stems (0.5–)1.5–4.5 dm, lengths 2–4(–5) times basal leaves. **Basal leaves** pinnate, 3–15(–20) cm; petiole 1–10(–15) cm, long hairs dense, appressed, 1.5–2.5 mm, usually stiff, short and crisped hairs usually absent, cottony hairs absent, glands sparse, often obscured; leaflets often conduplicate, lateral ones evenly paired, (3 or)4–6(or 7) per side on distal ⅓–⅔ of leaf axis, distal pairs ± decurrent, often confluent with terminal leaflet, larger leaflets narrowly cuneate or oblanceolate to obovate, 1–3(–4) × 0.2–0.8(–1) cm, distal ¼–½(–⅔) or less of margin incised ± ¼ or less to midvein, teeth (0 or)1–5(–9) per side, 1–2 mm, surfaces ± similar to ± dissimilar, abaxial silvery to greenish, long hairs usually dense (at least on primary veins), 1–2 mm, stiff, short-crisped hairs absent or sparse, cottony hairs usually absent, glands sparse to common, often obscured, adaxial ± green, long hairs sparse to common, sometimes absent, short, crisped, and cottony hairs absent, glands sparse. **Cauline leaves** 1–3(or 4). **Inflorescences** (5–) 10–30-flowered. **Pedicels** 0.5–2(–4) cm. **Flowers:** epicalyx bractlets lanceolate, 1.5–4.5 × 0.5(–1) mm,

½–⅔ as long as sepals, abaxial vestiture similar to or ± sparser than sepals, not glabrescent, straight hairs common, crisped or cottony hairs usually absent; hypanthium 2.5–5 mm diam.; sepals (3–)4–7 mm, apex acute to long acuminate; petals (3–)4.5–7.5(–8) × 4–6 mm; filaments 1–3 mm, anthers 0.6–1.1 mm; carpels 5–20, styles 1.6–2.6 mm. **Achenes** 1.4–1.7 mm, smooth or slightly rugose.

Flowering summer. Dry meadows, pygmy conifer, oak, aspen, or montane conifer woodlands; 2000–2600 m; Ariz., Colo., Nev., N.Mex., Utah.

Potentilla crinita occurs mainly in the upper foothills and mountains from southern Nevada to south-central Utah, northern Arizona, and northwestern New Mexico and is disjunct to southwestern Colorado (Archuleta County). It tends to grow on somewhat drier, rockier sites than co-occurring species of *Potentilla*. The often conduplicate leaflets, falcate in outline, bear relatively few, small teeth. Two varieties are sometimes recognized, based on leaflet and vestiture characters that do not reliably coincide.

Potentilla crinita can hybridize with *P. hippiana* where the two species overlap, in spite of ecological partitioning. N. H. Holmgren (1997b) noted the type of *P. crinita* may be such a hybrid. If correct, then *P. lemmonii* would be used for the species unless the name *P. crinita* were to be conserved with a conserved type.

8m. Potentilla Linnaeus sect. Subjugae (Rydberg) A. Nelson in J. M. Coulter and A. Nelson, New Man. Bot. Rocky Mt., 255. 1909 [E]

Barbara Ertter

Potentilla [unranked] *Subjugae* Rydberg, Bull. Torrey Bot. Club 23: 397. 1896

Perennials, ± tufted, not stoloniferous; taproots not fleshy-thickened; vestiture of long, crisped, and/or cottony hairs, glands sparse to common, sometimes red. **Stems** ascending to nearly erect, not flagelliform, not rooting at nodes, lateral to persistent basal rosettes, (0.2–)0.4–2.5(–3.5) dm, lengths 2–3 times basal leaves. **Leaves:** basal not 2-ranked; cauline 0–3; primary leaves palmate with additional lateral pairs to pinnate (with distal leaflets distinct), (1–)1.5–10(–14) cm; petiole: long hairs spreading to tightly appressed, weak to stiff, glands absent or sparse; leaflets 5–7(–9), on distal ⅒–⅓(–½) of leaf axis, strongly overlapping or not, oblanceolate-oblong to obovate, margins scarcely to ± revolute, ± whole length evenly incised ½–¾ to midvein, teeth 2–9 per side, surfaces ± to strongly dissimilar, abaxial white to grayish green, cottony hairs sparse to dense, adaxial green to grayish, not glaucous, long hairs ± stiff. **Inflorescences** (1–)2–20(–30) -flowered, usually cymose, open. **Pedicels** straight in fruit, 0.5–2(–3) cm, proximal ± longer than distal. **Flowers** 5-merous; hypanthium 3–5 mm diam.; petals yellow, ± obcordate, 4–8 mm, equal to or longer than sepals, apex retuse; stamens ca. 20; styles subapical, filiform to filiform-tapered or filiform-columnar, papillate-swollen in proximal less than ⅕ if at all, 0.8–2 mm. **Achenes** smooth.

Species 2 (2 in the flora): w North America.

Section *Subjugae* is used primarily to accommodate the unique leaf dissection and vestiture of *Potentilla subjuga*, which commonly has five palmately arranged leaflets and one, sometimes two, additional pairs of lateral leaflets. Sometimes leaves are simply pinnate with only three terminal leaflets, and sufficient intergradation occurs with *P. saximontana* that it is also included in the section. This circumscription coincides with the initial concept of Rydberg and Nelson and contrasts with that of B. C. Johnston (1980, 1985).

Section *Subjugae* is restricted to subalpine and alpine habitats in the Rocky Mountains and adjacent ranges of western North America centered in Colorado. Variation within the section is complex and poorly understood, such that the two species treated here represent only the best expressed extremes with existing names. Other named components of uncertain placement include *Potentilla lupina* Rydberg, *P. minutifolia* Rydberg, *P. rubricaulis* var. *nana* Clements & E. G. Clements, *P. rubripes* Rydberg [= *P. concinna* var. *rubripes* (Rydberg) C. L. Hitchcock], and *P. tenerrima* Rydberg.

1. Leaflets at tip of leaf axis (3–)5; styles 1.5–2 mm; stems (0.8–)1–2.5(–3.5) dm; inflorescences 3–20(–30)-flowered . . . . 34. *Potentilla subjuga*
1. Leaflets at tip of leaf axis 3; styles 0.8–1.5 mm; stems (0.2–)0.4–1(–1.5) dm; inflorescences (1 or)2–5(–10)-flowered. . . . . 35. *Potentilla saximontana*

34. Potentilla subjuga Rydberg, Bull. Torrey Bot. Club 23: 397, plate 274. 1896 • Colorado cinquefoil [E] [F]

Potentilla osterhoutiana Th. Wolf

Stems (0.8–)1–2.5(–3.5) dm. **Basal leaves** usually palmate with additional lateral leaflets, sometimes pinnate, 3–10(–14) cm; petiole 1.5–5 cm, vestiture seasonally dimorphic, long hairs abundant, spreading on first-formed leaves, tightly appressed to ascending on later-formed leaves, 1–2 mm, ± stiff (especially on later-formed leaves), cottony and crisped hairs usually absent, glands absent or sparse; leaflets (3–)5 at tip of leaf axis plus 1(or 2) additional pair(s) separated from tip by 3–20 mm, on distal 1/10–1/3(–1/2) of leaf axis, largest leaflets oblanceolate-oblong, (0.5–)1.5–2.5(–3) × 0.3–1 cm, ± whole margin incised 1/2–2/3(–3/4) to midvein, teeth (2–)4–9 per side, usually touching to strongly overlapping, sometimes separate, 2–6 mm, surfaces usually strongly dissimilar (less so on first-formed leaves), abaxial usually white, straight hairs ± abundant (mostly on veins), 1–2 mm, cottony or crisped/cottony hairs ± dense (sparser on first-formed leaves), glands absent or obscured, adaxial green (to grayish), straight hairs sparse to common, 0.5–1.5 mm, cottony and crisped hairs absent, glands sparse. **Cauline leaves** 1–3. **Inflorescences** 3–20(–30) -flowered. **Pedicels** 0.5–2 cm (proximal to 3 cm). **Flowers:** epicalyx bractlets narrowly to broadly lanceolate, 2–5(–6) × 1–1.5 mm; sepals 4–7 mm, apex acute to acuminate; petals 4–8 × 4–8 mm; filaments (0.5–)1–2 mm, anthers 0.3–0.8 mm; carpels 15–30, styles filiform to filiform-tapered, ± papillate-swollen in less than proximal 1/5, 1.5–2 mm. **Achenes** 1.2–1.6 mm.

Flowering summer. Alpine tundra and meadows, boulder piles, gravelly slopes, stabilized talus; 3400–4000 m; Alta.; Colo., N.Mex.

Potentilla subjuga is centered in the high mountains of Colorado and barely enters New Mexico in the Sangre de Cristo Mountains. Collections from Alberta also apparently belong to this species, but all known collections from Wyoming have been identified as different taxa, at least one currently undescribed. At its most distinctive, *P. subjuga* is easily recognized by its unique leaf division, with five palmate leaflets subtended by an additional pair (or two) of lateral leaflets. Southern populations, however, are more likely to have only three apical leaflets. The leaflets tend to be strongly bicolored with overlapping teeth, in contrast to most sympatric pinnate species. Petiole vestiture is also distinctive in being seasonally dimorphic, with long hairs on first-formed leaves spreading to ascending and those on later-formed leaves tightly appressed, as well as more conspicuously verrucose. Unresolved infraspecific variation exists, and field observations suggest that *P. subjuga* readily hybridizes with sympatric species, creating a swarm of intermediate specimens.

35. Potentilla saximontana Rydberg, Bull. Torrey Bot. Club 23: 399. 1896 • Rocky Mountains cinquefoil [E]

Stems (0.2–)0.4–1(–1.5) dm. **Basal leaves** subpalmate to pinnate, (1–)1.5–4(–7) cm; petiole 0.5–2(–4) cm, vestiture not or scarcely seasonally dimorphic, long hairs abundant to dense, appressed to spreading, 1–1.5(–2) mm, weak to ± stiff, cottony hairs usually absent, crisped hairs absent or sparse, glands sparse; leaflets 3 at tip of leaf axis plus 1 or 2(or 3) additional pairs separated from tip by 1–5 mm, on distal 1/10–1/3 of leaf axis, largest leaflets obovate to oblanceolate, 0.5–1.5 × 0.3–1.2 cm, ± whole margin incised 1/2–3/4 to midvein, teeth 2–4(or 5) per side, separate to overlapping, 1–4 mm, surfaces ± dissimilar, abaxial grayish green to white, straight hairs abundant, 1–1.5 mm, cottony and crisped hairs sparse to dense, rarely absent, glands sparse or obscured, adaxial green to grayish, straight hairs sparse to abundant, 1–2 mm, cottony hairs absent or rarely sparse, crisped hairs absent or sparse, glands ± sparse. **Cauline leaves** 0 or 1(or 2). **Inflorescences** (1 or)2–5(–10) -flowered. **Pedicels** 0.5–1.5 cm (proximal to 2.5 cm). **Flowers:** epicalyx bractlets ± ovate-elliptic, 2–4 × 1–1.5 mm; sepals 3.5–5 mm, apex bluntly acute; petals 4–6(–8) × 4–7 mm; filaments 0.5–2 mm, anthers 0.5 mm; carpels 10–20+, styles columnar-filiform, papillate-swollen at base, if at all, 0.8–1.5 mm. **Achenes** 1.5 mm.

Flowering summer. Alpine tundra and meadows, rocky slopes, talus; 3200–4100 m; Colo., Mont., Utah, Wyo.

Potentilla saximontana is a small plant of very high elevations in the mountains of Colorado, the La Sal and Uinta mountains of Utah, and the Absaroka and Beartooth mountains of Wyoming and Montana. Although sometimes confused with *P. rubricaulis* (for example, W. A. Weber and R. C. Wittmann 1996), the subpinnate leaves, open few-flowered inflorescences, and columnar-filiform styles make *P. saximontana* a distinct species.

The lectotype of *Potentilla nivea* Linnaeus var. *dissecta* S. Watson [= *P. saximontana* var. *dissecta* (S. Watson) Soják] has been tentatively included in *P. saximontana* (B. Ertter 2008), but *P. saximontana* is not otherwise known from the Canadian Rockies. See B. Ertter et al. (2013) for additional discussion.

The combination *Potentilla rubripes* Rydberg var. *saximontana* (Rydberg) Th. Wolf is incorrect, in that *P. saximontana* is the older name.

8n. Potentilla Linnaeus sect. Multijugae (Rydberg) A. Nelson in J. M. Coulter and A. Nelson, New Man. Bot. Rocky Mt., 255. 1909

Barbara Ertter

Potentilla [unranked] *Multijugae* Rydberg, Bull. Torrey Bot. Club 23: 432. 1896; *Potentilla* sect. *Candicantes* (Rydberg) O. Stevens

Perennials, rosetted to ± matted, not stoloniferous; taproots often fleshy-thickened; vestiture primarily of straight hairs, not differentiated into long and short, crisped hairs often on stems and inflorescence, cottony hairs only in *P. breweri*, glands usually absent or sparse, rarely common, rarely ± red. **Stems** usually prostrate to decumbent, sometimes ascending (especially in supporting vegetation), not flagelliform, not rooting at nodes, lateral to persistent basal rosettes, (0.3–)0.5–5(–7) dm, lengths 1–3(–5) times basal leaves. **Leaves:** basal rarely ± 2-ranked; cauline (0 or)1–3(or 4); primary leaves pinnate (with distal leaflets distinct or confluent), (1.5–)2–20 (–32) cm; petiole: long hairs soft to stiff, sometimes absent, spreading to ± appressed, glands usually absent or sparse; leaflets 7–21(–41), on distal (1/4–)1/3 to whole leaf axis, overlapping or not, oblanceolate to obovate, oblong, cuneate, flabellate, or elliptic, margins flat, distal (1/4–)1/2 to whole length evenly or unevenly incised ± 1/2 to completely to midvein, rarely entire, teeth (0 or) 1–5(–7) per side, surfaces ± similar, green to whitish, sometimes glaucous, long hairs, if present, soft to stiff, cottony hairs usually absent, sometimes sparse to dense (*P. breweri*). **Inflorescences** (1 or)2–15(–25)-flowered, usually ± cymose, sometimes racemiform when prostrate, ± compact (*P. arizonica*) to very open. **Pedicels** recurved or straight in fruit, (0.5–)1–4(–6) cm, proximal often not much longer than distal. **Flowers** 5-merous; hypanthium 2–6 mm diam.; petals yellow, narrowly to broadly obcordate, (3.5–)4–10(–12) mm, equal to or longer than sepals, apex usually ± retuse; stamens (15–)20; styles subapical, filiform, often papillate-swollen in proximal 1/5 or less, (1.5–)2–3.5 mm. **Achenes** smooth to faintly rugose (± rugose in *P. basaltica*).

Species 11 (10 in the flora): w North America, c Mexico.

Section *Multijugae* is used here for most North American species with fully pinnate leaves that have similar vestiture abaxially and adaxially, leaflets often so deeply divided and overlapping as to make the leaves nearly cylindric rather than planar, terminal leaflets often irregularly lobed and confluent with distalmost lateral leaflets, and relatively long styles somewhat glandular-papillate at the base. This circumscription includes the Mexican species *Potentilla candicans* Humboldt & Bonpland ex Nestler, which P. A. Rydberg (1898, 1908d) placed in its own group *Candicantes*.

This provisional circumscription of sect. *Multijugae* encompasses at least three species clusters. One cluster (species 40–45), largely coinciding with the original concept of group *Multijugae* Rydberg, tends to have fleshy-thickened taproots, prostrate stems that are usually no more than twice as long as the leaves, very open racemiform inflorescences with recurved pedicels, and caruncular achenes. This species cluster has evidently radiated in seasonally wet meadows in western North America, resulting in several highly localized endemics, two of which (*P. multijuga*, *P. uliginosa*) are presumed extinct (B. Ertter 1993; B. C. Johnston and Ertter 2010).

The second cluster consists of *Potentilla breweri* and *P. versicolor*, montane species with taproots that are not fleshy-thickened, stems usually more than twice as long as leaves, pedicels scarcely recurved in fruit, and no caruncle on the achenes. These differences from the previous species cluster are offset by the intermediacy of *P. ovina*.

Potentilla arizonica differs from both species clusters in its relatively congested and flat-topped inflorescences and anthers that are as long as or longer than the filaments. The species resembles the first species cluster in its fleshy-thickened taproots, but differs in that the stems are more likely to be decumbent to ascending, pedicels remain relatively straight in fruit, and achenes are not caruncular.

Since *Potentilla drummondii* (sect. *Graciles*), and *P. macounii* and *P. morefieldii* (sect. *Concinnae*) are sometimes identified as members of sect. *Multijugae*, they are included herein and key out in the third, tenth, and eleventh couplets, respectively.

1. Largest leaflets: teeth 2–5(–7) on distal ¼–½(–¾) of margin, narrowly cuneate-oblanceolate to obovate, green, hairs absent (except on margins) or sparse.
   2. Basal leaves 2–10(–13) cm; stem lengths (1.5–)2–3 times leaves; Rocky Mountains and e Great Basin . . . . 39. *Potentilla ovina* (in part)
   2. Basal leaves 10–22 cm; stem lengths 1–2 times leaves; coastal California . . . . 44. *Potentilla multijuga* (in part)
1. Largest leaflets: ultimate teeth or segments 2–10(–15) on distal ½ to whole margin, rarely 2(or 3)-fid or entire, oblanceolate-cuneate or -oblong to flabellate, rarely elliptic (*P. basaltica*), green to white, hairs usually sparse to dense, sometimes absent or nearly so.
   3. Leaflets 2–4 per side on distal ⅒–½(–¾) of leaf axis, incised ± ½ to midvein (often with additional incisions nearly to midvein), green on both surfaces . . . . 29. *Potentilla drummondii* (sect. *Graciles*)
   3. Leaflets usually more than 4 per side and/or on more than distal ⅕ of leaf axis, mostly incised more than ½ to midvein, green to grayish green or white on one or both surfaces.
      4. Largest leaflets (or at least distal ones in *P. ovina*) pinnately incised into linear to linear-oblanceolate teeth; distal leaflets ± distinct; Alberta to Arizona and e, barely entering British Columbia, Oregon, and Nevada.
         5. Anthers 1–1.5 mm, often as long as filaments; pedicels straight in fruit; leaflets on distal (½–)⅔–¾ of leaf axis; Coconino County, Arizona . . . . 36. *Potentilla arizonica*
         5. Anthers 0.4–1 mm, usually ± ½ as long as filaments; pedicels usually ± recurved, sometimes straight, in fruit; leaflets on distal (¼–)½–⅔(–¾) of leaf axis; not Coconino County, Arizona.

6. Leaflets: straight hairs 1–2 mm, loosely appressed, apical tufts on teeth 0.5–2 mm; fellfields, alpine ridges, rocky slopes, dry hillsides, 1500–3800 m . . . . 39. *Potentilla ovina* (in part)
6. Leaflets: straight hairs ± 0.5 mm, tightly appressed, apical tufts on teeth less than 0.5 mm; moist meadows, streamsides, reservoir margins, 300–2900 m . . . . 40. *Potentilla plattensis*

[4. Shifted to left margin.—Ed.]

4. Largest leaflets usually ± palmately or unevenly (± evenly in *P. multijuga*) incised into linear to broadly elliptic or oblanceolate, sometimes ± ovate, teeth or segments, sometimes 2(or 3) -fid or entire; distal leaflets ± confluent; California, Nevada, Oregon, Washington (or Alberta and Montana for *P. macounii*).
7. Primary lateral leaflets (5–)10–15(–20) per side on nearly whole leaf axis, entire or 2 (or 3)-fid to base (sometimes shallowly toothed as well), glaucous, straight hairs absent or sparse to common, 0.5 mm or less, tightly appressed; nw Nevada, ne California . . . . 45. *Potentilla basaltica*
7. Primary lateral leaflets 3–10(–13) per side on distal ⅓–¾(+) of leaf axis, incised ½ to completely to midvein into (1 or)2–10(–15) ultimate teeth or segments, not glaucous, straight hairs sparse to abundant, sometimes nearly absent, 0.5–2 mm, ± appressed to spreading; California, Nevada, Oregon, Washington.
8. Pedicels straight to slightly recurved in fruit; stem lengths 2–4(–5) times basal leaves; leaves green to white; achenes not carunculate; 1500–3600(–4000) m.
9. Leaflets: abaxial surfaces green to grayish green, cottony hairs absent, ultimate teeth or segments 2–5(–8); filaments (1–)1.5–2.5 mm; e Oregon, ne Nevada . . . . 38. *Potentilla versicolor*
9. Leaflets: abaxial surfaces grayish green to white, cottony hairs sparse to dense, ultimate teeth or segments 3–10; filaments (1–)2–4 mm.
10. Leaflets on distal ⅕–½ of leaf axis; Alberta, Montana . . . . .49. *Potentilla macounii* (sect. *Concinnae*)
10. Leaflets on distal (¼–)½–¾+ of leaf axis; California, Nevada, Oregon, Washington.
11. Leaflet surfaces ± similar, adaxial: cottony hairs sparse to dense, straight hairs soft; Sierra Nevada, Cascade Range, and n Great Basin . . . . .37. *Potentilla breweri*
11. Leaflet surfaces ± to strongly dissimilar, adaxial: cottony hairs absent, straight hairs stiff; White Mountains and adjacent Sierra Nevada, California . . . . 50. *Potentilla morefieldii* (sect. *Concinnae*)
8. Pedicels ± recurved in fruit; stem lengths 1–2 times basal leaves; leaves usually green, sometimes grayish green; achenes often carunculate; 0–2200 m.
12. Primary lateral leaflets incised ⅔ to completely to midvein, (3–)5–13 per side on distal ½–¾+ of leaf axis, ultimate teeth or segments (0.5–)1–2.5 mm wide; California, Nevada, Oregon.
13. Stems 0.4–2(–3) dm; basal leaves 2–15(–20) cm; leaflets on distal ⅔–¾+ of leaf axis; interior California, Oregon, Nevada, 700–2200 m . . . . 41. *Potentilla millefolia*
13. Stems 2.5–5.5 dm; basal leaves 15–32 cm; leaflets on distal ± ½ of leaf axis; Sonoma County, California, 30–40 m . . . . .43. *Potentilla uliginosa*
12. Primary lateral leaflets incised ½–⅔(+) to midvein, 3–8 per side on distal ⅓–⅔ of leaf axis, ultimate teeth or segments (1–)1.5–3 mm wide; coastal California.
14. Distal ½ to whole leaflet incised ½–⅔(+) to midvein; stems 0.5–2.5 dm; c California . . . . 42. *Potentilla hickmanii*
14. Distal ⅓–⅔ of leaflet incised ± ½ to midvein; stems 2–5(–7) dm; s California . . . . 44. *Potentilla multijuga* (in part)

36. Potentilla arizonica Greene, Pittonia 1: 104. 1887 • Garland Prairie cinquefoil [C] [E]

Ivesia pinnatifida S. Watson, Proc. Amer. Acad. Arts 20: 364. 1885, not *Potentilla pinnatifida* C. Presl 1822

Plants rosetted to tufted; taproots ± fleshy-thickened. **Stems** decumbent to ascending, sometimes prostrate, 0.6–2.2 dm, lengths 1.5–3 times basal leaves. **Basal leaves** pinnate with distal leaflets ± distinct, 4–10 × 1–2(–3) cm; petiole 1–3 cm, straight hairs dense, ± appressed, 1–2 mm, stiff, cottony hairs absent, glands absent or sparse; primary lateral leaflets 5–7(–9) per side (sometimes with additional interspersed leaflets), on distal (½–)⅔–¾ of leaf axis, ± overlapping, largest ones oblanceolate- to obovate-oblong, 0.5–1.5 × 0.4–0.8 cm, distal ¾ to whole margin pinnately incised nearly to midvein, teeth (5–)7–9, linear-oblanceolate, 2–6 × 0.5–1 mm, apical tufts 1 mm, surfaces grayish green, straight hairs sparse (adaxially) to common, ± appressed, 1–2 mm, stiff, cottony hairs absent, glands absent or obscured. **Cauline leaves** (1 or)2 or 3(or 4). **Inflorescences** 3–15-flowered, ± compactly cymose, opening in fruit. **Pedicels** 0.7–2(–2.5) cm, straight in fruit. **Flowers:** epicalyx bractlets narrowly elliptic, <sometimes doubled>, 2–3.5 × 1–1.5 mm; hypanthium 3–4 mm diam.; sepals 4–6 mm, apex ± acute; petals 4–5(–6) × 3–4(–5.5) mm; filaments 1–2 mm, anthers 1–1.5 mm, <often as long as filaments>; carpels 8–20, styles 2 mm. **Achenes** 1.8 mm, ± smooth, not carunculate.

Flowering late spring–early summer. Vernally wet clay of rocky basaltic meadows, openings in pine woodlands; of conservation concern; 1900–2100 m; Ariz.

Potentilla arizonica is known only from the Garland Prairie area in Coconino County. Although commonly included within *P. plattensis* (for example, T. H. Kearney and R. H. Peebles 1951; B. C. Johnston 1980; N. H. Holmgren 1997b), *P. arizonica* differs in its longer hairs, more erect habit, more condensed inflorescences, pedicels that remain straight in fruit, and significantly larger anthers.

37. Potentilla breweri S. Watson, Proc. Amer. Acad. Arts 8: 555. 1873 • Brewer's cinquefoil [C] [E]

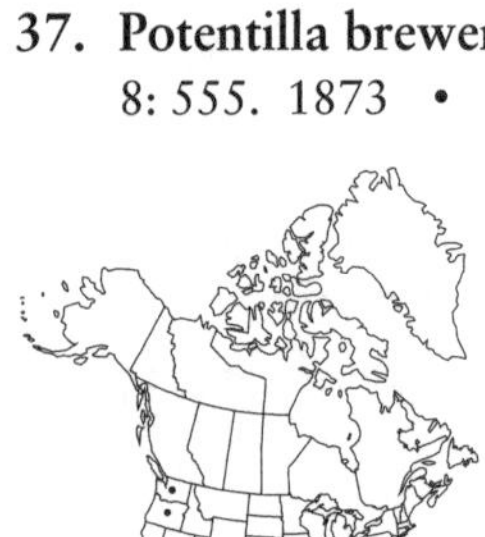

Potentilla drummondii Lehmann subsp. *breweri* (S. Watson) Ertter; *P. drummondii* var. *breweri* (S. Watson) N. H. Holmgren; *P. plattensis* Nuttall var. *leucophylla* Greene

Plants rosetted to ± matted; taproots not fleshy-thickened. **Stems** prostrate to ascending, (0.5–)1–3(–4.5) dm, lengths 2–4(–5) times basal leaves. **Basal leaves** pinnate with distal leaflets ± confluent, (2–)4–12(–17) × 1–2.5(–5) cm; petiole 1–3(–7) cm, straight hairs mostly absent, cottony hairs sparse to dense, glands absent or obscured to common; primary lateral leaflets 3–6 per side, on distal ½–⅔(–¾) of leaf axis, ± overlapping, largest ones cuneate-flabellate, 0.5–1.5(–2.5) × 0.4–2(–3) cm, distal ½ to whole margin unevenly incised ½ to completely to midvein (blade often medially split as well), ultimate teeth or segments 3–10, narrowly elliptic, 2–10(–15) × 1–3(–5) mm, apical tufts less than 1 mm, surfaces grayish green to white, not glaucous, straight hairs sparse to common (sparser adaxially), loosely appressed to ascending, 1–1.5 mm, soft, cottony hairs sparse to dense, glands sparse or obscured to common. **Cauline leaves** 1–3. **Inflorescences** 2–15(–25)-flowered, openly cymose. **Pedicels** (1–)1.5–2.5(–4) cm, straight in fruit. **Flowers:** epicalyx bractlets lanceolate to ovate, 2–5 × 1–1.5 mm; hypanthium (3–)4–5 mm diam.; sepals (3–)4–7 mm, apex acute; petals 5–9(–10) × 4–8(–10) mm; filaments (1–)2–4 mm, anthers (0.5–)1 mm; carpels 15–25, styles 2–3 mm. **Achenes** 1.8 mm, smooth, not carunculate. **2*n*** = 72, 73, 99, 100, 102.

Flowering summer. Rocky meadows, seasonally moist flats, rock crevices, often near streams and lakes; of conservation concern; 1500–3600 m; Calif., Nev., Oreg., Wash.

Potentilla breweri differs from other members of sect. *Multijugae* in its often dense, cottony vestiture. Density of cottony hairs and leaflet dissection can differ significantly between first-formed and mid- to late-season leaves of *P. breweri*; in the description above, leaves are those predominant at anthesis.

Potentilla breweri is most common in the Sierra Nevada of California, with sporadic occurrences to the Cascade Range in southern Washington. Disjunct populations occur in the Ruby and Snake ranges in Nevada and on Steens Mountain in Oregon, where often introgressed with *P. versicolor*. Reports from Utah are based on collections of *P. concinna* var. *proxima* or possible hybrids between *P. concinna* var. *proxima* and *P. ovina* var. *decurrens*.

J. Clausen et al. (1940) concluded that *Potentilla breweri* (with *P. versicolor* as synonym) belonged with *P. bruceae* and *P. drummondii* as members of a cenospecies that probably also included unnamed species (possibly *P. ovina* var. *decurrens*) from the mountains of Colorado, Utah, and Wyoming. Chromosome number varied within populations, and high numbers of univalents were present.

Recognition of *Potentilla breweri* as a distinct species differs from recent treatments in which it was considered to be a variety or subspecies of *P. drummondii*, parallel to the treatment of *P. bruceae*. The present continental perspective of the genus, as well as additional fieldwork, has resulted in the return of all three to species rank (B. Ertter and D. Mansfield 2007). Placement of

P. breweri in sect. *Multijugae*, separate from *P. bruceae* and *P. drummondii* in sect. *Graciles*, is based on the commonly prostrate habit, fully pinnate leaves, and deeply divided overlapping leaflets.

Although B. C. Johnston (1980) used *Potentilla breweri* var. *viridis* Jepson to accommodate greener-than-average sparsely cottony plants, including populations treated here as *P. versicolor*, the type is probably a sterile hybrid with *P. wheeleri* as one parent (B. Ertter 1992). The type of *P. millefolia* var. *algida* Jepson (included by Johnston in the synonymy of var. *viridis*) and comparable plants from north-central California combine features of *P. breweri* and *P. versicolor* but are more glandular than either. This variant, as well as exceptionally small plants from the Warner Mountains of California, may prove distinct.

38. Potentilla versicolor Rydberg in N. L. Britton et al., N. Amer. Fl. 22: 344. 1908 • Steens Mountain cinquefoil [E]

Plants rosetted to ± matted; taproots ± fleshy-thickened. **Stems** prostrate to ascending, (0.7–)1.5–2.5(–4) dm, lengths 2–4(–5) times basal leaves. **Basal leaves** pinnate with distal leaflets ± confluent, 4–12 × 1–2.5(–3.5) cm; petiole 1–3 cm, straight hairs absent (on early-season petioles) or sparse to common, spreading-ascending to loosely appressed, 1–2 mm, soft, cottony hairs usually absent, glands sparse; primary lateral leaflets 3–5 per side, on distal (⅓–)½–⅔ of leaf axis, overlapping, largest ones cuneate to flabellate, 0.5–1.5(–2) × 0.5–1.5 cm, distal ¾ to whole margin unevenly incised ⅔ to completely to midvein (blade often medially split as well), ultimate teeth or segments 2–5(–8), ± oblanceolate, 3–11 × 1–3 mm, apical tufts to 1 mm, surfaces green to grayish green, not glaucous, straight hairs sparse to common (sparser adaxially), sometimes absent (except on margins), loosely appressed to ascending, 1–2 mm, soft, cottony hairs absent, glands sparse to common. **Cauline leaves** 1 or 2(or 3). **Inflorescences** (1–)3–10-flowered, usually openly cymose. **Pedicels** 1–3(–5) cm, straight to slightly recurved in fruit. **Flowers:** epicalyx bractlets lanceolate to elliptic, rarely ovate, 2–5 × 1–2 mm, <sometimes apically toothed>; hypanthium 3–5 mm diam.; sepals 4–7 mm, apex acute; petals 4–7 × 3–5.5 mm; filaments (1–)1.5–2.5 mm, anthers 0.5–0.8 mm; carpels 10–25, styles 2 mm. **Achenes** 1.5–1.8 mm, smooth to faintly rugose, not caruculate.

Flowering summer. Rocky, alpine meadows, seasonally moist slopes, near streams or snowmelt; 2100–3200 m; Nev., Oreg.

Potentilla versicolor is most common on Steens Mountain in Harney County, Oregon. Collections are known from other mountain ranges in eastern Oregon (some possibly representing distinct taxa) and from near Island Lake in the Ruby Mountains of northeastern Nevada. The species often grows and, apparently, intergrades with *P. breweri*; it is usually distinct in its lack of cottony hairs. Petals, filaments, and styles tend to be somewhat shorter in *P. versicolor* than in *P. breweri*. Collections of *P. versicolor* have most often been identified as *P. breweri*, *P. millefolia*, or *P. ovina*.

39. Potentilla ovina J. M. Macoun, Canad. Rec. Sci. 6: 464. 1896 • Sheep cinquefoil [E]

Potentilla diversifolia Lehmann var. *pinnatisecta* S. Watson, Botany (Fortieth Parallel), 87. 1871; *P. pinnatisecta* (S. Watson) A. Nelson; *P. plattensis* Nuttall var. *pinnatisecta* (S. Watson) J. M. Macoun

Plants rosetted to ± matted; taproots sometimes ± fleshy-thickened. **Stems** prostrate to ascending, (0.3–)0.8–2 (–3.5) dm, lengths (1–)1.5–3 times basal leaves. **Basal leaves** pinnate with distal leaflets ± distinct, (1.5–) 2–10(–13) × 0.7–3.5(–5) cm; petiole 0.5–3.5(–5) cm, straight hairs sparse to abundant, sometimes absent (var. *decurrens*), ± appressed to ascending, 1 mm, ± stiff, cottony hairs absent, glands sparse, often obscured; primary lateral leaflets 3–6 per side (often with additional interspersed leaflets), on distal ½–⅔ (–¾) of leaf axis, ± separate to ± overlapping, largest ones narrowly cuneate-oblanceolate to ± obovate, (0.3–)0.5–2(–3.5) × (0.2–)0.3–0.8(–1) cm, distal ¼ to whole margin unevenly to pinnately (at least distal leaflets of var. *ovina*) incised ½–¾+ to midvein, ultimate teeth 2–9(–11), linear or oblong to ovate, 1–7(–9) × 1–2 mm, apical tufts 0.5–2 mm, surfaces green to grayish, not glaucous, straight hairs sparse to abundant (sparser adaxially), sometimes absent (except on margins), loosely appressed, 0.5–2 mm, ± stiff, cottony hairs absent, <crisped hairs sometimes sparse to common>, glands absent or inconspicuous. **Cauline leaves** 1 or 2. **Inflorescences** (1 or)2–11(–20)-flowered, usually openly cymose. **Pedicels** (0.8–)1–2(–4) cm, straight to ± recurved in fruit. **Flowers:** epicalyx bractlets linear-lanceolate to narrowly elliptic, <sometimes doubled>, 2–3.5(–5) × 1 mm; hypanthium 3–5 mm diam.; sepals 3.5–5.5(–7) mm, apex acute to obtuse; petals 4–7(–8) × 3.5–8 mm; filaments 1–2.5 mm, anthers 0.4–1 mm <usually ± ½ as long as filaments>; carpels 10–20, styles 2–3 mm. **Achenes** 1.5–2 mm, smooth, not caruculate.

Varieties 2 (2 in the flora): w North America.

Potentilla ovina is here accepted as an implicit new name by J. M. Macoun for *P. diversifolia* var. *pinnatisecta*, in agreement with N. H. Holmgren (1997b). In contrast, B. C. Johnston (1980) considered the names heterotypic. *Potentilla ovina* has priority at the species rank over *P. pinnatisecta* by one month.

1. Distal leaflets: ± whole margin pinnately incised ¾+ to midvein, teeth 4–9(–11), linear, surfaces grayish, rarely green, hairs sparse to dense . . . . 39a. *Potentilla ovina* var. *ovina*
1. Distal leaflets: distal ¼–½(–¾) of margin unevenly to pinnately incised ½–⅔(–¾) to midvein, teeth 2–5(–7), ovate to oblong, surfaces green, hairs absent (except on margins) or sparse. . . . . 39b. *Potentilla ovina* var. *decurrens*

39a. Potentilla ovina J. M. Macoun var. **ovina** [E]

Potentilla bipinnatifida Douglas var. *glabrata* (Lehmann) Kohli & Packer; *P. glabrella* Rydberg; *P. monidensis* A. Nelson; *P. pensylvanica* Linnaeus var. *glabrata* (Lehmann) S. Watson; *P. wyomingensis* A. Nelson

Stems: lengths (1–)1.5–2.5(–3) times basal leaves. **Basal leaves** (1.5–)2–8(–10) × 0.7–2 cm; petiole (0.5–)1–2(–3) cm; lateral leaflets on distal ½(–⅔) of leaf axis, ± overlapping, distal leaflets ± obovate, (0.3–)0.5–1.2(–2.5) cm, ± whole margin pinnately incised ¾+ to midvein, teeth 4–9(–11), linear, 2–6(–9) × 1 mm, surfaces grayish, rarely green, hairs sparse to abundant, 1–2 mm. **Achenes** 1.8–2 mm.

Flowering summer. Fellfields, alpine ridges, rocky slopes, dry hillsides; 1500–3800 m; Alta., B.C.; Colo., Idaho, Mont., Nev., Oreg., Utah, Wyo.

Typical var. *ovina* is a tufted, grayish plant with the leaflets divided (pinnately on distal leaflets, palmately or unevenly on proximal leaflets) nearly to the midvein into linear teeth with prominent apical hair tufts. Leaflets are often folded and closely overlapping, giving the leaves a distinctive silvery cylindric appearance especially when drought-stressed. Populations around Yellowstone National Park, in particular the type of *Potentilla wyomingensis*, are often larger and less hairy, with larger flowers in more open inflorescences. Some collections from southeastern Idaho and southwestern Wyoming are transitional between var. *ovina* and *P. multisecta* (sect. *Concinnae*); their optimal placement has yet to be determined.

The distribution of var. *ovina* given here largely matches that indicated by B. C. Johnston (1980), with the addition of southeastern British Columbia based on H. J. Scoggan (1978–1979, part 3). No record has been found to support the inclusion of Saskatchewan in the distribution, as done by C. L. Hitchcock and A. Cronquist (1961b). In Oregon, var. *ovina* occurs only as depauperate plants in the Wallowa Mountains; reports from Steens Mountain are mostly based on *Potentilla versicolor*. In Nevada, var. *ovina* is confirmed only from the Schell Creek Range, White Pine County. Not included here are collections from Alaska and Yukon that probably represent an undescribed taxon.

39b. Potentilla ovina J. M. Macoun var. **decurrens** (S. Watson) S. L. Welsh & B. C. Johnston, Great Basin Naturalist 42: 30. 1982 [E]

Potentilla dissecta Pursh var. *decurrens* S. Watson, Proc. Amer. Acad. Arts 8: 557. 1873; *P. decurrens* (S. Watson) Rydberg; *P. diversifolia* Lehmann var. *decurrens* (S. Watson) Th. Wolf; *P. nelsoniana* Rydberg

Stems: lengths (1.5–)2–3 times basal leaves. **Basal leaves** 2–10(–13) × 1–3.5(–5) cm; petiole 0.5–3.5(–5) cm; lateral leaflets on distal ½–⅔(–¾) of leaf axis, ± separate, distal leaflets narrowly cuneate-oblanceolate to obovate, (0.5–)1–2(–3.5) cm, distal ¼–½(–¾) of margin unevenly to pinnately incised ½–⅔(–¾) to midvein, teeth 2–5(–7), ovate to oblong, 1–7 × 1–2 mm, surfaces green, hairs absent (except on margins) or sparse, 0.5–1 mm. **Achenes** 1.5–1.8 mm.

Flowering summer. Dry meadows, stony slopes, exposed ridges and summits, alpine and subalpine; 2400–3800 m; Colo., Idaho, Mont., Nev., Utah, Wyo.

Variety *decurrens* at its most distinctive has elongate cuneate-oblanceolate, glabrate, apically toothed leaflets that are well spaced on the leaf axis. This extreme is most commonly found in the Uinta Mountains of Utah and the Medicine Bow Mountains of Wyoming. Collections from Colorado, Idaho, Montana, and Nevada are more equivocal and might represent hybrids or extreme forms of other (or undescribed) taxa.

Although B. C. Johnston (1980) is provisionally followed in associating var. *decurrens* with *Potentilla ovina*, comparisons can also be made with *P. drummondii* and *P. glaucophylla*, both placed here in sect. *Graciles*. Apparent intergradation exists in all directions.

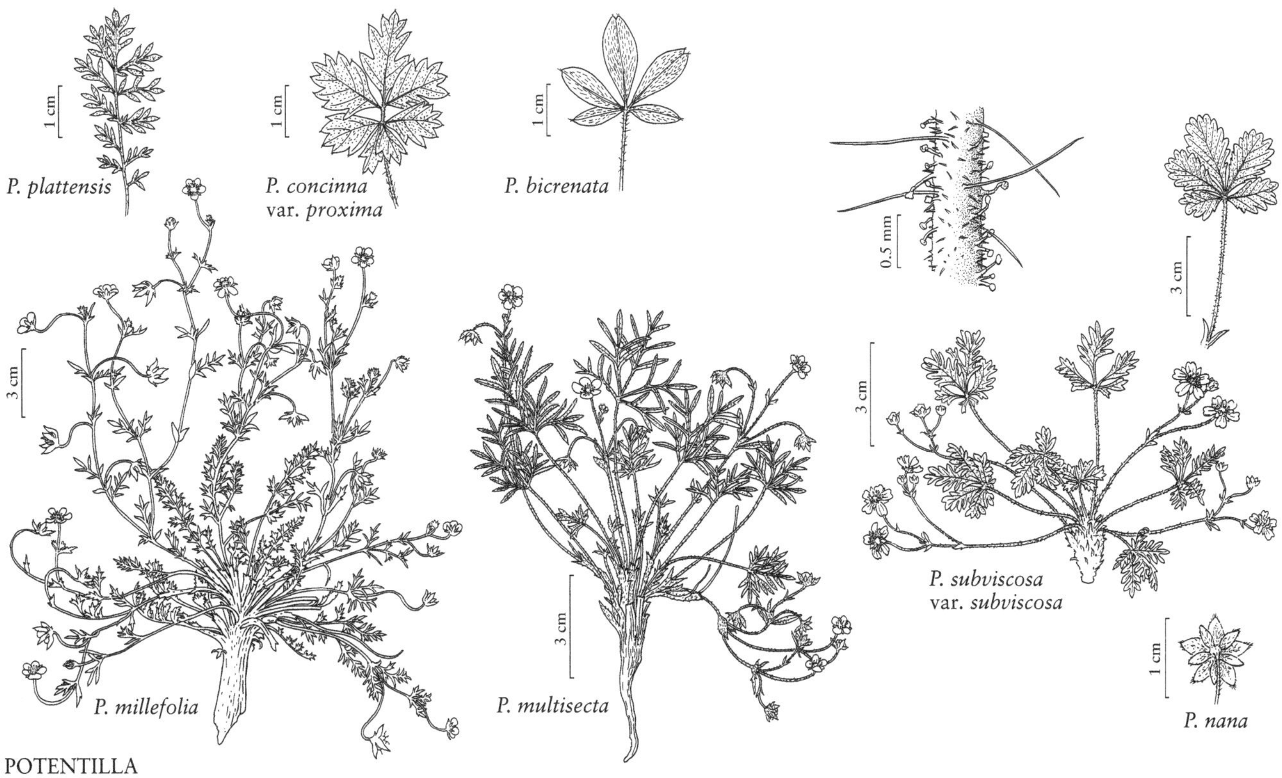

POTENTILLA

40. Potentilla plattensis Nuttall in J. Torrey and A. Gray, Fl. N. Amer. 1: 439. 1840 • Platte cinquefoil [E] [F]

Potentilla diversifolia Lehmann var. *madsenii* S. L. Welsh & N. D. Atwood; *P. plattensis* var. *pedicillata* A. Nelson

Plants rosetted to tufted; taproots fleshy-thickened. **Stems** initially decumbent to sometimes ascending, becoming prostrate or supported by vegetation, (0.3–)0.5–3.5(–4.5) dm, lengths 1.5–2.5(–4) times basal leaves. **Basal leaves** pinnate with distal leaflets ± distinct, 2–15(–20) × 1–3 cm; petiole 0.5–5 cm, straight hairs common, tightly appressed, 0.5(–1) mm, stiff, cottony hairs absent, glands absent or sparse; primary lateral leaflets (3 or)4–6(–8) per side, on distal (¼–)½–¾ of leaf axis, overlapping to separate, largest ones obovate, 0.5–1.5(–2) × 0.5–1(–1.3) cm, distal (½–)⅔ to whole margin pinnately incised ¾+ to midvein, ultimate teeth 5–10, linear-oblanceolate, 1.5–8(–10) × 1–2 mm, apical tufts less than 0.5 mm, surfaces green to grayish green, not glaucous, straight hairs sparse to common (sparser adaxially), tightly appressed, 0.5 mm, stiff, cottony hairs absent, glands absent or inconspicuous. **Cauline leaves** 1–3. **Inflorescences** (1–)3–15(–20)-flowered, loosely cymose, sometimes racemiform. **Pedicels** 1–4(–5) cm, ± recurved in fruit. **Flowers:** epicalyx bractlets narrowly elliptic to linear-lanceolate, 2–5(–6) × 0.5–2 mm; hypanthium 3–4 mm diam.; sepals 3–6 mm, apex acute; petals 4–7 × 3–6 mm; filaments 1–2.5 mm, anthers 0.5–1 mm, <usually ± ½ as long as filaments>; carpels 10–20, styles 1.5–2.5 mm. **Achenes** (1.3–)1.5–1.9 mm, smooth, often ± carunculate. $2n = 70$.

Flowering summer. Moist meadows, streamsides, reservoir margins; 300–2900 m; Alta., Man., N.W.T., Sask.; Ariz., Colo., Idaho, Mont., N.Mex., N.Dak., S.Dak., Utah, Wyo.

Potentilla plattensis occurs mostly east of the Continental Divide from the Canadian Prairies to the mountains of New Mexico. The species barely enters Idaho at Monida Pass (Clark County). Populations also exist in the White Mountains of east-central Arizona and the mountains of southern Utah. The species is relatively uniform throughout its range but there is often significant seasonal variation, such that plants can be compact and densely strigose in early summer but elongate and subglabrous later in the season.

Where the ranges of *Potentilla plattensis* and *P. ovina* overlap, the two are sometimes difficult to distinguish. The habitats are usually distinct, in that *P. plattensis* generally occurs in moist valley bottoms and *P. ovina* occurs in rocky uplands. Differences in vestiture are also diagnostic: hairs of *P. plattensis* are 0.5 mm and tightly appressed; those of *P. ovina* are longer and looser.

41. Potentilla millefolia Rydberg, Bull. Torrey Bot. Club 23: 433, plate 277, figs. 1–5. 1896 • Feather or many-leaved or Klamath cinquefoil [E] [F]

Potentilla klamathensis Rydberg; *P. millefolia* var. *klamathensis* (Rydberg) Jepson; *P. plattensis* Nuttall var. *klamathensis* (Rydberg) Jepson; *P. plattensis* var. *millefolia* (Rydberg) Jepson

Plants ± rosetted; taproots fleshy-thickened. **Stems** usually prostrate, sometimes ± decumbent, 0.4–2(–3) dm, lengths 1–2 times basal leaves. **Basal leaves** pinnate with distal leaflets ± confluent, 2–15(–20) × 1–3 cm; petiole 0.5–2(–3) cm, straight hairs sparse to abundant, appressed to spreading, 0.5–1.5 mm, stiff, cottony hairs absent, glands absent or sparse; primary lateral leaflets (3–)5–13 per side, on distal 2/3–3/4+ of leaf axis, separate to overlapping, largest ones cuneate to flabellate, 0.5–1.5(–2) × 0.5–2 cm, distal 2/3 to whole margin palmately or unevenly, rarely pinnately, incised 2/3 to completely to midvein, ultimate teeth or segments (1 or)2–10, linear to broadly oblanceolate, 2–10 × (0.5–)1–2 mm, apical tufts to 1 mm, surfaces green to grayish green, not glaucous, straight hairs sparse to abundant, appressed to spreading, 0.5–1.5(–2) mm, stiff, cottony hairs absent, glands sparse to common. **Cauline leaves** (0 or)1 or 2. **Inflorescences** 3–6(–10)-flowered, loosely cymose, sometimes racemiform. **Pedicels** (0.5–)1–2(–4.5) cm, ± recurved in fruit. **Flowers:** epicalyx bractlets ± elliptic, 2–4(–6) × 1–2(–2.5) mm; hypanthium 3–6 mm diam.; sepals 4–6(–8) mm, apex acute; petals 4–8(–10) × 3–7(–9) mm; filaments 2–3.5 mm, anthers 0.7–1 mm; carpels 10–30, styles (1.5–)2–3 mm. **Achenes** 1.5–2 mm, smooth, often ± carunculate.

Flowering spring–summer. Vernally to permanently wet meadows, moist openings in conifer forests and sagebrush, alkaline flats; 700–2200 m; Calif., Nev., Oreg.

Potentilla millefolia occurs from central Oregon to the east side of the Sierra Nevada of California, with a disjunct occurrence on the alkaline flats of Reese River Valley, Nevada. Significant variation occurs in vestiture type, leaflet dissection, and flower size, but with minimal geographic correlation. The most distinctive variant, represented by the type of *P. klamathensis*, has relatively long, slender, spreading, pustule-based hairs, often intermixed with shorter hairs. This vestiture type does not appear to be correlated with any other characters or geographic distribution and may vary within a population.

42. Potentilla hickmanii Eastwood, Bull. Torrey Bot. Club 29: 77. 1902 (as hickmani) • Hickman's cinquefoil [C] [E]

Plants ± rosetted; taproots fleshy-thickened. **Stems** usually prostrate to decumbent, sometimes ± ascending in supporting vegetation, 0.5–2.5 dm, lengths 1–2 times basal leaves. **Basal leaves** pinnate with distal leaflets ± confluent, 3–17 × 1–3.5 cm; petiole 1–5 cm, straight hairs sparse to common, appressed, 0.5–1.5(–2) mm, stiff, cottony hairs absent, glands absent or sparse; primary lateral leaflets 3–6 per side, on distal 1/2–2/3 of leaf axis, ± overlapping to nearly separate, largest ones cuneate to flabellate, 0.5–2 × 0.5–1.5 cm, distal 1/2 to whole margin ± palmately incised 1/2–2/3(+) to midvein, ultimate teeth or segments 2–5, narrowly elliptic to oblanceolate, 2–10(–15) × (1–)1.5–3 mm, apical tufts 0.5 mm, surfaces green, not glaucous, straight hairs sparse to common (often sparser to glabrate adaxially), appressed, 1–1.5 mm, stiff, cottony hairs absent, glands absent or inconspicuous. **Cauline leaves** 2. **Inflorescences** 2–5-flowered, very openly cymose, sometimes racemiform. **Pedicels** 1–2(–4) cm, ± recurved in fruit. **Flowers:** epicalyx bractlets ± elliptic, 3.5–6 × (1–)1.5–2.5 mm; hypanthium 3–5 mm diam.; sepals 3.5–7(–9) mm, apex acute; petals 6–12 × 5–9 mm; filaments (1.5–)2–3(–4) mm, anthers 0.7–1.2 mm; carpels (2–)5–15, styles 2.5–3.5 mm. **Achenes** 2 mm, smooth, ± carunculate.

Flowering spring–early summer. Vernally saturated coastal meadows, openings in Monterey Pine forests; of conservation concern; 30–100 m; Calif.

Potentilla hickmanii is known from only two extant populations in Monterey and San Mateo counties. The species is a federally listed endangered species (U.S. Fish and Wildlife Service, ecos.fws.gov/docs/recovery_plan/050617a.pdf). Plants from Sonoma County formerly included in *P. hickmanii* now constitute *P. uliginosa*. Although some morphologic overlap occurs with *P. millefolia*, *P. hickmanii* generally differs in having less dissected leaflets occupying less of the leaf axis, as well as somewhat larger flowers bearing fewer carpels.

43. Potentilla uliginosa B. C. Johnston & Ertter, J. Bot. Res. Inst. Texas 4: 14, fig. 1. 2010 • Cunningham Marsh cinquefoil [C] [E]

Plants rosetted to tufted; taproots not fleshy-thickened. **Stems** probably prostrate to decumbent, or ± ascending in supporting vegetation, 2.5–5.5 dm, lengths 1–2 times basal leaves. **Basal leaves** pinnate with distal leaflets ± confluent, 15–32 × 2–4 cm; petiole 5–10(–15) cm, straight hairs absent or sparse to common, appressed, 0.5–1.5 mm, stiff, cottony hairs absent, glands absent or sparse; primary lateral leaflets 6–10(–12) per side (irregularly paired), on distal ± ½ of leaf axis, loosely overlapping to proximally separate, largest ones cuneate to flabellate, 1.2–2.2 × 1–3 cm, ± whole margin unevenly incised ¾ to nearly to midvein, ultimate segments (3–)5–10(–15), linear to narrowly oblanceolate, 5–17 × 1–2.5 mm, apical tufts less than 0.5 mm, surfaces green, not glaucous, straight hairs sparse (often glabrate adaxially), appressed, 0.5–1 (–1.5) mm, stiff, cottony hairs absent, glands absent or sparse. **Cauline leaves** 2. **Inflorescences** 6–10-flowered, very openly cymose. **Pedicels** 1–6 cm, ± recurved in fruit. **Flowers:** epicalyx bractlets narrowly elliptic to oblong-ovate, 4–6 × 1–2.5 mm; hypanthium 5–6 mm diam.; sepals 4–7 mm, apex acute; petals 6–10 × 5–8 mm; filaments (1.5–)2–3 mm, anthers 0.7–1.2 mm; carpels ca. 10, styles 2.5–3.5 mm. **Achenes** 2–2.6 mm, smooth, ± caruncculate.

Flowering spring–summer. Permanent oligotrophic wetlands; of conservation concern; 30–40 m; Calif.

Potentilla uliginosa is known only from a handful of historical collections from Cunningham Marsh, one of three wetlands in southern Sonoma County that harbor a suite of disjunct and regionally rare species (B. C. Johnston and B. Ertter 2010). Although previously included in *P. hickmanii*, *P. uliginosa* differs in having woodier caudices, longer stems and leaves, and more deeply incised leaflets occupying less of the leaf axis. *Potentilla uliginosa* is presumed extinct; however, a morphologically comparable collection of the *P. millefolia* complex from Josephine County, Oregon (Deer Creek, near Selma, 14 June 1929, *Applegate 5735*, UC), opens the possibility of additional undiscovered populations in the mountains of northwestern California.

44. Potentilla multijuga Lehmann, Index Seminum (Hamburg) 1849: 6. 1849, name conserved • Ballona or lost cinquefoil [C] [E]

Plants ± rosetted; taproots ± fleshy-thickened. **Stems** prostrate to ascending, 2–5(–7) dm, lengths 1–2 times basal leaves. **Basal leaves** pinnate with distal leaflets ± confluent, 10–22 × 1.5–3.5 cm; petiole (2–) 4–10 cm, straight hairs absent or sparse, appressed, 0.5–1.5 mm, stiff, cottony hairs absent, glands absent; primary lateral leaflets 3–8 per side, on distal ⅓–⅔ of leaf axis, ± separate, largest ones cuneate to obovate, 1–2.2 × 0.8–1.5 cm, distal ⅓–⅔ of margin ± evenly incised ± ½ to midvein, ultimate teeth 3–6, ± ovate, 2–6 × 1.5–3 mm, apical tufts absent, surfaces green, not glaucous, straight hairs sparse to nearly absent (except on margins), ± appressed, 0.5–1 mm, stiff, cottony hairs absent, glands absent or sparse. **Cauline leaves** 1–3. **Inflorescences** 3–10(–20)-flowered, very openly cymose, sometimes racemiform. **Pedicels** 1.5–3.5(–5) cm, ± recurved in fruit. **Flowers:** epicalyx bractlets ovate-elliptic, 2.5–4(–5) × 1.5–3 mm, <sometimes toothed>; hypanthium 4–6 mm diam.; sepals 4–6(–7) mm, apex acute, rarely acuminate; petals 4.5–10 × 4–10 mm; filaments 1.5–3(–4.5) mm, anthers 0.7–1 mm; carpels 5–10, styles 2–3 mm. **Achenes** 1.8–2 mm, smooth to faintly rugose, possibly caruncculate.

Flowering spring–early summer. Brackish coastal meadows or marshes; of conservation concern; 0–10 m; Calif.

Potentilla multijuga is known only from a handful of collections made in the 1890s near the current site of Los Angeles International Airport. All efforts to locate extant plants have failed; the species is presumed extinct. Continued use of *P. multijuga* for this species required conservation of the name with a conserved type, because the type designated by Lehmann is a specimen of *Horkelia cuneata* Lindley var. *cuneata* (B. Ertter and J. L. Reveal 2008).

45. Potentilla basaltica Tiehm & Ertter, Brittonia 36: 228, fig. 1. 1984 • Soldier Meadow or Black Rock cinquefoil C E

Plants ± rosetted; taproots fleshy-thickened. **Stems** ± prostrate, sometimes ascending in supporting vegetation, 1.5–5 dm, lengths 2–3 times basal leaves. **Basal leaves** pinnate with distal leaflets ± confluent, 5–12(–18) × 1–1.5(–2) cm; petiole 0.5–1 (–1.5) cm, straight hairs absent, cottony hairs absent, glands absent or sparse; primary lateral leaflets (5–)10–15(–20) per side, on nearly whole leaf axis, ± overlapping, largest ones (or leaflet lobes) elliptic, 0.4–1(–2) × 0.2–0.5(–1) cm, simple and entire or 2(or 3)-fid to base (sometimes shallowly toothed as well), segments 1 or 2(or 3), narrowly to broadly elliptic, 4–10(–20) × (1–)2–3 mm, apical tufts absent or less than 0.5 mm, surfaces green, glaucous, straight hairs absent or sparse to common, tightly appressed, 0.5 mm or less, stiff, cottony hairs absent, glands absent or sparse. **Cauline leaves** 2–4. **Inflorescences** (5–)10–20-flowered, loosely cymose. **Pedicels** 0.8–3(–4) cm, straight to ± recurved in fruit. **Flowers:** epicalyx bractlets lanceolate-elliptic to ovate-elliptic, 1.5–2.5(–3) × 1 mm; hypanthium 2–5 mm diam.; sepals (3–)4–5(–6) mm, apex acute to acuminate; petals (3.5–)4.5–6.5 × 2.5–4 mm; filaments 1.5–3 mm, anthers 0.6–0.8 mm; carpels 3–10, styles 2–2.5 mm. **Achenes** 1.8–2.2 mm, smooth to ± rugose, not carunculate.

Flowering late spring–summer. Moist, subalkaline meadows in shrub steppe; of conservation concern; 1300–1600 m; Calif., Nev.

Potentilla basaltica is very distinctive in its glaucous, highly divided cylindric leaves with leaflets arranged in four ranks. The leaves strongly resemble those of some species of *Ivesia*, especially *I. kingii* S. Watson, but molecular analysis (B. Ertter et al. 1998) confirms a relationship with other members of sect. *Multijugae*.

Potentilla basaltica is known from only two localities, one in Lassen County, California, and the other in Humboldt County, Nevada. Both populations are highly localized, and *P. basaltica* is a candidate for federal listing as well as of conservation concern in both states.

The epithet *basaltica* was chosen as a reference to the Black Rock Desert, not habitat.

80. Potentilla Linnaeus sect. Concinnae (Rydberg) A. Nelson in J. M. Coulter and A. Nelson, New Man. Bot. Rocky Mt., 255. 1909

Barbara Ertter

Potentilla [unranked] *Concinnae* Rydberg, Bull. Torrey Bot. Club 23: 431. 1896

Perennials, usually rosetted to tufted, not stoloniferous; taproots sometimes thick, not fleshy; vestiture primarily of long and/or cottony hairs (and crisped hairs in inflorescence), glands absent or sparse to abundant, rarely reddish. **Stems** prostrate to ± decumbent, not flagelliform, not rooting at nodes, lateral to persistent basal rosettes, 0.2–1.5(–2.7) dm, lengths ½–3(–4) times basal leaves. **Leaves:** basal not in ranks; cauline 0–2; primary leaves palmate to subpalmate, sometimes subpinnate to pinnate (distal leaflets sometimes confluent), 1.5–12(–15) cm; petiole: long hairs ± appressed to ascending, stiff to weak, glands absent or sparse or obscured, rarely abundant; leaflets (3–)5–9(–11), at tip or to distal ¾+ of leaf axis, separate to strongly overlapping, narrowly oblanceolate or cuneate to obovate, margins flat, less than distal ⅕ to whole length evenly, sometimes unevenly, incised ¼–¾+ to midvein, sometimes medially cleft as well, rarely entire, teeth (0 or)1–5(–10) per side, surfaces similar to strongly dissimilar, abaxial green to white, cottony hairs absent or sparse to dense, adaxial green to grayish, not glaucous, long hairs usually stiff, rarely weak. **Inflorescences** (1 or)2–12(–15)-flowered, usually ± cymose, open, sometimes racemiform when prostrate. **Pedicels** often recurved in fruit, 0.7–3(–4.5) cm, proximal usually not much longer than distal. **Flowers** 5-merous; hypanthium 2.5–6 mm diam.; petals yellow, ± obcordate, (2.5–)3.5–7(–9) mm, longer than sepals, apex usually ± retuse; stamens ca. 20; styles subapical, filiform to filiform-tapered, not papillate-swollen or in proximal ⅕, (1–)1.5–3 mm. **Achenes** smooth to ± rugose.

Species ca. 10 (8 in the flora): w North America, n Mexico.

Section *Concinnae*, as here defined, accommodates a cluster of species that occur from the western Great Plains to California and to the mountains of northern Mexico. Stems are commonly prostrate or nearly so and are sometimes shorter than the basal leaves; pedicels tend to become recurved in fruit. Inflorescences often have abundant reddish papillae at the bases of hairs on hypanthia, epicalyx bractlets, and sepals. Leaves are most commonly palmate to subpalmate (except for *Potentilla angelliae*, *P. macounii*, and *P. morefieldii*), most often sparsely to densely cottony abaxially (except for *P. johnstonii*, *P. multisecta*, and *P. sierrae-blancae*), and strigose adaxially with stiff, verrucose hairs, which are sometimes yellowish. Petioles are usually sparsely to densely strigose with stiff, verrucose hairs.

Both P. A. Rydberg (1908d) and B. C. Johnston (1980, 1985) restricted sect. *Concinnae* to species with palmate, abaxially cottony leaves. Given the unifying or transitional nature of other morphologic and biogeographic features, the circumscription presented here seems more justified.

P. A. Rydberg (1908d) included New Mexico in the distribution of the otherwise Mexican species *Potentilla oblanceolata* Rydberg, repeated in subsequent floras (I. Tidestrom and T. Kittell 1941; W. C. Martin and C. R. Hutchins 1980). No specimens are currently known to confirm the presence of this species north of Mexico, and it is excluded here.

Since *Potentilla breweri* (sect. *Multijugae*) is sometimes identified as a member of sect. *Concinnae*, it is included herein and keys out in the ninth couplet.

1. Leaflet abaxial surfaces: cottony hairs absent.
    2. Teeth 2–5 per side on distal ½+ of central leaflets, 3–10(–14) mm. . . . . .51. *Potentilla multisecta*
    2. Teeth 1(–3) per side on distal ¼ or less of central leaflets, 1–2(–4) mm.
        3. Leaflets (3–)5–7(–9), straight hairs sparse to common; inflorescences 4–11-flowered; Nevada . . . . .52. *Potentilla johnstonii*
        3. Leaflets (3 or)4 or 5, straight hairs mostly absent (except on margins); inflorescences 1–3-flowered; New Mexico. . . . . 53. *Potentilla sierrae-blancae*
1. Leaflet abaxial surfaces: cottony hairs sparse to dense.
    4. Leaves palmate, leaflets on tip or at least less than distal ⅒ of leaf axis.
        5. Central leaflets: teeth (1 or)2–5(–10) per side on distal ¼ to whole margins . . . . .46. *Potentilla concinna* (in part)
        5. Central leaflets: teeth 0 or 1(–3) per side on less than distal ⅕(–⅓) of margins . . . . 47. *Potentilla bicrenata*
    4. Leaves subpalmate to pinnate, leaflets on distal (⅒–)⅕–¾+ of leaf axis.
        6. Leaflets 5(–7) on distal (⅒–)⅕–¼ of leaf axis, usually only 1 pair of leaflets separate from terminal leaflets.
            7. Central leaflets: teeth 2–5(or 6) per side on distal ⅓ to whole margins . . . . .46. *Potentilla concinna* (in part)
            7. Central leaflets: teeth (0 or)1(or 2) per side on distal ¼(–⅓) or less of margins . . . . .48. *Potentilla angelliae*
        6. Leaflets 5–9(–11) on distal ⅕–¾+ of leaf axis, usually at least 2 pairs of leaflets separate from terminal leaflets.
            8. Leaflets on distal ⅕–½ of leaf axis; Alberta, Montana . . . . 49. *Potentilla macounii*
            8. Leaflets on distal (¼–)½–¾+ of leaf axis; California, Nevada, Oregon, Washington.
                9. Leaflet surfaces ± similar, adaxial: cottony hairs sparse to dense, straight hairs soft; Sierra Nevada, Cascade Range, and n Great Basin . . . . .37. *Potentilla breweri* (sect. *Multijugae*)
                9. Leaflet surfaces ± to strongly dissimilar, adaxial: cottony hairs absent, straight hairs stiff; White Mountains and adjacent Sierra Nevada, California . . . . 50. *Potentilla morefieldii*

46. Potentilla concinna Richardson in J. Franklin, Narr. Journey Polar Sea, 739. 1823 • Early cinquefoil [E] [F]

Stems 0.2–1.6 dm, lengths ½–3(–4) times basal leaves. **Basal leaves** usually palmate to subpalmate, sometimes subpinnate, <1(or 2) pairs of leaflets separate from terminal leaflets, distal leaflets distinct>, 2–10(–15) cm; petiole 1–7(–10) cm, straight hairs ± abundant, ± appressed, 1–3 mm, stiff to weak, cottony hairs present, sometimes absent; leaflets 5(–7), on tip or to distal ¼ of leaf axis, separate to ± overlapping, proximal pair separated from others by 0–5(–10) mm of leaf axis, central leaflets narrowly oblanceolate to obovate, 1–3(–7) × 0.4–1.2 cm, petiolules 0–2(–4) mm, distal ¼ to whole margin incised ¼–¾+ to midvein, teeth (1 or)2–5(–10) per side, separate, 1–6 mm, surfaces strongly to ± dissimilar, abaxial grayish to white, straight hairs ± abundant, ± appressed, 0.5–2 mm, weak to stiff (especially on veins), cottony hairs ± dense, rarely sparse, glands sparse or obscured, adaxial green to grayish, straight hairs common to abundant, appressed, 0.5–2 mm, mostly stiff, sometimes weak or mixed, cottony hairs absent or sparse, rarely common, glands ± sparse. **Inflorescences** (1 or)2–12-flowered. **Pedicels** 1–2(–3) cm. **Flowers:** epicalyx bractlets lanceolate to narrowly ovate to linear-elliptic, <sometimes doubled>, (2–)2.5–5 × 1–1.5 mm; hypanthium 3–6 mm diam.; sepals 3.5–6 mm, apex acute; petals (2.5–)4–9 × (2–)3–7 mm; filaments 1–3 mm, anthers 0.5–1 mm; carpels (7–)10–30, styles (1–)1.5–2 mm. **Achenes** 1.5–2.5 mm, smooth to lightly rugose. $2n = 70$.

Varieties 3 (3 in the flora): w North America.

Potentilla concinna is a relatively widespread, low-growing, early-blooming species with palmate to subpalmate, abaxially cottony leaves. Flowers often appear before leaves are fully expanded. The three varieties treated here have some level of biogeographic identity and are, therefore, given provisional taxonomic recognition. Some treatments have also included *P. bicrenata*, *P. johnstonii*, and *P. macounii* as varieties of *P. concinna*. Two Mexican species, *P. leonina* Standley and *P. oblanceolata* Rydberg, have also been treated as varieties of *P. concinna* by J. Soják (2006); they are retained as distinct species here, pending further research on Mexican *Potentilla*. Two other erstwhile varieties of *P. concinna* are now placed in separate sections: var. *modesta* (Rydberg) S. L. Welsh & B. C. Johnston (misapplied) in sect. *Rubricaules* as *P. modesta*, and var. *rubripes* (Rydberg) C. L. Hitchcock in sect. *Subjugae*, intermediate between *P. saximontana* and *P. subjuga*. *Potentilla concinnaeformis* Rydberg, considered to be a close relative of *P. concinna* by I. W. Clokey (1939), is more likely a hybrid between *P. glaucophylla* (sect. *Graciles*) and *P. hippiana* (sect. *Leucophyllae*). Hybrids also occur with *P. multisecta*, and probably other species as well.

Potentilla concinna was first described in 1818 as *P. humifusa* Nuttall, a later homonym for *P. humifusa* Willdenow.

1. Distal (½–)¾ to full length of central leaflets incised ½–¾+ to midvein, teeth 2–6 mm; n Great Plains to c Utah. . . . . 46b. *Potentilla concinna* var. *divisa*
1. Distal ¼–¾ of central leaflets incised ¼–½ to midvein, teeth 1–3(–5) mm; mostly Rocky Mountains and west.
    2. Stem lengths ½–1.5(–2) times basal leaves; leaves palmate, proximal leaflet pair not separated from others . . . . 46a. *Potentilla concinna* var. *concinna*
    2. Stem lengths 2–3(–4) times basal leaves; leaves usually palmate to subpalmate, sometimes subpinnate, proximal leaflet pair often separated from others by 1–5(–10) mm of leaf axis . . . . 46c. *Potentilla concinna* var. *proxima*

46a. Potentilla concinna Richardson var. **concinna** [E]

Stems 0.2–0.8(–1) dm, lengths ½–1.5(–2) times basal leaves. **Basal leaves** palmate; leaflets on tip or at least less than distal 1/10 of leaf axis, separate to touching, proximal pair not separated from others; distal ¼–¾ of central leaflets incised ¼–½ to midvein, teeth (1 or)2–4(–10) per side, 1–3 mm. **Inflorescences** (1 or)2–6-flowered. **Petals** (4–)5–9 mm. **Achenes** 1.5–2 mm.

Flowering early summer (Apr–Jun), occasionally second flowering late summer (Jul–Sep). Meadows and vernally moist openings in conifer and aspen woodlands, sagebrush rangelands, prairie hillsides, rocky outcrops and ridges; 1300–3800 m; Alta., Sask.; Ariz., Colo., Mont., N.Mex., N.Dak., S.Dak., Utah, Wyo.

Variety *concinna* is the compact phase of *Potentilla concinna* with strictly palmate, short-toothed leaves that are often prominently bicolored. The variety occurs primarily in vernally wet meadows and similar situations in the Rocky Mountains of Colorado and New Mexico, and in forested areas of northern Arizona. Plants referable to var. *concinna* also extend into the primary range of var. *dissecta* (S. Watson) B. Boivin at least as far as Alberta and North Dakota. No vouchers have been located to confirm reports from Idaho (R. J. Davis 1952; B. C. Johnston 1980). The low elevation extreme is from Alberta; the highest elevations are in Colorado and Wyoming.

46b. Potentilla concinna Richardson var. **divisa** Rydberg, Bull. Torrey Bot. Club 23: 431. 1896 [E]

Potentilla divisa (Rydberg) Rydberg

Stems 0.2–1(–1.5) dm, lengths (½–)1–2 times basal leaves. **Basal leaves** palmate to subpalmate; leaflets on tip or to distal ¼ of leaf axis, ± overlapping, proximal pairs sometimes sep-arated from others by 1–3 mm of leaf axis; distal (½–)¾ to entire length of central leaflets incised ½–¾+ to midvein, teeth 3–5(or 6) per side, 2–6 mm. **Inflorescences** (1 or)2–6 -flowered. **Petals** 5–8 mm. **Achenes** 2–2.5 mm.

Flowering summer. High prairies, rocky slopes and ridges, gravelly foothills, open pine woodlands, often on limestone; 600–3300 m; Alta., Man., Sask.; Colo., Mont., N.Dak., Ohio, S.Dak., Utah, Wyo.

Variety *divisa* is the primary phase of the species in the northwestern prairies, Black Hills, and Rocky Mountain foothills. Plants differ from the typical variety in having more deeply toothed leaflets on average, as well as somewhat longer stems in relation to leaves, silkier vestiture, and occasional subpalmate leaves (possibly intergrading with *Potentilla macounii*). Also included in var. *divisa* are plants with deeply toothed leaflets from comparatively low elevations in northwestern Colorado, northeastern Utah, and Wyoming. These populations, which tend to have sparser, shaggier vestiture, were included by S. L. Welsh et al. (1993) in var. *modesta* (Rydberg) S. L. Welsh & B. C. Johnston (misapplied), along with some palmate-leaved collections of var. *proxima*. An anomalous collection from Stark County, Ohio (*T. Davis 876*, MO), which is exceptionally silky, is presumably introduced.

For the nomenclatural relation between var. *divisa* and *Potentilla nivea* var. *dissecta* S. Watson, see B. Ertter (2008).

46c. Potentilla concinna Richardson var. **proxima** (Rydberg) S. L. Welsh & B. C. Johnston, Great Basin Naturalist 42: 25. 1982 [C] [E] [F]

Potentilla proxima Rydberg in N. L. Britton et al., N. Amer. Fl. 22: 339. 1908; *P. beanii* Clokey; *P. diversifolia* Lehmann var. *proxima* (Rydberg) Soják

Stems (0.2–)0.6–1.6 dm, lengths 2–3(–4) times basal leaves. **Basal leaves** usually palmate to subpalmate, sometimes subpinnate; leaflets on tip or to distal ¼ of leaf axis, ± overlapping, proximal pairs <sometimes doubled> often separated from others by 1–5(–10) mm of leaf axis; distal ⅓–⅔ of central leaflets incised ¼–½ to midvein, teeth 2–4 per side, 1–3(–5) mm. **Inflorescences** (1–)3–12-flowered. **Petals** (2.5–)4–6 mm. **Achenes** 1.5–2 mm.

Flowering mid summer. Alpine tundra, windswept ridges, stony meadows; of conservation concern; 2600–3900 m; Ariz., Calif., Colo., Nev., N.Mex., Utah.

Variety *proxima* is the primary phase of *Potentilla concinna* in the mountains of southwestern Colorado, Nevada, and Utah. Some alpine collections from New Mexico are referable to var. *proxima*, and isolated populations also occur in the San Francisco Peaks of Arizona (at higher elevations than otherwise sympatric var. *concinna*) and the White Mountains of California. Specimens from Montana keying here are provisionally placed in *P. macounii*.

The most distinctive feature of var. *proxima* is the elongate flowering stems, which are usually over twice as long as the leaves. Leaves tend to be more sparsely cottony and less bicolored than in the typical variety, the central leaflet is more often petiolulate, and flowers are commonly somewhat smaller. Collections from southwestern Utah and adjacent Nevada, including the type of *Potentilla proxima*, commonly have subpalmate leaves; elsewhere leaves are rarely or scarcely subpalmate.

47. Potentilla bicrenata Rydberg, Bull. Torrey Bot. Club 23: 431. 1896 [E] [F]

Potentilla concinna Richardson var. *bicrenata* (Rydberg) S. L. Welsh & B. C. Johnston

Stems 0.2–0.8(–1.5) dm, lengths ½–1½ times basal leaves. **Basal leaves** palmate, (2–)4–10 (–15) cm; petiole (1–)2–6(–10) cm, straight hairs abundant, ± appressed to ascending, 1.5–2 mm, stiff, cottony hairs usually absent, glands sparse, often obscured; leaflets 5–7(–9), on tip or at least less than distal 1/10 of leaf axis, slightly overlapping, proximal pair separated from others by 0(–1) mm of leaf axis, central leaflets ± oblanceolate, 1–4(–5.5) × 0.5–1 cm, petiolules 0(–3) mm, less than distal 1/5 (–⅓) of margins incised ± ½ to midvein, teeth 0 or 1(–3) per side, separate, 0.5–2 mm, surfaces strongly to ± dissimilar, abaxial grayish white to white, straight hairs abundant, ± appressed, 1–1.5(–2) mm, weak to stiff (especially on veins), cottony hairs common to dense, glands absent or obscured, adaxial green to grayish green, straight hairs ± abundant, appressed, 0.5–1 mm, stiff, cottony hairs absent, rarely sparse, glands absent or obscured. **Inflorescences** 2–5(–8)-flowered. **Pedicels** 1–3(–4.5) cm. **Flowers:** epicalyx bractlets ± lanceolate, 2–4 × 1 mm; hypanthium 3–4 mm diam.; sepals (2.5–)

3–5.5 mm, apex ± acute; petals 3.5–7 × 2.5–5.5 mm; filaments 1–2 mm, anthers 0.5–1 mm; carpels 10–20, styles 2 mm. **Achenes** 2 mm, smooth to rugose.

Flowering early summer. Outcrops, dry flats, pine duff, in pine and/or juniper woodlands, sagebrush scrub; 1900–3300 m; Ariz., Colo., N.Mex., Utah, Wyo.

The entire to tridentate leaflets of *Potentilla bicrenata* generally serve to distinguish this species from *P. concinna*. Unequivocal *P. bicrenata* is most common in Utah and the southern Rocky Mountains from New Mexico to central Colorado; in Arizona, it is documented only from the Kaibab Plateau. Wyoming populations tend to be transitional to *P. concinna* var. *concinna* in leaflet toothing and/or petal size. Some collections from as far north as Waterton Lakes National Park, Alberta, vary in the direction of *P. bicrenata* but are retained here in *P. concinna*.

48. Potentilla angelliae N. H. Holmgren, Brittonia 39: 342, fig. 2. 1987 • Angell's or Boulder Mountain cinquefoil [C] [E]

Stems (0.4–)0.6–1.1 dm, lengths 2–3 times basal leaves. **Basal leaves** subpinnate, <1 pair of leaflets separate from terminal leaflets, distal leaflets distinct>, 1.5–5 cm; petiole 0.5–3(–4) cm, straight hairs abundant, loosely appressed to ascending, 1–2 mm, stiff, cottony hairs absent, glands sparse to abundant; leaflets 5, on distal (1/10–) 1/5–1/4 of leaf axis, separate, proximal pair separated from others by 1–3 mm of leaf axis, central leaflets narrowly cuneate to oblanceolate, 0.5–1.5 × 0.3–0.6 cm, petiolules 1 mm, distal 1/4(–1/3) or less of margins incised 1/2 or less to midvein, teeth (0 or)1(or 2) per side, ± separate, 1 mm, surfaces ± dissimilar, abaxial grayish to white, straight hairs ± abundant, appressed, 1–1.5 mm, ± stiff, cottony hairs abundant to sparse, glands ± sparse or obscured, adaxial greenish, straight hairs abundant, ± appressed, 1–1.5(–2) mm, stiff, cottony hairs absent, glands sparse to abundant. **Inflorescences** solitary flowers or 2- or 3-flowered. **Pedicels** (1–)1.5–3 cm. **Flowers:** epicalyx bractlets lanceolate-elliptic to ovate, 2–4 × 0.8–1.5 mm; hypanthium 3–5 mm diam.; sepals 3–5(–6) mm, apex ± acute; petals (4–)5–6.5 × 4–5.5 mm filaments 1.5–2.5 mm, anthers 0.7–1 mm; carpels 10–30, styles 2–2.5 mm. **Achenes** 1.5–2 mm, smooth to faintly rugose.

Flowering summer. Rocky subalpine meadows; of conservation concern; 3300–3400 m; Utah.

Potentilla angelliae is known only from Boulder Mountain on the Aquarius Plateau in Garfield and Wayne counties. Comparable plants, whose status has yet to be resolved, have been collected in Montezuma County, Colorado. S. L. Welsh et al. (1993) suggested that *P. angelliae* might best be treated as a variety of *P. concinna*; the combination of subpalmate leaves, tridentate leaflets, and relatively few-flowered inflorescences extending beyond the leaves makes *P. angelliae* a well-defined species with no evidence of intergradation.

49. Potentilla macounii Rydberg, Monogr. N. Amer. Potentilleae, 101, plate 41, figs. 1, 2. 1898 • Macoun's cinquefoil [C] [E]

Potentilla concinna Richardson var. *macounii* (Rydberg) C. L. Hitchcock

Stems 0.4–1.5(–2) dm, lengths 2 times basal leaves. **Basal leaves** subpinnate to pinnate, <usually at least 2 pairs of leaflets separate from terminal leaflets, distal leaflets often confluent>, 2–7 cm; petiole 1–4(–8) cm, straight hairs abundant, ± appressed to ascending, 1–2 mm, stiff to weak, cottony hairs sometimes present, glands absent or obscured; leaflets 5–9(–11), on distal 1/5–1/2 of leaf axis, strongly overlapping, proximal pair separate from others by 1–5 mm of leaf axis, central leaflets oblanceolate to obovate, <sometimes lobed>, (0.5–)1–1.5 × 0.5–1 cm, petiolules 0–1 mm, distal 2/3 of margins incised 1/2+ to midvein, teeth 2–5 per side, touching, 1–3 mm, surfaces ± dissimilar, abaxial white, straight hairs ± abundant, ± appressed, 1–1.5 mm, weak to stiff, cottony hairs ± dense, glands absent or obscured, adaxial greenish gray, straight hairs sparse to abundant, usually appressed, sometimes ascending, 1 mm, usually ± stiff, sometimes soft, cottony hairs sparse, glands obscured. **Inflorescences** (1–)3–6(–10)-flowered. **Pedicels** 0.8–2 cm. **Flowers:** epicalyx bractlets lanceolate to ovate-elliptic, 2.5–4 × 0.8–1.5 mm; hypanthium 3–4 mm diam.; sepals (3–)3.5–5.5(–6) mm, apex ± acute; petals 6–7 × 5–7 mm; filaments 1.5–3 mm, anthers 0.8–1 mm; carpels 15–20, styles 2–2.5 mm. **Achenes** 1.5–2 mm, smooth.

Flowering early summer. Rocky slopes, outcrops, gravel banks, prairies; of conservation concern; 1200–2200 m; Alta.; Mont.

Potentilla macounii differs from other species in sect. *Concinnae* in the combination of subpinnate leaves, with at least two proximal pairs of leaflets separated from the terminal leaflets, and in having somewhat softer hairs. It occurs on the eastern slopes of the Rocky Mountains in southwestern Alberta (where of conservation concern) and in western Montana. Some problematic collections from the Absaroka Range in western Wyoming are most comparable to *P. macounii*, but have softer vestiture and somewhat larger, less divided, more subpalmate leaves. If these Wyoming populations were included in *P. macounii*, the upper elevation range would be 3300 m.

50. Potentilla morefieldii Ertter, Brittonia 44: 432, fig. 1. 1992 • Morefield's cinquefoil [C] [E]

Stems 0.4–1.5(–1.7) dm, lengths 2–3 times basal leaves. **Basal leaves** pinnate, <usually at least 2 pairs of leaflets separate from terminal leaflets, distal leaflets usually distinct>, 2–6 cm; petiole 0.2–1.5 cm, straight hairs abundant, ± appressed, 1–1.5 mm, stiff, cottony hairs absent, glands absent or sparse; leaflets 5–9, on distal (¼–)½–¾+ of leaf axis, overlapping, proximal pair separated from others by 2–10 mm of leaf axis, central leaflets oblanceolate to narrowly obovate, (0.5–)1–2(–2.5) × 0.5–1.2 cm, petiolules 0–1 mm, distal ¾ of margins incised ¾ to midvein (sometimes medially cleft as well), teeth 3 or 4 per side, separate to slightly overlapping, (1–)2–5 mm, surfaces ± to strongly dissimilar, abaxial grayish to white, straight hairs ± abundant, ± appressed, 1 mm, weak to stiff (especially on veins), cottony hairs abundant to dense, glands absent or obscured, adaxial greenish to grayish, straight hairs ± abundant, appressed, 0.5–1.5 mm, stiff, cottony hairs absent, rarely sparse, glands sparse. **Inflorescences** 3–15-flowered. **Pedicels** 1–2.5(–3) cm. **Flowers:** epicalyx bractlets ± ovate-elliptic, 2–4 × 0.8–2 mm; hypanthium 2.5–5 mm diam.; sepals 3.5–5 mm, apex ± acute; petals 4–6 × 3–5 mm; filaments 1–2 mm, anthers 0.5–1 mm; carpels 15–20, styles 2 mm. **Achenes** 1.5–2 mm, smooth to faintly rugose.

Flowering summer. Alpine tundra and fellfields, mostly on dolomite substrates; of conservation concern; 3500–4000 m; Calif.

Potentilla morefieldii is known only from the White Mountains and adjacent Sierra Nevada. Plants with similar aspect but lacking strigose vestiture occur in White Pine County, Nevada; their disposition is uncertain. *Potentilla morefieldii* is sometimes confused with *P. breweri* (sect. *Multijugae*), but the latter differs in having more or less similar leaflet surfaces with cottony hairs nearly as abundant adaxially as abaxially. *Potentilla breweri* also lacks the stiff adaxial leaflet hairs that characterize most species in sect. *Concinnae*, including *P. morefieldii*.

W. L. Jepson (1909–1943, vol. 2) misapplied the name *Potentilla pseudosericea* var. *grandiflora* Th. Wolf, which is a synonym of an unresolved species from the Rocky Mountains (B. Ertter 1992; Ertter et al. 2013), to *P. morefieldii*.

51. Potentilla multisecta (S. Watson) Rydberg, Bull. Torrey Bot. Club 23: 397. 1896 • Great Basin cinquefoil [E] [F]

Potentilla diversifolia Lehmann var. *multisecta* S. Watson, Botany (Fortieth Parallel), 86. 1871

Stems 0.5–2(–2.7) dm, lengths 1–2(–3) times basal leaves. **Basal leaves** usually palmate, sometimes subpalmate, 3–12 (–15) cm; petiole 1–6(–13) cm, straight hairs ± abundant, ± appressed, 1–1.5 mm, stiff, cottony hairs absent, glands absent or sparse; leaflets (3–)5, usually on tip, sometimes to distal ¼ of leaf axis, ± overlapping, proximal pair separated from others by 0–10 mm of leaf axis, central leaflets ± obovate, 1–2.5(–4) × (0.5–)1–1.5(–2) cm, petiolule 0–2 mm, distal ½+ of margins incised ¾+ to midvein, teeth 2–5 per side, separate, 3–10(–14) mm, surfaces similar, green to grayish, straight hairs sparse to abundant, appressed, 1–1.5 mm, usually stiff, <short crisped hairs sometimes present>, cottony hairs absent, glands absent or sparse. **Inflorescences** (1 or)2–6(–10) -flowered. **Pedicels** 1–3(–4) cm. **Flowers:** epicalyx bractlets lanceolate-elliptic to narrowly ovate, 2–4 × 1–1.5 mm; hypanthium 3–5 mm diam.; sepals 3–5 (–6) mm, apex ± acute; petals (2.5–)5–7 × (2–)4–5 mm; filaments 1–2 mm, anthers 0.6–1 mm; carpels 5–20, styles 2–3 mm. **Achenes** 2–2.5 mm, smooth.

Flowering summer. Rocky slopes, dry meadows, in sagebrush scrub, pinyon-juniper woodlands, limber pine woodlands; 2100–3500 m; Idaho, Mont., Nev., Utah, Wyo.

The center of distribution of *Potentilla multisecta* is northeastern Nevada and northwestern Utah; the best disposition of more northern populations, which tend to be hairier and more compact, is unresolved. Populations in eastern Idaho, southwestern Montana, and western Wyoming are components of an unresolved zone of apparent intergradation that also includes *P. ovina* var. *ovina* (sect. *Multijugae*), *P. glaucophylla* var. *perdissecta* (sect. *Graciles*), and members of sect. *Subjugae*. Significant variation also exists among populations in Nevada and Utah. Not included here are some subpalmate collections from Saskatchewan that would key to *P. multisecta* (for example, *A. Breitung 4045*, MO, RM), whose optimum disposition has yet to be resolved.

52. Potentilla johnstonii Soják, Thaiszia 16: 97. 2007 • Sagebrush cinquefoil C E

Stems 0.4–2 dm, lengths 2 times basal leaves. **Basal leaves** palmate, 3–8(–13) cm; petiole 1.5–5(–8) cm, straight hairs abundant, ± appressed, 1–2 mm, stiff, cottony hairs absent, glands sparse to abundant; leaflets (3–)5–7(–9), on tip or at least less than distal 1/10 of leaf axis, separate to slightly overlapping, proximal pair separated from others by 0–1 mm of leaf axis, central leaflets narrowly cuneate to oblanceolate, 1–3(–4) × 0.3–0.7 cm, petiolule 1 mm, distal 1/4 or less of margins incised 1/3–2/3 to midvein, teeth 1(–3) per side, separate, 1–2(–4) mm, surfaces similar, green to grayish, straight hairs sparse to common, appressed, 1–1.5(–2) mm, stiff (sometimes softer and wavy), cottony hairs absent, glands sparse to abundant. **Inflorescences** 4–11 -flowered. **Pedicels** 0.7–2(–4) cm. **Flowers:** epicalyx bractlets ± lanceolate, 2–4 × 1 mm; hypanthium 3–4(–6) mm diam.; sepals 3–4.5 mm, apex acute to obtuse; petals 5–7 × 3–5 mm; filaments 1–2.5 mm, anthers 0.6–0.8 mm; carpels 3–10, styles 2–2.5 mm. **Achenes** 2.5 mm, smooth to slightly rugose.

Flowering summer. Rocky slopes in pinyon-juniper woodlands, with *Artemisia*; of conservation concern; 2300–2500 m; Nev.

Potentilla johnstonii occurs only in the Quinn Canyon Range in Nye County. Its distinctiveness was first noted by B. C. Johnston (1980), who treated it as a variety of *P. concinna* using the unpublished epithet *curvata*.

53. Potentilla sierrae-blancae Wooton & Rydberg in P. A. Rydberg, Monogr. N. Amer. Potentilleae, 57, plate 18, figs. 3–5. 1898 • Sierra Blanca cinquefoil C E

Stems 0.2–1 dm, lengths 1–1.5 times basal leaves. **Basal leaves** palmate, 2–8 cm; petiole 0.5–5 cm, straight hairs sparse to abundant, sometimes absent, ± appressed, 2–3 mm, usually ± stiff, cottony hairs absent, glands absent; leaflets (3 or)4 or 5, at tip of leaf axis, separate, proximal pair separated from others by 0(–1) mm of leaf axis, central leaflets narrowly cuneate-oblanceolate, 0.8–2.2 × 0.2–0.5 cm, petiolules 0(–1) mm, less than distal 1/4 of margins incised 1/2–3/4 to midvein, teeth 1(or 2) per side, ± separate, 1–2 mm, surfaces similar, green, straight hairs mostly absent (except on margins), appressed, 1–2 mm, stiff, cottony hairs absent, glands absent, rarely sparse. **Inflorescences** solitary flowers or 2- or 3-flowered. **Pedicels** 1–4 cm. **Flowers:** epicalyx bractlets narrowly to widely elliptic-ovate, 2–3.5 × 1–1.5 mm; hypanthium 3–5 mm diam.; sepals 3–5 mm, apex obtuse to acute; petals 5–7 × 4–6 mm; filaments 1–2 mm, anthers 0.4–0.8 mm; carpels 5–20, styles 2 mm. **Achenes** 1.5 mm, smooth.

Flowering summer. Windswept barren ridges, subalpine grasslands, rock outcrops; of conservation concern; 2400–3700 m; N.Mex.

Potentilla sierrae-blancae is restricted to the White Mountains in Lincoln and Otero counties. P. A. Rydberg (1898, 1908d) and B. C. Johnston (1985) placed *P. sierrae-blancae* in sect. (or group) *Aureae*; the species fits well in sect. *Concinnae*, combining the few-flowered inflorescence of *P. angelliae* with the tridentate leaflets of *P. bicrenata* and *P. johnstonii*.

8p. Potentilla Linnaeus sect. Subviscosae (Rydberg) O. Stevens in N. L. Britton et al., N. Amer. Fl. 22(7): 11. 1959

Barbara Ertter

Potentilla [unranked] *Subviscosae* Rydberg, Bull. Torrey Bot. Club 23: 429. 1896

Perennials, rosetted or tufted, not stoloniferous; taproots sometimes ± thickened; vestiture of long and short (rarely short-crisped) hairs, glands sparse to abundant, sometimes red. **Stems** ± prostrate to decumbent, sometimes pendent (on rock faces), not flagelliform, not rooting at nodes, lateral to persistent basal rosettes, 0.2–2(–3) dm, lengths 1–3(–5) times basal leaves. **Leaves:** basal not in ranks; cauline 0–2; primary leaves ternate or palmate, (1–)2–10(–18) cm; petiole: long hairs usually spreading to ascending, sometimes loosely appressed, weak to stiff, glands usually abundant, sometimes sparse; leaflets 3–5(–7), at tip of leaf axis, separate to slightly overlapping, oblanceolate or elliptic to broadly obovate or cuneate to flabellate, sometimes nearly orbiculate, margins flat, distal $\frac{1}{3}$–$\frac{3}{4}$ evenly to unevenly (*P. subviscosa*) incised $\frac{1}{5}$–$\frac{3}{4}$ to midvein, <sometimes deeply lobed as well (*P. subviscosa*)>, teeth (1 or)2–4(–9) per side, surfaces similar, green to gray-green, not glaucous, long hairs weak to stiff, cottony hairs absent. **Inflorescences** solitary flowers or 2–10(–20)-flowered, usually ± cymose, open, sometimes racemiform when prostrate. **Pedicels** usually recurved in fruit, 0.5–2(–4) cm, proximal usually not much longer than distal. **Flowers** 5-merous; hypanthium 2–6 mm diam.; petals pale to bright yellow, narrowly or ± obcordate, rarely broadly obovate, (2–)3–7(–8) mm, longer than sepals, apex usually ± retuse; stamens 15–20; styles subapical, usually filiform, sometimes tapered-columnar (*P. cottamii*), papillate-swollen in proximal $\frac{1}{10}$(–$\frac{1}{3}$ in *P. cottamii*), (1–)1.5–2.5(–3) mm. **Achenes** smooth to ± rugose.

Species ca. 10 (7 in the flora): sw United States, n Mexico.

Section *Subviscosae* consists of species that have radiated in the sky islands of southwestern North America and Mexico, mostly in rocky meadows or outcrops in forested montane areas (B. Ertter 2007c). Some species are chasmophytic, forming clumps on cliffs or steep rock outcrops, similar to species in *Ivesia* sect. *Setosae*. Mexican species are poorly known but include at least *Potentilla mexiae* Standley and possibly *P. rydbergiana* Rose, as well as undescribed taxa.

Section *Subviscosae* shares with sect. *Concinnae* leaves that are often palmate, prostrate stems, and slender pedicels that become recurved in fruit, as well as a similar biogeographic pattern. Section *Subviscosae* differs in that cottony and strongly verrucose hairs are lacking, leaves are never strongly bicolored, and petiole and stem hairs are spreading to loosely appressed. Vestiture consists of varying proportions of long hairs, short hairs, and glands; petals are often somewhat clawed, unlike the unclawed norm in the genus.

1. Basal leaves usually ternate, rarely palmate.
    2. Petals (2–)3–4 mm; styles 1–1.5 mm; ne Nevada, nw Utah . . . . 60. *Potentilla cottamii*
    2. Petals 3–8 mm; styles 2–3 mm; se Arizona.
        3. Petioles: long hairs 1–2(–3) mm; central leaflets with 2 or 3(or 4) teeth per side; Santa Rita, Huachuca, and Chiricahua ranges . . . . 58. *Potentilla rhyolitica*
        3. Petioles: long hairs 0.5–1(–1.5) mm; central leaflets with 3–5(or 6) teeth per side; Pinaleño Mountains, nearby Mogollon Rim, and Santa Catalina Range. . . . . . 59. *Potentilla albiflora*
1. Basal leaves usually palmate, rarely ternate.
    4. Petals nearly white abaxially, pale yellow adaxially, narrowly obcordate; central leaflets evenly to unevenly incised (sometimes deeply lobed as well), teeth 2–9 per side; short hairs well differentiated from long hairs, ± abundant to dense throughout . . . . . 54. *Potentilla subviscosa*
    4. Petals ± paler abaxially, bright yellow adaxially, ± obcordate; central leaflets evenly incised, teeth 2–4(or 5) per side; short hairs not well differentiated from long hairs, absent or sparse throughout.
        5. Plants rooted in ground; leaflets gray-green, long hairs abundant to dense; carpels ca. 20, styles (1.2–)1.5–2 mm; s Sierra Nevada and San Bernardino Mountains, California . . . . 55. *Potentilla wheeleri*
        5. Plants rooted in rock crevices, often on vertical surfaces; leaflets green, long hairs sparse to abundant; carpels 5–15(–20), styles 1.5–2.5 mm; s California, nw Arizona, and Baja California, Mexico.
            6. Central leaflets 1–3 cm, oblanceolate to broadly obovate-cuneate or nearly orbiculate, distal ± $\frac{1}{3}$ of margins incised ± $\frac{1}{4}$ to midvein; epicalyx bractlets lanceolate-elliptic, rarely linear; San Jacinto Mountains, California, and Baja California, Mexico . . . . 56. *Potentilla rimicola*
            6. Central leaflets 0.5–1.5(–2) cm, oblanceolate, distal ± $\frac{3}{4}$ of margins incised $\frac{1}{2}$–$\frac{3}{4}$ to midvein; epicalyx bractlets ovate-elliptic; Hualapai Mountains, Arizona . . . . 57. *Potentilla demotica*

54. Potentilla subviscosa Greene, Bull. Torrey Bot. Club 8: 97. 1881 • Mogollon cinquefoil [E] [F]

Short hairs well differentiated from long hairs, ± abundant to dense throughout. **Stems** 0.2–1.5(–2) dm. **Basal leaves** palmate, rarely ternate, (1–)2–7(–18) cm; petiole (0.5–)1–5(–11) cm, long hairs sparse to abundant, ± spreading (to ascending), 1–3(–4) mm, ± weak, glands ± abundant; leaflets (3–)5(–7), central flabellate to obovate-cuneate or oblanceolate, (0.5–)1–3(–8) × 0.5–2 cm, not or ± petiolulate, distal 2/3–3/4 of margins evenly to unevenly incised 1/5–3/4 to midvein, <sometimes deeply lobed as well>, teeth 2–9 per side (some secondarily toothed as well), surfaces green, long hairs sparse to common, 1–2 mm (late-season leaves and adaxial surfaces sometimes nearly glabrate), glands sparse to abundant. **Inflorescences** (1–)3–15-flowered. **Pedicels** 0.5–1.5(–2) cm. **Flowers:** epicalyx bractlets lanceolate-elliptic, 1.5–3(–5) × 0.5–1.5 mm; hypanthium 2.5–4 mm diam.; sepals (2.5–)3–5(–6) mm, apex ± acute; petals nearly white abaxially, pale yellow adaxially, narrowly obcordate, 3–6(–8) × 2.5–5 mm; filaments 1.5–2.5 mm, anthers 0.5–1 mm; carpels 4–12, styles 2–3 mm. **Achenes** 1.5–2 mm, ± rugose.

Varieties 2 (2 in the flora): w United States.

1. Leaves often notably seasonally dimorphic; leaflets: early-season evenly to unevenly incised 1/2–3/4 to midvein (sometimes deeply lobed as well), teeth 3–7 per side (sometimes secondarily toothed); late-season evenly incised 1/5–1/4 to midvein, teeth 6–9 per side; petioles: long hairs 1–1.5(–3) mm; n Arizona and New Mexico, barely entering Colorado .... 54a. *Potentilla subviscosa* var. *subviscosa*
1. Leaves not notably seasonally dimorphic; leaflets ± evenly incised 1/4–1/2 to midvein, teeth 2–4 (–6 in late-season) per side; petioles: long hairs 2–3(–4) mm; Santa Catalina and Rincon mountains and Sierra Ancha, se Arizona....... .... 54b. *Potentilla subviscosa* var. *ramulosa*

54a. Potentilla subviscosa Greene var. **subviscosa** [E] [F]

Stems 0.2–1(–1.5) dm. **Basal leaves** often notably seasonally dimorphic, (1–)2–6(–18 in late-season) cm; petiole (0.5–)1–4 (–10 in late-season) cm, long hairs 1–1.5(–3) mm (sparse in early-season), central leaflets obovate to flabellate (to oblanceolate in late-season), (0.5–)1–3(–8 in late-season) cm, early-season evenly to unevenly incised 1/2–3/4 to midvein (sometimes deeply lobed as well), teeth 3–7 per side (sometimes secondarily toothed), late-season evenly incised 1/5–1/4 to midvein, teeth 6–9 per side. **Flowers:** sepals 3–4 mm; petals (3–)4–6(–7) mm. **Achenes** 2 mm, strongly rugose, with large prominent scar.

Flowering Apr–Jun. Open pine woodlands, grassy flats, hillsides, sandstone ridges; 1200–2500 m; Ariz., Colo., N.Mex.

Variety *subviscosa* is widespread in the mountains of New Mexico and north of the Mogollon Rim in Arizona, barely entering Colorado in the Sangre de Cristo Range. The variety often has striking seasonal differences in leaf shape and petiole vestiture, best developed in Arizona populations. Leaves formed early in the season and coinciding with peak flowering have deeply divided leaflets, and the petiole vestiture often consists almost exclusively of glands. Later-formed leaves are sometimes significantly larger and have progressively less deeply divided leaflets and increasing densities of both short and long hairs.

54b. Potentilla subviscosa Greene var. **ramulosa** (Rydberg) Kearney & Peebles, J. Wash. Acad. Sci. 29: 481. 1939 [E]

Potentilla ramulosa Rydberg, Bull. Torrey Bot. Club 23: 430, plate 276. 1896

Stems (0.2–)0.4–1.5(–2) dm. **Basal leaves** not notably seasonally dimorphic, 2–7 (–15 in late-season) cm; petiole 1–5(–11 in late-season) cm, long hairs 2–3(–4) mm (abundant all seasons); central leaflets obovate-cuneate to lanceolate, 1–2.5(–5 in late-season) cm, ± evenly incised 1/4–1/2 to midvein, teeth 2–4(–6 in late-season) per side. **Flowers:** sepals (2.5–)3.5–5(–6) mm; petals 3–5(–8) mm. **Achenes** 1.5–2 mm, lightly rugose, with small scar.

Flowering Apr–Aug. Open conifer woodlands, edges of decomposing granite outcrops; 1800–2700 m; Ariz.

Variety *ramulosa* is the primary representative of sect. *Subviscosae* in the Rincon and Santa Catalina ranges in Pima County (the only other being *Potentilla albiflora*). Comparable plants have been collected on Aztec Peak in the Sierra Ancha of Gila County, where they overlap the range of and possibly intergrade with var. *subviscosa*. Intermediates also occur in the nearby Mazatzal Range and Fort Apache Indian Reservation. Historic collections of var. *ramulosa* exist from the Chiricahua, Huachuca, and Santa Rita ranges in Cochise and Santa Cruz counties, apparently from lower elevations than the known populations of *P. rhyolitica*.

55. Potentilla wheeleri S. Watson, Proc. Amer. Acad. Arts 11: 148. 1876 • Wheeler's cinquefoil [E]

Potentilla wheeleri var. *paupercula* Jepson

Short hairs not well differentiated from long hairs, absent or sparse throughout. **Stems** 0.2–2.5 dm. **Basal leaves** palmate, 1.5–9.5 cm; petiole 1–7 cm, long hairs abundant to dense, spreading to appressed, 1–2 mm, weak to stiff, glands ± abundant; leaflets 5, central cuneate-elliptic to obovate, 0.5–2.5 × 0.3–1.5 cm, scarcely to distinctly petiolulate, distal ⅓ of margins evenly incised ± ¼ to midvein, teeth 2–4(or 5) per side, surfaces gray-green, long hairs abundant to dense, 1–1.5 mm, glands ± abundant, sometimes obscured. **Inflorescences** 1–20-flowered. **Pedicels** 0.5–1.5 cm. **Flowers:** epicalyx bractlets lanceolate to ovate-elliptic, 1–3.5 × 1–1.5 mm; hypanthium 2–4 mm diam.; sepals 2–5 mm, apex ± acute; petals ± paler abaxially, bright yellow adaxially, ± obcordate, 3–6 × 2–5 mm; filaments 1–2 mm, anthers 0.5 mm; carpels ca. 20, styles (1.2–)1.5–2 mm. **Achenes** 1–1.5 mm, lightly rugose.

Flowering May–Sep. Sandy flats, streamsides, lake margins, open conifer woodlands, alpine fellfields; 1800–3500 m; Calif.

Potentilla wheeleri is found in the southern Sierra Nevada and San Bernardino Mountains of southern California. Compact plants on the summit of Mount San Gorgonio, described by Jepson as var. *paupercula*, show no consistent difference to justify their taxonomic segregation. Variety *viscidula* Rydberg has been misapplied to Arizona populations now called *P. rhyolitica*. Plants identified as *P. wheeleri* (excluding *P. rimicola*) from the Sierra San Pedro Mártir (Baja California) stand as a distinct species, *P. luteosericea* Rydberg (= *P. pinetorum* Wiggins).

56. Potentilla rimicola (Munz & I. M. Johnston) Ertter, Phytologia 71: 420. 1992 • Cliff cinquefoil [C]

Potentilla wheeleri S. Watson var. *rimicola* Munz & I. M. Johnston, Bull. S. Calif. Acad. Sci. 24: 19. 1925

Short hairs not well differentiated from long hairs, absent or sparse throughout. **Stems** 0.4–2(–3) dm. **Basal leaves** palmate, 3–10 cm; petiole 1–7 cm, long hairs abundant, spreading to ascending, 1–3 mm, weak, glands abundant; leaflets 5, central oblanceolate to broadly obovate-cuneate or nearly orbiculate, 1–3 × 0.8–2 cm, scarcely to distinctly petiolulate, distal ± ⅓ of margins evenly incised ± ¼ to midvein, teeth 2–4 per side, surfaces green, long hairs sparse to abundant, 1 mm, glands ± abundant. **Inflorescences** 3–20-flowered. **Pedicels** 0.5–2.5(–4) cm. **Flowers:** epicalyx bractlets lanceolate-elliptic, rarely linear, 1–2(–5) × 0.5–1.5 mm; hypanthium 2–3.5 mm diam.; sepals 2–4(–5) mm, apex ± acute; petals ± paler abaxially, bright yellow adaxially, ± obcordate, (3–)4–7 × 3–6 mm; filaments 1.5–2.5 mm, anthers 0.5–1 mm; carpels 5–20, styles 1.5–2.5 mm. **Achenes** 1.5 mm, smooth to faintly rugose.

Flowering May–Sep. Crevices on protected faces of granitic outcrops in conifer woodlands; of conservation concern; 2400–2900 m; Calif.; Mexico (Baja California).

Potentilla rimicola is known only from Tahquitz Peak in the San Jacinto Mountains, Riverside County, California, and the Sierra San Pedro Mártir in Baja California, Mexico. Plants are rooted in rock crevices, often on vertical surfaces, in contrast to *P. wheeleri*, which is rooted in the ground.

57. Potentilla demotica Ertter, J. Bot. Res. Inst. Texas 1: 53, fig. 2. 2007 • Hualapai cinquefoil [C] [E]

Short hairs not well differentiated from long hairs, absent or sparse throughout. **Stems** 0.2–1.5(–2) dm. **Basal leaves** usually palmate, sometimes ternate, 2–8 cm; petiole 1–7 cm, long hairs common, spreading to ascending, 1.5–2.5 mm, weak to ± stiff, glands abundant; leaflets (3–)5, central oblanceolate, 0.5–1.5(–2) × 0.4–0.8 cm, scarcely petiolulate, distal ± ¾ of margins ± evenly incised ½–¾ to midvein, teeth 2 or 3(or 4) per side, surfaces green, long hairs sparse to common, 0.5–1.5 mm, glands abundant. **Inflorescences** 1–7-flowered. **Pedicels** 0.5–1(–2) cm. **Flowers:** epicalyx bractlets ovate-elliptic, 1–2.5 × 1–1.5 mm; hypanthium 2–3 mm diam.; sepals 2.5–4 mm, apex obtuse; petals ± paler abaxially, bright yellow adaxially, ± obcordate, 3–7 × 2.5–4 mm; filaments 1.3–2 mm, anthers 0.5 mm; carpels 5–12, styles 2–2.5 mm. **Achenes** 1.5–1.8 mm, smooth to lightly rugose.

Flowering May–Jun. Crevices on granitic outcrops, in Yellow pine forests; of conservation concern; 2300–2400 m; Ariz.

Potentilla demotica is known from a single granitic knoll in the Hualapai Mountains, Mohave County. The species was reported as a western range extension of *P. subviscosa* by M. Butterwick et al. (1991); it differs in rock-dwelling habit, petal color, and epicalyx, among other characters.

58. Potentilla rhyolitica Ertter, J. Bot. Res. Inst. Texas 1: 50, fig. 1. 2007 [C] [E]

Short hairs not well differentiated from long hairs, absent or sparse throughout. **Stems** 0.2–2 dm. **Basal leaves** ternate, rarely palmate, 2–8(–11) cm; petiole 1–6(–8) cm, long hairs abundant to dense, spreading to loosely appressed, 1–2(–3) mm, weak to ± stiff, glands ± abundant; leaflets 3(–5), central obovate to broadly elliptic, 1–3 × 0.8–1.8 cm, often distinctly petiolulate, distal $\frac{1}{2}$–$\frac{2}{3}$ of margins evenly incised ± $\frac{1}{3}$ to midvein, teeth 2 or 3(or 4) per side, surfaces green to gray-green, long hairs common to dense, 1 mm, glands abundant. **Inflorescences** 1–10 -flowered. **Pedicels** 0.5–1(–2) cm. **Flowers:** epicalyx bractlets lanceolate-elliptic, 2–4 × 1 mm; hypanthium 3–5 mm diam.; sepals 3–5 mm, apex acute; petals ± paler abaxially, bright yellow adaxially, ± obcordate, 3–7 × 2–5 mm; filaments 1.5–3 mm, anthers 0.6–1 mm; carpels 5–15, styles 2–3 mm. **Achenes** 1.5–2 mm, smooth to lightly rugose.

Varieties 2 (2 in the flora): Arizona.

1. Leaves gray-green, hairs usually dense, sometimes common; styles 2–2.5 mm; achenes 1.5 mm, smooth to lightly rugose; Santa Rita and Huachuca ranges . . . . 58a. *Potentilla rhyolitica* var. *rhyolitica*
1. Leaves green, hairs common, rarely dense; styles (2–)2.5–3 mm; achenes 2 mm, smooth; Chiricahua Range . . . . . 58b. *Potentilla rhyolitica* var. *chiricahuensis*

58a. Potentilla rhyolitica Ertter var. **rhyolitica** • Huachuca cinquefoil [C] [E]

Stems 0.3–2 dm. **Leaves** gray-green; long hairs of petioles 1(–2) mm; leaflets: hairs usually dense, sometimes common. **Flowers** 2–10; filaments 1.5–2.5 mm; carpels 5–15, styles 2–2.5 mm. **Achenes** 1.5 mm, smooth to lightly rugose.

Flowering Apr–Jun. Rhyolite and quartzite outcrops, in open pine forests; of conservation concern; 2600–2900 m; Ariz.

Variety *rhyolitica* forms dense tufts in outcrops near the summits of the Huachuca and Santa Rita ranges (Cochise and Santa Cruz counties). The name *Potentilla viscidula* (Rydberg) Rydberg (= *P. wheeleri* var. *viscidula* Rydberg) was misapplied to these plants by T. H. Kearney and R. H. Peebles (1951); that name has been lectotypified on the hybrid of *P. rhyolitica* and *P. subviscosa* (B. Ertter 2007c).

58b. Potentilla rhyolitica Ertter var. **chiricahuensis** Ertter, J. Bot. Res. Inst. Texas 1: 52, fig. 1H. 2007 • Chiricahua cinquefoil [C] [E]

Stems 0.2–1(–1.5) dm. **Leaves** green; long hairs of petioles 1–2(–3) mm; leaflets: hairs common, rarely dense. **Flowers** 1–5; filaments 2–3 mm; carpels 5–10, styles (2–)2.5–3 mm. **Achenes** 2 mm, smooth.

Flowering May–Jun. Rocky openings, in mixed conifer forests; of conservation concern; 2700–2900 m; Ariz.

Variety *chiricahuensis* is known only from the Chiricahua Mountains (Cochise County). It differs from the typical variety in being less densely hairy, with somewhat longer and coarser hairs and larger achenes. It is also less likely to be found rooted in outcrops, favoring rocky flats.

59. Potentilla albiflora L. O. Williams, Bull. Torrey Bot. Club 61: 260. 1934 • Pinaleño cinquefoil [C] [E]

Short hairs not well differentiated from long hairs, absent or sparse throughout. **Stems** 0.2–1.5 dm. **Basal leaves** ternate, 1.5–10 cm; petiole 1–7 cm, long hairs ± abundant, spreading to ascending, 0.5–1 (–1.5) mm, weak to stiff, glands sparse to abundant; leaflets 3, central obovate, 0.5–2.5(–3.5) × 0.4–1.3(–2) cm, not to scarcely petiolulate, distal $\frac{1}{2}$–$\frac{2}{3}$ of margins evenly incised ± $\frac{1}{3}$ to midvein, teeth 3–5(or 6) per side, surfaces green, long hairs ± abundant, 0.5–1 mm, glands sparse to abundant. **Inflorescences** 1–5(–10)-flowered. **Pedicels** 0.5–2 cm. **Flowers:** epicalyx bractlets lanceolate-elliptic, (1.5–)2–3(–5) × 0.5–1.5 mm; hypanthium 3.5–6 mm diam.; sepals 3–6(–8) mm, apex acute; petals ± paler abaxially, bright yellow adaxially, narrowly obcordate, (3–)4–7(–8) × 3–5 mm; filaments 2–4 mm, anthers 0.5–1 mm; carpels 10–25, styles 2–3 mm. **Achenes** 1.5 mm, smooth.

Flowering May–Sep. Rocky slopes, open ground, in pine-oak and mixed conifer woodlands; of conservation concern; 1800–3200 m; Ariz.

Potentilla albiflora grows on the Pinaleño (Graham) Mountains in Graham County, with a few outlying collections known from the Mogollon Rim in Greenlee County and the Santa Catalina Mountains of Pima County.

Contrary to the specific epithet, petals of living plants are yellow, not white; like those of many *Potentilla* species, they tend to fade in pressed material.

60. **Potentilla cottamii** N. H. Holmgren, Brittonia 39: 340, fig. 1. 1987 • Cottam's cinquefoil C E

Short hairs not well differentiated from long hairs, absent or sparse throughout. **Stems** 0.4–1.7 dm. **Basal leaves** ternate, (1.5–)2–6(–8) cm; petiole 0.8–5(–7) cm, long hairs sparse to abundant, usually ± spreading, sometimes ascending, 0.5–1 mm, weak to ± stiff, glands sparse to common; leaflets 3, central obovate-cuneate, (0.2–)0.5–1(–1.5) × 0.2–1 cm, scarcely petiolulate, distal ½–⅔ of margins evenly incised ⅓–½ to midvein, teeth (1 or)2 or 3(or 4) per side, surfaces green, long hairs sparse to common, 0.5–1 mm, glands sparse to abundant. **Inflorescences** 1–4-flowered. **Pedicels** 0.5–2(–4) cm. **Flowers:** epicalyx bractlets <sometimes paired>, narrowly elliptic, 1.5–2(–2.5) × 0.5–1 mm <½–¾ as long as sepals>; hypanthium 2–3 mm diam.; sepals 2.5–4 mm, apex broadly acute; petals ± paler abaxially, bright yellow adaxially, ± obcordate to broadly obovate, (2–)3–4 × 2–3 mm; filaments 1–1.5 mm, anthers 0.3–0.5 mm; carpels 10–25, styles 1–1.5 mm. **Achenes** 1.3 mm, smooth to lightly rugose.

Flowering Jun–Aug. Protected rock crevices, near vertical faces of quartzite and granite; of conservation concern; 2200–3600 m; Nev., Utah.

Potentilla cottamii is known only from the Pilot Range, Elko County, Nevada, and in northwestern Utah from the Deep Creek, Raft River, and Stansbury mountains. Plants of *P. cottamii* can form dense clumps to 4 dm across. Except for the short style, which led Holmgren to associate the species with *P. hyparctica* and *P. robbinsiana* in sect. *Aureae*, *P. cottamii* fits well in sect. *Subviscosae*.

8q. POTENTILLA Linnaeus sect. BREVIFOLIAE (Rydberg) O. Stevens in N. L. Britton et al., N. Amer. Fl. 22(7): 11. 1959

Barbara Ertter

James L. Reveal

Potentilla [unranked] *Brevifoliae* Rydberg, Monogr. N. Amer. Potentilleae, 30. 1898

Perennials, tufted to openly matted, not stoloniferous; taproots not fleshy-thickened; vestiture of sparse long hairs, or eglandular hairs absent, glands common to abundant, red. **Stems** ± erect, not flagelliform, not rooting at nodes, arising laterally to persistent basal rosettes, (0.2–)0.5–1.5(–2.5) dm, lengths 1.5–3(–4) times basal leaves. **Leaves:** basal not in ranks; cauline 0–2; primary leaves ± pinnate to subpinnate (distal leaflets ± distinct), rarely ternate, 2–7 cm; petiole: long hairs absent or spreading, weak, glands common to abundant; leaflets (3–)5–7, on distal ¼ or less of leaf axis, usually overlapping, ± flabellate, margins flat, ± whole length unevenly incised ½ to nearly to midvein into 3+ lobes <these in turn incised ¼–½ to midvein>, primary teeth of each lobe usually 2 or 3, often secondarily toothed, surfaces similar, green, not glaucous, long hairs (if present) weak, cottony hairs absent. **Inflorescences** (1–)3–10-flowered, ± cymose, open, or solitary flowers. **Pedicels** straight in fruit, 0.1–1 cm, proximal not much longer than distal. **Flowers** 5-merous; hypanthium 2.5–4.5 mm diam.; petals yellow, ± obcordate, 3.5–6.5 mm, longer than sepals, apex retuse; stamens ca. 20; styles subapical, filiform, ± papillate-swollen at very base, if at all, 1.4–2 mm. **Achenes** smooth.

Species 3 (1 in the flora): nw United States, n Mexico.

Section *Brevifoliae* is primarily based on *Potentilla brevifolia*, distinctive in its abundant short-septate glands and the three to seven overlapping, more or less flabellate leaflets, most often in a subpinnate arrangement. Two Mexican species, *P. macdonaldii* B. L. Turner and *P. richardii* Lehmann, are included provisionally in the section. The rounded membranous stipules suggest a close relationship with sect. *Aureae*, especially *P. cristae*, which is also abundantly glandular.

61. Potentilla brevifolia Nuttall in J. Torrey and A. Gray, Fl. N. Amer. 1: 442. 1840 • Sparseleaf cinquefoil [E]

Basal leaves: petiole 1–4 cm, eglandular hairs absent or sparse, spreading, less than 1 mm, weak, glands abundant. **Flowers:** epicalyx bractlets lanceolate-elliptic to ovate, 2–4 × 0.8–2 mm; sepals 3–5 (–7) mm, apex broadly acute to obtuse; petals 3.5–6.5 × 3–5 mm; filaments 1–2.5 mm, anthers 0.5–0.6 mm; carpels 10–15. **Achenes** 1–1.5 mm.

Flowering summer. Moist rocky slopes and talus, alpine grassy areas, subalpine conifer woodlands; 2500–3600 m; Idaho, Mont., Nev., Oreg., Wyo.

Potentilla brevifolia is found mainly in alpine situations in the Pioneer, Sawtooth, Smoky, and White Cloud mountains of central Idaho; the Madison Range of western Montana; the Jarbidge Range of northeastern Nevada; the Blue, Steens, and Wallowa mountains of eastern Oregon; and the Teton Range of northwestern Wyoming. The elongate caudices are easily covered by moving soil and talus.

8r. Potentilla Linnaeus sect. Aureae (Rydberg) Juzepczuk in V. L. Komarov et al., Fl. URSS 10: 197. 1941

Reidar Elven Barbara Ertter

David F. Murray James L. Reveal

Potentilla [unranked] *Aureae* Rydberg, Bull. Torrey Bot. Club 23: 394. 1896; *Potentilla* ser. *Aureae* (Rydberg) T. T. Yu & C. L. Li; *Potentilla* sect. *Frigidae* (Rydberg) O. Stevens

Perennials, tufted to cushion-forming or matted, rarely rosetted, not stoloniferous (except sometimes *P. verna*); taproots not fleshy-thickened; vestiture primarily of straight hairs, not differentiated into long and short, glands absent or sparse to abundant, sometimes red (*P. hyparctica*). **Stems** usually ascending to erect, sometimes prostrate to decumbent or spreading, not flagelliform, not rooting at nodes (except *P. verna*), lateral to persistent basal rosettes, 0.1–2(–3) dm, lengths 1–3(–4) times basal leaves. **Leaves:** basal sometimes 2-ranked; cauline 0–2(or 3); primary leaves ternate or palmate, (0.5–)1–12(–20) cm; petiole: long hairs absent or spreading to ascending, weak to ± stiff, glands usually absent or sparse, sometimes abundant; leaflets 3–7, at tip of leaf axis, overlapping or not, broadly oblanceolate or obtriangular to flabellate, margins flat to revolute, distal ¼–¾+ evenly to unevenly incised ¼–½ to midvein, <sometimes 3–5-lobed ½–¾ to midvein as well>, primary teeth (1 or)2–5(–7) per side <or per lobe>, surfaces ± similar, abaxial pale to dark green, rarely grayish, brownish, or reddish, cottony hairs absent, adaxial green to dark green, rarely brownish or reddish, sometimes ± glaucous (*P. grayi*), long hairs (if present) soft to weak, sometimes stiff. **Inflorescences** 1–6(–12)-flowered, ± cymose, ± to very open, or solitary flowers. **Pedicels** usually straight in fruit (often recurved in *P. verna*), (0–)0.2–3(–9) cm, proximal usually not much longer than distal (sometimes longer in fruit). **Flowers** 5-merous; hypanthium 2–7 mm diam.; petals pale to dark yellow, mostly ± obcordate (to orbiculate in *P. elegans*), (2–)3–10(–12) mm, usually longer than sepals, apex usually ± retuse; stamens ca. 20; styles subapical, columnar, filiform, or tapered, not, scarcely, or ± papillate-swollen in proximal ⅕–½, 0.8–2.5 mm. **Achenes** ± smooth.

Species ca. 30 (10 in the flora): North America, Eurasia; reportedly introduced in s Australia.

The circumscription of sect. *Aureae* used here largely follows that of B. C. Johnston (1985), except that *Potentilla sierrae-blancae* is placed in sect. *Concinnae* and *P. rubella* and *P. stipularis* are in sect. *Chrysanthae*. Other treatments (for example, P. A. Rydberg 1908d; T. Wolf 1908) split the species included here in two groups and/or recombine them with species in other sections, notably sect. *Ranunculoides* (Th. Wolf) Juzepczuk.

Since *Potentilla glaucophylla* (sect. *Graciles*) and *P. cottamii* (sect. *Subviscosae*) are sometimes identified as members of sect. *Aureae*, they are included herein and key out in the third and twelfth couplets, respectively.

1. Basal leaflets (3 or)4–7.
    2. Basal leaves not in ranks; stems usually prostrate to decumbent, rarely weakly erect; caudex branches elongate and slender; pedicels often recurved . . . . 62. *Potentilla verna*
    2. Basal leaves 2-ranked; stems ascending to erect; caudex branches usually short and stout, sometimes elongate and slender; pedicels straight.
        3. Leaves glaucous, ± blue-green; styles 1.5–2.5(–3) mm; sepals: apex acute to short acuminate; w North America, Greenland . . . . .23. *Potentilla glaucophylla* (sect. *Graciles*)
        3. Leaves not glaucous, green; styles 1–1.6 mm; sepals: apex broadly acute; ne Canada, Greenland . . . . 63. *Potentilla crantzii* (in part)
1. Basal leaflets 3.
    4. Central leaflets deeply 3–5-lobed (sinuses extending ½ to nearly to midvein), lobes unevenly incised ¼–½ to midvein, teeth (1 or)2–5 per lobe, glands sparse to abundant.
        5. Styles 1.3–2 mm; stems 0.3–2 dm; sepals 3–5 mm; nw California. . . . . .66. *Potentilla cristae*
        5. Styles 0.7–0.9 mm; stems 0.1–0.5 dm; sepals 2.2–2.5 mm; arctic . . . . 71. *Potentilla elegans*
    4. Central leaflets not or shallowly lobed (sinuses extending less than ½ to midvein), evenly or unevenly incised ¼–½ to midvein, teeth 2–7 per side, glands absent or sparse to common.
        6. Leaflet and epicalyx bractlet margins revolute; British Columbia, Alaska.
            7. Petals 9–12 mm; caudex branches stout (diam. 1–2 cm, including old leaf-bases); filaments (2–)2.8–3.2 mm; styles 1.7–2 mm; 0–10 m . . . . 67. *Potentilla fragiformis*
            7. Petals 4–8 mm; caudex branches slender to ± stout (diam. 0.5–1 cm, including old leaf-bases); filaments 1.2–2 mm; styles 0.8–1.2 mm; 100–1100 m . . . . . . .68. *Potentilla nana*
        6. Leaflet and epicalyx bractlet margins flat, sometimes slightly or ± revolute; North America.
            8. Styles 1–2.5 mm; filaments 1–3 mm.
                9. Leaflet margins usually unevenly incised, teeth 3–7 per side, often secondarily toothed . . . . 64. *Potentilla flabellifolia*
                9. Leaflet margins evenly incised, teeth (2 or)3 or 4(or 5) per side, not secondarily toothed.
                    10. Petioles: long hairs 1–2.5 mm; central leaflets: petiolules 0–1 mm; Greenland, ne Canada. . . . . 63. *Potentilla crantzii* (in part)
                    10. Petioles: long hairs 0.5–1 mm; central leaflets: petiolules (1–)3–10 mm; California . . . . .65. *Potentilla grayi*
            8. Styles 0.6–1 mm; filaments 0.5–1(–1.5) mm.
                11. Epicalyx bractlets 3.5–7 mm; petals 4–9 mm. . . . . 69. *Potentilla hyparctica*
                11. Epicalyx bractlets (1.8–)2–2.5(–3) mm; petals 2–3 mm.
                    12. Epicalyx bractlets ½–¾ as long as sepals; inflorescences 1–4-flowered; Great Basin . . . . 60. *Potentilla cottamii* (sect. *Subviscosae*)
                    12. Epicalyx bractlets ± equal to sepals; inflorescences 1(or 2)-flowered; ne United States. . . . . 70. *Potentilla robbinsiana*

62. Potentilla verna Linnaeus, Sp. Pl. 1: 498. 1753, name conserved • Spring cinquefoil [I]

Potentilla tabernaemontani Ascherson

Plants openly matted; caudex branches elongate, slender, <sometimes rooting at nodes>. **Stems** prostrate to decumbent, rarely weakly erect, 0.3–1.5 dm, lengths 1–2(–3) times basal leaves. **Basal leaves** not in ranks, palmate, 2–5(–10) cm; stipules: apex acuminate to acute, rarely obtuse; petiole 1–6 cm, long hairs common, ascending to appressed, 1–2.5 mm, weak to ± stiff, glands absent or sparse; leaflets 5–7, central oblanceolate to obovate, (0.8–)1–3 × 0.4–1.5 cm, petiolule 0–2 mm, margins flat, not lobed, distal ¼–½ (–⅔) evenly incised ± ⅓ to midvein, teeth 2–4(or 5) per side, surfaces similar, green, hairs absent <especially adaxially> or sparse, 0.5–1.5(–2) mm, glands absent or sparse. **Inflorescences** 3–5(or 6)-flowered. **Pedicels** often recurved, 1–3(–4) cm, not much longer in fruit than in flower. **Flowers:** epicalyx bractlets lanceolate, 3.5–5 × 1–1.5 mm, margins flat; hypanthium 2–4 mm diam.; sepals 3–5 mm, apex broadly acute to obtuse; petals yellow, 5–8(–10) × 4–6 mm; filaments 0.8–2 mm, anthers 0.5–0.8 mm; carpels 40–60, styles ± columnar, not papillate-swollen proximally, 1 mm. **Achenes** 1.5–2 mm. **2*n*** = 28, 35, 42, 49, 56, 63, 70, 84 (Europe).

Flowering spring–early summer. Dry roadsides and lawns; 0–200 m; introduced; Ont.; Conn., Mich., Minn.; Europe.

The use of the name *Potentilla verna* here reflects the successful proposal by J. Soják (2009) to conserve *P. verna* with a conserved type. Alternatives in recent use include *P. neumanniana* Reichenbach (misapplied) and *P. tabernaemontani.* The species has an established horticultural presence, primarily as a ground cover, and is to be expected in North America beyond where reported here.

63. Potentilla crantzii (Crantz) Beck ex Fritsch, Excursionfl. Oesterreich 295. 1897 • Alpine cinquefoil, potentille de Crantz

Fragaria crantzii Crantz, Inst. Rei Herb. 2: 178. 1766; *Potentilla alpestris* Haller f.; *P. crantzii* var. *hirta* (Lange) Malte; *P. flabellifolia* Hooker ex Torrey & A. Gray var. *hirta* (Lange) B. Boivin; *P. langeana* Rydberg; *P. maculata* Pourret

Plants tufted to ± matted; caudex branches usually short, stout, sometimes elongate, slender. **Stems** ascending to erect, 0.5–2(–3) dm, lengths 1.5–3 times basal leaves. **Basal leaves** 2-ranked, usually palmate, sometimes ternate, 2–10(–15) cm; stipules: apex obtuse to truncate, rarely acute; petiole 0.5–6.5 (–10) cm, long hairs usually ± sparse, usually ± ascending, sometimes spreading or appressed, 1–2.5 mm, weak to ± stiff, glands absent or sparse; leaflets 3–5, central obovate, 1.5–3(–3.5) × 0.8–1.5(–2) cm, petiolule 0–1 mm, margins flat, not lobed, distal ½–⅔ evenly incised ⅓–½ to midvein, teeth 3 or 4(or 5) per side, <not secondarily toothed>, surfaces similar, green, <not glaucous>, hairs absent or sparse, rarely common, 0.8–1.5 mm, glands absent or sparse to abundant (fewer adaxially). **Inflorescences** (2 or)3–8(–12)-flowered. **Pedicels** straight, 1–3(–4) cm in flower, to 6 cm in fruit. **Flowers:** epicalyx bractlets ovate or narrowly ovate, (1.5–)2.5–4 × (0.8–) 1–1.3 mm, margins flat; hypanthium 3–4 mm diam.; sepals (3–)4–5(–6) mm, apex broadly acute; petals yellow, 4–7(–9) × 4–8 mm; filaments 1.8–2.2 mm, anthers 0.3–0.6 mm; carpels 30–40, styles columnar-filiform, not papillate-swollen proximally, 1–1.6 mm. **Achenes** 1.2 mm. **2*n*** = 42; 28, 35, 49 (Europe).

Flowering spring–summer. Herb meadows and slopes, usually on calcareous substrates, usually near coast; 0–1700 m; Greenland; Nfld. and Labr., Nunavut, Ont., Que.; Eurasia; reportedly introduced in s Australia.

Potentilla crantzii is primarily European, not restricted to coasts as in North America. The nomenclatural history is long and complex (A. Kurtto et al. in J. Jalas et al. 1972+, vol. 13), with *P. alpestris* and *P. maculata* in greatest historical use. The name *P. ×protea* Soják is available for presumed hybrids between *P. crantzii* and *P. hyparctica.*

The emphasis on two-ranked leaves to distinguish *Potentilla crantzii* from *P. verna* follows J. Soják (pers. comm.). Some collections of *P. crantzii* from Newfoundland approach *P. verna* in their more elongate caudex branches and shorter stature.

64. Potentilla flabellifolia Hooker ex Torrey & A. Gray, Fl. N. Amer. 1: 442. 1840 • Fan-leaved or fanfoil or Mount Rainier cinquefoil [E]

Plants ± openly matted; caudex branches short to elongate, ± stout. **Stems** ascending to erect, (0.3–)1–3 dm, lengths 2 times basal leaves. **Basal leaves** 2-ranked, ternate, (1–) 3–12(–20) cm; stipules: apex rounded; petiole (0.5–)1–15 (–18) cm, hairs and glands usually absent; leaflets 3, central widely obovate to flabellate, (0.5–)1–5 × (0.4–)1–3(–3.5) cm, petiolule 0–5(–10) mm, margins flat, not or shallowly lobed (sinuses extending less than ½ to midvein), distal ½–¾ usually unevenly incised ¼–½ to midvein, teeth 3–7 per side, <often secondarily toothed>, surfaces similar, green, hairs absent or sparse, rarely common, 0.5–1 mm, glands absent or sparse. **Inflorescences** 1–6-flowered. **Pedicels** straight, 0.3–4.5(–5) cm, not much longer in fruit than in flower. **Flowers:** epicalyx bractlets <often toothed or 2-lobed>, elliptic to oval, 3.5–7 × (1–)1.5–3 mm, margins flat; hypanthium 3–7 mm diam.; sepals 4–8 mm, apex acute to obtuse; petals dark yellow, (5–)6–10(–12) × (3–)5–10 mm; filaments 1.5–3 mm, anthers 0.6–1.5 mm; carpels 20–50, styles filiform, not or slightly papillate-swollen proximally, 1.6–2.5 mm. **Achenes** 1.2–1.5 mm. **2*n*** = 28.

Flowering summer. Moist to wet or swampy, sometimes dry, stream banks, lakeshores, meadows, in conifer woodlands, subalpine and alpine grassy, rocky slopes, near or above timberline; 1000–3700 m; Alta., B.C.; Calif., Idaho, Mont., Nev., Oreg., Wash., Wyo.

Potentilla flabellifolia is found in the mountains from southern British Columbia, southwestern Alberta, and southwestern Montana to the southern Sierra Nevada of California.

65. Potentilla grayi S. Watson, Proc. Amer. Acad. Arts 8: 560. 1873 • Gray's cinquefoil [E]

Potentilla flabellifolia Hooker ex Torrey & A. Gray var. *grayi* (S. Watson) Jepson

Plants rosetted or ± tufted; caudex branches absent or short to elongate, stout to slender. **Stems** ascending to erect, (0.4–)1–2 dm, lengths 2–4 times basal leaves. **Basal leaves** not in ranks, ternate, (1–)2–5(–7) cm; stipules: apex acute to rounded; petiole 1–4 cm, hairs absent or long hairs sparse to common, appressed, 0.5–1 mm, stiff, glands absent; leaflets 3, central obovate to flabellate, 1–2.5 × 0.5–2 cm, petiolule (1–)3–10 mm, margins flat, not lobed, distal ½–¾ ± evenly incised ¼–½ to midvein, teeth (2 or)3 (or 4) per side, <not secondarily toothed>, surfaces similar, green, <sometimes ± glaucous>, hairs absent or sparse to common, 0.5 mm, glands absent. **Inflorescences** 1–6-flowered. **Pedicels** straight, 0.5–2 cm in flower, to 5 cm in fruit. **Flowers:** epicalyx bractlets elliptic to oval, 2–4.5 × 1–2 mm, margins flat; hypanthium 2.5–4 mm diam.; sepals 3–5.5 mm, apex acute to obtuse; petals yellow, 4–7 × 4–6 mm; filaments 1–3 mm, anthers 0.5–0.9 mm; carpels 10–30, styles filiform, not or slightly papillate-swollen proximally, 1–2.5 mm. **Achenes** 1.2–1.5 mm.

Flowering summer. Moist stream banks, lakeshores, meadows, in conifer woodlands; 2000–2800 m; Calif.

Potentilla grayi occurs on the western slope of the Sierra Nevada from El Dorado County to Tulare County, well within the range of *P. flabellifolia*. Some collections are intermediate between the two. *Potentilla grayi* can be distinguished by its tendency to have glabrous or strigose leaves, distinctly petiolulate leaflets with simple teeth, paler yellow petals, and a less-branched caudex.

The report of *Potentilla grayi* in New Mexico (J. D. Garcia 1970) was based on a specimen of *P. norvegica* (sect. *Rivales*).

66. Potentilla cristae Ferlatte & Strother, Madroño 37: 190, fig. 1. 1990 • Crested cinquefoil [C] [E]

Plants tufted to densely matted; caudex branches usually short, stout, <sometimes embedded in old leaf bases>. **Stems** ascending to erect, 0.3–2 dm, lengths 2–3 times basal leaves. **Basal leaves** not in ranks, ternate, 1–5(–9) cm; stipules: apex acute to obtuse; petiole 1–4 cm, long hairs absent or sparse to common, spreading to ascending, 0.5–2 mm, ± weak, glands sparse to abundant; leaflets 3, central flabellate, 0.4–2 × 0.5–1.5 cm, petiolule 1–3 mm, margins flat, distal ¾+ deeply 3–5-lobed (sinuses extending ½ to nearly to midvein), lobes unevenly incised ¼–½ to midvein, teeth 2–5 per lobe, surfaces similar, dark green, hairs ± sparse, 0.5–1.5 mm, glands sparse to abundant. **Inflorescences** 1–7-flowered. **Pedicels** straight, 0.2–1.2(–2) cm, not much longer in fruit than in flower. **Flowers:** epicalyx bractlets broadly ovate, 2–4 × 2 mm, margins flat; hypanthium 3.5–6 mm diam.; sepals 3–5 mm, apex broadly acute to obtuse; petals yellow, 3–5.5 × 4–5 mm; filaments 0.7–1.7 mm, anthers 0.5–0.6 mm; carpels 25–40, styles tapered-filiform, ± papillate-swollen in proximal ⅕–⅓, 1.3–2 mm. **Achenes** 1–1.5 mm, <dorsally crested>. **2*n*** = 42.

Flowering summer. Rocky, open, serpentine slopes, in conifer woodlands; of conservation concern; 1800–2800 m; Calif.

Potentilla cristae is known only from Cory Peak, Mount Eddy, and the Marble Mountains in northwestern California. In addition to the characteristics in the key, the species is distinctive in having a low dorsal crest on the achenes, hence the specific epithet.

67. Potentilla fragiformis D. F. K. Schlechtendal, Ges. Naturf. Freunde Berlin Mag. Neuesten Entdeck. Gesammten Naturk. 7: 294. 1816

Plants tufted to cushion-forming; caudex branches short, stout, <diam. 1–1.5(–2) cm, including old leaf bases>. **Stems** ascending to erect, 0.3–1.5 (–2) dm, lengths 1–2 times basal leaves. **Basal leaves** not in ranks, ternate, (1–)4–10(–15) cm; stipules: apex acute to obtuse; petiole (0.3–)2–11(–13) cm, long hairs abundant, ascending to spreading, 1–2.3 mm, soft, glands sparse to common; leaflets 3, central obovate, 0.7–2.5 × 0.5–1.6 cm, petiolule 0–1 mm, margins revolute, not lobed, distal ½–⅔ evenly incised ± ⅓ to midvein, teeth 2 or 3(or 4) per side, surfaces ± similar, green (paler abaxially), hairs common, 1–1.5 mm, glands common (sparser adaxially). **Inflorescences** 1–3-flowered. **Pedicels** straight, 0.5–2 cm in flower, to 3.5 cm in fruit. **Flowers:** epicalyx bractlets oblong, 3–7 × 1.2–2.5 mm, margins revolute; hypanthium 4–6 mm diam.; sepals 4–6(–7) mm, apex broadly acute to obtuse; petals yellow, 9–12 × 8–10 mm; filaments (2–)2.8–3.2 mm, anthers 0.5–0.8 mm; carpels 40–50, styles filiform, not or scarcely papillate-swollen proximally, 1.7–2 mm. **Achenes** 1.3–1.5. **2*n*** = 42, 56 (Russian Far East).

Flowering late spring–summer. *Leymus-Potentilla* sand dunes, coastal gravel bars, beach ridges; 0–10 m; Alaska; e Asia (Russian Far East).

Recent collections of *Potentilla fragiformis* in Alaska on the coast of the Bering and Chukchi seas have confirmed historical reports (P. A. Rydberg 1908d) of this Beringian species in North America. Dwarf plants from St. Lawrence Island are somewhat intermediate to *P. nana* but have the larger petals and longer styles of *P. fragiformis*.

68. Potentilla nana D. F. K. Schlechtendal, Ges. Naturf. Freunde Berlin Mag. Neuesten Entdeck. Gesammten Naturk. 7: 296. 1816 • Dwarf cinquefoil [E] [F]

Potentilla emarginata Pursh subsp. *nana* (D. F. K. Schlechtendal) Hultén

Plants densely tufted; caudex branches short, slender to ± stout, <diam. 0.5–1 cm, including old leaf bases>. **Stems** ascending to erect, 0.1–0.5(–0.7) dm, lengths 1–2 times basal leaves. **Basal leaves** not in ranks, ternate, 1–5 cm; stipules: apex acute; petiole 0.5–3.5 cm, long hairs sparse to abundant, ascending to spreading, 0.5–2 mm, ± soft, glands sparse to common; leaflets 3, central obovate, 0.5–2 × 0.5–1 cm, petiolule 0–1 mm, margins revolute, not lobed, distal ½–⅔ evenly incised ± ½ to midvein, teeth 3 or 4(–6) per side, surfaces ± similar, green (paler abaxially), hairs sparse to abundant, 0.8–1.2 mm, glands sparse to common. **Inflorescences** 1(or 2)-flowered. **Pedicels** straight, 0.2–1.5 cm in flower, to 5 cm in fruit. **Flowers:** epicalyx bractlets oblong or ovate, 2.5–5 × 1.5–3.5 mm, margins revolute; hypanthium 3–3.5 mm diam.; sepals 2.5–5 mm, apex ± acute; petals pale yellow, 4–8 × 4–6 mm; filaments 1.2–2 mm, anthers 0.3–0.4 mm; carpels 40–50, styles ± columnar, not or scarcely papillate-swollen proximally, 0.8–1.2 mm. **Achenes** 1.4–1.6 mm.

Flowering summer. Alpine and coastal *Dryas-Salix-Empetrum* heaths, gravelly slopes, ridge crests, fellfields, scree and talus; 100–1100 m; B.C.; Alaska.

Potentilla nana has most often been considered a race of *P. hyparctica* or as the priority name for *P. hyparctica*. As now defined, *P. nana* occurs in an arc across southern Alaska, from the Aleutian and Pribilof islands to the Alaskan panhandle and adjacent British Columbia. T. Wolf (1908) treated it as a dwarf form of *P. fragiformis*; J. Soják (1996) interpreted *P. nana* as a hybrid species originating from cross(es) between *P. fragiformis* and *P. hyparctica*. Although the hybrid hypothesis has morphologic support, *P. nana* has a unique combination of morphologic features and has a distinct range from both *P. fragiformis* and *P. hyparctica*. The species are also ecologically segregated: *P. nana* is a coastal heath and gravel slope species, like *P. hyparctica*; *P. fragiformis* is strictly maritime coastal.

69. Potentilla hyparctica Malte, Rhodora 36: 177. 1934 • Arctic cinquefoil [F]

Potentilla robbinsiana (Lehmann) Oakes ex Rydberg subsp. *hyparctica* (Malte) D. Löve

Plants cushion-forming; caudex branches short to elongate, slender, <often embedded in old leaf bases>. **Stems** ascending to erect, 0.2–2(–2.5) dm, lengths 1–3(–4) times basal leaves. **Basal leaves** not in ranks, ternate, (1–)2–8(–10) cm; stipules: apex ± acute; petiole 1–6.5(–8.5) cm, long hairs sparse to common, spreading to subappressed, 0.8–2 mm, weak to ± stiff, glands sparse to common; leaflets 3, central obovate to broadly obovate, 0.5–2.5 (–2.8) × 0.3–2(–2.2) cm, petiolule 0–3 mm, margins flat or slightly revolute, not lobed, distal ½–⅔(–¾) evenly incised ± ½ to midvein, teeth (2 or)3–5 per side, surfaces ± similar, abaxial usually pale green, sometimes grayish, hairs sparse to abundant, 0.5–1.7 mm, adaxial darker green, hairs and glands sparse. **Inflorescences** 1–3 (–5)-flowered. **Pedicels** straight, 0.3–3 cm in flower, to 9 cm in fruit. **Flowers:** epicalyx bractlets narrowly to broadly oblong or ovate, 3.5–7 × 1.5–5 mm, margins flat; hypanthium 4–7 mm diam.; sepals 4–8 mm, apex subacute to rounded; petals pale or bright yellow, 4–9 × 4–6 mm; filaments 0.5–1.1 mm, anthers 0.2–0.4 mm; carpels 50–80, styles ± columnar, not or scarcely papillate-swollen proximally, 0.6–0.9 mm. **Achenes** 1.1–1.3 mm.

Subspecies 3 (2 in the flora): North America, n Eurasia; circumpolar.

Southern arctic and subarctic plants differ from the northern arctic ones in presumably independent features. Where the two forms are sympatric (western Greenland,

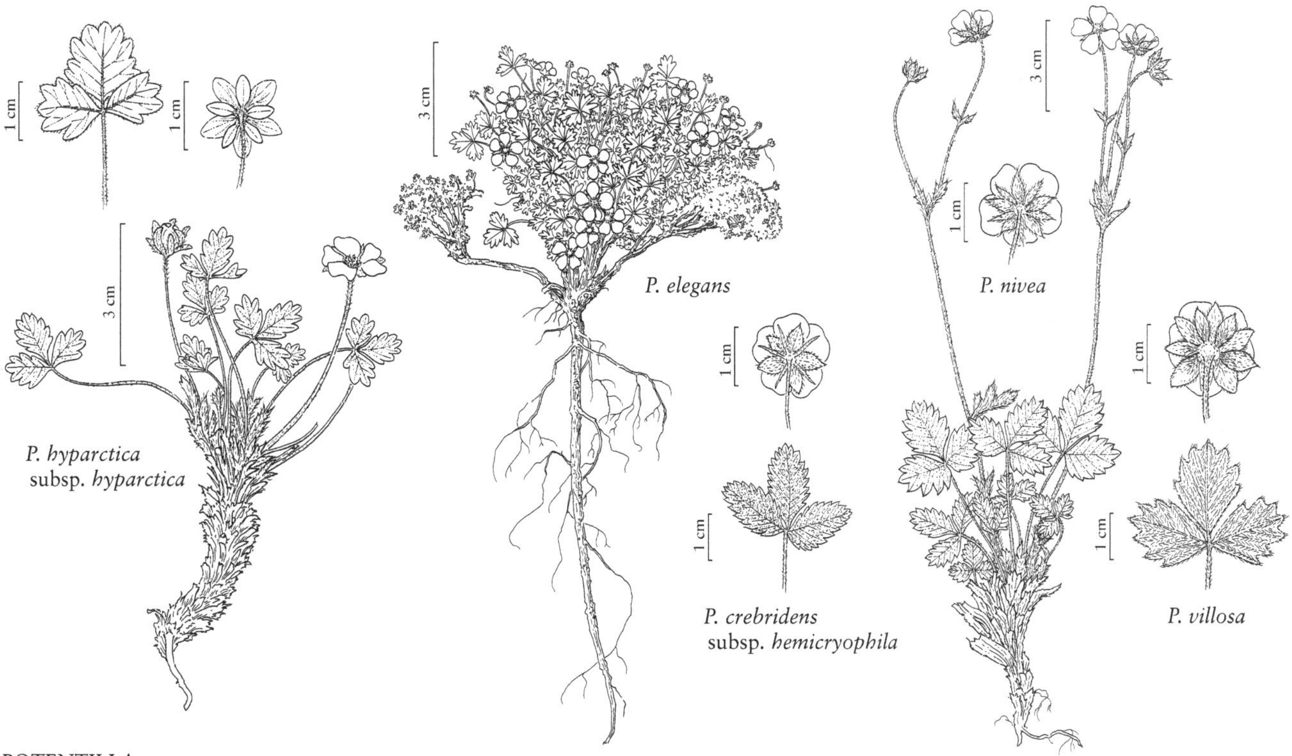

POTENTILLA

northern Quebec, Baffin Island), they appear to remain distinct even in mixed populations and are accordingly treated as subspecies.

Although some features of scattered plants in Alaska and Yukon approach subsp. *nivicola* Jurtzev & V. V. Petrovsky (described from northeastern Asia), such plants do not otherwise correspond fully with this taxon. Possible hybrids between *Potentilla hyparctica* and *P. pulchella* (sect. *Pensylvanicae*) are addressed in the discussion of sect. *Pensylvanicae*. Presumed hybrids with *P. villosa* from the Aleutian Islands, characterized by the combination of basally thickened styles and lack of cottony hairs on leaflet abaxial surfaces, have been named *P.* ×*aleutica* Soják.

1. Epicalyx bractlets narrowly to ± oblong or ovate, 1.5–2 mm wide; central leaflets: petiolules 0–2 mm, bases cuneate . . . . 69a. *Potentilla hyparctica* subsp. *hyparctica*
1. Epicalyx bractlets broadly oblong or ovate, 2–5 mm wide; central leaflets: petiolules (0–)2–3 (–5) mm, bases broadly cuneate to rounded . . . . 69b. *Potentilla hyparctica* subsp. *elatior*

69a. Potentilla hyparctica Malte subsp. **hyparctica** • Potentille subarctique [F]

Plants: hairs common to dense. **Stems** 0.2–1.2 dm. **Central leaflets** obovate, petiolule 0–2 mm, base cuneate, teeth narrowly triangular to oblong, tooth apex subacute to obtuse. **Inflorescences** 1(–3)-flowered. **Flowers:** epicalyx bractlets narrowly to ± oblong or ovate, 1.5–2 mm wide; sepals: apex subacute to obtuse. **2*n*** = 42.

Flowering late spring–summer. Polar deserts, moist tundra, meadows, boulder and gravel fields, rocky outcrops, talus, mainly acidic bedrock; 0–1500 m; Greenland; N.W.T., Nunavut, Que., Yukon; Alaska; Europe; Asia.

69b. Potentilla hyparctica Malte subsp. **elatior** (Abromeit) Elven & D. F. Murray, J. Bot. Res. Inst. Texas 1: 813. 2007 • Grande potentille subarctique

Potentilla emarginata Pursh var. *elatior* Abromeit, Biblioth. Bot. 42(2): 8. 1899; *P. flabellifolia* Hooker ex Torrey & A. Gray var. *emarginata* (Seringe) B. Boivin; *P. hyparctica* var. *elatior* (Abromeit) Fernald

Plants: hairs usually sparse, sometimes common to abundant. **Stems** 0.4–2(–2.5) dm. **Central leaflets** broadly obovate, petiolule (0–)2–3(–5) mm, base broadly cuneate to rounded, teeth broadly oblong, tooth apex obtuse to rounded. **Inflorescences** 1–3(–5)-flowered. **Flowers:** epicalyx bractlets broadly oblong or ovate, 2–5 mm wide; sepals: apex obtuse to rounded. $2n = 42$.

Flowering late spring–summer. Damp ericaceous heath, snow patch meadows, damp arctic and alpine tundra, rocky outcrops, solifluction slopes, talus, mainly acidic bedrock; 0–4000 m; Greenland; Alta., B.C., Nfld. and Labr. (Labr.), N.W.T., Nunavut, Que., Yukon; Alaska, Mont., Wash., Wyo.; Asia (Russian Far East).

Geographically isolated and morphologically deviant plants from alpine sites in Montana (Beartooth Mountains), Washington (Mount Baker), and Wyoming (Sublette County) are provisionally assigned to subsp. *elatior* but likely represent a segregate race. These plants are exceptionally compact and hairy, with leaflets usually lacking petiolules.

Potentilla emarginata Pursh (1813), sometimes used for this taxon, is an illegitimate later homonym of *P. emarginata* Desfontaines (1804).

70. Potentilla robbinsiana (Lehmann) Oakes ex Rydberg, Bull. Torrey Bot. Club 23: 304. 1896 • White Mountains or Robbins's cinquefoil [C] [E]

Potentilla minima Haller f. var. *robbinsiana* Lehmann, Revis. Potentill., 159. 1856; *P. hyparctica* Malte subsp. *robbinsiana* (Lehmann) Á. Löve & D. Löve

Plants tufted to densely matted; caudex branches short, ± slender, <often embedded in old leaf bases>. **Stems** spreading to erect, 0.1–0.4 dm, lengths 1–2 times basal leaves. **Basal leaves** not in ranks, ternate, 1–2 cm; stipules: apex broadly acute; petiole 0.7–3 cm, long hairs sparse to abundant, spreading to subappressed, 0.2–2 mm, weak, glands sparse to common; leaflets 3, central obovate, 0.5–1.3 × 0.2–1 cm, petiolule 0–1 mm, margins flat to ± revolute, not lobed, distal ± ⅔ evenly incised ± ½ to midvein, teeth 2–4(or 5) per side, surfaces similar, green, hairs ± abundant (or nearly absent adaxially), 0.5–1 mm, glands sparse to common. **Inflorescences** 1(or 2) -flowered. **Pedicels** straight, 0.5–3 cm, not much longer in fruit than in flower. **Flowers:** epicalyx bractlets oblong to elliptic, (1.8–)2–2.5(–3) × 0.7–1.2(–1.4) mm, <± equal to sepals>, margins flat to ± revolute; hypanthium (4–)5–7 mm diam.; sepals (1.8–)2–2.5(–3) mm, apex broadly acute; petals yellow, 2–3 × 2–3 mm; filaments 0.5–1 mm, anthers 0.4–0.6 mm; carpels 20–30, styles columnar-tapered, sometimes ± papillate-swollen in proximal ⅓–½, 0.8–1 mm. **Achenes** 1–1.3 mm. $2n = 49$.

Flowering summer. Moist rocky slopes and flats, in montane tundra; of conservation concern; 1400–1600 m; N.H., Vt.

Potentilla robbinsiana is known only from two sites in the White Mountains, Monroe County, New Hampshire, and is reported in Vermont. Additional populations have been established nearby as a result of transplant efforts. Listed in 1980 as a federally endangered species, an intense recovery program resulted in the species being delisted in 2002.

Some early specimens were distributed as *Potentilla frigida* Villars, a similar European species.

SELECTED REFERENCES Cogbill, C. V. 1968. The interplay of botanists and *Potentilla robbinsiana*: Discovery, systematics, collection, and stewardship of a rare species. Rhodora 95: 52–75. Graber, R. E. 1980. The life history and ecology of *Potentilla robbinsiana*. Rhodora 82: 131–140. Graber, R. E. and L. G. Brewer. 1985. Changes in the population of the rare and endangered plant *Potentilla robbinsiana* Oakes during the period 1973 to 1983. Rhodora 87: 449–457. Steele, F. L. 1964. *Potentilla robbinsiana* in the White Mountains of New Hampshire. Rhodora 66: 408–411.

71. Potentilla elegans Chamisso & Schlechtendal, Linnaea 2: 22. 1827 • Elegant cinquefoil [F]

Plants densely tufted to cushion-forming; caudex branches usually short, stout, <sometimes embedded in old leaves>. **Stems** erect, 0.1–0.5 dm, lengths 1–2 times leaves. **Basal leaves** not in ranks, ternate, 0.5–2.5(–4) cm; stipules: apex acute to obtuse; petiole 0.2–1.8(–2.5) cm, long hairs ± sparse, ± spreading, 0.2–1 mm, weak, glands sparse to abundant; leaflets 3, central obtriangular to flabellate, 0.5–0.7(–1) × 0.5–0.7 cm, petiolule 0–0.5 mm, margins flat, distal ¾+ deeply 3–5-lobed (sinuses extending ⅔–¾ to midvein), lobes unevenly incised ¼–½ to midvein, teeth 1–3(–5) per lobe, surfaces similar, green, brownish, or reddish, hairs absent or sparse, 0.2–0.5 mm, glands sparse to common (also densely punctate-glandular). **Inflorescences** 1-flowered. **Pedicels** straight or ± curved, 0–0.5(–2) cm in flower, to 5 cm in fruit. **Flowers:** epicalyx bractlets linear, elliptic, or obovate, 2–2.2 × 0.8–1.1 mm, margins flat; hypanthium 2.5–3.5 mm diam.; sepals 2.2–2.5 mm, apex broadly acute to obtuse; petals pale yellow, 3–3.5 × 3–3.5 mm; filaments 1–1.5 mm, anthers 0.3–0.4 mm; carpels 20–25, styles ± columnar, not papillate-swollen proximally, 0.7–0.9 mm. **Achenes** 1.2 mm. $2n$ = 14, 28 (Russian Far East).

Flowering spring–summer. Rock crevices, blocky scree, mountain summits, mostly on acidic bedrock; 100–1600 m; B.C., N.W.T., Yukon; Alaska; e Asia (Russian Far East).

For present purposes, *Potentilla elegans* is treated in sect. *Aureae*, as was done by B. C. Johnston (1985), largely on the basis of pragmatism. J. Soják (1994) places the species instead in sect. *Dumosae* Soják, which otherwise comprises several Himalayan species.

8s. POTENTILLA Linnaeus sect. NIVEAE (Rydberg) A. Nelson in J. M. Coulter and A. Nelson, New Man. Bot. Rocky Mt., 255. 1909

Reidar Elven

David F. Murray

Barbara Ertter

Potentilla [unranked] *Niveae* Rydberg, Bull. Torrey Bot. Club 23: 301. 1896

Perennials, ± tufted or cushion-forming, not stoloniferous; taproots not fleshy-thickened; vestiture of long, short, crisped, and/or cottony hairs, glands absent or sparse to common, sometimes red. **Stems** ascending to erect, not flagelliform, not rooting at nodes, lateral to persistent basal rosettes, 0.2–2.5(–4.5) dm, lengths 1.5–5 times basal leaves. **Leaves:** basal rarely 2-ranked; cauline 0–2(or 3); primary leaves usually ternate, rarely palmate on same plant, (0.5–)1–12(–20) cm; petiole: long hairs sometimes absent, ± appressed to spreading, soft to stiff, glands absent or sparse to common; leaflets 3(–5), at tip of leaf axis, separate to overlapping, usually obovate, sometimes elliptic, suborbiculate, or obtriangular, margins slightly to strongly revolute, rarely flat, distal (⅓–)½ to nearly whole length evenly incised ¼–¾ to midvein, teeth (1 or)2–6(–12) per side, surfaces ± to strongly dissimilar, abaxial gray to white, sometimes yellowish white or reddish, cottony-crisped hairs usually dense, <glands absent or obscured>, adaxial dark green to grayish white or yellowish white, sometimes reddish, not glaucous, long hairs soft to stiff. **Inflorescences** 1–7(–15)-flowered, cymose, ± to very open, or solitary flowers. **Pedicels** straight in fruit, 0.4–6(–10) cm, proximal often longer than distal. **Flowers** 5-merous; hypanthium (1.8–)2.5–7 mm diam.; petals pale yellow to yellow, ± obcordate, (3–)4–10(–15) mm, longer than sepals, apex retuse; stamens 15–30; styles subapical, narrowly columnar, tapered, or conic, usually papillate-swollen in proximal ⅕ or less, sometimes to proximal ⅓(–½) or not at all, 0.7–1.2(–1.5) mm. **Achenes** smooth or slightly rugose.

Species ca. 32 (11, including 1 hybrid, in the flora): North America, Eurasia.

Species of sect. *Niveae* are primarily trifoliate, with occasional supernumerary leaflets developing on otherwise trifoliate plants (B. Eriksen and J. Nyléhn 1999). Abaxial leaflet surfaces are covered with dense crisped and/or cottony hairs. This is perhaps the most taxonomically difficult section of panarctic *Potentilla*; the current treatment, which leans heavily on revisions of Russian and Beringian *Potentilla* by B. A. Jurtzev (1984, 1993, 2001) and J. Soják (1989, 2004), is decidedly provisional. Eastern North American and western montane variation in particular remains poorly known, with more taxa likely to be recognized pending further study. The disposition of some infraspecific names is currently unresolved, notably *P. nivea* var. *macrophylla* Seringe, *P. nivea* var. *tomentosa* Nilsson-Ehle ex Hultén, and *P. chamissonis* Hultén var. *umanakensis* Hultén.

North American species of sect. *Niveae* have traditionally been treated as *Potentilla nivea* in the broad sense (including *P. hookeriana* Lehmann, misapplied), *P. uniflora* Ledebour (or its illegitimate synonym *P. ledebouriana* A. E. Porsild), *P. vahliana*, and *P. villosa*. The finer taxonomy adopted here is discussed in greater detail elsewhere (R. Elven and D. F. Murray 2008, 2008b; Elven et al., nhm2.uio.no/paf). The species treated here mostly fall into three categories: the *P. nivea* group (*P. arenosa*, *P. crebridens*, *P. holmgrenii*, *P. nivea*), the *P. uniflora/villosa* group (*P. subvahliana*, *P. villosa*, *P. villosula*, *P. vulcanicola*), and species of presumed intrasectional hybrid origin (*P. subgorodkovii*, *P. vahliana*). The North American species comprising the *P. nivea* group are distinguished by relatively narrow epicalyx bractlets, multi-flowered inflorescences, leaflets with relatively numerous teeth, and either cottony or verrucose petiole hairs. The American taxa of the *P. uniflora/villosa* group are characterized by relatively broad epicalyx bractlets (often notched or split to base) and smooth, silky, straight hairs on petioles.

Most North American plants previously included in *Potentilla uniflora* are treated here as *P. subgorodkovii* or *P. vulcanicola*. The presence of *P. uniflora* in the narrow sense in North America is unconfirmed, with only one specimen from northwestern Alaska (Ogotoruk Creek, ALA) conforming in diagnostic features. This is about 400 km from the easternmost confirmed Asian occurrence of *P. uniflora*, which is widespread in northeastern Asia from Taimyr Peninsula eastward to the isthmus of the Chukchi Peninsula (B. A. Jurtzev 1984).

Also treated here is *Potentilla tikhomirovii*, a presumed hybrid species derived from sections *Aureae* and *Niveae*. For species assumed to have originated from crosses between species of sect. *Niveae* and sect. *Pensylvanicae*, see 8t. sect. *Rubricaules*. Hybrids that have not yet stabilized as recognizable taxa are not included in the key, even though putative hybrids are more frequent than the presumed parents in some regions, forming an unresolvable complex that also influences the parental species through back-crosses.

Plants interpreted as hybrids between *Potentilla arenosa* in the broad sense and *P. nivea* are particularly common throughout the sympatric ranges of these species. Among the names that have been applied to the resultant hybrids are *P.* ×*drymeja* Soják (type from Mongolia), *P.* ×*mischkinii* Juzepczuk (type from European Russia), *P.* ×*prostrata* Rottbøll (type from Greenland), *P.* ×*subquinata* (Lange) Rydberg (= *P. nivea* var. *subquinata* Lange; lectotype [J. Soják 1986] from Greenland), and *P.* ×*tomentulosa* Jurtzev (type from Siberia); *P.* ×*prostrata* is the priority name.

Hair types that are essential in defining species of sect. *Niveae* are summarized by B. Eriksen and B. A. Jurtzev (1999), modified for the current treatment as discussed in the introduction to the genus. Verrucose or finely warty hairs are most evident at high magnification (for example, 50×) but can usually be recognized by their dull, frosted appearance, in contrast to the smoother glassy appearance of non-verrucose hairs. Red-tipped glands, characteristic of *Potentilla tikhomirovii*, should not be confused with red tubercles at the base of some hairs.

SELECTED REFERENCES Elven, R. and D. F. Murray. 2008. The *Potentilla nivea* group in northwestern North America. Bot. Electronic News 389. Elven, R. and D. F. Murray. 2008b. The *Potentilla villosa-uniflora* group in northwestern North America. Bot. Electronic News 390. Eriksen, B. 1997. Morphometric analysis of Alaskan members of the genus *Potentilla* sect. *Niveae* (Rosaceae). Nordic J. Bot. 17: 621–630. Eriksen, B. and B. A. Jurtzev. 1999. Hair types in *Potentilla* sect. *Niveae* (Rosaceae) and related taxa—Terminology and systematic distribution. Skr. Norske Vidensk.-Akad. Mat.-Naturvidensk. Kl., n. s. 38: 201–221. Eriksen, B. and M. H. Töpel. 2006. Molecular phylogeography and hybridization in members of the circumpolar *Potentilla* sect. *Niveae* (Rosaceae). Amer. J. Bot. 93: 460–469. Hultén, E. 1945. Studies in the *Potentilla nivea* group. Bot. Not. 1945: 127–148. Soják, J. 1989. Notes on *Potentilla* (Rosaceae). VIII. *P. nivea* L. agg. Candollea 44: 741–762.

1. Epicalyx bractlets usually ½ or less as wide as sepals, margins usually flat; petioles: vestiture either primarily of cottony hairs or of ± stiff verrucose hairs; inflorescences usually more than 1-flowered; central leaflets: distal (½–)¾ to nearly whole length incised less than ½ to midvein, teeth (2 or)3–8(–12) per side.
   2. Petioles: long hairs sparse to abundant, usually stiff, spreading to ± ascending, verrucose, cottony hairs absent; central leaflets usually petiolulate, (petiolules to 5 mm) . . . . . .75. *Potentilla arenosa*
   2. Petioles: long hairs usually absent, sometimes sparse to common, usually soft, usually ± appressed, smooth, cottony hairs usually abundant to dense; central leaflets subsessile or short-petiolulate.
      3. Caudex branches columnar, at least partly sheathed with marcescent whole leaves; central leaflets 0.7–1.1 × 0.5–0.6 cm, teeth 2–4 per side, adaxial surfaces grayish green to grayish white, hairs abundant to dense; styles 1.2–1.4 mm, not or ± papillate-swollen in proximal ⅕ or less; mountains of ec Nevada and adjacent Utah . . . . 74. *Potentilla holmgrenii*
      3. Caudex branches not columnar, not sheathed with marcescent whole leaves; central leaflets 0.5–2(–4) × (0.2–)0.4–1.2(–2) cm, teeth (2 or)3–8(–12) per side, adaxial surfaces green to grayish green, hairs sparse to abundant; styles 0.7–1.2 mm, not or ± to strongly papillate-swollen in less than proximal ⅕, rarely in proximal ⅕–⅓; Alaska to Greenland, south to Arizona and New Mexico.
         4. Epicalyx bractlets linear, 0.2–0.6 mm wide; styles not or ± papillate-swollen in less than proximal ⅕; central leaflets: teeth (3–)5–8(–12) per side, adaxial surfaces green to grayish green; Alaska, Yukon .72. *Potentilla crebridens*
         4. Epicalyx bractlets narrowly to broadly lanceolate or elliptic, 0.6–1.7 mm wide; styles strongly papillate-swollen at very base, rarely in proximal ⅕–⅓; central leaflets: teeth (2 or)3–5(or 6) per side, adaxial surfaces usually green, sometimes grayish green; Alaska to Greenland, south to Arizona, New Mexico . . . . 73. *Potentilla nivea*
1. Epicalyx bractlets (½–)⅔ to ± as wide as sepals, margins often revolute; petioles: vestiture primarily of soft to weak smooth hairs (or stiff verrucose hairs in *P. tikhomirovii*); inflorescences often only 1-flowered; central leaflets: distal (⅓–)½–⅔(–¾) incised (⅓–)½–¾ to midvein (¼–½ to midvein in *P. villosa*), teeth (1 or)2 or 3(or 4) per side (3–7 in *P. villosa*).
   5. Epicalyx bractlets: red glands usually common; petioles: long hairs verrucose; central leaflet abaxial surfaces: cottony-crisped hairs common to dense . . . . .76. *Potentilla tikhomirovii*
   5. Epicalyx bractlets: red glands absent; petioles: long hairs smooth; central leaflet abaxial surfaces: cottony-crisped hairs ± dense.
      6. Carpels: apical hairs sparse to abundant (straight).
         7. Central leaflet surfaces ± dissimilar, adaxial grayish green, hairs abundant to dense; inflorescences (1 or)2 or 3(or 4)-flowered; Alaska and w Yukon, s in Canadian Rockies . . . . 78. *Potentilla villosula* (in part)
         7. Central leaflet surfaces strongly dissimilar, adaxial usually dark green, sometimes grayish green, hairs sparse to abundant; inflorescences 1 or 2(–5) -flowered; Alaska to Nunavut . . . . .79. *Potentilla vulcanicola*
      6. Carpels: apical hairs usually absent, rarely present (cottony).
         8. Inflorescences (1 or)2–7(–10)-flowered; plants ± to densely tufted; caudex branches sometimes ± columnar, not sheathed with marcescent whole leaves; central leaflet teeth (2 or)3–6(or 7) per side, ± approximate to ± distant; leaflets usually ± overlapping, bases mostly broadly cuneate to rounded, adaxial surfaces grayish green, hairs abundant to dense.
            9. Inflorescences (1 or)2–7(–10)-flowered; central leaflets incised ¼–½ to midvein, teeth 3–6(or 7) per side; filaments 1.8–2.1 mm; rocky habitats, coast and coastal mountains from Alaska to Oregon. . . . . .77. *Potentilla villosa*
            9. Inflorescences (1 or)2 or 3(or 4)-flowered; central leaflets incised ± ½ to midvein, teeth 2 or 3(or 4) per side; filaments 1.1–1.4 mm; various habitats, coast and inland from Alaska to w Yukon, s in Canadian Rockies . . . . 78. *Potentilla villosula* (in part)

[8. Shifted to left margin.—Ed.]

8. Inflorescences usually 1- or 2-flowered, rarely to 5-flowered; plants ± densely tufted to cushion-forming; caudex branches usually columnar, often sheathed with marcescent whole leaves; central leaflet teeth (1 or)2 or 3(or 4) per side, usually ± distant; leaflets separate to slightly overlapping, bases cuneate, adaxial surfaces dark green to greenish or yellowish gray, hairs sparse to ± abundant.
   10. Petioles: crisped/short-cottony hairs usually absent, sometimes sparse, long hairs ± weak, rarely stiff; plants usually cushion-forming; basal leaves 0.5–2.5(–3) cm; arctic . . . . 80. *Potentilla subvahliana*
   10. Petioles: crisped/short-cottony hairs usually sparse to abundant, long hairs soft to weak; plants ± densely tufted to cushion-forming; basal leaves 1–10(–15) cm; arctic, subarctic, and alpine.
      11. Epicalyx bractlets elliptic-lanceolate to ovate, (3–)4–6(–7) × (0.8–)1.2–2(–2.5) mm, (½–)⅔ to as wide as sepals; sepals 4–6(–7) mm; petals (5–)6–9 mm, lengths 1.5–1.8 times sepals; Alaska to Nunavut, s to British Columbia, Utah, and Colorado . . . . 81. *Potentilla subgorodkovii*
      11. Epicalyx bractlets broadly ovate, 2.5–4(–5) × 1.5–3 mm, ± as wide as sepals; sepals 2.5–5(–6) mm; petals 8–10 mm, lengths 1.8–2.1 times sepals; ne Canada and Greenland . . . . 82. *Potentilla vahliana*

72. Potentilla crebridens Juzepczuk, Bot. Mater. Gerb. Bot. Inst. Komarova Akad. Nauk S.S.S.R. 17: 218, fig. 2. 1955 [F]

Subspecies 2 (1 in the flora): nw North America, Asia.

72a. Potentilla crebridens Juzepczuk subsp. **hemicryophila** Jurtzev in A. I. Tolmatchew, Fl. Arct. URSS 9(1): 318. 1984 [F]

Potentilla matsuokana Makino subsp. *hemicryophila* (Jurtzev) Soják

Plants ± tufted. **Caudex branches** slender to stout, not columnar, not sheathed with marcescent whole leaves. **Stems** ascending to erect, 0.8–2 dm, lengths 2–3(–4) times basal leaves. **Basal leaves** 3–8 cm; petiole 0.5–5 cm, cottony hairs usually dense, sometimes sparse, other hair types and glands absent or obscured; leaflets ± overlapping, central obovate, 0.6–2 × 0.5–1.2 cm, subsessile, base cuneate, margins slightly revolute, distal ¾ to nearly whole length incised ¼–⅓(–½) to midvein, teeth (3–)5–8(–12) per side, approximate, surfaces ± to strongly dissimilar, abaxial ± white, long hairs 0.7–1.2 mm, cottony-crisped hairs usually dense, adaxial green to grayish green, long hairs sparse to abundant, short and/or crisped hairs sparse to abundant. **Cauline leaves** 0 or 1. **Inflorescences** 1–6 -flowered. **Pedicels** 1–1.5 cm in flower, to 3 cm in fruit. **Flowers:** epicalyx bractlets linear, 3–5 × 0.2–0.6 mm, usually less than ¼ as wide as sepals, margins flat, red glands absent; hypanthium 4–4.5 mm diam.; sepals 5–6 mm, apex acute; petals 5–7 × 5–8 mm, slightly longer than sepals; filaments 1 mm, anthers 0.4 mm; carpels 25–30, apical hairs absent, styles columnar, not or ± papillate-swollen on less than proximal ⅕, 0.9–1.1 mm. **Achenes** 1.1 mm. $2n = 28$ (Russian Far East).

Flowering summer. Dry alpine ridges and meadows, well-drained slopes, rocky outcrops, scree, on calcareous bedrock; 0–1500 m; Yukon; Alaska; Asia (Russian Far East, Siberia).

Nearly half of the Alaska and Yukon (Beringian) specimens previously assigned to *Potentilla nivea* belongs to *P. crebridens* subsp. *hemicryophila*, which is tetraploid; *P. nivea* is high polyploid. Intermediates were not seen, and the two species may be reproductively isolated by the ploidal difference. See B. A. Jurtzev (1984) for additional character differences.

73. Potentilla nivea Linnaeus, Sp. Pl. 1: 499. 1753, name conserved • Snow cinquefoil, potentille des neiges [F]

Potentilla prostrata Rottbøll subsp. *floccosa* Soják

Plants ± tufted. **Caudex branches** stout, not columnar, not sheathed with marcescent whole leaves. **Stems** ascending to erect, (0.3–)0.5–3(–4) dm, lengths 1.5–2.5(–4) times basal leaves. **Basal leaves** (1–)3–10 (–15) cm; petiole (0.5–)1–6(–10) cm, long hairs usually absent, sometimes sparse to common (less so than cottony hairs), ± appressed, 1–2 mm, soft, smooth, short-crisped hairs absent or obscured, cottony hairs abundant to dense <sometimes sparse with age>, glands absent, sparse, or obscured; leaflets overlapping, central obovate, 0.5–2(–4) × (0.2–)0.4–1.2(–2) cm, subsessile, base cuneate, margins slightly revolute, distal ± ¾ incised (¼–)⅓–½ to midvein, teeth (2 or)3–5(or 6) per side, ± approximate, surfaces dissimilar, often strongly so, abaxial ± white, long hairs 0.8–1.2 mm, cottony-crisped hairs dense, adaxial usually green, sometimes grayish green, long hairs sparse to abundant, short-crisped hairs sparse to common. **Cauline leaves** 0 or 1. **Inflorescences** 1–5(–7)-flowered. **Pedicels** 1–4 cm in flower, to 5 cm in fruit. **Flowers:** epicalyx bractlets narrowly to broadly lanceolate or elliptic, (2–)4–7 × 0.6–1.7 mm, usually ¼–½ as wide as sepals, margins flat, red glands usually absent, sometimes sparse, inconspicuous; hypanthium (2–)3–4 mm diam.; sepals (2.5–)4–8 mm, apex acute; petals (3–)4–8 × (3–)5–9 mm, slightly longer than sepals; filaments 0.9–1.2 mm, anthers 0.5 mm; carpels 20–40, apical hairs absent, styles narrowly columnar or columnar-tapered, strongly papillate-swollen at very base, rarely in proximal ⅕–⅓, 0.7–1.2 mm. **Achenes** 1.1–1.5 mm. **2*n*** = 56, 63; 28, 42, 49, 70 (Asia, Europe).

Flowering summer. Well-drained, exposed sites, ridge crests, coarse mineral soil, scree, usually on calcareous substrates; 400–3800 m; Greenland; Alta., B.C., Man., Nfld. and Labr., N.W.T., Nunavut, Que., Yukon; Alaska, Ariz., Colo., Mont., N.Mex., Utah, Wash., Wyo.; Eurasia.

Although now restricted to plants with exclusively (or at least predominantly) cottony hairs on the petioles, the name *Potentilla nivea* has a long history with an often wider application, sometimes including nearly all of sect. *Niveae*. As further confusion, J. Soják (1989) noted that the Linnaean type of *P. nivea* belonged to what is here treated as *P. arenosa*. Although historical usage of *P. nivea* has been re-established as a conserved name with a conserved type (B. Eriksen et al. 1999), from 1989 to 1999 the name *P. nivea* was applied to *P. arenosa*. During this period, *P. prostrata* subsp. *floccosa* was briefly adopted as the correct name for this species (for example, W. J. Cody 1996).

Molecular evidence (B. Eriksen and M. H. Töpel 2006) indicates that populations of *Potentilla nivea* in the Atlantic regions, including Greenland and eastern Canada, differ from those in the Beringian regions of northwestern North America, suggesting expansion from separate Pleistocene refugia. A comparable pattern was noted by R. Elven and S. G. Aiken (2007) based on morphologic characters. The conserved type of *P. nivea* is from northern Sweden (Eriksen et al. 1999) and belongs to the Atlantic morphologic group. The variation within each region is large, and racial recognition would accordingly be premature. Plants from sites south of the continental glaciation, which were not included in the analysis by Eriksen and Töpel, deviate in having acuminate leaflet teeth and epicalyx bractlets and in being generally more slender. Epicalyx bractlets of some Washington plants are nearly as narrow as those of *P. crebridens*.

Additional chromosome numbers have been reported for *Potentilla nivea*, but it is unknown whether these apply to this species, *P. crebridens*, some Asian relative, or hybrids.

74. Potentilla holmgrenii D. F. Murray & Elven, J. Bot. Res. Inst. Texas 1: 811, figs. 1, 2. 2007 • Holmgren's cinquefoil [C] [E]

Plants densely tufted to cushion-forming. **Caudex branches** stout, columnar, at least partly sheathed with marcescent whole leaves. **Stems** ascending, 0.2–0.8 dm, lengths 1.5–2.5 times basal leaves. **Basal leaves** 1.5–4 cm; petiole 0.5–2 cm, long hairs absent or sparse to common (less so than cottony hairs), ± appressed, 1–2 mm, soft, smooth, cottony hairs dense, other hairs and glands absent, sparse, or obscured; leaflets touching to overlapping, central broadly elliptic to obovate, 0.7–1.1 × 0.5–0.6 cm, sessile or short-petiolulate, <petiolule to 1 mm>, base cuneate, margins revolute, distal ½–¾ incised ± ½ to midvein, teeth 2–4 per side, approximate, surfaces ± dissimilar, abaxial white to grayish white, long hairs 1 mm, cottony-crisped hairs dense, adaxial grayish green to grayish white, long and short-crisped hairs abundant to dense. **Cauline leaves** 0 or 1. **Inflorescences** 1 or 2(or 3)-flowered. **Pedicels** 0.4–0.7 cm in flower, to 2 cm in fruit. **Flowers:** epicalyx bractlets lanceolate to narrowly elliptic, 2–2.2 × 0.6–0.9 mm, ¼–½ as wide as sepals, margins flat, red glands absent; hypanthium 1.8–2.2 mm diam.; sepals 3–3.5 mm, apex subacute; petals 4–5 × 2.5–4 mm, longer than sepals; filaments 1–1.5 mm, anthers 0.5 mm; carpels 30–40, apical hairs absent, styles narrowly columnar, not or ± papillate-swollen in proximal ⅕ or less, 1.2–1.4 mm. **Achenes** 1.1–1.4 mm.

Flowering summer. Wind-swept ridges, fellfields, rocky slopes; of conservation concern; 3300–4500 m; Nev., Utah.

Potentilla holmgrenii is known only from the Schell Creek and Snake ranges, Nevada, and Deep Creek Range, Utah. In addition to characteristics given in the key, lateral leaflets and leaflet teeth of *P. holmgrenii* are directed more forward than in *P. nivea*, leaflets are thicker and more densely hairy, and the terminal tooth is often orbiculate and frequently overlapped by the adjacent pair of lateral teeth.

75. Potentilla arenosa (Turczaninow) Juzepczuk in V. L. Komarov et al., Fl. URSS 10: 137. 1941

Potentilla nivea Linnaeus var. *arenosa* Turczaninow, Bull. Soc. Imp. Naturalistes Moscou 16: 607. 1843

Plants scarcely to ± tufted. **Caudex branches** thick, not columnar, not sheathed with marcescent whole leaves. **Stems** ascending to erect, (0.3–) 0.8–2.5(–4.5) dm, lengths (2–)3–5 times basal leaves. **Basal leaves** 1.5–12(–20) cm; petiole 1–7(–15) cm, long hairs sparse to abundant, spreading to ± ascending, rarely loosely appressed, 1–2(–2.5) mm, usually stiff, sometimes weak (subsp. *chamissonis*), verrucose, short and/or crisped hairs absent or sparse to abundant, cottony hairs absent, glands absent or sparse; leaflets separate to ± overlapping, central obovate, 1–3.5(–4.5) × 0.5–2(–3) cm, usually petiolulate, <petiolule to 5 mm>, base cuneate, margins slightly revolute, distal ± ¾ incised ± ½ to midvein, teeth (2 or)3 or 4(–6) per side, ± approximate to distant, surfaces dissimilar, often strongly so, abaxial white to gray, long hairs 0.5–1.8 mm, cottony-crisped hairs ± dense, adaxial green, sometimes grayish green, long hairs sparse to abundant, short-crisped hairs sparse to abundant. **Cauline leaves** (0 or)1 or 2. **Inflorescence** 1–7(–15)-flowered. **Pedicels** 1.5–5 cm in flower, to 6(–10) cm in fruit. **Flowers:** epicalyx bractlets linear-lanceolate to lanceolate-elliptic, 2–5(–7) × 0.4–1.2(–1.5) mm, ¼–½ as wide as sepals, margins usually flat, red glands absent or sparse and inconspicuous; hypanthium 3–5 mm diam.; sepals 3–6(–8) mm, apex acute; petals 4–7(–10) × 4–7(–9) mm, ± longer than sepals; filaments 0.8–1 mm, anthers 0.4 mm; carpels 28–40, apical hairs absent, styles conic-columnar, strongly papillate-swollen in proximal ⅕–⅓, 1–1.5 mm. **Achenes** 1.1 mm.

Subspecies 2 (2 in the flora): North America, Eurasia.

The name *Potentilla arenosa* is now used for most arctic and subarctic plants previously treated as *P. hookeriana* or *P. nivea* subsp. *hookeriana* (Lehmann) Hiitonen. As noted by J. Soják (1986), the type of *P. hookeriana* has quinate leaves; that name is now restricted to a Rocky Mountain species in sect. *Rubricaules*. The arctic and subarctic material was briefly (1989–1999) called *P. nivea*, as discussed under that species. Because the type of *P. nivea* var. *arenosa* and other northern Asian specimens correspond closely to the North American plants, the name *P. arenosa* is assigned here.

The two subspecies differ only in one character, the petiole hairs, but are largely allopatric. Subspecies *arenosa* occurs in western and northern Greenland, northern North America (very northern in the east), and northern Asia (and perhaps northeasternmost European Russia); subsp. *chamissonis* occurs in southern Greenland, northeastern North America (more southern than subsp. *arenosa*), and northern Europe at least east to the Urals.

1. Petioles with common to abundant short and/or stiff crisped hairs in addition to long verrucose hairs . . . . 75a. *Potentilla arenosa* subsp. *arenosa*
1. Petioles with sparse or no short and/or soft crisped hairs in addition to long verrucose hairs . . . . 75b. *Potentilla arenosa* subsp. *chamissonis*

75a. Potentilla arenosa (Turczaninow) Juzepczuk subsp. **arenosa**

Petioles with both long verrucose hairs and a lower layer of common to abundant short and/or stiff crisped hairs. **2*n*** = 28; 42, 49, 56, 70 (Asia).

Flowering spring–summer. Well-drained gravel, ridge tops, cliffs, dry soil banks, dry tundra; 0–1400 m; Greenland; Alta., B.C., Man., N.W.T., Nunavut, Sask., Yukon; Alaska; Asia (Siberia).

Plants of subsp. *arenosa* from Alaska and Yukon tend to be larger, with more open inflorescences and more distinctly petiolulate central leaflets, than plants in north-central Canada. Some collections from the range of *Potentilla furcata* (sect. *Rubricaules*) have red glands in the inflorescence, suggesting introgression between *P. arenosa* and *P. furcata*. Reported occurrences in Alberta and Saskatchewan (and the upper altitudinal limit given here) might actually be *P. hookeriana* as now circumscribed.

75b. Potentilla arenosa (Turczaninow) Juzepczuk subsp. **chamissonis** (Hultén) Elven & D. F. Murray, J. Bot. Res. Inst. Texas 1: 813. 2007 • Potentille de Chamisso

Potentilla chamissonis Hultén, Bot. Not. 1945: 140, fig. 5b. 1945; *P. hookeriana* Lehmann subsp. *chamissonis* (Hultén) Hultén; *P. nivea* Linnaeus subsp. *chamissonis* (Hultén) Hiitonen

Petioles with only or almost only long verrucose hairs, sometimes mixed with sparse short and/or soft crisped hairs.

Flowering summer. Cliffs, dry heath tundra, ridge tops; 0–1000 m; Greenland; Man., N.W.T., Nunavut, Que.; Europe.

Reports of subsp. *chamissonis* from Alaska or Yukon (E. Hultén 1968) have not been confirmed; they are unlikely.

The current use of subsp. *chamissonis* differs from that of J. Soják (1989, pers. comm.), who included *Potentilla chamissonis* among the hybrids of *P. arenosa* and *P. nivea*. He instead provided the combination *P. nivea* var. *nipharga* (Rydberg) Soják for the taxon currently under discussion, using *P. nivea* in its unconserved sense. However, he also noted that *P. nipharga* Rydberg might instead be based on an unstabilized hybrid between *P. arenosa* and *P. crebridens* or *P. nivea*. The latter interpretation is provisionally accepted here; if treated as conspecific with *P. arenosa*, then *P. nipharga* would have priority. The name has traditionally been applied in Greenland and northeastern Canada; the type from the Mackenzie River does not conform with how the name has been used.

76. Potentilla tikhomirovii Jurtzev in A. I. Tolmatchew, Fl. Arct. URSS 9(1): 318. 1984

Plants ± tufted. **Caudex branches** slender to stout, not columnar, not sheathed with marcescent whole leaves. **Stems** ± ascending, sometimes nearly erect, 0.5–1.5(–3) dm, lengths 2–4 times basal leaves. **Basal leaves** 2–4 cm; petiole 1–2.5 cm, long hairs common to abundant, spreading to ascending, rarely loosely appressed, 1–2 mm, ± stiff, verrucose, short-crisped hairs sparse to abundant, sometimes nearly absent, cottony hairs absent, glands sparse; leaflets ± overlapping, central elliptic to obovate, 1–1.5 × 0.6–1.1 cm, petiolulate, base broadly cuneate, margins flat or slightly revolute, distal ± 2/3 incised ± 1/2 to midvein, teeth (3 or)4 per side, ± approximate, surfaces ± dissimilar, abaxial reddish or greenish gray to white, long hairs 1 mm, cottony-crisped hairs common to dense, adaxial green or reddish to gray-green, long hairs sparse to abundant, short-crisped hairs absent or sparse to common. **Cauline leaves** 0 or 1. **Inflorescences** 2–5-flowered. **Pedicels** 1–2 cm in flower, to 3(–5) cm in fruit. **Flowers:** epicalyx bractlets narrowly elliptic to narrowly ovate, 4–6 × 1–2.5 mm, (1/2–)2/3 to as wide as sepals, margins revolute, sometimes flat, red glands usually common, conspicuous; hypanthium 3.5–4.5 mm diam.; sepals 5–7 mm, apex subacute; petals 6–8 × (4–)5–8 mm, longer than sepals; filaments 0.7–1.2 mm, anthers 0.4 mm; carpels 40–60, apical hairs absent, styles columnar, strongly papillate-swollen in proximal 1/5–1/3, 0.9–1 mm. **Achenes** 1.2–1.8 mm. **2*n*** = 42, 49 (Russian Far East).

Flowering summer. Dry tundra meadows, loamy soil banks, sedge-herb slopes; 0–1100 m; Greenland; N.W.T., Nunavut; Alaska; Europe (Russia); Asia.

Potentilla tikhomirovii is an assumed hybrid species combining characteristics from *P. arenosa* subsp. *arenosa* of sect. *Niveae* (abaxial leaflet surfaces with crisped-cottony hairs, adaxial surfaces with both long and short hairs, petioles often with both long and short hairs, and flower shape and size) and *P. hyparctica* of sect. *Aureae* (crisped-cottony hairs often not fully covering abaxial leaf surfaces, reddish color in most plant parts, sepals and epicalyx bractlets relatively large, broad, less acute, and with many and reddish glands, which are often also on petioles and stipules).

Potentilla tikhomirovii has an interrupted range and is not very coherent morphologically. It forms large populations locally, probably by agamic seed propagation, and is a significant part of the *Potentilla* variation in northern and northeastern Greenland, Ellesmere Island, and parts of the Brooks Range, northern Alaska. Both the distributional and the morphologic patterns indicate that it has evolved multiple times in different regions, such that it is a borderline case for status as an independent species. Its large populations and significant distribution in some regions support recognition, as also does the paucity of obvious back-crosses with the parents.

77. Potentilla villosa Pallas ex Pursh, Fl. Amer. Sept. 1: 353. 1813 • Villous cinquefoil [F]

Potentilla villosa var. *parviflora* C. L. Hitchcock

Plants ± to densely tufted. **Caudex branches** stout, sometimes short-columnar, not sheathed with marcescent whole leaves. **Stems** ascending, 0.5–2(–2.5) dm, lengths 1.5–3 times basal leaves. **Basal leaves** 2–12(–15) cm; petiole 1–9(–12) cm, long hairs ± abundant to dense, spreading to ascending, 1–2.5(–3) mm, soft to weak, smooth, crisped hairs absent or sparse, cottony hairs absent, glands sparse to common, sometimes obscured; leaflets usually ± overlapping, central broadly obovate to suborbiculate, (0.5–)1.5–2.5(–3) × (0.5–)1.5–2.6(–3.2) cm, sessile to subsessile, base cuneate to rounded, margins revolute, distal 1/2–2/3(–3/4) incised 1/4–1/2 to midvein, teeth 3–6(or 7) per side, ± approximate to ± distant, surfaces ± dissimilar, abaxial yellowish or grayish white, long hairs 1–2 mm, cottony-crisped hairs ± dense, adaxial grayish green, long hairs abundant to dense, crisped hairs absent, sparse, or obscured. **Cauline leaves** (0 or)1 or 2. **Inflorescences** (1 or)2–7(–10)-flowered. **Pedicels** 0.5–3(–4) cm in flower, to 4 cm in fruit. **Flowers:** epicalyx bractlets ovate to oval-elliptic, 3–8 × 2–5 mm, 2/3 to as wide as sepals, margins strongly revolute, red glands absent; hypanthium 5–7 mm diam.; sepals 4–8 mm, apex ± acute; petals (5–)7–15 × 7–16 mm, significantly longer than sepals; filaments 1.8–2.1 mm, anthers 0.7–0.8 mm; carpels 150–250, apical hairs usually absent, rarely present (cottony), styles narrowly conic to tapered, ± papillate-swollen on proximal 1/5–1/2, 0.9–1.1 mm. **Achenes** 0.9–1.3 mm. **2*n*** = 14.

Flowering spring–summer. Sea cliffs, gravel beaches, inland on alpine outcrops (primarily southern populations), pumice barrens, scree and rock ledges; 0–100(–1800) m; B.C.; Alaska, Oreg., Wash.; e Asia (Russian Far East).

Potentilla villosa is the only documented diploid species of sect. *Niveae* in North America and, almost certainly, it is fully sexual. It is a characteristic species of coarse-grained beaches and coastal cliffs from southwestern British Columbia to western Alaska and the Russian Far East, and also occurs on scattered mountains in the Olympic Peninsula and Cascade Range of Washington and Oregon (providing the higher elevational extreme). These southern populations, which have been called var. *parviflora*, tend to be smaller, more delicate, less hairy plants in general, with fewer and smaller flowers. At least some plants in some southern populations also have cottony hairs on the carpels.

78. Potentilla villosula Jurtzev in A. I. Tolmatchew, Fl. Arct. URSS 9(1): 319. 1984

Potentilla villosula subsp. *congesta* Jurtzev

Plants densely tufted. **Caudex branches** stout, sometimes ± columnar, not sheathed with marcescent whole leaves. **Stems** ascending to erect, 0.2–1.5(–2) dm, lengths 1.5–3(–4) times basal leaves. **Basal leaves** 1–5.5(–7) cm; petiole 0.5–3.5(–5) cm, long hairs common to dense, ascending to spreading, loosely appressed, sometimes retrorse, 1–2(–3) mm, soft, smooth, crisped/short-cottony hairs usually sparse, sometimes common, glands absent or sparse to common or obscured; leaflets overlapping, central broadly obovate to obtriangular, 0.8–2.5 × 0.6–2 cm, sessile to subsessile, base broadly cuneate, margins revolute, distal ½–⅔(–¾) incised ± ½ to midvein, teeth 2 or 3(or 4) per side, ± approximate to ± distant, surfaces ± dissimilar, abaxial grayish to white or yellowish white, long hairs 1.5–2.5 mm, cottony-crisped hairs ± dense, adaxial grayish green, long hairs abundant to dense, crisped hairs absent, sparse, or obscured. **Cauline leaves** (0 or) 1 or 2(or 3). **Inflorescences** (1 or)2 or 3(or 4)-flowered. **Pedicels** 0.5–3(–5) cm in flower, to 4(–6) cm in fruit. **Flowers:** epicalyx bractlets broadly lanceolate to narrowly ovate, 3–7(–8) × 1.5–3(–3.5) mm, ⅔ to as wide as sepals, margins ± revolute, red glands absent; hypanthium (3–)4–6 mm diam.; sepals 4–7(–8) mm, apex acute or rarely acuminate; petals 5–10 × 6–12 mm, significantly longer than sepals; filaments 1.1–1.4 mm, anthers 0.5–0.8 mm; carpels 40–70, apical hairs absent or sparse (straight), styles narrowly columnar to conic-tapered, papillate-swollen in proximal ⅕ or less, 1–1.2 mm. **Achenes** 0.9–2 mm. **2*n*** = 28 (Russian Far East).

Flowering late spring–summer. Rocky alpine heaths, outcrops, scree and talus, gravel outwash plains, dry tundra, coastal bluffs, stabilized sand dunes, mostly on acidic bedrock; 0–2900 m; Alta., B.C., Yukon; Alaska; e Asia (Russian Far East).

Potentilla villosula was included previously in *P. villosa*; the two overlap in the southern and western Alaskan coasts, where a transition is suspected to be associated with differing ploidy levels. The plants from this area with silky hairs and multiflowered inflorescences are assigned to *P. villosa*; *P. villosula* has thicker, stiffer, and less silky hairs, fewer teeth per leaflet, narrower bracts, fewer and smaller flowers with narrower epicalyx bractlets and sepals, and fewer achenes (sometimes with single apical hairs). While *P. villosa* is restricted to relatively rocky sites near the coast from western Alaska to Oregon, *P. villosula* often occurs on coastal sand dunes, farther inland, and/or farther to the north.

As circumscribed, *Potentilla villosula* is a major plant of Alaska, especially of the Bering Sea region. The species extends from Alaska and the Yukon to at least central British Columbia. Most of the range given by E. Hultén (1968) for *P. villosa* belongs to *P. villosula*. W. J. Cody (1996) accepted the name *P. villosula* for a common south-central Yukon plant, but the majority of the plants mapped by Cody probably belong to *P. subgorodkovii* (although the distinction between the two species needs more resolution). Provisionally included here are plants from southern and central Yukon south through the Canadian Rockies, possibly including the type of *P. nivea* subsp. *fallax* A. E. Porsild. Such plants tend to be significantly smaller overall than Alaskan *P. villosula*; at least some have straight hairs on carpel apices.

J. Soják (2004) believed that *Potentilla villosula* evolved from crosses between *P. villosa* and *P. vulcanicola*. The affinity with *P. villosa* is seen in the sericeous vestiture, number of leaflet teeth, and flower number and size, and with *P. vulcanicola* in the frequent occurrence of straight hairs on carpel apices in both species. Plants with columnar caudex branches, representing a significant part of the Alaskan material, have been called subsp. *congesta*. This morphology results from the persistence of marcescent whole leaves for several years (in typical *P. villosula* only sheaths and petioles are retained). This character suggests that *P. subvahliana*, with which the form called subsp. *congesta* is sympatric, may be part of its parentage. *Potentilla villosula* is accepted here in a fairly collective sense, possibly including several hybrid lineages, but with *P. villosa* a part of them all.

79. Potentilla vulcanicola Juzepczuk, Bot. Mater. Gerb. Bot. Inst. Komarova Akad. Nauk S.S.S.R. 17: 222, fig. 3. 1955

Plants densely tufted. **Caudex branches** stout, sometimes ± columnar, not sheathed with marcescent whole leaves. **Stems** ascending to erect, (0.2–)0.4–1.6(–2) dm, lengths 2–3 times basal leaves. **Basal leaves** 1–6(–11) cm; petiole 0.5–4(–9) cm, long hairs ± abundant, ascending to spreading, sometimes loosely appressed, 1–2.5 mm, soft to weak, smooth, crisped(/cottony) hairs absent or sparse, sometimes common, glands absent or sparse; leaflets overlapping, central broadly obovate, 0.8–2(–2.5) × 0.8–1.5(–2.5) cm, sessile or subsessile, base broadly cuneate, margins strongly revolute, distal ½–⅔ incised ± ½ to midvein, teeth 2 or 3(or 4) per side, distant, surfaces strongly dissimilar, abaxial grayish white, long hairs 1–2 mm, cottony-crisped hairs ± dense, adaxial usually dark green, sometimes grayish green, long hairs sparse to abundant, other hairs usually absent. **Cauline leaves** (0 or)1 or 2(or 3). **Inflorescences** 1 or 2(–5)-flowered. **Pedicels** 1–2 cm in flower, to 5 cm in fruit. **Flowers:** epicalyx bractlets ± oblong to ovate, 4–8(–10) × 1.8–3(–5) mm, ⅔ to as wide as sepals, margins revolute, sometimes flat, red glands absent; hypanthium 3.5–4.5 mm diam.; sepals 5–8 mm, apex ± acute; petals 7–9 × 7–10 mm, longer than sepals; filaments 0.8–1 mm, anthers 0.6–0.7 mm; carpels 20–30, apical hairs sparse to abundant (straight), styles conic-columnar, ± papillate-swollen in less than proximal ⅕, or sometimes to ⅕, 0.8–1.1 mm. **Achenes** 0.9–1.2 mm. $2n$ = 28 (Russian Far East).

Flowering late spring–summer. Dry alpine heaths, ridge crests, rock outcrops, herb slopes, scree and talus, stabilized sand dunes, mainly on calcareous bedrock; 0–3100 m; N.W.T., Nunavut, Yukon; Alaska; e Asia (Russian Far East).

Potentilla vulcanicola is close morphologically to *P. uniflora* in the narrow sense and largely replaces it in southern and eastern Chukotka, Alaska, and northwestern Canada, where it reaches east of Mackenzie River on the coastal mainland and the westernmost islands (Banks and Victoria islands). *Potentilla subgorodkovii* and *P. vulcanicola* account for a major part of North American records of *P. uniflora*.

80. Potentilla subvahliana Jurtzev in A. I. Tolmatchew, Fl. Arct. URSS 9(1): 319. 1984 [F]

Plants usually cushion-forming. **Caudex branches** stout, columnar, sheathed with marcescent whole leaves. **Stems** erect, 0.2–0.6(–1.1) dm, lengths 2–4 times basal leaves. **Basal leaves** 0.5–2.5(–3) cm; petiole 0.3–1.5(–2) cm, long hairs common to abundant, spreading to ascending, 1(–2) mm, ± weak, rarely stiff, smooth, crisped/short-cottony hairs usually absent, sometimes sparse, glands sparse to common; leaflets separate to slightly overlapping, central obovate, 0.5–1.2(–1.5) × 0.4–0.9(–1) cm, subsessile to short-petiolulate, base cuneate, margins revolute, distal ½ incised ½–¾ to midvein, teeth (1 or)2 or 3(or 4) per side, ± distant, surfaces ± dissimilar, abaxial grayish or yellowish white, long hairs 0.5–1 mm, cottony-crisped hairs ± dense (often obscured by long hairs), adaxial dark green to greenish gray, long hairs ± abundant, other hairs absent. **Cauline leaves** (0 or)1(or 2). **Inflorescences** 1(or 2)-flowered. **Pedicels** 1–2(–3) cm in flower, to 5.5 cm in fruit. **Flowers:** epicalyx bractlets lanceolate to elliptic-ovate, rarely linear, 3–6(–7) × 0.8–2(–2.3) mm, (½–)⅔ to nearly as wide as sepals, margins flat or ± revolute, red glands absent; hypanthium 2.5–4 mm diam.; sepals 3–5(–6) mm, apex obtuse to subacute; petals 4–8(–9) × 4–9 mm, longer than sepals; filaments 1–1.3 mm, anthers 0.3–0.5 mm; carpels 25–35, apical hairs absent, styles conic-columnar, ± papillate-swollen on less than proximal ⅕, 0.9–1.1 mm. **Achenes** 1.2–1.6 mm. $2n$ = 28 (Russian Far East).

Flowering late spring–summer. Rocky alpine heaths, rock outcrops and crevices, scree and talus, dry tundra, coastal bluffs, stabilized sand dunes, usually on calcareous bedrock; 0–1700 m; Greenland; Alta., B.C., N.W.T., Nunavut, Que., Yukon; Alaska; e Asia (Russian Far East).

The major part of the range assigned to *Potentilla vahliana* by E. Hultén (1968) belongs to *P. subvahliana*.

Two morphologic types are present within what is accepted here as *Potentilla subvahliana*. Plants corresponding to the type of *P. subvahliana* (Wrangel Island) are widespread throughout northeastern Asia (Chukotka) and arctic North America to northwestern Greenland. Some plants in alpine Alaska, Yukon, and east to Amundsen Gulf, Nunavut, are redder, more densely columnar, and less hairy, with smaller leaves having fewer and narrower lobes, more slender one-flowered stems, narrow and entire bracts, narrower sepals, and much narrower epicalyx bractlets. *Potentilla subvahliana* is tetraploid, perhaps an allopolyploid, and could consist of lineages from different parental combinations.

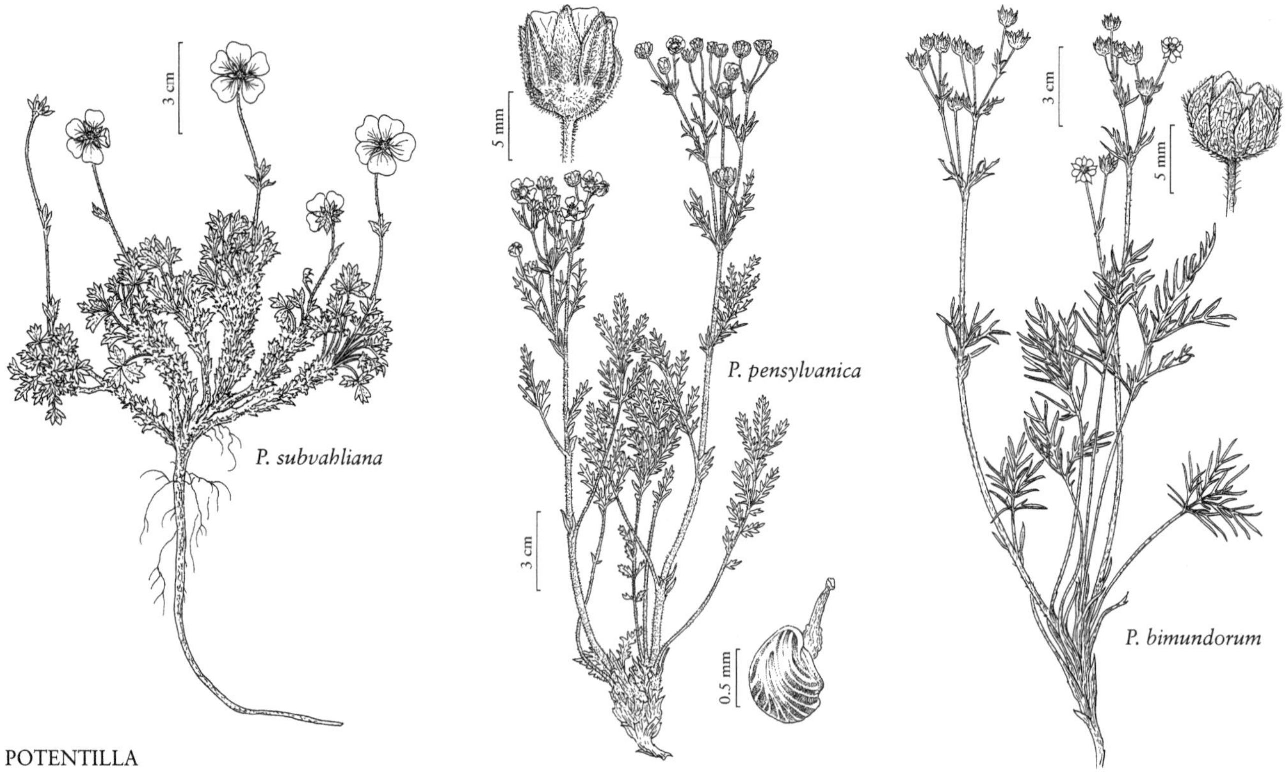

POTENTILLA

81. Potentilla subgorodkovii Jurtzev, Bot. Zhurn. (Moscow & Leningrad) 78(11): 83. 1993

Plants ± densely tufted. **Caudex branches** stout, usually columnar, sometimes sheathed with marcescent whole leaves. **Stems** erect, (0.2–)0.3–1.5(–2) dm, lengths 1.5–2.5(–3.5) times basal leaves. **Basal leaves** (1–)2–10(–15) cm; petiole (0.5–)1–5(–12) cm, long hairs common to dense, ± ascending to loosely appressed, sometimes spreading, 1–2 mm, ± soft, smooth, crisped/short-cottony hairs usually sparse, sometimes absent or common, glands absent, sparse, or obscured; leaflets separate to slightly overlapping, central obovate or obtriangular, (0.5–)1–2(–3) × (0.4–)0.8–1.5(–2) cm, sessile or subsessile, base cuneate, margins revolute, distal (⅓–)½–⅔(–¾) incised ½–⅔(–¾) to midvein, teeth 2 or 3(or 4) per side, usually ± distant, surfaces somewhat to more often strongly dissimilar, abaxial yellowish or grayish white to white, long hairs 0.8–1.5 mm, cottony-crisped hairs ± dense, adaxial green to grayish green, long hairs sparse to ± abundant, other hairs usually absent. **Cauline leaves** (0 or)1(or 2). **Inflorescences** usually 1(or 2)-flowered, rarely to 5-flowered. **Pedicels** (0.5–)2–4 cm in flower, to 5 cm in fruit. **Flowers:** epicalyx bractlets elliptic-lanceolate to ovate, (3–)4–6(–7) × (0.8–)1.2–2(–2.5) mm, (½–)⅔ to as wide as sepals, margins revolute, rarely flat, red glands absent; hypanthium 2.5–4 mm diam.; sepals 4–6(–7) mm, apex subacute; petals (5–)6–9 × (5–)7–9 mm, significantly longer than sepals; filaments 1–1.3 mm, anthers 0.4–0.6 mm; carpels 30–40, apical hairs absent, styles narrowly columnar to conic-tapered, papillate-swollen on proximal ⅕(–⅓), 0.9–1.1 mm. **Achenes** 1.2–2 mm. **2*n*** = 28, 42, 49, 56 (Russian Far East).

Flowering late spring–summer. Dry alpine heaths, exposed ridges and summits, rock outcrops, scree and talus, dry tundra, acidic and calcareous bedrock; 0–4300 m; Alta., B.C., N.W.T., Nunavut, Yukon; Alaska, Colo., Mont., Utah, Wyo.; Asia (Russian Far East, Sakha [Yakutia]).

The name *Potentilla subgorodkovii* was coined for a presumed hybrid species resulting from *P. crebridens* × *P. subvahliana*; here it is applied in a collective meaning for plants combining characteristics from multiple species of the *P. uniflora*/*villosa* and *P. nivea* groups. Morphologically, this collective entity is much closer to the *P. uniflora*/*villosa* group than to the *P. nivea* group and is clearly different from primary hybrids and clones, which are often observed.

Potentilla subgorodkovii constitutes about half of what previously has been considered *P. uniflora* in Alaska and Yukon and is the only such entity fully confirmed south of northern British Columbia. These southern populations occur outside the range of possible parental members of the *P. uniflora*/*villosa* group and do not support the hybrid hypothesis. They may instead represent a distinct species, not yet described.

82. Potentilla vahliana Lehmann, Monogr. Potentill., 172. 1820 • Vahl's cinquefoil, potentille de Vahl [E]

Potentilla hirsuta Vahl ex Hornemann in G. C. Oeder et al., Fl. Dan. 8(22): plate 1390. 1810, not Michaux 1803

Plants densely tufted to cushion-forming. **Caudex branches** usually stout, columnar, sheathed with marcescent whole leaves. **Stems** erect, (0.2–)0.3–0.8(–1) dm, lengths 2–4(–5) times basal leaves. **Basal leaves** 1–3 cm; petiole 0.5–2 cm, long hairs common to abundant, ascending to spreading, 0.5–1.5(–2) mm, soft to weak, smooth, short hairs absent, crisped/short-cottony hairs sparse to abundant, glands sparse to common; leaflets separate to slightly overlapping, central obovate, 0.8–1.2(–1.5) × 0.5–1(–1.5) cm, sessile or subsessile, base cuneate, margins revolute, distal ± ½ incised (⅓–)½–¾ to midvein, teeth 2 or 3 per side, ± approximate to distant, surfaces ± dissimilar, abaxial yellowish white or grayish, long hairs 1–2 mm, cottony-crisped hairs ± dense, adaxial green to yellowish gray, long(/short) hairs common to abundant, crisped hairs absent or sparse. **Cauline leaves** 0 or 1(or 2). **Inflorescences** 1 or 2(or 3)-flowered. **Pedicels** 0.5–1.5 cm in flower, to 3.5 cm in fruit. **Flowers:** epicalyx bractlets broadly ovate, 2.5–4(–5) × 1.5–3 mm, ± as wide as sepals, margins flat or revolute, red glands absent; hypanthium 2–4 mm diam.; sepals 2.5–5(–6) mm, apex subacute to acute; petals 8–10 × 7–12 mm, significantly longer than sepals; filaments 1–1.2 mm, anthers 0.4–0.6 mm; carpels 30–40, apical hairs absent, styles narrowly columnar, distinctly papillate-swollen at very base, 1–1.1 mm. **Achenes** 1.1–1.5 mm. $2n$ = 42, 49, 56.

Flowering summer. Gravelly dry tundra, dry ridges, gravel flats, herb slopes, acidic and calcareous bedrock; 0–700 m; Greenland; Nunavut, Que.

As circumscribed here, *Potentilla vahliana* is confined to northeastern arctic Canada (including northernmost Quebec) and western Greenland. Plants previously assigned to *P. vahliana* from northwestern Greenland, Yukon, most of arctic Canada, and Alaska are now relegated to *P. subvahliana*.

Plants from western Greenland, whence *Potentilla vahliana* was described and typified, have petioles with at least some short cottony or crisped hairs in addition to long, smooth hairs, and the styles have distinct papillae at the very base. These features are derived from a combination of the *P. uniflora* Ledebour and *P. nivea* groups. *Potentilla vahliana* is of a higher polyploidy than the species of the *P. uniflora* group. *Potentilla vahliana* is treated here as a geographically restricted species, most probably from cross(es) between *P. nivea* in the narrow sense (probably the Atlantic race, often $2n$ = 70) and *P. subvahliana* ($2n$ = 28); it shares several features with the latter, not least of which is leaf dissection and the columnar structure of the caudex.

In mainland Quebec, plants combine features from *Potentilla arenosa* and *P. vahliana*; that is, the straight hairs are slightly verrucose, not smooth as in *P. nivea*, *P. subvahliana*, and *P. vahliana* in the narrow sense. They form populations and probably should be described as a new hybrid species.

8t. POTENTILLA Linnaeus sect. RUBRICAULES (Rydberg) A. Nelson in J. M. Coulter and A. Nelson, New Man. Bot. Rocky Mt., 255. 1909

Reidar Elven

Barbara Ertter

Potentilla [unranked] *Rubricaules* Rydberg, Monogr. N. Amer. Potentilleae, 30. 1898; *Potentilla* ser. *Hookerianae* Soják

Perennials, tufted to ± matted, not stoloniferous; taproots not fleshy-thickened; vestiture of long, short, crisped, and/or cottony hairs, glands sparse to abundant, sometimes red. **Stems** decumbent to erect, not flagelliform, not rooting at nodes, lateral to persistent basal rosettes, (0.2–)0.5–3(–4) dm, lengths (1–)1.5–4(–6) times basal leaves. **Leaves:** basal not in ranks; cauline 0–3; primary leaves usually palmate to subpalmate, sometimes some (rarely all) ternate, 1–10 cm;

petiole: long hairs soft to stiff, ± appressed to spreading, glands sparse to abundant, sometimes absent; leaflets 3–5(–7) (usually more than 3 on at least some leaves), usually at tip, sometimes to distal ⅙(–¼) of leaf axis, usually ± overlapping, oblanceolate to obovate, oblong, or elliptic, margins slightly to strongly revolute, distal ⅔ to whole length evenly incised ½–¾+ to midvein, teeth 2–8 per side, surfaces ± to strongly dissimilar, abaxial usually grayish white to white, sometimes gray, cottony (or cottony-crisped) hairs abundant to dense, adaxial green to gray, not glaucous, long hairs soft to stiff. **Inflorescences** 1–12(–20)-flowered, ± cymose, congested to open, sometimes solitary flowers. **Pedicels** ± straight in fruit, 0.2–2(–5) cm, proximal often much longer than distal. **Flowers** 5-merous; hypanthium 2.5–6 mm diam.; petals pale yellow or yellow, usually ± obcordate, (2–)3–7 mm, equal to or distinctly longer than sepals, apex usually ± retuse; stamens ca. 20; styles subapical, ± columnar to tapered, papillate-swollen in proximal ⅕–½, rarely to nearly whole length, 0.7–1.2(–1.5 in *P. paucijuga*) mm. **Achenes** smooth to faintly rugose.

Species 35–40 (8 in the flora): North America, Eurasia.

Section *Rubricaules* is used here for independently propagating species that have at least some palmate to subpalmate basal leaves and styles less than 1.5 mm that are thickened and glandular-papillate basally. Such species purportedly originated as intersectional hybrids between ternate sect. *Niveae* and pinnate to subpalmate sect. *Pensylvanicae* (J. Soják 1986). The North American representatives of the section essentially coincide with *Potentilla rubricaulis* in the broadest sense. Although some early annotations and treatments by the authors (for example, R. Elven and S. G. Aiken 2007) reflect an inclusive *P. rubricaulis*, the current treatment recognizes several reasonably distinct species, based on characters and geographic ranges as presented here. These are only the best-defined elements, with further investigation needed to resolve poorly understood variants and transition zones, especially in western Canada. One additional segregated element, *P. saximontana*, is placed here in sect. *Subjugae*. Minor elements reported from North America but not otherwise addressed in the current treatment include *P. borealis* Soják, *P. insularis* Soják, *P. murrayi* Jurtzev, *P. petrovskyi* Soják, and *P. psychrophila* Soják.

The name *Potentilla nivea* var. *pentaphylla* Turczaninow (mistakenly attributed to Lehmann) and its replacement name *P. quinquefolia* Rydberg, although commonly used as a synonym of *P. rubricaulis*, do not apply to North American plants (B. Ertter et al. 2013). Independent of the name itself, references to *P. quinquefolia* from Oregon (C. L. Hitchcock and A. Cronquist 1961b) are based on specimens of *P. gracilis* var. *owyheensis*; the voucher for *P. quinquefolia* in Washington has been reidentified as *P. argentea*.

Potentilla nivea var. *subquinata* Lange [= *P. subquinata* (Lange) Rydberg, *P. nivea* subsp. *subquinata* (Lange) Hultén] is another name often used for members of this group. As discussed under sect. *Niveae*, however, the lectotype is here interpreted as the casual hybrid of *P. arenosa* and *P. nivea*. Plants of otherwise trifoliolate species in sect. *Niveae* will sometimes produce supernumerary leaflets; such plants might key to sect. *Rubricaules* and should be evaluated in a population context.

Further discussion and synonymy of all these species, and the section in general, are provided elsewhere (B. Ertter et al. 2013).

1. Inflorescences congested or ± elongating in fruit; pedicels 0.2–0.7(–1) cm, proximal to 1.5 cm; basal leaves usually palmate or subpalmate, rarely ternate on same plant.
    2. Leaves subpalmate, proximalmost leaflets separated by 1–5 mm; styles 1.2–1.5 mm; petals (4–)5–6 × 4.5–5.5 mm; hypanthia 3.5–5 mm diam.; La Sal Mountains, Utah . . . . 88. *Potentilla paucijuga*
    2. Leaves usually palmate, rarely subpalmate or ternate, proximalmost leaflets separated by 0–2 mm; styles 1–1.2 mm; petals 2–5 × 2–4 mm; hypanthia 3–4 mm diam.; Rocky Mountains to California.

3. Petioles: long hairs weak to ± stiff, ± ascending to almost spreading, 1–2 mm; sepals: glands abundant, usually not obscured; leaflets incised ½–¾ to midvein, teeth (1–)2–5 mm; Idaho and Montana to Utah and Colorado. . . . . 86. *Potentilla modesta*
3. Petioles: long hairs stiff, ± appressed, 0.5–1.2 mm; sepals: glands sparse, obscured; leaflets incised ¾+ to midvein, teeth (2–)3–6(–10) mm; California . . . . . . 87. *Potentilla pseudosericea*

1. Inflorescences ± open, sometimes solitary flowers; pedicels (0.5–)1–3 cm, proximal to 5 cm; basal leaves often both ternate and palmate or subpalmate on same plant.
   4. Inflorescence branch angles 5–30(–50)°; petals 3–5(–6) × 3–4(–5) mm, slightly longer than sepals, not overlapping; ne British Columbia, Yukon, interior Alaska . . . . 85. *Potentilla furcata*
   4. Inflorescence branch angles (10–)20–50°; petals 3–7 × 4–8(–9) mm, usually distinctly longer than sepals, sometimes overlapping; Alaska to Greenland, s to Nevada and Colorado.
      5. Petioles: long hairs ± weak to stiff, 0.5–1.5(–2) mm; adaxial leaflet surfaces: long hairs usually stiff; leaflet teeth: apical tufts 0.5–1 mm; proximalmost leaflets separated by 0(–1) mm; mostly subarctic and alpine; w North America.
         6. Stems 1.5–4 dm; basal leaflet teeth (4 or)5–8 per side; inflorescences 4–20 -flowered; hypanthia 4–6 mm diam.; ± subarctic regions of nw Canada, Alaska . . . . 83. *Potentilla rubricaulis*
         6. Stems (0.3–)0.5–2 dm; basal leaflet teeth 2–6 per side; inflorescences 1–6(–8) -flowered; hypanthia 3–4 mm diam.; mostly alpine regions, Rocky Mountains to eastern Great Basin . . . . 84. *Potentilla hookeriana*
      5. Petioles: long hairs soft to ± stiff, 1–2.5 mm; adaxial leaflet surfaces: long hairs soft to ± weak; leaflet teeth: apical tufts 1–1.5 mm; proximalmost leaflets separated by 0–2 mm; arctic, n Canada, Greenland.
         7. Inflorescences (1–)3–7-flowered; petioles: long hairs 1–2 mm, weak to ± stiff, verrucose; leaflet teeth: apical tufts ± 1 mm; adaxial leaflet surfaces: short (short-crisped) hairs absent or sparse, rarely common, cottony hairs absent; caudex branches not sheathed with marcescent whole leaves; petals usually not overlapping . . . . 89. *Potentilla pedersenii*
         7. Inflorescences 1–3(or 4)-flowered; petioles: long hairs (1–)1.5–2.5 mm, soft to ± weak, smooth; leaflet teeth: apical tufts 1–1.5 mm; adaxial leaflet surfaces: short/crisped/cottony hairs common to abundant; caudex branches often sheathed with marcescent whole leaves; petals often overlapping . . . . . . 90. *Potentilla uschakovii*

83. Potentilla rubricaulis Lehmann, Nov. Stirp. Pug. 2: 11. 1830 • Red-stemmed cinquefoil [E]

Potentilla dissecta Pursh var. *rubricaulis* (Lehmann) Rydberg; *P. nivea* Linnaeus subsp. *rubricaulis* (Lehmann) Hiitonen

Caudex branches not sheathed with marcescent whole leaves. **Stems** ascending to nearly erect, 1.5–4 dm. **Basal leaves** often both ternate and palmate on same plant, rarely subpalmate, 4–10 cm; petiole 2.5–7 cm, long hairs sparse to common, loosely appressed to ascending-spreading, 1–2 mm, ± weak to stiff, verrucose, short and/or ± crisped hairs common to abundant, cottony hairs absent, glands usually sparse; leaflets 3–5, proximalmost separated by 0(–1) mm, central oblong to obovate, 1.5–4 × 1–2.5 cm, petiolules 0–5 mm, distal ⅔–¾ of margin incised ½–¾(+), to midvein, teeth (4 or)5–8 per side, 4–5 mm, apical tufts 1 mm, abaxial surfaces gray to grayish white, long hairs abundant, cottony-crisped hairs usually dense, short hairs and glands absent or obscured, adaxial green to grayish green, long hairs sparse to common, 0.5–1.5 mm, stiff, short hairs absent or sparse, rarely common, crisped and cottony hairs absent, glands absent or sparse, rarely common. **Cauline leaves** 2 or 3. **Inflorescences** 4–20-flowered, open, branch angle (10–)20–45°. **Pedicels** 0.5–3 cm, proximal to 5 cm. **Flowers:** epicalyx bractlets linear to narrowly lanceolate, 3–4 × 0.8–1.2 mm; hypanthium 4–6 mm diam.; sepals 4–5 mm, apex subacute to acute, glands usually ± sparse, not obscured; petals pale yellow, not overlapping, 5–7 × (4–)5–6.5 mm, distinctly longer than sepals; filaments 0.5–1.5 mm, anthers 0.4 mm; carpels 30–60, styles 0.9–1.1 mm. **Achenes** 1.2 mm. $2n = 56$.

Flowering summer. Sandy lake and stream shores, open sandy forests, dry grassy slopes, sandy and loamy bluffs, rock crevices, scree; 0–1600 m; Alta., B.C., N.W.T., Sask., Yukon; Alaska.

As addressed by B. Ertter et al. (2013), the name *Potentilla rubricaulis* is here restricted to relatively large plants with open inflorescences occurring mainly in glaciated parts of subarctic northwestern Canada and Alaska. Plants from the Chugach Mountains of southern Alaska tend to be more conspicuously glandular than elsewhere.

The distinction between *Potentilla rubricaulis* and large forms of *P. arenosa* with supernumerary leaflets is problematic. Although both species have somewhat similar petiole vestiture (long, straight, verrucose hairs and a layer of short, stiff, or curly hairs), the latter species tends to have more stiffly spreading petiole hairs and prominently petiolulate central leaflets.

The octoploid chromosome count (P. M. Dansereau and E. Steiner 1956) from Great Bear Lake area, Northwest Territories, probably belongs to *Potentilla rubricaulis* in the narrow sense, since that is its type locality.

84. Potentilla hookeriana Lehmann, Index Seminum (Hamburg) 1849: 10. 1849 • Hooker's cinquefoil [E]

Potentilla nivea Linnaeus subsp. *hookeriana* (Lehmann) Hiitonen; *P. nivea* var. *hookeriana* (Lehmann) Th. Wolf

Caudex branches not sheathed with marcescent whole leaves. **Stems** ascending to nearly erect, (0.3–)0.5–2 dm. **Basal leaves** usually both ternate and palmate on same plant, rarely subpalmate, 1–9 cm; petiole 0.5–6 cm, long hairs common to abundant, ± appressed to ascending-spreading, (0.5–)1–1.5(–2) mm, usually stiff, rarely weak, verrucose, short-crisped hairs sparse to common, cottony hairs absent, glands ± sparse; leaflets 3–5, proximalmost separated by 0(–1) mm, central oblanceolate to obovate, 0.5–2.5 × 0.4–1.2 cm, petiolules 0–1 mm, distal ± ¾ of margin incised ½–¾(+), to midvein, teeth 2–6 per side, 1–5 mm, apical tufts 0.5–1 mm, abaxial surfaces grayish white to white, long hairs common to abundant, cottony-crisped hairs usually dense, short hairs and glands absent or obscured, adaxial green to grayish green, long hairs sparse to common, 0.5–1.5 mm, usually stiff, short and/or crisped hairs sparse to common, cottony hairs absent, glands sparse to common. **Cauline leaves** 1 or 2. **Inflorescences** 1–6(–8)-flowered, open, branch angle 20–50°. **Pedicels** (0.5–)1–2 cm, proximal to 3 cm. **Flowers:** epicalyx bractlets linear to oblong-lanceolate, 2–3 × 0.7–1.3 mm; hypanthium 3–4 mm diam.; sepals 3–5 mm, apex subacute, glands ± common, not obscured; petals yellow, ± overlapping, 3–6(–7) × 4–5 mm, usually longer than sepals; filaments 0.5–1.5 mm, anthers 0.3–0.5 mm; carpels 30–50, styles 0.8–1.2 mm. **Achenes** 1–1.3 mm.

Flowering (spring–)summer. Alpine ridges, fellfields, scree slopes, rocky canyons; 1400–3900 m; Alta., B.C.; Colo., Idaho, Mont., Nev., Utah, Wyo.

As here defined, *Potentilla hookeriana* consists of primarily alpine plants from the Rocky Mountains and adjacent desert mountains that share many of the same characters with *P. rubricaulis* but are smaller overall. Arctic plants previously included in *P. hookeriana* are now assigned to *P. arenosa* (J. Soják 1986; B. Ertter et al. 2013).

85. Potentilla furcata A. E. Porsild, Bull. Natl. Mus. Canada 121: 224, plate 18. 1951 • Forked cinquefoil [E]

Potentilla hookeriana Lehmann var. *furcata* (A. E. Porsild) Hultén; *P. rubricaulis* Lehmann var. *furcata* (A. E. Porsild) Soják

Caudex branches usually not sheathed with marcescent whole leaves. **Stems** ascending to erect, (0.5–)1–3 dm. **Basal leaves** often both ternate and palmate on same plant, sometimes subpalmate, 4–10 cm; petiole 2.5–8 cm, long hairs common to abundant, loosely appressed to ascending-spreading, 1–2 mm, weak to ± stiff, ± verrucose, short and/or crisped hairs sparse to abundant, cottony hairs absent, glands sparse to common; leaflets 3–5, proximalmost separated by 0(–2) mm, central oblong to obovate, 1.4–2.5(–5) × 0.6–1.3(–2) cm, petiolules 1–3(–5) mm, distal ⅔–¾ of margin incised ½–¾+ to midvein, teeth 3–5 per side, 3.5–6 mm, apical tufts 0.5–1 mm, abaxial surfaces grayish white to white, long hairs common to abundant, cottony-crisped hairs usually dense, short hairs and glands absent or obscured, adaxial green to grayish green, long hairs sparse to common, 0.5–1 (–1.5) mm, ± stiff, short and/or crisped hairs sparse to common, sometimes abundant, cottony hairs absent, glands sparse to abundant. **Cauline leaves** 1–3. **Inflorescences** 7–12(–20)-flowered, ± open, branch angle 5–30(–50)°. **Pedicels** (0.5–)1–2 cm, proximal to 3 cm. **Flowers:** epicalyx bractlets linear to narrowly lanceolate, 2–3.5 × 0.6–1 mm; hypanthium 2.5–3.5 mm diam.; sepals 2.5–4 mm, apex subacute to acute, glands ± common, usually not obscured; petals pale yellow, not overlapping, 3–5(–6) × 3–4(–5) mm, slightly longer than sepals; filaments 0.8–1.8 mm, anthers 0.4 mm; carpels 30–60, styles 0.7–0.9 mm. **Achenes** 1.2 mm.

Flowering spring–summer. Sandy bluffs, dry riverbanks, limestone outcrops, grassy openings in dry forests, dry mountain slopes; 100–2400 m; B.C., Yukon; Alaska.

Potentilla furcata is a characteristic species of the steppe bluffs of interior and south-central Alaska, Yukon, and northern British Columbia, mainly within the unglaciated Beringian region. It differs from *P. rubricaulis* in having less open inflorescences, smaller flowers, narrower petals, more papillae on styles, and denser glands on epicalyx and calyx.

86. Potentilla modesta Rydberg in N. L. Britton et al., N. Amer. Fl. 22: 331. 1908 • Modest cinquefoil [E]

Potentilla concinna Richardson var. *modesta* (Rydberg) S. L. Welsh & B. C. Johnston

Caudex branches usually not sheathed with marcescent whole leaves. **Stems** ascending to nearly erect, (0.3–)0.5–1.5(–2.5) dm. **Basal leaves** usually palmate, rarely also ternate on same plant or subpalmate, 2–6(–8) cm; petiole 0.5–3.5(–5) cm, long hairs common to abundant, ± ascending to almost spreading, 1–2 mm, weak to ± stiff, scarcely to ± verrucose, short and/or crisped hairs common to abundant, cottony hairs absent, glands sparse to common; leaflets (3–)5, proximalmost separated by 0–1 mm, central obovate to oblanceolate-elliptic, 0.5–2(–2.5) × 0.5–1(–1.5) cm, petiolules 1 mm, distal ¾ to nearly whole margin incised ½–¾ to midvein, teeth 2–5 per side, (1–)2–5 mm, apical tufts 0.5–1 mm, abaxial surfaces white, long hairs common to abundant, cottony-crisped hairs dense, short hairs and glands absent or obscured, adaxial grayish green, long hairs common to abundant, 0.5–1.5 mm, ± stiff, rarely soft, short and/or crisped hairs sparse to common, cottony hairs absent, glands sparse to common. **Cauline leaves** 1 or 2. **Inflorescences** (1–)3–10-flowered, congested <or ± elongating in fruit>, branch angle 20–35°. **Pedicels** 0.2–0.5 cm, proximal to 1.2 cm. **Flowers:** epicalyx bractlets narrowly elliptic, 1.5–3 × 0.5–1 mm; hypanthium 3–4 mm diam.; sepals 2.5–4 mm, apex ± acute, glands abundant, usually not obscured; petals yellow, ± overlapping, 3.5–5 × 3–4 mm, slightly longer than sepals; filaments 1 mm, anthers 0.3–0.7 mm; carpels 20–40, styles 1 mm. **Achenes** 1 mm.

Flowering summer. Alpine tundra, fellfields, talus slopes, cliffs, usually on limestone; 2500–3900 m; Colo., Idaho, Mont., Utah, Wyo.

Potentilla modesta is the dominant component of what has previously been called *P. rubricaulis* in the Intermountain Region (for example, N. H. Holmgren 1997b, including illustration). This and other species of alpine *Potentilla* often grow in mixed populations, which contributes to confusion and mistaken identifications. In general, *P. modesta* has a more compact inflorescence and more consistently palmate basal leaves than sympatric members of the section. Intergrades with *P. jepsonii* (sect. *Pensylvanicae*) can also be problematic, though leaves of the latter are generally subpinnate and abaxially grayish rather than white.

The epithet *modesta* is misapplied in the combination *Potentilla concinna* var. *modesta*, which S. L. Welsh et al. (1993) used for plants mostly placed here in *P. concinna* var. *divisa* (sect. *Concinnae*).

87. Potentilla pseudosericea Rydberg, Monogr. N. Amer. Potentilleae, 98, plate 36, figs. 1–5. 1898 • Mono or strigose cinquefoil [E]

Caudex branches sometimes sheathed with marcescent whole leaves. **Stems** decumbent to ascending, (0.2–)0.5–1.5(–2.2) dm. **Basal leaves** usually palmate, rarely also ternate on same plant or subpalmate, (1–)2–6(–8) cm; petiole (0.5–)1–2.5 cm, long hairs abundant to dense, ± appressed, 0.5–1.2 mm, stiff, verrucose, crisped hairs common, cottony hairs absent, glands sparse to abundant but usually obscured; leaflets (3–)5–7, proximalmost separated by 0–2 mm, central ± obovate, (0.5–)1–1.5(–2) × (0.3–)0.5–0.8(–1) cm, petiolules 0–1 mm, distal ¾ to nearly whole margin incised ¾+ to midvein, teeth 2–4(or 5) per side, (2–)3–6(–10) mm, apical tufts 0.5 mm, abaxial surfaces white, long hairs common to abundant, cottony-crisped hairs dense, short hairs and glands absent or obscured, adaxial grayish green, long hairs abundant, 0.5–1.5 mm, ± stiff, short-crisped (rarely ± cottony) hairs sparse to common, glands absent or sparse. **Cauline leaves** 1–3. **Inflorescences** (1–)3–10(–20)-flowered, congested <or ± elongating in fruit>, branch angle 5–30°. **Pedicels** 0.2–0.7 cm, proximal to 1.3 cm. **Flowers:** epicalyx bractlets narrowly elliptic, (1–)1.5–2.5(–3) × 0.5–1 mm; hypanthium 3–4 mm diam.; sepals 2–4(–5) mm, apex obtuse to bluntly acute, glands sparse, obscured; petals pale yellow, not overlapping, 2–4 × 2–3 mm, ± equal to sepals (slightly longer than in Inyo County); filaments 1–1.5 mm, anthers 0.3–0.5 mm; carpels 30–50, styles 1–1.2 mm. **Achenes** 1–1.4 mm.

Flowering summer. Alpine fellfields, moist gravel, talus, on dolomite and quartzite; 3200–3600 m; Calif.

Potentilla pseudosericea is currently confirmed only from the central Sierra Nevada and Sweetwater and White mountains of California (B. Ertter 2008). The species is similar to *P. modesta* but has a more tightly strigose, silvery vestiture and more numerous and deeply incised leaflets.

88. Potentilla paucijuga Rydberg in N. L. Britton et al., N. Amer. Fl. 22: 348. 1908 • La Sal cinquefoil [E]

Potentilla pensylvanica Linnaeus var. *paucijuga* (Rydberg) S. L. Welsh & B. C. Johnston; *P. rubricaulis* Lehmann var. *paucijuga* (Rydberg) Soják

Caudex branches not sheathed with marcescent whole leaves. **Stems** decumbent to ascending, 0.5–2(–2.5) dm. **Basal leaves** subpalmate, (2–)3–8 cm; petiole 1–4 cm, long hairs sparse to abundant, ascending to spreading, 0.5–2.5 mm, weak to ± stiff, scarcely to ± verrucose, short-crisped hairs abundant, cottony hairs absent, glands sparse to common; leaflets (3–)5(–7), proximalmost separated by 1–5 mm, central narrowly obovate-elliptic, (1–)1.5–2.5(–3) × 0.8–1.5 cm, petiolules 0–1 mm, distal 3/4 to whole margin incised 2/3–3/4 to midvein, teeth 5–7 per side, (2–)3–6 mm, apical tufts 0.5–1 mm, abaxial surfaces grayish white to white, long hairs abundant, cottony-crisped hairs usually dense, short hairs and glands absent or sparse, adaxial grayish green, long hairs common to abundant, 1–2 mm, ± stiff, short-crisped hairs abundant, cottony hairs absent, glands sparse to common. **Cauline leaves** 1 or 2. **Inflorescences** (3–)5–10-flowered, congested <or ± elongating in fruit>, branch angle 10–30°. **Pedicels** 0.2–1 cm, proximal to 1.5 cm. **Flowers:** epicalyx bractlets ± lanceolate, (2–)3–5(–6) × 0.5–1.5 mm; hypanthium 3.5–5 mm diam.; sepals 4–6 mm, apex narrowly acute, glands ± common, only partly obscured; petals yellow, ± overlapping, (4–)5–6 × 4.5–5.5 mm, longer than sepals; filaments 0.5–2 mm, anthers 0.7 mm; carpels 50–70, styles 1.2–1.5 mm. **Achenes** 1.3 mm.

Flowering summer. Alpine tundra, patches of turf in talus; 3300–3700 m; Utah.

Potentilla paucijuga is a distinctive species endemic to the La Sal Mountains in Grand and San Juan counties. Diagnostic features include silvery-silky subpalmate leaves, relatively showy flowers with elongate epicalyx bractlets and sepals, and styles that are nearly 1.5 mm. In Colorado (for example, W. A. Weber and R. C. Wittmann 1996), *P. pensylvanica* var. *paucijuga* has been applied to what is treated here as *P. jepsonii* (sect. *Pensylvanicae*).

89. Potentilla pedersenii (Rydberg) Rydberg in N. L. Britton et al., N. Amer. Fl. 22: 332. 1908

Potentilla subquinata (Lange) Rydberg var. *pedersenii* Rydberg, Bull. Torrey Bot. Club 28: 182. 1901; *P. tolmatchevii* Jurtzev & Soják

Caudex branches not sheathed with marcescent whole leaves. **Stems** ascending to nearly erect, 0.4–2 dm. **Basal leaves** often both ternate and palmate or subpalmate on same plant, 2.5–4 cm; petiole 1.5–2.5 cm, long hairs common to abundant, loosely appressed to ascending-spreading, 1–2 mm, weak to ± stiff, verrucose, crisped(/short) hairs absent or sparse to common, cottony hairs absent, glands sparse to common; leaflets 3–5, proximalmost separated by 0–2 mm, central broadly elliptic to obovate, 1–1.5 × 0.5–0.9 cm, petiolules 1–2 mm, distal 2/3–3/4 of margin incised 1/2–3/4 to midvein, teeth (2 or)3 or 4 per side, 4–6 mm, apical tufts ± 1 mm, abaxial surfaces grayish white to white, long hairs abundant (sometimes obscuring entire surface), cottony-crisped hairs abundant to dense, short hairs absent or obscured, glands sparse to common but usually obscured, adaxial grayish green to gray, long hairs sparse to abundant, 1–1.5(–2) mm, ± weak, short (short-crisped) hairs absent or sparse, rarely common, cottony hairs absent, glands absent or sparse, rarely common. **Cauline leaves** 0–2. **Inflorescences** (1–)3–7-flowered, open, branch angle 30–50°. **Pedicels** 1–2 cm, proximal to 4 cm. **Flowers:** epicalyx bractlets narrowly ovate to elliptic, 4–5 × 1–1.4 mm; hypanthium 3–4 mm diam.; sepals 4–6 mm, apex subacute to acute, glands sparse to common, usually not obscured; petals pale yellow, usually not overlapping, 6–7 × 4–8 mm, distinctly longer than sepals; filaments 1–2 mm, anthers ± 0.4 mm; carpels 40–80, styles 0.8–0.9 mm. **Achenes** 1.1–1.2 mm.

Flowering summer. Dry tundra, gravel and loam ridges, loam flats, rocky outcrops and crevices; 0–200 m; Greenland; N.W.T., Nunavut; ne Europe; n Asia.

Potentilla pedersenii and *P. uschakovii* account for the majority of arctic populations previously included in a broadly defined *P. rubricaulis*. The diagnostic morphological characters between the two species can be variable and overlapping; they are treated separately in part because of differences in presumed parental combinations. Whereas *P. pulchella* is the probable sect. *Pensylvanicae* parent for both species, the putative sect. *Niveae* parent for *P. pedersenii* is *P. arenosa* subsp. *arenosa*; that of *P. uschakovii* is *P. subvahliana*. Reflecting this parentage, *P. pedersenii* is distinguished by caudex branches with no marcescent whole leaves, verrucose long hairs on petioles, and inflorescences with usually several relatively small flowers. In contrast, *P. uschakovii* often has marcescent whole leaves sheathing the caudex branches, smooth long hairs on petioles, and one- or few-flowered inflorescences with mostly larger flowers. These generalities aside, there is much variation within both species, such that each island or population group may have its own features; it is probable that both *P. pedersenii* and *P. uschakovii* have evolved from multiple hybridization events.

90. Potentilla uschakovii Jurtzev, Bot. Zhurn. (Moscow & Leningrad) 73: 1613. 1988

Caudex branches often sheathed with marcescent whole leaves. **Stems** ascending to nearly erect, 0.3–1.5 dm. **Basal leaves** often both ternate and palmate or subpalmate on same plant, 1–5 cm; petiole 0.5–3 cm, long hairs sparse to dense, ± appressed to ascending, (1–)1.5–2.5 mm, soft to ± weak, smooth, short hairs absent or sparse, crisped(/cottony) hairs sparse to common, glands absent or sparse; leaflets 3–5, proximalmost separated by 0–2 mm, central broadly elliptic to obovate, 0.5–2 × 0.3–1.4 cm, petiolules ± 1 mm, distal ⅔ to nearly whole margin incised ⅔–⅗+ to midvein, teeth (2 or)3 or 4 per side, 3–5 mm, apical tufts 1–1.5 mm, abaxial surfaces grayish white to white, long hairs common to abundant, cottony(/crisped) hairs dense, short hairs and glands absent or obscured, adaxial grayish green to gray, long hairs abundant, 1–2 mm, soft (grading to cottony), short/crisped/cottony hairs common to abundant, glands sparse to common. **Cauline leaves** 0 or 1. **Inflorescences** 1–3(or 4)-flowered, open, branch angle 20–40°. **Pedicels** 1–1.5 cm, proximal to 3 cm. **Flowers:** epicalyx bractlets lanceolate to ovate, 3–4 × 1–2 mm; hypanthium 2.5–3.5 mm diam.; sepals 3.5–5 mm, apex obtuse to subacute, glands sparse to common, often ± obscured; petals pale yellow, often overlapping, 5–7 × 5–8(–9) mm, distinctly longer than sepals; filaments 0.5–1.5 mm, anthers ± 0.5 mm; carpels 30–60, styles 0.8–0.9 mm. **Achenes** 1–1.3 mm.

Flowering summer. Dry tundra meadows, gravel and loam ridges, loam flats, sandy bluffs; 0–200 m; Greenland; N.W.T., Nunavut; e Asia (Russian Far East).

For discussion of probable parentage and resultant diagnostic characters, see discussion of 89. *Potentilla pedersenii.*

8u. Potentilla Linnaeus sect. Pensylvanicae Poeverlein in P. F. A. Ascherson et al., Syn. Mitteleur. Fl. 6(1): 669. 1904 (as Pennsylvanicae)

Barbara Ertter

Reidar Elven

Potentilla [unranked] *Multifidae* Rydberg; *Potentilla* sect. *Multifidae* (Rydberg) A. Nelson; *Potentilla* ser. *Multifidae* (Rydberg) T. T. Yu & C. L. Li

Perennials, ± tufted, not stoloniferous; taproots not fleshy-thickened; vestiture of long, short, crisped, and/or cottony hairs, glands sparse to abundant, sometimes absent, not red. **Stems** decumbent to erect, not flagelliform, not rooting at nodes, lateral to persistent basal rosettes, (0.1–)0.5–4(–6) dm, lengths (1–)1.5–3(–4) times basal leaves. **Leaves:** basal not in ranks; cauline 1–4; primary leaves pinnate to subpinnate or subpalmate (with distal leaflets distinct), 1.5–25(–30) cm; petiole: long hairs spreading to appressed, soft to stiff, glands absent or sparse to abundant; leaflets (3–)5–13(–19), on distal ⅙–½+, rarely less, of leaf axis, separate to ± overlapping, narrowly oblanceolate or elliptic to obovate or broadly oblong, margins ± to strongly revolute, ± whole length evenly incised ± ½ to nearly completely to midvein, teeth (2 or)3–8(–12) per side, surfaces similar to strongly dissimilar, abaxial green to white, cottony hairs absent or sparse to dense, adaxial green to grayish, not glaucous, long hairs soft to ± stiff. **Inflorescences** (1 or)2–40(–100)-flowered, cymose, congested to very open. **Pedicels** ± straight in fruit, (0.1–)0.2–3(–5) cm, proximal often significantly longer than distal. **Flowers** 5-merous; hypanthium 2.5–10 mm diam.; petals pale to bright yellow, obovate to obcordate, (2–)3–5

(–10.5) mm, ± equal to or longer, sometimes shorter, than sepals, apex rounded or truncate to retuse; stamens (15–)20; styles subapical, conic to columnar, papillate-swollen in proximal ¼–¾+ or only at base, if at all, 0.8–1.2(–1.5) mm. **Achenes** smooth to rugose.

Species ca. 20 (8 in the flora): North America, ne Mexico, Eurasia, n Africa.

Section *Pensylvanicae*, as here presented, accommodates most perennial species with pinnate to subpinnate leaves and styles about 1 mm that are frequently papillate-swollen proximally. Plants are often notably glandular, leaflet margins are usually revolute, and petals are most commonly about the same length as the sepals and often somewhat erect at anthesis. Section *Multifidae* was commonly used for this group of species until J. Soják (1987b) noted that sect. *Pensylvanicae* was an older, validly published name.

About half of the North American members of sect. *Pensylvanicae* have been previously included in a broadly defined *Potentilla pensylvanica*, with or without infraspecific divisions (for example, C. L. Hitchcock and A. Cronquist 1961b; S. L. Welsh 1974; R. L. McGregor 1986b; H. A. Gleason and Cronquist 1991). In support of a more narrowly circumscribed *P. pensylvanica*, B. L. Kohli and J. G. Packer (1976) found that major morphologic types coincide with three different ploidy levels. Most plants with fully pinnate leaves (five to nine leaflets in remote pairs) and other characters of typical *P. pensylvanica* in the narrow sense are tetraploid; plants with subpinnate leaves (mostly five leaflets in approximate pairs) are octoploid (or sometimes tetraploid; Packer, pers. comm.). Diploid populations from the northern Great Plains, morphologically closest to *P. pensylvanica* in the narrow sense but with epicalyx bractlets usually longer than the sepals, were named *P. finitima* Kohli & Packer (a later synonym of *P. lasiodonta*).

The ploidy differences noted by B. L. Kohli and J. G. Packer (1976), in combination with morphology (vestiture, glandularity, leaf dissection, and features of the epicalyx) and ecogeographic correlates, serve collectively to delimit the species that are recognized here. The most distinctive vestiture type is a velvety covering of short (0.2 mm), straight, spreading hairs that generally characterizes *Potentilla lasiodonta* and *P. pensylvanica* in the narrow sense, although this feature is sometimes obscured by longer ascending hairs or is missing altogether. Leaves of *P. lasiodonta* and *P. pensylvanica* in the narrow sense are also distinctly pinnate, with the blade comprising at least half the total leaf length and being no more than about half as wide as long, usually bearing at least three pairs of lateral leaflets gradually reduced in size from the distalmost, and the leaflets themselves being pinnately toothed with four to 12 blunt teeth per side. Epicalyx bractlets in both species, and also in *P. litoralis*, tend to be nearly as long as the sepals (to twice as long in *P. lasiodonta*) and concave, at least in fruit, with revolute margins and prominent midveins. These diagnostic features are not consistently present and/or evident, especially in depauperate and immature specimens, but collectively they serve to identify most specimens with a reasonable level of confidence. Occasional hybrids are to be expected.

In addition to the species addressed here, *Potentilla lyngei* Jurtzev & Soják subsp. *lyngei* has been reported from northeastern Greenland. At least one purported voucher (annotated by Jurtzev in ALA) has been reidentified to *P. pedersenii* (sect. *Rubricaules*), leaving the North American presence of *P. lyngei* subsp. *lyngei* in question. This is not the case for *P. lyngei* subsp. *spissa* Soják, described from Greenland; as discussed elsewhere (B. Ertter et al. 2013), this name is probably based on hybrids between *P. pulchella* and *P. hyparctica* (sect. *Aureae*), for which *P.* ×*safronoviae* Jurtzev & Soják is the priority name.

Additional nomenclatural details of species in sect. *Pensylvanicae* were provided by M. L. Fernald (1935), J. Soják (1987c), and B. Ertter (2008). For species of presumed hybrid origin with sect. *Pensylvanicae* as one parent, see 8t. sect. *Rubricaules* and Ertter et al. (2013).

Since *Potentilla paucijuga* (sect. *Rubricaules*) is sometimes identified as a member of sect. *Pensylvanicae*, it is included herein and keys out in the sixth couplet.

SELECTED REFERENCE Kohli, B. L. and J. G. Packer. 1976. A contribution to the taxonomy of the *Potentilla pensylvanica* complex in North America. Canad. J. Bot. 54: 706–719.

1. Styles papillate-swollen at very base, if at all; inflorescences ± congested to very open.
    2. Sepal venation prominent at least proximally; inflorescences (5–)10–40+-flowered, ± congested to open; pedicels 0.3–1 cm (proximal to 3 cm). . . . . 97. *Potentilla bimundorum*
    2. Sepal venation indistinct; inflorescences 2–5-flowered, very open; pedicels 2–4 cm . . . . .98. *Potentilla anachoretica*
1. Styles papillate-swollen in proximal ¼–¾+; inflorescences usually congested, sometimes elongating in fruit (± open in *P. pulchella*).
    3. Basal leaves pinnate, short hairs usually abundant to dense, rarely absent, sometimes obscured (especially in Great Plains); leaflets (2 or)3–6(–9) per side, on distal (⅓–)½–⅗ of leaf axis; cottony and crisped hairs absent abaxially; epicalyx bractlets: lengths 1–2 times sepals, margins ± revolute.
        4. Leaflets: teeth 4–8(–10) per side, margins incised ± ¾ to midvein, leaving undivided medial blade 2–4(–8) mm wide; epicalyx bractlets equal to length of sepals; widespread . . . . .91. *Potentilla pensylvanica*
        4. Leaflets: teeth 8–12 per side, margins incised ± ½ to midvein, leaving undivided medial blade 5–9 mm wide; epicalyx bractlets 1–2 times length of sepals; sandy sites, n Great Plains. . . . . 92. *Potentilla lasiodonta*
    3. Basal leaves subpinnate to subpalmate, short hairs absent, sparse, or obscured; leaflets (1 or)2 or 3 per side, on distal ⅙–⅓(–½) of leaf axis; cottony and/or crisped hairs absent or sparse to dense abaxially; epicalyx bractlets: lengths ½–1 times sepals, margins revolute or not.
        5. Inflorescences ± open, (1 or)2–5(–9)-flowered; pedicels 0.5–5 cm; leaflet teeth 2–5 per side; Alaska, n Canada, Greenland . . . . .96. *Potentilla pulchella*
        5. Inflorescences congested, sometimes elongating in fruit, (1–)3–50(–100)-flowered; pedicels 0.2–1(–3) cm; leaflet teeth (2 or)3–8 per side; Alaska to Newfoundland, s to California, Colorado, and New Hampshire.
            6. Petals (4–)5–6 × 4.5–5.5 mm, longer than sepals; styles 1.2–1.5 mm; leaflets incised ⅔–¾ to midvein; stems 0.5–2(–2.5) dm; La Sal Mountains, Utah . . . . 88. *Potentilla paucijuga* (sect. *Rubricaules*)
            6. Petals 2–5 × 2–4 mm, ± equal to sepals; styles 0.8–1.2(–1.5) mm; leaflets incised ¾+ to midvein; stems (0.2–)0.5–5(–6) dm; widespread.
                7. Sepals: glands absent, sparse, or obscured; leaflets usually white abaxially, cottony (and sometimes crisped) hairs ± dense. . . . . .95. *Potentilla bipinnatifida*
                7. Sepals: glands abundant, evident at least distally; leaflets usually green to grayish abaxially, cottony hairs absent, crisped hairs absent or sparse to abundant.
                    8. Epicalyx bractlets: lengths ⅔–1 times sepals, margins ± revolute (at least in fruit); stems (0.4–)1–4.5(–6) dm; n Atlantic Coast and Hudson Bay across n prairies to n Rocky Mountains; 0–2200 m . . . . .93. *Potentilla litoralis*
                    8. Epicalyx bractlets: lengths ½–¾ times sepals, margins flat; stems (0.2–)0.5–2.6 dm; Great Basin and surrounding mountains; 2500–3700 m . . . . .94. *Potentilla jepsonii*

91. Potentilla pensylvanica Linnaeus, Mant. Pl. 1: 76. 1767 • Prairie cinquefoil, potentille de Pennsylvanie [F]

Potentilla atrovirens Rydberg; *P. brooksensis* Jurtzev; *P. pensylvanica* var. *arida* B. Boivin; *P. pensylvanica* var. *brevipila* Soják; *P. sibirica* Th. Wolf

Stems ascending to erect, (0.3–)1–4(–6) dm. **Basal leaves** pinnate, (2–)4–20(–25) cm; petiole 1–8 cm, long hairs sparse to abundant, rarely absent, spreading to ascending, 0.5–3 mm, weak to ± stiff, short hairs abundant to dense, rarely absent, cottony or crisped hairs absent, glands sparse to abundant, often obscured; leaflets (2 or) 3–5(–7) per side, on distal 1/2–3/5 of leaf axis, separate to scarcely overlapping, terminal ones narrowly to broadly oblanceolate, (1–)1.5–5(–6) × 0.5–1.5(–2) cm, margins strongly revolute, incised ± 3/4 to midvein, sometimes more or only ± 1/2, undivided medial blade 2–4(–8) mm wide, teeth 4–8(–10) per side, triangular-ovate to linear-elliptic, surfaces usually ± dissimilar, abaxial ± grayish to nearly white, rarely green, long hairs sparse to abundant (mostly on veins), rarely absent, 0.5–1.5 mm, ± weak, short hairs usually abundant to dense, rarely absent, crisped and cottony hairs absent, glands ± abundant, sometimes obscured, adaxial green to gray-green, long hairs sparse to abundant, rarely absent, spreading to ascending, ± 1 mm, short hairs ± abundant, sometimes obscured, rarely absent, cottony and crisped hairs absent, glands ± abundant. **Cauline leaves** 1–4. **Inflorescences** (1–)3–40-flowered, congested <or elongating in fruit>. **Pedicels** 0.2–1 cm (proximal to 3.5 cm). **Flowers:** epicalyx bractlets usually ± lanceolate-elliptic, rarely ± ovate-elliptic, 3–6(–8) × (1–)1.5–2(–2.5) mm, lengths ± equal to sepals, margins ± revolute; hypanthium 3–6 mm diam.; sepals (3–)4–6(–8) mm, apex acute to obtuse, abaxial surfaces: venation indistinct, glands abundant, ± evident; petals pale to bright yellow, (3–)4–7 × (3–)4–7 mm, lengths ± equal to sepals; filaments (0.5–)1–2 mm, anthers 0.5–0.8(–1) mm; carpels 50–100, styles papillate-swollen in proximal ± 1/2, ± 1 mm. **Achenes** 1.2 mm, ± to strongly rugose. **2*n*** = 28.

Flowering summer. Open prairies, steppe bluffs, shallow gravelly soil, rocky ridges and outcrops, dry stream channels, similar disturbed sites; 100–4000 m; Alta., B.C., Man., N.W.T., Ont., Que., Sask., Yukon; Alaska, Ariz., Calif., Colo., Idaho, Iowa, Minn., Mont., Nebr., Nev., N.Mex., N.Dak., S.Dak., Tex., Utah, Wyo.; Mexico (Tamaulipas); Eurasia.

Even in the reduced circumscription used here, *Potentilla pensylvanica* is exceptionally widespread, occurring in North America from Alaska to Hudson Bay, in the Great Plains and the North American Cordillera to northern Mexico. The species also grows across Siberia (where it has sometimes been called *P. sibirica*) and as isolated occurrences in the southwestern Alps and Spain (J. Soják 1987c; A. Kurtto et al. in J. Jalas et al. 1972+, vol. 13). One place *P. pensylvanica* does not occur is in Pennsylvania; the epithet was applied by Linnaeus to cultivated plants originating from Canada. Collections from Quebec are probably introductions (J. Cayouette, pers. comm.). The Oregon citation of *P. pensylvanica* (D. Mansfield 2000) refers to *P. jepsonii*.

Significant variation occurs within *Potentilla pensylvanica*, especially regarding vestiture and plant stature. J. Soják (1987c) has described several varieties, all Eurasian except var. *brevipila* (long hairs 0.6–1 mm); diminutive specimens from the Brooks Range of northern Alaska, isolated from other known populations of *P. pensylvanica*, have been described as *P. brooksensis*. Although much of the ecogeographic variation may merit recognition, at least at varietal level, doing so is postponed pending further analysis and clarification of nomenclature. Among the extremes are plants from the Great Plains, which have abundant long silky hairs overlying and obscuring the velvety layer of short hairs (resembling sympatric *P. litoralis*), and glabrate (except for glands) populations in Alaska. E. Hultén (1968) erroneously referred to the latter as *P. pensylvanica* var. *glabrata* (Lehmann) S. Watson; that name is a synonym of *P. ovina* var. *ovina* (sect. *Multijugae*). Most populations in Alaska also have larger than average flowers (petals to 7 × 7 mm, anthers more or less 1 mm).

Although *Potentilla pensylvanica* var. *strigosa* Pursh has been one of the most commonly used varieties in North America, the name has been effectively lectotypified by J. Soják (1987c) on a collection from Siberia. Soják identified this collection as the hybrid between *P. pensylvanica* and the Asian *P. sanguisorba* D. F. K. Schlechtendal; var. *strigosa* accordingly does not occur in North America.

92. Potentilla lasiodonta Rydberg in N. L. Britton et al., N. Amer. Fl. 22: 351. 1908 • Sandhills cinquefoil [E]

Potentilla finitima Kohli & Packer

Stems ± erect, 2–4 dm. **Basal leaves** pinnate, 12–20(–30) cm; petiole 5–10(–15) cm, long hairs ± abundant, spreading to ascending, 0.5–1 mm, weak to ± stiff, short hairs abundant to dense, cottony or crisped hairs absent, glands sparse to abundant, often obscured; leaflets 4–6(–9) per side, on distal (1/3–)1/2–3/5 of leaf axis, slightly overlapping, terminal ones oblanceolate to narrowly elliptic, 3–6

× 0.8–1.2(–1.5) cm, margins strongly revolute, incised ± ½ to midvein, undivided medial blade 5–9 mm wide, teeth 8–12 per side, narrowly triangular to lanceolate, surfaces ± dissimilar, abaxial grayish, long hairs ± sparse (or not differentiated from short hairs), 0.5–1.5 mm, weak, short hairs abundant to dense, cottony or crisped hairs absent, glands sparse or obscured, adaxial ± green, straight hairs (long and short not differentiated) abundant, spreading to ascending, 0.2–0.5 mm, cottony and crisped hairs absent, glands sparse to abundant, often obscured. **Cauline leaves** (1 or)2–4. **Inflorescences** 10–40-flowered, congested <or elongating in fruit>. **Pedicels** ± 0.1 cm (proximal to 1.2 cm). **Flowers:** epicalyx bractlets narrowly ovate-acuminate, (3–)4–8(–10) × 2–3 mm, lengths 1–2 times sepals, margins ± revolute; hypanthium (4–)5–10 mm diam.; sepals 3–5 mm, apex acute to obtuse, abaxial surfaces: venation moderate, glands ± abundant, obscured to evident; petals pale yellow, 3–5 × 2–4 mm, lengths ± equal to or shorter than sepals; filaments 1–1.5 mm, anthers 0.5–0.8 mm; carpels 80–90, styles papillate-swollen in proximal ¾+, 1–1.2 mm. **Achenes** ± 1 mm, ± rugose. **2*n*** = 14.

Flowering summer. Sandy sites in prairies; 300–1100 m; Alta., Man., Sask.; Minn., N.Dak.

Potentilla lasiodonta is a diploid relative of tetraploid *P. pensylvanica*, sharing similar velvety vestiture of dense short hairs, pinnate leaves, and epicalyx bractlets with revolute margins. The species differs in having larger epicalyx bractlets and less deeply incised leaflets. Plant height and leaf size are at the upper range of *P. pensylvanica*, and populations are evidently restricted to sandy substrates. Occurrence in Manitoba is based on B. L. Kohli and J. G. Packer (1976); no vouchers are in WIN. At least one collection from southeastern British Columbia (*Brown 779*, MO) approaches *P. lasiodonta*, but it is insufficient by itself to serve as a provincial record.

The fundamental ploidy and epicalyx distinctions of this species were established by B. L. Kohli and J. G. Packer (1976), who proposed the name *Potentilla finitima*. As noted by J. Soják (1994), the type of *P. lasiodonta* is the same entity. *Potentilla atrovirens* Rydberg and *P. pensylvanica* var. *arida* B. Boivin have sometimes been applied to this taxon; the types of both names fall within the circumscription of *P. pensylvanica* in the narrow sense (Packer, pers. comm.).

93. Potentilla litoralis Rydberg, Bull. Torrey Bot. Club 23: 264. 1896 • Coast cinquefoil, potentille du littoral [E]

Potentilla pensylvanica Linnaeus var. *litoralis* (Rydberg) B. Boivin; *P. pensylvanica* var. *pectinata* Lepage; *P. pensylvanica* var. *virgulata* (A. Nelson) Th. Wolf; *P. virgulata* A. Nelson

Stems decumbent to nearly erect, (0.4–)1–4.5(–6) dm. **Basal leaves** subpinnate to subpalmate, (3–)4–15(–25) cm; petiole (1–)2–8 (–15) cm, long hairs sparse to abundant, appressed to rarely ascending, ± 1 mm, ± stiff, short hairs absent or sparse, cottony or crisped hairs absent, glands ± sparse; leaflets 2 or 3 per side, on distal ⅙–⅓(–½) of leaf axis, ± overlapping, terminal ones narrowly lanceolate to narrowly obovate, (1.5–)2–5(–7) × 1–2.5 cm, margins revolute, incised ¾+ to midvein, undivided medial blade 2–10 mm wide, teeth (2–)5–8 per side, ± linear-elliptic, surfaces usually ± dissimilar, abaxial green to grayish, long hairs ± abundant especially on veins, 0.5–1.5 mm, weak, short hairs absent or sparse, cottony hairs absent, crisped hairs absent or sparse to abundant, glands ± abundant, adaxial ± green, long hairs sparse to common, ± appressed, 0.5–1.5 mm, short hairs absent or sparse, cottony hairs absent, crisped hairs absent or sparse, glands ± abundant. **Cauline leaves** (1 or)2–4. **Inflorescences** (1–)5–30-flowered, congested, <scarcely lengthening in fruit>. **Pedicels** 0.2–1 cm (proximal to 3 cm). **Flowers:** epicalyx bractlets narrowly, rarely broadly, elliptic to linear, 2.5–5(–6) × 1(–2) mm, lengths ± ⅔ to equal to sepals, margins ± revolute (at least in fruit); hypanthium 3–5 mm diam.; sepals 3–6 mm, apex ± acute, abaxial surfaces: venation moderate, glands abundant, evident at least distally; petals yellow, 3–5 × 2–4 mm, lengths ± equal to sepals; filaments (0.5–)1–1.5 mm, anthers 0.3–0.5 mm; carpels 50–100, styles papillate-swollen in proximal ½–¾+, 1–1.5 mm. **Achenes** ± 1 mm, smooth to faintly rugose. **2*n*** = 28, 56.

Flowering summer. Rocky and sandy shorelines, gravelly sites, open prairies, meadows, disturbed sites, talus, rocky outcrops; 0–2200 m; Alta., B.C., Man., Nfld. and Labr. (Nfld.), N.W.T., N.S., Nunavut, Ont., Que., Sask., Yukon; Alaska, Maine, Mich., Minn., Mont., Nev., N.H., N.Dak., S.Dak., Wyo.

Revolute epicalyx bractlets and visible glands on the sepal tips are the most reliable characteristics to distinguish *Potentilla litoralis* from *P. bipinnatifida* where their ranges overlap in the northern Great Plains. The former is also more likely to be decumbent-ascending and tends to darken more in drying. Intermediates occur, as well as possible hybrids with *P. bimundorum*, *P. hippiana*, and *P. pensylvanica*.

Rydberg described *Potentilla litoralis* to accommodate the coastal phase of this species, concentrated around the Gulf of St. Lawrence and along the north Atlantic Coast as far south as New Hampshire. *Potentilla pectinata* Rafinesque was briefly adopted as the older name (for example, Frère Marie-Victorin 1935); it is both superfluous and a later homonym (B. Boivin 1952). The homotypic synonyms *P. pensylvanica* var. *glabrata* (Lehmann) S. Watson, *P. bipinnatifida* var. *glabrata* (Lehmann) Kohli & Packer, and *P. glabrella* Rydberg have been misapplied to *P. litoralis*; the type is a specimen of *P. ovina* var. *ovina* (sect. *Multijugae*) (J. Soják 1996). The type of *P. dissecta* Pursh, traditionally applied to *P. glaucophylla* (sect. *Graciles*), is closest to this species but has palmate leaves; ambiguity has led to the name being rejected (B. Ertter et al. 2008).

The current circumscription of *Potentilla litoralis* includes prairie populations extending through the northern Great Plains and boreal regions into eastern Alaska; those plants have generally been called *P. virgulata* when not combined with *P. bipinnatifida*, *P. multifida* in the broad sense, or *P. pensylvanica*. Outlying populations occur at least as far south as Wyoming (including the type of *P. virgulata*), central Nevada (possibly as introductions), and, possibly, Colorado.

Other than the tendency of prairie plants to be somewhat larger and more erect than the coastal plants, no significant morphologic distinctions have been found to match the differences in habitat and distribution. Based on their distribution map, B. L. Kohli and J. G. Packer (1976) considered the coastal and prairie populations, as well as what is here separated as *Potentilla jepsonii*, to comprise a single taxon, which they called *P. bipinnatifida* var. *glabrata* (Lehmann) Kohli & Packer (misapplied).

If *Potentilla litoralis* is treated as a variety of another species, var. *virgulata* has priority at varietal rank.

Unequivocal *Potentilla litoralis* barely enters Alaska; the majority of collections from central, western, and northern Alaska previously placed in *P. litoralis* represent an undescribed taxon whose dimensions are not included in the preceding description. Significant differences include more deeply incised leaflets with fewer teeth, flat, linear epicalyx bractlets, and larger petals. Some collections from around Kotzebue with relatively wide sepals and sparse glands may indicate hybridization with *P. bimundorum*.

94. Potentilla jepsonii Ertter, J. Bot. Res. Inst. Texas 2: 202. 2008 • Jepson's potentilla [E]

Potentilla pensylvanica Linnaeus var. *ovium* Jepson, Fl. Calif. 2: 184. 1936 (as pennsylvanica); *P. litoralis* Rydberg var. *ovium* (Jepson) Soják

Stems decumbent to nearly erect, (0.2–)0.5–2.6 dm. **Basal leaves** subpinnate to subpalmate, 1.5–10(–17) cm; petiole 0.5–4(–8) cm, long hairs ± abundant, spreading to loosely appressed, 0.5–2 mm, weak to stiff, short hairs absent or sparse, cottony or crisped hairs absent, glands sparse to abundant; leaflets (1–)2–3 per side, on distal 1/6–1/3(–1/2) of leaf axis, separate to slightly overlapping, terminal ones oblanceolate-elliptic to narrowly obovate, (0.5–)1–2.5(–3.5) × 0.4–1(–1.3) cm, margins strongly revolute, incised 3/4+ to midvein, undivided medial blade 1.5–3(–5) mm wide, teeth (2–)3–5(–6) per side, ± linear-elliptic to rarely ovate or oblanceolate, surfaces ± dissimilar, abaxial ± grayish, rarely white, long hairs ± abundant especially on veins, 1 mm, weak, short hairs absent or sparse, cottony hairs absent, crisped hairs sparse to abundant, glands ± abundant, sometimes obscured, adaxial green to gray-green, long hairs sparse to abundant, loosely appressed, 0.5–1(–2) mm, short hairs absent or sparse, cottony and crisped hairs absent, glands ± abundant. **Cauline leaves** 1–2. **Inflorescences** (1–)3–12(–20)-flowered, congested, <scarcely opening in fruit>. **Pedicels** 0.2–0.5 cm (proximal to 1.2 cm). **Flowers:** epicalyx bractlets elliptic to linear, 1.5–3 × (0.3–)0.5–1.2 mm, lengths 1/2–3/4 times sepals, margins flat; hypanthium 2.5–5 mm diam.; sepals 2–5 mm, apex usually ± acute to obtuse, rarely acuminate, abaxial surfaces: venation faint, glands abundant, evident; petals pale yellow, 2–5 × 2–4 mm, lengths ± equal to sepals; filaments 0.5–2 mm, anthers 0.3–0.7 mm; carpels 20–60, styles papillate-swollen in proximal 1/2–3/4+, 0.8–1.2 mm. **Achenes** 1 mm, smooth to faintly rugose.

Flowering summer. Seasonally moist sites in alpine meadows, ridges, fellfields, rock ledges, talus slopes; 2500–3700 m; Alta.; Ariz., Calif., Colo., Idaho, Mont., Nev., Oreg., Utah, Wash.

Potentilla jepsonii (a replacement name for *P. pensylvanica* var. *ovium*) is widespread at higher elevations in the Intermountain Region, extending into the Sierra Nevada in California and the Rocky Mountains in Idaho and Wyoming. Disjunct occurrences are known on the north rim of the Grand Canyon and San Francisco Peaks, Arizona, and Steens Mountain, Oregon. Collections have most commonly been included in *P. pensylvanica* (for example, B. Ertter 1993; N. H. Holmgren 1997b); they differ in leaf division, vestiture, and epicalyx bractlets. J. Soják considered var. *ovium* to be a variety of *P. litoralis*, which has comparable leaves and vestiture; *P. jepsonii* lacks the enlarged, revolute epicalyx bractlets that distinguish *P. litoralis* and is for the most part well separated geographically and elevationally. The exception is the montane phase of *P. litoralis*, which, at the current stage of knowledge, is somewhat arbitrarily delimited from *P. jepsonii* in the northern Rocky Mountains; collections from Waterton Lakes National Park in Alberta and the Okanogan Mountains in Washington are tentatively placed here in *P. jepsonii*.

95. Potentilla bipinnatifida Douglas in W. J. Hooker, Fl. Bor.-Amer. 1: 188. 1832, name conserved • Potentille bipinnatifide [E]

Potentilla pensylvanica Linnaeus var. *bipinnatifida* (Douglas) Torrey & A. Gray

Stems ascending to erect, (1–)2–5 dm. **Basal leaves** subpinnate to subpalmate, (6–)10–25 cm; petiole (2–)5–15 cm, long hairs dense, appressed, 1–2 mm, soft to ± stiff, short hairs absent, crisped hairs sparse, glands absent, sparse, or obscured; leaflets 2 or 3 per side, on distal ⅙–⅓(–½) of leaf axis, separate to ± overlapping, terminal ones oblanceolate, (2–)3–6(–10) × 1–2(–3.5) cm, margins revolute, incised ¾+ to midvein, undivided medial blade 1.5–6 mm wide, teeth 5–8 per side, ± linear, surfaces ± to strongly dissimilar, abaxial usually white, rarely grayish, long hairs abundant especially on veins, 1–2 mm, ± weak, short hairs absent or obscured, cottony (and crisped) hairs ± dense, glands absent or obscured, adaxial green to grayish, long hairs sparse to abundant, loosely appressed, 0.5–1.5 mm, short hairs absent or sparse, crisped and/or cottony hairs sparse to common, glands sparse to common. **Cauline leaves** 2–4. **Inflorescences** (4–)10–50(–100)-flowered, congested <or elongating in fruit>. **Pedicels** 0.2–0.8 cm (proximal to 2 cm). **Flowers:** epicalyx bractlets narrowly elliptic to lanceolate, 2.5–6 mm, lengths ± ⅔ times sepals, margins flat; hypanthium 3–5 mm diam.; sepals 3–6 mm, apex ± acute, abaxial surfaces: venation indistinct, glands absent, sparse, or obscured; petals yellow, 3–5 × 3–4 mm, lengths ± equal to sepals; filaments 0.5–2 mm, anthers 0.3–0.5 mm; carpels 50–80, styles papillate-swollen in proximal ½–¾+, 1–1.2 mm. **Achenes** 1–1.2 mm, smooth to faintly rugose. **2*n*** = 56.

Flowering summer. Open shortgrass prairie, alkaline bottoms, streamsides in sagebrush, disturbed sites; 10–3400 m; Alta., Man., N.W.T., Ont., Que., Sask., Yukon; Alaska, Colo., Idaho, Mich., Minn., Mont., Nebr., N.Mex., N.Dak., S.Dak., Utah, Wyo.

Potentilla bipinnatifida is similar to *P. litoralis* in habit and leaf dissection but has flat, silky epicalyx bractlets and sepals with no evident glands. Vestiture is generally silkier, and the silvery to bicolor leaves are white-cottony abaxially. The two species are sympatric in the plains of Canada, with some intergradation; *P. bipinnatifida* is also common south to Colorado, where it is found in intermontane meadows and sagebrush flats. Outlying populations occur in Blaine and Custer counties, Idaho, and Duchesne and Piute counties, Utah. Eastern collections from disturbed sites might be adventive.

Potentilla missourica Hornemann ex Lindley and *P. normalis* Besser ex Sprengel are older names for this species; both were rejected against a conserved *P. bipinnatifida* with designated lectotypes (see J. Soják 2008b).

96. Potentilla pulchella R. Brown, Chlor. Melvill., 19. 1823 • Potentille jolie

Potentilla pulchella var. *elatior* Lange; *P. pulchella* subsp. *gracilicaulis* (A. E. Porsild) Jurtzev; *P. pulchella* var. *gracilicaulis* A. E. Porsild; *P. usticapensis* Fernald

Stems ascending, 0.1–1.5(–2) dm. **Basal leaves** subpinnate, 1.5–9 cm; petiole 0.7–4 cm, long hairs common to dense, ± appressed to ascending, 0.5–3 mm, soft, short hairs absent, crisped hairs sparse to abundant, glands absent or sparse; leaflets (1 or)2(or 3) per side, on distal ⅙–½ of leaf axis, separate to overlapping, terminal ones oblong or obovate, 0.4–3 × 0.2–1.2 cm, margins revolute, incised ⅔–¾+ to midvein, undivided medial blade 2–6 mm wide, teeth 2–5 per side, linear to narrowly oblong, surfaces dissimilar, abaxial white, long hairs ± abundant, 1–2 mm, soft, short hairs absent or obscured, crisped and/or cottony hairs dense, glands absent or obscured, adaxial grayish white to dark grayish green, long hairs sparse to dense, loosely appressed to spreading, 0.5–2 mm, short hairs absent, crisped hairs absent or sparse, glands absent. **Cauline leaves** 1 or 2. **Inflorescences** (1 or)2–5(–9)-flowered, ± open. **Pedicels** 0.5–1 cm (proximal to 5 cm). **Flowers:** epicalyx bractlets oblong to ovate, 2–4(–6) × 0.8–1.3(–2) mm, lengths ± ¾ times sepals, margins sometimes revolute; hypanthium 2.5–4 mm diam.; sepals 3–5 mm, apex obtuse to subacute, abaxial surfaces: venation indistinct, glands ± abundant, often obscured; petals pale yellow, 4–5 × 2–3 mm, lengths 1.5 times sepals; filaments 1–1.5 mm, anthers 0.4 mm; carpels 30–60, styles papillate-swollen in proximal ¼–⅓, 0.9–1.2 mm. **Achenes** 1.4–1.5 mm, smooth or rugose. **2*n*** = 28.

Flowering summer. Dry or damp, open tundra, silt flats, stream and seashore banks, bird cliff meadows, fine scree, on loam, clay, or calcareous substrates; 0–600 m; Greenland; Man., Nfld. and Labr., N.W.T., Nunavut, Ont., Que., Yukon; Alaska; Eurasia.

Potentilla pulchella is morphologically isolated within sect. *Pensylvanicae* and is very polymorphic. The proposed races (var. *elatior*, subsp./var. *gracilicaulis*) differ only in characters appearing throughout the range of the species (more slender and erect to ascending stems, less hairy leaves). Some different morphs are often found close together but in different habitats.

Some species with palmate or subpalmate leaves combine features of *Potentilla pulchella* and species from sect. *Niveae* (addressed under 8t. sect. *Rubricaules*); others are possible hybrids between *P. hyparctica* and *P. pulchella* (sect. *Aureae*).

97. **Potentilla bimundorum** Soják, Čas. Nár. Muz. Praze, Rada Přír. 141: 195. 1974 • Potentille des deux mondes [F]

Potentilla multifida Linnaeus subsp. *bimundorum* (Soják) Soják

Stems decumbent to ascending, rarely erect, 1–5 dm. **Basal leaves** subpinnate, 7–22(–25) cm; petiole 3–12(–15) cm, long hairs ± abundant, appressed to ascending, 0.5–3 mm, stiff, short hairs absent, cottony or crisped hairs absent, glands absent or sparse; leaflets 3(–4) per side, on distal 1/5–1/3(–1/2) of leaf axis, overlapping or not, terminal ones oblong to broadly oblong, 1.5–5 × 0.7–2 cm, margins strongly revolute, incised nearly to midvein, undivided medial blade 1–3 mm wide, teeth (3–)4–5 per side, linear, rarely narrowly oblong to narrowly lanceolate, surfaces strongly dissimilar, abaxial white or grayish white, long hairs abundant, ± 1 mm, stiff <verrucose> on veins, sparse and soft between veins, short hairs absent, crisped and cottony hairs dense, glands absent, adaxial dark green, long hairs sparse to common, tightly appressed, 0.5–1 mm, short hairs absent, cottony and crisped hairs absent, glands absent or sparse. **Cauline leaves** (1–)2–3. **Inflorescences** (5–)10–40+-flowered, ± congested to open. **Pedicels** 0.3–1 cm (proximal to 3 cm). **Flowers:** epicalyx bractlets linear to narrowly oblong-lanceolate, 3–5 × 0.9–1.3 mm, lengths ± 3/4 times sepals, margins flat; hypanthium 2.5–3.5 mm diam.; sepals 4–7(–8) mm, apex acute to obtuse, abaxial surfaces: venation prominent at least proximally, glands ± sparse, obscured to evident; petals pale yellow, 3.5–5 × 3–5 mm, lengths ± equal to sepals; filaments 0.5–1.5 mm, anthers 0.5–0.6 mm; carpels 40–60, styles papillate-swollen at very base, if at all, 0.8–0.9 mm. **Achenes** 1.3–1.5 mm, rugose. **2*n*** = 28.

Flowering summer. Road and railway verges, gravelly ruderal sites, acidic rocky outcrops, stream banks and gravel bars, sandy-gravelly flood plains, gravel ridges, sandy lakeshores, sandy steppe bluffs; 0–1000 m; Alta., Man., N.W.T., Nunavut, Ont., Que., Sask., Yukon; Alaska; e Asia (Russian Far East).

Potentilla bimundorum is the correct name for what has been treated as *P. multifida* Linnaeus in North America. It is the only American representative of the mainly Eurasian *P. multifida* group, which is sometimes treated as a distinct section. J. Soják (2005) presented a useful survey and key to this group, showing that its species, in spite of being merged within *P. multifida* in the broad sense by many authors, are each distinct and with their own range. *Potentilla multifida* in the narrow sense is restricted to eastern Europe and western Siberia.

98. **Potentilla anachoretica** Soják, Preslia 46: 70, fig. [p. 71]. 1974

Stems erect, 1–3 dm. **Basal leaves** pinnate, (4–)5–10 (–14) cm; petiole 3–7(–10) cm, long hairs abundant, subappressed to appressed, 0.5–2 mm, ± weak, short hairs absent, crisped or cottony hairs absent or sparse, glands absent or sparse; leaflets 2 or 3 per side, on distal 1/3–1/2 of leaf axis, separate, terminal ones elliptic to obovate, (1.8–)2–2.5(–3) × 0.8–1.3 cm, margins revolute, incised nearly to midvein, undivided medial blade 1–2 mm wide, teeth (4 or)5–7 per side, linear, surfaces ± dissimilar, abaxial grayish to white, long hairs ± abundant especially on veins, 1–1.5 mm, weak, short hairs absent, crisped hairs abundant to dense, glands absent or obscured, adaxial greenish gray, long hairs common, appressed, 0.5–1 mm, short hairs absent, crisped hairs absent or sparse, glands absent or obscured. **Cauline leaves** 1 or 2. **Inflorescences** 2–5 -flowered, very open. **Pedicels** 2–4 cm. **Flowers:** epicalyx bractlets narrowly elliptic to linear, 3–4 × 0.5–0.8(–1) mm, lengths 2/3–3/4 times sepals, margins flat; hypanthium 2.5–3 mm diam.; sepals 3.5–5 mm, apex subacute, abaxial surfaces: venation indistinct, glands sparse to common, often obscured; petals yellow, 7.5–10(–10.5) × 7.5–9(–10) mm, much exceeding sepals; filaments 1–2 mm, anthers 0.5–0.7 mm; carpels 50–60, styles papillate-swollen at very base, if at all, 0.8–0.9 mm. **Achenes** 1.4–1.6 mm, smooth to faintly rugose. **2*n*** = 28 (14, 42, 56) (Russian Far East).

Flowering summer. Dry ridges, rocky outcrops, south-facing slopes, mainly on basic substrates; 500 m; Alaska; e Asia (Russian Far East).

Potentilla anachoretica is isolated in sect. *Pensylvanicae*. B. A. Jurtzev (1993, pers. comm.) claimed indirect support for the presence of *P. anachoretica* in Alaska and Yukon by the presence of three hybrid species presumably involving *P. anachoretica* in their parentage: *P. borealis* Soják, *P. murrayi* Jurtzev, and *P. petrovskyi* Soják (sect. *Rubricaules*; discussed by B. Ertter et al. 2013). *Potentilla anachoretica* is confirmed from at least two sites in Brooks Range, northern Alaska.

9. IVESIA Torrey & A. Gray in War Department [U.S.], Pacif. Railr. Rep. 6(3): 72, plate 11. 1858 • [For Eli Ives, 1779–1861, professor of pediatrics, materia medica, and botany at Yale University]

Barbara Ertter

James L. Reveal

Horkelia Chamisso & Schlechtendal subg. *Ivesia* (Torrey & A. Gray) Rydberg; *Potentilla* Linnaeus sect. *Ivesia* (Torrey & A. Gray) Baillon; *Potentilla* subg. *Ivesia* (Torrey & A. Gray) Jepson

Herbs, perennial, rosetted, tufted, or matted, sometimes forming hanging clumps, <often aromatic>, (0.1–)0.3–6(–6.5) dm, sparsely to densely hirsute, sericeous, villous, or pilose, sometimes glabrate, sparsely to densely glandular-puberulent to stipitate-glandular; caudices simple or branched, branches usually relatively short, rarely elongate (*I. cryptocaulis*), taproot slender to stout, sometimes fusiform and fleshy. **Stems** (1 or)2–20+, prostrate or pendent to erect, <green, grayish, or reddish>. **Leaves** winter-marcescent, primarily basal, <cauline 0–10(–15), well developed to vestigial>, usually alternate (cauline opposite in *I. muirii*, *I. webberi*), odd-pinnate; stipules persistent, sometimes absent, basally adnate to petiole, linear to lanceolate or elliptic, margins usually entire, cauline sometimes lobed; petiole present; blade narrowly oblong to filiform, planar to cylindric, 0.5–20(–25) cm, foliaceous, leaflets (3–)7–161, <sometimes separate, more often overlapping at least distally, sometimes tightly imbricate, terminal distinct or, more often, confluent with distalmost lateral ones>, ovate or obovate to orbiculate or ± flabellate, <often oriented in 3-dimensions, leaflet arrangement appearing verticillate>, margins flat, palmately incised from ¼ to completely to base into teeth or linear or oblanceolate to oval or obovate lobes, venation palmate. **Inflorescences** terminal, (1–)3–100(–250)-flowered, ± cymes, open or of 1–several loose to capitate glomerules; bracts present, reduced; bracteoles absent. **Pedicels** present, <straight to sigmoid>. **Flowers** 4–15 mm diam.; epicalyx bractlets (0–)5; hypanthium patelliform to campanulate or cupulate to turbinate, 0.5–3(–4) mm; sepals 5 (usually 4 in *I. campestris*), spreading, lanceolate to broadly ovate or deltate; petals 5 (usually 4 in *I. campestris*), golden to pale yellow to white, sometimes pink-tinged (red in *I. multifoliolata*), linear or narrowly oblanceolate to obovate, sometimes obcordate; stamens 5–20(–40), usually shorter than petals; <filaments not forming tube>; torus ± flat to conic, or turbinate (in sect. *Comarella*), rarely elongate and stipelike (*I. arizonica*); carpels 1–20(–40), glabrous, styles subterminal, <usually ± filiform, sometimes rough-thickened proximally>; ovule 1. **Fruits** aggregated achenes, individually deciduous, 1–20+, obliquely ovoid to reniform, 0.8–3 mm, <sometimes rugose and/or carunculate>, glabrous; hypanthium persistent; sepals persistent, erect; styles tardily deciduous, <jointed>. $x = 7$.

Species 30 (30 in the flora): w United States, nw Mexico.

The often aromatic (resin-scented) ivesioid genera (*Horkelia*, *Horkeliella*, and *Ivesia*) have generally been assumed to be derived from *Potentilla* in the broad sense, an interpretation confirmed by molecular analyses (T. Eriksson et al. 1998, 2003; C. Dobeš and J. Paule 2010; M. H. Töpel et al. 2012). Continued recognition of the three genera (and *Duchesnea*) accordingly creates a paraphyletic *Potentilla*; the senior author does not accept the dictum that paraphyletic groups are inherently unnatural and finds it more useful to continue recognizing the ivesioid genera as here circumscribed.

In spite of the morphologic and ecologic extremes encompassed within *Ivesia*, spanning the continuum from planar-leaved desert chasmophytes to alpine species with mousetail-like leaves, molecular phylogenetic analyses of representatives from all sections of *Ivesia*, *Horkelia*, and *Horkeliella* result in a monophyletic near-polytomy, indicating rapid morphologic radiation not matched by molecular divergence (C. Dobeš and J. Paule 2010; M. H. Töpel et al. 2012). This interpretation is not fully supported by M. Töpel et al., but until the lack of concordance between molecular-based clades and even the most well-defined morphology-based groups is addressed, any conclusions based on molecular analysis alone are inconclusive.

The hypothesis on which the sequence of species used here is based is that the ivesiod genera represent a threshold-crossing radiation in response to the late Tertiary development of xeric conditions in western North America, with adaptations that allow drought-avoidance or minimize water loss. A recurrent trend is the progressive reduction in size and number of floral parts occurring independently in multiple lineages. The end product is a flower in which the nectariferous, patelliform hypanthium itself becomes the main pollinator attractor. Much of this evolutionary radiation has occurred in the various island-like habitats that characterize the arid West, with the result that the ivesioid genera include a high percentage of narrowly endemic species.

In the following descriptions, stems refer to flowering stems, with length including the inflorescence. For chasmophytic (growing on cliffs) species in sect. *Setosae*, terms such as decumbent are in reference to the vertical rock face. The inflorescence is composed of the branched portion of the flowering stem, including the proximalmost branched node. Cauline leaves are on the portion of the flowering stem proximal to the inflorescence. Planar leaves are those in which the leaflets are largely arranged in a single plane on both sides of the rachis; in cylindric leaves the lobes of deeply incised leaflets are arranged evenly around the rachis and thus appear verticillate. Well-developed blades of cauline leaves are similar to those of basal leaves, though progressively reduced distally; vestigial blades are reduced to a simple lobe, which, with the stipules, has the appearance of a 3-lobed bract. Leaf incisions are referred to as teeth if less than half to base, as lobes if more than half. Counts of leaflet lobes are of ultimate segments, including teeth, not just of primary segments incised to the base or nearly so. Inflorescence structure is described at peak anthesis; at early anthesis flowers are often more congested than indicated here and sometimes become more distant in fruit. Pedicels at proximal nodes are often significantly longer than others; those in most sections remain more or less straight in fruit or bend proximal to the flower, in contrast to those of sect. *Setosae* that generally become strongly curved or even sigmoid. Petal color is described at peak of flowering; yellow petals often fade afterward. Vestiture descriptions are primarily of non-glandular hairs; subsessile glands or evidently septate-glandular trichomes are universal in *Ivesia*, though often sparse or hidden by dense non-glandular vestiture.

SELECTED REFERENCE Ertter, B. 1989. Revisionary studies in *Ivesia* (Rosaceae). Syst. Bot. 14: 231–244.

1. Pedicels usually becoming ± curved, often sigmoid; basal leaves planar or loosely to tightly cylindric, leaflets separate to overlapping, toothed ¼–¾ or lobed to base; achenes often rugose, ± carunculate (except *I. cryptocaulis*); flowers arranged ± individually (or ± glomerate in *I. paniculata* and *I. rhypara*); plants usually ± petrophytic, often forming hanging clumps in vertical rock crevices . . . . 9a. *Ivesia* sect. *Setosae*, p. 221
1. Pedicels remaining ± straight; basal leaves loosely to tightly cylindric (sometimes weakly planar in *I. longibracteata*), leaflets ± overlapping, lobed to base (except *I. multifoliolata*); achenes smooth, not carunculate; flowers arranged individually or in 1–several loose to capitate glomerules; plants usually not petrophytic (except *I. longibracteata*), not forming hanging clumps.

[2. Shifted to left margin.—Ed.]

2. Stamens 5; petals not or scarcely clawed.
   3. Inflorescences usually ± congested, sometimes becoming open in fruit, flowers arranged in 1+ loose to capitate glomerules; hypanthia shallowly cupulate or campanulate, sometimes turbinate; petals usually narrowly oblanceolate or spatulate to broadly obovate, sometimes linear, not medially reflexed, usually yellow, sometimes white (*I. utahensis*); stems (0.2–)0.3–2(–4) dm . . . . 9b. *Ivesia* sect. *Ivesia* (in part), p. 230
   3. Inflorescences open, flowers arranged individually; hypanthia patelliform; petals linear to oblanceolate or narrowly elliptic, medially reflexed, yellow or red; stems (1.8–)2–6 (–6.5) dm . . . . .9d. *Ivesia* sect. *Comarella*, p. 243
2. Stamens 10–20; petals often clawed.
   4. Cauline leaves (0 or)1; stems 0.3–1(–1.5) dm; petals golden yellow. . .9b. *Ivesia* sect. *Ivesia* (in part), p. 230
   4. Cauline leaves (1 or)2–10(–15); stems (0.3–)1–4.5(–5.5) dm; petals light yellow to white, sometimes pink-tinged.
      5. Basal leaves not mousetail-like, loosely to tightly cylindric, individual leaflets distinguishable, sparsely to densely hairy or glabrate . . .9c. *Ivesia* sect. *Unguiculatae* (in part), p. 237
      5. Basal leaves mousetail-like, tightly to very tightly cylindric, individual leaflets not or scarcely distinguishable, densely hairy.
         6. Stamens 20; carpels (1 or)2–20; inflorescences: flowers arranged individually and/or in glomerules; anthers white to yellowish, ± as long as to longer than wide . . . . .9c. *Ivesia* sect. *Unguiculatae* (in part), p. 237
         6. Stamens 15; carpels 1; inflorescences: flowers arranged individually; anthers purple, shorter than wide . . . . .9e. *Ivesia* sect. *Stellariopsis*, p. 245

9a. IVESIA Torrey & A. Gray sect. SETOSAE (Rydberg) O. Stevens in N. L. Britton et al., N. Amer. Fl. 22(7): 8. 1959

Ivesia [unranked] *Setosae* Rydberg in N. L. Britton et al., N. Amer. Fl. 22: 290. 1908; *Horkelia* Chamisso & Schlechtendal [unranked] *Saxosae* Rydberg; *Ivesia* sect. *Saxosae* (Rydberg) Ertter & Reveal; *Potentilla* Linnaeus subg. *Purpusia* (Brandegee) J. T. Howell; *Potentilla* [unranked] *Saxosae* (Rydberg) Rydberg; *Potentilla* sect. *Saxosae* (Rydberg) O. Stevens; *Purpusia* Brandegee

Plants usually tufted or matted (diffusely so in *I. cryptocaulis*), sometimes rosetted, often forming hanging clumps in vertical rock crevices, ± aromatic; taproot slender or stout, not fusiform or fleshy. **Stems** (0.1–)0.2–2.5(–3) dm. **Basal leaves** planar or loosely to tightly cylindric; stipules present; leaflets separate to overlapping, usually individually distinguishable, toothed ¼–¾ or lobed to base, rarely entire, sparsely to densely hairy; terminal leaflets distinct or indistinct. **Cauline leaves** (0 or)1 or 2(–4), not paired; blade well developed to vestigial. **Inflorescences** open to congested, flowers arranged ± individually (or ± glomerulate in *I. paniculata* and *I. rhypara*). **Pedicels** usually becoming ± curved, often sigmoid. **Flowers:** hypanthium mostly patelliform, sometimes shallowly cupulate or (in *I. arizonica*) campanulate or turbinate; petals not medially reflexed, yellow to white, not clawed, apex obtuse to rounded; stamens 5–35(–40), anthers longer or shorter than wide, laterally dehiscent; carpels 1–20(–40). **Achenes** vertical, usually rugose, sometimes smooth, ± carunculate (except *I. cryptocaulis*).

Species 11 (11 in the flora): w United States, nw Mexico.

The planar-leaved, chasmophytic members of sect. *Setosae* found in and around the mountains centered on the Mojave Desert are hypothesized to represent the ancestral radiation of *Ivesia*, on the basis of morphology and distribution. A secondary radiation northward into the Great Basin is characterized by a complete sequence in size reduction, dissection, and increased number of leaflets, paralleled by a reduction in numbers of floral parts. *Ivesia saxosa* retains the most pleisiomorphies and could readily be accommodated within *Potentilla* in the strict sense, as has commonly been done. Gross morphology of *I. saxosa* is nearly identical to that of *I. arizonica*, which has been treated as a distinct genus, *Purpusia*, on the basis of its lack of an epicalyx, deep hypanthium, and stipelike torus. The narrow endemic *I. patellifera* is intermediate between these two species, and an identical habit is shared by *I. baileyi* and *I. jaegeri*, which have more dissected leaflets and more reduced flowers. All of these species are now placed in the same genus, either as *Ivesia* (B. Ertter 1989) or as *Potentilla* (J. T. Howell 1945).

Most species in sect. *Setosae* are more or less petrophytic (in contrast to other sections), often growing in crevices of vertical cliffs. The protection offered by these habitats often allows extended blooming periods and resultant indeterminate inflorescences that produce flowers as long as conditions are suitable, including at the height of summer when relatively few other desert species are in bloom. Stems can initially radiate in almost all directions relative to the substrate (vertical in chasmophytes, horizontal in others), but they generally succumb to gravity with increased number of flowers and become either pendent or prostrate. In particularly favorable years, blooming season and flower number can exceed the ranges given here. The small but often numerous flowers are visited by a wide diversity of potential pollinators, including ants, which are attracted to the glistening, nectariferous hypanthial disc. Most species of sect. *Setosae* have carunculate seeds, which may indicate that ants play a role in seed dispersal, a particularly useful adaptation for chasmophytic plants.

Since *Ivesia longibracteata* (sect. *Ivesia*) is sometimes identified as a member of sect. *Setosae*, it is included herein and keys out in the sixth couplet.

1. Basal leaves planar; lateral leaflets (2–)5–20(–25) mm, incised ¼–¾ to base into teeth or lobes.
    2. Stamens 15–35(–40) . . . . 1. *Ivesia saxosa*
    2. Stamens 5–10.
        3. Epicalyx bractlets 0(–3); Mojave Desert and n Arizona.
            4. Hypanthia patelliform, 0.5(–1) mm deep; Kingston Range, California . . . . 4. *Ivesia patellifera*
            4. Hypanthia turbinate or campanulate, 1.5–3(–5) mm deep; se California, s Nevada, and n Arizona . . . . 5. *Ivesia arizonica*
        3. Epicalyx bractlets 5; n Great Basin and adjacent mountains.
            5. Lateral leaflets incised ¼–¾ to base, lobes not setose apically; basal leaves: sheathing bases not or sparsely strigose abaxially; volcanic substrates . . . . 6. *Ivesia baileyi*
            5. Lateral leaflets incised ± ¾ to base, sometimes nearly to base, lobes usually ± setose apically; basal leaves: sheathing bases ± strigose abaxially; usually calcareous substrates . . . . 7. *Ivesia setosa* (in part)
1. Basal leaves ± cylindric to weakly planar; lateral leaflets 0.5–8 mm, incised to base or nearly so into separate lobes, sometimes entire.
    6. Epicalyx bractlets longer than sepals; inflorescences loosely subcapitate; pedicels remaining straight (Castle Crags, n California) . . . . 18. *Ivesia longibracteata* (sect. *Ivesia*)
    6. Epicalyx bractlets shorter than (or equal to) sepals; inflorescences open to congested, not subcapitate; pedicels usually ± curved in fruit.
        7. Stamens 20.
            8. Petals yellow, narrowly oblanceolate; lateral leaflet lobes 3–6; Spring Mountains, Nevada, and Clark Mountain, California . . . . 2. *Ivesia jaegeri*
            8. Petals white, obovate; lateral leaflet lobes (0–)2–4; San Jacinto Mountains, California . . . . 3. *Ivesia callida*
        7. Stamens 5.

[9. Shifted to left margin.—Ed.]

9. Plants grayish; lateral leaflet surfaces densely hirsute, cryptically glandular; inflorescences congested, 5–200-flowered; petals linear to narrowly oblanceolate, 1–1.5 mm, white to pale yellowish; 1300–1900 m.
    10. Leaflet lobes (0–)2–4(–9), elliptic to obovate or orbiculate, 0.5–3(–4) mm; n Nevada, se Oregon . . . . 10. *Ivesia rhypara*
    10. Leaflet lobes (0–)3–8(–15), elliptic to narrowly obovate, 0.5–2 mm; ne California. . . . 11. *Ivesia paniculata*
9. Plants green to grayish green; lateral leaflet surfaces ± sparsely hirsute or loosely long-strigose, evidently glandular; inflorescences ± open to ± congested, 1–20(–30)-flowered; petals oblanceolate to spatulate or narrowly obovate, (1–)1.5–3.2 mm, ± yellow; 1700–4000 m.
    11. Plants diffusely matted; lateral leaflet surfaces loosely long-strigose; Spring Mountains, s Nevada . . . . 9. *Ivesia cryptocaulis*
    11. Plants tufted to ± densely matted; lateral leaflet surfaces ± sparsely hirsute; Great Basin and n Sierra Nevada.
        12. Lateral leaflets incised ± ¾ to base, sometimes nearly to base, into ovate teeth to narrowly obovate lobes; usually calcareous substrates . . . . 7. *Ivesia setosa* (in part)
        12. Lateral leaflets incised to base or nearly so into oblanceolate to obovate or elliptic lobes; various substrates, seldom calcareous . . . . 8. *Ivesia shockleyi*

1. Ivesia saxosa (Lemmon ex Greene) Ertter, Syst. Bot. 14: 232. 1989 • Rock ivesia

Potentilla saxosa Lemmon ex Greene, Pittonia 1: 171. 1888; *P. acuminata* H. M. Hall; *P. saxosa* subsp. *sierrae* Munz

Plants green, ± tufted, often forming hanging clumps, sometimes rosetted. **Stems** pendent or prostrate to ascending, 0.4–2.6(–3) dm. **Basal leaves** planar, 2–15 cm; sheathing base not strigose abaxially; petiole 3–9 cm; lateral leaflets (1 or)2–4(–7) per side, separate to slightly overlapping, obovate to orbiculate or flabellate, (3–)5–15(–22) mm, incised ¼–¾ to base into 5–15 broadly ovate teeth or oblanceolate lobes, sometimes also medially split to base (Kern Plateau), apex not setose, surfaces ± sparsely short-pilose, prominently glandular; terminal leaflets ± distinct. **Cauline leaves** 1 or 2(–4); blade well developed. **Inflorescences** (1–)3–30(–60)-flowered, open, (0.5–)2–8 cm diam. **Pedicels** (5–)10–30 mm. **Flowers** 6–12 mm diam.; epicalyx bractlets 5, broadly lanceolate to oblong, 1–2.5 mm; hypanthium patelliform, 0.5–1.5 × 2–4(–4.5) mm; sepals 2–4.5 mm, acute; petals yellow, oblanceolate to obovate, (1.5–)2–4 mm; stamens 15–35(–40), filaments 0.3–1 mm, anthers yellow, subrotund, 0.2–0.3(–0.5) mm; carpels (3–)10–20(–40), styles 1–2 mm. **Achenes** greenish white to light tan, 1–1.8 mm, faintly rugose, ± carunculate.

Flowering late spring–summer. Dry, rocky outcrops of granitic or volcanic origin, usually crevices of more or less vertical protected cliffs or boulders, mainly in oak and conifer woodlands; 900–3300 m; Calif.; Mexico (Baja California).

Ivesia saxosa is encountered in the foothills and mountains of southeastern California from the White Mountains and adjacent eastern slopes of the Sierra Nevada of Mono County to the Transverse Ranges (including the Little San Bernardino Mountains) of Kern and San Bernardino counties. Populations on the Kern Plateau tend to have more dissected leaflets and may deserve recognition as a distinct variety. On the Peninsular Ranges, *I. saxosa* occurs in Riverside and San Diego counties and into northern Baja California, Mexico, as far south as the western slopes of the Sierra Juárez and Sierra San Pedro Mártir.

2. Ivesia jaegeri Munz & I. M. Johnston, Bull. Torrey Bot. Club 56: 165. 1929 • Jaeger's ivesia [C] [E]

Potentilla jaegeri (Munz & I. M. Johnston) L. C. Wheeler

Plants green, ± tufted to ± matted, often forming hanging clumps, sometimes rosetted. **Stems** pendent or prostrate to ascending, 0.3–1.5 dm. **Basal leaves** ± loosely cylindric to weakly planar, 2–8(–10) cm; sheathing base not or sparsely strigose abaxially; petiole 0.5–4 cm; lateral leaflets (1–)4–6(–8) per side, ± overlapping at least distally, ± flabellate, 2–6(–7) mm, incised nearly to base into 3–6 oblanceolate to narrowly obovate lobes, apex not or scarcely setose, surfaces ± sparsely short-pilose, ± glandular; terminal leaflets indistinct. **Cauline leaves** (1 or)2; blade well developed. **Inflorescences** (1–)3–15-flowered, open, (0.5–)1.5–4(–7) cm diam. **Pedicels** 5–30 mm. **Flowers** 5–11 mm diam.; epicalyx bractlets 5, lanceolate, 1–2 mm; hypanthium patelliform, 1 × 2–4 mm; sepals 2–3 mm, ± acute; petals yellow, narrowly oblanceolate, 1.5–2 mm; stamens 20, filaments 1–1.5(–1.8) mm, anthers yellow, subrotund, 0.2–0.3 mm; carpels 3–8, styles 1.5–2 mm. **Achenes** tan to light brown, 1–2 mm, ± rugose, ± carunculate.

Flowering summer. Dry, rocky outcrops of limestone origin, usually crevices of more or less vertical protected cliffs or boulders, mainly in conifer woodlands; of conservation concern; 1600–3600 m; Calif., Nev.

Ivesia jaegeri is known only from the Spring Mountains, Clark County, Nevada, and the Clark Mountains, San Bernardino County, California.

3. **Ivesia callida** (H. M. Hall) Rydberg in N. L. Britton et al., N. Amer. Fl. 22: 286. 1908 • Tahquitz ivesia [C] [E]

Potentilla callida H. M. Hall, Univ. Calif. Publ. Bot. 1: 86. 1902

Plants green, ± tufted to matted, often forming hanging clumps, sometimes rosetted. **Stems** usually pendent or prostrate to ascending, sometimes nearly erect, 0.2–1.5 dm. **Basal leaves** ± loosely cylindric to weakly planar, 1–7 cm; sheathing base sparsely strigose abaxially; petiole 0.3–3 cm; lateral leaflets (1–)5–8 per side, slightly overlapping, elliptic to flabellate, 2–6(–7) mm, incised nearly to base into (0–)2–4 oblanceolate to elliptic lobes, apex sometimes ± setose, surfaces loosely long-strigose, sparsely glandular; terminal leaflets indistinct. **Cauline leaves** (1 or)2; blade reduced. **Inflorescences** 1–10(–15) -flowered, open, (0.5–)2–5 cm diam. **Pedicels** 5–15 mm. **Flowers** 7–10 mm diam.; epicalyx bractlets 5, narrowly lanceolate, 1.5–3 mm; hypanthium patelliform, 0.5–1 × 2–3 mm; sepals 2–3.5 mm, acute; petals white, obovate, 2–3.5 mm; stamens 20, filaments 1.5–3 mm, anthers maroon, subrotund, 0.2–0.3 mm; carpels 4–8, styles 1.5–1.8 mm. **Achenes** greenish white to light tan, 1.5 mm, smooth to lightly rugose, ± carunculate.

Flowering summer. Crevices of dry, rocky outcrops of granite, sometimes more or less vertical protected cliffs, in montane conifer woodlands; of conservation concern; 2400–2500 m; Calif.

Ivesia callida occurs in the Tahquitz Peak area of the San Jacinto Wilderness Area, San Jacinto Mountains, Riverside County, where two populations are known (K. Berg 1983). Plants form tight mats in horizontal crevices and loose pendent clumps on vertical rock faces.

4. **Ivesia patellifera** (J. T. Howell) Ertter, Syst. Bot. 14: 232. 1989 • Kingston Mountains ivesia [C] [E]

Potentilla patellifera J. T. Howell, Leafl. W. Bot. 4: 173. 1945

Plants green, ± tufted, often forming hanging clumps, sometimes rosetted. **Stems** pendent or prostrate to ascending, (0.5–)1–2 dm. **Basal leaves** planar, (2–)5–15 cm; sheathing base not strigose abaxially; petiole 1–5 cm; lateral leaflets (1 or)2 or 3(or 4) per side, separate, obovate to orbiculate, 5–20 mm, incised ¼–½ to base into 5–9 broadly ovate teeth, apex not setose, surfaces short-pilose, ± glandular; terminal leaflets distinct. **Cauline leaves** (0–)2; blade well developed. **Inflorescences** (1–)3–20(–35)-flowered, open, (0.5–) 1.5–4(–6) cm diam. **Pedicels** 5–20(–30) mm. **Flowers** 7–10 cm diam.; epicalyx bractlets 0; hypanthium patelliform, 0.5(–1) × 2–3 mm; sepals 2–4 mm, broadly acute; petals yellow, narrowly oblanceolate, 2–3 mm; stamens 5–10, filaments 0.6–1.2(–1.5) mm, anthers yellow, oblong, 0.8–1 mm; carpels 4–10, styles 1.5–2 mm. **Achenes** greenish white to light tan, 1.5–2 mm, faintly rugose, ± carunculate.

Flowering summer. Dry, rocky outcrops of limestone, usually crevices of more or less vertical protected cliffs or boulders, in conifer woodlands; of conservation concern; 1400–2200 m; Calif.

Ivesia patellifera is confined to crevices in steep wash and canyon walls in the Kingston Mountains of San Bernardino County.

5. **Ivesia arizonica** (Eastwood ex J. T. Howell) Ertter, Syst. Bot. 14: 233. 1989 • Purpusia [C] [E] [F]

Purpusia arizonica Eastwood ex J. T. Howell, Madroño 2: 12. 1930

Plants green, ± tufted, often forming hanging clumps, sometimes rosetted. **Stems** pendent or prostrate to ascending, (0.2–)0.5–1.6(–3) dm. **Basal leaves** planar, 2–15(–20) cm; sheathing base usually not strigose abaxially; petiole 3–10 cm; lateral leaflets (2 or)3 or 4(or 5) per side, separate, broadly ovate or obovate to orbiculate, (2–)5–15 (–18) mm, incised ¼–¾ to base into (3–)7–11 ± ovate teeth, apex not setose, surfaces ± sparsely short-pilose, ± glandular; terminal leaflets ± distinct. **Cauline leaves** 1–3; blade well developed. **Inflorescences** (1–)5–30 (–150)-flowered, open, 0.5–14 cm diam. **Pedicels** 5–30 mm. **Flowers** 6–12 mm diam.; epicalyx bractlets 0(–3), lanceolate, 0.9–1.8 mm; hypanthium turbinate or campanulate, 1.5–3(–5) × 1–3(–4) mm; sepals 2–4(–5) mm,

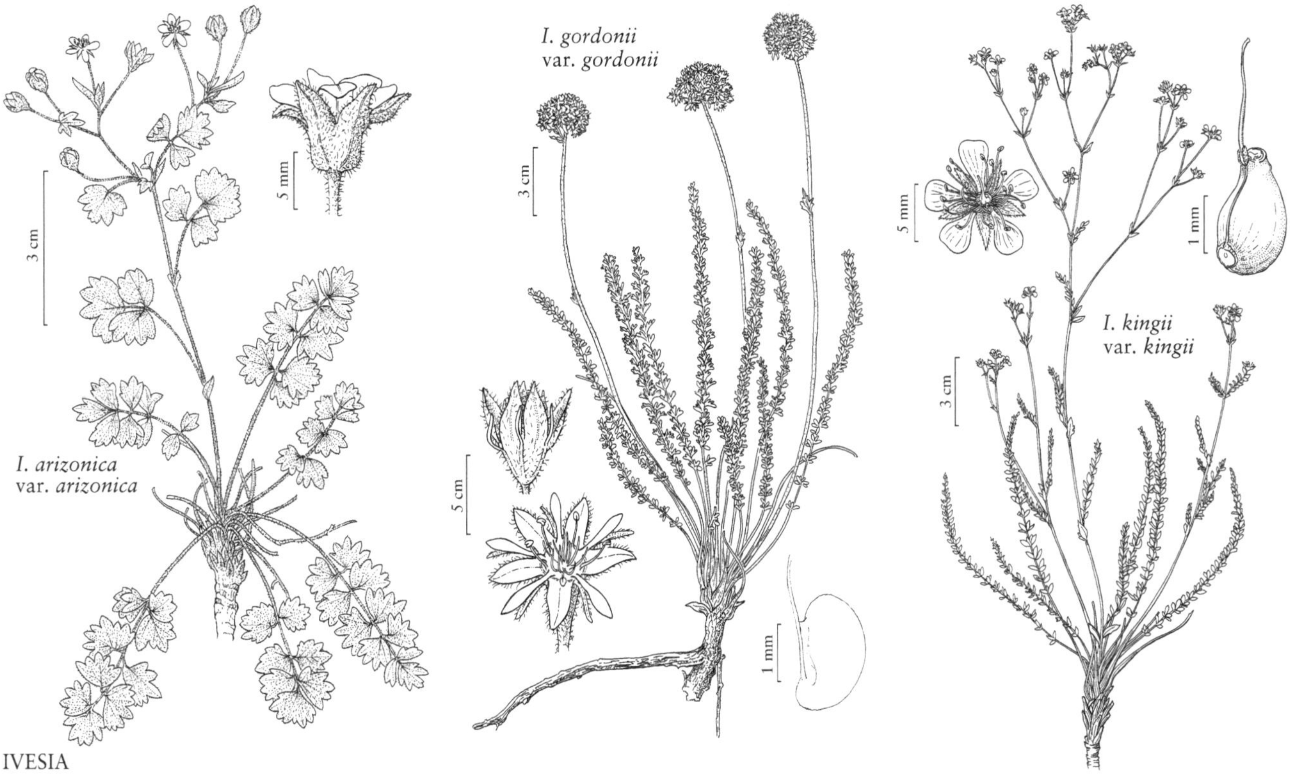

IVESIA

acute; petals yellow or white, oblanceolate to elliptic or obovate, (1.6–)2–4 mm; stamens 5, filaments 0.8–1.3 mm, anthers ± yellow, oblong to narrowly ovate, 0.6–1.5 mm; carpels (2–)6–10(–13), <atop a stipelike torus>, styles 0.9–1.8 mm. **Achenes** greenish white to light tan, 1.2–2 mm, faintly rugose, ± carunculate.

Varieties 2 (2 in the flora): sw United States.

Because the epithet *saxosa* was retained for *Potentilla saxosa* upon transfer to *Ivesia*, a different epithet (*arizonica*) was needed when *Purpusia* was likewise transferred in the same publication (B. Ertter 1989). The correct name for this species in *Potentilla* is *P. osterhoutii* (A. Nelson) J. T. Howell, due to the existence of *P. arizonica*.

1. Petals yellow; hypanthia campanulate. . . . . 5a. *Ivesia arizonica* var. *arizonica*
1. Petals white; hypanthia turbinate . . . . 5b. *Ivesia arizonica* var. *saxosa*

5a. Ivesia arizonica (Eastwood ex J. T. Howell) Ertter var. **arizonica** • Yellow purpusia [C] [E] [F]

Potentilla osterhoutii (A. Nelson) J. T. Howell; *Purpusia osterhoutii* A. Nelson

Inflorescences 1–20(–60)-flowered, 0.5–7 cm diam. **Flowers:** hypanthium campanulate, ± as deep as wide; petals yellow; anthers 0.6–1 mm.

Flowering summer. Dry, rocky outcrops of calcareous or volcanic origin, usually in crevices of more or less vertical protected cliffs or large boulders, in sagebrush communities, conifer woodlands; of conservation concern; 1200–2400(–3400) m; Ariz., Calif., Nev., Utah.

Variety *arizonica* occurs as isolated populations from the Funeral and Grapevine mountains, Inyo County, California, to the Kolob Plateau of Washington County, Utah, and on the walls of the Grand Canyon and in the Oak Creek Canyon area of northern Arizona. Significant variation in vestiture, hypanthium depth, and other features occurs among these widely scattered population clusters. High-elevation plants (3100–3400 m) from the Troy Peak area of the Grant Range in northeastern Nye County, Nevada, are notably depauperate (0.2–0.5 dm) with leaflets 2–5 mm and petals 1.6–2.5 mm. All other populations occur below 2400 m.

5b. Ivesia arizonica (Eastwood ex J. T. Howell) Ertter var. **saxosa** (Brandegee) Ertter, Syst. Bot. 14: 233. 1989 • Rock purpusia [C] [E]

Purpusia saxosa Brandegee, Bot. Gaz. 27: 447. 1899; *Potentilla osterhoutii* (A. Nelson) J. T. Howell var. *saxosa* (Brandegee) J. T. Howell

Inflorescences (1–)5–30(–150)-flowered, (0.5–)2–14 cm diam. **Flowers:** hypanthium turbinate, ± 2 times as deep as wide; petals white; anthers 1–1.5 mm.

Flowering summer. Dry, rocky outcrops of mainly volcanic origin, usually in crevices of more or less vertical protected cliffs or boulders, in sagebrush communities, pinyon-juniper woodlands; of conservation concern; 1500–2100 m; Nev.

Variety *saxosa* is known from scattered locations in the North and South Pahroc ranges, Lincoln County, and on Pahute Mesa, Nye County, Nevada. Reports of this variety in the Sheep Range of Clark County, Nevada (N. H. Holmgren 1997b; T. L. Ackerman et al. 2003), are probably based on var. *arizonica*.

6. Ivesia baileyi S. Watson, Botany (Fortieth Parallel), 90. 1871 [E]

Potentilla baileyi (S. Watson) Greene

Plants green, ± tufted, often forming hanging clumps, sometimes rosetted. **Stems** pendent or prostrate to ascending, 0.5–2(–2.5) dm. **Basal leaves** planar, 3–12(–18) cm; sheathing base not or sparsely strigose abaxially; petiole 1–8 cm; lateral leaflets 2–6(–10) per side, separate to slightly overlapping distally, ovate or obovate to flabellate, 4–15(–25) mm, incised ¼–¾ to base into 3–11(–15) ovate to oblanceolate teeth or lobes, apex not setose, surfaces ± sparsely short-pilose or hirsute, ± glandular; terminal leaflets ± distinct. **Cauline leaves** 1 or 2; blade reduced. **Inflorescences** (1–)5–40-flowered, open, (1–)1.5–8(–10) cm diam. **Pedicels** 2–15(–30) mm. **Flowers** 4–10 mm diam.; epicalyx bractlets 5, narrowly lanceolate to ovate, 0.8–2.5 mm; hypanthium patelliform, 0.5–2 × 2–4 mm; sepals (1.2–)1.5–4 mm, acute; petals white or pale yellow, oblanceolate to narrowly spatulate, 1.5–2.5 mm; stamens 5, filaments 0.5–1.1 mm, anthers yellow, sometimes with reddish margins, oblong, 0.4–0.7 mm; carpels (1–)3–8, styles 0.9–1.8 mm. **Achenes** greenish white to light tan, 1.5–2 mm, smooth or rugose, ± carunculate.

Varieties 2 (2 in the flora): w United States.

The two varieties accepted here were not recognized by D. D. Keck (1938), who instead treated *Ivesia setosa* as a variety of *I. baileyi*. Field investigations confirm that the three entities are reasonably distinct morphologically, with intergradation where their otherwise distinct ranges overlap in central Nevada.

1. Petals pale yellow; hypanthium interior yellow to golden; epicalyx bractlets ± ¾ to as long as sepals; achenes smooth . . . . 6a. *Ivesia baileyi* var. *baileyi*
1. Petals white; hypanthium interior pale green or cream to maroon; epicalyx bractlets usually less than ½ as long as sepals; achenes rugose . . . . 6b. *Ivesia baileyi* var. *beneolens*

6a. Ivesia baileyi S. Watson var. **baileyi** • Bailey's ivesia [C] [E]

Basal leaves: sheathing base densely glandular abaxially, sometimes sparsely strigose. **Pedicels** 2–5 mm at flowering, to 12 mm in fruit. **Flowers** (1–)5–30, 4–7 mm diam.; epicalyx bractlets broadly lanceolate to ovate, 1.2–2.5 mm, ± ¾ to as long as sepals; hypanthium interior yellow to golden; sepals (2–)2.5–4 mm; petals pale yellow; filaments yellow, anther margins not reddish. **Achenes** ± 1.5 mm, smooth.

Flowering summer. Crevices on north-facing cliffs or similarly protected sites in canyons and on rocky outcrops mainly of volcanic origin, in sagebrush communities, conifer woodlands; of conservation concern; 1300–2600 m; Calif., Nev.

Variety *baileyi* occurs from Lassen and Plumas counties, California, to Churchill, Humboldt, and Pershing counties, Nevada.

6b. Ivesia baileyi S. Watson var. **beneolens** (A. Nelson & J. F. Macbride) Ertter, Syst. Bot. 14: 236. 1989 • Owyhee ivesia [E]

Horkelia beneolens A. Nelson & J. F. Macbride, Bot. Gaz. 55: 374. 1913

Basal leaves: sheathing base sparsely glandular abaxially, otherwise glabrous. **Pedicels** 5–6+ mm at flowering, to 15(–30) mm in fruit. **Flowers** 5–40, 7–10 mm diam.; epicalyx bractlets narrowly lanceolate, 0.8–1 mm, usually less than ½ as long as sepals; hypanthium interior pale green or cream to maroon; sepals (1.2–)1.5–2.5 mm; petals white; filaments white, anther margins reddish. **Achenes** ± 2 mm, rugose. $2n = 28$.

Flowering summer. Crevices on north-facing cliffs or similarly protected sites in canyons and rocky outcrops mainly of volcanic origin, in sagebrush communities, conifer woodlands; 1000–2600 m; Calif., Idaho, Nev., Oreg.

Variety *beneolens* occurs from Harney and Malheur counties, Oregon, to Modoc County, California, and to Elmore, Owyhee, and Twin Falls counties, Idaho, and Elko and Humboldt counties, Nevada. Plants are particularly common on the vertical sides of the river canyons that cut through the Owyhee Plateau. The variety also barely enters the Idaho Batholith on volcanic intrusions along the South Fork of the Boise River.

7. Ivesia setosa (S. Watson) Rydberg in N. L. Britton et al., N. Amer. Fl. 22: 290. 1908 • Bristly ivesia [E]

Ivesia baileyi S. Watson var. *setosa* S. Watson, Botany (Fortieth Parallel), 91. 1871; *Potentilla baileyi* (S. Watson) Greene var. *setosa* (S. Watson) J. T. Howell

Plants green, tufted to ± densely matted. **Stems** ± ascending to nearly erect, 0.7–2.5(–2.8) dm. **Basal leaves** weakly planar to loosely cylindric, 5–9(–12) cm; sheathing base ± strigose abaxially; petiole 1–8 cm; lateral leaflets 5–10 per side, separate to overlapping distally, ± flabellate, 2–8 mm, incised ± ¾ to base, sometimes nearly to base, into (3–)7–11 ovate teeth to narrowly obovate lobes, apex usually ± setose, surfaces ± sparsely hirsute, conspicuously glandular; terminal leaflets ± indistinct. **Cauline leaves** (0 or)1; blade vestigial. **Inflorescences** (1–)5–15(–30)-flowered, ± open, 1–6(–12) cm diam. **Pedicels** 5–15(–20) mm. **Flowers** 7–10 mm diam.; epicalyx bractlets 5, lanceolate, 1.3–2.5(–3) mm; hypanthium patelliform, (1–)1.5–2 × 2–3.5(–4) mm; sepals (1.5–)2–3.5 mm, ± acute; petals yellow, oblanceolate to narrowly spatulate, 1.5–2.5 mm; stamens 5, filaments 0.8–1.5 mm, anthers yellow, oblong, 0.4–0.7 mm; carpels 2–8, styles 1.2–2 mm. **Achenes** greenish white to light tan, 1.7–2 mm, smooth, ± carunculate.

Flowering summer. Dry, rocky talus slopes, boulders and outcrops, most often but not always of calcareous origin, occasionally away from immediate outcrops, sagebrush communities, conifer woodlands; 1800–2600(–3100) m; Nev., Utah.

Ivesia setosa occurs mostly to the east and south of *I. baileyi*, from southeastern Humboldt and Churchill counties to Elko, White Pine, and northern Nye counties, Nevada, and in the Deep Creek Range of western Utah. It represents a transition between the planar-leaved, chasmophytic members of sect. *Setosae* and the cylindric-leaved, matted species of flatter sites. Stems of *I. setosa* are more generally erect than in other species in the section, and the usually calcareous substrate is noteworthy. The deeply incised (usually not quite to base) leaflets are somewhat intermediate between the toothed leaflets of *I. baileyi* and the leaflets of *I. shockleyi* that are incised to the base into separate lobes. The individual leaflets of both *I. baileyi* and *I. setosa* are more or less flat and distichously paired; in *I. shockleyi*, groups of leaflet lobes are folded over onto each other, giving a verticillate appearance to the leaflet arrangement.

8. Ivesia shockleyi S. Watson, Proc. Amer. Acad. Arts 23: 263. 1888 [C] [E]

Potentilla shockleyi (S. Watson) Jepson

Plants green to grayish green, ± densely matted. **Stems** prostrate to ascending, 0.2–1.5 dm. **Basal leaves** tightly to loosely cylindric, (1–)2–12(–14) cm; sheathing base strigose to hirsute abaxially; petiole 0.5–4 cm; lateral leaflets (3–)5–15(–18) per side, separate to overlapping at least distally, ± flabellate, 1–4(–6) mm, incised to base or nearly so into 2–10(–12) oblanceolate to obovate or elliptic lobes, apex sometimes setose, surfaces ± sparsely hirsute, ± glandular; terminal leaflets indistinct. **Cauline leaves** (0 or)1; blade vestigial or reduced. **Inflorescences** 2–20-flowered, ± open, 0.5–4(–6) cm diam. **Pedicels** 3–10(–12) mm. **Flowers** 5–10 mm diam.; epicalyx bractlets 5, elliptic or narrowly oblong to ovate, 0.7–2(–3) mm; hypanthium patelliform, 1–2 × 2.5–4.5(–5) mm; sepals 1.5–3.5(–4) mm, acute to obtuse; petals ± yellow, oblanceolate to spatulate or narrowly obovate, 1–3 mm; stamens 5, filaments 0.8–1.8 mm, anthers yellow, oblong, 0.4–0.9 mm; carpels 2–5(or 6), styles 1.8–2.5 mm. **Achenes** greenish cream to light brown, 1.5–2.2(–2.5) mm, smooth, prominently carunculate.

Varieties 2 (2 in the flora): w United States.

1. Lateral leaflets: lobes 2–5(–7); inflorescences 2–10-flowered, 0.5–3 cm diam.; e California, c, se Oregon, c, n Nevada .... 8a. *Ivesia shockleyi* var. *shockleyi*
1. Lateral leaflets: lobes (4–)6–10(–12); inflorescences 5–20-flowered, (1–)3–4(–6) cm diam.; wc Utah .... 8b. *Ivesia shockleyi* var. *ostleri*

8a. Ivesia shockleyi S. Watson var. **shockleyi** • Shockley's or sky ivesia [C] [E]

Stems 0.2–0.7(–1.1) dm. **Basal leaves** (1–)2–8(–10) cm; lateral leaflets (3–)5–10 per side, lobes 2–5(–7), oblanceolate to obovate. **Cauline leaves** 0.3–1 cm, blade vestigial, leaflets usually 0. **Inflorescences** 2–10-flowered, 0.5–3 cm diam.

Flowering spring–early summer. Gravelly to rocky slopes and ridges, mainly in high-elevation sagebrush communities, subalpine or alpine conifer woodlands, alpine tundra communities; of conservation concern; 1700–4000 m; Calif., Nev., Oreg.

Variety *shockleyi* occurs mainly on rocky slopes and talus near and above timberline from the crest of the northern and central Sierra Nevada in California and adjacent Nevada, to Lake and Malheur counties, Oregon, and on scattered sky islands as far as the Toquima Range (Nye County) and Jarbidge Mountains (Elko County) of Nevada. These isolated populations have developed variation that merits further analysis: plants from the northern Intermountain Region tend to have flowers and leaflets larger than those in the Sierra Nevada; central Nevada populations approach var. *ostleri* in leaflet number and inflorescence size. The Malheur County plants are unusual in occurring on canyon rimrock (West Little Owyhee River) 1000 m lower than any other known population, near *I. baileyi* var. *beneolens* on adjacent canyon walls.

8b. Ivesia shockleyi S. Watson var. **ostleri** Ertter, Syst. Bot. 14: 237, fig. 3. 1989 • Ostler's or Wah Wah ivesia [C] [E]

Stems (0.2–)0.5–1.5 dm. **Basal leaves** (2–)4–12(–14) cm; leaflets 7–15(–18) per side, lobes (4–)6–10(–12), oblanceolate to obovate or elliptic. **Cauline leaves** 0.5–2.5 cm, blade reduced pinnate, lateral leaflets 1–3 per side. **Inflorescences** 5–20-flowered, (1–)3–4(–6) cm diam. $2n = 28$.

Flowering spring–early summer. Crevices of quartzite outcrops, in pinyon-juniper and ponderosa pine woodlands; of conservation concern; 1900–2500 m; Utah.

Variety *ostleri* is known from the Needle Range and Wah Wah Mountains of Beaver County, Utah. A report from the Snake Range, White Pine County, Nevada (G. Clifton, unpubl.) remains to be verified.

9. Ivesia cryptocaulis (Clokey) D. D. Keck, Lloydia 1: 130. 1939 • Charleston Peak ivesia [C] [E]

Potentilla cryptocaulis Clokey, Bull. S. Calif. Acad. Sci. 37: 4, plate 1. 1938

Plants green, diffusely matted. **Stems** prostrate to decumbent, 0.2–0.6(–1) dm. **Basal leaves** tightly cylindric, 1–4(–5) cm; sheathing base mostly not strigose abaxially; petiole 0.3–2 cm, <hairs 0.5–1 mm>; lateral leaflets 5–10(–12) per side, ± overlapping, ± flabellate, 1–2.5 mm, incised to base or nearly so into 3–5 obovate to oval lobes, apex sometimes ± setose, surfaces loosely long-strigose, sparsely glandular; terminal leaflets indistinct. **Cauline leaves** 1(or 2); blade vestigial. **Inflorescences** 1–10-flowered, ± congested, 0.5–1.5 cm diam. **Pedicels** 3–7(–10) mm. **Flowers** 6–9 mm diam.; epicalyx bractlets 5, lanceolate to narrowly oblong or narrowly ovate, 0.8–1.3 mm; hypanthium shallowly cupulate, ± 1 × 2–3(–4) mm; sepals 1.5–2.3 mm, ± acute; petals yellow, oblanceolate to spatulate, 2–3.2 mm; stamens 5, filaments 0.9–1.2 mm, anthers yellow, oblong, 0.7–0.9 mm; carpels 6–10, styles 1–1.5 mm. **Achenes** greenish cream to light brown, 1.4–1.8(–2) mm, smooth, not carunculate.

Flowering summer. Dry to moist limestone gravel, scree, or talus slopes and outcrops, subalpine conifer woodlands, alpine tundra communities; of conservation concern; 2500–3700 m; Nev.

Ivesia cryptocaulis is known only from timberline and above in the Spring Mountains of Clark County. Plants differ from all others of the genus by having slender, elongate, diffuse caudex branches that ramify through its habitat of gravel, scree, and talus. The species is here provisionally included in sect. *Setosae* as a diffusely matted relative of *I. shockleyi* (as also suggested by D. D. Keck 1938); the ecarunculate seeds and relatively straight pedicels may indicate a greater similarity to *I. pygmaea* in sect. *Ivesia*.

10. Ivesia rhypara Ertter & Reveal, Madroño 24: 224, fig. 1. 1977 [C] [E]

Plants grayish, ± matted. **Stems** ± prostrate, (0.1–)0.2–1.5(–2) dm. **Basal leaves** tightly cylindric, 2–10 cm; sheathing base densely hairy abaxially; petiole 0.5–4(–10) cm; lateral leaflets 4–10(–15) per side, overlapping at least distally, ± flabellate, 0.5–3(–4) mm, incised to base or nearly so into (0–)2–4(–9) elliptic to obovate or orbiculate lobes, apex usually not setose, surfaces densely hirsute, cryptically glandular; terminal leaflets indistinct. **Cauline leaves** (0 or)1; blade vestigial.

Inflorescences 5–60(–100)-flowered, congested, 0.5–5(–7) cm diam. **Pedicels** 1.5–4(–8) mm. **Flowers** 4–8 mm diam.; epicalyx bractlets 5, linear to elliptic or narrowly ovate, 0.8–1.6(–2.1) mm; hypanthium patelliform to shallowly cupulate, 1–1.5 × (1.5–)2–3 (–3.5) mm; sepals 1.4–2.5(–2.8) mm, acute; petals white to pale yellowish, linear to narrowly oblanceolate, 1–1.5 mm; stamens 5, filaments 0.6–1.8 mm, anthers maroon or yellow with maroon margins, oblong, 0.3–0.6 mm; carpels 1 or 2(–4), styles 1–1.5 mm. **Achenes** brown, 1.2–1.6(–2) mm, smooth, prominently carunculate.

Varieties 2 (2 in the flora): w United States.

Ivesia rhypara forms densely hairy mats or mounds on isolated edaphic sites in the northern Great Basin. Only one of the widely disjunct population clusters is recognized as a distinct variety; the others have general differences in gestalt, but no significant distinctions have been determined (B. Ertter 1989). Isozyme analysis likewise does not show significant differences (T. Kaye et al., unpubl., BLM and Oregon Dept. of Agriculture, 1991). Reproductive studies by D. Wiens (pers. comm.) show low seed set (about 5% in var. *rhypara*, somewhat higher in var. *shellyi*), possibly indicating a deleterious genetic load.

1. Plants (4–)8–30 cm diam.; inflorescences (1–)2–5 (–7) cm diam.; scarcely or cryptically petrophytic in bedrock overlain by rubble and soil . . . . 10a. *Ivesia rhypara* var. *rhypara*
1. Plants 2–10(–15) cm diam.; inflorescences 0.5–2 (–5) cm diam.; overtly petrophytic in boulders and outcrops. . . . . 10b. *Ivesia rhypara* var. *shellyi*

10a. Ivesia rhypara Ertter & Reveal var. **rhypara** • Grimy or Grimes's ivesia [C] [E]

Plants (4–)8–30 cm diam. **Stems** (0.3–)0.4–1.5(–2) dm, usually exceeding leaves by more than 2 cm. **Inflorescences** 10–60 (–100)-flowered, (1–)2–5(–7) cm diam.

Flowering summer. Scarcely or cryptically petrophytic in ash tuff bedrock overlain by dry soil and pulverized rubble, in sagebrush communities, sometimes juniper woodlands; of conservation concern; 1400–1900 m; Nev., Oreg.

Variety *rhypara* is known from widely scattered locations in central Malheur and southeastern Lake counties, Oregon, and in northern Washoe, Humboldt, and Elko counties, Nevada. Although not overtly petrophytic, plants are generally associated with crevices in the underlying bedrock (E. M. Clark and W. H. Clark 2003). They might thereby benefit from a more favorable water relationship, allowing them to be in full bloom when most associated species are summer-dormant.

10b. Ivesia rhypara Ertter & Reveal var. **shellyi** Ertter, Syst. Bot. 14: 239, fig. 4. 1989 • Shelly's ivesia [C] [E]

Plants 2–10(–15) cm diam. **Stems** (0.1–)0.2–0.5(–0.9) dm, equal to or exceeding leaves by no more than 2(–7) cm. **Inflorescences** 5–30(–60)-flowered, 0.5–2(–5) cm diam. $2n = 28$.

Flowering summer. Overtly petrophytic in cracks and crevices of pumiceous welded ash-flow tuff boulders and outcrops, in sagebrush communities; of conservation concern; 1300–1600 m; Oreg.

In addition to their more overtly petrophytic habit, plants of var. *shellyi* are generally smaller and more compact than those of var. *rhypara*. It is known only from the Rehart and Venator canyons area along the border of Harney and Lake counties in southeastern Oregon.

11. Ivesia paniculata T. W. Nelson & J. P. Nelson, Brittonia 33: 165, fig. 1. 1981 • Ash Creek ivesia [C] [E]

Plants grayish, ± matted. **Stems** ± prostrate, 0.4–1.5(–2) dm. **Basal leaves** tightly cylindric, (1.5–)2–5(–7) cm; sheathing base densely hairy abaxially; petiole 0.5–4 cm; lateral leaflets (5–)8–15 per side, overlapping at least distally, ± flabellate, 0.5–2 mm, incised to base or nearly so into (0–)3–8(–15) elliptic to narrowly obovate lobes, apex not or obscurely setose, surfaces densely hirsute, cryptically glandular; terminal leaflets indistinct. **Cauline leaves** (0 or)1; blade reduced. **Inflorescences** 20–200-flowered, congested, (1–)1.5–6(–10) cm diam. **Pedicels** 1.5–6 mm. **Flowers** 4–6 mm diam.; epicalyx bractlets 5, lanceolate to elliptic, 0.6–1.5(–2) mm; hypanthium shallowly cupulate, 1 × 2–3 mm; sepals (1–)1.5–2.5(–3) mm, acute; petals white to pale yellowish, linear, 1 mm; stamens 5, filaments 0.3–1 mm, anthers yellow with maroon margins, ovate, 0.3–0.5 mm; carpels 1 or 2(or 3), styles 0.7–1.8 mm. **Achenes** brown, 0.8–1.5 mm, smooth, prominently carunculate.

Flowering summer. Dry shallow volcanic ash and cinders atop volcanic bedrock, open sagebrush communities, adjacent conifer woodlands; of conservation concern; 1500–1800 m; Calif.

Ivesia paniculata is known only from the Ash Valley area of Lassen County. The distinctions between *I. paniculata* and *I. rhypara* are perhaps on the same scale as variation among disjunct population clusters of *I. rhypara*, but no taxonomic adjustments are proposed at this time.

9b. IVESIA Torrey & A. Gray sect. IVESIA [E]

Horkelia Chamisso & Schlechtendal [unranked] *Lycopodioides* Rydberg; *Ivesia* [unranked] *Lycopodioides* (Rydberg) Rydberg; *Ivesia* sect. *Lycopodioides* (Rydberg) O. Stevens

Plants usually rosetted or tufted, rarely ± matted (*I. pygmaea*), not forming hanging clumps (not in vertical rock crevices except *I. longibracteata*), ± aromatic; taproot stout to fusiform and fleshy. **Stems** (0.2–)0.3–2(–4) dm. **Basal leaves** usually loosely to very tightly cylindric (mousetail-like in *I. muirii*), rarely weakly planar (*I. longibracteata*); stipules present; leaflets overlapping, individually distinguishable or not, lobed to base, sparsely to densely hairy, sometimes glabrous or glabrate; terminal leaflets indistinct. **Cauline leaves** (0 or)1–3, <sometimes paired>; blade usually ± reduced to vestigial, rarely well developed. **Inflorescences** usually ± congested, sometimes becoming open in fruit, flowers mostly arranged in 1–few(–several in *I. gordonii* var. *wasatchensis*) loose to capitate glomerules. **Pedicels** remaining straight. **Flowers:** hypanthium shallowly cupulate or campanulate, sometimes turbinate (*I. gordonii*); petals not medially reflexed, golden to pale yellow, sometimes white (*I. utahensis*) and then sometimes pink-tinged, not or scarcely clawed, apex acute to truncate, rounded, or emarginate; stamens 5 (10 in *I. pygmaea*), anthers longer than wide, laterally dehiscent; carpels (1 or)2–15 (–30 in *I. pygmaea*). **Achenes** vertical, smooth or nearly so, not caruncululate.

Species 8 (8 in the flora): w United States.

Species in sect. *Ivesia* tend to form compact rosettes or tufts in open montane to alpine areas, often where rocky but generally not growing in rock crevices (except *Ivesia longibracteata*). General characteristics of the section are an overall evident glandularity, loosely to tightly cylindric leaves (mousetail-like in *I. muirii*), relatively short-ascending to erect stems, straight pedicels, and flowers that are typically congested into one to few capitate clusters that sometimes become more open in fruit. The taproots of some species, especially *I. lycopodioides*, are markedly enlarged, an adaptation to alpine growing conditions. The inclusion of two anomalous species in the section, *I. longibracteata* and *I. webberi*, is tentative; it is also possible that *I. cryptocaulis* belongs here rather than in sect. *Setosae*.

Section *Ivesia* has its center of species diversity in the high Sierra Nevada of California, where *Ivesia pygmaea*, with its ten stamens and more open habit, is possibly transitional between sects. *Ivesia* and *Setosae*. *Ivesia gordonii* (the most widespread species), *I. tweedyi*, and *I. utahensis* form an arc around the northern Great Basin, as outliers from the core Sierran distribution of the section.

Some early floras treated *Ivesia pygmaea* and *I. lycopodioides* as varieties of *I. gordonii* (for example, W. H. Brewer et al. 1876–1880, vol. 1) or *Potentilla gordonii* (Hooker) Greene (W. L. Jepson [1923–1925], 1909–1943, vol. 2). Recent treatments follow D. D. Keck (1938[1939]) in recognizing all three as distinct species; some annotations and references are nevertheless carried over from the earlier expanded circumscription of *I. gordonii*.

Since *Ivesia cryptocaulis* (sect. *Setosae*) is sometimes identified as a member of sect. *Ivesia*, it is included herein and keys out in the seventh couplet.

1. Epicalyx bractlets longer than sepals, 2.5–5 mm; leaves weakly planar to loosely ± cylindric; leaflets 5 or 6 per side; Castle Crags, n California . . . . 18. *Ivesia longibracteata*
1. Epicalyx bractlets shorter than sepals, 0.5–3(–4 in *I. gordonii*) mm; leaves loosely to tightly cylindric; leaflets mostly 10–40 per side (4–10 in *I. webberi*); w United States.
    2. Plants silvery; leaves very tightly cylindric (mousetail-like, with leaflets scarcely distinguishable); leaflets 0.4–1 mm . . . . 17. *Ivesia muirii*
    2. Plants ± green; leaves loosely to tightly cylindric; leaflets distinguishable, (0.5–)2–13 (–18) mm.

[3. Shifted to left margin.—Ed.]

3. Cauline leaves 2, paired; basal leaves: leaflets 4–8(–10) per side; petiole hairs 2–4 mm .... 19. *Ivesia webberi*
3. Cauline leaves 0–2(or 3), not paired; basal leaves: leaflets (6–)10–35 per side; petiole hairs 0.2–1.5 mm.
    4. Hypanthia turbinate to campanulate, (1.5–)2–4(–4.5) mm (± as deep as wide); petals narrowly oblanceolate to narrowly spatulate .... 15. *Ivesia gordonii*
    4. Hypanthia shallowly cupulate or shallowly campanulate, 1–2 mm (wider than deep); petals oblanceolate or broadly elliptic to spatulate or broadly obovate.
        5. Carpels (1 or)2–6(–9); Idaho, Utah, Washington.
            6. Petals white, sometimes pink-tinged; stems prostrate to ascending; leaflets 2–4 mm; Utah .... 14. *Ivesia utahensis*
            6. Petals golden yellow; stems ascending to erect; leaflets 4–7(–10) mm; n Idaho, Washington .... 16. *Ivesia tweedyi*
        5. Carpels (5–)8–30; e California, Nevada.
            7. Plants diffusely matted; Spring Mountains, s Nevada....... 9. *Ivesia cryptocaulis* (sect. *Setosae*)
            7. Plants rosetted to tufted or ± tightly matted; Sierra Nevada, White Mountains, and Sweetwater Mountains of e California and adjacent w Nevada.
                8. Stamens 10; sheathing bases usually ± strigose abaxially; taproots stout, not fleshy... 12. *Ivesia pygmaea*
                8. Stamens 5; sheathing bases glabrous abaxially; taproots fusiform, fleshy .... 13. *Ivesia lycopodioides*

12. Ivesia pygmaea A. Gray, Proc. Amer. Acad. Arts 6: 531. 1865 • Dwarf ivesia [E]

Ivesia chaetophora (Rydberg) Rydberg; *I. gordonii* (Hooker) Torrey & A. Gray var. *pygmaea* (A. Gray) S. Watson; *Potentilla gordonii* (Hooker) Greene var. *chaetophora* (Rydberg) Jepson; *P. nubigena* Greene

Plants green, tufted to ± tightly matted; taproot stout, not fleshy. **Stems** decumbent to erect, 0.3–1(–1.5) dm. **Basal leaves** tightly cylindric, (0.5–)1–8(–12) cm; sheathing base usually ± strigose abaxially; petiole 0.5–4 cm, hairs 0.2–1 mm; leaflets 10–15(–20) per side, 0.5–4(–5) mm, sparsely to moderately short-villous, densely glandular, lobes 3–6(–8), oblanceolate to narrowly oblong or oval, apex setose. **Cauline leaves** (0 or)1, not paired. **Inflorescences** (1–)5–10-flowered, 0.8–3 cm diam.; glomerules usually 1. **Pedicels** 1.5–8(–10) mm. **Flowers** 9–11 mm diam.; epicalyx bractlets elliptic to oblong, 1.2–2 mm; hypanthium shallowly cupulate, 1–2 × 2.5–5 (–7) mm; sepals 2–3.5 mm, bluntly acute; petals golden yellow, broadly oblanceolate to spatulate or broadly obovate, (2–)2.5–3.5(–4) mm; stamens 10, filaments 0.6–1.8 mm, anthers yellow, 0.6–0.8 mm; carpels 10–30, styles 0.9–1.3 mm. **Achenes** greenish tan to light brown, 1.2–1.5 mm.

Flowering summer. Dry rocky slopes, sometimes in horizontal rock crevices, in high-elevation sagebrush communities, subalpine to alpine conifer woodlands, alpine tundra; 2700–4000 m; Calif.

Ivesia pygmaea is known only from the southern Sierra Nevada in California, where it overlaps the range of *I. lycopodioides* var. *megalopetala*. Stamen number is the only unequivocal way to distinguish between the two, though *I. pygmaea* is also more generally setose with somewhat shallower hypanthia. It is also more likely to have much-branched caudices and can form mats in rocky sites. Larger plants with relatively open cymes have sometimes been segregated as *I. chaetophora*; this probably represents ecological and phenotypic plasticity.

Early reports of *Ivesia pygmaea* from Nevada are based on the mistaken understanding of P. A. Rydberg (1898) of the type locality, which is actually in Tulare County, California. A reported occurrence from the northern Sierra Nevada in Nevada (J. T. Kartesz 1987) is presumably based on the same source as the unconfirmed report of *I. lycopodioides* from the same locality (D. D. Keck 1938).

The correct name for this species if treated as *Potentilla* is *P. nubigena*; *P. decipiens* Greene is a later homonym and illegitimate.

13. Ivesia lycopodioides A. Gray, Proc. Amer. Acad. Arts 6: 530. 1865 [E]

Potentilla lycopodioides (A. Gray) Baillon ex J. T. Howell

Plants green, usually rosetted, sometimes ± tufted; taproot fusiform, fleshy. **Stems** decumbent to erect, 0.3–3 dm. **Basal leaves** tightly to loosely cylindric, 1–15 cm; sheathing base glabrous abaxially; petiole 0.5–4 cm, hairs 0.2–1 mm; leaflets 10–35 per side, 1–8 mm, glabrous or short-hirsute, minutely glandular, lobes (2–)4–8(–10), linear to obovate or ± orbiculate, apex sometimes setose. **Cauline leaves** 0–2(or 3), not paired. **Inflorescences** 3–20(–25)-flowered, (0.5–) 1–2.5(–3.5) cm diam.; glomerules usually 1. **Pedicels** (0.5–)1–7(–11) mm. **Flowers** 6–12 mm diam.; epicalyx bractlets oblong to oval, 0.8–2.5(–3) mm; hypanthium shallowly cupulate, 1–2 × 2.5–5 mm; sepals (1.8–)2–4 (–4.5) mm, obtuse to acute; petals golden yellow, obovate, 2–5 mm; stamens 5, filaments 0.8–2 mm, anthers yellow, 0.6–0.8 mm; carpels (5–)8–15(–18), styles 1–3 mm. **Achenes** greenish tan to light brown, 1.2–1.5 mm.

Varieties 3 (3 in the flora): w United States.

The three varieties of *Ivesia lycopodioides* are for the most part readily distinguished, though intergradation is known. The high-elevation var. *lycopodioides* extends farthest north; var. *scandularis* is the only variety in the White Mountains. Variety *megalopetala* is found mostly at somewhat lower (subalpine) elevations and generally has a more southern range.

1. Leaflets short-hirsute, apical setae (0–)0.5–1(–2) mm; White Mountains and c Sierra Nevada . . . . 13b. *Ivesia lycopodioides* var. *scandularis*
1. Leaflets glabrous or sparsely short-hirsute, apical setae 0–0.5 mm; Sierra Nevada and Sweetwater Mountains.
    2. Leaflet lobes ± orbiculate, ± 1 mm; petals 2–3 × 1 mm . . . . 13a. *Ivesia lycopodioides* var. *lycopodioides*
    2. Leaflet lobes linear to oblanceolate, 2–8 mm; petals 3–5 × 2–4 mm . . . . 13c. *Ivesia lycopodioides* var. *megalopetala*

13a. Ivesia lycopodioides A. Gray var. **lycopodioides** • Clubmoss ivesia [E]

Stems decumbent to ascending, 0.3–1(–1.5) dm. **Basal leaves** 1–7 cm; leaflets tightly overlapping, glabrous or sparsely short-hirsute, lobes ± orbiculate, ± 1 mm, apical setae 0(–0.5) mm. **Cauline leaves** 0 or 1. **Inflorescences** usually ± capitate, 3–15(–20) -flowered, 0.5–1.5(–2) cm diam. **Flowers** 6–9 mm diam.; petals obovate, 2–3 × 1 mm; filaments 0.8–1.2 mm; styles 1–2 mm. $2n = 28$.

Flowering summer. Dry rocky flats or slopes, fellfields, in high-elevation sagebrush communities, subalpine to alpine conifer woodlands, alpine tundra; 3000–4000 m; Calif., Nev.

Variety *lycopodioides* occurs in the Sierra Nevada from El Dorado and Alpine to Fresno counties, and on the Sweetwater Mountains of Mono County, California. The voucher reported by D. D. Keck (1938[1939]) from the Carson Range of Washoe County, Nevada, has not been located for confirmation. Of the three varieties, var. *lycopodioides* tends to have plants with the smallest leaflet lobes (ca. 1 mm) which lack apical setae. Plants also have more consistently simple caudices bearing a single rosette of glabrous or sparsely hairy leaves atop an enlarged, fleshy taproot.

13b. Ivesia lycopodioides A. Gray var. **scandularis** (Rydberg) Ertter & Reveal, Novon 17: 324. 2007 • White Mountains ivesia [E]

Horkelia scandularis Rydberg, Monogr. N. Amer. Potentilleae, 150, plate 91, figs. 5–9. 1898; *Ivesia lycopodioides* subsp. *scandularis* (Rydberg) D. D. Keck; *Potentilla lycopodioides* (A. Gray) Baillon ex J. T. Howell var. *scandularis* (Rydberg) J. T. Howell

Stems decumbent to ascending, 0.5–1.5 dm. **Basal leaves** 3–8 cm; leaflets tightly overlapping, short-hirsute, lobes obovate, 1–3 mm, apical setae (0–)0.5–1(–2) mm. **Cauline leaves** 0 or 1. **Inflorescences** usually ± capitate, 3–15(–20)-flowered, 1–2(–3) cm diam. **Flowers** 6–12 mm diam.; petals broadly obovate, 3–5 × 2–3 mm; filaments 1.2–1.5 mm; styles 2–3 mm.

Flowering summer. Moist slopes, fellfields, in high-elevation sagebrush communities, subalpine to alpine conifer woodlands, alpine tundra; 3000–4000 m; Calif.

Variety *scandularis* is the only variety that occurs in the White Mountains of Inyo and Mono counties, California; scattered populations occur also on the eastern side of the Sierra Nevada. Leaflet lobes are intermediate in size between those of the other two varieties and usually are tipped with a single bristle to 2 mm.

13c. Ivesia lycopodioides A. Gray var. **megalopetala** (Rydberg) Ertter & Reveal, Novon 17: 324. 2007 • Rock Creek or big-petaled ivesia [E]

Horkelia gordonii Hooker var. *megalopetala* Rydberg, Monogr. N. Amer. Potentilleae, 152. 1898; *Ivesia lycopodioides* subsp. *megalopetala* (Rydberg) D. D. Keck; *Potentilla lycopodioides* (A. Gray) Baillon ex J. T. Howell var. *megalopetala* (Rydberg) J. T. Howell

Stems ascending to erect, 1–3 dm. **Basal leaves** 4–15 cm; leaflets loosely overlapping, ± glabrous or sparsely short-hirsute, lobes linear to oblanceolate, 2–8 mm, apical setae 0–0.5 mm. **Cauline leaves** 1 or 2(or 3). **Inflorescences** ± open to congested, sometimes subcapitate, 5–20(–25) -flowered, 1–2.5(–3.5) cm diam. **Flowers** 8–12 mm diam.; petals broadly obovate, 3–5 × 2–4 mm; filaments (1–)1.5–2 mm; styles 2.5–3 mm.

Flowering summer. Wet meadows, in sagebrush communities, subalpine conifer woodlands, alpine tundra; 2300–3700 m; Calif.

Variety *megalopetala* is known from the southern Sierra Nevada from Mono and Tuolumne to Tulare counties. It tends to be the largest of the three varieties, with plants having larger flowers and longer and narrower leaflet lobes, which commonly have apical setae to 0.5 mm. Variety *megalopetala* is also more likely to occur in wet meadows along subalpine streams.

14. Ivesia utahensis S. Watson, Proc. Amer. Acad. Arts 17: 371. 1882 • Utah ivesia [C] [E]

Potentilla utahensis (S. Watson) Greene

Plants green, rosetted to tufted; taproot proximally enlarged, not fleshy. **Stems** prostrate to ascending, 0.5–1.5(–2) dm. **Basal leaves** ± tightly cylindric, 2–7(–9) cm; sheathing base glabrous abaxially; petiole 0.5–2 cm, hairs 0.5–1 mm; leaflets 15–20 per side, 2–4 mm, glabrate or sparsely short-hirsute, sparsely to densely glandular, lobes (2 or)3–8, narrowly oblanceolate to obovate, apex not setose. **Cauline leaves** (0 or)1, not paired. **Inflorescences** (5–)10–30-flowered, 1–2.5(–5) cm diam.; glomerules 1–few. **Pedicels** 1–7 mm. **Flowers** 7–9 mm diam.; epicalyx bractlets narrowly oblong to oval, 0.8–2 mm; hypanthium shallowly cupulate, (1–)1.5–2 × 2.5–3.5 mm; sepals (1.8–)2–3 mm, acute; petals white, sometimes pink-tinged, oblanceolate to spatulate, 1.8–3 mm; stamens 5, filaments 1.3–1.8 mm, anthers orangish to reddish brown, 0.4–0.6 mm; carpels (1 or)2–4, styles 1.5–2 mm. **Achenes** yellowish green to light tan or gray-brown, 1.7–1.9 mm.

Flowering summer. Talus slopes, bare ridges, in high-elevation sagebrush communities, subalpine to alpine conifer woodlands; of conservation concern; 3200–3700 m; Utah.

Ivesia utahensis is known from the Wasatch and adjacent Uinta Mountains in northern Utah. The species stands as the white-petaled counterpart to the yellow-petaled Sierran *I. lycopodioides*, possibly indicating a common ancestral stock that was once more continuous across the Great Basin.

15. Ivesia gordonii (Hooker) Torrey & A. Gray in War Department [U.S.], Pacif. Railr. Rep. 6(3): 72. 1858 (as gordoni) [E] [F]

Horkelia gordonii Hooker, Hooker's J. Bot. Kew Gard. Misc. 5: 341, plate 12. 1853 (as gordoni); *Potentilla gordonii* (Hooker) Greene

Plants green, ± tufted, sometimes rosetted; taproot stout, not fleshy. **Stems** prostrate to erect, (0.2–)0.5–4 dm. **Basal leaves** tightly to loosely cylindric, (1–)3–20(–25) cm; sheathing base ± glandular abaxially, otherwise glabrous; petiole 0.5–8 cm, hairs 0.2–0.5 mm; leaflets (6–)10–25 per side, (1–)2–13(–18) mm, glabrous or ± hirsute or villous, glandular-puberulent or -pubescent, lobes (2–)4–8(–15), linear or narrowly oblanceolate to obovate, apex rarely setose. **Cauline leaves** 1(or 2), not paired. **Inflorescences** 5–50(–70)-flowered, 1–8(–11) cm diam.; glomerules 1–several. **Pedicels** 1–3(–5) mm. **Flowers** 5–12 mm diam.; epicalyx bractlets linear to narrowly elliptic, (0.5–)1–3.5(–4) mm; hypanthium turbinate to campanulate, (1.5–)2–4(–4.5) × 2–4(–5) mm; sepals (2–)2.5–5(–6) mm, obtuse to ± acute; petals yellow, narrowly oblanceolate to narrowly spatulate, (1–)1.5–3 mm; stamens 5, filaments 1.3–2.5 mm, anthers yellow, sometimes red-margined, 0.5–1 mm; carpels (1 or)2–4(–6), styles 2.5–4.5(–6) mm. **Achenes** grayish brown to mottled brown, ± 2 mm.

Varieties 4 (4 in the flora): w United States.

Ivesia gordonii is the most widespread species of the genus, occurring from Washington to Montana, south to central California and Colorado. The species can be distinguished from other members of sect. *Ivesia* by the relatively deep hypanthium, which is about as deep as wide. Four varieties are provisionally recognized here (B. Ertter and J. L. Reveal 2007), with the likelihood that future work may indicate additional and/or alternate circumscriptions. Populations that are difficult to assign to a variety can be found where the recognizable taxonomic units come together, for example, in northeastern Utah and western Wyoming, involving var. *gordonii* and var. *wasatchensis*, and in California and central Idaho where var. *alpicola* and var. *ursinorum* tend to merge.

1. Leaflets 7–13(–18) mm, glabrous or sparsely hirsute or villous marginally; stems (1–)1.5–4 dm; inflorescences usually branched, 2–8(–11) cm diam., glomerules (1 or)2–6(–10), ± capitate. . . . . .15d. *Ivesia gordonii* var. *wasatchensis*
1. Leaflets (1–)2–7(–9) mm, hairy or ± glabrous except marginally ciliate; stems (0.2–)0.5–2.5 dm; inflorescences simple or branched, 1–3(–5) cm diam., glomerules 1(–3), ± capitate to loosely congested.
    2. Stems usually prostrate to ascending, rarely nearly erect, usually dark reddish, minutely glandular or glandular-puberulent to -pubescent; leaves to 0.8 cm diam.; anthers usually red-margined . . . . .15c. *Ivesia gordonii* var. *ursinorum*
    2. Stems usually ascending to erect, sometimes decumbent, usually greenish, rarely reddish, hirsute to villous, glandular-puberulent to -pubescent; leaves to 1.5 cm diam.; anthers rarely red-margined.
        3. Stems usually hirsute to villous, sometimes densely so, glandular-pubescent or eglandular; basal leaves (3–)5–10 (–15) cm; flowers 7–12 mm diam.; e Idaho and w Montana to Utah, Wyoming, and w Colorado . . . 15a. *Ivesia gordonii* var. *gordonii*
        3. Stems not or sparsely hirsute to villous, glandular-puberulent or -pubescent; basal leaves 2–8(–10) cm; flowers 5–9 mm diam.; s Washington to California and e to w Montana. . . . 15b. *Ivesia gordonii* var. *alpicola*

15a. Ivesia gordonii (Hooker) Torrey & A. Gray var. **gordonii** • Gordon's ivesia [E] [F]

Stems usually greenish, ascending to erect, (0.5–)1–2.5 dm, usually hirsute to villous, sometimes densely so, glandular-pubescent or eglandular. **Basal leaves** (3–)5–10(–15) × 0.5–1.5 cm; leaflets (1–)3–7(–9) mm, hirsute to villous, sometimes hairy only marginally. **Inflorescences** 10–30 -flowered, simple or branched, (1–)2–3(–5) cm diam., glomerules 1(–3), ± capitate. **Flowers** 7–12 mm diam.; epicalyx bractlets (1.5–)2–3(–3.5) mm; hypanthium 2–4 × 2.5–4 mm; sepals (2.5–)3.5–5 mm, ± acute; anthers rarely red-margined.

Flowering summer. Gravelly to rocky flats, slopes, and outcrops, in grassland and sagebrush communities, montane and subalpine conifer woodlands; 2000–3600 m; Colo., Idaho, Mont., Utah, Wyo.

Variety *gordonii* occurs from western Montana and eastern Idaho through western and central Wyoming to western Colorado, and in the Uinta and Wasatch mountains of northeastern Utah. Plants have either villous stems, or stems that are both hirsute to villous and glandular-puberulent or -pubescent. Leaflets are usually densely hirsute, and the inflorescence is nearly always simply capitate. The variety is most distinctive at relatively low elevations in Montana south to the Uinta Mountains (Utah) and Moffat County, Colorado, where it is an erect, medium-sized plant with densely hairy leaves. Some populations here assigned to var. *gordonii* have features of var. *wasatchensis*, such as branched inflorescences in the Big Horn Mountains of Wyoming, or glabrate vestiture in the northern Colorado Rocky Mountains.

15b. Ivesia gordonii (Hooker) Torrey & A. Gray var. **alpicola** (Rydberg ex Howell) Ertter & Reveal, Novon 17: 322. 2007 • Howell's ivesia [E]

Ivesia alpicola Rydberg ex Howell, Fl. N.W. Amer., 182. 1898; *Horkelia gordonii* Hooker var. *alpicola* (Rydberg ex Howell) Rydberg

Stems usually greenish, rarely reddish, usually ascending to erect, sometimes decumbent, 0.3–2 dm, not or sparsely hirsute to villous, glandular-puberulent or -pubescent. **Basal leaves** 2–8(–10) × 0.5–1(–1.5) cm; leaflets 2–5(–9) mm, hirsute to pilose, rarely minutely glandular at least marginally. **Inflorescences** 5–25-flowered, usually simple, 1–3(–4) cm diam., glomerules usually 1, ± capitate. **Flowers** 5–9 mm diam.; epicalyx bractlets (1.5–)2–3(–4) mm; hypanthium (2–)2.5–3.5(–4) × 2.5–4 mm; sepals 2–4(–5) mm, acute; anthers rarely red-margined.

Flowering summer. Dry grassy areas on talus, or gravelly or rocky slopes and ridges, in subalpine and alpine montane conifer woodlands, alpine tundra; 2100–3300 m; Calif., Idaho, Mont., Oreg., Wash.

P. A. Rydberg (1898) used var. *alpicola* (as *Horkelia gordonii* var. *alpicola*) for compact, high-elevation plants throughout the range of *Ivesia gordonii*. As circumscribed here, var. *alpicola* is restricted to plants occurring disjunctly in an arc around the northwestern part of the species range. Populations occur in the northern Coast Ranges and central Sierra Nevada of California, sporadically to Mount Adams and the Wenatchee Mountains in Washington, and from there to the mountains of central Idaho and the Bitterroot Range of Montana. The variety is also disjunct on the Owyhee Mountains of southwestern Idaho.

Plants of var. *alpicola* average shorter than those of var. *gordonii*, which occurs farther east and south. In var. *alpicola*, stems with only glandular indumentum are common; such stems are rare in var. *gordonii*. Unlike those of var. *gordonii*, the leaflets of var. *alpicola* are frequently glandular with only marginal hairs.

15c. Ivesia gordonii (Hooker) Torrey & A. Gray var. **ursinorum** (Jepson) Ertter & Reveal, Novon 17: 323. 2007 • Spider ivesia [E]

Potentilla gordonii (Hooker) Greene var. *ursinorum* Jepson, Man. Fl. Pl. Calif., 492. 1925

Stems usually dark reddish, usually prostrate to ascending, rarely nearly erect, (0.2–) 0.5–1.5(–2.5) dm, usually not villous or hirsute, minutely glandular or glandular-puberulent to -pubescent. **Basal leaves** (1–)3–8(–15) × 0.3–0.8 cm; leaflets 1–4(–6) mm, glabrous or sparsely hirsute, at least marginally, rarely more densely hirsute. **Inflorescences** 5–20(–25)-flowered, usually simple, 1–2(–5) cm diam., glomerules usually 1, capitate to loosely congested. **Flowers** 6–9 mm diam.; epicalyx bractlets (0.5–)1.5–2(–3) mm; hypanthium (1.5–) 2–3(–4) × 2–3(–4) mm; sepals (2–)2.5–4(–4.5) mm, acute; anthers usually red-margined.

Flowering summer. Rocky slopes and outcrops, in high-elevation sagebrush communities, montane conifer woodlands; 1800–3500 m; Calif., Idaho, Nev., Oreg.

As circumscribed here, var. *ursinorum* encompasses "*tenuicaulis*" of N. H. Holmgren (1997b) but is more broadly defined to encompass generally prostrate to decumbent plants found at widely scattered locations in and adjacent to the northwestern Intermountain Region. Such plants are most common in the Independence and Jarbidge mountains (Elko County, Nevada) and on Steens Mountain and other ranges in southeastern Oregon. Similar plants dominate the mountains of central Idaho and the Wallowa Mountains of Oregon, mostly south of var. *alpicola*. In northern California, var. *ursinorum* is found on the Modoc, Scott, and Sweetwater mountains, and around Sonora Pass in the Sierra Nevada.

In general, var. *ursinorum* may be distinguished from var. *alpicola* by its more commonly prostrate to ascending habit, reddish stems, narrowly cylindric (bottlebrush-like) leaves, and rounder leaflet lobes. The inflorescence is also more prone to becoming lax at maturity. Plants are most often glandular-pubescent and minutely glandular but otherwise glabrate, the conspicuous exception being the type of var. *ursinorum* from northern California. Plants in the Wallowa Mountains approach var. *alpicola*, as do some populations in California and central Idaho, where they are sometimes only tentatively assigned to one variety or the other. Some central Idaho collections included here in var. *ursinorum* have more cupulate hypanthia and wider petals approaching those of *Ivesia lycopodioides*.

15d. Ivesia gordonii (Hooker) Torrey & A. Gray var. **wasatchensis** N. H. Holmgren ex Ertter & Reveal, Novon 17: 323. 2007 • Wasatch ivesia [E]

Stems usually greenish, ± erect, (1–)1.5–4 dm, sometimes sparsely villous, minutely glandular, often glandular-pubescent. **Basal leaves** 9–20 (–25) × 1–2(–3) cm; leaflets 7–13(–18) mm, glabrous or sparsely hirsute or villous marginally. **Inflorescences** 10–50(–70)-flowered, usually branched, 2–8(–11) cm diam., glomerules (1 or)2–6(–10), ± capitate. **Flowers** 9–12 mm diam.; epicalyx bractlets 2–4 mm; hypanthium 2.5–4(–4.5) × 2.5–4(–5) mm; sepals 3–5(–6) mm, acute to obtuse; anthers sometimes red-margined.

Flowering summer. Talus slopes and outcrops, in montane and subalpine conifer woodlands; 2000–3200 m; Idaho, Mont., Utah, Wyo.

Variety *wasatchensis* is more broadly defined here than the "*wasatchensis*" of N. H. Holmgren (1997b). The variety is most distinct in the greater Wasatch Range and in the western Uinta Mountains, where plants are generally larger than other varieties with larger leaves, more elongated leaflets, and multi-headed inflorescences. Less distinctive populations extend along the Idaho-Wyoming border to southwestern Montana. Variety *wasatchensis* merges with var. *gordonii* at low elevations in Montana and Wyoming, and with the high-elevation phase assigned here to var. *gordonii* in Utah.

16. Ivesia tweedyi Rydberg in N. L. Britton et al., N. Amer. Fl. 22: 288. 1908 • Tweedy's ivesia [C] [E]

Potentilla tweedyi (Rydberg) J. T. Howell

Plants green, ± rosetted to tufted; taproot stout, not fleshy. **Stems** ascending to erect, 0.4–2(–3.5) dm. **Basal leaves** loosely cylindric, (3–)4–12 (–17) cm; sheathing base glabrous abaxially; petiole 0.5–7 cm, hairs 1–1.5 mm; leaflets 10–16 per side, 4–7(–10) mm, glabrous or sparsely hirsute, glandular-puberulent or -pubescent, lobes (2–)5–15, linear to narrowly oblanceolate, apex not setose. **Cauline leaves**

(0 or)1 or 2, not paired. **Inflorescences** (5–)10–25(–35) -flowered, (1–)1.5–3(–4.5) cm diam.; glomerules 1–few. **Pedicels** 1–3(–5) mm. **Flowers** 9–12 mm diam.; epicalyx bractlets linear, 1–2 mm; hypanthium shallowly campanulate, 1–1.5 × 2–4(–5.5) mm; sepals 2–3.5 mm, acute; petals golden yellow, broadly elliptic to spatulate, 2.5–3.3 mm; stamens 5, filaments 1–1.7 mm, anthers yellow, 0.5–0.8 mm; carpels (2–)4–6(–9), styles 2–3 mm. **Achenes** olive green, 1.8–2 mm.

Flowering summer. Dry, gravelly to rocky flats, slopes, alpine ridges, often on serpentine, in subalpine conifer woodlands; of conservation concern; 1600–2300 m; Idaho, Mont., Wash.

Ivesia tweedyi is found from the Cascade Range in Washington to Boundary and Shoshone counties in Idaho, barely entering Montana in Mineral County.

17. Ivesia muirii A. Gray, Proc. Amer. Acad. Arts 8: 627. 1873 • Muir's ivesia, granite mousetail [E]

Potentilla muirii (A. Gray) Greene

Plants silvery, usually ± rosetted; taproot stout, sometimes fleshy. **Stems** usually ± erect, sometimes nearly prostrate, 0.5–1.5(–2) dm. **Basal leaves** very tightly cylindric (mousetail-like, with individual leaflets scarcely distinguishable), 2–5(–10) cm; sheathing base densely strigose abaxially; petiole 0.2–0.8(–1) cm, hairs 0.5–1.5 mm; leaflets 25–40 per side, 0.4–1 mm, densely sericeous, glands obscured, lobes 2–5, obovate or oval to orbiculate, apex not setose. **Cauline leaves** (0 or)1 or 2, paired if 2. **Inflorescences** 10–30-flowered, 1–2 (–3.5) cm diam.; glomerules usually 1. **Pedicels** 0.3–2(–3.5) mm. **Flowers** 5–6 mm diam.; epicalyx bractlets oblong to obovate, 0.5–1 mm; hypanthium shallowly cupulate, 0.5–1(–1.5) × 1.5–2.5 mm; sepals (1–)1.5–2.5 mm, acute; petals yellow, linear to oblanceolate or narrowly oblong, 1–2 mm; stamens 5, filaments 0.3–0.6 mm, anthers yellow, 0.4–0.6 mm; carpels 1–4, styles 0.7–1.2 mm. **Achenes** grayish brown, mottled with red, 1.6–2 mm.

Flowering summer. Dry rocky slopes, fellfields, mostly in alpine conifer woodlands and tundra; 2900–4000 m; Calif.

Ivesia muirii is known from alpine areas in the Sierra Nevada. It is one of the more distinctive species of the genus, in its silvery mousetail-like leaves and usually tightly capitate inflorescences. Putative hybrids are known with *I. lycopodioides* (D. D. Keck 1938[1939]) and *I. pygmaea* (Center Basin area of Tulare County).

18. Ivesia longibracteata Ertter, Syst. Bot. 14: 233, fig. 1. 1989 • Castle Crags ivesia [C] [E]

Plants green, ± tufted; taproot ± stout, not fleshy. **Stems** ascending to erect, 0.3–1.2 dm. **Basal leaves** weakly planar to loosely ± cylindric, (0.5–) 2–4(–6) cm; sheathing base glandular abaxially, otherwise glabrous; petiole 0.5–2 cm, hairs 0.2–1 mm; leaflets 5 or 6 per side, 2–6 mm, ± short-hirsute, glandular-puberulent, lobes 2–7, oblanceolate to spatulate or obovate, apex not setose. **Cauline leaves** 1–3, not paired. **Inflorescences** 3–14-flowered, 1–2.5 cm diam.; glomerules 1. **Pedicels** 1.5–6 mm. **Flowers** 8–10 mm diam.; epicalyx bractlets linear to narrowly lanceolate or elliptic, 2.5–5 mm, <longer than sepals>; hypanthium shallowly cupulate, 0.5–1 × 2–3 mm; sepals 1.5–2.5 mm, ± acute; petals pale yellow, linear to narrowly oblanceolate, 1.5–2.5 mm; stamens 5, filaments 0.7–1.3 mm, anthers yellow, 0.5–0.8 mm; carpels 6–11, styles 1–1.5 mm. **Achenes** cream to light tan, 1.2–1.5 mm.

Flowering summer. Dry crevices of granodioritic igneous rock, in mixed oak-conifer woodlands; of conservation concern; 1200–1400 m; Calif.

Ivesia longibracteata is known only from the Castle Crags area of Shasta County. The epithet alludes to a diagnostic characteristic unique in the genus: the epicalyx bractlets are longer than the sepals. The plants grow on vertical rock faces, a habitat more characteristic of sect. *Setosae*; however, the stems are ascending to erect and do not form hanging clumps, and the pedicels are not curved in fruit.

19. Ivesia webberi A. Gray, Proc. Amer. Acad. Arts 10: 71. 1874 • Webber's or wire ivesia [C] [E]

Potentilla webberi (A. Gray) Greene

Plants ± green, ± rosetted; taproot slender to ± stout, not fleshy. **Stems** decumbent to ascending, 0.5–1.5(–1.8) dm. **Basal leaves** loosely ± cylindric, 3–7(–10) cm; sheathing base ± strigose abaxially; petiole 0.5–5(–6) cm, hairs 2–4 mm; leaflets 4–8(–10) per side, (0.5–)3–8(–10) mm, loosely long-strigose or -villous and short-hirsute, ± glandular, lobes 2–5(–12), linear to lanceolate, apex not setose. **Cauline leaves** 2, paired. **Inflorescences** 5–15(–25)-flowered, 1.5–3(–6) cm diam.; glomerules 1. **Pedicels** (0.5–)1–8(–13) mm. **Flowers**

9–12 mm diam.; epicalyx bractlets linear, 1.2–3 mm; hypanthium cupulate, 1–2(–2.5) × 2.5–5 mm; sepals 2.5–4.5(–5.5) mm, acute; petals yellow, narrowly oblanceolate, 2–3(–4) mm; stamens 5, filaments 1.8–2.5(–3) mm, anthers yellow, (0.8–)1–1.6 mm; carpels 3–8, styles 1.8–2.2 mm. **Achenes** light brown, often mottled darker brown, 1.9–2.5 mm.

Flowering late spring–early summer. Dry flats and slopes, in sagebrush communities, conifer woodlands; of conservation concern; (1300–)1500–1900 m; Calif., Nev.

Ivesia webberi is known only from the eastern foothills of the northern Sierra Nevada and scattered ranges to the east in California and adjacent Nevada. It is among the more distinctive species in the genus and is only tentatively placed in sect. *Ivesia*. The leaflets are loosely incised into slender, sparsely villous segments, and the two cauline leaves are paired with dissected stipules. Previous reports of the stems and inflorescence branches being glandular-puberulent are due to a misinterpretation of the minute pustulose bases associated with the villous indumentum as being enlarged glands.

9c. Ivesia Torrey & A. Gray sect. Unguiculatae (Rydberg) O. Stevens in N. L. Britton et al., N. Amer. Fl. 22(7): 8. 1959

Horkelia Chamisso & Schlechtendal [unranked] *Unguiculatae* Rydberg, Monogr. N. Amer. Potentilleae, 124. 1898; *Horkelia* [unranked] *Eremicae* Rydberg; *Ivesia* [unranked] *Eremicae* (Rydberg) Rydberg; *Ivesia* sect. *Eremicae* (Rydberg) O. Stevens; *Ivesia* [unranked] *Unguiculatae* (Rydberg) Rydberg

Plants usually rosetted or tufted, rarely ± matted (*I. kingii* var. *eremica*), not forming hanging clumps, not aromatic; taproot stout to fusiform and fleshy. **Stems** (0.3–)1–4.5(–5.5) dm. **Basal leaves** loosely to tightly cylindric (± mousetail-like in *I. argyrocoma* and *I. kingii* var. *eremica*); stipules present or absent; leaflets loosely or, sometimes, tightly overlapping, individually distinguishable or not, lobed to base, sometimes entire; terminal leaflets indistinct, sparsely to densely hairy or glabrate. **Cauline leaves** (1 or)2–10(–15), not paired; blade ± well developed. **Inflorescences** open to congested, flowers arranged individually and/or in glomerules, these usually ± capitate. **Pedicels** remaining ± straight (rarely ± curved in *I. argyrocoma*). **Flowers:** hypanthium shallowly cupulate or campanulate to turbinate, rarely patelliform (*I. pityocharis*); petals not medially reflexed, light yellow to white, sometimes pink-tinged, ± clawed, apices rounded or truncate to emarginate; stamens usually 20 (10–15 in *I. unguiculata*, 12–20 in *I. campestris*), anthers ± as long as to longer than wide, laterally dehiscent; carpels (1 or)2–20. **Achenes** vertical, smooth, not carunculate.

Species 8 (8 in the flora): w United States, nw Mexico.

Section *Unguiculatae* encompasses a series of species that have radiated in montane meadows in California, from Siskiyou and Trinity counties through the Sierra Nevada to the San Bernardino Mountains, and in the Carson, Virginia, and Pine Nut ranges in adjacent Nevada; one variety of *Ivesia argyrocoma* is endemic to the Sierra San Pedro Mártir in Baja California, Mexico. The species occur in seasonally wet/dry meadows and flats, including some specific substrate types of localized occurrence that often support suites of endemic species. The most widespread species in the section, *I. kingii*, has adapted to alkali-crusted valley bottoms extending across the Great Basin to western Utah.

Members of sect. *Unguiculatae* are nearly glabrous to densely sericeous but lack the conspicuous glandularity and distinctive ivesioid smell that characterize sects. *Ivesia* and *Setosae*. The taproot is often elongate and fleshy-thickened, which is apparently an adaption for when the preferred habitat becomes seasonally desiccated. Leaves are cylindric with deeply lobed leaflets. Flowers are usually aggregated into multiple few- to many-flowered glomerules; sometimes they are individually arranged in diffuse inflorescences. Petals are white to pale yellow but never golden yellow and are often conspicuously clawed. Stamens are commonly 20 except in *Ivesia campestris* and *I. unguiculata*.

1. Filaments flattened; leaflets (1–)2–3.5 mm, tightly overlapping; plants silvery; stems prostrate to ascending, (0.3–)1–2.5(–3) dm; cauline leaves (1 or)2 or 3; San Bernardino Mountains, s California . . . . 27. *Ivesia argyrocoma*
1. Filaments filiform; leaflets (1.5–)2–15(–20) mm, loosely overlapping (tightly so in some *I. kingii*); plants green to grayish or silvery; stems prostrate-decumbent to erect, (0.5–)1–4(–5.5) dm; cauline leaves 2–15; w United States.
    2. Stamens 10–20, filaments 0.6–1.1 mm; leaflets 15–20(–25) per side; c, s Sierra Nevada, California.
        3. Petals 5, white, often tinged with pink; sepals heavily purple-mottled; inflorescences (15–)30–100(–200)-flowered; c Sierra Nevada . . . . 25. *Ivesia unguiculata*
        3. Petals 4(or 5), light yellow; sepals green; inflorescences 5–20(–40)-flowered; s Sierra Nevada . . . . 26. *Ivesia campestris*
    2. Stamens 20, filaments (1–)1.5–4 mm; leaflets 15–50(–60) per side; n California to Utah.
        4. Flowers arranged in tight glomerules of 5–20 flowers; pedicels 1–3(–15) mm.
            5. Petals white; petiole hairs usually spreading, 1–4 mm; hypanthia campanulate to shallowly turbinate, 1.5–3 mm, often nearly as deep as wide . . . . 22. *Ivesia sericoleuca*
            5. Petals light yellow; petiole hairs ascending, ± 1(–3) mm; hypanthia cupulate, 1–2 mm, ½–⅔ as deep as wide . . . . 23. *Ivesia aperta*
        4. Flowers arranged individually or in loose glomerules of 2–10 flowers; pedicels (1–)2–20(–25) mm.
            6. Petals oblanceolate to narrowly spatulate; hypanthia cupulate to turbinate, 1.5–3 mm, often nearly as deep as wide; on ultramafic-derived soil; n California . . . . 21. *Ivesia pickeringii*
            6. Petals spatulate or obovate to orbiculate; hypanthia shallowly cupulate to patelliform, 0.5–2.5 mm, ± ½ as deep as wide; on alkaline soil or in non-alkaline meadows; Great Basin and n Mojave Desert.
                7. Leaflets 15–60 per side, hairs absent or 0.2–0.5(–1) mm, ± appressed; carpels 2–9; alkali-crusted flats, e California, Nevada, sw Utah . . . . 20. *Ivesia kingii*
                7. Leaflets 5–25 per side, hairs 1–3 mm, spreading to ascending; carpels 8–20; meadows in sagebrush, Pine Nut Mountains, Nevada . . . . 24. *Ivesia pityocharis*

20. Ivesia kingii S. Watson, Botany (Fortieth Parallel), 91, 448. 1871 [C] [E] [F]

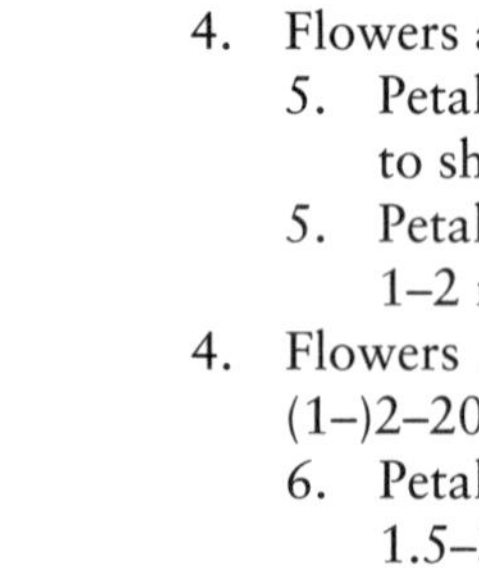

Potentilla kingii (S. Watson) Greene

Plants usually grayish green to silvery, often glaucous; glands sparse. **Stems** usually prostrate-decumbent to ascending, sometimes ± erect, 1–4(–5.5) dm. **Basal leaves** 5–20 cm; sheathing base glabrous or densely strigose abaxially; stipules absent or linear to lanceolate, 1–4 mm; petiole 0.2–1.2 cm, hairs absent or sparse to dense, appressed or ascending, 0.5–2 mm; leaflets 15–60 per side, loosely to tightly overlapping, (1.8–)2–6(–8) mm, lobes (0–)2–4, narrowly oblanceolate to obovate, hairs absent or sparse to dense, ± appressed, 0.2–0.5(–1) mm. **Cauline leaves** 4–15. **Inflorescences** 5–100-flowered, 1–12 cm diam., flowers arranged individually and/or in few to several loose glomerules of 2–10 flowers. **Pedicels** (1–)2–20(–25) mm. **Flowers** 8–12 mm diam.; epicalyx bractlets lanceolate to elliptic or ovate, (0.8–)1–2(–2.5) mm; hypanthium shallowly cupulate, 0.5–2 × 1.5–4 mm, ± ½ as deep as wide; sepals sometimes purple-suffused, 2–4(–5) mm, narrowly acute to acuminate; petals white, spatulate or obovate to orbiculate, (2.8–)3–5(–6) mm; stamens 20, filaments filiform, 1.5–2.5 mm, anthers white to cream, 0.3–0.6 mm; carpels 2–9, styles 2–2.5(–3) mm. **Achenes** light brown, 1.8–2.5 mm.

Varieties 2 (2 in the flora): w United States.

Significant variation in habit, indument, leaflet lobing, and compactness of inflorescence can be found among and within populations of *Ivesia kingii*; no taxonomic structure has yet been discerned beyond the varieties recognized here. The variation within widespread var. *kingii* suggests that var. *eremica* resulted from a southward, late Pleistocene migration of a hairy phase of *I. kingii* with compact leaflets out of the Great Basin onto the northern edge of the Mojave Desert (J. L. Reveal 1980). Although densely hairy plants are encountered widely in the Great Basin, only in the Ash Meadows area of southernmost Nye County, the home of var. *eremica*, is there a consistently hairy phase. The leaflets of var. *eremica* are so tightly compacted that individual leaflets cannot be distinguished; in var. *kingii*, leaflets are individually distinct, though plants from Mineral County sometimes approach var. *eremica* in this regard. In the field, shoots of var. *kingii* arise directly from a taproot with the spreading branches flowing outwardly from a single point of attachment. In var. *eremica*, older, mature plants form compact mats with short, spreading caudex branches.

1. Leaflets individually ± distinguishable, loosely to tightly overlapping; caudices simple or few branched; Great Basin, ec California, Nevada, w Utah . . . . 20a. *Ivesia kingii* var. *kingii*
1. Leaflets mostly not individually distinguishable, tightly overlapping; caudices often much branched; n Mojave Desert, Nevada . . . . 20b. *Ivesia kingii* var. *eremica*

20a. Ivesia kingii S. Watson var. **kingii** • Alkali or King's ivesia [C] [E] [F]

Plants usually grayish green to silvery; caudices simple or few branched. **Basal leaves:** petioles of inner leaves usually glabrous, sometimes with sparse to dense, appressed or ascending hairs 0.5–2 mm; leaflets 15–40 per side, loosely to tightly overlapping, individually ± distinguishable, glabrous or ± sericeous or villous. **Inflorescences** 10–100-flowered, 2–12 cm diam.

Flowering summer. Damp alkali-crusted valley bottoms, in saltbush and saltgrass communities; of conservation concern; 1200–2400 m; Calif., Nev., Utah.

Variety *kingii* is found across the Great Basin from eastern California to western Utah.

20b. Ivesia kingii S. Watson var. **eremica** (Coville) Ertter, Syst. Bot. 14: 244. 1989 • Ash Meadows ivesia [C] [E]

Potentilla eremica Coville, Proc. Biol. Soc. Wash. 7: 76. 1892; *Ivesia eremica* (Coville) Rydberg

Plants silvery; caudices often much branched. **Basal leaves:** petioles of inner leaves with dense, ascending hairs 1–1.5 mm; leaflets 30–60 per side, tightly overlapping, mostly not individually distinguishable, densely sericeous or villous. **Inflorescences** 5–30-flowered, 1–6 cm diam.

Flowering summer. Damp alkali-crusted flats, in saltbush communities; of conservation concern; 600–700 m; Nev.

Variety *eremica* is known only from Ash Meadows, Nye County.

21. Ivesia pickeringii Torrey ex A. Gray, Proc. Amer. Acad. Arts 6: 531. 1865 • Pickering's ivesia [C] [E]

Potentilla pickeringii (Torrey ex A. Gray) Greene

Plants grayish green; glands abundant. **Stems** ascending to erect, 3–5 dm. **Basal leaves** 8–20 cm; sheathing base ± strigose abaxially; stipules linear to narrowly lanceolate, 3–5 mm; petiole 1–3.5 cm, hairs abundant, ascending to spreading, 1–2 mm; leaflets 35–50 per side, loosely overlapping, 2–6 mm, lobes 3–5, oblanceolate to obovate or oval, hairs abundant, ± ascending, 1–2(–3) mm. **Cauline leaves** 5–10. **Inflorescences** 10–100-flowered, (1.5–)5–15 cm diam., flowers usually arranged individually, sometimes more congested. **Pedicels** (1–)2–10 mm. **Flowers** 8–13 mm diam.; epicalyx bractlets linear-lanceolate to lanceolate, (1.8–)2–2.5 mm; hypanthium cupulate to turbinate, 1.5–3 × 2.5–4 mm, often nearly as deep as wide; sepals often purple-suffused or -mottled, (2.5–)3–5 mm, acuminate; petals white, becoming pink-tinged with age, oblanceolate to narrowly spatulate, (2.8–)3–5(–6) mm; stamens 20, filaments filiform, 1.5–2.3 mm, anthers white to cream, 0.3–0.6 mm; carpels 2–4, styles 2.5–3.5 mm. **Achenes** dark brown, 2.5–3 mm.

Flowering summer. Moist, rocky, grassy meadows, mainly on ultramafic-derived clayey soil, in montane conifer woodlands; of conservation concern; 800–1500 m; Calif.

Ivesia pickeringii is known from Siskiyou and Trinity counties. Some early treatments (for example, P. A. Rydberg 1898, 1908–1918; W. L. Jepson [1923–1925], 1909–1943, vol. 2) conflated this species with *I. aperta*, hence reports of *I. pickeringii* in the Sierra Nevada. Historic collections from Edgewood, at the headwaters of the Shasta River in Siskiyou County, have inflorescences more glomerulate than elsewhere in the species range.

The chromosome count of $2n = 28$ (P. A. Munz 1959) needs confirmation. If based on *Kruckeberg 3665*, which was distributed as a voucher for that count, the determination of this collection has been changed to *Ivesia sericoleuca*.

22. Ivesia sericoleuca (Rydberg) Rydberg in N. L. Britton et al., N. Amer. Fl. 22: 284. 1908 • Plumas ivesia [C] [E]

Horkelia sericoleuca Rydberg, Monogr. N. Amer. Potentilleae, 144, plate 85. 1898; *Potentilla sericoleuca* (Rydberg) J. T. Howell

Plants silvery to grayish green; glands usually sparse, sometimes abundant. **Stems** decumbent to ascending, 1.5–4.5 dm. **Basal leaves** 10–20(–30) cm; sheathing base densely strigose abaxially; stipules absent; petiole 2–6(–10) cm, hairs abundant, usually spreading, 1–4 mm; leaflets 20–35 per side, loosely overlapping, 3–15 mm, lobes 0–4, oblanceolate to elliptic, hairs abundant, spreading to ascending, (0.5–)1–3(–4) mm. **Cauline leaves** 3–8(–10). **Inflorescences** 20–120 -flowered, (2–)4–14 cm diam., flowers mostly arranged in several to many tight glomerules of 5–10 flowers. **Pedicels** 1–3(–12) mm. **Flowers** 10–15 mm diam.; epicalyx bractlets narrowly lanceolate to narrowly elliptic, (1.5–)2–2.5(–3) mm; hypanthium campanulate to shallowly turbinate, 1.5–3 × 2.5–4.5(–5) mm, often nearly as deep as wide; sepals sometimes purple-suffused, 3–5.5 mm, acute to acuminate; petals white, broadly spatulate to broadly obovate or obcordate, 4–7 mm; stamens 20, filaments filiform, 1.5–3 mm, anthers white to cream, 0.5–0.7 mm; carpels 2–7, styles 2.5–4 mm. **Achenes** brown, 2–3 mm. $2n = 28$.

Flowering summer. Dry gravelly meadows, margins of seeps, usually on vernally saturated volcanic soil, in sagebrush and grass communities, conifer woodlands; of conservation concern; 1300–2300 m; Calif.

Ivesia sericoleuca is known from valleys and flats in the northern Sierra Nevada. Many historic collections were identified and distributed as *I. unguiculata*, and circumscriptions prior to 1962 include *I. aperta* (hence reports from Nevada).

Hairs are usually dense in plants of *Ivesia sericoleuca*, such that the leaves, and occasionally the stems and branches, are silvery gray, especially in Sierra Valley and the Feather River drainage. Plants in the Truckee River drainage tend to be less hairy with redder stems, less glomerate inflorescences, shallower hypanthia, and more conspicuous glandularity.

As mentioned by J. T. Howell (1962), the chromosome count given for *Ivesia sericoleuca* by P. A. Munz (1959) most likely was based on a collection of *I. aperta* var. *aperta*. The chromosome count given here is instead based on *Kruckeberg 3665*, originally distributed as *I. pickeringii*.

23. Ivesia aperta (J. T. Howell) Munz, Suppl. Calif. Fl., 111. 1968 [C] [E]

Potentilla aperta J. T. Howell, Leafl. W. Bot. 9: 239. 1962

Plants greenish gray to silvery; glands sparse. **Stems** decumbent to erect, 1.5–4.5 dm. **Basal leaves** (7–)10–20(–23) cm; sheathing base densely strigose abaxially; stipules absent; petiole 4–8 cm, hairs abundant, ascending, ± 1(–3) mm; leaflets 20–35 per side, loosely overlapping, 3–15(–20) mm, lobes 0–4(or 5), oblanceolate to elliptic, hairs abundant, ascending to appressed, ± 1(–2) mm. **Cauline leaves** 3–8. **Inflorescences** (10–)20–250 -flowered, 1–14 cm diam., flowers arranged in several to many tight glomerules of 5–20 flowers. **Pedicels** 1–3 (–15) mm. **Flowers** 5–12 mm diam.; epicalyx bractlets lanceolate, 1.5–2.5(–3.5) mm; hypanthium cupulate, 1–2 × 2.5–4(–5) mm, $\frac{1}{2}$–$\frac{2}{3}$ as deep as wide; sepals green, 2.5–4(–5.5) mm, acute to acuminate; petals light yellow, oblanceolate to broadly obovate, 2–7 mm; stamens 20, filaments filiform, 1–4 mm, anthers yellow, 0.4–0.6 mm; carpels 2–7, styles 2–4 mm. **Achenes** brown, 2–3 mm.

Varieties 2 (2 in the flora): w United States.

Many historic collections of *Ivesia aperta* were identified as *I. pickeringii*. D. D. Keck (1938) considered *I. aperta* to be no more than a yellow-flowered form of *I. sericoleuca*, a situation clarified and rectified by J. T. Howell (1962). The ranges of the two species overlap in Sierra Valley, California, with mixed populations occurring near Beckwourth and in the valleys north of Sierra Valley. In general, var. *aperta* occurs in the southeastern portions of the valley and *I. sericoleuca* occurs on the western side; the species are not otherwise sympatric. In addition to petal color, which fades in herbarium material, *I. aperta* differs from *I. sericoleuca* in having smaller flowers with shallower hypanthia. Hairs at the base of the stems and petioles of *I. aperta* are relatively short (to 2 mm) and generally ascending; those of *I. sericoleuca* are longer (to 4 mm) and spreading. Variety *canina* combines the petal color of var. *aperta* and floral dimensions of *I. sericoleuca*.

1. Petals 2–3 mm, oblanceolate; filaments 1–1.5 (–2) mm; Sierra Valley drainage, California, Carson and Virginia ranges, Nevada . . . . 23a. *Ivesia aperta* var. *aperta*
1. Petals usually 4–7 mm, narrowly to broadly obovate; filaments 2–4 mm; Dog Valley, California, Nevada . . . . . . 23b. *Ivesia aperta* var. *canina*

23a. Ivesia aperta (J. T. Howell) Munz var. **aperta** • Sierra Valley ivesia [C] [E]

Stems ascending to erect. **Flowers** (20–)30–250, usually more than 10 per glomerule, 5–10 mm diam.; hypanthium 2–3(–4) mm diam.; petals oblanceolate, 2–3 mm, shorter than sepals; filaments 1–1.5 (–2) mm; styles 2–3 mm. $2n = 28$.

Flowering summer. Dry meadows, edges of seeps, slopes, and flats, on vernally saturated volcanic soil, in sagebrush and grass communities, conifer woodlands; of conservation concern; 1300–2300 m; Calif., Nev.

Variety *aperta* is most abundant in Sierra Valley and its drainages in Plumas and Sierra counties, California, barely extending into Lassen County. Disjunct populations occur in the Carson Range in southern Washoe County and the Virginia Range of Storey County, Nevada.

The chromosome count given here is based on a collection originally identified as *Ivesia sericoleuca* (*D. D. Keck & A. Gustafsson 4901*).

23b. Ivesia aperta (J. T. Howell) Munz var. **canina** Ertter, Brittonia 40: 398, fig. 1. 1988 • Dog Valley ivesia [C] [E]

Stems decumbent to ascending. **Flowers** (10–)20–120, fewer than 10 per glomerule, 8–12 mm diam.; hypanthium (3–)3.5–5 mm diam.; petals narrowly to broadly obovate, usually 4–7 mm, ± equal to sepals; filaments 2–4 mm; styles 3–4 mm.

Flowering summer. Dry meadows, edges of seeps, slopes, and flats, on vernally saturated volcanic soil, in sagebrush and grass communities, conifer woodlands; of conservation concern; 1600–2000 m; Calif., Nev.

Variety *canina* is known only from the Dog Valley area, Sierra County, California, and adjacent Washoe County, Nevada.

24. Ivesia pityocharis Ertter, Syst. Bot. 14: 241, fig. 6. 1989 • Pine Nut Mountains ivesia [C] [E]

Plants grayish green to silvery, sometimes reddish tinged; glands sparse. **Stems** prostrate-decumbent to ascending, 0.5–2(–3) dm. **Basal leaves** 6–12(–15) cm; sheathing base weakly strigose abaxially; stipules absent; petiole (1–)1.5–2.5(–3) cm, hairs abundant, ascending to spreading, 1–4 mm; leaflets 15–25 per side, loosely overlapping, (1.5–)2–7 mm, lobes 0–4(–6), lanceolate or oblanceolate to elliptic, hairs abundant, spreading to ascending, 1–3 mm. **Cauline leaves** 2–5. **Inflorescences** (7–)15–50-flowered, 2–8 cm diam., flowers arranged individually and/or in several to many loose few-flowered glomerules. **Pedicels** (3–)6–15 (–22) mm. **Flowers** 8–13 mm diam.; epicalyx bractlets lanceolate to elliptic, 2–3 mm; hypanthium patelliform to shallowly cupulate, 1.5–2.5 × 3–4.5 mm, ± ½ as deep as wide; sepals often purple-suffused, 2.3–4(–5) mm, acute; petals white, broadly spatulate or obovate to orbiculate, (2.8–)3–5(–6) mm; stamens 20, filaments filiform, (1–)1.5–3 mm, anthers light pink, 0.3–0.5 mm; carpels 8–20, styles 2.2–3 mm. **Achenes** light brown, 1.2–1.3 mm.

Flowering summer. Vernally saturated meadows, in sagebrush communities; of conservation concern; 2100–2700 m; Nev.

Ivesia pityocharis is known only from the Pine Nut Mountains, Douglas County. Plants are somewhat intermediate between those of *I. kingii* and *I. sericoleuca* in having relatively shallow hypanthia and loosely sericeous vestiture. Hairs on petioles and stem bases of *I. pityocharis* are to 4 mm; those of *I. kingii* are 1 mm and generally appressed-ascending.

25. Ivesia unguiculata A. Gray, Proc. Amer. Acad. Arts 7: 339. 1868 • Yosemite ivesia [C] [E]

Potentilla unguiculata (A. Gray) Hooker f.

Plants ± grayish, often purple-tinged; glands sparse to abundant. **Stems** decumbent to ascending, 1–3.5 dm. **Basal leaves** (4–)7–15 cm; sheathing base ± glabrous abaxially; stipules linear, 3–6 mm; petiole (0.3–)0.5–4(–5) cm, hairs sparse to abundant, ascending to spreading, 1–2 mm; leaflets 15–20(–25) per side, loosely overlapping, 3–6 mm, lobes 3–8, linear to oblanceolate, hairs sparse to abundant, spreading, 1–2 mm. **Cauline leaves** 3–6. **Inflorescences** (15–)30–100(–200)-flowered, (1–)1.5–4(–8) cm diam., flowers mostly arranged in 1–several loose to tight glomerules of 5–10 flowers. **Pedicels** 1–3 mm. **Flowers** 6–9 mm diam.; epicalyx bractlets linear or narrowly lanceolate to elliptic or narrowly oblong, 1.2–2(–3) mm; hypanthium shallowly turbinate, 1.5–2.5 × 2–3(–3.5) mm, often nearly as deep as wide; sepals heavily purple-mottled, (1.5–)2–3(–3.5) mm, acute; petals white, often tinged with pink, oblanceolate to spatulate or obovate, 3–4 mm; stamens 10–15, filaments filiform, 0.6–1.1 mm, anthers maroon, 0.3–0.5 mm; carpels (1–)3–9, styles 1.4–2 mm. **Achenes** light brown, 1.2–1.5 mm.

Flowering summer. Moist meadows and slopes, in montane conifer woodlands; of conservation concern; 1500–2500 m; Calif.

Ivesia unguiculata is found in mid-elevation meadows of the central Sierra Nevada, mostly north of the Kings River. The distinctive deep red to purplish coloration of the inflorescence, and the plant in general, can make patches of this species conspicuous as a smoky purplish haze in meadows. The shape and color of the inflorescences are similar to those of the sympatric *Horkelia fusca* var. *parviflora*, suggesting shared pollinators.

The description and illustration by J. D. Hooker (1881), supposedly of *Potentilla* (*Ivesia*) *unguiculata*, were based actually on material grown from seed of *I. sericoleuca* due to initial confusion of the two species (W. H. Brewer et al. 1876–1880, vol. 1).

The type (*Kellogg s.n.*, CAS) of *Potentilla ciliata* Greene (not Rafinesque) is unquestionably this species; however, the purported locality (Owens Valley, Inyo County) is dubious and most likely an error in the labeling of the specimen by the collector.

26. Ivesia campestris (M. E. Jones) Rydberg in N. L. Britton et al., N. Amer. Fl. 22: 285. 1908 • Field or Kaweah ivesia [E]

Potentilla utahensis (S. Watson) Greene var. *campestris* M. E. Jones, Proc. Calif. Acad. Sci., ser. 2, 5: 679. 1895; *P. campestris* (M. E. Jones) Jepson

Plants green to grayish; glands sparse. **Stems** decumbent to ascending, 1–3.5 dm. **Basal leaves** (3–)5–15(–18) cm; sheathing base glabrous or sparsely strigose abaxially; stipules ± lanceolate, 2.5–5 mm; petiole (0.3–)0.5–6 (–8) cm, hairs sparse to abundant, appressed-ascending, 1–2 mm; leaflets 15–20 per side, loosely overlapping, 2–10 mm, lobes 2–5, oblanceolate, hairs sparse to abundant, spreading to ascending, 1–2 mm. **Cauline leaves** 3 or 4. **Inflorescences** 5–20(–40)-flowered, (1–)1.5–2.5(–4) cm diam., flowers arranged in 1–few ± tight glomerules of 10–15 flowers. **Pedicels** 1–4 mm. **Flowers** 7–10 mm diam.; epicalyx bractlets linear to oblong, 1–2(–2.5) mm; hypanthium campanulate, 1–2 × 2.5–4 mm, ± ½ as deep as wide; sepals green, 2–3 (–3.5) mm, acute; petals <4(or 5)>, light yellow, oblanceolate to spatulate or narrowly obovate, 3–4 mm; stamens 12–16 (4-merous flowers) or 16–20 (5-merous flowers), filaments filiform, 0.6–1.1 mm, anthers yellowish, 0.3–0.5 mm; carpels 4–20, styles 1.4–2 mm. **Achenes** light brown, 1–1.5 mm.

Flowering summer. Moist meadows and slopes, in montane to subalpine conifer woodlands; 2200–3400 m; Calif.

Ivesia campestris is found in the southern Sierra Nevada south of the Kings River, where it replaces *I. unguiculata* and often occurs at higher elevations.

Ivesia campestris is the only member of the genus that commonly has 4-merous flowers. It is sometimes misidentified as *I. unguiculata*, especially when the pale yellow petals have faded to creamy white; the glomerules of the inflorescences tend to be yellowish green to green rather than purplish. Also, the anthers of *I. campestris* are yellowish, rather than maroon.

27. Ivesia argyrocoma (Rydberg) Rydberg in N. L. Britton et al., N. Amer. Fl. 22: 284. 1908 [C]

Horkelia argyrocoma Rydberg, Monogr. N. Amer. Potentilleae, 144, plate 84. 1898; *Potentilla argyrocoma* (Rydberg) Jepson

Varieties 2 (1 in the flora): California, nw Mexico.

Plants from the central Sierra San Pedro Mártir in northern Baja California belong to var. *moranii* Ertter & Reveal, which may be distinguished by its compact stature (to 0.5 dm), densely villous basal stipules 2–4 mm, decumbent stems 2–7 cm that are stipitate-glandular under the villous indumentum, and solitary, loosely congested cyme. The petals are oblanceolate to narrowly obovate and 2.5–3.5 mm. The grayish, faintly mottled achenes are 1.3–2 mm.

27a. Ivesia argyrocoma (Rydberg) Rydberg var. **argyrocoma** • Silver-hair ivesia [C] [E]

Plants silvery; glands obscured. **Stems** prostrate to ascending, (0.3–)1–2.5(–3) dm. **Basal leaves** 4–8(–10) cm; sheathing base sparsely to densely strigose abaxially; stipules lanceolate, 1–2 mm; petiole (0.5–)1–3 cm, hairs abundant, spreading, (1–)3–5 mm; leaflets 25–35 per side, tightly overlapping, (1–)2–3.5 mm, lobes 2 or 3, elliptic to obovate or oval, hairs dense, appressed to ascending, 0.5–1.5 mm. **Cauline leaves** (1 or)2 or 3. **Inflorescences** (5–)10–30-flowered, (1–)1.5–4(–6) cm diam., flowers arranged in 1–several ± tight glomerules of 10–20(–30) flowers. **Pedicels** 1–3 mm. **Flowers** 7–9 mm diam.; epicalyx bractlets mostly linear to narrowly ovate, 0.8–2(–2.5) mm; hypanthium cupulate, (1–)1.5–2 × (2–)2.5–3(–3.5) mm, ½–⅔ as deep as wide; sepals sometimes purple-suffused, 2–3(–4) mm, acute; petals white, obovate, (2–)3–4(–4.5) mm; stamens 20, filaments flattened, 0.6–1.3 mm, anthers yellowish, 0.3–0.5 mm; carpels 4–8, styles 1.5–2.2 mm. **Achenes** brown, (1.5–)2–2.5 mm.

Flowering summer. Dry meadows, gravelly soil, in montane conifer woodlands; of conservation concern; 1400–2300 m; Calif.

Variety *argyrocoma* is found only on so-called pebble plains in the eastern San Bernardino Mountains, San Bernardino County. The filaments are significantly more flattened than those of other species of *Ivesia* and approach those that characterize *Horkelia*.

9d. Ivesia Torrey & A. Gray sect. Comarella (Rydberg) Ertter & Reveal, Novon 17: 316. 2007 [C] [E]

Comarella Rydberg, Monogr. N. Amer. Potentilleae, 156, plates 96, 97, figs. 1–5. 1898

Plants usually ± tufted, not forming hanging clumps, not aromatic; taproot stout, not fusiform or fleshy. **Stems** (1.8–)2–6(–6.5) dm. **Basal leaves** cylindric; stipules usually present, sometimes much-reduced; leaflets loosely to ± tightly overlapping, individually distinguishable at least distally, toothed ⅓–½ or lobed to base, sometimes entire, sparsely to ± densely hairy, sometimes glabrate; terminal leaflets indistinct. **Cauline leaves** 1–3, not paired; blade well developed. **Inflorescences** open, flowers arranged individually. **Pedicels** remaining straight. **Flowers:** hypanthium patelliform; petals medially reflexed, yellow or red, not clawed, apex acute to obtuse; stamens 5, anthers longer than wide, laterally dehiscent; carpels 1–5. **Achenes** horizontal, smooth, not carunculate.

Species 2 (2 in the flora): sw United States.

Rydberg established *Comarella* to accommodate two species with wheel-shaped centrally turbinate hypanthia, linear petals, five stamens, and few carpels that are nearly horizontal at maturity. He understood both species to have purple petals, but the petals of *Ivesia sabulosa* are actually bright yellow. Other than flower color and leaflet lobing, the two species are nearly identical, with diffuse inflorescences on erect stems emerging from tufts of erect, slender leaves with numerous pairs of leaflets. The medially reflexed petals are unique to these two species of *Ivesia*, though not always well developed. The stipules of the basal leaves are often little more than slightly rounded, winglike extensions of the petiole base, if they are present at all.

Comarella is the only section of the genus without representation in California. The plants occur at the southeastern corner of the range of *Ivesia* in the southern Great Basin, southwestern Colorado Plateau, and adjacent Mojave Desert. The Grand Canyon of the Colorado River divides the range of the two species in northern Arizona.

1. Petals brick- to blood-red; leaflets incised ⅓–½ to base; s of Colorado River, Arizona. . . . . . 28. *Ivesia multifoliolata*
1. Petals yellow; leaflets incised to base; n of Colorado River, Arizona, Nevada, Utah. . . . . . . 29. *Ivesia sabulosa*

28. Ivesia multifoliolata (Torrey) D. D. Keck, Lloydia 1: 125. 1939 • Coconino ivesia, red comarella C E

Horkelia multifoliolata Torrey in L. Sitgreaves, Rep. Exped. Zuni Colorado Rivers, 159. 1853; *Comarella multifoliolata* (Torrey) Rydberg; *Potentilla multifoliolata* (Torrey) Kearney & Peebles

Plants green. **Stems** ascending to erect, 2–6 dm. **Basal leaves** 6–25(–30) cm; sheathing base minutely glandular abaxially; petiole 0.5–8(–10) cm; leaflets 12–30 per side, obovate to broadly cuneate, 5–12(–15) mm, incised 1/3–1/2 to base into 3–5 oblong to ovate teeth, sometimes entire, sparsely short-strigose to -villous, sometimes glabrate. **Inflorescences** (5–) 8–60-flowered, 2–12 cm diam. **Pedicels** 5–30(–35) mm. **Flowers** 9–12 mm diam.; epicalyx bractlets narrowly lanceolate to lanceolate, 1–2 mm; hypanthium interior brick- to blood-red or golden, 0.5–1.5 × 3–5 mm; sepals 3–4.5 mm, base brick- to blood-red adaxially, apex acute to acuminate; petals brick- to blood-red, linear to oblanceolate or narrowly elliptic, 2–3.5 mm; stamens 5, filaments 1.3–2 mm, anthers maroon, oblong, 0.6–0.9 mm; carpels 1–5, styles 2–3 mm. **Achenes** brown, 2–2.5 mm. $2n = 28$.

Flowering summer. Dry flats and slopes, on gravelly volcanic or sandstone soil, often along seasonal stream courses, in oak and conifer woodlands; of conservation concern; 1800–2400 m; Ariz.

Ivesia multifoliolata is found between the Grand Canyon and Mogollon Rim in northern Arizona. It is both the only species of the genus with red petals and the only one that overlaps the range of the red-petaled *Potentilla* sect. *Rubrae*.

29. Ivesia sabulosa (M. E. Jones) D. D. Keck, Lloydia 1: 124. 1939 • Sevier ivesia, yellow comarella C E F

Potentilla sabulosa M. E. Jones, Proc. Calif. Acad. Sci., ser. 2, 5: 680. 1895; *Comarella sabulosa* (M. E. Jones) Rydberg

Plants ± grayish green. **Stems** ascending to erect, (1.8–)2–6 (–6.5) dm. **Basal leaves** 7–25 (–30) cm; sheathing base usually sparsely strigose abaxially; petiole 1–4(–5) cm; leaflets 15–40 per side, ± flabellate, 3–14 mm, usually incised to base into 2 or 3 oblanceolate lobes, ± densely short-hirsute to -villous. **Inflorescences** 10–60-flowered, 4–15 cm diam. **Pedicels** (1–)5–20 mm. **Flowers** 9–14 mm diam.; epicalyx bractlets lanceolate, 1–3(–3.3) mm; hypanthium interior golden, 1–2 × 3–5 mm; sepals (2.5–)3.5–6 mm, base golden adaxially, apex acute to acuminate; petals yellow, linear to narrowly oblanceolate, 2–4 mm; stamens 5, filaments 2–4 mm, anthers yellow, sometimes red-rimmed, oblong, 0.6–1.2 mm; carpels 1–5, styles 2–3 mm. **Achenes** brown, 1.7–2.2 mm.

Flowering summer. Dry flats and slopes, on gravelly volcanic or limestone soil, in sagebrush and other desert shrub communities, montane conifer woodlands; of conservation concern; 1500–2700 m; Ariz., Nev., Utah.

Ivesia sabulosa occurs from central Nevada and southwestern Utah south to Arizona north of the Grand Canyon.

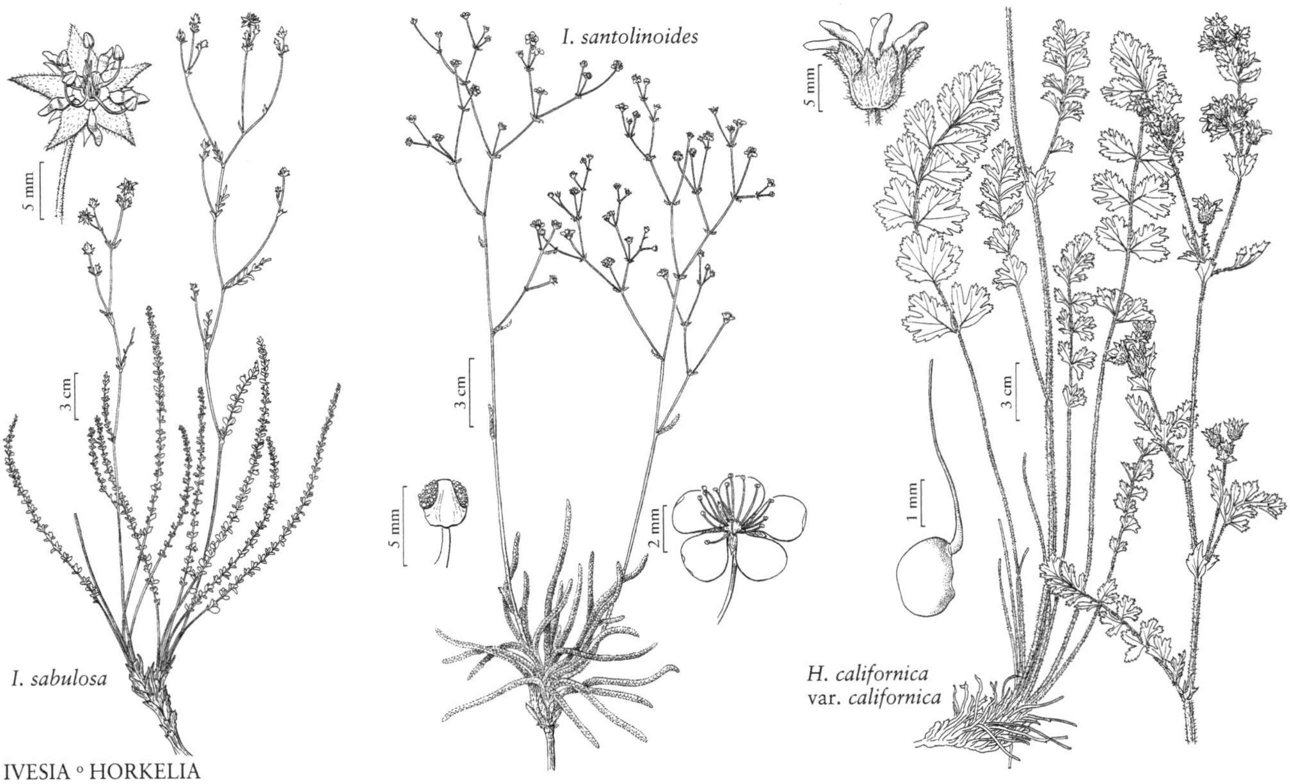

IVESIA ◦ HORKELIA

9e. IVESIA Torrey & A. Gray sect. STELLARIOPSIS (Baillon) Ertter & Reveal, Novon 17: 317. 2007 [E]

Potentilla Linnaeus sect. *Stellariopsis* Baillon, Hist. Pl. 1: 370. 1869; *Stellariopsis* (Baillon) Rydberg

Plants tufted, not forming hanging clumps, not aromatic; taproot ± stout, not fusiform or fleshy. **Stems** (1–)1.5–4 dm. **Basal leaves** very tightly cylindric (mousetail-like); stipules present; leaflets tightly overlapping, individually scarcely distinguishable, lobed to base, sometimes entire, densely hairy; terminal leaflets indistinct. **Cauline leaves** 1–3, not paired; blade relatively well developed. **Inflorescences** open, flowers arranged individually. **Pedicels** remaining ± straight. **Flowers:** hypanthium shallowly campanulate; petals not medially reflexed, white, short-clawed, apex rounded; stamens 15, anthers shorter than wide, dehiscing by subterminal pores to short lateral slits; carpels 1. **Achenes** vertical, smooth, not carunculate.

Species 1: California.

P. A. Rydberg (1898) recognized *Stellariopsis* at the generic level primarily on the basis of the obcordate anthers, which apparently open by terminal pores, as well as the single carpel, mousetail-like leaves, diffuse inflorescence, and flowers intermediate between *Ivesia* and *Potentilla*. J. Soják (2008) argued that the anther character alone justifies generic status. Examination of a series of flowers shows that porelike dehiscence is largely a result of the extreme shortness of the anthers relative to width, such that longer-than-average anthers approach a more typical dehiscence mode. The single carpel and mousetail-like leaves are consistent within the taxon, although both are found elsewhere in the genus (albeit not in the same species). The relatively large, maroon-tipped stipitate glands of the inflorescence branches, including the pedicel and hypanthium, are otherwise unknown in *Ivesia*.

30. Ivesia santolinoides A. Gray, Proc. Amer. Acad. Arts 6: 531. 1865 • Silver mousetail, stellariopsis [E] [F]

Potentilla santolinoides (A. Gray) Greene; *Stellariopsis santolinoides* (A. Gray) Rydberg

Plants ± grayish to silvery. **Stems** ascending to erect, (1–)1.5–4 dm. **Basal leaves** mousetail-like (individual leaflets scarcely distinguishable), 3–10 cm; sheathing base densely strigose abaxially; petiole 0.5–1.5 cm; leaflets 60–80 per side, 0.6–1.5 mm, lobes (0–)3–5, obovate to oval, densely villous. **Inflorescences** (10–)30–200-flowered, 3–8 (–30) cm diam. **Pedicels** 5–30 mm. **Flowers** 5–8 mm diam.; epicalyx bractlets oblong to broadly ovate or orbiculate, 0.2–0.5 mm; hypanthium 0.5–1.5 × 2–3 mm; sepals 1–2 mm, apex obtuse to acute; petals white, broadly obovate to orbiculate, 2–2.5 mm; stamens 15, filaments 1.2–1.8 mm, anthers purple, broadly obcordate to pouch-shaped, 0.3–0.4 mm; carpels 1, styles 2–3 mm. **Achenes** mottled grayish brown, 1.7–2 mm. $2n = 28$.

Flowering summer. Dry sandy granitic soil, decomposed granite accumulations, ledges and outcrops, in montane and subalpine conifer woodlands; 1500–3600 m; Calif.

Ivesia santolinoides is found on loose granitic substrates in the Sierra Nevada, Transverse Ranges, and San Jacinto Mountains from El Dorado to Riverside counties. The species is easily recognized by its silvery mousetail-like leaves and erect, diffuse inflorescences with small, plumlike flowers.

10. HORKELIA Chamisso & Schlechtendal, Linnaea 2: 26. 1827 • [For Johann Horkel, 1769–1846, German plant physiologist]

Barbara Ertter

James L. Reveal

Potentilla Linnaeus sect. *Horkelia* (Chamisso & Schlechtendal) Baillon; *Potentilla* subg. *Horkelia* (Chamisso & Schlechtendal) Jepson

Herbs, perennial, rosetted, tufted, or matted, <often resinously aromatic>, 0.3–10(–12) dm, sparsely to densely hairy, inconspicuously to conspicuously glandular; compactly to ± loosely rhizomatous. **Stems** 1–10+, decumbent to erect, rarely prostrate, <green, reddish, or stramineous>. **Leaves** marcescent or winter-persistent, primarily basal, <cauline 1–6(–10), gradually or abruptly reduced distally>, alternate, odd-pinnate; stipules persistent, basally adnate to petiole, linear to lanceolate or elliptic, margins entire or pinnately divided into linear to filiform segments; petiole present; blade oblanceolate-oblong to linear in outline, planar to cylindric, (1–)2–30(–40) cm, foliaceous, leaflets (3–)5–41, <separate or congested and overlapping, terminal usually ± confluent with distalmost lateral ones>, elliptic-oblong to ovate or flabellate in outline, margins flat, shallowly toothed to deeply divided into oblanceolate lobes, sometimes nearly entire except for 3-toothed apex, rarely completely entire, venation palmate to pinnate. **Inflorescences** terminal, 3–200-flowered, open to congested cymes, flowers arranged individually, in corymbiform clusters, and/or in ± globose glomerules, these sometimes capitate; bracts present at proximal nodes, ± leafy, becoming reduced distally; bracteoles absent. **Pedicels** present, <usually straight in fruit, sometimes reflexed or recurved>. **Flowers** 4–17 mm diam.; epicalyx bractlets 5; hypanthium ± cupulate with flattened bases, 1–5.5 mm, <less than ½ to nearly as deep as wide>, interior often with band of hairs; sepals 5, spreading to strongly reflexed, usually lanceolate, sometimes ovate to triangular; petals 5, usually white, sometimes pink- to rose-veined or cream, usually narrowly oblanceolate-elliptic to obovate, sometimes linear to round, <usually equal to or longer than sepals>; stamens 10, shorter than petals, <filaments usually ± erect, ± flattened, forming tube, anther thecae dehiscing introrsely>; torus conic; carpels 2–200(–220), glabrous, styles subterminal; ovule 1. **Fruits** aggregated achenes, (1 or)2–200(–220), obliquely ovoid to reniform, 0.8–3 mm, glabrous; hypanthium persistent; sepals persistent, erect; styles tardily deciduous, <jointed, usually ± rough-thickened basally, otherwise mostly filiform>. $x = 7$.

Species 20 (20 in the flora): w North America, nw Mexico.

As discussed under 9. *Ivesia*, recognition of the ivesioid genera including *Horkelia* results in a paraphyletic *Potentilla*, which the first author does not find sufficient to dictate generic circumscriptions. On the basis of morphology and biogeography, *Horkelia* itself is likely nested within *Ivesia* as the branch that has radiated primarily within the California Floristic Province (see discussions under *Ivesia* regarding discordance with M. Töpel et al. 2012, http://www.plosone.org/article/info%3Adoi%2F10.1371%2Fjournal.pone.0050358). *Ivesia* in the strict sense consists of an exceptionally diverse array of well-defined species; in contrast, *Horkelia* is more morphologically coherent as a genus but with species less sharply defined. The primary unifying features of the genus result from a flower specifically adapted for bee pollination: a relatively deep, flat-bottomed hypanthium containing abundant nectar, to which access is restricted by the columnar arrangement of ten erect stamens with flattened filaments, at least in early flowering. Anthers open introrsely by adaxial slits so that pollen is deposited on a visitor forcing its way into the staminal column.

Other diagnostic characteristics of *Horkelia* include a distinctive resinous aroma in most species (except sect. *Hispidulae* and sect. *Tridentatae*) and various levels of glandularity. The terminal leaflet is most often lobed and/or confluent with lateral leaflets, in contrast to *Drymocallis*, which is often confused with *Horkelia*. Leaves range from planar with distinct leaflets to cylindric with overlapping leaflets, the latter condition often a response to drought stress in species that are normally planar-leaved. Sepals are often strongly reflexed.

Stem lengths given below include inflorescences. Leaflet descriptions are based primarily on mid rachis leaflets of the larger basal leaves present at flowering; leaflets on the smaller outermost leaves formed early in the growing season are often less divided and less hairy. Counts of leaflet marginal segments include both shallow (teeth) and deep (lobes) ultimate divisions. Flowers are often aggregated into a single corymbiform cluster or one to many more globular glomerules, some of which are tightly capitate. Hypanthia interiors are glabrous in most species, except for the hairs surrounding the torus; some taxa are also more or less pilose on the internal hypanthial walls, especially just below the rim. Sepals are described as spreading or reflexed based on their orientation at anthesis. Larger and smaller anthers often are alternating; measurements are primarily of the larger filaments, with width measured at the base. See under 8. *Potentilla* and 9. *Ivesia* for additional discussion of special terminology.

1. Pedicels becoming reflexed or recurved, 3–15 mm; inflorescences open; Sierra Nevada foothills and San Bernardino Mountains, California . . . . 10c. *Horkelia* sect. *Parryae*, p. 257
1. Pedicels remaining straight, outermost sometimes ± reflexed in congested inflorescences, 1–30(–40) mm; inflorescences open to congested; w United States.
    2. Basal leaves: leaflet teeth (0–)3(–5), restricted to apex, leaflets 2–5 per side. . . . . 10e. *Horkelia* sect. *Tridentatae* (in part), p. 264
    2. Basal leaves: leaflet teeth or lobes (0–)4–30(–60), not restricted to apex, leaflets (1–)3–16(–20) per side.

[3. Shifted to left margin.—Ed.]

3. Plants not resinously aromatic, inconspicuously to moderately glandular; basal leaves ± cylindric to weakly planar; leaflets ± overlapping at least distally, divided ½–¾+ to midrib into (0–)2–8(–15) lobes.
   4. Plants forming dense mats; stems 0.3–2.5 dm; basal leaves: stipules entire; 2000–3400 m; White Mountains, s Sierra Nevada, and Siskiyou Mountains, California, w Nevada, sw Oregon .... 10a. *Horkelia* sect. *Hispidulae*, p. 248
   4. Plants forming rosettes or tufts, sometimes loose mats; stems 0.5–5 dm; basal leaves: stipules entire, forked, or pinnately divided into linear or filiform lobes; 60–2500 m; nw California, sw Oregon .... 10e. *Horkelia* sect. *Tridentatae* (in part), p. 264
3. Plants ± resinously aromatic, conspicuously to obscurely (and minutely) glandular; leaves usually ± planar, sometimes nearly cylindric; leaflets separate to ± overlapping especially distally, divided ± ⅙–¾+ to midrib into (3–)5–30(–60) teeth or lobes.
   5. Epicalyx bractlets narrowly elliptic-lanceolate to broadly ovate, 0.5–3 mm wide; pedicels 1–30(–40) mm; flowers arranged individually, in usually non-capitate glomerules, and/or in corymbiform clusters; Sierra Nevada foothills and w California .... 10b. *Horkelia* sect. *Horkelia*, p. 250
   5. Epicalyx bractlets linear, 0.2–0.3(–0.5) mm wide; pedicels 1–3(–10) mm; flowers usually arranged in ± capitate glomerules, rarely individually (in *H. fusca* var. *filicoides*); Sierra Nevada, California, to Washington and Wyoming .... 10d. *Horkelia* sect. *Capitatae*, p. 259

10a. Horkelia Chamisso & Schlechtendal sect. Hispidulae Ertter & Reveal, Novon 17: 316. 2007 [C] [E]

Plants forming dense mats, usually grayish green, sometimes reddish, inconspicuously glandular, not resinously aromatic. **Stems** decumbent to erect, 0.3–2.5 dm. **Basal leaves** ± cylindric to weakly planar; stipules entire; leaflets 4–14 per side, overlapping at least distally, divided ½–¾+ to midrib into 3–8 lobes not restricted to apex. **Inflorescences** ± congested, flowers arranged in dense corymbiform clusters. **Pedicels** remaining straight, 2–8(–12) mm. **Flowers:** epicalyx bractlets linear to lanceolate, 0.3–0.5(–0.8) mm wide, entire; hypanthium interior sparsely pilose to densely villous; sepals acute; petals white, sometimes pink-tinged, narrowly oblanceolate to narrowly elliptic or oblong, apex usually acute to rounded to truncate, sometimes slightly mucronate or emarginate; filaments white to pinkish, glabrous or sparsely pilose adaxially, anthers longer than wide; carpels 5–18(–20). **Achenes** 1.5–2.5 mm, smooth.

Species 3 (3 in the flora): w United States.

Section *Hispidulae* accommodates three widely separated, localized species, with plants all forming tight, densely hairy mats in rocky sites at elevations of 2000 to 3400 m in the White Mountains, southern Sierra Nevada, and Siskiyou Mountains of California, adjacent Nevada, and southwestern Oregon.

When describing *Horkelia tularensis* (as *Potentilla tularensis*), Howell compared it to *H. hispidula* with the speculation that these interior species bordering the Great Basin were relicts of an evolutionary line distinct from the more coastal species. Given the geographic proximity to *Horkeliella*, which serves as a morphologic bridge between *Horkelia* and *Ivesia*, it is possible that sect. *Hispidulae* represents the relictual ancestral radiation within the genus.

1. Sepals narrowly lanceolate, with silky hairs 1–1.5 mm; hypanthium interior densely villous; petals often pink-tinged; Siskiyou Mountains, California and Oregon .... 3. *Horkelia hendersonii*
1. Sepals broadly lanceolate, with stiff hairs 0.5–1 mm; hypanthium interior sparsely pilose; petals not pink-tinged; Sierra Nevada and White Mountains, California and Nevada.
   2. Basal leaves: leaflets 4–6(–10) per side; cauline leaves 1 or 2; carpels 5–12; s Sierra Nevada .... 1. *Horkelia tularensis*
   2. Basal leaves: leaflets (6–)10–14 per side; cauline leaves 3–7; carpels (10–)12–18(–20); White Mountains .... 2. *Horkelia hispidula*

1. Horkelia tularensis (J. T. Howell) Munz, Suppl. Calif. Fl., 110. 1968 • Kern Plateau horkelia [C] [E]

Potentilla tularensis J. T. Howell, Leafl. W. Bot. 10: 254. 1966

Plants 1–4 dm diam. **Stems** decumbent to erect, 0.3–1.8 (–2.5) dm. **Basal leaves** 2–8 (–10) × 0.5–1.3 cm; leaflets 4–6(–10) per side, overlapping, cuneate to flabellate, 3–8 mm, divided ¾+ to midrib into (3–)5–8 oblong to obovate or oblanceolate lobes, villous to pilose at least marginally or apically. **Cauline leaves** 1 or 2. **Pedicels** 2–7 mm. **Flowers** 3–15, 8–10 mm diam.; epicalyx bractlets linear-lanceolate, 1–2.5(–3) × 0.3–0.5(–0.8) mm, ± ⅔ length of sepals; hypanthium 1–1.5 × 2.5–4.5 mm, nearly ½ as deep as wide, interior sparsely pilose; sepals spreading to reflexed, broadly lanceolate, (2–)3–4.5(–5) mm, hairs stiff, 0.5–1 mm; petals not pink-tinged, narrowly oblanceolate to narrowly oblong, 2–4 mm, apex rounded to truncate, often slightly mucronate; filaments white, 1–2 × 0.3–0.5 mm, usually glabrous, rarely sparsely pilose, anthers 0.6–0.8 mm; carpels 5–12; styles 1.5–2 mm. **Achenes** light brown to brown, 2–2.5 mm.

Flowering summer. Dry, rocky metamorphic ridges, in subalpine conifer woodlands; of conservation concern; 2300–2900 m; Calif.

Horkelia tularensis is known only from the Kern Plateau in Tulare County. Greenish and reddish plants can grow intermixed in a single population.

2. Horkelia hispidula Rydberg in N. L. Britton et al., N. Amer. Fl. 22: 278. 1908 • White Mountains horkelia [C] [E]

Potentilla hispidula (Rydberg) Jepson

Plants 0.7–3 dm diam. **Stems** ascending to erect, 1–2.5 dm. **Basal leaves** 3–10 × 0.4–0.8 cm; leaflets (6–)10–14 per side, ± overlapping at least distally, cuneate to flabellate, 2.5–4 (–6) mm, divided ¾+ to midrib into 3–6 oblanceolate to obovate lobes, hispid. **Cauline leaves** 3–7. **Pedicels** 2–8(–12) mm. **Flowers** 3–15, 10 mm diam.; epicalyx bractlets linear to lanceolate, 1.5–3 × 0.2–0.5 mm, ± ⅔ length of sepals; hypanthium 1.8–3 × 3–4 mm, ± ½ as deep as wide, interior sparsely pilose; sepals reflexed, broadly lanceolate, 2.5–4(–5) mm, hairs stiff, 0.5 mm; petals not pink-tinged, oblanceolate to oblong or narrowly elliptic, 2.5–5 mm, apex rounded to truncate, sometimes slightly emarginate or mucronate; filaments white, 0.5–2 × 0.4–0.6 mm, glabrous or sparsely pilose adaxially, anthers 0.5–0.9 mm; carpels (10–)12–18(–20); styles 1.8–2.2 mm. **Achenes** brown to dark brown, 1.5–2 mm.

Flowering summer. Dry, rocky alpine flats, in subalpine conifer woodlands; of conservation concern; 3000–3400 m; Calif., Nev.

When describing *Horkelia hispidula*, Rydberg associated it with *H. sericata* in his group *Sericatae*. The species is known only from the White Mountains of California and adjacent Nevada.

3. Horkelia hendersonii Howell, Fl. N.W. Amer., 180. 1898 (as hendersoni) • Henderson's horkelia [C] [E]

Potentilla hendersonii (Howell) J. T. Howell

Plants 1.5–4 dm diam. **Stems** ascending to erect, 0.8–2 (–2.5) dm. **Basal leaves** 3–8 × 0.7–1.5 cm; leaflets (5 or) 6–12 per side, overlapping at least distally, cuneate to flabellate, 4–9(–10) mm, divided ½+ to midrib into 3–6 oblanceolate to obovate lobes (often medially notched more than ¾ to midrib as well), silky villous. **Cauline leaves** 2 or 3. **Pedicels** 2–8(–10) mm. **Flowers** (2–)4–15(–20), 10 mm diam.; epicalyx bractlets linear, 2.5–4.5 × 0.2–0.5 mm, ⅔ to nearly equal to length of sepals; hypanthium 1–2.5 × (2–)3–4(–5) mm, ± ½ as deep as wide, interior densely villous; sepals spreading, narrowly lanceolate, 3.5–6 mm, hairs silky, 1–1.5 mm; petals often pink-tinged, linear to narrowly oblanceolate, 2.5–5 × 1 mm, apex ± acute to narrowly rounded, sometimes slightly mucronate; filaments pinkish, 2 × 0.5–0.7 mm, glabrous or sparsely pilose adaxially, anthers 0.5–0.7 mm; carpels 8–17; styles (1.5–)2–3 mm. **Achenes** dark brown, 2 mm.

Flowering summer. Dry, metamorphosed granite and ultramafic talus slopes, in montane conifer woodlands; of conservation concern; 2000–2300 m; Calif., Oreg.

Horkelia hendersonii is one of the more attractive species of the genus, with its grayish mats and dusky pink inflorescences. P. A. Rydberg (1898) placed it initially in his group *Capitatae*, here restricted to *H. fusca*, and later (1908c) transferred it to his new group *Tenuilobae*, which otherwise comprised only *H. tenuiloba*. The species is known only from Jackson County, Oregon, and an isolated population in adjacent Siskiyou County, California. The populations are vulnerable due to their proximity to popular recreational sites.

10b. Horkelia Chamisso & Schlechtendal sect. Horkelia

Horkelia [unranked] *Californicae* Rydberg; *Horkelia* sect. *Californicae* (Rydberg) O. Stevens; *Horkelia* [unranked] *Cuneatae* Rydberg; *Horkelia* sect. *Cuneatae* (Rydberg) O. Stevens; *Horkelia* [unranked] *Tenuilobae* Rydberg; *Horkelia* sect. *Tenuilobae* (Rydberg) O. Stevens

Plants forming tufts or mats, green to grayish, obscurely (and minutely) glandular, resinously aromatic, often strongly so. **Stems** decumbent or ascending to erect, (0.5–)1–10(–12) dm. **Basal leaves** usually planar, sometimes ± cylindric; stipules usually entire, sometimes basally lobed; leaflets (1–)3–16(–20) per side, separate to overlapping, divided ± 1/6–3/4+ to midrib into 3–30 (–60) teeth or lobes not restricted to apex. **Inflorescences** open to congested, flowers arranged individually, in usually non-capitate glomerules, and/or in corymbiform clusters. **Pedicels** remaining straight, 1–30(–40) mm. **Flowers:** epicalyx bractlets narrowly elliptic-lanceolate to broadly ovate, 0.5–3 mm wide, usually entire, sometimes toothed; hypanthium interior pilose or glabrous; sepals acute; petals white, oblong-oblanceolate to round, apex obtuse to truncate to emarginate; filaments white, glabrous, anthers longer than wide; carpels 10–200(–220). **Achenes** 0.8–2 mm, usually smooth or slightly rugose, sometimes merely roughened.

Species 9 (9 in the flora): California, nw Mexico.

Section *Horkelia* encompasses the species that are most commonly encountered in heavily populated areas of California. Plants are notably glandular-viscid (unless obscured by dense vestiture) and have a distinctive resinous odor.

Previous revisions (for example, P. A. Rydberg 1908c; D. D. Keck 1938[1939]) have placed *Horkelia frondosa* (here treated as *H. californica* var. *frondosa*) at the beginning, implying that this is the least derived expression within the genus. Such an assumption is based on its gross resemblance to sympatric members of *Drymocallis*; molecular evidence (T. Eriksson et al. 1998; M. Lundberg et al. 2009; C. Dobeš and J. Paule 2010) confirms that this similarity is superficial. If, as speculated above, species composing sect. *Hispidulae* are relicts of the original radiation, then *H. californica* var. *frondosa* is actually one of the more derived members of the genus. Species within sect. *Horkelia* have been ordered here according to that interpretation.

SELECTED REFERENCE Ertter, B. 1993c. A re-evaluation of the *Horkelia bolanderi* (Rosaceae) complex, with the new species *Horkelia yadonii*. Syst. Bot. 18: 137–144.

1. Leaflets 1–3 per side, sparsely short-pilose to glabrate, (15–)20–30-toothed 1/6–1/4 to midrib; petals obovate to round, (3–)4–7 mm wide . . . . 11. *Horkelia truncata*
1. Leaflets (3–)5–16(–20) per side, hairy, 3–15(–30)-toothed or -lobed (collectively 10–60 -toothed in *H. californica*) 1/6–3/4+ to midrib; petals narrowly obovate or oblanceolate to elliptic or oblong, 1–3(–4) mm wide.
    2. Hypanthia 3–5.5 mm, 1/2 to nearly as deep as wide; epicalyx bractlets 4–6(–8) mm, ± equal to sepals, entire or toothed . . . . 12. *Horkelia californica*
    2. Hypanthia 1–2(–3) mm, less than 1/2 as deep as wide (except *H. clevelandii*); epicalyx bractlets 1–4(–5) mm, shorter than sepals, entire.
        3. Leaflets ± elliptic to obovate, divided less than 1/3 to midrib into (5–)10–15 teeth; epicalyx bractlets 1.5–3 mm wide . . . . 10. *Horkelia cuneata*
        3. Leaflets cuneate or flabellate to ovate or nearly round, divided (1/5–)1/3–3/4+ to midrib into 3–10(–30 in *H. yadonii*) teeth or lobes; epicalyx bractlets 0.5–2(–2.5) mm wide.
            4. Stems ± decumbent to ascending, 1–3.5(–4.5) dm; leaflets 5–10(–12) per side, cuneate, ± overlapping; inflorescences usually congested; plants matted; coastal . . . . 9. *Horkelia marinensis*
            4. Stems ascending to erect, (0.5–)1–7 dm; leaflets 6–16(–20) per side, cuneate to flabellate or nearly round, separate to ± overlapping; inflorescences open to ± congested; plants tufted to matted; coastal or interior.

[5. Shifted to left margin.—Ed.]

5. Epicalyx bractlets 1–2(–2.5) mm wide, lanceolate to ovate; anthers 0.8–1.2 mm; hypanthia 3–6.5 mm diam. . . . . 8. *Horkelia yadonii*
5. Epicalyx bractlets 0.5–1(–1.5) mm wide, narrowly elliptic to broadly lanceolate; anthers 0.4–1 mm; hypanthia 2–4.5(–6) mm diam.
    6. Stems: hairs ascending to appressed; plants usually grayish to grayish green; hypanthium interior pilose.
        7. Carpels 20–50(–120); styles 2–4 mm; basal leaves (4–)8–20(–30) cm; Transverse Ranges, s California . . . . 4. *Horkelia rydbergii*
        7. Carpels 10–20(–27); styles (1–)1.5–2 mm; basal leaves 3–8(–9) cm; Coast Ranges, n California . . . . 5. *Horkelia bolanderi*
    6. Stems: hairs ± spreading; plants green to grayish; hypanthium interior glabrous or pilose.
        8. Leaflets divided ½–¾+ to midrib into linear to narrowly oblanceolate or narrowly elliptic lobes; hypanthium interior pilose; n Coast Ranges, California . . . . 6. *Horkelia tenuiloba*
        8. Leaflets divided ⅓ to midrib into acute to obtuse teeth; hypanthium interior glabrous; Peninsular Ranges, s California . . . . 7. *Horkelia clevelandii*

4. Horkelia rydbergii Elmer, Bot. Gaz. 39: 50. 1905 • Rydberg's horkelia [E]

Horkelia bernardina Rydberg; *H. bolanderi* A. Gray subsp. *parryi* (S. Watson) D. D. Keck; *H. bolanderi* var. *parryi* S. Watson; *Potentilla bolanderi* (A. Gray) Greene var. *parryi* (S. Watson) Munz & I. M. Johnston

Plants tufted to ± matted, usually grayish to grayish green. **Stems** ascending to erect, (1–)2–7 dm, hairs ascending to appressed. **Basal leaves** ± planar, (4–)8–20(–30) × 0.8–2(–3) cm; stipules entire; leaflets 7–14 per side, separate to ± overlapping at least distally, cuneate to flabellate, (3–)5–15(–22) × 3–10(–15) mm, ± ½ to as wide as long, divided ± ⅓ to midrib into (3–)5–10 acute to obtuse teeth, densely (to sparsely) strigose or pilose. **Cauline leaves** (2 or)3–6(–8). **Inflorescences** ± open to congested, flowers arranged individually or in glomerules. **Pedicels** (1–)2–8 mm. **Flowers** 10–15 mm diam.; epicalyx bractlets lanceolate to narrowly elliptic, 2–3.5(–4) × 0.5–1 mm, ± ¾ length of sepals, entire; hypanthium 1–1.5(–2) × 2.5–4.5(–6) mm, less than ½ as deep as wide, interior pilose; sepals spreading to ± reflexed, lanceolate, (2.5–)3–5(–5.2) mm; petals oblong to oblanceolate, 4–5.5 × 1.2–2.5 mm, apex rounded to truncate or slightly emarginate; filaments 0.5–2 × 0.5–1 mm, anthers 0.6–1 mm; carpels 20–50 (–120); styles 2–4 mm. **Achenes** light to dark brown, 1–1.5 mm, smooth or roughened.

Flowering summer. Dry to moist meadows and stream banks, in conifer woodlands; 1200–2800 m; Calif.

Horkelia rydbergii occurs in the Transverse Ranges of Kern, Los Angeles, San Bernardino, and Ventura counties. Reports from farther north (as *H. bolanderi* var. *parryi*) are referable to *H. marinensis*, *H. yadonii*, or, possibly, *H. cuneata*. Although there has been occasional nomenclatural confusion between this taxon (as *H. bolanderi* var. *parryi*) and *H. parryi* Greene, the two are taxonomically and geographically distinct.

Plants in the Mount Pinos–Lockwood Valley area of Kern and Ventura counties, and on Frazier Mountain, are distinctly gray with an abundance of tightly curled hairs. The leaves tend to be relatively short (to 8 cm) with relatively small leaflets (3–6 mm) on the basal leaves. Stems are relatively short (to 3 dm), and the fruits are dark brown, distinctly rugose, and 1.3–1.5 mm. This phase (including the type of *Horkelia rydbergii*) abruptly gives way to a grayish green phase with a slightly less dense, pilose indument in the San Bernardino and San Gabriel mountains. Here the plants tend to have larger basal leaves (to 30 cm) and leaflets (5–12 mm), longer stems (to 7 dm), and fruits that are a lighter brown, mostly smooth, and 1–1.2 mm. This phase in turn grades into even less densely hairy plants near Bear Lake, where the type of *H. bolanderi* var. *parryi* was obtained.

5. Horkelia bolanderi A. Gray, Proc. Amer. Acad. Arts 7: 338. 1868 • Bolander's horkelia [C] [E]

Potentilla bolanderi (A. Gray) Greene

Plants openly matted, grayish. **Stems** ascending to erect, 1–3(–4.5) dm, hairs ascending to appressed. **Basal leaves** ± planar, 3–8(–9) × 0.7–1.5 cm; stipules entire; leaflets 6–9 per side, separate to ± overlapping at least distally, ± cuneate to obovate, 4–10 × 2–7 mm, ½ to nearly as wide as long, divided ⅓ to midrib into 3–6 acute, sometimes obtuse, teeth, densely strigulose, often villous marginally or at least apically. **Cauline leaves**

(2 or)3–6. **Inflorescences** ± congested to open, most flowers arranged in glomerules. **Pedicels** (1–)2–4(–6) mm. **Flowers** 10 mm diam.; epicalyx bractlets elliptic-lanceolate, (1.5–)1.8–3 × 0.5–1 mm, ± ⅔ length of sepals, entire; hypanthium 1–1.5 × 2–4 mm, less than ½ as deep as wide, interior slightly pilose; sepals spreading to ± reflexed, lanceolate, (2.5–)3–4 mm; petals oblong to oblanceolate, 3–5.5 × 1.5–2 mm, apex rounded to slightly emarginate; filaments 0.5–2 × 0.3–0.4 mm, anthers (0.4–)0.5–0.8 mm; carpels 10–20(–27); styles (1–)1.5–2 mm. **Achenes** brown to dark brown, 1–1.3 mm, usually faintly roughened.

Flowering summer. Edges of vernal lakes, meadows, seeps, in grasslands, conifer woodlands; of conservation concern; 400–1100 m; Calif.

W. L. Jepson (1923–1925) and P. A. Munz and I. M. Johnston (1925) independently adopted an expanded circumscription of *Potentilla bolanderi* that included four species of *Horkelia* recognized by P. A. Rydberg (1908c): *H. bernardina*, *H. bolanderi*, *H. clevelandii*, and *H. rydbergii*. These were treated as three varieties, which D. D. Keck (1938[1939]) converted to subspecies as *H. bolanderi* subsp. *bolanderi*, subsp. *clevelandii*, and subsp. *parryi*, the last encompassing *H. bernardina* and *H. rydbergii*. These subspecies were restored to species rank by B. Ertter (1993c) as *H. bolanderi*, *H. clevelandii*, and *H. rydbergii*, along with the establishment of *H. yadonii* to accommodate populations from Monterey County assigned to subsp. *parryi* by Keck.

As now circumscribed, *Horkelia bolanderi* is known from five locations in the northern Coast Ranges in Colusa, Lake, and Mendocino counties; records from more southerly localities would be referenced to one of the segregate species discussed above.

6. Horkelia tenuiloba (Torrey) A. Gray, Proc. Amer. Acad. Arts 6: 529. 1865 • Thin-lobed, Santa Rosa, or Sonoma horkelia [C] [E]

Horkelia fusca Lindley var. *tenuiloba* Torrey in War Department [U.S.], Pacif. Railr. Rep. 4(5): 84. 1857; *Potentilla micheneri* Greene; *P. stenoloba* Greene

Plants loosely matted, green. **Stems** ascending to erect, 1–4 dm, hairs ± spreading. **Basal leaves** weakly planar to ± cylindric, 5–15(–20) × 0.5–1.5 cm; stipules entire; leaflets 8–16(–20) per side, ± overlapping especially distally, cuneate to flabellate, 3–10 × 2–10 mm, ½ to nearly as wide as long, divided ½–¾+ to midrib into 3–8 linear to narrowly oblanceolate or narrowly elliptic lobes, sparsely villous at least marginally, often with a tuft of hairs apically. **Cauline leaves** 2–5. **Inflorescences** open to congested, flowers arranged individually and in glomerules, these sometimes subcapitate. **Pedicels** 1–6 mm. **Flowers** 10 mm diam.; epicalyx bractlets narrowly to broadly lanceolate, 1–3 × 0.5–1 mm, slightly shorter than sepals, entire; hypanthium 1–1.2 × 2.5–4.5 mm, less than ½ as deep as wide, interior pilose; sepals spreading to reflexed, lanceolate, 3–5 mm; petals oblanceolate, 2.5–4.5 × 1.5 mm, apex emarginate; filaments (1–)1.5–2 × 0.5 mm, anthers 0.4–0.6 mm; carpels 10–25; styles 1.8–2.2 mm. **Achenes** light brown, 1.5 mm, smooth or slightly rugose. $2n = 28$.

Flowering summer. Sandy soil, openings, in chaparral, oak woodlands; of conservation concern; 50–500 m; Calif.

Horkelia tenuiloba occurs on the western edges of the northern Coast Ranges in Marin, Mendocino, and Sonoma counties. Populations from San Luis Obispo formerly included in this species now are part of *H. yadonii*. A specimen (*M. K. C[urran]*, July 5, 1885, UC) unequivocally of *H. tenuiloba* purportedly from San Luis Obispo is in all likelihood mislabeled with respect to locality.

Horkelia tenuiloba is commonly associated with seral openings in chaparral and woodlands and might be dependent on periodic disturbance by fire.

W. L. Jepson (1909–1943, vol. 2) used *Potentilla stenoloba* (1895) for the species encompassing the types of *Horkelia tenuiloba* and *P. micheneri*. The epithet *micheneri* (1893) has priority at species rank within *Potentilla*, since *P. tenuiloba* (Torrey) Greene is a later homonym of *P. tenuiloba* Jordan.

7. Horkelia clevelandii (Greene) Rydberg, Bull. Torrey Bot. Club 25: 54. 1898

Potentilla clevelandii Greene, Pittonia 1: 102. 1887 (as clevelandi); *Horkelia bolanderi* A. Gray subsp. *clevelandii* (Greene) D. D. Keck; *P. bolanderi* (A. Gray) Greene var. *clevelandii* (Greene) Jepson

Varieties 2 (1 in the flora): California, nw Mexico.

The component of *Horkelia clevelandii* occurring in the Sierra San Pedro Mártir and most of the Sierra Juarez of northern Baja California, Mexico, is var. *brevibracteata* (Wiggins) Ertter & Reveal. It differs from var. *clevelandii* in tending to have a more compact and matted habit, shorter stems (1–3 dm), more numerous (11–16) pairs of smaller (4–8 mm) leaflets, shorter epicalyx bractlets (1.5–2 mm), and deeper hypanthia (2–2.5 mm).

7a. Horkelia clevelandii (Greene) Rydberg var. **clevelandii** • Cleveland's horkelia

Plants tufted, green to grayish. **Stems** ascending to erect, 1–5 dm, hairs spreading. **Basal leaves** planar, 5–15(–18) × 1–2.5 cm; stipules entire; leaflets 6–12 per side, ± separate, cuneate to nearly round, 5–12(–15) × 4–13 mm, nearly as wide as long, divided ⅓ to midrib into 5–10 acute to obtuse teeth, canescent and often short pilose, sparsely pilose on midveins and margins. **Cauline leaves** (2 or)3–6. **Inflorescences** mostly open, flowers arranged individually and in glomerules. **Pedicels** (1–)1.5–6 mm. **Flowers** 9–13 mm diam.; epicalyx bractlets ± lanceolate, 2–3.5 × 0.5–1.5 mm, ± ¾ length of sepals, entire; hypanthium 1–1.5 × 2–4 mm, ± ½ as deep as wide, interior glabrous; sepals ± reflexed, lanceolate, 2.5–4 mm; petals broadly oblanceolate, 3–6 × 2 mm, apex usually rounded; filaments 0.5–1.5 × 0.4–0.6 mm, anthers 0.5–1 mm; carpels (15–)20–40 (–60); styles 1.3–1.6 mm. **Achenes** brown, 1–1.3 mm, usually smooth.

Flowering summer. Dry meadows, in conifer woodlands; 1200–2500 m; Calif.; Mexico (Baja California).

Variety *clevelandii* occurs primarily on the Peninsular Ranges of Riverside and San Diego counties. Although most populations from Baja California belong to var. *brevibracteata*, historical collections from northern Baja California by C. R. Orcutt (GH, ND-G, UC), including a syntype of *Horkelia clevelandii*, have the diagnostic features of var. *clevelandii*. Populations in San Diego County serve as the sole larval host plant for the federally endangered Laguna Mountains Skipper, *Pyrgus ruralis lagunae* Scott.

8. Horkelia yadonii Ertter, Syst. Bot. 18: 139, fig. 2. 1993 • Yadon's or Santa Lucia horkelia [C] [E]

Plants tufted to matted, ± grayish green. **Stems** ascending to erect, (0.5–)2–6(–7) dm, hairs spreading. **Basal leaves** ± planar, (3–)6–20(–32) × 0.5–1.5(–2) cm; stipules entire; leaflets 7–16 per side, separate to ± overlapping especially distally, broadly cuneate to nearly round, (3–) 4–15(–17) × (2–)5–12 mm, ¾ to nearly as wide as long, divided ⅕–½ to midrib into (3–)8–15(–30) acute to obtuse teeth, <often medially notched as well ⅓–⅔ to midrib>, villous. **Cauline leaves** 3–5. **Inflorescences** open, flowers arranged individually and in glomerules. **Pedicels** 1–7(–20) mm. **Flowers** 8–12 mm diam.; epicalyx bractlets lanceolate to ovate, 1.5–4.5 × 1–2 (–2.5) mm, ½–¾ length of sepals, entire; hypanthium 1.5–3 × 3–6.5 mm, less than ½ as deep as wide, interior sparsely pilose; sepals spreading to ± reflexed, lanceolate, (3–)4–6.5 mm; petals broadly oblanceolate-elliptic, 3–5.5 × 2–3 mm, apex truncate; filaments 1–2 × 0.5–1 mm, anthers 0.8–1.2 mm; carpels 20–60; styles 2.5–3 mm. **Achenes** brown, 1.5 mm, smooth.

Flowering summer. Granitic sand, meadows, along stream banks, in chaparral, conifer woodlands; of conservation concern; 300–1900 m; Calif.

Horkelia yadonii is known only from localized populations in the La Panza, San Rafael, and Santa Lucia ranges of Monterey, San Luis Obispo, and Santa Barbara counties. The isolated populations often differ in such characters as general size, leaf proportions, and shape of epicalyx bractlets. The most distinctive extreme, in Monterey County, is characterized by relatively long, narrow leaves, small, few-toothed leaflets, long pedicels, few small flowers, and ovate bractlets. These differences, however, are not consistently or sharply defined.

9. Horkelia marinensis (Elmer) Crum, Lloydia 1: 91. 1939 • Point Reyes horkelia [C] [E]

Horkelia bolanderi A. Gray var. *marinensis* Elmer, Bot. Gaz. 41: 321. 1906; *Potentilla kelloggii* (Greene) Greene var. *marinensis* (Elmer) Jepson; *P. marinensis* (Elmer) J. T. Howell

Plants matted, green to grayish. **Stems** ± decumbent to ascending, 1–3.5(–4.5) dm, hairs spreading to ascending. **Basal leaves** ± planar to nearly cylindric, 4–12(–15) × 1–2.5(–3) cm; stipules entire or basally lobed; leaflets 5–10(–12) per side, ± overlapping especially distally, cuneate, 7–12 × 5–10 mm, nearly as wide as long, divided ⅓–½ to midrib into (3–)5–10 acute to obtuse teeth, villous. **Cauline leaves** 3–5. **Inflorescences** usually congested, flowers usually arranged in dense corymbiform clusters. **Pedicels** 1–6 mm. **Flowers** 10–15 mm diam.; epicalyx bractlets lanceolate, 2.5–4.5 × 1 mm, ± ⅔ length of sepals, entire; hypanthium 1.2–1.5 × 4–5 mm, less than ½ as deep as wide, interior pilose; sepals ± reflexed, lanceolate, 3–6 mm; petals narrowly oblanceolate to narrowly oblong, 4–7 × 2 mm, apex obtuse to rounded, rarely truncate; filaments 1–2.8 × 0.2–0.6 mm, anthers 0.6–0.9 mm; carpels 20–30; styles 2–4 mm. **Achenes** brown, 1.5–2 mm, smooth to slightly rugose. **2*n*** = 56.

Flowering summer. Sandy coastal flats, stabilized dunes; of conservation concern; 0–800 m; Calif.

Horkelia marinensis occurs in three population clusters along the coast of California: near Fort Bragg and Rockport in Mendocino County, on Point Reyes Peninsula in Marin County, and from Santa Cruz and (historically) San Mateo counties. The last population cluster has often been confused with *H. cuneata* var. *sericea*; plants of it have more deeply and apically toothed, cuneate leaflets and longer, silkier hairs. It is also the basis for the attribution by J. H. Thomas (1961) of *H. rydbergii* (as *H. bolanderi* subsp. *parryi*) to the Santa Cruz Mountains. The three population clusters differ somewhat in vestiture, evident glandularity, and leaflet shape and size; further studies may indicate that varietal status is justified. References to *H. marinensis* in Sonoma County (M. G. Barbour 1970) are based on specimens of *Acaena pinnatifida* var. *californica*; reports from southern Mendocino County (G. L. Smith and C. R. Wheeler 1990–1991) are based on misidentifications of *H. californica* and *H. tenuiloba*.

Horkelia marinensis is by far the most aromatic member of the genus; the odor, which has been compared to that of a beehive, often can be perceived even before the plants are located, especially on sunny days. Plants do relatively well in cultivation, forming spreading mats with flowers that are strongly attractive to both native bees and honeybees.

10. Horkelia cuneata Lindley, Edwards's Bot. Reg. 23: sub plate 1997. 1837 [E]

Potentilla kelloggii (Greene) Greene var. *cuneata* (Lindley) Hoover; *P. lindleyi* Greene

Plants matted or tufted, green to grayish. **Stems** decumbent to erect, (1–)2–6(–7.5) dm, hairs ascending to appressed, sometimes spreading. **Basal leaves** planar, (5–)10–30 × 1.5–4 (–5) cm; stipules entire or basally lobed; leaflets 5–12 per side, separate to slightly overlapping, ± elliptic to obovate or oblanceolate, 5–25(–30) × 5–15(–25) mm, ± ½ to nearly as wide as long, divided less than ⅓ to midrib into (5–)10–15 oblanceolate to obovate teeth, sparsely pilose to glabrate or sericeous. **Cauline leaves** 1–3(–5). **Inflorescences** ± open to congested, flowers arranged individually and/or in glomerules. **Pedicels** 1–30(–40) mm. **Flowers** 12–17 mm diam.; epicalyx bractlets lanceolate to ovate, 3–5 × 1.5–3 mm, ⅔–¾ length of sepals, entire; hypanthium 1.5–2 × 4–7 mm, less than ½ as deep as wide, interior glabrous or pilose in a narrow ring; sepals ± reflexed to spreading, lanceolate, 4–6.5 mm; petals oblanceolate to narrowly obovate, 4–8 × 1.5–4 mm, apex obtuse to rounded; filaments 1–3 × 0.5–2 mm, anthers 0.7–1.1 mm; carpels (30–) 40–60(–80); styles 2–3 mm. **Achenes** brown, 1.5–1.8 mm, smooth or minutely rugose.

Varieties 3 (3 in the flora): California.

As noted by D. D. Keck (1938) and B. Ertter (1997b), recognition of subunits within *Horkelia cuneata* is complicated by the abundance of intermediates, to the extent that a complete transition exists between the decumbent, densely hairy coastal extreme with a congested inflorescence, represented by var. *sericea*, and the erect, green-glandular interior extreme with an open inflorescence, represented by var. *puberula*. The traditional division into three infraspecific units is maintained here to emphasize the distinctiveness of the extremes, which would qualify as specifically distinct were they not ends of a continuum. The extremes are furthermore threatened by ongoing decimation of suitable habitats at the northern (San Francisco Bay Area) and southern (greater Los Angeles) ends of the species range.

Determining the correct name of *Horkelia cuneata* if treated as a species of *Potentilla* is complicated. *Potentilla cuneata* (Lindley) Baillon ex Munz & I. M. Johnston (1925) and *P. puberula* Greene are both later homonyms, while *P. multijuga* Lehmann has been conserved with a conserved type to maintain established usage for an unrelated species of *Potentilla* (B. Ertter and J. L. Reveal 2008). J. T. Howell (1945) adopted *P. lindleyi*; however, as noted by R. F. Hoover (1966), *P. kelloggii* has priority at species rank by a matter of months. The latter is accordingly the correct name in *Potentilla* for *H. cuneata*.

1. Plants: eglandular hairs sparse or absent, glandular hairs evident; hypanthium interior rim usually glabrous, sometimes sparsely pilose; pedicels (2–)5–10 mm, proximalmost to 40 mm; stems erect .... 10c. *Horkelia cuneata* var. *puberula*
1. Plants: eglandular hairs moderately abundant to dense, glandular hairs evident or obscured; hypanthium interior rim ± or densely pilose; pedicels 1–5 mm, proximalmost to 15 mm; stems decumbent to erect.
    2. Plants ± green, eglandular hairs moderately abundant, spreading to ascending, glandular hairs usually evident; stems ascending to erect........... 10a. *Horkelia cuneata* var. *cuneata*
    2. Plants grayish-sericeous, eglandular hairs dense, ascending to appressed, glandular hairs obscured; stems decumbent to ascending.... ....10b. *Horkelia cuneata* var. *sericea*

10a. Horkelia cuneata Lindley var. **cuneata** • Wedgeleaf horkelia [E]

Potentilla lindleyi Greene var. *lepida* Crum

Plants ± green; eglandular hairs moderately abundant, spreading to ascending, glandular hairs usually evident. **Stems** ascending to erect, (1–)2–5 dm. **Leaflets** 5–12 per side, elliptic to obovate or oblanceolate, 5–15 mm, distinctly pinnately veined. **Inflorescences** ± open; most flowers arranged in glomerules. **Pedicels** 1–5 mm, proximalmost to 15 mm. **Flowers:** hypanthium interior rim ± pilose; petals oblanceolate, 2.5–3.5 mm wide; filaments 1.5–3 × 0.5–1 mm. **2*n*** = 28.

Flowering spring–summer. Coastal sand dunes and hills, coastal scrub communities; 0–400 m; Calif.

Most extant populations of *Horkelia cuneata* are referable to var. *cuneata* or intermediates between it and the other two varieties, with the greatest concentrations around Monterey Bay and from Morro Bay to Santa Barbara. At its most distinctive, var. *cuneata* occurs from Santa Cruz to Los Angeles counties, often sharing habitat with var. *sericea* though also occurring somewhat more inland, but not as far inland as var. *puberula*. Although not as endangered as the extremes, unequivocal var. *cuneata* is far from abundant and shares some of the same threats to its prime coastal habitat.

10b. Horkelia cuneata Lindley var. **sericea** (A. Gray) Ertter & Reveal, Novon 17: 319. 2007 • Kellogg's or coast horkelia [C] [E]

Horkelia californica Chamisso & Schlechtendal var. *sericea* A. Gray, Proc. Amer. Acad. Arts 6: 529. 1865; *H. cuneata* subsp. *sericea* (A. Gray) D. D. Keck; *H. kelloggii* Greene; *H. sericea* (A. Gray) Rydberg; *Potentilla kelloggii* (Greene) Greene; *P. lindleyi* Greene var. *sericea* (A. Gray) J. T. Howell

Plants grayish-sericeous; eglandular hairs dense, ascending to appressed, glandular hairs obscured. **Stems** decumbent to ascending, 2–5(–7.5) dm. **Leaflets** 5–10 per side, obovate, (5–)10–25(–30) mm, not distinctly pinnately veined. **Inflorescences** usually ± congested; most flowers arranged in glomerules. **Pedicels** 1–3 mm, proximalmost to 12 mm. **Flowers:** hypanthium interior rim densely pilose; petals oblanceolate, 1.5–3 mm wide; filaments 1–3 × 0.5–1 mm. **2*n*** = 28.

Flowering summer. Coastal stabilized dunes and hills, coastal scrub communities; of conservation concern; 0–200 m; Calif.

Historic populations of var. *sericea* occurred along the immediate coast from Alameda and Marin counties south at least to Santa Barbara County, with some littoral collections of *Horkelia cuneata* from Los Angeles and San Diego counties having some *sericea* characteristics. Reports from farther north are all based on misidentifications, for example, of *H. californica* in Sonoma County. The most distinctive specimens are from the northern populations in Alameda, Marin, San Francisco, and San Mateo counties, all of which apparently no longer exist. Of the recently confirmed extant populations, those that come closest to justifying continued recognition of var. *sericea* are in Monterey, San Luis Obispo, and Santa Cruz counties.

10c. Horkelia cuneata Lindley var. **puberula** (Rydberg) Ertter & Reveal, Novon 17: 318. 2007 • Mesa horkelia [C] [E]

Horkelia puberula Rydberg, Bull. Torrey Bot. Club 25: 55. 1898, based on *Potentilla puberula* Greene, Pittonia 1: 102. 1887, not Krašan 1867; *H. cuneata* subsp. *puberula* (Rydberg) D. D. Keck; *P. kelloggii* (Greene) Greene var. *puberula* (Rydberg) Hoover; *P. lindleyi* Greene var. *puberula* (Rydberg) Jepson

Plants green; eglandular hairs sparse or absent, spreading, glandular hairs not obscured. **Stems** erect, 3–6 dm. **Leaflets** 5–12 per side, elliptic to obovate, 5–25 mm, distinctly pinnately veined. **Inflorescences** open, most flowers arranged individually. **Pedicels** (2–)5–10 mm, proximalmost to 40 mm. **Flowers:** hypanthium interior rim usually glabrous, sometimes sparsely pilose; petals oblanceolate to narrowly obovate, 2–4 mm wide; filaments 1–2.5 × 0.5–2 mm.

Flowering spring–summer. Coastal ranges away from the immediate coast, sandy to gravelly soil, coastal sage and chaparral communities; of conservation concern; 70–900 m; Calif.

At its most distinctive, var. *puberula* is represented by historic collections from the hills and plains of Los Angeles, western Riverside, extreme southwestern San Bernardino, and northwestern San Diego counties; these populations have now largely been destroyed by urbanization. Scattered populations still exist within this core area, to what extent and with what level of ensured continuity remains uncertain. In addition, multiple populations of *Horkelia cuneata* from as far north and west as San Luis Obispo County have characteristics of var. *puberula*; these tend to intergrade with var. *cuneata*, as do some historic collections from the core area.

11. Horkelia truncata Rydberg in N. L. Britton et al., N. Amer. Fl. 22: 274. 1908 • Ramona horkelia [C]

Potentilla truncata (Rydberg) Jepson

Plants tufted, green. **Stems** erect, 2–6(–8) dm, hairs spreading. **Basal leaves** planar, 4–13 × 2–3.5(–4) cm; stipules entire or basally lobed; leaflets 1–3 per side, separate, oblong to obovate, 10–30 × 5–12(–15) mm, ± 1/2 as wide as long, divided 1/6–1/4 to midrib into (15–)20–30 acute to obtuse teeth, sparsely short-pilose to glabrate, especially adaxially. **Cauline leaves** 4–7. **Inflorescences** open, flowers arranged individually. **Pedicels** 4–20(–30) mm. **Flowers** 12–17 mm diam.; epicalyx bractlets broadly ovate, 3–4.5 × 2–3 mm, ± 2/3 length of sepals, entire; hypanthium 1.5–2 × 5–8 mm, less than 1/2 as deep as wide, interior glabrous; sepals ± reflexed, broadly ovate, 3.5–5.5 mm; petals obovate to round, 5–7 × (3–)4–7 mm, apex ± rounded; filaments 1–2 × 0.5–2 mm, anthers 0.8–1 mm; carpels 50–80; styles 1.5–3 mm. **Achenes** brown, 1.3–1.5 mm, smooth.

Flowering summer. Dry clay soil, in chaparral communities, oak woodlands; of conservation concern; 400–1300 m; Calif.; Mexico (Baja California).

Horkelia truncata is confined to the Peninsular Ranges of San Diego County, California, and northern Baja California, Mexico. Plants of the species are unusual in *Horkelia* in that the terminal leaflet is usually notably larger than the lateral leaflets, unlobed, and distinct, rather than confluent with the distal lateral ones.

12. Horkelia californica Chamisso & Schlechtendal, Linnaea 2: 26. 1827 [E] [F]

Potentilla californica (Chamisso & Schlechtendal) Greene

Plants ± tufted, green. **Stems** ascending to erect, (1–)5–10(–12) dm, hairs ± spreading. **Basal leaves** planar, (5–)8–40(–50) × 1.5–11 cm; stipules entire; leaflets 3–9 per side, separate to slightly overlapping, ovate to round or broadly elliptic, 5–60 × (7–)10–30(–40) mm, 1/2 to as wide as long, unlobed or irregularly cleft 1/6–3/4+ to midrib into 5–15 linear or oblanceolate to obovate coarsely toothed lobes, collectively 10–60-toothed, pilose to villous. **Cauline leaves** 4–10. **Inflorescences** ± open, flowers arranged individually and in glomerules. **Pedicels** 1–20 mm. **Flowers** 8–15 mm diam.; epicalyx bractlets narrowly oblong to ovate, 4–6(–8) × 1–2.5 mm, ± equal to sepals, entire or toothed; hypanthium 3–5.5 × 4–10 mm, 1/2 to nearly as deep as wide, interior glabrous or pilose; sepals spreading, lanceolate, 4–6.5(–8) mm; petals oblanceolate to elliptic or oblong, 3–8 × 1–4 mm, apex obtuse to rounded; filaments 0.5–3 × 0.2–1.5 mm, anthers (0.8–)1.3–1.8 mm; carpels (50–)80–200(–220); styles 2–4 mm. **Achenes** brown, 0.8–1 mm, smooth or slightly rugose.

Varieties 3 (3 in the flora): California.

Plants of *Horkelia californica* have the largest hypanthia in the genus, and the hypanthia are significantly deeper than others that are equally wide. Other unique features include the often toothed bractlets and the slightly clawed petals that are often inwardly curved with undulate margins. These features also help to distinguish the species from *Drymocallis glandulosa*, with which it is often confused: both species have a similar gross morphology and overlapping ranges and habitats.

The three varieties recognized here were treated as distinct species by D. D. Keck (1938[1939]). The existence of intermediate populations makes varietal status the preferred option.

1. Lateral leaflets 3 or 4 per side, broadly elliptic to ovate, unlobed or cleft 1/6–1/4 to midrib . . . . 12c. *Horkelia californica* var. *frondosa*
1. Lateral leaflets 4–9 per side, ovate to round, cleft ± 1/2–3/4+ to midrib.
   2. Hypanthium interior ± pilose; sepals often red-mottled adaxially; lateral leaflets cleft ± 1/2(–3/4) to midrib; styles 3–4 mm . . . . 12a. *Horkelia californica* var. *californica*
   2. Hypanthium interior glabrous; sepals not red-mottled adaxially; lateral leaflets cleft 1/2–3/4+ to midrib; styles 2–3 mm . . . . 12b. *Horkelia californica* var. *elata*

12a. Horkelia californica Chamisso & Schlechtendal var. **californica** • California horkelia [E] [F]

Basal leaves 8–40 cm; sheathing bases ± strigose; lateral leaflets 4–7(–9) per side, ovate to round, 10–40 mm, cleft ± 1/2(–3/4) to midrib, ultimate teeth 10–50; terminal leaflet 10–40 mm. **Flowers:** epicalyx bractlets often toothed; hypanthium interior ± pilose; sepals often red-mottled adaxially; filaments 1.5–3 × 0.5–1 mm; carpels 80–200; styles 3–4 mm. $2n = 56$.

Flowering summer. Margins of coastal scrub; 0–400 m; Calif.

Variety *californica* is known from the immediate coast from Humboldt to San Luis Obispo counties.

12b. Horkelia californica Chamisso & Schlechtendal var. **elata** (Greene) Ertter & Reveal, Novon 17: 318. 2007 • Tall horkelia [E]

Potentilla californica (Chamisso & Schlechtendal) Greene var. *elata* Greene, Fl. Francisc. 1: 66. 1891, based on *P. elata* Greene, Pittonia 1: 100. 1887, not Salisbury 1796; *Horkelia californica* subsp. *dissita* (Crum) Ertter; *H. elata* (Greene) Rydberg; *P. elata* var. *dissita* Crum

Basal leaves (5–)10–25 cm; sheathing bases usually glabrous, sometimes sparsely strigose; lateral leaflets (5–)7–9 per side, ovate to round, 5–30 mm, cleft ½–¾+ to midrib, ultimate teeth 10–30; terminal leaflet 10–40 mm. **Flowers:** epicalyx bractlets usually entire; hypanthium interior glabrous; sepals not red-mottled adaxially; filaments 0.5–1.5 × 0.2–0.5 mm; carpels 50–100; styles 2–3 mm. $2n = 28$.

Flowering summer. Meadow edges, along streams, in chaparral; 50–1600 m; Calif.

Variety *elata* occurs away from the immediate coast in the northern coastal ranges (Contra Costa to Humboldt counties), western foothills of the Sierra Nevada (Butte to Tulare counties), and disjunctly in the San Bernardino Mountains. At its most distinctive, the variety is represented by relatively delicate plants with finely divided leaflets at higher elevations in the North Coast ranges. Populations in the Sierran foothills, including the type of *Potentilla elata* var. *dissita*, tend to be relatively coarse and densely villous. Intermediates with var. *californica* exist in interior mid elevation sites, including Mount Diablo in Contra Costa County and the type locality of *Horkelia glandulosa* Eastwood in Mendocino County.

12c. Horkelia californica Chamisso & Schlechtendal var. **frondosa** (Greene) Ertter & Reveal, Brittonia 17: 318. 2007 • Leafy horkelia [E]

Potentilla frondosa Greene, Pittonia 1: 300. 1889; *Horkelia californica* subsp. *frondosa* (Greene) Ertter; *H. frondosa* (Greene) Rydberg; *P. californica* (Chamisso & Schlechtendal) Greene var. *frondosa* (Greene) Jepson

Basal leaves 10–40(–50) cm; sheathing bases ± strigose; lateral leaflets 3 or 4 per side, broadly elliptic to ovate, 15–60 mm, unlobed or cleft ⅙–¼ to midrib, ultimate teeth 30–60; terminal leaflet 20–90 mm. **Flowers:** epicalyx bractlets entire or toothed; hypanthium interior glabrous; sepals not red-mottled adaxially; filaments 1–2 × 0.5–1.5 mm; carpels 80–200(–220); styles 2–3 mm.

Flowering summer. Coastal scrub, chaparral communities; 0–400 m; Calif.

Variety *frondosa* occurs mainly in the central coastal ranges from Alameda and Contra Costa counties south to Monterey and San Benito counties.

10c. Horkelia Chamisso & Schlechtendal sect. Parryae Ertter & Reveal, Novon 17: 316. 2007 [C] [E]

Plants forming rosettes or open mats, green, conspicuously glandular, ± resinously aromatic. **Stems** prostrate to erect, (0.5–)1–3(–3.5) dm. **Basal leaves** planar; stipules usually entire, rarely deeply 2-lobed; leaflets 3–7 per side, separate, divided ± ¼–¾ to midrib into 5–15 teeth or lobes not restricted to apex. **Inflorescences** open, flowers arranged individually. **Pedicels** becoming reflexed or recurved, 3–15 mm. **Flowers:** epicalyx bractlets linear-lanceolate to elliptic or ovate, 0.3–1.5 mm wide, entire; hypanthium interior glabrous; sepals acute; petals white, elliptic or oblong to obovate, apex obtuse to rounded or slightly emarginate; filaments white, glabrous, anthers at least slightly longer than wide; carpels 3 or 4, or (17–)20–50. **Achenes** 1.3–2.2 mm, finely reticulate or coarsely rugose.

Species 2 (2 in the flora): California.

Section *Parryae* accommodates two localized species with open inflorescences in which the slender pedicels become reflexed to recurved at or soon after flowering. P. A. Rydberg (1908c) included both species in his group *Cuneatae*.

1. Petals 4–7 mm; leaflets divided ± ¼ to midrib; carpels (17–)20–50; foothills of w Sierra Nevada . . . . 13. *Horkelia parryi*
1. Petals 2–3 mm; leaflets divided ½–¾ to midrib; carpels 3 or 4; San Bernardino Mountains . . . . 14. *Horkelia wilderae*

13. Horkelia parryi Greene, Bull. Calif. Acad. Sci. 2: 416. 1887 • Parry's or Chamise horkelia [C] [E]

Potentilla parryi (Greene) Greene

Plants forming open mats. **Stems** ascending to erect, 1–3(–3.5) dm. **Basal leaves** 4–10(–12) × 1–2 cm; leaflets 3–6 (or 7) per side, narrowly obovate, 5–12(–15) × 4–8(–10) mm, ½–¾ as wide as long, divided ± ¼ to midrib into 5–10 oblong to broadly obovate teeth, sparsely pilose especially on midveins and margins. **Cauline leaves** 1–4. **Pedicels** (3–)5–15 mm. **Flowers** 15 mm diam.; epicalyx bractlets linear-lanceolate to narrowly elliptic, 2–5 × 0.5–1.5 mm, ± ⅔ length of sepals; hypanthium 0.6–1 × 2.5–4 mm, less than ⅓ as deep as wide; sepals ± spreading to reflexed, lanceolate to narrowly elliptic, 3.5–6 mm; petals elliptic to obovate, 4–7 × 3 mm, apex obtuse to rounded to truncate or slightly emarginate; filaments 1–3 × 0.6–1.3 mm, anthers 0.6–1 mm; carpels (17–)20–50; styles 1.5–2.5 mm. **Achenes** grayish, 1.3–1.5 mm, finely reticulate. **2*n*** = 28.

Flowering late spring–early summer. Chaparral, pine-oak woodlands, primarily on Ione Formation, rarely on schist or limestone; of conservation concern; 80–900 m; Calif.

Horkelia parryi is known from the foothills of the western Sierra Nevada in Amador, Calaveras, and El Dorado counties, and is disjunct in Mariposa County.

14. Horkelia wilderae Parish, Bot. Gaz. 38: 460. 1904 • Barton Flats or Wilder's horkelia [C] [E]

Potentilla parryi (Greene) Greene var. *wilderae* (Parish) Jepson; *P. wilderae* (Parish) Munz & I. M. Johnston

Plants forming rosettes. **Stems** prostrate to decumbent, (0.5–)1–2.5 dm. **Basal leaves** (3–)4–8(–10) × 0.8–1.8 cm; leaflets (3 or)4–7 per side, narrowly to broadly obovate, 3–10 × 3–10 mm, ± as wide as long, divided ½–¾ to midrib into 5–15 narrowly oblong teeth or lobes, pilose especially on midveins and margins. **Cauline leaves** (0 or)1(or 2). **Pedicels** 3–15 mm. **Flowers** 5 mm diam.; epicalyx bractlets lanceolate to ovate, 0.8–1.7 × 0.3–0.7 mm, ½–⅔ length of sepals; hypanthium 1–1.5 × 1.5–3 mm, ± ½ as deep as wide; sepals reflexed, broadly elliptic to broadly ovate, 1.7–2.5 mm; petals oblanceolate to oblong, 2–3 × 1–2 mm, apex rounded; filaments 0.5–1 × 0.3–0.8 mm, anthers 0.4–0.5 mm; carpels 3 or 4; styles 1–2 mm. **Achenes** tan, 1.8–2.2 mm, coarsely rugose.

Flowering summer. Chaparral flats, hills, adjacent to montane conifer woodlands; of conservation concern; 1900–3000 m; Calif.

Horkelia wilderae is known from the Barton Flats area of the eastern San Bernardino Mountains, San Bernardino County.

Some morphological characteristics (small flowers, reflexed sepals) may indicate a closer relation to *Horkelia fusca* than to *H. parryi*.

10d. Horkelia Chamisso & Schlechtendal sect. Capitatae (Rydberg) O. Stevens in N. L. Britton et al., N. Amer. Fl. 22(7): 7. 1959 [E]

Horkelia [unranked] *Capitatae* Rydberg, Monogr. N. Amer. Potentilleae, 120, 122. 1898; *Potentilla* Linnaeus sect. *Capitatae* (Rydberg) Crum

Plants forming tufts or open mats, green or reddish, rarely grayish, conspicuously glandular, resinously aromatic. **Stems** ascending to erect, (0.6–)1–6(–9) dm. **Basal leaves** planar; stipules entire; leaflets 3–8(–15) per side, separate, sometimes ± overlapping, divided 1/5–3/4+ to midrib into (3–)5–15 teeth or lobes not restricted to apex. **Inflorescences** open to ± congested, flowers usually arranged in ± capitate glomerules (arranged individually in *H. fusca* var. *filicoides*). **Pedicels** remaining ± straight, outermost sometimes ± reflexed in congested inflorescences, 1–3 (–10) mm. **Flowers:** epicalyx bractlets linear, 0.2–0.3(–0.5) mm wide, entire; hypanthium interior glabrous; sepals acute to acuminate; petals white to pale pink, often veined with pink to rose, ± oblanceolate to cuneate, apex emarginate to truncate or rounded; filaments white, glabrous, anthers longer than wide; carpels 10–25. **Achenes** 1–1.8 mm, smooth.

Species 1: w United States.

Rydberg recognized seven species in his group *Capitatae*, treated here as intergrading variation within a single species. Plants of the section share the characteristic *Horkelia* odor, glandularity, and planar leaves of sect. *Horkelia*, but differ in the combination of relatively small, short-pedicelled flowers that (except for var. *filicoides*) are most commonly aggregated into one or more capitate, purple-suffused glomerules, with linear epicalyx bractlets, oblanceolate-cuneate petals that are often pink-tinged, and relatively short, broad filaments.

15. Horkelia fusca Lindley, Edwards's Bot. Reg. 23: plate 1997. 1837 • Dusky, pinewoods, or tawny horkelia [E] [F]

Potentilla douglasii Greene

Basal leaves (3–)4–20(–40) × (1–)1.5–4(–7) cm; leaflets narrowly cuneate to obovate to flabellate, 5–30(–35) × 2–20 (–30) mm, 1/3 as wide to wider than long, divided into linear or oblanceolate to obovate teeth or lobes, sparsely to ± densely short-villous or hirsute, sometimes glabrate. **Cauline leaves** 1–5(or 6). **Flowers** 5–12 mm diam.; epicalyx bractlets 1–3 mm, 1/2 length of to nearly equal to sepals; hypanthium 1–3 × 2–4 mm, 1/2 to nearly as deep as wide; sepals spreading to ± reflexed, narrowly to broadly lanceolate, (1.7–)2–4(–4.5) mm; petals 2–6(–6.5) mm; filaments 0.2–1.5 × (0.2–)0.4–0.6(–1) mm, anthers 0.4–0.6 mm; styles 0.9–1.5 mm. **Achenes** brown.

Varieties 7 (7 in the flora): w United States.

Horkelia fusca represents the primary radiation of the genus beyond the California Floristic Province. Within California, *H. fusca* occurs in the Sierra Nevada and mountains of northern California, and it is the only representative of the genus extending beyond California and Oregon into Washington, Idaho, Nevada [with the possible exception of *H. tridentata* (sect. *Tridentatae*) in Washoe County], and, questionably, Wyoming.

As here circumscribed, *Horkelia fusca* is the most diverse, most widely distributed species in the genus, with significant additional work needed to fully elucidate its variation patterns. The treatment presented here differs from that of D. D. Keck (1938[1939]) and B. Ertter (1993d) in using the rank variety instead of subspecies, circumscribing var. *capitata* more narrowly, and transferring the application of var. *pseudocapitata* from what is here called var. *brownii* to the bulk of what had been subsp. *capitata* (Lindley) D. D. Keck (Ertter and J. L. Reveal 2007).

There are two types of basal leaves in plants of *Horkelia fusca*. The ephemeral early-season leaves have leaflets that tend to be broadly cuneate-obovate, shallowly toothed, densely glandular but otherwise sparsely hairy, and deeply veined. The leaf features described below are drawn from the more persistent, mid season leaves that predominate at peak flowering and differ more strongly among varieties.

Petals of first-formed flowers are often larger than average; end-of-season petals can be smaller than average.

Although Montana is sometimes included in the range of *Horkelia fusca*, such references are based only on potential occurrence (W. E. Booth and J. C. Wright 1959).

1. Leaflets divided ± ½–¾+ to midrib.
    2. Inflorescences usually with flowers arranged individually; sw Oregon. . . . . 15b. *Horkelia fusca* var. *filicoides*
    2. Inflorescences composed of 3–20-flowered glomerules, (flowers rarely arranged individually in var. *fusca*); ne California, nw Oregon, sw Washington.
        3. Leaflets divided ± ½ to midrib, 5–18(–20) × 4–15 mm; petals 3–5 mm; filaments 0.5–1.3 mm, longer than wide; nw Oregon, sw Washington . . . . 15a. *Horkelia fusca* var. *fusca*
        3. Leaflets divided ¾+ to midrib, 5–10(–15) × 2–10(–20) mm; petals 2–3(–4) mm; filaments 0.2–0.5 mm, wider than long; ne California . . . . 15c. *Horkelia fusca* var. *tenella*
1. Leaflets divided less than ½ to midrib.
    4. Leaves grayish, surfaces often obscured, ± densely short-villous; leaflets narrowly to broadly cuneate, usually ⅓–½ as wide as long; ne California, adjacent Oregon and Nevada . . . . 15g. *Horkelia fusca* var. *brownii*
    4. Leaves green, surfaces not obscured, ± short-hirsute, villous, or glabrate; leaflets cuneate to broadly obovate, ½ to wider than long; California to Washington, Idaho, Wyoming.
        5. Inflorescences dark reddish purple, densely congested, bracts acuminate-lobed, obscuring pedicels and flowers even at maturity, glandular hairs usually conspicuously red-septate; epicalyx bractlets (2–)3.5–5(–6) mm; cauline leaves 3–6, leaflets of proximalmost 3–5 per side; ne Oregon, se Washington, n Idaho . . . . 15d. *Horkelia fusca* var. *capitata*
        5. Inflorescences green to reddish purple, congested to open, bracts acute-lobed, not obscuring pedicels and flowers at maturity, glandular hairs not or obscurely red-septate; epicalyx bractlets 1–3.5(–4) mm; cauline leaves 1–4(or 5), leaflets of proximalmost 2–4(or 5) per side; Cascade-Sierra axis of Oregon, California, e to Wyoming and n Nevada.
            6. Petals (3–)4–6 mm; leaflets 10–30 (–35) × (5–)10–20(–25) mm; epicalyx bractlets 2–3.5(–4) mm; mountains in and bordering n Intermountain Region, ne California, n Nevada, se Oregon, s Idaho . . . . 15e. *Horkelia fusca* var. *pseudocapitata*
            6. Petals 2–4.5(–6.5) mm; leaflets 5–15 (–20) × (2–)5–10(–15) mm; epicalyx bractlets 1–2(–3) mm; Sierra Nevada, California and Nevada, to Cascade Ranges, c Oregon, and mountains of c Idaho, adjacent Wyoming . . . . .15f. *Horkelia fusca* var. *parviflora*

15a. Horkelia fusca Lindley var. **fusca** • Dusky horkelia [E]

Horkelia tenuisecta Rydberg; *Potentilla douglasii* Greene var. *tenuisecta* (Rydberg) Crum

Stems 3–4(–5) dm. **Basal leaves** green, (4–)8–20 cm; leaflets 6–10 per side, cuneate-obovate, 5–18(–20) × 4–15 mm, ½ to nearly as wide as long, divided ± ½ to midrib into 5–8 teeth or lobes, surfaces not obscured, ± sparsely short-villous. **Cauline leaves** (2 or)3–5; leaflets of proximalmost 3–6 per side. **Inflorescences** reddish purple, congested to open, comprising less than ¼ of stem, usually composed of 10–20-flowered glomerules, rarely flowers arranged individually, glandular hairs often ± red-septate; bracts acute- to acuminate-lobed, not obscuring pedicels and flowers at maturity. **Flowers:** epicalyx bractlets 1.5–3 mm; hypanthium 1.5–3 × 2.5–4 mm; petals 3–5 mm; filaments 0.5–1.3 mm, longer than wide, anthers 0.5–0.6 mm; styles 1–1.5 mm. **Achenes** 1.6–1.8 mm.

Flowering summer. Moist meadows, in conifer woodlands, mainly on volcanic soil; 100–1500 m; Oreg., Wash.

Variety *fusca* occurs in the Cascade Ranges on both sides of the Columbia River Gorge, primarily around Mount Hood, Oregon, and Mount Adams, Washington. Pending further analysis, all collections of the species from the Cascade Range in Washington are treated here as var. *fusca*, including some with leaflets and/or petals in the size range of var. *parviflora*. Plants from lower elevations tend to have greener, more openly branched inflorescences and narrower, more finely divided leaflets; this expression provides the type of *Horkelia tenuisecta*.

15b. Horkelia fusca Lindley var. **filicoides** (Crum) M. Peck, Man. Pl. Oregon, 399. 1941 • Diamond Lake horkelia [E]

Potentilla douglasii Greene var. *filicoides* Crum, Leafl. W. Bot. 1: 100. 1934; *Horkelia fusca* subsp. *filicoides* (Crum) D. D. Keck

Stems (1–)1.5–3.5(–4) dm. **Basal leaves** green, 4–9(–12) cm; leaflets 4–8 per side, cuneate-obovate, 5–12 × 2–8 mm, ± ½ as wide as long, divided ± ½+ to midrib into 3–9 teeth or lobes, surfaces not obscured, sparsely short-villous to glabrate. **Cauline leaves** 2–4(–6); leaflets of proximalmost 2–5 per side. **Inflorescences** green, open, comprising ⅓–½ of stem, flowers usually arranged individually, glandular hairs sometimes red-septate; bracts acuminate-lobed, not obscuring pedicels and flowers at maturity. **Flowers:** epicalyx bractlets 1–2 mm; hypanthium 1–2 × 2.5–3.5 mm; petals 2–4(–5) mm; filaments 0.5–1 mm, ± as wide as long, anthers 0.5 mm; styles 1 mm. **Achenes** 1.2–1.5 mm.

Flowering summer. Open conifer woodlands, mainly on volcanic soil; 800–1600 m; Oreg.

Variety *filicoides* is known from the southern Cascade Range of southwestern Oregon, primarily in open lodgepole pine forests north and west of Crater Lake. This is the only variety in which the flowers are usually arranged individually rather than clustered into glomerules, at least in fully expanded inflorescences; some populations out of the core range have more congested inflorescences and are transitional to var. *parviflora* in this regard.

15c. Horkelia fusca Lindley var. **tenella** S. Watson in W. H. Brewer et al., Bot. California 1: 181. 1876 • Delicate horkelia [E]

Horkelia fusca subsp. *tenella* (S. Watson) D. D. Keck; *Potentilla douglasii* Greene var. *tenella* (S. Watson) Greene

Stems (0.6–)1–4(–5) dm. **Basal leaves** green, (3–)4–18(–27) cm; leaflets 8–15 per side, cuneate to flabellate, 5–10(–15) × 2–10 (–20) mm, ¾ as wide to wider than long, divided ¾+ to midrib into 5–15 lobes or teeth, surfaces not obscured, sparsely short-hirsute to glabrate. **Cauline leaves** 2–4(or 5); leaflets of proximalmost 3–6 per side. **Inflorescences** green to ± reddish, congested to open, comprising no more than ¼ of stem, composed of 3–10(–20)-flowered glomerules, glandular hairs sometimes red-septate; bracts acute- to acuminate-lobed, sometimes partly obscuring pedicels and flowers at maturity. **Flowers:** epicalyx bractlets 1.5–2 mm; hypanthium 1.5–2 × 2–3 mm; petals 2–3 (–4) mm; filaments 0.2–0.5 mm, wider than long, anthers 0.5 mm; styles 1 mm. **Achenes** 1.2–1.5 mm. $2n = 28$.

Flowering summer. Dry meadow edges, open conifer woodlands, mainly on volcanic soil; 1200–2200 m; Calif.

Plants of var. *tenella* have the most deeply and consistently divided leaflets, and the variety is the only one with such leaflets that occurs in California. It is most common in open lodgepole pine forests at the southern end of the Cascade Range in northeastern California. Populations outside this core area tend to intergrade with var. *brownii* or var. *parviflora*. A collection from west-central Tehama County (*Bracelin 405*, UC) matches this variety, but from an atypical habitat well outside the core range.

15d. Horkelia fusca Lindley var. **capitata** (Lindley) M. Peck, Man. Pl. Oregon, 398. 1941 • Palouse horkelia [E]

Horkelia capitata Lindley, Edwards's Bot. Reg. 23: sub plate 1997. 1837; *H. fusca* subsp. *capitata* (Lindley) D. D. Keck; *Potentilla capitata* (Lindley) Greene; *P. douglasii* Greene var. *capitata* (Lindley) J. T. Howell

Stems 4–9 dm. **Basal leaves** green, 10–20(–35) cm; leaflets 4–6(–8) per side, cuneate to broadly obovate, 10–30 (–35) × 10–25(–30) mm, ½ to as wide as long; divided ⅕–¼ to midrib into 8–15 teeth, surfaces not obscured, ± sparsely short-hirsute to glabrate. **Cauline leaves** 3–6; leaflets of proximalmost 3–5 per side. **Inflorescences** dark reddish purple, densely congested, comprising no more than ⅙ of stem, composed of 5–30-flowered glomerules, glandular hairs usually conspicuously red-septate; bracts acuminate-lobed, obscuring pedicels and flowers even at maturity. **Flowers:** epicalyx bractlets (2–)3.5–5(–6) mm; hypanthium 2–3 × 3–5 mm; petals 4–6.5 mm; filaments 0.5–1.5 mm, longer than wide, anthers 0.5–0.6 mm; styles 1–1.5 mm. **Achenes** 1.6–1.8 mm.

Flowering summer. Dry meadow edges, in conifer woodlands, mainly on volcanic or granitic soil; 100–2300 m; Idaho, Oreg., Wash.

As here circumscribed, var. *capitata* is found in meadow margins and grassy openings in the forested mountains surrounding the Palouse Prairie (in the broad sense) of northern Idaho, southeastern Washington, and northeastern Oregon. This is a significantly more limited range than accepted by D. D. Keck (1938[1939]) and in subsequent floras (for example, B. Ertter 1993d), with the bulk of what had been treated as var. (or subsp.) *capitata* here treated as var. *pseudocapitata*. The distinctions between the two varieties are more matters of degree than absolute differences, var. *capitata* generally having more congested, more purple, more glandular inflorescences in which the pedicels and hypanthia remain covered by larger involucral bracts divided into more elongate teeth. Plants of var. *capitata* can also be significantly larger, especially in the remnant prairie patches of northern Idaho, with more numerous cauline leaves having more pairs of larger leaflets. The extremes of the two varieties are reasonably distinct, with a significant biogeographic underpinning, but the differences merge in the Blue Mountains of Oregon.

Given the California focus of the published source (P. A. Munz 1959), the chromosome count of $2n = 28$ for var. *capitata* is most likely based on material here treated as var. *pseudocapitata*.

15e. Horkelia fusca Lindley var. **pseudocapitata** (Rydberg ex Howell) M. Peck, Man. Pl. Oregon, 398. 1941 (as pseudo-capitata) • Intermountain horkelia [E]

Horkelia pseudocapitata Rydberg ex Howell, Fl. N.W. Amer., 180. 1898; *H. fusca* subsp. *pseudocapitata* (Rydberg ex Howell) D. D. Keck; *Potentilla douglasii* Greene var. *pseudocapitata* (Rydberg ex Howell) Jepson

Stems 3–6 dm. **Basal leaves** green, 10–35(–40) cm; leaflets 4–6(or 7) per side, narrowly to broadly obovate, 10–30(–35) × (5–)10–20 (–25) mm, 1/2 to nearly as wide as long, divided 1/5–1/4 to midrib into 8–14 teeth, surfaces not obscured, ± sparsely hirsute to glabrate. **Cauline leaves** 1–4(or 5); leaflets of proximalmost 2 or 3(–5) per side. **Inflorescences** green to reddish purple, congested to open, comprising 1/6–1/3 of stem, composed of 10–30-flowered glomerules, glandular hairs not or obscurely red-septate; bracts acute-lobed, not obscuring pedicels and flowers at maturity. **Flowers:** epicalyx bractlets 2–3.5(–4) mm; hypanthium 2 × 2.5–4 mm; petals (3–)4–6 mm; filaments 0.5–1.5 mm, usually longer than wide, anthers 0.5–0.6 mm; styles 1–1.5 mm. **Achenes** 1.6–1.8 mm. **2*n*** = 28.

Flowering summer. Dry meadow edges, often with sagebrush, aspen, and/or willows, open conifer woodlands, mainly on volcanic or granitic soil; 900–2300 m; Calif., Idaho, Nev., Oreg.

The application of the epithet *pseudocapitata* here differs significantly from that of P. A. Rydberg (1908c), D. D. Keck (1938[1939]), and most floras, who used it for the taxon that is here called var. *brownii*. Instead, var. *pseudocapitata* encompasses the bulk of what had been called var. (or subsp.) *capitata*, except in mountains bordering the Palouse Prairie in Idaho, Oregon, and Washington. For nomenclatural details, see B. Ertter and J. L Reveal (2007).

As here circumscribed, var. *pseudocapitata* is a relatively large, big-petaled variety that grows in the mountains in and bordering the northern Intermountain Region in northeastern California (primarily the Warner Mountains), northern Nevada, southeastern Oregon, and southern Idaho, extending northward along valleys from Camas to Blaine counties. Intermediate plants are common where the range intersects those of var. *brownii*, var. *capitata*, and var. *parviflora*, though in the core of its range var. *pseudocapitata* is reasonably distinctive and uniform.

15f. Horkelia fusca Lindley var. **parviflora** (Nuttall ex Hooker & Arnott) Wawra, Itin. Princ. S. Coburgi 1: 17. 1883 • Small-flowered horkelia [E] [F]

Horkelia parviflora Nuttall ex Hooker & Arnott, Bot. Beechey Voy., 338. 1839; *H. fusca* subsp. *parviflora* (Nuttall ex Hooker & Arnott) D. D. Keck; *Potentilla andersonii* Greene; *P. douglasii* Greene var. *parviflora* (Nuttall ex Hooker & Arnott) J. T. Howell

Stems 1–4.5 dm. **Basal leaves** green, 4–15(–18) cm; leaflets 4–8 per side, cuneate to broadly obovate, 5–15 (–20) × (2–)5–10(–15) mm, 1/2 as wide to wider than long, divided 1/4–1/2 to midrib into 4–6(–10) teeth, surfaces not obscured, sparsely to moderately hirsute or villous, sometimes glabrate. **Cauline leaves** 1–3(or 4); leaflets of proximalmost 2–4(or 5) per side. **Inflorescences** green to purplish, congested to open, usually comprising less than 1/4 of stem, composed of 5–20-flowered glomerules, glandular hairs not red-septate; bracts acute-lobed, not obscuring pedicels and flowers at maturity. **Flowers:** epicalyx bractlets 1–2(–3) mm; hypanthium 1.5–2 × 2–3.5 mm; petals 2–4.5(–6.5) mm; filaments 0.2–1 mm, usually longer than wide, anthers 0.3–0.5 mm; styles 0.9–1.1 mm. **Achenes** 1–1.2 mm. **2*n*** = 28.

Flowering summer. Dry meadow edges, in conifer woodlands, mainly on volcanic or granitic soil; 800–3300 m; Calif., Idaho, Nev., Oreg., Wyo.

Variety *parviflora* is the most widespread and polymorphic in the species and might represent the repository for variation that is not accommodated by the other, more tightly circumscribed varieties. Population clusters occur in three discrete areas: the Sierra Nevada of eastern California and adjacent Nevada, the mountains of northern California extending north in the Cascade Range to central Oregon, and the mountains of central Idaho. Each of these three areas has plants that differ slightly from those of the other two.

The most distinct phase is found in the Cascade Range of Oregon, with larger-than-average petals and highly branched inflorescences. Plants from Lost Prairie in eastern Linn County are particularly anomalous, with petals to 6.5 mm; further studies may indicate that these deserve separate recognition.

Plants from the Sierra Nevada tend to have smaller petals, and those in the higher southern Sierra Nevada tend to have capitate or subcapitate inflorescences. From about Lake Tahoe northward, inflorescences tend to be somewhat more open but still congested compared to the Oregon phase.

The Idaho phase, represented by the type of var. *parviflora*, has petals that run the full range from 2–4 mm, with most individuals having a slightly branched inflorescence. Unlike those of the other two

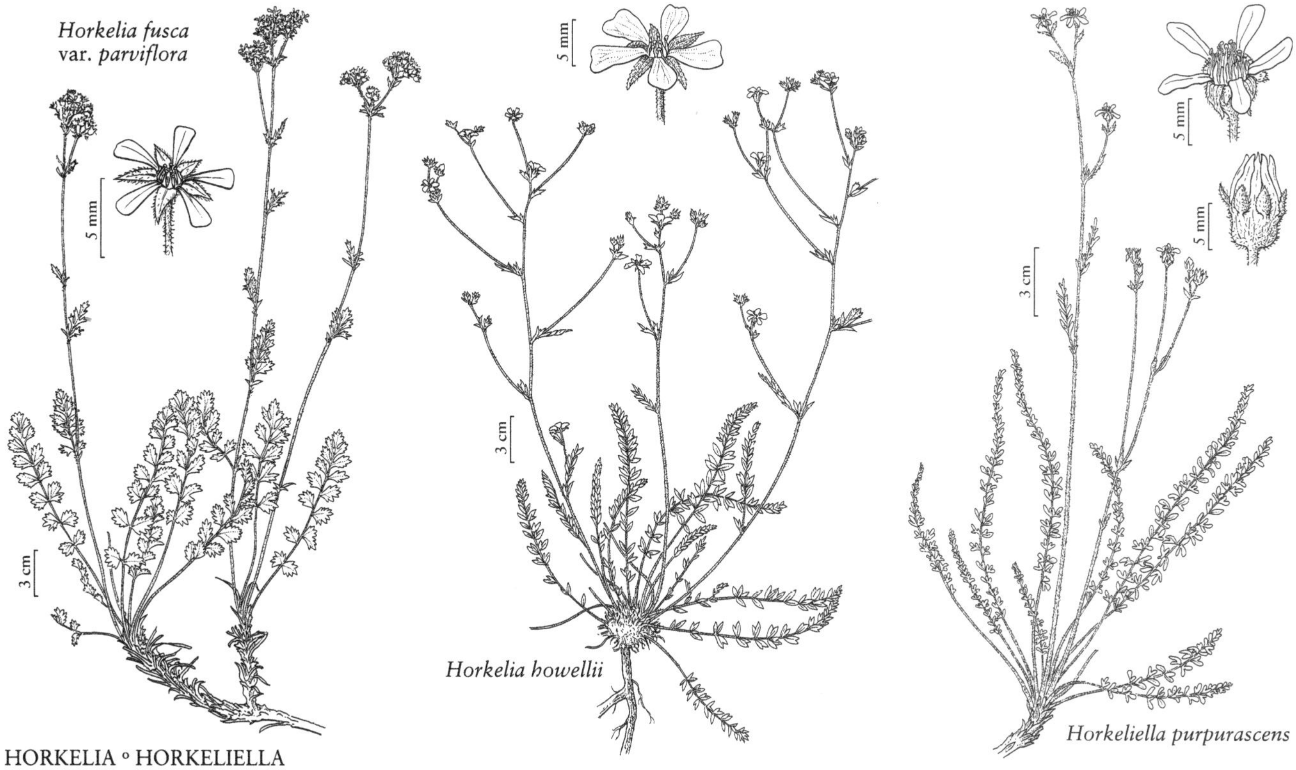

HORKELIA ° HORKELIELLA

phases, both the hypanthia and sepals are often purple; the hypanthium is otherwise typically greenish with only the sepals themselves a dark purple. The sole Wyoming record (*D. R. Goddard 1027*, F, UC) has yet to be confirmed by more recent collections in Yellowstone National Park. These plants have pale inflorescences, relatively blunt sepals, and short epicalyx bractlets, differing in these features from the closest populations in central Idaho. For reports from Montana, see species discussion.

15g. Horkelia fusca Lindley var. **brownii** (Rydberg) Ertter & Reveal, Novon 17: 319. 2007 • Shasta horkelia [E]

Horkelia brownii Rydberg in N. L. Britton et al., N. Amer. Fl. 22: 276. 1908

Stems 1–4.5 dm. **Basal leaves** grayish, 4–12(–18) cm; leaflets 3–7 per side, narrowly to broadly cuneate, 5–15(–17) × 2–8 (–12) mm, usually ⅓–½ as wide as long, divided ¼–½ to midrib into 4–6 teeth, surfaces often obscured, ± densely short-villous. **Cauline leaves** 2–4; leaflets of proximalmost 2–4 per side. **Inflorescences** grayish or purplish, usually ± congested, sometimes more open, usually comprising less than ⅙ of stem, composed of 5–20(–40)-flowered glomerules, glandular hairs usually not red-septate; bracts acuminate-lobed, often ± obscuring pedicels and flowers even at maturity. **Flowers:** epicalyx bractlets 1.5–2.5 mm; hypanthium 1.5–2.5 × 2.5–4 mm; petals 2–3(–4.5) mm; filaments (0.2–)0.6–1 mm, usually longer than wide, anthers 0.3–0.5 mm; styles 1 mm. **Achenes** 1–1.2 mm. **2*n*** = 28.

Flowering summer. Dry meadow edges in conifer woodlands, on pumice or sandy or clay soil; 900–2000 m; Calif., Nev., Oreg.

Variety *brownii* here takes the place of var./subsp. *pseudocapitata* as used in D. D. Keck (1938[1939]) and most floras; this variant was included within var. *parviflora* by B. Ertter (1993d). The diagnostic feature is the ash gray leaflets, which also tend to be more narrowly cuneate than in other varieties. The variety occurs from Mount Shasta in Siskiyou County east to southwestern Modoc County and southeast to Sierra Valley in Plumas and Sierra counties, California, skirting the range of var. *tenella*. Morphologically comparable plants occur as far south as the California-Nevada border on the south side of Lake Tahoe, and at least one collection from Nevada (Reno, *Hillman s.n.*, POM) is closest to this variety. Populations from the southeastern end of the range tend to be larger and less ash gray than those in the vicinity of Mount Shasta. Most references to var. *brownii* (as var./subsp. *pseudocapitata* misapplied) from Oregon are probably best attributed to the polymorphic var. *parviflora*, but at least one collection from Klamath Lake (*Newberry s.n.*, GH) fits well in var. *brownii*.

10e. Horkelia Chamisso & Schlechtendal sect. Tridentatae (Rydberg) O. Stevens in N. L. Britton et al., N. Amer. Fl. 22(7): 7. 1959 [E]

Horkelia [unranked] *Tridentatae* Rydberg, Monogr. N. Amer. Potentilleae, 122. 1898; *Horkelia* [unranked] *Ambiguae* A. Gray; *Horkelia* [unranked] *Hirsutae* Rydberg; *Horkelia* sect. *Hirsutae* (Rydberg) O. Stevens; *Horkelia* [unranked] *Sericatae* Rydberg; *Horkelia* sect. *Sericatae* (Rydberg) O. Stevens

Plants forming rosettes or tufts, sometimes loose mats, green to grayish or silvery, inconspicuously to moderately glandular, not resinously aromatic. **Stems** decumbent to erect, 0.5–5 dm. **Basal leaves** planar to ± cylindric; stipules entire, forked, or pinnately divided into linear to filiform lobes; leaflets 2–20 per side, separate to ± overlapping, divided either ½–¾+ to midrib into (0–)2–15 ultimate lobes or teeth not restricted to apex, or ⅒–¼ or less to midrib into (0–)3(–5) teeth restricted to apex. **Inflorescences** open to ± congested, flowers arranged individually, in corymbiform clusters, and/or in capitate or non-capitate glomerules. **Pedicels** remaining straight, outermost sometimes ± reflexed in congested inflorescences, 1–10(–20) mm. **Flowers**: epicalyx bractlets linear to lanceolate or narrowly elliptic, 0.2–0.5(–0.8) mm wide, entire; hypanthium interior usually glabrous, sometimes pilose; sepals obtuse to acute to slightly acuminate; petals white to pink or cream, sometimes drying yellowish, linear or cuneate to obovate or obcordate, apex usually acute to rounded to emarginate, sometimes truncate, rarely mucronate; filaments white, glabrous, anthers longer or shorter than wide; carpels 2–15(–20). **Achenes** 1.5–3 mm, smooth or rugose.

Species 5 (5 in the flora): w United States.

P. A. Rydberg (1908c) assigned the species here composing this section to three of his unranked groups: *Hirsutae*, *Sericatae* (also including *Horkelia hispidula*), and *Tridentatae*, validated inadvertently as sections by O. A. Stevens in the index to volume 22 of North American Flora. In addition to lacking the characteristic *Horkelia* odor, plants are often rosette-forming from simple caudices; stipules are sometimes highly divided; and leaflets are either few-toothed or deeply lobed. The section is centered in the Siskiyou-Klamath region of northwestern California and southwestern Oregon, with some affinity for ultramafic substrates. *Horkelia tridentata* extends south in the Sierra Nevada; *H. congesta* is a component of the Willamette Valley flora. Much of the variation in the section is highly localized, with a full series of intergrading morphologies, indicative of a relatively recent and active radiation. More segregates are recognized here than in some recent floras (M. E. Peck 1941; P. A. Munz 1959; B. Ertter 1993d), and additional variants are likely to warrant taxonomic recognition.

1. Leaflets 2–6 per side, separate to slightly overlapping, divided into (0–)3(–5) teeth restricted to apex; inflorescences: flowers usually arranged in ± capitate glomerules, sometimes in corymbiform clusters.
    2. Petals ± obovate, (1.5–)2–5 mm wide; stems: glands dense distally; Oregon, barely in nw California . . . . 19. *Horkelia congesta*
    2. Petals linear to broadly oblanceolate, 0.3–1.5 mm wide; stems: glands absent or sparse distally; California, barely in sw Oregon . . . . 20. *Horkelia tridentata*
1. Leaflets 5–20 per side, ± overlapping, divided into (0–)2–15 linear to elliptic or oblanceolate lobes or teeth not restricted to apex; inflorescences: flowers arranged individually, in corymbiform clusters, and/or in usually non-capitate glomerules.

[3. Shifted to left margin.—Ed.]

3. Leaflets 5–10 per side, 5–15(–25) mm, divided ¾+ to midrib into (0–)2–15 lobes; stems: hairs 2–3 mm proximally; inflorescences open to ± congested, flowers arranged individually or in ± corymbiform clusters; petals white to cream, often drying yellowish . . . . 18. *Horkelia daucifolia*
3. Leaflets (8–)10–20 per side, 2–10(–15) mm, divided ± ½–¾ to midrib into (0–)2–4 lobes; stems: hairs 1 mm proximally; inflorescences open, flowers arranged individually and/or in non-capitate glomerules; petals white to pink or red-veined.
    4. Plants silvery; basal leaves densely sericeous, 3–10 cm, stipules usually entire or forked, rarely pinnately divided. . . . . 16. *Horkelia sericata*
    4. Plants ± green; basal leaves usually villous to pilose, often glabrate, 5–15 cm, stipules deeply 2-lobed or pinnately divided into 3–5 lobes . . . . 17. *Horkelia howellii*

16. Horkelia sericata S. Watson, Proc. Amer. Acad. Arts 20: 364. 1885 • Silky horkelia [C] [E]

Potentilla sericata (S. Watson) Greene

Plants ± tufted, silvery. **Stems** ascending to erect, 1.5–4.5 dm, hairs 1 mm proximally, glands absent or sparse distally. **Basal leaves** ± cylindric to weakly planar, 3–10 × 0.3–1.2 cm, densely sericeous, often villous on margins apically; stipules usually entire or forked, rarely pinnately divided into linear lobes; leaflets (8–) 10–20 per side, ± overlapping, elliptic to flabellate, 2–8 × 1–4 mm, ½–⅔ as wide as long, divided ± ½+ to midrib into (0–)2–4 elliptic lobes 1–2 mm wide, these not restricted to apex. **Cauline leaves** (2 or)3–5; stipules 3–8 mm, entire or shallowly 1- or 2-toothed. **Inflorescences** open, flowers arranged individually and/or in non-capitate glomerules. **Pedicels** 1–4 mm. **Flowers** 10 mm diam.; epicalyx bractlets linear-lanceolate, 1.5–3 × 0.5 mm, ± ⅔ length of sepals; hypanthium 1–1.5 × 2–3 mm, ± ½ to as deep as wide, interior glabrous; sepals spreading to reflexed, abaxially green to reddish or purplish, 2–4 mm; petals white to pink or red-veined, narrowly obcordate, 3–4.5(–7) × 2–3 mm, apex ± emarginate; filaments 0.5–1.5 × 0.2–0.5 mm, anthers 0.4–0.6 mm; carpels 2–6; styles 1.5–2 mm. **Achenes** brown, 2–2.5 mm, smooth. **2*n*** = 28.

Flowering summer. Chaparral, oak-conifer woodlands, on serpentine-derived soil; of conservation concern; 100–1200 m; Calif., Oreg.

D. D. Keck (1938[1939]) expanded the circumscription of *Horkelia sericata* to encompass the plants segregated here as *H. howellii*, on the grounds that intergradation is too extensive to justify taxonomic recognition of the extremes. In this evaluation, however, most collections can be unequivocally divided between plants with compact silvery-sericeous leaves with entire or forked stipules (*H. sericata*), and plants with larger, greener leaves and pinnately divided stipules (*H. howellii*). It is not known if the chromosome count provided by P. A. Munz (1959) was derived from *H. howellii* or *H. sericata*.

As here circumscribed, *Horkelia sericata* is a localized taxon known only from Curry County, Oregon, and the Gasquet serpentine area in adjacent Del Norte County, California.

17. Horkelia howellii (Greene) Rydberg, Bull. Torrey Bot. Club 25: 55. 1898 • Howell's or Klamath horkelia [C] [E] [F]

Potentilla howellii Greene, Pittonia 1: 104. 1887

Plants tufted to openly matted, ± green. **Stems** ascending to erect, (0.9–)1.5–5 dm, hairs 1 mm proximally, glands absent or sparse distally. **Basal leaves** ± cylindric to weakly planar, 5–15 × 0.5–1.8(–2.5) cm, usually villous to pilose at least marginally, often glabrescent; stipules deeply 2-lobed or pinnately divided into 3–5 linear to filiform lobes that often form a tangled mass; leaflets 10–15 per side, ± overlapping, elliptic to flabellate, (3–)4–10(–15) × 2–8 mm, ½ to as wide as long, divided ½–¾ to midrib into (0–)2–4 elliptic to linear lobes 1–2 mm wide, these not restricted to apex. **Cauline leaves** (2 or)3–5; stipules 5–12(–15) mm, entire to deeply 2–4-lobed or -toothed (usually on 1 side). **Inflorescences** open, flowers arranged individually and/or in non-capitate glomerules. **Pedicels** 1–6 (–12) mm. **Flowers** 10 mm diam.; epicalyx bractlets linear-lanceolate, 1–3 × 0.5 mm, ± ⅔ length of sepals; hypanthium 1.1–1.5 × 2–3 mm, ± ½ to as deep as wide, interior glabrous; sepals spreading to reflexed, abaxially green to reddish or purplish, 2.5–4.5 mm; petals white to pink or red-veined, narrowly obcordate, 3–5 × 2–3 mm, apex truncate or emarginate; filaments 0.5–1.7 × 0.2–0.5 mm, anthers 0.4–0.6 mm; carpels 2–6; styles 1.5–2 mm. **Achenes** brown, 2–2.7 mm, smooth.

Flowering summer. Chaparral, oak-conifer woodlands, mainly on serpentine soil; of conservation concern; 60–1200 m; Calif., Oreg.

As noted above, *Horkelia howellii* comprises the bulk of what D. D. Keck (1938[1939]) and subsequent floras included in *H. sericata*, and the range of the latter is fully encompassed within that of the former. As here circumscribed, *H. howellii* occurs in northwestern California and southwestern Oregon. The handful of collections from Humboldt and Trinity counties, California, including the type of *Potentilla laxiflora* Drew, have leaves approaching *H. daucifolia* but the inflorescence of *H. howellii*; exact petal color is uncertain.

18. Horkelia daucifolia (Greene) Rydberg, Bull. Torrey Bot. Club 25: 55. 1898 [E]

Potentilla daucifolia Greene, Pittonia 1: 160. 1888

Plants rosette-forming to tufted, rarely ± matted, grayish or green. **Stems** ascending to erect, 1.5–3.5 dm, hairs 2–3 mm proximally, glands sparse to dense distally. **Basal leaves** weakly planar to loosely cylindric, (4–)5–12(–17) × 0.8–3(–4) cm, usually sericeous to villous, sometimes glabrate; stipules pinnately divided into 3–7 linear to filiform lobes; leaflets 5–10 per side, ± overlapping, obovate to broadly obcordate, 5–15(–25) × 5–20 mm, ± as wide as long, divided ¾+ to midrib into (0–)2–15 linear to oblanceolate lobes 0.4–2(–3) mm wide, these not restricted to apex. **Cauline leaves** 3–6; stipules 8–15 mm, deeply 3–7-lobed proximally, 1–3 -lobed distally. **Inflorescences** open to ± congested, flowers arranged individually or in ± corymbiform clusters. **Pedicels** 2–10(–20) mm. **Flowers** 10–15 mm diam.; epicalyx bractlets linear to linear-lanceolate, 2.5–3.5 × 0.5 mm, ⅔ to ± equal to length of sepals; hypanthium 1–1.5 × 3.5–5 mm, less than ½ as deep as wide, interior glabrous; sepals ± spreading, abaxially green, 3.5–7.5 mm; petals white to cream, often drying yellowish, cuneate to obcordate, (3.5–)4–8 × 2–8 mm, apex ± emarginate, sometimes rounded; filaments 1–3 × 0.5–1.5 mm, anthers 0.5–0.8 mm; carpels 5–15, styles 2–4 mm. **Achenes** dark brown, 2.4–3 mm, smooth to ± roughened.

Varieties 3 (3 in the flora): w United States.

D. D. Keck (1938[1939]) decided that the leaflet characters previously used to recognize segregates of *Horkelia daucifolia* were random variables that have not become localized into geographic subunits. At the same time, he established subsp. *latior* D. D. Keck to accommodate a single collection with leaves approaching those of *H. sericata*, with the speculation that the population was an amphiploid derivative of the two species. Subsequent collections from the same general area (Scott Mountain, Siskiyou County, California) provide a full range of variation between the extreme with exceptionally wide leaflet lobes and the typical form occurring at the base of the mountain. This variant of the species is accordingly not recognized here, but the extremes with narrower, more numerous leaflet lobes are recognized as varieties that coincide with major river drainages (B. Ertter and J. L. Reveal 2007).

In the descriptions below, the pedicels of all varieties are sparsely pilose in addition to being puberulent or not.

1. Leaflet lobes 2–5(–8), linear to oblanceolate, 1–2(–3) mm wide; petals cuneate to obovate, 2–4 mm wide. . . 18a. *Horkelia daucifolia* var. *daucifolia*
1. Leaflet lobes (2–)5–15, linear, 0.4–1 mm wide; petals broadly obovate to obcordate, 4–8 mm wide.
    2. Pedicels 2–7(–10) mm, puberulent, eglandular; Oregon . . . . . 18b. *Horkelia daucifolia* var. *caruifolia*
    2. Pedicels 5–10(–20) mm, not puberulent, minutely glandular; California. . . . . 18c. *Horkelia daucifolia* var. *indicta*

18a. Horkelia daucifolia (Greene) Rydberg var. **daucifolia** • Carrotleaf horkelia [E]

Horkelia daucifolia subsp. *latior* D. D. Keck; *Potentilla daucifolia* Greene var. *latior* (D. D. Keck) J. T. Howell

Plants mostly rosette-forming. **Basal leaves** grayish, 6–12(–17) cm; leaflets (5–)10–15(–25) mm, lobes 2–5(–8), linear to oblanceolate, 1–2(–3) mm wide. **Pedicels** (2–)3–7(–10) mm, puberulent, eglandular or sparsely glandular. **Flowers:** petals cuneate to obovate, 4–6 × 2–4 mm; filaments 1–2.5 mm.

Flowering summer. Open areas in conifer woodlands, mainly on serpentine soil; 500–1700 m; Calif., Oreg.

Variety *daucifolia* occurs mostly in valleys along the central Klamath River drainage in Shasta and Siskiyou counties of northern California; a single collection (*Applegate 2122*, DS and US) is known from extreme southwestern Klamath County, Oregon. Scattered populations occur also in the upper Trinity River drainage in Trinity County, California; some of them have plants with atypically short, 2-fid leaflets.

18b. Horkelia daucifolia (Greene) Rydberg var. **caruifolia** (Rydberg ex Howell) Ertter & Reveal, Novon 17: 319. 2007 • Rogue River horkelia [E]

Horkelia caruifolia Rydberg ex Howell, Fl. N.W. Amer., 181. 1898; *Potentilla caruifolia* (Rydberg ex Howell) Crum; *P. caruifolia* var. *divergens* Crum

Plants rosette-forming to ± matted. **Basal leaves** green to grayish, (4–)5–9(–15) cm; leaflets 5–10(–15) mm, lobes 4–15, linear, 0.4–1 mm wide. **Pedicels** 2–7(–10) mm, puberulent, eglandular. **Flowers:** petals broadly obcordate, (3.5–)5–8 × (4–)5–8 mm; filaments 2–3 mm.

Flowering summer. Grassy openings in mixed oak and conifer woodlands, on non-serpentine soil; 400–1100 m; Oreg.

Except for a single population in extreme southeastern Klamath County (noted under var. *daucifolia*), all populations of *Horkelia daucifolia* from Oregon are here referred to var. *caruifolia* (B. Ertter and J. L. Reveal 2007). The variety is found in the Rogue River drainage in Jackson and Josephine counties and might prove to be of conservation concern. Collections from near Ashland, including the type of *H. caruifolia*, are not as distinct from var. *daucifolia* as are populations from north of the Rogue River Valley, which have particularly broad petals and mat-forming capabilities. These northern populations include the type of *Potentilla caruifolia* var. *divergens*, which was described on the basis of its widely open inflorescence; subsequent collections from the same general area (near Trail) have the full range of inflorescence compactness found elsewhere in *H. daucifolia*.

18c. Horkelia daucifolia (Greene) Rydberg var. **indicta** (Jepson) Ertter & Reveal, Novon 17: 319. 2007 • Jepson's horkelia [C] [E]

Potentilla daucifolia Greene var. *indicta* Jepson, Man. Fl. Pl. Calif., 493. 1925

Plants rosette-forming to tufted. **Basal leaves** green, 7–11 cm; leaflets 5–10(–15) mm, lobes (2–)5–15, linear, 0.4–1 mm wide. **Pedicels** 5–10(–20) mm, not puberulent, minutely glandular. **Flowers:** petals broadly obovate to obcordate, 5–7.5 × 4–6 mm; filaments 2–3 mm.

Flowering summer. Vernally saturated openings in oak-conifer woodlands, mainly on clay soil; of conservation concern; 200–700 m; Calif.

Variety *indicta* was known only from the type in Tehama County until rediscovered in 1990 in western Shasta County, confirming that this is a distinct variety in the forested hills rimming the northern Sacramento Valley (B. Ertter and J. L. Reveal 2007). One other historic collection of *Horkelia daucifolia* from Shasta County (Montgomery Creek, *Smith s.n.*, CAS) is more similar to var. *daucifolia*.

19. Horkelia congesta Douglas ex Hooker, Bot. Mag. 56: plate 2880. 1829 [C] [E]

Potentilla congesta (Douglas ex Hooker) Baillon

Plants tufted or rosette-forming, green to grayish green. **Stems** ascending to erect, 1.5–5 dm, hairs 2–3 mm proximally, glands dense distally. **Basal leaves** ± planar, (3–)4–15 × 1–3 cm, usually villous at least abaxially, often densely so; stipules entire, deeply 2-lobed, or pinnately divided into 3–5 linear lobes; leaflets 2–6 per side, ± separate, narrowly cuneate or lanceolate-elliptic to oblong, 5–20 × (1–)2–5(–6) mm, 1/5–1/2 as wide as long, divided 1/6–1/4 or less to midrib into (0–)3(–5) ± ovate teeth 1–2 mm wide, these restricted to apex. **Cauline leaves** 2–7; stipules 5–25(–30) mm, divided 3/4+ into linear or linear-lanceolate lobes, or incompletely divided or divided no more than 2/3 into lanceolate, often acuminate teeth. **Inflorescences** open to congested, flowers usually arranged in ± capitate glomerules or corymbiform clusters. **Pedicels** 1–3(–6) mm. **Flowers** 8–16 mm diam.; epicalyx bractlets linear to lanceolate or narrowly elliptic, 1–3 × 0.3–0.8 mm, ± 2/3 length of sepals; hypanthium 2–2.5 × 3–4.5 mm, ± 1/2 to nearly as deep as wide, interior usually glabrous, sometimes sparsely pilose; sepals spreading, abaxially green, 2–4.5 mm; petals white to cream, ± obovate, (2.5–)3–6 × (1.5–)2–5 mm, apex ± rounded to ± emarginate; filaments (0.5–)1–2 × (0.2–)0.4–0.6 mm, anthers 0.4–0.7(–1) mm; carpels (6–)10–15(–20), styles 1.5–3 mm. **Achenes** brown, 1.6–2.2 mm, faintly rugose.

Varieties 2 (2 in the flora): w United States.

As discussed by D. D. Keck (1938[1939]) and analyzed by T. N. Kaye (unpubl., Oregon Dept. of Agriculture, 1995), *Horkelia congesta* comprises three population clusters that form a latitudinal gradient in Oregon west of the Cascade Range, transitioning to *H. tridentata* in the mountains of California. Populations of *H. congesta* in remnant prairies in the Willamette Valley (probably including the type of *H. hirsuta* Lindley) have relatively tall, erect stems, pectinate stipules, relatively numerous, narrow leaflets, openly branched inflorescences with flowers in corymbiform clusters, and clawed petals 5–6 × 4–5 mm. In contrast, populations centered in rocky serpentine flats within the Illinois River drainage in Josephine County, which were described by Keck as subsp. *nemorosa*, tend to have shorter ascending stems, less divided stipules, fewer and broader leaflets, more capitate inflorescences, and scarcely clawed petals 3–4 × 2–3 mm. The differences between these two extremes are blurred by populations from the Umpqua Valley in Douglas County (the probable type locality of *H. congesta*), as well as scattered populations in northern Jackson and Josephine counties. Based on morphometric analyses of 19 characters, Kaye concluded that the Umpqua material overlapped slightly with plants from the Willamette Valley, while the Josephine County populations clustered separately. He retained the Umpqua and Willamette valley populations as subsp. *congesta*, with mid stem stipule features of greatest diagnostic value. This taxonomic conclusion is adopted here, with a change of rank from subspecies to variety.

1. Cauline leaves: stipules divided ¾+ into linear to linear-lanceolate lobes; inflorescences composed of ± corymbiform clusters to capitate glomerules; petals (3–)4–6 × (2–)3–5 mm; remnant prairies, Willamette and Umpqua valleys, Oregon . . . . 19a. *Horkelia congesta* var. *congesta*
1. Cauline leaves: stipules incompletely divided or divided no more than ⅔ into lanceolate, often acuminate teeth; inflorescences usually composed of ± capitate glomerules; petals 2.5–3(–4) × (1.5–) 2–3 mm; serpentine flats, Illinois River drainage, Oregon, and Del Norte County, California . . . . 19b. *Horkelia congesta* var. *nemorosa*

19a. Horkelia congesta Douglas ex Hooker var. **congesta** • Shaggy horkelia [C] [E]

Horkelia hirsuta Lindley

Stems ± erect, usually greenish, rarely reddish, 3–5 dm, hairs 3 mm proximally. **Basal leaves** 6–15 × 1.5–3 cm; leaflets 3–6 per side, narrowly cuneate, 10–20 × 2–4(–6) mm, ⅕–⅓ as wide as long. **Cauline leaves** 3–7; stipules divided ¾+ into linear to linear-lanceolate lobes. **Inflorescences** composed of ± corymbiform clusters to capitate glomerules. **Flowers:** epicalyx bractlets linear to lanceolate, 2–3 mm; hypanthium ± ½ as deep as wide; petals (3–)4–6 × (2–)3–5 mm; styles 1.5–2 mm.

Flowering spring–summer. Wet to dry remnant prairies, generally near valley bottoms, or on balds of low hills in oak-conifer woodlands, generally on volcanic soil; of conservation concern; 80–700 m; Oreg.

Variety *congesta* is a component of the vanishing Willamette Prairie flora in western Oregon (T. N. Kaye, unpubl.). The variety has been extirpated from the northern part of its range in Marion, Polk, and Washington counties, and only isolated populations remain in the southern Willamette Valley, primarily in Lane County, and in the Umpqua Valley in Douglas County. A specimen from northern Josephine County (*Beach 509*, UC) is also referable to var. *congesta*, and two historic collections from the Rogue River Valley in Jackson County are intermediate between var. *congesta* and var. *nemorosa*.

19b. Horkelia congesta Douglas ex Hooker var. **nemorosa** (D. D. Keck) M. Peck, Man. Pl. Oregon, 399. 1941 • Josephine horkelia [C] [E]

Horkelia congesta subsp. *nemorosa* D. D. Keck, Lloydia 1: 108. 1939; *Potentilla congesta* (Douglas ex Hooker) Baillon var. *nemorosa* (D. D. Keck) J. T. Howell

Stems ± ascending, usually reddish, sometimes greenish, 1.5–3 dm, hairs 2 mm proximally. **Basal leaves** (3–)4–8 × 1–2.5 cm; leaflets 2–4 per side, lanceolate-elliptic to oblong, 5–12 × (1–) 3–5(–6) mm, (¼–)⅓–½ as wide as long. **Cauline leaves** 2–5; stipules incompletely divided or divided no more than ⅔ into lanceolate, often acuminate teeth. **Inflorescences** usually composed of ± capitate glomerules. **Flowers:** epicalyx bractlets linear to narrowly elliptic, 1–2(–3) mm; hypanthium ½ to nearly as deep as wide; petals 2.5–3(–4) × (1.5–)2–3 mm; styles 2–3 mm.

Flowering spring–summer. Serpentine flats, openings of conifer woodlands; of conservation concern; 200–800 m; Calif., Oreg.

Variety *nemorosa* occurs primarily in open ultramafic flats in the Illinois River Valley of southern Josephine County, Oregon; a single collection (*Van Deventer s.n.*, CAS) from Shelly Creek in Del Norte County, California, also belongs to this variety. Plants from the head of Babyfoot Creek in Curry County (*Leach 3006*, OSC) are referable to var. *nemorosa*, although transitional to *Horkelia tridentata* var. *flavescens*; a collection from southwest of Waldo in Josephine County (*Peck 8117*, WILLU) also combines features of the two taxa.

The specimen cited by D. D. Keck (1938[1939]) from Douglas County, Oregon, is here referred to *Horkelia tridentata* var. *tridentata*; specimens cited from Jackson County, Oregon, are all intermediate with var. *congesta*.

20. Horkelia tridentata Torrey in War Department [U.S.], Pacif. Railr. Rep. 4(5): 84, plate 6. 1857 [E]

Plants mostly rosette-forming, grayish. **Stems** decumbent to nearly erect, 0.5–4(–7) dm, hairs 1–3 mm proximally, glands absent or sparse distally. **Basal leaves** planar, 3–12 × 1–3(–4) cm, usually sericeous to villous, sometimes glabrate adaxially; stipules entire or rarely forked or pectinate; leaflets 2–5(or 6) per side, separate to slightly overlapping, elliptic to cuneate-oblong, (3–)5–20(–30) × 1.5–7(–10) mm, ($1/4$–)$1/3$–$1/2$ as wide as long, divided $1/10$–$1/6$ or less to midrib into (0–)3(or 4) triangular teeth 1 mm wide, these restricted to apex. **Cauline leaves** 1–5; stipules 3–12 mm, entire or deeply 3–5-lobed, lanceolate to ovate. **Inflorescences** open to congested, flowers arranged in ± capitate glomerules or in ± corymbiform clusters. **Pedicels** (1–)2–6(–10) mm. **Flowers** 4–10 mm diam.; epicalyx bractlets linear to narrowly elliptic, 0.8–2(–3) × 0.2–0.4 mm, $1/2$–$2/3$ length of sepals; hypanthium 1–2.5 × 2–5 mm, $1/2$ to nearly as deep as wide, interior glabrous or pilose; sepals moderately reflexed, abaxially green to reddish, 1.5–3 mm; petals white, sometimes tinged with pink, linear to broadly oblanceolate, (1–)1.5–4(–4.5) × 0.3–1.5 mm, apex acute to rounded or mucronate; filaments 0.5–2 × 0.1–0.4(–0.8) mm, anthers 0.2–0.5 mm; carpels 5–15; styles 1–2.5 mm. **Achenes** light brown, 1.5–2.5 mm, rugose.

Varieties 2 (2 in the flora): w United States.

Rydberg established *Horkelia flavescens* on the basis of its more congested inflorescence, purportedly yellowish petals, and broader filaments. As noted by D. D. Keck (1938[1939]), the petals are usually white, at least when fresh, and the other characters are highly variable; he reduced *H. flavescens* to a subspecies of *H. tridentata*, which is here treated at the rank of variety. In general, var. *flavescens* tends to be more decumbent with a single, loosely congested flower cluster and broader petals; these characters completely overlap and can vary independently. The presence or absence of hairs on the hypanthium interior is the single most reliably diagnostic character and is the primary determiner of the geographic distributions presented here. The exception is a cluster of populations from the extreme southern end of the coastal range of var. *flavescens* in Colusa, Lake, and Mendocino counties, California, which have glabrous hypanthia but otherwise have all the diagnostic characters of var. *flavescens*. Although provisionally included here within var. *flavescens*, these populations may be better treated as a separate variety.

1. Hypanthium interior glabrous; petals linear to oblanceolate, usually ± equal to or longer than sepals; inflorescences composed of 1+ ± capitate glomerules, these sometimes expanding with age . . . . 20a. *Horkelia tridentata* var. *tridentata*
1. Hypanthium interior pilose (except in some North Coast Ranges of California); petals ± broadly oblanceolate, often shorter than sepals; inflorescences composed of usually 1 ± corymbiform cluster. . . . . 20b. *Horkelia tridentata* var. *flavescens*

20a. Horkelia tridentata Torrey var. **tridentata** • Three-toothed or Sierra horkelia [F]

Potentilla tilingii (Regel) Greene

Stems decumbent to nearly erect, 1–4(–7) dm. **Inflorescences** composed of 1+ 5–40-flowered ± capitate glomerules, these sometimes expanding with age. **Flowers** 4–9 mm diam.; hypanthium 2–3.5(–5) mm diam., interior glabrous; petals linear to oblanceolate, (1.5–)2–3.5(–4.5) × 0.3–1 mm, usually ± equal to or longer than sepals, apex acute to obtuse; filaments 0.5–1(–2) × 0.1–0.2 mm, anthers 0.2–0.3(–0.4) mm; styles 1–2 mm. **Achenes** 1.5–2 mm, lightly rugose. $2n = 28$.

Flowering spring–summer. Open areas, primarily in sagebrush communities and conifer woodlands, mainly on granitic or volcanic soil; 300–2500 m; Calif., Oreg.

Variety *tridentata* occurs throughout the Cascade Mountains and Sierra Nevada of California, and barely extends into southern Oregon near Ashland, Josephine County, where the presence of distal cauline glands on some plants indicates possible introgression with *Horkelia congesta*. A specimen from Glendale, Oregon, (*Jones s.n.*, POM) that was annotated and cited by D. D. Keck (1938[1939]) as *H. congesta* subsp. *nemorosa* is fully comparable to the Ashland material of var. *tridentata*; the apparent disjunction in Douglas County needs to be confirmed.

20b. Horkelia tridentata Torrey var. **flavescens** (Rydberg) Ertter & Reveal, Novon 17: 320. 2007 • Hairycup horkelia [E]

Horkelia flavescens Rydberg, Monogr. N. Amer. Potentilleae, 138, plate 75. 1898; *H. tridentata* subsp. *flavescens* (Rydberg) D. D. Keck; *Potentilla tilingii* (Regel) Greene var. *flavescens* (Rydberg) J. T. Howell

Stems ± decumbent, 0.5–2.5(–4.5) dm. **Inflorescences** composed of usually 1 3–25-flowered ± corymbiform cluster. **Flowers** 6–10 mm diam.; hypanthium 2.5–5 mm diam., interior pilose (except in some North Coast Ranges populations); petals ± broadly oblanceolate, (1–)1.5–3(–4) × 0.5–1.5 mm, often shorter than sepals, apex obtuse to rounded or mucronate; filaments 1–2 × 0.2–0.4(–0.8) mm, anthers 0.3–0.5 mm; styles 1.5–2.5 mm. **Achenes** 2–2.5 mm, strongly rugose.

Flowering summer. Open areas, primarily in conifer woodlands, often on serpentine soil; 700–2000 m; Calif., Oreg.

Variety *flavescens* occurs primarily in the mountains of northwestern California, from Mendocino to Siskiyou counties. A single collection is confirmed from Curry County, Oregon (Bear Camp Trail, *Leach 4554*, OSC); other collections from Curry and Josephine counties may be hybrids between this variety and *Horkelia congesta* var. *nemorosa*. A second cluster of populations occurs in central Plumas and adjacent Lassen counties, California, and historic collections are known from near Truckee on the border of Nevada and Placer counties, California. D. D. Keck (1938[1939]) suggested that this bicentric distribution may have resulted from the independent coalescence of the diagnostic characteristics of var. *flavescens*. The locality of a collection purportedly from Washoe County, Nevada (80 miles north of Reno, *Brooks s.n.*, RENO) needs to be confirmed.

11. HORKELIELLA (Rydberg) Rydberg in N. L. Britton et al., N. Amer. Fl. 22: 282. 1908 • False horkelia [Genus *Horkelia* and Latin *-ella*, diminutive] [E]

Barbara Ertter

James L. Reveal

Horkelia Chamisso & Schlechtendal subg. *Horkeliella* Rydberg, Monogr. N. Amer. Potentilleae, 120. 1898; *Horkelia* [unranked] *Purpurascentes* Rydberg

Herbs, perennial, densely tufted to ± matted, <± aromatic>, 1.5–5(–9) dm, sparsely to densely villous and glandular-pubescent; short-rhizomatous. **Stems** 1–20+, ascending to erect, <green, stramineous, or reddish>. **Leaves** winter-marcescent, primarily basal, <cauline 2–7, reduced>, alternate, odd-pinnate; stipules persistent, basally adnate to petiole, linear to lanceolate or elliptic, margins entire, cauline sometimes lobed; petiole present; blade narrowly linear in outline, ± cylindric, 4–20(–35) cm, foliaceous, leaflets 30–70, <separate proximally, otherwise overlapping, sometimes imbricate, terminal confluent with distalmost lateral ones>, broadly obovate to reniform in outline, margins flat, palmately divided nearly to base into linear to oblanceolate lobes, venation palmate. **Inflorescences** terminal, 5–20-flowered, narrow cymes, ± open or ± congested; bracts present, reduced; bracteoles absent. **Pedicels** present, <straight>. **Flowers** 10–15 mm diam.; epicalyx bractlets 5; hypanthium bluntly campanulate, flat-bottomed, 2–3 × 4–6 mm; sepals 5, reflexed, <adaxially purple-streaked>, lanceolate to narrowly deltate; petals 5, white, sometimes pink-tinged, narrowly oblong to obovate or spatulate; stamens 20, shorter than petals, <filaments erect, ± flattened, forming tube, anther thecae dehiscing introrsely>; torus elongate; carpels 25–50, glabrous, styles subterminal, <slightly rough-thickened basally, otherwise mostly filiform>; ovule 1. **Fruits** aggregated achenes, individually deciduous, 25–50 or less, obliquely ovoid to reniform, 1–1.5 mm, glabrous; hypanthium persistent; sepals persistent, erect; styles tardily deciduous, <jointed>.

Species 2 (2 in the flora): California.

Rydberg established *Horkeliella* as a subgenus of *Horkelia*, at the same time treating *Ivesia* as another subgenus. The species resemble *Horkelia* in having flat-bottomed, campanulate hypanthia, reflexed, purple-streaked sepals, flattened, erect filaments that form a column around the styles, and inwardly dehiscing anthers. This floral morphology is evidently an adaptation to bee pollination. Unlike *Horkelia*, *Horkeliella* has 20 stamens, and the cylindric leaves with finely divided leaflets are most similar to those of *Ivesia* species, especially sect. *Comarella*. D. D. Keck (1938[1939]) accordingly decided that the taxa in question were better accommodated in *Ivesia* and also reduced the three species of Rydberg to two subspecies of a single species. Subsequent regional floras (for example, P. A. Munz 1959) followed this basic arrangement, although L. Abrams (1944) placed the bitypic species in *Horkelia* rather than *Ivesia*. In contrast, worldwide generic compilations (for example, J. Hutchinson 1964–1967) considered the treatment by Rydberg in 1908 of *Horkeliella* as a genus with three species, even while accepting the numerical circumscription of *Ivesia* by Keck.

Horkeliella was resurrected for regional floras by B. Ertter (1993b), comprising two species endemic to the southern Sierra Nevada of California. The genus may represent the transitional stage in the evolution of *Horkelia* from the ivesioid lineage.

1. Filaments opposite center of sepals longer than those on either side; herbage with non-glandular hairs sparse, not obscuring abundant stipitate-glandular ones; leaflets divided into 5–10 ultimate segments; e of Sierra Nevada crest, Inyo and Mono counties . . . . 1. *Horkeliella congdonis*
1. Filaments opposite center of sepals shorter than those on either side; herbage with non-glandular hairs sparse to dense, often obscuring stipitate-glandular ones; leaflets divided into 3–6 ultimate segments; w of Sierra Nevada crest, Kern and Tulare counties . . . 2. *Horkeliella purpurascens*

1. Horkeliella congdonis (Rydberg) Rydberg in N. L. Britton et al., N. Amer. Fl. 22: 283. 1908 • Congdon's horkeliella or false horkelia [E]

Horkelia congdonis Rydberg, Bull. Torrey Bot. Club 26: 543. 1899; *H. purpurascens* S. Watson subsp. *congdonis* (Rydberg) Abrams; *Ivesia purpurascens* (S. Watson) D. D. Keck subsp. *congdonis* (Rydberg) D. D. Keck; *Potentilla purpurascens* (S. Watson) Greene var. *congdonis* (Rydberg) Jepson

Plants green, non-glandular hairs sparse, not obscuring abundant stipitate-glandular ones. **Basal leaves** (5–)8–25(–27) cm; petiole 4–10 cm; leaflets 25–35 per side, 2–6 mm, divided into 5–10 ultimate segments, sparsely villous, more densely so marginally, minutely orangish sessile-glandular. **Cauline leaves** 2–7, 2–10 cm. **Inflorescences** ± congested. **Pedicels** usually 2 mm (proximalmost to 20 mm). **Flowers** 15 mm diam.; epicalyx bractlets lanceolate, 2.5–5(–6) mm; sepals 4–8(–9) mm, villous, short-stipitate-glandular; petals narrowly oblong-oblanceolate, 3–6(–6.5) mm; filaments 1.5–3 mm, filament opposite center of sepal longer than those on either side, anthers 1 mm; styles 2–5 mm.

Flowering late summer. Dry granitic meadows, in sagebrush communities, montane to subalpine conifer woodlands; 1500–3100 m; Calif.

Horkeliella congdonis is known from the eastern foothills and slopes of the Sierra Nevada northeast to the Sweetwater Mountains and the Glass Mountain area of Mono and northern Inyo counties. It generally forms dense clumps that are notably more glandular than *H. purpurascens*.

2. Horkeliella purpurascens (S. Watson) Rydberg in N. L. Britton et al., N. Amer. Fl. 22: 282. 1908 • Kern Plateau horkeliella or false horkelia [E] [F]

Horkelia purpurascens S. Watson, Proc. Amer. Acad. Arts 11: 148. 1876; *Ivesia purpurascens* (S. Watson) D. D. Keck; *Potentilla purpurascens* (S. Watson) Greene

Plants green or grayish green to gray, non-glandular hairs sparse to dense, often obscuring stipitate-glandular ones. **Basal leaves** (5–)7–17(–20) cm; petiole 1–5 cm; leaflets 15–30(–35) per side, (2–)3–10(–20) mm, divided into 3–6 ultimate segments, hirsute (sometimes densely so) and villous (especially marginally), infrequently viscid-stipitate-glandular, or nearly glabrous except for villous tuft at apex. **Cauline leaves** 2–4, 2–10 cm. **Inflorescences** ± open. **Pedicels** (1–)2–10(–15) mm. **Flowers** 10–15 mm diam., epicalyx bractlets linear to narrowly ovate, 1.5–3 mm; sepals 3–6 mm, hirsute, sparsely villous, sometimes minutely glandular or stipitate-glandular; petals oblanceolate to narrowly spatulate or oblong, 3–7 mm; filaments 1–2.5 mm, filament opposite center of sepal shorter than those on either side, anthers 0.5–0.8(–1) mm; styles 2.5–4 mm.

Flowering summer. Dry granitic meadows at edges of conifer woodlands; 1400–2900 m; Calif.

Typical *Horkeliella purpurascens* occurs on the Kern Plateau on the western slope of the southern Sierra Nevada in Kern and Tulare counties. More densely hairy populations farther south in the Piute, Scodie, and Tehachapi mountains of Kern County tend to have more congested inflorescences, shorter leaves, and fewer carpels; they may merit taxonomic recognition.

12. DUCHESNEA Smith, Trans. Linn. Soc. London 10: 372. 1811 • Mock- or Indian-strawberry [For Antoine Nicolas Duchesne, 1747–1827, French botanist and monographer of *Fragaria*] [I]

Barbara Ertter

James L. Reveal

Potentilla Linnaeus sect. *Duchesnea* (Smith) Dikshit & Panigrahi; *Potentilla* subg. *Duchesnea* (Smith) M. Shah & Wilcock

Herbs, perennial, ± matted, 0.5–3 dm, sparsely hairy; stoloniferous. **Stems:** number not recorded, prostrate or spreading to weakly erect, <stolon nodes bearing single leaves or condensed shoots, consisting of paired (1–)3-lobed green bractlets and 1+ leaves, developing into new plantlets, adventitious roots, and peduncles>. **Leaves** often winter-persistent, basal or cauline, alternate, ternate; stipules persistent, proximally adnate to base of petiole, sheathing, broadly lanceolate to narrowly ovate, margins entire; petiole present; blade ± cordate in outline, (1–)2–7 cm, foliaceous, leaflets 3, elliptic to obovate, margins flat, ± doubly serrate or crenate, venation pinnate, surfaces sparsely pilose or strigose (at least abaxially). **Inflorescences** axillary <at stolon nodes>, flowers solitary; bracts absent; bracteoles absent. **Pedicels** present, <straight>. **Flowers** [5–]15–20 mm diam.; epicalyx bractlets 5; hypanthium cupulate, 1–3 × 2.5–5 mm; sepals 5, spreading to erect, broadly lanceolate to ovate; petals 5, yellow, narrowly obcordate to obovate, <± equal to or shorter than sepals>; stamens 20, 25, or 30, shorter than petals, <filaments filiform, glabrous, thecae paired on sides of narrow connective, dehiscing by broad lateral slits>; torus turbinate to hemispheric; carpels 50–100, glabrous, styles subterminal, <filiform>; ovule 1. **Fruits** accessory; aggregated achenes, 50–100 or less, <borne on enlarged torus, superficial on flesh>, obliquely ovoid, [8–]10–15(–20) mm, glabrous; <torus reddish, compressed ovoid, [0.8–]1–1.5 cm, fleshy (not pulpy), apex rounded, glabrous>; hypanthium persistent; sepals persistent, spreading; styles tardily deciduous, <jointed>. $x = 7$.

Species 2 (1 in the flora): introduced; Asia; introduced also in South America, Europe, Africa.

Duchesnea is superficially similar to *Fragaria* in its fleshy red fruit and ternate leaves but differs in having yellow petals, doubly toothed leaflets, and protruding achenes, among other characteristics. The similarities result from convergence; phylogenetically, *Duchesnea* nests within *Potentilla* near sect. *Potentilla*. It is retained here as a distinct genus because of characteristics that would be anomalous in *Potentilla*, that is, condensed shoot structure at stolon nodes, enlarged 3-lobed epicalyx bractlets, and accessory fruit structure.

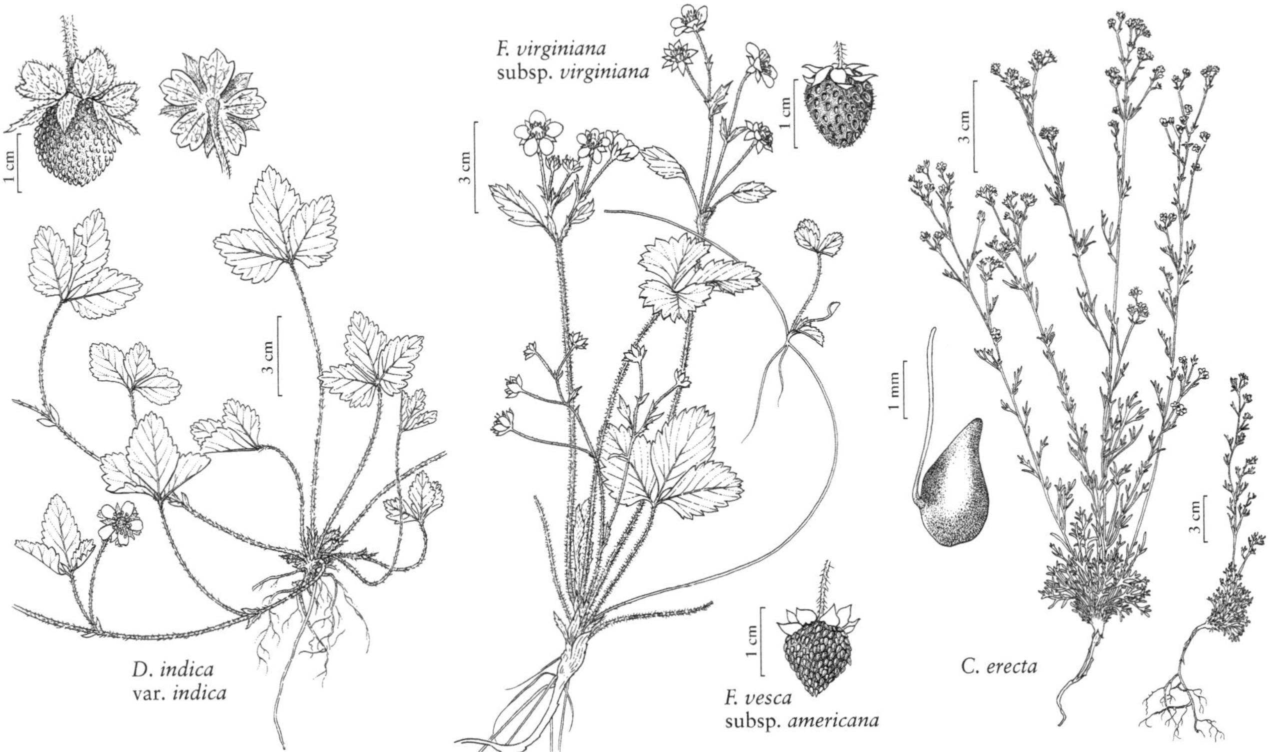

DUCHESNEA ◦ FRAGARIA ◦ CHAMAERHODOS

Our infrageneric taxonomy follows Li C. L. et al. (2003b), in which the stable diploid ($2n = 14$) *Duchesnea chrysantha* (Zollinger & Moritzi) Miquel is recognized as a distinct species in eastern Asia. In contrast, the most common phase of *D. indica*, which is the one naturalized in North America, appears to be a dodecaploid ($2n = 84$), although various other numbers have been reported ($2n = 14, 21, 28, 49, 56$), mainly from Europe. Morphological differences include achenes tuberculate in *D. chrysantha* versus smooth in *D. indica.*

SELECTED REFERENCE Hara, H. and S. Kurosawa. 1959. On the *Duchesnea indica* group. J. Jap. Bot. 34: 161–166.

1. Duchesnea indica (Andrews) Teschemacher, Hort. Reg. & Gard. Mag. 1: 460. 1835 (as Duchesnia) [F] [I]

Fragaria indica Andrews, Bot. Repos. 7: plate 479. 1807; *Potentilla indica* (Andrews) Th. Wolf

Varieties 2 (1 in the flora): introduced; Asia; introduced also in South America, Europe, Africa.

Authorship of *Fragaria indica* is sometimes ascribed to Jackson in Andrews, because it is likely that Jackson prepared the text that accompanied illustrations by Andrews (F. A. Stafleu and R. S. Cowan 1976–1988, vol. 1). Because this is not indicated in the work itself, authorship is here retained as Andrews.

Variety *microphylla* T. T. Yu & T. C. Ku of Tibet has smaller, densely villous leaves.

1a. Duchesnea indica (Andrews) Teschemacher var. **indica** • Indian-strawberry [F] [I]

Stolons (1–)3–10(–13) dm. **Leaves:** stipules 5–8 mm, sparsely villous; petiole 2–20 cm, pilose; leaflets petiolulate, (1–)3–5(–7) × (0.8–)1–3 cm, abaxial surface sparsely strigose, adaxial pilose or glabrous. **Petioles** 2–10(–13) cm. **Flowers:** epicalyx bractlets spreading to slightly reflexed, obovate, 4–12 × 4–10 mm, 3(–5)-toothed apically, foliaceous, sparsely villous at least on teeth and veins; sepals (3–)4–10 mm, margins entire; petals 4–8(–11) × 3–6(–8) mm, apex rounded to truncate or emarginate; styles 1–1.5 mm. **Achenes** red, 1–1.5 mm, smooth, glossy. **$2n$** = 84.

Flowering spring–fall. Roadsides, cultivated fields, lawns, gardens, waste places, disturbed areas, occasionally in moist places; 0–700[–3200] m; introduced; B.C., Ont.; Ala., Ariz., Ark., Calif., Conn., Del., D.C., Fla., Ga., Ill., Ind., Iowa, Kans., Ky., La., Md., Mich., Miss., Mo., Nebr., N.J., N.Y., N.C., Ohio, Okla., Oreg., Pa., S.C., Tenn., Tex., Va., Wash., W.Va., Wis.; Asia; introduced also in South America, Europe, Africa.

13. FRAGARIA Linnaeus, Sp. Pl. 1: 494. 1753; Gen. Pl. ed. 5, 218. 1754 • Strawberry, fraisier [Latin *fraga*, fragrance, and *aria*, possession, alluding to sweet-smelling strawberry fruit]

Günter Staudt†

Herbs, perennial, acaulescent, 0.2–2.8 dm; short-rhizomatous, stoloniferous. **Stems** 1–10 (crowns), procumbent, <sympodially branched, stolons not leafy, rooting and producing shoots at every 2d node>, glabrate to densely spreading-hairy or appressed-ascending hairy. **Leaves** deciduous, basal, alternate, ternate; stipules persistent, basally adnate to petiole, lanceolate, margins entire; petiole present; blade shallowly deltate to widely ovate or widely depressed ovate, 1–6 cm, thin to thick, sometimes leathery, leaflets 3, usually ovate to obovate, sometimes rhombic to roundish, margins flat, serrate to crenate, venation pinnate, abaxial surface sericeous (at least along veins), adaxial mostly glabrate. **Inflorescences** axillary <from leaf rosette>, 1–10-flowered, cymes, glabrate to densely spreading-hairy or appressed-ascending-hairy; bracts usually present; bracteoles present, caducous. **Pedicels** present. **Flowers** bisexual or unisexual (plants synoecious, gynodioecious, dioecious, or trioecious), 12–29.8 mm diam.; epicalyx bractlets 5; hypanthium saucer-shaped, 9.5–27.8 mm diam., hairy; sepals 5, spreading, broadly lanceolate; petals 5[–9], usually white, rarely pink, orbiculate to ovate or obovate, <base clawed>; <sterile staminodia in pistillate flowers>; stamens 15 in 3 whorls, shorter than petals; torus globose to conic; carpels 30–150, glabrous, styles basal; ovule 1. **Fruits** accessory; aggregated achenes, 30 (or less)–150, <borne on enlarged torus, superficial or embedded in torus flesh>, ovate or ± ovoid, 1–2 mm, glabrous; <torus usually red, wine red, or carmine, rarely white, globose or subglobose to oblong, conic, or oblate, 9–24(–37) mm, fleshy, apex sometimes pointed>; hypanthium persistent; sepals persistent, spreading, reflexed, or clasping tori; styles deciduous. $x = 7$.

Species 24 (3 in the flora): North America, Mexico, South America, Eurasia, Pacific Islands (Hawaii); introduced nearly worldwide.

The marginal teeth on terminal leaflets vary in size and number depending on length of the terminal leaflet in the species of *Fragaria*. The following ratios are used in the keys and descriptions: relative size of teeth (length in mm of terminal leaflet to number of teeth per side) and relative number of teeth (number of teeth per side to length in mm of terminal leaflet). The length to width ratio of leaflets is also given.

Fragaria belongs to subtribe Fragariinae (D. Potter et al. 2007), which is generally characterized by stamens with a single horseshoe-shaped theca (T. Eriksson et al. 2003), and includes, in the flora area, *Alchemilla*, *Aphanes*, *Chamaerhodos*, *Comarum*, *Dasiphora*, *Drymocallis*, *Sibbaldia*, and *Sibbaldiopsis*.

SELECTED REFERENCE Staudt, G. 1999. Systematics and geographic distribution of the American strawberry species. Taxonomic studies in the genus *Fragaria* (Rosaceae: Potentilleae). Univ. Calif. Publ. Bot. 81.

1. Leaves thick, leathery, strongly reticulately veined abaxially; achenes 1.4–2 mm, reddish brown to dark brown; coastal w North America . . . . 3. *Fragaria chiloensis*
1. Leaves thin, sometimes slightly leathery, not reticulately veined abaxially; achenes 1–1.8 mm, yellowish green to reddish brown; mostly inland North America, sometimes coastal.
    2. Leaves thin, bright green (sometimes darker green in subsp. *californica*), not glaucous; terminal tooth of terminal leaflet usually longer than or equal to adjacent teeth (sometimes equal to or shorter than adjacent teeth in subsp. *californica*); achenes 1–1.5 mm; plants hermaphroditic (sometimes pistillate in subsp. *bracteata*) . . . . 1. *Fragaria vesca*
    2. Leaves thin but ± stout (slightly thicker than *F. vesca*), sometimes slightly leathery, green to bluish green, sometimes bright green, sometimes glaucous; terminal tooth of terminal leaflet usually shorter than adjacent teeth; achenes 1.2–1.8 mm; plants hermaphroditic or unisexual . . . . 2. *Fragaria virginiana*

1. Fragaria vesca Linnaeus, Sp. Pl. 1: 494. 1753 • Wood strawberry [F]

Plants hermaphroditic (sometimes pistillate in subsp. *bracteata*). **Leaves** bright green (sometimes darker green in subsp. *californica*), not glaucous, thin, not leathery, not reticulately veined abaxially, terminal tooth of terminal leaflet longer than or equal to adjacent teeth (sometimes equal to or shorter than adjacent teeth in subsp. *californica*). **Flowers** bisexual or unisexual; hypanthium 11–21.2 mm diam.; petals 5, ± obovate to nearly orbiculate, margins overlapping or distinct. **Achenes** deeply embedded, in shallow pits, or superficial, yellowish green to reddish brown, 1–1.5 mm; bractlets and sepals clasping, spreading, or reflexed in fruit; torus easily or not easily separating from hypanthium.

Subspecies 4 (4 in the flora): North America, Mexico, Eurasia; introduced in West Indies, South America, Pacific Islands.

1. Fruiting tori easily separating from hypanthia, tori globose to ± oblong; achenes superficial or in shallow pits; leaves green to dark green, terminal leaflet blades ± round or slightly ovate-rhombic, length/width 0.9–1.3, teeth: relative number 0.4–0.9, relative size 1–2.3; bractlets and sepals spreading or reflexed in fruit . . . . 1d. *Fragaria vesca* subsp. *californica*
1. Fruiting tori easily or not easily separating from hypanthia, tori globose, subglobose, oblong-ovoid, or elongate-conic (slightly oblate or conic); achenes superficial, in shallow pits, or deeply embedded; leaves green to bright green, terminal leaflet blades ovate or obovate to slightly rhombic, length/width 1.1–1.8, teeth: relative number 0.2–0.6, relative size 1.5–4; bractlets and sepals clasping, spreading, or reflexed in fruit.

[2. Shifted to left margin.—Ed.]

2. Flowers usually bisexual, sometimes pistillate, 15.3–23.8 mm diam.; petals usually white, sometimes pinkish; bractlets clasping or spreading to slightly reflexed, sepals clasping in fruit; achenes in shallow pits or deeply embedded; tori usually not easily separating from hypanthia . . . . 1c. *Fragaria vesca* subsp. *bracteata*
2. Flowers bisexual, 12–20.5 mm diam.; petals white; bractlets and sepals spreading to reflexed in fruit; achenes superficial or in shallow pits; tori easily separating from hypanthia.
    3. Fruiting tori globose or subglobose (often slightly oblate or conic); leaves bright green; achenes superficial or in shallow pits . . . . 1a. *Fragaria vesca* subsp. *vesca*
    3. Fruiting tori elongate-conic (often somewhat pointed); leaves green to bright green; achenes superficial . . . . . 1b. *Fragaria vesca* subsp. *americana*

1a. Fragaria vesca Linnaeus subsp. **vesca** • Fraisier des bois [I]

Leaves bright green; terminal leaflet blade ovate or obovate to slightly rhombic, length/width 1.1–1.6, teeth: relative number 0.2–0.5, relative size 2–3.9, terminal tooth longer than adjacent teeth. **Flowers** bisexual, 13.2–20.5 mm diam.; hypanthium 11.2–17 mm diam.; petals white, <widely depressed>, obovate to nearly orbiculate, margins overlapping or distinct, sometimes distally crenate. **Achenes** superficial or in shallow pits; bractlets and sepals spreading to reflexed; torus carmine to wine red, shiny, globose or subglobose (often slightly oblate or conic), easily separating from hypanthium. $2n = 14$.

Flowering spring. Open forests, edges of woods, thickets, meadows; 50–1000 m; introduced; Nfld. and Labr. (Nfld.), Ont., Que.; Conn., Maine, Mass., N.Y., Ohio, Pa., Vt., Va., W.Va.; Eurasia; introduced also in Mexico, West Indies, South America, Pacific Islands (Hawaii).

Subspecies *vesca* is occasionally naturalized in North America.

1b. Fragaria vesca Linnaeus subsp. **americana** (Porter) Staudt, Canad. J. Bot. 40: 872. 1962 • Fraisier américain [E] [F]

Fragaria vesca var. *americana* Porter, Bull. Torrey Bot. Club 17: 15. 1890; *F. americana* (Porter) Britton

Leaves green to bright green; terminal leaflet blade ovate or obovate to slightly rhombic, length/width 1.3–1.6, teeth: relative number 0.2–0.6, relative size 1.5–4, terminal tooth longer than adjacent teeth. **Flowers** bisexual, 12–14.6 mm diam.; hypanthium 11–16.5 mm diam.; petals white, obovate to widely obovate, margins distinct, sometimes distally crenate. **Achenes** superficial; bractlets and sepals spreading to strongly reflexed; torus carmine, shiny, elongate-conic (often somewhat pointed), easily separating from hypanthium. **2*n*** = 14.

Flowering spring. Open forests, forest edges, clearings, along trails and roadsides, open grasslands, often on steep, rocky slopes and outcrops; 0–1900 m; Alta., B.C., Man., N.B., N.W.T., N.S., Ont., Que., Sask.; Conn., Ill., Iowa, Maine, Mass., Mich., Minn., Miss., Nebr., N.H., N.Y., N.C., N.Dak., Ohio, Pa., S.Dak., Vt., Va., W.Va., Wis.

It is assumed that after the maximum glaciation, this North American subspecies spread from an eastern refugium northeastward to the Gaspé and northwestward to the Mackenzie and Peace rivers (G. Staudt 1999).

1c. Fragaria vesca Linnaeus subsp. **bracteata** (A. Heller) Staudt, Canad. J. Bot. 40: 872. 1962

Fragaria bracteata A. Heller, Bull. Torrey Bot. Club 25: 194, plate 339. 1898; *F. crinita* Rydberg; *F. helleri* Holzinger; *F. vesca* var. *bracteata* (A. Heller) R. J. Davis; *F. vesca* var. *crinita* (Rydberg) C. L. Hitchcock

Leaves green; terminal leaflet blade obovate to slightly rhombic, length/width 1.2–1.8, teeth: relative number 0.2–0.6, relative size 1.5–3.7, terminal tooth longer than or equal to adjacent teeth. **Flowers** usually bisexual, sometimes pistillate (gynoecious plants), 15.3–23.8 mm diam.; hypanthium 11.8–21.2 mm diam.; petals usually white, sometimes pinkish, widely obovate to widely depressed-obovate, margins usually overlapping, sometimes distally crenate. **Achenes** in shallow pits or deeply embedded; bractlets clasping or spreading to slightly reflexed, sepals clasping; torus wine red, not shiny (skin slightly hirsute), usually oblong-ovoid, sometimes globose, usually not easily separating from hypanthium. **2*n*** = 14.

Flowering spring. Forest edges, open forests, meadows, stream banks, roadsides, usually shady and moderately damp; 50–3800 m; Alta., B.C.; Ariz., Calif., Colo., Idaho, Mont., N.Mex., Oreg., Tex., Utah, Wash., Wyo.; Mexico (Chiapas, Chihuahua, Coahuila, Distrito Federal, Durango, Michoacán, San Luis Potosí, Veracruz, Zacatecas).

Plants of subsp. *bracteata* with pinkish petals are relatively rare and have been called forma *helleri* (Holzinger) B. Boivin.

1d. Fragaria vesca Linnaeus subsp. **californica** (Chamisso & Schlechtendal) Staudt, Canad. J. Bot. 40: 872. 1962

Fragaria californica Chamisso & Schlechtendal, Linnaea 2: 20. 1827

Leaves green to dark green; terminal leaflet blade ± round or slightly ovate-rhombic, length/width 0.9–1.3, teeth: relative number 0.4–0.9, relative size 1–2.3, terminal tooth equal to or shorter than adjacent teeth. **Flowers** bisexual, 17–22 mm diam.; hypanthium 11–18.1 mm diam.; petals white, widely obovate to nearly orbiculate, margins distinct, sometimes crenate. **Achenes** superficial or in shallow pits; bractlets and sepals spreading or reflexed; torus carmine or red, shiny, globose to ± oblong, easily separating from hypanthium. **2*n*** = 14.

Flowering spring. Coastal strand, grassy coastal areas, coniferous forests; 0–100 m; Calif., Oreg.; Mexico (Baja California, Baja California Sur).

Whether subsp. *bracteata* has introgressed into subsp. *californica* not only in northern California and Oregon, but also in southern California, should be investigated.

2. Fragaria virginiana Miller, Gard. Dict. ed. 8, Fragaria no. 2. 1768 [E] [F]

Plants hermaphroditic or unisexual. **Leaves** usually green to bluish green, sometimes bright green, sometimes glaucous, thin but ± stout (slightly thicker than *F. vesca*), sometimes slightly leathery, not reticulately veined abaxially, terminal tooth of terminal leaflet usually shorter (often narrower) than adjacent teeth. **Flowers** bisexual or unisexual (plants dioecious, gynodioecious, or trioecious); hypanthium 9.5–27 mm diam.; petals 5, obovate to widely obovate, margins overlapping or distinct. **Achenes** usually deeply embedded, rarely in pits or ± superficial, yellowish green to reddish brown, 1.2–1.8 mm; bractlets clasping, spreading, or ± reflexed, sepals clasping; torus usually not easily separate from hypanthium.

Subspecies 4 (4 in the flora): North America.

It seems inappropriate to the author to follow the proposal of K. E. Hokanson et al. (2006) to reclassify the subspecies of *Fragaria virginiana* to forms.

1. Stolons, petioles, peduncles, and pedicels usually appressed ascending-hairy, sometimes almost glabrous . . . . 2b. *Fragaria virginiana* subsp. *glauca*
1. Petioles spreading-hairy; stolons, peduncles, and pedicels usually spreading- or appressed ascending-hairy, sometimes glabrous.
   2. Stolons and peduncles spreading- or appressed-ascending-hairy, proximally often spreading-hairy, distally ascending-hairy or glabrous; pedicels appressed-hairy or glabrous . . . . .2a. *Fragaria virginiana* subsp. *virginiana*
   2. Stolons, peduncles, and pedicels spreading-hairy throughout.
      3. Leaflet blades green to bright green, not glaucous, terminal leaflets slightly rhombic to broadly elliptic, margins sharply serrate throughout; bractlets slightly clasping or spreading to ± reflexed; e, sc United States . . . . 2c. *Fragaria virginiana* subsp. *grayana*
      3. Leaflet blades usually bluish green, slightly glaucous, sometimes dull, terminal leaflets ovate-obovate to cuneate or truncate, margins serrate or crenate only in distal ⅓; bractlets clasping or slightly spreading; w North America . . . . .2d. *Fragaria virginiana* subsp. *platypetala*

2a. Fragaria virginiana Miller subsp. **virginiana** • Scarlet or Virginia strawberry, fraisier des champs [E] [F]

Fragaria australis (Rydberg) Rydberg; *F. virginiana* var. *australis* Rydberg; *F. virginiana* var. *canadensis* Farwell

Stolons spreading-hairy proximally, sometimes ascending-hairy or glabrous distally. **Leaves:** petiole spreading-hairy; leaflet blade usually dark green to slightly bluish green, sometimes bright green, not glaucous, terminal leaflets oblong-ovate or slightly rhombic, not leathery, margins sharply serrate in distal ½ to throughout, teeth: relative number 0.2–0.8. **Peduncles** usually ± spreading-hairy proximally, appressed ascending-hairy or glabrous distally. **Pedicels** appressed-hairy or glabrous. **Flowers** 12.1–19.5 mm diam. (pistillate), 15.2–23.6 mm diam. (bisexual or staminate). **Fruits:** bractlets clasping, spreading, or ± reflexed. $2n = 56$.

Flowering spring. Moist to dry sites, open forests, forest edges, hedges, fields, roadsides, railroad embankments, abandoned fields, often ruderal; 0–1000 m; St. Pierre and Miquelon; Man., Nfld. and Labr. (Nfld.), N.S., Ont., Que.; Ala., Ark., Conn., Del., D.C., Fla., Ga., Ill., Ind., Iowa, Kans., Ky., La., Maine, Md., Mass., Mich., Minn., Miss., Mo., Nebr., N.H., N.J., N.Y., N.C., Ohio, Okla., Pa., R.I., S.C., S.Dak., Tenn., Tex., Vt., Va., W.Va., Wis.

2b. Fragaria virginiana Miller subsp. **glauca** (S. Watson) Staudt, Canad. J. Bot. 40: 881. 1962 • Fraisier glauque [E]

Fragaria virginiana var. *glauca* S. Watson, Botany (Fortieth Parallel), 85. 1871; *F. glauca* (S. Watson) Rydberg; *F. multicipita* Fernald; *F. terrae-novae* Rydberg; *F. virginiana* var. *terrae-novae* (Rydberg) Fernald & Wiegand

Stolons usually appressed ascending-hairy, sometimes almost glabrous. **Leaves:** petiole usually appressed-ascending-hairy, sometimes almost glabrous; leaflet blade dark green to bluish green, glaucous, terminal leaflets usually oblong-ovate to cuneate, sometimes roundish, not leathery, margins sharply serrate throughout, teeth: relative number 0.1–0.5. **Peduncles and pedicels** usually appressed ascending-hairy, sometimes almost glabrous. **Flowers** 11.5–17.7 mm diam. (pistillate), 16–25.5 mm diam. (bisexual and staminate). **Fruits:** bractlets clasping, spreading, or ± reflexed. $2n = 56$.

Flowering spring and fall. Moist to dry sites, open forests, forest edges, hedges, fields, roadsides, railroad embankments, often ruderal; 0–3400 m; Alta., B.C., Man., N.B., Nfld. and Labr., N.W.T., N.S., Nunavut, Ont., P.E.I., Que., Sask., Yukon; Alaska, Ariz., Calif., Colo., Idaho, Iowa, Maine, Mass., Minn., Mont., N.H., N.Mex., N.Y., N.Dak., R.I., S.Dak., Utah, Vt., Wis., Wyo.

It is assumed that after the maximum glaciation, recolonization of most glaciated areas by subsp. *glauca* spread from the Alaska-Yukon refugium eastward to Hudson Bay, Newfoundland, and the New England states and, perhaps from a putative Rocky Mountains refuge northward to the Yukon and eastward to the Great Lakes area (G. Staudt 1999).

2c. Fragaria virginiana Miller subsp. **grayana** (E. Vilmorin ex J. Gay) Staudt, Canad. J. Bot. 40: 881. 1962 [E]

Fragaria grayana E. Vilmorin ex J. Gay, Ann. Sci. Nat., Bot., sér. 4, 8: 202. 1857; *F. virginiana* var. *illinoensis* A. Gray

Stolons spreading-hairy. **Leaves:** petiole spreading-hairy (usually densely so); leaflet blade green to bright green, not glaucous, terminal leaflets slightly rhombic to broadly elliptic, not leathery, margins sharply serrate throughout, teeth: relative number 0.2–0.5. **Peduncles and pedicels** spreading-hairy (usually densely so). **Flowers** 14.8–25.5 mm diam. (pistillate), 19–29.8 mm diam. (bisexual and staminate). **Fruits:** bractlets slightly clasping or spreading to ± reflexed. **2*n*** = 56.

Flowering spring. Forest openings, forest edges, meadows; 200–1000 m; Ala., Ark., Ill., Ind., Kans., Ky., La., Miss., Mo., Nebr., N.Y., Ohio, Okla., Pa., Tenn., Tex., Va., Wis.

2d. Fragaria virginiana Miller subsp. **platypetala** (Rydberg) Staudt, Canad. J. Bot. 40: 881. 1962 [E]

Fragaria platypetala Rydberg, Monogr. N. Amer. Potentilleae, 177. 1898; *F. sibbaldifolia* Rydberg; *F. suksdorfii* Rydberg; *F. truncata* Rydberg; *F. virginiana* var. *platypetala* (Rydberg) H. M. Hall

Stolons usually densely spreading-hairy. **Leaves:** petiole usually densely spreading-hairy; leaflet blade usually bluish green, slightly glaucous, sometimes dull, terminal leaflets ovate-obovate to cuneate or truncate, sometimes slightly leathery, margins serrate or crenate only in distal ⅓, teeth: relative number 0.1–0.4. **Peduncles and pedicels** densely spreading-hairy. **Flowers** 15–18.6 mm diam. (pistillate), 16.3–26 mm diam. (bisexual and staminate). **Fruits:** bractlets clasping or slightly spreading. **2*n*** = 56.

Flowering spring. Moist to dry sites, prairies, savannas, open forests, roadsides, abandoned fields; 0–3700 m; Alta., B.C., Yukon; Calif., Colo., Idaho, Mont., Oreg., Utah, Wash., Wyo.

Plants of subsp. *platypetala* that are hypothesized to have resulted from a former introgression of subsp. *glauca* into *Fragaria chiloensis* on the Pacific coast have been grouped under subsp. *platypetala*. Plants growing away from coastal areas still show characteristics of *F. chiloensis*, such as having all stems densely spreading-hairy and relatively large hypanthia spreading or clasping the fruits.

3. Fragaria chiloensis (Linnaeus) Miller, Gard. Dict. ed. 8, Fragaria no. 4. 1768 • Chilean or beach or sand strawberry

Fragaria vesca Linnaeus var. *chiloensis* Linnaeus, Sp. Pl. 1: 495. 1753

Plants hermaphroditic or unisexual. **Leaves** dark or dull green, not glaucous, thick, leathery, strongly reticulately veined abaxially, terminal tooth of terminal leaflet usually shorter than adjacent teeth, <abaxial surface densely silky, adaxial glabrous, shiny>. **Flowers** bisexual, pistillate or staminate (plants dioecious or trioecious); hypanthium 14.5–27.8 mm diam.; petals 5(or 6), obovate to widely depressed obovate, margins distinct or overlapping. **Achenes** in shallow pits or partially embedded, reddish brown to dark brown, 1.4–2 mm; bractlets and sepals clasping; torus not easily separated from hypanthium.

Subspecies 4 (2 in the flora): w North America, South America, Pacific Islands (Hawaii).

Fragaria chiloensis was first described from cultivated plants that had been selected in Chile at least 400 years ago and brought to Europe in 1714. Subsequently, it was realized that wild-growing plants of this species exist in North America and South America. The typical subspecies is restricted to South America; subspp. *lucida* and *pacifica* are present in North America, and the latter in Hawaii as well.

The proposal by K. E. Hokanson et al. (2006) to classify subsp. *lucida* and subsp. *pacifica* as forms of one North American subspecies seems inappropriate.

1. Stolons, petioles, peduncles, and pedicels usually appressed ascending-hairy, sometimes almost glabrous . . . . 3a. *Fragaria chiloensis* subsp. *lucida*
1. Stolons, petioles, peduncles, and pedicels spreading-hairy (usually densely so) . . . . 3b. *Fragaria chiloensis* subsp. *pacifica*

3a. Fragaria chiloensis (Linnaeus) Miller subsp. **lucida** (E. Vilmorin ex J. Gay) Staudt, Canad. J. Bot. 40: 883. 1962 [E]

Fragaria lucida E. Vilmorin ex J. Gay, Ann. Sci. Nat., Bot., sér. 4, 8: 201. 1857

Stolons, petioles, peduncles, and pedicels usually appressed ascending-hairy, sometimes almost glabrous. **2*n*** = 56.

Flowering spring. Sandy beaches, dunes, rocky shores, usually in salt spray zone; 0–10 m; B.C.; Calif., Oreg., Wash.

3b. Fragaria chiloensis (Linnaeus) Miller subsp. **pacifica** Staudt, Canad. J. Bot. 40: 883. 1962 E

Stolons, petioles, peduncles, and pedicels spreading-hairy (usually densely so). $2n = 56$.

Flowering spring. Sandy beaches, dunes, rocky shores, ocean bluffs, usually in salt spray zone; 0–10 m; B.C.; Alaska, Calif., Oreg., Wash.

Hybrids:

Fragaria ×*ananassa* Duchesne ex Rozier subsp. *ananassa* [*Fragaria chiloensis* (Linnaeus) Miller var. *ananassa* (Duchesne ex Rozier) L. H. Bailey], the garden or dessert strawberry ($2n = 56$). *Fragaria* ×*ananassa* originated under cultivation in Europe from crossings of the cultivated *F. chiloensis* forma *chiloensis* and *F. virginiana*. It is grown today on all continents and occasionally escapes from gardens.

Fragaria ×*ananassa* Duchesne ex Rozier subsp. *cuneifolia* (Nuttall ex Howell) Staudt [*Fragaria cuneifolia* Nuttall ex Howell; *F.* ×*ananassa* var. *cuneifolia* (Nuttall ex Howell) Staudt]

Stolons spreading or appressed ascending-hairy. **Leaves** dark green to bluish green, slightly glaucous, sometimes dull, ± thin to ± thick, sometimes leathery, abaxial surface mostly not strongly reticulately veined, adaxial glabrous, often shiny; petiole usually densely spreading-hairy (hairs sometimes descending or appressed); terminal leaflets petiolulate, margins serrate to crenate in distal ⅓, terminal tooth shorter than adjacent teeth. **Flowers** 14–19 mm diam. (pistillate), 17.2–24.9 mm diam. (staminate or bisexual) (plants dioecious, gynodioecious, or trioecious); hypanthium 14.3–23.8 mm diam.; petals round to oval, margins distinct or overlapping. **Achenes** in shallow pits or embedded, dark reddish brown to dark brown, 1.4–2 mm; bractlets and sepals clasping; torus yellowish red to red, globose, flattened, or conic, not easily separated from hypanthium. $2n = 56$.

Flowering spring. Sandy beaches, coastal areas; 0–100 m; B.C.; Calif., Oreg., Wash.

Subspecies *cuneifolia* arose in nature from taxa in western North America (G. Staudt 1962). Wild-growing, reciprocal hybrid populations between *Fragaria chiloensis* subsp. *pacifica* and *F. virginiana* subsp. *glauca* occur in coastal areas between Vancouver Island and Point Sur, California. These plants were recognized by T. J. Howell (1897–1903) as *F. cuneifolia* Nuttall and by Staudt (1999) as *F.* ×*ananassa* subsp. *cuneifolia*.

Fragaria ×*bringhurstii* Staudt. Leaves mostly intermediate in texture and hairiness; flowers unisexual, intermediate, usually sterile, sometimes partially fertile; fruits not observed.

Fragaria ×*bringhurstii* is difficult to differentiate from *F. chiloensis* and *F.* ×*ananassa* subsp. *cuneifolia*. It has been found at eight localities on the coast of California, in Humboldt, Marin, Monterey, San Mateo, and Sonoma counties. Its origin as hybrids of *F. chiloensis* and *F. vesca* subsp. *californica* can be deduced from its chromosome numbers ($2n = 5x = 35$, $2n = 6x = 42$, and $2n = 9x = 63$) and morphology. Reduced and unreduced gametes were involved in both species. Partial fertility and abnormal gamete production in the hybrids may give rise to plants with new chromosome numbers hitherto not known in nature.

14. CHAMAERHODOS Bunge in C. F. von Ledebour, Fl. Altaica 1: 429. 1829 • Little rose [Greek *chamai*, dwarf, and *rhodon*, rose, alluding to appearance of plants]

Joshua C. Springer

Bruce D. Parfitt†

Herbs, biennial or short-lived [long-lived] perennial, [0.3–]0.6–3 dm, hispid-hirsute, soft-hairy, and stipitate-glandular [eglandular]; taproots woody, scaly. **Stems** 1(–10), <reddish tinged [green]>, simple or branched throughout, branches ascending or erect. **Leaves** basal winter-persistent, basal and cauline, rosulate, alternate, pinnately compound or simple and deeply pinnatifid; stipules absent; petiole present; blade obovate, 1.5–4 cm, herbaceous, 2–4-ternate, <cauline sessile or petioles to 2 mm>, blade 1- or 2-pinnate or -pinnatifid, lobes linear-oblong, margins flat, entire, venation pinnate, <1 vein per lobe (lateral veins not seen)>, surfaces hairy, stipitate- [or sessile] glandular. **Inflorescences** terminal or often axillary, 10–many-flowered, crowded [sparse], flat-topped panicles; <peduncles present>; bracts present; bracteoles absent.

Pedicels present or nearly absent [long-pedicellate]. **Flowers** 2–4[–4.5] mm diam.; epicalyx bractlets 5, reduced to bristles proximal to sepal bases; hypanthium campanulate, 1.5–2.5 [–3.5] mm, exterior shiny-bristled, adaxially sparsely villous at rim; sepals 5, erect-ascending [recurved], ovate-lanceolate [triangular], <bristle-tipped>; petals 5, white, obovate or cuneate-oblong, <apex emarginate>; stamens 5(or 6), shorter than petals; torus absent; carpels 5–10 (–20), glabrous, styles lateral; ovule 1. **Fruits** aggregated achenes, 1–6, <olivaceous to blackish [reddish to purplish at base]>, ovoid-pyriform, 1.1–1.4[–1.6] mm, glabrous; hypanthium persistent; sepals persistent, erect-ascending (recurved); styles persistent. $x = 7$.

Species 7 (1 in the flora): North America, Asia.

1. Chamaerhodos erecta (Linnaeus) Bunge in C. F. von Ledebour, Fl. Altaica 1: 430. 1829 [F]

Sibbaldia erecta Linnaeus, Sp. Pl. 1: 284. 1753; *Chamaerhodos erecta* subsp. *nuttallii* (Pickering ex Rydberg) Hultén; *C. erecta* var. *parviflora* (Nuttall) C. L. Hitchcock; *C. nuttallii* Pickering ex Rydberg

Leaves: basal petiole to 30 mm; lobes of cauline leaves 5–18 × 0.2–1 mm; surfaces white-hairy, stipitate-glandular, hairs rarely antrorsely barbed. **Flowers:** sepals 1.8–2.5 mm; petals 1.8–2.5 × 0.5–1.1 mm. $2n = 14$.

Flowering Jun–Aug; fruiting Jul–Sep. Prairies, dry hilltops, knolls, outcrops, buttes, rocky and sandy soil or gravel, exposed and disturbed sites; 300–3200 m; Alta., B.C., Man., N.W.T., Ont., Sask., Yukon; Alaska, Colo., Mich., Minn., Mont., N.Dak., S.Dak., Utah, Wyo.; Asia (Mongolia, Russian Far East, Siberia).

No clear distinction could be made among infraspecific taxa of *Chamaerhodos erecta*. Purportedly distinguishing characters, for example petal length relative to sepal length, vary throughout the range of the species. Stems may be in tufts, single or multiple (one North Dakota specimen exhibited leafy tufts with ten branches from the base). Most often, specimens are single-stemmed, branching distal to the middle, and producing flat-topped inflorescences.

15. DRYMOCALLIS Fourreau ex Rydberg, Monogr. N. Amer. Potentilleae, 190, plate 102, figs. 1–5; plates 103–111. 1898 • Wood beauty [Greek *drymos*, woods, and *kallos*, beauty]

Barbara Ertter

Potentilla Linnaeus [unranked] *Closterostyles* Torrey & A. Gray; *Potentilla* sect. *Closterostyles* (Torrey & A. Gray) Baillon; *Potentilla* subg. *Closterostyles* (Torrey & A. Gray) Juzepczuk; *Potentilla* sect. *Rupestres* Poeverlein

Herbs, perennial, 0.5–10 dm, sparsely to densely short eglandular-hairy, long septate-glandular, and peglike-glandular; tufted or openly rhizomatous. **Stems** 1–20+, ± erect, <green, reddish, or stramineous>. **Leaves** marcescent or winter-persistent, primarily basal, <cauline 0–4, gradually or abruptly reduced distally>, alternate, odd-pinnate; stipules persistent, basally adnate to petiole, narrowly to broadly triangular-ovate to rounded, margins entire to ± toothed; petiole present; blade oblanceolate-oblong in outline, 1–25 cm, foliaceous, leaflets 5–13(–21), <distinct, terminal not confluent with distalmost lateral pair, not or scarcely overlapping>, obovate or elliptic-ovate to flabellate or rhombic, margins flat, singly or doubly toothed, venation ± pinnate. **Inflorescences** terminal, 2–50(–60)-flowered, narrowly to widely cymose, open or congested; bracts present, leaflike or reduced; bracteoles absent. **Pedicels** present, <straight>. **Flowers** 3–25 mm diam. (smallest ones with erect petals); epicalyx bractlets 5; hypanthium shallowly cupulate, 0.5–3 × 2.5–7 mm; sepals 5, spreading to reflexed or erect, narrowly to broadly triangular-ovate; petals 5, white to pale or bright yellow, rarely red-tinged (*D. pseudorupestris* var. *crumiana*), narrowly to broadly ovate-elliptic or obovate to round; stamens 20–30(–40), shorter than petals, <anthers: connective broad, theca single, horseshoe-shaped, dehiscing by continuous slit>; torus hemispheric to conic; carpels 10–80, glabrous, styles sub-basal, <fusiform, medially rough-thickened>; ovule 1. **Fruits** aggregated achenes, individually deciduous, 10–80 or less, obliquely ovoid, 0.7–1.5 mm, glabrous; hypanthium persistent; sepals persistent, erect; styles tardily deciduous, <jointed>. $x = 7$.

Species 25–30 (15 in the flora): North America, nw Mexico, Eurasia; mostly temperate areas.

The convergence of morphologic (J. Soják 1985[1989]) and molecular (T. Eriksson et al. 1998; M. Lundberg et al. 2009; C. Dobeš and J. Paule 2010) evidence shows that the species formerly placed in *Potentilla* sect. *Rupestres* or subg. *Closterostyles* belong in Fragariinae, separate from *Potentilla* in the strict sense. *Drymocallis* has accordingly been resurrected for this complex of species, which has centers of radiation in western North America, southeastern Europe, and central Asia (A. Kurtto and Eriksson 2003; B. Ertter 2007; Soják 2006, 2011). Molecular studies by Lundberg et al. suggest that North American and Eurasian species form separate clades and hint at a possible hybrid origin of the genus.

In addition to anthers with a broad connective ringed by a single horseshoe shaped theca (shared with other subtribe Fragariinae), *Drymocallis* is characterized by fusiform styles attached near the bases of the achenes. Leaves are odd-pinnate with a distinct terminal leaflet, in contrast to some superficially similar species of *Horkelia*. Vestiture consists of various proportions of septate glands to 2 mm, subsessile peglike glands, short-spreading eglandular hairs, and, less frequently, rigid bristles to 1.5 mm.

The North American members of the genus, as *Potentilla glandulosa* and allies, were studied by J. Clausen et al. (1940) in their seminal biosystematic experiments. They determined that the complex consists of a wide diversity of often highly localized ecotypes differing from one another ecologically, physiologically, and morphologically. They also demonstrated that hybrids between those ecotypes are readily generated, aided by a uniform diploid chromosome number of $2n = 14$, and that intergradation zones where ecotypes intersect are the norm. Faced with this biosystematic complexity, D. D. Keck (in Clausen et al.) recognized *P. arguta*, *P. fissa*, and *P. glandulosa*, with additional ecotypic variation treated as subspecies of *P. arguta* and *P. glandulosa*.

During preparation of the present treatment, it became apparent that the outline by D. D. Keck did not adequately accommodate current evidence of variation in North American *Drymocallis*, especially beyond California. Subspecies accepted by Keck appeared no more closely related within a species than between species, and extremes of variation occurred beyond those addressed by Keck. The treatment here is an unabashedly provisional alternate framework, which may provide an improved foundation and incentive for much-needed further research on a relatively neglected genus.

The potentially most controversial aspect of this approach is the recognition of more of the variation within *Drymocallis* at the species level, with full acknowledgment that any attempt to recognize formal taxa in North American *Drymocallis* will be compromised by wide zones of intergradation and populations that defy placement. The alternative is to accept broadly defined taxa in which the extensive variation is disregarded, including variation of conservation significance and/or comparable to species currently recognized in Eurasia (A. Kurtto and T. Eriksson 2003; J. Soják 2011). Here, species are used for relatively cohesive core populations sharing multiple characteristics in a definable ecogeographic setting, even in the absence of sharp morphologic boundaries. Varieties are used where differences are less distinct and/or the intergradations more complex.

Among the more outstanding unresolved questions are plants from the Pacific Northwest that have been called *Drymocallis valida* (Greene) Piper [*Potentilla valida* Greene], which combine features of *D. convallaria* and *D. glabrata*. Variation in the Pacific Northwest is generally in critical need of careful analysis, as is that in other major zones of intergradation (for example, northern Great Basin, New Mexico, southern Utah). More collections with basal leaves, fully developed inflorescences, and careful note of petal orientation and color (which becomes unreliable upon drying) will help significantly in identification and future analyses.

In the descriptions, stems refer to flowering stems, with stem length including the inflorescence. Cauline leaves are on the unbranched portion of the stem. Counts of leaflet teeth include all primary teeth and, when double, significant secondary teeth. The inflorescence comprises the branched portion of the stem, including the proximalmost ramification; the fraction of total stem comprised of inflorescence is indicated in each species description. Leafy inflorescences are those in which bracts at the proximal nodes are well developed, that is, generally more than half as long as the subtended internode. Shorthand terms describing inflorescence architecture (for example, congested, narrow) are followed by more precise numeric counterparts. The diagnostic value of pedicel vestiture derives from the relative proportion of short (0.2 mm) simple eglandular hairs and longer septate glands. Epicalyx bractlets are usually simple but sometimes doubled, two-lobed, or otherwise toothed; shape and measurements are based on the simple form. Sepal and petal orientation (erect, spreading-reflexed) are at flowering; all tend to be erect in fruit. Numeric ranges given here (for example, stem and flower number), derived from wild-collected herbarium specimens, are sometimes greatly exceeded by the ranges reported by J. Clausen et al. (1940) for cultivated plants.

SELECTED REFERENCES Clausen, J., D. D. Keck, and W. M. Hiesey. 1940. Experimental studies on the nature of species. I. Effect of varied environments on western North American plants. Publ. Carnegie Inst. Wash. 520. Ertter, B. 2007. Generic realignments in tribe Potentilleae and revision of *Drymocallis* (Rosoideae: Rosaceae) in North America. J. Bot. Res. Inst. Texas 1: 31–46. Rydberg, P. A. 1908. *Drymocallis*. In: N. L. Britton et al., eds. 1905+. North American Flora.... 47+ vols. New York. Vol. 22, pp. 366–376.

1. Flowers opening narrowly (petals and sepals ± erect); styles slender, (1–)1.5–2.5 mm.
    2. Petals broadly obovate, 5–11 × (3–)5–8 mm; pedicels predominantly septate-glandular; c Oregon . . . . 13. *Drymocallis campanulata*
    2. Petals narrowly obovate, 2–5 × 1.5–3 mm; pedicels predominantly short-hairy; sw Oregon and California.
        3. Petals cream-white; epicalyx bractlets 2–3.5 × 1 mm; pedicels 1–5(–15) mm; nw California, sw Oregon. . . . . 14. *Drymocallis rhomboidea*
        3. Petals yellow; epicalyx bractlets 1–2 × 0.5 mm; pedicels 2–15(–30) mm; s California . . . . 15. *Drymocallis cuneifolia*
1. Flowers opening widely (petals and sepals ± spreading to reflexed); styles thickened, 0.8–1.5 mm.
    4. Basal leaves with (4 or)5 or 6(–10) pairs of leaflets, cauline with 4–6(–10) pairs; anthers (0.7–)1–1.4 mm; mostly e of Continental Divide. . . . . 1. *Drymocallis fissa*
    4. Basal leaves with (1 or)2–5 pairs of leaflets, cauline with 1–4 pairs; anthers 0.5–1 mm; w of Continental Divide (except *D. arguta*).
        5. Pedicels predominantly short-hairy, often velutinous, septate glands absent or sparsely to moderately abundant but fewer than short hairs.
            6. Petals shorter than sepals; inflorescences leafy; Utah, Wyoming.
                7. Petals 2–5 mm, cream-white; sepals 4–7(–9) mm; pedicels 1–5 mm . . . . 5. *Drymocallis micropetala*
                7. Petals (4–)6–10 mm, cream-white to light yellow; sepals (5–)6–12(–15) mm; pedicels 2–20 mm . . . . 11. *Drymocallis deseretica*

6. Petals usually longer than, sometimes equal to, sepals; inflorescences not leafy; California, Nevada, Oregon.
   8. Stem bases septate-glandular, (1–)2–4 mm diam.; leaflet teeth single to double . . . . 6. *Drymocallis hansenii* (in part)
   8. Stem bases usually not septate-glandular, 1–2.5 mm diam.; leaflet teeth usually single . . . . 7. *Drymocallis lactea* (in part)

[5. Shifted to left margin.—Ed.]

5. Pedicels predominantly septate-glandular, short hairs absent or sparsely to moderately abundant, fewer than septate glands.
   9. Inflorescences narrowly branched (branch angles 5–25°) or too congested to determine branch angles.
      10. Terminal leaflets obtuse to acute, usually densely hairy; teeth 15–30 per side (including 2° and 3° teeth); inflorescences usually less than ⅕ of stems; e of Continental Divide . . . . 3. *Drymocallis arguta*
      10. Terminal leaflets usually rounded to obtuse, sometimes acute (*D. arizonica*), moderately hairy to glabrate; teeth (4–)6–18(–20) per side; inflorescences often more than ⅕ of stems; w of Continental Divide.
          11. Petals narrowly obovate-elliptic, 1.5–2.5 mm wide, yellow; s California . . . . 12. *Drymocallis glandulosa* (in part)
          11. Petals broadly obovate to obovate-elliptic, 2.5–7 mm wide, most often cream-white to light yellow; widespread, not s California.
              12. Stems usually ± tufted, (1–)2–6 dm, base 1.5–3 mm diam.
                  13. Petals 3–6 × 2.5–5 mm, shorter than or equal to sepals; basal leaves: leaflet pairs (2 or)3 or 4(or 5); cauline leaves 1–4; Arizona, Utah . . . . 2. *Drymocallis arizonica*
                  13. Petals 5–9 × 4–7 mm, equal to or longer than sepals; basal leaves: leaflet pairs 2 or 3(or 4); cauline leaves 0–2; sw Oregon . . . . .8. *Drymocallis ashlandica* (in part)
              12. Stems usually solitary, sometimes tufted, (3–)4–9(–10) dm, base (1–)2–5 mm diam.
                  14. Inflorescences congested to open, usually less than ⅓ of stem; pedicels predominantly septate-glandular; sepals acute; widespread, not California . . . . 4. *Drymocallis convallaria*
                  14. Inflorescences open, usually more than ⅓ of stem; pedicels sparsely to moderately septate-glandular; sepals acute to acuminate; c California . . . . 6. *Drymocallis hansenii* (in part)
   9. Inflorescences widely branched (branch angles 15–60°), usually open.
      15. Petals shorter than or equal to sepals, narrowly to broadly obovate or ovate to nearly round, 2–6.5 × 1.5–5.5 mm, not overlapping; sepal apices usually broadly obtuse with mucronate tip; most common w of Sierran-Cascade axis. . . . . 12. *Drymocallis glandulosa* (in part)
      15. Petals usually longer than sepals, ± broadly obovate, (3–)4–12 × 3–11 mm, often overlapping; sepal apices usually ± acute, sometimes ± obtuse with mucronate tip; mostly on and e of Sierran-Cascade axis.
          16. Terminal leaflet apices obtuse to acute; inflorescences leafy; pedicels 5–40 (–45) mm; epicalyx bractlets linear-oblanceolate, 0.5–1 mm wide; petals usually bright, sometimes pale, yellow . . . . 10. *Drymocallis glabrata*
          16. Terminal leaflet apices rounded to truncate, sometimes ± obtuse; inflorescences not or ± leafy; pedicels 2–20(–40) mm; epicalyx bractlets linear to elliptic-ovate, (0.5–)1–2 mm wide; petals cream-white to bright yellow.
              17. Stems usually less than 2.5 dm, often loosely spaced from elongate caudex branches; petals narrowly to broadly obovate; usually in rocky habitats . . . . .9. *Drymocallis pseudorupestris* (in part)
              17. Stems usually more than 2 dm, usually ± tufted from short caudex branches (except variety of *D. pseudorupestris*, see couplet 19); petals broadly obovate; in various habitats.

[18. Shifted to left margin.—Ed.]

18. Inflorescences congested, less than ¼ of stem. . . . . .8. *Drymocallis ashlandica* (in part)
18. Inflorescences open, usually more than ¼ of stem.
    19. Terminal leaflet: teeth usually single, 4–10(–12) per side; petals (cream-white) pale to bright yellow; California, w Nevada, Oregon . . . . 7. *Drymocallis lactea* (in part)
    19. Terminal leaflet: teeth mostly double, (5–)8–15 per side; petals cream-white to pale yellow; n Idaho, Montana, Wyoming . . . . .9. *Drymocallis pseudorupestris* (in part)

1. Drymocallis fissa (Nuttall) Rydberg, Monogr. N. Amer. Potentilleae, 197. 1898 • Leafy drymocallis or wood beauty [E] [F]

Potentilla fissa Nuttall in J. Torrey and A. Gray, Fl. N. Amer. 1: 446. 1840

Caudex branches short to elongate. **Stems** tufted to loosely spaced, (1.2–)1.5–3.5(–4.5) dm; base (1.5–)2–3 mm diam., ± densely septate-glandular. **Leaves** sparsely to moderately hairy; basal (3–)7–19 cm, leaflet pairs (4 or)5 or 6(–10) (additional reduced leaflets sometimes interspersed) terminal leaflet usually broadly obovate-cuneate, sometimes elliptic, (1–)1.5–3.5(–5) × (1–)1.5–3(–3.5) cm, teeth single or double, 5–13 per side, apex rounded to obtuse; cauline 1–3, well developed, leaflet pairs 4–6 (–10). **Inflorescences** 5–15-flowered, leafy, congested to ± open, ⅙–½ of stem, narrow to wide, branch angles 15–30(–40)°. **Pedicels** 1–12 mm, short-hairy, septate-glandular. **Flowers** opening widely; epicalyx bractlets linear-oblanceolate, 3–7 × 1–2 mm; sepals spreading, 6–10 mm, apex acute to acuminate; petals overlapping, spreading, yellow, broadly obovate, 7–11 × 5–11 mm, equal to or exceeding sepals; filaments 1.5–4.5 mm, anthers (0.7–)1–1.4 mm; styles thickened, 1 mm. **Achenes** light brown, 1 mm.

Flowering May–Aug. Sagebrush slopes, open forests, stream banks, often in rocky or moderately disturbed sites; 1600–3000 m; Colo., S.Dak., Utah, Wyo.

Drymocallis fissa is distinctive in its relatively numerous leaflets (often with additional smaller ones), large flowers, and large, elongate anthers. It is most abundant in the Colorado Front Range, extending into the Medicine Bow Mountains of Wyoming. Outlying populations occur at least as far north as the Bighorn Mountains of northern Wyoming and the Black Hills of South Dakota. Tentatively included here are large-anthered populations from the eastern Uinta Mountains of Utah, though these often have fewer leaflets and smaller flowers of unknown color; they may represent a unique taxon worthy of separate recognition. Possible collections of *D. fissa* from New Mexico, including the type of *Potentilla fissa* var. *major* Torrey & A. Gray, are of uncertain placement in that they combine features of *D. arguta*, *D. arizonica*, and *D. fissa*. Reports from other states, including Montana, are probably all based on misidentified specimens.

2. Drymocallis arizonica Rydberg in N. L. Britton et al., N. Amer. Fl. 22: 373. 1908 • Arizona drymocallis or wood beauty [E]

Drymocallis glandulosa (Lindley) Rydberg subsp. *arizonica* (Rydberg) Soják; *Potentilla glandulosa* Lindley subsp. *arizonica* (Rydberg) D. D. Keck; *P. macdougalii* Tidestrom

Caudex branches short. **Stems** ± tufted, (1.5–)2.5–6 dm; base 1.5–2.5 mm diam., sparsely to densely septate-glandular. **Leaves** sparsely to moderately hairy; basal 5–22 cm, leaflet pairs (2 or)3 or 4(or 5); terminal leaflet broadly obovate-elliptic, 1–5 × 1–2.5(–4) cm, teeth double, 6–18 per side, apex rounded to acute; cauline 1–4, well developed, leaflet pairs 2 or 3. **Inflorescences** 5–30-flowered, not notably leafy, compact, ⅒–¼(–½) of stem, narrow, branch angles 5–25°. **Pedicels** 1–10 (proximal to 20) mm, sparsely to ± densely short-hairy, predominantly septate-glandular. **Flowers** opening widely; epicalyx bractlets linear-lanceolate to narrowly elliptic, 2–4(–7) × 1–2 mm; sepals spreading, 4–6(–11) mm, apex ± acute; petals not or scarcely overlapping, spreading, cream-white, ± obovate, 3–6 × 2.5–5 mm, shorter than or equal to sepals; filaments 1–3 mm, anthers 0.7–1 mm; styles thickened, 1 mm. **Achenes** light brown, 1 mm. $2n = 14$.

Flowering May–Jul. Streamsides, rocky sites, open forest floors, pine and aspen forests; 1800–3200 m; Ariz., Utah.

Drymocallis arizonica encompasses populations in Arizona north of the Mogollon Rim and in Utah as far north as Garfield and Sevier counties, including the Henry Mountains. It is most distinctive in northern Arizona, where plants commonly have basal leaves with four pairs of lateral leaflets and compact inflorescences. The species intergrades with *D. convallaria* but is in general shorter and more likely multistemmed. It also shares similarities with *D. pseudorupestris* but has more developed cauline leaves and a more compact inflorescence. The intergrade zone between all three species and *D. deseretica* in southwestern Utah is particularly problematic and perhaps indicative of yet additional taxa deserving recognition.

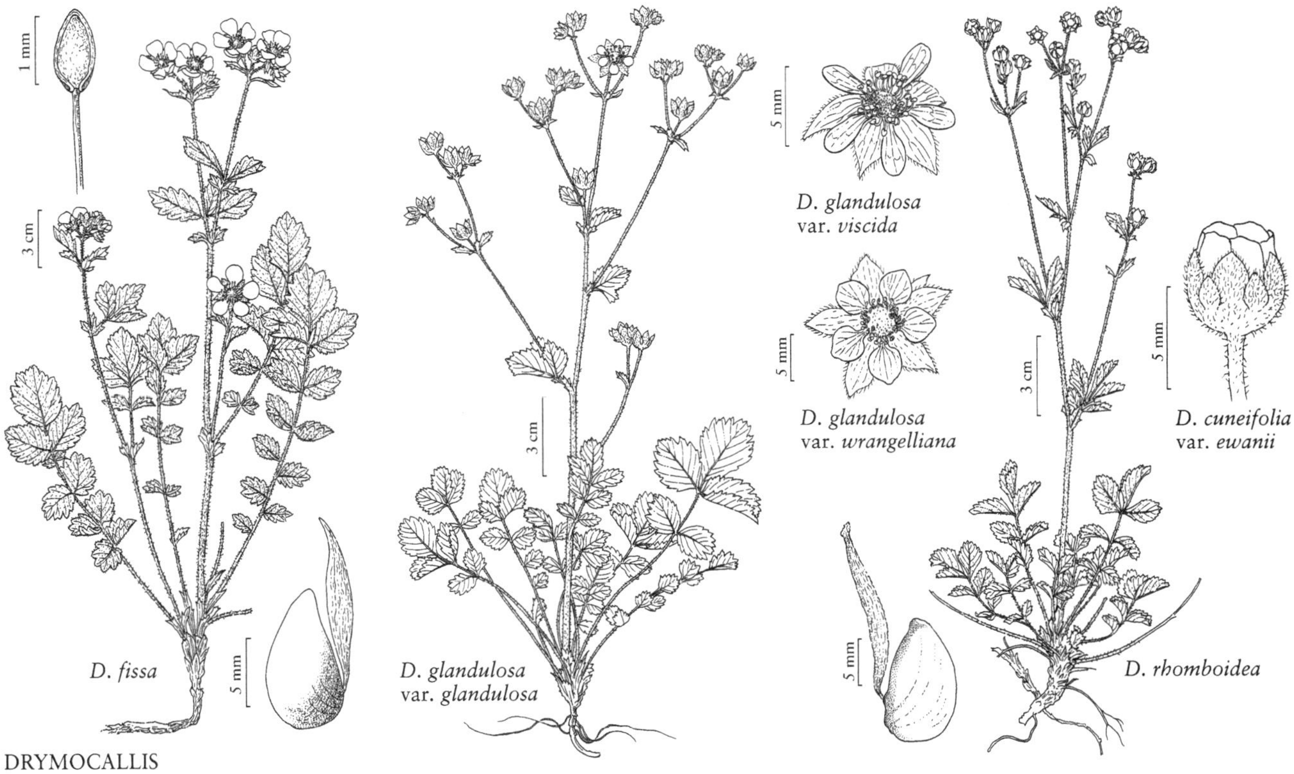

DRYMOCALLIS

3. **Drymocallis arguta** (Pursh) Rydberg, Monogr. N. Amer. Potentilleae, 192. 1898 • Tall drymocallis or wood beauty [E]

Potentilla arguta Pursh, Fl. Amer. Sept. 2: 736. 1813; *Drymocallis agrimonoides* (Pursh) Rydberg

Caudex branches short. **Stems** usually solitary, sometimes tufted, (1–)3–10 dm; base 3–6 mm diam., ± densely, sometimes sparsely, septate-glandular. **Leaves** usually moderately to densely hairy; basal (6–)12–40 cm, leaflet pairs (3 or)4 or 5; terminal leaflet broadly elliptic-obovate, (2–)4–10 × (1–)2–4.5 cm, teeth mostly double, 15–30+ per side, apex obtuse to acute; cauline 2 or 3, relatively well developed, leaflet pairs 2–4. **Inflorescences** 10–40 -flowered, not leafy, congested, sometimes more elongate in fruit, $^{1}/_{10}$–$^{1}/_{5}$(–$^{1}/_{3}$) of stem, narrow, branch angles 5–20°. **Pedicels** 1–5 (proximal to 15) mm, densely short-hairy, septate-glandular. **Flowers** opening widely; epicalyx bractlets ± elliptic, 4–6(–8) × 1–2 mm; sepals spreading, (5–)7–10 mm, apex obtuse to acute, apiculate; petals overlapping, spreading, cream-white to yellowish, broadly elliptic, (5–)7–9 × (5–)6–8 mm, ± equal to or slightly longer than sepals; filaments 1.5–2.5 mm, anthers 0.7–1 mm; styles thickened, 1 mm. **Achenes** light brown, 1 mm. $2n$ = 14.

Flowering May–Aug. Prairies, abandoned pastures, open woodlands, shallow soil and grassy openings on ridges, slopes, bluffs, and rocky barrens; 10–2300 m; Alta., Man., N.B., N.W.T., Ont., Que., Sask., Yukon; Ark., Colo., Conn., Ill., Ind., Iowa, Kans., Maine, Md., Mass., Mich., Minn., Mo., Mont., Nebr., N.H., N.J., N.Mex., N.Y., N.Dak., Ohio, Okla., Pa., S.Dak., Tenn., Vt., Va., W.Va., Wis., Wyo.

Drymocallis arguta is the only species of the genus occurring east of the Rocky Mountains, except for *D. fissa* in immediately adjacent prairies and the Black Hills of South Dakota. It is primarily a species of intact prairies and pasturelands throughout the Great Plains, extending as scattered populations in equivalent habitats throughout the northeastern states and adjacent provinces. Some of these states consider it a species of conservation concern (as *Potentilla arguta*, tall cinquefoil), and the sparsely hairy nature of some of these populations suggests that further taxonomic attention might be warranted. *Drymocallis arguta* occurs also in the Colorado Front Range, generally at elevations higher than *D. fissa*, with which it sometimes intergrades morphologically. It intergrades also with *D. convallaria* but tends to be larger, coarser, and more densely hairy with more acute leaflets.

4. Drymocallis convallaria (Rydberg) Rydberg, Monogr. N. Amer. Potentilleae, 193. 1898 • Cordilleran drymocallis or wood beauty [E]

Potentilla convallaria Rydberg, Bull. Torrey Bot. Club 24: 249, plate 306. 1897; *Drymocallis arguta* (Pursh) Rydberg subsp. *convallaria* (Rydberg) Soják; *P. arguta* Pursh subsp. *convallaria* (Rydberg) D. D. Keck; *P. arguta* var. *convallaria* (Rydberg) Th. Wolf

Caudex branches short. **Stems** usually solitary, sometimes tufted, 4–9(–10) dm; base 2–5 mm diam., ± densely septate-glandular. **Leaves** sparsely to moderately, rarely densely, hairy; basal (10–)15–25 (–30) cm, leaflet pairs (2 or)3 or 4; terminal leaflet usually obovate, sometimes elliptic, 3–5(–6) × 1.5–4.5 cm, teeth double, 9–16 per side, apex rounded-obtuse; cauline (1 or)2–4, well developed at least proximally, leaflet pairs usually 3. **Inflorescences** (5–)10–40-flowered, not leafy, congested to open, (⅛–)⅙–⅓(–½) of stem, narrow, branch angles 5–20°. **Pedicels** 1–5 (proximal to 25) mm, sparsely to ± densely short-hairy, predominantly septate-glandular. **Flowers** opening ± widely; epicalyx bractlets narrowly elliptic, (2–)4–6 × 0.7–1.5 mm; sepals ± spreading, (4–)5–8(–10) mm, apex acute; petals not or scarcely overlapping, ± spreading, cream-white to pale yellow, obovate-elliptic, sometimes broadly so, (3–)5–8 × (2–)4–6 mm, ± equal to sepals; filaments 1.5–2.5 mm, anthers 0.7–1 mm; styles thickened, 0.8 mm. **Achenes** light brown, 1 mm.

Flowering (May–)Jun–Aug. Dry to seasonally moist meadows, open forests, sagebrush and grassy rocky slopes; 100–3000 m; Alta., B.C., Yukon; Alaska, Ariz., Colo., Idaho, Mont., Nev., Oreg., Utah, Wash., Wyo.

Drymocallis convallaria is widespread in western North America, occurring mostly west of the Continental Divide from south-central Alaska to Arizona. It differs from most sympatric species in usually having a single thick stem, a narrow inflorescence, and cream-white to pale yellow petals, at most only slightly longer than the sepals.

Some collections from southern Idaho and northern Nevada that have very small petals in the size range of *Drymocallis micropetala* are included here on the basis of glandular pedicels. Excluded here are comparably small-petaled plants from the Warner Mountains of California and Oregon that have the aspect of *D. convallaria* but the blunter, redder achenes of *D. glandulosa*. Populations in northern Idaho and adjacent Washington that approach *D. arguta* in size and vestiture are provisionally treated as *D. convallaria* on the basis of leaflet shape. Specimens from New Mexico previously assigned to this taxon have largely been redetermined to *D. arguta* in the strict sense. See also discussions of *Potentilla fissa* var. *major* under 1. *D. fissa*, and of *D. valida* under the genus discussion. The illegitimate names *P. glutinosa* Nuttall ex Rydberg and *D. glutinosa* Rydberg have been applied most often to *D. convallaria*.

5. Drymocallis micropetala Rydberg in N. L. Britton et al., N. Amer. Fl. 22: 375. 1908 • Wasatch drymocallis or wood beauty [E]

Drymocallis glandulosa (Lindley) Rydberg subsp. *micropetala* (Rydberg) Soják; *Potentilla glandulosa* Lindley subsp. *micropetala* (Rydberg) D. D. Keck; *P. glandulosa* var. *micropetala* (Rydberg) S. L. Welsh & B. C. Johnston

Caudex branches short to elongate. **Stems** solitary or ± tufted, (2.5–)3–6 dm; base 2–4 mm diam., septate-glandular. **Leaves** moderately hairy; basal (6–)10–24 cm, leaflet pairs 2 or 3; terminal leaflet obovate, 2–5 × 1.5–4 cm, teeth double, (5–)8–12 (–15) per side, apex obtuse; cauline 1–3, well developed, leaflet pairs 2 or 3. **Inflorescences** 10–20(–30)-flowered, leafy, congested clusters terminating branches, ⅒–⅓(–½) of stem, narrow, branch angles 5–20°. **Pedicels** 1–5 mm, predominantly short-hairy, often velutinous, sparsely to moderately septate-glandular. **Flowers** opening widely; epicalyx bractlets linear to oblanceolate-elliptic, 2–4(–6) × (0.5–)1–1.5 mm; sepals spreading, 4–7(–9) mm, apex obtuse; petals not overlapping, spreading, cream-white, obovate-elliptic, 2–5 × 1.5–3.5 mm, shorter than sepals; filaments 1–2 mm, anthers 0.5–0.8 mm; styles thickened, 1 mm. **Achenes** reddish, 1.3 mm.

Flowering May–Jul. Mountain brush in canyons, pinyon-juniper woodlands; 1400–2600 m; Utah.

Drymocallis micropetala is poorly known; it combines the aspect and narrow inflorescence of *D. convallaria*, the small petals of *D. glandulosa*, and the predominantly short-hairy pedicels of *D. deseretica*. Plants with equally small petals that occur sporadically throughout the range of *D. convallaria* are predominantly septate-glandular rather than short-hairy on the petioles and are not included here. As circumscribed here, *D. micropetala* occurs only in the Wasatch Range of north-central Utah, from Juab to Rich counties; it is possibly of conservation concern.

6. Drymocallis hansenii (Greene) Rydberg, Monogr. N. Amer. Potentilleae, 200. 1898 (as hanseni) • Hansen's drymocallis or wood beauty [E]

Potentilla hansenii Greene, Pittonia 3: 20. 1896 (as hanseni); *Drymocallis glandulosa* (Lindley) Rydberg subsp. *hansenii* (Greene) Soják; *P. glandulosa* Lindley subsp. *hansenii* (Greene) D. D. Keck

Caudex branches short to elongate. **Stems** usually solitary, sometimes loosely tufted, (3–)4–9 dm; base (1–)2–4 mm diam., moderately to densely septate-glandular. **Leaves** sparsely to moderately hairy; basal (7–)10–25(–30) cm, leaflet pairs 3 or 4; terminal leaflet broadly obovate, 2–5(–6) × 1.5–3.5(–4) cm, teeth single or double, 6–11 per side, apex usually rounded, sometimes obtuse; cauline 1–3, proximally well developed, leaflet pairs 2–4. **Inflorescences** 10–20(–40)-flowered, not leafy, open, ¼–⅔ of stem, narrow, branch angles 10–30°. **Pedicels** 2–8 (proximal to 20) mm, predominantly short-hairy, sparsely to moderately septate-glandular. **Flowers** opening widely; epicalyx bractlets linear to narrowly elliptic, 2–4 × 0.5–1 mm; sepals spreading, 5–8 mm, apex acute to acuminate; petals overlapping or not, spreading, cream-white to pale yellow, broadly obovate, 4–6 × 3–6 mm, usually longer than, sometimes equal to, sepals; filaments 1.5–2.5(–3) mm, anthers 0.8–1 mm; styles thickened, 0.8–1.2 mm. **Achenes** light brown, 0.7–1 mm. $2n$ = 14.

Flowering Jun–Sep. Moist ground, meadows, open forests, streamsides; 1200–1900(–2200) m; Calif.

D. D. Keck (in J. Clausen et al. 1940) speculated that *Drymocallis hansenii* was the stabilized recombinant of *D. glandulosa* var. *reflexa* and *D. lactea*. Alternatively, it may represent the California counterpart of *D. convallaria*, because it tends to have tall, thick-based, single stems and narrow inflorescences. The species is centered in the west-central Sierra Nevada of California, usually occurring in moist meadows and equivalent habitats. Plants near Lake Tahoe, which provide the high-elevation extreme, combine the smaller stature of *D. lactea* and the glandular-septate stem bases of *D. hansenii*; their optimal taxonomic disposition is uncertain.

7. Drymocallis lactea (Greene) Rydberg in N. L. Britton et al., N. Amer. Fl. 22: 369. 1908 [E]

Potentilla glandulosa Lindley var. *lactea* Greene, Fl. Francisc., 65. 1891

Caudex branches short. **Stems** tufted, (0.3–)1.5–6(–6.5) dm; base 1–2.5 mm diam., not or sparsely, sometimes moderately, septate-glandular. **Leaves** usually sparsely to moderately hairy, sometimes glabrate; basal (2–)5–20(–30) cm, leaflet pairs (2 or)3 or 4(or 5); terminal leaflet broadly obovate to cuneate, 1–4 × 0.7–3.5 cm, teeth usually single, 4–10(–14) per side, apex usually rounded to obtuse, rarely acute; cauline 1 or 2, reduced, leaflet pairs 2 or 3. **Inflorescences** (2–)5–30(–50)-flowered, not leafy, open, (⅙–)⅕–½(–⅘) of stem, narrow to wide, branch angles 10–40(–50)°. **Pedicels** 2–10 (proximal to 30) mm, predominantly short-hairy, sometimes velutinous, not or sparsely, sometimes moderately, septate-glandular. **Flowers** opening widely; epicalyx bractlets linear-lanceolate to narrowly elliptic, 2–5 × 0.5–1.5 mm; sepals spreading, 3–8(–9) mm, apex usually acute, sometimes obtuse and apiculate; petals ± overlapping, spreading, cream-white to bright yellow, broadly obovate, (3–)4–8(–10) × 3–8 mm, usually longer than, rarely equal to, sepals; filaments 1–3(–3.5) mm, anthers 0.6–1 mm; styles thickened, 1 mm. **Achenes** light brown, 1 mm.

Varieties 2 (2 in the flora): w United States.

Drymocallis lactea has had a well-established identity as *Potentilla glandulosa* var. (or subsp.) *nevadensis*; however, the epithet *lactea* has priority at species rank and avoids possible confusion with *P. nevadensis* Boissier. D. D. Keck (in J. Clausen et al. 1940) based his concept of *P. glandulosa* subsp. *nevadensis* on white-petaled plants with narrow inflorescences, the common form in the central Sierra Nevada and south. His map and annotations, however, include populations with yellow petals and widely branched inflorescences from northern California and adjacent Oregon. The two extremes merge in the northern Sierra Nevada, and populations in the North Coast Ranges of California also are problematic; the extremes are accordingly treated here as varieties of a single species. The inclusion by Keck of Washington in the species range might have been based on specimens here assigned to *P. pseudorupestris*.

Drymocallis lactea can be distinguished from sympatric species by its relatively large, overlapping petals, non-leafy inflorescences, and singly toothed leaflets. Pedicels and distal stems are usually densely short-hairy with relatively few or no glandular hairs (in contrast to *D. pseudorupestris*), and septate glands are usually absent from stem bases (in contrast to *D. hansenii*). The distinction between *D. lactea* and *D. pseudorupestris* breaks down at the margin of the species range in Nevada and Oregon.

1. Inflorescences narrow, branch angles 10–20°; petals cream-white to pale yellowish; pedicels and stems not or sparsely septate-glandular. . . . . 7a. *Drymocallis lactea* var. *lactea*
1. Inflorescences wide, branch angles mostly 20–40°; petals usually pale to bright yellow; pedicels and stems sometimes moderately septate-glandular . . . . 7b. *Drymocallis lactea* var. *austiniae*

7a. Drymocallis lactea (Greene) Rydberg var. **lactea** • Sierran drymocallis or wood beauty [E]

Drymocallis glandulosa (Lindley) Rydberg subsp. *nevadensis* (S. Watson) Soják; *Potentilla glandulosa* Lindley subsp. *nevadensis* (S. Watson) D. D. Keck; *P. glandulosa* var. *nevadensis* S. Watson

Stems (0.3–)1–6 dm; base glabrate or sparsely to densely short-hairy, not or sparsely septate-glandular. **Inflorescences** 3–20-flowered, $\frac{1}{5}$–$\frac{2}{5}$(–$\frac{3}{4}$) of stem, narrow, branch angles 10–20°. **Pedicels** 5–10 mm, moderately to densely short-hairy, not or sparsely septate-glandular. **Petals** cream-white to pale yellowish. $2n$ = 14.

Flowering May–Aug. Rocky, seasonally moist places; 1600–3700 m; Calif., Nev., Oreg.

Variety *lactea* is the most common and conspicuous *Drymocallis* at higher elevations in the central and southern Sierra Nevada, where it is recognized by its large cream-white to yellowish petals and narrow inflorescences. It also occurs in the San Jacinto Mountains and Transverse Ranges of southern California, and extends into the Intermountain Region from the mountains of southeastern Oregon to central Nevada.

7b. Drymocallis lactea (Greene) Rydberg var. **austiniae** (Jepson) Ertter, J. Bot. Res. Inst. Texas 1: 36. 2007 • Austin's drymocallis or wood beauty [E]

Potentilla glandulosa Lindley var. *austiniae* Jepson, Fl. Calif. 2: 181. 1936 (as austinae)

Stems (0.4–)1.5–4.5(–6.5) dm; base sparsely to densely short-hairy, sometimes moderately septate-glandular. **Inflorescences** (2–)5–30(–50)-flowered, ($\frac{1}{6}$–)$\frac{1}{4}$–$\frac{1}{2}$(–$\frac{4}{5}$) of stem, wide, branch angles (10–)20–40(–50)°. **Pedicels** 2–20(–30) mm, sparsely to ± densely short-hairy, usually eglandular, sometimes sparsely to moderately septate-glandular. **Petals** (cream-white) pale to bright yellow.

Flowering May–Jul(–Aug). Rocky seasonally moist places, meadows; 900–2800 m; Calif., Nev., Oreg.

Variety *austiniae* was erroneously treated as *Potentilla glandulosa* subsp. *ashlandica* (Greene) D. D. Keck (B. Ertter 1993); it now is understood to be a separate entity. As here defined, it is the common large-petaled, open-flowered *Drymocallis* of northern California and adjacent Oregon and Nevada, differing from var. *lactea* in its more widely branched inflorescences and usually pale to bright yellow petals. The boundary between the two varieties is unclear, in part because petal color is uncertain on herbarium specimens. Some collections from Steens Mountain, Oregon, are transitional to *D. pseudorupestris* var. *saxicola*. Occasional populations in northern California that combine the morphology of var. *austiniae* with more glandular vestiture on pedicels and stems might indicate introgression with *D. ashlandica* or *D. pseudorupestris*.

8. Drymocallis ashlandica (Greene) Rydberg, Monogr. N. Amer. Potentilleae, 200. 1898 • Mazama drymocallis or wood beauty [E]

Potentilla ashlandica Greene, Pittonia 3: 248. 1897; *Drymocallis glandulosa* (Lindley) Rydberg subsp. *ashlandica* (Greene) Soják; *P. glandulosa* Lindley subsp. *ashlandica* (Greene) D. D. Keck

Caudex branches short. **Stems** tufted, (1–)2–5 dm; base 2–3 mm diam., ± densely septate-glandular. **Leaves** moderately to ± densely hairy; basal 6–16 cm, leaflet pairs 2 or 3(or 4); terminal leaflet broadly obovate, 1.5–4 × 1–3 cm, teeth ± double, 7–12 per side, apex rounded; cauline 0–2, moderately developed, leaflet pairs 2 or 3. **Inflorescences** 5–15(–20) -flowered, not leafy, congested to deeply branched, $\frac{1}{10}$–$\frac{2}{3}$ or less of stem, usually narrow, rarely ± wide, branch angles (5–)10–20(–30)°. **Pedicels** 2–10 (proximal to 30) mm, sparsely to moderately short-hairy, predominantly septate-glandular, often obscuring other vestiture. **Flowers** opening widely; epicalyx bractlets linear to broadly elliptic, 3–5 × 1–2 mm; sepals spreading, 5–7(–9) mm, apex acute; petals overlapping, spreading, light yellow, broadly obovate, 5–9 × 4–7 mm, equal to or longer than sepals; filaments 1.5–5 mm, anthers 0.7–1.2 mm; styles ± thickened, 1 mm. **Achenes** light brown, 1 mm.

Flowering Jun–Aug. Meadows, roadsides, sandy shorelines; 500–2000 m; Oreg.

The name *Potentilla glandulosa* subsp. *ashlandica* was misapplied to *Drymocallis lactea* var. *austiniae* (B. Ertter 1993); as circumscribed here, *D. ashlandica* is a poorly known species restricted to the Cascade and Siskiyou ranges of southwestern Oregon. It very likely also occurs in adjacent California; no collections from there are known. *Drymocallis ashlandica* differs from most sympatric populations of *D. lactea* var. *austiniae* and resembles *D. pseudorupestris* in its densely glandular inflorescences, which tend to be more densely congested and/or narrower than those of the other two species.

Drymocallis ashlandica was described by Greene as *Potentilla ashlandica*, and by Howell as *P. ciliata* (a later homonym), both based on a collection by Howell from the Siskiyou Mountains near Ashland, Oregon. The circumscription here, which matches that of D. D. Keck (in J. Clausen et al. 1940) and M. E. Peck (1961), encompasses populations from near Oregon Caves and Ashland in the Siskiyou Mountains to near McKenzie Bridge, Lane County, and Sparks Lake, Deschutes County. It is currently known from relatively few collections, mostly from the Crater Lake area, and may prove to be of conservation concern.

9. Drymocallis pseudorupestris (Rydberg) Rydberg, Monogr. N. Amer. Potentilleae, 194. 1898 [E]

Potentilla pseudorupestris Rydberg, Bull. Torrey Bot. Club 24: 250, plate 307. 1897; *Drymocallis glandulosa* (Lindley) Rydberg subsp. *pseudorupestris* (Rydberg) Soják; *P. glandulosa* Lindley subsp. *pseudorupestris* (Rydberg) D. D. Keck; *P. glandulosa* var. *pseudorupestris* (Rydberg) Breitung; *P. rupestris* Linnaeus var. *americana* Th. Wolf

Caudex branches elongate. **Stems** openly tufted to loosely spaced, (0.3–)0.6–4 dm; base 1–3 mm diam., ± densely septate-glandular. **Leaves** glabrate or sparsely to ± densely hairy; basal (2–)3–16 cm, leaflet pairs (2 or)3 or 4(or 5); terminal leaflet broadly obovate-cuneate to flabellate, 0.2–3(–4) × 0.5–3 cm, teeth single or double, 2–15 per side, apex usually rounded to truncate, sometimes obtuse; cauline 0–2, reduced, leaflet pairs 2 or 3. **Inflorescences** 2–40-flowered, not or ± leafy, open, 1/6–3/4(–4/5) of stem, ± wide, branch angles (10–)20–40(–50)°. **Pedicels** 3–20 (proximal to 40) mm, not or sparsely to moderately short-hairy, predominantly septate-glandular. **Flowers** opening widely; epicalyx bractlets linear to elliptic, 2–6 × 1–2 mm; sepals spreading, 4–7(–9) mm, apex acute to obtuse, apiculate; petals overlapping or not, spreading, cream-white to pale yellow (red-tinged in var. *crumiana*), narrowly to broadly obovate, 4–12 × 3–11 mm, longer than sepals; filaments 1–4 mm, anthers 0.7–1.2 mm; styles thickened, 1–1.5 mm. **Achenes** light brown, 1 mm.

Varieties 3 (3 in the flora): w North America.

Drymocallis pseudorupestris occurs from Alberta and Washington to California and Utah, mostly in montane habitats; it is the species most often associated with rocky habitats, including talus slopes, for which its relatively elongate caudex branches are an obvious adaptation. Vestiture is dominated by abundant septate glands on stems and in the inflorescences. Except for var. *pseudorupestris*, which occurs only in the northeastern part of the species range, plants are relatively short, usually less than 2.5 dm. Three intergrading varieties accommodate the extremes at the northeastern and southern ends of the range.

1. Stems (1–)2–4 dm, bases (1.5–)2–3 mm diam.; basal leaves (4–)7–16 cm; terminal leaflets (1–)2–3(–4) cm, teeth usually double, (5–)8–15 per side; flowers (5–)10–40; petals 6–12 × 5–11 mm, widely overlapping; filaments 2–4 mm . . . .
   . . . 9a. *Drymocallis pseudorupestris* var. *pseudorupestris*
1. Stems (0.3–)0.6–2.5 dm, bases 1–2(–3) mm diam.; basal leaves (2–)3–9(–15) cm; terminal leaflets 0.2–2(–4) cm, teeth single or ± double, 2–8 (–12) per side; flowers 2–12(–20); petals 4–8(–9) × 3–6(–8) mm, not or ± overlapping; filaments 1–2.5(–3) mm.
   2. Basal leaves: leaflet pairs (2 or)3(or 4); hypanthia and sepals not bristly or bristles less than 1 mm; short hairs sparse to moderately abundant on stems and pedicels (sometimes absent); styles usually golden brown, rarely reddish. . . . .
      . . . . . . 9b. *Drymocallis pseudorupestris* var. *saxicola*
   2. Basal leaves: leaflet pairs 3 or 4(or 5); hypanthia and sepals prominently bristly, bristles 1–1.5 mm; short hairs absent or sparse on stems and pedicels; styles usually dark red, rarely golden brown. . . . .
      . . . . . 9c. *Drymocallis pseudorupestris* var. *crumiana*

9a. Drymocallis pseudorupestris (Rydberg) Rydberg var. **pseudorupestris** • Montana drymocallis or wood beauty [E]

Stems (1–)2–4 dm, base (1.5–)2–3 mm diam., short hairs usually absent or sparse, sometimes moderately abundant. **Basal leaves** (4–)7–16 cm, sparsely to moderately hairy (hairs to 1 mm), peglike-glandular, not bristly; leaflet pairs (2 or)3(or 4); terminal leaflet broadly obovate, (1–)2–3(–4) × 1.5–3 cm, teeth usually double, (5–)8–15 per side. **Pedicels** 5–20 (proximal to 40) mm, not bristly, short hairs sparse to moderately abundant. **Flowers** (5–)10–40; hypanthia and sepals not bristly; epicalyx bractlets linear to elliptic, 3–6 × 1–2 mm; sepals 5–7(–9) mm; petals widely overlapping, not red-tinged, broadly obovate, 6–12 × 5–11 mm; filaments 2–4 mm; styles light reddish brown.

Flowering May–Aug. Rocky areas, exposed slopes, grasslands on bluffs; 1000–2900 m; Alta.; Idaho, Mont., Wyo.

Variety *pseudorupestris* is the tall, large-flowered extreme centered in the northern Rocky Mountains in Montana, extending into Alberta and to the Wind River Range in Wyoming. Some collections from northern Idaho are also tentatively included here. Unlike the other varieties, var. *pseudorupestris* occurs in grasslands as well as in more rocky habitats. It may intergrade with *Drymocallis fissa*, though petal color and leaflet number are usually distinct.

9b. Drymocallis pseudorupestris (Rydberg) Rydberg var. **saxicola** Ertter, J. Bot. Res. Inst. Texas 1: 37, figs. 1M–R. 2007 • Cliff drymocallis or wood beauty [E]

Stems 0.6–2.5 dm, base 1–2(–3) mm diam., short hairs sparse to moderately abundant, sometimes absent. **Basal leaves** 3–9(–15) cm, sparsely to densely hairy (hairs to 1.5 mm), usually densely peglike-glandular, not bristly; leaflet pairs (2 or)3(or 4); terminal leaflet broadly obovate-cuneate, 0.8–2(–4) × 0.7–2(–3) cm, teeth single or ± double, 3–8(–12) per side. **Pedicels** 3–15 (proximal to 20) mm, usually not bristly, short hairs sparse to moderately abundant (sometimes absent). **Flowers** (2 or)3–12(–20); hypanthia and sepals not bristly or bristles less than 1 mm; epicalyx bractlets linear-lanceolate to broadly elliptic, 2–5 × 1–1.5 mm; sepals 4–6(–7) mm; petals ± overlapping, not red-tinged, narrowly to broadly obovate, 4–8(–9) × 3–6(–8) mm; filaments 1–2.5 mm; styles usually golden brown, rarely reddish.

Flowering May–Aug. Cliffs, ledges, outcrops, ridges, talus slopes, lava beds, other rocky habitats; 1000–3400 m; Alta., B.C.; Calif., Idaho, Mont., Nev., Oreg., Utah, Wash., Wyo.

Variety *saxicola* accommodates the bulk of specimens previously placed in *Potentilla glandulosa* var. *pseudorupestris*, minus the extremes at the northeastern and southern ends of the species range. The circumscription here encompasses significant heterogeneity, which might be resolved into additional taxa upon further analysis [for example, the dwarfed high elevation form of *P. glandulosa* noted by N. H. Holmgren (1997b)]. Some collections from Alberta and Washington have petals as large as those of var. *pseudorupestris* but are here included in var. *saxicola* on the basis of stature. Plants from southwestern Idaho and southeastern Oregon combine features of *Drymocallis lactea* and *D. pseudorupestris*, some being atypically tall (to 5.5 dm) but with the vestiture and saxicolous preference of var. *saxicola*. In contrast, populations on Steens Mountain, Oregon, including the type of *D. pumila* Rydberg, have the typical habit of var. *saxicola* but the vestiture of *D. lactea*; their optimal placement is unresolved. In California, var. *saxicola* occurs in the Cascade Range (Mount Lassen, Mount Shasta) and extends sporadically south through the Sierra Nevada to Tulare County, intergrading with var. *crumiana* and *D. lactea*.

9c. Drymocallis pseudorupestris (Rydberg) Rydberg var. **crumiana** D. D. Keck ex Ertter, J. Bot. Res. Inst. Texas 1: 39, figs. 1G–L. 2007 • Crum's drymocallis or wood beauty [E]

Stems (0.3–)0.8–2(–2.5) dm, base 1–2 mm diam., short hairs absent or sparse. **Basal leaves** (2–)3–9(–15) cm, not to sparsely short-hairy, sparsely to moderately subsessile golden glandular, rigidly bristly (bristles 0.5–1 mm); leaflet pairs 3 or 4(or 5); terminal leaflet broadly obovate to flabellate, 0.2–1.2(–2) × 0.5–1.5 cm, teeth ± single, 2–5 per side. **Pedicels** 5–12 (proximal to 20) mm, bristly, short hairs absent or sparse. **Flowers** 2–8; hypanthia and sepals prominently bristly, bristles 1–1.5 mm; epicalyx bractlets elliptic-ovate to lanceolate, 2–3.5(–4) × 1–2 mm; sepals 4–6(–7) mm; petals not or scarcely overlapping, often red-tinged, narrowly to broadly obovate, 5–7(–9) × 3–5 mm; filaments 1.5–2(–3) mm; styles usually dark red, rarely golden brown.

Flowering Jul–Aug. Rocky slopes, talus, and ledges, on metamorphic, granitic, and volcanic substrates; 3200–3900 m; Calif.

The red-tinged petals of var. *crumiana*, initially noted by E. Crum (as summarized by B. Ertter 2007), are not reliably diagnostic; however, the distinctive leaflet aspect is not found elsewhere in *Drymocallis*, though approaching *D. fissa* in Colorado. The vestiture includes prominent golden glands and conspicuous bristles that are rarely found in other North American *Drymocallis*, and the stigmas are commonly (though not consistently) dark reddish, not the more common pale golden brown. The variety occurs in the southern Sierra Nevada in Fresno, Inyo, and Tulare counties, and in the White Mountains of Mono County.

10. Drymocallis glabrata Rydberg, Monogr. N. Amer. Potentilleae, 201, plate 109. 1898 • Idaho drymocallis or wood beauty [E]

Drymocallis glandulosa (Lindley) Rydberg subsp. *glabrata* (Rydberg) Soják; *Potentilla glandulosa* Lindley subsp. *glabrata* (Rydberg) D. D. Keck; *P. glandulosa* var. *incisa* Lindley; *P. glandulosa* var. *intermedia* (Rydberg) C. L. Hitchcock

Caudex branches short. **Stems** ± tufted, (2–)2.5–6(–8) dm; base 2–3 mm diam., moderately to densely septate-glandular. **Leaves** glabrate or sparsely hairy (hairier at northern margin of range); basal (8–)20–30 cm, leaflet pairs 2–4; terminal leaflet ± obovate, 2–6(–7) × (1–)1.5–3 cm, teeth ± double, 5–25 per side, apex obtuse to acute; cauline 0–2, well developed, leaflet pairs 1–3. **Inflorescences** (5–)10–30 (–60)-flowered, leafy, open, ⅙–⅔ of stem, wide, branch angles (20–)30–60°. **Pedicels** 5–40 (proximal to 45) mm, sparsely to moderately short-hairy, predominantly septate-glandular. **Flowers** opening widely; epicalyx bractlets linear-oblanceolate, 2–6 × 0.5–1 mm; sepals spreading, 4–7(–8) mm, apex usually acute, rarely obtuse; petals ± overlapping, spreading, usually bright, sometimes pale, yellow, broadly obovate to round, 5–12 × (4–)5–11 mm, usually longer, rarely shorter, than sepals; filaments 1.5–4 mm, anthers 0.7–1 mm; styles thickened, 1 mm. **Achenes** brown to reddish, 1.2 mm.

Flowering May–Aug. Montane meadows, open forests; 400–2800 m; Idaho, Mont., Nev., Oreg., Utah, Wash., Wyo.

Drymocallis glabrata is the common member of the genus in central Idaho and adjacent parts of neighboring states, characterized by acute, glabrate leaves, leafy, widely branched inflorescences, relatively long pedicels with slender, septate glands, linear-oblanceolate epicalyx bractlets, and relatively large, bright yellow petals. Plants from northern Idaho, at the lower elevational extreme, tend to be hairier than those elsewhere in the species range. The species enters Utah in Cache and Weber counties; most other Utah populations placed by S. L. Welsh et al. (1993) in this taxon (as *Potentilla glandulosa* var. *intermedia*) are treated here as *D. deseretica*.

11. Drymocallis deseretica Ertter, J. Bot. Res. Inst. Texas 1: 41, fig. 2. 2007 • Deseret drymocallis or wood beauty [E]

Caudex branches short. **Stems** ± tufted, (1.5–)2.5–6(–6.5) dm; base 2–3(–4) diam., sparsely to densely septate-glandular. **Leaves** sparsely hairy; basal (5–)7–20 cm, leaflet pairs (2 or)3; terminal leaflet broadly obovate to rhombic, (1.5–)2–4 × (1–)1.5–3 cm, teeth single or double, 5–9 per side, apex acute to obtuse; cauline 1 or 2, well developed, leaflet pairs 2 or 3. **Inflorescences** 3–15(–20)-flowered, leafy, ± compact, (⅙–)⅕–⅓(–½) of stem, narrow, branch angles 10–20°. **Pedicels** 2–15 (proximal to 20) mm, predominantly short-hairy, often velutinous, not or sparsely to moderately septate glandular. **Flowers** opening widely; epicalyx bractlets linear to lanceolate or narrowly elliptic, (2.5–)3–8 × 0.5–2(–3) mm; sepals spreading, (5–)6–12(–15) mm, apex usually acute, sometimes obtuse; petals rarely overlapping, spreading, cream-white to light yellow, narrowly to broadly obovate, (4–)6–10 × (2.5–)3.5 6(–7) mm, usually shorter than sepals; filaments 1–3 mm, anthers 0.8–1 mm; styles very thickened, 1 mm. **Achenes** light brown, 1.2–1.5 mm.

Flowering (May–)Jun–Sep. Openings among sagebrush, aspen, fir, and/or spruce forests, often where moist or rocky, below cliffs; 2000–3300 m; Utah, Wyo.

Drymocallis deseretica is common in the Wasatch and western Uinta mountains of northern and central Utah, where it has usually been treated as *Potentilla glandulosa* var. *intermedia* (= *D. glabrata*) or *P. glandulosa* var. *pseudorupestris* (= *D. pseudorupestris* var. *saxicola*). It differs from both taxa in its more compact inflorescences, predominantly short-hairy pedicels, and sepals that conspicuously enlarge in fruit, which create a distinctive aspect. Comparable plants with shorter, obtuse sepals occur in the Raft River Mountains; their optimal placement remains to be determined. Inclusion of Wyoming in the species range is based on *Goodding 1993* (UT) from Bridger Peak, Carbon County. *Drymocallis deseretica* intergrades with *D. glabrata* to the north and *D. arizonica* to the south, with the exact range yet to be determined.

12. Drymocallis glandulosa (Lindley) Rydberg, Monogr. N. Amer. Potentilleae, 198. 1898 [F]

Potentilla glandulosa Lindley, Edwards's Bot. Reg. 19: plate 1583. 1833

Caudex branches short. **Stems** tufted, (1–)1.5–7.5 dm; base 1–5 mm diam., moderately to densely septate-glandular. **Leaves** sparsely to moderately hairy; basal (3–)6–30 cm, leaflet pairs (1 or)2–4; terminal leaflet broadly obovate to rhombic-elliptic, (1–)1.5–7(–8) × 1–5 cm, teeth usually double (irregularly single in var. *viscida*), (4–)6–18(–20) per side, apex rounded to obtuse; cauline 0–3, developed or reduced, leaflet pairs 1–4. **Inflorescences** (2–)5–40 -flowered, leafy or not, open or of congested clusters terminating branches, $(^1/_{10}–)^1/_5–^3/_4$ of stem, usually wide (narrow in var. *viscida*), branch angles (15–)20–55°. **Pedicels** 1–10 (proximal to 30) mm, not or sparsely short-hairy, predominantly septate-glandular. **Flowers** opening widely; epicalyx bractlets linear-lanceolate to elliptic, 1.5–7(–8) × 0.5–2.5(–3) mm; sepals spreading or reflexed, (3–)4–11 mm, apex usually broadly obtuse with mucronate tip, sometimes acute; petals not overlapping, spreading or reflexed, cream-white to yellow, narrowly to broadly obovate or ovate to nearly round, 2–6.5 × 1.5–5.5 mm, shorter than or equal to sepals; filaments 0.5–2 mm, anthers 0.5–1 mm; styles thickened, 1–1.5 mm. **Achenes** reddish brown, 0.9–1.5 mm.

Varieties 4 (4 in the flora): w North America, nw Mexico.

Even with the removal of most of the subspecies of *Potentilla glandulosa* treated by Keck, *Drymocallis glandulosa* encompasses an inordinate range of variation. This is because of the apparently clinal variation between the coastal extreme (var. *wrangelliana*), with leafy inflorescences and broad cream-white petals, the montane var. *reflexa*, with less leafy inflorescences and narrow yellow petals, and the transitional interior var. *glandulosa*. Variety *reflexa*, in turn, intergrades in the mountains of southern California with var. *viscida*, which is anomalous in *D. glandulosa* with its narrow inflorescences and irregularly single-toothed leaflets. Variation outside of California is unclear, with no evident pattern to differences in petal size and color; these have largely been assigned to var. *glandulosa*.

At its most distinctive, *Drymocallis glandulosa* is a robust, densely glandular plant with widely branched leafy inflorescences (except var. *viscida*), relatively short, stout pedicels, broad sepals with rounded, mucronate tips (except var. *viscida* and sometimes var. *reflexa*) that are often significantly enlarged in fruit, and non-overlapping petals that are shorter than to scarcely equal to the sepals. The relatively large achenes are blunter and redder than those of most other species of *Drymocallis*, as discussed by J. Clausen et al. (1940).

1. Petals obovate-elliptic to broadly ovate to nearly round, (2–)3.5–5.5 mm wide, cream-white to ± yellow; sepals spreading, apices obtuse to rounded; inflorescences ± to very leafy; pedicels 1–5 (proximal to 30) mm.
   2. Petals (3–)3.5–5 mm, usually ± yellow, sometimes cream-white; inland . . . . 12a. *Drymocallis glandulosa* var. *glandulosa*
   2. Petals (4–)4.5–6.5 mm, usually cream-white, sometimes pale yellow; ± coastal . . . . 12b. *Drymocallis glandulosa* var. *wrangelliana*
1. Petals narrowly ovate to narrowly obovate or obovate-elliptic, 1.5–3 mm wide, yellow; sepals reflexed, apices acute to obtuse; inflorescences ± leafy or not; pedicels 2–10 (proximal to 20) mm.
   3. Inflorescences usually wide, branch angles (15–)25–55°; leaflet teeth usually double, rarely single . . . . 12c. *Drymocallis glandulosa* var. *reflexa*
   3. Inflorescences narrow, branch angles 15–30 (–40)°; leaflet teeth single or double. . . . . 12d. *Drymocallis glandulosa* var. *viscida*

12a. Drymocallis glandulosa (Lindley) Rydberg var. **glandulosa** • Douglas's drymocallis or wood beauty [E] [F]

Stems 2–6 dm, base 2–3(–4) mm diam. **Leaves:** basal (5–)10–25 cm, leaflet pairs (1–)2–3(–4); terminal leaflet obovate to nearly round, 2–6(–8) × (1–)2–5 cm, teeth double, 7–15(–20) per side, apex obtuse to rounded; cauline well developed, leaflet pairs 1–3. **Inflorescences** 5–40 -flowered, ± to very leafy, wide, branch angles (20–)30–50°. **Pedicels** 1–5 (proximal to 30) mm. **Flowers:** epicalyx bractlets linear-lanceolate to elliptic, 2–6(–8) × 0.5–2 mm; sepals spreading, 4.5–10 mm, apex ± obtuse, mucronate; petals spreading, usually ± yellow, sometimes cream-white, obovate-elliptic to nearly round, (3–)3.5–5 × (2–)3–4 mm; filaments (0.5–)1–2 mm. **Achenes** 1–1.4 mm. $2n = 14$.

Flowering (Mar–)Apr–Jul. Open slopes, stream banks, road banks, shrublands, open woodlands; 400–2500 m; B.C.; Ariz., Calif., Idaho, Mont., Nev., Oreg., Utah, Wash.

Variety *glandulosa* is the most widespread variety of *Drymocallis glandulosa* and the primary one occurring outside of California. It is most common away from the coast from southern British Columbia to central California, with scattered (and possibly introduced) collections known from Idaho, Montana, Nevada, and Utah. Disjunct, consistently white-petaled populations in Arizona south of the Mogollon rim may represent a distinct variety.

12b. Drymocallis glandulosa (Lindley) Rydberg var. **wrangelliana** (Fischer & Avé-Lallemant) Ertter, J. Bot. Res. Inst. Texas 1: 43. 2007 • Coast drymocallis or wood beauty [F]

Potentilla wrangelliana Fischer & Avé-Lallemant, Index Seminum (St. Petersburg) 7: 54. 1841; *Drymocallis albida* Rydberg; *P. glandulosa* Lindley var. *wrangelliana* (Fischer & Avé-Lallemant) Th. Wolf

Stems (1–)2–7 dm, base 2–5 mm diam. **Leaves:** basal (5–)10–30 cm, leaflet pairs (2 or)3 or 4; terminal leaflet broadly obovate, (2–)3–7 × 2–5 cm, teeth double, 9–17 per side, apex usually rounded, rarely obtuse; cauline well developed, leaflet pairs 2–4. **Inflorescences** (3–)5–30-flowered, ± to very leafy, wide, branch angles (15–)30–55°. **Pedicels** 1–5 (proximal to 30) mm. **Flowers:** epicalyx bractlets ± elliptic, (3–)4–7 × 1.5–2.5(–3) mm; sepals spreading, 5–11 mm, apex round-obtuse, mucronate; petals spreading, usually cream-white, sometimes pale yellow, broadly ovate or obovate, (4–)4.5–6.5 × 3–5.5 mm; filaments 1–2 mm. **Achenes** 1.2–1.5 mm. **2*n*** = 14.

Flowering Mar–Jun. Openings, moist and/or ± shaded places in coastal scrub and woodlands; 0–1000 m; Calif., Oreg., Wash.; Mexico (Baja California).

Variety *wrangelliana* is provisionally resurrected for the coastal extreme with exceptionally leafy inflorescences and relatively large, most commonly cream-white petals.

12c. Drymocallis glandulosa (Lindley) Rydberg var. **reflexa** (Greene) Ertter, J. Bot. Res. Inst. Texas 1: 43. 2007 • Greene's drymocallis or wood beauty

Potentilla glandulosa Lindley var. *reflexa* Greene, Fl. Francisc., 65. 1891; *Drymocallis glandulosa* (Lindley) Rydberg subsp. *reflexa* (Greene) Soják; *P. glandulosa* subsp. *reflexa* (Greene) D. D. Keck

Stems 1.5–7.5 dm, base 1.5–3 mm diam. **Leaves:** basal (3–)6–20(–30) cm, leaflet pairs 3 (or 4); terminal leaflet ± broadly obovate, (1–)1.5–5(–7) × 1–3 cm, teeth usually double, rarely single, 6–16 per side, apex rounded to obtuse; cauline well developed or reduced, leaflet pairs 2 or 3. **Inflorescences** (2–)10–30-flowered, ± leafy or not, usually wide, branch angles (15–)25–55°. **Pedicels** 2–10 (proximal to 20) mm. **Flowers:** epicalyx bractlets narrowly elliptic, 1.5–4 × 0.5–1(–1.5) mm; sepals reflexed, (3–)4–6(–8) mm, apex usually obtuse, sometimes acute; petals reflexed, yellow, narrowly ovate or obovate, 3–4(–5) × 1.5–3 mm; filaments 0.5–1.5(–2) mm. **Achenes** 0.9–1.3 mm. **2*n*** = 14.

Flowering May–Aug. Moist, somewhat shaded places, often somewhat disturbed; 400–2600 m; Calif., Nev., Oreg.; Mexico (Baja California).

At its morphologic extreme, var. *reflexa* is distinctive in its small, elliptic-oblanceolate, reflexed yellow petals, and widely branched inflorescences; however, in northern California and southern Oregon, var. *reflexa* and var. *glandulosa* completely intergrade. As circumscribed here, var. *reflexa* occurs at low to middle elevations in the mountains from northern Baja California, Mexico, and the Peninsular and Transverse ranges of southern California, through the Sierra Nevada to the mountains of northern California, barely entering adjacent Nevada and Oregon. Some disjunct collections from central Idaho approach this variety; their optimal disposition remains unclear.

12d. Drymocallis glandulosa (Lindley) Rydberg var. **viscida** (Parish) Ertter, J. Bot. Res. Inst. Texas 1: 43. 2007 • Parish's drymocallis or wood beauty [E] [F]

Drymocallis viscida Parish, Bot. Gaz. 38: 460. 1904

Stems (1.5–)2–6.5 dm, base 1–4 mm diam. **Leaves:** basal (3–)6–20 cm, leaflet pairs 2 or 3(or 4), terminal leaflet obovate to rhombic-elliptic, (1–)2–4 × 1–2.5 cm, teeth single or double, 4–9 per side, apex rounded to obtuse; cauline moderately developed, leaflet pairs 2 or 3. **Inflorescences** (2–)5–30-flowered, not leafy, narrow, branch angles 15–30(–40)°. **Pedicels** 2–10 (proximal to 20) mm. **Flowers:** epicalyx bractlets lanceolate-elliptic, (1.5–)2–3(–4.5) × 0.7–1.1(–1.5) mm; sepals reflexed, 4–6(–8) mm, apex obtuse to acute; petals reflexed, yellow, narrowly obovate-elliptic, 2–4(–5) × 1.5–2.5 mm; filaments 0.5–1.5(–2) mm. **Achenes** 0.9–1 mm.

Flowering Jun–Aug. Moist openings, montane forests; 1100–2500 m; Calif.

Variety *viscida* is provisionally resurrected to accommodate plants in the mountains of southern California that combine the flowers and vestiture of var. *reflexa* with the narrow inflorescences and frequently single-toothed leaflets of *Drymocallis lactea* var. *lactea*.

13. Drymocallis campanulata (C. L. Hitchcock) Ertter, J. Bot. Res. Inst. Texas 1: 43. 2007 • John Day drymocallis or wood beauty [E]

Potentilla glandulosa Lindley var. *campanulata* C. L. Hitchcock in C. L. Hitchcock et al., Vasc. Pl. Pacif. N.W. 1: 861. 1969

Caudex branches elongate. **Stems** openly tufted to loosely spaced, 1.5–4.5 dm; base 1.2–3.5 mm diam., ± densely septate-glandular. **Leaves** sparsely to ± densely hairy; basal 6–20 cm, leaflet pairs (2 or)3 or 4(or 5); terminal leaflet broadly obovate to nearly round, 1–4 × 1–3.5 cm, teeth usually ± double, 5–10 per side, apex rounded; cauline 0–2, well developed, leaflet pairs 2 or 3. **Inflorescences** 5–50-flowered, leafy, open, (1/5–)1/2(–4/5) of stem, wide, branch angles 25–50°. **Pedicels** 2–5 (proximal to 10) mm, sparsely to moderately short-hairy, predominantly septate-glandular. **Flowers** opening narrowly; epicalyx bractlets usually lanceolate, sometimes linear, 3–4 × 1–1.5 mm; sepals ± erect, 5–8 mm, apex acute to obtuse; petals overlapping, ± erect, light yellow, broadly obovate, 5–11 × (3–)5–8 mm, equal to or exceeding sepals; filaments 1.5–3 mm, anthers 1 mm; styles slender, (1–)1.5–2.5 mm. **Achenes** light brown, 1 mm.

Flowering May–Jul. Basaltic cliffs and talus above streams; 700–1400 m; Oreg.

Drymocallis campanulata is one of the more distinct species in the genus, with its campanulate flowers, relatively large butter yellow petals, and densely glandular pedicels. It occurs primarily in the canyons of the John Day River in Grant and Wheeler counties.

The invalidly published *Potentilla campanulata* D. D. Keck provides a full description and illustration for the subsequent basionym.

14. Drymocallis rhomboidea (Rydberg) Rydberg, Monogr. N. Amer. Potentilleae, 203. 1898 • Siskiyou or globose drymocallis or wood beauty [E] [F]

Potentilla rhomboidea Rydberg, Bull. Torrey Bot. Club 23: 248. 1896; *Drymocallis glandulosa* (Lindley) Rydberg subsp. *globosa* (D. D. Keck) Soják; *P. glandulosa* Lindley subsp. *globosa* D. D. Keck

Caudex branches short. **Stems** ± tufted, 1.7–4.7 dm; base 1–3 mm diam., usually not septate-glandular. **Leaves** moderately to densely hairy; basal 5–12 cm, leaflet pairs 3; terminal leaflet obovate-elliptic, 1–2.5(–3) × 1–2(–2.5) cm, teeth single or double, 6–10 per side, apex obtuse to rounded; cauline 0–2, developed or reduced, leaflet pairs 2 or 3. **Inflorescences** 5–35-flowered, ± leafy, usually ± open, 1/3–2/3 of stem, wide, branch angles 15–60°. **Pedicels** 1–5 (proximal to 15) mm, predominantly short-hairy, usually not or sparsely, sometimes moderately, septate-glandular. **Flowers** opening narrowly; epicalyx bractlets linear-oblanceolate, 2–3.5 × 1 mm; sepals ± erect, 4–6(–8) mm, apex obtuse; petals not overlapping, ± erect, cream-white, usually narrowly obovate, sometimes oblanceolate, 3–5 × 2–3 mm, usually ± shorter than sepals; filaments 1.5–3 mm, anthers 0.6–1 mm; styles slender, 1.5–2.5 mm. **Achenes** light reddish brown, 1–1.2 mm. **2*n*** = 14.

Flowering Jun–Jul(–Aug). Dry slopes and outcrops in open forests; 1200–2500 m; Calif., Oreg.

Drymocallis rhomboidea is centered in the Siskiyou Mountains. It differs from *D. campanulata* in its smaller, cream-white petals and globular flowers; in addition, the vestiture is usually predominantly short-hairy, except in scattered California populations. Plants from Mount Ashland, the type locality, tend to have more congested, leafier inflorescences than plants found elsewhere in the range.

15. Drymocallis cuneifolia Rydberg, Monogr. N. Amer. Potentilleae, 204, plate 111 [as cuneata]. 1898 [C] [E] [F]

Potentilla peirsonii Munz

Caudex branches elongate. **Stems** openly tufted to loosely spaced, 0.5–4.5 dm; base 1–2 mm diam., sparsely septate-glandular. **Leaves** sparsely to moderately hairy; basal 2–15 cm, leaflet pairs (2 or)3–5; terminal leaflet broadly cuneate to nearly round, 0.6–2(–3.5) × 0.5–2 cm, teeth irregularly single, 2–4 per side, apex rounded to truncate; cauline 0 or 1(or 2), reduced, leaflet pairs 1 or 2. **Inflorescences** 3–10(–15)-flowered, not leafy, open, 1/4–3/4 of stem, wide, branch angles 20–75°. **Pedicels** 2–15 (proximal to 30) mm, sparsely to moderately short-hairy, septate-glandular. **Flowers** opening narrowly; epicalyx bractlets linear to narrowly elliptic or lanceolate, 1–2 × 0.5 mm; sepals erect, 2–5 mm, apex broadly obtuse; petals scarcely overlapping, erect, yellow, narrowly obovate, 2–4 × 1.5–2.5 mm, ± equal to or slightly longer than sepals; filaments 1.5–2.5 mm, anthers 0.5–0.8 mm; styles slender, 1.5–2.5 mm. **Achenes** brown, 0.8–1.2 mm.

Varieties 2 (2 in the flora): California.

Drymocallis cuneifolia has a complex nomenclatural history, complicated by the extreme rarity of the typical variety. Rydberg described the species on the basis of a single collection from the San Bernardino Mountains of San Bernardino County, probably near Green Lead Mines. When comparable (though significantly smaller) plants were found in the San Gabriel Mountains of Los Angeles County, P. A. Munz and I. M. Johnston (1925) adopted *Potentilla cuneifolia* (Rydberg) Th. Wolf for both extremes; this name was later replaced with *P. peirsonii* because of the earlier *P. cuneifolia* Bertoloni. In the absence of comparable new collections from the San Bernardino Mountains, D. D. Keck (in J. Clausen et al. 1940) concluded that the type of *D. cuneifolia* was merely an immature specimen of *D. lactea* var. *lactea* and accordingly described *P. glandulosa* subsp. *ewanii* to accommodate populations in the San Gabriel Mountains.

The 2004 discovery of a small population of plants comparable to the type of *Drymocallis cuneifolia* from near the type locality (*Elvin 3555*, IRVC, UCR) confirms that *D. cuneifolia* is a valid taxon that shares most of the diagnostic features of populations in the San Gabriel Mountains. Plants in the San Bernardino Mountains tend to be larger than those in the San Gabriel Mountains, with fewer and somewhat differently shaped leaflets; they are accordingly treated here as varieties.

1. Stems 2–4.5 dm; basal leaves 5–15 cm; San Bernardino Mountains . . . . 15a. *Drymocallis cuneifolia* var. *cuneifolia*
1. Stems 0.5–2(–2.5) dm; basal leaves 2–10 cm; San Gabriel Mountains . . . . 15b. *Drymocallis cuneifolia* var. *ewanii*

15a. Drymocallis cuneifolia Rydberg var. **cuneifolia** • Wedgeleaf drymocallis or wood beauty C E

Stems 2–4.5 dm. **Basal leaves** 5–15 cm, leaflet pairs (2 or) 3 or 4(or 5); terminal leaflet broadly cuneate-flabellate, 1–2.5(–3.5) cm. **Epicalyx bractlets** linear to narrowly elliptic, 0.8–2 × 0.3–0.5 mm. **Achenes** 1–1.2 mm.

Flowering Jun–Aug(–Sep). Stream edges in riparian scrub; of conservation concern; 1800–2300 m; Calif.

Variety *cuneifolia* is currently known from only two populations on the north face of the San Bernardino Mountains, San Bernardino County.

15b. Drymocallis cuneifolia Rydberg var. **ewanii** (D. D. Keck) Ertter, J. Bot. Res. Inst. Texas 1: 44. 2007 • Ewan's drymocallis or wood beauty C E F

Potentilla glandulosa Lindley subsp. *ewanii* D. D. Keck, Publ. Carnegie Inst. Wash. 520: 47. 1940; *Drymocallis glandulosa* (Lindley) Rydberg subsp. *ewanii* (D. D. Keck) Soják

Stems 0.5–2(–2.5) dm. **Basal leaves** 2–10 cm, leaflet pairs 3–5; terminal leaflet broadly cuneate to nearly round, 0.6–1.5 cm. **Epicalyx bractlets** linear-elliptic to lanceolate, 1 × 0.5 mm. **Achenes** 0.8–1 mm.

Flowering May–Jul. Edges of seeps, springs, and waterways, in lower montane coniferous forests; of conservation concern; 1900–2400 m; Calif.

Variety *ewanii* is known from only four occurrences in the Dawson Saddle area of the San Gabriel Mountains, Los Angeles County.

16. DASIPHORA Rafinesque, Autik. Bot., 167. 1840 • Bush cinquefoil [Greek *dasys*, hairy, and *phoros*, bearing, alluding to villous hypanthium]

Barbara Ertter

James L. Reveal

Potentilla Linnaeus [unranked] *Comocarpa* Torrey & A. Gray; *Potentilla* sect. *Comocarpa* (Torrey & A. Gray) O. Stevens; *Potentilla* subg. *Dasiphora* (Rafinesque) Jepson; *Potentilla* sect. *Fruticosae* (Döll) Grenier & Godron; *Potentilla* sect. *Rhopalostylae* (Th. Wolf) Guşuleac

Shrubs, (0.2–)1–10(–16) dm; taprooted. **Stems:** number not recorded, prostrate to erect; bark reddish brown or grayish, <papery, shredding>; lateral short shoots often present (bearing fascicles of leaves); unarmed; villous first year, then glabrous. **Leaves** deciduous, sometimes persistent, cauline, alternate, usually ± odd-pinnate to subpalmate (jointed beyond stipules and at distal leaflet node), rarely ternate; stipules persistent, proximally adnate to base of petiole, broadly lanceolate to ovate, margins entire; petiole present; blade ± ovate in outline, 0.5–2.6 cm, foliaceous, leaflets (3–)5(–7)[–9], <proximal pairs confluent on rachis>, linear to narrowly elliptic or oblong, margins flat or revolute, entire, venation pinnate, surfaces glabrous or hairy. **Inflorescences** terminal and on lateral twigs, solitary flowers or 2–7(–30)-flowered, loosely cymose, glabrous or hairy; bracts present, ± leafy; bracteoles absent. **Pedicels** present, <straight>. **Flowers** occasionally functionally unisexual (plants monoecious), [10–]15–30 [–35] mm diam.; epicalyx bractlets 5; hypanthium patelliform to cupulate, 1–2 × 3.5–5 mm, villous; sepals 5, spreading, ovate [broadly elliptic]; petals 5, yellow [white], obovate to orbiculate, <longer than sepals>; stamens 20 or 25(–30), shorter than petals, <filaments filiform, glabrous, theca solitary, horseshoe-shaped, rimming broad connective, dehiscing by continuous marginal slit>; torus hemispheric; carpels 30–100, hirsute, styles sub-basal, <clavate>; ovule 1. **Fruits** aggregated achenes, individually deciduous, 30–100, obliquely ovoid, 1.4–1.8(–2) mm, hirsute; hypanthium persistent; sepals persistent, erect; styles tardily deciduous, <jointed>. $x = 7$.

Species ca. 12 (1 in the flora): North America, Eurasia.

Dasiphora comprises the shrubby cinquefoils, which are well known in the horticultural trade. All species occur in Asia; only *D. fruticosa* is also native to the flora area. Molecular analysis confirms the distinction of these species from *Potentilla* (T. Eriksson et al. 1998; M. Lundberg et al. 2009; C. Dobeš and J. Paule 2010), with morphological support provided by the shrubby habit, unique leaflet division, and anther structure (J. Soják 1989). *Pentaphylloides* Duhamel has been misapplied to this genus (for example, N. H. Holmgren 1997b); that name is based on *Potentilla* and accordingly not legitimate. Contrary to W. A. Weber and R. C. Wittmann (1996), *Dasiphora* is not nomenclaturally superfluous; P. A. Rydberg (1898) lectotypified it on *P. fruticosa*.

SELECTED REFERENCE Klackenberg, J. 1983. The holarctic complex *Potentilla fruticosa*. Nordic J. Bot. 3: 181–191.

1. **Dasiphora fruticosa** (Linnaeus) Rydberg, Monogr. N. Amer. Potentilleae, 188. 1898 • Shrubby cinquefoil, potentille frutescente [F]

Potentilla fruticosa Linnaeus, Sp. Pl. 1: 495. 1753; *Dasiphora floribunda* (Pursh) Rafinesque; *D. fruticosa* subsp. *floribunda* (Pursh) Kartesz; *D. fruticosa* var. *tenuifolia* (D. F. K. Schelchtendal) Rydberg; *P. floribunda* Pursh; *P. fruticosa* subsp. *floribunda* (Pursh) Elkington; *P. tenuifolia* D. F. K. Schelchtendal

Plants prostrate or spreading to weakly erect. **Leaves:** stipules often sheathing, 6–12 mm, scarious, glabrous; petiole 0.5–1(–1.5) cm; leaflets 0.5–2.5 × 0.2–1 cm, subglabrous to appressed silky-villous especially abaxially, sometimes merely villous on veins. **Pedicels** 5–15(–30) mm, elongating in fruit, thinly villous. **Flowers:** epicalyx bractlets narrowly lanceolate, 4–13 mm, usually equal to or shorter than sepals, sometimes revolute, villous; sepals 4–6(–9) mm, apex acuminate, villous; petals 5–15 mm; anthers 0.9–1.2 mm, connective reddish brown to blackish, theca yellowish; styles 0.8–1.5 mm. **Achenes** light brown. $2n = 14, 28$ (Europe).

Flowering summer. Sandy to gravelly shores, flats and rocky slopes, on ledges and bluffs in conifer and deciduous woodlands, meadows, or fens, alpine woodlands and meadows; 0–3600 m; St. Pierre and Miquelon; Alta., B.C., Man., N.B., Nfld. and Labr., N.W.T., N.S., Nunavut, Ont., Que., Sask., Yukon; Alaska, Ariz., Calif., Colo., Conn., Idaho, Ill., Ind., Iowa, Maine, Mass., Mich., Minn., Mont., Nev., N.H., N.J., N.Mex., N.Y., N.Dak., Ohio, Oreg., Pa., R.I., S.Dak., Utah, Vt., Wash., Wis., Wyo.; Eurasia.

Dasiphora fruticosa is circumboreal, encompassing significant variation, some of which has often been given taxonomic recognition on the basis of ploidy level, sexuality, vestiture, leaflet shape, and flower size. The phase that occurs in North America is monoecious; it is also found throughout much of Asia and portions of Europe. Although sometimes distinguished as *D. floribunda*, this phase is only weakly differentiated morphologically from the tetraploid, usually dioecious, phase (typical *D. fruticosa*) that occurs in the Old World, where the two phases often occur in mixed populations (J. Klackenberg 1983). Given this complexity, this taxon is conservatively treated here as a single species, as has been done in Europe by A. Kurtto and colleagues (in J. Jalas et al. 1972+, vol. 13).

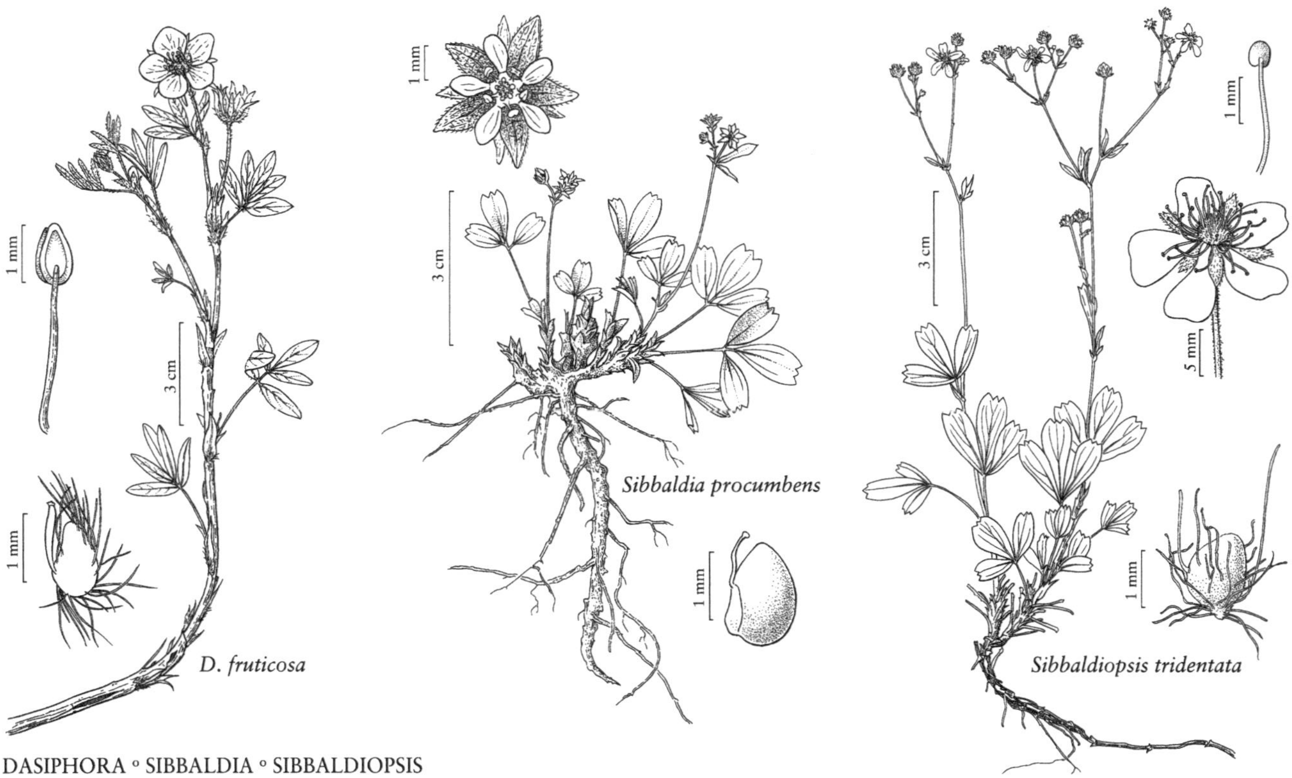

DASIPHORA ◦ SIBBALDIA ◦ SIBBALDIOPSIS

Both the American and European expressions are widely cultivated, with more than 150 cultivars, some of which differ greatly in petal color (for example, white, reddish orange).

Native Americans, mainly in Canada, occasionally brewed a tea from the leaves. Fruits are eaten occasionally by rodents and birds.

The names *Pentaphylloides floribunda* (Pursh) A. Löve and *P. fruticosa* (Linnaeus) O. Schwarz are not legitimate, the latter being widely used prior to 1998 (J. L. Reveal et al. 1999). The combination *P. fruticosa* var. *tenuifolia* (D. F. K. Schelchtendal) Lehmann is superfluous; its counterpart in *Dasiphora* (cited in synonymy) is valid and legitimate.

17. SIBBALDIA Linnaeus, Sp. Pl. 1: 284. 1753; Gen. Pl. ed. 5, 137. 1754 • [For Robert Sibbald, 1642–1722, professor of medicine at Edinburgh]

Barbara Ertter

James L. Reveal

Potentilla Linnaeus sect. *Sibbaldia* (Linnaeus) Hooker f.; *Potentilla* subg. *Sibbaldia* (Linnaeus) Syme

Herbs, perennial, [0.1–]0.2–1.5(–2)[–3] dm, [subglabrous or] sparsely strigose to subappressed pilose [to densely tomentose or sericeous]; horizontal rootstocks branched, compact, woody. **Stems** 1–10+, prostrate to ascending [or erect], <green to reddish>. **Leaves** deciduous or marcescent, mostly basal, alternate, ternate [or simple]; stipules persistent, basally adnate to petiole, sheathing, lanceolate to ovate, margins entire; petiole present; blade bluntly ovate to obovate [or cordate] in outline, 1–3[–4] cm, foliaceous, leaflets 3, cuneate or oblanceolate to broadly elliptic or obovate to nearly orbiculate, margins flat, shallowly (2 or)3(–5)[–7]-toothed apically, <central tooth usually shorter than others>, venation pinnate, surfaces often more densely strigose on veins and margins, sometimes glabrous. **Inflorescences** terminal, (1–)3–12 -flowered, loosely capitate or corymbose (or flowers solitary); bracts present, reduced; bracteoles absent. **Pedicels** present, <straight or curved>. **Flowers** 4–7[–10] mm diam.; epicalyx bractlets 5; hypanthium shallowly cupulate, 0.5–2 × 2–4[–5] mm, usually strigose, sometimes glabrous; sepals 5, spreading, broadly lanceolate to ovate; petals 5, ± yellow, oblanceolate [to obovate], <shorter [to slightly longer] than sepals>; stamens 5[rarely 10], shorter than petals, <filaments columnar, glabrous, anthers with theca solitary, ± horseshoe-shaped, rimming an indistinct connective, dehiscing by continuous marginal slit>; torus hemispheric; carpels 5–15[–30], glabrous, styles lateral, <narrowly clavate>; ovule 1. **Fruits** aggregated achenes, individually deciduous, 5–15[–30], obliquely ovoid, 1.2–1.7 mm, glabrous; hypanthium persistent; sepals persistent, erect; styles tardily deciduous, <jointed>. $x = 7$.

Species 6 (1 in the flora): North America, c Mexico, Eurasia.

When not included within *Potentilla*, *Sibbaldia* has been circumscribed to include 20 or more species, centered in the Himalayas (B. K. Dikshit and G. Panigrahi 1981; Li C. L. et al. 2003). This broad circumscription is not supported by molecular analyses (T. Eriksson et al. 1998, 2003; M. Lundberg et al. 2009; C. Dobeš and J. Paule 2010), in which *S. procumbens* (lectotype of the genus), *S. cuneata* Hornemann ex Kunze, *S. parviflora* Willdenow, and *S. semiglabra* C. A. Meyer form a clade in the Fragariinae; most other species nest within *Potentilla*. These molecular results coincide with morphological features emphasized by J. Soják (1985[1989], 2008), that is, each anther bearing a single horseshoe-shaped theca versus more typical paired thecae, and lateral versus subterminal attachment of narrowly clavate styles.

This citation of six species in *Sibbaldia* follows J. Soják (2008, pers. comm.), largely coinciding with sect. *Eusibbaldia* by O. A. Muravjeva (1936). Included are the four species mentioned above plus *S. aphanopetala* Handel-Mazzetti and *S. trullifolia* (Hooker f.) Chatterjee. The last is an enigmatic, simple-leaved species that is known only from the type; it has not been included in molecular analyses. More recently, Soják (2012) has proposed treating *S. aphanopetala*, *S. parviflora*, and *S. semiglabra* as subspecies of *S. procumbens*, but without validly publishing the corresponding combinations. The three species of *Sibbaldiopsis* have also been included in *Sibbaldia* (J. Paule and Soják 2009; Soják 2012), but too late for full incorporation in the current treatment. All other species traditionally placed in *Sibbaldia* are relegated to *Argentina* Hill, *Piletophyllum* (Soják) Soják, *Potentilla* in the strict sense, or *Sibbaldianthe* Juzepczuk.

SELECTED REFERENCE Muravjeva, O. A. 1936. The genus *Sibbaldia* L. and its species. Trudy Bot. Inst. Akad. Nauk S.S.S.R., Ser. 1, Sist. Vyssh. Rast. 2: 217–241.

1. Sibbaldia procumbens Linnaeus, Sp. Pl. 1: 284. 1753 • Sibbaldie couchée F

Potentilla sibbaldii Haller f.

Leaves: stipules 3–6 mm, membranous; petiole (0.5–) 1–9(–12) cm; leaflets dark green abaxially, light green adaxially, sessile or petiolulate, 0.5–2.5(–3) × 0.2–1.5(–2) cm, teeth roundish, mucronate, surfaces sparsely appressed-pilose. **Pedicels** 1–3(–5) mm, usually minutely stipitate-glandular. **Flowers:** epicalyx bractlets often becoming reddish with age, linear to narrowly lanceolate, 1–3.5 mm, shorter than sepals, sparsely hirsute or strigose; sepals often becoming reddish with age, 2.5–5 mm, strigose to sparsely pilose; anthers 0.3–0.5 mm; styles 0.4–0.7 mm. **Achenes** greenish brown to dark brown. $2n = 14$.

Flowering summer. Dry rocky to gravelly flats and slopes, often in tundra near ocean in the north to subalpine and alpine settings above timberline in the south, snow beds, damp, south-facing herb banks; 0–3800[–4200] m; Greenland; Alta., B.C., Nfld. and Labr., N.W.T., Nunavut, Que., Yukon; Alaska, Ariz., Calif., Colo., Idaho, Mont., Nev., N.H., N.Mex., Oreg., Utah, Wash., Wyo.; Mexico; Eurasia.

Sibbaldia procumbens is circumpolar and is common from southern Greenland (to near 72° N) to the Aleutian Islands (to 166° W) of westernmost Alaska.

It ranges southward in the Coast Ranges to the Siskiyou-Trinity Mountains, as well as Lassen Peak and Mt. Shasta of northern California, and farther inland to the Sierra-Cascade axis as far south as the San Bernardino Mountains of southern California, isolated mountain ranges in the central Great Basin, down the Rocky Mountains to the San Francisco Peaks in Arizona, and to the Sangre de Cristo Mountains of northern New Mexico. Farther south it reaches an elevation of more than 4200 m in the state of México. In eastern Canada and the northeastern United States, it is known from Labrador, the Long Range Mountains of Newfoundland, the Shickshock Mountains of the Gaspé, Quebec, and the White Mountains of New Hampshire.

18. SIBBALDIOPSIS Rydberg, Monogr. N. Amer. Potentilleae, 187, plate 100, figs. 6–10. 1898 • [Genus *Sibbaldia* and Greek *–opsis*, resembling] C

Barbara Ertter

James L. Reveal

Potentilla Linnaeus subg. *Sibbaldiopsis* (Rydberg) B. Boivin; *Potentilla* sect. *Tridentatae* (Th. Wolf) Dikshit & Panigrahi

Subshrubs to suffrutescent herbs, perennial, often ± matted, 0.2–3 dm, sparsely to moderately strigose to pilose; stoloniferous and rhizomatous (caudex diffusely branched). **Stems** 1–10+, ascending to erect, <reddish>. **Leaves** persistent, basal and proximally cauline [or evenly cauline], alternate, ternate; stipules persistent, basally adnate to petiole, sheathing, ovate, margins entire; petiole present; blade ± obovate [or broadly cordate] in outline, (0.5–)1–3(–4) cm, leathery, leaflets 3, oblanceolate to narrowly oblong, margins nearly flat to ± revolute, shallowly 3(–5) -toothed apically, venation pinnate, abaxial surface sparsely to moderately strigose to pilose, adaxial glabrous or sparsely strigose or pilose, rarely densely so. **Inflorescences** terminal, [1–]3–20-flowered, cymose, flat-topped; bracts present, reduced [or leafy]; bracteoles absent. **Pedicels** present, <± straight>. **Flowers** 10–20[–30] mm diam.; epicalyx bractlets 5; hypanthium patelliform, 0.5–1.5[–2] × 3–5 mm, villous; sepals 5, spreading, broadly ovate; petals 5, white, sometimes pink-tinged [or yellow], elliptic-obovate [to orbiculate], <longer than sepals>; stamens 20(–30), shorter than petals, <filaments filiform, not flattened, glabrous, anthers with theca horseshoe-shaped, rimming narrow to broad connective, dehiscing by a continuous, narrow, marginal slit>; torus ± hemispheric; carpels 20–50, hairy, styles attached proximal to middle of achene to sub-basal, <filiform>; ovule 1. **Fruits** aggregated achenes, individually deciduous, obliquely ovoid, 1.2–1.5 mm, villous; hypanthium persistent; sepals persistent, erect; styles tardily deciduous, <jointed>. $x = 7$.

Species 3 (1 in the flora): e North America, Asia.

Rydberg based *Sibbaldiopsis* solely on the North American *S. tridentata*, which T. Wolf (1908) placed in his Grex *Tridentatae* with the Himalayan *Potentilla ambigua* Cambessèdes (not Gaudin) and *P. miyabei* Makino of Hokkaido and the Kuril Islands. Both Asian species have leafier stems and inflorescence bracts than the type species, as well as fewer flowers with broader yellow (versus white) petals. The treatment presented here provisionally follows that by J. Soják (2004, 2008, 2009b) with the inclusion of all three species in *Sibbaldiopsis*, using *S. cuneifolia* (Bertoloni) Soják and *S. miyabei* (Makino) Soják for the Asian species. More recently, however, J. Paule and J. Soják (2009) have proposed to include *Sibbaldiopsis* in *Sibbaldia*, since the species are intermingled in molecular phylogenies. Whether *S. tridentata* itself can be retained in a distinct monospecific genus remains to be determined.

SELECTED REFERENCE Paule, J. and J. Soják. 2009. Taxonomic comments on the genus *Sibbaldiopsis* Rydberg (Rosaceae). Čas. Nár. Muz., Rada Přír. 178(4): 15–16.

1. **Sibbaldiopsis tridentata** (Aiton) Rydberg, Monogr. N. Amer. Potentilleae, 187. 1898 • Three-toothed cinquefoil, potentille tridentée C E F

Potentilla tridentata Aiton, Hort. Kew. 2: 216, plate 9. 1789; *Sibbaldia tridentata* (Aiton) Paule & Soják

Leaves: stipules 3–6 mm, margins villous-ciliate; petiole slender, (0.2–)0.5–6(–8) cm; leaflets dark green abaxially, light green adaxially, (0.5–)1–3(–4) × 0.4–1.5(–1.8) cm. **Pedicels** 2–35 mm, glabrous. **Flowers:** epicalyx bractlets lanceolate, 1–2 mm; sepals 2–3 mm, hirsute, glandular; petals white, sometimes pink-tinged, (4–)5–8[–11] mm; anthers reddish, 0.3–0.6 mm; styles 1–2.5(–3) mm. **Achenes** dark brown. **2*n*** = 14, 28.

Flowering summer. Dry rocky to gravelly shores or maritime habitats, rocky outcrops, montane balds and promontories in mixed conifer-hardwood woodlands, dry meadows of montane conifer communities, alpine tundra, often in acidic soil; of conservation concern; 0–1900 m; Greenland; St. Pierre and Miquelon; Alta., Man., N.B., Nfld. and Labr., N.W.T., N.S., Nunavut, Ont., P.E.I., Que., Sask.; Conn., Ga., Ill., Iowa, Maine, Md., Mass., Mich., Minn., N.H., N.J., N.Y., N.C., N.Dak., Pa., R.I., Tenn., Vt., Va., W.Va., Wis.

Sibbaldiopsis tridentata is uncommon to rare through most of its southern range from Connecticut south to Georgia and west to Iowa, Tennessee, and North Dakota. It is more common in Canada and around the Great Lakes. The inclusion of Scotland in the distribution of the species by P. A. Rydberg (1898) is erroneous (T. Wolf 1908).

19. COMARUM Linnaeus, Sp. Pl. 1: 502. 1753; Gen. Pl. ed. 5, 220. 1754 • Marshlocks [Greek *komaros*, strawberry-tree (*Arbutus unedo* Linnaeus), alluding to similarity of fruit]

Barbara Ertter

James L. Reveal

Potentilla Linnaeus sect. *Comarum* (Linnaeus) Tausch; *Potentilla* subg. *Comarum* (Linnaeus) Syme

Herbs, perennial, 1–6 dm; stoloniferous and rhizomatous. **Stems** 1–several, decumbent to ascending aerial stems arising from horizontal ground-level stems, horizontal stems often rooting at nodes, sometimes floating, aerial stems glabrous proximally, finely appressed-hairy and often purplish-glandular distally. **Leaves** usually marcescent, cauline, alternate, odd-pinnate to subpalmate; stipules marcescent, partially to completely adnate to base of petiole, linear-lanceolate to elliptic-ovate, usually entire, sometimes serrate apically; petiole present, <marcescent>; blade ± cordate in outline, 2–15 cm, foliaceous, leaflets (3–)5–7, lanceolate or oblanceolate to elliptic or obovate, margins flat, crenate to deeply serrate throughout or apically, venation pinnate, surfaces glabrous or finely appressed hairy. **Inflorescences** terminal, 1–10 -flowered, flowers solitary or loosely cymose, hairy and often purplish-glandular; bracts present, leafy, smaller distally; bracteoles absent. **Pedicels** present, <straight or nearly so>. **Flowers** 1–3 cm diam.; epicalyx bractlets 5; hypanthium slightly cupulate, 1–2 × (3–)4–8 mm, villous; sepals 5, spreading to reflexed, broadly ovate; petals 5, deep red to purple, rarely pink, ovate to spatulate, <shorter than sepals>; stamens (15–)20–25(–30), shorter than or equal to petals, <filaments filiform, not flattened, glabrous, thecae solitary, horseshoe-shaped, rimming broad connective, dehiscing by continuous, narrow, marginal slit>; torus hemispheric; carpels 30–100, glabrous, styles lateral to sub-basal, <filiform>; ovule 1. **Fruits** aggregated achenes, individually deciduous (sometimes tardily so), 30–100, obliquely ovoid, (1–)1.3–1.6 mm, glabrous; <torus enlarged, spongy, not fleshy, hairy>; hypanthium persistent; sepals persistent, reflexed; styles persistent, <jointed>. x = 14.

Species 1: North America, Eurasia.

In North America, *Comarum* differs from *Potentilla* by its strikingly deep red to purple petals that are shorter than the sepals and its generally sprawling habit with rooting nodes in distinctly wet habitats. The dark reddish or purplish color generally infuses other floral parts (calyces, anthers, styles). Petals in *Potentilla* sect. *Rubrae* (Rydberg) O. Stevens are similarly colored but more or less obcordate and longer than the sepals, and the plants are otherwise typical of that genus in habit, habitat, and palmate leaves.

The fruiting structure of *Comarum* is intermediate between that of *Fragaria* and those of most other Potentilleae. As in *Fragaria*, the torus becomes enlarged and spongy (but not fleshy) in fruit, albeit completely covered by ultimately deciduous achenes. As supported by molecular data (T. Eriksson et al. 1998, 2003), *Comarum* is more closely related to *Fragaria* than to *Potentilla* in the strict sense. It will hybridize with members of *Fragaria* (J. R. Ellis 1962), giving rise to commercially available pink-flowered strawberries.

This treatment follows J. Soják (2008) in treating *Comarum* as monospecific; the other species sometimes placed in this genus is *Farinopsis salesoviana* (Stephan) Chrtek & Soják.

SELECTED REFERENCES Fernald, M. L. and B. H. Long. 1914. The American variations of *Potentilla palustris*. Rhodora 16: 5–11. Tzvelev, N. N. 2007. On the genus *Comarum* (Rosaceae). Bot. Zhurn. (Moscow & Leningrad) 92: 894–900.

1. **Comarum palustre** Linnaeus, Sp. Pl. 1: 502. 1753 • Purple marshlocks, marsh potentilla, comaret des marais [F]

Comarum palustre subsp. *angustifolium* (Rafinesque) Tzvelev; *C. palustre* subsp. *arcticum* (Gandoger) Tzvelev; *Fragaria palustris* (Linnaeus) Crantz; *Potentilla palustris* (Linnaeus) Scopoli; *P. palustris* var. *parvifolia* (Rafinesque) Fernald & B. H. Long; *P. palustris* var. *villosa* (Persoon) Lehmann

Stems 1–10(–18) dm. **Leaves:** petiole 5–10(–15) cm; leaflets usually abaxially glaucous, adaxially green, sometimes silvery-silky, often aging reddish, 1.5–10(–12) × (0.8–)1–5 cm. **Pedicels** 5–25(–30) mm. **Flowers:** epicalyx bractlets narrowly lanceolate to lanceolate, 2–7(–9) mm, usually equal to or shorter than sepals; sepals greenish purple adaxially, (5–)7–12(–14) mm, apex acuminate; petals 2.5–6(–8) mm; anthers dark purple, 0.8–1.2 mm, connectives thickened, greenish purple to purple; styles 0.9–2 mm. **2*n*** = 28, (35)36, 42, (60, 62, 63)64.

Flowering summer. Shallow, often sphagnum-dominated ponds, bogs, wet meadows, marshes, stream banks, and lake margins, mainly in boreal and low arctic habitats in the north, subalpine to alpine sites in the south; 0–3100 m; Greenland; St. Pierre and Miquelon; Alta., B.C., Man., N.B., Nfld. and Labr., N.W.T., N.S., Nunavut, Ont., P.E.I., Que., Sask., Yukon; Alaska, Calif., Colo., Conn., Idaho, Ill., Ind., Iowa, Maine, Mass., Mich., Minn., Mont., N.H., N.J., N.Y., N.Dak., Ohio, Oreg., Pa., Utah, Vt., Wash., Wis., Wyo.; Eurasia.

Comarum palustre is circumboreal; it occurs in a broad band across the northernmost United States and Canada as far north as the low arctic. Disjunct populations occur well to the north, as on Baffin and Victoria islands, and in the south, as in Gunnison County, Colorado, and the Uinta Mountains of Utah. Appropriate habitats are uncommon and geographically restricted in the southern part of the range, for example, in Idaho, Montana, and North Dakota.

Comarum palustre is sometimes divided into subspecies, varieties, and/or forms on the basis of vestiture, leaflet dimensions, and flower number (for example, M. L. Fernald and B. H. Long 1914; N. N. Tzvelev 2007). Because there is no existing consensus, and a global evaluation of the situation is beyond the scope of this work, no infraspecific taxa are recognized here. Within North America, northern populations tend to have broader leaflets and inflorescences composed of one to relatively few flowers with relatively short sepals and petals, but these differences are not consistent; similar plants occur in subalpine settings in the mountains well to the south. Chromosome numbers also vary, apparently not correlated with morphology or geography; some populations occasionally include individuals of different disploid and polyploid levels.

Horizontal stems were used occasionally by Native Americans as a medicinal aid in the treatment of dysentery and stomach cramps (D. E. Moerman 1998). Aquatic birds in particular eat the leaves and achenes.

COMARUM ◦ ALCHEMILLA ◦ APHANES

20. ALCHEMILLA Linnaeus, Sp. Pl. 1: 123. 1753; Gen. Pl. ed. 5, 58. 1754 • Lady's mantle, alchémille [Arabic name *alkemelyeh*, perhaps alluding to alchemists' interest in reputed marvelous powers of its dew]

John McNeill

Martin Dubé

Peter Frost-Olsen

Herbs, perennial, 0.2–4(–8) dm; rhizomatous, often forming woody stock. **Stems** 1–6(–10), decumbent to ascending or erect, usually pubescent or puberulent, rarely glabrous. **Leaves** not persistent, basal in rosette, alternate, usually simple, sometimes palmately compound; stipules persistent, adnate to petiole, proximally connate opposite petiole (abpetiolar fusion), semiorbiculate to ovate, margins lobed or dentate; petiole present; blade (of simple leaves) reniform to orbiculate, sometimes reniform-orbiculate, 1–15 × 1.5–15 cm, usually palmately 7–9(–11)-lobed, sometimes palmately compound with 5–7 leaflets, margins flat or undulate, crenate to dentate, venation palmate, that of lobes or leaflets pinnate, surfaces usually pubescent or glabrous, sometimes sericeous; cauline: stipules relatively large, variously colored, becoming brownish at maturity or on drying, toothed or lobed, petiole shorter, lobes 3–7(–11), often shallow. **Inflorescences** terminal, multi-flowered, cymes, much-branched; flowers ± aggregated into clusters of 5–15, glabrous or densely pubescent; bracts usually absent; bracteoles absent. **Pedicels** present. **Flowers** 2–4 mm diam.; epicalyx bractlets 4; hypanthium urceolate, 1.5–2.5 mm, glabrous or sparsely to densely pubescent; sepals 4, usually erect to spreading, rarely reflexed (*A. glomerulans*), ovate, <glabrous or pubescent>; petals 0; stamens 4, <inserted outside nectar-secreting receptacular disc almost closing hypanthium, alternating with sepals, anthers

often poorly developed>; torus absent; carpel 1 (rarely, *A. filicaulis*, lowermost flower in a monochasium with 2), styles basal, <stigmas capitate, ± exserted from hypanthia>; ovules 1 (or 2). **Fruits** achenes, 1, ovoid, 2–2.5 mm, <smooth>; hypanthium persistent; sepals persistent, spreading or erect; styles persistent. $x = 8$.

Species 800–1000 (12 in the flora): North America, Eurasia, Africa.

The occurrence of an additional species, not included in the account below, has recently been recognized. This is *Alchemilla acutiloba* Opiz naturalized near St. John's, Newfoundland. In the key below, it would most likely be identified as *A. monticola*. It is most readily distinguished by its almost triangular leaf lobes with straight sides and acute triangular teeth. It has yellowish green to grass green leaves. Unlike *A. xanthochlora*, the adaxial leaf surface is hairy but much less so than that of *A. micans* or *A. monticola*. In addition, it differs from *A. micans* in the rounded hypanthium base and in the stipules being only slightly wine red suffused, and from *A. monticola* by the hypanthium being glabrous or rarely sparsely hairy proximally.

The majority of *Alchemilla* species, including all those in the flora area, are agamospermic, high polyploids (to 28-ploid). Most occurrences in North America are as naturalized introductions. Only four species, restricted to Greenland, St. Pierre and Miquelon, and Atlantic coastal regions of Canada, may be native. Some of the introduced species that occur as scattered meadow and hedgerow plants in their native range grow in eastern Canada in large patches as aggressive weeds.

The name *Alchemilla vulgaris* Linnaeus, used in some North American accounts, was applied in a broad sense to include all the species of sect. *Alchemilla*, those being all the species occurring in the flora area except *A. alpina* (sect. *Alpinae* Camus) and *A. mollis* and *A. venosa* (sect. *Erectae* Fröhner). The type of *A. vulgaris* represents a mixture of *A. glaucescens* and *A. wichurae* (both with very limited occurrences in North America) and the name now is either informally rejected as confused or applied, also informally, in an aggregate sense.

The commonly cultivated species of *Alchemilla* all fall within sect. *Erectae*, distinguished from sect. *Alchemilla* (*A. vulgaris* of authors) by the epicalyx bractlets being as long as the sepals and almost as wide. In addition to *A. venosa*, well established on Cape Breton Island and sporadically elsewhere in Atlantic Canada, *A. mollis* and *A. speciosa* Buser are in cultivation, the former widely so and recently reported as naturalized. These species are distinguished from *A. venosa* by their flat or only slightly undulate, densely pubescent leaves. *Alchemilla speciosa* differs from *A. mollis* by its pubescent pedicels and its epicalyx bractlets narrower than the sepals.

Aphanes, the members of which are readily distinguished by their annual habit, has sometimes been included in *Alchemilla* in North American floras and manuals. Over the past half century, the two genera have generally been maintained as distinct, a pattern set even earlier in North America by P. A. Rydberg (1908). As a result of molecular studies, R. Gehrke et al. (2008) have suggested that *Alchemilla* should be circumscribed more widely to include *Aphanes* and also *Lachemilla* (Focke) Rydberg, which occurs in Mexico, Central America, and South America. Both of these genera are shown to be monophyletic groups, as is *Alchemilla* apart from the African species currently placed in it that form a clade separate from the genus as represented in Eurasia and North America. The African species deserve more thorough investigation before incorporating within *Alchemilla* such a morphologically distinct genus as *Aphanes*.

M. J. M. Christenhusz and S. M. Väre (2012) have proposed 35 new names and combinations in *Potentilla* for those species of *Alchemilla* and *Aphanes* native or cultivated in the Nordic countries. This is said to be based on the molecular findings of C. Dobeš and J. Paule (2010), which support previous results by T. Eriksson et al. (1998). These studies note that when *Potentilla* is traditionally circumscribed (as for example by P. W. Ball et al. 1968), it is paraphyletic and *Alchemilla* is nested within it, leading Chistenhusz and Väre to adopt an inclusive *Potentilla* that Dobeš and Paule consider an impractical and unconventional generic circumscription, a view that the authors share. As laid out by B. Ertter (2007), in this flora the authors follow the alternate option taken by Dobeš and Paule and other authors (A. Kurtto and Eriksson 2003; J. Soják 2008) of regcognizing within the monophyletic subtribe Fragariinae not only *Alchemilla*, *Chamaerhodos*, and *Fragaria*, but also other genera previously segregated from *Potentilla* on morphological grounds, such as *Comarum*, *Dasiphora*, *Drymocallis*, *Farinopsis* Chrtek & Soják, *Sibbaldia*, and *Sibbaldiopsis*.

SELECTED REFERENCE Bradshaw, M. E., P. M. Dansereau, and D. H. Valentine. 1964. Notes on the genus *Alchemilla* in southeastern Canada. Canad. J. Bot. 42: 89–104.

1. Leaves palmately compound . . . . 1. *Alchemilla alpina*
1. Leaves palmately lobed.
   2. Epicalyx bractlet lengths equal to or longer than sepals.
      3. Leaf margins undulate, basal sinuses appearing closed, lobes overlapping, adaxial surfaces glabrous or hairy only on folds; hypanthia glabrous; leaves: middle lobes as long as or longer than their half-widths . . . . 11. *Alchemilla venosa*
      3. Leaf margins flat or slightly undulate, basal sinuses narrow or almost closed, lobes sometimes overlapping, adaxial surfaces hairy throughout; hypanthia usually densely, occasionally sparsely, hairy (in proximal ½); leaves: middle lobes shorter to as long as their half-widths . . . . 12. *Alchemilla mollis*
   2. Epicalyx bractlet lengths shorter than sepals.
      4. Stems glabrous or appressed-hairy.
         5. Adaxial leaf surfaces sparsely to densely appressed-hairy throughout or only on folds, (nerves glabrous); primary inflorescence branches densely appressed- to ascending-hairy . . . . 4. *Alchemilla glomerulans*
         5. Adaxial leaf surfaces glabrous or slightly hairy on margins only; primary inflorescence branches glabrous.
            6. Incisions absent from leaf sinuses; leaf teeth subacute or obtuse, strongly asymmetric (slightly concave near apex) . . . . 2. *Alchemilla glabra*
            6. Incisions present in leaf sinuses; leaf teeth acute, symmetric to slightly asymmetric . . . . 3. *Alchemilla wichurae*
      4. Stems spreading-, ascending-, or reflexed-pubescent or sericeous, sometimes glabrous (*A. micans*).
         7. Primary inflorescence branches glabrous . . . . 7. *Alchemilla filicaulis* (in part)
         7. Primary inflorescence branches sparsely to densely hairy.
            8. Pedicels (all or most) hairy.
               9. Stipules strongly wine red-tinged proximally; leaf blades usually reniform, sometimes orbiculate, basal sinuses usually wide, sometimes narrow . . . . 7. *Alchemilla filicaulis* (in part)
               9. Stipules translucent, colorless, sometimes flushed pale wine red proximally; leaf blades orbiculate, basal sinuses closed . . . . . . 10. *Alchemilla glaucescens*
            8. Pedicels usually glabrous or some of the proximal sparsely hairy.
               10. Leaves: basal sinuses closed, basal lobes overlapping. . . . . 6. *Alchemilla subcrenata*
               10. Leaves: basal sinuses wide or narrow, basal lobes not overlapping.
                   11. Adaxial leaf surfaces usually glabrous, rarely sparsely hairy on teeth, margins, and folds or on some folds only . . . . 5. *Alchemilla xanthochlora*
                   11. Adaxial leaf surfaces densely appressed-hairy or pubescent throughout.
                       12. Hypanthia attenuate at base, usually glabrous; stipules translucent, strongly wine red-tinged proximally; stems usually sparsely to densely spreading- to slightly ascending-hairy, sometimes glabrous in distal ½ . . . . 8. *Alchemilla micans*
                       12. Hypanthia rounded at base, usually sparsely to densely spreading-hairy; stipules translucent to pale green proximally, (apices green); stems usually densely spreading-hairy in distal ½ . . . . 9. *Alchemilla monticola*

1. Alchemilla alpina Linnaeus, Sp. Pl. 1: 123. 1753 • Alpine lady's mantle

Plants dwarfed, green to dark green, <carpet-forming, 5–20 cm, rarely taller>. **Stems** appressed- to ascending-hairy. **Leaves:** stipules translucent, quickly turning brownish; blade ± orbiculate, palmately compound, leaflets 5–7, narrowly elliptic to oblanceolate, apices with 7–11 teeth, margins flat, abaxial surface sericeous, adaxial shiny, glabrous. **Inflorescences:** primary branches densely appressed-hairy. **Pedicels** <usually shorter than hypanthia>, densely appressed-hairy. **Flowers:** epicalyx bractlet lengths 0.5 or less times sepals; hypanthium densely pubescent; <sepals erect after flowering>. **Achenes** not exserted.

Flowering mid Jun–mid Sep. Meadows, herb slopes, moist rock ledges; 0–500(–1000) m; Greenland; St. Pierre and Miquelon; Nfld. and Labr. (Nfld.); Europe.

In the flora area, *Alchemilla alpina* is possibly introduced except in Greenland.

2. Alchemilla glabra Neygenfind, Ench. Bot., 67. 1821 • Smooth lady's mantle, alchémille glabre [I]

Alchemilla alpestris (F. W. Schmidt) Opiz; *A. vulgaris* Linnaeus var. *alpestris* F. W. Schmidt; *A. vulgaris* var. *grandis* Blytt

Plants medium-sized to occasionally very large, yellowish grass green to dark green, often reddish brown on exposed parts of stems and petioles. **Stems** glabrous or appressed-hairy proximally, glabrous distally. **Leaves:** stipules translucent, usually suffused with red proximally, translucent to pale green distally, <lobes pale green; petiole glabrous or sparsely to densely appressed-hairy>; blade reniform-orbiculate, 7–9(–11) -lobed, margins flat or slightly undulate, basal sinuses narrow or wide, middle lobes as long as to longer than their half-widths, rounded to triangular-ovate; incisions absent; teeth ± connivent, slightly concave near apex, strongly asymmetric, apex subacute or obtuse, abaxial surface with nerves hairy throughout or in distal ½ only, internerve regions glabrous, adaxial glabrous or slightly hairy near margins, rarely (late-season leaves) hairy on folds. **Inflorescences:** primary branches glabrous. **Pedicels** glabrous. **Flowers:** epicalyx bractlet lengths 0.5 times sepals (narrower); hypanthium glabrous. **Achenes** exserted from discs.

Flowering Jul–Aug. Meadows; 0–100 m; introduced; N.B., Nfld. and Labr. (Nfld.), Que.; Alaska; Europe.

3. Alchemilla wichurae (Buser) Stefánsson, Fl. Íslands, 135. 1901

Alchemilla connivens Buser [unranked] *wichurae* Buser, Bull. Herb. Boissier 2: 110. 1894; *A. vulgaris* Linnaeus var. *wichurae* (Buser) B. Boivin

Plants medium-sized, often purple on stems, petioles, and margins of leaves and stem leaves, <usually thin and delicate, 20–30 cm>. **Stems** sparsely appressed-hairy throughout or glabrous in distal ½. **Leaves:** stipules translucent, colorless, usually flushed pale wine red proximally, <apex pale green, sometimes with pale wine red margins and veins>; blade reniform-orbiculate, 7–9 -lobed, margins flat, basal sinuses narrow or closed, middle lobes as long as and rounded to longer than their half-widths; incisions long or short, <2–5 mm>; teeth usually connivent, symmetric to slightly asymmetric, apex acute, abaxial surface with nerves hairy throughout or in distal ½ only, internerve regions usually glabrous (except sometimes on 2 basal lobes), adaxial glabrous or slightly hairy at margin. **Inflorescences:** primary branches glabrous. **Pedicels** glabrous. **Flowers:** epicalyx bractlet lengths 0.5 times to almost equal to sepals (always narrower); hypanthium glabrous. **Achenes** not exserted.

Flowering Jul. Moist herb slopes; 0–100 m; Greenland; Europe.

In the flora area, *Alchemilla wichurae* is restricted to a small area on the southeast coast of Greenland, mostly (or all) in Angmagssalik district.

4. Alchemilla glomerulans Buser, Bull. Herb. Boissier 1(app. 2): 30. 1893 • Clustered lady's mantle, alchémille à glomérules

Alchemilla obtusa Buser; *A. obtusa* var. *comosa* Brenner; *A. pseudomicans* Böcher; *A. vulgaris* Linnaeus var. *comosa* (Brenner) Fernald & Wiegand

Plants medium-sized, grass green or glaucous, often becoming reddish orange when young changing to dark brownish especially on margins of leaves and flowers (young flowers are yellowish), <often coarse, 30–40 cm>. **Stems** appressed-hairy throughout (hairs becoming looser and ± ascending distally). **Leaves:** stipules translucent to pale green, appearing brownish upon drying; <petiole thickly, usually densely appressed-hairy throughout, rarely glabrous or sparsely hairy (on spring leaves)>; blade reniform to orbiculate, 7–9-lobed, margins undulate, basal sinuses narrow, middle lobes equal to longer than their half-widths; incisions absent; teeth: sometimes proximal sides at least slightly connivent, slightly concave near apex, slightly asymmetric, apex subobtuse to acute, abaxial surface with nerves hairy throughout, internerve regions ± hairy throughout, adaxial <light to grass green, sometimes glaucous, margins and folds usually turning reddish orange>, sparsely to densely appressed-hairy throughout or only on folds. **Inflorescences:** primary branches densely appressed- to ascending-hairy; <peduncles appressed- to ascending-hairy or glabrous>. **Pedicels** mostly glabrous or some of the proximal hairy. **Flowers:** epicalyx bractlet lengths 0.5 times to almost equal to sepals (narrower); hypanthium glabrous or sparsely appressed-hairy (on proximalmost flowers). **Achenes** not exserted.

Flowering late Jun–Sep. Moist herb slopes, willow scrub; 0–500 m; Greenland; Nfld. and Labr. (Labr.), Que.; Europe.

In contrast to *Alchemilla wichurae*, *A. glomerulans* occurs throughout the southern (unglaciated) portion of Greenland.

5. Alchemilla xanthochlora Rothmaler, Repert. Spec. Nov. Regni Veg. 42: 167. 1937 • Yellow-green lady's mantle [I]

Plants large, yellowish green to green, rarely with reddish orange leaf margins, <robust, compact often with rather rigid stems, 40–60 cm>. **Stems** usually densely spreading-hairy, sometimes sparsely in distal ½. **Leaves:** stipules translucent, colorless proximally, <lobes green, quickly turning brownish; petiole thickly, densely pubescent (hairs patent or, when not fully developed, slightly ascending)>; blade reniform to orbiculate, 7–11-lobed, margins flat or slightly undulate, basal sinuses wide, <basal lobes not overlapping>, middle lobes equal to longer than their half-widths and rounded, or as long as wide and straight-sided; incisions usually absent, sometimes relatively short; teeth sometimes proximal sides at least slightly connivent, side slightly concave near apex, usually slightly, sometimes markedly, asymmetric, apex acute to subobtuse, abaxial surface with nerves hairy throughout, internerve regions uniformly or irregularly hairy throughout, adaxial <yellowish to light green>, usually glabrous, rarely sparsely hairy on teeth, margins, and folds or on some folds only. **Inflorescences:** primary branches densely hairy; <peduncles sparsely hairy or glabrous>. **Pedicels** glabrous or some of the proximal rarely sparsely hairy. **Flowers:** epicalyx bractlet lengths 0.5 times to almost equal to sepals (narrower); hypanthium usually glabrous or sparsely hairy (on proximalmost flowers). **Achenes** exserted ⅓ from discs.

Flowering late May–Sep. Moist meadows, herb slopes, willow scrub, ditches, lawns; 0–50 m; introduced; N.B., N.S., Que.; Europe.

Alchemilla xanthochlora was recognized by P. A. Rydberg (1908–1918) and M. L. Fernald (1950) as *A. pratensis* attributed to F. W. Schmidt, as typical *A. vulgaris* Linnaeus by M. L. Fernald and K. M. Wiegand (1912), and as *A. vulgaris* var. *vulgaris* by B. Boivin (1966b). Actually, Schmidt published a varietal name [*A. vulgaris* var. *pratensis* F. W. Schmidt, *A. pratensis* (F. W. Schmidt) Opiz], which does not apply to this species nor does *A. vulgaris*.

6. Alchemilla subcrenata Buser, Scrinia Fl. Select. 12: 285. 1893 • Round-toothed lady's mantle, alchémille subcrénelée [I]

Plants medium-sized, sometimes larger, yellowish green to dark green, often reddish brown especially on stems, petioles, and inflorescence branches, <to 50 cm>. **Stems** densely spreading- to slightly reflexed-pubescent or only sparsely so in distal ½. **Leaves:** stipules translucent, sometimes wine red-tinged proximally, <lobes green; petiole sparsely to densely spreading- to reflexed-hairy>; blade reniform to orbiculate, 7–9-lobed, margins strongly plicate to undulate, basal sinuses closed, <basal lobes overlapping (in plants from spring-flooded habitats, only cauline leaves with wide sinus may persist)>, middle lobes usually longer than their half-widths, as long as wide, longer than wide (with ± straight sides); incisions absent; teeth: proximal sides connivent or slightly so, sometimes slightly concave near apex, slightly to strongly asymmetric, apex subobtuse to obtuse, abaxial surface <grass green to dark green>, nerves hairy throughout, internerve regions irregularly or uniformly hairy, adaxial densely spreading-hairy throughout or only on margins and folds. **Inflorescences:** primary branches sparsely to densely hairy; <peduncles glabrous or sparsely hairy>. **Pedicels** glabrous. **Flowers** <green, often becoming reddish brown>; epicalyx bractlet lengths 0.5 times sepals (narrower); hypanthium glabrous. **Achenes** exserted to ⅓ from discs.

Flowering Jun–Sep. Moist grasslands, flood plains; 0–1400 m; introduced; B.C., Que.; Mont.; Europe.

7. Alchemilla filicaulis Buser, Bull. Herb. Boissier 1(app. 2): 22. 1893 [F]

Alchemilla vulgaris Linnaeus var. *filicaulis* (Buser) Fernald & Wiegand

Plants small to medium-sized, glaucous to grass green, young inflorescences yellowish, soon becoming reddish brown, <to 40 cm>. **Stems** densely spreading- to very sparsely hairy at least in proximal ½. **Leaves:** stipules translucent, usually strongly wine red-tinged proximally, <lobes green>; blade usually reniform, sometimes orbiculate, shallowly 7–9-lobed, margins flat or slightly, rarely distinctly, undulate, basal sinuses usually wide, sometimes narrow, middle lobes rounded, shorter than to equal to their half-widths, sometimes longer; incisions absent or relatively short to long; teeth sometimes slightly connivent, slightly to strongly asymmetric, apex usually acute to subobtuse, sometimes obtuse, abaxial surface glabrous or hairy, nerves glabrous proximally or hairy throughout, adaxial sparsely to densely hairy throughout or on folds only. **Inflorescences:** primary branches glabrous or densely pubescent; <peduncles glabrous or densely hairy>. **Pedicels** glabrous or hairy. **Flowers:** epicalyx bractlet lengths 0.5 times to almost equal to sepals (always narrower); hypanthium usually sparsely to densely hairy, sometimes glabrous. **Achenes** exserted.

Subspecies 2 (2 in the flora): North America, Europe.

1. Pedicels, distal ½ of stems, primary inflorescence branches, and peduncles usually glabrous; abaxial leaf surfaces usually glabrous, nerves usually glabrous proximally, sometimes hairy throughout, adaxial sparsely to densely hairy throughout or on folds only . . . . 7a. *Alchemilla filicaulis* subsp. *filicaulis*
1. All pedicels, distal ½ of stems, primary inflorescence branches, and peduncles densely hairy; abaxial leaf surfaces usually irregularly or uniformly hairy, sometimes glabrous, nerves hairy throughout, adaxial usually densely hairy throughout, rarely only on folds . . . . 7b. *Alchemilla filicaulis* subsp. *vestita*

7a. Alchemilla filicaulis Buser subsp. **filicaulis** • Thread-stemmed lady's mantle, alchémille filicaule

Stems densely hairy in proximal ½, glabrous or sparsely hairy in distal ½. **Leaf blades:** basal sinus incisions absent or, sometimes, relatively short, abaxial surface usually glabrous, nerves usually glabrous proximally, sometimes hairy throughout, adaxial sparsely to densely hairy throughout or on folds only. **Inflorescences:** primary branches and peduncles usually glabrous. **Pedicels** usually glabrous, rarely some of the proximal hairy. **Flowers:** hypanthium usually sparsely hairy, sometimes glabrous.

Flowering late Jun–Sep. Moist meadows, herb slopes, often near seashores; 0–700 m; Greenland; St. Pierre and Miquelon; N.B., Nfld. and Labr., N.S., Ont., Que.; Wyo.; Europe.

Subspecies *filicaulis* is introduced in New Brunswick, Nova Scotia, Ontario, Quebec, and Wyoming; it may be introduced also in parts of Newfoundland.

7b. Alchemilla filicaulis Buser subsp. **vestita** (Buser) M. E. Bradshaw, Watsonia 5: 305. 1963 • Alchémille vêtue [F]

Alchemilla filicaulis forma *vestita* Buser, Bull. Herb. Boissier 1(app. 2): 23. 1893; *A. filicaulis* var. *vestita* (Buser) H. J. Coste; *A. vestita* (Buser) Raunkiaer; *A. vulgaris* Linnaeus var. *vestita* (Buser) Fernald & Wiegand

Stems densely hairy. **Leaf blades:** basal sinus incisions relatively short or long, sometimes absent, abaxial surface usually irregularly or uniformly hairy, sometimes glabrous, nerves hairy throughout, adaxial usually densely hairy throughout, rarely only on folds. **Inflorescences:** primary branches densely hairy; peduncles hairy. **Pedicels,** or at least most, hairy. **Flowers:** hypanthium densely hairy.

Flowering late Jun–Aug. Moist to mesic meadows, herb slopes, sometimes near seashores; 0–500 m; Greenland; N.B., Nfld. and Labr., N.S., Que.; Europe.

Subspecies *vestita* was recognized by M. L. Fernald (1950) and A. E. Roland and E. C. Smith (1966) as *Alchemilla minor* Hudson, a misapplication of that name. The subspecies is probably introduced in New Brunswick and Nova Scotia.

8. Alchemilla micans Buser, Bull. Herb. Boissier 1(app. 2): 28. 1893 • Gleaming lady's mantle [I]

Plants medium-sized, dark green, often somewhat sericeous (abaxial surface of leaves), sometimes reddish brown, especially on exposed distal part of stems and inflorescences, <to 50 cm>. **Stems** usually densely spreading- to slightly ascending-hairy, usually glabrous in distal ½. **Leaves:** stipules translucent, strongly wine red-tinged proximally, <lobes pale green, sometimes suffused wine red; petiole densely spreading- or slightly ascending-hairy (especially in distal ⅓)>; blade usually reniform to orbiculate, 7–9-lobed, margins flat, sometimes slightly undulate, basal sinuses relatively wide or narrow, <basal lobes not overlapping>, middle lobes rounded, as long as to longer than their half-widths, to as long as wide and with straight sides; incisions usually absent or relatively short; teeth usually slightly connivent, almost symmetric to ± asymmetric, apex acute, abaxial surface with nerves hairy throughout, internerve regions uniformly or irregularly hairy, adaxial densely appressed-hairy throughout. **Inflorescences:** primary branches often sparsely, sometimes densely, ascending-hairy; <peduncles glabrous or hairy>. **Pedicels** glabrous or some of the proximal hairy. **Flowers** <dark green, often becoming reddish>; epicalyx bractlet lengths 0.5+ times sepals (narrower); hypanthium <attenuate at base>, usually glabrous, rarely sparsely hairy (in proximal flowers). **Achenes** not exserted.

Flowering late May–Sep. Meadows, moist sand; 0–400 m; introduced; Nfld. and Labr. (Nfld.), N.S., Que.; Maine, N.Y.; Europe.

Alchemilla micans has been widely known as *A. gracilis* Opiz; the type of that name is referable to *A. monticola*, a species that the original description also fits better.

9. Alchemilla monticola Opiz in F. Berchtold et al., Oekon.-techn. Fl. Böhm. 2(1): 13. 1838 • Mountain alchemilla, alchémille des montagnes [I]

Alchemilla pastoralis Buser; *A. vulgaris* Linnaeus var. *pastoralis* (Buser) B. Boivin

Plants medium-sized, grass green, young inflorescences yellowish, often becoming reddish brown, <25–30(–40) cm, often robust>. **Stems** densely spreading-hairy in distal ½. **Leaves:** stipules translucent to pale green proximally, <apex green>; blade reniform or reniform-orbiculate, rather shallowly 7–9-lobed, margins flat or slightly undulate, basal sinuses wide or narrow, <basal lobes not overlapping>, middle lobes rounded, equal to longer than their half-width; incisions usually relatively short, sometimes absent or relatively long; teeth usually slightly connivent, mostly slightly asymmetric but a few strongly so, apex acute to subobtuse, surfaces densely pubescent throughout, adaxial <gray green to dark green>. **Inflorescences:** primary branches densely hairy; <peduncles sparsely hairy or glabrous>. **Pedicels** glabrous or some of the proximal hairy. **Flowers** <green, often becoming reddish>; epicalyx bractlet lengths at least 0.5 times sepals (not longer, always narrower); hypanthium <rounded at base>, usually sparsely to densely spreading-hairy, rarely glabrous. **Achenes** exserted ⅕ from discs.

Flowering late May–Sep. Meadows, lawns, ditches, under trees; 0–3000 m; introduced; B.C., Nfld. and Labr. (Nfld.), N.S., Ont., Que.; Conn., Maine, Mass., Mont., N.H., N.Mex., N.Y., Vt., Wis.; Europe.

10. Alchemilla glaucescens Wallroth, Linnaea 14: 134. 1840 [I]

Plants small, gray-green, <to 20 cm>. **Stems** densely ascending-spreading-hairy, sometimes sericeous. **Leaves:** stipules translucent, colorless, sometimes flushed pale wine red proximally, green distally, <lobes green>; blade orbiculate, rather shallowly 7–9-lobed, margins flat or slightly undulate, basal sinuses closed, middle lobes rounded, as long as to slightly longer than their half-widths; incisions usually absent or relatively short, sometimes relatively long; teeth slightly connivent, slightly asymmetric, apex subacute or almost digitate, surfaces densely hairy. **Inflorescences:** primary branches sericeous; <peduncles sericeous>. **Pedicels** sericeous. **Flowers:** epicalyx bractlet lengths 0.5 times sepals (narrower); hypanthium sericeous. **Achenes** not or exserted to ¹⁄₂₀ from discs.

Flowering Jun–Aug. Waste ground; 0–50 m; introduced, Que.; Europe.

In the flora area, *Alchemilla glaucescens* is known only from Grosse-Ile in the St. Lawrence River.

11. Alchemilla venosa Juzepczuk in A. A. Grossheim, Fl. Kavkaza 4: 328. 1934 • Veined lady's mantle [I]

Plants medium-sized, darkish green, sometimes reddish tinged, <to 40 cm>. **Stems** ± appressed-hairy. **Leaves:** stipules green or slightly purplish tinged, turning brown; blade orbiculate, 7–9-lobed, margins undulate, basal sinuses appearing closed, <basal lobes overlapping>, middle lobes rounded, as long as or longer than their half-widths; incisions relatively short to long; teeth ± connivent, slightly asymmetric, apex acute, abaxial surface with nerves hairy throughout, internerve regions glabrous or densely hairy throughout, adaxial glabrous or hairy only on folds. **Inflorescences:** primary branches densely hairy; <peduncles hairy or glabrous>. **Pedicels** glabrous. **Flowers:** epicalyx bractlet lengths equal to slightly longer than sepals (usually almost as wide); <epicalyx segments and sepals patent after flowering, giving appearance of an 8-point star>; hypanthium <usually shorter than sepals>, glabrous. **Achenes** exserted from discs (distinctly longer than hypanthia).

Flowering late Jun–Aug. Grasslands, often near sea shores; 0–50 m; introduced; N.B., Nfld. and Labr. (Nfld.), N.S.; sw Asia (Caucasus, e Turkey).

12. Alchemilla mollis (Buser) Rothmaler, Repert. Spec. Nov. Regni Veg. 33: 347. 1934 • Garden lady's mantle [I]

Alchemilla acutiloba Steven var. *mollis* Buser, Bull. Herb. Boissier 4: 759. 1896

Plants large to very large, yellowish green, <to 80 cm, often very robust, densely hairy, hairs patent, soft>. **Stems** densely spreading-hairy (to inflorescences). **Leaves:** stipules translucent, usually pale pink proximally, <lobes turning brownish>; blade orbiculate or reniform-orbiculate, 9–11-lobed, margins flat or slightly undulate, basal sinuses narrow to almost closed, <basal lobes sometimes overlapping>, middle lobes <with lateral sides convex or slightly concave at base>, shorter than to as long as their half-widths; incisions absent; teeth slightly connivent or not, ± symmetric, apex acute or subacute, surfaces densely hairy throughout. **Inflorescences:** primary branches densely hairy; <peduncles pubescent or almost glabrous>. **Pedicels** glabrous. **Flowers:** epicalyx bractlet lengths 1 times sepals (as wide), <often with blunt teeth, glabrous; epicalyx bractlets and sepals patent after flowering, giving appearance of an 8-point star>; hypanthium usually densely, occasionally sparsely, hairy (in proximal ½). **Achenes** exserted from discs (distinctly longer than hypanthia).

Flowering late Jul–Sep. Densely vegetated lake shores; 0–100 m; introduced; B.C., Ont.; Europe (e Carpathians); w Asia (Caucasus, Turkey).

21. APHANES Linnaeus, Sp. Pl. 1: 123. 1753; Gen. Pl. ed. 5, 59. 1754 • Parsley-piert [Greek *aphanes*, unseen, invisible, alluding to inconspicuous nature of plants and/or flowers]

John McNeill

Barbara Ertter

Herbs, annual, prostrate to ascending, 0.1–3 dm, soft-hairy; taprooted. **Stems** 1–10+, erect, ascending, or spreading. **Leaves** not persistent, cauline, alternate, simple (deeply lobed); stipules persistent, free or distally adnate to petiole, asymmetric, ± orbiculate to ovate, margins lobed; petiole present, <short>; blade cuneate, 0.2–1 cm, herbaceous, deeply divided into 2 or 3(–5) segments, each segment (1 or)2 or 3(–6)-lobed, margins flat, entire, venation pinnate. **Inflorescences** lateral, 4–12-flowered, condensed cymes, often hidden by stipules; bracts absent; bracteoles absent. **Pedicels** present. **Flowers** 0.7–1.1 mm diam.; epicalyx bractlets (0–)4; hypanthium subglobose to ellipsoid or ovoid, 0.7–2 mm; sepals 4, connivent or erect to spreading, ovate to narrowly triangular; petals 0; stamens 1(or 2); torus absent or reduced; carpel 1, hairy, styles basal, <stigmas capitate>; ovule 1. **Fruits** achenes, 1, narrowly ovoid, 0.8–2.5 mm; hypanthium persistent; sepals persistent, erect; styles deciduous. $x = 8$.

Species ca. 20 (3 in the flora): North America, nw Mexico, Eurasia, n Africa; introduced in s South America, Pacific Islands, Australia.

On the basis of molecular studies, R. Gehrke et al. (2008) have suggested that *Aphanes* and the Central and South American *Lachemilla* (Focke) Rydberg should be included in a more broadly circumscribed *Alchemilla*. Both *Aphanes* and *Lachemilla* were shown to be monophyletic groups as is *Alchemilla*, apart from the African species currently placed in that genus; these form a clade separate from *Alchemilla* as represented in North America and Eurasia. The African species deserve more thorough investigation before such a morphologically distinct genus as *Aphanes* is abandoned.

The highly reduced nature of plants of *Aphanes*, coupled with their high dispersibility, complicates circumscription of species in the genus. The most recent treatments recognize a number of endemic species in South America, Europe, and North Africa, particularly in regions with Mediterranean climates. The extent to which these, and those recognizable in western North America, represent native radiations versus multiple introductions remains to be determined.

Measurements of flowers apply collectively to the length of the hypanthium and calyx measured in fruit.

1. Flowers 1.5–2.5 mm in fruit, sepals ± spreading to erect; leaf blades 4–10 mm; petioles free from stipules in proximal and distal leaves; stems 2–20(–30) cm; at least some hairs more than 1 mm; most stipule lobes on distal nodes ± triangular, 1–2(–4) times as long as wide . . . . 1. *Aphanes arvensis*
1. Flowers 1–1.5 mm in fruit, sepals connivent to spreading; leaf blades 2–5 mm; petioles adnate to stipules in distal leaves; stems 1–10(–22) cm; hairs less than 1 mm; most stipule lobes on distal nodes oblong or ovate, usually ⅔–1 times as long as wide.
    2. Inflorescences dense, flowers tending to remain hidden; pedicels usually less than 1 mm; sepals 0.2 mm, connivent; epicalyx bractlets minute (less than 0.1 mm) or absent; stipules overlapping, lobes oblong, almost equal to undivided portion; leaf blades 5 mm, divided into 3 segments . . . . 2. *Aphanes australis*
    2. Inflorescences open, flowers tending to become exposed; pedicels to 2 mm; sepals 0.2–0.6 mm, connivent to spreading; epicalyx bractlets 0–0.5 mm; stipules overlapping only distally, lobes ovate, ⅔–1 times undivided portion; leaf blades 2–5 mm, 4–6-lobed . . . . 3. *Aphanes occidentalis*

1. Aphanes arvensis Linnaeus, Sp. Pl. 1: 123. 1753 [I]

Alchemilla arvensis (Linnaeus) Scopoli

Plants slender or relatively robust, densely hairy throughout, hairs, at least some, more than 1 mm. **Stems** ± erect or spreading, usually branched from base, 2–20(–30) cm. **Leaves** 6–16 mm; stipules overlapping, 4–10 mm, divided ⅓–½ their length, lobes <8–12>, on distal nodes ± triangular, length 1–2(–4) times as long as width; petiole free from stipules in proximal and distal leaves; blade 4–10 mm, divided into 3 or 4 segments, each segment 3 or 4(or 5)-lobed. **Inflorescences** opposing leaves, dense, flowers not completely hidden. **Pedicels** less than 1 mm. **Flowers** 1.3–1.8 × 0.8–1 mm, 1.5–2.5 mm in fruit; epicalyx bractlets 0–0.1 mm; hypanthium ovoid to ellipsoid, contracted at apex, 8-ribbed, <spreading-erect hairy in proximal ⅔, glabrescent in distal ⅓>; sepals ± spreading to erect, 0.4–0.7 mm, long-ciliate. **Fruits** (1.6–)1.8–2.3(–2.5) mm. $2n = 48$.

Flowering Apr–May. Fields, usually on sandy soil, mossy and grassy sites, moist shade, gravelly roadsides, shady, bare patches in lawns, shallow soil pockets on rocky seashores, meadows, hillsides; 0–150 m; introduced; B.C., N.S.; Calif., Del., Ga., Idaho, N.J., Oreg., Wash.; Europe; sw Asia; n Africa; introduced also in South America (Chile), Pacific Islands (New Zealand), Australia.

2. Aphanes australis Rydberg in N. L. Britton et al., N. Amer. Fl. 22: 380. 1908 [I]

Aphanes inexspectata W. Lippert

Plants slender, appressed-hairy throughout, hairs usually less than 0.8 mm. **Stems** ascending or erect, simple or branched, (1–)3–10(–22) cm. **Leaves** 5–10 mm; stipules overlapping, (2.5–)4–6(–7) mm, divided ca. ½ their length, lobes <(5–)7–10(–12)>, on distal nodes oblong, 2–4(–5) times as long as wide; petiole free from stipules in proximal leaves, adnate in distal ones; blade 5 mm, prominently divided into 3 segments, each (1 or)2 or 3(or 4)-lobed. **Inflorescences** opposing leaves, dense, flowers tending to remain hidden. **Pedicels** less than 1 mm. **Flowers** (0.8–)1.2–1.3(–1.4) × 0.7–0.8 mm, 1–1.5 mm in fruit; epicalyx bractlets 0–0.1 mm; hypanthium subglobose to ellipsoid, not contracted at apex, not distinctly 8-ribbed, <hairy between ribs ⅔–¾ of length, sometimes glabrescent>; sepals connivent, 0.2 mm, long-ciliate. **Fruits** (0.9–)1.4–1.6(–1.7) mm. $2n = 16$.

Flowering Mar–May. Fields, sandy places; 0–300 m; introduced; B.C.; Ala., Ark., Del., D.C., Fla., Ga., Ky., La., Md., Miss., Mo., N.Y., N.C., Okla., Oreg., S.C., Tenn., Tex., Va., Wash.; Europe.

Plants of *Aphanes australis* have commonly been misidentified as *A. microcarpa* (Boisser & Reuter) Rothmaler (or *Alchemilla microcarpa* Boisser & Reuter), an endemic to the western Mediterranean region. Plants native to other parts of Europe but misidentified as *Aphanes microcarpa* were described in 1984 as a new species, *A. inexspectata*, but this taxon had previously been described by Rydberg as *A. australis* from plants introduced to the eastern United States (P. Frost-Olsen 1998). *Aphanes microcarpa* is not present in North America.

3. Aphanes occidentalis (Nuttall) Rydberg in N. L. Britton et al., N. Amer. Fl. 22: 380. 1908 [F]

Alchemilla occidentalis Nuttall in J. Torrey and A. Gray, Fl. N. Amer. 1: 432. 1840

Plants slender, sparsely to densely hairy, hairs less than 1 mm. **Stems** erect or ascending, simple or branched from base, 1–10(–15) cm. **Leaves** 3–9 mm; stipules overlapping only distally, 2–3(–4) mm, lobes on distal nodes ovate, length ⅔–1 times undivided portion; petiole free from stipules in proximal leaves, adnate in distal ones; blade 2–5 mm, divided ca. ⅓ length, each segment 4–6-lobed, <ovate, length 1.8–3 times width>. **Inflorescences** opposing leaves, open, flowers tending to become exposed. **Pedicels** to 2 mm. **Flowers** 0.5–2 × 0.7–1.1 mm, 1–1.5 mm in fruit; epicalyx bractlets 0–0.5 mm; hypanthium ovoid to globose, contracted or not at apex, faintly ribbed; sepals connivent to spreading, 0.2–0.6 mm, ciliate. **Fruits** 0.8–1 mm.

Flowering Mar–Jun. Vernally moist grasslands, chaparral, woodlands, clearings, rocky outcrops, cliffs, sandy beaches, waste ground, roadsides; 0–1200 m; B.C.; Calif., Oreg., Wash.; Mexico (Baja California).

Three races of *Aphanes occidentalis* are recognizable, apparently corresponding to previously described species. Some apparently intermediate plants exist, and not all specimens can be confidently assigned to these races, which are, therefore, not recognized as formal taxa here. The names under which they have been recognized and their distinguishing features and distributions are as follows:

Aphanes cuneifolia (Nuttall) Rydberg: hypanthium and calyx ovoid, 1 mm, smooth, often subglabrous; hypanthium contracted at apex, not densely hairy; epicalyx bractlets often present; sepals erect-connivent, 0.3–0.4 mm, ¼–⅓ flower length. California; Mexico (Baja California).

Aphanes macrosepala Rydberg: hypanthium not contracted at apex, usually densely hairy, hairs spreading; epicalyx bractlets present; sepals spreading erect, 0.5 mm, ca. ⅓ flower length. California, Oregon, Washington.

Aphanes occidentalis in the narrow sense: hypanthium and calyx ellipsoid, 1.5 mm, ± ribbed, sparsely short-hairy; hypanthium contracted at apex, not densely hairy; epicalyx bractlets usually absent; sepals usually connivent, 0.3–0.4 mm, ¼–⅓ flower length. British Columbia; California, Oregon, Washington.

a6. ROSACEAE Jussieu (subfam. ROSOIDEAE) tribe AGRIMONIEAE Lamarck & de Candolle, Syn. Pl. Fl. Gall., 333. 1806 (as Agrimoniaceae)

Luc Brouillet

Herbs, perennial (annual or biennial in *Poteridium*) [shrubs or trees]; unarmed (hypanthia armed in *Acaena*). **Leaves** alternate, odd-pinnately compound; stipules persistent (absent in *Acaena*), adnate to petiole (free in *Poterium*); venation pinnate. **Flowers:** perianth and androecium perigynous; epicalyx bractlets absent; hypanthium hemispheric, obconic, ovoid, urceolate, top-shaped, ellipsoid, nearly orbicular, or obtriangular; torus absent; carpels 1 or 2(or 3), rarely more, styles terminal, distinct; ovule 1, apical. **Fruits** achenes, enclosed within enlarged, often hardened, sometimes armed hypanthia; styles deciduous, not elongate.

Genera 12, species ca. 270 (5 genera, 17 species in the flora): North America, Mexico, West Indies (Hispaniola), South America, Eurasia, Africa, Atlantic Islands, Indian Ocean Islands, Pacific Islands (New Zealand), Australia.

The base chromosome number for Agrimonieae is $x = 7$. *Acaena* and *Sanguisorba* are host to *Phragmidium* rusts. The tribal name Agrimonieae has priority over Sanguisorbeae, used by D. Potter et al. (2007) and others. Agrimonieae also includes the genera *Aremonia* Necker ex Nestler (Europe), *Cliffortia* Linnaeus, *Hagenia* J. F. Gmelin and *Leucosidea* Ecklon & Zeyher (Africa), *Margyricarpus* Ruiz & Pavón and *Polylepis* Ruiz & Pavón (South America), and *Spenceria* Trimen (Asia) (Potter et al.).

SELECTED REFERENCES Kerr, M. S. 2004. A Phylogenetic and Biogeographic Analysis of Sanguisorbeae (Rosaceae), with Emphasis on the Pleistocene Radiation of the High Andean Genus *Polylepis*. Ph.D. dissertation. University of Maryland. Nordborg, G. 1966. *Sanguisorba* L., *Sarcopoterium* Spach, and *Bencomia* Webb et Berth.: Delimitation and subdivision of the genera. Opera Bot. 11: 1–103.

1. Stems creeping or suberect; hypanthia spiny . . . . 26. *Acaena*, p. 323
1. Stems ascending to erect; hypanthia not spiny (rim bristly in fruit in *Agrimonia*).
    2. Inflorescences racemes (simple or compound); leaf blades with minor leaflet pairs between major pairs; petals 5 (yellow); hypanthial rim with 2–5 rows of bristles . . . 22. *Agrimonia*, p. 312
    2. Inflorescences spikes (cylindric, globose, or headlike); leaf blades without minor leaflet pairs between major pairs; petals 0; hypanthial rim without rows of bristles.
        3. Leaflet margins pectinately pinnatisect . . . . 24. *Poteridium*, p. 319
        3. Leaflet margins crenate, serrate, or incised.
            4. Spikes headlike; flowers bisexual or pistillate (plants gynomonoecious); sepals distinct . . . . 23. *Poterium*, p. 318
            4. Spikes ellipsoid to cylindric; flowers bisexual; sepals basally connate . . . . 25. *Sanguisorba*, p. 321

22. AGRIMONIA Linnaeus, Sp. Pl. 1: 448. 1753; Gen. Pl. ed. 5, 206. 1754 • Agrimony, aigremoine [Greek *Argemone* from *argemos*, cataract of eye, alluding to supposed curative properties of plant for eye disease]

Genevieve J. Kline

Paul D. Sørensen

Herbs, perennial, 3–20 dm, <self-compatible>; rhizomatous, <roots slender, fibrous, sometimes with thickened, ± fusiform or ± narrowly oblong tubers>. **Stems** 1–5, ascending to erect. **Leaves** persistent or basal deciduous and cauline persistent, cauline (see discussion), alternate, odd pinnately compound; stipules persistent, adnate to petiole, ± ½-ovate, margins dentate or deeply incised, rarely entire; petiole present; blade ± narrowly obovate to ± narrowly elliptic, 1.5–22 cm, herbaceous, major leaflets 3–19, <leaflets along rachis in ± opposite pairs, minor leaflet pairs between each major pair 0–5, number usually increasing distally along rachis>, major leaflet blades ovate, lanceolate, elliptic, rhombic, and/or obovate, minor leaflet blades ovate, elliptic, or obovate, margins flat, serrate to dentate, rarely incised, or in minor leaflet blades entire or few-toothed, eglandular hairs of two types: soft and either straight or wavy (surfaces pilose, pubescent, or villous), and stiff and straight (surfaces hirsute); glandular hairs of two types: short-stipitate, and sessile, both ± glistening (indumentum description applies to all plant surfaces in *Agrimonia*). **Inflorescences** terminal and often axillary, 9–120-flowered, simple or compound racemes (flexible); peduncles present; bracts present; bracteoles present. **Pedicels** present. **Flowers** 5–10 mm diam.; hypanthium <stipitate, very short stipe adnate to hypanthium above bracteoles, reflexed at maturity, fruit abscising at base of stipe>, hemispheric to obconic, becoming indurate, 0.5–2 mm, <sulcate, throat occluded by annular disc>, bristles in 2–5 circumferential rows from rim; sepals 5, spreading, ± ovate, 1–3 mm; petals 5, yellow, obovate to ± oval; stamens 5–15, shorter than petals; carpels 2, rarely more. **Fruits** achenes, 1 (1 ovary aborts), (including hypanthia) top-shaped, 1–6.6 mm; hypanthium persistent, enclosing achenes; sepals persistent, erect, connivent, <bristles erect to reflexed, hooked>. $x = 7$.

Species ca. 20 (7 in the flora): North America, Mexico, West Indies, Central America, South America, Eurasia, s Africa; circumboreal.

Characters useful in separating *Agrimonia* species exhibit sufficient overlap of expressions among species, as well as variation within species, to confound identification. A suite of two or more characters is necessary to identify unequivocally any *Agrimonia* species. Because mature reflexed fruits, which provide the most consistent identifying characters, are frequently absent from material under examination, the key first presents the most useful vegetative characters. Hypanthium shape, size, indumentum, number of circumferential bristle rows, and position of the proximal row should be assessed on fully reflexed mature fruits. When mature fruits are not present, the hypanthium indumentum and number of bristle rows may be useful. Mid cauline leaves were selected to measure stipule shape, major leaflet shape and number, and minor leaflet number for comparison among the species, because most of those leaf characters vary in some way along the stem of a plant. The number of minor, or smaller, leaflets on each side of the rachis between any two major, or larger, leaflets also varies, increasing distally. The number of leaflets in the distal interval (the maximum number per leaf) can help separate species, for example, *A. gryposepala* and *A. rostellata*. Species with flowers more or less subopposite along the inflorescence axis have flowering inflorescences that are more compact in appearance, especially in the flowering portion. Presence of tuberous roots can separate *A. gryposepala* from *A. rostellata* and is useful in confirming identification.

Examining plant surfaces to determine the types of eglandular and glandular hairs is necessary and most accurately done with a 10× lens; frequently, glistening, sessile-glandular hairs can be seen with the naked eye. Despite the overlap in descriptions of the density of surface hairs (for example, sparsely versus scattered hirsute), these distinctions more completely describe the variation among species and between plant surfaces than merely describing them as hirsute.

No basal leaves occur in North American species of *Agrimonia*; before dormancy, a short section of stem is produced; it has the appearance of a basal rosette. The stem continues to elongate from that section in the spring.

SELECTED REFERENCES Bicknell, E. P. 1896b. The North American species of *Agrimonia*. Bull. Torrey Bot. Club 23: 508–523. Kline, G. J. and P. D. Sørensen. 2008. A revision of *Agrimonia* (Rosaceae) in North and Central America. Brittonia 60: 11–33. Skalický, V. 1973. Amerikanischer Arten Gattung *Agrimonia* L. ser. *Tuberosae* ser. nova. Folia Geobot. Phytotax. 8: 95–104.

1. Stems with glistening glandular hairs (short-stipitate, sometimes also sessile) and ± hirsute.
    2. Inflorescence axes ± ascending-hirsute (hairs 0.5–1 mm); minor leaflets 0 or 1(or 2) pairs; roots with fusiform tubers; fruiting hypanthia hemispheric, 1–3.1 mm . . . . .5. *Agrimonia rostellata*
    2. Inflorescence axes hirsute (hairs 2 mm); minor leaflets 1–4, rarely 0 or 1, pairs; roots without tubers; fruiting hypanthia turbinate to broadly campanulate, rarely broadly obconic, 2.3–5.8 mm. . . . . 6. *Agrimonia gryposepala*
1. Stems with glandular hairs sometimes glistening (short-stipitate and sessile) and pubescent to villous and hirsute.
    3. Abaxial leaflet surfaces (at 10×) rarely glistening with sessile-glandular hairs.
        4. Stem hairs 3–4 mm; mid cauline stipules deeply incised; minor leaflets 0 or 1 pair; fruiting hypanthia 2.2–4 × 3–4.6 mm, ridges rarely sparsely hirsute . . . . 3. *Agrimonia microcarpa*
        4. Stem hairs 2–3 mm; mid cauline stipules dentate (but not deeply so); minor leaflets 1 or 1–3 pairs; fruiting hypanthia 1.9–4.5 × 2–4.6 mm, ridges sparsely hirsute . . . . 4. *Agrimonia pubescens*
    3. Abaxial leaflet surfaces (at 10×, often with naked eye) glistening with sessile-glandular hairs.
        5. Major leaflets 5–7 on mid cauline leaves; mid cauline stipules proximally dentate or entire . . . . .7. *Agrimonia striata*
        5. Major leaflets 9–13 on mid cauline leaves; mid cauline stipules dentate.
            6. Major leaflet blades lanceolate to narrowly elliptic, rarely narrowly rhombic, margins serrate to dentate, apices acuminate or long-acuminate, rarely acute; flowers usually ± subopposite. . . . . 1. *Agrimonia parviflora*
            6. Major leaflet blades ± obovate, margins incised, apices obtuse to acute; flowers usually ± alternate . . . . 2. *Agrimonia incisa*

1. Agrimonia parviflora Aiton, Hort. Kew. 2: 130. 1789 [F]

Agrimonia polyphylla Urban; *A. serrifolia* Wallroth; *A. suaveolens* Pursh

Herbs, 5–20 dm. **Roots:** tubers absent. **Stems** with glistening glandular hairs (stipitate and sessile) and pubescent to villous and hirsute (hairs stiff, scattered, erect, 2–3 mm). **Leaves:** mid cauline stipules ± falcate, margins dentate, <apical lobes long-attenuate>; major leaflets 3–19 (mid cauline 9–13), minor 1–4 pairs; major leaflet blades lanceolate to narrowly elliptic, rarely narrowly rhombic, terminal largest, largest of these 3.4–8.5 × 1.2–2.4 cm, margins serrate to dentate, apex acuminate or long-acuminate, rarely acute, abaxial surface usually glistening with sessile-glandular hairs and pubescent to villous and hirsute (hairs stiff, ± scattered, 1–3 mm, densest along major veins). **Inflorescences:** axes glistening with sessile-glandular hairs and pubescent to villous and hirsute (hairs scattered, ± erect, 1–2 mm, usually absent distally). **Flowers** usually ± subopposite. **Fruiting hypanthia** broadly campanulate to broadly turbinate, rarely broadly obconic or hemispheric, 1.3–3 × 1.7–3.8 mm, moderately to shallowly sulcate, hooked bristles in 3 or 4 circumferential rows, proximal row spreading ± 90° to ± reflexed, usually glistening with sessile-glandular hairs, <ridges sometimes sparingly hirsute>. $2n = 28$.

Flowering mid Jul–early Sep. Edges, open spaces, thickets, deciduous or mixed woods and thickets in wet places; 0–1100 m; Ont.; Ala., Ark., Conn., Del., D.C., Ga., Ill., Ind., Iowa, Kans., Ky., La., Md., Mass., Mich., Miss., Mo., Nebr., N.J., N.Mex., N.Y., N.C., N.Dak., Ohio, Okla., Pa., S.C., S.Dak., Tenn., Tex., Va., W.Va., Wis.; West Indies (Dominican Republic); introduced in South America (Argentina, s Brazil).

Plants of *Agrimonia parviflora* with major leaflet blades narrowly elliptic and with acute apices are occasionally found throughout the range. *Eupatorium parviflorum* (Aiton) Nieuwland is an illegitimate name that pertains here.

2. Agrimonia incisa Torrey & A. Gray, Fl. N. Amer. 1: 431. 1840 [E] [F]

Herbs, 3–11 dm. **Roots:** tubers fusiform, thickened. **Stems** with glistening sessile-glandular hairs and ± matted-pubescent to villous and hirsute (hairs scattered, erect, stiff, 2–3 mm). **Leaves:** mid cauline stipules ± falcate, margins dentate; major leaflets 3–15 (mid cauline 9), minor 1–3 pairs; major leaflet blades ± obovate, terminal largest, largest of these 2.3–4.1 × 1–1.7 cm, margins incised, apex obtuse to acute, abaxial surfaces glistening with sessile-glandular hairs and pubescent to villous and sparsely hirsute (hairs stiff, 2 mm). **Inflorescences:** axes often with glistening sessile-glandular hairs and pubescent to villous and hirsute

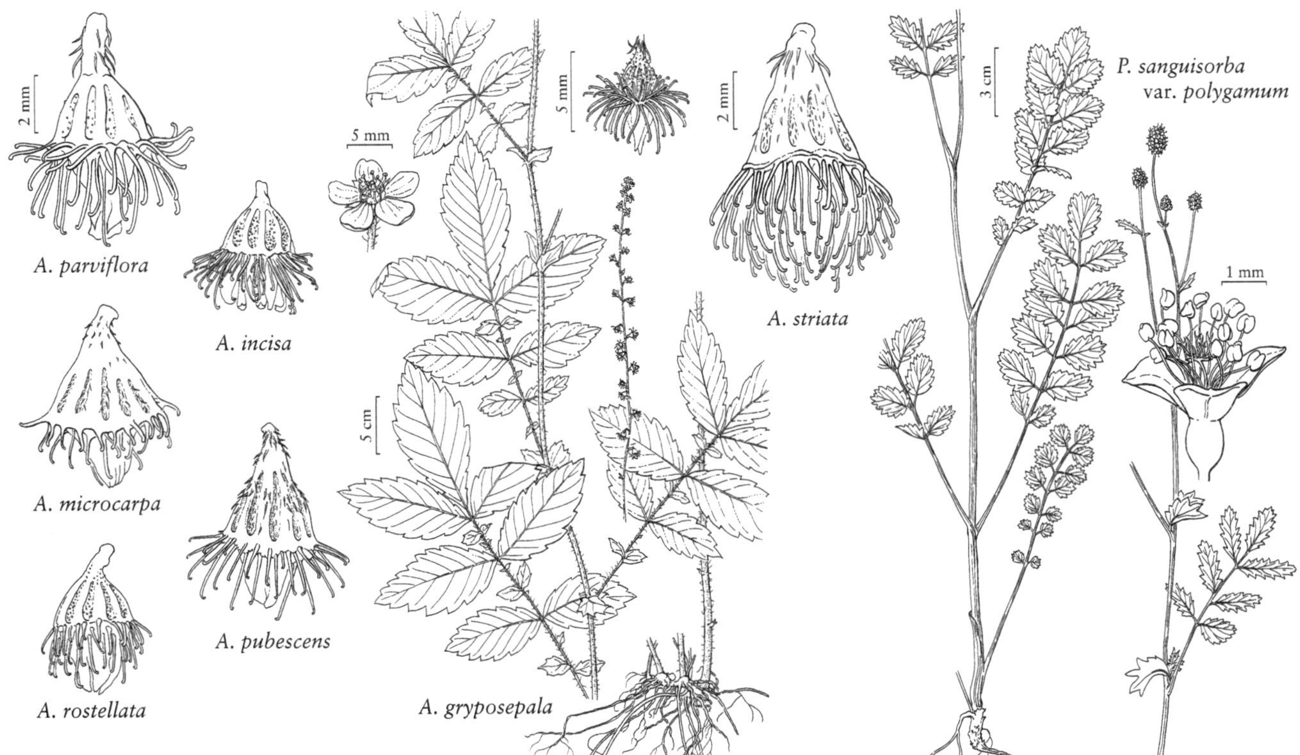

AGRIMONIA ∘ POTERIUM

(hairs stiff, erect, 2 mm). **Flowers** usually ± alternate. **Fruiting hypanthia** hemispheric to turbinate, 1.5–2.7 × 1.8–3.3 mm, obscurely sulcate, hooked bristles in 3 or 4 circumferential rows, proximal row spreading 45–90°, glistening with sessile-glandular hairs (often yellow).

Flowering late Jul–Nov. Dry to moist, longleaf pine-oak woods, oak-hickory slopes, roadsides, sand or shell maritime thickets; 0–200 m; Ala., Fla., Ga., La., Miss., Mo., S.C., Tex.

Populations of *Agrimonia incisa* are infrequent and usually widely scattered within the limited range (15–20 currently known). The plants are sometimes abundant where they occur.

3. Agrimonia microcarpa Wallroth, Beitr. Bot. 1: 39, plate 1, fig. 3. 1842, name conserved C E F

Agrimonia pubescens Wallroth var. *microcarpa* (Wallroth) H. E. Ahles; *A. pumila* Muhlenberg ex E. P. Bicknell

Herbs, 3–11 dm. **Roots:** tubers ± narrowly oblong, thickened. **Stems** with short-stipitate-glandular hairs and pubescent to villous and hirsute (hairs scattered, erect, 3–4 mm, stiff). **Leaves:** mid cauline stipules ± falcate to ± ½-round, margins deeply incised; major leaflets 3–9 (mid cauline 5–7), minor 0 or 1 pair; major leaflet blades obovate to elliptic, terminal largest, largest of these 2.9–7 × 1.7–3.7 cm, margins serrate to dentate, apex obtuse to acute, abaxial surface rarely with glistening sessile-glandular hairs and pubescent to pilose and hirsute (hairs stiff, scattered, 1–2 mm, densest along major veins). **Inflorescences:** axes pubescent to villous and hirsute (hairs stiff proximally, erect and 2 mm to ascending distally and 1 mm). **Flowers** usually ± alternate. **Fruiting hypanthia** turbinate to obconic, 2.2–4 × 3–4.6 mm, ± deeply sulcate, hooked bristles in 3 circumferential rows, proximal row spreading 45–90°, glandular-hairy (hairs short-stipitate), <grooves strigose, ridges rarely sparsely hirsute (sepal bases rarely pubescent along rim)>.

Flowering late Jul–Sep. Edges, open spaces and thickets, deciduous or mixed pine-deciduous woods; of conservation concern; 0–600 m; Ala., Ark., Fla., Ga., La., Miss., N.J., N.C., Ohio, Okla., Pa., S.C., Tenn., Tex., Va., W.Va.

The ranges of *Agrimonia microcarpa* and *A. pubescens* overlap from North Carolina through Arkansas. Only rarely do populations of either species occur in the range of the other species beyond that contact area. Plants occurring within the range of *A. pubescens* that resemble *A. microcarpa* but lack one or more of the above characteristics have been included in *A. pubescens.* Observations of plants of *A. pubescens* grown from seed in the garden at Northern Illinois University suggest that some, if not most, plants resembling *A. microcarpa* are young second-year plants or are under stress. Ambiguous identifications are most clearly resolved when mature fruits are present.

4. Agrimonia pubescens Wallroth, Beitr. Bot. 1: 45, plate 1, fig. 7. 1842 • Soft agrimony, aigremoine pubescente E F

Agrimonia bicknellii (Kearney) Rydberg; *A. eupatoria* Linnaeus var. *mollis* Torrey & A. Gray; *A. mollis* (Torrey & A. Gray) Britton; *A. mollis* var. *bicknellii* Kearney; *A. platycarpa* Wallroth

Herbs, 5–16 dm. **Roots:** tubers ± fusiform. **Stems** with short-stipitate-glandular hairs and pubescent to villous and hirsute (hairs scattered, erect, 2–3 mm, stiff). **Leaves:** mid cauline stipules ± broadly ½-ovate, margins dentate, <apical lobes sometimes attenuate>; major leaflets 3–13 (mid cauline 5–9), minor 1 or 1–3 pairs; major leaflet blades elliptic or narrowly elliptic to lanceolate, sometimes ± obovate (lateral sometimes ± falcate), terminal largest, largest of these 3.3–9.8 × 1.4–5.5 cm, margins serrate to dentate, apex obtuse to acute or acuminate, abaxial surface rarely with glistening glandular hairs, pubescent to pilose and sparsely hirsute (hairs stiff, 1–2 mm, usually densest along major veins). **Inflorescences:** axes with short-stipitate-glandular hairs to villous and hirsute (hairs stiff, ± scattered, usually erect proximally and 1–3 mm to ± ascending distally and 1 mm). **Flowers** usually ± alternate. **Fruiting hypanthia** turbinate to campanulate, 1.9–4.5 × 2–4.6 mm, deeply to shallowly sulcate, hooked bristles in 3 or 4 circumferential rows, proximal row spreading ± 90° (pressed upward on dried specimens), glandular-hairy, <grooves strigose, ridges sparsely hirsute (sepal bases usually pubescent along rim)>.

Flowering mid Jul–Sep. Edges, open spaces, thickets, deciduous or mixed deciduous woods; 0–1300 m; Ont., Que.; Ala., Ark., Conn., Del., D.C., Ga., Ill., Ind., Iowa, Kans., Ky., Md., Mass., Mich., Minn., Miss., Mo., Nebr., N.H., N.J., N.Y., N.C., Ohio, Okla., Pa., R.I., S.C., S.Dak., Tenn., Tex., Vt., Va., W.Va., Wis.

Agrimonia pubescens includes three recognizable morphs. The principal patterns are: leaves with major leaflets more or less obovate to more or less elliptic, 5–7 leaflets on mid cauline leaves, 1 minor pair; leaves with major leaflets more or less narrowly obovate to elliptic, 7 leaflets on mid cauline leaves, 0 or 1 or 1–3 minor pairs; and, leaves with major leaflets elliptic to narrowly elliptic or lanceolate, 7–9 leaflets on mid cauline leaves, 0 or 1 or 1–3 minor pairs. Lateral major leaflets are frequently more or less falcate in morphs two and three. Stipules with attenuate apical lobes are usual in morph three and frequent in morph two. No discontinuities in the variation occur within *A. pubescens*, and all patterns of variation are found throughout the range. Plants resembling *A. microcarpa* in major leaflet shape and in the number of major and minor leaflets can usually be identified as *A. pubescens* by stipule shape, the presence or absence of stiff hairs on mature hypanthia ridges, and/or the lengths of stem hairs. Additional study is needed to establish the phylogeny and perhaps a more precise taxonomy of this species as well as its relationship to *A. microcarpa*. *Eupatorium molle* (Torrey & A. Gray) Nieuwland, which pertains here, is illegitimate.

5. Agrimonia rostellata Wallroth, Beitr. Bot. 1: 42. 1842 E F

Eupatorium rostellatum (Wallroth) Nieuwland

Herbs, 4–10 dm. **Roots:** tubers fusiform. **Stems** with glistening stipitate-glandular hairs and hirsute (hairs stiff, sparse to scattered, erect, 1–2 mm). **Leaves:** mid cauline stipules ± falcate to ± ½-ovate, margins often deeply incised; major leaflets 3–11 (mid cauline 5–7), minor 0 or 1(or 2) pairs; major leaflet blades obovate to broadly elliptic to elliptic, terminal largest, largest of these 3.4–10.5 × 1.7–5.6 cm, margins serrate, apex acute to obtuse, abaxial surface glistening with sessile-glandular hairs (often shining) and hirsute (hairs sparse to scattered, 1 mm). **Inflorescences:** axes glistening with sessile-glandular hairs and ± ascending-hirsute (hairs stiff, sparse to scattered, 0.5–1 mm). **Flowers** ± alternate. **Fruiting hypanthia** hemispheric, 1–3.1 × 1.8–3.9 mm, obscurely sulcate, hooked bristles in 2 or 3 circumferential rows, proximal row spreading to ± 90°, usually glistening with sessile-glandular hairs.

Flowering Jul–mid Sep. Edges, open spaces, and thickets, deciduous or mixed deciduous woods; 0–1100 m; Ala., Ark., Conn., Del., D.C., Fla., Ga., Ill., Ind., Kans., Ky., La., Md., Mich., Miss., Mo., N.J., N.Y., N.C., Ohio, Okla., Pa., S.C., Tenn., Tex., Va., W.Va., Wis.

6. Agrimonia gryposepala Wallroth, Beitr. Bot. 1: 49, plate 1, fig. 8. 1842 • Common agrimony, aigremoine à sépales crochus F

Agrimonia eupatoria Linnaeus var. *parviflora* Hooker; *A. macrocarpa* (Focke) Rydberg; *A. parviflora* Aiton var. *macrocarpa* Focke; *Eupatorium gryposepalum* (Wallroth) Lunell

Herbs, 3.5–15 dm. **Roots:** tubers absent. **Stems** with glistening stipitate-glandular hairs and hirsute (hairs scattered, erect, 2 mm). **Leaves:** mid cauline stipules ± ½-ovate, margins incised; major leaflets 3–9 (mid cauline 5–7), minor 1–4, rarely 0 or 1, pairs; major leaflet blades ovate to elliptic or rhombic, terminal largest, largest of these 2.4–10.5 × 1.4–5.6 cm, margins serrate, apex acute, sometimes acuminate, abaxial surface usually with glistening stipitate-glandular hairs and hirsute (hairs scattered, 0.5–2 mm). **Inflorescences:** axes with glistening stipitate-glandular hairs and hirsute (hairs erect, 2 mm). **Flowers** usually ± alternate. **Fruiting hypanthia** turbinate to broadly campanulate, rarely broadly obconic, 2.3–5.8 × 2.8–6.2 mm, deeply to sometimes obscurely sulcate, hooked bristles in 4 or 5 circumferential rows, proximal row reflexed, with glistening stipitate-glandular hairs and sparsely hirsute. ***2n*** = 56.

Flowering Jul–mid Aug[–Mar]. Edges, open spaces, and thickets, deciduous or mixed pine-deciduous woods; 0–2000[–3000] m; B.C., Man., N.B., N.S., Ont., P.E.I., Que.; Ariz., Ark., Calif., Colo., Conn., Del., Ga., Ill., Ind., Iowa, Kans., Ky., Maine, Md., Mass., Mich., Minn., Mo., Mont., Nebr., N.H., N.J., N.Mex., N.Y., N.C., N.Dak., Ohio, Oreg., Pa., R.I., S.Dak., Tenn., Vt., Va., Wash., W.Va., Wis., Wyo.; Mexico (Chiapas, Coahuila, Nuevo León); Central America (Guatemala).

In the mountains of the western United States, Mexico, and Guatemala, populations of *Agrimonia gryposepala* are widely scattered. The species is absent from the Great Basin. Plants with stems and abaxial leaflet surfaces more densely hirsute and leaflets less than 7 cm are sometimes found in the mountains of the western United States and are the predominant morphs of *A. gryposepala* in Chiapas and Guatemala. Only three specimens from Coahuila and Nuevo León have been seen by the authors; the relative occurrence of the morphs there cannot be determined.

7. **Agrimonia striata** Michaux, Fl. Bor.-Amer. 1: 287. 1803 • Roadside agrimony, aigremoine striée [F]

Agrimonia brittoniana E. P. Bicknell; *A. striata* var. *campanulata* Fernald; *Eupatorium brittonianum* (Bicknell) Nieuwland & Lunell

Herbs, 4–13 dm. **Roots:** tubers absent. **Stems** with ± glistening stipitate-glandular hairs and pubescent to pilose (at least distally) and hirsute (hairs scattered, erect, 2 mm, stiff). **Leaves:** mid cauline stipules ± falcate, margins proximally dentate or entire; major leaflets 3–11 (mid cauline 5–7), minor 1–4 pairs; major leaflet blades elliptic to rhombic, terminal largest, largest of these 4.3–10.7 × 1.9–4.2 cm, margins serrate, apex acute to acuminate or long-acuminate, abaxial surface with glistening glandular hairs (stipitate and sessile) and pubescent to pilose and hirsute (hairs stiff, scattered, 1–2 mm, densest along major veins). **Inflorescences:** axes with stipitate-glandular hairs and pubescent to pilose and hirsute (hairs stiff, scattered, proximally ± erect and 1–2 mm to distally ± ascending and 1 mm). **Flowers** usually ± subopposite. **Fruiting hypanthia** obconic to ± campanulate, rarely turbinate, 2.1–6.6 × 2.4–5.2 mm, deeply sulcate, hooked bristles in 3 or 4 circumferential rows, proximal row spreading ± 90° (pressed upward on dried specimens), stipitate-glandular hairs, ± glistening, <grooves strigose, ridges usually sparingly hirsute (sepal bases usually pubescent along rim)>. ***2n*** = 56.

Flowering late Jun–Aug(–mid Sep). Edges, open spaces, thickets, in deciduous or mixed deciduous woods; 0–3700 m; St. Pierre and Miquelon; Alta., B.C., Man., N.B., Nfld. and Labr. (Nfld.), N.S., Ont., P.E.I., Que., Sask.; Ariz., Calif., Colo., Conn., Del., Ga., Ill., Iowa, Maine, Md., Mass., Mich., Minn., Mont., Nebr., Nev., N.H., N.J., N.Mex., N.Y., N.Dak., Ohio, Okla., Pa., R.I., S.Dak., Vt., W.Va., Wis., Wyo.; Mexico (Chihuahua, Distrito Federal, Hidalgo, Sinaloa, Sonora, Tamaulipas).

Plants of *Agrimonia striata* with fruiting hypanthia campanulate or obconic or both occur throughout the range. Plants with vestiture more like that of *A. gryposepala* are sometimes found, particularly in Quebec and the mountains of Arizona and New Mexico.

Excluded Species:

Agrimonia eupatoria Linnaeus

Agrimonia eupatoria has been sporadically cultivated in the flora area. No evidence has been found that this introduced European species, which rarely escapes cultivation, has become an established element of the flora. It is distinguished from *A. pubescens* by the absence of tuberous roots, inflorescences with closely spaced, more or less subopposite flowers, and pendent fruiting hypanthia, 7–10 mm, covered with appressed hairs.

Agrimonia repens Linnaeus

Agrimonia repens has been reported (as *A. odorata* Miller) as persisting from cultivation on Prince Edward Island. There is as yet no evidence that this species is established; it can be identified by major leaflets more or less overlapping, mid cauline stipules 4–5 cm, petals 5–7 mm, and fruiting hypanthia cylindric with disc projecting 1.5 mm, 5–7 rows of circumferential bristles, proximal row reflexed.

23. POTERIUM Linnaeus, Sp. Pl. 2: 994. 1753; Gen. Pl. ed. 5, 430. 1754 • Salad burnet [Greek *poterion*, goblet or cup, presumably alluding to shape of hypanthium] [I]

Alan S. Weakley

Herbs, perennial [unarmed or thorny shrubs], 1–8[–40] dm; rhizomatous [taprooted]. **Stems** 1–10+, ascending to erect, glabrescent to loosely villous [glabrous]. **Leaves** deciduous [persistent], basal and/or cauline; stipules persistent, adnate to petiole or free, lanceolate to ovate, margins entire or serrate; petiole present; blade oblanceolate to obovate [or elliptic], 1–15[–35] cm, herbaceous [subcoriaceous], leaflets 2–12 pairs, elliptic, orbiculate, or obovate, margins flat, crenate, serrate, or incised, surfaces glabrous or hairy <hairs unicellular to multicellular, acicular and/or gland-tipped>. **Inflorescences** terminal [or axillary] to distal leaves, [15–]20–100+-flowered, headlike spikes, glabrous or sparsely [to densely] pubescent; peduncles present, sometimes absent; bracts absent; bracteoles present. **Pedicels** absent. **Flowers** bisexual or pistillate (plants gynomonoecious [dioecious]), 2–4[–5] mm diam.; hypanthium <dry [fleshy, spongy]>, top-shaped, ellipsoid, or nearly orbicular, 2–9 mm, <angled, winged, and/or ornamented with irregular flanges and papillae, tan to brown>, hairy or glabrous; sepals 4, <distinct [basally connate]>, usually ascending or spreading, rarely slightly reflexed, ovate to elliptic; petals 0; stamens 0–30(–50), shorter or longer than sepals; carpels [1 or]2(or 3), glabrous, <styles repeatedly branched, brushlike>. **Fruits** achenes, 1 or 2[or 3], ellipsoid, 1.5–2 mm diam., glabrous (though reticulate and muricate); hypanthium persistent, enclosing achenes; sepals persistent, spreading to slightly ascending. $x = 7$.

Species 13 (1 in the flora): introduced; Europe (especially Mediterranean region), w Asia, n Africa, n Atlantic Islands (Macaronesia); introduced also in s South America, Australia.

Poterium is here circumscribed to include *Bencomia* Webb & Berthelot, *Dendriopoterium* Sventenius (all of Macaronesia), *Marcetella* Sventenius, and *Sarcopoterium* Spach (of the eastern Mediterranean) based on molecular phylogenetic analyses summarized by D. Potter et al. (2007). The inclusion of these woody Mediterranean and Macaronesian endemics significantly increases the morphological diversity in the genus; see also 25. *Sanguisorba*.

SELECTED REFERENCE Nordborg, G. 1967. The genus *Sanguisorba* section *Poterium*: Experimental studies and taxonomy. Opera Bot. 16: 1–166.

1. Poterium sanguisorba Linnaeus, Sp. Pl. 2: 994. 1753 [F] [I]

Sanguisorba minor Scopoli

Varieties 5 (1 in the flora): introduced; Europe, w Asia, n Africa; introduced also in s South America, Australia.

The *Poterium sanguisorba* complex has been treated in various and complex ways in Europe; a consensus has not emerged as to the number of species or infraspecific taxa to be recognized.

1a. Poterium sanguisorba Linnaeus var. **polygamum** (Waldstein & Kitaibel) Visiani, Fl. Dalmat. 3: 255. 1851 (as polygama) • Salad or fodder burnet, burnet bloodwort, sanguisorbe mineure [F] [I]

Poterium polygamum Waldstein & Kitaibel, Descr. Icon. Pl. Hung. 2: 217, plate 198. 1805; *P. balearicum* (Nyman) Bourgeau ex Porta; *P. sanguisorba* subsp. *muricatum* Bonnier & Layens; *Sanguisorba minor* Scopoli subsp. *balearica* (Nyman) Muñoz Garmendia & C. Navarro; *S. minor* subsp. *muricata* (Bonnier & Layens) Briquet

Leaves: stipules of basal leaves simple, lanceolate, margins entire, those of cauline lanceolate to ovate, coarsely serrate; petiole longer than blade in basal and proximal cauline leaves, to shorter than proximal rachis segment in distal cauline ones; leaflets 2–12 pairs, proximally alternate, subopposite, or opposite, distally subopposite or opposite, margins coarsely serrate, surfaces glabrous or sparsely pilose. **Spikes** subglobose to ovoid, 0.3–2.5 × 0.3–2 cm; peduncles glabrous or sparsely loosely villous; bracteoles ovate, equal to hypanthia, ciliate. **Flowers** generally gynomonoecious, bisexual flowers proximal and pistillate distal (rarely some spikes completely pistillate or completely bisexual); hypanthium broadly ellipsoid; sepals green to purple, margins scarious, apex acute to apiculate with tufts of trichomes; filaments filiform, equal to sepals. **Achenes** enclosed within broadly ellipsoid hypanthia 4 mm, 4-sided, angles winged, surfaces irregularly reticulately ridged, ridges sometimes also muricate. $2n$ = 28, 56.

Flowering May–Aug; fruiting Jul–Oct. Open disturbed areas, pastures, roadsides, woodland edges, along railroad rights-of-way, some preference for calcareous soil; 0–2700 m; introduced; B.C., N.S., Ont., Que.; Ala., Ariz., Calif., Colo., Conn., Del., Idaho, Ill., Ind., Md., Mass., Mich., Mont., Nev., N.J., N.Mex., N.Y., N.C., Ohio, Oreg., Pa., R.I., Tenn., Tex., Utah, Vt., Va., Wash., W.Va., Wyo.; Europe; w Asia; n Africa; introduced also in Australia.

All North American plants of *Poterium* belong to var. *polygamum*. This taxon has also been reported for New Brunswick; this is unconfirmed, but certainly plausible. Reports for North America of *Sanguisorba minor* or *P. sanguisorba* refer to that in a broad sense, not distinguishing the varieties or subspecies variously recognized, and do not imply the occurrence of the typical infrataxon in North America. It is possible that the typical variety (or other infrataxa, as defined in the European literature) is present in North America. The other infrataxa differ primarily in details of the ornamentation of the hypanthium, with either more or less pronounced wings, differences in the roundedness or acuteness of the wings, and variations in the reticulate patterns between the wings; refer to the European literature to identify material suspected of belonging to other infrataxa.

24. POTERIDIUM Spach, Ann. Sci. Nat., Bot., sér. 3, 5: 43. 1846 • American burnet [Genus *Poterium* and Latin *-idium*, diminutive] [E]

Alan S. Weakley

Herbs, annual or biennial, 1–10 dm; taprooted. **Stems** 1–10+, ascending to erect, <freely branched>, glabrous or sparsely pubescent. **Leaves** deciduous, basal and cauline (basal usually withering before or at flowering); stipules persistent, adnate to petiole, those of basal leaves simple, lanceolate, entire, of cauline pectinately divided; petiole present; blade oblanceolate to obovate, 2–7 cm, herbaceous, leaflets 3–8 pairs, <proximally alternate, subopposite, or opposite, distally subopposite or opposite>, broadly elliptic to obovate, margins flat, pectinately pinnatisect (segment margins nearly parallel), faces glabrous. **Inflorescences** terminal, also often axillary to distal leaves, 12–100+-flowered, headlike spikes, <cylindric, or subglobose when poorly developed>, glabrous or nearly so; peduncles present, sometimes absent; bracts absent; bracteoles present. **Pedicels** absent. **Flowers** 1–3 mm diam.; hypanthium urceolate, compressed at mouth, 1–3 mm, glabrous; sepals 4, ascending to spreading, ovate, <margins scarious>; petals 0; stamens 2 (and opposite inner sepals) or 4, equal to sepals; carpel 1, glabrous, <styles repeatedly branched, brushlike>. **Fruits** achenes, 1, globose, 1.5–2 mm diam., glabrous; hypanthium persistent, enclosing achenes; sepals persistent, ascending to spreading. x = 7.

Species 2 (2 in the flora): North America.

In North America, *Poteridium* has, with few exceptions (notably P. A. Rydberg 1908b), been included in *Sanguisorba* since its segregation by E. Spach. Molecular analyses clarify the distinction of *Poteridium* from *Sanguisorba* and bolster the morphologic distinctions (see 25. *Sanguisorba* for discussion).

SELECTED REFERENCE Rydberg, P. A. 1908b. *Poteridium.* In: N. L. Britton et al., eds. 1905+. North American Flora.... 47+ vols. New York. Vol. 22, p. 388.

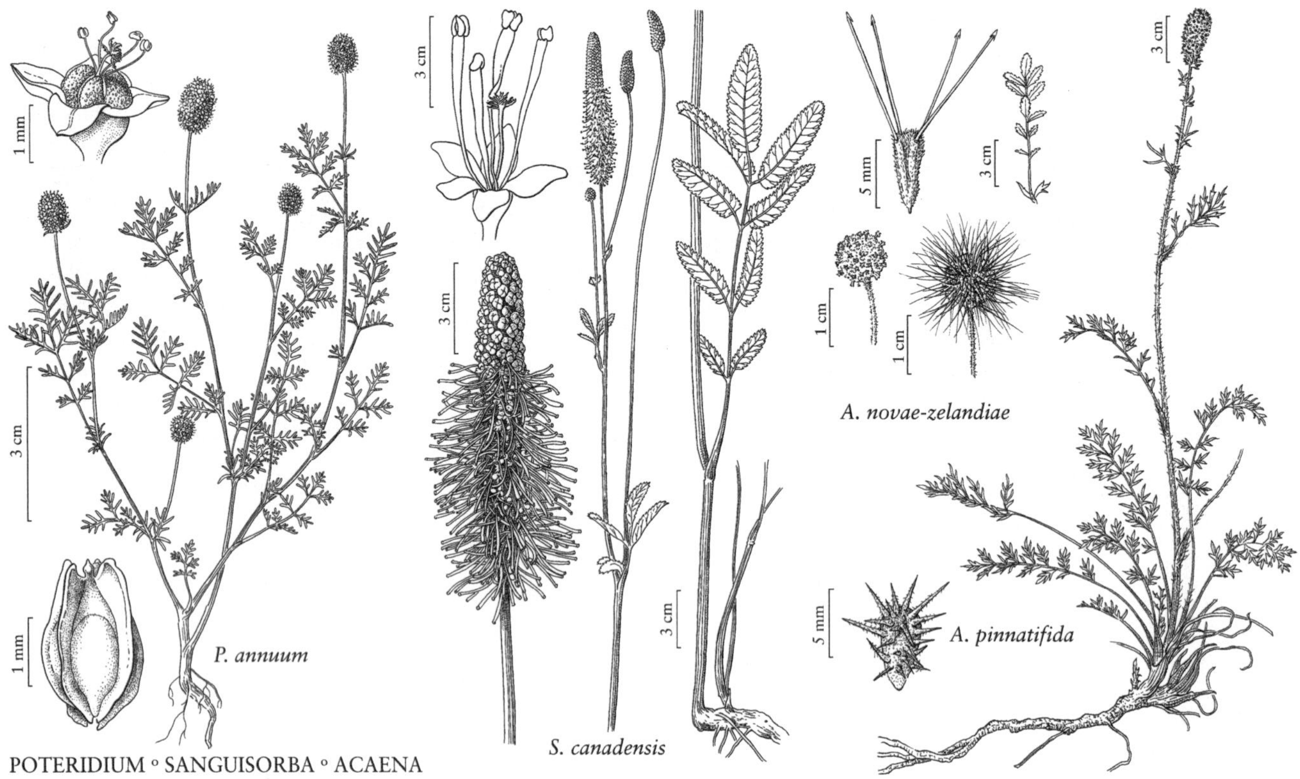

POTERIDIUM ◦ SANGUISORBA ◦ ACAENA

1. Fruits: hypanthia ridges extended into thin wings (0.5 mm wide); sepals tuberculose-thickened proximally; stamens (2 or)4; sc North America (adventive eastward) . . . . . . 1. *Poteridium annuum*
1. Fruits: hypanthia ridges rounded, thickened; sepals not thickened proximally; stamens 2 (or 4); w North America . . . . 2. *Poteridium occidentale*

1. Poteridium annuum (Nuttall) Spach, Ann. Sci. Nat., Bot., sér. 3, 5: 43. 1846 • Prairie burnet [E] [F]

Poterium annuum Nuttall in W. J. Hooker, Fl. Bor.-Amer. 1: 198. 1832; *Sanguisorba annua* (Nuttall) Nuttall

Plants winter-annual or biennial, glabrous; taproots 1–4(–6) dm. **Stamens** (2 or)4. **Fruits:** hypanthia ridges extended into thin wings, <0.5 mm wide>; sepals tuberculose-thickened proximally.

Flowering Mar–Jun; fruiting Apr–Jul. Sandy and rocky soil, prairies, oak savannas, oak woodlands, disturbed areas (pastures, roadsides), often locally moister sites; 0–500 m; Ont.; Ark., Iowa, Kans., Mass., N.Y., Okla., S.C., Tex.

The eastern records (Ontario, Iowa, Massachusetts, New York, and South Carolina) all appear to represent adventives or non-persistent waifs. A record from Maryland (J. B. S. Norton and R. G. Brown 1946) is plausible; no specimen is known to the author.

2. Poteridium occidentale (Nuttall) Rydberg in N. L. Britton et al., N. Amer. Fl. 22: 388. 1908 • Western burnet [E]

Sanguisorba occidentalis Nuttall in J. Torrey and A. Gray, Fl. N. Amer. 1: 429. 1840

Plants winter-annual, glabrous; taproots 3–10 dm. **Stamens** 2(or 4). **Fruits:** hypanthia ridges rounded, thickened; sepals not thickened proximally.

Flowering May–Jul; fruiting Jun–Aug. Prairies, sandy open ground, sagebrush flats, vernal pools, drawdown shorelines of streams and lakes, grassy clearings, roadsides, particularly with surficial or subterranean moisture; 10–2100 m; B.C.; Ariz., Calif., Idaho, Mont., Nev., N.Mex., Oreg., Wash., Wyo.

The taxonomic status of *Poteridium occidentale* has been controversial. Most floristicians of the twentieth century accepted it as distinct from the more eastern *P. annuum* (usually treating the two in *Sanguisorba*); but in recent decades a trend developed to merge the two taxa. While superficially similar, they seem to represent independent evolutionary lineages, as indicated by morphologic distinctions and allopatric distributions.

25. SANGUISORBA Linnaeus, Sp. Pl. 1: 116. 1753; Gen. Pl. ed. 5, 53. 1754 • Burnet, sanguisorbe [Latin *sanguis*, blood, and *sorbeo*, to absorb, apparently alluding to traditional European and Asian uses to stop external or internal bleeding, suggested by dark red flowers of some species and medieval doctrine of signatures]

Alan S. Weakley

Herbs, perennial, 2–20 dm; rhizomatous. **Stems** 1–10+, ascending to erect, glabrous or glabrescent. **Leaves** deciduous, basal and cauline; stipules persistent, adnate to petiole, free portion usually ovate, obovate, or suborbiculate, margins dentate; petiole present; blade oblong-elliptic, 5–55 cm (basal blades largest, cauline reduced distally), herbaceous, leaflets (3–)7–21, orbiculate, ovate, or oblong, <leafy stipels sometimes present on vigorous leaves>, margins flat, crenate or serrate, surfaces glabrous or sparsely hairy. **Inflorescences** terminal or axillary to distal leaves, 50–500+-flowered, spikes, <ellipsoid to cylindric>, glabrous or variously pubescent; peduncles present; bracts absent; bracteoles present. **Pedicels** absent. **Flowers** 2–5 mm diam.; hypanthium urceolate, 1–6 mm, <smooth, muricate, or winged>, glabrous; sepals 4, <basally connate>, spreading, petaloid, elliptic to ovate; petals 0; stamens [2–]4[–12], shorter or longer than sepals; carpels 1[or 2], glabrous, <styles repeatedly branched, brushlike; ovules pendulous>. **Fruits** achenes, 1, globose, 0.5–2 mm, glabrous; hypanthium persistent, enclosing achenes, <dry, hardened, top-shaped to ellipsoid, 2–5 mm, 4-angled or -winged, faces smooth between angles>; sepals persistent, spreading to ascending. $x = 14$.

Species ca. 15 (4 in the flora): North America, n Eurasia.

Sanguisorba is here circumscribed more narrowly than it has been traditionally (for example, H. A. Gleason and A. Cronquist 1991) (though exactly as it was circumscribed by P. A. Rydberg 1908–1918), based on recent molecular phylogenetic analyses summarized in D. Potter et al. (2007), following and refining initial findings by M. S. Kerr (2004). *Poteridium* and *Poterium* are in a clade separate from *Sanguisorba* and basal in Sanguisorbeae, with *Sanguisorba* then basal to the remainder of Sanguisorbeae (Potter et al.).

Acropetal flowering refers to flowers opening sequentially from the base toward the apex of the inflorescence. Basipetal flowering refers to flowers opening sequentially from the apex toward the base of the inflorescence.

1. Spikes flowering acropetal, 5–25 cm; calyx lobes white to greenish white, rarely cream to pinkish or slightly purple.
    2. Leaflets oblong-ovate to lance-oblong, lengths ca. 2–4 times widths, bases broadly cuneate, truncate, or rounded to slightly cordate; calyx lobe midveins thickened distally; plants 9–20 dm; c, e North America . . . . 1. *Sanguisorba canadensis*
    2. Leaflets ovate-oblong to ovate or suborbiculate, lengths ca. 1.2–2 times widths, bases usually deeply cordate, rarely truncate; calyx lobe midveins not thickened distally; plants 2–12 dm; nw North America . . . . 2. *Sanguisorba stipulata*
1. Spikes flowering basipetal, 1.5–7 cm; calyx lobes purple or dark purple.
    3. Stamen lengths 1.5–2.5 times calyx lobes, filaments flattened and dilated in distal ½; calyx lobes purple; leaflet lengths 1–2 times widths; spikes cylindric, 1.5–7 cm . . . . 3. *Sanguisorba menziesii*
    3 Stamen lengths ± equal to calyx lobes, filaments filiform throughout; calyx lobes dark purple; leaflet lengths 1.5–2.5 times widths; spikes ellipsoid or ovoid, 1.5–3 cm . . . . 4. *Sanguisorba officinalis*

1. Sanguisorba canadensis Linnaeus, Sp. Pl. 1: 117. 1753 • Canada burnet, sanguisorbe du Canada, herbe à pisser E F

Plants 9–20 dm. **Leaves:** blade 5–55 cm, leaflets 7–17, oblong-ovate to lance-oblong, 10 × 4 cm, lengths ca. 2–4 times widths, base broadly cuneate, truncate, or rounded to slightly cordate, sometimes with leafy stipels (especially on vigorous basal leaves). **Spikes** 250–500+-flowered, cylindric, main 10–25 cm, flowering acropetal. **Flowers:** calyx lobes white to greenish white, midveins thickened distally; stamen lengths 4–6 times calyx lobes, filaments 6–10 mm, distally flattened-clavate. $2n = 56$.

Flowering Jul–Sep; fruiting Aug–Oct. Fens, stream channels (also lower stream channels), bogs, seeps, swamps, stream shores; 0–1600 m; St. Pierre and Miquelon; N.B., Nfld. and Labr., N.S., Que.; Conn., Del., D.C., Ga., Ill., Ind., Ky., Maine, Md., Mass., Mich., N.H., N.J., N.Y., N.C., Ohio, Pa., R.I., S.C., Tenn., Vt., Va., W.Va.

Reports of *Sanguisorba canadensis* for Manitoba seem biogeographically unlikely and no definite documentation has been found; they are rejected pending additional information.

2. Sanguisorba stipulata Rafinesque, Herb. Raf., 47. 1833 • Sitka burnet

Sanguisorba canadensis Linnaeus subsp. *latifolia* (Hooker) Calder & Roy L. Taylor; *S. canadensis* var. *latifolia* Hooker; *S. canadensis* var. *sitchensis* (C. A. Meyer) Koidzumi; *S. sitchensis* C. A. Meyer

Plants 2–12 dm. **Leaves:** blade 5–35 cm, leaflets 9–15(–21), ovate-oblong to ovate or suborbiculate, to 10 × 7 cm, lengths ca. 1.2–2 times widths, base usually deeply cordate, rarely truncate, without stipels. **Spikes** 150–300+-flowered, cylindric, main 5–12 cm, flowering acropetal. **Flowers:** calyx lobes <distinct, petaloid>, white to greenish white, rarely cream to pinkish or slightly purple, midveins not thickened distally; stamen lengths 1–3 times calyx lobes, filaments 7–13 mm, distally flattened-clavate. $2n = 56$.

Flowering May–Sep; fruiting Jul–Sep. Fens, rich wet meadows, subalpine forests, avalanche tracks, gravel and scree, ravines, gorges, swamps, moist to wet meadows and forest glades, thickets, stream banks; 0–1800 m; B.C., Yukon; Alaska, Idaho, Oreg., Wash.; e Asia.

The taxonomic and nomenclatural treatment of *Sanguisorba stipulata* has varied. Plants from northwest North America and northeast Asia are not taxonomically distinguishable, a conclusion reached also by Li C. L. et al. (2003d). The name *S. sitchensis* has been customarily applied to North American plants; the oldest name available for a taxon bridging the Bering Strait is *S. stipulata*. An additional taxonomic question is the relationship of this taxon to eastern North American *S. canadensis*. While similar and undoubtedly closely related, the two differ in several morphologic characteristics, are strongly allopatric in distribution, and are here regarded as sibling species.

3. Sanguisorba menziesii Rydberg in N. L. Britton et al., N. Amer. Fl. 22: 387. 1908 • Menzies's burnet C E

Plants 3–10 dm. **Leaves:** blade 5–40 cm, leaflets 11–15, orbiculate to ovate, 6 × 4 cm, lengths 1–2 times widths, base usually deeply cordate, rarely truncate, without stipels. **Spikes** 100–300-flowered, cylindric, main 1.5–7 cm, flowering basipetal. **Flowers:** calyx lobes purple, midveins not thickened distally; stamen lengths 1.5–2.5 times hypanthium lobes, filaments 5–7 mm, flattened and dilated in distal ½.

Flowering May–Jul; fruiting Jun–Aug. Coastal bogs, marshes, upper brackish tidal marshes, wet meadows; of conservation concern; 0–1200 m; B.C.; Alaska, Wash.

Sanguisorba menziesii sometimes has been treated as a hybrid between *S. officinalis* and *S. stipulata*. Some of its characteristics do suggest intermediacy; it is possible that it arose via hybridization, but it seems best to regard it as a species, because it forms populations separate from its possible parents and is fully fertile (P. S. Holloway and G. E. M. Matheke 2003).

4. Sanguisorba officinalis Linnaeus, Sp. Pl. 1: 116. 1753 • Great burnet, burnet-bloodwort

Poterium officinale (Linnaeus) A. Gray; *Sanguisorba microcephala* C. Presl; *S. officinalis* subsp. *microcephala* (C. Presl) Calder & Roy L. Taylor

Plants 3–20 dm. **Leaves:** blade 5–40 cm, leaflets (7–)11–15, orbiculate to ovate, to 7 × 5 cm, lengths 1.5–2.5 times widths, base cuneate, truncate, rounded, or cordate, without stipels. **Spikes** 50–250-flowered, ellipsoid or ovoid, main 1.5–3 cm, flowering basipetal. **Flowers:** calyx lobes dark purple, midveins not thickened distally; stamen lengths ± equal to hypanthium lobes, filaments 1.5–2.5 mm, filiform throughout. $2n = 56, 112$.

Flowering (Jun–)Jul–Sep; fruiting Aug–Oct. Muskeg, fens, bogs, swamps, tundra, stream banks (especially in ice-scoured areas), often on serpentine rock; 0–1600 m; B.C., N.W.T., N.S., Yukon; Alaska, Calif., Maine, Mich., Oreg., Pa., Wash.; n Europe; n Asia.

Occurrences of *Sanguisorba officinalis* in eastern North America represent introductions from Eurasia; the species is native in western North America. Near the southern end of its distribution in western North America, the species is generally found over serpentine substrates. There seem to be no consistent differences upon which to base a taxonomic distinction at any level between native northwestern North American plants (*S. microcephala*) and Eurasian *S. officinalis* in the strict sense.

SELECTED REFERENCE Nordborg, G. 1963. Studies in *Sanguisorba officinalis* L. Bot. Not. 116: 267–288.

26. ACAENA Mutis ex Linnaeus, Mant. Pl., 145, 200. 1771 • [Greek *akaina*, thorn, alluding to barbed spines arising from wall of hypanthium]

Bryony Macmillan

Herbs, perennial, 1–15 dm diam.; from branched, woody caudex or stoloniferous. **Stems** 1–20, creeping and rooting or suberect, <basally woody, clad in old leaf bases>, glabrous or sparsely pilose. **Leaves** persistent, basal and/or cauline; stipules absent or persistent, adnate to petiole, free portion ovate to obovate, deeply 2–5-fid, or linear, margins entire; petiole present; blade spatulate, [1–]2–15[–18] cm, herbaceous, leaflets 4–10 per side, <opposite>, ovate, oblong, or elliptic, margins flat or revolute, incised or pinnatisect, dentate or entire, surfaces hairy or glabrous. **Inflorescences** terminal or axillary, [4–]15–100-flowered, spikes or globose heads; peduncles present or absent; bracts absent; bracteoles present. **Pedicels** present. **Flowers** 3.5–6 mm diam.; hypanthium obtriangular, ovoid, or obconic, 3.5–6 mm diam., <spiny [or winged or neither]>, tomentose or glabrous; sepals 4 or 5[or 6], spreading, elliptic or triangular; petals 0; stamens 2–4[–6], longer than sepals; carpels 1[or 2], <styles white or red, fimbriate>. **Fruits** achenes, 1[or 2], spindle-shaped; hypanthium persistent, enclosing achenes, <gray- to red-brown, ovoid, oblong, globose, or obconic, 3–6 mm diam., hardened, sometimes winged, distinctly or obscurely angled with a single spine at apex of each of 4 angles, or covered with 10–20[–50] spines [or without spines], spines usually bearing retrorse hairs or barbs at apices, pubescent or glabrous, eglandular or glandular>; sepals persistent, spreading. $x = 21$.

Species ca. 45 (3 in the flora): w United States, Mexico, South America (Argentina, Chile), s Africa (Cape Province), Atlantic Islands (Gough Island, South Georgia, Tristan da Cunha), Indian Ocean Islands (Kerguelen, Prince Edward Islands), Pacific Islands (Hawaii, New Guinea, New Zealand), Australia; introduced in Europe (England, Ireland).

SELECTED REFERENCE Bitter, F. A. G. [1910–]1911. Die Gattung *Acaena*. Biblioth. Bot. 74.

1. Inflorescences interrupted spikes; fruiting hypanthia bearing 10–20 spines 1–3 mm; plants decumbent to suberect; leaflets pinnatisect. . . . . 1. *Acaena pinnatifida*
1. Inflorescences globose heads; fruiting hypanthia 4-angled, each angle with a spine at apex (arising distally) 7–15 mm; plants creeping or stoloniferous; leaflets dentate.
    2. Fruiting hypanthia: spines 7–12 mm, without subsidiary spines on angles; stems 1.5–2 mm diam.; leaflet adaxial surfaces bright shiny green, smooth, abaxial surfaces glaucescent . . . . 2. *Acaena novae-zelandiae*
    2. Fruiting hypanthia: spines 9–15 mm, with subsidiary spines on angles; stems 2–3 mm diam.; leaflet adaxial surfaces glossy green, rugulose, abaxial surfaces whitish with uneven wax layer . . . . 3. *Acaena pallida*

1. Acaena pinnatifida Ruiz & Pavón, Fl. Peruv. 1: 68, plate 104, fig. b. 1798 [F]

Acaena californica Bitter; *A. pinnatifida* var. *californica* (Bitter) Jepson

Plants from branched woody caudex. **Stems** <decumbent, becoming suberect>, 1.5–5 dm (in fruit), 4 mm diam.; branches erect or leafy. **Leaves** (crowded at bases of stems); stipules absent; petiole to 4 cm; blade 3–15 cm, leaflets 6–10 per side, gray-green (at least on dried specimens), ovate, 4–10 mm, pinnatisect, lobes 3–7(–11), linear, margins revolute, <apex penicillate (having a little brush or tuft of hairs at tips)>, abaxial surface sericeous, adaxial sparsely pilose. **Scapes** terminal, to 50 cm, 1.5–3 mm diam., <sparsely pilose>, proximal scape and leaf sheaths with multicellular hairs among 1-celled hairs, with (0–)2 or 3 cauline leaves to 10 cm. **Inflorescences** 15–35-flowered, interrupted spikes, (flowers crowded distally, 2–15 cm in flower, elongating to 50 cm in fruit); bracteoles leaflike to linear, ciliate. **Flowers** 6 mm diam.; sepals 4–5, elliptic, 3 mm, abaxially pilose, adaxially glabrous; stamens usually 4, filaments 2–3 mm, anthers dark red; stigma rose, laciniate. **Fruiting hypanthia** ovoid, 3–4 mm diam., surfaces tomentose, obscurely 4-angled, bearing 10–20 spines in longitudinal ranks, to 3[–4] mm, spines with retrorse hairs or barbs.

Flowering Mar–Jun. Coastal grasslands, open rocky slopes, sand dunes, old roadsides; 50–400 m; Calif.; South America (Argentina, Chile).

Herbarium records indicate occurrences of *Acaena pinnatifida* between Santa Barbara and Sonoma counties.

2. Acaena novae-zelandiae Kirk, Trans. & Proc. New Zealand Inst. 3: 177. 1871 • Biddy biddy [F] [I] [W]

Plants creeping or stoloniferous. **Stems** to 15 dm, 1.5–2 mm diam.; branches erect, <short>, leafy. **Leaves:** stipules 3–5-fid; petiole 0.3–1 cm; blade 2–11 cm, leaflets 4–7 per side, bright shiny green, oblong or elliptic, 4–17 mm, margins crenate-dentate, teeth 8–15, <smooth adaxially>, abaxial surface glaucescent, usually pilose on veins and margins, adaxial sparsely pilose. **Scapes** axillary, 10–15 cm, 1–2 mm diam., <smooth>, cauline leaves and multicelled hairs absent. **Inflorescences** 80–100 -flowered, globose heads (9 mm diam. in flower, 20–35 mm diam. in fruit); bracteoles linear, pilose. **Flowers** 3.5 mm diam.; sepals 4, triangular, 1.5 mm, abaxially pilose, adaxially glabrous; stamens 2(or 3), filaments 1.5–2 mm, anthers white; stigma white, fimbriate. **Fruiting hypanthia** obtriangular, 3.5–5 mm diam., surfaces sericeous, 4-angled, each angle with a slender spine at apex, <red, sometimes lemon yellow>, 7–12 mm, spines with 4–10 translucent, retrorse barbs 0.5 mm at apex, <surfaces glabrous>. $2n = 42$ (New Zealand).

Flowering May–Jun. Roadsides and paths, forest edges, logged lands, lawns, among cultivated ground covers and waste grounds; 0–200 m; introduced; Calif., Oreg.; Pacific Islands (New Guinea, New Zealand); Australia; also introduced in Europe (England, Ireland).

Herbarium records of *Acaena novae-zelandiae* are from the coastal ranges in California, Humboldt and Sonoma counties, around San Francisco and Monterey, and in Oregon.

Acaena novae-zelandiae has noxious weed status in the United States. The name *A. anserinifolia* (J. R. Forster & G. Forster) Druce was misapplied to *A. novae-zelandiae* by P. A. Munz (1968), this being the first published record of the taxon from North America.

3. Acaena pallida (Kirk) Allan, Fl. New Zealand 1: 360. 1961 [I]

Acaena novae-zelandiae Kirk var. *pallida* Kirk, Stud. Fl. New Zealand, 134. 1899

Plants stoloniferous. **Stems** to 10 dm, 2–3 mm diam., <rough with protruding hair bases>; branches erect, leafy. **Leaves:** stipules 3- or 4-fid; petiole 1–1.5 cm; blade 2.5–12 cm, leaflets 4–7 per side, abaxially whitish with uneven wax layer, adaxially glossy green, oblong, 6–30 mm, margins dentate, teeth 4–10 per side, <rugulose adaxially>, abaxial surface pilose. **Scapes** axillary, 7–18 cm, 1.5–2 mm diam., cauline leaves and multicellular hairs absent. **Inflorescences** 80–90-flowered, globose heads (heads 10 mm diam. in flower, 2–4 cm diam. in fruit); bracteoles linear, pilose. **Flowers** 4 mm diam.; sepals 4, triangular, 2 mm, abaxially pilose, adaxially glabrous; stamens 2, filaments 3–4 mm, anthers white; stigma white, fimbriate. **Fruiting hypanthia** obconic, 4.5–6 mm diam., surfaces covered with appressed hairs, 4-angled, each angle with a spine at apex, 9–15 mm, spines with 4–10 translucent, retrorse barbs 0.5 mm at apex, usually with 1[–3] subsidiary spines on angles, <surfaces glabrous>. **2*n*** = 42 (New Zealand).

Flowering May–Jun. Coastal sands; 0–10 m; introduced, Calif.; Pacific Islands (New Zealand); Australia.

In the flora area, *Acaena pallida* is known from a single collection from Marin County.

b. Rosaceae Jussieu subfam. Dryadoideae Sweet, Brit. Fl. Gard., ser. 2, 1: sub plate 43. 1830 (as Dryadeae)

Luc Brouillet

Shrubs or trees. Leaves cauline; stipules present. **Flowers:** carpels 1–150, distinct, free, styles distinct; ovule 1. **Fruits** achenes or aggregated achenes; styles persistent, elongate, usually plumose.

The subfamily consists of a single tribe.

b1. Rosaceae Jussieu (subfam. Dryadoideae) tribe Dryadeae Lamarck & de Candolle, Syn. Pl. Fl. Gall., 334. 1806

Plants unarmed. **Leaves** alternate, simple, sometimes pinnately compound or basal leaflets 1(or 2); stipules deciduous or persistent, free or adnate to petiole; venation pinnate. **Flowers:** perianth and androecium perigynous; epicalyx bractlets absent; hypanthium saucer-shaped, obconic, funnelform-turbinate, or cylindric; torus absent or hemispheric; carpels 1(or 2) or 4–150, distinct, free, styles terminal, distinct; ovule 1, basal (without obturator). **Fruits** achenes or aggregated achenes; styles persistent, elongate.

Genera 4, species 31 (4 genera, 17 species in the flora): North America, Mexico, Eurasia.

Members of Dryadeae are nitrogen fixing in association with *Frankia*. Cyanogenic glycosides are present, and sorbitol is present in trace amounts. The base chromosome number for the tribe is $x = 9$ (D. Potter et al. 2007). Dryadeae is probably North American in origin.

The obturator is a placental (or funicular) outgrowth present at the micropylar end, which directs the passage of pollen tubes towards the ovule.

1. Leaves 2- or 3-odd-pinnately compound; inflorescences (1 or)2–7(–10)-flowered, corymbs or panicles . . . . 30. *Chamaebatia*, p. 343
1. Leaves simple or with 1(or 2) basal leaflets, (margins lobed, toothed, or entire); inflorescences 1–3(–18)-flowered clusters or flowers solitary.
    2. Petals 0; hypanthia narrowly cylindric; carpels 1(or 2); achenes solitary . . . . 28. *Cercocarpus*, p. 332
    2. Petals 5–10(–12); hypanthia saucer-shaped or obconic; carpels 1–31 or 60–150; achenes usually aggregated, sometimes solitary (*Purshia*).
        3. Shrubs 0.1–2.3 dm; hypanthia saucer-shaped; petals 8–10(–12) . . . . 27. *Dryas*, p. 326
        3. Shrubs 2–75 dm; hypanthia obconic; petals 5 . . . . 29. *Purshia*, p. 339

27. DRYAS Linnaeus, Sp. Pl. 1: 501. 1753; Gen. Pl. ed. 5, 220. 1754 • Mountain-avens, dryade [Greek *Dryas*, name for oak nymphs]

Joshua C. Springer

Bruce D. Parfitt†

Shrubs, <tufted or matted, scapose>, 0.1–2.3 dm; <caudices relatively long, woody>. **Stems** 3+, branched, <rooting freely>; bark green in 1st year, reddish to brown in 2d year, often appearing blackish, peeling; short and long shoots absent; <stoloniferous branches horizontal, relatively short, glabrous>. **Leaves** persistent, simple or (in *D. drummondii*) basal leaflets 1(or 2); stipules persistent, narrowly lanceolate, margins usually entire, sometimes 1- or 2-toothed; petiole present; blade elliptic, lanceolate, linear, oblong, oblong-elliptic, obovate, oval, or ovate, 0.2–3.9(–4.6) cm, smooth to rugose or plicate, margins usually flat, sometimes revolute, crenate, dentate, serrate, or entire, abaxial surface tomentose or woolly, adaxial usually glabrous, sometimes tomentose or sparsely hairy on proximal portion of midvein. **Inflorescences** terminal, flowers solitary; bracts absent; bracteoles absent. **Pedicels** present. **Flowers** 13–29 mm diam.; hypanthium saucer-shaped, 3–15 mm, exterior villous, stipitate-glandular; sepals 8–10, ascending to spreading, linear-oblong or linear-lanceolate to lanceolate, elliptic, or ovate; petals 8–10(–12), deciduous or persistent, usually white or cream, sometimes yellow, ovate to obovate, oblanceolate, or ovoid; stamens 50–130, shorter than petals; torus hemispheric; carpels 60–150, sessile, stigmas lateral. **Fruits** aggregated achenes, 20–40, linear to lanceolate, 0.8–3.5 mm, smooth <with many short white hairs>; hypanthium persistent; sepals persistent, erect-ascending. $x = 9$.

Species 15 (7 in the flora): North America, Eurasia; arctic, boreal, and alpine regions.

Hybridization is common in *Dryas* (E. Hultén 1959b), often rendering specimen identification difficult when based on morphological features. Molecular evidence indicates that there are multiple morphologically similar species within the broad concept of *D. octopetala* (R. Elven et al., unpubl.). A study of amplified fragment length polymorphism (AFLP) markers found *D. drummondii* to be sister to all other species of the genus (I. Skrede et al. 2006). The taxonomy in the present treatment is based on that in the Pan-arctic Flora (R. Elven et al., http://nhm2.uio.no/paf/) and reflects the findings by Skrede et al. that AFLP markers support various proposals of B. A. Jurtzev (1984) and refute some of those by Hultén.

Species of *Dryas* are among the important components of arctic and alpine tundra in terms of biomass (I. Skrede et al. 2006) and cover. *Dryas* has been placed in Dryadeae, which comprises four (*Frankia* association) actinorhizal genera, also including *Cercocarpus*, *Chamaebatia*, and *Purshia* (D. Potter et al. 2007). *Dryas octopetala* in the broad sense is characteristically a pioneer (T. T. Elkington 1971); it forms ectomycorrhizal associations with *Cenococcum geophilum* Fries (Ascomycetae, familial relationship unknown), *Hebeloma alpinum* (J. Favre) Bruchet (Cortinariaceae, Agaricales), and other fungi (M. Gardes and A. Dahlberg 1996). Nitrogen fixation occurs in root nodules of *D. drummondii*, *D. integrifolia*, and *D. octopetala* (in the broad sense) (D. B. Lawrence et al. 1967). The authors have observed root nodules on *D. ajanensis* subsp. *beringensis*. Soils in primary successional environments with *D. drummondii* have higher nitrogen and phosphorus, which facilitates growth of spruce seedlings (F. S. Chapin III et al. 1994). In late successional stages, *D. drummondii* is inhibited by alders by shading and from (possible) allelopathic effects (Chapin et al.).

In the key and descriptions below, feathery hairs, referred to by some authors as octopetala scales, consist of relatively thick brown axes bearing delicate white hairs, appearing featherlike (E. Hultén 1959b). The brown axes mentioned by Hultén are analogous to a feather rachis. Such hairs may be absent from rare individuals in species that normally have them. Normally glandular species may include rare eglandular individuals.

SELECTED REFERENCES Hultén, E. 1959b. Studies in the genus *Dryas*. Svensk Bot. Tidskr. 53: 507–542. Porsild, A. E. 1947. The genus *Dryas* in North America. Canad. Field-Naturalist 61: 175–192.

1. Petals yellow, erect; leaf blades: bases usually cuneate (sometimes truncate or cordate); basal leaflets 1(or 2); leaf surfaces tomentose; flowers nodding at flowering; filaments hairy .... 1. *Dryas drummondii*
1. Petals usually white or cream, sometimes yellow, spreading; leaf blades: bases usually truncate or cordate, sometimes cuneate or rounded; basal leaflets 0; leaf abaxial surfaces tomentose to woolly, adaxial glabrous or glabrate, sometimes tomentose; flowers erect at flowering; filaments glabrous.

[2. Shifted to left margin.—Ed.]

2. Leaf blades: margins usually entire or dentate to crenate, sometimes serrate in proximal ½; (plants eglandular) . . . . . 7. *Dryas integrifolia*
2. Leaf blades: margins serrate to crenate or dentate (sometimes almost pinnatifid); (plants usually glandular, sometimes eglandular).
   3. Feathery hairs usually present, especially abaxially on midveins.
      4. Feathery hairs and stipitate glands abundant along midveins abaxially; nw North America . . . . . 5. *Dryas ajanensis*
      4. Feathery hairs and stipitate glands scattered to rare along midveins abaxially; e Greenland . . . . . 6. *Dryas octopetala*
   3. Feathery hairs usually absent, rarely 1+ feathery hairs abaxially on midveins in *D. hookeriana* and *D. incisa*.
      5. Leaf blades: sinuses of margins 45–60% to midvein, apices rounded to obtuse, surfaces smooth to rugose, plicate, (midvein and lateral veins adaxially sunken into folds) . . . . . 3. *Dryas alaskensis*
      5. Leaf blades: sinuses of margins 5–30% to midvein, apices acute to slightly obtuse, surfaces smooth to rugulose or slightly plicate, (margins strongly revolute to flat, only midvein ± obscured adaxially within fold).
         6. Midveins and petioles not stipitate-glandular abaxially . . . . . 2. *Dryas incisa*
         6. Midveins and petioles stipitate-glandular abaxially . . . . . 4. *Dryas hookeriana*

1. Dryas drummondii Richardson ex Hooker, Bot. Mag. 57: plate 2972. 1830 • Drummond's or yellow mountain-avens, dryade de Drummond [E]

Dryas drummondii var. *eglandulosa* A. E. Porsild; *D. drummondii* var. *tomentosa* (Farr) L. O. Williams; *D. octopetala* Linnaeus var. *drummondii* (Richardson ex Hooker) S. Watson; *D. tomentosa* Farr

Plants 2–23 cm. **Leaf blades** oblong-elliptic to obovate, 4–38(–46) × 2–24 mm, base usually cuneate, sometimes truncate or cordate, margins <flat>, crenate, dentate, or ± serrate, sinuses 10–25% to midvein, apex rounded to obtuse, surfaces smooth, tomentose, feathery hairs and stipitate glands absent or on midveins and petioles; basal leaflets 1(or 2), 0.5–3 × 0.5–2 mm. **Peduncles** 40–263 mm. **Flowers** nodding at flowering; sepals ovate, 4–6 × 3–4 mm; petals 8–10(–12), erect, yellow, 8–12 × 4–7 mm; filaments hairy. **Achenes** 2.5–3 mm; styles 15–39 mm. **2*n*** = 18.

Flowering Jun–Aug(–Sep). Sandy and gravelly beaches, flood plains, stream bars, rocky streams, grassy areas, alpine, calcicolous; 0–3000 m; Alta., B.C., Nfld. and Labr. (Nfld.), N.W.T., Ont., Que., Sask., Yukon; Alaska, Mont., Oreg.

In some states and provinces, *Dryas drummondii* has a limited distribution: northwestern Saskatchewan, Lake Superior islands of Ontario (relatively rare elsewhere in Ontario), Gaspé Peninsula, Quebec, northern Montana (Glacier National Park and surrounding area), and northeastern Oregon.

Dryas ×*suendermannii* Kellerer ex Sündermann, a hybrid between *D. drummondii* and *D. hookeriana*, has been reported from Alberta (J. G. Packer 1994). The hybrid shares with *D. drummondii* abaxial median and lateral leaf veins covered by tomentum, not viscid, stipitate glands absent; adaxial surface not viscid, and punctate glands absent. The hybrid shares with *D. hookeriana* an adaxially shiny leaf surface. It is intermediate in feathery hairs occasionally present (absent in *D. hookeriana*).

2. Dryas incisa Juzepczuk, Izv. Glavn. Bot. Sada S.S.S.R. 28: 312, 323. 1929

Plants 1–9.5 cm. **Leaf blades** oblong-elliptic to lanceolate, 4–16(–25) × 1–6(–10) mm, base usually cuneate, sometimes truncate or cordate, margins <strongly revolute to entire>, usually serrate or crenate, sometimes dentate, sinuses 5–30% to midvein, apex acute to slightly obtuse, surfaces smooth to rugulose, <only midvein ± obscured adaxially within medial fold>, abaxial tomentose (obscuring lateral veins), adaxial usually glabrous, sometimes tomentose, feathery hairs usually absent, rarely 1–3 on midveins abaxially, midveins and petioles abaxially not stipitate-glandular, <sessile glands usually present, rarely absent, longer hairs on adaxial surface and petioles sparse>. **Peduncles** 17–150 mm. **Flowers** erect at flowering; sepals lanceolate to narrowly elliptic, 4–8 × 0.7–2.5 mm; petals 8, spreading, usually white or cream, sometimes yellow, 9–14 × 5–11 mm; filaments glabrous. **Achenes** 2.2–3.5 mm; styles 11–25 mm. **2*n*** = 18 (Siberia).

Flowering Jun–Aug. Alpine rocky limestone slopes, heath meadows, dry gravel, mesic tundra meadows, outwash plains; 0–1800 m; N.W.T., Yukon; Alaska; Asia (Russian Far East, Siberia).

Occasional plants without feathery hairs identified as *Dryas ajanensis* subsp. *beringensis* are likely *D. incisa*.

Dryas crenulata Juzepczuk and *D. integrifolia* subsp. *crenulata* (Juzepczuk) Scoggan have been misapplied to *D. incisa*, for example, by A. E. Porsild and W. J. Cody (1980) and Cody (2000). *Dryas crenulata* is of eastern Siberia (B. A. Jurtzev 1984).

3. **Dryas alaskensis** A. E. Porsild, Canad. Field-Naturalist 61: 187, plate 1, figs. 3–5. 1948 • Alaskan mountain-avens

Dryas octopetala Linnaeus subsp. *alaskensis* (A. E. Porsild) Hultén; *D. octopetala* var. *glabrata* Hultén; *D. punctata* Juzepczuk subsp. *alaskensis* (A. E. Porsild) Jurtzev

Plants 2–18 cm. **Leaf blades** obovate or oblong-elliptic to narrowly oblong, 4.5–39 × 2–19 mm, base usually cuneate, sometimes truncate or cordate, margins coarsely dentate to crenate, sinuses 45–60% to midvein, apex rounded to obtuse, surfaces smooth to rugose, plicate, <midvein and lateral veins adaxially sunken into folds>, abaxial sparsely hairy to tomentose, adaxial usually glabrous, sometimes sparsely hairy proximally on midvein, feathery hairs and stipitate glands absent. **Peduncles** 25–130 mm. **Flowers** erect at flowering; sepals lanceolate to narrowly elliptic, 5.5–9 × 1.5–2 mm; petals 8, spreading, usually white or cream, sometimes yellow, 9–14 × 5–11 mm; filaments glabrous. **Achenes** 2–3 mm; styles 12–27 mm. **2*n*** = 18.

Flowering Jun–Aug. Solifluction soil, sand-gravel beaches, sandy lakeshores, old beach ridges, wet meadows, stream banks, dry rocky knolls, rocky heath, ericaceous tundra; 100–1800 m; Yukon; Alaska; e Asia (Russian Far East: Chukotka).

Dryas punctata has been reported from North America as a synonym of *D. octopetala* var. *octopetala* (E. Hultén 1968); it is now considered to be confined to northeastern Europe and northern Asia (R. Elven et al., http://nhm2.uio.no/paf/). Hultén (1959b) reduced *D. octopetala* var. *glabrata* to the rank of forma and later to synonymy under *D. alaskensis* (Hultén 1968).

4. **Dryas hookeriana** Juzepczuk, Izv. Glavn. Bot. Sada S.S.S.R. 28: 325. 1929 • Hooker's mountain-avens [E]

Dryas octopetala Linnaeus var. *angustifolia* C. L. Hitchcock; *D. octopetala* subsp. *hookeriana* (Juzepczuk) Hultén; *D. octopetala* var. *hookeriana* (Juzepczuk) Breitung

Plants 1.5–9 cm. **Leaf blades** usually oblong-elliptic to lanceolate, sometimes ovate, 4–16(–25) × 1–6(–10) mm, base usually truncate or cordate, sometimes cuneate, margins <strongly revolute to flat>, coarsely dentate or serrate, sinuses 5–25(–30)% to midvein, apex acute to obtuse, surfaces smooth to slightly plicate, <only midvein ± obscured adaxially within medial fold>, abaxial tomentose to woolly (obscuring lateral veins), adaxial usually glabrous, sometimes sparsely hairy proximally on midvein, feathery hairs usually absent, rarely 1+ on midveins abaxially, midveins and petioles abaxially stipitate-glandular. **Peduncles** 10–30 mm. **Flowers** erect at flowering; sepals lanceolate to narrowly elliptic, 4–10 × 0.6–1.5 mm; petals 8, spreading, usually white or cream, sometimes yellow, 9–14 × 5–11 mm; filaments glabrous. **Achenes** 2.5–3 mm; styles 11–25 mm. **2*n*** = 18.

Flowering Jun–Aug. Alpine meadows, dry rocky slopes and ridges, alpine tundra; 1500–3900 m; Alta., B.C., N.W.T; Alaska, Colo., Idaho, Mont., Oreg., Utah, Wash., Wyo.

In some states and provinces, the distribution of *Dryas hookeriana* is restricted to northwestern and central Colorado, southeastern and northernmost Idaho, western and central Montana, northeastern Utah (Uinta and Wasatch mountains), eastern Oregon and Washington, southwestern and central Wyoming, and western Northwest Territories.

5. **Dryas ajanensis** Juzepczuk, Izv. Glavn. Bot. Sada S.S.S.R. 28: 318. 1929 [F]

Subspecies 3 (1 in the flora): North America, e Asia (Russian Far East).

5a. **Dryas ajanensis** Juzepczuk subsp. **beringensis** Jurtzev in A. I. Tolmatchew, Fl. Arct. SSSR 9(1): 322. 1984 • White dryad [F]

Dryas octopetala Linnaeus var. *kamtschatica* (Juzepczuk) Hultén; *D. octopetala* var. *luteola* Hultén; *D. octopetala* var. *viscida* Hultén

Plants 2.5–9 cm. **Leaf blades** oblong or ovate to elliptic, 3–10 × 0.5–4 mm, base usually truncate or cordate, sometimes

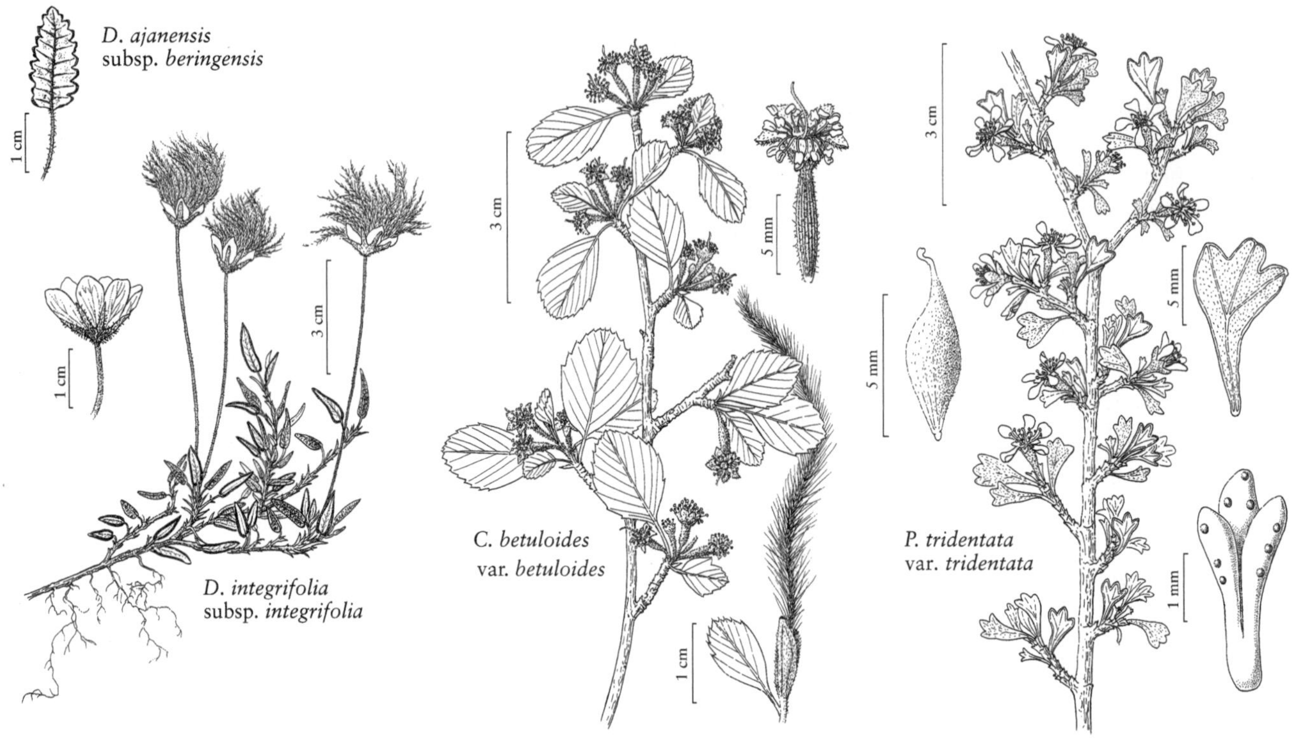

DRYAS ◦ CERCOCARPUS ◦ PURSHIA

cuneate, margins <flat>, coarsely dentate to serrate, sinuses 5–25% to midvein, apex acute to rounded, surfaces plicate along midvein and/or rugulose or rugose, <midvein and lateral veins sunken>, abaxial tomentose to woolly, adaxial usually glabrous, sometimes sparsely hairy proximally on midvein, feathery hairs abundant along midveins abaxially and on petioles, rarely scattered elsewhere, midveins and petioles abaxially usually stipitate-glandular. **Peduncles** 38–50 mm. **Flowers** erect at flowering; sepals linear-oblong, 4–6 × 1.5–2 mm; petals 8, spreading, white or cream, 9–14 × 5–11 mm; filaments glabrous. **Achenes** 2.5–3 mm; styles 16–38 mm. **2*n*** = 18 (Chukotka, Siberia).

Flowering Jun–Aug. Alpine heath, dry slopes, timberline, moraines, tundra; 0–1800 m; N.W.T., Yukon; Alaska; e Asia (Russian Far East: Chukotka).

Subspecies *beringensis* was formerly considered to be part of *Dryas octopetala* var. *octopetala* (R. Elven et al., http://nhm2.uio.no/paf/). This subspecies forms hybrid swarms with *D. alaskensis* and *D. integrifolia* in Alaska and eastern Siberia (Elven et al.). *Dryas incanescens* Juzepczuk and *D. ajanensis* var. *incanescens* (Juzepczuk) Kozhevnikov have been misapplied to subsp. *beringensis*.

6. Dryas octopetala Linnaeus, Sp. Pl. 1: 501. 1753 E

Plants 2.5–12 cm. **Leaf blades** oblong or ovate to elliptic, 3–16 × 0.5–6.5 mm, base usually truncate or cordate, sometimes cuneate, margins <flat>, coarsely dentate to serrate, sinuses 5–50(–65)% to midvein, apex acute to rounded, surfaces plicate along midvein, usually rugulose, sometimes rugose, <midvein and lateral veins sunken>, abaxial tomentose to woolly, adaxial usually glabrous, rarely tomentose or sparsely hairy proximally on midvein, feathery hairs scattered to rare along midveins abaxially and on petioles, scattered to rare elsewhere, midveins and petiole apices usually abaxially stipitate-glandular. **Peduncles** 11–125 mm. **Flowers** erect at flowering; sepals linear-oblong, 4–7 × 1.5–2.5 mm; petals 8, spreading, white or cream, 9–14 × 5–9 mm; filaments glabrous. **Achenes** 2.5–3 mm; styles 16–32 mm.

Flowering Jun–Aug. Tundra; 1000 m; Greenland.

On Greenland, a major zone of hybridization and introgression of *Dryas octopetala* with *D. integrifolia* occurs, mixing morphological characteristics so extensively as to nearly obliterate *D. octopetala* (E. Hultén 1971). M. Phillip and H. R. Siegismund (2003) found continuous variation, making a clear separation of the two species impossible. N. Tremblay and D. Schoen (1999) described a series of events isolating populations of *Dryas* species in refugia during glaciation followed by secondary contact between species after glacial retreat; such events could explain the large amount of variability seen on much of Greenland. Such hybridization and diversification are not uncommon in arctic species (Phillip and Siegismund) and were too recent to allow fixation of consistent characteristics throughout the range of taxa (I. Nordal et al. 1999).

Phenotypically, some plants of the eastern coast of Greenland seem to be outliers of *Dryas octopetala* in the strict sense (T. T. Elkington 1965). Introgressed plants on Greenland have been called *D. chamissonis* Sprengel ex Juzepczuk (formerly *D. chamissonis* Jurtzev).

7. Dryas integrifolia Vahl, Skr. Naturhist.-Selsk. 4(2): 171. 1798 • Entire-leaved mountain-avens, dryade à feuilles entières [F]

Plants 3–14 cm. **Leaf blades** linear to narrowly ovate or oblong to oval, 2–38 × 0.5–11 mm, base cordate to truncate or cuneate to rounded, margins usually entire or dentate to crenate, sometimes serrate in proximal 1/2, sinuses 5–15 (–20)% to midvein, apex usually obtuse or acute, sometimes acuminate, surfaces usually smooth, sometimes rugulose, abaxial slightly hairy to tomentose, adaxial glabrous or sparsely hairy proximally on midvein, feathery hairs and stipitate glands absent. **Peduncles** 10–150 mm. **Flowers** erect at flowering; sepals linear-lanceolate, 3.5–9 × 1.5–2.5 mm; petals 8 (or 9), spreading, usually white or cream, sometimes yellow, 9–14 × 5–11 mm; filaments glabrous. **Achenes** 0.8–3.5 mm; styles 8–29 mm.

Subspecies 2 (2 in the flora): North America, e Asia (Russian Far East).

Hybrid swarms between the two subspecies of *Dryas integrifolia* are common (E. Hultén 1968).

1. Leaf blades linear to narrowly ovate, margins usually strongly revolute, bases cordate to truncate, apices acute . . . . 7a. *Dryas integrifolia* subsp. *integrifolia*
1. Leaf blades oblong to oval, margins usually flat, sometimes narrowly revolute, bases cuneate to rounded, apices obtuse . . . . 7b. *Dryas integrifolia* subsp. *sylvatica*

7a. Dryas integrifolia Vahl subsp. **integrifolia** [F]

Dryas chamissonis Sprengel ex Juzepczuk; *D. integrifolia* var. *canescens* Simmons; *D. integrifolia* subsp. *chamissonis* (Sprengel ex Juzepczuk) Scoggan; *D. integrifolia* var. *subintegrifolia* Hultén

Leaf blades linear to narrowly ovate, base cordate to truncate, margins usually strongly revolute, entire or crenate to serrate or dentate in proximal 2/3–3/4, apex acute. $2n = 18$.

Flowering Jun–Aug. Tundra, heathlands, rocky mountain slopes, grasslands, sandy ridges and beaches, mesic and dry areas, solifluction soil; 0–2600 m; Greenland; Alta., B.C., Man., Nfld. and Labr., N.W.T., Nunavut, Ont., Que., Yukon; Alaska; e Asia (Russian Far East).

In some states and provinces, subsp. *integrifolia* has a limited distribution: central and northeastern Manitoba, northern Lake Superior and the Hudson Bay region of Ontario, and southeastern and northern Quebec.

In Greenland, subsp. *integrifolia* has widely introgressed with *Dryas octopetala*.

Dryas ×*lewinii* Rouleau (*D.* ×*wyssiana* Beauverd, name invalid) of Newfoundland is a putative hybrid between subsp. *integrifolia* and *D. drummondii*.

7b. Dryas integrifolia Vahl subsp. **sylvatica** (Hultén) Hultén, Svensk Bot. Tidskr. 53: 536. 1959

Dryas integrifolia var. *sylvatica* Hultén, Svensk Bot. Tidskr. 30: 527, figs. 2b,c. 1936; *D. sylvatica* (Hultén) A. E. Porsild

Leaf blades oblong to oval, base cuneate to rounded, margins usually flat, sometimes narrowly revolute, usually entire to serrate in proximal 2/3–3/4, rarely serrate throughout, apex obtuse. $2n = 18$.

Flowering Jun–Aug. Alpine turf slopes, wet spruce-birch forests, bogs, gravel beaches, stream bars, dry areas; 400–1600 m; N.W.T., Nunavut, Yukon; Alaska; e Asia (Russian Far East).

28. CERCOCARPUS Kunth in A. von Humboldt et al., Nov. Gen. Sp. 6(fol.): 183; 6(qto.): 232; plate 559. 1824 • Mountain-mahogany [Greek *kerkos*, tail, and *karpos*, fruit]

James Henrickson

Brian D. Vanden Heuvel

Shrubs or trees, (5–)10–85[–300+] dm. **Stems** 1–20, branched distally; bark silvery to dark gray or brownish black, often maroon-brown when young, smooth, roughened in older stems; long shoots straight or flexuous, short shoots short to elongate, branched or not, sometimes interspersed with long-shoot segments. **Leaves** persistent or drought or winter deciduous, <abscising distal to leaf bases and stipules, fascicled on short shoots>, simple; stipules caducous, lanceolate to ovate, margins entire; petiole present; blade narrowly to broadly obovate, oblanceolate, ovate, suborbiculate, elliptic, or lanceolate to linear, (0.3–)0.5–5(–7.8) cm, membranous to coriaceous, margins revolute or flat, serrate, dentate, crenate, or entire, abaxial surface hairier on veins, adaxial sericeous, pilose, hirsute, villous, glabrate, or glabrous. **Inflorescences** axillary, 1–3 (–18)-flowered clusters; bracts of highly reduced leaves or only of leaf bases; bracteoles of highly reduced leaves or only of leaf bases. **Pedicels** present. **Flowers** 6–15 mm diam.; hypanthium tube <greenish yellow or tinged reddish purple>, narrowly cylindric, <cups hemispheric-campanulate, thin, exterior hairy, interior glabrous, reticulately veined, circumscissially dehiscent from tube after flowering>, 1–10 mm, exterior hairy, eglandular; sepals (3–)5, erect to reflexed, broadly to narrowly deltate; petals 0; stamens 10–61; torus absent; carpels 1(or 2), stipitate, stigmas lateral. **Fruits** achenes, 1, cylindro-fusiform, 6–11(–13.5) mm, chartaceous-coriaceous, <strongly veined>; hypanthium tube persistent, loosely enclosing achenes, <cinnamon brown, 5–10-veined, longitudinally splitting distally as fruit enlarges>; sepals and hypanthial cup deciduous. $x = 9$.

Species 8 (4 in the flora): w, sc United States, Mexico.

Within *Cercocarpus*, nuclear DNA sequences and chloroplast DNA regions have been analyzed (B. D. Vanden Heuvel 2002) with support demonstrated for four clades: *C. betuloides* clade, including all the varieties of *C. betuloides*; *C. ledifolius* clade, including three varieties of *C. ledifolius*; *C. montanus* clade, including *C. breviflorus* and *C. montanus*; and, a clade including *C. fothergilloides* and other Mexican species. The data sets do not resolve the relationships among individual species or varieties within the four clades.

Cercocarpus species are heteroblastic, with new growth occurring in favorable seasons via long shoots, but with most leaves produced on flower-bearing short shoots. All but one species, *C. montanus*, are evergreen. Long-shoot leaves are larger than short-shoot leaves. Short-shoot leaves are variable in size with the proximal smallest, more ovate, the later-formed larger, often more oblong-obovate. Plants of mesic sites typically have larger, thinner leaves than those of more xeric sites, often with variation present within a mountain range; this is not considered taxonomically important. The small leaves of drier sites are more coriaceous and thicker; the veins have bundle sheath extensions, and the stomata are in depressions. Variation also occurs in vestiture within most taxa, with some plants sericeous-strigose or with hairs ascending to wavy (more or less villous) or erect, other plants having thicker, coiled hairs. These vestiture types may be geographically distributed (and useful taxonomically) or mixed within a population. Short shoots may develop few to many flowers in a cluster; this can be a useful character or be modified by the environment and result in exceptions.

SELECTED REFERENCES Lis, R. A. 1992. Leaf architectural survey of *Cercocarpus* (Rosaceae) and its systematic significance. Int. J. Pl. Sci. 153: 258–272. Martin, F. L. 1950. Revision of *Cercocarpus*. Brittonia 7: 91–111. Rydberg, P. A. 1913. *Cercocarpus*. In: N. L. Britton et al. 1905+. North American Flora.... 47+ vols. New York. Vol. 22, pp. 418–424. Vanden Heuvel, B. D. 2002. Molecular Systematics of *Cercocarpus* H.B.K. (Rosaceae). Ph.D. dissertation. University of Texas.

1. Leaf blades linear, linear-lanceolate, lanceolate, lance-elliptic, or elliptic-ovate, (l/w = 2.5–10), stiffly coriaceous, margins entire, weakly to strongly revolute; anthers glabrous . . . . 4. *Cercocarpus ledifolius*
1. Leaf blades usually ovate to broadly ovate, oblong-elliptic, oblong-obovate, or obovate to narrowly obovate, sometimes oblanceolate to narrowly oblanceolate, oblong-oblanceolate, rhombic, or suborbiculate, rarely lance-ovate, (l/w = 1–3), membranous, subcoriaceous, or coriaceous, margins serrulate, serrate, dentate, or crenate sometimes in distal 1/5–3/4, revolute or flat; anthers hirsute.
    2. Leaves winter deciduous, blades ± subcoriaceous (not stiff), ovate to broadly ovate, obovate to narrowly oblong-obovate, or oblanceolate, margins dentate to crenate or serrate (proximal teeth with convex margins, distal teeth smaller, more crowded); Arizona, Colorado, sw Idaho, w Nebraska, se Nevada, New Mexico, w South Dakota, w Oklahoma, nw Texas, Utah, e to s Wyoming . . . . 3. *Cercocarpus montanus*
    2. Leaves persistent or drought deciduous, blades membranous or subcoriaceous to coriaceous, oblong-obovate, rhombic, oblong-elliptic, ovate to broadly ovate, suborbiculate, narrowly oblanceolate, oblong-lanceolate, or narrowly obovate to obovate, rarely lance-ovate, margins dentate, serrulate, serrate, crenate (teeth apiculate), or entire in some *C. betuloides* var. *traskiae*; Arizona, California, New Mexico, Oregon, w Texas.
        3. Stipules (1.5–)3.5–6(–10) mm (1.5–2.5 mm in s California); achene styles: proximal setae 2.7–5 mm; flowering late Feb–Apr(–May); sw, sc Arizona, California, Oregon . . . . 1. *Cercocarpus betuloides*
        3. Stipules 1.5–2.5(–3) mm; achene styles: proximal setae 1.5–2.7 mm; flowering late Jun–Aug(–Sep); nc Arizona, New Mexico, Texas . . . . 2. *Cercocarpus breviflorus*

1. Cercocarpus betuloides Nuttall in J. Torrey and A. Gray, Fl. N. Amer. 1: 427. 1840 [F]

Shrubs or trees, 10–50(–80) dm, sparsely to moderately branched, <young stems, leaves, pedicels, and hypanthia sericeous (hairs straight, antrorse, appressed, or ± ascending), or hirsute-pilose (hairs erect, straight), or ± villous (hairs wavy or wavy-crinkled), or hairs tightly coiled or 1/2 coiled, or 1/2 coiled at base and erect-ascending, wavy distally, 0.2–0.8(–1.3) mm>. **Stems:** long-shoot internodes 3–20(–55) mm, ± sparsely hairy, glabrescent; short shoots 3–30(–55) × 1–2.5 mm. **Leaves** persistent or ± drought deciduous; stipules (1.5–)3.5–6(–10) mm, <1.5–2.5 mm in s California>; petiole 1.5–9(–16) mm; blade oblong-obovate, rhombic, suborbiculate to broadly ovate, rarely oblong-elliptic, (8–)12–50(–78) × (4–)8–35(–64) mm, membranous to coriaceous, base cuneate to rounded, rarely ± cordate, margins ± revolute or flat, dentate, serrate, or, in broader leaves, crenate in distal 1/5–3/4, apex rounded, sometimes acute, abaxial surface sericeous, pilose, villous, or with curled hairs on veins, areoles canescent or glabrous, adaxial sparsely hairy, sometimes glabrate. **Flowers** 1–18 per short shoot; hypanthial tubes sericeous, woolly, or villous or with coiled or spreading hairs; hypanthial cups 2–4 × 4–7 mm; sepals 5, broadly deltate, 1–2.5 mm, acute-acuminate; stamens 20–61, anthers 1–1.5 mm, hirsute. **Achenes** 7–11(–13.5) × 1.5–2.5(–2.9) mm; fruiting pedicels (1.5–)3–11(–21) mm; hypanthial tubes (5–)7–13(–17) mm; pedicel/tube ratio (15–)35–90(–126)%; fruit awns 5–8.5(–12.5) cm, proximal setae 2.7–5 mm.

Varieties 3 (3 in the flora): w United States, nw Mexico.

The western, spring-flowering *Cercocarpus betuloides* in the broad sense, centered in California, is well supported by molecular data. A total of five taxa have been recognized in recent floras; molecular data provide no support for these taxa, but instead support recognition of a single inclusive taxon.

1. Leaves, young stems, and hypanthial tubes appressed-sericeous or hypanthial tubes also with erect-ascending hairs (hairs not coiled); flowers 1–3(–10) per short shoot; transmontane s, c California, c Arizona, and sw Oregon . . . . 1a. *Cercocarpus betuloides* var. *betuloides*
1. Leaves, young stems, and hypanthial tubes commonly with tightly coiled hairs (except in some insular plants where sericeous); flowers (1–)3–6(–18) per short shoot; cismontane s California.
    2. Leaf blades sparsely to moderately hairy, subcoriaceous to coriaceous, margins serrate or dentate or crenate in distal 1/5–3/4; coastal s California and adjacent islands . . . . 1b. *Cercocarpus betuloides* var. *blancheae*
    2. Leaf blades abaxially densely woolly, strongly coriaceous, margins (often strongly revolute, usually obscuring teeth), crenate, dentate-serrate in distal 1/2–3/4; sw Santa Catalina Island. . . . . . 1c. *Cercocarpus betuloides* var. *traskiae*

1a. Cercocarpus betuloides Nuttall var. **betuloides** • Mountain-mahogany [F]

Cercocarpus betulifolius var. *minor* C. K. Schneider; *C. betuloides* subsp. *macrourus* (Rydberg) A. E. Murray; *C. betuloides* var. *macrourus* (Rydberg) Jepson; *C. douglasii* Rydberg; *C. macrourus* Rydberg; *C. minutiflorus* Abrams; *C. montanus* Rafinesque var. *glaber* (S. Watson) F. L. Martin; *C. montanus* var. *macrourus* (Rydberg) F. L. Martin; *C. montanus* var. *minutiflorus* (Abrams) F. L. Martin; *C. parvifolius* Nuttall ex Hooker & Arnott var. *glaber* S. Watson; *C. rotundifolius* Rydberg

Plants 10–50(–80) dm. **Young stems** sericeous or with ascending hairs. **Leaves:** stipules 1.5–5(–5.7) mm; blade narrowly to broadly obovate or rhombic to broadly ovate, (9–)15–30(–70) × (5–)10–18(–52) mm, subcoriaceous to coriaceous, margins serrate or dentate to crenate in distal 1/2–2/3(–3/4), abaxial veins usually sericeous or with ascending, rarely erect, hairs 0.2–0.5 mm, areoles canescent or some or all glabrous, adaxial surface mostly sericeous, or with ascending, erect hairs, or glabrate. **Flowers** 1–3(–10) per short shoot; hypanthial tubes sericeous or mixed sericeous with spreading hairs. **Achenes** 8–13.5 × 1.5–2.2(–2.9) mm; fruiting pedicels (1.5–)3–7(–21) mm; hypanthial tubes (5–)8–12(–17) mm; pedicel/tube ratio (15–)30–60(–126)%; fruit awns (3.5–)4.5–7.5(–12.5) cm.

Flowering Mar–May. Rocky slopes, outcrops, ravines, upper chaparral, oak woodlands, yellow pine forests; (50–)150–1800(–2200) m; Ariz., Calif., Oreg.; Mexico (Baja California).

Variety *betuloides* extends from transmontane southern California (the dry eastern side of the coastal ranges) except in San Diego and Orange counties (where it incorporates the small-leaved *Cercocarpus minutiflorus*), onto the western slopes of the Sierra Nevada and to central-northern California from Santa Barbara County to southwestern Oregon, where it incorporates what has been recognized as var. *macrourus*, extending into central Arizona, and into Baja California, Mexico. Leaves are similar to those of plants in coastal southern California in mesic sites, but are more narrowly obovate, thicker, more coriaceous, and cuneate-based in drier sites. The leaves are usually closely sericeous, less frequently the hairs are ascending to erect; they never have coiled hairs. The pedicels and hypanthial tubes can be sericeous throughout or with some to many spreading hairs. The plants also usually have few flowers per short-shoot cluster. Intermediates occur in zone of overlap with var. *blancheae*.

In Orange, Riverside, and San Diego counties of California, plants with small leaves, shorter stipules, and glabrous abaxial areoles have been recognized as *Cercocarpus minutiflorus*. Similar plants also occur from Santa Barbara to Trinity counties, many with small leaves, slightly longer stipules but with all (or only some) of the leaves with glabrous abaxial areoles indicating that the glabrous areole character is under developmental, not genetic, control, associated with thin, vernal leaves. Some leaves of var. *blancheae* also have glabrous areoles.

In Arizona, var. *betuloides* generally occurs at elevations below 1100(–1400) m, usually below populations of *Cercocarpus breviflorus*. In Arizona, var. *betuloides* is known from Gila, La Paz, Maricopa, Mojave, Pinal, and Yavapai counties; it can be distinguished by its silvery stems (versus red-brown in *C. breviflorus*) and spring versus late summer flowering.

Cercocarpus betulifolius Nuttall ex Hooker is an illegitimate name linked to this variety.

1b. Cercocarpus betuloides Nuttall var. **blancheae** (C. K. Schneider) Little, Phytologia 4: 308. 1953 • Coastal mountain mahogany

Cercocarpus betulifolius Nuttall ex Hooker var. *blancheae* C. K. Schneider, Mitt. Deutsch. Dendrol. Ges. 14: 127. 1905 (as belulaefolius); *C. alnifolius* Rydberg; *C. betuloides* subsp. *blancheae* (C. K. Schneider) Thorne; *C. betuloides* var. *multiflorus* Jepson; *C. montanus* Rafinesque var. *blancheae* (C. K. Schneider) F. L. Martin

Plants 10–50(–80) dm. **Young stems** with tightly coiled hairs. **Leaves:** stipules (3–)4–6(–10) mm; blade oblong-obovate or -elliptic, obovate, ovate, or suborbiculate, (8–)12–50(–78) × (4–)8–35(–64) mm, subcoriaceous to coriaceous, margins serrate or dentate or crenate in distal 1/5–3/4, abaxial veins with coiled, 1/2 coiled, or coiled-wavy hairs to 1 mm (or some appressed-sericeous hairs on off-coastal island plants), areoles canescent or glabrous, adaxial surface sparsely similarly hairy, glabrescent. **Flowers** (1–)3–8(–18) per short shoot; hypanthial tubes with coiled hairs (or some sericeous hairs on island plants). **Achenes** 8–11(–13) × 1.8–2.5(–2.9) mm; fruiting pedicels (1.5–)3–11(–14) mm; hypanthial tubes (6–)7–13(–16) mm; pedicel/tube ratio (20–)35–90(–120)%; fruit awns 5.3–7.5(–9.2) cm.

Flowering late Feb–Apr. Coastal chaparral; (10–)100–800 m; Calif.; Mexico (Baja California).

Variety *blancheae* grows primarily in chaparral in coastal southern California from San Luis Obispo to San Diego counties (rare in northern Baja California, Mexico) and on adjacent Santa Catalina, Santa Cruz (where frequent), and Santa Rosa (rare) islands. This treatment combines the coastal mainland and insular plants. Both tend to have multiple flowers on each short shoot and coiled hairs on young stems, leaves, hypanthia, and pedicels. This coiled hair type is consistent in mainland populations; some insular plants are sericeous throughout. Insular plants typically have larger leaves than mainland plants, except in very dry areas, and conversely, mainland coastal plants previously designated as var. *betuloides* have smaller leaves except in protected and shaded habitats. Recognition of these two as distinct, at any rank, is poorly supported, and the authors consider them phases of a single taxon. Plants of the more arid slopes in transmontane southern and central California are more sericeous throughout and have fewer flowers per short shoot (var. *betuloides*). Intermediates occur in zones of contact.

1c. Cercocarpus betuloides Nuttall var. **traskiae** (Eastwood) Dunkle, Bull. S. Calif. Acad. Sci. 39: 2. 1940 • Santa Catalina mountain mahogany C E

Cercocarpus traskiae Eastwood, Proc. Calif. Acad. Sci., ser. 3, 1: 136, plate 11, fig. 7. 1898; *C. betuloides* subsp. *traskiae* (Eastwood) A. E. Murray; *C. montanus* Rafinesque var. *traskiae* (Eastwood) F. L. Martin

Plants 10–30(–80) dm. **Young stems** woolly with tightly coiled hairs. **Leaves:** stipules 5–10 mm; blade ovate to broadly ovate, obovate, suborbiculate, or oblong-elliptic, (15–)25–50(–62) × (11–)17–35(–45) mm, strongly coriaceous, margins crenate, <often strongly revolute>, dentate-serrate in distal $\frac{1}{2}$–$\frac{3}{4}$, abaxial veins densely and ± persistently white villous to woolly with loosely curved, $\frac{1}{2}$ coiled at base, or coiled hairs 0.5–2 mm, areoles canescent, adaxial surface <shiny>, moderately hairy or glabrate. **Flowers** 3–8(–16) per short shoot; hypanthial tubes woolly with tightly coiled hairs. **Achenes** 7–11 × 1.8–2.3 mm; fruiting pedicels 6–10 mm; hypanthial tubes 9–11.5 mm; pedicel/tube ratio 52–69%; fruit awns 5–6 cm.

Flowering Mar–Apr. Metamorphosed volcanic soil; of conservation concern; 100–300 m; Calif.

Variety *traskiae* is known from a south-facing, steep canyon on Santa Catalina Island, where it grows on a highly mineralized metamorphic igneous-derived saussurite gabbro substrate, one not exposed elsewhere in southern California. The population now consists of 11 adult plants, down from 40–50 plants in 1898, the reduction attributed to herbivory by introduced mammals. In molecular studies summarized by L. H. Rieseberg and D. Gerber (1995), only six plants were considered pure *traskiae*; five were hybrids or introgressants with the insular phase of var. *blancheae*. The areas around several var. *traskiae* trees have been fenced to allow for recruitment.

Variety *traskiae* differs from var. *blancheae* in thicker, more coriaceous leaves with revolute margins that often obscure the marginal teeth, and thick, woolly tomentum of mostly coiled hairs on the abaxial leaf surface, pedicels, and hypanthial tubes. The thick vestiture can weather away over time.

Plants grown in gardens from cuttings of pure *traskiae* often lose the thick leaves, thick cuticles, and revolute margins that characterize the taxon in its native habitat, indicating that the diagnostic morphological traits in part may be a phenotypic response to its arid environment and perhaps its peculiar soil type.

2. Cercocarpus breviflorus A. Gray, Smithsonian Contr. Knowl. 5(6): 54. 1853 • Eastern mountain mahogany

Cercocarpus breviflorus var. *eximius* C. K. Schneider; *C. eximius* (C. K. Schneider) Rydberg; *C. montanus* Rafinesque var. *paucidentatus* (S. Watson) F. L. Martin; *C. parvifolius* Nuttall ex Hooker & Arnott var. *paucidentatus* S. Watson

Shrubs, (5–)10–30(–60) dm, sparsely to moderately branched. **Stems:** long-shoot internodes (3–)7–20(–37) mm, sericeous, pilose-hirsute or villous, glabrate; short shoots 5–60(–140) × 1.5–2.5 mm. **Leaves** persistent or drought deciduous; stipules 1.5–2.5(–3) mm; petiole (1–)2–5(–9) mm; blade narrowly oblanceolate, oblong-oblanceolate, oblong-elliptic, narrowly obovate to obovate, sometimes ovate, (3–)7–35(–52) × (1.3–)4–15(–28) mm, coriaceous to subcoriaceous, base narrowly to broadly cuneate, margins serrulate to crenate in distal $\frac{1}{3}$–$\frac{1}{2}$, <teeth acute-apiculate, antrorse or proximal divergent in large leaves>, or smaller blades entire, apex rounded to apiculate, abaxial surface sericeous, villous, or hirsute on veins, areoles canescent, rarely glabrous, adaxial moderately sericeous, hirsute, or villous. **Flowers** 1–3(–7) per short shoot; hypanthial tubes sericeous, hirsute-pilose to villous; hypanthial cups 1.5–2.5 × 2.5–4 mm; sepals 5, narrowly deltate, 1–2 mm, acute; stamens (15–)20–40, anthers 0.6–1.2 mm, hirsute. **Achenes** 6–9(–10.7) × 1.3–2 mm; fruiting pedicels (1–)2–5(–11.3) mm; hypanthial tubes (4–)6–9(–10.2) mm; pedicel/tube ratio (18–)30–50(–135)%; fruit awns (2.5–)3.5–5(–7.2) cm, proximal setae 1.5–2.7 mm.

Flowering late Jun–Aug(–Sep). Yellow pine, fir forests, pinyon, juniper, oak woodlands, chaparral, grasslands, desert mountains, rocky, limestone, sandstone or rhyolite substrates; 1500–2600 m; Ariz., N.Mex., Tex.; Mexico (Chihuahua, Coahuila, Durango, Hidalgo, Nuevo León, Querétaro, San Luis Potosí, Sonora, Tamaulipas, Zacatecas).

Cercocarpus breviflorus grows in the mountains of north-central Arizona, south-central New Mexico, and central and trans-Pecos Texas and in Mexico, over a 2000 km range. It has smaller flowers and shorter fruiting pedicels, hypanthial tubes, and stipules than *C. betuloides*. Flowering occurs after summer rains, providing reproductive isolation from *C. betuloides*.

In the mountains from east-central Arizona into north-central New Mexico, about 70–80% of plants of *Cercocarpus breviflorus* have pilose-hirsute to villous vestiture, and leaves tend to be thinner and more obovate with five to seven teeth in the rounded apical margins. In more arid montane regions of southern New Mexico, trans-Pecos Texas, and Mexico, leaves are more commonly sericeous (70%), often smaller, more coriaceous, with fewer teeth, in extreme being entire with a single terminal tooth. In strongly arid conditions, plants are smaller, more intricately branched, and thorny. These differences appear suitable for taxonomic recognition; a broad zone of intergradation occurs in southeastern Arizona and south-central New Mexico.

In the Big Bend region and Hill County of Texas and in the Chihuahuan Desert and mountains of eastern Mexico, plants of *Cercocarpus breviflorus* in more mesic chaparral habitats on north-facing slopes and drainages are taller with larger, sericeous, toothed leaves, often with the proximal teeth divergent. These plants blend completely into smaller-leaved plants in more arid sites on the same mountains. As in the *C. betuloides* complex, there are also plants with glabrous abaxial areoles.

3. Cercocarpus montanus Rafinesque, Atlantic J. 1: 146. 1832 • Alder-leaf mountain mahogany

Cercocarpus argenteus Rydberg; *C. montanus* var. *argenteus* (Rydberg) F. L. Martin; *C. parvifolius* Nuttall ex Hooker & Arnott; *C. parvifolius* var. *hirsutissimus* C. K. Schneider; *C. parvifolius* var. *minimus* C. K. Schneider

Shrubs, 10–20(–50) dm, moderately branched, <young stems, leaves, pedicels, and hypanthia sericeous or pilose-hirsute (hairs straight or crinkled, 0.7–1.2 mm)>. **Stems:** long-shoot internodes 10–25(–45) mm, sericeous to pilose-hirsute, soon glabrate; short shoots 2–25(–90) × 1.7–3 mm. **Leaves** winter deciduous; stipules 1.8–3.5 mm; petiole (1.5–)2–5(–9) mm; blade ovate to broadly ovate, obovate to narrowly oblong-obovate, or oblanceolate, (6.5–)10–30(–63) × (6.5–)8–20(–29) mm, ± subcoriaceous (not stiff), base narrowly cuneate to rounded, margins rarely ± revolute proximally, dentate to crenate or serrate, <teeth usually with convex margins>, apex rounded, abaxial surface sericeous or pilose on veins and veinlets, areoles canescent, adaxial sericeous, pilose, or villous. **Flowers** 1–3(–5) per short shoot; hypanthial tubes sericeous or villous to pilose; hypanthial cups 3–4 × 3–5 mm; sepals 5, oblong, 1(–2) mm, acute; stamens 25–45, anthers 0.8–1.2 mm, hirsute. **Achenes** 9.2–11 × 1.7–2.2 mm; fruiting pedicels (2.3–)3–9(–11) mm; hypanthial tubes (7–)8–12(–14) mm; pedicel/tube ratio 25–70(–98)%; fruit awns 4.5–6(–8.2) cm, proximal setae 2–2.5(–3) mm.

Flowering May–Jun. Ridges, rocky slopes, valley floors, sandstone, granite, limestone (Texas), grasslands, juniper-oak, pinyon-juniper, yellow pine woodlands, aspen, pine-fir forests; (800–)1200–2600(–3000) m; Ariz., Colo., Idaho, Nebr., Nev., N.Mex., Okla., S.Dak., Tex., Utah, Wyo.; Mexico (Coahuila).

Cercocarpus montanus is winter deciduous, with thin, usually ovate to obovate, strongly toothed leaf blades and teeth usually with convex margins. Later-formed leaves in some plants may be more narrowly obovate and tend to be more serrate. Vestiture ranges from sericeous or pilose to villous with straight, twisted, or variously crinkled hairs within populations.

Plants of *Cercocarpus montanus* in northwestern Texas and adjacent northeastern New Mexico tend to have narrowly obovoid, oblanceolate, narrowly cuneate-based, serrate-margined leaf blades, usually with sericeous surfaces and the tightly coiled hairs of abaxial areoles often extending well beyond the areole margins. Their long-shoot leaves are often relatively large. While distinctive at the extreme, areole vestiture differences are most often minor and difficult to evaluate, and leaf-shape differences are not diagnostic in that some other western Texas individuals have ovate leaves, and plants with oblanceolate leaves occur well within the range of *C. montanus*. These populations have been recognized as *C. argenteus* or *C. montanus* var. *argenteus*.

Putative hybrids between *Cercocarpus montanus* and *C. ledifolius* var. *intermontanus* have persistent, elliptic to narrowly obovate leaves, 20–30(–50) × (5–)8–12(–16) mm, strongly coriaceous, with revolute margins, and appressed or ascending hairs to 0.5 mm on abaxial veins and adaxial surfaces; they are sharply serrate in the distal 1/5–1/2, with mid and secondary veins raised on the abaxial surfaces and well impressed on the adaxial ones. These appear to be F_1 hybrids that may form wherever both parent taxa are present; they appear not to be reproductively viable.

4. Cercocarpus ledifolius Nuttall in J. Torrey and A. Gray, Fl. N. Amer. 1: 427. 1840 • Curl-leaf mountain mahogany

Shrubs or trees, (5–)10–85 dm, moderately to strongly branched distally. **Stems:** long-shoot internodes 4–22 mm, sericeous-villous to hirsute-pilose, rarely glabrous, glabrescent; short shoots (1–)2–55 × 1–3.4 mm. **Leaves** persistent or ± drought deciduous; stipules 1–3.8 mm; petiole 0.4–6 mm; blade linear, linear-lanceolate, lanceolate, lance-elliptic, or elliptic-ovate, (3–)5–48 × 1–11 mm, stiffly coriaceous, base cuneate, margins weakly to strongly revolute, entire, <thickened>, apex acute, apiculate-mucronate to obtuse-rounded, abaxial surface initially sericeous or villous (sometimes strongly so), sometimes glabrescent, areoles closely canescent, adaxial sericeous, hirsute, or villous, glabrescent. **Flowers** 1–10 per short shoot; hypanthial tubes densely to loosely sericeous, villous, or with indument of coiled ascending hairs; hypanthial cups 1.5–3 × 1.8–3.5 mm; sepals (3–)5, broadly deltate, 1–2.5 mm, acute-obtuse; stamens 10–25, anthers 0.7–1.2 mm, glabrous. **Achenes** 6–10.5 × 0.8–1.5 mm; fruiting pedicels 0.3–1.5(–2) mm; hypanthial tubes (2.6–)3–11 mm; pedicel/tube ratio (5–)8–18(–28)%; fruit awns 3–8.5 cm, proximal setae 1.6–3.5 mm.

Varieties 3 (3 in the flora): w United States, nw Mexico.

Characteristics of *Cercocarpus ledifolius* include entire leaf margins; adaxial epidermis-cuticle translucent showing underlying venation and bundle-sheath extensions; stipules and bracts more or less glutinous; pedicels relatively short; flowers with relatively small hypanthial cups and large, erect to spreading sepals, stamens set deep within hypanthial cup, with longer filaments and glabrous anthers.

1. Leaf blades (3–)5–13(–18) × 1–1.5(–2) mm, linear to linear-lanceolate (usually subterete), margins strongly revolute (central groove 0.1–0.9 mm wide on abaxial surface); shrubs . . . . 4c. *Cercocarpus ledifolius* var. *intricatus*
1. Leaf blades (7–)13–30(–48) × (1.5–)2–9(–11) mm, elliptic-ovate, lanceolate to linear, or lance-elliptic, margins ± revolute; shrubs or trees.

[2. Shifted to left margin.—Ed.]

2. Leaf blades narrowly lanceolate to linear, sometimes lance-elliptic, (1.5–)2–5(–7) mm wide, margins strongly revolute; abaxial leaf surfaces usually densely villous, sometimes sericeous . . . . 4a. *Cercocarpus ledifolius* var. *ledifolius*
2. Leaf blades lanceolate to lance-elliptic or elliptic-ovate, (2.8–)5–9(–11) mm wide, margins revolute; abaxial leaf surfaces sericeous or closely villous (hairs not obscuring surface), glabrescent . . . . 4b. *Cercocarpus ledifolius* var. *intermontanus*

4a. Cercocarpus ledifolius Nuttall var. **ledifolius** [E]

Cercocarpus hypoleucus Rydberg; *C. ledifolius* var. *intercedens* C. K. Schneider

Shrubs or trees, 10–30 dm. **Stems:** long-shoot internodes 10–17 mm, sparsely villous, sericeous, or hirsute; short shoots 1–15 × 1.7–3 mm. **Leaves:** petiole (1.5–)2–4.5 mm; blade narrowly lanceolate to linear, sometimes lance-elliptic, (7–)13–29(–48) × (1.5–)2–5(–7) mm, margins revolute (leaving much of abaxial surface exposed), abaxial surface usually densely villous, sometimes sericeous, adaxial usually moderately, often loosely, villous or sericeous, rarely pilose, usually glabrescent. **Flowers** 1–5 per short shoot; hypanthial tubes strongly villous; sepals 2–2.5 mm. **Achenes** 6.5–9 × 1.2–1.4 mm; fruit awns 4–7.5 cm, proximal setae 1.6–2.7 mm.

Flowering Apr–May. Rocky outcrops, talus slopes, with *Artemisia*, *Purshia*, Gambel oak, pinyon pine and juniper; (200–)1300–2300(–3000) m; Ariz., Calif., Colo., Idaho, Mont., Nev., Oreg., Utah, Wash., Wyo.

Variety *ledifolius* has narrowly lanceolate to linear leaves that are initially strongly villous adaxially and often strongly villous-woolly abaxially, obscuring the veins between the revolute margins. The adaxial surfaces are usually glabrescent; vestiture is not very dense in sericeous plants.

Here, var. *ledifolius* comprises intermediates between var. *intermontanus* and var. *intricatus* representing hybrids and introgressants that together form breeding populations in zones of parental overlap and in regions outside the range of either parent. Variety *ledifolius* is considered to be polyphyletic and is much more variable than either parent. Some specimens have relatively small leaves approaching those in var. *intricatus*. Some specimens have larger long-shoot leaves as in var. *intermontanus* (with diminished [glabrescent] vestiture and narrower margins) and short-shoot leaves as in var. *ledifolius*.

4b. Cercocarpus ledifolius Nuttall var. **intermontanus** N. H. Holmgren, Brittonia 39: 424, figs. 1A–D. 1987

Shrubs or trees, 20–70(–85) dm. **Stems:** long-shoot internodes 13–22 mm, sparsely sericeous-villous; short shoots 2–55 × 1.8–3.4 mm. **Leaves:** petiole (1.5–)2.5–6 mm; blade lanceolate to lance-elliptic or (when shorter) elliptic-ovate, (10–)16–30(–40) × (2.8–)5–9 (–11) mm, margins revolute (leaving much of abaxial surface exposed), abaxial surface sericeous or closely villous (hairs not obscuring surface), glabrescent, adaxial sericeous or sparsely villous <or hairs tightly coiled and/or ½ coiled>, glabrescent. **Flowers** 4–10 per short shoot; hypanthial tubes strongly sericeous or villous; sepals 2–2.5 mm. **Achenes** 6.5–10.5 × 1.3–1.5 mm; fruit awns 5–7(–8.5) cm, proximal setae 2.2–3.5 mm.

Flowering May–Jun. Rocky dry, often south- to west-facing slopes, canyons, with *Artemisia tridentata*, *Purshia tridentata*, Gambel oak, maple, juniper, pinyon, ponderosa pine, lodgepole pine, Douglas fir and white fir, subalpine zones, often forming pure stands; (500–)1300–3300 m; Ariz., Calif., Colo., Idaho, Nev., Oreg., Utah, Wash., Wyo.; Mexico (Baja California).

Variety *intermontanus* is sympatric with var. *ledifolius* in southern Idaho, southwestern to eastern Oregon, southeastern Washington, and western Wyoming, and with var. *intricatus* in northeastern Arizona, eastern California, Nevada, and Utah.

Leaves in individual specimens of var. *intermontanus* vary in size, with long-shoot leaves largest and the first-formed short-shoot leaves of a season more oblong-ovate and often smaller. Heartwood is hard, very dark, and valued for specialty items. Variety *intermontanus* has long been known as *Cercocarpus ledifolius*, or var. *ledifolius*; its type specimen is convarietal with types of var. *intercedens*.

4c. Cercocarpus ledifolius Nuttall var. **intricatus** (S. Watson) M. E. Jones, Bot. Gaz. 5: 154. 1880 • Littleleaf or dwarf mountain-mahogany [E]

Cercocarpus intricatus S. Watson, Proc. Amer. Acad. Arts 10: 346. 1875; *C. arizonicus* M. E. Jones; *C. intricatus* var. *villosus* C. K. Schneider

Shrubs, 5–25 dm. **Stems:** long-shoot internodes 4–13 mm, sparsely villous, sericeous, or pilose, glabrescent, rarely glabrous; short shoots 2–10(–17) × 1–1.5 mm. **Leaves:** petiole 0.4–1.4(–2) mm; blade linear to linear-lanceolate, (3–)5–13(–18) × 1–1.5(–2) mm, margins strongly revolute (central groove 0.1–0.9 mm wide on abaxial surface in dried leaves), abaxial surface (largely obscured by leaf margins except on raised midvein where moderately, rarely densely hairy) usually villous, sometimes sericeous or hirsute, adaxial villous, sericeous, or hirsute, sometimes glabrescent, or glabrous. **Flowers** 1–5 per short shoot; hypanthial tubes strongly villous or sericeous; sepals 1–1.5 mm. **Achenes** 6–7 × 0.8–1.3 mm; fruit awns 3–4.5 cm, proximal setae 1.8–2.5 mm.

Flowering late Mar–early Jun. Exposed, rocky slopes, ledges, bluffs with shallow soil, often cracks and crevices, alluvial slopes of limestone, dolomite, sandstone, or granite, with *Artemisia*, *Coleogyne*, *Ephedra*, *Nolina*, *Purshia*, pinyon and ponderosa pine, oaks, juniper woodlands; 1000–2400(–3100) m; Ariz., Calif., Colo., Nev., N.Mex., Utah, Wyo.

Variety *intricatus* is distinguished by rigid, moderately to strongly branched growth habit and stems densely clothed by linear, coriaceous, strongly revolute leaves with the abaxial surface exposed only along a narrow central groove. Long-shoot leaves are often larger than those of short shoots.

Variety *intricatus* is often treated as a distinct species; it is known to hybridize with var. *intermontanus* and var. *ledifolius*. Here, var. *intricatus* is restricted to plants with relatively small leaves. Variants with larger leaves, possible introgressants with the other taxa, are placed in var. *ledifolius*.

29. PURSHIA de Candolle ex Poiret in J. Lamarck et al., Encycl., Suppl. 4: 623. 1816 • [For Frederick Traugott Pursh, 1774–1820, German botanist, student of North American flora]

James Henrickson

Cowania D. Don

Shrubs, 2–75 dm. **Stems** 3–20+, branched; bark cinnamon brown, weathering gray, in thin papery sheets; short shoots present. **Leaves** persistent or drought deciduous, <abscising distal to bases, crowded on short-shoot spurs>, simple; stipules persistent, deltate, margins entire; petiole usually absent, sometimes present; blade linear or ovate to obovate, 0.3–1.5(–1.9) cm, coriaceous, sometimes thin, margins revolute, entire or toothed to lobed, abaxial surface usually densely hairy, adaxial glabrous or sparsely hairy. **Inflorescences** terminal <on short shoots>, flowers solitary; bracts present <resembling immature leaves>; bracteoles absent. **Pedicels** present. **Flowers** 10–15(–25) mm diam.; hypanthium obconic, sometimes hemispheric in fruit, 2.5–6.5 mm, exterior villous, stipitate-glandular or not, <interior glabrous, nectariferous>; sepals 5, spreading-reflexed, ovate to oblong-ovate, <abaxial surface pubescent-villous>; petals 5, caducous, white, cream-yellow, or yellow [rose red], broadly obovate or obovate-spatulate to spatulate, <base cuneate to clawed, apex spreading, rounded to emarginate>; stamens (17–) 20–80(–125), shorter than petals; torus absent; carpels 1–15(–31), stipitate, stigmas terminal. **Fruits** achenes, 1, obovoid-fusiform or oblanceoloid, 4–12 mm, leathery; hypanthium persistent; sepals persistent, spreading-reflexing. $x = 9$.

Species 6 (4 in the flora): sw United States, n Mexico.

This treatment combines *Cowania* (with 3–31 carpels that develop into wind-dispersed achenes with plumose awns) with *Purshia* (with 1–3 carpels that form into larger, leathery achenes with persistent style bases that are often dispersed by rodents that cache the fruits for later use). This is supported by the presence of intergeneric hybrids, by the similar vegetative and floral morphologies, except for carpel number and means of fruit dispersal, and by molecular data. The roots are often associated with the nitrogen-fixing actinomycete *Frankia*. Some species are important browse for ungulates.

SELECTED REFERENCES Stutz, H. C. and L. K. Thomas. 1963. Hybridization and introgression in *Cowania* and *Purshia*. Evolution 18: 183–195. Tiedemann, A. R. and K. L. Johnson, comps. 1983. Proceedings, Research and Management of Bitterbrush and Cliffrose in Western North America, Salt Lake City, Utah, April 13–15, 1982. Ogden. [U.S.D.A. Forest Serv., Gen. Techn. Rep. INT-152.]. Young, J. A. and C. D. Clements. 2002. *Purshia*: The Wild and Bitter Roses. Reno.

1. Carpels 1 or 2; achenes broadly obovoid-fusiform, 3.5–5.2 mm diam., persistent style 5–7 (–10) mm, densely hirtellous . . . . . 1. *Purshia tridentata*
1. Carpels (2–)5–15(–31); achenes oblanceoloid, 1.1–2.5 mm diam., persistent style (15–)20–59 mm, plumose.
    2. Leaf blades oblong to spatulate, margins irregularly lobed, lobes (3–)5(–7) . . . . . 2. *Purshia stansburyana*
    2. Leaf blades linear to oblanceolate or linear-subulate, margins entire or with 1 or 2 teeth or lobes.
        3. Leaf blades: margins revolute (not obscuring abaxial surfaces), entire or with 1 or 2 teeth or lobes; petals light yellowish cream; stamens (28–)34–54; carpels (2–) 4 or 5(or 6) . . . . . 3. *Purshia subintegra*
        3. Leaf blades: margins strongly revolute (nearly obscuring abaxial surfaces), entire; petals white or suffused with red; stamens (60–)80–120; carpels (12–)17–31 . . . . 4. *Purshia ericifolia*

1. Purshia tridentata (Pursh) de Candolle, Trans. Linn. Soc. London 12: 158. 1818 • Bitterbrush [F]

Tigarea tridentata Pursh, Fl. Amer. Sept. 1: 333, plate 15. 1813

Shrubs, evergreen or partially summer-winter deciduous, 2–25(–40) dm. **Stems:** young-stem internodes 4–12 mm, sparsely to densely arachnoid-villous, ± hirtellous, or glabrous, stipitate-glandular or not; short-shoot spurs simple or branched, 5–20(–60) × 1.5–2 mm. **Leaves:** blade of 1 or 2 types: (1) persistent, dark green adaxially, obovate to spatulate, 4–9 × (1–)3–6 mm, irregularly lobed in distal ⅓, lobes 3(–5), lobes entire, ± terete, central lobe deflexed, often apiculate, 2 lateral ascending, margins strongly revolute, abaxial surface arachnoid-villous, midvein glabrous, glabrate, or hirtellous, adaxial glabrous, sometimes arachnoid-villous or hirtellous, often glandular-punctate along margins; (2) winter deciduous, green-gray adaxially, flabelliform, 6–15+ × 3–7+ mm, margins weakly revolute, usually 3-toothed, sometimes 3-lobed, distally, abaxial surface arachnoid-villous to arachnoid-tomentose, veins ± glabrate, adaxial hirtellous and ± arachnoid villous, sometimes sessile or stipitate-glandular along margins. **Pedicels** 0.2–2.5 mm. **Flowers:** hypanthium obconic, 2.5–5 × 1.5–3 mm, 3.5–6 mm diam. in fruit, pubescent-sericeous, sessile or stipitate-glandular; petals usually lemon- to cream-yellow, sometimes nearly white, obovate-spatulate, 4–7+ mm; stamens 17–35; carpels 1 or 2. **Achenes** broadly obovoid-fusiform (circular in cross section or flattened on common side when paired), 7–12 × 3.5–5.2 mm, 15–22-veined, persistent style 5–7(–10) mm, densely hirtellous.

Varieties 2 (2 in the flora): w North America, nw Mexico.

1. Leaves dimorphic, blades flabelliform, thin, margins weakly revolute, usually 3-toothed, sometimes 3-lobed, apically, adaxial surfaces persistently arachnoid-villous and hirtellous, margins rarely with sessile or stipitate glands, these not strongly resinous, (at least proximalmost short-shoot leaves, smaller, thick, 3-lobed distally, margins with or without resinous glands); stems: young long shoots arachnoid-villous to pubescent, stipitate-glandular or not . . . . 1a. *Purshia tridentata* var. *tridentata*
1. Leaves monomorphic, blades obovate, coriaceous-thickened, 3(–5)-lobed, lobes oblong-linear, central lobes deflexed, 2 lateral ascending, margins strongly revolute, adaxial surfaces glabrous, sometimes weakly arachnoid-villous or hirtellous, margins often with large, resiniferous, punctate glands; stems: young long shoots glabrous or sparsely hirtellous, often stipitate-glandular . . . . 1b. *Purshia tridentata* var. *glandulosa*

1a. Purshia tridentata (Pursh) de Candolle var. **tridentata** • Antelope bitterbrush, deerbrush, buckbrush, quininebush, black sage, greasewood [E] [F]

Stems: young long shoots arachnoid-villous to pubescent <curved hairs 0.1–0.4 mm>, stipitate-glandular or not. **Leaves** dimorphic, long-shoot and most short-shoot leaves summer-winter deciduous; blade flabelliform, thin, margins weakly revolute, usually 3-toothed, sometimes 3-lobed, apically, not punctate or rarely with sessile or stipitate glands but these not strongly resinous, abaxial surface white arachnoid-villous to arachnoid-tomentose, adaxial persistently arachnoid-villous (hairs crinkled) and hirtellous; <late season short-shoot and sometimes first-formed (proximalmost) spring short-shoot leaves winter-persistent, remaining at base of new season's growth, smaller, thick, 3-lobed distally, margins with or without resinous punctate glands, glabrous or hirtellous>. $2n$ = 18.

Flowering Mar–Jun(–Jul). Mostly well-drained alluvial plains, sagebrush scrub, grasslands, pinyon-juniper woodlands, yellow pine, red fir forests, subalpine zones; (60–)1000–2500(–3400) m; B.C.; Ariz., Calif., Colo., Idaho, Mont., Nebr., Nev., N.Mex., Oreg., Utah, Wash., Wyo.

Plants of var. *tridentata* have dimorphic leaves. Short shoots typically produce thick glandular-margined, overwintering leaves in the later summer (and sometimes in early spring) and thinner, flabelliform winter-deciduous leaves in spring and summer or only on long shoots. Variation in persistent and deciduous leaves occurs through the range, with northern plants producing mostly thinner deciduous leaves and plants in more arid zones to the south producing more thick, persistent and fewer thin, deciduous leaves. At high elevations in the White Mountains of Inyo County, California, low vegetative plants have only thick, persistent leaves as in var. *glandulosa*, except that long-shoot leaves are of the thin, deciduous type characteristic of var. *tridentata*. In more arid regions plants produce only thick, persistent leaves on both long and short shoots, characteristic of var. *glandulosa* with a broad zone of transition between the taxa. The persistent leaves provide important, high-protein winter browse for ungulates. Development of stipitate glands on stems, leaves, and hypanthia is highly variable even within populations.

1b. Purshia tridentata (Pursh) de Candolle var. **glandulosa** (Curran) M. E. Jones, Proc. Calif. Acad. Sci., ser. 2, 5: 680. 1895 • Desert bitterbrush

Purshia glandulosa Curran, Bull. Calif. Acad. Sci. 1: 153. 1885

Stems: young long shoots glabrous or sparsely hirtellous, often stipitate-glandular. **Leaves** monomorphic, both long- and short-shoot leaves winter persistent; blade obovate to spatulate (when small), coriaceous-thickened, 3(–5)-lobed, lobes oblong-linear, narrow, central lobe deflexed, 2 lateral ascending, margins strongly revolute, often with large, resiniferous, punctate glands, abaxial surface arachnoid-villous, midvein glabrous, glabrate, or hirtellous, adaxial glabrous, sometimes weakly arachnoid-villous (this often obscured by broad, glabrous midvein in dried leaves) or hirtellous. $2n = 18$.

Flowering Apr–May. Juniper-pinyon, Joshua tree woodlands, desert blackbrush scrub, chaparral at desert margins; (600–)800–1500 m; Ariz., Calif., Nev., Utah; Mexico (Baja California).

Variety *glandulosa* is a derivative of var. *tridentata*. Some sterile specimens of var. *glandulosa* cannot be distinguished from *Purshia stansburyana*; both taxa sometimes have similar leaves, and hirtellous, more or less stipitate-glandular young stems.

G. L. Stebbins (1959) and H. C. Stutz and L. K. Thomas (1963) hypothesized that var. *glandulosa* is a product of hybridization and introgression between *Purshia tridentata* and *P. stansburyana*. Although introgression would explain certain similarities, *P. tridentata* stands as a distinct species strongly differentiated by ovary-fruit characteristics and dimorphic leaves, which allow it to adapt to a wide range of habitats. Hybrids with *P. stansburyana* are discussed under the latter species.

2. Purshia stansburyana (Torrey) Henrickson, Phytologia 60: 468. 1986 (as stansburiana) • Cliff rose

Cowania stansburyana Torrey in H. Stansbury, Exped. Great Salt Lake, 386, plate 3. 1852 (as stansburiana); *C. mexicana* D. Don var. *stansburyana* (Torrey) Jepson; *Purshia mexicana* (D. Don) S. L. Welsh var. *stansburyana* (Torrey) S. L. Welsh

Shrubs, evergreen, 10–40(–75) dm. **Stems:** young-stem internodes 5–14 mm, sparsely hirtellous or glabrous, usually stipitate-glandular; short-shoot spurs simple or branched, 10–20(–40) × 1.2–2.7 mm. **Leaves:** blade dark green adaxially, obovate, (3–)7–12(–19) × (2–)3–8(–13) mm, irregularly lobed in distal ½, lobes (3–)5(–7), proximal 2 lobes ascending, oblong to spatulate, <narrow, 2–6 mm>, distal 3 lobes deflexed, reduced, margins strongly revolute, usually entire, rarely toothed proximally, <apex rounded to truncate>, abaxial surface white-villous, midvein <greenish>, glabrate, <glandular-punctate>, adaxial sparsely hirtellous or arachnoid-villous, soon glabrate, ± glandular-punctate. **Pedicels** 2.5–4.5 mm. **Flowers:** hypanthium obconic, 4–6.5 × 1.5–3 mm, 3–5 mm diam. in fruit, arachnoid-villous to pubescent, usually stipitate-glandular; petals light to cream-yellow or -white, broadly obovate, 7–13(–15) mm; stamens (52–)60–90(–125); carpels (3 or)4–7(–10). **Achenes** oblanceoloid, 6–8 × 1.8–2.2 mm, 12-ribbed, persistent style 35–59 mm, plumose hairs 1.4–2.2 mm. $2n = 18$.

Flowering Apr–May(–Oct). Pinyon-juniper, Joshua tree, yellow pine woodlands, arid desert scrub, desert chaparral; (1000–)1200–2200(–2600) m; Ariz., Calif., Colo., Idaho, Nev., N.Mex., Utah; Mexico (Baja California, Sonora).

Purshia stansburyana is vegetatively similar to *P. tridentata* var. *glandulosa*, the latter usually having thicker leaves with narrower abaxial grooves and apiculate apices. The two sometimes can only be distinguished by flowers or fruits.

Purshia stansburyana forms hybrids with *P. tridentata* vars. *glandulosa* and *tridentata*. The hybrids easily can be recognized by their 2 or 3(–5) carpels and achenes with styles 12–28 mm, among other characteristics (H. C. Stutz and L. K. Thomas 1963). Hybrids between *P. stansburyana* and *P. tridentata* var. *glandulosa* have been given the names *Cowania mexicana* var. *dubia* Brandegee and *C. alba* Goodding. Introgression with *P. stansburyana* occurs in two populations of *P. subintegra*.

3. Purshia subintegra (Kearney) Henrickson, Phytologia 60: 468. 1986 • Arizona cliffrose [C] [E]

Cowania subintegra Kearney, Madroño 7: 15. 1943; *Purshia pinkavae* Schaack

Shrubs, evergreen or tardily drought deciduous, 2–8(–12) dm. **Stems:** young-stem internodes 2.5–9(–13) mm, closely arachnoid-villous, sometimes pubescent-hirtellous, rarely stipitate-glandular; short-shoot spurs simple, 1–10 × 1–1.3 mm. **Leaves:** blade green adaxially, linear to oblanceolate, (3–)4–8(–10)[–17] × 1–1.3(–2)[–4.2] mm, margins revolute (not obscuring abaxial surfaces), entire or with 1 or 2 teeth or lobes, <lobes ascending, oblong, 1(–2) mm>, margins strongly revolute, <apex obtuse-rounded, sometimes ± apiculate>, abaxial surface densely arachnoid-villous, midvein glabrous (when exposed), adaxial hirtellous (along midvein proximally), initially closely arachnoid-villous, glabrescent, eglandular. **Pedicels** 1.5–6(–12) mm. **Flowers:** hypanthium narrowly obconic, 3–6 × 1.5–3.2 mm, 4.5 mm diam. in fruit, arachnoid-villous, rarely stipitate-glandular; petals light yellowish cream, obovate-spatulate to spatulate, (4–)6–8 mm; stamens (28–)34–54; carpels (2–)4 or 5(or 6). **Achenes** oblanceoloid, 5–6 × 1.5–2.5 mm, 13–15 ribbed, persistent style 32–45 mm, plumose hairs 1.5–3 mm. **2*n*** = 18.

Flowering Mar–early Jun. Whitish, lacustrine-derived limy soil of Miocene-Pliocene age; of conservation concern; 600–1100 m; Ariz.

Purshia subintegra is known from four populations in Arizona: Mojave County (north of Burro Creek); Maricopa and adjacent Yavapai counties (Tonto Basin, western side of Horseshoe Reservoir on Verde River); Yavapai County (Verde Valley, south of Cottonwood, the largest population); and Graham County (San Carlos Basin northwest of Bylas). The first three populations consist of plants with both entire or lobed leaves (on individual plants) and some stipitate glands on pedicels and hypanthia. The Bylas population consists mainly of plants with mostly entire leaves and without stipitate glands, although some collections from that population have stipitate glands and others have lobed long-shoot leaves. The lobed leaves and stipitate glands have been considered markers of introgression with *P. stansburyana*, which has been confirmed by morphometric studies (F. W. Reichenbacher 1994) and by molecular markers (S. E. Travis et al. 2008) that provide evidence of introgression with *P. stansburyana* in all but the Bylas population.

Because of this variation and introgression, C. G. Schaack (1987) lectotypified *Purshia subintegra* on a specimen with lobed leaves, relegating that name to hybrid status. He named a new taxon (*P. pinkavae*) from the population near Bylas that mostly (but not entirely) lacks stipitate glands and lobed leaves; this seems unnecessary, as edaphic and morphologic continuity exists among known populations of *P. subintegra*, and all four populations share at least some traits with *P. stansburyana* as well as with *P. ericifolia*.

SELECTED REFERENCE Travis, S. E., J. E. Baggs, and J. Maschinski. 2008. Disentangling the role of hybridization in the evolution of the endangered Arizona cliffrose (*Purshia subintegra*; Rosaceae): A molecular and morphological analysis. Conservation Genet. 9: 1183–1194.

4. Purshia ericifolia (Torrey ex A. Gray) Henrickson, Phytologia 60: 468. 1986

Cowania ericifolia Torrey ex A. Gray, Smithsonian Contr. Knowl. 5(6): 106. 1853 (as ericaefolia)

Shrubs, evergreen, 5–15 (–25) dm. **Stems:** young-stem internodes 2–9 mm, closely white villous-tomentose, ± glabrate, eglandular; short-shoot spurs simple or branched, 5–20 × 1.5–2 mm. **Leaves:** blade dark green adaxially, linear-subulate, (2.5–)3–8(–10) × 0.7–1(–1.5) mm, margins strongly revolute (nearly obscuring abaxial surfaces), entire, <apex acuminate to cuspidate>, abaxial surface sparsely arachnoid-villous <abaxial groove 0.1–0.3 mm wide, broader at base, densely villous>, adaxial glabrate, eglandular. **Pedicels** 2–13(–20) mm. **Flowers:** hypanthium hemispheric, 2.5–4 × 2–3 mm, to 5.5 mm diam. in fruit, closely villous, sometimes stipitate-glandular; petals white or suffused with red, broadly obovate, 4–8(–11) mm; stamens (60–) 80–120; carpels (12–)17–31. **Achenes** oblanceoloid, 4–6 × 1.1–1.3 mm, weakly 2- or 3-ribbed, persistent style 15–30 mm, plumose hairs 1–2 mm. **2*n*** = 18.

Flowering Jun–Sep. Arid rocky limestone slopes, cliffs, Chihuahuan Desert; (600–)800–1600(–2100) m; Tex.; Mexico (Chihuahua, Coahuila).

Purshia ericifolia is known from Brewster and Presidio counties in trans-Pecos Texas. The species is distinctive in its subterete, arcuate leaves that are distichous in gradual spirals on short shoots. It flowers after rain.

30. CHAMAEBATIA Bentham, Pl. Hartw., 308. 1849 • Mountain misery [Greek *chamai*, low, and *batos*, bramble, alluding to habit]

James Henrickson

Shrubs, 1–12(–20) dm, <herbage strongly aromatic>. **Stems** 3–20+, sparsely branched; bark gray or reddish brown, smooth; long shoots present. **Leaves** persistent, 2- or 3-odd-pinnately compound; stipules persistent 1st year, deciduous with leaf base 2d year, linear-subulate, margins entire, <sometimes stipitate-glandular>; petiole present; blade obovate to obtrullate or lance-elliptic in outline, (2.8–)3.5–7(–9.4) cm, leathery, surfaces hirtellous; pinnae 8–17 per side, oblong to oblong-obovate, pinnately or 2-pinnately divided; ultimate segments oblong-obovate to obovate or obliquely obovoid, margins entire. **Inflorescences** terminal, (1 or)2–7 (–10)-flowered, corymbs or panicles; bracts present; bracteoles present. **Pedicels** present. **Flowers** 10–19 mm diam.; hypanthium funnelform-turbinate, 3–6 mm diam., exterior hairy, usually stipitate-glandular; sepals 5, reflexed, deltate; petals 5, caducous, white, orbiculate, <base short-clawed, margins erose-undulate, apex often retuse>; stamens 35–65(–76), shorter than petals; torus absent; carpels 1(or 2), sessile, stigmas lateral. **Fruits** achenes, 1, subglobose, 3–5.5 mm, smooth; hypanthium persistent; sepals persistent, erect. $x = 9$.

Species 2 (2 in the flora): California, nw Mexico.

Molecular data support inclusion of *Chamaebatia* with *Cercocarpus*, *Dryas*, and *Purshia* in subfam. Dryadoideae. *Chamaebatia* differs from the other genera in its sessile carpels and smooth, rounded fruit wall with a mesocarp of macrosclereids. Vegetative and floral similarities to *Chamaebatiaria* (subfam. Amygdaloideae) are perplexing.

1. Leaf blades obovate to obtrullate, 3-pinnately compound; terminal leaflet lobes each with terminal short-stipitate gland . . . . 1. *Chamaebatia foliolosa*
1. Leaf blades lance-elliptic, 2- or 3-pinnately compound; terminal leaf lobes each with terminal sessile or embedded gland . . . . 2. *Chamaebatia australis*

1. Chamaebatia foliolosa Bentham, Pl. Hartw., 308. 1849 E F

Plants 1–5(–7) dm, forming extensive colonies. **Stems:** internodes 5–10(–15) mm; bark reddish brown. **Leaves** 3-pinnately compound; petiole 10(–15) mm; blade obovate to obtrullate, (28–)45–70(–94) × (13–)20–45(–71) mm, surfaces moderately hirtellous, sparsely stipitate-glandular; rachises straight; pinnae (8–)10–12 per side, oblong to oblong-obovate, (3–)10–25(–40) × (1.8–)3–8(–13) mm; <pinnules (4–)7–10 per side, oblong, (1.2–)2.5–5(–10) × (0.6–)1.5–3(–5) mm>, each pinnately divided into (3–)5–11 pairs of segments, segments oblong-obovate to obovate, (0.3–)0.6–1.5 × (0.2–)0.5–1 mm, terminal largest, each with terminal short-stipitate gland, adaxial rachis and rachilla with additional segments between pinnae and pinnules. **Inflorescences** (3–)5–7(–10)-flowered, panicles, 35–60 (–80) mm; peduncles 2–5 mm. **Flowers:** hypanthium 4–6 mm; hypanthia and sepals pubescent, stipitate glands 0.8–1.5 mm; sepals oblong-lanceolate, 4.5–6(–7) mm, apex acute; petals 6–9 mm; stamens 55–65(–76); styles 5.5 mm. **Achenes** 4.5–5.5 mm. $2n = 18$.

Flowering Apr–Jun. Slopes, flats, coarse sand, yellow pine to montane conifer forests; (300–)600–1800 (–2300) m; Calif.

Chamaebatia foliolosa occurs on the western slope of the Sierra Nevada and into the Cascade Range, from Shasta to Inyo and Kern counties.

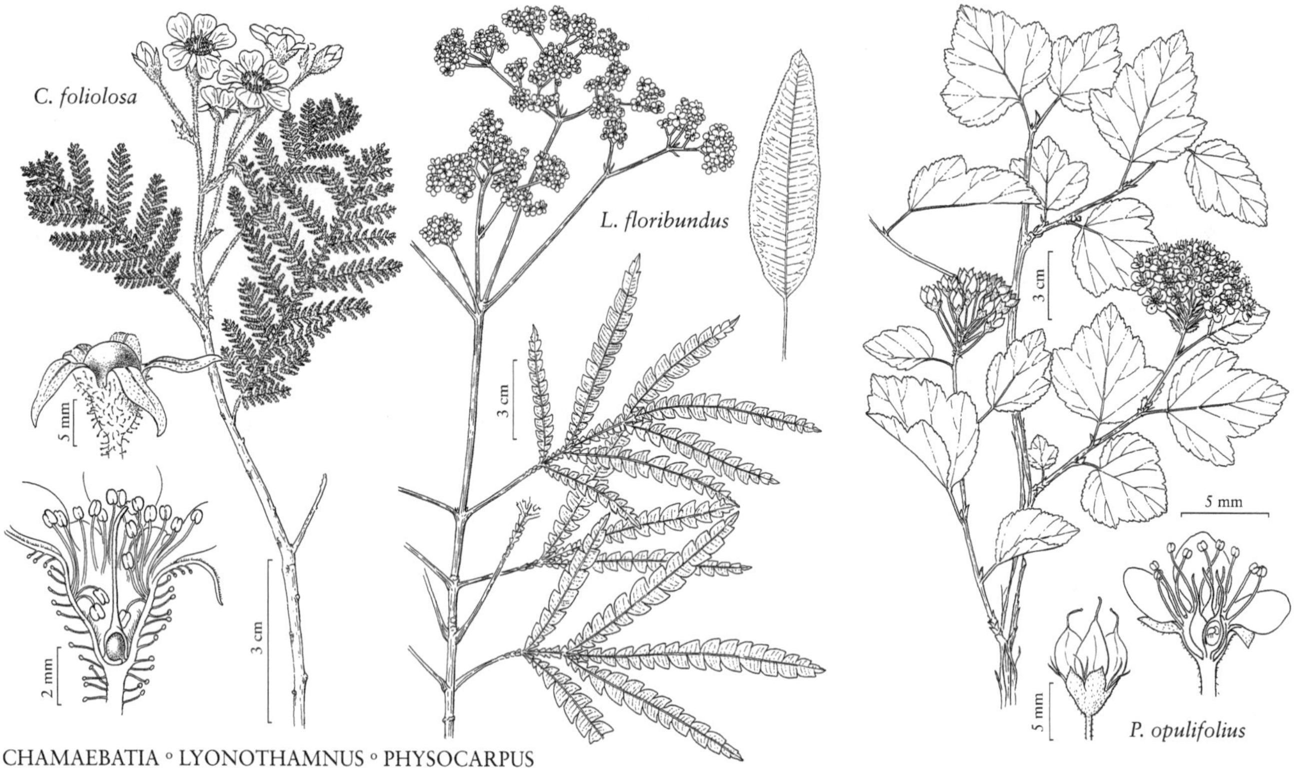

CHAMAEBATIA ◦ LYONOTHAMNUS ◦ PHYSOCARPUS

2. Chamaebatia australis (Brandegee) Abrams, Bull. Torrey Bot. Club 34: 263. 1907 • Southern mountain misery

Chamaebatia foliolosa Bentham var. *australis* Brandegee, Bot. Gaz. 27: 447. 1899

Plants 6–20 dm, separate, possibly colonial. **Stems:** internodes 2–10(–22) mm; bark gray. **Leaves** 2(or 3)-pinnately compound; petiole 2–5 mm; blade lance-elliptic, (30–)35–55 × 10–25 mm, surfaces hirtellous, stipitate-glandular; rachises straight to decurved; pinnae 13–17 per side, oblong, 2–13 × 1.3–4 mm, each divided into 3–9 pairs of segments, ultimate segments obliquely obovoid, 0.8–1.5 × 0.4–1 mm, terminal largest, each with terminal sessile or embedded gland, rachis (rachilla) with additional segments. **Inflorescences** (1 or)2–5-flowered, racemose corymbs, 40–65 mm; peduncles 3–15 mm. **Flowers:** hypanthium 3–4 mm diam.; hypanthia and sepals hirtellous, stipitate glands to 1 mm; sepals oblong, 3.5–4.5 mm; petals 5.5–6.5(–7.5) mm, apex cleft or not; stamens 35–40; styles 3.5–4 mm. **Achenes** 3–4 mm. $2n = 18$.

Flowering Apr–Jun. Dry slopes in chaparral, coarse granitic sand; (200–)300–1000(–1200) m; Calif.; Mexico (Baja California).

Chamaebatia australis occurs in San Diego County and adjacent northern Baja California.

c. ROSACEAE Jussieu subfam. AMYGDALOIDEAE Arnott, Botany, 107. 1832 (as Amygdaleae)

Luc Brouillet

Shrubs or trees, sometimes subshrubs or herbs. Leaves alternate, sometimes opposite, simple, sometimes pinnately compound; stipules present or absent. **Flowers:** torus absent or minute; carpels 1–5(–8), distinct or ± connate (Maleae), free or ± adnate to hypanthium (many Maleae), styles distinct or ± connate (some Maleae); ovules (1 or)2(–5+), collateral, clustered, or biseriate. **Fruits** follicles aggregated or not, capsules, drupes aggregated or not, aggregated drupelets, pomes, or aggregated nutlets, rarely achenes or aggregated achenes; styles persistent or deciduous, not elongate (elongate in Gillenieae). x = 8, 9, 15, 17.

Genera 55, species ca. 1300 (38 genera, 361 species, including 21 hybrids, in the flora): North America, Mexico, Central America, South America, Europe, Asia, Africa, Atlantic Islands (Madeira), Pacific Islands (Hawaii), Australia.

Each of the nine tribes recognized in subfam. Amygdaloideae has representative species found in the flora area.

Cyanogenic glycosides are usually present in Amygdaloideae; sorbitol is present.

The name Amygdaloideae Arnott (1832) has priority over Spiraeoideae Arnott (1832), used by D. Potter et al. (2007), because Amygdalaceae (1820) is an earlier conserved name.

SELECTED REFERENCES Campbell, C. S. et al. 1995. Phylogenetic relationships in Maloideae (Rosaceae): Evidence from sequences of the internal transcribed spacers of nuclear ribosomal DNA and its congruence with morphology. Amer. J. Bot. 82: 903–918. Campbell, C. S., C. W. Greene, and T. A. Dickinson. 1991. Reproductive biology in subfam. Maloideae (Rosaceae). Syst. Bot. 16: 333–349. Evans, R. C. and C. S. Campbell. 2002. The origin of the apple subfamily (Rosaceae: Maloideae) is clarified by DNA sequence data from duplicated GBSSI genes. Amer. J. Bot. 89: 1478–1484. Evans, R. C. and T. A. Dickinson. 1999. Floral ontogeny and morphology in subfamily Amygdaloideae T. & G. (Rosaceae). Int. J. Pl. Sci. 160: 955–979. Evans, R. C. and T. A. Dickinson. 1999b. Floral ontogeny and morphology in subfamily Spiraeoideae Endl. (Rosaceae). Int. J. Pl. Sci. 160: 981–1012. Iketani, H. and H. Ohashi. 1991. Anatomical structure of fruits and evolution of the tribe Sorbeae in the subfamily Maloideae (Rosaceae). J. Jap. Bot. 66: 319–351. Lee, S. and J. Wen. 2001. A phylogenetic analysis of *Prunus* and the Amygdaloideae (Rosaceae) using ITS sequences of nuclear ribosomal DNA. Amer. J. Bot. 88: 150–160. Phipps, J. B. et al. 1990. A checklist of the subfamily Maloideae (Rosaceae). Canad. J. Bot. 68: 2209–2269. Phipps, J. B. et al. 1991b. Origins and evolution of subfam. Maloideae (Rosaceae). Syst. Bot. 16: 303–332. Robertson, K. R. et al. 1991. A synopsis of genera in Maloideae (Rosaceae). Syst. Bot. 16: 376–394. Robertson, K. R., J. B. Phipps, and J. R. Rohrer. 1992. Summary of leaves in the genera of Maloideae (Rosaceae). Ann. Missouri Bot. Gard. 79: 81–94. Rohrer, J. R., K. R. Robertson, and J. B. Phipps. 1991. Variation in structure among fruits of Maloideae (Rosaceae). Amer. J. Bot. 78: 1617–1635. Rohrer, J. R., K. R. Robertson, and J. B. Phipps. 1994. Floral morphology of Maloideae (Rosaceae) and its systematic relevance. Amer. J. Bot. 81: 574–581. Sterling, C. 1966. Comparative morphology of the carpel in the Rosaceae. VIII. Spiraeoideae: Holodisceae, Neillieae, Spiraeeae, Ulmarieae. Amer. J. Bot. 53: 521–530.

c1. ROSACEAE Jussieu (subfam. AMYGDALOIDEAE) tribe LYONOTHAMNEAE Brouillet, J. Bot. Res. Inst. Texas 2: 385. 2008 [E]

Luc Brouillet

Trees, sometimes shrublike; unarmed. **Leaves** opposite, simple or pinnately compound; stipules caducous, free; venation pinnate. **Flowers:** perianth and androecium perigynous; epicalyx bractlets (1 or)2 or 3; hypanthium hemispheric; torus absent; carpels 2(or 3), distinct, free, styles terminal, distinct; ovules 2–4+, nearly apical, <pendulous>, clustered. **Fruits** aggregated follicles; styles tardily falling.

Species 1: California (Channel Islands).

The base chromosome number for Lyonothamneae is $x = 9$. Cyanogenic glycosides are absent from *Lyonothamnus*.

31. LYONOTHAMNUS A. Gray, Proc. Amer. Acad. Arts 20: 291. 1885 • Catalina or island ironwood [For William Scrugham Lyon, 1851–1916, botanist, nurseryman, plant collector in California and Philippines, and Greek *thamnos*, bush or shrub] [C] [E]

John L. Strother

Plants clonal, 50–100(–150) dm, herbage usually glabrous, glabrate, or tomentulose (glabrescent). **Stems** 1+; bark brown or gray; short shoots absent. **Leaves** tardily falling, cauline; stipules absent or lance-linear, margins entire; petiole present; blade simple (sometimes lobed), palmate, or pinnate, simple blades or leaflets elliptic, lance-linear, lanceolate, or oblong, 6–18 cm, leathery, leaflets 3–7, usually pinnately lobed, margins ± revolute, entire, crenulate, or serrate. **Inflorescences** terminal, 50–800+-flowered, compound corymbs or panicles; bracts present; bracteoles present. **Pedicels** present. **Flowers** produced after leaves, (5–)6–8 mm diam.; hypanthium 3–4(–5) mm (including sepals), exterior tomentulose, interior pilose; sepals 5, spreading, ovate to triangular; petals 5, white, obovate to orbiculate; stamens (12–)15+ in 1 series, shorter or longer than petals; carpels tomentulose. **Fruits** aggregated follicles, 2 (or 3), ± ovoid to lanceoloid, 3–4 mm (excluding styles), stipitate-glandular, dehiscent adaxially; hypanthium persistent; sepals persistent or tardily falling, ± spreading. **Seeds** 2–4+, ± ellipsoid to fusiform, firm, ends rounded or truncate, faces smooth or minutely papillate, seed coats tight. $x = 9$.

Species 1: California (Channel Islands).

Taxonomically, *Lyonothamnus* has been moved about within Rosaceae; here it is placed as an isolated taxon within Amygdaloideae. It is known from Miocene fossils from California, Nevada, and Oregon (D. M. Erwin and H. E. Shorn 2000).

SELECTED REFERENCES Brandegee, T. S. 1890. The plants of Santa Catalina Island. Zoë 1: 107–115. Bushakra, J. M. et al. 1999. The extent of clonality and genetic diversity in the Santa Cruz Island ironwood, *Lyonothamnus floribundus*. Molec. Ecol. 8: 471–475. Erwin, D. M. and H. E. Shorn. 2000. Revision of *Lyonothamnus* A. Gray (Rosaceae) from Neogene of western North America. Int. J. Pl. Sci. 161: 179–193.

1. Lyonothamnus floribundus A. Gray, Proc. Amer. Acad. Arts 20: 292. 1885 [C] [E] [F]

Lyonothamnus aspleniifolius Greene; *L. floribundus* subsp. *aspleniifolius* (Greene) P. H. Raven; *L. floribundus* var. *aspleniifolius* (Greene) Brandegee

Leaves: petiole 12–35 mm, often distally winged, wings often lobed, base attenuate to rounded; in lobed blades, lobes oblong-rounded to broadly falcate-acute or almost triangular, base broadly sessile along axes, margins near bases often overlapping, surfaces: abaxial tomentose, tardily glabrescent, adaxial tomentulose, soon glabrescent. **Flowers:** sepals 1.5–2 mm; stamens 2 mm, filaments unequal. $2n = 54$.

Flowering May–Jul. Canyons, ravines, ridges, and slopes in chaparral, oak woodlands, openings; of conservation concern; (20–)100–500+ m; Calif.

On Santa Catalina Island, most plants of *Lyonothamnus floribundus* have most leaves not divided into leaflets and have been called forma, subsp., or var. *floribundus*. Plants with most leaves palmately or pinnately divided, usually with 3–5(–7), pinnately lobed leaflets, have been called forma, subsp., or var. *aspleniifolius*; such plants are known from San Clemente, Santa Catalina, Santa Cruz, and Santa Rosa islands. Continuity of variation in leaf form in *L. floribundus* has been documented by T. S. Brandegee (1890) and others. J. M. Bushakra et al. (1999) noted genetic similarity among clones.

c2. Rosaceae Jussieu (subfam. Amygdaloideae) tribe Neillieae Maximowicz, Trudy Imp. S.-Peterburgsk. Bot. Sada 6: 164, 216. 1879

Luc Brouillet

Shrubs; unarmed. **Leaves** alternate, simple; stipules persistent or caducous, free; venation palmate. **Flowers:** perianth and androecium perigynous; epicalyx bractlets absent; hypanthium cup-shaped to campanulate; torus absent; carpels 1–5, basally connate (distinct when single), free, styles terminal, distinct; ovules 2–5[–10], marginal, <pendulous>, biseriate or clustered. **Fruits** solitary or aggregated follicles; styles persistent, not elongate.

Genera 2, species 23–30 (2 genera, 7 species in the flora): North America, n Mexico, Asia; introduced in Europe.

The base chromosome number for Neillieae is $x = 9$. Cyanogenic glycosides are absent in *Physocarpus*; there are no data on *Neillia*.

SELECTED REFERENCE Oh, S. H. and D. Potter. 2005. Molecular phylogenetic systematics and biogeography of tribe Neillieae (Rosaceae) using DNA sequences of cpDNA, rDNA, and *LEAFY*. Amer. J. Bot. 92: 179–192.

1. Inflorescences corymb- or umbel-like racemes; carpels 1–5; fruits usually aggregated follicles .... 32. *Physocarpus*, p. 347
1. Inflorescences racemes or panicles to corymbs; carpel 1; fruits solitary follicles .... 33. *Neillia*, p. 351

32. PHYSOCARPUS (Cambessèdes) Rafinesque, New Fl. 3: 73. 1838 (as Physocarpa), name and orthography conserved • Nine-bark, physocarpe [Greek *physa*, bladder, and *karpos*, fruit, alluding to inflated follicles of some species]

Crinan Alexander

Spiraea Linnaeus sect. *Physocarpus* Cambessèdes, Ann. Sci. Nat. (Paris) 1: 239, 385. 1824 (as Physocarpos)

Shrubs, (10–)30–60 dm, glabrous or stellate-hairy, mostly eglandular (except in inflorescence). **Stems** 5–20, ascending to erect, <branches erect proximally, somewhat sinuous>; bark buff to reddish brown or black, <usually peeling in narrow strips>; short shoots present; <buds superposed, each with ca. 5 exposed brown scales>. **Leaves** deciduous, cauline; stipules caducous, <decurrent on stem, leaving oblique, broadly ovate to rectangular stubs>, linear or oblong to narrowly ovate, obovate, or elliptic, sometimes rhombic, margins sometimes remotely glandular-dentate, <surfaces sparsely stellate-hairy (glabrous adaxially in *P. monogynus*)>; petiole present; blade broadly ovate to obovate or orbiculate, 0.5–10 cm, membranous, margins flat, crenate to doubly crenate or doubly dentate, 3–5(–7)-lobed, rarely almost unlobed, <3–5(–7)-veined>, surfaces glabrous or sparsely to densely stellate-hairy. **Inflorescences** terminal or axillary <on leafy short shoots>, 5–50-flowered, open to dense corymb- or umbel-like pedunculate racemes; bracts present; bracteoles absent. **Pedicels** present. **Flowers** 5–12 mm diam.; hypanthium cup-shaped to campanulate, 1.5–3 mm, exterior stellate-hairy; sepals 5, ascending to slightly reflexed, triangular to ovate; petals 5, white to pale pink, broadly obovate, elliptic, or orbiculate, <base short-clawed, sometimes irregularly toothed or notched>; stamens 20–40, shorter than, equal to, or exceeding petals; carpels 1–5; <styles slender, stigmas capitate>; ovules 2–5. **Fruits** aggregated follicles, sometimes solitary, 1–5, ovoid, often ± inflated, slightly flattened, size not recorded, splitting along both sutures; hypanthium persistent; sepals persistent, ascending to slightly reflexed. **Seeds** 1(–5), <shiny yellow, caruncuIate>. $x = 9$.

Species 8–10 (6 in the flora): North America, n Mexico, ne Asia; introduced in Europe.

Some species of *Physocarpus* are not well delimited. Various botanists have confused *Physocarpus* and *Neillia* (including *Stephanandra*); *Physocarpus* is distinguished by its rounded umbel-like racemes and bladderlike fruits splitting along both sutures; in *Neillia*, the inflorescences are longer and narrower, and the fruits split along the adaxial suture. As currently understood, *Neillia* is an Asian genus. Two or three species of *Physocarpus* are widely cultivated in North America and Europe; the others are less commonly seen, though they may be worthy of more attention.

Opulaster Medikus ex Kuntze is an illegitimate, superfluous name that pertains here.

1. Carpels 3–5, connate basally.
    2. Carpels densely stellate-hairy (sometimes only on sutures); follicles densely stellate-hairy (sometimes only on sutures) . . . . 2. *Physocarpus intermedius*
    2. Carpels mostly glabrous, sometimes hairy (on ventral sutures) or sparsely stellate-hairy, glabrescent; follicles glabrous or sparsely stellate-hairy, ± glabrescent.
        3. Follicle lengths ca. 2 times sepals; leaves usually longer than wide; racemes open . . . . 1. *Physocarpus opulifolius*
        3. Follicle lengths slightly exceeding sepals; leaves usually as wide as long; racemes dense. . . . . 3. *Physocarpus capitatus*
1. Carpels 1 or 2(or 3), if 2(or 3), connate ½ their lengths.
    4. Leaves 0.5–2 × 0.5–2 cm; carpels 1(or 2); stamens unequal (alternately long and short) . . . . 6. *Physocarpus alternans*
    4. Leaves 1.5–6 × 2–6 cm; carpels usually 2(or 3); stamens ± equal.
        5. Stipules oblong to elliptic or obovate, apices of some rounded or erose, 6 × 2.5 mm; follicles flattened, keeled apically . . . . 4. *Physocarpus malvaceus*
        5. Stipules linear to narrowly elliptic or subulate, apices acute, 4–5 × 1 mm; follicles inflated, not keeled apically. . . . . 5. *Physocarpus monogynus*

1. Physocarpus opulifolius (Linnaeus) Maximowicz, Trudy Imp. S.-Petersburgsk. Bot. Sada 6: 220. 1879 (as opulifolia), name conserved • Common or Atlantic nine-bark, physocarpe à feuilles d'obier [E] [F]

Spiraea opulifolia Linnaeus, Sp. Pl. 1: 489. 1753; *Opulaster alabamensis* Rydberg; *O. australis* Rydberg; *O. opulifolius* (Linnaeus) Kuntze; *O. stellatus* Rydberg ex Small

Shrubs, to 30 dm. **Stems** spreading to ascending, ± glabrous. **Leaves:** stipules narrowly ovate, 6–10 × 1.5–2.5 mm, <apex acute>; petiole 1–3 cm; blade ovate to obovate, 6–8.5(–10) × 4–7(–10) cm, usually longer than wide, base broadly cuneate to truncate, 3(–5)-lobed, margins irregularly serrate, apex obtuse to rounded, surfaces mostly glabrous, sometimes sparsely hairy abaxially. **Inflorescences** 30–50-flowered, open, hemispheric racemes, 5 cm diam.; bracts elliptic to spatulate or rhombic, 5 × 2.5 mm, apex acute to 3-fid or coarsely erose, <faces glandular>. **Pedicels** 1–2 cm, usually stellate-hairy. **Flowers** 7–10 mm diam.; hypanthium cup-shaped, 1.5–2 mm, glabrous or sparsely stellate-hairy; sepals <pale green to white, darker in center>, triangular, 1.5–2.5 mm, apex <mucronate>, gland-tipped, surfaces usually stellate-hairy; petals white to pale pink, broadly elliptic to orbiculate, 4–5 × 4–5 mm; stamens equal to or exceeding petals, <anthers purplish>; carpels 3–5, connate basally, sparsely stellate-hairy, glabrescent. **Follicles** 3–5, <connate basally, bright red to brownish red>, ovoid, 5–10 mm (lengths ca. 2 times sepals), sparsely stellate-hairy, ± glabrescent; styles 4 mm. **Seeds** 2(–5), <pyriform, 2 mm>. **2*n*** = 18.

Flowering May–Jun; fruiting Jul–Sep. Rocky stream banks, lake shores, moist woods, swampy ground; 0–1300 m; N.B., N.S., Ont., Que.; Ala., Ariz., Conn., Del., D.C., Fla., Ga., Ill., Ind., Iowa, Kans., Ky., Maine, Md., Mass., Mich., Minn., Miss., Nebr., N.H., N.J., N.Y., N.C., N.Dak., Ohio, Okla., Pa., R.I., S.C., S.Dak., Tenn., Vt., Va., W.Va., Wis.; introduced in Europe.

Physocarpus opulifolius is the most widespread species in the wild and is commonly cultivated in North America and Europe; it occasionally escapes from cultivation and has become established in parts of Europe.

2. Physocarpus intermedius (Rydberg) C. K. Schneider, Ill. Handb. Laubholzk. 1: 807. 1906 • Atlantic nine-bark

Opulaster intermedius Rydberg in N. L. Britton, Man. Fl. N. States, 492. 1901; *Physocarpus opulifolius* (Linnaeus) Maximowicz var. *intermedius* (Rydberg) B. L. Robinson

Shrubs, to 15 dm. **Stems** ascending to erect, ± glabrous. **Leaves:** stipules linear to narrowly ovate, 6 × 1.5 mm, <apex acute>; petiole 0.7–1.5 cm; blade orbiculate to broadly ovate, 3–6 × 2–4 cm, usually longer than wide, base broadly cuneate to truncate or slightly cordate, 3(–5)-lobed, margins deeply and irregularly crenate to doubly crenate, apex rounded to obtuse or acute, surfaces mostly glabrous, sometimes sparsely hairy abaxially. **Inflorescences** 15–20-flowered, dense, hemispheric racemes, 4 cm diam.; bracts elliptic to spatulate or rhombic, 5 × 2 mm, apex not recorded, <faces glandular-dentate>. **Pedicels** 1–2 cm, usually stellate-hairy. **Flowers** 10–12 mm diam.; hypanthium cup-shaped, 2 mm, moderately to densely stellate-hairy; sepals <pale green to white, darker in center>, triangular, 2.5–3.2 mm, apex <sometimes obtuse, mucronate>, gland-tipped, surfaces densely stellate-hairy; petals white, broadly elliptic to orbiculate, 5–6 × 5–6 mm; stamens shorter than or equal to petals; carpels 3–5, connate basally, densely stellate-hairy (sometimes only on sutures). **Follicles** 3–5, <connate basally>, ovoid, 5–8 mm (lengths ca. 2 times sepals), densely stellate-hairy (sometimes only on sutures); styles 4–5 mm. **Seeds** 2(–5), <pyriform>. $2n$ = 18.

Flowering Jun; fruiting Jul–Sep. Woods on bluffs and slopes, bluff prairies, forest openings and glades, abandoned fields, roadsides, thickets, stream banks and flood plains, wetland edges; 50–500[–2200] m; Ont.; Ark., Colo., Ill., Ind., Iowa, Kans., Md., Mich., Minn., Miss., Mo., Nebr., N.Y., N.Dak., Ohio, Okla., S.Dak., Wis.; Mexico (Coahuila, Nuevo León).

Physocarpus intermedius often has been treated as a variety of *P. opulifolius*; it can be distinguished by its densely hairy young carpels that remain hairy, at least on the sutures of the follicles. The species generally inhabits somewhat drier habitats than does *P. opulifolius*.

3. Physocarpus capitatus (Pursh) Kuntze, Revis. Gen. Pl. 1: 219. 1891 • Pacific nine-bark [E]

Spiraea capitata Pursh, Fl. Amer. Sept. 1: 342. 1813; *Physocarpus opulifolius* (Linnaeus) Maximowicz var. *tomentellus* (Seringe) B. Boivin

Shrubs, to 45(–60) dm. **Stems** erect, sometimes suckering, <angled>, glabrous or finely stellate-hairy. **Leaves:** stipules linear to narrowly elliptic, 4 × 0.5–2 mm; petiole 1–2(–3) cm; blade broadly ovate to obovate, (3–)4–8 cm, usually as wide as long, base rounded to truncate or slightly cordate, 3- or 5-lobed, margins irregularly crenate to doubly serrate, apex obtuse to acute, abaxial surface <paler>, glabrous or more densely stellate-hairy, adaxial glabrous or sparsely stellate-hairy. **Inflorescences** 30–50-flowered, dense, hemispheric racemes, 3 cm diam., <sometimes compound with some proximal pedicels becoming secondary peduncles>; bracts narrowly elliptic to spatulate, 4 × 2 mm, apex acute or erose-dentate, <faces glandular>. **Pedicels** 1–1.5 cm, densely stellate-hairy. **Flowers** 5–8 mm diam.; hypanthium cup-shaped, 2 mm, densely stellate-hairy; sepals <pale green to white, darker in center>, triangular, 2–3 mm, apex gland-tipped, surfaces densely stellate-hairy; petals white, broadly elliptic to orbiculate, 3–4 × 3–4 mm; stamens equal to or exceeding petals; carpels 3–5, connate basally, mostly glabrous, sometimes hairy (on ventral suture). **Follicles** 3–5, <connate basally, shiny brown>, ovoid, 5–7 mm (lengths slightly exceeding sepals), glabrous; styles 2.5–3 mm. **Seeds** 2(–5), <pyriform, 2.3–2.8 mm>. $2n$ = 18.

Flowering May–Jul; fruiting Jul–Aug. Open sunny slopes on clay (higher elevations), stream and swamp banks, lake margins in moist woods (lower elevations); 20–1300 m; B.C.; Alaska, Calif., Idaho, Oreg., Wash.

Physocarpus capitatus is commonly cultivated.

4. Physocarpus malvaceus (Greene) Kuntze, Revis. Gen. Pl. 1: 219. 1891 • Mallow-leaved or mallow nine-bark [E]

Neillia malvacea Greene, Pittonia 2: 30. 1889; *Opulaster pauciflorus* (Torrey & A. Gray) A. Heller; *Physocarpus pauciflorus* (Torrey & A. Gray) Piper

Shrubs, 20 dm. **Stems** spreading, <brown becoming grayish black>, glabrous or finely stellate-hairy. **Leaves:** stipules oblong to elliptic or obovate, 6 × 2.5 mm, <base broad, apex rounded and erose to apiculate or acute>; petiole 1–2.5(–3) cm; blade broadly ovate to orbiculate, 2–6 × 2.5–6 cm, sometimes wider than long, base truncate to slightly cordate, 3- or 5-lobed, <clefts between lobes may be shallow to deep>, margins doubly crenate to doubly serrate, apex rounded, obtuse, or acute, surfaces glabrous or sparsely stellate-hairy (more so on abaxial veins). **Inflorescences** 15–20-flowered, fairly dense, hemispheric racemes, 2.5–4 cm diam.; bracts narrowly obovate to spatulate, 4.5 mm, apex erose or acute. **Pedicels** 8–15 mm, densely stellate-hairy. **Flowers** 5–8 mm diam.; hypanthium campanulate, 1.5–2 mm, densely stellate-hairy; sepals triangular to ovate, 2–3 mm, apex gland-tipped, surfaces densely stellate-hairy; petals white, broadly elliptic to obovate or orbiculate, 4.5 × 4.5 mm; stamens <ca. 30>, equal to or slightly exceeding petals; carpels 2(or 3), connate at least ½ their lengths, densely stellate-hairy. **Follicles** 2(or 3), ovoid, flattened, 2.5 mm (lengths not exceeding sepals), <keeled apically>, densely stellate-hairy, <hairs white>; styles 2.5–3 mm. **Seeds** 1 or 2, <pyriform, 1.5 mm>.

Flowering May–Jul; fruiting Aug. Rocky canyon slopes and cliffs, open grassy slopes, dry open forests among *Pinus*, *Juniperus*, *Pseudotsuga*, *Picea*, and *Populus*; 500–2800 m; Alta., B.C.; Colo., Idaho, Mont., Nev., Oreg., Utah, Wash., Wyo.

5. **Physocarpus monogynus** (Torrey) J. M. Coulter, Contr. U.S. Natl. Herb. 2: 104. 1891 • Mountain nine-bark [E]

Spiraea monogyna Torrey, Ann. Lyceum Nat. Hist. New York 2: 194. 1827

Shrubs, 10 dm. **Stems** often decumbent, spreading, <reddish brown becoming dark gray>, glabrous or sparsely stellate-hairy. **Leaves:** stipules linear to narrowly elliptic or subulate, 4–5 × 1 mm, <base attenuate, apex acute>; petiole 0.5–2.2 cm; blade broadly ovate, 1.5–4 × 2–4.5 cm, sometimes wider than long, base usually truncate to cordate, rarely broadly cuneate, 0 or 1–5-lobed, margins doubly crenate to doubly serrate, apex rounded to obtuse, surfaces glabrous or sparsely stellate-hairy. **Inflorescences** 10(–15)–30-flowered, fairly dense, hemispheric racemes, 4 cm diam.; bracts linear to narrowly elliptic or narrowly obovate, 2–4 mm, apex acute, rarely 2-fid. **Pedicels** 8–18 mm, sparsely to moderately stellate-hairy. **Flowers** 8–10 mm diam.; hypanthium cup-shaped, 1.7–2 mm, sparsely to moderately stellate-hairy; sepals triangular to ovate, 3 mm, apex gland-tipped, surfaces sparsely to moderately stellate-hairy; petals white, broadly elliptic to orbiculate, 4.5 × 4 mm; stamens <20–40>, ± equal to petals; carpels (1 or)2(or 3), connate at least ½ their lengths, densely stellate-hairy. **Follicles** (1 or)2(or 3), inflated, 2.5 mm (lengths shorter than sepals), densely stellate-hairy, <hairs white>; styles 3.5 mm. **Seeds** 1–3, <pyriform, 2–2.5 mm>. $2n = 18$.

Flowering May–Aug; fruiting Jul–Oct. Open rocky wooded slopes, seepage ledges, canyons; 1800–2600 m; Ariz., Colo., Mont., Nev., N.Mex., Okla., S.Dak., Tex., Utah, Wyo.

6. **Physocarpus alternans** (M. E. Jones) J. T. Howell, Proc. Calif. Acad. Sci., ser. 4, 20: 130. 1931 • Nevada or dwarf nine-bark [E]

Neillia monogyna (Torrey) Greene var. *alternans* M. E. Jones, Zoë 4: 42. 1893; *Physocarpus alternans* subsp. *annulatus* J. T. Howell; *P. alternans* subsp. *panamintensis* J. T. Howell

Shrubs, 15 dm, <densely branched>. **Stems** orientation not recorded, <brown becoming gray>, stellate-hairy, glabrescent. **Leaves:** stipules oblong to narrowly elliptic or rhombic, 2–4 × 1.5–2 mm; petiole 2–10 mm; blade broadly ovate to orbiculate, 0.5–2 × 0.5–2 cm, slightly longer than wide, base rounded to truncate or slightly cordate, (0)3- or 5(7)-lobed, margins crenate to doubly crenate, apex rounded, abaxial surface more densely stellate-hairy (especially on veins), adaxial sparsely so. **Inflorescences** 5–10(–15)-flowered, open, umbel-like racemes, 3 cm diam.; bracts similar to stipules. **Pedicels** 0.6–1.5 cm, densely stellate-hairy. **Flowers** 7 mm diam.; hypanthium cup-shaped to campanulate, 1.5–3 mm, stellate-hairy; sepals <pale green to white, darker in center>, triangular to narrowly ovate, 2.5–3 mm, apex gland-tipped, surfaces stellate-hairy; petals white to pale pink, broadly obovate to orbiculate, 3 × 2.5 mm; stamens <20–30>, unequal (alternately long and short), shorter than petals, <anthers pink to dark brown>; carpels 1(or 2), if 2 then connate at least ½ their lengths, indumentum not recorded. **Follicles** 1(or 2), <pale buff>, shape not recorded, 3–6 mm (lengths equal to or slightly exceeding sepals), densely stellate-hairy; styles 2.5 mm <sometimes distinctly bent>. **Seeds** 1 or 2, <2 mm>.

Flowering Jun–Jul; fruiting Aug. Dry rocky canyon slopes, among pinyon pine, juniper, and *Cercocarpus* woodlands; 1500–3100 m; Calif., Colo., Idaho, Nev., Utah, Wyo.

The two subspecies described from California by Howell appear to be local variants with relatively small differences in leaf and hypanthium indumentum that probably do not merit formal recognition. Being drought tolerant, *Physocarpus alternans* should be considered more often for planting in dry habitats.

33. NEILLIA D. Don, Prodr. Fl. Nepal., 228. 1825 • [For Patrick Neill, 1776–1851, Scottish printer, naturalist, and secretary of the Caledonian Horticultural Society] [I]

Alan S. Weakley

Robert A. S. Wright

Stephanandra Siebold & Zuccarini

Shrubs, 2.5–25[–40] dm. **Stems** 1–20+, arching, sometimes rooting at tips; bark red-brown [to greenish]; glabrous or slightly hairy [to densely yellow-brown pubescent when young]. **Leaves** deciduous, cauline; stipules persistent [deciduous], ovate-lanceolate to elliptic, margins remotely to strongly serrate; petiole present; blade ovate, 2–6[–12] cm, herbaceous, margins flat, incised or lobed, lobes [3–]7–11, poorly or well developed, singly or doubly serrate, venation palmate, abaxial surface glabrous or sparsely hairy (especially on main veins), adaxial glabrous or sparsely hairy. **Inflorescences** terminal, sometimes also axillary, 3–70+-flowered, racemes or panicles to corymbs; bracts present, <deciduous>; bracteoles absent; <peduncles present>. **Pedicels** present. **Flowers** [3–]4–5[–10] mm diam.; hypanthium cup-shaped [obconic, campanulate], [2–]3–5[–12] mm, glabrous or sparsely hairy [glandular-pubescent, densely hairy]; sepals 5, ascending, triangular to oblong [elliptic]; petals 5, white [pink or reddish], obovate; stamens [5–]10[–30] in 1[or 2] whorl(s), shorter than [approximately equal to] petals; carpels 1[–5]; ovules 2[–10]. **Fruits** follicles [aggregated], 1, obliquely subglobose [ellipsoid or cylindric], 2–3[–5] mm diam., dehiscent from base, <enclosed in or protruding from hypanthium>, glabrous or sparsely to moderately pilose [glandular-pubescent, densely hairy]; hypanthium persistent; sepals persistent, ascending to erect. **Seeds** 1[–4]. $x = 9$.

Species 15–20 (1 in the flora): introduced, Virginia; Asia; introduced also in Europe.

Neillia has traditionally been considered to be one of three genera of Neillieae, the others being *Physocarpus* and *Stephanandra*. Few, if any, consistent morphologic differences exist between *Neillia* and *Stephanandra*, and a molecular phylogenetic analysis by Oh S. H. and D. Potter (2005) strongly suggested that *Stephanandra* originated from within *Neillia*, perhaps by hybridization of the two main lineages. The authors follow Oh (2006) in treating *Neillia* more broadly to include *Stephanandra*. Various members of *Neillia* are cultivated horticulturally in North America. Only *N. incisa* appears to have naturalized (and apparently only sparingly), but other species have the potential to do so.

SELECTED REFERENCE Oh, S. H. 2006. *Neillia* includes *Stephanandra*. Novon 16: 91–95.

1. Neillia incisa (Thunberg) S. H. Oh, Novon 16: 92. 2006 • Lace shrub, cutleaf stephanandra [F] [I]

Spiraea incisa Thunberg in J. A. Murray, Syst. Veg. ed. 14, 472. 1784; *Stephanandra incisa* (Thunberg) Siebold & Zuccarini ex Zabel

Leaves: stipules prominent. **Inflorescences:** peduncles glabrous or sparsely to densely pubescent. **Pedicels** 5–8 mm, pubescent. **Flowers:** sepals distinct or basally connate. **Follicles** pilose. **Seeds** ovoid, shiny. $2n = 18$ (Asia).

Flowering May; fruiting Jul–Sep. Moist forests, forest edges; 100–200 m; introduced, Va.; e Asia (Japan, Korea, Taiwan).

Neillia incisa is commonly cultivated and is promoted in the nursery trade as an attractive, deer-resistant, and easy-to-grow ornamental shrub. The establishment and spread of *N. incisa* in Richmond, Virginia, shows its ability to naturalize and occupy shaded sites in moist temperate forests of eastern North America. Field observations indicate local spread by proliferous tip sprouting. Despite the paucity of documented occurrences of its naturalization, it can be expected to naturalize more broadly in eastern North America and may prove to be locally invasive. 'Crispa' is one of the more popular cultivars and is shorter in stature (less than 1 m) than the rest of the species.

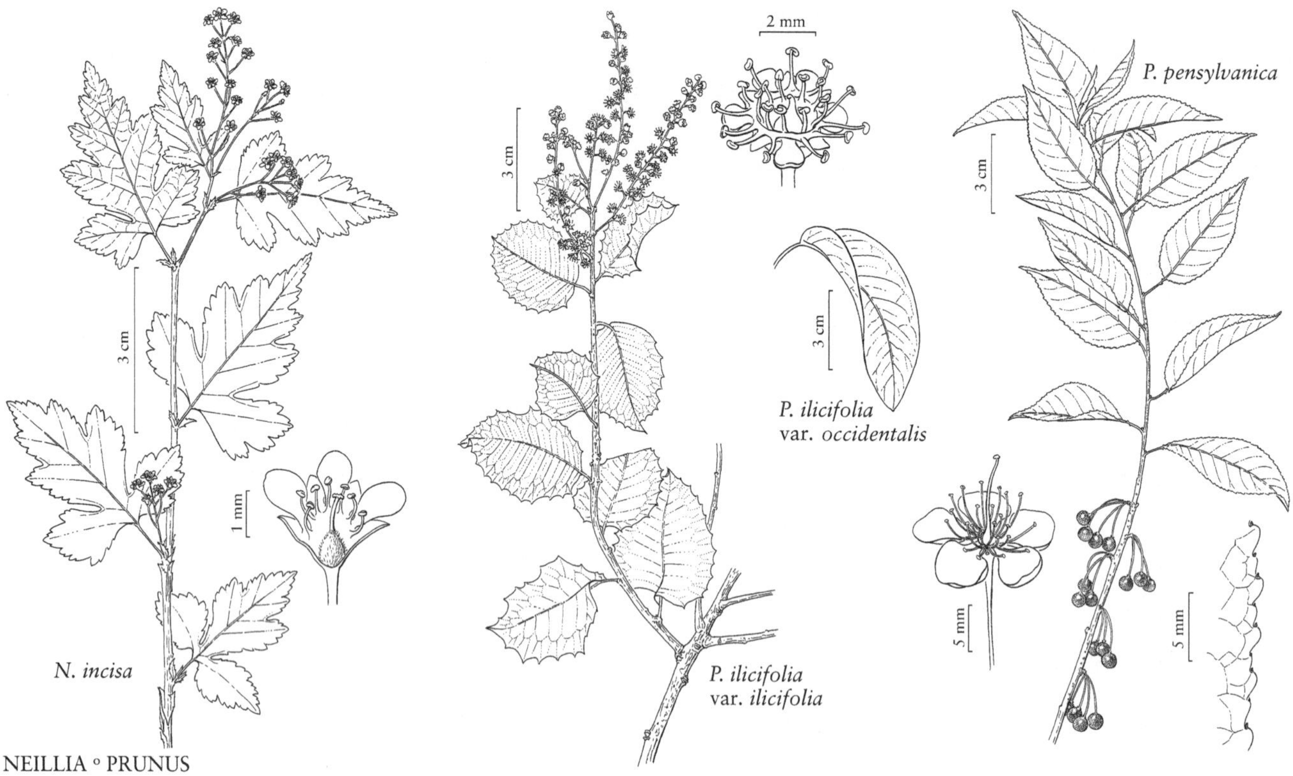

NEILLIA ◦ PRUNUS

c3. ROSACEAE Jussieu (subfam. AMYGDALOIDEAE) tribe AMYGDALEAE de Candolle in A. P. de Candolle and A. L. P. P. de Candolle, Prodr. 2: 529. 1825

Luc Brouillet

Shrubs or trees; sometimes armed. **Leaves** alternate, simple; stipules deciduous, free; venation pinnate. **Flowers:** perianth and androecium perigynous; epicalyx bractlets absent; hypanthium cup-shaped, obconic, campanulate, or tubular; torus absent; carpel 1, distinct, free, style terminal, distinct; ovules 2 (1 abortive), apical, collateral (obturator present). **Fruits** drupes; styles deciduous, not elongate.

Genus 1, species 200+ (44 in the flora): North America, Mexico, Central America, South America, Eurasia, Africa, Australia; widely introduced.

The base chromosome number for Amygdaleae is $x = 8$.

34. PRUNUS Linnaeus, Sp. Pl. 1: 473. 1753; Gen. Pl. ed. 5, 213. 1754 • Plum, cherry, prunier, cerisier [Greek *Prunum*, plum]

Joseph R. Rohrer

Amygdalus Linnaeus; *Armeniaca* Scopoli; *Cerasus* Miller; *Lauro-cerasus* Duhamel; *Padus* Miller; *Persica* Miller

Shrubs or trees, sometimes forming clonal thickets, 1–400 dm, glabrous or hairy. **Stems** 1–20+; bark reddish, reddish brown, gray-brown, or dark gray; long and short shoots usually present; thorns present or absent. **Leaves** deciduous or persistent, cauline; stipules caducous, linear to lanceolate, margins toothed to lobed, usually glandular; petiole present or absent, <usually glandular near blade>; blade elliptic, oblong, suborbiculate, ovate, lanceolate, linear, obovate, oblanceolate, spatulate, fan-shaped, or rhombic, seldom folded along midribs, 0.5–18 cm, membranous to leathery, margins flat, usually entire or toothed, sometimes undulate, <teeth usually glandular, sometimes eglandular>. **Inflorescences** terminal <on short shoots or from axils of previous year's leaves>, 1–64(–90)[–100]-flowered, racemes, corymbs, umbellate fascicles, 2-flowered fascicles, or solitary; bracts sometimes present; bracteoles present. **Pedicels** usually present, sometimes absent. **Flowers** usually bisexual, sometimes unisexual (then plants usually dioecious, sometimes andropolygamous), blooming before or at leaf emergence, 4–40 mm diam.; hypanthium 1.5–8 mm, exterior glabrous or hairy; sepals 5, erect to reflexed, usually triangular, semicircular, ovate, or oblong, rarely ovate-elliptic, lanceolate, or obovate; petals 5(–50+ in doubled ornamentals), usually white to pink or dark pink, sometimes yellowish, usually suborbiculate to elliptic or obovate, sometimes oblong, rarely ovate, oblanceolate, or rhombic, <base usually clawed>; stamens 10–30, usually shorter than or equal to petals, sometimes longer. **Drupes** 1, greenish yellow to yellowish or orange to bright or dark red, reddish brown, or dark purple to black, globose to ovoid, ovoid-oblong, ellipsoid, or obovoid, 5–30(–80) mm; hypanthium deciduous, rarely persistent in fruit; sepals falling with hypanthium; <mesocarps usually fleshy, sometimes leathery to dry, rarely splitting along suture to reveal stone; endocarps forming globose to ovoid or ellipsoid to fusiform stones, sometimes flattened laterally>. **Seed** 1. $x = 8$.

Species 200+ (44 in the flora): North America, Mexico, Central America, South America, Eurasia, Africa, Australia; most abundant in north-temperate regions.

Prunus is important economically; it includes almonds, apricots, cherries, nectarines, peaches, and plums. Most commercial species are of Old World origin; Native Americans made use of *Prunus* fruits, especially plums, long before the Eurasian species were introduced. North American *Prunus* was consumed fresh, dried, and prepared with meat to form pemmican. Even though native North American plums have never rivaled those of Eurasia for commercial significance, some species were sources for hundreds of named cultivars selected and developed by settlers, farmers, and horticulturists in the late nineteenth and early twentieth centuries (U. P. Hedrick 1911). Luther Burbank also made use of native plums in carrying out crosses with *P. salicina* Lindley that led to most of the commercial plum crop grown in California. Small-scale research into the commercial potential of native plums continues today (W. R. Okie 2001).

European and Asian species of *Prunus* have been introduced into North America for their ornamental beauty and are used in the landscape as specimen and accent trees, foundation shrubs, hedges, and screens. Their fleshy fruits are favorites of birds and other frugivores, which disperse the seeds widely. Some species escape cultivation and their seedlings appear nearby, especially in disturbed soil. Included in this treatment are introduced species that not only escape frequently but also have become naturalized in the flora area, reproducing on their own. The following species are cultivated in North America and have been reported as rarely escaped or persisting; they are excluded here because they do not yet appear to be naturalized in the flora area: *P. fruticosa* Pallas, *P. mume* Siebold & Zuccarini, and *P. triloba* Lindley.

The taxonomic history of *Prunus* is long and complicated, in part due to the economic value of its fruit crops and also the ease with which some species hybridize. Here, *Prunus* is circumscribed in its broad sense based on the argument that when viewed on a worldwide scale, the morphologic discontinuities among the segregate genera diminish, and they overlap with one another (R. McVaugh 1951). Included here are species that other botanists, from Linnaean times to the present (for example, Wu Z. and P. H. Raven 2003, vol. 9), have placed in *Amygdalus*, *Armeniaca*, *Cerasus*, *Lauro-cerasus*, *Padus*, and *Persica*. Systems of subgeneric classification have been proposed by those treating *Prunus* in the broad sense; as K. R. Robertson

(1974) concluded, no consensus was reached on how many subgenera or sections should be recognized, or which species should be placed in which taxa. The most widely followed system has been that of A. J. Rehder (1940), who recognized five subgenera: *Amygdalus*, *Cerasus*, *Lauro-cerasus*, *Padus*, and *Prunophora* [= *Prunus*]. Phylogenetic studies employing DNA sequences from chloroplast and nuclear genes (E. Bortiri et al. 2001, 2002; S. Lee and J. Wen 2001) have shown that the five subgenera accepted by Rehder and his placement of individual species within them are not supported. So far, an alternative infrageneric classification of *Prunus* that accounts for the majority of its species has yet to be proposed. Character reconstruction, or the mapping of morphological characters onto phylogenetic trees based in large part on molecular data, shows that there is considerable homoplasy in most of the characters previously used to delimit subgenera, sections, and species (J. Shaw and R. L. Small 2004; Bortiri et al. 2006). Few new morphological characters useful in writing a key to the clades resolved by molecular phylogenetics have emerged. *Prunus* is presented here without division into subgenera or sections.

At the species level, *Prunus* has been the object of the usual combining and splitting common among taxonomists with different philosophies and opinions. In particular, over-reliance on the indument of various vegetative and floral parts has led to the naming of numerous species and infraspecific taxa. Similarly, too much has been made of fruit color and palatability in naming taxa of *Prunus*. Fruits of some cherries and plums go through a progression of color changes as they mature. A typical series is green to red to purple to black. Often, an individual tree whose fruits would be nearly black when fully ripe has its fruits removed by birds when they are still red or purple. These same fruits, which may be highly astringent when red or purple, are often more palatable to humans when dark purple to black. In general, a conservative approach toward species circumscription is taken here, one that in many cases reflects current usage. Widespread hybridization among plums has made their taxonomy particularly troublesome. J. Shaw and R. L. Small (2005) have shown that chloroplast DNA data provide resolution for *P. subcordata* and *P. texana*, while the rest of the North American plums form three primary clades with little resolution within each clade and no resolution among them. Multiple accessions of most eastern species were scattered on more than one clade. Analysis of sequences from nuclear genes has not done much to resolve relationships among the plums (J. R. Rohrer et al. 2008). Surely, as more molecular and genetic data are analyzed and, more importantly, correlated with morphological data, circumscriptions will be redrawn and the number of North American plum species further reduced.

Measurements of leaf size are from leaves on both long and short shoots. The larger values, especially those in parentheses, are found only on long shoots, and exceptionally fast-growing long shoots may have leaves that occasionally exceed the dimensions given. In the majority of *Prunus*, a true terminal bud is formed at the end of each vegetative shoot. In subgenera *Emplectocladus* Torrey and *Prunus*, the apical meristem aborts and leaves a nib of stem tissue soon surpassed distally by an enlarged axillary bud that assumes the end bud position and role in continuing longitudinal growth of the shoot during the next growing season. Such axillary end buds are called false terminal buds, or the twigs are described as lacking terminal buds. For clarity, the descriptions state that twigs have either terminal end buds or axillary end buds. The lengths of racemes were measured from the base of the peduncle to the distal end of the rachis (= central axis of inflorescence). For racemes terminal on leafy shoots, the length was measured from the node of the most distal leaf not subtending a flower. The number of flowers per umbellate fascicle was counted as the number of flowers emerging from one bud, rather than from a node or spur. Reports of more than six flowers in an umbellate fascicle are probably the result of counting flowers from more than one bud. Fruit sizes given in the descriptions are for fresh fruits. As fruits dry, they shrink to little more than the size of the stone.

SELECTED REFERENCES Bortiri, E. et al. 2001. Phylogeny and systematics of *Prunus* (Rosaceae) as determined by sequence analysis of ITS and the chloroplast *trn*L-*trn*F spacer DNA. Syst. Bot. 26: 797–807. Bortiri, E. et al. 2002. The phylogenetic utility of nucleotide sequences of sorbitol 6-phosphate dehydrogenase in *Prunus* (Rosaceae). Amer. J. Bot. 89: 1697–1708. Bortiri, E., B. D. Vanden Heuvel, and D. Potter. 2006. Phylogenetic analysis of morphology in *Prunus* reveals extensive homoplasy. Pl. Syst. Evol. 259: 53–71. Hedrick, U. P. 1911. The Plums of New York. Albany. Lee, S. and J. Wen. 2001. A phylogenetic analysis of *Prunus* and the Amygdaloideae (Rosaceae) using ITS sequences of nuclear ribosomal DNA. Amer. J. Bot. 88: 150–160. Mason, S. C. 1913. The pubescent-fruited species of *Prunus* of the southwestern states. J. Agric. Res. 1: 147–178. McVaugh, R. 1951. A revision of the North American black cherries (*Prunus serotina* Ehrh., and relatives). Brittonia 7: 279–315. Shaw, J. and R. L. Small. 2005. Chloroplast DNA phylogeny and phylogeography of the North American plums (*Prunus* subgenus *Prunus* section *Prunocerasus*, Rosaceae). Amer. J. Bot. 92: 2011–2030. Wight, W. F. 1915. Native American species of *Prunus*. Bull. U.S.D.A. 179: 1–75.

1. Inflorescences 12–64(–90)-flowered, racemes (central axis lengths 4–25 times pedicels).
    2. Leaves deciduous; racemes leafy at bases.
        3. Leaf margins crenulate-serrulate to serrate, teeth incurved or appressed; lateral veins 15–30 per side, flush abaxially; sepal margins usually entire, rarely glandular-toothed; hypanthia persistent . . . . 6. *Prunus serotina*
        3. Leaf margins serrulate to serrate, teeth ascending to spreading; lateral veins 6–18 per side, raised abaxially; sepal margins usually toothed, erose; hypanthia deciduous (leaving discs at bases of drupes).
            4. Petals 2–5(–7) mm; sepals 0.7–1.4 mm (lengths equal to widths); stones ± smooth . . . . 7. *Prunus virginiana*
            4. Petals (5–)6–9 mm; sepals 1.2–2 mm (lengths greater than widths); stones rugulose . . . . 8. *Prunus padus*
    2. Leaves persistent; racemes leafless at bases.
        5. Leaf abaxial surfaces eglandular (if glands present, restricted to margins).
            6. Racemes: central axes 30–80 mm; leaf margins spinose-dentate to spinose-serrulate or entire; drupes 12–25 mm . . . . 3. *Prunus ilicifolia*
            6. Racemes: central axes 100–280 mm; leaf margins crenate-dentate; drupes 8–12 mm . . . . 5. *Prunus lusitanica*
        5. Leaf abaxial surfaces glandular, glands 1–several, proximal, flat, circular to oval.
            7. Leaf apices acute to acuminate, apicula obtuse; flowering Nov–Jan . . . . 2. *Prunus myrtifolia*
            7. Leaf apices usually acute to short-acuminate, sometimes obtuse-apiculate, apicula acute; flowering Feb–May.
                8. Racemes: central axes 13–30(–43) mm; petals 1–1.5 mm; leaf blade margins entire or spinose-serrate; drupes 9–12 mm . . . . 1. *Prunus caroliniana*
                8. Racemes: central axes (35–)55–130 mm; petals 3–5 mm; leaf blade margins remotely serrulate or nearly entire; drupes 13–17 mm . . . . . . 4. *Prunus laurocerasus*
1. Inflorescences 1–12-flowered, umbellate fascicles, corymbs, or racemes (central axis lengths 0–3.5 times pedicels).
    9. Specimens with leaves and fruits.
        10. Drupes hairy.
            11. Trees, 30–100 dm; drupes 25–80 mm.
                12. Leaf blades broadly ovate to suborbiculate, abaxial surfaces with tufts of hairs in vein axils; stones not pitted . . . . 27. *Prunus armeniaca* (in part)
                12. Leaf blades lanceolate to oblong, abaxial surfaces glabrous; stones pitted.
                    13. Petioles 5–10(–15) mm, not winged; leaf blades (50–)70–150 mm, folded along midribs; drupe mesocarps fleshy . . . . 21. *Prunus persica* (in part)
                    13. Petioles (8–)10–25 mm, usually winged distally; leaf blades 25–100 mm, not folded along midribs; drupe mesocarps leathery . . . . 22. *Prunus dulcis* (in part)
            11. Shrubs, 5–30 dm (–40 dm in *P. fremontii*, –60 dm in *P. subcordata*); drupes 7–25 mm.
                14. Drupe mesocarps fleshy.
                    15. Pedicels 5–15 mm; drupes 15–25 mm; petioles 4–18 mm, usually glandular distally. . . . . 32. *Prunus subcordata* (in part)
                    15. Pedicels 0–5 mm; drupes 8–15 mm; petioles 1–7 mm, eglandular.

16. Leaf blades 1.7–3.5(–4.7) cm wide, margins singly to doubly serrate, teeth acute, usually eglandular, sometimes glandular, glands spheric; drupes red, sparsely hairy . . . . .23. *Prunus tomentosa* (in part)
16. Leaf blades 0.4–1.1 cm wide, margins dentate, teeth obtuse, glandular, glands discoid; drupes usually yellow to greenish yellow, sometimes tinged with red, velutinous . . . . 33. *Prunus texana* (in part)

14. Drupe mesocarps leathery to dry.
17. Leaf bases long-attenuate; blades linear, narrowly elliptic, obovate, oblanceolate, or spatulate, (lengths 2.2–10 times widths).
18. Pedicels 0–4 mm; blades 0.1–0.2(–0.4) cm wide, margins nearly entire or obscurely and remotely serrulate in distal 1/3 . . . . .17. *Prunus fasciculata* (in part)
18. Pedicels (1–)4–12 mm; blades 0.2–0.6 cm wide, margins serrulate (sometimes obscurely so) . . . . .24. *Prunus andersonii* (in part)
17. Leaf bases cuneate, obtuse, rounded, truncate, or subcordate; blades elliptic, ovate, obovate, spatulate, suborbiculate, rhombic, or fan-shaped, (lengths 1–2.6 times widths).
19. Twigs glabrous; pedicels 2–12 mm, glabrous . . . . 25. *Prunus fremontii* (in part)
19. Twigs canescent or puberulent; pedicels 0–3 mm, puberulent.
20. Leaf blades rhombic, obovate, or fan-shaped, bases broadly obtuse or rounded to nearly truncate, margins serrate or dentate in distal 1/2 . . . . 19. *Prunus havardii* (in part)
20. Leaf blades elliptic, ovate, obovate, or spatulate, bases cuneate to obtuse, margins usually entire, sometimes irregularly serrulate (sometimes dentate on long shoots), or irregularly serrate.
21. Leaf blades and petioles glabrous. . . . . 18. *Prunus minutiflora* (in part)
21. Leaf blades and petioles hairy . . . . 20. *Prunus eremophila* (in part)

[10. Shifted to left margin.—Ed.]

10. Drupes glabrous.
22. Twigs with terminal end buds; plants not thorny; stones not or slightly flattened (without lateral edges).
23. Leaf blade margins toothed in distal 1/2–2/3 . . . . 26. *Prunus pumila* (in part)
23. Leaf blade margins toothed to base.
24. Leaf blade margins usually doubly, sometimes singly, serrate, teeth sharp to aristate.
25. Twigs, petioles, and leaf blades glabrous; leaf apices caudate. . . 14. *Prunus speciosa* (in part)
25. Twigs and petioles hairy, leaf blades hairy abaxially along midribs and veins; leaf apices acuminate.
26. Petioles 10–20 mm; leaf bases rounded; inflorescences corymbs . . . . 15. *Prunus yedoensis* (in part)
26. Petioles 5–10 mm; leaf bases obtuse; inflorescences umbellate fascicles . . . . 16. *Prunus subhirtella* (in part)
24. Leaf blade margins singly or doubly crenulate-serrulate to crenate-serrate, teeth blunt.
27. Leaf apices usually rounded to obtuse, rarely acute . . . . 11. *Prunus emarginata* (in part)
27. Leaf apices abruptly acuminate, acuminate, or acute.
28. Twigs densely puberulent; leaf blades 1.9–4.5 cm (lengths 1–1.5 times widths) . . . . 9. *Prunus mahaleb* (in part)
28. Twigs glabrous; leaf blades (2.5–)4.5–14 cm (lengths 1.5–4.3 times widths).
29. Petioles 1–6 mm; shrubs, 5–20 dm . . . . 31. *Prunus glandulosa* (in part)
29. Petioles 7–40 mm; shrubs or trees, 20–200 dm.
30. Leaf blades lanceolate, oblong-lanceolate, or elliptic; drupes 6–10 mm . . . . .10. *Prunus pensylvanica* (in part)
30. Leaf blades broadly elliptic, elliptic-obovate, oblong, ovate, or obovate; drupes 13–30 mm.

31. Leaf abaxial surfaces moderately hairy, especially along midribs and veins, blades (4–)7–14 cm; petioles (14–)20–40 mm, with 1–3 glands distally and/or glands on margins at bases of blades . . . . 12. *Prunus avium* (in part)
31. Leaf abaxial surfaces glabrous or glabrate, blades 4.4–6(–8) cm; petioles 10–24 mm, usually eglandular, sometimes with marginal glands at bases of blades . . . . 13. *Prunus cerasus* (in part)

[22. Shifted to left margin.—Ed.]

22. Twigs with axillary end buds; plants sometimes thorny; stones slightly to ± or strongly flattened (with lateral edges).
    32. Leaf blade margins: teeth sharp, usually eglandular, sometimes with glands (especially near bases of blades).
        33. Leaf blades: margins coarsely, doubly serrate, apices usually acuminate to abruptly acuminate, rarely acute; drupes 15–30 mm, stones strongly flattened.
            34. Leaf surfaces: abaxial glabrous or sparsely hairy along main veins; petioles usually hairy adaxially, rarely glabrous, usually eglandular, sometimes glandular distally . . . . 35. *Prunus americana* (in part)
            34. Leaf surfaces: abaxial moderately to densely hairy; petioles hairy on both surfaces, usually glandular distally (sometimes on margins at bases of blades).
                35. Shrubs or trees, suckering; twigs usually hairy, sometimes glabrous; leaf bases usually cuneate to obtuse, sometimes rounded, (blade lengths 1.9–2.6 times widths). . . . . 35. *Prunus americana* (in part)
                35. Trees, rarely suckering; twigs usually glabrous, sometimes hairy; leaf bases usually obtuse to rounded, sometimes subcordate, (blade lengths 1.6–2.1 times widths) . . . . 36. *Prunus mexicana* (in part)
        33. Leaf blades: margins finely and usually singly, sometimes doubly, serrulate, sometimes doubly serrate, apices usually obtuse to acute, sometimes short-acuminate, rarely rounded; drupes 9–18 mm, stones slightly to ± flattened.
            36. Twigs and pedicels usually glabrous, sometimes hairy; leaf apices usually acute, sometimes short-acuminate, (blade lengths 2–2.7 times widths); shrubs or trees, 10–60 dm; Michigan, Appalachian Mountains, se coastal plain and piedmont . . . . 40. *Prunus umbellata* (in part)
            36. Twigs and pedicels usually densely, sometimes hairy, glabrate; leaf apices usually acute to obtuse, rarely rounded, (blade lengths 1.6–2.1 times widths); shrubs, 3–25 dm; Atlantic seaboard or s Great Plains.
                37. Leaf adaxial surfaces glabrous; Atlantic seaboard . . . . 38. *Prunus maritima* (in part)
                37. Leaf adaxial surfaces finely hairy; s Great Plains . . . . 39. *Prunus gracilis* (in part)
    32. Leaf blade margins: teeth blunt, usually glandular (glands sometimes deciduous, each leaving scar on tooth).
        38. Twigs hairy.
            39. Leaf blades: largest 1–5(–6.5) cm, apices rounded to acute (if blades longer than 40 mm, apices obtuse or rounded).
                40. Pedicels 5–15 mm; petioles 4–18 mm. . . . . 32. *Prunus subcordata* (in part)
                40. Pedicels 0–5(–8) mm; petioles 3–7 mm.
                    41. Leaves: surfaces hairy (especially along midribs and veins), blades 10–22 mm wide; petioles hairy, eglandular; drupes bluish black, 10–15 mm. . . . . 29. *Prunus spinosa* (in part)
                    41. Leaves: surfaces glabrous, blades 4–13 mm wide; petioles usually hairy adaxially, sometimes with glands distally; drupes reddish, 12–25 mm . . . . 34. *Prunus geniculata* (in part)
            39. Leaf blades: largest 5–11 cm, apices usually acute to acuminate, sometimes obtuse (in *P. domestica*).
                42. Leaf blades lanceolate to elliptic or ovate, (lengths 2–3 times widths), usually folded along midribs; drupes 10–18 mm . . . . 44. *Prunus murrayana* (in part)
                42. Leaf blades broadly elliptic, elliptic, or obovate, (lengths 1.5–2 times widths), not folded along midribs; drupes 15–35 mm.

43. Leaf blades (2.5–)4–7(–9) cm, apices usually acute to abruptly acuminate, sometimes obtuse . . . . 30. *Prunus domestica* (in part)
43. Leaf blades (5–)7–11 cm, apices abruptly acuminate. . . . . 37. *Prunus nigra* (in part)

38. Twigs glabrous.
44. Leaf blades usually lanceolate, oblanceolate, oblong-lanceolate, or narrowly elliptic, sometimes elliptic or oblong-obovate, (lengths 2–3 times widths), apices acute to long-acuminate.
45. Petioles usually eglandular, sometimes with 1 or 2 glands distally; leaf blades 0.8–2 cm wide, marginal teeth with conspicuous, reddish orange glands . . . . 41. *Prunus angustifolia* (in part)
45. Petioles usually with 1–5 glands distally; leaf blades 1.5–5.5 cm wide, marginal teeth with inconspicuous, blackish glands.
46. Leaf blades folded along midribs; plants usually suckering . . . . 42. *Prunus rivularis* (in part)
46. Leaf blades not folded along midribs; plants rarely suckering . . . . 43. *Prunus hortulana* (in part)
44. Leaf blades broadly elliptic to elliptic, ovate, oblong-ovate, obovate, or suborbiculate, (lengths 1.5–2 times widths), apices rounded, obtuse, acute, or abruptly acuminate.
47. Leaf blades (5–)7–11 cm, apices abruptly acuminate . . . . 37. *Prunus nigra* (in part)
47. Leaf blades 2–7(–9) cm, apices usually acute, obtuse, or rounded, sometimes abruptly acuminate (in *P. domestica*).
48. Leaf blades elliptic, oblong-ovate, or suborbiculate, bases rounded or subcordate, apices usually obtuse to rounded, rarely acute . . . . 32. *Prunus subcordata* (in part)
48. Leaf blades ovate, elliptic, or obovate, bases cuneate to obtuse, apices acute, obtuse, or abruptly acuminate.
49. Leaves: abaxial surfaces hairy along midribs and veins; drupes purple-red to yellow . . . . 28. *Prunus cerasifera* (in part)
49. Leaves: abaxial surfaces hairy (especially along veins); drupes blue-black (green, yellow, or red in cultivars) . . . . 30. *Prunus domestica* (in part)

[9. Shifted to left margin.—Ed.]

9. Specimens with flowers (with or without leaves).
50. Ovaries hairy (or flowers staminate and petals 2–6 mm).
51. Ovaries sparsely hairy or glabrous proximally, villous distally.
52. Inflorescences (3–)6–12-flowered, corymbs or racemes; hypanthia obconic, 2.5–3 mm; sepal margins entire; petals 3–8 mm . . . . 11. *Prunus emarginata* (in part)
52. Inflorescences usually solitary flowers, sometimes 2-flowered fascicles; hypanthia tubular, 3.5–6 mm; sepal margins serrate; petals 9–13 mm . . . . 23. *Prunus tomentosa* (in part)
51. Ovaries usually densely hairy.
53. Trees, 30–100 dm, not thorny; twigs glabrous; hypanthia 4–7 mm.
54. Twigs with axillary end buds; sepals reflexed, margins remotely glandular-toothed, neither ciliate nor tomentose; petals 8–12 mm, white (pink in bud) . . . . 27. *Prunus armeniaca* (in part)
54. Twigs with terminal end buds; sepals erect-spreading to spreading, margins entire, ciliate or tomentose; petals 10–25 mm, dark pink or pink to nearly white.
55. Petals dark pink; sepals abaxially hairy, margins ciliate. . . . . 21. *Prunus persica* (in part)
55. Petals pink to nearly white; sepals abaxially glabrous, margins tomentose . . . . 22. *Prunus dulcis* (in part)
53. Shrubs, rarely trees (*P. subcordata*), (5–)10–40(–60) dm, sometimes thorny; twigs glabrous or hairy; hypanthia 1.5–4 mm.

56. Flowers unisexual; sepal margins entire.
57. Leaf blades 0.1–0.2(–0.4) cm wide, oblanceolate to linear; petioles 0 mm . . . . . 17. *Prunus fasciculata* (in part)
57. Leaf blades 0.2–1(–2.1) cm wide, ovate, elliptic, obovate, spatulate, rhombic, or fan-shaped; petioles 0.5–6 mm.
58. Hypanthial exterior and sepal abaxial surfaces hairy; California . . . . . 20. *Prunus eremophila* (in part)
58. Hypanthial exterior and sepal abaxial surfaces glabrous or glabrate; Texas.
59. Leaf blades elliptic or obovate, margins usually entire, sometimes irregularly serrulate, surfaces glabrous . . . . . 18. *Prunus minutiflora* (in part)
59. Leaf blades rhombic, obovate, or fan-shaped, margins serrate or dentate in distal ½, surfaces puberulent . . . . . 19. *Prunus havardii* (in part)
56. Flowers bisexual; sepal margins glandular-toothed.
60. Hypanthial exterior and sepal abaxial surfaces tomentose; emerging leaf blades tomentose; Texas . . . . . 33. *Prunus texana* (in part)
60. Hypanthial exterior and sepal abaxial surfaces glabrous or hairy; emerging leaf blades usually glabrous, sometimes puberulent; California, Nevada, Oregon.
61. Leaf blades narrowly elliptic, obovate, oblanceolate, or spatulate, bases long-attenuate; petals usually dark pink, sometimes nearly white. . . . . . 24. *Prunus andersonii* (in part)
61. Leaf blades elliptic, ovate, oblong-ovate, or suborbiculate, bases obtuse, rounded, subcordate, or truncate; petals usually white, sometimes pinkish rose.
62. Emerging leaf surfaces glabrous; sepals erect-spreading . . . . . 25. *Prunus fremontii* (in part)
62. Emerging leaf surfaces hairy; sepals broadly spreading to reflexed. . . . . . 32. *Prunus subcordata* (in part)

[50. Shifted to left margin.—Ed.]

50. Ovaries glabrous (or flowers lacking stamens and carpels, and petals 7–12 mm).
63. Inflorescences 2–12-flowered, racemes or corymbs (each with peduncle).
64. Hypanthia 4–8 mm (lengths greater than diams.); sepals 3–8 mm.
65. Pedicels glabrous or sparsely hairy; hypanthia and sepals glabrous; sepal margins entire or toothed, eglandular; plants blooming at leaf emergence . . . . . 14. *Prunus speciosa* (in part)
65. Pedicels, hypanthia, and sepals hairy; sepal margins glandular-toothed; plants blooming before leaf emergence . . . . . 15. *Prunus yedoensis* (in part)
64. Hypanthia 1.8–3 mm (lengths ± equal to diams.); sepals 1.2–2.8 mm.
66. Pedicels (8–)10–30 mm (subtended by minute bracts); central axes 0–8(–24) mm, (pedicels longer than central axes) . . . . . 10. *Prunus pensylvanica* (in part)
66. Pedicels 3–18 mm (subtended by leafy bracts); central axes (5–)8–40 mm, (pedicels shorter than or equal to central axes).
67. Leaves: apices abruptly acuminate, bases usually rounded to truncate, sometimes subcordate; twigs densely puberulent . . . . . 9. *Prunus mahaleb* (in part)
67. Leaves: apices usually rounded to obtuse, rarely acute, bases cuneate; twigs glabrous or hairy . . . . . 11. *Prunus emarginata* (in part)
63. Inflorescences 1–5(–8)-flowered, umbellate fascicles or solitary flowers.
68. Sepals entire or teeth eglandular.
69. Petals 2 mm; pedicels 0–3 mm . . . . . 34. *Prunus geniculata* (in part)
69. Petals 3–15 mm; pedicels 3–52 mm.
70. Sepals oblong, glabrous; twigs with terminal end buds; plants not thorny.
71. Petals 4–7 mm; hypanthia 1.8–3 mm. . . . . . 10. *Prunus pensylvanica* (in part)
71. Petals 12–15 mm; hypanthia 5–7 mm . . . . . 12. *Prunus avium* (in part)
70. Sepals lanceolate to oblong-ovate, ovate, or obovate, hairy (especially proximally on adaxial surfaces); twigs with axillary end buds; plants usually thorny.

[72. Shifted to left margin.—Ed.]

72. Sepals 2.2–4(–5) mm; hypanthia 2–3.5 mm (sepals longer than hypanthia) . . . 32. *Prunus subcordata* (in part)
72. Sepals 1–6 mm; hypanthia 1.5–7 mm (sepals equal to or shorter than hypanthia).
 73. Sepals broadly spreading to reflexed; petals 5–14 mm.
  74. Inflorescences usually solitary flowers, sometimes 2-flowered fascicles; emerging leaves: marginal teeth glandular . . . . 28. *Prunus cerasifera* (in part)
  74. Inflorescences 2–5-flowered (at least some buds 3- or 4-flowered); emerging leaves: marginal teeth eglandular.
   75. Hypanthial exterior and sepal abaxial surfaces glabrous . . . . 35. *Prunus americana* (in part)
   75. Hypanthial exterior (at least distally) and sepal abaxial surfaces hairy.
    76. Shrubs or trees, suckering; twigs usually hairy; pedicels hairy . . . . 35. *Prunus americana* (in part)
    76. Trees, rarely suckering; twigs usually glabrous; pedicels usually glabrous, sometimes sparsely hairy . . . . 36. *Prunus mexicana* (in part)
 73. Sepals erect to spreading; petals 3–8 mm.
  77. Emerging leaves: marginal teeth glandular, glands reddish orange, conic; sepals not 2-fid at apices . . . . 41. *Prunus angustifolia* (in part)
  77. Emerging leaves: marginal teeth eglandular or glandular, glands blackish, spheric; sepals sometimes 2-fid at apices.
   78. Twigs, pedicels, and hypanthial exteriors glabrous . . . . 40. *Prunus umbellata* (in part)
   78. Twigs, pedicels, and hypanthial exteriors hairy.
    79. Hypanthia tubular (often tubular-urceolate when dried); trees or shrubs, 10–60 dm . . . . 40. *Prunus umbellata* (in part)
    79. Hypanthia campanulate; shrubs, 3–25 dm.
     80. Emerging leaves: usually eglandular, rarely glandular; flowering Apr–Jun; Atlantic seaboard . . . . 38. *Prunus maritima* (in part)
     80. Emerging leaves: often with spheric glands on marginal teeth; flowering Mar–Apr; s Great Plains . . . . 39. *Prunus gracilis* (in part)

[68. Shifted to left margin.—Ed.]

68. Sepals glandular-toothed.
 81. Emerging leaves: marginal teeth eglandular; sepals irregularly or obscurely glandular-toothed.
  82. Hypanthial exterior and sepal abaxial surfaces glabrous. . . . . 35. *Prunus americana* (in part)
  82. Hypanthial exterior (at least distally) and sepal abaxial surfaces hairy.
   83. Shrubs or trees, suckering; twigs usually hairy; pedicels hairy . . . . 35. *Prunus americana* (in part)
   83. Trees, rarely suckering; twigs usually glabrous; pedicels usually glabrous, sometimes sparsely hairy . . . . 36. *Prunus mexicana* (in part)
 81. Emerging leaves: marginal teeth glandular; sepals regularly and conspicuously glandular-toothed.
  84. Inflorescences solitary flowers or 2-flowered fascicles.
   85. Petals 25+ (flowers lacking stamens and carpels) . . . . 31. *Prunus glandulosa* (in part)
   85. Petals 5 (flowers with stamens and carpels).
    86. Pedicels 0.5–5(–8) mm; hypanthia 1.5–2.5 mm; petals 4–8 mm; plants thorny . . . . 29. *Prunus spinosa* (in part)
    86. Pedicels (2–)10–20 mm; hypanthia 2–5 mm; petals 7–14 mm; plants not or slightly thorny.
     87. Twigs and pedicels glabrous; sepals 2–4 mm, margins eciliate . . . . 28. *Prunus cerasifera* (in part)
     87. Twigs and pedicels usually hairy; sepals 3.5–6 mm, margins ciliate . . . . 30. *Prunus domestica* (in part)
  84. Inflorescences 1–5-flowered (at least some buds on each plant 3- or 4-flowered).
   88. Twigs with terminal end buds; plants not thorny.
    89. Shrubs, 1–15(–25) dm; hypanthia 1.7–3 mm . . . . 26. *Prunus pumila* (in part)
    89. Shrubs or trees, 30–100 dm; hypanthia 4–7 mm.
     90. Twigs, pedicels, hypanthia glabrous; sepals reflexed . . . . 13. *Prunus cerasus* (in part)
     90. Twigs, pedicels, hypanthia hairy; sepals erect to spreading . . . . 16. *Prunus subhirtella* (in part)

[88. Shifted to left margin.—Ed.]

88. Twigs with axillary end buds; plants usually thorny.
   91. Sepals 2.2–4(–5) mm; hypanthia 2–3.5 mm (sepals longer than hypanthia); California, Oregon . . . . . 32. *Prunus subcordata* (in part)
   91. Sepals 1.5–4(–5) mm; hypanthia 2–4(–5) mm (sepals equal to or shorter than hypanthia); e of Rocky Mountains.
      92. Petals 8–13 mm; hypanthia 3–4(–5) mm, red-tinged . . . . . 37. *Prunus nigra* (in part)
      92. Petals 4–9 mm; hypanthia 2–3 mm, green.
         93. Twigs and pedicels hairy . . . . . 44. *Prunus murrayana* (in part)
         93. Twigs and pedicels glabrous.
            94. Leaf blades lanceolate, oblong-lanceolate, or elliptic to narrowly elliptic, (usually folded along midribs); plants usually suckering; shrubs or trees, 10–80 dm; pedicels 3–15 mm . . . . . 42. *Prunus rivularis* (in part)
            94. Leaf blades narrowly elliptic, lanceolate, oblanceolate, or oblong-obovate, (not folded along midribs); plants rarely suckering; trees, 40–100 dm; pedicels 8–20 mm . . . . . 43. *Prunus hortulana* (in part)

1. **Prunus caroliniana** (Miller) Aiton, Hort. Kew. 2: 163. 1789 • Carolina cherry laurel or laurel cherry, laurier amande [E]

Padus caroliniana Miller, Gard. Dict. ed. 8, Padus no. 6. 1768; *Lauro-cerasus caroliniana* (Miller) M. Roemer

Shrubs or trees, not suckering, 40–120 dm, not thorny. **Twigs** with terminal end buds, glabrous. **Leaves** persistent; petiole 5–8 mm, glabrous, eglandular; blade narrowly elliptic to elliptic or oblanceolate, 5–10 × 1.5–4 cm, base cuneate to obtuse, margins entire or spinose-serrate, sometimes undulate, teeth sharp, eglandular, apex usually acute to short-acuminate, sometimes obtuse-apiculate, <apicula acute>, surfaces glabrous, <abaxial glandular, glands 2, proximal, flat, circular to oval>. **Inflorescences** 12–30-flowered, racemes; central axes 13–30(–43) mm, <leafless at bases>. **Pedicels** 1–4 mm, glabrous. **Flowers** <usually bisexual, proximal sometimes staminate>, blooming before leaf emergence; hypanthium cupulate, 2.5–3 mm, glabrous externally; sepals spreading, semicircular, 0.5–1 mm, margins usually entire, sometimes glandular-toothed, surfaces glabrous; petals white, suborbiculate to elliptic, 1–1.5 mm; ovaries glabrous. **Drupes** black, ovoid, 9–12 mm, glabrous; mesocarps leathery; stones ovoid, not flattened, <usually splitting open>. $2n$ = 32.

Flowering Feb–Apr; fruiting May–Nov. Stream bottoms, thickets, wooded uplands, maritime forests, naturalizing in urban woodlands; 0–200 m; Ala., Ark., Fla., Ga., La., Miss., N.C., S.C., Tex.

Prunus caroliniana is a popular ornamental for screens and trimmed hedges and is widely planted in the southeastern United States because of its lustrous, dark green foliage persistent through the seasons. The species was probably common as a native plant on the southeastern barrier islands; most inland occurrences represent escapes from cultivation. It rarely escapes from cultivation in California.

2. **Prunus myrtifolia** (Linnaeus) Urban, Symb. Antill. 5: 93. 1904 • West Indies or myrtle laurel cherry

Celastrus myrtifolius Linnaeus, Sp. Pl. 1: 196. 1753; *Lauro-cerasus myrtifolia* (Linnaeus) Britton

Trees, not suckering, 60–120 dm, not thorny. **Twigs** with terminal end buds, glabrous. **Leaves** persistent; petiole 8–16 mm, glabrous, eglandular; blade elliptic to broadly elliptic, 4–10 × 2–4.5(–6.5) cm, base cuneate, obtuse, or nearly rounded, margins undulate, entire, apex acute to acuminate, <apicula obtuse>, surfaces glabrous, <abaxial glandular, glands 2, proximal, flat, circular to oval>. **Inflorescences** 12–30-flowered, racemes; central axes (11–)20–50 mm, <leafless at bases>. **Pedicels** (2–)3–6 mm, glabrous. **Flowers** <usually bisexual, proximal sometimes staminate>, blooming before leaf emergence; hypanthium cupulate, 1.5–2.5 mm, glabrous externally; sepals spreading, semicircular, 0.5–0.8 mm, margins usually entire, occassionally with a glandular tooth, surfaces glabrous; petals white, obovate to suborbiculate, 1.5 mm; ovaries glabrous. **Drupes** purple-black, globose to ± ovoid, 8–12 mm, glabrous; mesocarps leathery; stones subglobose, not flattened.

Flowering Nov–Jan; fruiting Mar–Jun. Hammocks, pinelands; 0–10 m; Fla.; Mexico; West Indies; Central America; South America.

Prunus myrtifolia, when compared with *P. caroliniana*, has flowers more widely spaced on longer rachises and pedicels, and flowers in the winter rather than the spring. The leaves of *P. myrtifolia* are broader on average, and their apices are blunt at the tip; the fruits are more rounded at the apices with smaller apicula.

3. Prunus ilicifolia (Nuttall ex Hooker & Arnott) D. Dietrich, Syn. Pl. 3: 43. 1842 • Hollyleaf cherry, islay [F]

Cerasus ilicifolia Nuttall ex Hooker & Arnott, Bot. Beechey Voy., 340. 1839

Shrubs or trees, sometimes suckering, 10–150 dm, not thorny. **Twigs** with terminal end buds, glabrous. **Leaves** persistent; petiole 3–25 mm, glabrous, eglandular; blade oblong-ovate to ovate-lanceolate or ovate to suborbiculate, 1.6–12 × 1.2–5(–7) cm, base usually broadly rounded to subcordate, sometimes obtuse, margins spinose-dentate to spinose-serrulate or entire, sometimes undulate, teeth sharp, callus-tipped, sometimes glandular at leaf base, apex acute to acuminate, surfaces glabrous, <abaxial eglandular, if glands present, restricted to margins>. **Inflorescences** 15–40-flowered, racemes; central axes 30–80 mm, <leafless at bases>. **Pedicels** 1–5 mm, glabrous. **Flowers** blooming before leaf emergence; hypanthium cupulate, 2–3 mm, glabrous externally; sepals erect to spreading, triangular, 0.7–1.2 mm, margins entire, surfaces usually glabrous, rarely with scattered hairs; petals white to yellowish, elliptic, obovate, or suborbiculate, 1–3 mm; ovaries glabrous. **Drupes** dark red to purple or blue-black, obovoid or ellipsoid to globose, 12–25 mm, glabrous; mesocarps leathery; stones subglobose to ellipsoid, ± flattened.

Varieties 2 (2 in the flora): California, nw Mexico.

1. Leaf blades ovate to suborbiculate, margins spinose-dentate to spinose-serrulate; petioles 3–10 mm. . . . . .3a. *Prunus ilicifolia* var. *ilicifolia*
1. Leaf blades oblong-ovate to ovate-lanceolate, margins usually entire, sometimes remotely spinose-serrulate; petioles 8–25 mm . . . . .3b. *Prunus ilicifolia* var. *occidentalis*

3a. Prunus ilicifolia (Nuttall ex Hooker & Arnott) D. Dietrich var. **ilicifolia** [F]

Shrubs or trees, 10–90 dm. **Leaves:** petiole 3–10 mm; blade ovate to suborbiculate, 1.6–6 × 1.2–4 cm, margins undulate, spinose-dentate to spinose-serrulate. **Drupes** dark red to purple, 12–18 mm.

Flowering Apr–May; fruiting Aug–Nov. Chaparral, woodlands, canyons, slopes, and ridges; 0–1600 m; Calif.; Mexico (Baja California).

Variety *ilicifolia* is widely cultivated in California because of its persistent, dark green, hollylike foliage.

3b. Prunus ilicifolia (Nuttall ex Hooker & Arnott) D. Dietrich var. **occidentalis** Brandegee, Proc. Calif. Acad. Sci., ser. 2, 1: 209. 1888 • Catalina cherry [F]

Prunus occidentalis W. S. Lyon, Bot. Gaz. 11: 202. 1886, not Swartz 1788; *P. ilicifolia* subsp. *lyonii* (Eastwood) P. H. Raven; *P. lyonii* (Eastwood) Sargent

Trees, 40–150 dm. **Leaves:** petiole 8–25 mm; blade oblong-ovate to ovate-lanceolate, 4–12 × 1.5–5.5(–7) cm, margins not or slightly undulate, usually entire, sometimes remotely spinose-serrulate. **Drupes** blue-black, 15–25 mm.

Flowering Mar–May; fruiting Aug–Nov. Chaparral, woodlands, canyons; 0–500 m; Calif.; Mexico (Baja California).

Variety *occidentalis* is native on the Channel Islands and occasionally escapes from cultivation on the mainland. Some botanists consider it a subspecies or variety of *Prunus ilicifolia* because the two taxa when grown together hybridize to form viable fertile offspring (P. H. Raven 1963), and since some specimens have more or less toothed leaves, intermediate between the extremes of the two taxa. Others prefer distinct species based on geographic isolation and differences in plant size and habit, leaf margin, petiole length, and fruit color.

Variety *integrifolia* Sudworth is an illegitimate and superfluous name that pertains here.

4. Prunus laurocerasus Linnaeus, Sp. Pl. 1: 474. 1753 • Common cherry laurel, laurel cherry [I]

Shrubs or trees, sometimes suckering, 20–60(–100) dm, not thorny. **Twigs** with terminal end buds, glabrous. **Leaves** persistent; petiole 5–15 mm, glabrous, eglandular; blade elliptic to obovate, 6–18 × 3–7 cm, base cuneate to obtuse, margins remotely serrulate or nearly entire, teeth blunt, glandular, apex abruptly short-acuminate, <apicula acute>, surfaces glabrous, <abaxial glandular, glands 1–several, proximal, flat, circular to oval>. **Inflorescences** 26–32-flowered, racemes; central axes (35–)55–130 mm, <leafless at bases>. **Pedicels** 1–5 mm, glabrous. **Flowers** blooming before leaf emergence; hypanthium cupulate, 3–4 mm, glabrous externally; sepals spreading, triangular, 0.7–1.2 mm, margins usually entire, sometimes with deciduous glands, <ciliate in spots>, surfaces glabrous or hairy; petals white, obovate or broadly elliptic to suborbiculate, 3–5 mm; ovaries glabrous. **Drupes** deep purple-red to nearly black, ovoid to conic-ovoid, 13–17 mm, glabrous; mesocarps fleshy to leathery; stones ovoid, not flattened.

Flowering Mar–May; fruiting Aug–Nov. Riparian thickets, shaded ravines, understory of urban and second-growth forests; 0–600 m; introduced; B.C.; Calif., Oreg., Wash.; Eurasia.

Flowering specimens of *Prunus laurocerasus* with small leaves and entire margins that otherwise resemble *P. caroliniana* can be identified by their larger hypanthia and longer petals. In fruit, the stone is much harder in *P. laurocerasus* and does not split open upon drying; the flesh around the stone is thicker and more succulent.

5. Prunus lusitanica Linnaeus, Sp. Pl. 1: 473. 1753 • Portugal laurel I

Shrubs or trees, not suckering, 30–80(–200) dm, not thorny. **Twigs** with terminal end buds, glabrous. **Leaves** persistent; petiole 14–20 mm, glabrous, eglandular; blade elliptic, oblong-ovate, or ovate-lanceolate, 6–13 × 2.5–7 cm, base obtuse to rounded, margins crenate-dentate, teeth blunt, glandular, apex acuminate, surfaces glabrous, <abaxial eglandular, if glands present, restricted to margins>. **Inflorescences** 18–60[–100] -flowered, racemes; central axes 100–280 mm, <leafless at bases>. **Pedicels** 7–15(–22) mm, glabrous. **Flowers** blooming after leaf emergence; hypanthium cupulate, 2.5–3.5 mm, glabrous externally; sepals spreading to reflexed, semicircular, 1–1.5 mm, margins entire, <± ciliate>, abaxial surface glabrous, adaxial hairy; petals white, suborbiculate to obovate, 4–7 mm; ovaries glabrous. **Drupes** dark purple, ovoid to conic-ovoid, 8–12 mm, glabrous; mesocarps fleshy to leathery; stones ovoid, not flattened. $2n$ = 32, 64.

Flowering May–Jul; fruiting Aug–Nov. Disturbed sites, thickets, urban forests; 0–400 m; introduced; B.C.; Calif., Oreg., Wash.; Europe.

6. Prunus serotina Ehrhart, Gartenkalender 3: 285. 1784, name conserved • Black or rum cherry, cerisier tardif or d'automne, merisier

Shrubs or trees, not suckering, 40–400 dm, not thorny. **Twigs** with terminal end buds, glabrous or hairy. **Leaves** deciduous; petiole 2–23(–30) mm, glabrous or sparsely to densely hairy, usually glandular distally or at petiole-blade junction, glands 1–6; blade usually narrowly elliptic, oblong-elliptic, or obovate, sometimes lanceolate, rarely ovate, 2–13.5 × 1.1–6.5 cm, base cuneate to rounded, margins crenulate-serrulate to serrate, teeth <incurved or appressed>, sharp or blunt, glandular or callus-tipped, apex usually acute to acuminate, sometimes obtuse, rounded to emarginate in var. *alabamensis*, <lateral veins 15–30 per side, flush abaxially>, abaxial surface usually densely hairy along midribs proximally, sometimes glabrous or sparsely hairy, adaxial glabrous. **Inflorescences** 18–55(–90) -flowered, racemes; central axes (25–)35–160 mm, <leafy at bases>. **Pedicels** 1–10 mm, glabrous or hairy. **Flowers** blooming after leaf emergence; hypanthium cupulate, 1.5–3 mm, glabrous externally; sepals erect-spreading to reflexed, semicircular, 0.5–1.5 mm, margins usually entire, rarely glandular-toothed, <rarely ciliate>, surfaces glabrous; petals white, obovate to suborbiculate, 2–4 mm; ovaries glabrous. **Drupes** dark purple to nearly black, globose, 5–10[–25] mm, glabrous; <hypanthium persistent>; mesocarps fleshy; stones subglobose, not flattened.

Varieties 4 (3 in the flora): North America, Mexico, Central America, South America; introduced in Europe.

Variety *capuli* (Cavanilles) Hatusima [subsp. *capuli* (Cavanilles) McVaugh] is known from Mexico (Chiapas, Distrito Federal, Jalisco, México, Puebla), Central America, and South America (Argentina, Bolivia, Colombia, Ecuador, Peru, Venezuela). It is distinguished from var. *serotina* by lanceolate to ovate-lanceolate leaves with decurrent bases (versus elliptic to obovate with obtuse to cuneate bases) and a long (to 2 cm), thin (1 mm) petiole; inflorescences tend to be subtended by 3 or 4 leaves [versus 2 (or 3), reduced in size], with a rachis relatively longer (greater than 15 cm), thicker, and flexuous compared to other varieties, and shorter pedicels (3–5 mm versus 5–10 mm, in var. *serotina*).

The characters given in the key usually allow easy separation of *Prunus serotina* from *P. virginiana*; some specimens from Arizona and southern Utah are difficult to determine. In that area, it is not uncommon for the sepals of *P. virginiana* flowers to lack the glandular teeth that are common elsewhere and thus mimic those of *P. serotina*. Variety *rufula* has thicker leaves that are also much shinier on both surfaces than those of *P. virginiana* and shorter petioles (2–15 mm versus 4–27 mm).

Compounding the confusion with chokecherry is a nomenclatural conundrum in which the epithet *virginiana* has been used for both chokecherry (the next species) and for black cherry (this species). The status of the two names *Prunus serotina* and *P. virginiana* was discussed by K. N. Gandhi et al. (2009), who proposed the conservation of both names; of the two only *P. serotina* has been conserved.

Of the native *Prunus* species only *P. serotina* grows large enough to produce commercial lumber, which is highly prized for its fine grain and rich, warm, reddish brown color.

1. Leaf blades elliptic to obovate, abaxial surfaces sparsely hairy, apices usually obtuse, rounded, or emarginate, sometimes abruptly acute or short-acuminate . . . . 6b. *Prunus serotina* var. *alabamensis*
1. Leaf blades usually elliptic, lanceolate, or oblong, rarely obovate or ovate, abaxial surfaces glabrous or midribs hairy, apices acute to acuminate.
   2. Leaf blades 4–13.5 cm, membranous or slightly leathery; petioles (7–)10–23(–30) mm; e North America . . . . 6a. *Prunus serotina* var. *serotina*
   2. Leaf blades 2–5.2(–7.4) cm, leathery; petioles (2–)4–10(–15) mm; Arizona to w Texas. . . . . 6c. *Prunus serotina* var. *rufula*

6a. Prunus serotina Ehrhart var. **serotina**

Prunus serotina subsp. *eximia* (Small) McVaugh; *P. serotina* var. *eximia* (Small) Little

Twigs usually glabrous, sometimes hairy, hairs rusty brown. **Leaves:** petiole (7–)10–23(–30) mm, glabrous, glandular distally, <glands 1–6>; blade usually narrowly elliptic to elliptic or lanceolate, rarely obovate, 4–13.5 × 1.7–5.8 cm, membranous or slightly leathery, apex acute to acuminate, sometimes abruptly so, abaxial surface glabrous or midribs sparsely to densely hairy proximally. **Inflorescences:** rachises 35–140 mm. $2n = 32$.

Flowering Mar–Jun; fruiting Jun–Sep. Roadsides, edges of fields, secondary forests, open woodlands, ravines, along streams; 0–1800 m; B.C., N.B., N.S., Ont., Que.; Ala., Ark., Conn., Del., D.C., Fla., Ga., Ill., Ind., Iowa, Kans., Ky., La., Maine, Md., Mass., Mich., Minn., Miss., Mo., Nebr., N.H., N.J., N.Y., N.C., Ohio, Okla., Pa., R.I., S.C., Tenn., Tex., Vt., Va., Wash., W.Va., Wis.; Mexico (Chiapas, Chihuahua, Coahuila, Colima [Socorro Island], Durango, Hidalgo, Jalisco, Michoacán, Nayarit, Nuevo León, Oaxaca, Puebla, San Luis Potosí, Tamaulipas, Veracruz); Central America (Guatemala).

Plants of *Prunus serotina* from the Edwards Plateau in central Texas have often been recognized as subsp. or var. *eximia*. If it were not for their geographic isolation from the rest of var. *serotina* by 200 km, it is doubtful that taxonomists would even think about segregating these outliers from the plants of eastern North America. All of the suggested distinctive characteristics of var. *eximia* (glabrous leaves, branchlets, and inflorescences; coarsely toothed leaf margins; long petioles) can be found both separately and combined in specimens from other areas within the main range of var. *serotina*. The trees of the Edwards Plateau seem to be merely a homogeneous subset of the wide-ranging, heterogeneous var. *serotina*.

6b. Prunus serotina Ehrhart var. **alabamensis** (C. Mohr) Little, Phytologia 4: 309. 1953 • Alabama black cherry [E]

Prunus alabamensis C. Mohr, Bull. Torrey Bot. Club 26: 118. 1899; *Padus alabamensis* (C. Mohr) Small; *P. australis* (Beadle) Beadle; *P. cuthbertii* (Small) Small; *Prunus serotina* subsp. *hirsuta* (Elliott) McVaugh

Twigs hairy, often densely so, hairs rusty brown. **Leaves:** petiole 2–12 mm, hairy to glabrate, glandular at petiole-blade junction, <glands 2>; blade elliptic to obovate, 3–12.5 × 2–6.5 cm, ± leathery, apex usually obtuse, rounded, or emarginate, sometimes abruptly acute or short-acuminate, abaxial surface sparsely hairy, especially along midribs and proximally. **Inflorescences:** rachises 50–160 mm.

Flowering Apr–May; fruiting Jun–Aug. Rocky slopes and ridges, sandhills, mixed oak-pine-hickory forests; 20–700 m; Ala., Fla., Ga., S.C.

Variety *alabamensis* has obovate leaf blades, much broader than var. *serotina*. The apices are usually obtuse, rounded, or rounded-emarginate, sometimes short-acute or short-acuminate. Branchlets and, sometimes, raceme axes are rusty-hairy; abaxial surfaces of leaf blades have scattered, rusty hairs. The trees are said to leaf out two weeks later than var. *serotina* and to hold their leaves much longer in the fall (C. L. Brown and L. K. Kirkman 1990).

6c. Prunus serotina Ehrhart var. **rufula** (Wooton & Standley) McVaugh, Brittonia 7: 307. 1951 • Southwestern black cherry

Padus rufula Wooton & Standley, Contr. U.S. Natl. Herb. 16: 132. 1913; *Prunus serotina* subsp. *virens* (Wooton & Standley) McVaugh; *P. serotina* var. *virens* (Wooton & Standley) McVaugh; *P. virens* (Wooton & Standley) Shreve

Twigs glabrous or hairy, hairs rusty brown or gray. **Leaves:** petiole (2–)4–10(–15) mm, glabrous or sparsely to densely hairy, sometimes glandular at petiole-blade junction; blade usually elliptic or oblong, rarely ovate or obovate, 2–5.2(–7.4 on long shoots) × 1.1–2.6(–3.8) cm, leathery, apex usually acute, rarely abruptly acute, abaxial surface usually densely hairy along midribs proximally, sometimes glabrous. **Inflorescences:** rachises (25–)35–70(–102) mm.

Flowering Mar–May; fruiting Jul–Aug. Along streams, moist slopes in canyons, mixed oak-pine-juniper woodlands; 1400–2500 m; Ariz., N.Mex., Tex.; Mexico (Baja California, Chihuahua, Coahuila, Durango, Guanajuato, Guerrero, Jalisco, San Luis Potosí, Sonora).

Black cherries of the semiarid southwestern mountains and canyons (var. *rufula*) have generally smaller, thicker leaf blades on shorter petioles than those of the eastern United States and southeastern Canada. The southwestern plants vary in the degree of indument (hairs often rusty, sometimes lighter on the branchlets and axes of the racemes, and on petioles and along midribs abaxially). The glabrous plants have been segregated by some botanists as var. *virens*; the variation in indument is nearly continuous, and the plants are otherwise similar to hairy specimens.

7. Prunus virginiana Linnaeus, Sp. Pl. 1: 473. 1753 • Chokecherry

Padus virginiana (Linnaeus) Miller

Shrubs or trees, often suckering, 10–60(–100) dm, not thorny. **Twigs** with terminal end buds, glabrous or hairy. **Leaves** deciduous; petiole 4–22(–27) mm, usually glabrous, sometimes hairy, glandular distally, glands 2, <discoid>; blade obovate or elliptic to ovate or oblanceolate, 2.5–9(–11) × 1.2–5 (–6.6) cm, base cuneate to rounded or subcordate, margins serrulate to serrate, teeth <ascending to spreading>, usually sharp, eglandular, sometimes callus-tipped, obscurely glandular, <lateral veins 6–13 per side, raised abaxially>, apex acuminate, acute, or obtuse, abaxial surface glabrous or hairy (especially along midribs), adaxial glabrous. **Inflorescences** 18–64 -flowered, racemes; central axes (18–)40–110(–130) mm, <leafy at bases>. **Pedicels** 2–8(–16) mm, usually glabrous, rarely hairy. **Flowers** blooming after leaf emergence; hypanthium cupulate, 1.5–3 mm, glabrous externally; sepals erect-spreading to reflexed, semicircular, 0.7–1.4 mm, margins erose, usually glandular-toothed, sometimes nearly eglandular, surfaces glabrous; petals white, obovate to suborbiculate, 2–5(–7) mm; ovaries glabrous. **Drupes** red, purple, dark purple, or black, globose, 6–14 mm, glabrous; <hypanthium deciduous, leaving discs at bases of drupes>; mesocarps fleshy; stones subglobose to ellipsoid, not flattened, <± smooth>.

Varieties 2 (2 in the flora): North America, Mexico.

Because of variation in the degree of hairiness on branches and leaves, thickness of leaves, leaf shape, sizes of racemes and flowers, and colors of fruits, *Prunus virginiana* has been split into forms, varieties, and segregate species. When viewed on a continental scale, most of the key characteristics used to separate these taxa intergrade enough so as to be useless.

The most useful distinction is between chokecherries found east of the Rocky Mountains and those found from the mountains westward. They differ in leaf shape and raceme size; the variation is imperfectly correlated. The eastern plants generally have their larger leaves obovate with lengths less than two times widths, racemes typically less than 7 cm, and petals to 4 mm. Plants of the Rockies and west have longer racemes and larger petals, and generally their larger leaves are elliptic to oblanceolate with lengths two or more times widths. Distinctions between the eastern and western varieties are least obvious from Montana to eastern Washington north into British Columbia, where the varietal ranges overlap. Plants of the High Plains show intergradation between those of the Rockies and those of the East. The taxonomic solution provided here is a single species of chokecherry in North America with an eastern and a western variety that overlap in range along a line from northeastern New Mexico to southwestern British Columbia.

1. Racemes (18–)40–70(–95) mm; petals (2–)2.5–4 mm; leaf blades usually obovate, lengths of larger less than 2 times widths .... 7a. *Prunus virginiana* var. *virginiana*
1. Racemes (30–)60–110(–130) mm; petals 4–5(–7) mm; leaf blades elliptic to oblanceolate, lengths of larger at least 2 times widths.............................. 7b. *Prunus virginiana* var. *demissa*

7a. Prunus virginiana Linnaeus var. **virginiana** • Eastern chokecherry, cerisier à grappes or de Virginie [E]

Padus nana (DuRoi) Borkhausen

Leaves: blade usually obovate, sometimes elliptic to ovate, lengths of larger less than 2 times widths. **Inflorescences:** racemes (18–)40–70(–95) mm. **Flowers:** hypanthium 1.5–2 mm; sepals 0.7–1 mm; petals (2–)2.5–4 mm.

Flowering Apr–Jun; fruiting Jul–Sep. Stream banks, fencerows, edge of woods, forests, roadsides, hillsides, canyons, thickets; 0–2600 m; St. Pierre and Miquelon; Alta., B.C., Man., N.B., Nfld. and Labr. (Nfld.), N.W.T., N.S., Ont., P.E.I., Que., Sask.; Ark., Colo., Conn., Del., D.C., Ga., Idaho, Ill., Ind., Iowa, Kans., Ky., Maine, Md., Mass., Mich., Minn., Mo., Mont., Nebr., N.H., N.J., N.Mex., N.Y., N.C., N.Dak., Ohio, Okla., Pa., R.I., S.Dak., Tenn., Tex., Vt., Va., Wash., W.Va., Wis., Wyo.

7b. Prunus virginiana Linnaeus var. **demissa** (Nuttall) Torrey in C. Wilkes et al., U.S. Expl. Exped. 17: 284. 1874 • Western chokecherry

Cerasus demissa Nuttall in J. Torrey and A. Gray, Fl. N. Amer. 1: 411. 1840; *C. demissa* var. *melanocarpa* A. Nelson; *Prunus demissa* (Nuttall) D. Dietrich; *P. demissa* var. *melanocarpa* (A. Nelson) A. Nelson; *P. melanocarpa* (A. Nelson) Rydberg; *P. virginiana* subsp. *demissa* (Nuttall) Roy L. Taylor & MacBryde; *P. virginiana* subsp. *melanocarpa* (A. Nelson) Roy L. Taylor & MacBryde; *P. virginiana* var. *melanocarpa* (A. Nelson) Sargent

Leaves: blade elliptic to oblanceolate, lengths of larger at least 2 times widths. **Inflorescences:** racemes (30–)60–110(–130) mm. **Flowers:** hypanthium 2–3 mm; sepals 0.7–1.4 mm; petals 4–5(–7) mm. $2n = 16$.

Flowering Apr–Jun; fruiting Jul–Sep. Canyons, gullies, mountain slopes, stream banks and terraces, chaparral, pine-oak forests, thickets, grassy rocky slopes, bluffs, forest edges, roadsides; 30–3000 m; Alta., B.C.; Ariz., Calif., Colo., Idaho, Mont., Nev., N.Mex., Oreg., Tex., Utah, Wash., Wyo.; Mexico (Chihuahua, Sonora).

8. Prunus padus Linnaeus, Sp. Pl. 1: 473. 1753 • European bird cherry [I]

Trees, not suckering, 50–150 dm, not thorny. **Twigs** with terminal end buds, glabrous or puberulent. **Leaves** deciduous; petiole 8–17 mm, glabrous or puberulent on adaxial surface or both surfaces, glandular distally, glands 1–4; blade elliptic to obovate, 5–10(–13) × 2.5–4.5(–7) cm, base obtuse to rounded or subcordate, margins serrate, teeth <ascending to spreading>, sharp, eglandular, apex acuminate to abruptly so, <lateral veins 10–18 per side, raised abaxially>, surfaces glabrous or abaxial hairy in vein axils or along midribs. **Inflorescences** 15–50-flowered, racemes; central axes 45–150 mm, <leafy at bases>. **Pedicels** 3–17 mm, glabrous. **Flowers** blooming after leaf emergence; hypanthium cupulate, 2–2.5 mm, glabrous externally; sepals broadly spreading to reflexed, oblong-ovate, 1.2–2 mm (lengths greater than widths), margins glandular-toothed, surfaces glabrous; petals white, obovate to suborbiculate, (5–)6–9 mm; ovaries glabrous. **Drupes** black, globose, 6–8 mm, glabrous; <hypanthium deciduous, leaving discs at bases of drupes>; mesocarps fleshy; stones subglobose, not flattened, <rugulose>. $2n = 32$.

Flowering Apr–May; fruiting Jul–Aug. Disturbed sites; 0–1700 m; introduced; N.B., Ont.; Alaska, Del., Ill., Mont., N.J., N.Y., Pa., Utah, Wash.; Eurasia; n Africa.

Prunus padus is cultivated in North America as an ornamental prized for its long, showy racemes of white flowers and its cold hardiness. Its value in the upper Midwest is often compromised by fungal infection causing black knot disease.

Prunus padus is difficult to distinguish from *P. virginiana* var. *demissa*, at least from herbarium material. Induments of hypanthium interior have been widely used in keys to separate the two: hairy in *P. padus* versus glabrous in *P. virginiana*. As noted by E. G. Voss (1972–1996, vol. 2), many of the specimens of *P. virginiana* have a pubescent hypanthium, especially basally. This is true throughout the range of *P. virginiana*, and also for some specimens of *P. serotina*. The hypanthia of *P. padus* are more hairy than those of *P. virginiana*. The petals of *P. padus* are longer than those of eastern chokecherry (*P. virginiana* var. *virginiana*), making it possible to distinguish flowering specimens from east of the Rocky Mountains. The petals are only slightly longer, on average, than those of *P. virginiana* var. *demissa*. Petals of *P. padus* are also a bit narrower and more elliptic. Shape of sepals may prove useful: 1.2–2 mm with lengths greater than widths in *P. padus* versus 0.7–1.4 mm with lengths equal to widths in *P. virginiana* var. *demissa*. In fruit, the difference between rugulose stones of *P. padus* and smooth stones of *P. virginiana* is subtle, perhaps too subtle for those unfamiliar with the range in variation of pit surfaces among these taxa.

9. Prunus mahaleb Linnaeus, Sp. Pl. 1: 474. 1753 • Mahaleb or perfumed or St. Lucie cherry [I]

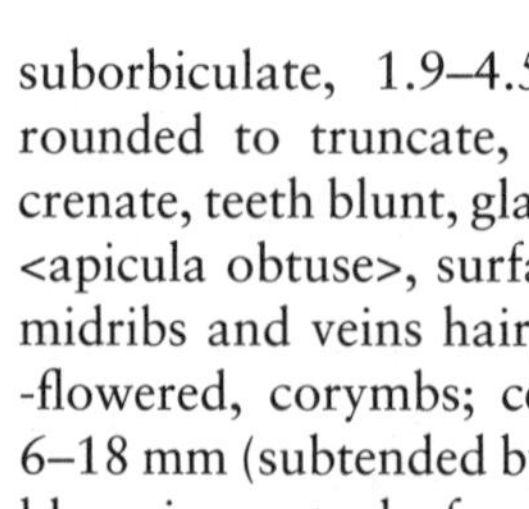

Shrubs or trees, not suckering, 30–150 dm, not thorny. **Twigs** with terminal end buds, densely puberulent. **Leaves** deciduous; petiole 4–20 mm, glabrous or ± puberulent on adaxial surface, sometimes glandular distally, glands 1 or 2, <discoid>; blade broadly ovate, oblong, or suborbiculate, 1.9–4.5 × 1.2–3.4 cm, base usually rounded to truncate, sometimes subcordate, margins crenate, teeth blunt, glandular, apex abruptly acuminate, <apicula obtuse>, surfaces usually glabrous, sometimes midribs and veins hairy abaxially. **Inflorescences** 4–10-flowered, corymbs; central axes 8–40 mm. **Pedicels** 6–18 mm (subtended by leafy bracts), glabrous. **Flowers** blooming at leaf emergence; hypanthium conic-campanulate, 2–3 mm, glabrous externally; sepals reflexed, oblong, 1.3–2 mm, margins entire, surfaces glabrous; petals white, elliptic to obovate, 6–7 mm; ovaries glabrous. **Drupes** dark red to black, ovoid, 6–10 mm, glabrous; mesocarps leathery; stones ellipsoid to subglobose, ± flattened. $2n = 16$.

Flowering Apr–May; fruiting Jun–Jul. Roadsides, stream banks, limestone bluffs and quarries, lowland thickets and woods, fencerows, chaparral; 0–2300 m; introduced; B.C., Ont.; Ala., Ark., Calif., Conn., Del., D.C., Idaho, Ill., Ind., Iowa, Kans., Ky., Md., Mass., Mich., Mo., Mont., N.J., N.Mex., N.Y., N.C., Ohio, Okla., Oreg., Pa., Tenn., Utah, Va., Wash., W.Va.; Eurasia.

Prunus mahaleb was introduced to North America as a rootstock for commercial cherries and is now sometimes cultivated for its attractive and fragrant flowers. At one time, the aromatic wood was a favorite for tobacco pipes.

10. Prunus pensylvanica Linnaeus f., Suppl. Pl., 252. 1782 • Pin or bird or fire cherry, cerisier de Pennsylvanie, petit merisier E F

Prunus corymbulosa Rydberg; *P. pensylvanica* subsp. *corymbulosa* (Rydberg) W. Wight; *P. pensylvanica* var. *saximontana* Rehder

Shrubs or trees, often suckering, 20–160 dm, not thorny. **Twigs** with terminal end buds, glabrous. **Leaves** deciduous; petiole (7–)9–20 mm, glabrous, glandular distally, glands 1–3; blade elliptic, oblong-lanceolate, or lanceolate, (2.5–)4.5–10(–14) × 1.5–5 cm, base cuneate to rounded, margins crenulate to crenate-serrate, teeth blunt, glandular, apex usually acuminate, sometimes acute (western specimens), surfaces glabrous. **Inflorescences** 2–5(–8)-flowered, umbellate fascicles or corymbs; central axes 0–8(–24) mm. **Pedicels** (8–)10–30 mm (subtended by minute bracts), glabrous. **Flowers** blooming at leaf emergence; hypanthium obconic, 1.8–3 mm, glabrous externally; sepals reflexed, oblong, 1.2–2.8 mm, margins entire, surfaces glabrous; petals white, elliptic, obovate, or suborbiculate, 4–7 mm; ovaries glabrous. **Drupes** bright red, ellipsoid, 6–10 mm, glabrous; mesocarps fleshy; stones ellipsoid, not flattened. **2*n*** = 16.

Flowering Apr–Jun; fruiting Jul–Aug. Forming thickets along streams and lakeshores, in clearings, roadsides, burned-over areas, disturbed sites, rocky hillsides, cliffs, open forests; 0–2800 m; St. Pierre and Miquelon; Alta., B.C., Man., N.B., Nfld. and Labr., N.W.T., N.S., Ont., P.E.I., Que., Sask.; Colo., Conn., Ga., Ill., Ind., Iowa, Maine, Md., Mass., Mich., Minn., Mont., N.H., N.J., N.Y., N.C., N.Dak., Ohio, Pa., R.I., S.Dak., Tenn., Vt., Va., W.Va., Wis., Wyo.

Throughout most of its range, *Prunus pensylvanica* appears distinct from *P. emarginata*. The leaves of *P. pensylvanica* are generally larger and lanceolate to ovate-lanceolate rather than oblanceolate to elliptic; the leaf apices are usually acuminate (sometimes acute) versus rounded to obtuse (rarely acute) in *P. emarginata*. Inflorescences of *P. pensylvanica* are corymbose to umbellate with central axes shorter than pedicels; in *P. emarginata* the inflorescences are corymbose to racemose with central axes longer than pedicels. Where their ranges overlap in British Columbia and western Montana, intermediates are found with the corymbose inflorescence of *P. emarginata* and leaves more comfortably accommodated within the variation of *P. pensylvanica*. Some of these specimens have been identified as *P. corymbulosa*, based on a type from Montana, here included within *P. pensylvanica*.

Along the eastern slopes of the Rockies and throughout the northwestern Great Plains, *Prunus pensylvanica* is shrubby and has smaller leaves (less than 60 mm) than it does farther east. Compared to leaves of eastern plants, those of western specimens are also broader in proportion to their length (1.7–2.3:1 versus 2–4.3:1), have acute rather than acuminate apices, and tend to be more coarsely toothed. Although sometimes segregated as var. *saximontana*, or subsp. *corymbulosa*, when having a corymbose inflorescence, these plants fit within the variation in plant habit, leaf size, leaf shape, and margin serration seen throughout the broad range of *P. pensylvanica*.

11. Prunus emarginata (Douglas) Eaton, Man. Bot. ed. 7, 463. 1836 • Bitter cherry

Cerasus emarginata Douglas in W. J. Hooker, Fl. Bor.-Amer. 1: 169. 1832; *Prunus emarginata* var. *crenulata* (Greene) Kearney & Peebles; *P. emarginata* var. *mollis* (Douglas) W. H. Brewer; *P. pensylvanica* Linnaeus f. var. *mollis* (Douglas) B. Boivin

Shrubs or trees, often suckering, 10–150 dm, not thorny. **Twigs** with terminal end buds, glabrous or hairy. **Leaves** deciduous; petiole 3–12 mm, glabrous or hairy, usually eglandular, rarely glandular distally, <glands discoid>; blade elliptic, oblong, oblanceolate, or obovate, (1.5–)2–6(–8) × 1–3(–4.2) cm, base cuneate, margins crenate to serrate, teeth blunt, glandular, apex usually rounded to obtuse, rarely acute, surfaces glabrous or downy, abaxial often hairier. **Inflorescences** (3–)6–12-flowered, corymbs or racemes; central axes (5–)9–30 mm. **Pedicels** 3–12(–18) mm (subtended by leafy bracts), glabrous or hairy. **Flowers** blooming at leaf emergence; hypanthium obconic, 2.5–3 mm, glabrous or hairy externally; sepals reflexed, oblong, 1.5–2 mm, margins entire, surfaces usually glabrous, sometimes hairy abaxially; petals white, elliptic to obovate, 3–8 mm; ovaries glabrous or with scattered hairs. **Drupes** bright red, globose to ovoid, 7–14 mm, glabrous; mesocarps fleshy; stones ellipsoid, not flattened.

Flowering Apr–Jul; fruiting Jun–Sep. Gravelly or sandy soil along streams, rocky mountain slopes, subalpine, thickets on exposed sites, cutover and burned areas, understory of conifer and oak forests; 0–3000 m; B.C.; Ariz., Calif., Idaho, Mont., Nev., N.Mex., Oreg., Utah, Wash.; Mexico (Baja California).

H. J. Scoggan (1978–1979, part 3) included *Prunus emarginata* within *P. pensylvanica*. Given that much of the problem of separating *P. emarginata* from *P. pensylvanica* arises from intermediate plants found in British Columbia and western Montana where the ranges of the two species meet, this makes some sense from a Canadian perspective. When viewed from a continental perspective, *P. emarginata* normally does not look much like *P. pensylvanica* (see notes on their differences under 10. *P. pensylvanica*).

Individuals of *Prunus emarginata* vary considerably in their habit, size and shape of leaves, and indument on branchlets, inflorescences, and leaves. Many segregate species have been described based on this variation. Most often, western botanists have recognized one species with two varieties based on habit or indument; those characters are not well correlated. Some keys use habit to separate the varieties (var. *emarginata*, a shrub, versus var. *mollis*, a tree, commonly used in Pacific Northwest floras); others have used indument (var. *emarginata*, glabrous, versus var. *mollis*, hairy, more commonly used in floras of the arid West). On the mountain slopes in the more arid portions of its range, *P. emarginata* is a thicket-forming shrub to about 6 m with the leaves expanding to no more than 50 mm; in the moister, milder climates of coastal Oregon to British Columbia, as well as the interior wetbelt of northern Idaho, northwestern Montana, and adjacent British Columbia, the plants are trees to 15 m with larger leaves expanding to 80 mm. Hairy plants occur throughout the range of *P. emarginata*, and glabrous plants are found from British Columbia to Arizona, generally not west of the Cascades.

Hybrids between *Prunus emarginata* and naturalized *P. avium* (*P.* ×*pugetensis* Jacobson & Zika) have been reported from the lowlands of western British Columbia, Oregon, and Washington.

12. Prunus avium (Linnaeus) Linnaeus, Fl. Suec. ed. 2, 165. 1755 • Sweet cherry, cerisier des oiseaux [I]

Prunus cerasus Linnaeus var. *avium* Linnaeus, Sp. Pl. 1: 474. 1753

Trees, not suckering, 60–200 dm, not thorny. **Twigs** with terminal end buds, glabrous. **Leaves** deciduous; petiole (14–)20–40 mm, nearly glabrous or sparsely hairy adaxially, glandular distally or on margins at bases of blades, glands 1–3, <discoid>; blade oblong, elliptic-obovate, or obovate, (4–)7–14 × (2.5–)4–8.5 cm, base obtuse to rounded, margins doubly crenate-serrate, teeth blunt, glandular, apex abruptly short-acuminate, abaxial surface moderately hairy (especially midribs and veins), adaxial glabrous. **Inflorescences** 1–3-flowered, umbellate fascicles. **Pedicels** (14–)25–52 mm, glabrous. **Flowers** blooming at leaf emergence; hypanthium tubular-urceolate, 5–7 mm, glabrous externally; sepals reflexed, oblong, 4–6 mm, margins entire or sparsely and irregularly toothed, surfaces glabrous; petals white, obovate to suborbiculate, 12–15 mm; ovaries glabrous. **Drupes** dark red to purplish black, globose, 13–30 mm, glabrous; mesocarps fleshy; stones subglobose, not flattened. $2n = 16$.

Flowering Mar–May; fruiting Jun–Jul. Roadsides, stream banks, forest edges, disturbed forests; 0–1600 m; introduced; B.C., N.B., N.S., Ont.; Ariz., Calif., Conn., Del., D.C., Idaho, Ill., Ind., Ky., Maine, Md., Mass., Mich., N.H., N.J., N.Y., N.C., Ohio, Oreg., Pa., R.I., S.C., Tenn., Utah, Va., Wash., W.Va., Wyo.; Eurasia.

Cultivars of *Prunus avium* are grown in North America both for their attractive flowers, especially where the Japanese ornamental cherries are not hardy, and for their sweet fruits. Most of the sweet cherry crop in North America is grown in the Pacific Coast states and in Michigan and Ontario. Most varieties are self-incompatible and require pollen from a variety in a different incompatibility group in order to set fruits.

13. Prunus cerasus Linnaeus, Sp. Pl. 1: 474. 1753 • Sour or pie cherry, cerisier acide [I]

Shrubs or trees, suckering, 30–50(–80) dm, not thorny. **Twigs** with terminal end buds, glabrous. **Leaves** deciduous; petiole 10–24 mm, glabrous, usually eglandular, sometimes with discoid marginal glands at bases of blades; blade broadly elliptic to ovate or obovate, 4.4–6(–8) × 2.8–4(–6) cm, base obtuse to rounded, margins doubly crenate-serrate, teeth blunt, glandular, apex acute to abruptly acuminate, abaxial surface glabrous or glabrate, adaxial glabrous. **Inflorescences** 1–4-flowered, umbellate fascicles. **Pedicels** 8–37 mm, glabrous. **Flowers** blooming at leaf emergence; hypanthium tubular-campanulate, 4–6 mm, glabrous externally; sepals reflexed, oblong, 4–7 mm, margins regularly glandular-toothed, surfaces glabrous; petals white, suborbiculate, 10–14 mm; ovaries glabrous. **Drupes** bright red, globose, 13–20 mm, glabrous; mesocarps fleshy; stones subglobose, not flattened. $2n = 32$.

Flowering Apr–May; fruiting Jun–Jul. Roadsides, thickets, woodland borders, abandoned fields; 0–1000 m; introduced; B.C., N.B., N.S., Ont., P.E.I., Que.; Alaska, Calif., Conn., Del., D.C., Ga., Idaho, Ill.,

Ind., Iowa, Kans., Ky., Maine, Md., Mass., Mich., Mo., Mont., Nebr., N.H., N.J., N.Mex., N.Y., N.C., Ohio, Oreg., Pa., R.I., Vt., Va., Wash., W.Va.; Eurasia.

Cultivars of *Prunus cerasus* with doubled flowers are used as landscape ornamentals. Most of the commercial crop of *P. cerasus* in North America comes from Michigan. Unlike *P. avium*, all widely grown varieties of *P. cerasus* are self-fertile. Bud scales at the bases of the pedicels on flowering and immature fruiting specimens of *P. cerasus* often have leaflike apices and the inner scales are erect; in *P. avium* the scales are not leaflike and the inner ones are reflexed or spreading.

14. Prunus speciosa (Koidzumi) Nakai, Bot. Mag. (Tokyo) 29: 139. 1915 • Japanese flowering or Oriental cherry I

Prunus jamasakura Siebold ex Koidzumi var. *speciosa* Koidzumi, Bot. Mag. (Tokyo) 25: 186. 1911; *P. serrulata* Lindley var. *lannesiana* (Carrière) Makino

Trees, not suckering, 60–100 (–250) dm, not thorny. **Twigs** with terminal end buds, glabrous. **Leaves** deciduous; petiole 7–45 mm, glabrous, glandular, glands 2–4, <discoid>; blade elliptic to obovate, 5–17 × 3–8 cm, base obtuse to rounded, margins singly to doubly serrate, teeth aristate, glandular, apex caudate, surfaces glabrous. **Inflorescences** (2 or)3–5(or 6)-flowered, corymbs; central axes 5–25(–60) mm. **Pedicels** 10–40 mm (subtended by leafy bracts), glabrous or sparsely hairy. **Flowers** blooming at leaf emergence; hypanthium tubular, 4–8 mm, glabrous externally; sepals spreading to reflexed, oblong-ovate to lanceolate, 3–8 mm, margins entire or toothed, eglandular, surfaces glabrous; petals white or pink, suborbiculate to oblong-obovate, 8–18 mm; ovaries glabrous. **Drupes** black, globose, 10–13 mm, glabrous; mesocarps fleshy; stones ellipsoid, slightly flattened. $2n = 16$ (Japan).

Flowering Apr–May; fruiting Jun–Jul. Disturbed sites, abandoned plantings; 0–200 m; introduced; Calif., Mass., N.C.; e Asia (Japan).

The name *Prunus serrulata* has been widely applied to *P. speciosa* by North American botanists and horticulturalists (for example, P. G. Russell 1934; A. J. Rehder 1940); some Japanese cherry experts now circumscribe *P. serrulata* more narrowly so that it includes only the white-petaled, double-flowered cultivars closely resembling the nomenclatural type. The single-flowered plants that are found escaping rarely and perhaps naturalizing in the flora area have been called *P. lannesiana* (Carrière) E. H. Wilson forma *albida* (Makino) E. H. Wilson or *P. speciosa*. Based on principal components analysis of 35 morphological characters from 468 individuals of the *P. serrulata* complex and related taxa, K. S. Chang et al. (2007) argued that forma *albida* is distinctive and separated from other taxa of the *P. serrulata* complex. H. Ohba (2001) recognized it at species rank as *Cerasus speciosa* (Koidzumi) H. Ohba. The classification and nomenclature of Japanese flowering cherries are complex, convoluted, and subject to varying interpretations, and no attempt is made to resolve them here. Centuries of selection and hybridization have blurred species distinctions, and it may be best to do as horticulturalists have and forsake botanical species names in favor of traditional and cultivar names. Whatever the name, these Japanese flowering cherries are widely grown as ornamentals where winters are not too cold nor summers too hot; they escape only rarely and have been found naturalizing only near planted specimens.

15. Prunus yedoensis Matsumura, Bot. Mag. (Tokyo) 15: 100. 1901 • Yoshino cherry I

Trees, not suckering, 40–80 (–160) dm, not thorny. **Twigs** with terminal end buds, sparsely hairy. **Leaves** deciduous; petiole 10–20 mm, hairy, sometimes glandular distally, glands 1 or 2; blade elliptic-ovate to obovate, 5–12 × 2.5–7 cm, base rounded, margins doubly serrate, teeth aristate, glandular, apex acuminate, abaxial surface hairy along midribs and veins, adaxial glabrous. **Inflorescences** 2–6-flowered, corymbs; central axes 4–8(–20) mm. **Pedicels** 15–25 mm (subtended by leafy bracts), hairy. **Flowers** blooming before leaf emergence; hypanthium tubular, 7–8 mm, hairy externally; sepals spreading, ovate-lanceolate, 4–5 mm, margins glandular-toothed, abaxial surface hairy, adaxial sparsely hairy; petals white or pink, broadly elliptic to obovate, 13–15 mm; ovaries glabrous. **Drupes** black, subglobose, 7–12 mm, glabrous; mesocarps fleshy; stones ellipsoid, not flattened. $2n = 16$ (Japan).

Flowering Mar–Apr; fruiting May. Abandoned plantings, disturbed sites; 0–200 m; introduced; Calif., D.C., Wash.; e Asia (Japan).

16. Prunus subhirtella Miquel, Ann. Mus. Bot. Lugduno-Batavi 2: 91. 1865 • Higan cherry I

Trees, not suckering, 30–100 dm, not thorny. **Twigs** with terminal end buds, hairy. **Leaves** deciduous; petiole 5–10 mm, hairy, glandular distally or on margins at bases of blades; blade elliptic, oblong-ovate, or ovate, 3–8 × 1.5–4 cm, base obtuse, margins doubly serrate, teeth sharp, glandular, apex acuminate, abaxial surface hairy along midribs and veins, adaxial glabrous or with appressed hairs along midrib. **Inflorescences**

2–5-flowered, umbellate fascicles. **Pedicels** 8–22 mm, hairy. **Flowers** blooming before leaf emergence; hypanthium tubular to tubular-urceolate, 4–7 mm, hairy externally; sepals erect to spreading, oblong-ovate, 2–5 mm, margins toothed, sometimes glandular, abaxial surface sparsely hairy, adaxial glabrous; petals pale pink, oblong to obovate, 8–12 mm; ovaries glabrous. **Drupes** black, subglobose, 8 mm, glabrous; mesocarps fleshy; stones ellipsoid, not flattened. $2n$ = 24 (Japan).

Flowering Mar–Apr; fruiting Jun. Disturbed sites, abandoned plantings; 0–200 m; introduced; D.C., Ohio, Va.; e Asia (Japan).

17. Prunus fasciculata (Torrey) A. Gray, Proc. Amer. Acad. Arts 10: 70. 1874 • Desert almond

Emplectocladus fasciculatus Torrey, Proc. Amer. Assoc. Advancem. Sci. 4: 192. 1851

Shrubs, suckering, <much branched>, 10–20(–30) dm, thorny. **Twigs** with axillary end buds, glabrous or canescent. **Leaves** deciduous; sessile; blade oblanceolate to linear, 0.5–2 × 0.1–0.2(–0.4) cm, base long-attenuate, margins nearly entire or obscurely and remotely serrulate in distal ⅓, teeth blunt to sharp, sometimes glandular, apex rounded to acute, surfaces puberulent or glabrous or low-papillate (var. *punctata*). **Inflorescences** solitary flowers or 2-flowered fascicles. **Pedicels** 0–4 mm, glabrous. **Flowers** <unisexual, plants dioecious>, blooming at leaf emergence; hypanthium campanulate, 1.5–3 mm, glabrous externally; sepals erect-spreading, triangular, 0.7–1 mm, margins entire, surfaces glabrous; petals white to yellowish, elliptic, obovate, or suborbiculate, 1.4–2.5(–4) mm; ovaries hairy. **Drupes** gray to red-brown, ovoid, <± compressed>, 7–15 mm, densely puberulent; <hypanthium tardily deciduous>; mesocarps leathery to dry; stones ovoid, ± flattened.

Varieties 2 (2 in the flora): sw United States, nw Mexico.

1. Leaf blades sparsely to densely puberulent, not papillate . . . . 17a. *Prunus fasciculata* var. *fasciculata*
1. Leaf blades glabrous, sometimes papillate. . . . . 17b. *Prunus fasciculata* var. *punctata*

17a. Prunus fasciculata (Torrey) A. Gray var. **fasciculata**

Leaf blades sparsely to densely puberulent, not papillate.

Flowering Mar–May; fruiting May–Jul. Sandy or gravelly washes, dry mountain slopes, desert scrub, pinyon-juniper-Joshua tree woodlands; 200–2100 m; Ariz., Calif., Nev., Utah; Mexico (Baja California).

17b. Prunus fasciculata (Torrey) A. Gray var. **punctata** Jepson, Fl. Calif. 2: 230. 1936 [E]

Leaf blades glabrous, sometimes papillate.

Flowering Mar–Apr; fruiting Apr–Jun. Sandy or grassy slopes and mesas, coastal scrub, oak woodlands; 10–200 m; Calif.

Variety *punctata* occupies a narrow range along the coast of California in San Luis Obispo and Santa Barbara counties.

18. Prunus minutiflora Engelmann ex A. Gray, Boston J. Nat. Hist. 6: 185. 1850 • Texas almond [E]

Shrubs, suckering, <much branched>, 10–20 dm, weakly thorny. **Twigs** with axillary end buds, canescent. **Leaves** deciduous; petiole 1–2(–6) mm, glabrous, eglandular; blade elliptic or obovate, 0.5–1.6 (–3.5) × 0.3–0.8(–2.1) cm, base cuneate, margins usually entire, sometimes irregularly serrulate (sometimes dentate on long shoots), teeth sharp to blunt, eglandular, some callus-tipped, apex usually obtuse to rounded, sometimes apiculate, surfaces glabrous. **Inflorescences** solitary flowers. **Pedicels** 0–2 mm, puberulent. **Flowers** <unisexual, plants dioecious>, blooming at leaf emergence; hypanthium campanulate, 2–3 mm, glabrous externally; sepals spreading, triangular, 0.7–1.5 mm, margins entire, surfaces glabrous; petals white, obovate, 2–3.5 mm; ovaries hairy. **Drupes** reddish brown, globose to ovoid, 9–12 mm, puberulent; <hypanthium tardily deciduous>; mesocarps leathery to dry (slightly splitting); stones ovoid to subglobose, not flattened.

Flowering Feb–Mar; fruiting May–Jun. Dry rocky streambeds and uplands, limestone hills, ledges; 100–700 m; Tex.

Prunus minutiflora is a rare species limited to central Texas around the Edwards Plateau.

19. Prunus havardii (W. Wight) S. C. Mason, J. Agric. Res. 1: 176. 1913 • Havard's almond

Amygdalus havardii W. Wight in W. R. Dudley et al., Dudley Mem. Vol., 133. 1913 (as harvardii)

Shrubs, suckering unknown, <much branched>, 10–20 dm, thorny. **Twigs** with axillary end buds, puberulent. **Leaves** deciduous; petiole 1–3 mm, glabrous or puberulent, eglandular; blade rhombic, obovate, or fan-shaped, 0.5–1.6(–2) × 0.2–0.8(–1.4) cm, base broadly obtuse or rounded to nearly truncate, margins serrate or dentate in distal ½, teeth blunt to sharp, some callus-tipped, rarely glandular, apex rounded to obtuse, surfaces puberulent. **Inflorescences** solitary flowers. **Pedicels** 0 mm. **Flowers** <unisexual, plants dioecious>, blooming at leaf emergence; hypanthium campanulate, 2.5–3 mm, glabrous externally; sepals spreading to reflexed, triangular, 0.7–1 mm, margins entire, <sparsely ciliate>, surfaces glabrate; petals white, obovate, 2 mm; ovaries hairy. **Drupes** reddish brown, ovoid, 8–11 mm, puberulent; <hypanthium tardily deciduous>; mesocarps leathery to dry (splitting); stones ovoid, slightly flattened.

Flowering Apr–Jun; fruiting Jun–Aug. Draws, dry rocky slopes of canyons, limestone soil, igneous rock; 700–1700 m; Tex.; Mexico (Chihuahua).

Prunus havardii is endemic to the Chihuahuan Desert of trans-Pecos Texas and across the Rio Grande in Mexico, with most collections from the Big Bend area.

20. Prunus eremophila Prigge, Madroño 49: 285, figs. 1, 2. 2003 [C] [E] [F]

Shrubs, suckering unknown, <much branched>, 10–25 dm, weakly thorny. **Twigs** with axillary end buds, canescent. **Leaves** deciduous; petiole 0.5–3(–5) mm, hairy, eglandular; blade ovate, obovate, or spatulate, 0.5–2(–3) × 0.2–1(–2) cm, base cuneate to obtuse, margins irregularly serrate, teeth usually sharp, eglandular, sometimes blunt, obscurely glandular, apex obtuse to rounded, often mucronate, surfaces hairy. **Inflorescences** solitary flowers or 2-flowered fascicles. **Pedicels** 0–3 mm, puberulent. **Flowers** <unisexual, plants dioecious>, blooming at leaf emergence; hypanthium campanulate, 2–4 mm, hairy externally; sepals erect, triangular, 1–2 mm, margins entire, <sparsely ciliate>, abaxial surface densely hairy, adaxial glabrous or slightly hairy; petals white, elliptic, rhombic, ovate, or suborbiculate, 2.5–6 mm, <abaxial surfaces hairy>; ovaries hairy. **Drupes** yellowish orange, obovoid to ovoid, 9–16 mm, velutinous; mesocarps leathery to dry; stones subglobose to ovoid, slightly flattened.

Flowering Mar–Apr; fruiting May–Jun. Desert washes, rocky slopes; of conservation concern; 900–1200 m; Calif.

Prunus eremophila is endemic to the southern Mojave Desert and has so far been collected only from the East Mojave Natural Preserve in eastern San Bernardino County.

21. Prunus persica (Linnaeus) Batsch, Beytr. Entw. Gewächsreich, 30. 1801 • Peach, pêcher [I]

Amygdalus persica Linnaeus, Sp. Pl. 1: 472. 1753; *Persica vulgaris* Miller

Trees, not suckering, 30–100 dm, not thorny. **Twigs** with terminal end buds, glabrous. **Leaves** deciduous; petiole 5–10(–15) mm, <not winged>, glabrous, sometimes glandular distally, glands 1–4, <discoid>; blade oblong to lanceolate, <folded along midribs, often falcate>, (5–)7–15 × 2–4.5 cm, base cuneate to obtuse, margins crenulate-serrulate, teeth blunt, glandular, apex acuminate, surfaces glabrous. **Inflorescences** usually solitary flowers, sometimes 2-flowered fascicles. **Pedicels** 0–3 mm, glabrous. **Flowers** blooming before leaf emergence; hypanthium cupulate, 4–5 mm, glabrous externally; sepals spreading, oblong-ovate, 3.5–5 mm, margins entire, <ciliate>, abaxial surface hairy (especially along margins), adaxial glabrous; petals dark pink, obovate to suborbiculate, 10–17 mm; ovaries hairy. **Drupes** yellow- to orange-tinged with red, globose, 40–80 mm, velutinous (glabrous in nectarines); mesocarps fleshy; stones ellipsoid, strongly flattened, <deeply pitted, furrowed>. $2n = 16$.

Flowering Feb–Apr; fruiting Jun–Aug. Roadsides, fencerows, abandoned farms, streamsides, canyons; 0–2300 m; introduced; N.S., Ont.; Ala., Ariz., Ark., Calif., Conn., Del., D.C., Fla., Ga., Idaho, Ill., Ind., Iowa, Kans., Ky., La., Maine, Md., Mass., Mich., Miss., Mo., N.J., N.Y., N.C., Ohio, Okla., Oreg., Pa., R.I., S.C., Tenn., Tex., Utah, Va., W.Va., Wis.; e Asia.

Prunus persica is cultivated throughout much of North America. Commercial production is diffuse, occurring in over half of the lower 48 states as well as in southern British Columbia and southern Ontario. Freestone peaches for fresh eating come largely from California, Georgia, New Jersey, North Carolina, Pennsylvania, and South Carolina; most clingstones are used for canning and are grown in California. Nectarines are a variety with hairless skin. The fruits are popular on picnics, and peach saplings are commonly encountered anywhere

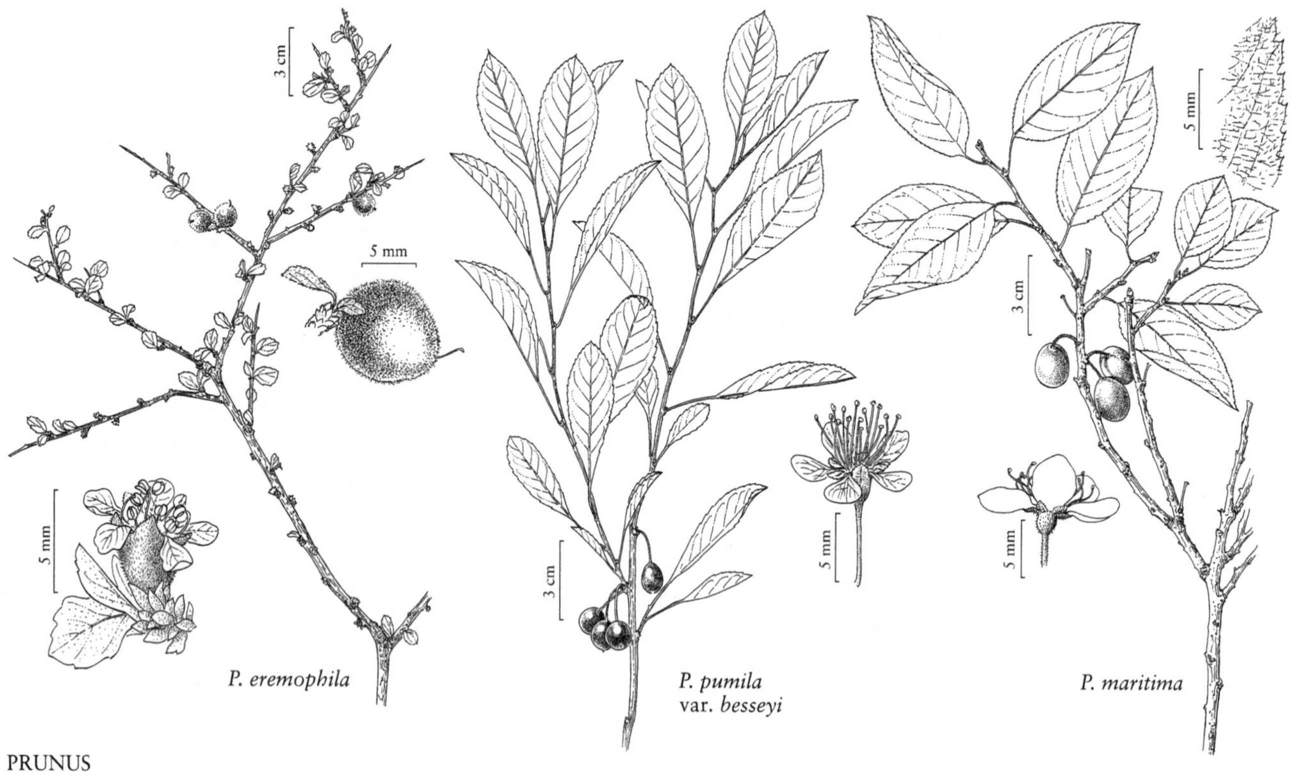

PRUNUS

pits are discarded. Escapes are usually short-lived; some escapes form naturalized populations. Double-flowered cultivars are used ornamentally.

22. Prunus dulcis (Miller) D. A. Webb, Feddes Repert. 74: 24. 1967 • Sweet almond, amandier [I]

Amygdalus dulcis Miller, Gard. Dict. ed. 8, Amygdalus no. 2. 1768; *A. communis* Linnaeus; *Prunus amygdalus* Batsch

Trees, not suckering, 50–80 dm, not thorny. **Twigs** with terminal end buds, glabrous. **Leaves** deciduous; petiole (8–)10–25 mm, <usually winged distally>, glabrous, usually glandular distally or on margins at bases of blades; blade oblong to lanceolate, 2.5–10 × 1–3 cm, base obtuse, margins crenulate-serrulate to crenate-serrate, teeth blunt, glandular, apex acuminate, surfaces glabrous. **Inflorescences** solitary flowers or 2-flowered fascicles. **Pedicels** 1–5 mm, glabrous. **Flowers** blooming before leaf emergence; hypanthium cupulate, 4–7 mm, glabrous externally; sepals erect-spreading to spreading, oblong-ovate, 4–8 mm, margins entire, <tomentose>, surfaces glabrous; petals pink to nearly white, obovate, elliptic, or suborbiculate, 12–25 mm; ovaries hairy. **Drupes** gray-green, ovoid-oblong, <compressed>, 25–40 mm,' velutinous; mesocarps leathery (splitting); stones ellipsoid, strongly flattened, <pitted>. **2*n*** = 16.

Flowering Feb–Mar; fruiting Jul–Sep. Roadsides, canyons, grasslands; 20–500 m; introduced; Calif., Idaho, Wash.; w Asia; n Africa.

The United States now dominates world almond production with over 40% of the annual crop, all of it grown in or near the Central Valley of California. The in-shell so-called nuts sold in stores are the pits of drupes with the leathery mesocarp removed. *Prunus dulcis* is among the earliest blossoming trees and one of the first signs of spring in areas where it is grown.

23. Prunus tomentosa Thunberg in J. A. Murray, Syst. Veg. ed. 14, 464. 1784 • Nanking or Manchu cherry [I]

Shrubs, not suckering, 10–20 dm, not thorny. **Twigs** with terminal end buds, densely hairy. **Leaves** deciduous; petiole 2–7 mm, hairy, eglandular; blade obovate to oblong-elliptic, 3–6(–7.5) × 1.7–3.5(–4.7) cm, base obtuse to rounded, margins coarsely, singly to doubly serrate, teeth sharp, usually eglandular, sometimes glandular, <glands dark, spheric>, apex abruptly short-acuminate to acute, abaxial surface tomentose, adaxial <rugose>, hairy, sometimes sparsely so. **Inflorescences** usually solitary flowers, sometimes 2-flowered fascicles. **Pedicels** 0–5 mm, hairy. **Flowers** blooming before or at leaf emergence; hypanthium tubular, 3.5–6 mm, glabrous or sparsely hairy (especially at bases) externally; sepals spreading, ovate, 2.5–3.5 mm, margins serrate, sometimes glandular, abaxial surface hairy, adaxial glabrous; petals white to pale pink (pink in bud), obovate or elliptic to suborbiculate, 9–13 mm; ovaries glabrous proximally, villous distally. **Drupes** red, globose, 8–15 mm, sparsely hairy; mesocarps fleshy; stones ellipsoid, not flattened. 2*n* = 16.

Flowering Apr–May; fruiting Jun–Jul. Roadsides, fencerows, vacant lots, riparian woods; 50–500 m; introduced; Man., Ont., Sask.; Ill., Iowa, Md., Mich., Minn., Nebr., N.Y., Ohio, Pa., S.Dak., Utah; Asia.

Prunus tomentosa is cultivated as a landscape plant in hedges and as a background shrub because of its spreading habit and pale pink to white petals. It produces edible fruits that are said to make excellent jelly and juice.

24. Prunus andersonii A. Gray, Proc. Amer. Acad. Arts 7: 337. 1868 • Desert peach [E]

Shrubs, suckering unknown, <much branched>, 10–20(–30) dm, thorny. **Twigs** with axillary end buds, usually glabrous, rarely puberulent. **Leaves** deciduous; ± sessile; blade narrowly elliptic, obovate, oblanceolate, or spatulate, 1–3 × 0.2–0.6 cm, base long-attenuate, margins usually serrulate, sometimes obscurely, teeth blunt, inconspicuously glandular, apex usually acute, sometimes obtuse, surfaces usually glabrous, sometimes puberulent. **Inflorescences** solitary flowers or 2-flowered fascicles. **Pedicels** (1–)4–12 mm, glabrous. **Flowers** blooming at leaf emergence; hypanthium campanulate, 3–4 mm, usually glabrous, rarely puberulent externally; sepals spreading, triangular, 1.5–2.5 mm, margins sparsely glandular-toothed, <ciliate>, surfaces glabrate; petals usually dark pink, sometimes nearly white, elliptic, obovate, or suborbiculate, (5–)8–11 mm; ovaries hairy. **Drupes** greenish yellow to red-orange, globose to asymmetrically obovoid, <compressed>, 10–18 mm, <base cuneate-stipitate, apex mucronate>, densely puberulent; <hypanthium persistent>; mesocarps leathery to dry (often splitting); stones ellipsoid to subglobose, ± flattened.

Flowering Mar–May; fruiting Jun–Aug. Dry rocky slopes, washes, canyons, sagebrush scrub, pinyon-juniper woodlands; 900–2600 m; Calif., Nev.

Prunus andersonii inhabits the Great Basin Desert region of central and western Nevada, ranging westward into California to the eastern slopes of the Sierra Nevada from Modoc County south to Inyo County.

25. Prunus fremontii S. Watson in W. H. Brewer et al., Bot. California 2: 442. 1880 (as fremonti) • Desert apricot

Prunus eriogyna S. C. Mason

Shrubs, suckering unknown, <much branched>, 10–40 dm, thorny. **Twigs** with axillary end buds, glabrous. **Leaves** deciduous; petiole 1–7 mm, glabrous, eglandular; blade elliptic, ovate, or suborbiculate, 0.6–3 × 0.5–2 cm, base obtuse to rounded, subcordate, or truncate, margins obscurely crenulate, crenulate-serrulate, or serrate, teeth blunt, glandular, apex usually obtuse to rounded, sometimes emarginate, surfaces glabrous. **Inflorescences** 1–3-flowered, umbellate fascicles. **Pedicels** 2–12 mm, glabrous. **Flowers** blooming at leaf emergence; hypanthium campanulate, 2–4 mm, glabrous externally; sepals erect-spreading, semicircular to ovate, 1.2–4 mm, margins glandular-toothed, <ciliate>, abaxial surface glabrous, adaxial hairy; petals usually white, sometimes pinkish rose, elliptic, obovate, or suborbiculate, 3–10 mm; ovaries hairy. **Drupes** yellowish, ellipsoid-ovoid, 8–15 mm, densely puberulent; mesocarps leathery to dry (splitting); stones ovoid, ± flattened.

Flowering Jan–Mar; fruiting Apr–Jun. Dry, sandy or rocky slopes, canyons, desert, chaparral, pinyon-juniper woodlands; 200–1500 m; Calif.; Mexico (Baja California).

Prunus fremontii is known only from the western edge of the Sonoran Desert.

26. Prunus pumila Linnaeus, Mant. Pl. 1: 75. 1767 • Sandcherry, cerisier des sables [E] [F]

Shrubs, sometimes suckering, 1–15(–25) dm, not thorny. **Twigs** with terminal end buds, glabrous or sparsely to densely puberulent (var. *susquehanae*). **Leaves** deciduous; petiole 2–10(–13) mm, glabrous or hairy only when young, sometimes glandular distally or on margins at bases of blades, glands 1 or 2; blade elliptic, oblanceolate, or obovate, 2.5–8 × 0.8–3 cm, base obtuse, cuneate, or long-attenuate, margins crenulate-serrulate to serrate in distal ½–⅔, teeth sharp or blunt, callus-tipped, sometimes glandular, apex short-acuminate to rounded, surfaces glabrous. **Inflorescences** 2–5-flowered, umbellate fascicles. **Pedicels** 3–19 mm, glabrous. **Flowers** blooming before leaf emergence; hypanthium campanulate, 1.7–3 mm, glabrous externally; sepals erect to reflexed, semicircular, 1.3–2.8

mm, margins glandular-toothed, surfaces glabrous; petals white, oblanceolate, oblong, or suborbiculate, 3–9 mm; ovaries glabrous. **Drupes** dark purple to nearly black, subglobose or broadly ellipsoid, 6–12 mm, glabrous; mesocarps fleshy; stones subglobose, ovoid, or fusiform, not flattened.

Varieties 4 (4 in the flora): North America.

Opinion has varied as to whether *Prunus pumila* is best treated as one variable species (for example, H. Groh and H. A. Senn 1940; H. A. Gleason 1952; J. R. Rohrer 2000) or as two, three, or four separate species (for example, W. F. Wight 1915; M. L. Fernald 1923b; P. M. Catling et al. 1999). The plants vary in stem posture, twig indument, leaf shape, fruit size and taste, pit size and shape, and ecologic preference. Even though the morphologic characters show almost continuous variation, four varieties are recognized here based largely on differences in habitat and geographic range.

SELECTED REFERENCE Catling, P. M., S. M. McKay-Kuja, and G. Mitrow. 1999. Rank and typification in North American dwarf cherries, and a key to the taxa. Taxon 48: 483–488.

1. Twigs densely puberulent (10× magnification); leaf apices usually obtuse, sometimes acute or rounded, blade lengths ca. 2.6 times widths; sandy pine-oak woods or barrens with open canopy, or adjacent fields and lakeshores . . . . 26d. *Prunus pumila* var. *susquehanae*
1. Twigs usually glabrous, sometimes sparsely puberulent; leaf apices short-acuminate, acute, or obtuse, blade lengths ca. 2.9–3.7 times widths; usually in open habitats.
    2. Stems prostrate to decumbent; leaf blades oblanceolate, lengths ca. 3.7 times widths; stones ovoid to fusiform, 4.5–5 mm wide . . . . 26c. *Prunus pumila* var. *depressa*
    2. Stems usually erect-ascending, sometimes decumbent or sprawling; leaf blades elliptic, obovate, or oblanceolate, lengths ca. 2.9–3.3 times widths; stones subglobose to ovoid, 5–7 mm wide.
        3. Leaf blades usually oblanceolate, lengths ca. 3.3 times widths; shores of the Great Lakes on sandy, gravelly, or rocky beaches, dunes, and interdunal flats . . . . 26a. *Prunus pumila* var. *pumila*
        3. Leaf blades elliptic, oblanceolate, or obovate, lengths ca. 2.9 times widths; sandy prairies, oak savannas, rock outcrops . . . . 26b. *Prunus pumila* var. *besseyi*

26a. Prunus pumila Linnaeus var. **pumila** • Great Lakes sandcherry [E]

Stems usually erect-ascending, sometimes decumbent, 3–15 (–25) dm; twigs glabrous. **Leaf blades** usually oblanceolate, sometimes elliptic or obovate, 3.1–5.8(–7) × 0.8–1.9(–2.3) cm, lengths ca. 3.3 times widths, base cuneate to attenuate, apex usually acute or short-acuminate, sometimes obtuse. **Drupes** subglobose, 8–12 × 5–10 mm; stones subglobose to ovoid, 8–9.5 × 6–7 mm.

Flowering May–Jun; fruiting Aug–Sep. Shores of the Great Lakes on sandy, gravelly, or rocky beaches, dunes, interdunal flats; 70–200 m; Ont.; Ill., Ind., Mich., Minn., N.Y., Ohio, Pa., Wis.

26b. Prunus pumila Linnaeus var. **besseyi** (L. H. Bailey) Waugh, Rep. (Annual) Vermont Agric. Exp. Sta. 12: 239. 1899 • Western or Bessey's sandcherry [E] [F]

Prunus besseyi L. H. Bailey, Cornell Univ. Exp. Sta. Bull. 70: 261, plate 1. 1894

Stems usually erect-ascending, sometimes sprawling with ascending tips, 1–7(–15) dm; twigs usually glabrous, sometimes sparsely puberulent. **Leaf blades** elliptic, oblanceolate, or obovate, 2.5–8 × 0.8–2.2(–2.9) cm, lengths ca. 2.9 times widths, base cuneate, apex short-acuminate, acute, or obtuse. **Drupes** subglobose, 8–12 × 5–12 mm; stones subglobose to ovoid, 7–9 × 5–7 mm.

Flowering Apr–May; fruiting Jul–Aug. Sandy prairies, oak savannas, rock outcrops; 200–1700 m; Man., Ont., Sask.; Colo., Ill., Ind., Iowa, Kans., Minn., Mont., Nebr., N.Dak., Ohio, Oreg., S.Dak., Utah, Wis., Wyo.

Variety *besseyi* has been introduced west of its native range as an ornamental and fruit bearer for the home garden, where its adaptation to cold winters and hot, dry summers is a valuable asset. Cultivars were developed by breeders at the Morden Experimental Station in Manitoba and at the University of South Dakota.

26c. Prunus pumila Linnaeus var. **depressa** (Pursh) Bean, Trees Shrubs Brit. Isles 2: 249. 1914 • Prostrate or eastern sandcherry, cerisier déprimé [E]

Prunus depressa Pursh, Fl. Amer. Sept. 1: 332. 1813

Stems prostrate to decumbent, 1–5 dm; twigs glabrous. **Leaf blades** oblanceolate, 3.9–7.5 × 1–2(–2.3) cm, lengths ca. 3.7 times widths, base long-attenuate, apex obtuse to acute. **Drupes** broadly ellipsoid, 6–10 × 3–7 mm; stones ovoid to fusiform, 6–8.5 × 4.5–5 mm.

Flowering May–Jun; fruiting Aug. Rocky, gravelly, or sandy shores or bars of streams and lakes; 10–200 m; N.B., Ont., Que.; Maine, Mass., N.H., N.J., N.Y., Pa., Vt.

Variety *depressa* always has prostrate to decumbent stems, and the plants form mats on the rocky or sandy substrate along streams in eastern North America. All of the other varieties sometimes have decumbent stems; specimens are sometimes misidentified as var. *depressa*, but nearly always some of the stems on these plants are ascending or erect. Stems of vars. *besseyi* and *pumila* are often partially buried in shifting sands on the dunes in the sandhills of Nebraska and on the dunes along the shores of the Great Lakes.

26d. Prunus pumila Linnaeus var. **susquehanae** (Willdenow) H. Jaeger, Ziergeh. Gärt. Parkanl., 400. 1865 • Appalachian or Susquehanna sandcherry, cerisier de la Susquehanna [E]

Prunus susquehanae Willdenow, Enum. Pl., 519. 1809; *P. cuneata* Rafinesque; *P. pumila* var. *cuneata* (Rafinesque) L. H. Bailey; *P. pumila* subsp. *susquehanae* (Willdenow) R. T. Clausen

Stems usually erect-ascending, sometimes decumbent, 3–10 (–20) dm; twigs densely puberulent (10× magnification). **Leaf blades** elliptic, broadly elliptic, or obovate, 3–5.8(–7.2) × 1.1–2.5(–3) cm, lengths ca. 2.6 times widths, base obtuse to cuneate, apex usually obtuse, sometimes acute or rounded. **Drupes** subglobose, 8–12 × 5–10 mm; stones subglobose to ovoid, 6–8 × 5–6 mm. **2*n*** = 16.

Flowering Apr–Jun; fruiting Jul–Aug. Sandy pine-oak woods or barrens with open canopy, or adjacent fields and lakeshores; 30–500 m; Man., Ont., Que.; Ark., Conn., Del., Ind., Maine, Md., Mass., Mich., Minn., N.H., N.J., N.Y., N.C., Ohio, Pa., R.I., Tenn., Vt., Va., W.Va., Wis.

27. Prunus armeniaca Linnaeus, Sp. Pl. 1: 474. 1753 • Apricot [I]

Armeniaca vulgaris Lamarck

Trees, not suckering, 50–100 dm, not thorny. **Twigs** with axillary end buds, glabrous. **Leaves** deciduous; petiole (12–)20–45 mm, glabrous, glandular distally or on margins at bases of blades, glands 1–5; blade broadly ovate to suborbiculate, (3–)5–9 × (2–)4–8 cm, base usually obtuse to rounded, sometimes truncate or subcordate, margins singly to doubly crenate-serrate, teeth blunt, glandular, apex abruptly short-acuminate, abaxial surface with tufts of hairs in vein axils, adaxial glabrous. **Inflorescences** solitary flowers. **Pedicels** 1–3 mm, hairy. **Flowers** blooming before leaf emergence; hypanthium tubular-campanulate, 4–6 mm, glabrous or sparsely hairy externally; sepals reflexed, oblong-ovate, 4–6 mm, margins remotely glandular-toothed, surfaces sparsely hairy; petals white (pink in bud), broadly elliptic to suborbiculate, 8–12 mm; ovaries hairy. **Drupes** yellow to orange, often tinged with red, ellipsoid to globose, <laterally compressed>, 25–60 mm, velutinous; mesocarps fleshy; stones ellipsoid to subglobose, strongly flattened, <not pitted>. **2*n*** = 16.

Flowering Mar–Apr; fruiting May–Jul. Roadsides, abandoned plantings; 20–1600 m; introduced; Calif., Colo., Idaho, Iowa, Kans., Mich., Mo., Mont., N.Mex., Oreg., Pa., Utah, Va., Wash.; Asia (China).

Commercial production of apricots in North America is in the western United States, mostly in the San Joaquin Valley of California. There is little market for fresh apricots because of their extremely short shelf life; most fruits are preserved by drying.

28. Prunus cerasifera Ehrhart, Gartenkalender 4: 192. 1784 • Cherry plum, myrobalan [I]

Trees, sometimes suckering, 40–80 dm, not or slightly thorny. **Twigs** with axillary end buds, glabrous. **Leaves** deciduous; petiole 5–20 mm, glabrous except for a few hairs on adaxial surface, eglandular; blade ovate, elliptic, or obovate, 3–7 × 1.5–3.5 cm, base obtuse, margins singly to doubly crenate-serrate, teeth blunt, glandular, apex obtuse to acute, abaxial surface hairy along midribs and veins, adaxial glabrous. **Inflorescences** usually solitary flowers, sometimes 2-flowered fascicles. **Pedicels** (4–)10–18 mm, glabrous. **Flowers** blooming before leaf emergence; hypanthium campanulate, 2–4 mm, glabrous externally; sepals reflexed to spreading, oblong-ovate, 2–4 mm, margins glandular-toothed to nearly entire, <eciliate>, abaxial surface glabrous, adaxial hairy at bases; petals white (reddish pink in cultivars), elliptic to suborbiculate, 7–14 mm; ovaries glabrous. **Drupes** purple-red to yellow, <sometimes glaucous>, ovoid, ellipsoid, or globose, 15–30 mm, glabrous; mesocarps fleshy; stones ellipsoid to ovoid, ± to strongly flattened. **2*n*** = 16.

Flowering Feb–Apr; fruiting Jun–Aug. Roadsides, stream banks, canyons, chaparral; 0–900 m; introduced; B.C., Ont.; Calif., Idaho, Md., Mass., N.H., N.Y., Ohio, Oreg., Pa., Wash.; se Europe.

The purple-leaved, pink-flowered cultivars of *Prunus cerasifera* are especially popular for ornamental use. The earliest purple form was introduced into European gardens about 1880 by M. Pissard, gardener to the Shah of Iran. *Prunus cerasifera* is widely used as a rootstock for commercial plums.

29. Prunus spinosa Linnaeus, Sp. Pl. 1: 475. 1753 • Blackthorn, sloe I

Shrubs, suckering, 10–40 dm, thorny. **Twigs** with axillary end buds, hairy. **Leaves** deciduous; petiole 4–7 mm, hairy, eglandular; blade elliptic to obovate, 1.5–4 × 1–2.2 cm, base obtuse to rounded, margins crenulate-serrulate, teeth blunt, often glandular, apex acute to obtuse, abaxial surface hairy (especially along midribs and veins), adaxial glabrate. **Inflorescences** usually solitary flowers, sometimes 2-flowered fascicles. **Pedicels** 0.5–5(–8) mm, usually glabrous, rarely hairy. **Flowers** blooming before leaf emergence; hypanthium cupulate, 1.5–2.5 mm, glabrous externally; sepals spreading, oblong, 1.5–2.5 mm, margins glandular-toothed, surfaces glabrous or adaxially hairy at bases; petals white, elliptic, 4–8 mm; ovaries glabrous. **Drupes** bluish black, globose, 10–15 mm, glabrous; mesocarps fleshy; stones subglobose, ± flattened. **2*n*** = 32.

Flowering Mar–May; fruiting Aug–Sep. Roadsides; 0–1000 m; introduced; B.C., N.S., Ont.; Conn., Idaho, Maine, Mass., Mich., N.Y., Oreg., Wash.; Eurasia; n Africa.

As past flora writers (C. L. Hitchcock et al. 1955–1969; E. G. Voss 1972–1996) have noted, the distinctions between *Prunus spinosa* and *P. domestica* are not clear. Some researchers consider the hexaploid *P. domestica* to have been derived from the tetraploid *P. spinosa*, often in a scenario involving hybridization with *P. cerasifera*. It should not be surprising that some of the characters used in keys to separate these three taxa (spininess, indument, leaf size, pedicel length, numbers of flowers per bud) are subject to variation within each species and overlap among the species.

30. Prunus domestica Linnaeus, Sp. Pl. 1: 475. 1753 • European plum, prunier damas I

Prunus domestica subsp. *insititia* (Linnaeus) C. K. Schneider; *P. domestica* var. *insititia* (Linnaeus) Fiori & Paoletti; *P. insititia* Linnaeus

Shrubs or trees, sometimes suckering, 20–60(–100) dm, not or slightly thorny. **Twigs** with axillary end buds, usually hairy, sometimes glabrous. **Leaves** deciduous; petiole 6–20 mm, glabrous or hairy on adaxial surface or both surfaces, eglandular or glandular distally or on margins at bases of blades, glands 1–3; blade elliptic to obovate, (2.5–)4–7(–9) × 1.5–5 cm, base cuneate to obtuse, margins singly to doubly crenate-serrate, teeth blunt, glandular, apex usually acute to abruptly acuminate, sometimes obtuse, abaxial surface hairy (especially along veins), adaxial glabrous or midribs hairy. **Inflorescences** solitary flowers or 2-flowered fascicles. **Pedicels** (2–)10–20 mm, glabrous or hairy. **Flowers** blooming before or at leaf emergence; hypanthium cupulate, 3–5 mm, glabrous externally; sepals spreading to reflexed, ovate-elliptic to lanceolate, 3.5–6 mm, margins glandular-toothed, <ciliate>, surfaces glabrous or hairy; petals white, oblong to suborbiculate, 7–14 mm; ovaries glabrous. **Drupes** blue-black (green, yellow, or red in cultivars), ellipsoid to globose, 15–35 mm, glabrous; mesocarps fleshy; stones ovoid to ellipsoid, strongly flattened. **2*n*** = 48.

Flowering Apr–May; fruiting Aug–Sep. Roadsides, fencerows, abandoned homesites; 0–1500 m; introduced; N.B., N.S., Ont., Que.; Calif., Conn., Del., Idaho, La., Maine, Md., Mass., Mich., Nev., N.H., N.J., N.Y., Ohio, Oreg., Pa., R.I., Utah, Vt., Va., Wash.; Eurasia.

The distinctions given in the literature between *Prunus domestica* and *P. insititia*, and which cultivated varieties are derived from each of them, are contradictory and confusing. After examining plants from North America, it seems best to treat these Eurasian introductions as one variable species without designation of infraspecific taxa.

European plums are grown along the West Coast, primarily in California, for prunes and other processed food. They are also grown in the Great Lakes region of Ontario, Michigan, and New York for both prunes and the local fresh market. Most of the commercial fresh plum market is supplied by fruits of the Japanese plum, *Prunus salicina*, and its hybrid derivatives, which include genetic material from native American species, for example, *P. americana*, *P. angustifolia*, and *P. hortulana*.

31. Prunus glandulosa Thunberg in J. A. Murray, Syst. Veg. ed. 14, 463. 1784 • Dwarf flowering almond [I]

Shrubs, suckering, 5–20 dm, not thorny. **Twigs** with terminal end buds, glabrous. **Leaves** deciduous; petiole 1–6 mm, hairy on adaxial surface, eglandular; blade narrowly elliptic to oblong-lanceolate, 2.5–9 × 0.9–2.5 cm, base cuneate to obtuse, margins crenulate-serrulate, teeth blunt, glandular, apex acute to acuminate, abaxial surface glabrous or slightly hairy along midribs and veins, adaxial glabrous. **Inflorescences** solitary flowers or 2-flowered fascicles. **Pedicels** 6–15 mm, glabrous or sparsely puberulent. **Flowers** blooming at leaf emergence; hypanthium campanulate, 2–3 mm, glabrous or sparsely puberulent externally; sepals spreading to reflexed, ovate, 2.5–5 mm, margins glandular-toothed, <± ciliate>, surfaces glabrous or sparsely puberulent; petals <25+>, white or pink, elliptic to narrowly elliptic, 8–11 mm; ovaries glabrous. **Drupes** reddish, subglobose, 10–15 mm, glabrous; mesocarps fleshy; stones subglobose, not flattened.

Flowering Mar–May. Roadsides, fencerows, abandoned plantings; 0–400 m; introduced; Ont.; Iowa, Mich., N.C.; e Asia (China, Japan).

Double-flowered cultivars of *Prunus glandulosa*, which lack stamens and carpels, are commonly planted as foundation shrubs and in borders. Although the plants are attractive in bloom, some horticulturists consider them of little ornamental value otherwise.

32. Prunus subcordata Bentham, Pl. Hartw., 308. 1849 • Sierra or Klamath or Pacific plum [E]

Prunus subcordata var. *kelloggii* Lemmon; *P. subcordata* var. *oregana* (Greene) W. Wight; *P. subcordata* var. *rubicunda* Jepson

Shrubs or trees, often suckering, 10–30(–60) dm, sometimes thorny. **Twigs** with axillary end buds, glabrous or hairy. **Leaves** deciduous; petiole 4–18 mm, glabrous or hairy, usually glandular distally, glands 1–3, <discoid>; blade elliptic, oblong-ovate, or suborbiculate, 2–5(–6.5) × 1.3–3.5(–4.5) cm, base rounded or subcordate, margins singly to doubly serrulate, teeth blunt, glandular, apex usually obtuse to rounded, rarely acute, abaxial surface glabrous or hairy, adaxial glabrous or appressed-hairy. **Inflorescences** 2–5-flowered, umbellate fascicles. **Pedicels** 5–15 mm, glabrous or hairy. **Flowers** blooming at leaf emergence; hypanthium campanulate, 2–3.5 mm, glabrous or hairy externally; sepals broadly spreading to reflexed, obovate, 2.2–4(–5) mm, margins glandular-toothed, sometimes obscurely so, <sometimes ciliate>, surfaces glabrous or hairy; petals white, oblong-obovate, 5–10 mm; ovaries usually glabrous, rarely hairy. **Drupes** usually red to reddish purple, sometimes yellow, globose to ellipsoid, 15–25 mm, usually glabrous, rarely puberulent; mesocarps fleshy; stones ovoid, ± flattened.

Flowering Mar–May; fruiting Aug–Sep. Stream banks, dry rocky slopes, chaparral, pine-oak and juniper-oak woodlands; 100–1900 m; Calif., Oreg.

Prunus subcordata shows variation in color, size, indument, and palatability of fruits, which has been the basis for recognition of taxonomic segregates. Variety *kelloggii* was described as a larger plant with less hairy foliage than in typical *P. subcordata*, and its fruits are yellow, larger, sweeter, and more pulpy. Variety *rubicunda* is a smaller shrub with bright red, bitter fruits. Most noteworthy is var. *oregana* with densely hairy ovaries and puberulent fruits, recorded only from the sparsely inhabited border region between northeastern California and adjacent Oregon. Specimens in herbaria are few and recent collections are lacking. The assignment of *P. texana* with hairy fruits to the plums (J. Shaw and R. L. Small 2005) makes this variant all the more interesting. Field study and collection followed by morphologic, genetic, and molecular study are needed to clarify its systematic position and significance.

33. Prunus texana D. Dietrich, Syn. Pl. 3: 45. 1842 • Peachbush, Texas wild peach [E]

Amygdalus glandulosa Hooker, Icon. Pl. 3: plate 288. 1840, not *Prunus glandulosa* Thunberg 1784

Shrubs, sometimes suckering, <much branched>, 5–15 dm, sometimes weakly thorny. **Twigs** with axillary end buds, tomentose. **Leaves** deciduous; petiole 1–4 mm, tomentose, eglandular; blade elliptic to oblong-elliptic, 1.1–4 × 0.4–1.1 cm, base cuneate to obtuse, margins dentate, teeth blunt, glandular, <glands discoid>, apex acute, obtuse, or rounded, abaxial surface hairy to canescent, adaxial hairy. **Inflorescences** solitary flowers or 2-flowered fascicles. **Pedicels** 1–5 mm, tomentose. **Flowers** blooming before or at leaf emergence; hypanthium campanulate, 2 mm, tomentose externally; sepals reflexed, triangular, 1–1.7 mm, margins glandular-toothed, surfaces tomentose; petals white, elliptic, 3–5 mm; ovaries hairy. **Drupes** usually yellow to greenish yellow, sometimes tinged with red, ovoid, <compressed>, 8–15 mm, velutinous; <hypanthium tardily deciduous>; mesocarps fleshy; stones ovoid, ± flattened.

Flowering Feb–Mar; fruiting Apr–Jun. Deep sand, plains and sand hills, grasslands, oak woods; 0–200 m; Tex.

Prunus texana is endemic to south-central Texas from the Edwards Plateau southeast to the coastal plain.

Despite the peachlike fruits of *Prunus texana*, DNA evidence supports its placement among the native American plums (J. Shaw and R. L. Small 2005). The leaf margins look like those of no other North American species of *Prunus*. The teeth project perpendicular to the margins and are capped with disc-shaped glands.

34. Prunus geniculata R. M. Harper, Torreya 11: 67. 1911 • Scrub plum [C] [E]

Shrubs, sometimes suckering, <much branched>, 5–10(–20) dm, thorny. **Twigs** with axillary end buds, hairy. **Leaves** deciduous; petiole 3–6 mm, hairy on adaxial surface, sometimes glandular distally, glands 1 or 2, <discoid>; blade elliptic, 0.8–2.5 × 0.4–1.3 cm, base obtuse to rounded, margins crenulate-serrulate in distal ½, nearly entire on smaller leaves, teeth blunt, glandular, apex obtuse to rounded, mucronate, surfaces glabrous. **Inflorescences** solitary flowers. **Pedicels** 0–3 mm, glabrous. **Flowers** <usually bisexual with some staminate flowers on same plant>, blooming before leaf emergence; hypanthium campanulate, 2–3 mm, glabrous externally; sepals erect to spreading, ovate, 1–1.5 mm, margins entire, <ciliate>, abaxial surface glabrous, adaxial hairy; petals white, elliptic, 2 mm; ovaries glabrous. **Drupes** reddish, ovoid, 12–25 mm, glabrous; mesocarps fleshy; stones ovoid, ± flattened.

Flowering Jan–Feb; fruiting Mar–May. Longleaf pine-turkey oak sandhills, pine-evergreen oak scrub; of conservation concern; 10–50 m; Fla.

Prunus geniculata is known only from the Lake Wales Ridge in central Florida. Much of its original habitat has been lost to residential, recreational, and agricultural development; it received federal listing as an endangered species in 1987.

35. Prunus americana Marshall, Arbust. Amer., 111. 1785 • Wild or American plum, prunier d'Amérique [E]

Shrubs or trees, suckering, 15–80 dm, moderately thorny. **Twigs** with axillary end buds, usually hairy, sometimes glabrous. **Leaves** deciduous; petiole 4–19 mm, usually hairy on adaxial surface, sometimes on both surfaces, rarely glabrous, usually eglandular, sometimes glandular distally, glands 1 or 2, <discoid>; blade usually elliptic, broadly elliptic, or obovate, rarely ovate, 5–11 × 2–5.5 cm, base usually cuneate to obtuse, sometimes rounded, margins <coarsely>, doubly serrate, teeth sharp, eglandular, apex usually abruptly acuminate, rarely acute, surfaces glabrous or sparsely hairy along main veins. **Inflorescences** 2–5-flowered, umbellate fascicles. **Pedicels** (4–)8–20 mm, usually glabrous, sometimes hairy. **Flowers** blooming before or at leaf emergence; hypanthium obconic, 2.5–5 mm, usually glabrous, sometimes hairy, externally; sepals broadly spreading to reflexed, ovate to lanceolate, 2–3.5(–5) mm, margins entire or irregularly or obscurely glandular-toothed, sometimes 2-fid at apices, <ciliate>, abaxial surface glabrous or hairy, adaxial tomentose; petals white, oval to oblong-obovate, 7–12 mm; ovaries glabrous. **Drupes** red, orange, or yellowish, <glaucous>, subglobose to ellipsoid, 15–30 mm, glabrous; mesocarps fleshy; stones ovoid, strongly flattened. $2n = 16$.

Flowering Apr–Jun; fruiting Jul–Sep. Thickets, moist soil, roadsides, fence rows, margins of woods, stream banks; 10–2100 m; Man., N.B., Ont., Que., Sask.; Ala., Ark., Colo., Conn., Del., D.C., Fla., Ga., Idaho, Ill., Ind., Iowa, Kans., Ky., La., Maine, Md., Mass., Mich., Minn., Miss., Mo., Mont., Nebr., N.H., N.J., N.Mex., N.Y., N.C., N.Dak., Ohio, Okla., Oreg., Pa., R.I., S.C., S.Dak., Tenn., Utah, Vt., Va., Wash., W.Va., Wis., Wyo.

Of the native plum species, *Prunus americana* is the most widely distributed and abundant in terms of individual plants. As might be expected, it is variable. U. P. Hedrick (1911) reported that about 260 horticultural varieties had been derived from it, far more than from any other North American plum. Not infrequently, individual specimens possess characteristics intermediate between *P. americana* and other plum species, probably the result of gene exchange, which seems to be common among American plums (J. Shaw and R. L. Small 2005).

The most obvious variation in *Prunus americana* is the presence or absence of indument on twigs, pedicels, hypanthia, sepals, petioles, and abaxial surfaces of leaves. The hairy form has been known by the epithets *mollis* and *lanata*, both as varieties of *P. americana* and at species rank. The type of *P. mollis*, upon which all of these names are ultimately based, is conspecific with that of *P. nigra*, and thus *mollis* and *lanata* are misapplied as infraspecific epithets of *P. americana* (L. H. Shinners 1956b). The hairy form is most abundant in the western part of the species range, where it is at least as prevalent as the glabrous form; hairy individuals are found throughout the range of the species. As in other species of *Prunus*, variation in indument is an easily noticed difference among individuals but not one of taxonomic significance.

In Missouri, Arkansas, and neighboring states, hairy *Prunus americana* intergrades with *P. mexicana*. In this zone of sympatry, they hybridize and produce intermediate forms, making determination of some herbarium specimens from this region difficult, if not arbitrary. One solution, not chosen here, would be to combine *P. americana* and *P. mexicana* into a single species. Tradition and morphologic distinctions argue for keeping them separate until more data can clarify the entire North American plum complex. Typically, *P. americana* is a shrub or tree to 80 dm tall and the plants sucker to form thickets; *P. mexicana* is a single tree 30–120 dm tall, rarely suckering. The leaves of *P. americana* are longer in proportion to their width than those of *P. mexicana*, and the leaf bases are cuneate in *P. americana* versus rounded or subcordate in *P. mexicana*. In hairy individuals of *P. americana*, the form most likely to be confused with *P. mexicana*, twigs are almost always hairy; about 70% of all *P. mexicana* specimens examined have glabrous twigs.

The following names have been misapplied to *Prunus americana*, though the types belong with that of *P. nigra*: *Prunus americana* var. *lanata* Sudworth, *P. americana* var. *mollis* (Torrey) Torrey & A. Gray, *P. lanata* (Sudworth) Mackenzie & Bush, and *P. mollis* Torrey.

36. Prunus mexicana S. Watson, Proc. Amer. Acad. Arts 17: 353. 1882 • Mexican or bigtree plum

Trees, rarely suckering, 30–120 dm, sparsely thorny. **Twigs** with axillary end buds, usually glabrous, sometimes hairy. **Leaves** deciduous; petiole 4–18 mm, hairy, rarely only adaxially, usually glandular distally, glands 1 or 2(–4); blade usually elliptic, sometimes broadly elliptic, obovate, ovate, or oblong, 6–12 × 3–7 cm, base usually obtuse to broadly rounded, sometimes subcordate, margins <coarsely>, doubly serrate, teeth sharp, eglandular, apex usually abruptly acuminate to acuminate, rarely acute, abaxial surface densely hairy, adaxial <rugose>, glabrous or hairy. **Inflorescences** 2–5 -flowered, umbellate fascicles. **Pedicels** 4–20 mm, usually glabrous, sometimes sparsely hairy. **Flowers** blooming before leaf emergence; hypanthium obconic, 2–4.5 mm, glabrous or hairy externally; sepals reflexed, ovate to lanceolate, 1.5–4 mm, margins entire or irregularly or obscurely glandular-toothed, sometimes 2-fid at apices, <ciliate>, surfaces hairy; petals white, sometimes turning pink, elliptic to obovate, 5–10 mm; ovaries glabrous. **Drupes** purplish red to dark blue, <glaucous>, subglobose to ellipsoid, 15–30 mm, glabrous; mesocarps fleshy; stones ovoid-ellipsoid, strongly flattened.

Flowering Mar–Apr; fruiting Aug–Oct. Stream bottoms, open woods, edges of woods; 10–400 m; Ala., Ark., Ill., Ind., Iowa, Kans., Ky., La., Miss., Mo., Okla., Tenn., Tex.; Mexico (Coahuila, Nuevo León, San Luis Potosí, Tamaulipas).

As L. H. Shinners (1956b) pointed out, plums from Texas determined as *Prunus americana* var. *lanata* Sudworth are not *P. americana* but *P. mexicana*. This was extrapolated by others to mean that all hairy *americana*-type plums, which had been called var. *lanata*, are *P. mexicana*. As a consequence, *P. mexicana* is often listed from states far from where it truly occurs (for example, North Carolina, Ohio, Wisconsin) and herbarium sheets of hairy *P. americana* can often be found filed under *P. mexicana*.

37. Prunus nigra Aiton, Hort. Kew. 2: 165. 1789 • Canada plum, prunier noir [E]

Shrubs or trees, sometimes suckering, 30–90 dm, moderately thorny. **Twigs** with axillary end buds, usually glabrous, rarely hairy. **Leaves** deciduous; petiole 8–22 mm, glabrate with hairs adaxially, sometimes glandular distally, glands 1–3, <discoid>; blade broadly elliptic to obovate, (5–)7–11 × 3–6.5 cm, base obtuse to rounded or subcordate, margins doubly crenate-serrate, teeth blunt, glandular, apex abruptly acuminate, abaxial surface hairy along midribs and major veins, adaxial glabrous. **Inflorescences** 2–4 -flowered, umbellate fascicles. **Pedicels** 7–20 mm, usually glabrous, rarely hairy. **Flowers** blooming before or at leaf emergence; hypanthium <red-tinged>, obconic, 3–4(–5) mm, usually glabrous, rarely hairy externally; sepals broadly spreading to reflexed, ovate, 2–4(–5) mm, margins glandular-toothed, abaxial surface usually glabrous, rarely sparsely hairy, adaxial glabrous or hairy; petals white, often fading to pink, suborbiculate to oblong-obovate, 8–13 mm; ovaries glabrous. **Drupes** red, orange, or yellowish, <barely glaucous>, globose to ellipsoid, 15–30 mm, glabrous; mesocarps fleshy; stones ovoid-ellipsoid, strongly flattened. $2n = 16$.

Flowering Apr–Jun; fruiting Aug–Sep. Borders of deciduous woods, bottomland forests, roadside thickets; 10–800 m; Man., N.B., N.S., Ont., Que.; Conn., Ill., Ind., Iowa, Maine, Mass., Mich., Minn., N.H., N.Y., N.Dak., Ohio, Vt., Wis.

In the northeastern United States and adjacent Canada, *Prunus nigra* co-occurs with, and is sometimes confused with, *P. americana*, despite being distinct in both flower and leaf. The red-tinged hypanthia and sepals of *P. nigra* give the entire inflorescence a pinkish coloration even when the petals are pure white, and the sepals bear obvious glandular teeth along their margins; in *P. americana* the hypanthia and sepals are green, and the sepals bear relatively few glandular teeth or are eglandular. The leaf marginal teeth of *P. nigra* are rounded and bear glands at their tips or have callus scars where the glands fell off; those of *P. americana* are acute and eglandular; some may have a callused tip.

38. Prunus maritima Marshall, Arbust. Amer., 112. 1785 • Beach plum E F

Prunus gravesii Small; *P. maritima* var. *gravesii* (Small) G. J. Anderson

Shrubs, suckering, 10–25 dm, not thorny. **Twigs** with axillary end buds, densely hairy to glabrate. **Leaves** deciduous; petiole 3–11 mm, usually densely hairy, rarely hairy only on adaxial surface, usually eglandular, sometimes glandular distally, glands 1 or 2, <discoid>; blade elliptic, broadly elliptic, or obovate (orbiculate in *gravesii* clone), 3.5–7 × 2–4 cm, base cuneate to obtuse (rounded in *gravesii* clone), margins <finely>, usually singly, rarely doubly, serrulate, teeth sharp, usually eglandular, rarely glandular, apex acute to obtuse (rounded in *gravesii* clone), abaxial surface usually hairy, sometimes only on veins or glabrescent, adaxial <rugose>, glabrous. **Inflorescences** 2–4-flowered, umbellate fascicles. **Pedicels** 5–15 mm, hairy. **Flowers** blooming before leaf emergence; hypanthium campanulate, 2.5–3.5 mm, hairy externally; sepals erect to spreading, ovate, 1.5–3.5 mm, margins entire, sometimes 2-fid at apices, <ciliate>, surfaces hairy; petals white, oblong-obovate, 3–8 mm; ovaries glabrous. **Drupes** red, yellow, dark blue, or purplish black, <glaucous>, globose, 11–18 mm, glabrous; mesocarps fleshy; stones ovoid, slightly flattened. $2n = 16$.

Flowering Apr–Jun; fruiting Aug–Oct. Sand dunes, beaches, sandy fields, roadsides, coastal barrens; 0–30 m; Conn., Del., Maine, Md., Mass., N.H., N.J., N.Y., Pa., R.I., Va.

Prunus gravesii has been known from only a single locality since its description in 1897, and there may never have been more than a single, multistemmed individual. G. J. Anderson (1980) concluded that it is a mutant derivative of *P. maritima*. In the study by Anderson, it differed with statistical significance from other specimens of *P. maritima* for only 4 of 38 morphological features measured, and even for those four the range in values between the two taxa overlapped. Their leaf flavonoid chromatograms were identical. From a conservation standpoint it is of value to designate the orbiculate-leaved *gravesii* clone as a distinct taxonomic variety, for which the name *P. maritima* var. *gravesii* is available. Once listed by the state of Connecticut as endangered, it is now believed to be extinct in the wild. Throughout much of its original range *P. maritima* is threatened by oceanside development.

39. Prunus gracilis Engelmann & A. Gray, Boston J. Nat. Hist. 5: 243. 1845 • Oklahoma plum E

Shrubs, suckering, 3–15 dm, not thorny. **Twigs** with axillary end buds, densely hairy. **Leaves** deciduous; petiole 3–8 mm, densely hairy, usually eglandular, sometimes glandular distally, glands 1–3, <discoid>; blade usually ovate or elliptic, rarely obovate, 2–5(–7) × 1–2.5(–3.6) cm, base obtuse to cuneate, margins <finely>, singly or doubly serrulate, teeth sharp, usually eglandular, sometimes glandular, <glands blackish, spheric>, apex acute to obtuse, abaxial surface densely hairy, adaxial sparsely hairy. **Inflorescences** 2–4(–6)-flowered, umbellate fascicles. **Pedicels** 4–15 mm, hairy. **Flowers** blooming before or at leaf emergence; hypanthium campanulate, 2–3 mm, hairy externally; sepals erect to spreading, ovate-oblong, 1.5–2.5 mm, margins usually entire, sometimes glandular-toothed or eglandular, <sometimes ciliate>, surfaces hairy; petals white, oblong to obovate, 4–7 mm; ovaries glabrous. **Drupes** yellow, orange, or red, <slightly glaucous>, globose to ellipsoid, 9–18 mm, glabrous; mesocarps fleshy; stones subglobose to ellipsoid, ± flattened.

Flowering Mar–Apr; fruiting Jun–Aug. Sandy roadsides, upland thickets, open woods, waste places; 100–1300 m; Ark., Colo., Kans., La., N.Mex., Okla., Tex.

Typically, sepals of *Prunus gracilis* are eglandular; in some specimens, glands occur along the margins. Similar glands are sometimes found on the tips of teeth along the leaf margins. The teeth are always sharp-pointed rather than blunt or rounded as they are in those plums whose leaves consistently have glandular teeth. Presence of glands on the sepals is not correlated with glands on leaf marginal teeth. *Prunus gracilis*, native to the southern Great Plains, is very similar to *P. maritima*, a plum of the eastern seaboard over 1600 km distant.

40. Prunus umbellata Elliott, Sketch Bot. S. Carolina 1: 541. 1821 • Hog or flatwoods or Allegheny plum, sloe E

Prunus alleghaniensis Porter; *P. alleghaniensis* var. *davisii* W. Wight; *P. injucunda* Small; *P. mitis* Beadle; *P. umbellata* var. *injucunda* (Small) Sargent

Shrubs or trees, sometimes suckering, 10–60 dm, moderately thorny. **Twigs** with axillary end buds, usually glabrous, sometimes hairy. **Leaves** deciduous; petiole 3–14 mm, hairy, usually eglandular, sometimes hairy only adaxially, glandular distally, glands 1–3, <discoid>; blade usually elliptic to broadly elliptic, sometimes oblanceolate to obovate, 3.5–8 × 1.5–4 cm, base usually cuneate to obtuse, rarely rounded, margins <finely>, usually singly serrulate, sometimes doubly serrate, teeth sharp, usually eglandular, sometimes glandular, <glands blackish, spheric>, apex usually acute, sometimes short-acuminate, abaxial surface hairy to glabrate, adaxial glabrous. **Inflorescences** 2–4(–6)-flowered, umbellate fascicles. **Pedicels** 5–22 mm, usually glabrous, sometimes hairy. **Flowers** blooming before or at leaf emergence; hypanthium tubular (often tubular-urceolate when dried), 2–4 mm, glabrous or hairy externally; sepals erect-spreading, ovate-oblong, 1.5–2.5 mm, margins usually entire, sometimes 2-fid at apices, <ciliate>, abaxial surface hairy or glabrous, adaxial hairy; petals white, sometimes turning pink, obovate to suborbiculate, 3–8 mm; ovaries glabrous. **Drupes** red, yellow, dark blue, or nearly black, <glaucous>, globose, 10–15 mm, glabrous; mesocarps fleshy; stones ovoid, slightly to ± flattened.

Flowering Feb–May; fruiting Jul–Sep. Sandy pine or oak woods, sandy barrens, shale ridges, limestone bluffs, rocky upland woods, old fields, roadsides; 10–800 m; Ala., Ark., Conn., Fla., Ga., La., Md., Mass., Mich., Miss., N.C., Pa., S.C., Tenn., Tex., Va., W.Va.

Traditionally treated as distinct species, *Prunus alleghaniensis* from the Appalachians and *P. umbellata* from southeastern United States are very similar; their ranges overlap in North Carolina and Tennessee. Similar plants disjunct in Michigan have been called *P. alleghaniensis* var. *davisii*. Morphological characters (petiole length, shape of blade along with its base and apex, degree of suckering) that have been used to separate these taxa vary as much within each taxon as among them. Since they can be separated only by geographic distribution, they are combined in this treatment.

Some specimens from Connecticut with hairy twigs, petioles, pedicels, and hypanthia were determined as *Prunus alleghaniensis* by Eames a century ago and do seem to fit within *P. umbellata*. Other Connecticut specimens determined as *P. alleghaniensis* (for example, *Eames 121*, MICH) with scattered glandular teeth on the sepals seem better placed in *P. americana*. The sole specimen known from Massachusetts (*Pease 10,005*, NEBC) has singly serrate leaves evenly tapered at both ends as in *P. umbellata*, but larger petals and reflexed sepals like those of *P. americana*.

In the southeastern United States, hairy forms have been called *Prunus injucunda* or *P. mitis;* the degree of indument on the twigs, petioles, leaf surfaces, pedicels, hypanthia, and sepals is subject to much trivial variation throughout the southeast, the Appalachians, and in Michigan.

41. **Prunus angustifolia** Marshall, Arbust. Amer., 111. 1785 • Chickasaw plum [E]

Prunus angustifolia subsp. *varians* W. Wight & Hedrick; *P. angustifolia* var. *watsonii* (Sargent) Waugh

Shrubs or trees, often suckering, 10–50 dm, thorny. **Twigs** with axillary end buds, glabrous. **Leaves** deciduous; petiole 2–14 mm, usually sparsely hairy on adaxial surface, rarely glabrous or hairy on both surfaces, usually eglandular, sometimes glandular distally, glands 1 or 2; blade lanceolate to narrowly elliptic, <usually folded along midribs, often ± falcate>, 1.5–6 × 0.8–2 cm, base cuneate to obtuse, margins crenulate-serrulate, teeth blunt, glandular, <glands reddish orange, conic>, apex acute, abaxial surface glabrate with hairs along midribs and major veins, adaxial glabrous. **Inflorescences** 2–4-flowered, umbellate fascicles. **Pedicels** 3–10 mm, glabrous. **Flowers** blooming before, sometimes at, leaf emergence; hypanthium campanulate, 1.5–3 mm, usually glabrous, rarely hairy, externally; sepals erect to spreading, ovate, 1–2 mm, margins entire, <sparsely ciliate>, abaxial surface glabrous, adaxial hairy; petals white, suborbiculate to obovate, 3–6 mm; ovaries glabrous. **Drupes** red to yellow, <lightly glaucous>, globose to ellipsoid, 15–20 mm, glabrous; mesocarps fleshy; stones ovoid, ± flattened. $2n = 16$.

Flowering Feb–Apr; fruiting May–Aug. Thickets, upland sandy soil, open woods, sand dunes, fence rows, pastures, roadsides, stream bottoms; 0–600 m; Ala., Ark., Colo., Del., D.C., Fla., Ga., Ill., Ind., Kans., Ky., La., Md., Miss., Mo., Nebr., N.J., N.Mex., N.C., Okla., Pa., S.C., Tenn., Tex., Va., W.Va.

A naturalized population of *Prunus angustifolia* in Santa Barbara County, California, has been extirpated.

Prunus angustifolia is one of the more distinctive plum species in North America. The leaves are relatively small, usually folded along their midribs, and have relatively large, reddish orange, conic glands along the margins. When it hybridizes with other plums, the hybrids often have distinctive characteristics and have been named as species more often than hybrids not involving *P. angustifolia*: *P.* ×*orthosepala* Koehne (*angustifolia* × *americana*), *P.* ×*slavinii* E. J. Palmer ex Rehder (*angustifolia* × *gracilis*), *P.* ×*utahensis* Dieck ex Koehne (*angustifolia* × *pumila* var. *besseyi*); see also discussion under 42. *P. rivularis*.

42. Prunus rivularis Scheele, Linnaea 21: 594. 1848 • Creek or hog plum E

Prunus munsoniana W. Wight & Hedrick; *P. reverchonii* Sargent

Shrubs or trees, usually suckering, 10–80 dm, thorny. **Twigs** with axillary end buds, glabrous. **Leaves** deciduous; petiole 7–21 mm, glabrous or sparsely hairy on adaxial surface, glandular distally, glands 1–4; blade lanceolate, oblong-lanceolate, or elliptic to narrowly elliptic, <usually folded along midribs>, 4–11 × 1.5–5 cm, base obtuse to rounded, margins singly to doubly crenate-serrulate, teeth blunt, glandular, <glands inconspicuous, blackish>, apex usually acuminate, sometimes acute, abaxial surface glabrous or ± hairy along midribs and veins, adaxial usually glabrous, rarely midribs hairy. **Inflorescences** 2–4-flowered, umbellate fascicles. **Pedicels** 3–15 mm, glabrous. **Flowers** blooming before or at leaf emergence; hypanthium campanulate, 2–3 mm, usually glabrous, rarely glabrate, externally; sepals broadly spreading to reflexed, oblong-ovate, 1.5–2.5 mm, margins glandular-toothed, abaxial surface glabrous or sparsely hairy, adaxial densely hairy at bases; petals white, obovate, 4–7 mm; ovaries glabrous. **Drupes** usually red, sometimes yellowish orange, with white dots, <not or only slightly glaucous>, globose, 12–25 mm, glabrous; mesocarps fleshy; stones ovoid to subglobose, ± flattened.

Flowering Mar–Apr; fruiting Jul–Sep. Stream banks, roadside thickets, prairie hillsides, borders of woods; 200–1000 m; Ala., Ark., Calif., Colo., Ga., Ill., Ind., Kans., Ky., La., Md., Miss., Mo., N.J., N.C., Ohio, Okla., Pa., Tenn., Tex.

Prunus munsoniana is here placed in synonymy with *P. rivularis*, greatly expanding the range of the latter compared to its traditional treatment. Descriptions of *P. munsoniana* and *P. rivularis* differ mostly in size of the plants, leaves, and floral structures. As suggested by G. M. Diggs et al. (1999), *P. munsoniana* is simply a larger version of *P. rivularis*.

Evidence from analysis of DNA sequences of two nuclear genes in native plums (J. R. Rohrer et al. 2008) indicates that *Prunus rivularis* may be an allopolyploid. Three plants sequenced from Texas contained alleles seemingly derived from *P. angustifolia* on the one hand and *P. hortulana* (or *P. murrayana*) on the other. Further, two individuals determined as *P. munsoniana*, one each from Pennsylvania and Texas, had sequences very similar to those of *P. rivularis*.

Other specimens in herbaria labeled as *Prunus munsoniana* include hybrids between *P. americana* and *P. angustifolia*, as well as individuals of *P. hortulana* that bloom before the leaves emerge or that sucker freely. *Prunus rivularis* is introduced in California.

43. Prunus hortulana L. H. Bailey, Gard. & Forest 5: 90. 1892 • Hortulan or wild goose plum E F

Prunus hortulana var. *mineri* L. H. Bailey

Trees, rarely suckering, 40–100 dm, moderately thorny. **Twigs** with axillary end buds, glabrous. **Leaves** deciduous; petiole 6–20 mm, hairy on adaxial surface, usually glandular distally, glands 1–5; blade narrowly elliptic, lanceolate, oblanceolate, or oblong-obovate, (5–)7–11(–13) × (2–)3–5.5 cm, base obtuse to rounded, margins singly to doubly crenate-serrate, teeth blunt, glandular, <glands inconspicuous, blackish>, apex long-acuminate, abaxial surface ± hairy along midribs and veins, adaxial glabrous or with hairs along midribs. **Inflorescences** 2–4-flowered, umbellate fascicles. **Pedicels** 8–20 mm, glabrous. **Flowers** blooming before or at leaf emergence; hypanthium campanulate, 2–3 mm, glabrous externally; sepals erect to reflexed, ovate, 1.5–3 mm, margins glandular-toothed, abaxial surface glabrous or sparsely hairy, adaxial densely hairy at bases; petals white, obovate, 4–9 mm; ovaries glabrous. **Drupes** red to yellowish with white dots, <not or only slightly glaucous>, globose, 20–30(–40) mm, glabrous; mesocarps fleshy; stones ovoid-ellipsoid, ± flattened. $2n = 16$.

Flowering Apr–May; fruiting Aug–Sep. Roadside thickets, flood plains, open woodlands; 50–500 m; Ark., Ill., Ind., Iowa, Kans., Ky., Mass., Mich., Mo., Nebr., Ohio, Okla., Tenn., Va., W.Va.

44. Prunus murrayana E. J. Palmer, J. Arnold Arbor. 10: 38. 1929 • Murray's plum E

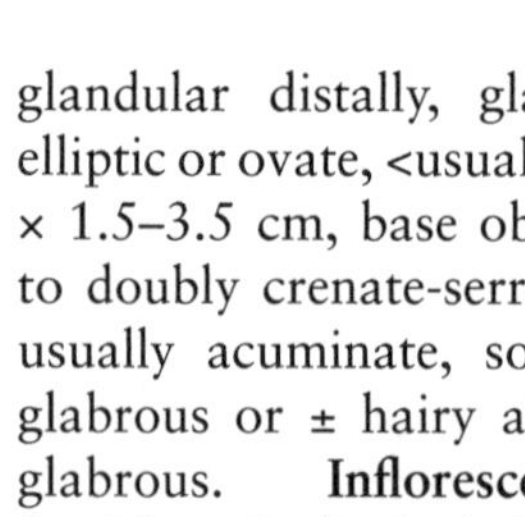

Prunus rivularis Scheele var. *pubescens* Enquist

Shrubs, usually suckering, 5–30(–50) dm, ± thorny. **Twigs** with axillary end buds, hairy. **Leaves** deciduous; petiole 8–16 mm, usually hairy on both surfaces, sometimes only on adaxial surface, usually glandular distally, glands 1–4; blade lanceolate to elliptic or ovate, <usually folded along midribs>, 3.5–7.5 × 1.5–3.5 cm, base obtuse to rounded, margins singly to doubly crenate-serrate, teeth blunt, glandular, apex usually acuminate, sometimes acute, abaxial surface glabrous or ± hairy along midribs and veins, adaxial glabrous. **Inflorescences** 1–4-flowered, umbellate fascicles. **Pedicels** 4–14 mm, hairy. **Flowers** blooming at leaf emergence; hypanthium campanulate, 2–2.5 mm, hairy externally; sepals erect to reflexed, ovate, 1.5–2 mm, margins glandular-toothed, abaxial surface

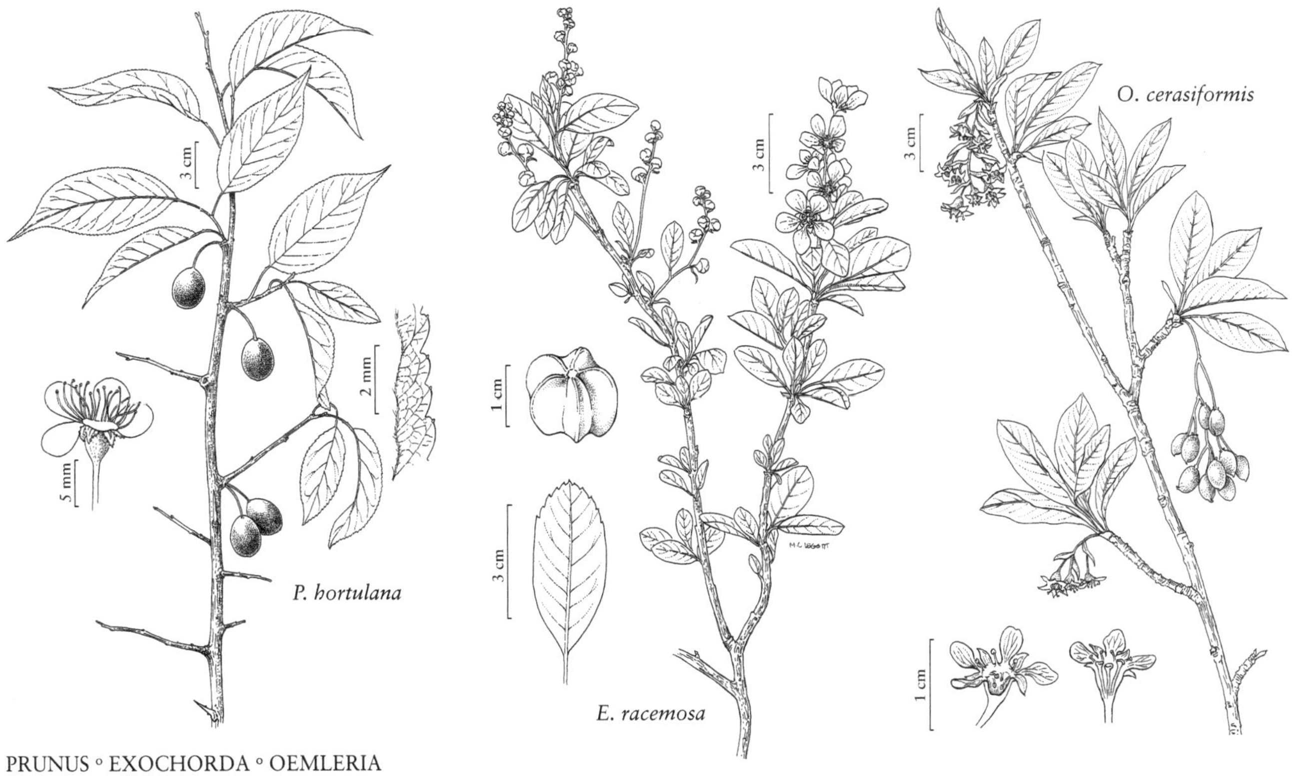

PRUNUS ◦ EXOCHORDA ◦ OEMLERIA

glabrate or hairy, adaxial densely hairy at bases; petals white, obovate, 4–8 mm; ovaries glabrous. **Drupes** red with white dots, <slightly to quite glaucous>, globose, 10–18 mm, glabrous; mesocarps fleshy; stones ovoid, ± flattened.

Flowering Mar–Apr; fruiting Jul–Aug. Rocky stream banks, canyons, dry washes, fence rows; 500–1500 m; Tex.

Prunus murrayana was long known only from trans-Pecos Texas. M. F. Enquist (1997) expanded the range by recognizing the similarity of these plants to plants from the Edwards Plateau, which he described as a hairy variety of *P. rivularis* and placed *P. murrayana* in synonymy. Sequence analysis of DNA from some of the same hairy plants collected by Enquist indicate that these plants are distinct from *P. rivularis* and are most similar to *P. hortulana* (J. R. Rohrer et al. 2008). *Prunus murrayana* is a smaller plant than *P. hortulana*, with smaller leaves and fruits. Also, *P. murrayana* suckers to form shrubby thickets; *P. hortulana* generally forms single trees. The twigs, pedicels, and hypanthium exteriors are hairy in *P. murrayana* and glabrous in *P. hortulana*.

c4. ROSACEAE Jussieu (subfam. AMYGDALOIDEAE) tribe EXOCHORDEAE Schulze-Menz ex Reveal, J. Bot. Res. Inst. Texas 4: 215. 2010

Luc Brouillet

Shrubs or trees; unarmed. **Leaves** alternate, simple; stipules absent or caducous (*Oemleria*), (rudimentary) free; venation pinnate. **Flowers:** perianth and androecium perigynous; epicalyx bractlets absent; hypanthium campanulate to turbinate or obturbinate; torus absent or minute; carpels [1–]5, distinct or connate, free, styles terminal, distinct; ovules 1 or 2, apical, collateral. **Fruits** capsules or aggregated drupes; styles not persistent, not elongate.

Genera 3, species 5 (2 genera, 2 species in the flora): North America, Asia.

The base chromosome number for Exochordeae is $x = 8$. *Prinsepia* Royle is Asian and is not present in North America. The tribal name Osmaronieae Rydberg, which pertains here, is illegitimate.

1. Plants synoecious; stipules usually absent; leaf blades membranous, margins entire or serrate in distal ½; carpels connate; fruits capsules . . . . 35. *Exochorda*, p. 384
1. Plants usually dioecious, rarely monoecious; stipules caducous; leaf blades coriaceous, margins entire; carpels distinct; fruits drupes . . . . 36. *Oemleria*, p. 385

35. EXOCHORDA Lindley, Gard. Chron. 1858: 925. 1858 • Pearlbrush [Greek *exo*-, outside, and *chorde*, string, alluding to free placentary cords external to carpels] [I]

Arthur Haines

Shrubs, to 30(–50) dm. **Stems** 1–20+, spreading to erect; bark brown to gray-brown on branches, <not exfoliating>; short shoots absent; glabrous or sparsely hairy in early season. **Leaves** deciduous, cauline, alternate; stipules usually absent; petiole present, <sometimes relatively short>; blade elliptic or oblong to oblong-obovate, 1.5–4.4(–6)[–9] cm, membranous, margins flat, entire or serrate in distal ½, surfaces glabrous or hairy. **Inflorescences** terminal, 4–10-flowered, racemes, axes glabrous or sparsely hairy prior to expansion of flowers; bracts present; bracteoles present. **Pedicels** present. **Flowers** bisexual (plants synoecious), appearing with leaves, 25–50 mm diam.; hypanthium shallowly campanulate to obturbinate, 3.5–4.2 mm, glabrous; sepals 5, ascending, triangular to broad-triangular or semiorbiculate; petals 5, white, obovate to suborbiculate [oblong], <base clawed>; stamens 15–20[–30], shorter than petals; torus absent; carpels 5, connate, glabrous, styles 5; ovules 1 or 2. **Fruits** capsules, semicircular in outline, shorter than wide or as tall as wide in side view, 6–10[–15] mm, glabrous; hypanthium not persistent; sepals not persistent. **Seeds** 1 or 2 per capsule, winged. $x = 8$.

Species 4 (1 in the flora): introduced; Asia; introduced also in Europe, Australia.

Species of *Exochorda* are attractive, spring-flowering shrubs with racemes of showy, white flowers, and are commonly planted.

1. Exochorda racemosa (Lindley) Rehder in C. S. Sargent, Pl. Wilson. 1: 456. 1913 • Common pearlbrush [F] [I]

Amelanchier racemosa Lindley, Edwards's Bot. Reg. 33: sub plate 38. 1847; *Exochorda grandiflora* Lindley

Branchlets red-brown, turning brown in following year. **Leaves:** petiole 4–7 mm; blade apex rounded to obtuse, usually minutely apiculate. **Bracteoles** 2, deciduous, linear to oblong, small, petaloid. **Pedicels** 3–11 mm, proximal bract leaflike, other reduced. **Flowers:** sepals 2.1–3 mm; petals 12–20 mm; anthers 0.8–1.4 mm.

Flowering spring. Forest borders, exotic-dominated woods, disturbed sites, areas of cultivation; 10–200 m; introduced; Ark., Conn., Del., Ga., La., Mass., Minn., Miss., Mo., N.Y., N.C., Ohio, Pa., S.C., Tenn., Va.; e Asia; introduced also in Europe, Australia.

36. OEMLERIA Reichenbach, Deut. Bot. Herb.-Buch, syn. red., 236. 1841 • Indianplum, osoberry [For Augustus Gottlieb Oemler, 1773–1852, Savannah pharmacist and entomologist] [E]

William J. Hess

Nuttallia Torrey & A. Gray ex Hooker & Arnott, Bot. Beechey Voy., 336. 1839, not Rafinesque 1817 [Loasaceae]

Shrubs or trees, 15–60 dm. **Stems** (1–)5–20, arching; bark purplish brown, <pith chambered>; long and short shoots present; glabrous. **Leaves** deciduous, cauline, alternate; stipules caducous, free, distinct, linear, margins entire; petiole present; blade oblong-lanceolate, elliptic, or oblong, 5–10(–15) cm, coriaceous, margins flat, entire, abaxial surface sometimes puberulent, glabrescent, adaxial glabrous. **Inflorescences** axillary <on lateral, pendulous branches>, 5–10 -flowered, determinate racemes, glabrous, glabrate, or hairy; <peduncles present>; bracts present; bracteoles present. **Pedicels** present. **Flowers** usually unisexual, sometimes bisexual (plants usually dioecious, rarely monoecious), 10–15 mm diam.; hypanthium turbinate to campanulate, 3–5 mm, glabrous, glabrate, or hairy; sepals 5, spreading (staminate flowers) or erect (pistillate), triangular; petals 5, greenish white, elliptic-ovate or elliptic-obovate, <base clawed; staminodia in pistillate flowers>; stamens 15 in 3 series, shorter than petals; torus absent or minute; carpels usually 5 (not evident or aborted in staminate flowers), distinct, glabrous, styles 5, <deciduous, included within hypanthia; stigmas 5, capitate>; ovules 2. **Fruits** aggregated drupes, (1 or)2–5, purple-black, elliptic-obovate to ovate, 5–15 mm, <pulp thin>, glabrous; hypanthium persistent (staminate) or mostly deciduous (pistillate); sepals not persistent. **Seeds** 1. $x = 8$.

Species 1: w North America.

The relationship of *Oemleria* within Rosaceae has vacillated traditionally between Maloideae (Maleae here) and Amygdaloideae (Amygdaleae here); recent work by R. C. Evans and T. A. Dickinson (1999) has suggested that its relationship is nearer Amygdaloideae. Even though *Oemleria* has five carpels per flower, a characteristic of Maloideae, Evans and Dickinson believed that its floral ontogeny and mature morphology support its inclusion in Amygdaloideae. They stated that they found the development of the gynoecium in *Oemleria* to be similar in some ways to that of Maloideae. Molecular phylogenetic studies (D. R. Morgan et al. 1994; S. Lee and J. Wen 2001; D. Potter et al. 2002) have recovered a well-supported clade of *Exochorda*, *Oemleria*, and *Prinsepia* Royle. Potter et al. (2007) have shown that this clade, Osmaronieae Rydberg, belongs to Amygdaloideae (as newly circumscribed) and is sister to Kerrieae, within a grade of tribes located between Amygdaleae and Maleae.

1. Oemleria cerasiformis (Torrey & A. Gray) J. W. Landon, Taxon 24: 200. 1975 [E] [F]

Nuttallia cerasiformis Torrey & A. Gray in W. J. Hooker and G. A. W. Arnott, Bot. Beechey Voy., 337. 1839; *Osmaronia cerasiformis* (Torrey & A. Gray) Greene; *O. cerasiformis* var. *lancifolia* Greene

Leaves: petiole 5–12(–15) mm; blade abaxially paler, 5–10(–15) × 1.5–4 cm, base cuneate, apex acute to obtuse. **Inflorescences** appearing with leaves, 3–10 cm. **Flowers:** sepals 1.5–2 mm, glabrous, glabrate, or hairy; petals (3–)4–6 mm (smaller on pistillate flowers); stamens 10 within hypanthium, 5 exserted from rim; carpels 3–4(–4.5) mm. **Drupes** glaucous, taste bitter. $2n = 16$.

Flowering Feb–Apr. Canyons, roadsides, stream banks, lowland wet woods, dry open woods, chaparral; 0–1900 m; B.C.; Calif., Oreg., Wash.

Oemleria cerasiformis is found from the Pacific coast to the western slopes of the Cascades and the northern Sierra Nevada. It is one of the earlier flowering woody plants in the Pacific Northwest and provides good cover and food for birds and other animals. The flowers are foul smelling, part of a fly-pollination syndrome.

c5. ROSACEAE Jussieu (subfam. AMYGDALOIDEAE) tribe KERRIEAE Focke in H. G. A. Engler and K. Prantl, Nat. Pflanzenfam. 24[III,3]: 12, 27. 1888

Luc Brouillet

Shrubs; unarmed (appearing ± thorny in *Coleogyne*). **Leaves** alternate or opposite, simple; stipules persistent or deciduous, free; venation pinnate. **Flowers:** perianth and androecium perigynous; epicalyx bractlets absent (present in *Rhodotypos*); hypanthium flat, shallowly cupped, saucer-shaped, or campanulate; torus absent, inconspicuous, or a rim or tubelike sheath surrounding carpels; carpels 2–5(–8) (usually 1 in *Coleogyne*), distinct, free, styles lateral or subapical (*Neviusia*), distinct; ovules 1 or 2, marginal, collateral when 2. **Fruits** aggregated nutlets or achenes; styles persistent, not elongate.

Genera 4, species 5 (4 genera, 5 species in the flora): North America, Asia; introduced in Europe.

The base chromosome number for Kerrieae is $x = 8$ or 9. C. Kalkman (2004) described the fruits of Kerrieae as drupaceous achenes, with slightly fleshy mesocarps that dry at maturity and woody endocarps. Such fruits might easily be interpreted as nutlets or achenes. The fruits of *Coleogyne*, unplaced as to tribe in his treatment, have a coriaceous pericarp and can be considered achenes.

1. Leaves opposite; sepals usually 4; tori surrounding ovary.
   2. Leaf blades ovate, membranous, margins doubly serrulate; stipules deciduous; epicalyx bractlets slender; petals 4, white; stamens 30–60; tori surrounding and overtopping ovaries, leaving styles free; carpels 4; fruits black, ovoid-globose drupelike nutlets . . . . 37. *Rhodotypos*, p. 386
   2. Leaf blades oblanceolate-spatulate, leathery, margins entire; stipules persistent; epicalyx bractlets absent; petals usually 0; stamens (16–)20–25(–30); tori extending into tubelike sheath surrounding ovary and proximal style; carpels 1 (rarely 2); fruits brown, ovoid-reniform achenes . . . . 40. *Coleogyne*, p. 390
1. Leaves alternate; sepals usually 5; tori absent or inconspicuous.
   3. Stems: bark proximally grayish, distally reddish brown; stipules persistent; inflorescences 2–10-flowered, subumbellate corymbs; petals usually 0, rarely 1 or 2, white . . . . 38. *Neviusia*, p. 387
   3. Stems: bark (periderm) not forming, epidermis green; stipules caducous; inflorescences: flowers solitary; petals usually (4 or)5(or 6), strong yellow to slightly orangish or yellow-cream . . . . 39. *Kerria*, p. 389

37. RHODOTYPOS Siebold & Zuccarini, Fl. Jap. 1: 185, plate 99. 1841 • Black jetbead [Greek *rhodon*, rose, and *typos*, pattern, alluding to resemblance to genus *Rosa*] ☐

James Henrickson

Shrubs, <spreading>, rounded, 5–20(–30) dm. **Stems** 5–20+, widely arcuate; bark gray; long shoots moderately branched; unarmed; glabrous. **Leaves** deciduous, cauline, opposite; stipules deciduous, linear, margins entire; petiole present; blade ovate, 4.5–10 cm, membranous, margins flat, sharply doubly serrulate, <teeth sometimes stipitate-glandular along distal margins>, surfaces sparsely sericeous. **Inflorescences** terminal <on vernal, leaf-bearing branches>, flowers solitary; bracts absent. **Pedicels** present. **Flowers** 30–45 mm diam.; epicalyx bractlets 4, <slender>; hypanthium shallowly cupped, 4–6 mm diam., exterior sericeous-strigose; sepals 4, spreading, ovate to oblong-ovate; petals 4, <spreading-ascending>, white, ovate-orbiculate, <base short-

clawed, apex rounded>; stamens 30–60 (± in 4 clusters), much shorter than petals; torus thickened, dome-shaped, surrounding and overtopping ovaries, constricted below styles, leaving styles free, distally crenulate or 4-toothed; carpels 4, glabrous, styles lateral; ovules 2. **Fruits** aggregated nutlets (appearing as dry drupelets), 1–4, <black>, ovoid-globose, compressed, 6–9 mm; <exocarp thin; mesocarp dry, thin; endocarp thick, cartilaginous>; hypanthium persistent; sepals persistent, reflexed. x = 9.

Species 1: introduced; e Asia.

Rhodotypos is an Asian mesophytic shrub that shares several characteristics with the western North American desert shrub *Coleogyne*, including opposite leaves, four-merous, terminal, solitary flowers, ovaries with lateral (adaxial) styles, and a distinctive inner, rimlike or sheathlike torus extension that surrounds the ovaries and occasionally bears some of the stamens. The fruit of *Rhodotypos* is drupelike during development, but dries at maturity while retaining a shiny, smooth exocarp; it can be considered a modified drupelet.

Along with *Coleogyne*, *Kerria*, and *Neviusia*, *Rhodotypos* form Kerrieae of subfam. Amygdaloideae, an alliance recognized by G. K. W. Schulze-Menz (1964) but in Rosoideae.

1. Rhodotypos scandens (Thunberg) Makino, Bot. Mag. (Tokyo) 27: 126. 1913 [F] [I]

Corchorus scandens Thunberg, Trans. Linn. Soc. 2: 335. 1794; *Kerria tetrapetala* Siebold; *Rhodotypos kerrioides* Siebold & Zuccarini; *R. tetrapetala* (Siebold) Makino

Stems reddish brown. **Leaves:** stipules 2–3.5 mm, membranous, margins strongly sericeous; petiole 3–10 mm; blade weakly bicolored, narrowly to broadly ovate, 4.5–7.5(–10) × 2–4(–6.5) cm, base rounded to broadly cuneate, plicate between impressed veins, apex acuminate, abaxial surface more strongly sericeous, hairs 0.1–0.2 mm. **Pedicels** 6–13 mm (to 30 mm and expanding in fruit). **Flowers:** epicalyx bractlets linear-lanceolate, leafy, 2.5–5 × 1 mm; sepals 8–15(–18) mm, thickened basally, distally leafy, fimbriate-serrate, acuminate, sericeous-strigose; petals spreading, 17–22 mm, margins entire-erose; stamens in several series mostly at sepal bases, rarely onto torus, filaments white, 4–6 mm, anthers yellow, 0.5–0.7 mm; torus 2–2.5 mm, interior densely sericeous-strigose, exterior glabrous; styles linear. **2*n*** = 18.

Flowering Apr–May; fruiting Jun–Sep. Mesic disturbed woods and thickets, especially in suburban areas; 0–300 m; introduced; Ont.; Ala., Conn., Del., D.C., Ga., Ill., Ind., Kans., Ky., Mass., Mich., Miss., Mo., Nebr., N.H., N.J., N.Y., N.C., Ohio, Pa., S.C., Vt., Va., W.Va., Wis.; Asia (China, Japan, Korea).

Rhodotypos scandens is expected elsewhere in North America. It is widely cultivated in Japan and infrequently in North America and Europe.

38. NEVIUSIA A. Gray, Mem. Amer. Acad. Arts, n. s. 6: 374, plate 30. 1859 • Snow-wreath [For Reuben Denton Nevius, 1827–1913, clergyman and amateur botanist] [C] [E]

Alice L. Heikens

Shrubs, 6–32 dm, ± strigose. **Stems** 1–6+, erect, <slender>; bark proximally grayish, distally reddish brown, <± exfoliating in strips near base, decurrently ridged>; short shoots absent; unarmed; <young stems reddish>. **Leaves** winter-persistent, deciduous, cauline, alternate; stipules persistent, linear to awl-shaped, margins serrate; petiole present; blade <bright green adaxially>, ovate or ovate-lanceolate to cordate, 2–9 cm, membranous, margins flat, shallowly lobed and coarsely serrate or doubly serrate, abaxial surface subsericeous to strigose, adaxial sparsely strigose to subsericeous. **Inflorescences** terminal, 2–10-flowered, subumbellate corymbs; bracts absent; bracteoles present. **Pedicels** present. **Flowers** appearing as leaves expand, <odorless>, 2.5–4 mm diam.; epicalyx bractlets 0; hypanthium flat, 1–3 mm diam., glabrous or glabrate;

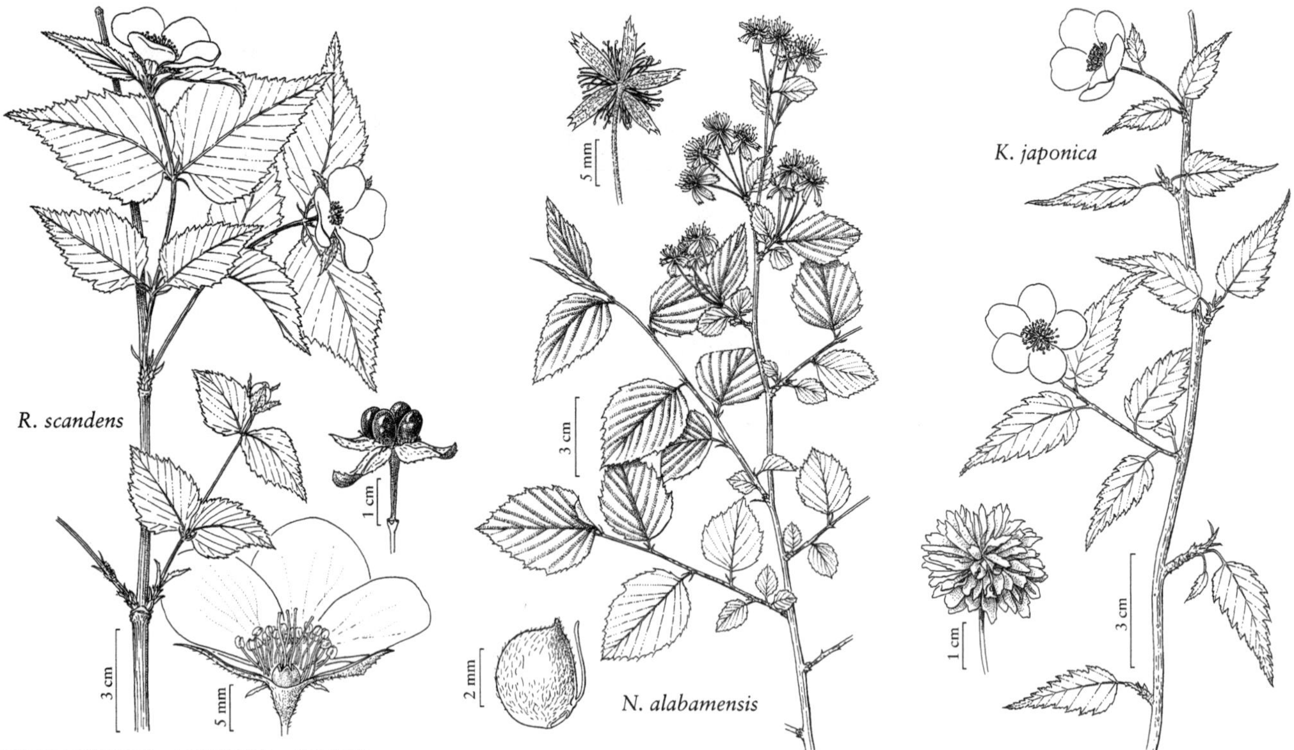

RHODOTYPOS ◦ NEVIUSIA ◦ KERRIA

sepals 5, spreading to reflexed, elliptic to obovate or lanceolate, <margins serrate (incised above middle)>; petals usually 0, rarely 1 or 2, white, oblanceolate; stamens 50–100+, equal to sepals (inner often shorter); torus inconspicuous; carpels 2–6, densely strigose, styles subapical, <elongate, stigmas minute>; ovule 1. **Fruits** aggregated achenes, 2–6, eccentrically ovoid, 3–4 mm, densely strigose; hypanthium persistent; sepals persistent, orientation in fruit unknown.

Species 2 (2 in the flora): w, se United States.

1. Leaf margins doubly serrate, some distal leaves only finely serrate; sepals 5–9(–12) mm; petals 0; stamens 100+; styles 4.5–6 mm; se United States . . . . 1. *Neviusia alabamensis*
1. Leaf margins shallowly lobed and coarsely serrate; sepals 3.5–6 mm; petals usually 0, rarely 1 or 2; stamens 50+; styles 3 mm; California . . . . 2. *Neviusia cliftonii*

1. Neviusia alabamensis A. Gray, Mem. Amer. Acad. Arts, n. s. 6: 374, plate 30. 1859 • Alabama snow-wreath [C] [E] [F]

Leaves: petiole 2–9 mm; blade ovate to ovate-lanceolate, 3–7 × 2–4.5 cm, base attenuate to obtuse, margins doubly serrate, some distal leaves only finely serrate, apex acuminate to acute, surfaces subsericeous. **Pedicels** 13–32 mm. **Flowers:** sepals ovate to lanceolate or elliptic, 5–9(–12) × 3–5 mm; petals 0; stamens 100+; carpels 2–5, styles 4.5–6 mm. **2*n*** = 18.

Flowering early–mid spring; fruiting late spring. Steep limestone, sandstone, and shale slopes with little soil; of conservation concern; 100–500 m; Ala., Ark., Ga., Miss., Tenn.

Neviusia alabamensis has been recommended for federal listing under the Endangered Species Act. The species is known currently from five southeastern states; a population known earlier from Poplar Bluff, Missouri, has been extirpated (D. M. Moore 1956).

The disjunct populations appear to be relicts from a former more widespread distribution and perhaps were associated with mesophytic forests during the Tertiary (A. A. Long 1989). It also has been suggested that *Neviusia alabamensis* is an epibiotic taxon that survived the Mississippi embayment since the late Paleozoic or early Mesozoic (D. M. Moore 1956; D. D. Horn and P. Somers 1981).

Neviusia alabamensis has been cultivated as far north as Boston and Chicago since its discovery (J. R. Shevock et al. 1992). It has been misidentified as *Physocarpus opulifolius* (D. M. Moore 1956).

2. Neviusia cliftonii Shevock, Ertter & D. W. Taylor, Novon 2: 286, fig. 1. 1992 • Shasta snow-wreath [C] [E]

Leaves: petiole 4–10 mm; blade ovate to cordate, 2–9 × 1.5–7 cm, base obtuse to truncate, margins shallowly lobed and coarsely serrate, apex acuminate to acute, surfaces strigose. **Pedicels** 1–3 cm. **Flowers:** sepals obovate, 3.5–6 × 2.5–4 mm; petals usually 0, rarely 1 or 2; stamens 50+; carpels 3–6, styles 3 mm.

Flowering late spring; fruiting late spring–early summer. Limestone substrates, shaded canyons along streams; of conservation concern; 300–500 m; Calif.

Neviusia cliftonii, known only from the Lake Shasta region in northern California, has been recommended for federal listing under the Endangered Species Act. The discovery of fossil leaves of the genus in British Columbia (S. M. Moore et al. 2002) substantiates the hypothesis that the disjunct populations are relicts from a former more widespread distribution (J. R. Shevock et al. 1992). Members of the California Native Plant Society currently are cultivating *N. cliftonii* (B. Ertter and J. R. Shevock 1993).

39. KERRIA de Candolle, Trans. Linn. Soc. London 12: 156. 1818 • [For William Kerr, d. 1814, collector in the Far East, sponsored by Royal Botanic Gardens, Kew, and superintendent of Botanic Garden, Peradinaya, Sri Lanka] [I]

James Henrickson

Alan S. Weakley

Shrubs, <spreading>, open, rounded, 10–20(–30) dm; <rhizomatous, suckering freely>. **Stems** 1–30+, widely arcuate, <sparingly branched>; bark (periderm) not forming, epidermis green, <striate, glabrous>; short shoots absent; unarmed; <bud scales imbricate>. **Leaves** winter-deciduous, cauline, alternate; stipules caducous, linear-subulate, <thin>, margins entire <± strigose-ciliate>; petiole present; blade ovate to lance-ovate, 2–8.5 cm, membranous, margins flat, coarsely doubly serrate, abaxial surface sparsely sericeous-strigose along veins. **Inflorescences** terminal <on vernal, leaf-bearing lateral branches>, flowers solitary, sparsely strigose or glabrous; bracts absent; bracteoles absent or present, <leaflike>. **Pedicels** present. **Flowers** 20–50 mm diam.; epicalyx bractlets 0; hypanthium saucer-shaped, 3–4 mm diam., glabrous; sepals (4 or)5(or 6), spreading to reflexed, oblong-ovate to ± orbiculate; petals (4 or)5(or 6), to 34 in multi-petaled cultivars, <spreading>, strong yellow to slightly orangish or yellow-cream (some petals white in multi-petaled cultivars), oblong-ovate to orbiculate, <base short-clawed, apex rounded to emarginate>; stamens 80–130 (0 or few in multi-petaled cultivars), shorter than petals; torus absent; carpels (4 or)5(–8), <globose>, glabrous, styles lateral, <linear, ± exceeding filaments>; ovules 2. **Fruits** aggregated nutlets, (1–)5 in 1 whorl, tan, obliquely obovoid-globose, 4.5–5 mm, <with distinct adaxial ridge>, glabrous except for coarse basal hypanthial hairs; hypanthium persistent; sepals persistent (withering), reflexed; <exocarp thin, wrinkled, mesocarp thin, dry, endocarp ± cartilaginous>. $x = 9$.

Species 1: introduced; e Asia (China, Japan); introduced also in Europe.

Kerria is a distinctive mesophytic shrub, native to montane forests and valleys at 200–3000 m in China and Japan, now widely cultivated. Morphologically, it stands closest to *Neviusia*.

1. **Kerria japonica** (Linnaeus) de Candolle, Trans. Linn. Soc. London 12: 157. 1818 • Japanese- or Easter-rose, corête du Japon F I

Rubus japonicus Linnaeus, Mant. Pl. 2: 245. 1771; *Corchorus japonicus* (Linnaeus) Houttuyn

Stems moderately flexuous. **Leaves:** stipules 1.5–2.5 mm; petiole channeled, 3–6(–9) mm; blade bicolored, (2–)3–6(–8.5) × (1–)1.5–3(–3.7) cm, base broadly cuneate to subcordate, proximal margins finely gland-tipped serrulate, ± plicate between impressed veins, apex long-acuminate, hairs 0.3–0.5 mm. **Pedicels** 5–22 mm, glabrous. **Flowers:** sepal margins entire, scarious, sometimes serrulate or glandular-fimbriate, apex acuminate, surfaces glabrous; petals (9–)17–25 × (2–)12–15 mm, claws 1–2 mm (innermost in multi-petaled cultivars showing transition toward stamens, 10 × 2–3 mm); filaments erect-ascending, straight, yellow, slender, 4–5 mm, anthers yellow, 0.2–0.3 mm. **Nutlets** 4–5 × 3.5–4 mm. $2n = 18$.

Flowering Mar–Jun; fruiting Jun–Aug. Moist suburban forests and woodlands, roadsides; 0–700 m; introduced; N.S., Ont.; Ala., Conn., Ill., Mich., N.H., N.Y., N.C., Ohio, Pa., R.I., S.C., Tenn., Utah, Va.; e Asia (China, Japan); introduced also in Europe.

Kerria japonica is cultivated in central Europe and North America, particularly in the east, as a shade-tolerant ornamental shrub. The species persists long after cultivation and spreads mostly vegetatively, apparently only rarely spreading by seed; it is expected elsewhere. The cultivar 'Pleniflora' has multiple petals appearing as a yellow-orange rose.

40. COLEOGYNE Torrey, Proc. Amer. Assoc. Advancem. Sci. 4: 192. 1851 • Blackbrush [Greek *koleos*, sheath, and *gyne*, female, alluding to thin staminal tubelike sheath surrounding ovary and style] E

James Henrickson

Bruce D. Parfitt†

Shrubs, 2–6(–10) dm, herbage strigose, with appressed medifixed hairs. **Stems** 15–150, stiffly erect; bark dark russet, becoming gray, striate; long shoots divaricately and oppositely branched, short shoots present; unarmed (stiff dead branches resembling thorns). **Leaves** persistent (drought deciduous), cauline, opposite and clustered on short shoots; stipules persistent on distal leaf-base margins, deltate-lanceolate, margins entire; petiole obscure; blade oblanceolate-spatulate, 0.6–1.2 cm, thickened, leathery, margins thick, rounded (not revolute), entire. **Inflorescences** terminal <on long- and short-shoot spurs>, flowers solitary; bracts present, <reduced, leaflike>. **Pedicels** reduced. **Flowers** 11–14 mm diam.; epicalyx bractlets 0; hypanthium campanulate, 1–2 mm; sepals 4 (rarely 5), weakly spreading, <abaxially yellow>, outer narrowly ovate, inner broadly ovate, <margins thin>; petals usually 0, infrequently 1–4(or 5) on scattered flowers, <caducous>, translucent yellow, spatulate; stamens (16–)20–25(–30), equal to sepals; torus extending into tubelike sheath surrounding ovary and proximal style (interior sericeous, exterior bearing some stamens); carpels 1 (rarely 2), glabrous except at style base, styles lateral, <exserted beyond sheath, sericeous most of length>; ovule 1, <lateral>. **Fruits** achenes, 1 (or 2), <brown>, ovoid-reniform, distally narrower, 5.3–5.8 mm, <shiny>, glabrous; hypanthium persistent; sepals persistent, ascending. $x = 8$.

Species 1: sw United States.

Coleogyne is a xeromorphic desert shrub with floral features unusual within Rosaceae: a peculiar sheathlike torus surrounds the carpel and most of the style that bears some of the stamens, four sepals, and petals commonly absent but the inner sepal surface strongly yellow; vegetatively, *Coleogyne* is further distinguished by its opposite leaves and uniformly oriented medifixed trichomes. The single-ovuled, solitary ovary has a lateral style, suggesting relationship with multi-ovuled ancestors. Molecular data support relationships in the Kerrieae in subfam. Amygdaloideae (D. Potter et al. 2007).

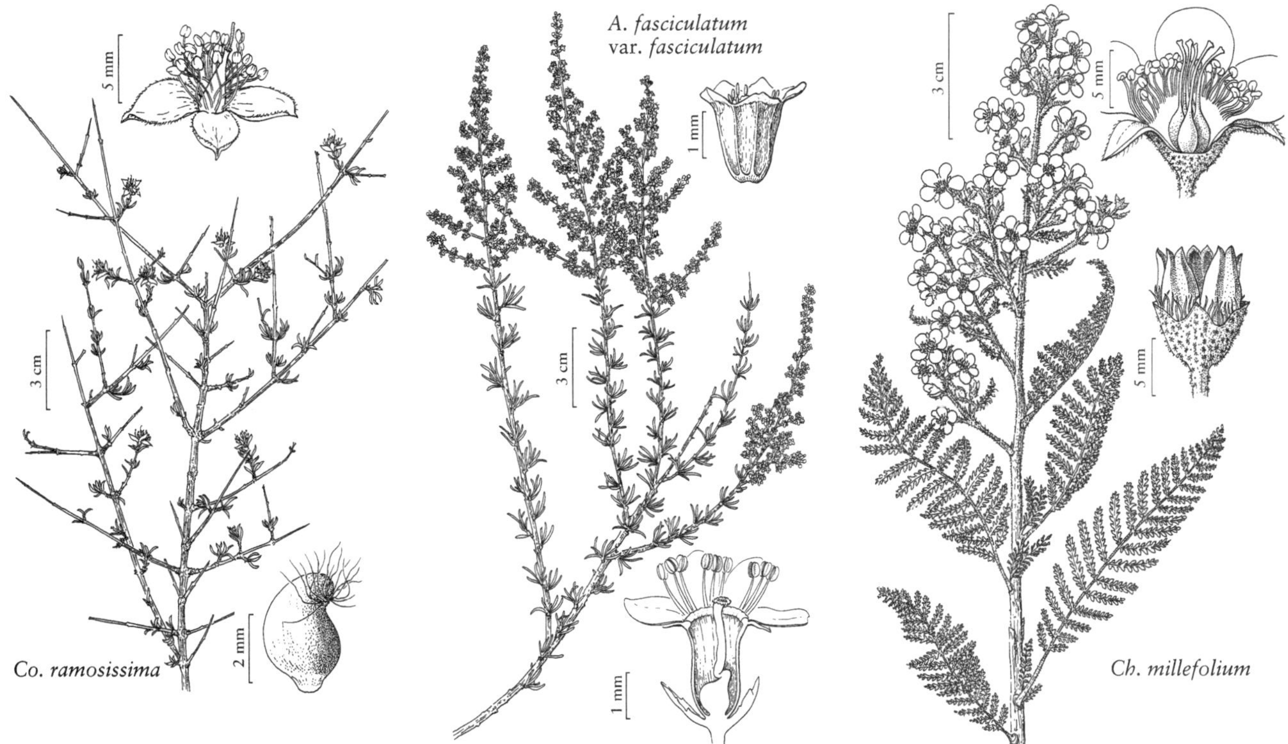

COLEOGYNE ◦ ADENOSTOMA ◦ CHAMAEBATIARIA

1. Colcogyne ramosissima Torrey, Smithsonian Contr. Knowl. 6(2)[as 5(1)]: 8, plate 4. 1853 [E] [F]

Stems strigose, medifixed hairs to 0.5 mm; internodes 1–2 cm, short-shoot spurs to 7 mm, 1 mm thick. **Leaves:** 0.4–1 (–1.2) mm, base clasping; stipules acute, free portion 0.2–0.5 mm; blade 3-angled in cross section, 0.6–1.4 mm wide, abaxially 5-ribbed, apex acute, surfaces strigose. **Flowers:** sepals ovate to broadly ovate or obovate, concave-convex, 5–7 mm, outer pair acuminate, inner cuspidate, abaxially red-green, strigose, adaxially yellow to orange-yellow, glabrous; petals, if present, 5–8 mm; stamens borne between sepals and on proximal ½ of torus sheath, filaments yellow, 5–8 mm; torus sheath light yellow, tubular, narrowly obconic, 4–6(–9) mm, thick-membranous, exterior glabrous, interior long-sericeous proximally, hair tips slightly protruding from orifice of sheath; styles 5–6 mm, slender, stigmas 3–5 mm. **Achenes** red-brown. $2n = 16$.

Flowering Apr–May; fruiting Jun–Oct. Arid scrub zones, Great Basin and Mojave deserts, pinyon-juniper, Joshua tree woodlands; (300–)900–1800(–2400) m; Ariz., Calif., Colo., Nev.

In the Great Basin and Mojave deserts, *Coleogyne ramosissima* is the dominant shrub in areas of shallow soils underlain by a caliche layer (S. A. Lei and L. R. Walker 1997). In deeper soils, grasses and other shrubs prevent dominance by *C. ramosissima.* The species has a notably long life span; seedling establishment is rare in arid sites but is more likely at higher elevations and in cooler portions of its range (R. H. Webb et al. 1987; S. E. Meyer and B. K. Pendleton 2005).

In some areas *Coleogyne ramosissima* provides important cover and winter browse for mule deer and desert bighorn sheep (D. L. Jeffries and J. M. Klopatek 1987). It is used to a much lesser extent by domestic livestock. B. K. Pendleton and R. L. Pendleton (1998) show the species is wind pollinated.

c6. ROSACEAE Jussieu (subfam. AMYGDALOIDEAE) tribe SORBARIEAE Rydberg in N. L. Britton et al., N. Amer. Fl. 22: 239, 256. 1908

Luc Brouillet

Shrubs; unarmed. **Leaves** alternate, sometimes opposite, pinnately compound (mature simple in *Adenostoma*); stipules persistent, free or adnate to petiole base, or reduced or absent (*Adenostoma*); venation pinnate. **Flowers:** perianth and androecium perigynous; epicalyx bractlets absent; hypanthium funnelform, cup-shaped, or campanulate to cylindric; torus absent; carpels 1–5, basally connate or distinct (sometimes appearing connivent at base), free, styles terminal or lateral, distinct; ovules 1, basal, or 2 and collateral, 4–8, or 10–12 in 2 series, declining. **Fruits** achenes or aggregated follicles; styles persistent or not, not elongate.

Genera 4, species 8 (3 genera, 5 species in the flora): North America, nw Mexico, Eurasia.

The base chromosome number for Sorbarieae is $x = 9$. *Spiraeanthus* (Fischer & C. A. Meyer) Maximowicz, with one species native to central Asia, also belongs here.

1. Leaves simple (juvenile pinnatifid); stamens 10 or 15; fruits achenes within hypanthium . . . . 41. *Adenostoma,* p. 392
1. Leaves pinnately compound; stamens 17–110; fruits aggregated follicles.
    2. Leaves persistent, 1(or 2)-pinnate-pinnatifid, leaflet margins entire or secondarily toothed-lobed; plants 6–20 dm . . . . 42. *Chamaebatiaria,* p. 394
    2. Leaves deciduous, odd-pinnately compound, leaflet margins usually doubly serrate; plants 10–70 dm . . . . 43. *Sorbaria,* p. 396

41. ADENOSTOMA Hooker & Arnott, Bot. Beechey Voy., 139. 1832 • Chamise [Greek *aden*, gland, and *stoma*, mouth, alluding to gland at rim of hypanthium]

William V. Jones

James Henrickson

Shrubs, erect to rounded-spreading, 8–60 dm; developing enlarged woody lignotubers. **Stems** 1–10+, erect to arching or spreading, <branched, terete or decurrently grooved>; bark of red brownish papery layers, weathering gray; short shoots present or absent; hirtellous, pilose, or glabrous. **Leaves** persistent, cauline, alternate, sometimes opposite, juvenile 1- or 2-pinnatifid, mature simple; stipules reduced or absent; petiole present or absent; blade linear to terete or linear-oblanceoloid, (0.2–)0.3–1.4(–2.6) mm, leathery or not, <base persistent, clasping, truncate>, margins revolute, entire, abaxial surface glabrous or sessile-glandular. **Inflorescences** terminal, 2–400+-flowered, panicles <well branched, pyramidal> or racemes or spikes <reduced to cylindric>, glabrous or hirtellous; bracts present; bracteoles present. **Pedicels** absent or relatively short. **Flowers** 2–6 mm diam.; hypanthium cylindric to campanulate, <inner rim annular gland thickened, undulate>, 0.6–2.4 mm, glabrous or hirtellous; sepals 5, erect to reflexed, <valvate>, broadly ovate to semiorbiculate; petals 5, white becoming rusty brown, obovate-orbiculate or elliptic to ovate, <base clawed>; stamens 10 or 15 in groups of 2 or 3 alternate with petals, shorter than petals; carpels 1(or 2), distinct, <obliquely truncate distally>, glabrous or sericeous, styles lateral, <each bending over ovary, erect distally, stigmas capitate to truncate>; ovules 1(or 2). **Fruits** achenes, 1(or 2), obovoid or ellipsoid, 1.1–1.8(–2.4) mm, hirtellous distally (exposed part); hypanthium persistent; sepals persistent, ± spreading; styles not persistent. $x = 9$.

Species 2 (2 in the flora): California, nw Mexico.

Adenostoma is a Madro-Tertiary derivative of a lineage with compound leaves. Seedlings and sucker shoots have 1- or 2-pinnatifid leaves; mature leaves are simple, undivided. Both species occur in chaparral and resprout after fires from basal lignotubers. The single achenes are derived from ancestral 5-follicled fruits, each carpel retaining the ancestral marginal style. Stamens are arranged in groups of two or three alternate with petals; filaments expand sequentially, first five (one opposite each sepal), followed by the second five and third five (when 15).

1. Leaves crowded on short shoots, blades dark green, adaxial surfaces glabrous; stipules present on long-shoot leaves; subtending flower bracts 3-lobed, medial lobe largest; bracteoles 2, subulate; stamens (10–)15 . . . . 1. *Adenostoma fasciculatum*
1. Leaves on slender long shoots (short-shoot fascicles absent), blades yellow-green, adaxial surfaces sparsely hirtellous and sessile-glandular; stipules absent; subtending flower bracts linear-subulate; bracteoles 5 or 6, outermost subulate, inner oblong-obovate; stamens 10(–12) . . . . 2. *Adenostoma sparsifolium*

1. Adenostoma fasciculatum Hooker & Arnott, Bot. Beechey Voy., 139. 1832 (as fasciculata) • Greasewood [F]

Plants decumbent or erect-ascending, 8–20 dm. **Stems** erect to arching; young stem internodes brown, 2–6.5(–12) mm, glabrous or hirtellous-pilose, rarely glandular. **Leaves** crowded on short shoots; stipules green, 0.5–1.5 mm on long-shoot leaves, smaller on fascicled leaves; blade dark green, linear-oblanceolate, (2–)3–10(–13) × 0.6–1(–1.7) mm, apex acute, apiculate to obtuse, abaxial surface glabrous or hirtellous or villous, adaxial glabrous. **Inflorescences** of alternately branched pyramidal panicles, 3.5–10(–17) × 1–9(–15) cm or reduced to cylindric racemes or spikes, glabrous; subtending flower bracts 3-lobed, medial lobe largest; bracteoles 2, subulate. **Pedicels** present, short. **Flowers** in clusters of 1–3(–7); hypanthium green, 1.5–2.4 × 1–1.6(–2) mm, <obscurely 10-veined, conspicuously so when dry>, glabrous or weakly hirtellous, inner rims whitish; sepals broadly ovate to semiorbiculate, 0.5–0.9 mm; petals tardily falling, white, turning rusty, obovate-orbiculate, to 1–2 mm diam.; stamens (10–)15, filaments 1–1.5 mm, anthers 0.4–0.8 mm; ovaries obconic; styles curved over ovary, erect distally, white, rarely pinkish distally, 1–1.4 mm. **Achenes** obovoid, 1.2–1.8(–2.4) mm, contained within hypanthium.

Varieties 3 (3 in the flora): California, nw Mexico.

1. Stems decumbent (plants low, mounded); insular . . . . 1c. *Adenostoma fasciculatum* var. *prostratum*
1. Stems erect-ascending, spreading (plants tall); mainland and insular.
    2. Young stems glabrous, sometimes ± hirtellous; leaves oblanceolate to linear, (2.5–)5–13 mm, apices acute-apiculate . . . . 1a. *Adenostoma fasciculatum* var. *fasciculatum*
    2. Young stems pubescent to villous; leaves oblanceolate-clavate, 4–6.5 mm, apices usually obtuse . . . . 1b. *Adenostoma fasciculatum* var. *obtusifolium*

1a. Adenostoma fasciculatum Hooker & Arnott var. **fasciculatum** [F]

Stems erect-ascending, spreading (plants tall); internodes 2.5–7(–13) mm; young stems glabrous, sometimes ± hirtellous. **Leaves** oblanceolate to linear, semi-terete, (2.5–)5–13 mm (not or slightly expanded distally), apex acute-apiculate. **2*n*** = 18.

Flowering (Apr–)May–Jun(–Aug). Dry slopes, ridges, lower chaparral and coastal sage scrub; 10–1800(–2100) m; Calif.; Mexico (Baja California).

Variety *fasciculatum* is a major component of lower chaparral, upper coastal sage scrub from northern, central, and southern cismontane California and northern Baja California.

1b. Adenostoma fasciculatum Hooker & Arnott var. **obtusifolium** S. Watson in W. H. Brewer et al., Bot. California 1: 184. 1876

Stems erect-ascending, spreading; internodes 1–7 mm; young stems pubescent to villous (hairs curled or retrorse). **Leaves** oblanceolate-clavate, obscurely semiterete, 4–6.5 mm (expanded distally), apex usually obtuse.

Flowering (Apr–)May–Jun (–Aug). Mesas, ravines, coastal sage scrub-chaparral; 20–1100(–2100) m; Calif.; Mexico (Baja California).

Variety *obtusifolium* is known from coastal southern California (southern Orange and San Diego counties) and northern Baja California, where it introgresses with var. *fasciculatum*, resulting in shorter, more strongly hairy stems, and leaves less apiculate or mixed obtuse-acute in some interior populations.

1c. Adenostoma fasciculatum Hooker & Arnott var. **prostratum** Dunkle, Bull. S. Calif. Acad. Sci. 40: 109. 1941 [E]

Stems decumbent (plants low, mounded, mostly to 5 dm); internodes 1.5–4.5 mm; young stems glabrous. **Leaves** clavate, semiterete, 2–6.3 mm, apex acute.

Flowering (Apr–)May–Jun (–Aug). Maritime chaparral; 0–200(–500) m; Calif.

Variety *prostratum* is known from the Channel Islands (San Nicolas, Santa Catalina, Santa Cruz, Santa Rosa islands) and the adjacent mainland (San Luis Obispo County) in southern California, where it often co-occurs with var. *fasciculatum*. The plants maintain their low-mounded habit in cultivation.

2. Adenostoma sparsifolium Torrey in W. H. Emory, Not. Milit. Reconn., 140. 1848 (as sparsifolia) • Red shank, ribbon bush

Plants erect, 20–60 dm. **Stems** erect to spreading; <bark distinctively cinnamon-colored, papery-layered>; young stem internodes yellowish green, 1.5–7(–15) mm (marked with slender paired decurrent grooves), finely hirtellous, sessile-glandular. **Leaves** on slender long shoots, <proximal 1 or 2 opposite, distal subalternate or alternate>; stipules absent; blade yellow-green, linear to terete, (2.5–)5–14(–26) × 0.4–0.6(–0.8) mm, apex recurved-apiculate, surfaces sparsely hirtellous and sessile-glandular. **Inflorescences** paniculate or reduced to spicate racemes (terminating all distal, long shoots of season), 0.8–5(–7.5) cm, hirtellous; subtending flower bracts linear-subulate; bracteoles 5 or 6, outermost subulate, inner oblong-ovate, concave. **Pedicels** present or absent. **Flowers** in sessile clusters of 2 or 3 (or single and pedicellate); hypanthium greenish, 0.6–2 mm, hirtellous, inner rims yellowish; sepals broadly ovate, 0.6–1 mm; petals persistent, whitish, turning rusty brown, elliptic to ovate, 1.5–2.2 mm; stamens 10(–12), filaments 1–1.5 mm, anthers 0.4–0.6 mm; ovaries obliquely obovoid; styles erect, pinkish, 0.5–1 mm. **Achenes** ellipsoid, 1.1–1.6 mm, exserted from hypanthium. $2n = 18$.

Flowering Jul–Aug(–Sep). Dry granitic slopes, mesas, ravines, chaparral, pinyon woodlands; 200–2000 m; Calif.; Mexico (Baja California).

Adenostoma sparsifolium is known in southern California from San Luis Obispo and Santa Barbara to Riverside and San Diego counties, and in northern Baja California.

42. CHAMAEBATIARIA (Porter ex W. H. Brewer & S. Watson) Maximowicz, Trudy Imp. S.-Peterburgsk. Bot. Sada 6: 225. 1879 • Fern bush, desert sweet [Genus *Chamaebatia* and Latin *-aria*, connection, alluding to resemblance] [E]

James Henrickson

Spiraea Linnaeus sect. *Chamaebatiaria* Porter ex W. H. Brewer & S. Watson in W. H. Brewer et al., Bot. California 1: 170. 1876

Shrubs, erect to widely spreading, 6–20 dm, <herbage strongly aromatic>. **Stems** 1–10+, ascending-spreading, <strongly branched from base and above>; bark russet, with age gray, <smooth>; long shoots present; strongly stipitate-glandular, stipitate-stellate, hirtellous, sometimes strongly glutinous. **Leaves** persistent, cauline, alternate, 1(or 2)-pinnate-pinnatifid; stipules free, subulate to oblanceolate, margins entire; petiole present; blade oblong-ovate, 2.6–7(–9.6) cm, herbaceous, pinnae 13–25 per side, <ascending, opposite to alternate on rachises>, narrow, deeply divided into obovate segments, each segment decurrent on rachilla on proximal base, margins entire or secondarily toothed-lobed, surfaces glabrous or stellate-hairy, glandular. **Inflorescences** terminal <on long shoots of season>, 20–400-flowered, panicles <narrowly ovoid, lateral branches leafy at bases>; bracts present, <proximal bracts of reduced leaves, distal bracts lanceolate, pinnate-lacerate, or entire>; bracteoles present. **Pedicels** present. **Flowers** 10–13 mm diam.; hypanthium funnelform, to 3–4 mm diam.; sepals 5, spreading, ovate-deltate; petals 5, white, orbiculate-reniform, <base short-clawed>; stamens 70–110 in 2 or 3 series, shorter than petals; carpels (4 or)5, distinct, appearing connivent at base, sericeous, styles <caducous>, terminal, <slender>, <stigmas broadened>; ovules 10–12 in 2 series. **Fruits** aggregations of (4 or)5 follicles, each obliquely lanceoloid, 3–5 mm, hirtellous, dehiscent to bases on ventral sutures and halfway along dorsal sutures; hypanthium persistent; sepals persistent, erect to spreading; styles persistent or not. **Seeds** 3–8, narrowly fusiform, thin walled. $x = 9$.

Species 1: w United States.

Vestiture on young stems, inflorescences, leaves, hypanthia, and sepals of *Chamaebatiaria* is complex, consisting of short, straight or curved, sessile hairs; sessile or thick-stalked stellate clusters of flattened, more or less twisted hairs; and sessile, subsessile, or thick-stalked stipitate glands. In some plants, glands will exude material rendering the stems and leaves glutinous. They may have hairs on the gland tip or on the stalk below the gland.

Molecular data indicate relationships of *Chamaebatiaria* with the Sorbarieae, with the Asian *Sorbaria* and *Spiraeanthus* (Fischer & C. A. Meyer) Maximowicz, and the North American *Adenostoma* (D. Potter et al. 2007). The similarity to *Chamaebatia* is striking. J. A. Wolfe and W. C. Wehr (1988) presented interesting fossil evidence of leaf structure development.

1. **Chamaebatiaria millefolium** (Torrey) Maximowicz, Trudy Imp. S.-Peterburgsk. Bot. Sada 6: 225. 1879 E F

Spiraea millefolium Torrey in War Department [U.S.], Pacif. Railr. Rep. 4(5): 83, plate 5. 1857

Stems: internodes (1–)6–12 mm. **Leaves:** stipules 2–4 mm, green on leaf base margins; blade 2.6–7(–9.6) × 0.9–2(–3) cm, 1(or 2)-pinnate-pinnatifid into 13–25 pairs of pinnae; pinnae oblong-lanceolate, (4–)6–14(–16) × 2–4 mm, divided into 6–10 opposite or alternate pairs of obovate to ovate segments 0.8–2 mm; segments obtuse-rounded, entire or secondarily lobed on proximal margins, proximal margins decurrent along rachillae extending to subtending divisions (pinnae then pinnatifid and leaves then 2-pinnate), pinnae margins of proximalmost lobes decurrent along adaxial leaf rachises, abaxial surfaces of rachises, rachillae, and petioles hairy, adaxial glabrous. **Inflorescences** 3–15(–23) × 2–8(–12) cm, leafy proximally, flowers crowded along distal branches; lobes of distal bracts and bracteoles acute. **Flowers:** sepals valvate, 2.5–4 mm, acute, abaxial surface hairy, glandular; petals spreading, margins erose, glabrous; filaments spreading, inner more erect, 3–4 mm, glabrous, anthers orbicular, introrse, versatile, 0.4–0.5 mm; styles 3–5 mm. **Follicles** reddish brown, leathery, hairy. **Seeds** yellowish, 2.5–3.5 mm. $2n = 18$.

Flowering Jul–Sep. Dry, rocky sagebrush scrub, pinyon-juniper woodlands, montane coniferous forests to subalpine woodlands; 900–3400 m; Ariz., Calif., Idaho, Nev., Oreg., Utah.

Chamaebatiaria millefolium grows in southern Oregon, northern California, and the eastern slope of the Sierra Nevada to southern Idaho, Nevada, north-central Arizona, and western Utah.

43. SORBARIA (Seringe) A. Braun in P. F. A. Ascherson, Fl. Brandenburg 1: 177. 1860, name conserved • False spiraea, sorbaire [Genus *Sorbus* and Latin *-aria*, possession, alluding to similar leaves] [I]

James Henrickson

Spiraea Linnaeus sect. *Sorbaria* Seringe in A. P. de Candolle and A. L. P. P. de Candolle, Prodr. 2: 545. 1825

Shrubs, erect to spreading, [2–]10–70 dm, vestiture simple, uniseriate or stellate hairs, often stipitate-glandular; rhizomatous, suckering. **Stems** 1–10+, widely arching; bark gray; long shoots sparsely branched; glabrous or strongly pubescent and/or stellate, <axillary buds ovoid to cylindric, bud scales exposed, alternate, glabrous or pubescent or stipitate-stellate>. **Leaves** deciduous, cauline, alternate, odd-pinnately compound; stipules obliquely adnate to petiole base, linear, lanceolate to ovate, margins entire or serrate, <hairy ciliate>; petiole present; blade oblong-ovate, 14–35 cm, herbaceous, leaflets (9–)11–19[–25], <opposite along rachis, sessile>, oblong-ovate to elliptic, margins flat, usually doubly serrate, with 0–6 teeth on each major tooth, surfaces glabrous or hairy. **Inflorescences** terminal, 15–1100+-flowered, panicles <becoming pendulous when large>, glabrous or pubescent and glandular; bracts present; bracteoles present. **Pedicels** present, short. **Flowers** (5–)7–13(–14) mm diam.; hypanthium shallowly cup-shaped, 1.5–4.6 mm diam., glabrous or stellate-hairy; sepals 5, strongly reflexed, ovate-rounded to oblong-ovate; petals 5, white, ovate to orbicular, <base short-clawed>; stamens 17–35[–50] usually in 1 series, much exceeding petals, those directly opposite petals sometimes much shorter, <filaments glabrous>; carpels 5, <whorled, opposite sepals>, basally and adaxially connate, laterally distinct, glabrous or pubescent, styles adaxial; ovules 4–8. **Fruits** aggregated follicles, 5, oblong, 3–6 mm, <leathery, united to hypanthium at base, weakly so on adaxial margins>, glabrous or pubescent, dehiscent along distal adaxial margin, <adaxial margin often elongating and deflecting style to abaxial edge>; hypanthium persistent; sepals persistent, reflexed, withering; styles not persistent. **Seeds** 4–8, fusiform, 4 mm. $x = 9$.

Species 4 (2 in the flora): introduced; Asia (Afghanistan, China, n India, Japan, Pakistan, e Russia); introduced also in Europe.

Three species of *Sorbaria* are cultivated in North America, two of which are also escaping. *Sorbaria tomentosa* (Lindley) Rehder is recorded as cultivated in the United States and may naturalize; it differs from *S. kirilowii* in having fruiting styles borne near the fruit tip and simple, straight hairs scattered all along the midveins of abaxial leaf surfaces (not just on primary-secondary vein axils). It has two varieties: var. *tomentosa*, with leaflet margins doubly serrate, and var. *angustifolia* (Wenzig) Rahn (= *S. aitchisonii* Hemsley) with leaflets narrow (5–12 mm wide), abaxially glabrous, with the margins mostly only serrate, and vegetative branches purplish. They are native to eastern Afghanistan, northern India, and Pakistan.

SELECTED REFERENCE Rahn, K. 1989. A survey of the genus *Sorbaria* (Rosaceae). Nordic J. Bot. 8: 557–563.

1. Follicles and ovaries sericeous; sepals ovate to oblong-ovate, margins often glandular-serrate; stamens 20–35[–50], 2–6.5 mm (of variable length) . . . . 1. *Sorbaria sorbifolia*
1. Follicles and ovaries glabrous or nearly so; sepals ovate-rounded, margins entire; stamens 20–25, those opposite petals 1–1.5 mm, others 1.9–2.5(–4.5) mm. . . . . 2. *Sorbaria kirilowii*

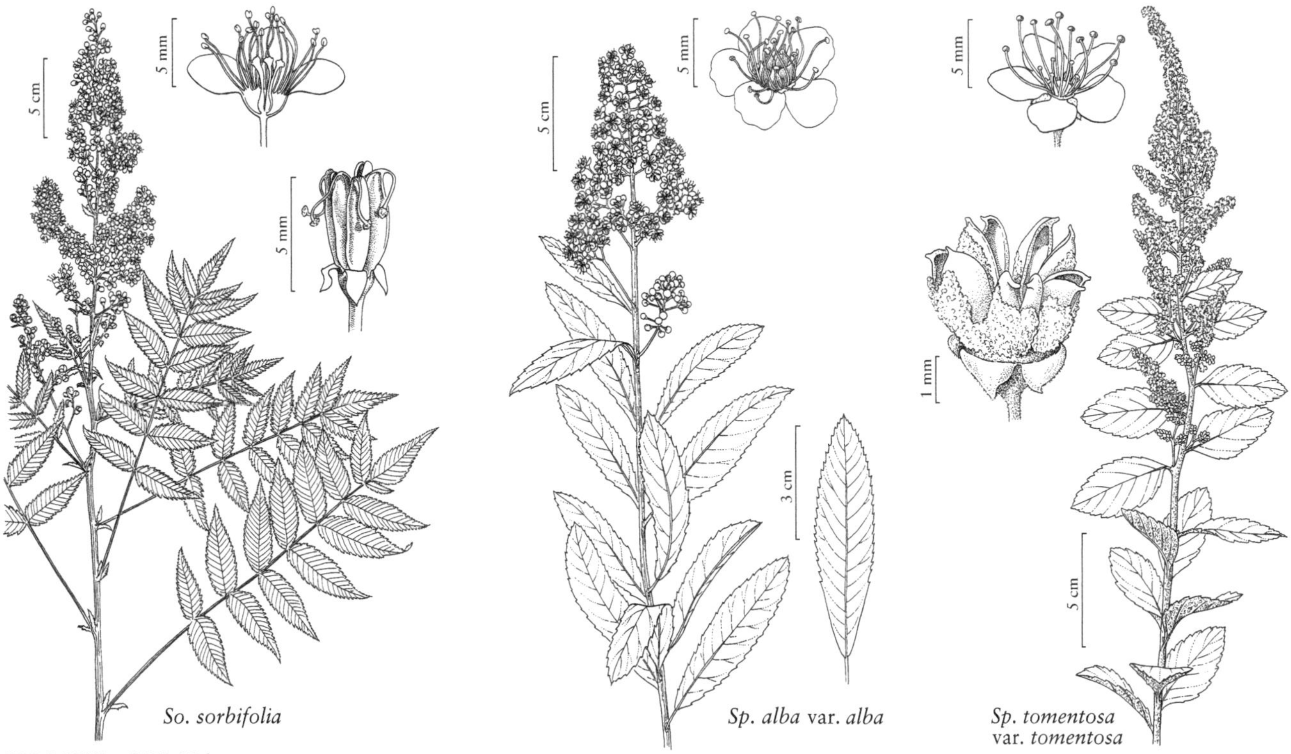

SORBARIA ◦ SPIRAEA

1. **Sorbaria sorbifolia** (Linnaeus) A. Braun in P. F. A. Ascherson, Fl. Brandenburg 1: 177. 1860 • Sorbaire à feuilles de sorbier [F] [I]

Spiraea sorbifolia Linnaeus, Sp. Pl. 1: 490. 1753; *Sorbaria sorbifolia* var. *stellipila* Maximowicz; *S. stellipila* (Maximowicz) C. K. Schneider

Plants 10–30 dm. **Leaves:** blade 14–30 × 5–17 cm; leaflets (9–)11–21(–29), oblong-ovate to elliptic, (25–)35–75(–95) × (8–)12–20(–25) mm, abaxial surface glabrous or ± sparsely stipitate-stellate, adaxial with some simple hairs near margins, otherwise glabrous. **Inflorescences** (7–)10–15(–34) × (3–)4–7(–14) cm. **Pedicels** (and axes) usually puberulent, <stipitate-glandular, less often stipitate-stellate>. **Flowers** 10–14 mm diam. (anther tip to tip); hypanthium puberulent, hirtellous, stellate, or glabrous; sepals ovate to oblong-ovate, margins often glandular-serrate; petals ovate to orbiculate, 2.7–4.3 × 2.1–3.4 mm; stamens 20–35[–50], 2–6.5 mm (of variable length); ovaries sericeous, styles 1.3–3.5 mm. **Follicles** 4.5–6 mm, sericeous. **2*n*** = 36.

Flowering Jun–Jul. Roadsides, old fields, waste areas, overgrown forest margins; 10–500 m; introduced; Alta., Man., N.B., N.S., Ont., P.E.I., Que., Sask., Yukon; Conn., Ind., Iowa, Maine, Mass., Mich., Minn., N.H., N.J., N.Y., Ohio, Pa., R.I., Tenn., Vt., Va., Wash., W.Va., Wis.; Asia (n China, Japan, Korea, Manchuria, e Siberia); introduced also in Europe.

Sorbaria sorbifolia is cultivated in North America and Europe. The plants are always colonial and are capable of becoming adventive.

2. **Sorbaria kirilowii** (Regel) Maximowicz, Trudy Imp. S.-Petersburgsk. Bot. Sada 6: 225. 1879 (as kirilowi) • Giant false spiraea [I]

Spiraea kirilowii Regel in E. Regel and H. S. T. Tiling, Fl. Ajan., 81. 1858; *Sorbaria arborea* C. K. Schneider

Plants 15–70 dm. **Leaves:** blade 14–35 × 8–17 cm; leaflets 13–21, oblong-ovate, (30–)60–90(–130) × (10–)13–30(–40) mm, abaxial surface with simple hairs to 0.8 mm in vein axils, sometimes also stipitate-stellate, adaxial glabrous. **Inflorescences** (7–)20–30(–42) × (5–)10–17(–33) cm. **Pedicels** (and axes) usually glabrous. **Flowers** 5–9 mm diam. (anther tip to tip); hypanthium glabrous; sepals ovate-rounded, margins entire; petals orbicular, 1.9–3.5 mm diam.; stamens 20–25, those opposite petals 1–1.5 mm, others 1.9–2.5(–4.5) mm; ovaries glabrous or nearly so, styles 1–1.2 mm. **Follicles** 3–5.3 mm, glabrous or nearly so.

Flowering Jun–Jul. Roadsides; 10–400 m; introduced, Wash.; Asia (China, Japan, Korea, Tibet); introduced also in Europe.

Sorbaria kirilowii is cultivated in the United States and Europe, persisting and becoming naturalized in northwestern Washington.

c7. ROSACEAE Jussieu (subfam. AMYGDALOIDEAE) tribe SPIRAEEAE de Candolle in A. P de Candolle and A. L. P. P. de Candolle, Prodr. 2: 541. 1825 (as Spiraeaceae)

Luc Brouillet

Shrubs, subshrubs, or herbs, perennial (*Aruncus*); unarmed. **Leaves** alternate, simple or pinnately compound; stipules absent; venation pinnate, sometimes palmate. **Flowers:** perianth and androecium perigynous; epicalyx bractlets absent or present; hypanthium shallowly bowl-shaped, hemispheric, campanulate, patelliform, ± crateriform, or turbinate; torus absent, minute, or thickened; carpels 3–5(or 6)[–8], distinct, free or adnate to hypanthium base, styles ± terminal, distinct; ovules 2–5, apical, collateral or clustered. **Fruits** aggregated follicles or achenes (*Holodiscus*); styles deciduous or persistent, not elongate.

Genera 8, species ca. 130 (6 genera, 25 species, including 3 hybrids, in the flora): North America, Mexico, Central America, South America, Eurasia; introduced widely.

The two genera not present in North America north of Mexico are the southeastern European-western Asian *Sibiraea* Maximowicz (five species) and the Mexican *Xerospiraea* Henrickson (one species).

The base chromosome number for Spiraeeae is $x = 9$.

SELECTED REFERENCE Potter, D. et al. 2007b. Phylogenetic relationships in tribe Spiraeeae (Rosaceae) inferred from nucleotide sequence data. Pl. Syst. Evol. 266: 105–118.

1. Herbs, perennial; flowers unisexual; hypanthia shallowly bowl-shaped . . . . 49. *Aruncus*, p. 422
1. Shrubs or subshrubs; flowers bisexual; hypanthia hemispheric, campanulate, patelliform, ± crateriform, or turbinate.
    2. Subshrubs; leaves 2 or 3(–5)-ternate, primarily crowded basally . . . . 48. *Luetkea*, p. 421
    2. Shrubs; leaves simple, cauline (tightly clustered in *Petrophytum*).
        3. Leaves deciduous, herbaceous, membranous, chartaceous, rarely coriaceous; stems erect, arching, or ascending, sometimes spreading to prostrate in *Spiraea*.
            4. Fruits aggregated follicles; inflorescences panicles or corymbiform or racemiform; carpels free; petals greenish, yellowish, white, pink, or purple . . . 44. *Spiraea*, p. 398
            4. Fruits aggregated achenes; inflorescences panicles; carpels adnate to hypanthium base; petals usually white, sometimes pink-tinged, rarely pink . . . . 47. *Holodiscus*, p. 415
        3. Leaves persistent, often marcescent, coriaceous; stems prostrate (mat-forming), sometimes ascending or erect in *Petrophytum*.
            5. Inflorescences panicles; petals white; stamens 20–40 . . . . 45. *Petrophytum*, p. 411
            5. Inflorescences: flowers solitary; petals pink to pinkish, often purple-tinged; stamens 7–12 . . . . 46. *Kelseya*, p. 414

44. SPIRAEA Linnaeus, Sp. Pl. 1: 489. 1753; Gen. Pl. ed. 5, 216. 1754 • Spirée [Greek *speira*, to become spiral, alluding to flexile branches being suitable for wreathing into garlands]

Richard A. Lis

Shrubs, 1–40 dm; usually rhizomatous. **Stems** 5–20+, usually erect to ascending or arching, sometimes spreading, prostrate, or decumbent; bark reddish to dark brown, gray, or gray-black, <exfoliating or not>; long and short shoots present; young stems <tan or reddish brown to brown>, glabrous or villous. **Leaves** deciduous (tardily so in *S. thunbergii*, partly persistent in *S. cantoniensis*), cauline, alternate, <dimorphic with shoot type>, simple; petiole present;

blade obovate to oblanceolate, rhombic, elliptic, or linear to lanceolate, ovate, or suborbiculate, 1–10 cm, membranous, chartaceous, or coriaceous, margins flat, serrate to serrulate, dentate, or entire, venation pinnate, surfaces glabrous or hairy. **Inflorescences** mostly terminal or mostly axillary, (2 or)3–1000+-flowered, panicles or corymbiform or racemiform, hairy or glabrous; bracts present or absent; bracteoles present or absent. **Pedicels** present. **Flowers** opening before or after full foliation, 2–15 mm diam.; epicalyx bractlets 1 or 2; hypanthium usually hemispheric or campanulate, sometimes turbinate or crateriform, 0.5–2(–5) mm, ± densely hairy or glabrous; sepals 5, usually erect, spreading, or reflexed, sometimes ascending, deltate, triangular, ovate-triangular, or ovate; petals 5, greenish, yellowish, chalky, or translucent white, white, pink [rose], or purple, ovate to obovate, suborbiculate, or orbiculate, sometimes elliptic; <staminodes present or reduced to annulus, nectariferous>; stamens 10–50 in 2–4 series, shorter or longer than petals; torus absent or reduced; carpels [3 or]4 or 5[–8], free, glabrate to tomentose, styles terminal or subterminal, <stigmas capitate or discoid>; ovules 2[–4]. **Fruits** aggregated follicles, 4 or 5, cymbiform, ellipsoid, falcate, fusiform, or oblanceoloid, (0.5–)1.5–4 mm, <coriaceous>, glabrous or tomentose; hypanthium persistent; sepals persistent or deciduous, spreading, ascending, erect, or reflexed; styles persistent or deciduous. **Seeds** 2–4 per follicle, fusiform to oblong, <2–4 mm>. $x = 9$.

Species 100–120 (17, including 3 hybrids, in the flora): North America, Europe, Asia; introduced widely, especially in the northern hemisphere, where often cultivated.

Spiraea has been a popular decorative plant in North America, Europe, and Asia. Horticultural hybrids have been named. Hybridization also occurs in natural settings, complicating the identification of the species and varieties. For specimens that are suspected of being naturalized or of hybrid origin, additional references for identification include horticultural works such as L. H. Bailey et al. (1949), W. J. Bean (1970–1988, vol. 4), A. J. Rehder (1927), H. S. Maxwell and S. G. Knees (1989), and A. Huxley et al. (1992, vol. 4), or the major floristic treatments by A. I. Pojarkova (1971) and Lu L. T. and C. Alexander (2003), and references therein.

The position of *Spiraea* within the Rosaceae has been the subject of considerable speculation; treatments by D. Potter et al. (2002, 2007b) are particularly useful. The floral ontogeny and morphology of *S. alba* and *S. trilobata* Linnaeus were examined and compared to other members of the Amygdaloideae by R. C. Evans and T. A. Dickinson (1999b).

Spiraea hypericifolia Linnaeus has been found naturalized in eastern Texas (J. N. Mink et al. 2011) and has been reported in Mississippi (plants.usda.gov). The species is native to central and southwestern Asia and southeastern Europe; it has been widely cultivated in Europe and North America, but there are few reports of its escape and naturalization. The invasive potential of *S. hypericifolia* is uncertain and will likely vary with local site conditions. In its native habitat, it can form thickets on dry slopes and in riparian zones and wet meadows.

A description of *Spiraea hypericifolia* follows: Shrubs 5–15 dm, usually rhizomatous. Stems erect to ascending. Leaves: petiole 1–4 mm; blade oblong-obovate or obovate-lanceolate, 1.5–2 × 0.3–0.8 cm, length 2–4 times width, base acute, margins entire or with a few reduced teeth at apex, apex acute to obtuse, surfaces usually glabrous. Inflorescences axillary, 4–12-flowered, 0.75–2.5 × 1–3 cm. Flowers 5–8 mm diam.; hypanthia campanulate, 2–4 mm; sepals triangular, 1.5–2 mm; petals white, obovate to suborbiculate, 3–4 mm; staminodes 10–12; stamens 16–22, 0.8 times petal length. Follicles oblanceoloid, 1.4–2 mm, glabrous.

In North America, *Spiraea hypericifolia* would most resemble *S. prunifolia* or *S. thunbergii* and can be distinguished from these two species in that they have a leaf blade margin that is sharply serrulate, whereas in *S. hypericifolia* the margin is entire or with only a few teeth at the apex. Leaf shape is another distinguishing character; see the descriptions of 12. *S. prunifolia* and 13. *S. thunbergii* for comparative details.

1. Inflorescences racemiform, corymbiform, or panicles, sometimes reduced or simple fascicles (*S. prunifolia*), usually mostly axillary.
    2. Leaves: venation pinnate simple craspedodromous . . . . 14. *Spiraea chamaedryfolia*
    2. Leaves: venation pinnate cladodromous, eucamptodromous, simple craspedodromous, or suprabasal actinodromous.
        3. Leaves: venation suprabasal actinodromous . . . . 17. *Spiraea* ×*vanhouttei*
        3. Leaves: venation pinnate cladodromous, pinnate eucamptodromous, or pinnate simple craspedodromous.
            4. Leaf blades linear to lanceolate, venation pinnate eucamptodromous. . . . . 13. *Spiraea thunbergii*
            4. Leaf blades ovate, elliptic, oblanceolate, or rhombic-lanceolate, venation pinnate cladodromous or pinnate simple craspedodromous.
                5. Leaf blade margins serrulate, usually from near base to apex, sometimes only distally from middle, rarely nearly entire with relatively few teeth apically, venation pinnate cladodromous . . . . .12. *Spiraea prunifolia*
                5. Leaf blade margins coarsely serrate or irregularly 3-fid to slightly lobed distally, venation pinnate simple craspedodromous . . . . 15. *Spiraea cantoniensis*
1. Inflorescences corymbiform or conic, cylindric, hemispheric, obconic, pyramidal, or pyriform panicles, mostly terminal.
    6. Petals usually pink, sometimes white or purple; leaves: secondary veins prominent.
        7. Stamens 15–20, 1 times petal length; follicles tomentose to arachnoid . . . . 3. *Spiraea tomentosa*
        7. Stamens (10–)20–40, (1–)2–3 times petal length; follicles glabrous or adaxial sutures sometimes sparsely ciliate.
            8. Inflorescences corymbiform or elongate-conic to elongate-pyriform or hemispheric panicles, height 0.5–1(–2) times diam.; stamens 10–40, 1–2 times petal length; leaf abaxial surfaces glabrous or puberulent.
                9. Stamens 35–40; leaf abaxial surfaces glabrous or puberulent. . . . . 8. *Spiraea splendens*
                9. Stamens 10–20; leaf abaxial surfaces puberulent . . . . 9. *Spiraea* ×*hitchcockii*
            8. Inflorescences conic or hemispheric to obconic panicles, height 1–4 times diam.; stamens 20–35, 2–3 times petal length; leaf abaxial surfaces lanate to tomentose or glabrous or puberulent to pubescent.
                10. Panicles conic, height 2–4 times diam.; stamens 25–35; staminodes 0 . . . . 10. *Spiraea douglasii*
                10. Panicles hemispheric to obconic, height 1–1.5 times diam.; stamens 20–30; staminodes 10–15 . . . . 11. *Spiraea* ×*pyramidata*
    6. Petals usually greenish, yellowish, or translucent white, white, or chalky white, sometimes pink-tinged to pink; leaves: secondary veins not prominent.
        11. Leaves: number of primary and secondary serrations 3–5 times number of secondary veins (excluding inter-secondary veins) . . . . .16. *Spiraea japonica*
        11. Leaves: primary and secondary serrations 0–1 times number of secondary veins (excluding inter-secondary veins).
            12. Shrubs 1–10(–15) dm; inflorescences corymbiform, height usually 0.4–1.1 times diam.; inflorescence branches rarely in axils of leaves.
                13. Leaves coriaceous, margins irregularly, coarsely and sharply doubly serrate, teeth acute and mucronate. . . . . .4. *Spiraea corymbosa*
                13. Leaves membranous, margins usually regularly to irregularly serrate, serrulate, or crenate, rarely entire, teeth rounded.
                    14. Stems erect to ascending, often dying to ground annually; leaf bases acute . . . . .6. *Spiraea lucida*
                    14. Stems ascending to prostrate, rarely dying to ground, (rarely branched); leaf bases obtuse . . . . .7. *Spiraea stevenii*
            12. Shrubs 10–40 dm; inflorescences usually conic, cylindric, obconic, or pyramidal panicles, height 1.4–5 times diam., sometimes corymbiform, then ± hemispheric, height 0.5–1 times diam.; inflorescence branches usually in axils of leaves.
                15. Leaves: blade length 3 times width, margins entire or dentate, (teeth 1–4 in distal ¼ of blade). . . . . .5. *Spiraea virginiana*
                15. Leaves: blade length 3–5 times width, margins serrulate to serrate.

[16. Shifted to left margin.—Ed.]

16. Leaf blades narrowly lanceolate to oblanceolate or broadly oblanceolate to obovate, secondary veins irregularly terminating in primary teeth, inter-secondary veins usually 8–12+ per leaf; inflorescence branches puberulent to pubescent. . . . . 1. *Spiraea alba*
16. Leaf blades narrowly rhombic to rhombic or lanceolate to narrowly elliptic, secondary veins regularly terminating in primary teeth, inter-secondary veins usually 1–4 per leaf; inflorescence branches puberulent or glabrous. . . . . 2. *Spiraea salicifolia*

1. Spiraea alba Du Roi, Harbk. Baumz. 2: 430. 1772 • Narrow-leaf, eastern, or white meadowsweet, spirée blanche [E] [F] [W]

Shrubs, 10–20 dm. **Stems** erect, unbranched. **Leaves:** petiole 2–8 mm, <puberulent or sparsely hairy>; blade narrowly lanceolate to oblanceolate or broadly oblanceolate to obovate, 2–9 × 0.5–3 cm, <length 3–5 times width>, chartaceous or membranous, base cuneate to rounded, margins finely to coarsely, sharply serrate to serrulate (sometimes doubly so on long shoot leaves), <number of primary and secondary serrations 0.5–1.1 times number of secondary veins (excluding inter-secondary veins)>, venation pinnate craspedodromous, secondary veins not prominent, <irregularly terminating in primary teeth, inter-secondary veins usually 8–12+ per leaf>, apex acute to obtuse, abaxial surface mostly glabrous, adaxial glabrous. **Inflorescences** mostly terminal, narrowly conic to open, pyramidal panicles, 5–20 × 3–10 cm <height 1.4–3.5 times diam.>; branches <usually in axils of leaves>, puberulent to pubescent. **Pedicels** 1–2(–3) mm, glabrous or glabrate. **Flowers** 3–8 mm diam.; hypanthia hemispheric, 0.6–0.8 mm, abaxial surface usually puberulent to sparsely strigose, sometimes glabrate or glabrous, adaxial glabrous; sepals triangular, 0.8–1.5 mm; petals usually white, sometimes pink-tinged (in bud), suborbiculate, 1.3–2(–3) mm; staminodes 0–4; stamens 30–50, 1–2 times petal length. **Follicles** oblanceoloid, 3–4 mm, <shiny>, glabrous.

Varieties 2 (2 in the flora): North America; introduced in Europe.

Varieties *alba* and *latifolia* have regions of hybridization, primarily around the Great Lakes, that produce intermediate forms that may be difficult to key. Recognition of these two taxa as either species or varieties has been problematic because the taxa are quite distinct at the extremes of their range. A. R. Kugel (1958) recognized them as distinct species, and her work illustrates the regional zone of hybridization. H. A. Gleason (1952) and Gleason and A. Cronquist (1963) also recognized the two taxa as separate species; later Gleason and Cronquist (1991) recognized them as varieties that frequently intergrade.

The European *Spiraea salicifolia* has a leaf morphology that is similar to that of var. *alba* and the two taxa have been treated as conspecific. With the interest in *Spiraea* as an ornamental, *S. salicifolia* was imported to North America, resulting in this species or hybrids of it becoming naturalized. Specimens that are difficult to identify may be *S. salicifolia*, as escapes from homesteads and gardens that became established or may have hybridized with native taxa.

Spiraea alba has become locally naturalized in western and central Europe; in the British Isles, its hybrids are commonly naturalized as *S.* ×*rosalba* Dippel (*S. alba* × *S. salicifolia*), or as *S.* ×*billardii* Hortus ex K. Koch (*S. alba* × *S. douglasii*) (A. J. Silverside 1990).

1. Leaves: lengths 3–4 times widths, margins finely serrate to serrulate; inflorescences narrowly conic . . . . 1a. *Spiraea alba* var. *alba*
1. Leaves: lengths 2–3 times widths, margins coarsely serrate; inflorescences open, pyramidal . . . . 1b. *Spiraea alba* var. *latifolia*

1a. Spiraea alba Du Roi var. **alba** [E] [F]

Spiraea ciliata Rafinesque; *S. cuneifolia* Borkhausen; *S. lancifolia* Hoffmannsegg; *S. salicifolia* Linnaeus var. *lanceolata* Torrey & A. Gray; *S. salicifolia* var. *linearis* Wenzig; *S. salicifolia* var. *paniculata* Aiton; *S. tomentosa* Linnaeus var. *alba* Marshall

Leaves: petiole 2–8 mm, puberulent; blade narrowly lanceolate to oblanceolate, 2–9 × 0.5–2.5 cm, length 3–4 times width, chartaceous, base cuneate, margins finely serrate to serrulate, apex acute. **Panicles** narrowly conic, 5–15 × 3–6 cm. **Pedicels** 1–2(–3) mm. **Flowers** 5–8 mm diam.; hypanthia abaxial surface usually puberulent to sparsely strigose, sometimes glabrous or glabrate; sepals 0.8–1 mm; petals 1.7–2 mm; stamens 40–50, 1 times petal length. $2n = 36$.

Flowering Jun–Sep; fruiting Jun–Nov. Wetlands, fens, riparian areas, moist open sites; 0–300 m; Alta., Man., Ont., Que., Sask.; Ill., Ind., Iowa, Ky., Mass., Mich., Minn., Mo., N.Y., N.C., N.Dak., Ohio, Pa., S.Dak., Tenn., Vt., Va., W.Va., Wis.; introduced in Europe.

Variety *alba* is of conservation concern in Kentucky and Tennessee.

1b. Spiraea alba Du Roi var. **latifolia** (Aiton) Dippel, Handb. Laubholzk. 3: 484. 1893 • Broad-leaved or northern or alpine or mountain meadowsweet, spirée à larges feuilles E W

Spiraea salicifolia Linnaeus var. *latifolia* Aiton, Hort. Kew. 2: 198. 1789; *S. alba* var. *septentrionalis* (Fernald) Fosberg; *S. bethlehemensis* Hortus ex K. Koch; *S. carpinifolia* Willdenow; *S. heterophylla* Rafinesque; *S. latifolia* (Aiton) Borkhausen; *S. latifolia* var. *septentrionalis* Fernald; *S. obovata* Rafinesque; *S. ovata* Rafinesque; *S. septentrionalis* (Fernald) Á. Löve & D. Löve

Leaves: petiole 2–4 mm, sparsely hairy; blade broadly oblanceolate to obovate, 2–8 × 1–3 cm, length 2–3 times width, membranous, base cuneate to rounded, margins coarsely serrate, apex acute to obtuse. **Panicles** open, pyramidal, 5–20 × 3–10 cm. **Pedicels** 2–3 mm. **Flowers** 3–5 mm diam.; hypanthia abaxial surface glabrous or glabrate; sepals 0.8–1.5 mm; petals 1.3–1.5(–3) mm; stamens 30–35, 2 times petal length. $2n$ = 36, 54.

Flowering Jun–Sep; fruiting Jun–Nov. Moist or dry open sites, meadows, fields, forest margins; 0–1500 m; St. Pierre and Miquelon; N.B., Nfld. and Labr., N.S., Ont., P.E.I., Que.; Conn., Del., Maine, Md., Mass., Mich., N.H., N.J., N.Y., N.C., Pa., R.I., Vt., Va., W.Va.

Among synonyms of var. *latifolia,* var. *septentrionalis* was first described by Fernald, who found it occupying a number of unique boreal and subalpine areas of the northeastern United States and southeastern Canada; it was raised to specific rank by Á. Löve and D. Löve based upon its chromosome number ($2n$ = 54). Populations of var. *latifolia* that have been ascribed to var. *septentrionalis* are cited as differing in being alpine, subalpine, and boreal. In New Hampshire, D. D. Sperduto (2005) characterized it as an alpine and subalpine obligate that occurs in ravines and snowbed communities. A. Haines (pers. comm.) could not find any unique floral morphological characteristic to distinguish populations of var. *septentrionalis* from var. *latifolia.* Additional work is needed to determine if the polyploid populations (identified as var. *septentrionalis*) have unique ecological tolerances and if other physiological or morphological characteristics will allow consistent distinction from var. *latifolia.*

2. Spiraea salicifolia Linnaeus, Sp. Pl. 1: 489. 1753 • Bridewort, willow-leaved meadowsweet I

Spiraea amena Rafinesque

Shrubs, 10–20 dm, <thicket forming>. **Stems** erect to ascending or spreading, rarely branched. **Leaves:** petiole 2–6 mm; blade narrowly rhombic to rhombic or lanceolate to narrowly elliptic, <usually widest at middle>, 3–7 × 1–3 cm, <length 3–5 times width>, chartaceous, base acute, margins sharply serrate to serrulate nearly to base, <number of primary and secondary serrations 1 times number of secondary veins (excluding inter-secondary veins)>, venation pinnate craspedodromous, secondary veins not prominent, <regularly terminating in primary teeth, inter-secondary veins usually 1–4 per leaf>, apex acute, abaxial surface mostly puberulent, adaxial glabrous. **Inflorescences** mostly terminal, cylindric to obconic panicles, 5–10 × 2–4 cm <height 2–5 times diam.>; branches <sometimes in axils of leaves>, puberulent or glabrous. **Pedicels** 3–6 mm, puberulent or glabrous. **Flowers** 4–7(–10) mm diam.; hypanthia hemispheric, 0.8–1 mm, abaxial surface glabrous or sparsely puberulent, adaxial glabrous; sepals ovate, 0.8–1 mm; petals pink, elliptic to widely obovate, 1.8–2 mm; staminodes 0; stamens 28–32, 2 times petal length. **Follicles** oblanceoloid, 4 mm, glabrous, <adaxial suture glabrous or ciliate>.

Flowering Jun–Aug; fruiting Jun–Sep. Old homesteads in moist areas, flood plains; 0–300 m; introduced; Ont.; Ga., Ky., Mich., Miss., N.C., Vt., Va.; e Europe; e Asia; introduced also in w, c Europe.

Spiraea salicifolia is escaped from cultivation in eastern North America. K. Sax (1936) reported a specimen of *S. salicifolia* as tetraploid; it is possible that it was not a European specimen, so this count may be unreliable.

The name *Spiraea salicifolia* is often misapplied to individuals of *S. alba* var. *alba* or *S. alba* var. *latifolia* (see discussion above). In addition, being a cultivated species that has been popular in gardens and likely introduced and escaped in various hybridized forms, the intermediates and potential hybrids with native taxa can make identification difficult. For problematic specimens that appear to be hybrids, see A. J. Silverside (1988, 1990).

3. Spiraea tomentosa Linnaeus, Sp. Pl. 1: 489. 1753 • Hardhack, steeplebush, spirée tomenteuse [E] [F] [W]

Shrubs, 3–15 dm. **Stems** erect to ascending, rarely branched. **Leaves:** petiole 2–3 mm; blade <bicolorous>, ovate-lanceolate, 3–6 × 1–2.5 cm, coriaceous, base cuneate, margins coarsely serrate to crenate on distal 3/4, (secondary teeth on large and long shoot leaves), venation pinnate craspedodromous, secondary veins prominent, apex acute or rounded, abaxial surface densely <white to gray, tan, or rusty>, tomentose, adaxial glabrous or puberulent. **Inflorescences** predominantly terminal, narrow, conic panicles, <150–15,000+-flowered>, 5–30 × 3–20 cm, <branching varying from compact to wide and spreading>; branches tomentose. **Pedicels** 0.1–1.5 mm, tomentose. **Flowers** 4–6 mm diam.; hypanthia hemispheric, 0.5–1 mm, abaxial surface tomentose, adaxial glabrous; sepals triangular, 0.7–1.1 mm; petals usually light to dark pink, sometimes white or purple, ovate to orbiculate, 1–1.5 mm; staminodes 0; stamens 15–20, 1 times petal length. **Follicles** cymbiform, 2–2.5 mm, tomentose to arachnoid. **2*n*** = 24, 36.

Varieties 2 (2 in the flora): c, e North America; introduced in Europe.

P. J. Salamun (1951) conducted a comprehensive study of variation of the inflorescence across much of the range of *Spiraea tomentosa* and concluded that the two varieties could be segregated based upon the number of flowers or fruits per centimeter (density) of a lateral branch of the inflorescence. Intermediate populations have been identified, particularly in western New York and Pennsylvania. A. Gille (1949) discussed the overall distribution of *S. tomentosa* without evaluating varieties, and conducted a detailed study of its ecology in Quebec. The occurrences of this species (var. *rosea*) in Oregon are as an introduced weed in cranberry bogs; it has the potential to spread to natural wetlands. The species is reported as invasive in Belgium and Poland.

1. Pedicels usually not visible, 0.1–0.5 mm; flowers or fruits 12–20 per cm of branches .... 3a. *Spiraea tomentosa* var. *tomentosa*
1. Pedicels easily visible, 0.5–1.5 mm; flowers or fruits 6–11 per cm of branches .... 3b. *Spiraea tomentosa* var. *rosea*

3a. Spiraea tomentosa Linnaeus var. **tomentosa** [E] [F] [W]

Spiraea glomerata Rafinesque; *S. parvifolia* Rafinesque

Inflorescences 300–15,000+-flowered, flowers or fruits 12–20 per cm of branches. **Pedicels** 0.1–0.5 mm. **2*n*** = 36.

Flowering Jul–Sep; fruiting Aug–Oct. Wetlands, woodland borders with moist acidic soil; 0–400 m; N.B., N.S., Ont., P.E.I., Que.; Conn., Del., Maine, Md., Mass., N.H., N.J., N.Y., Pa., R.I., Vt., Va., W.Va.; introduced in n, c Europe.

Forms with white petals have been identified as forma *albiflora* J. F. Macbride; plants with rose petals are assigned to forma *tomentosa*. The chromosome count of 2*n* = 36 by J. T. Baldwin (1951) was for forma *albiflora*.

3b. Spiraea tomentosa Linnaeus var. **rosea** (Rafinesque) Fernald, Rhodora 14: 190. 1912 [E]

Spiraea rosea Rafinesque, New Fl. 3: 62. 1838 (as Spirea); *S. ferruginea* Rafinesque

Inflorescences 150–10,000-flowered, flowers or fruits 6–11 per cm of branches. **Pedicels** 0.5–1.5 mm.

Flowering Jul–Sep; fruiting late Aug–Oct. Wetlands, poor fens, wet meadows, bogs, woodland borders with moist acidic soil; 0–1500 m; N.B., Ont., P.E.I., Que.; Ala., Ark., Ga., Ill., Ind., Kans., Ky., La., Md., Mich., Minn., Miss., Mo., N.J., N.Y., N.C., Ohio, Oreg., Pa., S.C., Tenn., Vt., Va., W.Va., Wis.

4. Spiraea corymbosa Rafinesque, Précis Découv. Somiol., 36. 1814 (as Spirea) • Shinyleaf meadowsweet, dwarf spiraea [E]

Spiraea betulifolia Pallas var. *corymbosa* (Rafinesque) S. Watson; *S. ostryfolia* Rafinesque; *S. repens* Rafinesque; *S. sororia* Penny ex G. Don

Shrubs, 3–10(–15) dm. **Stems** erect to arching, branched. **Leaves:** petiole 3–8 mm, <sparsely hairy>; blade ovate to elliptic or suborbiculate, 2–10 × 1–2.5 cm, coriaceous, base obtuse, rounded, or subcordate, margins irregularly, coarsely and sharply doubly serrate from midpoint to apex, <teeth acute and mucronate, number of primary and secondary serrations 1 times number of secondary

veins (excluding inter-secondary veins)>, venation pinnate cladodromous, secondary veins not prominent, apex acute, obtuse, or rounded, abaxial surface mostly glabrous, adaxial glabrous. **Inflorescences** mostly terminal, corymbiform, 2–5 × 3–10 cm <height 0.4–1.1 times diam.>; branches <rarely in axils of leaves>, glabrous or glabrate. **Pedicels** 1–3 mm, glabrous or glabrate. **Flowers** 4–7 mm diam.; hypanthia hemispheric, 0.8–1 mm, abaxial surface glabrous or pubescent, adaxial glabrous; sepals triangular, 0.5–1 mm; petals chalky white to pink, orbiculate, 1.3–1.5 mm; staminodes 5–15 <reduced to serrations>; stamens 15–20, 2 times petal length. **Follicles** nearly fusiform, 2–3 mm, <shiny>, glabrous. $2n = 36$.

Flowering May–Oct; fruiting Jun–Nov. Open rocky soil, rocky, lightly wooded sites, dry or fast draining slopes, rocky edges of woods; 0–500 m; N.S., Ont.; Ga., Md., Mass., N.J., N.C., Pa., Tenn., Va., W.Va.; introduced in e Europe.

H. A. Gleason and A. Cronquist (1963) and L. J. Uttal (1974) considered *Spiraea corymbosa* to be a variety of *S. betulifolia*. K. Sax (1936) found *S. corymbosa* to be a triploid with complete pollen sterility and hypothesized that it must exist as a diploid, or form viable egg cells, because it is involved in hybrids. If so, these cytological differences may be correlated with some morphological variation.

5. Spiraea virginiana Britton, Bull. Torrey Bot. Club 17: 314. 1890 • Virginia or West Virginia spiraea C E

Shrubs, 30–40 dm. **Stems** erect to arching or prostrate, densely branched. **Leaves:** petiole 2–5 mm; blade ovate to oblanceolate, 3–5 × 1–1.5 cm (length 3 times width), membranous, base acute, margins entire or dentate <teeth 1–4 in distal 1/4 of blade, mucronate, number of primary and secondary serrations 1 times number of secondary veins (excluding inter-secondary veins)>, venation pinnate cladodromous, secondary veins not prominent, apex acute to acuminate, abaxial surface <glaucous>, mostly glabrous, adaxial glabrous. **Inflorescences** mostly terminal, corymbiform or hemispheric panicles, 2–8 × 5–22 cm <height 0.5–1 times diam.>; branches <frequently in axils of leaves>, usually glabrous or pubescent, sometimes villous. **Pedicels** 1–3 mm, usually glabrous or pubescent, sometimes villous. **Flowers** 5–7 mm diam.; hypanthia turbinate to campanulate, 1.2–1.8 mm, surfaces glabrous or pubescent; sepals triangular, 1–1.2 mm; petals yellowish white, greenish white, or translucent white, ovate to orbiculate, 1.8–2 mm; staminodes 0; stamens 25–45, 2 times petal length. **Follicles** oblanceoloid, 1.5–2 mm, <shiny>, glabrous.

Flowering May–Jul; fruiting Jun–Sep. Riparian zones of rocky stream reaches with high flows capable of scour and deposition of sediments; of conservation concern; 200–1000 m; Ala., Ga., Ky., N.C., Ohio, Pa., Tenn., Va., W.Va.

Spiraea virginiana is endemic to the southern Blue Ridge and Appalachian Plateau physiographic provinces. In stream and river reaches, where there is little variation in flow volume and velocity, available habitat for *S. virginiana* would decline over time without scour and redeposition of sediments. The species reproduces primarily through asexual propagation: the stems break during high flows, are transported and deposited in sediments, and take root, establishing plants (D. W. Ogle 1991b; C. M. Anders and Z. E. Murrell 2001). Ogle (1991, 1991b) conducted a comprehensive analysis of the species and found viable seed to be rare on individuals found within local populations, confined to a single stream corridor. When plants from different populations were brought together in cultivation, abundant viable seed was produced (Ogle 1991b).

All populations of *Spiraea virginiana* examined by J. R. Brzyski and T. M. Culley (2011) had higher levels of genetic differentiation among populations, and lower values of genotypic richness within populations, than would be expected for a clonal organism.

Spiraea virginiana is in the Center for Plant Conservation's National Collection of Endangered Plants.

SELECTED REFERENCES Anders, C. M. and Z. E. Murrell. 2001. Morphological, molecular, and biogeographical variation within the imperiled Virginia *Spiraea*. Castanea 66: 24–41. Brzyski, J. R. and T. M. Culley. 2011. Genetic variation and clonal structure of the rare, riparian shrub *Spiraea virginiana* (Rosaceae). Conservation Genet. 12: 1323–1332. Clarkson, R. B. 1959. The West Virginia *Spiraea*. Castanea 24: 143–146. Ogle, D. W. 1991. *Spiraea virginiana* Britton: I. Delineation and distribution. Castanea 56: 287–296. Ogle, D. W. 1991b. *Spiraea virginiana* Britton: II. Ecology and species biology. Castanea 56: 297–303.

6. Spiraea lucida Douglas ex Greene, Pittonia 2: 221. 1892 • Shiny-leaf spiraea E

Spiraea betulifolia Pallas var. *lucida* (Douglas ex Greene) C. L. Hitchcock; *S. corymbosa* Rafinesque var. *lucida* (Douglas ex Greene) A. Zobel

Shrubs, 3–10 dm. **Stems** erect to ascending, <often dying to ground annually>, rarely branched. **Leaves:** petiole 4–15 mm; blade obovate to ovate, 2–8 × 1–3(–5) cm, membranous, base acute, margins usually coarsely serrate in distal 1/4–1/2 with secondary and tertiary teeth, <teeth rounded or nearly entire with 1 or 2 teeth near apex, number of primary and secondary serrations 0 or 1 times number of secondary veins (excluding inter-secondary

veins)>, venation pinnate cladodromous, secondary veins not prominent, apex acute to obtuse, surfaces glabrous. **Inflorescences** mostly terminal, corymbiform, 5–8 × 3–10 cm <height 0.5–1 times diam.>; branches <rarely in axils of leaves>, glabrous or sparsely pubescent. **Pedicels** 0.5–4 mm, glabrous or sparsely pubescent. **Flowers** 4–6 mm diam.; hypanthia hemispheric, 0.5–1 mm, abaxial surface glabrous, adaxial glabrous or glabrate; sepals triangular, 0.5–1.1 mm; petals white, orbiculate, 1–2 mm; staminodes 10–12; stamens 18–24, 2 times petal length. **Follicles** oblanceoloid to ellipsoid, 2–2.5 mm, glabrous or sparsely pubescent.

Flowering May–Aug; fruiting Jun–Nov. Open woods, meadows, stream banks, often on rocky scree slopes; 10–3300 m; Alta., B.C., Sask.; Idaho, Mont., N.Dak., Oreg., S.Dak., Wash., Wyo.

Spiraea lucida has been considered a variety of *S. betulifolia* based upon: hypanthial vestiture, inflorescence shape, extent of leaf serration, and leaf texture (L. J. Uttal 1974); or, corymbiform inflorescences, white petals, plant nearly glabrous throughout (C. L. Hitchcock et al. 1955–1969, vol. 3; H. J. Scoggan 1978–1979, part 3; A. Cronquist et al. 1997). Although these character states occur in the eastern Asian *S. betulifolia*, it has very different leaf architecture showing much greater organization of the tertiary venation, which appears to be unique and is not found in any North American taxa. *Spiraea lucida* has an extremely variable leaf morphology that seems related to the tendency for stems to die back annually and the consequent annual production of long shoots. C. Sterling (1966) did not examine all taxa of *Spiraea*; the character states of carpel morphology of 7. *S. stevenii* (see discussion there under) and *S. lucida* were reported to be similar and to differ slightly from those of *S. betulifolia* (the provenance of his specimen is not clear from the text).

7. Spiraea stevenii (C. K. Schneider) Rydberg in N. L. Britton et al., N. Amer. Fl. 22: 247. 1908 (as steveni) • Alaska or Beauverd's spiraea

Spiraea beauverdiana C. K. Schneider var. *stevenii* C. K. Schneider, Bull. Herb. Boissier, sér. 2, 5: 348. 1905 (as steveni); *S. chamaedryfolia* Linnaeus var. *media* Pursh

Shrubs, 1–10 dm. **Stems** ascending to prostrate, <rarely dying to ground>, rarely branched. **Leaves:** petiole 0.5–6 mm; blade ovate to elliptic, 1–6 × 0.5–2.5 cm, membranous, base obtuse, margins regularly to irregularly serrulate on distal ¼–½ to crenate or entire (long shoot leaves sometimes serrate on full blade), <teeth rounded, number of primary and secondary serrations 0–1 times number of secondary veins (excluding inter-secondary veins)>, venation pinnate cladodromous, secondary veins not prominent, apex obtuse, surfaces glabrous. **Inflorescences** mostly terminal, corymbiform (dense), 1–3[–5] × 0.5–3[–5] cm <height usually 1 times diam.>; branches <rarely in axils of leaves>, usually glabrous or puberulent, sometimes villous. **Pedicels** 1–5(–8) mm, usually glabrous or puberulent, sometimes villous. **Flowers** 4–6 mm diam.; hypanthia hemispheric, 0.7–1 mm, abaxial surface puberulent to pubescent, adaxial glabrate to pubescent; sepals triangular, 1–1.5 mm; petals white, orbiculate, 1.5–2.5 mm; staminodes 10–12; stamens 18–22, 2 times petal length. **Follicles** oblanceoloid, 2.5–3.5 mm, puberulent.

Flowering Jun–Sep; fruiting Jul–Nov. Meadows, tundra, open woods, thickets, stream banks, lake edges, alpine areas; 0–2000 m; B.C., N.W.T., Yukon; Alaska; e Asia.

Spiraea stevenii has been frequently referred to as *S. beauverdiana*; the nomenclatural history of this was well summarized by L. J. Uttal (1973), who noted that the type for *S. beauverdiana* is a specimen that belongs in the circumscription of *S. aemeliana*. A study of the pollen morphology of *S. betulifolia* and *S. stevenii* (*S. beauverdiana*) (T. A. Poljakova and G. N. Gataulina 2008) showed substantial differences between these two taxa. Examination of the carpel anatomy of various species of *Spiraea* (C. Sterling 1966) showed that *S. stevenii* differs from other species except *S. lucida*.

8. Spiraea splendens É. N. Baumann ex K. Koch, Monatsschr. Vereines Beförd. Gartenbaues Königl. Preuss. Staaten 18: 294. 1875 [E]

Shrubs, 2–10 dm. **Stems** spreading to ascending, branched. **Leaves:** petiole 1–4 mm; blade ovate to elliptic, 1–4 × 1–2(–3) cm, chartaceous, base usually obtuse, sometimes acute, margins crenate to serrate distally from middle, venation pinnate craspedodromous, secondary veins prominent, apex usually obtuse, sometimes acute, abaxial surface glabrous or puberulent, adaxial glabrous. **Inflorescences** mostly terminal, corymbiform or hemispheric panicles, <100–1000+ -flowered>, 2–4 × 2–4 cm <height 0.8–1 times diam.>; branches glabrous or puberulent. **Pedicels** 1–3(–5) mm, glabrous or puberulent. **Flowers** 2–4 mm diam.; hypanthia hemispheric, 0.5–1.1 mm, abaxial surface glabrous or pubescent, adaxial pubescent; sepals triangular, 0.8–1 mm; petals light to dark pink, ovate to obovate, 1–2.5 mm, <veins prominent>; staminodes 5–10, <reduced, fused, appearing as scalloped ring>; stamens 35–40, 2 times petal length. **Follicles** oblanceoloid, 2–3 mm, <shiny>, glabrous, <adaxial suture sparsely ciliate>.

Varieties 2 (2 in the flora): w North America.

The name *Spiraea densiflora* Nuttall ex Rydberg subsp. *splendens* (É. N. Baumann ex K. Koch) Abrams, which pertains here, is incorrect in relation to the priority of *S. splendens* (1875) over *S. densiflora*, which is an illegitimate name with no priority.

1. Young stems, leaf abaxial surfaces, and pedicels and bracteoles glabrous or glabrate. . . . . 8a. *Spiraea splendens* var. *splendens*
1. Young stems, leaf abaxial surfaces, and pedicels and bracteoles puberulent . . . . 8b. *Spiraea splendens* var. *rosea*

8a. Spiraea splendens É. N. Baumann ex K. Koch var. **splendens** [E]

Young stems, leaf abaxial surface, and pedicels and bracteoles glabrous or glabrate. **Flowers:** hypanthial abaxial surface glabrous or glabrate, adaxial pubescent.

Flowering Jun–Aug; fruiting Jul–Nov. Along streams, lakeshores, meadows, forest openings and edges, wooded to open slopes; 500–3500 m; Alta., B.C.; Calif., Idaho, Mont., Oreg., Wash., Wyo.

8b. Spiraea splendens É. N. Baumann ex K. Koch var. **rosea** (A. Gray) Kartesz & Gandhi, Phytologia 67: 464. 1989 [E]

Spiraea betulifolia Pallas var. *rosea* A. Gray, Proc. Amer. Acad. Arts 8: 381. 1872 (as betulaefolia); *S. arbuscula* Greene; *S. helleri* Rydberg; *S. lucida* Douglas ex Greene var. *rosea* (A. Gray) Greene

Young stems, leaf abaxial surface, and pedicels and bracteoles puberulent. **Flowers:** hypanthial surfaces puberulent.

Flowering Jun–Aug; fruiting Jul–Nov. Along streams, lakeshores, margins of woods, meadows, open rocky slopes; 200–3500 m; Alta.; Calif., Idaho, Mont., Oreg., Wash.

Spiraea densiflora Nuttall ex Greenman (1898) is an illegitimate name, sometimes found in synonymy with *S. splendens* var. *rosea*.

Spiraea densiflora Nuttall ex Rydberg is an illegitimate name that pertains here.

9. Spiraea ×hitchcockii W. J. Hess & Stoynoff, Sida 18: 827. 1999 • Hitchcock's spiraea [E]

Shrubs, 10–12 dm. **Stems** erect to arching, or decumbent, branched. **Leaves:** petiole 2–5 mm; blade narrowly elliptic to elliptic, obovate, 2–5(–6) × 1–2.8 cm, membranous, base acute to obtuse, margins irregularly serrulate to irregularly erose, venation pinnate, simple craspedodromous or semicraspedodromous, secondary veins prominent, apex usually obtuse, sometimes acute, abaxial surface puberulent, adaxial glabrescent. **Inflorescences** mostly terminal, irregularly shaped, usually elongate-conic to elongate-pyriform panicles, 1.5–6(–9) × 1.5–3(–4.5) cm <height 0.5–1(–2) times diam.>; branches puberulent. **Pedicels** 2–5 mm, puberulent. **Flowers** 3–8 mm diam.; hypanthia crateriform, 0.7–1.1 mm, abaxial surface puberulent, adaxial puberulent; sepals deltate, 0.5–1 mm; petals pink to dark pink, elliptic to widely elliptic, 1.3–1.7 mm; staminodes 5–12; stamens 10–20, 1–2 times petal length. **Follicles** asymmetrically falcate, 0.8–1.3 mm, <not shiny>, glabrous. $2n = 54$.

Flowering Jun–Sep; fruiting Jul–Nov. Wet meadows, aspen meadows, lake margins, moist woodland margins, wetlands, along creeks; 300–2000 m; Oreg.

Hess and Stoynoff described *Spiraea* ×*hitchcockii* as a naturally occurring hybrid of *S. douglasii* var. *menziesii* and *S. splendens* var. *rosea*; it has been found within the Cascades of Oregon and is very likely to be found in other areas of the Pacific Northwest, where the two species have overlapping ranges.

10. Spiraea douglasii Hooker, Fl. Bor.-Amer. 1: 172. 1832 • Douglas's or rose spiraea, western hardhack [E] [F]

Shrubs, 10–30 dm. **Stems** erect to ascending, branched. **Leaves:** petiole 3–5 mm; blade <bicolorous>, elliptic to obovate, 3–10 × 0.5–4 cm, chartaceous, base cuneate, margins serrate distally from middle, venation pinnate craspedodromous, secondary veins prominent, apex acute to obtuse, abaxial surface lanate to tomentose, glabrous, or pubescent, adaxial glabrous. **Inflorescences** mostly terminal, conic panicles, [5–]8–25 × 3–6[–10] cm <height 2–4 times diam.>; branches glabrous or glabrescent to tomentose. **Pedicels** 1–3 mm, glabrous or tomentose. **Flowers** 2–4 mm diam.; hypanthia hemispheric, 1 mm, abaxial surface tomentose, adaxial glabrous; sepals ovate, 1 mm; petals light to dark pink, ovate to obovate, 1–2 mm; staminodes 0; stamens 25–35, 2–3 times petal length. **Follicles** oblanceoloid, 2.5–3 mm, <shiny>, glabrous, <adaxial suture sometimes ciliate>.

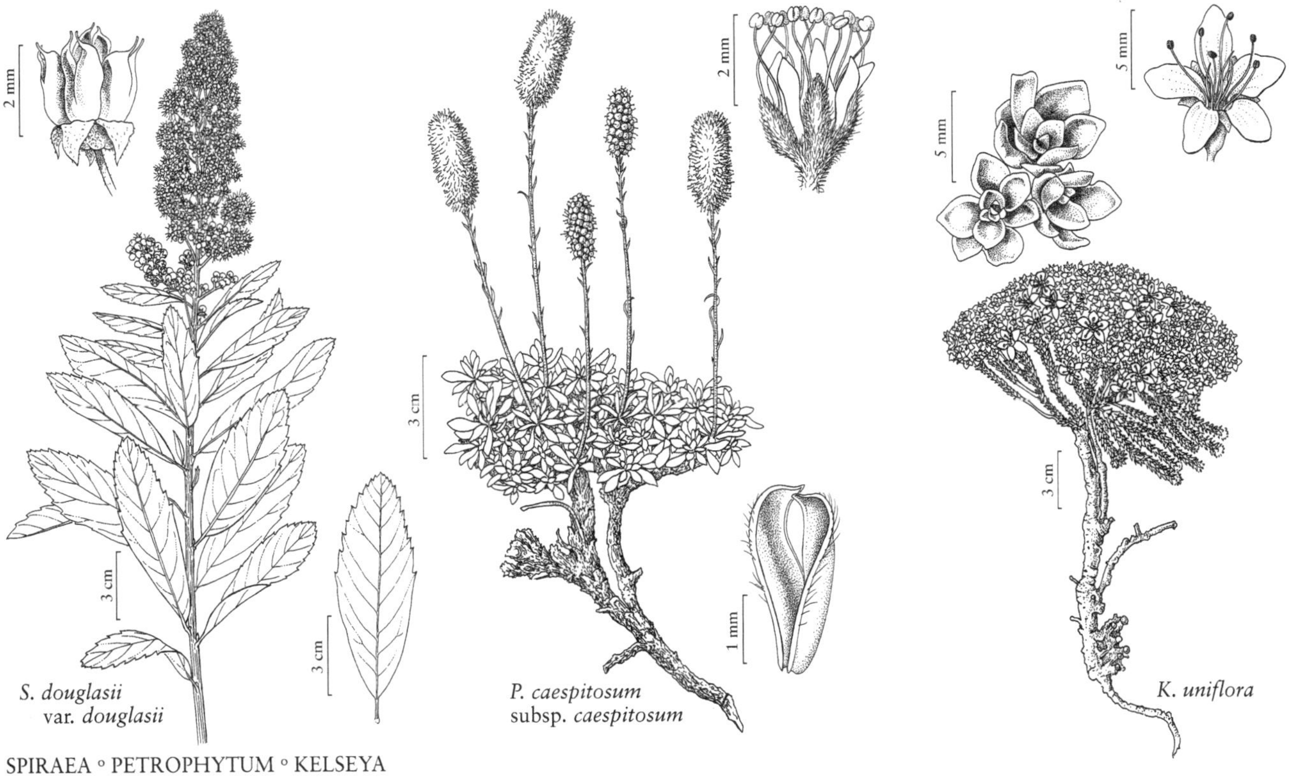

SPIRAEA ◦ PETROPHYTUM ◦ KELSEYA

Varieties 3 (3 in the flora): w North America; introduced in Europe.

Spiraea douglasii has been popular in the garden trade and was used to develop a number of hybrids. In North America, the hybrid most likely to be encountered is *S.* ×*billardii* Hortus ex K. Koch, because the parental species are both native, although *S. salicifolia* was widely imported from Europe and thus *S.* ×*pseudosalicifolia* Silverside would also likely be encountered.

1. Leaf abaxial surfaces lanate to tomentose . . . . 10a. *Spiraea douglasii* var. *douglasii*
1. Leaf abaxial surfaces glabrous or pubescent.
   2. Pedicels, hypanthia, and sepals tomentulose to sparsely pubescent . . . . 10b. *Spiraea douglasii* var. *menziesii*
   2. Pedicels, hypanthia, and sepals glabrescent or glabrous. . . . . .10c. *Spiraea douglasii* var. *roseata*

10a. Spiraea douglasii Hooker var. **douglasii** [E] [F]

Leaf abaxial surfaces <whitish to grayish>, lanate to tomentose. **Inflorescence branches** tomentose. **Pedicels** tomentose. **Flowers:** hypanthia and sepals tomentose.

Flowering May–Aug; fruiting Jul–Nov. Stream banks, moist woodland margins, wet meadows, fens, swamps, roadsides; 0–2500 m; B.C.; Calif., Oreg., Wash.; introduced in Europe.

H. J. Scoggan (1978–1979, part 3) identified two forms of var. *douglasii* based upon petal color: forma *alba* J. K. Henry (white) and forma *douglasii* (pink to deep rose). See discussion under 10b. var. *menziesii* for information that may also apply to var. *douglasii.*

10b. Spiraea douglasii Hooker var. **menziesii** (Hooker) C. Presl, Abh. Königl. Böhm. Ges. Wiss., ser. 5, 6: 555. 1851 [E]

Spiraea menziesii Hooker, Fl. Bor.-Amer. 1: 173. 1832; *S. douglasii* subsp. *menziesii* (Hooker) Calder & Roy L. Taylor; *S. subvillosa* Rydberg

Leaf abaxial surfaces glabrous or pubescent. **Inflorescence branches** tomentulose. **Pedicels** tomentulose to sparsely pubescent. **Flowers:** hypanthia and sepals tomentulose to sparsely pubescent.

Flowering Jun–Aug; fruiting Jul–Nov. Stream banks, flood plains, moist open forests, woodland margins, wet meadows and fields, wetlands, roadside ditches; 0–2500 m; B.C.; Alaska, Calif., Colo., Idaho, Mont., Oreg., Wash.; introduced in Europe.

H. J. Scoggan (1978–1979, part 3) identified two forms of var. *menziesii* based upon petal color: forma *menziesii* (Hooker) Voss (pink to deep rose) and forma *pseudosalicifolia* B. Boivin (white). These appear to be local forms that may be stable; petal color can be variable within varieties and species in *Spiraea*, and the stability of these forms, within the variety and species as a whole, has not been fully assessed. A. R. Kruckeberg (1967) tested two strains of var. *menziesii* and found that one showed adaptation to soils derived from ultramafic rocks (serpentinite, peridotite, dunite) and one did not. Although not discussed by Kruckeberg, variation of flower color within the variety may be correlated with substrate.

10c. Spiraea douglasii Hooker var. **roseata** (Rydberg) C. L. Hitchcock in C. L. Hitchcock et al., Vasc. Pl. Pacif. N.W. 3: 192. 1961 [E]

Spiraea roseata Rydberg in N. L. Britton et al., N. Amer. Fl. 22: 250. 1908; *S. idahoensis* A. Nelson

Leaf abaxial surfaces glabrous or pubescent. **Inflorescence branches** glabrescent or glabrous. **Pedicels** glabrescent or glabrous. **Flowers:** hypanthia and sepals glabrescent or glabrous.

Flowering Jun–Aug; fruiting Jul–Nov. Stream banks, woodland margins, meadows, roadsides; 300–2500 m; Idaho, Wash.

P. A. Rydberg (1917) considered *Spiraea roseata* to be a hybrid of *S. densiflora* and *S. menziesii* (here *S. douglasii* var. *menziesii*), and considered *S. idahoensis* to be a distinct species. At this time, it is unclear whether it is indeed a hybrid or a geographic variant of *S. douglasii*. The occurrence of specimens in the Columbia River region suggests a hybrid origin.

11. Spiraea ×pyramidata Greene, Pittonia 2: 221. 1892 (as species) [E]

Spiraea menziesii Hooker var. *pyramidata* (Greene) Piper; *S. tomentulosa* Rydberg

Shrubs, 3–15 dm. **Stems** erect to spreading, branched. **Leaves:** petiole 2–5 mm; blade elliptic to obovate, 2–6(–9) × 0.5–4 cm, chartaceous, base cuneate, margins doubly serrate distally from middle, venation pinnate craspedodromous, secondary veins prominent, apex acute to obtuse, abaxial surface puberulent to pubescent, adaxial glabrous. **Inflorescences** mostly terminal, hemispheric to obconic panicles, 5–15 × 3–10 cm <height 1–1.5 times diam.>; branches pubescent. **Pedicels** 1–3 mm, pubescent. **Flowers** 2–4 mm diam.; hypanthia hemispheric, 1 mm, abaxial surface tomentose, adaxial pubescent; sepals ovate, 1.5–2.5 mm; petals white to pink, ovate to obovate, 1.5–2.5 mm; staminodes 10–15; stamens 20–30, 2 times petal length. **Follicles** oblanceoloid, 1.5–2 mm, <shiny>, glabrous. **2*n*** = 36.

Flowering May–Jul; fruiting Jun–Nov. Disturbed sites, riparian zones, flood plains, scree slopes, dry canyon slopes, moist open forest and woodland edges, clearings, roadsides, railways, campsites, home or farm sites; 200–1500 m; B.C.; Idaho, Mont., Oreg., Wash.

Examining morphological characters of the leaf and flower, tests of pollen viability, cytological study of chromosomes, cultivation studies of seeds, cuttings, and live plants, assessment of pollinators, and controlled hybrid experiments led W. J. Hess (1969) to conclude that *Spiraea* ×*pyramidata* is a natural hybrid of *S. douglasii* var. *menziesii* and *S. lucida*. Hess found also that populations of *S.* ×*pyramidata* in Washington are more distinct and no introgression appeared to occur from the parental taxa; populations in British Columbia were less distinct and had more mixed populations with hybrids, parents, and probable introgressants. Hess determined that *S.* ×*pyramidata* is particularly adapted to colonization of open disturbed sites such as roadsides, railways, forest camps, and homesteads that occur in proximity to both parents. Once established, the hybrid spreads extensively by rhizomes.

SELECTED REFERENCE Hess, W. J. 1969. A taxonomic study of *Spiraea pyramidata* Greene (Rosaceae). Sida 3: 298–308.

12. Spiraea prunifolia Siebold & Zuccarini, Fl. Jap. 1: 131, plate 70. 1840 • Bridal-wreath spiraea [I]

Shrubs, 10–30 dm. **Stems** erect, <virgate>, branched. **Leaves:** petiole 1–4 mm, <pubescent>; blade usually ovate to elliptic, sometimes oblanceolate, 1–4(–5) × 1–2 cm, membranous, base obtuse, margins serrulate, usually from near base to apex, sometimes only distally from middle, rarely nearly entire with few teeth apically, venation pinnate cladodromous, secondary veins not prominent, apex acute to obtuse, abaxial surface lanate to glabrescent, adaxial glabrate. **Inflorescences** mostly axillary, usually hemispheric panicles, sometimes simple fascicles, <3–6-flowered, sessile>, 2–3 × 1–3 cm; <bractlets 3–7 × 3–5 mm>; branches pubescent. **Pedicels** 10–30 mm, pubescent. **Flowers** 5–15 mm diam.; hypanthia campanulate, 1–1.2 mm, abaxial surface glabrous or sparsely hairy, adaxial glabrous; sepals ovate to ovate-triangular, 1–3 mm; petals white to cream, <often with more than a single whorl of petals in most commonly escaped form>, ovate to obovate, 2–10 mm; staminodes 10–16, <irregularly fused>; stamens 20–25, 0.5 times petal length. **Follicles** oblanceoloid, 1.5–2 mm, glabrous, <adaxial suture sparsely hairy>. **2*n*** = 18.

Flowering Apr–Aug; fruiting May–Nov. Abandoned homesteads, roadsides, meadows, riparian zones; 0–300 m; introduced; N.S., Ont.; Ala., Ark., Calif., Conn., Del., D.C., Ga., Ill., Ky., La., Md., Mass., Miss., Mo., N.J., N.Y., N.C., Ohio, Pa., Tenn., Va., W.Va.; Asia (China, Japan, Korea).

Spiraea prunifolia is an ornamental shrub that can be found established in many areas. The double-flowered variant (most common in cultivation) may be assigned to var. *prunifolia*; single-flowered forms may be assigned to three other varieties [see Lu L. T. and C. Alexander (2003) for a key to native varieties in China]. L. H. Bailey et al. (1949), W. J. Bean (1970–1988, vol. 4), H. S. Maxwell and S. G. Knees (1989), A. Huxley et al. (1992, vol. 4), and references therein provide useful additional information on variation in *S. prunifolia*. M. De Cleene and J. De Ley (1981) noted that *S. prunifolia* and 17. *S.* ×*vanhouttei* (see discussion) are hosts to infectious hairy-root.

13. Spiraea thunbergii Siebold ex Blume, Bijdr. Fl. Ned. Ind. 17: 1115. 1827 • Thunberg's meadowsweet [I]

Shrubs, 10–20 dm. **Stems** spreading to arching, densely branched. **Leaves** <sometimes tardily deciduous>; petiole 1–2 mm or indistinct; blade linear to lanceolate, 2.5–4 × 0.3–1 cm, membranous, base acute, margins irregularly, sharply serrulate, venation pinnate eucamptodromous, secondary veins not prominent, apex acuminate, surfaces glabrous. **Inflorescences** mostly axillary, hemispheric panicles, <(2 or)3–7-flowered, sessile on leafless stems in spring>, 2–6 × 2–6 cm; branches glabrous. **Pedicels** 5–8 mm, glabrous. **Flowers** 6–8 mm diam.; hypanthia crateriform, 2–4 mm, abaxial surface glabrous, adaxial glabrous or puberulent; sepals ovate-triangular, 1.3–2 mm; petals white, obovate to suborbiculate, 2–4 mm; staminodes 10–14; stamens 18–20, 0.3 times petal length. **Follicles** oblanceoloid, 1.5–2 mm, glabrous. $2n = 18$.

Flowering Mar–May; fruiting Apr–Oct. Abandoned homesteads, meadows, roadsides, pastures; 0–500 m; introduced; Ont.; Ala., Conn., Ga., Kans., La., Md., Miss., Nebr., N.Y., N.C., Pa.; Asia.

Spiraea thunbergii flowers very early in the spring and may be easily observed due to the obvious flowers on branches lacking leaves; because it can thrive in a variety of soils, it will often persist as the native vegetation returns. Although established in a number of areas, it does not appear to spread aggressively or compete strongly with native species.

14. Spiraea chamaedryfolia Linnaeus, Sp. Pl. 1: 489. 1753 • Germander meadowsweet [I]

Shrubs, 10–20 dm. **Stems** erect, densely branched. **Leaves:** petiole 5–10 mm; blade ovate to ovate-lanceolate, 4–8 × 1.5–4 cm, membranous, base obtuse, margins usually coarsely, irregularly, sharply serrate to serrulate in distal ⅔, sometimes 2-serrate, venation pinnate simple craspedodromous, secondary veins not prominent, apex acute, surfaces <glaucous>, glabrous. **Inflorescences** mostly axillary, racemiform, corymbiform, or hemispheric panicles, (2–)3–10 × 2–5 cm; branches glabrous. **Pedicels** 5–10 mm, glabrous. **Flowers** 5–12 mm diam.; hypanthia campanulate, 1.5–2 mm, abaxial surface glabrous, adaxial glabrous or puberulent; sepals ovate-triangular, 2.5–3 mm; petals white, broadly ovate, 4–6 mm, <apex notched or erose>; staminodes 0; stamens 35–50, 2 times petal length. **Follicles** asymmetrical fusiform, 2–3 mm, glabrous or pubescent. $2n = 18$.

Flowering May–Jul; fruiting Jun–Oct. Old pastures, homestead sites; 0–700 m; introduced; Ont.; Conn., N.Y.; se Europe; Asia (China, Japan, Russia); introduced also in w Europe.

Two frequently mentioned varieties of *Spiraea chamaedryfolia* are var. *ulmifolia* (Scopoli) Maximowicz, which has primarily ovate leaves with an obtuse base and a frequently doubly-serrate margin and inflorescences 7–10 cm diam., treated as a species by Rydberg, and var. *flexuosa* (Fischer ex Cambessèdes) Maximowicz, which has ovate-lanceolate leaves with cuneate base and entire or singly-serrate margin with fewer teeth and inflorescences 4–6 cm diam. Variation in the leaf and inflorescence characters is inconsistent in plants established in North America. C. Sterling (1966) described the carpel anatomy of this species but did not examine enough taxa from Europe or Asia for comparison. W. J. Bean (1970–1988, vol. 4) and H. S. Maxwell and S. G. Knees (1989) are useful references for additional information on varieties and hybrids of this Eurasian native.

15. Spiraea cantoniensis Loureiro, Fl. Cochinch. 1: 322. 1790 • Reeve's spiraea or meadowsweet [I]

Shrubs, 3–20 dm, <not rhizomatous>. **Stems** arching, branched. **Leaves** <sometimes partly persistent>; petiole 3–10 mm; blade rhombic-lanceolate, 2–7 × 0.5–2 cm, membranous to chartaceous, base cuneate to obtuse, margins coarsely serrate or irregularly 3-fid to slightly lobed distally, venation pinnate simple craspedodromous, secondary veins not prominent, apex acute, abaxial surface <blue-green, glaucous>, glabrous, adaxial <dark green>, glabrous. **Inflorescences** mostly axillary, corymbiform or hemispheric panicles, 2–5 × 2–5 cm; <filiform bractlets sometimes present at base>; branches glabrous or pubescent. **Pedicels** 10–20 mm, glabrous or pubescent. **Flowers** <often double-flowered, usually distinct>, 10–15 mm diam.; hypanthia campanulate to turbinate, 3–5 mm, abaxial surface glabrous or sparsely hairy, adaxial glabrous; sepals deltate, 1–1.5 mm; petals white, elliptic to orbiculate, 2–4 mm; staminodes 10–12; stamens 20–28, 0.5–1 times petal length. **Follicles** inflated cymbiform, 0.5–2 mm, glabrous.

Flowering May–Jun; fruiting Jun–Sep. Meadows, woodland edges, abandoned homesteads; 0–300 m; introduced; Ala., Ark., La., N.Y., N.C., Ohio, Va.; Asia (China); introduced also in Mexico, Central America (Costa Rica, Panama), South America (Bolivia, Brazil, Colombia, Ecuador), s Africa, Pacific Islands (New Zealand), Australia.

Spiraea cantoniensis as found in China has a five-petaled flower, as is typical of the genus; the native species is rarely found or cultivated in North America. Instead, a cultivar (lanceata, flore pleno) with twice or more the number of petals, frequently referred to as "double-flowered," is the representative of *S. cantoniensis* most commonly established in the flora area; it does not appear to be invasive. Other references that may be helpful in identifying this taxon include L. H. Bailey et al. (1949), H. S. Maxwell and S. G. Knees (1989), A. Huxley et al. (1992, vol. 4), and Lu L. T. and C. Alexander (2003).

16. Spiraea japonica Linnaeus f., Suppl. Pl., 262. 1782 • Japanese spiraea [I] [W]

Spiraea callosa Thunberg; *S. fortunei* Planchon; *S. japonica* var. *fortunei* (Planchon) Rehder

Shrubs, 10–15 dm, <not rhizomatous>. **Stems** erect, branched. **Leaves:** petiole 2–7 mm; blade ovate to lanceolate, 5–10 × 2–4 cm, membranous, base cuneate, margins sharply doubly serrate, <number of primary and secondary serrations 3–5 times number of secondary veins (excluding inter-secondary veins)>, venation pinnate simple craspedodromous, secondary veins not prominent, apex acute, abaxial surface pubescent, adaxial glabrous. **Inflorescences** mostly terminal, corymbiform, 5–8 × 5–15 cm; branches finely villous. **Pedicels** 4–6 mm, finely villous. **Flowers** 4–8 mm diam.; hypanthia campanulate, 1–1.2 mm, abaxial surface strigose, adaxial pubescent; sepals triangular, 1–1.3 mm; petals light to dark pink, obovate, 1–1.5 mm; staminodes 0; stamens 25–30, 2 times petal length. **Follicles** oblanceoloid, 2–3 mm, glabrous, <adaxial suture glabrous or ciliate>. $2n = 36$.

Flowering May–Jun; fruiting Jun–Sep. Wet meadows, riparian zones; 0–1500 m; introduced; N.S., Ont.; Calif., Conn., Del., Ga., Ill., Ind., Ky., Maine, Md., Mass., Mich., Mo., N.H., N.Y., N.C., Ohio, Pa., R.I., S.C., Tenn., Utah, Va., W.Va., Wis.; Asia (China, Japan, Korea); introduced also in w, c, Europe.

Spiraea japonica is currently recognized as having eight varieties (Zhang Zhao Y. et al. 2002, 2006; Lu L. T. and C. Alexander 2003) and has been the source of many hybrids and cultivars that have been popular for gardens in temperate zones around the world.

Various varieties or cultivars of *Spiraea japonica* have become noxious weeds, primarily in wet sites and riparian areas. In Tennessee, *S. japonica* is ranked as a significant threat to native species by the Tennessee Exotic Pest Plant Council (2009); it is noted as a threat to *S. virginiana* through competition in the riparian zone (D. W. Ogle 1992).

One cultivar that is noted as having established populations is *Spiraea* ×*bumalda* Burvenich, which has been considered to be a hybrid of *S. albiflora* Miquel and *S. japonica* (B. A. E. Koehne 1893; A. J. Rehder 1927; L. H. Bailey et al. 1949; A. Huxley et al. 1992).

17. Spiraea ×vanhouttei (Briot) Carrière, Rev. Hort. 48: 260. 1876 (as species) • Bridal-wreath [I]

Spiraea aquilegiifolia Pallas var. *vanhouttei* Briot, Rev. Hort. 37: 269. 1866

Shrubs, 10–20(–25) dm, <not rhizomatous>. **Stems** arching, <virgate>, branched. **Leaves:** petiole 2–7 mm; blade ovate or rhombic-ovate to obovate, 3–5 × 2–3 cm, membranous, base cuneate to obtuse, margins crenate to coarsely serrate on distal ½–¾, irregularly doubly serrate, or, sometimes, 3–5 lobed (especially long shoot leaves), venation suprabasal actinodromous, secondary veins not prominent, apex acute, surfaces glabrous. **Inflorescences** mostly axillary (borne profusely along branches), corymbiform or hemispheric panicles, 3–6 × 3–6 cm; branches glabrous. **Pedicels** 5–10 mm, glabrous. **Flowers** 7–10 mm diam.; hypanthia campanulate, 1–1.5 mm, surfaces glabrous; sepals triangular, 1–1.5 mm; petals white to cream, orbiculate, 2–3 mm; staminodes 10–16, <mostly distinct>; stamens 20–25, 0.5 times petal length. **Follicles** falcate, 2–3 mm, glabrous.

Flowering May–Jun; fruiting Jun–Oct. Old homesteads, pastures, roadsides, railways; 0–2000 m; introduced; N.S., Ont.; Ark., Colo., Conn., Ill., Ky., La., Md., Mass., Mich., Minn., Mo., Mont., Nebr., N.Mex., N.Y., N.C., Ohio, Pa., Utah, Va., Wash., W.Va.; Asia (China); introduced also in Europe.

Spiraea ×*vanhouttei* is widely planted as an ornamental near homes and along streets. Concerning the potential for threat as an invasive or weedy species, R. B. Kaul (pers. comm.) has observed that it may persist for decades in old hedges and ornamental plantings; he has not seen any evidence of it volunteering or naturalizing and, despite repeated searches, has never seen a seed or seedling.

J. A. Harris (1917) questioned whether *Spiraea* ×*vanhouttei* had been proven to be the hybrid of *S. cantoniensis* with *S. trilobata*. C. Sterling (1966) described the carpel anatomy of *S.* ×*vanhouttei*; but he did not examine the postulated parents. M. De Cleene and J. De Ley (1981) attempted to define the host range of infectious hairy-root (*Agrobacterium rhizogenes*) and correlate hosts with taxonomic relationships; they verified that *S. prunifolia* and *S.* ×*vanhouttei* are hosts to this disease but did not test other species of the genus.

45. PETROPHYTUM (Nuttall ex Torrey & A. Gray) Rydberg, Mem. New York Bot. Gard. 1: 206. 1900 (as Petrophyton) • Rock-spiraea [Greek *petros*, rock, and *phyton*, plant, alluding to habitat]

Richard A. Lis

Spiraea Linnaeus [unranked] *Petrophytum* Nuttall ex Torrey & A. Gray, Fl. N. Amer. 1: 418. 1840

Shrubs, cespitose, densely matted, 0.1–10 dm. **Stems** 1+, prostrate, decumbent, erect, or ascending; bark brown to dark brown, aging to gray; short and long shoots present; short shoots glabrous. **Leaves** persistent, marcescent, cauline (tightly clustered), alternate, simple; petiole absent; blade oblanceolate to narrowly obtrullate, 0.2–2.5(–3) cm, coriaceous, margins flat, entire, venation indistinct, masked by hairs, or palmately 1–3-veined, surfaces glabrate, strigose, canescent, pilose, or sericeous. **Inflorescences** terminal, 10–100-flowered, panicles narrow or widely branched, flowers aggregated, dense, compact, puberulent, canescent to sericeous; bracts present, sometimes absent; bracteoles present. **Pedicels** present. **Flowers** 2–6 mm diam.; epicalyx bractlets 0; hypanthium hemispheric, turbinate, 0.5–1 mm, canescent, pilose, or sericeous, <sometimes glandular>; sepals 5, erect or reflexed, ovate to lanceolate; petals 5, persistent, withering, white, oval to oblong, obovate, or oblanceolate; stamens 20–40, equal to or longer than petals; torus thickened basally, <margins crenulate or entire>; carpels (3–)5 (or 6)[or 7], abaxially or adaxially connate or free, hirsute to pilose adaxially, styles terminal, <stigmas minute>; ovules 2 or 3[or 4]. **Fruits** aggregated follicles, (3–)5(or 6), lanceoloid, 1.5–2 mm, <coriaceous>, glabrous, glabrate, or sparsely pilose, dehiscent along abaxial and adaxial sutures; hypanthium persistent; sepals persistent, erect to reflexed; styles deciduous or persistent. **Seeds** 1 or 2, fusiform to terete. $x = 9$.

Species 3 (3 in the flora): w, sc United States, ne Mexico.

Taxa of *Petrophytum* are widely distributed in the mountains of western United States and northeastern Mexico, with *P. caespitosum* subsp. *caespitosum* the most widely distributed. The three other taxa recognized in this treatment, *P. caespitosum* subsp. *acuminatum*, *P. cinerascens*, and *P. hendersonii*, are isolated endemics with small ranges. With a single widely distributed taxon and three local endemics, it seems probable that taxa with closer geographical proximity would be likely to have greater morphological similarity. When C. Sterling (1966) examined carpel morphology in *P. caespitosum*, *P. cinerascens*, *P. elatius* [= *P. caespitosum* subsp. *caespitosum*], and *P. hendersonii*, he found that *P. elatius* and *P. hendersonii* had similar distributions of character states in five anatomical characters, yet differed as to whether carpels were fused (*P. hendersonii*) by their ventral bundles or distinct (*P. elatius*); *P. caespitosum* and *P. cinerascens* differed in only one character: carpels ventrally fused in *P. caespitosum* and distinct in *P. cinerascens*. Such incongruence of character states is seen also in leaves, flowers, and seeds; it is complicated by the reduction of these plants.

In molecular phylogenetic analyses of Rosaceae (D. Potter et al. 2007), *Petrophytum* was sister to *Kelseya*, with *Spiraea* basal to the pair. In a more detailed study of Spiraeeae (Potter et al. 2007b), *Kelseya* was sister to *Petrophytum* and *Spiraea*; however, the three topologies were not well supported, and additional work may show *Kelseya* and *Petrophytum* to be sister taxa.

1. Pedicel bracteoles rarely extending to base of sepals; leaf abaxial surfaces glabrate, sparsely strigose, pilose to pilose-sericeous on lamina, sparsely strigose along veins, venation visible through hairs (palmately 3-veined) . . . . 1. *Petrophytum hendersonii*
1. Pedicel bracteoles extending from middle to beyond apex of sepals; leaf abaxial surfaces minutely canescent to strigose, cinereous, or pilose to sericeous, venation usually not visible through hairs [3 veins may be visible through hairs (*P. cinerascens*) or 3+ veins may be visible in long shoot leaves (some plants of *P. caespitosum*)].
   2. Pedicel bracteoles extending from middle to apex of sepals, rarely beyond; leaf blades: apices obtuse, abaxial surfaces minutely canescent to strigose or cinereous, venation sometimes visible through hairs, palmately 3-veined; petal apices acute or rounded . . . . 2. *Petrophytum cinerascens*
   2. Pedicel bracteoles usually extending well beyond apex of sepals; leaf blades: apices acute, abaxial surfaces pilose to sericeous or sparsely strigose, venation rarely visible except on long-shoot leaves, 1(–3+)-veined; petal apices obtuse to slightly cleft or acute to acuminate . . . . 3. *Petrophytum caespitosum*

1. Petrophytum hendersonii (Canby) Rydberg in N. L. Britton et al., N. Amer. Fl. 22: 253. 1908 (as hendersoni) • Olympic mountain rockmat [E]

Eriogynia hendersonii Canby, Bot. Gaz. 16: 236. 1891 (as hendersoni); *Luetkea hendersonii* (Canby) Greene; *Spiraea hendersonii* (Canby) Piper

Shrubs, 2–9 dm diam. **Stems** decumbent to ascending, 1–10 cm, internodes 0.5–1.5 cm. **Leaves:** blade narrowly obtrullate, sometimes oblanceolate to spatulate-obovate, 1–2(–3) × 0.2–0.6 cm, palmately 3-veined, venation visible through hairs, apex acute, abaxial surface sparsely strigose, pilose to pilose-sericeous, sometimes glabrate, sparsely strigose along veins. **Panicles** densely racemose, 2–5 × 1–2 cm, sparsely sericeous; bracts linear, 2–4(–6) mm, pilose. **Pedicels** 2–5 mm; bracteoles 1, rarely extending to base of sepals. **Flowers** 3–5 mm diam.; hypanthium 0.5–1 mm, sparsely pilose; sepals reflexed, oblong to obtuse, 1–1.5 mm, margins ciliate, abaxial surface sparsely pilose; petals oval, oblong, or obovate, 2–2.5 mm, apex obtuse, sometimes 2-lobate; stamens 25–40, lengths 1–1.3 times petals (2 times sepals); carpels (4 or)5(or 6), abaxially connate. **Follicles** 1.5–2 mm. $2n = 18$.

Flowering Jul–Sep. Rock outcrops, cliffs, talus slopes; 1000–2300 m; Wash.

Petrophytum hendersonii is known only from the Olympic Mountains of northwestern Washington.

2. Petrophytum cinerascens (Piper) Rydberg in N. L. Britton et al., N. Amer. Fl. 22: 253. 1908 • Chelan rockmat [C] [E]

Spiraea cinerascens Piper, Erythea 7: 171. 1899; *Luetkea cinerascens* (Piper) A. Heller

Shrubs, 1–5 dm diam. **Stems** prostrate to decumbent, 2–8 cm, internodes (0.1–)1(–2) cm. **Leaves:** blade oblanceolate to obtuse, 1–2.5(–3) × 0.2–0.4 (–0.5) cm, palmately 3-veined, venation sometimes visible through hairs, apex obtuse, abaxial surface minutely canescent to strigose or cinereous, <adaxial sometimes glandular>. **Panicles** sometimes branched, 2–8(–15) × 1–5 cm, canescent to puberulent; bracts linear to oblanceolate, 5–10 mm, pilose. **Pedicels** 0.5–2(–4) mm; bracteoles 1(or 2), extending from middle to apex of sepals, rarely beyond. **Flowers** 2–4(–6) mm diam.; hypanthium 1 mm, canescent; sepals erect, ovate or lanceolate, 1–1.5 mm, margins ciliate, abaxial surface pubescent, <glandular>; petals oblanceolate or narrowly ovate, 1–2.5 mm, apex acute or rounded; stamens 20–25, lengths 1.3 times petals (1.3–1.5 times sepals); carpels (3–)5(or 6), distinct. **Follicles** 2 mm.

Flowering Jun–Aug. Crevices and ledges of outcrops, gneiss, schist, or granite; of conservation concern; 200–600 m; Wash.

Petrophytum cinerascens grows in crevices and ledges of outcrops above the Columbia River where there is little or no soil; it was found only on rocky outcrops of gneiss, schist, and granite between Chelan and Wenatchee in central Washington (D. J. Moore et al. 1998); C. L. Hitchcock et al. (1955–1969, vol. 3) erroneously cited its habitat as basaltic cliffs, which is the more common rock type on the Columbia River plateau. Moore et al. evaluated the ability of *P. cinerascens* for photosynthetic acclimation to increased growth temperature and drought stress under short-term experimental conditions, and concluded that it could not acclimate to such changes; they suggested that this endemic species might be at risk of extinction if warmer, drier local conditions result from projected climate changes.

3. Petrophytum caespitosum (Nuttall) Rydberg, Mem. New York Bot. Gard. 1: 206. 1900 (as Petrophyton) [F]

Spiraea caespitosa Nuttall in J. Torrey and A. Gray, Fl. N. Amer. 1: 418. 1840; *Eriogynia caespitosa* (Nuttall) S. Watson; *Luetkea caespitosa* (Nuttall) Kuntze

Shrubs, 1–10+ dm diam. **Stems** prostrate or decumbent, <loosely intertwined>, 1–4+ cm, and erect or ascending, <tightly coalesced>, 0.2–0.5 cm, internodes 0.1–1 cm. **Leaves:** blade spatulate, 0.4–1.8 × 0.2–0.4 cm, venation rarely visible except on long-shoot leaves, 1(–3)-veined, apex acute, abaxial surface pilose to sericeous, rarely sparsely strigose on lamina and veins. **Panicles** widely branched to compact and racemose, [0.5–]1–15[–20] × 0.5–6 cm, sericeous; bracts subulate to narrowly obtrullate, 3–8 mm, pilose to sericeous. **Pedicels** 0.5–2.5 mm; bracteoles 1, extending from middle to well beyond apex of sepals. **Flowers** 3–6 mm diam.; hypanthium 1 mm, densely sericeous; sepals erect, ovate-lanceolate, 1–1.5 mm, margins sericeous, abaxial surface sericeous to tomentose; petals narrowly oblanceolate, 1–2.5 mm, apex obtuse to slightly cleft or acute to acuminate; stamens 20, lengths 1.5–2(–2.5) times petals (1.5–3 times sepals); carpels (3–)5(or 6), adaxially connate. **Follicles** 2 mm. $2n = 18$.

Subspecies 2 (2 in the flora): w, sc United States, ne Mexico.

Petrophytum caespitosum primarily inhabits arid rocky outcrops and talus slopes at high elevations in mountain ranges of the western United States and northeastern Mexico.

1. Leaf blades 0.4–1.4 cm, abaxial surfaces pilose to sericeous; petal apices obtuse to slightly cleft . . . . .3a. *Petrophytum caespitosum* subsp. *caespitosum*
1. Leaf blades 1–1.8 cm, abaxial surfaces sparsely strigose; petal apices acute to acuminate . . . . 3b. *Petrophytum caespitosum* subsp. *acuminatum*

3a. Petrophytum caespitosum (Nuttall) Rydberg subsp. **caespitosum** [F]

Petrophytum caespitosum var. *elatius* (S. Watson) Tidestrom; *P. elatius* (S. Watson) A. Heller; *Spiraea caespitosa* Nuttall var. *elatior* S. Watson

Short shoots usually compressed into tight rosettes, sometimes internodes extended and rosette structure lost. **Leaf blades** 0.4–1.4 cm, abaxial surface pilose to sericeous. **Inflorescences** compressed panicles, proximal ½ of panicle varying from unbranched with scattered bracts along peduncles to ± branched with branches 1–3(–4+) cm, distal ½ of panicle often appearing racemose with no or relatively few branches. **Flowers:** petal apex obtuse to slightly cleft.

Flowering May–Oct. Dry rock ledges, rocky outcrops, talus slopes; 1100–3000 m; Ariz., Calif., Colo., Idaho, Mont., Nev., N.Mex., S.Dak., Tex., Utah, Wash.; Mexico (Coahuila, Tamaulipas).

Subspecies *caespitosum* is frequently found with relatively small leaves that are pilose to densely sericeous. Sometimes plants growing in moister conditions will have larger leaves and reduced indument, and they may develop visible primary and secondary venation.

For this treatment, var. *elatius* has been subsumed in subsp. *caespitosum*, as the author is uncertain as to whether it should be recognized as a variety within this subspecies or merits elevation to subspecies status. Variety *elatius* has short shoots with extended internodes (1–3 cm), larger panicles (10–20 × 8–15 cm), widely branched with 4–10 cm branches, with bracteoles linear to narrowly oblanceolate, 8–16 mm; it is found on dry rock ledges and outcrops, often on limestone, between 1100 and 2800 m in Arizona, Nevada, and Utah.

Variety *elatius* appears to intergrade with subsp. *caespitosum* (freely intergrades; T. H. Kearney and R. H. Peebles 1942, 1960) but does not appear to be merely an ecological variant of it. Morphological work on the carpel of *Petrophytum* by C. Sterling (1966) indicated that, for the carpel characters examined, the character state distribution of *P. elatius* (here cited as var. *elatius*) was most similar to that of *P. hendersonii*. Variety *elatius* was dissimilar to both *P. caespitosum* (infraspecific taxon not identified, most likely subsp. *caespitosum*) and *P. cinerascens*. If var. *elatius* was only an ecological variant of *P. caespitosum*, it would be expected that carpel morphology and anatomy would be more similar than found by Sterling. Variety *elatius* may merit taxonomic recognition.

3b. Petrophytum caespitosum (Nuttall) Rydberg subsp. **acuminatum** (Rydberg) Munz, Aliso 4: 92. 1958 C E

Petrophytum acuminatum Rydberg in N. L. Britton et al., N. Amer. Fl. 22: 253. 1908

Short shoots compressed into tight rosettes. **Leaf blades** 1–1.8 cm, abaxial surface sparsely strigose. **Inflorescences** compressed panicles, usually appearing racemose, with scattered bracts along peduncle, branches uncommon, 0.5–1 cm. **Flowers:** petal apex acute to acuminate.

Flowering Jun–Sep. Limestone cliffs; of conservation concern; 1200–2500 m; Calif.

Subspecies *acuminatum* is known only from the southern Sierra Nevada in Fresno and Tulare counties.

46. KELSEYA Rydberg in N. L. Britton et al., N. Amer. Fl. 22: 254. 1908 • [For Francis Duncan Kelsey, 1849–1905, Montana botanist] E

Richard A. Lis

Shrubs, densely cespitose, pulvinate, mat-forming, 0.3–1 dm; caudices woody. **Stems** 1–20+, semierect to prostrate and decumbent with very short internodes, <densely, intricately branched, leafy>; bark brown to gray or dark gray in age; short shoots usually present, long shoots rarely present; glabrous. **Leaves** persistent, cauline, alternate, tightly crowded and imbricate, simple; petiole absent; blade oblanceolate to spatulate-obovate, 0.2–0.4 cm, coriaceous, margins flat, entire, venation not visible, surfaces sericeous. **Inflorescences** terminal or pseudoterminal, flowers solitary, hirsute to villous; bracts absent; bracteoles absent. **Pedicels** absent. **Flowers** 3–6(–8) mm diam.; epicalyx bractlets 0; hypanthium hemispheric, 1–2 mm, sericeous; sepals 5, erect to slightly reflexed, oblanceolate; petals 5, pink to pinkish, often purple-tinged, elliptic to oblong; stamens 7–12, slightly longer than petals; torus annular to weakly developed and inconspicuous; carpels 3 or 4(or 5), free, glabrous or glabrate, sparsely villous on adaxial margin, styles terminal, <adherent to hypanthium>; ovules 3–5. **Fruits** aggregated follicles, 3–5, ellipsoid, 2–3 mm, glabrate or sparsely hirsute, sparsely villous on adaxial margin; hypanthium persistent; sepals persistent, erect to slightly reflexed; styles deciduous. **Seeds** 1 or 2(–4), elongate, fusiform. $x = 9$.

Species 1: nw United States.

The phylogenetic work on Rosaceae by D. Potter et al. (2007) and the nucleotide sequence data analysis of Spiraeeae by Potter et al. (2007b) identified a clade containing *Kelseya*, *Petrophytum*, and *Spiraea*. A cladistic and numerical taxonomic study using morphologic characters ascertained that *Kelseya* and *Petrophytum* were nearest neighbors (J. Henrickson 1985). The work of Potter et al. (2007b) suggested that the reduced growth habit in Spiraeeae evolved at least twice. The carpels of *Kelseya* are adherent to the hypanthium and each has an independent vascular supply of three bundles per carpel (C. Sterling 1966). Using *S. trilobata* as a model (R. C. Evans and T. A. Dickinson 1999b) for gynoecial development in *Kelseya*, the five primordia are initiated separately and grow independently, yet simultaneously, with the intercalary growth of the hypanthium that occurs between the point of insertion of the perianth and the androecium. As the carpels of *Kelseya* mature, they become adherent to the hypanthium (Sterling); the carpels of *S. trilobata* remain free from the hypanthium (Evans and Dickinson). In contrast, in

Maleae (J. R. Rohrer et al. 1991, 1994) and in *Sorbaria* and *Vauquelinia* (Evans and Dickinson), the region of hypanthial intercalary growth extends proximally from the gynoecium, and as the hypanthium and carpels develop, the abaxial margins of the carpels become adnate to the growing hypanthium. The adherence of the carpels displayed by *Kelseya*, found more typically in other genera of Maleae and Amygdaleae, appears to be an unusual morphological feature that may be a result of the reduction in plant stature and habit possibly associated with its xeric alpine habitat.

1. **Kelseya uniflora** (S. Watson) Rydberg in N. L. Britton et al., N. Amer. Fl. 22: 254. 1908 • One-flower kelseya or spiraea, alpine laurel E F

Eriogynia uniflora S. Watson, Proc. Amer. Acad. Arts 25: 130. 1890

Shrubs 3–6 dm diam. **Leaves** light green to grayish green, becoming brownish, marcescent, hardened. **Flowers:** sepals persistent, inflating radially with fruit maturity. **Follicles** brown, coriaceous, opening first along adaxial suture, later on abaxial suture. $2n = 18$.

Flowering Apr–Aug. Cracks of limestone rock outcrops; 1800–3500 m; Idaho, Mont., Wyo.

Kelseya uniflora can form dense carpets or mats that may cover areas of cliff faces. The solitary flower is unique in Spiraeeae. This species is of interest as an unusual rock-garden plant; it is difficult to cultivate.

47. HOLODISCUS (K. Koch) Maximowicz, Trudy Imp. S.-Peterburgsk. Bot. Sada 6: 253. 1879, name conserved • [Greek *holos*, whole, and *diskos*, disc, alluding to entire floral disc]

Richard A. Lis

Spiraea Linnaeus [unranked] *Holodiscus* K. Koch, Dendrologie 1: 309. 1869

Shrubs, spreading to densely compact, (2–)3–30(–60) dm. **Stems** 1–10+, erect to arching; bark gray, <exfoliating, periderm reddish to nearly black>; shoots 3 types: long <strongly angled or pleated 1st year>, short, and water-sprout; glabrous, glabrate, downy, canescent, or pilose-tomentose. **Leaves** deciduous, cauline, alternate, <dimorphic with shoot type, those of short shoots in fascicles of 3–8>, simple; petiole absent or present, <distinct, sometimes obscured by decurrent leaf base>; blade ± ovate, rhombic, trullate to broadly trullate, broadly obovate, or obtrullate, 1–10 cm, membranaceous to chartaceous, sometimes coriaceous, margins flat, usually serrate, sometimes doubly serrate, venation pinnate, simple craspedodromous, abaxial surface glabrous, glabrate, pilose to villous, tomentose, or sericeous, <sometimes sessile- or stipitate-glandular, granular deposits present or absent>, adaxial glabrous, glabrate, puberulent, pubescent, short-hirsute, velutinous, villous, <sessile- or stipitate-glandular or eglandular>. **Inflorescences** terminal <on long shoots, erect or pendulous>, 10–100+-flowered, panicles, glabrate, puberulent, villous, or tomentose <hairs golden, tan, or white, stipitate- or sessile-glandular or absent>; bracts absent or subtending inflorescence branches; bracteoles present. **Pedicels** present. **Flowers** appearing after leaves, 2–6 mm diam.; epicalyx bractlets 0; hypanthium patelliform to shallowly crateriform, 1.5–5 mm diam.; sepals 5, erect to spreading or slightly reflexed, triangular- to elliptic-ovate or deltate; petals 5, usually white, sometimes pink-tinged, rarely pink, ovate to elliptic; stamens 15–20, equal to or longer than petals, <with prominent,

raised nectar disc proximal to inner rim>, exterior puberulent to tomentose, <sometimes sessile-glandular>; torus inconspicuous; carpels 5, adnate to hypanthium base, hirsute and hispid, styles terminal, <stigmas 2-lobed>; ovules 2 (1 aborted). **Fruits** aggregated achenes, 5, laterally flattened, inverted dolabriform, with abaxial edge strongly convex, adaxial straight, 1–1.5 mm, hispid, hirsute, and often sessile- or stipitate-glandular; hypanthium persistent; sepals persistent, erect; styles deciduous or persistent. $x = 9$.

Species 5 (2 in the flora): North America, Mexico, Central America (Costa Rica, Guatemala, Honduras), South America (Colombia, Venezuela).

Holodiscus contains two species complexes: the *argenteus* and the *discolor* complexes (R. A. Lis 1990). The *discolor* complex comprises *H. discolor* and *H. microphyllus* and ranges from British Columbia and Montana through the western United States and south through Mexico along the Sierra Madre Occidental and Oriental, with the southern terminus near the Valley of Mexico. The *argenteus* complex comprises *H. argenteus* (Linnaeus f.) Maximowicz, *H. fissus* (Lindley) C. K. Schneider, and *H. orizabae* F. A. Ley and ranges from the Trans-Mexican Volcanic Belt south to Colombia and Venezuela.

A number of species and varieties have been described from the western United States that appear locally unique, and many names have been applied at the species and varietal levels. The leaves have provided the primary characters for circumscribing taxa in the genus, as the flowers and inflorescences show a narrow range of variation. *Holodiscus* can exhibit phenotypic plasticity in leaf morphology, particularly in response to changes in relative amounts of moisture and light.

Each plant has three shoot types that bear leaves of morphologies slightly to widely divergent among types: short shoots, long shoots, and water-sprouts. The short shoots bear leaves in fascicles of 3–8, the most diagnostic leaves for the taxa. Multiple seasons of leaf development on a short shoot result in a short woody stem with leaf scars in close succession in a phyllotactic spiral.

The long shoots develop in the spring, grow rapidly, and may attain a length of 1–2 m in a month or two; they often have a strongly angled or pleated appearance; leaves are alternate along the stem. The long-shoot leaves have rapid growth and expansion and differ from short-shoot leaves in characteristics such as petiole condition, blade size, shape, serration, and base and apex shapes. Long-shoot leaves are quite responsive to local environmental conditions and the season's water regime under which they developed; consequently, they are more variable morphologically than short-shoot leaves. Long shoots are often terminated by an inflorescence and are more frequently observed and collected. Long shoots may arise from anywhere on the plant, and in some plants, populations, or taxa, they may be relatively short (sometimes as short as 2–6 cm), but maintain the particular morphology of the long shoot.

Water-sprout shoots typically arise from the crown of the plant and appear to occur for two reasons: trauma to the plant by severe clipping, mowing, or intense grazing; or unusually high moisture during the spring and early summer growth phase. Water-sprout leaves are the least diagnostic because they have a similar morphology across taxa of the *discolor* complex. They are quite striking when found and can be dramatically different from short-shoot leaves. Water-sprout leaves are typically thin and delicate, strongly serrate with primary and secondary teeth, and glabrous or glabrescent to very sparsely pubescent, hence appearing very green.

Holodiscus has been segregated from the Spiraeeae, with a follicle fruit type, and placed in its own tribe (Holodisceae Focke) based upon the presence of achenes (W. O. Focke 1888–1891; P. A. Rydberg 1908–1918; C. Kalkman 1988; A. L. Takhtajan 1997). Recent molecular work by D. R. Morgan et al. (1994) and D. Potter et al. (2007, 2007b) confirms that *Holodiscus* belongs in Spiraeeae. The investigation into floral ontogeny and morphology in Amygdaloideae by R. C. Evans and T. A. Dickinson (1999b) found that *Aruncus*, *Holodiscus*, and *Spiraea* share three character states of ovule morphology, further supporting the inclusion of *Holodiscus* in Spiraeeae.

Two other generic names have been used for taxa in *Holodiscus*: *Schizonotus* Rafinesque, an illegitimate name, and *Sericotheca* Rafinesque, a rejected name.

Holodiscus was of particular horticultural interest in the nineteenth century, which led to the description of many species and varieties. The panicle of flowers can be attractive, and a shrub arched and sagging with relatively many panicles is eye-catching.

SELECTED REFERENCES Ley, F. A. 1943. A taxonomic revision of the genus *Holodiscus* (Rosaceae). Bull. Torrey Bot. Club 70: 275–288. Lis, R. A. 1990. A Taxonomic Revision of *Holodiscus* Maxim. (Spiraeoideae: Rosaceae) Based upon Numerical Analyses of Leaf Morphometric Variation. Ph.D. dissertation. University of California, Berkeley.

1. Leaves: long shoots usually predominant, blades ± ovate, rhombic, or ± trullate, 1–10 cm, margins frequently with secondary teeth; short shoots usually not predominant, blades (0.5–)2–8 cm, ± ovate, bases truncate, apices obtuse, or slightly ovate to obovate or rhombic, bases acute, decurrent, apices acute; blade maximal width usually proximal to middle . . . . 1. *Holodiscus discolor*
1. Leaves: long shoots usually not predominant, blades obtrullate or broadly obovate to ± ovate, 1–3 cm, margins infrequently with secondary teeth; short shoots usually predominant, blades 0.3–1.5(–3) cm, obtrullate, bases acute, decurrent, cuneate, apices obtuse; blade maximal width usually from middle distally. . . . . 2. *Holodiscus microphyllus*

1. Holodiscus discolor (Pursh) Maximowicz, Trudy Imp. S.-Peterburgsk. Bot. Sada 6: 254. 1879 F

Spiraea discolor Pursh, Fl. Amer. Sept. 1: 342. 1813

Shrubs, erect to arching, 5–50 (–60) dm. **Stems:** long shoots (predominant in moist or shaded conditions) (4–)10–200 cm, 1st year strongly angled or pleated, green to reddish, glabrate to <off-white or gray> downy or tomentose, rarely sessile- or stipitate-glandular, 2d or 3d year epidermis exfoliating, periderm shiny maroon or reddish, aging into light to dark gray bark; short shoots (not predominant) 0.5–2 cm, <developing more frequently in xeric or exposed conditions, rarely in moist or shaded ones>; water-sprout shoots uncommon, 20–150 cm. **Leaves:** petiole 0.5–3 cm, distinct, sometimes ± obscured by decurrent leaf base; blade maximal width usually proximal to middle, membranaceous to chartaceous, rarely coriaceous, <margins serrate>; long shoots (usually predominant): blade ± ovate, rhombic, or ± trullate, 1–10 × 0.5–8 cm, membranaceous to chartaceous, base truncate to cuneate, margins serrate, secondary teeth frequent, primary and secondary veins slightly to strongly protruding abaxially, venation percurrent, apex acute or obtuse, abaxial surfaces pilose to villous, sometimes sessile-glandular, adaxial glabrate or pubescent to hirsute, eglandular, rarely sessile-glandular; short shoots (usually not predominant): in fascicles of 3–8, blade ± ovate, base truncate, apex obtuse, or slightly ovate to obovate or rhombic, base cuneate, decurrent, apex acute, 1–8 × (0.5–)1–7 cm, margins serrate, secondary teeth frequent or rare, sharply acute, small, 1–3 mm, primary and secondary veins usually strongly protruding abaxially, venation percurrent, abaxial surfaces glabrate to densely tomentose or villous, sometimes sessile-glandular, adaxial glabrate to thinly puberulent, rarely sessile-glandular; water-sprout shoots: blade (3–)4–8 × 3–7 cm. **Inflorescences** 10–100+-flowered, 2–20 × 2–15 cm; branches usually tomentose to villous, sometimes glabrate, sometimes sessile-glandular; bracts absent, sometimes subtending branches (more frequent on basal branches, reducing distally), rarely subtending peduncles; bracteoles 0.5–1.5 mm. **Flowers** 2–6 mm diam.; hypanthium 3–5 mm diam.; sepals triangular-ovate to elliptic-ovate, 1–2 mm, apex acute, abaxially puberulent to tomentulose, usually eglandular; petals white, sometimes pink-tinged, rarely pink, ± ovate, 1.5–2 mm; stamens equal to or longer than petals, 2–3.5 mm; carpels 1–1.5 mm. **Achenes** beige to white, sparsely to densely hirsute, edges and remnant style usually hispid-ciliate, glands 0–15 per face, sessile or stipitate.

Varieties 2 (2 in the flora): w, sc North America, n, c Mexico.

Holodiscus discolor is uncommon in the Great Basin, where *H. microphyllus* is primarily found.

Holodiscus discolor displays the widest variation of leaf morphology in the genus, often with a strong geographic correlation. Two different treatments of the specific and subspecific taxa proposed by P. A. Rydberg (1908–1918; 14 species) and F. A. Ley (1943; 8 species and 13 varieties) were correlated with geographic areas. These circumscriptions were problematic because in many cases a single plant may exhibit leaves that would fit the descriptions of two or three taxa. There are strong tendencies of covariation among leaf morphology, geography, and local environmental conditions of moisture availability.

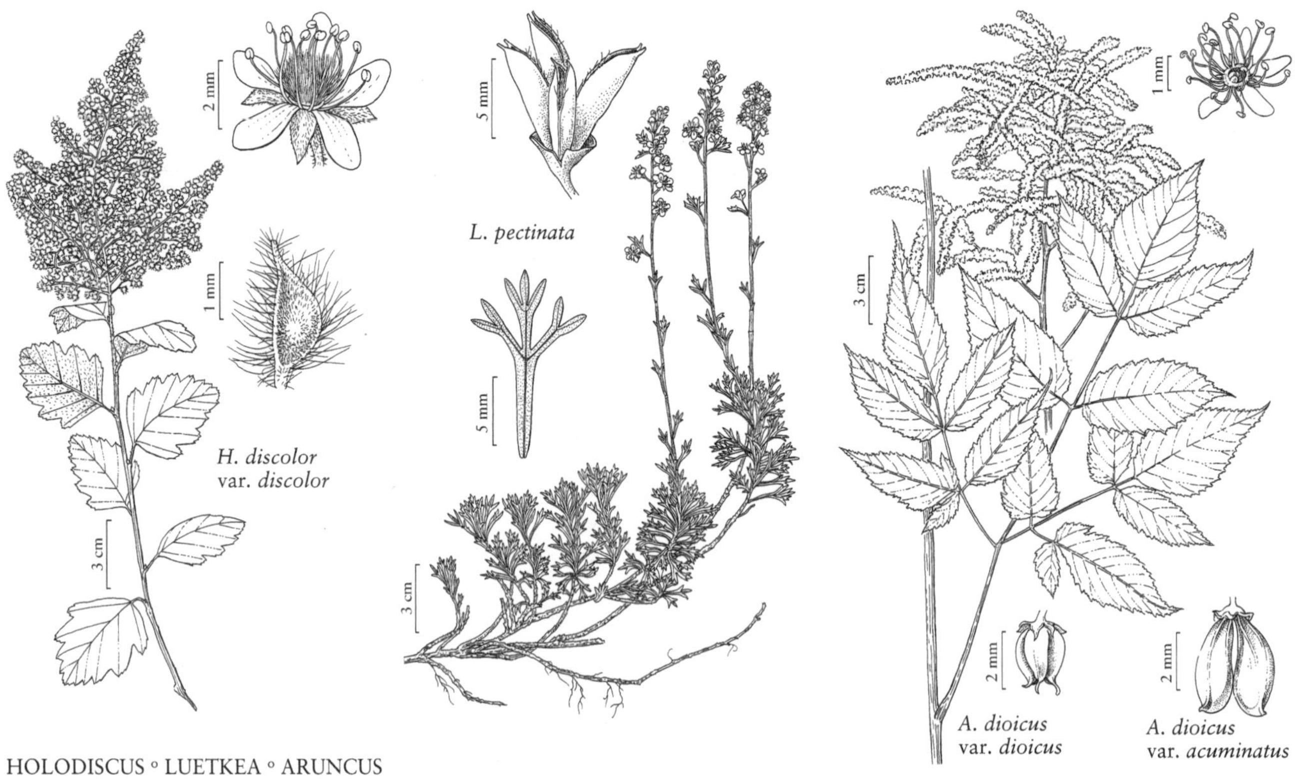

HOLODISCUS ◦ LUETKEA ◦ ARUNCUS

Some overlap in morphology occurs between the two varieties recognized here, particularly when populations occupy sites that are outside the normal habitat range. The basic morphology and habitats occupied by average populations of var. *discolor* and var. *dumosus* are sufficiently distinct that recognition at the varietal level is warranted. The physiological and ecological tolerances appear to be quite different between the two varieties, allowing their survival in different habitats.

R. A. Lis (1990) did not previously recognize varieties *discolor* or *dumosus*; that decision was based upon multivariate statistical analysis of leaf morphometric characters, which were unable to resolve any patterns or clusters within *Holodiscus discolor* with the methods of analysis employed. To consider *H. discolor* as an extremely variable species with a wide range of ecological habitat preferences would be feasible if, as a single highly variable species, it displayed the same morphological character state patterns in similar habitats; however, further work on the morphological variability in conjunction with ecological considerations and geographic distribution of populations has led the author to conclude that this is not the case. There are patterns of morphological characters and ecological adaptations that appear to have evolved in particular geographic subunits of the species range. Variety *discolor* likely includes some populations that will need to be recognized separately; this is not currently possible.

1. Leaves: long shoots predominant, petioles distinct, blades rarely decurrent, maximal width usually in proximal 1/4, broadly trullate or ovate to slightly rhombic, 2–10 cm, bases obtuse and truncate, sometimes to slightly acute, apices obtuse; short shoots absent or not predominant. . . . . 1a. *Holodiscus discolor* var. *discolor*
1. Leaves: long shoots not predominant, petioles indistinct, blades decurrent, maximal width usually at mid blade, rhombic to slightly ovate, 1–4(–6) cm, bases acute, decurrent, sometimes truncate and rounded, apices acute; short shoots predominant . . . . . 1b. *Holodiscus discolor* var. *dumosus*

1a. Holodiscus discolor (Pursh) Maximowicz var. **discolor** • Ocean-spray, cream- or foam-bush F

Holodiscus australis A. Heller; *H. boursieri* (Carrière) Rehder; *H. discolor* var. *delnortensis* F. A. Ley; *H. discolor* var. *franciscanus* (Rydberg) Jepson; *H. dumosus* (S. Watson) A. Heller var. *australis* (A. Heller) F. A. Ley; *H. dumosus* subsp. *saxicola* (A. Heller) Abrams; *Spiraea ariifolia* Smith

Stems 1–6 m. **Leaves:** long shoots (predominant): petiole distinct, rarely with decurrent blades; blade broadly trullate or ovate to slightly rhombic, maximal width usually in proximal 1/4 of blade, 2–10 × 1–8 cm, base obtuse and truncate, sometimes to slightly acute, apex obtuse; short shoots (absent or not predominant): petiole usually distinct, blade ovate to slightly rhombic, maximal width proximal to midpoint, 2–8 × 1–7 cm, base obtuse and truncate, sometimes acute, margins usually with secondary teeth, apex obtuse. **Inflorescences** (2–)4–20 × (2–)4–15 cm, multi-branched, dense. $2n$ = 36.

Flowering Apr–Aug. Open woodlands, forest edges, thickets, dry to moist, often partially shaded, rocky slopes; 0–3000 m; B.C.; Ariz., Calif., Colo., Idaho, Mont., Nev., N.Mex., Oreg., Tex., Utah, Wash., Wyo.; Mexico (Baja California, Chihuahua, Distrito Federal, Durango, Guanajuato, Nuevo León, Querétaro, San Luis Potosí, Sonora, Zacatecas).

As treated here, var. *discolor* covers wide elevational and latitudinal ranges and displays significant variation. The variety includes populations that might best be segregated to produce a classification within the species that would more precisely reflect the morphologic, ecologic, and geographic patterns observed. In a detailed study of leaf variation and carpel morphology in northeastern Oregon and Washington, C. J. Antieau (1985) found that: phenotype is a reliable indicator of general patterns of genetic divergence; for the characters examined, inherent variation was non-random; and, morphologies observed in field collections were being maintained in cultivation. Although the study by Antieau was on a restricted region of the range of *Holodiscus discolor*, his observations may be applicable to the species as a whole; infraspecific taxa with consistent phenotypic expressions and genetic bases might be segregated after further study.

1b. Holodiscus discolor (Pursh) Maximowicz var. **dumosus** (S. Watson) Maximowicz ex J. M. Coulter, Man. Bot. Rocky Mt., 79. 1885 • Rock-spiraea

Spiraea discolor Pursh var. *dumosa* S. Watson in W. H. Brewer et al., Bot. California 1: 170. 1876; *Holodiscus dumosus* (S. Watson) A. Heller

Stems 0.3–2 m. **Leaves:** long shoots (not predominant): petiole indistinct with decurrent blades; blade rhombic to slightly ovate, maximal width usually at mid blade, sometimes in proximal 1/4 of blade, 1–4(–6) × 0.5–2(–3) cm, base cuneate to decurrent, sometimes truncate and rounded, apex acute; short shoots (predominant): petiole obscured by decurrent blade; blade rhombic to ovate, maximal width approximately mid blade, 1–2(–3) × 0.5–1(–2) cm, base acute, decurrent, margins rarely with secondary teeth, apex acute. **Inflorescences** 3–12 × 3–8 cm, irregularly branched, loose. $2n$ = 36.

Flowering Jun–Aug. Dry habitats, mixed shrublands, ponderosa pine woodlands, rock outcrops, edges of talus slopes; 1400–2800 m; Ariz., Colo., Nev., N.Mex., Utah, Wyo.; Mexico (Chihuahua, Durango, San Luis Potosí).

Variety *dumosus* may be distinguished from var. *discolor* by the following features: leaf blade decurrent on the petiole on long- and short-shoot leaves, short-shoot leaves rhombic to slightly ovate with a narrower length to width ratio (2.1–3.1:1), and maximal width of blade near mid blade or slightly below. The inflorescences are generally shorter and narrower than in var. *discolor*. The shrub is generally shorter (1–2 m) than in var. *discolor* (1–6 m) with a more open habit. Variety *dumosus* occupies more arid, rocky, fast-draining sites such as scree slopes and rock outcrops. Its primary distribution is in the central Rocky Mountains, centered on Colorado and Wyoming, with western extensions into central and southern Utah and central Nevada, and southern extensions to northern Arizona, New Mexico, and Chihuahua, Durango, and San Luis Potosí in Mexico. The western extensions and Wyoming populations show indications of possible hybridization with *Holodiscus microphyllus*. The southern populations show some hybridization with populations of var. *discolor* that occupy higher elevations, in cooler, moister habitats. The long-shoot leaves of some plants of var. *dumosus* can be very similar to those of var. *discolor*, particularly when growing in moister sites.

2. Holodiscus microphyllus Rydberg, Bull. Torrey Bot. Club 31: 559. 1904

Holodiscus discolor (Pursh) Maximowicz subsp. *microphyllus* (Rydberg) A. E. Murray; *H. discolor* var. *microphyllus* (Rydberg) Jepson

Shrubs, erect to spreading, <often tangled and tightly condensed where exposed or with open growth where protected>, 3–10(–20) dm. **Stems:** long shoots (not predominant, depending upon seasonal conditions) 2–10(–20) cm, 1st year slightly angled or round, green to straw yellow, glabrous or canescent, pilose, or tomentose, sessile- and stipitate-glandular, 2d or 3d year epidermis exfoliating, periderm shiny maroon or dark red, aging into light to dark gray or nearly black bark; short shoots (usually predominant) 0.2–3 cm; water-sprout shoots very rare, 5–15 cm. **Leaves:** petiole often absent, sometimes obscured by decurrent leaf base; blade maximal width usually from middle distally, chartaceous to coriaceous; long shoots (not predominant): blade obtrullate or broadly obovate to ± ovate, 1–3 × 0.5–1.5 cm, coriaceous, base acute, cuneate or decurrent, margins entire proximally, serrate in distal 1/4–1/2, secondary teeth infrequent, primary and secondary veins slightly protruding abaxially, venation weakly percurrent, apex obtuse, sometimes acute, abaxial surfaces glabrous, glabrate, or pilose, tomentose to villous, sericeous, sessile- or stipitate-glandular, adaxial glabrous or glabrate, puberulent, or short-hirsute, ± sessile-glandular; short shoots (predominant): in fascicles of 3–8, blade obtrullate, 0.3–1.5(–3) × 0.2–1 cm, base cuneate, margins serrate, secondary teeth absent, primary and secondary veins usually protruding abaxially, venation very weakly percurrent, apex obtuse, abaxial surfaces <whitish (waxes and cutin)>, glabrous or strigose on veins and tomentose to villous or sericeous (in intercostal regions) to densely <long, white> sericeous, ± sessile- and stipitate-glandular, adaxial glabrous, glabrate, puberulent, velutinous, villous, with scattered longer hairs, usually stipitate- or sessile-glandular; water-sprout shoots: blade 2–3 × 1–2 cm. **Inflorescences** 10–50-flowered, 2–10 × 1–7 cm; branches puberulent to tomentose, ± stipitate- or sessile-glandular; bracts absent, rarely subtending basal branches, absent distally, usually subtending peduncles, sometimes absent; bracteoles 0.5–2(–3) mm. **Flowers** 3–4 mm diam.; hypanthium 1.5–2.5 mm diam.; sepals triangular to deltate, 1.3–2.5 mm, apex acute to acuminate, abaxially glabrate, sparsely sessile-glandular or densely villous, glandular; petals white, ovate to elliptic, 1.5–2.5 mm; stamens longer than petals, 2.5–3 mm; carpels 1.3–2 mm. **Achenes** beige to gray, appearing white from hairs, densely hirsute on edges, sometimes on faces, glands 20–40 per face, sessile.

Varieties 3 (3 in the flora): w United States, nw Mexico.

The distribution of *Holodiscus microphyllus* is centered in the Great Basin of the western United States, with expansions into arid rocky habitat of adjacent mountain ranges; it is adapted to xeric conditions on rocky, infertile sites. Short shoots are predominant in this species, and as the percentage of short shoots increases, the architecture of the plant becomes increasingly tangled and dense. In populations with a wider spreading habit, long shoots develop more frequently. Depending upon the moisture conditions of the growth season, the long-shoot leaves often differ strongly from the short-shoot ones.

Populations of *Holodiscus microphyllus* are usually small (1–20 plants); and in a number of locations they appear to consist of clones, and the identity of individuals cannot be determined with certainty. Individual populations often are on mountaintops, isolated rock outcrops, or rimrock areas that often are distant from the nearest adjacent population. The potential for genetic exchange between populations may be very limited.

The varieties of *Holodiscus microphyllus* are quite distinct and have strong geographic correlations. Fieldwork on intra-populational variation indicated no clinal variation in the indumentum features that separate varieties. No hybrid was found between varieties of *H. discolor* and *H. microphyllus*.

1. Short-shoot leaf abaxial surfaces glabrous or glabrate, sessile- or stipitate-glandular; veins, intercostal regions, and margins not obscured by hairs . . . . . 2c. *Holodiscus microphyllus* var. *glabrescens*
1. Short-shoot leaf abaxial surfaces tomentose to villous or densely sericeous, sessile- and/or stipitate-glandular (sometimes obscured by hairs); veins, intercostal regions, and margins frequently obscured by hairs.
    2. Short-shoot leaf abaxial surfaces gray to off-white, tomentose to villous, adaxial puberulent, short-hirsute, velutinous to villous . . . . 2a. *Holodiscus microphyllus* var. *microphyllus*
    2. Short-shoot leaf abaxial surfaces white, sometimes silvery white, densely sericeous, adaxial white-short-villous. . . . . 2b. *Holodiscus microphyllus* var. *sericeus*

2a. Holodiscus microphyllus Rydberg var. **microphyllus** • Rock-spiraea

Short-shoot leaf surfaces sessile- and/or stipitate-glandular, abaxial gray to off-white, veins, intercostal regions, and margins frequently obscured by hairs, tomentose to villous, often with white to gray granular deposits, adaxial puberulent, short-hirsute, velutinous to villous, sometimes hairs very short and surface appearing green to gray-green. **Inflorescences** 2–9 × 1–7 cm. **Pedicels** 2–3 mm. **Flowers:** nectar disc slightly pilose to puberulent, sometimes glabrous; sepals densely pilose to tomentose, usually eglandular; petal base abaxially with few, scattered long hairs.

Flowering Apr–Aug. Rock outcrops, rocky sites in sagebrush and bitterbrush shrublands, edges of talus slopes; 500–3100 m; Ariz., Calif., Colo., Idaho, Nev., Oreg., Utah, Wyo.; Mexico (Baja California).

Variety *microphyllus* has the widest range of the varieties. Individual populations are generally small and isolated.

2b. Holodiscus microphyllus Rydberg var. **sericeus** F. A. Ley, Bull. Torrey Bot. Club 70: 283. 1943

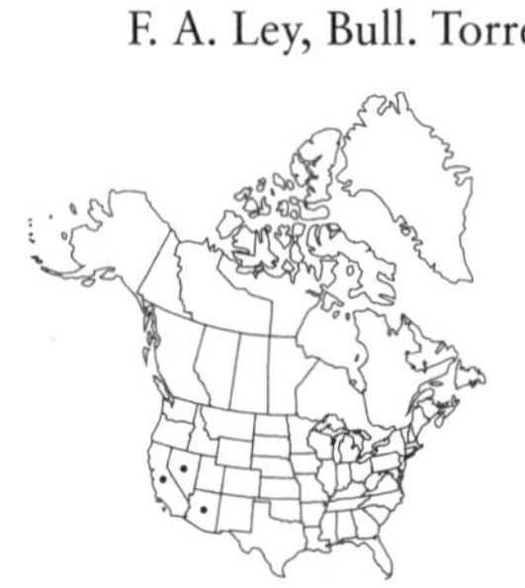

Short-shoot leaf surfaces sessile- and/or stipitate-glandular (often obscured by hairs), abaxial white, sometimes silvery white, veins, intercostal regions, and margins obscured by hairs, densely sericeous, white granular deposits sparse, adaxial white short-villous. **Inflorescences** 2–6 × 1–4 cm. **Pedicels** 2–3 mm. **Flowers:** nectar disc sparsely white-hirsute; sepals densely white-tomentose, usually eglandular; petal base abaxially with few, scattered long hairs.

Flowering Apr–Jun. Rock outcrops, often limestone, sandy loam soil derived from diorite, margins of ponderosa pine woodlands, Jeffrey pine forests, pinyon pine-juniper woodlands; 1500–3000 m; Ariz., Calif., Nev.; Mexico (Baja California).

The unusually dense, white to silvery white, sericeous indumentum that characterizes var. *sericeus* is found only in northwestern Arizona, extreme southern Nevada, some desert mountains of southern California, and northern Baja California, Mexico. In this region, var. *microphyllus* also occurs but its populations are usually disjunct. The primary problem in distinguishing var. *sericeus* arises when specimens actually belong to var. *microphyllus* but are abaxially off-white to light gray and tomentose to villous.

2c. Holodiscus microphyllus Rydberg var. **glabrescens** (Greenman) F. A. Ley, Bull. Torrey Bot. Club 70: 284. 1943 [E]

Spiraea discolor Pursh var. *glabrescens* Greenman, Erythea 7: 116. 1899; *Holodiscus discolor* (Pursh) Maximowicz var. *glabrescens* (Greenman) Jepson; *H. dumosus* (S. Watson) A. Heller var. *glabrescens* (Greenman) C. L. Hitchcock; *H. glabrescens* (Greenman) A. Heller

Short-shoot leaf surfaces: abaxial light green, veins, intercostal regions, and margins not obscured by hairs, glabrous or glabrate, sessile- or stipitate-glandular, without granular deposits, adaxial sparsely glabrate or glabrous, sessile-glandular, <sometimes margins and veins sparsely hairy>. **Inflorescences** 2–10 × 1–7 cm. **Pedicels** 1–2 mm. **Flowers:** nectar disc glabrous; sepals glabrous or puberulent, sessile-glandular; petal base glabrous or abaxial midvein with a few long hairs.

Flowering May–Sep. Rock outcrops, talus slopes, mountaintops in alpine and subalpine habitats; 1000–3000 m; Calif., Idaho, Nev., Oreg., Utah.

Specimens that belong to var. *microphyllus* are frequently misidentified as var. *glabrescens* when the vestiture is short and dense, such that the leaf surfaces appear to be light green or gray and insufficient magnification is used to see the trichomes. Careful examination is required and in some cases 10× magnification is insufficient due to the short, dense indumentum that occurs in some populations of var. *microphyllus*. Variety *glabrescens* is glabrous or glabrate with sessile or stipitate glands alone on the smooth green epidermis. The variety inhabits the northern Great Basin region, in areas that recently have been geologically active, or on mountains adjacent to this region that are currently or were recently tectonically active. Individual populations often comprise a small number of individuals that occupy a narrow ecologic zone on mountain peaks.

48. LUETKEA Bongard, Mém. Acad. Imp. Sci. St.-Pétersbourg, Sér. 6, Sci. Math. 2: 130, plate 2. 1832 • [For Friedrich Benjamin Lütke (later russified to Count Fyodor Petrovich Litke), 1797–1882, Russian sea captain and Arctic explorer] [E]

Richard A. Lis

Subshrubs, mat-forming, 1–1.5(–3) dm; rhizomatous, stoloniferous. **Stems** 1–5(–12), prostrate, glabrous. **Leaves** persistent, often marcescent, cauline, alternate (internodes elongating apically), primarily crowded basally, simple; petiole present, <narrowly winged>; blade obtrullate, 0.5–1.5 cm, pergamentaceous, 2 or 3(–5)-ternately lobed, margins flat, venation suprabasal, actinodromous, surfaces glabrous. **Inflorescences** terminal, 5–20(–30)-flowered, racemes, glabrescent to villous; <peduncles present>; bracts reducing acropetally from 2-ternate to simple, linear; bracteoles absent. **Pedicels** present. **Flowers** 3–6 mm diam.; epicalyx bractlets 0; hypanthium hemispheric, 0.5–1 mm, glabrous or villous; sepals 5, spreading to erect, deltate; petals 5, white, spatulate to obovate; stamens 20, shorter than petals, <filaments basally connate>; torus thickened, <10-lobed>; carpels (4 or)5(or 6), free, adaxially sparsely villous, styles terminal; ovules 3–5. **Fruits** aggregated follicles, (4 or)5(or 6), falcate, truncated at proximal end, 2–3 mm, sericeous adaxially; hypanthium persistent; sepals persistent, spreading to erect; styles deciduous. **Seeds** 1–4, elongate-fusiform. $x = 9$.

Species 1: nw North America.

D. Potter et al. (2007b), using nucleotide sequence data, grouped *Aruncus* and *Luetkea* into a subclade that was a sister group to the *Holodiscus-Xerospiraea* subclade; these two subclades composed a clade with strong support. A cladistic and numerical taxonomic study of Amygdaloideae (J. Henrickson 1985) using morphologic characters determined that *Kelseya* and *Petrophytum* were sister groups on both phenogram and cladogram; *Luetkea* and *Xerospiraea* exchanged positions depending upon the method used. In his evaluation of carpel anatomy in *Aruncus* (two species), *Holodiscus* (eight taxa), *L. pectinata*, and *Petrophytum* (four species), C. Sterling (1966) found that the combination of character states found in *Luetkea* was unlike those of any other of these taxa and similar only to those of some species of *Spiraea*. R. J. Hebda and C. C. Chinnappa (1990b) proposed two evolutionary lineages of Rosaceae based upon pollen perforation types, with *Luetkea* and 13 other genera in one lineage; *Aruncus* and four other genera formed the other. If the phylogeny proposed by Potter is correct (*Aruncus* and *Luetkea* are sister taxa), then their divergence is probably not an evolutionarily recent event; they are quite distinct in carpel anatomy and overall morphology.

1. Luetkea pectinata (Pursh) Kuntze, Revis. Gen. Pl. 1: 217. 1891 • Partridge foot [E] [F]

Saxifraga pectinata Pursh, Fl. Amer. Sept. 1: 312. 1813; *Eriogynia pectinata* (Pursh) Hooker; *Luetkea sibbaldioides* Bongard; *Spiraea pectinata* (Pursh) Torrey & A. Gray

Leaves marcescent; petiole 5–10 mm, narrowly winged; blade green to grayish green, veins abaxially ribbed, adaxially grooved. **Flowers:** sepals glabrous or sparsely pilose; stamens inserted outside edge of disc. **Follicles** opening along both sutures, coriaceous. $2n = 18$.

Flowering Jun–Sep. Rocky substrates, tundra, mesic to wet forest openings, heaths, meadows, seepage areas; 0–3500 m; Alta., B.C., N.W.T., Yukon; Alaska, Calif., Idaho, Mont., Oreg., Wash.

In an ecological study of competition in subalpine meadows in Deer Park, Olympic National Park, Washington, R. del Moral (1983) found that *Luetkea pectinata* occurred in colder sites, with lower relative drought stress, and moderately low net above-ground productivity (115 and 185 $g/m^2/yr$). Although not focused on *L. pectinata*, detailed work by del Moral evaluated microsite and microclimatic variation where *L. pectinata* is able to survive and may be useful in evaluating other sites where *L. pectinata* forms clonal carpets or where it is a component of a plant association. In contrast to the subalpine studies by del Moral, B. McAvoy (1931) presented a detailed study of plant associations in the Bella Coola Valley, British Columbia, from tidal flats to the alpine zone. She described some of the associations in which *L. pectinata* occurs, and the range of habitats in which it can occur within a limited geographic region.

At higher latitudes (48–65°), *Luetkea pectinata* ranges from sea level on rocky substrates to open tundra, high elevation alpine tundra, heaths, and other rocky moist habitats with well-drained thin soils. At lower latitudes (40–50°), the southern end of its range, it occurs at or above timberline in mesic to wet forest openings and edges of tree clumps, meadows, heathlands, open rocky slopes and mossy seepage areas, and in subalpine and alpine zones, often forming dense carpets or mats. Plants of *L. pectinata* appear to be shade intolerant.

49. ARUNCUS Linnaeus, Opera Var., 259. 1758 • Goat's-beard, bride's-feathers, barbe-de-bouc [Greek *arunkos*, goat's beard, alluding to showy fingerlike clusters forming feathery flowers]

T. Lawrence Mellichamp

Herbs, perennial, clump-forming, 5–15(–20) dm, herbage glabrous or sparsely to densely hairy, <hairs simple, straight to curved, sometimes multicellular or glandular>; rhizomatous, <rhizomes thick, branching, woody; roots coarse, fibrous>. **Stems** 1–5, erect, <stout, simple, sometimes glaucous>. **Leaves** deciduous, cauline, alternate, 2- or 3-irregularly ternate to odd-pinnate, equivalent to 2- or 3-pinnate; petiole present, <relatively long, base ½ sheathing stems>; blade broadly ovate, <cauline reduced in size and complexity distally>, 15–60 cm, thick to thin, leaflets 9–45, <long-petiolulate>, broadly ovate to narrowly lanceolate, margins flat, [serrate to] doubly serrate, venation pinnate. **Inflorescences** terminal, 500–2500-flowered, racemes forming panicles; bracts present; bracteoles present <hairs simple, glandular>. **Pedicels** present. **Flowers** unisexual (plants dioecious, rarely polygamous), 3–4 mm diam.; epicalyx bractlets 0; hypanthium shallowly bowl-shaped, 0.5–1 mm, <nectary ring inside with simple, glandular hairs>; sepals persistent, 5, spreading, triangular, <margins entire>; petals 5, white, obovate to elliptic, <base cuneate>; staminate: stamens (15–)20(–30) in 1 series (between calyx and nectar ring), longer than petals; torus minute; pistillate: nectar ring scarcely developed, stamens rudimentary, carpels 3, free, glabrous, styles terminal; ovules 2–4. **Fruits** follicles, 3(or 4), <pedicels recurving, fruits inverted, retrorse>, ovate-elongate, 1.5–5 mm, <shining>, glabrous, dehiscent along whole adaxial suture; hypanthium persistent; sepals persistent, spreading to appressed; styles persistent, <forming a beak>. **Seeds** 2–4, elongate, <membranous, caudate at both ends, scalariform-reticulate, seed coat loose>. x = 7, 9.

Species 1: North America, Eurasia; temperate regions.

SELECTED REFERENCE Mellichamp, T. L. 1976. A Comparative Study of *Aruncus* (Rosaceae) and *Astilbe* (Saxifragaceae), and the Problem of Their Relationships. Ph.D. dissertation. University of Michigan.

1. Aruncus dioicus (Walter) Fernald, Rhodora 41: 423. 1939 [F]

Actaea dioica Walter, Fl. Carol., 152. 1788

Leaves: leaflets green to gray-green, 6–15 cm, base subcordate to attenuate, apex acute to long-acuminate. **Flowers:** sepals 1 mm; petals 0.5–1.5 mm, apex acute to obtuse; stamens 1.5–2 mm, filaments slender, ± equal, anthers versatile; styles [0.2–]0.3–0.8 mm. **Seeds** 1.5–2(–2.5) mm.

Varieties 14 (4 in the flora): North America, Eurasia.

Aruncus dioicus is superficially similar to *Astilbe biternata* (Saxifragaceae), false goat's-beard, which is endemic to the southern Appalachian Mountains. Both species produce tall, dioecious plants with biternately to bipinnately compound leaves and plumelike terminal panicles with many, relatively small white, unisexual flowers, and may grow side by side. *Aruncus dioicus* is more widespread and may be readily distinguished by its unlobed rather than three-lobed terminal leaflets, flowers with 20 rather than 10 stamens, and three carpels reflexed in fruit rather than two erect carpels. See T. L. Mellichamp (1976) for an analysis of this remarkable example of convergent evolution.

Aruncus dioicus is striking and is grown in gardens throughout North America. It is variable throughout the temperate and subarctic regions of the Northern Hemisphere, especially in China and Japan. H. Hara (1955) recognized the following varieties that are found in the flora area.

1. Follicles 1.5–2.4 mm; styles 0.3–0.8 mm; seeds 1.5–2 mm.
   2. Follicles 1.5–2 mm; leaf surfaces glabrous or slightly hairy; s Appalachians and eastward . . . . 1a. *Aruncus dioicus* var. *dioicus*
   2. Follicles 1.6–2.4 mm; leaf surfaces densely hairy; w of Appalachians . . . . 1b. *Aruncus dioicus* var. *pubescens*
1. Follicles 2.5–5 mm; styles 0.3–0.5 mm; seeds 1.8–2.5 mm.
   3. Follicles 2.5–5 mm; Alaska and Yukon to n California . . . . . 1c. *Aruncus dioicus* var. *acuminatus*
   3. Follicles 2.5–3 mm; ne United States . . . . 1d. *Aruncus dioicus* var. *vulgaris*

1a. Aruncus dioicus (Walter) Fernald var. **dioicus** [E] [F]

Aruncus allegheniensis Rydberg

Leaf surfaces glabrous or slightly hairy. **Styles** 0.3–0.8 mm. **Follicles** somewhat plump, 1.5–2 mm. **Seeds** 1.5–2 mm. $2n$ = 18.

Flowering early summer. Rich woods, stream banks, north-facing roadsides, seeps, ditches; 10–1500 m; Ala., Ga., Ky., Md., Miss., N.C., Ohio, Pa., S.C., Tenn., Va., W.Va.

Variety *dioicus* is abundant in the southern Appalachians and into the Piedmont, where it typically occurs on mesic north-facing wooded slopes and roadsides.

1b. Aruncus dioicus (Walter) Fernald var. **pubescens** (Rydberg) Fernald, Rhodora 41: 423. 1939 E

Aruncus pubescens Rydberg in N. L. Britton et al., N. Amer. Fl. 22: 256. 1908

Leaf surfaces densely hairy. **Styles** 0.3–0.8 mm. **Follicles** slender, 1.6–2.4 mm. **Seeds** 1.5–2 mm. **2*n*** = 18.

Flowering early summer. Rich woods, limestone bluffs; 50–700 m; Ark., Ill., Ind., Iowa, Ky., Miss., Mo., Ohio, Okla., Tenn.

Variety *dioicus* gradually gives way to var. *pubescens*, which has more consistently thicker and hairier leaves and occurs west of the Appalachian Mountains; there is variation in indument throughout the eastern range of var. *dioicus*.

1c. Aruncus dioicus (Walter) Fernald var. **acuminatus** (Rydberg) H. Hara, J. Jap. Bot. 30: 68. 1955 E F

Aruncus acuminatus Rydberg in N. L. Britton et al., N. Amer. Fl. 22: 255. 1908

Leaf surfaces mostly glabrous. **Styles** 0.3–0.5 mm. **Follicles** slender, 2.5–5 mm. **Seeds** 1.8–2.5 mm. **2*n*** = 18.

Flowering early summer. Moist to wet ravines, rocky ledges, avalanche tracks, stream banks, forest edges and openings; 0–4800 m; Alta., B.C., Yukon; Alaska, Calif., Oreg., Wash.

1d. Aruncus dioicus (Walter) Fernald var. **vulgaris** (Maximowicz) H. Hara, J. Jap. Bot. 30: 68. 1955 I

Aruncus sylvester Kosteletzky ex Maximowicz var. *vulgaris* Maximowicz, Trudy Imp. S.-Petersburgsk. Bot. Sada 6: 170. 1879; *Spiraea aruncus* Linnaeus

Leaf surfaces usually glabrous. **Styles** 0.3–0.5 mm. **Follicles** slender, 2.5–3 mm. **Seeds** 1.8–2.5 mm. **2*n*** = 14, 18.

Flowering summer. Rich woods; 10–500 m; introduced; N.S., Ont., Que.; Conn., Del., Maine, Mich., N.J., N.Y., Wis.; c Europe; Asia (Caucasus).

Variety *vulgaris* seems to be introduced and locally naturalized in North America. The variety differs very little from var. *acuminatus* and may prove to be identical. The longer follicles of var. *acuminatus* seem to be the only difference, though H. Hara (1955) said that the petals of var. *vulgaris* are larger than those of the other varieties. One may wonder whether this European-Asian taxon is a westward circum-Pacific extension of the American var. *acuminatus*, or whether the latter is an eastward extension of var. *vulgaris*. *Aruncus dioicus* in the broad sense displays the classic Arcto-Tertiary Mesoflora distribution pattern, and it is impossible to say where it originated, though the greatest diversity of varieties (ten) is in eastern Asia (Hara). The names *A. aruncus* (Linnaeus) H. Karsten and *A. vulgaris* Rafinesque are invalid and linked to this variety.

c8. ROSACEAE Jussieu (subfam. AMYGDALOIDEAE) tribe GILLENIEAE Maximowicz, Trudy Imp. S.-Pterburgsk. Bot. Sada 6: 164, 222. 1879 E

Luc Brouillet

Herbs, perennial; unarmed. **Leaves** alternate, pinnately compound (ternate); stipules persistent or deciduous, free; venation pinnate. **Flowers:** perianth and androecium perigynous; epicalyx bractlets absent; hypanthium tubular to campanulate; torus absent; carpels 5, basally connate (becoming distinct at maturity), adnate to hypanthium, styles terminal, distinct; ovules 2–4, basal, collateral (obturator present). **Fruits** aggregated follicles; styles tardily deciduous, elongate.

Genus 1, species 2 (2 in the flora): North America.

The base chromosome number for Gillenieae is $x = 9$. *Gillenia* is host to *Gymnosporangium* and *Phragmidium* rusts.

50. GILLENIA Moench, Suppl. Meth., 286. 1802 • Indian physic [For Arnold Gillen, seventeenth-century German botanist and physician] [E]

Guy L. Nesom

Herbs, perennial, 4–10 dm, glabrous or hairy, <hairs simple and glandular>; rhizomatous, <rhizomes horizontal, woody>. **Stems** 1–5, <proximal to inflorescences>, erect to ascending, <usually red when young>, usually simple. **Leaves** persistent, basal and cauline, ternate; stipules depressed-ovate or linear, margins serrate, doubly serrate, shallowly laciniate, incised, lobed, or dissected; petiole present or absent; blade trullate to widely trullate, 3–9 cm, herbaceous, leaflets 3, lanceolate to ovate-oblong, <4–8 cm>, margins flat, incised, doubly serrate, <teeth gland-tipped>, variably glandular and hairy. **Inflorescences** terminal, 3(–9)-flowered, loose, leafy panicles; bracts present; bracteoles absent. **Pedicels** present, <1–7 on each branch>. **Flowers** 10–20 mm diam.; hypanthium <often red or purple>, 4.5–8 mm, glabrous; sepals 5, erect, triangular-ovate; petals 5, white to pale rose, linear-lanceolate, <unequal>; stamens 20 (in 3 whorls), shorter than petals; <styles filiform, glabrous, stigmas capitate>. **Fruits** aggregated follicles, 5 (1–5 maturing), broadly ovoid to broadly elliptic-ovoid, slightly flattened, 4.5–8 mm, strigose to villous-strigose, glabrous, or glabrate, partially dehiscent abaxially and partially adaxially; hypanthium persistent; sepals persistent, erect to loose spreading at apices. **Seeds** 2 per follicle, light brown to reddish, longitudinally rugulose, glabrous. $x = 9$.

Species 2 (2 in the flora): North America.

Porteranthus Britton, a superfluous and illegitimate name, was proposed by N. L. Britton in 1894 as a substitute for *Gillenia* because the similar name *Gillena* had been used by Adanson in 1763. Both names refer to Arnold Gillen. A proposal to conserve *Gillenia* (D. R. Hunt 1982) was opposed by K. R. Robertson (1983); a vote (E. G. Voss 1986) indicated that the two names are not confusable.

SELECTED REFERENCE Evans, R. C. and T. A. Dickinson. 2005. Floral ontogeny and morphology in *Gillenia* ("Spiraeoideae") and subfamily Maloideae C. Weber (Rosaceae). Int. J. Pl. Sci. 166: 427–447.

1. Stipules depressed-ovate, persistent, 10–20(–25) × 10–20 mm; leaflets: abaxial surfaces densely glandular; follicles glabrous or glabrate . . . . 1. *Gillenia stipulata*
1. Stipules linear, deciduous, 3–8 × 1–3 mm; leaflets: abaxial surfaces eglandular or sparsely glandular; follicles strigose to villous-strigose . . . . 2. *Gillenia trifoliata*

1. Gillenia stipulata (Muhlenberg ex Willdenow) Nuttall in W. P. C. Barton, Veg. Mater. Med. U.S. 1: 71. 1817 (as stipulacea) • American ipecac, Indian physic [E] [F]

Spiraea stipulata Muhlenberg ex Willdenow, Enum. Pl., 542. 1809; *Porteranthus stipulatus* (Muhlenberg ex Willdenow) Britton

Leaves: stipules persistent, depressed-ovate, 10–20(–25) × 10–20 mm (foliaceous), margins serrate to doubly serrate; leaflet blade ovate-oblong, those of proximal leaves laciniately lobed or dissected, abaxial surface densely sessile- or stipitate-glandular, otherwise hirsute to hirsutulous or rarely glabrous, adaxial hirsutulous, eglandular or sparsely sessile-glandular. **Hypanthia** 4.5–5.5 mm. **Petals** 10–13 mm. **Follicles** long-beaked, slightly to prominently rugulose, glabrous or glabrate. $2n = 18$.

Flowering (Apr–)May–Jul. Rich woods, post oak woodlands, moist places in second-growth pine hills, low woods, ravines, stream valleys, prairie thickets, dry, rocky slopes; 100–1000 m; Ala., Ark., Ga., Ill., Ind., Kans., Ky., La., Md., Mich., Miss., Mo., N.Y., N.C., Ohio, Okla., Pa., Tenn., Tex., Va., W.Va.

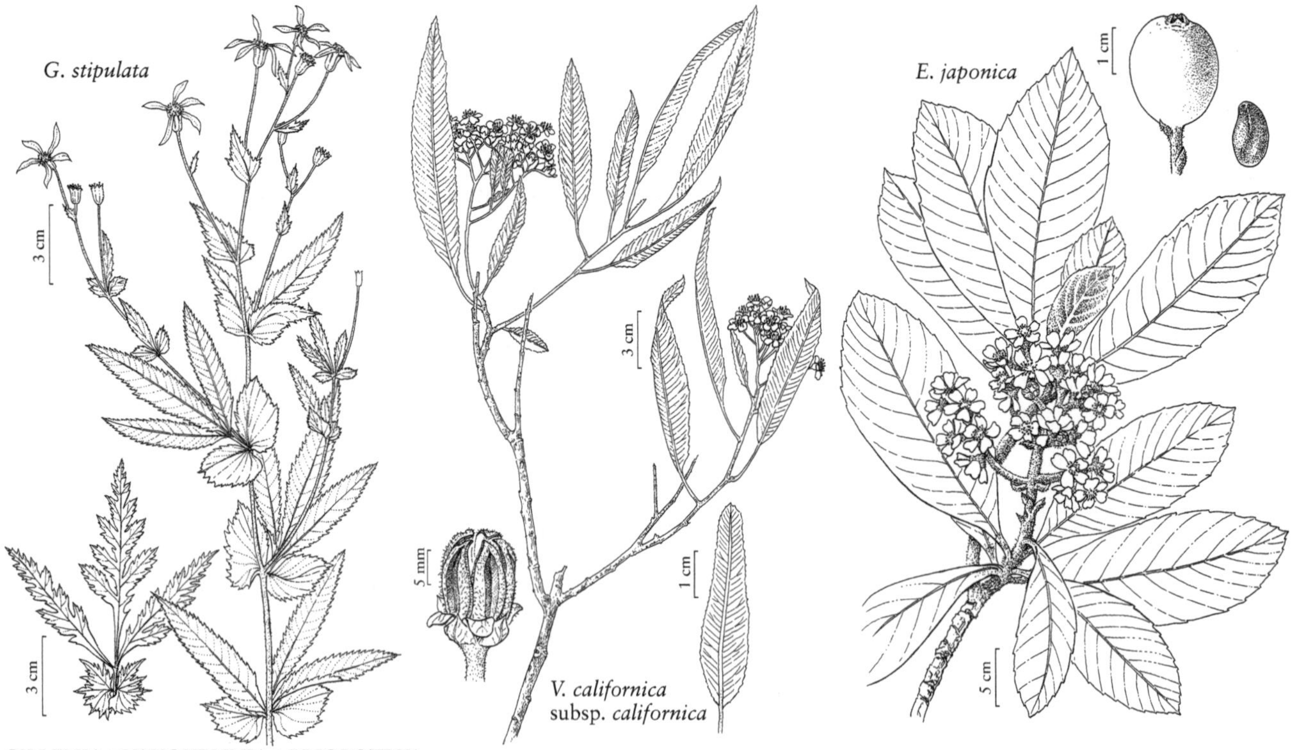

GILLENIA ◦ VAUQUELINIA ◦ ERIOBOTRYA

2. **Gillenia trifoliata** (Linnaeus) Moench, Suppl. Meth., 286. 1802 • Bowman's root [E]

Spiraea trifoliata Linnaeus, Sp. Pl. 1: 490. 1753; *Porteranthus trifoliatus* (Linnaeus) Britton

Leaves: stipules deciduous, linear, 3–8 × 1–3 mm (often with basal lobes), margins serrate to doubly serrate or shallowly laciniate; leaflet blade lanceolate, abaxial surface eglandular or sparsely sessile- or stipitate-glandular, otherwise glabrous or hairy, adaxial glabrous or sparsely to moderately hirsute. **Hypanthia** 5.5–8 mm. **Petals** 12–22 mm. **Follicles** short-beaked, smooth or slightly rugulose, loosely strigose to villous-strigose. **2*n*** = 18.

Flowering Apr–Jul. Rich woods, open woods, commonly on rocky slopes, road banks and cuts; 100–1200 m; Ont.; Ala., Ark., Del., D.C., Ga., Ill., Ind., Ky., Md., Mass., Mich., Mo., N.J., N.Y., N.C., Ohio, Pa., R.I., S.C., Tenn., Va., W.Va.

Occurrences in Massachusetts and Rhode Island are escapes from cultivation. M. Oldham (pers. comm.) noted that *Gillenia trifoliata* has not been seen in Ontario for several decades and may be extirpated there.

c9. ROSACEAE Jussieu (subfam. AMYGDALOIDEAE) tribe MALEAE Small, Man. S.E. Fl., 632. 1933

Luc Brouillet

Pyreae Baillon

Trees or shrubs; armed or unarmed. **Leaves** alternate, simple or pinnately compound; stipules persistent, deciduous, or absent, free, sometimes adnate or short-adnate to petiole (and base of blade in *Peraphyllum*); venation pinnate. **Flowers:** perianth and androecium epigynous (perigynous in *Vauquelinia*); epicalyx bractlets absent; hypanthium hemispheric, campanulate, cupulate, funnelform, or obconic, sometimes urceolate, cylindric, or saucer-shaped; torus absent (present in *Vauquelinia*); carpels 1–5, ± connate or distinct, adnate more than ½ to hypanthium (free in *Vauquelinia*, [*Dichotomanthes*]), styles terminal, sometimes subterminal or lateral, distinct or ± connate basally; ovules (1 or)2(or 3), basal and collateral, or 2–20+, marginal and biseriate (with funicular obturators). **Fruits** pomes or woody capsules surrounded by hypanthium and splitting into 5 follicles (coccetum) (*Vauquelinia*); styles persistent or deciduous, not elongate.

Genera 29, species 550–840+ (18 genera, 270 species, including 18 hybrids, in the flora): North America, Mexico, Central America, Eurasia, Africa, Atlantic Islands (Madeira), Pacific Islands (Hawaii); introduced in temperate southern hemisphere.

The range in the number of species of Maleae reflects the taxonomic difficulty of delimiting species in many genera. The base chromosome number for the tribe is mostly $x = 17$ ($x = 15$ in *Vauquelinia*). Cyanogenic glycosides are absent in some genera. Maleae include *Kageneckia* Ruiz & Pavón (South America), *Lindleya* Kunth (Mexico), and *Vauquelinia* with woody capsular or follicular fruits. These genera collectively are sister to subtribe Pyrinae (D. Potter et al. 2007), formerly recognized as subfam. Maloideae, which have hypanthia adnate to at least the proximal half of the carpels (except in the Chinese *Dichotomanthes* Kurz where they are free) and produce pomes. C. S. Campbell et al. (2007) produced a molecular phylogeny of subtribe Pyrinae.

The family name Malaceae Small (1903) is a conserved name, with *Malus* as its type genus. In contrast, the family name Pyraceae Vest (1818), with *Pyrus* as its type, is not a conserved name. Although Maleae was published later than Pyreae (1869), a Rosaceae tribe that includes both *Malus* and *Pyrus* is to be called Maleae (see Melbourne Code, Article 19.5, Example 5).

SELECTED REFERENCES Campbell, C. S. et al. 2007. Phylogeny of subtribe Pyrinae (formerly the Maloideae, Rosaceae): Limited resolution of a complex evolutionary history. Pl. Syst. Evol. 266: 119–145. Lo, E. Y. Y., S. Stefanovic, and T. A. Dickinson. 2007. Molecular reappraisal of relationships between *Crataegus* and *Mespilus*. Syst. Bot. 32: 596–616. Talent, N. and T. A. Dickinson. 2005. Polyploidy in *Crataegus* and *Mespilus* (Rosaceae, Maloideae): Evolutionary inferences from flow cytometry of nuclear DNA amounts. Canad. J. Bot. 83: 1268–1304.

1. Leaf margins usually horny; carpels free; flowers: perianth and androecium perigynous; fruits woody capsules surrounded by a hypanthium, splitting into 5 follicles; seeds winged . . . . 51. *Vauquelinia*, p. 429
1. Leaf margins not horny; carpels ± adnate to hypanthium; flowers: perianth and androecium epigynous; fruits pomes; seeds not winged or pyrenes.
    2. Fruiting carpels woody or bony.
        3. Leaf margins entire; stipules short-adnate to petiole; stems unarmed; sepals erect in flower; petal base clawed . . . . 56. *Cotoneaster*, p. 448
        3. Leaf margins ± serrate, crenate, serrulate, or crenulate, sometimes entire; stipules free; stems usually armed (sometimes with compound thorns), sometimes unarmed; sepals spreading in flower; petal base slightly or barely clawed.

4. Leaves persistent or late-deciduous; flowers 3–10(–12) mm diam., hypanthium campanulate; pomes 3–8 mm diam. . . . . 57. *Pyracantha*, p. 468
4. Leaves deciduous (sometimes winter-persistent in south); flowers 8–35 mm diam., hypanthium ± obconic; pomes 6–40 mm diam.
   5. Flowers 8–25 mm diam., stamens 5–20 (rarely 30–45); pomes yellow to red or purplish to black mature, 6–20(–25) mm diam.; pyrenes 1–5; short shoots present; inflorescences 1–50-flowered, domed panicles, corymbose, or flowers solitary . . . . 64. *Crataegus*, p. 491
   5. Flowers 25–35 mm diam., stamens 25–35(–40); pomes brownish, 15–40 mm diam.; pyrenes 5; short shoots rare or absent; inflorescences 1(or 2) -flowered . . . . 65. *Mespilus*, p. 643

[2. Shifted to left margin.—Ed.]

2. Fruiting carpels cartilaginous.
   6. Stems armed (thorns present).
      7. Stipules persistent; pedicels short or absent; styles basally connate ⅓ of length; pome flesh without stones; stamens 40–60; fruiting sepals deciduous . . . . . . 60. *Chaenomeles*, p. 484
      7. Stipules usually deciduous or caducous; pedicels present; styles distinct or basally connate; pome flesh with stones (at least near carpels and epidermis); stamens 15–50; fruiting sepals persistent or deciduous.
         8. Pome flesh with stone cells adjacent to carpels and epidermis; styles basally connate . . . . 58. *Malus*, p. 472
         8. Pome flesh with abundant stone or grit cells; styles distinct . . . . 59. *Pyrus* (in part), p. 479
   6. Stems unarmed.
      9. Inflorescences: flowers solitary or 1–5-flowered, corymbs or cymes.
         10. Pomes yellow; ovules (seeds) many; inflorescences: flowers solitary.
            11. Leaf margins entire, abaxial surfaces densely villous; buds ovoid, apices obtuse or acuminate, tomentose; young branches tomentose, glabrescent; stipules caducous; flowers 40–50 mm diam., petals white or light pink, suborbiculate, ovate, or obovate, stamens equal to or slightly longer than petals; pomes pyriform or subglobose, 30–50 mm. . . . . 61. *Cydonia*, p. 486
            11. Leaf margins sharply serrate, abaxial surfaces yellowish white-tomentose when young, glabrescent; buds triangular-ovoid, apices obtuse, glabrous or scale margins pubescent; young branches hairy; stipules persistent; flowers 25–30 mm diam., petals light to deep pink, obovate, stamens ½ length of petals; pomes ellipsoid or narrowly ellipsoid, 100–150 mm. . . . . . 62. *Pseudocydonia*, p. 487
         10. Pomes pink, yellow-orange, purple, purplish or bluish black, brownish, or nearly black; ovules (seeds) (1 or)2; inflorescences 1–5-flowered, cymes or corymbs.
            12. Pomes yellow-orange; stipules adnate to petiole and base of blade; petioles short or absent; leaf blades elliptic to oblanceolate or linear . . . . . . .68. *Peraphyllum*, p. 662
            12. Pomes pink, bluish or purplish black, purple, brownish, or nearly black; stipules free; petioles present; leaf blades elliptic, elliptic-oblong, or oblong-ovate to orbiculate.
               13. Leaves leathery, drought-deciduous or persistent; sepals nearly orbiculate (inner broadly deltate), petals round or kidney-shaped; carpels barely connate or distinct, styles lateral; pomes translucent, vivid pink, drying purplish black. . . . . 66. *Malacomeles*, p. 645
               13. Leaves membranous to coriaceous (not leathery), deciduous; sepals triangular to lanceolate, petals linear to orbiculate; carpels connate, styles terminal; pomes bluish or purplish to nearly black, pinkish or maroon-purple, dark purple-blue, or brownish . . . . 67. *Amelanchier* (in part), p. 646
      9. Inflorescences (4 or)5–400+-flowered, panicles, sometimes racemes, corymbs, or subumbellate.

[14. Shifted to left margin.—Ed.]

14. Leaves persistent, leathery; carpels basally adnate to hypanthium.
    15. Leaf margins flat; flowers 15–20 mm diam.; pedicels short or nearly absent; hypanthia usually tomentose; stamens 20; carpels connate, styles (2–)5; pomes soft apricot yellow, 20–30 mm (diam.). . . . . 52. *Eriobotrya*, p. 432
    15. Leaf margins revolute; flowers 10 mm diam.; pedicels present; hypanthia glabrous or weakly floccose; stamens 10; carpels distinct, styles 2 or 3; pomes usually bright red, sometimes yellow, 5–10 mm . . . . 55. *Heteromeles*, p. 447
14. Leaves usually deciduous, sometimes semipersistent or persistent (then margin entire), membranous to ± leathery; carpels adnate to all or $\frac{1}{3}$–$\frac{1}{2}$ of hypanthium.
    16. Pome flesh with stones or sclereids.
        17. Inflorescences terminal, 6–400+-flowered flat-topped or rounded panicles; flowers opening after leaf expansion, 5–17 mm diam.; sepals erect or ascending; leaves pinnately divided, sometimes simple or lobed . . . . 53. *Sorbus*, p. 433
        17. Inflorescences terminal on short shoots, 4–9-flowered racemes or simple corymbs, umbel-like; flowers developing with or before leaves, 15–45 mm diam.; sepals reflexed; leaves simple. . . . . 59. *Pyrus* (in part), p. 479
    16. Pome flesh without stones.
        18. Stipules adnate to petiole, persistent . . . . 54. *Aronia*, p. 445
        18. Stipules free, caducous or deciduous.
            19. Leaves persistent or deciduous; inflorescences corymbose or subumbellate; pomes red or black. . . . . 63. *Photinia*, p. 488
            19. Leaves deciduous; inflorescences racemes; pomes bluish or purplish to nearly black, pinkish or maroon-purple, dark purple-blue, or brownish. 67. *Amelanchier* (in part), p. 646

51. VAUQUELINIA Corrêa ex Bonpland in A. von Humboldt and A. J. Bonpland, Pl. Aequinoct. 1: 140, plate 40. 1807 • [For Nicolas Louis Vauquelin, 1763–1829, French chemist and pharmacist]

William J. Hess

Shrubs or trees, (10–)15–80(–100) dm. **Stems** 1–10+, orientation unknown; bark gray to dark gray, <smooth, older plaited>; short shoots absent; unarmed; tomentulose to villous-canescent, <hairs white, short, tightly crinkled>, often tardily to soon glabrescent. **Leaves** persistent, cauline, <erect-ascending to spreading>, simple; stipules tardily deciduous, free, subulate to narrowly deltate, margins entire, glandular; petiole present; blade oblong-elliptic or oblong-lanceolate to linear-lanceolate or linear to linear-oblong, (2.2–)3–13(–18.5) cm, leathery, margins flat, <usually horny>, serrate, serrulate, or crenulate, sometimes doubly serrate, rarely entire, venation pinnate and semicraspedodromous, surfaces tomentulose, sometimes glabrate or puberulent. **Inflorescences** terminal or axillary, 15–25+-flowered, compound corymbs, puberulent to tomentulose; bracts present; bracteoles present. **Pedicels** present. **Flowers:** perianth and androecium perigynous, 5–10 mm diam.; hypanthium hemispheric, 1.5–2.5(–3) mm, <leathery>, sericeous, glabrescent, <interior proximal surface nectariferous>; sepals 5, erect, broadly ovate; petals 5, white, oblong-ovate to oblong-obovate, <base clawed, apex rounded to emarginate>; stamens 18–20, shorter than petals; <torus thickened>; carpels 5, connate, free, strigose, styles terminal, distinct; ovules 2. **Fruits** capsules, broadly ovoid, 4.5–7.5 mm, <woody>, sericeous, ventrally (fully) and dorsally (in distal $\frac{1}{2}$) dehiscent, splitting into 5 follicles; hypanthium persistent; sepals persistent, erect; styles persistent. **Seeds** 2 per follicle, <winged>. $x = 15$.

Species 3 (2 in the flora): sw United States, Mexico.

Vauquelinia species are xerophytic. The third species in the genus, *V. australis* Standley, is known from Oaxaca and Puebla, Mexico.

SELECTED REFERENCE Hess, W. J. and J. Henrickson. 1987. A taxonomic revision of *Vauquelinia* (Rosaceae). Sida 12: 101–163.

1. Leaf margins serrulate or crenulate, teeth 10–35(–50) per 5 cm; sepal margins eglandular .... 1. *Vauquelinia californica*
1. Leaf margins usually serrate, sometimes partly doubly serrate, rarely entire, teeth (3–)5–10(–14) per 5 cm; sepal margins glandular............ 2. *Vauquelinia corymbosa*

1. Vauquelinia californica (Torrey) Sargent, Gard. & Forest 2: 400. 1889 • Arizona rosewood [F]

Spiraea californica Torrey in W. H. Emory, Not. Milit. Reconn., 140. 1848

Shrubs or trees, <rounded to ovoid>, 10–50(–80) dm, usually multistemmed. **Leaves:** petiole (1.5–)4–16(–26) mm; blade usually bicolor, sometimes green or yellow-green, oblong-elliptic or oblong-lanceolate to linear-lanceolate or linear, (2.2–)3–11(–15) × (0.6–)0.7–2(–3.2) cm, base obliquely cuneate to rounded, margins serrulate or crenulate, teeth 10–35(–50) per 5 cm, 0.1–0.5(–1.5) mm, apex acute, mostly mucronate, surfaces closely villous-tomentulose, abaxially more densely so, or glabrate or tardily glabrescent (some remaining crinkled hairs). **Corymbs** 1.5–5(–12) × 1.7–8(–13) cm, tomentulose, sometimes sparsely hairy or glabrate. **Flowers:** hypanthium 1.5–2.5(–3) × 2–3.7(–4.5) mm, tomentose to sparsely tomentulose or glabrate; sepals 1.1–2.2 × 1.4–2.3 mm, margins eglandular; petals oblong-ovate, 3.4–5.4 × 2.4–3.5 mm; filaments 2.5–6 mm. **Capsules** 4.5–6.5 × 3.5–4.5 mm. **Seeds** 3.5–5 × 0.9–1.4 mm.

Subspecies 4 (3 in the flora): sw United States, n Mexico.

Subspecies *retherfordii* (I. M. Johnston) W. J. Hess & Henrickson is known from Coahuila and Durango in northern Mexico.

SELECTED REFERENCE Williams, K. B. 1971. Ecological and Morphological Variations of *Vauquelinia californica* (Torr.) Sarg. Populations in Arizona. Ph.D. dissertation. University of Arizona.

1. Leaves green or yellow-green, glabrate or puberulent along midveins .... 1c. *Vauquelinia californica* subsp. *pauciflora*
1. Leaves bicolor, abaxially white-puberulent or villous-tomentulose (or tardily glabrescent in plants of c Arizona), adaxially green.
    2. Leaf blades lanceolate to oblong-lanceolate or narrowly elliptic, (0.7–)1–2(–3.2) cm wide, usually abaxially sparsely white-puberulent, sometimes glabrate or glabrous .... 1a. *Vauquelinia californica* subsp. *californica*
    2. Leaf blades linear to linear-lanceolate, (0.6–)0.7–1.2(–1.4) cm wide, abaxially villous-tomentulose, soon or tardily glabrescent except for hairy midveins............ 1b. *Vauquelinia californica* subsp. *sonorensis*

1a. Vauquelinia californica (Torrey) Sargent subsp. **californica** [F]

Young stems initially tomentulose, becoming canescent or tardily glabrate. **Leaves:** petiole (4–)7–15(–26) mm; blade bicolor, abaxially white, adaxially green and lustrous, lanceolate to oblong-lanceolate or narrowly elliptic, (3.5–)4–9(–15) × (0.7–)1–2(–3.2) cm, surfaces initially villous-tomentulose, abaxially usually sparsely white-puberulent, <hairs slightly coiled>, sometimes glabrate or glabrous, adaxially mostly glabrate or glabrous. **Corymbs** 3–5(–12) × 3–8(–13) cm. **Flowers:** hypanthium 2–2.5(–3) × 2–3.7(–4.5) mm, exterior tomentulose or tardily glabrate, interior glabrous; sepals 1.4–2.1 × 1.5–2.3 mm, puberulent-villous; petals 4–5 × 2.5–3.5 mm; filaments 3–5.5 mm. **Capsules** 5–6.5 × 3.5–4.5 mm. **Seeds** 4–5 × 1.2–1.3 mm. **2*n*** = 30.

Flowering spring. Limestone and granite substrates, shrublands of upper Sonoran Desert from above desert plains to lower pinyon-juniper zones; 700–1800 m; Ariz.; Mexico (Baja California, Baja California Sur).

Subspecies *californica* shows considerable variation among populations, particularly with regard to vestiture, leaf size and shape, petiole length, and number of marginal teeth. Often populations within the same mountain ranges in Arizona share the same pattern of variation. Most of the populations have distinctly bicolor leaves, at least when young, or abaxially become weakly puberulent and persistently, closely tomentulose to canescent with tightly coiled hairs.

The name *Vauquelinia torreyi* S. Watson, which is illegitimate, pertains here.

1b. Vauquelinia californica (Torrey) Sargent subsp. **sonorensis** W. J. Hess & Henrickson, Sida 12: 130, figs. 11a–c. 1987

Young stems densely white-tomentulose, becoming canescent. **Leaves:** petiole (4–)6–16(–22) mm; blade bicolor, abaxially white, adaxially green and nonlustrous, linear to linear-lanceolate, (2.5–)5–11(–15) × (0.6–)0.7–1.2(–1.4) cm, surfaces villous-tomentulose, soon or tardily glabrescent except for hairy midveins. **Corymbs** 1.5–4.5 × 2–6.5 cm, <villous to tomentulose>. **Flowers:** hypanthium 2–2.5 × 3–3.5 mm, exterior white villous-tomentulose, interior glabrous except at base; sepals 1.3–1.8 × 1.6–2 mm, villous-tomentulose; petals 4–5 × 2.4–3 mm; filaments 3–5 mm. **Capsules** 5–6 × 3.5–4 mm. **Seeds** 3.5–4 × 0.9–1.2 mm.

Flowering spring. Canyon margins and hillsides of the Sonoran Desert; 700–1500 m; Ariz.; Mexico (Baja California).

Subspecies *sonorensis* is known from the Ajo Mountains in Pima County. Some plants from the Baboquivari Mountains of Pima County have somewhat similar narrow leaves but lack the dense vestiture on stems, inflorescences, and abaxial leaf surfaces.

1c. Vauquelinia californica (Torrey) Sargent subsp. **pauciflora** (Standley) W. J. Hess & Henrickson, Sida 12: 135. 1987

Vauquelinia pauciflora Standley, Proc. Biol. Soc. Wash. 31: 132. 1918

Young stems loosely tomentulose, tardily glabrescent. **Leaves:** petiole (1.5–)4–16(–22) mm; blade green or yellow-green, lanceolate or oblong-lanceolate to elliptic or oblong-elliptic, sometimes oblong-ovate, (2.2–)3–7.5(–9) × (0.6–)0.8–1.4(–2) cm, surfaces glabrate or puberulent along midveins. **Corymbs** 1.5–5 × 1.7–7 cm, <puberulent to glabrate>. **Flowers:** hypanthium 1.5–2.5 × 2.5–3.3 mm, exterior puberulent, interior glabrate; sepals 1.1–2.2 × 1.4–2 mm, abaxially puberulent to glabrate; petals 3.4–5.4 × 2.4–3.4 mm; filaments (2.5–)3–6 mm. **Capsules** (4.5–)5–6 × 3.5–4(–4.5) mm. **Seeds** 3.8–5 × 1.1–1.4 mm.

Flowering spring. Limestone substrates in arid chaparral-desert scrub; 1400–2300 m; Ariz., N.Mex.; Mexico (Chihuahua, Coahuila, Durango).

Subspecies *pauciflora* is known from Cochise County, Arizona, and Hidalgo County, New Mexico. These plants have relatively small (to 5.5 mm) and thickened leaves with short petioles (to 7 mm), and small inflorescences.

2. Vauquelinia corymbosa Bonpland in A. von Humboldt and A. J. Bonpland, Pl. Aequinoct. 1: 140, plate 40. 1807

Subspecies 6 (1 in the flora): Texas, Mexico.

2a. Vauquelinia corymbosa Bonpland subsp. **angustifolia** (Rydberg) W. J. Hess & Henrickson, Sida 12: 148. 1987

Vauquelinia angustifolia Rydberg in N. L. Britton et al., N. Amer. Fl. 22: 260. 1908

Shrubs [trees], 15–70(–100) dm, multistemmed [single-stemmed]. **Leaves:** petiole (6–)9–28(–40) mm; blade green or yellow-green, linear to linear-oblong, (3.5–)6–13(–18.5) × (0.3–)0.5–0.9(–1.3) cm, base obliquely cuneate to truncate-subcordate, margins usually serrate, sometimes partly doubly serrate, rarely entire, teeth (3–)5–10(–14) per 5 cm, 0.2–1.5(–2.5) mm, apex acute to rounded, mucronate, <lustrous>, surfaces tomentulose, tardily glabrescent. **Corymbs** 2.5–5.5 × (3–)3.5–6(–7) cm, mostly glabrate [tomentulose or canescent]. **Flowers:** hypanthium 1.7–2.5 × 2.5–3.8 mm, floccose or sparsely villous to glabrate; sepals (1.2–)1.5–2.4 × 1.5–2.3 mm, margins glandular; petals oblong-ovate to oblong-obovate, 2.4–4.5(–5.6) × 2–3.5 mm; filaments 2.7–5 mm. **Capsules** 5–7.5 × 3.5–5.4 mm. **Seeds** 4.4–4.8 × 1–1.3 mm. $2n$ = 30 (Mexico).

Flowering spring. Limestone, rarely igneous rock slopes and canyons, and gypsum flats, Chihuahuan Desert scrub and chaparral; 1000–1500 m; Tex.; Mexico (Chihuahua, Coahuila).

Subspecies *angustifolia* is found in the flora area only in the mountains of Brewster and Presidio counties. Its leaves are typically 20 times longer than wide, with marginal serrations approximately one per centimeter or rarely entire; its inflorescences are mostly glabrate.

52. ERIOBOTRYA Lindley, Trans. Linn. Soc. London 13: 96, 102. 1821 • Loquat [Greek *erion*, wool, and *botrys*, bunch of grapes, alluding to woolly inflorescences] [I]

James B. Phipps

Shrubs or trees, 40–100[–200] dm. **Stems** ca. 1, erect; bark gray-brown; short shoots absent; unarmed; hairy. **Leaves** persistent, cauline, simple; stipules deciduous or ± persistent, free, linear, small, margins entire; petiole present; blade ± elliptic to oblong-lanceolate, 2–40 cm, leathery, margins flat, dentate, venation pinnate (craspedodromous), abaxial surface tomentose, adaxial glabrous [hairy]. **Inflorescences** terminal, 20–40-flowered, panicles, usually ± tomentose; bracts present at proximal nodes, <leafy>; bracteoles present. **Pedicels** present, short, or nearly absent. **Flowers:** perianth and androecium epigynous, 15–20 mm diam.; hypanthium urceolate, 3–4 mm, usually tomentose; sepals 5, suberect, triangular; petals 5, white, obovate to ± oblong; stamens [10–15]20, shorter than petals; carpels (2–)5, connate, basally adnate to hypanthium, indumentum not recorded, styles <(2–)5>, terminal, basally connate; ovules 2. **Fruits** pomes, soft apricot yellow [yellow to reddish or blackish], ellipsoid to subglobose, [10–]20–30 mm diam., softly short-hairy; hypanthium persistent; sepals usually persistent, covering hypanthial opening or reflexed; carpel walls thin; styles not persistent. **Seeds** 1–5 per fruit. $x = 17$.

Species ca. 30 (1 in the flora): introduced; s, se Asia; introduced also in Mexico, West Indies, Central America, South America, Europe, n, s Africa, Pacific Islands (New Zealand), Australia.

Eriobotrya is distinctive in Maleae; among other attributes it contains the largest plants in the subfamily with forest trees to 20 m and has pomes that do not have cores but have large seeds (J. R. Rohrer et al. 1991). Important regional revisions are by J. E. Vidal (1965) and Gu C. Z. and S. A. Spongberg (2003c). The closest relation is *Rhaphiolepis* Lindley (D. Potter et al. 2007), in conformity with morphologic evidence. *Eriobotrya japonica* provides a delectable fruit. In addition to this species cultivated and escaped in the southern United States, *E. deflexa* (Hemsley) Nakai may be encountered in arboreta in the same area.

SELECTED REFERENCES Gu, C. and S. A. Spongberg. 2003c. *Eriobotrya*. In: Wu Z. and P. H. Raven, eds. 1994+. Flora of China. 20+ vols. Vol. 9, pp. 138–141. Beijing and St. Louis. Vidal, J. E. 1965. Révision du genre *Eriobotrya* (Pomoideae). Adansonia, n. s. 5: 537–580.

1. Eriobotrya japonica (Thunberg) Lindley, Trans. Linn. Soc. London 13: 102. 1821 [F] [I]

Mespilus japonica Thunberg, Nova Acta Regiae Soc. Sci. Upsal. 3: 208. 1780

Stems: bark ± smooth. **Leaves:** petiole 6–10 mm; blade margins dentate in distal ½, lateral veins 15–25 per side, apex acute. **Inflorescences:** branches stiff, densely rufous-tomentose, with 1–3 barely reduced leaflike bracts, flowers ± sessile; bracteoles deciduous, narrowly triangular, margins entire, rufous-tomentose. **Flowers:** sepals 3 × 3 mm; petals ± spreading, often notched, 8–10 mm. **Pomes:** flesh sweet. **Seeds** 3–5, black, ovoid, shiny. $2n = 34$.

Flowering spring. Redwood forests, suburban and urban woodlots; 0–100 m; introduced; Calif., Fla., Ga., La.; Asia (China); introduced also in Mexico, West Indies, Central America, South America, Europe, n, s Africa, Pacific Islands (New Zealand), Australia.

Naturalized plants of *Eriobotrya japonica* are only sporadically found in North America. The species is apparently native to east-central China (Gu C. and S. A. Spongberg 2003c), but it has long been cultivated and is now spontaneous in a much larger Asian area. The species is cultivated widely for its fruit in warm temperate and subtropical regions.

53. SORBUS Linnaeus, Sp. Pl. 1: 477. 1753; Gen. Pl. ed. 5, 213. 1754 • Mountain ash, rowan, whitebeam, sorbier, alisier [Latin name for service tree, *Sorbus domestica* Linnaeus]

Peter F. Zika

Stéphane M. Bailleul

Shrubs or trees, 10–300 dm. **Stems** 1–8, erect or ascending; bark gray to brown or bronze, <usually smooth, becoming scaly with age, with conspicuous horizontal lenticels>; long and short shoots present; unarmed; <twigs with smell and taste of bitter almond (cyanogenic glycosides)>; glabrous or hairy. **Leaves** deciduous, cauline, simple or odd (rarely even) pinnately lobed or divided; stipules usually early deciduous, sometimes persistent, free or short-adnate to petiole, linear, ovate to lanceolate, or flabellate, margins entire or dentate to laciniate; petiole present; blade ovate to oblong, 5–33 cm, membranous to slightly leathery, leaflets 0 or 7–17(–19), <usually opposite, rarely alternate>, elliptic, ovate, oblong to lanceolate or oblanceolate, margins flat, sharply serrate or nearly entire, venation pinnate, surfaces glabrous or hairy, sometimes glaucous. **Inflorescences** terminal, 6–400+-flowered, panicles, <flat-topped or rounded>, glabrous, glaucous, or hairy; bracts present; bracteoles absent. **Pedicels** present. **Flowers** opening after leaf expansion, perianth and androecium epigynous or ½ epigynous, <odor strong, often considered unpleasant or rancid>, 5–17 mm diam.; hypanthium <green to red>, obconic, 2–6 mm, glabrous or hairy; sepals 5, erect or ascending, ovate or triangular; petals 5, white or pink [red], suborbiculate or broadly obovate to broadly ovate, <base clawed or not, claw often ± villous>; stamens [10–]14–20[–44] in 2 or 3 series, usually slightly longer than petals; carpels 2–5, distinct, partially or wholly connate and adnate to all or proximal ½ of hypanthium, usually apically woolly, styles <2–5>, terminal, distinct or connate ½ of length; ovules 2 or 3[or 4] (all but 1 usually aborting). **Fruits** pomes, usually orange or red, rarely brown or yellow [green, white, or pink], globose to ovoid, obovoid, ellipsoid, or oblong, [pyriform], 4–19 mm, <smooth or with lenticels, shiny, sometimes glaucous>, hairy or glabrous; <flesh usually with sclereids>; hypanthium persistent; sepals usually persistent, rarely deciduous, usually incurved, <fleshy>; carpels cartilaginous; styles often persistent. **Seeds** 3–5, <brown to reddish brown or yellowish, darkening with maturity, ovoid to lanceoloid, slightly asymmetric and flattened>. $x = 17$.

Species ca. 130 (10 in the flora): North America, Eurasia, n Africa, Atlantic Islands, Pacific Islands.

The taxonomy of *Sorbus* is complicated by apomixis, polyploidy, and hybridization among sections and genera, especially in Eurasia. Hybrids with other genera in Maleae are generally distinguished by their incompletely divided, deeply lobed leaves, and include ×*Amelasorbus* Rehder (*Sorbus* × *Amelanchier*), ×*Crataegosorbus* Makino (*Sorbus* × *Crataegus*), ×*Sorbaronia* C. K. Schneider (*Sorbus* × *Aronia*), ×*Sorbocotoneaster* Pojarkova (*Sorbus* × *Cotoneaster*), ×*Sorbopyrus* C. K. Schneider (*Sorbus* × *Pyrus*), and ×*Tormimalus* Holub (as *Torminalis* Medikus [= *Sorbus* subg. *Torminaria*] × *Malus*). Only ×*Amelasorbus* and ×*Sorbaronia* are known to occur naturally in North America. Parentages of some reported ×*Sorbaronia* nothospecies are difficult to verify, especially for densely hairy hybrids involving either *Aronia arbutifolia*, *A.* ×*prunifolia* (Marshall) Rehder, or *Sorbus aucuparia*; all these possibilities are claimed. In addition to their partially pinnate leaves, some ×*Sorbaronia* hybrids may be recognized by the presence of pink to red anthers and black or purple fruits; these are reported also to be somewhat sterile.

Diploid *Sorbus* are often obligate outcrossers; some apomictic triploids and tetraploids are self-compatible (C. S. Campbell et al. 1991; H. A. McAllister 2005). Some species have morphologically indistinguishable diploid and tetraploid races, with sexual and apomictic individuals. North American species are in need of biosystematic review.

Sorbus is treated here in the broad sense, as a provisional arrangement to accommodate more than 70 natural hybrids between simple- and pinnate-leaved subgenera (J. B. Phipps et al. 1990; E. B. Nelson-Jones et al. 2002; C. Kalkman 2004; A. Robertson et al. 2004, 2004b; J. J. Aldasoro et al. 2004, 2005). Additional hybrid taxa continue to be discovered in Europe (G. Aas et al. 1994; M. F. Fay et al. 2002; T. C. G. Rich and L. Houston 2006; A. Robertson and C. Sydes 2006). Nonetheless, the argument has been made to revive generic ranking for the simple-leaved subgenera *Aria* and *Torminaria*, based on flower and fruit characters (K. R. Robertson et al. 1991), with some support from molecular investigations (C. S. Campbell et al. 1995, 2007; D. Potter et al. 2007). Subgenus *Torminaria* is anomalous in *Sorbus*, having a unique flavonoid chemistry, leaves with relatively few broad lobes, and scented brown fruits, thick-skinned as in subg. *Cormus* (Spach) Duchartre and some *Pyrus*, and being dispersed by carnivorous mammals as well as birds (C. M. Herrera 1987, 1989). Eurasian hybrids between subgenera frequently involve *S. torminalis* of subg. *Torminaria*. Subgenera *Aria* and *Torminaria* are likely to be recognized at generic rank once molecular studies can consistently resolve their placement within the subtribe Pyrinae, overcoming current difficulties with interfertility, reticulate relationships, rapid radiation, and small samples (Campbell et al. 2007; Potter et al.).

Some species of *Sorbus* are cultivated as ornamentals. The colorful fruits of native and introduced species persist after the leaves drop and are consumed by birds across North America (A. C. Martin et al. 1951). One exotic species disseminated by birds is invasive (*S. aucuparia*). The pomes of *S. americana*, collected after a frost, can be sweetened for preserves. The pulp of some strains of *S. aucuparia* is reported to be less bitter. It may be eaten fresh or dried or used as flour. The fruits of a number of Eurasian species are made into vinegar, spirits, or medications. *Sorbus domestica* and *S. torminalis* pomes are sometimes eaten after being bletted (allowed to autodigest and become soft and mushy in texture). The fine-grained wood of *Sorbus* is used for carving or furniture in China. The bark of some species is antiseptic, or used to tan leather. *Sorbus sambucifolia* pomes show some potential in early testing for treating tumors (Y. Yoshizawa et al. 2000, 2000b).

SELECTED REFERENCES Aldasoro, J. J. et al. 2004. Revision of *Sorbus* subgenera *Aria* and *Torminaria* (Rosaceae–Maloideae). Syst. Bot. Monogr. 69: 1–148. Jones, G. N. 1939. A synopsis of the North American species of *Sorbus*. J. Arnold Arbor. 20: 1–43. McAllister, H. A. 2005. The Genus *Sorbus*: Mountain Ashes and Other Rowans. Kew.

1. Leaves pinnately compound at least proximally, petiole nodes 5-lacunate; pome flesh without tanniferous cells . . . . 53c. *Sorbus* subg. *Sorbus*, p. 437
1. Leaves simple, sometimes lobed, petiole nodes 3-lacunate; pome flesh with tanniferous cells.
    2. Blade margins relatively deeply lobed; pome flesh composed entirely of tanniferous cells; styles connate ½ their lengths . . . . 53a. *Sorbus* subg. *Torminaria*, p. 436
    2. Blade margins entire, toothed, or relatively shallowly lobed; pome flesh with tanniferous cells in groups; styles distinct. . . . . 53b. *Sorbus* subg. *Aria*, p. 436

In order to identify intergeneric hybrids involving *Sorbus* as one of the parents, the following key includes leads for ×*Amelasorbus* and ×*Sorbaronia*.

1. Leaves simple, lobed, or proximally pinnately compound (escapes and hybrids).
   2. Leaves simple, without leaflets; styles 2.
      3. Leaves thinly hairy to glabrate abaxially, subpalmately lobed, lobes ± triangular, (1–)1.5–2.5 cm wide, apices acute to acuminate; pomes brown, narrowly obovoid, lenticels abundant; sepal margins prominently glandular (and villous), glands often relatively thick . . . . 1. *Sorbus torminalis*
      3. Leaves tomentose abaxially, proximally pinnately lobed, lobes ± oblong, 1–1.5 (–1.8) cm wide, apices acute to obtuse; pomes bright red, ellipsoid, sometimes narrowly obovoid, lenticels few; sepal margins eglandular (and villous), rarely with inconspicuous glands . . . . 2. *Sorbus intermedia*
   2. Leaves, at least some, proximally pinnate, with 1–3(–5) basal pairs of distinct leaflets; styles 2–5.
      4. Leaves tomentose abaxially; terminal leaflets 7–10-lobed; styles 2 or 3; pomes red. . . 4. *Sorbus hybrida*
      4. Leaves glabrous, glabrate, or sparsely to densely villous abaxially; terminal leaflets 1–3-lobed; styles 3–5; pomes dark purple or red.
         5. Terminal leaflets coarsely serrate; pomes subglobose; sepals erect or spreading-ascending, prominent . . . . ×*Amelasorbus*
         5. Terminal leaflets finely serrate; pomes ellipsoid to ovoid; sepals incurved, inconspicuous. . . . . ×*Sorbaronia*
1. Leaves pinnately compound (mostly native species).
   6. Leaflets shiny adaxially (visible on herbarium specimens with strong lighting), green to dark green, never glaucous or blue-green; w North America.
      7. Pomes: sepals erect, prominent; flowering hypanthium plus sepals (4.5–)5–6 mm; petals white to pinkish; w Aleutians (Alaska) . . . . 9. *Sorbus sambucifolia*
      7. Pomes: sepals incurved, inconspicuous; flowering hypanthium plus sepals 3–4 mm; petals white; w North America (except w Aleutians).
         8. Indument primarily rufous on winter buds, leaflet axils, and inflorescences; fruiting pedicels essentially glabrous; leaflets 2.5–4.1(–4.5) cm . . . . .7. *Sorbus californica*
         8. Indument primarily whitish on winter buds, leaflet axils, and inflorescences; fruiting pedicels sparsely to densely villous; leaflets (3–)4–6.5(–8.7) cm . . . . . 8. *Sorbus scopulina*
   6. Leaflets dull adaxially, green to blue-green or yellowish green, sometimes slightly glaucous; w, e North America.
      9. Winter buds densely villous, hairs whitish, rarely rufous; hypanthia densely villous, hairs whitish . . . . 3. *Sorbus aucuparia*
      9. Winter buds glabrous or sparsely to densely villous, hairs primarily rufous; hypanthia glabrous or sparsely villous proximally, hairs whitish or rufous.
         10. Leaflet length/width ratios 3.4–5, apices gradually acuminate to long-acuminate; pomes 4–7 mm diam., not glaucous when fresh or dried . . . . .5. *Sorbus americana*
         10. Leaflet length/width ratios 1.9–3.5(–3.6), apices abruptly short-acuminate, cuspidate, or obtuse to acute; pomes (5–)7–13 mm diam., often glaucous (sometimes only when dried).
             11. Winter buds glutinous, shiny; leaflets (11–)13–17; e North America . . . . . . .6. *Sorbus decora*
             11. Winter buds not glutinous, dull; leaflets 7–13; w North America. . . . . . 10. *Sorbus sitchensis*

53a. Sorbus Linnaeus subg. Torminaria (de Candolle) K. Koch, Hort. Dendrol., 178. 1853 [I]

Pyrus Linnaeus sect. *Torminaria* de Candolle in A. P. de Candolle and A. L. P. P. de Candolle, Prodr. 2: 636. 1825; *Torminalis* Medikus

Trees, to 120[–200] dm. **Winter buds** glutinous [or not]. **Leaves** simple; petiole nodes 3-lacunate; blade margins relatively deeply lobed. **Flowers:** styles connate ½ their lengths. **Pomes:** epidermis multilayered; flesh composed entirely of tanniferous cells; core present.

Species 3 (1 in the flora): introduced, Washington; Eurasia, n Africa.

The two other species of this subgenus are of apparent hybrid origin, possibly with subg. *Aria*, and exhibit less deeply lobed leaves and pome flesh with tanniferous cells in groups.

1. Sorbus torminalis (Linnaeus) Crantz, Stirp. Austr. Fasc. 2: 45. 1763 • Wild service tree [I]

Crataegus torminalis Linnaeus, Sp. Pl. 1: 476. 1753

Trees, to 120[–200] dm. **Stems** 1; bark gray; winter buds green to brown, ovoid to ovoid-oblong, 5–10 mm, <scales dark-margined>, shiny, glutinous, glabrous, ciliate, or villous. **Leaves** simple; stipules early deciduous, whitish-villous; blade slightly shiny, green to dark green adaxially, ovate to broadly elliptic, 5.5–10.5 × 3–8 cm, margins deeply 1–4 subpalmately lobed, lobes <± triangular>, (1–)1.5–2.5 cm wide, basal sinuses deepest, margins finely serrate, secondary veins in 4–6 main pairs, apex acute to acuminate, abaxial surface thinly white-hairy to glabrate, adaxial glabrous. **Panicles** 15–60+-flowered, rounded, 3–17.5 cm diam.; peduncles whitish-villous. **Pedicels** whitish-villous. **Flowers** 10–17 mm diam.; hypanthium densely villous, hypanthium plus sepals 5.5–7.5 mm; sepals 1.5–3.5 mm, margins villous and prominently glandular, glands often relatively thick; petals white, elliptic, broadly elliptic, or broadly ovate, 4–7 mm; stamens 20; carpels adnate to hypanthium, apex conic, styles 2, 3.5–4 mm. **Infructescences** sparsely whitish-villous. **Pomes** brown, narrowly obovoid, 15–19 × 10–13 mm, dull, not glaucous; <lenticels abundant>; sepals inconspicuous, incurved. **Seeds** brown, oblong-lanceoloid, 8.1 × 3.4 mm, slightly asymmetric, slightly flattened. **2*n*** = 34, 68 (Europe).

Flowering spring; fruiting fall. Disturbed conifer forests; 0–50 m; introduced, Wash.; Europe; sw Asia; nw Africa.

In European gardens, one or more clones of intergeneric hybrids between *Pyrus communis* and *Sorbus* are called ×*Sorbopyrus auricularis* (Knoop) C. K. Schneider. J. Wiśniewska et al. (1969) suggested that *S. torminalis* may be involved in the parentage of at least one clone.

53b. Sorbus Linnaeus subg. Aria Persoon, Syn. Pl. 2: 38. 1806 [I]

Aria (Persoon) Host

Trees, to 70[–300] dm. **Winter buds** glutinous [or not]. **Leaves** simple; petiole nodes 3-lacunate; blade margins relatively shallowly lobed [entire or toothed]. **Flowers:** styles distinct. **Pomes:** epidermis single-layered [rarely multilayered]; flesh with tanniferous cells in groups; core present [absent].

Species ca. 40 (1 in the flora): introduced; Europe, Asia, n Africa.

Some species of apparent hybrid origin between subg. *Aria* and other subgenera exhibit more deeply lobed leaves and/or pome flesh composed entirely of tanniferous cells.

2. Sorbus intermedia (Ehrhart) Persoon, Syn. Pl. 2: 38. 1806 • Swedish whitebeam [I]

Pyrus intermedia Ehrhart, Gartenkalender 4: 197. 1784; *Aria intermedia* (Ehrhart) Schur

Trees, to 70[–200] dm. **Stems** 1; bark gray; winter buds green to red-brown, ovoid, 5–12 mm, shiny, slightly glutinous, whitish-villous. **Leaves** simple; stipules early deciduous, densely whitish-villous; blade shiny, green to dark green adaxially, ovate to elliptic, 6–13 × 3.5–8.5 cm, margins proximally 4–7 pinnately lobed, lobes <± oblong>, 1–1.5(–1.8) cm wide, basal sinuses deepest, margins distally coarsely serrate, main pairs of secondary veins 8–10, apex acute to obtuse, abaxial surface white-tomentose, adaxial glabrous or sparsely villous, indument usually persistent. **Panicles** 18–85-flowered, rounded, 6–17 cm diam.; peduncles white-tomentose. **Pedicels** white-tomentose. **Flowers** 14–16 mm diam.; hypanthium whitish-tomentose, hypanthium plus sepals 5–6 mm; sepals 1.5–2.5 mm, margins villous and eglandular, rarely with inconspicuous glands; petals white, elliptic to broadly ovate, 5–7 mm; stamens 20; carpels distinct, apex conic, styles 2, 2.5–4 mm. **Infructescences** thinly white-tomentose to glabrate. **Pomes** bright red, ellipsoid, sometimes narrowly obovoid, 7–16 × 6–11 mm, shiny, faintly glaucous; <lenticels few>; sepals inconspicuous, erect, <villous toward apex>. **Seeds** red-brown, narrowly ovoid, 6–6.5 × 2.5–3.1 mm, slightly asymmetric, slightly flattened. **2*n*** = 68 (Europe).

Flowering spring; fruiting fall. Thickets, roadsides, waste ground, disturbed conifer forests; 0–100 m; introduced; B.C., Ont.; Mass., Wash.; Europe.

Sorbus aria (Linnaeus) Crantz of Europe has been reported from San Juan County, Washington (S. Atkinson and F. Sharpe 1993; J. T. Kartesz and C. A. Meacham 2003). Those populations are *S. intermedia*, with more deeply lobed leaves and broader winter buds.

53c. Sorbus Linnaeus subg. Sorbus

Shrubs or trees, to 150 dm. **Winter buds** glutinous or not. **Leaves** pinnately compound at least proximally; petiole nodes 5-lacunate; leaflet margins usually serrate, rarely almost entire. **Flowers:** styles distinct. **Pomes:** epidermis single-layered; flesh without tanniferous cells; core absent.

Species 72 (8 in the flora): North America, Eurasia, n Africa, Atlantic Islands; introduced in Pacific Islands (New Zealand).

Four of the 11 recognized sections of subg. *Sorbus* occur in the flora area.

1. Leaflets minutely, densely papillose abaxially (at 70×); carpels in fruit distinct in distal ½, apices convex; winter buds densely hairy, hairs usually whitish, rarely rufous, not glutinous . . . . 53c.1. *Sorbus* sect. *Sorbus*, p. 437
1. Leaflets not papillose abaxially; carpels in fruit connate or distinct, apices flat or convex; winter buds glabrous, glutinous, or hairy, hairs usually rufous, sometimes whitish.
    2. Winter buds mostly glutinous; carpel apices flat or convex . . . . 53c.2. *Sorbus* sect. *Commixtae*, p. 439
    2. Winter buds never strongly glutinous; carpel apices convex.
        3. Leaflets not glaucous adaxially; flowers: perianth and androecium epigynous; carpels connate entire lengths; pomes red, not glaucous . . . . 53c.3. *Sorbus* sect. *Sambucifoliae*, p. 443
        3. Leaflets somewhat glaucous adaxially; flowers: perianth and androecium semiepigynous; carpels distinct most of lengths; pomes pinkish red, glaucous. . . . . 53c.4. *Sorbus* sect. *Tianshanicae*, p. 443

53c.1. Sorbus Linnaeus (subg. Sorbus) sect. Sorbus [I]

Winter buds ovoid to ovoid-conic, dull, not glaucous, not glutinous, densely hairy, hairs usually whitish, rarely rufous. **Leaflets** minutely, densely papillose abaxially (at 70×), not glaucous adaxially. **Flowers:** perianth and androecium semiepigynous. **Pomes** red to orange; sepals persistent; carpels distinct in distal ½, apex convex.

Species 4 (2 in the flora): introduced; Eurasia, n Africa, Atlantic Islands (Iceland, Madeira); introduced also in Pacific Islands (New Zealand).

3. Sorbus aucuparia Linnaeus, Sp. Pl. 1: 477. 1753 • Rowan, European mountain ash, sorbier des oiseleurs, sorbier des oiseaux [I]

Pyrus aucuparia (Linnaeus) Gaertner; *Sorbus subvestita* Greene

Trees, 60–150 dm. **Stems** 1–3(or 4); bark gray to bronze; winter buds purple, ovoid to ovoid-conic, 5–15 mm, dull, not glutinous, usually densely villous, hairs whitish, rarely rufous, <outer scales sometimes merely ciliate>. **Leaves** pinnately compound; stipules deciduous or persistent, whitish-villous or glabrate; blade paler or often whitish abaxially, dull green to bluish green adaxially, leaflets 11–17(–19), opposite, oblong to oblong-lanceolate or oblanceolate, (2.5–)3–6(–7.5) × 1.5–2 cm, l/w ratio 2.4–3.7, margins serrate at least in distal ½, often almost to base, apex acute to obtuse, abaxial surface white-tomentose to villous at flowering, indument often persisting, at least along midveins, leaflet axils, and petiole bases, hairs usually whitish, sometimes rufous. **Panicles** 75–200+-flowered, flat-topped or rounded, 6–18 cm diam.; peduncles ± whitish-villous. **Pedicels** densely whitish-villous. **Flowers** 8–11 mm diam.; hypanthium densely villous, <hairs whitish>, hypanthium plus sepals 2.9–3.2 mm; sepals 0.5–1 mm, margins often glandular; petals white, orbiculate, (3–)4–5 mm; stamens 15–20; carpels ½ adnate to hypanthium, apex conic, styles 3 or 4, 1.5–3 mm. **Infructescences** glabrous or densely villous. **Pomes** yellow or orange-red to red, globose to subglobose, 8–12 mm diam., shiny or dull, <slightly glaucous or not>; sepals inconspicuous, incurved. **Seeds** brown, ovoid to ovoid-lanceoloid, 3–4.5 × 1.5–2 mm, asymmetric, slightly flattened. $2n = 34$.

Flowering spring; fruiting fall. Thickets, roadsides, woods, other habitats near towns, scattered in remote natural settings; 0–2100 m; introduced; St. Pierre and Miquelon; Alta., B.C., N.B., Nfld. and Labr. (Nfld.), N.S., Ont., P.E.I., Que., Sask.; Alaska, Calif., Colo., Conn., D.C., Idaho, Ill., Ind., Iowa, Maine, Md., Mass., Mich., Minn., Mont., Nebr., N.H., N.J., N.Y., N.Dak., Ohio, Oreg., Pa., R.I., S.C., S.Dak., Utah, Vt., Wash., W.Va., Wis., Wyo.; Eurasia; n Africa; Atlantic Islands (Iceland, Madeira); introduced also in Pacific Islands (New Zealand).

Sorbus aucuparia is a commonly planted ornamental tree that is widely bird-disseminated. The species is often confused with *S. decora*, from which it differs by its ovoid and usually densely whitish-villous and nonglutinous winter buds and by its densely villous flowering pedicels and hypanthia. The leaflets of *S. aucuparia* are often tomentose abaxially at flowering; those of *S. decora* are glabrous or sparsely villous. Examination of leaflet serration is also useful; *S. decora* exhibits a prolonged apical tooth; it is no longer than the lateral ones in *S. aucuparia*. The lateral teeth are relatively finer and sharper in *S. decora* compared to those of *S. aucuparia*. At a magnification of at least 70×, leaflets of *S. aucuparia* are reportedly minutely papillose abaxially; those of *S. decora* are not papillose (H. A. McAllister 2005). Natural hybrids with *S. decora* have been reported by G. N. Jones (1939) (one fruiting specimen found near Opeongo Lake, Algonquin Park, Ontario), with *Aronia arbutifolia* [×*Sorbaronia hybrida* (Moench) C. K. Schneider; synonyms *A. hybrida* (Moench) Zabel, *Sorbus spuria* Persoon] in Truro, Nova Scotia, and also in Maine, New Hampshire, and Vermont (D. W. Magee and H. E. Ahles 1999), and with *A. melanocarpa* (×*Sorbaronia fallax* C. K. Schneider) in Maine and Massachusetts. Other nothospecies names for putative intergeneric hybrids involving *S. aucuparia* include: ×*Crataegosorbus miczurinii* Pojarkova (*Crataegus sanguinea* Pallas × *S. aucuparia*), and ×*Sorbocotoneaster pozdnjakovii* Pojarkova [*Cotoneaster melanocarpus* (Ledebour) Loddiges, G. Loddiges & W. Loddiges ex M. Roemer × *S. aucuparia* (as *S. sibirica* Hedlund)]. Subspecies of *S. aucuparia* were recognized by McAllister; the authors made no attempt to identify specimens infraspecifically. *Sorbus subvestita* Greene is variably considered to be a synonym of *S. aucuparia* (Jones) or *S. decora* (M. L. Fernald 1950; H. A. Gleason and A. Cronquist 1991). However, Jones explicitly cited the isotype of *S. subvestita* in his specimen citation for *S. aucuparia* along with an acknowledgement of the confusion with *S. decora*.

4. Sorbus hybrida Linnaeus, Sp. Pl. ed. 2, 1: 684. 1762 • Oak-leaf mountain ash [I]

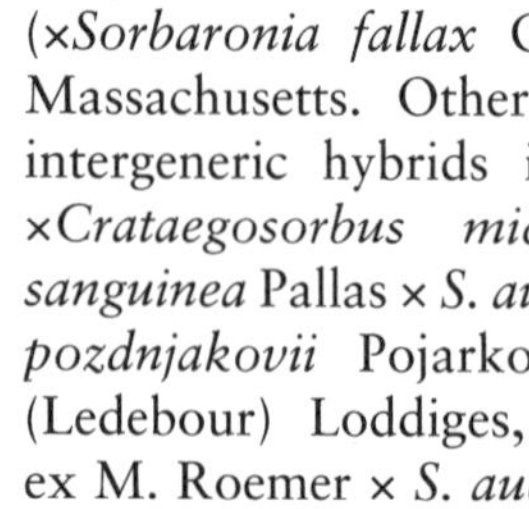

Trees, 30–120 dm. **Stems** single or multi-stemmed; bark gray; winter buds brownish to purple, ovoid to ovoid-conic, 4–10 mm, dull, not glutinous, densely white-villous, at least distally and on scale margins. **Leaves** proximally pinnately compound, lobed distally; stipules deciduous or persistent, often white-villous; blade paler green abaxially, dull green adaxially, ovate to oblong-ovate, rarely oblong, 7–13 × 5–11 cm, pinnately lobed with (1 or)2(or 3) pairs of sessile or decurrent, free, oblong leaflets, terminal leaflets 7–10-lobed, margins serrate at least distally, <lobe and leaflet apex acute or obtuse, lateral veins 7–10 pairs>, abaxial surface whitish-tomentose. **Panicles** 25–75-flowered, flat-topped or rounded, 4.5–10 cm diam.; peduncles white-villous. **Pedicels** white-villous. **Flowers** 10–15 mm diam.; hypanthium tomentose, hypanthium plus sepals 5–7.5 mm; sepals 1.5–4 mm, margins usually entire, rarely with inconspicuous glands; petals white, suborbiculate, broadly obovate, broadly ovate, or broadly elliptic, 5–7 mm; stamens 20; carpels ½ adnate to hypanthium, apex conic, styles 2 or 3, 2–3 mm. **Infructescences** glabrate to villous. **Pomes** red, globose to subglobose, 8–15 mm diam., shiny, not glaucous; sepals inconspicuous, incurved. **Seeds** red-brown, ovoid, 5–6 × 2 mm, asymmetric, slightly flattened. **2*n*** = 68.

Flowering spring; fruiting fall. Woods, rocky slopes, disturbed ground and edges near towns; 0–1000 m; introduced; N.B.; Utah, Vt., Wash.; n Europe.

Sorbus hybrida is considered an apomictic tetraploid producing fully fertile pollen and seed and subject to dispersal. An almost undistinguishable European taxon, *S.* ×*thuringiaca* (Ilse) Fritsch, often confused with *S. hybrida*, has been reported for New Hampshire and Vermont. Described as an unstable diploid hybrid between *S. aria* and *S. aucuparia*, *S.* ×*thuringiaca* has oblong leaves, one to three (to five) pairs of free leaflets, ten to 12 (or 13) pairs of lateral veins, and produces sterile pollen and relatively few viable seeds. Uncritical reports of escaped *S.* ×*thuringiaca* in North America should be considered doubtful; they probably refer to the more likely *S. hybrida*. The authors have not seen a specimen to document a report of *S. hybrida* from Montana.

Sorbus hybrida is placed in subg. *Sorbus* following J. J. Aldasoro et al. (2004), based on leaf division and pome characteristics, including the lack of tanniferous cells in parenchyma, lack of starch, and relatively small sclereid groups.

53c.2. SORBUS Linnaeus (subg. SORBUS) sect. COMMIXTAE McAllister, Gen. Sorbus, 236. 2005

Winter buds ovoid to conic, usually shiny, occasionally glaucous, mostly glutinous, glabrous or hairy, hairs whitish or rufous. **Leaflets** not papillose abaxially nor glaucous adaxially. **Flowers:** perianth and androecium semiepigynous. **Pomes** red to orange; sepals persistent; carpels distinct in distal ½, apex flat or convex.

Species 6 (4 in the flora): North America, e Asia (Japan, Russian Far East [Kuril Islands, Sakhalin], Taiwan).

5. **Sorbus americana** Marshall, Arbust. Amer., 145. 1785 • American mountain ash, sorbier d'Amérique E F

Pyrus americana (Marshall) Sprengel

Shrubs or trees, 40–100 dm. **Stems** 1–4+; bark gray to bronze; winter buds green to purple, shiny, ovoid to conic, 7–20 mm, glutinous, glabrous or hairy along scale margins and at apex, hairs primarily rufous. **Leaves** pinnately compound; stipules early deciduous, rufous-ciliate, <margins sometimes glandular>; blade paler abaxially, dull green to yellowish green adaxially, leaflets 11–17, opposite, lanceolate to oblong-lanceolate, 5–10 × 1–2.5 cm, l/w ratio 3.4–5, margins serrulate to serrate, at least in apical ½ and often almost to base, apex acuminate to long acuminate, surfaces glabrous or glabrate at flowering, some hairs occasionally persisting abaxially along midvein; leaflet axils and petiole bases glabrous or with rufous and/or whitish hairs adaxially. **Panicles** 125–400+-flowered, flat-topped or rounded, 6–15 cm diam.; peduncles glabrous or sparsely villous. **Pedicels** glabrous or sparsely villous. **Flowers** 5–7.5 (–8.5) mm diam.; hypanthium glabrous, hypanthium plus sepals 2–2.5 mm; sepals 0.5–1 mm, margins entire, often with a few thick glands; petals white, orbiculate to obovate, (2–)2.5–3.5(–4) mm; stamens 14–20; carpels ½ adnate to hypanthium, apex conic, styles 3 or 4, 1.5–2 mm. **Infructescences** glabrous or sparsely villous. **Pomes** bright red to orange-red, globose to subglobose, 4–7 mm diam., shiny, not glaucous; sepals inconspicuous, incurved. **Seeds** brown, yellowish when fresh or immature, ovoid to ovoid-lanceoloid, 2.5–3.3 × 1.5–2 mm, asymmetric, slightly flattened. **2*n*** = 34.

Flowering spring; fruiting fall. Cool, moist woods, lake and stream shores, rocky hillsides, thickets; 0–1300 m; St. Pierre and Miquelon; N.B., Nfld. and Labr. (Nfld.), N.S., Ont., P.E.I., Que.; Conn., Ga., Ill., Maine, Md., Mass., Mich., Minn., N.H., N.J., N.Y., N.C., Pa., R.I., S.C., Tenn., Vt., Va., W.Va., Wis.

Sorbus americana reputedly flowers about one week earlier than *S. decora*. Some authors have described *S. americana* as having generally thinner leaflets relative to thicker and firmer ones for *S. decora*; examination of herbarium specimens appears to support this, though the authors made no systematic study of blade thickness. No confirmed putative hybrids with *S. decora* are reported, although hybridization may occur. The hybrid ×*Sorbaronia jackii* Rehder [*Aronia* ×*prunifolia* (Marshall) Rehder × *S. americana*; synonym *Pyrus* ×*jackii* (Rehder) Fernald] has been reported from New Brunswick as a fastigiate shrub 20 dm tall with leaves intermediate in appearance between the putative parents.

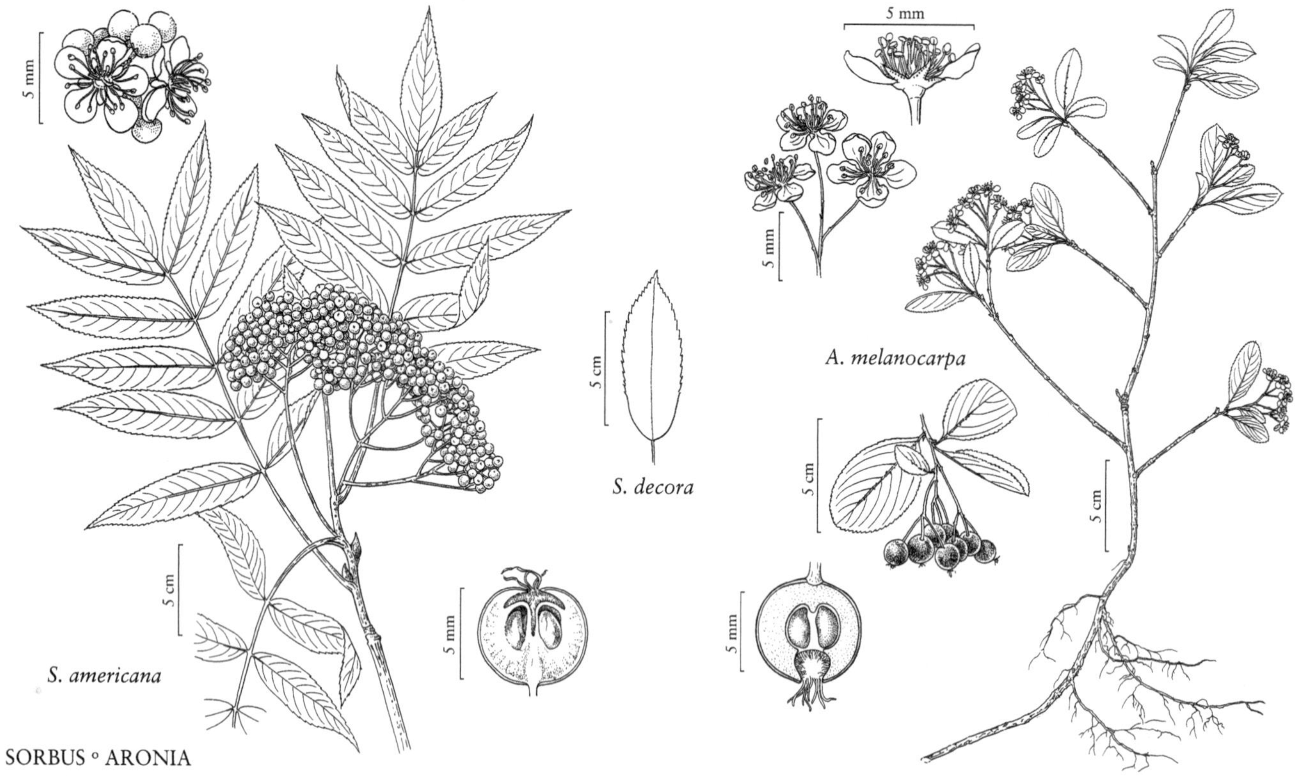

SORBUS ◦ ARONIA

Garden hybrids involving *S. americana* are occasionally cultivated in North America; none has been reported as escaped. These include: *Sorbus* ×*splendida* Hedlund (a hybrid with *S. aucuparia*), *S.* ×*plantierensis* Simon-Louis ex C. K. Schneider (with *S. aria*), ×*Sorbaronia monstrosa* (Zabel) C. K. Schneider (with *A. arbutifolia*), and ×*Sorbaronia sorbifolia* (Poiret) C. K. Schneider (with *A. melanocarpa*); synonyms of ×*Sorbaronia sorbifolia* include *A. sargentii* (Dippel) Zabel, *A. watsoniana* M. Roemer, and *P.* ×*mixta* Fernald.

6. **Sorbus decora** (Sargent) C. K. Schneider, Bull. Herb. Boissier, sér. 2, 6: 313. 1906 • Northern mountain ash, sorbier plaisant [E] [F]

Pyrus americana (Marshall) Sprengel var. *decora* Sargent, Silva 14: 101. 1902; *P. decora* (Sargent) Hyland; *P. decora* var. *groenlandica* (C. K. Schneider) Fernald; *Sorbus decora* var. *groenlandica* (C. K. Schneider) G. N. Jones; *S. groenlandica* (C. K. Schneider) Á. Löve & D. Löve

Shrubs or trees, 30–150 dm. **Stems** 1–4+; bark gray to bronze; winter buds purplish at maturity, shiny, conic, 10–20 mm, glutinous, scales ciliate to moderately (densely) villous, hairs primarily rufous. **Leaves** pinnately compound; stipules deciduous, sometimes persistent, glabrous or with whitish and/or rufous hairs; blade paler abaxially, dull green to bluish green adaxially, leaflets (11–)13–17, opposite, elliptic, oblong to oblong-ovate, 4–7(–8) × 1.5–2.5 cm, l/w ratio 2.4–3.2(–3.6), margins serrate, sometimes serrulate, at least in distal ½ and often almost to base, apex abruptly short-acuminate, cuspidate, acute, or obtuse, surfaces essentially glabrous or sparsely (moderately) villous at flowering, usually glabrescent thereafter, sometimes persistently villous along midveins abaxially; leaflet axils and petiole bases glabrous or with some rufous and/or whitish hairs adaxially. **Panicles** 75–400+-flowered, flat-topped or rounded, 6–15 cm diam.; peduncles sparsely to moderately (densely) villous. **Pedicels** sparsely to moderately (densely) villous. **Flowers** (7–)8–12 mm diam.; hypanthium glabrous or sparsely villous proximally, hairs whitish or rufous, hypanthium plus sepals 2.5–3.5 mm; sepals 1–1.5(–2) mm, margins entire, often with a few thick glands; petals white, orbiculate to obovate, (3–)3.5–5 mm; stamens 15–20; carpels ½ adnate to hypanthium, apex conic, styles 3 or 4, 1.5–2.5 mm. **Infructescences** sparsely to moderately villous, rarely glabrous. **Pomes** bright red, globose to subglobose, (5–)7–11 mm diam., shiny or dull, often glaucous when dried; sepals inconspicuous, incurved. **Seeds** brown, lanceoloid, 4.5–5 × 1.5–2 mm, slightly asymmetric, slightly flattened. **2*n*** = 68.

Flowering spring; fruiting fall. Moist or dry woods, montane woods, rocky slopes, lake and stream shores, thickets; 0–1300 m; Greenland; St. Pierre and Miquelon; Man., N.B., Nfld. and Labr., N.S., Nunavut, Ont., P.E.I., Que., Sask.; Conn., Ill., Ind., Iowa, Maine, Mass., Mich., Minn., N.H., N.Y., Ohio, Pa., Vt., Wis.

Sorbus decora reputedly flowers about one week later than *S. americana*. The two are sometimes confused when one is not familiar with the smaller flowers and fruits, and the distinctly longer and notably acuminate leaflets of *S. americana*. Leaflet length/width ratios greater than 3.2 are sometimes observed on narrower leaflets of *S. decora*, and a range of representative leaflets need to be sampled before determination. Another species frequently confused with *S. decora* is *S. aucuparia*, characterized by ovoid, usually densely whitish-villous and not glutinous winter buds as well as densely villous flowering pedicels and hypanthia, and often abaxially tomentose leaflets at flowering. The teeth on margins of leaves in *S. decora* are more finely pointed and sharper than those of *S. aucuparia*, and the apical tooth is often distinctly prolonged in *S. decora*; it is no more conspicuous than the lateral ones in *S. aucuparia* (E. G. Voss 1972–1996, vol. 2). The leaflets of *S. aucuparia* are reported to be minutely papillose abaxially (visible at 100×); those of *S. decora* and other members of sect. *Commixtae* are not papillose (H. A. McAllister 2005). *Sorbus decora* is reported to be extirpated from Indiana.

Chromosome counts of 2*n* = 34 were reported by Á. Löve and D. Löve (1965, 1982b) for *Sorbus decora*. Doubt exists about the identity of the material that they sampled, as the only voucher specimen retraced (*Á. & D. Löve 6559*, COLO, WIN), although only vegetative and juvenile, is nonetheless referable to *S. aucuparia*. These doubtful chromosome counts also formed the basis for the recognition of *S. groenlandica* (Löve and Löve 1965b), reputedly of Greenland, coastal Labrador, and alpine and subalpine regions of northeastern North America, which apparently differed from *S. decora* with a chromosome count of 2*n* = 68 (T. W. Böcher and K. Larsen 1950; K. Holmen in C. A. Jørgensen et al. 1958). Reliable *S. decora* chromosome counts by H. A. McAllister (2005) of three wild collections (Dundas, Ontario; St. Anne de Beaupré, Quebec; Narsarsuaq, Greenland) and two of cultivated material rendered this proposition untenable with consistent counts of 2*n* = 68. Further, critical examination of herbarium specimens from Greenland yielded no consistent morphologic character to separate *S. decora* from *S. groenlandica*, though leaflet and fruit size in the latter tended to be relatively smaller.

A naturally occurring putative intergeneric hybrid, ×*Sorbaronia arsenei* (Britton ex L. Arsène) G. N. Jones, synonym *Pyrus* ×*arsenei* (Britton ex L. Arsène) L. Arsène [= *Aronia* ×*prunifolia* (Marshall) Rehder × *Sorbus decora*], is known from St. Pierre and Miquelon, New Brunswick, Nova Scotia, Prince Edward Island, Quebec, and, possibly, Newfoundland. The hybrid is a 10–25 dm shrub with finely serrulate, partially pinnate leaves containing (zero or) one to three pairs of reduced, distinct leaflets proximally and a much larger, one- to three-lobed terminal segment. The midvein on the adaxial surface of the terminal leaflet may be irregularly glandular or eglandular. A possible variant of this hybrid (parentage unconfirmed), much closer to *Sorbus* in appearance, is found in southwestern Newfoundland. It possesses finely serrulate, almost fully pinnate leaves, often with four or five pairs of distinct leaflets, the distal first or second leaflet pairs more or less pinnatifid or partially fused to the terminal leaflet. The hypanthium and abaxial surface of the sepals are villous, the anthers are pink or red, and the pomes are usually red, though some are reported to be blackish in age, often with some tomentum persisting on the abaxial surface of the persistent sepals.

7. **Sorbus californica** Greene, Pittonia 4: 131. 1900 • California mountain ash [E]

Sorbus sitchensis M. Roemer subsp. *californica* (Greene) Abrams

Shrubs, 10–20(–40) dm. **Stems** 1–8; bark dark gray; winter buds red to brown, shiny, conic, 5–12 mm, slightly glutinous, sparsely hairy except along scale margins and apex, hairs rufous. **Leaves** pinnately compound; stipules often persistent, margins rufous-hairy; blade pale abaxially, shiny, green adaxially, leaflets 7–9(–11), opposite or subopposite, oblong to elliptic, sometimes narrowly elliptic, 2.5–4.1(–4.5) × 0.9–2 cm, l/w ratio 2.1–2.6, margins finely to coarsely serrate, sometimes doubly serrate, apex obtuse to acute, surfaces soon glabrous, leaflet axils and petiole bases hairy, hairs rufous. **Panicles** 25–120+-flowered, rounded, 3–11 cm diam.; peduncles glabrous or sparsely hairy. **Pedicels** glabrous or sparsely hairy; <fruiting pedicels essentially glabrous>. **Flowers** 9–10 mm diam.; hypanthium glabrous, hypanthium plus sepals 3 mm; sepals 0.7–1.5 mm, margins ciliolate, hairs whitish, irregularly glandular; petals white, broadly ovate, 3–4 mm; stamens 20; carpels distinct, apex conic, styles 3 or 4, 1.5–2 mm. **Infructescences** glabrous or nearly so. **Pomes** bright red, globose to subglobose, 6–9 mm diam., shiny if glaucous layer is rubbed off; sepals inconspicuous, incurved. **Seeds** red-brown, lanceolate, 4 × 2 mm, slightly asymmetric, slightly flattened.

Flowering spring; fruiting fall. Mountain slopes, meadows, stream banks, lakeshores, sun or shade; 1500–3400 m; Calif., Nev., Oreg.

Sorbus californica is recognized here in the narrow sense, occurring north to Crater Lake National Park, Oregon (P. F. Zika 2003). The concept in H. A. McAllister (2005) differs, including plants of British Columbia and Washington treated here as *S. scopulina*. *Sorbus californica* is distinct in its shiny leaflets less than 4 cm with rufous hairs in the axils. Where their ranges overlap, *S. scopulina* has hairier inflorescences, leaflets usually more than 4.5 cm, and whitish axillary hairs. The small leaflets and essentially glabrous inflorescences help distinguish *S. californica* from putative hybrids between *S. scopulina* and *S. sitchensis* found farther north, which are otherwise similar with their somewhat shiny leaflets, slightly glaucous fruits, and red axillary hairs. Some collections of *S. californica* from California are intermediate with either *S. sitchensis* or *S. scopulina* and may represent hybrids with those two species.

8. **Sorbus scopulina** Greene, Pittonia 4: 130. 1900 • Rocky Mountain or Cascade or Greene's mountain ash [E]

Sorbus andersonii G. N. Jones; *S. angustifolia* Rydberg; *S. cascadensis* G. N. Jones; *S. dumosa* Greene; *S. scopulina* var. *cascadensis* (G. N. Jones) C. L. Hitchcock

Shrubs, 10–50(–80) dm. **Stems** 1–8; bark gray, sometimes yellowish or reddish purple, or grayish red; winter buds olive brown to red-brown, conic, 8–14 mm, shiny, slightly glutinous, glabrous or sparsely or densely whitish-villous. **Leaves** pinnately compound; stipules deciduous or persistent, hairs whitish; blade paler abaxially, shiny, green to dark green adaxially, leaflets 7–13(–15), opposite or subopposite, lanceolate, oblong, narrowly ovate, oblanceolate, or obovate, (3–)4–6.5(–8.7) × 1.5–2.7 cm, l/w ratio 2.1–4.4, margins finely serrate, apex obtuse, acute, or acuminate, surfaces glabrous, leaf and leaflet axils hairy, hairs whitish. **Panicles** 40–200+-flowered, flat-topped to rounded, 5–15 cm diam.; peduncles sparsely to densely whitish-villous. **Pedicels** sparsely to densely whitish-villous (in flower and fruit). **Flowers** 11–13 mm diam.; hypanthium nearly glabrous or densely villous, hypanthium plus sepals 3–4 mm; sepals 0.8–2.3 mm, margins lightly to densely whitish-villous, glands absent or sparse and usually inconspicuous; petals white, ovate, 4–6 mm; stamens (14–)20; carpels distinct, apex conic, styles 3 or 4, 2–2.5 mm. **Infructescences** whitish-villous, rarely glabrous with age. **Pomes** bright orange to reddish orange, globose, subglobose, broadly obovoid, or broadly elliptic, 8–12.1 × 7.5–12.8 mm, shiny, sometimes lightly glaucous; sepals inconspicuous, incurved. **Seeds** brown, narrowly ovoid, 3.8–5.4 × 1.7–2.8 mm, slightly asymmetric, slightly flattened. **2*n*** = 34, 68.

Flowering spring; fruiting fall. Mountain slopes, open forests, forest edges, riparian zones, lakeshores; 0–3300 m; Alta., B.C., N.W.T., Sask., Yukon; Alaska, Ariz., Calif., Colo., Idaho, Mont., Nev., N.Mex., Oreg., S.Dak., Utah, Wash., Wyo.

Sorbus scopulina is variable, especially in leaflet shape, number, and indument. Plants with narrower leaflets were separated as *S. angustifolia*; plants with broader leaflets were segregated as *S. andersonii* and *S. cascadensis*. Densely hairy forms were named *S. dumosa*. All represent points on a morphologic continuum, united by their shiny leaflets, whitish indument, and western range. Sparsely hairy forms approach the more southern *S. californica* but have whitish axillary hairs and larger leaflets than that species, which has rufous axillary hairs and leaflets less than 4 cm. Most *S. scopulina* have shiny hypanthia and fruits. Occasional collections, all within the range of *S. sitchensis*, are glaucous-fruited. This may be natural variation, or it could represent introgression from *S. sitchensis*. H. A. McAllister (2005, pers. comm.) found that individuals of *S. scopulina* from cordilleran Idaho and Arizona were sexual diploids; plants from the Cascade and Coast ranges of British Columbia and Washington were tetraploid and presumably apomictic. Here, these entities are synonymized, awaiting a larger study that might correlate distinct morphologic characteristics with the ploidy levels. J. J. Aldasoro et al. (1998, 2004) noted that some European *Sorbus* species are both diploid and tetraploid, suggesting that sexual species occasionally produce spontaneous agamospermous individuals.

The intergeneric hybrid ×*Amelasorbus jackii* Rehder [*Amelanchier alnifolia* (Nuttall) Nuttall ex M. Roemer × *Sorbus scopulina*] has been collected in Idaho and Oregon. Foliage of the Idaho collection is very similar to that of *Amelanchier*, with basal pinnae inconspicuous and infrequent, suggesting that this hybrid may be overlooked.

The illegitimate later homonym, *Sorbus alaskana* G. N. Jones, pertains here.

53c.3. Sorbus Linnaeus (subg. Sorbus) sect. Sambucifoliae McAllister, Gen. Sorbus, 237. 2005

Winter buds conic, shiny, not glaucous, glutinous, sparsely hairy, hairs primarily rufous. **Leaflets** not papillose abaxially nor glaucous adaxially. **Flowers:** perianth and androecium epigynous. **Pomes** red, <not glaucous>; sepals persistent; carpels connate entire length, apex convex.

Species 1: Alaska, e Asia (Japan, Russian Far East [Commander Islands, Kamchatka, Kuril Islands, Sakhalin]).

9. Sorbus sambucifolia (Chamisso & Schlechtendal) M. Roemer, Fam. Nat. Syn. Monogr. 3: 139. 1847 • Aleutian mountain ash

Pyrus sambucifolia Chamisso & Schlechtendal, Linnaea 2: 36. 1827

Shrubs, 10–20 dm. **Stems** 1–8; bark gray; winter buds blackish red to black, conic, 7–18 mm, shiny, slightly glutinous, sparsely villous distally, hairs primarily rufous. **Leaves** pinnately compound; stipules sometimes persistent, margins and surface hairy, hairs rufous; blade paler abaxially, shiny, green to dark green adaxially, leaflets 7–11, opposite or subopposite, lanceolate, sometimes narrowly elliptic, elliptic, oblong, or oblanceolate, 2.5–6.5 × 0.9–2.5 cm, l/w ratio 1.9–3.8, margins serrate, apex usually acute to acuminate or attenuate, sometimes obtuse, surfaces soon glabrous; leaf and leaflet axils hairy, hairs rufous. **Panicles** 6–23-flowered, rounded, 2.5–7 cm diam.; peduncles sparsely rufous-villous. **Pedicels** sparsely rufous-villous. **Flowers** 10–15 mm diam.; hypanthium glabrous, hypanthium plus sepals (4.5–)5–6 mm; sepals 1.5–2.9 mm, margins usually villous, hairs white and reddish mixed [white], rarely sparsely glandular; petals white to pinkish, orbiculate or ovate, 4–5 mm; stamens 20; carpels almost completely adnate to hypanthium, apex truncate or slightly conic, styles 4 or 5, 1.5–3.5 mm. **Infructescences** glabrous. **Pomes** red, ellipsoid, 10–12 × 7–9 mm, shiny, not glaucous; sepals prominent, erect. **Seeds** dark brown, lanceoloid, 4 × 2 mm, symmetric, not flattened. **2*n*** = 34 (Asia).

Flowering spring; fruiting fall. Slopes and meadows; 0–100 m; Alaska; e Asia (Japan, Russian Far East [Commander Islands, Kamchatka, Kuril Islands, Sakhalin]).

In North America, *Sorbus sambucifolia* is known only from the four westernmost islands of the Aleutians, where it is allopatric from all other *Sorbus*. *Sorbus scopulina* is known from farther east in the Aleutians; sterile specimens differ in their whitish, not rufous indument, and the less deeply sunken reticulate venation on the adaxial surfaces of the leaflets.

53c.4. Sorbus Linnaeus (subg. Sorbus) sect. Tianshanicae (Komarov ex T. T. Yu) McAllister, Gen. Sorbus, 237. 2005

Sorbus ser. *Tianshanicae* Komarov ex T. T. Yu, Acta Phytotax. Sin. 8: 222. 1963

Winter buds ovoid to conic, dull, often glaucous, not glutinous, sparsely or densely hairy, hairs primarily rufous. **Leaflets** not papillose abaxially [exceptionally papillose], ± glaucous adaxially. **Flowers:** perianth and androecium semiepigynous. **Pomes** pinkish red, <glaucous [or not]>; sepals persistent; carpels distinct most of length, apex convex.

Species 3 (1 in the flora): w North America, Asia.

10. Sorbus sitchensis M. Roemer, Fam. Nat. Syn. Monogr. 3: 139. 1847 [E]

Shrubs, 10–40 dm. **Stems** 1–8; bark initially reddish purple to brownish gray, becoming grayish red; winter buds red-purple to red-brown, conic to ovoid, 8–13 mm, dull, <slightly glaucous>, not glutinous, sparsely or densely villous, hairs primarily rufous. **Leaves** pinnately compound; stipules persistent, sometimes deciduous, hairs rufous; blade paler abaxially, dull blue-green, usually slightly glaucous, rarely faintly shiny adaxially, leaflets 7–13, opposite, sometimes subopposite, oblong to narrowly elliptic, sometimes oblanceolate, ovate, or obovate, 1.8–6 × 0.6–2.5 cm, l/w ratio 1.9–3.5, margins entire or finely to coarsely serrate, sometimes doubly serrate, apex obtuse or acute, surfaces soon glabrous, leaf and leaflet axils hairy, hairs rufous. **Panicles** 12–80-flowered, rounded, 2–8 cm diam.; peduncles glabrous, glaucous, or sparsely to densely villous, hairs primarily rufous. **Pedicels** glabrous, glaucous, or sparsely to densely villous, hairs primarily rufous. **Flowers** 11–17 mm diam.; hypanthium glabrous, <glaucous>, hypanthium plus sepals 3–4 mm; sepals 1–2.2 mm, margins lightly to moderately ciliate, <hairs rufous or whitish>, infrequently glandular; petals white, rarely pinkish, rhombic to ovate, 4–7.5 mm; stamens 15–20; carpels distinct, apex concave, depressed, or slightly conic and truncate, styles 3–5, 1.5–3.2 mm. **Infructescences** glabrous, glaucous, or villous. **Pomes** pinkish red or red (often appearing slightly purplish), subglobose to broadly elliptic or obovoid, 7–13 × 7–13 mm, dull, glaucous; sepals inconspicuous, incurved. **Seeds** red-brown, ovoid to ovoid-lanceolate, oblong or elliptic, 3.4–5 × 2.2–2.9 mm, symmetric or trigonous, slightly flattened.

Varieties 2 (2 in the flora): w North America.

The other western *Sorbus* with rufous indument in the leaf and leaflet axils is *S. californica*, which differs in its smaller, shinier leaflets, less hairy inflorescences, and shiny winter buds. The winter buds of *S. sitchensis* are hairier and less scaly at the summit, also glaucous and never shiny or glutinous. In the field, the fresh fruits of *S. sitchensis* are often pinkish red; fruits of *S. californica* and *S. scopulina* are generally orange-red.

1. Leaflet margins toothed apically, serrate for more than ½ their length; petals white . . . . 10a. *Sorbus sitchensis* var. *sitchensis*
1. Leaflet margins entire or toothed apically, serrate for less than ½ their length; petals white or pinkish . . . . 10b. *Sorbus sitchensis* var. *grayi*

10a. Sorbus sitchensis M. Roemer var. **sitchensis** • Sitka mountain ash [E]

Leaflets: l/w ratio 2.2–3.1, margins toothed apically, serrate for more than ½ their length, serrations fine to coarse, sometimes doubly serrate. **Flowers:** petals white, pink or white in bud, claw glabrous or villous; styles 1.5–2.2(–3) mm. $2n = 68$.

Flowering spring; fruiting fall. Open conifer woods, forest edges, mountain slopes, meadows, shores of lakes and streams; 0–2400 m; Alta., B.C., Yukon; Alaska, Calif., Idaho, Mont., Oreg., Wash.

10b. Sorbus sitchensis M. Roemer var. **grayi** (Wenzig) C. L. Hitchcock in C. L. Hitchcock et al., Vasc. Pl. Pacif. N.W. 3: 189. 1961 • Dwarf or western mountain ash [E]

Sorbus sambucifolia (Chamisso & Schlechtendal) M. Roemer var. *grayi* Wenzig, Bot. Centralbl. 35: 342. 1888; *Pyrus occidentalis* S. Watson; *S. occidentalis* (S. Watson) Greene; *S. sitchensis* subsp. *grayi* (Wenzig) Calder & Roy L. Taylor

Leaflets: l/w ratio 1.9–3.5, margins entire or toothed apically, serrate for less than ½ their length, serrations fine to coarse. **Flowers:** petals white or pinkish, white or pink in bud, claw villous; styles (2–)2.4–3.2 mm. $2n = 34$.

Flowering spring (late summer); fruiting fall. Mountain slopes, forest edges, rockslides, edges of avalanche tracks, peatlands, muskegs, meadows; 0–3000 m; B.C.; Alaska, Calif., Oreg., Wash.

Variety *grayi* often has smaller leaflets than var. *sitchensis*, and plants near timberline frequently are dwarfed by exposure. It is usually found at higher elevations or on boggier soils than var. *sitchensis*. The rank is controversial; it was treated by G. N. Jones (1939) and H. A. McAllister (2005) as a species (*Sorbus occidentalis*). The only reliable character to segregate it is the extent of toothing on the leaf margins, which is variable on some individuals, with nearly entire leaflets on shoots exposed to sun and with leaflet margins to 50% toothed on shaded basal shoots.

Variety *grayi* occurs south to Crater Lake National Park in Klamath County, Oregon (P. F. Zika 2003). Most collections and reports from California (T. J. Rosatti 1993) were misidentified, except for a single sheet from Siskiyou County in the Klamath Mountains (*R. White 207*, HSC). One of the synonyms, *Pyrus occidentalis*, was described with several syntypes from California, which are not var. *grayi* but *Sorbus californica*. G. N. Jones (1939, p. 40) effectively lectotypified *P. occidentalis* with a syntype from the Canadian border (*Lyall s.n.*, GH), which has nearly entire leaflets and is var. *grayi*.

54. ARONIA Medikus, Philos. Bot. 1: 140, 155. 1789, name conserved • Chokeberry [Greek *Aria*, name for whitebeam (formerly a species of *Sorbus*), alluding to resemblance to chokeberry fruit] [E]

Richard J. Pankhurst†

Adenorachis (de Candolle) Nieuwland; *Pyrus* Linnaeus sect. *Adenorachis* de Candolle; *Sorbus* Linnaeus sect. *Aronia* (Medikus) C. K. Schneider

Shrubs, 8–20 dm; <suckering>. **Stems** 1–20+, erect; bark gray or brown, <smooth>; short shoots absent; unarmed; appressed-pilose, glabrous, or glabrescent. **Leaves** deciduous, cauline, simple; stipules persistent, adnate to petiole, narrowly triangular, margins glandular; petiole present; blade elliptic to obovate, 2.5–7.5(–18) cm, membranous, margins flat, glandular serrulate-dentate, venation pinnate, surfaces glabrous or glabrescent to pilose (or villous). **Inflorescences** lateral and apparently terminal, 5–12(–20)-flowered, corymbose, appressed pilose; bracts reduced to glands; bracteoles reduced to glands. **Pedicels** present. **Flowers:** perianth and androecium epigynous, 12–20 mm diam.; hypanthium campanulate, 1–2 mm, glabrous or villous; sepals 5, erect, triangular; petals 5, white to pale pink, elliptic to orbiculate, <base clawed>; stamens 16–22, equal to petals; carpels 5, connate proximally, adnate to hypanthium, hairy, styles terminal, distinct; ovules 2. **Fruits** pomes, red or black, obovoid or subglobose, 6–9(–11) mm, glabrous or pilose; hypanthium persistent; sepals persistent, ± appressed; carpels cartilaginous; styles <and often filaments> persistent. **Seeds** 1–8 per pome, <2–3 mm>. $x = 17$.

Species 2 (2 in the flora): e North America.

Aronia has been included in *Photinia* (K. R. Robertson et al. 1991) on morphologic evidence, but C. Kalkman (2004) doubted this conclusion; a phylogenetic analysis by C. S. Campbell et al. (2007), using chloroplast and nuclear DNA sequence data, did not find a close relationship between *A. arbutifolia* and *P. villosa*. Historically, species of *Aronia* have been assigned variously to *Adenorachis*, *Crataegus*, *Halmia* M. Roemer, *Malus*, *Mespilus*, *Pyrus*, and *Sorbus*. *Aronia latifolia* Riddell from Kentucky appears to be a form of *Amelanchier canadensis*. *Aronia* is cultivated for food (juice, wine, and jam, and as a soft drink flavoring) and as an ornamental for its leaf color, for example, in the former Soviet Union (as *A. mitschurinii* A. K. Skvortsov & Maitulina), Sweden (H. A. Persson Hovmalm et al. 2004), and in North America.

Experiments by J. W. Hardin (1973) suggested that species of *Aronia* are variously outbreeding, self-compatible, or apomictic. They can also hybridize with *Sorbus*, forming the intergeneric hybrid ×*Sorbaronia* C. K. Schneider (see 53. *Sorbus*). The primary pollinators are thought to be small bees.

Varieties have been described for each species, but they are not recognized here as they appear to represent merely extremes of variation.

Aronia ×*prunifolia* (Marshall) Rehder [*Mespilus prunifolia* Marshall; *Adenorachis atropurpurea* (Britton) Nieuwland; *Aronia atropurpurea* Britton; *A. floribunda* (Lindley) Sweet; *Photinia floribunda* (Lindley) K. R. Robertson & J. B. Phipps; *Pyrus floribunda* Lindley], the purple chokeberry, is intermediate between the two species in indumentum but has purple pomes. It is found in St. Pierre and Miquelon, eastern Canada (New Brunswick, Newfoundland, Nova Scotia, Ontario, Prince Edward Island, and Quebec), and the eastern United States (Connecticut, Delaware, District of Columbia, Florida, Illinois, Indiana, Maine, Maryland, Massachusetts, Michigan, Minnesota, New Hampshire, New Jersey, New York, North Carolina, Ohio, Pennsylvania, Rhode Island, South Carolina, Tennessee, Vermont, Virginia, West Virginia, and Wisconsin).

J. W. Hardin (1973) concluded that the two species are fairly distinct but that *Aronia* ×*prunifolia* tends to obscure the boundary between them, making meaningful identification difficult. The fact that the putative hybrid tends to make apparently normal fruit could be the result of apomixis. It could also explain why it has been able to spread beyond the range limits of at least one of its putative parents.

SELECTED REFERENCE Hardin, J. W. 1973. The enigmatic chokeberries. Bull. Torrey Bot. Club 100: 178–184.

1. Leaves shiny adaxially, glabrous or glabrescent; hypanthia glabrous; pomes black . . . . 1. *Aronia melanocarpa*
1. Leaves dull adaxially, abaxially pilose (except for glabrous forms); hypanthia villous, especially proximally; pomes red . . . . 2. *Aronia arbutifolia*

1. Aronia melanocarpa (Michaux) Elliott, Sketch Bot. S. Carolina 1: 557. 1821 • Black chokeberry E F

Mespilus arbutifolia Linnaeus var. *melanocarpa* Michaux, Fl. Bor.-Amer. 1: 292. 1803; *Aronia nigra* (Willdenow) Koehne; *Photinia melanocarpa* (Michaux) K. R. Robertson & J. B. Phipps; *Pyrus arbutifolia* (Linnaeus) Linnaeus f. var. *nigra* Willdenow; *P. melanocarpa* (Michaux) Willdenow; *Sorbus melanocarpa* (Michaux) Heynhold

Stems glabrous or glabrescent. **Leaves** pale green abaxially, dark green and shiny adaxially, becoming scarlet; blade 2.5–7 × 2.5–3.5 cm, apex subacute to acuminate or apiculate, surfaces glabrous or glabrescent, adaxial midrib stipitate-glandular. **Flowers** sweet-scented; hypanthium glabrous; sepal margins glabrous; anthers yellow to purplish red. **Pomes** black, glabrous, taste acid and bitter. $2n = 34, 68$.

Flowering Mar–Jun; fruiting Sep–Nov. Swamps, bogs, wet thickets, margins of ponds and lakes, beaver ponds, woods, moist high-elevation forests, rock outcrops; 0–2000 m; N.B., Nfld. and Labr. (Nfld.), N.S., Ont., P.E.I., Que.; Ala., Ark., Conn., Del., D.C., Fla., Ga., Ill., Ind., Iowa, Ky., Maine, Md., Mass., Mich., Minn., Mo., N.H., N.J., N.Y., N.C., Ohio, Pa., R.I., S.C., Tenn., Vt., Va., W.Va., Wis.

2. Aronia arbutifolia (Linnaeus) Persoon, Syn. Pl. 2: 39. 1806 • Red chokeberry E

Mespilus arbutifolia Linnaeus, Sp. Pl. 1: 478. 1753; *Aronia glabrescens* Spach; *A. pubens* Spach; *A. pumila* M. Roemer; *A. pyrifolia* (Lamarck) Persoon; *Crataegus pyrifolia* Lamarck; *Halmia tomentosa* M. Roemer var. *pyrifolia* (Lamarck) M. Roemer; *Photinia pyrifolia* (Lamarck) K. R. Robertson & J. B. Phipps; *Pyrus arbutifolia* (Linnaeus) Linnaeus f.; *Sorbus arbutifolia* (Linnaeus) Heinhold

Stems appressed-pilose. **Leaves** pale green abaxially, dark green and dull adaxially, coloring in autumn; blade 3–7.5 × 2–3.5 cm, apex subacute to acuminate, abaxial surface pilose, adaxial glabrous or glabrescent, midrib abaxially densely gray-hairy, adaxially glandular-hairy. **Flowers** sweet-scented; hypanthium villous, especially proximally; sepal margins villous proximally; anthers yellow to purplish red. **Pomes** red, pilose, taste acid and bitter. $2n = 34, 68$.

Flowering Feb–May; fruiting Sep–Nov. Swamps, wet thickets, peatland pocosins, bogs, fens, wet pine flatwoods, margins of freshwater wetlands, beaver ponds, mixed loblolly pine, *Magnolia virginiana*-*Rhus vernix* swamps, black gum swamps; 0–2000 m; N.B., N.S., Ont., P.E.I., Que.; Ala., Ark., Conn., Del., D.C., Fla., Ga., Ky., La., Maine, Md., Mass., Mich., Miss., N.H., N.J., N.Y., N.C., Ohio, Okla., Pa., R.I., S.C., Tenn., Tex., Va., W.Va.

The presence of *Aronia arbutifolia* in Maine is uncertain; it has been reported from there, but no specimens are known (A. Haines, pers. comm.).

55. HETEROMELES M. Roemer, Fam. Nat. Syn. Monogr. 3: 100, 105. 1847, name conserved • Christmas berry, toyon [Greek *heteros*, different, and *melon*, apple, alluding to low stamen number]

James B. Phipps

Shrubs or trees, <canopies dense>, 20–100 dm. **Stems** usually 1; bark (trunk) grayish, with fine tan to dark gray striations, <± smooth>; short shoots absent; unarmed; young stems puberulent. **Leaves** persistent, cauline, simple; stipules variably persistent, free, minute, margins unknown, <apex gland-tipped>; petiole present; blade ± elliptic to narrowly elliptic or narrowly oblong, 5–15(–20) cm, leathery, margins revolute, sharply, remotely serrate, venation pinnate (craspedodromous-brochidodromous), surfaces tomentose when young, glabrescent. **Inflorescences** terminal, 20–150-flowered, panicles, <± dome-shaped>, white-tomentose; bracts present on proximal nodes, <leaflike, plus numerous, scalelike appendages on axes>; bracteoles present <caducous, delicate>. **Pedicels** present. **Flowers:** perianth and androecium epigynous, 10 mm diam.; hypanthium urceolate, 2–4 mm, glabrous or weakly floccose; sepals 5, suberect, triangular; petals 5, white, irregularly round, <base weakly clawed>; stamens 10, shorter than petals; carpels 2 or 3, distinct, basally adnate to hypanthium, styles <2 or 3>, lateral, distinct; ovules 2. **Fruits** pomes, usually bright red, sometimes yellow, ellipsoid, 5–10 mm, glabrous or glabrate; <flesh mealy>; hypanthium persistent; sepals persistent, accrescent over hypanthial opening. **Pyrenes** 2 or 3 per fruit, <carpel walls thin, soft; styles not persistent; seeds 1 per pyrene, large>.

Species 1: California, nw Mexico.

Heteromeles resembles *Aronia* and *Photinia*; it differs by having ten stamens instead of 20, nearly free carpels (versus fully connate), and fruit flesh lacking stone cells (versus often possessing some). Its fruit is of the coreless type (J. R. Rohrer et al. 1991). Molecular evidence (D. Potter et al. 2007) shows a relationship to the Asiatic *Eriobotrya* and *Rhaphiolepis* Lindley miniclade; fruit type, notched petals, and other characteristics confirm this conclusion.

SELECTED REFERENCE Phipps, J. B. 1992. *Heteromeles* and *Photinia* (Rosaceae subfam. Maloideae) of Mexico and Central America. Canad. J. Bot. 70: 2138–2162.

1. **Heteromeles arbutifolia** (Lindley) M. Roemer, Fam. Nat. Syn. Monogr. 3: 105. 1847 [F]

Photinia arbutifolia Lindley, Bot. Reg. 6: plate 491. 1820, based on *Crataegus arbutifolia* W. T. Aiton in W. Aiton and W. T. Aiton, Hortus Kew. 3: 202. 1811, not Lamarck 1783; *Heteromeles arbutifolia* var. *cerina* (Jepson) A. E. Murray; *H. arbutifolia* var. *macrocarpa* (Munz) Munz; *H. fremontiana* Decaisne; *H. salicifolia* (C. Presl) Abrams; *P. salicifolia* C. Presl

Leaves: petiole 1–2 cm, tomentose; blade margins cartilaginous, veins 10–20 per side. **Inflorescences:** bracts persistent, 1–3; appendages persistent, hard; bracteoles subulate, 4–10 mm, chartaceous, margins ± revolute, teeth hard, gland-tipped, abaxial surfaces hairy. **Flowers:** sepals 1–2 mm. **Pyrenes** brown, smooth.

Flowering Apr–Aug. Chaparral, oak woodlands, mixed-evergreen forests; 0–1300 m; Calif.; Mexico (Baja California).

Heteromeles arbutifolia is widely distributed in the Coast Ranges north to Humboldt County and in a more scattered manner in the Sierra Nevada to Shasta County. Plants are used for Christmas decoration in the flora area and are sometimes cultivated for ornament where the species prospers.

Two varieties have been named, var. *cerina* (= *Photinia arbutifolia* var. *cerina*), which represents yellow-fruited forms, and var. *macrocarpa* (= *P. arbutifolia* var. *macrocarpa*), with more floriferous inflorescences and larger fruits (8–10 mm). The first is a sporadic variant worthy only of forma status; the variety named by Munz is more substantial, seems usually to have shorter and relatively broader leaves with barely toothed, revolute margins, and is said to be most common on the Channel Islands of San Clemente and Santa Catalina. This variety seems to intergrade with var. *arbutifolia*, which represents the bulk of the species.

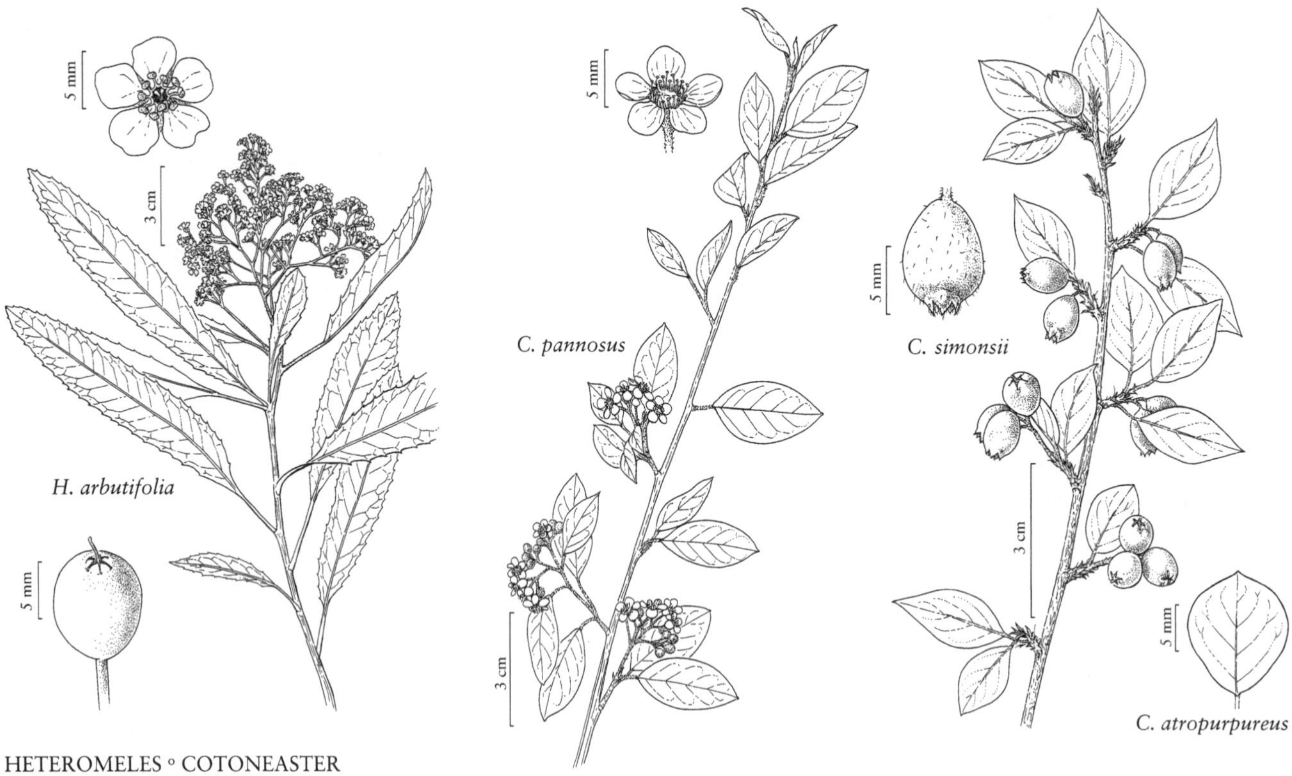

HETEROMELES ◦ COTONEASTER

56. COTONEASTER Medikus, Philos. Bot. 1: 154. 1789 • [Latin *cotoneum*, quince, and *-aster*, incomplete resemblance, alluding to similarity of leaves in some species] I

Jeanette Fryer

Bertil Hylmö†

Peter F. Zika

Ostinia Clairville

Shrubs, subshrubs, or trees, 0–150[–180] dm. **Stems** 1–25, erect, ascending, spreading to prostrate; bark usually dark gray, <smooth>; short shoots present or absent; unarmed; glabrous or tomentose to strigose; <branches irregular, distichous, divaricate, or spiraled>. **Leaves** persistent, semipersistent, or deciduous, cauline, simple; stipules mostly early deciduous, short-adnate to petiole, linear or narrowly lanceolate, inconspicuous, margins entire; petiole present; blade elliptic to linear, orbiculate, lanceolate, or ovate, [0.3–]0.4–15(–21) cm, leathery to membranous, margins flat or revolute, entire, venation pinnate, <2–14 pairs>, surfaces often hairy, rarely glandular. **Inflorescences** terminal <on lateral shoots>, 1–150[–200]-flowered, <domed or flattened> compound panicles, racemes, or flowers solitary, glabrous or hairy; bracts usually present; bracteoles absent. **Pedicels** usually present. **Flowers:** perianth and androecium epigynous, 4–15 mm diam.; hypanthium funnelform, cupulate, or campanulate, 1.5–5 mm, glabrous or pilose to tomentose; sepals 5, erect, triangular, <often fleshy>; petals 5, <spreading>, white [rarely pale pink], or <erect>, pink to red or blackish red, spatulate or ± orbiculate, <base

clawed>; stamens (8–)10–20(–23) in 1 or 2 series, shorter than petals; carpels 1–5, distinct, adnate to proximal ½ of hypanthium, glabrous or strigose to tomentose, styles <1–5>, terminal or lateral, distinct; ovules 2 (only 1 maturing). **Fruits** pomes, orange to red or purple to black, globose to obovoid or oblong, [3–]4–14[–15] mm, often hairy distally; <fleshy, flesh usually yellow, sclereids absent>; hypanthium persistent; sepals mostly persistent, erect to incurved or flat; carpels bony; style remnants terminal and projecting from pyrene apex, or lateral, on distal ¼–½ of central adaxial keel. **Pyrenes** 1–5, <without intervening flesh, rarely connate, brown, planoconvex or trigonous, bony>. $x = 17$.

Species ca. 400 (34, including 1 hybrid, in the flora): introduced; Eurasia, n Africa; introduced also in Europe, Pacific Islands, Australia.

A marked concentration of *Cotoneaster* species is found in the Himalaya and western China. The majority of species (ca. 90%) are apomictic, tetraploid (G. H. Kroon 1975; I. V. Bartish et al. 2001), and breed true when raised from seed.

Cotoneaster was long placed near *Crataegus*. Based on a reevaluation of flower and fruit characters, it is now believed to be more closely related to *Heteromeles* and *Pyracantha* (K. R. Robertson et al. 1991; J. R. Rohrer et al. 1991), but lacks thorns and toothed leaf margins. Recent molecular studies also distance *Crataegus* from *Cotoneaster*; as of yet, there is no consensus about the closest allies of *Cotoneaster* (C. S. Campbell et al. 1995, 2007; R. C. Evans and Campbell 2002). D. Potter et al. (2007) grouped *Cotoneaster* with *Chamaemeles* Lindley and *Malus*, but further study is needed to resolve its placement. The genera of Maleae are difficult to organize in a phylogeny, due in part to interfertility, lack of divergence, and reticulate evolution (Campbell et al. 2007).

Plants of *Cotoneaster* are used by apiarists, as the flowers have abundant nectar and are much loved by bees; the resulting honey is pale golden with a delicate flavor. In Asia, leaves and fruits are used in tea, and the wood for implements. *Cotoneaster* is used in India and Iran as the source of a sweet mannalike substance high in dextrose. The fruit of *C. integerrimus* Medikus has been used in the treatment of diarrhea. Many species are ornamentals, with brilliant fall foliage or colorful fruit persisting throughout the winter. The genus is bird-disseminated, mainly by American robins (*Turdus migratorius* Linnaeus) and cedar waxwings (*Bombycilla cedrorum* Vieillot) in North America. A few orange- or red-fruited species (*C. franchetii*, *C. lacteus*, *C. pannosus*, and *C. simonsii*) are weedy on the Pacific coast (J. M. Randall and J. Marinelli 1996; C. C. Bossard et al. 2000). Only *C. divaricatus* and the black-fruited *C. lucidus* are widely escaped in the interior.

More than 70 species of *Cotoneaster* are cultivated in North America, and additional escapes might be expected. Reports of several species in the wild appear to be unfounded. These include *C. acutifolius* Turczaninow and *C. apiculatus* Rehder & E. H. Wilson from Pennsylvania (A. F. Rhoads and W. M. Klein 1993; Rhoads and T. A. Block 2000), *C. bullatus* Bois from British Columbia (J. Pojar 1999), *C. hupehensis* Rehder & E. H. Wilson from New York (R. S. Mitchell and G. C. Tucker 1997), and *C. multiflorus* Bunge and *C. racemiflorus* (Desfontaines) K. Koch from Quebec (J. Cayouette et al. 1983). Without reference specimens it is often difficult to name adventives, given the lack of North American literature.

As with many plants of Maleae, *Cotoneaster* foliage is variable, yet mature leaf size, shape, texture, and luster on vigorous sterile shoots are usually diagnostic. The leaves of fertile shoots are usually smaller and less typical than those on sterile ones. Shaded plants have larger leaves than individuals in full sun. Deciduous species in flower have smaller leaves than when fruiting. A few species, such as *C. gamblei*, often hold leaves into January (at least on vigorous sheltered shoots) and key as evergreen shrubs, although naturalized populations in North America will eventually drop most leaves in late winter, especially where exposed. Several species are reported in the literature as semievergreen shrubs (for example, *C. horizontalis*, *C. simonsii*), but most of their leaves color and drop by midwinter where they are naturalized on the Pacific coast, so they are keyed as deciduous here. Shrubs with tomentose leaves can be sorted into evergreen and deciduous species by checking for the presence of darkened and weathered tomentum on the abaxial surfaces of overwintered leaves. Sunken (versus superficial) veins are mentioned in the keys and descriptions, always referring to the adaxial leaf surfaces. Collectors should observe this on fresh material and note it on herbarium labels. Sunken veins and wrinkling or bulging between the lateral veins can be obscured when pressing thin-leaved species. The number of veins in the descriptions represents the major lateral veins on one side of the leaf or midvein, most easily observed on the adaxial surface.

The pyrenes are usually pubescent at the apex, if exposed by the navel between the sepals on the summit of the pome. Extracted pyrenes are planoconvex or trigonous (like orange segments). Potentially useful taxonomic characters such as pyrene size, outline, and surface texture are not described here but warrant investigation. The style remnant is usually visible as a small peg or scar on the distal half of the inner surface or central keel of the pyrene; its location is easy to assess if fresh material is available. At least five pomes should be opened to determine the number of pyrenes present.

Identification of most naturalized species requires fruit characters and mature foliage; unfortunately, floral characters are also necessary to separate some species pairs. Floral and fruit dimensions are from fresh material and will be smaller on dried specimens. Floral diameter is provided throughout in subg. *Chaenopetalum* (Koehne) G. Klotz (species 1–16), which has rotate blossoms with spreading petals. For subg. *Cotoneaster* (species 17–34), the cupulate flowers are described by their floral length, including hypanthium and the erect petals.

Cotoneaster is properly treated as a masculine name (Yu T. T. 1954), although horticultural works still can be found with feminine endings applied to species epithets.

SELECTED REFERENCES Flinck, K. E. and B. Hylmö. 1966. A list of series and species in the genus *Cotoneaster*. Bot. Not. 119: 445–463. Fryer, J. and B. Hylmö. 2009. Cotoneasters: A Comprehensive Guide to Shrubs for Flowers, Fruit, and Foliage. Portland. Klotz, G. 1957. Uebersicht ueber die in Kulture befindlichen *Cotoneaster*-Arten und -Formen. Wiss. Z. Martin-Luther-Univ. Halle-Wittenberg, Math.-Naturwiss. Reihe 6: 945–982. Klotz, G. and T. Krügel. 1983. Zur zytologischen Struktur der Gattung *Cotoneaster* Medik. II. Wiss. Z. Friedrich-Schiller-Univ. Jena, Math.-Naturwiss. Reihe 32: 901–907. Rehder, A. J. and E. H. Wilson. 1913. *Cotoneaster*. In: C. S. Sargent, ed. 1913–1917. Plantae Wilsonianae.... 3 vols. in parts. Cambridge, Mass. Vol. 1, pp. 154–177.

Key to Groups of *Cotoneaster*

1. Shrubs, prostrate or long-trailing, rarely a few shoots to 30 cm tall. . . . . Group A
1. Shrubs or trees, erect, more than 30 cm tall.
    2. Mature leaves tomentose abaxially (hairs obscuring veins and most or all of surfaces).
        3. Leaves deciduous . . . . Group B
        3. Leaves persistent. . . . . Group C
    2. Mature leaves glabrous or slightly hairy abaxially (surfaces easily visible).
        4. Leaves persistent. . . . . Group D
        4. Leaves deciduous.
            5. Pomes black or dark purple; styles and pyrenes 1–3. . . . . Group E
            5. Pomes orange to red, dark red, or red-purple; styles and pyrenes (1 or)2–5 . . . . Group F

Key to Species by Group

Group A. Shrubs, prostrate or long-trailing, rarely a few shoots to 30 cm tall; leaves deciduous or persistent, not tomentose.

1. Leaves deciduous, blades chartaceous, margins slightly undulate, not revolute; petals erect, dark red . . . . 18. *Cotoneaster adpressus*
1. Leaves persistent, blades coriaceous, margins slightly revolute, not undulate; petals spreading, white.
   2. Styles and pyrenes 2(or 3); leaf blades 5–14 mm . . . . 14. *Cotoneaster cochleatus*
   2. Styles and pyrenes 4 or 5; leaf blades 13–43 mm . . . . 15. *Cotoneaster dammeri*

Group B. Leaves deciduous, abaxial surfaces tomentose.

1. Pomes purple-black, dark red, or ruby to maroon.
   2. Pomes dark red or ruby to maroon; petals spreading, white; anthers purple to blackish purple; styles and pyrenes 1(or 2) . . . . 9. *Cotoneaster monopyrenus* (in part)
   2. Pomes purple-black; petals erect, greenish white with pink or red; anthers white; styles and pyrenes 2–4 . . . . 34. *Cotoneaster melanocarpus* (in part)
1. Pomes bright red or orange-red.
   3. Filaments pink or pale pink; leaf blades: abaxial surfaces villous-strigose or sparsely to moderately tomentose . . . . 27. *Cotoneaster tengyuehensis* (in part)
   3. Filaments red or dark red; leaf blades: abaxial surfaces densely tomentose.
      4. Pomes orange-red . . . . 29. *Cotoneaster qungbixiensis* (in part)
      4. Pomes bright red . . . . 31. *Cotoneaster dielsianus*

Group C. Shrubs or trees, erect; leaves persistent, abaxial surfaces tomentose.

1. Shrubs, 0.5(–1.5) m; leaf blades 5–11 mm . . . . 12. *Cotoneaster hodjingensis*
1. Shrubs or trees, 1–9 m; leaf blades 15–120 mm.
   2. Leaves: lateral veins ± superficial, rarely faintly sunken, adaxial surfaces flat between lateral veins; styles and pyrenes (1 or)2.
      3. Pomes slightly glaucous, sepals flat to incurved; flowers (7–)10–11 mm diam. . . . . 2. *Cotoneaster crispii* (in part)
      3. Pomes not glaucous, sepals suberect; flowers 6–10 mm diam.
         4. Leaf blades 15–35(–39) × 10–19(–26) mm; flowers 8.5–10 mm diam. . . . . 4. *Cotoneaster pannosus*
         4. Leaf blades 35–77 × 15–35 mm; flowers 6–6.5(–9) mm diam. . . . . 5. *Cotoneaster vestitus*
   2. Leaves: lateral veins sunken, adaxial surfaces bulging or flat between lateral veins; styles and pyrenes 2–5.
      5. Petals spreading, creamy white; anthers red-purple; styles and pyrenes 2; leaf blades obovate to elliptic, rarely broadly elliptic . . . . 3. *Cotoneaster lacteus* (in part)
      5. Petals erect, pink or red; anthers white, pink, or purplish pink; styles and pyrenes (2 or)3–5 (always some with more than 2); leaf blades ovate to elliptic.
         6. Anthers white.
            7. Filaments pink or pale pink, whitish distally; pomes bright red; leaves sparsely to moderately tomentose abaxially . . . . 27. *Cotoneaster tengyuehensis* (in part)
            7. Filaments dark red, at least proximally; pomes orange-red; leaves densely tomentose abaxially . . . . 29. *Cotoneaster qungbixiensis* (in part)
         6. Anthers pink to purplish pink or mostly white with pink-tinged sutures.
            8. Filaments pink, distally white; anthers pink to purplish pink; leaves usually flat adaxially between lateral veins; pomes obovoid, bases rounded or cuneate to obconic, straight-sided . . . . 28. *Cotoneaster franchetii*
            8. Filaments red; anthers white with pink-tinged sutures; leaves usually bulging slightly adaxially between lateral veins (less obvious on shaded shoots); pomes subglobose to depressed-globose or globose, bases rounded . . . . 30. *Cotoneaster sternianus*

Group D. Stems erect; leaves persistent, glabrous or sparsely hairy abaxially.

1. Shrubs, usually less than 0.8 m; leaf blade margins revolute.
   2. Leaf blades usually oblanceolate or oblong; pomes dull or slightly shiny, glaucous .... 13. *Cotoneaster integrifolius*
   2. Leaf blades usually elliptic to narrowly elliptic; pomes shiny, not glaucous.
      3. Leaf blades 6–12(–20) × 2–6(–8) mm, adaxial surfaces grayish green, dull to slightly shiny; styles and pyrenes 2(or 3) .... 11. *Cotoneaster conspicuus*
      3. Leaf blades 10–27 × 4–12 mm, adaxial surfaces dark green, shiny; styles and pyrenes 2–4(or 5) (always some with more than 2) .... 16. *Cotoneaster* ×*suecicus*
1. Shrubs or trees, 1–10 m; leaf blade margins flat or revolute.
   4. Leaf blades: adaxial surfaces flat between lateral veins, veins superficial or essentially so; styles and pyrenes (1 or)2.
      5. Leaf blades 25–46 mm, adaxial surfaces dull, slightly glaucous; pomes 6–8 mm, globose or subglobose .... 2. *Cotoneaster crispii* (in part)
      5. Leaf blades (25–)50–127 mm, adaxial surfaces shiny, not glaucous; pomes 8.5–12.8 mm, obovoid, rarely subglobose .... 7. *Cotoneaster gamblei* (in part)
   4. Leaf blades: adaxial surfaces wrinkled or bulging between lateral veins, or veins deeply sunken; styles and pyrenes 2–5.
      6. Styles and pyrenes usually 3–5; petals erect, pink to red; anthers white .... 27. *Cotoneaster tengyuehensis* (in part)
      6. Styles and pyrenes usually 2 or 3; petals spreading, white or creamy white; anthers red-purple.
         7. Leaf blades shiny adaxially, narrowly elliptic to elliptic-lanceolate, 3–4 times as long as wide, margins strongly revolute, apices gradually tapered, acute or acuminate; styles and pyrenes 2 or 3(or 4); pomes maturing in October .... 1. *Cotoneaster salicifolius*
         7. Leaf blades dull or slightly shiny adaxially, obovate to elliptic, rarely broadly elliptic, 2 times as long as wide, margins flat or slightly revolute, apices short-tapered, acute, acuminate, or obtuse; styles and pyrenes 2; pomes maturing in November .... 3. *Cotoneaster lacteus* (in part)

Group E. Stems erect; leaves deciduous, glabrous abaxially or hairy, not densely tomentose; pomes black or purple-black.

1. Leaves usually bulging between lateral veins, veins sunken (easily obscured on pressed plants).
   2. Vigorous shoots with leaf apices short-acuminate or acute; pomes glaucous .... 24. *Cotoneaster lucidus*
   2. Vigorous shoots with leaf apices attenuate to long-acuminate; pomes not glaucous .... 25. *Cotoneaster villosulus*
1. Leaf blades: adaxial surfaces flat between lateral veins, veins superficial or slightly sunken.
   3. Styles and pyrenes 1 or 2; petals white, spreading, usually with hair tuft.
      4. Leaves of vigorous shoots 1.5 times as long as wide; pomes: sepal margins sparsely hairy .... 9. *Cotoneaster monopyrenus* (in part)
      4. Leaves of vigorous shoots 2 times as long as wide; pomes: sepal margins glabrous .... 10. *Cotoneaster transens* (in part)
   3. Styles and pyrenes 2–4; petals pink to red or purple, erect, glabrous.
      5. Leaf blades 8–28 mm; fall leaves turning bright red and gold; styles and pyrenes 2(or 3) .... 23. *Cotoneaster nitens*
      5. Leaf blades 33–45 mm; fall leaves without notable color; styles and pyrenes 2–4 .... 34. *Cotoneaster melanocarpus* (in part)

Group F. Leaves deciduous, abaxial surfaces glabrous or hairy, not densely tomentose; pomes orange or red, dark red, or red-purple.

1. Adaxial surfaces of fresh leaves strongly wrinkled or bulging between lateral veins (easily obscured when pressed); styles and pyrenes (4 or)5 . . . . 26. *Cotoneaster rehderi*
1. Adaxial surfaces of fresh leaves flat between lateral veins, sometimes faintly bulging (some taxa with faintly sunken midvein or wavy margins); styles and pyrenes (1 or)2–5.
    2. Leaf blades 4–30(–36) mm.
        3. Shrubs 1.5–4(–6) m; pomes usually cylindric to obovoid (proportionally narrower).
            4. Plants ± strict or ascending, lateral branches mostly suppressed, short, ± straight; stamens 20 (found under calyx in fruit, with careful dissection); pomes bright orange to orange-red, usually obovoid; styles and pyrenes (2 or)3–5; leaf blades on vigorous shoots usually broadly elliptic to ovate . . . . 17. *Cotoneaster simonsii* (in part)
            4. Plants spreading with rounded shape, lateral branches well developed, usually long, arching; stamens 10–15; pomes dark red to ruby (finally blackish red), cylindric, oblong-ellipsoid, or narrowly obovoid; styles and pyrenes (1 or)2 (or 3); leaf blades on vigorous shoots usually elliptic or broadly elliptic . . . . 20. *Cotoneaster divaricatus*
        3. Shrubs 0.5–1 m; pomes usually broadly obovoid.
            5. Leaf blades on vigorous shoots usually broadly elliptic . . . . 22. *Cotoneaster horizontalis*
            5. Leaf blades on vigorous shoots usually broadly obovate to orbiculate.
                6. Leaf blades on vigorous shoots obovate-orbiculate; pomes bright red; sepals sparsely pilose . . . . .19. *Cotoneaster atropurpureus*
                6. Leaf blades on vigorous shoots orbiculate, sometimes broadly obovate; pomes orange-red; sepals glabrous. . . . . .21. *Cotoneaster hjelmqvistii*
    2. Leaf blades (21–)30–150 mm.
        7. Leaf blades (25–)50–150 mm; anthers purple, red-purple, or black; pomes red to purplish red or maroon to purple-black.
            8. Pomes maroon to purple-black, sepals sparsely villous, margins glabrous . . . . 10. *Cotoneaster transens* (in part)
            8. Pomes red, sepals densely villous, margins tomentose.
                9. Pomes 4–6 × 4–5 mm, sepals erect or ascending; pedicels (and peduncles) densely tomentose . . . . .6. *Cotoneaster frigidus*
                9. Pomes 8.5–12.8 × 8–12 mm, sepals flat; pedicels (and peduncles) sparsely villous-strigose. . . . . .7. *Cotoneaster gamblei* (in part)
        7. Leaf blades usually less than 50 mm; anthers white; pomes orange to dark red.
            10. Petals spreading; flowers 12–14 mm diam.; petioles 5–8 mm . . . . 8. *Cotoneaster magnificus*
            10. Petals erect or erect-incurved; flowers 4–10 mm diam.; petioles 1–5 mm.
                11. Styles and pyrenes (2 or)3–5.
                    12. Pomes orange to orange-red; sepals erect or ascending, rarely nearly flat (not meeting over and obscuring pome summit), navel open; flowers pendent. . . . . 17. *Cotoneaster simonsii* (in part)
                    12. Pomes bright red; sepals flat (meeting over and obscuring pome summit), navel closed; flowers erect or ascending . . . . 27. *Cotoneaster tengyuehensis* (in part)
                11. Styles and pyrenes 2(or 3).
                    13. Pomes bright to dark red; leaves on vigorous shoots elliptic, broadly elliptic, or ovate, sometimes suborbiculate . . . . 32. *Cotoneaster fangianus*
                    13. Pomes bright orange-red; leaves on vigorous shoots elliptic to narrowly ovate (usually proportionally narrower) . . . . 33. *Cotoneaster miniatus*

1. Cotoneaster salicifolius Franchet, Nouv. Arch. Mus. Hist. Nat., sér. 2, 8: 225. 1885 (as salicifolia) • Willow-leaved cotoneaster [I]

Shrubs or trees, 1–8 m. **Stems** erect or arching [prostrate]; branches spiraled or distichous, <arching>, maroon, <slender>, initially tomentose-strigose. **Leaves** persistent; petiole 3–8 mm, tomentose-strigose; blade narrowly elliptic to elliptic-lanceolate, 30–90 × 8–23 mm, coriaceous, base cuneate, margins strongly revolute, veins 7–12, deeply sunken, apex gradually tapered, acute or acuminate, abaxial surfaces grayish green, initially densely villous, becoming glabrous or glabrate, adaxial dark green, shiny, not glaucous, bulging between lateral veins, sparsely pilose initially. **Inflorescences** on fertile shoots 20–60 mm with 3 or 4 leaves, 10–50 -flowered, compact. **Pedicels** 0.5–3(–4) mm, densely strigose-villous. **Flowers** (5–)6–9 mm diam.; buds white; hypanthium funnelform, pilose; sepals: margins sparsely villous, apex acute, surfaces sparsely pilose; petals spreading, white, glabrous; stamens 20, filaments white, anthers red-purple; styles 2 or 3(or 4). **Pomes** bright red (or with hint of crimson), broadly obovoid to globose, (5–)6–8.5 × (5–)6–8 mm, shiny, not glaucous, sparsely pilose; sepals depressed or flat, sparsely pilose; navel slightly open; style remnants at apex. **Pyrenes** 2 or 3 (or 4). $2n$ = 34 (Germany).

Flowering May–Jun; fruiting Oct–Apr. Thickets, mossy cliffs, cracks in walls, lakeshores, urban waste ground; 0–200 m; introduced; B.C.; Wash.; Asia (China: Sichuan, Xizang); introduced also in Europe.

L. Lingdi and A. R. Brach (2003) included as varieties of *Cotoneaster salicifolius* several taxa that here are considered full species, such as *C. henryanus* (C. K. Schneider) Rehder & E. H. Wilson, *C. hylmoei* Flinck & J. Fryer, and *C. rugosus* E. Pritzel.

A horticultural hybrid, presumably between *Cotoneaster dammeri* and *C. salicifolius*, is known in gardens as the cultivar *Cotoneaster* 'Hybridus Pendulus' and was recently collected as a local escape from cultivation in King County, Washington. This hybrid would key to *C.* ×*suecicus* but differs in its limp branching, much larger leaves (20–80 × 10–30 mm) and more numerous and smaller diameter flowers (9–11 mm wide). The foliage of the hybrid closely resembles *C. salicifolius*, but it has fewer (four to 13) and larger flowers and is less than 1 m in height.

2. Cotoneaster crispii Exell, Gard. Chron., ser. 3, 83: 44. 1928, as hybrid • Crisp's cotoneaster [I]

Shrubs or trees, 1.5–6 m. **Stems** narrowly erect, arching; branches spiraled, purple-black, initially silky tomentose. **Leaves** persistent, rarely semipersistent (in harsh winters); petiole 5–7 mm, tomentose; blade elliptic, sometimes ovate, 25–46 × 11–28 mm, subcoriaceous or coriaceous, base cuneate, margins flat, veins 4–7, superficial, rarely faintly sunken, apex acute, abaxial surfaces initially whitish-tomentose, becoming glabrate, adaxial green to dark green, dull, slightly glaucous, flat between lateral veins, initially sparsely pilose. **Inflorescences** on fertile shoots 40–70 mm with 4 leaves, 7–25-flowered, compact to ± lax. **Pedicels** 1–7 mm, silky tomentose. **Flowers** (7–)10–11 mm diam.; buds white; hypanthium funnelform, silky tomentose; sepals: margins villous, apex cuspidate or acuminate, surfaces silky tomentose; petals spreading, white, rarely with hair tuft; stamens 20, filaments white, anthers reddish purple; styles (1 or)2. **Pomes** red to dark red, globose or subglobose, 6–8 × 6–8 mm, dull or slightly shiny, slightly glaucous, glabrescent; sepals flat, depressed, or incurved, densely villous; navel open; style remnants at apex. **Pyrenes** (1 or)2. $2n$ = 51 (Germany).

Flowering Jun–Jul; fruiting Oct–Feb. Thickets, building edges, cracks in pavement; 0–200 m; introduced, B.C.; Asia (China).

Cotoneaster crispii was collected recently wild in Yunnan. The species is often treated as a hybrid of garden origin between *C. frigidus* and *C. pannosus*. Plants of it differ from those of *C. frigidus* in their smaller and persistent leaves, and it has larger leaves and fruits than *C. pannosus*. The species was overlooked in the *Flora of China* (L. Lingdi and A. R. Brach 2003).

3. Cotoneaster lacteus W. W. Smith, Notes Roy. Bot. Gard. Edinburgh 10: 23. 1917 (as lactea) • Cream-flowered or late or Parney's cotoneaster [I]

Shrubs or trees, 2–9 m. **Stems** erect, arching, spreading, <slender>; branches spiraled or distichous, purple-black, initially yellowish tomentose-villous. **Leaves** persistent; petiole 5–8 mm, initially tomentose; blade obovate to elliptic, rarely broadly elliptic, (26–)40–85(–120) × (13–)19–45(–60) mm, coriaceous, base cuneate, margins flat or slightly revolute, veins 7–9, deeply sunken, apex short-tapered, acute, acuminate, or obtuse, abaxial surfaces initially yellowish tomentose-villous, adaxial dark green, dull or slightly shiny, not glaucous, bulging slightly between lateral veins, initially sparsely pilose. **Inflorescences** on fertile shoots 45–100 mm with 4 leaves, 20–150-flowered, lax. **Pedicels** 1–5 mm, tomentose-strigose. **Flowers** 6–9 mm diam.; buds creamy white, sometimes slightly pinkish; hypanthium funnelform, tomentose-strigose; sepals: margins villous, apex cuspidate, surfaces tomentose; petals spreading, creamy white, sometimes with hair tuft; stamens 20, filaments white, anthers red-purple; styles 2. **Pomes** crimson, broadly obovoid, subglobose, or globose, 6.5–8.6 × 6.5–7.6 mm, shiny, not glaucous, sparsely villous; sepals depressed or slightly ascending, tomentose; navel slightly open or obscured by sepals; style remnants near apex. **Pyrenes** 2. $2n = 68$ (Germany).

Flowering May–Jul; fruiting Nov–Apr. Thickets, disturbed forests, edges, urban waste ground, stonework, grassy bluffs, canyons, pond shores; 0–700 m; introduced; B.C.; Calif., Oreg., Wash.; Asia (China); introduced also in Europe, Pacific Islands (New Zealand).

Cotoneaster lacteus sometimes is synonymized with *C. coriaceus* Franchet (for example, L. Lingdi and A. R. Brach 2003). As treated here, plants of this species are identified by their fertile shoots to 10 cm, scarcely shiny leaves to 120 × 60 mm with adaxial surfaces bulging between lateral veins, flowers to 150 in lax inflorescences, 6–9 mm diam. with cream petals, and larger, later-ripening fruits. Plants of *C. coriaceus* have fertile shoots to 5 cm, shiny leaves to 52 × 24 mm with flat adaxial surfaces, flowers to 50 in compact inflorescences, 9–10 mm diam. with white petals, and smaller, earlier-ripening fruits.

4. **Cotoneaster pannosus** Franchet, Pl. Delavay., 223. 1890 (as pannosa) • Silverleaf cotoneaster [F] [I]

Shrubs, 1–5 m. **Stems** erect, wide-spreading, arching, pendulous, <slender>; branches spiraled, purple-black, tomentose-villous. **Leaves** persistent; petiole 4–7 mm, tomentose; blade elliptic, sometimes broadly elliptic, 15–35(–39) × 10–19(–26) mm, mid-coriaceous, base cuneate, margins flat, veins 4–6, usually superficial, sometimes slightly sunken, apex acute or obtuse, abaxial surfaces whitish tomentose, adaxial green to blue-green, dull, not glaucous, usually flat between lateral veins, initially sparsely pilose. **Inflorescences** on fertile shoots 20–45 mm, usually with 4 leaves, (3–)5–15(–25)-flowered, compact. **Pedicels** 1–5 mm, tomentose. **Flowers** [7–]8.5–10 mm diam.; buds white; hypanthium funnelform or cupulate, silky-tomentose; sepals: margins villous, <borders white, narrow, membranous>, apex <red>, cuspidate, sometimes acute or obtuse, surfaces tomentose; petals spreading, white, sometimes with hair tuft; stamens 16–20, filaments white, anthers dark pink-purple to purple; styles (1 or)2. **Pomes** red, globose, depressed-globose, subglobose, or broadly obovoid, 5–9.2 × 5.5–9.5 mm, shiny, not glaucous, villous; sepals suberect, tomentose; navel open; style remnants at apex. **Pyrenes** (1 or)2. $2n = 68$ (Germany).

Flowering May–Aug; fruiting Oct–May. Thickets, meadows, sea cliffs, canyons, coastal chaparral, springs, edges, waste ground, pastures, disturbed forests; 0–800 m; introduced; Calif., Oreg.; Asia (China); introduced also in Europe, Africa (South Africa), Pacific Islands (Hawaii, New Zealand), Australia.

5. **Cotoneaster vestitus** (W. W. Smith) Flinck & B. Hylmö, Bot. Not. 119: 460. 1966 • Vested or woolly cotoneaster [I]

Cotoneaster glaucophyllus Franchet var. *vestitus* W. W. Smith, Notes Roy. Bot. Gard. Edinburgh 10: 21. 1917 (as glaucophylla var. vestita)

Shrubs, 2–5 m. **Stems** erect, arching to pendulous; branches spiraled or distichous, purple-black, <slender>, initially tomentose-villous. **Leaves** persistent; petiole 3–7 mm, tomentose; blade elliptic, 35–77 × 15–35 mm, subcoriaceous or coriaceous, base cuneate or obtuse, margins flat, veins 5–9, ± superficial, apex acute or obtuse, abaxial surfaces gray-tomentose, adaxial green, dull, slightly glaucous, flat between lateral veins, initially sparsely pilose. **Inflorescences** on fertile shoots 30–60 mm, usually with 3 leaves, 15–40-flowered, compact. **Pedicels** 1–5 mm, tomentose. **Flowers** 6–6.5(–9) mm diam.; buds white; hypanthium cupulate, tomentose; sepals: margins villous, apex cuspidate, acute or obtuse, surfaces tomentose; petals spreading, creamy white, glabrous; stamens 15–17(–20), filaments white, anthers pink-purple to dark purple; styles [1 or]2. **Pomes** red [rarely maroon], globose, 6–7 mm, shiny, not glaucous, sparsely villous; sepals suberect, <margins villous>, tomentose; navel open; style remnants at apex. **Pyrenes** [1 or]2.

Flowering Jun–Jul; fruiting Nov–Mar. Urban waste ground; 0–50 m; introduced, Wash.; Asia (China).

6. Cotoneaster frigidus Wallich ex Lindley, Edwards's Bot. Reg. 15: plate 1229. 1829 (as frigida) • Tree cotoneaster [I]

Shrubs or trees, 5–15 m. **Stems** loosely erect, spreading; branches spiraled, dark brown, initially downy. **Leaves** deciduous; petiole 5–8 mm, densely villous to tomentose; blade narrowly elliptic (to lance-elliptic or narrowly obovate), 55–150 × 19–55 mm, membranous, base cuneate, margins flat, veins 6–10, superficial or slightly sunken, apex acute, abaxial surfaces color not recorded, villous at least near midvein, adaxial green, dull, coating not recorded, flat between lateral veins, glabrous except villous midrib. **Inflorescences** on fertile shoots 60–80 mm with 2–4 leaves, 20–60-flowered, lax. **Pedicels** length not recorded, densely tomentose. **Flowers** 6–7 mm diam.; buds white; hypanthium cupulate [subcylindric], tomentose; sepals: margins tomentose, apex acute, surfaces densely villous; petals spreading, white, glabrous or with hair tuft; stamens 20, filaments white, anthers purple; styles 2. **Pomes** bright red, subglobose to ellipsoid, 4–6 × 4–5 mm, shiny, not glaucous, sparsely villous; sepals erect or ascending, <margins tomentose>, densely villous; navel hidden; style remnants near apex. **Pyrenes** 2. **2*n*** = 34 (India).

Flowering May–Jun; fruiting Sep–Oct. North-facing canyon slopes; 0–500 m; introduced, Calif.; Asia (Himalaya Mountains); introduced also in Europe.

Reports of *Cotoneaster frigidus* from Washington (F. C. Weinmann 2000; A. L. Jacobson 2001; J. T. Kartesz 2006) are here referred to *C. gamblei*, which are plants having larger fruits tending to be obovoid, less densely pubescent inflorescences, and thicker, more persistent leaves.

7. Cotoneaster gamblei G. Klotz, Wiss. Z. Martin-Luther-Univ. Halle-Wittenberg, Math.-Naturwiss. Reihe 15: 530. 1966 • Gamble's cotoneaster [I]

Shrubs or trees, 4–10 m. **Stems** loosely erect, spreading and arching; branches spiraled, brown or maroon-purple, initially tomentose-strigose. **Leaves** usually semipersistent (persisting to April if sheltered), sometimes deciduous; petiole 4–10 mm, densely villous-strigose; blade elliptic, (25–)50–127 × 13–49[–62 in cultivation] mm, chartaceous or subcoriaceous, base cuneate or obtuse, margins flat, veins 7 or 8, superficial, apex acute or obtuse, abaxial surfaces pale gray-green, initially villous, later glabrate, adaxial dark green, shiny, not glaucous, flat between lateral veins, sometimes becoming slightly rugose, glabrate; <fall leaves pale yellowish green>. **Inflorescences** on fertile shoots 40–70 mm with 2 or 3 leaves, 5–34-flowered, lax. **Pedicels** 1–8 mm, sparsely villous-strigose. **Flowers** 8–12.5 mm diam.; buds white; hypanthium cupulate, densely villous; sepals: margins <maroon>, tomentose, apex acute, surfaces densely villous; petals spreading, white, sometimes with hair tuft; stamens 18–20, filaments white, anthers red-purple or dark red-purple; styles (1 or)2. **Pomes** red, often tinged with pink, green, brown, or purple, obovoid, rarely subglobose, 8.5–12.8 × 8–12 mm, dull, slightly glaucous, sparsely villous; sepals flat, <margins tomentose>, densely villous; navel slightly open; style remnants near apex. **Pyrenes** [1 or]2.

Flowering May; fruiting Sep–Nov. Mesic thickets, flood plains, disturbed forests; 0–50 m; introduced, Wash.; Asia (Bhutan, India).

8. Cotoneaster magnificus J. Fryer & B. Hylmö, New Plantsman 5: 138, fig. p. 140 [lower left]. 1998 • Magnificent cotoneaster [I]

Shrubs, 3–5 m. **Stems** erect, spreading, arching; branches spiraled and distichous, maroon, initially tomentose-pilose. **Leaves** deciduous; petiole 5–8 mm, pilose-villous; blade suborbiculate, broadly elliptic, ovate, or rhombic, 34–40 × 20–32 mm, subcoriaceous, base obtuse or cuneate, margins flat, veins 5 or 6, superficial, apex acute [acuminate or apiculate], abaxial surfaces pale green, initially densely pilose, soon glabrous, adaxial brownish or coppery green, soon light green, slightly shiny, not glaucous, flat between lateral veins, glabrescent; <fall leaves wihtout notable color>. **Inflorescences** on fertile shoots 25–40 mm with 3 or 4 leaves, 5–12-flowered, lax. **Pedicels** 3–12 mm, glabrate. **Flowers** 12–14 mm diam.; buds white; hypanthium cupulate, glabrous; sepals: margins sparsely villous, apex acute or obtuse, surfaces glabrous; petals spreading, white, with hair tuft; stamens (16–)20, filaments white, anthers white; styles (1 or)2. **Pomes** red, dark red, red-purple, crimson, or rich red with cherry, globose or depressed-globose, 10–13 × 12–13 mm, slightly shiny, not glaucous, glabrous; sepals suberect, indumentum not recorded; navel semiopen; style remnants at or near apex. **Pyrenes** (1 or)2. **2*n*** = 68.

Flowering May–Jun; fruiting Sep–Nov. Thickets, hedgerows, edges; 0–300 m; introduced; Ill., Iowa; Asia (China).

Cotoneaster magnificus was synonymized with *C. multiflorus* Bunge by L. Lingdi and A. R. Brach (2003). Flowers of the latter are 9–10 mm wide in clusters of ten to 20, the flowering pedicels are to 7 mm, and the leaves are smaller, thinner, and tend to be ovate. Plants of *C. magnificus* have flowers 12–14 mm wide in clusters of five to 12, flowering pedicels are to 12 mm, and the leaves are larger, thicker, and mostly elliptic. Reports of wild *C. multiflorus* from Illinois (F. Swink and G. S. Wilhelm 1994; J. T. Kartesz and C. A. Meacham 2003) and Iowa (W. R. Norris et al. 2001; Kartesz and Meacham) are referred to *C. magnificus*.

9. **Cotoneaster monopyrenus** (W. W. Smith) Flinck & B. Hylmö, Bot. Not. 119: 459. 1966 • One-stoned cotoneaster [I]

Cotoneaster hebephyllus Diels var. *monopyrenus* W. W. Smith, Notes Roy. Bot. Gard. Edinburgh 10: 23. 1917 (as hebephylla var. monopyrena)

Shrubs or trees, 3–5 m. **Stems** loosely erect, arching, spreading; branches spiraled, maroon, <lenticellate>, initially pilose-strigose. **Leaves** deciduous; petiole 5–9 mm, pilose-strigose; blade elliptic, broadly elliptic, or broadly obovate, sometimes suborbiculate, 25–58 × 15–40 mm, subcoriaceous, base obtuse or cuneate, margins flat, veins 4–6, slightly sunken, apex obtuse or truncate, seldom acute, abaxial surfaces color not recorded, <reticulate>, tomentose-pilose, later thinning, adaxial dark green, dull, coating not recorded, flat between lateral veins, glabrescent; <fall leaves pale yellowish green>. **Inflorescences** on fertile shoots 30–50 mm with 3 or 4 leaves, 7–20-flowered, lax. **Pedicels** 2–5 mm, pilose-strigose. **Flowers** 10–12 mm diam., buds white, hypanthium campanulate to cupulate, sparsely pilose-strigose; sepals: margins <reddish>, villous, <borders maroon-tipped, membranous>, apex acute, acuminate, or obtuse, surfaces sparsely pilose-strigose; petals spreading, white, sometimes with hair tuft; stamens (15–)20, filaments white; anthers purple to blackish purple; styles 1(or 2). **Pomes** dark red or ruby to maroon, maturing dark purple or purple-black, obovoid, rarely globose or ellipsoid, 9–12.5 × 8.5–11.5 mm, dull, glaucous, glabrous; sepals flat, <margins sparsely hairy>, sparsely hairy to glabrate; navel open; style remnants at apex. **Pyrenes** 1(or 2).

Flowering May–Jun; fruiting Jul–Nov. Thickets; 0–200 m; introduced, Wash.; Asia (China); introduced also in Europe.

L. Lingdi and A. R. Brach (2003) synonymized *Cotoneaster monopyrenus* with *C. hebephyllus* (as *C. hebephyllus* var. *hebephyllus*).

10. **Cotoneaster transens** G. Klotz, Wiss. Z. Friedrich-Schiller-Univ. Jena, Math.-Naturwiss. Reihe 17(3): 337, fig. 4. 1968 • McLaren's cotoneaster [I]

Shrubs or trees, 4–6 m. **Stems** loosely erect, spreading, <stiff>; branches spiraled, maroon, <minutely verrucose>, initially pilose-strigose. **Leaves** deciduous; petiole 4–6 mm, densely pilose-strigose; blade on sterile shoots elliptic, (35–)50–80(–103) × (13–)17–37(–42) mm, chartaceous, margins flat, base cuneate, veins 6–10, superficial or lightly sunken, apex acute or obtuse, abaxial surfaces color not recorded, villous-strigose, adaxial brownish green, soon green, dull, coating not recorded, flat between lateral veins, glabrate; <fall leaves yellow>. **Inflorescences** on fertile shoots 40–50 mm with 2–4 leaves, 10–25-flowered, lax. **Pedicels** 3–5 mm, densely pilose-strigose. **Flowers** 10–12 mm diam.; buds white; hypanthium cupulate, villous-strigose; sepals: margins villous, apex acute, surfaces sparsely villous; petals spreading, white, with large hair tuft; stamens (16–)20, filaments white, anthers black, styles 1 or 2. **Pomes** maroon to purple-black, obovoid to elliptic-obovoid, 9–12 × 9–10 mm, slightly shiny, glaucous, sometimes sparsely villous; sepals flat, <margins glabrous>, sparsely villous; navel slightly open; style remnants 9/10 from base. **Pyrenes** 1 or 2.

Flowering Jun; fruiting Oct–Nov. Thickets; 0–50 m; introduced, Wash.; Asia (China); introduced also in Europe.

L. Lingdi and A. R. Brach (2003) synonymized *Cotoneaster transens* with *C. glaucophyllus* Franchet (as *C. glaucophyllus* var. *glaucophyllus*). *Cotoneaster transens* is a deciduous species with purple-black pomes and is not closely related to *C. glaucophyllus*, an evergreen species with smaller leaves, smaller red pomes, and smaller flowers, among other differences.

11. **Cotoneaster conspicuus** (Messel) Messel, J. Roy. Hort. Soc. 59: 303. 1934 (as conspicua) • Necklace or winter-green cotoneaster [I]

Cotoneaster microphyllus Wallich ex Lindley var. *conspicuus* Messel, Gard. Chron., ser. 3, 94: 299. 1933 (as microphylla var. conspicua); *C. conspicuus* var. *decorus* P. G. Russell; *C. conspicuus* var. *nanus* G. Klotz; *C. nanus* (G. Klotz) G. Klotz; *C. permutatus* G. Klotz

Shrubs, 0.5–1[–2.5] m. **Stems** weakly ascending or mound-forming [prostrate]; branches spiraled or distichous, <dense>, maroon, initially strigose. **Leaves**

persistent; petiole 1–3 mm, pilose-strigose; blade elliptic to narrowly elliptic, rarely lanceolate, 6–12(–20) × 2–6(–8) mm, subcoriaceous or coriaceous, base cuneate, margins revolute, veins 3–5, superficial, apex obtuse or acute, abaxial surfaces grayish green, <reticulate>, initially pilose-strigose, adaxial grayish green [rarely green], dull to slightly shiny, coating not recorded, slightly rugose, sparsely pilose or glabrous. **Inflorescences** on fertile shoots 8–12 mm with 3 or 4 leaves, 1-flowered, <2–4-flowered on shoot tips>. **Pedicels** 1–3 mm, pilose-strigose. **Flowers** 9–13 mm diam.; buds pink; hypanthium cupulate, pilose-strigose; sepals: margins villous, <borders mostly reddish, membranous>, apex acute or obtuse, surfaces pilose-strigose; petals spreading, white; stamens 17–21, filaments white, anthers purple-black; styles 2(or 3). **Pomes** orange-red to red, depressed-globose, 8–10 × 9–11 mm, shiny, not glaucous, glabrous; sepals suberect, pilose; navel not recorded; style remnants at elongated apex. **Pyrenes** 2(or 3). **2*n*** = 34 (Germany).

Flowering Apr–May; fruiting Oct–May. Brush edges in urban areas; 0–50 m; introduced, Wash.; Asia (China); introduced also in Europe.

G. Klotz (1963) emended the diagnosis of *Cotoneaster conspicuus* to recognize erect plants as var. *conspicuus* and more decumbent plants as var. *decorus*. This variation in habit is not significant taxonomically and not unusual for an outcrossing diploid, so the emended description by Klotz of the species is superfluous.

SELECTED REFERENCE Klotz, G. 1978. Neue oder kritische *Cotoneaster*-Arten. VIII. Wiss. Z. Friedrich-Schiller-Univ. Jena, Math.-Naturwiss. Reihe 27: 19–26.

12. Cotoneaster hodjingensis G. Klotz, Wiss. Z. Martin-Luther-Univ. Halle-Wittenberg, Math.-Naturwiss. Reihe 12: 774. 1963 • Earthquake cotoneaster [I]

Shrubs, <sprawling>, 0.5(–1.5) m. **Stems** erect, becoming suberect, ascending, wide-spreading; branches spiraled, <dense>, maroon, <stiff, minutely verruculose>, initially densely pilose or villous-strigose, later sparsely so. **Leaves** persistent; petiole 1–2 mm, tomentose; blade narrowly elliptic, sometimes lance-elliptic, rarely lanceolate, 5–11 × 2–6 mm, coriaceous, base narrowly cuneate, margins slightly revolute, veins 2–4, slightly sunken, apex acute or obtuse <with long mucro>, abaxial surfaces grayish white-tomentose, adaxial dark green, shiny, not glaucous, minutely rugose and verruculose, flat between lateral veins, initially pilose-strigose, soon glabrate. **Inflorescences** on fertile shoots 10–20 mm with 4 leaves, 1(–3)-flowered. **Pedicels** 1–3 mm, tomentose. **Flowers** <erect>, 7–8 mm diam.; buds white; hypanthium funnelform, densely villous, becoming glabrate; sepals: margins villous, apex cuspidate or acuminate, surfaces tomentose; petals spreading, white, glabrous or hairy; stamens 20, filaments white, anthers purple; styles 2. **Pomes** bright red, sometimes tinged crimson or cherry, globose, subglobose, or depressed-globose, 6–7 × 5–8 mm, shiny, not glaucous, sparsely villous; sepals suberect or erect as coronet, villous to sparsely so; navel open <with pyrenes exserted>; style remnants at apex. **Pyrenes** 2. **2*n*** = 68 (Germany).

Flowering Jun; fruiting Oct–Nov. Banks, disturbed ground; 0–50 m; introduced; Calif., Wash.; Asia (China).

Cotoneaster hodjingensis sometimes is synonymized with *C. buxifolius* Wallich ex Lindley (Zhou L. H. and Wu Z. Y. 2001; L. Lingdi and A. R. Brach 2003), a species from southern India with elliptic to ovate elliptic, usually obtuse (rarely short-acute) leaves to 17 mm, petioles to 4 mm, one to nine flowers per inflorescence (to 20 on end shoots), and a tomentose hypanthium. Here, *C. hodjingensis* is treated as separable by its normally narrower leaves no longer than 11 mm, usually acute, with petioles to 2 mm, solitary flowers (two or three flowers per inflorescence on end shoots), and villous hypanthium becoming glabrate (G. Klotz 1963).

13. Cotoneaster integrifolius (Roxburgh) G. Klotz, Wiss. Z. Martin-Luther-Univ. Halle-Wittenberg, Math.-Naturwiss. Reihe 12: 779. 1963 • Entire-leaved cotoneaster [I]

Crataegus integrifolius Roxburgh, Fl. Ind. ed. 1832, 2: 509. 1832 (as Crategus integrifolia)

Shrubs, <wide-spreading>, 0.5–1.5 m. **Stems** weakly ascending, suberect, or decumbent; branches spiraled, purple-black, sometimes green and brown striate, initially strigose. **Leaves** persistent, petiole 1–5 mm, strigose; blade oblanceolate or oblong, rarely lanceolate, 8–17 × 3–8 mm, coriaceous, base cuneate, margins revolute, veins 2–4, superficial, apex rotund or obtuse, rarely emarginate or acute, abaxial surfaces grayish white, <faintly reticulate>, initially densely strigose or strigose-pilose, later glabrate, adaxial dark green, sometimes blue-green, shiny, not glaucous, not bulging between lateral veins, initially strigose, soon glabrate. **Inflorescences** on fertile shoots 5–12 mm with 2 or 3 leaves, 1(or 2)-flowered, <subsessile>. **Pedicels** <erect>, densely strigose. **Flowers** 7–15 mm diam.; buds pink; hypanthium narrowly cupulate, densely strigose; sepals: margins densely villous-strigose, <borders membranous>, apex <often red>, acute or obtuse, sometimes shortly mucronate, surfaces often densely strigose-villous;

petals spreading, white, glabrous; stamens 20, filaments white, anthers red-purple; styles 2[or 3]. **Pomes** dark red with crimson, sometimes red, depressed-globose, 7–9 × 7–10 mm, dull or slightly shiny, glaucous, sparsely villous; sepals suberect or erect, villous; navel open; style remnants at apex. **Pyrenes** 2[or 3]. **2*n*** = 68 (Germany).

Flowering Apr–Jun; fruiting Aug–Mar. Openings in conifer forests; 0–200 m; introduced, Calif.; Asia; introduced also in Europe.

Cotoneaster integrifolius was treated as *C. microphyllus* Wallich ex Lindley var. *thymifolius* (Baker) Koehne by L. Lingdi and A. R. Brach (2003).

14. Cotoneaster cochleatus (Franchet) G. Klotz, Wiss. Z. Martin-Luther-Univ. Halle-Wittenberg, Math.-Naturwiss. Reihe 6: 952. 1957 • Yunnan cotoneaster [I]

Cotoneaster buxifolius Wallich ex Lindley forma *cochleatus* Franchet, Pl. Delavay., 224. 1890 (as buxifolia forma cochleata); *C. microphyllus* Wallich ex Lindley var. *cochleatus* (Franchet) Rehder & E. H. Wilson; *C. thymifolius* Baker var. *cochleatus* (Franchet) Franchet

Shrubs, to 0.4 m. **Stems** prostrate or nearly so, carpeting, <rooting>; branches spiraled and distichous, <dense>, red to purple-black, <slender>, initially yellow-green strigose. **Leaves** persistent; petiole 1–4 mm, strigose; blade obovate to broadly obovate, rarely suborbiculate, 5–14 × 3–9 mm, coriaceous, base obtuse or broadly cuneate, margins slightly revolute, veins 2 or 3, superficial, apex obtuse, sometimes emarginate, abaxial surfaces grayish, <reticulate>, initially densely strigose-villous, adaxial dark green, shiny, not glaucous, sometimes lightly rugose, glabrescent. **Inflorescences** on fertile shoots 8–15 mm, usually with 4 leaves, 1(–3) -flowered. **Pedicels** 3–5 mm, strigose. **Flowers** 7–10 mm diam.; buds pinkish; hypanthium cupulate, strigose or pilose-strigose; sepals: margins villous, apex obtuse or acute, surfaces initially sparsely pilose-strigose; petals spreading, white, glabrous; stamens (15–)20, filaments white, anthers dark purple; styles 2(or 3). **Pomes** bright red to crimson, subglobose, 7–9 × 8–10 mm, slightly shiny, not glaucous, sparsely pilose; sepals suberect, sparsely strigose; navel slightly open; style remnants at apex on small projection. **Pyrenes** 2(or 3). **2*n*** = 68 (Germany).

Flowering May–Jun; fruiting Sep–Nov. Forest edges; 0–50 m; introduced, N.Y.; Asia (China); introduced also in Europe.

Cotoneaster cochleatus was treated as a variety of *C. microphyllus* by L. Lingdi and A. R. Brach (2003); here the two are distinguished at species rank, following H. Nybom et al. (2005). Plants of *C. microphyllus* have a suberect habit, usually elliptic leaves (rarely broadly obovate) with acute apices, and pomes 6 mm wide; *C. cochleatus* is always prostrate and has usually obovate leaves (rarely suborbiculate) with blunt apices, and pomes 8–10 mm wide.

15. Cotoneaster dammeri C. K. Schneider, Ill. Handb. Laubholzk. 1: 761, figs. 429h–k. 1906 • Bearberry cotoneaster [I]

Shrubs, to 0.2 m. **Stems** carpeting or procumbent, <rooting, to 1.5 m, pliant>; branches distichous or spiraled, greenish to light brown, <densely lenticellate>, initially pilose-strigose. **Leaves** persistent; petiole 2–9 mm, villous-strigose; blade elliptic or obovate, rarely suborbiculate, 13–43 × 6–26 mm, coriaceous, base cuneate or obtuse, margins revolute, veins 5–8, sunken, apex obtuse or acute, rarely retuse, abaxial surfaces grayish green, <reticulate>, initially villous, adaxial light green to green, intensely shiny, not glaucous, reticulate-rugose, not bulging between lateral veins, often single hairs on midrib. **Inflorescences** on fertile shoots 20–40 mm, usually with 4 leaves, (1 or)2 or 3(or 4)-flowered, compact. **Pedicels** 4–15 mm, thin, villous-strigose. **Flowers** <slightly pendent>, 10–12 mm diam.; buds white; hypanthium cupulate, sparsely pilose-strigose; sepals: margins sparsely villous, apex obtuse or acute, surfaces sparsely pilose-strigose; petals spreading, white, glabrous; stamens 20, filaments white, anthers purple-black; styles 4 or 5. **Pomes** bright red, obovoid to broadly obovoid, rarely globose, 6–10 × 6–8 mm, shiny, not glaucous, glabrous; sepals suberect, glabrous; navel open; style remnants at apex. **Pyrenes** 4 or 5. **2*n*** = 34 (Germany).

Flowering May–Jun; fruiting Sep–Apr. Disturbed forests, grassy banks, rock roadcuts, urban waste ground; 0–200 m; introduced; B.C.; Wash.; Asia (China); introduced also in Europe.

SELECTED REFERENCE Zhou L. H., Yin Q., and Wu Z. Y. 2000. Taxonomic studies on *Cotoneaster dammeri* (Rosaceae). Acta Bot. Yunnan. 22: 379–382.

16. Cotoneaster ×suecicus G. Klotz, Beitr. Phytotax. 10: 47. 1982 • Swedish cotoneaster [I]

Shrubs, 0.4–0.6 m. **Stems** ± erect or ± procumbent, arching, <ends rooting>; branches spiraled, green proximally, purple-black distally, initially densely strigose. **Leaves** persistent; petiole <often red>, 1–3 mm, strigose; blade elliptic, 10–27 × 4–12 mm, coriaceous, base cuneate, margins revolute, veins 4–7, superficial, apex obtuse or acute, rarely emarginate, abaxial surfaces gray-green, initially strigose-villous, <midrib often red>, adaxial dark green, shiny, not glaucous, flat between lateral veins, initially sparsely long-haired pilose. **Inflorescences** on fertile shoots 8–40 mm with 3 or 4 leaves, 1 or 2(–6)-flowered, compact. **Pedicels** 1–3.5 mm, initially sparsely pilose. **Flowers** 12–13 mm diam.; buds white; hypanthium cupulate, initially sparsely pilose; sepals: margins villous, apex acute, surfaces villous; petals spreading, white, glabrous; stamens 15–20, filaments white, anthers dark red-purple; styles 2–4(or 5). **Pomes** red-orange to scarlet, obovoid to subglobose, [4–]7–11 × 7–10 mm, shiny, not glaucous, glabrous or sparsely villous; sepals depressed to slightly ascending, glabrous or sparsely villous; navel open; style remnants at apex. **Pyrenes** 2–4(or 5) (always some with more than 2). $2n = 34$ (Germany).

Flowering May–Jun; fruiting Oct–May. Hedges, edges, paths, urban waste ground; 0–200 m; introduced; B.C., Que.; Oreg., Wash.; Europe.

Cotoneaster ×*suecicus* is of garden origin, presumably a cross between *C. conspicuus* and *C. dammeri*.

17. Cotoneaster simonsii Baker in W. W. Saunders, Refug. Bot. 1: plate 55. 1869 • Khasiberry or Simon's cotoneaster [F] [I]

Shrubs, 2–4(–6) m; <± strict or ascending, crown narrow or ± columnar>. **Stems** erect, <stiff>; branches spiraled, divaricate, <lateral branches ± straight, mostly suppressed, short>, grayish brown, initially densely strigose. **Leaves** deciduous (sometimes tardily so on vigorous sprouts); petiole 3–5 mm, strigose; blade on vigorous shoots broadly elliptic to ovate, sometimes elliptic, rarely suborbiculate, 17–36 × 10–25 mm, chartaceous or subcoriaceous, base obtuse or cuneate, margins flat, veins 4 or 5, superficial or slightly sunken, apex acuminate or acute, abaxial surfaces light green, initially strigose, usually becoming sparsely so, adaxial green to dark green, shiny, not glaucous, flat or faintly bulging between lateral veins, or slightly rugose, sparsely strigose; <fall leaves yellow to bright orange or fiery red>. **Inflorescences** on fertile shoots 15–30 mm with 4 leaves, 2–6-flowered, compact. **Pedicels** 2–5 mm, strigose. **Flowers** <pendent>, (5–)8–10 mm, open; hypanthium campanulate, strigose; sepals: margins villous, <borders narrow, glabrous>, apex <recurved>, long acuminate or cuspidate, surfaces strigose; petals erect-incurved, dark red with white border; stamens 20, filaments pink, anthers white; styles (2 or)3–5. **Pomes** bright orange to orange-red, usually obovoid, rarely cylindric or broadly obovoid, 7–14 × 6–10 mm, shiny, not glaucous, glabrous or sparsely villous; sepals erect or ascending, rarely nearly flat, <margins villous>, glabrous or sparsely villous; navel open; style remnants ⅔ from base. **Pyrenes** (2 or)3–5.

Flowering May–Jun; fruiting Oct–Apr. Shores, clearings, disturbed or open forests, thickets; 0–200 m; introduced; B.C.; Calif., Oreg., Tenn., Wash.; Asia (Bhutan, India); introduced also in Europe, Pacific Islands (New Zealand).

Reports of *Cotoneaster simonsii* from Ohio (M. A. Vincent and A. W. Cusick 1998; T. S. Cooperrider et al. 2001) are here referred to *C. divaricatus*. The binomial *C. symondsii* T. Moore was proposed as an earlier name for *C. simonsii* (A. Kumar and G. Panigrahi 1992; W. B. Dickoré and G. Kasperek 2010), but J. Fryer and P. F. Zika (2014) typified *C. symondsii* so that it is a synonym of the high Himalayan species *C. marginatus* Lindley ex Loudon.

18. Cotoneaster adpressus Bois, Bull. Soc. Bot. France 51(sess. extraord.): cxlix, fig. [p. cl.] 1907 (as adpressa) • Creeping cotoneaster [I]

Shrubs, <mound-forming>, to 0.3 m. **Stems** <divaricate>, prostrate to procumbent, <rooting, to 0.8 m long, stiff>; branches distichous, maroon, initially strigose-villous. **Leaves** deciduous; petiole 1–2 mm, glabrescent; blade elliptic or ovate, rarely suborbiculate, 8–15 × 5–12 mm, chartaceous, base rounded, margins slightly undulate, not revolute, veins 2 or 3, superficial, apex acute or obtuse, abaxial surfaces green, glabrate, <midrib strigose>, adaxial green, dull to slightly shiny, not glaucous, flat between lateral veins, glabrous. **Inflorescences** on fertile shoots 6–10 mm with 2 or 3 leaves, 1(or 2)-flowered, <subsessile>. **Pedicels** 0–1.5 mm, glabrate. **Flowers** 4–7 mm; hypanthium funnelform, <maroon>, glabrate; sepals <sometimes recurved>, margins villous, apex lingulate or acute, surfaces glabrous; petals erect, dark red; stamens 10(–13), filaments pink to red, anthers white; styles 2(or 3).

Pomes bright red, broadly obovoid to globose, 6–7 mm, shiny, not glaucous, <succulent>, glabrous; sepals suberect, glabrous; navel open; style remnants 4/5 from base. **Pyrenes** 2(or 3). $2n = 34$ (Sweden).

Flowering Apr–May; fruiting Sep. Disturbed ground, gravel pits; 0–400 m; introduced, Mich.; Asia (China); introduced also in Europe.

19. Cotoneaster atropurpureus Flinck & B. Hylmö, Watsonia 18: 311. 1991 • Prostrate or purple-flowered cotoneaster [F] [I]

Shrubs, 0.5–1 m. **Stems** ascending to erect, spreading horizontally, arching; branches divaricate, sometimes distichous, red-purple, initially densely yellowish strigose-villous. **Leaves** deciduous; petiole 1–2 mm, yellow-villous; blade elliptic, on vigorous shoots obovate-orbiculate, 9–14 × 8–12 mm, chartaceous, base obtuse or acute, margins flat or slightly undulate, veins 2 or 3, superficial or slightly sunken, apex obtuse or truncate, mucronulate, abaxial surfaces <shiny>, sparsely golden yellow-pilose, adaxial green to dark green, shiny, not glaucous, flat or scarcely bulging between lateral veins, glabrous; <fall leaves ruby red>. **Inflorescences** on fertile shoots 9–11 mm with 3 leaves, (1 or)2 or 3-flowered, compact. **Pedicels** 0.2–1 mm, pilose. **Flowers** 6 mm, closed or nearly so; hypanthium funnelform, sparsely yellow-pilose-strigose; sepals: margins yellow-villous, apex <green and red-purple>, acuminate or acute, surfaces sparsely pilose; petals erect-incurved, dark red, base purple-black, margins narrowly white; stamens 10–14, filaments dark red to dark purple-red, white distally, anthers white [pink-tinged]; styles 2 or 3(or 4). **Pomes** bright red, usually broadly obovoid, sometimes obovoid, 7–8 × 7–8 mm, shiny, not glaucous, glabrate; sepals suberect, <margins villous>, sparsely pilose; navel open; style remnants 3/4 from base. **Pyrenes** 2 or 3(or 4).

Flowering May; fruiting Sep–Jan. Thickets, rock roadcuts, paths, edges of disturbed forests; 0–400 m; introduced; N.Y., Wash.; Asia (China); introduced also in Europe.

Although treated as synonymous with *Cotoneaster horizontalis* by L. Lingdi and A. R. Brach (2003), *C. atropurpureus* is separated by having somewhat thinner leaves on vigorous shoots that tend to be obovate instead of broadly elliptic, with obtuse to truncate apices and sometimes slightly wavy margins. It also tends towards less regular branching and somewhat darker petals and filaments.

20. Cotoneaster divaricatus Rehder & E. H. Wilson in C. S. Sargent, Pl. Wilson. 1: 157. 1912 (as divaricata) • Spreading or stretch-fruit cotoneaster [F] [I]

Shrubs, 1.5–2 m; <crown spreading, rounded>. **Stems** erect; branches divaricate or spiraled, <spreading, lateral branches well developed, usually arching, long>, brownish violet, initially strigose. **Leaves** deciduous; petiole 3–4 mm, strigose; blade elliptic, rarely ovate or suborbiculate, on sterile shoots usually elliptic or broadly elliptic, 10–30 × 7–21 mm, chartaceous, base obtuse, margins flat, not revolute, sometimes undulate, veins 3 or 4, superficial or slightly sunken, apex acute, rarely acuminate, abaxial surfaces pale green, initially strigose, adaxial dark green, shiny, not glaucous, flat between lateral veins, initially sparsely strigose; <fall leaves orange, red, or reddish purple>. **Inflorescences** on fertile shoots 10–20 mm with 4 leaves, 1–4-flowered, compact. **Pedicels** 1–3 mm, strigose. **Flowers** 6–8 mm, opening small; hypanthium cupulate, strigose; sepals: margins villous, apex acute or acuminate, surfaces strigose; petals erect-incurved, pale red, base dark red, border white; stamens 10–15, filaments dark red, whitish distally, anthers white; styles (1 or)2(or 3). **Pomes** dark red to ruby (finally blackish red), cylindric, oblong-ellipsoid, or narrowly obovoid, 8–11 × 4–8 mm, shiny, not glaucous, sparsely pilose; sepals suberect, sparsely pilose; navel open; style remnants 2/3 from base. **Pyrenes** (1 or)2(or 3) [rarely 4].

Flowering Apr–May; fruiting Sep–Dec. Thickets, edges, disturbed forests, flood plains, rarely epiphytic on maple; 0–700 m; introduced; B.C., Ont.; Ill., Ind., Ky., Mass., Mich., N.Y., Ohio, Oreg., Pa., Wash., Wis.; Asia (China); introduced also in Europe.

Although rarely collected, *Cotoneaster divaricatus* is the most widely escaped cotoneaster in eastern North America. Misidentified specimens of it have been reported from New York as *C. hupehensis* Rehder & E. H. Wilson (R. S. Mitchell and G. C. Tucker 1997), from Illinois as *C. acutifolius* (J. T. Kartesz and C. A. Meacham 2003), from Indiana and Ontario as *C. horizontalis* (Kartesz and Meacham), from Kentucky as *C. apiculatus* Rehder & E. H. Wilson (Kartesz and Meacham), and from Wisconsin as *C. lucidus* (Kartesz and Meacham).

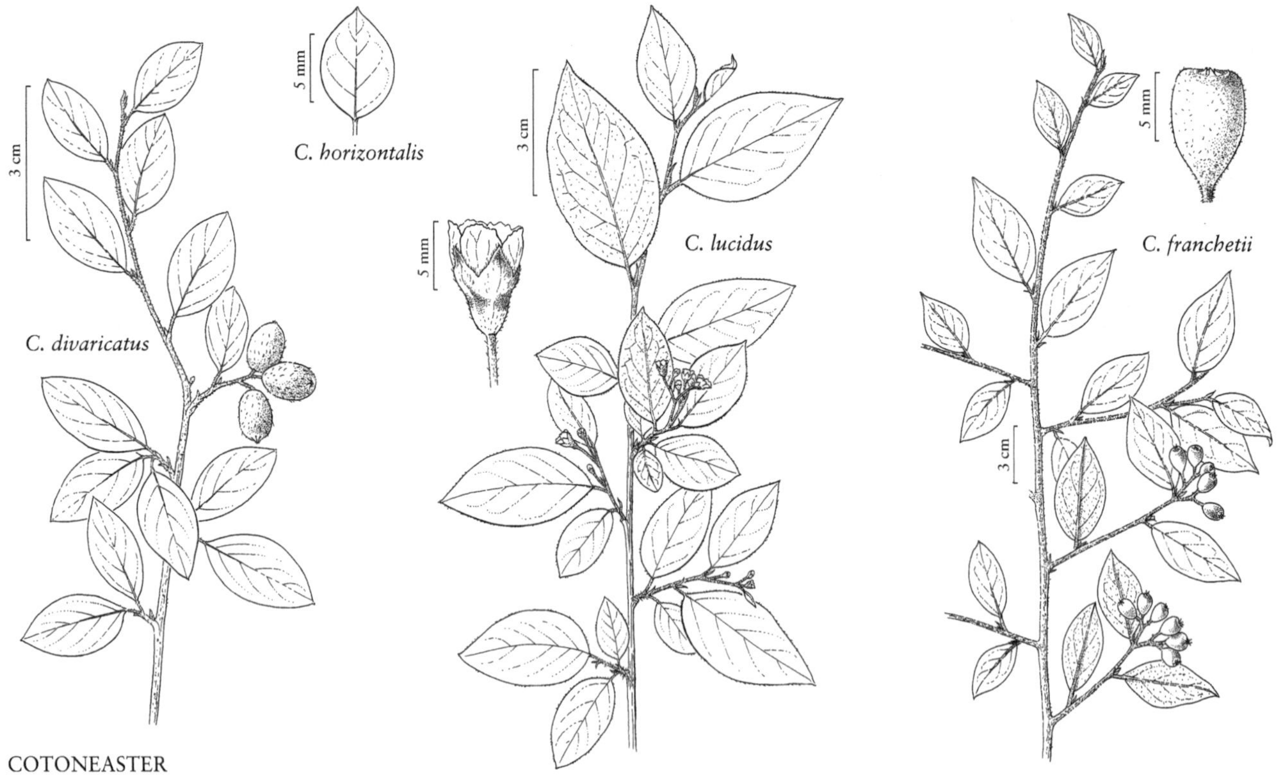

COTONEASTER

21. Cotoneaster hjelmqvistii Flinck & B. Hylmö, Watsonia 18: 312. 1991 • Hjelmqvist's cotoneaster [I]

Shrubs, 0.5–1 m. **Stems** ascending to erect, spreading horizontally, arching; branches usually distichous, brownish red, initially densely strigose. **Leaves** deciduous; petiole 2–4 mm, pilose; blade broadly elliptic, on vigorous shoots orbiculate, sometimes broadly obovate, <concave>, 10–18(–25) × 9–15(–25) mm, chartaceous, base obtuse, margins flat, veins 3 or 4 (or 5), superficial or slightly sunken, apex obtuse, mucronulate, abaxial surfaces pale green, yellowish pilose-strigose, adaxial vivid light green, shiny, not glaucous, flat between lateral veins, glabrous; <fall leaves red to intense red-purple>. **Inflorescences** on fertile shoots 5–25 mm with 3 or 4 leaves, 1–3(or 4) -flowered, compact. **Pedicels** 0.5–2(–3) mm, glabrescent. **Flowers** 5–6 mm, open; hypanthium cupulate, glabrescent; sepals: margins villous, apex <green and reddish purple>, acuminate or acute, surfaces glabrous; petals erect-incurved, dark pink, base red, margins pink; stamens 10–12(–16), filaments red and pale crimson, anthers white; styles 2 or 3. **Pomes** orange-red, usually broadly obovoid, sometimes subglobose or obovoid, 5.7–6.8[–8] × 4.9–7.2[–8] mm, shiny, not glaucous, glabrous; sepals suberect, <margins villous>, glabrous; navel open; style remnants 4/5 from base. **Pyrenes** 2 or 3, <distinct>.

Flowering May; fruiting Oct–Nov. Thickets, wet prairie remnants; 0–200 m; introduced, Oreg.; Asia (China); introduced also in Europe.

Cotoneaster hjelmqvistii was overlooked in the *Flora of China* (L. Lingdi and A. R. Brach 2003).

22. Cotoneaster horizontalis Decaisne, Ann. Gén. Hort. 22: 168. 1877 • Herringbone cotoneaster [F] [I]

Shrubs, 0.5–1 m. **Stems** ascending to erect, spreading horizontally; branches distichous, maroon, initially densely yellow-strigose. **Leaves** deciduous (tardily so on a few vigorous shoots); petiole 2–3 mm, strigose; blade elliptic or broadly elliptic, on vigorous shoots usually broadly elliptic, sometimes suborbiculate, orbiculate, or broadly obovate, 4–14 × 5–9 mm, subcoriaceous, base obtuse or cuneate, margins flat, veins 2–4, superficial, apex acute, apiculate or obtuse, mucronulate, abaxial surfaces green, sparsely long-strigose, adaxial dark green, shiny, not glaucous, flat between lateral veins, smooth to slightly rugose, glabrescent; <fall leaves orange to rich red>. **Inflorescences** on fertile shoots 8–20 mm with 3 or 4 leaves, 1–3-flowered, compact. **Pedicels** 1–2 mm, strigose. **Flowers** 5–7 mm, closed; hypanthium funnelform, sparsely pilose-strigose; sepals: margins villous, apex acute or acuminate, surfaces sparsely strigose; petals erect-incurved, dark pink to dark red, base blackish red, margins pale crimson; stamens (8–)10(–14), filaments dark red, anthers white; styles 2 or 3(or 4). **Pomes** bright red, usually broadly obovoid, sometimes obovoid, rarely subglobose, 5–8.5 × 4.5–7.7 mm, shiny, not glaucous, glabrate; sepals suberect, sparsely strigose or glabrate; navel open; style remnants ½–¾ from base. **Pyrenes** 2 or 3(or 4), <distinct>. $2n$ = 68 (Germany).

Flowering Apr–May; fruiting Sep–Mar. Thickets, disturbed forests, paths, cliffs, pilings, rockwork; 0–500 m; introduced; B.C.; Calif., Oreg., Wash.; Asia (China); introduced also in Europe.

L. Lingdi and A. R. Brach (2003) included within *Cotoneaster horizontalis* several taxa recognized here as distinct species, including *C. flinckii* J. Fryer & B. Hylmö and *C. perpusillus* (C. K. Schneider) Flinck & B. Hylmö. Reports of *C. horizontalis* from Ontario (J. K. Morton and J. M. Venn 1990) are here referred to *C. divaricatus*. Reports of *C. horizontalis* from Indiana are not supported by specimens.

23. Cotoneaster nitens Rehder & E. H. Wilson in C. S. Sargent, Pl. Wilson. 1: 156. 1912 • Few-flowered or Kokonor cotoneaster [I]

Shrubs, 1.5–3 m. **Stems** erect, becoming pendulous; branches divaricate, red-brown, initially strigose. **Leaves** deciduous; petiole 2–3(–4) mm, pilose-strigose; blade elliptic or obovate, sometimes broadly elliptic, broadly ovate, or suborbiculate, 8–28 × 9–23 mm, chartaceous, base rounded, margins flat, veins 3–5, superficial, apex obtuse or acute and mucronulate, abaxial surfaces pale green, initially densely pilose-strigose, adaxial initially reddish, soon bright green, dull to slightly shiny, not glaucous, flat between lateral veins, glabrous; <fall leaves turning bright red and gold>. **Inflorescences** on fertile shoots 15–30 mm with 3 or 4 (or 5) leaves, 1–3(or 4)-flowered, compact. **Pedicels** 2–4 mm, glabrescent. **Flowers** 6–7 mm, open; hypanthium widely cupulate, sparsely pilose; sepals: margins tomentose, apex obtuse, acute, or apiculate, surfaces sparsely pilose; petals erect-incurved, pink-tinged with purple-maroon or pink-red, base drying to purple-black, margins white, glabrous; stamens 10–14(or 15), filaments pink, sometimes red-purple, distally dark pink, anthers pink or white, <margins pale crimson, drying to dark red>; styles 2(or 3). **Pomes** purple-black, ellipsoid or obovoid to globose, (7–)8–9.5 × 6.5–8 mm, shiny, slightly glaucous, sparsely pilose to glabrate; sepals erect or suberect, sparsely pilose; navel open; style remnants ¾ from base. **Pyrenes** 2(or 3). $2n$ = 68 (Germany).

Flowering Mar–May; fruiting Jul–Nov. Thickets, edges, disturbed forests; 0–200 m; introduced; Oreg., Wash.; Asia (China); introduced also in Europe.

24. Cotoneaster lucidus Schlechtendal, Linnaea 27: 541. 1854 (as lucida) • Hedge or shiny cotoneaster, cotonéaster à feuilles aiguës [F] [I]

Cotoneaster acutifolius Turczaninow var. *lucidus* (Schlechtendal) L. T. Lu

Shrubs, 1.5–2.5 m. **Stems** erect, arching, or spreading; branches spiraled or distichous, <dense>, greenish gray to maroon, <not lenticellate>, initially strigose. **Leaves** deciduous; petiole 4–8 mm, strigose; blade elliptic or ovate, sometimes obovate, 28–50(–65) × 17–32(–40) mm, chartaceous, base cuneate or rounded, margins flat, not revolute, veins 3–5, sunken, apex short-acuminate or acute, abaxial surfaces green, sparsely pilose-strigose, adaxial dark green, intensely shiny, not glaucous, flat or scarcely bulging between lateral veins, slightly rugose, initially sparsely strigose; <fall leaves intense dark red, yellow, or orange>. **Inflorescences** on fertile shoots 25–50 mm with 4 or 5 leaves, 2–6(–15)-flowered, lax. **Pedicels** 5–12 mm, sparsely pilose. **Flowers** 5.5–8 mm, largely open; hypanthium cupulate, <base sparsely pilose>, glabrescent; sepals: margins villous, apex acute or acuminate, surfaces glabrous; petals erect, pinkish white, sometimes green-tinged, base dark pink or reddish; stamens 18–20, filaments pink, <base dark pink>, anthers white; styles 2 or 3. **Pomes** black, broadly obovoid to obovoid, rarely ellipsoid or globose, (8–)9.5–12.1 × (7.5–)8.7–11.8 mm, shiny, glaucous, sparsely pilose; sepals flat, <margins villous>, sparsely pilose; navel open; style remnants ⅔ from base. **Pyrenes** 2 or 3. $2n$ = 68 (Germany).

Flowering Mar–Jul (sometimes reflowering in Sep); fruiting Jul–Oct. Deciduous or coniferous forests, ravines, lakeshores, forested dunes, thickets, edges; 0–2300 m; introduced; Alta., B.C., Man., Ont., Que., Sask.; Colo., Idaho, Ind., Minn., Wash., Wyo.; Asia (Mongolia, Siberian Russia); introduced also in Europe.

Leaf glossiness is often retained on herbarium specimens and is visible with strong lighting and magnification.

Plants of *Cotoneaster acutifolius* of China and Mongolia are seldom seen in cultivation. Reports of it naturalized in North America are here referred to the commonly cultivated *C. lucidus*, which tolerates continental winters. No specimens have been seen to support reports of the latter from Montana and Oregon. Some reports of *C. melanocarpus* from Alberta and Manitoba (H. J. Scoggan 1978–1979, part 3) are here referred to *C. lucidus*.

25. Cotoneaster villosulus (Rehder & E. H. Wilson) Flinck & B. Hylmö, Bot. Not. 115: 383. 1962 • Taiping cotoneaster [I]

Cotoneaster acutifolius Turczaninow var. *villosulus* Rehder & E. H. Wilson in C. S. Sargent, Pl. Wilson 1: 158. 1912 (as acutifolia var. villosula)

Shrubs or trees, 3–5 m. **Stems** erect, arching, spreading, <coarse>; branches distichous or spiraled, brown, initially densely villous-strigose. **Leaves** deciduous; petiole 2–5 mm, strigose; blade elliptic or ovate, 32–72(–104) × 18–37(–46) mm, subcoriaceous to chartaceous, base obtuse or cuneate, margins flat, not revolute, veins 5–8, lightly sunken, apex attenuate to long-acuminate, often acute on short shoots, abaxial surfaces green, <shiny>, sparsely pilose-strigose, adaxial dark reddish green, shiny, not glaucous, soon dark green, dullish, bulging between lateral veins, rugose, pilose-strigose; <fall leaves purple and rich dark red>. **Inflorescences** on fertile shoots 25–50 mm with 4 leaves, 3–5(–7)-flowered, mostly compact. **Pedicels** 3–7 mm, densely strigose. **Flowers** 6–8 mm, opening small; hypanthium cupulate, tomentose-strigose; sepals: margins tomentose, <borders brown, broad, glabrous>, apex acute, obtuse, or apiculate, surfaces strigose; petals erect-incurved, pink or red, base dark red, margins white; stamens 18–20, filaments pale pink, white distally, anthers white; styles 2 or 3. **Pomes** purple-black, broadly ellipsoid to broadly obovoid, rarely cylindric, 9–11 × 8–11 mm, slightly shiny, not or slightly glaucous, villous; sepals flat or ascending, <margins tomentose>, strigose; forming star over open navel; style remnants ⅘ from base. **Pyrenes** 2 or 3. $2n = 68$ (Germany).

Flowering May (often reflowering Sep); fruiting Aug–Nov. Thickets, open forests; 0–50 m; introduced, Wash.; Asia (China); introduced also in Europe.

The difficulties of typification and taxonomy for black-fruited species were discussed by J. Fryer and B. Hylmö (1997). Some collections of *Cotoneaster villosulus* from eastern Washington need study. They approach the very similar *C. hsingshangensis* J. Fryer & B. Hylmö, a Chinese species with larger, more wrinkled leaves, more globose pomes, and less densely hairy flowers.

26. Cotoneaster rehderi Pojarkova, Bot. Mater. Gerb. Bot. Inst. Komarova Akad. Nauk S.S.S.R. 17: 184. 1955 • Bullate cotoneaster [I]

Cotoneaster bullatus Bois var. *macrophyllus* Rehder & E. H. Wilson in C. S. Sargent, Pl. Wilson. 1: 164. 1912 (as bullata var. macrophylla)

Shrubs, 2–5 m. **Stems** erect, arching; branches distichous or spiraled, maroon to brown, <lenticellate>, initially pilose-strigose. **Leaves** deciduous; petiole 0–3 mm, pilose-strigose; blade elliptic, sometimes ovate-elliptic, rarely obovate, (50–)70–124(–210) × (24–)30–55(–90) mm, subcoriaceous, base cuneate, obtuse, truncate, or auriculate, margins flat, not revolute, veins 8–11, deeply sunken, apex long-acuminate, sometimes acute on short shoots, abaxial surfaces light green, initially yellowish strigose, adaxial green, shiny, not glaucous, strongly wrinkled or bulging between lateral veins (bullate), initially sparsely pilose; <fall leaves intense butter yellow and reddish purple>. **Inflorescences** on fertile shoots 60–120 mm with 4 leaves, 8–32-flowered, lax. **Pedicels** 2–4 mm, pilose-strigose. **Flowers** (5–)7–9 mm, closed; hypanthium shallowly cupulate, <shiny>, sparsely pilose-strigose; sepals: margins villous, apex acute, surfaces <shiny>, glabrous; petals erect-incurved, pink, red, greenish pink, or maroon, margins pink; stamens 20–23, filaments pink, white distally, anthers white; styles (4 or)5. **Pomes** bright to dark red, obovoid, broadly obovoid, or obconic, rarely globose or depressed-globose, 8–12.4 × 7–11.3 mm, shiny, not glaucous, sparsely pilose; sepals flat, glabrous; navel closed; style remnants ¾ from base. **Pyrenes** (4 or)5.

Flowering May–Jun; fruiting Sep–Dec. Thickets, disturbed forests, flood plains, lakeshores; 0–200 m; introduced; B.C.; Alaska, Wash.; Asia (China); introduced also in Europe.

Reports of *Cotoneaster bullatus* (treated as *C. bullatus* var. *bullatus* by L. Lingdi and A. R. Brach 2003) from British Columbia (J. Pojar 1999) are here referred to *C. rehderi*.

27. Cotoneaster tengyuehensis J. Fryer & B. Hylmö, Watsonia 21: 338. 1997 • Tengyueh cotoneaster [I]

Shrubs, 2–2.5 m; <crown ± rounded>. **Stems** erect, arching, <long>; branches distichous, grayish maroon, initially densely strigose. **Leaves** deciduous (sometimes tardily so on vigorous shoots); petiole 1–4 mm, strigose; blade ovate or elliptic, (25–)30–50(–61) × 12–32 mm, chartaceous, base cuneate or rounded, margins flat, veins 3–5, deeply sunken, apex acuminate or acute, abaxial surfaces grayish green, villous-strigose or sparsely to moderately gray-tomentose, adaxial green to dark green, shiny, not glaucous, flat or faintly bulging between lateral veins, pilose-strigose; <fall leaves yellow to orange>. **Inflorescences** on fertile shoots 25–40 mm with (3 or)4 leaves, 3–7(–9)-flowered, compact. **Pedicels** 1–4 mm, densely strigose. **Flowers** <erect or ascending>, [5–]6–8 mm, opening small; hypanthium cupulate, strigose; sepals: margins villous, <borders reddish brown, glabrous>, apex acuminate or cuspidate, surfaces strigose; petals erect-incurved, pink to red, base dark pink or red, rarely dark red or maroon, margins white; stamens 20 or 21, filaments pink or pale pink, whitish distally, anthers white; styles (2 or)3–5. **Pomes** <spreading or pendent>, bright red, broadly obovoid or subglobose, rarely oblong-obovoid, 6.3–10.4 × 5.8–9.1 mm, shiny, not glaucous, sparsely strigose; sepals flat or suberect, strigose; navel closed; style remnants 2/3 from base. **Pyrenes** (2 or)3–5. $2n$ = 68 (Germany).

Flowering May–Jun; fruiting Sep–Feb. Thickets, creek banks, open forests, edges; 0–50 m; introduced, Wash.; Asia (China); introduced also in Europe.

L. Lingdi and A. R. Brach (2003) confused *Cotoneaster tengyuehensis* with *C. glomerulatus* W. W. Smith, a species with small fruits 5 mm in diameter, with five pyrenes and styles, and slightly smaller leaves to 40–50 mm. In the flora area, *C. tengyuehensis* is more likely to be mistaken for *C. franchetii*, which is a true evergreen with orange-red, obovate to obconic fruits, and pink to purple stamens. Naturalized shrubs of *C. tengyuehensis* drop their foliage in midwinter.

28. Cotoneaster franchetii Bois, Rev. Hort. 74: 380, figs. 159–161, 164. 1902 (as francheti) • Franchet's cotoneaster [F] [I]

Shrubs, 2–3 m. **Stems** erect, arching, <lax, slender>; branches distichous, maroon, initially densely pilose-strigose. **Leaves** persistent, <often 30–40 mm apart>; petiole 2–4 mm, tomentose-pilose; blade ovate or elliptic, 22–58 × 12–29 mm, coriaceous, base cuneate or obtuse, margins flat, veins 4 or 5, sunken, apex acute or acuminate, mucronate, abaxial surfaces silvery-tomentose, adaxial grayish green, slightly shiny, not glaucous, flat, rarely slightly bulging between lateral veins, drying slightly rugose, pilose. **Inflorescences** on fertile shoots 30–60 mm with 4 leaves, 5–15(–25) -flowered, compact. **Pedicels** 1–4 mm, densely long-silky hairy. **Flowers** (5–)7–8(–10) mm, opening small; hypanthium cupulate, densely long-silky hairy; sepals: margins villous, <borders purple, glabrous>, apex cuspidate or acuminate, surfaces silky hairy; petals erect-incurved, pink or red, base dark pink to dark red, margins off-white; stamens 16–21, filaments pink, distally white, anthers pink or purplish pink; styles 2 or 3(or 4). **Pomes** orange-red, obovoid <base rounded> or obconic <base cuneate, straight-sided>, 8–12 × 6–9 mm, <apex flat>, shiny, not glaucous, sparsely pilose; sepals flat, <often asymmetric with 2 sepals longer than others>, tomentose; navel slightly open; style remnants 3/4 from base. **Pyrenes** 2 or 3(or 4). $2n$ = 68 (Germany).

Flowering May–Jul; fruiting Oct–Apr. Edges, thickets, disturbed forests, flood plains, urban waste ground, mossy ledges, wet prairie remnants; 0–700 m; introduced; B.C.; Calif., Oreg., Wash.; Asia (China); introduced also in Europe, Pacific Islands (New Zealand), Australia.

Reports of *Cotoneaster franchetii* from Colorado (J. Ells 2006) are here referred to *C. lucidus*. A few specimens escaped in Washington have red filaments, not pink, but otherwise agree with *C. franchetii*. Other escapes in Washington suggest the closely related *C. wardii* W. W. Smith, distinguished by semievergreen habit, longer fertile shoots, more lax and elongate inflorescences, and white anthers. The entire complex needs biosystematic review.

29. Cotoneaster qungbixiensis J. Fryer & B. Hylmö, J. Bot. Res. Inst. Texas 2: 53, fig. 1. 2008 • Truncate-leaved or Cangshan cotoneaster [I]

Shrubs, 2–3 m. **Stems** erect, arching, spreading; branches distichous or spiraled, reddish brown, initially strigose-pilose. **Leaves** deciduous (sometimes tardily so on vigorous basal sprouts); petiole <often red>, 3–4 mm, strigose-pilose; blade elliptic or narrowly elliptic to sometimes elliptic-ovate or narrowly ovate, 20–48(–61) × (10–)15–29 mm, chartaceous, base cuneate or obtuse, margins flat, veins 2–4, sunken, apex minutely truncate, rarely acute, mucronate, abaxial surfaces densely grayish tomentose, adaxial dark green, shiny, not glaucous, flat between lateral veins, slightly rugose, initially pilose; <fall leaves orange>. **Inflorescences** on fertile shoots 30–50 mm with 4–6 leaves, 3–9(–15)-flowered, compact. **Pedicels** 1–3 mm, strigose-pilose. **Flowers** 6.5–9 mm, open; hypanthium funnelform, silky tomentose; sepals: margins tomentose, apex cuspidate or acuminate, surfaces tomentose; petals erect-incurved, pink or reddish, base dark red, margins yellowish or off-white; stamens 17–20, filaments dark red, at least proximally, anthers white; styles (2 or)3 or 4[or 5]. **Pomes** orange-red, obovoid to broadly obovoid or subglobose, rarely globose, (7–)8.2–10.1 × 6.6–7.6 mm, slightly shiny, not glaucous, pilose; sepals flat, tomentose; navel closed; style remnants ⅔ from base. **Pyrenes** (2 or)3 or 4[or 5]. $\mathbf{2n} = 68$ (England).

Flowering Jun–Jul; fruiting Oct–Dec. Thickets, disturbed forests; 0–50 m; introduced, B.C.; Asia (China).

30. Cotoneaster sternianus (Turrill) Boom, Jaarb. Ned. Dendrol. Ver. 20: 81. 1957 • Stern's cotoneaster [I]

Cotoneaster franchetii Bois var. *sternianus* Turrill, Bot. Mag. 167: plate 130. 1950 (as sterniana)

Shrubs, 1.5–4 m. **Stems** erect, arching, <stiff>; branches distichous, maroon, whitish tomentose-pilose. **Leaves** persistent; petiole 4–6 mm, tomentose-pilose; blade elliptic to broadly ovate, (18–)25–52(–56) × (12–)16–28 (–34) mm, coriaceous, base obtuse or cuneate, margins recurved, veins 4–6, deeply sunken, apex acute or acuminate, abaxial surfaces whitish tomentose, adaxial dark green, slightly shiny, not glaucous, on vigorous shoots usually bulging slightly between lateral veins (unless shaded), very rugose, pilose. **Inflorescences** on fertile shoots 25–70 mm with 3–5 leaves, 7–20 -flowered, compact. **Pedicels** 1–4 mm, tomentose-pilose. **Flowers** (5–)6–7.5 mm, opening small; hypanthium cupulate, silky tomentose; sepals: margins villous, <borders purple, glabrous>, apex acute or acuminate, surfaces silky tomentose; petals erect-incurved, pink to red, base dark red (with minute black dots), margins erect-incurved, pink or white; stamens 20, filaments red, distally pink or whitish, anthers white, <sutures pink-tinged>; styles (2 or)3 or 4(or 5). **Pomes** bright orange, eventually orange-red, subglobose to depressed-globose or globose, 6–10 × 6–10 mm, <base rounded>, shiny, not glaucous, pilose; sepals flat, tomentose; forming 5-pointed star over closed navel; style remnants ¾ from base. **Pyrenes** (2 or)3 or 4(or 5).

Flowering Jun–Jul; fruiting Oct–Jan. Mossy cliffs, thickets, hedges, urban waste ground; 0–200 m; introduced, B.C.; Asia (China, Myanmar); introduced also in Europe, Pacific Islands (New Zealand).

Cotoneaster sternianus is similar to *C. franchetii*, a species with more elongated pomes, uniformly pink or purple anthers, and less textured leaves.

31. Cotoneaster dielsianus E. Pritz ex Diels, Bot. Jahrb. Syst. 29: 385. 1900 • Diels' cotoneaster [I]

Shrubs, 2–4 m. **Stems** narrowly erect, spreading; branches distichous, grayish brown, initially tomentose-pilose. **Leaves** deciduous; petiole 2–5 mm, tomentose-pilose; blade ovate to elliptic or broadly ovate, 9–38 × 7–23 mm, coriaceous or subcoriaceous, base obtuse or cuneate, margins flat, not revolute, veins 2–4, lightly sunken, apex acute or acuminate, abaxial surfaces densely grayish tomentose, adaxial dark green to grayish green, shiny, not glaucous, flat, rarely slightly bulging, between lateral veins, slightly rugose, pilose; <fall leaves bright yellow to orange or red>. **Inflorescences** on fertile shoots 10–35 mm with 4 leaves, 2–7(–10) -flowered, compact. **Pedicels** 1–3 mm, tomentose-strigose. **Flowers** <erect>, 6–7.5 mm, closed; hypanthium funnelform, proximally tomentose-strigose, distally tomentose-pilose; sepals: margins reddish tomentose, <borders purple, broad, glabrous>, apex cuspidate or apiculate, surfaces pilose; petals erect-incurved, dark pink to red, base dark red, margins off-white; stamens 14–20, filaments red to dark red, anthers white; styles 3 or 4(or 5). **Pomes** bright red, broadly obovoid, sometimes globose or depressed-globose, 5–9.1 × 5.1–8.8 mm, slightly shiny, not glaucous, pilose; sepals flat, tomentose; navel open; style remnants ½–⅔ from base. **Pyrenes** 3 or 4(or 5).

Flowering May–Jun; fruiting Aug–Mar. Edges of woods, disturbed forests, flood plains, thickets, waste ground, cliffs, meadows, brushy wet prairie remnants; 0–500 m; introduced; B.C.; Oreg., Wash.; Asia (China); introduced also in Europe.

L. Lingdi and A. R. Brach (2003) included within *Cotoneaster dielsianus* several taxa recognized here as distinct species, including *C. bradyi* J. Fryer & E. C. Nelson and *C. splendens* Flinck & B. Hylmö. Some collections from Oregon approach an undescribed species from China but differ in fruit shape.

32. Cotoneaster fangianus T. T. Yu, Acta Phytotax. Sin. 8: 219. 1963 • Fang's cotoneaster [I]

Shrubs, 2–2.5 m. **Stems** erect, arching, spreading; branches distichous, maroon, purple, or brown, initially tomentose-pilose. **Leaves** deciduous; petiole 2–4 mm, tomentose; blade elliptic, broadly elliptic, or ovate, rarely suborbiculate, 21–56 × 13–30 mm, subcoriaceous, base obtuse or truncate, margins flat, veins 4–6, slightly to deeply sunken, apex obtuse or acute, abaxial surfaces grayish, initially moderately to densely villous, adaxial dark green, dull, not glaucous, flat between lateral veins, slightly rugose, pilose; <fall leaves yellowish>. **Inflorescences** on fertile shoots 20–30 mm, usually with 2 or 3 leaves, 3–15-flowered, compact. **Pedicels** 2–5 mm, tomentose-pilose. **Flowers** <erect>, 4–6 mm; hypanthium cupulate, densely villous; sepals: margins villous, apex acute, surfaces villous, sparsely hairy submarginally; petals erect, pink or fading whitish; stamens (16–)19 or 20, filaments pale pink or white, anthers white; styles [1 or]2(or 3). **Pomes** <pendent>, bright to dark red, obovoid to obconic, 8–10 × 8–9 mm, <apex nearly flat>, reflectance and coating not recorded, <succulent>, villous; sepals flat, <margins villous>, villous; navel slightly open; style remnants ⅔ from base. **Pyrenes** [1 or]2(or 3).

Flowering May; fruiting Oct–Nov. Disturbed mesic forests; 0–200 m; introduced, Oreg.; Asia (China); introduced also in Europe.

Reports of *Cotoneaster mucronatus* Franchet escaped in Oregon (P. F. Zika and E. R. Alverson 2005) are here referred to *C. fangianus*.

33. Cotoneaster miniatus (Rehder & E. H. Wilson) Flinck & B. Hylmö, Bot. Not. 119: 456. 1966 • Orange-drop cotoneaster [I]

Cotoneaster zabelii C. K. Schneider var. *miniatus* Rehder & E. H. Wilson in C. S. Sargent, Pl. Wilson. 3: 430. 1917 (as miniata)

Shrubs, 1–1.5(–2) m. **Stems** erect, spreading, arching, <slender, weak>; branches distichous, divaricate, light brown, initially tomentose-pilose. **Leaves** deciduous; petiole 3–5 mm, densely pilose; blade elliptic to narrowly ovate, (24–)30–50(–68) × 12–25(–32) mm, chartaceous, base cuneate or obtuse, margins flat, not revolute, veins 3–5, slightly sunken, apex acute or obtuse, abaxial surfaces pale, <reticulate>, pilose, adaxial yellowish green, dull, not glaucous, flat between lateral veins, pilose; <fall leaves intensely yellow>. **Inflorescences** on fertile shoots 20–45 mm with 4 leaves, 8–18-flowered, lax. **Pedicels** <slender>, 5–10 mm, pilose. **Flowers** <spreading>, 4–5.5 mm, open; hypanthium cupulate, villous-pilose; sepals: margins villous, apex obtuse, surfaces villous; petals erect, pink, base pink to dark pink or red-tinged; stamens 19–23, filaments white or pale pink, anthers white; styles 2(or 3). **Pomes** <pendent>, bright orange-red, broadly ellipsoid, broadly obovoid, or globose, rarely depressed-globose, (4–)6–9.3(–11) × 6–9.7 mm, slightly shiny, not glaucous, sparsely villous; sepals flat, villous; forming star over slightly open navel; style remnants ¾ from base. **Pyrenes** 2(or 3).

Flowering May–Jul; fruiting Sep–Nov. Thickets; 0–100 m; introduced, Wash.; Asia (China).

34. Cotoneaster melanocarpus (Ledebour) Loddiges, G. Loddiges & W. Loddiges ex M. Roemer, Fam. Nat. Syn. Monogr. 3: 223. 1847 (as melanocarpa) • Dark cotoneaster [I]

Cotoneaster vulgaris Lindley var. *melanocarpus* Ledebour, Fl. Altaica 2: 219. 1830 (as melanocarpa)

Shrubs, 2–2.5 m. **Stems** loosely erect; branches spiraled, yellow brown to red brown, <lenticellate, shiny>, initially densely pilose. **Leaves** deciduous; petiole 4–7 mm, tomentose-villous; blade elliptic to ovate, 33–45 × 20–32 mm, chartaceous, base rounded, margins flat, veins 5–7, superficial, apex acute or obtuse, abaxial surfaces densely silvery-pilose-villous, adaxial green to dark green, dull to slightly shiny, not glaucous, flat between lateral veins, rugose, sparsely pilose; <fall

leaves without notable color>. **Inflorescences** on fertile shoots 25–30 mm with 3 or 4 leaves, 5–13-flowered, <pendent>, lax. **Pedicels** 3–8 mm, sometimes sparsely villous. **Flowers** <erect>, 6–7 mm, open; hypanthium cupulate, <dark reddish brown>, glabrous; sepals: margins erose, apex acute or obtuse, surfaces glabrous, often apically villous; petals erect, greenish white with pink, red, base slightly darkened, margins white, glabrous; stamens 20(–22), filaments white, anthers white; styles 2–4. **Pomes** purple-black, obovoid or globose, 7–9 × 7–9 mm, dull, glaucous with blue tinge, glabrous; sepals suberect, glabrous or apically villous; navel open; style remnants $\frac{2}{3}$ from base. **Pyrenes** 2–4. $2n$ = 48–52, 68 (Russia).

Flowering Apr–May; fruiting Jul–Aug. Forested ravines; 300–500 m; introduced, Man.; Eurasia (Russia, Ukraine).

Cotoneaster melanocarpus is said to hybridize spontaneously with *Sorbus aucuparia* in Siberia, forming ×*Sorbocotoneaster* Pojarkova. The actual *Cotoneaster* parent is more likely to be a diploid and needs investigation.

57. PYRACANTHA M. Roemer, Fam. Nat. Syn. Monogr. 3: 104, 219. 1847 • Firethorn [Greek *pyr*, fire, and *acanthos*, thorn, alluding to fiery thorns] [I]

Ronald W. Lance

Peter F. Zika

Sportella Hance

Shrubs, <spreading or partially reclining>, 5–60 dm. **Stems** 1–3+, erect or divergent, straight or geniculate; bark grayish, <thin, smooth, becoming slightly fissured>; long and short shoots present, <leafy>; thorns present (commonly on short shoots); glabrous or tomentose. **Leaves** persistent or late-deciduous, cauline, <fascicled on short shoots>, simple; stipules caducous, free, lanceolate, <membranous>, margins remotely serrate, <glabrous or tomentose>; petiole present; blade obovate, ovate, oblong-obovate, oblong, or elliptic to lanceolate or oblanceolate, 1–7 cm, leathery, margins flat or slightly revolute, serrulate, crenulate, or entire, venation pinnate, surfaces glabrous or sparsely hairy to densely tomentose when young, often glabrescent later. **Inflorescences** terminal, 6–40-flowered, <flat-topped> panicles, glabrous or tomentose; bracts present, <caducous, lanceolate>; bracteoles present. **Pedicels** present. **Flowers:** perianth and androecium epigynous, 3–10(–12) mm diam.; hypanthium campanulate, 2–4 mm diam., exterior glabrous or densely tomentose; sepals 5, spreading, broadly elliptic or triangular; petals 5, <spreading>, white, usually suborbiculate, <base slightly clawed>; stamens 15–20, shorter than petals, <filaments 2–4 mm>; carpels 5, distinct, adnate to proximal $\frac{1}{2}$ of hypanthium, externally glabrous or tomentose, styles terminal, distinct, <about equal to stamens>; ovules 2. **Fruits** pomes, red or orange-red, rarely yellow, globose, 3–8 mm diam., glabrate; <flesh soft>; hypanthium persistent; sepals persistent, erect or incurved; carpels woody; styles usually persistent. **Pyrenes** 5. x = 17.

Species 10 (6 in the flora): introduced; Europe, Asia; introduced also in Africa, Pacific Islands, Australia.

Morphologic studies align *Pyracantha* near *Cotoneaster* or *Heteromeles* (J. R. Rohrer et al. 1991; K. R. Robertson et al. 1991, 1992; C. S. Campbell et al. 2007). Molecular studies to date are inconclusive and have failed to provide strong or consistent support for these or alternative placements of *Pyracantha* (R. C. Evans and Campbell 2002; D. Potter et al. 2007; Campbell et al.). The difficulty in resolving the generic phylogeny of Maleae may be due in part to interfertility, lack of divergence, and reticulate evolution (Campbell et al.).

Species of *Pyracantha* are widely cultivated as ornamentals, and over 100 cultivars (D. R. Egolf and A. O. Andrick 1995) and hybrids are known. Of the ten species traditionally recognized, nine are native to China, some showing sufficient overlap of distribution and morphology that the taxonomy is not well defined. The plants as a group have been assigned in earlier treatments to the genera *Cotoneaster*, *Crataegus*, *Mespilus*, *Sportella*, and *Timbalia* Clos. Reports of additional naturalizing taxa could increase the number of our six sporadically naturalized species; validating specimens were not seen. Seedlings of naturalized *Pyracantha* most frequently appear in ruderal areas, along roads, forest edges, thickets, and in proximity to plantings. Spontaneous intermediates occasionally appear where cultivated species or hybrids grow in proximity, as no barrier to hybridization is apparent. The pomes are eaten by birds, most notably American robins (*Turdus migratorius* Linnaeus) and cedar waxwings (*Bombycilla cedrorum* Vieillot).

1. Leaf margins crenulate, crenulate-serrulate, or serrulate (at least ½ length of blades).
    2. Leaf blades oblanceolate or obovate, apices obtuse, emarginate, or short-apiculate . . . . 1. *Pyracantha fortuneana*
    2. Leaf blades lanceolate, oblong, oblanceolate, ovate-lanceolate, elliptic, or ovate, apices usually acute or short-apiculate.
        3. Young twigs gray-hairy, glabrescent later; hypanthia finely hairy . . . . 2. *Pyracantha coccinea*
        3. Young twigs brown-hairy, glabrescent later; hypanthia glabrous . . . . 3. *Pyracantha crenulata*
1. Leaf margins usually entire, rarely remotely serrulate distally or with minute teeth near apices.
    4. Calyces and leaf abaxial surfaces persistently gray-tomentose . . . . 4. *Pyracantha angustifolia*
    4. Calyces and leaf abaxial surfaces brown- or yellowish brown-puberulent when young, glabrescent.
        5. Leaf blades elliptic, oblong, or oblong-obovate (usually widest near middle), apices obtuse, apiculate, or aristate, abaxial surfaces glaucescent . . . . 5. *Pyracantha atalantioides*
        5. Leaf blades oblanceolate or narrowly obovate (usually widest distal to middle), apices usually truncate or retuse, abaxial surfaces pale green but not glaucescent . . . . 6. *Pyracantha koidzumii*

1. Pyracantha fortuneana (Maximowicz) H. L. Li, J. Arnold Arbor. 25: 420. 1944 • Chinese firethorn [I]

Photinia fortuneana Maximowicz, Bull. Acad. Imp. Sci. Saint-Pétersbourg 19: 179. 1873; *Pyracantha crenatoserrata* (Hance) Rehder; *P. crenulata* (D. Don) M. Roemer var. *rogersiana* A. B. Jackson; *P. rogersiana* (A. B. Jackson) Bean; *P. yunnanensis* (M. Vilmorin ex Mottet) Chittenden

Plants 10–30 dm. **Stems:** thorns sparse to abundant; young twigs rusty-hairy (gray-hairy in 1 variation), glabrescent. **Leaves:** stipules 3–8 mm; petiole 2–5 mm; blade oblanceolate or obovate, 1.5–6 × 0.5–2.5 cm, base cuneate, margins remotely serrulate or crenulate, often entire proximally, apex obtuse, emarginate, or short-apiculate, surfaces glabrate. **Inflorescences** 3–4 cm diam. **Pedicels** 2–8 mm, usually glabrate, rarely sparsely appressed brown- or gray-hairy. **Flowers** 4–12 mm diam.; hypanthium glabrate, rarely slightly hairy; sepals triangular, 1–1.5 mm, apex obtuse; petals suborbiculate, 3–4 mm, apex rounded. **Pomes** orange-red to dark red, 3–6 mm diam.; pedicels 2–10 mm. $2n = 34$ (China).

Flowering Feb–May; fruiting Sep–Mar. Disturbed ground, roadsides, canyons, woodland edges, open forests, riparian areas; 0–1500 m; introduced; Ala., Calif., Fla., Ga., La., N.C., Oreg., S.C., Tex., Wash.; Asia (China); introduced also in Europe (England), Pacific Islands (Hawaii, New Zealand), Australia.

Pyracantha fortuneana is native to central and western China. It is variable in the amount of toothing on the leaf blades and color of the hairs on new growth. Plants having grayish hairs on young twigs rather than the usual rusty or brownish hairs have been attributed to the taxon *P. rogersiana*, which has also been interpreted as a variety of *P. crenulata*. It is insufficiently distinct to separate from *P. fortuneana* among the specimens examined.

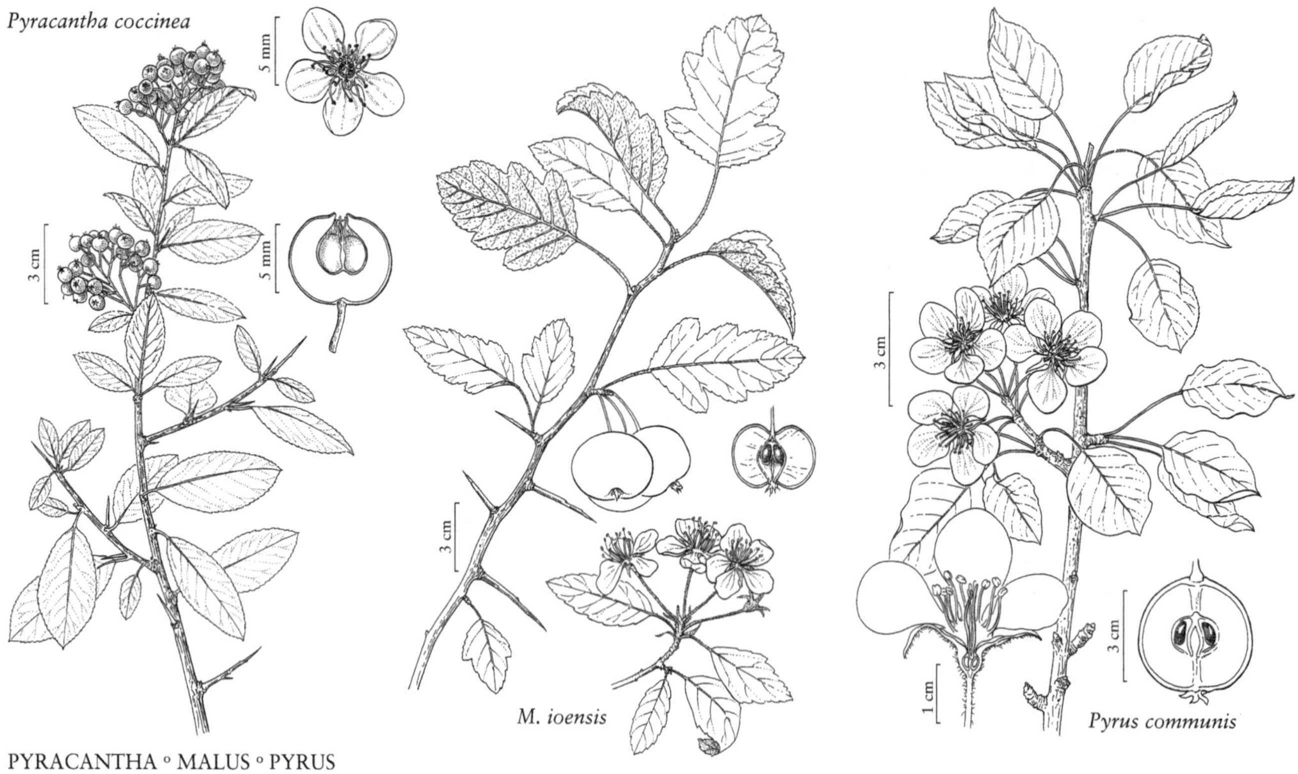

PYRACANTHA ◦ MALUS ◦ PYRUS

2. **Pyracantha coccinea** M. Roemer, Fam. Nat. Syn. Monogr. 3: 219. 1847, name conserved • Scarlet firethorn, buisson ardent [F] [I]

Mespilus pyracantha Linnaeus, Sp. Pl. 1: 478. 1753

Plants: <crowns spreading>, 10–60 dm. **Stems:** young twigs gray-hairy, glabrescent later. **Leaves:** stipules 4–8 mm; petiole 2–5 mm; blade elliptic or ovate to lanceolate or oblanceolate, 2–4 × 0.7–1.5 cm, base cuneate, margins finely crenulate-serrulate, apex acute, rarely obtuse, abaxial surface slightly hairy when young, glabrescent, adaxial glabrous. **Inflorescences** 3–4 cm diam.; <bracteoles in distal axils 4–10 × 3–4 mm>. **Pedicels** 3–10 mm, with coarse pale hairs. **Flowers** 6–8 mm diam.; hypanthium finely hairy; sepals triangular, 1.5–2 mm, apex acute; petals suborbiculate, 3–5 mm, apex rounded. **Pomes** bright red, 5–8 mm diam.; pedicels 5–12 mm. **2*n*** = 34 (Europe).

Flowering Mar–May; fruiting Aug–Jan. Mixed conifer forests, disturbed ground, fields, thickets, roadsides; 0–1500 m; introduced; B.C., Ont.; Ala., Ariz., Ark., Calif., Ga., Ind., Kans., La., Miss., Mo., N.Mex., N.Y., N.C., Ohio, Okla., Oreg., Pa., S.C., Tenn., Tex., Utah, Va., Wash.; s Europe; w Asia; introduced also in w Europe (England), e Asia (Japan), Africa (South Africa), Australia.

Pyracantha coccinea has been extensively planted throughout most of temperate North America. The species is native from southern Europe east to the Caucasus and western Asia.

3. **Pyracantha crenulata** (D. Don) M. Roemer, Fam. Nat. Syn. Monogr. 3: 220. 1847 [I]

Mespilus crenulata D. Don, Prodr. Fl. Nepal., 238. 1825; *Cotoneaster crenulatus* (D. Don) K. Koch; *Crataegus crenulata* Roxburgh; *C. pyracantha* (Linnaeus) Medikus var. *crenulata* (D. Don) Loudon; *Pyracantha chinensis* M. Roemer; *P. crenulata* var. *kansuensis* Rehder

Plants 10–50 dm. **Stems:** thorns usually abundant; young twigs brown-hairy, glabrescent later. **Leaves:** stipules 2–4 mm; petiole 2–6 mm; blade elliptic, oblong, or oblanceolate, sometimes ovate-lanceolate, 1–7 × 0.4–1.8 cm, base rounded to cuneate, margins crenulate to sparsely serrulate, apex acute or short-apiculate, sometimes obtuse, surfaces glabrous or nearly so. **Inflorescences** 2–5 cm diam.; <peduncles brown-hairy or glabrous>. **Pedicels** 4–10 mm, glabrous. **Flowers** 6–10 mm diam.; hypanthium glabrous; sepals triangular, 1 mm, apex acute; petals orbiculate, 3–5 mm, apex rounded. **Pomes** orange-yellow, orange, or red, <subglobose>, 3–8 mm diam.; pedicels 5–12 mm.

Flowering Mar–May; fruiting Aug–Oct. Canyons; 0–1500 m; introduced, Calif.; Asia; introduced also in Europe, e Asia (Japan), Africa (South Africa), Pacific Islands (New Zealand), Australia.

Pyracantha crenulata is native to southwestern China, India, Kashmir, and Nepal. A dwarf variant has been called var. *kansuensis* and is reported as native to China (A. J. Rehder 1949; Gu C. Z. and S. A. Spongberg 2003b); it is distinguished by its short stature (rarely exceeding 2 m), abundantly thorny branches, small leaf blades (1–2.5 × 0.4–0.8 cm), and small (3–5 mm diam.) red pomes. Its presence among the naturalized records of *P. crenulata* in California is suggested, but its morphology is not clearly distinctive. G. L. Nesom (2010) suggested *P. fortuneana* was not distinct from *P. crenulata*.

4. Pyracantha angustifolia (Franchet) C. K. Schneider, Ill. Handb. Laubholzk. 1: 761. 1906 [I]

Cotoneaster angustifolius Franchet, Pl. Delavay., 221. 1890 (as angustifolia)

Plants 20–40 dm. **Stems:** thorns sparse; young twigs tomentose, hairs dense grayish or yellowish (glabrescent 2d year). **Leaves:** stipules 4–8 mm; petiole 1–2.5 mm, <tomentose>; blade narrowly oblong to slightly oblanceolate, 1.5–5 × 0.4–1 cm, base rounded to cuneate, margins entire (sometimes revolute), apex apiculate, obtuse, or retuse, surfaces gray-tomentose (abaxial remaining gray-tomentose, adaxial glabrescent). **Inflorescences** 2–4 cm diam. **Pedicels** 1–3 mm, gray-tomentose. **Flowers** 3–8 mm diam.; hypanthium gray-tomentose; <calyx persistently gray-tomentose>, sepals triangular, 1 mm, apex acute; petals suborbiculate, 3–4 mm, apex rounded. **Pomes** red, <depressed globose>, 4–8 mm diam.; pedicels 2–5 mm.

Flowering Feb–May; fruiting Nov–Mar. Disturbed areas, fencerows, abandoned fields, roadsides; 0–200 m; introduced, Calif.; Asia (China); introduced also in Africa (South Africa), Pacific Islands (Hawaii, New Zealand), Australia.

Pyracantha angustifolia is native to southwestern China. It is widely cultivated in North America but apparently naturalized only within California. The dense abaxial leaf indument can thin with age, but the distinctive dense hairs are found also on the sepals, even in fruit. The narrowly oblong to slightly oblanceolate, entire leaves are also diagnostic.

5. Pyracantha atalantioides (Hance) Stapf, Bot. Mag. 151: sub plate 9099. 1926 [I]

Sportella atalantioides Hance, J. Bot. 15: 207. 1877; *Pyracantha discolor* Rehder

Plants: <crowns spreading or partially reclining>, 10–60 dm. **Stems:** thorns abundant; young twigs yellowish brown-hairy, glabrescent. **Leaves:** stipules 3–7 mm; petiole 2–6 mm; blade elliptic, oblong, or oblong-obovate, 1.5–5 × 1–2 cm, base rounded to cuneate, margins entire or remotely serrulate distally (on vigorous shoots), apex obtuse, apiculate, or aristate, surfaces yellowish brown-hairy, glabrescent (abaxial becoming slightly glaucescent). **Inflorescences** 3–4 cm diam. **Pedicels** 5–10 mm, yellowish brown-hairy. **Flowers** 8–10 mm diam.; hypanthium hairy; <calyx brown- or yellowish brown-puberulent when young, glabrescent>, sepals broadly elliptic or triangular, 1–1.5 mm, apex obtuse; petals ovate, 4–5 mm, apex slightly apiculate. **Pomes** bright red, 4–7 mm diam.; pedicels 5–12 mm.

Flowering Feb–May; fruiting Oct–Jan. Roadsides, thickets, disturbed ground, edges; 0–300 m; introduced; Calif., Fla., Ga., La.; Asia (China).

Pyracantha atalantioides is native to southeastern and western China. The species has been extensively planted in North America; it is apparently naturalized only sporadically in milder climatic areas of the continent.

6. Pyracantha koidzumii (Hayata) Rehder, J. Arnold Arbor. 1: 261. 1920 • Taiwan firethorn [I]

Cotoneaster koidzumii Hayata, J. Coll. Sci. Imp. Univ. Tokyo 30(1): 101. 1911 (as koizumii)

Plants 5–40 dm. **Stems:** thorns sparse to abundant; young twigs reddish brown-hairy, glabrescent. **Leaves:** stipules 3–8 mm; petiole 1–3 mm; blade oblanceolate or narrowly obovate, 2.5–4.5 × 0.5–1.2 cm, base cuneate, margins entire, rarely with minute teeth near apices, apex truncate or retuse, rarely rounded, abaxial surface brown-hairy, becoming glabrate and pale green, not glaucescent, adaxial glabrous. **Inflorescences** 3–4 cm diam. **Pedicels** 5–12 mm, sparsely brown-hairy. **Flowers** 8–10 mm diam.; hypanthium densely hairy; <calyx brown- or yellowish brown-puberulent when young, glabrescent>, sepals triangular, 1–1.3 mm, apex acute; petals suborbiculate, 3–4 mm, apex slightly emarginate. **Pomes** orange-red, 4–7 mm diam.; pedicels 5–13 mm.

Flowering Feb–May; fruiting Jul–Feb. Disturbed forests, beach bluffs, riparian areas, thickets, roadsides; 0–1500 m; introduced; Ala., Ariz., Ark., Calif., Fla., Ga., La., Miss., Okla., S.C., Tex.; e Asia (Taiwan); introduced also in Pacific Islands (Hawaii), Australia.

Occasional specimens of *Pyracantha fortuneana* with sparsely toothed leaf margins can be difficult to distinguish from *P. koidzumii*, and some plants in California may be of hybrid origin or represent escapes of horticultural selections.

58. MALUS Miller, Gard. Dict. Abr. ed. 4, vol. 2. 1754 • Apple, pommier [Latin *malus*, apple tree]

Elizabeth E. Dickson

Shrubs or trees, 2–200 dm. **Stems** 1+ (derived from root shoots), erect; bark dark brown, reddish brown, or gray, <firm, platy or scaly>; long and short shoots present; thorns present (modified short shoots); glabrous, glabrescent, villous, densely puberulent, or tomentose. **Leaves** deciduous [semipersistent], cauline, simple, <shoot dimorphism, long- and juvenile-shoot leaves usually larger and more deeply serrate (lobed) than short-shoot leaves>; stipules deciduous (or persistent on vigorous shoot leaves in *M. baccata*), basally adnate to petiole, linear-lanceolate to lanceolate, sometimes filiform, <membranous, sometimes herbaceous>, margins entire, serrulate, glandular-serrate, sparsely glandular-denticulate, or white-ciliate; petiole present; blade elliptic, ovate, obovate, triangular-ovate, oval, lanceolate, ovate-oblong, or oblong, (2–)4–12 cm, membranous or leathery, margins flat, sometimes lobed, dentate, serrulate, serrate, doubly serrate, crenate, or sometimes entire, venation pinnate (craspedodromous when lobed, camptodromous when unlobed), surfaces glabrous or tomentose. **Inflorescences** terminal <on short shoots>, 2–12 -flowered, <flat-topped> panicles, glabrous or tomentose; bracts present (absent in *M. fusca*), <caducous, ovate, linear-lanceolate, membranous, glabrous>; bracteoles present. **Pedicels** present. **Flowers** bisexual (occasionally andromonoecious in *M. halliana*), opening with leaves, perianth and androecium epigynous, 15–50 mm diam.; hypanthium campanulate, size not recorded, glabrous or tomentose; sepals 5, reflexed to wide spreading, triangular, triangular-lanceolate, triangular-ovate, or lanceolate; petals 5 (or more in *M. halliana*), white, pink, or red, suborbiculate, obovate, narrowly or oblong-obovate, ovate, or ± elliptic, <base clawed>; stamens 15–50, <unequal>, usually shorter than, rarely equal to, petals; carpels 3–5, connate, adnate to hypanthium, styles <3–5>, attachment unknown, basally connate, <glabrous or tomentose basally>; ovules 2. **Fruits** pomes, green, yellow, or red, globose, depressed-globose, obovoid, or oblong, 6–50(–70) mm diam., <glaucous, waxy>, punctate; <flesh homogeneous, stone cells adjacent to carpels and epidermis>; hypanthium persistent; sepals persistent or deciduous, erect or reflexed; carpels cartilaginous; styles persistent or deciduous. **Seeds** 2 per carpel, <light, dark, or reddish brown, smooth>. $x = 17$.

Species 25–55 (10 in the flora): North America, Eurasia; introduced widely, especially in temperate regions.

Malus has great economic value; species are widely cultivated throughout the world for their edible fruit (not all species), ornamental fruit, or flowers. Mammals and birds eat cultivated and wild apples, which can result in the spread of seed and naturalization of species. The hard wood is occasionally used to make furniture and tools, and as fuel. The apple has symbolic significance in western cultures, including Greek, Roman, and Judeo-Christian mythologies. The apple often represents desire, temptation, and sin in art and literature (*malum* is Latin for evil), symbolizing sexuality, love, and the forbidden fruit of the Garden of Eden.

Cross-compatibility among species is common. Hybridization can occur naturally in botanical gardens and in the wild, or artificially through breeding. Polyploid forms and asexual seed production (apomixis) occur in some species.

The taxonomy of *Malus* has been revised at least three times, with some authors placing species within the genus *Pyrus*. Recent morphologic work has suggested that *Malus* be retained as a separate genus based on fully adnate carpels and deeply inferior ovaries (K. R. Robertson et al. 1991), with molecular evidence providing support (C. S. Campbell et al. 1995). Cultivation, hybridization, and introgression have led to hundreds of species names within *Malus*. The genus requires a comprehensive worldwide revision.

The eastern North American apples stand apart from other members of the genus by having unique floral and fruit traits, which define sect. *Chloromeles* (Decaisne) Rehder. Although distinctive morphologic extremes are obvious, continuous variation within this overall homogeneous group blurs species boundaries. Historically, taxonomic confusion has resulted from names being applied to slight variations in: leaf lobing and serration; patterns of leaf venation; leaf, calyx, and pedicel indument; and fruit shape. Leaves of vigorous long shoots often differ in shape and hairiness from those of short flowering shoots on the same tree. Polyploidy and apomixis may contribute to taxonomic difficulties by locking suites of traits together into morphologic forms. Morphologic patterns within the group can shift. Studies indicate that diploids within natural populations occasionally produce triploid and tetraploid progeny; triploids can give rise to tetraploids (E. E. Dickson 1995). Preliminary genetic studies do not clarify species boundaries and suggest a single monotypic species, *Malus coronaria* (Dickson et al. 1991; Dickson 1995). This treatment takes a conservative approach by recognizing three morphologic extremes as species. Further work is warranted to clarify the taxonomy of sect. *Chloromeles*.

The description of *Malus prunifolia* is based on Y. Asami (1927) and Gu C. Z. and S. A. Spongberg (2003); *M. baccata* and *M. halliana* are based on Gu and Spongberg; *M. hupehensis* is based on Gu and Spongberg and C. A. Huckins (1972); and *M. toringo* (as *M. sieboldii*) is based on Asami, Huckins, and Gu and Spongberg.

SELECTED REFERENCES Dickson, E. E. 1995. Systematic Studies of *Malus* sect. *Chloromeles*. Ph.D. dissertation. Cornell University. Dickson, E. E., S. Kresovich, and N. F. Weeden. 1991. Isozymes in North American *Malus* (Rosaceae): Hybridization and species differentiation. Syst. Bot. 16: 363–375. Huckins, C. A. 1972. A Revision of the Sections of the Genus *Malus* Miller. Ph.D. dissertation. Cornell University.

1. Sepals deciduous in fruit.
    2. Leaves usually lobed.
        3. Styles glabrous; pomes yellow to purplish red, oblong, sometimes ovoid or obovoid . . . . 4. *Malus fusca*
        3. Styles proximally villous or lanate; pomes red or brownish yellow, subglobose . . . . 10. *Malus toringo*
    2. Leaves unlobed.
        4. Sepals lanceolate, longer than tube; anthers yellow before dehiscence. . . . . 7. *Malus baccata*
        4. Sepals triangular-ovate, equal to or shorter than tube; anthers white before dehiscence.
            5. Leaves obtusely serrulate; sepal apices obtuse; petals usually more than 5; styles 4 or 5; pomes pyriform or obovoid. . . . . 8. *Malus halliana*
            5. Leaves sharply serrulate; sepal apices acute or acuminate; petals 5; styles 3(or 4); pomes ellipsoid or subglobose . . . . 9. *Malus hupehensis*
1. Sepals persistent in fruit.
    6. Cores of pomes enclosed at apex, sclereids absent or sparse surrounding core; anthers yellow before dehiscence; leaf blades of vigorous shoots unlobed.
        7. Leaf margins obtusely serrate, sometimes serrate-crenate; pomes globose or depressed-globose, sepals not swollen at base (cultivated apple) . . . . 5. *Malus pumila*
        7. Leaf margins acutely serrate or serrulate, sometimes doubly serrate; pomes ovoid or oblong, sepals swollen at base . . . . 6. *Malus prunifolia*

[6. Shifted to left margin.—Ed.]

6. Cores of pomes not enclosed at apex, sclereids abundant surrounding core; anthers rose, pink, salmon, or purple before dehiscence; leaf blades of vigorous shoots often lobed.
   8. Sepals hoary-tomentose; leaves abaxially tomentose . . . . 1. *Malus ioensis*
   8. Sepals abaxially glabrous, sometimes glabrescent; leaves abaxially glabrous (villous only on veins).
      9. Flowering shoot leaves usually elliptic or oblong, bases usually cuneate, margins crenate, crenate-serrate, or entire, apices rounded (with point or acute) . . . . 2. *Malus angustifolia*
      9. Flowering shoot leaves usually ovate, triangular-ovate, or lanceolate, bases usually rounded or cordate, margins usually serrate, apices acute or broadly acute (rounded with point or rounded) . . . . 3. *Malus coronaria*

1. Malus ioensis (Alph. Wood) Britton in N. L. Britton and A. Brown, Ill. Fl. N. U.S. 2: 235. 1897 • Prairie or Iowa or western crabapple [E] [F]

Pyrus coronaria Linnaeus var. *ioensis* Alph. Wood, Class-book Bot. ed. s.n.(b), 333. 1861; *Malus angustifolia* (Aiton) Michaux var. *spinosa* (Rehder) C. F. Reed; *M. coronaria* (Linnaeus) Miller subsp. *ioensis* (Alph. Wood) Likhonos; *M. coronaria* var. *ioensis* (Alph. Wood) C. K. Schneider; *M. ioensis* var. *bushii* Rehder; *M. ioensis* var. *creniserrata* Rehder; *M. ioensis* var. *palmeri* Rehder; *M. ioensis* var. *spinosa* Rehder; *M. ioensis* var. *texana* Rehder; *P. ioensis* (Alph. Wood) L. H. Bailey

Trees, sometimes shrubs, (20–)40–60(–100) dm. **Stems** 6–45 cm diam.; bark dark reddish brown to gray, with narrow reddish brown scales, sometimes glaucous when young; young branches reddish gray, tomentose, becoming reddish brown or dark brown and glabrous or slightly puberulous; flowering shoots becoming spurs or spiny thorns, 10–60 mm. **Buds** reddish brown, ovoid, 3–4 mm, scale margins tomentose. **Leaves** conduplicate in bud; heteromorphic; stipules deciduous, narrowly lanceolate, 3–8 mm, apex acute; vigorous shoot leaves: petiole 15–30 mm, tomentose; blade ovate, sometimes triangular-ovate or oval, 3–9(–12) × 1.5–4(–6) cm, base cuneate to rounded, margins usually lobed, serrate or crenate-serrate, sometimes doubly serrate, apex acute, broadly acute, or rounded, abaxial surface tomentose (villous only on veins), adaxial glabrous or slightly villous; flowering shoot leaves: petiole 5–20(–25) mm, tomentose, sometimes puberulent; blade elliptic, oval, or ovate, 2–5 × 1–2 cm, base cuneate or rounded, sometimes truncate-rounded, margins lobed or unlobed, serrate, doubly serrate, or crenate-serrate, apex acute, broadly acute, or rounded, abaxial surface usually tomentose, adaxial glabrous. **Panicles** corymblike; peduncles absent; bracteoles sometimes persistent, filiform, 3–10 mm. **Pedicels** 15–30 mm, tomentose, sometimes puberulent or sparsely strigose. **Flowers** <fragrant>, 35–50 mm diam.; hypanthium tomentose; sepals triangular, 3–7 mm, equal to tube, apex acuminate, surfaces hoary-tomentose; petals pink, sometimes fading white, oblong to narrowly obovate, (11–)14–20(–22) mm, claws (2–)3–4 mm, margins entire, sinuate, or fimbriate, apex rounded; stamens 20, 8–16 mm, anthers rose before dehiscence; styles 5, connate basally, 9–15 mm, ± equal to stamens, tomentose in proximal 1/3. **Pomes** green, sometimes yellow, depressed-globose, sometimes globose, 20–30 mm diam., <greasy, waxy>, cores not enclosed at apex; sepals persistent, erect; sclereids abundant surrounding core. **Seeds** dark brown. $2n$ = 34, 51, 64.

Flowering Apr–May; fruiting Sep–Oct. Woods, thickets, barrens, hillsides, stream and lake banks, flood plains, roadsides, open fields, fence lines; 100–300 m; Ark., Ill., Ind., Iowa, Kans., Ky., La., Mich., Minn., Mo., Nebr., Ohio, Okla., S.Dak., Tex., Wis.

The fruit of *Malus ioensis* is extremely acrid in taste, and it has a high pectin content that allows it to be made into a clear yellow jelly. The beautiful, fragrant flowers provide a pleasant show in spring. Double-flowered forms have been selected for cultivation: Bechtel's crab and forma *fimbriata* A. D. Slavin. This species flowers later than the cultivated apple in most regions; artificial hybrids have been documented. *Malus* ×*soulardii* (L. H. Bailey) Britton (*Pyrus* ×*soulardii* L. H. Bailey) is a hybrid between *M. ioensis* and *M. pumila*.

2. Malus angustifolia (Aiton) Michaux, Fl. Bor.-Amer. 1: 292. 1803 • Southern or narrow-leaved crabapple [E]

Pyrus angustifolia Aiton, Hort. Kew. 2: 176. 1789; *Malus coronaria* (Linnaeus) Miller var. *angustifolia* (Aiton) Ponomarenko; *P. coronaria* Linnaeus var. *angustifolia* (Aiton) Wenzig

Trees, sometimes shrubs, 10–50(–100) dm. **Stems** 5–25 cm diam.; bark reddish brown to gray, longitudinally fissured with platelike scales; young branches reddish brown, sparsely puberulent, becoming gray or grayish brown, glabrescent; flowering shoots developing as short spurs or spiny thorns, (5–)10–40(–60) mm. **Buds** reddish brown, ovoid, 1.6 mm, scale margins tomentose. **Leaves** conduplicate in bud;

heteromorphic; stipules deciduous, linear-lanceolate, 2–4(–5) mm, apex acuminate; vigorous shoot leaves: petiole (10–)15–25 mm, villous, sometimes tomentose; blade elliptic, oval, or ovate, sometimes triangular-ovate, (3.5–)4–6(–8) × (1.5–)3–4(–5) cm, base cuneate, sometimes rounded, margins sometimes slightly lobed, crenate, crenate-serrate, serrate, or entire, sometimes doubly serrate, apex rounded, broadly acute, sometimes apiculate, abaxial surface glabrous (villous only on veins), adaxial glabrous; flowering shoot leaves: petiole 3–10(–25) mm, villous, sometimes glabrous; blade elliptic or oblong, sometimes ovate, obovate, or lanceolate, (9–)15–50(–65) × (7–)10–20(–30) mm, base cuneate, sometimes rounded, margins unlobed, crenate, crenate-serrate, or entire, apex rounded (with point or acute), abaxial surface glabrous (veins villous), adaxial glabrous. **Panicles** corymblike; peduncles absent; bracteoles sometimes persistent, filiform, (1–)3–4(–6) mm. **Pedicels** (10–)20–30(–40) mm, glabrous, sometimes slightly villous. **Flowers** 20–30 mm diam.; hypanthium glabrous, rarely slightly villous; sepals triangular, (2–)3–4(–5) mm, equal to tube, apex acuminate, abaxial surface glabrous, adaxial hoary-tomentose; petals pink, sometimes fading white, oblong to narrowly obovate, (10–)12–16(–22) mm, claws (2–)3 mm, margins entire, sinuate, or fimbriate, apex rounded; stamens 20, (7–)9–12(–14) mm, anthers pink or rose before dehiscence; styles 5, connate basally, (6–)9–12(–15) mm, usually slightly longer than stamens, tomentose in proximal ⅓. **Pomes** green or yellow-green, subglobose, 10–20(–30) mm diam., cores not enclosed at apex; sepals persistent, erect; sclereids abundant surrounding core. **Seeds** dark brown. $2n$ = 34 (68).

Flowering (Feb–)Mar–May; fruiting Aug–Sep. Open woods, thickets, hillsides, stream and lake banks, flood plains, terraces, roadsides, open fields, sandy soil, clay soil; 10–700 m; Ala., Ark., Del., D.C., Fla., Ga., Ill., Ky., La., Md., Miss., Mo., N.J., N.C., Ohio, Pa., S.C., Tenn., Va., W.Va.

The attractive flowers of *Malus angustifolia* have a pleasant, violetlike odor.

3. **Malus coronaria** (Linnaeus) Miller, Gard. Dict. ed. 8, Malus no. 2. 1768 • American or sweet or fragrant crabapple, garland tree [E]

Pyrus coronaria Linnaeus, Sp. Pl. 1: 480. 1753; *Malus angustifolia* (Aiton) Michaux var. *puberula* (Rehder) Rehder; *M. bracteata* Rehder; *M. carolinensis* Ashe; *M. coronaria* var. *bracteata* (Rehder) Likhonos; *M. coronaria* var. *dasycalyx* Rehder; *M. coronaria* var. *elongata* (Rehder) Rehder; *M. coronaria* var. *glabrata* (Rehder) Likhonos; *M. coronaria* var. *glaucescens* (Rehder) Likhonos; *M. coronaria* var. *hoopesii* Rehder; *M. coronaria* var. *lancifolia* (Rehder) C. F. Reed; *M. coronaria* var. *platycarpa* (Rehder) Likhonos; *M. coronaria* var. *plena* C. K. Schneider; *M. coronaria* var. *puberula* Rehder; *M. elongata* (Rehder) Ashe; *M. fragrans* Rehder; *M. glabrata* Rehder; *M. glaucescens* Rehder; *M. lancifolia* Rehder; *M. platycarpa* Rehder; *P. bracteata* (Rehder) L. H. Bailey; *P. coronaria* var. *dasycalyx* (Rehder) Fernald; *P. coronaria* var. *elongata* (Rehder) L. H. Bailey; *P. coronaria* var. *lancifolia* (Rehder) Fernald

Trees, sometimes shrubs, 25–75(–100) dm. **Stems** 1–45 cm diam.; bark reddish brown to gray, longitudinally fissured with platelike scales; young branches reddish brown to dark brown, with orange lenticels, pubescent, glabrescent; flowering shoots becoming spurs or spiny thorns, (10–)20–40(–100) mm. **Buds** reddish brown, ovoid, 1–6 mm, scale margins tomentose. **Leaves** conduplicate in bud; heteromorphic; stipules deciduous, linear-lanceolate, (3–)4–6(–7) mm, apex acuminate; vigorous shoot leaves: petiole (7–)10–25(–30) mm, glabrous or villous, sometimes puberulent; blade ovate or triangular-ovate, sometimes lanceolate, (2.5–)4–8(–10) × (1.5–)4–6(–8) cm, base rounded or cordate-rounded, sometimes cuneate, margins ± lobed, sometimes unlobed, serrate, sometimes doubly serrate or crenate-serrate, apex broadly acute or acute, sometimes rounded, apiculate, abaxial surface glabrous (villous only on veins), adaxial glabrous; flowering shoot leaves: petiole 5–15(–25) mm, glabrous or villous; blade ovate, triangular-ovate, or lanceolate, sometimes oval or elliptic, (15–)20–60 (–85) × (10–)15–45(–60) mm, base rounded or cordate, sometimes cuneate-rounded or cuneate, margins lobed or unlobed, serrate, sometimes crenate-serrate, doubly serrate, or entire, apex acute or broadly acute (rounded with point or rounded), abaxial surface glabrous (veins villous), adaxial glabrous. **Panicles** corymblike; peduncles absent; bracteoles sometimes persistent, filiform, 3–5(–10) mm. **Pedicels** (10–)20–30(–40) mm, glabrous, sometimes villous. **Flowers** 25–40 mm diam.; hypanthium glabrous, sometimes puberulent; sepals triangular, (3–)4–6(–7) mm, equal to slightly longer than tube, apex acute or acuminate, abaxial surface glabrous, sometimes glabrescent, adaxial hoary-tomentose; petals pink, sometimes fading white, oblong-obovate, ovate, or oblong, (11–)12–18(–20) mm, claws (2–)3–4 mm, margins entire, sinuate, or fimbriate, apex rounded; stamens 20, (8–)10–13(–14) mm, anthers pink, rose, salmon, or purple before dehiscence; styles 5, connate basally, (8–)10–12(–16) mm, equal to or slightly longer than stamens, villous in proximal ½. **Pomes** green or yellow-green, depressed-globose, 15–30(–55) mm diam., cores not enclosed at apex; sepals persistent, erect; sclereids abundant surrounding core. **Seeds** dark brown. $2n$ = 34, 51, 68.

Flowering (Apr–)May(–Jun); fruiting Sep–Oct. Open woods, forest edges, thickets, stream banks, fields, fencerows, roadsides; 50–1000 m; Ont.; Ala., Ark., Del., D.C., Ga., Ill., Ind., Kans., Ky., La., Md., Mich., Mo., N.J., N.Y., N.C., Ohio, Pa., S.C., Tenn., Va., W.Va., Wis.

Leaves of *Malus coronaria* are variable in shape and lobing, leading Rehder in the early 1900s to recognize a number of eastern taxa. For examples, *M. lancifolia* was described as having ovate-lanceolate to oblong-lanceolate leaves; *M. glabrata* and *M. glaucescens* were described as having triangular-ovate or ovate leaves that are distinctly lobed. Specimens with relatively large fruit and only slightly lobed leaves were named *M. platycarpa*.

Double-flowered forms have been found in nature and selected for cultivation: forma *charlottae* Rehder and forma *nieuwlandiana* A. D. Slavin.

4. Malus fusca (Rafinesque) C. K. Schneider, Ill. Handb. Laubholzk. 1: 723. 1906 • Oregon or western or Pacific crabapple [E]

Pyrus fusca Rafinesque, Med. Fl. 2: 254. 1830; *Malus diversifolia* (Bongard) M. Roemer; *M. fusca* var. *diversifolia* (Bongard) C. K. Schneider; *M. fusca* var. *levipes* (Nuttall) C. K. Schneider; *M. macounii* Greene; *M. rivularis* (Douglas) M. Roemer; *P. diversifolia* Bongard; *P. rivularis* Douglas

Trees, sometimes shrubs, 50–100(–200) dm. **Stems** (12–)20–30(–40) cm diam.; bark gray to reddish brown, smooth when young, scaly and deeply fissured at maturity; young branches reddish, puberulent when young, becoming reddish brown or gray and glabrous; lateral flowering shoots becoming spurs, 10–30(–50) mm. **Buds** red-brown, ovoid, 1.5–4 mm, scale margins ciliate. **Leaves** conduplicate in bud; isomorphic; stipules deciduous, narrowly lanceolate, 1–5 mm, apex acuminate; petiole 10–30 mm, tomentose, glabrescent; blade ovate, sometimes oval, elliptic, or lanceolate, 3–9(–11) × 1–4 cm, base rounded to cuneate, margins unlobed or 3-lobed, serrate to doubly serrate, sometimes serrulate, apex acute or acuminate, abaxial surface glabrous, adaxial puberulent. **Panicles** corymblike; peduncles absent; bracts absent; bracteoles sometimes persistent, filiform, 1–4 mm. **Pedicels** 15–40 mm, villous or glabrous. **Flowers** 15–20 mm diam.; hypanthium glabrous or tomentose; sepals triangular, 3–6 mm, shorter than tube, apex apiculate, abaxial surface glabrous, adaxial hoary-tomentose; petals white, sometimes pink, orbiculate to obovate, (6–) 10–15 mm, claws 1.5–2 mm, margins erose or undulate, apex rounded; stamens ca. 20, 4–6 mm, anthers white before dehiscence; styles 3(or 4), connate in proximal ⅓, 6–7 mm, longer than stamens, glabrous. **Pomes** yellow to purplish red, oblong, sometimes ovoid or obovoid, 6–10(–13) mm diam., cores enclosed at apex; sepals deciduous, sometimes tardily; sclereids abundant surrounding core. **Seeds** reddish brown. **2*n*** = 34.

Flowering Apr–early Jul; fruiting Jun–Nov. Moist, open places, open Sitka spruce forests, forest edges, beach edges, sea cliffs, swamps, bogs, brackish marshes, clearings; 0–600 m; B.C.; Alaska, Calif., Oreg., Wash.

Malus fusca is tolerant of occasional exposure to salt water. The dry flesh of the fruit is edible, and while tart, becomes sweeter with time. The bark was prepared as medicine by western coastal Native Americans.

Malus fusca is a western North American apple species that is morphologically and phylogenetically closer to Asian species than to eastern North American ones (E. E. Dickson 1995). The species hybridizes in Oregon and Washington with the cultivated apple, *M. pumila* (P. F. Zika 2004), and the hybrid is *M.* ×*dawsoniana* Rehder. Similar in habit and bark to *M. fusca*, the hybrid differs by its larger flowers and larger elliptic-oblong or obovate-oblong, (four- or) five-loculed fruit with a persistent calyx. It differs from *M. prunifolia* by its corymblike inflorescences and the pome lacking swollen sepals.

5. Malus pumila Miller, Gard. Dict. ed. 8, Malus no. 3. 1768 • Orchard or common or paradise crabapple, pommier commun [I]

Pyrus malus Linnaeus var. *paradisiaca* Linnaeus, Sp. Pl. 1: 479. 1753; *M. sylvestris* Miller; *P. malus* Linnaeus

Trees, 20–100(–150) dm. **Stems** 5–30 cm diam.; bark dark gray or gray-brown, scaly; young branches dark brown or reddish brown, densely tomentose, becoming glabrous; flowering shoots becoming spurs, 40–150 mm. **Buds** dark red or purple, ovoid, 3–4(–5) mm, scale margins densely puberulous. **Leaves** convolute in bud; isomorphic; stipules deciduous, lanceolate, 3–5 mm, apex acuminate; petiole 10–35 mm, tomentose to sparsely pubescent; blade elliptic, ovate, or broadly elliptic, (2–)5–10 × (1–)3–6.5 cm, base broadly cuneate or rounded, margins unlobed, obtusely serrate, sometimes serrate-crenate, apex acute, abaxial surface glabrescent, adaxial densely puberulent. **Panicles** umbel-like; peduncles absent; bracteoles deciduous, filiform, 5–7 mm. **Pedicels** 10–25 mm, puberulous. **Flowers** 30–40 mm diam.; hypanthium tomentose; sepals <reflexed at flowering>, triangular-lanceolate or triangular-ovate, 6–8 mm, equal to or longer than tube, apex acuminate, surfaces tomentose; petals (rose in bud) white, sometimes pink, obovate, 15–25 mm, claws 1 mm, margins entire, apex rounded; stamens 20, 9–10 mm, anthers yellow before dehiscence; styles 5, basally connate less than ½ length, 9–10 mm, slightly longer than stamens, basally gray-tomentose; <stigmas green>. **Pomes** green, yellow, or red, <pure, striped, or blushed>, globose or depressed-globose, 20–50(–70) mm diam., <skin with bloom or wax, sometimes russetted or dotted>, cores enclosed at apex; sepals persistent, erect; sclereids absent. **Seeds** light brown. **2*n*** = 34 (51, 68).

Flowering Apr–May; fruiting Jul–Oct. Abandoned or naturalized in thickets, forests, fields, fence edges, shores, roadsides; 0–1600 m; introduced; St. Pierre and Miquelon; B.C., Man., N.B., Nfld. and Labr. (Nfld.), N.S., Ont., P.E.I., Que.; Ala., Alaska, Ark., Calif., Colo., Conn., Del., D.C., Ga., Ill., Ind., Iowa, Kans., Ky., La., Maine, Md., Mass., Mich., Minn., Miss., Mo., Mont., Nebr., Nev., N.H., N.J., N.Mex., N.Y., N.C., N.Dak., Ohio, Oreg., Pa., R.I., S.C., Tenn., Utah, Vt., Va., Wash., W.Va., Wis., Wyo.; c Asia; introduced also in Mexico, Central America (El Salvador, Guatemala), South America (Argentina), Europe, e Asia, Africa (South Africa), Pacific Islands (New Zealand, Tristan da Cunha), Australia.

Malus pumila is cultivated for its edible fruit. Historically, the nomenclature of the orchard apple has been confusing, with many species names applied to cultivar groups. Based on extensive nomenclatural research, D. J. Mabberley et al. (2001) concluded that *M. pumila* is the correct binomial for the species of central Asia that is ancestral to orchard apples. Commonly used binomials, such as *M. domestica* Borkhausen (an illegitimate name), *M. paradisiaca* (Linnaeus) Medikus, and *M. sylvestris*, have been placed in synonymy. Quian G. Z. et al. (2010) have proposed to conserve *M. domestica* and reject *M. pumila*. Their proposal is still under consideration by the Nomenclature Committee for Vascular Plants.

Some trees, although originally planted, may appear naturalized when found in old, overgrown areas. Naturalized trees are derived from seeds distributed by birds and mammals and from apples discarded by people. Trees grown from seeds often produce small, bitter, and sour fruit.

6. Malus prunifolia (Willdenow) Borkhausen, Theor. Prakt. Handb. Forstbot. 2: 1278. 1803 • Plumleaf or pearleaf crabapple, Chinese apple [I]

Pyrus prunifolia Willdenow, Phytographia, 8. 1794

Trees, 30–80(–100) dm. **Stems** 5–20 cm diam.; bark purplish brown or gray-brown, peeling irregularly; young branches purplish brown and densely puberulous, becoming grayish purple or grayish brown and glabrous at maturity; flowering shoots producing spurs, (5–)10–20(–35) mm. **Buds** purplish brown, ovoid, 5–7 mm, scale margins sparsely pubescent. **Leaves** convolute in bud; isomorphic; stipules deciduous, lanceolate, 4–5 mm, apex acuminate; petiole 10–50 mm, tomentose; blade elliptic or ovate, (3.5–)5–9.5(–11) × (2–)4–5(–8) cm, base broadly cuneate, margins unlobed, acutely serrate or serrulate, sometimes doubly serrate, apex acuminate or acute, sometimes cuspidate, abaxial surface glabrous, adaxial puberulent or subglabrous. **Panicles** umbel-like; peduncles 0–2 mm; bracteoles deciduous, rare, <± middle of pedicel>, filiform, 3–4 mm. **Pedicels** 20–35 mm, puberulous. **Flowers** (30–) 40–50 mm diam.; hypanthium puberulent; sepals triangular-lanceolate or lanceolate, 7–9 mm, longer than tube, apex acuminate, surfaces puberulent; petals white, obovate or elliptic, 25–30 mm, claws 1–2 mm, margins entire, apex rounded; stamens ca. 20, 8–10 mm, anthers yellow before dehiscence; styles 4 or 5, basally connate ⅓ length, 9–11 mm, longer than stamens, proximally tomentose. **Pomes** yellow to dark red, ovoid or oblong, 20–25(–34) mm diam., cores enclosed at apex; sepals persistent, reflexed (erect), <raised, swollen at base>; sclereids sparse surrounding core. **Seeds** dark brown. $2n$ = 34, 51.

Flowering Apr–Jun; fruiting Aug–Oct. Hillsides, flats; 0–1300 m; introduced; N.B., N.S.; Conn., D.C., Ill., Maine, Mass., Minn., N.H., N.Y., Pa., R.I., S.C., Tenn., Vt., Va., Wash., Wis.; ne Asia.

Malus prunifolia is closely related to the cultivated apple, *M. pumila*, and sometimes is used as rootstock for the latter because of its extensive root system. *Malus prunifolia* differs from *M. pumila* in having more sharply serrate leaves, glabrate or glabrous abaxial surfaces of mature leaves, relatively long pedicels, and pomes with raised, swollen, fleshy-based sepals.

Malus prunifolia is cultivated worldwide and utilized as grafting rootstock, as a fruit tree, and as an ornamental. The species is potentially a genetic source of cold tolerance and disease resistance for apple-breeding programs. See discussion under 4. *M. fusca* for distinguishing *M. prunifolia* from the hybrid *M.* ×*dawsoniana*.

Malus ×*scheideckeri* Späth ex Zabel is considered a hybrid of *M. floribunda* and *M. prunifolia* (see discussion of the latter under 8. *M. halliana*). The 'Scheidecker crabapple' is cultivated as an ornamental tree for its semidouble pale pink flowers and yellow to orange globose pomes.

7. Malus baccata (Linnaeus) Borkhausen, Theor. Prakt. Handb. Forstbot. 2: 1280. 1803 • Siberian crabapple, pommier de Sibérie [I]

Pyrus baccata Linnaeus, Syst. Nat. ed. 12, 2: 344. 1767; Mant. Pl. 1: 75. 1767

Trees, 100–140 dm. **Stems** to 30 cm diam.; bark dark gray with platelike scales; young branches reddish orange and glabrous, becoming reddish brown; flowering shoots becoming spurs, 3–10(–25) mm. **Buds** reddish brown, ovoid, 3–4 mm, scale margins tomentose. **Leaves** convolute in bud; isomorphic; stipules deciduous or persistent on vigorous shoot leaves, lanceolate, sometimes filiform, 3 mm, apex acuminate; petiole 20–50 mm, glabrous; blade elliptic or ovate, 3–8 × 2–3.5 cm, base cuneate or rounded, margins unlobed, serrate, apex long-acuminate, sometimes caudate, surfaces glabrous or slightly puberulent when young. **Panicles** umbel-like; peduncles absent; bracteoles absent. **Pedicels** 15–40 mm, glabrous. **Flowers** <not fragrant>, 30–35 mm diam.; hypanthium <constricted distal to ovaries>, glabrous; sepals lanceolate, 5–7 mm, longer than tube, apex acuminate, abaxial surface glabrous, adaxial tomentose; petals white, obovate, 20–30 mm, claws 1–2 mm, margins entire, apex obtuse rounded; stamens 15–20, 7–9 mm, anthers yellow before dehiscence; styles (4 or)5, basally connate to ½ length, 8–10 mm, longer than stamens, densely villous basally. **Pomes** yellow to red, subglobose, 8–10 mm diam., cores enclosed at apex; sepals deciduous; sclereids sparse surrounding core. **Seeds** light reddish brown. $2n = 34$.

Flowering Apr–Jun; fruiting Sep–Oct. Roadsides, pastures, mixed woods, thickets; 0–1500 m; introduced; N.B., N.S., Ont.; Conn., Ill., Ky., Maine, Mass., Mich., Minn., Mo., N.H., N.J., N.Y., N.C., Ohio, Pa., R.I., S.C., Tenn., Va., W.Va., Wis., Wyo.; Asia; introduced also in Europe.

Malus baccata is grown ornamentally for flowers and fruit (red or yellow) and occasionally escapes from cultivation. The species is used as a rootstock for grafting other apple cultivars and is a possible genetic source of disease resistance and cold hardiness. The species is possibly naturalized in Honduras.

8. Malus halliana Koehne, Gatt. Pomac., 27. 1890 • Hall crabapple [I]

Malus domestica Borkhausen var. *halliana* (Koehne) Likhonos; *Pyrus halliana* (Koehne) Voss

Trees, sometimes shrubs, 30–50(–80) dm. **Stems** to 35 cm diam.; bark gray, longitudinally fissured at maturity; young branches green and sparsely puberulent, becoming dark purplish brown and glabrous; flowering shoots developing into spurs, 5–10(–15) mm. **Buds** purplish brown, ovoid, 4–5 mm, scale margins glabrous or ciliate. **Leaves** convolute in bud; isomorphic; stipules deciduous, lanceolate, 4–6 mm, apex acuminate; petiole 5–25 mm, glabrescent; blade ovate, elliptic, or narrowly elliptic, 3.5–8 × 2.5–4.5 cm, base cuneate or subrounded, margins unlobed, obtusely serrulate, apex long-acuminate, surfaces glabrous or sometimes puberulous along midveins. **Panicles** umbel-like; peduncles absent; bracteoles absent. **Pedicels** 20–40 mm, sparsely puberulent. **Flowers** <some trees reportedly andromonoecious, inflorescences with 1 or more staminate flowers or flowers all bisexual>, 30–35 mm diam.; hypanthium glabrous; sepals triangular-ovate, 3–5 mm, equal to or shorter than tube, apex obtuse, abaxial surface glabrous, adaxial tomentose; petals <usually more than 5>, pink, obovate, 15 mm, claws 1 mm, margins entire, apex rounded; stamens 20–25, 4–6(–8) mm, anthers white before dehiscence; styles 4 or 5, basally connate, length unknown, slightly longer than stamens, long-tomentose basally. **Pomes** dark red or purple, pyriform or obovoid, 6–8 mm diam., cores enclosed at apex; sepals deciduous; sclereids sparse to moderate surrounding core. **Seeds** brown. $2n = 34$.

Flowering Mar–May; fruiting Sep–Oct. Along streams or slopes; 0–1200 m; introduced, N.C.; Asia.

In Japan, cultivated plants of *Malus halliana* have larger flowers (3.5–5 cm), with sepals sometimes acute at the apex, to 40 stamens, and styles slightly shorter than the stamens (Y. Asami 1927). Further study is required to define the morphologic variation and species boundaries of this taxon.

Malus floribunda Siebold ex Van Houtte is similar to *M. halliana*; it differs by its anthers yellow before anthesis, leaves sharply serrulate and conduplicate in bud, and pomes impressed at both ends. The geographic origin of *M. floribunda* is unknown; wild Asian populations have not been documented. It may be of hybrid origin and probably represents a cultivated selection, rather than a natural species.

9. Malus hupehensis (Pampanini) Rehder, J. Arnold Arbor. 14: 207. 1933 • Tea or flowering tea crabapple [I]

Pyrus hupehensis Pampanini, Nuovo Giorn. Bot. Ital., n. s. 17: 291. 1910 (as Pirus); *Malus domestica* Borkhausen var. *hupehensis* (Pampanini) Likhonos; *M. theifera* Rehder

Trees, to 80 dm. **Stems** to 30 cm diam.; bark dark gray or dark brown, smooth or rough; young branches dark green and puberulous, becoming purplish or purplish brown and glabrous at maturity; flowering shoots developing as spurs, 5–20(–40) mm. **Buds** dark purple, ovoid, 3–4(–5) mm, scale margins sparsely ciliate. **Leaves** convolute in bud; isomorphic; stipules deciduous, linear-lanceolate, 5–6 mm, apex acuminate; petiole 10–30 mm, sparsely pubescent when young, glabrescent; blade ovate or ovate-oblong, 5–10 × 2.4–4 cm, base broadly cuneate, rarely rounded, margins unlobed, sharply serrulate, apex acuminate, surfaces sparsely puberulent when young, becoming glabrous. **Panicles** umbel-like; peduncles absent; bracteoles rarely persistent, lanceolate, 2 mm. **Pedicels**

30–40 mm, glabrous or slightly villous. **Flowers** 35–40 mm diam.; hypanthium <constricted distal to ovaries>, glabrous or sparsely villous; sepals triangular-ovate, 4–5 mm, equal to or shorter than tube, apex acute or acuminate, abaxial surface glabrous, adaxial hairy; petals white, sometimes pink-white, obovate, 15 mm, claws 1 mm, margins crenulate, apex rounded; stamens 20, 6–8 mm, anthers white before dehiscence; styles 3(or 4), basally connate to ½ length, 7–9 mm, slightly longer than stamens, proximally tomentose. **Pomes** greenish yellow with red blush, ellipsoid or subglobose, 10 mm diam., cores enclosed at apex; sepals deciduous; sclereids sparse to moderate surrounding core. **Seeds** brown. **2*n*** = 51, 68.

Flowering Apr–May; fruiting Aug–Sep. Slopes and valleys; 0–2900 m; introduced; Ill., Wash.; Asia; introduced also in Europe.

Malus hupehensis is possibly naturalized in Egypt.

10. Malus toringo (Siebold) Siebold ex de Vriese, Tuinb.-fl. 3: 368. 1856 • Japanese or Toringo or Siebold crabapple [I]

Sorbus toringo Siebold, Jaarb. Kon. Ned. Maatsch. Aanm. Tuinb. 1848: 47. 1848; *Crataegus cavaleriei* H. Léveillé & Vaniot; *C. taquetii* H. Léveillé; *Malus baccata* (Linnaeus) Borkhausen subsp. *toringo* (Siebold) Koidzumi; *M. sargentii* Rehder; *Photinia rubrolutea* H. Léveillé; *Pyrus esquirolii* H. Léveillé; *P. subcrataegifolia* H. Léveillé; *P. toringo* (Siebold) Siebold ex Miquel

Trees, sometimes shrubs, 30–80(–100) dm. **Stems** to 30 cm diam.; bark purplish brown, smooth, split irregularly when old; young branches orangish brown and puberulent, becoming dark purple or purplish brown and glabrous; flowering shoots developing as spurs, rarely as thorns, (5–)15–20(–35) mm. **Buds** purplish brown, ovoid, 20–30 mm, glabrous or only terminal scale margins puberulous. **Leaves** conduplicate in bud; isomorphic; stipules deciduous, lanceolate, 4–6 mm, apex acute or acuminate; petiole 10–25 mm, puberulous; blade narrowly elliptic, elliptic, or ovate, 3–7 × 2–4 cm, base broadly cuneate or rounded, margins usually 3-lobed, serrate, apex acute, abaxial surface puberulous when young, veins puberulous at maturity, adaxial glabrous. **Panicles** umbel-like; peduncles absent; bracteoles deciduous, filiform, 4–5 mm. **Pedicels** 20–25 mm, pubescent or subglabrous. **Flowers** 20–30 mm diam.; hypanthium subglabrous or pubescent; sepals lanceolate, 6–9 mm, to 2 times as long as tube, apex caudate-acuminate, abaxial surface glabrous, adaxial tomentose; petals white, sometimes pink-white, elliptic-obovate, 15–18 mm, claws 1–2 mm, margins entire, sometimes crenulate, apex rounded; stamens 20, 7–9 mm, anthers yellow before dehiscence; styles 3 or 4 (or 5), basally connate to ½ length, 8–10 mm, slightly longer than stamens, proximally villous or lanate. **Pomes** red or brownish yellow, subglobose, 6–8(–10) mm diam., cores enclosed at apex; sepals deciduous; sclereids absent or relatively few surrounding core. **Seeds** reddish brown. **2*n*** = 34, 51.

Flowering Apr–Jun; fruiting Aug–Sep. Slopes, mixed forests, or shrublands; 100–2000 m; introduced; Ont.; Ill., Kans., Md., Mass., N.Y., Ohio, Oreg., Pa., Va., Wash., Wis.; Asia.

Malus toringo is possibly naturalized in Europe.

Variety *sargentii* (Rehder) Asami is distinguished by its horizontally spreading branches, pure white, suborbiculate petals, and dark red fruits with a slight bloom; var. *zumi* (Matsumara) Asami is recognized by its ascending and spreading branches (Asami).

Malus ×*zumi* (Matsumura) Rehder has been reported for Ohio. The hybrid is morphologically similar to *M. toringo*, but differs by its larger flowers and fruits and the vigorous shoot having sparsely distributed, only slightly lobed leaves. It has been considered by some to be a variety of *M. toringo* and by others to be a hybrid between *M. baccata* and *M. toringo*.

Malus sieboldii Rehder and *Pyrus sieboldii* Regel are illegitimate names that pertain here.

59. PYRUS Linnaeus, Sp. Pl. 1: 479. 1753; Gen. Pl. ed. 5, 214. 1754 • Pear, poirier [Latin *pirum*, pear] [I]

Paul M. Catling

Gisèle Mitrow

Trees, <often pyramidal or oval in silhouette>, (50–)80–300 dm. **Stems** 1–many, erect; bark gray, reddish or orange-brown, <smooth, older with gray or blackish rectangular plates>; short shoots sometimes present; thorns present or absent; young branches glabrous or tomentose; <buds

ovoid, narrowly ovoid, or triangular-ovoid, apex obtuse or acuminate, glabrous or sparsely to densely tomentose>. **Leaves** deciduous, cauline, simple; stipules sometimes caducous, free, lanceolate, margins serrate or entire; petiole present; blade suborbiculate, ± ovate or ovate-lanceolate to elliptic, oblong, or obovate, (2–)5–9(–12) cm, firm or leathery, margins flat, serrate, crenate, serrulate, crenulate, or entire, venation pinnate, surfaces hairy or glabrous. **Inflorescences** terminal <on short shoots>, 4–9-flowered, racemes or simple corymbs, umbel-like, <with short persistent axis>, glabrous or tomentose; bracts absent; bracteoles absent. **Pedicels** present. **Flowers** developing with or before leaves, perianth and androecium epigynous, 15–45 mm diam.; hypanthium campanulate or cupular, 3–5 mm diam., interior usually pubescent, exterior hairy or glabrous; sepals 5, reflexed, lanceolate or triangular-lanceolate to triangular or triangular-ovate; petals 5, white or creamy, ovate or ovate-elliptic to obovate or obovate-elliptic, <base short-clawed, apex rounded>; stamens 20, equal to or slightly shorter than petals; carpels 2–5, mostly connate, fully adnate to hypanthium, glabrous, styles terminal, distinct (constricted basally by hypanthium), <± equal to stamens>; ovules 2. **Fruits** pomes, brown, red, green, or yellow, often dotted, globose, subglobose, ovoid, obovoid, or pyriform, 8–120 mm diam., glabrous; <fleshy with abundant stone or grit cells>; hypanthium persistent; sepals persistent or deciduous, reflexed; carpels cartilaginous; styles persistent or not. **Seeds** 2–5(–10) per fruit. x = 34.

Species ca. 25 (6 in the flora): introduced; Europe, Asia, n Africa; introduced also in Mexico, South America, Pacific Islands (New Zealand), Australia.

Pyrus communis is widely cultivated and naturalized in temperate and subtropical regions around the world. The evergreen pear, *P. kawakamii* Hayata, is widely cultivated, especially in California. This taxon has recently been included in *P. calleryana* (G. Cuizhi and S. A. Spongberg 2003), which is escaped in California. Some of the ornamental pear cultivars available in North America are selections of the oriental *P. betulifolia* Bunge, with fruits less than 1 cm diameter and leaves tomentose or at least slightly hairy proximally. *Pyrus betulifolia* is frequently used as a grafting stock. The willow-leaved pear, *P. salicifolia* Pallas, with densely silvery-pubescent leaves, pendent branches, and brown, pyriform pomes, is rarely cultivated in North America and is not known to escape. The Ussurian pear, *P. ussuriensis* Maximowicz, is less often cultivated and not known in North America as an escape; it is distinguished by persistent sepals on the fruit and spinulose-serrate leaf margins.

Pyrus may be divided into three groups (J. S. Challice and M. N. Westwood 1973; R. L. Bell 1990; K. Browicz 1993). *Pyrus calleryana* belongs to the Asian pea pear group. *Pyrus pyrifolia* belongs to the medium- to large-fruited Asian pears. Both *P. communis* and *P. nivalis* belong to the group of European species.

Although the cultivated fruits of apples and pears are easily distinguished, the wild pears and wild apples are sometimes quite similar. Pears differ from apples (*Malus*) by their hard glossy leaves, leaf margins with callused serrations, flower clusters with a central axis, and fruit with abundant stone or grit cells. The closely related quince (*Cydonia oblonga*) differs in its basally subcordate leaves that are abaxially tomentose and have entire margins, and its tomentose, pyriform pomes.

The identification of wild pears in North America is difficult. For more information see 4. *Pyrus calleryana*. The genus requires more study to resolve relationships, taxa, and hybrid origins.

SELECTED REFERENCES Browicz, K. 1993. Conspect and chorology of the genus *Pyrus* L. Abor. Kórnickie 38: 17–33. Challice, J. S. and M. N. Westwood. 1973. Numerical taxonomic studies of the genus *Pyrus* using both chemical and botanical characters. Bot. J. Linn. Soc. 67: 121–148. Lee, S. H. 1948. A taxonomic survey of the oriental pears. Proc. Amer. Soc. Hort. Sci. 51: 152–156. Rubtsov, G. A. 1944. Geographical distribution of the genus *Pyrus* and trends and factors in its evolution. Amer. Midl. Naturalist 78: 358–366. Terpó, A. 1985. Studies in the taxonomy and grouping of *Pyrus* species. Feddes Repert. 96: 73–87.

1. Branches white-tomentose when young; leaf blades: abaxial surfaces densely whitish gray pubescent . . . . . 1. *Pyrus nivalis*
1. Branches and leaf blades glabrous or glabrescent.
   2. Leaf margins spinulose-serrate, surfaces brown-lanate when young; flowers 25–35 mm diam.; pomes subglobose . . . . . 2. *Pyrus pyrifolia*
   2. Leaf margins entire or crenulate, serrulate, crenate-dentate, obtusely serrate, or crenate, surfaces glabrous or brown-tomentose abaxially when young; flowers 15–35 mm diam.; pomes globose, subglobose, obovoid, ovoid, or pyriform.
      3. Styles 2 or 3(or 4); pomes blackish brown or dark brown, brown, or yellow-brown (often with pale dots), subglobose, globose, or obovoid, 10–15(–22) mm diam.
         4. Sepals mostly persistent; pomes subglobose or obovoid, 15–22 mm diam.; styles 3(or 4) . . . . . 3. *Pyrus serrulata*
         4. Sepals usually deciduous; pomes globose, 10–15 mm diam.; styles 2 or 3 . . . . 4. *Pyrus calleryana*
      3. Styles 3(–5); pomes yellowish, green, brown, or red (with or without pale dots), globose, subglobose, ovoid, or pyriform, 8–160 × 8–120 mm.
         5. Leaf margins densely ciliate when young; sepals persistent; petals (12–)13–15 mm; leaf blades ovate or suborbiculate to elliptic, bases broadly cuneate to almost rounded . . . . . 5. *Pyrus communis*
         5. Leaf margins eciliate or sparsely ciliate when young; sepals deciduous, sometimes persistent; petals 6–12 mm; leaf blades ovate, ovate-lanceolate, or suborbiculate, bases usually cordate, sometimes broadly rounded . . . . . 6. *Pyrus cordata*

1. Pyrus nivalis Jacquin, Fl. Austriac. 2: 4, plate 107. 1774 • Snow pear [I]

Pyrus austriaca A. Kerner; *P. eriopleura* Reichenbach; *P. salviifolia* de Candolle

Plants 50–200 dm. **Branches** white-tomentose when young, later blackish; thorns usually absent. **Leaves:** petiole 1–2 cm, tomentose; blade obovate, 5–9 × 3–4 cm, base cuneate, decurrent, margins entire or slightly crenulate, apex acute or short-acuminate, abaxial surface densely whitish gray-pubescent, adaxial sparsely pubescent. **Pedicels** 3–6 cm, tomentose-lanate. **Flowers** 35–45 mm diam.; sepals triangular, 6–8 × 3–4 mm, apex acuminate; petals white, obovate-elliptic, 14–16 × 12–14 mm; ovaries 5-locular; styles 5. **Pomes** yellowish green with purple dots, globose, 30–50 mm diam.; sepals persistent.

Flowering Apr. Old fields; 50–200 m; introduced; Md., Wash.; Europe.

In the flora area, *Pyrus nivalis* possibly is only persisting after cultivation. The species is native to central and south-central Europe. Some specimens of common pears are called snow pears because of the abundant white flowers; the true snow pear also has whitish foliage as a result of the persistent indument on the abaxial surface of the leaf.

Pyrus elaeagnifolia Pallas, differing from *P. nivalis* in styles villous to the middle instead of only at the base, smaller fruit 2–3 cm, and often narrower lanceolate or narrowly elliptic leaves, would key here. It is occasionally cultivated but not yet definitely known as an escape.

2. Pyrus pyrifolia (Burman f.) Nakai, Bot. Mag. (Tokyo) 40: 564. 1926 • Chinese pear [I]

Ficus pyrifolia Burman f., Fl. Indica, 226. 1768; *Pyrus serotina* Rehder

Plants 70–150 dm. **Branches** purplish brown or dark brown, tawny villous or tomentose when young, glabrescent; thorns usually absent. **Leaves:** petiole 3–4.5 cm, tomentose when young; blade ovate or ovate-elliptic, 7–12 × 4–6.5 cm, base rounded or subcordate, margins spinulose-serrate, apex acute, surfaces glabrous or brown-lanate when young. **Pedicels** 3.5–5 cm, pubescent when young. **Flowers** 25–35 mm diam.; sepals triangular-ovate, 5 mm, apex acuminate; petals white, ovate, 15–17 mm; ovaries 5-locular; styles 5. **Pomes** brownish with pale dots, subglobose, 20–25 mm diam.; sepals deciduous. $2n = 34$.

Flowering Apr. Old fields; 50–400 m; introduced; Ill., Va., W.Va.; Asia (China, Laos, Vietnam).

The type collection of the species is from Japan, but the plant is doubtfully native there.

3. Pyrus serrulata Rehder, Proc. Amer. Acad. Arts 50: 234. 1915 • Serrulate pear [I]

Plants 80–100 dm. **Branches** purplish brown or dark brown, brown-tomentose when young, glabrescent; thorns not recorded. **Leaves:** petiole 3.5–7.5 cm, brown-tomentose when young; blade ovate or narrowly ovate, 5–11 × 3.5–7.5 cm, base rounded or broadly cuneate, margins serrulate, apex acuminate, surfaces glabrous or brown-tomentose when young. **Pedicels** 3–5 cm, pubescent when young. **Flowers** 20–30 mm diam.; sepals triangular-ovate, 3 mm, apex acute or acuminate; petals white, broadly ovate, 10–12 mm; ovaries 3(or 4)-locular; styles 3(or 4). **Pomes** dark brown, subglobose or obovoid, 15–22 mm diam.; sepals mostly persistent. 2*n* = 34.

Flowering Apr. Fields; 300 m; introduced, Mo.; Asia (China).

The tree nursery industry has associated the common name "red spire" with this species but that name is frequently applied also to *Pyrus calleryana*. *Pyrus serrulata* has been considered a hybrid, *P. calleryana* × *P. pyrifolia*, but recent taxonomic work has accorded it specific rank. The only report of this species as an escape in North America is based on a fruiting specimen collected in Howell County, Missouri. Potential hybrids involving *P. calleryana* may resemble this species.

4. Pyrus calleryana Decaisne, Jard. Fruit. 1: sub plate 8. 1872 (as Pirus) • Callery or Bradford pear [I]

Plants 50–200 dm. **Branches** reddish brown when young, becoming grayish brown with age, initially tomentose, usually soon glabrescent; short shoots of young plants often thorn-tipped (some escaped cultivars thornless). **Leaves:** petiole 2–4.5 cm, glabrous; blade ovate, broadly ovate, or oblong-lanceolate, 4–9 × 3.5–6 cm, base broadly cuneate to rounded, margins serrate or entire, apex acuminate, surfaces glabrous. **Pedicels** 1–3 cm, glabrous. **Flowers** 15–25 mm diam.; sepals lanceolate, 5 mm, apex acuminate; petals white, obovate, 6–7(–13) × 6–7(–13) mm; ovaries 2 or 3(or 4)-locular; styles 2 or 3. **Pomes** blackish brown, brown, or yellow-brown with white or tan dots, globose, 10–15 mm diam.; sepals deciduous.

Flowering late Feb–early May (sometimes partial second flowering Sep–Oct). Woodland edges, bottomland forests, old field fencerows; 0–400 m; introduced; Ala., Ark., Calif., Fla., Ga., Ill., Ind., Kans., Ky., La., Md., Mich., Miss., Mo., N.J., N.C., Ohio, Okla., Pa., S.C., Tenn., Tex., Va., W.Va.; Asia (China, Japan, Vietnam).

The presence of *Pyrus calleryana* in Oklahoma is not confirmed.

The densely hairy buds are a useful identification characteristic, and the leaves are often so deltate or acuminate as to be confused with poplar.

Four varieties are recognized within the native range of *Pyrus calleryana* (G. Cuizhi and S. A. Spongberg 2003). These varieties may all be found in North America throughout the range of the species. They are separated principally by differences in the leaf shape. Variety *calleryana* has ovate to broadly ovate leaf blades, with a rounded to broadly cuneate base and serrate margins (obtuse teeth). Variety *integrifolia* T. T. Yu has ovate leaf blades with a rounded base and entire margins. Variety *lanceata* Rehder (synonym *P. kawakamii* Hayata) has ovate- to oblong-lanceolate leaf blades with a rounded to broadly cuneate base and shallowly serrate or nearly entire margins. These three varieties have two- or three-locular ovaries, while the following has three- or four-locular ovaries. Variety *koehnei* (C. K. Schneider) T. T. Yu (synonym *P. koehnei* C. K. Schneider) has ovate leaf blades with a cuneate base and serrate margins.

The cultivated forms vary markedly in leaf shape, and recognition of variation on this basis, within what has become a complex cultivated species, may be unwarranted. Most escaped plants in the flora area are referable to var. *calleryana*.

The hard, close-grained wood of *Pyrus calleryana* has been used for making furniture and tools. This species is also frequently used as a stock for grafting *P. pyrifolia*; it is most well known as an ornamental, attractive not only during flowering but also with dramatic autumn foliage varying from maroon to burgundy or, occasionally, yellow, orange, or red. It has been deemed a cultivated nuisance since some of the cultivars split apart when adult size is achieved and cultivation increases the threat of invasion of natural habitats to the detriment of native species. It has rapidly become invasive in its horticultural range.

The Bradford pear is actually the first cultivar of the Callery pear. It was selected at the Plant Introduction Station in Glenn Dale, Maryland, in 1963, and it clearly demonstrated the potential ornamental value of this species. The cultivar name, which honors F. C. Bradford, former horticulturist at Glenn Dale, has become commonly used for the species. Some of the escaped individuals resembling the Bradford pear appear to be of hybrid origin, suggested by larger round fruits (to 2.5 cm diam.), glabrous buds, sharply serrate leaf margins, and persistent calyces (M. A. Vincent 2005). Because *Pyrus communis* cultivars have been found to be incompatible with the Bradford pear (*P. calleryana*), these hybrids are believed to involve *P. betulifolia* Bunge or *P. bretschneideri* Rehder rather than *P. communis* (Vincent). Some support for this idea exists; cultivars have been produced from crosses of *P. calleryana* and *P. betulifolia*. Some of these putative hybrids appear within the range of *P. communis*, which includes small, round-fruited wild pears as treated by J. S. Challice and M. N. Westwood (1973). Furthermore, some successful crosses of wild *P. calleryana* and *P. communis* have been reported (Westwood and H. O. Bjornstad 1971). Wild pears similar to *P. calleryana* in their rather small, round fruits that are, nevertheless, larger than 1.5 cm diam. may be referable to *P. communis* or to hybrids involving *P. betulifolia*, *P. bretschneideri*, or *P. communis*. Although the majority of wild pears can be placed with either *P. calleryana* or *P. communis*, plants of uncertain identity are widespread (Vincent). Such material has been seen or reported from Alabama, Arkansas, Florida, Georgia, Kentucky, Louisiana, Michigan, Mississippi, North Carolina, Ohio, South Carolina, Tennessee, and Texas. Some plants appear very close to *P. calleryana* but have larger fruit (1.3–1.5 cm diam.) and three or four styles. These may be referable to *P. fauriei* C. K. Schneider, or even to some other taxon such as *P. cordata*.

Pyrus calleryana and its putative hybrids are well suited to invasive behavior, having broad ecologic tolerance, relatively few pests, rapid growth, and early sexual maturity (three years). Seeds are produced in large quantity and may be dispersed by birds; control has been recommended (M. A. Vincent 2005). Although individual cultivars cannot self-pollinate and are, therefore, not invasive, different cultivars are widely planted, and crossing between these or between the scion and rootstock of cultivated individuals can produce invasive plants (T. M. Culley and N. A. Hardiman 2007). The rate of invasion is likely to be increased by the introduction of new cultivars into the mixture already present. Interestingly, thornless cultivars appear to retain genes for thorniness that recombine in their progeny, producing thorny invasives.

SELECTED REFERENCES Culley, T. M. and N. A. Hardiman. 2007. The beginning of a new invasive plant: A history of the ornamental Callery pear in the United States. BioScience 57: 956–964. Vincent, M. A. 2005. On the spread and current distribution of *Pyrus calleryana* in the United States. Castanea 70: 20–31.

5. **Pyrus communis** Linnaeus, Sp. Pl. 1: 479; 2: 1200. 1753 • Common pear [F] [I]

Pyrus ×*amphigenea* Domin ex Dostálek; *P. caucasica* Fedorov; *P. pyraster* (Linnaeus) Burgsdorff

Plants 50–150(–300) dm. **Branches** grayish brown or dark reddish brown, glabrous; short shoots of young plants often thorn-tipped. **Leaves:** petiole 1.5–5 cm, slightly pubescent when young; blade ovate or suborbiculate to elliptic, 2–5(–7) × 1.5–2.5 cm, base broadly cuneate to almost rounded, margins obtusely serrate, serrulate, or crenulate, sometimes entire, <densely ciliate when young>, apex acute or shortly acuminate, surfaces pubescent when young, glabrescent. **Pedicels** 2–3.5 cm, pubescent or glabrate. **Flowers** 25–35 mm diam.; sepals triangular-lanceolate, 5–9 × 3–4 mm, apex acuminate; petals white, obovate, (12–)13–15 × 10–13 mm; ovaries 5-locular; styles (3–)5. **Pomes** green, yellowish, or reddish green, globose, subglobose, ovoid, or pyriform, 30–160 × 15–120 mm; sepals persistent. $2n = 34$.

Flowering Mar–May; fruiting Jul–Aug. Open woods, old fields, clearings, fencerows; 0–1000+ m; introduced; B.C., N.S., Ont.; Ala., Ark., Calif., Conn., Del., Fla., Ga., Idaho, Ill., Ind., Iowa, Kans., Ky., La., Maine, Md., Mass., Mich., Miss., Mo., Mont., N.H., N.J., N.Mex., N.Y., N.C., Ohio, Okla., Oreg., Pa., R.I., S.C., Tenn., Tex., Utah, Vt., Va., Wash., W.Va., Wis.; Eurasia; introduced also in Mexico.

Pyrus communis is frequently cultivated and escapes to adjacent habitats, where it is sometimes common or even dominant. It includes more than a thousand cultivars, most of these being fruit (pomological) varieties. Ornamental forms with variegated leaves, dissected leaves, and unusual growth habits also occur. The taxon is of hybrid origin, and its parentage is thought to involve multiple species, including *P. austriaca* K. Kerner, *P. nivalis* Jacquin, *P. pyraster*, and *P. syriaca* Boissier. *Pyrus pyraster* is often treated as a wild species with smaller, rounder, and more sour fruit, yet it seems likely that wild pears fitting this description have originated from cultivated stock referable to *P. communis*, just as self-sown apples (*Malus pumila*) often have smaller, sour fruits. Recent authors are followed here in treating *P. pyraster* as a synonym.

The leaf margins of *Pyrus communis* are usually densely ciliate in flowering material, and this vestiture is retained to some extent in fruiting specimens. This characteristic, along with the broader fruiting pedicels, assists in distinguishing this species from *P. cordata*.

The high eating-quality pears were developed in the eighteenth or nineteenth century in northern Europe, although pears had been cultivated for centuries prior to that time. Currently, much of the cultivation in North America is in the Great Lakes area and on the Pacific coast. The highest production is in Europe (France, Germany, Switzerland), where most of the crop is used for making pear cider (perry); large-scale cultivation occurs also in Argentina, Australia, and Japan. The trees sometimes attain great age, and some in eastern North America, the French mission pears, persist today where planted 300 years ago on the sites of frontier forts in what was then a wilderness. Some of these old pears are of a type now no longer cultivated and have been identified as a significant germplasm resource.

6. Pyrus cordata Desvaux, Observ. Pl. Angers, 152. 1818 • Plymouth pear [I]

Pyrus boissieriana Buhse

Plants 50–80(–100) dm. **Branches** mostly purplish brown or dark reddish brown, glabrous; thorns present at end of twigs. **Leaves:** petiole 2–5 cm, glabrous when young; blade ovate, ovate-lanceolate, or suborbiculate, 2–5.5 × 1.5–3.5 cm, base usually cordate, sometimes broadly rounded, margins crenate-dentate or serrulate, <eciliate or sparsely ciliate when young>, apex acute or shortly acuminate, surfaces glabrous. **Pedicels** 2–3.5 cm, pubescent or glabrate. **Flowers** 20–25 mm diam.; sepals triangular-lanceolate, 2–3(–5.5) × 1–1.5 mm, apex acuminate; petals white or creamy, ovate-elliptic or obovate, 6–12 × 5–7 mm; ovaries 5-locular; styles (3–)5. **Pomes** red or reddish brown or reddish green, densely pale dotted, usually globose, sometimes subglobose, 8–15(–20) × 8–15 mm; sepals deciduous, sometimes persistent. $2n = 34$.

Flowering late Apr–early May; fruiting Aug–Sep. Moist soil at edge of wetlands; 200–300 m; introduced, Mich.; Europe; w Asia; n Africa.

Occasional plants that are considered possible hybrids involving either *Pyrus calleryana* or *P. communis* may actually belong here; in most cases, sufficient information is lacking to identify them with certainty. Report of this taxon for Michigan is based on both flowering and fruiting material from the same stand and on extensive notes and photographs.

The concept used here for *Pyrus cordata* is based largely on J. J. Aldasoro et al. (1996), who distinguished it from *P. communis* by its narrower fruit pedicels, 0.8–1.6 mm diam. versus 1.5–2.8 mm diam.; the authors have observed much variation in pedicel diameter in wild plants referable to *P. communis*, in which the pedicels of dry specimens are often less than 1.5 mm diam. Among the species occurring in Europe and North Africa, *P. cordata* is considered to be quite distinctive (Aldasoro et al.; J. S. Challice and M. N. Westwood 1973). The flowers are reported to have a particularly unpleasant odor.

60. CHAENOMELES Lindley, Trans. Linn. Soc. London 13: 97. 1821 (as Choenomeles), name and orthography conserved • Flowering quince [Greek *chaino*, open, and *melon*, apple, alluding to mistakenly presumed splitting of fruit] [I]

Paul M. Catling

Gisèle Mitrow

Shrubs [or trees], (0.2–)10–20 dm. **Stems** few to many, erect or spreading; bark purplish brown, blackish brown, purplish black, or purple, with scattered pale brown lenticels; long and short shoots present; thorns present; glabrous or hairy young, <smooth older; buds triangular-ovoid, apex obtuse or acute, scale margins glabrous or hairy>. **Leaves** deciduous or semipersistent, cauline, simple; stipules persistent, free, reniform or suborbiculate, rarely ovate, <leaflike>, margins serrate or crenate-serrate; petiole present; blade spatulate, obovate, elliptic, or ovate, 3–9 cm, firm or leathery, margins flat, serrate or crenate-serrate, venation pinnate, surfaces glabrous, sometimes midvein abaxially. **Inflorescences** terminal <on short branches, appearing lateral on branch as a whole>, [2 or]3–5[–10]-flowered, fascicles, glabrous or hairy; bracts present or absent; bracteoles present or absent. **Pedicels** present, short, or absent. **Flowers** opening before or with leaves, perianth and androecium epigynous, 25–50 mm diam.; hypanthium campanulate, ± constricted at mouth, 4–7 mm diam., exterior glabrous; sepals 5, reflexed or ascending, suborbiculate or ovate, <abaxial surface glabrous, adaxial hairy>; petals 5, white, pink, or red, obovate or ovate to suborbiculate, <base short-clawed, apex rounded>; stamens 40–60, equal to or ½ length petals; carpels 5, connate, adnate to hypanthium, indumentum not recorded, styles <2–5>, terminal, basally connate ⅓ of length, <nearly equal to stamens>; ovules 2. **Fruits** pomes, <sessile>, yellow or yellowish green, globose, subglobose, or ovoid, 23–60 mm diam., <5-locular>, glabrous; <fleshy>; hypanthium persistent; sepals deciduous; carpels cartilaginous; styles deciduous. **Seeds** 10 per locule. x = 17.

Species 4 (2 in the flora): introduced; Europe, Asia (China, Japan).

The flowering quinces are widely cultivated as ornamental shrubs for their attractive and abundant pink, red, or white flowers. Other species differ from those in the flora area in their entire leaf margins and tomentose leaves.

1. Branches smooth (not verrucose with age); leaf margins serrate; pomes 40–60 mm diam. . . . . 1. *Chaenomeles speciosa*
1. Branches scabrous (verrucose with age); leaf margins crenate-serrate; pomes 23–40 mm diam. . . . . 2. *Chaenomeles japonica*

1. Chaenomeles speciosa (Sweet) Nakai, Jap. J. Bot. 4: 331. 1929 (as Choenomeles) • Common flowering-quince F I

Cydonia speciosa Sweet, Hort. Suburb. Lond., 113. 1818

Shrubs, 2–20 dm. **Branches** purplish brown or blackish brown, smooth (not verrucose with age). **Leaves:** stipules of vegetative branches usually reniform or suborbiculate, rarely ovate, to 1 cm, margins serrate, <teeth triangular>, apex acute; petiole 5–10 mm; blade ovate, elliptic, or narrowly elliptic, 3–9 × 1.5–5 cm, base cuneate to broadly cuneate, margins serrate, apex obtuse to acute, abaxial surface glabrous or hairy on veins. **Flowers** 30–50 mm diam.; sepals suborbiculate, rarely ovate, 3–4 mm; petals scarlet red, rarely pinkish or white, ovate or suborbiculate, 13–24 mm; stamens 40–50, equal to petals. **Pomes** yellow or yellowish green, globose or ovoid, 40–60 mm diam. $2n$ = 34 (China).

Flowering Apr–May; fruiting Aug–Oct. Vacant lots, old fields, fencerows, wastelands; 0–500 m; introduced; Conn., D.C., Ky., La., Mass., Mo., N.Y., N.C., Ohio, Pa., Tenn., W.Va., Wis.; Asia (China); introduced also in Europe.

Chaenomeles speciosa is cultivated as an ornamental for its showy spring flowers and, occasionally, for medicinal use. Putative hybrids of *C. japonica* and *C. speciosa* are called *C. ×superba* (Frahm) Rehder. They also are cultivated and may be expected to spread occasionally through the dumping of garden waste. The putative hybrids are difficult to distinguish from *C. speciosa*.

2. Chaenomeles japonica (Thunberg) Lindley ex Spach, Hist. Nat. Vég. 2: 159. 1834 (as Choenomeles) • Japanese flowering quince I

Pyrus japonica Thunberg, Nova Acta Regiae Soc. Sci. Upsal. 3: 208. 1780; *Cydonia japonica* (Thunberg) Persoon

Shrubs, 0.2–10 dm. **Branches** purple, scabrous (and tomentose when young, becoming blackish brown, verrucose, glabrous with age). **Leaves:** stipules of vegetative branches reniform, to 1 cm, margins crenate-serrate, apex usually obtuse; petiole 4–6 mm; blade obovate, spatulate, or broadly ovate, 3–5 × 2–3 cm, base cuneate to broadly cuneate, margins crenate-serrate, apex obtuse to acute, abaxial surface glabrous. **Flowers** 25–40 mm diam.; sepals ovate, rarely suborbiculate, 4–5 mm; petals dark red, obovate or suborbiculate, 15–25 mm; stamens 40–60, ½ length of petals. **Pomes** yellow, subglobose, 23–40 mm diam.

Flowering Apr–May; fruiting Aug–Oct. Vacant lots, old fields, fencerows, wastelands; 10–600 m; introduced; Ill., Miss., N.Y., Pa., Vt., W.Va.; e Asia (Japan); introduced also in Europe.

Chaenomeles japonica is cultivated as an ornamental for its showy spring flowers.

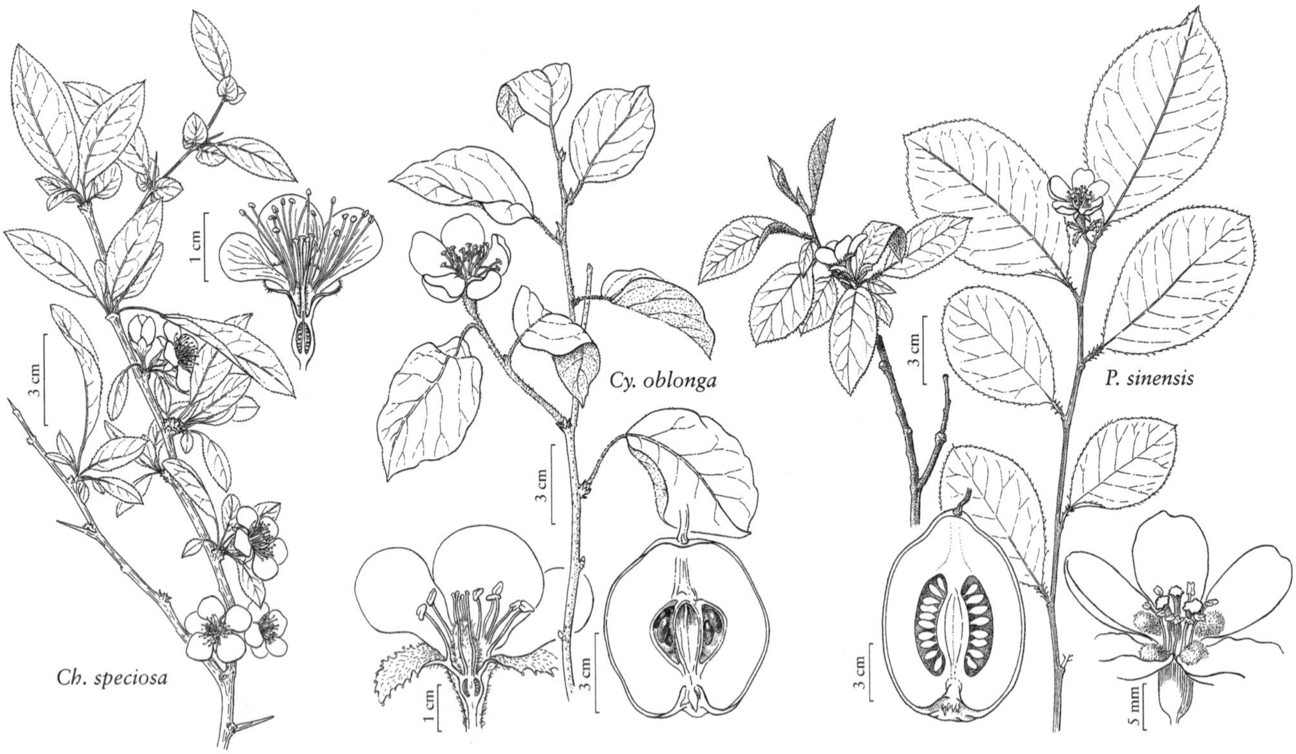

CHAENOMELES ◦ CYDONIA ◦ PSEUDOCYDONIA

61. CYDONIA Miller, Gard. Dict. Abr. ed. 4, vol. 1. 1754 • Quince [Greek *Kydonia*, alluding to nativity in Kydon, ancient city-state in Crete] [I]

Paul M. Catling

Gisèle Mitrow

Trees, <crowns rounded>, (30–)80 dm. **Stems** 1, sometimes several, erect or spreading; bark: older trunks gray, <fissures dividing bark into rectangular plates>; short shoots absent; unarmed; young branches tomentose, glabrescent; <buds ovoid, apex obtuse or acuminate, tomentose>. **Leaves** deciduous, cauline, simple; stipules caducous, free, ovate, margins glandular-serrate; petiole present; blade broadly ovate to oblong, 5–10 cm, firm or leathery, margins flat, entire, venation pinnate, abaxial surface densely villous, adaxial glabrous. **Inflorescences** terminal <on leafy branches>, flowers solitary, densely hairy (tomentose) within, otherwise glabrous; bracts present <similar to stipules>; bracteoles absent. **Pedicels** present. **Flowers** developing with leaves, perianth and androecium epigynous, 40–50 mm diam.; hypanthium campanulate, ± constricted, 4–6 mm diam., inner rims hairy; sepals 5, reflexed, ovate or broadly lanceolate; petals 5, white or light pink, suborbiculate, ovate, or obovate, <base short-clawed, apex rounded>; stamens 20, equal to or slightly longer than petals; carpels 5, connate, adnate to hypanthium, <ovaries 5-locular>, tomentose, styles <5>, terminal, distinct, <± equal to or 3/4 length of stamens, protruding through pit in top of hypanthium>; ovules many. **Fruits** pomes, <sessile or pedicellate>, yellow, pyriform or subglobose, 30–50 mm, tomentose (sometimes tomentum ± rubbed off, with abundant, grouped stone or grit cells); <fleshy>; hypanthium persistent; sepals persistent, reflexed; carpels cartilaginous; styles ± persistent. **Seeds** many.

Species 1: introduced; Asia; introduced also in South America, Europe, Africa.

Cydonia is native from Iran to Turkestan; it is introduced and naturalized throughout much of temperate Asia and the Mediterranean region, where it has been cultivated for its edible fruit since ancient times.

The species of 'flowering quince' (*Chaenomeles* spp.) are often called quince, but this latter common name is best reserved for *Cydonia oblonga*.

1. Cydonia oblonga Miller, Gard. Dict. ed. 8, Cydonia no. 1. 1768 [F] [I]

Branches purplish red, densely tomentose young, purplish brown, glabrous with age. **Leaves:** petiole 3–10 mm, tomentose young; blade 5–10 × 3–5 cm, base rounded or subcordate, apex acute or emarginate, lustrous or semi-lustrous. **Pedicels** 0–5 mm. **Flowers:** sepals 5–6 mm, apex acute; petals 15–20 mm. $2n = 34$.

Flowering Apr–May; fruiting Sep–Oct. Old fields, fencerows, wastelands; 0–500 m; introduced; Ont.; Calif., Conn., Ill., Md., Mass., N.Y., Ohio, Pa., Vt.; Asia; introduced also in South America, Europe, Africa.

Cydonia oblonga is cultivated in the United States, particularly California and New York, for its fruit and less often throughout temperate North America as an ornamental and a curiosity. It is often used as a dwarfing rootstock for pears. In the Mediterranean regions of Europe, South America, China, and elsewhere, it is cultivated for the golden yellow aromatic fruit that is rich in pectin, making fine jellies and preserves. The fruit is also used in pies, turning deep red when cooked.

62. PSEUDOCYDONIA (C. K. Schneider) C. K. Schneider, Repert. Spec. Nov. Regni Veg. 3: 180. 1906 • Chinese quince [Greek *pseudo-*, false, and genus *Cydonia*, alluding to resemblance] [I]

Paul M. Catling

Gisèle Mitrow

Chaenomeles Lindley sect. *Pseudocydonia* C. K. Schneider, Ill. Handb. Laubholzk. 1: 729, figs. 405a–g. 1906

Shrubs or trees, 50–100 dm. **Stems** 1–several, erect; bark purplish brown or purplish black with scattered pale brown lenticels; short shoots absent; unarmed; young branches hairy; <buds triangular-ovoid, apex obtuse, glabrous or scale margins pubescent>. **Leaves** deciduous or semipersistent, cauline, simple; stipules persistent, free, (of vegetative branches) ovate-oblong, rhombic, or lanceolate, <leaflike>, margins serrate; petiole present, <short>; blade elliptic-ovate or elliptic-oblong, rarely obovate, 5–8 cm, firm, margins flat, sharply serrate, venation pinnate, abaxial surface yellowish white-tomentose when young, glabrescent. **Inflorescences** terminal <on leafy branches>, flowers solitary, internally tomentose; bracts absent; bracteoles present or absent. **Pedicels** present. **Flowers** developing before or with leaves, perianth and androecium epigynous, 25–30 mm diam.; hypanthium campanulate, ± constricted, size not recorded, interior brown tomentose, exterior glabrous; sepals 5, reflexed, triangular-lanceolate; petals 5, light to deep pink, obovate; stamens 25, ½ length of petals; carpels 5, connate, adnate to hypanthium, indumentum not recorded, styles <3–5>, terminal, basally connate ⅓ of length, <as long as stamens>; ovules many. **Fruits** pomes, <sessile or short-pedicellate>, dark yellow, ellipsoid or narrowly ellipsoid, 100–150 mm, glabrous; hypanthium persistent; sepals deciduous; carpels cartilaginous; styles persistence not recorded. **Seeds** many.

Species 1: introduced, California; Asia (China); introduced also in Europe.

Pseudocydonia sinensis is morphologically intermediate between *Cydonia* and *Chaenomeles*, having the basally fused styles and 25 or fewer stamens of the former and the serrated leaves, glabrous fruits, and deciduous sepals of the latter. It has been placed in both those genera but is anomalous in both.

1. **Pseudocydonia sinensis** (Thouin) C. K. Schneider, Repert. Spec. Nov. Regni Veg. 3: 181. 1906 [F] [I]

Cydonia sinensis Thouin, Ann. Mus. Natl. Hist. Nat. 19: 145, plates 8, 9. 1812; *Chaenomeles sinensis* (Thouin) Koehne

Leaves: stipules 5–12 mm, herbaceous, apex acute; petiole 5–10 mm, pubescent; blade 5–8 × 3.5–5.5 cm, base broadly cuneate or round, apex acute. **Pedicels** 5–10 mm. **Flowers:** sepals 6–10 mm; petals 10–15 mm, base short-clawed, apex rounded. **2*n*** = 32.

Flowering Mar–Apr; fruiting Aug–Oct. Swampy waste areas; elevation unknown; introduced, Calif.; Asia (China); introduced also in Europe.

Pseudocydonia sinensis is cultivated as an ornamental and for its medicinal value. The species was reported from Mendocino County in 1979 (F. Hrusa et al. 2002) but is likely to have a much broader range of habitat in North America.

63. PHOTINIA Lindley, Bot. Reg. 6: plate 491. 1820, name conserved • [Greek *photeinos*, shining, alluding to leaves] [I]

Guy L. Nesom

Stranvaesia Lindley

Shrubs or trees, 10–120 dm. **Stems** 1–15+, erect; bark gray; short shoots absent; unarmed; glabrous. **Leaves** persistent or deciduous, cauline, simple; stipules deciduous, free, usually subulate, margins dentate or entire; petiole present; blade oblong to elliptic or elliptic-obovate, 3–20 cm, coriaceous or herbaceous, margins flat, serrate to serrulate, rarely entire, venation pinnate, surfaces glabrous at least at maturity. **Inflorescences** terminal, <erect or pendulous in fruit>, 150–300-flowered, corymbose (compound-corymboid) or subumbellate [compound-racemose], glabrous or glabrate; bracts present; bracteoles present. **Pedicels** present. **Flowers:** perianth and androecium epigynous, 6–12[–15] mm diam.; hypanthium cupulate to campanulate or cylindric, 1–2 mm, usually glabrous; sepals 5, incurved, triangular; petals 5, white, oblong to elliptic, obovate, or orbiculate to suborbiculate, <base clawed>; stamens (16–)20(–25), ± as long as or slightly shorter than petals; carpels (1 or)2–5, connate, adnate to proximal ⅓–½ of hypanthium, free apically, glabrous or apically pilose, styles <[1 or]2–4[or 5]>, terminal, distinct or ± basally connate, <stigmas truncate or capitate>; ovules 1 or 2. **Fruits** pomes, red or black, globose to ovoid or ellipsoid, sometimes apically depressed, 4–8[–12] mm, usually glabrous; <somewhat fleshy; carpels capped by hard-shelled dome that rises above hypanthium>; hypanthium persistent; sepals persistent, spreading (adnate to hypanthium); carpels cartilaginous; styles deciduous. **Seeds** (1 or)2 per carpel. ***x*** = 17.

Species 40–60 (4 in the flora): introduced; Mexico, Central America, Asia; introduced also in Europe, Pacific Islands (Hawaii, New Zealand), Australia.

Photinia is widely introduced in cultivation and naturalizing, at least in the continental United States. Species are native mostly to warm temperate Asia, from the Himalayas east to Japan and south to India and Thailand. Molecular data suggest that *Heteromeles* originated through hybridization, with a species of *Photinia* as pollen parent (Guo W. et al. 2011).

Five species from Mexico and Central America were included within *Photinia* by J. B. Phipps (1992), and molecular evidence corroborates their placement there (Guo W. et al. 2011).

The Asian *Stranvaesia* has sometimes been treated as distinct from *Photinia*, but botanists have noted the apparent artificiality of a primary character used to differentiate the two genera, namely an endocarp dehiscent in *Stranvaesia*, indehiscent in *Photinia* (C. Kalkman 1973; J. B. Phipps et al. 1991b; K. R. Robertson et al. 1991; J. R. Rohrer et al. 1991, 1994). Kalkman observed that the putative dehiscence of *Stranvaesia* carpels is an artifact of specimen preparation. Rohrer et al. (1991, 1994) found that *Photinia* and *Stranvaesia* do not differ in connation of the carpels or in the adnation of the carpels to the hypanthium. Molecular phylogenetic analyses (Guo W. et al. 2011) show that the *Stranvaesia* species have arisen within the cladistic topology of *Photinia*.

The primarily eastern North American genus *Aronia* (for example, *A. floribunda*, *A. melanocarpa*) has been included in *Photinia* in some classifications (K. R. Robertson et al. 1991) but is here treated as distinct.

Many species of *Photinia* are ornamental trees and shrubs with large lustrous leaves and masses of white flowers in the spring followed by red fruits in the fall. The flowers last about two weeks and commonly have an unpleasant smell. The wood is hard and heavy and has been used for furniture and small articles such as axe handles.

Photinias have been widely and abundantly planted for hedging in the southeastern United States. The rapid spread since about the mid 1980s of a fungal leaf spot, *Entomosporium maculatum*, has been lethal to many such hedges. *Photinia* ×*fraseri* ('red tip'; see under 2. *P. serratifolia*) and *P. glabra* are severely affected; *P. serratifolia* is said to be partially resistant.

1. Leaf margins entire . . . . 4. *Photinia davidiana*
1. Leaf margins usually serrate or crenate-serrulate.
    2. Leaves deciduous, blades herbaceous; inflorescences 3–6 cm diam.; pedicels with warty lenticels . . . . 3. *Photinia villosa*
    2. Leaves persistent, blades coriaceous; inflorescences 4–18 cm diam.; pedicels without lenticels.
        3. Leaf blades 5–9 cm, lateral veins 10–15(–18) pairs, margins crenate-serrulate; inflorescences 4–10 cm diam. . . . . 1. *Photinia glabra*
        3. Leaf blades (6–)9–20 cm, lateral veins 20–30 pairs, margins sharply serrate to inconspicuously toothed or entire; inflorescences 10–18 cm diam. . . . . 2. *Photinia serratifolia*

1. Photinia glabra (Thunberg) Franchet & Savatier, Enum. Pl. Jap. 1: 141. 1873 • Japanese photinia [I]

Crataegus glabra Thunberg in J. A. Murray, Syst. Veg. ed. 14, 465. 1784; Fl. Jap., 205. 1784

Plants 20–50 dm. **Leaves** persistent; petiole 10–30 mm, glabrous; blade elliptic to elliptic-obovate or elliptic-oblong, 5–9 × 2.5–5 cm, coriaceous, base cuneate, margins crenate-serrulate, lateral veins 10–15(–18) pairs, apex abruptly acuminate, surfaces glabrous. **Inflorescences** 4–10 cm diam. **Pedicels** without lenticels. **Flowers** 7–8 mm diam.; petals obovate, 2–3 mm, adaxially white-tomentose proximally.

Flowering late Mar–Apr. Fencerows, thickets, disturbed sites; 20–100 m; introduced, La.; Asia (China, Japan, Myanmar, Thailand); Pacific Islands (New Zealand); Australia.

As noted by C. Kalkman (1973), Lindley cited *Crataegus glabra* as a synonym of his proposed new species *Photinia serrulata*. The concept of *P. serrulata* by Lindley was not the same as that of *C. glabra* by Thunberg, and *P. serrulata* Lindley has persisted as a name in use for *P. serratifolia*. In any case, the publication of *P. serrulata* was superfluous and illegitimate.

Photinia ×*fraseri* Dress, or 'red tip,' is a popular hybrid cultivar between *P. glabra* and *P. serrulata*, valued especially for its brilliant burgundy red young leaves, which mature to green over weeks through shades of copper red. Shrubs are 3–4(–4.5) m; leaf blades elliptic to oblong-obovate or elliptic-obovate, 7–12 × 3–4(–4.5)

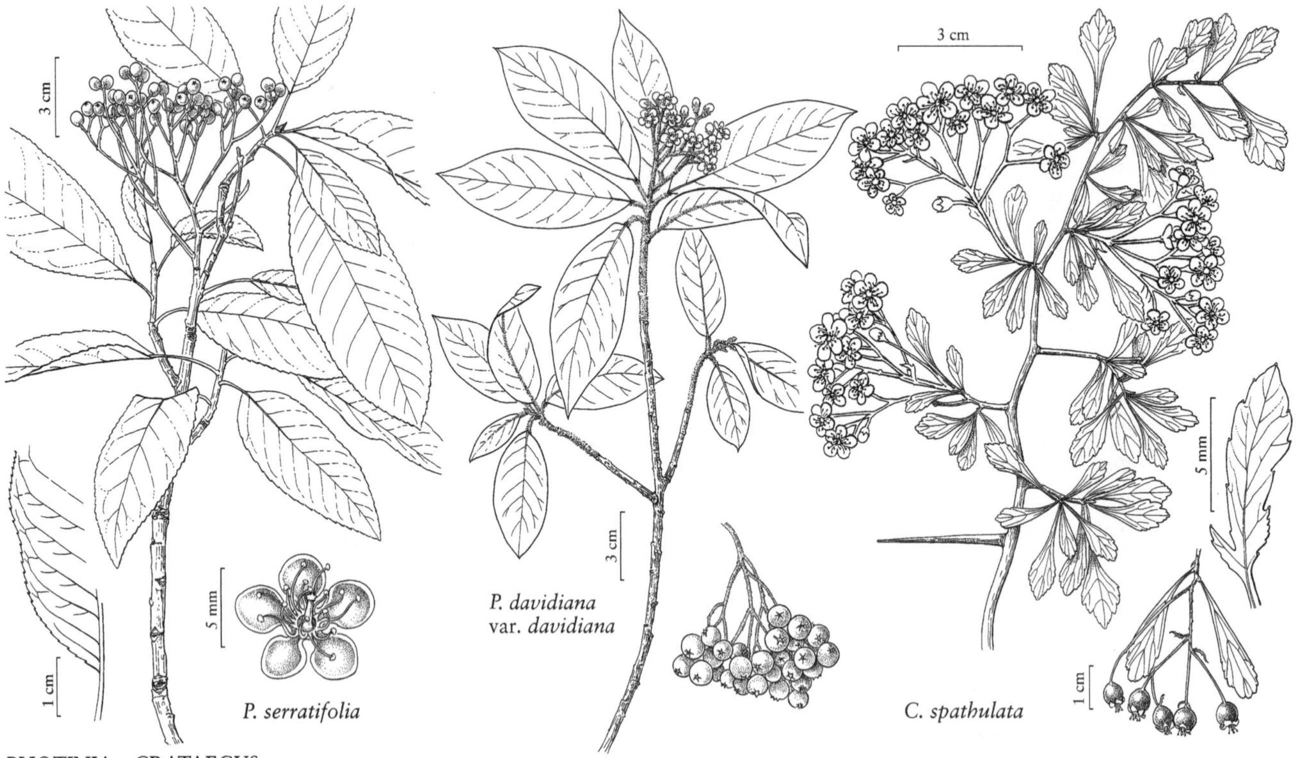

PHOTINIA ∘ CRATAEGUS

cm, basally cuneate, with petioles 1–1.5(–2.3) cm. The leaves are smaller than those of *P. serratifolia* but tend to be longer and slightly narrower, with sharper teeth, than those of *P. glabra*. Apparently, fruit set is very low and the plants have not been recorded as escaping cultivation. In Texas, *P.* ×*fraseri* produces an abundant flush of new leaves in late October through November, when neither of the parents is doing so.

2. Photinia serratifolia (Desfontaines) Kalkman, Blumea 21: 424. 1973 • Chinese photinia or hawthorn [F] [I]

Crataegus serratifolia Desfontaines, Tabl. École Bot. ed. 3, 288, 408. 1829

Plants 40–60(–120) dm. **Leaves** persistent; petiole 20–40 mm, villous adaxially when young, glabrescent; blade narrowly elliptic to oblong, obovate-elliptic, or narrowly obovate, (6–)9–20 × 3–6.5 cm, coriaceous, base rounded or broadly cuneate, margins sharply serrate to rarely inconspicuously toothed or entire, lateral veins 20–30 pairs, apex acuminate, abaxial surfaces slightly villous along veins when young, quickly glabrescent. **Inflorescences** 10–18 cm diam. **Pedicels** without lenticels. **Flowers** 6–8 mm diam.; petals suborbiculate, 3–4 mm, glabrous or villous.

Flowering late Mar–Apr. Fencerows, thickets, disturbed sites; 20–200 m; introduced; Ala., Ga., La., Miss., Tex.; Asia.

Four varieties of *Photinia serratifolia* are recorded in China; var. *serratifolia* occurs in the flora area and has leaf margins prominently serrate, entire only near the base (versus entire to shallowly and inconspicuously serrate in other varieties). Cultivars have been developed, however, differing in growth form, coloration, and leaf margins, and a formal identification to variety is problematic. Chinese photinia can grow larger than *P.* ×*fraseri* ('red tip') or Japanese photinia, and flowers appear before those of either.

3. Photinia villosa (Thunberg) de Candolle in A. P. de Candolle and A. L. P. P. de Candolle, Prodr. 2: 631. 1825 • Smooth oriental photinia [I]

Crataegus villosa Thunberg in J. A. Murray, Syst. Veg. ed. 14, 465. 1784; Fl. Jap., 204. 1784; *Pourthiaea villosa* (Thunberg) Decaisne

Plants 20–50 dm. **Leaves** deciduous; petiole 2–6 mm, villous; blade obovate, oblong-obovate, or elliptic, 3–8 × 2–4 cm, herbaceous, base cuneate, margins densely sharply serrate distally, lateral veins 5–7 pairs, apex caudate, surfaces initially white-villous, at maturity abaxial surfaces villous only along veins, adaxial glabrescent or glabrate. **Inflorescences** 3–6 cm diam. **Pedicels** with warty lenticels. **Flowers** 7–12 mm diam.; petals suborbiculate, 4–5 mm, adaxially pilose near base.

Flowering late Mar–Apr. Fencerows, thickets, disturbed sites; 20–200 m; introduced; Del., D.C., N.H., N.J., N.Y., Ohio, Pa., Wash.; Asia (China, Japan, Korea).

About five varieties of *Photinia villosa* are recognized in Asia, differing mainly in persistence of vestiture, leaf shape, and number of flowers per inflorescence. At least some of the plants offered for sale in the United States apparently are var. *sinica* Rehder & E. H. Wilson, but cultivars have been developed, and a formal identification at varietal rank would be speculative, based on characters apparently of overlapping morphology.

4. **Photinia davidiana** (Decaisne) Cardot, Bull. Mus. Natl. Hist. Nat. 25: 399. 1919 [F] [I]

Stranvaesia davidiana Decaisne, Nouv. Arch. Mus. Hist. Nat. 10: 179. 1874

Varieties 3 (1 in the flora): introduced, Washington; Asia; introduced also in Europe, Pacific Islands.

Stranvaesia davidiana is cultivated in California and was reported by Eric Walther to be commonly spontaneous in the Strybing Arboretum in Golden Gate Park (J. T. Howell et al. 1958), but current accounts of the California flora have not included the species as naturalized in that state; California botanists believe that it is not established there. The species is now considered naturalized in Washington, as well as in Hawaii and New Zealand (P. F. Zika 2002b).

4a. **Photinia davidiana** Decaisne var. **davidiana** • Chinese photinia or stranvaesia [F] [I]

Plants 10–80 dm. **Leaves** persistent; petiole 12–20 mm, pilose, gradually glabrescent; blade elliptic to oblong or oblong-lanceolate, (3–)5–12 × 2–4.5 cm, coriaceous, base rounded to cuneate, margins entire, lateral veins 8–16 pairs, inconspicuous, apex acute or abruptly pointed, surfaces villous along abaxial midvein, otherwise glabrous. **Inflorescences** <with 8 laterals of 1st order>, 5–10 cm diam., <pendulous in fruit>. **Pedicels** without lenticels. **Flowers** 5–10 mm diam.; petals suborbiculate, 4–7 mm, glabrous.

Flowering Mar–Jun; fruiting Nov–Apr. Bird-sown weed in thickets, open or disturbed forests, beaches, rarely epiphytic; 0–150 m; introduced, Wash.; Asia (China, Malaysia, North Vietnam, Taiwan); introduced also in Europe, Pacific Islands (Hawaii, New Zealand).

64. CRATAEGUS Linnaeus, Sp. Pl. 1: 475. 1753; Gen. Pl. ed. 5, 213. 1754 • Hawthorn, aubépine [Greek *Krataigon*, thorn, from Greek *kratos*, strength, and *akis*, sharp tip, alluding to thorns of some species]

James B. Phipps

Shrubs or trees, <usually main trunk dominant>, 3–120 dm. **Stems** usually 1 in larger plants, more in smaller plants, outer spreading; bark usually checked into rough, exfoliating plates, when usually dark, but freshly exposed ± russet, sometimes deeply corrugated and dark gray-brown, or smooth, thin and exfoliating, when usually pale, rarely with prominent horizontal lenticels; extension (long) and short shoots present; compound thorns on trunk abundant or absent; <twigs ± thorny, thorns 10–60(–100) mm, growth determinate (indeterminate in sect. *Crataegus*)>; glabrous or tomentose in first year. **Leaves** deciduous (sometimes winter-persistent in south), cauline, alternate, <scattered on extension shoots, often crowded on short shoots>, simple or deeply lobed; stipules deciduous with leaves or sometimes earlier, free, circinate or falcate, margins serrate to crenate, glandular; petiole present, <often glandular>; blade ± ovate to narrowly elliptic or obovate, (1.2–)2–8(–12) cm, <wider leaves shallowly to deeply incised (rarely pinnately compound), the narrower less incised>, margins flat, serrate, <young teeth often gland-tipped>, venation pinnate, usually craspedodromous, surfaces glabrous to tomentose. **Inflorescences** terminal <on few-leaved annual short shoots that usually arise from a subterminal bud on a woody short shoot, rarely arise laterally direct from an extension

shoot>, 1–50-flowered, panicles <domed, monopodial>, corymbose, or flowers solitary; bracts sometimes present, <leafy>; bracteoles present, <usually caducous>. **Pedicels** present. **Flowers:** perianth and androecium epigynous, 8–25 mm diam.; hypanthium ± obconic, constricted at disc except for opening, 2–6 mm wide; sepals 5, spreading, triangular; petals 5, white, sometimes pale cream (ser. *Montaninsulae*) or pale pink (ser. *Aestivales*, ser. *Molles*), ± circular, sometimes elliptic (ser. *Lacrimatae*) or ± elliptic (ser. *Apiifoliae*, ser. *Triflorae*), <base barely clawed, margins ± entire>; stamens 5–20 (30–45 in *C. triflora*), usually shorter than petals; torus absent; carpels 1–5, distinct, laterally touching, partially adnate to hypanthium, styles <1–5>, lateral, distinct, <adnate most of length, exsert>; ovules 2, <superposed>. **Fruits** pomes, yellow to red or purplish to black mature, suborbicular to ellipsoid or pyriform, 6–20(–25) mm diam. (larger in some cultivars), <flesh soft, sometimes hard>; hypanthium persistent; sepals often persistent, appressed to erect; carpels woody; styles usually persistent. **Pyrenes** 1–5, <dorsally grooved, sides plane or eroded, excavated, or pitted>. x = 17.

Species ca. 230 (169, including 17 hybrids, in the flora): North America, Mexico, Europe, Asia, n Africa; introduced in temperate South America (including equatorial Andes), s Africa, Pacific Islands (New Zealand), Australia.

Species of *Crataegus* vary from heliophiles to moderately shade tolerant. Most hawthorns are mesophytes and soil moisture regime appears to be a more important habitat determinant than soil type. Hawthorns are mostly found on topland, mesic, fine-textured soils, usually reflective of a reasonable rainfall. None are strict calciphiles or acidophiles. In drier parts of the Great Plains and intermontane regions, suitable habitats are often restricted to the vicinity of watercourses, or marsh edges, seeps, ditches, and other similar habitats. Relatively few taxa grow in drier (for example, ser. *Lacrimatae*) or moister (ser. *Aestivales*) soils than the norm.

Often abundant, species of *Crataegus* can be important components of some ecosystems, acting as sources of shelter and protection for animal life, nesting habitat, source of food (pomes), mainly for medium-sized passerine birds, and source of nectar and pollen. Hawthorns may be nurse plants to deciduous trees and play an important role in succession. *Gymnosporangium* R. Hedwig rusts may be serious pests on hawthorns where junipers are common. Other, locally serious threats to native hawthorns include aggressive introduced woody competitors such as *Lonicera japonica* in the southeast and *Rhamnus cathartica* in the northeast (P. M. Catling and G. Mitrow 2012b).

Species of *Crataegus* flower in spring during (rarely before) leaf expansion. They are mass-flowerers such that the anthesis season for a single plant lasts seven to ten days, the timing principally dependent on accumulated heat. Phenological separation is an important evolutionary driver and makes knowledge of flowering sequence very helpful in identification. The length of local flowering season may be as long as six weeks in localities with multiple hawthorns. Pollination is primarily by Diptera and Hymenoptera (N. Power 1980).

Seven species, including *Crataegus azarolus* Linnaeus, *C. mexicana* Mociño & Sessé ex de Candolle, *C. pinnatifida* Bunge, and *C. scabrifolia* (Franchet) Rehder as well as those of ser. *Aestivales* of the flora area, are minor commercial fruit crops. Hawthorn extracts are used to treat hypertension in alternative medicine. In addition to *hawthorn*, the names *haw*, *thorn*, and *thorn apple* are used colloquially (see J. B. Phipps et al. 2003).

Three ploidy levels (2*x*, 3*x*, and 4*x*) have been demonstrated in a wide range of North American hawthorns both cytologically (various authors) and by flow cytometry (N. Talent and T. A. Dickinson 2005). Diploids are usually sexual outbreeders; triploids are obligate apomicts; tetraploids are self-compatible facultative apomicts; and apomicts are pseudogamous. The polyploid groups are most difficult taxonomically.

Hybridization is routinely touted as a principal cause of taxonomic difficulty in *Crataegus*, and among closely related species, it may be overlooked as it is hard to detect. However, there is now groundbreaking work demonstrating hybridization in the *C. douglasii* complex by the Dickinson group (for example, E. Y. Y. Lo et al. 2010). Putative hybrids between species in different series, often resulting in plants receiving hybrid species rank, are easier to suspect and have been proposed in numerous instances, but on investigation nearly all cases proved non-persistent and only rarely re-created (J. B. Phipps 2005). Series *Anomalae*, which has some widespread and not uncommon species, may be one exception. Hybrids between species in different sections are nearly unknown today in North America, except for hybrid swarms between the introduced sexual compilospecies *C. monogyna* and several native diploids (R. Love and M. Feigen 1979; T. C. Wells and Phipps 1989). Ancient, wide hybridization is proposed for the origin of three common, taxonomically isolated, southeastern species (Lo et al. 2009). Hybridization, as contributing to taxonomic difficulty, appears to be restricted to situations involving closely related species or within particular species-complexes.

This treatment of *Crataegus* is the first attempted for the whole flora area since the work of J. Torrey and A. Gray (1838–1843). It is mainly conservatively based on E. J. Palmer (1950, 1952, 1960), except for series *Apricae* and *Lacrimatae*, and new species, mainly western, not known to Palmer.

The taxonomic complexity of North American *Crataegus* has resulted in extremely narrow species (C. S. Sargent and others), middling species (E. J. Palmer), and very broad ones (A. Cronquist). Here, a morphological species concept is used. In general, middling positions are favored, and the number of species treated proves similar to what might be extrapolated from work by Palmer except for some southeastern series and western taxa unknown to Palmer. As used here, the morphological species concept assumes that distinguishing suites of character states are adaptive.

A principal objective of this treatment has been to draw attention to the often great, but little discussed, infraspecific variation. This has been possible due to extensive fieldwork by the author and large quantities of available herbarium material. Because of evolutionary implications, infraspecific variation is given considerable attention. Levels of variation in some cases exceed those of woody sexual species of comparable abundance and range yet may not lead to clear dissection into several less variable species. Where high levels of variation occur, varieties are not necessarily recognized, either due to insufficient study or simply because the variation appears chaotic. For example, in the broadly treated *Crataegus macrosperma*, the infrataxa earlier recognized are not formally treated herein, but detailed discussion of variation is provided. When varieties are used, these themselves may also be variable, as with *C. pruinosa* and *C. viridis*. An interesting feature revealed for North American *Crataegus* is the frequency of occurrence of sympatry in infraspecific variation, this also shown by E. Y. Y. Lo et al. (2009b) in an in-depth study of the *C. douglasii* complex.

A frequent difficulty for North American *Crataegus* derives from the co-existence of sympatric, significantly atypical, named entities with otherwise well-defined taxa. They are often uncommon, might be hybrid or simply particularly divergent forms; usually, no one knows. Their treatment could be as species (greatly inflating the text), as synonyms (which seems misleading), or dismissed from the flora altogether (which would cripple proper understanding of North American *Crataegus* variation). Here, they are diagnosed very briefly in the discussion of the most relevant taxon.

Higher level variation is arranged into six sections, these divided into 32 series, mostly following the circumscriptions of E. J. Palmer (1925). Treatment of series may run into similar problems to species delimitation, namely, the existence of atypical forms that obscure taxon limits. Also, some species (for example, *Crataegus flabellata* [ser. *Tenuifoliae*]), may act as a connecting link to another series, here ser. *Rotundifoliae*. Both kinds of issues can necessitate repeated keying out of some series. However, rather than amalgamating the series recognized into larger, even more amorphous, ill-defined entities to mitigate these problems, it is preferred to maintain the series in a familiar Palmerian sense preserving the time-tested utility of their core elements.

Species that are similar in flower may often be readily distinguished in fruit and vice versa; users of this text are encouraged to verify identification from the same plant at both seasons.

Variation occurs in the thorns. Compound thorns, which are adventitious and indeterminate, occur on the trunks of many species. Thorns on twigs, which develop with the new growth, persist a few years and are of two kinds. Determinate thorns (aphyllous thorns of K. I. Christensen 1992) are lateral and are the only type found in North American native species. They have diagnostic value, varying in abundance, color in their second year, length, curvature, and stoutness. Indeterminate thorns, as found in Eurasian species of sect. *Crataegus*, commence as lateral or terminal growths and may develop into thorn-tipped shoot systems.

Leaves supply useful characters for *Crataegus* taxonomy, but variation within a plant may be confusing. Median leaves of short shoots are the normal standard for descriptions; they are relatively invariant in size and form. Size at anthesis varies according to species from 0% to 100% mature size. Lobing, here often expressed as Leaf Incision Index, or LII, varies from 0% (not incised) to 100% (incised to midvein). As LII typically diminishes towards the leaf apex, maximum LII is usually used. Leaves on extension shoots are often larger, more deeply incised, sometimes of a completely different shape from short-shoot leaves, and may be more variable on one plant. Although apparently diagnostic in some cases, they are insufficiently known and too difficult to describe to be generally useful for description.

Hawthorns may possess both leaflike bracts and bracteoles in the inflorescence, the former being either relatively few or absent. Bracteoles are often relatively numerous and can usually be found throughout young inflorescences; they are sparse in some species. Typical bracteoles of North American hawthorns are caducous, more or less linear-oblong to linear, bilaterally symmetric, and membranous with glandular margins. Some North American hawthorns have larger, harder, more or less herbaceous and more persistent bracteoles. A different bracteole type found in some North American hawthorns (for example, *Crataegus brachyacantha*) is smaller than any of the above and more or less lacking marginal glands.

Fruit color change during development is a valuable distinguishing character in some cases. In comparing two sympatric species, it is critical to compare them at the same time and a good example is found with *Crataegus okennonii* versus *C. douglasii*. Although pomes both finish black, pomes of *C. douglasii* are usually already black when those of *C. okennonii* are brownish or reddish, as illustrated in J. B. Phipps et al. (2003).

Crataegus is best treated as sister to *Mespilus* (E. Y. Y. Lo and M. J. Donoghue 2012; Li Q. Y. et al. 2012); other relationships in Maleae are less close. Molecular evidence (C. S. Campbell et al. 2007; D. Potter et al. 2007) shows that the *Amelanchier* clade is sister to *Crataegus-Mespilus*, but *Amelanchier* is different in many morphologic features.

The following names could not be accounted for in the current treatment: *Crataegus ardua* Sargent, *C. brachyphylla* Sargent, *C. evansiana* Sargent, *C. faxonii* Sargent var. *praetermissa* (Sargent) E. J. Palmer, *C. glareosa* Ashe, *C. harryi* Sargent, *C. immanis* Ashe, *C. lecta* Sargent, *C. littoralis* Sargent, *C. menandiana* Sargent, *C. mercerensis* Sargent, *C. pilosa* Sargent, *C. pinguis* Sargent, *C. populifolia* Walter, *C. sublobulata* Sargent, and *C. whitakeri* Sargent.

SELECTED REFERENCES Lo, E. Y. Y. et al. 2009. Evidence for the genetic association between East Asian and western North American *Crataegus* L. (Rosaceae) and rapid divergence of the eastern North American lineages based on multiple DNA sequences. Molec. Phylogen. Evol. 51: 157–168. Palmer, E. J. 1925. Synopsis of North American *Crataegi*. J. Arnold Arbor. 6: 5–128. Palmer, E. J. 1950. *Crataegus*. In: M. L. Fernald. 1950. Gray's Manual of Botany, ed. 8. New York. Pp. 767–801. Palmer, E. J. 1952. *Crataegus*. In: H. A. Gleason. 1952. The New Britton and Brown Illustrated Flora of the Northeastern United States and Adjacent Canada. 3 vols. New York. Vol. 2, pp. 338–375. Palmer, E. J. 1960. *Crataegus*. Key to series and species. In: R. A. Vines. 1960. Trees, Shrubs, and Woody Vines of the Southwest. Austin. Pp. 329–334. Phipps, J. B. 1998. Synopsis of *Crataegus* series *Apiifoliae*, *Cordatae*, *Microcarpae*, and *Brachyacanthae* (Rosaceae subfam. Maloideae). Ann. Missouri Bot. Gard. 85: 475–491. Phipps, J. B. 2005. Review of hybridization in *Crataegus*—Another look at "The *Crataegus* Problem." Ann. Missouri Bot. Gard. 92: 113–127. Phipps, J. B., R. J. O'Kennon, and R. W. Lance. 2003. Hawthorns and Medlars. Portland.

1. Leaves: evidently lobed (except in some small-leaved specimens of *C. spathulata*, ser. *Microcarpae*), veins to both lobes and deeper sinuses.
    2. Inflorescences: falcate bracteoles subtending lower branches absent, branches glabrous or pubescent; pomes vermillion to red or orange-red; leaf blades 1.4–6 cm; styles 1–3; pyrene sides usually plane (except grooved in ser. *Crataegus*) . . . . . 64a. *Crataegus* sect. *Crataegus*, p. 495
    2. Inflorescences: falcate bracteoles subtending lower branches present, branches glabrous; pomes yellowish to tan; leaf blades 5–9 cm; styles 4 or 5; pyrene sides strongly pitted . . . . 64d. *Crataegus* sect. *Sanguineae* (in part), p. 514
1. Leaves: lobed or not, veins absent to sinuses.
    3. Inflorescences: falcate bracteoles subtending lower branches present, branches glabrous; mature pomes yellowish to tan; leaf blades 5–9 cm; pyrene sides strongly pitted. . . . . 64d. *Crataegus* sect. *Sanguineae* (in part), p. 514
    3. Inflorescences: falcate bracteoles subtending lower branches absent, branches glabrous or pubescent to tomentose; mature pomes usually red to black, rarely yellow to orange; leaf blades 2–8(–12) cm; pyrene sides plane or concave, eroded, or pitted.
        4. Pomes red or bright red at full maturity, or colors other than deep red to black.
            5. Mature pomes orange to red; pyrene sides ± pitted or eroded . . . . 64e. *Crataegus* sect. *Macracanthae*, p. 516
            5. Mature pomes usually red, sometimes orange, yellow, or pink-mauve (sometimes ± pruinose), sometimes remaining green until late, or green-blotched; pyrene sides plane to concave . . . . 64f. *Crataegus* sect. *Coccineae* (in part), p. 525
        4. Pomes deep red to purple or black at full maturity, sometimes bright red when immature.
            6. Pyrene sides usually pitted, sometimes only shallowly . . . . 64c. *Crataegus* sect. *Douglasia*, p. 502
            6. Pyrene sides ± plane, sometimes shallowly concave (*C. ursopedensis*, ser. *Rotundifoliae*).
                7. Leaves glossy; thorns on twigs 1–1.5 cm, recurved; post-mature petals ± orange; anthers cream to orange; Louisiana and adjacent states . . . . 64b. *Crataegus* sect. *Brevispinae*, p. 500
                7. Leaves matte; thorns on twigs 2.5–4 cm, ± straight; post-mature petals turning pale brown; anthers pale pink to pink, pink-purple, or ivory; n Great Plains, ne United States, and adjacent Canada.
                    8. Leaf blades: lobes absent or very short, length (1.2–)1.6–2(–2.4) times width, veins 6–8 per side; stamens 20, anthers pink-purple or ivory; ne United States and adjacent Canada . . . . . 64f. *Crataegus* sect. *Coccineae* (in part), p. 525
                    8. Leaf blades: lobed, length 1.2–1.5 times width, veins (3–)5–8 per side; stamens 10, anthers pale pink to pink; Cypress Hills, w Canada only . . . . 64g. *Crataegus* sect. unassigned (ser. *Montaninsulae*), p. 633

64a. CRATAEGUS Linnaeus sect. CRATAEGUS

Shrubs or trees, 20–100[–120] dm. **Stems:** trunks 1–few, ± erect, bark flattened-scaly or exfoliating; compound thorns on trunks present or absent; thorns on twigs determinate or indeterminate, straight or slightly curved, ± slender unless very short. **Leaves:** blade ± ovate,

1.2–6 cm, ± thin, lobes (0 or)1–4 per side, sinuses shallow or deep, veins 2–8 per side, to lobes and sinuses, <sometimes glossy>. **Inflorescences:** branches glabrous or pubescent; symmetric bracteoles present, basal stipuliform, falcate bracteoles absent. **Flowers:** post-mature petals pale paper brown; stamens [10–]20, anthers ivory, pale yellow, pink to purple, or reddish. **Pomes** vermillion to red or orange-red [black]; pyrene sides plane or grooved.

Species 28 (5 in the flora): introduced; w Eurasia, n Africa; introduced also in South America (Argentina, Chile), s Africa, Pacific Islands (New Zealand), Australia.

All four of the series recognized in sect. *Crataegus* are represented in the flora area.

Section *Crataegus* is extended here beyond ser. *Crataegus* to include three related North American series that have only determinate thorns and that have been suggested by E. Y. Y. Lo et al. (2009) to represent a separate subclade of hybrid origin.

1. Veins to all sinuses; styles and pyrenes 1–3.
    2. Pomes bright to dark red, 6–14 mm diam., ± orbicular to ± cylindric; inflorescence branches glabrous; thorns on twigs indeterminate (when appearing terminal on often leafy twigs), sometimes determinate (aphyllous and lateral); trunk bark flattened-scaly, exfoliating in elongate strips; pyrene sides grooved . . . . 64a.1. *Crataegus* ser. *Crataegus*, p. 496
    2. Pomes bright, glossy red, 4–6 mm diam., ellipsoid to sometimes orbicular; inflorescence branches densely pubescent; thorns on twigs determinate; trunk bark smooth, thin, exfoliating in irregular flakes; pyrene sides plane. . . . . 64a.2. *Crataegus* ser. *Apiifoliae*, p. 498
1. Veins only to deeper sinuses; styles and pyrenes 3–5.
    3. Trunk bark flattened-scaly in elongate strips; leaf blades 3–6 cm, broad . . . . .64a.3. *Crataegus* ser. *Cordatae*, p. 498
    3. Trunk bark exfoliating with smooth, thin flakes; leaf blades 1.5–3 cm, narrow . . . . 64a.4. *Crataegus* ser. *Microcarpae*, p. 499

64a.1. CRATAEGUS Linnaeus (sect. CRATAEGUS) ser. CRATAEGUS [I]

Shrubs or trees, 40–100[–120] dm. **Stems:** trunk bark freshly exposed ± orange, older gray-brown, flattened-scaly, exfoliating in elongate strips; compound thorns on trunks usually absent; thorns on twigs determinate or indeterminate. **Leaves** <deciduous>; petiole length 30–50% blade; blade broadly elliptic to obovate or ovate to deltate, lobes 1–3 per side, sinuses shallow or deep, venation craspedodromous, veins to lobes and sinuses; <blades on extension shoots sometimes fully pinnate>. **Inflorescences** 3–15-flowered; branches glabrous [to lanate]; bracteoles caducous, ± linear, small, membranous, margins entire or denticulate or with small protuberances or pegs bearing terminal, small glands. **Flowers** 8–22 mm diam.; sepals narrowly triangular; stamens [10–]20, <filaments long>, anthers pink to purple; styles 1–3. **Pomes** bright to dark red [black], ± orbicular to ± cylindric, 6–14 mm diam., <glabrous [to tomentose]>; <flesh yellowish>; sepal remnants persistent; pyrenes 1–3, dorsally shallowly grooved, sides grooved [plane].

Species ca. 25 (2 in the flora): introduced; Europe, w Asia, n Africa; introduced also in South America (Argentina, Chile), s Africa, Pacific Islands (New Zealand), Australia.

Crataegus oxyacantha Linnaeus is a rejected name of this series that is replaced by *C. rhipidophylla* Gandoger. There are several instances of its misapplication in North America to *C. monogyna*.

1. Leaf lobe apices obtuse, sinuses ± shallow; styles 2 or 3. . . . . 1. *Crataegus laevigata*
1. Leaf lobe apices acute, sinuses usually deep; style 1 . . . . 2. *Crataegus monogyna*

1. Crataegus laevigata (Poiret) de Candolle in A. P. de Candolle and A. L. P. P. de Candolle, Prodr. 2: 630. 1825 • Woodland hawthorn, aubépine épineuse, aubépine à deux styles [I]

Mespilus laevigata Poiret in J. Lamarck et al., Encycl. 4: 439. 1798; *Crataegus oxyacanthoides* Thuillier

Shrubs or trees, 80 dm. **Stems:** twigs: new growth sparsely villous or glabrous; thorns on twigs few to numerous, determinate or indeterminate, stout, 1 cm. **Leaves:** petiole slender, 0.8–2 cm, glabrous, eglandular; blade broadly elliptic to obovate, 1.4–5.8 cm, base cuneate, lobes 1 or 2 per side, sinuses ± shallow, lobe apex obtuse, margins serrate except proximally, veins 3–7 per side, apex obtuse, abaxial surface pale green, ± pilose along veins and in axils, adaxial ± lustrous dark green, pilose on veins. **Inflorescences** 3–11-flowered, lax; branches glabrous; bracteole margins irregularly denticulate, bearing few to numerous glands. **Flowers** 12–22 mm diam.; hypanthium glabrous [to lanate]; sepals 1.5–2.5 mm, margins entire, apex acute; stamens 20, anthers purple; styles 2 or 3. **Pomes** bright red, orbicular or ± cylindric, 6–14 mm diam.; sepals recurved or spreading; pyrenes 2 or 3. $2n = 34$.

Flowering Apr–May; fruiting Sep–Oct. Woodlands, scrub, hedges; 10–50 m; introduced, Wash.; Europe.

In the flora area, *Crataegus laevigata* is known only from the San Juan Islands; alleged wild records from elsewhere in North America proved to be misidentifications. The species is widespread in central Europe and scarce beyond that. Woodland hawthorn is more shade tolerant than many other species, hence the common name.

Crataegus laevigata has more shallowly incised leaves and more obtuse lobes than other species in ser. *Crataegus*.

Pink- or red-petaled cultivated forms attributed to *Crataegus laevigata* are important ornamentals; they are usually cultivars of *C.* ×*media* Bechstein. Similar -colored cultivars of true *C. laevigata* are scarce in North America.

2. Crataegus monogyna Jacquin, Fl. Austriac. 3: 50, plate 292, fig. 1. 1775 • English hawthorn, one-seeded hawthorn, aubépine monogyne [I]

Varieties 2 (1 in the flora): introduced; Europe, w Asia, n Africa; introduced also in South America, s Africa, Pacific Islands, Australia.

Crataegus monogyna was introduced early to North America to be used for agricultural hedges, a practice now discontinued on this continent.

Cultivar 'Crimson Cloud,' often under an incorrect species name, is widely cultivated as an ornamental.

2a. Crataegus monogyna Jacquin var. **monogyna** [I]

Shrubs or trees, 40–100[–120] dm. **Stems:** twigs: new growth glabrous; <stipules of flowering shoots ± entire>; thorns on twigs few to numerous, usually indeterminate and terminal, 0.7–1 cm, when determinate and lateral ± stout, to 2.4 cm. **Leaves:** petiole slender, 1–3 cm, glabrous, eglandular; blade ovate to deltate, 1.2–5.5 cm, base cuneate to truncate, lobes 1–3 per side, sinuses usually deep, lobe apex acute, margins weakly and often few-toothed at lobe apices, veins 2–5 per side, apex acute, abaxial surface paler green, villous on major veins, adaxial ± lustrous dark green, glabrous or mainly villous. **Inflorescences** 5–15-flowered, ± lax; branches glabrous [to villous]; bracteole margins entire or bearing few to numerous glands on small irregular protuberances or on small pegs. **Flowers** 8–16 mm diam.; hypanthium glabrous [to lanate]; sepals 1.5–2.5 mm, margins entire, apex acute or obtuse; stamens 15–20, anthers pink-purple; style 1. **Pomes** sometimes bright, usually dark red, orbicular to ± cylindric, 6–11 mm diam.; sepals reflexed; pyrene 1. $2n = 34, 51$.

Flowering Apr–May; fruiting Sep–Oct. Brush, woodland edges and old pastures in medium rainfall areas; 10–800 m; introduced; B.C., N.B., N.S., Ont., P.E.I., Que.; Calif., Colo., Conn., Del., Ill., Ind., Maine, Mass., Mich., Minn., N.H., N.J., N.Y., Ohio, Oreg., Pa., R.I., Vt., Wash., Wis.; Europe; sw Asia; n Africa; introduced also in South America (Argentina, Chile), s Africa, Pacific Islands (New Zealand), Australia.

Variety *monogyna* differs from var. *lasiocarpa* (Lange) K. I. Christensen in the absence of villous abaxial leaf surfaces and inflorescences. Variety *monogyna* is distributed throughout the area of the species; var. *lasiocarpa* primarily occurs in the Balkan, Iranian, and Mediterranean areas, and is most common in Greece.

Variety *monogyna* is locally common in the Pacific Northwest and adjacent Canada, as well as from the Great Lakes to the Atlantic. The variety hybridizes in North America with the diploid species *Crataegus suksdorfii* and rarely with *C. punctata*. The hybrid with *C. laevigata*, *C.* ×*media* Bechstein, is well known in the wild in Europe; only its cultivars are seen in North America.

64a.2. Crataegus Linnaeus (sect. Crataegus) ser. Apiifoliae (Loudon) Rehder, Man. Cult. Trees ed. 2, 367. 1940 [E]

Crataegus sect. *Apiifoliae* Loudon, Arbor. Frutic. Brit. 2: 824. 1838

Shrubs or trees, 20–80 dm <often with crown of ± layered branches>. **Stems:** trunk bark parti-colored, smooth, thin, exfoliating in irregular flakes; compound thorns on trunks absent; thorns on twigs determinate. **Leaves:** petiole length 80–150% blade; blade broadly ovate to deltate, lobes 3 per side <at 45–70° angle with midvein>, sinuses deep, venation craspedodromous, veins to major sinuses. **Inflorescences** 4–10-flowered; branches densely pubescent; bracteoles caducous, linear, small, membranous, margins irregularly toothed, teeth gland-tipped. **Flowers** 12–17 mm diam.; sepals narrowly triangular; <petals elliptic>; stamens 20, anthers reddish; styles 1 or 2(or 3). **Pomes** bright red, ellipsoid to sometimes orbicular, 4–6 mm diam., <glossy>; <flesh mealy when ripe>; sepal remnants persistent; pyrenes 1 or 2(or 3), dorsally convex, sides plane.

Species 1: se United States.

Series *Apiifoliae* is widespread and common and has obvious similarities to European ser. *Crataegus* in the distinctive leaf shape, relatively small flowers, long filaments, and small, few-pyrened fruit; it differs from the latter series in possessing only determinate thorns.

3. Crataegus marshallii Eggleston in N. L. Britton and J. A. Shafer, N. Amer. Trees, 473. 1908 • Parsley haw [E]

Mespilus apiifolia Marshall, Arbust. Amer., 89. 1785, not *Crataegus apiifolia* Medikus 1793

Shrubs or trees, 20–80 dm. **Stems** 1-trunked; 1-year old twigs dark purple-brown, older dark, dull gray; thorns on twigs straight or slightly recurved, 1.5–3 cm, 3-years old dark and shiny, slender, 2–3 cm. **Leaves:** petiole long, pubescent; blade broadly ovate to deltate, 1.5–3 cm, base broadly cuneate to ± truncate, lobes 3 per side, narrow, margins serrate, particularly toward lobe apices, veins 7 or 8 per side (including those to sinuses), apex acute, surfaces hairy, especially along veins abaxially. **Inflorescences** 4–10-flowered; branches densely pubescent; bracteole margins short-stipitate-glandular. **Flowers** 12–17 mm diam.; hypanthium hairy; sepals narrowly triangular, margins slightly incised; stamens 20, anthers deep rose to red; styles 1 or 2(or 3). **2*n*** = 34, 51.

Flowering Mar–Apr; fruiting Sep–Nov. Alluvial and other woodlands, light shade, open areas, calcareous or mafic rock, uplands; 10–500 m; Ala., Ark., Fla., Ga., Ill., Ky., La., Miss., Mo., N.C., Okla., S.C., Tenn., Tex., Va.

Crataegus marshallii is common in the southeastern United States; reports from Kansas are unconfirmed. The exfoliating bark is an attractive ornamental characteristic.

Crataegus apiifolia (Marshall) Michaux is an illegitimate name.

64a.3. Crataegus Linnaeus (sect. Crataegus) ser. Cordatae (Beadle) Rehder, Man. Cult. Trees ed. 2, 367. 1940 [E]

Crataegus [unranked] *Cordatae* Beadle in J. K. Small, Fl. S.E. U.S., 532. 1903; *Crataegus* sect. *Cordatae* (Beadle) Eggleston

Trees, 40–100 dm. **Stems:** trunk bark freshly exposed ± orange, older dark gray brown, flattened-scaly in elongate strips; compound thorns on trunks present; thorns on twigs determinate. **Leaves:** petiole length 40–60% blade; blade <brilliantly bronze and purple in fall, glossy>, ± deltate, <broad, 3–6 cm>, lobes palmate or 2 or 3(or 4) per side, sinuses deep, <triangular>, venation craspedodromous, veins to lobes and deeper sinuses. **Inflorescences** 15–30-flowered; branches glabrous; bracteoles caducous, linear, small, membranous, margins stipitate-glandular. **Flowers** 10–12 mm diam.; sepals broadly triangular; stamens 20, anthers ivory; styles 3(or 4). **Pomes** vermillion, ± orbicular, 5–8 mm diam., <shiny>; sepal remnants persistent or deciduous; pyrenes 3, dorsally grooved, sides plane.

Species 1: c, se United States; introduced elsewhere in the flora area.

Series *Cordatae* occurs in a wide band from Missouri to North Carolina with outliers in all states southward except Alabama. Anthesis occurs last among sympatric hawthorns.

4. Crataegus phaenopyrum (Linnaeus f.) Medikus, Gesch. Bot., 84. 1793 • Washington thorn E

Mespilus phaenopyrum Linnaeus f., Suppl. Pl., 254. 1782; *Crataegus youngii* Sargent

Trees, 40–100 dm. **Stems:** 3-year old twigs shiny, deep reddish brown, older dark gray; thorns on twigs ± straight, 2–5 cm. **Leaves:** petiole slender, sometimes both abaxially and adaxially short rusty-tomentose, eglandular; blade broadly to narrowly deltate, 3–6 cm, base truncate to rarely cuneate or cordate, sometimes very shallowly so, lobes palmately or pinnately 2(or 3) per side, margins serrulate, veins 6 or 7(or 8) per side (to sinuses), apex acute, surfaces glabrous. **Inflorescences** 15–30-flowered, flowers strong-smelling; branches glabrous; bracteole margins stipitate-glandular. **Flowers** 10–12 mm diam.; hypanthium glabrous; sepals 2 mm; stamens 20, anthers ivory; styles 3(or 4). $2n$ = 51, 68.

Flowering Jun; fruiting Sep–Nov. Woodland edges, gaps, scrub, damp locations; 10–300 m; Ont.; Ark., Del., Fla., Ga., Ill., Ind., Ky., La., Md., Mich., Miss., Mo., N.C., Ohio, Oreg., Pa., S.C., Tenn., Va., Wash., W.Va.

Crataegus phaenopyrum has a wide native range across the middle latitudes of the United States from Missouri to Maryland and southward but is not particularly abundant. In the northern and western states and Ontario, it is introduced. All alleged specimens for Alabama proved to be wrongly identified (J. B. Phipps 1998).

The flowering season of the Washington thorn is the latest of all North American hawthorns. The species is one of the most widely planted ornamentals in the genus, being elegantly treelike and possessing beautiful, remarkably rust-resistant foliage at all seasons as well as attractive flowers and fruit. Several cultivars are known.

Crataegus cordata (Miller) Aiton 1789 (= *Mespilus cordata* Miller 1757) was commonly used for *C. phaenopyrum* until the early 20th century (J. B. Phipps 1999).

64a.4. CRATAEGUS Linnaeus (sect. CRATAEGUS) ser. MICROCARPAE (Loudon) Rehder, Man. Cult. Trees ed. 2, 367. 1940 E

Crataegus sect. *Microcarpae* Loudon, Arbor. Frutic. Brit. 2: 825. 1838

Shrubs or trees, 30–70 dm. **Stems:** trunk bark cream to cinnamon brown or light gray-brown in patches, in shade darker patches pale tan or olive green, exfoliating with smooth, thin flakes; compound thorns on trunks sometimes present; thorns on twigs determinate. **Leaves** <long-persistent>: petiole length 5–30% blade; blade narrowly or broadly spatulate, <narrow, 1.5–3 cm, coriaceous>, lobes 0 or 1 per side, sinuses ± shallow to deep, venation craspedodromous, veins to lobes and deeper sinuses; <blades on extension shoots usually more diverse in shape, larger, more deeply incised>. **Inflorescences** 20–30-flowered; branches glabrous; bracteoles ± caducous, linear, small, membranous to herbaceous, margins usually eglandular, sometimes ± stipuliform. **Flowers** 10 mm diam.; sepals triangular; stamens 20, anthers pale yellow; styles 3–5. **Pomes** bright orange-red, orbicular, 3–5 mm diam., <matte to glossy>; sepal remnants persistent; pyrenes 3–5, dorsally obscurely grooved, sides plane <apical quarters exposed>.

Species 1: sc, se United States.

5. Crataegus spathulata Michaux, Fl. Bor.-Amer. 1: 288. 1803 • Littlehip hawthorn [E] [F]

Crataegus microcarpa Lindley

Shrubs or trees, 30–70 dm, branching often tabulate when open grown. **Stems:** 3-year old twigs deep purple-brown, older dark gray; 3-year old thorns on twigs blackish, ± straight, 3–4(–5) cm. **Leaves:** petiole short; blade usually dark green, <sometimes ± glaucous>, narrowly or broadly spatulate, 1.5–3 cm, base cuneate, lobes 0 or 1 per side, lobe apex acute to rounded, margins subentire, crenate, or serrate, veins 3 per side (extending to sinuses and lobes), apex subacute to acute, surfaces usually glabrous, marginal and scattered abaxial hairs, adaxial midvein with long hairs (± dense young). **Inflorescences** 20–30-flowered; branches glabrous; largely ebracteolate, sometimes a few curved, green stipular bractlets; bracteole margins eglandular or with very few, very small glands. **Flowers** strong-smelling, 10 mm diam.; hypanthium glabrous; stamens 20, anthers pale yellow; styles 3–5. $2n$ = 34, 51.

Flowering Apr; fruiting Sep–Nov. Brush, calcareous substrates; 10–500 m; Ala., Ark., Fla., Ga., Ky., La., Miss., Mo., N.C., Okla., S.C., Tenn., Tex., Va.

Crataegus spathulata is locally common across the southern states from southern Missouri to Virginia south to eastern Texas to Florida.

Mature *Crataegus spathulata* has very attractive, often honey-colored, exfoliating bark. The illustration herein depicts an atypically large, unusually shaped, herbaceous bracteole.

64b. CRATAEGUS Linnaeus sect. BREVISPINAE (Beadle) C. K. Schneider, Ill. Handb. Laubholzk. 1: 791. 1906 [E]

Crataegus [unranked] *Brevispinae* Beadle in J. K. Small, Fl. S.E. U.S., 532, 534. 1903

Species 1: sc, se United States.

64b.1. CRATAEGUS Linnaeus (sect. BREVISPINAE) ser. BREVISPINAE (Beadle) Rehder, Man. Cult. Trees ed. 2, 366. 1940 [E]

Crataegus [unranked] *Brevispinae* Beadle in J. K. Small, Fl. S.E. U.S., 532, 534. 1903

Shrubs or trees, 60–100(–150) dm. **Stems:** trunks 1–few, erect, bark plated; compound thorns on trunks present when older; thorns on twigs absent or determinate, recurved, short, <1–1.5 cm>. **Leaves:** blade elliptic, 2–3 cm, coriaceous, lobes 0, veins 5 or 6(–8) per side, absent to sinuses, <glossy>; <blades on extension shoots often much larger (more than 6 cm) and variously lobed, to deeper sinuses>. **Inflorescences:** branches glabrous; symmetric bracteoles present, basal stipuliform, falcate bracteoles absent. **Flowers:** post-mature petals ± orange; stamens 20, anthers cream to orange. **Pomes** black to bluish black, <heavily pruinose mature>; pyrene sides plane.

Species 1: sc, se United States.

Series *Brevispinae* is centered in Louisiana and occurs in neighboring states, a distribution unique among hawthorns; the series is disjunct in Georgia. Formerly, *Crataegus saligna* (ser. *Cerrones*) was placed in ser. *Brevispinae* due to similarities of fruit, flower, and foliage. Section *Brevispinae* is distinct because of its short, recurved thorns, relatively large plant size (in some specimens), narrow, glossy leaves with lobes absent, more or less orange post-mature petals, and black fruit with plane-sided pyrenes. The names *Brachyacanthae* Sargent and sect. *Brachyacanthae* E. J. Palmer, which pertain here, were not validly published.

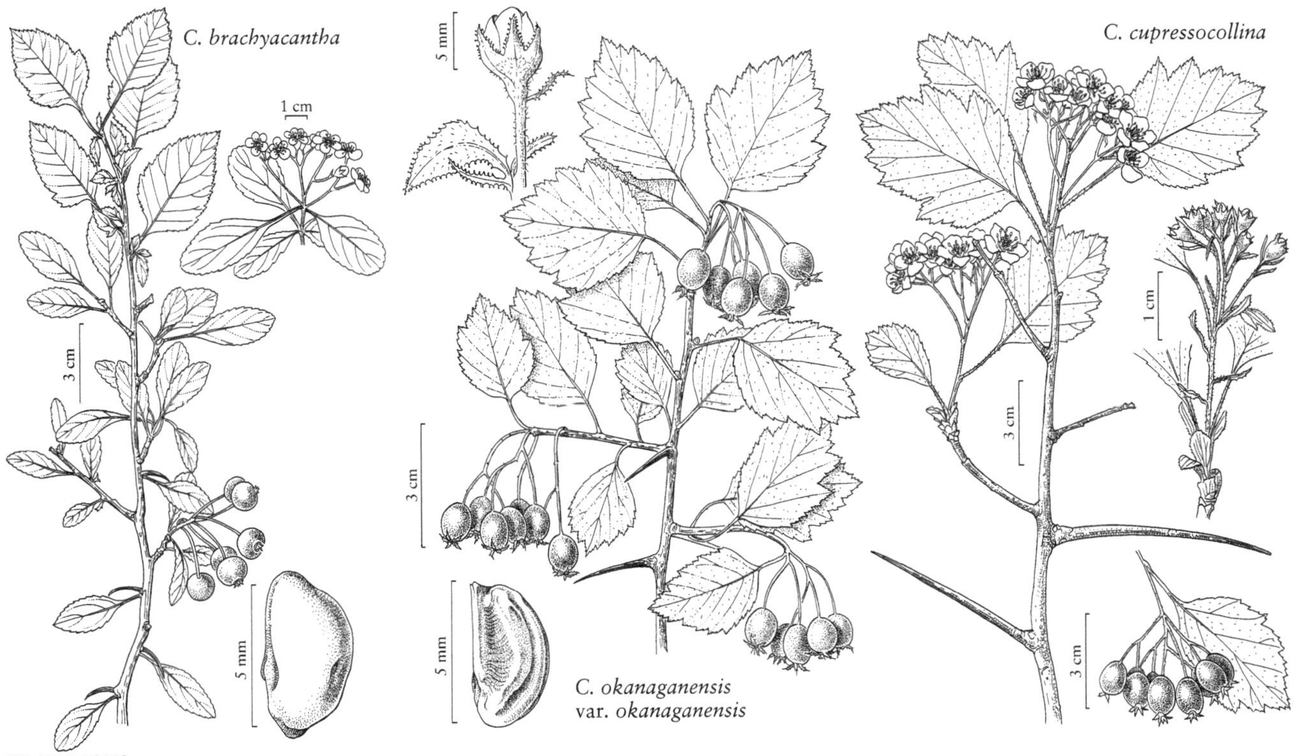

CRATAEGUS

6. **Crataegus brachyacantha** Sargent & Engelmann, Bot. Gaz. 7: 128. 1882 • Blueberry haw, hoghaw, pomette bleue [E] [F]

Shrubs or trees, 60–100(–150) dm. **Stems:** trunk bark dark gray-brown, plated; thorns on twigs absent or present, recurved, short, to 1.5 cm. **Leaves:** petiole length 15–30% blade; blade elliptic, 2–3 cm, coriaceous, shiny, lobes 0, margins crenate, venation camptodromous, veins 5 or 6(–8) per side, abaxial surface glabrous, adaxial sparingly appressed-hairy, hairy on midvein, sometimes to sinuses. **Inflorescences** 15–25-flowered; branches glabrous; bracteoles caducous, narrow, small, membranous, margins eglandular, nearly glabrous. **Flowers** 12 mm diam.; hypanthium glabrous; sepals triangular, 1.5 mm; stamens 20, anthers cream to orange, 0.5 mm; styles 4 or 5. **Pomes** black to bluish black, pruinose, oblate-orbiculate, 8–14 mm diam.; flesh mealy; fruiting calyx suberect, on small collar, sepal tips often reflexed; pyrenes 4 or 5, dorsally slightly grooved, sides plane. $2n$ = 34, 51.

Flowering Apr; fruiting Sep–Nov. Wet prairies, alluvial flats, well-drained mesic sites, woodland margins; 10–200 m; Ark., Ga., La., Miss., Okla., Tex.

Crataegus brachyacantha occurs throughout Louisiana, where it is locally common, and in the adjacent parts of all surrounding states. A disjunct population was known in Georgia but cannot be found there now. The species is seemingly more shade tolerant than many other hawthorns.

Crataegus brachyacantha is among the taller hawthorns in North America; its petals turn orange with age or on drying. The short, recurved thorns and bitter, oblate-orbiculate, black fruit also are distinctive and help to distinguish *C. brachyacantha* from *C. saligna* (ser. *Cerrones*) of Colorado and Utah, which is similar in foliage and flower.

The names blueberry haw and pomette bleue may have a special appropriateness in late August when the waxy covering of the fruit is still thick and the underlying color is a dark purple. At maturity, the flesh is thin and bitter and the skin usually black. The fall foliage, brilliant lustrous orange, bronze, and red, suggests potential ornamental use. Forma *leucocarpa* Sargent is a white-fruited form, which was collected by E. J. Palmer at Natchitoches, Louisiana, September 1915; it is the only recorded white-fruited hawthorn.

64c. CRATAEGUS Linnaeus sect. DOUGLASIA Loudon, Arbor. Frutic. Brit. 2: 823. 1838 (as Douglasii) [E]

Shrubs or trees, (20–)30–100(–140) dm. **Stems:** trunks 1–few, usually ± erect, bark plated, ± exfoliating; compound thorns on trunks sometimes present; thorns on twigs determinate, <usually numerous>, straight to ± recurved, slender to stout. **Leaves:** blade ± rhombic to ± elliptic, elliptic-oblong, or ovate-rhombic, sometimes ± obovate, oblanceolate, or suborbiculate, 3–8 cm, ± coriaceous to ± thin, lobes 0 or 1–4 per side, obscure to evident and sinuses shallow to deep, veins (3 or)4–9(–12) per side, absent to sinuses. **Inflorescences:** branches glabrous or pubescent; symmetric bracteoles present, basal stipuliform, falcate bracteoles absent. **Flowers:** post-mature petals pale paper brown; stamens 5 or 10–20, anthers white to ivory, pink, or purple, sometimes puce. **Pomes** bright or deep red to purple or black (late Aug), reddish purple to black (late Sep); pyrene sides usually pitted, sometimes plane.

Species 16 (16 in the flora): w, nc North America.

Section *Douglasia* comprises all the North American hawthorns (except *Crataegus brachyacantha*) with purple (burgundy in two) to black fully mature fruit, usually fairly short (1.5–3.5 cm) thorns, and laterally eroded pyrenes. The lateral erosion or pitting on the pyrenes is not always very deep or as conspicuous as in *C. macracantha* and close allies.

SELECTED REFERENCE Dickinson, T. A. et al. 1996. American black-fruited hawthorns. I. Variation in floral construction, breeding system correlates and their possible evolutionary significance in *Crataegus* sect. *Douglasii* Loudon. Folia Geobot. Phytotax. 31: 355–371.

1. Leaves: blades narrow, more than 2:1, lobes 0, veins 6–8(–12) per side; flowers 10–13 mm diam.; stamens 20, anthers cream; bark on younger 2–5 cm thick branches often shining coppery, horizontal lenticels present .... 64c.2. *Crataegus* ser. *Cerrones* (in part), p. 508
1. Leaves: blades usually broader, less than 1.6:1, lobes 0 or 1–4 per side, veins (3 or)4 or 5(–7) per side; flowers (10–)12–22 mm diam.; stamens 5 or 10(–20), anthers white to ivory, pink, or purple, sometimes puce; bark on younger 2–5 cm thick branches usually dark gray-brown or orange, conspicuous horizontal lenticels absent.
   2. Stamens 10, anthers pink or pink-purple to purple; flowers 14–18 mm diam.; thorns on twigs slender .... 64c.2. *Crataegus* ser. *Cerrones* (in part), p. 508
   2. Stamens 10(–20), anthers pink or ivory to white, sometimes puce; flowers (10–) 12–18(–22) mm diam.; thorns on twigs usually ± stout, sometimes slender.
      3. Thorns 2–5(–7) cm, slender to stout; inflorescence branches usually sparsely villous, sometimes glabrate; sepal margins denticulate to finely, sometimes strongly, glandular-serrate; pomes bright red to purple (late Aug), reddish purple to deep purple, often nearly black (late Sep), sepals conspicuous unless eroded, usually erect-patent, sometimes ± reflexed, usually accrescent (except in *C. williamsii*) .... 64c.1. *Crataegus* ser. *Purpureofructus*, p. 502
      3. Thorns 1–3.5 cm, ± stout; inflorescence branches glabrous or pubescent; sepal margins subentire, finely glandular distally; pomes black or dark purple mature, sepals short, reflexed or patent-recurved, not accrescent . . .64c.3. *Crataegus* ser. *Douglasianae*, p. 511

64c.1. CRATAEGUS Linnaeus (sect. DOUGLASIA) ser. PURPUREOFRUCTUS J. B. Phipps & O'Kennon, Sida 18: 184. 1998 (as Purpureofructi) [E]

Shrubs or trees, (20–)30–60(–100) dm. **Stems:** twigs: new growth glabrous or densely pubescent, 1-year old deep reddish brown, brown, or purple-brown, older gray to dark gray, sometimes orange-brown overlaid with gray; bark on younger 2–5 cm thick branches mainly dark gray-brown, conspicuous horizontal lenticels absent; thorns on twigs straight to recurved, 2-years old dark brown, red-brown, or blackish brown, shiny, slender to stout, 2–5(–7) cm. **Leaves:**

petiole 1–3 cm, glabrous or sparsely hairy young, eglandular or glandular, <glands black>; blade ± rhombic to elliptic-oblong or ovate-rhombic, sometimes suborbiculate, 3–8 cm, coriaceous or thin, base cuneate to ± rounded, lobes (0 or)1–4 per side, obscure to evident, sinuses shallow to deep, lobe apex acute or ± subacute to ± rounded, margins serrate except at base, <teeth with gland-dots>, venation craspedodromous, veins (3 or)4–6(or 7) per side, <± impressed in coriaceous-leaved plants>, apex usually acute, sometimes subacute or obtuse, abaxial surface glabrous or glabrate, sometimes subglabrous or glabrescent, <veins hairy or not>, adaxial scabrous-hairy young, sometimes glabrescent. **Inflorescences** (5–)8–25-flowered; branches usually sparsely villous, sometimes glabrate; bracteoles caducous, linear, membranous, margins glandular, sometimes nearly eglandular. **Flowers** (10–)12–18(–22) mm diam.; hypanthium glabrous or pubescent; sepals (2.5–)4–7 mm, margins denticulate to finely, sometimes strongly, glandular-serrate; stamens 5 or 10(–20), anthers white to ivory, cream, or pink; styles 2–5. **Pomes** bright red to purple (late Aug), reddish purple to deep purple, often nearly black (late Sep), ± orbicular to ellipsoid, ellipsoid-urceolate, ± oblong, or ovoid, (7–)8–10(–15) mm diam., sparsely hairy or glabrous; sepals <prominent, usually accrescent (except in *C. williamsii*)>, erect-patent, sometimes ± reflexed; pyrenes 2–4(or 5), sides plane to deeply pitted.

Species 8 (8 in the flora): w North America.

Members of ser. *Purpureofructus* occur from the Okanagan Valley of British Columbia and Washington east to Montana and southern Saskatchewan. Most species are locally common.

Plants of ser. *Purpureofructus* share characteristics with those of ser. *Douglasianae*, but the thorns are usually considerably longer, the partly ripe fruit color is more likely to be red, even bright red, in late August, and the fruiting sepals are usually accrescent. *Crataegus williamsii* is an exception (see comments, p. 504). Several species of ser. *Purpureofructus* have been shown to have alloploid origin between members of ser. *Douglasianae* and sect. *Coccineae* or *Macracanthae* (M. Zarrei and T. A. Dickinson, pers. comm.). In late summer and fall, the usually abundant fruit of ser. *Purpureofructus* may turn from bright red to deep purple, and the leaves in some manifest brilliant coloring in bronze, burgundy, and yellow.

1. Pomes usually short-ellipsoid, pale red young, bright red (late Aug), deepening to oxblood red or darker mature, sepals spreading to reflexed, not thickened or enlarged; anthers pink; nw Montana . . . . 7. *Crataegus williamsii*
1. Pomes ± orbicular to ellipsoid or nearly ovoid, dull reddish, brick red, bright red, or purplish young (mid Aug), bright red to purple (late Aug), deepening to deep burgundy or purple to black mature (Sep), sepals suberect and sometimes incurved to erect-patent, spreading to erect, ± accrescent; anthers pink, white, or ivory; mainly of the Okanagan or Cypress Hills, only *C. phippsii* extending to nw Montana.
    2. Leaf blades of a wider form, broadly elliptic to suborbiculate or ovate to broadly deltate-ovate, lobe apices ± subacute or rounded to broadly angled.
        3. Leaf blades broadly elliptic to suborbiculate, lobes short-triangular, (pale green at anthesis); inflorescences sparsely villous, sometimes glabrate distally; pomes 7–10 mm diam., ± oblong, brick red (late Aug) becoming deep burgundy (Sep) . . . 13. *Crataegus orbicularis*
        3. Leaf blades broadly elliptic to ovate or broadly deltate-ovate, lobes rounded to broadly angled, (mid green at anthesis); inflorescences densely short-tomentose; pomes 12 mm diam., ± orbicular to broadly ovoid, red to plum purple (Aug) becoming nearly black (Sep) . . . . 14. *Crataegus phippsii*
    2. Leaf blades narrowly ovate to elliptic-rhombic or elliptic, or ovate to ovate-rhombic, lobe apices usually acute.

[4. Shifted to left margin.—Ed.]

4. Thorns on twigs usually ± stout, 2–4(–5) cm; leaf blades ± ovate to ovate-rhombic or ± oblong, ± coriaceous; anthers cream or ivory, sometimes very pale pink; sepals in fruit erect-patent to reflexed; pomes bright to deeper red (late Aug) becoming dark purple (Sep).
    5. Leaf blades 3.5–6 cm, ± coriaceous, venation ± impressed adaxially, (flattish to strongly concave, mid green until fall) . . . . . 8. *Crataegus okanaganensis*
    5. Leaf blades 4–8 cm, softly coriaceous, venation impressed adaxially, (flattish to convex, deep green until fall) . . . . . 9. *Crataegus enderbyensis*
4. Thorns on twigs usually ± slender, 3–5(–7) cm; leaf blades broadly elliptic to narrowly ovate or trullate-ovate to broadly elliptic-rhombic, thin or coriaceous; anthers pink, sometimes white; sepals in fruit suberect, erect-recurved, or spreading, (herbaceous); pomes deep red to purple (late Aug), deep burgundy to nearly black (Sep).
    6. Leaves matte, lobes small, max LII ca. 10%; flowers 10–14 mm diam.; pomes dull; n Okanagan . . . . . 11. *Crataegus atrovirens*
    6. Leaves shiny, lobes larger, max LII 15–25%; flowers 15–18 mm diam.; pomes dull or glossy; Cypress Hills and Bear's Paw Mountains.
        7. Anthers pale pink to pink; bark on 2–5 cm thick branches gray-brown; leaves ± coriaceous, lobe apices acute, max LII 15–25%; pomes glabrate or moderately to densely pubescent, dull . . . . . 10. *Crataegus aquacervensis*
        7. Anthers white; bark on 2–5 cm thick branches orange-brown; leaves thin, lobe apices acute to ± rounded, max LII ca. 15%; pomes glabrous, glossy . . . 12. *Crataegus cupressocollina*

7. Crataegus williamsii Eggleston, Bull. Torrey Bot. Club 36: 641. 1909 • Williams hawthorn [E]

Shrubs, 40–60(–100) dm. **Stems:** <expanding bud-scales conspicuous, coral red>; twigs: new growth glabrous, 1-year old deep reddish brown; thorns on twigs ± straight to slightly recurved, <black, sometimes dark purple young, older paler, shiny>, ± slender, 3–4 cm. **Leaves:** petiole 2–3 cm, ± hairy, eglandular or glandular; blade rhombic to elliptic, 4–6 cm, thin, base cuneate, lobes 2 or 3 per side, short, lobe apex acute, margins ± evenly sharp-serrate except at base, <teeth ± inconspicuously gland-tipped>, veins 4 or 5(or 6) per side, apex acute, abaxial surface glabrous, veins hairy, adaxial shortly and densely appressed-hairy. **Inflorescences** 8–20-flowered; branches sparsely villous; bracteole margins sessile-glandular. **Flowers** 16 mm diam.; hypanthium glabrous or densely villous externally; sepals ± narrowly triangular, 4–7 mm, margins finely glandular-serrate, abaxially glabrous; stamens 10, anthers pink; styles 3 or 4(or 5). **Pomes** usually pale red young, bright red (late Aug), deepening to oxblood red or darker mature (Sep), usually short-ellipsoid, 8 mm diam., <longer than wide>, sparsely villous; sepals <not thickened or enlarged>, spreading to reflexed, narrow, 4–7 mm; pyrenes 3 or 4(or 5), sides variably pitted.

Flowering May–Jun; fruiting Sep–Oct. Scrub; 700–1000 m; Mont.

Crataegus williamsii occurs in the Clark Fork and Flathead drainage basins of northwestern Montana; possible records for Idaho require more study. The species is notable for its extremely late anthesis, after sympatric *C. macracantha*; it is the last hawthorn to flower in its area of occurrence. The immature fruit is bright red, in contrast to its later duller color. The fall color of the foliage is predominantly crimson burgundy or a pallid version thereof. *Crataegus williamsii* is the most distinct member of ser. *Purpureofructus* because of its relatively small and delicate, spreading to reflexed fruiting sepals (versus accrescent, long, erect-patent) and its lack of purple- to black-colored, fully mature fruit. See also comments under 26. *C. macracantha*.

8. Crataegus okanaganensis J. B. Phipps & O'Kennon, Sida 18: 178, figs. 5, 9d–f. 1998 • Okanagan hawthorn [E] [F]

Shrubs, <often multi-trunked>, 30–60 dm. **Stems:** twigs: new growth sparsely pubescent, 1-year old mid to dark brown; thorns on twigs recurved, <deep brown, shiny young>, usually stout, 2–4(–5) cm. **Leaves:** petiole 1–2 cm, sparsely hairy, eglandular or glandular when young; blade <glossy especially when young, expanding leaves often reddish, glabrescent>, ± ovate to ovate-rhombic or ± oblong, 3.5–6 cm, ± coriaceous, base rounded to cuneate, lobes 0–4 per side (3 or 4 per side, well defined in ± ovate to ovate-rhombic leaves; 0–3 per side, distinct in apical ½ only in ± oblong leaves), lobe apex acute, margins finely, doubly serrate, veins 4 or 5 per side, <± impressed adaxially>, apex acute, abaxial surface glabrous, veins hairy, adaxial densely appressed-hairy. **Inflorescences** 10–20-flowered; branches ± villous, <gland-dotted>; bracteole margins glandular (glands near base stipitate only in larger bracteoles). **Flowers** <notably cup-shaped young>, 12–15 mm diam.; hypanthium villous (at base) or glabrous; sepals narrowly triangular, 4–6 mm, margins finely glandular-serrate, abaxially sparsely hairy; stamens 5 or 10, anthers ivory, rarely very pale pink; styles 2–4. **Pomes** bright red young (mid late Aug), older usually burgundy to deep purple, sometimes almost black (mid Sep), ovoid and 8 mm diam., or orbicular and 10 mm diam., glabrous or hairy; sepals reflexed, sometimes erect-patent, narrow, 5–7 mm; pyrenes 2–4, sides ± eroded.

Varieties 2 (2 in the flora): w North America.

Crataegus okanaganensis is widely distributed and common from the Okanagan valley of southern British Columbia and northern Washington to northern Idaho.

1. Leaf blades ± ovate to ovate-rhombic, lobes well defined; leaves ± bronze in fall; pomes pendent, ovoid . . . . . . 8a. *Crataegus okanaganensis* var. *okanaganensis*
1. Leaf blades ± oblong, lobes absent or short and only in distal ½; leaves crimson or pink burgundy in fall; pomes ± erect, orbicular. . . . . 8b. *Crataegus okanaganensis* var. *wellsii*

8a. Crataegus okanaganensis J. B. Phipps & O'Kennon var. **okanaganensis** E F

Leaves ± bronze in fall; blade ± ovate to ovate-rhombic, lobes well defined. **Pomes** pendent, ovoid.

Flowering May; fruiting late Aug–Sep. Natural hedgerows and mesic brush, mainly in valley bottoms; 400–700 m; B.C.; Idaho, Wash.

Variety *okanaganensis* is somewhat like *Crataegus williamsii* of northwestern Montana both in leaf shape and fruit shape, but it has glossier, more coriaceous leaves with the color often yellowish green in mid season, the blades proportionately wider, usually with less acute lobes, stouter thorns, different anther color, and a different fruiting calyx. The young fruit is brilliant red when *C. williamsii* is usually a pale to deeper dull red; mature fruits are deep purple instead of blood or deep red. Northern Idaho specimens of *C. okanaganensis* flower about a week before *C. williamsii* in northwestern Montana. Apparent intermediates with var. *wellsii* occur.

8b. Crataegus okanaganensis J. B. Phipps & O'Kennon var. **wellsii** J. B. Phipps & O'Kennon, Sida 20: 132, figs. 6, 7. 2002 E

Leaves crimson or pink burgundy in fall; blade ± oblong, ± lobes absent or short and only in distal ½, <sinuses shallow>. **Pomes** ± erect, orbicular.

Flowering May; fruiting late Aug–Sep. Natural hedgerows and mesic brush, mainly in valley bottoms; 400–700 m; B.C.; Wash.

Variety *wellsii* is known from the Okanagan drainage, where it is usually, but not always, local in occurrence; it is less common than var. *okanaganensis*. Variety *wellsii* has been occasionally recorded with pink anthers; it is very distinct with its unusual, differently lobed leaves; it also differs in fruit shape and presentation, fewer pyrenes, and fall foliage color crimson or pink burgundy versus copper bronze in var. *okanaganensis*. Pomes tend to be duller or deeper red than those of var. *okanaganensis* at the same date in late August; in mid August they are also brilliant scarlet. The most extreme form of this variety, virtually thornless, is found at Palmer Lake, northern Washington.

9. Crataegus enderbyensis J. B. Phipps & O'Kennon, Sida 20: 136, fig. 8. 2002 • Enderby hawthorn E

Shrubs or trees, 50–60 dm. **Stems:** twigs: new growth pubescence not recorded, 1-year old purple-brown, older dark gray; thorns on twigs recurved, <mahogany brown, shiny young>, stout, 2.5 cm. **Leaves:** petiole 1–2 cm, glabrous, glandularity not recorded; blade ovate-rhombic, 4–8 cm, softly coriaceous, base broadly cuneate to rounded, lobes 3 per side, short, lobe apex acute, margins serrulate, veins 4 per side, <distinctly impressed adaxially>, apex acute or subacute, abaxial surface glabrous, veins hairy, adaxial appressed-hairy. **Inflorescences** 12–25-flowered; branches villous; bracteole margins glandular. **Flowers** 15 mm diam.; hypanthium glabrous or pubescent; sepals narrowly triangular, 5–7 mm, margins finely glandular-serrate, abaxially glabrous; stamens 10, anthers cream; styles 3. **Pomes** <often in large, ± pendent infructescences>, cream young (mid Aug), turning salmon, then red (late Aug), purple mature (Sep), ellipsoid-urceolate, 8–10 mm diam., glabrous; sepals erect-patent to reflexed, narrow, 5–7 mm, <hairy>; pyrenes 3, sides plane or shallowly eroded. $2n = 68$.

Flowering May; fruiting Sep–Oct. Brush, natural hedgerows, on mesic sites; 300–400 m; B.C.

Crataegus enderbyensis is known from the Armstrong-Enderby region of the upper Okanagan. The species is easily recognized in the field by its dark green, softly coriaceous, somewhat drooping, convex leaves, with deeply impressed venation, as well as its usually large drooping corymbs with very pale fruit (at first), becoming red and finally turning purple-black. In flower, it is very similar to *C. okanaganensis*. The indumentum difference from *C. okanaganensis* originally noted has been found not to hold up after further collections. The differences as now understood between the two species reside only in unique characteristics of the *C. enderbyensis* leaf and its early pome color.

10. Crataegus aquacervensis J. B. Phipps & O'Kennon, J. Bot. Res. Inst. Texas 1: 1061, plate 7.2c, fig. 11. 2007 • Elkwater hawthorn [E]

Shrubs, <densely branched>, 20–40 dm. **Stems:** twigs: new growth <anthocyanin-rich>, nearly glabrous, 1-year old medium to darker brown, older gray to dark gray; <bark on 2–5 cm thick branches gray-brown>; thorns on twigs usually recurved, ± stout, 3–6.5 cm, 2-years old dark red-brown. **Leaves** <deciduous>; petiole length 30–40% blade, hairy (only in adaxial sulcus), sessile-glandular; blade ovate or trullate-ovate to broadly elliptic, 3–3.5 cm at anthesis, 4–7 cm mature, ± coriaceous, base cuneate to broadly cuneate, lobes (2 or)3 or 4 per side, pronounced, <max LII 15–25%>, lobe apex acute, margins serrate, <teeth numerous, small, acute to subacute>, veins (3 or)4 or 5 per side, <distinctly impressed adaxially>, apex acute, <shiny>, abaxial surface glabrous, veins sometimes densely finely hairy along sides, adaxial ± densely appressed-scabrous hairy. **Inflorescences** 5–12-flowered; branches ± sparsely to moderately or densely pubescent; bracteoles <light brown>, margins glandular. **Flowers** 15–18 mm diam.; hypanthium glabrate or sparsely to moderately pubescent; sepals <pale with some anthocyanin or light brown>, triangular, 2.5 mm, margins denticulate to glandular-serrate, abaxially glabrous; stamens 10, anthers pale pink or pink; styles 3 or 4. <**Fruiting pedicels** glabrate or sparsely to moderately or densely hairy>. **Pomes** reddish plum early (late Aug), dull purple mature (mid Sep), ellipsoid to broadly ellipsoid or ovoid, 10–12 mm diam., glabrate or moderately to densely pubescent; sepals usually erect-recurved, wide, 2.5–4 mm; pyrenes 3 or 4(or 5), sides ± eroded, often shallowly.

Flowering late May–early Jun; fruiting Aug–Sep. Thickets, light shade of *Populus* spp.; 800–1200 m; Alta., Sask.

Crataegus aquacervensis is widely distributed in the northern part of the Cypress Hills westward from the Adams Creek area and Fort Walsh. The species is superficially similar to the sympatric *C. rivuloadamensis* (ser. *Montaninsulae*) and *C. chrysocarpa* (ser. *Rotundifoliae*), differing from both by the more or less eroded lateral faces of the pyrenes and from the latter by the larger, more venous leaves, prominent fruiting calyx, and ripe fruit color.

11. Crataegus atrovirens J. B. Phipps & O'Kennon, Sida 20: 141, fig. 10. 2002 [E]

Shrubs, 20–50 dm. **Stems** <multiple, suberect to erect>; twigs: new growth pubescent, 1-year old glossy brown, older dark gray; <bark on 2–5 cm thick branches gray-brown>; thorns on twigs ± straight to very slightly curved, ± slender, 3–5 cm, 2-years old dark brown, black at tip, older deep gray. **Leaves:** petiole length 15–20% blade, sparsely or densely (in adaxial sulcus mature) hairy, eglandular or sparsely small glandular mainly distally; blade <dark green (in fall turning bronze-brown), matte>, narrowly ovate to broadly elliptic-rhombic or broadly elliptic, 4–6(–7) cm, subcoriaceous, base cuneate, lobes 3 or 4 per side, small, <max LII ca. 10%>, lobe apex acute, margins serrulate, <minutely glandular at apices>, veins 3–5 per side, apex acute to subacute, <matte>, abaxial surface glabrate, vein pubescence not recorded, adaxial scabrous hairy, glabrescent. **Inflorescences** 12–20-flowered; branches densely spreading-pubescent; bracteoles <narrow but widened distally>, margins densely glandular. **Flowers** 10–14 mm diam.; hypanthium pubescent; sepals narrowly triangular, 4–5 mm, margins strongly glandular-serrate, abaxial pubescence not recorded; stamens 10(–20), anthers pink; styles 2 or 3. **Pomes** burgundy (late Aug), blackish purple (Sep), dull, ± ovoid, 13–15 × 10 mm, hairy; <flesh soft orange>; sepals suberect, wide, 4–6 mm; pyrenes 2 or 3, sides pitted.

Flowering May; fruiting Sep–Oct. Natural hedgerows, brush in valleys, dry habitats; 300–500 m; B.C.

Crataegus atrovirens is known from the northern Okanagan, where it is locally common to abundant. One of its most notable characteristics is the unusually large glands on the leaf teeth. It also has smallish flowers for its series. Of species known to date from this region, *C. shuswapensis* is the one most likely to be confused with *C. atrovirens*. The mature fruit and the growth habit of *C. shuswapensis* are similar, but that species has leaves more sharply lobed, glabrous inflorescences, sometimes 15–18 stamens per flower, redder fruit in the fourth week of August, much smaller, less prominent fruiting sepals, whose margins, rather than being serrate-laciniate, are minutely glandular. *Crataegus atrovirens* also has similarities to *C. okanaganensis*; it is easily differentiated in the field by the usually much longer thorns, smaller flowers with pink anthers, and much darker fruit in late August. Compared to *C. orbicularis*, the fruit of *C. atrovirens* is also conspicuously larger; it is soft and ripe in the fourth week of August when the former is hard but not ripe. The erect, multi-trunked habit is also somewhat distinctive. It is a dull hawthorn, somber in appearance, contrasting markedly with more colorful congeners such as *C. okanaganensis*.

12. Crataegus cupressocollina J. B. Phipps & O'Kennon, J. Bot. Res. Inst. Texas 1: 1056, plates 6.2a, 7a,b, fig. 6. 2007 • Cypress Hills hawthorn E F

Shrubs or trees, <± erect>, 25–60 dm. **Stems:** twigs: new growth <greenish red young>, sparsely pilose, 1-year old deep to reddish brown, shiny, older orange-brown overlaid with gray; <bark on 2–5 cm thick branches orange-brown>; thorns on twigs <few to numerous>, ± straight to strongly recurved, slender, 3–7 cm, 2-years old blackish brown. **Leaves:** petiole 1–2 cm, sparsely hairy young, sparsely glandular; blade ± ovate to broadly elliptic, 3–6 cm, thin, base broadly cuneate, lobes 3 or 4 per side, <max LII ca. 15%>, lobe apex acute to ± rounded, margins serrate, <teeth numerous, small, gland-tipped when young>, veins 3 or 4 per side, apex acute, <shiny>, abaxial surface glabrate, some axils of midvein pilose, adaxial appressed-hairy. **Inflorescences** 5–15-flowered; branches sparsely to moderately or densely pilose; bracteoles <pale brownish>, margins nearly eglandular, minutely short, delicately stipulate-glandular. **Flowers** 15–18 mm diam.; hypanthium glabrous; sepals <± anthocyanic>, triangular, length not recorded, margins finely glandular-serrate, abaxially glabrous, <adaxially pilose>; stamens 10, anthers white; styles 3–5. **Pomes** <on pendulous branches>, burgundy (Aug–early Sep), black (late Sep), glossy, ellipsoid, 7–10 mm diam., glabrous; sepals spreading, wide, 3–4 mm, <apex ± acuminate>; pyrenes 3–5, sides ± irregularly pitted or scarred.

Flowering late May–early Jun; fruiting Aug–Sep. Thickets, grasslands, light shade of *Populus* spp.; 800–1200 m; Alta., Sask.; Mont.

Crataegus cupressocollina is found through much of the Cypress Hills, where it may be common, and in a restricted area in the Bear's Paw Mountains, Montana. The species is distinctive in its tall, willowy, upright habit, leaf form, relatively long, slender thorns, and two to five year-old orange-brown bark. The flowers differ from those in ser. *Douglasianae* in their white anthers. The ripe fruit of *C. cupressocollina* is glabrous and plum red to purple-brown in late August, after which it ripens to a glossy purple-black. The large, pendulous clusters of fruit at once draw attention, as do the reddish tips of the sepals and the soft red-brown of the often nearly eglandular bracteoles when in flower. The colorful expanding bud scales are also rather striking. *Crataegus cupressocollina* is the first to flower of sympatric species.

13. Crataegus orbicularis J. B. Phipps & O'Kennon, Sida 20: 138, fig. 9. 2002 C E

Shrubs, <wide-spreading>, 20–40 dm. **Stems:** twigs: new growth sparsely pubescent, 1-year old strong mid brown, older dull gray; thorns on twigs straight or more often recurved, slender, 3–5 cm. **Leaves:** petiole 1.2–2 cm, adaxially short-hairy, usually eglandular; blade broadly elliptic to suborbiculate, 5–8 cm, thin, base curved in wider blades, ± attenuate in narrower ones, lobes 3 or 4 per side, short-triangular, margins serrate, <teeth very small, except near base>, veins 5 or 6(or 7) per side, apex ± subacute, abaxial surface glabrate, veins hairy, adaxial short-appressed-hairy. **Inflorescences** 10–18-flowered; branches sparsely villous throughout, sometimes glabrate distally; bracteole margins sessile-glandular. **Flowers** 18 mm diam.; hypanthium glabrate; sepals narrowly triangular, 5–6 mm, margins finely glandular-serrate, abaxial pubescence not recorded; stamens 10, anthers pink; styles 3 or 4. **Pomes** <in showy infructescences>, brick red (mid to late Aug) becoming deep burgundy (Sep), ± oblong, 7–10 mm diam., glabrate; sepals ± erect, wide, length not recorded; pyrenes 3 or 4, sides shallowly pitted or almost plane.

Flowering May; fruiting Sep–Oct. Brush, natural hedgerows; of conservation concern; 300–400 m; B.C.

Crataegus orbicularis is known only from within a few miles of Enderby. The species is readily distinguished from sympatric taxa by the unusual pale yellowish green color of the foliage when young, leaf shape, large flowers, and small glabrous pomes with prominent erect sepals.

14. Crataegus phippsii O'Kennon, Sida 18: 185, figs. 7, 9a–c. 1998 [E]

Shrubs or trees, 50–60(–70) dm. **Stems:** <main trunk usually present, sometimes suckering>; twigs: new growth densely short-pubescent, 1-year old dark purple-brown, older dark gray; thorns on twigs <few>, slightly recurved, <deep brown to black>, ± slender, 2–4 cm. **Leaves:** petiole 1.5–2 cm, hairy, glandularity not recorded; blade broadly elliptic to ovate or broadly deltate-ovate, 4–8 cm (–10 cm on extension shoots), becoming ± coriaceous, base cuneate to very broadly cuneate, lobes 3 per side, sinuses ± shallow, lobe apex rounded to broadly angled, margins short-serrate, veins 4(or 5) per side, apex obtuse or subacute, abaxial surface subglabrous or glabrescent, veins persistently short-hairy, adaxial short-appressed-hairy. **Inflorescences** 6–12-flowered; branches densely short-tomentose; bracteole margins stipitate- or sessile-glandular. **Flowers** 15–22 mm diam.; hypanthium densely tomentose, <disc interior glabrous, bristly at center>; sepals broadly triangular, 4–5 mm, margins glandular-serrate, <apex subacute>, abaxially pubescent (especially near base); stamens 10, anthers pale pink; styles 2–4, <stigmas capitate>. **Pomes** red to plum purple (Aug) to nearly black (Sep), ± orbicular to broadly ovoid, 12 mm diam., short-pubescent; sepals erect-patent to patent-incurved, 4–5 mm; pyrenes 3, sides irregularly eroded or ± plane. $2n = 68$.

Flowering May; fruiting Sep–Oct. Brush, natural hedgerows on mesic soil; 300–700 m; B.C.; Mont., Wash.

Crataegus phippsii is found from the Okanagan Valley, southern British Columbia and northwestern Washington, to near Flathead Lake, Montana.

Crataegus phippsii is one of the more distinctive North American hawthorns and does not appear to be closely similar to any other known in the west. The plants are large, somewhat reminiscent of *C. mollis* in foliage, habit, and pubescence; in its fruit (red changing to purple to black) there is a color change similar to those in other species in this series. The bracteoles are smaller than in the *mollis* group, caducous and membranous, and bright crimson when fresh. The light gray bark flakes off in long narrow strips, which is unusual in *Crataegus*.

64c.2. CRATAEGUS Linnaeus (sect. DOUGLASIA) ser. CERRONES J. B. Phipps, Canad. J. Bot. 76: 1872. 1999 [E]

Shrubs or trees, 30–50 dm. **Stems:** twigs: new growth glabrous or sparsely hairy, 1-year old reddish tan, reddish mahogany, or red-purple, older gray or copper colored; bark on younger 2–5 cm thick branches mainly dark gray-brown or copper colored, horizontal lenticels present; thorns on twigs straight or recurved, 2-years old black or purple-black, glossy or shiny, slender, 1–4 cm. **Leaves:** petiole <slender>, 0.8–2.5 cm, pubescence and glandularity not recorded; blade ± elliptic to rhombic-elliptic, 3–8 cm, thin to coriaceous, base cuneate, lobes 0, or 3 or 4 per side, sinuses shallow, lobe apex acute, margins shallowly crenate or serrate, venation craspedodromous to semicamptodromous, veins 4 or 5(–12) per side, apex usually acute, sometimes obtuse, abaxial surface glabrate or glabrous mature. **Inflorescences** 5–12-flowered; branches glabrous, <usually glandular-punctate>; bracteoles caducous, linear, membranous, margins usually sessile-glandular, sometimes eglandular. **Flowers** (10–)14–18 mm diam.; hypanthium glabrous; sepals (1.5–)3–4(–8) mm, margins subentire or entire; stamens 10(or 20), anthers pink or pink-purple to purple, sometimes cream; styles 3–5. **Pomes** deep red to vinous purple or black, orbicular, (8–)10 mm diam., glabrous; sepals suberect to reflexed; pyrenes 3–5, sides excavated.

Species 3 (3 in the flora): w United States.

Series *Cerrones* was initially treated among the red-fruited series (J. B. Phipps 1999b) as it was then understood only from its type species, *Crataegus erythropoda*, which has the least typical fruit color for the series. Series *Cerrones* contains two closely related species of intermontane hawthorn to which *C. saligna*, formerly placed in ser. *Brachyacanthae*, has been added. Series *Cerrones* is characterized by copper-colored mature bark, sometimes displaying horizontal lenticels, fine and short thorns, early caducous, few gland-margined bracteoles, burgundy to black mature fruit, and pyrenes with lateral excavations.

SELECTED REFERENCE Phipps, J. B. 1999c. The relationships of the American black-fruited hawthorns *Crataegus erythropoda*, *C. rivularis*, *C. saligna* and *C. brachyacantha* to C. ser. *Douglasianae* (Rosaceae). Sida 18: 647–620.

1. Leaf blades: veins 6–9(–12) per side, margins crenate; flowers 10–13 mm diam., stamens 20, anthers cream . . . . 15. *Crataegus saligna*
1. Leaf blades: veins 4 or 5 per side, margins serrate; flowers 14–18 mm diam., stamens 10, anthers usually pink to pink-purple or purple, sometimes ivory.
   2. Leaf blades rhombic-elliptic, length/width = 1.6, lobes 3 or 4 per side; pomes deep red to vinous purple mature . . . . 16. *Crataegus erythropoda*
   2. Leaf blades elliptic to narrowly elliptic, at least 2 times as long as wide, lobes 0 or with small apiculi at ends of some vein tips; pomes black or blackish purple mature . . . 17. *Crataegus rivularis*

15. Crataegus saligna Greene, Pittonia 3: 99. 1896 • Willow hawthorn [E]

Crataegus douglasii Lindley var. *duchesnensis* S. L. Welsh

Shrubs, 30–50 dm, <often clonal>. **Stems:** twigs: new growth green, glabrous, 1-year old reddish tan, older deep gray; bark on younger 2–5 cm thick branches mainly dark (with horizontal lenticels); thorns on twigs straight (often needlelike) or recurved, 2-years old black, shiny, ± fine, 1–3 cm (sometimes bigger and stouter). **Leaves:** petiole 0.8–1.8 cm; blade narrowly elliptic to bluntly elliptic-lanceolate, 3–5 cm, ± coriaceous, base narrowly cuneate, lobes 0, margins crenate, teeth numerous, small, obtuse, venation semicamptodromous, veins 6–9(–12) per side, apex usually obtuse, sometimes acute, abaxial surface glabrous, adaxial appressed pubescent young, glabrescent. **Inflorescences** 12-flowered; branches not glandular-punctate; bracteoles early caducous, few, margins eglandular or sessile-glandular. **Flowers** 10–13 mm diam.; sepals proximally broadly triangular with short, narrower limbs, 1.5–2 mm, margins not recorded, apex acute, pubescence not recorded; stamens 20, anthers cream; styles 5. **Pomes** black mature, orbicular, 8 mm diam.; sepals suberect, 1.5–2 mm, <tips reflexed>; pyrenes 3–5, sides irregular. $2n$ = 34, 51.

Flowering May–Jun; fruiting Sep–Oct. Along streams, flood plains; 1600–2400 m; Colo., Utah.

Crataegus saligna is restricted to intermontane areas of northeastern Utah and adjacent Colorado; it may be locally plentiful, and the plants flower early. The species may be confused with *C. rivularis*; it differs by its smaller flowers, fruit, and foliage, 20 stamens, cream anthers, and leaves relatively narrower, more venous, and with smaller, more obtuse teeth. It is more similar in flower and foliage to *C. brachyacantha* but has very different thorns and bark color, as well as irregular lateral faces of pyrenes.

Crataegus saligna is a striking plant with its suckering habit, its glowing copper-tan bark, long arching twigs and branches, dark green foliage, narrow and more or less lanceolate leaves, often beautifully reddish bronze foliage in fall, and black fruit.

16. Crataegus erythropoda Ashe, Bull. North Carolina Agric. Exp. Sta. 175: 113. 1900 • Cerro hawthorn [E]

Crataegus cerronis A. Nelson

Shrubs or trees, 50 dm. **Stems:** twigs: new growth greenish, glabrous, 1-year old dark reddish mahogany; bark on younger 2–5 cm thick branches dark gray-brown, sometimes copper colored; thorns on twigs straight or slightly recurved, 2-years old black, shiny, moderately stout, 2–4 cm. **Leaves:** petiole 1–2 cm; blade rhombic-elliptic, 3–5 cm, <length 1.6 times width>, ± coriaceous-shining, base cuneate, lobes 3 or 4 per side, lobe apex acute, margins serrate, teeth very short, venation craspedodromous, veins 4 or 5 per side, apex acute, abaxial surface glabrate, adaxial sparsely pilose young. **Inflorescences** 5–10-flowered; branches glandular-punctate; bracteoles few or absent, margins sessile-glandular. **Flowers** 14–18 mm diam.; sepals narrowly triangular, 3–4 mm, margins glandular-serrate, teeth small, apex acute, glabrous; stamens 10, anthers pink-purple to purple; styles 4 or 5. **Pomes** deep red to vinous purple mature, orbicular, 10 mm diam.; sepals reflexed, 4 mm; pyrenes 3–5, sides excavated.

Flowering May–Jun; fruiting Sep–Oct. Sagebrush, pastures, usually along streams; 1700–2600 m; Colo., N.Mex., Wyo.

Crataegus erythropoda is found mainly in Colorado westward from the foothills of the Front Range; it also extends north and south into adjacent states. N. H. Holmgren (1997b) recorded it from Arizona and Utah; no authenticating specimen has been encountered. The species is locally common and is early-flowering.

Crataegus erythropoda varies little; it has striking deep red to vinous fruit color and is usually easily recognized, although the black-fruited *C. rivularis* may also show a similar color when not fully ripe. Paucity of observed bracteoles may indicate they are very early caducous. The specific epithet alludes to the pedicels, which may turn red in the fall or under particularly dry conditions, which is neither constant in nor unique to *C. erythropoda*.

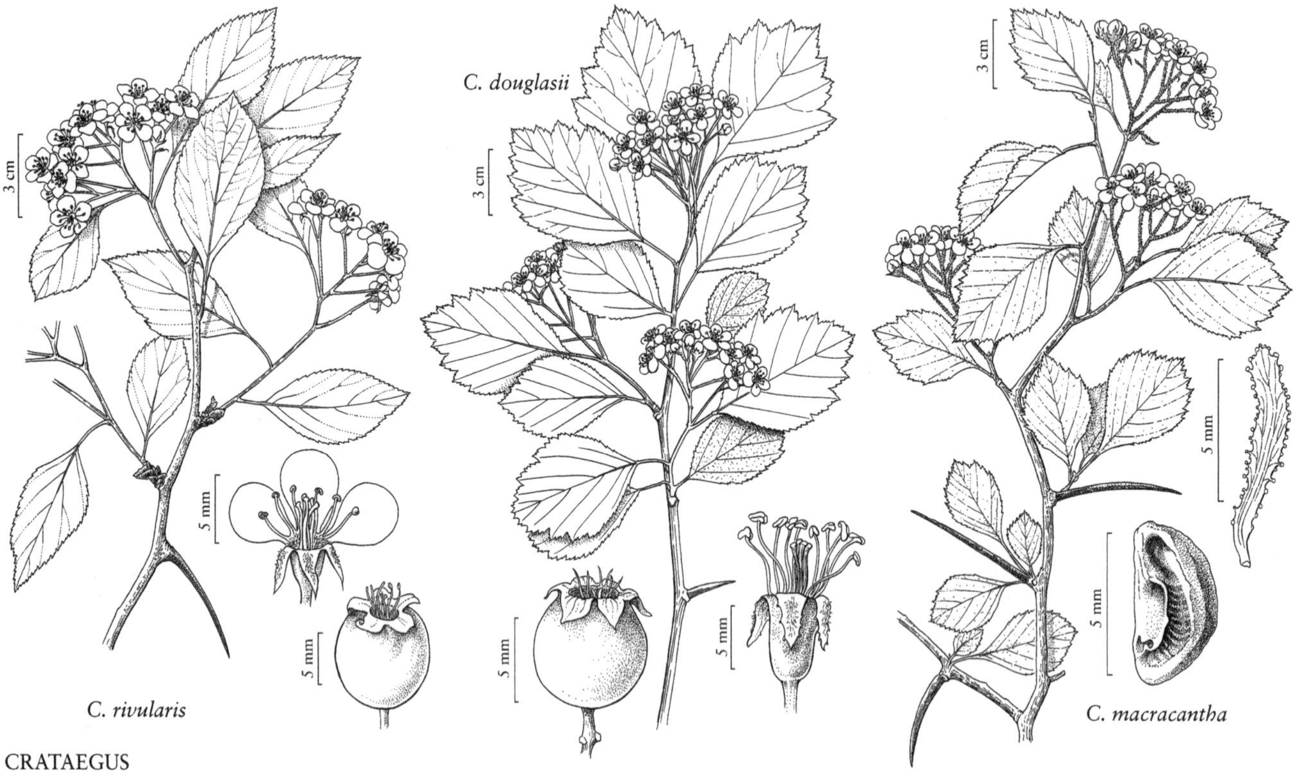

CRATAEGUS

17. Crataegus rivularis Nuttall in J. Torrey and A. Gray, Fl. N. Amer. 1: 464. 1840 • Mountain river hawthorn [E] [F]

Shrubs or trees, 30–50 dm. **Stems:** twigs: new growth greenish, sparsely pubescent, early glabrescent, 1–2-years old often red-purple; bark on younger 2–5 cm thick branches dark gray-brown, sometimes copper colored; thorns on twigs straight or ± recurved, 2-years old black or purple-black, glossy, fine, 1.5–4 cm. **Leaves:** petiole 1–2.5 cm, <glabrous>; blade elliptic to narrowly elliptic, 3–8 cm, <length 2+ times width>, thin, base cuneate, lobes 0 or small apiculi at ends of some vein tips, margins serrate, teeth numerous, usually acute, fine to coarse, venation semicamptodromous, veins 4 or 5 per side, <larger often branched proximal to ends>, apex acute, abaxial surface glabrous, adaxial appressed-strigose young. **Inflorescences** 6–12-flowered; branches glandular-punctate; bracteoles early caducous, usually present, margins sessile-glandular. **Flowers** 14–17 mm diam.; sepals broadly deltate, 6–8 mm, margins entire, apex narrow, glabrous; stamens 10, anthers pink, sometimes ivory; styles 3–5. **Pomes** deep red (mid Aug), black or blackish purple mature, suborbicular, 10 mm diam.; sepals ± erect to strongly outcurved or ± patent, 2–3 mm; pyrenes 3 or 4, sides eroded.

Flowering mid May–early Jun; fruiting Aug–Sep. Intermontane streamsides, ditches, flood plains; 1300–2300 m; Ariz., Colo., Idaho, Nev., N.Mex., Utah, Wyo.

Crataegus rivularis is most abundant in intermontane Colorado and Utah and is one of two hawthorns found in Arizona; it is rare in Nevada. There are reports from northwestern Texas (for example, D. S. Correll and M. C. Johnston 1970); they have not been verified. *Crataegus rivularis* is one of the more common woody species of mesic, valley bottom intermontane habitats. The species flowers early.

Crataegus rivularis is superficially similar to the more or less sympatric, less common, *C. saligna*; it has larger leaves with fewer veins, larger flowers with ten stamens, and larger pomes. It differs from *C. erythropoda* little other than in leaf shape and ripe fruit color.

64c.3. Crataegus Linnaeus (sect. Douglasia) ser. Douglasianae Rehder, Man. Cult. Trees ed. 2, 368. 1940 [E]

Shrubs or trees, 25–100(–140) dm. **Stems:** twigs: new growth glabrous, sometimes pubescent, 1-year old usually tan to deep mahogany; bark on younger 2–5 cm thick branches mainly dark gray-brown or orange, conspicuous horizontal lenticels absent; thorns on twigs recurved or straight, dark brown, shiny young, usually ± stout, 1–3.5 cm. **Leaves:** petiole 0.7–1.5(–2.2) cm, glabrous or pubescent, eglandular or glandular; blade broadly elliptic or elliptic to ± obovate, rhombic, oblanceolate, or ovate, 3–5(–8) cm, chartaceous to coriaceous, base usually cuneate, lobes 0 or 2–4 per side, sinuses shallow, lobe apex obtuse to subacute, margins serrate, <teeth apices glandular>, venation craspedodromous to semicamptodromous (in some forms with lobes absent), veins (3 or)4 or 5 per side, apex usually acute, sometimes obtuse, abaxial surface usually glabrous except on veins, sometimes sparsely pubescent, adaxial usually appressed-pubescent. **Inflorescences** (7–)12–20(–25)-flowered; branches glabrous or pubescent; bracteoles caducous, linear, <small>, membranous, margins stipitate- or sessile-glandular. **Flowers** (10–)12–15(–20) mm diam.; hypanthium pubescent or glabrous; sepals 3–4(–5) mm, margins subentire, finely glandular distally; stamens 10(–20), anthers pink, sometimes puce; styles 3–5. **Pomes** black or dark purple mature, ellipsoid to suborbicular, 6–10(–12) mm diam., glabrous or sparsely pilose; sepals reflexed or patent-recurved; pyrenes 3–5, sides plane to pitted.

Species 5 (5 in the flora): w, c North America.

Series *Douglasianae* comprises mainly western, widespread, black-fruited hawthorns, one of which, *Crataegus douglasii*, extends disjunctly to central Ontario. Its species are notable for their usually laterally pitted pyrenes, relatively short thorns, short and often blunt-fruiting sepals, and dark purple to black fruits at full maturity.

1. Stamens 20; thorns 1–2(–2.5) cm; leaf blades: lobes 0, or 2 or 3 per side, lobe apices acute to obtuse; pomes suborbicular to broadly ovoid .... 18. *Crataegus gaylussacia*
1. Stamens 10(–18); thorns (1.5–)2–3.5 cm; leaf blades: lobes 2–4 per side, lobe apices subobtuse to acute; pomes ellipsoid to suborbicular, ± oblate, or ± ampulliform-orbicular.
   2. Inflorescence branches and hypanthia sparsely to densely pubescent, sometimes glabrous; thorns very few to many, some paired or in triads; pomes orbicular or ± oblate (recessed at junction with pedicel) .... 20. *Crataegus castlegarensis*
   2. Inflorescence branches and hypanthia glabrous, sometimes ± sparsely pubescent; thorns single; pomes suborbicular to ellipsoid.
      3. Thorns 2 cm; flowers 15–20 mm diam.; pomes ± ampulliform-orbicular, red-burgundy to chestnut or reddish brown (late Aug), often shiny and usually with little bloom when ripe... 21. *Crataegus okennonii*
      3. Thorns (1.5–)2–3.5 cm; flowers 10–15 mm diam.; pomes usually ellipsoid, rarely suborbicular, usually already very deep purple or dull black mature (late Aug), often still glaucous when ripe.
         4. Pyrene sides excavated, sometimes only shallowly; leaf blades elliptic to broadly elliptic or subrhombic, lobe sinuses usually shallow, lobe apices subobtuse to acute; thorns (1.5–)2–3.5 cm; pomes usually ellipsoid, 6–8 mm diam., vinous and usually very waxy glaucous young (mid Aug), dull black mature; sepal remnants reflexed, apices obtuse; stamens 10...................... 19. *Crataegus douglasii*
         4. Pyrene sides plane to ± concave, grooved, or irregularly scarred; leaf blades rhombic, lobe sinuses deep, lobe apices acute; thorns 2–3 cm; pomes suborbicular to ± broadly ellipsoid, 12 mm diam., brick red young (mid Aug), deep purple mature, ± glaucous; sepal remnants patent-recurved, apices acute; stamens 10–18 .... 22. *Crataegus shuswapensis*

18. Crataegus gaylussacia A. Heller, Bull. S. Calif. Acad. Sci. 2: 69. 1903 • Huckleberry or Suksdorf hawthorn [E]

Crataegus douglasii Lindley var. *suksdorfii* Sargent; *C. punctata* Jacquin var. *brevispina* Douglas ex Hooker; *C. suksdorfii* (Sargent) Kruschke

Shrubs, 30–100 dm. **Stems:** 1-year old twigs tan, older grayish; thorns on twigs recurved or straight, dark brown young, 1–2(–2.5) cm. **Leaves:** petiole 0.7–1.5 cm, glabrous, eglandular, sometimes glandular (Montana); blade ± rhombic-elliptic to elliptic-obovate or broadly ovate, 3–8 cm, lobes 0, or 2 or 3 per side, lobe apex acute to obtuse, margins serrate, <teeth glandular in Montana>, venation ± semicamptodromous, veins 3 or 4 per side, apex obtuse, abaxial surface ± glabrate, adaxial appressed-pubescent, <hairy near veins in Oregon>. **Inflorescences** 7–15-flowered; branches <punctate>, glabrous, sometimes densely pubescent (coastal Washington); bracteole margins stipitate- or sessile-glandular. **Flowers** 15–17 mm diam.; hypanthium glabrous; sepals triangular, 4 mm, margins minutely and sparsely glandular-serrate, <glabrous>; stamens 20, anthers pink; styles 4 or 5. **Pomes** dull vinous purple becoming black mature, suborbicular to broadly ovoid, 8–10 mm diam., glabrous; sepals reflexed, apex obtuse; pyrenes 4 or 5, sides eroded. $2n$ = 34, 51, 68.

Flowering Apr–May; fruiting Sep–Oct. Brush, especially near groundwater; 10–400 m; B.C.; Alaska, Calif., Idaho, Mont., Oreg., Wash.

Crataegus gaylussacia occurs from north of San Francisco to coastal mainland British Columbia, Vancouver Island, the Queen Charlotte Islands, and coastal Alaska around Anchorage. The species ranges somewhat sporadically to Montana; it is the most northerly occurring *Crataegus* in North America, reaching 61° N.

Crataegus gaylussacia was an ignored name that has priority over *C. suksdorfii*. The plants have considerable variation in leaf shape, extremes of which are the type of *C. gaylussacia* from the southern edge of the species range, with 3–4 cm, more or less unlobed, elliptic to ovate leaf blades, and more or less acute teeth (not unlike syntype material of *C. suksdorfii*), and a contrasted northern Californian type with large (8 cm), deeply incised leaves. T. A. Dickinson and coworkers (pers. comm.) restrict *C. gaylussacia* to the autotriplod from California and Oregon, calling the remainder *C. suksdorfii*.

Crataegus gaylussacia differs from *C. douglasii* by stamen number. Also, the thorns are usually shorter and the fruit is usually smaller and more orbicular than in *C. douglasii*, as well as being of a different early color. In addition, *C. gaylussacia* usually has more or less camptodromous venation, unlike *C. douglasii*. Hybrids with the introduced European *C. monogyna* were first noticed by R. Love and M. Feigen (1979) in the Willamette Valley, Oregon, and are now locally common.

SELECTED REFERENCE Brunsfeld, S. J. and F. D. Johnston. 1990. Cytological, morphological, ecological and phenological support for specific status of *Crataegus suksdorfii* (Rosaceae). Madroño 37: 274–282.

19. Crataegus douglasii Lindley, Edwards's Bot. Reg. 21: plate 1810. 1835 • Douglas's or black hawthorn [E] [F]

Crataegus columbiana Howell; *C. tennowana* A. Nelson

Shrubs, 40–80 dm. **Stems:** 1-year old twigs deep tan to dark mahogany, ± shiny young, older gray to dark gray, often showing brown or alternatively tan-brown eastward; thorns on twigs straight or slightly recurved, deep brown young, (1.5–)2–3.5 cm. **Leaves:** petiole 0.7–1.5 cm, pubescent young, slightly glandular; blade usually elliptic to broadly elliptic or subrhombic (elliptic-obovate when lobes very small), 4–7 cm, lobes 2–4 per side, <sinuses shallow, LII 10–20%>, lobe apex subobtuse to acute, margins coarsely, sharply serrate, <teeth small, gland-tipped young>, venation craspedodromous, veins 4 or 5 per side, apex acute to obtuse, abaxial surface usually glabrous except on veins, adaxial densely short-appressed-pubescent. **Inflorescences** 10–25-flowered; branches glabrous; bracteole margins stipitate- or sessile-glandular. **Flowers** 10–15 mm diam.; hypanthium glabrous; sepals broadly triangular, 3–4 mm, margins sparsely glandular, <adaxially glabrous>; stamens 10, anthers pink; styles 3 or 4. **Pomes** vinous and usually waxy glaucous young, dull black mature, usually ellipsoid, 6–8 mm diam., glabrous; sepal remnants reflexed, apex obtuse; pyrenes 3 or 4, sides excavated, sometimes only shallowly. $2n$ = 68.

Flowering May–Jun; fruiting Sep–Oct. Brush, old fields, fencerows, woodland edges; 50–1600 m; Alta., B.C., Ont., Sask.; Alaska, Calif., Idaho, Mich., Minn., Mont., Oreg., Wash., Wis.

Crataegus douglasii occurs from southern and western British Columbia to the panhandle of Alaska and ranges to northern California and the Rocky Mountains of Idaho, Montana, and southwestern Alberta. Disjunct populations occur farther east in the Cypress Hills (Alberta and Saskatchewan), the Montana montane islands, and around the northern Great Lakes. In western North America, *C. douglasii* is found mainly near water in drier areas; in more mesic regions, it may occur in open woodlands. In the Great Lakes area, it mainly occurs in old pastures and on fencerows.

Crataegus douglasii is distinguished from *C. gaylussacia* by having ten stamens and is rather variable in leaf shape and size. The species is distinguished from *C. okennonii* by the latter usually being taller and typically more erect as well as having a usually straight, single trunk, shorter thorns, ampulliform-orbicular fruit, and often purplish crimson fall foliage. *Crataegus castlegarensis* is the most similar species, and beyond possessing relatively few to many multiple thorns, it is distinguished from *C. douglasii* by its hairy hypanthia, pedicels, and, often, fruit, which is orbicular, plump, crimson to purple in late August, ripening to deep purple (sometimes black) at a time when sympatric *C. douglasii* is fully black.

20. Crataegus castlegarensis J. B. Phipps & O'Kennon, Sida 20: 121, figs. 3, 4. 2002 • Castlegar hawthorn [E]

Shrubs, 25–50 dm. **Stems** <erect; branches spreading>; 1-year old twigs brown; thorns on twigs <usually branched, some paired or in triads>, straight to slightly recurved, dark brown with blackish tip young, 2–3 cm. **Leaves:** petiole 0.7–1.5 cm, pubescent, eglandular; blade oblanceolate to ovate-rhombic, 3.5–6 cm, lobes 3 or 4 per side, <sinuses shallow to deep>, lobe apex usually acute, margins serrate, <teeth apices finally glandular young>, venation craspedodromous, veins 4 or 5 per side, apex broadly triangular, abaxial surface sparsely hairy or glabrous except on veins, adaxial conspicuously appressed-pubescent young, glabrescent except on midvein. **Inflorescences** 12–20-flowered; branches sparsely to densely pubescent; bracteole margins stipitate-glandular. **Flowers** 12 mm diam.; hypanthium pubescent or glabrous; sepals triangular, 3 mm, margins remotely glandular-serrate; stamens 10, anthers pink; styles 3 or 4. **Pomes** crimson (mid Aug) turning to reddish plum or, ultimately, blackish purple, orbicular, ± oblate (recessed at junction with pedicel), 10 mm diam., sparsely pilose; sepals reflexed, apex obtuse; pyrenes 3 or 4, sides usually pitted. $2n = 68$.

Flowering May–Jun; fruiting Sep–Oct. Mesic brush; 300–1200 m; Alta., B.C., Sask.; Calif., Idaho, Mont., Oreg., Utah, Wash., Wyo.

Crataegus castlegarensis occurs from around the northern Okanagan, British Columbia, to the northern California Coast Ranges, to northwestern Montana, and the Rocky Mountains to northeast of Salt Lake City, Utah. The species occurs also in the Cypress Hills of Alberta and Saskatchewan; it is found in habitats similar to those of *C. douglasii* and is at least as abundant as that species in a number of parts of its range.

Crataegus castlegarensis is readily recognized by a combination of hairy inflorescence branches, pomes more or less orbicular, crimson or burgundy (with irregular earlier ripening) around the third week of August, soon becoming purple, often when nearby *C. douglasii* is already black, as well as a tendency to possess thorns on the young twigs branched at the base to become double, triple, or even sometimes quadruple. Such multiple thorns, though sometimes abundant on a bush, are more often few and may require searching for. Inflorescence pubescence, as in other species with this characteristic, may become sparse by fruiting. *Crataegus castlegarensis* is most similar to *C. douglasii*; its fruit is usually more orbicular, even oblately so, than is normal in that species, sometimes even with a recessed junction to the pedicel like an apple.

21. Crataegus okennonii J. B. Phipps, Sida 18: 170, figs. 2, 4a–c, f. 1998 • O'Kennon hawthorn [E]

Shrubs or trees, 30–100(–140) dm. **Stems** <usually single-trunked>; 1-year old twigs deep mahogany, shiny, older gray-brown; thorns on twigs <single>, slightly recurved, usually brown young, 2 cm. **Leaves:** (fall color plum to crimson or light crimson); petiole 1–2 cm, pubescence and glandularity not recorded; blade elliptic to broadly elliptic, sometimes ± narrowly ovate or rhombic, 4–6 cm, lobes 2 or 3 per side, lobe apex acute to ± obtuse, margins serrate, venation craspedodromous, veins 4 per side, apex subacute to obtuse, abaxial surface glabrous except on veins, adaxial finely appressed-pubescent. **Inflorescences** 12–20-flowered; branches glabrous, sometimes ± sparsely pubescent; bracteole margins stipitate-glandular. **Flowers** 15–20 mm diam.; hypanthium glabrous; sepals broadly triangular, 3 mm, margins obscurely glandular; stamens 10–12, anthers pale pink; styles 3 or 4. **Pomes** red-burgundy to chestnut or reddish brown (late Aug), deep purple to black at full maturity, ± ampulliform-orbicular, 8–10 mm diam., glabrous; sepal remnants reflexed, apex obtuse; pyrenes 3 or 4, sides ± eroded, sometimes plane. $2n = 68$.

Flowering May; fruiting Sep–Oct. Mesic thickets, natural hedgerows, scrub; 300–700 m; B.C.; Idaho, Mont., Wash.

Crataegus okennonii is common in southern interior British Columbia and the eastern half of northern Washington, extending to Idaho and northwestern Montana, usually along streams or other water sources.

Although *Crataegus okennonii* is treated as a microspecies of *C. douglasii* by T. A. Dickinson et al. (2008), the ease with which most specimens can be identified and its relatively wide range merits taxonomic recognition. The plants are often distinctively tall, usually with a straight trunk.

While *Crataegus okennonii* fits towards an extreme of *C. douglasii* on individual characters, for example, relatively short thorns, broad, shallowly incised leaves, and larger flowers, the ensemble is unique. Further, putatively later anthesis, crimson fall foliage, and the peculiar color of the subripe fruit, red, reddish brown, or deep tan, are unknown for *C. douglasii*.

22. Crataegus shuswapensis J. B. Phipps & O'Kennon, Sida 20: 128, fig. 5. 2002 [C] [E]

Shrubs, 55 dm. **Stems:** 1-year old twigs tan to deep brown; thorns on twigs straight or slightly recurved, very dark brown young, 2–3 cm. **Leaves:** petiole 1.5–2.2 cm, pubescent, sparsely gland-dotted; blade rhombic, 5–7 cm, lobes 4 per side, <sinuses deep>, lobe apex acute, margins remotely serrate, venation craspedodromous, veins 4 or 5 per side, apex acute, abaxial surface glabrous, adaxial sparsely appressed-pubescent. **Inflorescences** 12–18-flowered; branches glabrous; bracteole margins glandular. **Flowers** 15 mm diam.; hypanthium glabrous; sepals triangular, 4–5 mm, margins minutely glandular; stamens 10–18, anthers pale to bright pink or puce; styles 3 or 4. **Pomes** brick red young, deep purple mature, somewhat glaucous, suborbicular to ± broadly ellipsoid, 12 mm diam., glabrous; sepal remnants patent-recurved, apex acute; pyrenes (3 or)4, sides plane to ± concave, grooved, or irregularly scarred. $2n = 68$.

Flowering May; fruiting Sep–Oct. Thickets, natural hedgerows, shade of riverine *Populus*; of conservation concern; 300–500 m; B.C.

Crataegus shuswapensis is locally common near Enderby in the northern Okanagan and southwestern Shuswap drainage. Flowering plants from Lake County, Montana, are indistinguishable; fruiting plants from these populations have not been collected.

Crataegus shuswapensis is a distinctive member of ser. *Douglasianae* in the Okanagan because of its relatively long, stout thorns, large, wide, rhombic, and deeply and sharply incised leaf blades, pink, sometimes puce anthers, somewhat low style and pyrene numbers, and fairly large for the series, suborbicular to broadly ellipsoid purple fruit. Also, this species has a greater frequency of plane-sided pyrenes than has been noted in *C. douglasii*.

Crataegus shuswapensis is unusual in the genus for its unstable stamen number; in its other characteristics, it is more uniform. The majority of specimens that have been assigned to this species have about ten stamens; these are not dissimilar to an extreme form of *C. douglasii* that is locally common in northwest Montana and has been noted occasionally elsewhere. This latter variant of *C. shuswapensis* may be a single apomictic race.

64d. CRATAEGUS Linnaeus sect. SANGUINEAE (Zabel ex Rehder) C. K. Schneider, Ill. Handb. Laubholzk. 1: 771. 1906 [I]

Crataegus [unranked] *Sanguineae* Zabel ex Rehder in M. Vilmorin and D. G. J. M. Bois, Frutic. Vilmor., 111. 1904

Trees [or shrubs], 30–60[–80] dm. **Stems:** trunks 1–few, ± erect, bark not recorded; compound thorns on trunks not recorded; thorns on twigs determinate, ± straight and conic, stout. **Leaves:** blade ± ovate, [3–]5–9 cm, ± coriaceous, lobes 3–5 per side, sinuses shallow to deep (if deep, usually proximal 1 or 2 lobes only with deep sinuses, with veins to sinuses), veins 4 or 5[–12] per side, to lobes and deeper sinuses. **Inflorescences:** branches glabrous [or pubescent]; symmetric bracteoles present, basal bracteoles present, falcate, stipuliform. **Flowers:** post-mature petals pale paper brown; stamens 20, anthers ivory [pink to purple]. **Pomes** yellow or orange [red to reddish purple or black]; pyrene sides strongly pitted.

Species ca. 20 (1 in the flora): introduced, Washington; c, e Europe, Asia.

Section *Sanguineae* composes three series ranging continuously from eastern Russia across the whole of southern Siberia and Mongolia to the Pacific, then southward to Afghanistan and the mountains of Pakistan as well as most of eastern China. The section is largely absent from the dry interior tablelands and is disjunct in the Danube basin of central Europe, represented there by *Crataegus nigra* Waldstein & Kitaibel.

Some of the most diagnostic features of this section are the usually short, stout conic thorns, very dark, one- to two-year twigs, presence of basal falcate bracteoles, long filaments, and laterally erose pyrenes.

64d.1. CRATAEGUS Linnaeus (sect. SANGUINEAE) ser. ALTAICAE J. B. Phipps, J. Bot. Res. Inst. Texas 3: 243. 2009 [I]

Trees [or shrubs], 30–60 dm. **Stems:** 1-year old twigs deep red- or purple-brown, shining, older dull or dark gray. **Leaves** deciduous; petiole length $1/5$–$1/3$ blade; blade ± ovate, 5–9 cm, base truncate to broadly cuneate, lobes 3–5 per side, sinuses shallow to deep (if deep, usually only proximal 1 or 2), margins sharply serrate, venation craspedodromous, veins 4 or 5 per side, to lobes and deeper sinuses, apex acute. **Inflorescences** 20–50-flowered; <branches often warty>; bracteoles absent or present, caducous, linear to elliptic, membranous, margins glandular; basal branches subtended by persistent, chlorophyllous, falcate, stipuliform, glandular bracteoles. **Flowers** 12–15 mm diam.; sepals triangular-ovate, short, entire or 1- or 2-toothed; filament length usually 2 times petals; styles 4 or 5. **Pomes** yellow or orange [to red or reddish], suborbicular; sepals reflexed; pyrenes 4 or 5.

Species 2 or 3 (1 in the flora): introduced, Washington; c Asia.

Series *Altaicae* is closely related to ser. *Sanguineae*; its glabrous inflorescences, unusual fruit color, and distribution combine to distinguish it.

23. Crataegus wattiana Hemsley & Lace, J. Linn. Soc., Bot. 28: 323, plate 40. 1891 • Altai hawthorn [I]

Varieties 2 (1 in the flora): introduced, Washington; c Asia.

Crataegus wattiana is the only Asiatic hawthorn known to have escaped from cultivation in North America. The species is one of the relatively few usually yellowish-fruited hawthorns.

The depth of sinuses of the leaf blades in *Crataegus wattiana* is variable, ranging between three or four, more or less equally shallow to moderately deep lobes per side to deeply incised at the proximal two sinuses.

23a. Crataegus wattiana Hemsley & Lace var. wattiana [I]

Crataegus altaica (Loudon) Lange; *C. korolkowii* L. Henry; *C. purpurea* Bosc ex de Candolle var. *altaica* Loudon

Trees, 30–60 dm. **Stems:** 1-year old twigs deep red- or purple-brown, older dull or dark gray; thorns on twigs absent or sparse, stout, 1–4 cm. **Leaves** deciduous; petiole 2–3.5 cm, adaxially glabrous or sparsely pubescent; blade pale green abaxially, dull bluish green adaxially, ± ovate to triangular-ovate, 5–9 cm, base truncate to broadly cuneate, lobes 3–5 per side, evenly spaced, sinuses deep, margins sharply serrate, veins 4 or 5 per side, to lobes and deeper sinuses, apex acute, surfaces glabrous. **Inflorescences** 20–50 -flowered; proximal bracteoles falcate, stipuliform, 2 cm, herbaceous, distal caducous, linear, membranous, margins sparsely sessile-glandular. **Flowers** 12–15 mm diam.; hypanthium glabrous; sepals triangular-ovate, 2–4 mm, entire or 1- or 2-toothed; styles 4 or 5. **Pomes** yellow to bright orange or orange-brown, suborbicular, 8–12 mm diam.; sepals persistent, reflexed; pyrenes 4 or 5. $2n$ = 51, 68.

Flowering Apr–Jun; fruiting Sep–Oct. Scrub; 300–2500 m; introduced, Wash.; c Asia.

Variety *wattiana* has become naturalized in the Washington Park Arboretum, Seattle. The variety has leaves shallowly incised throughout; those of var. *incisa* C. K. Schneider are pinnatisect basally.

64e. Crataegus Linnaeus sect. Macracanthae Loudon, Arbor. Frutic. Brit. 2: 819. 1838 [E]

Shrubs or trees, 20–70(–80) dm. **Stems:** trunks 1–several, ± erect to oblique, bark flattened-scaly; compound thorns on trunks present or absent; thorns on twigs determinate, recurved, usually stout, sometimes ± slender. **Leaves:** blade ± ovate or rhombic to rhombic-elliptic or broadly elliptic, 3–8(–12) cm, coriaceous or thin, lobes 0 or 3–5(–8) per side, sinuses shallow, max LII usually 20%, veins (3–)5–8 per side, absent to sinuses. **Inflorescences:** branches usually pubescent; symmetric bracteoles present, basal bracteoles absent, falcate bracteoles absent. **Flowers:** post-mature petals pale paper brown; stamens 10 or 20, anthers red, pink, purple, sometimes white, ivory, or cream. **Pomes** orange to red; pyrenes dorsally grooved, sides ± pitted or eroded, sometimes plane.

Species 8 (8 in the flora): North America.

Section *Macracanthae* is similar to some species of sect. *Coccineae*; it differs in possessing more or less pitted pyrenes.

1. Pyrenes 2 or 3, (obtuse at ends), sides eroded; petioles usually eglandular . . . . 64e.1. *Crataegus* ser. *Macracanthae* (in part), p. 516
1. Pyrenes 2–5, (obtuse or acute at ends), sides ± smooth to erose or pitted; petioles usually glandular.
   2. Pyrene sides eroded, (always obtuse at ends); Great Plains . . . . 64e.1. *Crataegus* ser. *Macracanthae* (in part), p. 516
   2. Pyrene sides ± smooth to erose or pitted, (sometimes acute at ends); Great Lakes to New England . . . . 64e.2. *Crataegus* ser. *Anomalae*, p. 522

64e.1. Crataegus Linnaeus (sect. Macracanthae) ser. Macracanthae (Loudon) Rehder, Man. Cult. Trees ed. 2, 368. 1940 [E]

Crataegus sect. *Macracanthae* Loudon, Arbor. Frutic. Brit. 2: 819. 1838

Shrubs or trees, 20–70(–80) dm. **Stems:** 1–2-year old twigs usually very dark; thorns on twigs ± stout. **Leaves:** petiole glandular or eglandular; blade usually rhombic to rhombic-elliptic or broadly elliptic to ovate, lobes 0 or 3–5(–8) per side, lobe apex acute to obtuse, venation craspedodromous, veins (3–)6–8 per side, adaxial surface scabrous-pubescent young, glabrescent. **Inflorescences** (5–)12–30-flowered; bracteoles caducous, membranous, margins sessile- or stipitate-glandular. **Flowers** (12–)13–19(–22) mm diam.; sepal margins glandular-denticulate to glandular-laciniate, rarely nearly entire; stamens 7–10 or 20; styles 2 or 3(–5). **Pomes** orange to red, broadly ellipsoid to suborbicular, glabrous or hairy; pyrenes 2 or 3(–5), sides eroded.

Species 5 (5 in the flora): North America.

Three common core species comprise ser. *Macracanthae*: the similar *Crataegus macracantha* and *C. succulenta* and the somewhat dissimilar *C. calpodendron*. These are notable for their mid late to very late anthesis compared to sympatric hawthorns, stout thorns, and eglandular petioles. *Crataegus sheridana* and *C. rubribracteolata* are placed here for convenience, may not be closely related, and perhaps would better be placed within ser. *Anomalae*. Series *Macracanthae* is one of only two series of red-fruited hawthorns in North America that has eroded pyrenes.

Members of ser. *Macracanthae* form putative hybrids with members of ser. *Crus-galli* (*Crataegus persimilis*), ser. *Parvifoliae* (*C.* ×*vailiae*), ser. *Pruinosae* (*C. chadsfordiana* Sargent), ser. *Punctatae* (*C. florifera*), and ser. *Rotundifoliae* (*C. laurentiana*), at least. Some are treated here as interserial hybrid species numbers 154–169; *C. laurentiana*, *C. persimilis*, and *C.* ×*vailiae* key out in the eighth, first, and second couplets, respectively.

1. Leaves: adaxial surfaces glabrous or glabrate, glossy, lobes 0, or 3 or 4 per side . . . . 166. *Crataegus persimilis*, p. 641
1. Leaves: adaxial surfaces pubescent young, matte, lobes 0 or (1–)3–5(–8) per side.
   2. Inflorescences 2–8(–20)-flowered; sepals ± equal to petals, margins glandular-laciniate . . . . 169. *Crataegus* ×*vailiae*, p. 643
   2. Inflorescences (5–)12–30-flowered; sepals shorter than petals, margins usually glandular-denticulate to glandular-laciniate.
      3. Petioles eglandular, rarely glandular on distal wing.
         4. New growth pubescent; thorns on twigs absent or numerous; leaves (often yellowish green, without impressed venation), abaxial surfaces persistently appressed-pubescent; sepal margins nearly entire or shallowly glandular-serrate; (anthesis 1–3 weeks after sympatric species of this series); stamens 20 . . . . 24. *Crataegus calpodendron*
         4. New growth glabrous; thorns on twigs usually numerous; leaves (often blue-green, with impressed venation), abaxial surfaces glabrous (except the *pertomentosa* form of *C. macrantha*); sepal margins glandular-serrate to glandular-laciniate; (anthesis 1–3 weeks before *C. calpodendron* in same locality); stamens 10 or 20.
            5. Stamens 20; anthers 0.5–0.7 mm . . . . 25. *Crataegus succulenta*
            5. Stamens 10; anthers 0.9–1.2 mm . . . . 26. *Crataegus macracantha* (in part)
      3. Petioles ± sessile-glandular.
         6. Pyrene sides pitted or ± deeply concave.
            7. Leaf blade sinuses obscure to well-marked, LII 5–15%, lobe apices usually obtuse; pyrene sides pitted; (anthesis late among sympatric hawthorns); bracteoles in young inflorescences ± pallid; anthers white or pink; nw Montana . . . . 26. *Crataegus macracantha* (in part)
            7. Leaf blade sinuses moderately deep, max LII 25–33%, lobe apices acute to acuminate; pyrene sides ± deeply concave; (anthesis one of two earliest sympatric hawthorns); bracteoles in young inflorescences deep orange-red or fading; anthers ivory or cream; Cypress Hills, Alberta and Saskatchewan . . . . 27. *Crataegus rubribracteolata*
         6. Pyrene sides ± roughened with irregular pitting or diagonal scarring.
            8. Leaf blade lobe apices acute and sinuses shallow, LII 5–15%, or lobe apices subacute, sinuses deeper, LII 15–30%; anthers ivory; Great Plains. . . 28. *Crataegus sheridana*
            8. Leaf blade lobe apices acute and sinuses moderately deep, LII 20–33%; anthers cream or pale pink; Maine, Michigan, Minnesota, Wisconsin, to Quebec and Newfoundland . . . . 164. *Crataegus laurentiana*, p. 640

24. Crataegus calpodendron (Ehrhart) Medikus, Gesch. Bot., 83. 1793 • Late or pear hawthorn [E]

Mespilus calpodendron Ehrhart, Beitr. Naturk. 2: 67. 1788; *Crataegus calpodendron* var. *globosa* (Sargent) E. J. Palmer; *C. calpodendron* var. *hispidula* (Sargent) E. J. Palmer; *C. calpodendron* var. *microcarpa* (Chapman) E. J. Palmer; *C. calpodendron* var. *mollicula* (Sargent) E. J. Palmer; *C. globosa* Sargent; *C. hispidula* Sargent; *C. insperata* Sargent; *C. mollicula* Sargent; *C. mollita* Sargent; *C. obscura* Sargent; *C. scabera* Sargent; *C. spinulosa* Sargent; *C. tomentosa* Linnaeus var. *microcarpa* Chapman

Shrubs or trees, <usually multi-stemmed>, 50–70 dm. **Stems:** twigs: new growth golden brown, appressed-pubescent, 1-year old brown to grayish brown, 2-years old gray-brown, older ± pale gray; thorns on twigs absent or numerous, recurved (if longer), 1-year old shiny, dark brown, older turning gray, slender, 2.5–6 cm. **Leaves:** petiole length 25–30% blade, <narrowly winged most of length>, finely pubescent, eglandular; blade ± rhombic or rhombic-narrowly ovate, sometimes suborbiculate, 3–12 cm, thin, base cuneate, lobes 0 or 5–8 per side, sinuses shallow, lobe apex subacute to acute, margins serrate except in proximal 1/4, veins 8 per side, apex usually acute, abaxial surface persistently appressed-pubescent, especially on veins, adaxial scabrate young, glabrous mature. **Inflorescences** 12–25-flowered; branches densely appressed-pubescent; bracteoles linear, margins stipitate-glandular. **Flowers**

13–15(–22) mm diam.; hypanthium pubescent; sepals narrowly triangular, length not recorded, margins nearly entire or shallowly glandular-serrate, abaxially pubescent, adaxial pubescence not recorded; stamens 20, anthers usually red, pink, or purple, sometimes white, <small>; styles 2 or 3. **Pomes** orange-red to scarlet, ellipsoid to sometimes orbicular, (5–)7–9(–10) mm diam., pubescent; <flesh soft>; sepals reflexed or spreading; pyrenes 2 or 3, sides pitted. $2n = 34, 51$.

Flowering May–Jun; fruiting Sep–Oct. Open woods, brush, field margins; 10–800 m; Ont.; Ala., Ark., Ga., Ill., Ind., Iowa, Kans., Ky., Md., Mich., Minn., Miss., Mo., Nebr., N.Y., Ohio, Okla., Pa., Tenn., Va., W.Va., Wis.

Crataegus calpodendron is most common in the northern and Appalachian part of its range, where it is more likely to grow in the open (shrubs to 50 dm); in the southern part of its range, it is often in shaded habitats (trees to 70 dm). The distinctive vase-shaped branching pattern of well-grown northern *C. calpodendron* makes it easily distinguished even in winter, the lack of a central dominant trunk and the arching side trunks being characteristic. In open areas in the northern part of its range, mature leaves of *C. calpodendron* are often somewhat coriaceous and keeled along the midvein. In the south, the leaves remain thin until maturity. It is the second latest native hawthorn to flower, usually mid June (north) and early May (south, for example, central Alabama).

Vegetatively, distinction between *Crataegus calpodendron* and *C. succulenta* or *C. macracantha* is best supported by the more strongly impressed leaf venation, usually glabrous abaxial leaf surfaces, dark purple-brown, shiny, one-year old twigs, and often longer and stouter thorns in the latter.

Crataegus calpodendron is variable in branching pattern, indumentum, and leaf and fruit shapes. Examination of a large number of specimens does not reveal discontinuities between leaf shapes, neither do these shapes correlate with the varieties hitherto recognized primarily on fruit characteristics. In general, *C. calpodendron* in the narrow sense, *C. globosa*, *C. mollicula*, and *C. mollita* represent forms with narrower, rhombic-elliptic, less deeply incised leaves with somewhat blunter lobes and teeth; *C. hispidula*, *C. insperata*, and *C. obscura* represent forms with nearly isodiametric leaves, the deepest and sharpest sinuses, and the biggest and sharpest teeth; *C. scabera* and *C. spinulosa* are intermediate. Smaller leaf and pome sizes are more common in drier or hotter terrain.

Crataegus ×*rupicola* Sargent (synonym of *C. harveyana* Sargent), from Missouri, represents a form with glands at the base of the leaf and on the petiole and perhaps is not pure *calpodendron*. *Crataegus warneri* Sargent from east-central Texas is provisionally placed here, although it could be a hybrid or hybrid derivative of *C. calpodendron* × *C. crus-galli* or an odd form of *C. ashei* with ten stamens and red anthers. *Crataegus warneri* has broadly ovate leaves 3.5–5 cm and appears to be somewhat xeromorphic; it is distinguishable from other forms of *C. calpodendron* with similar leaves only by its ten stamens and lack of pitted pyrenes. *Crataegus* ×*simulata* Sargent, from Missouri, is a probable hybrid of *C. calpodendron* with *C. reverchonii* var. *palmeri* (as *C. palmeri*) according to E. J. Palmer (1952). *Crataegus* ×*vailiae* is the most convincing hybrid of *C. calpodendron* (with *C. uniflora*; see Interserial Hybrids, p. 643).

25. Crataegus succulenta Schrader ex Link, Handbuch 2: 78. 1831 • Succulent hawthorn, aubépine succulente [E]

Shrubs or trees, 40–80 dm. **Stems:** <older trunks usually bearing compound thorns>; twigs: new growth reddish green, glabrous, 1-year old dark, shiny red-brown, 2–3-years old becoming dark gray, older ± paler gray; thorns on twigs numerous, usually recurved, shiny, 1-year old dark blackish brown, stout, 3–6(–8) cm. **Leaves:** petiole 1–2 cm, <narrowly winged distally>, glabrous, eglandular; blade rhombic-elliptic to broadly rhombic-ovate or elliptic, 4–7 cm <widest near middle>, subcoriaceous mature (then often blue-green), base cuneate (constricted), lobes 3–5 per side, obscure to well-marked, sinuses shallow, lobe apex usually subacute to obtuse, margins serrate except proximally, veins 6–8 per side, <impressed>, apex acute to subacute, rarely obtuse, abaxial surface glabrous, adaxial scabrate-pubescent young. **Inflorescences** 15–30-flowered; branches pubescent or glabrous; bracteoles linear, <1.7 cm>, margins glandular. **Flowers** 12–17 mm diam.; hypanthium glabrous or pubescent; sepals narrowly triangular, 4–6 mm, margins glandular-serrate to glandular-laciniate, abaxially glabrous, adaxial pubescence not recorded; stamens 20, anthers usually red or pink, rarely white, <0.5–0.7 mm>; styles 2 or 3. **Pomes** bright or deep red, lustrous, suborbicular, (4–)7–10(–14) mm diam., glabrous, rarely pubescent; <flesh mealy or succulent mature>; sepals spreading-reflexed; pyrenes 2 or 3, sides pitted. $2n = 51$.

Varieties 5 (5 in the flora): North America.

Crataegus succulenta ranges through the southern Great Lakes area to the middle St. Lawrence and southern New England, to Minnesota, Iowa, Missouri (very rare), and Ohio, and the Appalachians to North Carolina. An outlier has recently been recognized in Manitoba.

The dark twig colors of *Crataegus succulenta* are dramatic in winter and the coral red expanding bud scales are conspicuous in spring, more so than in most other species of hawthorn except *C. macracantha*. In summer, its commonly bluish green leaves, eglandular petioles, and impressed venation combine with thorn and twig characteristics to make this and *C. macracantha* usually instantly recognizable.

Crataegus succulenta often forms suckering thickets in the north.

1. Leaf blades proportionately narrow (1.6–2:1), elliptic to rhombic-elliptic or narrowly ovate . . . . 25e. *Crataegus succulenta* var. *neofluvialis*
1. Leaf blades proportionately wide (1.3–1.6:1), ovate to broadly rhombic-ovate or broadly elliptic.
   2. Pomes 4–6 mm diam.. . . . . 25d. *Crataegus succulenta* var. *pisifera*
   2. Pomes 7–14 mm diam.
      3. Pomes 12–14 mm diam. . . . . .25c. *Crataegus succulenta* var. *gemmosa*
      3. Pomes 7–10 mm diam.
         4. Anthers red or pink . . . . .25a. *Crataegus succulenta* var. *succulenta*
         4. Anthers white . . . . .25b. *Crataegus succulenta* var. *michiganensis*

25a. Crataegus succulenta Schrader ex Link var. **succulenta** [E]

Crataegus membranacea Sargent; *C. vaga* Sargent

Leaves: blade ovate to broadly elliptic, proportionately wide (1.3–1.6:1). **Inflorescence branches** pubescent. **Anthers** red or pink. **Pomes** 7–10 mm diam.

Flowering Apr–Jun; fruiting Sep–Oct. Brush, field margins, open woodlands; 20–800 m; Man., Ont., Que.; Conn., Ill., Ind., Iowa, Maine, Md., Mass., Mich., Minn., Mo., N.H., N.Y., N.C., Ohio, Pa., R.I., Tenn., Vt., Va., W.Va., Wis.

Variety *succulenta* is found throughout the range of the species; it is less common southwards.

25b. Crataegus succulenta Schrader ex Link var. **michiganensis** (Ashe) E. J. Palmer, Brittonia 5: 489. 1946 [E]

Crataegus michiganensis Ashe, Bull. North Carolina Agric. Exp. Sta. 175: 111. 1900

Leaves: blade ovate to broadly elliptic, proportionately wide (1.3–1.6:1). **Inflorescence branches** pubescent. **Anthers** white. **Pomes** 7–10 mm diam.

Flowering May; fruiting Sep–Oct. Brush, field margins, open woodlands; 200–300 m; Ont.; Mich.

Variety *michiganensis* resembles var. *succulenta* but has white anthers. The similar but glabrate *Crataegus tanuphylla* Sargent was rediscovered in the 1990s in central Pennsylvania. *Crataegus laxiflora* Sargent [*C. succulenta* var. *laxiflora* (Sargent) Kruschke] is similar to var. *michiganensis* but has unusually lax inflorescences.

25c. Crataegus succulenta Schrader ex Link var. **gemmosa** (Sargent) Kruschke, Publ. Bot. Milwaukee Public Mus. 3: 159. 1965 [E]

Crataegus gemmosa Sargent, Bot. Gaz. 33: 119. 1902

Leaves: blade narrowly ovate to broadly elliptic, proportionately wide (1.3–1.5:1). **Inflorescence branches** pubescent. **Anthers** red or pink. **Pomes** 12–14 mm diam.

Flowering May; fruiting Sep–Oct. Brush, field margins, open woodlands; 100–300 m; Mich.

Variety *gemmosa* resembles var. *succulenta* except for larger fruit size.

25d. Crataegus succulenta Schrader ex Link var. **pisifera** (Sargent) Kruschke, Publ. Bot. Milwaukee Public Mus. 3: 159. 1965 [E]

Crataegus pisifera Sargent, Rhodora 7: 163. 1905

Leaves: blade ovate to broadly elliptic, proportionately wide (1.3–1.6:1). **Inflorescence branches** glabrous. **Anthers** red or pink. **Pomes** 4–6 mm diam.

Flowering May–Jun; fruiting Sep–Oct. Brush, field margins, open woodlands; 100–300 m; Vt.

Variety *pisifera* is local and scarce. The variety is remarkable for its very small fruit and glabrous inflorescences.

25e. Crataegus succulenta Schrader ex Link var. **neofluvialis** (Ashe) E. J. Palmer in E. J. Dole, Fl. Vermont ed. 3, 155. 1937 [E]

Crataegus neofluvialis Ashe, J. Elisha Mitchell Sci. Soc. 16: 71. 1900 (as neo-fluvialis)

Leaves: blade elliptic to rhombic-elliptic or narrowly ovate, proportionately narrow (1.6–2:1). **Inflorescence branches** glabrate. **Anthers** red or pink. **Pomes** 7–10 mm diam.

Flowering Apr–May; fruiting Sep–Oct. Brush, field margins, open woodlands; 100–1600 m; Mo., N.C., Vt.

Variety *neofluvialis* is sporadic in occurrence. It is an extreme form of *Crataegus succulenta* with regard to leaf proportions.

26. Crataegus macracantha Loddiges ex Loudon, Arbor. Frutic. Brit. 2: 819, figs. 572, 573. 1838 • Large-thorn hawthorn, aubépine à épines longues [E] [F]

Crataegus divida Sargent; *C. laneyi* Sargent; *C. macracantha* var. *occidentalis* (Britton) Eggleston; *C. macracantha* var. *pertomentosa* (Ashe) Kruschke; *C. occidentalis* Britton; *C. pertomentosa* Ashe; *C. succulenta* Schrader ex Link var. *macracantha* (Loddiges ex Loudon) Eggleston; *C. succulenta* var. *occidentalis* (Britton) E. J. Palmer; *C. succulenta* var. *pertomentosa* (Ashe) E. J. Palmer

Shrubs, 40–60(–80) dm. **Stems:** <older trunks usually bearing compound thorns>; twigs: new growth reddish green, usually glabrous, 1-year old dark, shiny reddish brown, 2–3-years old becoming dark gray, older ± paler gray; thorns on twigs usually numerous, recurved, shiny, dark blackish brown, older gray, stout, 3–7(–11) cm. **Leaves:** petiole 1–2 cm, <narrowly winged distally>, glabrous, eglandular, rarely glandular (on distal wing); blade narrowly rhombic-elliptic to broadly rhombic-ovate (sometimes narrowly obovate), 4–7(–10) cm <widest near middle>, subcoriaceous mature (then often blue-green), base cuneate (constricted into winged distal part of petioles), lobes 3–5 per side, sinuses obscure to well-marked, LII 5–15%, lobe apex usually obtuse, margins serrate except near bases, veins 6–8 per side, <impressed>, apex acute to obtuse, abaxial surface glabrous (except in var. *pertomentosa*), adaxial scabrous-pubescent young. **Inflorescences** 15–30 -flowered; branches pubescent or glabrous; bracteoles <± pallid>, linear, margins glandular, sometimes long-stipitate-glandular. **Flowers** 13–19 mm diam.; hypanthium glabrous or sparsely to densely pubescent; sepals narrowly triangular, 4–6 mm, margins glandular-serrate to glandular-laciniate, abaxial pubescence not recorded, adaxially glabrous; stamens 10, anthers white or pink, <0.9–1.2 mm>; styles 2 or 3. **Pomes:** brilliant scarlet (west) to full crimson (east), suborbicular, 7–12 mm diam., <succulent mature>, glabrous or pubescent; sepals spreading-reflexed; pyrenes 2 or 3, sides pitted. $2n$ = 68.

Flowering Apr–Jun; fruiting Sep–Oct. Brush; 20–1600 m; Alta., B.C., Man., N.B., Ont., Que., Sask.; Ariz., Colo., Conn., Ill., Ind., Iowa, Kans., Md., Mass., Mich., Minn., Mo., Mont., Nebr., N.H., N.Mex., N.Y., N.Dak., Ohio, Oreg., Pa., R.I., S.Dak., Utah, Vt., Va., Wash., W.Va., Wis., Wyo.

Crataegus macracantha is the most widespread North American hawthorn and one of the more variable. Variation is particularly great in Quebec and Montana. Attempts to organize the variation into named taxa have hitherto run into the problems that the type is an illustration of inadequate quality and that authors have been insufficiently aware of the total variation. Two common eastern forms are similar, white-anthered and pink-anthered with abaxially glabrous, broadly elliptic-ovate leaves, shallowly and sharply lobed, and fruit ripening directly from green to red; these have the longest thorns in the species, and both have been treated as if typical.

Crataegus occidentalis, a white-anthered, inter-montane form with leaves similar to the foregoing usually has shorter thorns and fruit ripening to red through a characteristic orange or salmon stage. Pink-anthered *C. pertomentosa*, with leaves of a similar shape but abaxially pubescent to tomentose, grows on the Great Plains. The last two forms have been treated as varieties. Other striking variants have brilliant red bracteoles or strongly convex leaves (both from Montana) and entire-margined sepals (*C. integriloba* Sargent from Quebec). Some rare forms have glabrous inflorescences. Some populations in northwestern Montana, otherwise much like a common pink-anthered sympatric form, have glandular petioles and may represent introgression from another species.

Crataegus macracantha often forms suckering thickets in the east.

27. Crataegus rubribracteolata J. B. Phipps & O'Kennon, J. Bot. Res. Inst. Texas 1: 1073, plates 6.1c, 7.3c, fig. 20. 2007 • Red bracteole hawthorn [E]

Shrubs, <erect, multi-stemmed in open, wide branching in shade>, 20–30 dm. **Stems:** <expanding bud-scales bright rufous-red>; twigs: new growth color not recorded, appressed-pubescent, 1-year old mid brown, 2-years old dark red-brown, older gray or dark gray; thorns on twigs usually numerous, straight to slightly curved, 1-year old dark blackish brown, ± stout, 3.5–5.5 cm. **Leaves:** petiole length 30–35% blade, densely pubescent in adaxial sulcus, ± sessile-glandular; blade ± ovate to ovate-trullate, 4–7 cm, ± thin, base cuneate to broadly cuneate, lobes (4 or)5 or 6 per side, sinuses moderately deep, max LII 25–33%, lobe apex acute to acuminate, margins with regular, acute teeth, often subacute mature, veins (5 or)6 or 7 per side, apex acute to acuminate, abaxial surface glabrous except for sparse hairs along midvein proximally, adaxial ± densely scabrous-pubescent young (variably persistent to maturity). **Inflorescences** (6–)8–12-flowered; branches ± tomentose; bracteoles <abundant, deep orange-red or

fading>, narrow, margins densely sessile-glandular. **Flowers** 12–15 mm diam.; hypanthium ± tomentose or densely pubescent; sepals <greenish or pale green>, triangular, 3–5 mm, margins glandular-denticulate, <apex ± red or paler red>, abaxially glabrous, adaxially pubescent; stamens 10, anthers ivory or cream; styles 3 or 4. **Pomes** <3–10 in ± compact infructescences>, red young, bright red or orange-scarlet in shade mature, ± ellipsoid to suborbicular, 9–13 mm tall, densely to sometimes sparsely pubescent; sepals appressed, <3–5 mm>; pyrenes 3 or 4(or 5), sides ± deeply concave.

Flowering late May; fruiting late Aug–Sep. Thickets, sometimes under aspen; 700–1100 m; Alta., Sask.

Crataegus rubribracteolata is restricted to the Cypress Hills and is common in suitable habitats. The record from Montana (Bear's Paw Mountains, Hill County) is now considered dubious.

Crataegus rubribracteolata can be recognized by its upright habit when grown in full light. Surprisingly, it is most easily confused with 101. *C. ursopedensis* (see key characters), not with *C. sheridana*. To *C. ursopedensis*, *C. rubrobracteolata* is similar in leaf shape, inflorescence, and indumentum; the mature leaves of *C. rubribracteolata* average smaller, except in some shade forms; both inflorescence form and indumentum are denser. It lacks the strikingly flabellate young leaves of *C. ursopedensis*, flowers significantly earlier, and has smaller flowers, as well as different colored bracteoles. The two species are easily confused when these characteristics are not available; in fruit, the technical characteristics of pyrene and sepal orientation need careful observation.

28. Crataegus sheridana A. Nelson, Bot. Gaz. 34: 370. 1902 • Great Plains hawthorn [E]

Shrubs, 20–35 dm. **Stems:** twigs: new growth color not recorded, pubescent, 1-year old light brown to olive-tan or mid brown, 2-years old dark gray or reddish brown overlaid dark gray, older gray; thorns on twigs usually numerous, usually recurved, 2-years old very dark, shiny red-brown, slender to moderately thick, (2.5–)3–5(–7) cm. **Leaves:** petiole length 40–50% blade, pubescent (especially in adaxial sulcus), ± densely sessile-glandular; blade sometimes broadly ovate or suborbiculate, sometimes ± ovate, elliptic-ovate to ovate, 3.5–5(–6) cm, <nearly full grown at flowering>, ± thin, base cuneate, broadly cuneate, or subtruncate to subcordate, lobes (3 or)4(or 5) per side, sinuses shallow, LII 5–15%, lobe apex acute (except where blades elliptic-ovate to ovate, then lobes deeper, LII 15–30%, lobe apex subacute), margins with subacute teeth, veins 3 or 4 per side, apex acute to obtuse, abaxial surface glabrous, veins glabrous or pubescent young, adaxial densely appressed-scabrous-pubescent young (variably persistent). **Inflorescences** 5–15-flowered; branches ± densely pubescent; bracteoles <pallid>, narrow, margins sessile-glandular. **Flowers** 14–16 mm diam.; hypanthium proximally ± tomentose, medially densely short-pubescent; sepals <greenish>, triangular, length not recorded, margins glandular-serrate, abaxially glabrous, adaxially pubescent; stamens 10, anthers ivory; styles 4 or 5. **Pomes** <1–6 in infructescences>, bright red, broadly ellipsoid to suborbicular, 9–11 mm tall, pubescent, rarely glabrate; sepals ± spreading to reflexed; pyrenes 3 or 4, sides ± roughened with shallow, irregular pitting or diagonal scarring.

Flowering May–Jun; fruiting Sep. Thickets, brush; 300–1200 m; Alta., Sask.; Minn., Mont., N.Dak., S.Dak., Wyo.

Crataegus sheridana occurs only east of the Rocky Mountains and is abundant in the Cypress Hills of Alberta and Saskatchewan. Collections from the montane islands of Montana are also known. Until confusion with the superficially similar *C. chrysocarpa* of the Great Plains and Prairies is cleared up in all the relevant herbaria, there will be an incomplete knowledge of the distribution of *C. sheridana*; however, it is widespread in the Great Plains region. Its plants are usually taller than sympatric *C. chrysocarpa* and also *C. macracantha*.

Crataegus sheridana has superficial resemblances to the *C. chrysocarpa* group; it is distinguished by its pitted pyrenes and usually larger leaves. In the Cypress Hills, where this species has the shortest thorns of any hawthorn in the red-fruited group, it is further distinguished from most specimens of the *C. chrysocarpa* group by the densely hairy to tomentose inflorescence branches and hypanthia and by its leaf shape. Collections from the Cypress Hills indicate that form of *C. sheridana* to be peculiarly round-leaved with pallid bracteoles; Montana forms have longer thorns, longer, more acutely lobed leaves, more elliptic-ovate to ovate blades, and red-brown rather than pallid bracteoles.

In all likelihood, *Crataegus sheridana* represents a series of hybrid descendants between *C. chrysocarpa* and *C. macracantha*. The reasons for ignoring its existence over a long period were discussed by J. B. Phipps et al. (2007).

64e.2. CRATAEGUS Linnaeus (sect. MACRACANTHAE) ser. ANOMALAE (Sargent ex Eggleston) Rehder, Man. Cult. Trees ed. 2, 367. 1940 [E]

Crataegus sect. *Anomalae* Sargent ex Eggleston in B. L. Robinson and M. L. Fernald, Manual ed. 7, 477. 1908; *Crataegus* ser. *Brainerdianae* Eggleston ex E. J. Palmer

Shrubs or trees, 30–70 dm. **Stems:** 2-year old twigs <shiny>, dark or reddish brown; thorns on twigs stout to ± slender. **Leaves:** petiole eglandular or glandular; blade ± ovate, coriaceous to thin, lobes 3–5 per side, lobe apex ± acute, venation craspedodromous, veins 5 or 6(–8) per side, adaxial surface glabrous or appressed-pubescent. **Inflorescences** 6–12-flowered; bracteoles caducous, membranous, margins sessile-glandular. **Flowers** 15–20 mm diam.; sepal margins subentire or glandular-serrate; stamens 7 or 8(–10) or 20; styles 2–4(or 5). **Pomes** reddish or orange-red, usually suborbicular, sometimes ± oblong-orbicular, glabrous or hairy; pyrenes 2–4, sides ± plane to erose or pitted.

Species 3 (3 in the flora): North America.

Members of ser. *Anomalae* are locally common hawthorns found from west of Lake Michigan to the Atlantic in Massachusetts and New Hampshire. Some members of the series are plausibly of hybrid origin, and if so, a member of ser. *Macracanthae* is always one parent, the other perhaps being either ser. *Punctatae* (*Crataegus florifera*) or ser. *Tenuifoliae* (most taxa, for example, *C. oreophila* Lance). *Crataegus* ×*vailiae*, usually treated with ser. *Macracanthae*, may also key here. Other (very rare or even extinct) putative hybrids that might be referred to ser. *Anomalae* include: *C. chadsfordiana* Sargent and *C. spatiosa* Sargent (ser. *Pruinosae* × ser. *Macracanthae*), *C. desueta* Sargent (*C. punctata* × ser. *Anomalae*), *C. dunbarii* Sargent and *C. ideae* Sargent (ser. *Anomalae* × ser. *Rotundifoliae*), *C. harryi* Sargent and *C. shirleyensis* Sargent (ser. *Anomalae* × ser. *Punctatae*), *C. kennedyi* Sargent (*C. jesupii* × ser. *Anomalae*), *C. randiana* Sargent and *C. rubrocarnea* Sargent (ser. *Anomalae* × *C. macrosperma*), and *C. websteri* Sargent (ser. *Anomalae* × ser. *Macracanthae*), without precise parentage suggested [putative parents as suggested by E. J. Palmer (1952) are given parenthetically]. The very local and perhaps extinct *C. kingstonensis* Sargent is perhaps a hybrid of ser. *Anomalae* × *C. pruinosa*.

The interserial hybrids *Crataegus* ×*bicknellii* and *C.* ×*vailiae* key out in the second couplet; *C.* ×*coleae* keys out in the fifth couplet.

1. Sepal margins glandular-laciniate.
    2. Twigs: new growths and inflorescence branches glabrous; stamens 10 . . . . 155. *Crataegus* ×*bicknellii*, p. 636
    2. Twigs: new growths and inflorescence branches ± densely pubescent; stamens 20. . . . . 169. *Crataegus* ×*vailiae*, p. 643
1. Sepal margins subentire or glandular-serrate.
    3. Stamens 7 or 8(–10) . . . . 31. *Crataegus scabrida*
    3. Stamens 20.
        4. Inflorescence branches villous; adaxial leaf surfaces pubescent (young) . . . . . . 29. *Crataegus florifera*
        4. Inflorescence branches glabrous; adaxial leaf surfaces glabrous, glabrescent, or glabrate.
            5. Fruiting sepals ± sessile. . . . . 30. *Crataegus brainerdii*
            5. Fruiting sepals elevated on collar . . . . 156. *Crataegus* ×*coleae*, p. 636

29. Crataegus florifera Sargent, Ontario Nat. Sci. Bull. 4: 14. 1908 [E]

Crataegus celsa Sargent

Shrubs or trees, 70 dm. **Stems:** twigs: new growth ± reddish young, soon glabrescent, 1-year old glossy red-brown, 2-years old duller, older dull gray; thorns on twigs at 1-year old shiny, black, stout to ± slender, 4–6 cm. **Leaves:** petiole length 30–70% blade, slightly pubescent young, glabrescent, eglandular; blade suborbiculate or broadly ovate to obovate, 4–7 cm, base narrowly to broadly cuneate or rounded, lobes 3 or 4 per side, <obscure to well-marked>, margins serrulate, veins 5 or 6 per side, apex acute, abaxial surface sparsely hairy, densely pubescent along veins, adaxial pubescent (late persistent). **Inflorescences** 6–10-flowered; branches villous; <bracteoles: anthesis late mid season>. **Flowers** 17–20 mm diam.; hypanthium sparsely pilose; sepals narrowly triangular, margins glandular-serrate; stamens 20, anthers rose to dark red; styles 2–4. **Pomes** deep red, suborbicular, 8–13 mm diam., glabrous; sepals erose or reflexed; pyrenes 2–4, sides plane to erose. $2n = 68$.

Flowering May; fruiting Sep–Oct. Successional fields, pastures with woody invaders, fencerows; 100–300 m; Ont.; N.Y.

Crataegus florifera is frequent in the western part of southern Ontario.

There appear to be two main entities here: typical *florifera* (= *Crataegus celsa*), plausibly of hybrid origin between *C. punctata* and *C. succulenta* (E. J. Palmer 1952), with narrower, somewhat obovate leaves and petiole length 30–40% blade, and a *sylvestris* type, named for the later homonym *C. sylvestris* [Anonymous], with much broader leaves and petiole length often 50–70% blade.

30. Crataegus brainerdii Sargent, Rhodora 3: 27. 1901 (as brainerdi) • Brainerd hawthorn, aubépine de Brainerd [E]

Shrubs or trees, 40–50 dm. **Stems:** twigs: new growth green, glabrous, 1–2-years old shiny, dark brown, older dark gray; thorns on twigs at 1-year old very dark, shiny, stout, 3–4.5 cm. **Leaves:** petiole length 40% blade, glabrous, glandular; blade ovate to narrowly ovate, 4–7 cm, base cuneate to rounded, lobes 3 or 4 per side, margins serrate, teeth numerous, sharp, fine, gland-tipped, especially near base, veins 5 or 6 per side (even young), apex acute, abaxial surface glabrous except for tufts in vein axils, adaxial glabrate. **Inflorescences** 7–12 -flowered; branches glabrous; <bracteoles absent or few at anthesis>. **Flowers** 15–18 mm diam.; hypanthium glabrous; sepals narrowly triangular, margins glandular-serrate; stamens 20, anthers pink to red; styles 2 or 3(–5). **Pomes** reddish, ± oblong-orbicular, 8–9 mm diam., glabrous; sepals ± sessile; pyrenes 2 or 3, sides erose.

Flowering May–Jun; fruiting Sep–Oct. Successional fields, brush; 10–300 m; Ont., Que.; Maine, Mass., Mich., N.H., N.Y., Vt., Wis.

Crataegus brainerdii has a scattered distribution and seems nowhere common. E. J. Palmer (1950) recorded it from North Carolina but this is well out of range. This species is similar in leaf shape to *C. scabrida* var. *scabrida*; it differs in stamen number.

31. Crataegus scabrida Sargent, Rhodora 3: 29. 1901 • Rough hawthorn, aubépine scabride [E]

Crataegus brainerdii Sargent var. *scabrida* (Sargent) Eggleston

Shrubs, 30–50 dm. **Stems:** twigs: new growth reddish early, glabrous, 1-year old deep tan, 2-years old tan to deep reddish brown, older dull gray; thorns on twigs at 1-year old shiny, dark brown to nearly black, stout or slender, 4–6 cm. **Leaves:** petiole <slender>, length 40–60% blade, sparsely pubescent or glabrous, eglandular or sessile-glandular; blade short-ovate to ovate, broadly elliptic to narrowly rhombic, or ± trullate, 3–8 cm, base rounded to cuneate, lobes 3–5 per side, margins serrate, veins 5–7(or 8) per side, apex acute, abaxial surface glabrous, adaxial sparsely to densely appressed-pubescent young, glabrescent. **Inflorescences** 6–10 -flowered; branches glabrous or villous. **Flowers** 15–18 mm diam.; hypanthium glabrous or pubescent; sepals narrowly triangular, margins subentire or glandular-serrate; stamens 7 or 8(–10), anthers pink; styles 3 or 4. **Pomes** reddish or orange-red, suborbicular, 6–9 mm diam., <pruinose>, glabrous or slightly pubescent; sepals spreading to ± erose; pyrenes 2–4, sides often pitted. $2n = 51$.

Varieties 4 (4 in the flora): nc, ne North America.

Crataegus scabrida is variable in leaf shape and size, as well as in pubescence of adaxial leaf surfaces and pedicels. The species falls into four reasonably well-marked varieties, only the first two of which are frequent. The range of the species as a whole is somewhat poorly documented because of frequent incorrect identification.

1. Inflorescence branches glabrous.
   2. Leaf blades ovate to narrowly rhombic, 4–7 cm . . . . 31a. *Crataegus scabrida* var. *scabrida*
   2. Leaf blades broadly elliptic, 3–5 cm. . . . . .31d. *Crataegus scabrida* var. *cyclophylla*
1. Inflorescence branches villous.
   3. Leaf blades trullate or broadly trullate, 5–8 cm . . . . 31b. *Crataegus scabrida* var. *asperifolia*
   3. Leaf blades short-ovate to broadly elliptic, 4 cm. . . . . . . 31c. *Crataegus scabrida* var. *egglestonii*

31a. Crataegus scabrida Sargent var. **scabrida** E

Crataegus improvisa Sargent

Leaf blades ovate, 4–7 cm, adaxial surface variably pubescent young. **Inflorescence branches** glabrous. **Flowers:** hypanthium glabrous.

Flowering May–Jun; fruiting Sep–Oct. Brush, roadsides, woodland edges; 50–400 m; N.B., N.S., Ont., Que.; Maine, Mass., Mich., N.Y., Vt., Wis.

Variety *scabrida* is widely spread throughout the range of *Crataegus scabrida* from Wisconsin to New England and Nova Scotia; it is nearly identical to *C. brainerdii* except for stamen number. *Crataegus improvisa* is a form with narrowly rhombic leaves.

31b. Crataegus scabrida Sargent var. **asperifolia** (Sargent) Kruschke, Publ. Bot. Milwaukee Public Mus. 3: 170. 1965 E

Crataegus asperifolia Sargent, Rhodora 3: 31. 1901; *C. brainerdii* Sargent var. *asperifolia* (Sargent) Eggleston

Leaf blades trullate or broadly trullate, 5–8 cm, adaxial surface densely scabrous-pubescent, glabrescent. **Inflorescence branches** densely villous, variably glabrescent by anthesis. **Flowers:** hypanthium pubescent.

Flowering May–Jun; fruiting Sep–Oct. Brush, roadsides, woodland edges; 50–900 m; Ont.; Mass., Mich., N.H., N.Y., Vt., Va., Wis.

Variety *asperifolia*, like var. *scabrida*, has a wide range from the southern Great Lakes area to the Virginia Blue Ridge. It is particularly distinct, thus perhaps worthy of specific rank, and may be locally common.

31c. Crataegus scabrida Sargent var. **egglestonii** (Sargent) Kruschke, Publ. Bot. Milwaukee Public Mus. 3: 170. 1965 E

Crataegus egglestonii Sargent, Rhodora 3: 30. 1901 (as egglestoni); *C. brainerdii* Sargent var. *egglestonii* (Sargent) B. L. Robinson; *C. honesta* Sargent; *C. scabrida* var. *honesta* (Sargent) Kruschke

Leaf blades short-ovate to broadly elliptic, 4 cm, adaxial surface sparsely to densely appressed-pubescent young. **Inflorescence branches** sparsely to densely villous. **Flowers:** hypanthium pubescent.

Flowering May–Jun; fruiting Sep–Oct. Brush, roadsides, woodland edges; 30–200 m; Que.; Mass., Mich., N.H., N.Y., Vt.

Variety *egglestonii* is nearly restricted to New England, adjacent New York, and the St. Lawrence valley in Quebec. E. J. Palmer (1950) recorded the variety from Nova Scotia and, if so, it probably also occurs in Maine; it appears to be scarce.

31d. Crataegus scabrida Sargent var. **cyclophylla** (Sargent) Kruschke, Publ. Bot. Milwaukee Public Mus. 3: 170. 1965 C E

Crataegus cyclophylla Sargent, Rhodora 7: 210. 1905; *C. brainerdii* Sargent var. *cyclophylla* (Sargent) E. J. Palmer

Leaf blades broadly elliptic, 3–5 cm, adaxial surface sparsely to densely appressed-pubescent young, glabrescent. **Inflorescence branches** glabrous. **Flowers:** hypanthium glabrous.

Flowering May–Jun; fruiting Sep–Oct. Brush, roadsides, woodland edges; of conservation concern; 20–200 m; Ont.; Conn., Mass., Vt.

Variety *cyclophylla* grows in southern New England and eastern Ontario.

64f. Crataegus Linnaeus sect. Coccineae Loudon, Arbor. Frutic. Brit. 2: 816. 1838

Shrubs or trees, (5–)30–110(–150) dm. **Stems:** trunks 1–several, ± erect to oblique, bark usually flattened-scaly, sometimes corrugated or thin-exfoliating; compound thorns on trunks present or absent; thorns on twigs determinate, straight to recurved, slender to stout. **Leaves:** blade broadly lanceolate, elliptic, spatulate, obovate, obtrullate, rhombic-ovate, broadly ovate to suborbiculate, 1–8(–12) cm, <length 1.6–2 times width>, coriaceous to thin, lobes 0 or 1–4 or 5(–9), sinuses usually shallow, sometimes deep, veins 1–8(–10) per side, absent to sinuses, <matte>. **Inflorescences:** branches glabrous or pubescent; symmetric bracteoles present, basal stipuliform, falcate bracteoles absent. **Flowers:** post-mature petals pale paper brown; stamens (5–)10 or 20(–47), rarely ca. 15, anthers pink to purple or white to cream. **Pomes** usually red, sometimes yellow, orange, or pink mauve, sometimes remaining green until late, or green-blotched, <sometimes ± conspicuously pruinose>; pyrene sides plane (to shallowly concave in *C. ursopedensis*).

Species 128 (119 in the flora): North America, Mexico; introduced in Europe.

Section *Coccineae* includes 20 series and constitutes the main North American hawthorn diversification which comprises a single clade when sect. *Macracanthae* is included (E. Y. Y. Lo et al. 2009). Variation in habit, leaf shape, glandularity, and other characters permits recognition of more or less distinct series; the delimitation of some series may warrant further attention. Most groups with leaf lobes absent, for example, ser. *Aestivales*, *Crus-galli*, *Punctatae*, and *Madrenses*, are found here, as are series that tolerate particularly extreme conditions for *Crataegus*, for example, ser. *Aestivales* (southeastern wetlands) and ser. *Lacrimatae* (southeastern xeromorphs).

Series *Anomalae* and ser. *Macracanthae* are often included in sect. *Coccineae*; here they are placed in sect. *Macracanthae* because their pyrenes are pitted.

Series *Microcarpae* (sect. *Crataegus*) is included at couplet 44 below to handle common forms that key out there with little lobing and, thus, no veins to sinuses. The interserial hybrids *Crataegus persimilis* and *C.* ×*sicca* key out at couplets 16 and 20, respectively; *C. turnerorum* keys out in part at couplets 18 and 31.

SELECTED REFERENCE Phipps, J. B. et al. 2006. *Crataegus* series *Bracteatae* and *Triflorae* (Rosaceae). Sida 22: 1009–1025.

1. Inflorescences 2–6-flowered (± umbellate); flowering before or with young leaves, anthesis very early, before other sympatric hawthorns; pomes ripe May–late June; se United States, wetlands . . . . 64f.4. *Crataegus* ser. *Aestivales*, p. 546
1. Inflorescences usually 7–25-flowered (if 1–6-flowered, not ± umbellate); not flowering before or with young leaves, anthesis later; pomes ripe August–October; mostly in toplands, except few se United States species.
    2. Bracteoles, at least larger, usually ± herbaceous, usually ± persistent, others sometimes membranous but sometimes caducous; flowers 18–30 mm diam.
        3. Leaves and petioles eglandular or sparsely glandular.
            4. Pomes orange-red, bright or deep red, or bright yellow, not pruinose; flesh mealy; fruiting sepals erect or erect-patent to broadly spreading, rarely incurved, often ± accrescent; larger bracteoles narrowly oblong, abaxially villous . . . . 64f.7. *Crataegus* ser. *Molles* (in part), p. 551
            4. Pomes bright pink to crimson, glabrous or pubescent at ends; flesh ± hard; fruiting sepals spreading or ± reflexed, non-accrescent; larger bracteoles oblong to curved, abaxially glabrous . . . . 64f.12. *Crataegus* ser. *Dilatatae* (in part), p. 578
        3. Leaves and petioles gland-dotted or stipitate-glandular.

5. Flowers 25–30 mm diam.; stamens 10 or 30–45(–47); shrubs multi-stemmed, main trunk not dominant; leaf blades ± chartaceous; inflorescences frequently arising laterally from extension shoots . . . . 64f.18. *Crataegus* ser. *Triflorae*, p. 609
5. Flowers 18–25 mm diam.; stamens (5 or)10 or 20–25; shrubs or trees 1- to few-stemmed, main trunk dominant; leaf blades usually thick and ± chartaceous or thinner but subcoriaceous; inflorescences either arising solely subterminally from woody short shoots (ser. *Molles*, ser. *Dilatatae*), or sometimes arising laterally from extension shoots (ser. *Bracteatae*).
   6. Leaves: lobes 0 or 1–4 per side, merely apiculi or sinuses very shallow, deep green, subcoriaceous; petioles 15–25% length of blade; inflorescences 3–12-flowered; stamens 20(–25); larger bracteoles ligulate, semipersistent, margins strongly stipitate-glandular. . . . . . 64f.17. *Crataegus* ser. *Bracteatae* (in part), p. 607
   6. Leaves: lobes usually 1–6 per side, distinct, sinuses shallow to deep, green or bluish green, ± thick or chartaceous; petioles 30–50% length of blade; inflorescences 5–20-flowered; stamens (5 or)10 or 20; larger bracteoles narrow to linear, subherbaceous, margins sessile-glandular (ser. *Molles*), or larger bracteoles oblong to curved, abaxially glabrous, margins strongly stipitate-glandular (ser. *Dilatatae*).
      7. Pomes orange-red, bright or deep red, not pruinose; flesh mealy; fruiting sepals erect or erect-patent to broadly spreading, rarely incurved, often ± accrescent; larger bracteoles narrowly oblong, abaxially villous . . . . 64f.7. *Crataegus* ser. *Molles* (in part), p. 551
      7. Pomes bright pink to crimson, glabrous or pubescent at ends; flesh ± hard; fruiting sepals spreading or ± reflexed, not accrescent; larger bracteoles oblong to curved, abaxially glabrous . . . . 64f.12. *Crataegus* ser. *Dilatatae* (in part), p. 578

[2. Shifted to left margin.—Ed.]

2. Bracteoles usually ± membranous, sometimes ± herbaceous, usually caducous, sometimes fugacious or ± persistent, rarely deciduous; flowers 8–25(–26) mm diam.
   8. Petioles usually eglandular; if petioles ± glandular, sepal margins not glandular-denticulate, -serrate, or -laciniate.
      9. Leaves: lobes 0 or not much bigger than teeth.
         10. Bracteoles linear-filiform, margins eglandular or sparsely glandular; pomes 5–8 mm diam.. . . . . 64f.1. *Crataegus* ser. *Virides* (in part), p. 531
         10. Bracteoles narrowly elliptic, linear, or narrow, margins sparsely glandular; pomes 8–15 mm diam.
            11. Sepal margins entire or slightly glandular-serrate.
               12. Leaf blades usually ± thin or chartaceous, matte.
                  13. Inflorescence branches pubescent . . .64f.3. *Crataegus* ser. *Punctatae* (in part), p. 541
                  13. Inflorescence branches glabrous . . . . .64f.15. *Crataegus* ser. *Rotundifoliae* (in part), p. 593
               12. Leaf blades subcoriaceous, glossy.
                  14. Leaves and inflorescence branches glabrous . . . . .64f.2. *Crataegus* ser. *Crus-galli* (in part), p. 536
                  14. Leaves and inflorescence branches hairy, particularly young.
                     15. Styles and pyrenes 1–3; se United States . . . . .64f.2. *Crataegus* ser. *Crus-galli* (in part), p. 536
                     15. Styles and pyrenes 3–5; Texas. . . . . 64f.5. *Crataegus* ser. *Madrenses* (in part), p. 549
            11. Sepal margins glandular-serrate to glandular-laciniate or -pectinate.
               16. Pyrene sides pitted. . . . . 166. *Crataegus persimilis*, p. 641
               16. Pyrene sides plane.
                  17. Leaf blades ± isodiametric, blue-green mature, woolly . . . . 64f.7. *Crataegus* ser. *Molles* (in part), p. 551
                  17. Leaf blades longer than wide, green to dark green mature, appressed- or scabrous-pubescent, glabrous, or white-tomentose.

18. Leaves: veins 3 per side; stamens 15 . . . . 168. *Crataegus turnerorum* (in part), p. 642
18. Leaves: veins 4–7 per side; stamens 10 or 20.
    19. Petiole length 20–30% blade; blades matte, villous to tomentose . . . . 64f.6. *Crataegus* ser. *Greggianae* (in part), p. 550
    19. Petiole length usually 20–50% blade; blades glossy, glabrous or sparsely appressed-pubescent.
        20. Leaf blades scabrous, veins 5 or 6 per side; Texas . . . . 64f.5. *Crataegus* ser. *Madrenses* (in part), p. 549
        20. Leaf blades glabrous, veins 7 per side; n United States, adjacent Canada .167. *Crataegus* ×*sicca*, p. 641

[9. Shifted to left margin.—Ed.]

9. Leaves: lobes evident and distinct.
    21. Inflorescence branches pubescent.
        22. Bracteoles linear-filiform, margins eglandular or sparsely glandular . . . . 64f.1. *Crataegus* ser. *Virides* (in part), p. 531
        22. Bracteoles linear to oblong, margins stipitate- or sessile-glandular.
            23. Sepal margins glandular-laciniate to -serrate; leaf blades hard, chartaceous, or coriaceous; fruiting sepals erect or erect-patent, sometimes spreading.
                24. Leaf blades broadly ovate or broadly elliptic to rhombic-elliptic, 2.5–4(–5) cm; flowers 15 mm diam.; c Texas, ± xeromorphs . . . . 64f.6. *Crataegus* ser. *Greggianae* (in part), p. 550
                24. Leaf blades broadly elliptic to ovate, 4–8(–12) cm; flowers 18–22 mm diam.; e Texas to Alabama n to Minnesota and Quebec, mainly mesophytes . . . . 64f.7. *Crataegus* ser. *Molles* (in part), p. 551
            23. Sepal margins entire or subentire; leaf blades thin; fruiting sepals erect-patent to recurved.
                25. Flowers (10–)13–17(–26) mm diam. . . . . . .64f.9. *Crataegus* ser. *Tenuifoliae* (in part), p. 561
                25. Flowers 18–23 mm diam.
                    26. Fruiting sepals sessile. . . . . .64f.3. *Crataegus* ser. *Punctatae* (in part), p. 541
                    26. Fruiting sepals elevated . . . . .64f.11. *Crataegus* ser. *Pruinosae* (in part), p. 571
    21. Inflorescence branches glabrous.
        27. Inflorescences (10–)15–30+-flowered, bracteoles linear-filiform (length/width = 10–20:1), margins eglandular or sparsely glandular; pomes 5–8 mm diam.; leaves with tufts of hair in abaxial vein axils (not discernable in rare pubescent form); southeastern, damp ground . . . . 64f.1. *Crataegus* ser. *Virides* (in part), p. 531
        27. Inflorescences 3–15-flowered, bracteoles linear or narrowly elliptic, margins sessile- or stipitate-glandular; pomes 7–15(–20) mm diam.; leaves lacking tufts of hair in abaxial vein axils; topland species.
            28. Leaves: adaxial surfaces of young blades glabrous (except along veins in some taxa and on whole adaxial surface in *C. virella*); pomes pink, mauve, or pale green, sometimes scarlet, deep crimson, or purple, often strongly pruinose, flesh hard; fruiting sepals usually on prominent collar . . . . .64f.11. *Crataegus* ser. *Pruinosae* (in part), p. 571
            28. Leaves: adaxial surfaces of young blades glabrous or sparsely hairy; pomes yellow to reddish mature, not strongly pruinose, flesh hard or mellow; fruiting sepals usually sessile, sometimes slightly elevated.
                29. Twigs 1–2-years old dark brown, dark purple-brown, or reddish brown to blackish, often ± pruinose.
                    30. Leaves: veins 4–6 per side, blades 2.5–4(–5) cm; petioles eglandular, length 25–33% blade; stamens 10. . . . 64f.6. *Crataegus* ser. *Greggianae* (in part), p. 550
                    30. Leaves: veins 3–5(or 6) per side, blades 2–7 cm; petioles sparsely glandular, length 20–30% blade; stamens 10(or 20) . . . . 64f.10. *Crataegus* ser. *Populneae* (in part), p. 567
                29. Twigs 1–2-years old not both purple-brown and ± pruinose.

31. Stamens 15, anthers purple; c Texas . . . . . . 168. *Crataegus turnerorum* (in part), p. 642
31. Stamens 5–10 or 20, anthers pink to purple or ivory; e, n United States, adjacent Canada.
32. Leaves: lobes well defined, lobe apices usually acute, blades 3–6(–8) cm, apices acute; petioles glandular or eglandular; anthers pink to red or purple; pomes red, ellipsoid to suborbicular, sepals erect-patent to recurved or erose. . . 64f.9. *Crataegus* ser. *Tenuifoliae* (in part), p. 561
32. Leaves: lobe sinuses usually shallow, lobe apices subacute to obtuse, sometimes acuminate, blades 2–4(–5) cm, apices acuminate to obtuse; petioles eglandular or with 1 or 2 tiny glands; anthers ivory, sometimes pink; pomes usually bright to deep red or yellowish, sometimes orangish or burgundy, usually ± suborbicular, sometimes ellipsoid or oblong, sepals reflexed . . . . .64f.15. *Crataegus* ser. *Rotundifoliae* (in part), p. 593

[8. Shifted to left margin.—Ed.]

8. Petioles, at least some, glandular; sepal margins glandular-denticulate, -serrate, -laciniate, or -pectinate.
33. Petioles, at least some, stipitate-glandular; bracteole margins usually short-stipitate- to stipitate-glandular.
34. Leaf blades ± elliptic to narrowly obovate or ovate, lobes 0 or obscure, subcoriaceous, (shiny at maturity); pomes orange-red to red, sepals not elevated on collar . . . . 64f.17. *Crataegus* ser. *Bracteatae* (in part), p. 607
34. Leaf blades broadly elliptic to elliptic, ovate or deltate, broadly ovate to rhombic-elliptic or rhombic, lobes 0 or 1–5 (if lobes 0, not elliptic to narrowly obovate), if so, short to moderately deep, usually ± thin, (not particularly shiny at maturity); pomes yellow to red, sepals ± elevated on collar, sometimes sessile.
35. Pomes with sepals sessile; se United States.
36. Leaf blades 3–5 cm, elliptic to broadly elliptic or suborbicular, base rounded to broadly truncate, lobes 3 or 4 per side, distinct, veins 4 or 5 per side; inflorescence branches coarsely hairy. . . . . 64f.10. *Crataegus* ser. *Populneae* (in part), p. 567
36. Leaf blades 2.5–4(–5) cm, ovate to obovate or rhombic-ovate, base ± cuneate, lobes 0 or 1–3 per side, obscure, veins 5 or 6(or 7) per side; inflorescence branches appressed-scabrous. . . 64f.19. *Crataegus* ser. *Apricae* (in part), p. 610
35. Pomes with sepals ± elevated on collar; e or se United States.
37. Stamens 10; trunk bark fibrous, checked into longitudinal plates or corrugated; e United States to Missouri . . . . 64f.14. *Crataegus* ser. *Intricatae* (in part), p. 585
37. Stamens 20; trunk bark often corrugated; Texas to Georgia.
38. Inflorescence branches hairy . . . . 64f.14. *Crataegus* ser. *Intricatae* (in part), p. 585
38. Inflorescence branches glabrous.
39. Leaf blades 4–7 cm . . . . 64f.13. *Crataegus* ser. *Pulcherrimae* (in part), p. 579
39. Leaf blades 2–4 cm . . . . 64f.14. *Crataegus* ser. *Intricatae* (in part), p. 585
33. Petioles sessile-glandular; bracteole margins usually sessile-glandular to short-stipitate-glandular, except stipitate-glandular in ser. *Intricatae*, ser. *Pulcherrimae*, or *C. dodgei* group (larger bracteoles only, or some ser. *Lacrimatae*), rarely ± eglandular.
40. Twigs ± strongly flexuous, rarely slightly flexuous; leaf blades 1–4(–5) cm; (ultimate branches ± conspicuously pendulous, except in dwarf forms less than 10 dm).
41. Dwarf species usually 0.5–2 m; twigs ± strongly flexuous, ultimate branches not markedly pendulous. . . . . .64f.20. *Crataegus* ser. *Lacrimatae* (in part), p. 618
41. Larger species usually 2–7 m; twigs slightly to strongly flexuous, upper branches often ± pendulous.

42. Leaf blades predominantly narrow, usually elliptic to oblong or narrowly obovate, sometimes obtrullate, ovate, or suborbiculate, usually widest toward tip, veins 1–5 per side (exiting mainly or wholly in apical part of blade), apices obtuse to acute; (upper branches conspicuously drooping) . . . . .64f.20. *Crataegus* ser. *Lacrimatae* (in part), p. 618
42. Leaf blades suborbiculate to broadly ovate or rhombic-elliptic, veins 1–3 per side in small-leaved forms (1–2 cm), usually 3–5 per side in larger-leaved forms, terminating along leaf length, apices acute to obtuse; (upper branches gently to conspicuously drooping).
    43. Leaves, especially young, glabrous or scabrous adaxially; (upper branches often slightly drooping, at most slightly flexuous); bracteoles ± stipitate-glandular; stamens 10 or 20, anthers ivory, cream, or pink to purple or red . . . . 64f.19. *Crataegus* ser. *Apricae* (in part), p. 610
    43. Leaves, especially young, ± glabrous or woolly-tomentose adaxially; (upper branches drooping, clearly flexuous); bracteoles sessile- to short-stipitate-glandular; stamens 20, anthers ivory or cream . . . . .64f.20. *Crataegus* ser. *Lacrimatae* (in part), p. 618

[40. Shifted to left margin.—Ed.]

40. Twigs straight, only rarely flexuous; leaf blades 2–6(–10) cm; (ultimate branches not especially drooping).
    44. Trunk bark smooth and exfoliating; pomes 4–6 mm diam. . . . . .64a.4. *Crataegus* (sect. *Crataegus*) ser. *Microcarpae*, p. 499
    44. Trunk bark rough, scaly, fibrous, checked into longitudinal plates, or corrugated; pomes 8–14(–20) mm diam.
        45. Leaves: lobes 0 or shallow, sinuate; inflorescences 1–5-flowered; sepals ± equal to petals . . . . .64f.16. *Crataegus* ser. *Parvifoliae*, p. 604
        45. Leaves: lobes 0 or distinctly lobed; inflorescences 3–15-flowered; sepals shorter than petals.
            46. Stamens (10–)20; pomes pink, mauve, or pale green, sometimes scarlet, deep crimson, or purple, strongly pruinose, flesh ± hard. . . . . .64f.11. *Crataegus* ser. *Pruinosae* (in part), p. 571
            46. Stamens 10 or 20; pomes reddish or yellowish, seldom strongly pruinose, flesh ± hard to mealy.
                47. Stamens 13–18, anthers purple . . . . .64f.14. *Crataegus* ser. *Intricatae* (in part), p. 585
                47. Stamens 5–10 or ca. 20, anthers white to cream or pale pink to pink-purple or red.
                    48. Fruiting sepals elevated on short but distinct collar; glands of at least larger bracteoles clearly stipitate.
                        49. Stamens 10 .64f.14. *Crataegus* ser. *Intricatae* (in part), p. 585
                        49. Stamens 20.
                            50. Inflorescence branches hairy; leaves 4–9 cm (rarely glabrous or sparsely hairy; if so, leaves 3–4 cm and anthers rose) . . . . .64f.14. *Crataegus* ser. *Intricatae* (in part), p. 585
                            50. Inflorescence branches glabrous; leaves 4–9 cm (if 3–4 cm, anthers white to cream).
                                51. Fruiting sepals elevated on distinct collar; glands on all bracteoles stipitate; se United States. . . . . 64f.13. *Crataegus* ser. *Pulcherrimae* (in part), p. 579
                                51. Fruiting sepals elevated on short collar; glands only on larger, lower bracteoles stipitate; midwestern United States and adjacent Canada . . . . .64f.15. *Crataegus* ser. *Rotundifoliae* (in part), p. 593
                    48. Fruiting sepals sessile; glands of bracteoles sessile or stipitate.

[52. Shifted to left margin.—Ed.]

52. Anthers white to cream, rarely palest pink; bracteoles sessile- or stipitate-glandular.
    53. Bracteoles sessile- or very short-stipitate-glandular; stamens 10 or 20; inflorescence branches glabrous or hairy.
        54. Young leaves: adaxial surfaces glabrous; stamens 10; Delaware . . . . 64f.10. *Crataegus* ser. *Populneae* (in part), p. 567
        54. Young leaves: adaxial surfaces ± scabrous-hairy; stamens 10 or 20; n United States, s Canada. . . . . 64f.15. *Crataegus* ser. *Rotundifoliae* (in part), p. 593
    53. At least larger bracteoles stipitate-glandular; stamens 10; inflorescence branches glabrous.
        55. Shrubs or trees, 4–6 m; all bracteoles stipitate-glandular; Ozarks, possible for se United States. . . . . 64f.14. *Crataegus* ser. *Intricatae* (in part), p. 585
        55. Shrubs, 2–4 m; larger bracteoles stipitate-glandular; ne United States, adjacent Canada . . . . 64f.15. *Crataegus* ser. *Rotundifoliae* (in part), p. 593
52. Anthers pale pink or pink to red or purple; bracteoles sessile- or very short-stipitate-glandular.
    56. Sepals glandular-denticulate; stamens 10; narrower pyrenes often concave-sided; pomes bright red in late August, becoming deep red at maturity; s Alberta, Saskatchewan . . . . 64g.1. *Crataegus* (sect. unassigned) ser. *Montaninsulae*, p. 633
    56. Sepals usually glandular-serrate; stamens 10 or 20; all pyrenes plane-sided; pomes orange to vermillion or bright red, green, or dull red, becoming bright or deep red or burgundy at maturity; n, ne United States, s Canada.
        57. Anthers very pale pink or pale cream-pink; inflorescence branches glabrous.
            58. Anthers pale cream-pink; stamens 10; mature fruit orange to orange-brown or orange-red; Ozarks, possible for se United States . . . . 64f.14. *Crataegus* ser. *Intricatae* (in part), p. 585
            58. Anthers very pale pink; stamens 20; mature fruit burgundy; British Columbia . . . . 64f.15. *Crataegus* ser. *Rotundifoliae* (in part), p. 593
        57. Anthers pink to pink-purple or red; inflorescence branches glabrous or hairy.
            59. Inflorescence branches glabrous . . . . 64f.10. *Crataegus* ser. *Populneae* (in part), p. 567
            59. Inflorescence branches hairy.
                60. Shrubs 4–10 m; leaves 5–9 cm; lobes usually shallow, but if deeper, broad; flowers 16–22 mm diam.; fruits 10–14 mm; Great Lakes to New England s in Appalachians to North Carolina . . . . 64f.8. *Crataegus* ser. *Coccineae*, p. 557
                60. Shrubs 3–5 m; leaves 3–7 cm; lobes moderately deep, acute; flowers 13–17 mm diam.; fruits 8–10 mm; se Canada, ne United States.
                    61. Leaf blades usually thin, often glossy, apices acute, lobe apices acuminate; inflorescence branches villous; sepal margins usually glandular-serrate, rarely nearly entire and eglandular; styles 3–5; pomes ellipsoid . . . . 64f.9. *Crataegus* ser. *Tenuifoliae* (in part), p. 561
                    61. Leaf blades thin to chartaceous, matte, apices and lobe apices acute to subacute; inflorescence branches densely pubescent or villous; sepal margins glandular-serrate; styles 3 or 4(or 5); pomes suborbicular . . . . 64f.15. *Crataegus* ser. *Rotundifoliae* (in part), p. 593

64f.1. CRATAEGUS Linnaeus (sect. COCCINEAE) ser. VIRIDES (Gordon) Rehder, Man. Cult. Trees ed. 2, 364. 1940 [E]

Crataegus sect. *Virides* Gordon in J. C. Loudon, Arbor. Frutic. Brit. 2: 847. 1838; *Crataegus* [unranked] *Virides* (Gordon) Sargent

Shrubs or trees, 80–150 dm, main trunk dominant. **Stems:** mature trunk bark dark gray or black, plated or whitish to light gray, thin-exfoliating, freshly exposed bark not recorded; <compound thorns on trunks present>; twigs: new growth usually glabrous (not recorded in *C. nitida*), 1-year old gray to reddish brown or chestnut brown; thorns on twigs absent or few to abundant, usually straight, 1–2-years old blackish or chestnut brown, slender, 3–4 cm. **Leaves:** petiole <long>, length 33–40% blade, adaxially glabrous or pubescent, usually eglandular, sometimes sparsely glandular; blade narrowly elliptic to suborbiculate, narrowly rhombic to oblong and ovate, oblanceolate, or broadly lanceolate to narrowly rhombic-elliptic, (2–)4–7(–8) cm, subcoriaceous or thin, base cuneate to rounded or tapered, lobes 0 or 1–3 per side, sinuses vary in depth (deepest LII 20–25%), lobe apex acute (obtuse in *C. viridis* var. *ovata*), margins strongly serrate to obscurely crenate-serrate in distal ½ only, venation craspedodromous, sometimes semicamptodromous, veins 3–5(–7) per side, apex acute to obtuse or acuminate, surfaces usually glabrous, except with tufts of hair in abaxial vein axils. **Inflorescences** 10(–50) -flowered; branches usually glabrous (densely hairy young in *C. viridis* var. *velutina*); bracteoles fugacious (not seen in *C. nitida*), linear-filiform, membranous, margins eglandular or sparsely glandular (somewhat glandular in *C. viridis* var. *glabriuscula*). **Flowers** (10–)13–15(–18) mm diam.; hypanthium usually glabrous (hairy in *C. viridis* var. *velutina*); sepals narrowly triangular, shorter than petals, margins ± entire or glandular-serrate; stamens 20, anthers usually cream or ivory, very rarely pale pink; styles 3–5. **Pomes** orange to deep red or brick red to red, rarely yellow, suborbicular, 5–8(–12) mm diam., usually glabrous; flesh mealy or hard; sepals usually recurved, sometimes spreading; pyrenes 3–5.

Species 2 (2 in the flora): sc, se United States.

Series *Virides* is widespread in the United States from Texas to Maryland in the coastal plain and inland to southeast Kansas and southern Illinois; it is absent from Appalachia except in deep river valleys. Taxa of ser. *Virides* are usually found in or at the edges of marshes, swamps, ditches, and streams.

Series *Virides* consists of one species complex and a narrowly defined species. The series is normally easily recognized by its nearly glabrous nature except for young adaxial leaf surfaces and abaxial leaf pubescence in axillary tufts. It is also barely glandular, has tall stature, usually exfoliating bark, pale gray bark on twigs to 5 mm diam., 20 stamens, cream anthers, narrowly linear, eglandular bracteoles, and orange to red pomes. Petioles are proportionately longer in the broad-leaved forms. Less common forms have colored anthers, larger fruit, or hairy foliage and inflorescences. Series *Virides* is perhaps most closely related to ser. *Crus-galli*.

The interserial hybrid *Crataegus* ×*atrorubens* keys out in the first couplet.

1. One-year old twigs bright red-brown; sepal margins glandular-serrate; pomes dark red, 9–11 mm diam. . . . . 154. *Crataegus* ×*atrorubens*, p. 636
1. One-year old twigs gray to reddish brown or chestnut brown; sepal margins ± entire or glandular-denticulate; pomes orange to deep red or brick red to red, rarely yellow, 5–12 mm diam.
    2. Leaf blades 2–6(–7) cm, marginal teeth 0–2 mm; sepal margins ± entire; pomes 5–8 mm diam., not pruinose . . . . 32. *Crataegus viridis*
    2. Leaf blades 5–8 cm, marginal teeth 2–3 mm; sepal margins glandular-denticulate; pomes 8–12 mm diam., pruinose on drying. . . . . 33. *Crataegus nitida*

32. Crataegus viridis Linnaeus, Sp. Pl. 1: 476. 1753 • Green hawthorn [E] [F]

Shrubs or trees, 80–150 dm. **Stems:** mature trunk bark checked dark gray or black, rough or whitish to light gray, thin-exfoliating; twigs: new growth often reddish, glabrous, 1-year old gray to red-brown, older gray, smooth; thorns on twigs few to abundant, 2-years old blackish, 3–4 cm. **Leaves:** petiole 0.7–2.5 cm, length 33–40% blade, adaxially glabrous or pubescent, eglandular; blade narrowly elliptic to suborbiculate, narrowly rhombic to oblong and ovate, oblanceolate, 2–6(–7) cm, thin (except in vars. *glabriuscula* and *nitens*), base cuneate to rounded, lobes 0 or 1–3 per side, margins serrate (teeth 0–2 mm) to obscurely crenate-serrate in distal ½ only, venation craspedodromous, sometimes semicamptodromous, veins 3–5(–7) per side, apex ± acute to obtuse, surfaces usually glabrous, except with tufts of hair in abaxial vein axils. **Inflorescences** (3–)10(–50)-flowered; branches usually glabrous, sometimes densely hairy young; <bracteoles ± eglandular, except somewhat glandular in var. *glabriuscula*>. **Flowers** (10–)13–15(–18) mm diam.; hypanthium usually glabrous (hairy in var. *velutina*); sepal margins ± entire; anthers cream or ivory. **Pomes** orange to deep red, rarely yellow, not pruinose, orbicular, 5–8 mm diam.; sepals recurved. $2n$ = 34, 51.

Varieties 6 (6 in the flora): se United States.

Crataegus viridis ranges from southeastern Texas along the coastal plain and piedmont to central Florida and to tidewater Maryland; inland, it extends to near St. Louis, Missouri, and the Tennessee-Kentucky border. The species favors streamsides and moist, fertile, alluvial woodlands or agricultural derivatives of these.

Crataegus viridis is variable in leaf shape and size. The varieties intergrade somewhat, and most are sympatric; they are valuable for recognizing variation. Varieties *glabriuscula* and *velutina* are regional.

Beyond the recognized varieties, sporadic forms exist with yellow or gold fruit or anthocyanic anthers. *Crataegus arborescens* Elliott from northern Florida is similar to *C. viridis* but has rose-purple anthers and slightly hairy inflorescence branches. Other local forms with red or pink anthers, such as *C. antimima* Sargent, *C. enucleata* Sargent, and *C. pechiana* Sargent, are found. Selected forms with bright, rather deep green foliage, and heavy crops of red fruit, for example, cultivar Winter King, are ornamentals.

Rare putative interserial hybrids exist with ser. *Crus-galli* (for example, *Crataegus* ×*permixta* E. J. Palmer), ser. *Molles* (for example, *C.* ×*anamesa* Sargent, *C.* ×*poliophylla* Sargent, *C.* ×*stenosepala* Sargent), and ser. *Punctatae* (J. B. Phipps 2005).

1. Young leaf blades, inflorescence branches, and hypanthia densely hairy . . . . 32f. *Crataegus viridis* var. *velutina*
1. Young leaf blades, inflorescence branches, and hypanthia glabrous except for tufts of hair in vein axils of abaxial leaf surfaces.
   2. Leaf blades broadly lanceolate to narrowly rhombic, lobes distinct, 2 or 3 per side, sinuses shallow to moderately deep, max LII 15–40%, veins 4 or 5 per side. . . . . 32a. *Crataegus viridis* var. *viridis*
   2. Leaf blades ± lanceolate, narrowly to broadly elliptic or rhombic-elliptic to narrowly obovate or oblong to ovate or ± orbiculate, lobes mostly obscure or short, 0 or 1(–3) per side, max LII 5(–20)%, veins 3–7 per side.
      3. Leaf blades broadly elliptic to ovate, 3–4 cm, lobes 0 or 1–3 per side, sinuses shallow to deep, coriaceous; 1-year old twigs reddish brown . . . . .32e. *Crataegus viridis* var. *nitens*
      3. Leaf blades ± lanceolate, narrowly to broadly elliptic or rhombic-elliptic, narrowly obovate or oblong to oblong-ovate, or ± orbiculate, 2–6.3 cm, lobes 0, or 1 or 2 per side and very short or rarely 1 per side and longer, but then not on most leaves, sinuses shallow, thin to chartaceous; 1-year old twigs gray, dark gray, or tan to reddish brown.
         4. Leaf blades ± lanceolate, narrowly elliptic to narrowly obovate or oblong, veins 3–7 per side . . . . 32b. *Crataegus viridis* var. *lanceolata*
         4. Leaf blades oblong, broadly elliptic, rhombic-elliptic or oblong-ovate to ± orbiculate, veins 3–5 per side.
            5. Leaf blades broadly elliptic to oblong-ovate, sometimes suborbiculate, 2.5–6.3 cm, marginal teeth 1 mm, veins 4 or 5 per side . . . . 32c. *Crataegus viridis* var. *ovata*
            5. Leaf blades ± orbiculate to narrowly ovate, oblong, rhombic-elliptic, or narrowly obovate, 2–4 cm, marginal teeth 1.5–2 mm, veins 3(or 4) per side. . . . . .32d. *Crataegus viridis* var. *glabriuscula*

32a. Crataegus viridis Linnaeus var. **viridis** [E] [F]

Stems: trunk bark light gray, ± exfoliating; 1-year old twigs gray-brown. **Leaves:** blade broadly lanceolate to narrowly rhombic, 4–7 cm, thin, base usually broadly cuneate, lobes 2 or 3 per side, distinct, sinuses shallow to moderately deep, max LII 15–40%, <lobe apex acute to subacute>, margins crenate-serrate, teeth 1 mm,

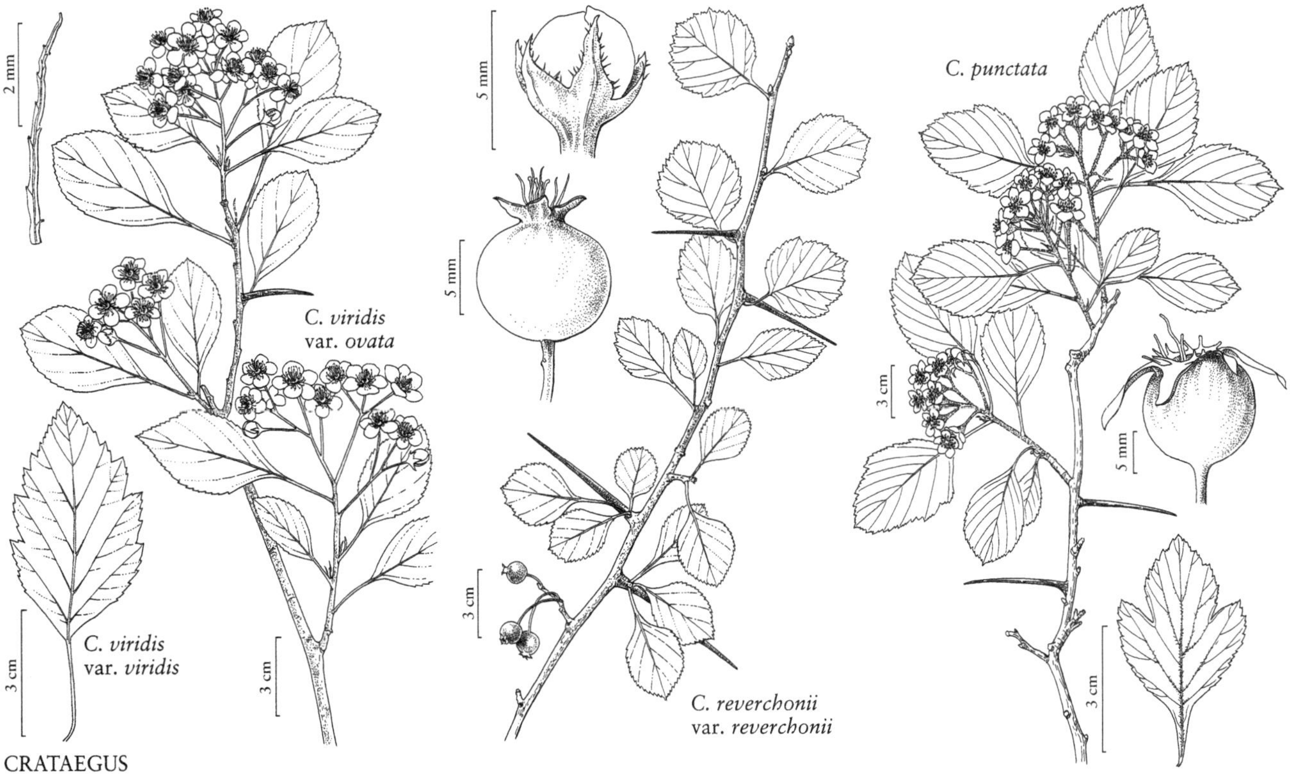

CRATAEGUS

venation craspedodromous, veins 4 or 5 per side, apex acute, surfaces glabrous young except abaxially with tufts of hair in vein axils, glabrescent. **Inflorescences:** branches glabrous. **Flowers:** hypanthium glabrous.

Flowering Apr–May; fruiting Sep–Nov. Moist, fertile, alluvial woodlands, agricultural derivatives of these; 10–200 m; Ala., Ark., Fla., Ga., Ill., Ind., Ky., La., Md., Miss., Mo., N.C., S.C., Tenn., Va., W.Va.

Variety *viridis* is widespread and sympatric with the other varieties except var. *glabriuscula*, with which it shares a diffuse boundary. Reports from Kansas, Oklahoma, and Texas need confirmation. Variety *viridis* grades into var. *glabriuscula*, var. *lanceolata*, var. *ovata*, and *Crataegus nitida*.

32b. Crataegus viridis Linnaeus var. **lanceolata** (Sargent) E. J. Palmer, Ann. Missouri Bot. Gard. 22: 561. 1935 E

Crataegus lanceolata Sargent, Trees & Shrubs 2: 65, plate 130. 1908; *C. interior* Beadle

Stems: trunk bark light gray, ± exfoliating; 1-year old twigs gray. **Leaves:** blade ± lanceolate, narrowly elliptic to narrowly obovate or oblong, 2.5–6 cm, thin, base cuneate, lobes 0 or obscure else rarely 1 per side, distinct, but not on most leaves, sinuses shallow, max LII 5(–20)%, margins serrulate-crenate, mainly in distal ⅔, sometimes entire, teeth to 1 mm, venation craspedodromous, sometimes semicamptodromous, veins 3–7 per side (often dividing before margin), apex acute to acuminate, surfaces glabrous except abaxially with tufts of hair in vein axils. **Inflorescences:** branches glabrous. **Flowers:** hypanthium glabrous.

Flowering Apr; fruiting Sep–Nov. Moist, fertile, alluvial woodlands, agricultural derivatives of these; 10–200 m; Ala., Ark., Fla., Ga., La., Miss., Mo., S.C., Tex.

The range of var. *lanceolata* is poorly documented; it appears to be rather common in Louisiana and extends into Missouri, into eastern Texas, and through Alabama to South Carolina. In its most extreme expression, it is readily distinguished; it intergrades with both var. *ovata* and var. *viridis*. The foliage, with its distinctive venation, and the usually small- and few-flowered inflorescences are characteristic. Particularly large, oblong leaf forms of var. *lanceolata* have been referred to *Crataegus interior*.

32c. Crataegus viridis Linnaeus var. **ovata** (Sargent) E. J. Palmer, Ann. Missouri Bot. Gard. 22: 561. 1935 E F

Crataegus ovata Sargent, Man. Trees, 402, fig. 321. 1905; *C. blanda* Sargent; *C. larga* Sargent

Stems: trunk bark gray, ± exfoliating, smooth; 1-year old twigs reddish brown. **Leaves:** blade broadly elliptic to oblong-ovate, sometimes suborbiculate, 2.5–6.3 cm, thin, base broadly or acutely concave-cuneate, lobes 0 or 1 per side, very short, else rarely 1 per side, distinct, but not on most leaves, usually very short, sinuses shallow, max LII 5 (-20)%, <lobe apex obtuse>, margins finely crenate-serrate, rarely serrate, except base entire, teeth 1 mm, venation craspedodromous, sometimes semicamptodromous, veins 4 or 5 per side, <midvein slender>, apex subacute to acute, surfaces glabrous except abaxially with persistent axillary tufts of hairs, adaxially with scattered hairs near base of midvein. **Inflorescences:** branches glabrous. **Flowers:** hypanthium glabrous.

Flowering early May; fruiting Oct. Moist, fertile, alluvial woodlands, agricultural derivatives of these; 10–200 m; Ark., La., Miss., Mo., Tex.

The range of var. *ovata* is not precisely documented and may be wider than noted. The variety appears to be well represented in Arkansas and Louisiana. The type specimen is of a rare form with particularly large leaves, some distinctively lobed, that is restricted to Missouri. A form with smaller and almost always unlobed leaves, also included here, is the most abundant of this variety. Intermediates with var. *lanceolata* and var. *viridis* are common. Forms with black scaly bark, relatively large flowers (said to be 20 mm diam.), thornless, and upland habitat were referred to *Crataegus blanda*, though specimens from the type tree have smaller (14–15 mm diam.) flowers. As the bark of *C. viridis* characteristically exfoliates in large flakes and is usually whitish or pale gray, the topland habitat of *C. blanda*, as well as the original collector's comments on its bark, and its type, which cannot be differentiated from var. *ovata* in the herbarium, suggest that the nature of the mature bark may need to be taken into account more seriously in the taxonomy of this species. Rare *ovata*-like forms with pink anthers and growing in atypically dry habitats in northeastern Missouri, perhaps hybrid, may be referred to *C. davisii* Sargent and *C. pechiana* Sargent and perhaps no longer exist.

32d. Crataegus viridis Linnaeus var. **glabriuscula** (Sargent) J. B. Phipps, Novon 16: 386. 2006 E

Crataegus glabriuscula Sargent, Bot. Gaz. 31: 235. 1901; *C. sutherlandensis* Sargent

Stems: trunk bark whitish, exfoliating, smooth; <compound thorns on trunks sometimes present>; 1-year old twigs tan to reddish brown. **Leaves:** blade ± orbiculate to narrowly ovate, oblong, rhombic-elliptic, or narrowly obovate, 2–4 cm, thin to chartaceous, base cuneate, lobes 0, or 1 or 2 per side, if present, very short, <lobe apices acute>, margins usually sharply serrate, teeth 1.5–2 mm, venation craspedodromous, veins 3(or 4) per side, apex acute to obtuse, surfaces glabrous except abaxially with tufts of hair in vein axils, adaxially short-pubescent young, glabrescent. **Inflorescences** <3–12-flowered>; branches glabrous. **Flowers:** hypanthium glabrous.

Flowering Apr; fruiting Sep–Nov. Rich moist soil near streams, blackland soil; 10–200 m; Ark., Kans., Okla., Tex.

Variety *glabriuscula* is common in the eastern half of Texas and Oklahoma and extends to southeastern Kansas, to western Arkansas, possibly to Louisiana, and probably to southern Missouri.

Variety *glabriuscula* is weakly distinguished in the north of its range from var. *viridis* because of intermediates such as *Crataegus furcata* Sargent and *C. lutensis* Sargent [= *C. viridis* var. *lutensis* (Sargent) E. J. Palmer]. The variety is also somewhat distinguished by smaller leaves from forms of var. *ovata*, which, however, are almost perfectly allopatric. The mature bark of some specimens is described as the smoothest of the *viridis* group, nearly white, with finely exfoliating large flakes. Bark characteristics (and habitat) are also credited with helping to distinguish *C. blanda* (see under 32c. var. *ovata*). Variety *glabriuscula* represents a distinct southwestern regional variant of the species.

Specimens with fruit diameter in the 8–10 mm range sometimes found in the Dallas area may be hybrids with *Crataegus reverchonii*. The rare *C. desertorum* Sargent, known from Bandera and Uvalde counties, Texas, is distinguished by the more scrublike nature of the plant, zigzag twigs, greater thorniness, frequently ovate leaves that are distinctly and bluntly lobed, persistent, dense pubescence on the petioles, and semipersistent pilosity on the abaxial leaf surface.

32e. Crataegus viridis Linnaeus var. **nitens** (Sargent) J. B. Phipps, Novon 16: 387. 2006 [E]

Crataegus nitens Sargent, Rep. (Annual) Missouri Bot. Gard. 22: 76. 1912

Stems: trunk bark dark gray or black, rough, thinly exfoliating whitish to light gray areas; 1-year old twigs reddish brown. **Leaves:** blade broadly elliptic to ovate, 3–4 cm, coriaceous, base ± rounded, lobes 0 or 1–3 per side, sinuses shallow to deep, max LII 5(–20)%, margins crenate, teeth 1 mm, venation craspedodromous, veins 5 per side, apex subacute, surfaces glabrous except abaxially with tufts of hair in vein axils. **Inflorescences:** branches glabrous. **Flowers:** hypanthium glabrous.

Flowering Apr; fruiting Sep–Oct. Stream bluffs; 100–200 m; Mo.

Variety *nitens* has been found near the northern limits of the *Crataegus viridis* complex near the Mississippi River at Hannibal. The leaves of var. *nitens* are more coriaceous than usual in *C. viridis*, the one-year old twigs are usually a reddish brown, and the two-year old twigs a darker gray. The variety may represent a hybrid with a member of ser. *Pruinosae*.

32f. Crataegus viridis Linnaeus var. **velutina** (Sargent) J. B. Phipps, J. Bot. Res. Inst. Texas 1: 1009. 2007 [E]

Crataegus velutina Sargent, Trees & Shrubs 2: 238. 1913; *C. micrantha* Sargent

Stems: trunk bark black, rough; 1-year old twigs gray. **Leaves:** blade ± lanceolate to narrowly ovate, broadly elliptic, or oblanceolate, 3–5 cm, thin, base rounded-cuneate, lobes 0 or obscure, margins serrulate-crenate, mainly in distal ⅔, sometimes entire, teeth 1 mm, venation craspedodromous, sometimes semicamptodromous, veins 3–7 per side (often dividing before margin), apex acute to subacute, surfaces silky young, variably glabrescent, except abaxially with hair tufts in vein axils. **Inflorescences:** branches densely pubescent young, sparsely scabrous-pubescent by anthesis. **Flowers:** hypanthium hairy.

Flowering Apr; fruiting Sep–Oct. Moist, fertile, alluvial woodlands, agricultural derivatives of these; 10–200 m; Ark., La., Tex.

Variety *velutina* is concentrated in western Arkansas, western Louisiana, and eastern Texas. Similar forms that are less velutinous when young are found in Florida.

Variety *velutina* is easily recognized in young foliage and at anthesis; it has virtually no intermediates with the glabrous varieties of *Crataegus viridis*. By anthesis, the adaxial leaf surfaces are usually only sparsely hairy and by fruiting they are usually glabrous. The strong tendency to glabrescence may explain the relatively small percentage of fruiting specimens noted for this variety.

33. Crataegus nitida (Engelmann ex Britton & A. Brown) Sargent, Bot. Gaz. 31: 231. 1901 [C] [E]

Crataegus viridis Linnaeus var. *nitida* Engelmann ex Britton & A. Brown, Ill. Fl. N. U.S. 2: 242. 1897

Trees, 100 dm. **Stems:** mature trunk bark broken into gray-brown platelike scales; twigs: new growth not recorded, 1-year old chestnut brown, subcoriaceous; thorns on twigs absent or few, 1–2 -years old chestnut brown, 2.5–4 cm. **Leaves:** petiole 1.5–2.5 cm, length 33% blade, adaxially pubescent young, sparsely glandular; blade broadly lanceolate to narrowly rhombic-elliptic, 5–8 cm, subcoriaceous, base tapered, lobes 0, or 1 or 2(or 3) per side, <sinuses shallow, max LII (0–)5–10(–20)%>, margins strongly serrate (teeth 2–3 mm), venation craspedodromous, veins 5 or 6 per side, apex acute to acuminate, surfaces glabrous. **Inflorescences** 10–20-flowered, <dome-shaped>; branches glabrous. **Flowers** 15–18 mm diam.; hypanthium glabrous; sepals <7 mm>, margins glandular-denticulate, <abaxially glabrous>; anthers cream or pale pink. **Pomes** brick red to red, pruinose on drying, suborbicular, 8–12 mm diam.; sepals spreading. $2n = 51$.

Flowering Apr; fruiting Sep–Oct. Flood plains; of conservation concern; 10–200 m; Ark., Ill., Mo.

Crataegus nitida is found from southern Illinois to eastern Arkansas along the Mississippi River flood plain; it appears to be rare. The species differs only slightly from *C. viridis* var. *viridis*, mainly in the larger fruit, more pronounced leaf marginal teeth, and more coriaceous mature leaves. Horticultural plants labeled *C. nitida* are usually a *C. crus-galli* × *C. viridis* cross.

64f.2. CRATAEGUS Linnaeus (sect. COCCINEAE) ser. CRUS-GALLI (Loudon) Rehder, Man. Cult. Trees ed. 2, 364. 1940

Crataegus sect. *Crus-galli* Loudon, Arbor. Frutic. Brit. 2: 820. 1838

Shrubs or trees, 10–100 dm, main trunk dominant (more commonly in taller plants). **Stems:** trunk bark older gray-ochre, usually fibrous, checked into longitudinal plates, freshly exposed bark not recorded; <compound thorns on trunks (of larger individuals) usually present>; twigs: new growth usually glabrous, sometimes pilose, 1-year old dark gray, pale brown, tan, brown, or orange-brown, 2-years old usually medium pale gray; thorns on twigs numerous, straight to recurved, sometimes reflexed, 2-years old dark gray to blackish or chestnut brown, medium to stout, (2–)3–6.5 cm. **Leaves:** petiole <usually short>, length 0–18(–31)% blade, usually glabrous, sometimes villous or glabrescent, usually eglandular, sometimes glandular; blade <paler green abaxially, bright, glossy, deep green adaxially>, narrowly oblanceolate to obovate, sometimes elliptic or ± rhombic, 2–7 cm (usually much longer than wide), usually coriaceous, sometimes thin and firm, base cuneate, tapered, or narrowed, lobes 0 or 1–3 per side, sinuses shallow, lobe apex acute, margins usually finely and evenly serrate, sometimes crenate, except near bases, or toothed only in distal part, venation semicamptodromous to craspedodromous, veins 4–6(–9) per side, apex acute to obtuse, rarely rounded, short-pointed, surfaces glabrous or hairy. **Inflorescences** 5–20-flowered; branches glabrous or hairy, <sparsely glandular-punctate>; bracteoles caducous (not seen in *C. schizophylla*), narrow, membranous, margins sparsely glandular. **Flowers** (8–)10–20 mm diam.; hypanthium glabrous, pilose, or villous; sepals narrowly triangular, much shorter than petals, margins entire or irregularly glandular-serrate; stamens 10–20, anthers ivory, cream, pink, dark purple, or red; styles 1–5. **Pomes** rose to deep red, sometimes orange-red, rarely yellow, suborbicular to ± oblong, (7–)8–15 mm diam., glabrous or pubescent; flesh hard, becoming mealy; sepals usually spreading to reflexed; pyrenes 1–5.

Species 7 (5 in the flora): c, e North America, Mexico.

Members of ser. *Crus-galli* are common hawthorns to the east of a line from central Texas to southern Minnesota and have a southeastern limit in north-central Florida. Members of ser. *Crus-galli* are less shade tolerant than most American hawthorns of the southeast, rarely succeeding in woodland understory.

Members of ser. *Crus-galli* are usually recognized by their rather narrow leaves more or less coriaceous, shiny, fairly dark green, lobes absent, usually short-petiolate, strongly tapered to the base, and rather late anthesis relative to sympatric hawthorns. The caducous bracteoles resemble those of ser. *Virides* in being especially narrow and often sparsely glandular. Sporadic yellow-fruited variants, some of which have been named, are well known in this series.

Series *Crus-galli* constitutes one of the most distinctive groups of hawthorns, but taxonomy of the species is difficult. A large range of variation is subsumed in *Crataegus crus-galli*, less in species like *C. fecunda*. E. J. Palmer (1925) cited well over 100 names for the series (*C. persimilis*, which keys out in the second couplet below, is an interserial hybrid). Stamen number and anther color in the more common species as interpreted here are rather inconstant compared to those in many *Crataegus* series. Rare putative hybrids with ser. *Punctatae* and ser. *Virides* are known. The distinctive *C.* ×*permixta* E. J. Palmer, from Illinois and Missouri, was placed in ser. *Crus-galli* by Palmer (1952), but its very broad and somewhat lobed leaves suggest hybridity with ser. *Virides*.

1. Sepal margins subentire or shallowly to deeply glandular-serrate; petiole length 25+% blade.
    2. Young leaves adaxially scabrous; leaf apices acute; sepal margins subentire or glandular-serrate; pyrene sides plane; 1-year old twigs orange-brown . . . . 38. *Crataegus fecunda*
    2. Young leaves adaxially glabrous except for midvein; leaf apices broadly acute; sepal margins deeply glandular-serrate; pyrene sides plane to deeply excavated; 1-year old twigs purplish brown . . . . 166. *Crataegus persimilis*, p. 641
1. Sepal margins ± entire; petiole length to 20% blade.
    3. Leaves hairy adaxially young, glabrescent or becoming scabrous; inflorescence branches densely pubescent . . . . 37. *Crataegus berberifolia*
    3. Leaves glabrous or sparsely hairy adaxially young, except sometimes for hairy midveins; inflorescence branches usually glabrous.
        4. Most short-shoot leaf lobes 1–3 per side. . . . . 36. *Crataegus schizophylla*
        4. Short-shoot leaf lobes 0.
            5. Leaf blades ± orbiculate to broadly elliptic; styles and pyrenes (1–)3–5 . . . . 34. *Crataegus reverchonii*
            5. Leaf blades broadly spatulate to narrowly oblanceolate or narrowly elliptic; styles and pyrenes 1 or 2(or 3) . . . . 35. *Crataegus crus-galli*

34. Crataegus reverchonii Sargent, Trees & Shrubs 1: 55, plate 28. 1903 (as reverchoni) • Reverchon hawthorn [E] [F]

Shrubs, 10–30(–80) dm. **Stems:** twigs: new growth glabrous, 1-year old dark gray; thorns on twigs slightly reflexed, 2-years old blackish, usually stout, 3–6 cm. **Leaves:** petiole 2–5 mm, length 7–11% blade, glabrous, eglandular; blade (on extension shoots) ± orbiculate to broadly elliptic, (on short shoots) ± orbiculate to broadly oblong-elliptic, 2–6 cm, very coriaceous, base broadly cuneate, lobes 0, margins serrulate in distal ½, venation semicamptodromous, veins 5 or 6 per side, apex subacute to obtuse, surfaces sparsely hairy young, glabrescent. **Inflorescences** 8–15-flowered; branches usually glabrous; bracteoles linear, <small>, margins short-stipitate-glandular. **Flowers** 8–12 mm diam.; hypanthium glabrous, sometimes slightly pilose; sepals 4 mm, margins subentire, abaxially glabrous; stamens 10–15 (–20), anthers cream, rose, or red; styles (2 or)3–5. **Pomes** red to orange-red, suborbicular, 10 mm diam., glabrous; sepals <persistent>, patent or reflexed; pyrenes (1–)3–5.

Varieties 3 (3 in the flora): sc, se United States.

Crataegus reverchonii is differentiated from *C. crus-galli* by a greater number of styles and pyrenes (except for var. *mohrii*) and extension-shoot leaves that tend to be suborbiculate. Distinction from a small-leaved form of *C. crus-galli* from the Edwards Plateau, Texas, needs elucidation. *Crataegus stevensiana* is sometimes separated out as a variety or species; however, the complex, unresolved variation makes this distinction moot at the moment.

The principal variation within *Crataegus reverchonii* is between smaller-statured, smaller-leaved, smaller-fruited, more-styled, xeromorphic plants of the southwest (var. *reverchonii*), and larger-leaved, larger-flowered, fewer-styled plants of the mid latitude, eastern United States (var. *palmeri*). Variety *mohrii* is intermediate.

1. Short-shoot leaf blades 4–6 cm; flowers 10–12 mm diam.; styles (2 or)3 . . . . 34c. *Crataegus reverchonii* var. *palmeri*
1. Short-shoot leaf blades 2–3.5 cm; flowers 8–12 mm diam.; styles 3–5.
    2. Flowers 8–10 mm diam.; inflorescences usually glabrous; w of Mississippi River . . . . 34a. *Crataegus reverchonii* var. *reverchonii*
    2. Flowers 12 mm diam.; inflorescences often pilose; e of Mississippi River . . . . 34b. *Crataegus reverchonii* var. *mohrii*

34a. Crataegus reverchonii Sargent var. **reverchonii** [E] [F]

Crataegus acutifolia Sargent var. *insignis* (Sargent) E. J. Palmer; *C. discolor* Sargent; *C. insignis* Sargent; *C. reverchonii* var. *discolor* (Sargent) E. J. Palmer; *C. reverchonii* var. *stevensiana* (Sargent) E. J. Palmer; *C. rubrisepala* Sargent; *C. stevensiana* Sargent

Plants 10–30 dm. **Short-shoot leaf blades** 2–3.5 cm, not usually drying dark. **Inflorescences** usually glabrous. **Flowers** 8–10 mm diam.; styles 3–5.

Flowering Apr; fruiting Sep–Oct. Mesic areas in grasslands, by streams, brush; 20–200 m; Ark., Kans., La., Mo., Okla., Tex.

Variety *reverchonii* is common in parts of the Edwards Plateau, Texas, and extends to southwestern Missouri and northwestern Louisiana. It is the most abundant variety and, with its small flowers, three to five styles and pyrenes, the most distinctive; in other respects, it is far from uniform. The type form, with ten stamens and pink anthers, is recorded from Texas; *Crataegus stevensiana*, with ten stamens and cream anthers, occurs through most of the range. Less common variants, which may have slightly larger leaves and flowers and all occur in Missouri are: *C. jasperensis* Sargent (20 stamens and yellow anthers), *C. parciflora* Sargent and *C. rotunda* Sargent (both with 20 stamens and rose anthers), and *C. discolor* (ten stamens and rose anthers). These are in some respects intermediate with var. *palmeri*.

34b. Crataegus reverchonii Sargent var. **mohrii** (Beadle) J. B. Phipps, J. Bot. Res. Inst. Texas 1: 1006. 2007 E

Crataegus mohrii Beadle, Bot. Gaz. 28: 416. 1899 (as mohri); *C. denaria* Beadle

Plants 10–30 dm. **Short-shoot leaf blades** 2–3.5 cm, not usually drying dark. **Inflorescences** often pilose. **Flowers** 12 mm diam.; styles 3–5.

Flowering Apr; fruiting Sep–Oct. Habitat not recorded, probably brush; 10–200 m; Ala., Ga., Miss., Tenn.

Variety *mohrii* is recorded with certainty only east of the Mississippi River; it is perhaps very rare. In the co-type, *Beadle 236A*, the inflorescences may be glabrous or pilose. The relatively small, short leaves, small flowers, three to five styles, and broadly elliptic to suborbiculate extension-shoot leaves clearly mark var. *mohrii* as a member of the *reverchonii* group. Variety *mohrii* is notable for its strongly and sharply serrate leaf margins.

34c. Crataegus reverchonii Sargent var. **palmeri** (Sargent) J. B. Phipps, Novon 16: 386. 2006 E

Crataegus palmeri Sargent, Trees & Shrubs 1: 57, plate 29. 1903; *C. canbyi* Sargent

Plants 80 dm. **Short-shoot leaf blades** 4–6 cm, usually drying dark. **Inflorescences** usually glabrous. **Flowers** 10–12 mm diam.; styles (2 or)3.

Flowering Apr; fruiting Sep–Oct. Brush; 10–200 m; Ark., Del., Kans., Mo., Tenn.

Variety *palmeri* ranges from the Ozark region to Delaware. *Crataegus pennypackeri* Sargent from Delaware, sometimes treated as distinct, is perhaps no more than a form of *C. canbyi*.

35. Crataegus crus-galli Linnaeus, Sp. Pl. 1: 476. 1753 (as Crus galli) • Cockspur hawthorn, aubépine ergot-de-coq E

Shrubs or trees, 50–100 dm, <often with tabulate branching>. **Stems:** twigs: new growth glabrous, 1-year old pale brown, older gray; thorns on twigs often ± recurved, 2-years old dark gray to blackish, ± stout, 3–6 cm. **Leaves:** petiole obsolescent or 5 mm, length 0–7% blade, glabrous, eglandular; blade <usually deep green, very shiny adaxially>, broadly spatulate to narrowly oblanceolate or narrowly elliptic, 2–7 cm, coriaceous, base tapered, lobes 0, margins serrulate to finely crenate, <teeth 1 mm>, venation semicamptodromous, veins 4–7 per side, apex acute to obtuse, adaxial surface glabrous, sometimes sparsely hairy young, glabrescent. **Inflorescences** 8–20-flowered; branches usually glabrous; bracteoles linear, <small>, margins sessile-glandular. **Flowers** 12–17 mm diam.; hypanthium glabrous; sepals 5–6 mm, margins ± entire, adaxially glabrous; stamens 10 or 20, anthers ivory to pink to red; styles 1 or 2(or 3). **Pomes** usually ruddy to crimson, sometimes bright yellow, ± oblong to suborbicular, 8–15 mm diam., <often pruinose>, glabrous; sepals patent-reflexed; pyrenes 1 or 2(or 3). **2*n*** = 34, 51, 68.

Varieties 2 (2 in the flora): e North America; introduced in Europe.

Crataegus crus-galli is one of the more abundant North American hawthorns and has been reported as naturalized in The Netherlands and other parts of Europe. It may occur in more or less pure stands sometimes many hectares in size. This species is variable, a situation attested to by its synonymy, most of which belongs to var. *crus-galli*. Some forms are locally distinct and may represent apomictic clones. In this treatment, only one variety other than var. *crus-galli* is recognized, this on the basis of leaf shape.

SELECTED REFERENCES Dickinson, T. A. 1983. *Crataegus crus-galli* L., sens. lat. in Southern Ontario. Ph.D. thesis. University of Western Ontario. Phipps, J. B. 1988. Typification of *Crataegus crus-galli* L. (Rosaceae) and nomenclatural notes on some related species of hawthorn. Bot. J. Linn. Soc. 96: 359–369.

1. Leaf blades relatively wide (1.5–2.5:1) . . . . 35a. *Crataegus crus-galli* var. *crus-galli*
1. Leaf blades extremely narrow (greater than 2.75:1) . . . 35b. *Crataegus crus-galli* var. *pyracanthifolia*

35a. Crataegus crus-galli Linnaeus var. **crus-galli** [E]

Crataegus acutifolia Sargent; *C. albanthera* Sargent; *C. arborea* Beadle; *C. barrettiana* Sargent; *C. calophylla* Sargent; *C. candens* Sargent; *C. cherokeenis* Sargent; *C. consueta* Sargent; *C. crus-galli* var. *capillata* Sargent; *C. crus-galli* var. *exigua* (Sargent) Eggleston; *C. crus-galli* var. *leptophylla* (Sargent) E. J. Palmer; *C. crus-galli* var. *macra* (Beadle) E. J. Palmer; *C. crus-galli* var. *oblongata* Sargent; *C. crus-galli* var. *pachyphylla* (Sargent) E. J. Palmer; *C. hamata* Sargent; *C. hannibalensis* E. J. Palmer; *C. infera* Sargent; *C. leptophylla* Sargent; *C. ludoviciensis* Sargent; *C. monosperma* Sargent; *C. pachyphylla* Sargent; *C. paradoxa* Sargent; *C. parkiae* Sargent; *C. permera* Sargent; *C. phaneroneura* Sargent; *C. polyclada* Sargent; *C. pyracanthoides* Beadle var. *arborea* (Beadle) E. J. Palmer; *C. regalis* Beadle; *C. rubrifolia* Sargent; *C. rudis* Sargent; *C. severa* Sargent; *C. strongylophylla* Sargent; *C. tantula* Sargent; *C. tardiflora* Sargent; *C. tenax* Ashe; *C. tenuispina* Sargent; *C. truncata* Sargent

Leaf blades relatively wide (1.5–2.5:1).

Flowering Apr–Jun; fruiting Sep–Nov. Brush, sandy soil, dense shade; 10–300 m; Ont., Que.; Ala., Ark., Conn., Del., D.C., Fla., Ga., Ill., Ind., Iowa, Kans., Ky., La., Md., Mass., Mich., Minn., Miss., Mo., N.J., N.Y., N.C., Ohio, Okla., Pa., R.I., S.C., Tenn., Tex., Va., W.Va., Wis.

Attempts to divide this complex into taxa based on stamen number and anther color have been unsatisfactory. *Crataegus arborea* (cream anthers) and *C. tenax* (pink anthers) appear to be the earliest names for 20-stamen forms, which E. J. Palmer and others wrongly called *C. fontanesiana* (Spach) Steudel [= *C. calpodendron* (see J. B. Phipps 1987)]. *Crataegus crus-galli* in the strict sense would be restricted to the ten-stamen, white-anthered form. Such a circumscription is not followed here because the collective variation in all ten- and 20-stamen forms, not only in stamen number and anther color, is much greater than the differences between them. T. A. Dickinson (1983) showed that in southern Ontario, clear phenological and morphological differences occur between ten- and 20-stamen forms.

Rare forms with three to five styles, for example, *Crataegus leptophylla*, may be hybrids with *C. reverchonii* var. *reverchonii*. The widely recognized var. *capillata* Sargent with inflorescence branches sparsely villous is here synonymized as that is all that distinguishes it.

35b. Crataegus crus-galli Linnaeus var. **pyracanthifolia** Aiton, Hort. Kew. 2: 170. 1789 [E]

Crataegus limnophila Sargent; *C. pyracanthoides* Beadle

Leaf blades extremely narrow (greater than 2.75:1).

Flowering Apr; fruiting Sep–Nov. Brush; 20–300 m; Ala., Ark., Del., Fla., Ga., La., Md., Miss., Mo., N.C., S.C., Tenn., Tex., Va.

Variety *pyracanthifolia* has a range to Missouri and Delaware and is especially common in Louisiana and parts of Florida, where it appears to be primarily a bottomland species.

Variety *pyracanthifolia* is a complex of narrowly leaved forms with glabrous inflorescences that is usually easily recognized. The typical form of var. *pyracanthifolia* has particularly narrow leaves with very fine teeth; it is the dominant type south of the Mason-Dixon line. North of the Mason-Dixon line is a sporadic morphotype differing by more acute-tipped, narrowly elliptic leaves with more distinct teeth. A form from northern Florida with hairy pedicels (*Crataegus limnophila* Sargent) is synonymized, as, *contra* the protologue, and has zero to three styles. An undescribed form from the Gulf Coast, and perhaps Tamaulipas, is differentiated by more strikingly serrate leaves than any known in *C. crus-galli*, as well as yellow to orange fruit. Similar, but with red fruit, is *C. sabineana* Ashe from Louisiana and Texas.

36. Crataegus schizophylla Eggleston, Bull. Torrey Bot. Club 38: 243. 1911 [C] [E]

Shrubs, 20–40 dm. **Stems:** twigs: new growth <flexuous>, glabrous, 1-years old tan, older gray; thorns on twigs ± straight, 1-year old shiny black, 3–5 cm. **Leaves:** petiole 3–7 mm, length 12–15% blade, glabrous, eglandular; blade <yellow-green, shiny>, elliptic to narrowly ovate, 2.5–5 cm, coriaceous, base cuneate, lobes 0 or 1–3 per side, margins serrulate, venation craspedodromous, veins 4–6 per side, apex acute to obtuse, surfaces glabrous. **Inflorescences** 5–20-flowered; branches sparsely hairy; bracteoles not seen. **Flowers** 12–15 mm diam.; hypanthium glabrous; sepals 3–4 mm, margins ± entire, adaxially glabrous; stamens 10, anthers pink; styles 1–3. **Pomes** bright red, ± oblong-obovoid, 7–9 mm diam., glabrous; sepals patent-reflexed; pyrenes 1–3.

Flowering May–Jun; fruiting Oct. Thickets and banks on sandy soil; of conservation concern; 10–20 m; Mass.

Crataegus schizophylla is known from Martha's Vineyard, Duke County. The extension-shoot leaves are larger, relatively wider, with acute lobes. Except for its distinctive leaf form, unique in this series, it closely resembles *C. crus-galli.*

37. Crataegus berberifolia Torrey & A. Gray, Fl. N. Amer. 1: 469. 1840 • Barberry-leaved hawthorn [E]

Shrubs or trees, 60 dm. **Stems:** twigs: new growth <orange-brown or green-tinged with red>, ± pubescent, 1-year old brown, older gray; thorns on twigs ± straight or recurved, 2-years old shiny black to chestnut brown, fine or stouter, (2–)3–4(–6.5) cm. **Leaves:** petiole 4–6 mm, length 13–18% blade, glabrescent, eglandular; blade narrowly obovate to oblanceolate, (2.5–)3(–4) cm, coriaceous, base narrowly cuneate, lobes 0, margins finely crenate or serrate except at base, or only beyond widest part, venation craspedodromous, veins 4–6 per side, apex subacute to obtuse, <lustrous>, abaxial surface ± densely pilose on veins, sometimes pubescent on surface, adaxial hairy young, glabrescent or becoming scabrous. **Inflorescences** 8–12-flowered; branches densely pubescent; bracteoles linear, margins glandular. **Flowers** 10–20 mm diam.; hypanthium villous or glabrous; sepals 3–5 mm, margins entire, abaxially glabrous; stamens 10 or 20, anthers cream or pink; styles 2 or 3. **Pomes** reddish to yellow, suborbicular, 8–10 mm diam., glabrous; sepals erose or patent; pyrenes 2 or 3.

Varieties 2 (2 in the flora): sc, se United States.

Crataegus berberifolia is widespread from Texas to Missouri, Florida, and Virginia; it is particularly abundant in Louisiana.

Crataegus berberifolia is little differentiated from some forms of *C. crus-galli*, except in indumentum; it has relatively small and less variably shaped leaves. Its distribution is quite different. Intermediates with hairy leaves and glabrous inflorescences or nearly glabrous leaves and hairy inflorescences may represent hybrids with *C. crus-galli* (where they might be reached in the key). Such a situation is found in *C. araioclada.* Abrasion of the adaxial leaf pubescence may occur, rendering identification more difficult with fruiting material. *Crataegus berberifolia* has a plethora of yellow- and orange-fruited forms, particularly from southern Louisiana. *Crataegus fera* and *C. tersa* are red-fruited, *C. crocina* yellow. The fruit color in the type is unknown; E. J. Palmer (in specimen annotation) called it yellow-orange. A form with exceptionally white-tomentose leaves near Copenhagen, Louisiana, is probably this species. *Crataegus regalis* var. *paradoxa* (Sargent) E. J. Palmer, from Missouri and adjacent Kansas and Arkansas, is probably a hybrid between the deeply serrated '*regalis*' leaf form of *C. crus-galli* and a form of *C. berberifolia.* Two common forms of *C. berberifolia* occur, treated here as varieties: var. *engelmannii* with ten pink anthers and var. *berberifolia* with 20 cream anthers. Forms with 20 pink or ten cream anthers also occur sporadically.

1. Stamens 20, anthers usually cream; flowers 16–20 mm diam. . . . 37a. *Crataegus berberifolia* var. *berberifolia*
1. Stamens 10, anthers usually pink; flowers 10–15 mm diam. . . . 37b. *Crataegus berberifolia* var. *engelmannii*

37a. Crataegus berberifolia Torrey & A. Gray var. **berberifolia** [E]

Crataegus araioclada Sargent; *C. crocina* Beadle; *C. edita* Sargent; *C. edura* Beadle; *C. fera* Beadle; *C. tersa* Beadle; *C. villiflora* Sargent

Flowers 16–20 mm diam.; stamens 20, anthers usually cream.

Flowering Apr; fruiting Sep–Nov. Brushy subxeric to xeric communities, mixed pine-oak forests; 10–200 m; Ala., Ark., Fla., Ga., La., Miss., Mo., N.C., S.C., Tenn.

Most commonly occurring in Louisiana, var. *berberifolia* extends to southern Missouri, where it is fairly abundant, and to Florida. This variety contains the yellow-fruited forms. There is not much difference from *Crataegus rosei* [subsp. *rosei*] var. *amoena* from Coahuila, Mexico.

37b. Crataegus berberifolia Torrey & A. Gray var. **engelmannii** (Sargent) Eggleston, Rhodora 10: 76. 1908 (as engelmanni) [E]

Crataegus engelmannii Sargent, Bot. Gaz. 31: 2. 1901 (as engelmanni); *C. albicera* Beadle; *C. bellica* Sargent; *C. crus-galli* Linnaeus var. *bellica* (Sargent) E. J. Palmer; *C. engelmannii* var. *sinistra* (Beadle) E. J. Palmer; *C. hirtella* Sargent; *C. munita* Sargent; *C. pilifera* Sargent; *C. setosa* Sargent; *C. signata* Beadle; *C. sinistra* Beadle; *C. tenuis* Sargent; *C. triumphalis* Sargent; *C. vallicola* Sargent

Flowers 10–15 mm diam.; stamens 10, anthers usually pink.

Flowering Apr; fruiting Sep–Nov. Brush; 10–200 m; Ala., Ark., Kans., La., Miss., Mo.

Variety *engelmannii* is red-fruited and is the most common variety in Missouri; it extends without break from Kansas to Louisiana and Alabama. *Crataegus signata* possibly does not belong; it has three to five styles, and Beadle compares it to *C. ashei* (ser. *Bracteatae*) in his protologue.

38. Crataegus fecunda Sargent, Bot. Gaz. 33: 111. 1902 • St. Clair or fecund hawthorn C E

Trees, 60–80 dm. **Stems:** <trunks to 3 dm diam.>; twigs: new growth <dark green turning orange, lustrous>, pilose, 1-year old orange-brown, older ashy gray; thorns on twigs ± straight to recurved, 1-year old chestnut or blackish brown, slender, 5–6.5 cm. **Leaves:** <stipules stipitate, linear to semilunate, 1.3–2 cm, coarsely glandular-serrate>; petiole <± winged apically>, 13–20 mm, <stout>, length 26–31% blade, adaxially <grooved>, villous, often scattered dark glandular; blade <pale yellow-green abaxially, dark green when mature, red in fall>, oblong-obovate to ovate or broadly ovate, 5–6.5 cm, thin and firm, base gradually or abruptly narrowed, <obscurely ciliate>, lobes 0, margins usually doubly, coarsely serrate (except proximally), <teeth spreading, 2–3 mm, glandular>, venation craspedodromous, <midvein stout>, veins 7–9 per side, <remote, only slightly impressed adaxially>, apex acute, rarely rounded, short-pointed, <lustrous>, adaxial scabrous young, scabrous or glabrous mature, midveins and primary veins villous <with occasional white hairs>. **Inflorescences** 6–17-flowered; branches sparsely villous; bracteoles linear to oblong-obovate, margins sparsely glandular-serrate. **Flowers** 18 mm diam.; hypanthium ± villous; sepals <spreading after anthesis, elongated>, shorter than petals, margins subentire or shallowly to deeply glandular-serrate, stipitate-glandular, <apex acuminate>, abaxially villous; stamens 10(–15), anthers dark purple, <small>; styles 2 or 3. **Pomes** <borne in wide, drooping, multi-fruited glabrous clusters>, dull orange-red, <with dark lenticels>, short-oblong to suborbicular <ends full, rounded>, 13 mm diam., sparsely hairy until nearly fully grown; <flesh pale green, thick, firm, hard, dry, sweet>; sepals erect, incurved, <linear-lanceolate, coarsely glandular-serrate above middle>; pyrenes 2 or 3, <rounded, 8 mm>.

Flowering Apr; fruiting Sep–Oct. Ravines, brush, rich woodlands, near streams; of conservation concern; 50–200 m; Ark., Ill., Ky., Mo.

Crataegus fecunda was formerly found in the central United States, mainly near the Mississippi and Ohio rivers. It has apparently not been seen in the wild since the 1930s, an example of the decline of previously well-documented taxa, some of which are seemingly extinct.

Crataegus fecunda is similar to glabrous-inflorescence forms of *C. persimilis*; it always has sharper and more prominent leaf marginal teeth, usually proportionately narrower leaves with more acute apices, slightly hairy inflorescences, and plane-sided pyrenes. According to Sargent, it had fruit 22–25 mm diam.

Crataegus shinnersii Kruschke, known from a handful of specimens in Wisconsin, is similar and would key here but has cream or ivory anthers and glabrous inflorescences; its precise taxonomic disposition is uncertain.

64f.3. Crataegus Linnaeus (sect. Coccineae) ser. Punctatae (Loudon) Rehder, Man. Cult. Trees ed. 2, 365. 1940

Crataegus sect. *Punctatae* Loudon, Arbor. Frutic. Brit. 2: 818. 1838; *Crataegus* sect. *Mexicanae* Loudon; *Crataegus* ser. *Mexicanae* (Loudon) Rehder

Shrubs or trees, (30–)50–80 dm, main trunk usually dominant. **Stems:** trunk bark buff to gray-brown, fibrous, checked into longitudinal plates, freshly exposed bark orange-brown; <compound thorns on trunks and larger branches often present>; twigs: new growth pubescent, 1-year old pale gray or grayish brown to tan or reddish brown, older pale gray or gray; thorns on twigs absent or few to numerous, ± straight to slightly recurved, 2-years old gray to dark gray, rarely ± shiny dark brown or black, older gray, usually slender, sometimes stout, 2–6 cm. **Leaves:** petiole <short>, length 10–25(–50)% blade, pubescent, usually eglandular, sometimes sparsely sessile-glandular; blade broadly elliptic to narrowly obovate, oblanceolate (broader in *C. jonesiae*), or rhombic-elliptic, 3–6(–8) cm, ± thin, base usually cuneate, lobes 0 or 1–4(–9) per side, sinuses usually shallow, lobe apex usually acute, rarely obtuse, margins serrate to crenate except subentire base, venation craspedodromous, veins (4 or)5–10 per side, apex subacute to obtuse, abaxial surface hairy, veins pubescent or not, adaxial hairy, especially young. **Inflorescences** 5–25-flowered; branches usually pubescent; bracteoles caducous, linear, membranous or ± herbaceous, margins glandular. **Flowers** (10–)14–23 mm diam.; hypanthium usually pubescent, sometimes glabrous; sepals narrowly triangular, much shorter than petals, margins entire or glandular-serrate; stamens (5–)10–20, anthers rose, pink-purple, cream, white, or yellow; styles 2–5. **Pomes** red, yellow, orange, scarlet, or burgundy, usually suborbicular, sometimes oblong or obovoid, (8–)10–16[–25] mm diam., <usually punctate>, glabrous or pubescent; flesh mealy; sepals persistent, reflexed or erect-patent; pyrenes 2–5.

Species 5 (4 in the flora): c, e North America, Mexico.

Series *Punctatae* consists of three widespread species (*Crataegus collina*, *C. mexicana* de Candolle, and *C. punctata*) plus two local ones (*C. spes-aestatum* and *C. jonesiae*), all found in typical hawthorn habitats.

The common species are narrow-leaved with leaf lobes absent or obscure, usually 20 stamens, and three to five styles and pyrenes. Apparently no meaningful distinction exists between ser. *Punctatae* and ser. *Mexicanae*, and the two have been united here. *Crataegus jonesiae* appears intermediate between ser. *Punctatae* and *C. chrysocarpa* (ser. *Rotundifoliae*).

1. Flowers 18–23 mm diam.; leaf blades 4–7(–8) cm, broadly ovate to broadly rhombic-elliptic or nearly circular, equal to or less than 1.5 times as long as wide; pomes oblong or obovoid; stamens 10. . . . . 42. *Crataegus jonesiae*
1. Flowers (10–)12–20(–22) mm diam.; leaf blades 3–7(–7.5) cm, narrowly obovate to oblanceolate or broadly elliptic, sometimes ± obtrullate or rhombic-obovate or -elliptic, 2–4 times as long as wide; pomes suborbicular or orbicular; stamens (5–)20.
    2. Leaves: veins 7–10 per side; inflorescences 10–25-flowered; sepal margins entire or subentire; pomes usually deep burgundy, sometimes scarlet, orange, or yellow; ne United States and adjacent Canada, in Appalachia at 1000–1800 m . . . . 39. *Crataegus punctata*
    2. Leaves: veins 4–7 per side; inflorescences 5–12-flowered; sepal margins usually glandular-serrate; pomes usually orange, sometimes deep red to ruddy; se United States, in Appalachia only at lower elevations.
        3. Leaf blades 3–5 cm (except to 7.5 cm in var. *lettermanii*), rhombic-obovate or broadly elliptic, sometimes ± narrowly obovate, ± obtrullate, or broadly rhombic-elliptic, apices subacute to obtuse, lobes 0, or 1 or 2(or 3) per side, lobe apices acute to subacute, rarely obtuse; flowers (10–)13–18(–20) mm diam. . . . . 40. *Crataegus collina*
        3. Leaf blades 4–6 cm, broadly rhombic-elliptic, apices acuminate, lobes 3 or 4 per side, lobe apices acute; flowers 20–22 mm diam. .41. *Crataegus spes-aestatum*

39. Crataegus punctata Jacquin, Hort. Bot. Vindob. 1: 10, plate 28. 1770 • Dotted or white haw, aubépine ponctuée [E] [F]

Crataegus punctata var. *microphylla* Sargent

Shrubs or trees, 40–80 dm. **Stems:** <compound thorns on trunks present>; twigs: new growth pubescent, 1-year old pale gray; thorns on twigs usually numerous, ± straight to ± recurved, 1-year old gray, slender, 2–5 cm. **Leaves:** petiole 0.5–1.5 cm, length 12–20% blade, <winged distally>, pubescent young, eglandular; blade narrowly obovate to oblanceolate or broadly elliptic, 4–7 cm, equal to or less than 1.5 times as long as wide, base narrowly cuneate, lobes 0 or 5–7 per side <distal to widest part of leaf>, lobe apex acute, margins sharply serrate to sometimes crenate distally, <teeth frequency decreasing proximally>, veins 7–10 per side, <impressed young>, apex subacute to obtuse, abaxial surface sparsely short-pilose except on veins, adaxial <matte>, sparsely short-pilose, glabrate mature. **Inflorescences** 10–25-flowered, <compact>; branches appressed-hairy; bracteoles membranous, margins sessile- or short-stipitate-glandular. **Flowers** 14–19 mm diam.; hypanthium densely pubescent; sepals <6–8 mm>, margins entire or subentire, adaxially pubescent proximally; stamens (10–)20, anthers cream, rose, or pink-purple; styles 3–5. **Pomes** usually deep burgundy, sometimes scarlet, orange, or yellow, suborbicular, 11–16(–20) mm diam., <punctate>, traces of pubescence remaining; sepals reflexed; pyrenes 3–5. **2*n*** = 34.

Flowering May–Jun; fruiting Sep–Oct. Open sites, fencerows, successional fields, brush, light shade in woodlands, edges of mesic hardwood forests, brushy secondary forests; 30–1800 m; N.B., Ont., Que.; Conn., Ill., Ind., Iowa, Ky., Maine, Mass., Mich., Minn., N.H., N.J., N.Y., N.C., Ohio, Pa., R.I., Tenn., Vt., Va., W.Va., Wis.

Crataegus punctata is a common northeastern North American hawthorn, ranging across the northeast quarter of the United States and adjacent Canada to New Brunswick, to North Carolina and Tennessee, occurring mainly above 1000 m in the southern Appalachians. South of New Jersey, it is not found near the coast. The species may occur in nearly pure populations of up to several thousand individuals. The best growth is in sunny sites near streams.

Crataegus punctata is recognized among sympatric species by its matte green leaves (unlike the somewhat similar *C. crus-galli*) with impressed veins, tabulate branching when grown in the open, usually burgundy fruits with pale dots; the whitish or ashy gray twigs present a distinctive feature in winter (J. B. Phipps et al. 2003, plate 55), responsible for the common name white haw. The similar 40. *C. collina*, with an allopatric distribution and much earlier flowering season, is often confused with it but is readily differentiated (see key and discussion for that species). Extension-shoot leaves of *C. punctata* may be very deeply incised with veins to the sinuses, a characteristic not found in *C. collina*.

Some forms of *Crataegus punctata* bear scarlet or yellow (var. *aurea* Aiton) fruit. Hairier forms, especially hairier leaves, have been named var. *canescens* Britton; very small-leaved forms have been called var. *microphylla* Sargent. Nearly glabrous forms, all with somewhat glossy foliage, less coarse leaf toothing than *C. punctata*, and ten stamens, occur sporadically and have been named *C. grandis* Ashe and var. *pausiaca* (Ashe) E. J. Palmer. They are probably hybrids with *C. crus-galli*, as is the similar *C.* ×*disperma*, treated under interserial hybrids. The very rare *C.* ×*puberis* Sargent is perhaps a hybrid with *C. macrosperma*, *C.* ×*kellermanii* Sargent a hybrid with *C. pruinosa*, and *C.* ×*neobaxteri* Sargent a hybrid with a species of ser. *Rotundifoliae*.

40. Crataegus collina Chapman, Fl. S.E. U.S. ed. 2 repr. 2, 684. 1892 • Hillside hawthorn [E]

Shrubs, 60–80 dm. **Stems:** twigs: new growth pubescent, 1–2-years old grayish brown, older gray; thorns on twigs few to numerous, ± straight, 1-year old dark gray, usually ± slender, 2–4 cm. **Leaves:** petiole 1 cm, length 10–20% blade, <unwinged part often short or absent>, pubescent, eglandular; blade rhombic-obovate or broadly elliptic, sometimes ± narrowly obovate, ± obtrullate, or broadly rhombic-elliptic, 3–5(–7.5) cm, 1.4–2.5 times as long as wide, base tapering into petioles, lobes 0, or 1 or 2(or 3) per side, lobe apex acute to subacute, rarely obtuse, margins serrate in distal ¾, veins 5–7 per side, apex subacute to obtuse, abaxial surface scabrous-pubescent, especially on veins, adaxial scabrous-pubescent young, glabrescent. **Inflorescences** 5–10-flowered; branches appressed-pubescent; bracteoles membranous, margins sessile- or short-stipitate-glandular; <flowering usually early, often before full leaf expansion>. **Flowers** (10–)13–18(–20) mm diam.; hypanthium glabrous or ± sparsely pubescent; sepal margins usually glandular-serrate, abaxially glabrous; stamens (5–)20, anthers white, cream, or rose; styles (3–)5. **Pomes** orange or ruddy to deep red, orbicular, 8–14 mm diam., <punctate>, usually glabrous; sepals patent-reflexed; pyrenes (3–)5. **2*n*** = 34, 51, 68.

Varieties 5 (5 in the flora): c, se United States.

Crataegus collina is found from Oklahoma and Kansas to the Atlantic. It is absent from middle and higher elevations in the Appalachians, where it is replaced by *C. punctata*.

Crataegus collina is similar to its almost allopatric counterpart *C. punctata*, differing in its less tabulate growth habit, leaves on extension shoots not deeply incised, fewer-flowered inflorescences, fewer (five to seven versus seven to ten) and less impressed veins, often wider leaf shape (especially in var. *hirtiflora*), somewhat darker twig color, sepal margins more or less glandular-serrate, and sometimes different stamen number. *Crataegus collina* is one of the first hawthorns to flower in its area of distribution, unlike *C. punctata*, which is late mid season. *Crataegus* ×*disperma* (see Interserial Hybrids, p. 637) is somewhat similar to *C. collina*; it is allopatric and usually has more lateral leaf veins and fewer pyrenes. *Crataegus collina* is apparently intolerant of drought and is often leafless by late August in dry summers.

Crataegus collina is variable and has been divided into varieties. The varieties have been worked out in Missouri but not systematically recorded elsewhere; the varietal distributions provided here may be incomplete. Three putative hybrids between *C. collina* or *C. punctata* (*C.* ×*collicola*, *C.* ×*disperma*, and *C.* ×*incaedua*) are included in the following key.

1. Styles and pyrenes 2 or 3; stamens 10–15; leaves: adaxial surfaces sparsely hairy or subglabrous young.
   2. Inflorescence branches glabrous; anthers pink or red-purple . . . . . 158. *Crataegus* ×*disperma*, p. 637
   2. Inflorescence branches puberulent; anthers cream. . . . . 161. *Crataegus* ×*incaedua*, p. 638
1. Styles and pyrenes (3–)5; stamens (5–)15–20; leaves: adaxial surfaces short-pubescent, puberulent, or pubescent young (except var. *sordida* and *C.* ×*collicola*).
   3. Leaf blades ⅔ to nearly full grown at anthesis, length/width = 1.3, lobes, if present, usually obtuse.
      4. Stamens 15–20; leaf blades 3–5 cm, lobes 0 or obscure . . . . 40c. *Crataegus collina* var. *hirtiflora*
      4. Stamens 5–15(–20); leaf blades 5–7.5 cm, lobes 1 or 2 per side . . . . 40d. *Crataegus collina* var. *lettermanii*

[3. Shifted to left margin.—Ed.]

3. Leaf blades usually 1/3–1/2 full grown at anthesis, length/width = 1.4–2.25+, lobes, if present, usually acute.
    5. Anthers white or cream.
        6. Leaf blades at anthesis: length/width = 1.4–1.7; flowers 14–18 mm diam. . . . . 40a. *Crataegus collina* var. *collina*
        6. Leaf blades at anthesis: length/width = 1.75–2.25; flowers 10–12 mm diam. . . . . 40b. *Crataegus collina* var. *succincta* (in part)
    5. Anthers rose.
        7. Leaf blades at anthesis: length/width = 1.75–2.25 . . . . 40b. *Crataegus collina* var. *succincta* (in part)
        7. Leaf blades at anthesis: length/width = 1.4–1.5.
            8. Leaf margins serrate distally, teeth 1.5 mm . . . . . . 40e. *Crataegus collina* var. *sordida*
            8. Leaf margins serrulate, teeth 0.5–0.7 mm . . . . 157. *Crataegus* ×*collicola*, p. 637

40a. Crataegus collina Chapman var. **collina** [E]

Crataegus sucida Sargent; *C. vicina* Sargent

Leaf blades rhombic-obovate or broadly rhombic-elliptic to narrowly obovate, 3–6 cm, 1/3–1/2 full grown at anthesis, length/width = 1.4–1.7, membranous (young) to chartaceous or subcoriaceous (mature), lobes 0, or 1 or 2 per side, max LII 10%, lobe apex acute, margins crenate to serrate, teeth 0.5–0.8 mm, apex subacute to obtuse, adaxial surface matte or lustrous, densely short-pubescent young. **Inflorescence branches** hairy. **Flowers** 14–18 mm diam.; stamens 20, anthers white or cream.

Flowering Apr–May; fruiting Sep–Oct. Pastures, fencerows, scrub, open woodlands; 10–500 m; Ala., Ark., Ga., Ill., Ind., Kans., Ky., La., Miss., Mo., N.C., S.C., Tenn., Va.

Crataegus collina is recorded from Oklahoma without variety, probably var. *collina*.

Variety *collina* comprises three forms. The two predominant forms are thin-leaved. The first, most resembling the type, has leaves more obovate with lobes absent and marginal teeth smaller and obtuse; the second has narrower leaves, more like var. *succincta* but somewhat broader, with two or three acute lobes per side and marginal teeth sharper. Both are widespread. The third, *Crataegus sucida*, from Missouri, is distinguished by leaves that are more or less obovate, subcoriaceous, with lobes absent, sharply and evenly toothed with teeth 1 mm.

40b. Crataegus collina Chapman var. **succincta** (Sargent) E. J. Palmer, Ann. Missouri Bot. Gard. 22: 560. 1935 [E]

Crataegus succincta Sargent, Crataegus Missouri, 78. 1908; *C. angustata* Sargent

Leaf blades narrowly obovate, 3–5 cm, 1/3–1/2 grown at anthesis, length/width = 1.75–2.25, membranous (young) to chartaceous (mature), lobes 0 or 1–3 per side, evident or obscure, max LII 10%, lobe apex acute, margins serrate, teeth 0.5 mm, apex acute, adaxial surface matte, densely puberulent young. **Inflorescence branches** pubescent. **Flowers** 10–12 mm diam.; stamens 20, anthers white, cream, or rose.

Flowering Apr; fruiting Sep–Oct. Pastures, fencerows, scrub, open woodlands; 10–200 m; Mo.

Variety *succincta* is uncommon; *Crataegus angustata* is the rose-anthered form. The leaf blades are particularly narrow, and small at anthesis.

40c. Crataegus collina Chapman var. **hirtiflora** (Sargent) J. B. Phipps, Novon 16: 385. 2006 [E]

Crataegus hirtiflora Sargent, Crataegus Missouri, 82. 1908 (as hertiflora); *C. macropoda* Sargent

Leaf blades broadly elliptic, 3–5 cm, 2/3 to nearly full grown at anthesis, length/width = 1.3, thin to chartaceous, lobes 0 or 1 per side, obscure, max LII 10%, lobe apex usually obtuse, margins serrate, teeth 0.5 mm, apex usually subacute, adaxial surface matte, short-pubescent young. **Inflorescence branches** short-pubescent. **Flowers** 15–18 mm diam.; stamens 15–20, anthers white or cream.

Flowering Apr; fruiting Sep–Oct. Pastures, fencerows, scrub, open woodlands; 10–200 m; Ark., La., Mo.

Variety *hirtiflora* is common west of the Mississippi River.

40d. Crataegus collina Chapman var. **lettermanii** (Sargent) Eggleston, Rhodora 10: 76. 1908 (as lettermani) [E]

Crataegus lettermanii Sargent, Bot. Gaz. 31: 220. 1901 (as lettermani); *C. collina* var. *secta* (Sargent) E. J. Palmer; *C. secta* Sargent

Leaf blades broadly elliptic, 5–7.5 cm, nearly full grown at anthesis, length/width = 1.3, chartaceous, lobes 1 or 2 per side, evident, max LII 20%, lobe apex obtuse, margins serrate, teeth 1 mm, apex subacute to obtuse, adaxial surface matte, pubescent young. **Inflorescence branches** pubescent. **Flowers** 14–18 mm diam.; stamens 5–15 (–20), anthers white or cream.

Flowering Apr; fruiting Sep–Oct. Pastures, fencerows, scrub, open woodlands; 10–500 m; Mo.

Variety *lettermanii* has leaves rather large for *Crataegus collina*; they are also relatively wide and have well-developed, obtuse lobes, making it like a larger var. *hirtiflora*.

40e. Crataegus collina Chapman var. **sordida** (Sargent) Eggleston, Rhodora 10: 76. 1908 [C] [E]

Crataegus sordida Sargent, Bot. Gaz. 33: 114. 1902

Leaf blades ± obtrullate, 3.5–5 cm, ½ grown at anthesis, length/width = 1.4–1.5, subcoriaceous, lobes 0 or 1–3 per side, max LII 15%, lobe apex acute, margins serrate distally, teeth 1.5 mm, apex subacute to acute, adaxial surface matte to sublustrous young, with short, scattered, caducous hairs. **Inflorescence branches** sparsely pubescent. **Flowers** 15(–20) mm diam.; stamens 20, anthers rose.

Flowering Apr; fruiting Sep–Nov. Brush, open woodlands; of conservation concern; 100–200 m; Mo.

Variety *sordida* is rare in southern Missouri.

B. F. Bush noted on a herbarium label that var. *sordida* flowered later and had larger flowers than sympatric *Crataegus collina*, assumed to be var. *collina*. The bracteoles of var. *sordida* are particularly narrow (20:1). Variety *sordida* has not been recorded since the early 1900s.

41. Crataegus spes-aestatum J. B. Phipps, Novon 16: 382, fig. 1. 2006 • Hope-of-summers [C] [E]

Shrubs or trees, 30–60 dm. **Stems:** twigs: new growth canescent, 1-year old reddish brown, older ± pale gray; thorns on twigs absent or few, straight or slightly curved, 1-year old silver-black, slender, 2.5–6 cm. **Leaves:** petiole 0.7–1.5 cm, length 18–25% blade, densely pubescent young, glabrescent, sparsely sessile-glandular; blade rhombic-elliptic, 4–6 cm (3–4 cm at anthesis), length/width = 1.5–2, base cuneate, lobes 3 or 4 per side, lobe apex acute, margins serrate, <teeth gland-tipped>, veins 4 or 5 per side, apex acuminate, abaxial surface sparsely pubescent, densely pubescent on veins young, adaxial scabrous-pubescent young. **Inflorescences** 6–12 -flowered; branches densely canescent; bracteoles membranous or ± herbaceous, margins glandular. **Flowers** 20–22 mm diam.; hypanthium canescent; sepal margins glandular-serrate, abaxially sparsely pubescent; stamens 20, anthers yellow or cream; styles 5. **Pomes** red, suborbicular, 10 mm diam. (dry), glabrous, short-pubescent at base and apex; sepals ± eroded (in specimen seen); pyrenes 3–5.

Flowering Apr; fruiting Sep–Oct. Thickets, hills, dry open woods, alluvial stream banks; of conservation concern; 100–200 m; Ill., Mo.

Crataegus spes-aestatum is known only from southern Illinois and counties bordering the Mississippi River in southern Missouri. It is similar to *C. collina* and has not been seen with certainty since the 1930s.

42. Crataegus jonesiae Sargent, Bot. Gaz. 31: 14. 1901 (as jonesae) • Jones hawthorn [C] [E]

Shrubs or trees, 60–70 dm. **Stems:** twigs: new growth ± densely appressed-pubescent, 1-year old pale tan, 2-years old olive gray, older gray; thorns on twigs: frequency not recorded, straight to recurved, 1-year old chestnut brown, 2-years old shiny dark brown or black, slender to stout, 3–6 cm. **Leaves:** petiole 1.5–3.5 cm, length 35–50% blade, pubescent, eglandular; blade broadly ovate to broadly rhombic-elliptic or nearly circular, 4–7(–8) cm, length/width = 1–1.5, base tapered, wedge-shaped to subtruncate, lobes (4 or)5–9 per side, obscure, lobe apex usually acute, margins doubly serrate, veins 5–10 per side, apex acute, abaxial surface densely appressed-pubescent on veins, adaxial appressed-pubescent young, sparsely pubescent or glabrescent later. **Inflorescences** 8–15(–20)-flowered, <lax>; branches with antrorse pubescence; bracteoles membranous, margins glandular. **Flowers** 18–23 mm diam.; hypanthium densely pubescent; sepals <6–8 mm>, margins ± entire, abaxially pubescent; stamens 10, anthers pink; styles 2 or 3(or 4). **Pomes** bright red, oblong or obovoid, 10–15 mm diam., pubescent, glabrescent; sepals erect-patent; pyrenes 2 or 3(or 4).

Flowering Jun; fruiting Sep–Oct. Woodland glades, stream borders, rocky banks; of conservation concern; 10–200 m; N.B., N.S., P.E.I.; Maine.

Reports of *Crataegus jonesiae* from Quebec have not been substantiated; vegetatively similar plants from Matapédia and Matane counties have shallowly, laterally scarred pyrenes; sepals are gland-margined and are barely denticulate.

Crataegus jonesiae is usually placed in ser. *Anomalae*. However, it is a particularly distinct species that has smooth pyrenes and other similarities to *C. punctata*, and the possibility of a hybrid origin should be considered; its leaves range from nearly entire, prototypically punctate, to sharply lobed as in *C. chrysocarpa*.

64f.4. CRATAEGUS Linnaeus (sect. COCCINEAE) ser. AESTIVALES (Sargent) Rehder, Man. Cult. Trees ed. 2, 366. 1940 [E]

Crataegus [unranked] *Aestivales* Sargent, Silva 13: 35. 1902; *Crataegus* sect. *Aestivales* (Sargent) C. K. Schneider

Shrubs or trees, 30–120 dm, main trunk dominant. **Stems:** trunk bark dark gray, scaling; freshly exposed bark gray; twigs: new growth glabrous or rufous-tomentose, 1-year old dark chestnut brown to dark gray, older gray; thorns on twigs few to numerous (less abundant on older shoots in *C. aestivalis*), straight, 2-years old dark gray to black, stout, 1–5 cm. **Leaves:** petiole <short>, length 10–16% blade, glabrous or rufous-tomentose, eglandular; blade elliptic to narrowly oblanceolate, lance-elliptic, or ovate, 2.5–7 cm, somewhat coriaceous, base cuneate, lobes 0 or sinuate, sinuses shallow, lobe apex obtuse, margins finely crenate-serrate to obscurely serrulate or entire, <glandular, sometimes eglandular>, venation ± craspedodromous, veins (3 or)4–9 (or 10) per side, apex acute to obtuse, <glossy to matte>, surfaces glabrous or hairy. **Inflorescences** <appearing before or with young leaves>, 2–6-flowered, ± umbellate; branches glabrous or hairy; bracteoles often caducous, narrow or oblong-linear, membranous, margins glandular. **Flowers** 12–25(–30) mm diam.; hypanthium glabrous, rarely rufous-tomentose; sepals triangular, much shorter than petals, margins entire or glandular-serrate; <petals white to sometimes pale pink in *C. opaca*>; stamens 20, anthers pink to reddish or rose; styles 4 or 5. **Pomes** <ripe May–late June>, usually red, sometimes yellow, suborbicular, 8–15(–20) mm diam., glabrous; flesh succulent; sepals usually ± recurved, <very short, accrescent>; pyrenes 4 or 5.

Species 3 (3 in the flora): se United States.

Members of ser. *Aestivales* are found in wetlands from eastern Texas to North Carolina. Two species are largely allopatric (*Crataegus aestivalis* and *C. opaca*) and a third (*C.* ×*rufula*) is more or less intermediate between the first two and occurs where the ranges of the other two overlap. The mayhaws are distinct in precocious flowering, habitat, exceptionally early fruit ripening, and more or less umbellate inflorescences. Also, uniquely in sect. *Coccineae*, it appears that some thorns may be indeterminate.

The fruit is widely consumed in foods such as pies, jellies, jams, and wine. The species are cultivated in home gardens and are also, on a small scale, commercially grown and processed.

SELECTED REFERENCES Croft, B. R., G. Melcher, and E. Langston. 1996. Mayhaws: A Guide to Orchard Production and Propagation. Kearney. Phipps, J. B. 1988b. *Crataegus* (Maloideae, Rosaceae) of the southeastern United States. I. Introduction and series *Aestivales*. J. Arnold Arbor. 69: 401–431. Sargent, C. S. 1920. *Crataegus* ser. *Aestivales*. In: Notes on North American Trees. VI. J. Arnold Arbor. 1: 248–252.

1. Leaf blades elliptic to oblanceolate or narrowly so, margins eglandular (except sometimes on teeth), entire proximally, serrulate to crenate in distal ½, abaxial surfaces glabrous except for tufts of usually pale gray hair in vein axils, adaxial ± shiny . . . . 43. *Crataegus aestivalis*
1. Leaf blades ± elliptic to lance-elliptic, sometimes broadly elliptic to ovate, margins gland-dotted, sinuate, abaxial surfaces rufous-tomentose, adaxial usually dull.
   2. Leaf blades ± elliptic to lance-elliptic, sometimes broadly elliptic, 5–7 cm, veins 5–9 (or 10) per side; inflorescence branches absent or glabrous . . . . .44. *Crataegus opaca*
   2. Leaf blades elliptic to ovate, 2.5–4.5 cm, veins 3–5 per side; inflorescence branches glabrous or rufous-tomentose . . . . .45. *Crataegus ×rufula*

43. Crataegus aestivalis (Walter) Torrey & A. Gray, Fl. N. Amer. 1: 468. 1840 • Eastern mayhaw E

Mespilus aestivalis Walter, Fl. Carol., 148. 1788; *Crataegus cerasoides* Sargent; *C. fruticosa* Sargent; *C. luculenta* Sargent; *C. maloides* Sargent; *C. monantha* Sargent

Shrubs or trees, 30–120 dm. **Stems:** twigs: new growth <reddish>, glabrous, 1-year old dark brown (in late summer); thorns on twigs 1-year old dark, ± shiny, older grayish, 2–4 cm. **Leaves:** <stipules broadly circinate, margins glandular>; petiole <3–8 mm>, length 10–16% blade, <winged distally>; blade <± dark green>, elliptic to oblanceolate or narrowly so, 3–5 cm (sometimes much larger on extension shoots, then sometimes few-lobed, sometimes deeply so, even almost tripartite), lobes 0, margins entire proximally, serrulate to crenate in distal ½, usually eglandular, teeth tips sometimes with black glands, veins 4 or 5 per side, apex acute to subobtuse, abaxial surface glabrous except for tufts of usually pale gray (sometimes ± rufous) hair in lateral vein axils, sometimes also along midvein, adaxial ± shiny, scabrous or glabrous. **Inflorescences** 2–4 -flowered umbels; branches glabrous or sparsely long-pilose; bracteoles narrow, margins sessile-glandular. **Flowers** 12–30 mm diam.; hypanthium glabrous; sepal margins ± entire, <glabrous>; anthers pink; styles 4 or 5. **Pomes** red, sometimes yellow, 8–20 mm diam., <shiny; flesh edible>; <sepals obtuse>; pyrenes 4 or 5.

Flowering Feb–Mar (later northward); fruiting May–Jul. Seasonally inundated depressions, ditches, sink holes, streamsides; 10–100 m; Ala., Fla., Ga., Miss., N.C., S.C.

Crataegus aestivalis is common in northern Florida and southern Georgia and continues up the coastal plain to New Bern, North Carolina, and west to Mississippi; it is scarce in the northern portion of its range.

Crataegus aestivalis is variable in flower and fruit size. *Crataegus maloides* has unusually large, and *C. luculenta* unusually small, flowers; *C. cerasoides* has large, late-ripening fruit. *Crataegus fruticosa* is distinguished by late anthesis, at nearly full leaf expansion, and could be mistaken for a form of *C. crus-galli*. *Crataegus monantha* is a dwarf form from northern Florida and might prove useful as a dwarfing stock for the mayhaw industry.

Crataegus aestivalis is widely used in the southern states for conserves.

The description of *Crataegus aestivalis* by Torrey and Gray, together with the distribution they ascribed to it, indicates that they included *C. opaca*. The Pursh collection of *C. elliptica* Aiton (reputedly at BM or OXF) has not been located, and the type of *C. lucida* Elliott (at CHARL) has not been accessed; these two names may belong here. C. S. Sargent (1920) thoroughly discussed the typification of *C. aestivalis*, suggesting that *C. elliptica* and *C. lucida* represent the same species.

44. Crataegus opaca Hooker & Arnott, Compan. Bot. Mag. 1: 25. 1835 • Western mayhaw E F

Trees, 80+ dm. **Stems:** <trunks 0.3 m diam.>; twigs: new growth rufous-tomentose, 1-year old medium to dark gray; thorns on twigs 2-years old dark gray to black, 2–4 cm. **Leaves:** petiole <4–7 mm>, length 10–15% blade, <short rufous-tomentose>; blade ± elliptic to lance-elliptic, sometimes broadly elliptic, 5–7 cm, lobes 0 or sinuate (sometimes more deeply and irregularly lobed on vegetative shoots, then lobes often wider), margins entire or obscurely or shallowly crenate distally, teeth gland-dotted, veins 5–9(or 10) per side, apex subacute to acute, abaxial surface rufous-tomentose, especially along veins, sometimes glabrescent, adaxial usually dull, scabrous, especially young. **Inflorescences** ± sessile or in 3–6-flowered umbels; branches absent or short, glabrous; bracteoles <few>, oblong-linear, margins glandular. **Flowers** <precocious>, 12.5–17.5 mm diam.; hypanthium glabrous; sepals <4 mm>, margins entire or slightly glandular-serrate; <petals white to sometimes pale pink, 7 mm>; anthers reddish or rose, <1 mm>; styles 4 or 5. **Pomes** red, sometimes yellow, (8–)12–15 mm diam.; pyrenes 4 or 5. $2n = 34$.

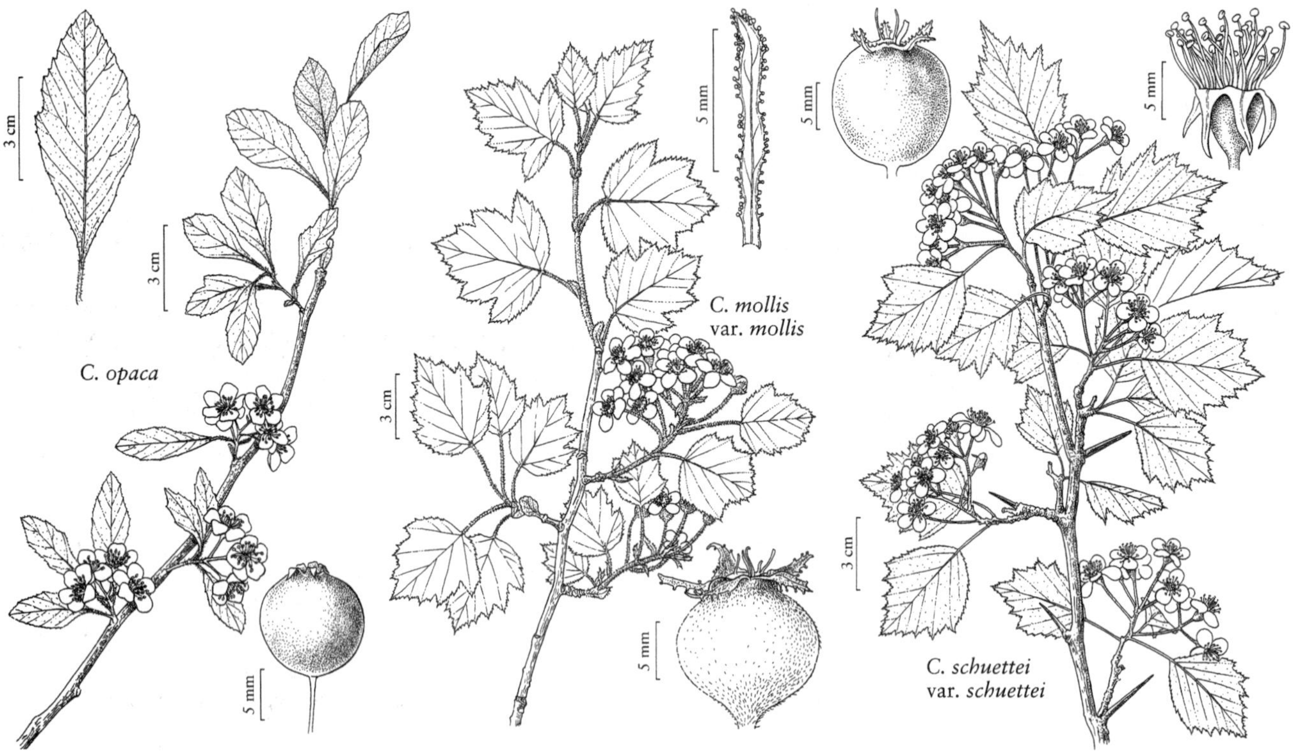

CRATAEGUS

Flowering Feb–Mar; fruiting May–Jun. Seasonally inundated depressions, ditches, sink holes, relict interdune swales, streamsides; 10–100 m; Ala., La., Miss., Tex.

Crataegus opaca occurs from eastern Texas to Alabama, is most common in Louisiana, and is apparently rare east of the Pearl River, Mississippi. The fruit is edible and used locally for conserves.

45. Crataegus ×rufula Sargent, J. Arnold Arbor. 1: 251. 1920, as species • Rufous mayhaw [E]

Shrubs or trees, 30–50 dm. **Stems:** twigs: new growth rusty-tomentose, 1-year old chestnut brown; thorns on twigs 2-years old dark gray to black, 1–3 (–5) cm. **Leaves:** petiole length 10–15% blade; blade elliptic to ovate, 2.5–4.5 cm, lobes 0 or sinuate (those on rapidly elongating shoots larger, usually proportionately wider than those on short shoots, sinuately lobed), margins entire or barely serrate or crenate in distal 1/3 or 1/2, teeth gland-tipped, veins 3–5 per side, apex subacute to obtuse, abaxial surface densely rufous-tomentose, especially on veins, adaxial densely white- or rufous-tomentose young, scabrous mature. **Inflorescences** 2–5 -flowered umbels; branches glabrous or rufous-tomentose; bracteoles narrow, margins glandular. **Flowers** <± precocious>, 15–27.5 mm diam.; hypanthium glabrous or rufous-tomentose; sepal margins entire or slightly glandular-serrate; <petals 12 mm>; anthers red; styles (4 or)5. **Pomes** red, 10 mm diam.; <sepals short-triangular>; pyrenes 5. $2n$ = 34, 51.

Flowering Feb–Mar; fruiting May–Jun. Seasonally inundated depressions, ditches, sink holes, streamsides; 10–100 m; Ala., Fla., Ga.

Crataegus ×rufula is restricted to the Florida panhandle and adjacent Georgia and southeastern Alabama.

Crataegus ×rufula is a presumed hybrid between the other two mayhaws or descendants thereof (J. B. Phipps 1988b) as shown by morphometric analyses. Although in some characteristics (for example, the usually sinuate leaf margin) *C. ×rufula* is like *C. opaca*, it lacks the characteristic elongate leaves of the latter. *Crataegus ×rufula* also tends to intergrade with *C. aestivalis*, with which it is fully sympatric. The frequency of *C. aestivalis*-like intermediates could well be accounted for by pure *C. aestivalis* being reasonably common through the *C. ×rufula* range and *C. opaca* being scarce there. Some, not all, individuals of *C. ×rufula* have relatively large flowers. *Crataegus ×rufula* is locally common and conveniently treated as a species.

64f.5. CRATAEGUS Linnaeus (sect. COCCINEAE) ser. MADRENSES J. B. Phipps, Monogr. N. Mex. Crataegus, 14. 1997

Shrubs or trees, 40–80 dm, main trunk dominant. **Stems:** trunk bark not recorded; twigs: new growth pubescent at least initially, 1-year old brown, older dull gray; thorns on twigs numerous, straight to slightly recurved, 2-years old dark gray, ± shiny, ± slender, 2.5–5 cm. **Leaves:** petiole length 10–25% blade, pubescent, glandularity not recorded; blade broadly elliptic or elliptic-rhombic, rarely narrowly oblanceolate, 3–4 cm, subcoriaceous, base cuneate, lobes 0 or sinuses shallow, lobe apex acute, margins serrate except proximally, venation craspedodromous, veins 5 or 6 per side, apex acute to obtuse, abaxial surface glabrate except on veins, adaxial appressed-pilose, usually persisting. **Inflorescences** 6–15-flowered; branches canescent; bracteoles caducous, linear, membranous, margins stipitate-glandular. **Flowers** 15–17 mm diam.; hypanthium tomentose especially proximally; sepals narrowly triangular, $\frac{2}{3}$ petal length, margins glandular-serrate to glandular-pectinate; stamens 10–20, anthers pink; styles 3–5. **Pomes** bright red, suborbicular, 8–10(–20) mm diam., pubescent; flesh not recorded; sepals patent-reflexed; pyrenes 3.

Species 2 (1 in the flora): Texas, ne Mexico.

The two species of ser. *Madrenses* occur in northeastern Mexico. *Crataegus tracyi* extends to west-central Texas; *C. aurescens* J. B. Phipps is endemic to northeastern Mexico. Species of ser. *Madrenses* resemble those of ser. *Crus-galli* but have more styles and pyrenes (three to five) than is usual in the latter series, as well as strongly glandular-serrate to glandular-laciniate sepal margins.

46. Crataegus tracyi Ashe ex Eggleston, Bull. Torrey Bot. Club 36: 639. 1909 • Tracy hawthorn

Crataegus montivaga Sargent

Shrubs or trees, 40–80 dm. **Stems:** twigs: new growth densely pubescent, 1-year old brown, older dull gray; thorns on twigs straight to slightly recurved, 2-years old dark gray, ± shiny, ± slender, 2.5–5 cm. **Leaves:** petiole 3–10 mm, length 10–25% blade; blade dark green, broadly elliptic or elliptic-rhombic, rarely narrowly oblanceolate, 3–4 cm, subcoriaceous, base cuneate, lobes 0 or sinuses shallow, margins serrate except proximally, veins 5 or 6 per side, apex acute to obtuse, glossy, abaxial surface glabrate except on veins, adaxial appressed-pubescent, usually persisting. **Inflorescences** 6–15-flowered; branches densely pubescent; bracteoles caducous, linear, membranous, margins strikingly stipitate-glandular. **Flowers** 15–17 mm diam.; hypanthium tomentose especially proximally; sepals narrowly triangular, margins glandular-serrate to glandular-pectinate, adaxially hairy in central part; stamens 10–20, anthers pink; styles 3–5. **Pomes** bright red, suborbicular, 8–10(–20) mm diam., pubescent; sepals patent-reflexed; pyrenes 3.

Flowering Apr–May; fruiting Sep–Oct. Streamsides, sometimes under shade; 300–1700 m; Tex.; Mexico (Coahuila, Nuevo León).

Crataegus tracyi is common in parts of Texas (Davis Mountains and Edwards Plateau) and in the mountains of Coahuila and Nuevo León in Mexico.

Variation in Texas is considerable, much of which was documented by J. B. Phipps (1997), with the typical form the most abundant in the state. The typical form has its widest extension-shoot leaves longer than broad, thorns straight, fine, 2–5 cm, and three styles. Forms resembling the common Mexican variety, var. *madrensis* J. B. Phipps, which has its widest extension-shoot leaves as broad as long, thorns short and stout, somewhat recurved, 2–3 cm, and four or five styles, but with ten stamens, have recently been collected in Texas. Variation in sepal margin dissection and inflorescence indumentum is also significant, much of the variation being geographically partitioned between the Davis Mountains and Edwards Plateau. *Crataegus tracyi* is a striking plant, having rich glossy green foliage during summer, coloring burgundy in the fall. It is hardy to USDA Zone 5, and thus has good potential in ornamental horticulture.

64f.6. CRATAEGUS Linnaeus (sect. COCCINEAE) ser. GREGGIANAE J. B. Phipps, Monogr. N. Mex. Crataegus, 14. 1997 [C]

Shrubs or trees, 40–80 dm, usually main trunk dominant. **Stems:** trunk bark buff to pale gray-brown, fibrous, checked into longitudinal plates, freshly exposed bark orange-brown; <branches spreading, ± straight toward tips>; twigs: new growth densely appressed-pubescent, 1-year old gray, sometimes ± shiny, 2-years old dull gray; thorns on twigs numerous, ± recurved, 2-years old shiny black, ± slender, 2.5–5 cm. **Leaves:** petiole length 25–33% blade, densely pubescent, eglandular; blade broadly ovate or broadly elliptic to rhombic-elliptic, 2.5–4(–5) cm, chartaceous, base ± truncate or cuneate, lobes obscure or 1–3 per side, sinuses very shallow, lobe apex acute, margins serrate, venation craspedodromous, veins 4–6 per side, apex acute to subacute, abaxial surface glabrate to tomentose, pubescent on veins, adaxial pubescent young. **Inflorescences** 3–8 -flowered, convex panicles; branches pubescent; bracteoles caducous, linear, membranous, margins stipitate-glandular. **Flowers** 15 mm diam.; hypanthium tomentose; sepals narrowly triangular, much shorter than petals, margins glandular-serrate; stamens 10[–20], anthers pink or rose [or white]; styles [3–]5. **Pomes** brick red to bright red, suborbicular, 7–11 mm diam., pubescent; flesh mealy; sepals sometimes eroded, spreading when retained, <non-accrescent>; pyrenes [2–]5.

Species 2 (1 in the flora): Texas, Mexico.

Series *Greggianae* is restricted to the northern Sierra Madre Oriental, Mexico, and Edwards Plateau, Texas; it is found in dry and mesic scrub. The strictly Mexican species in the series is *Crataegus serratissima* J. B. Phipps.

Members of ser. *Greggianae* are similar to ser. *Molles* but are smaller leaved, smaller flowered, have less prominent bracteoles, and are more xeromorphic.

47. Crataegus greggiana Eggleston, Bull. Torrey Bot. Club 36: 511. 1909 • Gregg hawthorn [C]

Crataegus uvaldensis Sargent

Shrubs or trees, 40–80 dm. **Stems:** twigs: new growth densely appressed-pubescent, 1-year old gray, sometimes ± shiny, 2-years old and older dull gray; thorns on twigs numerous, ± recurved, 2-years old shiny black, 2.5–5 cm. **Leaves:** petiole 6–12 mm, length 25–33% blade, densely pubescent, eglandular; blade mid green, broadly ovate (base ± truncate) or broadly elliptic to rhombic-elliptic (base cuneate), 2.5–4(–5) cm, lobes obscure or 1–3 per side, sinuses very shallow, lobe apex acute, margins serrate, teeth tips glandular, especially proximally, veins 4–6 per side, abaxial surface glabrate to tomentose, pubescent on veins, adaxial appressed-hairy young, becoming scabrous or glabrescent. **Inflorescences** 3–8-flowered; branches tomentose; bracteoles caducous, linear, membranous, margins stipitate-glandular, pubescent. **Flowers** 15 mm diam.; hypanthium tomentose; sepals narrowly triangular, 4–6 mm, margins glandular-serrate; stamens 10[–20], anthers pink or rose; styles 5. **Pomes** brick red to bright red, suborbicular, 7–11 mm diam., pubescent; sepals spreading or erose; pyrenes 5. $2n = 68$.

Flowering Apr; fruiting Sep–Oct. Deciduous scrub, pastures, pinyon woodlands; of conservation concern; 300–700 m; Tex.; Mexico (Coahuila, Nuevo León).

Crataegus greggiana is a scarce plant of the Edwards Plateau of Texas; it is more common and variable in the Mexican Sierra Madre Oriental. It usually occurs in more xeric habitats than most *Crataegus*.

The stamens are always ten in Texas but are often 20 in Mexico. *Crataegus uvaldensis* is a particularly tomentose form. A somewhat similar local species of the Edwards Plateau is 168. *C. turnerorum* (see there for differences).

64f.7. CRATAEGUS Linnaeus (sect. COCCINEAE) ser. MOLLES (Sargent) Rehder, Man. Cult. Trees ed. 2, 361. 1940 [E]

Crataegus [unranked] *Molles* Sargent, Silva 13: 33. 1902; *Crataegus* sect. *Molles* (Sargent) C. K. Schneider

Shrubs or trees, (20–)40–100 dm, main trunk dominant. **Stems:** trunk bark buff to gray-brown, fibrous, checked into longitudinal plates, freshly exposed bark orange-brown; <branches spreading; twigs ± straight>, new growth pubescent, sometimes glabrescent, 1-year old tan or grayish tan to brown, 2-years old pale gray-brown, older gray; thorns on twigs absent (particularly mature) or few to numerous, ± straight to ± recurved, 2-years old dark brown to black, ± stout, 3–7(–9) cm. **Leaves:** petiole length (25–)30–50% blade, pubescent to tomentose, sometimes glabrescent, usually eglandular; blade ± elliptic to ± ovate, ± rhombic, or rhombic-ovate, sometimes suborbiculate or ovate-deltate, 4–8(–12) cm, thin to ± firmly chartaceous, base broadly cuneate to subtruncate or rounded, lobes 0 or 1–6 per side (1 or 2 apiculi per side in *C. brazoria*), sinuses shallow to deep, lobe apex obtuse to very acute, margins usually strongly serrate, venation craspedodromous, veins (3 or)4–7(–9) per side, apex acute to obtuse, abaxial surface hairy to glabrate except on veins, adaxial hairy young. **Inflorescences** (5–)10–20-flowered, convex panicles, <arising subterminally from woody short shoots of season>; branches densely pubescent to tomentose; bracteoles (smaller) caducous, linear, membranous, margins sessile- or stipitate-glandular, (larger) ± persistent, narrowly oblong, ± herbaceous, margins sessile- or stipitate-glandular, abaxially villous. **Flowers** (14–)18–25 mm diam.; hypanthium ± densely pubescent to tomentose; sepals narrowly triangular, much shorter than petals, margins glandular-laciniate to glandular-serrate, rarely nearly entire; <petals white, sometimes pink at late anthesis in *C. mollis*>; stamens 5 or 6, 10, or 20, anthers usually ivory or rose, sometimes cream to pale pink or salmon, red, or purple; styles (3–)5. **Pomes** orange-red, bright or deep red, or bright yellow, usually suborbicular, 8–15(–25) mm diam., hairy (at least base and apex); flesh mealy; sepals ± retained, erect or erect-patent to broadly spreading, rarely incurved, <± accrescent>; pyrenes (3–)5.

Species 6 (6 in the flora): North America; introduced in Europe.

Crataegus mollis and immediate relatives (with 20 stamens except in rare forms of *C. mollis*) mostly occur from the Great Plains to the Appalachians south to the Gulf Coast; *C. pennsylvanica* and *C. submollis* (with ten stamens) range from Wisconsin to New England, to Delaware, and North Carolina (*C. pennsylvanica*).

As defined here, and following E. J. Palmer (1925) except for *Crataegus greggiana*, ser. *Molles* comprises four wide-ranging and two local species.

Plants in ser. *Molles* comprise relatively large-leaved, large-flowered, rather early-flowering hawthorns that are shrubs or trees with stout thorns, relatively large, usually red fruit, glandular-serrate sepals, and plant parts that are usually very hairy, especially when young; similar characteristics are found in ser. *Coccineae*. Compared to ser. *Coccineae*, specimens of ser. *Molles* are usually more hairy, particularly on the abaxial leaf surfaces, in the inflorescences, and in the density and persistence of indumentum. Other distinctions lie in the larger bracteoles in ser. *Molles* being broader and subherbaceous (versus narrower and membranous in ser. *Coccineae*) and in the fruiting sepals in ser. *Molles* being accrescent and tending to be suberect (versus non-accrescent and spreading in ser. *Coccineae*). The characteristic bracteoles of ser. *Molles* taxa are very evident in figures for J. B. Phipps (2012). Additionally, at least in their region of sympatry, the filament lengths of ser. *Molles* are shorter relative to length (50–60% versus 70–90%). Because of difficulties in measuring these delicate organs, this information is not used in the key.

Members of ser. *Molles* may also be confused with the somewhat similar ser. *Punctatae* but differ in their much more tomentose indumentum, usually larger and relatively wider leaves, wider and more persistent bracteoles, and, in most species, more deeply dissected sepals. *Crataegus phippsii*, a purple-fruited species of ser. *Purpureofructus* from the Pacific Northwest, is also similar in aspect and can be confused in flower. The rare local species *C.* ×*dispessa*, *C.* ×*kelloggii*, and *C.* ×*latebrosa*, previously assigned to the series, represent hybrids or hybrid derivatives with species of other series and key out in the eleventh, second, and fifth couplets, respectively; they possess more membranous, less persistent bracteoles, are usually not tomentose in the same way, have thinner leaves, and are treated under interserial hybrids.

SELECTED REFERENCE Phipps, J. B. 2012. Critical taxa in series *Molles* (*Crataegus*, Rosaceae): Typifications, new combinations, and taxonomic review. Phytoneuron 2012-78: 1–23.

1. Pomes yellow, gold, or orange.
    2. Inflorescences: branches sparsely villous; hypanthia hairy proximally; fruiting pedicels glabrous; Missouri . . . . 162. *Crataegus* ×*kelloggii*, p. 639
    2. Inflorescences: branches tomentose; hypanthia tomentose; fruiting pedicels pubescent; Texas.
        3. Leaf blades ± broadly ovate, lobes 2–4 per side, sinuses shallow, abaxially rough-hairy mature; anthers ivory-cream; pomes 8–14(–20) mm diam. . . . . . . 48. *Crataegus mollis* (in part)
        3. Leaf blades usually ± elliptic, lobes 0, abaxially glabrescent mature; anthers rose; pomes 8–10 mm diam. . . . . 51. *Crataegus brazoria* (in part)
1. Pomes orange-red to deep red.
    4. Stamens 5–10.
        5. Thorns on twigs usually numerous; anthers rose; sw Missouri, n Arkansas . . . . 163. *Crataegus* ×*latebrosa*, p. 639
        5. Thorns on twigs usually absent or few, sometimes numerous; anthers ivory or cream to pale pink or salmon; ne Great Plains to Missouri, nw United States, adjacent Canada.
            6. Anthers ivory, stamens 10; thorns on twigs numerous, 2-years old shiny blackish; leaf blades: bases cuneate . . . . 52. *Crataegus submollis* (in part)
            6. Anthers cream to pale pink or salmon, stamens 5 or 6, or 10; thorns on twigs absent or few, 2-years old shiny dark brown; leaf blades: bases broadly cuneate to truncate . . . . 53. *Crataegus pennsylvanica*
    4. Stamens 20.
        7. Anthers rose, rose-purple, red, or purple.
            8. Leaf blades ovate to suborbiculate, 4–5 cm, lobes 1 or 2 per side, bluish mature; shrubs 20–80 dm; thorns on twigs numerous; sw Missouri . . . . 50. *Crataegus lanuginosa*
            8. Leaf blades ± rhombic or ovate, 4–8 cm, lobes 1–4 per side, sinuses deep to shallow, or ± elliptic, lobes 0, or 1 or 2 apicula per side, green mature; shrubs 80–100 dm; thorns on twigs absent or few, sometimes frequent; Missouri to Texas.
                9. Leaf blades ± ovate, lobes 1–4 per side, sinuses shallow to deep; Missouri to Texas . . . . 49. *Crataegus texana*
                9. Leaf blades ± elliptic, lobes 0, or 1 or 2 apicula per side; sc, se Texas . . . . 51. *Crataegus brazoria* (in part)
        7. Anthers ivory or cream, rarely pink.
            10. Petioles glandular . . . . 52. *Crataegus submollis* (in part)
            10. Petioles eglandular.
                11. Leaf blades ovate or suborbiculate, lobes 0 or 2–6 per side; larger bracteoles subherbaceous, narrowly oblong . . . . 48. *Crataegus mollis* (in part)
                11. Leaf blades elliptic to suborbiculate, lobes 0 or 1–3 per side; all bracteoles ± membranous, linear . . . . 159. *Crataegus* ×*dispessa*, p. 638

48. Crataegus mollis (Torrey & A. Gray) Scheele, Linnaea 21: 569. 1848 • Red haw, downy hawthorn [E] [F]

Crataegus coccinea Linnaeus var. *mollis* Torrey & A. Gray, Fl. N. Amer. 1: 465. 1840

Shrubs or trees, 40–100 dm. **Stems:** <trunk bark fibrous, checked into longitudinal plates, ± exfoliating or rough-corrugated; compound thorns on trunks frequent in some populations>; twigs: new growth tomentose young, glabrescent, 1-year old tan, older pale gray; thorns on twigs absent or frequent, rarely plentiful, ± straight or ± recurved, 2-years old blackish brown, graying with age, usually ± stout, 3–6 cm. **Leaves:** petiole length 40–50% blade, tomentose young, tomentose to glabrate mature, usually eglandular; blade ± elliptic to ± ovate or suborbiculate, 4–8(–12) cm, thin to chartaceous, base truncate to cuneate, lobes 0 or 2–6 per side, sinuses shallow to deep, lobe apex acute to obtuse, margins strongly serrate, <teeth eglandular>, veins 4–7(–9) per side, apex acute to obtuse, abaxial surface pubescent young, densely so on veins, adaxial densely scabrous-pubescent young, densely scabrous-pubescent or glabrous mature. **Inflorescences** 5–15-flowered; branches tomentose; bracteoles caducous, <greenish>, linear-lanceolate or narrowly oblong, thin, larger ones subherbaceous, narrowly oblong, membranous to subherbaceous, margins short-stipitate or sessile-glandular, <multi-veined, abaxially pubescent>. **Flowers** 20–25 mm diam.; hypanthium densely tomentose; sepals narrowly triangular, 6 mm, margins glandular-serrate to glandular-laciniate, abaxially pubescent; <petals white, sometimes pink at late anthesis>; stamens (10 or)20, anthers ivory-cream; styles (3–)5. <**Infructescences:** fruiting pedicels pubescent>. **Pomes** bright red, sometimes orange-red or deep red, rarely yellow, suborbicular, <often tapered at base>, 8–14 mm diam., ± tomentose; sepals obsolescent or erect-patent; pyrenes (3–)5.

Varieties 5 (5 in the flora): c, e North America.

Crataegus mollis is common and widespread west of the Appalachians from eastern South Dakota to extreme southwestern Ontario, to southern Texas and central Alabama. It is rare in the southeastern part of its range. Sporadic records to the east of this are misidentifications or recent introductions.

Crataegus mollis has 20, or, in one unusual form, ten, cream anthers and is distinguished from the superficially similar ten-stamen *C. submollis* by lack of pedicel glandularity and stamen number and from pink-anthered taxa of restricted range such as *C. texana* varieties by anther color. Whether *C. mollis* is truly distinct from the *C. texana* complex will require biosystematic work. *Crataegus mollis* is less glandular than the allopatric *C. submollis*, from which it differs in further ways (see 52. *C. submollis*). *Crataegus canadensis*, from Quebec, but with glandular petioles, could key out here and is treated as a form of *C. submollis*.

Crataegus mollis is variable in leaf shape and size; other characters are more constant. Northern plants are more floriferous, and their trunk bark is more exfoliating than southern ones, which have a more rough-corrugated bark. The leaves are usually shortest and proportionately widest in Texas. Larger and proportionately narrower leaves are found northeastward. The most superficially distinct form is var. *incisifolia*, with acute lobes and deep sinuses. The lobing varies from very small to absent in var. *dumetosa* and *C. invisa* Sargent. *Crataegus meridionalis* from Alabama and Mississippi is somewhat similar to var. *dumetosa* but has longer, elliptic to narrowly ovate, sharply serrate, rather coriaceous leaves.

1. Thorns on twigs few to numerous; twigs ± strongly flexuous; leaf blades elliptic to narrowly ovate .... 48e. *Crataegus mollis* var. *meridionalis*
1. Thorns on twigs absent or few; twigs ± straight; leaf blades ± ovate or suborbiculate to broadly elliptic.
    2. Leaf blades: lobes 4–6 per side, sinuses deep (LII 20–40%)...48b. *Crataegus mollis* var. *incisifolia*
    2. Leaf blades: lobes 0 or obscure, or 2–4 per side, sinuses ± shallow (LII 0–20%).
        3. Pomes yellow .... 48d. *Crataegus mollis* var. *viburnifolia*
        3. Pomes red.
            4. Leaf blades: lobes 2–4 per side, sinuses shallow (LII 8–20%) ....48a. *Crataegus mollis* var. *mollis*
            4. Leaf blades: lobes 0 or obscure, sinuses shallow (LII 0–8%) .... 48c. *Crataegus mollis* var. *dumetosa*

48a. Crataegus mollis (Torrey & A. Gray) Scheele var. **mollis** [E] [F]

Crataegus mollis var. *sera* (Sargent) Eggleston; *C. gravida* Beadle; *C. limaria* Sargent; *C. sera* Sargent; *C. transmississippiensis* Sargent

Stems: twigs ± straight; thorns on twigs absent or few. **Leaf blades** ovate to broadly elliptic, lobes 2–4 per side, sinuses shallow (LII 8–20%), lobe apex subacute to obtuse. **Stamens** (10 or)20. **Pomes** red. $2n = 34$.

Flowering Mar–May; fruiting Sep–Nov. Brush, woodland edges, open woodlands, mesic habitats in grasslands; 50–400 m; Ont.; Ala., Ark., Ill., Ind., Iowa, Kans., Ky., La., Mich., Minn., Miss., Mo., Nebr., N.Dak., Ohio, Okla., S.Dak., Tenn., Tex., Wis.

Forms of var. *mollis* with smaller leaf blades (4–6 cm) of varying shape, characterized by subacute apices, often more or less cuneate bases, triangular lobes, thinner texture, and usually smaller flowers (15 mm diam.) match the type of *Crataegus gravida* (the form chosen for the *C. mollis* illustration) and appear to constitute a more or less distinct race of open or light woodlands in Alabama and Tennessee. Forms described as *C. sera* are similar to typical *mollis* but have thinner leaves. A form similar to the Ozarkian *C. transmississippiensis*, also having ten stamens, and commonly three styles, has been found in Minnesota, where it was called *C. submollis* (W. R. Smith 2008).

48b. Crataegus mollis (Torrey & A. Gray) Scheele var. **incisifolia** Kruschke, J. Bot. Res. Inst. Texas 2: 473. 2008 [E]

Crataegus arkansana Sargent

Stems: twigs ± straight; thorns on twigs absent or few. **Leaf blades** broadly elliptic to ovate, lobes 4–6 per side, sinuses deep (LII 20–40%), lobe apex acute. **Stamens** 20. **Pomes** red.

Flowering Mar–May; fruiting Sep–Nov. Brush, open woodlands, flood plains; 50–300 m; Ont.; Ill., Mo., Tex., Wis.

48c. Crataegus mollis (Torrey & A. Gray) Scheele var. **dumetosa** (Sargent) Kruschke, Publ. Bot. Milwaukee Public Mus. 3: 188. 1965 [C] [E]

Crataegus dumetosa Sargent, Crataegus Missouri, 109. 1908

Stems: twigs ± straight; thorns on twigs absent or few. **Leaf blades** broadly elliptic to suborbiculate, lobes 0 or obscure, sinuses shallow (LII 0–8%). **Stamens** 20. **Pomes** red.

Flowering Mar–Apr; fruiting Sep–Oct. Wooded sites; of conservation concern; 150–300 m; Mo., Tex.

Variety *dumetosa* is recorded only from Neck City, Jasper County in Missouri. *Crataegus invisa* Sargent from Arkansas and Oklahoma, which is intermediate with var. *mollis*, may have some leaves without lobes but most are usually shallowly incised and larger than in var. *dumetosa*.

48d. Crataegus mollis (Torrey & A. Gray) Scheele var. **viburnifolia** (Sargent) Lance, Phytoneuron 2011-3: 6. 2011 [C] [E]

Crataegus viburnifolia Sargent, Trees & Shrubs 2: 145, plate 159. 1911; *C. brazoria* Sargent var. *viburnifolia* (Sargent) J. B. Phipps

Stems: twigs ± straight; thorns on twigs absent or few. **Leaf blades** usually ± broadly ovate, lobes 2–4 per side, sinuses shallow (LII 8–15%), lobe apex subacute to obtuse. **Stamens** 20. **Pomes** yellow, <to 25 mm diam.>. $2n = 51$.

Flowering Mar; fruiting Sep–Oct. Rich alluvial soil, brushy habitats; of conservation concern; 10–50 m; Tex.

Variety *viburnifolia* occurs in southeastern Texas. The pure form of the variety is occasionally encountered. Yellow-fruited plants similar to var. *viburnifolia* with deeply incised leaves and salmon anthers are perhaps hybrids with *Crataegus texana*.

48e. Crataegus mollis (Torrey & A. Gray) Scheele var. **meridionalis** (Sargent) Lance, Phytoneuron 2011-3: 6. 2011 [E]

Crataegus meridionalis Sargent, J. Arnold Arbor. 1: 252. 1920

Stems: twigs ± strongly flexuous; thorns on twigs few to numerous. **Leaf blades** elliptic to narrowly ovate, lobes 2–4 per side, sinuses shallow (LII 8–15%), lobe apex subacute to obtuse. **Stamens** 20. **Pomes** red.

Flowering Mar–Apr; fruiting Sep–Oct. Blackland soil, brushy habits; 30–100 m; Ala., Miss., Tenn.

Variety *meridionalis* has strongly toothed leaves that are very tomentose abaxially at anthesis.

49. Crataegus texana Buckley, Proc. Acad. Nat. Sci. Philadelphia 13: 454. 1862 • Texas red haw [E]

Crataegus mollis (Torrey & A. Gray) Scheele var. *texana* (Buckley) Lance

Shrubs or trees, 100 dm. **Stems:** <older trunk bark nearly black, grooved, younger gray, fibrous, checked into longitudinal plates>; twigs: new growth lanate, 1-year old pale grayish tan, older gray; thorns on twigs absent or frequent, straight, 2-years old ± bright chestnut brown or shiny black, fine, 4–5 cm. **Leaves:** petiole length 37–43% blade, tomentose young, glabrescent, eglandular; blade broadly elliptic, narrowly rhombic, rhombic, rhombic-ovate, or broadly ovate, 4–7 cm, thin, base broadly cuneate to rounded, lobes 1–4 per side, sinuses shallow to deep, lobe apex obtuse to subacute, margins strongly serrate except proximally, <teeth 2–3 mm>, veins 5 per side, apex acute to subacute or obtuse, abaxial surface sparsely to densely white-tomentose young, less dense mature, veins densely hairy, adaxial densely scabrous young, glabrescent. **Inflorescences** 7–12-flowered; branches tomentose, sometimes glabrescent; bracteoles caducous, ± linear, (larger) herbaceous to (smaller) membranous, margins glandular. **Flowers** 14–22 mm diam.; hypanthium tomentose; sepals narrowly triangular, 4–6 mm, margins glandular-serrate to glandular-laciniate, abaxially densely pubescent; stamens 20, anthers rose, rose-purple, red, or purple; styles 4 or 5. **Pomes** red, sometimes red-orange, suborbicular, 9–14(–25) mm diam., with remnant tomentum; sepals often eroded or broken, spreading; pyrenes 4 or 5.

Varieties 2 (2 in the flora): sc United States.

Crataegus texana is distributed through much of the eastern half of Texas and into Oklahoma, Arkansas, and southwestern Missouri.

Crataegus texana is provisionally treated as a complex of pink- to purple-anthered, 20-stamened, red-fruited forms.

1. Leaf blades: lobes 2–4 per side, sinuses deep, lobe apices ± angled . . . . . . 49a. *Crataegus texana* var. *texana*
1. Leaf blades: lobes 1 or 2(or 3) per side, sinuses shallow, lobe apices ± obtuse to subacute . . . . 49b. *Crataegus texana* var. *dasyphylla*

49a. Crataegus texana Buckley var. **texana** [E]

Crataegus quercina Ashe

Leaf blades: lobes 2–4 per side, sinuses deep, lobe apex ± angled.

Flowering Mar–Apr; fruiting Sep–Nov. Brush; 10–100 m; Tex.

Variety *texana* is apparently restricted to south-central Texas, where it is locally common. Sporadic forms from southeastern Texas with rhombic-ovate, deeply incised leaves and yellowish fruit likely represent introgression with *Crataegus mollis* var. *viburnifolia*.

49b. Crataegus texana Buckley var. **dasyphylla** (Sargent) J. B. Phipps, Phytoneuron 2012–78: 5. 2012 [E]

Crataegus dasyphylla Sargent, Rep. (Annual) Missouri Bot. Gard. 22: 80. 1912; *C. induta* Ashe

Leaf blades: lobes 1 or 2(or 3) per side, sinuses shallow, lobe apex ± obtuse to subacute.

Flowering Mar–Apr; fruiting Sep–Nov. Brush; 100–200 m; Ark., Mo., Okla.

Crataegus induta is a small-anthered and particularly large-fruited variant, locally known as turkey haw. *Crataegus brachyphylla* Sargent is very similar to var. *dasyphylla*, but with only three styles and pyrenes; it is from dry hills in southwestern Arkansas.

50. Crataegus lanuginosa Sargent, Trees & Shrubs 1: 113, plate 57. 1903 [C] [E]

Crataegus mollis (Torrey & A. Gray) Scheele var. *lanuginosa* (Sargent) Lance

Shrubs or arborescent shrubs, 20–80 dm. **Stems:** <trunk bark pale, stout>; twigs: new growth <light green>, villous, <at end of 1st year dull red-brown, sparsely villous or pubescent>, <strongly flexuous>, 1-year old gray-brown, <stout>; thorns on twigs numerous, straight, shiny purplish black, ultimately ashy gray, stout, (3–)7(–9) cm. **Leaves:** <stipules caducous, oblong, foliaceous, acute, villous, minutely red-glandular>; petiole <stout>, length 25–30% blade, <slightly grooved>, densely tomentose, glandularity not recorded; blade <bluish>, short-ovate to suborbiculate, 4–5 cm, <less than ½ grown when flowers open>, subcoriaceous, base broadly cuneate to rounded, lobes 1 or 2 per side, short, wide, lobe apex acute, margins coarsely, sharply doubly serrate, <teeth glandular>, veins 3–5 per side, apex acute or rounded and short-pointed, abaxial surface yellow-green, tomentose, midvein prominent, adaxial <dark or bluish green, lustrous>, densely hoary tomentose, later scabrous. **Inflorescences** 5–10-flowered, <compact>; branches hoary-tomentose; bracteoles persistent, oblong-obovate, (smaller) membranous or (larger) herbaceous, margins finely glandular-serrate, <apex acute or rounded, villous>. **Flowers** 20 mm diam.; hypanthium densely covered with long, matted, white hairs; sepals broad, short, margins stipitate-glandular, abaxially densely villous; stamens 20, anthers rose; styles 5, <base surrounded by large tufts of white hairs>. **Pomes** <in clusters on erect, tomentose pedicels>, dark crimson, sometimes with pale dots, orbicular to suborbicular or short-oblong, 10–12 mm diam., slightly hairy at ends; <flesh thin, orange, dry, mealy; calyces sessile, enlarged, with wide deep cavity>; sepals persistent, usually erect and spreading or incurved, <coarsely serrate, villous>; pyrenes 5, <dorsally irregularly grooved, sides smooth, acute at ends>.

Flowering late Apr; fruiting late Oct. Dry, gravelly hills; of conservation concern; 100–200 m; Mo.

Crataegus lanuginosa was recorded as common near Webb City, Jasper County, in the early part of the 20th century; it has not been collected since 1957. The species is distinguished from others of ser. *Molles* by its relatively small leaves of a distinctly bluish color; the dark crimson, hard fruits and remarkable development of the thorns are also unusual in ser. *Molles*, as is the habitat on dry gravelly hills (when 20–30 dm tall), although it also occurs in more mesic sites (when 40–80 dm tall).

51. Crataegus brazoria Sargent, Bot. Gaz. 31: 233. 1901 • Brazos hawthorn [C] [E]

Shrubs or trees, 80–100 dm. **Stems:** <older trunk bark gray and black, scaly>; twigs: new growth densely white-hairy, 1-year old shiny brown, older pale gray; thorns on twigs absent or few, ± straight, 2-years old black, slender, 4–7 cm. **Leaves:** petiole length 35–40% blade, densely pubescent young, becoming sparsely to densely pubescent, eglandular; blade elliptic to narrowly ovate, 4–7 cm, thin, base curved-cuneate, lobes 0, or 1 or 2 apicula per side, margins serrate, <teeth 1.5–2.5 mm>, veins 4–6 per side, apex acute, abaxial surface woolly pubescent, glabrescent, adaxial scabrous young, variably glabrescent. **Inflorescences** 7–12-flowered; branches tomentose; bracteoles caducous, linear, subherbaceous, margins glandular, <parallel veined, abaxially pubescent>. **Flowers** 15–20 mm diam.; hypanthium tomentose; sepals narrowly triangular, 5–6 mm, margins glandular-laciniate to nearly entire, abaxially densely pubescent; stamens 20, anthers rose; styles 4 or 5. <**Infructescences:** fruiting pedicels pubescent>. **Pomes** usually red, sometimes bright yellow to golden or orange, suborbicular, 8–12 mm diam., often bearing residual hairs, particularly toward base or apex; sepals often broken or absent, broadly spreading; pyrenes 4 or 5.

Flowering Mar–Apr; fruiting Sep–Oct. Rich alluvial soil, brush; of conservation concern; 10–100 m; Tex.

Crataegus brazoria is restricted to south-central and southeastern Texas.

Crataegus brazoria was originally described as yellow-fruited with narrow leaves that more or less lack lobes. Such forms are very rare. The species is here expanded to include more abundant, red-fruited forms that extend into central Texas and are generally similar to *C. texana* except in leaf shape. It is plausible, because of sympatry and fruit color, that yellow-fruited *C. brazoria* represents introgression between its red-fruited form and *C. mollis* var. *viburnifolia*.

52. Crataegus submollis Sargent, Bot. Gaz. 31: 7. 1901 • Northern downy hawthorn, northern red haw, aubépine subsoyeuse [E]

Crataegus arnoldiana Sargent; *C. canadensis* Sargent; *C. champlainensis* Sargent

Shrubs, sometimes trees, 60–70 dm. **Stems:** twigs: new growth appressed-pubescent, 1-year old tan to brown, older gray; thorns on twigs usually numerous, straight to curved, 2-years old shiny blackish, ± thick, 4–7 cm. **Leaves:** petiole length 30–50% blade, densely pubescent, ± glandular, <glands black>; blade ± broadly ovate to rhombic-elliptic, 4–9 (–11) cm, chartaceous, base cuneate to almost truncate, lobes 3–5 per side, <obscure on smaller leaves>, sinuses shallow to moderately deep, lobe apex acuminate, margins sharply, doubly serrate, <on larger leaves teeth 2–3 mm, gland-tipped at least initially>, veins 5–7 per side, apex acute to subobtuse, abaxial surface sparsely appressed-pubescent, sometimes glabrescent, veins abaxially pubescent, adaxial densely scabrous-pubescent young, variably glabrescent. **Inflorescences** 5–15-flowered; branches densely pubescent; bracteoles caducous, linear, membranous, margins sessile-glandular, larger similar but late-deciduous, more herbaceous. **Flowers** 20–22 mm diam.; hypanthium tomentose; sepals narrowly triangular, 6–7 mm, margins deeply glandular-serrate, abaxially pubescent; stamens 10(or 20), anthers ivory; styles (3–)5. **Pomes** bright red, suborbicular to pyriform or ellipsoid, 10–15 mm diam., hairy, especially proximally and distally; sepals prominent, erect-patent, <stiff>; pyrenes 3–5. $2n = 68$.

Flowering May–Jun; fruiting Sep–Oct. Brush, natural hedgerows, field margins, apparently calciphile; 20–400 m; N.B., N.S., Ont., Que.; Maine, Mass., Mich., N.H., N.Y., Ohio, Pa., Vt., Wis.; introduced in Europe.

Crataegus submollis ranges from eastern Wisconsin and northern Michigan through the limestone areas of southern Ontario into Quebec, and New York to Maine and southern Nova Scotia. It is occasionally found in Pennsylvania and is plausible for northern West Virginia. It is allopatric with *C. mollis*. *Crataegus submollis* has also been recorded wild in Europe, including France and northern Italy.

Crataegus submollis, though quite similar to *C. mollis*, differs in its distribution, stature, petiole glandularity, stamen number, and greater thorniness.

The typical form of *Crataegus submollis* has proportionately broader leaves that are broadly cuneate to subcordate basally. By contrast, *C. champlainensis* is a form with proportionately narrower leaves more or less truncate to subcordate at the base. Intermediate forms are common in Quebec and New England. Placing the 20-stamen *C. canadensis*, a distinctive local form from the Montreal area, with *C. submollis* is based on leaf shape, glandular petioles, and disjunction from the range of *C. mollis*. *Crataegus arnoldiana* is a form with particularly large fruit.

Possession of glandular petioles and similar flowering time means that *Crataegus submollis* can sometimes show surprising similarity to large specimens of *C. chrysocarpa*. Then, the large fruiting calyces and the tendency to a more herbaceous larger bracteole type of *C. submollis*, typical for ser. *Molles*, differentiate the two.

53. **Crataegus pennsylvanica** Ashe, Ann. Carnegie Mus. 1: 394. 1902 • Pennsylvania thorn [E]

Crataegus tatnalliana Sargent

Shrubs or trees, 70–80 dm. **Stems:** <trunk bark newly exposed orange-brown, older buff, fibrous, checked into longitudinal plates>; twigs: new growth densely pubescent, 1-year old fawn, older gray; thorns on twigs absent or few, usually recurved, 2-years old shiny dark brown to blackish, stout to more slender, 3–5 cm. **Leaves:** <stipules axillary, circinate, herbaceous, gland-bordered>; petiole <relatively slender, stouter than *C. coccinea*>, length 30–35% blade, densely pubescent young, sparsely mature, often sparsely stipitate-glandular; blade broadly ovate to ovate-deltate, 6–10 cm, slightly coriaceous, base broadly cuneate to truncate, lobes 4–6 per side, sinuses shallow, lobe apex acute to obscure, margins sharply serrate, <teeth at base often stipitate-glandular>, veins 5–7 per side, apex acute, abaxial surface sparsely scabrous-pubescent young, glabrescent, veins hairy mature, adaxial densely appressed-scabrous young. **Inflorescences** 8–20-flowered; branches tomentose; bracteoles variably persistent, linear-lanceolate, membranous to (larger) ± herbaceous, margins glandular. **Flowers** 17–21 mm diam.; hypanthium densely tomentose; sepals narrowly triangular, 7–8 mm, margins glandular-laciniate, abaxially glabrate; stamens 5 or 6, or 10, anthers cream to pale pink or salmon; styles 4 or 5. **Pomes** bright red, suborbicular, 10–12 mm diam., densely villous; sepals prominent, erect-patent; pyrenes 4 or 5.

Flowering May–Jun; fruiting Sep–Oct. Brush, successional fields, open woodlands, fencerows; 10–200 m; Ont.; Del., N.Y., N.C., Ohio, Pa.

Crataegus pennsylvanica ranges mainly to the south of *C. submollis* and may be locally common. A highly disjunct record from North Carolina (Buncombe County) appears unequivocal.

Crataegus pennsylvanica has been confused with *C. mollis* and various members of ser. *Coccineae*. Except for a small overlap in northeastern Ohio, its range lies outside that of *C. mollis*. *Crataegus pennsylvanica* is more similar to *C. submollis*, from which it differs by its less variable leaves (usually broadly ovate, more or less truncate, relatively large), by its much less thorny nature, by its indumentum on young parts being tomentose rather than densely pubescent, and by its being more treelike (*C. submollis* is usually a shrub). Thorns are absent or sparse on older trees.

64f.8. CRATAEGUS Linnaeus (sect. COCCINEAE) ser. COCCINEAE (Loudon) Rehder, Man. Cult. Trees ed. 2, 362. 1940 [E]

Crataegus sect. *Coccineae* Loudon, Arbor. Frutic. Brit. 2: 816. 1838; *Crataegus* [unranked] *Lobulatae* Sargent

Shrubs or trees, (40–)70–120 dm, main trunk dominant. **Stems:** trunk bark older buff, newly exposed orange, exfoliating in fibrous plates, freshly exposed bark orange-brown (not recorded for some species); <branches spreading; twigs ± straight>, new growth glabrous or hairy, 1-year old golden to pale tan, older light gray; thorns on twigs few to numerous, straight to recurved, 2-years old dark blackish brown, shiny, ± stout, sometimes slender, 2–6 cm. **Leaves:** petiole length 30–40% blade, glabrate to hairy, eglandular or at least some glandular; blade <bright green>, broadly elliptic to elliptic-ovate or ± ovate to deltate-ovate, (4–)5–8(–10) cm, thin, base cuneate to subtruncate, rarely slightly cordate, lobes 4 or 5 per side, sinuses shallow to

± deep, lobe apex acute, sometimes acuminate, margins serrate, <teeth numerous>, venation craspedodromous, veins 5–7(or 8) per side, apex acute to subacute, abaxial surface usually glabrous, veins pubescent, adaxial appressed-pubescent. **Inflorescences** 8–15-flowered, convex panicles; branches usually pubescent, rarely glabrous or tomentose; bracteoles caducous, linear, membranous, margins sessile- or short-stipitate-glandular. **Flowers** 16–22 mm diam.; hypanthium usually pubescent, sometimes glabrous; sepals narrowly triangular, much shorter than petals, margins glandular-serrate; stamens 5–8, rarely to 10 or 20, anthers pink to rose-purple, sometimes crimson; styles (3 or)4 or 5. **Pomes** bright to deep red, suborbicular to oblong or ± ovoid, sometimes ellipsoid to broadly pyriform, 10–14 mm tall, glabrous or villous; flesh ± mealy, mellow; sepals variably persistent, ± spreading or erose, <non-accrescent>; pyrenes (3 or)4 or 5.

Species 3 (3 in the flora): c, e North America; introduced in Europe.

As treated here, ser. *Coccineae* has three species, largely following J. A. Macklin (2001). They occur from extreme southeastern Minnesota and Illinois eastward through the southern Great Lakes to southern New England and south down the Appalachians to North Carolina.

Members of ser. *Coccineae* are easily confused with those of ser. *Molles*; *Crataegus pennsylvanica* from that series keys out in the first couplet. A detailed comparison is presented with ser. *Molles*. The best distinctions concern bracteoles (hairier and more herbaceous in ser. *Molles*) and indumentum (more likely tomentose in ser. *Molles*). Rare putative hybrids with various species are recorded, including ser. *Dilatatae* (*C. putnamiana* Sargent), ser. *Punctatae* (*C. mansfieldensis* Sargent), ser. *Tenuifoliae* (*C. merita* Sargent), ser. *Populneae* (*C. xanthophylla* Sargent); see also the section on Interserial Hybrids (p. 635).

Confusion about the name of the series was caused by C. S. Sargent (1901), who suggested that *Crataegus coccinea* was the correct name for *C. chrysocarpa* (at that time called *C. rotundifolia*). Consequently, until his correction in 1909, the group *Coccineae* by Sargent was approximately the current ser. *Rotundifoliae*. His original view was followed by various other workers, which added to the confusion.

Members of ser. *Coccineae* are shrubs or trees of forest edges and gaps as well as successional sites on agricultural land. They can be exceptionally attractive throughout the growing season and are among the most dramatic fall-fruiting shrubs. Anthesis is early midseason relative to sympatric species.

SELECTED REFERENCES Macklin, J. A. 2001. Systematics of *Crataegus* Series *Coccineae* (Rosaceae). Ph.D. thesis. University of Western Ontario. Macklin, J. A. and J. B. Phipps. 2002. Sargent's names in *Crataegus* series *Coccineae* (Rosaceae). Harvard Pap. Bot. 7: 25–35. Macklin, J. A. and J. B. Phipps. 2006. Studies in *Crataegus* ser. *Coccineae*. I. Canad. J. Bot. 84: 70–86. Sargent, C. S. 1901. Notes on *Crataegus* in the Champlain Valley. Rhodora 3: 19–31.

1. Stamens 5 or 6, or 10, anthers cream to pale pink or salmon; inflorescence branches tomentose; pomes densely villous; sepals in fruit erect-patent, ± accrescent . . . . 53. *Crataegus pennsylvanica* (ser. *Molles*), p. 557
1. Stamens 5–8, rarely 10 or 20, anthers pink to rose-purple, sometimes crimson; inflorescence branches sparsely to densely pubescent, rarely glabrous or tomentose; pomes glabrous or pubescent; sepals in fruit ± spreading, non-accrescent, sometimes erose.
    2. Stamens 20.
        3. Leaf blades ovate to elliptic-ovate, bases ± cuneate . . . . 54. *Crataegus coccinea* (in part)
        3. Leaf blades ovate to deltate-ovate, bases broadly ± cordate or rounded to truncate . . . . 55. *Crataegus magniflora*
    2. Stamens 5–10.
        4. Leaf blades ovate or broadly elliptic, bases broadly cuneate to subtruncate, rarely slightly cordate; twigs: new growth greenish; pomes usually bright red, suborbicular to oblong; thorns on twigs ± stout, 2–4 cm . . . . 54. *Crataegus coccinea* (in part)
        4. Leaf blades ± narrowly ovate, bases cuneate or ± rounded; twigs: new growth greenish to reddish; pomes bright to deep red, ellipsoid to broadly pyriform; thorns on twigs often slender, 3–6 cm . . . . 56. *Crataegus holmesiana*

54. Crataegus coccinea Linnaeus, Sp. Pl. 1: 476. 1753 [E]

Shrubs or trees, 70–80(–120) dm. **Stems:** twigs: new growth greenish, glabrous or slightly hairy; thorns on twigs straight to recurved, ± stout, 2–4 cm. **Leaves:** petiole length 30–40% blade, glabrate to densely hairy, glandular or eglandular; blade ovate or broadly elliptic to elliptic-ovate, (4–)5–8 cm, base broadly cuneate to subtruncate, rarely slightly cordate, lobe apex acute, margins serrate, sometimes doubly serrate, <teeth 2 mm>, adaxial surface usually densely scabrous young. **Inflorescences:** branches sparsely to densely pubescent. **Flowers:** hypanthium glabrous or densely pubescent; stamens (5–)8–10(–20), anthers pink to rose-purple. **Pomes** usually bright red, suborbicular to oblong, 10–14 mm, often sparsely pubescent (especially at ends); sepals spreading or missing.

Varieties 3 (3 in the flora): e North America; introduced in Europe.

Crataegus coccinea is found from extreme southeastern Minnesota, Wisconsin, and Illinois through the southern Great Lakes area to coastal Maine, and in the Appalachians at higher altitudes to North Carolina; it is common in the north of its range.

Crataegus coccinea varies in leaf shape and planeness and in fruit shape. In some areas, more or less pure populations of the different variants occur. It is easiest to group the variation around two relatively strongly marked varieties with ten or fewer stamens, vars. *coccinea* and *pringlei*, between which intermediates are common, and the 20-stamen var. *fulleriana*.

1. Leaf blades (elongate), ovate (length/width = 1.4–1.5), 5–8 cm (widest in proximal ⅓), plane, lobe sinuses deep, bases broadly cuneate to truncate, rarely slightly cordate; pomes usually suborbicular . . . . . 54a. *Crataegus coccinea* var. *coccinea*
1. Leaf blades broadly elliptic to elliptic-ovate (length/width = 1.2), 4–7 cm, plane or concavo-convex, lobe sinuses often shallow, bases rounded to broadly cuneate; pomes usually oblong or suborbicular.
   2. Stamens 8–10; leaf blades conspicuously concavo-convex; pomes usually oblong . . . . 54b. *Crataegus coccinea* var. *pringlei*
   2. Stamens 20; leaf blades usually plane, rarely concavo-convex; pomes usually suborbicular . . . . 54c. *Crataegus coccinea* var. *fulleriana*

54a. Crataegus coccinea Linnaeus var. **coccinea** • Scarlet hawthorn, aubépine écarlate [E]

Crataegus acclivis Sargent; *C. albicans* Ashe; *C. delecta* Sargent; *C. digna* Sargent; *C. ellwangeriana* Sargent; *C. lobulata* Sargent; *C. pedicellata* Sargent; *C. pedicellata* var. *albicans* (Ashe) E. J. Palmer; *C. pedicellata* var. *ellwangeriana* (Sargent) Eggelston; *C. pringlei* Sargent var. *lobulata* (Sargent) Eggelston; *C. uticaensis* Sargent

Leaf blades ovate, 5–8 cm <widest in proximal ⅓> (length/width = 1.4–1.5), <elongate>, plane, base broadly cuneate to truncate, rarely slightly cordate, lobe sinuses deep. **Stamens** (5–)8–10. **Pomes** usually suborbicular. **2*n*** = 51, 68.

Flowering May; fruiting Sep–Oct. Brush, woodland margins, fencerows, overgrown pastures; 10–1300 m; Ont.; Conn., Ill., Ind., Maine, Md., Mass., Mich., N.H., N.Y., N.C., Ohio, Pa., R.I., Vt., Va., W.Va., Wis.; introduced in Europe.

Variety *coccinea* is found throughout the range of the species except for southern Quebec; it is the only variety found in the Appalachians south of Pennsylvania. An old report for Minnesota has not been confirmed. *Crataegus albicans* is a variant with particularly broad leaves and nearly glabrous inflorescences. *Crataegus aulica* Sargent perhaps belongs here. According to J. A. Macklin and J. B. Phipps (2002), much of the material named *C. ellwangeriana* [*C. pedicellata* var. *ellwangeriana*] belongs to *C. pennsylvanica* Ashe.

54b. Crataegus coccinea Linnaeus var. **pringlei** (Sargent) J. A. Macklin & J. B. Phipps, J. Bot. Res. Inst. Texas 1: 1005. 2007 • Pringle hawthorn, aubépine de Pringle [E]

Crataegus pringlei Sargent, Rhodora 3: 21. 1901; *C. exclusa* Sargent; *C. gloriosa* Sargent; *C. pringlei* var. *exclusa* (Sargent) Eggleston

Leaf blades broadly elliptic to elliptic-ovate, 4–7 cm (length/width = 1.2), conspicuously concavo-convex, base rounded to broadly cuneate, lobe sinuses often shallow. **Stamens** 8–10. **Pomes** usually oblong. **2*n*** = 51, 68.

Flowering May; fruiting Sep–Oct. Brush, woodland margins, fencerows, overgrown pastures; 10–300 m; Ont., Que.; Conn., Ill., Ind., Mass., Mich., Minn., N.H., N.Y., Ohio, Pa., R.I., Vt., Wis.

Variety *pringlei* is found throughout much of the range of the species, just reaching Quebec, but is absent in Virginia and North Carolina. Extreme forms are very distinct in appearance with their strikingly convex, broadly elliptic, usually shallowly incised leaves and often oblong fruit. Intermediates with var. *coccinea* are common.

54c. Crataegus coccinea Linnaeus var. **fulleriana** (Sargent) J. A. Macklin & J. B. Phipps, Phytoneuron 2011–29: 1. 2011 E

Crataegus fulleriana Sargent, Proc. Rochester Acad. Sci. 4: 111. 1903; *C. chippewaensis* Sargent; *C. holmesiana* Ashe var. *chippewaensis* (Sargent) E. J. Palmer; *C. illecebrosa* Sargent

Leaf blades broadly elliptic to elliptic-ovate, 4–7 cm (length/width = 1.2), usually plane, rarely concavo-convex, base rounded to broadly cuneate, lobe sinuses often shallow. **Stamens** 20. **Pomes** usually suborbicular.

Flowering May; fruiting Sep–Oct. Brush, woodland margins, fencerows, overgrown pastures; 10–300 m; Ont.; Mich., Wis.

The leaf shape of var. *fulleriana* is somewhat intermediate between var. *coccinea*, var. *pringlei*, and *Crataegus holmesiana*. In J. A. Macklin (2001), the placement of var. *fulleriana* indicates it is not simply a 20-stamen variant of one of the five- to ten-stamen entities. *Crataegus confragosa* Sargent may represent a glabrate form. Very rare convex-leaved forms might also be interpreted as mutants of var. *pringlei*.

55. Crataegus magniflora Sargent, Bot. Gaz. 35: 383. 1903 E

Crataegus dayana Sargent; *C. hillii* Sargent; *C. macounii* Sargent

Shrubs or trees, 40–70 dm. **Stems:** twigs: new growth greenish, glabrous or sparsely hairy; thorns on twigs straight to recurved, usually ± stout, 2–4 cm. **Leaves:** petiole length 30–40% blade, nearly glabrate to hairy, glandular or eglandular; blade ovate to deltate-ovate, 6–10 cm, base broadly ± cordate or rounded to truncate, lobe apex acute, margins serrate, sometimes doubly serrate, <teeth 2 mm>, adaxial usually densely scabrous young. **Inflorescences:** branches sparsely to densely pubescent. **Flowers:** hypanthium glabrous or densely pubescent; stamens 20, anthers pink to rose-purple. **Pomes** usually bright red, suborbicular to ± ovoid, 10–14 mm, often sparsely pubescent (especially at ends); sepals ± spreading or missing. $2n$ = 68 (as *C. corusca*).

Flowering May; fruiting Sep–Oct. Brush, woodland margins, fencerows, overgrown pastures; 20–300 m; Ont.; Ill., Mass., Mich., N.Y., Pa., Wis.

Crataegus magniflora ranges through the southern Great Lakes area from Illinois and Wisconsin to Pennsylvania and Massachusetts. Two forms occur, one with a more nearly truncate leaf base, which is more common in Ontario, and a more *fulleriana*-like form, which is most common westward.

The leaf of *Crataegus magniflora* is most similar to that of *C. coccinea* var. *coccinea* but differs in usually being a little larger, often relatively wider (length/width sometimes equal to 1.1), and in the base being more truncate to more or less cordate than is usual in var. *coccinea*. *Crataegus magniflora* was often identified as *C. corusca* Sargent before it was recognized that the latter represented a putative interserial hybrid of ser. *Coccineae* × ser. *Molles* origin.

56. Crataegus holmesiana Ashe, J. Elisha Mitchell Sci. Soc. 16: 78. 1900 • Holmes' hawthorn, aubépine de Holmes E

Crataegus anomala Sargent; *C. eamesii* Sargent; *C. elongata* Sargent; *C. holmesiana* var. *villipes* Ashe; *C. pura* Sargent; *C. sertata* Sargent; *C. tardipes* Sargent

Shrubs or trees, 40–70(–100) dm, <single-trunked>. **Stems:** twigs: new growth greenish to reddish, glabrous or pubescent; thorns on twigs recurved, often slender, 3–6 cm. **Leaves:** petiole length 30–35% blade, glabrate to densely hairy, often sparsely glandular; blade ± narrowly ovate, 6–9 cm, base cuneate or ± rounded, lobe apex acuminate, margins serrate or doubly serrate, <teeth 2 mm>, adaxial usually densely scabrous-pubescent young. **Inflorescences:** branches usually pubescent, rarely glabrous or tomentose. **Flowers:** hypanthium usually pubescent; stamens 5–8(–10), anthers pink to rose-purple, sometimes crimson. **Pomes** bright to deep red, ellipsoid to broadly pyriform, 12–14 mm, usually glabrous; sepals wide-spreading or erose. $2n$ = 34, 51.

Flowering May; fruiting Sep–Oct. Woodland edges, old pastures, fencerows; 10–300 m; Ont., Que.; Conn., Ill., Ind., Maine, Mass., Mich., Minn., N.H., N.Y., Ohio, Pa., R.I., Vt., Wis.

Crataegus holmesiana ranges from Wisconsin and Illinois through the southern Great Lakes region to southern Quebec and Maine, and to Pennsylvania; it is common in much of its range and should be expected in West Virginia.

In its usual form, *Crataegus holmesiana* is readily distinguished from *C. coccinea* vars. *coccinea* and *pringlei*, though occasional intermediates occur. *Crataegus robesoniana* Sargent [*C. pedicellata* var. *robesoniana* (Sargent) E. J. Palmer] is one and perhaps *C. amicta* Ashe [*C. holmesiana* var. *amicta* (Ashe) E. J. Palmer] is another. *Crataegus holmesiana* is routinely interpreted as a typical member of ser. *Coccineae*, thus, a villous inflorescence taxon with serrate sepals. However, the protologue by Ashe (authentic material not located) states that the inflorescences are glabrous and the sepals are nearly entire, causing confusion with *C. tenuifolia* Britton (ser. *Tenuifoliae*). Current usage for *C. holmesiana* became fixed after Ashe erected a var. *villosa* with serrate sepals and villous inflorescences.

64f.9. Crataegus Linnaeus (sect. Coccineae) ser. Tenuifoliae (Sargent) Rehder, Man. Cult. Trees ed. 2, 362. 1940 [E]

Crataegus [unranked] *Tenuifoliae* Sargent, Silva 13: 34. 1902; *Crataegus* sect. *Tenuifoliae* (Sargent) Eggleston

Shrubs or trees, 20–70(–80) dm, main trunk dominant. **Stems:** trunk bark buff to gray-brown, fibrous, checked into longitudinal plates, freshly exposed bark orange-brown; <branches spreading; twigs ± straight>, new growth usually glabrous, sometimes sparsely pubescent, 1-year old deep reddish brown, older gray; thorns on twigs usually numerous, usually recurved, 1-year old shiny, dark brown to black, stout to slender, 2.5–6 cm. **Leaves:** petiole <1–3 cm>, length 30–60% blade, glabrous, sessile-glandular or eglandular; blade <pale green abaxially, mid dark green adaxially>, elliptic to ovate or ovate-deltate, 3–6(–8) cm, thin to chartaceous, base cuneate to rounded or truncate to subcordate, lobes 0 or 3–6 per side, sinuses shallow to deep, lobe apex usually acute, margins serrate, venation craspedodromous, veins (3 or)4–7 per side, apex usually acute, sometimes subacute or acuminate, abaxial surface glabrous, veins pilose young in some, adaxial appressed-scabrous-pubescent young, glabrescent. **Inflorescences** (4 or)5–10(–20)-flowered, convex panicles; branches glabrous or villous; bracteoles caducous, <absent or few to several>, narrow, membranous, margins short-stipitate-glandular. **Flowers** (10–)13–17(–26) mm diam.; hypanthium usually glabrous, sometimes villous at base; sepals narrowly triangular, much shorter than petals, margins subentire or glandular-serrate, rarely nearly eglandular; stamens 5–10 or 20, anthers pink to red or purple; styles 3–5. **Pomes** red, rarely with orange or orange-red spots, ellipsoid to suborbicular, sometimes broadly pyriform, 8–15(–20) mm diam., glabrous; flesh mellow; sepals erect-patent to recurved or erose, <non-accrescent>; pyrenes (2 or)3–5.

Species 7 (7 in the flora): North America.

Members of ser. *Tenuifoliae* range through nearly the whole of the eastern United States (except Florida) and also in adjacent Canada. *Crataegus wootoniana* is a disjunct species in the mountains of New Mexico.

Relatively small and thin leaves, entire or weakly serrate sepals, and pink to reddish anthers characterize most ser. *Tenuifoliae*. Bracteole frequency in this series appears to correlate positively with inflorescence hairiness; the species flower relatively early among sympatric congeners and are often strikingly tinged with anthocyanins at leaf expansion.

Members of ser. *Tenuifoliae* are most similar to those of ser. *Coccineae* (the latter usually being larger in all parts and usually more hairy), to ser. *Rotundifoliae* via *Crataegus flabellata*, and to some members of ser. *Populneae*, this last series a grouping of taxa intermediate between ser. *Pruinosae* and ser. *Tenuifoliae*.

The interserial hybrids *Crataegus* ×*lucorum* and *C.* ×*fretalis* key out in the fourth and eighth couplets, respectively.

1. Stamens ca. 20.
    2. Inflorescence branches glabrous.
        3. Leaf marginal teeth 1 mm; leaf blades on extension shoots ovate . . . .61. *Crataegus schuettei* (in part)
        3. Leaf marginal teeth 1.5–3 mm; leaf blades on extension shoots suborbiculate or broadly deltate . . . . 62. *Crataegus roribacca*
    2. Inflorescence branches hairy.
        4. Leaf lobe apices acute, lobe sinuses shallow to moderately deep; anthers purple; adaxial leaf surfaces glabrous . . . . .165. *Crataegus* ×*lucorum*, p. 640
        4. Leaf lobe apices acute to acuminate, lobe sinuses deep; anthers usually pink or red; adaxial leaf surfaces appressed-scabrous young.
            5. Leaf blades narrowly ovate; inflorescence branches sparsely pubescent . . . . .61. *Crataegus schuettei* (in part)
            5. Leaf blades broadly elliptic to ovate; inflorescence branches villous . . . . 63. *Crataegus flabellata* (in part)
1. Stamens 5–10.
    6. Inflorescence branches hairy.
        7. Leaf blades not glossy, mid green, lobes usually acute, sinuses shallow to moderately deep; sepal margins usually eglandular, subentire in c and e of range; Minnesota to Quebec and Maine, uncommon in area of sympatry with *C. flabellata* . . . . . . 60. *Crataegus fluviatilis*
        7. Leaf blades glossy, dark green, lobes acuminate, sinuses deep; sepal margins usually glandular, often serrate; Ontario to New England and Nova Scotia . . . . 63. *Crataegus flabellata* (in part)
    6. Inflorescence branches glabrous.
        8. Sepal margins coarsely glandular-serrate. . . . . 160. *Crataegus* ×*fretalis*, p. 638
        8. Sepal margins usually subentire.
            9. Leaves 40–60% mature size at anthesis; Louisiana to South Carolina piedmont north to se Virginia coastal plain . . . . 58. *Crataegus iracunda*
            9. Leaves 70–100% mature size at anthesis; New Mexico or ne and midwestern United States and adjacent Canada south in the Appalachians at high altitudes to Georgia.
                10. Flowers 13–17(–22) mm diam.; leaf sinuses moderately deep; petioles sparsely to densely glandular; e North America . . . . .57. *Crataegus macrosperma*
                10. Flowers 10–12 mm diam.; leaf sinuses shallow; petioles eglandular or sparsely glandular; New Mexico . . . . 59. *Crataegus wootoniana*

57. Crataegus macrosperma Ashe, J. Elisha Mitchell Sci. Soc. 16: 73. 1900 • Eastern hawthorn, aubépine à lobes aigus [E]

Crataegus acutiloba Sargent; *C. cyanophylla* Sargent; *C. demissa* Sargent; *C. iracunda* Beadle var. *brumalis* (Ashe) Kruschke; *C. macrosperma* var. *acutiloba* (Sargent) Eggleston; *C. macrosperma* var. *demissa* (Sargent) Eggleston; *C. macrosperma* var. *eganii* (Ashe) Kruschke; *C. macrosperma* var. *matura* (Sargent) Eggleston; *C. macrosperma* var. *pentandra* (Sargent) Eggleston; *C. macrosperma* var. *roanensis* (Sargent) E. J. Palmer; *C. matura* Sargent; *C. pastorum* Sargent; *C. pentandra* Sargent; *C. roanensis* Ashe; *C. tenella* Ashe

Shrubs or trees, <multi-stemmed>, 30–70 dm. **Stems:** twigs: new growth golden green to strongly reddish tinged, glabrous, 1-year old usually shiny, dark brown, older gray; thorns on twigs 1-year old shiny, deep brown to blackish, often black-tipped, stout to slender, 2.5–6 cm. **Leaves:** petiole length 30–60% blade, sparsely to densely glandular; blade ovate, 3–6(–9) cm, <70–100% mature size at anthesis>, base subcordate, truncate, or rounded to cuneate, lobes 3–6 per side, sinuses moderately deep, lobe apex acute to acuminate, margins serrate, <teeth numerous, small, larger ones conspicuously glandular, particularly at base, smaller ones inconspicuously gland-tipped>, veins 4–7 per side, apex acute, adaxial surface densely appressed-scabrous-pubescent young, glabrescent, rarely glabrate. **Inflorescences** 5–12-flowered; branches glabrous; bracteoles usually absent (even young), sometimes few to several, <often red-tinged>, ± linear. **Flowers** 13–17 (–22) mm diam.; hypanthium glabrous; sepals 4–5 mm, margins subentire, abaxially glabrous; stamens 5–10, anthers pink to reddish purple; styles 3–5. **Pomes** bright to deep red, suborbicular to ellipsoid, 8–15 mm diam.; sepals erect-patent or erose; pyrenes 3–5, dorsally shallowly grooved. **2*n*** = 68.

Flowering Apr–Jun; fruiting Sep–Nov. Old fields, fencerows, brush, open woods, montane balds and rocky slopes; 0–1600 m; N.B., Nfld. and Labr. (Nfld.), N.S., Ont., P.E.I., Que.; Ala., Ark., Conn., Ill., Ind., Ky., Maine, Md., Mass., Mich., Minn., N.H., N.J., N.Y., N.C., Ohio, Pa., R.I., Tenn., Vt., Va., W.Va., Wis.

Crataegus macrosperma ranges from north-central Minnesota through the central Great Lakes region and the St. Lawrence Valley to southern Newfoundland south to Arkansas, and in the Appalachians, to northeastern Alabama. Recent observations show that the species had been cultivated at the Arnold Arboretum from an alleged southern Missouri provenance, which links with the discovery of specimens of wild *C. macrosperma* and *C. schuettei* from the Ozarks of northern Arkansas. These populations are relictual, and the southwestern limits of the species are poorly known.

Crataegus macrosperma is one of the earliest hawthorns to flower, preceded only by *C. collina* in the south, *C. mollis* in the Midwest, and *C. chrysocarpa* and allies in the north.

Crataegus macrosperma is most similar to *C. fluviatilis*, the latter differing only by the hairy inflorescences. It is similar also to *C. wootoniana* of New Mexico, the latter differing mainly from the *roanensis* form of *C. macrosperma* by its frailer nature, smaller flowers, and perhaps thinner, smaller, but proportionately wider leaves, as well as its disjunct distribution. *Crataegus schuettei* is another similar but usually stouter species with 20 stamens; *C. flabellata*, often confused with *C. macrosperma* in the literature, should be readily distinguished by its glandular to glandular-serrate sepal margins and its hairy inflorescences.

In *Crataegus macrosperma*, leaves vary in size, shape of base, number, depth and sharpness of lobes, and number of lateral veins. The fruit is usually more or less oblong; forms with suborbicular fruits are found nearly throughout. Recent typification (J. B. Phipps 2009) shows that some of the varieties were based on wrong material; critical modern study of their variation is lacking, and none are formally recognized here. Finally, the stamen number (five to ten) of *C. macrosperma* is rather variable.

Suberect shrubby forms with small foliage (3–4 cm), cuneate leaf bases, and ellipsoid fruits represent *Crataegus roanensis* and are common to the Great Lakes region to Pennsylvania extending southward to the species southern limits, where they become large shrubs or small trees. Plants of similar range, stature, and leaf size, but with a more nearly truncate leaf base and suborbicular fruit, represent *C. macrosperma* in the strict sense, but intermediates are quite common.

Crataegus pastorum is similar to the typical form. *Crataegus demissa* is a scarce, poorly understood form with the most proportionately wide, broad-based, but small leaves. Forms with particularly large leaves include *C. acutiloba* and *C. matura*. These usually have the most numerous lateral veins and lobes, which are particularly acuminate, and are most common in Quebec, New England, and the Canadian Maritime provinces. *Crataegus acutiloba* is distinguished by rather oblong and large leaves (to 8 cm) with cuneate to broadly cuneate bases, acute and usually reflexed lobes, and narrow, oblong-pyriform fruits. *Crataegus matura* has the largest leaves in the species (to 9 cm) with broad-cuneate to subtruncate bases, particularly deeply incised lobes, but with suborbiculate to broadly oblong fruits. A puzzling observation is that the two large-leaved forms may have shallowly concave pyrene sides. *Crataegus pentandra* has long and quite narrow leaves, with a cuneate to rounded, rather than truncate, base, usually rather shallow lobes as well as narrow fruits.

The recently investigated Minnesota populations of *Crataegus macrosperma* barely differ from the *C. apiomorpha* form of *C. fluviatilis* except in lacking inflorescence pubescence; they are further discussed under the latter. *Crataegus cyanophylla* from Illinois is probably conspecific with *C. macrosperma*. The identity of *C. tenuifolia* Britton, usually placed here, requires investigation (see comments under 56. *C. holmesiana*). *Crataegus tenella*, from southeastern Pennsylvania and Delaware, represents intergradation with the southeastern Virginia populations of *C. iracunda*. Forms with nearly glabrous young adaxial leaf surfaces and slender blackish thorns found not uncommonly in the southeast may represent introgression with *C. pruinosa* or a species in ser. *Populneae*.

58. Crataegus iracunda Beadle, Biltmore Bot. Stud. 1: 124. 1902 • Passionate hawthorn [E]

Crataegus drymophila Sargent; *C. iracunda* var. *silvicola* E. J. Palmer

Shrubs or trees, 30–80 dm. **Stems:** twigs: new growth greenish, glabrous, 1-year old deep reddish brown, 2-years old grayer; thorns on twigs 2-years old blackish or dark gray, <shiny>, slender, 2.5–4 cm. **Leaves:** petiole length 40–50% blade, eglandular; blade ovate to ovate-deltate, 2–4(–5) cm, <1.2–1.4 times as long as wide, 40–60% mature size at anthesis>, base broadly rounded to subtruncate or subcordate, lobes

4 per side, sinuses moderately shallow, lobe apex ca. 90° at tip, often less, margins serrate, <teeth regular, 0.5–1 mm, minute, caducous gland-tipped>, veins 3 or 4(or 5) per side, apex acute, <abaxial surface glabrous except along veins>, adaxial finely appressed-pubescent young. **Inflorescences** 4–10-flowered; branches glabrous; bracteoles few, ± linear. **Flowers** 12–15 mm diam.; hypanthium glabrous; sepals 4–5 mm, margins ± entire or slightly glandular-serrate, abaxially glabrous; stamens 10, anthers pink to purple; styles 3–5. **Pomes** orange to red, or blotched green, suborbicular, 8–10 mm diam.; sepals spreading; pyrenes 3–5, dorsally grooved.

Flowering Apr; fruiting Sep–Oct. Open woodlands, brush, fencerows, cutovers; 20–300 m; Ala., Ga., La., Miss., N.C., S.C., Va.

Crataegus iracunda ranges from Louisiana to South Carolina and Georgia, and to Virginia, but only at low altitudes. Nearly all records of *C. iracunda* north of the listed distribution are referable to *C. macrosperma* or to various species in ser. *Populneae*.

The distinguishing features of *Crataegus iracunda* are most evident when it is in flower. The species can be construed as a smaller, more southern, allopatric relative of *C. macrosperma* distinguished by smaller leaves (particularly at anthesis), more slender thorns, and harder, drier pomes. Its distribution is almost wholly to the south of or in Appalachia, at lower elevations than *C. macrosperma*. The nearest records of *C. macrosperma* to the Louisiana populations of *C. iracunda* are from the Arkansas Ozarks.

Crataegus iracunda has been confused with *C. gattingeri* (ser. *Pruinosae*) but is readily distinguished in flower by adaxial leaf pubescence, stamen number, and by lacking the usually attenuate terminal leaf lobes of the latter. However, in fruit, unless the filament bases can be counted, one is left with the less reliable feature of the terminal lobe shape. Confusion with the larger-leaved *C. populnea* (ser. *Populneae*) is perhaps possible, but the two species are essentially allopatric and their leaves differ markedly in size and texture.

Earlier attempts to segregate varieties based on leaf size are not taken up here but may have merit as the relatively numerous Louisiana (*Crataegus drymophila*) form has much smaller leaves than North Carolina specimens.

Variety *brumalis* (*Crataegus brumalis* Ashe) with syntype material from near Pittsburgh, Pennsylvania, is a different entity. Compared to *C. iracunda*, it has larger and differently shaped (often more or less truncate-based) leaves, proportionately larger at anthesis, and adaxially subglabrous, conspicuously glandular petioles, and larger flowers. This taxon (as a variety of *C. iracunda*) is primarily responsible for the northward extension of the range of the latter species. In fact, north of the Mason-Dixon line, most of the specimens attributed to *C. brumalis* seen by the author appear to be forms of *C. macrosperma*.

59. Crataegus wootoniana Eggleston, Torreya 7: 236. 1908 • Wooton's hawthorn [C] [E]

Shrubs, 20 dm. **Stems** <little branched, thorny>; twigs: new growth not recorded, glabrous, 1-year old reddish brown, older gray; thorns on twigs 1-year old reddish brown, ± stout, 4 cm. **Leaves:** petiole length 40–50% blade, eglandular or sparsely glandular; blade ovate, 3–4(–5) cm, <70–100% mature size at anthesis>, base truncate to broadly cuneate, lobes 3 or 4 per side, sinuses shallow, lobe apex obtuse to cuspidate, margins serrate, veins 4 or 5 per side, apex subacute, adaxial surface short-pubescent young, later glabrate. **Inflorescences** 5–10-flowered; branches glabrous; bracteoles absent or few, shape not recorded; <anthesis in spring>. **Flowers** 10–12 mm diam.; hypanthium glabrous; sepals 2–3 mm, margins subentire, abaxially glabrous; stamens 10, anthers pink; styles 3. **Pomes** shiny red, ellipsoid, 8 mm; sepal remnants erect-patent; pyrenes 2 or 3, dorsally grooved.

Flowering May; fruiting Sep. Grassy areas near streams; of conservation concern; 2000–3000 m; N.Mex.

No authentic records of *Crataegus wootoniana* for Arizona are known; all alleged examples seen proved to be *C. macracantha*.

Crataegus wootoniana is similar to the widespread 57. *C. macrosperma* (see discussion thereunder), from which it differs little.

60. Crataegus fluviatilis Sargent, Rhodora 5: 117. 1903 [E]

Crataegus apiomorpha Sargent

Shrubs or trees, <multi-stemmed>, 30–40(–70) dm. **Stems:** twigs: new growth golden green to strongly reddish tinged, sparsely pubescent or glabrous, 1-year old usually shiny, dark brown, older gray; thorns on twigs 1-year old shiny, deep brown to blackish, often black-tipped, stout to slender, 2.5–5 cm. **Leaves:** petiole length 30–60% blade, sparsely to densely glandular; blade ovate, 3–6 cm, base subcordate, truncate, or rounded to cuneate, lobes 3–6 per side, sinuses shallow to moderately deep, lobe apex acute to acuminate, margins serrate, <teeth numerous, small, larger ones conspicuously glandular, particularly at base, smaller ones inconspicuously gland-tipped>, veins 4–6 per side, apex acute, adaxial surface densely appressed-scabrous-pubescent young, glabrescent. **Inflorescences** 5–12-flowered; branches pubescent; bracteoles usually absent (even young), sometimes few to several, <often red-tinged>, ± linear. **Flowers** 12–17 mm diam.; hypanthium glabrous; sepals 4–5 mm, margins subentire to weakly serrate, usually eglandular, abaxially glabrous or pubescent; stamens 5–10, anthers pink to reddish purple; styles 3–5. **Pomes** bright to deep red, suborbicular to ellipsoid, 8–15 mm diam.; sepals erose or erect-patent; pyrenes 3–5, dorsally shallowly grooved.

Flowering May–Jun; fruiting Sep–Oct. Old fields, fencerows, brush, woodland margins, open woods; 0–500 m; Ont., Que.; Conn., Ill., Maine, Mass., Mich., Minn., N.H., N.Y., R.I., Vt., Wis.

Crataegus fluviatilis includes a range of intermediates between *C. flabellata* and *C. macrosperma*. There are three main forms. In the typical form, sporadic in Quebec and New England, the glandular-serration of the sepal margins is less pronounced, and the inflorescence branches at anthesis are less densely hairy than in *C. flabellata*, but the leaves are more similar to those of *C. macrosperma*. They could represent rare hybrid derivatives of these two species. In the central Great Lakes region (Illinois to Ontario), more frequent forms occur attributable to *C. apiomorpha*, similar to sympatric *C. macrosperma*, differing mainly in their soft-hairy inflorescences. More complex variation occurs in Minnesota, where forms with densely pubescent inflorescences occur in local populations, often intermixed with otherwise identical *C. macrosperma* (glabrous inflorescences). Both the latter are notable for their serrate sepal margins, not the norm for *C. macrosperma*. Into the mix are identical forms with 20 stamens, which perhaps are *C. asperata* Sargent. This Minnesota complex is notable for smallish leaves (4–6 cm), is constant for characters other than those mentioned above, and suggests active introgression.

Confusion between the Minnesota form of *Crataegus fluviatilis* and sympatric *C. chrysocarpa* might be possible; a helpful field character at anthesis is the pink anthers of the former; at other times, it is the flexible petioles that allow the leaves of *C. fluviatilis* to flutter in the wind, unlike those of *C. chrysocarpa*, which have stiffer petioles.

61. Crataegus schuettei Ashe, J. Elisha Mitchell Sci. Soc. 17: 7. 1901 • Schuette or royal hawthorn, aubépine de Schuette [E] [F]

Shrubs or trees, 40–70 dm. **Stems:** twigs: new growth reddish, glabrous, 1-year old shiny reddish brown, older dull gray; thorns on twigs 1-year old glossy, dark brown to blackish, ± stout, 3–6 cm. **Leaves:** petiole length 50–60% blade, usually glandular; blade ovate to oblong-ovate or narrowly ovate (ovate on extension shoots), 3–6 cm, base truncate, rounded, or cuneate, lobes 4 or 5 per side, sinuses moderately deep, lobe apex acute to acuminate, margins serrate, <teeth 1 mm>, veins 5 or 6 per side, apex acute, adaxial surface sparsely to densely scabrous-pubescent young, early glabrescent. **Inflorescences** 4–10-flowered; branches usually glabrous, rarely sparsely pubescent; bracteoles absent or few, linear. **Flowers** 15–26 mm diam.; hypanthium glabrous; sepals 3–4 mm, margins subentire or glandular-serrate, abaxially glabrous; stamens 20, anthers red; styles 3–5. **Pomes** red, ellipsoid to suborbicular, 8–20 mm diam., <succulent>; sepals ± erose or erect-patent; pyrenes 3–5, dorsally grooved. $2n = 51$.

Varieties 4 (4 in the flora): c, e North America.

Crataegus schuettei ranges from Wisconsin through the southern Great Lakes to southern New England, to North Carolina. Recent discoveries add the Ozarks and may add Minnesota (see discussion under 60. *C. fluviatilis*) to the documented range. Variety *schuettei* closely resembles *C. macrosperma*, differing principally in its often larger flowers, 20 stamens, usually glandular-serrate sepal margins, and larger fruit but is not nearly so variable as *C. macrosperma*; the other varieties are very rare. Still, the four varieties are fairly distinct. *Crataegus basilica* appears to be the southern form of the type variety and is very like sympatric *C. macrosperma* except for stamen number.

1. Inflorescence branches sparsely pubescent; leaf blades narrowly ovate, bases ± narrowly cuneate . . . . 61d. *Crataegus schuettei* var. *cuneata*
1. Inflorescence branches glabrous; leaf blades ± ovate to ovate-oblong, bases subtruncate or broadly cuneate to rounded.
    2. Flowers 24–26 mm diam.; pomes 15–20 mm diam. . . . . .61b. *Crataegus schuettei* var. *gigantea*
    2. Flowers 15–20 mm diam.; pomes 8–15 mm diam.
        3. Leaf blades ± ovate, sinuses shallow (max LII 15–20%), proximal lobes almost never most deeply separated, spreading at 45–60° . . . 61a. *Crataegus schuettei* var. *schuettei*
        3. Leaf blades ovate to ovate-oblong, proximal sinuses deep (max LII 30–40%), proximal lobes the most deeply separated, spreading at 60–80° . . . . .61c. *Crataegus schuettei* var. *ferrissii*

61a. Crataegus schuettei Ashe var. **schuettei** E F

Crataegus basilica Beadle

Leaf blades ± ovate, 3–6 cm, base subtruncate to rounded, sinuses shallow (max LII 15–20%), proximal lobe almost never most deeply separated, spreading at 45–60°. **Inflorescence branches** glabrous. **Flowers** 15–20 mm diam. **Pomes** 8–15 mm diam. **2*n*** = 51.

Flowering Apr–May; fruiting Sep–Oct. Brush, field margins, woodland edges; 50–1500 m; Ont., Que.; Ark., Conn., Ill., Ind., Ky., Maine, Md., Mass., Mich., Mo., N.H., N.J., N.Y., N.C., Ohio, Pa., R.I., Tenn., Vt., Va., W.Va., Wis.

Variety *schuettei* represents the bulk of the species and is locally common throughout much of its range.

61b. Crataegus schuettei Ashe var. **gigantea** Kruschke, J. Bot. Res. Inst. Texas 2: 473. 2008 C E

Leaf blades ± ovate, 3–6 cm, base rounded to broadly cuneate, sinuses shallow (max LII 15–20%). **Inflorescence branches** glabrous. **Flowers** 24–26 mm diam. **Pomes** 15–20 mm diam.

Flowering May; fruiting Sep–Oct. Brush, field margins, woodland edges; of conservation concern; 300 m; Wis.

Variety *gigantea* is known only from Wisconsin and is hardly differentiated from var. *schuettei* except by its much larger flowers and larger fruit.

61c. Crataegus schuettei Ashe var. **ferrissii** (Ashe) Kruschke, Publ. Bot. Milwaukee Public Mus. 3: 176. 1965 C E

Crataegus ferrissii Ashe, J. Elisha Mitchell Sci. Soc. 17: 11. 1901 (as ferrissi)

Leaf blades ovate to ovate-oblong, 3–6 cm, base rounded to subtruncate, sinuses deep (max LII 25–40%), proximal lobe most deeply separated, spreading at 60–80°. **Inflorescence branches** glabrous. **Flowers** 15–20 mm diam. **Pomes** 9–12 mm diam.

Flowering May; fruiting Sep–Oct. Brush, field margins, woodland edges; of conservation concern; 300 m; Ill.

Variety *ferrissii* is a very distinct form from northern Illinois that has recently been rediscovered near Tinley Park; it should perhaps be referred to another species.

61d. Crataegus schuettei Ashe var. **cuneata** Krushke ex J. B. Phipps, Phytoneuron 2014–39: 2. 2014 C E

Leaf blades narrowly ovate, 3–4 cm, base ± narrowly cuneate, sinuses deep (max LII 15–20%). **Inflorescence branches** sparsely pubescent. **Flowers** 15–20 mm diam. **Pomes** 8–15 mm diam.

Flowering Apr–May; fruiting Sep–Oct. Brush, field margins, woodland edges; of conservation concern; 200–300 m; Minn., Wis.

Variety *cuneata* could perhaps be considered a 20 -stamen variant of *Crataegus fluviatilis* or a particularly narrowly leaved form of *C. flabellata* var. *grayana*.

62. Crataegus roribacca Ashe, J. Elisha Mitchell Sci. Soc. 17: 4. 1900 C E

Trees, 40–70 dm. **Stems:** <trunk bark dark gray, thin, scaly>; twigs: new growth not recorded, 1-year old mid brown, ± glossy, older dark gray; thorns on twigs <few>, 1-year old color not recorded, 3–5 cm. **Leaves:** petiole length 50–60% blade, glandularity not recorded; blade ovate to broadly elliptic (suborbiculate or broadly deltate on extension shoots), 4–5.5 cm (larger on extension shoots), base broadly cuneate to truncate (truncate to cordate on extension shoots), lobes 0, or 3 or 4 per side, sinuses shallow, lobes often mere apiculi, margins serrate, <teeth 1.5–3 mm>, veins 6 or 7 per side, apex acuminate, adaxial surface densely crisp-pubescent young, glabrescent. **Inflorescences** 8–20-flowered; branches glabrous; bracteoles few, linear. **Flowers** 13–15 mm diam., hypanthium glabrous; sepal length not recorded, margins entire, abaxially glabrous; stamens 20, anthers purple; styles 3. **Pomes** dark red, sometimes with orange or orange-red spots, broadly pyriform, 14–17 mm diam., <smooth>; <flesh pale yellow>; sepals recurved; pyrenes 3, dorsally deeply grooved.

Flowering Apr; fruiting Sep. Habitat not recorded; of conservation concern; 600–700 m; N.C.

Crataegus roribacca was found by Ashe in Swain County; no recent collection is known. The specimen studied (CM 289601) is in excellent condition and has all the hallmarks of ser. *Tenuifoliae*, although E. J. Palmer (1925) left it unassigned. The leaf differs from those of other members of the series in shape and size of marginal teeth, and from most in vein number. *Crataegus roribacca* does not seem to be a hybrid, so it is perhaps only an unusual form of *C. schuettei*.

63. Crataegus flabellata (Bosc ex Spach) K. Koch, Verh. Vereins Beförd. Gartenbaues Königl. Preuss. Staaten, ser. 2, 1: 240. 1853 • Fan-leaved hawthorn, aubépine flabelliforme [E]

Mespilus flabellata Bosc ex Spach, Hist. Nat. Vég. 2: 63. 1834; *Crataegus crudelis* Sargent; *C. flabellata* var. *densiflora* Kruschke; *C. flabellata* var. *grayana* (Eggleston) E. J. Palmer; *C. grayana* Eggleston

Shrubs or trees, 40–50 dm. **Stems:** twigs: new growth reddish at first, usually sparsely pubescent, 1-year old deep chestnut brown, older dull gray; thorns on twigs 1-year old shiny, dark brown, stout to slender, 4–6 cm. **Leaves:** petiole <1–2 mm wide>, length 40–50% blade, glandular; blade <glossy, dark green>, broadly elliptic to ovate, 4–7 cm, base cuneate to slightly rounded, lobes 4 or 5 per side, sinuses deep, lobe apex acuminate, margins sharply serrate, <teeth gland-tipped>, veins 5 or 6 per side, apex acute, <abaxial surface glabrous except along veins>, adaxial densely appressed-scabrous young, glabrescent. **Inflorescences** 5–10-flowered; branches villous; bracteole frequency not recorded, linear. **Flowers** 13–17 mm diam.; hypanthium glabrous or villous at base; sepals 4–5 mm, margins entire, glandular to glandular-serrate, rarely nearly eglandular, <apex acute>, abaxial pubescence not recorded; stamens 5–10 or 20, anthers usually pink; styles 3–5. **Pomes** crimson, ± ellipsoid, 8–10 mm diam., <sometimes ± pruinose>; sepals spreading or usually erose, <not or obscurely elevated>; pyrenes 3–5, dorsally deeply grooved. 2*n* = 51.

Flowering May–Jun; fruiting Sep–Oct. Brush, successional fields, fencelines, woodland edges, open woodlands; 10–200 m; N.B., N.S., Ont., Que.; Conn., Maine, Mass., N.H., N.Y., Vt.

Crataegus flabellata occurs along the Niagara Escarpment in Ontario, on both sides of Lake Ontario and the St. Lawrence, to New Brunswick and Nova Scotia, and to Massachusetts. Reports for Newfoundland have not been substantiated.

Crataegus flabellata is distinctive with its leaf blades glossy dark green, ovate to broadly elliptic, cuneate at the base, and with deep acuminate lobes. Specimens with less glandularity or serration of the sepal margins may resemble *C. fluviatilis*. In spite of the obvious differences in morphology and range, *C. flabellata* and *C. macrosperma* often have been combined by floristicians. From *C. chrysocarpa*, *C. flabellata* is distinguished by its shinier, ellipsoid, glabrous fruit, by the usually deeper, acuminate lobes of its glossy, dark green leaves, and by longer, slightly more acute sepals. Typically, the inflorescences and hypanthia of eastern *C. chrysocarpa* are densely hairy, practically tomentose, unlike *C. flabellata* with its more sparsely hairy pedicels and glabrous distal hypanthia. It is conceivable that, due to morphological intermediacy and the sympatry of all three species within the range of *C. flabellata*, this species represents the result of ancient hybridization between *C. chrysocarpa* and *C. macrosperma*.

Crataegus flabellata is usually divided into two varieties: var. *flabellata* (ten stamens) and var. *grayana* (20 stamens), the latter being particularly common in southern Quebec. There appears to be little, if any, other constant differentiating feature. *Crataegus crudelis*, also from Quebec, is a form recognized by long thorns (to 10 cm).

The name *Crataegus densiflora* Sargent, which pertains to *C. flabellata*, is illegitimate.

64f.10. Crataegus Linnaeus (sect. Coccineae) ser. Populneae J. B. Phipps, Phytoneuron 2014–39: 1. 2014 [E]

Shrubs or trees, (10–)20–70(–80) dm, main trunk dominant. **Stems:** trunk bark buff to gray-brown, fibrous, checked into longitudinal plates, freshly exposed bark orange-brown; <branches spreading; twigs ± straight>, new growth usually glabrous, sometimes pubescent, 1-year old reddish brown or dark reddish brown, sometimes tan, 2-years old ± gray, sometimes gray-brown or blackish, older paler gray; thorns on twigs moderately frequent, straight to recurved, 1-year old usually dark reddish brown to blackish, usually shiny, usually slender, 2.5–5 cm. **Leaves:** petiole length (16–)30–55% blade, glabrous, sometimes pubescent young, eglandular or sparsely glandular, usually sessile-glandular; blade <mid to dark green adaxially>, broadly elliptic to ovate or deltate, rarely suborbiculate or rhombic, 2–7(–8) cm, chartaceous to subcoriaceous, base broadly cuneate or rounded to truncate or subcordate, lobes 3–5 per side, sinuses shallow to deep, lobe apex acute to acuminate, margins serrate, <teeth numerous>, venation craspedodromous, veins 3–5(or 6) per side, apex subacute to acuminate, abaxial surface glabrous except sometimes veins, adaxial pubescent young. **Inflorescences** 4–10(–12)-flowered, convex panicles; branches usually glabrous, sometimes pubescent; bracteoles caducous, <absent or few>, linear, membranous, margins sessile- or short-stipitate-glandular. **Flowers** 14–18 mm diam.; hypanthium usually glabrous, sometimes pubescent; sepals triangular or lanceolate, much shorter than petals, margins subentire or shallowly to moderately glandular-serrate; stamens (6–)10 or 20, anthers usually pink or rose to purple, rarely yellow; styles 3–5. **Pomes** usually bright red, sometimes orange-red, suborbicular to oblong or pyriform, 8–12(–13) mm diam., <usually pruinose>, glabrous; flesh usually hard, sometimes mellow; sepals erose or persistent, <sessile>, patent to reflexed, <non-accrescent>; pyrenes 3–5.

Species 8 (8 in the flora): c, e North America.

There is only one widespread member of this series, *Crataegus populnea*; *C. aemula* is locally common, and the others are sporadic in occurrence. Some species of ser. *Populneae* are difficult to separate from *C. macrosperma* in ser. *Tenuifoliae*, which usually has more fugacious bracteoles. Other species are closer to ser. *Pruinosae*. Some members of ser. *Populneae* may be of hybrid origin between these two series. The very rare and poorly known *C. haemacarpa* Ashe perhaps belongs here.

1. Stamens 20 . . . . 69. *Crataegus beata*
1. Stamens 6–10, rarely 20 (*C. jesupii*).
    2. Pomes oblong or pyriform-oblong to pyriform.
        3. Leaf blades subcoriaceous, sinuses shallow.
            4. Leaves blades ± ovate, length/width = 1.5 . . . . 66. *Crataegus populnea* (in part)
            4. Leaf blades deltate-ovate to ovate, length/width = 1.1–1.2. . . . . .68. *Crataegus prona*
        3. Leaf blades thin, sinuses deep.
            5. Leaf blades trullate-ovate, lobe apices often ± aligned . . . . 70. *Crataegus jesupii*
            5. Leaf blades ovate, lobe apices (at least proximal) not aligned. . . . . 71. *Crataegus levis*
    2. Pomes suborbicular.
        6. Leaf blades: length/width = 1.8–2; anthers pale yellow. . . . . 64. *Crataegus delawarensis*
        6. Leaf blades: length/width = to 1.5; anthers pink to purple.
            7. Leaf blades ± thin or chartaceous; petioles stipitate-glandular; inflorescence branches hairy . . . . .65. *Crataegus aemula*
            7. Leaf blades subcoriaceous; petioles eglandular or sessile-glandular; inflorescence branches glabrous.
                8. Leaf blades 4–7 cm; extension-shoot leaf blades widest at 2/5 distance from base, 1.3–1.4 times as long as wide, sinuses not deep (LII less than 25%), bases broadly cuneate, rarely truncate; pomes: flesh hard . . .66. *Crataegus populnea* (in part)
                8. Leaf blades 4–5 cm; extension-shoot leaf blades widest in basal 1/4, nearly as long as wide, sinuses deep (LII 25–50%), bases nearly truncate; pomes: flesh mellow when ripe . . . . 67. *Crataegus stolonifera*

64. Crataegus delawarensis Sargent, Bot. Gaz. 35: 102. 1903 [C] [E]

Crataegus pruinosa (H. L. Wendland) K. Koch var. *delawarensis* (Sargent) E. J. Palmer

Shrubs, 10–20 dm. **Stems:** twigs: new growth light green at first, pubescence not recorded, 1-year old dull red-brown, <marked by occasional large, oblong lenticels>, 2-years old light gray-brown; thorns on twigs straight or slightly curved, 1-year old dark red-brown, slender, 3–4 cm. **Leaves:** <stipules linear, acuminate, coarsely glandular-serrate>; petiole length not recorded, <distally slightly wing-margined, grooved>, pubescence not recorded, glandular, <glands deciduous, few, dark red, small>; blade <abaxial paler than adaxial, light yellow-green adaxially>, rhombic to ovate, 5–6 cm <length/width = 1.8–2>, membranous, base narrowed from near mid blade, or base broadly cuneate and entire on vigorous shoots, lobes 3 or 4 per side distally, short, lobe apex acute, margins finely, sometimes doubly, serrate, <teeth gland-tipped>, veins 3 or 4 per side, apex acute, surfaces glabrous. **Inflorescences** 5–8-flowered;

branches glabrous; bracteoles <reddish, acute>, margins glandular. **Flowers** 15 mm diam.; hypanthium usually glabrous, sometimes hairy; sepals broadly triangular, margins coarsely glandular-serrate, <apex acuminate>; stamens 10, anthers pale yellow; styles 3 or 4, <arising from rough tomentum>. **Pomes** dark red, suborbicular, 11–12 mm diam. <wider than tall, not pruinose>; sepals closely appressed; pyrenes 3 or 4, dorsally prominently grooved.

Flowering Apr; fruiting Sep–Oct. Brush; of conservation concern; 20–100 m; Del.

Crataegus delawarensis is known only from northern Delaware; it differs from other members of the series by its narrow leaf blades and yellow anthers.

65. Crataegus aemula Beadle, Biltmore Bot. Stud. 1: 53. 1902 E

Shrubs or trees, 30–50 dm. **Stems:** <trunks short, slender, 5–7 cm diam., bark smooth to slightly fissured or scaly; compound thorns on trunks sometimes present>; twigs: new growth reddish green, pubescent, 1-year old pale tan, becoming gray-brown; thorns on twigs ± straight, chestnut brown, gray later, ± slender, 3–5 cm. **Leaves:** petiole 0.5–1.5 cm, pubescent at least young, stipitate-glandular; blade elliptic-ovate or broadly elliptic to suborbicular, 3–5 cm <length/width = to 1.5>, ± thin, base rounded or broadly truncate, lobes 3 or 4 per side, sinuses shallow, lobe apex acute, margins serrate, <teeth regular>, veins 4 or 5 per side, apex acute, abaxial surface glabrous, veins sparsely hairy, adaxial sparsely pubescent young, glabrescent. **Inflorescences** 5–10-flowered; branches coarsely hairy; bracteole margins densely glandular. **Flowers** 14–18 mm diam.; hypanthium sparsely pubescent; sepals lanceolate, <4 mm>, margins glandular-serrate, abaxial pubescence not recorded; stamens 10, anthers pink-purple; styles 3–5. **Pomes** bright red, suborbicular, 10–13 mm diam., <shiny>; sepals spreading-reflexed or often erose; pyrenes 3–5, dorsally shallowly grooved. $2n$ = 51, 68.

Flowering Apr; fruiting Sep–Oct. Oak flatwoods, brush; 50–600 m; Ala., Ga., Miss.

Crataegus aemula grows from northwestern Georgia to central Mississippi; it is locally common. The species has escaped notice in floras since Beadle in Small (1903). It is distinguished from sympatric species of the series by twig color, new growth and inflorescence indumentum, leaf shape, stipitate-glandular petioles, and bright red pomes. Remarkably, it seems most similar to *C. frugiferens*, provisionally placed in ser. *Apricae*, with which it shares stipitate-glandular petioles but differs in larger (3.5–5 cm) leaves of a proportionately narrower (somewhat narrowly ovate to broadly elliptic) shape with cuneate bases and larger marginal teeth, and two or three distinct lobes per side. *Crataegus frugiferens* also has a similar distribution.

66. Crataegus populnea Ashe, Ann. Carnegie Mus. 1: 395. 1902 • Poplar hawthorn E

Crataegus compta Sargent; *C. iracunda* Beadle var. *populnea* (Ashe) Kruschke; *C. luxuriosa* Sargent

Shrubs or trees, 60 dm. **Stems:** twigs: new growth reddish green, glabrous, 1-year old reddish brown, sometimes dark, older mid gray; thorns on twigs recurved, 2-years old dark blackish red, ± shiny, ± slender, 3–5 cm. **Leaves:** petiole 2–4 cm, glabrous, eglandular or sparsely glandular; blade ± ovate, 4–7 cm <length/width = to 1.5>, subcoriaceous, base broadly rounded to broadly cuneate, rarely truncate, lobes 3–5 per side, sinuses usually shallow, lobe apex acute to acuminate, margins strongly serrate, veins 5 or 6 per side, apex acuminate, adaxial surface glabrate to appressed-hairy young, glabrescent; <extension-shoot leaf blades widest 2/5 from base, length/width = 1.3–1.4, sinuses not deep, LII less than 25%, base broadly cuneate, rarely truncate>. **Inflorescences** 5–12-flowered; branches glabrous; bracteoles <few, very narrow>, margins glandular. **Flowers** 16–18 mm diam.; hypanthium glabrous; sepals narrowly triangular, <5–6 mm>, margins subentire, <abaxially glabrous>; stamens 6–10, anthers red to purple; styles 3 or 4. **Pomes** reddish orange to bright red, suborbicular to oblong, 8–11 mm diam.; <flesh hard>; sepals commonly erose, patent-reflexed; pyrenes 3 or 4, dorsally grooved. $2n$ = 68.

Flowering Apr–May; fruiting Sep–Oct. Brush, fencerows; 10–400 m; Ont.; Mich., N.Y., Ohio, Pa., Va., W.Va., Wis.

Crataegus populnea occurs from Wisconsin to New York, to Ohio, and Virginia; it is common at least from southern Ontario to Pennsylvania. Dubious specimens from Georgia and South Carolina have small leaves and are only differentiated from *C. iracunda* (ser. *Tenuifoliae*) by their more coriaceous nature. Being out of range for *C. populnea*, they may represent an odd form of *C. iracunda*.

Crataegus compta is a variant with more or less pyriform-oblong pomes. *Crataegus populnea* might be mistaken for *C. macrosperma*, but it has more coriaceous leaves, usually significantly larger in the area of sympatry, and a harder fruit.

67. Crataegus stolonifera Sargent, Bot. Gaz. 35: 109. 1903 [E]

Crataegus iracunda Beadle var. *stolonifera* (Sargent) Kruschke

Shrubs, 25 dm. **Stems:** twigs: new growth golden green, glabrous, 1-year old reddish brown, 2-years old gray; thorns on twigs usually recurved, 2-years old shiny reddish brown or darker, ± slender, 3–5 cm. **Leaves:** petiole 1.5–2.5 cm, glabrous, sparsely glandular; blade ovate, 4–5 cm, <length/width = to 1.5>, subcoriaceous, base cuneate to broadly cuneate (leaves on extension shoots broadly ovate to ovate-deltate or deltate, base nearly truncate), lobes 4 per side, sinuses moderately deep, lobe apex acute, margins serrate, veins 4 or 5 per side, apex acute to subacute, adaxial surface finely appressed-pubescent young, glabrescent; <extension-shoot leaf blades broadly ovate to ovate-deltate or deltate, widest in basal ¼, nearly as long as wide, sinuses deep, LII 25–50%, base nearly truncate>. **Inflorescences** 5–10-flowered; branches glabrous; bracteoles <absent or few>, margins glandular. **Flowers** 15–18 mm diam.; hypanthium glabrous; sepals broadly to narrowly triangular, <6–7 mm>, margins glandular-serrate; stamens 10, anthers rose-purple; styles 3 or 4. **Pomes** red, suborbicular, 9–12 mm diam.; <flesh mellow when ripe>; sepals reflexed; pyrenes 3 or 4, dorsally grooved.

Flowering Apr–May; fruiting Sep–Oct. Brush, thickets, roadsides; 10–400 m; Ont.; Del., N.Y., Ohio, Pa., Wis.

Crataegus stolonifera has been recorded sporadically from Wisconsin to southern Ontario and New York, to Pennsylvania and Delaware. The species is similar to *C. populnea*, differing primarily in the deltate, deeply incised extension-shoot leaves and more mellow fruit.

68. Crataegus prona Ashe, J. Elisha Mitchell Sci. Soc. 19: 17. 1903 [E]

Crataegus gravis Ashe

Shrubs or trees, 60 dm. **Stems:** twigs: new growth reddish green, glabrous, 1-year old reddish brown, sometimes very dark, older mid gray; thorns on twigs recurved, 2-years old dark blackish red, ± shiny, ± slender, 3–5 cm. **Leaves:** petiole 2–4 cm, glabrous, eglandular or sparsely glandular; blade deltate-ovate to ovate, 4–7 cm, <length/width = 1.1–1.2>, subcoriaceous, base broadly rounded to ± cuneate, lobes 3–5 per side, sinuses usually shallow, lobe apex acute, margins strongly serrate, veins 5 or 6 per side, apex acute to acuminate, surfaces glabrous except adaxial appressed-hairy young. **Inflorescences** 5–12-flowered; branches glabrous; bracteoles <few>, margins glandular. **Flowers** 16–18 mm diam.; hypanthium glabrous; sepals narrowly triangular, <5–6 mm>, margins subentire, <abaxially glabrous>; stamens 6–10, anthers red to purple; styles 3 or 4. **Pomes** reddish orange to bright red, suborbicular to oblong, 8–11 mm diam.; sepals patent-reflexed or commonly erose; pyrenes 3 or 4, dorsally grooved.

Flowering May; fruiting Sep–Oct. Brush; 100–300 m; Ont.; Ind., Mich., N.Y., Pa., Wis.

Crataegus prona ranges from Wisconsin and Indiana through Michigan to New York and Pennsylvania.

Crataegus prona is similar to *C. populnea* but with proportionately wider leaves and narrow pomes. When better known, it may prove the same as *C. populnea*. *Crataegus gravis* is a form with more oblong fruit and smaller leaves with sharper lobes.

69. Crataegus beata Sargent, Proc. Rochester Acad. Sci. 4: 97. 1903 [E]

Shrubs, 60 dm. **Stems:** twigs: new growth dark orange-green-tinged with red, glabrous, 1-year old bright reddish brown, older mid gray; thorns on twigs straight to slightly recurved, 2-years old dark, shiny reddish brown, stout, 3.5–4.5 cm. **Leaves:** petiole 2.5–4 cm, hairy young, glabrescent, ± densely glandular; blade broadly ovate to deltate-ovate, 5–8 cm, subcoriaceous, base broadly rounded to subtruncate, sometimes broadly cuneate, lobes 4 or 5 per side, <± triangular>, sinuses not recorded, lobe apex acute to acuminate, margins strongly serrate, veins 5 per side, apex acuminate, surfaces glabrous except adaxial ± appressed-hairy young. **Inflorescences** 5–12-flowered; branches glabrous; bracteoles <few>, margins glandular. **Flowers** 16–18 mm diam.; hypanthium glabrous; sepals narrowly triangular, margins subentire or slightly glandular, <abaxially glabrous>; stamens 20, anthers deep maroon; styles 3 or 4. **Pomes** crimson, oblong to oblong-obovoid, 9–10 mm diam.; sepals spreading; pyrenes 4 or 5, dorsally grooved.

Flowering May; fruiting Sep–Oct. Brush; 100–300 m; Ont.; Mich., N.Y., Ohio, Wis.

Crataegus beata is a rare, variable taxon that here comprises all the 20-stamen forms of ser. *Populneae* and may be more widespread than indicated. Syntype material has resemblances to *C. schuettei* and *C. stolonifera*. The very small amount of material available to the author is characterized by tan or very dark (brownish to blackish) one-year old wood and one-year thorns, sharply dissected, narrower,

C. levis-like leaves, or more shallowly incised (except in extension shoots) *C. populnea*-like coriaceous foliage and glandular-serrate sepals. In these, the anthers are pink to red and the pomes large, oblong-orbicular, and red. The young leaves are densely short-scabrous hairy adaxially when young.

70. Crataegus jesupii Sargent, Rhodora 5: 61. 1903 (as jesupi) • Jesup hawthorn [C] [E]

Crataegus bellula Sargent; *C. filipes* Ashe

Shrubs or trees, 70 dm. **Stems:** twigs: new growth reddish, glabrous, 1-year old dark brown, <often strongly pruinose>, 2-years old blackish, older dark gray, becoming paler later; thorns on twigs recurved to ± straight, 2-years old blackish, ± slender, 3–5 cm. **Leaves:** petiole 2–4 cm, glabrous, glandular; blade <sometimes ± blue-green mature>, trullate-ovate, 4–7 cm, thin but firm, base ± broadly cuneate, lobes 3 or 4 per side, sinuses deep, <angled>, lobe apex <often ± aligned>, acuminate, margins serrate, <proximal teeth gland-tipped>, veins 4 or 5 per side, apex acute, abaxial surface glabrous, adaxial short-pubescent young, soon glabrescent. **Inflorescences** 4–8-flowered; branches glabrous; bracteoles <few>, margins glandular. **Flowers** 15–18 mm diam.; hypanthium glabrous; sepals ± triangular, <4–6 mm>, margins entire or glandular-serrate; stamens 10(or 20), anthers pink or rose; styles 4. **Pomes** red, pyriform-oblong, 8–10 mm diam., <slightly pruinose>; sepals spreading; pyrenes 3–5, dorsally grooved.

Flowering May; fruiting Sep–Oct. Borders of woods, thickets; of conservation concern; 20–300 m; Ont.; Conn., Mich., N.Y., Ohio, Pa., Wis.

Crataegus jesupii ranges from Wisconsin to Connecticut, Ohio, and Pennsylvania.

In Michigan, *Crataegus jesupii* is notable for its stipitate-glandular bracteoles. Forms with wider leaves, about as wide as long, are represented by *C. filipes*. An otherwise identical Ontario specimen has 20 stamens and would key out to *C. beata*.

71. Crataegus levis Sargent, Rhodora 7: 198. 1905 [E]

Crataegus laetans Sargent; *C. milleri* Sargent

Shrubs, 30–50 dm. **Stems:** twigs: new growth reddish, glabrous, 1-year old brown, older dark to paler gray; thorns on twigs straight to recurved, 2-years old shiny dark brown, ± slender, 3–5 cm. **Leaves:** petiole 1.5–3 cm, pubescence not recorded, eglandular or sparsely glandular; blade ovate, 3–6 cm, very thin, base rounded or broadly cuneate, lobes 3 or 4(or 5) per side, sinuses deep, <angled>, lobe apex <at least proximal, not aligned>, usually acuminate, margins serrate, <proximal teeth gland-tipped>, veins 4 or 5 per side, apex acute, abaxial surface glabrous, adaxial short-pubescent young, soon glabrescent. **Inflorescences** 5–8-flowered; branches glabrous; bracteoles <absent or few>, margins glandular. **Flowers** 15–17 mm diam.; hypanthium glabrous; sepals basally broadly triangular (lamina nearly linear), <6 mm>, margins entire, <abaxially glabrous>; stamens 10, anthers rose-red; styles 4. **Pomes** red, ± pyriform, 8–10 mm diam., <flesh hard>; sepals spreading; pyrenes 4, dorsally grooved.

Flowering Apr–May; fruiting Sep–Oct. Open woods, brush; 50–300 m; Mass., N.Y., N.C., Ohio, Pa.

Crataegus levis ranges from the southern Great Lakes area to southern New England, to Ohio, Pennsylvania, and in mountains to North Carolina.

Crataegus levis is often confused with the *C. compta* form of *C. populnea*; that taxon usually has slightly larger, more coriaceous leaves, with less acute lobes than *C. levis*. *Crataegus levis* is possibly a hybrid of *C. macrosperma* × *C. pruinosa*.

64f.11. Crataegus Linnaeus (sect. Coccineae) ser. Pruinosae (Sargent) Rehder, Man. Cult. Trees ed. 2, 363. 1940 • Frosted thorns [E]

Crataegus [unranked] *Pruinosae* Sargent, Silva 13: 32. 1902

Shrubs or trees, (20–)30–80 dm, main trunk dominant. **Stems:** trunk bark gray-brown, flattened-scaly, freshly exposed bark often orange-brown; <branches spreading; twigs ± straight>, new growth glabrous, 1-year old usually purple-brown, sometimes tan or reddish brown, 2-years old grayish; thorns on twigs usually numerous, straight to recurved, 2-years old usually black or blackish, shiny, sometimes reddish brown or purple, ± fine to slender or moderately thick, 1.5–5 cm. **Leaves:** petiole length (28–)50–100% blade, glabrous, usually glandular, sometimes eglandular; blade <mid to dark green or ± blue-green>, ± ovate or ± trullate to rhombic or deltate, sometimes suborbiculate, (2–)3–6(–7) cm, thin to subcoriaceous, base cuneate to truncate, sometimes cordate or rounded, lobes (1–)3 or 4 per side, sinuses shallow to moderately deep, lobe apex ± acute, sometimes ± obtuse, margins finely to coarsely serrate, venation craspedodromous, veins 4–7 per side, apex acute to obtuse, surfaces usually glabrous, sometimes adaxial sparsely hairy young. **Inflorescences** 4–10-flowered, convex panicles; branches usually glabrous, sometimes sparsely villous; bracteoles caducous, <few to numerous, rarely absent>, linear, membranous, margins sessile- or short-stipitate-glandular. **Flowers** 15–30 mm diam.; hypanthium glabrous; sepals triangular, length ½ petal, margins usually subentire, sometimes entire, rarely glandular-serrate; stamens (10 or)20, anthers white to pink or red, sometimes cream; styles 3–5. **Pomes** pink, mauve, or pale green, sometimes scarlet, deep crimson, or purple, suborbicular, (6–)7–15(–20) mm diam., <strongly pruinose>, glabrous; flesh hard (until overripe); sepals <usually on prominent collar>, spreading, <non-accrescent>; pyrenes (2 or)3–5, <often angular>.

Species 6 (6 in the flora): c, e North America.

Members of ser. *Pruinosae* are common from Wisconsin through the southern Great Lakes to southern Quebec and southern New England, to Arkansas and Louisiana to Georgia but absent from the southern coastal plain. In the north, they are found mainly in the open; southward they are more often found in woodland shade.

The core of ser. *Pruinosae* is a natural group of very thorny hawthorns with glabrous, somewhat blue-green, medium-sized to fairly large leaves, mid season anthesis relative to congeners, usually entire sepal margins, and hard, pruinose, often pink or mauve pomes, most with strikingly elevated fruiting calyx. All taxa except *Crataegus pruinosa* var. *dissona* have 20 stamens. *Crataegus pruinosa* var. *virella*, with the adaxial surface of younger leaves scabrous, might key to ser. *Tenuifoliae* when in flower. *Crataegus gattingeri* resembles *C. iracunda* (ser. *Tenuifoliae*) in fruit, though differing in leaf adaxial surface indumentum, leaf shape, stamen number, and anther size (see discussion below and under 58. *C. iracunda*). Members of ser. *Pruinosae* are similar to members of ser. *Intricatae* in many particulars, including range; they differ in bracteole form, fruit color, and leaf indumentum and often in stamen number.

The interserial hybrid *Crataegus* ×*coleae* keys out in the sixth couplet.

SELECTED REFERENCE Sinnott, Q. P. and J. B. Phipps. 1983. Variation patterns in *Crataegus* series *Pruinosae* (Rosaceae) in southern Ontario. Syst. Bot. 8: 59–70.

1. Anthers 0.3–0.4 mm.
    2. Leaf blades trullate to rhombic-ovate, sinuses shallow, lobe apices subacute, bases cuneate, marginal teeth 1 mm . . . . . 72. *Crataegus compacta*
    2. Leaf blades ± deltate to narrowly ovate, sinuses shallow, lobe apices acute, bases rounded to broadly cuneate, marginal teeth 2 mm . . . . . 73. *Crataegus gattingeri*
1. Anthers 0.6–0.8 mm.
    3. Leaf blades deltate, broadly elliptic, rhombic, or suborbiculate (widest near middle).
        4. Anthers cream, pink, or red; flowers 20–24 mm diam. . . . . . 74. *Crataegus suborbiculata*
        4. Anthers white; flowers 22–30 mm diam.. . . . . . 77. *Crataegus formosa*
    3. Leaf blades ± ovate to oval, trullate, or broadly rhombic-elliptic (widest in basal ⅓).
        5. Anthers white . . . . . 76. *Crataegus cognata*
        5. Anthers pink to bright rose or dull purple, sometimes cream.
            6. Pomes greenish with pink or mauve areas, sometimes bright crimson or scarlet, highly pruinose, not punctate; pyrene sides plane. . . . . . 75. *Crataegus pruinosa*
            6. Pomes bright orange-red, at most slightly pruinose, punctate; pyrene sides each linearly scarred. . . . . . 156. *Crataegus* ×*coleae*, p. 636

72. Crataegus compacta Sargent, Crataegus S. Michigan, 522. 1907 • Thames hawthorn [E]

Shrubs, <erect>, 60 dm. **Stems:** twigs: new growth reddish green, 1-year old tan, 2-years old dull gray; thorns on twigs <± numerous>, ± straight to slightly recurved, 2-years old shiny blackish, medium thickness, 1.5–3 cm. **Leaves:** petiole length 50–75% blade, sparsely glandular; blade trullate to rhombic-ovate, 3–5 cm, thin, base cuneate, lobes 2 or 3 per side, sinuses shallow, lobe apex subacute, margins serrate, <teeth 1 mm>, veins 4 or 5 per side, apex acute, surfaces glabrous, sometimes adaxial sparsely hairy young. **Inflorescences** 4–7-flowered; branches glabrous, <sparsely punctate>; bracteoles <sometimes greenish, large>, margins short-stipitate-glandular. **Flowers** 15–18 mm diam.; sepals narrowly triangular, 4–5 mm, margins subentire, abaxial pubescence not recorded; stamens 20, anthers pink to red, 0.3 mm; styles 3 or 4. **Pomes** <usually borne singly>, greenish to pinkish or pale purple, or dull reddish purple, 8–11 mm diam., moderately pruinose (sometimes bracteolate); sepals on collar, spreading, sometimes erose; pyrenes 2 or 3. $2n$ = 34, 51, 68.

Flowering May; fruiting Sep–Oct. Scrub, woodland edges, pastures, fencerows; 100–300 m; Ont.; Mich., N.Y., Ohio, Pa., Va., W.Va.

Crataegus compacta is common in the Carolinian zone of Ontario and extends to Michigan, New York, and Pennsylvania (where infrequent), and to the Virginia Blue Ridge (reported for Ohio by Palmer).

Crataegus compacta is most similar to *C. gattingeri*, one of the two most widespread species in the series, but the allopatry of the two, as well as consistent differences in leaf shape, underlines their distinction. The pedicels and hypanthia in young flowers are conspicuously reddish. Similar forms from Pennsylvania, though with ivory anthers, are *C. repentina* Sargent with ten to 20 stamens and minute anthers, and *C. varians* Sargent with ten to 15 stamens.

73. Crataegus gattingeri Ashe, J. Elisha Mitchell Sci. Soc. 17: 12. 1900 • Gattinger's hawthorn [E]

Crataegus bona Sargent; *C. callida* Beadle; *C. georgiana* Sargent; *C. pruinosa* (H. L. Wendland) K. Koch var. *gattingeri* (Ashe) Lance; *C. rustica* Beadle

Shrubs or trees, 80 dm. **Stems:** twigs: new growth reddish, 1-year old deep reddish brown, 2-years old dark gray-brown, older deep gray; thorns on twigs straight to recurved, 2-years old very dark reddish brown or blackish, ± slender, 2.5–3.5 cm. **Leaves:** petiole length 50% blade, eglandular or with a few small glands; blade ± deltate to narrowly ovate, 2.5–5 cm, thin, base rounded to broadly cuneate, lobes 3 or 4 per side, sinuses shallow, lobe apex acute, margins serrate, <teeth numerous, 2 mm>, veins 5 per side, apex acute, surfaces glabrous, adaxial sparsely hairy young, especially midvein. **Inflorescences** 4–8 -flowered; branches glabrous; bracteoles <numerous, small>, margins sessile- or stipitate-glandular. **Flowers** 15 mm diam.; sepals triangular, 4 mm, margins entire, pubescence not recorded; stamens 20, anthers pink to crimson, 0.3–0.4 mm; styles 4. **Pomes** greenish or greenish with pale blush young, ruddy to dull dark red when ripe, 6–10 mm diam., ± pruinose; sepals conspicuous, spreading; pyrenes 4.

Flowering Apr–May; fruiting Sep–Oct. Open woodlands, brush; 50–150 m; Ala., Ga., Ky., La., Miss., Mo., N.C., Ohio, Pa., Tenn., Va., W.Va.

Crataegus gattingeri extends from Missouri to Pennsylvania south to Louisiana, Alabama, and Georgia; it should be sought in southern Illinois and Indiana.

Crataegus gattingeri is similar to the variably hairy *C. pruinosa* var. *virella*, which has normal-sized anthers. Considerable similarity occurs between *C. compacta* and *C. gattingeri*; both may have stipitate-glandular bracteoles, the leaves and flowers are about the same size, and the thorn and twig colors are similar. In the latter species a greater range of leaf shape occurs, from narrowly ovate to practically deltate, and bases rounded to broadly cuneate. The lobes and apices of the leaves of *C. gattingeri* are usually much sharper. *Crataegus gattingeri* is also superficially similar to small-leaved forms of ser. *Tenuifoliae* such as *C. iracunda*, which also usually has particularly small leaves at anthesis. It differs from the latter primarily in its 20 stamens, small anthers, much less hairy adaxial leaf surfaces at anthesis, and more glandular-bracteolate inflorescences. These two may be difficult to differentiate in fruit. The author has not encountered *C. gattingeri* var. *rigida* E. J. Palmer, from Indiana and Kentucky, which apparently differs from the widespread type variety in its stouter and flexuous twigs, stouter thorns, and smaller flowers. It would not seem to be part of *C. gattingeri*. *Crataegus arcana* Beadle represents a somewhat similar form of possible conspecificity.

74. Crataegus suborbiculata Sargent, Rhodora 3: 72. 1901 [C] [E]

Shrubs or trees, 80 dm. **Stems:** twigs: new growth reddish, 1-year old tan, 2-years old light gray; thorns on twigs straight, 2-years old blackish, usually shiny, ± slender, 2.5–4.5 cm. **Leaves:** petiole length 50–100% blade, eglandular or with 1 or 2(or 3) glands; blade broadly elliptic or rhombic to suborbiculate, 4–6 cm, thin, base cuneate to broadly cuneate, lobes 3 or 4 per side, sinuses shallow, lobe apex acute, margins serrate, <eglandular>, veins 5 or 6 per side, apex acute to obtuse, surfaces glabrous, adaxial sparsely hairy young. **Inflorescences** 5–10-flowered; branches glabrous; bracteoles <sometimes absent>, margins glandular. **Flowers** 20–24 mm diam.; sepals narrowly triangular, 5–6 mm, margins subentire, abaxially glabrous; stamens 20, anthers cream, pink, or red, 0.7 mm; styles 4. **Pomes** pink mauve, ripening to orange-red or purplish, 10–14 mm diam., pruinose at first; sepals not on collar, spreading; pyrenes 3 or 4. $2n = 68$.

Flowering May–Jun; fruiting Sep–Oct. Woodland margins, brush; of conservation concern; 50–300 m; Ont., Que.; Conn., Mich., N.Y.

Crataegus suborbiculata is uncommon but is found in a number of locations in southern Ontario, and extends east to Quebec, New York, and Connecticut, and west to Michigan. Fruiting specimens, seemingly of this species but in poor condition, have been collected along the Skyline Drive of Virginia.

Very rare forms with slight inflorescence pilosity may occur, and then *Crataegus suborbiculata* may look very much like some forms *of C. florifera* (ser. *Anomalae*), which is distinguished by its pyrenes. One large Ontario population is rather similar to *C. compacta* except that it is larger in all its parts and has proportionately slightly wider leaves. Very rare forms with narrower leaves (elliptic-rhombic) and small anthers may be referred to *C. nitidula* Sargent, known from southern Ontario, Michigan, Pennsylvania, and Wisconsin. Forms named *C. umbratilis* Sargent from Connecticut and Pennsylvania have slightly larger, more venous leaves with larger marginal teeth than the above and an elevated calyx in fruit. The poorly understood *C. durobrivensis* Sargent was suggested by E. J. Palmer (1952) to be a hybrid with *C. pruinosa*.

75. Crataegus pruinosa (H. L. Wendland) K. Koch, Hort. Dendrol., 168. 1853 • Frosted hawthorn [E] [F]

Mespilus pruinosa H. L. Wendland, Flora 6: 701. 1823

Shrubs or trees, <dense, often suberect>, 20–70 dm. **Stems:** <compound thorns on trunks present>; twigs: new growth reddish brown, 1-year old dull purple-brown, 2-years old dull gray, older paler; thorns on twigs straight to slightly recurved, 2-years old deep purple to shiny black, usually ± fine, 3–5 cm. **Leaves:** petiole length 50–66% blade, sparsely glandular; blade trullate to ovate, ovate-oblong, or broadly rhombic, rarely ± deltate, 2–6(–7) cm, subcoriaceous, base broadly cuneate to subtruncate to weakly subcordate, lobes (1–)3 or 4 per side, sinuses shallow to deeper, lobe apex acute, sometimes ± obtuse, margins serrate, veins 5 or 6 per side (except smaller leaves), apex acute, surfaces glabrous (except var. *virella*). **Inflorescences** 5–10 -flowered; branches glabrous, sometimes sparsely villous; bracteoles caducous, <usually few>, margins short-stipitate-glandular. **Flowers** 15–25 mm diam.; sepals narrowly triangular, 5–6 mm, margins usually entire or subentire, rarely glandular-serrate, abaxially glabrous; stamens (10 or)20, anthers pale pink to bright rose or dull purple, sometimes cream, 0.6–0.8 mm; styles 3–5. **Pomes** greenish with pink or mauve areas, sometimes bright crimson or scarlet, <often rather angular>, 10–20 mm diam., highly pruinose, <not punctate>; <flesh hard>; sepals on collar, spreading; pyrenes 3–5.

Varieties 6 (6 in the flora): c, e North America.

Crataegus pruinosa extends from Arkansas to Wisconsin, through the southern Great Lakes to southern New England, and, in the south, mainly in the Appalachians to northern Georgia.

In the north of its range, *Crataegus pruinosa* is mainly a shrub of open successional habitats but in the south may commonly be a taller tree of open or thin woodlands.

Many hawthorns have a little waxy bloom on their pomes; it is particularly prominent on *Crataegus pruinosa* and *C. cognata* compared to others. Some authors include in *C. pruinosa* their white-anthered counterparts, here assigned to *C. cognata*. Whereas *C. pruinosa* characteristically has entire or subentire sepal margins, some forms in the southwest of the range of the species (for example, *C. callianthe* Sargent, *C. seducta* Sargent) may have glandular-serrate sepal margins. They may represent introgression with *C. coccinioides*. The varieties of *C. pruinosa* are weakly differentiated from each other, most of them on leaf shape and size characters. The more widespread varieties constitute a range of morphotypes held together by common traits. *Crataegus gaudens* Sargent is a strikingly distinct form from Pennsylvania that has more or less elliptic leaves with lobes absent; it is clearly related to *C. pruinosa*. Note that 156. *Crataegus* ×*coleae*, a Michigan endemic, will key out here if its laterally scarred pyrenes are missed.

1. Stamens 10 . . . . . . 75b. *Crataegus pruinosa* var. *dissona*
1. Stamens 20.
   2. Leaf surfaces sparsely short-hairy adaxially young, usually glabrescent; inflorescence branches sometimes sparsely villous. . . . . 75e. *Crataegus pruinosa* var. *virella*
   2. Leaf surfaces glabrous adaxially young, sometimes sparsely appressed-hairy along veins; inflorescence branches glabrous.
      3. Leaf blades 5–7 cm, ovate to ovate-deltate, lobes 0 or 1–4 per side, obscure, lobe apices ± obtuse, bases broadly cuneate to subtruncate; anthers cream. 75f. *Crataegus pruinosa* var. *magnifolia*
      3. Leaf blades 2–7 cm, ovate or ovate-oblong to ± deltate or broadly rhombic, lobes 2–4 per side, distinct, lobe apices acute, bases broadly cuneate to subtruncate or truncate to weakly subcordate; anthers pale pink.
         4. Leaf blades 3–7 cm, broadly ovate to deltate, length/width = 1–1.2, bases very broadly cuneate or truncate to weakly subcordate. . . . . .75d. *Crataegus pruinosa* var. *rugosa*
         4. Leaf blades 2–6 cm, broadly rhombic, ovate, or ovate-oblong to narrowly deltate, length/width = 1.5, bases broadly cuneate to subtruncate.
            5. Leaf blades ovate or ovate-oblong to narrowly deltate, 4–6 cm, bases broadly cuneate to subtruncate . . . . 75a. *Crataegus pruinosa* var. *pruinosa*
            5. Leaf blades broadly rhombic, 2–4 cm, bases broadly cuneate . . . . 75c. *Crataegus pruinosa* var. *parvula*

75a. Crataegus pruinosa (H. L. Wendland) K. Koch var. **pruinosa** [E] [F]

Crataegus bracteata Sargent; *C. mackenziei* Sargent ex Mackenzie var. *bracteata* (Sargent) E. J. Palmer; *C. tumida* Sargent

Shrubs or trees, 20–70 dm, <usually branched to base>. **Leaves:** blade ovate or ovate-oblong to narrowly deltate, 4–6 cm, length/width = 1.5, lobes 2–4 per side, lobe apex acute, base broadly cuneate to subtruncate, adaxial surface glabrous, sometimes sparsely appressed-hairy along veins. **Inflorescence branches** glabrous. **Flowers** 15–22(–25) mm diam.; stamens 20, anthers pale pink, rarely red.

Flowering Apr–May; fruiting Sep–Oct. Open scrub, light woodland shade; 20–1300 m; Ont., Que.; Conn., Ill., Ind., Iowa, Kans., Ky., Maine, Mass., Mich., Mo., N.H., N.J., N.Y., N.C., Ohio, Tenn., Vt., Va., W.Va., Wis.

Variety *pruinosa* is common through the northern range of the species, mainly north of southern Missouri and the Ohio River; it is found southward at middle and higher elevations in the Appalachians. *Crataegus bracteata* from Missouri is somewhat atypical; the adaxial leaf surfaces are briefly soft-pubescent, and the abaxial midrib permanently hairy. With its somewhat large leaves (5–6 cm) and slightly glandular sepal margins, as well as its very conspicuous bracteoles, it probably represents introgression with *C. coccinioides*. A red-anthered variant of var. *pruinosa* with a more deltate-ovate leaf shape, perhaps matching *C. tumida*, has been collected from New York.

75b. Crataegus pruinosa (H. L. Wendland) K. Koch var. **dissona** (Sargent) Eggleston, Bull. New York State Mus. Nat. Hist. 254: 422. 1924 [E]

Crataegus dissona Sargent, Rhodora 5: 60. 1903; *C. brachypoda* Sargent; *C. disjuncta* Sargent; *C. pruinosa* var. *brachypoda* (Sargent) E. J. Palmer; *C. rigida* Sargent

Shrubs, 20–40 dm. **Leaves:** blade trullate to ovate or ovate-oblong, 3–7 cm, length/width = 1.5, lobes 4 per side, lobe apex acute, base broadly cuneate to truncate, adaxial surface glabrous, sometimes sparsely appressed-hairy along veins. **Inflorescence branches** glabrous. **Flowers** 15–22 mm diam.; stamens 10, anthers bright rose to dull purple, sometimes cream. $2n = 51$.

Flowering Apr–May; fruiting Sep–Nov. Brush, open woodlands; 20–300 m; Ont.; Conn., Ga., Ind., Ky., Mass., Mo., N.Y., Ohio, Wis.

Variety *dissona*, found from Missouri to Wisconsin and through the Great Lakes to New England, then south to Georgia, is widely scattered throughout the range of var. *pruinosa* and an unnamed southern form (see 75d. var. *rugosa*). Its detailed distribution is not well known, but it is almost certainly more widespread than recorded. The variation in leaf shape of var. *dissona* closely parallels that of var. *pruinosa*. *Crataegus disjuncta* is a similar form with particularly large leaves; *C. brachypoda* has cream anthers.

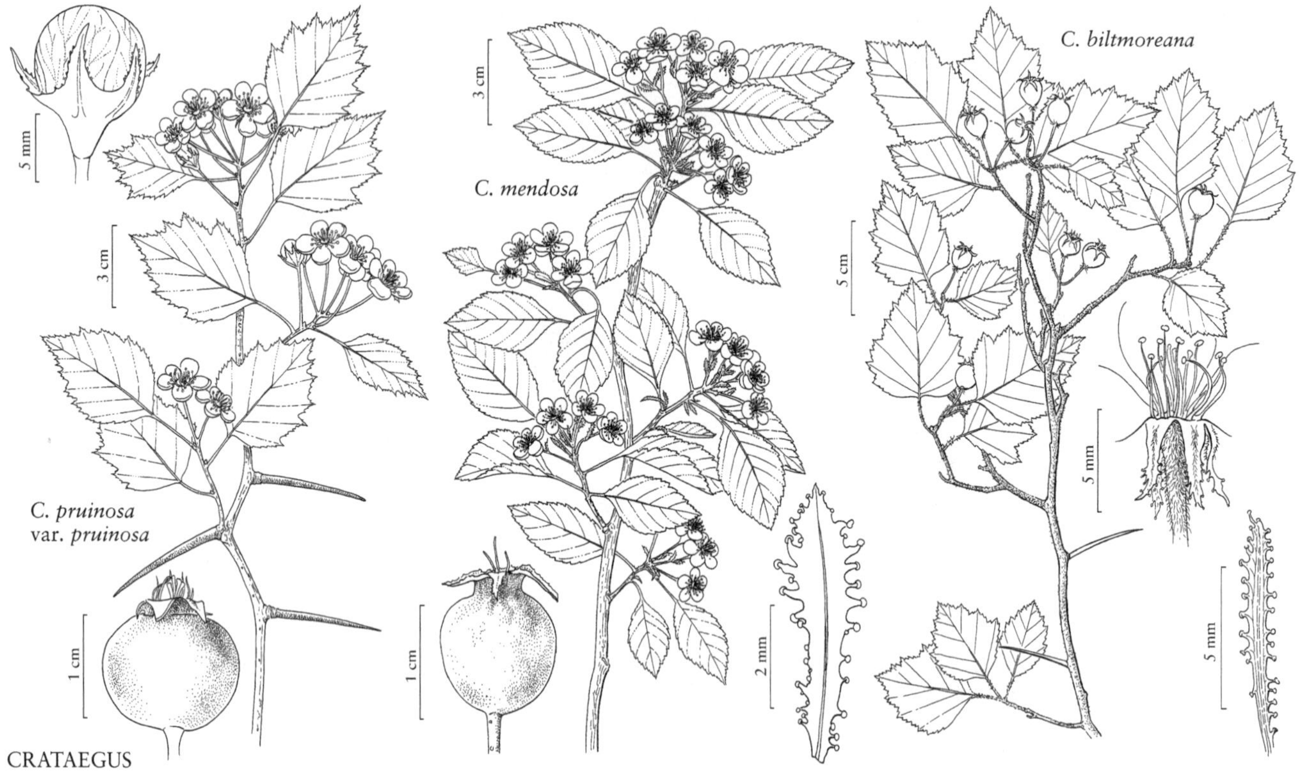

CRATAEGUS

75c. Crataegus pruinosa (H. L. Wendland) K. Koch var. **parvula** (Sargent) J. B. Phipps, Canad. J. Bot. 58: 1684. 1980 [E]

Crataegus parvula Sargent, Crataegus S. Michigan, 527. 1907

Shrubs, 20–30 dm, <usually branched to base>. **Leaves:** blade broadly rhombic, 2–4 cm <widest in middle>, length/width = 1.5, lobes 2–4 per side, <sinuses shallow>, lobe apex acute, base broadly cuneate, adaxial surface glabrous, sometimes sparsely appressed-hairy along veins. **Inflorescence branches** glabrous. **Flowers** 15–22 mm diam.; stamens 20, anthers pale pink.

Flowering May; fruiting Sep–Oct. Brush, successional fields, unmanaged pastures; 20–300 m; Ont.; Mass., Mich., N.Y., N.C., Va.

Variety *parvula* is scattered through the northern range of the species and at higher elevations south into the Appalachians; it is characterized by smaller leaves that usually are very shallowly incised.

75d. Crataegus pruinosa (H. L. Wendland) K. Koch var. **rugosa** (Ashe) Kruschke, Publ. Bot. Milwaukee Public Mus. 3: 181. 1965 [E]

Crataegus rugosa Ashe, J. Elisha Mitchell Sci. Soc. 17: 5. 1900; *C. leiophylla* Sargent; *C. mackenziei* Sargent ex Mackenzie; *C. rubicundula* Sargent; *C. seclusa* Sargent; *C. seducta* Sargent

Shrubs or trees, 20–40(–70) dm, <usually branched to base>. **Leaves:** blade broadly ovate to deltate, 3–7 cm, length/width = 1–1.2, lobes 2–4 per side, distinct, lobe apex acute, base very broadly cuneate or truncate to weakly subcordate, adaxial surface glabrous, sometimes sparsely appressed-hairy along veins. **Inflorescence branches** glabrous. **Flowers** 15–22 mm diam.; stamens 20, anthers pale pink.

Flowering Apr–May; fruiting Sep–Nov. Open scrub, light woodland shade; 50–300 m; Ont.; Ill., Ind., Iowa, Ky., Mich., Mo., N.Y., N.C., Ohio, Pa., Tenn., Wis.

Variety *rugosa* is fairly common from Missouri to the mountains of North Carolina, then northward through the range of the species. The variety often has the most proportionately wide and most deeply incised leaves of the *Crataegus pruinosa* group. The leaves also characteristically have sharper and longer marginal teeth than other varieties. Extremes are *C. mackenziei* with acute lobes and *C. seducta* with rather obtuse lobes.

Unnamed plants of the southern Appalachian piedmont, very likely worthy of at least varietal rank, will key here and are differentiated from var. *rugosa* as follows: leaf blade 4–6 cm (3–7 cm in var. *rugosa*), laminas thin (finally chartaceous to somewhat coriaceous in var. *rugosa*), basal corners of leaf rounded, bases often subcordate (not conspicuously rounded, bases rarely subcordate in var. *rugosa*).

75e. Crataegus pruinosa (H. L. Wendland) K. Koch var. **virella** (Ashe) Kruschke, Publ. Bot. Milwaukee Public Mus. 3: 181. 1965 [E]

Crataegus virella Ashe, Ann. Carnegie Mus. 1: 396. 1902

Shrubs or trees, 30–70 dm, <larger often single-trunked>. **Leaves:** blade short-ovate to nearly deltate, sometimes ± ovate, 3–4(–5) cm <widest proximal to middle, sometimes near base>, length/width = 1.1–1.2, lobes 3 or 4 per side, <sinuses shallow>, lobe apex acute, base truncate to weakly subcordate (basal corners of leaf rounded), adaxial surface sparsely short-hairy young, usually glabrescent. **Inflorescence branches** glabrous, sometimes sparsely villous. **Flowers** 15–22 mm diam.; stamens 20, anthers pale pink.

Flowering Apr–May; fruiting Sep–Nov. Brush, open woodlands; 50–300 m; Ont.; Ark., Mich., Mo., N.Y., Ohio, Pa., Wis.

Variety *virella* is scattered from Arkansas and Missouri to Wisconsin and to Pennsylvania and New York.

The adaxial leaf surface of var. *virella* is hairy, at least when young. This variety sometimes has hairy pedicels while the leaf blades are among the smallest of the *Crataegus pruinosa* group; its most commonly found form in recent years, from Missouri to Wisconsin, has proportionately very wide leaves. These are the typical form. The most extreme type possibly belonging to var. *virella*, the rare *C. aspera* Sargent [*C. mackenziei* var. *aspera* (Sargent) E. J. Palmer] from Missouri, differs by its hairy pedicels, a scabrous hairy adaxial leaf surface as well as permanently densely hairy petioles and veins on abaxial leaf surfaces. It is so densely hairy that it might well pass for *C. biltmoreana* (ser. *Intricatae*). However, *C. mackenziei* var. *mackenziei* and *C. decorata* Sargent, usually placed here, possess only scattered hairs on the adaxial surfaces of the young leaves plus some more persistent hairs along the major veins of adaxial surfaces; they are synonymized under var. *rugosa* and var. *pruinosa*, respectively.

75f. Crataegus pruinosa (H. L. Wendland) K. Koch var. **magnifolia** (Sargent) J. B. Phipps, Novon 16: 386. 2006 [C] [E]

Crataegus magnifolia Sargent, Trees & Shrubs 2: 69, plate 132. 1908

Shrubs or trees, height not recorded. **Leaves:** blade ovate to ovate-deltate, 5–7 cm, length/width = 1.3–1.5, lobes 0 or 1–4 per side, <obscure>, lobe apex ± obtuse, base broadly cuneate to subtruncate, adaxial surface glabrous, sometimes sparsely appressed-hairy along veins. **Inflorescence branches** glabrous. **Flowers** 18 mm diam.; stamens 20, anthers cream.

Flowering Apr; fruiting Sep–Oct. Chalk barrens, along rocky watercourses; of conservation concern; 100–200 m; Mo.

Variety *magnifolia*, a local variant from Jasper County in southern Missouri, not seen for many years, is distinct for its remarkable leaf shape and cream anthers. It would be more consistent with the logic of this treatment to place var. *magnifolia* with *Crataegus cognata*, but the combination has not been made.

76. Crataegus cognata Sargent, Rhodora 5: 58. 1903 [E]

Crataegus aperta Sargent; *C. patrum* Sargent; *C. pruinosa* (H. L. Wendland) K. Koch var. *cognata* (Sargent) J. B. Phipps; *C. pruinosa* var. *latisepala* (Ashe) Eggleston

Shrubs, 20–40 dm. **Stems:** <main branches often suberect>; twigs: new growth reddish brown, 1-year old purple-brown, older dark gray; thorns on twigs <numerous>, straight to slightly recurved, 2-years old black, shiny, ± slender, 3–5 cm. **Leaves:** petiole length 28–36% blade, eglandular or with 1 or 2 glands distally; blade ovate, 3.5–5.5 cm, subcoriaceous, base broadly cuneate to truncate, lobes 3(or 4) per side, sinuses shallow, lobe apex subacute, margins crenate-serrate, veins 5–7 per side, apex subacute, surfaces glabrous. **Inflorescences** 4- or 5-flowered; branches glabrous; bracteole margins short-stipitate-glandular. **Flowers** 17–19 mm diam.; hypanthium glabrous; sepals narrowly triangular, 5 mm, <base broad>, margins subentire, abaxially glabrous; stamens 20, anthers white, 0.6–0.8 mm; styles 3 or 4. **Pomes** green with pink or purple pruinosity, becoming orange-red or crimson, 8–10 mm diam.; <flesh thin>; sepals on collar, spreading-reflexed; pyrenes 3 or 4. $2n = 51$.

Flowering Apr–May; fruiting Sep–Oct. Successional fields, fencerows, brush, open wood margins; 50–300 m; Ont.; Mass., Mich., Mo., N.Y.

Crataegus cognata in the strict sense occurs from Michigan to Massachusetts with the southern limits poorly understood. Specimens from Missouri with large (5–7 cm) leaf blades and no or only weakly developed lobes should be checked against *C. pruinosa* var. *magnifolia*.

Crataegus perjucunda Sargent, native to Middlesex County, Ontario, is a distinctive local species similar to *C. cognata* but with shorter (2–3 cm) and finer thorns as well as relatively wider leaves (length/width = 1.2).

77. **Crataegus formosa** Sargent, Proc. Rochester Acad. Sci. 4: 101. 1903 [C] [E]

Shrubs or trees, 50 dm. **Stems:** twigs: new growth reddish, 1-year old dark purple-brown, 2-years old gray-tan, older gray; thorns on twigs recurved, 2-years old dark brown, shiny, older charcoal black, then gray, ± slender, 2.5–5 cm. **Leaves:** petiole length 30–36% blade, eglandular or slightly glandular; blade deltate (mainly on shoots with terminal inflorescence) to broadly rhombic-elliptic, 4–7 cm, subcoriaceous, base cuneate to nearly truncate, lobes 3 or 4 per side (larger leaves), sinuses moderately deep, lobe apex acute, margins serrate, veins 5 or 6 per side, apex acute, surfaces glabrous. **Inflorescences** 4–7-flowered; branches glabrous; bracteoles caducous, <absent or several>, margins glandular. **Flowers** 22–30 mm diam.; sepals triangular, 6 mm, margins subentire, abaxially glabrous; stamens 20, anthers white, 0.6–0.8 mm; styles 4 or 5. **Pomes** red, <often ± angular>, 10 mm diam., ± pruinose; sepals on collar, spreading; pyrenes 4 or 5.

Flowering May; fruiting Sep–Oct. Brush; of conservation concern; 100–300 m; Ont.; N.Y., Ohio, Wis.

The known distribution of *Crataegus formosa* is very scattered. The species is one of the most beautiful hawthorns, and one specimen set the *Crataegus* record for flower size with a diameter of 38 mm (E. P. Kruschke 1965). *Crataegus porteri* Britton [including *C. porteri* var. *caerulescens* (Sargent) E. J. Palmer], from Pennsylvania, is similar but with smaller leaves and may be conspecific.

64f.12. CRATAEGUS Linnaeus (sect. COCCINEAE) ser. DILATATAE (Sargent) Rehder, Man. Cult. Trees ed. 2, 362. 1940 [E]

Crataegus [unranked] *Dilatatae* Sargent, Silva 13: 33. 1902

Shrubs or trees, 40–70 dm, main trunk dominant. **Stems:** trunk bark buff to darker, fibrous, checked into longitudinal plates, freshly exposed bark orange-brown; <branches spreading; twigs ± straight>, new growth glabrous or pilose, 1-year old shiny, dark brown to sometimes tan, 2-years old deep gray, older pale; thorns on twigs numerous, usually ± recurved, 2-years old blackish, shiny, moderately thick, 3–6 cm. **Leaves:** petiole length 30–50% blade, glabrous or sparsely hairy, usually sparsely glandular, sometimes eglandular; blade <mid to dark green>, ovate to broadly ovate or ovate-deltate, (4–)5–9(–11) cm, subcoriaceous, base subcordate to truncate or broadly cuneate, lobes 3–5 per side, sinuses shallow, lobe apex acute, margins ± serrate, <teeth large>, venation craspedodromous, veins 5–7 per side, apex acute, abaxial surface glabrous, adaxial sometimes appressed-pubescent young. **Inflorescences** 5–12-flowered, convex panicles; branches glabrous or pilose; bracteoles caducous (larger ± persistent), <hyaline to red-tinged>, (larger) oblong to curved, membranous to semiherbaceous, margins stipitate-glandular. **Flowers** 20–25 mm diam.; hypanthium glabrous or pilose; sepals triangular, much shorter than petals, margins glandular-laciniate; stamens 20; anthers white to rose-red; styles (4 or)5. **Pomes** bright pink to crimson, suborbicular, 10–20(–25) mm diam., glabrous or pubescent at ends; flesh ± hard (until overripe); sepals spreading or ± reflexed, <non-accrescent>; pyrenes 5.

Species 1: c, e North America.

Series *Dilatatae* appears to be closely related to ser. *Pruinosae*, from which it is distinguished by often hairy adaxial leaf surfaces and inflorescence branches, at least some bracteoles more or less hard and persistent, more deeply dissected sepals, usually larger, non-angular pomes lacking elevated calyces, and by all the parts being equal to or greater than the normal upper-size range for ser. *Pruinosae*.

78. **Crataegus coccinioides** Ashe, J. Elisha Mitchell Sci. Soc. 16: 74. 1900 • Kansas hawthorn [E]

Crataegus callicarpa Sargent; *C. conspecta* Sargent; *C. declivitatis* Sargent; *C. dilatata* Sargent; *C. locuples* Sargent; *C. speciosa* Sargent

Shrubs or trees, 40–70 dm. **Stems:** twigs: new growth ± reddish at first, glabrous or pilose, 1-year old shiny, dark brown to sometimes tan, 2-years old deep gray, older paler; thorns numerous, usually ± recurved, 2-years old blackish, shiny, 3–6 cm. **Leaves:** petiole length 30–50% blade, glabrous or sparsely hairy, usually sparsely glandular, sometimes eglandular; blade usually ovate, sometimes very broadly ovate, to ovate-deltate, (4–)5–9(–11) cm, base broadly cuneate, ± truncate, or subcordate (broader leaves), lobes 3–5 per side, sinuses shallow, lobe apex acute, margins serrate or sharply serrate, veins 5–7 per side, apex acute, abaxial surface glabrous, veins glabrous or tomentose, adaxial glabrous or pubescent. **Inflorescences** 5–12-flowered; branches glabrous or densely pubescent; bracteoles hyaline to red-tinged, oblong to curved, membranous to semiherbaceous, margins glandular. **Flowers** 20–25 mm diam.; hypanthium glabrous or pubescent; sepals 6–8 mm; stamens 20, anthers white, rose, or red; styles (4 or)5. **Pomes** bright pink to crimson, suborbicular, 10–20(–25) mm diam., often ± pruinose; sepals spreading or ± reflexed, not on collar; pyrenes 5. **2*n*** = 51.

Flowering Apr–May; fruiting Sep–Oct. Brush, successional fields, pastures, woodland margins; 50–300 m; Ont., Que.; Ark., Ill., Ind., Kans., Ky., Mass., Mo., N.Y., Ohio, Pa.

Crataegus coccinioides occurs from the southwestern Midwest to New England and southern Quebec. Records from Vermont and Oklahoma are unconfirmed.

Crataegus coccinioides, in typical plants, has 20 rose-colored anthers and more or less glabrous parts. The opposite characteristics (white anthers, very hairy parts) are found in *C. declivitatis*. Another white-anthered form is *C. conspecta*. Somewhat similar to the latter is *C. locuples*, while *C. dilatata* is more of an intermediate type. Combinations of these varying characteristics can be found both in Missouri and the Niagara Peninsula of Ontario in various mixes. When in fruit, pruinosity and the pink-crimson color differentiate *C. coccinioides* from members of ser. *Coccineae*, which are scarlet or deeper red. *Crataegus coccinioides* also has somewhat more coriaceous leaf blades than ser. *Coccineae*, with the toothing more continuous, almost to the petiole, and much more conspicuous bracteoles.

64f.13. CRATAEGUS Linnaeus (sect. COCCINEAE) ser. PULCHERRIMAE (Beadle) K. R. Robertson, J. Arnold Arbor. 55: 628. 1974 [E]

Crataegus [unranked] *Pulcherrimae* Beadle in J. K. Small, Fl. S.E. U.S., 532, 535. 1903

Shrubs or trees, 20–70 dm, usually main trunk dominant. **Stems:** trunk bark gray, often rough or corrugated; <branches spreading; twigs ± straight>, new growth glabrous, 1-year old reddish brown to blackish brown, often shiny, 2-years old deep gray, sometimes grayish red or gray-brown; thorns on twigs few to numerous, straight to slightly recurved, 2-years old very dark brown to black, ± slender, (1.5–)2–3(–5) cm. **Leaves:** petiole length 20–50(–55)% blade, glabrous, sparsely to densely sessile-glandular, sometimes stipitate-glandular; blade <mid to dark green>, narrowly elliptic and lanceolate to trullate or deltate, (2.5–)4–7(–8) cm, thin or chartaceous, base ± cuneate to rounded, lobes 0 or (1–)3 or 4(or 5) per side, sinuses usually shallow, lobe apex obtuse to acute, sometimes ± cuspidate, margins crenate to serrate, venation craspedodromous, veins 4–9 per side, apex acuminate to obtuse, surfaces glabrous. **Inflorescences** (2–)5–10 (or 11)-flowered, slightly domed panicles; branches usually glabrous, sometimes sparsely pilose;

bracteoles usually caducous, sometimes ± persistent, <usually numerous>, linear to narrowly oblong or very narrowly obovate, membranous to subherbaceous, margins stipitate-glandular. **Flowers** (13–)14–18(–25) mm diam.; hypanthium glabrous; sepals narrowly triangular or triangular, much shorter than petals, margins ± glandular-serrate; stamens 20, anthers usually pink to purple, sometimes ivory to cream; styles 3–5. **Pomes** yellow, orange, ruddy, red, red-purple, or green, broadly ellipsoid to suborbicular, 5–13(–14) mm diam., glabrous; flesh ± hard to mealy; sepals <on collar>, usually reflexed, sometimes spreading, or erose, <non-accrescent>; pyrenes 2–5.

Species 10 (10 in the flora): se United States.

Series *Pulcherrimae* has a southern limit along the Gulf Coast of eastern Texas to northern Florida. Its species are fairly common over much of Mississippi and Alabama, the panhandle of Florida, and the western part of Georgia. There are also sporadic records east to South Carolina, one record in southern Tennessee, and one locality for North Carolina in Buncombe County. Anthesis is in April through much of Alabama, Georgia, and Mississippi; in the panhandle of Florida, it may start late March, while in North Carolina or Tennessee, it is in May. Fruit is ripe anywhere from late August to early October.

Species of ser. *Pulcherrimae* are mesophytes of woodland edge or woodlands, and if the latter, usually under oak or oak-mixed canopy but not in denser shade. They are recorded mainly on finer-grained soils.

The deeply corrugated bark of several species, found in few other series, glabrous or nearly glabrous plant parts, about 20 stamens, more or less glandular-bracteolate inflorescences, usually reflexed, somewhat elevated sepals in fruit, and sometimes large numbers of lateral veins in the leaf make ser. *Pulcherrimae* distinctive. Series *Intricatae* differs in usually possessing ten stamens, having some very hairy species, northern limits much farther north and southern limits somewhat to the north. As in ser. *Intricatae*, yellowish-fruited forms are at least as numerous as red-fruited ones. Because of the relative scarcity of many species, taxonomic limits are sometimes somewhat difficult to be certain about and may be resolved differently in the future. *Crataegus pearsonii* Ashe, a poorly understood member of ser. *Lacrimatae*, quite often has been used by Louisiana and Texas authors for species of ser. *Pulcherrimae* from those states.

SELECTED REFERENCE Phipps, J. B. and K. A. Dvorsky. 2006b. Review of *Crataegus* series *Pulcherrimae* (Rosaceae). Sida 22: 973–1007.

1. Leaf blades: lobes 0 or obscure, sinuses very shallow, LII 0–10%.
    2. Leaf blades ovate-trullate . . . . 79. *Crataegus sargentii* (in part)
    2. Leaf blades broadly elliptic to narrowly ovate or broadly lanceolate.
        3. Leaf blades narrowly ovate to broadly lanceolate, apices acute to acuminate, veins 5–7(–9) per side, margins serrulate; pomes greenish yellow to yellow . . . . 80. *Crataegus gilva* (in part)
        3. Leaf blades broadly oblong to elliptic or narrowly ovate, apices subacute to obtuse, veins 8 or 9 per side, margins crenate or crenate-serrate; pomes yellow-green with pink or salmon blush, or red . . . . 81. *Crataegus mendosa*
1. Leaf blades: lobes (1 or)2–4(or 5) per side, sinuses shallow to deep, LII 10–50%.
    4. Leaf blades narrower (length/width = 1.5–2).
        5. Lobes 3 or 4 per side, sinuses deep (max LII 25–50%), lobe apices usually ± narrowly cuspidate . . . . 87. *Crataegus incilis*
        5. Lobes 1–4 per side, sinuses shallow (max LII 10–30%), lobe apices subacute to obtuse.
            6. Leaf blade margins with extremely small teeth; pomes dull red; petioles sometimes densely glandular . . . . 83. *Crataegus pinetorum*
            6. Leaf blade margins with average-size to large teeth; pomes greenish yellow to deep yellow, sometimes flushed red or purplish red; petioles of low to average glandularity.

7. Lobes 1–3 per side, lobe apex acute distally, sides ± straight; pomes greenish yellow to yellow. . . . . 80. *Crataegus gilva* (in part)
7. Lobes 2–4 per side, lobe apex obtuse to subacute, sides ± bowed; pomes deep yellow, sometimes flushed red or purplish red . . . . . 82. *Crataegus pulcherrima* (in part)

[4. Shifted to left margin.—Ed.]

4. Leaf blades broader (length/width = 1.2–1.5).
 8. Leaf blades ovate-trullate, distal ⅓ often unlobed, terminal ½ ± straight-sided across lobe tips, 4–6(–7) cm, lobes 1–3 per side, sinuses shallow, lobe apices acute; anthers pale pink to pale purple; pomes yellow or flushed pink, sometimes reddish . . . . 79. *Crataegus sargentii* (in part)
 8. Leaf blades of quite different form, if ± trullate-ovate, then terminal ½ ± rounded across lobe apices, 2.5–8 cm, lobes 0 or 1–4(or 5) per side, sinuses shallow to deep, lobe apices obtuse to acute; anthers ivory to cream or pink to purple; pomes greenish to deep yellow, sometimes flushed red, purplish red, or red.
  9. Leaf blades: sinuses deep (LII 15–30%).
   10. Leaf blades 4.5–8 cm, ovate-rhombic to ovate-trullate, lobes angled; anthers cream to pale or deep purple; pomes red; flowering mid May, flowers 22–25 mm diam.. . . . . 88. *Crataegus eximia*
   10. Leaf blades 3–6 cm, ovate to broadly ovate or ± deltate, trullate, or rhombic-ovate, lobes angled or not; anthers cream or pink to pale or deep purple; pomes yellow to green or red, sometimes ivory to cream; flowering early to mid April, flowers 16–19 mm diam.
    11. Lobe apices ± obtuse to subacute. . . . . .85. *Crataegus opima* (in part)
    11. Lobe apices ± acute . . . . .86. *Crataegus tecta*
  9. Leaf blades: sinuses shallow (LII 10–15%).
   12. Lobe apices acute to acuminate young.
    13. Leaf blades narrowly ovate, lobe apices acute distally, (straight, ± unchanging with age); angle from widest lobe to tip ca. 60° . . . . 80. *Crataegus gilva* (in part)
    13. Leaf blades broadly oblong to ovate, lobe apices acuminate young, (sometimes reflexed), obtuse later; angle from widest lobe to tip ca. 90° . . . . 84. *Crataegus venusta*
   12. Lobe apices obtuse to subacute young.
    14. Leaf blades 5–8 cm, lobes 1–3 per side, sinuses shallow, base cuneate, veins 5–7 per side . . . . *Crataegus* species (see discussion under 81. *C. mendosa*)
    14. Leaf blades to 5 cm, lobes 2–4 per side, sinuses shallow, base cuneate to broadly cuneate, veins 4 or 5(or 6) per side.
     15. Leaf blades: length/width = 1.5, bases cuneate. . . . . . 82. *Crataegus pulcherrima* (in part)
     15. Leaf blades: length/width less than 1.5, bases cuneate to broadly cuneate . . . . .85. *Crataegus opima* (in part)

79. Crataegus sargentii Beadle, Bot. Gaz. 28: 407. 1899 (as sargenti) • Sargent hawthorn [E]

Shrubs or trees, 30–50 dm. **Stems:** twigs: new growth reddish, 1-year old shiny, deep reddish brown, 2-years old grayish red, older gray; thorns on twigs <few to numerous>, straight to slightly recurved, 2-years old black, slender, 2–5 cm. **Leaves:** petiole length 25–40% blade, sparsely sessile-glandular; blade ovate-trullate, 4–6(–7) cm, <terminal ½ ± straight-sided across lobe tips>, base rounded to broadly cuneate, lobes 0 or 1–3 per side, barely present or sinuses shallow, <LII 0–20%>, lobe apex acute, margins serrulate, <teeth to 1 mm>, veins 5–7 per side, apex acute to acuminate, abaxial veins sparsely hairy young. **Inflorescences** 4–10(or 11)-flowered; branches glabrous; bracteoles caducous, <numerous, often ± curved, green>, linear, membranous to subherbaceous, margins glandular. **Flowers** 15–18 mm diam.; sepals narrowly triangular, 3 mm, <base flaring to wide>, margins entire or glandular-denticulate, <apex ± obtuse, abaxially glabrous>; anthers pale pink to pale purple; styles 3 or 4. **Pomes** yellow or flushed pink, sometimes reddish, suborbicular, 8–10(–12) mm diam.; sepals <on collar>, reflexed; pyrenes 3 or 4. 2*n* = 34, 51.

Flowering late Mar–mid Apr; fruiting Sep–Nov. Woodland edges, cut-over woodlands, other open woodlands; 20–300 m; Ala., Fla., Ga., Miss.

Crataegus sargentii is mainly from Alabama, extending west to central Mississippi, eastwards with scattered records from Georgia, and south to the panhandle of Florida; it is locally common. Anthesis is three weeks later in the Alabama canyonlands than on the Gulf Coast. The species is most similar to *C. gilva*; the latter has narrower leaves and smaller pomes. Occasional more deeply incised forms may resemble other species to some extent; they will usually retain the characteristic overall leaf shape of the species.

80. Crataegus gilva Beadle, Biltmore Bot. Stud. 1: 60. 1902 [E]

Crataegus sargentii Beadle var. *gilva* (Beadle) Lance

Shrubs or trees, 20–50 dm. **Stems:** twigs: new growth reddish, 1-year old deep reddish brown, older dark, dull gray; thorns on twigs straight or slightly recurved, 2-years old dull purplish brown, fine, 2–4 cm. **Leaves:** petiole length 20–30% blade, glandular; blade broadly lanceolate to narrowly elliptic or very narrowly ovate to narrowly ovate or narrowly ovate-rhombic, 3.5–4.5 cm, base curved-cuneate, lobes 0 or 1–3 per side, <sides ± straight>, sinuses shallow, <LII 0–20%>, lobe apex acute distally, margins serrulate, <teeth sometimes large>, veins 5–7(–9) per side (except smaller leaves), apex acute to acuminate <angle from widest lobe to tip ca. 60°>, adaxial veins sparsely hairy especially young. **Inflorescences** 3–7-flowered; branches glabrous; bracteoles caducous, <dense>, narrowly oblong to very narrowly obovate, membranous to subherbaceous, margins densely glandular. **Flowers** (13–)18–20 mm diam.; sepals narrowly triangular, 6–7 mm, margins glandular-serrate; anthers pale pink; styles 3 or 4. **Pomes** greenish yellow to yellow, orbicular, 5–8 mm diam.; sepals ± reflexed; pyrenes 3 or 4.

Flowering Mar–Apr; fruiting Sep–Oct. Woodland edges and gaps, clear-cuts, pine-oak woodlands; 0–200 m; Ala., Fla., Ga., Miss.

Crataegus gilva occurs from Mississippi to Georgia and northern Florida. The leaves are rather like those of *C. mendosa*, but with an acuminate leaf apex, much sharper lobes (when present) and marginal teeth, and somewhat resemble a less clearly lobed *C. pinetorum*. The leaves are also usually somewhat like those of *C. sargentii* but narrower. In the flowering type, they are narrowly elliptic, with cuneate bases, virtually devoid of lobes, with particularly large marginal teeth and acuminate apices, this being the most extreme specimen of the species seen. The suspicion that *C. gilva* may represent an extreme form of *C. sargentii* is heightened by the almost complete lack of known fruiting specimens of the species, the fruiting co-type excepted (*Harbison 4374/2*); nevertheless, that specimen suggests the possibility that the leaves are often wider after anthesis. *Crataegus agrestina* Beadle, known only from a syntype specimen, is similar.

81. Crataegus mendosa Beadle, Biltmore Bot. Stud. 1: 65. 1902 [E] [F]

Shrubs or trees, 30–60 dm. **Stems:** <trunk bark not recorded>; twigs: new growth reddish green, 1-year old shiny reddish or purplish brown, older dark or dull gray; thorns on twigs <absent or few to numerous>, straight or slightly recurved, 1-year old reddish black, 2-years old blackish, fine, 2–5 cm. **Leaves:** petiole length 30–50% blade, sparsely glandular; blade <adaxially dark green>, broadly oblong to elliptic or narrowly ovate, 4–7 cm, base cuneate, lobes 0 or 1–3, obscure, <max LII 10%>, margins crenate or crenate-serrate, veins 8 or 9 per side (fewer on smaller leaves), apex subacute (obtuse on some larger leaves), abaxial midvein sparsely hairy young. **Inflorescences** 4–7 -flowered; branches glabrous; bracteoles caducous, <dense>, narrow, semiherbaceous (and green) to membranous, margins glandular. **Flowers** 14–17 mm diam.; sepals triangular, much shorter than petals, margins glandular-denticulate; anthers pink or pink-purple; styles 3–5. **Pomes** yellow-green with pink or salmon blush, or red, suborbicular to broadly ellipsoid, 8–10 mm diam.; sepals reflexed; pyrenes 3–5.

Flowering Mar–Apr; fruiting Sep–Nov. Woodland edges, cut-over woodlands, brush; 20–200 m; Ala., Fla., Ga., La., Miss., S.C., Tex.

Crataegus mendosa is scattered from southeastern Texas to western Georgia and northern Florida with a disjunct record in South Carolina. The species is one of the more distinctive members of the series, although it may be approached in form by *C. gilva*, which has an acuminate leaf apex, much sharper lobes (when present), and marginal teeth. The extension-shoot leaves of *C. mendosa* are ovate to broadly ovate, usually quite wide, with three or four subacute to acute lobes, which can be confusing if only few typical short-shoot leaves are present.

Plants from Sardis, Alabama, with similarities to *Crataegus mendosa*, *C. opima*, and *C. pulcherrima*, may key out here; they have leaf blades to 8 cm with shallow (LII 10–15%) sinuses and blunt to subacute lobe apices.

82. Crataegus pulcherrima Ashe, J. Elisha Mitchell Sci. Soc. 16: 77. 1900 • Beautiful hawthorn [E]

Shrubs, 20–40 dm. **Stems:** twigs: young growth green or reddish tinged, glabrous, 1-year old reddish black, 2-years old dark gray; thorns on twigs ± straight, 2-years old blackish, fine, 2–3.5 cm. **Leaves:** petiole length 25–40% blade, sessile- or stipitate-glandular; blade narrowly ovate or ± oblong to narrowly obovate or narrowly rhombic, 2.5–5 cm, <length/width = 1.5, thin>, base cuneate, lobes 2–4 per side, <sides ± bowed>, sinuses shallow, <max LII 10–30%>, lobe apex obtuse to subacute, margins serrulate, ± distant from one another for their size, veins 4 or 5(or 6) per side, apex acute, adaxial veins pilose young. **Inflorescences** 4–7-flowered (central axis very short); branches glabrous; bracteoles caducous, <numerous, small>, narrowly oblong, membranous, margins stipitate-glandular. **Flowers** 15–20 mm diam.; sepals narrowly triangular, 3–4 mm, margins glandular-serrate, <teeth average size for series, apex obtuse, abaxially glabrous>; anthers pink, <small>; styles 3. **Pomes** <in clusters of 2–5>, deep yellow, sometimes flushed red or purplish red, suborbicular, 6–7 mm diam.; sepals reflexed; pyrenes 3.

Flowering mid Apr; fruiting Sep–Oct. Open or rocky woodlands, swamp borders; 0–200 m; Ala., Fla., Ga., La., Miss.

Crataegus pulcherrima occurs from Louisiana to Georgia and northern Florida. *Crataegus robur* Beadle is similar and perhaps the same.

83. Crataegus pinetorum Beadle, Bot. Gaz. 30: 343. 1900 [E]

Shrubs or trees, 20–60 dm. **Stems:** twigs: new growth strongly reddish, 1-year old shiny reddish brown, 2-years old dull gray-brown, older dull gray; thorns on twigs <usually present>, ± straight, 1-year old shiny black or reddish black, 2-years old becoming black, fine, 2–3.5 cm. **Leaves:** petiole length 25–30% blade, sparsely to densely glandular or stipitate-glandular; blade narrowly ovate to narrowly elliptic, 4–6 cm (nearly mature at anthesis), base cuneate, lobes 2 or 3 per side, sinuses shallow, <max LII 10–30%>, lobe apex acute, margins with extremely small teeth, veins 5 or 6 per side, apex acute, vein pubescence not recorded. **Inflorescences** 2–4-flowered; branches glabrous; bracteoles ± persistent, <numerous>, narrowly obovate, chartaceous to subherbaceous, margins densely glandular. **Flowers** at anthesis not recorded; sepals narrowly triangular, 6 mm, margins finely glandular-serrate, <abaxially glabrous>; anther color not recorded; style number not recorded. **Pomes** dull red, suborbicular, 12–14 mm diam.; sepals erose; pyrenes 2–4. $2n = 51$.

Flowering Apr; fruiting Sep–Oct. Woodland edges and gaps; 100–200 m; Ala., Ga., Miss.

Crataegus pinetorum is found from south-central Mississippi through Alabama to northern Georgia; it appears to be scarce. The species is most similar to *C. pulcherrima*, differing in its sharply, rather than obtusely, lobed leaves, often densely glandular petioles, exceptionally small leaf marginal teeth, and red pomes. The species also somewhat resembles the more deeply incised forms of *C. gilva*.

84. Crataegus venusta Beadle, Bot. Gaz. 30: 338. 1900 [E]

Shrubs or trees, 40–70 dm. **Stems:** twigs: new growth reddish green, 1-year old usually deep chestnut; thorns on twigs ± straight, 2-years old very dark, usually slender, 2–5 cm. **Leaves:** petiole length 25–30% blade, sparsely sessile-glandular; blade broadly oblong to ovate, 5–8 cm, base cuneate to broadly cuneate, lobes 0, or 1 or 2 per side, sinuses shallow, lobe apex acuminate young, obtuse later, margins dentate, <teeth acute to acuminate>, veins 5–7 per side, apex acute to acuminate <angle from widest lobe to tip ca. 90°>, adaxial midvein sparsely long-hairy young. **Inflorescences** 5–8-flowered; branches glabrous; bracteoles caducous, <numerous, hyaline to greenish>, narrowly oblanceolate to linear, membranous, margins densely glandular. **Flowers** 15–20 mm diam.; sepals narrowly triangular, length not recorded, margins finely glandular, <abaxially glabrous>; anthers pink to purplish, sometimes cream; styles 3 or 4. **Pomes** greenish yellow to yellow, sometimes reddish, suborbicular, 8–10 mm diam.; sepals spreading or erose; pyrenes 3–5.

Flowering early–mid Apr; fruiting Sep–Nov. Brushy forest regrowth, forest gaps and margins; 20–200 m; Ala., Fla., Ga., La., Miss., Tex.

Crataegus venusta is most abundant in Alabama. When flowering, it is one of the most striking members of ser. *Pulcherrimae*, with typical oblong leaf blades, their margins with lobes more or less absent and bearing numerous acuminate teeth. At maturity this characteristic is diminished, and typical larger short-shoot leaves then have several shallow lobes per side, these more or less subacute. *Crataegus venusta* was described by Beadle as possessing reddish pomes and cream anthers, but all material where color determination was possible had yellowish pomes and pink to purple anthers.

The most distinctive form possibly synonymous with *Crataegus venusta* is *C. pallens* Beadle [*C. venusta* Beadle var. *pallens* (Beadle) Lance], a morphotype with a very sparse and scattered distribution that is found from its type locality in western North Carolina to the type locality of its synonym *C. alma* Beadle in Mississippi. Compared to typical *C. venusta*, it usually has smaller leaves (blades typically less than 5 cm), differently shaped lobes that are wider than those of *C. venusta* with the leading edge usually more or less perpendicular to the midvein, and less sharp teeth. Definite intermediates with typical *C. venusta* are not known.

85. Crataegus opima Beadle, Biltmore Bot. Stud. 1: 40. 1901 [E]

Crataegus abstrusa Beadle; *C. pulcherrima* Ashe var. *opima* (Beadle) Lance

Shrubs or trees, 70 dm. **Stems:** twigs: new growth reddish brown, 1-year old blackish brown, older gray; thorns on twigs <absent or few to numerous>, straight to slightly recurved, 2-years old dark brown to blackish, slender, 2.5–4 cm. **Leaves:** petiole length 20–30% blade, sparsely short-stipitate-glandular; blade broadly ovate to ± deltate, 3–5 cm, <length/width less than 1.5>, base cuneate to broadly cuneate, lobes 3 or 4 per side, sinuses shallow in common form, lobe apex usually ± obtuse, <angled or not>, margins serrate, <teeth subacute>, veins 4 or 5 per side, apex subacute to obtuse, adaxial midvein sparsely pilose young. **Inflorescences** 4–7 -flowered; branches glabrous or sparsely pilose; bracteoles caducous, <sometimes curved>, narrowly oblong, membranous to subherbaceous (and green), margins glandular. **Flowers** 16–18 mm diam.; sepals triangular, 4 mm, distal margins glandular-serrate; anthers cream to pale or deep purple; styles 3–5. **Pomes** yellow or red, suborbicular, 7–12 mm diam.; sepals <on slight collar>, erose or reflexed; pyrenes 3–5.

Flowering early–mid Apr; fruiting Sep–Oct. Open woods, woodland margins; 20–100 m; Ala., Ga., Miss., Tex.

Crataegus opima is found from southeastern Texas to Georgia.

Crataegus opima comprises the smaller-leaved members of the series with proportionately broad leaves and more or less obtuse lobes that are few and shallow. Both red- and yellow-fruited forms are known, as are forms with cream or purple anthers. It is possible that, as elsewhere in the series, these forms should be recognized as distinct taxa, but more information would be required to decide this.

A distinctive local form of *Crataegus opima* is known with ovate to ovate-trullate leaves 5–8 cm, having five to seven (or eight) veins per side, and five to eight usually short and sometimes subacute, deeper lobes per side.

86. Crataegus tecta Beadle, Biltmore Bot. Stud. 1: 26. 1901 [E]

Shrubs or trees, 40–60 dm. **Stems:** <trunk bark not recorded>; twigs: new growth reddish green, 1-year old shiny, dark reddish black, older dull gray; thorns on twigs ± straight, 2-years old shiny reddish black, ± fine, 2.5–4 cm. **Leaves:** petiole length 30–50% blade, sparsely sessile-glandular; blade trullate to rhombic-ovate or ovate, sometimes broadly so, 3–6 cm, base broadly cuneate to cuneate, lobes 3 or 4 per side, sinuses usually ± deep, lobe apex acute or acuminate to ± cuspidate (larger ones sharper), <angled or not>, margins with small acuminate teeth except at base, veins 4 or 5 per side, apex subacute, adaxial midvein hairy young. **Inflorescences** 3–6-flowered; branches usually glabrous; bracteoles caducous, <numerous, sometimes curved>, sublinear (often broader near apex), membranous to subherbaceous, margins glandular. **Flowers** 17–19 mm diam.; sepals narrowly triangular, 7 mm, margins subentire or glandular-denticulate; anthers pale pink to purple, sometimes ivory to cream; styles 3–5. **Pomes** red or green to yellow, suborbicular, 7 mm diam.; sepals reflexed; pyrenes 3–5.

Flowering Apr; fruiting Sep–Oct. Woodland gaps and edges, brush; 10–100 m; Ala., Fla., Ga., Miss., Tenn.

Crataegus tecta occurs from Mississippi to western Georgia and northern Florida. The species is fairly common in southern Alabama, and there is one somewhat atypical example for Tennessee, the second most northerly record for the series.

The typical form of *Crataegus tecta* is a distinctive morphotype in ser. *Pulcherrimae*. Compared to the somewhat similar *C. incilis*, the leaves are proportionately wider and have fewer lateral veins, and the anthers are cream or ivory. One specimen of *C. tecta* has a relatively densely pubescent inflorescence and, if correctly identified, this is the only such example for the series. Some specimens with conspicuously acuminate and slightly reflexed lobes are provisionally referred here rather than to *C. venusta*. *Crataegus opima* is usually the species most similar to *C. tecta*, lobe size and shape being the main discriminators. *Crataegus austrina* Beadle, with a narrower leaf form, is possibly the same.

87. Crataegus incilis Beadle, Biltmore Bot. Stud. 1: 41. 1901 [E]

Crataegus pulcherrima Ashe var. *incilis* (Beadle) Lance

Shrubs or trees, 40–70 dm. **Stems:** twigs: new growth reddish green, 1-year old dark red-brown, 2-years old dark gray; thorns on twigs ± straight, 2-years old dark brown to blackish, fine, 1.5–3 cm. **Leaves:** petiole length 40–55% blade, sparsely to densely sessile-glandular; blade trullate to narrowly ovate, 3.5–5.5(–7.5) cm, base cuneate (tapering into distal winged part of petiole), lobes 3 or 4 per side, sinuses deep, <max LII 25–50%>, lobe apex usually ± narrowly cuspidate, margins dentate, <teeth ± distant, subacute>, veins 5–7 per side, apex subacute, vein pubescence not recorded. **Inflorescences** 3–7-flowered; branches glabrous; bracteoles caducous, <numerous>, linear, membranous, margins glandular. **Flowers** 14–18 mm diam.; sepals narrowly triangular, 7 mm, margins subentire or glandular-denticulate; anthers pink; styles 3–5. **Pomes** red or yellow, orbicular, 5–8(–10) mm diam.; sepals wide-spreading; pyrenes 3–5.

Flowering Apr; fruiting Sep–Oct. Open woods, woodland margins; 20–200 m; Ala., Fla., Ga., Miss.

The deeply cut trullate leaves of *Crataegus incilis*, with the acute tips of their triangular lobes in a more or less straight line in the more extreme forms, are very distinctive. The type, and most specimens seen, do not have quite straight-sided lobes as seen in the illustration in J. B. Phipps and K. A. Dvorsky (2006b), and their lobes may therefore be described as more or less narrowly cuspidate. Forms with somewhat more rounded lobes and yellow pomes that might key out here perhaps represent *C. concinna* Beadle.

88. Crataegus eximia Beadle, Biltmore Bot. Stud. 1: 62. 1902 [C] [E]

Shrubs or trees, 30–40 dm. **Stems:** twigs: new growth reddish green, 1-year old reddish brown, older gray; thorns on twigs slightly curved, 2-years old blackish gray, usually fine, 3–4 cm. **Leaves:** petiole length 25–40% blade, densely stipitate-glandular (at least young); blade ovate-rhombic to ovate-trullate, 4.5–6.5 cm (flowering) to 8 cm (mature), base cuneate to broadly cuneate, lobes 3 or 4(or 5) per side, sinuses deep, lobe apex subacute to acute, <angled>, margins finely serrate, veins 5 or 6 per side, apex acute, adaxial midvein scabrous young. **Inflorescences** 4–6-flowered; branches glabrous; bracteoles caducous, very narrowly obovate, ± membranous, margins glandular. **Flowers** 22–25 mm diam.; sepals narrowly triangular, 4–5 mm, margins finely glandular-serrate; anthers cream; styles 3 or 4. **Pomes** red, orbicular, 8–12 mm diam.; sepals spreading; pyrenes 3–5.

Flowering late May; fruiting Sep–Oct. Rocky woods and glades; of conservation concern; 200 m; Ala.

Known only from the flowering type specimen and the protologue by Beadle, *Crataegus eximia* seems one of the most distinct entities in the series on the basis of its large rhomboid-ovate leaves with deep sinuses and angled lobes, large flowers, and late flowering date (23 May in 1901).

64f.14. CRATAEGUS Linnaeus (sect. COCCINEAE) ser. INTRICATAE (Sargent) Rehder, Man. Cult. Trees ed. 2, 363. 1940 [E]

Crataegus [unranked] *Intricatae* Sargent, Silva 13: 32. 1902; *Crataegus* sect. *Flavae* Loudon; *Crataegus* ser. *Flavae* (Loudon) Rehder

Shrubs or trees, 10–60(–80) dm, main trunk dominant. **Stems:** trunk bark gray and corrugated or buff to gray-brown and fibrous, checked into longitudinal plates, freshly exposed bark orange-brown; <branches spreading; twigs ± straight, very rarely slightly flexuous>, new growth usually glabrous, sometimes pubescent to tomentose, 1–2-years old gray to purple, purple-brown, reddish black or brown to very dark gray; thorns on twigs usually frequent, straight to recurved, 2-years old usually dark gray, almost black, sometimes purple- or chestnut-brown, usually slender, (1–)2–4(–6) cm. **Leaves:** petiole length 20–66% blade, glabrous or pubescent, usually sessile-, sometimes stipitate-, glandular (often conspicuously so); blade <mid to dark green>, ovate, rhombic, elliptic, or broadly elliptic, (1.5–)3–6(–9) cm, thin to chartaceous, base cuneate to rounded, sometimes truncate to subcordate, lobes 0 or 1–4(or 5) per side, sinuses shallow to ± deep, lobe apex usually acute, margins strongly serrate, <often glandular, particularly toward base>, venation craspedodromous, veins 3–5(–7) per side, apex acuminate to obtuse, surfaces glabrous or pubescent. **Inflorescences** (2 or)3–8-flowered, convex panicles; branches glabrous or hairy; bracteoles (not seen in *C. stonei* and *C. ouachitensis*), caducous, <conspicuous>, linear (oblong-linear in *C. biltmoreana*), membranous (to nearly herbaceous in *C. intricata*), margins stipitate-glandular, rarely shortly so. **Flowers** (12–)14–20(–22) mm diam.; hypanthium glabrous or hairy; sepals narrowly triangular (lanceolate in *C. nananixonii* and *C. biltmoreana*), much shorter than petals, margins glandular-serrate to glandular-pectinate; stamens (5–)10(–20), anthers ivory, cream, or pink to pale purple; styles 2–5. **Pomes** dull yellow to orange, ruddy, russet, or scarlet to red, sometimes green, suborbicular, pyriform, obovoid, or oblong, sometimes broadly ellipsoid, (7–)8–12(–15) mm diam., glabrous or hairy; flesh ± hard to mealy; sepals <usually on collar>, patent to reflexed, <non-accrescent>; pyrenes 2–5.

Species 12 (12 in the flora): c, e North America.

Members of ser. *Intricatae* are among the most characteristic hawthorns of the Appalachian foothills and valleys and extend beyond these areas less frequently. They are found in brush, open woodlands, and woodland edge habitats. The series includes the very rare *Crataegus flava* (the type of ser. *Flavae*), but ser. *Intricatae* is chosen for the name of the amalgamated series because, while both names have equal priority, the majority of ser. *Flavae* as previously understood are excluded from the combined series, and *C. flava* is now suspected of hybrid origin. Otherwise, ser. *Intricatae* is treated here in its traditional sense. Series *Pulcherrimae*, though always with 20 stamens, is similar to the glabrous inflorescence, southern elements of ser. *Intricatae*. This is seen particularly in the sometimes stipitate-glandular petioles, the usually densely bracteolate inflorescence, the bracteolar margins usually having relatively large, short-stipitate glands, and the fruiting calyx being usually elevated on a collar. Most varieties of *C. chrysocarpa* (ser. *Rotundifoliae*) also have similarities to hairy inflorescence members of ser. *Intricatae* but differ in lacking stipitate-glandular petioles, having less markedly glandular bracteoles, and having barely elevated fruiting calyces.

The most widespread members of ser. *Intricatae* are the glabrate *Crataegus intricata* and the hairy *C. biltmoreana*, while extremes of variation are toward the nearly entire-leaved *C. padifolia* among the glabrous group and the diminutive *C. nananixonii* among the hairy group.

1. Inflorescence branches pubescent to villous or tomentose.
    2. Leaf blades 1.5–3.5 cm; shrubs 10–15 dm; stamens 10, anthers rose-purple; e Texas . . . . 95. *Crataegus nananixonii*
    2. Leaf blades (2–)3–8 cm; shrubs 10–60 dm; stamens (5–)10 or 20, anthers ivory to pale yellow, pink to purple; Louisiana to Florida and Missouri to Great Lakes and New England.
        3. Stamens 5–10.
            4. Anthers cream; leaf blades narrowly to broadly ovate, lobes 3–5 per side, distinct; New England to Georgia and Arkansas . . . . 96. *Crataegus biltmoreana*
            4. Anthers pink; leaf blades narrowly ovate, lobes 0 or 1–3 per side, sinuses shallow; Massachusetts, New York, Pennsylvania . . . . 97. *Crataegus stonei*
        3. Stamens 20.
            5. Anthers pink or rose to rose-purple; leaf blades ovate to deltate-ovate, (2–)3–4 cm; petioles 0.5 mm wide; leaf blade lobes 1–3 per side; pedicels ± tomentose young; Arkansas . . . . 98. *Crataegus ouachitensis* (in part)
            5. Anthers pale yellow or cream; leaf blades ovate to broadly elliptic, 4–8 cm; petioles 1–2 mm wide; leaf blade lobes 3 or 4 per side; pedicels pubescent; Appalachia . . . . 99. *Crataegus craytonii*

1. Inflorescence branches glabrous or sparsely villous.
   6. Inflorescence branches glabrous or sparsely villous; stamens 13–16 . . . . 89. *Crataegus flava* (in part)
   6. Inflorescence branches glabrous; stamens 10–20.
      7. Stamens 20; leaf blades ovate to deltate-ovate, 3–4 cm; petioles 0.5 mm wide . . . . 98. *Crataegus ouachitensis* (in part)
      7. Stamens 10(or 13–16); leaf blades ovate to elliptic, 2.5–8(–9) cm; petioles 1 mm wide.
         8. Leaf blades ± elliptic, rhombic-elliptic, or ± obtrullate to ± ovate, lobes 0 or mere apiculi, sinuses very shallow or margins sinuous.
            9. Leaf blades rhombic-elliptic to ± obtrullate, marginal teeth strong; fruiting sepals on collar; pomes red to orange-red; e United States . . . . 93. *Crataegus rubella* (in part)
            9. Leaf blades elliptic to broadly elliptic, marginal teeth fine; fruiting sepals sessile; pomes dull orange-brown to ± shiny or red-orange to red; Ozarks . . . . 100. *Crataegus padifolia* (in part)
         8. Leaf blades usually narrowly to broadly ovate, sometimes obovate or broadly elliptic, lobes evident, 1–5 per side, sinuses ± deep, margins not sinuous.
            10. Anthers ivory or cream.
                11. Shrubs 10–60 dm; leaf blades broadly elliptic to broadly ovate or oblong, lobe apices acute to acuminate, often reflexed; sepal margins glandular-serrate; pomes dull yellow to orange or russet; fruiting sepals on collar . . . . 90. *Crataegus intricata*
                11. Shrubs 40–60 dm; leaf blades ovate, lobe apices ± obtuse; sepal margins finely glandular-serrate distally; pomes red-orange to red; fruiting sepals sessile . . . . 100. *Crataegus padifolia* (in part)
            10. Anthers rose to purple.
                12. Leaf blades with sinuous lobes . . . . 89. *Crataegus flava* (in part)
                12. Leaf blades with acute or subobtuse lobes.
                    13. Leaf blades usually ± ovate to broadly elliptic, sinuses shallow or moderately deep; pomes suborbicular.
                        14. Shrubs, 20–40 dm; leaf blades trullate-ovate, bases ± cuneate, lobe apices acute; Ozarks, rare in se United States . . . . . 91. *Crataegus neobushii*
                        14. Shrubs or trees, 10–80 dm; leaf blades broadly elliptic to ovate or broadly ovate, bases broadly cuneate, lobe apices subacute; s Appalachia. . . . . 92. *Crataegus buckleyi*
                    13. Leaf blades ± elliptic or rhombic-elliptic, sometimes ± obtrullate, sinuses shallow to non-existent; pomes usually oblong or short-oblong or obovoid, sometimes suborbicular.
                        15. Leaf blades ± elliptic or rhombic-elliptic, sometimes ± obtrullate, lobes 0, or 1 or 2 per side, sinuses shallow, lobe apices acute; pomes red to orange-red . . . . 93. *Crataegus rubella* (in part)
                        15. Leaf blades broadly elliptic, lobes 0, or 2 or 3 per side, reduced to apiculi, lobe apices subobtuse; pomes yellow-green to dull orange, orange-red, or red . . . . 94. *Crataegus communis*

89. Crataegus flava Aiton, Hort. Kew. 2: 169. 1789 • Yellow hawthorn C E

Shrubs or trees, 50–60 dm. **Stems:** twigs <straight or slightly flexuous>, new growth color not recorded, glabrous, 1-year old purple-brown, older dark gray; thorns on twigs straight to slightly recurved, 2-years old purple-brown, slender, 2–3 cm. **Leaves:** petiole <1 mm wide>, length 30–45% blade, <winged distally>, slightly pubescent, glandular; blade rhombic-obovate, 5–8 cm (in type material), base cuneate, lobes 1–3 per side, <sinuous>, sinuses shallow, lobe apex obtuse to acute, margins crenate to crenate-serrate, <teeth gland-tipped>, veins 3 or 4 per side, apex subacute to obtuse, adaxial surface pubescent young, glabrescent. **Inflorescences** 4–6-flowered; branches glabrous or sparsely villous; bracteole margins glandular. **Flowers** 16–18 mm diam.; hypanthium glabrous; sepals 4–5 mm, margins glandular; stamens 13–16, anthers purple; styles 3–5. **Pomes** dull orange, ± pyriform-oblong, 8–12 mm diam., glabrous; sepals on collar, ± reflexed; pyrenes 3–5.

Flowering early Apr; fruiting Sep–Oct. Deep, sandy soil; of conservation concern; 10–100 m; Fla., Ga., S.C.

Crataegus flava ranges from southeastern Georgia to South Carolina and northern Florida. Wild specimens of the typical form have not been seen since the early 1900s; vegetative specimens resembling this species were collected from the Florida panhandle in the early part of the present century.

Crataegus flava has been widely confused with members of both ser. *Apricae* and ser. *Lacrimatae* (J. B. Phipps et al. 2007). Its stamen number has been wrongly credited as ten or 20; stamen number in the type has recently been verified as 13 to 16 by S. Landrein (Kew). Recent collections with 13 to 16 stamens are unknown. The correct stamen counts are unusual for *Crataegus*, which are mainly narrowly dispersed around modes of ten or 20, so this perhaps suggests, together with its somewhat flexuous twigs, and rarity, hybrid origin. *Crataegus flava* is similar to *C. rubella*, particularly in fruit shape and color as well as general glandularity, so its placement with members of sect. *Intricatae* seems appropriate. Lack of recent collections similar to the type suggests that modern attributions to *C. flava* may represent hybrids involving members of ser. *Apricae*.

90. Crataegus intricata Lange, Bot. Tidsskr. 19: 264. 1895 • Intricate hawthorn E

Crataegus boyntonii Beadle; *C. foetida* Ashe; *C. intricata* var. *boyntonii* (Beadle) Kruschke

Shrubs, 10–60 dm. **Stems:** twigs: new growth reddish, glabrous, 1–2-years old dark reddish black, older grayer; thorns on twigs <absent or frequent>, straight to recurved, 2-years old blackish, slender, 2.5–5 cm. **Leaves:** petiole <1 mm wide>, length 40–50% blade, usually glabrous, sometimes hairy, glandular; blade broadly elliptic to broadly ovate or oblong, 4–8(–9) cm, base broadly cuneate to ± rounded, lobes 3–5 per side, sinuses ± shallow to moderately deep, <angled to rounded>, lobe apex acute to acuminate, <often reflexed>, margins ± serrate, <teeth gland-tipped at least young>, veins 4–6 per side, apex acute, surfaces glabrous. **Inflorescences** 3–8 -flowered; branches glabrous; bracteoles <membranous to nearly herbaceous>, margins short-stipitate-glandular. **Flowers** 16–20 mm diam.; hypanthium glabrous; sepals 6 mm, margins glandular-serrate; stamens 10, anthers ivory or cream; styles 3–5. **Pomes** dull yellow to orange or russet, sometimes green mature, orbicular to ± pyriform, 8–13 mm diam., glabrous; sepals on collar, patent-reflexed; pyrenes 3–5. $2n = 51$.

Flowering late Apr–early Jun; fruiting Sep–Oct. Open forests, forest gaps, open places; 10–300 m; Ont.; Ala., Conn., Del., Ga., Ill., Ind., Ky., Md., Mass., Mich., Mo., N.J., N.Y., N.C., Ohio, Pa., R.I., S.C., Tenn., Vt., Va., W.Va.

Crataegus intricata ranges in Appalachia from southern New England to Georgia and Alabama. The distribution given above, relative to similar taxa, may be somewhat over-represented because the species has been the default identification for doubtful (particularly fruiting) material of much ser. *Intricatae* with glabrous leaves and inflorescences.

Crataegus intricata is the common, widespread member of this series with lobed leaves, ten stamens, white or cream anthers, and glabrous inflorescences. Depth and sharpness of lobes and location of widest part of leaf are somewhat variable, as is plant stature, some forms appearing mature at 1–2 m.

W. W. Eggleston collected a distinctive form in the *Crataegus intricata* complex in September 1908 from North Carolina, in which the short-shoot leaves are broadly ovate to deltate with shallow obtuse lobes. This has a printed label naming it *C. virgata* Ashe. Though type material of *C. virgata* could not be located, the protologue by Ashe could fit. It is possible that all the southeastern forms from the *intricata* complex with shallow and obtuse lobes, proportionately wide leaves, and cream anthers represent *C. virgata*.

91. Crataegus neobushii Sargent, Trees & Shrubs 2: 9, plate 105. 1907 (as neo-bushii) • Bush's hawthorn [E]

Crataegus intricata Lange var. *neobushii* (Sargent) Kruschke; *C. leioclada* Sargent

Shrubs, 20–40 dm. **Stems:** twigs: new growth reddish young, glabrous, 1-year old brownish purple, older dark gray; thorns on twigs straight to reflexed, 2-years old dark, sometimes blackish, ± slender, 3–5 cm. **Leaves:** petiole <1 mm wide>, length 33–50% blade, glabrous, sometimes stipitate-glandular; blade trullate-ovate, 4–6 cm, <length/width = 1.2–1.6, thin>, base ± cuneate, lobes 4 per side, sinuses shallow, lobe apex acute, margins serrate except proximally, <teeth 1 mm, gland-tipped>, veins 4–6 per side, apex acute, abaxial surface glabrous, adaxial sparsely pubescent or glabrous young, glabrescent, veins persistently pilose. **Inflorescences** 3–7-flowered; branches glabrous; bracteoles <initially abundant, often reddish>, margins short-stipitate-glandular. **Flowers** 18–22 mm diam.; hypanthium glabrous; sepal length not recorded, margins glandular-pectinate, at least distally; stamens 10, anthers rose; styles 3–5. **Pomes** yellowish to orangish, suborbicular, 10 mm diam., glabrous, <sometimes gland-dotted>; sepals on collar, patent; pyrenes 3–5.

Flowering Apr; fruiting Sep–Oct. Open woodland understories; 20–300 m; Ala., Ark., Ga., Mo., Tenn., Va.

Crataegus neobushii is sparsely distributed from Missouri and Arkansas to Virginia and Georgia.

In *Crataegus neobushii*, the length-width ratio of the leaves is variable; the more or less straight edge from near the widest part of the leaf along the lobe apices to the tip of the leaf, forming a trullate to deltate overall shape, is distinctive. The difference between *C. neobushii* and *C. buckleyi* may seem slight but the latter has a different leaf shape as indicated in the key. Specimens of *C. neobushii* with leaves shallowly incised may be hard to distinguish from *C. rubella*. *Crataegus neobushii* is similar to the inadequately studied *C. cuprea* Sargent (a prior name), and this requires resolution.

92. Crataegus buckleyi Beadle, Biltmore Bot. Stud. 1: 25. 1901 [E]

Shrubs or trees, 10–80 dm. **Stems:** <compound thorns on larger trunks present>; twigs: new growth reddish, glabrous, 1-year old dark reddish brown, 2-years old dark ashy gray; thorns on twigs straight to slightly recurved, 2-years old dark chestnut brown, slender, 2–4 cm. **Leaves:** petiole <1 mm wide>, length 33–60% blade, glabrous, glandular; blade broadly elliptic to ovate or broadly ovate, 3–5 cm, <subcoriaceous>, base broadly cuneate, lobes 2 or 3 per side, sinuses moderately deep, lobe apex subacute, margins irregularly serrate, veins 3 or 4 per side, apex acute, surfaces glabrous. **Inflorescences** 3–7-flowered; branches glabrous; bracteole margins short-stipitate-glandular. **Flowers** 18–20 mm diam.; hypanthium glabrous; sepals 5 mm, margins glandular-pectinate proximally, glandular-serrate distally, <abaxially glabrous>; stamens 10, anthers purplish; styles 3–5. **Pomes** russet to red, suborbicular, <angled>, 8–12 mm diam., glabrous; sepals on collar, patent-reflexed; pyrenes 3–5.

Flowering Apr; fruiting Sep–Oct. Pine, oak, hickory woodlands; 50–300 m; Ala., Ga., N.C.

Crataegus buckleyi is pink-anthered like *C. neobushii*, but its leaf form is more like that of the common forms of sympatric *C. intricata* (cream anthers). Beadle suggested on some herbarium labels that *C. buckleyi* is usually a larger plant than *C. neobushii*, has relatively later anthesis, and stipitate glands along the proximal margins of the calyx. *Crataegus communis* is a related form with a more broadly elliptic leaf and stems only 10–20 dm.

93. Crataegus rubella Beadle, Bot. Gaz. 30: 344. 1900 [E]

Crataegus apposita Sargent; *C. intricata* Lange var. *rubella* (Beadle) Kruschke; *C. intricata* var. *straminea* (Beadle) E. J. Palmer; *C. straminea* Beadle

Shrubs, 10–40 dm. **Stems:** twigs: new growth reddish green, glabrous, 1-year old dull reddish brown, 2-years old dark gray, older ashy gray; thorns on twigs <absent or few to several>, straight, 2-years old nearly black, fine, 3 cm. **Leaves:** petiole <1 mm wide>, length 60% blade, glabrous, glandular; blade ± elliptic or rhombic-elliptic, sometimes ± obtrullate, 2.5–5 cm, <thin>, base cuneate, lobes 0, or 1 or 2 per side <in distal ½>, sinuses very shallow, lobe apex acute, margins serrate, <teeth gland-tipped, especially near base>, veins 4 per side, apex acuminate, surfaces glabrous, abaxial veins sparsely hairy young. **Inflorescences** 3–8-flowered; branches glabrous; bracteole margins short-stipitate-glandular. **Flowers** 15–18 mm diam.; hypanthium glabrous; sepals 5 mm, margins glandular-serrate; stamens 10, anthers pale purple; styles 2–4. **Pomes** red to orange-red, obovoid or oblong, 10–12 mm diam., glabrous; sepals on collar, ± reflexed; pyrenes 2–4.

Flowering late Apr–early May; fruiting Sep–Oct. Woods and thickets; 0–300 m; Ala., Del., Ga., Ky., La., Md., Mo., N.C., S.C., Tenn., Va.

Crataegus rubella is wide-ranging and common in the southeastern United States. The species is one of the more distinct taxa of ser. *Intricatae*, with its somewhat narrow leaves with distinctively shaped lobes. *Crataegus straminea*, somewhat arbitrarily synonymized here, represents uncommon intermediates with *C. communis* and shares the broader leaf shape with the latter but has two to three lobes per side of the form of *C. rubella*.

94. Crataegus communis Beadle, Biltmore Bot. Stud. 1: 58. 1902 [C] [E]

Shrubs, 15–20 dm. **Stems:** twigs: new growth reddish, glabrous, 1-year old dull reddish brown, 2-years old dark gray, older paler gray; thorns on twigs: course not recorded, 2-years old dark, slender, short. **Leaves:** petiole <1 mm wide>, length 30–50% blade, pubescence not recorded, glandular; blade broadly elliptic, 3–8 cm, <thin to subcoriaceous>, base cuneate, lobes 0, or 2 or 3 per side, <obscure or reduced to apiculi>, lobe apex subobtuse, margins strongly serrate, <teeth gland-tipped, especially toward base>, veins 4 or 5 per side, apex acute, surfaces hairy young, glabrescent. **Inflorescences** 3–8-flowered; branches glabrous; bracteole margins short-stipitate-glandular. **Flowers** 13–18 mm diam.; hypanthium glabrous; sepals 5–7 mm, margins glandular-serrate, <abaxially glabrous>; stamens 10, anthers rose; styles 2–4(or 5). **Pomes** yellow-green to dull orange, orange-red, or red, suborbicular to short-oblong or obovoid, 9–12 mm diam., glabrous; sepals on collar, patent-reflexed; pyrenes 2–4(or 5).

Flowering Apr; fruiting Sep–Oct. Brush, successional fields, open woodlands; of conservation concern; 100–400 m; Ala., N.C., Tenn.

Crataegus communis is uncommon and often wrongly annotated as *C. straminea*. The species has broadly elliptic leaves with small, very obscure or lacking, subobtuse lobes.

95. Crataegus nananixonii J. B. Phipps & O'Kennon, Sida 17: 569, fig. 1. 1997 • Nixon's dwarf hawthorn [E]

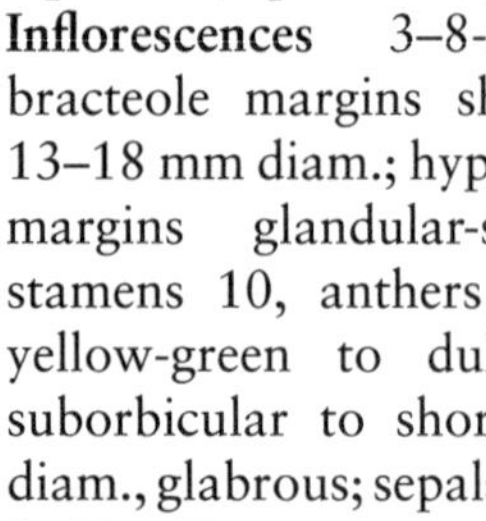

Shrubs, 10–15 dm. **Stems** <intricately branched>; twigs: new growth reddish green, pubescent, 1-year old gray to gray-brown, older gray; thorns on twigs straight to slightly recurved, 2-years old dark dull gray, fine, 1–3 cm. **Leaves:** petiole length 20–40% blade, pubescence not recorded, gland-dotted; blade rhombic-ovate, 1.5–3.5 cm, base broadly cuneate, lobes 0 or with sinuses very shallow, margins ± irregularly serrate, <some teeth gland-tipped, especially proximally>, veins 3 per side, apex acute, abaxial surface glabrous, veins sparsely pubescent, adaxial sparsely scabrous young, glabrescent. **Inflorescences** 3–5-flowered; branches pubescent; bracteole margins short-stipitate-glandular. **Flowers** 12–15 mm diam.; hypanthium glabrous; sepals <lanceolate>, 4–5 mm, margins glandular, <abaxially glabrous>; <petals slightly clawed>; stamens 10, anthers rose-purple; styles 3–5. **Pomes** coppery red, suborbicular, 10 mm diam., glabrous; sepal <remnants present or not>, on collar, patent-reflexed; pyrenes 3–5.

Flowering Apr; fruiting Sep–Oct. Open sandy sites; 200 m; Tex.

Crataegus nananixonii is restricted to open sandy sites in Nacogdoches and nearby counties of eastern Texas. The species is a very dwarf and distinctive member of ser. *Intricatae*, similar to *C. biltmoreana* but with much smaller parts.

96. Crataegus biltmoreana Beadle, Bot. Gaz. 28: 406. 1899 • Biltmore hawthorn [E] [F]

Crataegus confusa Sargent; *C. intricata* Lange var. *biltmoreana* (Beadle) Lance; *C. villicarpa* Sargent

Shrubs, 20–30 dm. **Stems:** twigs: new growth reddish, tomentose, 1-year old dark gray, 2-years old gray; thorns on twigs straight to slightly recurved, 2-years old shiny black, slender, 3–5 cm. **Leaves:** petiole length 40–45% blade, <sometimes winged>, pubescent, strongly glandular; blade narrowly to broadly ovate, 4–7 cm, base cuneate, lobes 3–5 per side, distinct, lobe apex acute, margins strongly serrate, <teeth gland-tipped>, veins 4 or 5 per side, apex acute, surfaces pilose (abaxial more sparsely) young, veins pilose. **Inflorescences** 2–7-flowered; branches tomentose; bracteoles <oblong-linear>, margins short-stipitate-glandular. **Flowers** 15–20 mm diam.; hypanthium usually densely tomentose; sepals <lanceolate>, 7 mm, margins deeply glandular-serrate, <abaxially pilose>; stamens 5–10, anthers cream; styles 3–5. **Pomes** yellow to orange, often with red cheek, suborbicular, 9–14 mm diam., hairy; sepals on collar, reflexed-patent, <8–9 mm>; pyrenes 3–5. **2*n*** = 51.

Flowering Apr–May; fruiting Sep–Oct. Understory of open woodlands, montane coves, ridge tops; 0–300 m; Ala., Ark., Conn., Ga., Ind., Md., Mass., Mo., N.Y., N.C., Ohio, Pa., Tenn., Vt., Va., W.Va.

Crataegus biltmoreana is wide-ranging but somewhat scarce; it is known from Vermont to Georgia and west to Missouri and Arkansas, but is most abundant in Appalachia.

Crataegus biltmoreana is distinctive with its very hairy young twigs, leaves, and inflorescence branches, low stature, long thorns, large flowers, and yellow to orange pomes. Indumentum is the only consistent difference from *C. intricata*; specific rank is maintained because wide-ranging glabrous/hairy variants of one species are so unusual in North American *Crataegus*. Eventually, it may prove better to accept *C. biltmoreana* as a variety of that species. Rare forms with almost unlobed leaves, the sinuses shallow and lobes obtuse when present, may be the same.

97. Crataegus stonei Sargent, Rhodora 5: 62. 1903 • Stone's hawthorn C E

Crataegus biltmoreana Beadle var. *stonei* (Sargent) Kruschke

Shrubs, 10–20 dm. **Stems:** twigs: new growth reddish green, glabrous or villous, 1-year old dull reddish brown, eventually gray; thorns on twigs ± straight, 2-years old blackish, ultimately gray, ± slender, 4–6 cm. **Leaves:** petiole length 30–40% blade, <winged distally>, villous, glandular; blade narrowly ovate, 7–8 cm, base cuneate, lobes 0 or 1–3 per side, sinuses shallow, lobe apex acute, margins serrate, <teeth gland-tipped>, veins 4 or 5 per side, apex acuminate, abaxial surface sparsely villous along veins, adaxial sparsely villous. **Inflorescences** 4–6 -flowered; branches villous; bracteoles not seen. **Flowers** 18–20 mm diam.; hypanthium densely pubescent; sepal length not recorded, margins coarsely glandular-serrate, <acuminate, abaxially pubescent>; stamens 5–10, anthers pink; styles 3 or 4. **Pomes** dull yellow to russet, broadly ellipsoid, 12–14 mm diam., hairy; sepals on collar, ± reflexed; pyrenes 3 or 4.

Flowering late May–early Jun; fruiting Sep–Oct. Woods, openings; of conservation concern; 20–300 m; Mass., N.Y., Pa.

Crataegus stonei is another species related to *C. biltmoreana*, from which it differs mainly in its pink anthers, proportionately narrower and more elongate leaves, and lower stature.

98. Crataegus ouachitensis E. J. Palmer, J. Arnold Arbor. 7: 124. 1926 • Ouachita hawthorn C E

Shrubs or trees, 40–60 dm. **Stems:** twigs: new growth not recorded, 1-year old dark purplish, older gray; thorns on twigs ± straight, 2-years old dark purple, fine, 2–4 cm. **Leaves:** petiole <very slender, 0.5 mm wide>, length ca. equal to or 50–66% blade, sparsely hairy or glabrous, slightly glandular or eglandular; blade ovate to deltate-ovate, (2–)3–4 cm, <thin>, base broadly rounded to nearly truncate or subcordate, lobes 1–3 per side <obscure>, sinuses shallow, lobe apex subacute, margins sharply serrate nearly to base, veins 3 per side, apex acuminate, abaxial surface sparsely to ± densely hairy on veins, adaxial variable scabrous, glabrescent. **Inflorescences** 3–8-flowered; branches glabrous or sparsely villous; bracteoles not recorded. **<Pedicels** glabrous or ± tomentose young>. **Flowers** 15–18 mm diam.; hypanthium not seen; sepal length not recorded, margins glandular-serrate or subentire; stamens 20, anthers pink or rose to rose-purple; styles 3 or 4. **Pomes** green until late, becoming bright red, obovoid to pyriform, 8–15 mm diam., glabrous; <flesh hard>; sepals on collar, strongly reflexed; pyrenes 3 or 4.

Varieties 2: Arkansas.

Crataegus ouachitensis is endemic to the Ouachita Mountains. Among ser. *Intricatae*, the species is distinct because of its small leaves, only slightly glandular nature, 20 stamens, and bright red pomes. The delicate leaves and pome color recall *C. wootoniana* (ser. *Tenuifoliae*). *Crataegus ouachitensis* is also similar to Louisianan *C. iracunda*, differing primarily in having 20 stamens and conspicuously glandular-serrate sepal margins in the type form. *Crataegus ouachitensis* is very rare and likely of interserial hybrid origin, though it could also plausibly be referred to ser. *Pulcherrimae*. *Crataegus thermopegea* E. J. Palmer, from the same area, is perhaps related; it has laterally pitted pyrenes and probably ten stamens, according to Palmer.

1. Petioles 50–66% length of blade; inflorescence branches glabrous; pedicels glabrous; sepal margins glandular-serrate ....98a. *Crataegus ouachitensis* var. *ouachitensis*
1. Petioles ca. equal to length of blade; inflorescence branches sparsely villous; pedicels ± tomentose young; sepal margins subentire ....98b. *Crataegus ouachitensis* var. *minor*

98a. Crataegus ouachitensis E. J. Palmer var. **ouachitensis** C E

Leaves: petioles 50–66% length of blade; blade 3–4 cm, longer than wide to as long as wide, lobes 1–3 per side, obscure or shallow, margins sharply serrate, abaxial surface sparsely hairy on veins, adaxial scabrous young. **Inflorescences:** branches glabrous. **Pedicels** glabrous. **Flowers:** sepal margins glandular-serrate.

Flowering Apr; fruiting Sep–Oct. Thickets along streams; of conservation concern; 100–300 m; Ark.

98b. Crataegus ouachitensis E. J. Palmer var. **minor** E. J. Palmer, J. Arnold Arbor. 7: 124. 1926 [C] [E]

Leaves: petioles ca. equal to length of blade; blade 2–3.5 cm, longer than wide to as long as wide, lobes somewhat more deeply incised, margins sharply serrate (more so than var. *ouachitensis*), abaxial surface more densely hairy on veins, adaxial more densely scabrous young. **Inflorescences:** branches sparsely villous. **Pedicels** ± tomentose young. **Flowers:** sepal margins subentire.

Flowering Apr; fruiting Sep–Oct. Thickets along streams; of conservation concern; 100–300 m; Ark.

Variety *minor* is very different from the typical variety. The entire plant is more or less glabrescent by maturity. It is quite possible that var. *minor* is a hybrid of *Crataegus marshallii*, as originally suggested on the type material, in which case it is not a form of *C. ouachitensis*.

99. Crataegus craytonii Beadle, Biltmore Bot. Stud. 1: 121. 1902 (as craytoni) • Crayton hawthorn [C] [E]

Shrubs, 20–30 dm. **Stems:** twigs: new growth reddish green, densely hairy, 1–2-years old dark gray-brown, older gray; thorns on twigs ± straight, 2-years old blackish, very slender, 2–3 cm. **Leaves:** petiole <1–2 mm wide>, length 25–40% blade, <winged distally>, pubescent, glandular; blade ovate to broadly elliptic, 4–8 cm, base cuneate to truncate (suddenly narrowed to petiole), lobes 3 or 4 per side, <often obscure>, sinuses sometimes quite deep, lobe apex acute, margins strongly serrate, <teeth gland-tipped>, veins 5 per side, apex acute, surfaces rough-hairy. **Inflorescences** 3–7-flowered; branches pubescent; bracteole margins stipitate-glandular. <**Pedicels** pubescent>. **Flowers** 20 mm diam.; hypanthium densely pubescent; sepals 7 mm, margins glandular-pectinate to glandular-serrate, <abaxially pubescent>; stamens 20, anthers pale yellow or cream; styles 3–5. **Pomes** red to ruddy, broadly ellipsoid, 7–10 mm diam., pubescent; sepals on collar, ± reflexed; pyrenes 3–5.

Flowering Apr; fruiting Sep–Oct. Wooded areas; of conservation concern; 100–300 m; Ala., Ga., N.C., Va.

Crataegus craytonii is a rare shrub from a few scattered localities. The species resembles *C. biltmoreana* in its hairiness but has usually larger, differently shaped leaves, 20 stamens, and red to ruddy pomes.

100. Crataegus padifolia Sargent, Trees & Shrubs 2: 75, plate 135. 1908 • Bird-cherry hawthorn [C] [E]

Crataegus intricata Lange var. *padifolia* (Sargent) Lance

Shrubs or trees, 40–60 dm. **Stems:** twigs: new growth reddish green, glabrous, 1-year old deep red-brown, 2-years old dark gray-black, older grayish; thorns on twigs <sometimes absent>, straight or slightly recurved, 2-years old blackish, slender, 2–4 cm. **Leaves:** petiole <1 mm wide>, length 33–40% blade, glabrous, sparsely glandular; blade <deep glossy green>, ± elliptic (var. *padifolia*) or ovate (var. *incarnata*), 3–5 cm, base cuneate or broadly cuneate to rounded or subtruncate, lobes 0, or 3 or 4 per side (enlarged toothlike apiculi in var. *padifolia* or larger in var. *incarnata*), sinuses shallow, lobe apex very short and acute to ± obtuse, margins serrate at least young <teeth gland-tipped>, veins 4–7 per side, apex subacute to obtuse, surfaces glabrous, adaxial larger veins sparsely short-hairy. **Inflorescences** 3–6-flowered; branches glabrous; bracteole margins stipitate-glandular. **Flowers** 15–18 mm diam.; hypanthium glabrous; sepals 4 mm, margins finely glandular-serrate distally, <abaxially glabrous>; stamens 10, anthers cream, flushed pink; styles 2 or 3. **Pomes** dull orange-brown to ± shiny, red-orange to red, ± oblong, 12–14 mm diam., glabrous; sepals sessile, patent-reflexed; pyrenes 2 or 3. $2n = 51$.

Varieties 2 (2 in the flora): c United States.

Crataegus padifolia is rare, confirmed only for the Ozark region of Arkansas, southern Missouri, and eastern Oklahoma. Occasional specimens from southern Appalachia may prove to belong here.

Crataegus padifolia is one of the less typical taxa usually placed in ser. *Intricatae* as it lacks the characteristic elevated fruiting calyx. The two varieties, at least in their more extreme forms, could pass for different species, but too little material is available on which to justify a change of status. *Crataegus fortunata* Sargent from Pennsylvania has yellow fruit but is otherwise similar and may belong here.

1. Leaf blades: bases cuneate, length/width = 1.4–2, lobes 0 or usually at most short, acute apiculi, veins 5–7 per side . . . . 100a. *Crataegus padifolia* var. *padifolia*
1. Leaf blades: bases broadly cuneate to rounded or subtruncate, length/width = 1–1.4, lobes 3 or 4 per side, lobe apices ± obtuse, sinuses shallow, veins 4–6 per side . . . . 100b. *Crataegus padifolia* var. *incarnata*

100a. Crataegus padifolia Sargent var. **padifolia** C E

Leaf blades: base cuneate, length/width = 1.4–2, lobes 0 or usually at most short, acute apiculi, veins 5–7 per side. **Pomes** dull orange-brown.

Flowering Apr; fruiting Sep–Oct. Open woodlands; of conservation concern; 100–300 m; Ark., Mo., Okla.

100b. Crataegus padifolia Sargent var. **incarnata** Sargent, J. Arnold Arbor. 6: 4. 1925 C E

Leaf blades: base broadly cuneate to rounded or subtruncate, length/width = 1–1.4, lobes 3 or 4 per side, sinuses shallow, lobe apex ± obtuse, veins 4–6 per side. **Pomes** red.

Flowering Apr; fruiting Sep–Oct. Open woodlands; of conservation concern; 100–300 m; Ark., Mo.

64f.15. Crataegus Linnaeus (sect. Coccineae) ser. Rotundifoliae (Eggleston) Rehder, Man. Cult. Trees ed. 2, 363. 1940 E

Crataegus sect. *Rotundifoliae* Eggleston in B. L. Robinson and M. L. Fernald, Manual ed. 7, 467. 1908 (as Rotundifolia)

Shrubs, sometimes trees, (15–)20–60(–70) dm, often multi-stemmed. **Stems:** trunk bark buff to darker, fibrous, checked into longitudinal plates, freshly exposed bark orange-brown; <branches spreading; twigs ± straight>, new growth <green to reddish>, pubescent or glabrous, 1-year old usually golden green to tan, sometimes deep reddish brown or gray-brown, older graying; thorns on twigs usually numerous, usually ± recurved, sometimes straight, 1-year old usually shiny, dark brown or blackish, older graying, ± stout to ± slender, 2.5–5(–7) cm. **Leaves:** petiole length 28–60% blade, <winged>, glabrous or pubescent, distally variably sessile-glandular; blade <mid to dark green or blue-green>, matte, usually ± rhombic, sometimes ovate, elliptic-oblong, or suborbiculate, 2–6(–9.5) cm, thin to chartaceous, sometimes subcoriaceous, base cuneate to rounded, truncate, or broadly ovate, lobes 0 or 1–5(or 6) per side, sinuses shallow to deep, lobe apex subacute to obtuse, sometimes acuminate, margins serrate, sometimes crenate, <teeth gland-tipped>, venation craspedodromous, veins (3–)5 or 6(–8) per side, apex acuminate to obtuse, abaxial surface usually glabrous young, sometimes persistently hairy, adaxial usually scabrous, glabrescent, sometimes persistently scabrous or glabrate. **Inflorescences** (1–)4–15 (–20)-flowered, convex panicles; branches densely pubescent, sometimes glabrous; bracteoles caducous, <usually numerous>, linear or narrowly elliptic, usually membranous, sometimes subherbaceous, margins sessile-glandular, sometimes stipitate-glandular <larger bracteoles>. **Flowers** (8–)12–20 mm diam.; hypanthium <obconic>, densely pubescent, sometimes glabrous; sepals narrowly triangular, $\frac{1}{2}$–$\frac{2}{3}$ petal length, margins entire or slightly glandular-serrate, <apex usually obtuse, sometimes acute, abaxially pubescent or glabrous>; stamens (5–)10 or (18–)20, anthers ivory, sometimes pink; styles 2–4(or 5). **Pomes** usually bright to deep red or yellowish, sometimes orangish or burgundy, usually ± suborbicular, sometimes ellipsoid or oblong, (7–) 8–12(–14) mm diam., glabrous or pubescent; flesh mealy; sepals <sessile except in *C. margarettae*>, remnants erect-patent to reflexed, <non-accrescent>; pyrenes 2–4(or 5), <sides plane, sometimes shallowly concave in *C. ursopedensis*, dorsally grooved>.

Species 8 (8 in the flora): North America.

Series *Rotundifoliae* ranges from southern interior British Columbia to Newfoundland, south to Oregon, Colorado, the northern Great Plains, the Midwest, the southern Great Lakes, New England, and mid-latitude Appalachians. East of the Rocky Mountains, ser. *Rotundifoliae* is the most northerly series of *Crataegus*, and only ser. *Macracanthae* has as great an east-west range. The most wide-ranging species are *C. chrysocarpa*, ranging from nearly the Pacific to the Atlantic, *C. margarettae*, midwestern, and *C. dodgei*, common mainly from the Great Lakes to New England.

Plants of ser. *Rotundifoliae* are relatively small shrubs that are almost never treelike; their inflorescences are quite copiously glandular-bracteolate, at least when young. Putative hybrids with series other than ser. *Macracanthae* are rare.

Series *Rotundifoliae* comprises two main groups plus the somewhat atypical *Crataegus ursopedensis*. The first group comprises the *C. chrysocarpa* complex, the second, *C. margarettae* and *C. dodgei*. Members of the *C. chrysocarpa* complex have relatively small, more or less ovate to rhombic-ovate, acute-tipped, sharply lobed leaves with glandular petioles and gland-tipped teeth. They also tend to have bracteoles with sessile glands and stouter thorns than members of the second group. The group occurs mostly in northern intermontane, Great Plains, and deciduous forest-conifer transition zones. It is also more xeromorphic, the plants typically being shrubs of dryish areas in the west and often on thin or rocky soils in the east. They usually have pubescent inflorescences and red mature pomes. Members of the second group, around *C. margarettae* and *C. dodgei*, are less glandular and often have more or less obtuse-tipped leaves, finer and often fewer thorns, larger, glabrous inflorescences, basal bracteoles that are often prominently stipitate-glandular, and a strong tendency to bear yellowish pomes. This group is more southerly.

Members of the chrysocarpoid group of ser. *Rotundifoliae* can be difficult to distinguish from members of ser. *Intricatae*. Nearly all members of ser. *Intricatae* have a somewhat elevated calyx in fruit; their distribution is almost entirely south of *Crataegus chrysocarpa* in eastern North America, and they often have stipitate-glandular bracteoles. *Crataegus submollis* (ser. *Molles*), apart from being larger in all its parts and having more persistent bracteoles, is also sometimes difficult to differentiate from *C. chrysocarpa* and might key out under ser. *Rotundifoliae*, as might also *C. flabellata* (ser. *Tenuifoliae*). The inter-relationships of the taxa mentioned require deeper investigation; consequently, the key to species includes *C. padifolia* (ser. *Intricatae*) and *C. submollis* (ser. *Molles*) in the fourth and thirteenth couplets, respectively. The interserial hybrid *C. ×sicca* keys out in part in the fifth and fifteenth couplets, and *C. laurentiana* keys out in the eighth couplet.

1. Inflorescence branches glabrous or sparsely pilose young.
    2. Stamens (18–)20.
        3. Petioles sparsely to densely persistently glandular; leaf blades ± rhombic, lobes 2–4 per side, max LII 20%, lobe apices acute, apices acuminate, veins 4 or 5 per side, green mature; sw Canada . . . . 102. *Crataegus sheila-phippsiae*
        3. Petioles usually with small glands at first, becoming eglandular; leaf blades suborbiculate, ovate, or obovate, rarely elliptic-rhombic to elliptic or lance-elliptic, lobes 1–3(or 4) per side, sinuses usually shallow, sometimes deep, lobe apices obtuse to angled, apices subacute to obtuse, veins 3 or 4 per side, green to bluish green; c, e North America . . . . 106. *Crataegus margarettae* (in part)
    2. Stamens (5–)10(or 20).
        4. Petioles sparsely glandular; bracteole margins stipitate-glandular; pomes ± oblong, 12–14 mm diam.; (Ozarks). 100. *Crataegus padifolia* (ser. *Intricatae*), p. 592
        4. Petioles eglandular or with few, very tiny glands; at most, larger bracteoles only sometimes with margins stipitate-glandular; pomes usually suborbicular, 7–12 mm diam.
            5. Flowers 10–14 mm diam.; styles and pyrenes 3–5 . . . . . . .167. *Crataegus ×sicca* (in part), p. 641
            5. Flowers 13–20 mm diam.; styles and pyrenes 2–4(or 5).

6. Leaf blades broadly rhombic-ovate, 4–5 cm; bracteoles: larger with margins sessile-glandular; sepal margins glandular-serrate; pomes red . . . . 103. *Crataegus chrysocarpa* (in part)
6. Leaf blades suborbiculate, ± ovate, broadly rhombic to rhombic-ovate, sometimes broadly elliptic, 2–5 cm; bracteoles: some with margins stipitate-glandular; sepal margins entire or ± glandular-denticulate; pomes yellow to yellow-orange, dull orange, or ruddy to red. . . . . 107. *Crataegus dodgei*

1. Inflorescence branches pubescent, sparsely to densely villous, or pilose.
 7. Anthers pink or rose.
  8. Leaf surfaces: adaxial appressed-scabrous, glabrescent, or short-appressed-pubescent; petiolar glands persistent; stamens 10 or 20; styles 3 or 4(or 5); pomes suborbicular; pyrene sides plane . . . . 103. *Crataegus chrysocarpa* (in part)
  8. Leaf surfaces: adaxial short-appressed pubescent young; petiolar glands caducous, (dark); stamens 10; styles 3–5; pomes oblong to pyriform; pyrene sides irregularly eroded or pitted . . . . 164. *Crataegus laurentiana*, p. 640
 7. Anthers white to cream, sometimes pale yellow.
  9. Stamens 20.
   10. Petioles usually with small glands at first, becoming eglandular; leaf blades suborbiculate, ovate, or obovate, rarely elliptic-rhombic to elliptic or lance-elliptic, apices subacute to obtuse, lobe apices obtuse to angled, veins 3 or 4 per side, green to bluish green; sepal margins entire or subentire; pomes yellow, red, or dull burgundy; c, e North America . . . . 106. *Crataegus margarettae* (in part)
   10. Petioles persistently glandular; leaf blades broadly rhombic-ovate to broadly ovate to oblong-ovate or broadly elliptic, apices acute, lobe apices acute or triangular, veins 3–8 per side, green; sepal margins glandular-serrate; pomes red; n United States, s Canada.
    11. Leaf blades ovate-oblong to oblong, (4–)5–7(–8) cm, (on extension shoots similar, larger), lobes 5 or 6 per side, veins 5–7 per side. . . . . 105. *Crataegus irrasa*
    11. Leaf blades rhombic-ovate to ovate or broadly ovate, oblong-ovate, or broadly elliptic, 2–5(–7.5) cm, (on extension shoots similar or suborbiculate, larger), lobes 2–5(or 6) per side, veins 3–5(–8) per side.
     12. Leaf blades rhombic-ovate to ovate, (on extension shoots similar shape, larger), lobes 2–4(–6) per side, sinuses moderately deep, veins 3–5(–7) per side. . . . . 103. *Crataegus chrysocarpa* (in part)
     12. Leaf blades broadly ovate to oblong-ovate or broadly elliptic, (on extension shoots usually suborbiculate, larger), lobes 4 or 5 per side, sinuses shallow, veins 6–8 per side. . . . . .104. *Crataegus oakesiana*
  9. Stamens 10.
   13. Plants 60–70 dm; petioles strongly glandular; larger bracteoles subherbaceous; flowers 20–22 mm diam.; styles and pyrenes 3–5; pomes 10–15 mm diam.; fruiting sepals stiff, erect-patent; Great Lakes to New England . . . . 52. *Crataegus submollis* (ser. *Molles*), p. 556
   13. Plants 20–50 dm; petioles ± glandular or eglandular; larger bracteoles membranous; flowers 13–20 mm diam.; styles and pyrenes 2–4(or 5); pomes 7–12 mm diam.; fruiting sepals thin, spreading to reflexed or patent to ± appressed.
    14. Leaves at early anthesis strikingly flabellate, only 1–2 cm, 5.5–9.5 cm, glossy and subcoriaceous mature; thorns (3–)4–7 cm; bracteoles cream to pale green, often suffused light red especially distally. . . . . 101. *Crataegus ursopedensis*
    14. Leaves at early anthesis slightly or not flabellate, much larger than an open flower, 2–5(–7.5) cm, matte and thin to chartaceous mature; thorns 2.5–6(–7) cm; bracteoles usually hyaline.
     15. Leaves 2–3 cm; bracteoles reddish; flowers 18–20 mm diam.; pomes pale orange, strongly pruinose. . . . . .167. *Crataegus* ×*sicca* (in part), p. 641
     15. Leaves 2–5(–7.5) cm; bracteoles usually hyaline; flowers 13–20 mm diam.; pomes yellowish or orangish to ruddy, scarlet, or red, not pruinose.

[16. Shifted to left margin.—Ed.]

16. Bracteole margins sessile-glandular; sepals 4 mm, triangular, margins glandular-serrate; pomes scarlet to deep red . . . . 103. *Crataegus chrysocarpa* (in part)
16. Bracteole margins of larger ones ± stipitate-glandular; sepals 5 mm, narrowly triangular, margins entire, glandular; pomes yellowish, dull orange, or reddish . . . . 108. *Crataegus lumaria*

101. Crataegus ursopedensis J. B. Phipps & O'Kennon, J. Bot. Res. Inst. Texas 1: 1081, plates 6.1d, 7.4d, fig. 27. 2007 • Bear's Paw hawthorn [E]

Shrubs, 20–30(–40) dm. **Stems:** <expanding bud-scales orange-tan or paler>; twigs: new growth pubescent, 1-year old shiny red-brown, 2-years old red-brown with white flakes, older gray; thorns on twigs <usually numerous>, ± straight to recurved, 2-years old shiny, very dark red-brown, ± stout, (3–)4–7 cm. **Leaves:** petiole length 30–45% blade, glabrous (adaxial sulcus hairy mature), densely sessile-glandular; blade ovate to ovate-trullate, 5.5–9.5 cm (flabellate, 2–3 cm at early anthesis), subcoriaceous, base cuneate to narrowly cuneate, lobes 4 or 5 per side, sinuses deep (max LII 25–35%), lobe apex acute to acuminate, margins ± serrate, <teeth numerous, small>, veins 5 per side, apex acuminate, <glossy>, abaxial surface glabrous, main veins sometimes sparsely to moderately pubescent young, adaxial densely scabrous young, glabrate mature. **Inflorescences** (1–)5–12-flowered, <widely spread at full anthesis>; branches moderately pubescent; bracteoles <cream to pale green, often suffused light red especially distally>, narrowly elliptic, membranous, margins densely sessile-glandular, <appearing to grade into bud scales>. **Flowers** 15–18 mm diam.; hypanthium exterior pubescent, interior glabrous; sepals <pale green to scarious>, triangular, 4–5 mm, margins glandular or glandular-denticulate, <abaxially glabrous, adaxially sparsely hairy>; stamens 10, anthers ivory; styles 3 or 4. **Pomes** <1–8 mm; branches pubescent, sometimes sparsely>; orange-red young, becoming orange-red to bright red, ellipsoid or oblong, 8–11 mm tall, pubescent; sepals spreading to ± appressed, <3–4 mm>; pyrenes 3 or 4, <sides plane to shallowly concave>.

Flowering May–Jun; fruiting not recorded. Thickets, shade of aspen or poplar; 800–1300 m; Alta., Sask.; Mont.

Crataegus ursopedensis is restricted to and locally common in the Bear's Paw Mountains (Montana) and the Cypress Hills (Alberta and Saskatchewan).

Crataegus ursopedensis can have strikingly large, deeply incised and sharply lobed, short-shoot leaves; they are often smaller and may then be difficult to differentiate from those of *C. rubribracteolata*. Compared to the latter, the thorns are often longer, the inflorescences more open, the petioles proportionately longer, the bracteoles paler and, significantly, the pyrene sides more nearly plane. *Crataegus ursopedensis* is strikingly distinct within ser. *Rotundifoliae* and perhaps does not belong there.

102. Crataegus sheila-phippsiae J. B. Phipps & O'Kennon, Sida 21: 67, fig. 2. 2004 [E]

Shrubs, arborescent, 30–50 dm, or thicket-forming, 15–30 dm. **Stems:** twigs: new growth <green>, glabrous, 1-year old shiny, brown, older gray; thorns on twigs straight or slightly recurved, 1-year old shiny, dark blackish brown, older deep gray, moderately slender to stout, 2–6 cm. **Leaves:** petiole length 28–40% blade, pubescent, sparsely to densely persistently glandular; blade <green mature>, ± rhombic, (2–)3.5–5(–6) cm (2.5–5 cm at flowering), thin-chartaceous, base cuneate, lobes 2–4 per side, sinuses: max LII 20%, lobe apex acute, margins serrate, <teeth small>, veins 4 or 5 per side, apex acuminate, abaxial surface glabrous, hairy in axils and parts of main veins, adaxial scabrous. **Inflorescences** 3–18-flowered; branches glabrous or sparsely pilose young; bracteoles linear, membranous, margins sessile-glandular. **Flowers** 14–20 mm diam.; hypanthium glabrous; sepals triangular, length not recorded, margins barely glandular-serrate; stamens 18–20, anthers white to pale pink, sometimes medium pink or puce; styles 4 or 5 (arising from small tuft of hairs). **Pomes** brilliant orange-red or crimson scarlet early, becoming bright red or dark burgundy, suborbicular, 10–12 mm diam., glabrous; sepals spreading; pyrenes 4 or 5.

Varieties 2 (2 in the flora): sw Canada.

1. Plants arborescent, 30–50 dm; anthers white to pale pink; thorns on twigs 2–4 cm; leaf blades 4–6 cm, at early anthesis larger than flowers; pomes burgundy, sometimes darker; British Columbia . . . . 102a. *Crataegus sheila-phippsiae* var. *sheila-phippsiae*
1. Plants shrubby, thicket-forming, 15–30 dm; anthers white to cream; thorns on twigs 3–6 cm; leaf blades 2–4(–5) cm, at early anthesis little larger than flowers; pomes orange-red to bright red; s Saskatchewan . . . . 102b. *Crataegus sheila-phippsiae* var. *saskatchewanensis*

102a. Crataegus sheila-phippsiae J. B. Phipps & O'Kennon var. **sheila-phippsiae** [C] [E]

Plants arborescent, 30–50 dm. **Stems:** thorns on twigs 2–4 cm. **Leaf blades** 4–6 cm, at early anthesis larger than flowers. **Flowers:** anthers white to pale pink. **Pomes** burgundy, sometimes darker.

Flowering May–Jun; fruiting Sep–Oct. Open, dry brush; of conservation concern; 300–400 m; B.C.

Variety *sheila-phippsiae* is locally frequent in the northern Okanagan of British Columbia.

102b. Crataegus sheila-phippsiae J. B. Phipps & O'Kennon var. **saskatchewanensis** J. B. Phipps & O'Kennon, J. Bot. Res. Inst. Texas 1: 1079, plate 7.4c, fig. 25. 2007 • Saskatchewan hawthorn [E]

Plants shrubby, thicket-forming, 15–30 dm. **Stems:** thorns on twigs 3–6 cm. **Leaf blades** 2–4(–5) cm, at early anthesis little larger than flowers. **Flowers:** anthers white to cream. **Pomes** orange-red to bright red.

Flowering May–Jun; fruiting Aug–Sep. Thickets; 700–1000 m; Sask.

The main range of var. *saskatchewanensis* is the Cypress Hills; records are known also from the prairie regions of southern Saskatchewan. The differences from the type variety are considerable and constant, and var. *saskatchewanensis* is perhaps more similar to *Crataegus chrysocarpa* var. *chrysocarpa*, although where sympatric its fruit is larger.

103. Crataegus chrysocarpa Ashe, Bull. North Carolina Agric. Exp. Sta. 175: 110. 1900 • Fireberry hawthorn, aubépine dorée [E]

Crataegus columbiana Howell var. *chrysocarpa* (Ashe) Dorn; *C. rotundifolia* Moench var. *chrysocarpa* (Ashe) Eggleston

Shrubs, 20–35 dm, <sometimes clonal>. **Stems:** twigs: new growth usually appressed-pubescent, 1-year old usually dull yellowish to greenish brown or gray-brown to light or dark tan; thorns on twigs <variable>, straight to slightly recurved, 1-year old shiny, dark mahogany or black, ± slender to ± stout, 3–6 cm. **Leaves:** petiole length 50% blade, <usually narrowly winged distally>, pubescent, glandular young, often persisting; blade <yellow or purplish in very dry spots (fall)>, ± rhombic to rhombic-ovate, rhombic-obovate, or ovate to broadly elliptic, <on extension shoots similar, larger>, 2–5(–7.5) cm, thin to chartaceous, base ± cuneate, sometimes rounded to broadly ovate (at early anthesis not strikingly flabellate), lobes 2–4(–6) per side, sinuses moderately deep, lobe apex acute to subacute, margins serrate, <teeth glands caducous, numerous, small>, veins 3–5(–7) per side, apex acute or subacute, abaxial surface glabrous, sparsely pilose-glabrescent, or villous, veins usually pubescent, adaxial appressed-scabrous, glabrescent or short-appressed-pubescent. **Inflorescences** 5–10-flowered, <compact>; branches sparsely to densely villous, rarely glabrous; bracteoles <usually hyaline>, linear, membranous, margins sessile- or short-stipitate-glandular. **Flowers** 15–20 mm diam.; hypanthium usually villous, sometimes glabrous; sepals triangular, 4 mm, margins glandular-serrate, <apex subacute to obtuse, abaxially pubescent>; stamens (5–)10(or 20), anthers cream or ivory, sometimes pink; styles 3 or 4(or 5). **Pomes** scarlet to deep red, only darkened to burgundy if dried and shriveled (except in var. *vernonensis*), usually suborbicular to broadly ellipsoid or oblong, 8–10(–12) mm diam., pubescent or glabrous; sepal <remnants present>, spreading to reflexed, <not or rarely obscurely elevated>; pyrenes 3 or 4(or 5). $2n = 68$.

Varieties 9 (9 in the flora): North America.

Crataegus chrysocarpa is one of the most wide-ranging North American species of the genus, occurring from inland of the Cascades and Coastal Ranges of Oregon, Washington, and British Columbia to the Atlantic from Newfoundland to New England; it is correspondingly variable. The type by Ashe from the Rocky Mountain foothills in northern Colorado is from the south of the range. East of the Rocky Mountains, *C. chrysocarpa* is the most northerly and cold-adapted species of hawthorn in North America and nearly reaches Hudson Bay in Ontario. Rare northeastern forms are arborescent (to 70 dm).

Variety *chrysocarpa* is by far the most common form through most of the species range; numerous well-marked to only slightly different forms, many accepted here at varietal level, occur in different parts of the range. The main regional variant is var. *piperi*, a well-marked intermontane form.

One- to two-year old twigs of var. *chrysocarpa* are much darker in the east (reddish tan or darker) than in the west (light tan to gray-brown).

1. Stamens 20.
   2. Anthers cream or ivory . . . . 103g. *Crataegus chrysocarpa* var. *vigintistamina*
   2. Anthers pink . . . . 103i. *Crataegus chrysocarpa* var. *blanchardii*
1. Stamens 5–10.

[3. Shifted to left margin.—Ed.]

3. Inflorescence branches lanate; anthers pale pink; fully ripe pomes burgundy . . . . 103d. *Crataegus chrysocarpa* var. *vernonensis*
3. Inflorescence branches glabrous or pubescent, sparsely to densely villous, or thin-pilose; anthers cream to ivory, sometimes pink; fully ripe pomes red.
    4. Abaxial leaf surfaces persistently hairy (rough-pubescent or with scattered long hairs).
        5. Leaf blades ovate to broadly ovate or broadly elliptic, bases rounded to broadly cuneate, lobe apices subacute, max LII 5–10%, veins 6 or 7 per side; Great Lakes region and eastwards . . . . 103b. *Crataegus chrysocarpa* var. *faxonii*
        5. Leaf blades rhombic to narrowly rhombic or rhombic-elliptic, bases ± cuneate, lobe apices acute, max LII 15–20%, veins 5 or 6 per side; Pacific Northwest and adjacent Canada. . . . . 103c. *Crataegus chrysocarpa* var. *piperi*
    4. Abaxial leaf surfaces glabrous, veins sometimes pilose.
        6. Anthers pink. . . . . 103h. *Crataegus chrysocarpa* var. *praecox*
        6. Anthers cream or ivory.
            7. Inflorescence branches glabrous; leaf surfaces: adaxial glabrous young. . . . . .103f. *Crataegus chrysocarpa* var. *phoeniceoides*
            7. Inflorescence branches sparsely to densely villous or thin-pilose; leaf surfaces: adaxial appressed-short-pubescent or appressed-pubescent young, glabrescent or becoming short scabrous.
                8. Hypanthia densely villous . . . . 103a. *Crataegus chrysocarpa* var. *chrysocarpa*
                8. Hypanthia glabrous . . . . 103e. *Crataegus chrysocarpa* var. *subrotundifolia*

103a. Crataegus chrysocarpa Ashe var. **chrysocarpa** E

Crataegus coloradoides Ramaley; *C. doddsii* Ramaley; *C. rotundifolia* Moench var. *pubera* Sargent

Shrubs, 20–30 dm. **Leaves:** blade ± ovate to rhombic-ovate, base ± cuneate, sinuses: max LII 15–20%, lobe apex acute, veins 3–5 per side, abaxial surface glabrous, sometimes hairy on veins or hairy young, glabrescent, adaxial appressed-short-pubescent, glabrescent. **Inflorescences:** branches sparsely to densely villous. **Flowers** 15–20 mm diam.; hypanthium densely villous; stamens 10, anthers cream or ivory. **Pomes** red (darkened to burgundy only if dried and shriveled), suborbicular to broadly ellipsoid, 8–10 mm diam., hairy.

Flowering May–Jun; fruiting Sep–Oct. Well-drained mesic sites with high light intensity, under aspen; 0–2200 m; Alta., B.C., Man., N.B., Nfld. and Labr. (Nfld.), N.S., Ont., P.E.I., Que., Sask.; Colo., Conn., Iowa, Maine, Mass., Mich., Minn., Mont., N.H., N.Y., N.Dak., Oreg., R.I., S.Dak., Utah, Vt., Wash., Wis., Wyo.

Variety *chrysocarpa* occurs through nearly the whole range of the species. Through most of its range the leaves are 2–4(–7.5) cm and ovate to rhombic-ovate. A variant with smaller leaves, glabrous surfaces, densely tomentose inflorescence branches, and early salmon-orange, eventually crimson, ellipsoid pomes has been found scattered in the valleys of southern British Columbia and into Alberta. Its range is not understood because it can be difficult to determine fruit shape from herbarium specimens. Forms with larger leaves (to 6–8 cm) are not infrequent in Newfoundland, Quebec, and northern New England, paralleling a similar situation in *Crataegus macrosperma*; occasional narrowly elliptic-leaved forms occur in Prince Edward Island. Taller and more treelike variants are known from Maine. Some specimens, particularly from Minnesota, have a thin, non-persistent abaxial surficial indumentum. Such specimens are abundant in some areas and, except for leaf shape, seem to represent intergradation with var. *faxonii*.

103b. Crataegus chrysocarpa Ashe var. **faxonii** (Sargent) Eggleston, Bull. New York State Mus. Nat. Hist. 243–244: 64. 1923 (as faxoni) E

Crataegus faxonii Sargent, Rhodora 5: 161. 1903 (as faxoni); *C. rotundifolia* Moench var. *faxonii* (Sargent) Eggleston

Shrubs, 30–35 dm. **Leaves:** blade ovate to broadly ovate or broadly elliptic, base rounded to broadly cuneate, sinuses: max LII 5–10%, lobe apex subacute, veins 6 or 7 per side, abaxial surface sparsely tomentose, veins densely tomentose, adaxial with long, soft, white hairs until after flowering, later with many long, not very conspicuous, ± appressed hairs. **Inflorescences:** branches densely villous. **Flowers:** 15–20 mm diam.; hypanthium densely villous; stamens 5(–10), anthers cream or ivory. **Pomes** dark crimson, oblong, <full and rounded at ends>, 8–10 mm diam., pubescent.

Flowering May; fruiting late Sep. Brush, successional fields, unmanaged pastures; 0–300 m; Que.; Ill., N.H., Vt., Wis.

The typical form of the variety may have broadly ovate to broadly elliptic leaf blades 4.5–6 cm with six or seven veins per side. With far more densely hairy parts than typical var. *chrysocarpa* and the indumentum persistent on the abaxial surfaces of the leaves, var. *faxonii* is an eastern counterpart of var. *piperi*, from which it differs in details of leaf.

103c. Crataegus chrysocarpa Ashe var. **piperi** (Britton) Kruschke, Publ. Bot. Milwaukee Public Mus. 3: 163. 1965 • Piper's hawthorn [E]

Crataegus piperi Britton, Torreya 1: 55. 1901; *C. columbiana* Howell var. *piperi* (Britton) Eggleston

Shrubs, 20–35 dm. **Leaves:** blade rhombic to narrowly rhombic or rhombic-elliptic, base ± cuneate, sinuses: max LII 15–20%, lobe apex acute, veins 5 or 6 per side, abaxial surface persistently rough-pubescent, adaxial appressed-pubescent, glabrescent. **Inflorescences:** branches ± villous. **Flowers:** 15–20 mm diam.; hypanthium villous; stamens 10, anthers cream to ivory, sometimes pink. **Pomes** <salmon-orange young>, bright red, suborbicular, 10 mm diam., pubescent. $2n = 68$.

Flowering May–Jun; fruiting Sep–Oct. Streamsides and washes, sagebrush, pine forests; 300–800 m; B.C.; Idaho, Mont., Oreg., Wash.

Variety *piperi* occurs in the intermontane region from the southern Coastal Range of British Columbia and the Cascades of Washington, to the Rocky Mountains of British Columbia and Montana, where it is widespread and locally common in suitable valley habitats.

Variety *piperi* has been treated frequently in the Pacific Northwest literature as *Crataegus columbiana*; when that species was divided by floristicians into var. *columbiana* (glabrous) and var. *piperi* (hairy), the former could not be found. *Crataegus columbiana* was based on two different species, and its recent typification, which makes it a synonym of *C. douglasii*, reflects this anomaly.

103d. Crataegus chrysocarpa Ashe var. **vernonensis** J. B. Phipps & O'Kennon, Sida 21: 73, fig. 3. 2004 [C] [E]

Shrubs, 20–35 dm. **Leaves:** blade ovate to rhombic-ovate, base ± cuneate, sinuses: max LII 10–15%, lobe apex subacute, veins 3 or 4 per side, abaxial surface glabrous, veins pilose, adaxial appressed-scabrous-pubescent, persistent. **Inflorescences:** branches lanate. **Flowers** 18–20 mm diam.; hypanthium lanate proximally, glabrous distally; stamens 8–10, anthers pale pink. **Pomes** burgundy to nearly black, suborbicular, 10–12 mm diam., pubescent.

Flowering May; fruiting late Aug. Xeric habitats, sometimes sandy soil; of conservation concern; 300–400 m; B.C.

Variety *vernonensis* is locally common in the northern Okanagan valley between Vernon and Salmon Arm. Because of its extreme habitat for a hawthorn and its early ripening, pomes are apt to shrivel by September, when pomes of other hawthorns in the area are usually still plump. Possibly that is why it is under-collected. For the same reason, the variety is seldom found growing among the larger, more mesomorphic sympatric hawthorns. The variety is among the earliest to flower of the hawthorns of its area.

Variety *vernonensis* is remarkably uniform in its diagnostic characteristics. The multi-stemmed, relatively little-branching, erect habit is conspicuous in well-developed plants, as are the lanate inflorescences, pale pink anthers, and pomes already ripe and burgundy in late August. These characters help to distinguish the taxon from other members of ser. *Rotundifoliae*, while the tendency to subacute lobes in the leaf helps to distinguish it in midsummer from rare lanate forms of sympatric var. *chrysocarpa*.

103e. Crataegus chrysocarpa Ashe var. **subrotundifolia** (Sargent) J. B. Phipps, J. Bot. Res. Inst. Texas 3: 240. 2009 [E]

Crataegus subrotundifolia Sargent, Bot. Gaz. 35: 394. 1903; *C. coccinata* Sargent

Shrubs, 20–30 dm. **Leaves:** blade ovate to rhombic-obovate, base ± cuneate, sinuses: max LII 15–20%, lobe apex acute, veins 3–5 per side, abaxial surface glabrous, adaxial appressed-pubescent young, glabrescent or becoming short-scabrous. **Inflorescences:** branches thin-pilose. **Flowers** 15–20 mm diam.; hypanthium glabrous; stamens 10, anthers cream or ivory. **Pomes** red, suborbicular, 8–10 mm diam., pubescent or glabrous.

Flowering May–Jun; fruiting Sep–Oct. Brush; 0–400 m; Alta., Ont., Que., Sask.; Mass., Mich., N.Y., Vt., Wis.

Nearly all Cypress Hills (Alberta and Saskatchewan) specimens of *Crataegus chrysocarpa* match the characteristics of var. *subrotundifolia*, but they are evidently different from the eastern form representing the type.

Variety *subrotundifolia* is a weakly differentiated variety originally held to be restricted to the southern section of the eastern part of the range of *Crataegus chrysocarpa*, where it is locally common. In plants from this area, the leaves are more coriaceous and a darker green, and the fruit is a deeper, richer red than those of sympatric var. *chrysocarpa*.

103f. Crataegus chrysocarpa Ashe var. **phoeniceoides** J. B. Phipps & Sennikov, Willdenowia 43: 39. 2013 [E]

Shrubs, 20–30 dm. **Leaves:** blade ovate to rhombic-obovate, base ± cuneate, sinuses: max LII 15–20%, lobe apex acute, veins 3–5 per side, abaxial surface glabrous, adaxial glabrous young. **Inflorescences:** branches glabrous. **Flowers** 15–20 mm diam.; hypanthium <externally> glabrous; stamens 10, anthers cream or ivory. **Pomes** red, suborbicular, 8–10 mm diam., glabrous.

Flowering May–Jun; fruiting Sep–Oct. Open brush; 50–400 m; Ont., Que.; Conn., Maine, Mass., N.Y., Vt., Wis.

Because of confusion with similar, insufficiently studied, glabrous-inflorescence forms, the distribution given is inclusive.

Variety *phoeniceoides* in the strict sense is, in general appearance, like typical forms of var. *chrysocarpa* but is completely glabrous and, uniquely for this species, has the larger bracteoles stipitate-glandular like those in the *Crataegus dodgei* group.

103g. Crataegus chrysocarpa Ashe var. **vigintistamina** J. B. Phipps, J. Bot. Res. Inst. Texas 1: 1008. 2007 [E]

Shrubs, 20–30 dm. **Leaves:** blade ovate to rhombic-ovate, base ± cuneate, sinuses: max LII 15–20%, lobe apex acute, veins 3–5 per side, abaxial surface glabrous or lamina sparsely hairy at anthesis, veins hairy, adaxial appressed-scabrous, glabrescent. **Inflorescences:** branches pubescent. **Flowers** 15–20 mm diam.; hypanthium hairy; stamens 20, anthers cream or ivory. **Pomes** red, suborbicular, 8–10 mm diam., glabrous.

Flowering May; fruiting Sep–Oct. Open brush; 100–400 m; Ont.; Minn., N.Y.

Variety *vigintistamina* is similar to var. *chrysocarpa* except with 20 stamens; it is frequent in parts of Minnesota, where it has the intermediate indumentum mentioned under var. *chrysocarpa* and var. *faxonii*. *Crataegus divergens* (Peck) Sargent, a prior name, with ten to 18 stamens according to Sargent, is known from New York; its type (*C. H. Peck 70*) has ten stamens, preventing the name from being used for the current taxon.

103h. Crataegus chrysocarpa Ashe var. **praecox** (Sargent) J. B. Phipps, J. Bot. Res. Inst. Texas 1: 1009. 2007 [E]

Crataegus praecox Sargent, Rhodora 3: 27. 1901; *C. lemingtonensis* Sargent

Shrubs, 20–30 dm. **Leaves:** blade ovate to rhombic-ovate, base ± cuneate, sinuses: max LII 15–20%, lobe apex acute, veins 3–5 per side, abaxial surface glabrous young or lamina sparsely hairy at anthesis, veins hairy, adaxial appressed-scabrous, glabrescent. **Inflorescences:** branches pubescent. **Flowers** 15–20 mm diam.; hypanthium pubescent; stamens 10, anthers pink. **Pomes** red, suborbicular, 8–10 mm diam., glabrous.

Flowering May; fruiting mid Sep. Open brush, rocky hillsides; 100–300 m; N.H., N.Y., Vt.

Variety *praecox* is similar to var. *chrysocarpa* except for its pink anthers and is sparsely scattered through the eastern part of the range of the species. Reports from Quebec and Wisconsin are unconfirmed. Forms attributed to *Crataegus lemingtonensis* in Vermont are said to be locally common and represent two variants, the more common like *C. praecox* and a second with a distinctive leaf shape and relatively slender and long petiole characteristic of members of ser. *Tenuifoliae*. The latter represents typical *C. lemingtonensis* and is perhaps a hybrid.

103i. Crataegus chrysocarpa Ashe var. **blanchardii** (Sargent) J. B. Phipps, J. Bot. Res. Inst. Texas 1: 1009. 2007 [E]

Crataegus blanchardii Sargent, Rhodora 7: 218. 1905 (as blanchardi); *C. irrasa* Sargent var. *blanchardii* (Sargent) Eggleston

Shrubs, 20–30 dm. **Leaves:** blade ovate to rhombic-ovate, base ± cuneate, sinuses: max LII 15–20%, lobe apex acute, veins 3–5 per side, abaxial surface glabrous or veins hairy, adaxial appressed-scabrous, glabrescent. **Inflorescences:** branches pubescent. **Flowers** 15–20 mm diam.; hypanthium pubescent; stamens 20, anthers pink. **Pomes** red, suborbicular, 8–10 mm diam., glabrous.

Flowering May; fruiting Sep–Oct. Open brush; 10–200 m; Que.; Minn., N.Y., Vt.

Variety *blanchardii* is usually like var. *chrysocarpa*, but with 20 pink anthers; it ranges from Minnesota and Quebec to Vermont and New York. Eggleston perhaps combined it under *Crataegus irrasa* because of its leaves being (in specimens seen by him) more like *C. irrasa* in shape and size than those of *C. chrysocarpa*; however, the type material is like *C. chrysocarpa* in these features.

104. Crataegus oakesiana Eggleston, Torreya 7: 35. 1907 C E

Shrubs, 60 dm. **Stems:** twigs: new growth <green>, glabrous, 1-year old shiny, mid brown, older deep gray; thorns on twigs <frequent>, recurved, 1-year old shiny reddish to blackish brown, slender to moderately thick, 2–4.5 cm. **Leaves:** petiole length 40% blade, glabrous, densely glandular; blade broadly ovate to oblong-ovate or broadly elliptic, smaller sometimes ± rhombic, 3–5 cm, <on extension shoots larger, often suborbiculate>, thin to chartaceous, base cuneate to broadly cuneate or ± rounded, lobes 4 or 5 per side, sinuses shallow, lobe apex triangular, margins serrate <teeth numerous, small>, veins 6–8 per side, apex acute, abaxial surface glabrous, adaxial densely appressed-scabrous young, glabrescent. **Inflorescences** 6–10-flowered; branches pilose; bracteoles linear, membranous, margins sessile-glandular. **Flowers** 16 mm diam.; hypanthium pubescent; sepals narrowly triangular, 6 mm, margins glandular, sometimes barely glandular-serrate, <abaxially glabrous>; stamens 20, anthers cream; styles 3 or 4. **Pomes** reddish, broadly ellipsoid to suborbicular, 11–14 mm diam., <smooth>, glabrous; sepals spreading, often erose; pyrenes 3 or 4.

Flowering May; fruiting Sep–Oct. Rocky outcrops; of conservation concern; 100–200 m; Vt.

Crataegus oakesiana is not very different from *C. chrysocarpa* var. *vigintistamina*, differing primarily in its suborbiculate extension-shoot leaves.

105. Crataegus irrasa Sargent, Rhodora 5: 116. 1903 • Aubépine pubérulente E

Crataegus noveboracensis Sargent

Shrubs, <spreading into broad thickets>, 30–60 dm. **Stems** <numerous>; twigs: new growth <olive- or orange-green, slender, marked by large, oblong, pale lenticels>, glabrous, 1-year old light, grayish, or deep reddish brown, at end of 1d year often deep red-brown, 2-years old gray or light brown-tinged with red, older ashy gray; thorns on twigs <numerous>, straight, 1-year old shiny, deep red-brown, older dark gray, ultimately ashy gray, stout, 3–6 cm. **Leaves:** <stipules oblong-obovate, to falcate and acuminate on extension shoots, glandular-serrate, acuminate, villous>; petiole length 33–40% blade, <slightly winged distally>, adaxially villous young, glandular, <glands minute, dark red>; blade ovate-oblong to oblong, (4–)5–7(–8) cm (more than ½ grown at anthesis), <on extension shoots similar, larger>, thin to chartaceous, base cuneate to nearly truncate, lobes 5 or 6 per side, sinuses moderately deep (max LII 15%), lobe apex acute, ± triangular, margins serrulate, veins 5–7 per side, apex acute, abaxial surface glabrous, adaxial appressed-scabrous, glabrescent. **Inflorescences** 9–12 -flowered; branches villous; bracteoles linear, <small>, membranous, margins sessile-glandular. **Flowers** 13–15 mm diam.; hypanthium glabrous or pubescent; sepals <reflexed after flowering>, tapering from broad bases, short, margins minutely <bright red> glandular, <apex acute, abaxially villous>; stamens 20, anthers pale yellow, <small>; styles 4 or 5. **Pomes** <in broad, multi-fruited clusters, drooping or erect on stout branches>; reddish, suborbicular to oblong, 12–14 mm diam., glabrous or pubescent; sepals <persistent, sometimes until following spring, prominent>, appressed to spreading; pyrenes 4 or 5, <full and rounded at ends>.

Flowering May–Jun; fruiting Sep. Brush, old fields; 0–300 m; Nfld. and Labr. (Nfld.), Ont., Que.; Conn., N.Y., Wis.

The proven range of *Crataegus irrasa* is from eastern Ontario and nearby New York to Newfoundland. This species is to be expected in New England.

Crataegus irrasa is readily distinguished from both *C. chrysocarpa* var. *vigintistamina* and *C. oakesiana* by its larger, more oblong, and more venous leaves. Dubious records from Minnesota have more rhombic leaves. The most distinctive form, *C. noveboracensis*, which flourishes in Muskoka, Ontario, has the largest, most venous, and most oblong leaves; similar plants are known also from Newfoundland. *Crataegus irrasa* is distinguished from the somewhat similar *C. flabellata* var. *grayana* by its wider and thicker leaves, much hairier corymbs and hypanthia, pale yellow anthers, and the color of the flesh of the ripe pomes.

106. Crataegus margarettae Ashe, J. Elisha Mitchell Sci. Soc. 16: 72. 1900 (as margaretta) • Margaret's hawthorn E

Shrubs, 60–70 dm. **Stems:** twigs: new growth <becoming shiny, brownish tan by fall in 1st year>, glabrous, 1-year old brownish gray, older dull gray; thorns on twigs <few to numerous>, straight to recurved, 1-year old blackish, ± shiny, slender, 2.5–5 cm. **Leaves:** petiole length 30–60% blade, glabrous, usually with small glands at first, becoming eglandular, <glands

± caducous, small>; blade <green to bluish green>, suborbiculate, ovate, or obovate, rarely elliptic-rhombic to elliptic or lance-elliptic, 3–5 cm (less than 2.5 in var. *meiophylla*), subcoriaceous, base cuneate to broadly cuneate, lobes 1–3(or 4) per side, sinuses usually shallow, sometimes deep, lobe apex obtuse to angled, margins crenate to finely, sharply serrate, veins 3 or 4 per side, apex subacute to obtuse, abaxial surface glabrous, adaxial ± sparsely appressed-short-pubescent young, glabrescent. **Inflorescences** 6–15-flowered; branches glabrous, rarely pubescent; bracteoles <late caducous>, linear, subherbaceous, margins glandular. **Flowers** (8–)12–17 mm diam.; hypanthium glabrous; sepals narrowly triangular, 4–5 mm, margins entire or subentire, glandular, <abaxially glabrous>; stamens 20, anthers ivory; styles 2 or 3(or 4). **Pomes** yellow, orange-red, red, or dull burgundy, broadly ellipsoid to suborbicular, 7–12 mm diam., glabrous; sepals reflexed, <often on collar>; pyrenes 2–4. $2n = 34$.

Varieties 4 (4 in the flora): c, e North America.

Crataegus margarettae is widespread and often common, extending east from Missouri and Iowa to Wisconsin, southern Michigan, southwestern Ontario (where it was always rare and is now perhaps lost), western Pennsylvania, and western Virginia.

Leaves of *Crataegus margarettae* are often somewhat blue-green at maturity, and the thorns are usually finer than found on its close relatives *C. dodgei* and *C. lumaria*. The variation in fruit color (yellow to burgundy) and leaf shape (suborbiculate to narrowly elliptic) is considerable; there are four varieties but only the typical one is common. The other varieties are extreme forms of which perhaps only var. *angustifolia* has a genetic basis; var. *brownii* and var. *meiophylla* are possibly products of dry seasons or situations.

1. Leaf blades proportionately narrow, lance-elliptic or oblong-lanceolate, bases cuneate, lobes (0 or obscure), sinuses shallow, apices acute ....106b. *Crataegus margarettae* var. *angustifolia*
1. Leaf blades proportionately wide, broadly elliptic, obovate to ovate, suborbiculate, or round, bases broadly cuneate, lobe sinuses ± shallow to ± deep, apices obtuse to subacute.
   2. Flowers 8–10 mm diam.; pomes ellipsoid ... .106c. *Crataegus margarettae* var. *brownii*
   2. Flowers 12–17 mm diam.; pomes broadly ellipsoid to suborbicular.
      3. Leaf blades 3–5 cm, broadly elliptic to suborbiculate .... 106a. *Crataegus margarettae* var. *margarettae*
      3. Leaf blades less than 2.5 cm, round .... 106d. *Crataegus margarettae* var. *meiophylla*

106a. Crataegus margarettae Ashe var. **margarettae** [E]

Leaves: blade broadly elliptic to suborbiculate, 3–5 cm, base broadly cuneate, lobe sinuses ± shallow to ± deep, apex obtuse to subacute. **Flowers** 12–17 mm diam. **Pomes** broadly ellipsoid to suborbicular.

Flowering Apr–May; fruiting Sep–Oct. Hedgerows, thickets, wood margins, dry areas; 100–300 m; Ont.; Ill., Ind., Iowa, Ky., Mich., Mo., Ohio, Pa., Va., W.Va., Wis.

Variety *margarettae* is the most common variety, occurring nearly throughout the range of the species. It has an uncommon form with distinctly pubescent inflorescences.

106b. Crataegus margarettae Ashe var. **angustifolia** E. J. Palmer in C. C. Deam, Fl. Indiana, 543. 1940 (as margaretta) [C] [E]

Leaves: blade lance-elliptic or oblong-lanceolate, 3–5 cm, base cuneate, lobes <0 or obscure>, sinuses shallow, apex acute. **Flowers** 12–17 mm diam. **Pomes** broadly ellipsoid to suborbicular.

Flowering May; fruiting Sep–Oct. Hedgerows, thickets, wood margins, dry areas; of conservation concern; 100–300 m; Ont.; Ind., Mich.

Variety *angustifolia* is very sporadic in its occurrence but has a rather limited range.

106c. Crataegus margarettae Ashe var. **brownii** (Britton) Sargent, J. Arnold Arbor. 3: 199. 1922 (as margaretta) [E]

Crataegus brownii Britton, Bull. New York Bot. Gard. 1: 447. 1900

Leaves: blade broadly elliptic, obovate to ovate, or suborbiculate, 3–5 cm, base broadly cuneate, lobe sinuses shallow, apex obtuse to subacute. **Flowers** 8–10 mm diam. **Pomes** ellipsoid.

Flowering Apr–May; fruiting Sep–Oct. Hedgerows, thickets, wood margins, dry areas; 100–300 m; Ky., Mo., Va.

Variety *brownii* is sporadic in occurrence. It is notable for having small flowers. Reports from Indiana and West Virginia need validation.

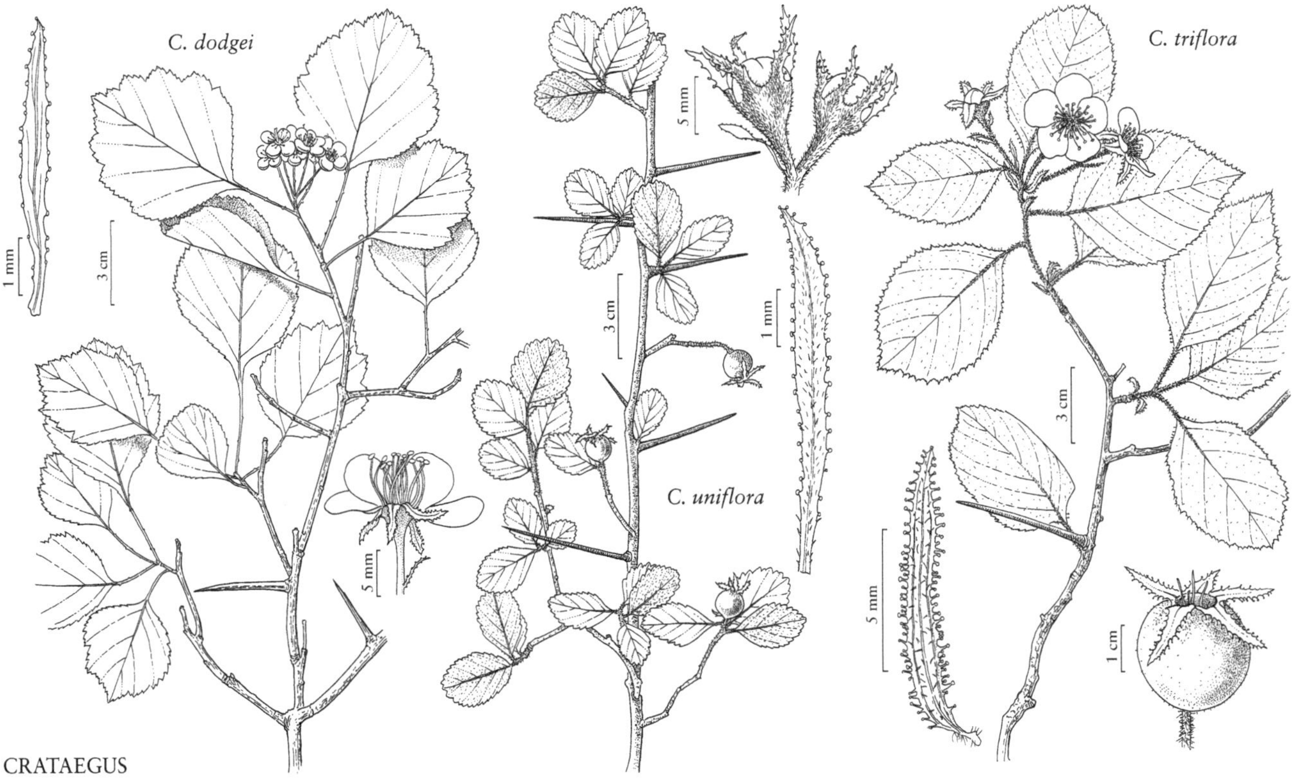

CRATAEGUS

106d. Crataegus margarettae Ashe var. **meiophylla** (Sargent) E. J. Palmer, Brittonia 5: 485. 1946 (as margaretta) [E]

Crataegus meiophylla Sargent, J. Arnold Arbor. 3: 198. 1922

Leaves: blade round, less than 2.5 cm, base broadly cuneate, lobe sinuses shallow to ± deep, apex obtuse to subacute. **Flowers** 12–17 mm diam. **Pomes** broadly ellipsoid to suborbicular.

Flowering Apr–May; fruiting Sep–Oct. Hedgerows, thickets, margins of woods, often in somewhat dry areas; 70–200 m; Ohio.

Variety *meiophylla* is a rare plant notable for its small leaves.

107. Crataegus dodgei Ashe, J. Elisha Mitchell Sci. Soc. 19: 26. 1903 • Dodge hawthorn [E] [F]

Crataegus chrysocarpa Ashe var. *dodgei* (Ashe) E. J. Palmer; *C. crassifolia* Sargent; *C. delosii* Sargent; *C. dodgei* var. *flavida* (Sargent) P. G. Smith & J. B. Phipps; *C. dodgei* var. *rotundata* (Sargent) Kruschke; *C. flavida* Sargent; *C. gravesii* Sargent; *C. rotundata* Sargent

Shrubs or trees, 50 dm. **Stems:** twigs: new growth <often ± reddish young>, glabrous, 1-year old gray-brown, older dull gray; thorns on twigs <few to plentiful>, straight to ± recurved, 1-year old shiny, dark brown or black, older gray, slender, 2.5–7 cm. **Leaves:** petiole length 40–60% blade, glabrous, sparsely to densely minutely glandular; blade <green>, suborbiculate, ± ovate, broadly rhombic to rhombic-ovate, sometimes broadly elliptic, 2–5 cm, thin, base rounded to cuneate, lobes 0 or 1–4 per side, sinuses shallow to moderately deep, lobe apex acute to obtuse, margins crenate-serrate to serrate, veins 4–6 per side, apex subacute to obtuse, abaxial surface glabrous, adaxial glabrate young, soon glabrescent. **Inflorescences** 6–15-flowered; branches glabrous; bracteoles linear, membranous, margins glandular, larger proximally stipitate-glandular. **Flowers** 13–18 mm diam.; hypanthium glabrous; sepals narrowly triangular,

5 mm, margins entire or ± glandular-denticulate; stamens 10, anthers ivory; styles 2 or 3. **Pomes** yellow to yellow-orange, dull orange, or ruddy to red, suborbicular, 7–12 mm diam., glabrous; sepals reflexed; pyrenes 2 or 3. **2*n*** = 51, 68.

Flowering Apr; fruiting Sep–Oct. Thickets, successional fields, brush; 0–900 m; Ont., Que.; Conn., Ill., Ind., Mass., Mich., N.Y., N.C., Ohio, Pa., Va., W.Va., Wis.

Crataegus dodgei ranges from Wisconsin and northern Illinois through the southern Great Lakes to Connecticut, and south to West Virginia and the North Carolina Blue Ridge. The southern limits are poorly understood.

Crataegus dodgei is unusually variable in leaf shape, and also variable in plant size, thorn length, and bracteole glandularity. The typical form resembles a ten-stamen *C. margarettae*, though often with larger and more veiny leaves. It occurs throughout the range of *C. dodgei* and is usually easily recognizable. An opposite tendency is found in the *C. flavida* form, which is also common but where the plants are smaller with usually finer thorns and particularly small, usually acutely lobed leaves. Rather scarce forms with larger, proportionately broad, often suborbiculate leaves and obtuse lobes represent the *C. delosii* form. Still other variants are known.

108. Crataegus lumaria Ashe, J. Elisha Mitchell Sci. Soc. 19: 25. 1903 E

Crataegus dodgei Ashe var. *lumaria* (Ashe) Sargent; *C. jackii* Sargent

Shrubs or trees, 50 dm. **Stems:** twigs: new growth <often ± reddish young>, glabrous, 1-year old gray-brown, older dull gray; thorns on twigs ± recurved, 1-year old shiny, dark brown or black, older gray, slender, 2.5–7 cm. **Leaves:** petiole length 40–60% blade, glabrous, sparsely to densely minutely glandular; blade <green>, broadly rhombic to rhombic-ovate, or narrowly ovate to broadly elliptic, 2–4(–5) cm, thin, base rounded to cuneate, lobes 3 or 4 per side, sinuses shallow, lobe apex acute, margins serrate, veins 4–6 per side, apex acute to obtuse, abaxial surface glabrous, adaxial glabrate young, soon glabrescent. **Inflorescences** 6–15-flowered; branches pilose; bracteoles <usually hyaline>, linear, membranous, margins of proximal bracteoles usually stipitate-glandular. **Flowers** 13–18 mm diam.; hypanthium glabrous; sepals narrowly triangular, 5 mm, margins entire, glandular; stamens 10, anthers ivory; styles 2 or 3. **Pomes** yellowish, dull orange, or reddish, suborbicular, 7–12 mm diam., glabrous; sepals reflexed; pyrenes 2 or 3. **2*n*** = 68.

Flowering May; fruiting Sep–Oct. Brush, fallow pastures; 30–300 m; Ont., Que.; Mich., N.Y., Wis.

Crataegus lumaria is sporadic through much of its range; it is locally common in southern Ontario.

Crataegus lumaria is much like a hairy-inflorescence form of *C. dodgei.* The disjunct *C. jackii* is a form with particularly hairy, suborbiculate leaves and larger flowers (to 20 mm).

64f.16. CRATAEGUS Linnaeus (sect. COCCINEAE) ser. PARVIFOLIAE (Loudon) Rehder, Man. Cult. Trees ed. 2, 366. 1940

Crataegus sect. *Parvifoliae* Loudon, Arbor. Frutic. Brit. 2: 841. 1838

Shrubs, 10–50 dm, <± xeromorphic>, multi-stemmed. **Stems:** trunk bark gray-brown, rough; <branches spreading; twigs ± straight>, new growth pubescent, 1-year old gray-brown or brown, older gray; thorns on twigs usually numerous, sometimes absent, straight, rarely slightly recurved, 1-year old blackish, fine, 3–5(–8) cm. **Leaves:** petiole nearly absent, sessile-glandular or eglandular; blade <dark green>, elliptic to narrowly obovate or similar, sometimes suborbiculate, (1–)2–3.5(–6) cm (small, usually longer than wide), coriaceous, base ± cuneate, lobes 0 or 1(or 2) per side, sinuses shallow, lobe apex obtuse, margins crenate or crenate-serrate, <teeth often gland-tipped, glands black>, venation ± craspedodromous to semicamptodromous, veins 3 or 4(or 5) per side, apex broadly acute to rounded, surfaces pubescent. **Inflorescences** 1–3(–6)-flowered, reduced corymbs; branches moderately to densely pubescent; bracteoles of two types: not occurring in all specimens, ± persistent, green, narrowly elliptic, herbaceous, margins strikingly toothed, glandular, or occurring in all specimens, ± caducous, linear, small, membranous to herbaceous, margins stipitate-glandular. **Flowers** 10–18 mm diam.; hypanthium densely tomentose; sepals narrowly triangular, equal to or longer than petals, margins usually glandular-serrate, often deeply so (except in *C. uniflora*); stamens 20, anthers white to cream or ivory; styles (3–)5. **Pomes** yellowish or red to ruddy, sometimes dull brown, suborbicular, 8–12(–14) mm diam., pubescent to tomentose; flesh mealy or not recorded; sepals patent or reflexed, <glandular-serrate, often deeply so, non-accrescent>; pyrenes (3 or)4 or 5.

Species 2 (2 in the flora): se United States, ne Mexico.

Series *Parvifoliae* contains two species: *Crataegus uniflora*, widespread and often common in the coastal plain from Texas to Long Island, New York, more sparse inland to Missouri and Illinois and in Tamaulipas, Mexico; and *C. brittonii*, much rarer, from North Carolina and adjacent states. *Crataegus choriophylla* Sargent may not belong in this series; it is a rare, perhaps extinct, tree with distinctly petiolate, thin, strongly toothed, often slightly incised leaves, 3–5-flowered inflorescences, cream anthers, and orange-red pomes. *Crataegus* ×*vailiae*, a putative interserial hybrid with ser. *Macracanthae*, keys out in the second couplet because of its long foliaceous sepals.

Features that best characterize ser. *Parvifoliae* are relatively small stature, rather glandular plants, relatively small, more or less unlobed leaves, uniflorous to few-flowered inflorescences, relatively small flowers, more or less foliaceous sepals equal to or longer than the petals, and ivory to cream anthers.

Members of ser. *Parvifoliae* share characteristics with members of ser. *Triflorae*, but their plant parts are so much smaller that even casual confusion is not possible. At least in some specimens, the series shares the characteristic with ser. *Triflorae* that some flowering shoots are leafy shoots of the season borne directly on the extension shoots and not subterminally on woody short shoots. Confusion is quite frequent with smaller members of ser. *Lacrimatae*. Members of both series contain small- to medium-sized xeromorphs that are rather glandular, often hairy, and often have yellowish pomes. Members of ser. *Lacrimatae*, though, may be readily distinguished by their strongly flexuous twigs, entire or glandular-serrate sepals much shorter than petals, and fewer styles and pyrenes.

The interserial hybrid *Crataegus* ×*vailiae* keys out in part in the second couplet.

SELECTED REFERENCE Phipps, J. B. and K. A. Dvorsky. 2006. *Crataegus* ser. *Parvifoliae* and its putative hybrids in the southeastern United States. Sida 22: 423–445.

1. Shrubs 10–20(–50) dm; leaf blades narrowly elliptic, elliptic, obtrullate, spatulate, oblanceolate, or narrowly rhombic-elliptic, sometimes suborbiculate, length/width = usually 1.5 or narrower, lobes 0; anthers white to cream . . . . 109. *Crataegus uniflora*
1. Shrubs 20–40(–50) dm; leaf blades elliptic to rhombic-elliptic, length/width = 1.3 or wider, lobes 0 or 1(or 2) per side, sinuses shallow; anthers ivory or cream, pink, or red.
   2. Leaves: lobe apices obtuse, margins crenate-serrate, teeth small; veins 3 or 4 per side; pyrene sides plane . . . . 110. *Crataegus brittonii*
   2. Leaves: lobe apices subacute, margins serrate to ± crenate-serrate, teeth larger; veins 4–6 per side; pyrene sides usually shallowly eroded. . . . . 169. *Crataegus* ×*vailiae*, p. 643

109. Crataegus uniflora Münchhausen, Hausvater 5: 147. 1770 • One-flowered hawthorn F

Crataegus parvifolia Aiton; *C. trianthophora* Sargent

Shrubs, 10–20(–50) dm. **Stems:** twigs: new growth densely appressed-pubescent, 1-year old brown, older grayish; thorns on twigs usually numerous, sometimes absent, ± straight, 1-year old black, needlelike, 3–5(–8) cm; <young thorns soft, rapidly growing, needle-pointed shoots with linear bracts, often curved initially, straightening>. **Leaves:** petiole very short or absent, pubescent, glandular or not; blade narrowly elliptic, elliptic, obtrullate, spatulate, oblanceolate, or narrowly rhombic-elliptic, sometimes suborbiculate, (1–)1.5–3(–6) cm, length/width = 1.5 or narrower, base cuneate, lobes 0, margins crenate to serrate, teeth variably caducous, black eglandular or gland-tipped, venation semicamptodromous to camptodromous, veins 3 or 4(or 5) per side, <± impressed>, apex obtuse to broadly acute, <often somewhat glossy>, abaxial surface sparsely to moderately pubescent, veins ± densely pubescent, adaxial densely scabrous, especially young. **Inflorescences** 1–3(–5)-flowered; branches pubescent; bracteoles of two types: ± persistent, green, herbaceous, margins serrate, teeth glandular, or caducous, linear, membranous to herbaceous, margins glandular. **Flowers** 10–15 mm diam.; hypanthium strongly tomentose, (sometimes bearing a caducous, linear, membranous to herbaceous, gland-margined bracteole); sepals foliaceous, usually longer than petals, <2–4 mm wide>, surfaces sparsely pubescent; anthers white to cream; styles 5. **Pomes** yellowish to ruddy, 8–10(–12) mm diam., ± tomentose; <flesh firm, dry, or mealy>; sepals <persistent>, patent, <7 mm>; pyrenes 4 or 5. $2n$ = 51, 68.

Flowering Apr–May; fruiting Sep–Oct. Sand barrens, sandhills, open scrubby woodlands, often associated with *Pinus clausa*, *P. palustris*, *P. rigida* and xerophytic species of *Quercus* and *Carya*; 10–200 m; Ala., Ark., Del., Fla., Ga., Ill., Ind., Ky., La., Md., Miss., Mo., N.J., N.Y., N.C., Ohio, Okla., Pa., S.C., Tenn., Tex., Va., W.Va.; Mexico (Tamaulipas).

Crataegus uniflora is widespread in the southeastern United States from Long Island (New York) to eastern Texas; it is most abundant in the coastal plain and lower piedmont. A disjunction in the Mississippi embayment between trans-Mississippian and cis-Mississipian populations mirrors a similar pattern in *C. spathulata* (J. B. Phipps 1998).

Crataegus uniflora is usually easy to identify because of its small size, leaves small and coriaceous with lobes absent, numerous straight, fine thorns, inflorescences uniflorous or few-flowered, sepals glandular-pectinate and equal to or longer than the petals, anthers white to cream, and pomes yellowish to ruddy. Possible confusion with some members of ser. *Lacrimatae* is treated under the series discussion. Sterile plants might be confused with small, sterile ones of *C. berberifolia* (ser. *Crus-galli*) except for the smaller number of lateral veins on the leaves. An interesting feature of *C. uniflora* is the occasional deep lobing of extension-shoot leaves, even narrow ones, with veins to sinuses.

Crataegus uniflora is somewhat variable in plant size, leaf shape, number of flowers to an inflorescence (though commonly one), and anther color (usually ivory to cream). The typical form, which is common, has one- or two-flowered inflorescences, and is usually a small shrub with more or less spatulate to narrowly obovate leaves. Some particularly dwarf forms of this, on laterite soil in Georgia, have been observed with over 100 flowers when less than 3 dm.

Forms of *Crataegus uniflora* with three- to six-flowered inflorescences are also common and are represented by *C. parvifolia* and *C. trianthophora*. These are connected to the typical uniflorous kind by numerous intermediates and may be distinguished from the related *C. brittonii* by their smaller and narrower leaves with lobes absent.

Three anomalous variants, very similar or identical to *Crataegus uniflora*, except for possessing small, more or less unlobed but glandular-serrate sepals, are known. These are: *Crataegus croomiana* Sargent from northern Florida, *C. bisulcata* Ashe from North Carolina, and *C. grossiserrata* Ashe from Florida. The first, also best known, is further distinguished by pink anthers, the second by broader leaves, and the third by often orbiculate (unusual for ser. *Parvifoliae*) extension-shoot leaves with a tendency to deep lobing sometimes with veins to sinuses, and particularly strong and sharp leaf teeth. All suggest hybrid origin with ser. *Lacrimatae*.

110. Crataegus brittonii Eggleston, Bull. Torrey Bot. Club 36: 640. 1909 • Britton hawthorn C E

Crataegus uniflora Munchhausen var. *brittonii* (Eggleston) Lance

Shrubs, <slender>, 30–40 dm. **Stems:** twigs: new growth pubescent, 1-year old gray-brown, ± shiny, <glabrous>, older gray; thorns on twigs often numerous, ± straight to slightly recurved, 1-year old blackish, ± slender, 3–5 cm. **Leaves:** petiole 3–6 mm,

<flared distally>, pubescent, glandularity not recorded; blade elliptic-rhombic to broadly elliptic-rhombic, 2–3.5 cm (usually widest near middle), length/width = 1.3 or wider, base broadly cuneate, lobes 0 or 1(or 2) per side, margins crenate-serrate, teeth caducous, gland-tipped, venation ± craspedodromous, veins 3 or 4 per side, apex rounded to subacute, abaxial surface <pale green>, soft pubescent, indumentum variably persistent with age, adaxial <dark green>, appressed-pubescent. **Inflorescences** 2–6-flowered; branches densely pubescent; bracteoles type 2. **Flowers** 14–18 mm diam.; hypanthium densely pubescent; sepals foliaceous, at least as long as petals, adaxially pubescent; anthers ivory; styles 3–5. **Pomes** red or ruddy, sometimes becoming dull brown, 14 mm, pubescent; sepals reflexed; pyrenes 3–5.

Flowering Apr–May; fruiting Sep–Nov. Brush; of conservation concern; 0–200 m; Ala., Ga., N.C., Tenn.

Crataegus brittonii is a rare species with a scattered distribution.

For a discussion of the possibly synonymous but untypified pink-anthered *Crataegus rhodella* Ashe and *C. arenicola* Ashe see J. B. Phipps and K. A. Dvorsky (2006).

64f.17. Crataegus Linnaeus (sect. Coccineae) ser. Bracteatae (Sargent) Rehder, Man. Cult. Trees ed. 2, 363. 1940 [E]

Crataegus [unranked] *Bracteatae* Sargent, Silva 13: 34. 1902

Shrubs or trees, 30–80 dm, usually main trunk dominant, <1–few-stemmed>. **Stems:** trunk bark not recorded; <branches spreading; twigs ± straight>, new growth pubescent, 1-year old pale to mid, sometimes dark, brown, older gray; thorns on twigs few to numerous, ± straight, 2-years old ± shiny, dark brown to black, ± stout to ± slender, 2–4 cm. **Leaves:** petiole length 15–25% blade, <winged distally>, pubescent, gland-dotted or stipitate-glandular; blade <deep green>, elliptic to narrowly obovate or broadly elliptic to ovate, 3–7 cm, subcoriaceous, base ± narrowly cuneate, lobes 0 or 1–4 per side, sometimes apiculi, sinuses very shallow, margins regularly, strongly glandular-serrate, <proximal bracteoles usually stipitate-glandular>, venation craspedodromous, veins 5–7 per side, apex acute to subacute, <glossy>, abaxial surface hairy mainly on veins, adaxial appressed-pubescent young, sometimes glabrescent. **Inflorescences** 3–12-flowered, convex panicles, <usually subterminal from a woody short shoot, sometimes lateral to extension shoots>; branches densely villous or pilose; bracteoles: larger ones semipersistent, ligulate, subherbaceous, margins glandular-pectinate to glandular-serrate, smaller ones caducous, linear, herbaceous to membranous, margins stipitate-glandular. **Flowers** (15–)20–25 mm diam.; hypanthium tomentose; sepals narrowly triangular or lanceolate, length 66–75% petals, margins glandular-pectinate or glandular-laciniate; stamens 20(–25), anthers usually ivory; styles 3–5. **Pomes** <1–few>, orange-red to red, suborbicular, 10–14(–22) mm diam., hairy; flesh mealy to ± hard; sepals patent-reflexed or -incurved, <non-accrescent>; pyrenes 3–5.

Species 2 (2 in the flora): se United States.

Series *Bracteatae* comprises one fairly common species (*Crataegus ashei*) and one rare species (*C. harbisonii*) together occurring from Louisiana to Alabama in the piedmont, and north into Tennessee and the Virginia Blue Ridge. These species are found in brush, including prairie margins and open woodlands.

Series *Bracteatae* are characterized by relatively few- and large-flowered inflorescences and by very glandular petioles, leaf margins, pedicels, and sepal margins, which are also usually very hairy. They share these characteristics with ser. *Triflorae*, to which they appear to be related. The leaves of ser. *Bracteatae* have longer petioles than those of the somewhat similar ser. *Crus-galli*. *Crataegus harbisonii* may bear its inflorescences on side shoots of the season. All are tetraploid.

1. Leaf blades: lobes 0 or obscure, marginal teeth to 1.5 mm; 2-year old thorns ± slender; inflorescences arising subterminally from perennial short shoots; bracteoles caducous, linear, herbaceous to membranous . . . . 111. *Crataegus ashei*
1. Leaf blades: lobes 0 or 1–4 per side, sinuses shallow, marginal teeth 2 mm; 2-year old thorns stout; inflorescences arising subterminally from perennial short shoots, also frequently arising laterally from extension shoots; bracteoles semipersistent, ligulate, subherbaceous . . . . 112. *Crataegus harbisonii*

111. Crataegus ashei Beadle, Bot. Gaz. 30: 339. 1900 • Ashe hawthorn [E]

Shrubs, 30–50 dm. **Stems:** twigs: new growth densely pubescent, 1-year old tan to chestnut, <± shiny>, older gray; thorns on twigs few to numerous, 2-years old shiny, very dark brown to black, ± slender, 2–3 cm. **Leaves:** petiole length 25% blade, densely pubescent, stipitate-glandular; blade mid green, elliptic to narrowly obovate, 3–6 cm, base cuneate, lobes 0 or obscure, margins glandular-serrate, teeth to 1.5 mm, <proximal teeth stipitate-glandular>, veins 5–7 per side, apex acute to subacute, abaxial surface sparsely pubescent or glabrous, veins hairy, adaxial <shiny>, scabrous. **Inflorescences** 3–10-flowered, arising proleptically (subterminal on perennial short shoots); branches densely villous; bracteoles caducous, linear, 4–8 mm, herbaceous to membranous, margins stipitate-glandular. **Flowers** (15–)20–23 mm diam.; hypanthium tomentose; sepals narrowly triangular, <6–7 mm>, margins glandular-pectinate, <abaxially pilose>; stamens 20 (–25), anthers cream, sometimes pink; styles 3(–5) (with bristly hairs at bases). **Pomes** orange-red to deep red, 10–14 mm diam., densely short-pubescent; sepals patent-incurved; pyrenes 3(–5). **2*n*** = 68.

Flowering Apr; fruiting Sep–Nov. Brush, open woodlands; 0–300 m; Ala., La., Miss., Tenn.

Crataegus ashei occurs from Louisiana to Alabama and is apparently scarce.

Crataegus ashei has often been confused with both *C. triflora* (ser. *Triflorae*) and *C. harbisonii*; it differs from *C. triflora* in its different growth habit with branching more or less layered and leaves smallish, flowers smaller, stamens 20–24, and leaves usually more coriaceous and shiny with lobes absent. *Crataegus ashei* is more similar to *C. harbisonii*, mainly differing in being smaller in its vegetative parts, invariably lacking reproductive side shoots of the season, and by having entire leaves.

112. Crataegus harbisonii Beadle, Bot. Gaz. 28: 413. 1899 (as harbisoni) • Harbison hawthorn [C] [E]

Shrubs or trees, 50–80 dm. **Stems:** twigs: new growth sparsely pilose, 1-year old chestnut brown, <becoming smooth>; thorns on twigs frequent, 2-years old glossy, dark brown or nearly black, stout, 3–4 cm. **Leaves:** petiole length 15–20% blade, sparsely pilose, gland-dotted; blade <shiny> green, broadly elliptic to ovate, 4–7 cm, base cuneate, lobes 0 or 1–4 per side, sometimes apiculi, sinuses very shallow, margins doubly serrate, <5 teeth per cm>, teeth 2 mm, <gland-tipped early>, veins 6 or 7 per side, apex acute, abaxial surface pubescent on veins mature, adaxial appressed-pubescent young. **Inflorescences** 5–12-flowered, arising subterminally from perennial short shoots, also frequently on leafy side shoots of season lateral to extension shoots; branches pilose; bracteoles semipersistent, ligulate, 7–18 mm, subherbaceous, margins glandular-pectinate to glandular-serrate. **Flowers** 20–25 mm diam.; hypanthium pubescent; sepals lanceolate, margins glandular-laciniate; stamens 20, anthers cream to light yellow; styles 3–5. **Pomes** orange-red to red, 12–22 mm diam., pubescent; sepals patent-reflexed; pyrenes 3–5. **2*n*** = 68.

Flowering early May; fruiting Sep–Nov. Wooded hills, creekside brush; of conservation concern; 0–200 m; Ala., Tenn.

Crataegus harbisonii is rare and is similar both to *C. ashei* and *C. triflora* and could perhaps be thought of as a particularly robust form of *C. ashei*; however, it is perhaps more likely of hybrid origin between these species. The most significant similarity to *C. triflora* (ser. *Triflorae*) is an ability to produce inflorescences on shoots of the season lateral to extension shoots. Nevertheless, it is easily distinguished from *C. triflora* by characteristics of growth habit, by its smaller flowers (except in the Louisiana and some Mississippi forms of *C. triflora*) with fewer stamens, by the more coriaceous leaves with larger and more distant teeth, and by having among the most persistent bracteoles in the genus, a feature that led to the creation of ser. *Bracteatae*.

SELECTED REFERENCE Lance, R. and J. B. Phipps. 2000. *Crataegus harbisonii* rediscovered and amplified. Castanea 85: 291–296.

64f.18. CRATAEGUS Linnaeus (sect. COCCINEAE) ser. TRIFLORAE (Beadle) Rehder, Man. Cult. Trees ed. 2, 364. 1940 [E]

Crataegus [unranked] *Triflorae* Beadle in J. K. Small, Fl. S.E. U.S., 533, 537. 1903; *Crataegus* sect. *Triflorae* (Beadle) C. K. Schneider

Shrubs, 20–50 dm, <trunks fasciculate>, multi-stemmed. **Stems:** trunk bark younger grayish, smooth, sometimes exfoliating in irregular papery patches, mature rough, gray-brown; <branches spreading; twigs ± straight>, new growth pubescent, 1-year old mid brown, older gray; thorns on twigs sparse to moderately frequent, ± straight, 2-years old dark gray or blackish, ± thin, 2–4 cm. **Leaves:** petiole length 33% blade, <winged distally>, densely pubescent, stipitate-glandular; blade <mid to dark green>, broadly ovate to broadly elliptic or nearly suborbiculate, 3–8(–10) cm, ± chartaceous, base broadly cuneate to rounded or truncate, lobes 0 or 1–3 (or 4) per side, sinuses shallow, lobe apex subacute to acute, margins regularly, sharply glandular-serrate, <proximally usually stipitate-glandular>, venation craspedodromous, veins 5–7 per side, apex acute to subacute, abaxial surface mainly hairy on veins, adaxial hairy until maturity. **Inflorescences** 2–6(–12)-flowered, reduced or larger convex panicles, <usually borne on leafy short shoots of season lateral to extension shoots, sometimes also on woody short shoots>; branches densely tomentose (intermixed with stipitate glands); bracteoles caducous to semipersistent, <ligulate>, linear, <large>, membranous to herbaceous, margins stipitate-glandular. **Flowers** 25–30 mm diam.; hypanthium densely tomentose; sepals <foliaceous>, triangular, wide, $\frac{2}{3}$–$\frac{3}{4}$ length of petals, <herbaceous>, margins glandular-pectinate, <abaxially pubescent>; <petals broadly elliptic in *C. austromontana*>; stamens 10 or 30–45(–47), anthers cream; styles 3–5. **Pomes** <1–few per infructescence>, red to ruddy, suborbicular to obovoid, 12–15 mm diam., hairy; flesh mealy to ± hard; sepals <persistent>, patent-reflexed, <long, non-accrescent>; pyrenes 3–5.

Species 2 (2 in the flora): se United States.

Series *Triflorae* comprises one relatively common (*Crataegus triflora*) and one rare (*C. austromontana*) species, occurring from Louisiana to Georgia in the piedmont and north into Tennessee, with an outlier in Arkansas. The species are found in brush, including prairie margins and open woodlands; they are also able to tolerate considerable shade.

Species in ser. *Triflorae* are characterized by few- and large-flowered inflorescences, often borne on lateral leafy short shoots of the season, as well as by very glandular petioles, leaf margins, pedicels, and sepal margins, which are also usually very hairy. Also characteristic are the multiple stems lacking main trunk dominance. The most similar series is ser. *Bracteatae*. The 30 to 45 stamens of *Crataegus triflora* are unique in the genus.

1. Leaf blades broadly ovate to suborbiculate, lobes 3(or 4) per side; stamens 10 .... 113. *Crataegus austromontana*
1. Leaf blades broadly elliptic, entire or lobes 1–3(or 4) per side; stamens 30–45(–47) ... 114. *Crataegus triflora*

113. Crataegus austromontana Beadle, Bot. Gaz. 28: 412. 1899 • Valley Head hawthorn [C] [E]

Shrubs, 40–50 dm. **Stems:** twigs: new growth densely pubescent, 1–2-years old brown; thorns on twigs not recorded. **Leaves:** blade broadly ovate to suborbiculate, 4–7 cm, base cuneate, lobes 3(or 4) per side, sinuses shallow, lobe apex subacute, margins sharply glandular-serrate throughout, veins 5 per side, apex subacute, abaxial surface sparsely pubescent except on veins, adaxial appressed-hairy young, especially on veins, surfaces usually glabrescent. **Inflorescences** 3-flowered, on lateral leafy short shoots of season; bracteoles caducous, membranous. **Flowers** 25 mm diam.; sepals narrowly triangular, 7 mm; <petals broadly elliptic>; stamens 10; styles 3–5. **Pomes** red, suborbicular.

Flowering early May; fruiting Sep–Nov. Wooded hills, brush; of conservation concern; 70–200 m; Ala., Tenn.

Crataegus austromontana is a possibly extinct species; it was last collected in 1916. Beadle reported that it occurred throughout the Sand Mountain region of Alabama and had also been collected in the Cumberland Mountains and hill country of eastern and middle Tennessee. The species is similar to *C. triflora* but has wider, usually larger and more deeply incised leaves, and only ten stamens.

114. Crataegus triflora Chapman, Fl. South. U.S. ed. 2 repr. 2, 684. 1892 • Three-flower hawthorn [E] [F]

Shrubs, 20–50 dm, <multi-stemmed>. **Stems:** <compound thorns on trunks present>; twigs: new growth appressed-pubescent, 1–2-years old brown, older gray; thorns on twigs ± straight, 2-years old dark gray or blackish, ± thin, 2–4 cm. **Leaves:** blade broadly elliptic, 3–8(–10) cm, base cuneate to rounded or truncate, entire or lobes 1–3(or 4) per side, sinuses shallow, lobe apex subacute to acute, margins serrulate <teeth 1 mm, 10 per cm>, gland-tipped, veins 5–7 per side, apex acute, abaxial surface pubescent young, only abaxial veins pubescent when old, adaxial appressed-pubescent young. **Inflorescences** 2–6(–12)-flowered, on leafy short shoots of season, arising lateral to extension shoots, also from woody short shoots; bracteoles caducous to semipersistent, membranous to herbaceous. **Flowers** 25–30 mm diam.; sepals triangular, 8–10 mm; stamens 30–45(–47); styles 4. **Pomes** ruddy to deep or bright red, sometimes coral red, suborbicular to obovoid. $2n = 34, 51, 68$.

Flowering late Apr–early May; fruiting Sep–Nov. Hardwood gaps, pine forests, cattle-grazed scrub on blackland soil, prairie margins; 0–200 m; Ala., Ark., Ga., La., Miss., N.C., Tenn.

Records of *Crataegus triflora* from Virginia probably refer to *C.* ×*vailiae*.

Crataegus triflora is usually an open-grown, multi-stemmed, rather lax shrub. A fairly rare, quite distinctive morph from central Mississippi has more or less ovate leaves that sometimes have nearly truncate bases and larger marginal teeth than the typical form. Tennessee plants usually have proportionately wide leaves with unusually large marginal teeth. Anomalous populations from Caldwell Parish, Louisiana, have much smaller leaves and flowers as in *C. ashei* but retain the large stamen number. Vegetative specimens with smallish leaves with lobes absent may resemble *C. collina* but are easily distinguished by their highly glandular leaf margins and petioles.

Crataegus triflora often produces inflorescences on 2–5 cm peduncles of the season that arise laterally from extension shoots, instead of from woody short shoots as is usual among hawthorn species. Such a feature is also found in ×*Crataemespilus canescens* (J. B. Phipps) J. B. Phipps. *Crataegus triflora* flowers are among the most spectacular in the genus, their size reminiscent of a wild rose, and the inflorescences may be much more floriferous then the name suggests. Ploidy level determinations by N. Talent and T. A. Dickinson (2005) suggested that tetraploidy and triploidy are common but that diploidy is rare in *C. triflora*.

64f.19. CRATAEGUS Linnaeus (sect. COCCINEAE) ser. APRICAE J. B. Phipps, J. Bot. Res. Inst. Texas 1: 173. 2007 [E]

Shrubs or trees, 20–70(–120) dm, often with main trunk dominant. **Stems:** trunk bark seldom recorded, black, dark gray, or brownish, ± rimose; <branches spreading rather than pendent; twigs slightly flexuous (± straight in *C. frugiferens*)>, new growth glabrous or pubescent, 1-year old reddish to dark or very dark brown, rarely tan to chestnut, older gray or gray-brown, sometimes dark purple-brown; thorns on twigs usually few, straight to slightly recurved, 1–2 -years old brown, purple-brown, deep chestnut brown to shiny, dark brown, black, or dark gray-brown, slender or ± fine, 1.5–4 cm. **Leaves:** petiole length 15–50% blade, glabrous or pubescent, usually sessile-glandular (sometimes stipitate-glandular in *C. frugiferens*); blade <light to mid green>, broadly ovate or rhombic-elliptic to obtrullate or obovate or ± orbiculate, 1.5–5 cm, thin to coriaceous, base ± cuneate, lobes 0 or 1–3 per side, sinuses shallow (sometimes apiculi) or obscure, more rarely moderately deep, lobe apex obtuse to acute, margins finely glandular-crenate to glandular-serrate, venation craspedodromous, veins 3–5 (5–7 in *C. frugiferens*) per side, apex acute to obtuse, abaxial surface glabrous, rarely pilose young, veins sparsely hairy, adaxial usually hairy young, glabrescent, sometimes glabrous. **Inflorescences** (1 or)2–6(or 7) -flowered, panicles, sometimes highly reduced, larger convex; branches glabrous or ± pubescent; bracteoles caducous (persistent in *C. ignava*), linear to narrowly oblong, <small>, membranous (to herbaceous in *C. ignava*), margins sessile-glandular, sometimes stipitate-glandular. **Flowers** 13–25 mm diam.; hypanthium glabrous or ± pubescent; sepals narrowly triangular, much shorter than petals, margins glandular-serrate; stamens 10 or 20 (12–15), anthers pink to purple, sometimes ivory or cream; styles (2 or)3–5. **Pomes** orange-red to red, sometimes green-mottled, suborbicular to orbicular, sometimes pyriform or broadly ellipsoid, 8–15(–18) mm diam., glabrous or sparsely hairy; flesh ± mealy; sepals <sessile>, usually reflexed, sometimes recurved or spreading, <non-accrescent>; pyrenes (2 or)3–5.

Species 12 (12 in the flora): se United States.

Series *Apricae* is essentially southern Appalachian and adjacent piedmont in distribution, with some extension into the coastal plain in northern Florida and South Carolina.

Series *Apricae* is one of the two series to emerge from the remnants of ser. *Flavae* (Loudon) Rehder and results from the loss of the type species of the latter to ser. *Intricatae*. The species included here belong to groups *Euflavae* (minus *C. flava*), *Ignavae*, *Segnes*, *Sororiae*, and *Visendae* of Beadle. Series *Apricae* differs from the other *Flavae*-descendent series, *Lacrimatae*, in fruiting later (thus not a 'summer haw'), by lacking a conspicuously weeping habit, in the bracteoles generally being stipitate-glandular (compared to sessile-glandular), and in many taxa having anthocyanic anthers (unknown in ser. *Lacrimatae* except in sporadic individuals).

In spite of the generally unambiguous nature of ser. *Apricae*, many species are neither particularly common nor well-marked, which makes the taxonomy difficult. It is possible that many of the taxa in this series arose by hybridization, perhaps with a ser. *Lacrimatae* parent when they would have flexuous twigs. *Crataegus ignava* and *C. frugiferens* are the most distinct species. Interesting, yet rare, forms include some not seen recently such as unnamed members of a rich *Crataegus* flora collected by Curtiss from the Augusta, Georgia, sandhills and *C. calva* Beadle from Alabama. Little is known about breeding systems and chromosome numbers, there being only one triploid count (N. Talent and T. A. Dickinson 2005), that from a cultivated specimen of uncertain identity.

Since *Crataegus flava* (ser. *Intricatae*) is sometimes identified as a member of ser. *Apricae*, it is included herein and keys out in the first couplet.

SELECTED REFERENCE Phipps, J. B. and K. A. Dvorsky. 2007. Review of *Crataegus* series *Apricae*, ser. nov., and *C. flava* (Rosaceae). J. Bot. Res. Inst. Texas 1: 171–202.

1. Twigs straight or slightly flexuous; leaf blades 5–8 cm; stamens 13–16, anthers purple; pomes dull orange, ± pyriform-oblong. . . . . 89. *Crataegus flava* (ser. *Intricatae*), p. 588
1. Twigs ± flexuous (± straight in *C. frugiferens*); leaf blades 1.5–5 cm; stamens 10 or 20 (12–15), anthers pink, purple, or red, sometimes ivory or cream; pomes orange, reddish orange, russet, yellow, or orange-red to red, suborbicular to orbicular, sometimes pyriform or broadly ellipsoid.

[2. Shifted to left margin.—Ed.]

2. Twigs ± straight; leaf blades ovate to obovate or rhombic-ovate, margins serrate; petioles sometimes stipitate-glandular; inflorescence branches appressed-scabrous . . . . 124. *Crataegus frugiferens*
2. Twigs ± flexuous; leaf blades elliptic-ovate or ovate to rhombic-ovate or trullate, obovate to obtrullate or rhombic-obovate, ± elliptic-rhombic to rhombic or oblong to orbiculate, margins finely serrate, crenate-serrate, finely or obscurely crenate-serrate, or ± crenate; petioles sessile-glandular; inflorescence branches glabrous, glabrate, pubescent, ± pilose (hairy), or tomentose.
    3. Leaf blades broadly elliptic to orbiculate or broadly rhombic-elliptic to ± suborbiculate, lobes 1–3 per side, small, neat, and regular, sinuses shallow, lobe apices ± acute.
        4. Flowers 15 mm diam.; inflorescence branches tomentose; anthers ivory, pink-purple, or red . . . . 119. *Crataegus sororia*
        4. Flowers 20–25 mm diam.; inflorescence branches glabrous or ± pilose; anthers cream or purple . . . . 120. *Crataegus mira*
    3. Leaf blades elliptic-ovate or ovate to rhombic-ovate or trullate, obovate to obtrullate or rhombic-obovate, ± elliptic-rhombic to rhombic or broadly elliptic to oblong, lobes 0 or 1–3 per side, sometimes obscure, small to large, sinuses shallow to moderately deep, lobe apices ± obtuse, acute, or acuminate.
        5. Stamens 10 or 12–15(or 20).
            6. Leaf blades 1.5–4 cm, widest near midpoint, rhombic-elliptic or broadly elliptic, lobes 0 or apiculi, margins crenate or serrate; inflorescence branches ± densely pilose; flowers 13–15 mm diam.; anthers ivory to cream. . . . . . . 118. *Crataegus aprica*
            6. Leaf blades (2–)3–5 cm, widest towards base, elliptic-ovate or rhombic-ovate, lobes 2 or 3 per side, short and acute or obscure, margins serrulate; inflorescence branches glabrous or pilose; flowers 16–20 mm diam.; anthers pink or dark purple.
                7. Lobes 3 per side, sinuses short to moderately deep, lobe apices acute to acuminate; inflorescence branches glabrous; stamens 10 . . . . . . 123. *Crataegus alleghaniensis*
                7. Lobes 2 or 3 per side, obscure to short, sinuses ± shallow, lobe apices acute; inflorescence branches pilose; stamens 12–15(or 20) . . . . 125. *Crataegus extraria*
        5. Stamens 20.
            8. Inflorescence branches pubescent; anthers ivory . . . . 126. *Crataegus annosa*
            8. Inflorescence branches glabrous, pubescent, or glabrate to ± pilose or hairy; anthers pink to purple.
                9. Leaf blades ± ovate or ovate-rhombic, 2.5–3.5 cm, lobes 2 or 3 per side, sinuses moderately deep, lobe apices acute; inflorescence branches glabrous; 1-year old twigs tan to shiny chestnut; 1–2-year old thorns black . . . . 122. *Crataegus ignava*
                9. Leaf blades ± rhombic or rhombic-elliptic to broadly elliptic, oblong, ± ovate, or ± obovate, 1.5–5 cm, lobes 0, or 1 or 2 per side, sometimes with only a series of 1–3 subterminal apiculi per side, sinuses shallow to moderately deep, lobe apices acute, subacute, or ± obtuse; inflorescence branches glabrate, pubescent to moderately pilose, hairy, or pubescent; 1-year old twigs dark brown; 1–2-year old thorns purple-brown to very dark brown or dark gray-brown.
                    10. Leaf blades rhombic, lobes 1(or 2) per side, well defined, lobe apices acute to ± obtuse; sepal margins subentire to glandular on barely discernible teeth; pomes slightly pyriform . . . . 115. *Crataegus egregia*
                    10. Leaf blades broadly elliptic or oblong to elliptic-rhombic, rhombic or narrowly to broadly obovate, trullate to narrowly ovate or rhombic-ovate, lobes 0, or 1 or 2 per side, obscure, sometimes with only a series of 1–3 subterminal apiculi per side, lobe apices ± acute to obtuse; sepal margins glandular-serrate; pomes usually suborbicular, sometimes pyriform.

[11. Shifted to left margin.—Ed.]

11. Leaf blades broadly elliptic or oblong to obovate, apices subacute to acute, often ± cuspidate, lobes 0, sometimes with only a series of 1–3 subterminal apiculi per side . . . . .117. *Crataegus segnis*
11. Leaf blades ± rhombic, trullate to ± obovate, or narrowly ovate to rhombic-ovate or broadly ovate, sometimes oblong, apices acute to subobtuse, lobes 0, or 1 or 2 per side.
    12. Leaf blades rhombic-elliptic to rhombic-obovate or trullate, 1.5–3 cm . . . . 116. *Crataegus visenda*
    12. Leaf blades narrowly ovate to rhombic-ovate or broadly ovate, sometimes oblong or ± obovate, 2.5–5 cm . . . . 121. *Crataegus leonensis*

115. Crataegus egregia Beadle, Biltmore Bot. Stud. 1: 82. 1902 [E]

Trees, 40–60 dm. **Stems:** <trunk bark dark, very rough>; twigs ± flexuous, new growth pubescent, 1-year old brown, older grayish; thorns on twigs <absent or several>, straight, 1–2-years old dark, slender, 3–4 cm. **Leaves:** petiole <slender>, length 30–50% blade, pubescent, very sessile-glandular; blade rhombic, 1.5–2.5 cm at anthesis, 2–3 cm mature, thin, base cuneate, lobes 1(or 2) per side, sinuses moderately deep, well defined, lobe apex acute to ± obtuse, margins shallowly crenate, teeth gland-tipped, veins 3 per side, apex acute to subacute, abaxial surface glabrate, adaxial sparsely hairy young, glabrescent. **Inflorescences** 3-flowered; branches hairy; bracteoles caducous, <greenish>, narrowly oblong, margins glandular. **Flowers** 15 mm diam.; hypanthium sparsely pilose; sepals narrowly triangular from wide bases, margins subentire to glandular on barely discernible teeth, <abaxially glabrous>; stamens 20, anthers bright purple; styles 3 or 4. **Pomes** red, slightly pyriform, 9–12 mm diam., glabrous; sepals reflexed; pyrenes 3 or 4.

Flowering Apr; fruiting Sep–Oct. Open brush, open woodlands; 0–100 m; Ala., Fla., Ga., S.C.

Crataegus egregia was described from Bristol, Florida, and is understood primarily from its type. It is in some ways like *C. egens* (ser. *Lacrimatae*) but with larger, more rhombic leaf blades, purple anthers, and where present, longer thorns. South Carolina specimens are thorny; the type material is thornless. Material represented by about six specimens, with similar leaves but more or less tomentose pedicels and much smaller pomes, is also known.

116. Crataegus visenda Beadle, Biltmore Bot. Stud. 1: 79. 1902 [E]

Crataegus arrogans Beadle; *C. sororia* Beadle var. *visenda* (Beadle) Lance; *C. tristis* Beadle

Shrubs or trees, 100 dm. **Stems:** <trunk bark dark gray or brownish, rough>; twigs ± flexuous, new growth pubescent, 1-year old dark brown; thorns on twigs <few to numerous>, straight, 1-year old very dark, slender, 1.5–3 cm. **Leaves:** petiole <slender>, length 25–40% blade, <winged distally>, pilose, sessile-glandular; blade rhombic-elliptic to rhombic-obovate or trullate, 1.5–3 cm, thin, base ± broadly cuneate, lobes 0, or 1 or 2 per side, sinuses shallow, lobe apex subacute, margins obscurely crenate <teeth sharper young>, veins 3 or 4 (or 5) per side, apex acute to subobtuse, abaxial surface sparsely pilose young, adaxial conspicuously pilose on veins, otherwise glabrate. **Inflorescences** 2–4-flowered; branches glabrate to pilose; bracteoles caducous, linear, margins glandular. **Flowers** 14–18 mm diam.; hypanthium glabrous or pilose; sepals triangular, margins irregularly glandular-serrate; stamens 20, anthers pale to bright purple; styles (2 or)3–5. **Pomes** orange or orange flushed red to red, suborbicular to pyriform, 10–12 mm diam., glabrous; sepals recurved; pyrenes (2 or)3–5.

Flowering Apr; fruiting Sep–Oct. Dry woods, gravelly ridges, sandy soil; 20–100 m; Ala., Fla., Ga.

C. D. Beadle (in J. K. Small 1903) recognized ten species in his group *Visendae*, from which *Crataegus furtiva* and *C. egens* have been moved to ser. *Lacrimatae* in this treatment.

117. Crataegus segnis Beadle, Biltmore Bot. Stud. 1: 32. 1901 [E]

Crataegus consanguinea Beadle; *C. sororia* Beadle var. *segnis* (Beadle) Lance

Shrubs or trees, 40–70 dm. **Stems:** twigs ± flexuous, new growth pubescent, 1-year old dark brown (partly covered with abraded cutin), older pale to mid gray; thorns on twigs <absent or numerous>, straight, 1-year old purple-brown, older gray, slender, 1.5–2.5 cm. **Leaves:** petiole <slender>, length 30–45% blade, pubescent, sessile-glandular; blade broadly elliptic or oblong to obovate, 2–3.5 cm, thin, base ± abruptly narrowed, lobes 0, sometimes with only a series of 1–3 subterminal apiculi per side, sinuses lacking or extremely shallow, margins crenate-serrate, teeth gland-tipped, veins 3–5 per side (except smaller leaves), apex subacute to acute, <often ± cuspidate>, abaxial surface glabrate young, adaxial sparsely hairy young, especially on midvein, glabrescent. **Inflorescences** 1–5-flowered; branches finely pubescent to pilose; bracteoles caducous, narrowly oblong, margins glandular. **Flowers** 15–20(–25) mm diam.; hypanthium sparsely pilose; sepals triangular, margins irregularly, strongly glandular-serrate, <abaxially glabrous>; stamens 20, anthers pale purple; styles 3–5. **Pomes** orange flushed red to red, suborbicular, 10–15 mm diam., glabrous; sepals recurved; pyrenes 3–5.

Flowering Apr; fruiting Sep–Oct. Brush; 20–150 m; Ala., Fla., Ga., N.C., S.C.

Crataegus segnis is treated broadly to include all members of the *C. visenda* group with leaves that lack lobes or are terminally apiculate. The species has larger leaves than *C. visenda*. Suborbiculate leaves are reported on extension shoots.

118. Crataegus aprica Beadle, Bot. Gaz. 30: 335. 1900 • Sunny hawthorn [E]

Shrubs, 20–30 dm. **Stems:** twigs ± flexuous, new growth <olive green>, sparsely pubescent, 1-year old reddish brown, <pubescent>, older dark gray-brown, <glabrous>; thorns on twigs straight or recurved, 2-years old dark gray-brown, slender, 3–4 cm. **Leaves:** petiole length 20–25% blade, winged distally, pubescent, sessile-glandular; blade rhombic-elliptic or broadly elliptic, 1.5–4 cm, <widest near midpoint>, ± coriaceous, base sharply constricted, lobes 0 or apiculi, sinuses very shallow, lobe apex acute, margins crenate or serrate, teeth obtuse gland-tipped, veins 3 or 4 per side, apex obtuse, abaxial surface glabrous, adaxial pilose young, glabrescent. **Inflorescences** 3–6-flowered; branches ± densely pilose; bracteoles caducous, linear, margins glandular. **Flowers** 13–15 mm diam.; hypanthium pilose (at least proximally); sepals narrowly triangular, <4 mm>, margins glandular-serrate, <midvein sometimes prominent, abaxially sparsely pubescent>; stamens 10, anthers ivory to cream; styles 3–5. **Pomes** red or reddish orange, orbicular, 9–15 mm diam., sparsely hairy; sepals patent-reflexed; pyrenes 3–5.

Flowering Apr–May; fruiting Sep–Oct. Open brush; 20–800 m; Ala., Ga., N.C., S.C., Tenn., Va.

Crataegus aprica is found at the southern end of the Appalachians and upper piedmont, where it may be quite common. Specimens from the Tuskegee National Forest, Macon County, Alabama, have narrower (ca. 2:1), more pointed leaves with four or five veins per side.

119. Crataegus sororia Beadle, Bot. Gaz. 30: 336. 1900 [E]

Shrubs, 20–30 dm. **Stems:** twigs ± flexuous, new growth <olive green>, pubescent, 1-year old reddish brown, <pubescent>, older dark gray-brown, <glabrous>; thorns on twigs straight or recurved, 2-years old dark gray-brown, slender, 3–4 cm. **Leaves:** petiole length 20–25% blade, <winged distally>, pubescent, sessile-glandular; blade broadly elliptic to orbiculate, 1.5–4 cm, ± coriaceous, base attenuate, lobes 1 or 2 per side, <small, neat, and regular>, sinuses shallow, lobe apex acute, margins serrulate, teeth gland-tipped, veins 3 or 4 per side, apex obtuse, abaxial surface glabrous, adaxial pilose young, glabrescent. **Inflorescences** 3–6-flowered; branches tomentose; bracteoles caducous, linear, margins glandular. **Flowers** 15 mm diam.; hypanthium densely pilose; sepals narrowly triangular, <4 mm>, margins glandular-serrate, <midvein sometimes prominent, abaxially sparsely pubescent>; stamens 20, anthers ivory, pink-purple, or red; styles 4 or 5. **Pomes** reddish orange, suborbicular, 12–18 mm diam., sparsely hairy; sepals patent-reflexed; pyrenes 4 or 5.

Flowering Apr; fruiting Sep–Oct. Open scrub; 50–200 m; Ala., Ga., S.C.

Crataegus sororia occurs around Rome, Georgia, where it is common, in adjacent areas of Alabama, and in Aiken County, South Carolina.

Most similar to *Crataegus aprica*, *C. sororia* differs in leaf shape (proportionately wider), stamen number, and anther color. Beadle said that *C. sororia* is found south to Florida; no specimen from there was seen. The more or less tomentose inflorescence is a good recognition character.

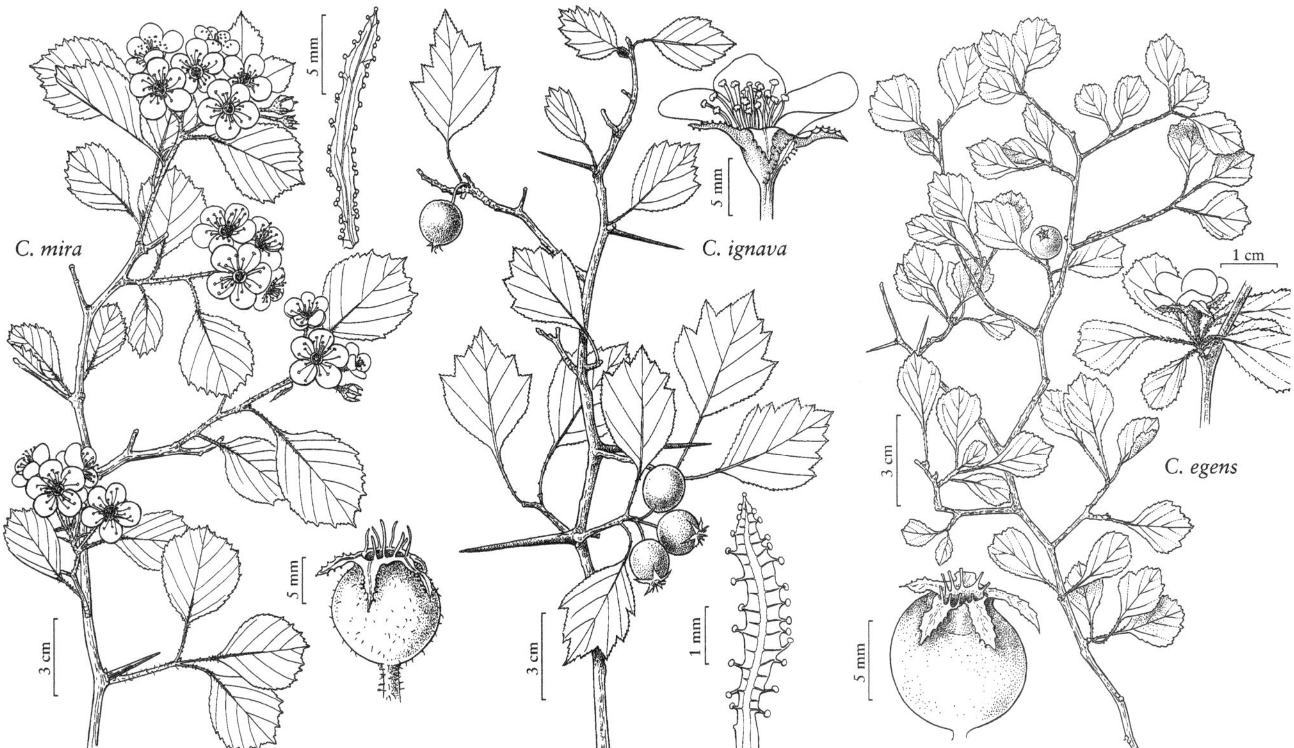

CRATAEGUS

120. Crataegus mira Beadle, Biltmore Bot. Stud. 1: 78. 1902 [E] [F]

Crataegus alleghaniensis Beadle var. *mira* (Beadle) Lance

Shrubs, 20–30 dm. **Stems:** twigs ± flexuous, new growth <olive green>, pubescent, 1-year old reddish brown, <pubescent>, older dark gray-brown, <glabrous>; thorns on twigs straight or recurved, 2-years old dark gray-brown, slender, 3–4 cm. **Leaves:** petiole length 15–25% blade, winged distally, pubescent, sessile-glandular; blade broadly rhombic-elliptic to ± suborbiculate (at least some, usually), 2–3.5 cm, ± coriaceous, base attenuate, lobes 1–3 per side, more prominent young (anthesis), <small, neat, and regular>, sinuses shallow, lobe apex ± acute, margins crenate-serrate to serrate, teeth gland-tipped (prominent young), veins 3 or 4 per side, apex acute, abaxial surface glabrous, adaxial pilose young, glabrescent. **Inflorescences** 3–6 -flowered; branches glabrous or ± pilose; bracteoles caducous, linear, margins usually stipitate-glandular. **Flowers** 20–25 mm diam.; hypanthium sparsely pilose or glabrous; sepals narrowly triangular, <5 mm>, margins glandular-serrate, <midvein sometimes prominent, abaxially sparsely pubescent>; stamens 10 or 20, anthers cream or purple; styles 4 or 5. **Pomes** red or reddish orange, suborbicular, 9–15 mm diam., sparsely hairy; sepals patent-reflexed; pyrenes 4 or 5.

Flowering Apr; fruiting Sep–Oct. Open brush; 20–200 m; Ala., Fla., Ga., Miss., N.C., S.C.

The main range of *Crataegus mira* is the Florida panhandle to central Georgia, but it extends to the Carolinas and Alabama; the species is locally common.

Crataegus mira is most similar to *C. aprica*, *C. sororia*, and especially *C. leonensis*, but differs from them in its wider and often larger leaves, thinner inflorescence indumentum (than *C. sororia*), and larger flower size. The illustration depicts an unusually formed sessile-glandular bracteole.

121. Crataegus leonensis E. J. Palmer, J. Arnold Arbor. 13: 422, fig. 1. 1932 [E]

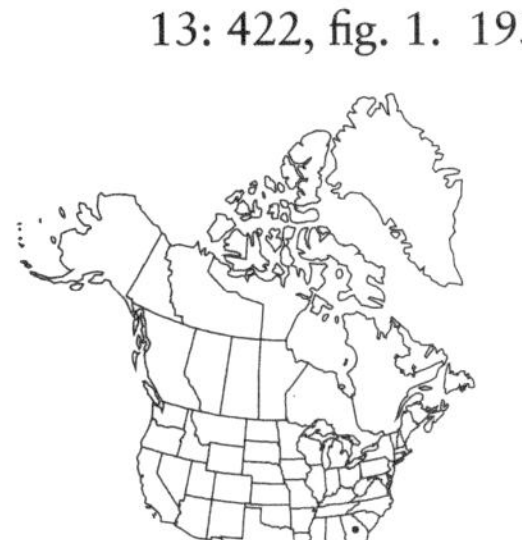

Trees, 100–120 dm. **Stems:** <trunk bark dark gray to nearly black, thick, ridged; branches broadly spreading, intricate>; twigs ± flexuous, new growth <olive green>, glabrous, 1-year old very dark brown, older gray; thorns on twigs <few>, straight, 1-year old very dark, older graying, fine, 3.5 cm. **Leaves:** petiole length 40–50% blade, pubescent in sulcus, densely sessile-glandular; blade narrowly ovate to rhombic-ovate or broadly ovate, sometimes oblong or ± obovate, 2.5–5 cm, ± thin, base cuneate, lobes 0, or 1 or 2 per side, obscure, sinuses shallow, lobe apex obtuse or acute, margins finely to obscurely crenate-serrate, teeth gland-tipped, veins 3 or

4 per side, apex acute, abaxial surface glabrous, adaxial glabrate, main veins sparsely pilose. **Inflorescences** 3–7 -flowered; branches glabrate to moderately pilose; bracteoles caducous, narrowly oblong, margins glandular. **Flowers** 16–20 mm diam.; hypanthium glabrous; sepals narrowly triangular, margins glandular-serrate; stamens 20, anthers pink to pale purple; styles 2–5. **Pomes** orange-red to russet, often green-mottled, suborbicular, 9–12 mm diam., glabrous; sepals <prominent, ± elevated>, spreading; pyrenes (2 or)3 or 4(or 5).

Flowering Apr; fruiting Sep–Oct. Sandy upland woods; 20–50 m; Fla., Ga.

Most specimens of *Crataegus leonensis* are from Leon County, Florida, but a few come from southern Georgia.

Crataegus leonensis is most similar to *C. mira*, of which it may prove to be an extreme form; it has smaller flowers and differently shaped, narrower leaves. An unnamed entity from northern Florida and southeastern Alabama to North Carolina, which could key to *C. leonensis*, is held to be closer to *C. annosa*. It differs from *C. leonensis* by larger leaves (3–5 cm) and fruit (12–15 mm), somewhat different leaf shape (broadly elliptic to rhombic, sharply lobed), always dense inflorescence indumentum, and cream anthers. *Crataegus subflavida* Murrill may be synonymous but has larger yellow pomes.

122. Crataegus ignava Beadle, Biltmore Bot. Stud. 1: 31. 1901 [E] [F]

Crataegus alleghaniensis Beadle var. *ignava* (Beadle) Lance

Shrubs, 20–40(–50) dm. **Stems:** twigs ± flexuous, new growth <reddish green>, glabrous, 1-year old tan to shiny chestnut, older ± dark gray, often reddish tinged; thorns on twigs <± numerous>, straight, 1–2-years old black, ± fine, 2–3 cm (3–5 cm at 2-years old). **Leaves:** petiole length 30–40% blade, glabrous, sessile-glandular, <glands black>; blade ± ovate or ovate-rhombic, 2.5–3.5 cm, coriaceous, base broadly cuneate or ± rounded, lobes 2 or 3 per side, sinuses moderately deep, lobe apex acute, margins obscurely crenate-serrate, teeth gland-dotted, veins 4 or 5 per side, apex acute, abaxial surface glabrous, adaxial sparsely pilose young, soon glabrescent. **Inflorescences** 2–5-flowered; branches glabrous; bracteoles mainly persistent through petal fall, <numerous>, linear, <herbaceous to membranous>, margins usually stipitate-glandular. **Flowers** 15 mm diam.; hypanthium glabrous; sepals narrowly triangular, <6 mm>, margins glandular-serrate, <abaxially glabrous>; stamens 20, anthers pale pink, sometimes pale purple; styles 3–5. **Pomes** orange-red, suborbicular, 10–17 mm diam., glabrous; sepals spreading-recurved; pyrenes 3–5.

Flowering Apr; fruiting Sep–Oct. Brush, rocky hills; 50–200 m; Ala., Ga.

Crataegus ignava is locally common in east-central Alabama and north-central and central Georgia; it is one of the more distinct species in the series with its often rhombic-ovate, quite deeply lobed leaves and large orange-red fruit.

123. Crataegus alleghaniensis Beadle, Bot. Gaz. 30: 337. 1900 [E]

Shrubs, 20–40(–50) dm. **Stems:** twigs ± flexuous, new growth <reddish>, glabrous, 1-year old reddish gray, older gray; thorns on twigs straight or slightly recurved, 2-years old deep chestnut brown to blackish, fine, 1.5–4 cm. **Leaves:** petiole length 25–30% blade, glabrous, densely sessile-glandular; blade elliptic-ovate to rhombic-ovate, (2–)3–5 cm, <widest towards base>, thin, base broadly cuneate, lobes 3 per side, sinuses short to moderately deep, lobe apex acute to acuminate, margins serrulate, teeth with small glands, veins 5 per side, apex acute, abaxial surface glabrous, adaxial pilose young, soon glabrescent. **Inflorescences** 2–4-flowered; branches glabrous; bracteoles caducous, linear, margins glandular. **Flowers** 20 mm diam.; hypanthium glabrous; sepals narrowly triangular, <5 mm>, margins glandular-serrate, <abaxially glabrous>; stamens 10, anthers pink; styles 4. **Pomes** red, orbicular-pyriform, 8–12 mm diam., glabrous; sepals spreading; pyrenes 2–5.

Flowering Apr; fruiting Sep–Oct. Rocky hills, brush; 50–200 m; Ala., Ga., Tenn.

Crataegus alleghaniensis is known from a few locations in northeastern Alabama, Georgia, and Tennessee, and is apparently scarce.

124. Crataegus frugiferens Beadle, Biltmore Bot. Stud. 1: 30. 1901 [E]

Shrubs, 30–60 dm. **Stems:** twigs ± straight, new growth <reddish young>, ± appressed-pubescent, 1-year old ± shiny tan, older gray; thorns on twigs straight to slightly recurved, 2-years old shiny, dark brown or blackish, older gray, ± fine, 2–5 cm. **Leaves:** petiole <slender>, length 25–40% blade, sparsely pubescent (more in adaxial groove), sometimes stipitate-glandular; blade ovate to obovate or rhombic-ovate, 2.5–4(–5) cm, thin, base cuneate to broadly cuneate, lobes 0 or 1–3 per side, sometimes little more than apiculi, sinuses shallow, margins serrate, teeth gland-tipped basally, glands sometimes stipitate, veins 5 or 6(or 7) per side, apex acute, abaxial surface glabrous, adaxial appressed-scabrous young, glabrescent. **Inflorescences** 3–7 -flowered; branches appressed-scabrous; bracteoles caducous, <few>, linear, margins glandular. **Flowers** 15–16 mm diam.; hypanthium appressed-scabrous, at least proximally; sepals narrowly triangular, <4–5 mm>, margins glandular-serrate; stamens 10 or 20, anthers cream or pink; styles 3 or 4. **Pomes** red, suborbicular or slightly pyriform, 9–12 mm diam., indumentum not recorded; sepals reflexed; pyrenes 3–5.

Flowering Apr; fruiting Sep–Oct. Open woodlands, woodland edges, brush; 50–200 m; Ala., Ga., Miss.

The main range of *Crataegus frugiferens* is north-central Alabama and adjacent Georgia; in the center of its range, the species is locally common.

Crataegus frugiferens is one of the more distinct species of ser. *Apricae*, even to being superficially similar to *C. collina*, from which it differs in its glandular, often stipitately so, petioles and leaf bases as well as leaf shape, which is proportionately wider, and with much more broadly cuneate bases. The species also has more strongly glandular-serrate sepals and, usually, deeper red, softer pomes than most in the series. The leaves at flowering are no larger than the flowers, as is common in *C. collina*. Its thin leaves make it less xeromorphic than other members of the series. Specimens of it have been annotated as *C. rigens* Beadle. It is possible that its true affinity lies with *C. aemula* (ser. *Populneae*), from which it differs by smaller (2.5–4 cm) leaves of a proportionately wider (± ovate) shape with rounded to broadly cuneate bases, finer marginal teeth, and zero to three obscure lobes per side.

125. Crataegus extraria Beadle, Biltmore Bot. Stud. 1: 73. 1902 [E]

Crataegus alleghaniensis Beadle var. *extraria* (Beadle) Lance

Shrubs, 25 dm. **Stems:** twigs ± flexuous, new growth not recorded, 1-year old reddish brown, older dark purple-brown; thorns on twigs <numerous>, straight, 2-years old blackish, fine, 2–3 cm. **Leaves:** petiole length 33–40% blade, sparsely pubescent, sessile-glandular; blade rhombic-ovate, 3–4 cm, <widest towards base>, thin, base cuneate, lobes 2 or 3 per side, obscure to short, sinuses ± shallow, lobe apex acute, margins serrulate, teeth gland-tipped, veins 4 or 5 per side, apex acuminate, abaxial surface sparsely pubescent, adaxial glabrous. **Inflorescences** 3–5-flowered; branches pilose; bracteoles not seen. **Flowers** 16–20 mm diam.; hypanthium pilose; sepals narrowly triangular, margins glandular-serrate; stamens 12–15(or 20), anthers dark purple; styles 2 or 3(or 4). **Pomes** red (yellow-orange immature), suborbicular, 9–12 mm diam., glabrous; sepals <on collar>, spreading; pyrenes 4.

Flowering Apr–May; fruiting Sep–Oct. Brush; 50–500 m; Ala., Ga., N.C., Va.

Crataegus extraria has always been very rare, occurring in scattered localities in North Carolina and Virginia.

As treated here, there is considerable extension to forms beyond the typical that may include hybrids.

126. Crataegus annosa Beadle, Biltmore Bot. Stud. 1: 83. 1902 [E]

Trees, 80 dm. **Stems:** <trunk bark dark, rimose>; twigs not recorded; thorns on twigs ± straight, 1-year old chestnut brown or gray, slender, 2–3 cm. **Leaves:** petiole <slender>, length 25–50% blade, pubescent (at least young), sessile-glandular; blade rhombic-elliptic to obovate or obtrullate, 2–4 cm, texture and base not recorded, lobes small across apical part, sinuses, lobe apex, and margins not recorded, veins 2 or 3 per side <extension shoots with 3 or 4 pairs>, apex not recorded, surfaces sparsely pubescent. **Inflorescences** 3–5-flowered; branches densely pubescent; bracteoles not recorded. **Flowers** 15–20 mm diam.; hypanthium pubescent; sepals <3–4.5 mm>, margins glandular-serrate; stamens 20, anthers ivory; styles 3–5. **Pomes** reddish orange to yellow, washed red, suborbicular to broadly ellipsoid, 10–12 mm diam., pubescence not recorded; sepals reflexed; pyrenes 3–5.

Flowering Apr; fruiting Sep. Brush; 50 m; Ala.

Crataegus annosa is typical for ser. *Apricae* in foliage and fruit, but in anther number and color, and in the narrower short-shoot leaves, it is typical of ser. *Lacrimatae*. It is known only from syntype specimens and is only doubtfully placed in ser. *Apricae*.

64f.20. Crataegus Linnaeus (sect. Coccineae) ser. Lacrimatae J. B. Phipps, Taxon 37: 113. 1988 (as Larimatae) [E]

Shrubs or trees, 5–70(–100) dm, usually main trunk dominant. **Stems:** trunk bark usually dark brown to dark gray or black, sometimes ashy gray, highly corrugated; <distal branches usually pendulous (particularly in the slender-twigged species), rarely ± spreading (in dwarf taxa and *C. exilis*), ± strongly flexuous (slightly so in *C. teres* and *C. exilis*)>; twigs: new growth usually pubescent, sometimes glabrate or glabrous, 1-year old purple-brown, older gray or dark gray, sometimes gray-brown; thorns on twigs absent or sparse to numerous, usually ± straight, sometimes ± recurved, 1-year old very dark, usually slender, 1–5 cm. **Leaves:** petiole length (0–)10–40(–60)% blade, pubescent (young), sessile-glandular (young); blade <light to mid green abaxially>, usually elliptic to oblong or narrowly obovate, sometimes obtrullate, ovate, or suborbiculate, (0.6–)1–4(–5) cm, thin to coriaceous, <firm to floppy>, base narrowly cuneate to rounded, lobes 0 or 1–3 per side, sinuses usually shallow, lobe apex acute to obtuse, margins entire or serrate, <very glandular>, venation craspedodromous (semicamptodromous in *C. attrita*), veins 1–6 per side (exiting mainly or wholly in apical part of blade), apex obtuse to acute, surfaces ± woolly-tomentose (young). **Inflorescences** 1–4(–7)-flowered, usually highly reduced panicles; branches usually appressed-white-pubescent to canescent or tomentose, sometimes pilose or glabrous; bracteoles usually deciduous, rarely ± caducous (in *C. lacrimata*), sometimes semipersistent, linear or oblong-linear, <small>, membranous, margins usually glandular, rarely eglandular (in *C. lacrimata*), adaxially sometimes short-pubescent. **Flowers** 10–20(–25) mm diam.; hypanthium usually densely hairy, sometimes glabrous; sepals ± narrowly triangular, much shorter than petals, margins entire or glandular-serrate; <petals elliptic or ± circular in *C. crocea*, *C. quaesita*>; stamens 20, anthers white to cream; styles (2 or)3–5. **Pomes** yellow to orange and copper to red (1 face often colored most brightly), usually suborbicular, sometimes pyriform or ellipsoid, (5–)6–12(–15) mm diam., glabrous or sparsely hairy; flesh texture not recorded; sepals <± sessile>, usually reflexed, <sometimes circumscissile, non-accrescent>; pyrenes (2 or)3–5.

Species 24 (24 in the flora): se United States.

Series *Lacrimatae* constitutes a characteristic and often abundant element of the sand plains, scrubby areas, and pinelands of the southeastern United States coastal plain and piedmont from Louisiana east of the Mississippi to central Florida and on both sides of the Appalachians to northern Alabama and southeastern Virginia.

Crataegus lacrimata, with conspicuously weeping branches, supplies the basis for the serial name. The widely used vernacular name, summer haw, is particularly apt as in most species the fruit ripens in August. Distinctions from the segregate ser. *Apricae* are given with that series. Nearly all the 24 species derive from Beadle who recognized 54 species. Those recognized here range from narrowly defined to variable entities lacking clear internal discontinuities. Regional authors after Beadle usually favored a few-species taxonomy, which tended to confuse ser. *Lacrimatae* with ser. *Apricae*. The present treatment reflects the substantial variation in stature, habit, leaf size, shape and venosity, indumentum, inflorescence size, and fruit size, color and time of ripening first noticed by Beadle. Series *Lacrimatae* contains dwarf shrubs as well as larger shrubs and trees, which mostly arrange straightforwardly into slender, very weeping types (subser. *Tenues* J. B. Phipps) or more robust, less weeping ones (subser. *Robustae* J. B. Phipps) (see J. B. Phipps and K. A. Dvorsky 2008); the bark is highly corrugated and presumably fire-resistant, unlike that of most other *Crataegus* series.

The taxonomy of ser. *Lacrimatae* and the often subsumed ser. *Apricae* has been confused by the use of the name *Crataegus michauxii* Persoon (1806), first by C. D. Beadle, followed by several regional floras, as the descriptions of the floristic authors were largely imaginary. The type of *C. michauxii* is a vegetative fragment at Paris (P), which cannot confidently be aligned with a particular species, or even series, without molecular study, though D. B. Ward (2009) ambitiously tried to equate it to *C. condigna*.

SELECTED REFERENCE Phipps, J. B. and K. A. Dvorsky. 2008. A taxonomic revision of *Crataegus* series *Lacrimatae* (Rosaceae). J. Bot. Res. Inst. Texas 2: 1101–1162.

1. Shrubs, dense, twiggy, 5–20(–25) dm, branches not weeping except in larger specimens; leaf blades: lobes 0 or very obscure.
    2. Leaves deciduous, thin to subchartaceous, blades narrowly obovate to cuneate or oblanceolate; Mississippi to s Virginia . . . . 130. *Crataegus munda*
    2. Leaves deciduous or semipersistent, subcoriaceous or coriaceous, blades broadly obovate to ± isodiametric or obovate-cuneate; c Florida to South Carolina.
        3. Leaf blades broadly obovate to ± isodiametric; petiole lengths 15–30% blade; thorns on twigs 0.6–2 cm . . . . 127. *Crataegus lepida*
        3. Leaf blades obovate-cuneate; petiole lengths 5–15% blade; thorns on twigs 3–5 cm . . . . 135. *Crataegus invicta*
1. Shrubs or trees, (10–)20–70(–100) dm, branches usually ± weeping; leaf blades: lobes 0 or conspicuous.
    4. Branches strongly weeping; inflorescence branches and hypanthia glabrous; bracteole margins eglandular or nearly so . . . . 129. *Crataegus lacrimata*
    4. Branches ± to strongly weeping; inflorescence branches and hypanthia pubescent to tomentose; bracteole margins glandular.
        5. Mature branches strongly weeping (moving in slight wind); habit usually rather slender; twigs slender (1.5 mm or less thick, at 5 cm from tip); thorns on twigs fine, 1–4 cm; leaf blades 1–4 cm, drooping or fluttering in many species; pomes 6–8(–12) mm diam., usually yellow with crimson cheek (subser. *Tenues*).
            6. Leaf blades: lobes 0 or obscure.
                7. Leaf blades usually spatulate or narrowly oblong to narrowly oblanceolate or narrowly elliptic, margins entire, sometimes obscurely crenate-serrate distally . . . . 131. *Crataegus crocea*
                7. Leaf blades narrowly obovate or ± spatulate to obtrullate, margins shallowly serrate to crenate-serrate except near base . . . . 132. *Crataegus condigna* (in part)
            6. Leaf blades: lobes usually 1 or 2 per side, usually quite evident.
                8. Leaf blades 1–2(–3) cm, short-obtrullate to broadly obovate (length/width mainly less than 1.4:1) . . . . 128. *Crataegus egens*
                8. Leaf blades 1.5–4 cm, narrowly to broadly cuneate to obtrullate or spatulate (length/width = 1.4–1.8:1).
                    9. Leaf blades 1.5–3 cm, narrowly to broadly cuneate to obtrullate, ± stiff to ± floppy mature; petiole lengths 15–20% blade . . . . 133. *Crataegus quaesita*
                    9. Leaf blades (1.5–)3–4 cm (usually some shorter on same specimen), spatulate to obtrullate, readily fluttering in wind; petiole lengths 25–50% blade . . . . 134. *Crataegus floridana*
        5. Mature branches ± weeping; habit more robust and stocky; twigs usually relatively stout (ca. 1.5–3 mm diam., at 5 cm from tip); thorns on twigs absent or slender to stout, sometimes fine, (1–)2–5(–7) cm; leaf blades (1.5–)2.5–4.5 cm, rather thick and stiff or coriaceous in most species, others thin or subcoriaceous, mostly not drooping or fluttering (except in *C. lassa* and *C. lanata*); pomes (5–)8–15 mm diam., orange-red to red, often copper-colored mature (subser. *Robustae*, except *C. condigna*, *C. exilis*).

[10. Shifted to left margin.—Ed.]

10. Shrubs, 15–30 dm; thorns on twigs usually numerous, 2–5(–7) cm; leaf blades 1.5–3 cm.
    11. Leaf blades: lobes 0, margins basally entire or subentire, distally ± denticulate; thorns on twigs 2–4.5 cm . . . . . 138. *Crataegus colonica*
    11. Leaf blades: lobes usually obscure, sinuses shallow, sometimes lobes 0, margins strongly crenate-serrate most of length; thorns on twigs 3–5(–7) cm. . . . . 139. *Crataegus pexa*
10. Shrubs or trees, (10–)20–70(–100) dm; thorns on twigs usually absent or few, sometimes moderately frequent, 1–4(–5) cm; leaf blades 1.5–5 cm.
    12. Leaf blades: usually distinctly lobed, only shallowly, margins often toothed.
        13. Leaf blades 2–5 cm mature; most leaves ± narrow, length/width = 1.75+, ± obovate or narrowly to broadly obtrullate, sometimes oblanceolate or oblong-spatulate to cuneate or narrowly obdeltate.
            14. Leaf blades 3–5 cm mature, floppy, older remaining pubescent; inflorescences 1–4-flowered; flowers 17–20 mm diam. . . . . . 144. *Crataegus lanata*
            14. Leaf blades 2–3 cm mature, ± floppy, older becoming glabrous; inflorescences 3–7-flowered; flowers (10–)14–16(–20) mm diam.
                15. Leaf blades usually narrowly obovate to obtrullate, sometimes oblanceolate, lobes 0, or 1 or 2(or 3) per side distally, lobe apices subacute, margins weakly crenate-serrate, veins 1–3 per side . . . . . 147. *Crataegus lancei*
                15. Leaf blades oblong-spatulate to cuneate or narrowly obdeltate, lobes 1 or 2 per side distally, lobe apices acute, margins strongly crenate-serrate almost to bases, veins (2 or)3 or 4(or 5) per side . . . . . 148. *Crataegus senta*
        13. Leaf blades 1.5–4 cm mature; ⅔+ of leaves relatively broad, length/width = 1.4–1.6, broadly obovate to obdeltate or obtrullate, obovate-spatulate, or rhombic-elliptic.
            16. Leaf blades often persistently lanate, lobes 1 or 2 per side, obscure, lobe apices obtuse . . . . . 145. *Crataegus furtiva* (in part)
            16. Leaf blades not persistently lanate, lobes 1–3 per side, prominent, lobe apices obtuse to acute.
                17. Upper twigs suberect, others ± spreading; inflorescences 3–7-flowered, branches pilose-pubescent; leaves glabrescent; pomes 5–7 mm diam.. . . 150. *Crataegus exilis*
                17. Most twigs ± pendent; inflorescences 2–4-flowered, branches appressed-canescent or densely appressed-pubescent; leaves remaining pubescent or glabrescent; pomes 8–12 mm diam.
                    18. Leaf blades narrowly to broadly cuneate to obtrullate, lobes 1–3 per side, margins crenate-serrate; petiole lengths 30–50% blade. . . 146. *Crataegus meridiana*
                    18. Leaf blades ± obovate-spatulate, lobes 2 per side, margins glandular-serrate; petiole lengths 20–35% blade . . . . . 149. *Crataegus dispar*
    12. Leaf blades: lobes 0, sometimes present, few, subterminal, obscure, very short-triangular or margins merely sinuous, toothed or not.
        19. Leaf margins glandular-crenate or glandular-denticulate to subentire.
            20. Leaf blades 2–4 cm (length/width = 1.6–2.1), obovate-cuneate to narrowly obovate, sides evenly tapered to base, apices often cuspidate; petiole lengths 25–33% blade . . . . . 143. *Crataegus lassa*
            20. Leaf blades 1.5–4 cm (length/width = 1–1.7), ± broadly obtrullate to obdeltate or ± obovate, sometimes suborbiculate, alternatively sometimes obscurely sinuously lobed around broadest point, sides variably tapered, apices acute to ± flat or truncate-obtuse, sometimes cuspidate; petiole lengths 10–25% blade.
                21. Leaf blades obovate-cuneate to obovate or suborbiculate, apices ± acute to weakly cuspidate, surfaces pubescent young; lobes 0 or obscure . . . . . 137. *Crataegus integra*
                21. Leaf blades broadly obtrullate or obdeltate to obtrullate, apices acute or rounded to subacute, sometimes cuspidate, surfaces densely tomentose young; lobes 1–3 per side, often evident, occasionally obscure. . . . . 145. *Crataegus furtiva* (in part)
        19. Leaf margins crenate, serrate, or glandular-serrate.

[22. Shifted to left margin.—Ed.]

22. Hypanthia and inflorescence branches glabrous, pilose, or glabrate.
 23. Leaf blade margins finely serrate, veins (2 or)3 or 4 per side; twigs slightly flexuous; inflorescence branches glabrate; styles and pyrenes 2 or 3 . . . . 140. *Crataegus teres*
 23. Leaf blade margins coarsely and/or sharply serrate, veins 1–4 per side; twigs usually flexuous; inflorescence branches ± pilose or glabrous; styles and pyrenes 3–5.
  24. Leaf blades 2–4.5 cm, abaxial vein axils glabrous, venation craspedodromous, veins (1 or)2–4 per side; inflorescence branches sparsely to moderately pilose . . . . 141. *Crataegus florens*
  24. Leaf blades 1.5–3.5 cm, abaxial vein axils with tufts of hair, venation semicamptodromous, veins 1 or 2 per side; inflorescence branches glabrous or very sparsely pilose . . . . 142. *Crataegus attrita*

22. Hypanthia and inflorescence branches tomentose.
 25. Leaf blade margins coarsely, irregularly serrate, lobes 0 or 1 per side, sinuses shallow . . . . *Crataegus pulla* (see discussion under 144. *C. lanata*)
 25. Leaf blade margins regularly, finely serrate to crenate-serrate, lobes 0 or obscure.
  26. Leaf blades narrowly obovate or ± spatulate to obtrullate, apices acute to subacute, often cuspidate; twigs usually slender, (1.5 mm diam. at 5 cm from tip); flowers 12–16 mm diam.; pomes 8–12 mm diam., yellow-orange to reddish, sometimes with flushed cheeks . . . . 132. *Crataegus condigna* (in part)
  26. Leaf blades broadly oblong to cuneate, apices ± flattened to slightly cuspidate; twigs ± stout, (2–3 mm diam. at 5 cm from tip); flowers 20–25 mm diam.; pomes 10–15 mm diam., red . . . . 136. *Crataegus alabamensis*

127. Crataegus lepida Beadle, Biltmore Bot. Stud. 1: 36. 1901 • Dwarf hawthorn C E

Shrubs, 5–10(–20) dm, <dense and twiggy>, branches not weeping except in larger specimens, <capable of flowering at 3 dm>. **Stems:** <trunk bark dark gray, scaly>; twigs: new growth appressed-pubescent, 1-year old brown, <slender>; thorns on twigs straight, 1–2 -years old color not recorded, fine, 0.6–2 cm. **Leaves** <late deciduous to wintergreen; stipules herbaceous, falcate, margins glandular>; petiole length 15–30% blade, <winged distally>, pubescent, glandular; blade <deep green>, broadly obovate to ± isodiametric, 0.6–1.5 cm, coriaceous, base abruptly contracted, lobes 0 or 1 per side, <subterminal, barely bumps>, sinuses extremely shallow, lobe apex obtuse, margins minutely glandular-crenate to sometimes with conspicuous, glandular-scalloped teeth distally, veins 1 or 2 per side, apex obtuse or with slight apiculus, <± glossy>, abaxial surface pubescent, especially on veins, adaxial very sparsely appressed-pilose young, particularly so on veins. **Inflorescences** 1–3-flowered; branches appressed-pubescent; bracteoles linear, gland-margined. **Flowers** 12 mm diam.; hypanthium pubescent; sepals triangular, margins ± entire, abaxially pubescent; anthers ivory; styles 3 or 4. **Pomes** yellow-green to yellow or orange, broadly ellipsoid to suborbicular, 8–13 mm diam., <black gland-dotted>, pubescent; sepals broken off or recurved; pyrenes 3. $2n$ = 51, 68.

Flowering Apr–May; fruiting Jul–Aug. Dwarf scrub, sandy areas including inland dunes; of conservation concern; 0–100 m; Fla., Ga., S.C.

Crataegus lepida is a scarce, distinctive species occurring at scattered localities from central Florida to southern Georgia and South Carolina. A sterile specimen from Louisiana (*Thomas 57507*) may be this species.

In central Florida, plants of *Crataegus lepida* start their extension growth and flower in May, long after growth of other nearby plants has started, while further north anthesis about the second week of April is usual. The southern populations are also the most wintergreen in the species. Thus, the central Florida plants appear to be ecophysiologically distinct and should be investigated further. Particularly small-leaved specimens with numerous, small, sharp marginal teeth may represent *C. garrettii* of Murrill.

Crataegus lepida is a dwarf semi-evergreen shrub with dark, glossy, often suborbiculate leaves that are striking in the field. Larger specimens might be confused with *C. egens*.

128. Crataegus egens Beadle, Biltmore Bot. Stud. 1: 85. 1902 [E] [F]

Crataegus quaesita Beadle var. *egens* (Beadle) Lance

Shrubs or trees, to 50 dm, branches strongly weeping (moving in slight wind). **Stems:** <plants seldom flowering at less than 1 m>; twigs: new growth white-canescent, 1-year old purple-brown, older darker or grayer, <slender>; thorns on twigs sometimes absent, straight or slightly recurved, 1–2-years old blackish to deep gray, fine, 1–2 cm. **Leaves:** petiole <very slender>, length 25–50% blade, <winged distally>, pubescent, glandular; blade short obtrullate to broadly obovate, 1–2(–3) cm (widest in distal ½, length-width mostly less than 1.4:1), thin, base cuneate to slightly rounded, lobes 1 or 2 per side distally, short, lobe apex usually subacute, margins obscurely crenate, sometimes crenate-serrate in distal ½, teeth glandular, veins 1 or 2 per side (exiting at or beyond at widest part), apex subacute, <not glossy>, surfaces sparsely hairy young, abaxial surface soon glabrate except on veins. **Inflorescences** 1- or 2-flowered; branches densely appressed white-pubescent; bracteoles <few>, linear, margins sessile-glandular, <adaxially thin short-pubescent>. **Flowers** 12–16 mm diam.; hypanthium white-tomentose; sepals triangular, <2–3 mm>, margins glandular, abaxially pubescent; anthers ivory; styles 3. **Pomes** deep, dull yellow, sometimes with red cheek, suborbicular, 8–10 mm diam., glabrous; sepals appressed, sometimes late circumscissile; pyrenes 3.

Flowering late Feb–early Apr; fruiting Jul–Aug. Pine barrens, brush, usually on sandy soil; 10–100 m; Ala., Fla., Ga., S.C.

Crataegus egens is locally common in north-central Florida and Georgia with outliers occurring in Clark and Henry counties, Alabama, and Jasper County, South Carolina.

The long-petiolate leaves of *Crataegus egens* are very small and short-obtrullate to broadly obovate, thus proportionally wide. There is some variation in leaf lobing from acute to obtuse and even obscure. An additional characteristic of this species is the way in which extension-shoot leaves may flare outward toward the apex, their obtuse and large terminal end being quite deeply and somewhat irregularly lobed. Specimens with more elongated (1.4–1.6:1) leaves but more or less typically shaped distal ends may be referred to *C. cirrata* Beadle and are somewhat intermediate with *C. floridana*. Similar forms with leaves obovate-cuneate, lobes virtually absent, very obtuse and strongly crenate terminally but of uncertain taxonomic status, may key here. Uncommon short specimens of *C. egens* may be distinguished from *C. lepida* by leaves of that species being coriaceous and shiny with lobes almost or entirely absent.

129. Crataegus lacrimata Small, Torreya 1: 97. 1901 • Weeping hawthorn [E]

Shrubs or trees, 50(–80) dm, branches strongly weeping. **Stems:** twigs: new growth pubescent, 1–2-years old gray or purple-gray, <slender>; thorns on twigs straight, 1–2 -years old purple-gray, fine, 1.5–3 cm. **Leaves:** petiole <slender>, length 20–30% blade, pubescent, glandularity not recorded; blade oblong or ± narrowly obtrullate, sometimes elliptic, 1–2 cm, thin, base cuneate, lobes 0, sometimes very slightly lobed subterminally, margins subentire or finely serrate in distal ½, veins 1 or 2 per side (exiting in distal ½ of leaf), apex usually truncate to obtuse, sometimes cuspidately subacute, surfaces glabrous. **Inflorescences** 1–3 -flowered; branches glabrous; bracteoles <± caducous>, linear, margins eglandular or nearly so, <glabrous>. **Flowers** 15 mm diam.; hypanthium glabrous; sepals triangular, <2–3 mm>, margins ± entire, abaxially glabrous; anthers cream; styles 3 or 4. **Pomes** yellow, yellow blushed red, or red, suborbicular, 8 mm diam., glabrous; sepal remnants patent-reflexed; pyrenes 3 or 4.

Flowering Mar–Apr; fruiting Jul–Aug. Pinewoods, open scrub, sandy soil; 10–100 m; Ala., Fla., Ga., S.C.

Crataegus lacrimata is found abundantly over parts of the Florida panhandle and through adjacent Alabama and Georgia to South Carolina.

Crataegus lacrimata is an upright, usually single-stemmed shrub with slender, more or less weeping branches and small, narrow, glabrous, bright green leaves. The plants are easily recognized among ser. *Lacrimatae*, the whole plant being almost completely glabrous, having narrow, nearly unlobed leaves, and being the only member of the series to possess more or less eglandular bracteoles. Only with poor material might the pubescent inflorescence of *C. munda* (dwarf, non-lacrimate) or *C. crocea* (tall, lacrimate) be confused. A form similar to *C. lacrimata* and with the same range, but apparently discontinuously larger, has longer leaves (blades 2.2–4 cm versus 1–1.7 cm) and larger flowers (petals 7–9 mm versus 5–7 mm) and more orange-colored ripe fruit.

130. Crataegus munda Beadle, Biltmore Bot. Stud. 1: 38. 1901 [E]

Shrubs, <stiff, compact>, 5–20 dm, branches not weeping except in larger specimens. **Stems:** twigs: new growth <green>, densely appressed-pubescent, 1-year old purple brown-gray, <slender>; thorns on twigs <numerous>, ± straight, 2-years old purple-brown, slender, 1–3 cm. **Leaves** <deciduous; stipules approximately obtriangular, thin>; petiole length 0–28% blade, <winged distally>, pubescent only young, glandular; blade <dark green young>, narrowly obovate to cuneate or oblanceolate, 0.8–1.8 cm, thin to subchartaceous, base narrowly cuneate, lobes 0 or 3 per side distally, obscure, margins serrulate-crenate (distal ½), teeth glandular, veins 1 or 2 per side (exiting beyond widest part of leaf), apex usually obtuse, rarely acute, <shiny mature>, surfaces sparsely pubescent, abaxial ± densely pubescent near vein axils. **Inflorescences** 1- or 2(or 3)-flowered; branches ± sparsely appressed-pubescent; bracteoles linear, margins sessile-glandular. **Flowers** 12–15 mm diam.; hypanthium pubescent; sepals triangular, margins glandular, abaxially pubescent; anthers ivory, rarely light purple; styles 3–5. **Pomes** red, suborbicular, 7–8 mm diam., glabrate; sepal remnants appressed; pyrenes 3–5. $2n = 68$.

Flowering Mar–Apr; fruiting Jul–Aug. Sandy soil, open brush or among pines; 0–200 m; Ala., Fla., Ga., Miss., N.C., S.C., Va.

Crataegus munda is one of the most widespread species in the series; it occurs from central Florida and southeastern Alabama to North Carolina with scarce outliers in Mississippi (Forest and Jones counties) and Virginia (Nansemond County).

Crataegus munda, like *C. lepida*, has a stiff, compact habit and usually lacks the weeping branches typical of the series; both are among the few species of hawthorn that regularly flower at under 1 m tall. Herbarium specimens of *C. munda* somewhat resemble *C. lacrimata*, but have shorter petioles, hairy young pedicels, and are more likely to have acute leaves with more pronounced marginal teeth. Some similarities may also exist with narrower-leaved forms of *C. crocea*, which has more pendant leaves with, at most, denticulate margins, usually much longer petioles, and nearly tomentose inflorescences and hypanthia.

A similar form, perhaps *Crataegus geniculata* Ashe, with usually larger, more sharply toothed leaves, three or four veins per side, larger flowers with less hairy pedicels, and subpyriform fruit, has been found in northeastern Georgia.

131. Crataegus crocea Beadle, Biltmore Bot. Stud. 1: 113. 1902 [E] [F]

Crataegus curva Beadle; *C. dapsilis* Beadle; *C. frugalis* Beadle; *C. incana* Beadle; *C. lassa* Beadle var. *recurva* (Beadle) Lance; *C. rava* Beadle; *C. recurva* Beadle; *C. villaris* Beadle

Shrubs or trees, 20–60 dm, branches strongly weeping (moving in slight wind). **Stems:** twigs: new growth densely appressed-white-hairy, 1-year old gray-brown to purple-brown, older gray, <slender>; thorns on twigs <absent or numerous>, straight, 2-years old purple-brown to gray, ± fine, 1–4 cm. **Leaves:** petiole length 20–25% blade, densely pubescent, glandular at least young; blade usually spatulate or narrowly oblong to narrowly oblanceolate or narrowly elliptic, 1–2.5 cm, thin (drooping or fluttering in wind), base cuneate, lobes 0 or obscure, margins entire, sometimes obscurely crenate-serrate distally, gland-dotted young, veins 1–4 per side (narrowly diverging, exiting beyond widest part of leaf), apex subacute, obtuse, or truncate, usually tapered sharply, abaxial surface slightly pubescent, veins conspicuously pubescent, glabrescent, adaxial often densely pubescent young. **Inflorescences** 1–3-flowered; branches densely appressed-white-pubescent; bracteoles linear, margins glandular, <adaxially short-pubescent>. **Flowers** 16–18 mm diam.; hypanthium densely white-pubescent; sepals narrowly triangular, <4 mm>, margins glandular, abaxially sparsely pubescent; <petals elliptic or ± circular>; anthers cream; styles 3. **Pomes** yellow to copper-orange or reddish orange, suborbicular, 8 mm diam., pubescent to glabrescent; sepal remnants none or reflexed, sometimes circumscissile; pyrenes 3.

Flowering late Feb–Apr; fruiting Jul–Aug. Sand plains, open scrub; 0–200 m; Ala., Fla., Ga., N.C., S.C.

Crataegus crocea ranges from southwestern Alabama to central Florida into North Carolina; it is particularly abundant and locally common in Alachua and Marion counties, Florida.

A wide view is taken of *Crataegus crocea*. The typical form, which represents one pole of variation, has narrowly elliptic leaves perfectly smooth in outline, acute at the apices, the distal margins sometimes finely denticulate and larger leaves slightly incised distally. It is the most distinctive form of the species and is locally common at least in northern Florida. Other forms with similar leaves may be referred to *C. frugalis* and, if with larger leaves at maturity sometimes slightly incised distally, to *C. recurva*. *Crataegus curva* is a form that is somewhat similar to the type but is wider and shorter and has shorter petioles. Often, leaves of

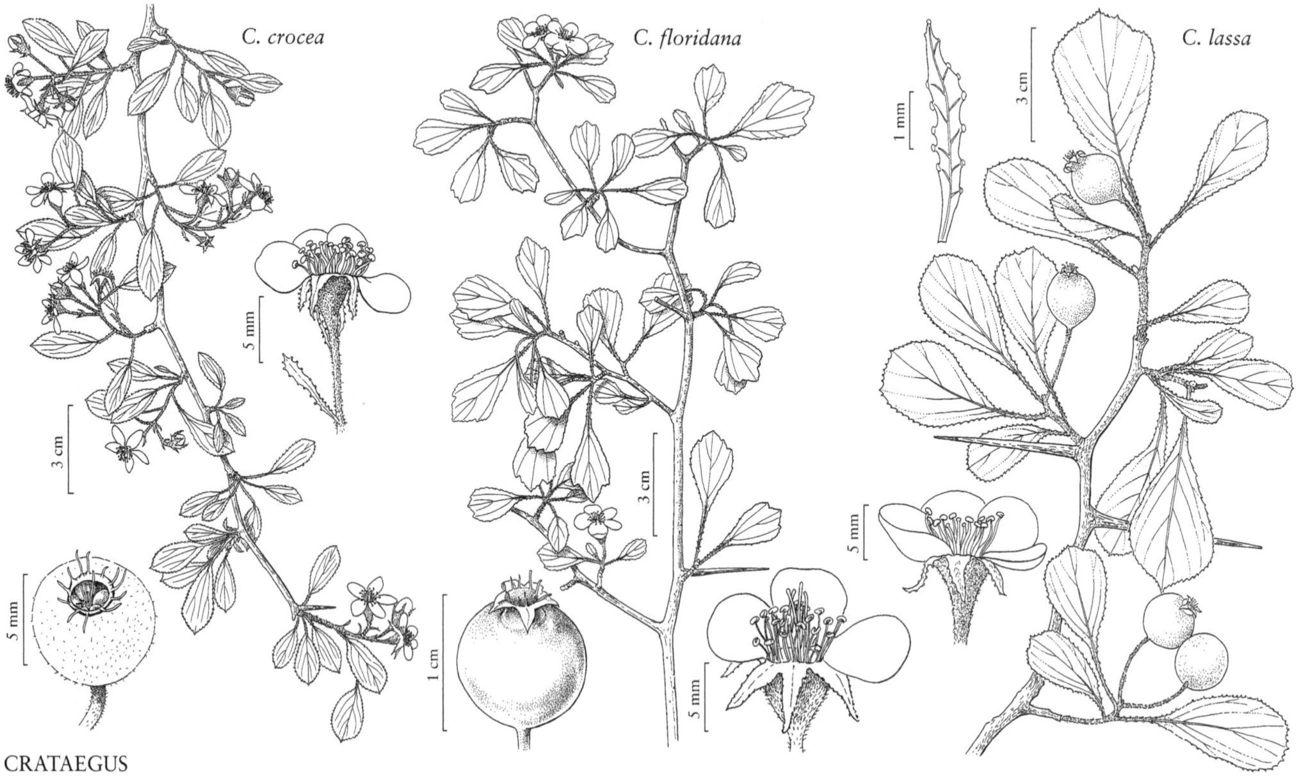

CRATAEGUS

C. crocea are wider than the type form and somewhat spatulate; these represent *C. villaris* and *C. incana*, the latter with particularly dense tomentum when young. All the foregoing forms intergrade. Forms such as *C. adusta* Beadle, with more or less spatulate leaves often with strongly crenulate to obscurely lobed leaf apices, are intermediate with *C. condigna*. They are larger in all their parts than typical *C. crocea*.

In spite of its variability, there should usually be no difficulty separating *Crataegus crocea* either from *C. munda* or from the small-leaved *C. condigna*, both subser. *Tenues*, both of which have obviously toothed leaf margins, nor from *C. lacrimata* with its glabrous inflorescences and more or less eglandular bracteoles. Larger-leaved forms of *C. crocea*, with spatulate-cuneate blades obscurely crenate-serrate distally, may resemble the *C. illudens* form of *C. lassa* (subser. *Robustae*) from Selma, Alabama; they retain the smaller fruit of *C. crocea*.

A related, unnamed form known from northern Florida has particularly long petioles, narrowly rhombic-obovate leaf blades (that is, with the lateral vertices nearer the distal end), and more or less entire margins.

Crataegus crocea can form an attractive, delicate shrub, which should be very suitable for xeriscaping.

132. Crataegus condigna Beadle, Biltmore Bot. Stud. 1: 35. 1901 [E]

Crataegus alabamensis Beadle var. *ravenelii* (Sargent) Lance; *C. arguta* Beadle; *C. clara* Beadle; *C. compitalis* Beadle; *C. insidiosa* Beadle; *C. ravenelii* Sargent; *C. rimosa* Beadle; *C. vicana* Beadle

Shrubs or trees, <habit ± intricate in smaller specimens, ± lacrimate in larger ones>, 10–60(–100) dm, branches ± to strongly weeping (moving in slight wind). **Stems:** twigs: new growth densely appressed-pubescent, 1-year old purple-brown, older dark gray, <usually slender>; thorns on twigs <usually numerous>, ± straight, 2-years old blackish gray, fine, 1–4 cm. **Leaves:** petiole length 20–35% blade, densely pubescent, glandular; blade <bright light green young>, narrowly obovate or ± spatulate to obtrullate, <on extension shoots suborbiculate>, 1.5–4 cm, coriaceous, base tapered, lobes 0 or slight protuberances distally, margins shallowly serrate to crenate-serrate except near base, teeth gland-tipped, veins 1–4 per side (exiting in distal part), apex acute to subacute, often cuspidate, <± shiny mature>, abaxial surface scabrous, main veins tomentose, adaxial pubescent young. **Inflorescences** 2–4-flowered; branches densely appressed-pubescent; bracteoles linear, margins sessile-glandular. **Flowers** 12–16 mm diam.; hypanthium white-canescent; sepals narrowly triangular, <3 mm>, margins glandular-serrate, abaxially pubescent; anthers cream; styles 3–5. **Pomes** yellow-orange to reddish, sometimes with flushed cheeks, suborbicular to sometimes ellipsoid, 8–12 mm diam., pubescent; sepals spreading to reflexed; pyrenes 3–5.

Flowering Mar–Apr; fruiting Jul–Aug. Open areas, open woodlands, sandy soil; 20–200 m; Ala., Fla., Ga., N.C., S.C.

Crataegus condigna extends from southern Alabama to North Carolina, being most abundant in northern Florida and South Carolina. The species is abundant and variable, characterized by leaves more or less obovate, lobes absent, and margins finely toothed. Extreme variation in stature suggests the existence of at least two forms; in the herbarium they cannot be distinguished. The typical form, with smaller leaves and more delicate branching, is best distinguished from the more robust, wider-leaved form of *C. crocea* by better defined leaf-marginal teeth. Larger-leaved forms of *C. condigna* may resemble narrower-leaved forms of *C. florens* (subser. *Robustae*) but differ by their smaller leaf marginal teeth and denser indumentum. Both have circular extension-shoot leaves. *Crataegus vicana* is a form with particularly tapered leaf bases and long petioles.

133. Crataegus quaesita Beadle, Biltmore Bot. Stud. 1: 33. 1901 [E]

Crataegus resima Beadle; *C. viaria* Beadle

Shrubs or trees, 20–50 dm, branches strongly weeping (moving in slight wind). **Stems:** twigs: new growth ± densely appressed-white-pubescent, 1-year old gray-brown to purple-brown, older gray, <± slender>; thorns on twigs <usually present>, straight, 1–2-years old purple-brown to gray, ± fine, 1–2 cm. **Leaves:** petiole length 15–20% blade, densely pubescent, glandular at least young; blade narrowly to broadly cuneate to obtrullate, 1.5–3 cm, ± stiff or ± floppy, base ± cuneate, lobes 1 or 2 per side, <subterminal>, small, lobe apex acute, <on extension shoots often deeply incised>, margins entire, sometimes obscurely crenate-serrate apically, gland-dotted young, veins 1–4 per side (narrowly diverging, exiting beyond widest part of leaf), apex acute to subacute, abaxial surface slightly pubescent, glabrescent, veins conspicuously pubescent, adaxial often densely pubescent at first, glabrescent, veins sparsely pubescent. **Inflorescences** 1–3(or 4)-flowered; branches densely appressed-white-pubescent; bracteoles linear, margins sessile-glandular, <adaxially short-pubescent>. **Flowers** 10–13 mm diam.; hypanthium white-pubescent; sepals narrowly triangular, <4 mm>, margins glandular, abaxially sparsely pubescent; <petals elliptic or ± circular>; anthers cream; styles 3 or 4. **Pomes** usually reddish orange to red, suborbicular, 8–10 mm diam., glabrescent to pubescent; sepal remnants none or reflexed; pyrenes 3 or 4(or 5).

Flowering Mar–Apr; fruiting Jul–Aug. Scrub on sandy soil; 0–100 m; Fla., Ga.

Crataegus quaesita is common in north-central Florida. The protologue by Beadle describes the anthers as purple, not seen since. The species is distinguished by narrow, more or less obtrullate leaves with several more or less sharp lobes distally. The *C. viaria* form, somewhat like a very elongated *C. egens*, has less sharp lobes. *Crataegus laxa*, a form with very little terminal lobing, forms a link with the *C. vicana* form of *C. condigna*. *Crataegus meridiana* (subser. *Robustae*) is another similar form but more robust and with larger flowers.

A striking local form, mainly from near Jacksonville, Florida, and from adjacent Georgia, is probably distinct. Its mature leaves, 2–3 cm, are much more sharply lobed and sharp-tipped than in *Crataegus quaesita*; the leaves at anthesis are unusually small, only 1 cm, because they are very little expanded (in *C. quaesita* the leaves then are nearly full size). The plant is very thorny, the extension-shoot leaves are 3 cm wide, more or less isodiametric, with deep, narrow sinuses, presenting a *C. marshallii*-like appearance.

134. Crataegus floridana Sargent, Bot. Gaz. 33: 124. 1902 [E] [F]

Crataegus anisophylla Beadle; *C. quaesita* Beadle var. *floridana* (Sargent) Lance; *C. versuta* Beadle

Shrubs, 60 dm, branches strongly weeping (moving in slight wind). **Stems:** <trunk bark dark brown, rimose>; twigs: new growth pubescent, 1–2-years old dark purple-brown, <slender>; thorns on twigs ± straight, 1–2-years old dark purple-brown, fine, 2 cm. **Leaves:** petiole <slender>, length 25–50% blade, pubescent, glandularity not recorded; blade spatulate to obtrullate, (1.5–)3–4 cm, thin (readily fluttering in wind), base cuneate, lobes 1 or 2 per side, ± obscure to ± distinct, lobe apex subacute, margins entire proximally, crenate to serrate distally, strongly gland-dotted young, veins 2 or 3 per side, apex subacute to acute, surfaces ± persistently pubescent. **Inflorescences** 2–4-flowered; branches canescent-tomentose; bracteoles linear, margins glandular. **Flowers** 12–15 mm diam.; hypanthium tomentose; sepals narrowly triangular, margins ± entire, abaxially tomentose; <petals 5>;

anthers cream; styles 3–5. **Pomes** orange, ellipsoid to suborbicular, 6–8 mm diam., glabrate or hairy; sepals spreading; pyrenes 3–5.

Flowering Mar–Apr; fruiting Jul–Aug. Dry brush; 0–200 m; Fla., Ga., N.C., S.C.

Crataegus floridana is frequent in north-central Florida and extends through much of Georgia with single records from Kershaw County, South Carolina, and Brunswick County, North Carolina.

Crataegus floridana resembles *C. quaesita* in its more or less obtrullate, but distally lobed leaves (which can always be observed fluttering in the wind) and longer petioles. The largest leaves in subser. *Tenues* may be found in *C. floridana*; there is considerable size variation. The depth and sharpness of the lobes is variable, with a maximum in the proportionately broadest leaves, these having been called *C. anisophylla*. Differences from 146. *C. meridiana* are noted there.

135. Crataegus invicta Beadle, Biltmore Bot. Stud. 1: 37. 1901 C E

Shrubs, 10–25 dm, branches not weeping except in larger specimens. **Stems:** twigs: new growth densely canescent (rapidly becoming dark underneath indumentum), 1-year old purple-brown, <stiff>; thorns on twigs <numerous>, straight, 1-year old dark purple-brown, becoming gray, fine, 3–5 cm. **Leaves** <deciduous or semipersistent>, petiole length 5–15% blade, pubescent, glandular; blade obovate-cuneate, 1.5–2 cm, stiff, subcoriaceous, base cuneate, lobes apical, sinuses shallow, obscurely so in narrower or smaller leaves, margins obscurely crenate or crenate-serrate, teeth glandular, veins 1–3 per side (exiting at or beyond widest part), apex subacute to obtuse, <adaxially very glossy>, surfaces (late spring to early summer) moderately crisped-hairy. **Inflorescences** 1- or 2(or 3) -flowered; branches very densely white-pubescent; bracteoles <few, semipersistent>, linear, margins glandular. **Flowers** 12–15 mm diam.; hypanthium ± densely pubescent; sepals narrowly triangular, margins glandular-serrate, abaxially sparsely pubescent; anther color not recorded; styles 3–5. **Pomes** color not recorded, ± pyriform, 6–8 mm diam., ripe fruit not recorded.

Flowering Mar–early Apr; fruiting Jul–Aug. Dry sandy soil; of conservation concern; 0–100 m; Ga., S.C.

Crataegus invicta is restricted to southeastern Georgia and adjacent South Carolina, where it is scarce.

Crataegus invicta is distinctive and is capable of flowering at under 1.5 m, as in the type collection. Somewhat similar to the *C. cirrata* form of *C. egens*, it differs in its relatively shorter petioles, more thorny habit, and blunter and shinier, apparently more coriaceous leaves.

136. Crataegus alabamensis Beadle, Bot. Gaz. 30: 342. 1900 E

Crataegus adunca Beadle

Shrubs or trees, 50–60 dm, branches ± weeping. **Stems:** twigs: new growth densely tomentose, 1-year old dark gray, older gray, <± stout>; thorns on twigs <usually sparse or absent>, ± recurved, 1-year old dark blackish gray, slender, 2–3 cm. **Leaves:** petiole <slender>, length 20–30% blade, pubescent, glandular; blade broadly oblong to cuneate, 2–3 cm (at anthesis, larger mature), ± thick, firm, base evenly tapered, lobes 0, margins crenate-serrate, veins 3–5 per side, <adaxially shiny mature>, apex ± flattened to slightly cuspidate, abaxial surface pubescent only on veins, adaxial glabrate young. **Inflorescences** 3–6 -flowered; branches tomentose; bracteoles linear, margins sessile-glandular, <adaxially short-pubescent>. **Flowers** 20–25 mm diam.; hypanthium tomentose; sepals narrowly triangular, margins glandular-serrate, abaxially appressed-white-pubescent in center of laminae, edges glabrate; anthers yellow; styles 3–5. **Pomes** red <often very glaucous on drying>, suborbicular to pyriform, 10–15 mm diam., glabrate; sepals spreading; pyrenes 3–5.

Flowering Mar–Apr; fruiting Jul–Aug. Open woods; 0–200 m; Ala., Fla.

Usually scarce, *Crataegus alabamensis* occurs in south-central Alabama and northern Florida near Tallahassee. The type location is recorded as being on clay soil near Montgomery, Alabama, a somewhat unusual habitat for this series.

Crataegus alabamensis, with its crenate-serrate leaves, resembles the larger forms of *C. condigna* (subser. *Tenues*); it is a more robust plant with much larger flowers and fruits. The type form has leaves particularly wide relative to length. The rare *C. adunca* has violet anthers, which is unusual for this series.

137. Crataegus integra (Nash) Beadle, Biltmore Bot. Stud. 1: 87. 1902 E

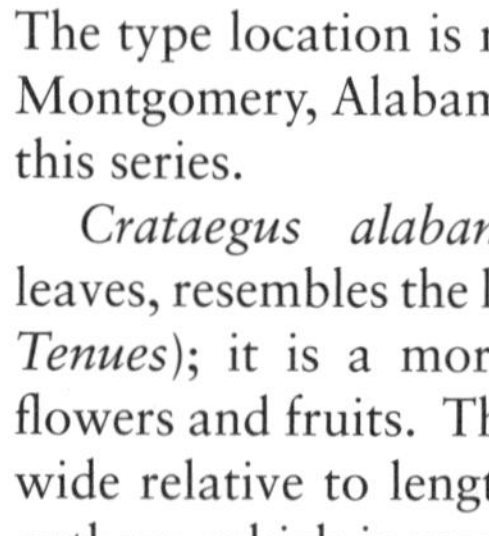

Crataegus flava Aiton var. *integra* Nash, Bull. Torrey Bot. Club 22: 150. 1895; *C. dolosa* Beadle; *C. lassa* Beadle var. *integra* (Nash) Lance; *C. sodalis* Beadle

Shrubs or trees, 20–50 dm, branches ± weeping. **Stems:** <trunk bark ashy gray, rough or scaly>; twigs: new growth densely white-canescent, 1-year old purple-brown or blackish, older dark gray, <± stout>; thorns on twigs <usually absent>, slightly recurved, 1-year old blackish or purple-brown, ± fine, 2–3 cm. **Leaves:** petiole length

10–25% blade, <winged distally>, densely pubescent, sometimes densely glandular; blade obovate-cuneate to obovate or suborbiculate, 1.5–4 cm (length/width = 1.2–1.7), stiff, ± coriaceous, base gradually tapered, lobes 0 or obscure, margins glandular-denticulate to subentire, veins 2–4(or 5) per side (except smaller leaves), apex ± acute to weakly cuspidate, surfaces pubescent at anthesis, particularly abaxially on veins, glabrescent. **Inflorescences** 2–4-flowered; branches tomentose; bracteoles oblong-linear, margins sessile-glandular, <adaxially thin short-pubescent>. **Flowers** 15–20 mm diam.; hypanthium tomentose; sepals narrowly triangular, <4 mm>, margins finely glandular-serrate, abaxially ± tomentose; anthers cream or ivory; styles 3–5. **Pomes** yellow blushed red to orange-red, suborbicular, (8–)10–12 mm diam., pubescent; sepals spreading; pyrenes 3–5.

Flowering mid Mar–mid Apr; fruiting Jul–Aug. Sand plains, dry scrub, open woodlands; 0–200 m; Ala., Fla., Ga., S.C.

Crataegus integra extends from Alabama to South Carolina and is concentrated in the northern half of peninsular Florida. The species is one of the more common members of ser. *Lacrimatae* in its main area of distribution.

Crataegus integra is relatively variable in leaf shape and is notable for its almost entire leaf blades, the margins of which often have numerous, barely detectible teeth. This species often displays a considerable variety of leaf size even on one short shoot and has more venous leaves than many species of ser. *Lacrimatae*. One of the most different entities in the *C. integra* complex is *C. sodalis* (= *C. dolosa*). It has densely glandular petioles, denticulate leaf margins, short-shoot leaves broadly elliptic to broadly obovate, and rather large suborbiculate extension-shoot leaves, and perhaps should be considered part of *C. alabamensis*. It is concentrated in south-central Alabama. *Crataegus integra* differs from the *C. illudens* form of *C. lassa* in its proportionately narrower and usually shorter leaves, lacking the distinctive cuneate-cuspidate shape of *C. lassa* but also sometimes obscurely subterminally lobed. *Crataegus integra* is particularly common in the northern half of peninsular Florida, where *C. lassa* is rare. *Crataegus constans* Beadle, with less copious inflorescence tomentum, longer petioles, and proportionately larger and thinner leaves, is perhaps a shade form of *C. integra*.

138. Crataegus colonica Beadle, Biltmore Bot. Stud. 1: 104. 1902 [E]

Crataegus lassa Beadle var. *colonica* (Beadle) Lance

Shrubs, 15–30 dm, branches ± weeping. **Stems:** <trunk bark ashy gray, rough or scaly>; twigs: new growth densely white-canescent, 1-year old purple-brown or blackish, older dark gray, <± stout>; thorns on twigs <usually numerous>, straight or slightly recurved, 1-year old blackish or purple-brown, moderately stout, 2–4.5 cm. **Leaves:** petiole length 15% blade, <winged distally>, densely pubescent, glandularity not recorded; blade obovate, 1.5–2.5(–3) cm, stiff, ± coriaceous, base gradually tapered, lobes 0, margins basally entire or subentire, distally ± denticulate, veins 2–4(or 5) per side (except smaller leaves), apex acute and ± cuspidate to ± obtuse, abaxial surface white-tomentose at anthesis, particularly on veins, adaxial glabrescent. **Inflorescences** 2–4-flowered; branches tomentose; bracteoles oblong-linear, margins sessile-glandular, <adaxially thin pubescent>. **Flowers** 12–15 mm diam.; hypanthium tomentose; sepals narrowly triangular, <4 mm>, margins finely glandular-serrate, abaxially ± tomentose; anthers cream or ivory; styles 3–5. **Pomes** green turning red or blushed red, pyriform or suborbicular, 6–9 mm diam., pubescent; sepals spreading; pyrenes 3–5.

Flowering early–mid Apr; fruiting Aug. Sand hills, dry sandy woodlands, sandy roadsides; 0–100 m; N.C., S.C.

Crataegus colonica is particularly abundant in south-central North Carolina. Very thorny species are unusual in ser. *Lacrimatae*, and *C. colonica* shares this, as well as habit, small leaves, and distribution with *C. pexa*. From the latter, it is distinguished by a distinctive leaf shape reminiscent of forms of *C. crus-galli*, relatively small, more or less pyriform fruits and shorter and less often numerous thorns. *Crataegus colonica* and *C. pexa* could represent two poles of the same species.

139. Crataegus pexa Beadle, Biltmore Bot. Stud. 1: 116. 1902 [E]

Crataegus munda Beadle var. *pexa* (Beadle) Lance

Shrubs, small (on only specimen with indication of stature), branches ± weeping. **Stems:** twigs: new growth densely white-tomentose, 1-year old dark gray, older gray, <± stout>; thorns on twigs <usually numerous>, slightly to strongly recurved,

1-year old purple-black, older blackish, slender, 3–5(–7) cm. **Leaves:** petiole length 10–20% blade, tomentose, glandular young; blade <dark green mature>, obovate to narrowly obovate-rhombic, 1.5–3 cm, stiff and ± coriaceous, base cuneate, lobes 0 or very short and subterminal, sinuses shallow, lobe apex subacute, margins strongly crenate-serrate, veins 2 or 3 per side, apex acute to obtuse, surfaces white-tomentose young, glabrescent. **Inflorescences** 1–3-flowered; branches tomentose; bracteoles oblong-linear, margins glandular. **Flowers** 12 mm diam.; hypanthium densely tomentose; sepals ± linear, margins glandular-serrate, abaxially tomentose; anther color not recorded; styles 4. **Pomes** orange, suborbicular, 8–10 mm diam., tomentose; sepals reflexed; pyrenes 4.

Flowering Mar–Apr; fruiting Aug. Dry woodlands; 20–200 m; Ga., N.C., S.C.

Crataegus pexa is widely distributed in North Carolina with a few records from South Carolina and Georgia. The similarities of 138. *C. colonica* to *C. pexa* have been noted under that species. *Crataegus pexa* differs in its often slightly incised and more prominently toothed leaf margins, generally lower plant stature, relatively wider leaf blades, suborbicular fruits, longer and abundant thorns, and strongly flexuous twigs.

140. Crataegus teres Beadle, Biltmore Bot. Stud. 1: 43. 1901 [E]

Crataegus alabamensis Beadle var. *teres* (Beadle) Lance

Shrubs or trees, 30–60 dm, branches ± weeping. **Stems:** twigs <slightly flexuous>, new growth glabrate, 1-year old tawny to reddish brown, 2-years old dark gray; thorns on twigs ± straight, 2-years old shiny blackish gray, ± fine to ± stout, 2–3 cm. **Leaves:** petiole length 15–20% blade, ± pilose young, glandular; blade <dark green mature adaxially>, narrowly obovate, 2–3.5 cm, thin, slightly coriaceous, base ± evenly tapered, lobes 0, margins finely glandular-serrate, veins (2 or)3 or 4 per side, apex flat-triangular to rounded or ± truncate, often minutely cuspidate, <adaxially shiny mature>, abaxial surface glabrous, adaxial veins sparsely hairy; <on extension shoots broadly elliptic to suborbiculate, larger, often slightly incised, strongly serrate>. **Inflorescences** 2–6-flowered; branches glabrate; bracteoles linear, margins glandular. **Flowers** 16 mm diam.; hypanthium glabrous; sepals narrowly triangular, <4–5 mm>, margins glandular-serrate, adaxially sparsely hirsute; anthers pale yellow; styles 2 or 3. **Pomes** red, ± ellipsoid, 8–10 mm diam., glabrous; sepals spreading, <7 mm>; pyrenes 2 or 3.

Flowering Mar; fruiting Aug–Sep. Pine woods, sandy fields; 0–100 m; Ala.

Crataegus teres, *C. attrita*, and *C. florens* appear to form a natural group and may be difficult to distinguish. They share more or less unlobed, rather narrow, subcoriaceous, bright to deep green leaves with more or less sharply denticulate margins, subglabrous herbage, and inflorescence branches subglabrous to pilose, lacking the dense tomentum characteristic of many members of ser. *Lacrimatae*. They are not part of the *C. alabamensis* complex. *Crataegus teres*, the least common of the three, has perhaps the most individual characteristics, including manifesting a merely slightly flexuous twig, having the smoothest leaf tips, and with *C. florens*, the most venous leaves (usually three or four per side), also subglabrous or pilose inflorescence branches, and two or three styles, an ensemble of characteristics not unlike what might be found in ser. *Crus-galli*.

141. Crataegus florens Beadle, Biltmore Bot. Stud. 1: 94. 1902 [E]

Crataegus alabamensis Beadle var. *florens* (Beadle) Lance

Shrubs or trees, 30–60 dm, branches ± weeping. **Stems:** twigs: new growth sparsely to densely spreading-pubescent, 1-year old tawny to reddish brown, 2-years old dark gray; thorns on twigs <usually numerous>, straight to slightly decurved, 2-years old shiny blackish gray, ± fine to ± stout, 2–4.5 cm. **Leaves:** petiole length 25–33% blade, ± pilose young, glandular; blade <dark green mature adaxially>, narrowly obovate to obovate, 2–4.5 cm, thin, slightly coriaceous, base abruptly tapered, lobes 0, or 1 or 2 per side distally, obscure, lobe apex obtuse to subacute, margins ± coarsely glandular-serrate, veins (1 or)2–4 per side, apex often ± cuspidate, <adaxially shiny mature>, surfaces glabrous, veins sparsely pubescent; <on extension shoots broadly elliptic to ± orbiculate, larger, often slightly incised, strongly serrate>. **Inflorescences** 3–6-flowered; branches sparsely to moderately pilose; bracteoles <caducous>, linear, margins glandular. **Flowers** 15–18(–25) mm diam.; hypanthium sparsely pubescent; sepals narrowly triangular, <4–5 mm>, margins glandular-serrate, adaxially sparsely hirsute; anthers pale yellow; styles 3(–5). **Pomes** orange-red to red, suborbicular, 8–15 mm diam., glabrous; sepals spreading, <7 mm>; pyrenes 3(–5).

Flowering Mar–Apr; fruiting Aug–Sep. Brush on sand plains; 0–200 m; Ala., Fla., Ga., La., Miss., N.C., S.C.

Crataegus florens occurs mainly from central Mississippi to the Carolinas and extreme northern Florida; it is common in central Alabama and is disjunct in West Feliciana Parish, Louisiana. The species may occur in more mesic habitats than other members of ser. *Lacrimatae*.

The type form of *Crataegus florens* has the smallest, least strongly serrate leaves, veins two or three per side, and largest flowers, but the circumscription is here expanded to include forms of smaller flower dimensions and with larger, more strongly serrate leaves, including those with three or four veins per side. As such, *C. florens* is a fairly common and striking plant with its glossy, deep green leaves, thus superficially resembling members of ser. *Crus-galli* except for the strongly flexuous twigs. Except for its type form, which is uncommon, it flowers before most other sympatric members of ser. *Lacrimatae*. Differences from 142. *C. attrita*, to which it is most similar, and 140. *C. teres* are given in the key and discussed under those species.

142. Crataegus attrita Beadle, Biltmore Bot. Stud. 1: 98. 1902 [E]

Shrubs or trees, 20–50 dm, branches ± weeping. **Stems:** <trunk bark rough or rimose>; twigs: new growth <dark greenish red>, glabrous, 1-year old dull gray; thorns on twigs <numerous>, ± straight, 1-year old dark grayish black, ± fine, 2 cm. **Leaves:** petiole <slender>, length 30% blade, sulcus short-pubescent, sparsely sessile-glandular; blade cuneiform, 1.5–3.5 cm, thin, stiff, base tapered, lobes 0 or 1 per side, <subterminal>, sinuses shallow, margins serrate (coarsely distally), <venation semicamptodromous>, veins 1 or 2 per side (originating in proximal part of blade, ending in widest, distal $1/3$), apex rounded, surfaces glabrous, midveins ± tomentose near proximal end, abaxial main vein axils with tufts of hair. **Inflorescences** 2- or 3-flowered; branches glabrous or very sparsely pilose; bracteoles <deciduous>, linear, margins glandular. **Flowers** 18–22 mm diam.; hypanthium glabrous or very sparsely pubescent; sepals ± ligulate from wide bases, <4–5 mm>, margins ± entire, ± glandular, abaxially glabrous; anthers cream; styles 3–5. **Pomes** yellow blushed red, suborbicular, 10–14 mm diam., glabrous; sepals not seen; pyrenes 3–5.

Flowering Mar; fruiting Aug. Brush; 0–100 m; Ala., Fla., Miss.

Crataegus attrita is known only from southeastern Mississippi (Lowndes County), southeastern Alabama (Dale and Greene counties), and adjacent Florida (Washington County). It is most similar to *C. florens* and *C. teres* and is more or less intermediate between the two; its dense tufts of hair in the abaxial vein axils are reminiscent of *C. aestivalis*.

143. Crataegus lassa Beadle, Biltmore Bot. Stud. 1: 29. 1901 [E] [F]

Crataegus illudens Beadle; *C. panda* Beadle

Shrubs or trees, 20–50 dm, branches ± weeping. **Stems:** <trunk bark ashy gray, rough or scaly>; twigs: new growth densely white-canescent, 1-year old purple-brown or blackish, older dark gray; thorns on twigs <absent or rare>, straight or slightly recurved, 1-year old blackish or purple-brown, ± stout, 3–4.5 cm. **Leaves:** petiole length 25–33% blade, <winged distally>, densely pubescent, glandular; blade obovate-cuneate to narrowly obovate, 2–4 cm (length/width = 1.6–2.1), ± coriaceous (floppy in wind), base gradually tapered, lobes 0 or 2–4 per side, <subterminal, toothlike>, margins glandular-denticulate to subentire, veins 2–4(or 5) per side (except smaller leaves), apex often strikingly cuspidate, surfaces ± white-pubescent young, especially on abaxial veins, glabrescent. **Inflorescences** 3–5-flowered; branches tomentose; bracteoles linear, margins glandular. **Flowers** 15–20 mm diam.; hypanthium tomentose; sepals narrowly triangular, <6–8 mm>, margins finely glandular-serrate, abaxially ± tomentose, especially proximally; anthers cream or ivory; styles 3–5. **Pomes** yellow blushed red to orange-red, suborbicular, (8–)10–12 mm diam., pubescent; sepals spreading-recurved or lost; pyrenes 3–5.

Flowering Mar–mid Apr; fruiting Jul–Aug. Sand plains, dry scrub, open woodlands; 0–100 m; Ala., Fla., Ga., S.C.

Crataegus lassa is particularly abundant around its type locality of Selma, Alabama. The species normally has big, floppy leaves. The type material has rather shorter leaves than most specimens seen commonly near Selma today and is notable for having a few small terminal lobes. *Crataegus illudens* is barely different but has somewhat longer leaves and leaf tips more often entire. *Crataegus panda* is a form with a slightly shorter and relatively wider leaf, minutely crenate on the margins; also see the discussion under 144. *C. lanata*.

144. Crataegus lanata Beadle, Biltmore Bot. Stud. 1: 86. 1902 [E]

Crataegus inops Beadle; *C. lassa* Beadle var. *lanata* (Beadle) Lance

Shrubs, <widely spreading mature>, 40–60 dm, branches ± weeping. **Stems:** twigs: new growth densely appressed white-pubescent, 1-year old brown to purple-brown, <glabrous>; thorns on twigs <absent or numerous>, straight, 1-year old purple-brown, slender, 1–4 cm. **Leaves:** petiole length 20–25% blade, pubescent young, densely glandular; blade usually narrowly to broadly obtrullate, sometimes ± obovate, 3–5 cm, thick, floppy, base usually gradually tapered, lobes 1 or 2 (or 3) per side, <subterminal>, short, well defined, margins subentire or obscurely glandular crenate-serrate, veins 2 or 3 per side (exiting beyond widest part of leaf), apex acute to subacute, surfaces pubescent. **Inflorescences** 1–4-flowered; branches densely white-canescent; bracteoles <deciduous>, linear, margins sessile-glandular, <adaxially short-pubescent>. **Flowers** 17–20 mm diam.; hypanthium densely canescent; sepals narrowly triangular, <5 mm>, margins glandular-serrate, abaxially pubescent; anthers ivory; styles 4 or 5. **Pomes** usually orange or ruddy to crimson, suborbicular to ± pyriform, 9 mm diam., pubescent; sepals reflexed; pyrenes 4 or 5.

Flowering Mar–Apr; fruiting Jul–Aug. Pine barrens, open scrub, on sand; 0–200 m; Ala., Fla., Ga., N.C., S.C., Va.

Crataegus lanata occurs from Alabama and northern Florida to the Carolinas; there is one record from Chesterfield County, Virginia.

Crataegus lanata is one of the most abundant species in ser. *Lacrimatae*; it is most like *C. lassa* and is noted for its very large (for ser. *Lacrimatae*), usually floppy, distinctly but shallowly incised, kite-shaped leaves, persistently somewhat white-hairy. The fruit of *C. lanata* ripens copper to red.

Variants of *Crataegus lanata* include some intermediates with the typical form of *C. lassa* and forms with wider, more obovate leaves, some of which may be the same as *C. amica* Beadle. *Crataegus pulla* Beadle, described from sandy flats along the Tombigbee River, Mississippi, is a form little-known apart from its type and protologue. *Crataegus pulla* has similarities to *C. lanata*, but differs by its more or less coarsely and rather irregularly toothed, obovate leaves often with a single, relatively small, somewhat obscure lobe on each side; the leaves are at first more or less tomentose and then variably somewhat glabrescent; flowers are smaller (14–17 mm wide); and fruits are larger (9–12 mm wide).

145. Crataegus furtiva Beadle, Biltmore Bot. Stud. 1: 81. 1902 [E]

Shrubs or trees, 30–60 dm, branches ± weeping. **Stems:** twigs: new growth densely appressed white-pubescent, 1-year old purple-brown, older deep gray, <relatively stout>; thorns on twigs <absent or few>, straight, 1-year old purple-brown, ± stout, 2–3 cm. **Leaves:** petiole length 20–25% blade, <broadly winged distally>, pubescent, glandular; blade broadly obtrullate or obdeltate to obtrullate, 1.5–2.5 cm (length/width = 1–1.6), thick, stiff, base abruptly tapered, lobes 1–3 per side, <subterminal or lateral>, sinuses shallow, lobe apex obtuse, margins obscurely glandular-crenate, veins 2 or 3 per side (diverging at narrow angle from midvein), apex usually rounded to subacute, sometimes sharply cuspidate, or acute (in narrower-leaved forms), surfaces densely tomentose young, variably glabrescent, <often persistently lanate>. **Inflorescences** 2–4-flowered; branches densely tomentose-canescent; bracteoles <deciduous>, ± linear, margins sessile-glandular, <adaxially short-pubescent>. **Flowers** 14–16 mm diam.; hypanthium densely white appressed-pubescent; sepals narrowly triangular, <5 mm>, margins glandular-serrate, abaxially appressed-white-pubescent; anthers cream; styles 3. **Pomes** orange to red, broadly ellipsoid to suborbicular, 8–15 mm diam., glabrous or sparsely pilose; sepals reflexed; pyrenes 3.

Flowering Mar–Apr; fruiting Jul–Aug. Open brush, sandy soil, among sparse pines; 0–200 m; Ala., Fla., Ga., N.C., S.C.

Crataegus furtiva is concentrated in Georgia and South Carolina with records from Alabama, Florida, and North Carolina. The species, as it is usually seen, has among the smallest leaves of the thick-twigged group (subser. *Robustae*) and is one of the most densely hairy of all members of ser. *Lacrimatae* when young. Its usually wide, obtrullate leaves are distinctive. The type form, which is also common, represents a narrower-leaved form. In spite of some variability in leaf shape and fruit size, shape, and residual indumentum, *C. furtiva* is usually easily recognized.

146. Crataegus meridiana Beadle, Biltmore Bot. Stud. 1: 115. 1902 [E]

Crataegus tristis Beadle

Shrubs or trees, 30–70 dm, branches ± weeping. **Stems:** twigs: new growth appressed-pubescent, 1-year old gray-brown to blackish gray, <slender to relatively stout>; thorns on twigs <absent or frequent>, ± straight, 1-year old purple-brown, becoming gray, slender, 2–3.5 cm. **Leaves:** petiole length 30–50% blade, persistently pubescent, glandular; blade narrowly to broadly cuneate to obtrullate, 2–4 cm (1/3 smaller young), thin, becoming floppy, lobes 1–3 per side, <subterminal>, sinuses shallow, lobe apex obtuse to acute, margins glandular crenate-serrate, veins 2 or 3 per side, apex rounded to subacute, surfaces pubescent, glabrescent adaxially, abaxial veins densely tomentose. **Inflorescences** 2–4-flowered; branches appressed-canescent; bracteoles linear, margins glandular. **Flowers** 18–20 mm diam.; hypanthium densely canescent; sepals narrowly triangular, <4–6 mm>, margins glandular-serrate, abaxially pubescent; anthers ivory; styles 2–4 (or 5). **Pomes** orange-yellow to ruddy, suborbicular, 9–12 mm diam., pubescent; sepals <on collar>, reflexed; pyrenes 2–4(or 5).

Flowering Mar–Apr; fruiting Jul–Aug. Brush, open woodlands; 0–200 m; Ala., Fla., Ga., S.C.

Crataegus meridiana is a poorly understood complex of forms with long-petiolate, terminally lobed, more or less narrowly obtrullate leaves, with similarities to *C. quaesita* and *C. senta*. The general form of the leaf is similar to that of *C. quaesita* but is more shortly and obscurely lobed than in *C. senta*. Unlike *C. quaesita*, extension-shoot leaves are more likely to be more or less isodiametric. More delicate specimens of *C. meridiana* might key to *C. floridana*, but the fruit is larger.

147. Crataegus lancei J. B. Phipps, J. Bot. Res. Inst. Texas 2: 1143, fig. 30 [top]. 2008 [E]

Shrubs or trees, 50–80 dm, branches ± weeping. **Stems:** twigs: new growth densely pubescent to tomentose, 1-year old purple-brown to blackish, older dark gray; thorns on twigs <absent or numerous>, ± straight, 1-year old purple-brown to blackish, slender, 3–4 cm. **Leaves:** petiole length 15–25% blade, pubescent, glandular; blade usually narrowly obovate to obtrullate, sometimes oblanceolate, 2–3 cm, thin to subcoriaceous, slightly floppy, base ± contracted, lobes 0 or 1 or 2(or 3) per side distally, short, lobe apex subacute, margins glandular, weakly crenate-serrate, veins 1–3 per side (terminating beyond widest part of leaf), apex acute to obtuse, surfaces canescent to woolly only near midvein young, becoming glabrate, abaxial main vein axils with tufts of hair. **Inflorescences** 3–7-flowered; branches pubescent to densely pubescent; bracteoles <deciduous>, linear, margins short-stipitate-glandular, <adaxially short-pubescent>. **Flowers** 10–12 mm diam.; hypanthium densely pubescent; sepals triangular, <3–4 mm>, margins glandular-serrate, abaxially pubescent; anthers cream; styles 3 or 4. **Pomes** often copper to deep red, suborbicular, 10 mm diam., pubescent; sepals patent-reflexed; pyrenes 3 or 4.

Flowering Apr–Sep; fruiting Aug–Sep. Open scrub, woodlands; 0–800 m; Ala., Fla., Ga., N.C., S.C.

Crataegus lancei is fairly common from Alabama to the Carolinas and northern Florida. In Buncombe County, North Carolina, *C. lancei* is found near the highest elevations for members of ser. *Lacrimatae*. The species is one of the narrow-leaved members of the thick-twigged group. Its crenate-margined short-shoot leaves vary from more or less elliptic-oblong, lobes absent, to narrowly obovate (rarely narrowly elliptic-rhombic) with a single well-defined, short, subterminal, toothlike lobe on each side. In this it differs from *C. senta*, in which short-shoot leaves are oblong-spatulate to cuneate with several short but acute, subterminal lobes. Extension-shoot leaves also differ markedly (see 148. *C. senta*). When the narrower type of short-shoot leaves, which are broadest near the middle, are plentiful, *C. lancei* is one of the most distinctive species in the series. Other striking features include inflorescences more floriferous than usual for the series and often dark orange to deep red, cherrylike fruits.

148. Crataegus senta Beadle, Bot. Gaz. 30: 341. 1900 [E]

Shrubs or trees, 60 dm, branches ± weeping. **Stems:** twigs: new growth <green>, pubescent, 1-year old purplish brown under exfoliating wax, older deep dull gray; thorns on twigs <absent or numerous>, straight to slightly recurved, 1-year old often copper to deep red, 1.5–5 cm. **Leaves:** petiole length 30–40% blade, roughly pubescent, black-glandular; blade <dull green, abaxially paler>, oblong-spatulate to cuneate or narrowly obdeltate, 3 cm, thin, ± floppy, base cuneate, lobes 1 or 2 per side distally, well defined, lobe apex acute, margins strongly crenate-serrate almost to bases, densely glandular, <glands black>, veins (2 or)3 or 4(or 5) per side, apex usually cuspidate, sometimes acute, adaxial surface sparsely pubescent young, main veins and axils hairy; <on extension shoots larger, relatively wider, more

deeply incised>. **Inflorescences** 3–7-flowered; branches densely short-canescent; bracteoles <deciduous>, linear, margins short-stipitate-glandular, <adaxially short-pubescent>. **Flowers** 15–20 mm diam.; hypanthium tomentose; sepals narrowly triangular, margins glandular-serrate, abaxially pubescent; anther color not recorded; styles 3–5. **Pomes** <1–3 per infructescence>, deep red, suborbicular, 10 mm diam., <punctate>, glabrous; sepals patent to reflexed; pyrenes 3–5.

Flowering Apr–early May; fruiting Aug–Sep. Dry pine woods, open scrub, sand plains; 100–800 m; Ala., Fla., Ga., La., N.C., S.C.

Crataegus senta is widespread from western North Carolina and South Carolina (where it is most common) to northern Florida and Georgia, with a disjunct record from West Feliciana Parish, Louisiana. Habitats are typical for the series.

Crataegus senta most resembles *C. dispar* but is more often confused with *C. lancei*. Short-shoot leaves of *C. senta* are narrower and are not as lanate as in *C. dispar*. Smaller leaves of *C. senta* have fewer veins (one or two per side); their shape is more like that of *C. lancei*, but the differences of lobes and leaf teeth are striking (see key and 147. *C. lancei* discussion). In *C. senta*, extension-shoot leaves differ little from those of the short shoots, having multiple, relatively short, acute lobes across the subterminal part. By contrast, the extension-shoot leaves of *C. lancei* are of a long-petiolate, narrowly rhombic form with one to three acute to subacute lobes per side. The large, red, cherrylike fruit, similar to that of *C. lancei*, is very striking.

149. Crataegus dispar Beadle, Biltmore Bot. Stud. 1: 28. 1901 [E]

Shrubs or trees, 50 dm, branches ± weeping. **Stems:** <ultimate twigs relatively stout>; twigs: new growth densely appressed-canescent, 2–3-years old dark gray or gray-brown; thorns on twigs <absent or numerous>, straight, 2-years old purple-brown, slender, 3–4 cm. **Leaves:** petiole length 20–35% blade, densely appressed-hairy young, glandular; blade ± obovate-spatulate, 2.3–3.5 cm, thin, base gradually tapered, lobes 2 per side, <subterminal>, sinuses shallow (LII 15–20%), lobe apex acute, margins glandular-serrate, veins 2 or 3 per side, apex acuminate, abaxial surface ± persistently pubescent, adaxial densely tomentose, ± glabrate. **Inflorescences** 2–4-flowered; branches densely appressed-pubescent; bracteoles <deciduous>, linear, margins glandular, <adaxially short-pubescent>. **Flowers** 13–16 mm diam.; hypanthium densely pubescent; sepals narrow, <3 mm>, margins glandular-serrate, abaxially pubescent; anthers ivory; styles 4 or 5. **Pomes** bright orange-red or reddish, orbicular, 8–12 mm diam., glabrate to slightly pubescent; sepals reflexed; pyrenes 4 or 5.

Flowering Mar–Apr; fruiting Jul–Aug. Brush, sandy soil; 30–100 m; Ala., Fla., Ga., S.C.

Crataegus dispar is a somewhat scarce species, concentrated in South Carolina and with scattered records from Alabama, Florida, and Georgia. The author notes the species in South Carolina grows in extremely dry conditions.

Crataegus dispar is one of the most easily recognized members of ser. *Lacrimatae*, even identifiable in specimens lacking reproductive parts and sometimes from extension shoots alone. Of note are the wide extension-shoot leaves, often deeply incised into wide-spreading, sharp segments lobed nearly to the sinuses, rather as in *C. marshallii*.

150. Crataegus exilis Beadle, Biltmore Bot. Stud. 1: 76. 1902 [C] [E]

Shrubs, 20–40 dm, branches ± weeping. **Stems:** <trunk bark dark gray or brownish, rough; upper twigs suberect, others ± spreading>; twigs <slightly flexuous>, new growth pubescent, 1-year old light gray-brown, older gray-brown; thorns on twigs <few>, straight, dark, 1–1.5 cm. **Leaves:** petiole length 25–30% blade, <winged distally>, pubescent young, glandularity not recorded; blade broadly obovate to broadly rhombic-elliptic, the smaller ± oblong-cuneate, 1.5–2.5(–3) cm, firm to subcoriaceous, base tapered, lobes 0, or 1(or 2) per side, <subterminal>, sinuses shallow (LII to 10%), lobe apex subacute, margins crenate-serrate (distal ½), veins 3 per side, apex subacute, surfaces pubescent young, particularly abaxially on veins, glabrescent; <on extension shoots deeply incised>. **Inflorescences** 3–7-flowered; branches pilose-pubescent; bracteoles oblong-linear, margins glandular. **Flowers** 15 mm diam.; hypanthium pilose; sepals narrowly triangular, <3–4 mm>, margins subentire or finely glandular-serrate, abaxial pubescence not recorded; anthers cream or ivory; styles 3–5. **Pomes** red, orbicular, 5–7 mm diam., glabrous; sepals usually erose; pyrenes 3–5.

Flowering Mar–Apr; fruiting Aug–Sep. Sandy soil, stream banks; of conservation concern; 60 m; Ga.

Crataegus exilis is only known from the type locality in south-central Georgia. Its appearance is distinctive; unfortunately, it is known only from flowering and fruiting syntypes. The species is somewhat anomalous in ser. *Lacrimatae* and slightly resembles a larger *C. calva* (ser. *Apricae*) but has pilose inflorescence branches, slightly different leaf shape, and somewhat different leaf margins.

64g. CRATAEGUS Linnaeus sect. unassigned

64g.1. CRATAEGUS Linnaeus (sect. unassigned) ser. MONTANINSULAE J. B. Phipps & O'Kennon, J. Bot. Res. Inst. Texas 1: 1065. 2007 [E]

Shrubs or trees, 20–50 dm. **Stems:** trunk bark dark, rough; twigs ± straight, new growth deep reddish green, glabrous or pubescent at first, 1-year old shiny, dark red-brown or glossy, brown to deep purple-brown, 2-years old dark reddish and gray, mid brown, or gray; thorns on twigs usually numerous, ± straight to recurved, 1-year old shiny, dark red-brown, deep brown, or black, slender to stout, 1.5–4(–5) cm. **Leaves** deciduous; petiole length 25–40% blade, slender, eglandular or minutely sessile-glandular, glands black, sometimes stipitate-glandular distally, adaxial sulcus pubescent; blade broadly elliptic to elliptic-oblong, or elliptic-rhombic, 3.5–8 cm, length/width = 1.2–1.5, ± coriaceous to thin, base narrowly cuneate to cuneate-rounded, lobes 2–4 per side, sinuses moderately deep, max LII (0–)15–25%, lobe apex subacute to acute, margins crenate-serrate or serrate, venation craspedodromous, veins (3–)5–8 per side, absent to sinuses, apex acute to subacute, matte, abaxial surface glabrous, veins glabrous or scabrous, adaxial ± densely scabrous, sometimes glabrescent. **Inflorescences** 4–15-flowered; branches glabrous or densely pubescent to tomentose; bracteoles caducous, linear, membranous, margins sessile-glandular. **Flowers** 13–21 mm diam.; hypanthium glabrous or tomentose; sepals triangular to broadly triangular, 3–4 mm, margins usually finely glandular-denticulate; <petals pale cream in *C. rivuloadamensis*>, post-mature petals pale brown; stamens 10, anthers pale pink to pink; styles 3–5. **Pomes** ripening from bright to deep red, deep red to purple (often nearly black) or pale reddish purple, turning through deep red to reddish plum and burgundy, suborbicular to ellipsoid or turbinate, 8–12 mm diam.; sepals sessile, prominent, ± patent to recurved or subappressed; pyrenes 3–5, sides plane to slightly concave.

Species 3 (3 in the flora): sw Canada.

Series *Montaninsulae* is restricted to the Cypress Hills and is erected to accommodate a suite of species with predominantly purple fully ripe fruit and plane-sided pyrenes. The series shares characteristics with sect. *Coccineae* and sect. *Douglasia* and, pending further study, remains unassigned to a section. It is plausible that ser. *Montaninsulae* represents three different hybrid derivatives between the two sections; each of the species has distinctive characteristics and can be found repeatedly over county-sized areas in suitable habitats in the Cypress Hills. The two most similar species, *Crataegus purpurella* and *C. rivulopugnensis*, differ primarily in fruit color, leaf shape, size, and venation, the latter being the red-fruited member of the group; *C. rivuloadamensis*, much hairier, with the largest fruit and often short, stout, more or less conic thorns, is the most distinct species.

1. Leaf blades 4–8 cm mature; inflorescence branches ± densely tomentose; flowers 17–21 mm diam., hypanthia densely pubescent except sometimes distally; pomes usually ± densely hairy . . . . 151. *Crataegus rivuloadamensis*
1. Leaf blades 5–6 cm mature; inflorescence branches glabrate or ± pilose; flowers 13–16 mm diam., hypanthia glabrous or ± pilose, sometimes proximally only; pomes glabrous or sparsely pubescent.
    2. Twigs: thorns 2–3.5 cm; leaf blades broadly rhombic to broadly elliptic or oblong, 3.5–4.5(–5) cm mature, veins 3 or 4(or 5) per side, barely impressed; inflorescence branches glabrate to sparsely pilose; pomes deep red to red-burgundy younger, deep purple fully ripe. 152. *Crataegus purpurella*
    2. Twigs: thorns 2.5–4(–5) cm; leaf blades broadly elliptic to ovate, 4–6 cm mature, veins 5–8 per side, slightly impressed; inflorescence branches ± pilose; pomes red, becoming bright to deep red . . . . 153. *Crataegus rivulopugnensis*

151. Crataegus rivuloadamensis J. B. Phipps & O'Kennon, J. Bot. Res. Inst. Texas 1: 1065, plate 7.3a, fig. 13. 2007 • Adams Creek hawthorn [E]

Shrubs or trees, 25–40(–50) dm, wide-branching. **Stems:** twigs: new growth dark reddish green, pubescent, 1-year old shiny, dark red-brown, 2-years old dark reddish and gray; thorns on twigs ± straight, 1-year old glossy, very dark red-brown, older dark gray, <shorter conic>, stout, 1.5–2.5(–3) cm. **Leaves:** petiole length 25–35% blade, eglandular or sparsely sessile-glandular, stipitate-glandular at leaf base, adaxial sulcus densely pubescent; blade ± elliptic or broadly elliptic to rhombic-elliptic, 2–4 cm at anthesis, 4–8 cm mature, base cuneate to narrowly cuneate, lobes 2–4 per side, max LII (0–) 5–20(–25)%, lobe apex subacute to broadly cuspidate, margins finely crenate-serrate or serrate, veins 4–6 per side, hardly impressed, apex acute, abaxial surface glabrous or sparsely hairy young, veins appressed-scabrous, adaxial ± densely appressed-scabrous (indumentum persisting). **Inflorescences** 4–10-flowered; branches ± densely tomentose; bracteoles ± reddish, narrow, margins densely sessile-glandular. **Flowers** 17–21 mm diam.; hypanthium densely pubescent except sometimes distally; sepal color not recorded, broadly triangular, 3 mm, <margins greenish>, abaxial surface glabrous, adaxial pubescent; <petals pale cream>; anthers pink; styles 3–5. **Pomes** <1–5(or 6) per infructescence>, pale reddish purple, becoming deep red to reddish plum and burgundy, suborbicular to broadly turbinate, 10–12 mm, usually ± densely hairy; sepals ± patent to recurved, <4–5 mm>; pyrenes 3–5, dorsally strongly grooved, sides plane to very shallowly concave.

Flowering late May–early Jun; fruiting Aug–Sep. Thickets, under light shade of aspens; 800–1200 m; Alta., Sask.

Crataegus rivuloadamensis is known only from the western half of the Cypress Hills and is most abundant at elevations above 1150 m. The extremely dark young wood and short and conic thorns show strong resemblance to species of the Eurasian sect. *Sanguineae* Zabel ex C. K. Schneider. The unpitted pyrenes are a marked difference. The petals appear to be pale cream rather than pure white.

152. Crataegus purpurella J. B. Phipps & O'Kennon, J. Bot. Res. Inst. Texas 1: 1066, plates 6.2b, 7.2b, fig. 15. 2007 • Loch Lomond hawthorn [E]

Shrubs or trees, 20–35 dm, dense and twiggy. **Stems:** twigs: new growth ± red, glabrous, 1-year old ± glossy, medium to deep brown, 2-years old gray; thorns on twigs <numerous>, straight to slightly curved, 1-year old deep brown, 2-years old nearly black, slender to ± stout, 2–3.5 cm. **Leaves:** petiole length 30–40% blade, sessile-glandular, glands 1–few, small, adaxial sulcus pubescent; blade broadly rhombic to broadly elliptic or oblong, <length/width = 0.7–1>, 2.5–4 cm at anthesis, 3.5–4.5(–5) cm mature, base cuneate, lobes 2 or 3 per side, max LII 15–25%, lobe apex acute, margins serrulate or crenate-serrate, veins 3 or 4(or 5) per side, barely impressed, apex acute to subacute, abaxial surface glabrous, adaxial appressed-scabrous (young), often glabrescent. **Inflorescences** 6–15-flowered; branches glabrate to sparsely pilose; bracteoles pale brown, very narrow, margins glandular. **Flowers** 13–16 mm diam.; hypanthium glabrous or sparsely pilose; sepals usually pale green, triangular, 3 mm, abaxial surface glabrous, adaxial pubescent; anthers pink; styles 3 or 4. **Pomes** deep red to red-burgundy younger, deep purple fully ripe, ellipsoid to suborbicular, (9–)11–12 mm, glabrous; sepals eroded or spreading, <4–4.5 mm, pointed>; pyrenes 3 or 4, dorsally shallowly or more deeply grooved, sides longitudinally grooved or very shallowly concave.

Flowering late May–early Jun; fruiting Aug–Sep. Thickets, light shade of aspens; 800–1100 m; Sask.

Crataegus purpurella is known only from the eastern half of the Cypress Hills; it is common in thickets along north-slope draws. The species is most similar to *C. rivulopugnensis*, with which it is allopatric. In the open, it is a much denser shrub than *C. rivulopugnensis*; both have short, slender thorns. *Crataegus purpurella* also differs from *C. rivulopugnensis* in its usually smaller, proportionately wider, fewer-veined leaves and in the deep purple, rather than red, fruit at maturity.

153. Crataegus rivulopugnensis J. B. Phipps & O'Kennon, J. Bot. Res. Inst. Texas 1: 1070, plates 5c, 7.2d, fig. 17. 2007 • Battle Creek hawthorn [E]

Shrubs or trees, 20–50 dm, slender unless in full sun. **Stems:** twigs: new growth color not recorded, sparsely pubescent young, glabrescent, 1-year old deep reddish or deep purple-brown, 2-years old mid brown; thorns on twigs <moderate to very frequent>, straight to recurved, 1-year old bright, shiny, dark red-brown, slender, 2.5–4(–5) cm. **Leaves:** petiole length 25–30% blade, sessile-glandular, glands few to several, small, adaxial sulcus pubescent; blade broadly elliptic to ovate, 4–6 cm mature, base cuneate to broadly cuneate or rounded, lobes 3 or 4 per side, max LII 15–25%, lobe apex acute, margins serrulate or crenate-serrate, veins 5–8 per side, slightly impressed, apex acute, abaxial surface glabrous young, vein axils sometimes with axillary tufts, adaxial sparsely to moderately scabrous young, glabrate to sparsely scabrous mature. **Inflorescences** 5–15-flowered; branches sparsely to moderately pilose; bracteoles <few, even young>, light brown, very narrow, margins glandular. **Flowers** 16 mm diam.; hypanthium glabrous or ± pilose proximally; sepals green with pale edges, triangular, 4 mm, abaxial surface glabrous, adaxial sparsely pilose; anthers pale pink to pink; styles 4. **Pomes** red, becoming bright to deep red, ellipsoid to suborbicular, 8–10(–12) mm, glabrous or sparsely pubescent; sepals spreading-reflexed to recurved triangular, <glandular-serrate>; pyrenes 3–5, dorsally grooved, sides ± plane to shallowly and smoothly concave <concavity deepest on narrowest pyrenes>.

Flowering late May–early Jun; fruiting Aug–Sep. Thickets, light shade of aspens; 800–1200 m; Alta., Sask.

Crataegus rivulopugnensis, known only from the Cypress Hills, is widely distributed in the western part. The species is most similar to *C. purpurella*, though the color of the ripe fruit in the two species is quite different (red versus deep purple). They differ also in an almost allopatric distribution, habit when open-grown, and leaf size and numbers of veins per side (four or five versus three or four). Both species have less hairy adaxial leaf surfaces than most other sympatric hawthorns, often being glabrate adaxially at maturity.

Interserial hybrids and species of hybrid origin:

Numerous putative interserial hybrids among North American hawthorns have been given binomials (J. B. Phipps 2005). Nearly all the named putative hybrids have proven to be rare or even apparently extinct. All of the 16 treated here satisfy at least two of the following criteria: there is a high degree of confidence that they are of interserial hybrid origin; they do not fit into any accepted series; they continue to be found long after their initial description, indicating either apomictic perpetuation, repeated creation, or both; or, they have appeared regularly as species in the literature, notably in works by E. J. Palmer. It is notable that most remain uncommon, suggesting that these are not particularly fit biotypes. Nevertheless, persisting populations of some (for example, *Crataegus persimilis*) may be readily found. The fifteenth example (*C. turnerorum*) is a well-collected, obviously persistent, local species in Texas.

The putative interserial hybrid, *Crataegus florifera*, however, is treated in the main text (in this case under ser. *Anomalae*) because it fits the serial description adequately and has significant and continuous ranges. Other examples included in the main text are some putative ser. *Punctatae* × ser. *Crus-galli* crosses that will be found associated with *C. collina*. The best studied interserial hybrids are two without names (though this will change with a forthcoming paper by the Dickinson group), both hybrids of the introduced *C. monogyna*. Here they are treated in the main text with their native parental species (*C. punctata* and *C. suksdorfii*). Neither of these hybrids is considered especially rare, and both appear to be increasing in frequency, but so far there is no evidence of a fixed hybrid establishing.

Hybrids among North American native species have probably arisen multiple times (J. B. Phipps 2005); however, there was little evidence for the creation of successful new biotypes until some recent papers by the Dickinson group demonstrating modern derivatives of ancient alloploid of ser. *Crataegus* (E. Y. Y. Lo et al. 2009) and perhaps more recent alloploid derivatives from ser. *Douglasianae* (T. A. Dickinson, pers. comm.). Hybridization among more closely related apomicts may be more frequent than previously thought (Dickinson et al. 2008) but, with more cryptic characteristics, may more easily escape detection.

154. Crataegus ×atrorubens Ashe, J. Elisha Mitchell Sci. Soc. 16: 78. 1900, as species E

Shrubs or trees, 40–60 dm. **Stems:** twigs: new growth not recorded, 1-year old bright red-brown, older grayish; thorns on twigs not recorded. **Leaves:** petiole slender, length 30–50% blade, adaxial sulcus pubescent; blade ovate, (3–)4–7 cm, base rounded to very broadly cuneate, lobes 3 or 4 per side, sinuses very shallow to ± deep, max LII 5–20%, lobe apex acute, margins serrulate, venation camptodromous, veins 4 or 5(or 6) per side, apex acute, abaxial surface with axillary hair tufts extending ± on basal part of leaf, adaxial sparsely appressed-hairy young, glabrescent. **Inflorescences** 6–14-flowered; branches sparsely pubescent; bracteoles usually persistent, linear-spatulate, margins glandular, sometimes caducous, linear, membranous, margins eglandular. **Flowers** 15–18 mm diam.; hypanthium densely canescent; sepals narrow, margins glandular-serrate; stamens 20, anther color not recorded; styles 4 or 5. **Pomes** dark red, orbicular, 9–11 mm diam., glabrous; sepals erose or ± reflexed; pyrenes 4 or 5, dorsally grooved, sides plane.

Flowering Apr; fruiting Sep–Oct. Riverine flats; 200 m; Mo.

Crataegus ×atrorubens is probably a hybrid of *C. viridis* × *C. mollis*, originating near St. Louis. The *C. viridis* parentage is apparent in the axillary tufts of hairs on the abaxial leaf surfaces and in the occasional presence of eglandular bracteoles; the *C. mollis* parentage is apparent in leaf size and shape. This handsome tree is much favored by arboreta; a quite different *C. crus-galli* hybrid is often misidentified as it.

155. Crataegus ×bicknellii (Eggleston) Eggleston, Bull. Torrey Bot. Club 38: 244. 1911, as species E

Crataegus rotundifolia Moench var. *bicknellii* Eggleston, Rhodora 10: 79. 1908; *C. chrysocarpa* Ashe var. *bicknellii* (Eggleston) E. J. Palmer

Shrubs or trees, 30–80 dm. **Stems:** twigs: new growth greenish, glabrous, 1-year old tan, older darker or grayer; thorns on twigs few to numerous, 1-year old dark brown, ± stout, 2–6 cm. **Leaves** deciduous; petiole slender, length 50–70% blade, adaxial sulcus sparsely hairy, eglandular to barely glandular; blade broadly elliptic, 4–6 cm, base cuneate, lobes 4 or 5 per side, short, LII 10–15%, lobe apex acute, margins basally entire, otherwise serrate, veins 5 or 6 per side, adaxially impressed, apex acuminate, abaxial surface glabrous, adaxial sparsely scabrous young, glabrescent. **Inflorescences** 7–15-flowered; branches glabrous; bracteoles conspicuous, 7–10 mm, membranous, linear, margins glandular-serrate, proximal ones sometimes very long. **Flowers** 16–23 mm diam.; hypanthium glabrous; sepals 7 mm, margins glandular-laciniate; stamens 10, anthers dull purple; styles 3–5. **Pomes** red, ± orbicular, 10–12 mm diam., glabrous; sepals persistent, reflexed; pyrenes 4 or 5, dorsally grooved, sides plane or roughened with obscure cavities.

Flowering Jun; fruiting Sep–Oct. Brush; 0–10 m; Mass.

Crataegus ×bicknellii is known only from Nantucket Island; it is a good example of a small population of a rare hawthorn known to have persisted about 100 years. Palmer assigned *C. ×bicknellii* to *C. chrysocarpa*; it is not much like var. *chrysocarpa* in leaf shape, indumentum, or glandularity, nor in its striking laciniate sepals, nor purple anthers. *Crataegus ×bicknellii* also flowers much later than the *C. chrysocarpa* group. The author agrees with E. P. Kruschke (1965) that it is closest to *C. sylvestris*. The specimens seen by the author have glabrous pedicels in contrast to the slightly villous pedicels reported by Eggleston and Palmer. The flowers in recently collected material are fewer per inflorescence and much larger than described in the protologue.

156. Crataegus ×coleae Sargent, Trees & Shrubs 1: 7, plate 4. 1902, as species E

Shrubs, 40–50 dm, broad. **Stems:** trunks numerous, spreading, stout, bark close dark; twigs nearly straight or slightly flexuous, slender, new growth light yellow-green, glabrous, marked by oblong pale lenticels, at end of 1st year bright red-brown, very lustrous, 1-year old dark gray-brown; thorns on twigs numerous, straight, stout, 2.5–4.5 cm, lustrous, often retrorse. **Leaves:** petiole nearly terete, slender, 1.2–1.5 cm; blade ovate to oval, 4.5–5 × 3–3.5 cm, base broadly cuneate, lobes 4 or 5 per side, sinuses shallow, margins sharply serrate, teeth minutely glandular, glands dark, venation craspedodromous, veins 5 per side, apex acute, abaxial surface pale, adaxial dark green, lustrous, glabrate young, glabrescent. **Inflorescences** 8–12-flowered; branches glabrous; bracteoles caducous, linear, membranous, margins sessile- or stipitate-glandular. **Flowers** 18–20 mm diam.; hypanthium glabrous; sepals narrowly triangular, margins glandular-serrate; stamens 20, anthers pink, 0.6–0.8 mm; styles 2–4(or 5). **Pomes** bright orange-red, at most slightly pruinose, suborbicular, 8–10 mm diam., punctate, glabrous; sepals reflexed, elevated on collar; pyrenes 2–4(or 5), dorsally grooved, sides each linearly scarred.

Flowering May; fruiting late Sep. Successional fields, brush; 200–300 m; Mich.

Crataegus ×*coleae* is rare and known from near Grand Rapids from the late 1800s to the 1940s; it is a presumptive interserial hybrid *C. pruinosa* (ser. *Pruinosae*) × *C. scabrida* (ser. *Anomalae*). The somewhat unstable leaf form and scarred lateral pyrene faces, reminiscent of *C. scabrida*, and elevated fruiting calyx, 20 stamens, and glabrous parts, reminiscent of *C. pruinosa*, are good markers.

157. Crataegus ×collicola Ashe, J. Elisha Mitchell Sci. Soc. 16: 75. 1900, as species [E]

Crataegus collina Chapman var. *collicola* (Ashe) E. J. Palmer; *C.* ×*verruculosa* Sargent

Shrubs or trees, 50–80 dm. **Stems:** twigs: new growth glabrous or densely pubescent young, 1-year old gray-brown, older grayish; thorns on twigs numerous, dark chestnut brown, older gray, slender, 2–8 cm. **Leaves:** petiole length 20–25% blade, pubescent, glandular; blade narrowly obovate, 3.5–5 cm, base narrowly cuneate, lobes 0, or 2 or 3 per side, sinuses shallow, lobe apex acute, margins serrulate, teeth 0.5–0.7 mm, venation camptodromous, veins 3–6 per side, apex acute, ± lustrous, abaxial surface villous on veins, adaxial soft-white-hairy young, becoming sparsely scabrous. **Inflorescences** 6–14 -flowered; branches glabrous or sparsely pubescent; bracteoles caducous, numerous, reddish, very narrow, membranous, margins glandular. **Flowers** 18 mm diam.; hypanthium densely villous; sepals narrowly triangular, margins entire or sparsely glandular-serrate; stamens 20, anthers rose; styles 3–5. **Pomes** orange to dull or dark red, depressed orbicular to orbicular, 8–13 mm diam., pubescent young; sepals small, erose or ± patent; pyrenes 3–5, dorsally grooved, sides plane.

Flowering May; fruiting Sep–Oct. Fertile ground along streams, fields, dry pastures in mountains; 100–1000 m; Ark., Ky., Mo., N.C., Va.

Crataegus ×*collicola* and *C.* ×*verruculosa* are two different putative hybrids, both of which are expected to possess parentage suggested by E. J. Palmer (1952) of *C. collina* and *C. crus-galli*. *Crataegus* ×*collicola* has thorns 4–8 mm, young twigs glabrous, leaf blade broadly elliptic to obovate, two or three lobes, and four to six veins per side, glabrous inflorescence branches, entire sepal margins, and pomes orange to dull red, depressed orbicular, and 8–11 mm diam. *Crataegus* ×*verruculosa* has thorns 2–2.5 mm, young twigs hairy, leaf blades narrowly obovate, unlobed, veins three to five per side, hairy inflorescence branches, sparingly glandular-serrate sepal margins, and pomes dark red, orbicular, 10–13 mm diam. If these are considered to be the same hybrid species, separate nothovariety names should be erected to maintain distinction.

158. Crataegus ×disperma Ashe, J. Elisha Mitchell Sci. Soc. 17: 14. 1900, as species [E]

Crataegus grandis Ashe; *C. pausiaca* Ashe; *C. peoriensis* Sargent; *C. punctata* Jacquin var. *pausiaca* (Ashe) E. J. Palmer

Shrubs or trees, 50–80 dm. **Stems:** twigs: new growth reddish young, glabrous or pubescent, 1-year old shiny brown, glabrous, older dark gray; thorns on twigs 2-years old shiny, dark brown, fine, 3–4 cm. **Leaves** deciduous; petiole 0.7–1.5 cm, winged nearly to base, glabrous; blade dark green, obovate or elliptic, 2.5–5 cm, subcoriaceous, base broadly to narrowly cuneate, lobes 0 or few, sinuses very shallow, margins serrate, venation craspedodromous, veins 5–7 per side, apex acute, surfaces glabrous. **Inflorescences** 5–12-flowered; branches glabrous; bracteoles caducous, linear, membranous, margins glandular. **Flowers** 13–18 mm diam.; hypanthium glabrous; sepals narrow, margins ± entire or subserrate; stamens 10, anthers pink or red-purple; styles 2(or 3). **Pomes** deep red, ± oblong, 12–14 × 9–11 mm; sepals spreading, subsessile; pyrenes 2, dorsally grooved, sides plane.

Flowering May; fruiting Sep–Oct. Brush, along streams; 50–200 m; Ill., Ind., Iowa, Ohio, Pa., Va.

Crataegus ×*disperma* has been found sporadically from Iowa east to Virginia and Pennsylvania, where the ranges of *C. punctata* and *C. crus-galli* overlap.

Crataegus ×*disperma* belongs to a group of taxa intermediate between *C. crus-galli* and *C. punctata*; it may key to *C. collina*, although it has glossier leaves and usually fewer styles and pyrenes. Some specimens are more hairy, or with more strongly glandular-serrate sepal margins, than in the description, which is based on Ashe. *Crataegus danielsii* E. J. Palmer from Missouri, with glossy, more or less coriaceous, but lobed leaves is another problematic *C. crus-galli* × *C. collina* hybrid according to Palmer (1952). The pink-anthered *C.* ×*disperma* and cream-anthered *C.* ×*incaedua* are similar to *C. collina* var. *sordida* except for the number of styles and pyrenes (two or three).

159. Crataegus ×dispessa Ashe, J. Elisha Mitchell Sci. Soc. 19: 17. 1903, as species [E]

Crataegus pyriformis Britton, Bull. New York Bot. Gard. 1: 449. 1900, not Jacques 1858; *C. treleasei* Sargent

Trees, 55 dm. **Stems:** twigs: new growth canescent, 1-year old dark brown, older gray; thorns on twigs absent or few, straight or slightly curved, 1-year old dark purple, 3 cm. **Leaves** deciduous; petiole slender, length 33–43% blade, distally winged, densely pubescent, eglandular; blade elliptic to suborbiculate, 3–7 cm, base cuneate, lobes 0 or 1–3 per side, apiculi, margins sharply serrate, venation craspedodromous, veins 5 per side, sharply impressed, apex acute, surfaces glabrate, glabrescent, abaxial veins densely pubescent, adaxial appressed-pubescent young. **Inflorescences** 10–20-flowered; branches tomentose; bracteoles caducous, numerous, linear, ± membranous, margins glandular. **Flowers** 20–25 mm diam.; hypanthium canescent; sepals narrowly triangular, margins glandular-serrate to glandular-laciniate, abaxial surface densely pubescent; stamens 20, anthers cream, sometimes pink; styles 4 or 5. **Pomes** crimson, suborbicular, 12–17 mm diam., pubescent; sepals erose or patent; pyrenes 4 or 5, dorsally grooved, sides plane.

Flowering May; fruiting Sep–Oct. Brush, along streams and woodland margins; 100–200 m; Mo.

Crataegus ×*dispessa* is found in southern Missouri to Boone County and was last collected in Dallas County.

Crataegus ×*dispessa* is probably of ser. *Molles* × *C. collina* var. *hirtiflora* parentage, the bracteole type not being characteristic of ser. *Molles*.

160. Crataegus ×fretalis Sargent, Rhodora 5: 112. 1903, as species [E]

Shrubs or trees, 60–70 dm. **Stems:** trunk bark grayish to grayish brown, checked; twigs straight or slightly flexuous, new growth green, tinged red, at end of 1st year bright reddish brown, lustrous, 1–2-years old darker red-brown; thorns on twigs few, ± straight, 1-year old shiny red-brown, stout, 2–4 cm. **Leaves** deciduous; petiole slender, length 33–60% blade, glabrous, minutely red-glandular; blade yellow-green, ovate or oval, 5–8 × 4.5–7 cm, thin but firm, base rounded, lobes 3–5 per side, small, lobe apex acute to acuminate (or larger and sometimes with proximal lobes enlarged and deeply cleft), margins sharply serrate, venation craspedodromous, veins 4–6 per side, apex acute to acuminate, abaxial surface glabrous, adaxial short-appressed-pubescent young, glabrescent. **Inflorescences** 5–12 -flowered; branches glabrous; bracteoles caducous, linear, membranous, margins glandular. **Flowers** 20 mm diam.; hypanthium glabrous; sepals lanceolate, margins coarsely glandular-serrate; stamens 5(–8), anthers red, small; styles 3. **Pomes** bright red, oblong or obovate, 10–12 × 9–10 mm, glabrous; flesh slightly succulent; sepals often lost, ± erect, incurved; pyrenes 3, dorsally grooved, sides plane.

Flowering May; fruiting Sep–Oct. Brush, open woods, rocky pastures; 20–100 m; Conn., Maine, Mass.

Crataegus ×*fretalis* is probably a hybrid between *C. macrosperma* and a species of ser. *Coccineae*. *Crataegus tortilis* Ashe from northern Illinois and Wisconsin is similar but has 20 stamens and usually five pyrenes. *Crataegus vittata* Ashe, from Pennsylvania, is also similar, but has truncate leaf bases. *Crataegus knieskerniana* Sargent is similar in foliage, flowers, and fruit, but it has pyrenes depressed or pitted on the inner faces and occurs near East Herkimer, New York.

161. Crataegus ×incaedua Sargent, Trees & Shrubs 2: 3, plate 102. 1907, as species [E]

Shrubs or trees, 50–70 dm. **Stems:** twigs: new growth pubescent, 1-year old light gray-brown, older grayish; thorns on twigs numerous, 1-year old bright chestnut brown, slender, 4–7 cm. **Leaves:** petiole length 20% blade, villous early, glabrescent; blade narrowly obovate, 5–6 cm, base narrowly cuneate, lobes 0, margins serrulate, venation camptodromous, veins 4–6 per side, apex acute, abaxial surface veins villous, adaxial soft-white-hairy young, glabrescent. **Inflorescences** 12–20-flowered; branches puberulent; bracteoles ± persistent, numerous, very narrow, membranous, margins glandular. **Flowers** 14–15 mm diam.; hypanthium densely villous; sepals narrowly triangular, margins glandular-serrate; stamens 10–15, anthers cream; styles 2 or 3. **Pomes** yellowish red, suborbicular, 9–10 mm diam., punctate, glabrous; sepals narrow, ± reflexed; pyrenes 2, dorsally grooved, sides plane.

Flowering May; fruiting Sep–Oct. Gravelly stream beds; 150–200 m; Ark., Ind., Mo.

Crataegus ×*incaedua* is similar to *C. collina* var. *sordida* except for the cream anthers and two or three styles and pyrenes; it probably is a hybrid of *C. collina* with *C. calpodendron*, according to E. J. Palmer (1952).

162. Crataegus ×kelloggii Sargent, Trees & Shrubs 1: 117, plate 59. 1903, as species [E]

Trees, 60–70 dm. **Stems:** trunks nearly black, 10 cm diam., bark deeply grooved; twigs: new growth nearly straight, dark brown-tinged with red, slender, slightly villous young, at end of 1st year bright red-brown, lustrous, marked by many small pale lenticels, 1-year old dark, dull reddish brown; thorns on twigs absent or present, nearly straight, shiny, 1-year old bright chestnut brown, slender, 3 cm. **Leaves** deciduous; stipules caducous, turning reddish, linear to linear-obovate, apex acuminate, glandular; petiole slender, 2–4 cm, slightly winged apically, white villous young, glabrescent, grooved; blade dark yellow-green, broadly ovate to suborbiculate, 5–6.5 × 4.5–6.5 cm, often wider than long, ½ grown when flowers open, thin but firm, base rounded, broadly cuneate, or truncate, lobes 4–6 per side, short, wide, lobe apex subacute to acute, margins coarsely serrate almost to base, teeth straight, gland-tipped, venation craspedodromous, veins 4–6 per side, prominent, slender, apex rounded, often short-pointed, abaxial surface paler, veins hairy, glabrescent except few hairs near base of thin yellow midvein, adaxial short-pale-hairy young, glabrescent. **Inflorescences** 5–10-flowered, compact; branches sparsely villous; bracteoles persistent until anthesis, oblong-obovate to linear, apex acuminate, margins glandular. **Flowers** 16 mm diam.; hypanthium hairy proximally; sepals slender, margins entire or minute, dark red stipitate-glandular, apex acuminate, abaxial surface glabrous, adaxial sparsely villous; stamens 20, anthers pale rose; styles 4 or 5. **Pomes** drooping on long, slender, glabrous pedicels, bright yellow, suborbicular to short-ovate, 12–18 mm diam., lenticels numerous, small, pale; flesh yellow, thin, dry, mealy; sepals often deciduous from ripe fruit, small, with narrowly shallow cavity, slightly villous distally; pyrenes 4 or 5, acute at ends, rounded, 8 × 6 mm, dorsally very slightly grooved on back, sides plane.

Flowering Apr; fruiting Sep–Oct. Brush, open woods, rocky pastures; 100–200 m; Mo.

Crataegus ×*kelloggii* was originally collected at the mouth of River des Peres, St. Louis, and was last found in Hannibal, near the Mississippi River; always rare, it is unknown in the wild today.

In all probability, *Crataegus* ×*kelloggii* is *C. mollis* × *C. margarettae*, as suggested by E. J. Palmer (1952).

163. Crataegus ×latebrosa Sargent, Rep. (Annual) Missouri Bot. Gard. 22: 73. 1912, as species [E]

Crataegus noelensis Sargent

Trees, 60–70 dm. **Stems:** trunks 15 cm diam.; compound thorns sometimes present on old trunks; branches spreading, forming wide flat or round-topped head; twigs: new growth with matted white hairs, later in 1st year becoming reddish brown, pubescent or puberulous, strongly flexuous, 1-year old gray, stout; thorns on twigs few or numerous, straight, 1-year old purple, slender, 2–7 cm, lustrous. **Leaves** deciduous; petiole slender, length 20–37% blade, slightly winged apically, hoary tomentose early, glabrescent; blade dark yellow-green, ovate to oval, 5–8 × 3–6 cm, more than ½ grown at anthesis, base narrowly or broadly cuneate, lobes 0, margins coarsely doubly serrate, teeth straight, venation craspedodromous, midvein prominent, veins thin, conspicuous, apex acute or acuminate, rarely rounded, abaxial surface villous, adaxial densely villous glabrescent; leaves at end of vigorous extension shoots broadly ovate, often 9 × 7–8 cm, base broadly rounded or cuneate, lobes usually short, lobe apex broadly acuminate, margins more coarsely serrate, apex acuminate. **Inflorescences** 5–10 -flowered, compact; branches long-matted white-hairy; bracteoles caducous, linear, membranous, margins glandular. **Flowers** 20–25 mm diam.; hypanthium pubescent; sepals slender, margins minutely glandular-serrate, apex long-acuminate, abaxial surface slightly villous; stamens (5–)10, anthers rose; styles 3–5, base surrounded by wide ring of pale tomentum. **Pomes** on slender, drooping, pubescent pedicels, orange-red, suborbicular, 10 mm diam., hairy; flesh yellow, thin, soft; sepals prominent, with short tube, spreading or closely appressed; pyrenes 3–5, 5–6 mm, base rounded, apex narrowed and rounded, dorsally slightly grooved, sides plane.

Flowering Apr; fruiting Sep–Oct. Brush, open woods, rocky pastures; 100–200 m; Ark., Mo.

Crataegus ×*latebrosa* differs little from *C.* ×*dispessa* except in the lower stamen number (five to ten) and wider range of style number (three to five). The only record of fruit color says orange-red; reports for *C.* ×*dispessa* uniformly declare crimson. The very narrow, caducous bracteoles and the leaf shape of *C.* ×*latebrosa* suggest a *C. collina* × ser. *Molles* ancestry.

164. Crataegus laurentiana Sargent, Rhodora 3: 77. 1901 • Laurentian hawthorn [E]

Crataegus brunetiana Sargent; *C. brunetiana* var. *fernaldii* (Sargent) E. J. Palmer; *C. fernaldii* Sargent; *C. laurentiana* var. *brunetiana* (Sargent) Kruschke

Shrubs, 30–50 dm. **Stems:** trunks much-branched, thick, stout; twigs strongly flexuous, stout, new growth light orange-green to dark green, tomentose, soon glabrescent, 1-year old bright orange-brown, very lustrous, older gray; thorns on twigs often retrorse, 1-year old lustrous chestnut brown to very dark red-brown, very stout, 5–7.5 cm. **Leaves** deciduous; stipules bright red before falling, lanceolate to oblanceolate, 8–15 mm, finely glandular-serrate; petiole often dark red after mid summer, stout to slender, length 40–50% blade, distally ± broadly winged, villous young, glabrescent, deeply grooved, sessile-glandular young, glands caducous, dark, small; blade in early spring yellow-green, dark green, abaxially paler, oblong to oblong-obovate or ovate to rhombic (suborbiculate in var. *dissimilifolia*), 4–6 cm, thin to subcoriaceous, base gradually or abruptly narrowed from near middle, lobes 4 or 5 per side, sinuses ± deep (except in var. *dissimilifolia*), LII 20–33%, narrow, lobe apex acute, margins sharply glandular-serrate except proximally, venation craspedodromous, veins 5 per side, apex acute or acuminate, matte, abaxial surface with stout midvein, glabrous, rarely sparsely hairy, veins sparsely white villous, adaxial pale, short-appressed-hairy. **Inflorescences** 6–15-flowered, lax; branches hairy; bracteoles caducous, pallid bright red, linear, membranous, margins finely glandular-serrate. **Flowers** 15(–20) mm diam.; hypanthium externally pale, long-matted hairy at base, internally glabrate to puberulous; sepals reflexed, shorter than petals, narrow, margins conspicuously glandular-serrate, apex acuminate, abaxial surface glabrate, adaxial villous; stamens 10, anthers cream or pale pink, fading purple; styles 3–5. **Pomes** in drooping or erect infructescences, branches slightly villous, red to dark crimson, oblong to pyriform, 8–15 mm; flesh yellow, thin, becoming sweet and succulent; sepals prominent, appressed, with deep, wide cavity; pyrenes (3 or)4 or 5, thick, wide, 6 mm, dorsally prominently grooved, sides ± roughened, irregularly eroded or pitted.

Flowering Jun; fruiting Sep–Oct. Brush, open woods, rocky pastures; 20–200 m; Nfld. and Labr. (Nfld.), Que.; Maine, Mich., Minn., Wis.

Crataegus laurentiana was considered a sporadic entity of the northeast and Wisconsin until the recent discovery of a large population in Minnesota.

Hybrid origin for *Crataegus laurentiana* between a ser. *Macracanthae* species and *C. chrysocarpa* in the broad sense, or *C. jonesiae*, is supported by intermediacy and sympatry of the suspected parents. Chrysocarpoid characteristics include leaf shape and glandular petioles; macracanthoid traits include late anthesis, pink anthers, and variably eroded pyrene sides.

Specimens from Newfoundland and Quebec are var. *laurentiana* and are characterized by larger but proportionately narrow leaves, thicker and paler twigs, and pink, normal-sized (1 mm) anthers. They may be distinguished from the partly sympatric var. *brunetiana* by the smaller and proportionately broader leaves of the latter, and cream anthers, more like *Crataegus chrysocarpa*. In both forms, the leaves abaxially have conspicuously appressed-pilose veins. A Maine variant is atypically arborescent.

The rare var. *dissimilifolia* Kruschke from northern Wisconsin has white-cream anthers but larger and more nearly isodiametric, shallowly incised leaves. Similar, also rare, plants from southern Manitoba have pink anthers. Stature and leaf size and shape in both of these place them outside the range of *Crataegus chrysocarpa*; perhaps *C. submollis* is a parent.

Unnamed populations from a large area in northeastern Minnesota are the most biologically interesting. They are self-perpetuating and, compared to other forms, have more slender twigs that are generally very dark in their second year, like *Crataegus macracantha*, smaller leaves, somewhat impressed leaf venation, small (0.6 mm), pink anthers, and ellipsoid fruits.

Crataegus fernaldii is a rare form from Maine and Quebec with lax, elongated, extremely villous corymbs, large flowers (20 mm diam.), pink anthers, and large, pear-shaped fruits gracefully drooping in wide clusters on their long stems.

165. Crataegus ×lucorum Sargent, Bot. Gaz. 31: 227. 1901, as species [E]

Shrubs or trees, 50–70 dm. **Stems:** twigs: new growth slightly villous, 1-year old brown, older dark gray-brown; thorns on twigs ± straight to ± recurved, 2-years old bright red-brown, slender to moderately stout, 4–5 cm. **Leaves** deciduous; petiole very slender, length 50–60% blade, glandular; blade ovate to obovate, 4–5 cm, thin, base rounded to broadly cuneate, lobes 3 or 4 per side, sinuses shallow to moderately deep, lobe apex acute, margins serrate, venation craspedodromous, veins 4 or 5 per side, apex acute, surfaces glabrous. **Inflorescences** 5–10-flowered; branches sparsely villous; bracteoles caducous, ± linear, membranous, margins glandular-serrate. **Flowers** 18 mm diam.; hypanthium glabrous; sepal margins glandular-serrate; stamens 20, anthers purple; styles 4 or 5. **Pomes** red, ± obovoid, 12–16 mm; flesh succulent; pyrenes dorsally grooved, sides plane.

Flowering May; fruiting Sep–Oct. Brush, borders of woods, stream banks; 200–300 m; Ont.; Ill., Ind., N.Y., Ohio.

Crataegus ×*lucorum* is scattered from Illinois to New York, exhibiting a mixture of characteristics of ser. *Coccineae* and ser. *Tenuifoliae*.

166. Crataegus persimilis Sargent, Proc. Rochester Acad. Sci. 4: 94. 1903 • Plumleaf hawthorn [E]

Crataegus cerasina Sargent; *C. crus-galli* Linnaeus var. *prunifolia* Torrey & A. Gray; *C. helderbergensis* Sargent; *C. laetifica* Sargent; *C. livoniana* Sargent; *C. nuda* Sargent; *C. robusta* Sargent

Shrubs or trees, 50–60 dm. **Stems:** twigs: new growth glabrous, 1-year old purplish brown, older dull gray; thorns on twigs straight to ± recurved, 2-years old shiny blackish purple, slender to moderately stout, 4–7 cm. **Leaves** deciduous; petiole length 20–30% blade, glabrous; blade ± narrowly obovate to broadly elliptic or rhombic-elliptic, 4–5 cm, coriaceous, base cuneate, lobes 0, or 3 or 4 per side, sinuses very shallow, lobe apex acute, margins sharply serrate, teeth 1–2 mm, venation craspedodromous, veins 7 per side, apex broadly acute, glossy, surfaces glabrate or glabrous, adaxial midvein hairy young. **Inflorescences** 8–18-flowered; branches punctate, sparsely to densely pubescent; bracteoles caducous, linear, membranous, margins glandular. **Flowers** 12–17 mm diam.; hypanthium glabrous or villous at base; sepals narrowly triangular, 4–5 mm, margins deeply glandular-serrate, adaxial surface glabrous; stamens 10–15(–20), anthers cream or pink; styles (1 or)2(or 3). **Pomes** bright scarlet, glossy, suborbicular to ± ellipsoid, 10 mm diam., glabrous; flesh mealy, ± thin; sepals patent; pyrenes (1 or)2(or 3), dorsally grooved, sides plane to deeply excavated.

Flowering May; fruiting Sep–Oct. Brush, open woods, rocky pastures; 20–100 m; Ont.; Ky., N.Y., Ohio, Va., Wis.

Crataegus persimilis has a large but sporadic distribution; some populations appear to be self-perpetuating. The species is widely cultivated.

Crataegus persimilis is plausibly a hybrid complex of *C. crus-galli* in the broad sense with *C. succulenta* in the broad sense. This is suggested by the glossy and subcoriaceous leaves like *C. crus-galli*, with larger size and broader proportions as in *C. succulenta*, and also dark and glossy one-year old twigs, very dark, more or less stout one-year old thorns, deeply incised sepal margins, and commonly pitted pyrenes (the sides usually excavated with a slit).

Included here is a substantial range of forms among which there is considerable variation in leaf shape, inflorescence indumentum, stamen number, anther color, pyrene excavation as well as other characters. The type form from near Rochester, New York, has the narrowest leaves (length/width = 1.5–2), most stamens (ten to 20), pink anthers, sparsely pubescent inflorescence branches, and deeply pitted pyrenes. The commonly cultivated '*prunifolia*' form is similar but has a much broader leaf (length/width = 1.3–1.5) as well as densely pubescent or nearly tomentose inflorescence branches, and ten pink anthers. *Crataegus helderbergensis* resembles the last but has ten to 14 white anthers and no lateral excavation to the pyrenes. Other similar taxa that are candidates for inclusion are: *C. robusta*, with wide leaves and nearly glabrous inflorescence branches, larger flowers (18 mm diam.), ten to 20 pink anthers, plane-sided pyrenes in the type though excavated in syntype material; and *C. cerasina*, similar, with laterally pitted pyrenes and smaller flowers (12 mm diam.). *Crataegus livoniana* Sargent has leaves more like *C. crus-galli* but barely eroded pyrenes and may not belong. *Crataegus laetifica* of Pennsylvania and Ohio is another similar form.

The cultivated plant is often called *Crataegus prunifolia* Persoon in the horticultural literature; this name is invalid because it is a later homonym of *C. prunifolia* (Marshall) Baumgarten, a synonym of *Aronia prunifolia*, the chokeberry. *Crataegus persimilis*, in its '*prunifolia*' form is one of the more popular cultivated hawthorns. It is a handsome ornamental when well grown, with glossy foliage, striking scarlet fruit and fall color of foliage.

167. Crataegus ×sicca Sargent, Crataegus Missouri, 101. 1908, as species [E]

Shrubs, 20–50 dm. **Stems:** twigs: new growth glabrate, 1-year old purple-brown, 2-years old dark gray-brown; thorns on twigs straight, 2-years old black, 3–7 cm. **Leaves:** petiole length 50% blade, winged distally, glandular; blade green, rhombic to rhombic-ovate, 2–4 cm, base cuneate, lobes 2–4 per side, sinuses shallow, lobe apex acute or rounded, margins finely serrate, venation craspedodromous, veins 5 per side, apex acute, veins abaxially sparsely pilose young, persistent, adaxial glabrous or sparsely appressed-pilose young, glabrescent. **Inflorescences** 4–6-flowered; branches glabrous or sparsely, sometimes densely, villous; bracteoles caducous, numerous, reddish, linear, membranous, margins glandular. **Flowers** 10–20 mm diam.; hypanthium glabrous; sepals narrowly triangular, margins glandular-serrate, abaxial surface glabrous; stamens 10, anthers cream, sometimes pale pink; styles 3–5. **Pomes** pale orange, strongly pruinose, or deep red, suborbicular, 10 mm diam., glabrous; flesh soft when mature; sepals patent-reflexed; pyrenes 3–5, dorsally grooved, sides plane.

Nothovarieties 2 (2 in the flora): c United States.

Crataegus ×*sicca* is a poorly understood taxon from scattered locations in southern Missouri and southern Ohio. It is likely an interserial hybrid between *C. margarettae* and one or more entities in the *C. pruinosa* complex.

1. Flowers 18–20 mm diam.; inflorescence branches and leaves villous . . . . 167a. *Crataegus* ×*sicca* nothovar. *sicca*
1. Flowers 10–14 mm diam.; inflorescence branches and leaves glabrous. . . . . 167b. *Crataegus* ×*sicca* nothovar. *glabrifolia*

167a. Crataegus ×sicca Sargent nothovar. **sicca** E

Shrubs, 20–30 dm. **Stems:** thorns 3–5 cm. **Leaf blades:** lobe apex acute, surfaces sparsely villous young, glabrescent. **Inflorescence branches** sparsely, sometimes densely, villous. **Flowers** 18–20 mm diam. **Pomes** pale orange, <strongly pruinose>.

Flowering Apr; fruiting Sep–Oct. Brush; 100–200 m; Mo.

Nothovariety *sicca* is scattered throughout southern Missouri.

The small size of plant and leaves, with a certain amount of villosity, suggests *Crataegus margarettae* as one parent; the thorns, flowers, pomes, and leaf shape suggest *C. pruinosa* var. *pruinosa* for the other.

167b. Crataegus ×sicca Sargent nothovar. **glabrifolia** (Sargent) E. J. Palmer, Ohio J. Sci. 56: 213. 1956, as variety E

Crataegus glabrifolia Sargent, Rep. (Annual) Missouri Bot. Gard. 22: 74. 1912

Shrubs, 40–50 dm. **Stems:** thorns 4–7 cm. **Leaf blades:** lobe apex rounded, surfaces glabrous. **Inflorescence branches** glabrous. **Flowers** 10–14 mm diam. **Pomes** deep red.

Flowering Apr; fruiting Sep–Oct. Brush; 100–200 m; Mo., Ohio.

Nothovariety *glabrifolia* is scattered over southern Missouri and southern Ohio.

In spite of the differences from the typical nothovariety, a similar origin is likely.

168. Crataegus turnerorum Enquist, Phytologia 68: 298, fig. 1. 1990 • Scalp Creek hawthorn E

Crataegus secreta J. B. Phipps

Shrubs or trees, 20–60 dm. **Stems:** trunk bark gray, plated; twigs straight or slightly flexuous, new growth glabrous, 1-year old light brown, 2-years old shiny, older pale ashy gray; thorns on twigs numerous, straight to slightly recurved, 2-years old black, slender, 2–5 cm. **Leaves** deciduous; petiole 25–33% length of blade, pubescent, eglandular; blade broadly elliptic to ovate or narrowly obovate, 2–3 cm, thin, base broadly cuneate, lobes 2 or 3 per side, sinuses very shallow, lobe apex subacute to broadly acute, margins serrate except near base, teeth small, acute or subacute, eglandular, venation craspedodromous, veins 3 or 4 per side, apex broadly acute, abaxial surface glabrous, adaxial pubescent young, glabrescent. **Inflorescences** 3–12-flowered, flattish, convex panicles; branches sparsely pubescent or glabrous; bracteoles caducous, few, very narrowly elliptic, membranous, margins glandular. **Flowers** 15 mm diam.; hypanthium glabrous; sepals narrowly triangular, ca. ⅔ length of petals, margins ± entire with some stipitate glands to glandular-serrate, adaxial surface pubescent; stamens 15, anthers purple; styles 3–5. **Pomes** red, suborbicular (slightly oblate), 10–13 mm diam., glabrous; flesh mealy; sepals or sepal remnants spreading; pyrenes 3–5, dorsally grooved, sides plane.

Flowering Apr; fruiting Sep–Oct. Brush, dwarf oak scrub, stream banks; 300–600 m; Tex.

Crataegus turnerorum is a distinctive and little-varying taxon recorded from at least seven counties in the Edwards Plateau of Texas. It is placed here as it is unassigned to series and may perhaps be a hybrid between the partly sympatric *C. greggiana*, than which it is much less hairy, with a more or less glabrous small-leaved taxon such as the partly sympatric *C. viridis* var. *desertorum* or even a member of ser. *Tenuifoliae* as discussed in J. B. Phipps (1990). *Crataegus turnerorum* sometimes has slightly flexuous twigs, which could credibly be inherited from a *C. viridis* var. *desertorum* ancestor. However, the nearest existing locality for the ser. *Tenuifoliae* option is montane New Mexico, even though the Edwards Plateau harbors many woody species with such typically eastern-northeastern ranges as ser. *Tenuifoliae*.

SELECTED REFERENCE Phipps, J. B. 1990. *Crataegus secreta* (Rosaceae), a new species of hawthorn from the Edwards Plateau, Texas. Sida 14: 13–19.

169. Crataegus ×vailiae Britton, Bull. Torrey Bot. Club 24: 53. 1897, as species • Vail's hawthorn [E]

Crataegus conjungens Sargent; *C. missouriensis* Ashe; *C. palliata* Sargent

Shrubs, 20–40(–50) dm. **Stems:** twigs: new growth densely pubescent, 1-year old sparsely pubescent, 2-years old shiny, dark blackish brown to dark gray, glabrous, older dark gray; thorns on twigs absent or few, straight to slightly recurved, 2-years old black, ± narrow, 3 cm. **Leaves** deciduous; petiole length 5–15% blade, pubescent, sometimes glandular; blade elliptic to rhombic-elliptic, 4–5(–6) cm, ± coriaceous, base cuneate, lobes 1–3 per side, sinuses extremely shallow, lobe apex subacute, margins toothed, teeth large, acute or subacute except near base, sharp or obtuse, eglandular, tapered to bases, venation craspedodromous, veins 4–6 per side, apex subacute, adaxial surfaces matte, appressed-pubescent young, abaxial very sparsely pubescent, glabrescent, veins persistently densely pubescent. **Inflorescences** 2–8(–20)-flowered; branches very densely appressed-pubescent; bracteoles caducous, few, linear, membranous, margins glandular, sometimes larger (15 × 3 mm), more herbaceous in basal parts. **Flowers** 15–20 mm diam.; hypanthium densely pubescent; sepals narrowly triangular, 6–8 mm, ± equal to petals, ± foliaceous, margins glandular-laciniate, abaxial surface slightly pubescent; stamens 20, anthers cream, pink, or red; styles 3–5. **Pomes** straw yellow to ruddy, suborbicular, 7–9 mm diam., hairy; sepals reflexed; pyrenes 3–5, dorsally grooved, sides usually shallowly eroded.

Flowering Apr–May; fruiting Sep–Oct. Brush, stream banks, borders of woods; 150–700 m; Ark., Ga., Mo., N.C., Tenn., Va.

Crataegus ×vailiae, with its similarities to *C. uniflora*, may be a hybrid between that species and a member of ser. *Macracanthae*, most likely *C. calpodendron*. Fruit color is recorded as in cultivation at Kew; B. F. Bush (on herbarium label) noted it as deep red and succulent on wild plants in Missouri. *Crataegus conjungens* appears to be a smaller form of *C. ×vailiae*.

65. MESPILUS Linnaeus, Sp. Pl. 1: 478. 1753; Gen. Pl. ed. 5, 214. 1754 • Medlar [Greek *mesos*, half, and *pilos*, felt or ball, perhaps alluding to shape of medlar fruit resembling half a ball] [I]

James B. Phipps

Crataegus Linnaeus sect. *Mespilus* (Linnaeus) T. A. Dickinson & E. Y. Y. Lo

Shrubs or trees, 40–70 dm. **Stems** 1–few, ± erect; bark grayish, <smooth, with numerous small horizontal lenticels, older rough, dark gray-brown>; short shoots rare or present; armed or unarmed, <compound thorns on trunk absent>; extending twigs canescent. **Leaves** deciduous, cauline, simple; stipules caducous, free, ± elliptic, <herbaceous>, margins entire or lobed, <asymmetric, ± pilose>; petiole present; blade narrowly oblong or narrowly elliptic to oblanceolate, rarely oblong-ovate, 5–12 cm, thin, margins flat, entire or finely serrate distally, venation camptodromous, surfaces ± pubescent. **Inflorescences** terminal <on leafy short shoots, these terminal on woody short shoots or arising directly from extension shoots>, 1 (or 2)-flowered, axes canescent; bracts present; bracteoles present, <caducous or persistent, ± linear, 5–15 mm, membranous to coriaceous, abaxially ± hairy, eglandular>. **Pedicels** present, short. **Flowers:** perianth and androecium epigynous, 25–35 mm diam.; hypanthium ± obconic, constricted apically, size not recorded, externally canescent; sepals 5, spreading, triangular; petals 5, white, usually suborbiculate, <base barely clawed, apex ± notched>; stamens 25–35(–40), shorter than petals; carpels 5, connate, adnate to hypanthium, styles <5>, subterminal, distinct; ovules 2, <superposed>. **Fruits** pomes, brownish, turbinate to subspheric, 15–40 mm diam. (larger in some cultivars), <rough, punctate>, glabrous or pubescent; hypanthium persistent; sepals persistent, erect to ± connivent, <bases not touching>; carpels woody; styles ± persistent, <emerging through disc>. **Pyrenes** 5, <sides plane, covered by disc>. $x = 17$.

Species 1: introduced; Eurasia.

The Eurasian *Mespilus germanica* is a minor fruit, occasionally cultivated and naturalized in North America. *Mespilus* is retained here in the traditional concept of *M. germanica*, because it is distinct on a variety of characteristics and has a sister position to *Crataegus* (C. S. Campbell et al. 2007; E. Y. Y. Lo et al. 2007; D. Potter et al. 2007; Li Q. Y. et al. 2012; Lo and M. J. Donoghue 2012). Reproductive woody short shoots with close internodes, normal in *Crataegus*, occur frequently, though not universally; in the latter case, the inflorescence is borne on an annual short shoot lateral to or terminal on an extension shoot. Other notable characteristics of *Mespilus* that distinguish it from *Crataegus* include brownish fruit indicating mammal dispersal (C. M. Herrera 1989); large, hairy, eglandular bracteoles; extremely wide hypanthial opening in fruit; pyrenes covered by tissue of disc; and erect to connivent foliaceous fruiting sepals with bases distinct. *Mespilus* characters rarely seen in *Crataegus* are uniflorous inflorescences (found in *C. uniflora*); inflorescences borne on annual short shoots as noted above; very large flowers; and stamens 25 to 35 (found in *C. triflora*).

The triploid ×*Crataemespilus canescens* (J. B. Phipps) J. B. Phipps [*Mespilus canescens* J. B. Phipps and *Crataegus* ×*canescens* (J. B. Phipps) T. A. Dickinson & E. Y. Y. Lo], is a plant with great horticultural potential. It is a large bush with numerous fasciculate stems; exfoliating bark; simple, nearly entire leaves; 2–6-flowered, semipendant, usually racemose, inflorescences that are commonly borne on short shoots of the season lateral to an extension shoot; flowers very like *M. germanica* in size and general appearance, but with fewer stamens; small bright red, hawthornlike fruit; and large, abaxially hairy bracteoles with scattered very large marginal glands. A substantial colony has existed at one site in Arkansas.

SELECTED REFERENCES Baird, J. R. and J. W. Thieret. 1989. The medlar (*Mespilus germanica*, Rosaceae) from antiquity to obscurity. Econ. Bot. 43: 328–372. Li, Q. Y. et al. 2012. Generic limits of Pyrinae: Insights from ribosomal DNA sequences. Bot. Stud. 53: 151–164. Lo, E. Y. Y. and M. J. Donoghue. 2012. Expanded phylogenetic and dating analyses of the apples and their relatives (Pyreae, Rosaceae). Molec. Phylogen. Evol. 63: 230–243. Phipps, J. B. et al. 1991. Isozyme evidence for the naturalness of *Mespilus* L. (Rosaceae, subfam. Maloideae). Syst. Bot. 16: 546–552. Phipps, J. B. et al. 2003. Hawthorns and Medlars. Portland.

1. Mespilus germanica Linnaeus, Sp. Pl. 1: 478. 1753 • Common medlar F I

Stems not fascicled; twigs sometimes thorny, thorns 1–2 cm. **Leaves** closely spaced on twigs; petiole 1–4 mm; blade: lateral veins 8–12 per side. **Inflorescences** 1(or 2)-flowered; bracteoles 5–15 mm, adaxially hairy, marginal glands absent. **Flowers:** anthers red (cream). $2n = 34$.

Flowering May. Scrubby habitats; 700 m; introduced, Wash.; w Eurasia.

Mespilus germanica is native to western Eurasia, primarily to the east and south of the Black Sea. The species was previously widely cultivated as a minor fruit in the Old World; it is hardly known in North America outside arboreta. The brownish fruits are related to a relatively unusual mechanism of fruit dispersal by carnivorous animals (C. M. Herrera 1989) found also in other Maleae species. The only wild plants known from the flora area are in Washington; there are other suitable areas in North America where this species might become naturalized.

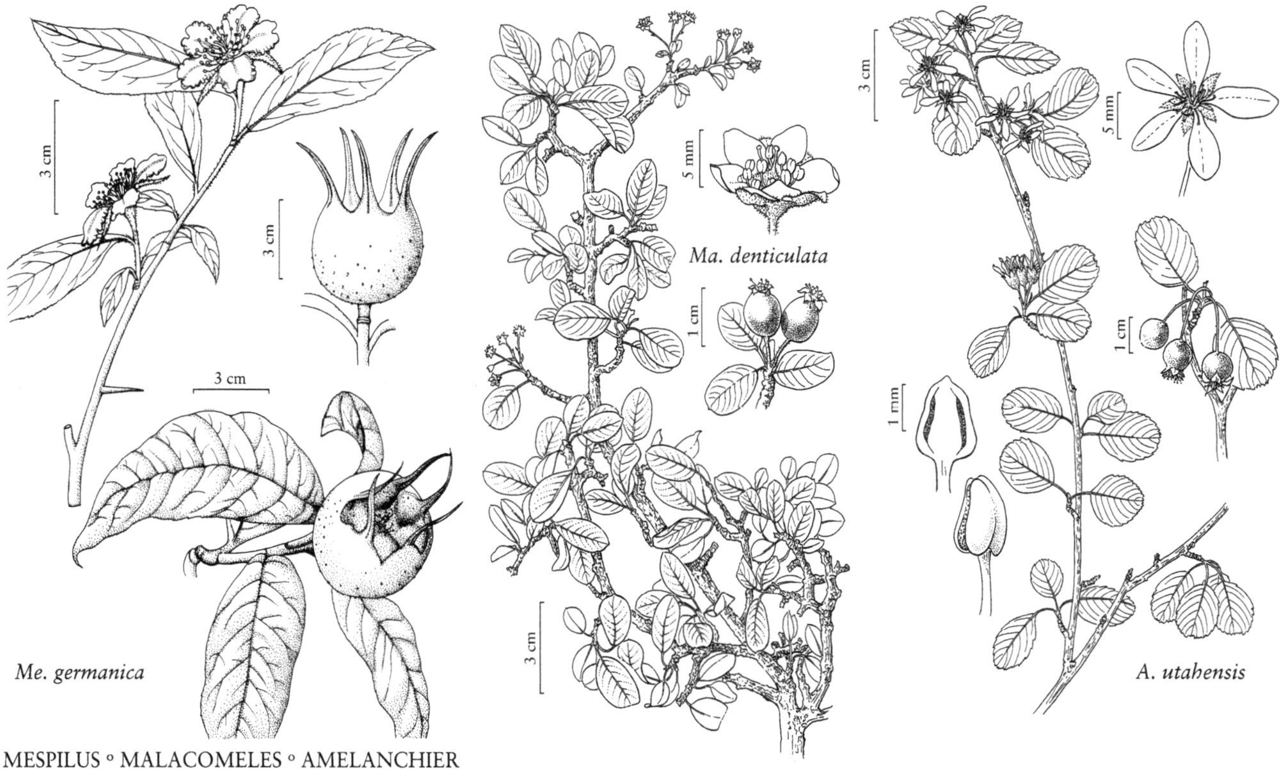

MESPILUS ◦ MALACOMELES ◦ AMELANCHIER

66. MALACOMELES (Decaisne) Decaisne, Ann. Gén. Hort. 23: 156. 1882 • [Greek *malakos*, soft, and *melon*, apple]

Christopher S. Campbell

Cotoneaster Medikus sect. *Malacomeles* Decaisne, Nouv. Arch. Mus. Hist. Nat. 10: 177. 1874, based on *Nagelia* Lindley, Edwards's Bot. Reg. 31: Misc. 40. 1845, not *Naegelia* Rabenhorst 1844 [Fungi]

Shrubs, 10–30[–50] dm. **Stems** 1–20, ascending, <much branched, often ± twisted, young stems terete, closely villous, smooth>; bark brown or gray, <young stems not hairy>; long shoots present, short shoots sometimes present; unarmed; grayish- or whitish-tomentose, sometimes glabrate; <overwintering buds tomentose>. **Leaves** drought-deciduous or persistent, cauline, simple, <slightly conduplicate>; stipules tardily deciduous or ± persistent, free, lanceolate, <glandlike>, margins entire; petiole present; blade elliptic or oblong-ovate to orbiculate, 0.5–1.5[–3.1] cm, leathery, margins flat, denticulate or entire, <teeth gland-tipped>, venation pinnate, abaxial surface tomentose, adaxial ± glabrous. **Inflorescences** terminal <on leafy shoots of season>, 1–5[–7]-flowered, corymbs [panicles], axes glabrous or hairy; bracts absent; bracteoles sometimes present. **Pedicels** present. **Flowers:** perianth and androecium epigynous, 10 mm diam.; hypanthium campanulate [urceolate], (2–)3–4 mm, distally copiously white-tomentose; sepals 5, spreading, nearly orbiculate (inner broadly deltate); petals 5, white, round or kidney-shaped, <base short-clawed>; stamens [8–]14–20[–22], longer or shorter than petals; carpels 2 or 3[–5], barely connate or distinct, adnate to proximal $\frac{1}{2}$–$\frac{4}{5}$ of hypanthium, <ovary appearing 4–10-loculed by false partitions>, densely hairy, styles <2 or 3>, lateral, distinct;

ovules 2 (separated by false partition). **Fruits** berrylike pomes, translucent, vivid pink, drying purplish black, globose, 6–10 mm, glabrous; <flesh whitish>; hypanthium persistent; sepals persistent, erect or recurving; carpels cartilaginous; styles persistent. **Seeds** 4–6(–10).

Species 2 (1 in the flora): Texas, Mexico, Central America.

Malacomeles shares with *Amelanchier* and *Peraphyllum* the feature of false partitions within the carpels; the partitions partially divide the locules and make the fruit appear to have twice as many locules as the number of carpels (G. N. Jones 1945; K. R. Robertson et al. 1991; J. R. Rohrer et al. 1991). Sequences from multiple chloroplast and nuclear genes indicate that *Malacomeles* is closely related to *Amelanchier* and *Peraphyllum* (C. S. Campbell et al. 2007).

The other species in the genus, *Malacomeles nervosa* (Decaisne) G. N. Jones, is known from Nuevo León, Mexico south to Guatemala.

SELECTED REFERENCES Hunt, D. R. and C. King. 1978. *Amelanchier denticulata* (Rosaceae). Bot. Mag. 182: 20–23. Jones, G. N. 1945. *Malacomeles*, a genus of Mexican and Guatemalan shrubs. Madroño 8: 33–39.

1. Malacomeles denticulata (Kunth) Decaisne, Fl. Serres Jard. Paris 23: 156. 1882 • Big Bend or southern false serviceberry [F]

Cotoneaster denticulatus Kunth in A. von Humboldt et al., Nov. Gen. Sp. 6(fol.): 169; 6(qto.): 214; plate 556. 1824 (as denticulata); *Amelanchier denticulata* (Kunth) K. Koch

Leaves: stipules 1 mm; petiole 2–4(–10) mm; blade 0.5–1.5 × 0.3–1 mm, base cuneate to rounded or truncate, margins: teeth 4–8 per side, apex truncate to rounded or acute to mucronate. **Bracts** persistent, lanceolate, ciliate. **Pedicels** 2–7(–9) mm. **Flowers:** sepals 1.6 × 2.8 mm, ciliate; petals 2.5–5.5 mm. **Pomes** edible, insipid.

Flowering spring; fruiting summer. Talus slopes, canyons, on limestone and igneous substrates; (500–)800–1700 m; Tex.; Mexico; Central America (Guatemala).

Malacomeles denticulata is a xerophytic shrub with small leaves and flowers. The species is found in the flora area only in trans-Pecos Texas.

67. AMELANCHIER Medikus, Philos. Bot. 1: 135, 155. 1789 • Shadbush, serviceberry, sarvisberry, sarvis, Juneberry, saskatoon, shadblow, shadwood, sugarplum, wild-plum [Old Savoy name for *Amelanchier ovalis* Medikus]

Christopher S. Campbell Michael B. Burgess
Kevin R. Cushman Eric T. Doucette
Alison C. Dibble Christopher T. Frye

Shrubs or trees, <sometimes fastigiate>, 2–250 dm; <suckering and/or rhizomatous, sometimes forming colonies>. **Stems** 1–150, erect to ascending; bark usually gray, sometimes brown or salmon colored, <smooth or fissuring in older trees>; short shoots present or absent; unarmed; young stems glabrous or hairy. **Leaves** deciduous, cauline, simple; stipules caducous, free, linear, margins entire; petiole present; blade elliptic, elliptic-oblong, elliptic-oval, ovate, oval, oblanceolate, oblong, oblong-ovate, obovate, quadrangular, suborbiculate, or orbiculate, 0.2–10 × 0.2–6 cm, membranous to coriaceous, <base tapering or truncate to rounded or cordate>, margins flat, entire or serrate, sometimes doubly serrate, or dentate distally or throughout, venation pinnate, <apex acuminate to truncate or retuse, or mucronate to apiculate>, abaxial surface glabrous or sparsely to densely hairy by flowering, glabrous or ± hairy at maturity, <usually paler than adaxial surface>. **Inflorescences** terminal, <arching, ascending, erect, or drooping, sometimes spreading or nodding>, 1–4-flowered clusters or 4–17-flowered racemes (panicles); bracts absent; bracteoles sometimes present. **Pedicels** present, <proximal 1–3 subtended by leaves>. **Flowers** opening during leaf expansion, perianth and androecium epigynous, 10–55 mm diam.; hypanthium campanulate or funnelform to saucer-shaped, 3–9 mm diam., hairy or glabrous; sepals 5, erect to reflexed, triangular to lanceolate; petals 5, usually white, sometimes ivory, rarely ± pink, linear to orbiculate; stamens 7–28, shorter than or equal to petals; carpels 2–5, connate, adnate to hypanthium, <ovary 6–10-loculed by false partitions>, apex densely hairy or glabrous, styles <2–6>, terminal, distinct or ± connate; ovules 2. **Fruits** pomes, bluish or purplish to nearly black, pinkish or maroon-purple, dark purple-blue, or brownish, globose (pear-shaped in *A. bartramiana*), 6–15 mm diam.; <taste insipid to sweet>; hypanthium persistent; sepals persistent, erect to strongly reflexed; carpels cartilaginous; styles sometimes persistent. **Seeds** absent or few. $x = 17$.

Species ca. 20 (18 in the flora): North America, nw Mexico, Europe, e, w Asia, n Africa.

Amelanchier plants grow well in full sun and tend to form larger populations in more or less disturbed habitats and woodland margins.

Identification from herbarium specimens is often inconclusive. Identification is best undertaken in the field, with visits during flowering and fruiting seasons, and observations of habitat, habit, presence of congeners, and flowering time relative to sympatric congeners.

No consensus exists regarding the enumeration of North American *Amelanchier* taxa. J. B. Phipps et al. (1990) listed about 33 species in the North Temperate Zone and about six in the Old World. In eastern North America, most floras and regional treatments have roughly the corresponding number (although sometimes not the same species) as G. N. Jones (1946); often substantial differences of opinion exist regarding circumscriptions of individual taxa. Disagreement is deeper for western North America, for which Jones had seven species and P. Landry (1975) had but one. W. H. Blanchard (1907) observed that diverging and intergrading forms are abundant in the genus, and M. L. Fernald (1946e) remarked that no genus in North America, except *Crataegus* and *Rubus*, has offered so much perplexity and such contradictory treatment as *Amelanchier*.

Identification of *Amelanchier* taxa is difficult because the genus has relatively few taxonomically informative morphologic characters, and some of these are variable. Leaves vary in shape between and within individuals, depending on whether they are on reproductive or sterile branches, on rapidly or slowly growing shoots, or at the tip or near the base of a branch. Habit also varies within species. *Amelanchier arborea* can develop into a single-stemmed tree and can flower when stems are less than one meter. Adverse conditions can constrain arboreal species to less than a meter in height for many years. *Amelanchier* is taxonomically complex also because some species have not diverged much from one another genetically. C. S. Campbell et al. (1997) reported low levels of divergence among North American species in sequences from the internal transcribed spacer (ITS) of nuclear ribosomal DNA, and R. D. Overath and J. L. Hamrick (1998) found little divergence in allozymes between *A. arborea* and *A. laevis*.

Interspecific hybridization is frequent among eastern North American *Amelanchier* taxa (K. M. Wiegand 1912; E. L. Nielsen 1939; M. L. Fernald 1950; J. E. Cruise 1964; L. Cinq-Mars 1971; S. M. McKay 1973; P. Landry 1975; W. A. Robinson 1982; E. G. Voss 1972–1996, vol. 2; C. S. Campbell and W. A. Wright 1996; A. C. Dibble et al. 1998; A. Haines and T. F. Vining 1998; H. R. Hinds 2000). *Amelanchier bartramiana*, for example, has been reported to hybridize with all 11 other species with which it is sympatric. G. N. Jones (1946) considered hybridization to be insignificant.

Hybridization is important and has varying consequences in *Amelanchier*, one of which is illustrated by *A. bartramiana* and *A. laevis*. These two species usually hybridize when they grow together, but at one well-sampled site, the F_1, *A. bartramiana* × *A. laevis* (often referred to by the invalid name *A.* ×*neglecta* Eggleston), apparently does not reproduce (J. E. Weber and C. S. Campbell 1989). Hybrids of other taxa have reproduced and spawned hybrid swarms (J. E. Cruise 1964; Campbell and W. A. Wright 1996). Wide dispersal of hybrids can be inferred from the distributions of *A. arborea* and *A. laevis*, both of which include allopolyploid plants (R. D. Overath and J. L. Hamrick 1998) found over much of eastern North America. These two species, like many others in *Amelanchier*, consist of both diploids and tetraploids (Cruise; W. A. Robinson and C. R. Partanen 1980; Campbell et al. 1985; A. C. Dibble et al. 1998; M. B. Burgess et al., unpubl.).

Ploidy levels noted for each taxon in the treatment below ($2n = 2x$ indicates diploidy, $3n = 3x$ triploidy, and $2n = 4x$ tetraploidy) were determined by flow cytometry by M. B. Burgess et al. (unpubl.) or are from publications to be cited by Burgess et al.

The interplay of hybridization and agamospermy has the potential for generating considerable taxonomic complexity in *Amelanchier*. Agamospermy has been documented in five species, one informally named taxon ["*A. erecta*" (not *A. erecta* Blanchard)], and two putative hybrids (C. S. Campbell et al. 1985, 1987; J. E. Weber and Campbell 1989; Campbell and W. A. Wright 1996; A. C. Dibble et al. 1998; M. B. Burgess et al., unpubl.). Given that agamospermy is strongly associated with polyploidy in *Amelanchier* and close relatives (Campbell et al. 1991) and that polyploidy predominates in North American *Amelanchier*, agamospermy is assumed to be widespread in North American amelanchiers. *Amelanchier* agamosperms that have been studied are facultative; they can produce seed both sexually and asexually. What is most pertinent to the taxonomy of the group is that the phenotypes of agamosperms can be perpetuated and, if a particular phenotype disperses over some appreciable geographic range, could be recognized taxonomically. Phenotypes that are more or less distinct and have a limited geographic range can be called microspecies (Campbell and Wright). Examples of well-characterized microspecies are presented in the treatments of *A. laevis* and *A. spicata*.

Species of *Amelanchier* are used in horticulture, and cultivars of *A. alnifolia*, *A. canadensis*, *A. laevis*, and *A. lamarckii* Schroeder are especially popular. *Amelanchier lamarckii* was named from plants grown in European gardens from plants originally from eastern North America (F.-G. Schroeder 1968). It appears that plants resembling *A. lamarckii* have escaped from cultivation in North America. Cultivars of *A. alnifolia* are grown as fruit crops. *Amelanchier* fruits are important to birds and other wildlife because of their size, availability, sweetness, and early ripening.

For additional information about this genus, see http://biology.umaine.edu/Amelanchier/.

SELECTED REFERENCES Blanchard, W. H. 1907. Our eastern shadwoods. Torreya 7: 97–102. Campbell, C. S. and W. A. Wright. 1996. Apomixis, hybridization, and taxonomic complexity in eastern North American *Amelanchier* (Rosaceae). Folia Geobot. Phytotax. 31: 345–354. Cruise, J. E. 1964. Studies of natural hybrids in *Amelanchier*. Canad. J. Bot. 42: 651–663. Dibble, A. C. 1995. Conservation Biology of Shadbush, *Amelanchier* (Rosaceae): Evidence from Systematics, Population Structure, and Reproductive Ecology. Ph.D. thesis. University of Maine. Jones, G. N. 1946. American Species *of Amelanchier*. Urbana. Landry, P. 1975. Le concept d'espèce et la taxonomie du genre *Amelanchier* (Rosacées). Bull. Soc. Bot. France 122: 43–252. Robinson, W. A. 1982. Experimental taxonomy in the genus *Amelanchier*. II. Do the taxa in *Amelanchier* form an agamic complex? Rhodora 84: 85–100. Robinson, W. A. and C. R. Partanen. 1980. Experimental taxonomy in the genus *Amelanchier*. I. A new look at the chromosome numbers of the *Amelanchier* species growing in the northeastern United States. Rhodora 82: 483–493. Wiegand, K. M. 1912. The genus *Amelanchier* in eastern North America. Rhodora 14: 117–161. Wiegand, K. M. 1920. Additional notes on *Amelanchier*. Rhodora 22: 146–151.

1. Inflorescences (1 or)2 or 3(or 4)-flowered; pomes pear-shaped. . . . . 14. *Amelanchier bartramiana*
1. Inflorescences (3–)5–12(–17)-flowered; pomes globose.
    2. Petals (2.2–)3–4.5(–6.8) mm (sometimes bearing 1 or 2 tiny pollen sacs near margins on adaxial surfaces). . . . . 13. *Amelanchier nantucketensis*
    2. Petals (4–)5–20(–23.2) mm (not bearing pollen sacs).
        3. Leaves: each margin with (0–)4–9(–20) teeth in distalmost cm.
            4. Leaves: abaxial surfaces densely (moderately) hairy by flowering.
                5. Petals (8–)10–15(–19) mm; leaf apices acute to acuminate; sepals soon reflexed after flowering; shrubs or trees . . . . 17. *Amelanchier arborea*
                5. Petals (4–)6–10.2(–15) mm; leaf apices rounded or acute to obtuse and mucronate; sepals erect, spreading, or recurved after flowering; shrubs.
                    6. Leaf blades oval to orbiculate; sepals recurved after flowering; ovary apices densely hairy (or glabrous) . . . . 12. *Amelanchier spicata* (in part)
                    6. Leaf blades elliptic or oval to oblong or obovate; sepals erect, ascending, or spreading after flowering; ovary apices glabrous (or moderately hairy) . . . . 15. *Amelanchier canadensis*
            4. Leaves: abaxial surfaces glabrous or sparsely hairy by flowering.
                7. Ovary apices usually densely (moderately) hairy.
                    8. Shrubs, 0.5–1 m (rhizomatous); stems 1–50, usually forming colonies . . . . 9. *Amelanchier fernaldii*
                    8. Shrubs or trees, 1–10 m; stems 1–10, forming colonies or solitary . . . . 10. *Amelanchier interior*
                7. Ovary apices glabrous (or sparsely hairy).
                    9. Leaves: abaxial surfaces sparsely hairy by flowering . . . . 16. *Amelanchier intermedia*
                    9. Leaves: abaxial surfaces glabrous (or sparsely hairy) by flowering . . . . 18. *Amelanchier laevis*
        3. Leaves: each margin with (0–)2–6(–11) teeth in distalmost cm.
            10. Styles (2 or)3 or 4.
                11. Proximalmost pedicels (8–)12–18(–25) mm. . . . . 1. *Amelanchier pallida*
                11. Proximalmost pedicels (2–)6–16(–25) mm.
                    12. Leaf blades (6–)16–24(–31) mm; petioles (2.5–)3.4–6(–8) mm . . . .2. *Amelanchier nitens*
                    12. Leaf blades (14–)21–36(–63) mm; petioles (3–)6–13(–22) mm . . . . 3. *Amelanchier utahensis* (in part)
            10. Styles (2–)4 or 5(or 6).
                13. Petals 4.3–14 mm, usually less than 9 mm.
                    14. Styles (2.2–)3–3.8(–4.6) mm . . . . 12. *Amelanchier spicata* (in part)
                    14. Styles (1.7–)2.3–3(–3.9) mm.
                        15. Sepals (1.6–)2.6–4.2(–6.5) mm . . . . 3. *Amelanchier utahensis* (in part)
                        15. Sepals (1–)1.6–2.7(–3.6) mm . . . . .11. *Amelanchier humilis*
                13. Petals 4.9–23.2 mm, usually more than 9 mm.
                    16. Leaves: each margin with 0–3(–11) teeth on proximal ½.
                        17. Petals usually less than 9.8 × 3.1 mm . . . . 3. *Amelanchier utahensis* (in part)
                        17. Petals usually more than 9.5 × 3.3 mm.
                            18. Petals (5.7–)9–14(–18.8) mm . . . . .4. *Amelanchier alnifolia*
                            18. Petals (10.8–)13.3–20.1(–23.2) mm . . . . .5. *Amelanchier cusickii* (in part)
                    16. Leaves: each margin with (0–)4–10(–16) teeth on proximal ½.
                        19. Ovary apices glabrous or sparsely hairy . . . . .5. *Amelanchier cusickii* (in part)
                        19. Ovary apices densely (moderately) hairy.
                            20. Petals (15–)16.6–19.7(–22) mm .7. *Amelanchier amabilis*
                            20. Petals (8–)9.7–16.1(–20) mm.
                                21. Petals (8–)10.7–16.1(–20) mm. . . . . .6. *Amelanchier sanguinea*
                                21. Petals (8–)9.7–13.5(–15) mm. . . . . 8. *Amelanchier gaspensis*

1. Amelanchier pallida Greene, Fl. Francisc., 53. 1891 • Pale serviceberry or shadbush E

Shrubs, 0.5–6 m. **Stems** 1–50, often densely colonial, much branched. **Leaves** fully unfolded; petiole (1–)4.5–8.5(–18) mm; blade suborbiculate to oval or obovate, (15–)20–29(–44) × (10–)14–20(–28) mm, base rounded to subcordate, sometimes cuneate, each margin with 0(–2) teeth on proximal ½ and (0–)2–5(–9) teeth in distalmost cm, largest teeth less than 1 mm, apex rounded, truncate, or emarginate to acute and mucronate, abaxial surface moderately (sparsely) hairy by flowering, sparsely to moderately hairy later, adaxial sparsely hairy later. **Inflorescences** (6 or)7–11(–17)-flowered, (15–)21–29(–38) mm. **Pedicels:** (0 or)1 or 2(or 3) subtended by a leaf, proximalmost (8–)12–18(–25) mm. **Flowers:** sepals usually erect to recurved after flowering, (1.3–)1.6–3(–4.5) mm; petals elliptic to broadly oval, (4–)5.2–8(–9.5) × (2.2–)3.2–4.6(–6.6) mm; stamens (12–)17–21(–23); styles (2 or)3 or 4(or 5), (1.4–)2.1–2.7(–3.3) mm; ovary apex moderately to densely (sparsely) hairy. **Pomes** often brownish, 6–10 mm diam. **2*n*** = 2*x*, 4*x*.

Flowering Apr–Jun; fruiting Jul–Sep. Dry rocky slopes, canyons, chaparral, mountainsides; 1000–2000 m; Calif., Oreg.

Amelanchier pallida is known from northern California and southern Oregon. The species is distinctive for its often broadly oval petals, relatively long proximalmost pedicels, leaves with teeth that are relatively small and few in number, and usually absent in proximal half, and much-branched habit.

2. Amelanchier nitens Tidestrom, Proc. Biol. Soc. Wash. 36: 182. 1923 • Shining shadbush E

Amelanchier alnifolia (Nuttall) Nuttall ex M. Roemer var. *nitens* (Tidestrom) Munz

Shrubs, 2–6 m. **Stems** 20–150, in colonies. **Leaves** fully unfolded; petiole (2.5–)3.4–6(–8) mm; blade ovate to obovate, (6–)16–24(–31) × (7–)11–16(–19) mm, base cuneate to rounded, each margin with 0–3(–6) teeth on proximal ½ and (0–)2–6(–9) teeth in distalmost cm, largest teeth less than 1 mm, apex acute to rounded, abaxial surface sparsely to moderately hairy (or glabrous) by flowering, surfaces moderately to densely hairy (or glabrous) later. **Inflorescences** (3 or)4–7(–12)-flowered, (8–)10–18(–27) mm. **Pedicels:** (0 or)1 or 2(or 3) subtended by a leaf, proximalmost (4–)6–12(–19) mm. **Flowers:** sepals spreading to recurved after flowering, (1.3–)2.2–3.3(–4.3) mm; petals linear to oblanceolate, (5.2–)6.7–9.1(–12.2) × (1.7–)2.4–3.5(–4.6) mm; stamens (7–)12–19(–21); styles 2–4(or 5), (1.8–)2.3–3.4(–4.7) mm; ovary apex glabrous or sparsely to moderately (densely) hairy. **Pomes** pinkish purple, 8–14 mm diam. **2*n*** = 2*x*.

Flowering Apr–Jun; fruiting Jun–Jul. Dry rocky slopes, canyons, stream banks, mountainsides, foothills, high deserts; 1500–2400 m; Ariz., Nev.

Amelanchier nitens is rarely found outside of pinyon pine-juniper community types; it often grows adjacent to seasonal streams and washes. The species is distinguished by its lustrous and coriaceous leaves (particularly in lower elevation populations), moderately hairy twigs, usually salmon-colored bark on older branches and trunks, relatively short inflorescences, and (typically) three styles. Petioles on many live specimens are a bright, lustrous red, a character state that is less noticeable on herbarium specimens. Stems are usually highly contorted, with relatively short internodes, abundant short shoots, and, often, divaricately branching ultimate branches. Seeds of *A. nitens* are much larger than congenerics (a sample of 25 averaged 7.3 times heavier than 25 seeds from an average eastern North American *Amelanchier*).

The authors have observed incomplete herbarium specimens conforming to *Amelanchier nitens* (1923) morphology from Colorado, New Mexico, and Utah, including the type specimen of *A. rubescens* Greene (1900), which may prove to be the correct name for this species.

3. Amelanchier utahensis Koehne, Gatt. Pomac., 25, plate 2, fig. 20e. 1890 • Utah serviceberry or shadbush F

Amelanchier covillei Standley; *A. glabra* Greene; *A. gracilis* A. Heller; *A. prunifolia* Greene; *A. utahensis* subsp. *covillei* (Standley) Clokey; *A. utahensis* var. *covillei* (Standley) N. H. Holmgren

Shrubs, 0.5–5 m. **Stems** 1–100, often colonial, much branched. **Leaves** mostly or fully unfolded; petiole (3–)6–13(–22) mm; blade suborbiculate to oval or obovate, (14–)21–36(–63) × (9–)16–32(–54) mm, base usually rounded to subcordate, sometimes cuneate, each margin with 0–3(–6) teeth on proximal ½ and (0–)3–5(–7) teeth in distalmost cm, largest teeth more than 1 mm, apex usually rounded to truncate or emarginate, sometimes acute and mucronate, abaxial surface moderately (sparsely or densely) hairy by flowering, sparsely to moderately hairy later, adaxial glabrous or sparsely (moderately) hairy later. **Inflorescences** (4–)6–10(–13)-flowered, (8–)16–30(–43) mm. **Pedicels:**

(0 or)1 or 2(or 3) subtended by a leaf, proximalmost (2–)7–16(–25) mm. **Flowers:** sepals usually recurved after flowering, (1.6–)2.6–4.2(–6.5) mm; petals oblanceolate to oblong, (4.9–)7.2–9.8(–14) × (1.6–)2.6–3.1(–5.3) mm; stamens (9–)13–19(–20); styles (2 or) 3 or 4, (1.7–)2.3–3(–3.9) mm; ovary apex moderately to densely hairy (or glabrous). **Pomes** purplish black, 6–10 mm diam. $2n = 4x$.

Flowering Apr–May; fruiting Jul–Sep. Dry rocky slopes, canyons, stream banks, mountainsides, foothills, deserts; 900–3500 m; Ariz., Calif., Colo., Idaho, Mont., Nev., N.Mex., Oreg., Tex., Utah, Wash., Wyo.; Mexico (Baja California).

The sparsely to moderately hairy mature leaves and twigs, rounded to truncate or emarginate leaf apices, relatively short petals, and reduced numbers of stamens and styles are distinctive characteristics for the wide-ranging and common *Amelanchier utahensis*. Within 25 years of its publication, about a dozen names were published that G. N. Jones (1946) considered synonyms of *A. utahensis*. Some recent floras have recognized some of these synonymized taxa, including *A. bakeri* Greene, *A. covillei*, *A. mormonica* C. K. Schneider, *A. oreophila* A. Nelson, and *A. venulosa* Greene.

4. Amelanchier alnifolia (Nuttall) Nuttall ex M. Roemer, Fam. Nat. Syn. Monogr. 3: 147. 1847 • Alder-leaved shadbush, saskatoon, saskatoon berry, amélanchier à feuilles d'aulne [E]

Aronia alnifolia Nuttall, Gen. N. Amer. Pl. 1: 306. 1818

Shrubs, 1–12 m. **Stems** 1–50, solitary or colonial. **Leaves** mostly unfolded; petiole (3–)6.8–19.1(–28) mm; blade usually elliptic to oval to suborbiculate, sometimes quadrangular, (14–)24–47(–67) × (7–)17–36(–55) mm, base usually subcordate to truncate, sometimes ± tapering or ± cuneate, each margin with 0–3(–9) teeth on proximal ½ and (0–)3–5(–8) teeth in distalmost cm, largest teeth more than 1 mm, apex rounded to truncate or occasionally acute or mucronate, abaxial surface sparsely to densely hairy (or glabrous) by flowering, sparsely to moderately hairy (or glabrous) later, adaxial glabrous or sparsely (moderately) hairy later. **Inflorescences** (4–)6–11(–16)-flowered, (8–)14–43 (–62) mm. **Pedicels:** (0 or)1 or 2(or 3) subtended by a leaf, proximalmost (2–)3–20(–29) mm. **Flowers:** sepals erect to recurved after flowering, (1.4–)2.2–4(–4.9) mm; petals oblanceolate to oval or obovate to elliptic, (5.7–)9.5–14(–18.8) × (2.2–)3.3–5.2(–6.6) mm; stamens (10–)15–21(or 22); styles (3 or)4 or 5(or 6), (1.3–)2–2.9 (–3.9) mm; ovary apex moderately to densely hairy (or glabrous). **Pomes** black or purple, 8–15 mm diam.

Varieties 3 (3 in the flora): w North America.

Amelanchier alnifolia is widespread and polymorphic, and its taxonomic and geographic limits have been viewed differently. L. Cinq-Mars (1971) considered this species to range eastward to the Gaspé Peninsula. Disagreements about the boundary between *A. alnifolia* and *A. humilis* are evident in herbarium specimen annotations. The distinctness of the varieties of *A. alnifolia* has also been questioned. G. N. Jones (1946) treated the three varieties recognized here as distinct species and noted that the leaves of var. *alnifolia* and var. *semiintegrifolia* (*A. florida*) are virtually indistinguishable, and, although petal lengths of the two do not overlap, occasional larger-flowered var. *alnifolia* and occasional smaller-flowered var. *semiintegrifolia* occur, so the petal length distinction is not an absolute one. Geographically these two varieties are largely separate, with var. *alnifolia* occurring in the Great Plains and Rocky Mountains and var. *semiintegrifolia* on the Pacific slopes of mountains from Alaska to northern California.

Amelanchier alnifolia is thought to hybridize with *Sorbus scopulina* (×*Amelasorbus jackii* Rehder).

1. Ovary apices glabrous (or sparsely hairy); shrubs 1–2(–4) m . . . . 4c. *Amelanchier alnifolia* var. *pumila*
1. Ovary apices moderately to densely hairy (or glabrous); shrubs or trees, 1–12 m.
    2. Inflorescences (8–)26–43(–62) mm; proximalmost pedicels (5–)8–16(–29) mm . . . . 4a. *Amelanchier alnifolia* var. *alnifolia*
    2. Inflorescences (8–)14–26(–35) mm; proximalmost pedicels (2 or)3–8(–13) mm. . . . . .4b. *Amelanchier alnifolia* var. *semiintegrifolia*

4a. Amelanchier alnifolia (Nuttall) Nuttall ex M. Roemer var. **alnifolia** [E]

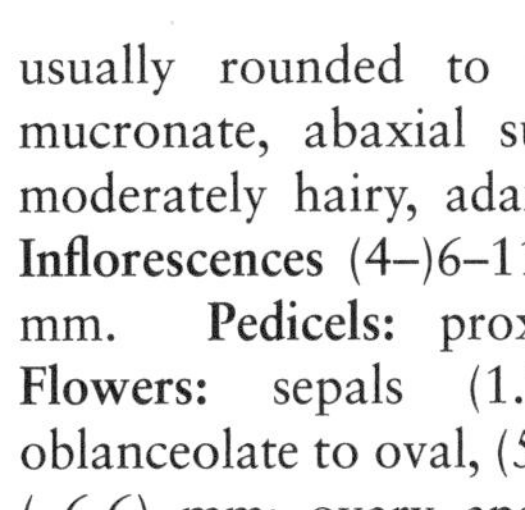

Shrubs, 1–8 m. **Stems** 1–20, often colonial. **Leaves:** petiole (5–)11.7–17.5(–28) mm; blade usually elliptic to oval to suborbiculate, sometimes quadrangular, (20–)33–47(–67) × (14–)24–36(–55) mm, base usually subcordate to truncate, sometimes ± tapering, apex usually rounded to truncate, sometimes acute or mucronate, abaxial surface glabrous or sparsely to moderately hairy, adaxial glabrous or sparsely hairy. **Inflorescences** (4–)6–11(–16)-flowered, (8–)26–43(–62) mm. **Pedicels:** proximalmost (5–)8–16(–29) mm. **Flowers:** sepals (1.7–)2.6–3.6(–4.9) mm; petals oblanceolate to oval, (5.7–)9–14(–18.8) × (2.2–)3.7–5.1 (–6.6) mm; ovary apex moderately to densely hairy (or glabrous). **Pomes** 10–15 mm diam. $2n = 4x$.

Flowering May–Jul; fruiting Jul–Aug. Stream banks and shores, lake shores, mountainsides, dry rocky and grassy slopes (northern shrub-steppe), hillsides, woods, thickets, shaded canyons, moist roadsides; 30–2900 m; Alta., B.C., Man., N.W.T., Ont., Sask., Yukon; Colo., Idaho, Iowa, Kans., Minn., Mont., Nebr., Nev., N.Mex., N.Dak., Oreg., S.Dak., Tex., Utah, Wash., Wyo.

Variety *alnifolia* extends much farther east than any other western shadbush; it is distinguished by its rounded to truncate leaves, hairy ovary apices, and relatively large fruits. The fruits are grown commercially and are an ingredient of the native American food pemmican.

4b. Amelanchier alnifolia (Nuttall) Nuttall ex M. Roemer var. **semiintegrifolia** (Hooker) C. L. Hitchcock in C. L. Hitchcock et al., Vasc. Pl. Pacif. N.W. 3: 94. 1961 • Many-flowered shadbush [E]

Amelanchier ovalis Medikus var. *semiintegrifolia* Hooker, Fl. Bor.-Amer. 1: 202. 1832; *A. florida* Lindley var. *humptulipensis* G. N. Jones

Shrubs or trees, 1–12 m. **Stems** 1–20, solitary or in colonies. **Leaves:** petiole (3–)6.8–12.6(–20) mm; blade elliptic to oval to orbiculate, (16–)24–34(–49) × (7–)17–26(–37) mm, base rounded or subcordate, apex subtruncate to rounded or acute, abaxial surface sparsely to moderately hairy, adaxial glabrous or sparsely (moderately) hairy. **Inflorescences** (4–)6–9(–12)-flowered, (8–)14–26(–35) mm. **Pedicels:** proximalmost (2–)3–8(–13) mm. **Flowers:** sepals (1.4–)2.2–3.2(–4) mm; petals oblanceolate to oval, (7.6–)9.9–13.9(–18.2) × (2.6–)3.3–4.7(–6.6) mm; ovary apex moderately to densely hairy (or glabrous). **Pomes** 10–13 mm diam. **2*n*** = 4*x*.

Flowering Mar–Jun; fruiting Jun–Aug. Woods, hillsides, roadsides, mountainsides; 0–2100 m; B.C.; Alaska, Calif., Oreg., Wash.

Variety *semiintegrifolia* is most similar to and intergrades with var. *alnifolia*; inflorescences and lowest pedicel lengths distinguish the two.

4c. Amelanchier alnifolia (Nuttall) Nuttall ex M. Roemer var. **pumila** (Torrey & A. Gray) C. K. Schneider, Ill. Handb. Laubholzk. 1: 739. 1906 • Dwarf shadbush [E]

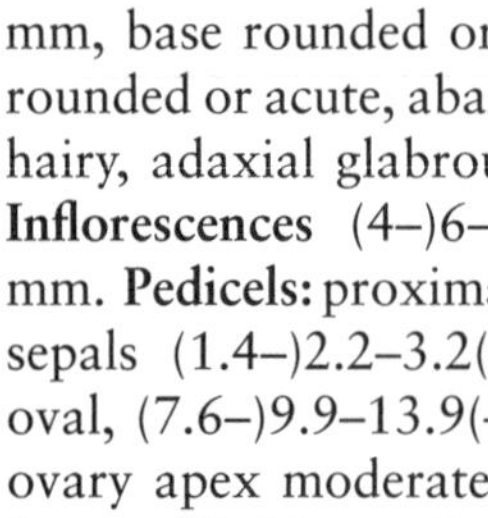

Amelanchier canadensis (Linnaeus) Medikus var. *pumila* Torrey & A. Gray, Fl. N. Amer. 1: 474. 1840

Shrubs, 1–2(–4) m. **Stems** 1–50, solitary or in colonies. **Leaves:** petiole (7–)11.9–19.1(–27) mm; blade orbiculate to oval, (14–)26–37(–50) × (18–)23–36(–43) mm, base usually cordate to truncate, sometimes ± cuneate, apex rounded or subtruncate, surfaces glabrous. **Inflorescences** (6 or) 7–9(–12)-flowered, (24–)29–40(–47) mm. **Pedicels** proximalmost (8–)10–20(–27) mm. **Flowers:** sepals (2.7–)3.2–4(–4.5) mm; petals obovate to elliptic, (9–)10.9–11.5(–12.6) × (3–)3.8–5.2(–5.9) mm; ovary apex glabrous (or sparsely hairy with a ring of hairs at base of styles). **Pomes** 8–9 mm diam. **2*n*** = 4*x*.

Flowering May–Jun; fruiting Jul–Sep. Mountain slopes; 2000–3400 m; Ariz., Calif., Colo., Utah, Wyo.

The most distinctive features of var. *pumila* are that its leaves, pedicels, sepals, and ovary apices are glabrous (with the exception of a few hairs on emerging leaves and a ring of hairs around the base of the styles in some plants), and its leaf teeth are relatively long and sharp-pointed. This variety appears to be restricted to higher elevations.

5. Amelanchier cusickii Fernald, Erythea 7: 121. 1899 • Cusick's shadbush [E]

Amelanchier alnifolia (Nuttall) Nuttall ex M. Roemer var. *cusickii* (Fernald) C. L. Hitchcock; *A. basalticola* Piper

Shrubs, 2–8 m. **Stems** 1–50, solitary or in colonies. **Leaves** fully unfolded; petiole (9–)12.8–19(–28) mm; blade elliptic to suborbiculate, (22–)33–46(–61) × (13–)25–32(–50) mm, base subcordate to truncate, each margin with (0 or)1–6(–11) teeth on proximal ½ and (2–)4 or 5(–7) teeth in distalmost cm, largest teeth 1 mm, apex usually truncate, sometimes mucronate to acute, abaxial surface glabrous or sparsely hairy by flowering, glabrous (or sparsely hairy) later, adaxial glabrous later. **Inflorescences** (4–)6–8(–10)-flowered, (17–)26–36(–47) mm. **Pedicels:** (0 or)1 or 2 subtended by a leaf, proximalmost (4–)9–14(–18) mm. **Flowers:** sepals erect to recurved after flowering, (2.6–)3.3–4.3(–5.3) mm; petals obovate to oblanceolate, (10.8–)13.3–20.1(–23.2) × (2.3–)3.3–6.3(–7.5) mm; stamens (14–)18–20(or 21); styles (4 or)5, (1.7–)2.6–4(–4.8) mm; ovary apex glabrous or sparsely hairy (or with a ring of hairs at base of styles). **Pomes** bluish black, 10 mm diam. **2*n*** = 4*x*.

Flowering Mar–Jul; fruiting May–Aug. Basaltic ledges, cliffs, and bluffs along streams, stony soil, streambeds, stream banks, copses, mountainsides, roadsides; 600–2300 m; B.C.; Idaho, Mont., Oreg., Wash.

Amelanchier cusickii has distinctively long petals. Leaves that are glabrous upon expanding and glabrous or sparsely hairy ovary apices are also useful for identification. The species flowers before *A. alnifolia*, which suggests that the two are genetically distinct. G. N. Jones (1946) noted that *A. alnifolia* and *A. cusickii* frequently grow together and that there is no evidence of hybridization; the authors have observed hybridization between *A. alnifolia* var. *semiintegrifolia* and *A. cusickii*.

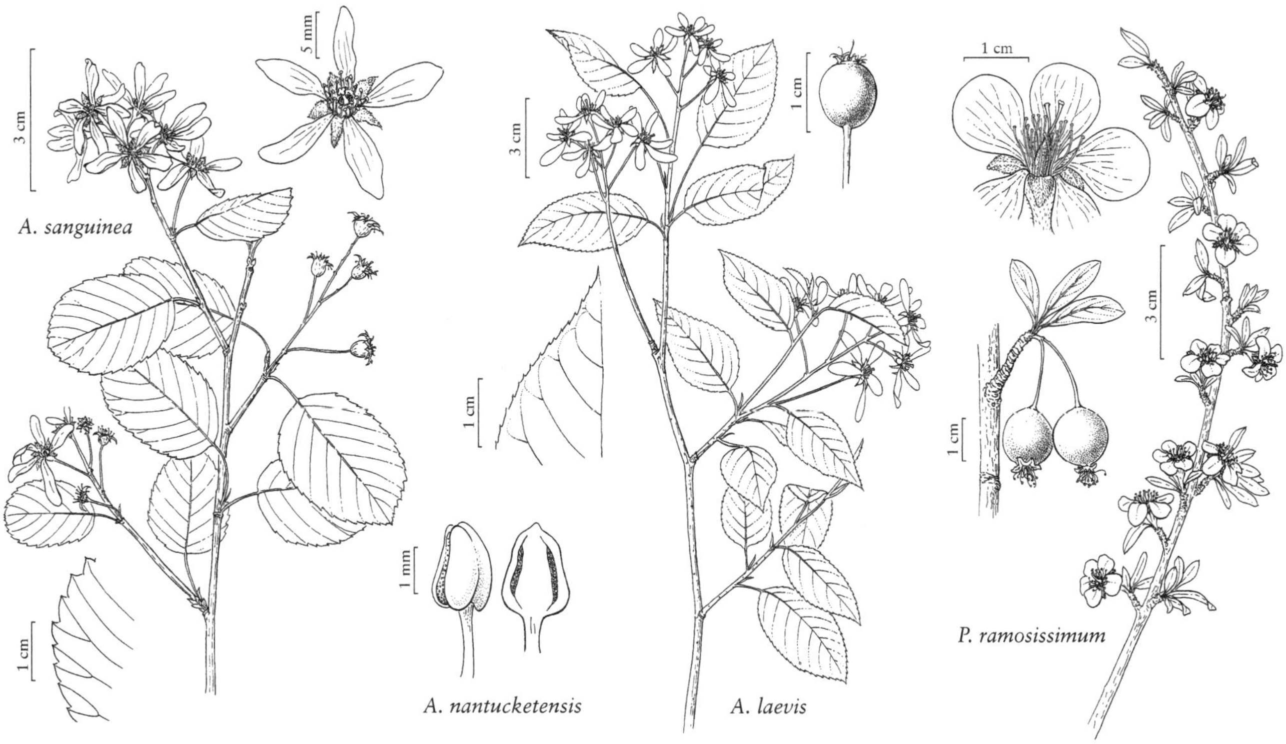

AMELANCHIER ◦ PERAPHYLLUM

6. **Amelanchier sanguinea** (Pursh) de Candolle in A. P. de Candolle and A. L. P. P. de Candolle, Prodr. 2: 633. 1825 • Red-twigged shadbush, amélanchier sanguin E F

Pyrus sanguinea Pursh, Fl. Amer. Sept. 1: 340. 1813; *Amelanchier huronensis* Wiegand

Shrubs or small trees, 1–7 m. **Stems** 1–20, ± colonial. **Leaves** fully unfolded or nearly so; petiole (8–)11.3–18.1(25) mm; blade elliptic-oblong to suborbiculate or obovate, (27–)38–59(–78) × (19–)26–43(–55) mm, base rounded to subcordate, each margin with (0–)4–11(–16) teeth on proximal ½ and (0–)3–6(–8) teeth in distalmost cm, largest teeth more than 1 mm, apex rounded and mucronate to subacute, abaxial surface densely (sparsely) hairy by flowering, surfaces sparsely (moderately) hairy (or glabrous) later. **Inflorescences** (4–)7–10(–13)-flowered, (20–)30–51(–68) mm. **Pedicels:** 0–2 subtended by a leaf, proximalmost (8–)11–23(–36) mm. **Flowers:** sepals recurved or spreading after flowering, (1.8–)2.5–4.1(–6) mm; petals linear to narrowly spatulate, (8–)10.7–16.1(–20) × (2–)3.5–5.6(–7) mm; stamens (16–)19–21(–24); styles (4 or)5, (1.6–)2.5–3.4(–4.6) mm; ovary apex densely (moderately) hairy. **Pomes** dark purple or almost black, 10 mm diam. **2*n*** = 2*x*, 3*x*, 4*x*.

Flowering May–Jun; fruiting Jul–Aug. Margins of woods, river ledges, shorelines, rocky slopes, crevices of open rock faces and cliffs, noncalcareous to slightly calcareous sites; 0–1000 m; N.B., Ont., Que., Sask.; Ala., Conn., Ga., Ill., Iowa, Ky., Maine, Md., Mass., Mich., Minn., N.H., N.J., N.Y., N.C., N.Dak., Ohio, Pa., Tenn., Vt., Va., W.Va., Wis.

This treatment follows those of G. N. Jones (1946) and E. G. Voss (1972–1996, vol. 2) with the inclusion of *Amelanchier huronensis* in *A. sanguinea*. The former was considered to differ from *A. sanguinea* in having longer proximalmost pedicels, sepals, and petals, but overlap occurs in the ranges of these characters. Moreover, *A. sanguinea* has been seen far from the range of *A. huronensis* (southern Ontario, Michigan, Minnesota, and Wisconsin) with petals as long as the longest reported for *A. huronensis*. Occurrences of *A. sanguinea* in Kentucky and North Dakota are questionable, as no specimens were actually seen to confirm the presence of this species from these states.

Amelanchier sanguinea has shorter petals than *A. amabilis*. *Amelanchier humilis* differs from *A. amabilis* and *A. sanguinea* in having shorter proximalmost pedicels, erect inflorescences, and shorter petals. *Amelanchier amabilis*, *A. sanguinea*, and, to a lesser extent, *A. humilis* resemble western North American congeners.

Amelanchier sanguinea flowers ten to 14 days after *A. arborea* and *A. laevis*, according to

K. M. Wiegand (1912), and at the same time as *A. amabilis*. Hybrids between *A. sanguinea* and *A. bartramiana*, *A. canadensis*, *A. laevis*, and *A. spicata* have been reported (M. L. Fernald 1950; P. Landry 1975). The authors have observed *A. sanguinea* flowering with *A. amabilis*, but no putative hybrid was observed.

7. **Amelanchier amabilis** Wiegand, Rhodora 23: 48. 1921 • Lovely shadbush, amélanchier gracieux [E]

Amelanchier sanguinea (Pursh) de Candolle forma *grandiflora* Wiegand, Rhodora 14: 139. 1912, not *A.* ×*grandiflora* Rehder 1920; *A. sanguinea* var. *grandiflora* (Wiegand) Rehder

Shrubs, 1–5 m. **Stems** 1–20, forming colonies. **Leaves** fully unfolded; petiole (9–)12.1–19.9(–28) mm; blade ovate-oval, (33–)41–58(–72) × (24–)30–46(–57) mm, base rounded to subcordate, each margin with (2–)6–10(–15) teeth on proximal ½ and (1 or)3 or 4(or 5) teeth in distalmost cm, largest teeth more than 1 mm, apex subacute, abaxial surface densely (moderately) hairy by flowering, sparsely (moderately) hairy (or glabrous) later, adaxial sparsely hairy (or glabrous) later. **Inflorescences** (4 or)5–8(–11)-flowered, (26–)38–57(–75) mm. **Pedicels:** (0 or)1 or 2(or 3) subtended by a leaf, proximalmost (9–)18–26(–33) mm. **Flowers:** sepals spreading to recurved after flowering, (2.4–)3.5–4.9(–6.5) mm; petals linear to narrowly spatulate, (15–)16.6–19.7(–22) × (4–)5.3–7.3(–8.6) mm; stamens (17–)19–21(–23); styles (3–)5, (2.3–)2.8–3.6(–4.2) mm; ovary apex densely hairy. **Pomes** dark purple or almost black, 10 mm diam. $2n = 4x$.

Flowering May–Jun; fruiting Jul–Aug. Open woods, rocky banks, shores, calcareous sites; 0–300 m; Ont., Que.; N.Y.

Amelanchier amabilis makes a conspicuous display with its relatively large flowers; in eastern North America, it resembles most closely *A. sanguinea*. Leaves with three or four teeth on distal cm of margin and largest teeth more than 1 mm are also useful for identification.

Amelanchier amabilis flowers at the same time as *A. humilis* and *A. sanguinea*. The authors have observed *A. amabilis* growing with *A. sanguinea* without the occurrence of putative hybrids; putative hybrids with *A. arborea* and *A. humilis* have been observed in the field.

8. **Amelanchier gaspensis** (Wiegand) Fernald & Weatherby, Rhodora 33: 235. 1931 • Gaspé shadbush, amélanchier de Gaspésie [E]

Amelanchier sanguinea (Pursh) de Candolle var. *gaspensis* Wiegand, Rhodora 14: 139. 1912

Shrubs, 0.5–6 m. **Stems** 1–40, suckering and densely colonial. **Leaves** mostly unfolded; petiole (8–)12.8–20.7(–25) mm; blade broadly oblong to suborbiculate, (32–)41–57(–63) × (26–)30–40(–45) mm, base cordate or rounded, each margin with (0–)3–9(–12) teeth on proximal ½ and 3–5 teeth in distalmost cm, largest teeth more than 1 mm, apex rounded to subtruncate or short-pointed, abaxial surface moderately (sparsely or densely) hairy by flowering, surfaces sparsely hairy (or glabrous) later. **Inflorescences** (3–)5–10(–12)-flowered, (20–)35–50(–55) mm. **Pedicels:** 1 or 2 subtended by a leaf, proximalmost (9–)12–19(–26) mm. **Flowers:** sepals recurved after flowering, (2.6–)3–4.5(–5.9) mm; petals oblanceolate, (8–)9.7–13.5(–15) × (3–)4.1–6.2(–7) mm; stamens (17 or)18–20(or 21); styles (4 or)5, (2.3–)2.6–3.5(–4.1) mm; ovary apex densely hairy. **Pomes** blackish purple, 10 mm diam. $2n = 4x$.

Flowering May–Jun; fruiting Jul–Sep. Calcareous shores, gravel beaches, cliffs, ledges, alluvial woods, roadsides; 0–300 m; Que.

K. M. Wiegand (1912) considered *Amelanchier gaspensis* to be perplexing because it varies much in stature and habit, as well as in leaf outline and dentition. The leaves suggest an intermediate condition between *A. sanguinea* and *A. humilis*, especially in the venation.

G. N. Jones (1946) considered the range of *Amelanchier gaspensis* to extend off the Gaspé Peninsula in Quebec and into Ontario and Michigan; he listed 13 specimens away from the Gaspé, and uncertainty exists about the identity of some of these specimens. For some specimens, Jones first annotated them as *A. spicata* and later changed his annotation to *A. gaspensis*. Wiegand annotated one of the specimens as "*A. humilis* x?," another specimen as "*A. humilis*," and another as "*A. florida*" (*A. alnifolia* var. *semiintegrifolia*). E. G. Voss (1972–1996, vol. 2) included *A. gaspensis* in *A. sanguinea*. M. L. Fernald (1950) reported *A. gaspensis* from northern Maine; specimens from there assigned to *A. gaspensis* do not match it in all respects. The occurrence of *A. gaspensis* away from the Gaspé remains questionable.

9. Amelanchier fernaldii Wiegand, Rhodora 22: 149. 1920 • Fernald's shadbush, amélanchier de Fernald [E]

Shrubs, 0.5–1 m, <rhizomatous>. **Stems** 1–50, usually forming colonies. **Leaves** mostly or fully unfolded; petiole 1–25 mm; blade elliptic-oblong to oblong-obovate, 50–80 × 15–45 mm, base rounded to subcordate, each margin with 4–18 teeth on proximal ½ and 5–10 teeth in distalmost cm, largest teeth less than 1 mm, apex subacute to rounded or mucronate, abaxial surface glabrous (or sparsely hairy) by flowering, surfaces glabrous later. **Inflorescences** 3–8-flowered, 20–40 mm. **Pedicels:** usually 1 subtended by a leaf, proximalmost 15–35 mm. **Flowers:** sepals ascending or spreading after flowering, 3–5 mm; petals oval to broadly oblanceolate, 8–15 × 3–6 mm; stamens 20; styles 5, length unknown; ovary apex densely hairy. **Pomes** dark purple or almost black, 6–10 mm diam.

Flowering Jun–Jul; fruiting Jul–Aug. Ravines, thickets, barrens, rocky banks, shores, sea cliffs, swamps, wet woods, often calcareous sites; 0–100 m; N.B., Nfld. and Labr. (Nfld.), N.S., P.E.I., Que.

The rhizomatous habit, leaves that are glabrous upon expanding, and densely hairy ovary apices in *Amelanchier fernaldii* are a distinctive combination of characteristics. Its relationships to congeners are unclear. Access to flowering material clearly assignable to *A. fernaldii* has not been available, and the size of floral parts was taken from M. L. Fernald (1950).

10. Amelanchier interior E. L. Nielsen, Amer. Midl. Naturalist 22: 185, plate 13. 1939 • Wiegand's shadbush, amélanchier de l'intérieur [E]

Amelanchier wiegandii E. L. Nielsen

Shrubs or trees, 1–10 m. **Stems** 1–10, forming colonies or solitary. **Leaves** not fully unfolded; petiole 10–30 mm; blade broadly ovate to elliptic, 30–70 × 20–50 mm, base rounded to subcordate, each margin with 3–15 teeth on proximal ½ and (2–)4 or 5(–7) teeth in distalmost cm, largest teeth less than 1 mm, apex short acuminate to apiculate, abaxial surface glabrous or sparsely hairy by flowering, surfaces glabrous later. **Inflorescences** 4–12-flowered, 30–75 mm. **Pedicels:** 1 or 2 subtended by a leaf, proximalmost 10–45 mm. **Flowers:** sepals recurved after flowering, 2–5 mm; petals obovate, 6–15 × 4–5 mm; stamens 20; styles 5, 3–5 mm; ovary apex densely (moderately) hairy (or glabrous). **Pomes** purple-black, 6–8 mm diam. $2n = 4x$.

Flowering May–Jun; fruiting Jul–Aug. Dry woods, bluffs, rocky areas and slopes, stream banks, fields, thickets, sandy areas, sometimes wetlands; 0–300 m; N.B., Nfld. and Labr. (Nfld.), N.S., Ont., P.E.I., Que.; Ill., Iowa, Mich., Minn., Ohio, S.Dak., Wis.

Amelanchier interior is distinguished by its capacity to grow into a tree to ten meters and by having sparsely hairy young leaves that are often reddish and densely hairy ovary apices. E. L. Nielsen (1939) differentiated *A. interior* and *A. wiegandii* on the basis of leaves being carinate in *A. wiegandii* as opposed to flat in *A. interior*, and leaf sinuses rounded versus acute. G. N. Jones (1946) considered those differences to be slight, and he included *A. wiegandii* in *A. interior*. The authors follow Jones, but retain the common name Wiegand's shadbush to commemorate Wiegand's early insights about the taxonomy of *Amelanchier*. M. L. Fernald (1950) considered *A. wiegandii* as suggesting a small-leaved *A. laevis* but with shorter buds, fewer teeth, fewer veins, and summit of ovary heavily tomentose. P. Landry (1975) thought *A. wiegandii* to be the hybrid of *A. arborea* and *A. sanguinea*, and E. G. Voss (1972–1996, vol. 2) reasoned that *A. interior* is a hybrid swarm involving *A. laevis* (or sometimes *A. arborea*) and plants of the *A. spicata* and/or *A. sanguinea* complex. A hybrid origin of *A. interior* from *A. laevis* and *A. sanguinea* is reasonable given that stem height, the number of leaf teeth, and petal length are more or less intermediate between those two species. DNA sequences from ITS region indicate that *A. interior* (*A. wiegandii*) is a possible later-generation hybrid involving a member of the western North American ITS clade (which also includes *A. humilis* and *A. sanguinea* of eastern North America) and some eastern North American taxon (C. S. Campbell et al. 1997). Multiple hybrid origins, possibly from different species, may explain the variability of *A. interior*. *Amelanchier interior* has unusually large ranges of lengths for proximal pedicels, sepals, and petals. The leaves of *A. wiegandii* were described by Nielsen as bronze at flowering; Jones described the leaves of *A. interior* as green when young. The authors assume that *A. interior* as they interpret it is polymorphic for the color of young leaves.

11. Amelanchier humilis Wiegand, Rhodora 14: 141, plate 95, figs. 2A–G. 1912 • Low shadbush, amélanchier bas [E]

Amelanchier humilis var. *campestris* E. L. Nielsen; *A. humilis* var. *compacta* E. L. Nielsen; *A. humilis* var. *exserrata* E. L. Nielsen; *A. mucronata* E. L. Nielsen

Shrubs, to 1.5 m (sometimes to 8 m in shaded conditions). **Stems** 1–50, rhizomatous or suckering and forming scattered colonies. **Leaves** partly to fully unfolded; petiole (5–)9.3–16.7(–24) mm; blade elliptic or oblong to ovate or obovate, (20–)31–47(–62) × (12–)21–35(–45) cm, base rounded to cordate, each margin with (0 or)1–6(–13) teeth on proximal ½ and (0–)3–6(–9) teeth in distalmost cm, largest teeth more than 1 mm, apex broadly subacute to rounded and mucronate, abaxial surface densely hairy by flowering, surfaces sparsely (moderately) hairy (or glabrous) later. **Inflorescences** (5–)7–11(–14)-flowered, (12–)18–32 (–51) mm. **Pedicels:** (0 or)1(or 2) subtended by a leaf, proximalmost (4–)6–11(–16) mm. **Flowers:** sepals spreading to recurved after flowering, (1–)1.6–2.7(–3.6) mm; petals ivory, obovate-oblong, (4.3–)5.9–8.7 (–11.3) × (1.4–)2.3–3.9(–5) mm; stamens (14–)18–21 (or 22); styles (3–)5, (1.8–)2.3–3(–3.6) mm; ovary apex densely hairy. **Pomes** almost black, 6–8 mm diam. $2n = 2x, 3x, 4x$.

Flowering Apr–May; fruiting Jul–Aug. Dry, open sites with rocky, gravelly, or sandy soil, calcareous soil; 0–500 m; Man., Ont., Que., Sask.; Ill., Ind., Iowa, Kans., Md., Mich., Minn., Mo., Nebr., N.J., N.Y., N.Dak., Ohio, Pa., S.Dak., Vt., W.Va., Wis.

Amelanchier humilis is distinctive in its strongly rhizomatous or suckering habit, erect inflorescences, leaf margins with three to six teeth on distal cm, and preference for basic or neutral soils. *Amelanchier humilis* was considered to be conspecific with *A. spicata* by G. N. Jones (1946). Occurrence of *A. humilis* in Saskatchewan, Kansas, and Nebraska is questionable, as no specimens were seen to confirm its presence there.

M. L. Fernald (1950) reported hybrids between *Amelanchier humilis* and *A. arborea*, *A. bartramiana*, and *A. spicata*. The authors have observed putative hybrids with *A. amabilis*.

12. Amelanchier spicata (Lamarck) K. Koch, Dendrologie 1: 182. 1869 • Thicket shadbush, amélanchier en épis [E]

Crataegus spicata Lamarck in J. Lamarck et al., Encycl. 1: 84. 1783; *Amelanchier arborea* (F. Michaux) Fernald var. *austromontana* (Ashe) H. E. Ahles; *A. austromontana* Ashe; *A. lucida* (Fernald) Fernald; *A. stolonifera* Wiegand

Shrubs, 0.3–1.2(–2) m. **Stems** 1–100, rhizomatous and in scattered colonies. **Leaves** half-unfolded; petiole (9.5–)11.5–16(–19) mm; blade oval to orbiculate, (22–)34–45(–61) × (13–)21–33(–52) mm, base subcordate or rounded, each margin with (0–) 3–9(–14) teeth on proximal ½ and (3 or)4–7(–11) teeth in distalmost cm, largest teeth less than 1 mm, apex acute to obtuse and sometimes mucronate, abaxial surface densely (moderately) hairy by flowering, surfaces glabrous or sparsely hairy later. **Inflorescences** (5 or) 6–9(–14)-flowered, (11–)22–37(–52) mm. **Pedicels:** (0 or)1 subtended by a leaf, proximalmost 7–16(–22) mm. **Flowers:** sepals recurved after flowering, (1.7–) 2–3(–4.4) mm; petals linear-oblong, (5.5–)6.5–9(–11) × (2–)2.7–4.4(–6.5) mm; stamens (10–)20; styles (4 or)5, (2.2–)3–3.8(–4.6) mm; ovary apex densely hairy (or glabrous). **Pomes** purple-black, 7–12 mm diam. $2n = 3x, 4x$.

Flowering Apr–Jun; fruiting Jul–Aug. Summits and cliffs of low mountains, open woods, woodland clearings, rocky soil, crevices, shores, fields, roadsides, peaty, sandy, or gravelly and, typically, acidic soil; 0–1200 m; N.B., Nfld. and Labr. (Nfld.), N.S., Ont., P.E.I., Que.; Ala., Conn., Ga., Ill., Iowa, Maine, Md., Mass., Mich., Minn., N.H., N.J., N.Y., N.C., N.Dak., Ohio, Pa., R.I., S.C., Vt., Va., W.Va., Wis.

Amelanchier spicata is strongly rhizomatous and has finely toothed leaves and a glabrous or densely hairy ovary apex. The species is similar to *A. humilis* in habit and vestiture of the ovary apices; it differs in leaf teeth, style length, style fusion, and fruit diameter. *Amelanchier spicata* prefers acidic soil; *A. humilis* is a calciphile. How *A. spicata* and 13. *A. nantucketensis* differ is discussed under the latter.

Amelanchier oblongifolia var. *micropetala* B. L. Robinson was transferred to *A. stolonifera* as forma *micropetala* (B. L. Robinson) Rehder. The type of this name has petals that fall within the size range of those of *A. nantucketensis*, and the authors consider it a synonym of the latter.

P. M. Catling (2006) analyzed the morphology, including flowers, of *Amelanchier lucida* and concluded that it is distinct from *A. spicata* because of its shiny leaves and erect orientation of the sepals at flowering. *Amelanchier lucida* closely resembles *A. spicata* in overall habit, leaves, inflorescences, and fruits, and the authors have observed somewhat lustrous leaves in *A. spicata*; therefore, *A. lucida* is included as a synonym.

Informally recognized, *Amelanchier* "*maritima*" is a well-studied microspecies restricted to the Atlantic Coast from Massachusetts to mid Maine. This microspecies resembles *A. spicata* in habit and mature leaf characters and differs significantly with longer inflorescences, more flowers per inflorescence, and longer and wider petals.

M. L. Fernald (1950) and L. Cinq-Mars (1971) reported hybrids between *Amelanchier spicata* and *A. arborea*, *A. bartramiana*, *A. intermedia*, *A. laevis*, and *A. sanguinea*. Plants determined to be apomictic and attributed to this species by C. S. Campbell et al. (1987) were actually *A. nantucketensis*.

13. Amelanchier nantucketensis E. P. Bicknell, Bull. Torrey Bot. Club 38: 453. 1911 (as nantucketense) • Nantucket shadbush [E] [F]

Amelanchier oblongifolia (Torrey & A. Gray) M. Roemer var. *micropetala* B. L. Robinson

Shrubs, 0.3–2.5 m. **Stems** 1–70, rhizomatous, suckering and forming colonies. **Leaves** less than half-unfolded; petiole (6–)9–15(–22) mm; blade elliptic to oblanceolate or oblong-elliptic, (11–)29–41(–55) × (13–)19–29(–42) mm, base rounded to cuneate, each margin with 0–4(–10) teeth on proximal ½ and (0–)3–8(–12) teeth in distalmost cm, largest teeth less than 1 mm, apex subacute to rounded and mucronate, abaxial surface densely (moderately) hairy by flowering, surfaces sparsely hairy (or glabrous) later. **Inflorescences** (4–)6–8(–11)-flowered, (12–)24–36 (–53) mm. **Pedicels:** (0 or)1(or 2) subtended by a leaf, proximalmost (4–)7–14(–38) mm. **Flowers:** sepals irregularly spreading or recurved after flowering, (1.3–) 2–3(–3.9) mm; petals spatulate to oblong, (2.2–)3–4.5 (–6.8) × (0.6–)1–2(–3) mm, <sometimes bearing 1 or 2 tiny pollen sacs near margins on adaxial surfaces>; stamens (12–)18–20; styles (4 or)5, (1.7–)2.6–3.7(–4.3) mm; ovary apex glabrous or sparsely to densely hairy. **Pomes** dark purple-blue, 7.5–10 mm diam. **2*n*** = 4*x*.

Flowering Mar–May; fruiting Jun–Jul. Fields, sand-plain grasslands, heaths, glacial outwash plains, forest openings, disturbed sites, stream shores, among rocks or sand, dry habitats, ditches, swales; 0–400 m; N.S.; Conn., Maine, Md., Mass., N.H., N.J., N.Y., R.I., S.C., Va.

In the absence of flowering material, identification of *Amelanchier nantucketensis* is tentative; it is easily confused with *A. spicata*. K. M. Wiegand (1912) thought that *A. nantucketensis* originated as a hybrid between *A. spicata* (which he called *A. stolonifera*) and *A. canadensis* (as *A. oblongifolia*).

Amelanchier nantucketensis can reach 2.5 m but is usually much shorter, with spindly, straight, pale gray stems. Some petals of *A. nantucketensis* bear one or two tiny pollen sacs near margins on adaxial surfaces. Petal-borne pollen, which is highly unusual within the plant kingdom, is viable in *A. nantucketensis* and associated with a unique pollinator guild of native bees (A. C. Dibble et al. 1997). Conservation challenges for this apomictic tetraploid were discussed by Dibble and C. S. Campbell (1995).

Amelanchier nantucketensis has been documented to be self-compatible and able to produce seed asexually (C. S. Campbell et al. 1987).

14. Amelanchier bartramiana (Tausch) M. Roemer, Fam. Nat. Syn. Monogr. 3: 145. 1847 • Mountain shadbush or serviceberry, amélanchier de Bartram [E]

Pyrus bartramiana Tausch, Flora 21: 715. 1838

Shrubs, 0.3–2.5(–5) m. **Stems** 1–50, fastigiate, suckering and forming ± dense colonies. **Leaves** half-unfolded; petiole (2–)4.5–10.5(–25) mm; blade narrowly elliptic-oval to oblong to broadly oval, (26–)37–51 (–74) × (12–)20–29(–48) mm, base usually cuneate, each margin with (2–)8–14(–27) teeth on proximal ½ and (2–)7–12(–21) teeth in distalmost cm, largest teeth less than 1 mm, apex acute to rounded, abaxial surface sparsely (moderately) hairy (or glabrous) by flowering, glabrous or sparsely hairy later, adaxial glabrous (or sparsely hairy) later. **Inflorescences** (1 or)2 or 3(or 4) -flowered, (6–)13–25(–38) mm. **Pedicels:** (0 or)1(–3) subtended by a leaf, proximalmost (4–)11–21(–35) mm. **Flowers:** sepals ascending to recurved after flowering, (1.7–)2.7–3.9(–6) mm; petals oblong-oval to broadly elliptic, (5.5–)7.1–8.7(–16.9) × (2.6–)3.9–5.3(–7) mm; stamens (8–)18–21(–25); styles (3 or)4 or 5, (2.7–)3.8–5.2 (–6.1) mm; ovary apex densely hairy (or glabrous). **Pomes** dark purple, <pear-shaped>, 10–15 mm diam. **2*n*** = 2*x*, 3*x*, 4*x*.

Flowering May–Aug; fruiting Jul–Sep. Cool woods, mountain slopes, summits, bogs, poor fens, conifer swamps, acidic soil, sandy lake shores, stream banks, rocky ridges, roadside thickets; 0–1500 m; St. Pierre and Miquelon; N.B., Nfld. and Labr., N.S., Ont., P.E.I., Que.; Maine, Mass., Mich., Minn., N.H., N.Y., Pa., Vt., W.Va., Wis.

Amelanchier bartramiana is the only North American *Amelanchier* with leaves that are imbricate in bud, usually fewer than four flowers per inflorescence, and conic ovary apices (and hence fruits that are more or less pear-shaped rather than globose as in other members of the genus). Because of these differences, P. Landry (1975) placed *A. bartramiana* in its own subgenus, and all other members of the genus in another. W. H. Blanchard (1907), W. A. Robinson and C. R. Partanen (1980), and Robinson (1982) also recognized the distinctness of this species. *Amelanchier bartramiana* grows farther north than any other shadbush in eastern North America, and more than other shadbushes, it occupies relatively undisturbed habitats, such as peatlands and natural breaks in mature forests. Some plants in this species produce seed sexually (C. S. Campbell et al. 1987). Sexual plants of *A. bartramiana* are self-incompatible diploids; a tetraploid individual has been reported (A. C. Dibble et al. 1998); it had relatively large petals and might have been an autotetraploid.

Amelanchier bartramiana usually flowers with *A. laevis*, and it frequently hybridizes with other members of the genus. M. L. Fernald (1950) and L. Cinq-Mars (1971) reported hybrids with *A. arborea* (*A.* ×*quinti-martii* Louis-Marie), *A. canadensis*, *A. fernaldii*, *A. gaspensis*, *A. humilis*, *A. intermedia*, *A. laevis*, *A. sanguinea*, *A. spicata*, and *A. interior*. The hybrid with *A. laevis* can usually be found when these two species grow together (J. E. Weber and C. S. Campbell 1989). The authors have documented a hybrid between *A. arborea* and *A. bartramiana* in eastern Pennsylvania (M. B. Burgess et al., unpubl.).

15. Amelanchier canadensis (Linnaeus) Medikus, Gesch. Bot., 79. 1793 • Eastern shadbush, amélanchier du Canada [E]

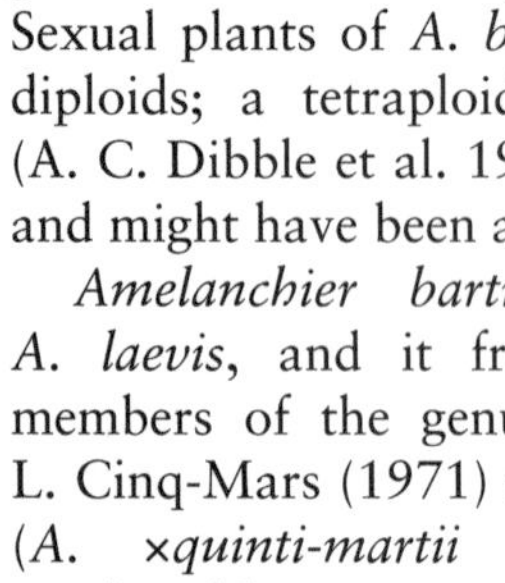

Mespilus canadensis Linnaeus, Sp. Pl. 1: 478. 1753

Shrubs, 0.2–8 m. **Stems** 1–100, in open to dense colonies, evidently stoloniferous in var. *obovalis*. **Leaves** less than half-unfolded; petiole (5–)10–15 (–25) mm; blade elliptic or oval to oblong or obovate, (24–) 34–48(–67) × (12–)18–27(–42) mm, base subcordate, rounded, or cuneate, each margin with (0–)6–15(–29) teeth on proximal ½ and (0–)6–12(–17) teeth in distalmost cm, largest teeth less than 1 mm, apex subacute or obtuse to rounded and mucronate, abaxial surface densely hairy by flowering, glabrous or sparsely to moderately hairy later, adaxial glabrous or sparsely (moderately) hairy later. **Inflorescences** (3–)7–10(–15) -flowered, (6–)15–37(–74) mm. **Pedicels:** 0 or 1(or 2) subtended by a leaf, proximalmost (1–)5–14(–17) mm. **Flowers:** sepals erect, ascending, or spreading after flowering, (0.3–)1.8–3.1(–4.6) mm; petals linear, elliptic, or oblong, (4–)6–10.2(–15) × (1.8–)2.6–4(–5.3) mm; stamens (15–)19–21(–28); styles (3–)5(or 6), (1.5–) 3.4–4.7(–5.8) mm; ovary apex glabrous (or moderately hairy). **Pomes** purplish black or maroon-purple, 6–10 mm diam.

Varieties 2 (2 in the flora): e North America.

The type specimen of *Amelanchier canadensis*, from Virginia or Canada, includes a single stem with five leaves and the remnants of an infructescence with three fruits. Much of the difficulty in determining the identities of specimens and species distributions owes to the limited diagnostic value of the type. The authors include *A. obovalis* as a variety of *A. canadensis* because of ecologic and morphologic similarities. W. W. Ashe (1903) described *A. obovalis* as a shrub of swamps and loose soils, 9–15 dm or a tree to 4 m with inflorescences 3–5 cm and proximalmost pedicels 3–9 mm at flowering and 20–30 mm in fruit. K. M. Wiegand (1912) described *A. canadensis* (as *A. oblongifolia*) as a shrub with slender, erect stems in dense fastigiate colonies with inflorescences 25–60 mm, the proximalmost pedicels 8–18 mm and scarcely longer in fruit. Wiegand was uncertain about placement of specimens of *A. obovalis* from the southeastern United States and stated that they appeared to be this species. The circumscription by Wiegand of *A. oblongifolia* and the circumscription by Ashe of *A. obovalis* overlap considerably. The two species co-occur in similar habitats and have overlapping phenologies, and individual populations may contain specimens with intermediate morphologic characters, especially at early successional sites. Experimental crosses between plants representing the morphologic and ecologic extremes result in formation of mature fruit and seed (C. T. Frye, unpubl.). Plants growing among rocky outcrops in the Potomac River Gorge in Maryland and Virginia suggest *A. spicata* in growth form and habitat but differ in having glabrous ovary apices and the finely toothed leaves of *A. canadensis*. These plants are also unusual in that most individuals in the population rarely produce fruits suggesting the presence of locally dominant self-incompatibility allele(s).

1. Petals (4–)7–10.2(–15) mm; inflorescences (12–) 21–37(–74) mm; proximalmost pedicels (6–)8–14 (–17) mm; shrubs 0.5–8 m; stems in open to dense colonies. . . 15a. *Amelanchier canadensis* var. *canadensis*
1. Petals (5–)6–8(–10.4) mm; inflorescences (6–) 15–27(–45) mm; proximalmost pedicels (1–)5–10 (–15) mm; shrubs 0.2–2 m; stems in dense colonies, often evidently stoloniferous. . . . . 15b. *Amelanchier canadensis* var. *obovalis*

15a. Amelanchier canadensis (Linnaeus) Medikus var. **canadensis** [E]

Amelanchier canadensis var. *subintegra* Fernald; *A. oblongifolia* (Torrey & A. Gray) M. Roemer

Shrubs, 0.5–8 m. **Stems** 1–25, in open to dense colonies. **Leaves:** blade oblong to oblong-elliptic or narrowly oblong-obovate, (24–)35–48(–67) × (12–)20–27 (–42) mm, base rounded, apex subacute to rounded and mucronate, abaxial surface glabrous to moderately hairy, adaxial glabrous or sparsely (moderately) hairy. **Inflorescences** (3–)7–10 (–15)-flowered, usually loose, (12–)21–37(–74) mm. **Pedicels:** 0 or 1(or 2) subtended by a leaf, proximalmost (6–)8–14(–17) mm. **Flowers:** sepals erect or spreading after flowering, (0.3–)2.1–3.1(–4.6) mm; petals linear to oblong, (4–)7–10.2(–15) × (1.8–)3–4(–5.3) mm; stamens (15–)19–20(or 21); styles (3–)5. **Pomes** maroon-purple, 7–10 mm diam. **2*n*** = 2*x*, 4*x*.

Flowering Mar–May; fruiting Jun–Jul. Damp soil of swamps, wet streamheads, bogs, moist to wet thickets, woods; 0–200 m; N.B., N.S., P.E.I., Que.; Conn., Del., D.C., Ga., Maine, Md., Mass., N.H., N.J., N.Y., N.C., Pa., R.I., S.C., Tenn., Vt., Va.

The dense colonies of larger plants, erect inflorescences, usually glabrous ovary apices, and sepals that are erect or ascending after flowering are, in combination, distinctive for var. *canadensis*. Unusual features are its restriction mostly to the Atlantic coastal plain and tolerance of hydric soils. The variety is abundant in forested wetlands and ecotones with fresh-tidal marshes in the Chesapeake Bay region. K. M. Wiegand (1912) and C. T. Frye (2006) noted that the inflorescence of var. *canadensis* does not elongate much following flowering.

Variety *canadensis* is remarkably plastic in habit, flowering at heights of less than 0.5 m in thin soils and growing to 8 m in better conditions. New Brunswick plus eastern Maine and Maryland plus Virginia have dwarf variants that do not appear to be edaphic. Fernald distinguished var. *subintegra*, which grows in low pine barrens of southeastern Virginia, by leaf blades that are entire or toothed only near the apex.

Variety *canadensis* is morphologically close to three species of eastern North America that have finely toothed leaves and glabrous ovary apices, and often grow into trees: *Amelanchier arborea*, *A. intermedia*, and *A. laevis*.

Variety *canadensis* generally flowers at about the same time as *Amelanchier laevis*. M. L. Fernald (1950) and J. E. Cruise (1964) reported that it hybridizes with *A. arborea*, *A. intermedia*, *A. laevis*, and *A. spicata*. Hybrids with *A. arborea* appear to be common, and some microspecies may result. C. T. Frye (2006) reported morphologic evidence of introgression with *A. obovalis*. Tetraploid var. *canadensis* is self-compatible and produces seeds asexually (C. S. Campbell et al. 1987); diploids are self-incompatible and sexual (C. T. Frye, unpubl.).

15b. Amelanchier canadensis (Linnaeus) Medikus var. **obovalis** (Michaux) Britton, Sterns & Poggenburg, Prelim. Cat., 17. 1888 [E]

Mespilus canadensis Linnaeus var. *obovalis* Michaux, Fl. Bor.-Amer. 1: 291. 1803; *Amelanchier canadensis* (Linnaeus) Medikus subsp. *obovalis* (Michaux) P. Landry

Shrubs, 0.2–2 m. **Stems** 1–100, forming dense colonies and often evidently stoloniferous. **Leaves:** blade elliptic or oval to oblong or obovate, (24–)34–40 (–62) × (12–)18–27(–38) mm, base subcordate to cuneate, apex acute or obtuse and mucronate, surfaces glabrous (or sparsely hairy). **Inflorescences** (5–)7–9(–11)-flowered, usually compact, (6–)15–27(–45) mm. **Pedicels:** (0 or)1 subtended by a leaf, proximalmost (1–)5–10(–15) mm. **Flowers:** sepals ascending or spreading after flowering, (0.8–)1.8–2.7(–4) mm; petals elliptic, (5–)6–8(–10.4) × (1.9–)2.6–3.5(–4.3) mm; stamens (15–)19–21(–28); styles (3 or)4 or 5(or 6). **Pomes** purplish black, 6–8 mm diam. **2*n*** = 2*x*, 3*x*.

Flowering Mar–May; fruiting May–Jun. Pine savannas, sandhills, roadsides, dry openings in woods, sandy soil; 0–200 m; Del., Ga., Md., N.J., N.C., Pa., S.C., Va.

C. T. Frye (2006) found morphologic evidence of introgression between var. *canadensis* and var. *obovalis*; the introgression may have been facilitated by land clearing and other anthropogenic disturbances. Variety *obovalis* represents a geographic trend towards smaller petal and inflorescence dimensions in the southern range; there is much overlap range-wide. Additionally, characters such as plant height and extent of lateral spread that have been used in major couplets of keys are variable and may largely depend on site disturbance history, for example, mowing, burning. Variety *obovalis* is self-incompatible and apparently sexual (C. T. Frye, unpubl.) and diploid throughout its range (M. B. Burgess et al., unpubl.). The triploid count (2*n* = 51) is represented by a single specimen determined by W. A. Robinson and C. R. Partenen (1980).

16. Amelanchier intermedia Spach, Hist. Nat. Vég. 2: 85. 1834 • Intermediate shadbush, amélanchier intermédiaire [E]

Shrubs or trees, 2–7 m. **Stems** 1–50, fastigiate, solitary or in colonies. **Leaves** half-unfolded; petiole (14–)16–21.8(–24) mm; blade elliptic to slightly obovate, (39–)46–65(–77) × (25–)26–38 (–45) mm, base subcordate or rounded, each margin with (5–)13–19(or 20) teeth on proximal ½ and (5–)13–19(or 20) teeth in distalmost cm, largest teeth less than 1 mm, apex acute to short-acuminate, abaxial surface sparsely hairy by flowering, surfaces glabrous later. **Inflorescences** (6 or)7–10 -flowered, 35–59(–77) mm. **Pedicels:** 0 or 1 subtended by a leaf, proximalmost (13–)14–25(–33) mm. **Flowers:** sepals ascending to recurved after flowering, (2–)3.3–4.5 (–5.2) mm; petals oblong, (10–)12.7–17.7(–18.5) × (3–) 3.9–5.3(–5.7) mm; stamens (18–)20(or 21); styles (4 or)5, (3.3–)3.8–4.8(–5) mm; ovary apex glabrous (or sparsely hairy). **Pomes** dark purple, 7–12 mm diam. $2n = 4x$.

Flowering May–Jun; fruiting Jun–Aug. Swamps, bogs, thickets, shores; 0–1000 m; N.B., Nfld. and Labr. (Nfld.), N.S., Ont., P.E.I., Que.; Conn., Maine, Md., Mich., Minn., N.H., N.J., N.Y., N.C., Pa., S.C., Vt., Va.

Amelanchier intermedia is morphologically closest to *A. laevis*. In a study of morphologic variation involving *A. canadensis*, *A. intermedia*, *A. laevis*, and other species of *Amelanchier*, A. C. Dibble et al. (1998) found that the only two species that overlap are *A. intermedia* and *A. laevis*. P. Landry (1975) considered *A. intermedia* to be the hybrid of *A. arborea* (including *A. laevis*) and *A. canadensis*, and the authors have data that suggest *A. intermedia* is a hybrid of *A. canadensis* and *A. laevis*.

Amelanchier intermedia hybridizes with *A. bartramiana*, *A. humilis*, and *A. spicata* (L. Cinq-Mars 1971).

A. C. Dibble et al. (1998) reported that two individuals of *Amelanchier intermedia*, both tetraploids, were apomictic.

17. Amelanchier arborea (F. Michaux) Fernald, Rhodora 43: 563. 1941 • Downy shadbush [E]

Mespilus arborea F. Michaux, Hist. Arbr. Forest. 3: 68, plate 11. 1812; *Amelanchier arborea* var. *alabamensis* (Britton) G. N. Jones

Shrubs or trees, 2–20 m. **Stems** 1–20, solitary or in colonies. **Leaves** less than half-unfolded; petiole (5.5–)12–20.6(–29.2) mm; blade ovate to obovate, (30–)47–68(–93) × (16–)26–40(–56) cm, base cordate to rounded, each margin with (0–)11–21(–30) teeth on proximal ½ and (3–)5–9(–13) teeth in distalmost cm, largest teeth less than 1 mm, apex acute to acuminate, abaxial surface densely (moderately) hairy by flowering, glabrous or sparsely to moderately hairy later, adaxial glabrous or sparsely (moderately) hairy later. **Inflorescences** (3–)6–12(–15)-flowered, (19–)30–55(–79) mm. **Pedicels:** 0 or 1 subtended by a leaf, proximalmost (5–)10–20(–32) mm. **Flowers:** sepals soon reflexed after flowering, (1.6–)2.3–3.7(–5.3) mm; petals linear to oblong, (8–)10–15(–19) × (2–)3.1–5(–6.8) mm; stamens (16–)20(or 21); styles (3–)5(or 6), (2.1–)3.2–4.2(–5.5) mm; ovary apex glabrous or sparsely (densely) hairy. **Pomes** maroon-purple, 6–10 mm diam. $2n = 2x, 4x$.

Flowering Feb–May; fruiting May–Jul. Dry to moist woods, mesic mixed hardwoods and pine-hardwoods, fields, thickets, roadsides, circumneutral soil, especially northward; 0–1500 m; N.B., N.S., Ont., Que.; Ala., Ark., Conn., Del., D.C., Fla., Ga., Ill., Ind., Iowa, Kans., Ky., La., Maine, Md., Mass., Mich., Minn., Miss., Mo., Nebr., N.H., N.J., N.Y., N.C., Ohio, Okla., Pa., R.I., S.C., Tenn., Tex., Vt., Va., W.Va., Wis.

Amelanchier arborea is distinctive for its arborescent habit, generally precocious flowering, leaf margins each with (3–)5–9(–13) teeth on distal cm, densely hairy abaxial surface of leaf blades, and strongly reflexed sepals at flowering. Leaves typically have subcordate to strongly cordate bases, acuminate apices, and margins sometimes finely double serrate.

The name *Amelanchier canadensis* was applied by many botanists to *A. arborea* and *A. laevis* until M. L. Fernald (1941) clarified the issue; this usage appears in some older manuals. S. M. McKay transferred *A. laevis* to subspecific status under *A. arborea*, and the two species are generally considered to be closely related. Allozyme data show that, with one exception, populations of each are more closely related to a nearby population of the other than they are to conspecific populations (R. D. Overath and J. L. Hamrick 1998). At flowering, *A. arborea* and *A. laevis* can be distinguished by leaf color, which is typically reddish green or reddish brown in *A. laevis* and green to greenish white in *A. arborea*. In addition, the leaves of *A. laevis* are at least half-expanded and unfolding as well as glabrous (or sparsely hairy) by flowering, and its sepals are spreading to reflexed.

Amelanchier arborea exhibits considerable variation; infraspecific taxa have been named. Variety *alabamensis* differs in its densely hairy ovary apices. Variation in ovary summit indument in *A. arborea* does not appear to be associated with any other taxonomically informative variation; individuals with varying amounts of ovary indument occur sporadically in populations; this does not appear ecologically significant. The authors conclude that this variety does not warrant taxonomic recognition; they have observed groups of populations that differ in the number of flowers per inflorescence, color of the expanding leaves, size of the teeth, and other features.

Amelanchier arborea usually flowers a week or more before sympatric congeners. The species has been reported (M. L. Fernald 1950) to hybridize with *A. bartramiana*, *A. canadensis*, *A. humilis*, and *A. laevis*. The authors have collected putative hybrids with *A. amabilis* in New York and have observed putative hybrids between *A. arborea* and other eastern North American species of *Amelanchier*. Hybrid swarms in New Jersey between this species and *A. laevis* but not *A. canadensis* have been reported (J. E. Cruise 1964).

18. Amelanchier laevis Wiegand, Rhodora 14: 154, plate 96, figs. 7A–G. 1912 • Smooth shadbush, amélanchier glabre E F

Amelanchier arborea (F. Michaux) Fernald subsp. *laevis* (Wiegand) S. M. McKay ex P. Landry; *A. arborea* var. *laevis* (Wiegand) H. E. Ahles; *A. laevis* var. *nitida* (Wiegand) Fernald

Shrubs or trees, 2–25 m. **Stems** 1–20, fastigiate, solitary or in colonies. **Leaves** less than half-unfolded; petiole (7–)15–26(–36) mm; blade elliptic to ovate to oblong or obovate, (32–)48–75(–100) × (19–)29–42(–60) mm, base subcordate to rounded, each margin with (0–)10–23(–38) teeth on proximal ½ and (1–)5–9(–13) teeth in distalmost cm, largest teeth less than 1 mm, apex acute to acuminate, abaxial surface glabrous (or sparsely hairy) by flowering, surfaces glabrous later. **Inflorescences** (4–)7–11(–14)-flowered, (25–)43–65(–85) mm. **Pedicels:** 0 or 1(or 2) subtended by a leaf, proximalmost (7–)15–28(–41) mm. **Flowers:** sepals spreading to recurved after flowering, (1.9–)2.8–4(–5.3) mm; petals linear-oblong, (8–)12.5–17.3(–22.5) × (2.9–)3.9–5.7(–7.5) mm; stamens (14–)19–21(–24); styles (4 or)5(or 6), (2.5–)3.3–4.4(–5.3) mm; ovary apex glabrous (or sparsely hairy). **Pomes** dark purple, 8–15 mm diam. $2n = 2x, 4x$.

Flowering Apr–Jun; fruiting Jun–Jul. Dry to moist, deciduous, mixed, and coniferous forests, fields, thickets, roadsides; 0–2000 m; St. Pierre and Miquelon; N.B., Nfld. and Labr. (Nfld.), N.S., Ont., P.E.I., Que.; Ala., Conn., Del., D.C., Ga., Ill., Ind., Iowa, Ky., Maine, Md., Mass., Mich., Minn., N.H., N.J., N.Y., N.C., Ohio, Pa., R.I., S.C., Tenn., Vt., Va., W.Va., Wis.

Amelanchier laevis is common through much of its range and readily identified by its typically arborescent habit, leaves that are reddish and glabrous by flowering, and relatively long inflorescences, pedicels, and petals. Southeastern United States populations tend to be restricted to high elevations. The relationship to its closest relative, 17. *A. arborea*, is discussed under the latter.

Amelanchier laevis frequently hybridizes with congeners, including *A. arborea*, *A. bartramiana*, *A. canadensis*, *A. humilis*, *A. interior*, *A. sanguinea*, and *A. spicata* (M. L. Fernald 1950; L. Cinq-Mars 1971). J. E. Cruise (1964) documented hybrid swarms between *A. laevis* and both *A. arborea* and *A. canadensis* in New Jersey. The hybrid with *A. bartramiana* can usually be found when these two species grow together (J. E. Weber and C. S. Campbell 1989). The hybrid with *A. arborea*, *A.* ×*grandiflora* Rehder, is used ornamentally.

A. C. Dibble et al. (1998) concluded that *Amelanchier laevis* is possibly one of the parents of *A.* "rubra," an entity that is morphologically distinct. This entity is a tetraploid shrub with stems to 3.5 m and usually colonial, leaves that are reddish and glabrous at flowering, and petals that are often faintly reddish and slightly twisted. Ovary apices are mostly sparsely hairy, but may also be densely hairy or glabrous. Quantitative analysis places *A.* "rubra" between the cluster of *A. laevis* plus *A. intermedia* on the one hand and *A. nantucketensis* or *A. spicata* on the other (Dibble et al.). It is possible that this entity could be a hybrid with one of these arborescent species (*A. laevis* and *A. intermedia*) plus one of the shrub species (*A. nantucketensis* and *A. spicata*). The authors have located populations of *A.* "rubra" on Mount Desert Island and eastward for about 120 km along the Maine coast. The authors have observed plants that appear to be F_1s of this entity and later-generation hybrids with *A. bartramiana* as one of the parents.

Amelanchier laevis has been documented to be self-compatible and to produce seeds asexually (C. S. Campbell et al. 1985; A. C. Dibble et al. 1998).

68. PERAPHYLLUM Nuttall in J. Torrey and A. Gray, Fl. N. Amer. 1: 474. 1840 • Wild crab apple, squaw apple [Genus *Pera* and Greek *phyllon*, leaf, alluding to resemblance to leaves of *P. arborea*] [E]

Christopher S. Campbell

Shrubs, to 20 dm. **Stems** 1–50, ascending, forming clumps, <much branched>; bark gray, <smooth>; long and short shoots present; unarmed; slightly hairy. **Leaves** deciduous, cauline, simple; stipules soon deciduous, adnate to petiole and base of blade, triangular, margins entire or serrulate; petiole short or absent; blade elliptic to oblanceolate or linear, 1–6 cm, coriaceous, margins flat, entire or serrulate, venation pinnate, camptodromous, <apex with deciduous gland>, surfaces sparsely pubescent. **Inflorescences** terminal <usually on short shoots>, 1–3 -flowered, cymes, sparsely to moderately hairy; bracts absent; bracteoles absent. **Pedicels** present. **Flowers** developing after leaves, perianth and androecium epigynous, 10–20 mm diam.; hypanthium funnel-shaped, 2–4 mm diam., glabrous; sepals 5, ascending to spreading, triangular; petals 5, white or pinkish, obovate to orbiculate, <base clawed>; stamens 15–20, shorter than petals; carpels 2 or 3, connate, adnate to hypanthium, <appearing 4–6-loculed by false partitions>, styles <2 or 3>, terminal, connate; ovules 2. **Fruits** berrylike pomes, yellow-orange, globose to ellipsoid, 8–18 mm, glabrous; <bitter tasting>; hypanthium persistent; sepals persistent, erect to recurved; carpels cartilaginous; styles persistent. **Seeds** 1–6. $x = 17$.

Species 1: w United States.

Peraphyllum shares with *Amelanchier* and *Malacomeles* the feature of false partitions within the carpels that partially divide the locules and make the fruit appear to have twice as many locules as the number of carpels (G. N. Jones 1945; K. R. Robertson et al. 1991; J. R. Rohrer et al. 1991). Sequences from multiple chloroplast and nuclear genes indicate that *Amelanchier* and *Peraphyllum* are sister taxa (C. S. Campbell et al. 2007).

1. **Peraphyllum ramosissimum** Nuttall in J. Torrey and A. Gray, Fl. N. Amer. 1: 474. 1840 [E] [F]

Leaves crowded toward tips of short shoots; stipules minute; blade 4–12 mm wide. **Pedicels** 1–2 cm. **Flowers:** sepals 2.5–5 mm, adaxially pubescent; petals 4–8 mm. $2n = 34$.

Flowering spring; fruiting summer. Dry hillsides, mostly in full sun, pinyon pine-juniper, ponderosa pine, oak-sagebrush, and other communities; 500–3000 m; Calif., Colo., Idaho, Nev., N.Mex., Oreg., Utah.

Peraphyllum ramosissimum is distinctive for its narrow leaves, small flowers, and yellow-orange pomes.

Literature Cited

Robert W. Kiger, Editor

This is a consolidated list of all works cited in volume 9, whether as selected references, in text, or in nomenclatural contexts. In citations of articles, both here and in the taxonomic treatments, and also in nomenclatural citations, the titles of serials are rendered in the forms recommended in G. D. R. Bridson and E. R. Smith (1991), Bridson (2004), and Bridson and D. W. Brown (http://fmhibd.library.cmu.edu/fmi/iwp/cgi?-db=BPH_Online&-loadframes). When those forms are abbreviated, as most are, cross references to the corresponding full serial titles are interpolated here alphabetically by abbreviated form. In nomenclatural citations (only), book titles are rendered in the abbreviated forms recommended in F. A. Stafleu and R. S. Cowan (1976–1988) and Stafleu et al. (1992–2009). Here, those abbreviated forms are indicated parenthetically following the full citations of the corresponding works, and cross references to the full citations are interpolated in the list alphabetically by abbreviated form. Two or more works published in the same year by the same author or group of coauthors will be distinguished uniquely and consistently throughout all volumes of *Flora of North America* by lower-case letters (b, c, d, ...) suffixed to the date for the second and subsequent works in the set. The suffixes are assigned in order of editorial encounter and do not reflect chronological sequence of publication. The first work by any particular author or group from any given year carries the implicit date suffix "a"; thus, the sequence of explicit suffixes begins with "b". Works missing from any suffixed sequence here are ones cited elsewhere in the *Flora* that are not pertinent in this volume.

Aas, G. et al. 1994. Morphology, isozyme variation, cytology, and reproduction of hybrids between *Sorbus aria* (L.) Crantz and *S. torminalis* (L.) Crantz. Bot. Helv. 104: 195–214.

Abh. Königl. Böhm. Ges. Wiss. = Abhandlungen der Königlichen böhmischen Gesellschaft der Wissenschaften.

Abh. Naturwiss. Vereins Bremen = Abhandlungen herausgegeben vom Naturwissenschaftlichen Vereins zu Bremen.

Abh. Westfäl. Mus. Naturk. = Abhandlungen der Westfälischen Museum für Naturkunde.

Abrams, L. 1944. *Horkelia*. In: L. Abrams and R. S. Ferris. 1923–1960. Illustrated Flora of the Pacific States: Washington, Oregon, and California. 4 vols. Stanford. Vol. 2, pp. 416–421.

Abrams, L. and R. S. Ferris. 1923–1960. Illustrated Flora of the Pacific States: Washington, Oregon, and California. 4 vols. Stanford.

Academia Sinica. 1959+. Flora Reipublicae Popularis Sinicae. 80+ vols. Beijing.

Ackerman, T. L., J. Bair, and A. Tiehm. 2003. A flora of the Desert National Wildlife Refuge, Nevada. Mentzelia 7: 1–90.

Acta Bot. Brasil. = Acta Botanica Brasilica.

Acta Bot. Neerl. = Acta Botanica Neerlandica.

Acta Bot. Yunnan. = Acta Botanica Yunnanica. [Yunnan Zhiwu Yanjiu.]

Acta Hort. = Acta Horticulturae; Technical Communications of I S H S (International Society for Horticultural Science).

Acta Phytotax. Sin. = Acta Phytotaxonomica Sinica. [Chih Wu Fen Lei Hsüeh Pao.]

Acta Soc. Bot. Poloniae = Acta Societatis Botanicorum Poloniae.

Acta Univ. Agric. Suec., Agrar. = Acta Universitatis Agriculturae Suecicae. Agraria.

Adansonia = Adansonia; Recueil Périodique d'Observations Botaniques.

Aiken, S. G. et al. 2007. Flora of the Canadian Arctic Archipelago.... Ottawa. [CD-ROM.]

Aiton, W. 1789. Hortus Kewensis; or, a Catalogue of the Plants Cultivated in the Royal Botanic Garden at Kew. 3 vols. London. (Hort. Kew.)

Aiton, W. and W. T. Aiton. 1810–1813. Hortus Kewensis; or a Catalogue of the Plants Cultivated in the Royal Botanic Garden at Kew. 5 vols. London. (Hortus Kew.)

Aldasoro, J. J. et al. 1998. The genus *Sorbus* (Maloideae, Rosaceae) in Europe and North Africa: Morphological analysis and systematics. Syst. Bot. 23: 189–208.

Aldasoro, J. J. et al. 2004. Revision of *Sorbus* subgenera *Aria* and *Torminaria* (Rosaceae—Maloideae). Syst. Bot. Monogr. 69: 1–148.

Aldasoro, J. J., C. Aedo, and F. Muñoz Garmendia. 1996. The genus *Pyrus* L. (Rosaceae) in south-west Europe and North Africa. Bot. J. Linn. Soc. 121: 143–158.

Aldasoro, J. J., C. Aedo, and C. Navarro. 2005. Phylogenetic and phytogeographical relationships in Maloideae (Rosaceae) based on morphological and anatomical characters. Blumea 50: 3–32.

Alice, L. A. and C. S. Campbell. 1999. Phylogeny of *Rubus* (Rosaceae) based on nuclear ribosomal DNA internal transcribed spacer region sequences. Amer. J. Bot. 86: 81–97.

Alice, L. A., T. M. Dodson, and B. L. Sutherland. 2008. Diversity and relationships of Bhutanese *Rubus*. Acta Hort. 777: 63–70.

Allan, H. H. et al. 1961–2000. Flora of New Zealand. 5 vols. Wellington, Christchurch, and Lincoln. (Fl. New Zealand)

Ambrose, K. V. 2006. Clarification of Basal Relationships in *Rubus* (Rosaceae) and the Origin of *Rubus chamaemorus*. M.S. thesis. Western Kentucky University.

Amer. Anthropol. = American Anthropologist.

Amer. J. Bot. = American Journal of Botany.

Amer. J. Chin. Med. = The American Journal of Chinese Medicine.

Amer. Midl. Naturalist = American Midland Naturalist; Devoted to Natural History, Primarily That of the Prairie States.

Anal. Fam. Pl.—See: B. C. J. Dumortier 1829

Anders, C. M. and Z. E. Murrell. 2001. Morphological, molecular, and biogeographical variation within the imperiled Virginia *Spiraea*. Castanea 66: 24–41.

Anderson, G. J. 1980. The status of the very rare *Prunus gravesii* Small. Rhodora 82: 113–129.

Angiosperm Phylogeny Group. 2003. An update of the Angiosperm Phylogeny Group classification for the orders and families of flowering plants: APG II. Bot. J. Linn. Soc. 141: 399–436.

Ann. Bot. Fenn. = Annales Botanici Fennici.

Ann. Carnegie Mus. = Annals of the Carnegie Museum.

Ann. Gén. Hort. = Annales Générales d'Horticulture.

Ann. Lyceum Nat. Hist. New York = Annals of the Lyceum of Natural History of New York.

Ann. Missouri Bot. Gard. = Annals of the Missouri Botanical Garden.

Ann. Mus. Bot. Lugduno-Batavi = Annales Musei Botanici Lugduno-Batavi.

Ann. Mus. Natl. Hist. Nat. = Annales du Muséum National d'Histoire Naturelle. ["National" dropped after vol. 5.]

Ann. Naturhist. Mus. Wien, B = Annalen des Naturhistorischen Museums in Wien. Serie B, Botanik und Zoologie.

Ann. Sci. Nat., Bot. = Annales des Sciences Naturelles. Botanique.

Ann. Sci. Nat. (Paris) = Annales des Sciences Naturelles. (Paris.)

Anticancer Res. = Anticancer Research.

Antieau, C. J. 1985. Patterns of Natural Variation in Oceanspray, *Holodiscus discolor* (Pursh) Maxim. (Rosaceae). M.S. thesis. University of Washington.

Arbor. Frutic. Brit.—See: J. C. Loudon [1835–]1838

Arbor. Kórnickie = Arboretum Kórnickie.

Arbust. Amer.—See: H. Marshall 1785

Arnott, G. A. W. 1832. Botany. Edinburgh. [Preprinted from M. Napier, ed. 1830–1842. Encyclopaedia Britannica..., ed. 7. 21 vols. Edinburgh. Vol. 5.] (Botany)

Asami, Y. 1927. The Crabapples and Nectarines of Japan.... Tokyo.

Ascherson, P. F. A. 1859–1864. Flora der Provinz Brandenburg.... 3 vols. Berlin. (Fl. Brandenburg)

Ascherson, P. F. A. et al. 1896–1939. Synopsis der mitteleuropäischen Flora. 12 vols., some in parts. Leipzig. (Syn. Mitteleur. Fl.)

Ashe, W. W. 1903. *Amelanchier obovalis*. Bot. Gaz. 35: 434.

Atkinson, S. and F. Sharpe. 1993. Wild Plants of the San Juan Islands, ed. 2. Seattle.

Atlantic J. = Atlantic Journal, and Friend of Knowledge.

Austral. Syst. Bot. = Australian Systematic Botany.

Autik. Bot.—See: C. S. Rafinesque 1840

Bailey, L. H. 1941–1945. Species batorum. The genus *Rubus* in North America (north of Mexico) I–X. Gentes Herbarum 5: 1–918.

Bailey, L. H. and Bailey Hortorium Staff. 1949. Manual of Cultivated Plants..., rev. ed. New York. [Reprinted 1951, New York.]

Baillon, H. E. 1866–1895. Histoire des Plantes. 13 vols. Paris, London, and Leipzig. (Hist. Pl.)

Baird, J. R. and J. W. Thieret. 1989. The medlar (*Mespilus germanica*, Rosaceae) from antiquity to obscurity. Econ. Bot. 43: 328–372.

Baldwin, J. T. 1951. Chromosomes of *Spiraea* and of certain other genera of Rosaceae. Rhodora 53: 203–206.

Ball, P. W., B. Pawloski, and S. M. Walters. 1968. *Potentilla*. In: T. G. Tutin et al., eds. 1964–1980. Flora Europaea. 5 vols. Cambridge. Vol. 2, pp. 36–47.

Bammi, R. K. and H. P. Olmo. 1966. Cytogenetics of *Rubus*. V. Natural hybridization between *R. procerus* P. J. Muell. and *R. laciniatus* Willd. Evolution 20: 617–633.

Banasiak, S. E. and S. J. Meiners. 2009. Long term dynamics of *Rosa multiflora* in a successional system. Biol. Invasions 11: 215–224.

Barbour, M. G. 1970. The flora and plant communities of Bodega Head, California. Madroño 20: 289–336.

Barnaulov, O. D. and P. P. Denisenko. 1980. Anti-ulcer action of a decoction of the flowers of the dropwort, *Filipendula ulmaria* (L.) Maxim. [In Russian.] Farmakol. Toksikol. 43: 700–705.

Bartish, I. V., B. Hylmö, and H. Nybom. 2001. RAPD analysis of intraspecific relationships in presumably apomictic *Cotoneaster* species. Euphytica 120: 273–280.

Barton, W. P. C. 1817–1819. Vegetable Materia Medica of the United States; or Medical Botany, Containing a Botanical, General, and Medical History, of Medicinal Plants Indigenous to the United States. 2 vols. in 4 parts each. Philadelphia. [Parts numbered consecutively throughout, vols. paged independently.] (Veg. Mater. Med. U.S.)

Batsch, A. J. G. K. 1801. Beyträge und Entwürfe zur pragmatischen Geschichte der drey Natur-Reiche... Gewächsreich. Erster Theil.... 1 fasc. only. Weimar. (Beytr. Entw. Gewächsreich)

Bean, R. C. et al. 1967. Seventeenth report of the Committee on Plant Distribution. Rhodora 69: 306–318.

Bean, W. J. 1914–1933. Trees and Shrubs Hardy in the British Isles. 3 vols. London. (Trees Shrubs Brit. Isles)

Beitr. Bot.—See: C. F. W. Wallroth 1842–1844

Beitr. Naturk.—See: J. F. Ehrhart 1787–1792

Beitr. Phytotax. = Beiträge zur Phytotaxonomie. [Some vols. form parts of Wiss. Z. Friedrich-Schiller-Univ. Jena, Math.-Naturwiss. Reihe.]

Bell, R. L. 1990. Pears *(Pyrus)*. In: J. N. Moore and J. R. Ballington Jr., eds. 1990. Genetic Resources of Temperate Fruit and Nut Crops. 2 vols. Wageningen. Vol. 2, pp. 655–697.

Bentham, G. 1839[–1857]. Plantas Hartwegianas Imprimis Mexicanas.... London. [Issued by gatherings with consecutive signatures and pagination.] (Pl. Hartw.)

Berchtold, F. et al. 1836–1843. Oekonomisch-technische Flora Böhmens.... 3 vols. in 6 parts. Prague. (Oekon.-techn. Fl. Böhm.)

Berg, K. 1983. Searching for rare plants. I. The Tahquitz ivesia, presumed extinct. Fremontia 2: 13–15.

Bergner, P. 1997. Herbs and dysmenorrhea. Med. Herbalism 9(3): 5–13.

Bernard, J.-P. and R. Gauthier. 1986. Observations sur le *Geum urbanum* L. dans la région de Québec et description de deux hybrides. Naturaliste Canad. 113: 317–324.

Beytr. Entw. Gewächsreich—See: A. J. G. K. Batsch 1801

Biblioth. Bot. = Bibliotheca Botanica; original Abhandlungen aus dem Gesammtgebiete der Botanik.

Bicknell, E. P. 1896b. The North American species of *Agrimonia*. Bull. Torrey Bot. Club 23: 508–523.

Bigelow, J. 1824. Florula Bostoniensis. A Collection of Plants of Boston and Its Vicinity..., ed. 2. Boston. (Fl. Boston. ed. 2)

Bijdr. Fl. Ned. Ind.—See: C. L. Blume 1825–1826[–1827]

Biltmore Bot. Stud. = Biltmore Botanical Studies; a Journal of Botany Embracing Papers by the Director and Associates of the Biltmore Herbarium.

Biochem. Syst. & Ecol. = Biochemical Systematics and Ecology.

Biol. Invasions = Biological Invasions.

Biol. J. Linn. Soc. = Biological Journal of the Linnean Society.

Biol. Pharm. Bull. = Biological and Pharmaceutical Bulletin.

Biol. Skr. = Biologiske Skrifter.

Bitter, F. A. G. [1910–]1911. Die Gattung *Acaena*. Biblioth. Bot. 74.

Blanchard, W. H. 1907. Our eastern shadwoods. Torreya 7: 97–102.

Blue Jay = Blue Jay; Bulletin of the Yorktown Natural History Society.

Bluff, M. J., C. A. Fingerhuth, and C. F. W. Wallroth. 1825–1833. Compendium Florae Germaniae. 4 vols. Nuremberg. (Comp. Fl. German.)

Blume, C. L. 1825–1826[–1827]. Bijdragen tot de Flora van Nederlandsch Indië. 17 parts. Batavia. [Parts paged consecutively.] (Bijdr. Fl. Ned. Ind.)

Blumea = Blumea; Tidjschrift voor die Systematiek en die Geografie der Planten (A Journal of Plant Taxonomy and Plant Geography).

Blytt, A. G. 1906. Haandbog i Norges Flora. Oslo.

Bocek, B. R. 1984. Ethnobotany of Costanoan Indians, based on collections by John P. Harrington. Econ. Bot. 38: 240–255.

Böcher, T. W. 1938. Biological distributional types in the flora of Greenland: A study on the flora and plant-geography of South Greenland and East Greenland between Cape Farewell and Scoresby Sound [6. og 7. Thule-Expedition til Sydøstgrønland 1931–1933. Leader: Knud Rasmussen]. Meddel. Grønland 106(2): 1–339.

Böcher, T. W. and K. Larsen. 1950. Chromosome numbers of some arctic or boreal flowering plants. Meddel. Grønland 147(6): 1–32.

Boivin, B. 1952. Pugillus *Potentillarum*. Phytologia 4: 89–93.

Boivin, B. 1966b. Énumération des plantes du Canada, 2. Lignidées (suite). Naturaliste Canad. 93: 371–437.

Boivin, B. 1967–1979. Flora of the prairie provinces. Part I(–IV). Phytologia 15–18, 22–23, 42–43: passim. [Reprinted as Provancheria 2–5 with auxiliary pagination.]

Bolle, F. 1933. Eine Übersicht über die Gattung *Geum* L. und die ihr nahestehenden Gattungen. Repert. Spec. Nov. Regni Veg. Beih. 72: 1–119.

Booth, W. E. and J. C. Wright. 1959. Flora of Montana. Part II. Dicotyledons. Bozeman.

Borkhausen, M. B. 1800–1803. Theoretisches-praktisches Handbuch der Forstbotanik und Forsttechnologie. 2 vols. Giessen and Darmstadt. [Vols. paged consecutively.] (Theor. Prakt. Handb. Forstbot.)

Bortiri, E. et al. 2001. Phylogeny and systematics of *Prunus* (Rosaceae) as determined by sequence analysis of ITS and the chloroplast *trn*L-*trn*F spacer DNA. Syst. Bot. 26: 797–807.

Bortiri, E. et al. 2002. The phylogenetic utility of nucleotide sequences of sorbitol 6-phosphate dehydrogenase in *Prunus* (Rosaceae). Amer. J. Bot. 89: 1697–1708.

Bortiri, E., B. D. Vanden Heuvel, and D. Potter. 2006. Phylogenetic analysis of morphology in *Prunus* reveals extensive homoplasy. Pl. Syst. Evol. 259: 53–71.

Bossard, C. C., J. M. Randall, and M. C. Hoshovsky, eds. 2000. Invasive Plants of California's Wildlands. Berkeley.

Boston J. Nat. Hist. = Boston Journal of Natural History.

Bot. Beechey Voy.—See: W. J. Hooker and G. A. W. Arnott [1830–]1841

Bot. Beob.—See: J. C. Wendland 1798

Bot. California—See: W. H. Brewer et al. 1876–1880

Bot. Centralbl. = Botanisches Centralblatt.

Bot. Electronic News = Botanical Electronic News. [Electronic journal.]

Bot. Gaz. = Botanical Gazette; Paper of Botanical Notes.

Bot. Helv. = Botanica Helvetica.

Bot. J. Linn. Soc. = Botanical Journal of the Linnean Society.

Bot. Jahrb. Syst. = Botanische Jahrbücher für Systematik, Pflanzengeschichte und Pflanzengeographie.

Bot. Mag. = Botanical Magazine; or, Flower-garden Displayed.... [Edited by Wm. Curtis.] [With vol. 15, 1801, title became Curtis's Botanical Magazine; or....]

Bot. Mag. (Tokyo) = Botanical Magazine. [Shokubutsu-gaku Zasshi.] (Tokyo.)

Bot. Mater. Gerb. Bot. Inst. Komarova Akad. Nauk S.S.S.R. = Botanicheskie Materialy Gerbariya Botanicheskogo Instituti Imeni V. L. Komarova, Akademii Nauk S S S R.

Bot. Not. = Botaniska Notiser.

Bot. Reg. = Botanical Register....

Bot. Repos. = Botanists' Repository, for New, and Rare Plants.

Bot. Rev. (Lancaster) = Botanical Review, Interpreting Botanical Progress.

Bot. Stud. = Botanische Studien.

Bot. Tidsskr. = Botanisk Tidsskrift.

Bot. Zhurn. (Moscow & Leningrad) = Botanicheskii Zhurnal. (Moscow and Leningrad.)

Botany—See: G. A. W. Arnott 1832

Botany (Fortieth Parallel)—See: S. Watson 1871

Bradshaw, M. E., P. M. Dansereau, and D. H. Valentine. 1964. Notes on the genus *Alchemilla* in southeastern Canada. Canad. J. Bot. 42: 89–104.

Brandegee, T. S. 1890. The plants of Santa Catalina Island. Zoë 1: 107–115.

Brewer, W. H. et al. 1876–1880. Geological Survey of California.... Botany.... 2 vols. Cambridge, Mass. (Bot. California)

Bridson, G. D. R. 2004. BPH-2: Periodicals with Botanical Content. 2 vols. Pittsburgh.

Bridson, G. D. R. and E. R. Smith. 1991. B-P-H/S. Botanico-Periodicum-Huntianum/Supplementum. Pittsburgh.

Brit. Fl. Gard.—See: R. Sweet 1823–1837

Britton, N. L. 1901. Manual of the Flora of the Northern States and Canada. New York. (Man. Fl. N. States)

Britton, N. L. et al., eds. 1905+. North American Flora.... 47+ vols. New York. [Vols. 1–34, 1905–1957; ser. 2, parts 1–13+, 1954+.] (N. Amer. Fl.)

Britton, N. L. and A. Brown. 1896–1898. An Illustrated Flora of the Northern United States, Canada and the British Possessions from Newfoundland to the Parallel of the Southern Boundary of Virginia, and from the Atlantic Ocean Westward to the 102d Meridian.... 3 vols. New York. (Ill. Fl. N. U.S.)

Britton, N. L. and J. A. Shafer. 1908. North American Trees.... New York. (N. Amer. Trees)

Britton, N. L., E. E. Sterns, J. F. Poggenburg, et al. 1888. Preliminary Catalogue of Anthophyta and Pteridophyta Reported As Growing Spontaneously within One Hundred Miles of New York City. New York. [Authorship often attributed as B.S.P. in nomenclatural contexts.] (Prelim. Cat.)

Brittonia = Brittonia; a Journal of Systematic Botany....

Browicz, K. 1993. Conspect and chorology of the genus *Pyrus* L. Arbor. Kórnickie 38: 17–33.

Brown, C. L. and L. K. Kirkman. 1990. Trees of Georgia and Adjacent States. Portland.

Brown, R. 1823. Chloris Melvilliana. A List of Plants Collected in Melville Island...in the Year 1820.... London. [Preprint with independent pagination from W. E. Parry. 1824. A Supplement to the Appendix of Captain Parry's Voyage.... London.] (Chlor. Melvill.)

Brown, R. G. and M. L. Brown. 1972. Woody Plants of Maryland. College Park.

Brown, S. W. 1943. The origin and variability in the Pacific Coast blackberries (*Rubus ursinus* Cham. & Schlecht. and *R. lemurum* sp. nov.). Amer. J. Bot. 30: 686–697.

Brummitt, R. K. and C. E. Powell, eds. 1992. Authors of Plant Names. A List of Authors of Scientific Names of Plants, with Recommended Standard Forms of Their Names, Including Abbreviations. Kew.

Bruneau, A., J. R. Starr, and S. Joly. 2007. Phylogenetic relationships in the genus *Rosa:* New evidence from chloroplast DNA sequences and an appraisal of current knowledge. Syst. Bot. 32: 366–378.

Brunsfeld, S. J. and F. D. Johnston. 1990. Cytological, morphological, ecological and phenological support for specific status of *Crataegus suksdorfii* (Rosaceae). Madroño 37: 274–282.

Brzyski, J. R. and T. M. Culley. 2011. Genetic variation and clonal structure of the rare, riparian shrub *Spiraea virginiana* (Rosaceae). Conservation Genet. 12: 1323–1332.

Bull. Acad. Imp. Sci. Saint-Pétersbourg = Bulletin de l'Académie Impériale des Sciences de Saint Pétersbourg.

Bull. Acad. Polon. Sci., Cl. 2, Sér. Sci. Biol. = Bulletin de l'Académie Polonaise des Sciences. Classe II, Série des Sciences Biologiques.

Bull. Bot. Surv. India = Bulletin of the Botanical Survey of India.

Bull. Brit. Mus. (Nat. Hist.), Bot. = Bulletin of the British Museum (Natural History). Botany.

Bull. Calif. Acad. Sci. = Bulletin of the California Academy of Sciences.

Bull. Canad. Bot. Assoc. = Bulletin, Canadian Botanical Association.

Bull. Herb. Boissier = Bulletin de l'Herbier Boissier.

Bull. Mus. Natl. Hist. Nat. = Bulletin du Muséum National d'Histoire Naturelle.

Bull. Natl. Mus. Canada = Bulletin of the National Museum of Canada.

Bull. New York Bot. Gard. = Bulletin of the New York Botanical Garden.

Bull. New York State Mus. Nat. Hist. = Bulletin of the New York State Museum of Natural History.

Bull. New York State Mus. Sci. Serv. = Bulletin of the New York State Museum and Science Service.

Bull. North Carolina Agric. Exp. Sta. = Bulletin, North Carolina Agricultural Experiment Station.

Bull. Public Mus. Milwaukee = Bulletin of the Public Museum of Milwaukee.

Bull. S. Calif. Acad. Sci. = Bulletin of the Southern California Academy of Sciences.

Bull. Sci. Soc. Philom. Paris = Bulletin des Sciences, par la Société Philomatique (de Paris).

Bull. Soc. Bot. France = Bulletin de la Société Botanique de France.

Bull. Soc. Imp. Naturalistes Moscou = Bulletin de la Société Impériale des Naturalistes de Moscou.

Bull. Soc. Roy. Bot. Belgique = Bulletin de la Société Royale de Botanique de Belgique.

Bull. Torrey Bot. Club = Bulletin of the Torrey Botanical Club.

Bull. U.S.D.A. = Bulletin of the U S Department of Agriculture.

Bull. Univ. Mus. Univ. Tokyo = Bulletin, University Museum, University of Tokyo. [Sogo Kenkyu Shiryokan.]

Burman, N. L. 1768. Flora Indica ... Nec Non Prodromus Florae Capensis. Leiden and Amsterdam. (Fl. Indica)

Bushakra, J. M. et al. 1999. The extent of clonality and genetic diversity in the Santa Cruz Island ironwood, *Lyonothamnus floribundus*. Molec. Ecol. 8: 471–475.

Butterwick, M. L., B. D. Parfitt, and D. Hillyard. 1991. Vascular plants of the northern Hualapai Mountains, Arizona. J. Arizona-Nevada Acad. Sci. 24/25: 31–49.

Callihan, R. H., S. L. Carson, and R. T. Dobbins. 1995. NAWEEDS, Computer-aided Weed Identification for North America. Illustrated User's Guide plus Computer Floppy Disk. Moscow, Idaho.

Campbell, C. S. et al. 1985. Apomixis in *Amelanchier laevis,* shadbush (Rosaceae, Maloideae). Amer. J. Bot. 72: 1397–1403.

Campbell, C. S. et al. 1995. Phylogenetic relationships in Maloideae (Rosaceae): Evidence from sequences of the internal transcribed spacers of nuclear ribosomal DNA and its congruence with morphology. Amer. J. Bot. 82: 903–918.

Campbell, C. S. et al. 1997. Persistent nuclear ribosomal DNA sequence polymorphism in the *Amelanchier* agamic complex (Rosaceae). Molec. Biol. Evol. 14: 81–90.

Campbell, C. S. et al. 2007. Phylogeny of subtribe Pyrinae (formerly the Maloideae, Rosaceae): Limited resolution of a complex evolutionary history. Pl. Syst. Evol. 266: 119–145.

Campbell, C. S., C. W. Greene, and S. E. Bergquist. 1987. Apomixis and sexuality in three species of *Amelanchier,* shadbush (Rosaceae, Maloideae). Amer. J. Bot. 74: 321–328.

Campbell, C. S., C. W. Greene, and T. A. Dickinson. 1991. Reproductive biology in subfam. Maloideae (Rosaceae). Syst. Bot. 16: 333–349.

Campbell, C. S. and W. A. Wright. 1996. Apomixis, hybridization, and taxonomic complexity in eastern North American *Amelanchier* (Rosaceae). Folia Geobot. Phytotax. 31: 345–354.

Canad. J. Pl. Sci. = Canadian Journal of Plant Science.

Canad. J. Res., Sect. C, Bot. Sci. = The Canadian Journal of Research. Section C. Botanical Sciences.

Canad. Rec. Sci. = The Canadian Record of Science; Including the Proceedings of the Natural History Society of Montreal [etc.].

Candolle, A. P. de. 1813b. Catalogus Plantarum Horti Botanici Monspeliensis.... Montpellier, Paris, and Strasbourg. (Cat. Pl. Hort. Monsp.)

Candolle, A. P. de and A. L. P. P. de Candolle, eds. 1823–1873. Prodromus Systematis Naturalis Regni Vegetabilis.... 17 vols. Paris etc. [Vols. 1–7 edited by A. P. de Candolle, vols. 8–17 by A. L. P. P. de Candolle.] (Prodr.)

Candollea = Candollea; Organe du Conservatoire et du Jardin Botaniques de la Ville de Genève.

Caryologia = Caryologia; Giornale di Citologia, Citosistematica e Citogenetica.

Čas. Nár. Muz. Praze, Rada Přír. = Časopis Národního Muzea v Praze. Rada Přírodovědna.

Čas. Nár. Muz., Rada Přír. = Časopis Národního Muzea. Rada Přírodovědna.

Castanea = Castanea; Journal of the Southern Appalachian Botanical Club.

Castroviejo, S. et al., eds. 1986+. Flora Ibérica: Plantas Vasculares de la Peninsula Ibérica e Islas Baleares. 17+ vols. Madrid.

Cat. Pl. Hort. Monsp.—See: A. P. de Candolle 1813b

Catling, P. M. 2006. Characteristics of *Amelanchier lucida* based on an examination of the holotype and plants from the type locality. Rhodora 108: 347–355.

Catling, P. M., S. M. McKay-Kuja, and G. Mitrow. 1999. Rank and typification in North American dwarf cherries, and a key to the taxa. Taxon 48: 483–488.

Catling, P. M. and G. Mitrow. 2012b. Major invasive aliens in Canada. 5. Common buckthorn. Bull. Canad. Bot. Assoc. 45(3): 22–27.

Cayouette, J. et al. 1983. Plantes vasculaires nouvelles pour le Québec: Additions, échappées de culture et éphémérophytes. Naturaliste Canad. 110: 293–312.

Cayouette, R. 1966. *Potentilla hippiana* Lehm. et *Potentilla thuringiaca* Bernh. au Québec. Naturaliste Canad. 93: 889–894.

Challice, J. S. and M. N. Westwood. 1973. Numerical taxonomic studies of the genus *Pyrus* using both chemical and botanical characters. Bot. J. Linn. Soc. 67: 121–148.

Chang, K. S. et al. 2007. Reconsideration of the *Prunus serrulata* complex (Rosaceae) and related taxa in eastern Asia. Bot. J. Linn. Soc. 154: 35–54.

Chapin, F. S. III et al. 1994. Mechanisms of primary succession following deglaciation at Glacier Bay, Alaska. Ecol. Monogr. 64: 149–175.

Chapman, A. W. 1892. Flora of the Southern United States..., ed. 2 reprint 2. New York, Cincinnati, and Chicago. (Fl. South. U.S. ed. 2 repr. 2)

Chlor. Melvill.—See: R. Brown 1823

Cho, E. J. et al. 2004. *Rosa rugosa* attenuates diabetic oxidative stress in rats with streptozotocin-induced diabetes. Amer. J. Chin. Med. 32: 487–496.

Christenhusz, M. J. M. and H. Väre. 2012. New combinations in *Potentilla* (Rosaceae) for the Nordic Flora. Phytotaxa 57: 1–5.

Christensen, K. I. 1992. Revision of *Crataegus* sect. *Crataegus* in the Old World. Syst. Bot. Monogr. 35: 1–199.

Chrubasik, C., R. K. Duke, and S. Chrubasik. 2006. The evidence for clinical efficacy of rose hip and seed: A systematic review. Phytotherapy Res. 20: 1–3.

Cinq-Mars, L. 1971. Le genre *Amelanchier* au Québec. Naturaliste Canad. 98: 329–346.

Clark, E. M. and W. H. Clark. 2003. The distribution of *Ivesia rhypara* (Rosaceae) along bedrock lineaments at the type locality, Leslie Gulch, Malheur County, Oregon. J. Idaho Acad. Sci. 39: 1–5.

Clark, J. R. et al. 2007. Blackberry breeding and genetics. Pl. Breed. Rev. 29: 19–144.

Clark, L. V., K. J. Evans, and M. Jasieniuk. 2013. Origins and distribution of invasive *Rubus fruticosus* L. agg. (Rosaceae) clones in the western United States. Biol. Invasions 15: 1331–1342.

Clark, L. V. and M. Jasieniuk. 2012. Spontaneous hybrids between native and exotic *Rubus* in the western United States produce offspring both by apomixis and by sexual recombination. Heredity 109: 320–328.

Clarkson, R. B. 1959. The West Virginia *Spiraea.* Castanea 24: 143–146.

Class-book Bot. ed. s.n.(b)—See: A. Wood 1861

Clausen, J., D. D. Keck, and W. M. Hiesey. 1940. Experimental studies on the nature of species. I. Effect of varied environments on western American plants. Publ. Carnegie Inst. Wash. 520.

Clokey, I. W. 1939. Notes on the flora of the Charleston Mountains, Clark County, Nevada—III. Bull. S. Calif. Acad. Sci. 38: 1–7.

Cody, W. J. 1996. Flora of the Yukon Territory. Ottawa.

Cody, W. J. 2000. Flora of the Yukon Territory, ed. 2. Ottawa.

Cogbill, C. V. 1968. The interplay of botanists and *Potentilla robbinsiana:* Discovery, systematics, collection, and stewardship of a rare species. Rhodora 95: 52–75.

Collicutt, L. M. 1992. 'Prairie Joy' rose. HortScience 27: 478–479.

Collicutt, L. M. 1992b. 'Winnipeg Parks' rose. HortScience 27: 1140–1141.

Comp. Fl. German.—See: M. J. Bluff et al. 1825–1833

Compan. Bot. Mag. = Companion to the Botanical Magazine....

Conservation Biol. = Conservation Biology; Journal of the Society for Conservation Biology.

Conservation Genet. = Conservation Genetics.

Contemp. Probl. Ecol. = Contemporary Problems of Ecology.

Contr. U.S. Natl. Herb. = Contributions from the United States National Herbarium.

Contr. W. Bot. = Contributions to Western Botany.

Cooperrider, T. S., A. W. Cusick, and J. T. Kartesz, eds. 2001. Seventh Catalog of the Vascular Plants of Ohio. Columbus.

Cornell Univ. Agric. Exp. Sta. Bull. = Cornell University Agricultural Experiment Station Bulletin.

Correll, D. S. and M. C. Johnston. 1970. Manual of the Vascular Plants of Texas. Renner, Tex.

Coulter, J. M. [1885.] Manual of the Botany (Phaenogamia and Pteridophyta) of the Rocky Mountain Region.... New York and Chicago. (Man. Bot. Rocky Mt.)

Coulter, J. M. and A. Nelson. 1909. New Manual of Botany of the Central Rocky Mountains (Vascular Plants). New York. (New Man. Bot. Rocky Mt.)

Crantz, H. J. N. von. 1762–1767. Stirpium Austriarum Fasciculus I [–III]. 3 fascs. Vienna and Leipzig. (Stirp. Austr. Fasc.)

Crantz, H. J. N. von. 1766. Institutiones Rei Herbariae.... 2 vols. Vienna. (Inst. Rei Herb.)

Crataegus Missouri—See: C. S. Sargent 1908

Crataegus S. Michigan—See: C. S. Sargent 1907

Croft, B. R., G. Melcher, and E. Langston. 1996. Mayhaws: A Guide to Orchard Production and Propagation. Kearney.

Cronquist, A. 1981. An Integrated System of Classification of Flowering Plants. New York.

Cronquist, A. 1982. Floristic provinces of North America. Brittonia 34: 143–145.

Cronquist, A. et al. 1972+. Intermountain Flora. Vascular Plants of the Intermountain West, U.S.A. 6+ vols. in 7+. New York and London.

Cronquist, A. and N. H. Holmgren. 1997. *Rosa.* In: A. Cronquist et al. 1972+. Intermountain Flora. Vascular Plants of the Intermountain West, U.S.A. 6+ vols. in 7+. New York and London. Vol. 3, part A, pp. 134–140.

Cronquist, A., N. H. Holmgren, and P. K. Holmgren. 1997. *Spiraea.* In: A. Cronquist et al. 1972+. Intermountain Flora. Vascular Plants of the Intermountain West, U.S.A. 6+ vols. in 7+. New York and London. Vol. 3, part A, pp. 72–73.

Cruise, J. E. 1964. Studies of natural hybrids in *Amelanchier.* Canad. J. Bot. 42: 651–663.

Crum, E. 1936. A note on *Potentilla newberryi.* Leafl. W. Bot. 1: 195–196.

Cuizhi, G. and S. A. Spongberg. 2003. *Pyrus.* In: Wu Z. and P. H. Raven, eds. 1994+. Flora of China. 20+ vols. Beijing and St. Louis. Vol. 9, pp. 173–179.

Culley, T. M. and N. A. Hardiman. 2007. The beginning of a new invasive plant: A history of the ornamental Callery pear in the United States. BioScience 57: 956–964.

Cycl.—See: A. Rees [1802–]1819–1820

Dansereau, P. M. and E. Steiner. 1956. Studies in *Potentillae* of high latitudes and altitudes. II. Central Baffin Island populations. Bull. Torrey Bot. Club 83: 113–135.

Davies, H. 1813. Welsh Botanology.... 2 parts. London. [Parts paged consecutively.] (Welsh Botanol.)

Davis, H. A. 1990. Studies in *Rubus.* Castanea 55: 22–30.

Davis, H. A., A. M. Fuller, and T. Davis. 1967–1970. Contributions towards the revision of eubati of eastern North America. Castanea 32: 20–37; 33: 50–76, 206–241; 34: 157–179, 235–266; 35: 176–194.

Davis, R. J. 1952. Flora of Idaho. Dubuque.

De Cleene, M. and J. De Ley. 1981. The host range of infectious hairy-root. Bot. Rev. (Lancaster) 47: 147–194.

Deam, C. C. 1940. Flora of Indiana. Indianapolis. (Fl. Indiana)

Decaisne, J. [1858–]1862–1875. Le Jardin Fruitier du Muséum.... 9 vols. Paris. (Jard. Fruit.)

del Moral, R. 1983. Competition as a control mechanism in subalpine meadows. Amer. J. Bot. 70: 232–245.

Dendrologie—See: K. H. E. Koch 1869–1873

Descr. Icon. Pl. Hung.—See: F. Waldstein and P. Kitaibel [1799–]1802–1812

Desfontaines, R. L. 1829[–1832]. Tableau de l'École Botanique du Muséum d'Histoire Naturelle, ed. 3. 1 vol. + Add. Paris. [Title: Catalogus Plantarum Horti Regii Parisiensis....] (Tabl. École Bot. ed. 3)

Desvaux, N. A. 1818. Observations sur les Plantes des Environs d'Angers.... Angers and Paris. (Observ. Pl. Angers)

Deut. Bot. Herb.-Buch—See: H. G. L. Reichenbach 1841

Dibble, A. C. 1995. Conservation Biology of Shadbush, *Amelanchier* (Rosaceae): Evidence from Systematics, Population Structure, and Reproductive Ecology. Ph.D. thesis. University of Maine.

Dibble, A. C. et al. 1998. Quantitative morphologic analyses at a taxonomically complex site in the *Amelanchier* agamic complex (Rosaceae). Syst. Bot. 23: 31–40.

Dibble, A. C. and C. S. Campbell. 1995. Distribution and conservation of Nantucket shadbush, *Amelanchier nantucketensis* (Rosaceae). Rhodora 97: 339–349.

Dibble, A. C., F. A. Drummond, and W. E. LaBerge. 1997. Floral syndrome in *Amelanchier nantucketensis* (Rosaceae). II. Bee preference and diversity associated with andropetaly. Canad. J. Bot. 75: 1860–1867.

Dickinson, T. A. 1983. *Crataegus crus-galli* L., sens. lat. in Southern Ontario. Ph.D. thesis. University of Western Ontario.

Dickinson, T. A. et al. 1996. American black-fruited hawthorns. I. Variation in floral construction, breeding system correlates and their possible evolutionary significance in *Crataegus* sect. *Douglasii* Loudon. Folia Geobot. Phytotax. 31: 355-371.

Dickinson, T. A. et al. 2008. Black-fruited hawthorns of western North America, one or two agamic complexes. Botany 86: 846–865.

Dickison, W. C. 1986. Floral morphology and anatomy of Staphyleaceae. Bot. Gaz. 147: 312–326.

Dickison, W. C. 1987. Leaf and nodal anatomy and systematics of Staphyleaceae. Bot. Gaz. 148: 475–489.

Dickoré, W. B. and G. Kasperek. 2010. Species of *Cotoneaster* (Rosaceae, Maloideae) indigenous to, naturalizing or commonly cultivated in central Europe. Willdenowia 40: 13–45.

Dickson, E. E. 1995. Systematic Studies of *Malus* Sect. *Chloromeles*. Ph.D. dissertation. Cornell University.

Dickson, E. E., S. Kresovich, and N. F. Weeden. 1991. Isozymes in North American *Malus* (Rosaceae): Hybridization and species differentiation. Syst. Bot. 16: 363–375.

Dietrich, D. N. F. 1839–1852. Synopsis Plantarum.... 5 vols. Weimar. (Syn. Pl.)

Dietrich, W., W. L. Wagner, and P. H. Raven. 1997. Systematics of *Oenothera* section *Oenothera* subsection *Oenothera*. Syst. Bot. Monogr. 50.

Diggs, G. M., B. L. Lipscomb, and R. J. O'Kennon. 1999. Shinners' and Mahler's Illustrated Flora of North Central Texas. Fort Worth. [Sida Bot. Misc. 16.]

Dikshit, B. K. and G. Panigrahi. 1981. Revision of the genus *Sibbaldia* L. (Rosaceae) in India. Proc. Indian Acad. Sci., Pl. Sci. 90: 253–272.

Dippel, L. 1889–1893. Handbuch der Laubholzkunde. 3 vols. Berlin. (Handb. Laubholzk.)

Dobeš, C. and J. Paule. 2010. A comprehensive chloroplast DNA-based phylogeny of the genus *Potentilla* (Rosaceae): Implications for its geographic origin, phylogeography and generic circumscription. Molec. Phylogen. Evol. 56: 156–175.

Dole, E. J., ed. 1937. The Flora of Vermont..., ed. 3. Burlington. (Fl. Vermont ed. 3)

Döll, J. C. 1843. Rheinische Flora. Frankfurt am Main. (Rhein. Fl.)

Don, D. 1825. Prodromus Florae Nepalensis.... London. (Prodr. Fl. Nepal.)

Douglas, G. W. et al. 1998–2002. Illustrated Flora of British Columbia. 8 vols. Victoria.

Drobney, P. M. and M. P. Widrlechner. 2012. Japanese raspberry (*Rubus parvifolius* L.): An invasive species threat in savanna and prairie. In: Dave Williams et al., eds. 2012. Proceedings of the 22nd North American Prairie Conference. Cedar Falls. Pp. 148–152.

Du Roi, J. P. 1771–1772. Die Harbkesche wilde Baumzucht.... 2 vols. Braunschweig. (Harbk. Baumz.)

Dudley Mem. Vol.—See: W. R. Dudley et al. 1913

Dudley, W. R. et al. 1913. Dudley Memorial Volume.... Stanford. (Dudley Mem. Vol.)

Duke, J. A. and E. S. Ayensu. 1985. Medicinal Plants of China. 2 vols. Algonac, Mich.

Dumortier, B. C. J. 1829. Analyse des Familles des Plantes.... Tournay. (Anal. Fam. Pl.)

Eaton, A. 1836b. A Manual of Botany for the Northern States, ed. 7. Albany. (Man. Bot. ed. 7)

Ecol. Monogr. = Ecological Monographs.

Ecology = Ecology, a Quarterly Journal Devoted to All Phases of Ecological Biology.

Econ. Bot. = Economic Botany; Devoted to Applied Botany and Plant Utilization.

Edinburgh J. Bot. = Edinburgh Journal of Botany.

Edwards's Bot. Reg. = Edwards's Botanical Register....

Egolf, D. R. and A. O. Andrick. 1995. A Checklist of *Pyracantha* Cultivars. Washington. [U.S.D.A. Agric. Res. Serv., Natl. Arbor. Contr. 8.]

Ehrhart, J. F. 1787–1792. Beiträge zur Naturkunde.... 7 vols. Hannover and Osnabrück. (Beitr. Naturk.)

Elkington, T. T. 1965. Studies on the variation of the genus *Dryas* in Greenland. Meddel. Grønland 178: 1–56.

Elkington, T. T. 1971. *Dryas octopetala*. J. Ecol. 59: 887–905.

Elliott, S. [1816–]1821–1824. A Sketch of the Botany of South-Carolina and Georgia. 2 vols. in 13 parts. Charleston. (Sketch Bot. S. Carolina)

Ellis, J. R. 1962. *Fragaria–Potentilla* intergeneric hybridization and evolution of *Fragaria*. Proc. Linn. Soc. London 173: 99–106.

Ells, J. 2006. Rocky Mountain Flora. Golden.

Elven, R. et al. 1999. An operational species concept for the Panarctic Flora. Skr. Norske Vidensk.-Akad. Oslo, Mat.-Naturvidensk. Kl., n. s. 38: 23–32.

Elven, R. and S. G. Aiken. 2007. Rosaceae. In: S. G. Aiken et al. 2007. Flora of the Canadian Arctic Archipelago.... Ottawa. CD-ROM.

Elven, R. and D. F. Murray. 2008. The *Potentilla nivea* group in northwestern North America. Bot. Electronic News 389.

Elven, R. and D. F. Murray. 2008b. The *Potentilla villosa-uniflora* group in northwestern North America. Bot. Electronic News 390.

Emory, W. H. 1848. Notes of a Military Reconnoissance, from Fort Leavenworth, in Missouri, to San Diego, in California, Including Part of the Arkansas, Del Norte, and Gila Rivers.... Made in 1846–7, with the Advanced Guard of the "Army of the West." Washington. (Not. Milit. Reconn.)

Ench. Bot.—See: F. W. Neygenfind 1821

Encycl.—See: J. Lamarck et al. 1783–1817

Endlicher, S. L. 1836–1840[–1850] Genera Plantarum Secundum Ordines Naturales Disposita. 18 parts + 5 suppls. in 6 parts. Vienna. [Paged consecutively through suppl. 1(2); suppls. 2–5 paged independently.] (Gen. Pl.)

Engl. Bot.—See: J. E. Smith et al. 1790–1866

Engler, H. G. A. and K. Prantl, eds. 1887–1915. Die natürlichen Pflanzenfamilien.... 254 fascs. Leipzig. [Sequence of fasc. (Lieferung) numbers (order of publication) is independent of the sequence of division (Teil) and subdivision (Abteilung) numbers (taxonomic order).] (Nat. Pflanzenfam.)

Enquist, M. F. 1997. A new variety of *Prunus rivularis* (Rosaceae) for Texas. Phytologia 82: 289–299.

Ensign, M. R. 1942. A revision of the celastraceous genus *Forsellesia (Glossopetalon)*. Amer. Midl. Naturalist 27: 501–511.

Enum. Hort. Berol. Alt.—See: J. H. F. Link 1821–1822

Enum. Pl.—See: C. L. Willdenow 1809–1813[–1814]

Enum. Pl. Jap.—See: A. R. Franchet and P. A. L. Savatier [1873–]1875–1879

Eriksen, B. 1996. Mating systems in two species of *Potentilla* from Alaska. Folia Geobot. Phytotax. 31: 333–344.

Eriksen, B. 1997. Morphometric analysis of Alaskan members of the genus *Potentilla* sect. *Niveae* (Rosaceae). Nordic J. Bot. 17: 621–630.

Eriksen, B., B. Jonsell, and Ö. Nilsson. 1999. Proposal to conserve the name *Potentilla nivea* (Rosaceae) with a conserved type. Taxon 48: 165–166.

Eriksen, B. and B. A. Jurtzev. 1999. Hair types in *Potentilla* sect. *Niveae* (Rosaceae) and related taxa—Terminology and systematic distribution. Skr. Norske Vidensk.-Akad. Oslo, Mat.-Naturvidensk. Kl., n. s. 38: 201–221.

Eriksen, B. and J. Nyléhn. 1999. Cases of phenotypic plasticity in leaves of *Potentilla* L. (Rosaceae). Skr. Norske Vidensk.-Akad. Oslo, Mat.-Naturvidensk. Kl., n. s. 38: 191–199.

Eriksen, B. and M. H. Töpel. 2006. Molecular phylogeography and hybridization in members of the circumpolar *Potentilla* sect. *Niveae* (Rosaceae). Amer. J. Bot. 93: 460–469.

Eriksson, T. et al. 2003. The phylogeny of Rosoideae (Rosaceae) based on sequences of the internal transcribed spacers (ITS) of nuclear ribosomal DNA and the *trn*L/F region of chloroplast DNA. Int. J. Pl. Sci. 164: 197–211.

Eriksson, T., M. J. Donoghue, and M. S. Hibbs. 1998. Phylogenetic analysis of *Potentilla* using DNA sequences of nuclear ribosomal internal transcribed spacers (ITS), and implications for the classification of Rosoideae (Rosaceae). Pl. Syst. Evol. 211: 155–179.

Erlandsson, S. 1942. *Potentilla egedii* Wormsk. and *P. pacifica* Howell in the flora of Alaska and Yukon. Förh. Kungl. Fysiogr. Sällsk. Lund 12: 203–214.

Erlandsson, S. 1942b. Cytological studies in *Potentilla anserina* L. and closely allied Scandinavian species. Hereditas (Lund) 28: 503–504.

Erlanson, E. W. 1931. A group of tetraploid roses in central Oregon. Bot. Gaz. 91: 55–64.

Erlanson, E. W. 1934. Experimental data for a revision of the North American wild roses. Bot. Gaz. 96: 197–259.

Ertter, B. 1989. Revisionary studies in *Ivesia* (Rosaceae). Syst. Bot. 14: 231–244.

Ertter, B. 1992. A re-evaluation of *Potentilla drummondii* and *P. breweri* (Rosaceae), with the new species *Potentilla morefieldii*. Brittonia 44: 429–435.

Ertter, B. 1993. *Potentilla*. In: J. C. Hickman, ed. 1993. The Jepson Manual. Higher Plants of California. Berkeley, Los Angeles, and London. Pp. 964–969.

Ertter, B. 1993b. *Horkeliella*. In: J. C. Hickman, ed. 1993. The Jepson Manual. Higher Plants of California. Berkeley, Los Angeles, and London. P. 957.

Ertter, B. 1993c. A re-evaluation of the *Horkelia bolanderi* (Rosaceae) complex, with the new species *Horkelia yadonii*. Syst. Bot. 18:137–144.

Ertter, B. 1993d. *Horkelia*. In: J. C. Hickman, ed. 1993. The Jepson Manual. Higher Plants of California. Berkeley, Los Angeles, and London. Pp. 953–957.

Ertter, B. 1993e. *Rosa*. In: J. C. Hickman, ed. 1993. The Jepson Manual. Higher Plants of California. Berkeley, Los Angeles, and London. Pp. 972–973.

Ertter, B. 1997b. Taxonomic grays vs. black and white expectations: Implications for conservation management of diversity. In: T. N. Kaye et al., eds. 1997. Conservation of Native Plants and Fungi. Corvallis. Pp. 11–13.

Ertter, B. 2007. Generic realignments in tribe Potentilleae and revision of *Drymocallis* (Rosoideae: Rosaceae) in North America. J. Bot. Res. Inst. Texas 1: 31–46.

Ertter, B. 2007b. Proposal to conserve the name *Rosa bridgesii* against *R. yainacensis, R. calvaris, R. covillei, R. crenulata,* and *R. myriadenia* (Rosaceae). Taxon 56: 962–964.

Ertter, B. 2007c. Lectotypifications and new taxa in *Potentilla* sect. *Subviscosa* (Rosaceae) in Arizona. J. Bot. Res. Inst. Texas 1: 47–57.

Ertter, B. 2008. Nomenclatural notes in North American *Potentilla* (Rosaceae). J. Bot. Res. Inst. Texas 2: 201–205.

Ertter, B. et al. 2008. Proposal to reject the name *Potentilla dissecta* (Rosaceae). Taxon 57: 1010–1011.

Ertter, B., C. Baysdorfer, and D. Alonso. 1998. Monophyly of the ivesioid Potentilleae (Rosaceae) based on ITS sequence data. [Abstract.] Amer. J. Bot. 85(6, suppl.): 126–127.

Ertter, B., R. Elven, and D. F. Murray. 2013. Changes to *Potentilla rubricaulis* s.l., *P. hookeriana,* (Rosaceae), and erstwhile synonyms in Flora of North America North of Mexico. J. Bot. Res. Inst. Texas 7: 703–711.

Ertter, B. and W. H. Lewis. 2008. New *Rosa* (Rosaceae) in California and Oregon. Madroño 55: 170–177.

Ertter, B. and D. Mansfield. 2007. A new variety of *Potentilla gracilis* (Rosaceae) and re-evaluation of the *Potentilla drummondii* complex. Madroño 54: 180–186.

Ertter, B. and J. L. Reveal. 2007. New sections, combinations, and varieties in Rosaceae, Potentilleae. Novon 17: 315–325.

Ertter, B. and J. L. Reveal. 2008. Proposal to conserve the name *Potentilla multijuga* (Rosaceae) with a conserved type. Taxon 57: 312–313.

Ertter, B. and J. R. Shevock. 1993. Snow-wreath and its relatives in the garden. Fremontia 22: 10–11.

Erwin, D. M. and H. E. Shorn. 2000. Revision of *Lyonothamnus* A. Gray (Rosaceae) from Neogene of western North America. Int. J. Pl. Sci. 161: 179–193.

Erythea = Erythea; a Journal of Botany, West American and General.

Euphytica = Euphytica. Netherlands Journal of Plant Breeding.

Evans, K. J. et al. 2007. Systematics of the *Rubus fruticosus* aggregate (Rosaceae) and other exotic *Rubus* taxa in Australia. Austral. Syst. Bot. 20: 187–251.

Evans, R. C. and C. S. Campbell. 2002. The origin of the apple subfamily (Rosaceae: Maloideae) is clarified by DNA sequence data from duplicated GBSSI genes. Amer. J. Bot. 89: 1478–1484.

Evans, R. C. and T. A. Dickinson. 1999. Floral ontogeny and morphology in subfamily Amygdaloideae T. & G. (Rosaceae). Int. J. Pl. Sci. 160: 955–979.

Evans, R. C. and T. A. Dickinson. 1999b. Floral ontogeny and morphology in subfamily Spiraeoideae Endl. (Rosaceae). Int. J. Pl. Sci. 160: 981–1012.

Evans, R. C. and T. A. Dickinson. 2005. Floral ontogeny and morphology in *Gillenia* ("Spiraeoideae") and subfamily Maloideae C. Weber (Rosaceae). Int. J. Pl. Sci. 166: 427–447.

Evolution = Evolution; International Journal of Organic Evolution.

Excursionsfl. Oesterreich—See: K. Fritsch 1897

Exped. Great Salt Lake—See: H. Stansbury 1852

Fam. Nat. Syn. Monogr.—See: M. J. Roemer 1846–1847

Farmakol. Toksikol. = Farmakologiya i Toksikologiya.

Fassett, N. C. 1928. Notes from the herbarium of the University of Wisconsin—III. Rhodora 30: 205–207.

Fassett, N. C. 1941. Mass collections: *Rubus odoratus* and *R. parviflorus*. Ann. Missouri Bot. Gard. 28: 299–375.

Fay, M. F. et al. 2002. Parentage of an unknown member of the *Sorbus latifolia* (Lam.) Pers. group. Watsonia 24: 91–100.

Feddes Repert. = Feddes Repertorium.

Fedorov, A. A. and Z. T. Artjushenko. 1979. Organographia Illustrata Plantarum Vascularium: Inflorescentia. [In Russian.] Leningrad.

Fernald, M. L. 1900. *Rubus idaeus* and its variety *anomalus* in America. Rhodora 2: 195–200.

Fernald, M. L. 1919b. *Rubus idaeus* and some of its variations in North America. Rhodora 21: 89–98.

Fernald, M. L. 1923b. The identities of the sand cherries of eastern America. Rhodora 25: 69–74.

Fernald, M. L. 1931. *Potentilla canadensis* and *P. simplex*. Rhodora 33: 180–191.

Fernald, M. L. 1935. Critical plants of the upper Great Lakes region of Ontario and Michigan. Rhodora 37: 197–222, 238–262, 272–301, 324–341.

Fernald, M. L. 1941. Another century of additions to the flora of Virginia. Rhodora 53: 559–567.

Fernald, M. L. 1946e. A monograph of *Amelanchier*. [Review of G. N. Jones 1946.] Rhodora 48: 129–144.

Fernald, M. L. 1950. Gray's Manual of Botany, ed. 8. New York.

Fernald, M. L. and B. H. Long. 1914. The American variations of *Potentilla palustris*. Rhodora 16: 5–11.

Fernald, M. L. and C. A. Weatherby. 1922. Varieties of *Geum canadense*. Rhodora 24: 47–50.

Fernald, M. L. and K. M. Wiegand. 1912. *Alchemilla alpina* and *A. vulgaris* in North America. Rhodora 14: 229–234.

Fl. Alaska Yukon—See: E. Hultén 1941–1950

Fl. Altaica—See: C. F. von Ledebour 1829–1833

Fl. Amer. Sept.—See: F. Pursh [1813]1814

Fl. Arct. URSS—See: A. I. Tolmatchew 1960–1987

Fl. Austriac.—See: N. J. Jacquin 1773–1778

Fl. Bor.-Amer.—See: W. J. Hooker [1829–]1833–1840; A. Michaux 1803

Fl. Boston. ed. 2—See: J. Bigelow 1824

Fl. Brandenburg—See: P. F. A. Ascherson 1859–1864

Fl. Brit.—See: J. E. Smith 1800–1804

Fl. Calif.—See: W. L. Jepson 1909–1943

Fl. Carol.—See: T. Walter 1788

Fl. Cochinch.—See: J. de Loureiro 1790

Fl. Dalmat.—See: R. de Visiani 1842–1852

Fl. Dan.—See: G. C. Oeder et al. [1761–]1764–1883

Fl. France—See: J. C. M. Grenier and D. A. Godron [1847–]1848–1856

Fl. Francisc.—See: E. L. Greene 1891–1897

Fl. Ind. ed. 1832—See: W. Roxburgh 1832

Fl. Ind. Occid.—See: O. P. Swartz 1797–1806

Fl. Indiana—See: C. C. Deam 1940

Fl. Indica—See: N. L. Burman 1768

Fl. Íslands—See: S. Stefánsson 1901

Fl. Jap.—See: P. F. von Siebold and J. G. Zuccarini 1835[–1870]

Fl. Kavkaza—See: A. A. Grossheim 1928–1934

Fl. N. Amer.—See: J. Torrey and A. Gray 1838–1843

Fl. N. Mitt.-Deutschland ed. 4—See: C. A. F. Garcke 1858

Fl. N.W. Amer.—See: T. J. Howell 1897–1903

Fl. New Zealand—See: H. H. Allan et al. 1961–2000

Fl. Peruv.—See: H. Ruiz López and J. A. Pavon 1798–1802

Fl. Rocky Mts.—See: P. A. Rydberg 1917

Fl. S.E. U.S.—See: J. K. Small 1903

Fl. S.E. Washington—See: H. St. John 1937

Fl. South. U.S. ed. 2 repr. 2—See: A. W. Chapman 1892

Fl. Suec. ed. 2—See: C. Linnaeus 1755b

Fl. URSS—See: V. L. Komarov et al. 1934–1964

Fl. Vermont ed. 3—See: E. J. Dole 1937

Flexner, S. B. and L. C. Hauck, eds. 1987. The Random House Dictionary of the English Language, ed. 2 unabridged. New York.

Flinck, K. E. and B. Hylmö. 1966. A list of series and species in the genus *Cotoneaster*. Bot. Not. 119: 445–463.

Flora = Flora; oder (allgemeine) botanische Zeitung. [Vols. 1–16, 1818–1833, include "Beilage" and "Ergänzungsblätter"; vols. 17–25, 1834–1842, include "Beiblatt" and "Intelligenzblatt."]

Focke, W. O. 1888–1891. Rosaceae. In: H. G. A. Engler and K. Prantl, eds. 1887–1915. Die natürlichen Pflanzenfamilien.... 254 fascs. Leipzig. Fascs. 24, 59[III,3], pp. 1–61.

Focke, W. O. 1910. Species *Ruborum*. Monographie generis *Rubi* prodromus. Biblioth. Bot. 72.

Folia Geobot. Phytotax. = Folia Geobotanica et Phytotaxonomica.

Forest Ecol. Managem. = Forest Ecology and Management.

Förh. Kungl. Fysiogr. Sällsk. Lund = Förhandlingar, Kungliga Fysiografiska Sällskapet i Lund.

Foster, S. and J. A. Duke. 1990. A Field Guide to Medicinal Plants: Eastern and Central North America. Boston.

Franchet, A. R. 1889[–1890]. Plantae Delavayanae; Plantes de Chine Recueillis au Yun-nan par l'Abbé Delavay.... 3 parts. Paris. [Parts paged consecutively.] (Pl. Delavay.)

Franchet, A. R. and P. A. L. Savatier. [1873–]1875–1879. Enumeratio Plantarum in Japonia Sponte Crescentium.... 2 vols. in 5 parts. Paris. [Parts paged consecutively within vols.] (Enum. Pl. Jap.)

Franklin, J. et al. 1823. Narrative of a Journey to the Shores of the Polar Sea, in the Years 1819, 20, 21 and 22. London. [Richardson: Appendix VII. Botanical appendix, pp. [729]–768, incl. bryophytes by Schwägrichen, algae and lichens by Hooker.] (Narr. Journey Polar Sea)

Fremontia = Fremontia; Journal of the California Native Plant Society.

Fritsch, K. 1897. Excursionsflora für Oesterreich.... Vienna. (Excursionsfl. Oesterreich)

Frost-Olsen, P. 1998. *Aphanes*. In: S. Castroviejo et al., eds. 1986+. Flora Ibérica: Plantas Vasculares de la Peninsula Ibérica e Islas Baleares. 17+ vols. Madrid. Vol. 6, pp. 357–369.

Frutic. Vilmor.—See: M. Vilmorin and D. G. J. M. Bois 1904

Frye, C. T. 2006. Morphometric analysis of an *Amelanchier* (Rosaceae: Maloideae) complex on the Delmarva Peninsula (Delaware, Maryland and eastern Virginia) resolves the taxonomic identities of *Amelanchier obovalis* and *A. canadensis*. Sida 22: 1027–1048.

Fryer, J. and B. Hylmö. 1997. Five new species of *Cotoneaster* Medik. (Rosaceae) naturalized in Britain. Watsonia 21: 335–340.

Fryer, J. and B. Hylmö. 2009. Cotoneasters: A Comprehensive Guide to Shrubs for Flowers, Fruit, and Foliage. Portland.

Fryer, J. and P. F. Zika. 2014. The typification of *Cotoneaster symondsii* (Rosaceae). Phytotaxa 164: 149–153.

Gajewski, W. 1955. Cytogenetic relations of *Geum macrophyllum* Willd. with *G. perincisum* Rydb. and *G. oregonense* Rydb. Acta Soc. Bot. Poloniae 24: 311–334.

Gajewski, W. 1957. A cytogenetic study on the genus *Geum* L. Monogr. Bot. 4: 1–416.

Gandhi, K. N., J. L. Reveal, and L. Brouillet. 2009. Proposals to conserve the names *Prunus serotina* and *P. virginiana* (Rosaceae) with conserved types. Taxon 58: 312–314.

Garcia, J. D. 1970. A Study of the Genus *Potentilla* in New Mexico. M.S. thesis. University of New Mexico.

Garcke, C. A. F. 1858. Flora von Nord- und Mittel-Deutschland..., ed. 4. Berlin (Fl. N. Mitt.-Deutschland ed. 4)

Gard. & Forest = Garden and Forest; a Journal of Horticulture, Landscape Art and Forestry.

Gard. Chron. = Gardener's Chronicle.

Gard. Dict. Abr. ed. 4—See: P. Miller 1754

Gard. Dict. ed. 8—See: P. Miller 1768

Gardes, M. and A. Dahlberg. 1996. Mycorrhizal diversity in arctic and alpine tundra: An open question. New Phytol. 133: 147–157.

Gartenflora = Gartenflora; Monatsschrift für deutsche und schweizerische Garten- und Blumenkunde.

Gatt. Pomac.—See: B. A. E. Koehne 1890

Gehrke, R. et al. 2008. Molecular phylogenetics of *Alchemilla, Aphanes* and *Lachemilla* (Rosaceae) inferred from plastid and nuclear intron and spacer DNA sequences with comments on generic classification. Molec. Phylogen. Evol. 47: 1030–1044.

Gen. N. Amer. Pl.—See: T. Nuttall 1818

Gen. Pl.—See: S. L. Endlicher 1836–1840[–1850]

Gen. Pl. ed. 5—See: C. Linnaeus 1754

Gen. Sorbus—See: H. A. McAllister 2005

Gentes Herbarum = Gentes Herbarum; Occasional Papers on the Kinds of Plants.

Georgia J. Sci. = Georgia Journal of Science.

Ges. Naturf. Freunde Berlin Mag. Neuesten Entdeck. Gesammten Naturk. = Der Gesellschaft naturforschender Freunde zu Berlin Magazin für die neuesten Entdeckungen in der gesammten Naturkunde.

Ges. Naturf. Freunde Berlin Neue Schriften = Der Gesellschaft naturforschender Freunde zu Berlin, neue Schriften.

Gesch. Bot.—See: F. K. Medikus 1793

Gille, A. 1949. Le *Spiraea tomentosa* L. dans la région de Granby (Comté de Shefford, Québec, Canada): Étude écologique et phytosociologique. Vegetatio 2: 166–196.

Gilman, A. V. 2012. *Rosa sherardii* Davies, an overlooked species in Vermont, and new to North America. Phytoneuron 2012-31: 1–6.

Gleason, H. A. 1952. The New Britton and Brown Illustrated Flora of the Northeastern United States and Adjacent Canada. 3 vols. New York.

Gleason, H. A. and A. Cronquist. 1963. Manual of Vascular Plants of Northeastern United States and Adjacent Canada. Princeton.

Gleason, H. A. and A. Cronquist. 1991. Manual of Vascular Plants of Northeastern United States and Adjacent Canada, ed. 2. Bronx.

Godfrey, R. K. 1988. Trees, Shrubs, and Woody Vines of Northern Florida and Adjacent Georgia and Alabama. Athens, Ga.

Godfrey, R. K. and J. W. Wooten. 1981. Aquatic and Wetland Plants of Southeastern United States: Dicotyledons. Athens, Ga.

Graber, R. E. 1980. The life history and ecology of *Potentilla robbinsiana*. Rhodora 82: 131–140.

Graber, R. E. and L. G. Brewer. 1985. Changes in the population of the rare and endangered plant *Potentilla robbinsiana* Oakes during the period 1973 to 1983. Rhodora 87: 449–457.

Gray, A. 1854. Plantae Novae Thurberianae.... Cambridge, Mass. [Preprinted from Mem. Amer. Acad. Arts, n. s. 5: 297–328. 1855.] (Pl. Nov. Thurb.)

Gray, A., S. Watson, B. L. Robinson, et al. 1878–1897. Synoptical Flora of North America. 2 vols. in parts and fascs. New York etc. [Vol. 1(1,1), 1895; vol. 1(1,2), 1897; vol. 1(2), 1884; vol. 2(1), 1878.] (Syn. Fl. N. Amer.)

Great Plains Flora Association. 1986. Flora of the Great Plains. Lawrence, Kans.
Greene, E. L. 1891–1897. Flora Franciscana. An Attempt to Classify and Describe the Vascular Plants of Middle California. 4 parts. San Francisco. [Parts paged consecutively.] (Fl. Francisc.)
Greene, E. L. 1906. Segregates from *Sieversia*. Leafl. Bot. Observ. Crit. 1: 174–179.
Greene, E. L. 1906c. The genus *Tridophyllum*. Leafl. Bot. Observ. Crit. 1: 188–189.
Gregor, T., V. Poncet, and J.-M. Tison. 2009. Proposal to conserve the name *Potentilla inclinata* against *P. assurgens* (Rosaceae). Taxon 58: 651–652.
Grenier, J. C. M. and D. A. Godron. [1847–]1848–1856. Flore de France.... 3 vols in 2 parts each. Paris etc. [Parts paged consecutively within vols.] (Fl. France)
Groh, H. and H. A. Senn. 1940. *Prunus* in eastern Canada. Canad. J. Res., Sect. C, Bot. Sci. 18: 318–346.
Grossheim, A. A. 1928–1934. Flora Kavkaza.... 4 vols. Tiflis. (Fl. Kavkaza)
Grossi, C., O. Raymond, and M. Jay. 1998. Flavonoid and enzyme polymorphisms and taxonomic organization of *Rosa* sections: *Carolinae*, *Cinnamomeae*, *Pimpinellifoliae*, and *Systylae*. Biochem. Syst. & Ecol. 26: 857–871.
Gu, C. Z. and S. A. Spongberg. 2003. *Malus*. In: Wu Z. and P. H. Raven, eds. 1994+. Flora of China. 20+ vols. Beijing and St. Louis. Vol. 9, pp. 179–189.
Gu, C. Z. and S. A. Spongberg. 2003b. *Pyracantha*. In: Wu Z. and P. H. Raven, eds. 1994+. Flora of China. 20+ vols. Beijing and St. Louis. Vol. 9, pp. 108–111.
Gu, C. Z. and S. A. Spongberg. 2003c. *Eriobotrya*. In: Wu Z. and P. H. Raven, eds. 1994+. Flora of China. 20+ vols. Beijing and St. Louis. Vol. 9, pp. 138–141.
Guillard, J. C. 1857. Idée générale de l'inflorescence. Bull. Soc. Bot. France 4: 29–41, 116–124.
Guo, W. et al. 2011. A phylogeny of *Photinia* sensu lato (Rosaceae) and related genera based on nrITS and cpDNA analysis. Pl. Syst. Evol. 291: 91–102.
Haines, A. 1998. Flora of Maine.... Bar Harbor.
Handb. Laubholzk.—See: L. Dippel 1889–1893
Handbuch—See: J. H. F. Link 1829–1833
Hara, H. 1955. Critical notes on some type specimens of east-Asiatic plants in foreign herbaria (2). J. Jap. Bot. 30: 65–72.
Hara, H. and S. Kurosawa. 1959. On the *Duchesnea indica* group. J. Jap. Bot. 34: 161–166.
Harbk. Baumz.—See: J. P. Du Roi 1771–1772
Hardin, J. W. 1973. The enigmatic chokeberries. Bull. Torrey Bot. Club 100: 178–184.
Harkness, P. 2003. The Rose. London.
Harmon, P. J., D. I. Ford, and W. Grafton. 2006. Checklist and Atlas of the Vascular Flora of West Virginia. Elkins.
Harms, V. L. 1974. The native roses of Saskatchewan. Blue Jay 32: 131–139.
Harris, J. A. 1917. On the distribution of abnormalities in the inflorescence of *Spiraea vanhouttei*. Amer. J. Bot. 4: 624–636.
Harvard Pap. Bot. = Harvard Papers in Botany.
Haston, E. M. et al. 2007. A linear sequence of Angiosperm Phylogeny Group II families. Taxon 56: 7–12.
Hausvater—See: O. von Münchhausen 1765–1773
Heath, P. V. 1992. Preliminary notes on British roses. Calyx 1: 141–196.
Hebda, R. J. and C. C. Chinnappa. 1990. Pollen morphology of the Rosaceae of western Canada. III. Canad. J. Bot. 68: 1369–1378.
Hebda, R. J. and C. C. Chinnappa. 1990b. Studies on pollen morphology of Rosaceae in Canada. Rev. Palaeobot. Palynol. 64: 103–108.
Hedrick, U. P. 1911. The Plums of New York. Albany.
Henderson, L. 1995. Plant Invaders of Southern Africa. A Pocket Guide to the Identification of 161 of the Most Important and Potentially Important Alien Species. Pretoria.
Henrickson, J. 1985. *Xerospiraea*, a generic segregate of *Spiraea* (Rosaceae) from Mexico. Aliso 11: 199–211.
Henrickson, J. 2001. Systematics and relationships of *Fallugia* (Rosoideae–Rosaceae). Aliso 20: 1–15.
Herb. Raf.—See: C. S. Rafinesque 1833
Heredity = Heredity; an International Journal of Genetics.
Herrera, C. M. 1987. Vertebrate-dispersed plants of the Iberian Peninsula: A study of fruit characteristics. Ecol. Monogr. 57: 305–331.
Herrera, C. M. 1989. Frugivory and seed dispersal by carnivorous animals, and associated fire characteristics, in undisturbed Mediterranean habitats. Oikos 55: 250–262.
Herrick, J. W. 1977. Iroquois Medical Botany. Ph.D. dissertation. State University of New York, Albany.
Heslop-Harrison, Y. 1968. *Rubus*. In: T. G. Tutin et al., eds. 1964–1980. Flora Europaea. 5 vols. Cambridge. Vol. 2, pp. 7–25.
Hess, W. J. 1969. A taxonomic study of *Spiraea pyramidata* Greene (Rosaceae). Sida 3: 298–308.
Hess, W. J. and J. Henrickson. 1987. A taxonomic revision of *Vauquelinia* (Rosaceae). Sida 12: 101–163.
Hickman, J. C., ed. 1993. The Jepson Manual. Higher Plants of California. Berkeley, Los Angeles, and London.
Hill, J. 1768. Hortus Kewensis. London. (Hort. Kew.)
Hinds, H. R. 2000. Flora of New Brunswick..., ed. 2. Fredericton.
Hist. & Mém. Acad. Roy. Sci. Toulouse = Histoire et Mémoires de l'Académie Royale des Sciences, Inscriptions et Belles Lettres de Toulouse.
Hist. Arbr. Forest.—See: F. Michaux 1810–1813
Hist. Nat. Vég.—See: É. Spach 1834–1848
Hist. Pl.—See: H. E. Baillon 1866–1895
Hist. Pl. Dauphiné—See: D. Villars 1786–1789
Hitchcock, C. L. et al. 1955–1969. Vascular Plants of the Pacific Northwest. 5 vols. Seattle. [Univ. Wash. Publ. Biol. 17.] (Vasc. Pl. Pacif. N.W.)
Hitchcock, C. L. and A. Cronquist. 1961. *Petrophytum*. In: C. L. Hitchcock et al. 1955–1969. Vascular Plants of the Pacific Northwest. 5 vols. Seattle. Vol. 3, pp. 124–125.
Hitchcock, C. L. and A. Cronquist. 1961b. Rosaceae. In: C. L. Hitchcock et al. 1955–1969. Vascular Plants of the Pacific Northwest. 5 vols. Seattle. Vol. 3, pp. 89–194.
Hoffman, D. 2003. Medical Herbalism: The Science and Practice of Herbal Medicine. Rochester, Vt.

Hokanson, K. E. et al. 2006. Relationships among subspecies of New World octoploid strawberry species, *Fragaria virginiana* and *Fragaria chiloensis,* based on simple sequence repeat marker analysis. Canad. J. Bot. 84: 1829–1841.

Holloway, P. and G. Matheke. 2003. Germination of three Alaska species of *Sanguisorba.* Native Pl. J. 4: 95–99.

Holmgren, N. H. 1988. *Glossopetalon* (Crossosomataceae) and a new variety of *G. spinescens* from the Great Basin, U.S.A. Brittonia 40: 269–274.

Holmgren, N. H. 1997. Crossosomataceae. In: A. Cronquist et al. 1972+. Intermountain Flora. Vascular Plants of the Intermountain West, U.S.A. 6+ vols. in 7+. New York and London. Vol. 3, part A, pp. 158–161.

Holmgren, N. H. 1997b. Rosaceae In: A. Cronquist et al. 1972+. Intermountain Flora. Vascular Plants of the Intermountain West, U.S.A. 6+ vols. in 7+. New York and London. Vol. 3, part A, pp. 64–158.

Hooker, J. D. 1881. *Potentilla (Ivesia) unguiculata.* Bot. Mag., ser. 3, 37: plate 6560.

Hooker, W. J. [1829–]1833–1840. Flora Boreali-Americana; or, the Botany of the Northern Parts of British America.... 2 vols. in 12 parts. London, Paris, and Strasbourg. (Fl. Bor.-Amer.)

Hooker, W. J. and G. A. W. Arnott. [1830–]1841. The Botany of Captain Beechey's Voyage; Comprising an Account of the Plants Collected by Messrs Lay and Collie, and Other Officers of the Expedition, during the Voyage to the Pacific and Bering's Strait, Performed in His Majesty's Ship Blossom, under the Command of Captain F. W. Beechey...in the Years 1825, 26, 27, and 28. 10 parts. London. [Parts paged and plates numbered consecutively.] (Bot. Beechey Voy.)

Hooker's J. Bot. Kew Gard. Misc. = Hooker's Journal of Botany and Kew Garden Miscellany.

Hoover, R. F. 1966. Miscellaneous new names for California plants. Leafl. W. Bot. 10: 337–350.

Horn, D. D. and P. Somers. 1981. *Neviusia alabamensis* in Tennessee. Sida 9: 90–91.

Hort. Berol.—See: C. L. Willdenow 1803–1816

Hort. Bot. Vindob.—See: N. J. Jacquin 1770–1776

Hort. Canal.—See: I. F. Tausch 1823–1825

Hort. Dendrol.—See: K. H. E. Koch 1853–1854

Hort. Kew.—See: W. Aiton 1789; J. Hill 1768

Hort. Reg. & Gard. Mag. = Horticultural Register, and Gardener's Magazine.

Hort. Suburb. Lond.—See: R. Sweet 1818

Hortus Kew.—See: W. Aiton and W. T. Aiton 1810–1813

Howell, J. T. 1945. Studies in Rosaceae, tribe Potentilleae. Leafl. W. Bot. 4: 171–176.

Howell, J. T. 1962. Five days to Reno: A botanical motorlogue. Leafl. W. Bot. 9: 233–242.

Howell, J. T., P. H. Raven, and P. Rubtzoff. 1958. A flora of San Francisco, California. Wasmann J. Biol. 16: 1–157.

Howell, T. J. 1897–1903. A Flora of Northwest America. 1 vol. in 8 fascs. Portland. [Fascs. 1–7 (text) paged consecutively, fasc. 8 (index) independently.] (Fl. N.W. Amer.)

Hrusa, F. et al. 2002. Catalogue of non-native vascular plants occurring spontaneously in California beyond those addressed in The Jepson Manual—Part I. Madroño 49: 61–98.

Huckins, C. A. 1972. A Revision of the Sections of the Genus *Malus* Miller. Ph.D. dissertation. Cornell University.

Hultén, E. 1941–1950. Flora of Alaska and Yukon. 10 vols. Lund and Leipzig. [Vols. paged consecutively and designated as simultaneous numbers of Lunds Univ. Årsskr. (= Acta Univ. Lund.) and Kungl. Fysiogr. Sällsk. Handl.] (Fl. Alaska Yukon)

Hultén, E. 1945. Studies in the *Potentilla nivea* group. Bot. Not. 1945: 127–148.

Hultén, E. 1959b. Studies in the genus *Dryas.* Svensk Bot. Tidskr. 53: 507–542.

Hultén, E. 1968. Flora of Alaska and Neighboring Territories: A Manual of the Vascular Plants. Stanford.

Hultén, E. 1971. The circumpolar plants. II. Dicotyledons. Kongl. Svenska Vetensk. Acad. Handl., ser. 4, 13: 1–463.

Hultén, E. and M. Fries. 1986. Atlas of North European Vascular Plants North of the Tropic of Cancer. 3 vols. Königstein.

Humboldt, A. von and A. J. Bonpland. [1805–]1808–1809 [–1817]. Plantae Aequinoctiales.... 2 vols. in 17 parts. Paris and Tübingen. [Parts numbered consecutively.] (Pl. Aequinoct.)

Humboldt, A. von, A. J. Bonpland, and C. S. Kunth. 1815[1816]–1825. Nova Genera et Species Plantarum Quas in Peregrinatione Orbis Novi Collegerunt, Descripserunt.... 7 vols. in 36 parts. Paris. (Nov. Gen. Sp.)

Hunt, D. R. 1982. Proposal to conserve 3325 *Gillenia* Moench (1802) [Rosaceae] against *Gillena* Adans. (1763) [Clethraceae]. Taxon 31: 568.

Hunt, D. R. and C. King. 1978. *Amelanchier denticulata* (Rosaceae). Bot. Mag. 182: 20–23, plate 753.

Hutchinson, J. 1964–1967. The Genera of Flowering Plants (Angiospermae).... 2 vols. Oxford.

Hutton, E. E. and R. B. Clarkson. 1961. Two plants new for North America and some new or otherwise interesting plants in West Virginia. Castanea 26: 84–88.

Huxley, A., M. Griffiths, and M. Levy, eds. 1992. The New Royal Horticultural Society Dictionary of Gardening. 4 vols. London and New York.

Icon. Pl. Rar.—See: N. J. Jacquin 1781–1793[–1795]

Iketani, H. and H. Ohashi. 1991. Anatomical structure of fruits and evolution of the tribe Sorbeae in the subfamily Maloideae (Rosaceae). J. Jap. Bot. 66: 319–351.

Ill. Fl. N. U.S.—See: N. L. Britton and A. Brown 1896–1898

Ill. Handb. Laubholzk.—See: C. K. Schneider [1904–]1906–1912

Index Seminum (Hamburg) = Semina in Horto Botanico Hamburgensi...Collecta Quae pro Mutua Commutatione Offeruntur. [1823–1840 for years 1822–1840. Title after 1829: Delectus Seminum Quae in Horto Hamburgensium Botanico e Collectioni Anni...Mutuae Commutatione Offeruntur.]

Index Seminum (St. Petersburg) = Index Seminum, Quae Hortus Botanicus Imperialis Petropolitanus pro Mutua Commutatione Offert.

Inst. Rei Herb.—See: H. J. N. von Crantz 1766

Int. J. Food Sci. Nutr. = International Journal of Food Sciences and Nutrition.

Int. J. Hort. Sci. = International Journal of Horticultural Science.

Int. J. Pl. Sci. = International Journal of Plant Sciences.

Isis (Oken) = Isis, oder encyclopädische Zeitung (vorzüglich für Naturgeschichte, vergleichende Anatomie und Physiologie). [Oken ed.]

Itin. Princ. S. Coburgi—See: H. Wawra 1883–1888

Iwatsuki, K. et al., eds. 1993+. Flora of Japan. 3+ vols. in 6+. Tokyo.

Izv. Akad. Nauk S.S.S.R., Ser. Biol. = Izvestiya Akademii Nauk S S S R. Seriya Biologicheskaya.

Izv. Glavn. Bot. Sada S.S.S.R. = Izvestiya Glavnogo Botanicheskogo Sada S S S R.

J. Agric. Food Chem. = Journal of Agricultural and Food Chemistry.

J. Agric. Res. = Journal of Agricultural Research.

J. Amer. Soc. Hort. Sci. = Journal of the American Society for Horticultural Science.

J. Arizona-Nevada Acad. Sci. = Journal of the Arizona-Nevada Academy of Science.

J. Arnold Arbor. = Journal of the Arnold Arboretum.

J. Bot. = Journal of Botany, British and Foreign.

J. Bot. Res. Inst. Texas = Journal of the Botanical Research Institute of Texas.

J. Coll. Sci. Imp. Univ. Tokyo = Journal of the College of Science, Imperial University of Tokyo.

J. Ecol. = Journal of Ecology.

J. Elisha Mitchell Sci. Soc. = Journal of the Elisha Mitchell Scientific Society.

J. Ethnopharmacol. = Journal of Ethnopharmacology; Interdisciplinary Journal Devoted to Bioscientific Research on Indigenous Drugs.

J. Hort. Sci. Biotechnol. = Journal of Horticultural Science and Biotechnology.

J. Idaho Acad. Sci. = Journal of the Idaho Academy of Science.

J. Iowa Acad. Sci. = Journal of the Iowa Academy of Science.

J. Jap. Bot. = Journal of Japanese Botany.

J. Linn. Soc., Bot. = Journal of the Linnean Society. Botany.

J. Nat. Prod. (Lloydia) = Journal of Natural Products (Lloydia).

J. Pharm. Pharmacol. = Journal of Pharmacy and Pharmacology.

J. Range Managem. = Journal of Range Management.

J. Roy. Hort. Soc. = Journal of the Royal Horticultural Society.

J. Wash. Acad. Sci. = Journal of the Washington Academy of Sciences.

Jaarb. Kon. Ned. Maatsch. Aanm. Tuinb. = Jaarboek van de Koninklijke Nederlandsche Maatschappij tot Aanmoediging van den Tuinbouw.

Jaarb. Ned. Dendrol. Ver. = Jaarboek, Nederlandsche Dendrologische Vereeniging.

Jacobson, A. L. 2001. Wild Plants of Greater Seattle. Seattle.

Jacquin, N. J. 1770–1776. Hortus Botanicus Vindobonensis.... 3 vols. Vienna. (Hort. Bot. Vindob.)

Jacquin, N. J. 1773–1778. Florae Austriaceae.... 5 vols. Vienna. (Fl. Austriac.)

Jacquin, N. J. 1781–1793[–1795]. Icones Plantarum Rariorum. 3 vols. in fascs. Vienna etc. [Vols. paged independently, plates numbered consecutively.] (Icon. Pl. Rar.)

Jaeger, H. 1865. Die Ziergehölze der Gärten und Parkanlagen. Weimar. (Ziergeh. Gärt. Parkanl.)

Jalas, J. et al., eds. 1972+. Atlas Florae Europaeae: Distribution of Vascular Plants in Europe. 14+ vols. Helsinki.

Jap. J. Bot. = Japanese Journal of Botany.

Jard. Fruit.—See: J. Decaisne [1858–]1862–1875

Jeffries, D. L. and J. M. Klopatek. 1987. Effects of grazing on the vegetation of the blackbrush association. J. Range Managem. 40: 390–392.

Jensen, D. B. and K. D. Christensen, eds. 2003. The Biodiversity of Greenland—A Country Study. Pinngortitalerifik.

Jepson, W. L. 1909–1943. A Flora of California.... 3 vols. in 12 parts. San Francisco etc. [Pagination consecutive within each vol.; vol. 1 page sequence independent of part number sequence (chronological); part 8 of vol. 1 (pp. 1–32, 579–index) never published.] (Fl. Calif.)

Jepson, W. L. [1923–1925.] A Manual of the Flowering Plants of California.... Berkeley. (Man. Fl. Pl. Calif.)

Johnson, K. L. and A. R. Tiedemann, eds. 1983. Proceedings: Research and Management of Bitterbush and Cliffrose in Western North America. Ogden. [U.S.D.A. Forest Serv., Gen. Techn. Rep. INT-152.]

Johnston, B. C. 1980. Studies of Population Variability Leading to a New Classification of *Potentilla* sect. *Multijugae*. Ph.D. dissertation. University of Colorado.

Johnston, B. C. 1985. Studies in *Potentilla*. I. Key to North American sections. Phytologia 57: 292–302.

Johnston, B. C. and B. Ertter. 2010. *Potentilla uliginosa* (Rosaceae: Rosoideae), a new presumed extinct species from Sonoma County, California. J. Bot. Res. Inst. Texas 4: 13–18.

Joly, S. et al. 2006. Polyploid and hybrid evolution in roses east of the Rocky Mountains. Amer. J. Bot. 93: 412–425.

Joly, S. and A. Bruneau. 2006. Incorporating allelic variation for reconstructing the evolutionary history of organisms from multiple genes: An example from *Rosa* in North America. Syst. Biol. 55: 623–636.

Joly, S. and A. Bruneau. 2007. Delimiting species boundaries in *Rosa* sect. *Cinnamomeae* (Rosaceae) in eastern North America. Syst. Bot. 32: 819–836.

Jones, G. N. 1939. A synopsis of the North American species of *Sorbus*. J. Arnold Arbor. 20: 1–43.

Jones, G. N. 1945. *Malacomeles,* a genus of Mexican and Guatemalan shrubs. Madroño 8: 33–39.

Jones, G. N. 1946. American Species of *Amelanchier.* Urbana.

Jones, R. L. 2005. Plant Life of Kentucky: An Illustrated Guide to the Vascular Flora. Lexington, Ky.

Jørgensen, C. A., T. J. Sørensen, and M. Westergaard. 1958. The flowering plants of Greenland. A taxonomical and cytological survey. Biol. Skr. 9(4): 1–172.

Jurtzev, B. A. 1984. Rosaceae. In: A. I. Tolmatchew, ed. 1960–1987. Flora Arctica URSS. 10 vols. in 12. Moscow and Leningrad. Vol. 9(1), pp. 102–323.

Jurtzev, B. A. 1993. New taxa of the genus *Potentilla* (Rosaceae) from arctic Alaska. Bot. Zhurn. (Moscow & Leningrad) 78(11): 78–87.

Jurtzev, B. A. 2001. Annotated check-list of the genus *Potentilla* (Rosaceae) in the circumpolar Arctic. Bot. Zhurn. (Moscow & Leningrad) 86: 131–143.

Kalkman, C. 1973. The Malesian species of the subfamily Maloideae (Rosaceae). Blumea 21: 413–442.

Kalkman, C. 1988. The phylogeny of the Rosaceae. Bot. J. Linn. Soc. 98: 37–59.

Kalkman, C. 2004. Rosaceae. In: K. Kubitzki et al., eds. 1990+. The Families and Genera of Vascular Plants. 10+ vols. Berlin etc. Vol. 6, pp. 343–386.

Kartesz, J. T. 1987. A Flora of Nevada. Ph.D. thesis. 3 vols. University of Nevada, Reno.

Kartesz, J. T. 2006. Synthesis of the North American Flora, ver. 2.1. Chapel Hill. [CD-ROM.]

Kartesz, J. T. and C. A. Meacham. 2003. Synthesis of the North American Flora, ver. 2.0. Chapel Hill. [CD-ROM.]

Kashiwada, Y. et al. 1998. Anti-AIDS agents. 30. Anti-HIV activity of oleanolic acid, pomolic acid, and structurally related triterpenoids. J. Nat. Prod. (Lloydia) 61: 1090–1095.

Kaye, T. N. et al., eds. 1997. Conservation of Native Plants and Fungi. Corvallis.

Kearney, T. H. and R. H. Peebles. 1942. Flowering Plants and Ferns of Arizona.... Washington.

Kearney, T. H. and R. H. Peebles. 1951. Arizona Flora. Berkeley.

Kearney, T. H. and R. H. Peebles. 1960. Arizona Flora, ed. 2. Berkeley.

Keating, W. H. 1824. Narrative of an Expedition to the Source of St. Peter's River...under the Command of Stephen H. Long.... 2 vols. Philadelphia. (Narrat. Exp. St. Peter's River)

Keck, D. D. 1938[1939]. Revision of *Horkelia* and *Ivesia*. Lloydia 1: 75–142.

Kemp, J. R. et al. 1993. Morphological differences and changes of the gynoecium in short-lived flowers of *Rosa setigera* Michaux and their relationship to dioecy. Int. J. Pl. Sci. 154: 550–556.

Kemp, J. R. et al. 1993b. Floral development of *Rosa setigera*. Canad. J. Bot. 71: 74–86.

Kerr, M. S. 2004. A Phylogenetic and Biogeographic Analysis of Sanguisorbeae (Rosaceae), with Emphasis on the Pleistocene Radiation of the High Andean Genus *Polylepis*. Ph.D. dissertation. University of Maryland.

Kevan, P. G. et al. 1990. Cryptic dioecy and insect pollination in *Rosa setigera* Michx. (Rosaceae), a rare plant of Carolinian Canada. Biol. J. Linn. Soc. 40: 229–243.

Kew Bull. = Kew Bulletin.

Kiger, R. W. and D. M. Porter. 2001. Categorical Glossary for the Flora of North America Project. Pittsburgh.

Kirk, T. [1899.] The Students' Flora of New Zealand and the Outlying Islands.... Wellington. (Stud. Fl. New Zealand)

Klackenberg, J. 1983. The holarctic complex *Potentilla fruticosa*. Nordic J. Bot. 3: 181–191.

Klášterský, I. 1968. *Rosa*. In: T. G. Tutin et al., eds. 1964–1980. Flora Europaea. 5 vols. Cambridge. Vol. 2, pp. 25–32.

Klein, H. 2002. Legislation Regarding Harmful Plants in South Africa. Pretoria.

Kline, G. J. and P. D. Sørensen. 2008. A revision of *Agrimonia* (Rosaceae) in North and Central America. Brittonia 60: 11–33.

Klotz, G. 1957. Übersicht ueber die in Kulture befindlichen *Cotoneaster*-Arten und -Formen. Wiss. Z. Martin-Luther-Univ. Halle-Wittenberg, Math.-Naturwiss. Reihe 6: 945–982.

Klotz, G. 1963. Neue oder kritische *Cotoneaster*-Arten II. Wiss. Z. Martin-Luther-Univ. Halle-Wittenberg, Math.-Naturwiss. Reihe 12: 769–786.

Klotz, G. 1978. Neue oder kritische *Cotoneaster*-Arten VIII. Wiss. Z. Friedrich-Schiller-Univ. Jena, Math.-Naturwiss. Reihe 27: 19–26.

Klotz, G. and T. Krügel. 1983. Zur zytologischen Struktur der Gattung *Cotoneaster* Medik. II. Wiss. Z. Friedrich-Schiller-Univ. Jena, Math.-Naturwiss. Reihe 32: 901–907.

Koch, K. H. E. 1853–1854. Hortus Dendrologicus. 2 sects. Berlin. [Sections paged consecutively.] (Hort. Dendrol.)

Koch, K. H. E. 1869–1873. Dendrologie. Bäume, Sträucher und Halbsträucher, welche in Mittel- und Nord-Europa im Freien kultivirt werden. 2 vols. in 3. Erlangen. (Dendrologie)

Koehne, B. A. E. 1890. Die Gattungen der Pomaceen.... Berlin. [Wissenschaftliche Beilage zum Program des Falk-Realgymnasiums zu Berlin 95.] (Gatt. Pomac.)

Koehne, B. A. E. 1893. Deutsche Dendrologie. Stuttgart.

Kohli, B. L. and J. G. Packer. 1976. A contribution to the taxonomy of the *Potentilla pensylvanica* complex in North America. Canad. J. Bot. 54: 706–719.

Kollmann, J., T. Steinger, and B. A. Roy. 2000. Evidence of sexuality in European *Rubus* (Rosaceae) species based on AFLP and allozyme analysis. Amer. J. Bot. 87: 1592–1598.

Komarov, V. L. et al., eds. 1963+. Flora of the U.S.S.R. (Flora SSSR). Translated from Russian. 30+ vols. Jerusalem etc.

Komarov, V. L., B. K. Schischkin, and E. Bobrov, eds. 1934–1964. Flora URSS.... 30 vols. Leningrad. (Fl. URSS)

Kongl. Svenska Vetensk. Acad. Handl. = Kongl[iga]. Svenska Vetenskaps Academiens Handlingar.

Kroon, G. H. 1975. Polyploidy in *Cotoneaster* II. Acta Bot. Neerl. 24: 417–420.

Kruckeberg, A. R. 1967. Ecotypic response to ultramafic soils by some plant species of northwestern United States. Brittonia 19: 133–151.

Kruschke, E. P. 1965. Contributions to the taxonomy of *Crataegus*. Publ. Bot. Milwaukee Public Mus. 3: 11–273.

Kubitzki, K. et al., eds. 1990+. The Families and Genera of Vascular Plants. 10+ vols. Berlin etc.

Kudriashov, B. A. et al. 1991. Heparin from the meadowsweet *(Filipendula ulmaris)* and its properties. [In Russian.] Izv. Akad. Nauk S.S.S.R., Ser. Biol. 6: 939–943.

Kudriashov, B. A., L. A. Liapina, and L. D. Azieva. 1990. The content of a heparin-like anticoagulant in the flowers of the meadowsweet *(Filipendula ulmaria)*. [In Russian.] Farmakol. Toksikol. 53(4): 39–41.

Kugel, A. R. 1958. Variation in the *Spiraea alba-latifolia* Complex. Ph.D. dissertation. University of Michigan.

Kumar, A. and G. Panigrahi. 1989[1992]. Nomenclatural notes on *Cotoneaster symondsii* Th. Moore (Rosaceae). Bull. Bot. Surv. India 31: 168–170.

Kuntze, O. 1891–1898. Revisio Generum Plantarum Vascularium Omnium atque Cellularium Multarum.... 3 vols. Leipzig etc. [Vol. 3 in 3 parts paged independently; parts 1 and 3 unnumbered.] (Revis. Gen. Pl.)

Kurtto, A. and T. Eriksson. 2003. Atlas Florae Europaea notes. 15. Generic delimitations and nomenclatural adjustments in Potentilleae. Ann. Bot. Fenn. 40: 135–141.

Laicharding, J. N. 1790–1791. Vegetabilia Europaea.... 2 vols. Innsbruck. (Veg. Europ.)

Lamarck, J. et al. 1783–1817. Encyclopédie Méthodique. Botanique.... 13 vols. Paris and Liège. [Vols. 1–8, suppls. 1–5.] (Encycl.)

Lamarck, J. and A. P. de Candolle. 1806. Synopsis Plantarum in Florâ Gallicâ Descriptarum. Paris. (Syn. Pl. Fl. Gall.)

Lance, R. and J. B. Phipps. 2000. *Crataegus harbisonii* rediscovered and amplified. Castanea 85: 291–296.

Landry, P. 1975. Le concept d'espèce et la taxonomie du genre *Amelanchier* (Rosacées). Bull. Soc. Bot. France 122: 43–252.

Lawrence, D. B. et al. 1967. The role of *Dryas drummondii* in vegetation development following ice recession at Glacier Bay, Alaska, with special reference to its nitrogen fixation by root nodules. J. Ecol. 55: 793–813.

Leafl. Bot. Observ. Crit. = Leaflets of Botanical Observation and Criticism.

Leafl. W. Bot. = Leaflets of Western Botany.

Ledebour, C. F. von. 1829–1833. Flora Altaica. 4 vols. Berlin. (Fl. Altaica)

Lee, S. and J. Wen. 2001. A phylogenetic analysis of *Prunus* and the Amygdaloideae (Rosaceae) using ITS sequences of nuclear ribosomal DNA. Amer. J. Bot. 88: 150–160.

Lee, S. H. 1948. A taxonomic survey of the oriental pears. Proc. Amer. Soc. Hort. Sci. 51: 152–156.

Lehmann, J. G. C. 1820. Monographis Generis *Potentillarum*.... Paris and London. (Monogr. Potentill.)

Lehmann, J. G. C. 1828–1857. Novarum et Minus Cognitarum Stirpium Pugillus I–X.... 10 vols. Hamburg. (Nov. Stirp. Pug.)

Lehmann, J. G. C. 1856. Revisionem *Potentillarum*.... Breslau and Bonn. [Nov. Actorum Acad. Caes. Leop.-Carol. Nat. Cur. 23(suppl.): 1–230.] (Revis. Potentill.)

Lei, S. A. and L. R. Walker. 1997. Biotic and abiotic factors influencing the distribution of *Coleogyne* communities in southern Nevada. Great Basin Naturalist 57: 163–171.

Levin, G. A. 1986. *Rosa minutifolia* Engelm. in Parry (Rosaceae). Madroño 33: 150.

Lewis, W. H. 1958. A monograph of the genus *Rosa* in North America. I. *R. acicularis.* Brittonia 11: 1–24.

Lewis, W. H. 1959. A monograph of the genus *Rosa* in North America. II. *R. foliolosa.* SouthW. Naturalist 3: 145–153.

Lewis, W. H. 1959b. A monograph of the genus *Rosa* in North America. III. *R. setigera.* SouthW. Naturalist 3: 154–174.

Lewis, W. H. 1962. Monograph of *Rosa* in North America. IV. *R.* ×*dulcissima.* Brittonia 14: 65–71.

Lewis, W. H. 1965. Monograph of *Rosa* in North America. V. Subgenus *Hesperhodos.* Ann. Missouri Bot. Gard. 52: 99–113.

Lewis, W. H. 2008. *Rosa carolina* (Rosaceae) subspecies and hybrids in eastern and midwestern United States, Canada, and Mexico. Novon 18: 192–198.

Lewis, W. H. 2008b. Proposal to conserve the name *Rosa virginiana* Mill. against *R. virginiana* Herrm. (Rosaceae). Taxon 57: 1363–1364.

Lewis, W. H. and M. P. F. Elvin-Lewis. 2003. Medical Botany: Plants Affecting Human Health, ed. 2. Hoboken.

Lewis, W. H. and B. Ertter. 2007. Subspecies of *Rosa nutkana* and *R. woodsii* (Rosaceae) in western North America. Novon 17: 341–353.

Lewis, W. H. and B. Ertter. 2010. *Rosa woodsii* subsp. *puberulenta* and var. *ertterae* (Rosaceae), new to western North America. Novon 20: 47–52.

Ley, F. A. 1943. A taxonomic revision of the genus *Holodiscus* (Rosaceae). Bull. Torrey Bot. Club 70: 275–288.

Li, C. L., H. Ikeda, and H. Ohba. 2003. *Sibbaldia.* In: Wu Z. and P. H. Raven, eds. 1994+. Flora of China. 20+ vols. Beijing and St. Louis. Vol. 9, pp. 329–333.

Li, C. L., H. Ikeda, and H. Ohba. 2003b. *Duchesnea.* In: Wu Z. and P. H. Raven, eds. 1994+. Flora of China. 20+ vols. Beijing and St. Louis. Vol. 9, pp. 338–339.

Li, C. L., H. Ikeda, and H. Ohba. 2003c. *Potentilla.* In: Wu Z. and P. H. Raven, eds. 1994+. Flora of China. 20+ vols. Beijing and St. Louis. Vol. 9, pp. 291–328.

Li, C. L., H. Ikeda, and H. Ohba. 2003d. *Sanguisorba.* In: Wu Z. and P. H. Raven, eds. 1994+. Flora of China. 20+ vols. Beijing and St. Louis. Vol. 9, pp. 384–388.

Li, Q. Y. et al. 2012. Generic limits of Pyrinae: Insights from ribosomal DNA sequences. Bot. Stud. 53: 151–164.

Lindley, J. 1820. Rosarum Monographia; or, a Botanical History of Roses. London. (Ros. Monogr.)

Lindley, J. 1838. Flora Medica.... London.

Lingdi, L. and D. E. Boufford. 2003. *Rubus.* In: Wu Z. and P. H. Raven, eds. 1994+. Flora of China. 20+ vols. Beijing and St. Louis. Vol. 9, pp. 195–286.

Lingdi, L. and A. R. Brach. 2003. *Cotoneaster.* In: Wu Z. and P. H. Raven, eds. 1994+. Flora of China. 20+ vols. Beijing and St. Louis. Vol. 9, pp. 85–108.

Link, J. H. F. 1821–1822. Enumeratio Plantarum Horti Regii Berolinensis Altera.... 2 parts. Berlin. (Enum. Hort. Berol. Alt.)

Link, J. H. F. 1829–1833. Handbuch zur Erkennung der nutzbarsten und am häufigsten vorkommenden Gewächse.... 3 vols. Berlin. (Handbuch)

Linnaea = Linnaea; ein Journal für die Botanik in ihrem ganzen Umfange.

Linnaeus, C. 1753. Species Plantarum.... 2 vols. Stockholm. (Sp. Pl.)

Linnaeus, C. 1754. Genera Plantarum, ed. 5. Stockholm. (Gen. Pl. ed. 5)

Linnaeus, C. 1755b. Flora Suecica..., ed. 2. Stockholm. (Fl. Suec. ed. 2)

Linnaeus, C. 1758. Opera Varia...Fundamenta Botanica, Sponsalia Plantarum, et Systema Naturae.... Lucca. (Opera Var.)

Linnaeus, C. 1762–1763. Species Plantarum..., ed. 2. 2 vols. Stockholm. (Sp. Pl. ed. 2)

Linnaeus, C. 1766–1768. Systema Naturae per Regna Tria Naturae..., ed. 12. 3 vols. Stockholm. (Syst. Nat. ed. 12)

Linnaeus, C. 1767[–1771]. Mantissa Plantarum. 2 parts. Stockholm. [Mantissa [1] and Mantissa [2] Altera paged consecutively.] (Mant. Pl.)

Linnaeus, C. f. 1781[1782]. Supplementum Plantarum Systematis Vegetabilium Editionis Decimae Tertiae, Generum Plantarum Editionis Sextae, et Specierum Plantarum Editionis Secundae. Braunschweig. (Suppl. Pl.)

Lis, R. A. 1990. A Taxonomic Revision of *Holodiscus* Maxim. (Spiraeoideae: Rosaceae) Based upon Numerical Analyses of Leaf Morphometric Variation. Ph.D. dissertation. University of California, Berkeley.

Lis, R. A. 1992. Leaf architectural survey of *Cercocarpus* (Rosaceae) and its systematic significance. Int. J. Pl. Sci. 153: 258–272.

Lloydia = Lloydia; a Quarterly Journal of Biological Science.

Lo, E. Y. Y. et al. 2009. Evidence for the genetic association between East Asian and western North American *Crataegus* L. (Rosaceae) and rapid divergence of the eastern North American lineages based on multiple DNA sequences. Molec. Phylogen. Evol. 51: 157–168.

Lo, E. Y. Y. et al. 2009b. Population genetic structure of diploid sexual and polyploid apomictic hawthorns (*Crataegus;* Rosaceae) in the Pacific Northwest. Molec. Ecol. 18: 1145–1160.

Lo, E. Y. Y. and M. J. Donoghue. 2012. Expanded phylogenetic and dating analyses of the apples and their relatives (Pyreae, Rosaceae). Molec. Phylogen. Evol. 63: 230–243.

Lo, E. Y. Y., S. Stefanovic, and T. A. Dickinson. 2007. Molecular reappraisal of relationships between *Crataegus* and *Mespilus* (Rosaceae, Pyreae)—Two genera or one? Syst. Bot. 32: 596–616.

Lo, E. Y. Y., S. Stefanovic, and T. A. Dickinson. 2010. Reconstructing reticulation history in a phylogenetic framework and the potential of allopatric speciation driven by polyploidy in an agamic complex in *Crataegus* (Rosaceae). Evolution 64: 3593–3608.

Long, A. A. 1989. Disjunct populations of the rare shrub, *Neviusia alabamensis* Gray (Rosaceae). Castanea 54: 29–39.

Loudon, J. C. [1835–]1838. Arboretum et Fruticetum Britannicum; or, the Trees and Shrubs of Britain, Native and Foreign.... 8 vols. London. (Arbor. Frutic. Brit.)

Loureiro, J. de. 1790. Flora Cochinchinensis.... 2 vols. Lisbon. [Vols. paged consecutively.] (Fl. Cochinch.)

Löve, Á. 1954. Cytotaxonomical remarks on some American species of circumpolar taxa. Svensk Bot. Tidskr. 48: 211–232.

Löve, Á. and D. Löve. 1965. Taxonomic remarks on some American alpine plants. Univ. Colorado Stud., Ser. Biol. 17: 1–43.

Löve, Á. and D. Löve. 1966. Cytotaxonomy of the alpine vascular plants of Mount Washington. Univ. Colorado Stud., Ser. Biol. 24: 1–74.

Löve, Á. and D. Löve. 1975b. Cytotaxonomical Atlas of the Arctic Flora. Vaduz.

Löve, Á. and D. Löve. 1982b. In: IOPB chromosome number reports LXXV. Taxon 31: 342–368.

Love, R. and M. Feigen. 1979. Interspecific hybridization between native and naturalized *Crataegus* (Rosaceae) in western Oregon. Madroño 25: 211–216.

Lu, L. T. and C. Alexander. 2003. *Spiraea.* In: Wu Z. and P. H. Raven, eds. 1994+. Flora of China. 20+ vols. Beijing and St. Louis. Vol. 9, pp. 47–73.

Ludwig, C. G. 1760. Definitiones Generum Plantarum..., ed. 3. Leipzig.

Lundberg, M. et al. 2009. Allopolyploidy in Fragariinae (Rosaceae): Comparing four DNA sequence regions, with comments on classification. Molec. Phylogen. Evol. 51: 269–280.

Mabberley, D. J., C. E. Jarvis, and B. E. Juniper. 2001. The name of the apple. Telopea 9: 421–430.

Macklin, J. A. 2001. Systematics of *Crataegus* Series *Coccineae* (Rosaceae). Ph.D. thesis. University of Western Ontario.

Macklin, J. A. and J. B. Phipps. 2002. Sargent's names in *Crataegus* series *Coccineae* (Rosaceae). Harvard Pap. Bot. 7: 25–35.

Macklin, J. A. and J. B. Phipps. 2006. Studies in *Crataegus* ser. *Coccineae.* I. Canad. J. Bot. 84: 70–86.

Madroño = Madroño; Journal of the California Botanical Society [from vol. 3: a West American Journal of Botany].

Magee, D. W. and H. E. Ahles. 1999. Flora of the Northeast: A Manual of the Vascular Flora of New England and Adjacent New York. Amherst.

Mallah, T. S. 1954. Natural Hybridization Between *Rubus ursinus* Cham. and Schlecht. and Other *Rubi* in California. Ph.D. dissertation. University of California, Davis.

Man. Bot. ed. 7—See: A. Eaton 1836b

Man. Bot. Rocky Mt.—See: J. M. Coulter [1885]

Man. Cult. Trees ed. 2—See: A. J. Rehder 1940

Man. Fl. N. States—See: N. L. Britton 1901

Man. Fl. Pl. Calif.—See: W. L. Jepson [1923–1925]

Man. Pl. Oregon—See: M. E. Peck 1941

Man. S.E. Fl.—See: J. K. Small 1933

Man. Trees—See: C. S. Sargent 1905

Mansfield, D. 2000. Flora of Steens Mountain. Corvallis.

Mant. Pl.—See: C. Linnaeus 1767[–1771]

Manual ed. 7—See: B. L. Robinson and M. L. Fernald 1908

Marie-Victorin, Frère. 1935. Flore Laurentienne. Montreal.

Marshall, H. 1785. Arbustrum Americanum: The American Grove.... Philadelphia. (Arbust. Amer.)

Martin, A. C., H. S. Zim, and A. L. Nelson. 1951. American Wildlife & Plants.... New York. [Reprinted 1961, New York.]

Martin, F. L. 1950. Revision of *Cercocarpus.* Brittonia 7: 91–111.

Martin, W. C. and C. R. Hutchins. 1980. A Flora of New Mexico. 2 vols. Vaduz.

Mason, S. C. 1913. The pubescent-fruited species of *Prunus* of the southwestern states. J. Agric. Res. 1: 147–178.

Matfield, B. and J. R. Ellis. 1972. The allopolyploid origin and genomic constitution of *Potentilla anglica.* Heredity 29: 315–327.

Matthews, M. L. and P. K. Endress. 2005. Comparative floral structure and systematics in Crossosomatales (Crossosomataceae, Stachyuraceae, Staphyleaceae, Aphloiaceae, Geissolomataceae, Ixerbaceae, Strasburgeriaceae). Bot. J. Linn. Soc. 147: 1–46.

Mavi, A. 2004. Antioxidant properties of some medicinal plants. Biol. Pharm. Bull. 27: 702–705.

Maximowicz, C. J. 1879. Adnotationes de *Spiraeaceis*. Trudy Glavn. Bot. Sada 6: 105–261.

Maxwell, H. S. and S. G. Knees. 1989. *Spiraea*. In: S. M. Walters et al., eds. 1984+. The European Garden Flora.... 6+ vols. Cambridge etc. Vol. 3, pp. 331–341.

McAllister, H. A. 2005. The Genus *Sorbus:* Mountain Ashes and Other Rowans. Kew. (Gen. Sorbus)

McArthur, E. D., H. C. Stutz, and S. C. Sanderson. 1983. Taxonomy, distribution, evaluation, and cytogenetics of *Purshia, Cowania,* and *Fallugia* (Rosoideae, Rosaceae). In: K. L. Johnson and A. R. Tiedemann, eds. 1983. Proceedings: Research and Management of Bitterbush and Cliffrose in Western North America. Ogden. Pp. 4–24.

McAvoy, B. 1931. Ecological survey of the Bella Coola Region. Bot. Gaz. 92: 141–171.

McCutcheon, A. R. et al. 1995. Antiviral screening of British Columbian medicinal plants. J. Ethnopharmacol. 49: 101–110.

McFarland, J. H. 1938. Roses of the World in Color. Boston and New York.

McGregor, R. L. 1986b. Rosaceae. In: Great Plains Flora Association. 1986. Flora of the Great Plains. Lawrence, Kans. Pp. 364–406.

McKay, S. M. 1973. A Biosystematic Study of the Genus *Amelanchier* in Ontario. M.Sc. thesis. University of Toronto.

McLaughlin, S. P. 2007. Tundra to Tropics: The Floristic Plant Geography of North America. Fort Worth. [Sida Bot. Misc. 3.]

McNeill, J. et al., eds. 2006. International Code of Botanical Nomenclature (Vienna Code): Adopted by the Seventeenth International Botanical Congress, Vienna, Austria, July 2005. Königstein. [Regnum Veg. 146.]

McVaugh, R. 1951. A revision of the North American black cherries (*Prunus serotina* Ehrh., and relatives). Brittonia 7: 279–315.

Med. Fl.—See: C. S. Rafinesque 1828[–1830]

Med. Herbalism = Medical Herbalism; a Journal for the Clinical Practitioner.

Med. Repos. = Medical Repository.

Meddel. Grønland = Meddelelser om Grønland, af Kommissionen for Ledelsen af de Geologiske og Geografiske Undersølgeser i Grønland.

Medikus, F. K. 1789–1791. Philosophische Botanik.... 2 vols. Mannheim. (Philos. Bot.)

Medikus, F. K. 1793. Geschichte der Botanik unserer Zeiten. Mannheim. (Gesch. Bot.)

Melchior, H., ed. 1964. A. Engler's Syllabus der Pflanzenfamilien..., ed. 12. 2 vols. Berlin.

Mellichamp, T. L. 1976. A Comparative Study of *Aruncus* (Rosaceae) and *Astilbe* (Saxifragaceae), and the Problem of Their Relationships. Ph.D. dissertation. University of Michigan.

Mém. Acad. Imp. Sci. St.-Pétersbourg, Sér. 6, Sci. Math. = Mémoires de l'Académie Impériale des Sciences de St.-Pétersbourg. Sixième Série. Sciences Mathématiques, Physiques et Naturelles.

Mem. Amer. Acad. Arts = Memoirs of the American Academy of Arts and Science.

Mem. New York Bot. Gard. = Memoirs of the New York Botanical Garden.

Mém. Soc. Imp. Naturalistes Moscou = Mémoires de la Société Impériale des Naturalistes de Moscou.

Mentzelia = Mentzelia; Journal of the Northern Nevada Native Plant Society.

Mercure, M. and A. Bruneau. 2008. Hybridization between the escaped *Rosa rugosa* (Rosaceae) and native *R. blanda* in eastern North America. Amer. J. Bot. 95: 597–607.

Merriam-Webster. 1988. Webster's New Geographical Dictionary. Springfield, Mass.

Methodus—See: C. Moench 1794

Meyer, S. E. and B. K. Pendleton. 2005. Factors affecting seed germination and seedling establishment of a long-lived desert shrub (*Coleogyne ramosissima:* Rosaceae). Pl. Ecol. 178: 171–187.

Michaux, A. 1803. Flora Boreali-Americana.... 2 vols. Paris and Strasbourg. (Fl. Bor.-Amer.)

Michaux, F. 1810–1813. Histoire des Arbres Forestiers de l'Amérique Septentrionale.... 3 vols. in parts and fascs. Paris. [Pagination independent by vols., plate numbering independent by parts.] (Hist. Arbr. Forest.)

Michigan Bot. = Michigan Botanist.

Miller, P. 1754. The Gardeners Dictionary.... Abridged..., ed. 4. 3 vols. London. (Gard. Dict. Abr. ed. 4)

Miller, P. 1768. The Gardeners Dictionary..., ed. 8. London. (Gard. Dict. ed. 8)

Mink, J. N., J. R. Singhurst, and W. C. Holmes. 2011. Discovery of *Spiraea hypericifolia* naturalized in Texas. Phytoneuron 2011-63: 1–2.

Mitchell, R. S. and G. C. Tucker. 1997. Revised checklist of New York State plants. Bull. New York State Mus. Sci. Serv. 490.

Mitt. Deutsch. Dendrol. Ges. = Mitteilungen der Deutschen dendrologischen Gesellschaft.

Moench, C. 1794. Methodus Plantas Horti Botanici et Agri Marburgensis.... Marburg. (Methodus)

Moench, C. 1802. Supplementum ad Methodum Plantas.... Marburg. (Suppl. Meth.)

Moerman, D. E. 1998. Native American Ethnobotany. Portland.

Molec. Biol. Evol. = Molecular Biology and Evolution.

Molec. Ecol. = Molecular Ecology.

Molec. Phylogen. Evol. = Molecular Phylogenetics and Evolution.

Monasterio-Huelin, E. and H. E. Weber. 1996. Taxonomy and nomenclature of *Rubus ulmifolius* and *Rubus sanctus* (Rosaceae). Edinburgh J. Bot. 53: 311–322.

Monatsschr. Vereines Beförd. Gartenbaues Königl. Preuss. Staaten = Monatsschrift des Vereines zur Beförderung des Gartenbaues in den Königlich Preussischen Staaten für Gärtnerei und Pflanzenkunde.

Monogr. Bot. = Monographiae Botanicae.

Monogr. N. Amer. Potentilleae—See: P. A. Rydberg 1898

Monogr. N. Mex. Crataegus—See: J. B. Phipps 1997

Monogr. Potentill.—See: J. G. C. Lehmann 1820

Moore, D. J., R. S. Nowak, and C. L. Nowak. 1998. Photosynthetic acclimation to temperature and drought in the endemic chelan rockmat, *Petrophytum cinerascens* (Rosaceae). Amer. Midl. Naturalist 139: 374–382.

Moore, D. M. 1956. *Neviusia alabamensis* in Arkansas. Rhodora 58: 187–191.

Moore, J. N. and J. R. Ballington Jr., eds. 1990. Genetic Resources of Temperate Fruit and Nut Crops. 2 vols. Wageningen. [Acta Hort. 290.]

Moore, S. M. et al. 2002. Fossil *Neviusia* (Rosaceae) from the Middle Eocene of Princeton, BC, Canada. [Abstract.] Georgia J. Sci. 60: 43.

Morgan, D. R., D. E. Soltis, and K. R. Robertson. 1994. Systematic and evolutionary implications of *rbc*L sequence variation in Rosaceae. Amer. J. Bot. 81: 890–903.

Morton, J. F. 1987. Fruits of Warm Climates. Miami.

Morton, J. K. and J. M. Venn. 1990. A Checklist of the Flora of Ontario: Vascular Plants. Waterloo.

Muhlenbergia = Muhlenbergia; a Journal of Botany.

Münchhausen, O. von. 1765–1773. Der Hausvater. 6 vols. in parts. Hannover. (Hausvater)

Müntzing, A. 1958. Further studies on intraspecific polyploidy in *Potentilla argentea* (coll.). Bot. Not. 111: 209–227.

Müntzing, A. and G. Müntzing. 1945. The mode of reproduction of hybrids between sexual and apomictic *Potentilla argentea*. Bot. Not. 98: 49–71.

Munz, P. A. 1959. A California Flora. Berkeley and Los Angeles.

Munz, P. A. 1968. Supplement to A California Flora. Berkeley and Los Angeles. (Suppl. Calif. Fl.)

Munz, P. A. and I. M. Johnston. 1925. The potentillas of southern California. Bull. S. Calif. Acad. Sci. 24: 5–25.

Muravjeva, O. A. 1936. The genus *Sibbaldia* L. and its species. Trudy Bot. Inst. Akad. Nauk S.S.S.R, Ser. 1, Fl. Sist. Vyssh. Rast. 2: 217–241.

Murray, J. A. 1784. Caroli à Linné Equitis Systema Vegetabilium.... Editio Decima Quarta.... Göttingen. (Syst. Veg. ed. 14)

Mus. Helv. Bot.—See: N. C. Seringe 1818–1823

N. Amer. Fl.—See: N. L. Britton et al. 1905+

N. Amer. Trees—See: N. L. Britton and J. A. Shafer 1908

Narr. Journey Polar Sea—See: J. Franklin et al. 1823

Narrat. Exp. St. Peter's River—See: W. H. Keating 1824

Naruhashi, N. and H. Ikeda. 1999. Variation in nectar guides in Himalayan *Potentilla* (Rosaceae). Bull. Univ. Mus. Univ. Tokyo 39: 23–29.

Nat. Pflanzenfam.—See: H. G. A. Engler and K. Prantl 1887–1915

Native Pl. J. = Native Plants Journal.

Naturaliste Canad. = Naturaliste Canadien. Bulletin de Recherches, Observations et Découvertes se Rapportant à l'Histoire Naturelle du Canada.

Nature = Nature; a Weekly Illustrated Journal of Science.

Nelson-Jones, E. B., D. Briggs, and A. G. Smith. 2002. The origin of intermediate species of the genus *Sorbus*. Theor. Appl. Genet. 105: 953–963.

Neogenyton—See: C. S. Rafinesque 1825

Nesom, G. L. 2010. *Pyracantha* (Rosaceae) naturalized in Texas and the southeastern United States. Phytoneuron 2010-2: 1–6.

New Fl.—See: C. S. Rafinesque 1836[–1838]

New Man. Bot. Rocky Mt.—See: J. M. Coulter and A. Nelson 1909

New Phytol. = New Phytologist; a British Botanical Journal.

Neygenfind, F. W. 1821. Enchiridium Botanicum.... Meissen. (Ench. Bot.)

Ng, T. B. et al. 2004. A gallic acid derivative and polysaccharides with antioxidative activity from rose *(Rosa rugosa)* flowers. J. Pharm. Pharmacol. 56: 537–545.

Nielsen, E. L. 1939. A taxonomic study of the genus *Amelanchier* in Minnesota. Amer. Midl. Naturalist 22: 160–208.

Nomencl. Bot.—See: E. A. Raeuschel 1797

Nordal, I., B. A. Jurtzev, and M. Philipp. 1999. The *Dryas* workshop. Skr. Norske Vidensk.-Akad. Oslo, Mat.-Naturvidensk. Kl., n. s. 38: 345–350.

Nordborg, G. 1963. Studies in *Sanguisorba officinalis* L. Bot. Not. 116: 267–288.

Nordborg, G. 1966. *Sanguisorba* L., *Sarcopoterium* Spach, and *Bencomia* Webb et Berth.: Delimitation and subdivision of the genera. Opera Bot. 11: 1–103.

Nordborg, G. 1967. The genus *Sanguisorba* section *Poterium:* Experimental studies and taxonomy. Opera Bot. 16: 1–153.

Nordic J. Bot. = Nordic Journal of Botany.

Norris, W. R. et al. 2001. Lessons from an inventory of the Ames, Iowa, flora (1859–2000). J. Iowa Acad. Sci. 108: 34–63.

Norton, J. B. S. and R. G. Brown. 1946. A catalog of the vascular plants of Maryland. Castanea 11: 1–50.

Not. Milit. Reconn.—See: W. H. Emory 1848

Notes Roy. Bot. Gard. Edinburgh = Notes from the Royal Botanic Garden, Edinburgh.

Nouv. Arch. Mus. Hist. Nat. = Nouvelles Archives du Muséum d'Histoire Naturelle.

Nov. Gen. Sp.—See: A. von Humboldt et al. 1815[1816]–1825

Nov. Stirp. Pug.—See: J. G. C. Lehmann 1828–1857

Nova Acta Regiae Soc. Sci. Upsal. = Nova Acta Regiae Societatis Scientiarum Upsaliensis.

Novi Comment. Soc. Regiae Sci. Gott. = Novi Commentarii Societatis Regiae Scientiarum Gottingensis.

Novon = Novon; a Journal for Botanical Nomenclature.

Nuovo Giorn. Bot. Ital. = Nuovo Giornale Botanico Italiano.

Nuttall, T. 1818. The Genera of North American Plants, and Catalogue of the Species, to the Year 1817.... 2 vols. Philadelphia. (Gen. N. Amer. Pl.)

Nybom, H., I. V. Bartish, and J. Fryer. 2005. Genetic relatedness among Asian *Cotoneaster* species investigated with DNA marker analysis. Int. J. Hort. Sci. 11: 43–46.

Nybom, H., G. Werlemark, and A. M. E. Olsson. 2000. Genetic variation in hemisexual dogrose species, *Rosa* sect. *Caninae*. Acta Hort. 538: 531–534.

Observ. Pl. Angers—See: N. A. Desvaux 1818

Oeder, G. C. et al., eds. [1761–]1764–1883. Icones Plantarum... Florae Danicae Nomine Inscriptum. 17 vols. in 51 fascs. Copenhagen. [Fascs. paged independently and numbered consecutively throughout vols.] (Fl. Dan.)

Oekon.-techn. Fl. Böhm.—See: F. Berchtold et al. 1836–1843

Ogle, D. W. 1991. *Spiraea virginiana* Britton: I. Delineation and distribution. Castanea 56: 287–296.

Ogle, D. W. 1991b. *Spiraea virginiana* Britton: II. Ecology and species biology. Castanea 56: 297–303.

Ogle, D. W. 1992. Virginia Spiraea (*Spiraea virginiana* Britton) Recovery Plan. Newton Corner, Mass.

Oh, S. H. 2006. *Neillia* includes *Stephandra*. Novon 16: 91–95.

Oh, S. H. and D. Potter. 2005. Molecular phylogenetic systematics and biogeography of tribe Neillieae (Rosaceae) using DNA sequences of cpDNA, rDNA, and *LEAFY*. Amer. J. Bot. 92: 179–192.

Oh, S. H. and D. Potter. 2006. Phylogenetic position of a new angiosperm family, Guamatelaceae, inferred from chloroplast *rbc*L, *atp*B, and *mat*K sequences. Syst. Bot. 31: 730–738.

Ohba, H. 2001. *Cerasus.* In: K. Iwatsuki et al., eds. 1993+. Flora of Japan. 3+ vols. in 6+. Tokyo. Vol. 2b, pp. 128–144.
Ohio J. Sci. = Ohio Journal of Science.
Ohwi, J. 1965. Flora of Japan (in English).... A Combined, Much Revised, and Extended Translation by the Author of His...Flora of Japan (1953) and...Flora of Japan—Pteridophyta (1957). Edited by Frederick G. Meyer... and Egbert H. Walker.... Washington. [Reprinted 1984, Washington.]
Oikos = Oikos; Acta Oecologica Scandinavica.
Okie, W. R. 2001. Plum crazy: Rediscovering our lost *Prunus* resources. HortScience 36: 209–213.
Olkers, T. and M. P. Hill, eds. 1999. Biological Control of Weeds in South Africa. A Pocket Guide to the Identification of 161 of the Most Important and Potentially Important Alien Species (1990–1998). Pretoria.
Ontario Nat. Sci. Bull. = Ontario Natural Science Bulletin; Journal of the Wellington Field Naturalists' Club.
Opera Bot. = Opera Botanica a Societate Botanice Lundensi.
Opera Var.—See: C. Linnaeus 1758
Overath, R. D. and J. L. Hamrick. 1998. Allozyme diversity in *Amelanchier arborea* and *A. laevis* (Rosaceae). Rhodora 100: 276–292.
Pacif. Railr. Rep.—See: War Department 1855–1860
Packer, J. G. 1994. A mountain-avens, *Dryas* ×*sundermannii* Kellerer ex Sundermann, in Alberta. Canad. Field-Naturalist 108: 77–79.
Palmer, E. J. 1925. Synopsis of North American *Crataegi.* J. Arnold Arbor. 6: 5–128.
Palmer, E. J. 1950. *Crataegus.* In: M. L. Fernald. 1950. Gray's Manual of Botany, ed. 8. New York. Pp. 767–801.
Palmer, E. J. 1952. *Crataegus.* In: H. A. Gleason. 1952. The New Britton and Brown Illustrated Flora of the Northeastern United States and Adjacent Canada. 3 vols. New York. Vol. 2, pp. 338–375.
Palmer, E. J. 1960. *Crataegus.* Key to series and species. In: R. A. Vines. 1960. Trees, Shrubs, and Woody Vines of the Southwest. Austin. Pp. 329–334.
Paterson, I. G. and M. Snyder. 1999. Genetic evidence supporting the taxonomy of *Geum peckii* (Rosaceae) and *G. radiatum* as separate species. Rhodora 101: 325–340.
Patterson, D. T. et al. 1989. Composite List of Weeds. Champaign.
Paule, J., T. F. Sharbel, and C. Dobeš. 2011. Apomictic and sexual lineages of the *Potentilla argentea* L. group (Rosaceae): Cytotype and molecular genetic differentiation. Taxon 60: 721–732.
Paule, J. and J. Soják. 2009. Taxonomic comments on the genus *Sibbaldiopsis* Rydberg (Rosaceae). Čas. Nár. Muz., Rada Přír. 178(4): 15–16.
Peck, M. E. 1941. A Manual of the Higher Plants of Oregon. Portland. (Man. Pl. Oregon)
Peck, M. E. 1961. A Manual of the Higher Plants of Oregon, ed. 2. Corvallis.
Pendleton, B. K. and R. L. Pendleton. 1998. Pollination biology of *Coleogyne ramosissima* (Rosaceae). SouthW. Naturalist 43: 376–380.
Persoon, C. H. 1805–1807. Synopsis Plantarum.... 2 vols. Paris and Tubingen. (Syn. Pl.)
Persson Hovmalm, H. A. et al. 2004. RAPD analysis of diploid and tetraploid populations of *Aronia* point to different reproductive strategies within the genus. Hereditas (Lund) 141: 301–312.
Philipp, M. and H. R. Siegismund. 2003. What can morphology and isozymes tell us about the history of the *Dryas integrifolia-octopetala* complex? Molec. Ecol. 12: 2231–2242.
Phillips, A. M. 1992. A new subspecies of *Rosa stellata* (Rosaceae) from northwestern Arizona. Madroño 39: 31–35.
Philos. Bot.—See: F. K. Medikus 1789–1791
Phipps, J. B. 1987. What is *Crataegus fontanesiana* (Spach) Steud.? Taxon 36: 641–644.
Phipps, J. B. 1988. Typification of *Crataegus crus-galli* L. (Rosaceae) and nomenclatural notes on some related species of hawthorn. Bot. J. Linn. Soc. 96: 359–369.
Phipps, J. B. 1988b. *Crataegus* (Maloideae, Rosaceae) of the southeastern United States. I. Introduction and series *Aestivales.* J. Arnold Arbor. 69: 401–431.
Phipps, J. B. 1990. *Crataegus secreta* (Rosaceae), a new species of hawthorn from the Edwards Plateau, Texas. Sida 14: 13–19.
Phipps, J. B. 1992. *Heteromeles* and *Photinia* (Rosaceae subfam. Maloideae) of Mexico and Central America. Canad. J. Bot. 70: 2138–2162.
Phipps, J. B. 1997. Monograph of Northern Mexican *Crataegus* (Rosaceae, Subfam. Maloideae). Fort Worth. (Monogr. N. Mex. Crataegus)
Phipps, J. B. 1998. Synopsis of *Crataegus* series *Apiifoliae, Cordatae, Microcarpae,* and *Brachyacanthae* (Rosaceae subfam. Maloideae). Ann. Missouri Bot. Gard. 85: 475–491.
Phipps, J. B. 1999. Proposal to reject *Mespilus cordata* (Rosaceae), and typification of *Mespilus phaenopyrum.* Taxon 48: 831–832.
Phipps, J. B. 1999b. Introduction to the red-fruited hawthorns of western North America. Canad. J. Bot. 76: 1863–1899.
Phipps, J. B. 1999c. The relationships of the American black-fruited hawthorns *Crataegus erythropoda, C. rivularis, C. saligna* and *C. brachyacantha* to C. ser. *Douglasianae* (Rosaceae). Sida 18: 647–660.
Phipps, J. B. 2005. Review of hybridization in *Crataegus*—Another look at "The *Crataegus* Problem." Ann. Missouri Bot. Gard. 92: 113–127.
Phipps, J. B. 2009. Miscellaneous typifications, one new series, and one new varietal combination in *Crataegus* (Rosaceae). J. Bot. Res. Inst. Texas 3: 239–243.
Phipps, J. B. 2012. Critical taxa in series *Molles* (*Crataegus,* Rosaceae): Typifications, new combinations, and taxonomic review. Phytoneuron 2012-78: 1–23.
Phipps, J. B. et al. 1990. A checklist of the subfamily Maloideae (Rosaceae). Canad. J. Bot. 68: 2209–2269.
Phipps, J. B. et al. 1991. Isozyme evidence for the naturalness of *Mespilus* L. (Rosaceae, subfam. Maloideae). Syst. Bot. 16: 546–552.
Phipps, J. B. et al. 1991b. Origins and evolution of subfam. Maloideae (Rosaceae). Syst. Bot. 16: 303–332.
Phipps, J. B. et al. 2006. *Crataegus* series *Bracteatae* and *Triflorae* (Rosaceae). Sida 22: 1009–1025.

Phipps, J. B. and K. A. Dvorsky. 2006. *Crataegus* ser. *Parvifoliae* and its putative hybrids in the southeastern United States. Sida 22: 423–445.

Phipps, J. B. and K. A. Dvorsky. 2006b. Review of *Crataegus* series *Pulcherrimae* (Rosaceae). Sida 22: 973–1007.

Phipps, J. B. and K. A. Dvorsky. 2007. Review of *Crataegus* series *Apricae,* ser. nov., and *C. flava* (Rosaceae). J. Bot. Res. Inst. Texas 1: 171–202.

Phipps, J. B. and K. A. Dvorsky. 2008. A taxonomic revision of *Crataegus* series *Lacrimatae* (Rosaceae). J. Bot. Res. Inst. Texas 2: 1101–1162.

Phipps, J. B. and R. J. O'Kennon. 2007. The hawthorns of the Cypress Hills, Alberta and Saskatchewan. J. Bot. Res. Inst. Texas 1: 1031–1090.

Phipps, J. B., R. J. O'Kennon, and R. W. Lance. 2003. Hawthorns and Medlars. Portland.

Phytographia—See: C. L. Willdenow 1794

Phytologia = Phytologia; Designed to Expedite Botanical Publication.

Phytotherapy Res. = Phytotherapy Research; an International Journal Devoted to Medical and Scientific Research on Plants and Plant Products.

Pl. Aequinoct.—See: A. von Humboldt and A. J. Bonpland [1805–]1808–1809[–1817]

Pl. Breed. Rev. = Plant Breeding Reviews.

Pl. Delavay.—See: A. R. Franchet 1889[–1890]

Pl. Diversity Evol. = Plant Diversity and Evolution.

Pl. Ecol. = Plant Ecology.

Pl. Hartw.—See: G. Bentham 1839[–1857]

Pl. Icon. Ined.—See: J. E. Smith 1789–1791

Pl. Nov. Thurb.—See: A. Gray 1854

Pl. Syst. Evol. = Plant Systematics and Evolution.

Pl. Wilson.—See: C. S. Sargent 1913–1917

PLOS Currents Tree Life = PLOS Currents Tree of Life. [Electronic journal.]

Pojar, J. 1999. Rosaceae. In: G. W. Douglas et al. 1998–2002. Illustrated Flora of British Columbia. 8 vols. Victoria. Vol. 4, pp. 258–370.

Pojarkova, A. I. 1971. *Spiraea.* In: V. L. Komarov et al., eds. 1963+. Flora of the U.S.S.R. (Flora SSSR). Translated from Russian. 30+ vols. Jerusalem etc. Vol. 9, pp. 218–235.

Poljakova, T. A. and G. N. Gataulina. 2008. Morphology and variability of pollen of the genus *Spiraea* L. (Rosaceae) in Siberia and the Far East. Contemp. Probl. Ecol. 1: 420–424.

Porsild, A. E. 1947. The genus *Dryas* in North America. Canad. Field-Naturalist 61: 175–192.

Porsild, A. E. 1964. Illustrated flora of the Canadian Arctic Archipelago, ed. 2. Bull. Natl. Mus. Canada 146: 1–218.

Porsild, A. E. and W. J. Cody. 1980. Vascular Plants of Continental Northwest Territories, Canada. Ottawa.

Porter, T. C. and J. M. Coulter. 1874. Synopsis of the Flora of Colorado.... Washington. (Syn. Fl. Colorado)

Potter, D. et al. 2002. Phylogenetic relationships in Rosaceae inferred from chloroplast *mat*K and *trn*L-*trn*F nucleotide sequence data. Pl. Syst. Evol. 231: 77–89.

Potter, D. et al. 2007. Phylogeny and classification of Rosaceae. Pl. Syst. Evol. 266: 5–43.

Potter, D. et al. 2007b. Phylogenetic relationships in tribe Spiraeeae (Rosaceae) inferred from nucleotide sequence data. Pl. Syst. Evol. 266: 105–118.

Power, N. 1980. *Crataegus*-Insect Relationships. Honors thesis. University of Western Ontario.

Précis Découv. Somiol.—See: C. S. Rafinesque 1814

Prelim. Cat.—See: N. L. Britton et al. 1888

Preslia = Preslia. Věstník (Časopis) Československé Botanické Společnosti.

Proc. Acad. Nat. Sci. Philadelphia = Proceedings of the Academy of Natural Sciences of Philadelphia.

Proc. Amer. Acad. Arts = Proceedings of the American Academy of Arts and Sciences.

Proc. Amer. Assoc. Advancem. Sci. = Proceedings of the American Association for the Advancement of Science.

Proc. Amer. Philos. Soc. = Proceedings of the American Philosophical Society.

Proc. Amer. Soc. Hort. Sci. = Proceedings of the American Society for Horticultural Science.

Proc. Biol. Soc. Wash. = Proceedings of the Biological Society of Washington.

Proc. Calif. Acad. Sci. = Proceedings of the California Academy of Sciences.

Proc. Indian Acad. Sci., Pl. Sci. = Proceedings, Indian Academy of Sciences. Plant Sciences.

Proc. Linn. Soc. London = Proceedings of the Linnean Society of London.

Proc. Rochester Acad. Sci. = Proceedings of the Rochester Academy of Science.

Prodr.—See: A. P. de Candolle and A. L. P. P. de Candolle 1823–1873; O. P. Swartz 1788

Prodr. Fl. Nepal.—See: D. Don 1825

Prodr. Monogr. Rosier—See: C. A. Thory 1820

Publ. Bot. Milwaukee Public Mus. = Publications in Botany, Milwaukee Public Museum.

Publ. Carnegie Inst. Wash. = Publications of the Carnegie Institution of Washington.

Pursh, F. [1813]1814. Flora Americae Septentrionalis; or, a Systematic Arrangement and Description of the Plants of North America. 2 vols. London. (Fl. Amer. Sept.)

Qian, G. Z., Liu L. F., and Tang G. G. 2010. Proposal to conserve the name *Malus domestica* against *M. pumila, M. communis, M. frutescens* and *Pyrus dioica* (Rosaceae). Taxon 59: 650–652.

Quest-Ritson, C. and B. Quest-Ritson. 2003. The Royal Horticultural Society Encyclopedia of Roses. London.

Radford, A. E., H. E. Ahles, and C. R. Bell. 1968. Manual of the Vascular Flora of the Carolinas. Chapel Hill.

Raeuschel, E. A. 1797. Nomenclator Botanicus.... Leipzig. (Nomencl. Bot.)

Rafinesque, C. S. 1814. Précis des Découvertes et Travaux Somiologiques.... Palermo. (Précis Découv. Somiol.)

Rafinesque, C. S. 1825. Neogenyton, or Indication of Sixty-six New Genera of Plants of North America. [Lexington, Ky.] (Neogenyton)

Rafinesque, C. S. 1828[–1830]. Medical Flora; or, Manual of the Medical Botany of the United States of North America. 2 vols. Philadelphia. (Med. Fl.)

Rafinesque, C. S. 1833. Herbarium Rafinesquianum. .... Herbals; or Botanical Collections of C. S. Rafinesque... the Labor of a Whole Life! 3 parts. Philadelphia. [Parts designated as extras of Atlantic J., paged consecutively.] (Herb. Raf.)

Rafinesque, C. S. 1836[–1838]. New Flora and Botany of North America.... 4 parts. Philadelphia. [Parts paged independently.] (New Fl.)

Rafinesque, C. S. 1840. Autikon Botanikon. 3 parts. Philadelphia. [Parts paged consecutively.] (Autik. Bot.)
Rahn, K. 1989. A survey of the genus *Sorbaria* (Rosaceae). Nordic J. Bot. 8: 557–563.
Randall, J. M. and J. Marinelli, eds. 1996. Invasive Plants: Weeds of the Global Garden. Brooklyn.
Raup, H. M. 1931. The genus *Geum* in the Athabasca-Great Slave Lake region. Rhodora 33: 172–176.
Raven, P. H. 1963. A flora of San Clemente Island, California. Aliso 5: 289–347.
Raynor, L. A. 1952. Cytotaxonomic studies of *Geum*. Amer. J. Bot. 39: 713–719.
Reagan, A. B. 1936. Plants used by the Hoh and Quileute Indians. Trans. Kansas Acad. Sci. 37: 55–70.
Redouté, P. J. 1817–1824. Les Roses. 3 vols. Paris. (Roses)
Rees, A. [1802–]1819–1820. The Cyclopaedia; or, Universal Dictionary of Arts, Sciences, and Literature.... 39 vols. in 79 parts. London. [Pages unnumbered.] (Cycl.)
Refug. Bot.—See: W. W. Saunders [1868–]1869–1873
Regel, E. and H. S. T. Tiling. 1858. Florula Ajanensis.... Moscow.
Rehder, A. J. 1927. Manual of Cultivated Trees and Shrubs Hardy in North America.... New York.
Rehder, A. J. 1940. Manual of Cultivated Trees and Shrubs Hardy in North America..., ed. 2. New York. (Man. Cult. Trees ed. 2)
Rehder, A. J. 1949. Bibliography of Cultivated Trees and Shrubs Hardy in the Cooler Temperate Regions of the Northern Hemisphere. Jamaica Plain.
Rehder, A. J. and E. H. Wilson. 1913. *Cotoneaster.* In: C. S. Sargent, ed. 1913–1917. Plantae Wilsonianae: An Enumeration of the Woody Plants Collected in Western China for the Arnold Arboretum...During the Years 1907, 1908, and 1910 by E. H. Wilson. 3 vols. in parts. Cambridge, Mass. Vol. 1, pp. 154–177.
Reichenbach, H. G. L. 1841. Der deutsche Botaniker.... Erster Band. Das Herbarienbuch. Dresden and Leipzig. [Alt. title: Repertorium Herbarii....] (Deut. Bot. Herb.-Buch)
Reichenbacher, F. W. 1994. Identification of *Purshia subintegra* (Rosaceae). Great Basin Naturalist 54: 256–271.
Rep. (Annual) Missouri Bot. Gard. = Report (Annual) of the Missouri Botanical Garden.
Rep. (Annual) Vermont Agric. Exp. Sta. = Report (Annual), Vermont Agricultural Experiment Station.
Rep. Exped. Zuni Colorado Rivers—See: L. Sitgreaves 1853
Repert. Spec. Nov. Regni Veg. = Repertorium Specierum Novarum Regni Vegetabilis.
Repert. Spec. Nov. Regni Veg. Beih. = Repertorium Specierum Novarum Regni Vegetabilis. Beihefte.
Rev. Hort. = Revue Horticole; Journal d'Horticulture Pratique.
Rev. Palaeobot. Palynol. = Review of Palaeobotany and Palynology; an International Journal.
Reveal, J. L. 1980. Intermountain biogeography—a speculative appraisal. Mentzelia 4: 1–92.
Reveal, J. L., G. E. Moulton, and A. E. Schuyler. 1999. The Lewis and Clark collections of vascular plants: Names, types and comments. Proc. Acad. Nat. Sci. Philadelphia 149: 1–64.
Revis. Gen. Pl.—See: O. Kuntze 1891–1898
Revis. Potentill.—See: J. G. C. Lehmann 1856
Rhein. Fl.—See: J. C. Döll 1843
Rhoads, A. F. and T. A. Block. 2000. The Plants of Pennsylvania: An Illustrated Manual. Philadelphia.
Rhoads, A. F. and W. M. Klein. 1993. The Vascular Flora of Pennsylvania: Annotated Checklist and Atlas. Philadelphia.
Rhodora = Rhodora; Journal of the New England Botanical Club.
Rich, T. C. G. and L. Houston. 2006. *Sorbus whiteana* (Rosaceae), a new endemic tree from Britain. Watsonia 26: 1–7.
Rich, T. C. G. and M. D. B. Rich. 1988. Plant Crib. London.
Rickett, H. W. 1944. The classification of inflorescences. Bot. Rev. (Lancaster) 10: 187–231.
Rieseberg, L. H. and D. Gerber. 1995. Hybridization in the Catalina Island mountain mahogany *(Cercocarpus traskiae):* RAPD evidence. Conservation Biol. 9: 199–203.
Robertson, A., A. C. Newton, and R. A. Ennos. 2004. Multiple hybrid origins, genetic diversity and population genetic structure of two endemic *Sorbus* taxa on the Isle of Arran, Scotland. Molec. Ecol. 13: 123–143.
Robertson, A., A. C. Newton, and R. A. Ennos. 2004b. Breeding systems and continuing evolution in the endemic *Sorbus* taxa on Arran. Heredity 93: 487–495.
Robertson, A. and C. Sydes. 2006. *Sorbus pseudomeinichii,* a new endemic *Sorbus* (Rosaceae) microspecies from Arran, Scotland. Watsonia 26: 9–14.
Robertson, K. R. 1974. The genera of Rosaceae in the southeastern United States. J. Arnold Arbor. 55: 303–332, 344–401, 611–662.
Robertson, K. R. 1983. Comment opposed to (670) proposal to conserve *Gillenia* (Rosaceae). Taxon 32: 306.
Robertson, K. R. et al. 1991. A synopsis of genera in Maloideae (Rosaceae). Syst. Bot. 16: 376–394.
Robertson, K. R., J. B. Phipps, and J. R. Rohrer. 1992. Summary of leaves in the genera of Maloideae (Rosaceae). Ann. Missouri Bot. Gard. 79: 81–94.
Robinson, B. L. and M. L. Fernald. 1908. Gray's New Manual of Botany: A Handbook of the Flowering Plants and Ferns of the Central and Northeastern United States and Adjacent Canada, ed. 7. New York, Cincinnati, and Chicago. (Manual ed. 7)
Robinson, W. A. 1982. Experimental taxonomy in the genus *Amelanchier.* II. Do the taxa in *Amelanchier* form an agamic complex? Rhodora 84: 85–100.
Robinson, W. A. and C. R. Partanen. 1980. Experimental taxonomy in the genus *Amelanchier.* I. A new look at the chromosome numbers of the *Amelanchier* species growing in the northeastern United States. Rhodora 82: 483–493.
Roemer, M. J. 1846–1847. Familiarum Naturalium Regni Vegetabilis Synopses Monographicae.... 4 fascs. Weimar. (Fam. Nat. Syn. Monogr.)
Rohrer, J. R. 2000. The sand cherry in Wisconsin and neighboring states. Michigan Bot. 39: 59–69.
Rohrer, J. R., M. A. O'Brien, and J. A. Anderson. 2008. Phylogenetic analysis of North American plums (*Prunus* sect. *Prunocerasus:* Rosaceae) based on nuclear *LEAFY* and *S6PDH* sequences. J. Bot. Res. Inst. Texas 2: 401–414.

Rohrer, J. R., K. R. Robertson, and J. B. Phipps. 1991. Variation in structure among fruits of Maloideae (Rosaceae). Amer. J. Bot. 78: 1617–1635.

Rohrer, J. R., K. R. Robertson, and J. B. Phipps. 1994. Floral morphology of Maloideae (Rosaceae) and its systematic relevance. Amer. J. Bot. 81: 574–581.

Roland, A. E. and E. C. Smith. 1966. The Flora of Nova Scotia. 2 vols. Halifax.

Ros. Monogr.—See: J. Lindley 1820

Rosac. Monogr.—See: L. Trattinnick 1823–1824

Rosatti, T. J. 1993. *Sorbus.* In: J. C. Hickman, ed. 1993. The Jepson Manual. Higher Plants of California. Berkeley, Los Angeles, and London. P. 976.

Roses—See: P. J. Redouté 1817–1824

Rousi, A. 1965. Biosystematic studies on the species aggregate *Potentilla anserina* L. Ann. Bot. Fenn. 2: 47–112.

Roxburgh, W. 1832. Flora Indica; or, Descriptions of Indian Plants. 3 vols. Serampore. (Fl. Ind. ed. 1832)

Rubo—See: C. P. Thunberg 1813

Rubtsov, G. A. 1944. Geographical distribution of the genus *Pyrus* and trends and factors in its evolution. Amer. Midl. Naturalist 78: 358–366.

Ruiz López, H. and J. A. Pavon. 1798–1802. Flora Peruviana, et Chilensis, sive Descripciones, et Icones Plantarum Peruvianarum, et Chilensium.... 3 vols. [Madrid.] (Fl. Peruv.)

Russell, P. G. 1934. The Oriental Flowering Cherries. Washington. [U.S.D.A., Circ. 313.]

Rydberg, P. A. 1896. Notes on *Potentilla.* I. Bull. Torrey Bot. Club 23: 244–248.

Rydberg, P. A. 1897. Notes on *Potentilla.* VI. Bull. Torrey Bot. Club 24: 1–13.

Rydberg, P. A. 1898. A Monograph of the North American Potentilleae. New York. [Mem. Dept. Bot. Columbia Coll. 2.] (Monogr. N. Amer. Potentilleae)

Rydberg, P. A. 1908. *Drymocallis.* In: N. L. Britton et al., eds. 1905+. North American Flora.... 47+ vols. New York. Vol. 22, pp. 366–376.

Rydberg, P. A. 1908b. *Poteridium.* In: N. L. Britton et al., eds. 1905+. North American Flora.... 47+ vols. New York. Vol. 22, p. 388.

Rydberg, P. A. 1908c. *Horkelia.* In: N. L. Britton et al., eds. 1905+. North American Flora.... 47+ vols. New York. Vol. 22, pp. 269–282.

Rydberg, P. A. 1908d. *Potentilla.* In: N. L. Britton et al., eds. 1905+. North American Flora.... 47+ vols. New York. Vol. 22, pp. 293–388.

Rydberg, P. A. 1908–1918. Rosaceae. In: N. L. Britton et al., eds. 1905+. North American Flora.... 47+ vols. New York. Vol. 22, pp. 239–533.

Rydberg, P. A. 1913. *Cercocarpus.* In: N. L. Britton et al., eds. 1905+. North American Flora.... 47+ vols. New York. Vol. 22, pp. 418–424.

Rydberg, P. A. 1913b. *Geum.* In: N. L. Britton et al., eds. 1905+. North American Flora.... 47+ vols. New York. Vol. 22, pp. 401–407.

Rydberg, P. A. 1913c. *Rubus.* In: N. L. Britton et al., eds. 1905+. North American Flora.... 47+ vols. New York. Vol. 22, pp. 428–480.

Rydberg, P. A. 1917. Flora of the Rocky Mountains and Adjacent Plains. New York. (Fl. Rocky Mts.)

Rydberg, P. A. 1918. *Rosa.* In: N. L. Britton et al., eds. 1905+. North American Flora.... 47+ vols. New York. Vol. 22, pp. 482–533.

Salamun, P. J. 1951. A population study of the variation in the inflorescence of *Spiraea tomentosa.* Rhodora 53: 280–292.

Sanso, A. M. and M. N. Seo. 2005. Chromosomes of some Argentine angiosperms and their taxonomic significance. Caryologia 58: 171–177.

Sargent, C. S. 1890–1902. The Silva of North America.... 14 vols. Boston and New York. (Silva)

Sargent, C. S. 1901. Notes on *Crataegus* in the Champlain Valley. Rhodora 3: 19–31.

Sargent, C. S. [1902–]1905–1913. Trees and Shrubs.... 2 vols. in parts. Boston and New York. (Trees & Shrubs)

Sargent, C. S. 1905. Manual of the Trees of North America (Exclusive of Mexico).... Boston and New York. (Man. Trees)

Sargent, C. S. 1907. *Crataegus* in Southern Michigan. Lansing. ["Published by the State Board of Geological Survey as a part of the Report for 1906 and as a contribution to the Biological Survey of the state authorized by Act 250, Session, 1905."] (Crataegus S. Michigan)

Sargent, C. S. 1908. *Crataegus* in Missouri.... St. Louis. [Preprinted from Rep. (Annual) Missouri Bot. Gard. 19: 35–126. 1908.] (Crataegus Missouri)

Sargent, C. S., ed. 1913–1917. Plantae Wilsonianae: An Enumeration of the Woody Plants Collected in Western China for the Arnold Arboretum...During the Years 1907, 1908, and 1910 by E. H. Wilson. 3 vols. in parts. Cambridge, Mass. [Parts paged consecutively within vols.] (Pl. Wilson.)

Sargent, C. S. 1920. *Crataegus* ser. *Aestivales.* In: Notes on North American trees. J. Arnold Arbor. 1: 248–252.

Saunders, W. W., ed. [1868–]1869–1873. Refugium Botanicum.... 5 vols. London. (Refug. Bot.)

Savile, D. B. O. 1968. Parasite relationships and disposition of *Filipendula.* Brittonia 20: 230–231.

Savolainen, V. et al. 2000b. Phylogeny of the eudicots: A nearly complete familial analysis based on *rbc*L gene sequences. Kew Bull. 55: 257–309.

Sax, K. 1936. Polyploidy and geographic distribution in *Spiraea.* J. Arnold Arbor. 17: 352–356.

Scand. J. Rheumatol. = Scandinavian Journal of Rheumatology.

Schaack, C. G. 1987. Lectotypification of *Cowania subintegra* Kearney, basionym of *Purshia subintegra* (Kearney) Henrickson (Rosaceae). Taxon 36: 452–454.

Schanzer, I. A. 1994. Taxonomic revision of the genus *Filipendula* Mill. (Rosaceae). J. Jap. Bot. 69: 290–319.

Schneider, C. K. [1904–]1906–1912. Illustriertes Handbuch der Laubholzkunde.... 2 vols. in 12 fascs. Jena. (Ill. Handb. Laubholzk.)

Schroeder, F.-G. 1968. Zur Nomenklatur in der Gattung *Amelanchier* (Rosaceae). Taxon 17: 633–634.

Schulze-Menz, G. K. W. 1964. Rosaceae. In: H. Melchior, ed. 1964. A. Engler's Syllabus der Pflanzenfamilien..., ed. 12. 2 vols. Berlin. Vol. 2, pp. 209–218.

Scoggan, H. J. 1978–1979. The Flora of Canada. 4 parts. Ottawa. [Natl. Mus. Nat. Sci. Publ. Bot. 7.]

Scrinia Fl. Select. = Scrinia Florae Selectae.

Seringe, N. C. 1818–1823. Musée Helvétique d'Histoire Naturelle. (Partie Botanique.) 4 parts in 3. Geneva. [Parts paged and plates numbered consecutively.] (Mus. Helv. Bot.)

Seymour, F. C. 1969. The Flora of New England: A Manual for the Identification of All Vascular Plants...Growing without Cultivation.... Rutland, Vt.

Shaw, J. and R. L. Small. 2004. Addressing the "hardest puzzle in American pomology": Phylogeny of *Prunus* sect. *Prunocerasus* (Rosaceae) based on seven noncoding chloroplast DNA regions. Amer. J. Bot. 91: 985–996.

Shaw, J. and R. L. Small. 2005. Chloroplast DNA phylogeny and phylogeography of the North American plums (*Prunus* subgenus *Prunus* section *Prunocerasus,* Rosaceae). Amer. J. Bot. 92: 2011–2030.

Shevock, J. R., B. Ertter, and D. W. Taylor. 1992. *Neviusia cliftonii* (Rosaceae: Kerrieae), an intriguing new relict species from California. Novon 2: 285–289.

Shinners, L. H. 1956b. *Prunus americana* var. *lanata,* a synonym of *P. nigra.* Rhodora 58: 330–331.

Shorthouse, J. D. 1991. An unusual population of galls of *Diplolepis polita* (Hymenoptera, Cynipidae) in the Cypress Hills of southeastern Alberta. Canad. Field-Naturalist 105: 542–549.

Sida = Sida; Contributions to Botany.

Siebold, P. F. von and J. G. Zuccarini. 1835[–1870]. Flora Japonica.... 2 vols. in 30 parts. Leiden. [Vols. paged and parted independently, plates numbered consecutively.] (Fl. Jap.)

Silva—See: C. S. Sargent 1890–1902

Silverside, A. J. 1988. *Spiraea.* In: T. C. G. Rich and M. D. B. Rich, comps. 1988. Plant Crib. London. Pp. 46–48.

Silverside, A. J. 1990. The nomenclature of some hybrids of the *Spiraea salicifolia* group naturalized in Britain. Watsonia 18: 147–151.

Simmons, S. L. 2007. Staphyleaceae. In: K. Kubitzki et al., eds. 1990+. The Families and Genera of Vascular Plants. 10+ vols. Berlin etc. Vol. 9, pp. 440–445.

Sinnott, Q. P. and J. B. Phipps. 1983. Variation patterns in *Crataegus* series *Pruinosae* (Rosaceae) in southern Ontario. Syst. Bot. 8: 59–70.

Sitgreaves, L. 1853. Report of an Expedition down the Zuni and Colorado Rivers.... Washington. [Botany by J. Torrey, pp. 153–178, plates 1–21.] (Rep. Exped. Zuni Colorado Rivers)

Skalický, V. 1973. Amerikanischer Arten der Gattung *Agrimonia* L. ser. *Tuberosae* ser. nova. Folia Geobot. Phytotax. 8: 95–104.

Sketch Bot. S. Carolina—See: S. Elliott [1816–]1821–1824

Skr. Naturhist.-Selsk. = Skrivter af Naturhistorie-Selskabet.

Skr. Norske Vidensk.-Akad. Oslo, Mat.-Naturvidensk. Kl. = Skrifter Utgitt av det Norske Videnskaps-Akademi i Oslo. Matematisk-naturvidenskapelig Klasse.

Skrede, I. et al. 2006. Refugia, differentiation and postglacial migration in arctic-alpine Eurasia, exemplified by the mountain-avens (*Dryas octopetala* L.). Molec. Ecol. 15: 1827–1840.

Small, J. K. 1903. Flora of the Southeastern United States.... New York. (Fl. S.E. U.S.)

Small, J. K. 1933. Manual of the Southeastern Flora, Being Descriptions of the Seed Plants Growing Naturally in Florida, Alabama, Mississippi, Eastern Louisiana, Tennessee, North Carolina, South Carolina and Georgia. New York. (Man. S.E. Fl.)

Smedmark, J. E. E. 2006. Recircumscription of *Geum* (Colurieae: Rosaceae). Bot. Jahrb. Syst. 126: 409–417.

Smedmark, J. E. E. et al. 2003. Ancient allopolyploid speciation in Geinae (Rosaceae): Evidence from nuclear granule-bound starch synthase (GBSSI) gene sequences. Syst. Biol. 52: 374–385.

Smedmark, J. E. E. and T. Eriksson. 2002. Phylogenetic relationships of *Geum* (Rosaceae) and relatives inferred from the nrITS and *trn*L-*trn*F regions. Syst. Bot. 27: 303–317.

Smedmark, J. E. E. and T. Eriksson. 2006. Early stages of development shed light on fruit evolution in allopolyploid species of *Geum* (Rosaceae). Int. J. Pl. Sci. 167: 791–803.

Smith, E. B. 1994. Keys to the Flora of Arkansas. Fayetteville, Ark.

Smith, G. L. and C. R. Wheeler. 1990–1991. A flora of the vascular plants of Mendocino County, California. Wasmann J. Biol. 48–49(1–2).

Smith, Hu. H. 1923. Ethnobotany of the Menomini Indians. Bull. Public Mus. Milwaukee 4: 1–82.

Smith, Hu. H. 1928. Ethnobotany of the Meskwaki Indians. Bull. Public Mus. Milwaukee 4: 189–274.

Smith, Hu. H. 1932. Ethnobotany of the Ojibwe Indians. Bull. Public Mus. Milwaukee 4: 348–433.

Smith, Hu. H. 1933. Ethnobotany of the Forest Potawatomi Indians. Bull. Public Mus. Milwaukee 7: 32–1272.

Smith, J. E. 1789–1791. Plantarum Icones hactenus Ineditae.... 3 fascs. London. [Fascicles paged independently, plates numbered consecutively.] (Pl. Icon. Ined.)

Smith, J. E. 1800–1804. Flora Britannica. 3 vols. London. [Vols. paged consecutively. Vols. 1 and 2, 1800; vol. 3, 1804.] (Fl. Brit.)

Smith, J. E. et al. 1790–1866. English Botany; or, Coloured Figures of British Plants.... 36 vols. + 5 suppls. London. (Engl. Bot.)

Smith, W. R. 2008. Trees and Shrubs of Minnesota. Minneapolis.

Smithsonian Contr. Knowl. = Smithsonian Contributions to Knowledge.

Soják, J. 1969. Nomenklatorische Anmerkungen zur Gattung *Potentilla.* Folia Geobot. Phytotax. 4: 205–209.

Soják, J. 1985[1989]. Generická problematika *Potentilla* s.l. Čas. Nár. Muz. Praze, Rada Přír. 154: 117–118.

Soják, J. 1986. Notes on *Potentilla.* I. Hybridogenous species derived from intersectional hybrids of sect. *Niveae* × sect. *Multifidae.* Bot. Jahrb. Syst. 106: 145–210.

Soják, J. 1987. Notes on *Potentilla paradoxa* and *P. supina.* Preslia 59: 271–272.

Soják, J. 1987b. Notes on *Potentilla.* IV. Classification of Wolf's group *"Potentilla trichocarpae."* Candollea 42: 491–500.

Soják, J. 1987c. Notes on *Potentilla.* V. *Potentilla pensylvanica* group in the Old World. Preslia 59: 289–305.

Soják, J. 1989. Notes on *Potentilla* (Rosaceae) VIII. *P. nivea* agg. Candollea 44: 741–762.

Soják, J. 1993. Taxonomische Bemerkungen zu einigen mediterranean *Potentilla*-Sippen. Preslia 65: 117–118.

Soják, J. 1994. Notes on *Potentilla* (Rosaceae). X. The section *Dumosae.* XI. The *P. microphylla* and *P. stenophylla* groups (sect *Pentaphylloides*). XII. Key to the taxa of *P.* sect. *Petaphylloides (Anserina).* Bot. Jahrb. Syst. 116: 11–81.

Soják, J. 1996. Notes on *Potentilla* (Rosaceae). XIV. Type specimens in the Lehmann herbarium. Preslia 68: 97–124.

Soják, J. 2004. *Potentilla* L. (Rosaceae) and related genera in the former USSR (identification key, checklist and figures). Notes on *Potentilla* XVI. Bot. Jahrb. Syst. 125: 253–340.

Soják, J. 2005. *Potentilla* L. s.l. (Rosaceae) in flora Europae orientalis (notes on *Potentilla* XVIII). Candollea 60: 59–78.

Soják, J. 2006. New infraspecific nomenclatural combinations in twelve American species of *Drymocallis* and *Potentilla* (Rosaceae). Thaiszia 16: 47–50.

Soják, J. 2007. *Potentilla* (Rosaceae) in China. Notes on *Potentilla* XIX. Harvard Pap. Bot. 12: 285–324.

Soják, J. 2008. Notes on *Potentilla* XXI. A new division of the tribe Potentilleae (Rosaceae) and notes on generic delimitations. Bot. Jahrb. Syst. 127: 349–358.

Soják, J. 2008b. Proposals to conserve the name *Potentilla bipinnatifida* against *P. normalis* and *P. missourica* and the name *P. stolonifera* against *P. sprengeliana* (Rosaceae). Taxon 57: 1361–1362.

Soják, J. 2009. Proposal to conserve the name *Potentilla verna* (Rosaceae) with a conserved type. Taxon 58: 652–653.

Soják, J. 2009b. *Potentilla* L. (Rosaceae) in the former USSR; second part: Comments. Notes on *Potentilla* XXIV. Feddes Repert. 120: 185–217.

Soják, J. 2010. *Argentina* Hill, a genus distinct from *Potentilla* (Rosaceae). Thaiszia 20: 91–97.

Soják, J. 2011. Synopsis of *Drymocallis* Fourr. ex Rydb. (Rosaceae-Potentilleae) in the Old World. Ann. Naturhist. Mus. Wien, B 112: 319–328.

Soják, J. 2012. *Potentilla* L. (Rosaceae) and related genera in Asia (excluding the former USSR), Africa and New Guinea. Notes on *Potentilla* XXVIII. Pl. Diversity Evol. 130: 7–157.

Soltis, D. E. et al. 2000. Angiosperm phylogeny inferred from 18S rDNA, *rbc*L, and *atp*B sequences. Bot. J. Linn. Soc. 133: 381–461.

Sorrie, B. A. 2005. Alien vascular plants in Massachusetts. Rhodora 107: 284–329.

Sosa, V. and M. W. Chase. 2003. Phylogenetics of Crossosomataceae based on *rbc*L sequence data. Syst. Bot. 28: 96–105.

SouthW. Naturalist = Southwestern Naturalist.

Sp. Pl.—See: C. Linnaeus 1753

Sp. Pl. ed. 2—See: C. Linnaeus 1762–1763

Spach, É. 1834–1848. Histoire Naturelle des Végétaux. Phanérogames.... 14 vols., atlas. Paris. (Hist. Nat. Vég.)

Sperduto, D. D. 2005. Natural Community Systems of New Hampshire. Report from the New Hampshire Natural Heritage Bureau and The Nature Conservancy. Concord, N.H.

Spongberg, S. A. 1971b. The Staphyleaceae in the southeastern United States. J. Arnold Arbor. 52: 196–203.

St. John, H. 1937. Flora of Southeastern Washington and of Adjacent Idaho. Pullman. (Fl. S.E. Washington)

St. John, H. 1942. Nomenclatorial changes in *Glossopetalon* (Celastraceae). Proc. Biol. Soc. Wash. 55: 109–112.

Stafleu, F. A. et al. 1992–2009. Taxonomic Literature: A Selective Guide to Botanical Publications and Collections with Dates, Commentaries and Types. Supplement. 8 vols. Königstein.

Stafleu, F. A. and R. S. Cowan. 1976–1988. Taxonomic Literature: A Selective Guide to Botanical Publications and Collections with Dates, Commentaries and Types, ed. 2. 7 vols. Utrecht, Antwerp, The Hague, and Boston.

Stansbury, H. 1852. An Expedition to the Valley of the Great Salt Lake of Utah.... Philadelphia. [Botanical appendix by J. Torrey, pp. 381–397, plates 1–9.] (Exped. Great Salt Lake)

Starostenko, N. T. and V. N. Starostenko. 1971. Therapeutic effect of the silver-leaved cinquefoil in cirrhosis of the liver in the edema-ascitic stage. Zdravookhranenie (Kishinev) 14(6): 15–19.

Staudt, G. 1962. Taxonomic studies in the genus *Fragaria*. Typification of *Fragaria* species known at the time of Linnaeus. Canad. J. Bot. 40: 869–886.

Staudt, G. 1999. Systematics and geographic distribution of the American strawberry species. Taxonomic studies in the genus *Fragaria* (Rosaceae: Potentilleae). Univ. Calif. Publ. Bot. 81.

Stebbins, G. L. 1959. The role of hybridization in evolution. Proc. Amer. Philos. Soc. 103: 231–251.

Steele, F. L. 1964. *Potentilla robbinsiana* in the White Mountains of New Hampshire. Rhodora 66: 408–411.

Stefánsson, S. 1901. Flóra Íslands.... Kaupmannahöfn. (Fl. Íslands)

Sterling, C. 1966. Comparative morphology of the carpel in the Rosaceae. VIII. Spiraeoideae: Holodisceae, Neillieae, Spiraeeae, Ulmarieae. Amer. J. Bot. 53: 521–530.

Stevens, O. A. 1959. Index. In: N. L. Britton et al., eds. 1905+. North American Flora.... 47+ vols. New York. Vol. 22, part 7, pp. 1–22.

Steyermärk. Z. = Steyermärkische Zeitschrift.

Stirp. Austr. Fasc.—See: H. J. N. von Crantz 1762–1767

Strausbaugh, P. D. and E. L. Core. 1978. Flora of West Virginia, ed. 2 corr. Grantsville, W.Va.

Stud. Fl. New Zealand—See: T. Kirk [1899]

Stutz, H. C. and L. K. Thomas. 1963. Hybridization and introgression in *Cowania* and *Purshia*. Evolution 18: 183–195.

Suppl. Calif. Fl.—See: P. A. Munz 1968

Suppl. Meth.—See: C. Moench 1802

Suppl. Pl.—See: C. Linnaeus f. 1781[1782]

Svensk Bot. Tidskr. = Svensk Botanisk Tidskrift Utgifven af Svenska Botaniska Föreningen.

Swartz, O. P. 1788. Nova Genera & Species Plantarum seu Prodromus.... Stockholm, Uppsala, and Åbo. (Prodr.)

Swartz, O. P. 1797–1806. Flora Indiae Occidentalis.... 3 vols. Erlangen and London. (Fl. Ind. Occid.)

Sweet, R. 1818. Hortus Suburbanus Londinensis.... London. (Hort. Suburb. Lond.)

Sweet, R. 1823–1837. The British Flower Garden.... 7 vols. London. [Vols. 4–7 also issued as vols. 1–4 of series 2.] (Brit. Fl. Gard.)

Swink, F. and G. S. Wilhelm. 1994. Plants of the Chicago Region: An Annotated Checklist of the Vascular Flora..., ed. 4. Indianapolis.

Symb. Antill.—See: I. Urban 1898–1928
Syn. Fl. Colorado—See: T. C. Porter and J. M. Coulter 1874
Syn. Fl. N. Amer.—See: A. Gray et al. 1878–1897
Syn. Mitteleur. Fl.—See: P. F. A. Ascherson et al. 1896–1939
Syn. Pl.—See: D. N. F. Dietrich 1839–1852; C. H. Persoon 1805–1807
Syn. Pl. Fl. Gall.—See: J. Lamarck and A. P. de Candolle 1806
Syst. Biol. = Systematic Biology.
Syst. Bot. = Systematic Botany; Quarterly Journal of the American Society of Plant Taxonomists.
Syst. Bot. Monogr. = Systematic Botany Monographs; Monographic Series of the American Society of Plant Taxonomists.
Syst. Nat. ed. 12—See: C. Linnaeus 1766[–1768]
Syst. Veg. ed. 14—See: J. A. Murray 1784
Tabl. École Bot. ed. 3—See: R. L. Desfontaines 1829[–1832]
Takhtajan, A. L. 1997. Diversity and Classification of Flowering Plants. New York.
Talent, N. and T. A. Dickinson. 2005. Polyploidy in *Crataegus* and *Mespilus* (Rosaceae, Maloideae): Evolutionary inferences from flow cytometry of nuclear DNA amounts. Canad. J. Bot. 83: 1268–1304.
Tausch, I. F. 1823–1825. Hortus Canalius.... 2 parts. Prague. (Hort. Canal.)
Taxon = Taxon; Journal of the International Association for Plant Taxonomy.
Tennessee Exotic Plant Pest Council. 2009. Invasive exotic pest plants in Tennessee—2009. Wildland Weeds 13: 13–16.
Teppner, H. 1968. Zur Kenntnis der Gattung *Waldsteinia*. Graz.
Terpó, A. 1985. Studies in the taxonomy and grouping of *Pyrus* species. Feddes Repert. 96: 73–87.
Thaiszia = Thaiszia; Journal of Botany.
Theor. Appl. Genet. = Theoretical and Applied Genetics; International Journal of Breeding Research and Cell Genetics.
Theor. Prakt. Handb. Forstbot.—See: M. B. Borkhausen 1800–1803
Thomas, J. H. 1961. Flora of the Santa Cruz Mountains of California, a Manual of the Vascular Plants. Stanford.
Thomas, W. W. 1988. A conspectus of Mexican and Central American *Picramnia* (Simaroubaceae). Brittonia 40: 89–105.
Thomas, W. W. 1990. The American genera of Simaroubaceae and their distribution. Acta Bot. Brasil. 4: 11–18.
Thompson, M. M. 1997. Survey of chromosome numbers in *Rubus* (Rosaceae: Rosoideae). Ann. Missouri Bot. Gard. 84: 128–164.
Thorne, R. F. and R. Scoggin. 1978. *Forsellesia* Greene (*Glossopetalon* Gray), a third genus in the Crossosomataceae, Rosineae, Rosales. Aliso 9: 171–178.
Thory, C. A. 1820. Prodrome de la Monographie des Espèces et Variétés Connues du Genre Rosier.... Paris. (Prodr. Monogr. Rosier)
Thunberg, C. P. 1813. Dissertationem Botanico-medicam de *Rubo*. Uppsala. (Rubo)
Tidestrom, I. and T. Kittell. 1941. A Flora of Arizona and New Mexico.... Washington.
Tiedemann, A. R. and K. L. Johnson, comps. 1983. Proceedings, Research and Management of Bitterbrush and Cliffrose in Western North America, Salt Lake City, Utah, April 13–15, 1982. Ogden.
Tolmatchew, A. I., ed. 1960–1987. Flora Arctica URSS. 10 vols. in 12. Moscow and Leningrad. (Fl. Arct. URSS)
Töpel, M. H. et al. 2011. Molecular data and ploidal levels indicate several putative allopolyploidization events in the genus *Potentilla* (Rosaceae). PLOS Currents Tree Life 18 May. Ed. 1. DOI: 10.1371/currents.RRN1237.
Töpel, M. H. et al. 2012. Past climate change and plant evolution in western North America: A case study in Rosaceae. PLOS ONE 7 Dec: DOI 10.1371/journal.pone.0050358.
Torrey, J. and A. Gray. 1838–1843. A Flora of North America.... 2 vols. in 7 parts. New York, London, and Paris. (Fl. N. Amer.)
Torreya = Torreya; a Monthly Journal of Botanical Notes and News.
Trans. & Proc. New Zealand Inst. = Transactions and Proceedings of the New Zealand Institute.
Trans. Kansas Acad. Sci. = Transactions of the Kansas Academy of Science.
Trans. Linn. Soc. = Transactions of the Linnean Society.
Trans. Linn. Soc. London = Transactions of the Linnean Society of London.
Trattinnick, L. 1823–1824. Rosacearum Monographia.... 4 vols. Vienna. (Rosac. Monogr.)
Travis, S. E., J. E. Baggs, and J. Maschinski. 2008. Disentangling the role of hybridization in the evolution of the endangered Arizona cliffrose (*Purshia subintegra;* Rosaceae): A molecular and morphological analysis. Conservation Genet. 9: 1183–1194.
Trees & Shrubs—See: C. S. Sargent [1902–]1905–1913
Trees Shrubs Brit. Isles—See: W. J. Bean 1914–1933
Tremblay, N. and D. Schoen. 1999. Molecular phylogeography of *Dryas integrifolia:* Glacial refugia and postglacial recolonization. Molec. Ecol. 8: 1187–1198.
Trudy Bot. Inst. Akad. Nauk S.S.S.R, Ser. 1, Fl. Sist. Vyssh. Rast. = Trudy Botanicheskogo Instituta Akademii Nauk S S S R. Ser. 1, Flora i Sistematika Vysshikh Rastenii.
Trudy Glavn. Bot. Sada = Trudy Glavnago Botanicheskago Sada.
Trudy Imp. S.-Peterburgsk. Bot. Sada = Trudy Imperatorskago S.-Peterburgskago Botanicheskago Sada.
Tuinb.-fl.—See: W. H. de Vriese 1855–1856[–1857]
Turner, N. J. 1988. "The importance of a rose": Evaluating the cultural significance of plants in Thompson and Lillooet Interior Salish. Amer. Anthropol. 90: 272–290.
Tutin, T. G. et al., eds. 1964–1980. Flora Europaea. 5 vols. Cambridge.
Tzvelev, N. N. 2007. On the genus *Comarum* (Rosaceae). Bot. Zhurn. (Moscow & Leningrad) 92: 894–900.
U.S. Expl. Exped.—See: C. Wilkes et al. 1854–1876
Univ. Calif. Publ. Bot. = University of California Publications in Botany.
Univ. Colorado Stud., Ser. Biol. = University of Colorado Studies. Series in Biology.
University of Chicago Press. 1993. The Chicago Manual of Style, ed. 14. Chicago.

Urban, I., ed. 1898–1928. Symbolae Antillanae seu Fundamenta Florae Indiae Occidentalis.... 9 vols. Berlin etc. (Symb. Antill.)

Uttal, L. J. 1973. The scientific name of the Alaska *Spiraea.* Bull. Torrey Bot. Club 100: 236–237.

Uttal, L. J. 1974. The varieties of *Spiraea betulifolia.* Bull. Torrey Bot. Club 101: 35–36.

Vanden Heuvel, B. D. 2002. Molecular Systematics of *Cercocarpus* H.B.K. (Rosaceae). Ph.D. dissertation. University of Texas.

Vasc. Pl. Pacif. N.W.—See: C. L. Hitchcock et al. 1955–1969

Veg. Europ.—See: J. N. Laicharding 1790–1791

Veg. Mater. Med. U.S.—See: W. P. C. Barton 1817–1819

Vegetatio = Vegetatio; Acta Geobotanica.

Verh. Vereins Beförd. Gartenbaues Königl. Preuss. Staaten = Verhandlungen des Vereins zur Beförderung des Gartenbaues in den Königlich preussischen Staaten.

Vidal, J. E. 1965. Révision du genre *Eriobotrya* (Pomoideae). Adansonia 5: 537–580.

Vidensk. Meddel. Naturhist. Foren. Kjøbenhavn = Videnskabelige Meddelelser fra den Naturhistoriske Forening i Kjøbenhavn.

Villars, D. 1786–1789. Histoire des Plantes de Dauphiné. 3 vols. Grenoble, Lyon, and Paris. (Hist. Pl. Dauphiné)

Vilmorin, M. and D. G. J. M. Bois. 1904. Fruticetum Vilmorinianum. Paris. (Frutic. Vilmor.)

Vincent, M. A. 2005. On the spread and current distribution of *Pyrus calleryana* in the United States. Castanea 70: 20–31.

Vincent, M. A. and A. W. Cusick. 1998. New records of alien species in the Ohio vascular flora. Ohio J. Sci. 98(2): 10–17.

Vines, R. A. 1960. Trees, Shrubs, and Woody Vines of the Southwest. Austin.

Visiani, R. de. 1842–1852. Flora Dalmatica.... 3 vols. Leipzig. (Fl. Dalmat.)

Voss, E. G. 1972–1996. Michigan Flora.... 3 vols. Bloomfield Hills and Ann Arbor.

Voss, E. G. 1986. General Committee report 1982–1985. Taxon 35: 551–552.

Vriese, W. H. de. 1855–1856[–1857]. Tuinbouw-flora van Nederland.... 3 vols. Leiden. (Tuinb.-fl.)

Waldstein, F. and P. Kitaibel. [1799–]1802–1812. Descriptiones et Icones Plantarum Rariorum Hungariae. 3 vols. in parts. Vienna. (Descr. Icon. Pl. Hung.)

Walker, C. A. 1996. The Cherokee Rose in the Southeastern United States: A Historical Perspective. Ph.D. dissertation. North Carolina State University.

Walker, C. A. and D. J. Werner. 1997. Isozyme and randomly amplified polymorphic DNA (RAPD) analyses of Cherokee rose and its putative hybrids 'Silver Moon' and 'Anemone'. J. Amer. Soc. Hort. Sci. 122: 659–664.

Wallroth, C. F. W. 1842–1844. Beiträge zur Botanik. 1 vol. in 2 parts. Leipzig. [Parts paged consecutively.] (Beitr. Bot.)

Walter, T. 1788. Flora Caroliniana, Secundum Systema Vegetabilium Perillustris Linnaei Digesta.... London. (Fl. Carol.)

Walters, S. M. et al., eds. 1984+. The European Garden Flora.... 6+ vols. Cambridge etc.

War Department [U.S.]. 1855–1860. Reports of Explorations and Surveys, to Ascertain the Most Practicable and Economical Route for a Railroad from the Mississippi River to the Pacific Ocean. Made under the Direction of the Secretary of War, in 1853[–1856].... 12 vols. in 13. Washington. (Pacif. Railr. Rep.)

Ward, D. B. 2009. Keys to the flora of Florida. 21. *Crataegus* (Rosaceae). Phytologia 91: 18–25.

Wasmann J. Biol. = Wasmann Journal of Biology.

Watson, S. 1871. United States Geological Expolration [sic] of the Fortieth Parallel. Clarence King, Geologist-in-charge. [Vol. 5] Botany. By Sereno Watson.... Washington. [Botanical portion of larger work by C. King.] [Botany (Fortieth Parallel)]

Watsonia = Watsonia; Journal of the Botanical Society of the British Isles.

Wawra, H. 1883–1888. Itinera Principum S. Coburgi. 2 vols. Vienna. (Itin. Princ. S. Coburgi)

Weakley, A. S. 2012. Flora of the Southern and Mid-Atlantic States. Working Draft of 30 November 2012. Chapel Hill.

Weakley, A. S., J. C. Ludwig, and J. F. Townsend. 2012. Flora of Virginia. Fort Worth.

Webb, R. H., J. W. Steiger, and R. M. Turner. 1987. Dynamics of Mojave Desert shrub assemblages in the Panamint Mountains, California. Ecology 68: 478–490.

Weber, H. E. 1985. *Rubi* Westfälici. Abh. Westfäl. Mus. Naturk. 47(3).

Weber, J. E. and C. S. Campbell. 1989. Breeding system of a hybrid between a sexual and an apomictic species of *Amelanchier,* shadbush (Rosaceae, Maloideae). Amer. J. Bot. 73: 341–347.

Weber, W. A. and R. C. Wittmann. 1996. Colorado Flora: Western Slope, ed. 2. Niwot, Colo.

Weinmann, F. C. 2000. King County Checklist of Vascular Plants. Issaquah.

Wells, T. C. and J. B. Phipps. 1989. Studies in *Crataegus* (Rosaceae: Maloideae), XX. Interserial hybridization between *C. monogyna* Jacq. (series *Oxyacanthae*) and *C. punctata* Jacq. (series *Punctatae*). Canad. J. Bot. 56: 2465–2472.

Welsh Botanol.—See: H. Davies 1813

Welsh, S. L. 1974. Anderson's Flora of Alaska and Adjacent Parts of Canada. Provo.

Welsh, S. L. et al., eds. 1993. A Utah Flora, ed. 2. Provo.

Wendland, J. C. 1798. Botanische Beobachtungen.... Hannover. (Bot. Beob.)

Werlemark, G. 2000. Genetic variability and reproductive strategies in Nordic dogroses, *Rosa* section *Caninae.* Acta Univ. Agric. Suec., Agrar. 257.

Werner, P. A. and J. D. Soule. 1976. The biology of Canadian weeds. 18. *Potentilla recta* L., *P. norvegica* L., and *P. argentea* L. Canad. J. Pl. Sci. 56: 591–603.

Westwood, M. N. and H. O. Bjornstad. 1971. Some fruit characteristics of interspecific hybrids and self fertility in *Pyrus.* Bull. Torrey Bot. Club 98: 22–24.

Widrlechner, M. P. 1998. The genus *Rubus* in Iowa. Castanea 63: 415–465.

Wiegand, K. M. 1912. The genus *Amelanchier* in eastern North America. Rhodora 14: 117–161.

Wiegand, K. M. 1920. Additional notes on *Amelanchier.* Rhodora 22: 146–151.

Wight, W. F. 1915. Native American species of *Prunus.* Bull. U.S.D.A. 179: 1–75.

Wildland Weeds = Wildland Weeds; a Quarterly Publication of the Florida Exotic Pest Plant Council.

Wilkes, C. et al. 1854–1876. United States Exploring Expedition. During the years 1838, 1839, 1840, 1841, 1842. Under the Command of Charles Wilkes, U.S.N..... 18 vols. (1–17, 19). Philadelphia. [Vol. 15: Botany, Phanerogamia (A. Gray), 1854; Atlas, 1856. Vol. 16: Botany, Cryptogamia, Filices (W. D. Brackenridge), 1854; Atlas, 1855. Vol. 17: incl. Phanerogamia of Pacific North America (J. Torrey), 1874. Vol. 19 (2 parts): Geographical Distribution of Animals and Plants (C. Pickering), 1854. Vol. 18: Botany, Phanerogamia, part 2 (A. Gray) not published.] (U.S. Expl. Exped.)

Willdenow, C. L. 1794. Phytographia seu Descriptio Rariorum Minus Cognitarum Plantarum. 1 fasc. only. Erlangen. (Phytographia)

Willdenow, C. L. 1803–1816. Hortus Berolinensis.... 2 vols. in 10 fascs. Berlin. [Fascs. and plates numbered consecutively.] (Hort. Berol.)

Willdenow, C. L. 1809–1813[–1814]. Enumeratio Plantarum Horti Regii Botanici Berolinensis.... 2 parts + suppl. Berlin. [Parts paged consecutively.] (Enum. Pl.)

Williams, Dave, B. Butler, and Daryl Smith, eds. 2012. Proceedings of the 22nd North American Prairie Conference. Cedar Falls.

Williams, K., L. J. Westrick, and B. J. Williams. 2006. Effects of blackberry *(Rubus discolor)* invasion on oak population dynamics in a California savanna. Forest Ecol. Managem. 228: 187–196.

Williams, K. B. 1971. Ecological and Morphological Variations of *Vauquelinia californica* (Torr.) Sarg. Populations in Arizona. Ph.D. dissertation. University of Arizona.

Winther, K., K. Apel, and G. Thamsborg. 2005. A powder made from seeds and shells of a rose-hip subspecies *(Rosa canina)* reduces symptoms of knee and hip osteoarthritis: A randomized, double-blind, placebo-controlled clinical trial. Scand. J. Rheumatol. 34: 302–308.

Wiśniewska, J., O. Kucewicz, and B. Machnik. 1969. *Sorbopirus* fruit structure in light of comparative study. Bull. Acad. Polon. Sci., Cl. 2, Sér. Sci. Biol. 17: 617–622.

Wiss. Z. Friedrich-Schiller-Univ. Jena, Math.-Naturwiss. Reihe = Wissenschaftliche Zeitschrift der Friedrich-Schiller-Universität Jena/Thüringen. Mathematisch-naturwissenschaftliche Reihe.

Wiss. Z. Martin-Luther-Univ. Halle-Wittenberg, Math.-Naturwiss. Reihe = Wissenschaftliche Zeitschrift der Martin-Luther-Universität Halle-Wittenberg. Mathematisch-naturwissenschaftliche Reihe.

Wissemann, V. and C. M. Ritz. 2005. The genus *Rosa* (Rosoideae, Rosaceae) revisited: Molecular analysis of nrITS-1 and *atp*B-*rbc*L intergenic spacer (IGS) versus conventional taxonomy. Bot. J. Linn. Soc. 147: 275–290.

Wolf, T. 1908. Monographie der Gattung *Potentilla.* Biblioth. Bot. 71.

Wolfe, J. A. and W. C. Wehr. 1988. Rosaceous *Chamaebatiaria*-like foliage from the paleogene of western North America. Aliso 12: 177–200.

Wood, A. 1861. A Class-book of Botany.... New York. (Class-book Bot. ed. s.n.(b))

Wu, S. et al. 2001. Phylogenetic analysis of Japanese *Rosa* species using DNA sequence of nuclear ribosomal internal transcribed spacers (ITS). J. Hort. Sci. Biotechnol. 76: 127–132.

Wu, Z. and P. H. Raven, eds. 1994+. Flora of China. 20+ vols. Beijing and St. Louis.

Wunderlin, R. P. 1998. Guide to the Vascular Plants of Florida. Gainesville.

Yi, O. et al. 2007. Antioxidant and antimicrobial activities of native *Rosa* sp. from British Columbia, Canada. Int. J. Food Sci. Nutr. 58: 178–189.

Yoshizawa, Y. et al. 2000. Differentiation-inducing effects of small fruit juices on HL-60 leukemic cells. J. Agric. Food Chem. 48: 3177–3182.

Yoshizawa, Y. et al. 2000b. Antiproliferative effects of small fruit juices on several cancer cell lines. Anticancer Res. 20: 4285–4289.

Young, J. A. and C. D. Clements. 2002. *Purshia:* The Wild and Bitter Roses. Reno.

Yu, T. T. 1954. Cotoneasters from the eastern Himalaya. Bull. Brit. Mus. (Nat. Hist.), Bot. 1: 125–141.

Zhang, Zhao Y. et al. 2006. Alkaloid polymorphisms and ITS sequence variation in the *Spiraea japonica* complex (Rosaceae) in China: Traces of the biological effects of the Himalayan-Tibet plateau uplift. Amer. J. Bot. 93: 762–769.

Zhang, Zhao Y., Sun H., and Gu Z. J. 2002. Karyomorphological study of the *Spiraea japonica* complex (Rosaceae). Brittonia 54: 168–174.

Zhou, L. H. and Wu Z. Y. 2001. Taxonomic revision on series *Buxifolii* in genus *Cotoneaster* (Rosaceae). Acta Bot. Yunnan. 23: 29–36.

Zhou, L. H., Yin Q., and Wu Z. Y. 2000. Taxonomic studies on *Cotoneaster dammeri* (Rosaceae). Acta Bot. Yunnan. 22: 379–382.

Ziergeh. Gärt. Parkanl.—See: H. Jaeger 1865

Zika, P. F. 2002b. Noteworthy collections, Oregon and Washington. Madroño 49: 195–197.

Zika, P. F. 2003. A Crater Lake National Park Vascular Plant Checklist. Crater Lake.

Zika, P. F. 2004. Noteworthy collections from Oregon and Washington. Madroño 51: 395–397.

Zika, P. F. and E. R. Alverson. 2005. Noteworthy collections from Oregon. Madroño 52: 273–274.

Ziola, B. and J. R. Dugle. 1970. A Biosystematic Study of Manitoba Roses. Pinawa, Man.

Zoë = Zoë; a Biological Journal.

Index

Names in *italics* are synonyms, casually mentioned hybrids, or plants not established in the flora. Page numbers in **boldface** indicate the primary entry for a taxon. Page numbers in *italics* indicate an illustration. Roman type is used for all other entries, including author names, vernacular names, and accepted scientific names for plants treated as established members of the flora.

Aas, G., 434
Abrams, L., 117, 271
Acaena, 19, 312, **323**
anserinifolia, 324
californica, 324
novae-zelandiae, *320*, **324**
pallida, 325
pallida, 324, **325**
pinnatifida, *320*, **324**
californica, 254, 324
Acer
glabrum, 36
Acicular rose, 114
Ackerman, T. L., 226
Acomastylis, 58
rossii, 63
Actaea
dioica, 423
Adams Creek hawthorn, 634
Adenorachis, 445
atropurpurea, 445
Adenostoma, **392**, 393, 395
fasciculatum, *391*, **393**
fasciculatum, *391*, **393**, **394**
obtusifolium, 393, **394**
prostratum, 393, **394**
sparsifolium, 393, **394**
Agaricales, 327
Agrimonia, 19, **312**, 313
bicknellii, 316
brittoniana, 317
eupatoria, 317
mollis, 316
parviflora, 316
gryposepala, 313, 314, *315*, **316**, 317
incisa, **314**, *315*
macrocarpa, 316
microcarpa, 314, *315*, 316
mollis, 316
bicknellii, 316
odorata, 317
parviflora, **314**, *315*
macrocarpa, 316
platycarpa, 316
polyphylla, 314
pubescens, 314, *315*, **316**, 317
microcarpa, 315
pumila, 315
repens, 317
rostellata, 313, 314, *315*, **316**
serrifolia, 314
striata, 314, *315*, **317**
campanulata, 317
suaveolens, 314
Agrimonieae, 22, **312**
Agrimony, 312
Agrobacterium
rhizogenes, 411
Ahart's rose, 110
Ahles, H. E., 29, 438
Aigremoine, 312
Aigremoine à sépales crochus, 316
Aigremoine pubescente, 316
Aigremoine striée, 317
Aiken, S. G., 199, 206
Alabama black cherry, 364
Alabama snow-wreath, 388
Alaska spiraea, 405
Alaskan mountain-avens, 329
Alba, 132
Alchemilla, 20, 29, 119, 120, 274, **302**, 303, 304, 310
sect. *Alchemilla*, 303
sect. *Alpinae*, 303
sect. *Erectae*, 303
acutiloba
mollis, 309
alpestris, 305
alpina, 303, 304, **305**
arvensis, 310
connivens
wichurae, 305
filicaulis, *302*, 303, 304, **307**
filicaulis, **307**
vestita, *302*, **307**
glabra, 304, **305**
glaucescens, 303, 304, **308**
glomerulans, 302, 304, **305**, 306
gracilis, 308
micans, 304, **308**
microcarpa, 311
minor, 307
mollis, 303, 304, **309**
monticola, 304, **308**
obtusa, 305
comosa, 305
occidentalis, 311
pastoralis, 308
pratensis, 306
pseudomicans, 305
speciosa, 303
subcrenata, 304, **306**
venosa, 303, 304, **309**
vestita, 307
vulgaris, 303, 306
alpestris, 305
comosa, 305
filicaulis, 307
grandis, 305
pastoralis, 308
pratensis, 306
vestita, 307
vulgaris, 306
wichurae, 305
wichurae, 303, 304, **305**, 306
xanthochlora, 304, **306**
Alchémille, 302
Alchémille à glomérules, 305
Alchémille des montagnes, 308
Alchémille filicaule, 307
Alchémille glabre, 305
Alchémille subcrénelée, 306
Alchémille vêtue, 307
Aldasoro, J. J., 434, 439, 442, 484
Alder-leaf mountain mahogany, 336
Alder-leaved shadbush, 651
Aleutian avens, 58
Aleutian mountain ash, 443
Alexander, C., 347, 399, 409, 410
Alice, L. A., 28, 29, 46, 52
Alisier, 433
Alkali ivesia, 239
Allegheny blackberry, 34
Allegheny plum, 380
Almond, 19, 353, 372
Alpine cinquefoil, 190
Alpine lady's mantle, 305
Alpine laurel, 415
Alpine meadowsweet, 402
Altai hawthorn, 515
Alvaradoa, 3, **4**
amorphoides, **4**, *5*
amorphoides, 4
psilophylla, **4**, *5*
psilophylla, 4
subovata, 4
Alverson, E. R., 467
Amandier, 372

Amber raspberry, 30, 42
Ambrose, K. V., 38, 52
Amelanchier, 19, 20, 428, 429, 433, 442, **646**, 647, 648, 650, 655, 658, 660, 661, 662
alnifolia, 442, 648, 649, **651**, 652
alnifolia, **651**, 652
cusickii, 652
nitens, 650
pumila, 651, **652**
semiintegrifolia, 651, **652**, 654
amabilis, 649, 653, **654**, 656, 661
arborea, 647, 648, 649, 653, 654, 655, 656, 657, 658, 659, **660**, 661
alabamensis, 660, 661
austromontana, 656
laevis, 661
austromontana, 656
bakeri, 651
bartramiana, 647, 648, 649, 654, 656, **657**, 658, 660, 661
basalticola, 652
canadensis, 445, 648, 649, 654, 657, **658**, 660, 661
canadensis, 658, **659**
obovalis, 658, **659**
pumila, 652
subintegra, 659
covillei, 650, 651
cusickii, 649, **652**
denticulata, 646
erecta, 648
fernaldii, 649, **655**, 658
florida, 651, 654
humptulipensis, 652
gaspensis, 650, **654**, 658
glabra, 650
gracilis, 650
×*grandiflora*, 654, 661
humilis, 649, 651, 653, 654, 655, **656**, 658, 660, 661
compacta, 656
exserrata, 656
huronensis, 653
interior, 649, **655**, 658, 661
intermedia, 649, 657, 658, 659, **660**, 661
laevis, 647, 648, 649, *653*, 654, 655, 657, 658, 659, 660, **661**
nitida, 661
lamarckii, 648
lucida, 656, 657
mormonica, 651
mucronata, 656
nantucketensis, 649, *653*, 656, **657**, 661
×*neglecta*, 648
nitens, 649, **650**
oblongifolia, 657, 658, 659
micropetala, 656, 657
obovalis, 658, 659
oreophila, 651
ovalis
semiintegrifolia, 652
pallida, 649, **650**
prunifolia, 650
×*quinti-martii*, 658
racemosa, 384
rubescens, 650
rubra, 661
sanguinea, 649, ***653***, 654, 655, 657, 658, 661
gaspensis, 654
grandiflora, 654
spicata, 648, 649, 654, 655, **656**, 657, 658, 659, 660, 661
stolonifera, 656, 657
utahensis, *645*, 649, **650**, 651
covillei, 650
venulosa, 651
wiegandii, 655
Amélanchier à feuilles d'aulne, 651
Amélanchier bas, 656
Amélanchier de Bartram, 657
Amélanchier de Fernald, 655
Amélanchier de Gaspésie, 654
Amélanchier de l'intérieur, 655
Amélanchier du Canada, 658
Amélanchier en épis, 656
Amélanchier glabre, 661
Amélanchier gracieux, 654
Amélanchier intermédiaire, 660
Amélanchier sanguin, 653
×*Amelasorbus*, 433, 435
jackii, 442, 651
American bladdernut, 8
American burnet, 319
American crabapple, 475
American ipecac, 425
American mountain ash, 439
American plum, 378, 379
American red raspberry, 43
American robin, 449, 469
Amygdalaceae, 345
Amygdaleae, 20, 21, **352**, 385, 415
Amygdaloideae, 20, 21, 343, **345**, 346, 347, 352, 383, 385, 386, 387, 390, 392, 398, 399, 416, 422, 424, 427
Amygdalus, 352, 353, 354
communis, 372
dulcis, 372
glandulosa, 377
havardii, 371
persica, 371
Anders, C. M., 404
Anderson, D. G., 164
Anderson, G. J., 380
Andrick, A. O., 469
Angell's cinquefoil, 180
Antelope bitterbrush, 340
Antieau, C. J., 419
Apache plume, 73
Apacheria, 9, **10**
chiricahuensis, ***11***
Aphanes, 119, 120, 274, 303, **309**, 310
arvensis, **310**
australis, 310, **311**
cuneifolia, 311
inexspectata, 311
macrosepala, 311
microcarpa, 311
occidentalis, *302*, 310, **311**
Appalachian avens, 64
Appalachian sandcherry, 375
Apple, 472, 473, 474, 476, 477, 478, 480, 483
Apricot, 19, 353, 375
Arctic cinquefoil, 192
Arctic raspberry, 35
Aremonia, 312
Argentina, 126, 127, 298
anserina, 127
egedei, 130
groenlandica, 130
argentea, 128
egedei, 130
occidentalis, 129
pacifica, 129
Aria, 436
intermedia, 437
Arizona cliffrose, 342
Arizona drymocallis, 284
Arizona rose, 108
Arizona rosewood, 430
Arizona wood beauty, 284
Armeniaca, 352, 353
vulgaris, 375
Aronia, 429, 433, **445**, 447, 489
alnifolia, 651
arbutifolia, 433, 438, 440, 445, **446**
atropurpurea, 445
floribunda, 445, 489
glabrescens, 446
hybrida, 438
latifolia, 445
melanocarpa, 438, *440*, **446**, 489
mitschurinii, 445
nigra, 446
×*prunifolia*, 433, 439, 441, 445, 446
pubens, 446
pumila, 446
pyrifolia, 446
sargentii, 440
watsoniana, 440
Artemisia, 182
Artjuschenko, Z., 24
Aruncus, 398, 416, **422**
acuminatus, 424
allegheniensis, 423
dioicus, *418*, **423**, 424
acuminatus, *418*, 423, **424**
dioicus, *418*, **423**, 424
pubescens, 423, **424**
vulgaris, 423, **424**
pubescens, 424
sylvester
vulgaris, 424
Asami, Y., 473, 478
Ascomycetae, 327
Ash Creek ivesia, 229
Ash Meadows ivesia, 239
Ashe hawthorn, 608
Ashy cinquefoil, 145
Asian pea pear, 480
Astilbe
biternata, 423
Atkinson, S., 437
Atlantic nine-bark, 348, 349
Aubépine, 491
Aubépine à deux styles, 497
Aubépine à épines longues, 520
Aubépine à lobes aigus, 562
Aubépine de Brainerd, 523
Aubépine de Holmes, 560
Aubépine de Pringle, 559
Aubépine de Schuette, 565
Aubépine dorée, 597
Aubépine écarlate, 559
Aubépine épineuse, 497
Aubépine ergot-de-coq, 538
Aubépine flabelliforme, 567
Aubépine monogyne, 497
Aubépine ponctuée, 542
Aubépine pubérulente, 601
Aubépine scabride, 523
Aubépine subsoyeuse, 556
Aubépine succulente, 518
Austin's drymocallis, 288
Austin's wood beauty, 288
Australian bramble, 48
Australian raspberry, 48
Austrian brier rose, 94
Avens, 58, 68
Ayensu, E. S., 87

Bailey, L. H., 29, 36, 42, 43, 50, 54, 279, 399, **409**, 410
Bailey's ivesia, 226
Bailleul, S. M., 433
Baked-apple berry, 29, 38
Baldhip rose, 116
Ball, P. W., 142, 303
Ballona cinquefoil, 175
Bammi, R. K., 37, 44
Banasiak, S. E., 85
Barba de chivo, 73
Barbe-de-bouc, 422
Barberry-leaved hawthorn, 540
Barbour, M. G., 254
Barnaulov, O. D., 24, 27
Barren-strawberry, 71, 132
Bartish, I. V., 449
Barton Flats horkelia, 258
Barton's raspberry, 36
Battle Creek hawthorn, 635
Beach plum, 380
Beach strawberry, 278
Beadle, C. D., 613, 614
Bean, R. C., 145
Bean, W. J., 89, 90, 93, 399, 409
Bearberry cotoneaster, 459
Bearded cinquefoil, 164
Bear's Paw hawthorn, 597
Beautiful cinquefoil, 158
Beautiful hawthorn, 583
Beauverd's spiraea, 405
Bell, R. L., 480
Bencomia, 318
Benoîte, 58
Benoîte à grandes feuilles, 66
Benoîte commune, 70

Benoîte d'Alep, 67
Benoîte des ruisseaux, 65
Benoîte du Canada, 68
Benoîte lacinicée, 68
Bent avens, 65
Bergner, P., 47
Bessey's sandcherry, 374
Biddy biddy, 324
Biennial cinquefoil, 140
Big Bend serviceberry, 646
Big-petaled ivesia, 233
Bigtree plum, 379
Biltmore hawthorn, 590
Bird cherry, 367
Bird-cherry hawthorn, 592
Bitter cherry, 367
Bitterbrush, 340
Bitterbush, 4
Bitterbush Family, 3
Bjornstad, H. O., 483
Black cherry, 363, 365
Black chokeberry, 446
Black hawthorn, 512
Black jetbead, 386
Black raspberry, 30, 46
Black Rock cinquefoil, 176
Black sage, 340
Blackberry, 19, 29, 30, 39, 44, 46
Blackbrush, 390
Blackthorn, 376
Bladdernut, 7
Bladdernut Family, 6
Blanchard, W. H., 647, 658
Block, T. A., 449
Blueberry haw, 501
Blueleaf cinquefoil, 152
Blytt, A. G., 127
Bocek, B. R., 112
Böcher, T. W., 52, 441
Boehmer, G. R., 127
Boivin, B., 142, 148, 158, 161, 216, 306
Bolander's horkelia, 251
Bolle, F., 59, 63
Bombycilla
cedrorum, 449, 469
Bortiri, E., 354
Bossard, C. C., 449
Boufford, D. E., 30
Boulder Mountain cinquefoil, 180
Bowman's root, 426
Boysenberry, 30, 56
Brach, A. R., 454, 455, 457, 458, 459, 461, 462, 463, 464, 465, 467
Brachyacanthae, 500
Bradford pear, 482
Brainerd hawthorn, 523
Bramble, 28
Branched cinquefoil, 163
Brandegee, T. S., 346
Brassicales, 6
Brazos hawthorn, 556
Brewer, W. H., 230, 242
Brewer's cinquefoil, 170
Bridal-wreath, 410
Bridal-wreath spiraea, 408
Bride's-feathers, 422
Bridewort, 402
Brier, 75
Bristly blackberry, 53
Bristly dewberry, 41
Bristly ivesia, 227
Britton hawthorn, 606
Britton, N. L., 425
Broad-leaved meadowsweet, 402
Brook cinquefoil, 139
Brouillet, L., 6, 21, 23, 27, 57, 74, 119, 312, 326, 345, 347, 352, 383, 386, 392, 398, 424, 427
Browicz, K., 480
Brown, C. L., 364
Brown, M. L., 29
Brown, R. G., 29, 320
Brown, S. W., 56
Bruce's cinquefoil, 159
Bruneau, A., 75, 78, 88, 95, 105, 116, 117, 119
Brzyski, J. R., 404
Buckbrush, 340
Buisson ardent, 470
Bullate cotoneaster, 464
Burgess, M. B., 646
Burnet, 321
Burnet bloodwort, 318
Burnet rose, 94
Burnet-bloodwort, 323
Bush cinquefoil, 295
Bush's hawthorn, 589
Bushakra, J. M., 346
Bushy cinquefoil, 139
Butterwick, M., 185

Cabbage rose, 88
California blackberry, 55
California horkelia, 256
California mountain ash, 441
California rose, 111
Callery pear, 482
Caltha-leaved avens, 64
Campbell, C. S., 29, 46, 52, 427, 434, 445, 449, 468, 473, 494, 644, 645, 646, 647, 648, 655, 657, 658, 661, 662
Canada burnet, 322
Canada plum, 379
Canadian highbush blackberry, 37
Cangshan cotoneaster, 466
Cannabis, 143
Carolina cherry laurel, 361
Carolina laurel cherry, 361
Carolina rose, 101
Carrotleaf horkelia, 266
Carya, 606
Cascade mountain ash, 442
Castle Crags ivesia, 236
Castlegar hawthorn, 513
Catalina cherry, 362
Catalina ironwood, 346
Catalina rockflower, 12
Catling, P. M., 374, 479, 484, 486, 487, 492, 657
Cayouette, J., 449
Cayouette, R., 147
Cedar waxwing, 449, 469
Celastraceae, 9
Celastrus
myrtifolius, 361
Cenococcum
geophilum, 327
Cerasus, 352, 353, 354
demissa, 366
melanocarpa, 366
emarginata, 367
ilicifolia, 362
speciosa, 369
Cercocarpus, 19, 73, 326, 327, **332**, 343
alnifolius, 334
argenteus, 336
arizonicus, 338
betulifolius, 334
blancheae, 334
minor, 334
betuloides, *330*, 332, **333**, 336
betuloides, *330*, 333, **334**, 335
blancheae, 333, **334**, 335
macrourus, 334
multiflorus, 334
traskiae, 333, **335**
breviflorus, 332, 333, 334, **335**, 336
eximius, 335
douglasii, 334
eximius, 335
fothergilloides, 332
hypoleucus, 337
intricatus, 338
villosus, 338
ledifolius, 332, 333, **337**, 338
intercedens, 337, 338
intermontanus, 336, 337, **338**
intricatus, 337, **338**
ledifolius, **337**, 338
macrourus, 334
minutiflorus, 334
montanus, 332, 333, **336**
argenteus, 336
blancheae, 334
glaber, 334
macrourus, 334
minutiflorus, 334
paucidentatus, 335
traskiae, 335
parvifolius, 336
glaber, 334
hirsutissimus, 336
minimus, 336
paucidentatus, 335
rotundifolius, 334
traskiae, 335
Cerisier, 352
Cerisier à grappes, 365
Cerisier acide, 368
Cerisier d'automne, 363
Cerisier de la Susquehanna, 375
Cerisier de Pennsylvanie, 367
Cerisier de Virginie, 365
Cerisier déprimé, 375
Cerisier des oiseaux, 368
Cerisier des sables, 373
Cerisier tardif, 363
Cerro hawthorn, 509
Chaenomeles, 20, 428, **484**, 487, 488
sect. *Pseudocydonia*, 487
japonica, **485**
sinensis, 488
speciosa, **485**, *486*
×*superba*, 485
Challice, J. S., 480, 483, 484
Chamaebatia, 326, 327, **343**, 344, 395
australis, 343, **344**
foliolosa, **343**, *344*
australis, 344
Chamaebatiaria, 343, 392, **394**, 395
millefolium, *391*, **395**
Chamaemeles, 449
Chamaemorus
anglica, 38
norwegica, 38
Chamaerhodos, 119, 120, 274, **279**, 304
erecta, *273*, **280**
nuttallii, 280
parviflora, 280
nuttallii, 280
Chamise, 392
Chamise horkelia, 258
Chang, K. S., 369
Chapin, F. S. III, 327
Charleston Peak ivesia, 228
Chase, M. W., 6, 10
Chelan rockmat, 412
Cherokee rose, 87
Cherry, 19, 352, 353, 354, 367, 368
Cherry plum, 375
Chesapeake blackberry, 49
Chickasaw plum, 381
Chicouté, 38
Chilean strawberry, 278
China rose, 77
Chinese apple, 477
Chinese firethorn, 469
Chinese hawthorn, 490
Chinese pear, 481
Chinese photinia, 490, 491
Chinese quince, 487
Chinese stranvaesia, 491
Chinnappa, C. C., 64, 422
Chiricahua cinquefoil, 186
Chiricahua rockflower, 11
Cho, E. J., 99
Chocolate-root, 65
Chokeberry, 445
Chokecherry, 363, 365
Christenhusz, M. J. M., 303, 304
Christensen, K. D., 52
Christensen, K. I., 494
Christmas berry, 447
Chrubasik, C., 90
Chrysobalanaceae, 20
Cinnamon rose, 97
Cinq-Mars, L., 647, 651, 657, 658, 660, 661
Cinquefoil, 121
Clark, E. M., 229

Clark, J. R., 30
Clark, L. V., 37, 50, 56
Clark, W. H., 229
Clarkson, R. B., 30
Clausen, J., 122, 150, 151, 159, 160, 281, 282, 287, 289, 292, 295
Cleveland's horkelia, 253
Cliff cinquefoil, 185
Cliff drymocallis, 290
Cliff rose, 341
Cliff wood beauty, 290
Cliffortia, 312
Climbing prairie rose, 83
Clokey, I. W., 178
Clokey's greasebush, 14
Cloudberry, 29, 38
Clover-root, 70
Clubmoss ivesia, 232
Cluster rose, 110
Clustered lady's mantle, 305
Coast cinquefoil, 215
Coast drymocallis, 293
Coast ground rose, 118
Coast horkelia, 255
Coast wood beauty, 293
Coastal mountain mahogany, 334
Cockspur hawthorn, 538
Coconino ivesia, 244
Cody, W. J., 199, 202, 329
Coleogyne, 21, 386, 387, **390**
  ramosissima, ***391***
Collicutt, L. M., 104
Colorado cinquefoil, 166
Coluria, 59
Colurieae, 21, 22, 57, 58, 73
Comarella, 243
  multifoliolata, 244
  sabulosa, 244
Comaret des marais, 301
Comarum, 119, 120, 122, 124, 274, **300**, 301, 304
  palustre, **301**, *302*
    angustifolium, 301
    arcticum, 301
Combleaf cinquefoil, 156
Common agrimony, 316
Common barren-strawberry, 71
Common blackberry, 34
Common cherry laurel, 362
Common crabapple, 476
Common flowering-quince, 485
Common medlar, 644
Common nine-bark, 348
Common pear, 483
Common pearlbrush, 384
Common silverweed, 128
Congdon's false horkelia, 271
Congdon's horkeliella, 271
Cooperrider, T. S., 29, 460
Corchorus
  japonicus, 390
  scandens, 387
Cordilleran drymocallis, 286
Cordilleran wood beauty, 286
Core, E. L., 29
Corête du Japon, 390
Correll, D. S., 510
Cortinariaceae, 327
Cotoneaster, 20, 427, 433, **448**, 449, 450, 454, 461, 468, 469
  sect. *Malacomeles*, 645
  subg. *Chaenopetalum*, 450
  subg. *Cotoneaster*, 450
  acutifolius, 449, 461, 464
    lucidus, 463
    villosulus, 464
  adpressus, 451, **460**
  angustifolius, 471
  apiculatus, 449, 461
  atropurpureus, *448*, 453, **461**
  bradyi, 467
  bullatus, 449, 464
    bullatus, 464
    macrophyllus, 464
  buxifolius, 458
    cochleatus, 459
  cochleatus, 451, **459**
  conspicuus, 452, **457**, 458, 460
    conspicuus, 458
    decorus, 457, 458
    nanus, 457
  coriaceus, 455
  crenulatus, 470
  crispii, 451, 452, **454**
  dammeri, 451, 454, **459**, 460
  denticulatus, 646
  dielsianus, 451, **466**, 467
  divaricatus, 449, 453, 460, **461**, 462, 463
  fangianus, 453, **467**
  flinckii, 463
  franchetii, 449, 451, *462*, **465**, 466
    sternianus, 466
  frigidus, 453, 454, **456**
  gamblei, 450, 452, 453, **456**
  glaucophyllus, 457
    glaucophyllus, 457
    vestitus, 455
  glomerulatus, 465
  hebephyllus, 457
    hebephyllus, 457
    monopyrenus, 457
  henryanus, 454
  hjelmqvistii, 453, **462**
  hodjingensis, 451, **458**
  horizontalis, 450, 453, 461, **462**, 463
  hsingshangensis, 464
  hupehensis, 449, 461
  hylmoei, 454
  integerrimus, 449
  integrifolius, 452, **458**, 459
  koidzumii, 471
  lacteus, 449, 451, 452, **454**, 455
  lucidus, 449, 452, 461, *462*, **463**, 464, 465
  magnificus, 453, **456**, 457
  marginatus, 460
  melanocarpus, 438, 451, 452, 464, **467**, 468
  microphyllus, 459
    cochleatus, 459
    conspicuus, 457
    thymifolius, 459
  miniatus, 454, **467**
  monopyrenus, 451, 452, **457**
  mucronatus, 467
  multiflorus, 449, 457
  nanus, 457
  nitens, 452, **463**
  pannosus, *448*, 449, 451, 454, **455**
  permutatus, 457
  perpusillus, 463
  qungbixiensis, 451, **466**
  racemiflorus, 449
  rehderi, 453, **464**
  rugosus, 454
  salicifolius, 452, **454**
  simonsii, *448*, 449, 450, 453, **460**
  splendens, 467
  sternianus, 451, **466**
  ×suecicus, 452, 454, **460**
  symondsii, 460
  tengyuehensis, 451, 452, 453, **465**
  thymifolius
    cochleatus, 459
  transens, 452, 453, **457**
  vestitus, 451, **455**
  villosulus, 452, **464**
  vulgaris
    melanocarpus, 467
  wardii, 465
  zabelii
    miniatus, 467
Cotonéaster à feuilles aiguës, 463
Cottam's cinquefoil, 187
Coulter, J. M., 122
Cowan, R. S., 273
Cowania, 73, 339
  alba, 341
  ericifolia, 342
  mexicana
    dubia, 341
    stansburyana, 341
  stansburyana, 341
  subintegra, 342
Crassulaceae, 20
×*Crataegosorbus*, 433
  miczurinii, 438
Crataegus, 19, 20, 428, 433, 445, 449, 469, **491**, 492, 493, 494, 508, 512, 525, 536, 550, 578, 581, 588, 591, 594, 611, 618, 644
  [unranked] *Aestivales*, 546
  [unranked] *Bracteatae*, 607
  [unranked] *Brevispinae*, 500
  [unranked] *Cordatae*, 498
  [unranked] *Dilatatae*, 578
  [unranked] *Intricatae*, 585
  [unranked] *Lobulatae*, 557
  [unranked] *Molles*, 551
  [unranked] *Pruinosae*, 571
  [unranked] *Pulcherrimae*, 579
  [unranked] *Sanguineae*, 514
  [unranked] *Tenuifoliae*, 561
  [unranked] *Triflorae*, 609
  [unranked] *Virides*, 531
  sect. *Aestivales*, 546
  sect. *Anomalae*, 522
  sect. *Apiifoliae*, 498
  sect. *Brachyacanthae*, 500
  sect. Brevispinae, 495, **500**
  sect. Coccineae, 495, 503, 516, **525**, 531, 536, 541, 546, 549, 550, 551, 557, 561, 567, 571, 578, 579, 585, 593, 604, 607, 609, 610, 618, 633
  sect. *Cordatae*, 498
  sect. Crataegus, 491, 494, **495**, 496, 498, 499, 525, 529
  sect. *Crus-galli*, 536
  sect. Douglasia, 495, **502**, 508, 511, 633
  sect. *Flavae*, 585
  sect. Macracanthae, 495, 503, 515, **516**, 522, 525
  sect. *Mexicanae*, 541
  sect. *Microcarpae*, 499
  sect. *Molles*, 551
  sect. *Parvifoliae*, 604
  sect. *Punctatae*, 541
  sect. *Rotundifoliae*, 593
  sect. Sanguineae, 495, **514**, 515, 634
  sect. *Tenuifoliae*, 561
  sect. *Triflorae*, 609
  sect. *Virides*, 531
  ser. Aestivales, 492, 525, **546**
  ser. Altaicae, **515**
  ser. Anomalae, 493, 516, **522**, 525, 546, 574, 635, 637
  ser. Apiifoliae, 492, 496, **498**
  ser. Apricae, 493, 528, 529, 569, 588, **610**, 611, 617, 618, 619, 632
  ser. *Brachyacanthae*, 508
  ser. Bracteatae, 526, 528, 540, **607**, 608, 609
  ser. *Brainerdianae*, 522
  ser. Brevispinae, **500**
  ser. Cerrones, 500, 501, 502, **508**
  ser. Coccineae, 530, 551, **557**, 558, 561, 579, 638, 641
  ser. Cordatae, 496, **498**, 499
  ser. Crataegus, 495, **496**, 497, 498, 636
  ser. Crus-galli, 516, 525, 526, 531, 532, **536**, 549, 606, 607, 628, 629
  ser. Dilatatae, 525, 526, 558, **578**, 579
  ser. Douglasianae, 502, 503, 507, **511**, 514, 636
  ser. *Flavae*, 585, 586, 611
  ser. Greggianae, 527, **550**
  ser. Intricatae, 528, 529, 530, 572, 577, 580, **585**, 586, 588, 590, 591, 592, 594, 611
  ser. Lacrimatae, 492, 493, 525, 528, 529, 580, 588, 605, 606, 611, 613, 617, **618**, 619, 622, 627, 628, 629, 630, 631, 632

Crataegus (*continued*)
ser. Macracanthae, **516**, 522, 525, 594, 605, 640, 643
ser. Madrenses, 525, 526, 527, **549**
ser. *Mexicanae*, 541, 542
ser. Microcarpae, 495, 496, **499**, 525, 529
ser. Molles, 492, 525, 526, 527, 532, 550, **551**, 552, 556, 557, 558, 594, 595, 638, 639
ser. Montaninsulae, 492, 495, 506, 530, **633**
ser. Parvifoliae, 516, 529, **604**, 605, 606
ser. Populneae, 527, 528, 530, 558, 561, 563, 564, **567**, 568, 570, 617
ser. Pruinosae, 516, 522, 527, 529, 535, 561, 564, 568, **571**, 572, 579, 637
ser. Pulcherrimae, 528, 529, **579**, 580, 583, 584, 586, 591
ser. Punctatae, 516, 522, 525, 526, 527, 532, 536, **541**, 542, 552, 558
ser. Purpureofructus, **502**, 503, 504, 552
ser. Rotundifoliae, 494, 495, 506, 516, 522, 526, 528, 529, 530, 542, 543, 558, 561, 586, **593**, 594, 596, 599
ser. *Sanguineae*, 515
ser. Tenuifoliae, 494, 522, 527, 528, 530, 558, **561**, 566, 568, 569, 572, 573, 591, 594, 600, 641, 642
ser. Triflorae, 492, 525, 605, 607, 608, **609**
ser. Virides, 526, 527, **531**, 536
subser. *Robustae*, 618, 619, 624, 625, 630
subser. *Tenues*, 618, 619, 624, 626
abstrusa, 584
acclivis, 559
acutifolia, 539
insignis, 537
acutiloba, 562, 563
adunca, 626
adusta, 624
aemula, 568, **569**, 617
aestivalis, 546, **547**, 548, 629
agrestina, 582
alabamensis, 621, **626**, 627, 628
florens, 628
ravenelii, 624
teres, 628
albanthera, 539
albicans, 559
albicera, 540
alleghaniensis, 612, **616**
extraria, 617
ignava, 616
mira, 615
alma, 584
altaica, 515
amica, 630
amicta, 561
×*anamesa*, 532
angustata, 544
anisophylla, 625, 626
annosa, 612, 616, **617**
anomala, 560
antimima, 532
aperta, 577
apiifolia, 498
apiomorpha, 563, 564, 565
apposita, 589
aprica, 612, **614**, 615
aquacervensis, 504, **506**
araioclada, 540
arborea, 539
arborescens, 532
arbutifolia, 447
arcana, 574
ardua, 494
arenicola, 607
arguta, 624
arkansana, 554
arnoldiana, 556, 557
arrogans, 613
ashei, 518, 540, 607, **608**, 610
aspera, 577
asperata, 565
asperifolia, 524
×atrorubens, 531, **636**
atrovirens, 504, **506**, 507
attrita, 618, 621, 628, **629**
aulica, 559
aurescens, 549
austrina, 584
austromontana, 609, **610**
azarolus, 492
barrettiana, 539
basilica, 565, 566
beata, 568, **570**, 571
bellica, 540
bellula, 571
berberifolia, 537, 539, **540**, 606
berberifolia, **540**
engelmannii, **540**
×bicknellii, 522, **636**
biltmoreana, 576, 577, 586, **590**, 591, 592
stonei, 591
bisulcata, 606
blanchardii, 600
blanda, 534
bona, 573
boyntonii, 588
brachyacantha, 494, ***501***, 502, 509
leucocarpa, 501
brachyphylla, 494, 555
brachypoda, 575
bracteata, 575
brainerdii, 522, **523**, 524
asperifolia, 524
cyclophylla, 524
egglestonii, 524
scabrida, 523
brazoria, 551, 552, **556**
viburnifolia, 554
brittonii, 605, **606**, 607
brownii, 602
brumalis, 564
brunetiana, 640
fernaldii, 640
buckleyi, 587, **589**
calliantha, 574
callicarpa, 579
callida, 573
calophylla, 539
calpodendron, 516, **517**, 518, 539, 639, 643
globosa, 517
hispidula, 517
microcarpa, 517
mollicula, 517
calva, 611, 632
canadensis, 553, 556, 557
canbyi, 538
candens, 539
×*canescens*, 644
castlegarensis, 511, **513**
cavaleriei, 479
celsa, 523
cerasina, 641
cerasoides, 547
cerronis, 509
chadsfordiana, 516, 522
champlainensis, 556, 557
cherokeenis, 539
chippewaensis, 560
choriophylla, 605
chrysocarpa, 506, 521, 542, 546, 557, 558, 563, 565, 567, 586, 594, 595, 596, **597**, 599, 601, 636, 640
bicknellii, 636
blanchardii, 597, **600**
chrysocarpa, 597, **598**, 599, 600
dodgei, 603
faxonii, **598**, 600
phoeniceoides, 598, **600**
piperi, 597, 598, **599**
praecox, 598, **600**
subrotundifolia, 598, **599**
vernonensis, 597, 598, **599**
vigintistamina, 597, **600**, 601
cirrata, 622, 626
clara, 624
coccinata, 599
coccinea, 557, 558, **559**
coccinea, **559**, 560, 561
fulleriana, 559, **560**
mollis, 553
pringlei, **559**, 560, 561
coccinioides, 574, 575, **579**
cognata, 572, 574, **577**, 578
×coleae, 522, 572, 575, **636**, 637
×collicola, 543, 544, **637**
collina, 542, **543**, 544, 545, 563, 610, 617, 635, 637, 638, 639
collicola, 637
collina, **544**, 545
hirtiflora, 543, **544**, 545, 638
lettermanii, 542, 543, **545**
secta, 545
sordida, 543, 544, **545**, 638, 639
succincta, **544**
colonica, 620, **627**, 628
coloradoides, 598
columbiana, 512, 599
chrysocarpa, 597
columbiana, 599
piperi, 599
communis, 587, 589, **590**
compacta, 572, **573**, 574
compitalis, 624
compta, 569, 571
concinna, 585
condigna, 619, 621, **624**, 625, 626
confragosa, 560
confusa, 590
conjungens, 643
consanguinea, 614
conspecta, 579
constans, 627
consueta, 539
cordata, 499
corusca, 560
crassifolia, 603
craytonii, 586, **592**
crenulata, 470
crocea, 618, 619, 622, **623**, 624, 625
crocina, 540
croomiana, 606
crudelis, 567
crus-galli, 518, 535, 536, 537, **538**, 539, 540, 543, 547, 627, 636, 637, 638, 641
bellica, 540
capillata, 539
crus-galli, 538, **539**
exigua, 539
leptophylla, 539
macra, 539
oblongata, 539
pachyphylla, 539
prunifolia, 641
pyracanthifolia, 538, **539**
cuprea, 589
cupressocollina, *501*, 504, **507**
curva, 623
cyanophylla, 562, 563
cyclophylla, 524
danielsii, 637
dapsilis, 623
dasyphylla, 555
davisii, 534
dayana, 560
declivitatis, 579
decorata, 577
delawarensis, **568**, 569
delecta, 559
delosii, 603, 604
demissa, 562, 563
denaria, 538
densiflora, 567
desertorum, 534
desueta, 522

digna, 559
dilatata, 579
discolor, 537, 538
disjuncta, 575
dispar, 620, **632**
×disperma, 543, **637**, 638
×dispessa, 552, **638**, 639
dissona, 575
divergens, 600
divida, 520
doddsii, 598
dodgei, 528, 594, 595, 600, 602, **603**, 604
flavida, 603
lumaria, 604
rotundata, 603
dolosa, 626, 627
douglasii, 493, 494, *510*, 511, **512**, 513, 514, 599
duchesnensis, 509
suksdorfii, 512
drymophila, 563, 564
dumetosa, 554
dunbarii, 522
durobrivensis, 574
eamesii, 560
edita, 540
edura, 540
egens, 613, *615*, 619, 621, **622**, 625, 626
egglestonii, 524
egregia, 612, **613**
elliptica, 547
ellwangeriana, 559
elongata, 560
enderbyensis, 504, **505**, 506
engelmannii, 540
sinistra, 540
enucleata, 532
erythropoda, 508, **509**, 510
evansiana, 494
exclusa, 559
exilis, 618, 619, 620, **632**
eximia, 581, **585**
extraria, 612, **617**
faxonii, 598
praetermissa, 494
fecunda, 536, 537, **541**
fera, 540
fernaldii, 640
ferrissii, 566
filipes, 571
flabellata, 494, 561, 562, 563, 565, **567**, 594
densiflora, 567
flabellata, 567
grayana, 566, 567, 601
flava, 586, 587, **588**, 611
integra, 626
flavida, 603, 604
florens, 621, 625, **628**, 629
floridana, 619, 622, *624*, **625**, 626, 631
florifera, 516, 522, **523**, 574, 635
fluviatilis, 562, 563, **564**, 565, 566, 567
foetida, 588
fontanesiana, 539
formosa, 572, **578**
fortunata, 592
×fretalis, 561, 562, **638**
frugalis, 623
frugiferens, 569, 610, 611, 612, **616**, 617
fruticosa, 547
fulleriana, 560
furcata, 534
furtiva, 613, 620, **630**
garrettii, 621
gattingeri, 564, 572, **573**, 574
rigida, 573
gaudens, 575
gaylussacia, 511, **512**, 513
gemmosa, 519
geniculata, 623
georgiana, 573
gilva, 581, **582**, 583
glabra, 489
glabrifolia, 642
glabriuscula, 534
glareosa, 494
globosa, 517, 518
gloriosa, 559
grandis, 543, 637
gravesii, 603
gravida, 553, 554
gravis, 570
grayana, 567
greggiana, **550**, 642
grossiserrata, 606
haemacarpa, 568
hamata, 539
hannibalensis, 539
harbisonii, 607, **608**
harryi, 494, 522
harveyana, 518
helderbergensis, 641
hillii, 560
hirtella, 540
hirtiflora, 544
hispidula, 517, 518
holmesiana, 558, **560**, 561, 563
amicta, 561
chippewaensis, 560
villipes, 560
villosa, 561
honesta, 524
ideae, 522
ignava, 611, 612, *615*, **616**
illecebrosa, 560
illudens, 624, 627, 629
immanis, 494
improvisa, 524
×incaedua, 543, **638**, 639
incana, 623, 624
incilis, 580, 584, **585**
induta, 555
infera, 539
inops, 630
insidiosa, 624
insignis, 537
insperata, 517, 518
integra, 620, **626**, 627
integrifolius, 458
integriloba, 520
interior, 533
intricata, 586, 587, **588**, 589, 591
biltmoreana, 590
boyntonii, 588
neobushii, 589
padifolia, 592
rubella, 589
straminea, 589
invicta, 619, **626**
invisa, 553, 554
iracunda, 562, **563**, 564, 569, 572, 573, 591
brumalis, 562, 564
populnea, 569
silvicola, 563
stolonifera, 570
irrasa, 595, **601**
blanchardii, 600
jackii, 604
jasperensis, 538
jesupii, 522, 568, **571**
jonesiae, 542, **545**, 546, 640
×*kellermanii*, 543
×kelloggii, 552, **639**
kennedyi, 522
kingstonensis, 522
knieskerniana, 638
korolkowii, 515
lacrimata, 618, 619, **622**, 623, 624
laetans, 571
laetifica, 641
laevigata, 496, **497**
lanata, 619, 620, 621, 629, **630**
lancei, 620, **631**, 632
lanceolata, 533
laneyi, 520
lanuginosa, 552, **555**, 556
larga, 534
lassa, 619, 620, *624*, 627, **629**, 630
colonica, 627
integra, 626
lanata, 630
recurva, 623
×latebrosa, 552, **639**
laurentiana, 516, 517, 594, 595, **640**
brunetiana, 640
dissimilifolia, 640
laurentiana, 640
laxa, 625
laxiflora, 519
lecta, 494
leioclada, 589
leiophylla, 576
lemingtonensis, 600
leonensis, 613, **615**, 616
lepida, 619, **621**, 622, 623
leptophylla, 539
lettermanii, 545
levis, 568, 570, **571**
limaria, 553
limnophila, 539
littoralis, 494
livoniana, 641
lobulata, 559
locuples, 579
lucida, 547
×lucorum, 561, 562, **640**, 641
luculenta, 547
ludoviciensis, 539
lumaria, 596, 602, **604**
lutensis, 534
luxuriosa, 569
mackenziei, 576
aspera, 577
bracteata, 575
mackenziei, 577
macounii, 560
macracantha, 502, 504, *510*, 516, 517, 518, 519, **520**, 521, 564, 640
occidentalis, 520
pertomentosa, 520
macropoda, 544
macrosperma, 493, 522, 543, **562**, 563, 564, 565, 567, 568, 569, 571, 598, 638
acutiloba, 562
demissa, 562
eganii, 562
matura, 562
pentandra, 562
roanensis, 562, 563
magniflora, 558, **560**
magnifolia, 577
maloides, 547
mansfieldensis, 558
margarettae, 593, 594, 595, **601**, 602, 604, 639, 642
angustifolia, **602**
brownii, **602**
margarettae, **602**
meiophylla, 602, **603**
marshallii, **498**, 592, 625, 632
matura, 562, 563
×*media*, 497
meiophylla, 603
membranacea, 519
menandiana, 494
mendosa, *576*, 580, 581, **582**
mercerensis, 494
meridiana, 620, 625, 626, **631**
meridionalis, 553, 554
merita, 558
mexicana, 492, 542
michauxii, 619
michiganensis, 519
micrantha, 535
microcarpa, 500
milleri, 571
mira, 612, **615**, 616
missouriensis, 643
mohrii, 538
mollicula, 517, 518
mollis, 508, *548*, 551, 552, 553, 556, 557, 563, 636, 639
dumetosa, 553, **554**
incisifolia, 553, **554**
lanuginosa, 555
meridionalis, 553, **554**
mollis, *548*, **553**, 554
sera, 553
texana, 554
viburnifolia, 553, **554**, 555, 556

Crataegus (*continued*)
mollita, 517, 518
monantha, 547
monogyna, 493, 496, **497**, 512, 635
  lasiocarpa, 497
  monogyna, **497**
monosperma, 539
montivaga, 549
munda, 619, 622, **623**, 624
  pexa, 627
munita, 540
nananixonii, 586, **590**
×*neobaxteri*, 543
neobushii, 587, **589**
neofluvialis, 519
nigra, 515
nitens, 535
nitida, 531, 533, **535**
nitidula, 574
noelensis, 639
noveboracensis, 601
nuda, 641
oakesiana, 595, **601**
obscura, 517, 518
occidentalis, 520
okanaganensis, *frontispiece*, *501*, **504**, 505, 506, 507
  okanaganensis, *frontispiece*, *501*, **505**
  wellsii, **505**
okennonii, 494, 511, **513**, 514
opaca, 546, **547**, *548*
opima, 581, 582, **584**
orbicularis, 503, **507**
oreophila, 522
ouachitensis, 586, 587, **591**, 592
  minor, 591, **592**
  ouachitensis, **591**, 592
ovata, 534
oxyacantha, 496
oxyacanthoides, 497
pachyphylla, 539
padifolia, 586, 587, **592**, 594
  incarnata, 592, **593**
  padifolia, 592, **593**
pallens, 584
palliata, 643
palmeri, 518, 538
panda, 629
paradoxa, 539
parciflora, 538
parkiae, 539
parvifolia, 606
parvula, 576
pastorum, 562, 563
patrum, 577
pausiaca, 637
pearsonii, 580
pechiana, 532, 534
pedicellata, 559
  albicans, 559
  ellwangeriana, 559
  robesoniana, 561
pennsylvanica, 551, 552, **557**, 558, 559
pennypackeri, 538
pentandra, 562, 563
peoriensis, 637
perjucunda, 578
permera, 539
×*permixta*, 532, 536
persimilis, 516, 525, 526, 536, 537, 541, 635, **641**
pertomentosa, 520
pexa, 620, **627**, 628
phaenopyrum, **499**
phaneroneura, 539
phippsii, 503, **508**, 552
pilifera, 540
pilosa, 494
pinetorum, 580, 582, **583**
pinguis, 494
pinnatifida, 492
piperi, 599
pisifera, 519
×*poliophylla*, 532
polyclada, 539
populifolia, 494
populnea, 564, 568, **569**, 570, 571
porteri, 578
  caerulescens, 578
praecox, 600
pringlei, 559
  exclusa, 559
  lobulata, 559
prona, 568, **570**
pruinosa, 493, 522, 543, 563, 571, 572, **574**, 575, 576, 577, 637, 642
  brachypoda, 575
  cognata, 577
  delawarensis, 568
  dissona, 572, **575**
  gattingeri, 573
  latisepala, 577
  magnifolia, 575, **577**, 578
  parvula, 575, **576**
  pruinosa, **575**, 576, 577, 642
  rugosa, 575, **576**, 577
  virella, 572, 573, 574, 575, **577**
prunifolia, 641
×*puberis*, 543
pulcherrima, 581, 582, **583**
  incilis, 585
  opima, 584
pulla, 621, 630
punctata, 497, 522, 523, *533*, **542**, 543, 546, 635, 637
  aurea, 543
  brevispina, 512
  canescens, 543
  microphylla, 542, 543
  pausiaca, 543, 637
pura, 560
purpurea
  altaica, 515
purpurella, 633, **634**, 635
putnamiana, 558
pyracantha
  crenulata, 470
pyracanthoides, 539
  arborea, 539
pyrifolia, 446
pyriformis, 638
quaesita, 618, 619, **625**, 626, 631
  egens, 622
  floridana, 625
quercina, 555
randiana, 522
rava, 623
ravenelii, 624
recurva, 623
regalis, 539
  paradoxa, 540
repentina, 573
resima, 625
reverchonii, *533*, 534, **537**
  discolor, 537
  mohrii, 537, **538**
  palmeri, 518, 537, **538**
  reverchonii, *533*, **537**, 538, 539
  stevensiana, 537
rhipidophylla, 496
rhodella, 607
rigens, 617
rigida, 575
rimosa, 624
rivularis, 509, ***510***
rivuloadamensis, 506, 633, **634**
rivulopugnensis, 633, 634, **635**
roanensis, 562, 563
robesoniana, 561
robur, 583
robusta, 641
roribacca, 562, **566**
rosei
  amoena, 540
  rosei, 540
rotunda, 538
rotundata, 603
rotundifolia, 558
  bicknellii, 636
  chrysocarpa, 597
  faxonii, 598
  pubera, 598
rubella, 587, 588, **589**, 590
rubicundula, 576
rubribracteolata, 516, 517, **520**, 521, 596
rubrifolia, 539
rubrisepala, 537
rubrocarnea, 522
rudis, 539
×rufula, 546, 547, **548**
rugosa, 576
×*rupicola*, 518
rustica, 573
sabineana, 539
saligna, 500, 501, 508, **509**, 510
sanguinea, 438
sargentii, 580, **581**, 582
  gilva, 582
scabera, 517, 518
scabrida, 522, **523**, 524, 637
  asperifolia, **524**
  cyclophylla, **524**
  egglestonii, **524**
  honesta, 524
  scabrida, 523, **524**
scabrifolia, 492
schizophylla, 536, 537, **539**
schuettei, *548*, 562, 563, **565**, 566, 570
  cuneata, 565, **566**
  ferrissii, 565, **566**
  gigantea, 565, **566**
  schuettei, *548*, 565, **566**
seclusa, 576
secreta, 642
secta, 545
seducta, 574, 576
segnis, 613, **614**
senta, 620, **631**, 632
sera, 553, 554
serratifolia, 490
serratissima, 550
sertata, 560
setosa, 540
severa, 539
sheila-phippsiae, 594, **596**
  saskatchewanensis, 596, **597**
  sheila-phippsiae, 596, **597**
sheridana, 516, 517, **521**
shinnersii, 541
shirleyensis, 522
shuswapensis, 506, 507, 511, **514**
×sicca, 525, 527, 594, 595, **641**, 642
  glabrifolia, **642**
  sicca, **642**
signata, 540
×*simulata*, 518
sinistra, 540
sodalis, 626, 627
sordida, 545
sororia, 612, **614**, 615
  segnis, 614
  visenda, 613
spathulata, *490*, 495, **500**, 606
spatiosa, 522
speciosa, 579
spes-aestatum, 542, **545**
spinulosa, 517, 518
×*stenosepala*, 532
stevensiana, 537, 538
stolonifera, 568, **570**
stonei, 586, **591**
straminea, 589, 590
strongylophylla, 539
subflavida, 616
sublobulata, 494
submollis, 551, 552, 553, 554, **556**, 557, 594, 595, 640
suborbiculata, 572, **574**
subrotundifolia, 599
succincta, 544
succulenta, 516, 517, **518**, 519, 520, 523, 641
  gemmosa, **519**
  laxiflora, 519
  macracantha, 520
  michiganensis, **519**
  neofluvialis, **519**, 520
  occidentalis, 520

pertomentosa, 520
pisifera, **519**
succulenta, **519**
sucida, 544
suksdorfii, 497, 512, 635
sutherlandensis, 534
sylvestris, 523, 636
tantula, 539
tanuphylla, 519
taquetii, 479
tardiflora, 539
tardipes, 560
tatnalliana, 557
tecta, 581, **584**
tenax, 539
tenella, 562, 563
tennowana, 512
tenuifolia, 561, 563
tenuis, 540
tenuispina, 539
teres, 618, 621, **628**, 629
tersa, 540
texana, 552, 553, **554**, 555, 556
dasyphylla, **555**
texana, **555**
thermopegea, 591
tomentosa
microcarpa, 517
torminalis, 436
tortilis, 638
tracyi, **549**
madrensis, 549
transmississippiensis, 553, 554
treleasei, 638
trianthophora, 606
triflora, 492, *603*, 608, 609, **610**, 644
tristis, 613, 631
triumphalis, 540
truncata, 539
tumida, 575
turnerorum, 525, 527, 528, 550, 635, **642**
umbratilis, 574
uniflora, 518, *603*, 605, **606**, 643, 644
brittonii, 606
ursopedensis, 495, 521, 525, 593, 594, 595, **596**
uticaensis, 559
uvaldensis, 550
vaga, 519
×vailiae, 516, 518, 522, 605, 610, **642**, 643
vallicola, 540
varians, 573
velutina, 535
venusta, 581, **583**, 584
pallens, 584
×*verruculosa*, 637
versuta, 625
viaria, 625
viburnifolia, 554
vicana, 624, 625
vicina, 544
villaris, 623, 624
villicarpa, 590
villiflora, 540
villosa, 490
virella, 527, 577
virgata, 588
viridis, 493, 531, **532**, *533*, 534, 535, 636
desertorum, 642
glabriuscula, 531, 532, 533, **534**
lanceolata, 532, **533**, 534
lutensis, 534
nitens, 532, **535**
nitida, 535
ovata, 531, 532, *533*, **534**
velutina, 531, 532, **535**
viridis, **532**, *533*, 534, 535
visenda, **613**, 614
vittata, 638
warneri, 518
wattiana, **515**
incisa, 515
wattiana, **515**
websteri, 522
whitakeri, 494
williamsii, 502, 503, **504**, *505*
wootoniana, 561, 562, 563, **564**, 591
xanthophylla, 558
youngii, 499
×*Crataemespilus*
canescens, 613, 644
Crayton hawthorn, 592
Cream avens, 69
Cream-bush, 418
Cream-flowered cotoneaster, 454
Creek plum, 382
Creeping blackberry, 55
Creeping cinquefoil, 135
Creeping cotoneaster, 460
Crepis, 29
Crested cinquefoil, 191
Crisp's cotoneaster, 454
Crispa, 351
Cronquist, A., 6, 9, 29, 36, 75, 104, 108, 109, 116, 132, 172, 206, 212, 321, 401, 404, 405, 493
Crossosoma, 10, **11**
bigelovii, *11*, **12**
glaucum, 12
parviflorum, 12
californicum, **12**
glaucum, 12
parviflorum, 12
Crossosoma Family, 9
Crossosomataceae, 6, **9**
Crossosomatales, 6, 10
Cruise, J. E., 647, 648, 659, 661
Crum, E., 137
Crum's drymocallis, 290
Crum's wood beauty, 290
Cuizhi, G., 480, 482
Culley, T. M., 404, 483
Cunningham Marsh cinquefoil, 175
Curl-leaf mountain mahogany, 337
Cushman, K. R., 646
Cusick, A. W., 460
Cusick's shadbush, 652
Cutleaf blackberry, 43
Cutleaf stephanandra, 351
Cydonia, 20, 428, **486**, 487, 488
japonica, 485
oblonga, 480, *486*, **487**
sinensis, 488
Cynipid wasp, 77
Cypress Hills hawthorn, 507

Dahlberg, A., 327
Dalibarda, 28, 29
fragarioides, 71
lobata, 72
repens, 29, 51, 52
Dalibarde rampante, 51
Dalrympelea, 6
Damask rose, 88
Dansereau, P. M., 208
Dark cotoneaster, 467
Dark raspberry, 45
Dasiphora, 120, 122, 131, 274, **295**, 296, 297, 304
floribunda, 296
fruticosa, **296**, 297
floribunda, 296
tenuifolia, 296
Davis, H. A., 29, 49
Davis, R. J., 110, 178
De Cleene, M., 409, 411
De Ley, J., 409, 411
Deerbrush, 340
del Moral, R., 422
Delicate horkelia, 261
Delicious raspberry, 39
Dendriopoterium, 318
Denisenko, P. P., 24, 27
Deseret drymocallis, 291
Deseret wood beauty, 291
Desert almond, 370
Desert apricot, 373
Desert bitterbrush, 341
Desert peach, 373
Desert rose, 80
Desert sweet, 394
Dessert strawberry, 279
Diamond Lake horkelia, 260
Dibble, A. C., 646, 647, 648, 657, 658, 660, 661
Dichotomanthes, 427
Dickinson, T. A., 385, 399, 414, 415, 416, 492, 514, 539, 610, 611, 636
Dickison, W. C., 6
Dickoré, W. B., 460
Dickson, E. E., 472, 473, 476
Diels' cotoneaster, 466
Dietrich, W., 29
Diggs, G. M., 382
Dikshit, B. K., 298
Dilleniales, 10
Diplolepis, 77
polita, 77
spinosa, 77
Dobeš, C., 122, 127, 131, 132, 133, 137, 138, 219, 220, 281, 296, 303, 304
Dodge hawthorn, 603
Dog rose, 90
Dog Valley ivesia, 241
Donoghue, M. J., 494, 644
Dorothy Perkins, 85
Dotted haw, 542
Doucette, E. T., 646
Douglas's drymocallis, 292
Douglas's hawthorn, 512
Douglas's spiraea, 406
Douglas's wood beauty, 292
Downy cinquefoil, 145
Downy hawthorn, 553
Downy shadbush, 660
Drobney, P. M., 48
Dropwort, 23
Drummond's cinquefoil, 159
Drummond's mountain-avens, 328
Dryade, 326
Dryade à feuilles entières, 331
Dryade de Drummond, 328
Dryadeae, 21, **326**, 327
Dryadoideae, 20, 21, **326**, 343
Dryas, 73, 131, **326**, 327, 331, 343
ajanensis, 328, **329**, *330*
beringensis, 327, **329**, *330*
incanescens, 330
alaskensis, 328, **329**, 330
chamissonis, 331
crenulata, 329
drummondii, 327, **328**, 331
eglandulosa, 328
tomentosa, 328
hookeriana, 328, **329**
incanescens, 330
incisa, **328**, 329
integrifolia, 327, 328, *330*, **331**
canescens, 331
chamissonis, 331
crenulata, 329
integrifolia, *330*, **331**
subintegrifolia, 331
sylvatica, **331**
×*lewinii*, 331
octopetala, 327, 328, **330**, 331
alaskensis, 329
angustifolia, 329
drummondii, 328
glabrata, 329
hookeriana, 329
kamtschatica, 329
luteola, 329
octopetala, 329, 330
viscida, 329
pentapetala, 58
punctata, 329
alaskensis, 329
×*sundermannii*, 328
sylvatica, 331
tomentosa, 328
×*wyssiana*, 331
Drymocallis, 120, 122, 247, 250, 274, **280**, 281, 288, 290, 292, 304
agrimonoides, 285
albida, 293
arguta, 282, 283, 284, **285**, 286

Drymocallis arguta (*continued*)
convallaria, 286
arizonica, 283, **284**, 291
ashlandica, 283, 284, **288**
campanulata, 282, **294**
convallaria, 282, 283, 284, 285, **286**, 287
cuneifolia, 282, *285*, **294**, 295
cuneifolia, **295**
ewanii, *285*, **295**
deseretica, 282, 284, 286, **291**
fissa, 282, **284**, *285*, 286, 290
glabrata, 282, 283, **291**
glandulosa, 256, 283, *285*, 286, **292**
arizonica, 284
ashlandica, 288
ewanii, 295
glabrata, 291
glandulosa, *285*, **292**, 293, 294
hansenii, 287
micropetala, 286
nevadensis, 288
pseudorupestris, 289
reflexa, 287, 292, **293**
viscida, *285*, 292, **293**
wrangelliana, *285*, 292, **293**
glutinosa, 286
hansenii, 282, 283, **287**
lactea, 283, 284, **287**, 290
austiniae, **288**, 289
lactea, **288**, 293, 295
micropetala, 282, **286**
pseudorupestris, 283, 284, 287, 288, **289**, 290
crumiana, 281, 289, **290**
pseudorupestris, **289**, 290
saxicola, 288, 289, **290**, 291
pumila, 290
rhomboidea, 282, *285*, **294**
valida, 282, 286
Dubé, M., 302
Duchesnea, 120, 122, 124, 133, 219, **272**
chrysantha, 273
indica, **273**
indica, **273**
microphylla, 273
Dugle, J. R., 95, 107
Duke, J. A., 24, 26, 27, 87
Dusky horkelia, 259, 260
Dutch rose, 99
Dvorsky, K. A., 585, 607, 618
Dwarf cinquefoil, 134, 192
Dwarf flowering almond, 377
Dwarf greasebush, 14
Dwarf hawthorn, 621
Dwarf ivesia, 231
Dwarf mountain ash, 444
Dwarf mountain-mahogany, 338
Dwarf nine-bark, 350
Dwarf raspberry, 29, 35
Dwarf red blackberry, 51
Dwarf red raspberry, 51
Dwarf shadbush, 652
Dwarf spiraea, 403

Early cinquefoil, 178
Earthquake cotoneaster, 458
Eastern chokecherry, 365, 366
Eastern hawthorn, 562
Eastern mayhaw, 547
Eastern meadowsweet, 401
Eastern mountain mahogany, 335
Eastern sandcherry, 375
Eastern shadbush, 658
Easter-rose, 390
Eglantine rose, 90
Egolf, D. R., 469
Elegant cinquefoil, 194
Elkington, T. T., 327, 331
Elkwater hawthorn, 506
Ellis, J. R., 135, 301
Ells, J., 465
Elmer's cinquefoil, 156
Elm-leaf blackberry, 54
Elven, R., 76, 121, 122, 126, 130, 146, 188, 195, 196, 199, 205, 206, 211
Elvin-Lewis, M. P. F., 77, 100, 106
Emplectocladus, 354
fasciculatus, 370
Enderby hawthorn, 505
Endress, P. K., 6, 10
English cinquefoil, 135
English hawthorn, 497
Enquist, M. F., 383
Ensenada rose, 79
Ensign, M. R., 15, 16, 17
Entire-leaved cotoneaster, 458
Entire-leaved mountain-avens, 331
Entomosporium maculatum, 489
Erect cinquefoil, 136
Eriksen, B., 122, 123, 195, 196, 199
Eriksson, T., 24, 57, 59, 122, 219, 250, 274, 281, 296, 298, 301, 303, 304
Eriobotrya, 19, 429, **432**, 447
deflexa, 432
japonica, 426, **432**
Eriogynia
caespitosa, 413
hendersonii, 412
pectinata, 422
uniflora, 415
Erlandsson, S., 128
Erlanson rose, 81
Erlanson, E. W., 75, 76, 108, 117, 118, 119
Ertter rose, 108
Ertter, B., 75, 76, 82, 106, 107, 110, 112, 119, 121, 122, 130, 131, 132, 136, 137, 141, 143, 146, 148, 150, 152, 159, 160, 165, 167, 168, 170, 171, 175, 176, 179, 181, 182, 183, 186, 187, 188, 195, 205, 206, 208, 209, 211, 212, 213, 216, 218, 219, 222, 225, 229, 234, 246, 252, 254, 259, 261, 262, 263, 264, 266, 267, 270, 271, 272, 280, 281, 288, 290, 295, 297, 299, 300, 304, 309, 389
Erwin, D. M., 346
Erythrocoma, 58
triflora, 62
Eurasian acicular rose, 114
European bird cherry, 366
European blackberry, 36, 55, 56
European dewberry, 37
European mountain ash, 438
European plum, 376
European raspberry, 42
Euscaphis, 6
Evans, K. J., 37
Evans, R. C., 385, 399, 414, 415, 416, 449, 468
Evergreen pear, 480
Evergreen rose, 82
Ewan's drymocallis, 295
Ewan's wood beauty, 295
Exochorda, **384**, 385
grandiflora, 384
racemosa, *383*, **384**
Exochordeae, 21, **383**, 384

Fallugia, 23, 57, **73**
micrantha, 73
paradoxa, **73**, *74*
acuminata, 73
False goat's-beard, 423
False horkelia, 270
False spiraea, 396
False violet, 51
Fanfoil cinquefoil, 190
Fang's cotoneaster, 467
Fan-leaved cinquefoil, 190
Fan-leaved hawthorn, 567
Farinopsis, 304
salesoviana, 301
Fassett, N. C., 48, 62
Faux-fraisier, 71
Fay, M. F., 434
Feather cinquefoil, 174
Fecund hawthorn, 541
Fedorov, A. A., 24
Feigen, M., 493, 512
Fern bush, 394
Fernald, M. L., 29, 36, 42, 43, 48, 68, 69, 71, 132, 133, 134, 135, 141, 154, 213, 301, 306, 307, 374, 647, 654, 655, 656, 657, 658, 659, 660, 661
Fernald's shadbush, 655
Ferris, R. S., 117
Few-flowered cotoneaster, 463
Ficus
pyrifolia, 481
Field ivesia, 242
Filipendula, 20, **23**, 24, 27
angustiloba, 25
camtschatica, 25
denudata, 26
hexapetala, 27
lobata, 25
occidentalis, 24, **25**
palmata, 25
×*purpurea*, 24
rubra, 24, **25**, *26*
stepposa, 27
ulmaria, 24, 25, **26**, 27
denudata, 26, 27
picbaueri, 27
ulmaria, 27
vulgaris, 24, **27**
Filipendule, 23
Filipendule rouge, 25
Filipendule vulgaire, 27
Fire cherry, 367
Fireberry hawthorn, 597
Firethorn, 468
Flagstaff cinquefoil, 149
Flatwoods plum, 380
Florida bitterbush, 5
Flowering quince, 484, 485, 487
Flowering raspberry, 29
Flowering tea crabapple, 478
Foam-bush, 418
Focke, W. O., 29, 45, 416
Fodder burnet, 318
Forked cinquefoil, 208
Forsellesia, 13
arida, 15
clokeyi, 14
meionandra, 16
nevadensis, 15
planitierum, 16
pungens, 14
glabra, 14
spinescens, 14
stipulifera, 15
texensis, 17
Foster, S., 24, 26, 27
Fragaria, 19, 120, 131, 132, 272, **274**, 301, 304
americana, 276
×*ananassa*, 279
ananassa, 279
cuneifolia, 279
australis, 277
bracteata, 276
×*bringhurstii*, 279
californica, 276
chiloensis, 275, **278**, 279
ananassa, 279
chiloensis, 279
lucida, **278**
pacifica, 278, **279**
crantzii, 190
crinita, 276
cuneifolia, 279
glauca, 277
grayana, 278
helleri, 276
indica, 273
lucida, 278
multicipita, 277
palustris, 301
platypetala, 278
sibbaldifolia, 278
sterilis, 132
suksdorfii, 278

terrae-novae, 277
truncata, 278
vesca, *273*, **275**
americana, *273*, 275, **276**
bracteata, 275, **276**
californica, 275, **276**, 279
chiloensis, 278
crinita, 276
helleri, 276
vesca, **275**
virginiana, *273*, 275, **276**, 277, 279
australis, 277
canadensis, 277
glauca, **277**, 278, 279
grayana, 277, **278**
illinoensis, 278
platypetala, 277, **278**
terrae-novae, 277
virginiana, *273*, **277**
Fragariinae, 127, 274, 281, 298, 304
Fragrant crabapple, 475
Fraisier, 274
Fraisier américain, 276
Fraisier des bois, 275
Fraisier des champs, 277
Fraisier glauque, 277
Framboisier, 42
Framboisier d'Europe, 42
Framboisier noir, 46
Framboisier sauvage, 43
Franchet's cotoneaster, 465
Frankia, 326, 327, 339
French mission pear, 484
French rose, 88
Fries, M., 132, 139
Frosted hawthorn, 574
Frosted thorns, 571
Frost-Olsen, P., 302, 311
Frye, C. T., 646, 659
Fryer, J., 448, 460, 464

Gajewski, W., 59, 63, 65, 66
Gallic rose, 88
Gamble's cotoneaster, 456
Garcia, J. D., 191
Garden lady's mantle, 309
Garden strawberry, 279
Gardes, M., 327
Garland Prairie cinquefoil, 170
Garland tree, 475
Gaspé shadbush, 654
Gasquet rose, 117
Gataulina, G. N., 405
Gattinger's hawthorn, 573
Gehrke, R., 303, 310
Gerber, D., 335
German cinquefoil, 146
Germander meadowsweet, 409
Geum, 19, 20, 23, 57, **58**, 59, 60, 62, 64, 69, 71, 73
album, 69
aleppicum, 60, 61, 65, 67, 68, 69
strictum, 67
×*aurantiacum*, 65, 68
calthifolium, 60, 63, **64**
camporum, 68
canadense, 61, **68**, 69, 70
brevipes, 68
camporum, 68
grimesii, 68
texanum, 68, 69
canescens, 63
×*catlingii*, 69, 70
ciliatum, 63
decurrens, 67
donianum, 71
flavum, 69
fragarioides, 71
geniculatum, 60, **65**
glaciale, 59, 60, **62**
hirsutum, 69
idahoense, 72
laciniatum, 60, 61, **68**, 69, 70
trichocarpum, 68
lobatum, 72
×*macneillii*, 68, 70
×*macranthum*, 63, 64
macrophyllum, 61, **66**, 67, 68
macrophyllum, 60, 65, **66**, 67
perincisum, 65, **66**, 67, 68
rydbergii, 66
oregonense, 66
peckii, 60, **64**
pentapetalum, 58
perincisum, 66
intermedium, 66
×*pervale*, 65, 67
×*pulchrum*, 65, 66
radiatum, 60, **64**, 65
rivale, 60, **65**, 66, 67, 68
rossii, 60, **63**, 64
depressum, 63
turbinatum, 63
schofieldii, 60, **63**, 64
strictum, 68
decurrens, 67
triflorum, *52*, 60, **62**
campanulatum, 63
canescens, 63
ciliatum, 60, 62, **63**
triflorum, *52*, **62**
turbinatum, 63
urbanum, 61, 68, 69, **70**
vernum, 61, 67, **70**
virginianum, 61, 68, **69**, 70
Giant cinquefoil, 164
Giant false spiraea, 397
Gille, A., 403
Gillena, 425
Gillenia, 19, 424, **425**
stipulata, **425**, *426*
trifoliata, 425, **426**
Gillenieae, 21, 345, **424**
Gilman, A. V., 92, 93, 99
Glacier avens, 62
Gleaming lady's mantle, 308
Gleason, H. A., 29, 36, 132, 212, 321, 374, 401, 404
Globose drymocallis, 294
Globose wood beauty, 294
Glossopetalon, 9, 10, **12**, 13
clokeyi, 13, **14**
meionandrum, 16
nevadense, 15
stipuliferum, 15
planitierum, 16
pungens, 13, **14**
glabrum, 14
spinescens, *11*, 13, **14**, 15, 16, 17
aridum, *11*, **15**, 16
meionandrum, 15, **16**
mexicanum, 15
microphyllum, 15, **16**
planitierum, 15, **16**
spinescens, 13, 14, **15**
stipuliferum, 15
texense, 13, **17**
Goat's-beard, 422
Godfrey, R. K., 29
Golden raspberry, 30
Goldman, D. H., 28
Gooseberry rose, 80
Gordon's ivesia, 234
Grand Canyon rose, 81
Grande potentille subarctique, 194
Granite mousetail, 236
Gray's cinquefoil, 191
Greasebush, 12
Greasewood, 340, 393
Great Basin cinquefoil, 181
Great burnet, 323
Great Lakes sandcherry, 374
Great Plains Flora Association, 29
Great Plains hawthorn, 521
Green hawthorn, 532
Greene, E. L., 59, 138
Greene's cinquefoil, 140
Greene's drymocallis, 293
Greene's mountain ash, 442
Greene's wood beauty, 293
Gregg hawthorn, 550
Grimes's ivesia, 229
Grimy ivesia, 229
Groh, H., 374
Grossi, C., 84
Gu, C. Z., 432, 471, 473
Guamatelaceae, 6
Guillard, J. C., 30
Guo, W., 488, 489
Gymnosporangium, 424, 492

Hagenia, 312
Haines, A., 384, 647
Hairycup horkelia, 270
Hall crabapple, 478
Hall's cinquefoil, 155
Halmia, 445
tomentosa
pyrifolia, 446
Hamrick, J. L., 647, 648, 661
Hansen's drymocallis, 287
Hansen's wood beauty, 287
Hara, H., 423, 424
Harbison hawthorn, 608
Hardhack, 403
Hardiman, N. A., 483
Hardin, J. W., 445, 446
Harison's yellow rose, 94
Harkness, P., 77
Harmon, P. J., 30
Harms, V. L., 107
Harris, J. A., 411
Harsh downy-rose, 92
Havard's almond, 371
Haw, 492
Hawthorn, 491, 492, 494, 499, 500, 501, 502, 504, 507, 508, 510, 511, 515, 516, 517, 518, 519, 520, 521, 522, 525, 536, 538, 542, 543, 551, 563, 572, 574, 578, 586, 597, 599, 610, 623, 635, 636, 641
Heath, P. V., 88
Hebda, R. J., 64, 422
Hebeloma
alpinum, 327
Hedge cotoneaster, 463
Hedrick, U. P., 353, 378
Heikens, A. L., 387
Henderson, L., 39
Henderson's horkelia, 249
Henrickson, J., 73, 332, 339, 343, 386, 389, 390, 392, 394, 396, 414, 422
Herb bennet, 70
Herbe à pisser, 322
Herrera, C. M., 434, 644
Herrick, J. W., 52
Herringbone cotoneaster, 462
Heslop-Harrison, Y., 36
Hesperhodos, 78
minutifolius, 79
stellatus, 80
mirificus, 81
stellatus, 80
Hess, W. J., 385, 408, 429
Heteromeles, 429, **447**, 449, 468, 488
arbutifolia, **447**, *448*
arbutifolia, 447
cerina, 447
macrocarpa, 447
fremontiana, 447
salicifolia, 447
Hickman's cinquefoil, 174
Hieracium, 29
Higan cherry, 369
Highbush blackberry, 38
Hill raspberry, 46
Hill, M. P., 39
Hillside hawthorn, 543
Himalayan blackberry, 36
Hinds, R. R., 647
Hipp's cinquefoil, 162
Hitchcock, C. L., 172, 206, 212, 376, 405, 413
Hitchcock's spiraea, 406
Hjelmqvist's cotoneaster, 462
Hoary cinquefoil, 144
Hoffman, D., 42
Hog plum, 380, 382
Hoghaw, 501
Hokanson, K. E., 277, 278
Holloway, P. S., 322
Hollyleaf cherry, 362
Holmen, K., 441
Holmes' hawthorn, 560
Holmgren, N. H., 15, 16, 75, 108, 109, 116, 155, 163, 165, 170, 172, 209, 216, 226, 235, 290, 296, 509

Holmgren's cinquefoil, 199
Holodisceae, 416
Holodiscus, 21, 398, **415**, 416, 417, 422
argenteus, 416
australis, 418
boursieri, 418
discolor, 416, **417**, *418*, 419, 420
delnortensis, 418
discolor, ***418***, 419
dumosus, 418, **419**
franciscanus, 418
glabrescens, 421
microphyllus, 419
dumosus, 419
australis, 418
glabrescens, 421
saxicola, 418
fissus, 416
glabrescens, 421
microphyllus, 416, 417, **419**, 420
glabrescens, 420, **421**
microphyllus, **420**, 421
sericeus, **420**, 421
orizabae, 416
Honeybee, 254
Hooker, J. D., 242
Hooker's cinquefoil, 208
Hooker's mountain-avens, 329
Hoover, R. F., 117, 254
Hope-of-summers, 545
Horkelia, 120, 121, 122, 133, 219, 220, 243, **246**, 247, 248, 252, 256, 259, 264, 271, 281
[group] *Capitatae*, 249, 259
[group] *Cuneatae*, 258
[group] *Hirsutae*, 264
[group] *Sericatae*, 249, 264
[group] *Tenuilobae*, 249
[group] *Tridentatae*, 264
[unranked] *Ambiguae*, 264
[unranked] *Californicae*, 250
[unranked] *Capitatae*, 259
[unranked] *Cuneatae*, 250
[unranked] *Eremicae*, 237
[unranked] *Hirsutae*, 264
[unranked] *Lycopodioides*, 230
[unranked] *Purpurascentes*, 270
[unranked] *Saxosae*, 221
[unranked] *Sericatae*, 264
[unranked] *Tenuilobae*, 250
[unranked] *Tridentatae*, 264
[unranked] *Unguiculatae*, 237
sect. *Californicae*, 250
sect. Capitatae, 248, **259**
sect. *Cuneatae*, 250
sect. *Hirsutae*, 264
sect. Hispidulae, 247, **248**, 250
sect. Horkelia, 248, **250**, 259
sect. Parryae, 247, **257**, 258
sect. *Sericatae*, 264
sect. *Tenuilobae*, 250
sect. Tridentatae, 247, 248, 259, **264**
subg. *Horkeliella*, 270, 271
subg. *Ivesia*, 219
argyrocoma, 242
beneolens, 226
bernardina, 251, 252
bolanderi, **251**, 252
bolanderi, 252
clevelandii, 252
marinensis, 253
parryi, 251, 252, 254
brownii, 263
californica, *245*, 250, 254, 255, **256**
californica, *245*, **256**, 257
dissita, 257
elata, 256, **257**
frondosa, 250, 256, **257**
sericea, 255
capitata, 261
caruifolia, 267
clevelandii, 250, 251, **252**, 253
brevibracteata, 252, 253
clevelandii, 252, **253**
congdonis, 271
congesta, 264, **267**, 268, 269
congesta, **268**
nemorosa, **268**, 269, 270
cuneata, 250, 251, **254**, 255
cuneata, 175, 254, **255**
puberula, 254, **255**
sericea, 254, **255**
daucifolia, 265, **266**, 267
caruifolia, 266, **267**
daucifolia, **266**, 267
indicta, 266, **267**
latior, 266
elata, 257
flavescens, 269, 270
frondosa, 250, 257
fusca, 249, 258, **259**, 260, *263*
brownii, 259, 260, 261, 262, **263**
capitata, 259, 260, **261**, 262
filicoides, 248, 259, **260**, 261
fusca, **260**
parviflora, 242, 260, 261, **262**, 263
pseudocapitata, 259, 260, 261, **262**, 263
tenella, 260, **261**, 263
tenuiloba, 252
glandulosa, 257
gordonii, 233
alpicola, 234
megalopetala, 233
hendersonii, 248, **249**
hirsuta, 267, 268
hispidula, 248, **249**, 264
howellii, *263*, **265**, 266
kelloggii, 255
marinensis, 250, 251, **253**, 254
multifoliolata, 244
parryi, 251, **258**
parviflora, 262
pseudocapitata, 262
puberula, 255
purpurascens, 271
congdonis, 271
rydbergii, **251**, 252, 254
scandularis, 232
sericata, 249, **265**, 266
sericea, 255
sericoleuca, 240
tenuiloba, 249, 251, **252**, 254
tenuisecta, 260
tridentata, 259, 264, 267, **269**
flavescens, 268, 269, **270**
tridentata, 268, **269**
truncata, 250, **256**
tularensis, 248, **249**
wilderae, **258**
yadonii, 250, 251, 252, **253**
Horkeliella, 120, 121, 122, 133, 219, 220, 248, **270**, 271
congdonis, **271**
purpurascens, *263*, **271**, 272
Horn, D. D., 388
Hortulan plum, 382
Houston, L., 434
Howell, J. T., 222, 240, 254, 491
Howell, T. J., 279
Howell's horkelia, 265
Howell's ivesia, 234
Huachuca cinquefoil, 186
Hualapai cinquefoil, 185
Huckins, C. A., 473
Huckleberry hawthorn, 512
Huertea, 6
Huerteales, 6
Hultén, E., 132, 139, 162, 201, 202, 203, 214, 327, 329, 331
Hunt, D. R., 425
Hutchins, C. R., 177
Hutchinson, J., 271
Hutton, E. E., 30
Huxley, A., 399, 409, 410
Hybrid tea, 77
Hybridus Pendulus, 454
Hylmö, B., 448, 464

Idaho barren-strawberry, 72
Idaho drymocallis, 291
Idaho wood beauty, 291
Ikeda, H., 123
Indian physic, 425
Indianplum, 385
Indian-strawberry, 272, 273
Interior rose, 109
Intermediate shadbush, 660
Intermountain horkelia, 262
Intricate hawthorn, 588
Iowa crabapple, 474
Island ironwood, 346
Islay, 362
Ivesia, 120, 121, 122, 133, 137, 176, **219**, 220, 221, 222, 225, 243, 245, 247, 248, 271
[unranked] *Eremicae*, 237
[unranked] *Lycopodioides*, 230
[unranked] *Setosae*, 221
[unranked] *Unguiculatae*, 237
sect. Comarella, 219, 221, **243**, 271
sect. *Eremicae*, 237
sect. Ivesia, 221, 222, 228, **230**, 233, 237
sect. *Lycopodioides*, 230
sect. *Saxosae*, 221
sect. Setosae, 183, 220, **221**, 222, 227, 228, 230, 231, 236, 238
sect. Stellariopsis, 221, **245**
sect. Unguiculatae, 221, **237**
subg. *Horkeliella*, 271
alpicola, 234
aperta, 238, 239, **240**
aperta, 240, **241**
canina, 240, **241**
argyrocoma, 237, 238, **242**
argyrocoma, **242**, 243
moranii, 242
arizonica, 219, 221, 222, **224**, *225*
arizonica, **225**, 226
saxosa, 225, **226**
baileyi, 222, **226**, 227
baileyi, **226**
beneolens, **226**, 227, 228
setosa, 227
callida, 223, **224**
campestris, 219, 237, 238, **242**
chaetophora, 231
cryptocaulis, 219, 220, 221, 223, **228**, 230, 231
eremica, 239
gordonii, *225*, 230, 231, **233**, 234
alpicola, **234**, 235
gordonii, *225*, **234**, 235
pygmaea, 231
ursinorum, 234, **235**
wasatchensis, 230, 234, **235**
gracilis, 137
jaegeri, 222, **223**, 224
kingii, 176, *225*, 237, **238**, 241
eremica, 237, 238, **239**
kingii, *225*, 238, **239**
lemmonii, 164
longibracteata, 220, 222, 230, **236**
lycopodioides, 230, 231, **232**, 233, 235, 236
lycopodioides, **232**
megalopetala, 231, 232, **233**
scandularis, **232**
muirii, 219, 230, **236**
multifoliolata, 124, 219, 220, 243, **244**
paniculata, 220, 221, 223, **229**
patellifera, 222, **224**
pickeringii, 238, **239**, 240

pinnatifida, 170
pityocharis, 237, 238, **241**
purpurascens, 271
congdonis, 271
pygmaea, 228, 230, **231**, 236
rhypara, 220, 221, 223, **228**, 229
rhypara, **229**
shellyi, **229**
sabulosa, 243, **244**, *245*
santolinoides, *245*, **246**
saxosa, 221, 222, **223**
sericoleuca, 238, 239, **240**, 241, 242
setosa, 222, 223, 226, **227**
shockleyi, 223, **227**, 228
ostleri, **228**
shockleyi, 227, **228**
tweedyi, 230, 231, **235**, 236
unguiculata, 237, 238, 240, **241**, 242
utahensis, 221, 230, 231, **233**
webberi, 219, 230, 231, **236**, 237

Jacobson, A. L., 456
Jaeger's ivesia, 223
Jalas, J., 128, 132, 135, 139, 141, 142, 143, 144, 190, 214, 296
Japanese bramble, 48
Japanese crabapple, 479
Japanese flowering cherry, 369
Japanese flowering quince, 485
Japanese photinia, 489, 490
Japanese rose, 84, 98
Japanese spiraea, 410
Japanese-rose, 390
Jasieniuk, M., 37, 50, 56
Jeffries, D. L., 391
Jensen, D. B., 52
Jepson, W. L., 117, 181, 230, 239, 252
Jepson's horkelia, 267
Jepson's potentilla, 216
Jesup hawthorn, 571
John Day drymocallis, 294
John Day wood beauty, 294
Johnston, B. C., 122, 127, 137, 150, 159, 160, 161, 162, 163, 165, 168, 170, 172, 175, 177, 178, 182, 189, 195
Johnston, I. M., 252, 295
Johnston, M. C., 510
Joly, S., 95, 101, 104
Jones hawthorn, 545
Jones, G. N., 438, 444, 445, 646, 647, 651, 652, 653, 654, 655, 656, 662
Jones, R. L., 29
Jones, W. V., 392
Jørgensen, C. A., 441
Josephine horkelia, 268
Juneberry, 646
Jurtzev, B. A., 123, 195, 196, 198, 218, 327, 329

Kageneckia, 427
Kalkman, C., 57, 386, 416, 434, 445, 489
Kansas hawthorn, 579
Kartesz, J. T., 231, 437, 456, 457, 461
Kashiwada, Y., 106
Kasperek, G., 460
Kaweah ivesia, 242
Kearney, T. H., 170, 186, 413
Keck, D. D., 122, 159, 226, 228, 230, 231, 232, 236, 240, 250, 252, 254, 259, 261, 262, 263, 265, 266, 267, 268, 269, 270, 271, 281, 287, 289, 295
Kellogg's horkelia, 255
Kelseya, 398, 412, **414**, 415, 422
uniflora, *407*, **415**
Kemp, J. R., 84
Kern Plateau false horkelia, 271
Kern Plateau horkelia, 249
Kern Plateau horkeliella, 271
Kerr, M. S., 321
Kerria, 386, 387, **389**
japonica, *388*, **390**
tetrapetala, 387
Kerrieae, 21, 385, **386**, 387, 390
Kevan, P. G., 84
Khasiberry cotoneaster, 460
Kiger, R. W., 29
King's ivesia, 239
Kingston Mountains ivesia, 224
Kirkman, L. K., 364
Kittell, T., 177
Klackenberg, J., 296
Klamath cinquefoil, 174
Klamath horkelia, 265
Klamath plum, 377
Klášterský, I., 98
Klein, H., 39
Klein, W. M., 29, 449
Kline, G. J., 312
Klopatek, J. M., 391
Klotz, G., 458
Knees, S. G., 399, 409, 410
Koehne, B. A. E., 410
Kohli, B. L., 212, 215, 216
Kokonor cotoneaster, 463
Kollmann, J., 37
Kroon, G. H., 449
Kruckeberg, A. R., 408
Kruschke, E. P., 578, 636
Kudriashov, B. A., 24, 27
Kugel, A. R., 401
Kumar, A., 460
Kurtto, A., 122, 128, 132, 135, 139, 141, 142, 143, 144, 190, 214, 281, 296, 304

La Sal cinquefoil, 210
Lace shrub, 351
Lachemilla, 303, 310
Lady's mantle, 302
Laguna Mountains Skipper, 253
Lance, R. W., 468
Landry, P., 647, 654, 655, 658, 660
Large-leaved avens, 66
Large-thorn hawthorn, 520
Larsen, K., 441
Late cotoneaster, 454
Late hawthorn, 517
Laurel cherry, 362
Laurentian hawthorn, 640
Laurier amande, 361
Lauro-cerasus, 352, 353, 354
caroliniana, 361
myrtifolia, 361
Lawrence, D. B., 327
Leafy drymocallis, 284
Leafy horkelia, 257
Leafy wood beauty, 284
Lee, S., 354, 385
Lei, S. A., 391
Lemmon's cinquefoil, 164
Leucosidea, 312
Levin, G. A., 80
Lewis, W. H., 75, 76, 77, 80, 81, 82, 83, 84, 95, 100, 101, 104, 106, 107, 110, 112
Ley, F. A., 417
Li, C. L., 131, 148, 273, 298, 322
Li, Q. Y., 494, 644
Lindley, J., 88
Lindleya, 427
Lingdi, L., 30, 454, 455, 457, 458, 459, 461, 462, 463, 464, 465, 467
Lis, R. A., 398, 411, 414, 415, 416, 418, 421
Little rose, 279
Littlehip hawthorn, 500
Littleleaf mountain-mahogany, 338
Lo, E. Y. Y., 493, 494, 496, 525, 636, 644
Loch Lomond hawthorn, 634
Loganberry, 30, 56
Long, A. A., 388
Long, B. H., 301
Lonicera
japonica, 492
Loquat, 19, 432
Lost cinquefoil, 175
Löve, Á., 127, 141, 441
Löve, D., 127, 141, 441
Love, R., 493, 512
Lovely shadbush, 654
Low avens, 58
Low shadbush, 656
Lu, L. T., 399, 409, 410
Lucie rose, 85
Ludwig, C. G., 127
Luetkea, 398, **421**, 422
caespitosa, 413
cinerascens, 412
hendersonii, 412
pectinata, *418*, **422**
sibbaldioides, 422
Lundberg, M., 281, 296
Lyonothamneae, 21, **345**
Lyonothamnus, 345, **346**
aspleniifolius, 346
floribundus, *344*, **346**
aspleniifolius, 346
floribundus, 346
Mabberley, D. J., 477
Macartney rose, 86
Macdougal's rose, 113
Macklin, J. A., 28, 558, 559, 560
Macmillan, B., 323
Macoun, J. M., 172
Macoun's cinquefoil, 180
Magee, D. W., 29, 438
Magnificent cotoneaster, 456
Mahaleb cherry, 366
Malaceae, 20, 427
Malacomeles, 428, **645**, 646, 662
denticulata, *645*, **646**
nervosa, 646
Maleae, 20, 21, 345, 385, 415, **427**, 432, 433, 449, 468, 494, 644
Mallah, T. S., 37, 50, 56
Mallow nine-bark, 349
Mallow-leaved nine-bark, 349
Maloideae, 385, 427
Malus, 19, 427, 428, 433, 445, 449, **472**, 473, 480
sect. *Chloromeles*, 473
angustifolia, **474**, 475
puberula, 475
spinosa, 474
baccata, 472, 473, **477**, 478, 479
toringo, 479
bracteata, 475
carolinensis, 475
coronaria, 473, 474, **475**, 476
angustifolia, 474
bracteata, 475
dasycalyx, 475
elongata, 475
glabrata, 475
glaucescens, 475
hoopesii, 475
ioensis, 474
lancifolia, 475
platycarpa, 475
plena, 475
puberula, 475
×*dawsoniana*, 476, 477
diversifolia, 476
domestica, 477
halliana, 478
hupehensis, 478
elongata, 475
floribunda, 477, 478
fragrans, 475
fusca, 472, 473, **476**, 477
diversifolia, 476
levipes, 476
glabrata, 475, 476
glaucescens, 475, 476
halliana, 472, 473, 477, **478**
hupehensis, 473, **478**, 479
ioensis, *470*, **474**
bushii, 474
creniserrata, 474
palmeri, 474
spinosa, 474
texana, 474
lancifolia, 475, 476

Malus (*continued*)
macounii, 476
paradisiaca, 477
platycarpa, 475, 476
prunifolia, 473, 476, **477**
pumila, 473, 474, **476**, 477, 483
rivularis, 476
sargentii, 479
×*scheideckeri*, 477
sieboldii, 473, 479
×*soulardi*, 474
sylvestris, 476, 477
theifera, 478
toringo, 473, **479**
sargentii, 479
zumi, 479
×*zumi*, 479
Malvales, 6
Manchu cherry, 372
Mancos rose, 108
Mansfield, D., 159, 170, 214
Many-flowered shadbush, 652
Many-leaved cinquefoil, 174
Marcetella, 318
Margaret's hawthorn, 601
Margyricarpus, 312
Marie-Victorin, F., 216
Marinelli, J., 449
Marsh potentilla, 301
Marshlocks, 300
Martin, A. C., 434
Martin, W. C., 177
Mason, C. T. Jr., 9
Matfield, B., 135
Matheke, G. E. M., 322
Matthews, M. L., 6, 10
Mavi, A., 94
Maximowicz, C. J., 24
Maxwell, H. S., 399, 409, 410
Mazama drymocallis, 288
Mazama wood beauty, 288
McAllister, H. A., 434, 438, 441, 442, 444
McArthur, E. D., 73
McAvoy, B., 422
McCutcheon, A. R., 113
McFarland, J. H., 77
McGregor, R. L., 140, 212
McKay, S. M., 647
McLaren's cotoneaster, 457
McLaughlin, S. P., 102, 103
McNeill, J., 77, 302, 309
McVaugh, R., 353
Meacham, C. A., 437, 457, 461
Meadow rose, 105
Meadowsweet, 23
Medlar, 643
Meiners, S. J., 85
Melanobatus
nigerrimus, 45
Mellichamp, T. L., 422, 423
Memorial rose, 85
Menzies' burnet, 322
Mercure, M., 105
Merisier, 363
Mesa horkelia, 255
Mespilus, 20, 428, 445, 469, 494, **643**
aestivalis, 547
apiifolia, 498
arborea, 660
arbutifolia, 446
melanocarpa, 446
calpodendron, 517
canadensis, 658
obovalis, 659
canescens, 644
crenulata, 470
flabellata, 567
germanica, 643, **644**, *645*
japonica, 432
laevigata, 497
phaenopyrum, 499
pruinosa, 574
prunifolia, 445
pyracantha, 470
Mexican alvaradoa, 4
Mexican plum, 379
Meyer, S. E., 391
Mink, J. N., 399
Mitchell, R. S., 29, 449, 461
Mitrow, G., 479, 484, 486, 487, 492
Mock-strawberry, 272
Modest cinquefoil, 209
Moerman, D. E., 25, 42, 77, 104, 110, 301
Mogollon cinquefoil, 184
Mojave rose, 109
Monasterio-Huelin, E., 55
Mono cinquefoil, 209
Montana drymocallis, 289
Montana wood beauty, 289
Moore, D. J., 413
Moore, D. M., 388
Moore, G., 28
Moore, S. M., 389
Moraceae, 20
Morefield's cinquefoil, 181
Morgan, D. R., 24, 73, 385, 416
Morton, J. F., 46
Morton, J. K., 463
Mount Rainier cinquefoil, 190
Mountain alchemilla, 308
Mountain ash, 433
Mountain meadowsweet, 402
Mountain misery, 343
Mountain nine-bark, 350
Mountain river hawthorn, 510
Mountain serviceberry, 657
Mountain shadbush, 657
Mountain-avens, 64, 326
Mountain-mahogany, 332, 334
Muir's ivesia, 236
Multiflora rose, 84
Müntzing, A., 144
Müntzing, G., 144
Munz, P. A., 117, 239, 240, 252, 261, 264, 271, 295, 324
Muravjeva, O. A., 298
Murray, D. F., 121, 126, 130, 188, 195, 196
Murray's plum, 382
Murrell, Z. E., 404
Musk rose, 83
Myrobalan, 375
Myrtle laurel cherry, 361
Mysore raspberry, 46

Nanking cherry, 372
Nanticoke blackberry, 49
Nantucket shadbush, 657
Narrow-leaf meadowsweet, 401
Narrow-leaved crabapple, 474
Naruhashi, N., 123
Necklace cotoneaster, 457
Nectarine, 353, 371
Neillia, 347, 348, **351**
incisa, **351**, *352*
malvacea, 349
monogyna
alternans, 350
Neillieae, 21, **347**, 351
Nelson, A., 122
Nelson-Jones, E. B., 434
Nesom, G. L., 425, 471, 488
Neuradaceae, 20
Nevada nine-bark, 350
Neviusia, 386, **387**, 389
alabamensis, ***388***
cliftonii, 388, **389**
New Mexico raspberry, 45
Newberry's cinquefoil, 137
Ng, T. B., 99
Nielsen, E. L., 647, 655
Nine-bark, 347
Nixon's dwarf hawthorn, 590
Nootka rose, 113
Nordal, I., 331
Norris, W. R., 457
Northern dewberry, 40
Northern downy hawthorn, 556
Northern meadowsweet, 402
Northern mountain ash, 440
Northern red haw, 556
Northwest raspberry, 45
Norton, J. B. S., 320
Norwegian cinquefoil, 140
Nothotalisia, 3
Novosieversia, 58
glacialis, 62
Nuttall's cinquefoil, 155
Nuttallia, 385
cerasiformis, 385
Nybom, H., 89, 459
Nyléhn, J., 195

Oak-leaf mountain ash, 438
Ocean-spray, 418
Oemleria, 383, 384, **385**
cerasiformis, *383*, **385**
Oenothera, 29
Ogle, D. W., 404, 410
Oh, S. H., 6, 351
Ohba, H., 369
Okanagan hawthorn, 504
O'Kennon hawthorn, 513
Okie, W. R., 353
Oklahoma plum, 380
Old man's whiskers, 62
Old-field cinquefoil, 134
Olkers, T., 39
Olmo, H. P., 37, 44
Olympic mountain rockmat, 412
One-flower kelseya, 415
One-flower spiraea, 415
One-flowered hawthorn, 606
One-seeded hawthorn, 497
One-stoned cotoneaster, 457
Opulaster, 348
alabamensis, 348
australis, 348
intermedius, 349
opulifolius, 348
pauciflorus, 349
stellatus, 348
Orange-drop cotoneaster, 467
Orchard apple, 477
Orchard crabapple, 476
Oregon crabapple, 476
Oriental cherry, 369
Osmaronia
cerasiformis, 385
lancifolia, 385
Osmaronieae, 384, 385
Osoberry, 385
Ostinia, 448
Ostler's ivesia, 228
Ouachita hawthorn, 591
Overath, R. D., 647, 648, 661
Owyhee cinquefoil, 156
Owyhee ivesia, 226

Pacific blackberry, 55
Pacific crabapple, 476
Pacific nine-bark, 349
Pacific plum, 377
Pacific silverweed, 129
Packer, J. G., 212, 215, 216, 328
Padus, 352, 353, 354
caroliniana, 361
nana, 365
Pale avens, 69
Pale serviceberry, 650
Pale shadbush, 650
Palmer, E. J., 493, 522, 523, 524, 536, 537, 540, 551, 566, 574, 637, 638, 639
Palouse horkelia, 261
Panigrahi, G., 298, 460
Pankhurst, R. J., 445
Paradise crabapple, 476
Parfitt, B. D., 73, 279, 326, 390
Parish's drymocallis, 293
Parish's wood beauty, 293
Parmena
menziesii, 56
Parney's cotoneaster, 454
Parry's horkelia, 258
Parsley haw, 498
Parsley-piert, 309
Partanen, C. R., 648, 658, 659
Partridge foot, 422
Passionate hawthorn, 563
Pasture rose, 101
Paterson, I. G., 64
Paule, J., 122, 127, 131, 132, 133, 137, 138, 144, 145, 219, 220, 281, 296, 298, 299, 303, 304
Pea rose, 110
Peach, 19, 353, 371
Peachbush, 377
Pear, 19, 479, 480, 483, 484

Pear hawthorn, 517
Pearlbrush, 384
Pearleaf crabapple, 477
Pêcher, 371
Peck, M. E., 45, 264, 289
Peebles, R. H., 170, 186, 413
Pendleton, B. K., 391
Pendleton, R. L., 391
Pennsylvania blackberry, 50
Pennsylvania thorn, 557
Pentaphylloides, 127, 296
floribunda, 297
fruticosa, 297
tenuifolia, 297
Peraphyllum, 427, 428, 646, **662**
ramosissimum, *653*, **662**
Perfumed cherry, 366
Persica, 352, 353
vulgaris, 371
Persson Hovmalm, H. A., 445
Petit merisier, 367
Petrophytum, 398, **411**, 412, 413, 414, 422
acuminatum, 414
caespitosum, *407*, 411, 412, **413**
acuminatum, 411, 413, **414**
caespitosum, *407*, 411, **413**
elatius, 413
cinerascens, 411, **412**, 413
elatius, 411, 413
hendersonii, 411, **412**, 413
Phillip, M., 331
Phillips, A. M., 81
Phipps, J. B., 18, 71, 432, 434, 447, 489, 491, 492, 493, 494, 499, 508, 521, 532, 539, 543, 548, 549, 551, 559, 563, 585, 588, 606, 607, 618, 635, 642, 643, 647
Photinia, 20, 429, 445, 447, **488**, 489
arbutifolia, 447
cerina, 447
macrocarpa, 447
davidiana, *490*, **491**
davidiana, *490*, **491**
floribunda, 445
fortuneana, 469
×*fraseri*, 489, 490
glabra, **489**, 490
melanocarpa, 446
pyrifolia, 446
rubrolutea, 479
salicifolia, 447
serratifolia, 489, ***490***
serratifolia, 490
serrulata, 489
villosa, 445, 489, **490**, 491
sinica, 491
Phragmidium, 28, 57, 120, 312, 424
Physocarpe, 347
Physocarpe à feuilles d'obier, 348
Physocarpus, 20, **347**, 348, 351
alternans, 348, **350**
annulatus, 350
panamintensis, 350
capitatus, 348, **349**
intermedius, 348, **349**
malvaceus, 348, **349**, 350
monogynus, 347, 348, **350**
opulifolius, *344*, **348**, 349, 388
intermedius, 349
tomentellus, 349
pauciflorus, 349
Pickering's ivesia, 239
Picramnia, 3, **4**, 5
pentandra, *5*
Picramnia Family, 3
Picramniaceae, **3**
Pie cherry, 368
Piedmont barren-strawberry, 72
Piletophyllum, 127, 298
Pin cherry, 367
Pinaleño cinquefoil, 186
Pine Nut Mountains ivesia, 241
Pine rose, 117
Pinewoods horkelia, 259
Pinus
clausa, 606
muricata, 117
palustris, 606
radiata, 117
rigida, 606
Piper's hawthorn, 599
Plains greasebush, 16
Plaquebière, 38
Plateau rose, 108
Platte cinquefoil, 173
Pleniflora, 390
Plum, 19, 352, 353, 354, 376, 377, 378, 379, 380, 381, 382
Plumas ivesia, 240
Plumboys, 29
Plumleaf crabapple, 477
Plumleaf hawthorn, 641
Plymouth pear, 484
Point Reyes horkelia, 253
Poirier, 479
Pojar, J., 449, 464
Pojarkova, A. I., 399
Poljakova, T. A., 405
Polylepis, 312
Pomette bleue, 501
Pommier, 472
Pommier commun, 476
Pommier de Sibérie, 477
Poplar hawthorn, 569
Populus, 506, 507, 514
tremuloides, 114
Porsild, A. E., 147, 329
Porter, D. M., 29
Porteranthus, 425
stipulatus, 425
trifoliatus, 426
Portugal laurel, 363
Potentilla, 19, 20, 120, **121**, 122, 123, 127, 131, 132, 133, 137, 138, 148, 150, 159, 165, 178, 186, 195, 201, 209, 219, 221, 222, 225, 245, 247, 252, 254, 272, 281, 296, 298, 300, 301, 303, 304
[group] *Candicantes*, 168
[group] *Multijugae*, 168
[unranked] *Arenicolae*, 136
[unranked] *Argenteae*, 143
[unranked] *Aureae*, 188
[unranked] *Biflorae*, 130
[unranked] *Brevifoliae*, 187
[unranked] *Candidae*, 150
[unranked] *Chrysanthae*, 146
[unranked] *Closterostyles*, 280
[unranked] *Comocarpa*, 295
[unranked] *Concinnae*, 176
[unranked] *Graciles*, 150
[unranked] *Haematochrus*, 148
[unranked] *Hippianae*, 160
[unranked] *Leucophyllae*, 160
[unranked] *Longipedunculatae*, 150
[unranked] *Multifidae*, 211
[unranked] *Multijugae*, 167
[unranked] *Niveae*, 195
[unranked] *Nuttallianae*, 150
[unranked] *Pectinisectae*, 150
[unranked] *Permolles*, 150
[unranked] *Rectae*, 141
[unranked] *Rubrae*, 148
[unranked] *Rubricaules*, 205
[unranked] *Saxosae*, 221
[unranked] *Subjugae*, 165
[unranked] *Subviscosae*, 182
[unranked] *Terminales*, 143
sect. *Anserina*, 126
sect. Arenicolae, 124, **136**
sect. *Argenteae*, 143
sect. Aureae, 125, 144, 147, 152, 182, 187, **188**, 189, 195, 196, 201, 212, 218
sect. Biflorae, 124, **130**, 131
sect. Brevifoliae, 126, **187**
sect. *Candicantes*, 167
sect. *Candidae*, 150
sect. *Capitatae*, 259
sect. *Chenopotentilla*, 127
sect. Chrysanthae, 125, **146**, 151, 189
sect. *Closterostyles*, 280
sect. *Comarum*, 300
sect. *Comocarpa*, 295
sect. Concinnae, 125, 168, 169, 172, **176**, 177, 180, 181, 182, 183, 189, 209
sect. *Duchesnea*, 272
sect. *Dumosae*, 195
sect. *Fragariastrum*, 131
sect. *Fragarioides*, 131
sect. *Fruticosae*, 295
sect. Graciles, 125, 126, 143, **150**, 151, 154, 157, 158, 159, 160, 161, 168, 171, 172, 178, 181, 189, 216
sect. *Haematochroae*, 148
sect. *Haematochrus*, 148
sect. *Heptaphyllae*, 150, 151
sect. *Hippianae*, 160
sect. *Horkelia*, 246
sect. *Ivesia*, 219
sect. *Leptostylae*, 127
sect. Leucophyllae, 126, 150, 158, **160**, 161, 162, 164, 178
sect. *Longipedunculatae*, 150
sect. Lupinoides, 124, **131**, 132
sect. *Multifidae*, 211, 212
sect. Multijugae, 125, 126, 153, 159, 160, 161, **167**, 168, 170, 171, 176, 177, 181, 214, 216
sect. Niveae, 123, 125, **195**, 196, 199, 201, 202, 206, 210, 218
sect. *Nuttallianae*, 150
sect. *Pectinisectae*, 150
sect. Pensylvanicae, 124, 125, 193, 196, 206, 209, 210, **211**, 212, 213, 217, 218
sect. Pentaphylloides, 121, 124, **126**, 127
sect. *Permolles*, 150
sect. Potentilla, 124, **132**, 133, 143, 272
sect. *Quinquefolium*, 131
sect. *Ranunculoides*, 150, 189
sect. Rectae, 124, **141**
sect. *Rhopalostylae*, 295
sect. Rivales, 124, **137**, 138, 191
sect. Rubrae, 124, **148**, 244, 301
sect. Rubricaules, 123, 125, 178, 196, 200, **205**, 206, 212, 213, 218
sect. *Rupestres*, 280, 281
sect. *Saxosae*, 221
sect. *Sibbaldia*, 297
sect. *Stellariopsis*, 245
sect. Subjugae, 124, 126, 161, **165**, 166, 178, 181, 206
sect. Subviscosae, 125, **182**, 183, 187, 189
sect. *Supinae*, 137
sect. Terminales, 124, **143**, 144, 147
sect. *Tormentilla*, 132
sect. *Tridentatae*, 299
ser. *Anserina*, 126
ser. *Aureae*, 188
ser. *Biflorae*, 130
ser. *Hookerianae*, 205
ser. *Multifidae*, 211
ser. *Supinae*, 137
subg. *Argentina*, 126
subg. *Closterostyles*, 280, 281
subg. *Comarum*, 300
subg. *Dasiphora*, 295
subg. *Duchesnea*, 272
subg. *Fragariastrum*, 131
subg. *Horkelia*, 246
subg. *Ivesia*, 219
subg. *Micropogon*, 130
subg. *Purpusia*, 221
subg. *Sibbaldia*, 297

Potentilla (*continued*)
  subg. *Sibbaldiopsis*, 299
  acuminata, 223
  alaskana, 154, 155
  alba, 132
  albiflora, 183, 184, **186**
  ×*aleutica*, 193
  alpestris, 190
  ambigens, 162, **164**
  ambigua, 299
  anachoretica, 213, **218**
  andersonii, 262
  angelliae, 177, **180**, 182
  anglica, 133, **135**
  angustata, 155
  anomalofolia, 159
  anserina, 122, **127**, *129*
    anserina, 127, **128**, 130
    egedei, 127, 130
    grandis, 129
    groenlandica, 127, 128, *129*, **130**
    pacifica, 127, 128, ***129***, 130
    sericea, 128
    yukonensis, 127, 128, ***129***
  aperta, 240
  arenosa, 196, 197, 199, **200**, 201, 205, 206, 208
    arenosa, **200**, 201, 210
    chamissonis, **200**, 201
  argentea, **144**, 145, 206
    pseudocalabra, 145
  arguta, 281, 285
    convallaria, 286
  argyrea, 161, 162
  argyrocoma, 242
  argyrophylla, 148
    atrosanguinea, 148
  arizonica, 167, 168, **170**, 225
  ashlandica, 288, 289
  assurgens, 145
  atrorubens, 149
  atrosanguinea, 148
  atrovirens, 214, 215
  baileyi, 226
    setosa, 227
  bakeri, 156
  basaltica, 167, 168, 169, **176**
  beanii, 179
  bicrenata, *173*, 177, 178, **179**, 180, 182
  biennis, 137, 138, 139, **140**
  biflora, *129*, **131**
    biflora, 131
  bimundorum, *204*, 213, 216, **218**
  bipinnatifida, 213, 216, **217**
    glabrata, 172, 216
  blaschkeana, 154, 155, 157
    permollis, 155
  bolanderi, 251, 252
    clevelandii, 252
    parryi, 251
  borealis, 206, 218
  brevifolia, 187, **188**
  breweri, 159, 160, 167, 168, 169, **170**, 171, 177, 181
    viridis, 171
  brooksensis, 214
  bruceae, 151, **159**, 160, 170, 171
  brunnescens, 152, **157**, 158
  calabra, 145
  californica, 256
    elata, 257
    frondosa, 257
  callida, 224
  campanulata, 294
  campestris, 242
  canadensis, **134**, 135
    ochroleuca, 134
    pumila, 134
    simplex, 134
    villosissima, 134
  candicans, 168
  canescens, 145
  capitata, 261
  caroliniana, 134
  caruifolia, 267
    divergens, 267
  cascadensis, 159, 160
  chamissonis, 200, 201
    umanakensis, 195
  chrysantha
    thuringiaca, 146
  ciliata, 242, 289
  clevelandii, 252
  collina, 144
  concinna, *173*, 177, **178**, 179, 180
    bicrenata, 179
    concinna, **178**, 179, 180
    curvata, 182
    dissecta, 178
    divisa, 178, **179**, 209
    macounii, 180
    modesta, 178, 179, 209
    proxima, 170, *173*, 178, **179**
    rubripes, 166, 178
  concinnaeformis, 178
  congesta, 267
    nemorosa, 268
  convallaria, 286
  cottamii, 183, **187**, 189
  crantzii, 152, 189, **190**
    hirta, 190
  crebridens, *193*, 196, 197, **198**, 199, 201, 204
    hemicryophila, *193*, **198**
  crinita, *142*, 161, 162, **164**, 165
    lemmonii, 164
  cristae, 187, 189, **191**
  cryptocaulis, 228
  cuneata, 254
  cuneifolia, 295
  daucifolia, 266
    indicta, 267
    latior, 266
  decipiens, 231
  decurrens, 172
  demotica, 184, **185**
  dissecta, 152, 216
    decurrens, 172
    rubricaulis, 207
  diversifolia, 152, 153, 160
    decurrens, 172
    glaucophylla, 152
    madsenii, 173
    multisecta, 181
    perdissecta, 153
    pinnatisecta, 171, 172
    proxima, 179
    ranunculus, 152
  divisa, 179
  douglasii, 259
    capitata, 261
    filicoides, 260
    parviflora, 262
    pseudocapitata, 262
    tenella, 261
    tenuisecta, 260
  drummondii, 151, **159**, 160, 168, 170, 171, 172
    breweri, 170
    bruceae, 159
    cascadensis, 159
  ×*drymeja*, 196
  effusa, *142*, 161, **162**, 163
    argyrea, 162
    ×*coloradensis*, 163
    effusa, *142*, **163**, 164
    filicaulis, 163
    rupincola, 163, **164**
  egedei, 130
    grandis, 129
    groenlandica, 130
    pacifica, 129
    yukonensis, 129
  elata
    dissita, 257
  elegans, 188, 189, *193*, **194**, 195
  elmeri, 156
  emarginata, 194
    elatior, 194
    nana, 192
  erecta, 133, 135, **136**, 143
  eremica, 239
  etomentosa, 155
    hallii, 155
  fastigiata, 155
    jucunda, 155
    permollis, 155
  filipes, 155, 158
  finitima, 212, 214, 215
  fissa, 281, 284
    major, 284, 286
  flabellifolia, 189, **190**, 191
    emarginata, 194
    grayi, 191
    hirta, 190
  flabelliformis, 157
    ctenophora, 157
    inyoensis, 156
  flexuosa, 140, 141
  floribunda, 296
  fragiformis, 189, **191**, 192
  frigida, 194
  frondosa, 257
  fruticosa, 122, 296
    floribunda, 296
  furcata, 200, 207, **208**
  glabrella, 172, 216
  glandulosa, 122, 281, 290, 292
    arizonica, 284
    ashlandica, 288
    austiniae, 288
    campanulata, 294
    ewanii, 295
    glabrata, 291
    globosa, 294
    hansenii, 287
    incisa, 291
    intermedia, 291
    lactea, 287
    micropetala, 286
    nevadensis, 287, 288
    pseudorupestris, 289, 290, 291
    reflexa, 293
    wrangelliana, 293
  glaucophylla, *142*, 151, **152**, 153, 155, 160, 172, 178, 189, 216
    glaucophylla, *142*, **152**, 153
    perdissecta, 152, **153**, 181
    ranunculus, 153
  glomerata, 154
  glutinosa, 286
  goldbachii, 146
  gordonii, 230, 233
    chaetophora, 231
    ursinorum, 235
  gracilis, *142*, 143, 150, 151, 152, **153**, 154, 155, 156, 157, 158
    brunnescens, 157, 158
    ctenophora, 157
    elmeri, *142*, 154, **156**, 158
    fastigiata, 154, **155**
    filipes, 158
    flabelliformis, 154, 155, **157**
    glabrata, 155
    gracilis, **154**, 155, 157, 158
    hallii, 155
    hippianoides, 159, 162
    nuttallii, 155
    owyheensis, 153, 154, **156**, 206
    permollis, 155
    pulcherrima, 158
    rigida, 155
  grayi, 189, **191**
  hansenii, 287
  heidenreichii, 145
  hendersonii, 249
  hickmanii, 169, **174**, 175
  hippiana, *142*, 150, 152, 153, 158, 159, 161, **162**, 163, 164, 165, 178, 216
    argyrea, 162
    diffusa, 162
    effusa, 162, 163
    filicaulis, 163
  hirsuta, 205
  hispidula, 249
  holmgrenii, 196, 197, **199**
  holopetala
    fastigiata, 155
  hookeriana, 196, 200, 207, **208**
    chamissonis, 200

furcata, 208
howellii, 265
humifusa, 178
hyparctica, 147, 187, 189, 190, **192**, *193*, 201, 212, 218
elatior, 193, **194**
hyparctica, ***193***
nivicola, 193
robbinsiana, 194
inclinata, 143, 144, **145**
indica, 273
indiges, 155
insularis, 206
intermedia, 143, 144, **145**, 147
canescens, 145
jaegeri, 223
jepsonii, 209, 210, 214, **216**, 217
johnstonii, 177, 178, **182**
jucunda, 155
kelloggii, 254, 255
cuneata, 254
marinensis, 253
puberula, 255
kingii, 238
klamathensis, 174
labradorica, 140, 141
langeana, 190
lasiodonta, 212, 213, **214**, 215
laxiflora, 266
ledebouriana, 196
lemmonii, 164, 165
leonina, 178
leptophylla, 150
leucocarpa, 140
leucophylla, 162
lindleyi, 254
puberula, 255
sericea, 255
litoralis, 212, 213, 214, **215**, 216, 217
ovium, 216
virgulata, 216
longipedunculata, 154
lupina, 166
luteosericea, 185
lycopodioides, 232
megalopetala, 233
scandularis, 232
lyngei
lyngei, 212
spissa, 212
macdonaldii, 187
macdougalii, 284
macounii, 168, 169, 177, 178, 179, **180**
macropetala, 154
maculata, 190
marinensis, 253
matsuokana
hemicryophila, 198
mexiae, 183
micheneri, 252
millefolia, 169, 171, *173*, **174**, 175
algida, 171
klamathensis, 174
millegrana, 139, 140
minima
robbinsiana, 194
minutifolia, 166
×*mischkinii*, 196
missourica, 217
×*mixta*, 135
miyabei, 299
modesta, 178, 207, **209**
monidensis, 172
monspeliensis, 140, 141
morefieldii, 168, 169, 177, **181**
muirii, 236
multifida, 216, 218
bimundorum, 218
multifoliolata, 244
multijuga, 168, 169, **175**, 254
multisecta, 172, *173*, 177, 178, **181**
murrayi, 206, 218
nana, *173*, 189, **192**
nelsoniana, 172
nepalensis, 135, 148
nestleriana, 146
neumanniana, 190
nevadensis, 287
newberryi, *129*, **137**
nicolletii, 139
nipharga, 201
nivea, *193*, 196, 197, **198**, 199, 200, 201, 205, 206
arenosa, 200
chamissonis, 200
dissecta, 167, 179
fallax, 202
hookeriana, 200, 208
macrophylla, 195
nipharga, 201
pentaphylla, 206
rubricaulis, 207
subquinata, 196, 206
tomentosa, 195
nivea group, 204, 205
normalis, 217
norvegica, 138, **140**, 141, 144, 191
hirsuta, 140, 141
labradorica, 140, 141
monspeliensis, 140
norvegica, 141
nubigena, 231
nuttallii, 155
glabrata, 155
oblanceolata, 177, 178
osterhoutiana, 166
osterhoutii, 225
saxosa, 226
ovina, 161, 168, 169, **171**, 172, 173
decurrens, 153, 170, 171, **172**
ovina, 171, **172**, 181, 214, 216
pacifica, 129
palustris, 122, 301
parvifolia, 301
villosa, 301
paradoxa, 139
parryi, 258
wilderae, 258
patellifera, 224
paucijuga, 206, **210**, 213
pectinata, 216
pectinisecta, 156
comosa, 156
pedersenii, 207, **210**, 211, 212
peirsonii, 294, 295
pensylvanica, 162, *204*, 212, 213, **214**, 215, 216
arida, 214, 215
bipinnatifida, 217
brevipila, 214
glabrata, 172, 214, 216
litoralis, 215
ovium, 216
paucijuga, 210
pectinata, 215
strigosa, 214
virgulata, 215
pentandra, 139
perdissecta, 153
petrovskyi, 206, 218
pickeringii, 239
pinetorum, 185
pinnatisecta, 171, 172
plattensis, 169, 170, ***173***
klamathensis, 174
leucophylla, 170
millefolia, 174
pedicillata, 173
pinnatisecta, 171
pratincola, 128
procumbens, 135
propinqua, 162
prostrata
floccosa, 198, 199
×*prostrata*, 196
×*protea*, 190
proxima, 179
pseudorupestris, 287, 289
pseudosericea, 207, **209**
grandiflora, 181
psychrophila, 206
pulchella, 193, 210, 212, 213, **217**, 218
elatior, 217
gracilicaulis, 217
pulcherrima, 150, 151, 154, 155, **158**, 159, 162
condensata, 158
filipes, 158
wardii, 158
pumila, 134
purpurascens, 271
congdonis, 271
quinquefolia, 206
ramulosa, 184
ranunculus, 152
recta, 138, **142**, 143, 144
obscura, 142
pilosa, 142
recta, 142
sulphurea, 142
reptans, 133, **135**
rhomboidea, 294
rhyolitica, 183, 184, 185, **186**
chiricahuensis, **186**
rhyolitica, **186**
richardii, 187
rimicola, 183, **185**
rivalis, 138, **139**, 140
millegrana, 139
milligrana, 140
pentandra, 139, 140
rivalis, 140
robbinsiana, 187, 189, **194**
hyparctica, 192
rolandii, 129
rubella, 146, **147**, 189
rubelloides, 147
rubida, 149
rubricaulis, 167, 206, **207**, 208, 209, 210
furcata, 208
nana, 166
paucijuga, 210
rubripes, 166
saximontana, 167
rupestris
americana, 289
rupincola, 164
rydbergiana, 183
sabulosa, 244
×*safronoviae*, 212
sanguinea, **149**, 150
sanguisorba, 214
santolinoides, 246
saximontana, 165, **166**, 167, 178, 206
dissecta, 167
saxosa, 223, 225
sierrae, 223
sericata, 265
sericoleuca, 240
shockleyi, 227
sibbaldii, 298
sibirica, 214
sierrae-blancae, 177, **182**, 189
simplex, **134**, 135
argyrisma, 134
calvescens, 134
stenoloba, 252
sterilis, **132**
stipularis, *142*, 146, **147**, 189
groenlandica, 147
×*suberecta*, 135
subgorodkovii, 196, 198, 202, 203, **204**
subjuga, *142*, 161, 165, **166**, 178
subquinata, 206
pedersenii, 210
×*subquinata*, 196
subvahliana, 196, 198, 202, **203**, *204*, 205, 210
subvillosa, 156
subviscosa, *173*, 183, **184**, 185, 186
ramulosa, **184**
subviscosa, *173*, **184**
sulphurea, 142
supina, *129*, 138, **139**
nicolletii, 139
paradoxa, *129*, **139**
tabernaemontani, 190

Potentilla (*continued*)
tenerrima, 166
tenuifolia, 296
tenuiloba, 252
thurberi, 148, **149**
atrorubens, **149**
sanguinea, 149
thurberi, **149**, 150
thuringiaca, **146**, 147
tikhomirovii, 196, 197, **201**
tilingii, 269
flavescens, 270
tolmatchevii, 210
×*tomentulosa*, 196
×*tonguei*, 135
tormentilla, 136
townsendii, *142*, 150, 152, **153**
tridentata, 122, 300
truncata, 256
tularensis, 248, 249
tweedyi, 235
uliginosa, 168, 169, 174, **175**
unguiculata, 241, 242
uniflora, 196, 203, 204, 205
uniflora/villosa group, 196, 204
uschakovii, 207, 210, **211**
usticapensis, 217
utahensis, 233
campestris, 242
vahliana, 196, 198, 203, **205**
valida, 282
verna, 188, 189, **190**
versicolor, 168, 169, 170, **171**, 172
villosa, *193*, 196, 197, **201**, 202
parviflora, 201, 202
villosula, 196, 197, **202**
congesta, 202
virgulata, 215, 216
viscidula, 186
vulcanicola, 196, 197, 202, **203**
webberi, 236
wheeleri, 171, 183, **185**
paupercula, 185
rimicola, 185
viscidula, 185, 186
wilderae, 258
wrangelliana, 293
wyomingensis, 172
yukonensis, 127, 129
Potentille, 121
Potentille ansérine, 128
Potentille argentée, 144
Potentille bipinnatifide, 217
Potentille d'Angleterre, 135
Potentille de Chamisso, 200
Potentille de Crantz, 190
Potentille de Hipp, 162
Potentille de Norvège, 140
Potentille de Nuttall, 155
Potentille de Pennsylvanie, 214
Potentille de Thuringe, 146
Potentille de Vahl, 205
Potentille des deux mondes, 218
Potentille des neiges, 198
Potentille dressée, 142
Potentille du littoral, 215
Potentille frutescente, 296
Potentille grêle, 153
Potentille inclinée, 145
Potentille intermédiaire, 145
Potentille jolie, 217
Potentille rampante, 135
Potentille simple, 134
Potentille subarctique, 193
Potentille tridentée, 300
Potentilleae, 22, 23, **119**, 120, 122, 301
Poteridium, 312, **319**, 321
annuum, *320*
occidentale, **320**
Poterium, 312, **318**, 319, 321
annuum, 320
balearicum, 318
officinale, 323
polygamum, 318
sanguisorba, *315*, **318**, 319
muricatum, 318
polygamum, *315*, **318**, 319
Potter, D., 6, 20, 23, 57, 73, 75, 120, 274, 312, 318, 321, 327, 345, 351, 385, 390, 395, 399, 412, 414, 416, 422, 427, 432, 434, 447, 449, 468, 494, 644
Pourthiaea
villosa, 490
Power, N., 492
Prairie burnet, 320
Prairie cinquefoil, 214
Prairie crabapple, 474
Prairie Joy, 104
Prairie rose, 104
Prairie smoke, 62
Pringle hawthorn, 559
Prinsepia, 384, 385
Prostrate cotoneaster, 461
Prostrate sandcherry, 375
Prunier, 352
Prunier damas, 376
Prunier d'Amérique, 378
Prunier noir, 379
Prunophora, 354
Prunus, 19, **352**, 353, 354, 363, 378
alabamensis, 364
alleghaniensis, 380, 381
davisii, 380, 381
americana, 357, 360, 376, **378**, 379, 381, 382
lanata, 379
mollis, 379
amygdalus, 372
andersonii, 356, 359, **373**
angustifolia, 358, 360, 376, **381**, 382
varians, 381
watsonii, 381
armeniaca, 355, 358, **375**
avium, 357, 359, **368**, 369
besseyi, 374
caroliniana, 355, **361**, 363
cerasifera, 358, 360, **375**, 376
cerasus, 357, 360, **368**, 369
avium, 368
corymbulosa, 367
cuneata, 375
demissa, 366
melanocarpa, 366
depressa, 375
domestica, 357, 358, 360, **376**
insititia, 376
dulcis, 355, 358, **372**
emarginata, 356, 358, 359, **367**, 368
crenulata, 367
emarginata, 368
mollis, 367, 368
eremophila, 356, 359, **371**, *372*
eriogyna, 373
fasciculata, 356, 359, **370**
fasciculata, **370**
punctata, **370**
fremontii, 359, **373**
fruticosa, 353
geniculata, 357, 359, **378**
glandulosa, 356, 360, **377**
gracilis, 357, 360, **380**, 381
gravesii, 380
havardii, 356, 359, **371**
hortulana, 358, 361, 376, **382**, *383*
mineri, 382
ilicifolia, *352*, 355, **362**
ilicifolia, *352*, **362**
integrifolia, 362
lyonii, 362
occidentalis, *352*, **362**
injucunda, 380, 381
insititia, 376
jamasakura
speciosa, 369
lanata, 378, 379
lannesiana, 369
laurocerasus, 355, **362**, 363
lusitanica, 355, **363**
lyonii, 362
mahaleb, 356, 359, **366**, 367
maritima, 357, 360, *372*, **380**
gravesii, 380
melanocarpa, 366
mexicana, 357, 360, 378, **379**
minutiflora, 356, 359, **370**
mitis, 380, 381
mollis, 378, 379
mume, 353
munsoniana, 382
murrayana, 357, 361, **382**, 383
myrtifolia, 355, **361**
nigra, 358, 361, 378, **379**
occidentalis, 362
×*orthosepala*, 381
padus, 355, **366**
pensylvanica, *352*, 356, 359, **367**, 368
corymbulosa, 367
mollis, 367
saximontana, 367
persica, 355, 358, **371**
×*pugetensis*, 368
pumila, 356, 360, *372*, **373**, 374
besseyi, *372*, **374**, 375, 381
cuneata, 375
depressa, 374, **375**
pumila, **374**, 375
susquehanae, 373, 374, **375**
reverchonii, 382
rivularis, 358, 361, 381, **382**, 383
pubescens, 382
salicina, 353, 376
serotina, 355, **363**, 366
alabamensis, 363, **364**
capuli, 363
eximia, 364
rufula, 363, **364**, 365
serotina, 363, **364**
virens, 364, 365
serrulata, 369
lannesiana, 369
×*slavinii*, 381
speciosa, 356, 359, **369**
spinosa, 357, 360, **376**
subcordata, 354, 355, 357, 358, 359, 360, 361, **377**
kelloggii, 377
oregana, 377
rubicunda, 377
subhirtella, 356, 360, **369**
susquehanae, 375
texana, 354, 356, 359, **377**, 378
tomentosa, 356, 358, **372**, 373
triloba, 353
umbellata, 357, 360, **380**, 381
injucunda, 380
×*utahensis*, 381
virginiana, 355, 363, **365**, 366
demissa, 365, **366**
melanocarpa, 366
virginiana, **365**, 366
yedoensis, 356, 359, **369**
Pseudocydonia, 428, **487**
sinensis, *486*, **488**
Purple avens, 65
Purple flowering raspberry, 47
Purple marshlocks, 301
Purple raspberry, 30, 43
Purple-flowered cotoneaster, 461
Purpusia, 221, 222, 225
arizonica, 224
osterhoutii, 225
saxosa, 226
Purpusia, 224
Purshia, 326, 327, **339**, 343
ericifolia, 340, **342**
glandulosa, 341
mexicana
stansburyana, 341
pinkavae, 342
stansburyana, 339, **341**, 342
subintegra, 339, 341, **342**
tridentata, *330*, 339, **340**, 341

glandulosa, 340, **341**
tridentata, *330*, **340**, 341
Pyracantha, 20, 428, 449, **468**, 469
angustifolia, 469, **471**
atalantioides, 469, **471**
chinensis, 470
coccinea, 469, ***470***
crenatoserrata, 469
crenulata, 469, **470**, 471
kansuensis, 470, 471
rogersiana, 469
fortuneana, **469**, 471, 472
koidzumii, 469, **471**, 472
rogersiana, 469
yunnanensis, 469
Pyraceae, 427
Pyreae, 427
Pyrgus
ruralis
lagunae, 253
Pyrinae, 427, 434
Pyrus, 19, 427, 428, 429, 433, 434, 445, 473, **479**, 480
sect. *Adenorachis*, 445
sect. *Torminaria*, 436
americana, 439
decora, 440
×*amphigenea*, 483
angustifolia, 474
arbutifolia, 446
aucuparia, 438
austriaca, 481, 483
baccata, 477
bartramiana, 657
betulifolia, 480, 482, 483
boissieriana, 484
bracteata, 475
bretschneideri, 483
calleryana, 480, 481, **482**, 483, 484
calleryana, 482
integrifolia, 482
koehnei, 482
lanceata, 482
caucasica, 483
communis, 436, *470*, 480, 481, 482, **483**, 484
cordata, 481, 483, **484**
coronaria, 475
angustifolia, 474
dasycalyx, 475
elongata, 475
ioensis, 474
lancifolia, 475
decora, 440
groenlandica, 440
diversifolia, 476
elaeagnifolia, 481
eriopleura, 481
esquirolii, 479
fauriei, 483
floribunda, 445
fusca, 476
halliana, 478
hupehensis, 478
intermedia, 437
ioensis, 474
×*jackii*, 439
kawakamii, 480, 482
koehnei, 482
malus, 476
paradisiaca, 476
melanocarpa, 446
nivalis, 480, **481**, 483
occidentalis, 444, 445
prunifolia, 477
pyraster, 483
pyrifolia, 480, **481**, 482
rivularis, 476
salicifolia, 480
salviifolia, 481
sambucifolia, 443
sanguinea, 653
serotina, 481
serrulata, 481, **482**
sieboldii, 479
soulardii, 474
subcrataegifolia, 479
syriaca, 483
toringo, 479
ussuriensis, 480

Queen Charlotte avens, 63
Queen of the forest, 25
Queen of the prairie, 25
Quercus, 606
Quest-Ritson, B., 77
Quest-Ritson, C., 77
Quian, G. Z., 477
Quince, 480, 486, 487
Quininebush, 340

Radford, A. E., 29
Ramona horkelia, 256
Randall, J. M., 449
Raspberry, 19, 29, 30, 43, 46, 47, 48
Raup, H. M., 66
Raven, P. H., 353, 362
Raynor, L. A., 69
Reagan, A. B., 46
Red bracteole hawthorn, 520
Red chokeberry, 446
Red comarella, 244
Red haw, 553
Red raspberry, 30, 42, 51
Red shank, 394
Red spire, 482
Red tip, 489, 490
Redleaf rose, 98
Red-spined rose, 103
Red-stemmed cinquefoil, 207
Red-twigged shadbush, 653
Reeve's meadowsweet, 409
Reeve's spiraea, 409
Rehder, A. J., 98, 354, 369, 399, 410, 471
Reichenbacher, F. W., 342
Reine-des-prés, 26
Reveal, J. L., 121, 131, 132, 136, 137, 141, 143, 146, 148, 150, 160, 175, 187, 188, 219, 234, 238, 246, 254, 259, 262, 266, 267, 270, 272, 295, 297, 299, 300
Reverchon hawthorn, 537
Rhamnaceae, 20
Rhamnus
cathartica, 492
Rhaphiolepis, 20, 432, 447
Rhoads, A. F., 29, 449
Rhodotypos, **386**, 387
kerrioides, 387
scandens, **387**, *388*
tetrapetala, 387
Ribbon bush, 394
Ribes, 36
Rich, T. C. G., 434
Rickett, H. W., 30
Rieseberg, L. H., 335
Ritz, C. M., 78
River cinquefoil, 139
Roadside agrimony, 317
Robbins's cinquefoil, 194
Robertson, A., 434
Robertson, K. R., 65, 69, 353, 425, 434, 445, 449, 468, 473, 489, 646, 662
Robin runaway, 51
Robinson, B. L., 68, 69
Robinson, W. A., 647, 648, 658, 659
Rock Creek ivesia, 233
Rock ivesia, 223
Rock purpusia, 226
Rock-spiraea, 411, 419, 420
Rockflower, 12
Rockloving cinquefoil, 164
Rocky Mountain mountain ash, 442
Rocky Mountain rose, 113
Rocky Mountains cinquefoil, 166
Rogue River horkelia, 267
Rohrer, J. R., 19, 57, 58, 352, 354, 374, 382, 383, 415, 432, 447, 449, 468, 489, 646, 662
Roland, A. E., 307
Ronce à flagelles, 40
Ronce acaule, 35
Ronce de Pennsylvanie, 50
Ronce des Alléghanys, 34
Ronce du Canada, 37
Ronce hispide, 41
Ronce odorante, 47
Ronce pubescente, 51
Ronce sétuleuse, 53
Rosa, 20, **75**, 76, 77, 78, 89, 95, 104, 108
sect. Bracteatae, 78, **86**
sect. Caninae, 76, 78, **89**, 98
sect. *Chinenses*, 77
sect. *Cinnamomeae*, 77, 84, 94
sect. Gallicae, 77, 78, **88**
sect. *Gymnocarpae*, 116
sect. Laevigatae, 78, **87**
sect. *Luteae*, 93
sect. Minutifoliae, 77, **79**
sect. Pimpinellifoliae, 78, 82, **93**
sect. Rosa, 77, 78, 84, **94**, 98, 116
sect. *Sericeae*, 93
sect. *Sinicae*, 87
sect. Systylae, 78, **82**, 84
subg. Hesperhodos, 77, **78**, 79
subg. *Hulthemia*, 78
subg. Rosa, 77, **82**, 86, 87, 88, 89, 93, 94
supertribe Rosodae, 75
abietorum, 116
acicularis, 96, 107, 112, **114**, *115*
acicularis, **114**
bourgeauiana, 115
sayi, 77, 107, 112, 114, ***115***
adenosepala, 107
alcea, 104
aldersonii, 111
aleutensis, 113
arizonica, 108
arkansana, 76, 95, 96, 101, *102*, **104**
suffulta, 104
aucuparia, 102
blanda, 77, 95, 100, *102*, 104, **105**
bracteata, *84*, **86**
bridgesii, 97, *115*, 116, 117, **118**, 119
brownii, 113
butleri, 115
californica, 94, 97, 109, ***111***, 112, 116
glabrata, 109
orthacantha, 112
ultramontana, 109
calvaria, 118, 119
canina, *84*, 89, **90**
corymbifera, 90
dumetorum, 90
carolina, 95, 96, **101**, *102*, 104
aculeata, 100
carolina, 101, ***102***
grandiflora, 102
mexicoensis, 101, *102*
subserrulata, 101, ***102***
villosa, 102
×*centifolia*, 77, 88
cherokeensis, 87
chinensis, 77
chrysocarpa, 108
cinnamomea, 77, **97**, 98
collaris, 95
conjuncta, 104
corymbifera, 90
covillei, 119
crenulata, 119
damascena, 88
demareei, 107
×*dulcissima*, 95
dumetorum, 90
durandii, 113
eglanteria, 91
×*engelmannii*, 112
fendleri, 106, 107
ferruginea, 98
floridana, 99
×*foetida*, 94
foliolosa, 96, *102*, **103**, 104
gallica, **88**
gemella, 99
glauca, 95, **98**

Rosa (*continued*)
granulata, 118
granulifera, 108
gratissima, 109
grosseserrata, 109
gymnocarpa, 94, 95, 96, 112, *115*, **116**, 117, 119
gymnocarpa, *115*, **116**, 117
pinetorum, 117
pubescens, 118
serpentina, 116, **117**
×*harisonii*, 94
helleri, 116
×*hollandica*, 96, 99
hugonis, 93
humilis, 102
ignota, 103
illinoiensis, 94
johannensis, 105
laevigata, *84*, **87**
lancifolia, 99
lapwaiensis, 109
leucopsis, 116
lucida, 100
lamprophylla, 100
lucieae, 83, **85**
lutescens, 94
lyonii, 102
macdougalii, 113
macounii, 106, 107
manca, 108
×*medioccidentis*, 101, 104
melina, 113
michiganensis, 100
micrantha, 91
minutifolia, *74*, **79**
mirifica, 81
mohavensis, 109
mollis, 90, 92, **93**
montezumae, 90
×*moschata*, 83, 88
multiflora, 83, ***84***, 85
myriadenia, 119
nanella, 100
nemoralis, 91
neomexicana, 108
nitida, 95, 96, *102*, **103**
×*novae-angliae*, 101
nutkana, 96, 97, 108, 109, *111*, **112**
hispida, 113
macdougalii, *111*, **113**, 114, 115
melina, 108, *111*, 112, **113**, 114
muriculata, 113
nutkana, *111*, 112, **113**
setosa, 113
obtusifolia, 89
obtusiuscula, 99
×*odorata*, 77
omissa, 92
oreophila, 113
palmeri, 102
×*palustriformis*, 100
palustris, *84*, 95, 96, **99**, 100, 103
dasistema, 99
pandorana, 113
parviflora, 102
persica, 78
pimpinellifolia, 94
pinetorum, 97, *115*, 116, **117**, 118, 119
piscatoria, 116
pisocarpa, 97, **109**, 110, *111*
ahartii, 97, **110**
gratissima, 109
pisocarpa, **110**, *111*
ultramontana, 109
puberulenta, 108
pyrifera, 109
rainierensis, 113
rivalis, 110
rotundata, 109
rousseauiorum, 105
rubifolia, 83
rubiginosa, **90**, 91
nemoralis, **91**
rubiginosa, **91**
rubrifolia, 98
rudiuscula, 102
rugosa, 94, 96, **98**, 99, 105, 112
salictorum, 109
sayi, 115
schuetteana, 100
sempervirens, 82
serrulata, 102
setigera, 74, 75, **83**, 84
elatior, 83
glabra, 83
tomentosa, 83
sherardii, 90, **92**
sonomensis, 118
spaldingii, 113
spinosissima, **94**, 104
spithamea, 97, *115*, 116, 117, **118**, 119
pinetorum, 117
solitaria, 118
sonomensis, 118
subinermis, 118
stellata, *74*, 78, 79, **80**
abyssa, 80, **81**
erlansoniae, 81
mirifica, 80, **81**
stellata, *74*, **80**, 81
suaveolens, 91
subblanda, 105
subglobosa, 92
subserrulata, 102
suffulta, 104
texarkana, 102
tomentosa, 90, **92**
ultramontana, 109
underwoodii, 113
vernonii, 81
villosa, 93
virginiana, *84*, 95, 96, **100**, 101, 103
wichuraiana, 85
williamsii, 105
woodsii, 95, 96, 97, 104, **105**, 106, *107*, 108, 110, 112
arizonica, 106, *107*, **108**
ertterae, 108
fendleri, 107
glabrata, 109
gratissima, 106, **109**
maderensis, 107
manca, 106, *107*, **108**, 114
mohavensis, 109
puberulenta, 106, **108**
terrens, 107
ultramontana, 95, 97, 106, *107*, **109**, 110, 115
woodsii, 106, ***107***
yainacensis, 118, 119
Rosaceae, 10, **18**, 19, 20, 21, 23, 24, 27, 28, 57, 74, 75, 119, 312, 345, 346, 347, 352, 383, 384, 385, 386, 390, 392, 399, 412, 414, 422, 424, 427
subfam. Amygdaloideae, 20, 21, 343, **345**, 346, 347, 352, 383, 385, 386, 387, 390, 392, 398, 399, 416, 422, 424, 427
subfam. Dryadoideae, 20, 21, **326**, 343
subfam. *Maloideae*, 385, 427
subfam. Rosoideae, 20, 21, **23**, 24, 27, 57, 73, 74, 75, 119, 312, 387
subfam. *Spiraeoideae*, 345
subtribe Fragariinae, 127, 274, 281, 298, 304
subtribe Pyrinae, 427, 434
tribe Agrimonieae, 22, **312**
tribe Amygdaleae, 20, 21, **352**, 385, 415
tribe Colurieae, 21, 22, **57**, 58, 73
tribe Dryadeae, 21, **326**, 327
tribe Exochordeae, 21, **383**, 384
tribe Gillenieae, 21, 345, **424**
tribe *Holodisceae*, 416
tribe Kerrieae, 21, 385, **386**, 387, 390
tribe Lyonothamneae, 21, **345**
tribe Maleae, 20, 21, 345, 385, 415, **427**, 432, 433, 449, 468, 494, 644
tribe Neillieae, 21, **347**, 351
tribe *Osmaronieae*, 384, 385
tribe Potentilleae, 22, 23, **119**, 120, 122, 301
tribe *Pyreae*, 427
tribe Roseae, 21, 22, 23, **74**
tribe Rubeae, 22, 23, **27**, 28
tribe *Sanguisorbeae*, 312, 321
tribe Sorbarieae, 21, **392**, 395
tribe Spiraeeae, 21, **398**, 412, 414, 415, 416
tribe Ulmarieae, 22, **23**
Rosales, 10, 20
Rosatti, T. J., 445
Rose, 75, 77, 85, 87, 100, 104, 106, 109, 111, 113
Rose Family, 18
Rose of Provence, 88
Rose spiraea, 406
Roseae, 21, 22, 23, **74**
Rosier, 75
Rosier aciculaire, 115
Rosier brillant, 103
Rosier canelle, 97
Rosier de Caroline, 101
Rosier de France, 88
Rosier de Virginie, 100
Rosier de Woods, 107
Rosier des chiens, 90
Rosier églantier, 90
Rosier glauque, 98
Rosier inerme, 105
Rosier multiflore, 84
Rosier palustre, 99
Rosier pimprenelle, 94
Rosier rugueux, 98
Rosodae, 75
Rosoideae, 20, 21, **23**, 24, 27, 57, 73, 74, 75, 119, 312, 387
Ross's avens, 63
Rough avens, 68
Rough cinquefoil, 140
Rough hawthorn, 523
Roughfruit berry, 44
Round-toothed lady's mantle, 306
Rousi, A., 127, 128, 129
Rowan, 433, 438
Royal hawthorn, 565
Rubacer
odoratum, 47
Rubeae, 22, 23, **27**, 28
Rubus, 19, 20, **28**, 29, 30, 35, 38, 48, 49, 52, 54, 55, 647
sect. *Arguti*, 50
sect. *Canadenses*, 38
sect. *Cuneifolii*, 49
sect. *Flagellares*, 40, 54
sect. *Persistentes*, 54
sect. *Procumbentes*, 40, 54
sect. *Verotriviales*, 54
ser. *Discolores*, 37
subg. *Anoplobatus*, 29
subg. *Chamaebatus*, 29, 46
subg. *Chamaemorus*, 29
subg. *Cylactis*, 29
subg. *Dalibarda*, 29
subg. *Idaeobatus*, 29
subg. *Micranthobatus*, 28
subg. *Rubus*, 29, 30, 46
abactus, 50
×*abbrevians*, 35, 53
ablatus, 50
aboriginum, 40
abundiflorus, 50
acadiensis, 34
acaulis, 35
×*acer*, 39, 50
subacer, 39, 50
×*aculiferus*, 35, 53
×*adenocaulis*, 35, 53
×*adjacens*, 42, 53
agilis, 54
×*akermanii*, 41, 50
alacer, 40
alaskensis, 35
aliceae, 43
alius, 40

allegheniensis, *26*, **34**, 35, 38, 41, 50, 53, 56
calycosus, 34
gravesii, 34
neoscoticus, 34
plausus, 34
populifolius, 34
almus, 40
x*alter*, 41, 53
alumnus, 34
x*amabilis*, 38, 50
x*ambigens*, 41, 42
x*amicalis*, 38, 50
amnicola, 50
amplificatus, 56
andrewsianus, 50
x*angustifoliatus*, 35, 53
apianus, 34
apogaeus, 40
apparatus, 53
aptatus, 40
arcticus, 31, 32, **35**
acaulis, **35**, 36, 51
acaulis, 35
arcticus, **35**
grandiflorus, 35
pentaphylloides, 35
stellatus, 35, **36**
x*arcuans*, 41, 53
arenicola, 40
x*argutinus*, 39, 50
argutus, 50
randii, 37
arizonensis, 40, 41
arizonicus, 43
armatus, 40
armeniacus, 36, 37
arundelanus, 40
jecklyanus, 40
arvensis, 50
x*ascendens*, 35, 53
ashei, 40
associus, 50
attractus, 34
x*atwoodii*, 35, 53
audax, 39
auroralis, 34
austrinus, 40
x*avipes*, 35, 50
baileyanus, 40
barbarus, 50
bartonianus, 31, **36**, 39, 45
beatus, 53
bellobatus, 50
benneri, 53
bernardinus, 45
besseyi, 37
betulifolius, 50
bicknellii, 53
x*biformispinus*, 41, 53
bifrons, 30, 32, 33, **36**, 37, 44, 49, 55, 56
x*bigelovianus*, 41, 53
blakei, 50
blanchardianus, 41
bollianus, 40
bonus, 40
boottianus, 53
botruosus, 40
x*boyntonii*, 35, 41
bractealis, 50
x*bracteolifer*, 41, 53
brainerdii, 50
bretonis, 40
brevipedalis, 40
burnhamii, 50
bushii, 50
cacaponensis, 40
caesius, 31, **37**, 41, 42, 56
campester, 34
camurus, 40
canadensis, 34, 35, **37**, 38, 41, 42, 50, 53
elegantulus, 38, 50
imus, 37
pergratus, 38, 50
cardianus, 50
carolinianus, 43
cathartium, 40
celer, 40
census, 40
centralis, 40
chamaemorus, 28, 31, **38**
chapmanii, 39
clairbrownii, 40
clandestinus, 40
clarus, 40
x*clausenii*, 35, 53
coloniatus, 40
complex, 40
compos, 41
conabilis, 40
concameratus, 34
condensiflorus, 50
condignus, 53
congruus, 50
connixus, 40
continentalis, 54
cordialis, 40
cordifrons, 40
x*crux*, 41, 50
cubitans, 41
cuneifolius, 33, 34, **39**, 42, 49, 50, 54
angustior, 39
austrifer, 39
spiniceps, 39
subellipticus, 39
cupressorum, 50
currulis, 40
curtipes, 40
dalibarda, 29
x*darlingtonii*, 41, 50
davisiorum, 41
deamii, 40
deaneanus, 53
decor, 40
defectionis, 50
deliciosus, 31, 36, **39**, 45
neomexicanus, 45
densipubens, 50
densissimus, 50
depavitus, 40
discolor, 37, 54
discretus, 53
dissensus, 53
dissimilis, 53
x*dissitiflorus*, 41, 50
x*distinctus*, 41, 42
dives, 40
duplaris, 54
eastwoodianus, 55
eflagellaris, 40
egglestonii, 43
electus, 53
x*elegantulus*, 38, 50
elongatus, 41, 42
x*elongatus*, 41, 42
x*emeritus*, 41, 42
enslenii, 40
eriocarpus, 46
exemptus, 40
exrubicundus, 45
exsularis, 40
exter, 53
facetus, 50
fandus, 40
fassettii, 41
fatuus, 50
fecundus, 40
felix, 40
ferrofluvius, 40
flagellaris, 30, 32, 35, 37, 38, **40**, 41, 42, 50, 53, 54, 56
almus, 40
occidualis, 40
x*flavianus*, 35, 53
x*floricomus*, 35, 50
x*floridensis*, 39, 50
floridus, 50
foliaceus, 40
foliofiorus, 40
x*forestalis*, 35, 38
franciscanus, 53
x*fraseri*, 47, 48
x*fraternalis*, 35, 41
fraternus, 35, 41
x*fraternus*, 35, 41
x*frondisentis*, 35, 53
frondosus, 50
fryei, 34
fulleri, 53
x*furtivus*, 41, 42
ganderi, 41
georgiensis, 39
glandicaulis, 34
glaucifolius, 31, **41**
ganderi, 41
gnarus, 50
x*grandidens*, 42, 53
greeneanus, 43
grimesii, 40
groutianus, 53
gulosus, 53
hancinianus, 40
x*harmonicus*, 42, 53
hawaiiensis, 54
hesperius, 45
heterophyllus, 50
hirsutus, 30
hispidoides, 53
hispidus, 26, 32, 38, 39, **41**, 42, 53
blanchardianus, 41
cupulifer, 41
obovalis, 41
suberectus, 53
housei, 40
humilior, 50
huttonii, 41
x*hypolasius*, 41, 50
ictus, 54
idaeus, 30, 31, **42**
aculeatissimus, 43
anomalus, 43
canadensis, 43
caudatus, 43
egglestonii, 43
eucyclus, 43
gracilipes, 43
heterolasius, 43
idaeus, 30, **42**
inermis, 42
melanolasius, 43
peramoenus, 43
sachalinensis, 43
strigosus, 30, 35, 42, **43**, 46
strigosus, 43
ignarus, 40
illecebrosus, 28, 33, **43**
illustris, 37
x*immanis*, 35, 38
impar, 50
imperiorum, 40
impos, 34
inclinis, 34
independens, 50
indianensis, 40
inermis, 55
x*inferior*, 39, 54
iniens, 40
injunctus, 40
inobvius, 40
insons, 50
insulanus, 50
interioris, 50
invisus, 40
irregularis, 34
ithacanus, 40
x*jacens*, 42, 53
specialis, 41, 53
jactus, 40
x*janssonii*, 41, 50
japonicus, 390
jaysmithii, 40
angustior, 40
jejunus, 53
jennisonii, 50
jugosus, 50
junceus, 53
junior, 53
x*kalamazoensis*, 41, 42
kelloggii, 50
kennedyanus, 37
kentuckiensis, 40
laciniatus, 32, 33, 37, **43**, 44
laetabilis, 37
x*laevior*, 41, 53
x*largus*, 41, 50
lasiococcus, 31, 32, **44**, 52
lassus, 40
latifoliolus, 50
laudatus, 50
lawrencei, 53
leggii, 50
x*lepagei*, 38, 41
leucodermis, 31, **44**, 46
bernardinus, 45

Rubus leucodermis (*continued*)
glaucifolius, 41
leucodermis, 44, **45**
nigerrimus, 44, **45**
trinitatis, 45
leviculus, 40
libratus, 50
×*licens*, 35, 41
licitus, 34
linkianus, 36
longii, 39
longipes, 41
louisianus, 50
lucidus, 54
macraei, 56
macropetalus, 55
macrophyllus, 56
macvaughii, 54
mainensis, 40
maltei, 40
maniseesensis, 40
mediocris, 53
medius, 39
melanolasius, 43
menziesii, 54, 56
meracus, 40
michiganensis, 40
minnesotanus, 40
mirus, 54
×*miscix*, 38, 53
missouricus, 50
mollior, 50
montanus, 35
montensis, 37
montpelierensis, 34
multifer, 40
×*multiformis*, 41, 53
×*multilicius*, 38, 50
multispinus, 50
mundus, 40
navus, 53
nefrens, 40
neglectus, 43, 46
×*neglectus*, 43
neomexicanus, 31, 36, 39, **45**
nessianus, 54
nigerrimus, 45
nigrobaccus, 34
nivalis, 31, 32, **45**, 46
niveus, 31, **46**
nocivus, 53
notatus, 53
×*novanglicus*, 38, 42
×*noveboracus*, 38, 50
nuperus, 34
obsessus, 40
obvius, 40
occidentalis, 30, 31, 43, 44, **46**, 47
leucodermis, 44
occidualis, 40
odoratus, 31, **47**, 48
albidus, 47
columbianus, 47
oklahomus, 50
oligospermus, 41
×*orarius*, 35, 50
originalis, 50
ortivus, 53
×*ostryifolius*, 35, 41
paganus, 41
paludivagus, 50
×*paracaulis*, 36, 51
×*parlinii*, 42, 53
particeps, 40
particularis, 40
parviflorus, 31, 47, **48**
grandiflorus, 48
velutinus, 48
parvifolius, 32, **48**
pascuus, 33, 36, 39, **49**
paulus, 34
pauper, 50
pauxillus, 50
peculiaris, 38, 53
×*peculiaris*, 38, 53
pedatus, 32, **49**
penetrans, 50
pensilvanicus, 34, 35, 38, 39, 41, 49, **50**, 53, 56
perfoliosus, 50
×*pergratus*, 38, 50
perinvisus, 53
permixtus
laevior, 41, 53
×*permixtus*, 41, 42
pernagaeus, 40
perpauper, 50
persistens, 50
×*perspicuus*, 35, 53
pervarius, 41
philadelphicus, 50
phoenicolasius, 33, **51**
pityophilus, 40
plexus, 40
plicatifolius, 40
plus, 41
porteri, 41
praepes, 50
pratensis, 50
prestonensis, 50
pringlei, 46
probabilis, 39
probativus, 39
procerus, 37, 44
profusiflorus, 40
pronus, 40
provincialis, 41
pubescens, 32, 36, 47, **51**
paracaulis, 36, 51
pilosifolius, 51
pubifolius, 50
pudens, 41
pugnax, 34
pumilus, 46
racemiger, 53
randolphiorum, 39
×*ravus*, 35, 53
recurvans, 50
×*recurvicaulis*, 41, 50
redundans, 40
regionalis, 53
repens, 29, 31, **51**, *52*
reravus, 53
rhodophyllus, 40
riograndis, 54
×*rixosus*, 35, 41
roezlii, 39
roribaccus, 40
rosa, 34
rosagnetis, 40
×*rosendahlii*, 41, 42
×*rossbergianus*, 41, 50
rotundior, 53
rowleei, 41
rubrisetus, 54
russeus, 40
saepescandens, 50
sailori, 40
saltuensis, 34
sanfordii, 40
satis, 40
sativus, 50
saxatilis, 32, **52**
canadensis, 51
scambens, 40
sceleratus, 34
schneideri, 53
schoolcraftianus, 40
×*segnis*, 41, 42
sejunctus, 39
semisetosus, 53
serissimus, 49
×*setospinosus*, 41, 53
setosus, 32, 33, 35, 38, 41, 42, 50, **53**
×*severus*, 41, 53
sewardianus, 40
signatus, 41
sirbenus, 55
sons, 54
spectabilis, 33, **53**, 54, 56
franciscanus, 48, 53
menziessii, 56
spectatus, 53
×*spiculosus*, 42, 53
steelei, 40
stellatus, 36
stipulatus, 53
strigosus, 43
acalyphaceus, 43
arizonicus, 43
subinnoxius, 40
subsolanus, 50
subtractus, 50
suus, 50
tallahasseanus, 54
×*tardatus*, 42, 53
temerarius, 40
tetricus, 40
×*tholiformis*, 42, 53
thyrsoideus, 37
tomentosus
canescens, 30
trifrons
pudens, 41
×*trifrons*, 42, 53
trilobus, 39
triphyllus, 48
trivialis, 32, 39, **54**
tygartensis, 50
ucetanus, 50
uliginosus, 49
ulmifolius, 31, 36, 37, **54**, 55, 56
anoplothyrsus, 54, 55
inermis, 55
uniformis, 53
univocus, 53
ursinus, 28, 30, 31, 32, 37, 50, 54, **55**, 56
eastwoodianus, 55
macropetalus, 55
menziesii, 56
sirbenus, 55
uvidus, 40
vagulus, 41
vagus, 41
×*varus*, 41, 42
vermontanus, 53
vestitus, 30, 33, 34, 36, 55, **56**
viburnifolius, 43
vigil, 41
×*vigoratus*, 41, 42
villosus
montanus, 34
viridifrons, 53
virilis, 50
vitifolius, 55
eastwoodianus, 55
ursinus, 55
vixalacer, 40
vixargutus, 50
wahlii, 50
whartoniae, 40
wheeleri, 53
wiegandii, 50
×*wisconsinensis*, 50, 53
×*zaplutus*, 42, 53
Rufous mayhaw, 548
Rugose rose, 98
Rum cherry, 363
Russell, P. G., 369
Rydberg, P. A., 29, 59, 63, 66, 75, 122, 123, 127, 137, 144, 150, 168, 177, 182, 189, 192, 231, 234, 239, 245, 249, 250, 252, 258, 262, 264, 271, 296, 303, 306, 319, 321, 408, 416, 417
Rydberg's horkelia, 251

Sagebrush cinquefoil, 182
Salad burnet, 318
Salamun, P. J., 403
Salmonberry, 53
San Diego raspberry, 41
Sand blackberry, 39
Sand strawberry, 278
Sandcherry, 373
Sandhills cinquefoil, 214
Sanguisorba, 312, 318, 319, 320, **321**
annua, 320
canadensis, *320*, 321, **322**
latifolia, 322
sitchensis, 322
menziesii, 321, **322**
microcephala, 323
minor, 318, 319
balearica, 318
muricata, 318
occidentalis, 320
officinalis, 321, 322, **323**
microcephala, 323
sitchensis, 322
stipulata, 321, **322**
Sanguisorbe, 321

Sanguisorbe du Canada, 322
Sanguisorbe mineure, 318
Sanguisorbeae, 312, 321
Sanso, A. M., 4
Santa Catalina mountain mahogany, 335
Santa Lucia horkelia, 253
Santa Rosa horkelia, 252
Sapindales, 6
Sarcopoterium, 318
Sargent hawthorn, 581
Sargent, C. S., 493, 547, 558
Sarvis, 646
Sarvisberry, 646
Saskatchewan hawthorn, 597
Saskatoon, 646, 651
Saskatoon berry, 651
Savile, D. B. O., 24
Savolainen, V., 6
Sax, K., 402, 404
Saxifraga
  pectinata, 422
Saxifragaceae, 20, 423
Say's acicular rose, 115
Scalp Creek hawthorn, 642
Scarlet cinquefoil, 149
Scarlet firethorn, 470
Scarlet hawthorn, 559
Scarlet strawberry, 277
Schaack, C. G., 342
Schanzer, I. A., 23, 27
Scheidecker crabapple, 477
Schizonotus, 417
Schmidt, F. W., 306
Schoen, D., 331
Schuette hawthorn, 565
Schulze-Menz, G. K. W., 387
Scoggan, H. J., 75, 91, 92, 148, 155, 172, 368, 405, 407, 464
Scoggin, R., 9
Scots rose, 94
Scrub plum, 378
Senn, H. A., 374
Seo, M. N., 4
Sericotheca, 417
Serrulate pear, 482
Serviceberry, 646
Sevier ivesia, 244
Seymour, F. C., 145
Shadblow, 646
Shadbush, 646, 658
Shadwood, 646
Shaggy horkelia, 268
Sharpe, F., 437
Shasta horkelia, 263
Shasta snow-wreath, 389
Shaw, J., 354, 377, 378
Sheep cinquefoil, 171
Shelly's ivesia, 229
Sherard's downy-rose, 92
Shevock, J. R., 388, 389
Shining rose, 103
Shining shadbush, 650
Shinners, L. H., 378, 379
Shiny cotoneaster, 463
Shinyleaf meadowsweet, 403
Shiny-leaf spiraea, 404
Shockley's ivesia, 228
Shorn, H. E., 346
Shorthouse, J. D., 77
Shrubby cinquefoil, 296
Sibbaldia, 120, 274, **297**, 298, 299, 304
  sect. *Eusibbaldia*, 298
  aphanopetala, 298
  cuneata, 298
  erecta, 280
  parviflora, 298
  procumbens, 297, **298**, 299
  semiglabra, 298
  tridentata, 300
  trullifolia, 298
Sibbaldianthe, 298
Sibbaldie couchée, 298
Sibbaldiopsis, 120, 122, 124, 274, 298, **299**, 304
  cuneifolia, 299
  miyabei, 299
  tridentata, 297, 299, **300**
Siberian crabapple, 477
Sibiraea, 398
Siebold crabapple, 479
Siegismund, H. R., 331
Sierra bladdernut, 8
Sierra Blanca cinquefoil, 182
Sierra horkelia, 269
Sierra plum, 377
Sierra Valley ivesia, 241
Sierran drymocallis, 288
Sierran ground rose, 118
Sierran wood beauty, 288
Sieversia, **57**, 58, 59
  campanulata, 63
  canescens, 63
  ciliata, 63
  glacialis, 62
  gracilipes, 63
  paradoxa, 73
  pentapetala, *52*, **58**
  pusilla, 58
  radiata, 64
  rossii, 63
  triflora, 62
Silky horkelia, 265
Silkyleaf cinquefoil, 164
Silver cinquefoil, 144
Silver mousetail, 246
Silver-hair ivesia, 242
Silverleaf cotoneaster, 455
Silverside, A. J., 401, 402
Silverweed, 127
Simmons, S. L., 6, 7
Simon's cotoneaster, 460
Simple cinquefoil, 134
Siskiyou drymocallis, 294
Siskiyou wood beauty, 294
Sitka burnet, 322
Sitka mountain ash, 444
Sitka rose, 98
Skrede, I., 327
Sky ivesia, 228
Slender cinquefoil, 153
Sloe, 376, 380
Small, J. K., 68, 69, 613
Small, R. L., 354, 377, 378
Small-flowered horkelia, 262
Small-leaved rose, 79
Small-petaled barren-strawberry, 72
Smedmark, J. E. E., 57, 58, 59, 71
Smith, E. B., 29
Smith, E. C., 307
Smith, G. L., 254
Smith, H. H., 35, 101
Smith, W. R., 554
Smooth highbush blackberry, 37
Smooth lady's mantle, 305
Smooth oriental photinia, 490
Smooth rose, 105
Smooth shadbush, 661
Snow cinquefoil, 198
Snow pear, 481
Snow raspberry, 29, 45
Snowpeaks raspberry, 46
Snow-wreath, 387
Snyder, M., 64
Soft agrimony, 316
Soft downy-rose, 93
Soják, J., 122, 127, 130, 139, 140, 141, 142, 152, 158, 161, 178, 190, 192, 195, 196, 199, 200, 202, 206, 208, 212, 213, 214, 215, 216, 217, 245, 281, 296, 298, 299, 301, 304
Soldier Meadow cinquefoil, 176
Soltis, D. E., 6
Somers, P., 388
Sonoma horkelia, 252
Sorbaire, 396
Sorbaire à feuilles de sorbier, 397
Sorbaria, 392, 395, **396**, 415
  aitchisonii, 396
  arborea, 397
  kirilowii, 396, **397**
  sorbifolia, 396, **397**
    stellipila, 397
  stellipila, 397
  tomentosa, 396
    angustifolia, 396
    tomentosa, 396
Sorbarieae, 21, **392**, 395
×*Sorbaronia*, 433, 434, 435, 445
  arsenii, 441
  fallax, 438
  hybrida, 438
  jackii, 439
  monstrosa, 440
  nothospecies, 433
  sorbifolia, 440
Sorbier, 433
Sorbier d'Amérique, 439
Sorbier des oiseaux, 438
Sorbier des oiseleurs, 438
Sorbier plaisant, 440
×*Sorbocotoneaster*, 433, 468
  pozdnjakovii, 438
×*Sorbopyrus*, 433
  auricularis, 436
Sorbus, 20, 429, **433**, 434, 435, 436, 441, 442, 443, 444, 445
  sect. *Aronia*, 445
  sect. Commixtae, 437, **439**, 441
  sect. Sambucifoliae, 437, **443**
  sect. Sorbus, **437**
  sect. Tianshanicae, 437, **443**
  ser. *Tianshanicae*, 443
  subg. Aria, 434, **436**
  subg. *Cormus*, 434
  subg. Sorbus, 434, **437**, 439, 443
  subg. Torminaria, 433, 434, **436**
  alaskana, 442
  americana, 434, 435, **439**, *440*, 441
  andersonii, 442
  angustifolia, 442
  arbutifolia, 446
  aria, 437, 439, 440
  aucuparia, 433, 434, 435, **438**, 439, 440, 441, 468
  californica, 435, **441**, 442, 444, 445
  cascadensis, 442
  decora, 435, 438, 439, ***440***, 441
    groenlandica, 440
  domestica, 434
  dumosa, 442
  groenlandica, 440, 441
  hybrida, 435, **438**, 439
  intermedia, 435, **437**
  melanocarpa, 446
  occidentalis, 444
  ×*plantierensis*, 440
  sambucifolia, 434, 435, **443**
    grayi, 444
  scopulina, 435, **442**, 443, 444, 651
    cascadensis, 442
  sibirica, 438
  sitchensis, 436, 442, **444**
    californica, 441
    grayi, **444**, 445
    sitchensis, **444**
  ×*splendida*, 440
  spuria, 438
  subvestita, 438
  ×*thuringiaca*, 439
  toringo, 479
  torminalis, 434, 435, **436**
Sørensen, P. D., 312
Sorrie, B. A., 132
Sosa, V., 6, 10
Soule, J. D., 141, 144
Sour cherry, 368
Southern crabapple, 474
Southern dewberry, 54
Southern false serviceberry, 646
Southern mountain misery, 344
Southern Rocky Mountain cinquefoil, 164
Southwestern black cherry, 364
Sow-teat berry, 34
Spach, E., 319
Spalding rose, 113
Sparseleaf cinquefoil, 188
Spenceria, 312
Sperduto, D. D., 402
Spider ivesia, 235
Spiny greasebush, 14
Spiraea, 20, 24, **398**, 399, 401,

Spiraea (*continued*) 405, 407, 412, 414, 416, 422
[unranked] *Holodiscus*, 415
[unranked] *Petrophytum*, 411
sect. *Chamaebatiaria*, 394
sect. *Physocarpus*, 347
sect. *Sorbaria*, 396
aemeliana, 405
alba, *397*, 399, **401**
  alba, *397*, **401**, 402
  latifolia, 401, **402**
  septentrionalis, 402
albiflora, 410
amena, 402
aquilegiifolia
  vanhouttei, 410
arbuscula, 406
ariifolia, 418
aruncus, 424
beauverdiana, 405
  stevenii, 405
bethlehemensis, 402
betulifolia, 404, 405
  corymbosa, 403
  lucida, 404
  rosea, 406
×*billardii*, 401, 407
×*bumalda*, 410
caespitosa, 413
  elatius, 413
californica, 430
callosa, 410
camtschatica
  occidentalis, 25
cantoniensis, 398, 400, **409**, 410, 411
capitata, 349
carpinifolia, 402
chamaedryfolia, 400, **409**
  flexuosa, 409
  media, 405
  ulmifolia, 409
ciliata, 401
cinerascens, 412
corymbosa, 400, **403**, 404
  lucida, 404
cuneifolia, 401
densiflora, 406, 408
  splendens, 406
denudata, 26
discolor, 417
  dumosa, 419
  glabrescens, 421
douglasii, 400, **406**, *407*, 408
  alba, 407
  douglasii, **407**
  menziesii, 406, **407**, 408
  pseudosalicifolia, 407
  roseata, 407, **408**
ferruginea, 403
filipendula, 27
fortunei, 410
glomerata, 403
helleri, 406
hendersonii, 412
heterophylla, 402
×hitchcockii, 400, **406**
hypericifolia, 399
idahoensis, 408
incisa, 351
japonica, 400, **410**
  fortunei, 410
kirilowii, 397
lancifolia, 401
latifolia, 402
  septentrionalis, 402
lucida, 400, **404**, 405, 408
  rosea, 406
menziesii, 407, 408
  pyramidata, 408
millefolium, 395
monogyna, 350
obovata, 402
occidentalis, 25
opulifolia, 348
ostryfolia, 403
ovata, 402
parvifolia, 403
pectinata, 422
prunifolia, 399, 400, **408**, 409, 411
  prunifolia, 409
×*pseudosalicifolia*, 407
×pyramidata, 400, **408**
repens, 403
×*rosalba*, 401
rosea, 403
roseata, 408
rubra, 25
salicifolia, 401, **402**, 407
  lanceolata, 401
  latifolia, 402
  linearis, 401
  paniculata, 401
septentrionalis, 402
sorbifolia, 397
sororia, 403
splendens, 400, **405**, 406
  rosea, **406**
  splendens, **406**
stevenii, 400, **405**
stipulata, 425
subvillosa, 407
thunbergii, 398, 399, 400, **409**
tomentosa, *397*, 400, **403**
  alba, 401
  albiflora, 403
  rosea, **403**
  tomentosa, *397*, **403**
tomentulosa, 408
trifoliata, 426
trilobata, 399, 411, 414
ulmaria, 26
×vanhouttei, 400, 409, **410**, 411
virginiana, 400, **404**, 410
Spiraeanthus, 392, 395
Spiraeeae, 21, **398**, 412, 414, 415, 416
Spiraeoideae, 345
Spirée, 398
Spirée à larges feuilles, 402
Spirée blanche, 401
Spirée tomenteuse, 403
Spongberg, S. A., 432, 471, 473, 480, 482
Sportella, 468, 469
  atalantioides, 471
Spreading avens, 64
Spreading cotoneaster, 461
Spring avens, 70
Spring cinquefoil, 190
Springer, J. C., 279, 326
Squaw apple, 662
St. Clair hawthorn, 541
St. John, H., 15
St. Lucie cherry, 366
Stachyuraceae, 6, 10
Stafleu, F. A., 273
Staphylea, 6, **7**
  bolanderi, 7, **8**
  pringlei, 8
  trifolia, *5*, 7, **8**
Staphyleaceae, **6**, 10
  subfam. *Staphyleoideae*, 6
  subfam. *Tapiscioideae*, 6
Staphyleoideae, 6
Staphylier, 7
Staphylier à trois folioles, 8
Star rose, 80
Starostenko, N. T., 145
Starostenko, V. N., 145
Staudt, G., 274, 276, 277, 279
Stebbins, G. L., 341
Steens Mountain cinquefoil, 171
Steeplebush, 403
Steiner, E., 208
Stellariopsis, 245, 246
  santolinoides, 246
Stellariopsis, 246
Stephanandra, 348, 351
  incisa, 351
Sterling, C., 405, 409, 411, 413, 414, 422
Stern's cotoneaster, 466
Stipulate cinquefoil, 147
Stone bramble, 52
Stone's hawthorn, 591
Stranvaesia, 489
  davidiana, 491
Strausbaugh, P. D., 29
Strawberry, 19, 274, 301
Strawberry leaved cinquefoil, 132
Strawberry raspberry, 43
Strawberryleaf raspberry, 49
Stretch-fruit cotoneaster, 461
Strigose cinquefoil, 209
Strother, J. L., 346
Stutz, H. C., 341
Stylypus, 58
  vernus, 70
Succulent hawthorn, 518
Sugarplum, 646
Suksdorf hawthorn, 512
Sulphur cinquefoil, 142
Summer haw, 611
Sunny hawthorn, 614
Susquehanna sandcherry, 375
Swamp dewberry, 41
Swamp rose, 99
Swedish cotoneaster, 460
Swedish whitebeam, 437
Sweet almond, 372
Sweet briar rose, 90
Sweet cherry, 368
Sweet crabapple, 475
Swink, F., 457
Sydes, C., 434

Tahquitz ivesia, 224
Taihangia, 59
Taiping cotoneaster, 464
Taiwan firethorn, 471
Takhtajan, A. L., 416
Talent, N., 492, 610, 611
Tall cinquefoil, 285
Tall drymocallis, 285
Tall horkelia, 257
Tall wood beauty, 285
Tapiscia, 6
Tapisciaceae, 6
Tapiscioideae, 6
Taraxacum, 29
Tawny horkelia, 259
Tea rose, 77
Tea crabapple, 478
Tehachapi rose, 109
Tengyueh cotoneaster, 465
Texas almond, 370
Texas greasebush, 17
Texas red haw, 554
Texas wild peach, 377
Thames hawthorn, 573
Thecanisia, 23
  angustifolia, 25
  lobata, 25
  purpurea, 25
  ulmaria, 26
Thicket shadbush, 656
Thimbleberry, 29, 48
Thin-lobed horkelia, 252
Thomas, J. H., 254
Thomas, L. K., 341
Thomas, W. W., 3
Thompson, M. M., 46
Thorn, 492
Thorn apple, 492
Thorne, R. F., 9
Thread-stemmed lady's mantle, 307
Three-flower hawthorn, 610
Three-sisters, 62
Three-toothed cinquefoil, 300
Three-toothed horkelia, 269
Thunberg's meadowsweet, 409
Thurber's cinquefoil, 149
Tidestrom, I., 177
Tigarea
  tridentata, 340
Timbalia, 469
Töpel, M. H., 122, 127, 131, 132, 133, 138, 199
Topsy, 49
Torchflower, 62
Toringo crabapple, 479
Tormentil, 136
Tormentilla, 132, 133
  erecta, 136
×*Tormimalus*, 433
Torminalis, 436
Townsend's cinquefoil, 153
Toyon, 447
Tracy hawthorn, 549
Trailing tormentil, 135
Tree cotoneaster, 456
Tremblay, N., 331

Tridophyllum, 137, 138
Truncate-leaved cotoneaster, 466
Tucker, G. C., 29, 449, 461
Turdus
migratorius, 449, 469
Turner, N. J., 112
Turpinia, 6
Tweedy's ivesia, 235
Two-flower cinquefoil, 131
Tylosperma, 127
Tzvelev, N. N., 301

Ulmaceae, 20
Ulmaria, 23
rubra, 25
Ulmarieae, 22, **23**
Ultramontane rose, 109
Urticaceae, 20
Ussurian pear, 480
Utah ivesia, 233
Utah serviceberry, 650
Utah shadbush, 650
Uttal, L. J., 404, 405

Vahl's cinquefoil, 205
Vail's hawthorn, 643
Valley Head hawthorn, 610
Vanden Heuvel, B. D., 332
Väre, S. M., 303, 304
Vari-leaved cinquefoil, 152
Vauquelinia, 21, 415, 427, **429**, 430
angustifolia, 431
australis, 430
californica, *426*, **430**
californica, *426*, **430**, 431
pauciflora, 430, **431**
retherfordii, 430
sonorensis, 430, **431**
corymbosa, 430, **431**
angustifolia, **431**
pauciflora, 431
torreyi, 431
Veined lady's mantle, 309
Velascoa, 9
recondita, 9
Venn, J. M., 463
Vested cotoneaster, 455
Vidal, J. E., 432
Villous cinquefoil, 201
Vincent, M. A., 460, 482, 483
Vining, T. F., 647
Virginia avens, 69
Virginia rose, 100
Virginia spiraea, 404
Virginia strawberry, 277
Voss, E. G., 29, 59, 366, 376, 425, 441, 647, 653, 654, 655

Wah Wah ivesia, 228
Waldsteinia, 57, 59, **71**
doniana, 71
fragarioides, 67, **71**, 72
doniana, 71
parviflora, 72
idahoensis, 71, **72**
lobata, 71, **72**
parviflora, 67, 71, **72**
Waldsteinie, 71
Walker, C. A., 87
Walker, L. R., 391
Ward, D. B., 619
Wasatch drymocallis, 286
Wasatch ivesia, 235
Wasatch wood beauty, 286
Washington thorn, 499
Water avens, 65
Weakley, A. S., 29, 318, 319, 321, 351, 389
Webb, R. H., 391
Webber's ivesia, 236
Weber, H. E., 37, 55
Weber, J. E., 648, 658, 661
Weber, W. A., 167, 210, 296
Wedgeleaf drymocallis, 295
Wedgeleaf horkelia, 255
Wedgeleaf wood beauty, 295
Weeping hawthorn, 622
Wehr, W. C., 395
Weinmann, F. C., 456
Wells, T. C., 493
Welsh, S. L., 179, 180, 212, 291
Wen, J., 354, 385
Werier, D., 99
Werlemark, G., 89
Werner, D. J., 87
Werner, P. A., 141, 144
West Indies laurel cherry, 361
West Virginia spiraea, 404
Western burnet, 320
Western chokecherry, 366
Western crabapple, 474, 476
Western hardhack, 406
Western mayhaw, 547
Western mountain ash, 444
Western sandcherry, 374
Westwood, M. N., 480, 483, 484
Wheeler, C. R., 254
Wheeler's cinquefoil, 185
White avens, 68
White dryad, 329
White haw, 542
White meadowsweet, 401
White Mountains cinquefoil, 194
White Mountains horkelia, 249
White Mountains ivesia, 232
White prairie rose, 103
Whitebark raspberry, 45
Whitebeam, 433
Widrlechner, M. P., 29, 37, 48
Wiegand, K. M., 306, 647, 654, 657, 658, 659
Wiegand's shadbush, 655
Wight, W. F., 374
Wild crab apple, 662
Wild goose plum, 382
Wild plum, 378
Wild service tree, 436
Wilder's horkelia, 258
Wild-plum, 646
Wilhelm, G. S., 457
Williams hawthorn, 504
Williams, K., 37
Willow hawthorn, 509
Willow-leaved cotoneaster, 454
Willow-leaved meadowsweet, 402
Willow-leaved pear, 480
Wineberry, 51
Winnipeg Parks, 104
Winter-green cotoneaster, 457
Winther, K., 90
Wire ivesia, 236
Wiśniewska, J., 436
Wissemann, V., 78
Wittmann, R. C., 167, 210, 296
Wolf, T., 122, 123, 137, 144, 189, 192, 299
Wolfe, J. A., 395
Wonderful rose, 81
Wood avens, 70
Wood beauty, 280
Wood cinquefoil, 135
Wood rose, 116
Wood strawberry, 275
Woodland hawthorn, 497
Woods's rose, 107
Woolly cinquefoil, 162
Woolly cotoneaster, 455
Wooten, J. W., 29
Wooton's hawthorn, 564
Wright, R. A. S., 351
Wright, W. A., 647, 648
Wu, S., 78
Wu, Z., 353
Wu, Z. Y., 458
Wunderlin, R. P., 29

Xerospiraea, 398, 422

Yadon's horkelia, 253
Yatskievych, G., 9
Yellow avens, 67
Yellow comarella, 244
Yellow hawthorn, 588
Yellow mountain-avens, 328
Yellow purpusia, 225
Yellow-green lady's mantle, 306
Yerba del pasmo, 73
Yi, O., 109
Yosemite ivesia, 241
Yoshino cherry, 369
Yoshizawa, Y., 434
Yu, T. T., 450
Yukon silverweed, 129
Yunnan cotoneaster, 459

Zhang Zhao, Y., 410
Zhou, L. H., 458
Zika, P. F., 433, 442, 445, 448, 460, 467, 468, 476, 491
Ziola, B., 95, 107

Political Map of North America North of Mexico

Canadian Provinces

Alta.	*Alberta*
B.C.	*British Columbia*
Man.	*Manitoba*
N.B.	*New Brunswick*
Nfld. and Labr.	*Newfoundland and Labrador*
N.W.T.	*Northwest Territories*
N.S.	*Nova Scotia*
Nunavut	
Ont.	*Ontario*
P.E.I.	*Prince Edward Island*
Que.	*Quebec*
Sask.	*Saskatchewan*
Yukon	

United States

Ala.	*Alabama*
Alaska	
Ariz.	*Arizona*
Ark.	*Arkansas*
Calif.	*California*
Colo.	*Colorado*
Conn.	*Connecticut*
Del.	*Delaware*
D.C.	*District of Columbia*
Fla.	*Florida*
Ga.	*Georgia*
Idaho	
Ill.	*Illinois*
Ind.	*Indiana*
Iowa	
Kans.	*Kansas*
Ky.	*Kentucky*
La.	*Louisiana*
Maine	
Md.	*Maryland*
Mass.	*Massachusetts*
Mich.	*Michigan*
Minn.	*Minnesota*
Miss.	*Mississippi*
Mo.	*Missouri*
Mont.	*Montana*
Nebr.	*Nebraska*
Nev.	*Nevada*
N.H.	*New Hampshire*
N.J.	*New Jersey*
N.Mex.	*New Mexico*
N.Y.	*New York*
N.C.	*North Carolina*
N.Dak.	*North Dakota*
Ohio	
Okla.	*Oklahoma*
Oreg.	*Oregon*
Pa.	*Pennsylvania*
R.I.	*Rhode Island*
S.C.	*South Carolina*
S.Dak.	*South Dakota*
Tenn.	*Tennessee*
Tex.	*Texas*
Utah	
Vt.	*Vermont*
Va.	*Virginia*
Wash.	*Washington*
W.Va.	*West Virginia*
Wis.	*Wisconsin*
Wyo.	*Wyoming*